PEARSON ALWAYS LEARNING

Michael Sullivan • Michael Sullivan III

Algebra & Trigonometry
Enhanced with Graphing Utilities

Third Custom Edition for Columbus State Community College

Taken from:
Algebra & Trigonometry: Enhanced with Graphing Utilities, Seventh Edition
by Michael Sullivan, Michael Sullivan III

Cover Art: Umedar3 and Umedar1, by Barry Cronin.

Taken from:

Algebra & Trigonometry: Enhanced with Graphing Utilities, Seventh Edition
by Michael Sullivan, Michael Sullivan III

This special edition published in cooperation with Pearson Education, Inc.

Pearson Education, Inc., 330 Hudson Street, New York, New York 10013
A Pearson Education Company
www.pearsoned.com

Printed in the United States of America

000200010272050685

GG

PEARSON

ISBN 10: 1-323-45549-3
ISBN 13: 978-1-323-45549-4

Contents

Prepare for Class "Read the Book"

Feature	Description	Benefit	Page(s)
Every Chapter Opener begins with ...			
Chapter-Opening Topic & Project	Each chapter begins with a discussion of a topic of current interest and ends with a related project.	In the concluding project, you will apply what you have learned to solve a problem related to the topic.	407, 511
Internet-Based Projects	These projects allow for the integration of spreadsheet technology that you will need to be a productive member of the workforce.	The projects give you an opportunity to collaborate and use mathematics to deal with issues of current interest.	407, 511
Every Section begins with ...			
Learning Objectives 2	Each section begins with a list of objectives. Individual objectives also appear in the text where they are covered.	These objectives focus your studying by emphasizing what's most important and where to find it.	428
Sections contain ...			
PREPARING FOR THIS SECTION	Most sections begin with a list of key concepts to review, with page numbers.	Ever forget what you've learned? This feature highlights previously learned material to be used in this section. Review it, and you'll always be prepared to move forward.	428
Now Work the 'Are You Prepared?' Problems	These problems assess whether you have the prerequisite knowledge for the upcoming section.	Not sure you need the Preparing for This Section review? Work the 'Are You Prepared?' problems. If you get one wrong, you'll know exactly what you need to review and where to review it!	428, 439
Now Work PROBLEMS	These follow most examples and direct you to a related exercise.	We learn best by doing. You'll solidify your understanding of examples if you try a similar problem right away, to be sure you understand what you've just read.	437
WARNING	Warnings are provided in the text.	These point out common mistakes and help you avoid them.	462
Explorations and Seeing the Concept	These graphing utility activities foreshadow a concept or reinforce a concept just presented.	You will obtain a deeper and more intuitive understanding of theorems and definitions.	377, 434
In Words	This feature provides alternative descriptions of select definitions and theorems.	Does math ever look foreign to you? This feature translates math into plain English.	430
Calculus	This symbol appears next to information essential for the study of calculus.	Pay attention–if you spend extra time now, you'll do better later!	236, 238, 373
SHOWCASE EXAMPLES	These examples provide "how to" instruction by offering a guided, step-by-step approach to solving a problem.	With each step presented on the left and the mathematics displayed on the right, you can immediately see how each step is employed.	342–343
Model It! Examples and Problems	These examples and problems require you to build a mathematical model from either a verbal description or data. The homework Model It! problems are marked by purple problem numbers.	It is rare for a problem to come in the form "*Solve the following equation.*" Rather, the equation must be developed based on an explanation of the problem. These problems require you to develop models that will enable you to describe the problem mathematically and suggest a solution to the problem.	453, 482

Practice "Work the Problems"

Feature	Description	Benefit	Page(s)
'Are You Prepared?' Problems	These problems assess your retention of the prerequisite material. Answers are given at the end of the section exercises. This feature is related to the Preparing for This Section feature.	Do you always remember what you've learned? Working these problems is the best way to find out. If you get one wrong, you'll know exactly what you need to review and where to review it!	428, 439
Concepts and Vocabulary	These short-answer questions, mainly fill-in-the-blank, multiple-choice, and true/false items, assess your understanding of key definitions and concepts in the current section.	It is difficult to learn math without knowing the language of mathematics. These problems test your understanding of the formulas and vocabulary.	440
Skill Building	Correlated with section examples, these problems provide straightforward practice.	It's important to dig in and develop your skills. These problems give you ample opportunity to do so.	440–442
Mixed Practice	These problems offer comprehensive assessment of the skills learned in the section by asking problems related to more than one concept or objective. These problems may also require you to utilize skills learned in previous sections.	Learning mathematics is a building process. Many concepts build on each other and are related. These problems help you see how mathematics builds on itself and how the concepts are linked together.	442
Applications and Extensions	These problems allow you to apply your skills to real-world problems. They also enable you to extend concepts learned in the section.	You will see that the material learned within the section has many uses in everyday life.	442–444
Explaining Concepts: Discussion and Writing	"Discussion and Writing" problems are colored red. They support class discussion, verbalization of mathematical ideas, and writing and research projects.	To verbalize an idea, or to describe it clearly in writing, shows real understanding. These problems nurture that understanding. Many are challenging, but you'll get out what you put in.	445
NEW! Retain Your Knowledge	These problems allow you to practice content learned earlier in the course.	Remembering how to solve all the different kinds of problems that you encounter throughout the course is difficult. This practice helps you remember previously learned skills.	445
Now Work PROBLEMS	Many examples refer you to a related homework problem. These related problems are marked by a pencil and orange numbers.	If you get stuck while working problems, look for the closest Now Work problem, and refer to the related example to see if it helps.	429, 437, 438, 441
Review Exercises	Every chapter concludes with a comprehensive list of exercises to practice. Use the list of objectives to determine what objective and examples correspond to each problem.	Work these problems to ensure that you understand all the skills and concepts employed in the chapter. Think of it as a comprehensive review of the chapter. All answers to Chapter Review problems appear in the back of the text.	506–509

Review "Study for Quizzes and Tests"

Feature	Description	Benefit	Page(s)
The Chapter Review at the end of each chapter contains ...			
Things to Know	A detailed list of important theorems, formulas, and definitions from the chapter.	Review these and you'll know the most important material in the chapter!	504–505
You Should Be Able to ...	A complete list of objectives by section and, for each, examples that illustrate the objective, and practice exercises that test your understanding of the objective.	Do the recommended exercises and you'll have mastered the key material. If you get something wrong, go back and work through the example listed, and try again.	505–506
Review Exercises	These provide comprehensive review and practice of key skills, matched to the Learning Objectives for each section.	Practice makes perfect. These problems combine exercises from all sections, giving you a comprehensive review in one place.	506–509
Chapter Test	About 15–20 problems that can be taken as a Chapter Test. Be sure to take the Chapter Test under test conditions—no notes!	Be prepared. Take the sample practice test under test conditions. This will get you ready for your instructor's test. If you get a problem wrong, you can watch the Chapter Test Prep Video.	509
Cumulative Review	These problem sets appear at the end of each chapter, beginning with Chapter 2. They combine problems from previous chapters, providing an ongoing cumulative review. When you use them in conjunction with the Retain Your Knowledge problems, you will be ready for the final exam.	These problem sets are really important. Completing them will ensure that you are not forgetting anything as you go. This will go a long way toward keeping you primed for the final exam.	510
Chapter Projects	The Chapter Projects apply to what you've learned in the chapter. Additional projects are available on the Instructor's Resource Center (IRC).	The Chapter Projects give you an opportunity to apply what you've learned in the chapter to the opening topic. If your instructor allows, these make excellent opportunities to work in a group, which is often the best way of learning math.	511
Internet-Based Projects	In selected chapters, a Web-based project is given.	These projects give you an opportunity to collaborate and use mathematics to deal with issues of current interest by using the Internet to research and collect data.	511

Three Distinct Series

Students have different goals, learning styles, and levels of preparation. Instructors have different teaching philosophies, styles, and techniques. Rather than write one series to fit all, the Sullivans have written three distinct series. All share the same goal—to develop a high level of mathematical understanding and an appreciation for the way mathematics can describe the world around us. The manner of reaching that goal, however, differs from series to series.

Enhanced with Graphing Utilities Series, Seventh Edition

This series provides a thorough integration of graphing utilities into topics, allowing students to explore mathematical concepts and encounter ideas usually studied in later courses. Using technology, the approach to solving certain problems differs from the Contemporary or Concepts through Functions Series, while the emphasis on understanding concepts and building strong skills does not: *College Algebra, Algebra & Trigonometry, Precalculus*.

Contemporary Series, Tenth Edition

The Contemporary Series is the most traditional in approach, yet modern in its treatment of precalculus mathematics. Graphing utility coverage is optional and can be included or excluded at the discretion of the instructor: *College Algebra, Algebra & Trigonometry, Trigonometry: A Unit Circle Approach, Precalculus.*

Concepts through Functions Series, Third Edition

This series differs from the others, utilizing a functions approach that serves as the organizing principle tying concepts together. Functions are introduced early in various formats. This approach supports the Rule of Four, which states that functions are represented symbolically, numerically, graphically, and verbally. Each chapter introduces a new type of function and then develops all concepts pertaining to that particular function. The solutions of equations and inequalities, instead of being developed as stand-alone topics, are developed in the context of the underlying functions. Graphing utility coverage is optional and can be included or excluded at the discretion of the instructor: *College Algebra; Precalculus, with a Unit Circle Approach to Trigonometry; Precalculus, with a Right Triangle Approach to Trigonometry*.

The Enhanced with Graphing Utilities Series

College Algebra

This text provides an approach to college algebra that completely integrates graphing technology without sacrificing mathematical analysis and conceptualization. The text has three chapters of review material preceding the chapters on functions. After completing this text, a student will be prepared for trigonometry, finite mathematics, and business calculus.

Algebra & Trigonometry

This text contains all the material in *College Algebra*, but it also develops the trigonometric functions using a right triangle approach and shows how that approach is related to the unit circle approach. Graphing techniques are emphasized, including a thorough discussion of polar coordinates, parametric equations, and conics using polar coordinates. Graphing calculator usage is integrated throughout. After completing this text, a student will be prepared for finite mathematics, business calculus, and engineering calculus.

Precalculus

This text contains one review chapter before covering the traditional precalculus topics of functions and their graphs, polynomial and rational functions, and exponential and logarithmic functions. The trigonometric functions are introduced using a unit circle approach and show how it is related to the right triangle approach. Graphing techniques are emphasized, including a thorough discussion of polar coordinates, parametric equations, and conics using polar coordinates. Graphing calculator usage is integrated throughout. The final chapter provides an introduction to calculus, with a discussion of the limit, the derivative, and the integral of a function. After completing this text, a student will be prepared for finite mathematics, business calculus, and engineering calculus.

Preface to the Instructor

As professors at an urban university and a community college, Michael Sullivan and Michael Sullivan III are aware of the varied needs of Algebra and Trigonometry students. Such students range from those who have little mathematical background and are fearful of mathematics courses, to those with a strong mathematical education and a high level of motivation. For some of your students, this will be their last course in mathematics, whereas others will further their mathematical education. We have written this text with both groups in mind.

As a teacher, and as an author of precalculus, engineering calculus, finite mathematics, and business calculus texts, Michael Sullivan understands what students must know if they are to be focused and successful in upper-level math courses. However, as a father of four, he also understands the realities of college life. As an author of a developmental mathematics series, Michael's son and co-author, Michael Sullivan III, understands the trepidations and skills that students bring to the Algebra and Trigonometry course. As the father of a current college student, Michael III realizes that today's college students demand a variety of media to support their education. This text addresses that demand by providing technology and video support that enhances understanding without sacrificing math skills. Together, both authors have taken great pains to ensure that the text offers solid, student-friendly examples and problems, as well as a clear and seamless writing style.

A tremendous benefit of authoring a successful series is the broad-based feedback we receive from teachers and students. We are sincerely grateful for their support. Virtually every change in this edition is the result of their thoughtful comments and suggestions. We are confident that, building on the success of the first six editions and incorporating many of these suggestions, we have made *Algebra & Trigonometry Enhanced with Graphing Utilities,* 7th Edition, an even better tool for learning and teaching. We continue to encourage you to share with us your experiences teaching from this text.

Features in the Seventh Edition

A descriptive list of the many special features of *Algebra & Trigonometry* can be found in the front of this text.

This list places the features in their proper context, as building blocks of an overall learning system that has been carefully crafted over the years to help students get the most out of the time they put into studying. Please take the time to review this and to discuss it with your students at the beginning of your course. When students utilize these features, they are more successful in the course.

New to the Seventh Edition

- **Retain Your Knowledge** This new category of problems in the exercise set is based on the article "To Retain New Learning, Do the Math" published in the *Edurati Review*. In this article, Kevin Washburn suggests that "the more students are required to recall new content or skills, the better their memory will be." It is frustrating when students cannot recall skills learned earlier in the course. To alleviate this recall problem, we have created "Retain Your Knowledge" problems. These are problems considered to be "final exam material" that students can use to maintain their skills. All the answers to these problems appear in the back of the text, and all are programmed in MyMathLab.
- **Guided Lecture Notes** Ideal for online, emporium/redesign courses, inverted classrooms, or traditional lecture classrooms. These lecture notes help students take thorough, organized, and understandable notes as they watch the Author in Action videos. They ask students to complete definitions, procedures, and examples based on the content of the videos and text. In addition, experience suggests that students learn by doing and understanding the why/how of the concept or property. Therefore, many sections have an exploration activity to motivate student learning. These explorations introduce the topic and/or connect it to either a real-world application or a previous section. For example, when the vertical-line test is discussed in Section 3.2, after the theorem statement, the notes ask the students to explain why the vertical-line test works by using the definition of a function. This challenge helps students process the information at a higher level of understanding.
- **Illustrations** Many of the figures now have captions to help connect the illustrations to the explanations in the body of the text.
- **TI Screen Shots** In this edition we have replaced all the screen shots from the sixth edition with screen shots using TI-84 Plus C. These updated screen shots help students visualize concepts clearly and help make stronger connections among equations, data, and graphs in full color.
- **Exercise Sets** All the exercises in the text have been reviewed and analyzed for this edition, some have been removed, and new ones have been added. All time-sensitive problems have been updated to the most recent information available. The problem sets remain classified according to purpose.

  The ***'Are You Prepared?'*** problems have been improved to better serve their purpose as a just-in-time review of concepts that the student will need to apply in the upcoming section.

  The ***Concepts and Vocabulary*** problems have been expanded and now include multiple-choice exercises. Together with the fill-in-the-blank and true/false problems, these exercises have been written to serve as reading quizzes.

Skill Building problems develop the student's computational skills with a large selection of exercises that are directly related to the objectives of the section. ***Mixed Practice*** problems offer a comprehensive assessment of skills that relate to more than one objective. Often these require skills learned earlier in the course.

Applications and Extensions problems have been updated. Further, many new application-type exercises have been added, especially ones involving information and data drawn from sources the student will recognize, to improve relevance and timeliness.

The ***Explaining Concepts: Discussion and Writing*** exercises have been improved and expanded to provide more opportunity for classroom discussion and group projects.

New to this edition, ***Retain Your Knowledge*** exercises consist of a collection of four problems in each exercise set that are based on material learned earlier in the course. They serve to keep information that has already been learned "fresh" in the mind of the student. Answers to all these problems appear in the Student Edition.

The ***Review Exercises*** in the Chapter Review have been streamlined, but they remain tied to the clearly expressed objectives of the chapter. Answers to all these problems appear in the Student Edition.

- **Annotated Instructor's Edition** As a guide, the author's suggestions for homework assignments are indicated by a blue underscore below the problem number. These problems are assignable in MyMathLab.

Content Changes in the Seventh Edition

- **Section 3.1** The objective Find the Difference Quotient of a Function has been added.
- **Section 5.2** The objective Use Descartes' Rule of Signs has been included.
- **Section 5.2** The theorem Bounds on the Zeros of a Polynomial Function is now based on the traditional method of using synthetic division.
- **Section 5.5** Content has been added that discusses the role of multiplicity of the zeros of the denominator of a rational function as it relates to the graph near a vertical asymptote.

Using the Seventh Edition Effectively with Your Syllabus

To meet the varied needs of diverse syllabi, this text contains more content than is likely to be covered in an *Algebra and Trigonometry* course. As the chart illustrates, this text has been organized with flexibility of use in mind. Within a given chapter, certain sections are optional (see the details that follow the accompanying figure) and can be omitted without loss of continuity.

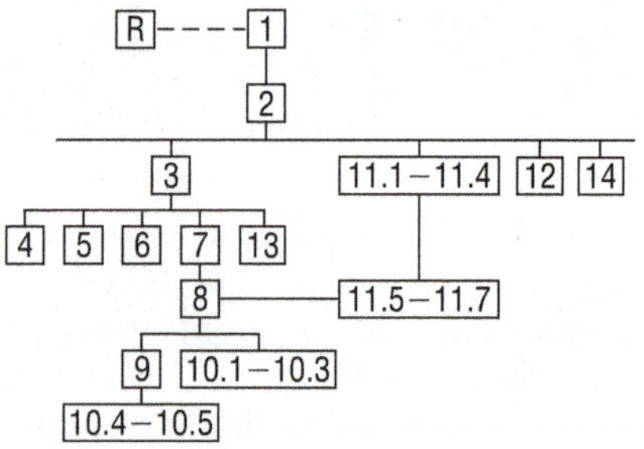

Chapter R Review
This chapter consists of review material. It may be used as the first part of the course or later as a just-in-time review when the content is required. Specific references to this chapter occur throughout the text to assist in the review process.

Chapter 1 Equations and Inequalities
Primarily a review of intermediate algebra topics, this material is a prerequisite for later topics. The coverage of complex numbers and quadratic equations with a negative discriminant is optional and may be postponed or skipped entirely without loss of continuity.

Chapter 2 Graphs
This chapter lays the foundation for functions. Section 2.4 is optional.

Chapter 3 Functions and Their Graphs
This is perhaps the most important chapter. Section 3.6 is optional.

Chapter 4 Linear and Quadratic Functions
Topic selection depends on your syllabus. Sections 4.2 and 4.4 may be omitted without loss of continuity.

Chapter 5 Polynomial and Rational Functions
Topic selection depends on your syllabus.

Chapter 6 Exponential and Logarithmic Functions
Sections 6.1–6.6 follow in sequence. Sections 6.7, 6.8, and 6.9 are optional.

Chapter 7 Trigonometric Functions
Section 7.8 may be omitted in a brief course.

Chapter 8 Analytic Trigonometry
Sections 8.2, 8.6, and 8.7 may be omitted in a brief course.

Chapter 9 Applications of Trigonometric Functions
Sections 9.4 and 9.5 may be omitted in a brief course.

Chapter 10 Polar Coordinates; Vectors
Sections 10.1–10.3 and Sections 10.4–10.5 are independent and may be covered separately.

Chapter 11 Analytic Geometry

Sections 11.1–11.4 follow in sequence. Sections 11.5, 11.6, and 11.7 are independent of each other, but each requires Sections 11.1–11.4.

Chapter 12 Systems of Equations and Inequalities

Sections 12.2–12.7 may be covered in any order, but each requires Section 12.1. Section 12.8 requires Section 12.7.

Chapter 13 Sequences; Induction; The Binomial Theorem

There are three independent parts: Sections 13.1–13.3, Section 13.4, and Section 13.5.

Chapter 14 Counting and Probability

The sections follow in sequence.

Acknowledgments

Texts are written by authors, but they evolve from idea to final form through the efforts of many people.

Thanks are due to the following people for their assistance and encouragement during the preparation of this edition:

- From Pearson Education: Anne Kelly for her substantial contributions, ideas, and enthusiasm; Dawn Murrin, for her unmatched talent at getting the details right; Joseph Colella for always getting the reviews and pages to us on time; Peggy McMahon for directing the always difficult production process; Rose Kernan for handling liaison between the compositor and author; Peggy Lucas for her genuine interest in marketing this text; Chris Hoag for her continued support and genuine interest; Paul Corey for his leadership and commitment to excellence; and the Pearson Math and Science Sales team for their continued confidence and personal support of our texts.
- Accuracy checkers: C. Brad Davis read the entire manuscript and checked the accuracy of answers. His attention to detail is amazing. Timothy Britt created the Solutions Manuals and accuracy-checked answers.
- Michael Sullivan III would like to thank his colleagues at Joliet Junior College for their support and feedback.

Finally, we offer our sincere thanks to the dedicated users and reviewers of our texts, whose collective insights form the backbone of each text revision.

The list of those to whom we are indebted continues to grow. If we've forgotten anyone, please accept our apology. Thank you to all.

Ryan Adams, Northwest Florida State College
James Africh, College of DuPage
Steve Agronsky, Cal Poly State University
Gererdo Aladro, Florida International University
Grant Alexander, Joliet Junior College
Dave Anderson, South Suburban College
Richard Andrews, Florida A&M University
Joby Milo Anthony, University of Central Florida
James E. Arnold, University of Wisconsin-Milwaukee
Adel Arshaghi, Center for Educational Merit
Carolyn Autray, University of West Georgia
Agnes Azzolino, Middlesex County College
Taoufik Bahadi, University of Tampa
Wilson P. Banks, Illinois State University
Scott Barnett, henry Ford Community College
Sudeshna Basu, Howard University
Dale R. Bedgood, East Texas State University
Beth Beno, South Suburban College
Carolyn Bernath, Tallahassee Community College
Rebecca Berthiaume, Edison State College
William H. Beyer, University of Akron
John Bialas, Joliet Junior College
Annette Blackwelder, Florida State University
Richelle Blair, Lakeland Community College
Linda Blanco, Joliet Junior College
Kevin Bodden, Lewis and Clark College
Jeffrey Boerner, University of Wisconsin-Stout
Barry Booten, Florida Atlantic University
Rebecca Bonk, Joliet Junior College
Larry Bouldin, Roane State Community College
Bob Bradshaw, Ohlone College
Trudy Bratten, Grossmont College
Martin Bredeck, Northern Virginia Community College (Annandale Campus)
Tim Bremer, Broome Community College
Tim Britt, Jackson State Community College
Michael Brook, University of Delaware
Joanne Brunner, Joliet Junior College
Warren Burch, Brevard Community College
Mary Butler, Lincoln Public Schools
Melanie Butler, West Virginia University
Jim Butterbach, Joliet Junior College
William J. Cable, University of Wisconsin-Stevens Point
Lois Calamia, Brookdale Community College
Jim Campbell, Lincoln Public Schools
Roger Carlsen, Moraine Valley Community College
Elena Catoiu, Joliet Junior College
Mathews Chakkanakuzhi, Palomar College
Tim Chappell, Penn Valley Community College
John Collado, South Suburban College
Alicia Collins, Mesa Community College
Nelson Collins, Joliet Junior College
Rebecca Connell, Troy University
Jim Cooper, Joliet Junior College
Denise Corbett, East Carolina University
Carlos C. Corona, San Antonio College
Theodore C. Coskey, South Seattle Community College
Rebecca Connell, Troy University
Donna Costello, Plano Senior High School
Paul Crittenden, University of Nebraska at Lincoln
John Davenport, East Texas State University
Faye Dang, Joliet Junior College
Antonio David, Del Mar College
Stephanie Deacon, Liberty University
Duane E. Deal, Ball State University
Jerry DeGroot, Purdue North Central
Timothy Deis, University of Wisconsin-Platteville
Joanna DelMonaco, Middlesex Community College
Vivian Dennis, Eastfield College
Deborah Dillon, R. L. Turner High School
Guesna Dohrman, Tallahassee Community College
Cheryl Doolittle, Iowa State University
Karen R. Dougan, University of Florida
Jerrett Dumouchel, Florida Community College at Jacksonville
Louise Dyson, Clark College
Paul D. East, Lexington Community College
Don Edmondson, University of Texas-Austin
Erica Egizio, Lewis University
Laura Egner, Joliet Junior College
Jason Eltrevoog, Joliet Junior College
Christopher Ennis, University of Minnesota
Kathy Eppler, Salt Lake Community College
Ralph Esparza Jr., Richland College
Garret J. Etgen, University of Houston
Scott Fallstrom, Shoreline Community College
Pete Falzone, Pensacola Junior College
Arash Farahmand, Skyline College
W.A. Ferguson, University of Illinois-Urbana/Champaign

Iris B. Fetta, Clemson University
Mason Flake, student at Edison Community College
Timothy W. Flood, Pittsburg State University
Robert Frank, Westmoreland County Community College
Merle Friel, Humboldt State University
Richard A. Fritz, Moraine Valley Community College
Dewey Furness, Ricks College
Mary Jule Gabiou, North Idaho College
Randy Gallaher, Lewis and Clark College
Tina Garn, University of Arizona
Dawit Getachew, Chicago State University
Wayne Gibson, Rancho Santiago College
Loran W. Gierhart, University of Texas at San Antonio and Palo Alto College
Robert Gill, University of Minnesota Duluth
Nina Girard, University of Pittsburgh at Johnstown
Sudhir Kumar Goel, Valdosta State University
Adrienne Goldstein, Miami Dade College, Kendall Campus
Joan Goliday, Sante Fe Community College
Lourdes Gonzalez, Miami Dade College, Kendall Campus
Frederic Gooding, Goucher College
Donald Goral, Northern Virginia Community College
Sue Graupner, Lincoln Public Schools
Mary Beth Grayson, Liberty University
Jennifer L. Grimsley, University of Charleston
Ken Gurganus, University of North Carolina
James E. Hall, University of Wisconsin-Madison
Judy Hall, West Virginia University
Edward R. Hancock, DeVry Institute of Technology
Julia Hassett, DeVry Institute, Dupage
Christopher Hay-Jahans, University of South Dakota
Michah Heibel, Lincoln Public Schools
LaRae Helliwell, San Jose City College
Celeste Hernandez, Richland College
Gloria P. Hernandez, Louisiana State University at Eunice
Brother Herron, Brother Rice High School
Robert Hoburg, Western Connecticut State University
Lynda Hollingsworth, Northwest Missouri State University
Deltrye Holt, Augusta State University
Charla Holzbog, Denison High School
Lee Hruby, Naperville North High School
Miles Hubbard, St. Cloud State University
Kim Hughes, California State College-San Bernardino
Stanislav, Jabuka, University of Nevada, Reno
Ron Jamison, Brigham Young University
Richard A. Jensen, Manatee Community College
Glenn Johnson, Middlesex Community College
Sandra G. Johnson, St. Cloud State University
Tuesday Johnson, New Mexico State University
Susitha Karunaratne, Purdue University North Central
Moana H. Karsteter, Tallahassee Community College
Donna Katula, Joliet Junior College
Arthur Kaufman, College of Staten Island
Thomas Kearns, North Kentucky University
Jack Keating, Massasoit Community College
Shelia Kellenbarger, Lincoln Public Schools
Rachael Kenney, North Carolina State University
John B. Klassen, North Idaho College
Debra Kopcso, Louisiana State University
Lynne Kowski, Raritan Valley Community College
Yelena Kravchuk, University of Alabama at Birmingham
Ray S. Kuan, Skyline College
Keith Kuchar, Manatee Community College
Tor Kwembe, Chicago State University
Linda J. Kyle, Tarrant Country Jr. College
H.E. Lacey, Texas A & M University
Harriet Lamm, Coastal Bend College
James Lapp, Fort Lewis College
Matt Larson, Lincoln Public Schools
Christopher Lattin, Oakton Community College
Julia Ledet, Lousiana State University
Adele LeGere, Oakton Community College
Kevin Leith, University of Houston
JoAnn Lewin, Edison College
Jeff Lewis, Johnson County Community College
Heidi Lyne, Joliet Junior College
Janice C. Lyon, Tallahassee Community College
Jean McArthur, Joliet Junior College
Virginia McCarthy, Iowa State University
Karla McCavit, Albion College
Michael McClendon, University of Central Oklahoma
Tom McCollow, DeVry Institute of Technology
Marilyn McCollum, North Carolina State University
Jill McGowan, Howard University
Will McGowant, Howard University
Dave McGuire, Joliet Junior College
Angela McNulty, Joliet Junior College
Laurence Maher, North Texas State University
Jay A. Malmstrom, Oklahoma City Community College
Rebecca Mann, Apollo High School
Lynn Marecek, Santa Ana College
Sherry Martina, Naperville North High School
Alec Matheson, Lamar University
Nancy Matthews, University of Oklahoma
James Maxwell, Oklahoma State University-Stillwater
Marsha May, Midwestern State University
James McLaughlin, West Chester University
Judy Meckley, Joliet Junior College
David Meel, Bowling Green State University
Carolyn Meitler, Concordia University
Samia Metwali, Erie Community College
Rich Meyers, Joliet Junior College
Matthew Michaelson, Glendale Community College
Eldon Miller, University of Mississippi
James Miller, West Virginia University
Michael Miller, Iowa State University
Kathleen Miranda, SUNY at Old Westbury
Chris Mirbaha, The Community College of Baltimore County
Val Mohanakumar, Hillsborough Community College
Thomas Monaghan, Naperville North High School
Miguel Montanez, Miami Dade College, Wolfson Campus
Maria Montoya, Our Lady of the Lake University
Susan Moosai, Florida Atlantic University
Craig Morse, Naperville North High School
Samad Mortabit, Metropolitan State University
Pat Mower, Washburn University
Tammy Muhs, University of Central Florida
A. Muhundan, Manatee Community College
Jane Murphy, Middlesex Community College
Richard Nadel, Florida International University
Gabriel Nagy, Kansas State University
Bill Naegele, South Suburban College
Karla Neal, Lousiana State University
Lawrence E. Newman, Holyoke Community College
Dwight Newsome, Pasco-Hernando Community College
Victoria Noddings, MiraCosta College
Denise Nunley, Maricopa Community Colleges
James Nymann, University of Texas-El Paso
Mark Omodt, Anoka-Ramsey Community College
Seth F. Oppenheimer, Mississippi State University
Leticia Oropesa, University of Miami
Linda Padilla, Joliet Junior College
Sanja Pantic, University of Illinois at Chicago
E. James Peake, Iowa State University
Kelly Pearson, Murray State University
Dashamir Petrela, Florida Atlantic University
Philip Pina, Florida Atlantic University
Charlotte Pisors, Baylor University
Michael Prophet, University of Northern Iowa
Laura Pyzdrowski, West Virginia University
Carrie Quesnell, Weber State University
Neal C. Raber, University of Akron
Thomas Radin, San Joaquin Delta College
Aibeng Serene Radulovic, Florida Atlantic University
Ken A. Rager, Metropolitan State College
Kenneth D. Reeves, San Antonio College
Elsi Reinhardt, Truckee Meadows Community College
Jose Remesar, Miami Dade College, Wolfson Campus
Jane Ringwald, Iowa State University
Douglas F. Robertson, University of Minnesota, MPLS
Stephen Rodi, Austin Community College
William Rogge, Lincoln Northeast High School
Howard L. Rolf, Baylor University
Mike Rosenthal, Florida International University
Phoebe Rouse, Lousiana State University
Edward Rozema, University of Tennessee at Chattanooga
David Ruffato, Joliet Junior College
Dennis C. Runde, Manatee Community College
Alan Saleski, Loyola University of Chicago
Susan Sandmeyer, Jamestown Community College
Brenda Santistevan, Salt Lake Community College
Linda Schmidt, Greenville Technical College
Ingrid Scott, Montgomery College

A.K. Shamma, University of West Florida
Zachery Sharon, University of Texas at San Antonio
Martin Sherry, Lower Columbia College
Carmen Shershin, Florida International University
Tatrana Shubin, San Jose State University
Anita Sikes, Delgado Community College
Timothy Sipka, Alma College
Charlotte Smedberg, University of Tampa
Lori Smellegar, Manatee Community College
Gayle Smith, Loyola Blakefield
Cindy Soderstrom, Salt Lake Community College
Leslie Soltis, Mercyhurst College
John Spellman, Southwest Texas State University
Karen Spike, University of North Carolina
Rajalakshmi Sriram, Okaloosa-Walton Community College
Katrina Staley, North Carolina Agricultural and Technical State University
Becky Stamper, Western Kentucky University
Judy Staver, Florida Community College-South
Robin Steinberg, Pima Community College
Neil Stephens, Hinsdale South High School
Sonya Stephens, Florida A&M Univeristy
Patrick Stevens, Joliet Junior College
Mary Stinnett, Umpqua Community College
John Sumner, University of Tampa
Matthew TenHuisen, University of North Carolina, Wilmington
Christopher Terry, Augusta State University
Diane Tesar, South Suburban College
Theresa Thompson, Tulsa Community College
Tommy Thompson, Brookhaven College
Martha K. Tietze, Shawnee Mission Northwest High School
Richard J. Tondra, Iowa State University
Florentina Tone, University of West Florida
Suzanne Topp, Salt Lake Community College
Marilyn Toscano, University of Wisconsin, Superior
Marvel Townsend, University of Florida
Jim Trudnowski, Carroll College
Robert Tuskey, Joliet Junior College
Mihaela Vajiac, Chapman University-Orange
Julia Varbalow, Thomas Nelson Community College-Leesville
Richard G. Vinson, University of South Alabama
Jorge Viola-Prioli, Florida Atlantic University
Mary Voxman, University of Idaho
Jennifer Walsh, Daytona Beach Community College
Donna Wandke, Naperville North High School
Timothy L. Warkentin, Cloud County Community College
Melissa J. Watts, Virginia State University
Hayat Weiss, Middlesex Community College
Kathryn Wetzel, Amarillo College
Darlene Whitkenack, Northern Illinois University
Suzanne Williams, Central Piedmont Community College
Larissa Williamson, University of Florida
Christine Wilson, West Virginia University
Brad Wind, Florida International University
Anna Wiodarczyk, Florida International University
Mary Wolyniak, Broome Community College
Canton Woods, Auburn University
Tamara S. Worner, Wayne State College
Terri Wright, New Hampshire Community Technical College, Manchester
Aletheia Zambesi, University of West Florida
George Zazi, Chicago State University
Steve Zuro, Joliet Junior College

Michael Sullivan
Chicago State University

Michael Sullivan III
Joliet Junior College

BREAKTHROUGH
To improving results

Resources for Success

MyMathLab® Online Course for the Enhanced with Graphing Utilities, Series, 7th ed., by Michael Sullivan and Michael Sullivan III

(access code required)

MyMathLab delivers proven results in helping individual students succeed. The author team, led by Michael Sullivan and Michael Sullivan III, has developed specific content in MyMathLab to ensure quality resources are available to help foster success in mathematics – and beyond! The MyMathLab features described here will help:

- Review math skills and forgotten concepts
- Retain new concepts while moving through the course
- Develop skills that will help with the transition to college

Supportive Exercise Sets

With *Getting Ready* content students refresh prerequisite topics through assignable skill review quizzes and personalized homework. *New video assessment questions* are tied to key Author in Action videos to check students' conceptual understanding of important math concepts. *Guided Visualizations* help students better understand the visual aspects of key concepts in figure format. The figures are included in MyMathLab as both a teaching and an assignable learning tool.

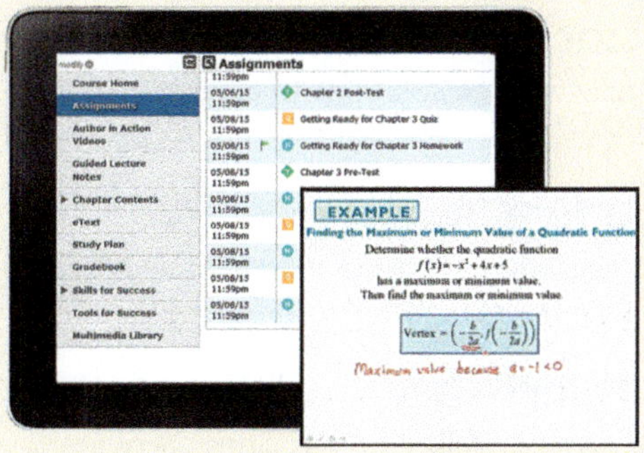

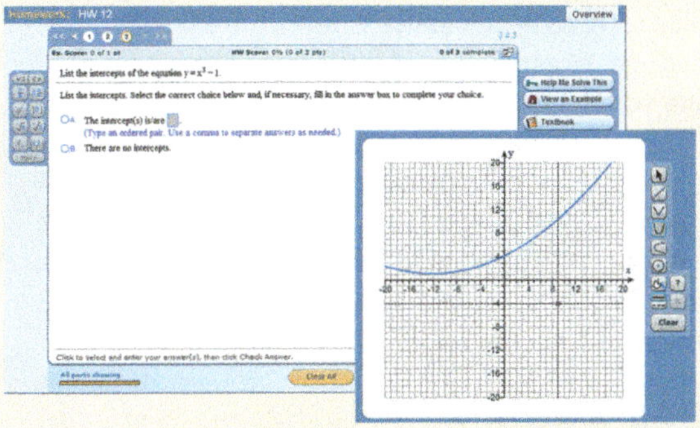

Encourage Retention

New *Retain Your Knowledge* quizzes promote ongoing review at the course level and help students maintain essential skills. New functionality within the graphing utility allows graphing of 3-point quadratic functions, 4-point cubic graphs, and transformations in exercises.

Boost Study Skills

Skills for Success Modules are integrated with MyMathLab courses to help students succeed in collegiate courses and prepare for future professions. Topics such as "Time Management," "Stress Management" and "Financial Literacy" are available for you to assign to your students.

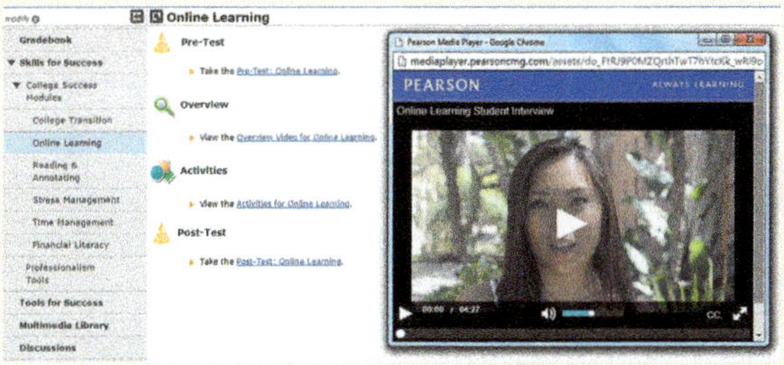

www.mymathlab.com

BREAKTHROUGH
To improving results

Resources for Success

Instructor Resources

Additional resources can be downloaded from **www.mymathlab.com** or **www.pearsonhighered.com** or hardcopy resources can be ordered from your sales representative.

Annotated Instructor Edition

Includes all answers to the exercises sets. Shorter answers are on the page beside the exercises, and longer answers are in the back of the text. Sample homework assignments are indicated by a blue underline within each end-of-section exercise set and may be assigned in MyMathLab.

Instructor's Solutions Manual

Includes fully worked solutions to all exercises in the text.

Mini Lecture Notes

This guide includes additional examples and helpful teaching tips, by section.

PowerPoint® Lecture Slides

These files contain fully editable slides correlated with the text.

Test Gen®

Test Gen® (**www.pearsoned.com/testgen**) enables instructor to build, edit, print, and administer tests using a computerized bank of question developed to cover all the objectives of the text.

Online Chapter Projects

Additional projects that give students an opportunity to apply what they learned in the chapter.

Student Resources

Additional resources to promote student success:

Lecture Videos

Author in Action videos are actual classroom lectures with fully worked-out examples presented by Michael Sullivan III. All video is assignable in MyMathLab.

Chapter Test Prep Videos

Students can watch instructors work through step-by-step solutions to all chapter test exercises from the text. These are available in MyMathLab and on YouTube.

Student Solutions Manual

Provides detailed worked-out solutions to odd-numbered exercises.

Guided Lecture Notes

These lecture notes assist students in taking thorough, organized, and understandable notes while watching Author in Action videos. Students actively participate in learning the how/why of important concepts through explorations and activities. The Guided Lecture Notes are available as pdfs and customizable Word files in MyMathLab. They can also be packaged with the text and MyMathLab access code.

Algebra Review

Four Chapters of Intermediate Algebra review. Perfect for a slower-paced course or for individual review.

Applications Index

To the Student

As you begin, you may feel anxious about the number of theorems, definitions, procedures, and equations you encounter. You may wonder if you can learn it all in time. Don't worry, your concerns are normal. This text was written with you in mind. If you attend class, work hard, and read and study effectively, you will build the knowledge and skills you need to be successful. Here's how you can use the text to your benefit.

Read Carefully

When you get busy, it's easy to skip reading and go right to the problems. Don't! The text provides a large number of examples and clear explanations to help you break down the mathematics into easy-to-understand steps. Reading will provide you with a clearer understanding, beyond simple memorization. Read before class (not after) so you can ask questions about anything you didn't understand. You'll be amazed at how much more you'll get out of class when you do this.

Use the Features

We use many different methods in the classroom to communicate. Those methods, when incorporated into the text, are called "features." The features serve many purposes, from supplying a timely review of material you learned before (just when you need it), to providing organized review sessions to help you prepare for quizzes and tests. Take advantage of the features and you will master the material.

To make this easier, we've provided a brief guide to getting the most from this book. Refer to the "Prepare for Class," "Practice," and "Review" guidelines on pages i–iii. Spend fifteen minutes reviewing the guide and familiarizing yourself with the features by flipping to the page numbers provided. Then, as you read, use them. This is the best way to make the most of your text.

Please do not hesitate to contact us, through Pearson Education, with any questions, comments, or suggestions about ways to improve this text. We look forward to hearing from you, and good luck with all of your studies.

Best Wishes!

Michael Sullivan
Michael Sullivan III

R Review

A Look Ahead •••

Chapter R, as the title states, contains review material. Your instructor may choose to cover all or part of it as a regular chapter at the beginning of your course or later as a just-in-time review when the content is required. Regardless, when information in this chapter is needed, a specific reference to this chapter will be made so you can review.

Outline

R.1 Real Numbers

PREPARING FOR THIS TEXT *Before getting started, read "To the Student" at the front of this text.*

OBJECTIVES
1. Work with Sets (p. 2)
2. Classify Numbers (p. 4)
3. Evaluate Numerical Expressions (p. 8)
4. Work with Properties of Real Numbers (p. 10)

1 Work with Sets

A **set** is a well-defined collection of distinct objects. The objects of a set are called its **elements**. By **well-defined**, we mean that there is a rule that enables us to determine whether a given object is an element of the set. If a set has no elements, it is called the **empty set**, or **null set**, and is denoted by the symbol $\varnothing$.

For example, the set of **digits** consists of the collection of numbers 0, 1, 2, 3, 4, 5, 6, 7, 8, and 9. If we use the symbol D to denote the set of digits, then we can write

$$D = \{0, 1, 2, 3, 4, 5, 6, 7, 8, 9\}$$

In this notation, the braces { } are used to enclose the objects, or **elements**, in the set. This method of denoting a set is called the **roster method**. A second way to denote a set is to use **set-builder notation**, where the set D of digits is written as

$$D = \{\; x \;|\; x \text{ is a digit}\}$$

Read as "*D* is the set of all *x* such that *x* is a digit."

EXAMPLE 1 Using Set-builder Notation and the Roster Method

(a) $E = \{x|x \text{ is an even digit}\} = \{0, 2, 4, 6, 8\}$

(b) $O = \{x|x \text{ is an odd digit}\} = \{1, 3, 5, 7, 9\}$

Because the elements of a set are distinct, we never repeat elements. For example, we would never write $\{1, 2, 3, 2\}$; the correct listing is $\{1, 2, 3\}$. Because a set is a collection, the order in which the elements are listed is immaterial. $\{1, 2, 3\}$, $\{1, 3, 2\}$, $\{2, 1, 3\}$, and so on, all represent the same set.

If every element of a set A is also an element of a set B, then A is a **subset** of B, which is denoted $A \subseteq B$. If two sets A and B have the same elements, then A **equals** B, which is denoted $A = B$.

For example, $\{1, 2, 3\} \subseteq \{1, 2, 3, 4, 5\}$ and $\{1, 2, 3\} = \{2, 3, 1\}$.

DEFINITION

If A and B are sets, the **intersection** of A with B, denoted $A \cap B$, is the set consisting of elements that belong to both A and B. The **union** of A with B, denoted $A \cup B$, is the set consisting of elements that belong to either A or B, or both.

EXAMPLE 2 Finding the Intersection and Union of Sets

Let $A = \{1, 3, 5, 8\}$, $B = \{3, 5, 7\}$, and $C = \{2, 4, 6, 8\}$. Find:

(a) $A \cap B$ (b) $A \cup B$ (c) $B \cap (A \cup C)$

Solution

(a) $A \cap B = \{1, 3, 5, 8\} \cap \{3, 5, 7\} = \{3, 5\}$

(b) $A \cup B = \{1, 3, 5, 8\} \cup \{3, 5, 7\} = \{1, 3, 5, 7, 8\}$

(c) $B \cap (A \cup C) = \{3, 5, 7\} \cap [\{1, 3, 5, 8\} \cup \{2, 4, 6, 8\}]$
$= \{3, 5, 7\} \cap \{1, 2, 3, 4, 5, 6, 8\} = \{3, 5\}$

Now Work PROBLEM 15

Usually, in working with sets, we designate a **universal set** U, the set consisting of all the elements that we wish to consider. Once a universal set has been designated, we can consider elements of the universal set not found in a given set.

DEFINITION

If A is a set, the **complement** of A, denoted $\overline{A}$, is the set consisting of all the elements in the universal set that are not in A.*

EXAMPLE 3 Finding the Complement of a Set

If the universal set is $U = \{1, 2, 3, 4, 5, 6, 7, 8, 9\}$ and if $A = \{1, 3, 5, 7, 9\}$, then $\overline{A} = \{2, 4, 6, 8\}$.

It follows from the definition of complement that $A \cup \overline{A} = U$ and $A \cap \overline{A} = \varnothing$. Do you see why?

Now Work PROBLEM 19

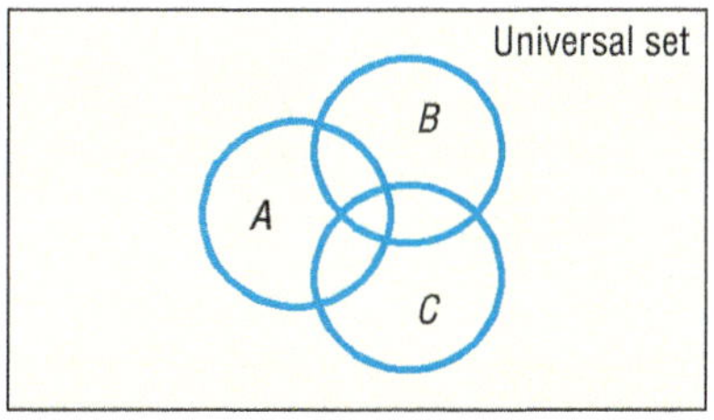

Figure 1 Venn diagram

It is often helpful to draw pictures of sets. Such pictures, called **Venn diagrams**, represent sets as circles enclosed in a rectangle, which represents the universal set. Such diagrams often help us to visualize various relationships among sets. See Figure 1.

If we know that $A \subseteq B$, we might use the Venn diagram in Figure 2(a). If we know that A and B have no elements in common—that is, if $A \cap B = \varnothing$—we might use the Venn diagram in Figure 2(b). The sets A and B in Figure 2(b) are said to be **disjoint**.

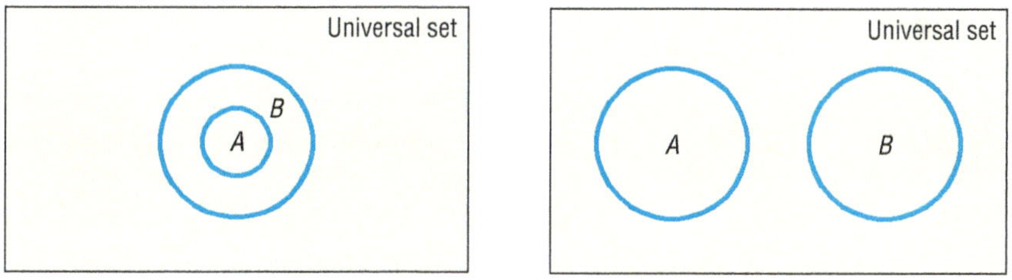

Figure 2 **(a)** $A \subseteq B$ subset **(b)** $A \cap B = \varnothing$ disjoint sets

Figures 3(a), 3(b), and 3(c) use Venn diagrams to illustrate the definitions of intersection, union, and complement, respectively.

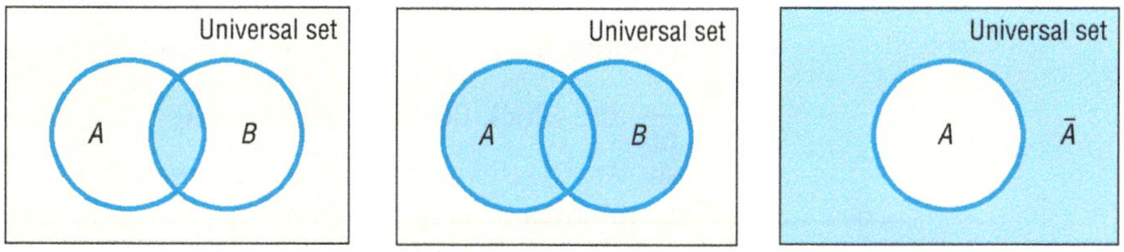

Figure 3 **(a)** $A \cap B$ intersection **(b)** $A \cup B$ union **(c)** $\overline{A}$ complement

*Some texts use the notation A' for the complement of A.

2 Classify Numbers

It is helpful to classify the various kinds of numbers that we deal with as sets. The **counting numbers**, or **natural numbers**, are the numbers in the set $\{1, 2, 3, 4, \ldots\}$. (The three dots, called an **ellipsis**, indicate that the pattern continues indefinitely.) As their name implies, these numbers are often used to count things. For example, there are 26 letters in our alphabet; there are 100 cents in a dollar. The **whole numbers** are the numbers in the set $\{0, 1, 2, 3, \ldots\}$—that is, the counting numbers together with 0. The set of counting numbers is a subset of the set of whole numbers.

DEFINITION

The **integers** are the set of numbers $\{\ldots, -3, -2, -1, 0, 1, 2, 3, \ldots\}$.

These numbers are useful in many situations. For example, if your checking account has \$10 in it and you write a check for \$15, you can represent the current balance as −\$5.

Each time we expand a number system, such as from the whole numbers to the integers, we do so in order to be able to handle new, and usually more complicated, problems. The integers enable us to solve problems requiring both positive and negative counting numbers, such as profit/loss, height above/below sea level, temperature above/below 0°F, and so on.

But integers alone are not sufficient for *all* problems. For example, they do not answer the question "What part of a dollar is 38 cents?" To answer such a question, we enlarge our number system to include *rational numbers*. For example, $\frac{38}{100}$ answers the question "What part of a dollar is 38 cents?"

DEFINITION

A **rational number** is a number that can be expressed as a quotient $\frac{a}{b}$ of two integers. The integer a is called the **numerator**, and the integer b, which cannot be 0, is called the **denominator**. The rational numbers are the numbers in the set $\left\{x \middle| x = \frac{a}{b}, \text{ where } a, b \text{ are integers and } b \neq 0\right\}$.

Examples of rational numbers are $\frac{3}{4}, \frac{5}{2}, \frac{0}{4}, -\frac{2}{3}$, and $\frac{100}{3}$. Since $\frac{a}{1} = a$ for any integer a, it follows that the set of integers is a subset of the set of rational numbers.

Rational numbers may be represented as **decimals**. For example, the rational numbers $\frac{3}{4}, \frac{5}{2}, -\frac{2}{3}$, and $\frac{7}{66}$ may be represented as decimals by merely carrying out the indicated division:

$$\frac{3}{4} = 0.75 \qquad \frac{5}{2} = 2.5 \qquad -\frac{2}{3} = -0.666\ldots = -0.\overline{6} \qquad \frac{7}{66} = 0.1060606\ldots = 0.1\overline{06}$$

Notice that the decimal representations of $\frac{3}{4}$ and $\frac{5}{2}$ terminate, or end. The decimal representations of $-\frac{2}{3}$ and $\frac{7}{66}$ do not terminate, but they do exhibit a pattern of repetition. For $-\frac{2}{3}$, the 6 repeats indefinitely, as indicated by the bar over the 6; for $\frac{7}{66}$, the block 06 repeats indefinitely, as indicated by the bar over the 06. It can be shown that every rational number may be represented by a decimal that either terminates or is nonterminating with a repeating block of digits, and vice versa.

On the other hand, some decimals do not fit into either of these categories. Such decimals represent **irrational numbers**. Every irrational number may be represented by a decimal that neither repeats nor terminates. In other words, irrational numbers cannot be written in the form $\frac{a}{b}$, where a, b are integers and $b \neq 0$.

Irrational numbers occur naturally. For example, consider the isosceles right triangle whose legs are each of length 1. See Figure 4. The length of the hypotenuse is $\sqrt{2}$, an irrational number.

Also, the number that equals the ratio of the circumference C to the diameter d of any circle, denoted by the symbol π (the Greek letter pi), is an irrational number. See Figure 5.

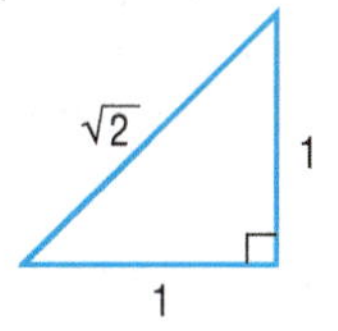

Figure 4

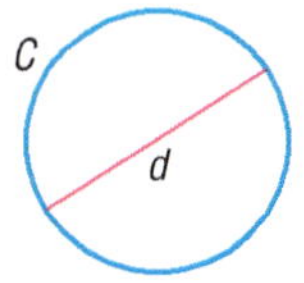

Figure 5 $\pi = \dfrac{C}{d}$

DEFINITION The set of **real numbers** is the union of the set of rational numbers with the set of irrational numbers.

Figure 6 shows the relationship of various types of numbers.*

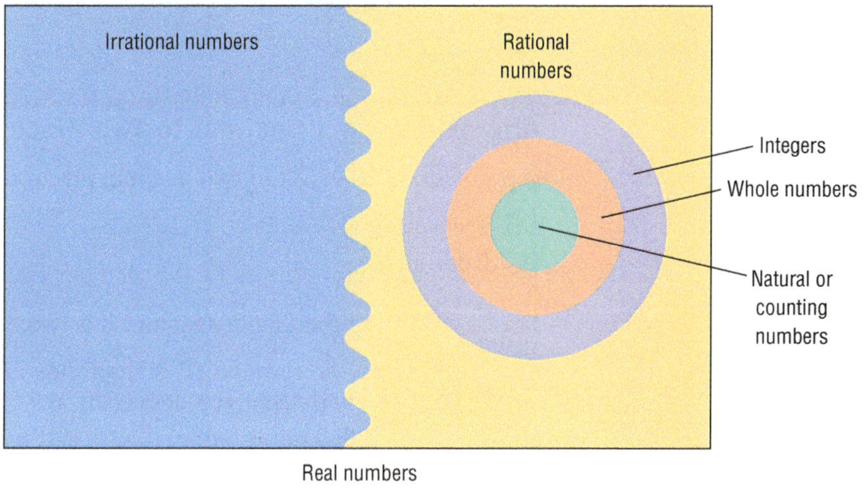

Figure 6

EXAMPLE 4 Classifying the Numbers in a Set

List the numbers in the set

$$\left\{-3, \frac{4}{3}, 0.12, \sqrt{2}, \pi, 10, 2.151515\ldots \text{(where the block 15 repeats)}\right\}$$

that are

(a) Natural numbers (b) Integers (c) Rational numbers
(d) Irrational numbers (e) Real numbers

Solution
(a) 10 is the only natural number.
(b) -3 and 10 are integers.
(c) $-3, 10, \frac{4}{3}, 0.12$, and $2.151515\ldots$ are rational numbers.
(d) $\sqrt{2}$ and π are irrational numbers.
(e) All the numbers listed are real numbers. ■

Now Work PROBLEM 25

*The set of real numbers is a subset of the set of complex numbers. We discuss complex numbers in Chapter 1, Section 1.4.

Approximations

Every decimal may be represented by a real number (either rational or irrational), and every real number may be represented by a decimal.

In practice, the decimal representation of an irrational number is given as an approximation. For example, using the symbol $\approx$ (read as "approximately equal to"), we can write

$$\sqrt{2} \approx 1.4142 \qquad \pi \approx 3.1416$$

In approximating decimals, we either *round off* or *truncate* to a given number of decimal places.* The number of places establishes the location of the *final digit* in the decimal approximation.

Truncation: Drop all of the digits that follow the specified final digit in the decimal.

Rounding: Identify the specified final digit in the decimal. If the next digit is 5 or more, add 1 to the final digit; if the next digit is 4 or less, leave the final digit as it is. Then truncate following the final digit.

EXAMPLE 5

Approximating a Decimal to Two Places

Approximate 20.98752 to two decimal places by

(a) Truncating
(b) Rounding

Solution For 20.98752, the final digit is 8, since it is two decimal places from the decimal point.

(a) To truncate, we remove all digits following the final digit 8. The truncation of 20.98752 to two decimal places is 20.98.
(b) The digit following the final digit 8 is the digit 7. Since 7 is 5 or more, we add 1 to the final digit 8 and truncate. The rounded form of 20.98752 to two decimal places is 20.99. ■

EXAMPLE 6

Approximating a Decimal to Two and Four Places

Number	Rounded to Two Decimal Places	Rounded to Four Decimal Places	Truncated to Two Decimal Places	Truncated to Four Decimal Places
(a) 3.14159	3.14	3.1416	3.14	3.1415
(b) 0.056128	0.06	0.0561	0.05	0.0561
(c) 893.46125	893.46	893.4613	893.46	893.4612

■

Now Work PROBLEM 29

Significant Digits

There are two types of numbers—*exact* and *approximate*. **Exact numbers** are numbers whose value is known with 100% certainty and accuracy. For example, there are 12 donuts in a dozen donuts, or there are 50 states in the United States.

*Sometimes we say "correct to a given number of decimal places" instead of "truncated."

Approximate numbers are numbers whose value is not known with 100% certainty or whose measurement is inexact. When values are determined from measurements they are typically approximate numbers because the exact measurement is limited by the accuracy of the measuring device and the skill of the individual obtaining the measurement. The **number of significant digits** in a number represents the level of accuracy of the measurement.

The following rules are used to determine the number of significant digits in approximate numbers.

The Number of Significant Digits

- Leading zeros are not significant. For example, 0.0034 has two significant digits.
- Embedded zeros are significant. For example, 208 has three significant digits.
- Trailing zeros are significant only if the decimal point is specified. For example, 2800 has two significant digits. However, if we specify the measurement is accurate to the ones digit, then 2800 has four significant digits.

When performing computations with approximate numbers, it is important not to report the result with more accuracy than the measurements used in the computation.

When performing computations using significant digits, proceed with the computation as you normally would, then round the final answer to the number of significant digits as the least accurately known number. For example, suppose we want to find the area of a rectangle whose width is 1.94 inches (three significant digits) and whose length is 2.7 inches (two significant digits). Because the length has two significant digits, we report the area to two significant digits. The area, (1.94 inches) (2.7 inches) $= 5.238$ square inches, can only be written to two significant digits and is reported as 5.2 square inches.

Calculators

Calculators are incapable of displaying decimals that contain a large number of digits. For example, some calculators are capable of displaying only eight digits. When a number requires more than eight digits, the calculator either truncates or rounds. To see how your calculator handles decimals, divide 2 by 3. How many digits do you see? Is the last digit a 6 or a 7? If it is a 6, your calculator truncates; if it is a 7, your calculator rounds.

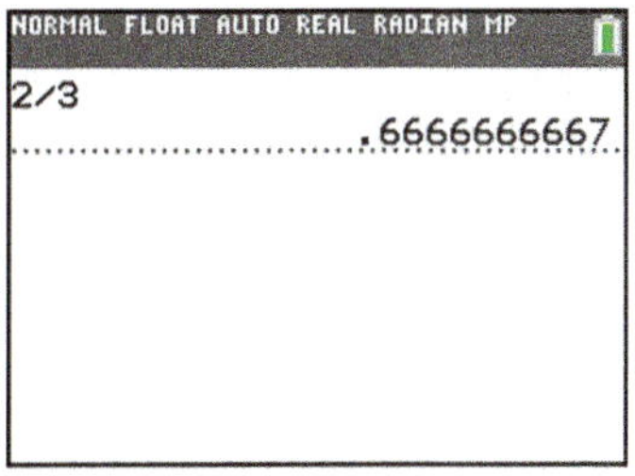

Figure 7

There are different kinds of calculators. An **arithmetic** calculator can only add, subtract, multiply, and divide numbers; therefore, this type is not adequate for this course. **Scientific** calculators have all the capabilities of arithmetic calculators and contain **function keys** labeled ln, log, sin, cos, tan, x^y, inv, and so on. **Graphing** calculators have all the capabilities of scientific calculators and contain a screen on which graphs can be displayed. As you proceed through this text, you will discover how to use many of the function keys.

Figure 7 shows $\frac{2}{3}$ on a TI-84 Plus C graphing calculator. How many digits are displayed? Does a TI-84 Plus C round or truncate? What does your calculator do?

Operations

In algebra, we use letters such as x, y, a, b, and c to represent numbers. The symbols used in algebra for the operations of addition, subtraction, multiplication, and division are $+$, $-$, $\cdot$, and $/$. The words used to describe the results of these operations are **sum**, **difference, product**, and **quotient**. Table 1 summarizes these ideas.

Table 1

Operation	Symbol	Words
Addition	$a + b$	Sum: a plus b
Subtraction	$a - b$	Difference: a minus b
Multiplication	$a \cdot b$, $(a) \cdot b$, $a \cdot (b)$, $(a) \cdot (b)$, ab, $(a)b$, $a(b)$, $(a)(b)$	Product: a times b
Division	a/b or $\frac{a}{b}$	Quotient: a divided by b

In algebra, we generally avoid using the multiplication sign $\times$ and the division sign $\div$ so familiar in arithmetic. Notice also that when two expressions are placed next to each other without an operation symbol, as in ab, or in parentheses, as in $(a)(b)$, it is understood that the expressions, called **factors**, are to be multiplied.

We also prefer not to use mixed numbers in algebra. When mixed numbers are used, addition is understood; for example, $2\frac{3}{4}$ means $2 + \frac{3}{4}$. In algebra, use of a mixed number may be confusing because the absence of an operation symbol between two terms is generally taken to mean multiplication. The expression $2\frac{3}{4}$ is therefore written instead as 2.75 or as $\frac{11}{4}$.

The symbol $=$, called an **equal sign** and read as "equals" or "is," is used to express the idea that the number or expression on the left of the equal sign is equivalent to the number or expression on the right.

EXAMPLE 7 **Writing Statements Using Symbols**

(a) The sum of 2 and 7 equals 9. In symbols, this statement is written as $2 + 7 = 9$.

(b) The product of 3 and 5 is 15. In symbols, this statement is written as $3 \cdot 5 = 15$.

Now Work PROBLEM 41

3 Evaluate Numerical Expressions

Consider the expression $2 + 3 \cdot 6$. It is not clear whether we should add 2 and 3 to get 5, and then multiply by 6 to get 30; or first multiply 3 and 6 to get 18, and then add 2 to get 20. To avoid this ambiguity, we have the following agreement.

In Words
Multiply first, then add.

We agree that whenever the two operations of addition and multiplication separate three numbers, the multiplication operation will always be performed first, followed by the addition operation.

For $2 + 3 \cdot 6$, then, we have

$$2 + 3 \cdot 6 = 2 + 18 = 20$$

EXAMPLE 8 **Finding the Value of an Expression**

Evaluate each expression.

(a) $3 + 4 \cdot 5$ (b) $8 \cdot 2 + 1$ (c) $2 + 2 \cdot 2$

Solution

(a) $3 + 4 \cdot 5 = 3 + 20 = 23$
↑ Multiply first.

(b) $8 \cdot 2 + 1 = 16 + 1 = 17$
↑ Multiply first.

(c) $2 + 2 \cdot 2 = 2 + 4 = 6$ ■

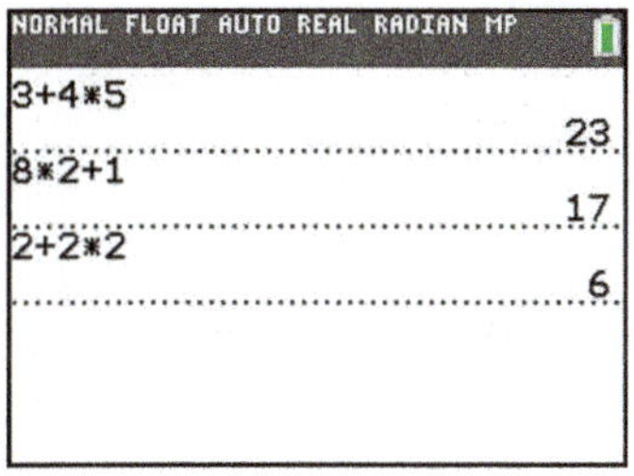

Figure 8

Figure 8 shows the solution to Example 8 using a TI-84 Plus C graphing calculator. Notice that the calculator follows the agreed order of operations.

Now Work PROBLEM 53

When we want to indicate adding 3 and 4 and then multiplying the result by 5, we use parentheses and write $(3 + 4) \cdot 5$. Whenever parentheses appear in an expression, it means "perform the operations within the parentheses first!"

EXAMPLE 9 Finding the Value of an Expression

(a) $(5 + 3) \cdot 4 = 8 \cdot 4 = 32$

(b) $(4 + 5) \cdot (8 - 2) = 9 \cdot 6 = 54$ ■

When we divide two expressions, as in

$$\frac{2+3}{4+8}$$

it is understood that the division bar acts like parentheses; that is,

$$\frac{2+3}{4+8} = \frac{(2+3)}{(4+8)}$$

Rules for the Order of Operations

1. Begin with the innermost parentheses and work outward. Remember that in dividing two expressions, we treat the numerator and denominator as if they were enclosed in parentheses.
2. Perform multiplications and divisions, working from left to right.
3. Perform additions and subtractions, working from left to right.

EXAMPLE 10 Finding the Value of an Expression

Evaluate each expression.

(a) $8 \cdot 2 + 3$

(b) $5 \cdot (3 + 4) + 2$

(c) $\dfrac{2+5}{2+4 \cdot 7}$

(d) $2 + [4 + 2 \cdot (10 + 6)]$

Solution

(a) $8 \cdot 2 + 3 = 16 + 3 = 19$
↑ Multiply first.

(b) $5 \cdot (3 + 4) + 2 = 5 \cdot 7 + 2 = 35 + 2 = 37$
↑ Parentheses first ↑ Multiply before adding.

(c) $\dfrac{2+5}{2+4\cdot 7} = \dfrac{2+5}{2+28} = \dfrac{7}{30}$

(d) $2 + [4 + 2\cdot(10+6)] = 2 + [4 + 2\cdot(16)]$

$= 2 + [4 + 32] = 2 + [36] = 38$ ■

Be careful if you use a calculator. For Example 10(c), you need to use parentheses. See Figure 9(a).* If you don't, the calculator will compute the expression

$$2 + \frac{5}{2} + 4\cdot 7 = 2 + 2.5 + 28 = 32.5$$

giving a wrong answer.

Another option, when using a TI-84 Plus C graphing calculator, is to use the fraction template. See Figure 9(b).

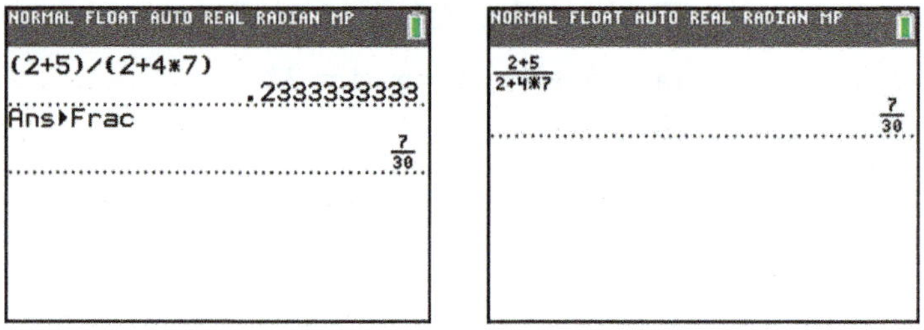

Figure 9 (a) (b)

Now Work PROBLEMS 59 AND 67

4 Work with Properties of Real Numbers

The equal sign is used to mean that one expression is equivalent to another. Four important properties of equality are listed next. In this list, a, b, and c represent real numbers.

1. The **reflexive property** states that a number equals itself; that is, $a = a$.
2. The **symmetric property** states that if $a = b$, then $b = a$.
3. The **transitive property** states that if $a = b$ and $b = c$, then $a = c$.
4. The **principle of substitution** states that if $a = b$, then we may substitute b for a in any expression containing a.

Now, let's consider some other properties of real numbers.

EXAMPLE 11 **Commutative Properties**

(a) $3 + 5 = 8$
$5 + 3 = 8$
$3 + 5 = 5 + 3$

(b) $2\cdot 3 = 6$
$3\cdot 2 = 6$
$2\cdot 3 = 3\cdot 2$ ■

This example illustrates the **commutative property** of real numbers, which states that the order in which addition or multiplication takes place does not affect the final result.

*Notice that we converted the decimal into its fraction form in Figure 9(a). Consult your manual to see how to enter such expressions on your calculator.

Commutative Properties

$$a + b = b + a \quad \textbf{(1a)}$$

$$a \cdot b = b \cdot a \quad \textbf{(1b)}$$

Here, and in the properties listed next and on pages 12–14, a, b, and c represent real numbers.

EXAMPLE 12 **Associative Properties**

(a) $2 + (3 + 4) = 2 + 7 = 9$
$(2 + 3) + 4 = 5 + 4 = 9$
$2 + (3 + 4) = (2 + 3) + 4$

(b) $2 \cdot (3 \cdot 4) = 2 \cdot 12 = 24$
$(2 \cdot 3) \cdot 4 = 6 \cdot 4 = 24$
$2 \cdot (3 \cdot 4) = (2 \cdot 3) \cdot 4$ ■

The way we add or multiply three real numbers does not affect the final result. Expressions such as $2 + 3 + 4$ and $3 \cdot 4 \cdot 5$ present no ambiguity, even though addition and multiplication are performed on one pair of numbers at a time. This property is called the **associative property**.

Associative Properties

$$a + (b + c) = (a + b) + c = a + b + c \quad \textbf{(2a)}$$

$$a \cdot (b \cdot c) = (a \cdot b) \cdot c = a \cdot b \cdot c \quad \textbf{(2b)}$$

Distributive Property

$$a \cdot (b + c) = a \cdot b + a \cdot c \quad \textbf{(3a)}$$

$$(a + b) \cdot c = a \cdot c + b \cdot c \quad \textbf{(3b)}$$

The **distributive property** may be used in two different ways.

EXAMPLE 13 **Distributive Property**

(a) $2 \cdot (x + 3) = 2 \cdot x + 2 \cdot 3 = 2x + 6$ Use to remove parentheses.

(b) $3x + 5x = (3 + 5)x = 8x$ Use to combine two expressions.

(c) $(x + 2)(x + 3) = x(x + 3) + 2(x + 3) = (x^2 + 3x) + (2x + 6)$
$= x^2 + (3x + 2x) + 6 = x^2 + 5x + 6$ ■

Now Work PROBLEM 89

The real numbers 0 and 1 have unique properties called the *identity properties*.

EXAMPLE 14 **Identity Properties**

(a) $4 + 0 = 0 + 4 = 4$ (b) $3 \cdot 1 = 1 \cdot 3 = 3$ ■

Identity Properties

$$0 + a = a + 0 = a \tag{4a}$$

$$a \cdot 1 = 1 \cdot a = a \tag{4b}$$

We call 0 the **additive identity** and 1 the **multiplicative identity**.

For each real number a, there is a real number $-a$, called the **additive inverse** of a, having the following property:

Additive Inverse Property

$$a + (-a) = -a + a = 0 \tag{5a}$$

EXAMPLE 15 **Finding an Additive Inverse**

(a) The additive inverse of 6 is -6, because $6 + (-6) = 0$.

(b) The additive inverse of -8 is $-(-8) = 8$, because $-8 + 8 = 0$.

The additive inverse of a, that is, $-a$, is often called the *negative* of a or the *opposite* of a. The use of such terms can be dangerous, because they suggest that the additive inverse is a negative number, which may not be the case. For example, the additive inverse of -3, or $-(-3)$, equals 3, a positive number.

For each *nonzero* real number a, there is a real number $\frac{1}{a}$, called the **multiplicative inverse** of a, having the following property:

Multiplicative Inverse Property

$$a \cdot \frac{1}{a} = \frac{1}{a} \cdot a = 1 \quad \text{if } a \neq 0 \tag{5b}$$

The multiplicative inverse $\frac{1}{a}$ of a nonzero real number a is also referred to as the **reciprocal** of a.

EXAMPLE 16 **Finding a Reciprocal**

(a) The reciprocal of 6 is $\frac{1}{6}$, because $6 \cdot \frac{1}{6} = 1$.

(b) The reciprocal of -3 is $\frac{1}{-3}$, because $-3 \cdot \frac{1}{-3} = 1$.

(c) The reciprocal of $\frac{2}{3}$ is $\frac{3}{2}$, because $\frac{2}{3} \cdot \frac{3}{2} = 1$.

With these properties for adding and multiplying real numbers, we can define the operations of subtraction and division as follows:

DEFINITION

The **difference** $a - b$, also read "a less b" or "a minus b," is defined as

$$a - b = a + (-b) \tag{6}$$

To subtract b from a, add the opposite of b to a.

DEFINITION

If b is a nonzero real number, the **quotient** $\frac{a}{b}$, also read as "a divided by b" or "the ratio of a to b," is defined as

$$\frac{a}{b} = a \cdot \frac{1}{b} \quad \text{if } b \neq 0 \qquad \textbf{(7)}$$

EXAMPLE 17 **Working with Differences and Quotients**

(a) $8 - 5 = 8 + (-5) = 3$

(b) $4 - 9 = 4 + (-9) = -5$

(c) $\frac{5}{8} = 5 \cdot \frac{1}{8}$

For any number a, the product of a times 0 is always 0; that is,

In Words
The result of multiplying by zero is zero.

Multiplication by Zero

$$a \cdot 0 = 0 \qquad \textbf{(8)}$$

For a nonzero number a,

Division Properties

$$\frac{0}{a} = 0 \qquad \frac{a}{a} = 1 \quad \text{if } a \neq 0 \qquad \textbf{(9)}$$

Note: Division by 0 is not defined. One reason is to avoid the following difficulty: $\frac{2}{0} = x$ means to find x such that $0 \cdot x = 2$. But $0 \cdot x$ equals 0 for all x, so there is no unique number x such that $\frac{2}{0} = x$.

Rules of Signs

$$a(-b) = -(ab) \qquad (-a)b = -(ab) \qquad (-a)(-b) = ab$$

$$-(-a) = a \qquad \frac{a}{-b} = \frac{-a}{b} = -\frac{a}{b} \qquad \frac{-a}{-b} = \frac{a}{b} \qquad \textbf{(10)}$$

EXAMPLE 18 **Applying the Rules of Signs**

(a) $2(-3) = -(2 \cdot 3) = -6$

(b) $(-3)(-5) = 3 \cdot 5 = 15$

(c) $\frac{3}{-2} = \frac{-3}{2} = -\frac{3}{2}$

(d) $\frac{-4}{-9} = \frac{4}{9}$

(e) $\frac{x}{-2} = \frac{1}{-2} \cdot x = -\frac{1}{2}x$

Reduction Properties

$$ac = bc \text{ implies } a = b \quad \text{if } c \neq 0$$

$$\frac{ac}{bc} = \frac{a}{b} \quad \text{if } b \neq 0, c \neq 0 \qquad \textbf{(11)}$$

EXAMPLE 19 **Using the Reduction Properties**

(a) If $2x = 6$, then

$$\begin{aligned} 2x &= 6 \\ 2x &= 2 \cdot 3 && \text{Factor 6.} \\ x &= 3 && \text{Divide out the 2's.} \end{aligned}$$

(b) $\frac{18}{12} = \frac{3 \cdot \cancel{6}}{2 \cdot \cancel{6}} = \frac{3}{2}$

↑ Divide out the 6's. ■

Note: We follow the common practice of using slash marks to indicate factors dividing out. ■

In Words
If a product equals 0, then one or both of the factors is 0.

Zero-Product Property

If $ab = 0$, then $a = 0$, or $b = 0$, or both. **(12)**

EXAMPLE 20 **Using the Zero-Product Property**

If $2x = 0$, then either $2 = 0$ or $x = 0$. Since $2 \neq 0$, it follows that $x = 0$. ■

Arithmetic of Quotients

$$\frac{a}{b} + \frac{c}{d} = \frac{ad}{bd} + \frac{bc}{bd} = \frac{ad + bc}{bd} \quad \text{if } b \neq 0, d \neq 0 \qquad \textbf{(13)}$$

$$\frac{a}{b} \cdot \frac{c}{d} = \frac{ac}{bd} \quad \text{if } b \neq 0, d \neq 0 \qquad \textbf{(14)}$$

$$\frac{\frac{a}{b}}{\frac{c}{d}} = \frac{a}{b} \cdot \frac{d}{c} = \frac{ad}{bc} \quad \text{if } b \neq 0, c \neq 0, d \neq 0 \qquad \textbf{(15)}$$

EXAMPLE 21 **Adding, Subtracting, Multiplying, and Dividing Quotients**

(a) $\frac{2}{3} + \frac{5}{2} = \frac{2 \cdot 2}{3 \cdot 2} + \frac{3 \cdot 5}{3 \cdot 2} = \frac{2 \cdot 2 + 3 \cdot 5}{3 \cdot 2} = \frac{4 + 15}{6} = \frac{19}{6}$

↑ By equation (13)

(b) $\frac{3}{5} - \frac{2}{3} = \frac{3}{5} + \left(-\frac{2}{3}\right) = \frac{3}{5} + \frac{-2}{3}$

↑ By equation (6) ↑ By equation (10)

$= \frac{3 \cdot 3 + 5 \cdot (-2)}{5 \cdot 3} = \frac{9 + (-10)}{15} = \frac{-1}{15} = -\frac{1}{15}$

↑ By equation (13)

Note: Slanting the slash marks in different directions for different factors, as shown here, is a good practice to follow, since it will help in checking for errors.

(c) $\frac{8}{3}\cdot\frac{15}{4} = \frac{8\cdot 15}{3\cdot 4} = \frac{2\cdot \not{4}\cdot \not{3}\cdot 5}{\not{3}\cdot \not{4}\cdot 1} = \frac{2\cdot 5}{1} = 10$

By equation (14) By equation (11)

(d) $\dfrac{\frac{3}{5}}{\frac{7}{9}} = \frac{3}{5}\cdot\frac{9}{7} = \frac{3\cdot 9}{5\cdot 7} = \frac{27}{35}$

By equation (15) By equation (14)

Note: In writing quotients, we shall follow the usual convention and write the quotient in lowest terms. That is, we write it so that any common factors of the numerator and the denominator have been removed using the Reduction Properties, equation (11). As examples,

$$\frac{90}{24} = \frac{15\cdot \not{6}}{4\cdot \not{6}} = \frac{15}{4}$$

$$\frac{24x^2}{18x} = \frac{4\cdot \not{6}\cdot x\cdot \not{x}}{3\cdot \not{6}\cdot \not{x}} = \frac{4x}{3} \qquad x \neq 0$$

Now Work PROBLEMS 69, 73, AND 83

Sometimes it is easier to add two fractions using *least common multiples* (LCM). The LCM of two numbers is the smallest number that each has as a common multiple.

EXAMPLE 22 Finding the Least Common Multiple of Two Numbers

Find the least common multiple of 15 and 12.

Solution To find the LCM of 15 and 12, we look at multiples of 15 and 12.

15, 30, 45, 60, 75, 90, 105, 120,...

12, 24, 36, 48, 60, 72, 84, 96, 108, 120,...

The *common* multiples are in blue. The *least* common multiple is 60.

EXAMPLE 23 Using the Least Common Multiple to Add Two Fractions

Find: $\frac{8}{15} + \frac{5}{12}$

Solution We use the LCM of the denominators of the fractions and rewrite each fraction using the LCM as a common denominator. The LCM of the denominators (12 and 15) is 60. Rewrite each fraction using 60 as the denominator.

$$\frac{8}{15} + \frac{5}{12} = \frac{8}{15}\cdot\frac{4}{4} + \frac{5}{12}\cdot\frac{5}{5}$$
$$= \frac{32}{60} + \frac{25}{60}$$
$$= \frac{32 + 25}{60}$$
$$= \frac{57}{60}$$
$$= \frac{19}{20}$$

Now Work PROBLEM 77

Historical Feature

The real number system has a history that stretches back at least to the ancient Babylonians (1800 BC). It is remarkable how much the ancient Babylonian attitudes resemble our own. As we stated in the text, the fundamental difficulty with irrational numbers is that they cannot be written as quotients of integers or, equivalently, as repeating or terminating decimals. The Babylonians wrote their numbers in a system based on 60 in the same way that we write ours based on 10. They would carry as many places for π as the accuracy of the problem demanded, just as we now use

$$\pi \approx 3\frac{1}{7} \quad \text{or} \quad \pi \approx 3.1416 \quad \text{or} \quad \pi \approx 3.14159$$
$$\text{or} \quad \pi \approx 3.14159265358979$$

depending on how accurate we need to be.

Things were very different for the Greeks, whose number system allowed only rational numbers. When it was discovered that $\sqrt{2}$ was not a rational number, this was regarded as a fundamental flaw in the number concept. So serious was the matter that the Pythagorean Brotherhood (an early mathematical society) is said to have drowned one of its members for revealing this terrible secret. Greek mathematicians then turned away from the number concept, expressing facts about whole numbers in terms of line segments.

In astronomy, however, Babylonian methods, including the Babylonian number system, continued to be used. Simon Stevin (1548–1620), probably using the Babylonian system as a model, invented the decimal system, complete with rules of calculation, in 1585. [Others, for example, al-Kashi of Samarkand (d. 1429), had made some progress in the same direction.] The decimal system so effectively conceals the difficulties that the need for more logical precision began to be felt only in the early 1800s. Around 1880, Georg Cantor (1845–1918) and Richard Dedekind (1831–1916) gave precise definitions of real numbers. Cantor's definition, although more abstract and precise, has its roots in the decimal (and hence Babylonian) numerical system.

Sets and set theory were a spin-off of the research that went into clarifying the foundations of the real number system. Set theory has developed into a large discipline of its own, and many mathematicians regard it as the foundation upon which modern mathematics is built. Cantor's discoveries that infinite sets can also be counted and that there are different sizes of infinite sets are among the most astounding results of modern mathematics.

R.1 Assess Your Understanding

Concepts and Vocabulary

1. The numbers in the set $\left\{x \middle| x = \frac{a}{b}, \text{ where } a, b \text{ are integers and } b \neq 0\right\}$ are called ________ numbers.

2. The value of the expression $4 + 5 \cdot 6 - 3$ is ___.

3. The fact that $2x + 3x = (2 + 3)x$ is a consequence of the __________ Property.

4. "The product of 5 and $x + 3$ equals 6" may be written as __________.

5. The intersection of sets A and B is denoted by which of the following?

(a) $A \cap B$ (b) $A \cup B$ (c) $A \subseteq B$ (d) $A \varnothing B$

6. Choose the correct name for the set of numbers $\{0, 1, 2, 3, \ldots\}$.

(a) Counting numbers (b) Whole numbers
(c) Integers (d) Irrational numbers

7. ***True or False*** Rational numbers have decimals that either terminate or are nonterminating with a repeating block of digits.

8. ***True or False*** The Zero-Product Property states that the product of any number and zero equals zero.

9. ***True or False*** The least common multiple of 12 and 18 is 6.

10. ***True or False*** No real number is both rational and irrational.

Skill Building

In Problems 11–22, use U = universal set = $\{0, 1, 2, 3, 4, 5, 6, 7, 8, 9\}$, $A = \{1, 3, 4, 5, 9\}$, $B = \{2, 4, 6, 7, 8\}$, and $C = \{1, 3, 4, 6\}$ to find each set.

11. $A \cup B$ **12.** $A \cup C$ **13.** $A \cap B$ **14.** $A \cap C$

15. $(A \cup B) \cap C$ **16.** $(A \cap B) \cup C$ **17.** $\overline{A}$ **18.** $\overline{C}$

19. $\overline{A \cap B}$ **20.** $\overline{B \cup C}$ **21.** $\overline{A} \cup \overline{B}$ **22.** $\overline{B} \cap \overline{C}$

In Problems 23–28, list the numbers in each set that are (a) Natural numbers, (b) Integers, (c) Rational numbers, (d) Irrational numbers, (e) Real numbers.

23. $A = \left\{-6, \frac{1}{2}, -1.333\ldots \text{(the 3's repeat)}, \pi, 2, 5\right\}$

24. $B = \left\{-\frac{5}{3}, 2.060606\ldots \text{(the block 06 repeats)}, 1.25, 0, 1, \sqrt{5}\right\}$

25. $C = \left\{0, 1, \frac{1}{2}, \frac{1}{3}, \frac{1}{4}\right\}$

26. $D = \{-1, -1.1, -1.2, -1.3\}$

27. $E = \left\{\sqrt{2}, \pi, \sqrt{2} + 1, \pi + \frac{1}{2}\right\}$

28. $F = \left\{-\sqrt{2}, \pi + \sqrt{2}, \frac{1}{2} + 10.3\right\}$

In Problems 29–40, approximate each number (a) rounded and (b) truncated to three decimal places.

29. 18.9526 **30.** 25.86134 **31.** 28.65319 **32.** 99.05249 **33.** 0.06291 **34.** 0.05388

35. 9.9985 **36.** 1.0006 **37.** $\frac{3}{7}$ **38.** $\frac{5}{9}$ **39.** $\frac{521}{15}$ **40.** $\frac{81}{5}$

In Problems 41–50, write each statement using symbols.

41. The sum of 3 and 2 equals 5.

42. The product of 5 and 2 equals 10.

43. The sum of x and 2 is the product of 3 and 4.

44. The sum of 3 and y is the sum of 2 and 2.

45. The product of 3 and y is the sum of 1 and 2.

46. The product of 2 and x is the product of 4 and 6.

47. The difference x less 2 equals 6.

48. The difference 2 less y equals 6.

49. The quotient x divided by 2 is 6.

50. The quotient 2 divided by x is 6.

In Problems 51–88, evaluate each expression.

51. $9 - 4 + 2$ **52.** $6 - 4 + 3$ **53.** $-6 + 4 \cdot 3$ **54.** $8 - 4 \cdot 2$

55. $4 + 5 - 8$ **56.** $8 - 3 - 4$ **57.** $4 + \frac{1}{3}$ **58.** $2 - \frac{1}{2}$

59. $6 - [3 \cdot 5 + 2 \cdot (3 - 2)]$ **60.** $2 \cdot [8 - 3(4 + 2)] - 3$ **61.** $2 \cdot (3 - 5) + 8 \cdot 2 - 1$ **62.** $1 - (4 \cdot 3 - 2 + 2)$

63. $10 - [6 - 2 \cdot 2 + (8 - 3)] \cdot 2$ **64.** $2 - 5 \cdot 4 - [6 \cdot (3 - 4)]$

65. $(5 - 3)\frac{1}{2}$ **66.** $(5 + 4)\frac{1}{3}$ **67.** $\frac{4 + 8}{5 - 3}$ **68.** $\frac{2 - 4}{5 - 3}$

69. $\frac{3}{5} \cdot \frac{10}{21}$ **70.** $\frac{5}{9} \cdot \frac{3}{10}$ **71.** $\frac{6}{25} \cdot \frac{10}{27}$ **72.** $\frac{21}{25} \cdot \frac{100}{3}$

73. $\frac{3}{4} + \frac{2}{5}$ **74.** $\frac{4}{3} + \frac{1}{2}$ **75.** $\frac{5}{6} + \frac{9}{5}$ **76.** $\frac{8}{9} + \frac{15}{2}$

77. $\frac{5}{18} + \frac{1}{12}$ **78.** $\frac{2}{15} + \frac{8}{9}$ **79.** $\frac{1}{30} - \frac{7}{18}$ **80.** $\frac{3}{14} - \frac{2}{21}$

81. $\frac{3}{20} - \frac{2}{15}$ **82.** $\frac{6}{35} - \frac{3}{14}$ **83.** $\dfrac{\frac{5}{18}}{\frac{11}{27}}$ **84.** $\dfrac{\frac{5}{21}}{\frac{2}{35}}$

85. $\frac{1}{2} \cdot \frac{3}{5} + \frac{7}{10}$ **86.** $\frac{2}{3} + \frac{4}{5} \cdot \frac{1}{6}$ **87.** $2 \cdot \frac{3}{4} + \frac{3}{8}$ **88.** $3 \cdot \frac{5}{6} - \frac{1}{2}$

In Problems 89–100, use the Distributive Property to remove the parentheses.

89. $6(x + 4)$ **90.** $4(2x - 1)$ **91.** $x(x - 4)$ **92.** $4x(x + 3)$

93. $2\left(\frac{3}{4}x - \frac{1}{2}\right)$ **94.** $3\left(\frac{2}{3}x + \frac{1}{6}\right)$ **95.** $(x + 2)(x + 4)$ **96.** $(x + 5)(x + 1)$

97. $(x - 2)(x + 1)$ **98.** $(x - 4)(x + 1)$ **99.** $(x - 8)(x - 2)$ **100.** $(x - 4)(x - 2)$

Explaining Concepts: Discussion and Writing

101. Explain to a friend how the Distributive Property is used to justify the fact that $2x + 3x = 5x$.

102. Explain to a friend why $2 + 3 \cdot 4 = 14$, whereas $(2 + 3) \cdot 4 = 20$.

103. Explain why $2(3 \cdot 4)$ is not equal to $(2 \cdot 3) \cdot (2 \cdot 4)$.

104. Explain why $\frac{4 + 3}{2 + 5}$ is not equal to $\frac{4}{2} + \frac{3}{5}$.

105. Is subtraction commutative? Support your conclusion with an example.

106. Is subtraction associative? Support your conclusion with an example.

107. Is division commutative? Support your conclusion with an example.

108. Is division associative? Support your conclusion with an example.

109. If $2 = x$, why does $x = 2$?

110. If $x = 5$, why does $x^2 + x = 30$?

111. Are there any real numbers that are both rational and irrational? Are there any real numbers that are neither? Explain your reasoning.

112. Explain why the sum of a rational number and an irrational number must be irrational.

113. A rational number is defined as the quotient of two integers. When written as a decimal, the decimal will either repeat or terminate. By looking at the denominator of the rational number, there is a way to tell in advance whether its decimal representation will repeat or terminate. Make a list of rational numbers and their decimals. See if you can discover the pattern. Confirm your conclusion by consulting books on number theory at the library. Write a brief essay on your findings.

114. The current time is 12 noon CST. What time (CST) will it be 12,997 hours from now?

115. Both $\frac{a}{0}(a \neq 0)$ and $\frac{0}{0}$ are undefined, but for different reasons. Write a paragraph or two explaining the different reasons.

R.2 Algebra Essentials

OBJECTIVES

The Real Number Line

Real numbers can be represented by points on a line called the **real number line**. There is a one-to-one correspondence between real numbers and points on a line. That is, every real number corresponds to a point on the line, and each point on the line has a unique real number associated with it.

Pick a point on a line somewhere in the center, and label it O. This point, called the **origin**, corresponds to the real number 0. See Figure 10. The point 1 unit to the right of O corresponds to the number 1. The distance between 0 and 1 determines the **scale** of the number line. For example, the point associated with the number 2 is twice as far from O as 1. Notice that an arrowhead on the right end of the line indicates the direction in which the numbers increase. Points to the left of the origin correspond to the real numbers $-1, -2$, and so on. Figure 10 also shows the points associated with the rational numbers $-\frac{1}{2}$ and $\frac{1}{2}$ and with the irrational numbers $\sqrt{2}$ and π.

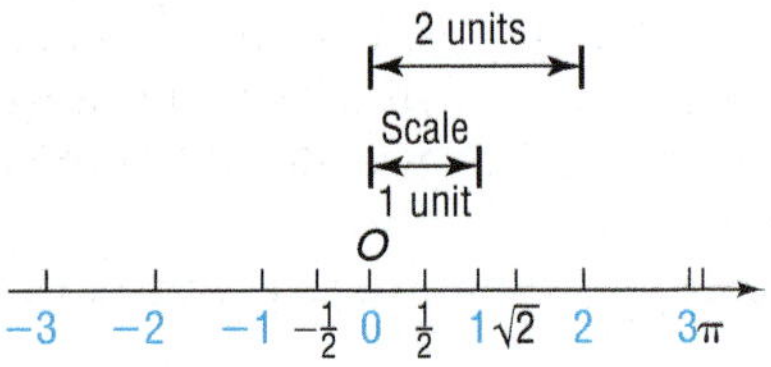

Figure 10 Real number line

DEFINITION

The real number associated with a point P is called the **coordinate** of P, and the line whose points have been assigned coordinates is called the **real number line**.

Now Work PROBLEM 13

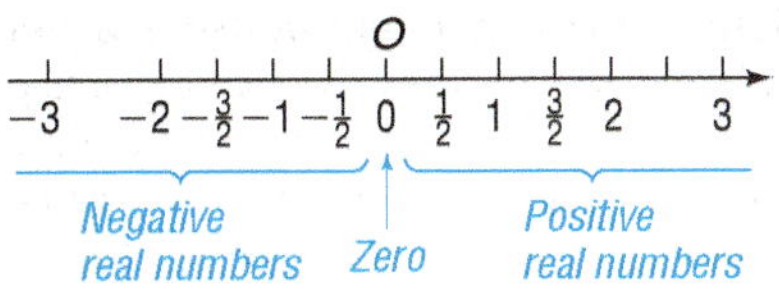

Figure 11

The real number line consists of three classes of real numbers, as shown in Figure 11.

1. The **negative real numbers** are the coordinates of points to the left of the origin O.
2. The real number **zero** is the coordinate of the origin O.
3. The **positive real numbers** are the coordinates of points to the right of the origin O.

Multiplication Properties of Positive and Negative Numbers

1. The product of two positive numbers is a positive number.
2. The product of two negative numbers is a positive number.
3. The product of a positive number and a negative number is a negative number.

1 Graph Inequalities

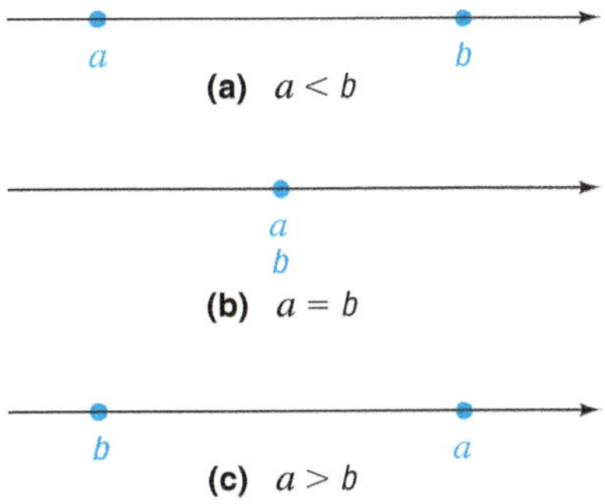

Figure 12

An important property of the real number line follows from the fact that, given two numbers (points) a and b, either a is to the left of b, or a is at the same location as b, or a is to the right of b. See Figure 12.

If a is to the left of b, then "a is less than b," which is written $a < b$. If a is to the right of b, then "a is greater than b," which is written $a > b$. If a is at the same location as b, then $a = b$. If a is either less than or equal to b, then $a \leq b$. Similarly, $a \geq b$ means that a is either greater than or equal to b. Collectively, the symbols $<, >, \leq,$ and $\geq$ are called **inequality symbols**.

Note that $a < b$ and $b > a$ mean the same thing. It does not matter whether we write $2 < 3$ or $3 > 2$.

Furthermore, if $a < b$ or if $b > a$, then the difference $b - a$ is positive. Do you see why?

EXAMPLE 1 **Using Inequality Symbols**

(a) $3 < 7$	(b) $-8 > -16$	(c) $-6 < 0$
(d) $-8 < -4$	(e) $4 > -1$	(f) $8 > 0$ ■

In Example 1(a), we conclude that $3 < 7$ either because 3 is to the left of 7 on the real number line or because the difference, $7 - 3 = 4$, is a positive real number.

Similarly, we conclude in Example 1(b) that $-8 > -16$ either because -8 lies to the right of -16 on the real number line or because the difference, $-8 - (-16) = -8 + 16 = 8$, is a positive real number.

Look again at Example 1. Note that the inequality symbol always points in the direction of the smaller number.

An **inequality** is a statement in which two expressions are related by an inequality symbol. The expressions are referred to as the **sides** of the inequality. Inequalities of the form $a < b$ or $b > a$ are called **strict inequalities**, whereas inequalities of the form $a \leq b$ or $b \geq a$ are called **nonstrict inequalities**.

Based on the discussion so far, we conclude that

$a > 0$	is equivalent to	a is positive
$a < 0$	is equivalent to	a is negative

We sometimes read $a > 0$ by saying that "a is positive." If $a \geq 0$, then either $a > 0$ or $a = 0$, and we may read this as "a is nonnegative."

Now Work PROBLEMS 17 AND 27

EXAMPLE 2

Graphing Inequalities

(a) On the real number line, graph all numbers x for which $x > 4$.
(b) On the real number line, graph all numbers x for which $x \leq 5$.

Solution

(a) See Figure 13. Notice that we use a left parenthesis to indicate that the number 4 is *not* part of the graph.
(b) See Figure 14. Notice that we use a right bracket to indicate that the number 5 *is* part of the graph. ■

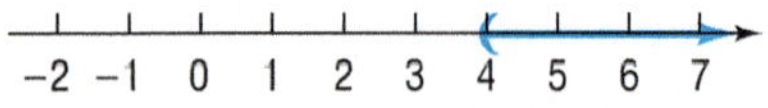

Figure 13 $x > 4$

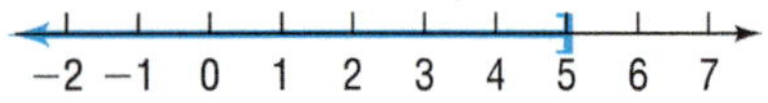

Figure 14 $x \leq 5$

Now Work PROBLEM 33

2 Find Distance on the Real Number Line

The *absolute value* of a number a is the distance from 0 to a on the number line. For example, -4 is 4 units from 0, and 3 is 3 units from 0. See Figure 15. That is, the absolute value of -4 is 4, and the absolute value of 3 is 3.

A more formal definition of absolute value is given next.

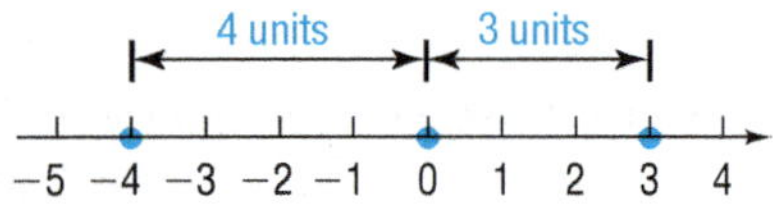

Figure 15

DEFINITION

The **absolute value** of a real number a, denoted by the symbol $|a|$, is defined by the rules

$$|a| = a \quad \text{if } a \geq 0 \qquad \text{and} \qquad |a| = -a \quad \text{if } a < 0$$

For example, because $-4 < 0$, the second rule must be used to get $|-4| = -(-4) = 4$.

EXAMPLE 3

Computing Absolute Value

(a) $|8| = 8$ (b) $|0| = 0$ (c) $|-15| = -(-15) = 15$ ■

Look again at Figure 15. The distance from -4 to 3 is 7 units. This distance is the difference $3 - (-4)$, obtained by subtracting the smaller coordinate from the larger. However, since $|3 - (-4)| = |7| = 7$ and $|-4 - 3| = |-7| = 7$, we can use absolute value to calculate the distance between two points without being concerned about which is smaller.

DEFINITION

If P and Q are two points on a real number line with coordinates a and b, respectively, the **distance between P and Q**, denoted by $d(P, Q)$, is

$$d(P, Q) = |b - a|$$

Since $|b - a| = |a - b|$, it follows that $d(P, Q) = d(Q, P)$.

EXAMPLE 4 **Finding Distance on a Number Line**

Let P, Q, and R be points on a real number line with coordinates -5, 7, and -3, respectively. Find the distance

(a) between P and Q (b) between Q and R

Solution See Figure 16.

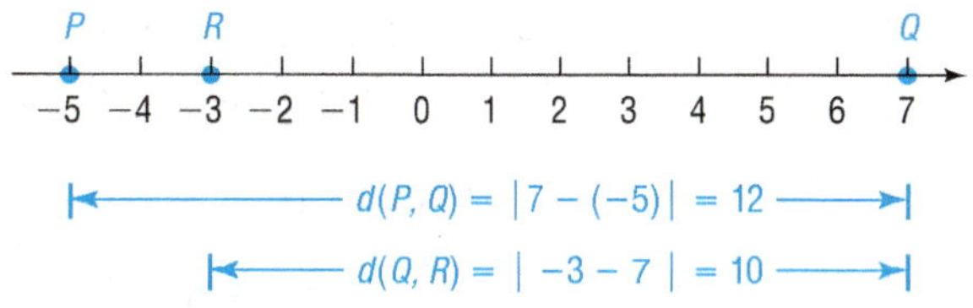

Figure 16

(a) $d(P, Q) = |7 - (-5)| = |12| = 12$
(b) $d(Q, R) = |-3 - 7| = |-10| = 10$ ■

Now Work PROBLEM 39

3 Evaluate Algebraic Expressions

Remember, in algebra we use letters such as x, y, a, b, and c to represent numbers. If the letter used is to represent *any* number from a given set of numbers, it is called a **variable**. A **constant** is either a fixed number, such as 5 or $\sqrt{3}$, or a letter that represents a fixed (possibly unspecified) number.

Constants and variables are combined using the operations of addition, subtraction, multiplication, and division to form *algebraic expressions*. Examples of algebraic expressions include

$$x + 3 \qquad \frac{3}{1 - t} \qquad 7x - 2y$$

To evaluate an algebraic expression, substitute a numerical value for each variable.

EXAMPLE 5 **Evaluating an Algebraic Expression**

Evaluate each expression if $x = 3$ and $y = -1$.

(a) $x + 3y$ (b) $5xy$ (c) $\dfrac{3y}{2 - 2x}$ (d) $|-4x + y|$

Solution (a) Substitute 3 for x and -1 for y in the expression $x + 3y$.

$$x + 3y = 3 + 3(-1) = 3 + (-3) = 0$$

$x = 3, y = -1$ (points to the first equals sign)

(b) If $x = 3$ and $y = -1$, then

$$5xy = 5(3)(-1) = -15$$

(c) If $x = 3$ and $y = -1$, then

$$\frac{3y}{2 - 2x} = \frac{3(-1)}{2 - 2(3)} = \frac{-3}{2 - 6} = \frac{-3}{-4} = \frac{3}{4}$$

(d) If $x = 3$ and $y = -1$, then

$$|-4x + y| = |-4(3) + (-1)| = |-12 + (-1)| = |-13| = 13$$ ■

A graphing calculator can be used to evaluate an algebraic expression. Figure 17 shows the results of Example 5 using a TI-84 Plus C.

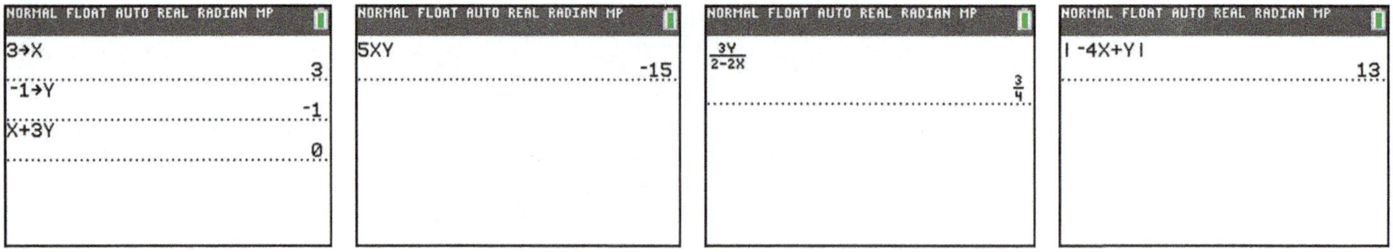

(a) (b) (c) (d)

Figure 17

Now Work PROBLEMS 41 AND 49

4 Determine the Domain of a Variable

In working with expressions or formulas involving variables, the variables may be allowed to take on values from only a certain set of numbers. For example, in the formula for the area A of a circle of radius r, $A = \pi r^2$, the variable r is necessarily restricted to the positive real numbers. In the expression $\frac{1}{x}$, the variable x cannot take on the value 0, since division by 0 is not defined.

DEFINITION The set of values that a variable may assume is called the **domain of the variable**.

EXAMPLE 6 Finding the Domain of a Variable

The domain of the variable x in the expression

$$\frac{5}{x-2}$$

is $\{x \mid x \neq 2\}$ since, if $x = 2$, the denominator becomes 0, which is not defined.

EXAMPLE 7 Circumference of a Circle

In the formula for the circumference C of a circle of radius r,

$$C = 2\pi r$$

the domain of the variable r, representing the radius of the circle, is the set of positive real numbers, $\{r \mid r > 0\}$. The domain of the variable C, representing the circumference of the circle, is also the set of positive real numbers, $\{C \mid C > 0\}$.

In describing the domain of a variable, we may use either set notation or words, whichever is more convenient.

Now Work PROBLEM 59

5 Use the Laws of Exponents

Integer exponents provide a shorthand notation for representing repeated multiplications of a real number. For example,

$$3^4 = 3 \cdot 3 \cdot 3 \cdot 3 = 81$$

Additionally, many formulas have exponents. For example,

- The formula for the horsepower rating H of an engine is

$$H = \frac{D^2 N}{2.5}$$

where D is the diameter of a cylinder and N is the number of cylinders.

- A formula for the resistance R of blood flowing in a blood vessel is

$$R = C\frac{L}{r^4}$$

where L is the length of the blood vessel, r is the radius, and C is a positive constant.

DEFINITION

If a is a real number and n is a positive integer, then the symbol $\boldsymbol{a^n}$ represents the product of n factors of a. That is,

$$a^n = \underbrace{a \cdot a \cdot \ldots \cdot a}_{n \text{ factors}} \qquad (1)$$

From this definition $a^1 = a$, $a^2 = a \cdot a$, $a^3 = a \cdot a \cdot a$, and so on. In the expression a^n, a is called the **base** and n is called the **exponent**, or **power**. We read a^n as "a raised to the power n" or as "a to the nth power." We usually read a^2 as "a squared" and a^3 as "a cubed."

WARNING Be careful with minus signs and exponents.

$$-2^4 = -1 \cdot 2^4 = -16$$

whereas

$$(-2)^4 = (-2)(-2)(-2)(-2) = 16$$

In working with exponents, the operation of *raising to a power* is performed before any other operation. As examples,

$$4 \cdot 3^2 = 4 \cdot 9 = 36 \qquad 2^2 + 3^2 = 4 + 9 = 13$$

$$-2^4 = -16 \qquad 5 \cdot 3^2 + 2 \cdot 4 = 5 \cdot 9 + 2 \cdot 4 = 45 + 8 = 53$$

Parentheses are used to indicate operations to be performed first. For example,

$$(-2)^4 = (-2)(-2)(-2)(-2) = 16 \qquad (2 + 3)^2 = 5^2 = 25$$

DEFINITION

If $a \neq 0$, then

$$a^0 = 1$$

DEFINITION

If $a \neq 0$ and if n is a positive integer, then

$$a^{-n} = \frac{1}{a^n}$$

Whenever you encounter a negative exponent, think "reciprocal."

EXAMPLE 8 **Evaluating Expressions Containing Negative Exponents**

(a) $2^{-3} = \frac{1}{2^3} = \frac{1}{8}$ (b) $x^{-4} = \frac{1}{x^4}$ (c) $\left(\frac{1}{5}\right)^{-2} = \frac{1}{\left(\frac{1}{5}\right)^2} = \frac{1}{\frac{1}{25}} = 25$

Now Work PROBLEMS 77 AND 97

The following properties, called the **Laws of Exponents**, can be proved using the preceding definitions. In the list, a and b are real numbers, and m and n are integers.

THEOREM

Laws of Exponents

$$a^m a^n = a^{m+n} \qquad (a^m)^n = a^{mn} \qquad (ab)^n = a^n b^n$$

$$\frac{a^m}{a^n} = a^{m-n} = \frac{1}{a^{n-m}} \quad \text{if } a \neq 0 \qquad \left(\frac{a}{b}\right)^n = \frac{a^n}{b^n} \quad \text{if } b \neq 0$$

EXAMPLE 9 **Using the Laws of Exponents**

(a) $x^{-3} \cdot x^5 = x^{-3+5} = x^2 \quad x \neq 0$

(b) $(x^{-3})^2 = x^{-3 \cdot 2} = x^{-6} = \dfrac{1}{x^6} \quad x \neq 0$

(c) $(2x)^3 = 2^3 \cdot x^3 = 8x^3$

(d) $\left(\dfrac{2}{3}\right)^4 = \dfrac{2^4}{3^4} = \dfrac{16}{81}$

(e) $\dfrac{x^{-2}}{x^{-5}} = x^{-2-(-5)} = x^3 \quad x \neq 0$ ■

Note: Always write the final answer using positive exponents. ■

Now Work PROBLEM 79

EXAMPLE 10 **Using the Laws of Exponents**

Write each expression so that all exponents are positive.

(a) $\dfrac{x^5 y^{-2}}{x^3 y} \quad x \neq 0, \quad y \neq 0$

(b) $\left(\dfrac{x^{-3}}{3y^{-1}}\right)^{-2} \quad x \neq 0, \quad y \neq 0$

Solution

(a) $\dfrac{x^5 y^{-2}}{x^3 y} = \dfrac{x^5}{x^3} \cdot \dfrac{y^{-2}}{y} = x^{5-3} \cdot y^{-2-1} = x^2 y^{-3} = x^2 \cdot \dfrac{1}{y^3} = \dfrac{x^2}{y^3}$

(b) $\left(\dfrac{x^{-3}}{3y^{-1}}\right)^{-2} = \dfrac{(x^{-3})^{-2}}{(3y^{-1})^{-2}} = \dfrac{x^6}{3^{-2}(y^{-1})^{-2}} = \dfrac{x^6}{\frac{1}{9}y^2} = \dfrac{9x^6}{y^2}$ ■

Now Work PROBLEM 89

6 Evaluate Square Roots

In Words
$\sqrt{36}$ means "give me the nonnegative number whose square is 36."

A real number is squared when it is raised to the power 2. The inverse of squaring is finding a **square root**. For example, since $6^2 = 36$ and $(-6)^2 = 36$, the numbers 6 and -6 are square roots of 36.

The symbol $\sqrt{}$, called a **radical sign**, is used to denote the **principal**, or nonnegative, square root. For example, $\sqrt{36} = 6$.

DEFINITION

If a is a nonnegative real number, the nonnegative number b such that $b^2 = a$ is the **principal square root** of a, and is denoted by $b = \sqrt{a}$.

The following comments are noteworthy:

1. Negative numbers do not have square roots (in the real number system), because the square of any real number is *nonnegative*. For example, $\sqrt{-4}$ is not a real number, because there is no real number whose square is -4.
2. The principal square root of 0 is 0, since $0^2 = 0$. That is, $\sqrt{0} = 0$.
3. The principal square root of a positive number is positive.
4. If $c \geq 0$, then $(\sqrt{c})^2 = c$. For example, $(\sqrt{2})^2 = 2$ and $(\sqrt{3})^2 = 3$.

EXAMPLE 11 **Evaluating Square Roots**

(a) $\sqrt{64} = 8$ (b) $\sqrt{\dfrac{1}{16}} = \dfrac{1}{4}$ (c) $(\sqrt{1.4})^2 = 1.4$ ■

Examples 11(a) and (b) are examples of square roots of perfect squares, since $64 = 8^2$ and $\dfrac{1}{16} = \left(\dfrac{1}{4}\right)^2$.

Consider the expression $\sqrt{a^2}$. Since $a^2 \geq 0$, the principal square root of a^2 is defined whether $a > 0$ or $a < 0$. However, since the principal square root is nonnegative, we need an absolute value to ensure the nonnegative result. That is,

$$\sqrt{a^2} = |a| \quad a \text{ any real number} \tag{2}$$

EXAMPLE 12

Using Equation (2)

(a) $\sqrt{(2.3)^2} = |2.3| = 2.3$ (b) $\sqrt{(-2.3)^2} = |-2.3| = 2.3$

(c) $\sqrt{x^2} = |x|$ ■

Now Work PROBLEM 85

7 Use a Calculator to Evaluate Exponents

Your calculator has either a caret key, $\boxed{\wedge}$, or an $\boxed{x^y}$ key, that is used for computations involving exponents.

EXAMPLE 13

Exponents on a Graphing Calculator

Evaluate: $(2.3)^5$

Solution Figure 18 shows the result using a TI-84 Plus C graphing calculator. ■

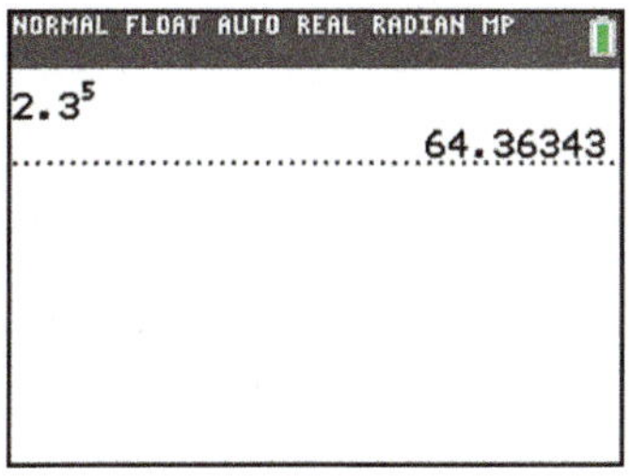

Figure 18

Now Work PROBLEM 115

8 Use Scientific Notation

Measurements of physical quantities can range from very small to very large. For example, the mass of a proton is approximately 0.00000000000000000000000000167 kilogram and the mass of Earth is about 5,980,000,000,000,000,000,000,000 kilograms. These numbers obviously are tedious to write down and difficult to read, so we use exponents to rewrite them.

DEFINITION

When a number has been written as the product of a number x, where $1 \leq x < 10$, times a power of 10, it is said to be written in **scientific notation**. ■

In scientific notation,

$$\text{Mass of a proton} = 1.67 \times 10^{-27} \text{ kilogram}$$

$$\text{Mass of Earth} = 5.98 \times 10^{24} \text{ kilograms}$$

Converting a Decimal into Scientific Notation

To change a positive number into scientific notation:

1. Count the number N of places that the decimal point must be moved to arrive at a number x, where $1 \leq x < 10$.
2. If the original number is greater than or equal to 1, the scientific notation is $x \times 10^N$. If the original number is between 0 and 1, the scientific notation is $x \times 10^{-N}$.

EXAMPLE 14 Using Scientific Notation

Write each number in scientific notation.

(a) 9582 (b) 1.245 (c) 0.285 (d) 0.000561

Solution (a) The decimal point in 9582 follows the 2. Count left from the decimal point

9 5 8 2 .
3 2 1

stopping after three moves, because 9.582 is a number between 1 and 10. Since 9582 is greater than 1, we write

$$9582 = 9.582 \times 10^3$$

(b) The decimal point in 1.245 is between the 1 and 2. Since the number is already between 1 and 10, the scientific notation for it is $1.245 \times 10^0 = 1.245$.

(c) The decimal point in 0.285 is between the 0 and the 2. We count

0 . 2 8 5
1

stopping after one move, because 2.85 is a number between 1 and 10. Since 0.285 is between 0 and 1, we write

$$0.285 = 2.85 \times 10^{-1}$$

(d) The decimal point in 0.000561 is moved as follows:

0 . 0 0 0 5 6 1
1 2 3 4

As a result,

$$0.000561 = 5.61 \times 10^{-4}$$

■

Now Work PROBLEM 121

EXAMPLE 15 Changing from Scientific Notation to Decimals

Write each number as a decimal.

(a) 2.1×10^4 (b) 3.26×10^{-5} (c) 1×10^{-2}

Solution

(a) $2.1 \times 10^4 = 2\ .\ 1\ 0\ 0\ 0 \times 10^4 = 21{,}000$
1 2 3 4

(b) $3.26 \times 10^{-5} = 0\ 0\ 0\ 0\ 0\ 3\ .\ 2\ 6 \times 10^{-5} = 0.0000326$
5 4 3 2 1

(c) $1 \times 10^{-2} = 0\ 0\ 1\ .\ 0 \times 10^{-2} = 0.01$
2 1

■

Now Work PROBLEM 129

EXAMPLE 16 **Using Scientific Notation**

(a) The diameter of the smallest living cell is only about 0.00001 centimeter (cm).* Express this number in scientific notation.

(b) The surface area of Earth is about 1.97×10^8 square miles.† Express the surface area as a whole number.

Solution (a) $0.00001 \text{ cm} = 1 \times 10^{-5} \text{ cm}$ because the decimal point is moved five places and the number is less than 1.

(b) 1.97×10^8 square miles = 197,000,000 square miles. ■

Now Work PROBLEM 155

COMMENT On a calculator, a number such as 3.615×10^{12} is usually displayed as 3.615E12. ■

* *Powers of Ten,* Philip and Phylis Morrison.

† *2011 Information Please Almanac.*

Historical Feature

The word *algebra* is derived from the Arabic word *al-jabr.* This word is a part of the title of a ninth-century work, "Hisâb al-jabr w'al-muqâbalah," written by Mohammed ibn Músâ al-Khwârizmî. The word *al-jabr* means "a restoration," a reference to the fact that if a number is added to one side of an equation, then it must also be added to the other side in order to "restore" the equality. The title of the work, freely translated, is "The Science of Reduction and Cancellation." Of course, today, algebra has come to mean a great deal more.

R.2 Assess Your Understanding

Concepts and Vocabulary

1. A(n) ________ is a letter used in algebra to represent any number from a given set of numbers.

2. On the real number line, the real number zero is the coordinate of the ______.

3. An inequality of the form $a > b$ is called a(n) ______ inequality.

4. In the expression 2^4, the number 2 is called the _____ and 4 is called the ________________.

5. In scientific notation, $1234.5678 =$ ________________.

6. ***True or False*** The product of two negative real numbers is always greater than zero.

7. ***True or False*** The distance between two distinct points on the real number line is always greater than zero.

8. ***True or False*** The absolute value of a real number is always greater than zero.

9. ***True or False*** When a number is expressed in scientific notation, it is expressed as the product of a number x, $0 \le x < 1$, and a power of 10.

10. ***True or False*** To multiply two expressions having the same base, retain the base and multiply the exponents.

11. If a is a nonnegative real number, then which inequality statement best describes a?

(a) $a < 0$ (b) $a > 0$ (c) $a \le 0$ (d) $a \ge 0$

12. Let a and b be non-zero real numbers and m and n be integers. Which of the following is not a law of exponents?

(a) $\left(\frac{a}{b}\right)^n = \frac{a^n}{b^n}$ (b) $(a^m)^n = a^{m+n}$

(c) $\frac{a^m}{a^n} = a^{m-n}$ (d) $(ab)^n = a^n b^n$

Skill Building

13. On the real number line, label the points with coordinates $0, 1, -1, \frac{5}{2}, -2.5, \frac{3}{4}$, and 0.25.

14. Repeat Problem 13 for the coordinates $0, -2, 2, -1.5, \frac{3}{2}, \frac{1}{3}$, and $\frac{2}{3}$.

In Problems 15–24, replace the question mark by <, >, or =, whichever is correct.

15. $\frac{1}{2} \ ? \ 0$

16. $5 \ ? \ 6$

17. $-1 \ ? \ -2$

18. $-3 \ ? \ -\frac{5}{2}$

19. $\pi \ ? \ 3.14$

20. $\sqrt{2} \ ? \ 1.41$

21. $\frac{1}{2} \ ? \ 0.5$

22. $\frac{1}{3} \ ? \ 0.33$

23. $\frac{2}{3} \ ? \ 0.67$

24. $\frac{1}{4} \ ? \ 0.25$

In Problems 25–30, write each statement as an inequality.

25. x is positive

26. z is negative

27. x is less than 2

28. y is greater than -5

29. x is less than or equal to 1

30. x is greater than or equal to 2

In Problems 31–34, graph the numbers x on the real number line.

31. $x \geq -2$

32. $x < 4$

33. $x > -1$

34. $x \leq 7$

In Problems 35–40, use the given real number line to compute each distance.

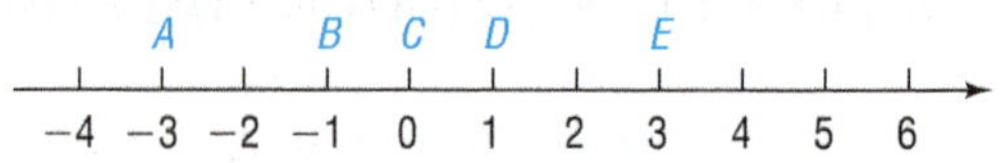

35. $d(C, D)$

36. $d(C, A)$

37. $d(D, E)$

38. $d(C, E)$

39. $d(A, E)$

40. $d(D, B)$

In Problems 41–48, evaluate each expression if $x = -2$ and $y = 3$.

41. $x + 2y$

42. $3x + y$

43. $5xy + 2$

44. $-2x + xy$

45. $\dfrac{2x}{x - y}$

46. $\dfrac{x + y}{x - y}$

47. $\dfrac{3x + 2y}{2 + y}$

48. $\dfrac{2x - 3}{y}$

In Problems 49–58, find the value of each expression if $x = 3$ and $y = -2$.

49. $|x + y|$

50. $|x - y|$

51. $|x| + |y|$

52. $|x| - |y|$

53. $\dfrac{|x|}{x}$

54. $\dfrac{|y|}{y}$

55. $|4x - 5y|$

56. $|3x + 2y|$

57. $||4x| - |5y||$

58. $3|x| + 2|y|$

In Problems 59–66, determine which of the values (a) through (d), if any, must be excluded from the domain of the variable in each expression.

(a) $x = 3$ (b) $x = 1$ (c) $x = 0$ (d) $x = -1$

59. $\dfrac{x^2 - 1}{x}$

60. $\dfrac{x^2 + 1}{x}$

61. $\dfrac{x}{x^2 - 9}$

62. $\dfrac{x}{x^2 + 9}$

63. $\dfrac{x^2}{x^2 + 1}$

64. $\dfrac{x^3}{x^2 - 1}$

65. $\dfrac{x^2 + 5x - 10}{x^3 - x}$

66. $\dfrac{-9x^2 - x + 1}{x^3 + x}$

In Problems 67–70, determine the domain of the variable x in each expression.

67. $\dfrac{4}{x - 5}$

68. $\dfrac{-6}{x + 4}$

69. $\dfrac{x}{x + 4}$

70. $\dfrac{x - 2}{x - 6}$

In Problems 71–74, use the formula $C = \frac{5}{9}(F - 32)$ for converting degrees Fahrenheit into degrees Celsius to find the Celsius measure of each Fahrenheit temperature.

71. $F = 32°$

72. $F = 212°$

73. $F = 77°$

74. $F = -4°$

In Problems 75–86, simplify each expression.

75. $(-4)^2$

76. -4^2

77. 4^{-2}

78. -4^{-2}

79. $3^{-6} \cdot 3^4$

80. $4^{-2} \cdot 4^3$

81. $(3^{-2})^{-1}$

82. $(2^{-1})^{-3}$

83. $\sqrt{25}$

84. $\sqrt{36}$

85. $\sqrt{(-4)^2}$

86. $\sqrt{(-3)^2}$

In Problems 87–96, simplify each expression. Express the answer so that all exponents are positive. Whenever an exponent is 0 or negative, we assume that the base is not 0.

87. $(8x^3)^2$

88. $(-4x^2)^{-1}$

89. $(x^2y^{-1})^2$

90. $(x^{-1}y)^3$

91. $\dfrac{x^2y^3}{xy^4}$

92. $\dfrac{x^{-2}y}{xy^2}$

93. $\dfrac{(-2)^3x^4(yz)^2}{3^2xy^3z}$

94. $\dfrac{4x^{-2}(yz)^{-1}}{2^3x^4y}$

95. $\left(\dfrac{3x^{-1}}{4y^{-1}}\right)^{-2}$

96. $\left(\dfrac{5x^{-2}}{6y^{-2}}\right)^{-3}$

In Problems 97–108, find the value of each expression if $x = 2$ and $y = -1$.

97. $2xy^{-1}$ **98.** $-3x^{-1}y$ **99.** $x^2 + y^2$ **100.** x^2y^2

101. $(xy)^2$ **102.** $(x+y)^2$ **103.** $\sqrt{x^2}$ **104.** $(\sqrt{x})^2$

105. $\sqrt{x^2 + y^2}$ **106.** $\sqrt{x^2} + \sqrt{y^2}$ **107.** x^y **108.** y^x

109. Find the value of the expression $2x^3 - 3x^2 + 5x - 4$ if $x = 2$. What is the value if $x = 1$?

110. Find the value of the expression $4x^3 + 3x^2 - x + 2$ if $x = 1$. What is the value if $x = 2$?

111. What is the value of $\dfrac{(666)^4}{(222)^4}$?

112. What is the value of $(0.1)^3(20)^3$?

In Problems 113–120, use a calculator to evaluate each expression. Round your answer to three decimal places.

113. $(8.2)^6$ **114.** $(3.7)^5$ **115.** $(6.1)^{-3}$ **116.** $(2.2)^{-5}$

117. $(-2.8)^6$ **118.** $-(2.8)^6$ **119.** $(-8.11)^{-4}$ **120.** $-(8.11)^{-4}$

In Problems 121–128, write each number in scientific notation.

121. 454.2 **122.** 32.14 **123.** 0.013 **124.** 0.00421

125. 32,155 **126.** 21,210 **127.** 0.000423 **128.** 0.0514

In Problems 129–136, write each number as a decimal.

129. 6.15×10^4 **130.** 9.7×10^3 **131.** 1.214×10^{-3} **132.** 9.88×10^{-4}

133. 1.1×10^8 **134.** 4.112×10^2 **135.** 8.1×10^{-2} **136.** 6.453×10^{-1}

Applications and Extensions

In Problems 137–146, express each statement as an equation involving the indicated variables.

137. Area of a Rectangle The area A of a rectangle is the product of its length l and its width w.

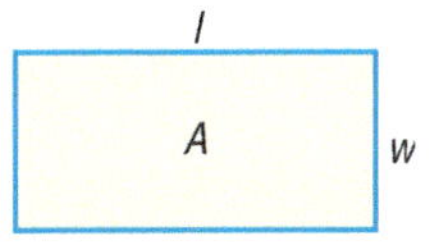

138. Perimeter of a Rectangle The perimeter P of a rectangle is twice the sum of its length l and its width w.

139. Circumference of a Circle The circumference C of a circle is the product of π and its diameter d.

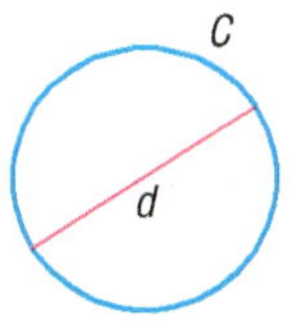

140. Area of a Triangle The area A of a triangle is one-half the product of its base b and its height h.

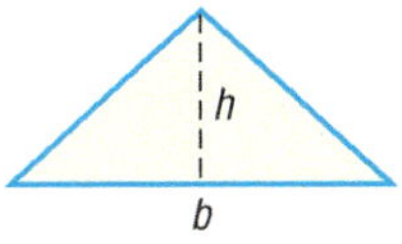

141. Area of an Equilateral Triangle The area A of an equilateral triangle is $\dfrac{\sqrt{3}}{4}$ times the square of the length x of one side.

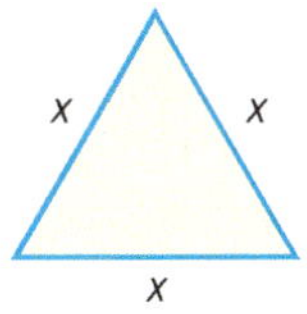

142. Perimeter of an Equilateral Triangle The perimeter P of an equilateral triangle is 3 times the length x of one side.

143. Volume of a Sphere The volume V of a sphere is $\dfrac{4}{3}$ times π times the cube of the radius r.

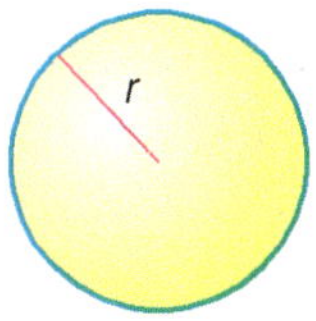

144. Surface Area of a Sphere The surface area S of a sphere is 4 times π times the square of the radius r.

145. Volume of a Cube The volume V of a cube is the cube of the length x of a side.

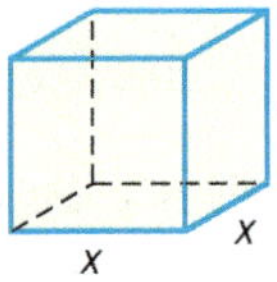

146. Surface Area of a Cube The surface area S of a cube is 6 times the square of the length x of a side.

147. Manufacturing Cost The weekly production cost C of manufacturing x watches is given by the formula $C = 4000 + 2x$, where the variable C is in dollars.
(a) What is the cost of producing 1000 watches?
(b) What is the cost of producing 2000 watches?

148. Balancing a Checkbook At the beginning of the month, Mike had a balance of \$210 in his checking account. During the next month, he deposited \$80, wrote a check for \$120, made another deposit of \$25, and wrote two checks: one for \$60 and the other for \$32. He was also assessed a monthly service charge of \$5. What was his balance at the end of the month?

In Problems 149 and 150, write an inequality using an absolute value to describe each statement.

149. x is at least 6 units from 4.

150. x is more than 5 units from 2.

151. U.S. Voltage In the United States, normal household voltage is 110 volts. It is acceptable for the actual voltage x to differ from normal by at most 5 volts. A formula that describes this is

$$|x - 110| \leq 5$$

(a) Show that a voltage of 108 volts is acceptable.
(b) Show that a voltage of 104 volts is not acceptable.

152. Foreign Voltage In other countries, normal household voltage is 220 volts. It is acceptable for the actual voltage x to differ from normal by at most 8 volts. A formula that describes this is

$$|x - 220| \leq 8$$

(a) Show that a voltage of 214 volts is acceptable.
(b) Show that a voltage of 209 volts is not acceptable.

153. Making Precision Ball Bearings The FireBall Company manufactures ball bearings for precision equipment. One of its products is a ball bearing with a stated radius of 3 centimeters (cm). Only ball bearings with a radius within 0.01 cm of this stated radius are acceptable. If x is the radius of a ball bearing, a formula describing this situation is

$$|x - 3| \leq 0.01$$

(a) Is a ball bearing of radius $x = 2.999$ acceptable?
(b) Is a ball bearing of radius $x = 2.89$ acceptable?

154. Body Temperature Normal human body temperature is 98.6°F. A temperature x that differs from normal by at least 1.5°F is considered unhealthy. A formula that describes this is

$$|x - 98.6| \geq 1.5$$

(a) Show that a temperature of 97°F is unhealthy.
(b) Show that a temperature of 100°F is not unhealthy.

155. Distance from Earth to Its Moon The distance from Earth to the Moon is about 4×10^8 meters.* Express this distance as a whole number.

156. Height of Mt. Everest The height of Mt. Everest is 8848 meters.* Express this height in scientific notation.

157. Wavelength of Visible Light The wavelength of visible light is about 5×10^{-7} meter.* Express this wavelength as a decimal.

158. Diameter of an Atom The diameter of an atom is about 1×10^{-10} meter.* Express this diameter as a decimal.

159. Diameter of Copper Wire The smallest commercial copper wire is about 0.0005 inch in diameter.† Express this diameter using scientific notation.

160. Smallest Motor The smallest motor ever made is less than 0.05 centimeter wide.† Express this width using scientific notation.

161. Astronomy One light-year is defined by astronomers to be the distance that a beam of light will travel in 1 year (365 days). If the speed of light is 186,000 miles per second, how many miles are in a light-year? Express your answer in scientific notation.

162. Astronomy How long does it take a beam of light to reach Earth from the Sun when the Sun is 93,000,000 miles from Earth? Express your answer in seconds, using scientific notation.

163. Does $\frac{1}{3}$ equal 0.333? If not, which is larger? By how much?

164. Does $\frac{2}{3}$ equal 0.666? If not, which is larger? By how much?

Explaining Concepts: Discussion and Writing

165. Is there a positive real number "closest" to 0?

166. **Number game** I'm thinking of a number! It lies between 1 and 10; its square is rational and lies between 1 and 10. The number is larger than π. Correct to two decimal places (that is, truncated to two decimal places), name the number. Now think of your own number, describe it, and challenge a fellow student to name it.

167. Write a brief paragraph that illustrates the similarities and differences between "less than" ($<$) and "less than or equal to" ($\leq$).

168. Give a reason why the statement $5 < 8$ is true.

* *Powers of Ten*, Philip and Phylis Morrison.
† *2011 Information Please Almanac.*

R.3 Geometry Essentials

OBJECTIVES
1 Use the Pythagorean Theorem and Its Converse (p. 31)
2 Know Geometry Formulas (p. 32)
3 Understand Congruent Triangles and Similar Triangles (p. 33)

1 Use the Pythagorean Theorem and Its Converse

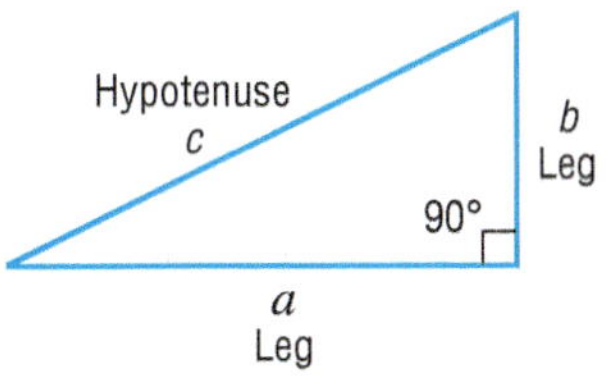

Figure 19 A right triangle

The *Pythagorean Theorem* is a statement about *right triangles*. A **right triangle** is one that contains a **right angle**—that is, an angle of 90°. The side of the triangle opposite the 90° angle is called the **hypotenuse**; the remaining two sides are called **legs**. In Figure 19 we have used c to represent the length of the hypotenuse and a and b to represent the lengths of the legs. Notice the use of the symbol ⊾ to show the 90° angle. We now state the Pythagorean Theorem.

PYTHAGOREAN THEOREM

In a right triangle, the square of the length of the hypotenuse is equal to the sum of the squares of the lengths of the legs. That is, in the right triangle shown in Figure 19,

$$c^2 = a^2 + b^2 \qquad \textbf{(1)}$$

A proof of the Pythagorean Theorem is given at the end of this section.

EXAMPLE 1 Finding the Hypotenuse of a Right Triangle

In a right triangle, one leg has length 4 and the other has length 3. What is the length of the hypotenuse?

Solution Since the triangle is a right triangle, we use the Pythagorean Theorem with $a = 4$ and $b = 3$ to find the length c of the hypotenuse. From equation (1),

$$c^2 = a^2 + b^2$$
$$c^2 = 4^2 + 3^2 = 16 + 9 = 25$$
$$c = \sqrt{25} = 5$$

Now Work PROBLEM 15

The converse of the Pythagorean Theorem is also true.

CONVERSE OF THE PYTHAGOREAN THEOREM

In a triangle, if the square of the length of one side equals the sum of the squares of the lengths of the other two sides, the triangle is a right triangle. The 90° angle is opposite the longest side.

A proof is given at the end of this section.

EXAMPLE 2 Verifying That a Triangle Is a Right Triangle

Show that a triangle whose sides are of lengths 5, 12, and 13 is a right triangle. Identify the hypotenuse.

Solution Square the lengths of the sides.

$$5^2 = 25 \qquad 12^2 = 144 \qquad 13^2 = 169$$

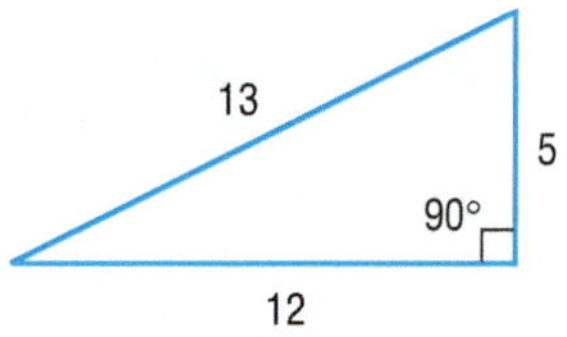

Figure 20

Notice that the sum of the first two squares (25 and 144) equals the third square (169). That is, because $5^2 + 12^2 = 13^2$, the triangle is a right triangle. The longest side, 13, is the hypotenuse. See Figure 20.

Now Work PROBLEM 23

EXAMPLE 3

Applying the Pythagorean Theorem

The tallest building in the world is Burj Khalifa in Dubai, United Arab Emirates, at 2717 feet and 163 floors. The observation deck is 1483 feet above ground level. How far can a person standing on the observation deck see (with the aid of a telescope)? Use 3960 miles for the radius of Earth.

Source: Council on Tall Buildings and Urban Habitat

Solution From the center of Earth, draw two radii: one through Burj Khalifa and the other to the farthest point a person can see from the observation deck. See Figure 21. Apply the Pythagorean Theorem to the right triangle.

Since 1 mile = 5280 feet, 1483 feet $= \frac{1483}{5280}$ mile. Then

$$d^2 + (3960)^2 = \left(3960 + \frac{1483}{5280}\right)^2$$

$$d^2 = \left(3960 + \frac{1483}{5280}\right)^2 - (3960)^2 \approx 2224.58$$

$$d \approx 47.17$$

A person can see more than 47 miles from the observation deck.

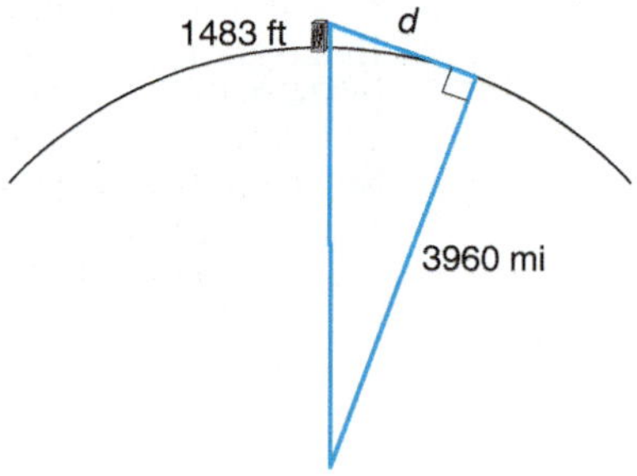

Figure 21

Now Work PROBLEM 55

2 Know Geometry Formulas

Certain formulas from geometry are useful in solving algebra problems.

For a rectangle of length l and width w,

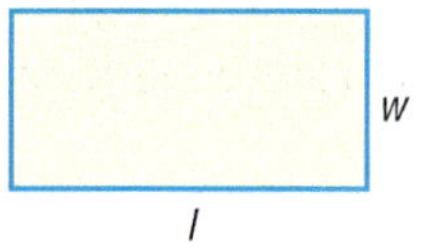

$$\text{Area} = lw \qquad \text{Perimeter} = 2l + 2w$$

For a triangle with base b and altitude h,

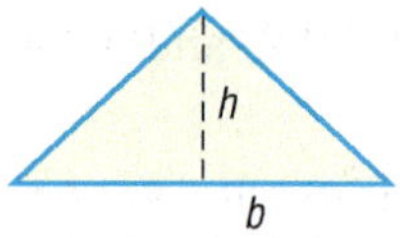

$$\text{Area} = \frac{1}{2}bh$$

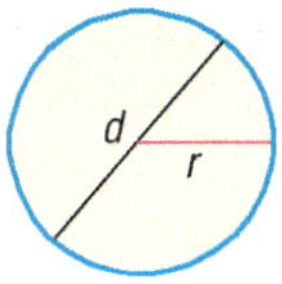

For a circle of radius r (diameter $d = 2r$),

$$\text{Area} = \pi r^2 \qquad \text{Circumference} = 2\pi r = \pi d$$

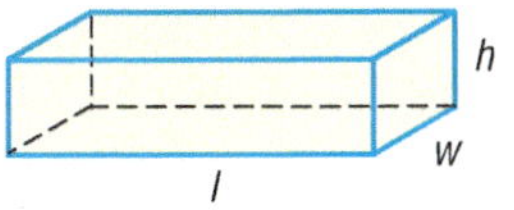

For a closed rectangular box of length l, width w, and height h,

$$\text{Volume} = lwh \qquad \text{Surface area} = 2lh + 2wh + 2lw$$

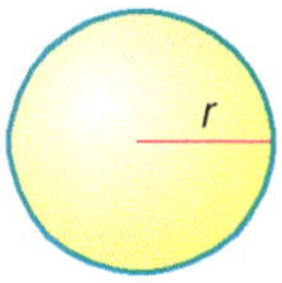

For a sphere of radius r,

$$\text{Volume} = \frac{4}{3}\pi r^3 \qquad \text{Surface area} = 4\pi r^2$$

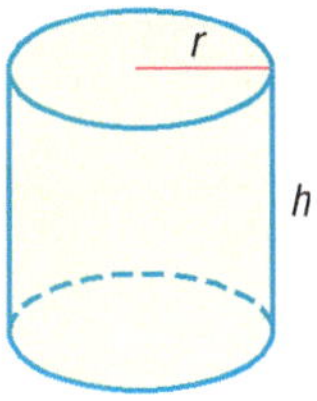

For a closed right circular cylinder of height h and radius r,

$$\text{Volume} = \pi r^2 h \qquad \text{Surface area} = 2\pi r^2 + 2\pi rh$$

Now Work PROBLEM 31

EXAMPLE 4 **Using Geometry Formulas**

A Christmas tree ornament is in the shape of a semicircle on top of a triangle. How many square centimeters (cm^2) of copper is required to make the ornament if the height of the triangle is 6 cm and the base is 4 cm?

Solution See Figure 22. The amount of copper required equals the shaded area. This area is the sum of the areas of the triangle and the semicircle. The triangle has height $h = 6$ and base $b = 4$. The semicircle has diameter $d = 4$, so its radius is $r = 2$.

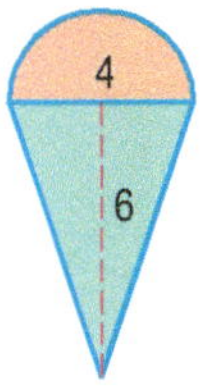

Figure 22

$$\begin{aligned}\text{Total area} &= \text{Area of triangle} + \text{Area of semicircle}\\ &= \frac{1}{2}bh + \frac{1}{2}\pi r^2 = \frac{1}{2}(4)(6) + \frac{1}{2}\pi \cdot 2^2 \qquad b = 4;\ h = 6;\ r = 2\\ &= 12 + 2\pi \approx 18.28\ \text{cm}^2\end{aligned}$$

About $18.28\ \text{cm}^2$ of copper is required. ■

Now Work PROBLEM 49

3 Understand Congruent Triangles and Similar Triangles

Throughout the text we will make reference to triangles. We begin with a discussion of *congruent* triangles. According to dictionary.com, the word **congruent** means "coinciding exactly when superimposed." For example, two angles are congruent if they have the same measure, and two line segments are congruent if they have the same length.

DEFINITION

Two triangles are **congruent** if each pair of corresponding angles have the same measure and each pair of corresponding sides are the same length. ■

In Words
Two triangles are congruent if they have the same size and shape.

In Figure 23 on the next page, corresponding angles are equal and the corresponding sides are equal in length: $a = d$, $b = e$, and $c = f$. As a result, these triangles are congruent.

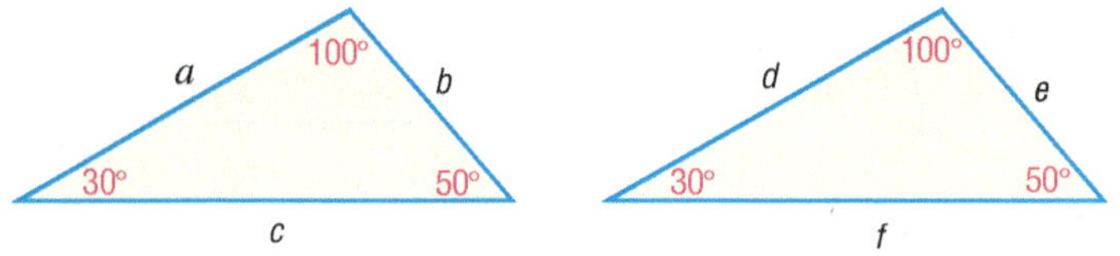

Figure 23 Congruent triangles

Actually, it is not necessary to verify that all three angles and all three sides are the same measure to determine whether two triangles are congruent.

Determining Congruent Triangles

1. **Angle–Side–Angle Case** Two triangles are congruent if two of the angles are equal and the lengths of the corresponding sides between the two angles are equal.
   For example, in Figure 24(a), the two triangles are congruent because two angles and the included side are equal.
2. **Side–Side–Side Case** Two triangles are congruent if the lengths of the corresponding sides of the triangles are equal.
   For example, in Figure 24(b), the two triangles are congruent because the three corresponding sides are all equal.
3. **Side–Angle–Side Case** Two triangles are congruent if the lengths of two corresponding sides are equal and the angles between the two sides are the same.
   For example, in Figure 24(c), the two triangles are congruent because two sides and the included angle are equal.

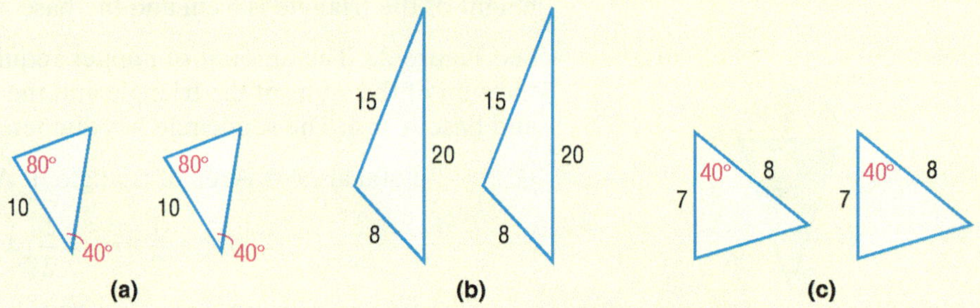

Figure 24

We contrast congruent triangles with *similar* triangles.

DEFINITION

Two triangles are **similar** if the corresponding angles are equal and the lengths of the corresponding sides are proportional.

In Words
Two triangles are similar if they have the same shape, but (possibly) different sizes.

For example, the triangles in Figure 25 are similar because the corresponding angles are equal. In addition, the lengths of the corresponding sides are proportional because each side in the triangle on the right is twice as long as each corresponding side in the triangle on the left. That is, the ratio of the corresponding sides is a constant: $\frac{d}{a} = \frac{e}{b} = \frac{f}{c} = 2$.

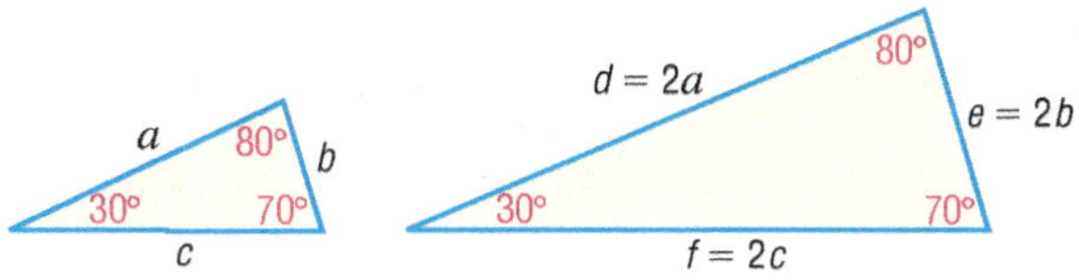

Figure 25 Similar triangles

It is not necessary to verify that all three angles are equal and all three sides are proportional to determine whether two triangles are similar.

Determining Similar Triangles

1. **Angle–Angle Case** Two triangles are similar if two of the corresponding angles are equal.

   For example, in Figure 26(a), the two triangles are similar because two angles are equal.

2. **Side–Side–Side Case** Two triangles are similar if the lengths of all three sides of each triangle are proportional.

   For example, in Figure 26(b), the two triangles are similar because

$$\frac{10}{30} = \frac{5}{15} = \frac{6}{18} = \frac{1}{3}$$

3. **Side–Angle–Side Case** Two triangles are similar if two corresponding sides are proportional and the angles between the two sides are equal.

   For example, in Figure 26(c), the two triangles are similar because $\frac{4}{6} = \frac{12}{18} = \frac{2}{3}$ and the angles between the sides are equal.

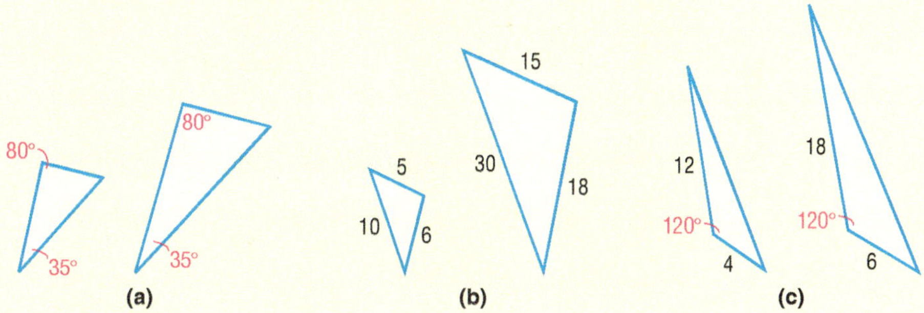

Figure 26

EXAMPLE 5 Using Similar Triangles

Given that the triangles in Figure 27 are similar, find the missing length x and the angles A, B, and C.

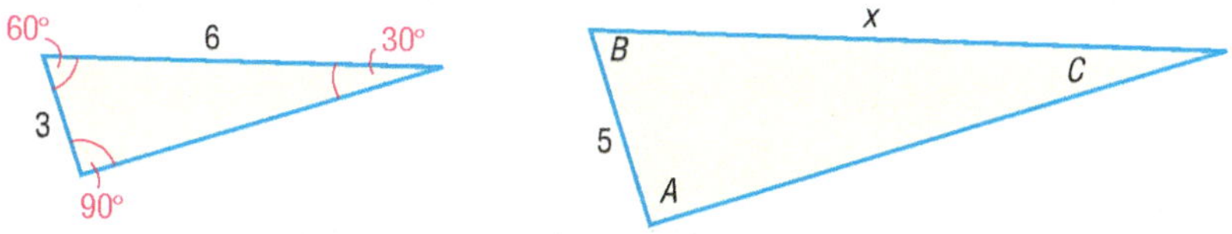

Figure 27

Solution Because the triangles are similar, corresponding angles are equal. So $A = 90°$, $B = 60°$, and $C = 30°$. Also, the corresponding sides are proportional. That is, $\frac{3}{5} = \frac{6}{x}$. We solve this equation for x.

$$\frac{3}{5} = \frac{6}{x}$$

$$5x \cdot \frac{3}{5} = 5x \cdot \frac{6}{x} \quad \text{Multiply both sides by } 5x.$$

$$3x = 30 \quad \text{Simplify.}$$

$$x = 10 \quad \text{Divide both sides by 3.}$$

The missing length is 10 units. ■

Now Work PROBLEM 43

Proof of the Pythagorean Theorem Begin with a square, each side of length $a + b$. In this square, form four right triangles, each having legs equal in length to a and b. See Figure 28. All these triangles are congruent (two sides and their included angle are equal). As a result, the hypotenuse of each is the same, say c, and the pink shading in Figure 28 indicates a square with an area equal to c^2.

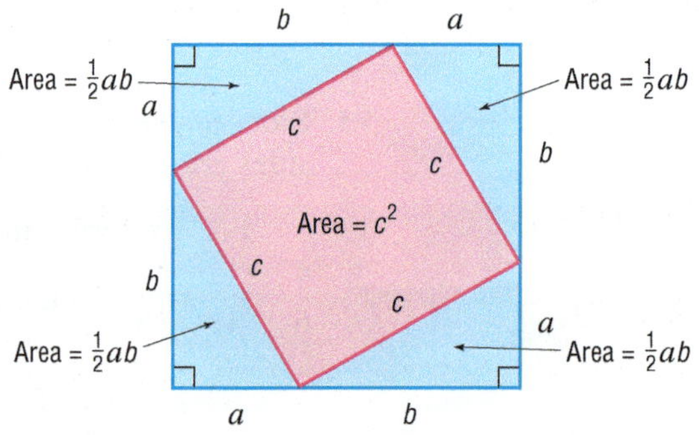

Figure 28

The area of the original square with sides $a + b$ equals the sum of the areas of the four triangles (each of area $\frac{1}{2}ab$) plus the area of the square with side c. That is,

$$(a + b)^2 = \frac{1}{2}ab + \frac{1}{2}ab + \frac{1}{2}ab + \frac{1}{2}ab + c^2$$

$$a^2 + 2ab + b^2 = 2ab + c^2$$

$$a^2 + b^2 = c^2$$

The proof is complete. ■

Proof of the Converse of the Pythagorean Theorem Begin with two triangles: one a right triangle with legs a and b and the other a triangle with sides a, b, and c for which $c^2 = a^2 + b^2$. See Figure 29. By the Pythagorean Theorem, the length x of the third side of the first triangle is

$$x^2 = a^2 + b^2$$

But $c^2 = a^2 + b^2$. Then,

$$x^2 = c^2$$

$$x = c$$

The two triangles have the same sides and are therefore congruent. This means corresponding angles are equal, so the angle opposite side c of the second triangle equals 90°.

The proof is complete. ■

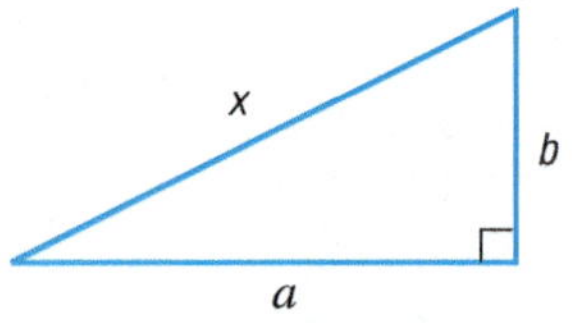

(a)

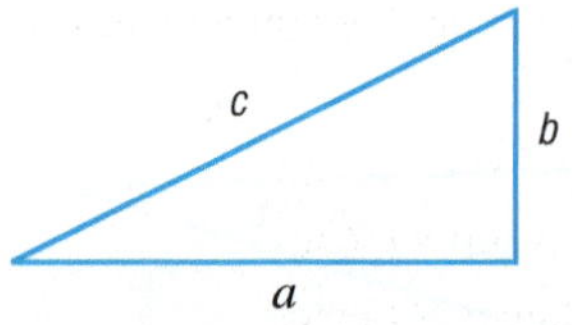

(b) $c^2 = a^2 + b^2$

Figure 29

R.3 Assess Your Understanding

Concepts and Vocabulary

1. A(n) ____ triangle is one that contains an angle of 90 degrees. The longest side is called the __________.

2. For a triangle with base b and altitude h, a formula for the area A is ________.

3. The formula for the circumference C of a circle of radius r is ________.

4. Two triangles are _______ if corresponding angles are equal and the lengths of the corresponding sides are proportional.

5. Which of the following is not a case for determining congruent triangles?

(a) Angle–Side–Angle (b) Side–Angle–Side

(c) Angle–Angle–Angle (d) Side-Side-Side

6. Choose the formula for the volume of a sphere of radius r.

(a) $\frac{4}{3}\pi r^2$ (b) $\frac{4}{3}\pi r^3$ (c) $4\pi r^3$ (d) $4\pi r^2$

7. ***True or False*** In a right triangle, the square of the length of the longest side equals the sum of the squares of the lengths of the other two sides.

8. ***True or False*** The triangle with sides of lengths 6, 8, and 10 is a right triangle.

9. ***True or False*** The surface area of a sphere of radius r is $\frac{4}{3}\pi r^2$.

10. ***True or False*** The triangles shown are congruent.

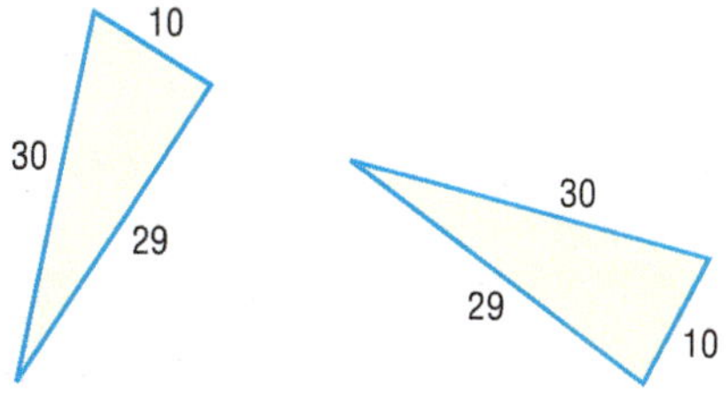

11. ***True or False*** The triangles shown are similar.

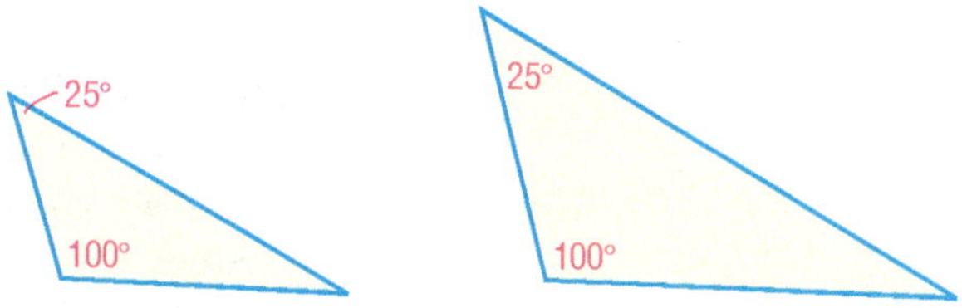

12. ***True or False*** The triangles shown are similar.

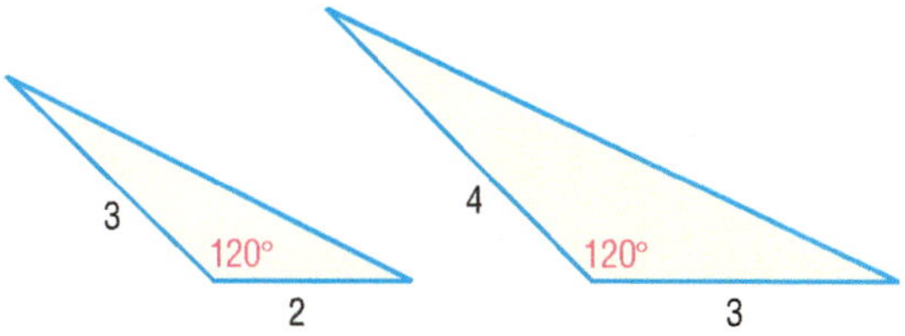

Skill Building

In Problems 13–18, the lengths of the legs of a right triangle are given. Find the hypotenuse.

13. $a = 5$, $b = 12$ **14.** $a = 6$, $b = 8$ **15.** $a = 10$, $b = 24$

16. $a = 4$, $b = 3$ **17.** $a = 7$, $b = 24$ **18.** $a = 14$, $b = 48$

In Problems 19–26, the lengths of the sides of a triangle are given. Determine which are right triangles. For those that are, identify the hypotenuse.

19. 3, 4, 5 **20.** 6, 8, 10 **21.** 4, 5, 6 **22.** 2, 2, 3

23. 7, 24, 25 **24.** 10, 24, 26 **25.** 6, 4, 3 **26.** 5, 4, 7

27. Find the area A of a rectangle with length 4 inches and width 2 inches.

28. Find the area A of a rectangle with length 9 centimeters and width 4 centimeters.

29. Find the area A of a triangle with height 4 inches and base 2 inches.

30. Find the area A of a triangle with height 9 centimeters and base 4 centimeters.

31. Find the area A and circumference C of a circle of radius 5 meters.

32. Find the area A and circumference C of a circle of radius 2 feet.

33. Find the volume V and surface area S of a closed rectangular box with length 8 feet, width 4 feet, and height 7 feet.

34. Find the volume V and surface area S of a closed rectangular box with length 9 inches, width 4 inches, and height 8 inches.

35. Find the volume V and surface area S of a sphere of radius 4 centimeters.

36. Find the volume V and surface area S of a sphere of radius 3 feet.

37. Find the volume V and surface area S of a closed right circular cylinder with radius 9 inches and height 8 inches.

38. Find the volume V and surface area S of a closed right circular cylinder with radius 8 inches and height 9 inches.

In Problems 39–42, find the area of the shaded region.

39.

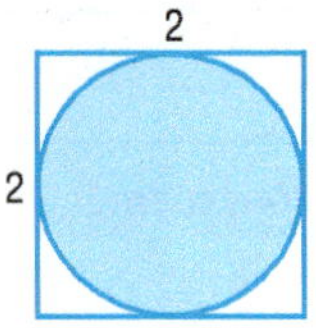

40.

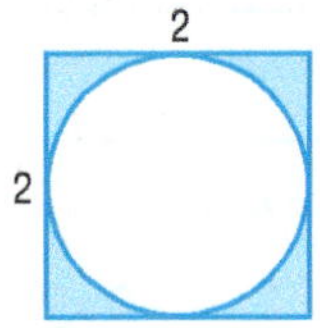

41.

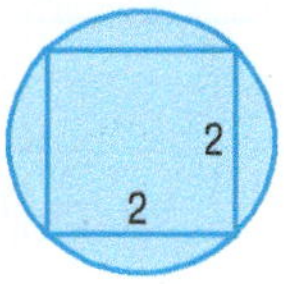

42.

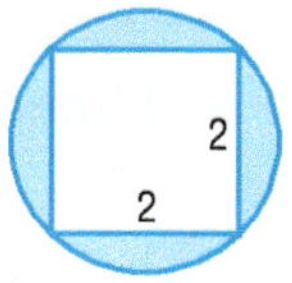

In Problems 43–46, the triangles in each pair are similar. Find the missing length x and the missing angles A, B, and C.

43.

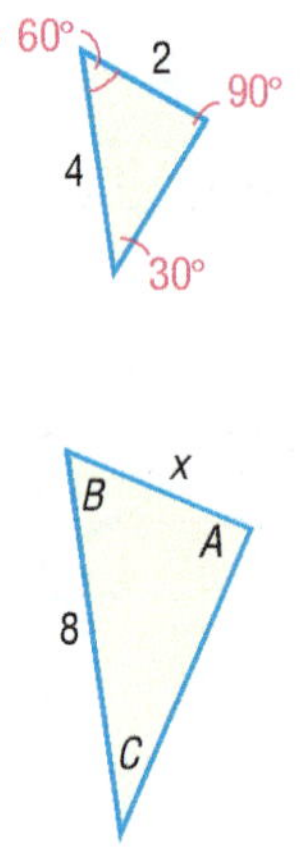

44.

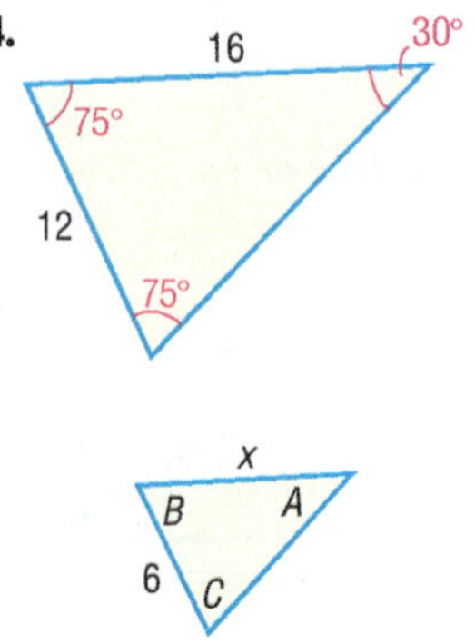

45.

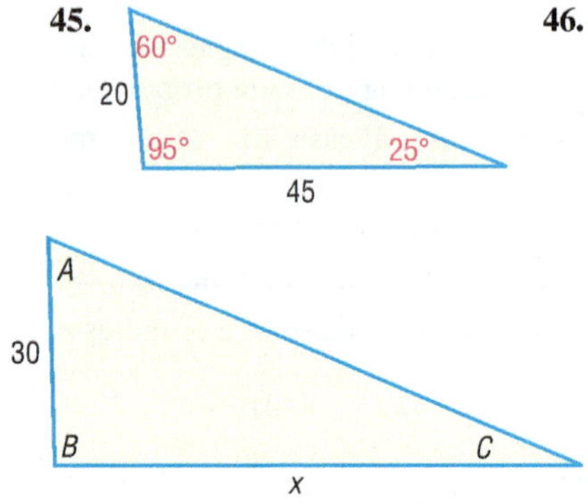

46.

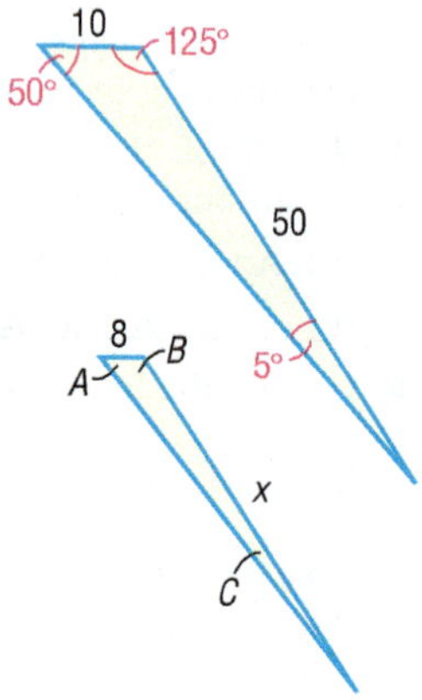

Applications and Extensions

47. How many feet has a wheel with a diameter of 16 inches traveled after four revolutions?

48. How many revolutions will a circular disk with a diameter of 4 feet have completed after it has rolled 20 feet?

49. In the figure shown, $ABCD$ is a square, with each side of length 6 feet. The width of the border (shaded portion) between the outer square $EFGH$ and $ABCD$ is 2 feet. Find the area of the border.

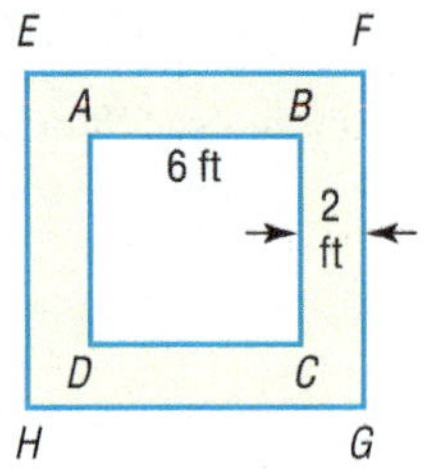

50. Refer to the figure. Square $ABCD$ has an area of 100 square feet; square $BEFG$ has an area of 16 square feet. What is the area of the triangle CGF?

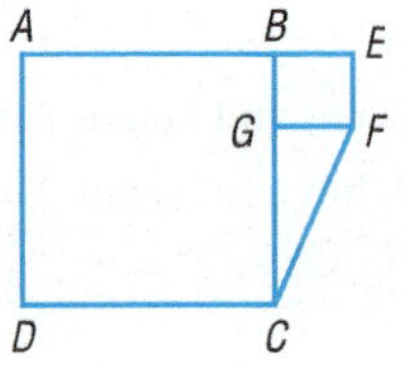

51. Architecture A Norman window consists of a rectangle surmounted by a semicircle. Find the area of the Norman window shown in the illustration. How much wood frame is needed to enclose the window?

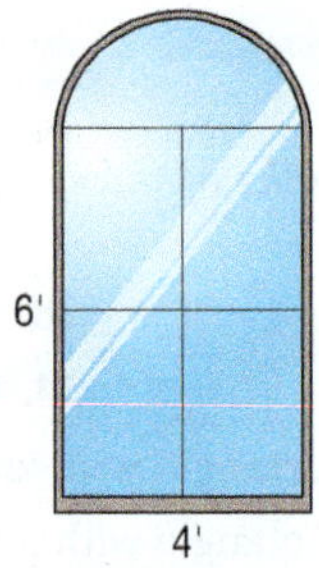

52. Construction A circular swimming pool that is 20 feet in diameter is enclosed by a wooden deck that is 3 feet wide. What is the area of the deck? How much fence is required to enclose the deck?

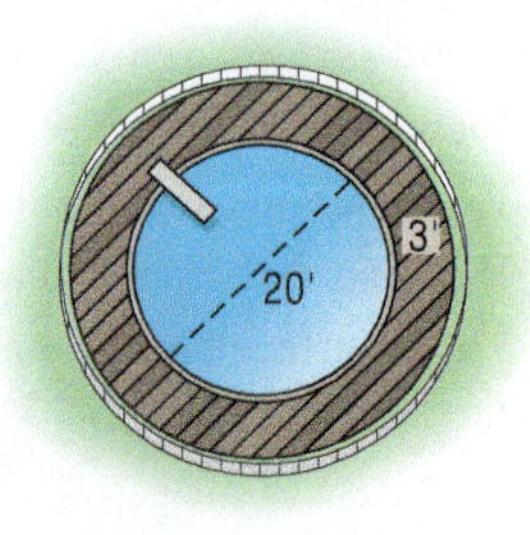

53. How Tall Is the Great Pyramid? The ancient Greek philosopher Thales of Miletus is reported on one occasion to have visited Egypt and calculated the height of the Great Pyramid of Cheops by means of shadow reckoning. Thales knew that each side of the base of the pyramid was 252 paces and that his own height was 2 paces. He measured the length of the pyramid's shadow to be 114 paces and determined the length of his shadow to be 3 paces. See the illustration. Using similar triangles, determine the height of the Great Pyramid in terms of the number of paces.

Source: Diggins, Julie E, *String Straightedge and Shadow: The Story of Geometry,* 2003, Whole Spirit Press, http://wholespiritpress.com.

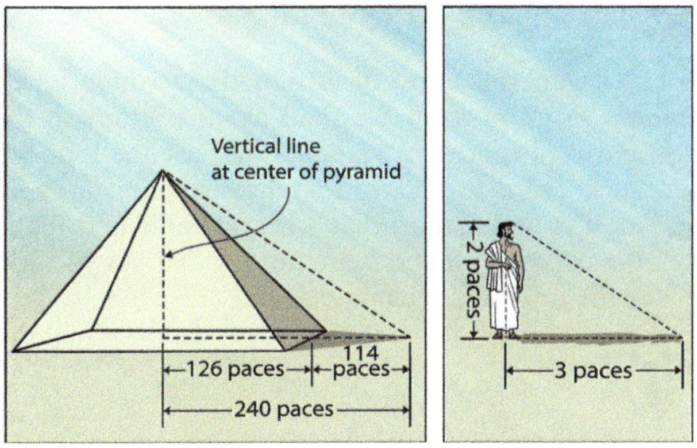

54. The Bermuda Triangle Karen is doing research on the Bermuda Triangle, which she defines roughly by Hamilton, Bermuda; San Juan, Puerto Rico; and Fort Lauderdale, Florida. On her atlas Karen measures the straight-line distances from Hamilton to Fort Lauderdale, Fort Lauderdale to San Juan, and San Juan to Hamilton to be approximately 57 millimeters (mm), 58 mm, and 53.5 mm respectively. If the actual distance from Fort Lauderdale to San Juan is 1046 miles, approximate the actual distances from San Juan to Hamilton and from Hamilton to Fort Lauderdale.

In Problems 55–57, use the facts that the radius of Earth is 3960 miles and 1 mile = 5280 feet.

55. How Far Can You See? The conning tower of the U.S.S. *Silversides,* a World War II submarine now permanently stationed in Muskegon, Michigan, is approximately 20 feet above sea level. How far can you see from the conning tower?

56. How Far Can You See? A person who is 6 feet tall is standing on the beach in Fort Lauderdale, Florida, and looks out onto the Atlantic Ocean. Suddenly, a ship appears on the horizon. How far is the ship from shore?

57. How Far Can You See? The deck of a destroyer is 100 feet above sea level. How far can a person see from the deck? How far can a person see from the bridge, which is 150 feet above sea level?

58. Suppose that m and n are positive integers with $m > n$. If $a = m^2 - n^2$, $b = 2mn$, and $c = m^2 + n^2$, show that a, b, and c are the lengths of the sides of a right triangle. (This formula can be used to find the sides of a right triangle that are integers, such as 3, 4, 5; 5, 12, 13; and so on. Such triplets of integers are called **Pythagorean triples**.)

Explaining Concepts: Discussion and Writing

59. You have 1000 feet of flexible pool siding and intend to construct a swimming pool. Experiment with rectangular-shaped pools with perimeters of 1000 feet. How do their areas vary? What is the shape of the rectangle with the largest area? Now compute the area enclosed by a circular pool with a perimeter (circumference) of 1000 feet. What would be your choice of shape for the pool? If rectangular, what is your preference for dimensions? Justify your choice. If your only consideration is to have a pool that encloses the most area, what shape should you use?

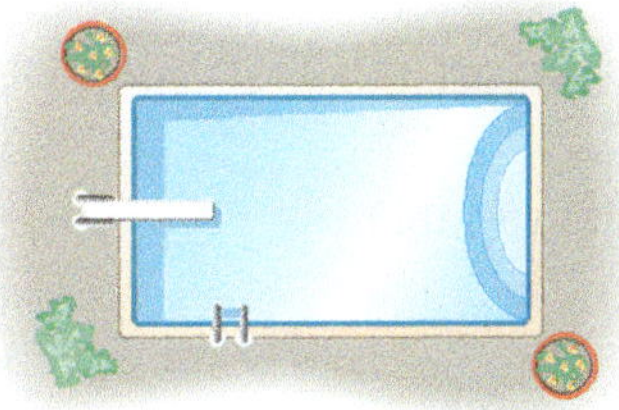

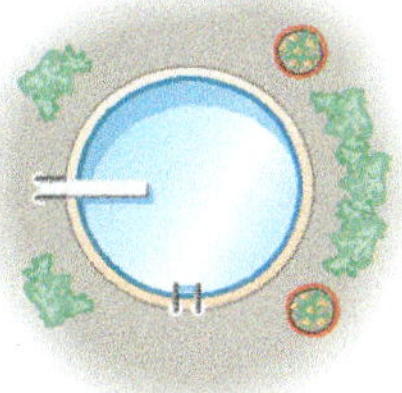

60. The Gibb's Hill Lighthouse, Southampton, Bermuda, in operation since 1846, stands 117 feet high on a hill 245 feet high, so its beam of light is 362 feet above sea level. A brochure states that the light itself can be seen on the horizon about 26 miles distant. Verify the accuracy of this information. The brochure further states that ships 40 miles away can see the light and that planes flying at 10,000 feet can see it 120 miles away. Verify the accuracy of these statements. What assumption did the brochure make about the height of the ship?

R.4 Polynomials

OBJECTIVES

We have described algebra as a generalization of arithmetic in which letters are used to represent real numbers. From now on, we shall use the letters at the end of the alphabet, such as x, y, and z, to represent variables and use the letters at the beginning of the alphabet, such as a, b, and c, to represent constants. In the expressions $3x + 5$ and $ax + b$, it is understood that x is a variable and that a and b are constants, even though the constants a and b are unspecified. As you will find out, the context usually makes the intended meaning clear.

1 Recognize Monomials

DEFINITION

Note: The nonnegative integers are the integers 0, 1, 2, 3,.... ■

A **monomial** in one variable is the product of a constant and a variable raised to a nonnegative integer power. A monomial is of the form

$$ax^k$$

where a is a constant, x is a variable, and $k \geq 0$ is an integer. The constant a is called the **coefficient** of the monomial. If $a \neq 0$, then k is called the **degree** of the monomial. ■

EXAMPLE 1 **Examples of Monomials**

Monomial	Coefficient	Degree	
(a) $6x^2$	6	2	
(b) $-\sqrt{2}x^3$	$-\sqrt{2}$	3	
(c) 3	3	0	Since $3 = 3 \cdot 1 = 3x^0, x \neq 0$
(d) $-5x$	-5	1	Since $-5x = -5x^1$
(e) x^4	1	4	Since $x^4 = 1 \cdot x^4$

■

EXAMPLE 2 **Examples of Nonmonomial Expressions**

(a) $3x^{1/2}$ is not a monomial, since the exponent of the variable x is $\frac{1}{2}$, and $\frac{1}{2}$ is not a nonnegative integer.

(b) $4x^{-3}$ is not a monomial, since the exponent of the variable x is -3, and -3 is not a nonnegative integer. ■

Now Work PROBLEM 9

2 Recognize Polynomials

Two monomials with the same variable raised to the same power are called **like terms**. For example, $2x^4$ and $-5x^4$ are like terms. In contrast, the monomials $2x^3$ and $2x^5$ are not like terms.

We can add or subtract like terms using the Distributive Property. For example,

$$2x^2 + 5x^2 = (2 + 5)x^2 = 7x^2 \quad \text{and} \quad 8x^3 - 5x^3 = (8 - 5)x^3 = 3x^3$$

The sum or difference of two monomials having different degrees is called a **binomial**. The sum or difference of three monomials with three different degrees is called a **trinomial**. For example,

$x^2 - 2$ is a binomial.
$x^3 - 3x + 5$ is a trinomial.
$2x^2 + 5x^2 + 2 = 7x^2 + 2$ is a binomial.

DEFINITION

A **polynomial** in one variable is an algebraic expression of the form

$$a_n x^n + a_{n-1}x^{n-1} + \cdots + a_1 x + a_0 \quad \textbf{(1)}$$

where $a_n, a_{n-1}, \ldots, a_1, a_0$ are constants,* called the **coefficients** of the polynomial, $n \geq 0$ is an integer, and x is a variable. If $a_n \neq 0$, it is called the **leading coefficient**, $a_n x^n$ is called the **leading term**, and n is the **degree** of the polynomial.

In Words
A polynomial is a sum of monomials.

The monomials that make up a polynomial are called its **terms**. If all of the coefficients are 0, the polynomial is called the **zero polynomial**, which has no degree.

Polynomials are usually written in **standard form**, beginning with the nonzero term of highest degree and continuing with terms in descending order according to degree. If a power of x is missing, it is because its coefficient is zero.

EXAMPLE 3 Examples of Polynomials

Polynomial	Coefficients	Degree
$-8x^3 + 4x^2 - 6x + 2$	$-8, 4, -6, 2$	3
$3x^2 - 5 = 3x^2 + 0 \cdot x + (-5)$	$3, 0, -5$	2
$8 - 2x + x^2 = 1 \cdot x^2 + (-2)x + 8$	$1, -2, 8$	2
$5x + \sqrt{2} = 5x^1 + \sqrt{2}$	$5, \sqrt{2}$	1
$3 = 3 \cdot 1 = 3 \cdot x^0$	3	0
0	0	No degree

Although we have been using x to represent the variable, letters such as y and z are also commonly used.

$3x^4 - x^2 + 2$ is a polynomial (in x) of degree 4.
$9y^3 - 2y^2 + y - 3$ is a polynomial (in y) of degree 3.
$z^5 + \pi$ is a polynomial (in z) of degree 5.

Algebraic expressions such as

$$\frac{1}{x} \quad \text{and} \quad \frac{x^2 + 1}{x + 5}$$

* The notation a_n is read as "a sub n." The number n is called a **subscript** and should not be confused with an exponent. We use subscripts to distinguish one constant from another when a large or undetermined number of constants are required.

are not polynomials. The first is not a polynomial because $\frac{1}{x} = x^{-1}$ has an exponent that is not a nonnegative integer. Although the second expression is the quotient of two polynomials, the polynomial in the denominator has degree greater than 0, so the expression cannot be a polynomial.

Now Work PROBLEM 19

3 Add and Subtract Polynomials

Polynomials are added and subtracted by combining like terms.

EXAMPLE 4 Adding Polynomials

Find the sum of the polynomials:

$$8x^3 - 2x^2 + 6x - 2 \quad \text{and} \quad 3x^4 - 2x^3 + x^2 + x$$

Solution We shall find the sum in two ways.

Horizontal Addition: The idea here is to group the like terms and then combine them.

$$\begin{aligned}&(8x^3 - 2x^2 + 6x - 2) + (3x^4 - 2x^3 + x^2 + x)\\ &\quad = 3x^4 + (8x^3 - 2x^3) + (-2x^2 + x^2) + (6x + x) - 2\\ &\quad = 3x^4 + 6x^3 - x^2 + 7x - 2\end{aligned}$$

Vertical Addition: The idea here is to vertically line up the like terms in each polynomial and then add the coefficients.

$$\begin{array}{rrrrrr} & x^4 & x^3 & x^2 & x^1 & x^0\\ & & 8x^3 & -\ 2x^2 & +\ 6x & -\ 2\\ + & 3x^4 & -\ 2x^3 & +\ x^2 & +\ x & \\ \hline & 3x^4 & +\ 6x^3 & -\ x^2 & +\ 7x & -\ 2\end{array}$$

We can subtract two polynomials horizontally or vertically as well.

EXAMPLE 5 Subtracting Polynomials

Find the difference: $(3x^4 - 4x^3 + 6x^2 - 1) - (2x^4 - 8x^2 - 6x + 5)$

Solution ***Horizontal Subtraction:***

$$\begin{aligned}&(3x^4 - 4x^3 + 6x^2 - 1) - (2x^4 - 8x^2 - 6x + 5)\\ &\quad = 3x^4 - 4x^3 + 6x^2 - 1 + \underbrace{(-2x^4 + 8x^2 + 6x - 5)}_{\text{Be sure to change the sign of each term in the second polynomial.}}\\ &\quad \underset{\text{Group like terms.}}{=} (3x^4 - 2x^4) + (-4x^3) + (6x^2 + 8x^2) + 6x + (-1 - 5)\\ &\quad = x^4 - 4x^3 + 14x^2 + 6x - 6\end{aligned}$$

COMMENT Vertical subtraction will be used when we divide polynomials. ■

Vertical Subtraction: We line up like terms, change the sign of each coefficient of the second polynomial, and add.

$$\begin{array}{rcr} \begin{array}{r} \scriptstyle x^4 \quad x^3 \quad x^2 \quad x^1 \quad x^0 \\ 3x^4 - 4x^3 + 6x^2 \qquad - 1 \\ -\ [2x^4 \qquad - 8x^2 - 6x + 5] \\ \hline \end{array} & \begin{array}{c} \\ = \\ = + \\ \\ \end{array} & \begin{array}{r} \scriptstyle x^4 \quad x^3 \quad x^2 \quad x^1 \quad x^0 \\ 3x^4 - 4x^3 + 6x^2 \qquad - 1 \\ -2x^4 \qquad + 8x^2 + 6x - 5 \\ \hline x^4 - 4x^3 + 14x^2 + 6x - 6 \end{array} \end{array}$$

■

Which method to use for adding and subtracting polynomials is up to you. To save space, we shall most often use the horizontal format.

Now Work PROBLEM 31

4 Multiply Polynomials

Two monomials may be multiplied using the Laws of Exponents and the Commutative and Associative Properties. For example,

$$(2x^3) \cdot (5x^4) = (2 \cdot 5) \cdot (x^3 \cdot x^4) = 10x^{3+4} = 10x^7$$

Products of polynomials are found by repeated use of the Distributive Property and the Laws of Exponents. Again, you have a choice of horizontal or vertical format.

EXAMPLE 6 **Multiplying Polynomials**

Find the product: $(2x + 5)(x^2 - x + 2)$

Solution ***Horizontal Multiplication:***

$$(2x + 5)(x^2 - x + 2) = 2x(x^2 - x + 2) + 5(x^2 - x + 2)$$

↑ Distributive Property

$$= (2x \cdot x^2 - 2x \cdot x + 2x \cdot 2) + (5 \cdot x^2 - 5 \cdot x + 5 \cdot 2)$$

↑ Distributive Property

$$= (2x^3 - 2x^2 + 4x) + (5x^2 - 5x + 10)$$

↑ Law of Exponents

$$= 2x^3 + 3x^2 - x + 10$$

↑ Combine like terms.

Vertical Multiplication: The idea here is very much like multiplying a two-digit number by a three-digit number.

$$\begin{array}{rrl} & x^2 - \ \ x + \ 2 & \\ & 2x + \ 5 & \\ \hline & 2x^3 - 2x^2 + 4x \qquad & \text{This line is } 2x(x^2 - x + 2). \\ (+) & 5x^2 - 5x + 10 & \text{This line is } 5(x^2 - x + 2). \\ \hline & 2x^3 + 3x^2 - \ \ x + 10 & \text{Sum of the above two lines} \end{array}$$

■

Now Work PROBLEM 47

5 Know Formulas for Special Products

Certain products, which we call **special products**, occur frequently in algebra. We can calculate them easily using the **FOIL** (*F*irst, *O*uter, *I*nner, *L*ast) method of multiplying two binomials.

Outer, First, Inner, Last

$$(ax + b)(cx + d) = ax(cx + d) + b(cx + d)$$

First Outer Inner Last

$$= ax \cdot cx + ax \cdot d + b \cdot cx + b \cdot d$$
$$= acx^2 + adx + bcx + bd$$
$$= acx^2 + (ad + bc)x + bd$$

EXAMPLE 7 Using FOIL

(a) $(x - 3)(x + 3) = x^2 + 3x - 3x - 9 = x^2 - 9$

F O I L

(b) $(x + 2)^2 = (x + 2)(x + 2) = x^2 + 2x + 2x + 4 = x^2 + 4x + 4$

(c) $(x - 3)^2 = (x - 3)(x - 3) = x^2 - 3x - 3x + 9 = x^2 - 6x + 9$

(d) $(x + 3)(x + 1) = x^2 + x + 3x + 3 = x^2 + 4x + 3$

(e) $(2x + 1)(3x + 4) = 6x^2 + 8x + 3x + 4 = 6x^2 + 11x + 4$ ■

Now Work PROBLEMS 49 AND 57

Some products have been given special names because of their form. The following special products are based on Examples 7(a), (b), and (c).

Difference of Two Squares

$$(x - a)(x + a) = x^2 - a^2 \quad \textbf{(2)}$$

Squares of Binomials, or Perfect Squares

$$(x + a)^2 = x^2 + 2ax + a^2 \quad \textbf{(3a)}$$
$$(x - a)^2 = x^2 - 2ax + a^2 \quad \textbf{(3b)}$$

EXAMPLE 8 Using Special Product Formulas

(a) $(x - 5)(x + 5) = x^2 - 5^2 = x^2 - 25$ — Difference of two squares

(b) $(x + 7)^2 = x^2 + 2 \cdot 7 \cdot x + 7^2 = x^2 + 14x + 49$ — Square of a binomial

(c) $(2x + 1)^2 = (2x)^2 + 2 \cdot 1 \cdot 2x + 1^2 = 4x^2 + 4x + 1$ — Notice that we used $2x$ in place of x in formula (3a).

(d) $(3x - 4)^2 = (3x)^2 - 2 \cdot 4 \cdot 3x + 4^2 = 9x^2 - 24x + 16$ — Replace x by $3x$ in formula (3b). ■

Now Work PROBLEMS 67, 69, AND 71

Let's look at some more examples that lead to general formulas.

EXAMPLE 9

Cubing a Binomial

(a) $(x + 2)^3 = (x + 2)(x + 2)^2 = (x + 2)(x^2 + 4x + 4)$ **Formula (3a)**

$$= (x^3 + 4x^2 + 4x) + (2x^2 + 8x + 8)$$
$$= x^3 + 6x^2 + 12x + 8$$

(b) $(x - 1)^3 = (x - 1)(x - 1)^2 = (x - 1)(x^2 - 2x + 1)$ **Formula (3b)**

$$= (x^3 - 2x^2 + x) - (x^2 - 2x + 1)$$
$$= x^3 - 3x^2 + 3x - 1$$

■

Cubes of Binomials, or Perfect Cubes

$$(x + a)^3 = x^3 + 3ax^2 + 3a^2x + a^3 \quad \textbf{(4a)}$$
$$(x - a)^3 = x^3 - 3ax^2 + 3a^2x - a^3 \quad \textbf{(4b)}$$

Now Work PROBLEM 87

EXAMPLE 10

Forming the Difference of Two Cubes

$$(x - 1)(x^2 + x + 1) = x(x^2 + x + 1) - 1(x^2 + x + 1)$$
$$= x^3 + x^2 + x - x^2 - x - 1$$
$$= x^3 - 1$$

■

EXAMPLE 11

Forming the Sum of Two Cubes

$$(x + 2)(x^2 - 2x + 4) = x(x^2 - 2x + 4) + 2(x^2 - 2x + 4)$$
$$= x^3 - 2x^2 + 4x + 2x^2 - 4x + 8$$
$$= x^3 + 8$$

■

Examples 10 and 11 lead to two more special products.

Difference of Two Cubes

$$(x - a)(x^2 + ax + a^2) = x^3 - a^3 \quad \textbf{(5)}$$

Sum of Two Cubes

$$(x + a)(x^2 - ax + a^2) = x^3 + a^3 \quad \textbf{(6)}$$

6 Divide Polynomials Using Long Division

The procedure for dividing two polynomials is similar to the procedure for dividing two integers.

EXAMPLE 12 Dividing Two Integers

Divide 842 by 15.

Solution

$$\begin{array}{rrl} & 56 & \leftarrow \textbf{Quotient} \\ \textbf{Divisor} \rightarrow & 15\overline{)842} & \leftarrow \textbf{Dividend} \\ & \underline{75} & \leftarrow \textbf{5} \cdot \textbf{15 (subtract)} \\ & 92 & \leftarrow \textbf{Bring down the 2.} \\ & \underline{90} & \leftarrow \textbf{6} \cdot \textbf{15 (subtract)} \\ & 2 & \leftarrow \textbf{Remainder} \end{array}$$

So, $\dfrac{842}{15} = 56 + \dfrac{2}{15}$. ■

In the long-division process detailed in Example 12, the number 15 is called the **divisor**, the number 842 is called the **dividend**, the number 56 is called the **quotient**, and the number 2 is called the **remainder**.

To check the answer obtained in a division problem, multiply the quotient by the divisor and add the remainder. The answer should be the dividend.

$$(\text{Quotient})(\text{Divisor}) + \text{Remainder} = \text{Dividend}$$

For example, we can check the results obtained in Example 12 as follows:

$$(56)(15) + 2 = 840 + 2 = 842$$

To divide two polynomials, we first must write each polynomial in standard form. The process then follows a pattern similar to that of Example 12. The next example illustrates the procedure.

EXAMPLE 13 Dividing Two Polynomials

Find the quotient and the remainder when

$$3x^3 + 4x^2 + x + 7 \quad \text{is divided by} \quad x^2 + 1$$

Solution Each polynomial is in standard form. The dividend is $3x^3 + 4x^2 + x + 7$, and the divisor is $x^2 + 1$.

Note: Remember, a polynomial is in standard form when its terms are written in descending powers of x. ■

STEP 1: Divide the leading term of the dividend, $3x^3$, by the leading term of the divisor, x^2. Enter the result, $3x$, over the term $3x^3$, as follows:

$$\begin{array}{r} 3x \\ x^2 + 1\overline{)3x^3 + 4x^2 + x + 7} \end{array}$$

STEP 2: Multiply $3x$ by $x^2 + 1$, and enter the result below the dividend.

$$\begin{array}{rl} \begin{array}{r} 3x \\ x^2 + 1\overline{)3x^3 + 4x^2 + x + 7} \\ \underline{3x^3 + 3x} \end{array} & \leftarrow \mathbf{3x \cdot (x^2 + 1) = 3x^3 + 3x} \end{array}$$

↑ **Align the $3x$ term under the x to make the next step easier.**

STEP 3: Subtract and bring down the remaining terms.

$$\begin{array}{rl} 3x & \\ x^2 + 1\overline{)3x^3 + 4x^2 + x + 7} & \\ \underline{3x^3 + 3x} & \leftarrow \textbf{Subtract (change the signs and add).} \\ 4x^2 - 2x + 7 & \leftarrow \textbf{Bring down the } \mathbf{4x^2} \textbf{ and the 7.} \end{array}$$

Step 4: Repeat Steps 1–3 using $4x^2 - 2x + 7$ as the dividend.

$$\begin{array}{r}
3x \quad + 4 \\
x^2 + 1 \overline{)3x^3 + 4x^2 + \quad x + 7} \\
3x^3 \qquad + 3x \quad \\
\hline
4x^2 - 2x + 7 \\
4x^2 \qquad + 4 \\
\hline
-2x + 3
\end{array}$$

← Divide $4x^2$ by x^2 to get 4.
← Multiply $x^2 + 1$ by 4; subtract.

COMMENT If the degree of the divisor is greater than the degree of the dividend, then the process ends. ■

Since x^2 does not divide $-2x$ evenly (that is, the result is not a monomial), the process ends. The quotient is $3x + 4$, and the remainder is $-2x + 3$.

✓**Check:** (Quotient) (Divisor) + Remainder

$$= (3x + 4)(x^2 + 1) + (-2x + 3)$$
$$= 3x^3 + 3x + 4x^2 + 4 + (-2x + 3)$$
$$= 3x^3 + 4x^2 + x + 7 = \text{Dividend}$$

Then

$$\frac{3x^3 + 4x^2 + x + 7}{x^2 + 1} = 3x + 4 + \frac{-2x + 3}{x^2 + 1}$$ ■

The next example combines the steps involved in long division.

EXAMPLE 14 Dividing Two Polynomials

Find the quotient and the remainder when

$$x^4 - 3x^3 + 2x - 5 \quad \text{is divided by} \quad x^2 - x + 1$$

Solution In setting up this division problem, it is necessary to leave a space for the missing x^2 term in the dividend.

$$\begin{array}{r}
x^2 - 2x \quad - 3 \\
x^2 - x + 1 \overline{)x^4 - 3x^3 \qquad\quad + 2x - 5} \\
x^4 - \ x^3 + \ x^2 \qquad\qquad\quad \\
\hline
-2x^3 - \ x^2 + 2x - 5 \\
-2x^3 + 2x^2 - 2x \quad\ \ \\
\hline
-3x^2 + 4x - 5 \\
-3x^2 + 3x - 3 \\
\hline
x - 2
\end{array}$$

Divisor → $x^2 - x + 1$; ← Quotient: $x^2 - 2x - 3$; ← Dividend: $x^4 - 3x^3 + 2x - 5$; Subtract → (each step); ← Remainder: $x - 2$

✓**Check:** (Quotient) (Divisor) + Remainder

$$= (x^2 - 2x - 3)(x^2 - x + 1) + (x - 2)$$
$$= x^4 - x^3 + x^2 - 2x^3 + 2x^2 - 2x - 3x^2 + 3x - 3 + x - 2$$
$$= x^4 - 3x^3 + 2x - 5 = \text{Dividend}$$

As a result,

$$\frac{x^4 - 3x^3 + 2x - 5}{x^2 - x + 1} = x^2 - 2x - 3 + \frac{x - 2}{x^2 - x + 1}$$ ■

The process of dividing two polynomials leads to the following result:

THEOREM

Let Q be a polynomial of positive degree, and let P be a polynomial whose degree is greater than or equal to the degree of Q. The remainder after dividing P by Q is either the zero polynomial or a polynomial whose degree is less than the degree of the divisor Q.

Now Work PROBLEM 95

7 Work with Polynomials in Two Variables

A **monomial in two variables** x and y has the form $ax^n y^m$, where a is a constant, x and y are variables, and n and m are nonnegative integers. The **degree** of a monomial is the sum of the powers of the variables.

For example,

$$2xy^3, \quad x^2y^2, \quad \text{and} \quad x^3y$$

are all monomials that have degree 4.

A **polynomial in two variables** x and y is the sum of one or more monomials in two variables. The **degree of a polynomial** in two variables is the highest degree of all the monomials with nonzero coefficients.

EXAMPLE 15 **Examples of Polynomials in Two Variables**

$3x^2 + 2x^3y + 5$	$\pi x^3 - y^2$	$x^4 + 4x^3y - xy^3 + y^4$
Two variables, degree is 4.	Two variables, degree is 3.	Two variables, degree is 4.

Multiplying polynomials in two variables is handled in the same way as multiplying polynomials in one variable.

EXAMPLE 16 **Using a Special Product Formula**

To multiply $(2x - y)^2$, use the Squares of Binomials formula (3b) with $2x$ instead of x and with y instead of a.

$$\begin{aligned}(2x - y)^2 &= (2x)^2 - 2 \cdot y \cdot 2x + y^2 \\ &= 4x^2 - 4xy + y^2\end{aligned}$$

Now Work PROBLEM 81

R.4 Assess Your Understanding

Concepts and Vocabulary

1. The polynomial $3x^4 - 2x^3 + 13x^2 - 5$ is of degree __. The leading coefficient is __.
2. $(x^2 - 4)(x^2 + 4) =$ _______.
3. $(x - 2)(x^2 + 2x + 4) =$ _______.
4. The monomials that make up a polynomial are called which of the following?
   (a) terms (b) variables (c) factors (d) coefficients
5. Choose the degree of the monomial $3x^4y^2$.
   (a) 3 (b) 8 (c) 6 (d) 2
6. ***True or False*** $4x^{-2}$ is a monomial of degree -2.
7. ***True or False*** The degree of the product of two nonzero polynomials equals the sum of their degrees.
8. ***True or False*** $(x + a)(x^2 + ax + a) = x^3 + a^3$.

Skill Building

In Problems 9–18, tell whether the expression is a monomial. If it is, name the variable(s) and the coefficient, and give the degree of the monomial. If it is not a monomial, state why not.

9. $2x^3$ **10.** $-4x^2$ **11.** $\frac{8}{x}$ **12.** $-2x^{-3}$ **13.** $-2xy^2$

14. $5x^2y^3$ **15.** $\frac{8x}{y}$ **16.** $-\frac{2x^2}{y^3}$ **17.** $x^2 + y^2$ **18.** $3x^2 + 4$

In Problems 19–28, tell whether the expression is a polynomial. If it is, give its degree. If it is not, state why not.

19. $3x^2 - 5$ **20.** $1 - 4x$ **21.** 5 **22.** $-\pi$ **23.** $3x^2 - \frac{5}{x}$

24. $\frac{3}{x} + 2$ **25.** $2y^3 - \sqrt{2}$ **26.** $10z^2 + z$ **27.** $\frac{x^2 + 5}{x^3 - 1}$ **28.** $\frac{3x^3 + 2x - 1}{x^2 + x + 1}$

In Problems 29–48, add, subtract, or multiply, as indicated. Express your answer as a single polynomial in standard form.

29. $(x^2 + 4x + 5) + (3x - 3)$ **30.** $(x^3 + 3x^2 + 2) + (x^2 - 4x + 4)$

31. $(x^3 - 2x^2 + 5x + 10) - (2x^2 - 4x + 3)$ **32.** $(x^2 - 3x - 4) - (x^3 - 3x^2 + x + 5)$

33. $(6x^5 + x^3 + x) + (5x^4 - x^3 + 3x^2)$ **34.** $(10x^5 - 8x^2) + (3x^3 - 2x^2 + 6)$

35. $(x^2 - 3x + 1) + 2(3x^2 + x - 4)$ **36.** $-2(x^2 + x + 1) + (-5x^2 - x + 2)$

37. $6(x^3 + x^2 - 3) - 4(2x^3 - 3x^2)$ **38.** $8(4x^3 - 3x^2 - 1) - 6(4x^3 + 8x - 2)$

39. $(x^2 - x + 2) + (2x^2 - 3x + 5) - (x^2 + 1)$ **40.** $(x^2 + 1) - (4x^2 + 5) + (x^2 + x - 2)$

41. $9(y^2 - 3y + 4) - 6(1 - y^2)$ **42.** $8(1 - y^3) + 4(1 + y + y^2 + y^3)$

43. $x(x^2 + x - 4)$ **44.** $4x^2(x^3 - x + 2)$

45. $-2x^2(4x^3 + 5)$ **46.** $5x^3(3x - 4)$

47. $(x + 1)(x^2 + 2x - 4)$ **48.** $(2x - 3)(x^2 + x + 1)$

In Problems 49–66, multiply the polynomials using the FOIL method. Express your answer as a single polynomial in standard form.

49. $(x + 2)(x + 4)$ **50.** $(x + 3)(x + 5)$ **51.** $(2x + 5)(x + 2)$

52. $(3x + 1)(2x + 1)$ **53.** $(x - 4)(x + 2)$ **54.** $(x + 4)(x - 2)$

55. $(x - 3)(x - 2)$ **56.** $(x - 5)(x - 1)$ **57.** $(2x + 3)(x - 2)$

58. $(2x - 4)(3x + 1)$ **59.** $(-2x + 3)(x - 4)$ **60.** $(-3x - 1)(x + 1)$

61. $(-x - 2)(-2x - 4)$ **62.** $(-2x - 3)(3 - x)$ **63.** $(x - 2y)(x + y)$

64. $(2x + 3y)(x - y)$ **65.** $(-2x - 3y)(3x + 2y)$ **66.** $(x - 3y)(-2x + y)$

In Problems 67–90, multiply the polynomials using the special product formulas. Express your answer as a single polynomial in standard form.

67. $(x - 7)(x + 7)$ **68.** $(x - 1)(x + 1)$ **69.** $(2x + 3)(2x - 3)$ **70.** $(3x + 2)(3x - 2)$

71. $(x + 4)^2$ **72.** $(x + 5)^2$ **73.** $(x - 4)^2$ **74.** $(x - 5)^2$

75. $(3x + 4)(3x - 4)$ **76.** $(5x - 3)(5x + 3)$ **77.** $(2x - 3)^2$ **78.** $(3x - 4)^2$

79. $(x + y)(x - y)$ **80.** $(x + 3y)(x - 3y)$ **81.** $(3x + y)(3x - y)$ **82.** $(3x + 4y)(3x - 4y)$

83. $(x + y)^2$ **84.** $(x - y)^2$ **85.** $(x - 2y)^2$ **86.** $(2x + 3y)^2$

87. $(x - 2)^3$ **88.** $(x + 1)^3$ **89.** $(2x + 1)^3$ **90.** $(3x - 2)^3$

In Problems 91–106, find the quotient and the remainder. Check your work by verifying that

$$(\textit{Quotient})(\textit{Divisor}) + \textit{Remainder} = \textit{Dividend}$$

91. $4x^3 - 3x^2 + x + 1$ divided by $x + 2$

92. $3x^3 - x^2 + x - 2$ divided by $x + 2$

93. $4x^3 - 3x^2 + x + 1$ divided by x^2

94. $3x^3 - x^2 + x - 2$ divided by x^2

95. $5x^4 - 3x^2 + x + 1$ divided by $x^2 + 2$

96. $5x^4 - x^2 + x - 2$ divided by $x^2 + 2$

97. $4x^5 - 3x^2 + x + 1$ divided by $2x^3 - 1$

98. $3x^5 - x^2 + x - 2$ divided by $3x^3 - 1$

99. $2x^4 - 3x^3 + x + 1$ divided by $2x^2 + x + 1$

100. $3x^4 - x^3 + x - 2$ divided by $3x^2 + x + 1$

101. $-4x^3 + x^2 - 4$ divided by $x - 1$

102. $-3x^4 - 2x - 1$ divided by $x - 1$

103. $1 - x^2 + x^4$ divided by $x^2 + x + 1$

104. $1 - x^2 + x^4$ divided by $x^2 - x + 1$

105. $x^3 - a^3$ divided by $x - a$

106. $x^5 - a^5$ divided by $x - a$

Explaining Concepts: Discussion and Writing

107. Explain why the degree of the product of two nonzero polynomials equals the sum of their degrees.

108. Explain why the degree of the sum of two polynomials of different degrees equals the larger of their degrees.

109. Give a careful statement about the degree of the sum of two polynomials of the same degree.

110. Do you prefer adding two polynomials using the horizontal method or the vertical method? Write a brief position paper defending your choice.

111. Do you prefer to memorize the rule for the square of a binomial $(x + a)^2$ or to use FOIL to obtain the product? Write a brief position paper defending your choice.

R.5 Factoring Polynomials

OBJECTIVES
1. Factor the Difference of Two Squares and the Sum and Difference of Two Cubes (p. 51)
2. Factor Perfect Squares (p. 52)
3. Factor a Second-Degree Polynomial: $x^2 + Bx + C$ (p. 53)
4. Factor by Grouping (p. 54)
5. Factor a Second-Degree Polynomial: $Ax^2 + Bx + C,\ A \neq 1$ (p. 55)
6. Complete the Square (p. 57)

Consider the following product:

$$(2x + 3)(x - 4) = 2x^2 - 5x - 12$$

The two polynomials on the left side are called **factors** of the polynomial on the right side. Expressing a given polynomial as a product of other polynomials—that is, finding the factors of a polynomial—is called **factoring**.

We shall restrict our discussion here to factoring polynomials in one variable into products of polynomials in one variable, where all coefficients are integers. We call this **factoring over the integers**.

Any polynomial can be written as the product of 1 times itself or as -1 times its additive inverse. If a polynomial cannot be written as the product of two other polynomials (excluding 1 and -1), then the polynomial is **prime**. When a polynomial has been written as a product consisting only of prime factors, it is **factored completely**. Examples of prime polynomials (over the integers) are

$$2, \quad 3, \quad 5, \quad x, \quad x+1, \quad x-1, \quad 3x+4, \quad x^2+4$$

COMMENT Over the real numbers, $3x + 4$ factors into $3\left(x + \frac{4}{3}\right)$. It is the noninteger $\frac{4}{3}$ that causes $3x + 4$ to be prime over the integers. In most instances, we will be factoring over the integers. ■

The first factor to look for in a factoring problem is a common monomial factor present in each term of the polynomial. If one is present, use the Distributive Property to factor it out. Continue factoring out monomial factors until none are left.

EXAMPLE 1

Identifying Common Monomial Factors

Polynomial	Common Monomial Factor	Remaining Factor	Factored Form
$2x+4$	2	$x+2$	$2x+4=2(x+2)$
$3x-6$	3	$x-2$	$3x-6=3(x-2)$
$2x^2-4x+8$	2	x^2-2x+4	$2x^2-4x+8=2(x^2-2x+4)$
$8x-12$	4	$2x-3$	$8x-12=4(2x-3)$
x^2+x	x	$x+1$	$x^2+x=x(x+1)$
x^3-3x^2	x^2	$x-3$	$x^3-3x^2=x^2(x-3)$
$6x^2+9x$	$3x$	$2x+3$	$6x^2+9x=3x(2x+3)$

■

Notice that once all common monomial factors have been removed from a polynomial, the remaining factor is either a prime polynomial of degree 1 or a polynomial of degree 2 or higher. (Do you see why?)

Now Work PROBLEM 9

1 Factor the Difference of Two Squares and the Sum and Difference of Two Cubes

When you factor a polynomial, first check for common monomial factors. Then see whether you can use one of the special formulas discussed in the previous section.

Difference of Two Squares	$x^2-a^2=(x-a)(x+a)$
Perfect Squares	$x^2+2ax+a^2=(x+a)^2$
	$x^2-2ax+a^2=(x-a)^2$
Sum of Two Cubes	$x^3+a^3=(x+a)(x^2-ax+a^2)$
Difference of Two Cubes	$x^3-a^3=(x-a)(x^2+ax+a^2)$

EXAMPLE 2

Factoring the Difference of Two Squares

Factor completely: x^2-4

Solution Note that x^2-4 is the difference of two squares, x^2 and 2^2.

$$x^2-4=(x-2)(x+2)$$

■

EXAMPLE 3 **Factoring the Difference of Two Cubes**

Factor completely: $x^3 - 1$

Solution Because $x^3 - 1$ is the difference of two cubes, x^3 and 1^3,

$$x^3 - 1 = (x - 1)(x^2 + x + 1)$$

EXAMPLE 4 **Factoring the Sum of Two Cubes**

Factor completely: $x^3 + 8$

Solution Because $x^3 + 8$ is the sum of two cubes, x^3 and 2^3,

$$x^3 + 8 = (x + 2)(x^2 - 2x + 4)$$

EXAMPLE 5 **Factoring the Difference of Two Squares**

Factor completely: $x^4 - 16$

Solution Because $x^4 - 16$ is the difference of two squares, $x^4 = (x^2)^2$ and $16 = 4^2$,

$$x^4 - 16 = (x^2 - 4)(x^2 + 4)$$

But $x^2 - 4$ is also the difference of two squares. Then,

$$x^4 - 16 = (x^2 - 4)(x^2 + 4) = (x - 2)(x + 2)(x^2 + 4)$$

Now Work PROBLEMS 19 AND 37

2 Factor Perfect Squares

When the first term and third term of a trinomial are both positive and are perfect squares, such as x^2, $9x^2$, 1, and 4, check to see whether the trinomial is a perfect square.

EXAMPLE 6 **Factoring a Perfect Square**

Factor completely: $x^2 + 6x + 9$

Solution The first term, x^2, and the third term, $9 = 3^2$, are perfect squares. Because the middle term, $6x$, is twice the product of x and 3, we have a perfect square.

$$x^2 + 6x + 9 = (x + 3)^2$$

EXAMPLE 7 **Factoring a Perfect Square**

Factor completely: $9x^2 - 6x + 1$

Solution The first term, $9x^2 = (3x)^2$, and the third term, $1 = 1^2$, are perfect squares. Because the middle term, $-6x$, is -2 times the product of $3x$ and 1, we have a perfect square.

$$9x^2 - 6x + 1 = (3x - 1)^2$$

EXAMPLE 8 **Factoring a Perfect Square**

Factor completely: $25x^2 + 30x + 9$

Solution The first term, $25x^2 = (5x)^2$, and the third term, $9 = 3^2$, are perfect squares. Because the middle term, $30x$, is twice the product of $5x$ and 3, we have a perfect square.

$$25x^2 + 30x + 9 = (5x + 3)^2$$

Now Work PROBLEMS 29 AND 103

If a trinomial is not a perfect square, it may be possible to factor it using the technique discussed next.

3 Factor a Second-Degree Polynomial: $x^2 + Bx + C$

The idea behind factoring a second-degree polynomial like $x^2 + Bx + C$ is to see whether it can be made equal to the product of two (possibly equal) first-degree polynomials.

For example, consider

$$(x + 3)(x + 4) = x^2 + 7x + 12$$

The factors of $x^2 + 7x + 12$ are $x + 3$ and $x + 4$. Notice the following:

$$x^2 + 7x + 12 = (x + 3)(x + 4)$$

12 is the product of 3 and 4.

7 is the sum of 3 and 4.

In general, if $x^2 + Bx + C = (x + a)(x + b) = x^2 + (a + b)x + ab$, then $ab = C$ and $a + b = B$.

To factor a second-degree polynomial $x^2 + Bx + C$, find integers whose product is C and whose sum is B. That is, if there are numbers a, b, where $ab = C$ and $a + b = B$, then

$$x^2 + Bx + C = (x + a)(x + b)$$

EXAMPLE 9 **Factoring a Trinomial**

Factor completely: $x^2 + 7x + 10$

Solution First determine all pairs of integers whose product is 10, and then compute their sums.

Integers whose product is 10	1, 10	−1, −10	2, 5	−2, −5
Sum	11	−11	7	−7

The integers 2 and 5 have a product of 10 and add up to 7, the coefficient of the middle term. As a result,

$$x^2 + 7x + 10 = (x + 2)(x + 5)$$

■

EXAMPLE 10 **Factoring a Trinomial**

Factor completely: $x^2 - 6x + 8$

Solution First determine all pairs of integers whose product is 8, and then compute each sum.

Integers whose product is 8	1, 8	−1, −8	2, 4	−2, −4
Sum	9	−9	6	−6

Since −6 is the coefficient of the middle term,

$$x^2 - 6x + 8 = (x - 2)(x - 4)$$

■

EXAMPLE 11

Factoring a Trinomial

Factor completely: $x^2 - x - 12$

Solution First determine all pairs of integers whose product is -12, and then compute each sum.

Integers whose product is -12	$1, -12$	$-1, 12$	$2, -6$	$-2, 6$	$3, -4$	$-3, 4$
Sum	-11	11	-4	4	-1	1

Since -1 is the coefficient of the middle term,

$$x^2 - x - 12 = (x + 3)(x - 4)$$

EXAMPLE 12

Factoring a Trinomial

Factor completely: $x^2 + 4x - 12$

Solution The integers -2 and 6 have a product of -12 and have the sum 4. So,

$$x^2 + 4x - 12 = (x - 2)(x + 6)$$

To avoid errors in factoring, always check your answer by multiplying it out to see whether the result equals the original expression.

When none of the possibilities works, the polynomial is prime.

EXAMPLE 13

Identifying a Prime Polynomial

Show that $x^2 + 9$ is prime.

Solution First list the pairs of integers whose product is 9, and then compute their sums.

Integers whose product is 9	$1, 9$	$-1, -9$	$3, 3$	$-3, -3$
Sum	10	-10	6	-6

Since the coefficient of the middle term in $x^2 + 9 = x^2 + 0x + 9$ is 0 and none of the sums equals 0, we conclude that $x^2 + 9$ is prime.

Example 13 demonstrates a more general result:

THEOREM Any polynomial of the form $x^2 + a^2$, a real, is prime.

Now Work PROBLEMS 43 AND 87

4 Factor by Grouping

Sometimes a common factor does not occur in every term of the polynomial but does occur in each of several groups of terms that together make up the polynomial. When this happens, the common factor can be factored out of each group by means of the Distributive Property. This technique is called **factoring by grouping**.

EXAMPLE 14

Factoring by Grouping

Factor completely by grouping: $(x^2 + 2)x + (x^2 + 2) \cdot 3$

Solution Notice the common factor $x^2 + 2$. Applying the Distributive Property yields

$$(x^2 + 2)x + (x^2 + 2) \cdot 3 = (x^2 + 2)(x + 3)$$

Since $x^2 + 2$ and $x + 3$ are prime, the factorization is complete.

The next example shows a factoring problem that occurs in calculus.

EXAMPLE 15 **Factoring by Grouping**

Factor completely by grouping: $3(x-1)^2(x+2)^4 + 4(x-1)^3(x+2)^3$

Solution Here, $(x-1)^2(x+2)^3$ is a common factor of both $3(x-1)^2(x+2)^4$ and $4(x-1)^3(x+2)^3$. As a result,

$$\begin{aligned}3(x-1)^2(x+2)^4 + 4(x-1)^3(x+2)^3 &= (x-1)^2(x+2)^3[3(x+2) + 4(x-1)]\\ &= (x-1)^2(x+2)^3[3x+6+4x-4]\\ &= (x-1)^2(x+2)^3(7x+2)\end{aligned}$$

■

EXAMPLE 16 **Factoring by Grouping**

Factor completely by grouping: $x^3 - 4x^2 + 2x - 8$

Solution To see whether factoring by grouping will work, group the first two terms and the last two terms. Then look for a common factor in each group. In this example, factor x^2 from $x^3 - 4x^2$ and 2 from $2x - 8$. The remaining factor in each case is the same, $x - 4$. This means that factoring by grouping will work, as follows:

$$\begin{aligned}x^3 - 4x^2 + 2x - 8 &= (x^3 - 4x^2) + (2x - 8)\\ &= x^2(x-4) + 2(x-4)\\ &= (x-4)(x^2+2)\end{aligned}$$

Since $x^2 + 2$ and $x - 4$ are prime, the factorization is complete. ■

Now Work PROBLEMS 55 AND 131

5 Factor a Second-Degree Polynomial: $Ax^2 + Bx + C$, $A \neq 1$

To factor a second-degree polynomial $Ax^2 + Bx + C$, when $A \neq 1$ and A, B, and C have no common factors, follow these steps:

Steps for Factoring $Ax^2 + Bx + C$, When $A \neq 1$ and A, B, and C Have No Common Factors

STEP 1: Find the value of AC.
STEP 2: Find a pair of integers whose product is AC and that add up to B. That is, find a and b such that $ab = AC$ and $a + b = B$.
STEP 3: Write $Ax^2 + Bx + C = Ax^2 + ax + bx + C$.
STEP 4: Factor this last expression by grouping.

EXAMPLE 17 **Factoring a Trinomial**

Factor completely: $2x^2 + 5x + 3$

Solution Comparing $2x^2 + 5x + 3$ to $Ax^2 + Bx + C$, we find that $A = 2$, $B = 5$, and $C = 3$.

STEP 1: The value of AC is $2 \cdot 3 = 6$.

STEP 2: Determine the pairs of integers whose product is $AC = 6$ and compute their sums.

Integers whose product is 6	1, 6	−1, −6	2, 3	−2, −3
Sum	7	−7	5	−5

Step 3: The integers whose product is 6 that add up to $B = 5$ are 2 and 3.

$$2x^2 + 5x + 3 = 2x^2 + 2x + 3x + 3$$

Step 4: Factor by grouping.

$$\begin{aligned} 2x^2 + 2x + 3x + 3 &= (2x^2 + 2x) + (3x + 3) \\ &= 2x(x + 1) + 3(x + 1) \\ &= (x + 1)(2x + 3) \end{aligned}$$

As a result,

$$2x^2 + 5x + 3 = (x + 1)(2x + 3)$$

EXAMPLE 18 **Factoring a Trinomial**

Factor completely: $2x^2 - x - 6$

Solution Comparing $2x^2 - x - 6$ to $Ax^2 + Bx + C$, we find that $A = 2, B = -1$, and $C = -6$.

Step 1: The value of AC is $2 \cdot (-6) = -12$.

Step 2: Determine the pairs of integers whose product is $AC = -12$ and compute their sums.

Integers whose product is −12	1, −12	−1, 12	2, −6	−2, 6	3, −4	−3, 4
Sum	−11	11	−4	4	−1	1

Step 3: The integers whose product is −12 that add up to $B = -1$ are −4 and 3.

$$2x^2 - x - 6 = 2x^2 - 4x + 3x - 6$$

Step 4: Factor by grouping.

$$\begin{aligned} 2x^2 - 4x + 3x - 6 &= (2x^2 - 4x) + (3x - 6) \\ &= 2x(x - 2) + 3(x - 2) \\ &= (x - 2)(2x + 3) \end{aligned}$$

As a result,

$$2x^2 - x - 6 = (x - 2)(2x + 3)$$

Now Work PROBLEM 61

SUMMARY

Type of Polynomial	Method	Example
Any polynomial	Look for common monomial factors. (Always do this first!)	$6x^2 + 9x = 3x(2x + 3)$
Binomials of degree 2 or higher	Check for a special product:	
	Difference of two squares, $x^2 - a^2$	$x^2 - 16 = (x - 4)(x + 4)$
	Difference of two cubes, $x^3 - a^3$	$x^3 - 64 = (x - 4)(x^2 + 4x + 16)$
	Sum of two cubes, $x^3 + a^3$	$x^3 + 27 = (x + 3)(x^2 - 3x + 9)$
Trinomials of degree 2	Check for a perfect square, $(x \pm a)^2$	$x^2 + 8x + 16 = (x + 4)^2$ $x^2 - 10x + 25 = (x - 5)^2$
	Factoring $x^2 + Bx + C$ (p. 53)	$x^2 - x - 2 = (x - 2)(x + 1)$
	Factoring $Ax^2 + Bx + C$ (p. 55)	$6x^2 + x - 1 = (2x + 1)(3x - 1)$
Four or more terms	Grouping	$2x^3 - 3x^2 + 4x - 6 = (2x - 3)(x^2 + 2)$

6 Complete the Square

The idea behind completing the square in one variable is to "adjust" an expression of the form $x^2 + bx$ to make it a perfect square. Perfect squares are trinomials of the form

$$x^2 + 2ax + a^2 = (x + a)^2 \text{ or } x^2 - 2ax + a^2 = (x - a)^2$$

For example, $x^2 + 6x + 9$ is a perfect square because $x^2 + 6x + 9 = (x + 3)^2$. And $p^2 - 12p + 36$ is a perfect square because $p^2 - 12p + 36 = (p - 6)^2$.

So how do we "adjust" $x^2 + bx$ to make it a perfect square? We do it by adding a number. For example, to make $x^2 + 6x$ a perfect square, add 9. But how do we know to add 9? If we divide the coefficient of the first-degree term, 6, by 2, and then square the result, we obtain 9. This approach works in general.

WARNING To use $\left(\frac{1}{2}b\right)^2$ to complete the square, the coefficient of the x^2 term must be 1. ■

Completing the Square of $x^2 + bx$

Identify the coefficient of the first-degree term. Multiply this coefficient by $\frac{1}{2}$ and then square the result. That is, determine the value of b in $x^2 + bx$ and compute $\left(\frac{1}{2}b\right)^2$.

EXAMPLE 19 Completing the Square

Determine the number that must be added to each expression to complete the square. Then factor the expression.

Start	Add	Result	Factored Form
$y^2 + 8y$	$\left(\frac{1}{2} \cdot 8\right)^2 = 16$	$y^2 + 8y + 16$	$(y + 4)^2$
$x^2 + 12x$	$\left(\frac{1}{2} \cdot 12\right)^2 = 36$	$x^2 + 12x + 36$	$(x + 6)^2$
$a^2 - 20a$	$\left(\frac{1}{2} \cdot (-20)\right)^2 = 100$	$a^2 - 20a + 100$	$(a - 10)^2$
$p^2 - 5p$	$\left(\frac{1}{2} \cdot (-5)\right)^2 = \frac{25}{4}$	$p^2 - 5p + \frac{25}{4}$	$\left(p - \frac{5}{2}\right)^2$

■

Notice that the factored form of a perfect square is either

$$x^2 + bx + \left(\frac{b}{2}\right)^2 = \left(x + \frac{b}{2}\right)^2 \quad \text{or} \quad x^2 - bx + \left(\frac{b}{2}\right)^2 = \left(x - \frac{b}{2}\right)^2$$

Now Work PROBLEM 73

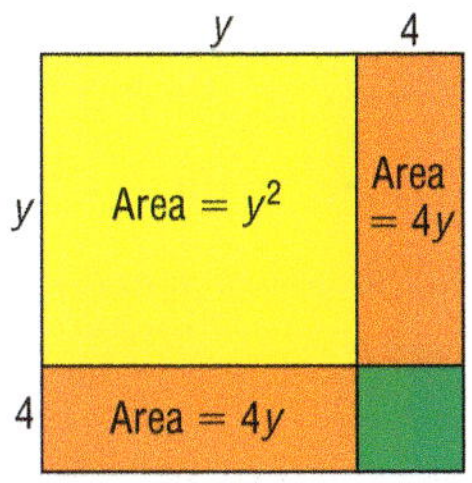

Figure 30

Are you wondering why we refer to making an expression a perfect square as "completing the square"? Look at the square in Figure 30. Its area is $(y + 4)^2$. The yellow area is y^2 and each orange area is $4y$ (for a total area of $8y$). The sum of these areas is $y^2 + 8y$. To complete the square, we need to add the area of the green region: $4 \cdot 4 = 16$. As a result, $y^2 + 8y + 16 = (y + 4)^2$.

R.5 Assess Your Understanding

Concepts and Vocabulary

1. If factored completely, $3x^3 - 12x =$ ______________.
2. If a polynomial cannot be written as the product of two other polynomials (excluding 1 and -1), then the polynomial is said to be ______.
3. For $x^2 + Bx + C = (x + a)(x + b)$, which of the following must be true?
   (a) $ab = B$ and $a + b = C$
   (b) $a + b = C$ and $a - b = B$
   (c) $ab = C$ and $a + b = B$
   (d) $ab = B$ and $a - b = C$
4. Choose the best description of $x^2 - 64$.
   (a) Prime (b) Difference of two squares
   (c) Difference of two cubes (d) Perfect Square
5. Choose the complete factorization of $4x^2 - 8x - 60$.
   (a) $2(x + 3)(x - 5)$ (b) $4(x^2 - 2x - 15)$
   (c) $(2x + 6)(2x - 10)$ (d) $4(x + 3)(x - 5)$
6. To complete the square of $x^2 + bx$, use which of the following?
   (a) $(2b)^2$ (b) $2b^2$ (c) $\left(\frac{1}{2}b\right)^2$ (d) $\frac{1}{2}b^2$
7. ***True or False*** The polynomial $x^2 + 4$ is prime.
8. ***True or False*** $3x^3 - 2x^2 - 6x + 4 = (3x - 2)(x^2 + 2)$.

Skill Building

In Problems 9–18, factor each polynomial by removing the common monomial factor.

9. $3x + 6$ 10. $7x - 14$ 11. $ax^2 + a$ 12. $ax - a$ 13. $x^3 + x^2 + x$
14. $x^3 - x^2 + x$ 15. $2x^2 - 2x$ 16. $3x^2 - 3x$ 17. $3x^2y - 6xy^2 + 12xy$ 18. $60x^2y - 48xy^2 + 72x^3y$

In Problems 19–26, factor the difference of two squares.

19. $x^2 - 1$ 20. $x^2 - 4$ 21. $4x^2 - 1$ 22. $9x^2 - 1$
23. $x^2 - 16$ 24. $x^2 - 25$ 25. $25x^2 - 4$ 26. $36x^2 - 9$

In Problems 27–36, factor the perfect squares.

27. $x^2 + 2x + 1$ 28. $x^2 - 4x + 4$ 29. $x^2 + 4x + 4$ 30. $x^2 - 2x + 1$
31. $x^2 - 10x + 25$ 32. $x^2 + 10x + 25$ 33. $4x^2 + 4x + 1$ 34. $9x^2 + 6x + 1$
35. $16x^2 + 8x + 1$ 36. $25x^2 + 10x + 1$

In Problems 37–42, factor the sum or difference of two cubes.

37. $x^3 - 27$ 38. $x^3 + 125$ 39. $x^3 + 27$ 40. $27 - 8x^3$ 41. $8x^3 + 27$ 42. $64 - 27x^3$

In Problems 43–54, factor each polynomial.

43. $x^2 + 5x + 6$ 44. $x^2 + 6x + 8$ 45. $x^2 + 7x + 6$ 46. $x^2 + 9x + 8$
47. $x^2 + 7x + 10$ 48. $x^2 + 11x + 10$ 49. $x^2 - 10x + 16$ 50. $x^2 - 17x + 16$
51. $x^2 - 7x - 8$ 52. $x^2 - 2x - 8$ 53. $x^2 + 7x - 8$ 54. $x^2 + 2x - 8$

In Problems 55–60, factor by grouping.

55. $2x^2 + 4x + 3x + 6$ 56. $3x^2 - 3x + 2x - 2$ 57. $2x^2 - 4x + x - 2$
58. $3x^2 + 6x - x - 2$ 59. $18x^2 + 27x + 12x + 18$ 60. $45x^3 - 30x^2 + 15x^2 - 10x$

In Problems 61–72, factor each polynomial.

61. $3x^2 + 4x + 1$ 62. $2x^2 + 3x + 1$ 63. $2z^2 + 5z + 3$ 64. $6z^2 + 5z + 1$
65. $3x^2 + 2x - 8$ 66. $3x^2 + 10x + 8$ 67. $3x^2 - 2x - 8$ 68. $3x^2 - 10x + 8$
69. $12x^4 + 56x^3 + 32x^2$ 70. $21x^2 - 98x + 56$ 71. $3x^2 + 10x - 8$ 72. $3x^2 - 10x - 8$

In Problems 73–78, determine what number should be added to complete the square of each expression. Then factor each expression.

73. $x^2 + 10x$ 74. $p^2 + 14p$ 75. $y^2 - 6y$
76. $x^2 - 4x$ 77. $x^2 - \frac{1}{2}x$ 78. $x^2 + \frac{1}{3}x$

Mixed Practice

In Problems 79–126, factor each polynomial completely. If the polynomial cannot be factored, say it is prime.

79. $x^2 - 36$

80. $x^2 - 9$

81. $2 - 8x^2$

82. $3 - 27x^2$

83. $8x^2 + 88x + 80$

84. $10x^3 + 50x^2 + 40x$

85. $x^2 - 10x + 21$

86. $x^2 - 6x + 8$

87. $4x^2 - 8x + 32$

88. $3x^2 - 12x + 15$

89. $x^2 + 4x + 16$

90. $x^2 + 12x + 36$

91. $15 + 2x - x^2$

92. $14 + 6x - x^2$

93. $3x^2 - 12x - 36$

94. $x^3 + 8x^2 - 20x$

95. $y^4 + 11y^3 + 30y^2$

96. $3y^3 - 18y^2 - 48y$

97. $8x^5 + 24x^4 + 18x^3$

98. $36x^6 - 48x^5 + 16x^4$

99. $6x^2 + 8x + 2$

100. $8x^2 + 6x - 2$

101. $x^4 - 81$

102. $x^4 - 1$

103. $x^6 - 2x^3 + 1$

104. $x^6 + 2x^3 + 1$

105. $x^7 - x^5$

106. $x^8 - x^5$

107. $16x^2 + 24x + 9$

108. $9x^2 - 24x + 16$

109. $5 + 16x - 16x^2$

110. $5 + 11x - 16x^2$

111. $4y^2 - 16y + 15$

112. $9y^2 + 9y - 4$

113. $1 - 8x^2 - 9x^4$

114. $4 - 14x^2 - 8x^4$

115. $x(x + 3) - 6(x + 3)$

116. $5(3x - 7) + x(3x - 7)$

117. $(x + 2)^2 - 5(x + 2)$

118. $(x - 1)^2 - 2(x - 1)$

119. $(3x - 2)^3 - 27$

120. $(5x + 1)^3 - 1$

121. $3(x^2 + 10x + 25) - 4(x + 5)$

122. $7(x^2 - 6x + 9) + 5(x - 3)$

123. $x^3 + 2x^2 - x - 2$

124. $x^3 - 3x^2 - x + 3$

125. $x^4 - x^3 + x - 1$

126. $x^4 + x^3 + x + 1$

Applications and Extensions

In Problems 127–136, expressions that occur in calculus are given. Factor each expression completely.

127. $2(3x + 4)^2 + (2x + 3) \cdot 2(3x + 4) \cdot 3$

128. $5(2x + 1)^2 + (5x - 6) \cdot 2(2x + 1) \cdot 2$

129. $2x(2x + 5) + x^2 \cdot 2$

130. $3x^2(8x - 3) + x^3 \cdot 8$

131. $2(x + 3)(x - 2)^3 + (x + 3)^2 \cdot 3(x - 2)^2$

132. $4(x + 5)^3(x - 1)^2 + (x + 5)^4 \cdot 2(x - 1)$

133. $(4x - 3)^2 + x \cdot 2(4x - 3) \cdot 4$

134. $3x^2(3x + 4)^2 + x^3 \cdot 2(3x + 4) \cdot 3$

135. $2(3x - 5) \cdot 3(2x + 1)^3 + (3x - 5)^2 \cdot 3(2x + 1)^2 \cdot 2$

136. $3(4x + 5)^2 \cdot 4(5x + 1)^2 + (4x + 5)^3 \cdot 2(5x + 1) \cdot 5$

137. Show that $x^2 + 4$ is prime.

138. Show that $x^2 + x + 1$ is prime.

Explaining Concepts: Discussion and Writing

139. Make up a polynomial that factors into a perfect square.

140. Explain to a fellow student what you look for first when presented with a factoring problem. What do you do next?

R.6 Synthetic Division

OBJECTIVE 1 Divide Polynomials Using Synthetic Division (p. 59)

1 Divide Polynomials Using Synthetic Division

To find the quotient as well as the remainder when a polynomial of degree 1 or higher is divided by $x - c$, a shortened version of long division, called **synthetic division**, makes the task simpler.

To see how synthetic division works, first consider long division for dividing the polynomial $2x^3 - x^2 + 3$ by $x - 3$.

$$
\begin{array}{r l}
 & \quad 2x^2 + 5x + 15 \qquad \leftarrow \textbf{Quotient} \\
x - 3 \overline{)} & 2x^3 - x^2 \qquad\qquad + 3 \\
 & \underline{2x^3 - 6x^2} \\
 & \qquad 5x^2 \\
 & \qquad \underline{5x^2 - 15x} \\
 & \qquad\qquad 15x + 3 \\
 & \qquad\qquad \underline{15x - 45} \\
 & \qquad\qquad\qquad 48 \leftarrow \textbf{Remainder}
\end{array}
$$

Check: (Divisor) · (Quotient) + Remainder

$$
\begin{aligned}
&= (x - 3)(2x^2 + 5x + 15) + 48 \\
&= 2x^3 + 5x^2 + 15x - 6x^2 - 15x - 45 + 48 \\
&= 2x^3 - x^2 + 3
\end{aligned}
$$

The process of synthetic division arises from rewriting the long division in a more compact form, using simpler notation. For example, in the long division above, the terms in blue are not really necessary because they are identical to the terms directly above them. With these terms removed, we have

$$
\begin{array}{r l}
 & \quad 2x^2 + 5x + 15 \\
x - 3 \overline{)} & 2x^3 - x^2 \qquad\qquad + 3 \\
 & \underline{\quad - 6x^2} \\
 & \qquad 5x^2 \\
 & \qquad\qquad \underline{- 15x} \\
 & \qquad\qquad 15x \\
 & \qquad\qquad\qquad \underline{- 45} \\
 & \qquad\qquad\qquad 48
\end{array}
$$

Most of the x's that appear in this process can also be removed, provided that we are careful about positioning each coefficient. In this regard, we will need to use 0 as the coefficient of x in the dividend, because that power of x is missing. Now we have

$$
\begin{array}{r rrrr}
 & & 2x^2 + & 5x + & 15 \\
x - 3 \overline{)} & 2 & -1 & 0 & 3 \\
 & & \underline{-6} & & \\
 & & 5 & & \\
 & & & \underline{-15} & \\
 & & & 15 & \\
 & & & & \underline{-45} \\
 & & & & 48
\end{array}
$$

We can make this display more compact by moving the lines up until the numbers in blue align horizontally.

$$
\begin{array}{r rrrr l}
 & & 2x^2 + & 5x + & 15 & \textbf{Row 1} \\
x - 3 \overline{)} & 2 & -1 & 0 & 3 & \textbf{Row 2} \\
 & & -6 & -15 & -45 & \textbf{Row 3} \\
\hline
 & \bigcirc & 5 & 15 & 48 & \textbf{Row 4}
\end{array}
$$

Because the leading coefficient of the divisor is always 1, the leading coefficient of the dividend will also be the leading coefficient of the quotient. So we place the leading coefficient of the quotient, 2, in the circled position. Now, the first three numbers in row 4 are precisely the coefficients of the quotient, and the last number in row 4 is the remainder. Since row 1 is not really needed, we can compress the process to three rows, where the bottom row contains both the coefficients of the quotient and the remainder.

$$\begin{array}{rrrrl} x-3\overline{)2} & -1 & 0 & 3 & \textbf{Row 1} \\ & -6 & -15 & -45 & \textbf{Row 2 (subtract)} \\ \hline 2 & 5 & 15 & 48 & \textbf{Row 3} \end{array}$$

Recall that the entries in row 3 are obtained by subtracting the entries in row 2 from those in row 1. Rather than subtracting the entries in row 2, we can change the sign of each entry and add. With this modification, our display will look like this:

$$\begin{array}{rrrrl} x-3\overline{)2} & -1 & 0 & 3 & \textbf{Row 1} \\ & 6 & 15 & 45 & \textbf{Row 2 (add)} \\ \hline 2 & 5 & 15 & 48 & \textbf{Row 3} \end{array}$$

Notice that the entries in row 2 are three times the prior entries in row 3. Our last modification to the display replaces the $x - 3$ by 3. The entries in row 3 give the quotient and the remainder, as shown next.

$$\begin{array}{rrrrl} 3\overline{)2} & -1 & 0 & 3 & \textbf{Row 1} \\ & 6 & 15 & 45 & \textbf{Row 2 (add)} \\ \hline 2 & 5 & 15 & 48 & \textbf{Row 3} \end{array}$$

Quotient: $2x^2 + 5x + 15$ **Remainder**: 48

Let's go through an example step by step.

EXAMPLE 1 Using Synthetic Division to Find the Quotient and Remainder

Use synthetic division to find the quotient and remainder when

$$x^3 - 4x^2 - 5 \quad \text{is divided by} \quad x - 3$$

Solution

STEP 1: Write the dividend in descending powers of x. Then copy the coefficients, remembering to insert a 0 for any missing powers of x.

$$\begin{array}{rrrrl} 1 & -4 & 0 & -5 & \textbf{Row 1} \end{array}$$

STEP 2: Insert the usual division symbol. In synthetic division, the divisor is of the form $x - c$, and c is the number placed to the left of the division symbol. Here, since the divisor is $x - 3$, insert 3 to the left of the division symbol.

$$\begin{array}{rrrrl} 3\overline{)1} & -4 & 0 & -5 & \textbf{Row 1} \end{array}$$

STEP 3: Bring the 1 down two rows, and enter it in row 3.

$$\begin{array}{rrrrl} 3\overline{)1} & -4 & 0 & -5 & \textbf{Row 1} \\ \downarrow & & & & \textbf{Row 2} \\ \hline 1 & & & & \textbf{Row 3} \end{array}$$

STEP 4: Multiply the latest entry in row 3 by 3, and place the result in row 2, one column over to the right.

$$\begin{array}{rrrrl} 3\overline{)1} & -4 & 0 & -5 & \textbf{Row 1} \\ & 3 & & & \textbf{Row 2} \\ \hline 1 & & & & \textbf{Row 3} \end{array}$$

STEP 5: Add the entry in row 2 to the entry above it in row 1, and enter the sum in row 3.

$$\begin{array}{rrrrl} 3\overline{)1} & -4 & 0 & -5 & \textbf{Row 1} \\ & 3 & & & \textbf{Row 2} \\ \hline 1 & -1 & & & \textbf{Row 3} \end{array}$$

Step 6: Repeat Steps 4 and 5 until no more entries are available in row 1.

$$\begin{array}{r|rrrrl} 3 & 1 & -4 & 0 & -5 & \text{Row 1} \\ & & 3 & -3 & -9 & \text{Row 2} \\ \hline & 1 & -1 & -3 & -14 & \text{Row 3} \end{array}$$

Step 7: The final entry in row 3, the -14, is the remainder; the other entries in row 3, the 1, -1, and -3, are the coefficients (in descending order) of a polynomial whose degree is 1 less than that of the dividend. This is the quotient. That is,

$$\text{Quotient} = x^2 - x - 3 \qquad \text{Remainder} = -14$$

✓**Check:** (Divisor) (Quotient) + Remainder

$$= (x - 3)(x^2 - x - 3) + (-14)$$
$$= (x^3 - x^2 - 3x - 3x^2 + 3x + 9) + (-14)$$
$$= x^3 - 4x^2 - 5 = \text{Dividend}$$

Let's do an example in which all seven steps are combined.

EXAMPLE 2 Using Synthetic Division to Verify a Factor

Use synthetic division to show that $x + 3$ is a factor of

$$2x^5 + 5x^4 - 2x^3 + 2x^2 - 2x + 3$$

Solution The divisor is $x + 3 = x - (-3)$, so place -3 to the left of the division symbol. Then the row 3 entries will be multiplied by -3, entered in row 2, and added to row 1.

$$\begin{array}{r|rrrrrrl} -3 & 2 & 5 & -2 & 2 & -2 & 3 & \text{Row 1} \\ & & -6 & 3 & -3 & 3 & -3 & \text{Row 2} \\ \hline & 2 & -1 & 1 & -1 & 1 & 0 & \text{Row 3} \end{array}$$

Because the remainder is 0, we have

(Divisor) (Quotient) + Remainder

$$= (x + 3)(2x^4 - x^3 + x^2 - x + 1) = 2x^5 + 5x^4 - 2x^3 + 2x^2 - 2x + 3$$

As we see, $x + 3$ is a factor of $2x^5 + 5x^4 - 2x^3 + 2x^2 - 2x + 3$.

As Example 2 illustrates, the remainder after division gives information about whether the divisor is, or is not, a factor. We shall have more to say about this in Chapter 5.

Now Work PROBLEMS 9 AND 19

R.6 Assess Your Understanding

Concepts and Vocabulary

1. To check division, the expression that is being divided, the dividend, should equal the product of the ________ and the ______ plus the _________.
2. To divide $2x^3 - 5x + 1$ by $x + 3$ using synthetic division, the first step is to write $\underline{\quad}\overline{)\qquad\qquad}$.
3. Choose the division problem that cannot be done using synthetic division.
   (a) $2x^3 - 4x^2 + 6x - 8$ is divided by $x - 8$
   (b) $x^4 - 3$ is divided by $x + 1$
   (c) $x^5 + 3x^2 - 9x + 2$ is divided by $x + 10$
   (d) $x^4 - 5x^3 + 3x^2 - 9x + 13$ is divided by $x^2 + 5$
4. Choose the correct conclusion based on the following synthetic division:
$$\begin{array}{r|rrrr} -5 & 2 & 3 & -38 & -15 \\ & & -10 & 35 & 15 \\ \hline & 2 & -7 & -3 & 0 \end{array}$$
   (a) $x + 5$ is a factor of $2x^3 + 3x^2 - 38x - 15$
   (b) $x - 5$ is a factor of $2x^3 + 3x^2 - 38x - 15$
   (c) $x + 5$ is not a factor of $2x^3 + 3x^2 - 38x - 15$
   (d) $x - 5$ is not a factor of $2x^3 + 3x^2 - 38x - 15$
5. ***True or False*** In using synthetic division, the divisor is always a polynomial of degree 1, whose leading coefficient is 1.
6. ***True or False***
$$\begin{array}{r|rrrr} -2 & 5 & 3 & 2 & 1 \\ & & -10 & 14 & -32 \\ \hline & 5 & -7 & 16 & -31 \end{array}$$
means $\dfrac{5x^3 + 3x^2 + 2x + 1}{x + 2} = 5x^2 - 7x + 16 + \dfrac{-31}{x + 2}$.

Skill Building

In Problems 7–18, use synthetic division to find the quotient and remainder when:

7. $x^3 - x^2 + 2x + 4$ is divided by $x - 2$
8. $x^3 + 2x^2 - 3x + 1$ is divided by $x + 1$
9. $3x^3 + 2x^2 - x + 3$ is divided by $x - 3$
10. $-4x^3 + 2x^2 - x + 1$ is divided by $x + 2$
11. $x^5 - 4x^3 + x$ is divided by $x + 3$
12. $x^4 + x^2 + 2$ is divided by $x - 2$
13. $4x^6 - 3x^4 + x^2 + 5$ is divided by $x - 1$
14. $x^5 + 5x^3 - 10$ is divided by $x + 1$
15. $0.1x^3 + 0.2x$ is divided by $x + 1.1$
16. $0.1x^2 - 0.2$ is divided by $x + 2.1$
17. $x^5 - 1$ is divided by $x - 1$
18. $x^5 + 1$ is divided by $x + 1$

In Problems 19–28, use synthetic division to determine whether $x - c$ is a factor of the given polynomial.

19. $4x^3 - 3x^2 - 8x + 4$; $x - 2$
20. $-4x^3 + 5x^2 + 8$; $x + 3$
21. $3x^4 - 6x^3 - 5x + 10$; $x - 2$
22. $4x^4 - 15x^2 - 4$; $x - 2$
23. $3x^6 + 82x^3 + 27$; $x + 3$
24. $2x^6 - 18x^4 + x^2 - 9$; $x + 3$
25. $4x^6 - 64x^4 + x^2 - 15$; $x + 4$
26. $x^6 - 16x^4 + x^2 - 16$; $x + 4$
27. $2x^4 - x^3 + 2x - 1$; $x - \frac{1}{2}$
28. $3x^4 + x^3 - 3x + 1$; $x + \frac{1}{3}$

Applications and Extensions

29. Find the sum of a, b, c, and d if

$$\frac{x^3 - 2x^2 + 3x + 5}{x + 2} = ax^2 + bx + c + \frac{d}{x + 2}$$

Explaining Concepts: Discussion and Writing

30. When dividing a polynomial by $x - c$, do you prefer to use long division or synthetic division? Does the value of c make a difference to you in choosing? Give reasons.

R.7 Rational Expressions

OBJECTIVES

1 Reduce a Rational Expression to Lowest Terms

If we form the quotient of two polynomials, the result is called a **rational expression**. Some examples of rational expressions are

$$\text{(a)}\ \frac{x^3 + 1}{x} \qquad \text{(b)}\ \frac{3x^2 + x - 2}{x^2 + 5} \qquad \text{(c)}\ \frac{x}{x^2 - 1} \qquad \text{(d)}\ \frac{xy^2}{(x - y)^2}$$

Expressions (a), (b), and (c) are rational expressions in one variable, x, whereas (d) is a rational expression in two variables, x and y.

Rational expressions are described in the same manner as rational numbers. In expression (a), the polynomial $x^3 + 1$ is the **numerator**, and x is the **denominator**. When the numerator and denominator of a rational expression contain no common factors (except 1 and -1), we say that the rational expression is **reduced to lowest terms**, or **simplified**.

The polynomial in the denominator of a rational expression cannot be equal to 0 because division by 0 is not defined. For example, for the expression $\frac{x^3 + 1}{x}$, x cannot take on the value 0. The domain of the variable x is $\{x \mid x \neq 0\}$.

A rational expression is reduced to lowest terms by factoring the numerator and the denominator completely and dividing out any common factors using the Reduction Property:

$$\frac{a\cancel{c}}{b\cancel{c}} = \frac{a}{b} \quad \text{if } b \neq 0, c \neq 0 \tag{1}$$

EXAMPLE 1 **Reducing a Rational Expression to Lowest Terms**

Reduce to lowest terms: $\dfrac{x^2 + 4x + 4}{x^2 + 3x + 2}$

Solution Begin by factoring the numerator and the denominator.

$$x^2 + 4x + 4 = (x + 2)(x + 2)$$
$$x^2 + 3x + 2 = (x + 2)(x + 1)$$

WARNING Apply the Reduction Property only to rational expressions written in factored form. Be sure to divide out only common factors, not common terms! ■

Since a common factor, $x + 2$, appears, the original expression is not in lowest terms. To reduce it to lowest terms, use the Reduction Property:

$$\frac{x^2 + 4x + 4}{x^2 + 3x + 2} = \frac{\cancel{(x + 2)}(x + 2)}{\cancel{(x + 2)}(x + 1)} = \frac{x + 2}{x + 1} \qquad x \neq -2, -1$$ ■

EXAMPLE 2 **Reducing Rational Expressions to Lowest Terms**

Reduce each rational expression to lowest terms.

(a) $\dfrac{x^3 - 8}{x^3 - 2x^2}$ (b) $\dfrac{8 - 2x}{x^2 - x - 12}$

Solution

(a) $\dfrac{x^3 - 8}{x^3 - 2x^2} = \dfrac{\cancel{(x - 2)}(x^2 + 2x + 4)}{x^2\cancel{(x - 2)}} = \dfrac{x^2 + 2x + 4}{x^2} \qquad x \neq 0, 2$

(b) $\dfrac{8 - 2x}{x^2 - x - 12} = \dfrac{2(4 - x)}{(x - 4)(x + 3)} = \dfrac{2(-1)\cancel{(x - 4)}}{\cancel{(x - 4)}(x + 3)} = \dfrac{-2}{x + 3} \qquad x \neq -3, 4$ ■

Now Work PROBLEM 7

2 Multiply and Divide Rational Expressions

The rules for multiplying and dividing rational expressions are the same as the rules for multiplying and dividing rational numbers. If $\frac{a}{b}$ and $\frac{c}{d}$, $b \neq 0$, $d \neq 0$, are two rational expressions, then

$$\frac{a}{b} \cdot \frac{c}{d} = \frac{ac}{bd} \quad \text{if } b \neq 0, d \neq 0 \tag{2}$$

$$\frac{\frac{a}{b}}{\frac{c}{d}} = \frac{a}{b} \cdot \frac{d}{c} = \frac{ad}{bc} \quad \text{if } b \neq 0, c \neq 0, d \neq 0 \tag{3}$$

In using equations (2) and (3) with rational expressions, be sure first to factor each polynomial completely so that common factors can be divided out. Leave your answer in factored form.

EXAMPLE 3 **Multiplying and Dividing Rational Expressions**

Perform the indicated operation and simplify the result. Leave your answer in factored form.

(a) $\dfrac{x^2 - 2x + 1}{x^3 + x} \cdot \dfrac{4x^2 + 4}{x^2 + x - 2}$ (b) $\dfrac{\dfrac{x + 3}{x^2 - 4}}{\dfrac{x^2 - x - 12}{x^3 - 8}}$

Solution (a)

$$\frac{x^2 - 2x + 1}{x^3 + x} \cdot \frac{4x^2 + 4}{x^2 + x - 2} = \frac{(x - 1)^2}{x(x^2 + 1)} \cdot \frac{4(x^2 + 1)}{(x + 2)(x - 1)}$$

$$= \frac{(x - 1)^{\cancel{2}}(4)\cancel{(x^2 + 1)}}{x\cancel{(x^2 + 1)}(x + 2)\cancel{(x - 1)}}$$

$$= \frac{4(x - 1)}{x(x + 2)} \qquad x \neq -2, 0, 1$$

(b)

$$\frac{\dfrac{x + 3}{x^2 - 4}}{\dfrac{x^2 - x - 12}{x^3 - 8}} = \frac{x + 3}{x^2 - 4} \cdot \frac{x^3 - 8}{x^2 - x - 12}$$

$$= \frac{x + 3}{(x - 2)(x + 2)} \cdot \frac{(x - 2)(x^2 + 2x + 4)}{(x - 4)(x + 3)}$$

$$= \frac{\cancel{(x + 3)}\cancel{(x - 2)}(x^2 + 2x + 4)}{\cancel{(x - 2)}(x + 2)(x - 4)\cancel{(x + 3)}}$$

$$= \frac{x^2 + 2x + 4}{(x + 2)(x - 4)} \qquad x \neq -3, -2, 2, 4$$

■

Now Work PROBLEMS 19 AND 27

3 Add and Subtract Rational Expressions

In Words
To add (or subtract) two rational expressions with the same denominator, keep the common denominator and add (or subtract) the numerators.

The rules for adding and subtracting rational expressions are the same as the rules for adding and subtracting rational numbers. If the denominators of two rational expressions to be added (or subtracted) are equal, then add (or subtract) the numerators and keep the common denominator.

If $\dfrac{a}{b}$ and $\dfrac{c}{b}$ are two rational expressions, then

$$\frac{a}{b} + \frac{c}{b} = \frac{a + c}{b} \qquad \frac{a}{b} - \frac{c}{b} = \frac{a - c}{b} \qquad \text{if } b \neq 0 \qquad \textbf{(4)}$$

EXAMPLE 4 **Adding and Subtracting Rational Expressions with Equal Denominators**

Perform the indicated operation and simplify the result. Leave your answer in factored form.

(a) $\dfrac{2x^2 - 4}{2x + 5} + \dfrac{x + 3}{2x + 5} \quad x \neq -\dfrac{5}{2}$ (b) $\dfrac{x}{x - 3} - \dfrac{3x + 2}{x - 3} \quad x \neq 3$

Solution (a)

$$\frac{2x^2 - 4}{2x + 5} + \frac{x + 3}{2x + 5} = \frac{(2x^2 - 4) + (x + 3)}{2x + 5}$$

$$= \frac{2x^2 + x - 1}{2x + 5} = \frac{(2x - 1)(x + 1)}{2x + 5}$$

$$\text{(b)}\ \frac{x}{x-3} - \frac{3x+2}{x-3} = \frac{x-(3x+2)}{x-3} = \frac{x-3x-2}{x-3}$$
$$= \frac{-2x-2}{x-3} = \frac{-2(x+1)}{x-3}$$

EXAMPLE 5 Adding Rational Expressions Whose Denominators Are Additive Inverses of Each Other

Perform the indicated operation and simplify the result. Leave your answer in factored form.

$$\frac{2x}{x-3} + \frac{5}{3-x} \qquad x \neq 3$$

Solution Notice that the denominators of the two rational expressions are different. However, the denominator of the second expression is the additive inverse of the denominator of the first. That is,

$$3 - x = -x + 3 = -1 \cdot (x-3) = -(x-3)$$

Then

$$\frac{2x}{x-3} + \frac{5}{3-x} \underset{3-x=-(x-3)}{=} \frac{2x}{x-3} + \frac{5}{-(x-3)} \underset{\frac{a}{-b}=\frac{-a}{b}}{=} \frac{2x}{x-3} + \frac{-5}{x-3}$$
$$= \frac{2x+(-5)}{x-3} = \frac{2x-5}{x-3}$$

Now Work PROBLEMS 39 AND 45

If the denominators of two rational expressions to be added or subtracted are not equal, we can use the general formulas for adding and subtracting rational expressions.

$$\frac{a}{b} + \frac{c}{d} = \frac{a \cdot d}{b \cdot d} + \frac{b \cdot c}{b \cdot d} = \frac{ad+bc}{bd} \qquad \text{if } b \neq 0, d \neq 0 \qquad \textbf{(5a)}$$
$$\frac{a}{b} - \frac{c}{d} = \frac{a \cdot d}{b \cdot d} - \frac{b \cdot c}{b \cdot d} = \frac{ad-bc}{bd} \qquad \text{if } b \neq 0, d \neq 0 \qquad \textbf{(5b)}$$

EXAMPLE 6 Adding and Subtracting Rational Expressions with Unequal Denominators

Perform the indicated operation and simplify the result. Leave your answer in factored form.

$$\text{(a)}\ \frac{x-3}{x+4} + \frac{x}{x-2} \quad x \neq -4, 2 \qquad \text{(b)}\ \frac{x^2}{x^2-4} - \frac{1}{x} \quad x \neq -2, 0, 2$$

Solution

$$\text{(a)}\ \frac{x-3}{x+4} + \frac{x}{x-2} \underset{(5a)}{=} \frac{x-3}{x+4} \cdot \frac{x-2}{x-2} + \frac{x+4}{x+4} \cdot \frac{x}{x-2}$$
$$= \frac{(x-3)(x-2) + (x+4)(x)}{(x+4)(x-2)}$$
$$= \frac{x^2 - 5x + 6 + x^2 + 4x}{(x+4)(x-2)} = \frac{2x^2 - x + 6}{(x+4)(x-2)}$$

(b) $$\frac{x^2}{x^2-4} - \frac{1}{x} \underset{\substack{\uparrow \\ (5b)}}{=} \frac{x^2}{x^2-4}\cdot\frac{x}{x} - \frac{x^2-4}{x^2-4}\cdot\frac{1}{x} = \frac{x^2(x) - (x^2-4)(1)}{(x^2-4)(x)}$$

$$= \frac{x^3 - x^2 + 4}{(x-2)(x+2)(x)}$$

Now Work PROBLEM 49

4 Use the Least Common Multiple Method

If the denominators of two rational expressions to be added (or subtracted) have common factors, we usually do not use the general rules given by equations (5a) and (5b). Just as with fractions, we apply the **least common multiple (LCM) method**. The LCM method uses the polynomial of least degree that has each denominator polynomial as a factor.

The LCM Method for Adding or Subtracting Rational Expressions

The Least Common Multiple (LCM) Method requires four steps:

STEP 1: Factor completely the polynomial in the denominator of each rational expression.

STEP 2: The LCM of the denominators is the product of each of these factors raised to a power equal to the greatest number of times that the factor occurs in the polynomials.

STEP 3: Write each rational expression using the LCM as the common denominator.

STEP 4: Add or subtract the rational expressions using equation (4).

We begin with an example that requires only Steps 1 and 2.

EXAMPLE 7 Finding the Least Common Multiple

Find the least common multiple of the following pair of polynomials:

$$x(x-1)^2(x+1) \quad \text{and} \quad 4(x-1)(x+1)^3$$

Solution

STEP 1: The polynomials are already factored completely as

$$x(x-1)^2(x+1) \quad \text{and} \quad 4(x-1)(x+1)^3$$

STEP 2: Start by writing the factors of the left-hand polynomial. (Or you could start with the one on the right.)

$$x(x-1)^2(x+1)$$

Now look at the right-hand polynomial. Its first factor, 4, does not appear in our list, so we insert it.

$$4x(x-1)^2(x+1)$$

The next factor, $x - 1$, is already in our list, so no change is necessary. The final factor is $(x+1)^3$. Since our list has $x + 1$ to the first power only, we replace $x + 1$ in the list by $(x+1)^3$. The LCM is

$$4x(x-1)^2(x+1)^3$$

Notice that the LCM is, in fact, the polynomial of least degree that contains $x(x-1)^2(x+1)$ and $4(x-1)(x+1)^3$ as factors. ■

Now Work PROBLEM 55

EXAMPLE 8 Using the Least Common Multiple to Add Rational Expressions

Perform the indicated operation and simplify the result. Leave your answer in factored form.

$$\frac{x}{x^2+3x+2}+\frac{2x-3}{x^2-1} \qquad x \neq -2, -1, 1$$

Solution **STEP 1:** Factor completely the polynomials in the denominators.

$$x^2+3x+2=(x+2)(x+1)$$
$$x^2-1=(x-1)(x+1)$$

STEP 2: The LCM is $(x+2)(x+1)(x-1)$. Do you see why?

STEP 3: Write each rational expression using the LCM as the denominator.

$$\frac{x}{x^2+3x+2}=\frac{x}{(x+2)(x+1)}=\frac{x}{(x+2)(x+1)}\cdot\frac{x-1}{x-1}=\frac{x(x-1)}{(x+2)(x+1)(x-1)}$$

Multiply numerator and denominator by $x-1$ to get the LCM in the denominator.

$$\frac{2x-3}{x^2-1}=\frac{2x-3}{(x-1)(x+1)}=\frac{2x-3}{(x-1)(x+1)}\cdot\frac{x+2}{x+2}=\frac{(2x-3)(x+2)}{(x-1)(x+1)(x+2)}$$

Multiply numerator and denominator by $x+2$ to get the LCM in the denominator.

STEP 4: Now add by using equation (4).

$$\frac{x}{x^2+3x+2}+\frac{2x-3}{x^2-1}=\frac{x(x-1)}{(x+2)(x+1)(x-1)}+\frac{(2x-3)(x+2)}{(x+2)(x+1)(x-1)}$$
$$=\frac{(x^2-x)+(2x^2+x-6)}{(x+2)(x+1)(x-1)}$$
$$=\frac{3x^2-6}{(x+2)(x+1)(x-1)}=\frac{3(x^2-2)}{(x+2)(x+1)(x-1)}$$

■

EXAMPLE 9 Using the Least Common Multiple to Subtract Rational Expressions

Perform the indicated operation and simplify the result. Leave your answer in factored form.

$$\frac{3}{x^2+x}-\frac{x+4}{x^2+2x+1} \qquad x \neq -1, 0$$

Solution **STEP 1:** Factor completely the polynomials in the denominators.

$$x^2+x=x(x+1)$$
$$x^2+2x+1=(x+1)^2$$

STEP 2: The LCM is $x(x+1)^2$.

STEP 3: Write each rational expression using the LCM as the denominator.

$$\frac{3}{x^2+x} = \frac{3}{x(x+1)} = \frac{3}{x(x+1)} \cdot \frac{x+1}{x+1} = \frac{3(x+1)}{x(x+1)^2}$$

$$\frac{x+4}{x^2+2x+1} = \frac{x+4}{(x+1)^2} = \frac{x+4}{(x+1)^2} \cdot \frac{x}{x} = \frac{x(x+4)}{x(x+1)^2}$$

STEP 4: Subtract, using equation (4).

$$\begin{aligned}\frac{3}{x^2+x} - \frac{x+4}{x^2+2x+1} &= \frac{3(x+1)}{x(x+1)^2} - \frac{x(x+4)}{x(x+1)^2}\\ &= \frac{3(x+1) - x(x+4)}{x(x+1)^2}\\ &= \frac{3x+3-x^2-4x}{x(x+1)^2}\\ &= \frac{-x^2-x+3}{x(x+1)^2}\end{aligned}$$

■

Now Work PROBLEM 65

5 Simplify Complex Rational Expressions

When sums and/or differences of rational expressions appear as the numerator and/or denominator of a quotient, the quotient is called a **complex rational expression**.* For example,

$$\frac{1+\dfrac{1}{x}}{1-\dfrac{1}{x}} \quad \text{and} \quad \frac{\dfrac{x^2}{x^2-4}-3}{\dfrac{x-3}{x+2}-1}$$

are complex rational expressions. To **simplify** a complex rational expression means to write it as a rational expression reduced to lowest terms. This can be accomplished in either of two ways.

Simplifying a Complex Rational Expression

Option 1: Treat the numerator and denominator of the complex rational expression separately, performing whatever operations are indicated and simplifying the results. Follow this by simplifying the resulting rational expression.

Option 2: Find the LCM of the denominators of all rational expressions that appear in the complex rational expression. Multiply the numerator and denominator of the complex rational expression by the LCM and simplify the result.

We use both options in the next example. By carefully studying each option, you can discover situations in which one may be easier to use than the other.

EXAMPLE 10 **Simplifying a Complex Rational Expression**

Simplify: $\dfrac{\dfrac{1}{2}+\dfrac{3}{x}}{\dfrac{x+3}{4}} \qquad x \neq -3, 0$

* Some texts use the term **complex fraction**.

Solution ***Option 1:*** First, we perform the indicated operation in the numerator, and then we divide.

$$\frac{\frac{1}{2}+\frac{3}{x}}{\frac{x+3}{4}} = \frac{\frac{1\cdot x + 2\cdot 3}{2\cdot x}}{\frac{x+3}{4}} = \frac{\frac{x+6}{2x}}{\frac{x+3}{4}} = \frac{x+6}{2x}\cdot\frac{4}{x+3}$$

Rule for adding quotients; Rule for dividing quotients

$$= \frac{(x+6)\cdot 4}{2\cdot x\cdot(x+3)} = \frac{\cancel{2}\cdot 2\cdot(x+6)}{\cancel{2}\cdot x\cdot(x+3)} = \frac{2(x+6)}{x(x+3)}$$

Rule for multiplying quotients

Option 2: The rational expressions that appear in the complex rational expression are

$$\frac{1}{2},\ \frac{3}{x},\ \frac{x+3}{4}$$

The LCM of their denominators is $4x$. We multiply the numerator and denominator of the complex rational expression by $4x$ and then simplify.

$$\frac{\frac{1}{2}+\frac{3}{x}}{\frac{x+3}{4}} = \frac{4x\cdot\left(\frac{1}{2}+\frac{3}{x}\right)}{4x\cdot\left(\frac{x+3}{4}\right)} = \frac{4x\cdot\frac{1}{2}+4x\cdot\frac{3}{x}}{\frac{4x\cdot(x+3)}{4}}$$

Multiply the numerator and denominator by $4x$. Use the Distributive Property in the numerator.

$$= \frac{\cancel{2}\cdot 2x\cdot\frac{1}{\cancel{2}}+4\cancel{x}\cdot\frac{3}{\cancel{x}}}{\frac{\cancel{4}x\cdot(x+3)}{\cancel{4}}} = \frac{2x+12}{x(x+3)} = \frac{2(x+6)}{x(x+3)}$$

Simplify. Factor. ■

EXAMPLE 11 Simplifying a Complex Rational Expression

Simplify: $\dfrac{\frac{x^2}{x-4}+2}{\frac{2x-2}{x}-1} \qquad x \neq 0, 2, 4$

Solution We will use Option 1.

$$\frac{\frac{x^2}{x-4}+2}{\frac{2x-2}{x}-1} = \frac{\frac{x^2}{x-4}+\frac{2(x-4)}{x-4}}{\frac{2x-2}{x}-\frac{x}{x}} = \frac{\frac{x^2+2x-8}{x-4}}{\frac{2x-2-x}{x}}$$

$$= \frac{\frac{(x+4)(x-2)}{x-4}}{\frac{x-2}{x}} = \frac{(x+4)\cancel{(x-2)}}{x-4}\cdot\frac{x}{\cancel{x-2}}$$

$$= \frac{(x+4)\cdot x}{x-4}$$ ■

Now Work PROBLEM 75

Application

EXAMPLE 12 **Solving an Application in Electricity**

An electrical circuit contains two resistors connected in parallel, as shown in Figure 31. If these two resistors provide resistance of R_1 and R_2 ohms, respectively, their combined resistance R is given by the formula

$$R = \frac{1}{\frac{1}{R_1} + \frac{1}{R_2}}$$

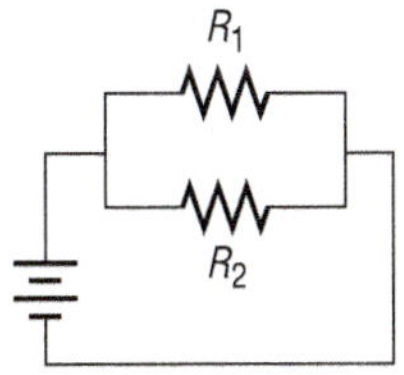

Figure 31

Express R as a rational expression; that is, simplify the right-hand side of this formula. Evaluate the rational expression if $R_1 = 6$ ohms and $R_2 = 10$ ohms.

Solution We will use Option 2. If we consider 1 as the fraction $\frac{1}{1}$, the rational expressions in the complex rational expression are

$$\frac{1}{1}, \quad \frac{1}{R_1}, \quad \frac{1}{R_2}$$

The LCM of the denominators is $R_1 R_2$. We multiply the numerator and denominator of the complex rational expression by $R_1 R_2$ and simplify.

$$\frac{1}{\frac{1}{R_1} + \frac{1}{R_2}} = \frac{1 \cdot R_1 R_2}{\left(\frac{1}{R_1} + \frac{1}{R_2}\right) \cdot R_1 R_2} = \frac{R_1 R_2}{\frac{1}{R_1} \cdot R_1 R_2 + \frac{1}{R_2} \cdot R_1 R_2} = \frac{R_1 R_2}{R_2 + R_1}$$

So,

$$R = \frac{R_1 R_2}{R_2 + R_1}$$

If $R_1 = 6$ and $R_2 = 10$, then

$$R = \frac{6 \cdot 10}{10 + 6} = \frac{60}{16} = \frac{15}{4} \text{ ohms}$$

■

R.7 Assess Your Understanding

Concepts and Vocabulary

1. When the numerator and denominator of a rational expression contain no common factors (except 1 and -1), the rational expression is in _______ _______.
2. LCM is an abbreviation for _____ _________ _________.
3. Choose the statement that is not true. Assume $b \neq 0, c \neq 0$, and $d \neq 0$ as necessary.

   (a) $\frac{ac}{bc} = \frac{a}{b}$ (b) $\frac{a}{b} + \frac{c}{b} = \frac{a+c}{b}$

   (c) $\frac{a}{b} - \frac{c}{d} = \frac{ad - bc}{bd}$ (d) $\frac{\frac{a}{b}}{\frac{c}{d}} = \frac{ac}{bd}$
4. Choose the rational expression that simplifies to -1.

   (a) $\frac{a-b}{b-a}$ (b) $\frac{a-b}{a-b}$

   (c) $\frac{a+b}{a-b}$ (d) $\frac{b-a}{b+a}$
5. ***True or False*** The rational expression $\frac{2x^3 - 4x}{x - 2}$ is reduced to lowest terms.
6. ***True or False*** The LCM of $2x^3 + 6x^2$ and $6x^4 + 4x^3$ is $4x^3(x + 1)$.

Skill Building

In Problems 7–18, reduce each rational expression to lowest terms.

7. $\frac{3x + 9}{x^2 - 9}$
8. $\frac{4x^2 + 8x}{12x + 24}$
9. $\frac{x^2 - 2x}{3x - 6}$
10. $\frac{15x^2 + 24x}{3x^2}$
11. $\frac{24x^2}{12x^2 - 6x}$
12. $\frac{x^2 + 4x + 4}{x^2 - 4}$
13. $\frac{y^2 - 25}{2y^2 - 8y - 10}$
14. $\frac{3y^2 - y - 2}{3y^2 + 5y + 2}$
15. $\frac{x^2 + 4x - 5}{x^2 - 2x + 1}$
16. $\frac{x - x^2}{x^2 + x - 2}$
17. $\frac{x^2 + 5x - 14}{2 - x}$
18. $\frac{2x^2 + 5x - 3}{1 - 2x}$

In Problems 19–36, perform the indicated operation and simplify the result. Leave your answer in factored form.

19. $\dfrac{3x+6}{5x^2}\cdot\dfrac{x}{x^2-4}$

20. $\dfrac{3}{2x}\cdot\dfrac{x^2}{6x+10}$

21. $\dfrac{4x^2}{x^2-16}\cdot\dfrac{x^3-64}{2x}$

22. $\dfrac{12}{x^2+x}\cdot\dfrac{x^3+1}{4x-2}$

23. $\dfrac{4x-8}{-3x}\cdot\dfrac{12}{12-6x}$

24. $\dfrac{6x-27}{5x}\cdot\dfrac{2}{4x-18}$

25. $\dfrac{x^2-3x-10}{x^2+2x-35}\cdot\dfrac{x^2+4x-21}{x^2+9x+14}$

26. $\dfrac{x^2+x-6}{x^2+4x-5}\cdot\dfrac{x^2-25}{x^2+2x-15}$

27. $\dfrac{\dfrac{6x}{x^2-4}}{\dfrac{3x-9}{2x+4}}$

28. $\dfrac{\dfrac{12x}{5x+20}}{\dfrac{4x^2}{x^2-16}}$

29. $\dfrac{\dfrac{8x}{x^2-1}}{\dfrac{10x}{x+1}}$

30. $\dfrac{\dfrac{x-2}{4x}}{\dfrac{x^2-4x+4}{12x}}$

31. $\dfrac{\dfrac{4-x}{4+x}}{\dfrac{4x}{x^2-16}}$

32. $\dfrac{\dfrac{3+x}{3-x}}{\dfrac{x^2-9}{9x^3}}$

33. $\dfrac{\dfrac{x^2+7x+12}{x^2-7x+12}}{\dfrac{x^2+x-12}{x^2-x-12}}$

34. $\dfrac{\dfrac{x^2+7x+6}{x^2+x-6}}{\dfrac{x^2+5x-6}{x^2+5x+6}}$

35. $\dfrac{\dfrac{2x^2-x-28}{3x^2-x-2}}{\dfrac{4x^2+16x+7}{3x^2+11x+6}}$

36. $\dfrac{\dfrac{9x^2+3x-2}{12x^2+5x-2}}{\dfrac{9x^2-6x+1}{8x^2-10x-3}}$

In Problems 37–54, perform the indicated operation and simplify the result. Leave your answer in factored form.

37. $\dfrac{x}{2}+\dfrac{5}{2}$

38. $\dfrac{3}{x}-\dfrac{6}{x}$

39. $\dfrac{x^2}{2x-3}-\dfrac{4}{2x-3}$

40. $\dfrac{3x^2}{2x-1}-\dfrac{9}{2x-1}$

41. $\dfrac{x+1}{x-3}+\dfrac{2x-3}{x-3}$

42. $\dfrac{2x-5}{3x+2}+\dfrac{x+4}{3x+2}$

43. $\dfrac{3x+5}{2x-1}-\dfrac{2x-4}{2x-1}$

44. $\dfrac{5x-4}{3x+4}-\dfrac{x+1}{3x+4}$

45. $\dfrac{4}{x-2}+\dfrac{x}{2-x}$

46. $\dfrac{6}{x-1}-\dfrac{x}{1-x}$

47. $\dfrac{4}{x-1}-\dfrac{2}{x+2}$

48. $\dfrac{2}{x+5}-\dfrac{5}{x-5}$

49. $\dfrac{x}{x+1}+\dfrac{2x-3}{x-1}$

50. $\dfrac{3x}{x-4}+\dfrac{2x}{x+3}$

51. $\dfrac{x-3}{x+2}-\dfrac{x+4}{x-2}$

52. $\dfrac{2x-3}{x-1}-\dfrac{2x+1}{x+1}$

53. $\dfrac{x}{x^2-4}+\dfrac{1}{x}$

54. $\dfrac{x-1}{x^3}+\dfrac{x}{x^2+1}$

In Problems 55–62, find the LCM of the given polynomials.

55. $x^2-4,\;\; x^2-x-2$

56. $x^2-x-12,\;\; x^2-8x+16$

57. $x^3-x,\;\; x^2-x$

58. $3x^2-27,\;\; 2x^2-x-15$

59. $4x^3-4x^2+x,\;\; 2x^3-x^2,\;\; x^3$

60. $x-3,\;\; x^2+3x,\;\; x^3-9x$

61. $x^3-x,\;\; x^3-2x^2+x,\;\; x^3-1$

62. $x^2+4x+4,\;\; x^3+2x^2,\;\; (x+2)^3$

In Problems 63–74, perform the indicated operations and simplify the result. Leave your answer in factored form.

63. $\dfrac{x}{x^2-7x+6}-\dfrac{x}{x^2-2x-24}$

64. $\dfrac{x}{x-3}-\dfrac{x+1}{x^2+5x-24}$

65. $\dfrac{4x}{x^2-4}-\dfrac{2}{x^2+x-6}$

66. $\dfrac{3x}{x-1}-\dfrac{x-4}{x^2-2x+1}$

67. $\dfrac{3}{(x-1)^2(x+1)}+\dfrac{2}{(x-1)(x+1)^2}$

68. $\dfrac{2}{(x+2)^2(x-1)}-\dfrac{6}{(x+2)(x-1)^2}$

69. $\dfrac{x+4}{x^2-x-2}-\dfrac{2x+3}{x^2+2x-8}$

70. $\dfrac{2x-3}{x^2+8x+7}-\dfrac{x-2}{(x+1)^2}$

71. $\dfrac{1}{x}-\dfrac{2}{x^2+x}+\dfrac{3}{x^3-x^2}$

72. $\dfrac{x}{(x-1)^2}+\dfrac{2}{x}-\dfrac{x+1}{x^3-x^2}$

73. $\dfrac{1}{h}\left(\dfrac{1}{x+h}-\dfrac{1}{x}\right)$

74. $\dfrac{1}{h}\left[\dfrac{1}{(x+h)^2}-\dfrac{1}{x^2}\right]$

In Problems 75–86, perform the indicated operations and simplify the result. Leave your answer in factored form.

75. $\dfrac{1+\dfrac{1}{x}}{1-\dfrac{1}{x}}$

76. $\dfrac{4+\dfrac{1}{x^2}}{3-\dfrac{1}{x^2}}$

77. $\dfrac{2-\dfrac{x+1}{x}}{3+\dfrac{x-1}{x+1}}$

78. $\dfrac{1-\dfrac{x}{x+1}}{2-\dfrac{x-1}{x}}$

79. $\dfrac{\dfrac{x+4}{x-2}-\dfrac{x-3}{x+1}}{x+1}$

80. $\dfrac{\dfrac{x-2}{x+1}-\dfrac{x}{x-2}}{x+3}$

81. $\dfrac{\dfrac{x-2}{x+2}+\dfrac{x-1}{x+1}}{\dfrac{x}{x+1}-\dfrac{2x-3}{x}}$

82. $\dfrac{\dfrac{2x+5}{x}-\dfrac{x}{x-3}}{\dfrac{x^2}{x-3}-\dfrac{(x+1)^2}{x+3}}$

83. $1-\dfrac{1}{1-\dfrac{1}{x}}$

84. $1-\dfrac{1}{1-\dfrac{1}{1-x}}$

85. $\dfrac{2(x-1)^{-1}+3}{3(x-1)^{-1}+2}$

86. $\dfrac{4(x+2)^{-1}-3}{3(x+2)^{-1}-1}$

In Problems 87–94, expressions that occur in calculus are given. Reduce each expression to lowest terms.

87. $\dfrac{(2x+3)\cdot 3-(3x-5)\cdot 2}{(3x-5)^2}$

88. $\dfrac{(4x+1)\cdot 5-(5x-2)\cdot 4}{(5x-2)^2}$

89. $\dfrac{x\cdot 2x-(x^2+1)\cdot 1}{(x^2+1)^2}$

90. $\dfrac{x\cdot 2x-(x^2-4)\cdot 1}{(x^2-4)^2}$

91. $\dfrac{(3x+1)\cdot 2x-x^2\cdot 3}{(3x+1)^2}$

92. $\dfrac{(2x-5)\cdot 3x^2-x^3\cdot 2}{(2x-5)^2}$

93. $\dfrac{(x^2+1)\cdot 3-(3x+4)\cdot 2x}{(x^2+1)^2}$

94. $\dfrac{(x^2+9)\cdot 2-(2x-5)\cdot 2x}{(x^2+9)^2}$

Applications and Extensions

95. **The Lensmaker's Equation** The focal length f of a lens with index of refraction n is

$$\frac{1}{f}=(n-1)\left[\frac{1}{R_1}+\frac{1}{R_2}\right]$$

where R_1 and R_2 are the radii of curvature of the front and back surfaces of the lens. Express f as a rational expression. Evaluate the rational expression for $n = 1.5$, $R_1 = 0.1$ meter, and $R_2 = 0.2$ meter.

96. **Electrical Circuits** An electrical circuit contains three resistors connected in parallel. If these three resistors provide resistance of R_1, R_2, and R_3 ohms, respectively, their combined resistance R is given by the formula

$$\frac{1}{R}=\frac{1}{R_1}+\frac{1}{R_2}+\frac{1}{R_3}$$

Express R as a rational expression. Evaluate R for $R_1 = 5$ ohms, $R_2 = 4$ ohms, and $R_3 = 10$ ohms.

Explaining Concepts: Discussion and Writing

97. The following expressions are called **continued fractions**:

$$1+\frac{1}{x},\quad 1+\frac{1}{1+\frac{1}{x}},\quad 1+\frac{1}{1+\frac{1}{1+\frac{1}{x}}},\quad 1+\frac{1}{1+\frac{1}{1+\frac{1}{1+\frac{1}{x}}}},\ldots$$

Each simplifies to an expression of the form

$$\frac{ax+b}{bx+c}$$

Trace the successive values of a, b, and c as you "continue" the fraction. Can you discover the patterns that these values follow? Go to the library and research Fibonacci numbers. Write a report on your findings.

98. Explain to a fellow student when you would use the LCM method to add two rational expressions. Give two examples of adding two rational expressions: one in which you use the LCM and the other in which you do not.

99. Which of the two options given in the text for simplifying complex rational expressions do you prefer? Write a brief paragraph stating the reasons for your choice.

R.8 *n*th Roots; Rational Exponents

PREPARING FOR THIS SECTION *Before getting started, review the following:*

- Exponents, Square Roots (Section R.2, pp. 22–25)

Now Work the **'Are You Prepared?'** problems on page 79.

OBJECTIVES
1. Work with *n*th Roots (p. 74)
2. Simplify Radicals (p. 75)
3. Rationalize Denominators (p. 76)
4. Simplify Expressions with Rational Exponents (p. 77)

1 Work with *n*th Roots

DEFINITION

The **principal *n*th root of a real number *a***, $n \geq 2$ an integer, symbolized by $\sqrt[n]{a}$, is defined as follows:

$$\sqrt[n]{a} = b \quad \text{means} \quad a = b^n$$

where $a \geq 0$ and $b \geq 0$ if n is even and a, b are any real numbers if n is odd.

In Words
The symbol $\sqrt[n]{a}$ means "give me the number that, when raised to the power n, equals a."

Notice that if a is negative and n is even, then $\sqrt[n]{a}$ is not defined. When it is defined, the principal nth root of a number is unique.

The symbol $\sqrt[n]{a}$ for the principal nth root of a is called a **radical**; the integer n is called the **index**, and a is called the **radicand**. If the index of a radical is 2, we call $\sqrt[2]{a}$ the **square root** of a and omit the index 2 by simply writing $\sqrt{a}$. If the index is 3, we call $\sqrt[3]{a}$ the **cube root** of a.

EXAMPLE 1 **Simplifying Principal *n*th Roots**

(a) $\sqrt[3]{8} = \sqrt[3]{2^3} = 2$

(b) $\sqrt[3]{-64} = \sqrt[3]{(-4)^3} = -4$

(c) $\sqrt[4]{\dfrac{1}{16}} = \sqrt[4]{\left(\dfrac{1}{2}\right)^4} = \dfrac{1}{2}$

(d) $\sqrt[6]{(-2)^6} = |-2| = 2$

These are examples of **perfect roots**, since each simplifies to a rational number. Notice the absolute value in Example 1(d). If n is even, then the principal nth root must be nonnegative.

In general, if $n \geq 2$ is an integer and a is a real number, we have

$$\sqrt[n]{a^n} = a \quad \text{if } n \geq 3 \text{ is odd} \qquad \textbf{(1a)}$$

$$\sqrt[n]{a^n} = |a| \quad \text{if } n \geq 2 \text{ is even} \qquad \textbf{(1b)}$$

Now Work PROBLEM 11

Radicals provide a way of representing many irrational real numbers. For example, it can be shown that there is no rational number whose square is 2. Using radicals, we can say that $\sqrt{2}$ is the positive number whose square is 2.

EXAMPLE 2

Using a Calculator to Approximate Roots

Use a calculator to approximate $\sqrt[5]{16}$.

Solution Figure 32 shows the result using a TI-84 Plus C graphing calculator. ■

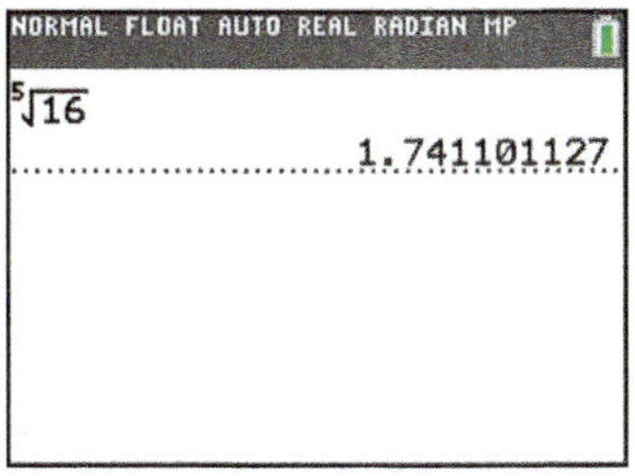

Figure 32

Now Work PROBLEM 119

2 Simplify Radicals

Let $n \geq 2$ and $m \geq 2$ denote integers, and let a and b represent real numbers. Assuming that all radicals are defined, we have the following properties:

Properties of Radicals

$$\sqrt[n]{ab} = \sqrt[n]{a}\sqrt[n]{b} \qquad \textbf{(2a)}$$

$$\sqrt[n]{\frac{a}{b}} = \frac{\sqrt[n]{a}}{\sqrt[n]{b}} \qquad b \neq 0 \qquad \textbf{(2b)}$$

$$\sqrt[n]{a^m} = \left(\sqrt[n]{a}\right)^m \qquad \textbf{(2c)}$$

When used in reference to radicals, the direction to "simplify" will mean to remove from the radicals any perfect roots that occur as factors.

EXAMPLE 3

Simplifying Radicals

(a) $\sqrt{32} = \sqrt{16 \cdot 2} = \sqrt{16} \cdot \sqrt{2} = 4\sqrt{2}$

Factor out 16, a perfect square. (2a)

(b) $\sqrt[3]{16} = \sqrt[3]{8 \cdot 2} = \sqrt[3]{8} \cdot \sqrt[3]{2} = \sqrt[3]{2^3} \cdot \sqrt[3]{2} = 2\sqrt[3]{2}$

Factor out 8, a perfect cube. (2a)

(c) $\sqrt[3]{-16x^4} = \sqrt[3]{-8 \cdot 2 \cdot x^3 \cdot x} = \sqrt[3]{(-8x^3)(2x)}$

Factor perfect cubes inside radical. Group perfect cubes.

$$= \sqrt[3]{(-2x)^3 \cdot 2x} = \sqrt[3]{(-2x)^3} \cdot \sqrt[3]{2x} = -2x\sqrt[3]{2x}$$

(2a)

(d) $\sqrt[4]{\frac{16x^5}{81}} = \sqrt[4]{\frac{2^4 x^4 x}{3^4}} = \sqrt[4]{\left(\frac{2x}{3}\right)^4 \cdot x} = \sqrt[4]{\left(\frac{2x}{3}\right)^4} \cdot \sqrt[4]{x} = \left|\frac{2x}{3}\right| \sqrt[4]{x}$ ■

Now Work PROBLEMS 15 AND 27

Two or more radicals can be combined, provided that they have the same index and the same radicand. Such radicals are called **like radicals**.

EXAMPLE 4

Combining Like Radicals

(a) $-8\sqrt{12} + \sqrt{3} = -8\sqrt{4 \cdot 3} + \sqrt{3}$

$$= -8 \cdot \sqrt{4}\sqrt{3} + \sqrt{3}$$

$$= -16\sqrt{3} + \sqrt{3} = -15\sqrt{3}$$

(b) $\sqrt[3]{8x^4} + \sqrt[3]{-x} + 4\sqrt[3]{27x} = \sqrt[3]{2^3x^3x} + \sqrt[3]{-1\cdot x} + 4\sqrt[3]{3^3x}$

$$= \sqrt[3]{(2x)^3}\cdot\sqrt[3]{x} + \sqrt[3]{-1}\cdot\sqrt[3]{x} + 4\sqrt[3]{3^3}\cdot\sqrt[3]{x}$$
$$= 2x\sqrt[3]{x} - 1\cdot\sqrt[3]{x} + 12\sqrt[3]{x}$$
$$= (2x + 11)\sqrt[3]{x}$$

Now Work PROBLEM 45

3 Rationalize Denominators

When radicals occur in quotients, it is customary to rewrite the quotient so that the new denominator contains no radicals. This process is referred to as **rationalizing the denominator**.

The idea is to multiply by an appropriate expression so that the new denominator contains no radicals. For example:

If a Denominator Contains the Factor	Multiply by	To Obtain a Denominator Free of Radicals
$\sqrt{3}$	$\sqrt{3}$	$(\sqrt{3})^2 = 3$
$\sqrt{3} + 1$	$\sqrt{3} - 1$	$(\sqrt{3})^2 - 1^2 = 3 - 1 = 2$
$\sqrt{2} - 3$	$\sqrt{2} + 3$	$(\sqrt{2})^2 - 3^2 = 2 - 9 = -7$
$\sqrt{5} - \sqrt{3}$	$\sqrt{5} + \sqrt{3}$	$(\sqrt{5})^2 - (\sqrt{3})^2 = 5 - 3 = 2$
$\sqrt[3]{4}$	$\sqrt[3]{2}$	$\sqrt[3]{4}\cdot\sqrt[3]{2} = \sqrt[3]{8} = 2$

In rationalizing the denominator of a quotient, be sure to multiply both the numerator and the denominator by the expression.

EXAMPLE 5 **Rationalizing Denominators**

Rationalize the denominator of each expression:

(a) $\dfrac{1}{\sqrt{3}}$ (b) $\dfrac{5}{4\sqrt{2}}$ (c) $\dfrac{\sqrt{2}}{\sqrt{3} - 3\sqrt{2}}$

Solution (a) The denominator contains the factor $\sqrt{3}$, so we multiply the numerator and denominator by $\sqrt{3}$ to obtain

$$\frac{1}{\sqrt{3}} = \frac{1}{\sqrt{3}}\cdot\frac{\sqrt{3}}{\sqrt{3}} = \frac{\sqrt{3}}{(\sqrt{3})^2} = \frac{\sqrt{3}}{3}$$

(b) The denominator contains the factor $\sqrt{2}$, so we multiply the numerator and denominator by $\sqrt{2}$ to obtain

$$\frac{5}{4\sqrt{2}} = \frac{5}{4\sqrt{2}}\cdot\frac{\sqrt{2}}{\sqrt{2}} = \frac{5\sqrt{2}}{4(\sqrt{2})^2} = \frac{5\sqrt{2}}{4\cdot 2} = \frac{5\sqrt{2}}{8}$$

(c) The denominator contains the factor $\sqrt{3} - 3\sqrt{2}$, so we multiply the numerator and denominator by $\sqrt{3} + 3\sqrt{2}$ to obtain

$$\frac{\sqrt{2}}{\sqrt{3} - 3\sqrt{2}} = \frac{\sqrt{2}}{\sqrt{3} - 3\sqrt{2}}\cdot\frac{\sqrt{3} + 3\sqrt{2}}{\sqrt{3} + 3\sqrt{2}} = \frac{\sqrt{2}(\sqrt{3} + 3\sqrt{2})}{(\sqrt{3})^2 - (3\sqrt{2})^2}$$
$$= \frac{\sqrt{2}\sqrt{3} + 3(\sqrt{2})^2}{3 - 18} = \frac{\sqrt{6} + 6}{-15} = -\frac{6 + \sqrt{6}}{15}$$

Now Work PROBLEM 59

4 Simplify Expressions with Rational Exponents

Radicals are used to define rational exponents.

DEFINITION

If a is a real number and $n \geq 2$ is an integer, then

$$a^{1/n} = \sqrt[n]{a} \quad \textbf{(3)}$$

provided that $\sqrt[n]{a}$ exists.

Note that if n is even and $a < 0$, then $\sqrt[n]{a}$ and $a^{1/n}$ do not exist.

EXAMPLE 6 **Writing Expressions Containing Fractional Exponents as Radicals**

(a) $4^{1/2} = \sqrt{4} = 2$

(b) $8^{1/2} = \sqrt{8} = 2\sqrt{2}$

(c) $(-27)^{1/3} = \sqrt[3]{-27} = -3$

(d) $16^{1/3} = \sqrt[3]{16} = 2\sqrt[3]{2}$

DEFINITION

If a is a real number and m and n are integers containing no common factors, with $n \geq 2$, then

$$a^{m/n} = \sqrt[n]{a^m} = (\sqrt[n]{a})^m \quad \textbf{(4)}$$

provided that $\sqrt[n]{a}$ exists.

We have two comments about equation (4):

1. The exponent $\dfrac{m}{n}$ must be in lowest terms, and $n \geq 2$ must be positive.
2. In simplifying the rational expression $a^{m/n}$, either $\sqrt[n]{a^m}$ or $(\sqrt[n]{a})^m$ may be used, the choice depending on which is easier to simplify. Generally, taking the root first, as in $(\sqrt[n]{a})^m$, is easier.

EXAMPLE 7 **Using Equation (4)**

(a) $4^{3/2} = (\sqrt{4})^3 = 2^3 = 8$

(b) $(-8)^{4/3} = (\sqrt[3]{-8})^4 = (-2)^4 = 16$

(c) $(32)^{-2/5} = (\sqrt[5]{32})^{-2} = 2^{-2} = \dfrac{1}{4}$

(d) $25^{6/4} = 25^{3/2} = (\sqrt{25})^3 = 5^3 = 125$

Now Work PROBLEM 69

It can be shown that the Laws of Exponents hold for rational exponents. The next example illustrates using the Laws of Exponents to simplify.

EXAMPLE 8 **Simplifying Expressions Containing Rational Exponents**

Simplify each expression. Express your answer so that only positive exponents occur. Assume that the variables are positive.

(a) $(x^{2/3}y)(x^{-2}y)^{1/2}$ (b) $\left(\dfrac{2x^{1/3}}{y^{2/3}}\right)^{-3}$ (c) $\left(\dfrac{9x^2y^{1/3}}{x^{1/3}y}\right)^{1/2}$

Solution (a) $(x^{2/3}y)(x^{-2}y)^{1/2} = (x^{2/3}y)[(x^{-2})^{1/2}y^{1/2}]$

$$= x^{2/3}yx^{-1}y^{1/2}$$
$$= (x^{2/3}\cdot x^{-1})(y\cdot y^{1/2})$$
$$= x^{-1/3}y^{3/2}$$
$$= \frac{y^{3/2}}{x^{1/3}}$$

(b) $\left(\frac{2x^{1/3}}{y^{2/3}}\right)^{-3} = \left(\frac{y^{2/3}}{2x^{1/3}}\right)^{3} = \frac{(y^{2/3})^3}{(2x^{1/3})^3} = \frac{y^2}{2^3(x^{1/3})^3} = \frac{y^2}{8x}$

(c) $\left(\frac{9x^2y^{1/3}}{x^{1/3}y}\right)^{1/2} = \left(\frac{9x^{2-(1/3)}}{y^{1-(1/3)}}\right)^{1/2} = \left(\frac{9x^{5/3}}{y^{2/3}}\right)^{1/2} = \frac{9^{1/2}(x^{5/3})^{1/2}}{(y^{2/3})^{1/2}} = \frac{3x^{5/6}}{y^{1/3}}$ ■

Now Work PROBLEM 89

The next two examples illustrate some algebra that you will need to know for certain calculus problems.

EXAMPLE 9 Writing an Expression as a Single Quotient

Write the following expression as a single quotient in which only positive exponents appear.

$$(x^2+1)^{1/2} + x\cdot\frac{1}{2}(x^2+1)^{-1/2}\cdot 2x$$

Solution

$$(x^2+1)^{1/2} + x\cdot\frac{1}{2}(x^2+1)^{-1/2}\cdot 2x = (x^2+1)^{1/2} + \frac{x^2}{(x^2+1)^{1/2}}$$
$$= \frac{(x^2+1)^{1/2}(x^2+1)^{1/2} + x^2}{(x^2+1)^{1/2}}$$
$$= \frac{(x^2+1) + x^2}{(x^2+1)^{1/2}}$$
$$= \frac{2x^2+1}{(x^2+1)^{1/2}}$$ ■

Now Work PROBLEM 95

EXAMPLE 10 Factoring an Expression Containing Rational Exponents

Factor and simplify: $\frac{4}{3}x^{1/3}(2x+1) + 2x^{4/3}$

Solution Begin by writing $2x^{4/3}$ as a fraction with 3 as the denominator.

$$\frac{4}{3}x^{1/3}(2x+1) + 2x^{4/3} = \frac{4x^{1/3}(2x+1)}{3} + \frac{6x^{4/3}}{3} = \frac{4x^{1/3}(2x+1) + 6x^{4/3}}{3}$$

↑ Add the two fractions

$$= \frac{2x^{1/3}[2(2x+1) + 3x]}{3} = \frac{2x^{1/3}(7x+2)}{3}$$

↑ 2 and $x^{1/3}$ are common factors ↑ Simplify ■

Now Work PROBLEM 107

Historical Note

The radical sign, $\sqrt{}$, was first used in print by Christoff Rudolff in 1525. It is thought to be the manuscript form of the letter *r* (for the Latin word *radix* = *root*), although this has not been quite conclusively confirmed. It took a long time for $\sqrt{}$ to become the standard symbol for a square root and much longer to standardize $\sqrt[3]{}$, $\sqrt[4]{}$, $\sqrt[5]{}$, and so on. The indexes of the root were placed in every conceivable position, with

$$\sqrt{\overset{3}{8}}, \quad \sqrt{③8}, \quad \text{and} \quad \underset{3}{\sqrt{8}}$$

all being variants for $\sqrt[3]{8}$. The notation $\sqrt{}\sqrt{16}$ was popular for $\sqrt[4]{16}$. By the 1700s, the index had settled where we now put it.

The bar on top of the present radical symbol, as follows,

$$\sqrt{a^2 + 2ab + b^2}$$

is the last survivor of the **vinculum**, a bar placed atop an expression to indicate what we would now indicate with parentheses. For example,

$$a\overline{b + c} = a(b + c)$$

R.8 Assess Your Understanding

'Are You Prepared?' *Answers are given at the end of these exercises. If you get a wrong answer, read the pages in red.*

1. $(-3)^2 =$ __; $-3^2 =$ ___ (pp. 22–25)

2. $\sqrt{16} =$ __; $\sqrt{(-4)^2} =$ __ (pp. 22–25)

Concepts and Vocabulary

3. In the symbol $\sqrt[n]{a}$, the integer n is called the ______.

4. We call $\sqrt[3]{a}$ the ______ ______ of a.

5. Let $n \geq 2$ and $m \geq 2$ be integers, and let a and b be real numbers. Which of the following is not a property of radicals? Assume all radicals are defined.

(a) $\sqrt[n]{\dfrac{a}{b}} = \dfrac{\sqrt[n]{a}}{\sqrt[n]{b}}$ (b) $\sqrt[n]{a+b} = \sqrt[n]{a} + \sqrt[n]{b}$

(c) $\sqrt[n]{ab} = \sqrt[n]{a}\sqrt[n]{b}$ (d) $\sqrt[n]{a^m} = (\sqrt[n]{a})^m$

6. If a is a real number and $n \geq 2$ is an integer, then which of the following expressions is equivalent to $\sqrt[n]{a}$, provided that it exists?

(a) a^{-n} (b) a^n (c) $\dfrac{1}{a^n}$ (d) $a^{1/n}$

7. Which of the following phrases best defines like radicals?
(a) Radical expressions that have the same index
(b) Radical expressions that have the same radicand
(c) Radical expressions that have the same index and the same radicand
(d) Radical expressions that have the same variable

8. To rationalize the denominator of the expression $\dfrac{\sqrt{2}}{1-\sqrt{3}}$, multiply both the numerator and the denominator by which of the following?

(a) $\sqrt{3}$ (b) $\sqrt{2}$ (c) $1+\sqrt{3}$ (d) $1-\sqrt{3}$

9. ***True or False*** $\sqrt[5]{-32} = -2$

10. ***True or False*** $\sqrt[4]{(-3)^4} = -3$

Skill Building

In Problems 11–54, simplify each expression. Assume that all variables are positive when they appear.

11. $\sqrt[3]{27}$ **12.** $\sqrt[4]{16}$ **13.** $\sqrt[3]{-8}$ **14.** $\sqrt[3]{-1}$

15. $\sqrt{8}$ **16.** $\sqrt{75}$ **17.** $\sqrt{700}$ **18.** $\sqrt{45x^3}$

19. $\sqrt[3]{32}$ **20.** $\sqrt[3]{54}$ **21.** $\sqrt[3]{-8x^4}$ **22.** $\sqrt[3]{192x^5}$

23. $\sqrt[4]{243}$ **24.** $\sqrt[4]{48x^5}$ **25.** $\sqrt[4]{x^{12}y^8}$ **26.** $\sqrt[5]{x^{10}y^5}$

27. $\sqrt[4]{\dfrac{x^9y^7}{xy^3}}$ **28.** $\sqrt[3]{\dfrac{3xy^2}{81x^4y^2}}$ **29.** $\sqrt{36x}$ **30.** $\sqrt{9x^5}$

31. $\sqrt[4]{162x^9y^{12}}$ **32.** $\sqrt[3]{-40x^{14}y^{10}}$ **33.** $\sqrt{3x^2}\sqrt{12x}$ **34.** $\sqrt{5x}\sqrt{20x^3}$

35. $(\sqrt{5}\sqrt[3]{9})^2$ **36.** $(\sqrt[3]{3}\sqrt{10})^4$ **37.** $(3\sqrt{6})(2\sqrt{2})$ **38.** $(5\sqrt{8})(-3\sqrt{3})$

39. $3\sqrt{2} + 4\sqrt{2}$ **40.** $6\sqrt{5} - 4\sqrt{5}$ **41.** $-\sqrt{18} + 2\sqrt{8}$ **42.** $2\sqrt{12} - 3\sqrt{27}$

43. $(\sqrt{3}+3)(\sqrt{3}-1)$ **44.** $(\sqrt{5}-2)(\sqrt{5}+3)$ **45.** $5\sqrt[3]{2} - 2\sqrt[3]{54}$ **46.** $9\sqrt[3]{24} - \sqrt[3]{81}$

47. $(\sqrt{x}-1)^2$ **48.** $(\sqrt{x}+\sqrt{5})^2$ **49.** $\sqrt[3]{16x^4} - \sqrt[3]{2x}$ **50.** $\sqrt[4]{32x} + \sqrt[4]{2x^5}$

51. $\sqrt{8x^3} - 3\sqrt{50x}$ **52.** $3x\sqrt{9y} + 4\sqrt{25y}$

53. $\sqrt[3]{16x^4y} - 3x\sqrt[3]{2xy} + 5\sqrt[3]{-2xy^4}$ **54.** $8xy - \sqrt{25x^2y^2} + \sqrt[3]{8x^3y^3}$

In Problems 55–68, rationalize the denominator of each expression. Assume that all variables are positive when they appear.

55. $\dfrac{1}{\sqrt{2}}$ **56.** $\dfrac{2}{\sqrt{3}}$ **57.** $\dfrac{-\sqrt{3}}{\sqrt{5}}$ **58.** $\dfrac{-\sqrt{3}}{\sqrt{8}}$

59. $\dfrac{\sqrt{3}}{5-\sqrt{2}}$ **60.** $\dfrac{\sqrt{2}}{\sqrt{7}+2}$ **61.** $\dfrac{2-\sqrt{5}}{2+3\sqrt{5}}$ **62.** $\dfrac{\sqrt{3}-1}{2\sqrt{3}+3}$

63. $\dfrac{5}{\sqrt{2}-1}$ **64.** $\dfrac{-3}{\sqrt{5}+4}$ **65.** $\dfrac{5}{\sqrt[3]{2}}$ **66.** $\dfrac{-2}{\sqrt[3]{9}}$

67. $\dfrac{\sqrt{x+h}-\sqrt{x}}{\sqrt{x+h}+\sqrt{x}}$ **68.** $\dfrac{\sqrt{x+h}+\sqrt{x-h}}{\sqrt{x+h}-\sqrt{x-h}}$

In Problems 69–84, simplify each expression.

69. $8^{2/3}$ **70.** $4^{3/2}$ **71.** $(-27)^{1/3}$ **72.** $16^{3/4}$

73. $16^{3/2}$ **74.** $25^{3/2}$ **75.** $9^{-3/2}$ **76.** $16^{-3/2}$

77. $\left(\dfrac{9}{8}\right)^{3/2}$ **78.** $\left(\dfrac{27}{8}\right)^{2/3}$ **79.** $\left(\dfrac{8}{9}\right)^{-3/2}$ **80.** $\left(\dfrac{8}{27}\right)^{-2/3}$

81. $(-1000)^{-1/3}$ **82.** $-25^{-1/2}$ **83.** $\left(-\dfrac{64}{125}\right)^{-2/3}$ **84.** $-81^{-3/4}$

In Problems 85–92, simplify each expression. Express your answer so that only positive exponents occur. Assume that the variables are positive.

85. $x^{3/4}x^{1/3}x^{-1/2}$ **86.** $x^{2/3}x^{1/2}x^{-1/4}$ **87.** $(x^3y^6)^{1/3}$ **88.** $(x^4y^8)^{3/4}$

89. $\dfrac{(x^2y)^{1/3}(xy^2)^{2/3}}{x^{2/3}y^{2/3}}$ **90.** $\dfrac{(xy)^{1/4}(x^2y^2)^{1/2}}{(x^2y)^{3/4}}$ **91.** $\dfrac{(16x^2y^{-1/3})^{3/4}}{(xy^2)^{1/4}}$ **92.** $\dfrac{(4x^{-1}y^{1/3})^{3/2}}{(xy)^{3/2}}$

Applications and Extensions

In Problems 93–106, expressions that occur in calculus are given. Write each expression as a single quotient in which only positive exponents and/or radicals appear.

93. $\dfrac{x}{(1+x)^{1/2}}+2(1+x)^{1/2} \quad x>-1$

94. $\dfrac{1+x}{2x^{1/2}}+x^{1/2} \quad x>0$

95. $2x(x^2+1)^{1/2}+x^2\cdot\dfrac{1}{2}(x^2+1)^{-1/2}\cdot 2x$

96. $(x+1)^{1/3}+x\cdot\dfrac{1}{3}(x+1)^{-2/3} \quad x\neq -1$

97. $\sqrt{4x+3}\cdot\dfrac{1}{2\sqrt{x-5}}+\sqrt{x-5}\cdot\dfrac{1}{5\sqrt{4x+3}} \quad x>5$

98. $\dfrac{\sqrt[3]{8x+1}}{3\sqrt[3]{(x-2)^2}}+\dfrac{\sqrt[3]{x-2}}{24\sqrt[3]{(8x+1)^2}} \quad x\neq 2, x\neq -\dfrac{1}{8}$

99. $\dfrac{\sqrt{1+x}-x\cdot\dfrac{1}{2\sqrt{1+x}}}{1+x} \quad x>-1$

100. $\dfrac{\sqrt{x^2+1}-x\cdot\dfrac{2x}{2\sqrt{x^2+1}}}{x^2+1}$

101. $\dfrac{(x+4)^{1/2}-2x(x+4)^{-1/2}}{x+4} \quad x>-4$

102. $\dfrac{(9-x^2)^{1/2}+x^2(9-x^2)^{-1/2}}{9-x^2} \quad -3<x<3$

103. $\dfrac{\dfrac{x^2}{(x^2-1)^{1/2}}-(x^2-1)^{1/2}}{x^2} \quad x<-1 \text{ or } x>1$

104. $\dfrac{(x^2+4)^{1/2}-x^2(x^2+4)^{-1/2}}{x^2+4}$

105. $\dfrac{\dfrac{1+x^2}{2\sqrt{x}}-2x\sqrt{x}}{(1+x^2)^2} \quad x>0$

106. $\dfrac{2x(1-x^2)^{1/3}+\dfrac{2}{3}x^3(1-x^2)^{-2/3}}{(1-x^2)^{2/3}} \quad x\neq -1, x\neq 1$

In Problems 107–116, expressions that occur in calculus are given. Factor each expression. Express your answer so that only positive exponents occur.

107. $(x+1)^{3/2} + x \cdot \frac{3}{2}(x+1)^{1/2} \quad x \geq -1$

108. $(x^2+4)^{4/3} + x \cdot \frac{4}{3}(x^2+4)^{1/3} \cdot 2x$

109. $6x^{1/2}(x^2+x) - 8x^{3/2} - 8x^{1/2} \quad x \geq 0$

110. $6x^{1/2}(2x+3) + x^{3/2} \cdot 8 \quad x \geq 0$

111. $3(x^2+4)^{4/3} + x \cdot 4(x^2+4)^{1/3} \cdot 2x$

112. $2x(3x+4)^{4/3} + x^2 \cdot 4(3x+4)^{1/3}$

113. $4(3x+5)^{1/3}(2x+3)^{3/2} + 3(3x+5)^{4/3}(2x+3)^{1/2} \quad x \geq -\frac{3}{2}$

114. $6(6x+1)^{1/3}(4x-3)^{3/2} + 6(6x+1)^{4/3}(4x-3)^{1/2} \quad x \geq \frac{3}{4}$

115. $3x^{-1/2} + \frac{3}{2}x^{1/2} \quad x > 0$

116. $8x^{1/3} - 4x^{-2/3} \quad x \neq 0$

In Problems 117–124, use a calculator to approximate each radical. Round your answer to two decimal places.

117. $\sqrt{2}$

118. $\sqrt{7}$

119. $\sqrt[3]{4}$

120. $\sqrt[3]{-5}$

121. $\dfrac{2+\sqrt{3}}{3-\sqrt{5}}$

122. $\dfrac{\sqrt{5}-2}{\sqrt{2}+4}$

123. $\dfrac{3\sqrt[3]{5}-\sqrt{2}}{\sqrt{3}}$

124. $\dfrac{2\sqrt{3}-\sqrt[3]{4}}{\sqrt{2}}$

125. Calculating the Amount of Gasoline in a Tank A Shell station stores its gasoline in underground tanks that are right circular cylinders lying on their sides. See the illustration. The volume V of gasoline in the tank (in gallons) is given by the formula

$$V = 40h^2\sqrt{\frac{96}{h} - 0.608}$$

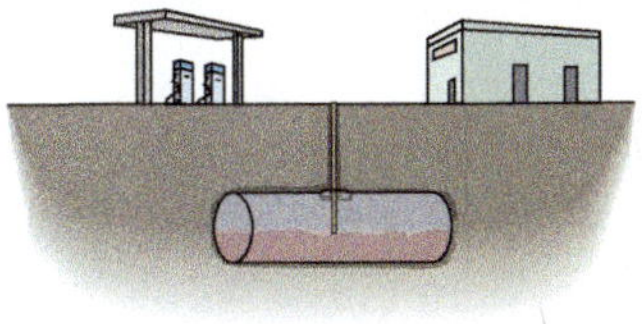

where h is the height of the gasoline (in inches) as measured on a depth stick.

(a) If $h = 12$ inches, how many gallons of gasoline are in the tank?
(b) If $h = 1$ inch, how many gallons of gasoline are in the tank?

126. Inclined Planes The final velocity v of an object in feet per second (ft/sec) after it slides down a frictionless inclined plane of height h feet is

$$v = \sqrt{64h + v_0^2}$$

where v_0 is the initial velocity (in ft/sec) of the object.

(a) What is the final velocity v of an object that slides down a frictionless inclined plane of height 4 feet? Assume that the initial velocity is 0.
(b) What is the final velocity v of an object that slides down a frictionless inclined plane of height 16 feet? Assume that the initial velocity is 0.
(c) What is the final velocity v of an object that slides down a frictionless inclined plane of height 2 feet with an initial velocity of 4 ft/sec?

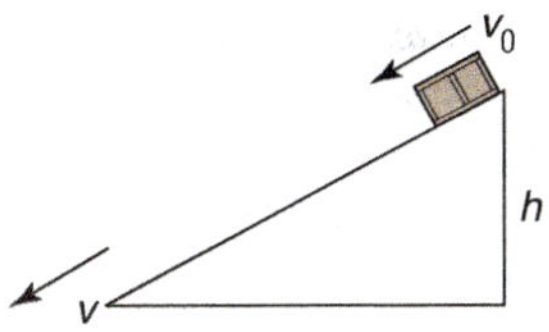

Problems 127 and 128 require the following information.

Period of a Pendulum *The period T, in seconds, of a pendulum of length l, in feet, may be approximated using the formula*

$$T = 2\pi\sqrt{\frac{l}{32}}$$

In Problems 127 and 128, express your answer both as a square root and as a decimal.

127. Find the period T of a pendulum whose length is 64 feet.

128. Find the period T of a pendulum whose length is 16 feet.

Explaining Concepts: Discussion and Writing

129. Give an example to show that $\sqrt{a^2}$ is not equal to a. Use it to explain why $\sqrt{a^2} = |a|$.

'Are You Prepared?' Answers

1. 9; −9 **2.** 4; 4

1 Graphs, Equations, and Inequalities

Financing a Purchase

Whenever we make a major purchase, such as an automobile or a house, we often need to finance the purchase by borrowing money from a lending institution, such as a bank. Have you ever wondered how the bank determines the monthly payment? How much total interest will be paid over the course of the loan? What roles do the rate of interest and the length of the loan play?

—See the Internet-based Chapter Project I—

Outline

A Look Ahead •••

In this chapter, we use the rectangular coordinate system to graph equations. Then, we algebraically solve equations and use the rectangular coordinate system to visualize the solution. The idea of using a system of rectangular coordinates dates back to ancient times, when such a system was used for surveying and city planning. Apollonius of Perga, in 200 BC, used a form of rectangular coordinates in his work on conics, although this use does not stand out as clearly as it does in modern treatments. Sporadic use of rectangular coordinates continued until the 1600s. By that time, algebra had developed sufficiently so that René Descartes (1596–1650) and Pierre de Fermat (1601–1665) could take the crucial step, which was the use of rectangular coordinates to translate geometry problems into algebra problems, and vice versa. This step allowed both geometers and algebraists to gain new insights into their subjects, which previously had been regarded as separate, but now were seen to be connected in many important ways.

With the advent of modern technology, in particular, graphing utilities, not only are we able to visualize the dual roles of algebra and geometry, but we are also able to solve many problems that required advanced methods before this technology.

1.1 The Distance and Midpoint Formulas; Graphing Utilities; Introduction to Graphing Equations

PREPARING FOR THIS SECTION *Before getting started, review the following:*

- Algebra Essentials (Section R.2, pp. 18–27)
- Geometry Essentials (Section R.3, pp. 31–36)

Now Work the **'Are You Prepared?'** problems on page 94.

OBJECTIVES
1 Use the Distance Formula (p. 85)
2 Use the Midpoint Formula (p. 87)
3 Graph Equations by Plotting Points (p. 88)
4 Graph Equations Using a Graphing Utility (p. 90)
5 Use a Graphing Utility to Create Tables (p. 92)
6 Find Intercepts from a Graph (p. 93)
7 Use a Graphing Utility to Approximate Intercepts (p. 93)

Rectangular Coordinates

We locate a point on the real number line by assigning it a single real number, called the *coordinate of the point*. For work in a two-dimensional plane, points are located by using two numbers.

Begin with two real number lines located in the same plane: one horizontal and the other vertical. The horizontal line is called the ***x*-axis**, the vertical line the ***y*-axis**, and the point of intersection the **origin *O***. See Figure 1. Assign coordinates to every point on these number lines using a convenient scale. Recall that the scale of a number line is the distance between 0 and 1. In mathematics, we usually use the same scale on each axis, but in applications, different scales appropriate to the application may be used.

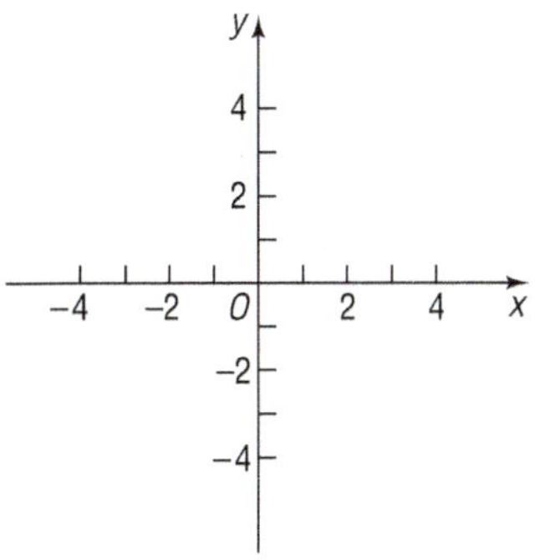

Figure 1 *xy*-Plane

The origin O has a value of 0 on both the x-axis and y-axis. Points on the x-axis to the right of O are associated with positive real numbers, and those to the left of O are associated with negative real numbers. Points on the y-axis above O are associated with positive real numbers, and those below O are associated with negative real numbers. In Figure 1, the x-axis and y-axis are labeled as x and y, respectively, and we have used an arrow at the end of each axis to denote the positive direction.

The coordinate system described here is called a **rectangular** or **Cartesian*** **coordinate system**. The plane formed by the x-axis and y-axis is sometimes called the ***xy*-plane**, and the x-axis and y-axis are referred to as the **coordinate axes**.

Any point P in the xy-plane can be located by using an **ordered pair** (x, y) of real numbers. Let x denote the signed distance of P from the y-axis (*signed* means that, if P is to the right of the y-axis, then $x > 0$, and if P is to the left of the y-axis, then $x < 0$); and let y denote the signed distance of P from the x-axis. The ordered pair (x, y), also called the **coordinates** of P, gives us enough information to locate the point P in the plane.

For example, to locate the point whose coordinates are $(-3, 1)$, go 3 units along the x-axis to the left of O and then go straight up 1 unit. We **plot** this point by placing a dot at this location. See Figure 2, in which the points with coordinates $(-3, 1)$, $(-2, -3)$, $(3, -2)$, and $(3, 2)$ are plotted.

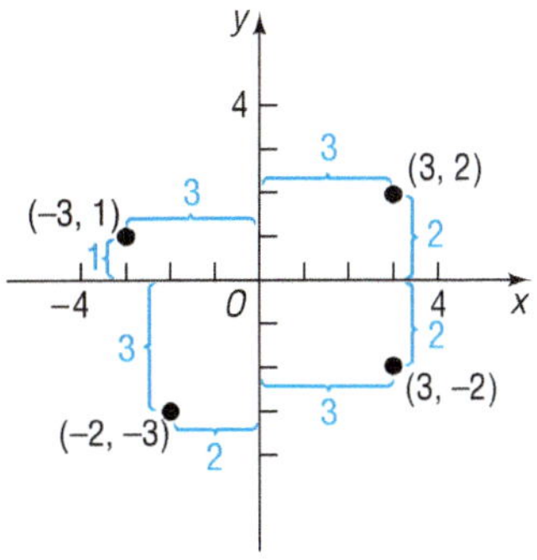

Figure 2

The origin has coordinates $(0, 0)$. Any point on the x-axis has coordinates of the form $(x, 0)$, and any point on the y-axis has coordinates of the form $(0, y)$.

If (x, y) are the coordinates of a point P, then x is called the ***x*-coordinate**, or **abscissa**, of P and y is the ***y*-coordinate**, or **ordinate**, of P. We identify the point P using its coordinates (x, y) by writing $P = (x, y)$. Usually, we will simply say "the point (x, y)" rather than "the point whose coordinates are (x, y)."

*Named after René Descartes (1596–1650), a French mathematician, philosopher, and theologian.

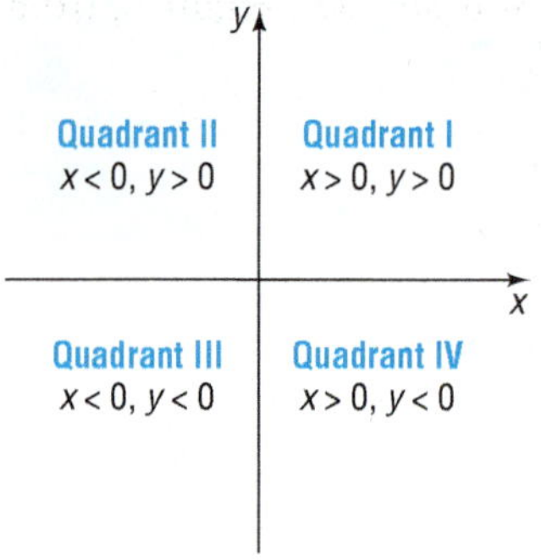

Figure 3

The coordinate axes divide the *xy*-plane into four sections called **quadrants**, as shown in Figure 3. In quadrant I, both the *x*-coordinate and the *y*-coordinate of all points are positive; in quadrant II, *x* is negative and *y* is positive; in quadrant III, both *x* and *y* are negative; and in quadrant IV, *x* is positive and *y* is negative. Points on the coordinate axes belong to no quadrant.

Now Work PROBLEM 15

Graphing Utilities

All graphing utilities (graphing calculators and computer software graphing packages) graph equations by plotting points on a screen. The screen itself actually consists of small rectangles, called **pixels**. The more pixels the screen has, the better the resolution. Most graphing calculators have 50 to 100 pixels per square inch; most smartphones have 300 to 450 pixels per square inch. When a point to be plotted lies inside a pixel, the pixel is turned on (lights up). The graph of an equation is a collection of lighted pixels. Figure 4 shows how the graph of $y = 2x$ looks on a TI-84 Plus C graphing calculator.

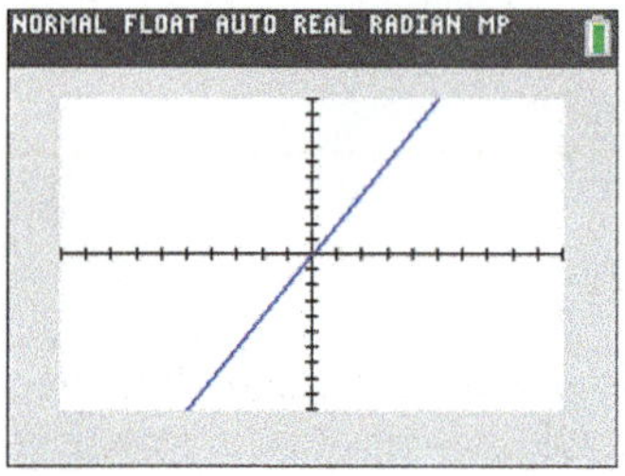

Figure 4 $Y_1 = 2x$

The screen of a graphing utility will display the coordinate axes of a rectangular coordinate system. However, the scale must be set on each axis. The smallest and largest values of *x* and *y* to be included in the graph must also be set. This is called **setting the viewing rectangle** or **viewing window**. Figure 5 illustrates a typical viewing window.

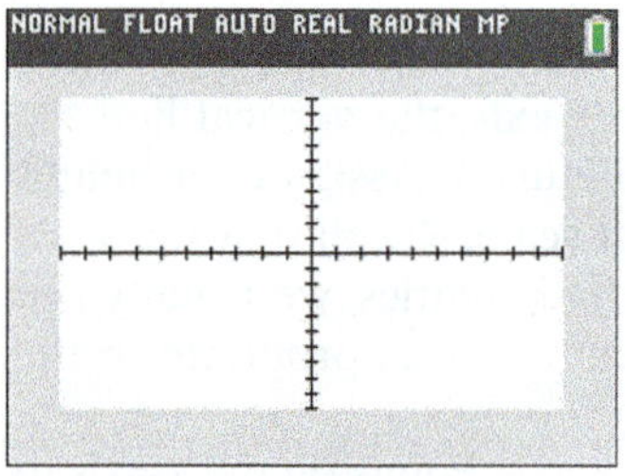

Figure 5 Viewing window

To set the viewing window, values must be given to the following expressions:

*X*min: the smallest value of *x* shown on the viewing window
*X*max: the largest value of *x* shown on the viewing window
*X*scl: the number of units per tick mark on the *x*-axis
*Y*min: the smallest value of *y* shown on the viewing window
*Y*max: the largest value of *y* shown on the viewing window
*Y*scl: the number of units per tick mark on the *y*-axis

Figure 6 illustrates these settings and their relation to the Cartesian coordinate system.

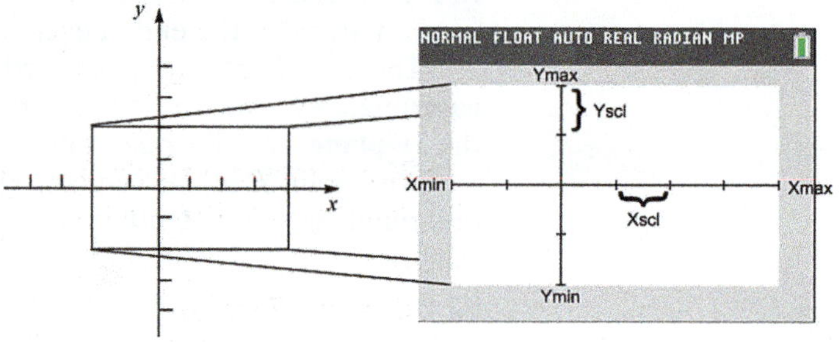

Figure 6

If the scale used on each axis is known, the minimum and maximum values of *x* and *y* shown on the screen can be determined by counting the tick marks. Look again at Figure 5. For a scale of 1 on each axis, the minimum and maximum values of *x* are -10 and 10, respectively; the minimum and maximum values of *y* are also -10 and 10. If the scale is 2 on each axis, then the minimum and maximum values of *x* are -20 and 20, respectively; and the minimum and maximum values of *y* are -20 and 20, respectively.

Conversely, if the minimum and maximum values of *x* and *y* are known, the scales can be determined by counting the tick marks displayed. This text follows the practice of showing the minimum and maximum values of *x* and *y* in illustrations

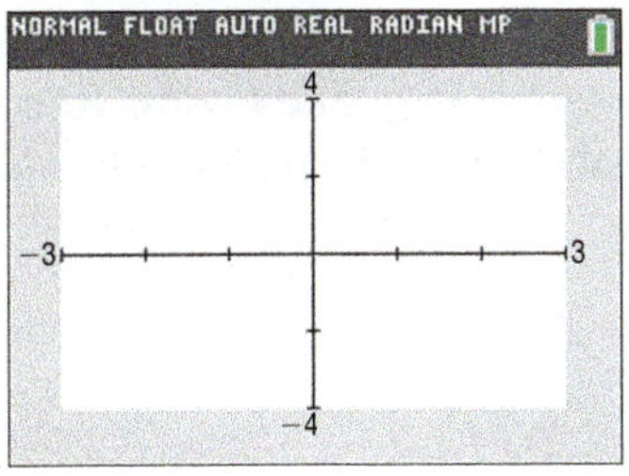

Figure 7

so that the reader will know how the viewing window was set. See Figure 7, from which the following window settings can be determined:

$$X\text{min} = -3 \qquad Y\text{min} = -4$$
$$X\text{max} = 3 \qquad Y\text{max} = 4$$
$$X\text{scl} = 1 \qquad Y\text{scl} = 2$$

Now Work PROBLEMS 19 AND 29

1 Use the Distance Formula

If the same units of measurement (such as inches, centimeters, and so on) are used for both the x-axis and y-axis, then all distances in the xy-plane can be measured using this unit of measurement.

EXAMPLE 1 **Finding the Distance between Two Points**

Find the distance d between the points $(1, 3)$ and $(5, 6)$.

Solution First plot the points $(1, 3)$ and $(5, 6)$ and connect them with a straight line. See Figure 8(a). To find the length d, begin by drawing a horizontal line from $(1, 3)$ to $(5, 3)$ and a vertical line from $(5, 3)$ to $(5, 6)$, forming a right triangle, as in Figure 8(b). One leg of the triangle is of length 4 (since $|5 - 1| = 4$) and the other is of length 3 (since $|6 - 3| = 3$). By the Pythagorean Theorem, the square of the distance d that we seek is

$$d^2 = 4^2 + 3^2 = 16 + 9 = 25$$
$$d = \sqrt{25} = 5$$

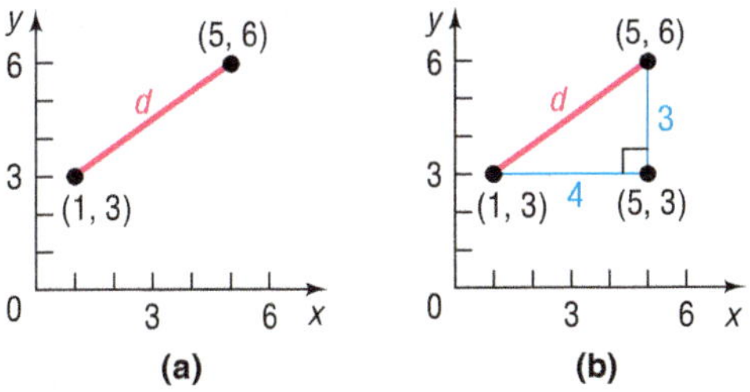

Figure 8

The **distance formula** provides a straightforward method for computing the distance between two points.

THEOREM

Distance Formula

The distance between two points $P_1 = (x_1, y_1)$ and $P_2 = (x_2, y_2)$, denoted by $d(P_1, P_2)$, is

$$d(P_1, P_2) = \sqrt{(x_2 - x_1)^2 + (y_2 - y_1)^2} \qquad \textbf{(1)}$$

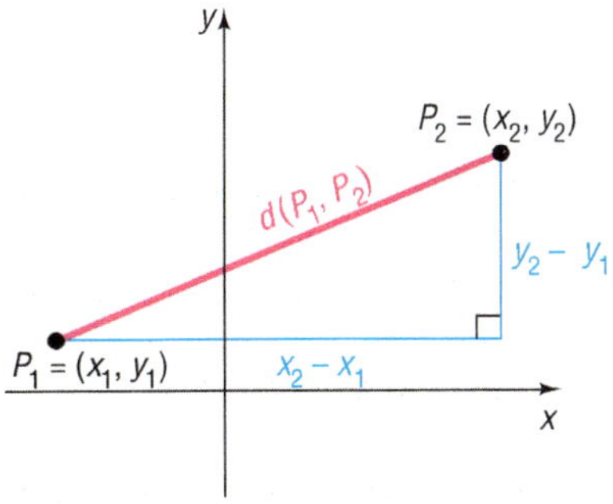

Figure 9 Illustration of the Distance Formula

Figure 9 illustrates the theorem.

Proof of the Distance Formula Let (x_1, y_1) denote the coordinates of point P_1, and let (x_2, y_2) denote the coordinates of point P_2. Assume that the line joining P_1 and P_2 is neither horizontal nor vertical. Refer to Figure 10(a). The coordinates of P_3 are (x_2, y_1). The horizontal distance from P_1 to P_3 is the absolute value of

> **In Words**
> To compute the distance between two points, find the difference of the *x*-coordinates, square it, and add this to the square of the difference of the *y*-coordinates. The square root of this sum is the distance.

the difference of the x-coordinates, $|x_2 - x_1|$. The vertical distance from P_3 to P_2 is the absolute value of the difference of the y-coordinates, $|y_2 - y_1|$. See Figure 10(b). The distance $d(P_1, P_2)$ is the length of the hypotenuse of the right triangle, so, by the Pythagorean Theorem, it follows that

$$\begin{aligned} [d(P_1, P_2)]^2 &= |x_2 - x_1|^2 + |y_2 - y_1|^2 \\ &= (x_2 - x_1)^2 + (y_2 - y_1)^2 \\ d(P_1, P_2) &= \sqrt{(x_2 - x_1)^2 + (y_2 - y_1)^2} \end{aligned}$$

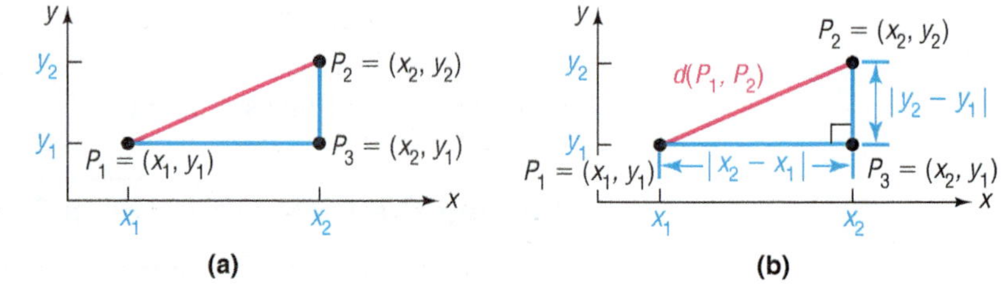

Figure 10

Now, if the line joining P_1 and P_2 is horizontal, then the y-coordinate of P_1 equals the y-coordinate of P_2; that is, $y_1 = y_2$. Refer to Figure 11(a). In this case, the distance formula (1) still works, because, for $y_1 = y_2$, it reduces to

$$d(P_1, P_2) = \sqrt{(x_2 - x_1)^2 + 0^2} = \sqrt{(x_2 - x_1)^2} = |x_2 - x_1|$$

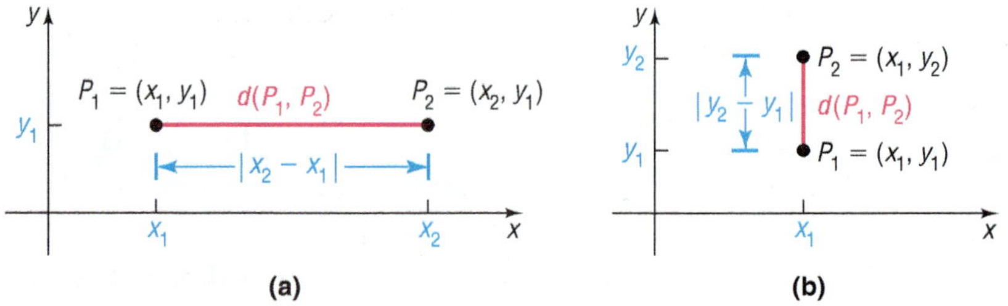

Figure 11

A similar argument holds if the line joining P_1 and P_2 is vertical. See Figure 11(b). ■

EXAMPLE 2 Finding the Length of a Line Segment

Find the length of the line segment shown in Figure 12.

Solution The length of the line segment is the distance between the points $P_1 = (x_1, y_1) = (-4, 5)$ and $P_2 = (x_2, y_2) = (3, 2)$. Using the distance formula (1) with $x_1 = -4$, $y_1 = 5$, $x_2 = 3$, and $y_2 = 2$, the length d is

$$\begin{aligned} d &= \sqrt{(x_2 - x_1)^2 + (y_2 - y_1)^2} = \sqrt{[3 - (-4)]^2 + (2 - 5)^2} \\ &= \sqrt{7^2 + (-3)^2} = \sqrt{49 + 9} = \sqrt{58} \approx 7.62 \end{aligned}$$

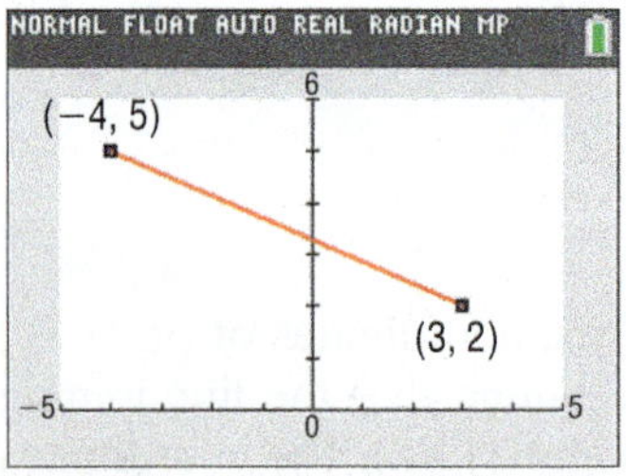

Figure 12

Now Work PROBLEM 35

The distance between two points $P_1 = (x_1, y_1)$ and $P_2 = (x_2, y_2)$ is never a negative number. Also, the distance between two points is 0 only when the points are identical—that is, when $x_1 = x_2$ and $y_1 = y_2$. Also, because $(x_2 - x_1)^2 = (x_1 - x_2)^2$ and $(y_2 - y_1)^2 = (y_1 - y_2)^2$, it makes no difference whether the distance is computed from P_1 to P_2 or from P_2 to P_1; that is, $d(P_1, P_2) = d(P_2, P_1)$.

The introduction to this chapter mentioned that rectangular coordinates enable us to translate geometry problems into algebra problems, and vice versa. The next example shows how algebra (the distance formula) can be used to solve geometry problems.

EXAMPLE 3 **Using Algebra to Solve Geometry Problems**

Consider the three points $A = (-2, 1)$, $B = (2, 3)$, and $C = (3, 1)$.

(a) Plot each point and form the triangle ABC.
(b) Find the length of each side of the triangle.
(c) Verify that the triangle is a right triangle.
(d) Find the area of the triangle.

Solution

(a) Figure 13 shows the points A, B, C and the triangle ABC.
(b) To find the length of each side of the triangle, use the distance formula (1).

$$d(A, B) = \sqrt{[2 - (-2)]^2 + (3 - 1)^2} = \sqrt{16 + 4} = \sqrt{20} = 2\sqrt{5}$$
$$d(B, C) = \sqrt{(3 - 2)^2 + (1 - 3)^2} = \sqrt{1 + 4} = \sqrt{5}$$
$$d(A, C) = \sqrt{[3 - (-2)]^2 + (1 - 1)^2} = \sqrt{25 + 0} = 5$$

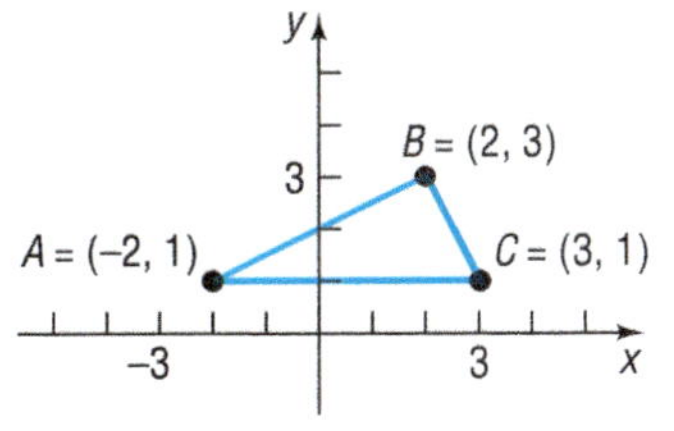

Figure 13

(c) If the sum of the squares of the lengths of two of the sides equals the square of the length of the third side, the triangle is a right triangle. (Why is this sufficient?) Looking at Figure 13, it seems reasonable to conjecture that the angle at vertex B might be a right angle. We shall check to see whether

$$[d(A, B)]^2 + [d(B, C)]^2 = [d(A, C)]^2$$

Using the results from part (b),

$$[d(A, B)]^2 + [d(B, C)]^2 = (2\sqrt{5})^2 + (\sqrt{5})^2$$
$$= 20 + 5 = 25 = [d(A, C)]^2$$

It follows from the converse of the Pythagorean Theorem that triangle ABC is a right triangle.

(d) Because the right angle is at vertex B, the sides AB and BC form the base and height of the triangle. Its area is

$$\text{Area} = \frac{1}{2}(\text{Base})(\text{Height}) = \frac{1}{2}(2\sqrt{5})(\sqrt{5}) = 5 \text{ square units}$$

Now Work PROBLEM 51

2 Use the Midpoint Formula

We now derive a formula for the coordinates of the **midpoint of a line segment**. Let $P_1 = (x_1, y_1)$ and $P_2 = (x_2, y_2)$ be the endpoints of a line segment, and let $M = (x, y)$ be the point on the line segment that is the same distance from P_1 as it is from P_2. See Figure 14. The triangles P_1AM and MBP_2 are congruent. [Do you see why? Angle AP_1M = angle BMP_2,* angle P_1MA = angle MP_2B, and $d(P_1, M) = d(M, P_2)$ is given. So, we have angle–side–angle.] Hence, corresponding sides are equal in length. That is,

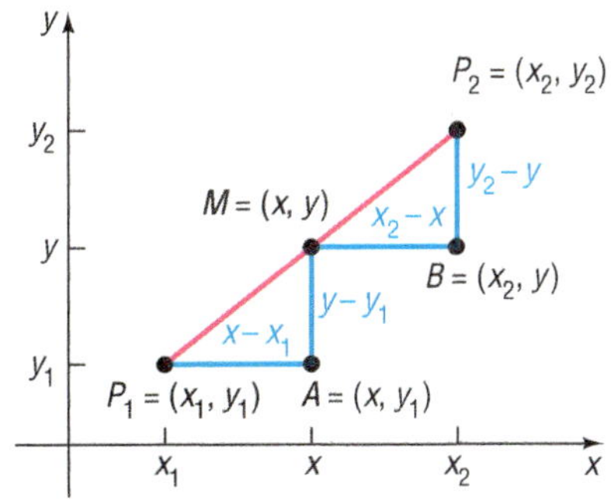

Figure 14 Illustration of midpoint

$$x - x_1 = x_2 - x \qquad \text{and} \qquad y - y_1 = y_2 - y$$
$$2x = x_1 + x_2 \qquad\qquad 2y = y_1 + y_2$$
$$x = \frac{x_1 + x_2}{2} \qquad\qquad y = \frac{y_1 + y_2}{2}$$

*A postulate from geometry states that the transversal $\overline{P_1P_2}$ forms congruent corresponding angles with the parallel line segments $\overline{P_1A}$ and $\overline{MB}$.

THEOREM

In Words
To find the midpoint of a line segment, average the x-coordinates of the endpoints, and average the y-coordinates of the endpoints.

Midpoint Formula

The midpoint $M = (x, y)$ of the line segment from $P_1 = (x_1, y_1)$ to $P_2 = (x_2, y_2)$ is

$$M = (x, y) = \left(\frac{x_1 + x_2}{2}, \frac{y_1 + y_2}{2}\right) \quad (2)$$

EXAMPLE 4

Finding the Midpoint of a Line Segment

Find the midpoint of a line segment from $P_1 = (-5, 5)$ to $P_2 = (3, 1)$. Plot the points P_1 and P_2 and their midpoint.

Solution Apply the midpoint formula (2) using $x_1 = -5, y_1 = 5, x_2 = 3$, and $y_2 = 1$. Then the coordinates (x, y) of the midpoint M are

$$x = \frac{x_1 + x_2}{2} = \frac{-5 + 3}{2} = -1 \quad \text{and} \quad y = \frac{y_1 + y_2}{2} = \frac{5 + 1}{2} = 3$$

That is, $M = (-1, 3)$. See Figure 15.

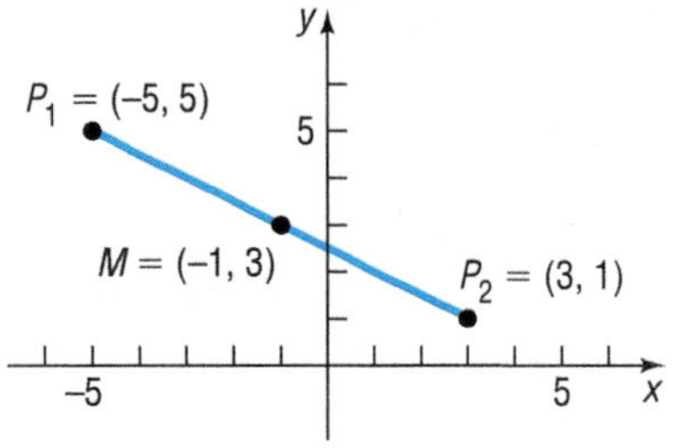

Figure 15

Now Work PROBLEM 57

3 Graph Equations by Plotting Points

An **equation in two variables**, say x and y, is a statement in which two expressions involving x and y are equal. The expressions are called the **sides** of the equation. Since an equation is a statement, it may be true or false, depending on the value of the variables. Any values of x and y that result in a true statement are said to **satisfy** the equation.

For example, the following are all equations in two variables x and y:

$$x^2 + y^2 = 5 \qquad 2x - y = 6 \qquad y = 2x + 5 \qquad x^2 = y$$

The first of these, $x^2 + y^2 = 5$, is satisfied for $x = 1, y = 2$, since $1^2 + 2^2 = 5$. Other choices of x and y, such as $x = -1, y = -2$, also satisfy this equation. It is not satisfied for $x = 2$ and $y = 3$, since $2^2 + 3^2 = 4 + 9 = 13 \neq 5$.

The **graph of an equation in two variables** x and y consists of the set of points in the xy-plane whose coordinates (x, y) satisfy the equation.

Graphs play an important role in helping us to visualize the relationships that exist between two variables or quantities. Figure 16 shows the relation between the

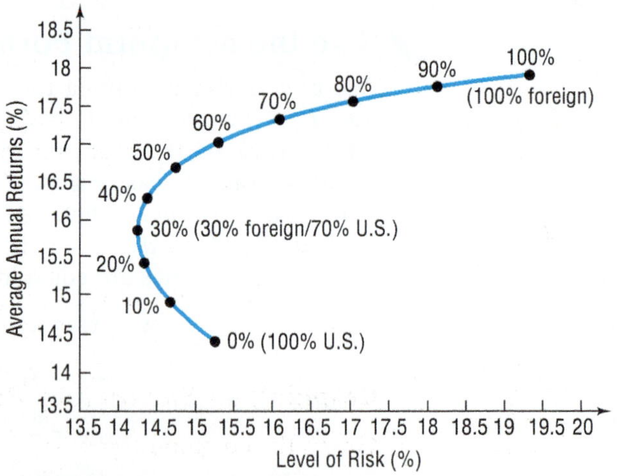

Figure 16
Source: T. Rowe Price

level of risk in a stock portfolio and the average annual rate of return. The graph shows that, when 30% of a portfolio of stocks is invested in foreign companies, risk is minimized.

EXAMPLE 5 Determining Whether a Point Is on the Graph of an Equation

Determine whether each of the following points is on the graph of the equation $2x - y = 6$.

(a) $(2, 3)$ (b) $(2, -2)$

Solution (a) For the point $(2, 3)$, check to see whether $x = 2$, $y = 3$ satisfies the equation $2x - y = 6$.

$$2x - y = 2(2) - 3 = 4 - 3 = 1 \neq 6$$

The equation is not satisfied, so the point $(2, 3)$ is not on the graph of $2x - y = 6$.

(b) For the point $(2, -2)$,

$$2x - y = 2(2) - (-2) = 4 + 2 = 6$$

The equation is satisfied, so the point $(2, -2)$ is on the graph of $2x - y = 6$. ■

Now Work PROBLEM 65

EXAMPLE 6 Graphing an Equation by Plotting Points

Graph the equation: $y = -2x + 3$

Step-by-Step Solution

Step 1: Find some points (x, y) that satisfy the equation. To determine these points, choose values of x and use the equation to find the corresponding values for y. See Table 1.

Table 1

x	$y = -2x + 3$	(x, y)
-2	$-2(-2) + 3 = 7$	$(-2, 7)$
-1	$-2(-1) + 3 = 5$	$(-1, 5)$
0	$-2(0) + 3 = 3$	$(0, 3)$
1	$-2(1) + 3 = 1$	$(1, 1)$
2	$-2(2) + 3 = -1$	$(2, -1)$

Step 2: Plot the points listed in the table as shown in Figure 17(a). Now connect the points to obtain the graph of the equation (a line), as shown in Figure 17(b).

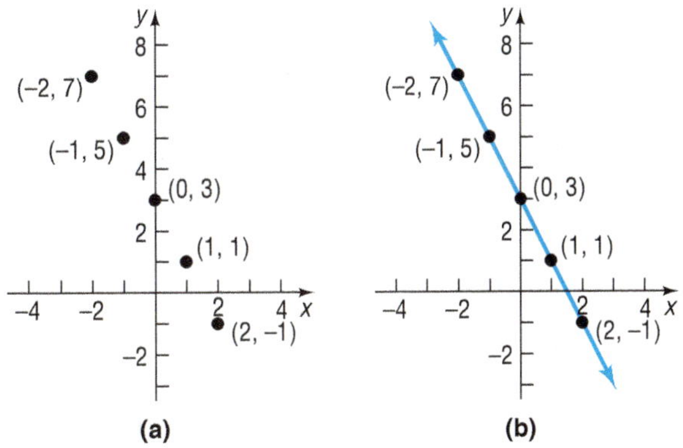

Figure 17 $y = -2x + 3$ ■

EXAMPLE 7 Graphing an Equation by Plotting Points

Graph the equation: $y = x^2$

Solution Table 2 on page 90 provides several points on the graph. Plotting these points and connecting them with a smooth curve gives the graph (a *parabola*) shown in Figure 18.

Table 2

x	$y = x^2$	(x, y)
−4	16	(−4, 16)
−3	9	(−3, 9)
−2	4	(−2, 4)
−1	1	(−1, 1)
0	0	(0, 0)
1	1	(1, 1)
2	4	(2, 4)
3	9	(3, 9)
4	16	(4, 16)

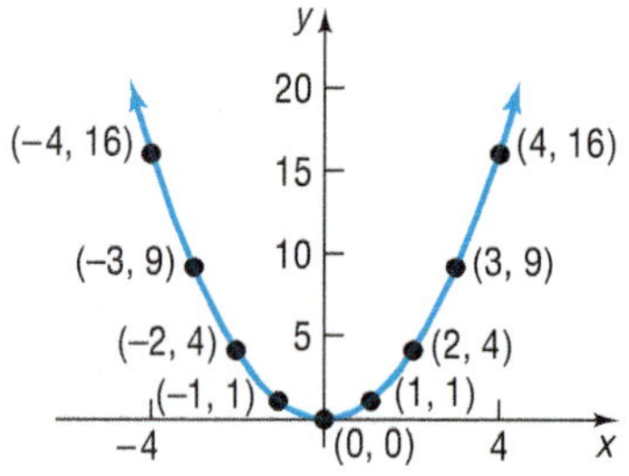

Figure 18 $y = x^2$

The graphs of the equations shown in Figures 17(b) and 18 do not show all the points that are on the graph. For example, in Figure 17(b), the point $(20, -37)$ is a part of the graph of $y = -2x + 3$, but it is not shown. Since the graph of $y = -2x + 3$ could be extended out as far as we please, we use arrows to indicate that the pattern shown continues. It is important when illustrating a graph to present enough of the graph so that any viewer of the illustration will "see" the rest of it as an obvious continuation of what is actually there. This is referred to as a **complete graph**.

One way to obtain a complete graph of an equation is to continue plotting points on the graph until a pattern becomes evident. Then these points are connected with a smooth curve following the suggested pattern. But how many points are sufficient? Sometimes knowledge about the equation tells us. For example, we will learn in the next chapter that if an equation is of the form $y = mx + b$, then its graph is a line. In this case, two points would suffice to obtain the graph.

One purpose of this text is to investigate the properties of equations in order to decide whether a graph is complete. Sometimes we shall graph equations by plotting points on the graph until a pattern becomes evident and then connect these points with a smooth curve, following the suggested pattern. (Shortly, we shall investigate various techniques that will enable us to graph an equation without plotting so many points.) Other times we shall graph equations using a graphing utility.

4 Graph Equations Using a Graphing Utility

From Examples 6 and 7, we see that a graph can be obtained by plotting points in a rectangular coordinate system and connecting them. Graphing utilities perform these same steps when graphing an equation. For example, the TI-84 Plus C determines 265 evenly spaced input values (starting at *X*min and ending at *X*max),* uses the equation to determine the output values, plots these points on the screen, and finally (if in the connected mode) draws a line between consecutive points.

To graph an equation in two variables x and y using a graphing utility requires that the equation be written in the form $y = \{\text{expression in } x\}$. If the original equation is not in this form, rewrite it using equivalent equations until the form $y = \{\text{expression in } x\}$ is obtained. In general, there are four ways to obtain equivalent equations.

Procedures That Result in Equivalent Equations

1. Interchange the two sides of the equation:

$$3x + 5 = y \quad \text{is equivalent to} \quad y = 3x + 5$$

2. Simplify the sides of the equation by combining like terms, eliminating parentheses, and so on:

$$2y + 2 + 6 = 2x + 5(x + 1) \quad \text{is equivalent to} \quad 2y + 8 = 7x + 5$$

*These input values depend on the values of *X*min and *X*max. For example, if *X*min $= -10$ and *X*max $= 10$, then the first input value will be -10 and the next input value will be $-10 + (10 - (-10))/264 = -9.9242$, and so on.

3. Add or subtract the same expression on both sides of the equation:

$$y + 3x - 5 = 4 \quad \text{is equivalent to} \quad y + 3x - 5 + 5 = 4 + 5$$

4. Multiply or divide both sides of the equation by the same nonzero expression:

$$3y = 6 - 2x \quad \text{is equivalent to} \quad \frac{1}{3} \cdot 3y = \frac{1}{3}(6 - 2x)$$

EXAMPLE 8 Expressing an Equation in the Form $y = \{\text{expression in } x\}$

Solve for y: $2y + 3x - 5 = 4$

Solution Replace the original equation by a succession of equivalent equations.

$$2y + 3x - 5 = 4$$

$$2y + 3x - 5 + 5 = 4 + 5 \qquad \text{Add 5 to both sides.}$$

$$2y + 3x = 9 \qquad \text{Simplify.}$$

$$2y + 3x - 3x = 9 - 3x \qquad \text{Subtract } 3x \text{ from both sides.}$$

$$2y = 9 - 3x \qquad \text{Simplify.}$$

$$\frac{2y}{2} = \frac{9 - 3x}{2} \qquad \text{Divide both sides by 2.}$$

$$y = \frac{9 - 3x}{2} \qquad \text{Simplify.}$$

WARNING Be careful when entering the expression $\frac{9 - 3x}{2}$. Use a fraction template $\left(\frac{\square}{\square}\right)$ or use parentheses as follows: $(9 - 3x)/2$.

We are now ready to graph equations using a graphing utility.

EXAMPLE 9 Graphing an Equation Using a Graphing Utility

Use a graphing utility to graph the equation: $6x^2 + 3y = 36$

Step-by-Step Solution

Step 1: Solve the equation for y in terms of x.

$$6x^2 + 3y = 36$$

$$3y = -6x^2 + 36 \qquad \text{Subtract } 6x^2 \text{ from both sides.}$$

$$y = -2x^2 + 12 \qquad \text{Divide both sides by 3 and simplify.}$$

Step 2: Enter the equation to be graphed into your graphing utility. Figure 19 shows the equation to be graphed entered on a TI-84 Plus C.

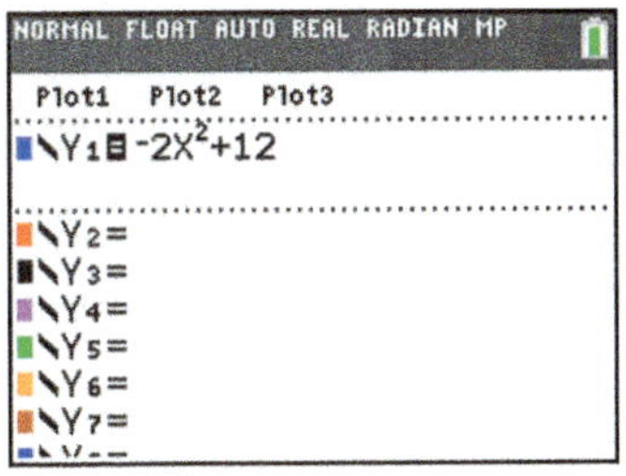

Figure 19

Step 3: Choose an initial viewing window. Without any knowledge about the behavior of the graph, it is common to choose the standard viewing window as the initial viewing window. The standard viewing window is

$X\min = -10$ $Y\min = -10$
$X\max = 10$ $Y\max = 10$
$X\text{scl} = 1$ $Y\text{scl} = 1$

See Figure 20.

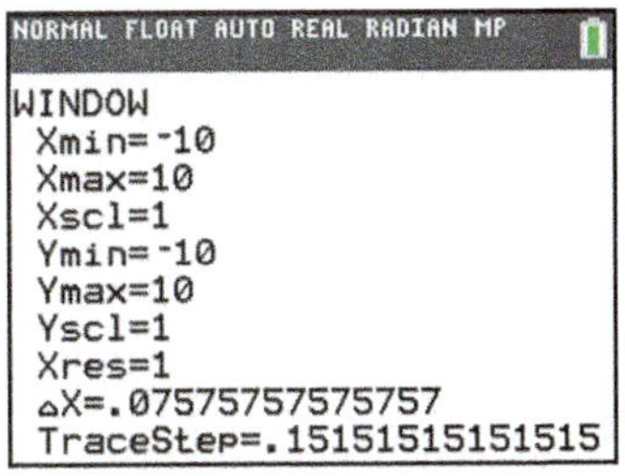

Figure 20 Standard viewing window

Step 4: Graph the equation. See Figure 21.

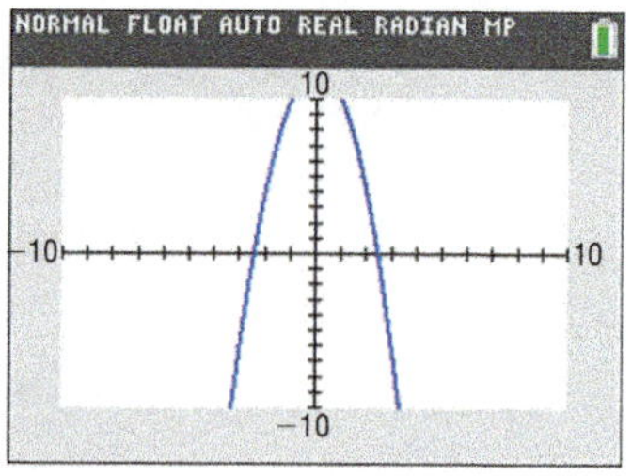

Figure 21 $Y_1 = -2x^2 + 12$

Step 5: Adjust the viewing window until a complete graph is obtained.

The graph of $y = -2x^2 + 12$ is not complete. The value of Ymax must be increased so that the top portion of the graph is visible. After increasing the value of Ymax to 12, we obtain the graph in Figure 22. The graph is now complete.

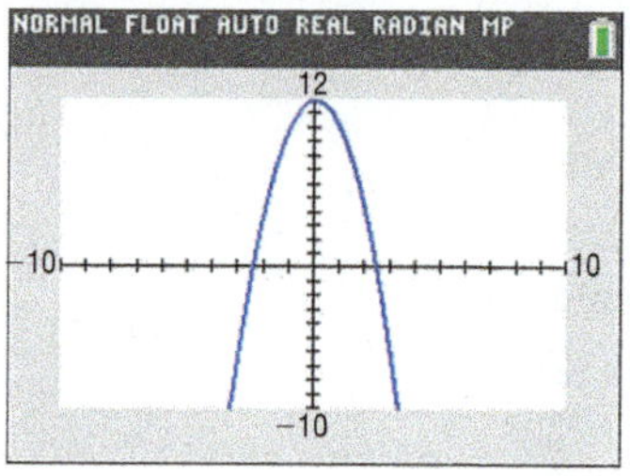

Figure 22 $Y_1 = -2x^2 + 12$

Note: Some graphing utilities have a ZOOM-STANDARD feature that automatically sets the viewing window to the standard viewing window. In addition, some graphing utilities have a ZOOM-FIT feature that determines the appropriate *Y*min and *Y*max for a given *X*min and *X*max. Consult your owner's manual for the appropriate keystrokes. ■

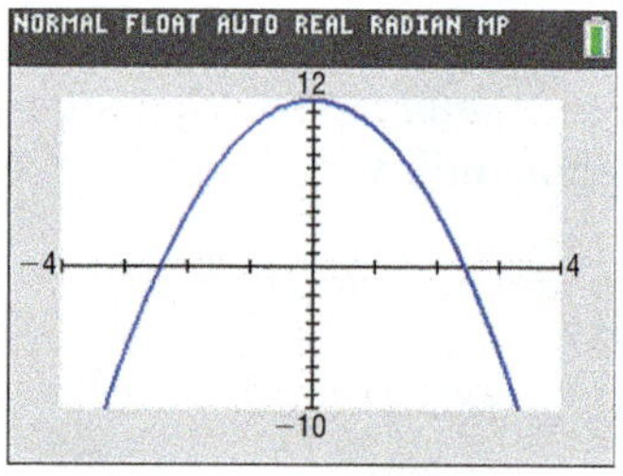

Figure 23 $Y_1 = -2x^2 + 12$

Now Work PROBLEM 83

Look again at Figure 22. Although a complete graph is shown, the graph might be improved by adjusting the values of Xmin and Xmax. Figure 23 shows the graph of $y = -2x^2 + 12$ using $X\text{min} = -4$ and $X\text{max} = 4$. Do you think this is a better choice for the viewing window?

5 Use a Graphing Utility to Create Tables

In addition to graphing equations, graphing utilities can also be used to create a table of values that satisfy the equation. This feature is especially useful in determining an appropriate viewing window when graphing an equation.

EXAMPLE 10 Creating a Table Using a Graphing Utility

Create a table that displays the points on the graph of $6x^2 + 3y = 36$ for $x = -3, -2, -1, 0, 1, 2,$ and 3.

Step-by-Step Solution

Step 1: Solve the equation for *y* in terms of *x*.

We solved the equation for y in terms of x in Example 9 and obtained $y = -2x^2 + 12$.

Step 2: Enter the expression in *x* following the *Y* = prompt of the graphing utility.

See Figure 19 on page 91.

Step 3: Set up the table. Graphing utilities typically have two modes for creating tables. In the AUTO mode, the user determines a starting point for the table (TblStart) and ΔTbl (pronounced "delta table"). The ΔTbl feature determines the increment for *x* in the table. The ASK mode requires the user to enter values of *x*, and then the utility determines the corresponding value of *y*.

Create a table using AUTO mode. The table we wish to create starts at −3, so TblStart = −3. The increment in x is 1, so ΔTbl = 1. See Figure 24.

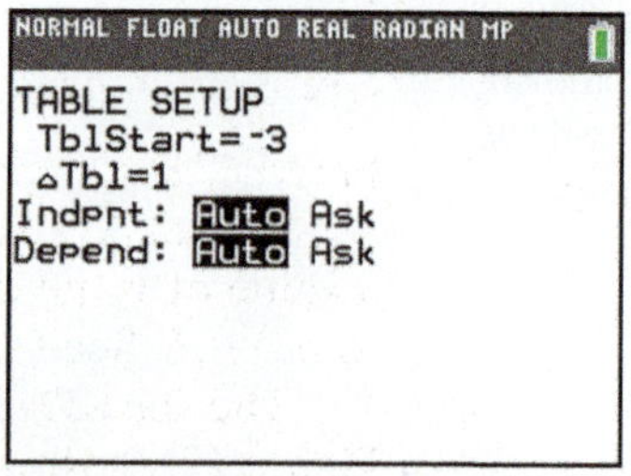

Figure 24 Table setup

Step 4: Create the table. See Table 3.

Table 3

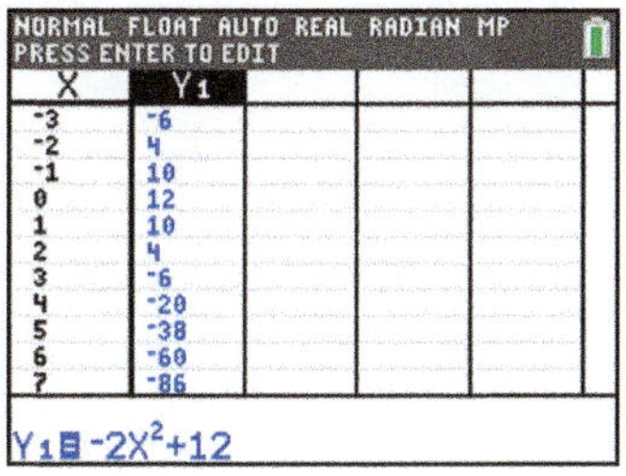
NORMAL FLOAT AUTO REAL RADIAN MP
PRESS ENTER TO EDIT

X	Y_1			
-3	-6			
-2	4			
-1	10			
0	12			
1	10			
2	4			
3	-6			
4	-20			
5	-38			
6	-60			
7	-86			

$Y_1 = -2X^2+12$

In AUTO mode, the user can scroll forward or backward within the table to find additional values. ■

In looking at Table 3, notice that $y = 12$ when $x = 0$. This information could have been used to help to create the initial viewing window by letting us know that Ymax needs to be at least 12 in order to get a complete graph.

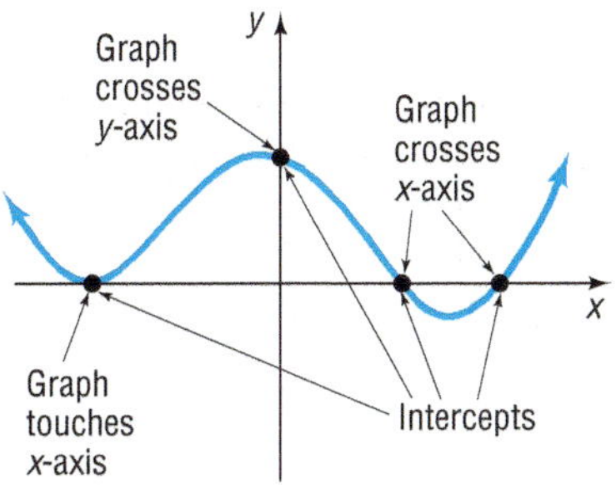

Figure 25 Intercepts

6 Find Intercepts from a Graph

The points, if any, at which a graph crosses or touches the coordinate axes are called the **intercepts**. See Figure 25. The x-coordinate of a point at which the graph crosses or touches the x-axis is an ***x*-intercept**, and the y-coordinate of a point at which the graph crosses or touches the y-axis is a ***y*-intercept**. For a graph to be complete, all its intercepts must be displayed.

EXAMPLE 11 **Finding Intercepts from a Graph**

Find the intercepts of the graph in Figure 26. What are its x-intercepts? What are its y-intercepts?

Solution The intercepts of the graph are the points

$$(-3, 0), \quad (0, 3), \quad \left(\frac{3}{2}, 0\right), \quad \left(0, -\frac{4}{3}\right), \quad (0, -3.5), \quad (4.5, 0)$$

The x-intercepts are -3, $\frac{3}{2}$, and 4.5; the y-intercepts are -3.5, $-\frac{4}{3}$, and 3. ■

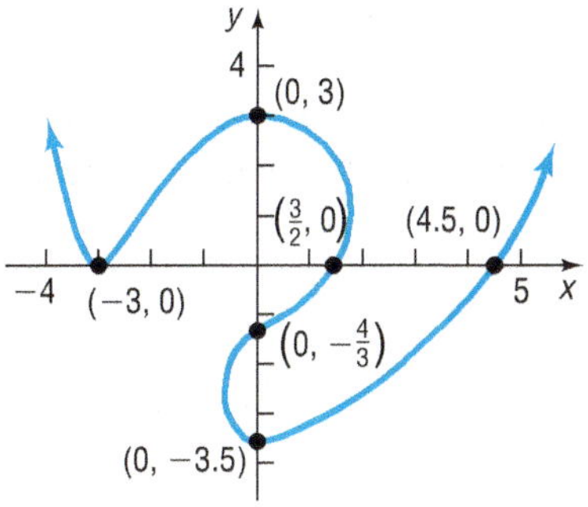

Figure 26

In Example 11 notice the following usage: If the type of intercept is not specified (x- versus y-), then report the intercept as an ordered pair. However, if the type of intercept is specified, then report the coordinate of the specified intercept. For x-intercepts, report the x-coordinate of the intercept; for y-intercepts, report the y-coordinate of the intercept.

Now Work PROBLEM 71

7 Use a Graphing Utility to Approximate Intercepts

We can use a graphing utility to approximate the intercepts of the graph of an equation.

EXAMPLE 12 **Approximating Intercepts Using a Graphing Utility**

Use a graphing utility to approximate the intercepts of the equation $y = x^3 - 16$.

Solution Figure 27(a) on page 94 shows the graph of $y = x^3 - 16$.

The eVALUEate feature of a TI-84 Plus C graphing calculator accepts as input a value of x and determines the value of y. If we let $x = 0$, the y-intercept is found to be -16. See Figure 27(b).

The ZERO feature of a TI-84 Plus C is used to find the x-intercept(s). See Figure 27(c). Rounded to two decimal places, the x-intercept is 2.52.

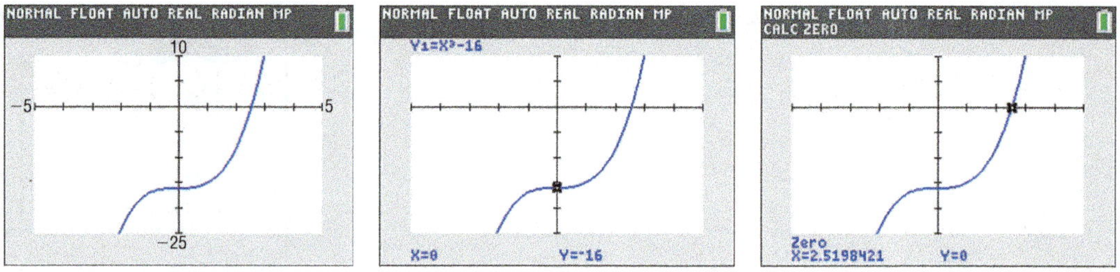

(a) (b) (c)

Figure 27

Now Work PROBLEM 93

To find the intercepts algebraically requires the ability to solve equations, the subject of the following four sections.

1.1 Assess Your Understanding

'Are You Prepared?' *Answers are given at the end of these exercises. If you get a wrong answer, read the pages listed in red.*

1. On a real number line the origin is assigned the number _______. (p. 18)

2. If -3 and 5 are the coordinates of two points on the real number line, the distance between these points is _______. (pp. 20–21)

3. If 3 and 4 are the legs of a right triangle, the hypotenuse is _______. (p. 31)

4. Use the converse of the Pythagorean Theorem to show that a triangle whose sides are of lengths 11, 60, and 61 is a right triangle. (p. 32)

5. The area of a triangle whose base is b and whose altitude is h is $A =$ _______. (p. 32)

6. ***True or False*** Two triangles are congruent if two angles and the included side of one equals two angles and the included side of the other. (pp. 33–34)

Concepts and Vocabulary

7. If (x, y) are the coordinates of a point P in the xy-plane, then x is called the _____________ of P and y is the _____________ of P.

8. The coordinate axes divide the xy-plane into four sections called ___________.

9. If three distinct points P, Q, and R all lie on a line and if $d(P, Q) = d(Q, R)$, then Q is called the __________ of the line segment from P to R.

10. ***True or False*** The distance between two points is sometimes a negative number.

11. ***True or False*** The point $(-1, 4)$ lies in quadrant IV of the Cartesian plane.

12. ***True or False*** The midpoint of a line segment is found by averaging the x-coordinates and averaging the y-coordinates of the endpoints.

13. Given that the intercepts of a graph are $(-4, 0)$ and $(0, 5)$, choose the statement that is true.
(a) The y-intercept is -4 and the x-intercept is 5.
(b) The y-intercepts are -4 and 5.
(c) The x-intercepts are -4 and 5.
(d) The x-intercept is -4 and the y-intercept is 5.

14. Which of the following points does not satisfy the equation $2x^2 - 5y = 20$?
(a) $(0, -4)$ (b) $(5, 6)$ (c) $(-5, -14)$ (d) $(10, 36)$

Skill Building

In Problems 15 and 16, plot each point in the xy-plane. Tell in which quadrant or on what coordinate axis each point lies.

15. (a) $A = (-3, 2)$ (b) $B = (6, 0)$ (c) $C = (-2, -2)$ (d) $D = (6, 5)$ (e) $E = (0, -3)$ (f) $F = (6, -3)$

16. (a) $A = (1, 4)$ (b) $B = (-3, -4)$ (c) $C = (-3, 4)$ (d) $D = (4, 1)$ (e) $E = (0, 1)$ (f) $F = (-3, 0)$

17. Plot the points $(2, 0)$, $(2, -3)$, $(2, 4)$, $(2, 1)$, and $(2, -1)$. Describe the set of all points of the form $(2, y)$, where y is a real number.

18. Plot the points $(0, 3)$, $(1, 3)$, $(-2, 3)$, $(5, 3)$, and $(-4, 3)$. Describe the set of all points of the form $(x, 3)$, where x is a real number.

In Problems 19–22, determine the coordinates of the points shown. Tell in which quadrant each point lies. Assume the coordinates are integers.

19. 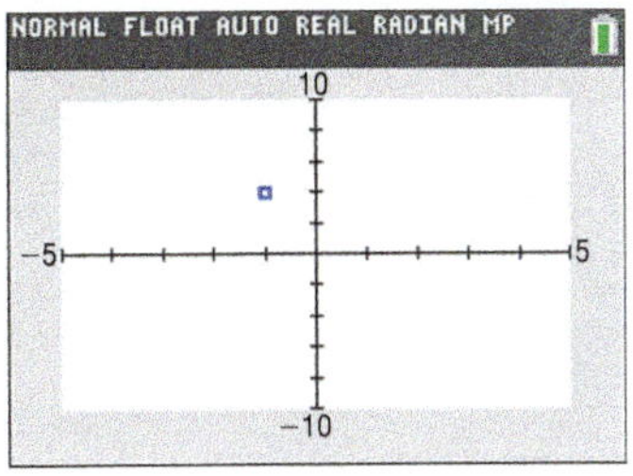

20. 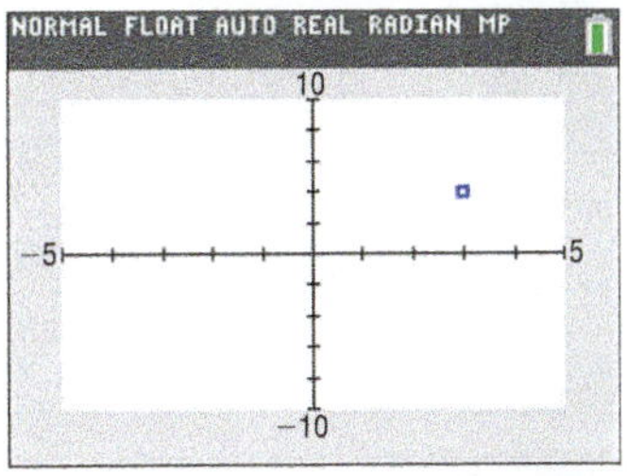

21. 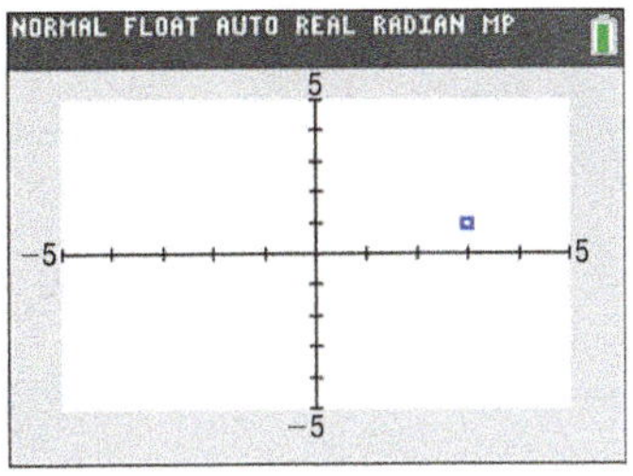

22. 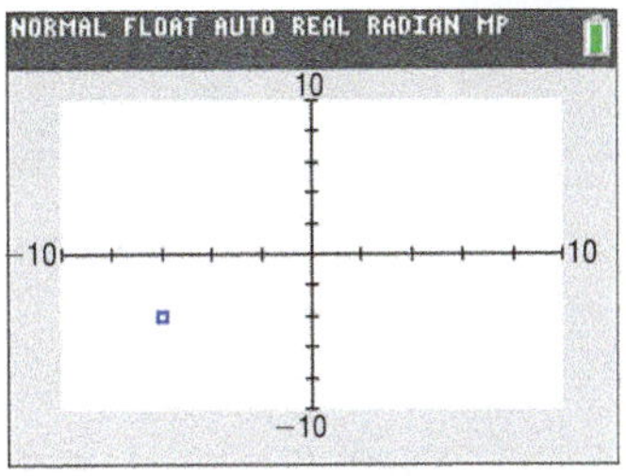

In Problems 23–28, select a setting so that each given point will lie within the viewing window.

23. $(-10, 5)$, $(3, -2)$, $(4, -1)$

24. $(5, 0)$, $(6, 8)$, $(-2, -3)$

25. $(40, 20)$, $(-20, -80)$, $(10, 40)$

26. $(-80, 60)$, $(20, -30)$, $(-20, -40)$

27. $(0, 0)$, $(100, 5)$, $(5, 150)$

28. $(0, -1)$, $(100, 50)$, $(-10, 30)$

In Problems 29–34, determine the viewing window used.

29. 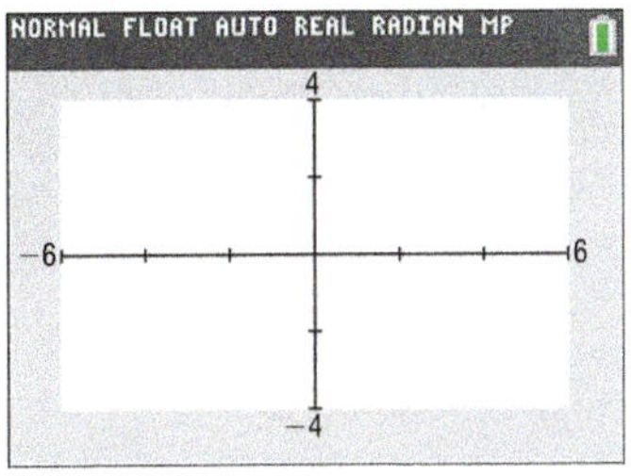

30. 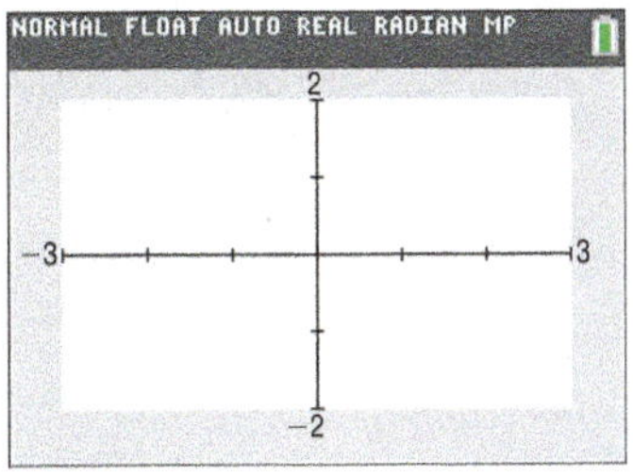

31. 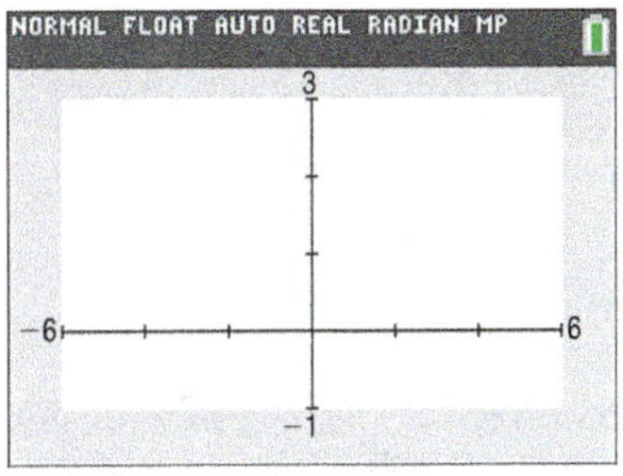

32. 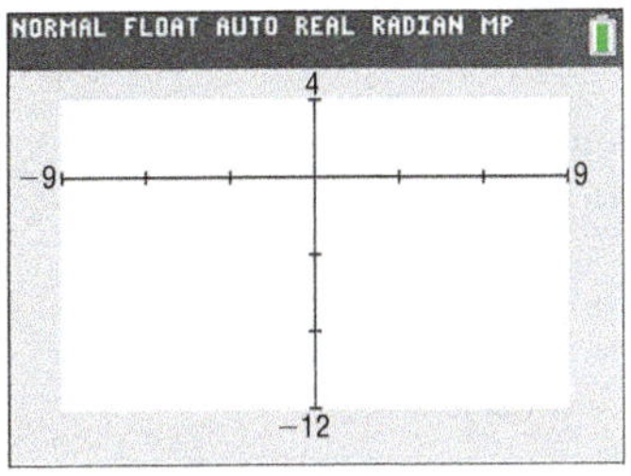

33. 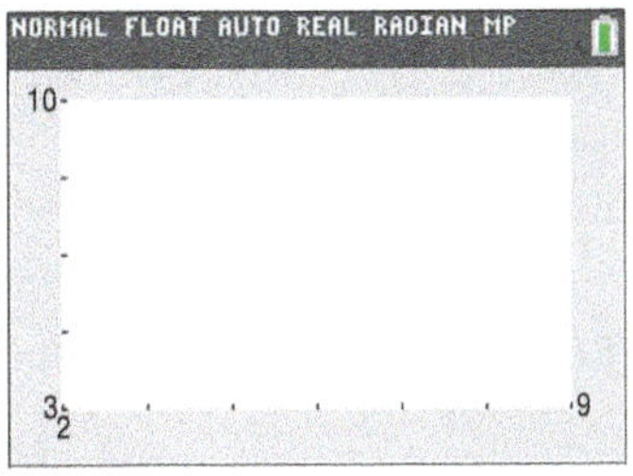

34. 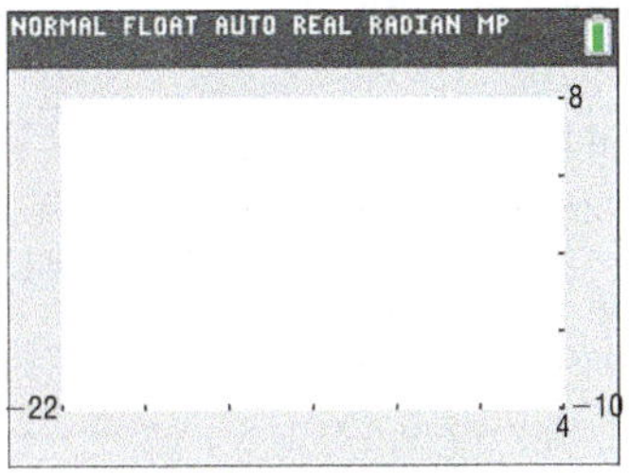

In Problems 35–46, find the distance $d(P_1, P_2)$ between the points P_1 and P_2.

35. 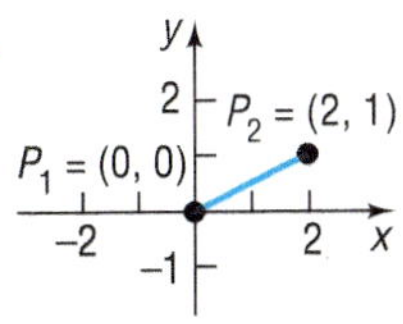

36. 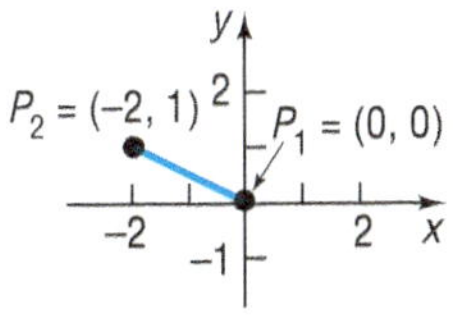

37. 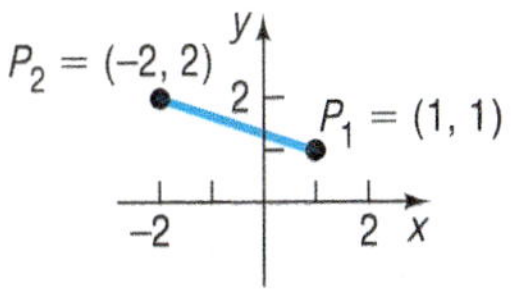

38. 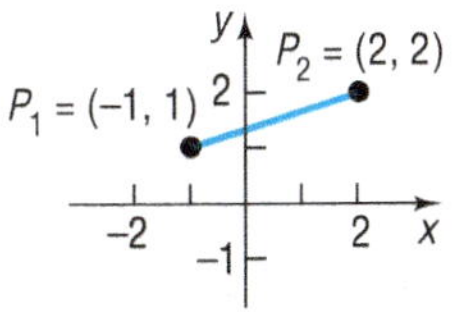

39. $P_1 = (3, -4)$; $P_2 = (5, 4)$

40. $P_1 = (-1, 0)$; $P_2 = (2, 4)$

41. $P_1 = (-5, -3)$; $P_2 = (11, 9)$

42. $P_1 = (2, -3)$; $P_2 = (10, 3)$

43. $P_1 = (4, -3)$; $P_2 = (6, 4)$

44. $P_1 = (-4, -3)$; $P_2 = (6, 2)$

45. $P_1 = (a, b)$; $P_2 = (0, 0)$

46. $P_1 = (a, a)$; $P_2 = (0, 0)$

In Problems 47–50, find the length of the line segment. Assume that the endpoints of each line segment have integer coordinates.

47. 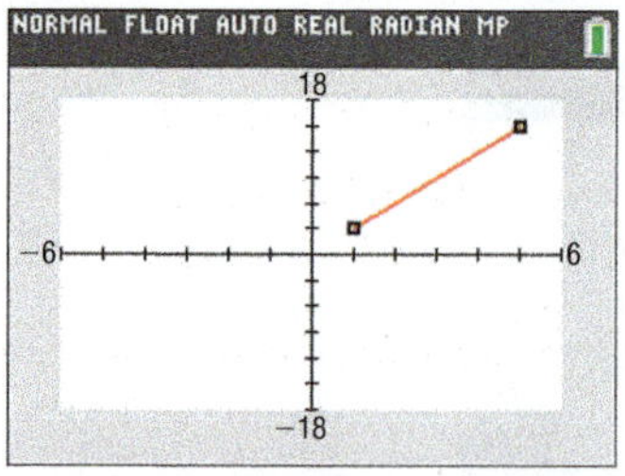

48. 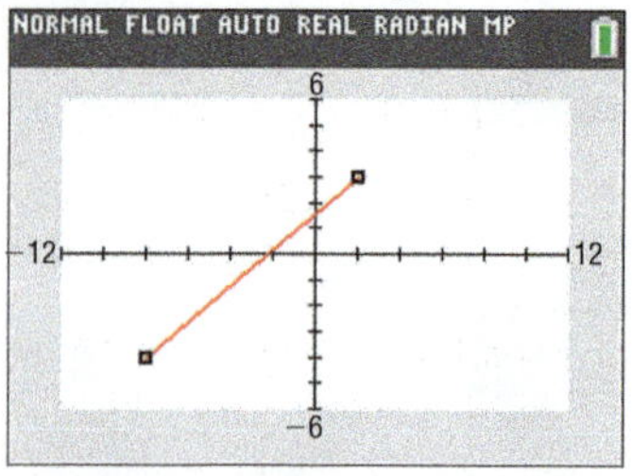

49. 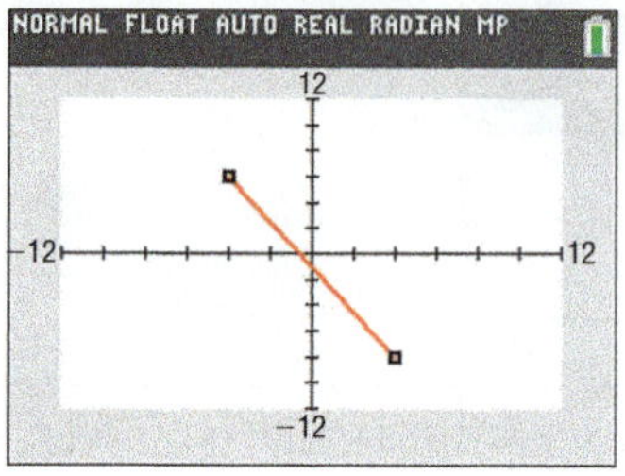

50. 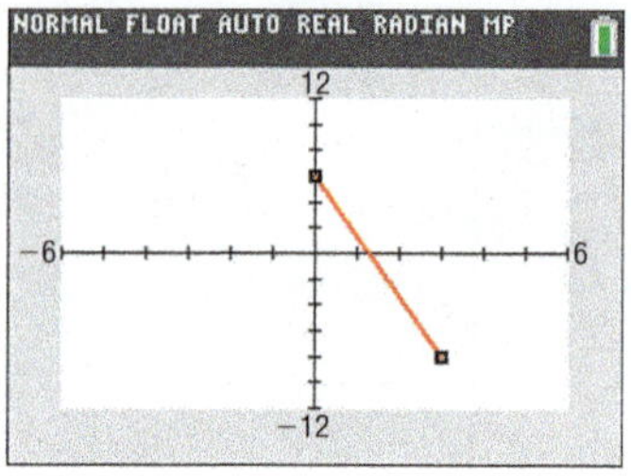

In Problems 51–56, plot each point and form the triangle ABC. Verify that the triangle is a right triangle. Find its area.

51. $A = (-2, 5)$; $B = (1, 3)$; $C = (-1, 0)$

52. $A = (-2, 5)$; $B = (12, 3)$; $C = (10, -11)$

53. $A = (-5, 3)$; $B = (6, 0)$; $C = (5, 5)$

54. $A = (-6, 3)$; $B = (3, -5)$; $C = (-1, 5)$

55. $A = (4, -3)$; $B = (0, -3)$; $C = (4, 2)$

56. $A = (4, -3)$; $B = (4, 1)$; $C = (2, 1)$

In Problems 57–64, find the midpoint of the line segment joining the points P_1 and P_2.

57. $P_1 = (3, -4)$; $P_2 = (5, 4)$

58. $P_1 = (-2, 0)$; $P_2 = (2, 4)$

59. $P_1 = (-5, -3)$; $P_2 = (11, 9)$

60. $P_1 = (2, -3)$; $P_2 = (10, 3)$

61. $P_1 = (4, -3)$; $P_2 = (6, 1)$

62. $P_1 = (-4, -3)$; $P_2 = (2, 2)$

63. $P_1 = (a, b)$; $P_2 = (0, 0)$

64. $P_1 = (a, a)$; $P_2 = (0, 0)$

In Problems 65–70, tell whether the given points are on the graph of the equation.

65. Equation: $y = x^4 - \sqrt{x}$
Points: $(0, 0)$; $(1, 1)$; $(-1, 0)$

66. Equation: $y = x^3 - 2\sqrt{x}$
Points: $(0, 0)$; $(1, 1)$; $(1, -1)$

67. Equation: $y^2 = x^2 + 9$
Points: $(0, 3)$; $(3, 0)$; $(-3, 0)$

68. Equation: $y^3 = x + 1$
Points: $(1, 2)$; $(0, 1)$; $(-1, 0)$

69. Equation: $x^2 + y^2 = 4$
Points: $(0, 2)$; $(-2, 2)$; $(\sqrt{2}, \sqrt{2})$

70. Equation: $x^2 + 4y^2 = 4$
Points: $(0, 1)$; $(2, 0)$; $\left(2, \frac{1}{2}\right)$

In Problems 71–78, the graph of an equation is given. List the intercepts of the graph.

71. 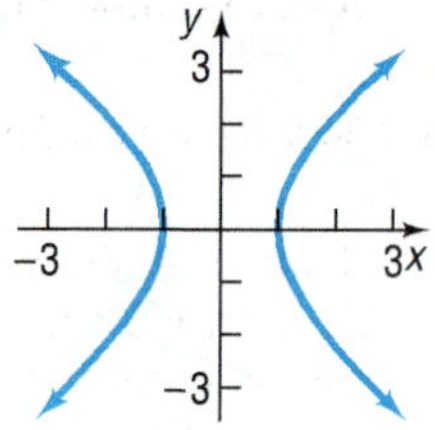

72. 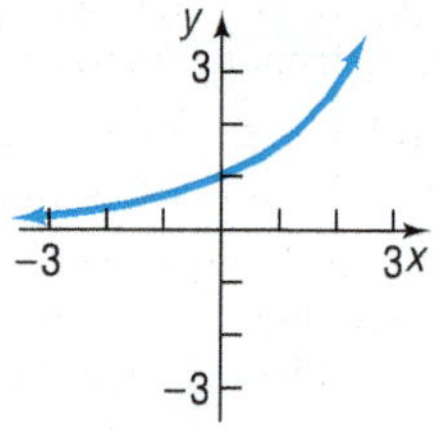

73. 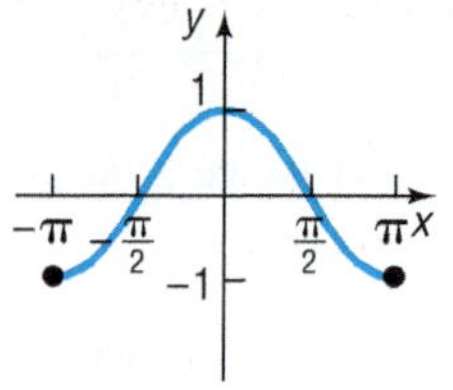

74. 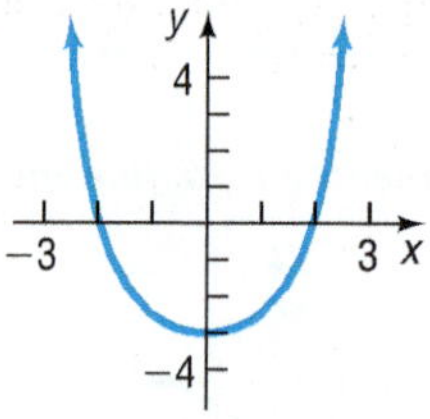

75. 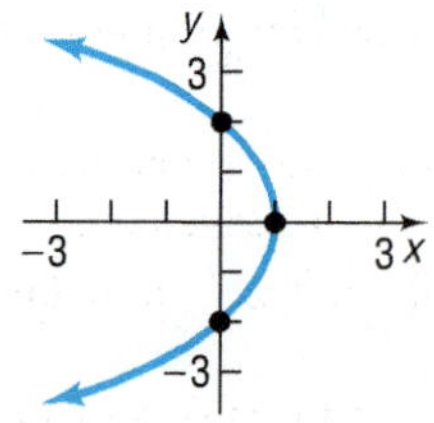

76. 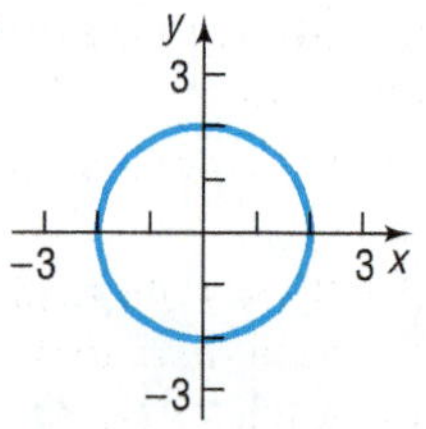

77. 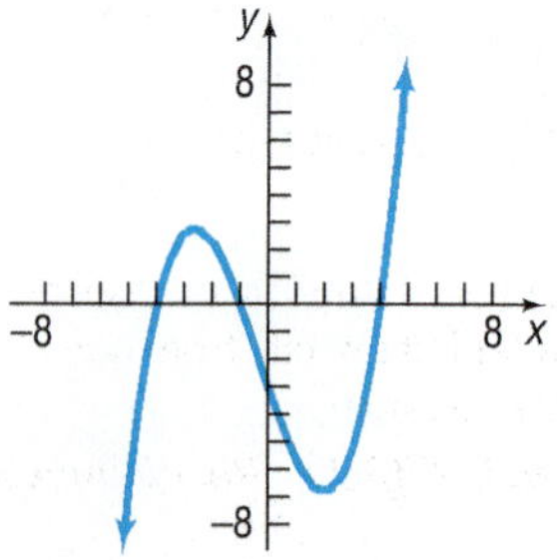

78. 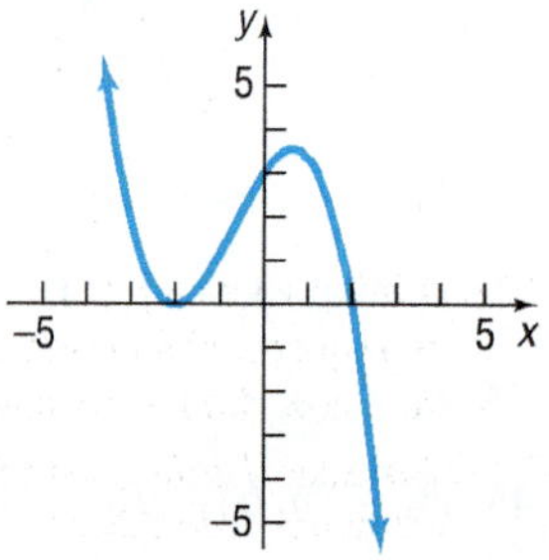

In Problems 79–90, graph each equation by plotting points. Verify your results using a graphing utility.

79. $y = x + 2$
80. $y = x - 6$
81. $y = 2x + 8$
82. $y = 3x - 9$
83. $y = x^2 - 1$
84. $y = x^2 - 9$
85. $y = -x^2 + 4$
86. $y = -x^2 + 1$
87. $2x + 3y = 6$
88. $5x + 2y = 10$
89. $9x^2 + 4y = 36$
90. $4x^2 + y = 4$

In Problems 91–98, graph each equation using a graphing utility. Use a graphing utility to approximate the intercepts rounded to two decimal places. Use the TABLE feature to help to establish the viewing window.

91. $y = 2x - 13$
92. $y = -3x + 14$
93. $y = 2x^2 - 15$
94. $y = -3x^2 + 19$
95. $3x - 2y = 43$
96. $4x + 5y = 82$
97. $5x^2 + 3y = 37$
98. $2x^2 - 3y = 35$

99. If the point $(2, 5)$ is shifted 3 units right and 2 units down, what are its new coordinates?

100. If the point $(-1, 6)$ is shifted 2 units left and 4 units up, what are its new coordinates?

Applications and Extensions

101. The **medians** of a triangle are the line segments from each vertex to the midpoint of the opposite side (see the figure). Find the lengths of the medians of the triangle with vertices at $A = (0, 0)$, $B = (6, 0)$, and $C = (4, 4)$.

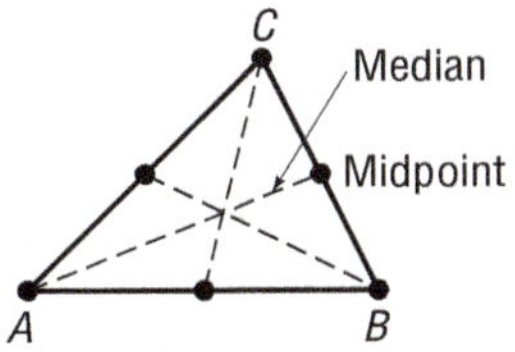

102. An **equilateral triangle** is one in which all three sides are of equal length. If two vertices of an equilateral triangle are $(0, 4)$ and $(0, 0)$, find the third vertex. How many of these triangles are possible?

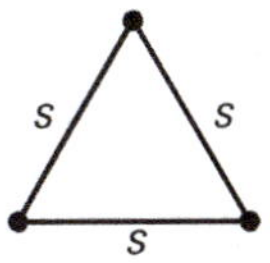

In Problems 103–106, find the length of each side of the triangle determined by the three points P_1, P_2, and P_3. State whether the triangle is an isosceles triangle, a right triangle, neither of these, or both. (An ***isosceles triangle*** *is one in which at least two of the sides are of equal length.)*

103. $P_1 = (2, 1);\ P_2 = (-4, 1);\ P_3 = (-4, -3)$
104. $P_1 = (-1, 4);\ P_2 = (6, 2);\ P_3 = (4, -5)$
105. $P_1 = (-2, -1);\ P_2 = (0, 7);\ P_3 = (3, 2)$
106. $P_1 = (7, 2);\ P_2 = (-4, 0);\ P_3 = (4, 6)$

107. Completing a Line Segment Plot the points $A = (-1, 8)$ and $M = (2, 3)$ in the xy-plane. If M is the midpoint of a line segment AB, find the coordinates of B.

108. Completing a Line Segment Plot the points $M = (5, -4)$ and $B = (7, -2)$ in the xy-plane. If M is the midpoint of a line segment AB, find the coordinates of A.

109. Baseball A major league baseball "diamond" is actually a square, 90 feet on a side (see the figure). What is the distance directly from home plate to second base (the diagonal of the square)?

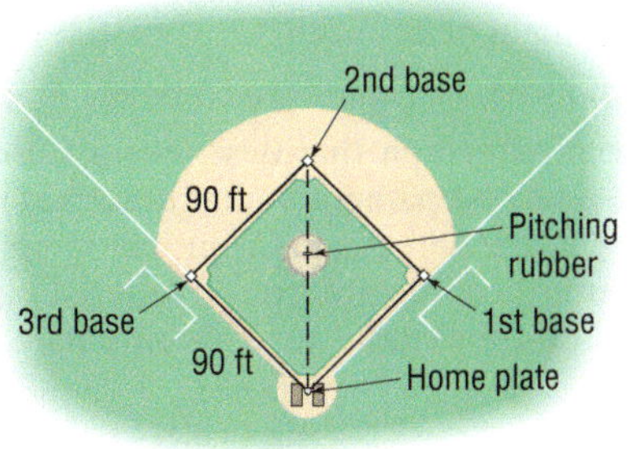

110. Little League Baseball The layout of a Little League playing field is a square, 60 feet on a side. How far is it directly from home plate to second base (the diagonal of the square)?

Source: *Little League Baseball, Official Regulations and Playing Rules, 2014.*

111. Baseball Refer to Problem 109. Overlay a rectangular coordinate system on a major league baseball diamond so that the origin is at home plate, the positive x-axis lies in the direction from home plate to first base, and the positive y-axis lies in the direction from home plate to third base.

(a) What are the coordinates of first base, second base, and third base? Use feet as the unit of measurement.

(b) If the right fielder is located at $(310, 15)$, how far is it from there to second base?

(c) If the center fielder is located at $(300, 300)$, how far is it from there to third base?

112. Little League Baseball Refer to Problem 110. Overlay a rectangular coordinate system on a Little League baseball diamond so that the origin is at home plate, the positive x-axis lies in the direction from home plate to first base, and the positive y-axis lies in the direction from home plate to third base.

(a) What are the coordinates of first base, second base, and third base? Use feet as the unit of measurement.

(b) If the right fielder is located at $(180, 20)$, how far is it from there to second base?

(c) If the center fielder is located at $(220, 220)$, how far is it from there to third base?

113. Distance between Moving Objects A Ford Focus and a Freightliner truck leave an intersection at the same time. The Focus heads east at an average speed of 30 miles per hour, while the truck heads south at an average speed of 40 miles per hour. Find an expression for their distance apart d (in miles) at the end of t hours.

114. Distance of a Moving Object from a Fixed Point A hot-air balloon, headed due east at an average speed of 15 miles per hour at a constant altitude of 100 feet, passes over an intersection (see the figure). Find an expression for its distance d (measured in feet) from the intersection t seconds later.

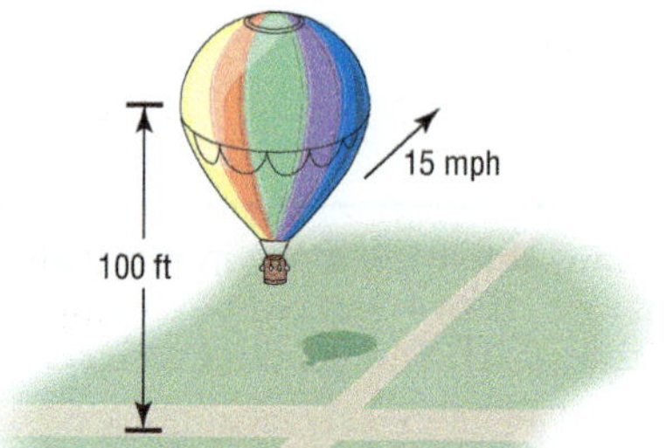

115. Drafting Error When a draftsman draws three lines that are to intersect at one point, the lines may not intersect as intended and subsequently will form an **error triangle.** If this error triangle is long and thin, one estimate for the location of the desired point is the midpoint of the shortest side. The figure shows one such error triangle.

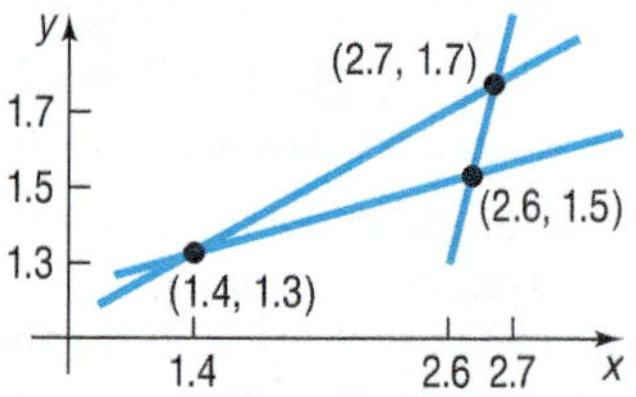

(a) Find an estimate for the desired intersection point.

(b) Find the length of the median for the midpoint found in part (a). See Problem 101.

116. Net Sales The figure illustrates how net sales of Wal-Mart Stores, Inc., have grown from 2011 through 2015. Use the midpoint formula to estimate the net sales of Wal-Mart Stores, Inc., in 2013. How does your result compare to the reported value of \$469 billion?

Source: *Wal-Mart Stores, Inc., 2015 Annual Report*

117. Poverty Threshold Poverty thresholds are determined by the U.S. Census Bureau. A poverty threshold represents the minimum annual household income for a family not to be considered poor. In 2004, the poverty threshold for a family of four with two children under the age of 18 years was \$19,157. In 2014, the poverty threshold for a family of four with two children under the age of 18 years was \$24,008. Assuming poverty thresholds increase in a straight-line fashion, use the midpoint formula to estimate the poverty threshold of a family of four with two children under the age of 18 in 2009. How does your result compare to the actual poverty threshold in 2009 of \$21,756?

Source: *U.S. Census Bureau*

Explaining Concepts: Discussion and Writing

In Problem 118, you may use a graphing utility, but it is not required.

118. (a) Graph $y = \sqrt{x^2}, y = x, y = |x|$, and $y = (\sqrt{x})^2$, noting which graphs are the same.

(b) Explain why the graphs of $y = \sqrt{x^2}$ and $y = |x|$ are the same.

(c) Explain why the graphs of $y = x$ and $y = (\sqrt{x})^2$ are not the same.

(d) Explain why the graphs of $y = \sqrt{x^2}$ and $y = x$ are not the same.

119. Make up an equation satisfied by the ordered pairs $(2, 0)$, $(4, 0)$, and $(0, 1)$. Compare your equation with a friend's equation. Comment on any similarities.

120. Draw a graph that contains the points $(-2, -1)$, $(0, 1)$, $(1, 3)$, and $(3, 5)$. Compare your graph with those of other students. Are most of the graphs almost straight lines? How many are "curved"? Discuss the various ways that these points might be connected.

121. Explain what is meant by a complete graph.

122. Write a paragraph that describes a Cartesian plane. Then write a second paragraph that describes how to plot points in the Cartesian plane. Your paragraphs should include the terms "coordinate axes," "ordered pair," "coordinates," "plot," "x-coordinate," and "y-coordinate."

'Are You Prepared?' Answers

1. 0 **2.** 8 **3.** 5 **4.** $11^2 + 60^2 = 61^2$ **5.** $\frac{1}{2}bh$ **6.** True

1.2 Solving Equations Using a Graphing Utility; Linear and Rational Equations

PREPARING FOR THIS SECTION *Before getting started, review the following:*

- Properties of Real Numbers (Section R.1, pp. 10–15)
- Domain of a Variable (Section R.2, p. 22)
- Evaluate Algebraic Expressions (Section R.2, pp. 21–22)
- Rational Expressions (Section R.7, pp. 63–71)

Now Work the **'Are You Prepared?'** problems on page 107.

OBJECTIVES
1. Solve Equations Using a Graphing Utility (p. 100)
2. Solve Linear Equations (p. 102)
3. Solve Rational Equations (p. 103)
4. Solve Problems That Can Be Modeled by Linear Equations (p. 105)

An **equation in one variable** is a statement in which two expressions, at least one containing the variable, are set equal. The expressions are called the **sides** of the equation. Since an equation is a statement, it may be true or false, depending on the value of the variable. Unless otherwise restricted, the admissible values of the variable are those in the domain of the variable. Those admissible values of the variable, if any, that result in a true statement are called **solutions**, or **roots**, of the equation. To **solve an equation** means to find all the solutions of the equation.

For example, the following are all equations in one variable, x:

$$x + 5 = 9 \qquad x^2 + 5x = 2x - 2 \qquad \frac{x^2 - 4}{x + 1} = 0 \qquad x^2 + 9 = 5$$

The first of these statements, $x + 5 = 9$, is true when $x = 4$ and false for any other choice of x. That is, 4 is a solution of the equation $x + 5 = 9$. We also say that 4 **satisfies** the equation $x + 5 = 9$, because, when 4 is substituted for x, a true statement results.

Sometimes an equation will have more than one solution. For example, the equation

$$\frac{x^2 - 4}{x + 1} = 0$$

has $x = -2$ and $x = 2$ as solutions.

Usually, we will write the solutions of an equation in set notation. This set is called the **solution set** of the equation. For example, the solution set of the equation $x^2 - 9 = 0$ is $\{-3, 3\}$.

Unless indicated otherwise, we will limit ourselves to real solutions—that is, solutions that are real numbers. Some equations have no real solution. For example, $x^2 + 9 = 5$ has no real solution, because there is no real number whose square, when added to 9, equals 5.

An equation that is satisfied for every choice of the variable for which both sides are defined is called an **identity**. For example, the equation

$$3x + 5 = x + 3 + 2x + 2$$

is an identity, because this statement is true for any real number x.

In this text, we present two methods for solving equations: algebraic and graphical. We shall see that some equations can be solved using algebraic techniques to obtain *exact* solutions. For other equations, however, there are no algebraic techniques that lead to an exact solution. For such equations, a graphing utility can often be used to investigate possible solutions.

One goal of this text is to determine when equations can be solved algebraically. If an algebraic method for solving an equation exists, we shall use it to obtain an exact solution. A graphing utility can then be used to support the algebraic result. However, if no algebraic techniques are available to solve an equation, a graphing utility will be used to obtain approximate solutions.

1 Solve Equations Using a Graphing Utility

When a graphing utility is used to solve an equation, usually *approximate* solutions are obtained. Unless otherwise stated, we shall follow the practice of giving approximate solutions as decimals *rounded to two decimal places*.

The ZERO (or ROOT) feature of a graphing utility can be used to find the solutions of an equation when one side of the equation is 0. In using this feature to solve equations, make use of the fact that when the graph of an equation in two variables, x and y, crosses or touches the x-axis then $y = 0$. For this reason, any value of x for which $y = 0$ will be a solution to the equation. That is, solving an equation for x when one side of the equation is 0 is equivalent to finding where the graph of the corresponding equation in two variables crosses or touches the x-axis.

EXAMPLE 1 **Using ZERO (or ROOT) to Approximate Solutions of an Equation**

Find the solution(s) of the equation $x^3 - x + 1 = 0$. Round answers to two decimal places.

Solution The solutions of the equation $x^3 - x + 1 = 0$ are the same as the x-intercepts of the graph of $Y_1 = x^3 - x + 1$. Begin by graphing Y_1. Figure 28 shows the graph.

From the graph there appears to be one x-intercept (solution to the equation) between -2 and -1. Using the ZERO (or ROOT) feature of a graphing utility, determine that the x-intercept, and thus the solution to the equation, is $x = -1.32$ rounded to two decimal places. See Figure 29.

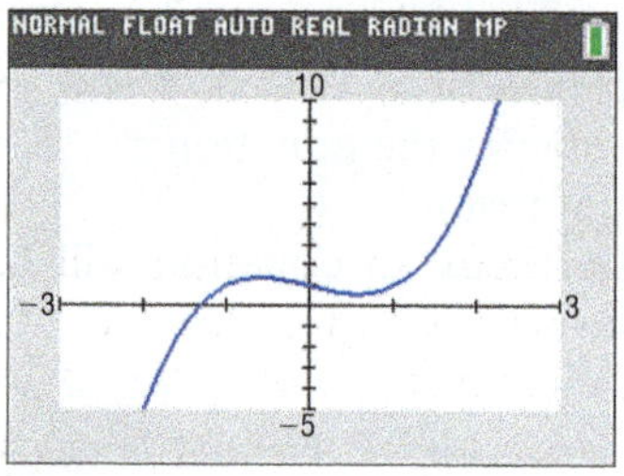

Figure 28 $Y_1 = x^3 - x + 1$

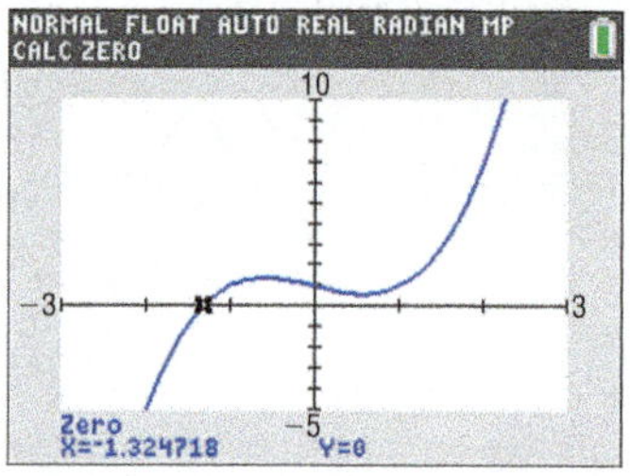

Figure 29

Note: Graphing utilities use a process in which they search for a solution until the answer is found within a certain tolerance level (such as within 0.0001). Therefore, the y-coordinate may sometimes be a nonzero value such as 1.1527 E-8, which is 1.1527×10^{-8}, very close to zero.

Now Work PROBLEM 19

A second method for solving equations using a graphing utility involves the INTERSECT feature of the graphing utility. This feature is used most effectively when neither side of the equation is 0.

EXAMPLE 2 **Using INTERSECT to Approximate Solutions of an Equation**

Find the solution(s) to the equation $4x^4 - 3 = 2x + 1$. Round answers to two decimal places.

Solution Begin by graphing each side of the equation as follows: graph $Y_1 = 4x^4 - 3$ and $Y_2 = 2x + 1$. See Figure 30.

At a point of intersection of the graphs, the value of the y-coordinate is the same for Y_1 and Y_2. Thus, the x-coordinate of the point of intersection represents a solution to the equation. Do you see why? The INTERSECT feature on a graphing utility determines a point of intersection of the graphs. Using this feature, find that the graphs intersect at $(-0.87, -0.73)$ and $(1.12, 3.23)$ rounded to two decimal places. See Figure 31(a) and (b). The solutions of the equation are $x = -0.87$ and $x = 1.12$ rounded to two decimal places.

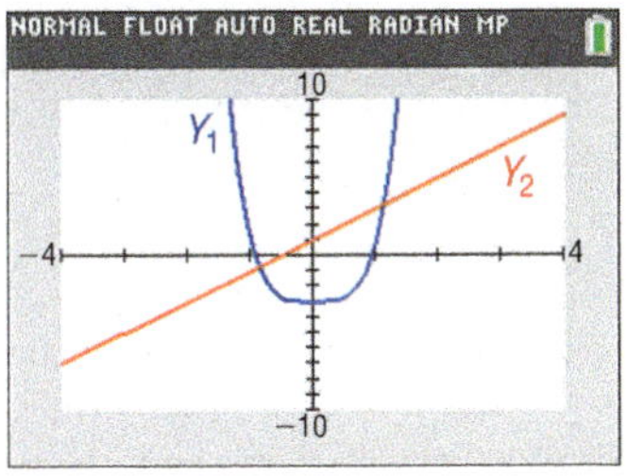

Figure 30 $Y_1 = 4x^4 - 3$; $Y_2 = 2x + 1$

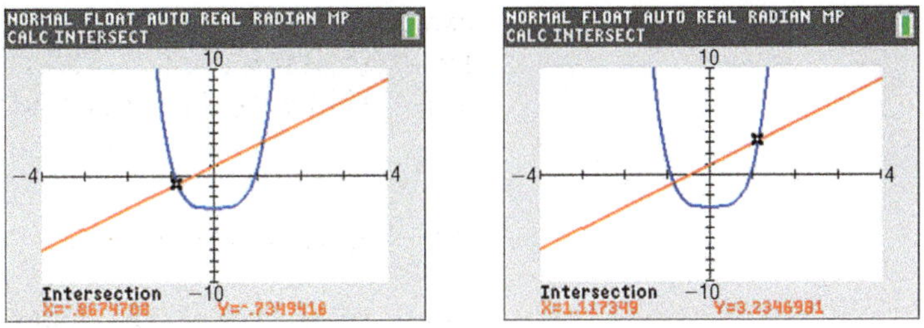

(a) (b)

Figure 31

Now Work PROBLEM 21

SUMMARY

Steps for Approximating Solutions of Equations Using ZERO (or ROOT)

STEP 1: Write the equation in the form {expression in x} = 0.

STEP 2: Graph Y_1 = {expression in x}. Be sure that the graph is complete. That is, be sure that all the x-intercepts are shown on the screen.

STEP 3: Use ZERO (or ROOT) to determine each x-intercept of the graph.

Steps for Approximating Solutions of Equations Using INTERSECT

STEP 1: Graph Y_1 = {expression in x on the left side of the equation}
Y_2 = {expression in x on the right side of the equation}

Be sure that the graphs are complete. That is, be sure that all the points of intersection are shown on the screen.

STEP 2: Use INTERSECT to determine the x-coordinate of each point of intersection.

Solving Equations Algebraically

One method for solving equations algebraically requires that a series of *equivalent equations* be developed from the original equation until an obvious solution results.

For example, consider the following succession of equivalent equations:

$$2x + 3 = 13$$
$$2x = 10$$
$$x = 5$$

We conclude that the solution set of the original equation is {5}.

How are equivalent equations obtained? In general, there are five ways to do so.

Procedures That Result in Equivalent Equations

1. Interchange the two sides of the equation:

$$3 = x \quad \text{is equivalent to} \quad x = 3$$

2. Simplify the sides of the equation by combining like terms, eliminating parentheses, and so on:

$$x + 2 + 6 = 2x + 3(x + 1)$$

$$\text{is equivalent to } x + 8 = 5x + 3$$

3. Add or subtract the same expression on both sides of the equation:

$$3x - 5 = 4$$

$$\text{is equivalent to } (3x - 5) + 5 = 4 + 5$$

4. Multiply or divide both sides of the equation by the same nonzero expression:

$$\frac{3x}{x-1} = \frac{6}{x-1} \qquad x \neq 1$$

$$\text{is equivalent to } \frac{3x}{x-1} \cdot (x-1) = \frac{6}{x-1} \cdot (x-1)$$

5. If one side of the equation is 0 and the other side can be factored, then we may use the Zero-Product Property* and set each factor equal to 0:

$$x(x-3) = 0$$

$$\text{is equivalent to } x = 0 \quad \text{or} \quad x - 3 = 0$$

WARNING Squaring both sides of an equation does not necessarily lead to an equivalent equation. For example, $x = 3$ has one solution, but $x^2 = 9$ has two solutions, $x = -3$ and $x = 3$. ■

Whenever it is possible to solve an equation in your head, do so. For example:

The solution of $2x = 8$ is $x = 4$.

The solution of $3x - 15 = 0$ is $x = 5$.

Now Work PROBLEM 11

2 Solve Linear Equations

Linear equations are equations such as

$$3x + 12 = 0 \qquad \frac{3}{4}x - \frac{1}{5} = 0 \qquad 0.62x - 0.3 = 0$$

A **linear equation in one variable** is equivalent to an equation of the form

$$ax + b = 0$$

where a and b are real numbers and $a \neq 0$.

Sometimes a linear equation is called a **first-degree equation**, because the left side is a polynomial in x of degree 1.

EXAMPLE 3 Solving a Linear Equation

Solve the equation: $3(x-2) = 5(x-1)$

Algebraic Solution

$3(x-2) = 5(x-1)$	
$3x - 6 = 5x - 5$	Use the Distributive Property.
$3x - 6 - 5x = 5x - 5 - 5x$	Subtract $5x$ from each side.
$-2x - 6 = -5$	Simplify.
$-2x - 6 + 6 = -5 + 6$	Add 6 to each side.
$-2x = 1$	Simplify.
$\frac{-2x}{-2} = \frac{1}{-2}$	Divide each side by -2.
$x = -\frac{1}{2}$	Simplify.

Graphing Solution

Graph $Y_1 = 3(x-2)$ and $Y_2 = 5(x-1)$. See Figure 32. Using INTERSECT, the point of intersection is found to be $(-0.5, -7.5)$. The solution of the equation is $x = -0.5$.

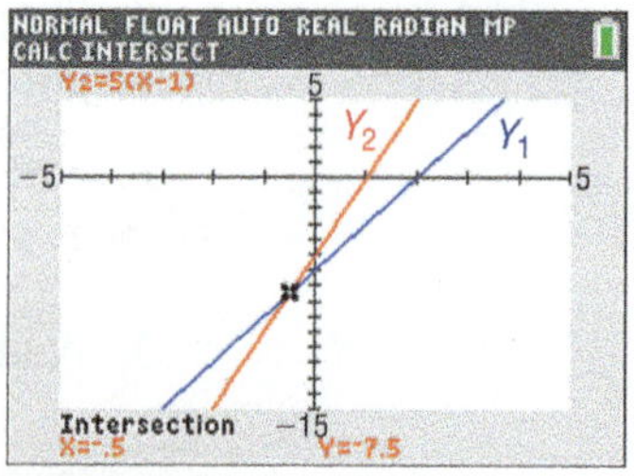

Figure 32 $Y_1 = 3(x-2)$; $Y_2 = 5(x-1)$

*The Zero-Product Property says that if $ab = 0$, then $a = 0$, $b = 0$, or both equal 0.

Check: Let $x = -\frac{1}{2}$ in the expression in x on the left side of the equation and simplify. Let $x = -\frac{1}{2}$ in the expression in x on the right side of the equation and simplify. If the two expressions are equal, the solution checks.

$$3(x-2) = 3\left(-\frac{1}{2} - 2\right) = 3\left(-\frac{5}{2}\right) = -\frac{15}{2}$$
$$5(x-1) = 5\left(-\frac{1}{2} - 1\right) = 5\left(-\frac{3}{2}\right) = -\frac{15}{2}$$

Since the two expressions are equal, the solution $x = -\frac{1}{2}$ checks.

The solution set is $\left\{-\frac{1}{2}\right\}$. ■

Now Work PROBLEM 37

EXAMPLE 4 Solving an Equation That Leads to a Linear Equation

Solve the equation: $(2x-1)(x-1) = (x-5)(2x-5)$

Algebraic Solution

$(2x-1)(x-1) = (x-5)(2x-5)$	
$2x^2 - 3x + 1 = 2x^2 - 15x + 25$	Multiply and combine like terms.
$2x^2 - 3x + 1 - 2x^2 = 2x^2 - 15x + 25 - 2x^2$	Subtract $2x^2$ from each side.
$-3x + 1 = -15x + 25$	Simplify.
$-3x + 1 - 1 = -15x + 25 - 1$	Subtract 1 from each side.
$-3x = -15x + 24$	Simplify.
$-3x + 15x = -15x + 24 + 15x$	Add $15x$ to each side.
$12x = 24$	Simplify.
$\frac{12x}{12} = \frac{24}{12}$	Divide each side by 12.
$x = 2$	Simplify.

Graphing Solution

Graph $Y_1 = (2x-1)(x-1)$ and $Y_2 = (x-5)(2x-5)$. See Figure 33. Using INTERSECT, the point of intersection is found to be $(2, 3)$. The solution of the equation is $x = 2$.

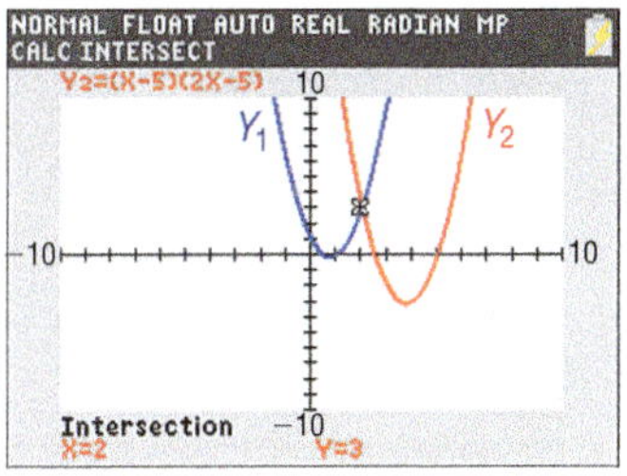

Figure 33 $Y_1 = (2x-1)(x-1)$; $Y_2 = (x-5)(2x-5)$

Check: $(2x-1)(x-1) = (2\cdot 2 - 1)(2-1) = (3)(1) = 3$

$(x-5)(2x-5) = (2-5)(2\cdot 2 - 5) = (-3)(-1) = 3$

Since the two expressions are equal, the solution checks. The solution set is $\{2\}$. ■

Now Work PROBLEM 55

3 Solve Rational Equations

A **rational equation** is an equation that contains a rational expression. Examples of rational equations are

$$\frac{3}{x+1} = \frac{2}{x-1} + 7 \quad \text{and} \quad \frac{x-5}{x-4} = \frac{3}{x+2}$$

To solve a rational equation, multiply both sides of the equation by the least common multiple of the denominators of the rational expressions that make up the rational equation. We shall only show the algebraic solution to rational equations. We leave it to you to verify the results using a graphing utility.

EXAMPLE 5 Solving a Rational Equation

Solve the equation: $\dfrac{3}{x-2} = \dfrac{1}{x-1} + \dfrac{7}{(x-1)(x-2)}$

Solution The domain of the variable is $\{x \mid x \neq 1, x \neq 2\}$. Clear the equation of rational expressions by multiplying both sides by the least common multiple of the denominators of the three rational expressions, $(x-1)(x-2)$.

$$\frac{3}{x-2} = \frac{1}{x-1} + \frac{7}{(x-1)(x-2)}$$

$$(x-1)\cancel{(x-2)}\frac{3}{\cancel{x-2}} = (x-1)(x-2)\left[\frac{1}{x-1} + \frac{7}{(x-1)(x-2)}\right]$$

Multiply both sides by $(x-1)(x-2)$. Divide out common factors on the left.

$$3x - 3 = \cancel{(x-1)}(x-2)\frac{1}{\cancel{x-1}} + \cancel{(x-1)(x-2)}\frac{7}{\cancel{(x-1)(x-2)}}$$

Use the Distributive Property on each side; divide out common factors on the right.

$$3x - 3 = (x-2) + 7$$

$$3x - 3 = x + 5$$

Combine like terms.

$$2x = 8$$

Add 3 to each side; subtract x from each side.

$$x = 4$$

Divide each side by 2.

Check: $\dfrac{3}{x-2} = \dfrac{3}{4-2} = \dfrac{3}{2}$

$$\frac{1}{x-1} + \frac{7}{(x-1)(x-2)} = \frac{1}{4-1} + \frac{7}{(4-1)(4-2)} = \frac{1}{3} + \frac{7}{3 \cdot 2} = \frac{2}{6} + \frac{7}{6} = \frac{9}{6} = \frac{3}{2}$$

Since the two expressions are equal, the solution checks. The solution set is $\{4\}$. ■

Now Work PROBLEM 71

Sometimes the process of creating equivalent equations leads to apparent solutions that are not solutions of the original equation. These are called **extraneous solutions**.

EXAMPLE 6 A Rational Equation with No Solution

Solve the equation: $\dfrac{3x}{x-1} + 2 = \dfrac{3}{x-1}$

Solution The domain of the variable is $\{x \mid x \neq 1\}$. Since the two rational expressions in the equation have the same denominator, $x - 1$, simplify by multiplying both sides by $x - 1$. The resulting equation is equivalent to the original equation, since we are multiplying by $x - 1$, which is not 0. (Remember, $x \neq 1$.)

$$\frac{3x}{x-1} + 2 = \frac{3}{x-1}$$

$$\left(\frac{3x}{x-1} + 2\right) \cdot (x-1) = \frac{3}{\cancel{x-1}} \cdot \cancel{(x-1)}$$

Multiply both sides by $x - 1$; divide out common factors on the right.

$$\frac{3x}{\cancel{x-1}} \cdot \cancel{(x-1)} + 2 \cdot (x-1) = 3$$

Use the Distributive Property and divide out common factors on the left side.

$$3x + 2x - 2 = 3$$

Simplify.

$$5x - 2 = 3$$

Combine like terms.

$$5x = 5$$

Add 2 to each side.

$$x = 1$$

Divide both sides by 5.

The solution appears to be 1. But recall that $x = 1$ is not in the domain of the variable, so $x = 1$ is an extraneous solution. The equation has no solution. The solution set is $\varnothing$. ■

Now Work PROBLEM 61

4 Solve Problems That Can Be Modeled by Linear Equations

Although each situation has unique features, we can provide an outline of the steps to follow when solving applied problems.

Note: The icon is a *Model It!* icon. It indicates that the discussion or problem involves modeling. ■

Note: It is a good practice to choose a variable that reminds you of the unknown. For example, use t for time. ■

Steps for Solving Applied Problems

STEP 1: Read the problem carefully, perhaps two or three times. Pay particular attention to the question being asked in order to identify what you are looking for. Identify any relevant formulas you may need ($d = rt$, $A = \pi r^2$, etc.). If you can, determine realistic possibilities for the answer.

STEP 2: Assign a letter (variable) to represent what you are looking for, and, if necessary, express any remaining unknown quantities in terms of this variable.

STEP 3: Make a list of all the known facts, and translate them into mathematical expressions. These may take the form of an equation (or, later, an inequality) involving the variable. The equation (or inequality) is called the **model**. If possible, draw an appropriately labeled diagram to assist you. Sometimes a table or chart helps.

STEP 4: Solve the equation for the variable, and then answer the question using a complete sentence.

STEP 5: Check the answer with the facts in the problem. If it agrees, congratulations! If it does not agree, try again.

EXAMPLE 7

Solving an Applied Problem: Investments

A total of \$18,000 is invested, some in stocks and some in bonds. If the amount invested in bonds is half that invested in stocks, how much is invested in each category?

Step-by-Step Solution

Step 1: Determine what you are looking for.

We are being asked to find the amount of two investments. These amounts must total \$18,000, because a total of \$18,000 is invested.

Step 2: Assign a variable to represent what you are looking for. If necessary, express any remaining unknown quantities in terms of this variable.

Let s represent the amount invested in stocks. Now, for example, if \$10,000 were invested in stocks, $\$18{,}000 - \$10{,}000 = \$8000$ would be left to invest in bonds. In general, if s represents the amount in stocks, then $18{,}000 - s$ is the amount invested in bonds.

Step 3: Translate the English into mathematical statements. It may be helpful to draw a figure that represents the situation. Also, a table can be used to organize the information. Use the information to build your model.

Set up a table.

Amount in Stocks	Amount in Bonds	Reason
s	$18{,}000 - s$	Total invested is \$18,000

We also know that:

Total amount invested in bonds is one-half that in stocks

$$18{,}000 - s = \frac{1}{2}(s) \qquad \text{The Model}$$

Step 4: Solve the equation and answer the original question.

$$18{,}000 - s = \frac{1}{2}s$$

$$18{,}000 = s + \frac{1}{2}s \qquad \text{Add } s \text{ to both sides.}$$

$$18{,}000 = \frac{3}{2}s \qquad \text{Simplify.}$$

$$\left(\frac{2}{3}\right)18{,}000 = \left(\frac{2}{3}\right)\left(\frac{3}{2}s\right) \quad \text{Multiply both sides by } \frac{2}{3}.$$

$$12{,}000 = s \quad \text{Simplify.}$$

So \$12,000 is invested in stocks and \$18,000 − \$12,000 = \$6000 is invested in bonds.

Step 5: Check your answer with the facts presented in the problem.

The total invested is \$12,000 + \$6000 = \$18,000, and the amount invested in bonds, \$6000, is half the amount invested in stocks, \$12,000. ■

Now Work PROBLEM 95

EXAMPLE 8 **Determining an Hourly Wage**

Shannon grossed \$725 one week by working 52 hours. Her employer pays time-and-a-half for all hours worked in excess of 40 hours. With this information, can you determine Shannon's regular hourly wage?

Solution

Step 1: We are looking for an hourly wage. Our answer will be in dollars per hour.

Step 2: Let w represent the regular hourly wage, measured in dollars per hour. Then $1.5w$ is the overtime hourly wage.

Step 3: Set up a table:

	Hours Worked	Hourly Wage	Salary
Regular	40	w	$40w$
Overtime	12	$1.5w$	$12(1.5w) = 18w$

The sum of regular salary plus overtime salary will equal \$725. From the table, $40w + 18w = 725$. This is our model.

Step 4:

$$40w + 18w = 725$$
$$58w = 725$$
$$w = 12.50$$

Shannon's regular hourly wage is \$12.50 per hour.

Step 5: Forty hours yields a salary of $40(12.50) = \$500$, and 12 hours of overtime yields a salary of $12(1.5)(12.50) = \$225$, for a total of \$725. ■

Now Work PROBLEM 99

Historical Feature

Solving equations is among the oldest of mathematical activities, and efforts to systematize this activity determined much of the shape of modern mathematics.

Consider the following problem and its solution using only words: Solve the problem of how many apples Jim has, given that

"Bob's five apples and Jim's apples together make twelve apples" by thinking,
"Jim's apples are all twelve apples less Bob's five apples" and then concluding,
"Jim has seven apples."

The mental steps translated into algebra are

$$5 + x = 12$$
$$x = 12 - 5$$
$$= 7$$

The solution of this problem using only words is the earliest form of algebra. Such problems were solved exactly this way in Babylonia in 1800 BC. We know almost nothing of mathematical work before this date, although most authorities believe the sophistication of the earliest known texts indicates that a long period of previous development must have occurred. The method of writing out equations in words persisted for thousands of years, and although it now seems extremely cumbersome, it was used very effectively by many generations of mathematicians. The Arabs developed a good deal of the theory of cubic equations while writing out all the equations in words. About AD 1500, the tendency to abbreviate words into written equations began to lead in the direction of modern notation; for example, the Latin word *et* (meaning *and*) developed into the plus sign, +. Although the occasional use of letters to represent variables dates back to AD 1200, the practice did not become common until about AD 1600. Development thereafter was rapid, and by 1635 algebraic notation did not differ essentially from what we use now.

1.2 Assess Your Understanding

'Are You Prepared?' *Answers are given at the end of these exercises. If you get a wrong answer, read the pages listed in red.*

1. Simplify: $-3(x-5)$ (p. 11)

2. Evaluate: $2x - 3 - 5(x+1)$ for $x = 3$. (pp. 21–22)

3. Is $x = 4$ in the domain of $\dfrac{3}{x-4}$? (p. 22)

4. Find the least common multiple of the denominators of $\dfrac{1}{x^2-1}$ and $\dfrac{3}{x^2+4x+3}$. (pp. 67–68)

Concepts and Vocabulary

5. Multiplying both sides of an equation by any number except ________ results in an equivalent equation.
(a) -1 (b) 0 (c) 1 (d) π

6. An equation that is satisfied for every value of the variable for which both sides are defined is called a(n) ____________.

7. An equation of the form $ax + b = 0$ is called a(n) ___________ equation or a(n) _________-degree equation.

8. An admissible value for the variable that makes the equation a true statement is called a(n) ________ of the equation.
(a) degree (b) identity (c) model (d) solution

9. ***True or False*** Some equations have no solution.

10. ***True or False*** When solving equations using a graphing utility, the solutions are always exact.

Skill Building

In Problems 11–18, mentally solve each equation.

11. $7x = 21$

12. $6x = -24$

13. $3x + 15 = 0$

14. $6x + 18 = 0$

15. $2x - 3 = 0$

16. $3x + 4 = 0$

17. $\dfrac{1}{3}x = \dfrac{5}{12}$

18. $\dfrac{2}{3}x = \dfrac{9}{2}$

In Problems 19–30, use a graphing utility to approximate the real solutions, if any, of each equation rounded to two decimal places. All solutions lie between −10 and 10.

19. $x^3 - 4x + 2 = 0$

20. $x^3 - 8x + 1 = 0$

21. $-2x^4 + 5 = 3x - 2$

22. $-x^4 + 1 = 2x^2 - 3$

23. $x^4 - 2x^3 + 3x - 1 = 0$

24. $3x^4 - x^3 + 4x^2 - 5 = 0$

25. $-x^3 - \dfrac{5}{3}x^2 + \dfrac{7}{2}x + 2 = 0$

26. $-x^4 + 3x^3 + \dfrac{7}{3}x^2 - \dfrac{15}{2}x + 2 = 0$

27. $-\dfrac{2}{3}x^4 - 2x^3 + \dfrac{5}{2}x = -\dfrac{2}{3}x^2 + \dfrac{1}{2}$

28. $\dfrac{1}{4}x^3 - 5x = \dfrac{1}{5}x^2 - 4$

29. $x^4 - 5x^2 + 2x + 11 = 0$

30. $-3x^4 + 8x^2 - 2x - 9 = 0$

In Problems 31–76, solve each equation algebraically. Verify your results using a graphing utility.

31. $3x + 4 = x$

32. $2x + 9 = 5x$

33. $2t - 6 = 3 - t$

34. $5y + 6 = -18 - y$

35. $6 - x = 2x + 9$

36. $3 - 2x = 2 - x$

37. $2(3 + 2x) = 3(x - 4)$

38. $3(2 - x) = 2x - 1$

39. $8x - (3x + 2) = 3x - 10$

40. $7 - (2x - 1) = 10$

41. $2(3x - 5) + 6(x - 3) = -3(4 - 5x) + 5x - 6$

42. $5(x - 2) - 2(3x + 1) = 4(1 - 2x) + x$

43. $\dfrac{3}{2}x + 2 = \dfrac{1}{2} - \dfrac{1}{2}x$

44. $\dfrac{1}{3}x = 2 - \dfrac{2}{3}x$

45. $\dfrac{2}{3}p = \dfrac{1}{2}p - \dfrac{1}{3}$

46. $\dfrac{1}{2} - \dfrac{1}{3}p = \dfrac{4}{3}$

47. $0.9t = 0.4 + 0.1t$

48. $0.9t = 1 + t$

49. $\dfrac{x+1}{3} + \dfrac{x+2}{7} = 2$

50. $\dfrac{2x+1}{3} + 16 = 3x$

51. $\dfrac{2}{y} + \dfrac{4}{y} = 3$

52. $\dfrac{4}{y} - 5 = \dfrac{5}{2y}$

53. $\dfrac{1}{2} + \dfrac{2}{x} = \dfrac{3}{4}$

54. $\dfrac{3}{x} - \dfrac{1}{3} = \dfrac{1}{6}$

55. $(x + 7)(x - 1) = (x + 1)^2$

56. $(x + 2)(x - 3) = (x + 3)^2$

57. $x(2x - 3) = (2x + 1)(x - 4)$

58. $x(1 + 2x) = (2x - 1)(x - 2)$

59. $z(z^2 + 1) = 3 + z^3$

60. $w(4 - w^2) = 8 - w^3$

61. $\dfrac{x}{x-2} + 3 = \dfrac{2}{x-2}$

62. $\dfrac{2x}{x+3} = \dfrac{-6}{x+3} - 2$

63. $\dfrac{2x}{x^2-4} = \dfrac{4}{x^2-4} - \dfrac{3}{x+2}$

64. $\dfrac{x}{x^2-9} + \dfrac{4}{x+3} = \dfrac{3}{x^2-9}$

65. $\dfrac{x}{x+2} = \dfrac{3}{2}$

66. $\dfrac{3x}{x-1} = 2$

67. $\dfrac{5}{2x-3} = \dfrac{3}{x+5}$

68. $\dfrac{-4}{x+4} = \dfrac{-3}{x+6}$

69. $\dfrac{6t+7}{4t-1} = \dfrac{3t+8}{2t-4}$

70. $\dfrac{8w+5}{10w-7} = \dfrac{4w-3}{5w+7}$

71. $\dfrac{4}{x-2} = \dfrac{-3}{x+5} + \dfrac{7}{(x+5)(x-2)}$

72. $\dfrac{-4}{2x+3} + \dfrac{1}{x-1} = \dfrac{1}{(2x+3)(x-1)}$

73. $\dfrac{2}{y+3} + \dfrac{3}{y-4} = \dfrac{5}{y+6}$

74. $\dfrac{5}{5z-11} + \dfrac{4}{2z-3} = \dfrac{-3}{5-z}$

75. $\dfrac{x}{x^2-1} - \dfrac{x+3}{x^2-x} = \dfrac{-3}{x^2+x}$

76. $\dfrac{x+1}{x^2+2x} - \dfrac{x+4}{x^2+x} = \dfrac{-3}{x^2+3x+2}$

Applications and Extensions

77. If $(a, 2)$ is a point on the graph of $y = 5x + 4$, what is a?

78. If $(2, b)$ is a point on the graph of $y = x^2 + 3x$, what is b?

79. If (a, b) is a point on the graph of $2x + 3y = 6$, write an equation that relates a to b.

80. If $(2, 0)$ and $(0, 5)$ are points on the graph of $y = mx + b$, what are m and b?

In Problems 81–86, solve each equation. The letters a, b, and c are constants.

81. $ax - b = c, \quad a \neq 0$

82. $1 - ax = b, \quad a \neq 0$

83. $\dfrac{x}{a} + \dfrac{x}{b} = c, \quad a \neq 0, b \neq 0, a \neq -b$

84. $\dfrac{a}{x} + \dfrac{b}{x} = c, \quad c \neq 0, \quad a + b \neq 0$

85. $\dfrac{1}{x-a} + \dfrac{1}{x+a} = \dfrac{2}{x-1}$

86. $\dfrac{b+c}{x+a} = \dfrac{b-c}{x-a}, \quad c \neq 0, a \neq 0$

87. Find the number a for which $x = 4$ is a solution of the equation

$$x + 2a = 16 + ax - 6a$$

88. Find the number b for which $x = 2$ is a solution of the equation

$$x + 2b = x - 4 + 2bx$$

Problems 89–94 list some formulas that occur in applications. Solve each formula for the indicated variable.

89. Electricity $\dfrac{1}{R} = \dfrac{1}{R_1} + \dfrac{1}{R_2}$ for R

90. Finance $A = P(1 + rt)$ for r

91. Mechanics $F = \dfrac{mv^2}{R}$ for R

92. Chemistry $PV = nRT$ for T

93. Mathematics $S = \dfrac{a}{1-r}$ for r

94. Mechanics $v = -gt + v_0$ for t

95. Finance A total of \$20,000 is to be invested, some in bonds and some in certificates of deposit (CDs). If the amount invested in bonds is to exceed that in CDs by \$3000, how much will be invested in each type of investment?

96. Finance A total of \$10,000 is to be divided between Sean and George, with George to receive \$3000 less than Sean. How much will each receive?

97. Finance An inheritance of \$900,000 is to be divided among Scott, Alice, and Tricia in the following manner: Alice is to receive $\frac{3}{4}$ of what Scott gets, while Tricia gets $\frac{1}{2}$ of what Scott gets. How much does each receive?

98. Sharing the Cost of a Pizza Judy and Tom agree to share the cost of an \$18 pizza based on how much each ate. If Tom ate $\frac{2}{3}$ the amount that Judy ate, how much should each pay? [**Hint:** Some pizza may be left.]

99. Computing Hourly Wages Sandra, who is paid time-and-a-half for hours worked in excess of 40 hours, had gross weekly wages of \$598 for 48 hours worked. What is her regular hourly rate?

100. Computing Hourly Wages Leigh is paid time-and-a-half for hours worked in excess of 40 hours and double-time for hours worked on Sunday. If Leigh had gross weekly wages of \$798 for working 50 hours, 4 of which were on Sunday, what is her regular hourly rate?

101. **Computing Grades** Going into the final exam, which will count as two tests, Brooke has test scores of 80, 83, 71, 61, and 95. What score does Brooke need on the final in order to have an average score of 80?

102. **Computing Grades** Going into the final exam, which will count as two-thirds of the final grade, Mike has test scores of 86, 80, 84, and 90. What minimum score does Mike need on the final in order to earn a B, which requires an average score of 80? What does he need to earn an A, which requires an average of 90?

103. **Business: Discount Pricing** A builder of homes reduced the price of a model by 15%. If the new price is \$170,000, what was its original price? How much can be saved by purchasing the model?

104. **Business: Discount Pricing** At a year-end clearance, a car dealer reduces the list price of last year's models by 12%. If a certain four-door model has a discounted price of \$28,160, what was its list price? How much can be saved by purchasing last year's model?

105. **Personal Finance: Concession Markup** A movie theater marks up the candy it sells by 275%. If a box of candy sells for \$3.00 at the theater, how much did the theater pay for the box?

106. **Personal Finance: Cost of a Car** The suggested list price of a new car is \$34,000. The dealer's cost is 85% of list. How much will you pay if the dealer is willing to accept \$100 over cost for the car?

107. **Business: Theater Attendance** The manager of the Coral Theater wants to know whether the majority of its patrons is adults or children. During a week in July, 5200 tickets were sold and the receipts totaled \$32,200. The adult admission is \$8.50, and the children's admission is \$6.00. How many adult patrons were there?

108. **Business: Discount Pricing** A wool suit, discounted by 30% for a clearance sale, has a price tag of \$399. What was the suit's original price?

109. **Geometry** The perimeter of a rectangle is 60 feet. Find its length and width if the length is 8 feet longer than the width.

110. **Geometry** The perimeter of a rectangle is 42 meters. Find its length and width if the length is twice the width.

111. **Smartphones** In January, 2015, there were 97.89 million people in the United States who owned a smartphone that ran the Google Android operating system (OS). The Google Android OS was used on 53.2% of all smartphones. How many people in the United States owned a smartphone in January, 2015?

Source: *comScore Networks*

112. **Social Networking** A September, 2014, survey indicated that 53% of U.S. adults aged 18–29 used the social media platform Instagram. If this was 21 percentage points less than twice the percent that used Twitter, what percent of U.S. adults aged 18–29 used Twitter?

Discussion and Writing

113. One step in the following list contains an error. Identify it and explain what is wrong.

$$x = 2 \quad (1)$$
$$3x - 2x = 2 \quad (2)$$
$$3x = 2x + 2 \quad (3)$$
$$x^2 + 3x = x^2 + 2x + 2 \quad (4)$$
$$x^2 + 3x - 10 = x^2 + 2x - 8 \quad (5)$$
$$(x - 2)(x + 5) = (x - 2)(x + 4) \quad (6)$$
$$x + 5 = x + 4 \quad (7)$$
$$5 = 4 \quad (8)$$

114. The equation

$$\frac{5}{x+3} + 3 = \frac{8+x}{x+3}$$

has no solution, yet when we go through the process of solving it we obtain $x = -3$. Write a paragraph to explain what causes this to happen.

115. Make up an equation that has no solution and give it to a fellow student to solve. Ask the fellow student to write a critique of your equation.

116. Explain the difference between the directions "solve," "evaluate," and "simplify." Write an example using each direction with the expression $3(x + 2) - x$.

'Are You Prepared?' Answers

1. $-3x + 15$ **2.** -17 **3.** No **4.** $(x + 1)(x - 1)(x + 3)$

1.3 Quadratic Equations

PREPARING FOR THIS SECTION *Before getting started, review the following:*

- Zero-Product Property (Section R.1, p. 14)
- Factoring (Section R.5, pp. 50–56)
- Square Roots (Section R.2, pp. 24–25)
- Completing the Square (Section R.5, p. 57)

Now Work the 'Are You Prepared?' problems on page 117.

OBJECTIVES
1 Solve Quadratic Equations by Factoring (p. 110)
2 Solve Quadratic Equations Using the Square Root Method (p. 112)
3 Solve Quadratic Equations by Completing the Square (p. 113)
4 Solve Quadratic Equations Using the Quadratic Formula (p. 113)
5 Solve Problems That Can Be Modeled by Quadratic Equations (p. 116)

Quadratic equations are equations such as

$$2x^2 + x + 8 = 0 \qquad 3x^2 - 5x = 0 \qquad x^2 - 9 = 0$$

A **quadratic equation** is an equation equivalent to one of the form

$$ax^2 + bx + c = 0 \tag{1}$$

where a, b, and c are real numbers and $a \neq 0$.

A quadratic equation written in the form $ax^2 + bx + c = 0$ is in **standard form**.

Sometimes, a quadratic equation is called a **second-degree equation** because, when it is in standard form, the left side is a polynomial of degree 2. We shall discuss four algebraic ways of solving quadratic equations: by factoring, by the square root method, by completing the square, and by using the quadratic formula.

1 Solve Quadratic Equations by Factoring

When a quadratic equation is written in standard form, it may be possible to factor the expression on the left side into the product of two first-degree polynomials. The Zero-Product Property can then be used by setting each factor equal to 0 and solving the resulting linear equations. With this approach we obtain the *exact* solutions of the quadratic equation. This approach leads us to a basic premise in mathematics. Whenever a problem is encountered, use techniques that reduce the problem to one you already know how to solve. In this instance, we are reducing quadratic equations to linear equations using the technique of factoring.

Let's look at an example.

EXAMPLE 1 Solving a Quadratic Equation by Factoring and by Graphing

Solve the equation: $2x^2 - x - 3 = 0$

Algebraic Solution

The equation is in standard form. The left side may be factored as

$$2x^2 - x - 3 = 0$$

$$(2x - 3)(x + 1) = 0 \quad \text{Factor.}$$

Graphing Solution

Graph $Y_1 = 2x^2 - x - 3$. See Figure 34(a). From the graph it appears there are two solutions to the equation (since the graph crosses the x-axis in two places). Using ZERO, the x-intercepts, and therefore solutions to the equation, are -1 and 1.5. See Figures 34(b) and (c). The solution set is $\{-1, 1.5\}$.

Use the Zero-Product Property and set each factor equal to zero.

$$2x - 3 = 0 \quad \text{or} \quad x + 1 = 0$$
$$2x = 3 \quad \text{or} \quad x = -1$$
$$x = \frac{3}{2}$$

The solution set is $\left\{-1, \frac{3}{2}\right\}$. ■

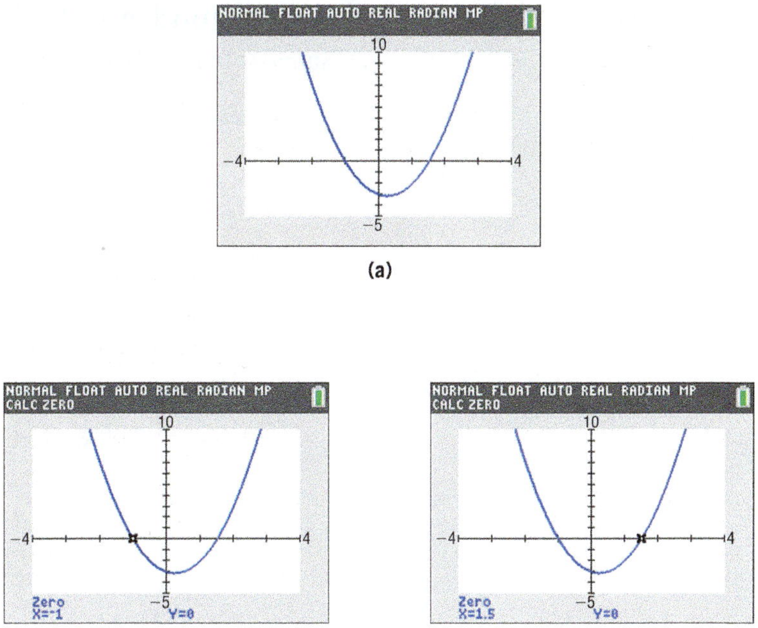

(a)

(b) (c)

Figure 34 ■

Now Work PROBLEM 13

When the left side factors into two linear equations with the same solution, the quadratic equation is said to have a **repeated solution.** This solution is also called a **root of multiplicity 2**, or a **double root**.

EXAMPLE 2 Solving a Quadratic Equation by Factoring and by Graphing

Solve the equation: $9x^2 + 1 = 6x$

Algebraic Solution

Put the equation in standard form by subtracting $6x$ from each side.

$$9x^2 + 1 = 6x$$
$$9x^2 - 6x + 1 = 0$$

Factor the left side of the equation.

$$(3x - 1)(3x - 1) = 0$$
$$3x - 1 = 0 \quad \text{or} \quad 3x - 1 = 0 \qquad \text{Zero-Product Property}$$
$$x = \frac{1}{3} \quad \text{or} \quad x = \frac{1}{3}$$

The equation has only the repeated solution $\frac{1}{3}$.

The solution set is $\left\{\frac{1}{3}\right\}$. ■

Graphing Solution

Graph $Y_1 = 9x^2 + 1$ and $Y_2 = 6x$. See Figure 35. Using INTERSECT, the only point of intersection is (0.33, 2), so the solution of the equation is $x = 0.33$, rounded to two decimal places. The solution set is $\{0.33\}$. This solution is approximate.

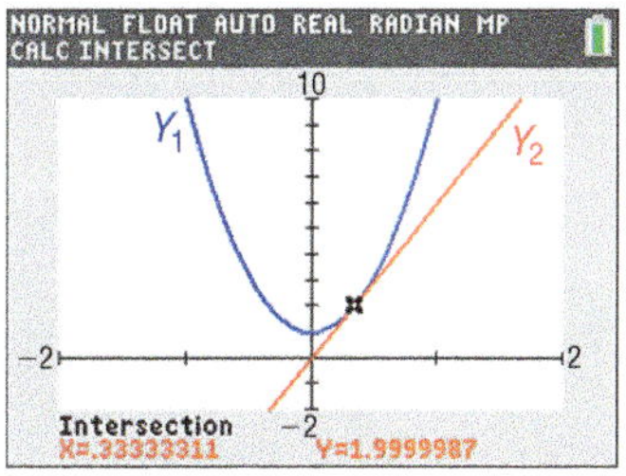

Figure 35 $Y_1 = 9x^2 + 1$; $Y_2 = 6x$ ■

Now Work PROBLEM 23

2 Solve Quadratic Equations Using the Square Root Method

Suppose that we wish to solve the quadratic equation

$$x^2 = p \qquad \textbf{(2)}$$

where p is a nonnegative number. Proceeding as in the earlier examples,

$$x^2 - p = 0 \quad \text{Put in standard form.}$$
$$(x - \sqrt{p})(x + \sqrt{p}) = 0 \quad \text{Factor (over the real numbers).}$$
$$x = \sqrt{p} \quad \text{or} \quad x = -\sqrt{p} \quad \text{Solve.}$$

we have the following result:

$$\text{If } x^2 = p \text{ and } p \geq 0, \text{ then } x = \sqrt{p} \text{ or } x = -\sqrt{p}. \qquad \textbf{(3)}$$

When statement (3) is used, it is called the **Square Root Method**. Note that in statement (3), if $p > 0$ the equation $x^2 = p$ has two solutions, $x = \sqrt{p}$ and $x = -\sqrt{p}$. We usually abbreviate these solutions as $x = \pm\sqrt{p}$, which is read as "x equals plus or minus the square root of p."

For example, the two solutions of the equation

$$x^2 = 4$$

are

$$x = \pm\sqrt{4} \quad \text{Use the Square Root Method.}$$

and, since $\sqrt{4} = 2$, we have

$$x = \pm 2$$

The solution set is $\{-2, 2\}$.

EXAMPLE 3 **Solving Quadratic Equations Using the Square Root Method**

Solve each equation: (a) $x^2 = 5$ (b) $(x - 2)^2 = 16$

Solution (a) $x^2 = 5$

$$x = \pm\sqrt{5} \quad \text{Use the Square Root Method.}$$
$$x = \sqrt{5} \quad \text{or} \quad x = -\sqrt{5}$$

The solution set is $\{-\sqrt{5}, \sqrt{5}\}$.

(b) $(x - 2)^2 = 16$

$$x - 2 = \pm\sqrt{16} \quad \text{Use the Square Root Method.}$$
$$x - 2 = \pm 4 \quad \sqrt{16} = 4$$
$$x - 2 = 4 \quad \text{or} \quad x - 2 = -4$$
$$x = 6 \quad \text{or} \quad x = -2$$

The solution set is $\{-2, 6\}$.

Check: Verify the solutions using a graphing utility. Are the solutions provided by the utility exact?

Now Work PROBLEM 33

3 Solve Quadratic Equations by Completing the Square

EXAMPLE 4 **Solving a Quadratic Equation by Completing the Square**

Solve by completing the square: $x^2 + 5x + 4 = 0$

Solution Always begin this procedure by rearranging the equation so that the constant is on the right side.

$$x^2 + 5x + 4 = 0$$
$$x^2 + 5x = -4$$

Since the coefficient of x^2 is 1, we can complete the square on the left side by adding $\left(\frac{1}{2} \cdot 5\right)^2 = \frac{25}{4}$. Remember, in an equation, whatever is added to the left side must also be added to the right side. So add $\frac{25}{4}$ to *both* sides.

$$x^2 + 5x + \frac{25}{4} = -4 + \frac{25}{4} \qquad \text{Add } \frac{25}{4} \text{ to both sides.}$$
$$\left(x + \frac{5}{2}\right)^2 = \frac{9}{4} \qquad \text{Factor; simplify.}$$
$$x + \frac{5}{2} = \pm\sqrt{\frac{9}{4}} \qquad \text{Use the Square Root Method.}$$
$$x + \frac{5}{2} = \pm\frac{3}{2} \qquad \sqrt{\frac{9}{4}} = \frac{\sqrt{9}}{\sqrt{4}} = \frac{3}{2}$$
$$x = -\frac{5}{2} \pm \frac{3}{2}$$
$$x = -\frac{5}{2} + \frac{3}{2} = -1 \quad \text{or} \quad x = -\frac{5}{2} - \frac{3}{2} = -4$$

The solution set is $\{-4, -1\}$.

Check: Verify the solutions using a graphing utility. ■

Note: If the coefficient of the square term is not 1, divide through by the coefficient of the square term before attempting to complete the square. For example, to solve $2x^2 - 8x = 5$ by completing the square, divide both sides of the equation by 2 and obtain $x^2 - 4x = \frac{5}{2}$. ■

The solution of the equation in Example 4 can also be obtained by factoring. Rework Example 4 using factoring.

Now Work PROBLEM 37

4 Solve Quadratic Equations Using the Quadratic Formula

The method of completing the square can be used to obtain a general formula for solving the quadratic equation

$$ax^2 + bx + c = 0, \qquad a > 0$$

Note: There is no loss in generality to assume that $a > 0$, since if $a < 0$ we can multiply both sides by -1 to obtain an equivalent equation with a positive leading coefficient. ■

As in Example 4, rearrange the equation as

$$ax^2 + bx = -c$$

Since $a > 0$, divide both sides by a to get

$$x^2 + \frac{b}{a}x = -\frac{c}{a}$$

Now the coefficient of x^2 is 1. To complete the square on the left side, add the square of $\frac{1}{2}$ of the coefficient of x; that is, add

$$\left(\frac{1}{2} \cdot \frac{b}{a}\right)^2 = \frac{b^2}{4a^2}$$

to each side. Then

$$x^2 + \frac{b}{a}x + \frac{b^2}{4a^2} = \frac{b^2}{4a^2} - \frac{c}{a}$$

$$\left(x + \frac{b}{2a}\right)^2 = \frac{b^2 - 4ac}{4a^2} \qquad \frac{b^2}{4a^2} - \frac{c}{a} = \frac{b^2}{4a^2} - \frac{4ac}{4a^2} = \frac{b^2 - 4ac}{4a^2} \qquad \textbf{(4)}$$

Provided that $b^2 - 4ac \geq 0$, we now can use the Square Root Method to get

$$x + \frac{b}{2a} = \pm\sqrt{\frac{b^2 - 4ac}{4a^2}}$$

$$x + \frac{b}{2a} = \frac{\pm\sqrt{b^2 - 4ac}}{2a} \qquad \text{The square root of a quotient equals the quotient of the square roots. Also, } \sqrt{4a^2} = 2a \text{ since } a > 0.$$

$$x = -\frac{b}{2a} \pm \frac{\sqrt{b^2 - 4ac}}{2a} \qquad \text{Add } -\frac{b}{2a} \text{ to both sides.}$$

$$= \frac{-b \pm \sqrt{b^2 - 4ac}}{2a} \qquad \text{Combine the quotients on the right.}$$

What if $b^2 - 4ac$ is negative? Then equation (4) states that the left expression (a real number squared) equals the right expression (a negative number). Since this is impossible for real numbers, we conclude that if $b^2 - 4ac < 0$, the quadratic equation has no *real* solution.*

THEOREM

Quadratic Formula

Consider the quadratic equation

$$ax^2 + bx + c = 0 \qquad a \neq 0$$

If $b^2 - 4ac < 0$, this equation has no real solution.

If $b^2 - 4ac \geq 0$, the real solution(s) of this equation is (are) given by the **quadratic formula**

$$x = \frac{-b \pm \sqrt{b^2 - 4ac}}{2a} \qquad \textbf{(5)}$$

The quantity $\boldsymbol{b^2 - 4ac}$ is called the **discriminant** of the quadratic equation, because its value tells us whether the equation has real solutions. In fact, it also tells us how many solutions to expect.

Discriminant of a Quadratic Equation

For a quadratic equation $ax^2 + bx + c = 0$:

1. If $b^2 - 4ac > 0$, there are two unequal real solutions.
2. If $b^2 - 4ac = 0$, there is a repeated real solution, a root of multiplicity 2.
3. If $b^2 - 4ac < 0$, there is no real solution.

When asked to find the real solutions, if any, of a quadratic equation, always evaluate the discriminant first to see if there are any real solutions.

*We consider quadratic equations where $b^2 - 4ac$ is negative in the next section.

EXAMPLE 5 Solving a Quadratic Equation Using the Quadratic Formula

Find the real solutions, if any, of the equation $3x^2 - 5x + 1 = 0$.

Algebraic Solution

The equation is in standard form, so compare it to $ax^2 + bx + c = 0$ to find a, b, and c.

$$3x^2 - 5x + 1 = 0$$

$$ax^2 + bx + c = 0 \qquad a = 3, b = -5, c = 1$$

With $a = 3$, $b = -5$, and $c = 1$, evaluate the discriminant $b^2 - 4ac$.

$$b^2 - 4ac = (-5)^2 - 4(3)(1) = 25 - 12 = 13$$

Since $b^2 - 4ac > 0$, there are two real solutions.

Use the quadratic formula with $a = 3, b = -5, c = 1$, and $b^2 - 4ac = 13$.

$$x = \frac{-b \pm \sqrt{b^2 - 4ac}}{2a} = \frac{-(-5) \pm \sqrt{13}}{2(3)} = \frac{5 \pm \sqrt{13}}{6}$$

The solution set is $\left\{\frac{5 - \sqrt{13}}{6}, \frac{5 + \sqrt{13}}{6}\right\}$. These solutions are exact. ■

Graphing Solution

Figure 36 shows the graph of the equation

$$Y_1 = 3x^2 - 5x + 1$$

There are two x-intercepts: one between 0 and 1, the other between 1 and 2. Using ZERO (or ROOT), we find the solutions to the equation are 0.23 and 1.43, rounded to two decimal places. These solutions are approximate.

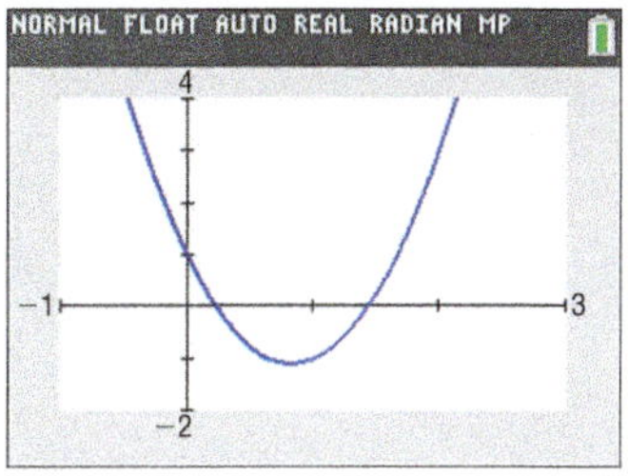

Figure 36 $Y_1 = 3x^2 - 5x + 1$ ■

Now Work PROBLEM 43

EXAMPLE 6 Solving a Quadratic Equation Using the Quadratic Formula

Find the real solutions, if any, of the equation $3x^2 + 2 = 4x$.

Algebraic Solution

The equation, as given, is not in standard form.

$$3x^2 + 2 = 4x$$

$$3x^2 - 4x + 2 = 0 \qquad \text{Subtract } 4x \text{ from both sides to put the equation in standard form.}$$

$$ax^2 + bx + c = 0 \qquad \text{Compare to standard form.}$$

With $a = 3$, $b = -4$, and $c = 2$, the discriminant is

$$b^2 - 4ac = (-4)^2 - 4(3)(2) = 16 - 24$$
$$= -8$$

Since $b^2 - 4ac < 0$, the equation has no real solution. ■

Graphing Solution

Use the standard form of the equation and graph

$$Y_1 = 3x^2 - 4x + 2$$

See Figure 37. Notice that there are no x-intercepts, so the equation has no real solution, as expected based on the value of the discriminant.

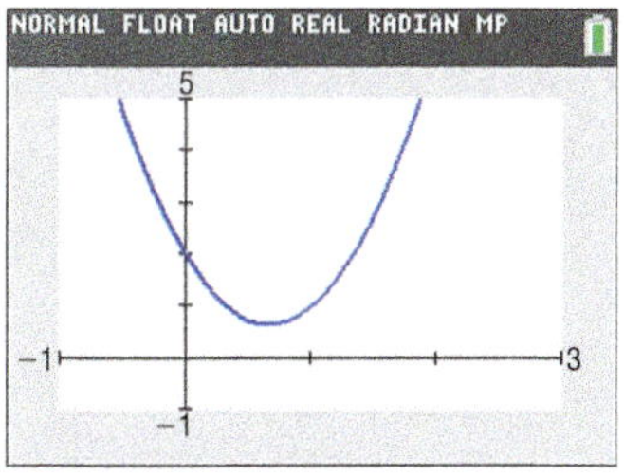

Figure 37 $Y_1 = 3x^2 - 4x + 2$ ■

Now Work PROBLEM 49

SUMMARY

Procedure for Solving a Quadratic Equation Algebraically

To solve a quadratic equation algebraically, first put it in standard form:

$$ax^2 + bx + c = 0$$

Then

STEP 1: Identify a, b, and c.

STEP 2: Evaluate the discriminant, $b^2 - 4ac$.

STEP 3: (a) If the discriminant is negative, the equation has no real solution.

(b) If the discriminant is zero, the equation has one real solution, a repeated root.

(c) If the discriminant is positive, the equation has two distinct real solutions.

If you can easily spot factors, use the factoring method to solve the equation. Otherwise, use the quadratic formula or the method of completing the square.

5 Solve Problems That Can Be Modeled by Quadratic Equations

Many applied problems require the solution of a quadratic equation. Let's look at one that you will probably see again in a slightly different form if you study calculus.

EXAMPLE 7 Constructing a Box

From each corner of a square piece of sheet metal, remove a square of side 9 centimeters. Turn up the edges to form an open box. If the box is to hold 144 cubic centimeters (cm^3), what should be the dimensions of the piece of sheet metal?

Solution Use Figure 38 as a guide. We have labeled the length of a side of the square piece of sheet metal, x. The box will be of height 9 centimeters, and its square base will have $x - 18$ as the length of each side. The volume V (Length $\times$ Width $\times$ Height) of the box is therefore

$$V = (x - 18)(x - 18) \cdot 9 = 9(x - 18)^2 \quad \textbf{The Model}$$

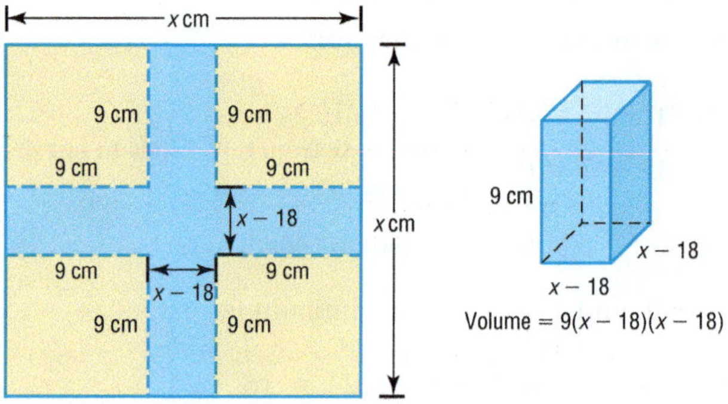

Figure 38

Since the volume of the box is to be 144 cm^3, we have

$$9(x - 18)^2 = 144 \quad V = 144$$

$$(x - 18)^2 = 16 \quad \text{Divide each side by 9.}$$

$$x - 18 = \pm 4 \quad \text{Use the Square Root Method.}$$

$$x = 18 \pm 4$$

$$x = 22 \quad \text{or} \quad x = 14$$

Discard the solution $x = 14$ (do you see why?) and conclude that the sheet metal should be 22 centimeters by 22 centimeters.

Check: If we take a piece of sheet metal that measures 22 centimeters by 22 centimeters, cut out a 9-centimeter square from each corner, and fold up the edges, we get a box whose dimensions are 9 cm by 4 cm by 4 cm, with volume $9 \text{ cm} \times 4 \text{ cm} \times 4 \text{ cm} = 144 \text{ cm}^3$, as required. ■

Now Work PROBLEM 95

Historical Feature

Problems using quadratic equations are found in the oldest known mathematical literature. Babylonians and Egyptians were solving such problems before 1800 BC. Euclid solved quadratic equations geometrically in his *Data* (300 BC), and the Hindus and Arabs gave rules for solving any quadratic equation with real roots. Because negative numbers were not freely used before AD 1500, there were several different types of quadratic equations, each with its own rule. Thomas Harriot (1560–1621) introduced the method of factoring to obtain solutions, and François Viète (1540–1603) introduced a method that is essentially completing the square.

Until modern times it was usual to neglect the negative roots (if there were any), and equations involving square roots of negative quantities were regarded as unsolvable until the 1500s.

Historical Problems

1. *One solution of al-Khwārizmī* Solve $x^2 + 12x = 85$ by drawing the square shown. The area of the four white rectangles and the yellow square is $x^2 + 12x$. We then set this expression equal to 85 to get the equation $x^2 + 12x = 85$. If we add the four blue squares, we will have a larger square of known area. Complete the solution.

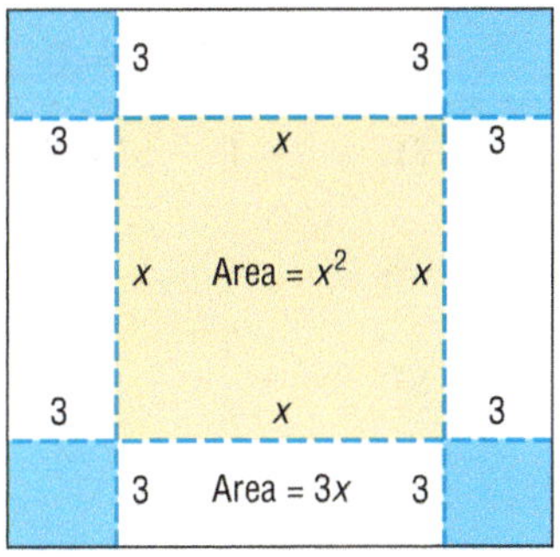

2. *Viète's method* Solve $x^2 + 12x - 85 = 0$ by letting $x = u + z$.

   Then

$$(u + z)^2 + 12(u + z) - 85 = 0$$
$$u^2 + (2z + 12)u + (z^2 + 12z - 85) = 0$$

   Now select z so that $2z + 12 = 0$ and finish the solution.

3. *Another method to get the quadratic formula* Look at equation (4) on page 114. Rewrite the right side as $\left(\dfrac{\sqrt{b^2 - 4ac}}{2a}\right)^2$ and then subtract it from each side. The right side is now 0 and the left side is a difference of two squares. If you factor this difference of two squares, you will easily be able to get the quadratic formula and, moreover, the quadratic expression is factored, which is sometimes useful.

1.3 Assess Your Understanding

'Are You Prepared?' *Answers are given at the end of these exercises. If you get a wrong answer, read the pages listed in red.*

1. Factor: $x^2 - 5x - 6$ (pp. 53–54)
2. Factor: $2x^2 - x - 3$ (pp. 55–56)
3. The solution set of the equation $(x - 3)(3x + 5) = 0$ is ________. (p. 14)
4. Simplify: $\sqrt{8^2 - 4 \cdot 2 \cdot 3}$ (pp. 75–76)
5. Complete the square of the expression $x^2 + 5x$. Factor the new expression. (p. 57)

Concepts and Vocabulary

6. When a quadratic equation has a repeated solution, it is called a(n) ____________ root or a root of ________________.
7. The quantity $b^2 - 4ac$ is called the ______________ of a quadratic equation. If it is __________, the equation has no real solution.
8. ***True or False*** Quadratic equations always have two real solutions.
9. A quadratic equation is sometimes called a ________ equation.
   (a) first-degree (b) second-degree
   (c) third-degree (d) fourth-degree
10. Which of the following quadratic equations is in standard form?
    (a) $x^2 - 7x = 5$ (b) $9 = x^2$
    (c) $(x + 5)(x - 4) = 0$ (d) $0 = 5x^2 - 6x - 1$

Skill Building

In Problems 11–30, solve each equation by factoring. Verify your solution using a graphing utility.

11. $x^2 - 9x = 0$

12. $x^2 + 4x = 0$

13. $x^2 - 25 = 0$

14. $x^2 - 9 = 0$

15. $z^2 + z - 6 = 0$

16. $v^2 + 7v + 6 = 0$

17. $2x^2 - 5x - 3 = 0$

18. $3x^2 + 5x + 2 = 0$

19. $3t^2 - 48 = 0$

20. $2y^2 - 50 = 0$

21. $x(x - 8) + 12 = 0$

22. $x(x + 4) = 12$

23. $4x^2 + 9 = 12x$

24. $25x^2 + 16 = 40x$

25. $6(p^2 - 1) = 5p$

26. $2(2u^2 - 4u) + 3 = 0$

27. $6x - 5 = \frac{6}{x}$

28. $x + \frac{12}{x} = 7$

29. $\frac{4(x - 2)}{x - 3} + \frac{3}{x} = \frac{-3}{x(x - 3)}$

30. $\frac{5}{x + 4} = 4 + \frac{3}{x - 2}$

In Problems 31–36, solve each equation by the Square Root Method. Verify your solution using a graphing utility.

31. $x^2 = 25$

32. $x^2 = 36$

33. $(x - 1)^2 = 4$

34. $(x + 2)^2 = 1$

35. $(2y + 3)^2 = 9$

36. $(3z - 2)^2 = 4$

In Problems 37–42, solve each equation by completing the square. Verify your solution using a graphing utility.

37. $x^2 + 4x = 21$

38. $x^2 - 6x = 13$

39. $x^2 - \frac{1}{2}x - \frac{3}{16} = 0$

40. $x^2 + \frac{2}{3}x - \frac{1}{3} = 0$

41. $3x^2 + x - \frac{1}{2} = 0$

42. $2x^2 - 3x - 1 = 0$

In Problems 43–66, find the real solutions, if any, of each equation. Use the quadratic formula. Verify your solution using a graphing utility.

43. $x^2 - 4x + 2 = 0$

44. $x^2 + 4x + 2 = 0$

45. $x^2 - 4x - 1 = 0$

46. $x^2 + 6x + 1 = 0$

47. $2x^2 - 5x + 3 = 0$

48. $2x^2 + 5x + 3 = 0$

49. $4y^2 - y + 2 = 0$

50. $4t^2 + t + 1 = 0$

51. $4x^2 = 1 - 2x$

52. $2x^2 = 1 - 2x$

53. $4x^2 = 9x$

54. $5x = 4x^2$

55. $9t^2 - 6t + 1 = 0$

56. $4u^2 - 6u + 9 = 0$

57. $\frac{3}{4}x^2 - \frac{1}{4}x - \frac{1}{2} = 0$

58. $\frac{2}{3}x^2 - x - 3 = 0$

59. $\frac{5}{3}x^2 - x = \frac{1}{3}$

60. $\frac{3}{5}x^2 - x = \frac{1}{5}$

61. $2x(x + 2) = 3$

62. $3x(x + 2) = 1$

63. $4 - \frac{1}{x} - \frac{2}{x^2} = 0$

64. $4 + \frac{1}{x} - \frac{1}{x^2} = 0$

65. $\frac{3x}{x - 2} + \frac{1}{x} = 4$

66. $\frac{2x}{x - 3} + \frac{1}{x} = 4$

In Problems 67–72, use the discriminant to determine whether each quadratic equation has two unequal real solutions, a repeated real solution (a double root), or no real solution, without solving the equation.

67. $2x^2 - 6x + 7 = 0$

68. $x^2 + 4x + 7 = 0$

69. $9x^2 - 30x + 25 = 0$

70. $25x^2 - 20x + 4 = 0$

71. $3x^2 + 5x - 8 = 0$

72. $2x^2 - 3x - 7 = 0$

Mixed Practice

In Problems 73–88, find the real solutions, if any, of each equation. Use any method. Verify your solution using a graphing utility.

73. $x^2 - 5 = 0$

74. $x^2 - 6 = 0$

75. $16x^2 - 8x + 1 = 0$

76. $9x^2 - 12x + 4 = 0$

77. $10x^2 - 19x - 15 = 0$

78. $6x^2 + 7x - 20 = 0$

79. $2 + z = 6z^2$

80. $2 = y + 6y^2$

81. $x^2 + \sqrt{2}x = \frac{1}{2}$

82. $\frac{1}{2}x^2 = \sqrt{2}x + 1$

83. $x^2 + x = 4$

84. $x^2 + x = 1$

85. $5x(x - 1) = -7x^2 + 2$

86. $10x(x + 2) = -3x + 5$

87. $\frac{x}{x-2} + \frac{2}{x+1} = \frac{7x+1}{x^2 - x - 2}$

88. $\frac{3x}{x+2} + \frac{1}{x-1} = \frac{4-7x}{x^2 + x - 2}$

Applications and Extensions

89. Pythagorean Theorem How many right triangles have a hypotenuse that measures $2x + 3$ meters and legs that measure $2x - 5$ meters and $x + 7$ meters? What are the dimensions of the triangle(s)?

90. Pythagorean Theorem How many right triangles have a hypotenuse that measures $4x + 5$ inches and legs that measure $3x + 13$ inches and x inches? What are the dimensions of the triangle(s)?

91. Dimensions of a Window The area of the opening of a rectangular window is to be 143 square feet. If the length is to be 2 feet more than the width, what are the dimensions?

92. Dimensions of a Window The area of a rectangular window is to be 306 square centimeters. If the length exceeds the width by 1 centimeter, what are the dimensions?

93. Geometry Find the dimensions of a rectangle whose perimeter is 26 meters and whose area is 40 square meters.

94. Watering a Field An adjustable water sprinkler that sprays water in a circular pattern is placed at the center of a square field whose area is 1250 square feet (see the figure). What is the shortest radius setting that can be used if the field is to be completely enclosed within the circle?

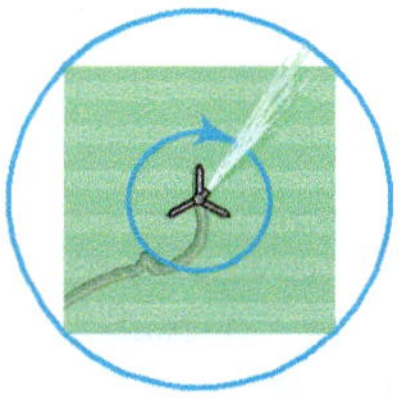

95. Constructing a Box An open box is to be constructed from a square piece of sheet metal by removing a square of side 1 foot from each corner and turning up the edges. If the box is to hold 4 cubic feet, what should be the dimensions of the sheet metal?

96. Constructing a Box Rework Problem 95 if the piece of sheet metal is a rectangle whose length is twice its width.

97. Physics A ball is thrown vertically upward from the top of a building 96 feet tall with an initial velocity of 80 feet per second. The distance s (in feet) of the ball from the ground after t seconds is $s = 96 + 80t - 16t^2$.

(a) After how many seconds does the ball strike the ground?

(b) After how many seconds will the ball pass the top of the building on its way down?

98. Physics An object is propelled vertically upward with an initial velocity of 20 meters per second. The distance s (in meters) of the object from the ground after t seconds is $s = -4.9t^2 + 20t$.

(a) When will the object be 15 meters above the ground?

(b) When will it strike the ground?

(c) Will the object reach a height of 100 meters?

99. Reducing the Size of a Candy Bar A jumbo chocolate bar with a rectangular shape measures 12 centimeters in length, 7 centimeters in width, and 3 centimeters in thickness. Due to escalating costs of cocoa, management decides to reduce the volume of the bar by 10%. To accomplish this reduction, management decides that the new bar should have the same 3 centimeter thickness, but the length and width of each should be reduced an equal number of centimeters. What should be the dimensions of the new candy bar?

100. Reducing the Size of a Candy Bar Rework Problem 99 if the reduction is to be 20%.

101. Constructing a Border around a Pool A circular pool measures 10 feet across. One cubic yard of concrete is to be used to create a circular border of uniform width around the pool. If the border is to have a depth of 3 inches, how wide will the border be? (1 cubic yard = 27 cubic feet) See the illustration.

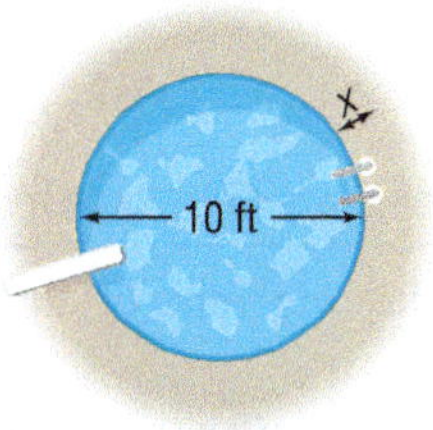

102. Constructing a Border around a Pool Rework Problem 101 if the depth of the border is 4 inches.

103. Constructing a Border around a Garden A landscaper, who just completed a rectangular flower garden measuring 6 feet by 10 feet, orders 1 cubic yard of premixed cement, all of which is to be used to create a border of uniform width around the garden. If the border is to have a depth of 3 inches, how wide will the border be? (1 cubic yard = 27 cubic feet)

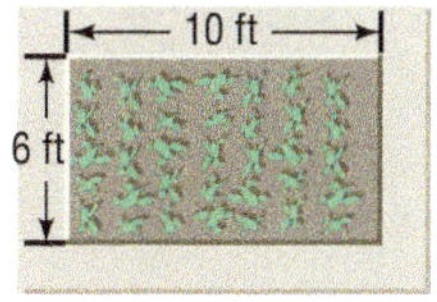

104. Dimensions of a Patio A contractor orders 8 cubic yards of premixed cement, all of which is to be used to pour a patio that will be 4 inches thick. If the length of the patio is specified to be twice the width, what will be the patio dimensions? (1 cubic yard = 27 cubic feet)

105. Comparing Tablets The screen size of a tablet is determined by the length of the diagonal of the rectangular screen. The 9.7-inch iPad Air™ comes in a 4:3 format, which means that the ratio of the length to the width of the rectangular screen is 4:3. What is the area of the iPad's screen? What is the area of a 10-inch Google Nexus™ if its screen is in a 16:10 format? Which screen is larger? (**Hint:** If x is the length of a 4:3 format screen, then $\frac{3}{4}x$ is the width.)

iPad mini 4:3

Google Nexus 16:10

106. Comparing Tablets Refer to Problem 105. Find the screen area of a 7.9-inch iPad mini with Retina™ in a 4:3 format, and compare it with an 8-inch Dell Venue Pro™ if its screen is in a 16:9 format. Which screen is larger?

107. Field Design A football field is sloped from the center toward the sides for drainage. The height h, in feet, of the field, x feet from the side, is given by $h = -0.00025x^2 + 0.04x$. Find the height of the field a distance of 35 feet from the side. Round to the nearest tenth of a foot.

108. College Value The difference, d, in median earnings, in \$1000s, between high school graduates and college graduates can be approximated by $d = -0.002x^2 + 0.319x + 7.512$, where x is the number of years after 1965. Based on this model, estimate to the nearest year when the difference in median earnings was \$15,000. (***Source:*** *Current Population Survey*)

109. Student Working A study found that a student's GPA, g, is related to the number of hours worked each week, h, by the equation $g = -0.0006h^2 + 0.015h + 3.04$. Estimate the number of hours worked each week for a student with a GPA of 2.97. Round to the nearest whole hour.

110. Fraternity Purchase A fraternity wants to buy a new LED Smart TV that costs \$1470. If 7 members of the fraternity are not able to contribute, the share for the remaining members increases by \$5. How many members are in the fraternity?

111. The sum of the consecutive integers 1, 2, 3, . . . n is given by the formula $\frac{1}{2}n(n+1)$. How many consecutive integers, starting with 1, must be added to get a sum of 666?

112. Geometry If a polygon of n sides has $\frac{1}{2}n(n-3)$ diagonals, how many sides will a polygon with 65 diagonals have? Is there a polygon with 80 diagonals?

113. Show that the sum of the roots of a quadratic equation is $-\frac{b}{a}$.

114. Show that the product of the roots of a quadratic equation is $\frac{c}{a}$.

115. Find k such that the equation $kx^2 + x + k = 0$ has a repeated real solution.

116. Find k such that the equation $x^2 - kx + 4 = 0$ has a repeated real solution.

117. Show that the real solutions of the equation $ax^2 + bx + c = 0$ are the negatives of the real solutions of the equation $ax^2 - bx + c = 0$. Assume that $b^2 - 4ac \geq 0$.

118. Show that the real solutions of the equation $ax^2 + bx + c = 0$ are the reciprocals of the real solutions of the equation $cx^2 + bx + a = 0$. Assume that $b^2 - 4ac \geq 0$.

Explaining Concepts: Discussion and Writing

119. Which of the following pairs of equations are equivalent? Explain.

(a) $x^2 = 9$; $x = 3$ (b) $x = \sqrt{9}$; $x = 3$

(c) $(x-1)(x-2) = (x-1)^2$; $x - 2 = x - 1$

120. Describe three ways that you might solve a quadratic equation. State your preferred method; explain why you chose it.

121. Explain the benefits of evaluating the discriminant of a quadratic equation before attempting to solve it.

122. Create three quadratic equations: one having two distinct solutions, one having no real solution, and one having exactly one real solution.

123. The word *quadratic* seems to imply four (*quad*), yet a quadratic equation is an equation that involves a polynomial of degree 2. Investigate the origin of the term *quadratic* as it is used in the expression *quadratic equation*. Write a brief essay on your findings.

'Are You Prepared?' Answers

1. $(x-6)(x+1)$ **2.** $(2x-3)(x+1)$ **3.** $\left\{-\frac{5}{3}, 3\right\}$ **4.** $2\sqrt{10}$ **5.** $x^2 + 5x + \frac{25}{4} = \left(x + \frac{5}{2}\right)^2$

1.4 Complex Numbers; Quadratic Equations in the Complex Number System*

PREPARING FOR THIS SECTION *Before getting started, review the following:*

- Classification of Numbers (Section R.1, pp. 4–5)
- Rationalizing Denominators (Section R.8, p. 76)

Now Work the 'Are You Prepared?' problems on page 128.

OBJECTIVES 1 Add, Subtract, Multiply, and Divide Complex Numbers (p. 122)
2 Solve Quadratic Equations in the Complex Number System (p. 125)

Complex Numbers

One property of a real number is that its square is nonnegative (greater than or equal to 0). For example, there is no real number x for which

$$x^2 = -1$$

To remedy this situation, we introduce a new number called the *imaginary unit.*

DEFINITION

The **imaginary unit**, which we denote by i, is the number whose square is -1. That is,

$$i^2 = -1$$

This should not surprise you. If our universe were to consist only of integers, there would be no number x for which $2x = 1$. This was remedied by introducing numbers such as $\frac{1}{2}$ and $\frac{2}{3}$, the *rational numbers*. If our universe were to consist only of rational numbers, there would be no x whose square equals 2. That is, there would be no number x for which $x^2 = 2$. To remedy this, we introduced numbers such as $\sqrt{2}$ and $\sqrt[3]{5}$, the *irrational numbers*. Recall that the *real numbers* consist of the rational numbers and the irrational numbers. Now, if our universe were to consist only of real numbers, then there would be no number x whose square is -1. To remedy this, we introduced a number i, whose square is -1.

In the progression outlined, each time we encountered a situation that was unsuitable, a new number system was introduced to remedy this situation. And each new number system contained the earlier number system as a subset. The number system that results from introducing the number i is called the **complex number system.**

DEFINITION

Complex numbers are numbers of the form $\boldsymbol{a + bi}$, where a and b are real numbers. The real number a is called the **real part** of the number $a + bi$; the real number b is called the **imaginary part** of $a + bi$; and i is the imaginary unit, so $i^2 = -1$.

For example, the complex number $-5 + 6i$ has the real part -5 and the imaginary part 6.

When a complex number is written in the form $a + bi$, where a and b are real numbers, it is in **standard form.** However, if the imaginary part of a complex number is negative, such as in the complex number $3 + (-2)i$, we agree to write it instead in the form $3 - 2i$.

Also, the complex number $a + 0i$ is usually written simply as a. This serves to remind us that the real numbers are a subset of the complex numbers. The complex number $0 + bi$ is usually written as bi. Sometimes the complex number bi is called a **pure imaginary number.**

*This section may be omitted without any loss of continuity.

1 Add, Subtract, Multiply, and Divide Complex Numbers

Equality, addition, subtraction, and multiplication of complex numbers are defined so as to preserve the familiar rules of algebra for real numbers. Two complex numbers are equal if and only if their real parts are equal and their imaginary parts are equal.

Equality of Complex Numbers

$$a + bi = c + di \quad \text{if and only if} \quad a = c \text{ and } b = d \qquad \textbf{(1)}$$

Two complex numbers are added by forming the complex number whose real part is the sum of the real parts and whose imaginary part is the sum of the imaginary parts.

Sum of Complex Numbers

$$(a + bi) + (c + di) = (a + c) + (b + d)i \qquad \textbf{(2)}$$

To subtract two complex numbers, use this rule:

Difference of Complex Numbers

$$(a + bi) - (c + di) = (a - c) + (b - d)i \qquad \textbf{(3)}$$

EXAMPLE 1 Adding and Subtracting Complex Numbers

(a) $(3 + 5i) + (-2 + 3i) = [3 + (-2)] + (5 + 3)i = 1 + 8i$

(b) $(6 + 4i) - (3 + 6i) = (6 - 3) + (4 - 6)i = 3 + (-2)i = 3 - 2i$ ■

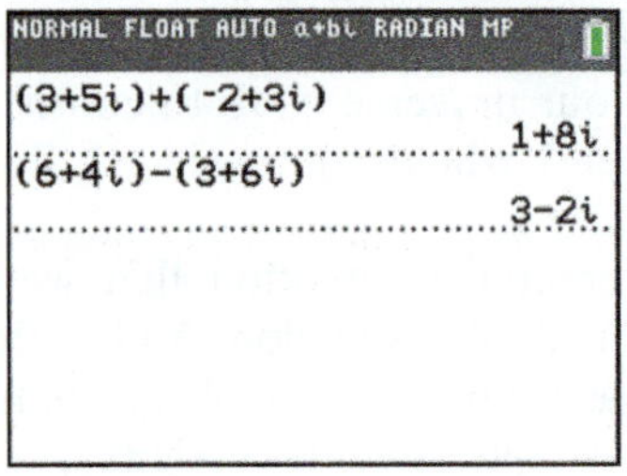

Figure 39

Some graphing calculators have the capability of handling complex numbers. For example, Figure 39 shows the results of Example 1 using a TI-84 Plus C graphing calculator.

Now Work PROBLEM 15

Products of complex numbers are calculated as illustrated in Example 2.

EXAMPLE 2 Multiplying Complex Numbers

$$\begin{aligned}(5 + 3i) \cdot (2 + 7i) &= 5 \cdot (2 + 7i) + 3i(2 + 7i) && \text{Distributive Property}\\ &= 10 + 35i + 6i + 21i^2 && \text{Distributive Property}\\ &= 10 + 41i + 21(-1) && i^2 = -1\\ &= -11 + 41i\end{aligned}$$

■

Based on the procedure of Example 2, the **product** of two complex numbers is defined as follows:

Product of Complex Numbers

$$(a + bi) \cdot (c + di) = (ac - bd) + (ad + bc)i \qquad \textbf{(4)}$$

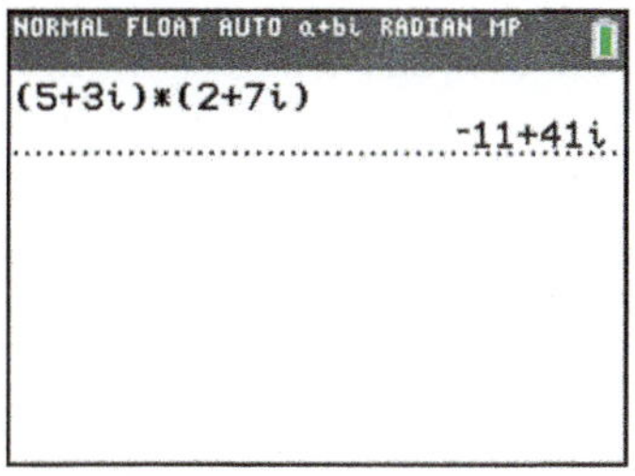

Figure 40

Do not bother to memorize formula (4). Instead, whenever it is necessary to multiply two complex numbers, follow the usual rules for multiplying two binomials, as in Example 2, remembering that $i^2 = -1$. For example,

$$(2i)(2i) = 4i^2 = 4(-1) = -4$$
$$(2 + i)(1 - i) = 2 - 2i + i - i^2 = 3 - i$$

Graphing calculators may also be used to multiply complex numbers. Figure 40 shows the result obtained in Example 2 using a TI-84 Plus C graphing calculator.

Now Work PROBLEM 21

Algebraic properties for addition and multiplication, such as the Commutative, Associative, and Distributive Properties, hold for complex numbers. However, the property that every nonzero complex number has a multiplicative inverse, or reciprocal, requires a closer look.

Conjugates

DEFINITION

If $z = a + bi$ is a complex number, then its **conjugate**, denoted by $\bar{z}$, is defined as

$$\bar{z} = \overline{a + bi} = a - bi$$

Note: The conjugate of a complex number can be found by changing the sign of the imaginary part. ■

For example, $\overline{2 + 3i} = 2 - 3i$ and $\overline{-6 - 2i} = -6 + 2i$.

EXAMPLE 3 **Multiplying a Complex Number by Its Conjugate**

Find the product of the complex number $z = 3 + 4i$ and its conjugate $\bar{z}$.

Solution Since $\bar{z} = 3 - 4i$, we have

$$z\bar{z} = (3 + 4i)(3 - 4i) = 9 - 12i + 12i - 16i^2 = 9 + 16 = 25$$ ■

The result obtained in Example 3 has an important generalization.

THEOREM

The product of a complex number and its conjugate is a nonnegative real number. That is, if $z = a + bi$, then

$$z\bar{z} = a^2 + b^2 \qquad (5)$$

■

Proof If $z = a + bi$, then

$$z\bar{z} = (a + bi)(a - bi) = a^2 - abi + abi - (bi)^2 = a^2 - b^2i^2 = a^2 + b^2$$ ■

To express the reciprocal of a nonzero complex number z in standard form, multiply the numerator and denominator of $\frac{1}{z}$ by its conjugate $\bar{z}$. That is, if $z = a + bi$ is a nonzero complex number, then

$$\frac{1}{a + bi} = \frac{1}{z} = \frac{1}{z} \cdot \frac{\bar{z}}{\bar{z}} = \frac{\bar{z}}{z\bar{z}} \underset{\uparrow \text{ Use (5).}}{=} \frac{a - bi}{a^2 + b^2} = \frac{a}{a^2 + b^2} - \frac{b}{a^2 + b^2}i$$

EXAMPLE 4 **Writing the Reciprocal of a Complex Number in Standard Form**

Write $\frac{1}{3 + 4i}$ in standard form $a + bi$; that is, find the reciprocal of $3 + 4i$.

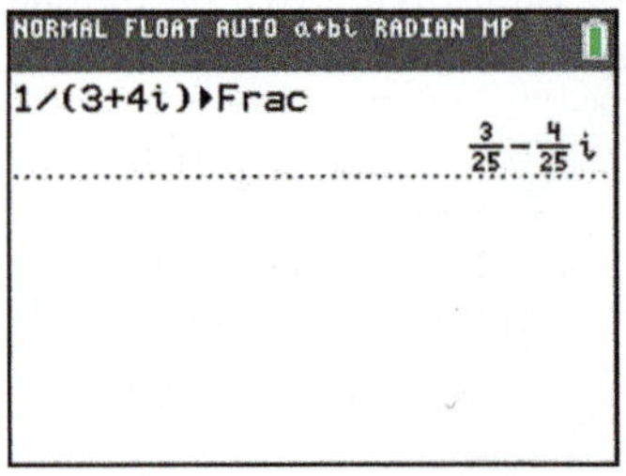

Figure 41

Solution Multiply the numerator and denominator of $\dfrac{1}{3+4i}$ by the conjugate of $3+4i$, the complex number $3-4i$. The result is

$$\frac{1}{3+4i}=\frac{1}{3+4i}\cdot\frac{3-4i}{3-4i}=\frac{3-4i}{9+16}=\frac{3}{25}-\frac{4}{25}i$$

A graphing calculator can be used to verify the result of Example 4. See Figure 41.

To express the quotient of two complex numbers in standard form, multiply the numerator and denominator of the quotient by the conjugate of the denominator.

EXAMPLE 5 Writing Quotients of Complex Numbers in Standard Form

Write each of the following in standard form.

(a) $\dfrac{1+4i}{5-12i}$ (b) $\dfrac{2-3i}{4-3i}$

Solution

(a) $$\frac{1+4i}{5-12i}=\frac{1+4i}{5-12i}\cdot\frac{5+12i}{5+12i}=\frac{5+12i+20i+48i^2}{25+144}$$
$$=\frac{-43+32i}{169}=-\frac{43}{169}+\frac{32}{169}i$$

(b) $$\frac{2-3i}{4-3i}=\frac{2-3i}{4-3i}\cdot\frac{4+3i}{4+3i}=\frac{8+6i-12i-9i^2}{16+9}$$
$$=\frac{17-6i}{25}=\frac{17}{25}-\frac{6}{25}i$$

Now Work PROBLEM 29

EXAMPLE 6 Writing Other Expressions in Standard Form

If $z=2-3i$ and $w=5+2i$, write each of the following expressions in standard form.

(a) $\dfrac{z}{w}$ (b) $\overline{z+w}$ (c) $z+\overline{z}$

Solution

(a) $$\frac{z}{w}=\frac{z\cdot\overline{w}}{w\cdot\overline{w}}=\frac{(2-3i)(5-2i)}{(5+2i)(5-2i)}=\frac{10-4i-15i+6i^2}{25+4}$$
$$=\frac{4-19i}{29}=\frac{4}{29}-\frac{19}{29}i$$

(b) $\overline{z+w}=\overline{(2-3i)+(5+2i)}=\overline{7-i}=7+i$

(c) $z+\overline{z}=(2-3i)+(2+3i)=4$

The conjugate of a complex number has certain general properties that will be useful later.

For a real number $a=a+0i$, the conjugate is $\overline{a}=\overline{a+0i}=a-0i=a$.

THEOREM

The conjugate of a real number is the real number itself.

Other properties that are direct consequences of the definition of the conjugate are given next. In each statement, z and w represent complex numbers.

THEOREM

The conjugate of the conjugate of a complex number is the complex number itself.

$$\overline{(\overline{z})} = z \quad \textbf{(6)}$$

The conjugate of the sum of two complex numbers equals the sum of their conjugates.

$$\overline{z + w} = \overline{z} + \overline{w} \quad \textbf{(7)}$$

The conjugate of the product of two complex numbers equals the product of their conjugates.

$$\overline{z \cdot w} = \overline{z} \cdot \overline{w} \quad \textbf{(8)}$$

The proofs of equations (6), (7), and (8) are left as exercises. See Problems 94–96.

Powers of *i*

The powers of i follow a pattern that is useful to know.

$$\begin{aligned} i^1 &= i & i^5 &= i^4 \cdot i = 1 \cdot i = i \\ i^2 &= -1 & i^6 &= i^4 \cdot i^2 = -1 \\ i^3 &= i^2 \cdot i = -i & i^7 &= i^4 \cdot i^3 = -i \\ i^4 &= i^2 \cdot i^2 = (-1)(-1) = 1 & i^8 &= i^4 \cdot i^4 = 1 \end{aligned}$$

And so on. The powers of i repeat with every fourth power.

EXAMPLE 7 **Evaluating Powers of *i***

(a) $i^{27} = i^{24} \cdot i^3 = (i^4)^6 \cdot i^3 = 1^6 \cdot i^3 = -i$

(b) $i^{101} = i^{100} \cdot i^1 = (i^4)^{25} \cdot i = 1^{25} \cdot i = i$ ■

EXAMPLE 8 **Writing the Power of a Complex Number in Standard Form**

Write $(2 + i)^3$ in standard form.

Solution Use the special product formula for $(x + a)^3$.

$$(x + a)^3 = x^3 + 3ax^2 + 3a^2x + a^3$$

Using this special product formula,

$$\begin{aligned} (2 + i)^3 &= 2^3 + 3 \cdot i \cdot 2^2 + 3 \cdot i^2 \cdot 2 + i^3 \\ &= 8 + 12i + 6(-1) + (-i) \\ &= 2 + 11i \end{aligned}$$ ■

Note: Another way to find $(2 + i)^3$ is to multiply out $(2 + i)^2 (2 + i)$. ■

Now Work PROBLEMS 35 AND 43

2 Solve Quadratic Equations in the Complex Number System

Quadratic equations with a negative discriminant have no real number solution. However, if we extend our number system to allow complex numbers, quadratic equations will always have a solution. Since the solution to a quadratic equation involves the square root of the discriminant, we begin with a discussion of square roots of negative numbers.

DEFINITION

If N is a positive real number, we define the **principal square root of $-N$**, denoted by $\sqrt{-N}$, as

$$\sqrt{-N} = \sqrt{N}i$$

where i is the imaginary unit and $i^2 = -1$.

EXAMPLE 9 **Evaluating the Square Root of a Negative Number**

(a) $\sqrt{-1} = \sqrt{1}i = i$ (b) $\sqrt{-16} = \sqrt{16}i = 4i$

(c) $\sqrt{-8} = \sqrt{8}i = 2\sqrt{2}i$

Now Work PROBLEM 51

EXAMPLE 10 **Using the Square Root Method in the Complex Number System**

Solve each equation in the complex number system.

(a) $x^2 = 4$ (b) $x^2 = -9$

Solution (a) $x^2 = 4$

$$x = \pm\sqrt{4} = \pm 2$$

The equation has the solution set $\{-2, 2\}$.

(b) $x^2 = -9$

$$x = \pm\sqrt{-9} = \pm\sqrt{9}i = \pm 3i$$

The equation has the solution set $\{-3i, 3i\}$.

WARNING When working with square roots of negative numbers, do not set the square root of a product equal to the product of the square roots (which can be done with positive numbers). To see why, look at this calculation: We know that $\sqrt{100} = 10$. However, it is also true that $100 = (-25)(-4)$, so

$$\begin{aligned} 10 &= \sqrt{100} \\ &= \sqrt{(-25)(-4)} \\ &\neq \sqrt{-25}\sqrt{-4} \end{aligned}$$

because

$$\begin{aligned} &\sqrt{-25} \cdot \sqrt{-4} \\ &= (\sqrt{25}i)(\sqrt{4}i) \\ &= (5i)(2i) \\ &= 10i^2 = -10 \end{aligned}$$

Now Work PROBLEM 59

Because we have defined the square root of a negative number, we can now restate the quadratic formula without restriction.

THEOREM Quadratic Formula

In the complex number system, the solutions of the quadratic equation $ax^2 + bx + c = 0$, where a, b, and c are real numbers and $a \neq 0$, are given by the formula

$$x = \frac{-b \pm \sqrt{b^2 - 4ac}}{2a} \qquad \textbf{(9)}$$

EXAMPLE 11 **Solving a Quadratic Equation in the Complex Number System**

Solve the equation $x^2 - 4x + 8 = 0$ in the complex number system.

Solution Here $a = 1$, $b = -4$, $c = 8$, and $b^2 - 4ac = 16 - 4(1)(8) = -16$. Using equation (9),

$$x = \frac{-(-4) \pm \sqrt{-16}}{2(1)} = \frac{4 \pm \sqrt{16}i}{2} = \frac{4 \pm 4i}{2} = 2 \pm 2i$$

The equation has the solution set $\{2 - 2i, 2 + 2i\}$.

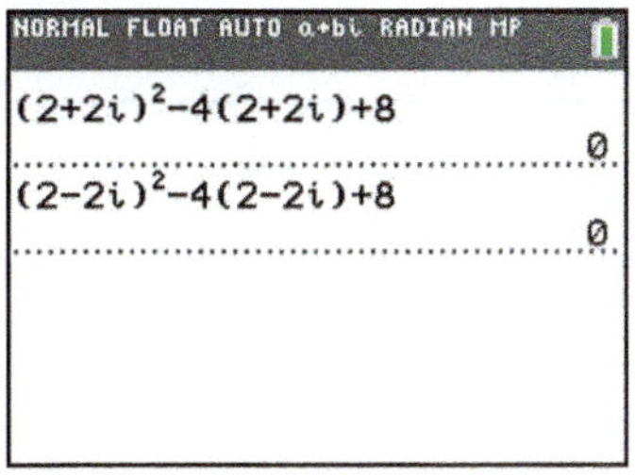

Figure 42

Check:

$$2 + 2i: \quad (2+2i)^2 - 4(2+2i) + 8 = 4 + \cancel{8i} + 4i^2 - \cancel{8} - \cancel{8i} + \cancel{8} = 4 - 4 = 0$$

$$2 - 2i: \quad (2-2i)^2 - 4(2-2i) + 8 = 4 - \cancel{8i} + 4i^2 - \cancel{8} + \cancel{8i} + \cancel{8} = 4 - 4 = 0$$

Figure 42 shows the check of the solution using a TI-84 Plus C graphing calculator. Graph $Y_1 = x^2 - 4x + 8$. How many x-intercepts are there?

Now Work PROBLEM 65

The discriminant $b^2 - 4ac$ of a quadratic equation still serves as a way to determine the character of the solutions.

Character of the Solutions of a Quadratic Equation

In the complex number system, consider a quadratic equation $ax^2 + bx + c = 0$ with real coefficients.

1. If $b^2 - 4ac > 0$, the equation has two unequal real solutions.
2. If $b^2 - 4ac = 0$, the equation has a repeated real solution, a double root.
3. If $b^2 - 4ac < 0$, the equation has two complex solutions that are not real. The solutions are conjugates of each other.

The third conclusion above is a consequence of the fact that if $b^2 - 4ac = -N < 0$, then by the quadratic formula, the solutions are

$$x = \frac{-b + \sqrt{b^2 - 4ac}}{2a} = \frac{-b + \sqrt{-N}}{2a} = \frac{-b + \sqrt{N}i}{2a} = \frac{-b}{2a} + \frac{\sqrt{N}}{2a}i$$

and

$$x = \frac{-b - \sqrt{b^2 - 4ac}}{2a} = \frac{-b - \sqrt{-N}}{2a} = \frac{-b - \sqrt{N}i}{2a} = \frac{-b}{2a} - \frac{\sqrt{N}}{2a}i$$

which are conjugates of each other.

EXAMPLE 12 Determining the Character of the Solutions of a Quadratic Equation

Without solving, determine the character of the solutions of each equation.

(a) $3x^2 + 4x + 5 = 0$ (b) $2x^2 + 4x + 1 = 0$

(c) $9x^2 - 6x + 1 = 0$

Solution

(a) Here $a = 3, b = 4,$ and $c = 5$, so $b^2 - 4ac = 4^2 - 4(3)(5) = -44$. The solutions are two complex numbers that are not real and are conjugates of each other.

(b) Here $a = 2, b = 4,$ and $c = 1$, so $b^2 - 4ac = 4^2 - 4(2)(1) = 8$. The solutions are two unequal real numbers.

(c) Here $a = 9, b = -6,$ and $c = 1$, so $b^2 - 4ac = (-6)^2 - 4(9)(1) = 0$. The solution is a repeated real number—that is, a double root.

Now Work PROBLEM 79

1.4 Assess Your Understanding

'Are You Prepared?' *Answers are given at the end of these exercises. If you get a wrong answer, read the pages listed in red.*

1. Name the integers and the rational numbers in the set $\left\{-3, 0, \sqrt{2}, \frac{6}{5}, \pi\right\}$. (pp. 4–5)

2. ***True or False*** Rational numbers and irrational numbers are in the set of real numbers. (pp. 4–5)

3. Rationalize the denominator of $\frac{3}{2+\sqrt{3}}$. (p. 76)

Concepts and Vocabulary

4. In the complex number $5 + 2i$, the number 5 is called the ______ part; the number 2 is called the __________ part; the number i is called the __________ ______.

5. ***True or False*** The conjugate of $2 + 5i$ is $-2 - 5i$.

6. ***True or False*** All real numbers are complex numbers.

7. ***True or False*** If $2 - 3i$ is a solution of a quadratic equation with real coefficients, then $-2 + 3i$ is also a solution.

8. Which of the following is the principal square root of -4?
(a) $-2i$ (b) $2i$ (c) -2 (d) 2

9. Which operation involving complex numbers requires the use of a conjugate?
(a) division (b) multiplication
(c) subtraction (d) addition

10. Powers of i repeat every ______ power.
(a) second (b) third (c) fourth (d) fifth

Skill Building

In Problems 11–48, write each expression in the standard form a + bi. Verify your results using a graphing utility.

11. $(2 - 3i) + (6 + 8i)$ **12.** $(4 + 5i) + (-8 + 2i)$ **13.** $(-3 + 2i) - (4 - 4i)$ **14.** $(3 - 4i) - (-3 - 4i)$

15. $(2 - 5i) - (8 + 6i)$ **16.** $(-8 + 4i) - (2 - 2i)$ **17.** $3(2 - 6i)$ **18.** $-4(2 + 8i)$

19. $2i(2 - 3i)$ **20.** $3i(-3 + 4i)$ **21.** $(3 - 4i)(2 + i)$ **22.** $(5 + 3i)(2 - i)$

23. $(-6 + i)(-6 - i)$ **24.** $(-3 + i)(3 + i)$ **25.** $\frac{10}{3 - 4i}$ **26.** $\frac{13}{5 - 12i}$

27. $\frac{2 + i}{i}$ **28.** $\frac{2 - i}{-2i}$ **29.** $\frac{6 - i}{1 + i}$ **30.** $\frac{2 + 3i}{1 - i}$

31. $\left(\frac{1}{2} + \frac{\sqrt{3}}{2}i\right)^2$ **32.** $\left(\frac{\sqrt{3}}{2} - \frac{1}{2}i\right)^2$ **33.** $(1 + i)^2$ **34.** $(1 - i)^2$

35. i^{23} **36.** i^{14} **37.** i^{-15} **38.** i^{-23}

39. $i^6 - 5$ **40.** $4 + i^3$ **41.** $6i^3 - 4i^5$ **42.** $4i^3 - 2i^2 + 1$

43. $(1 + i)^3$ **44.** $(3i)^4 + 1$ **45.** $i^7(1 + i^2)$ **46.** $2i^4(1 + i^2)$

47. $i^6 + i^4 + i^2 + 1$ **48.** $i^7 + i^5 + i^3 + i$

In Problems 49–58, perform the indicated operations and express your answer in the form a + bi.

49. $\sqrt{-4}$ **50.** $\sqrt{-9}$ **51.** $\sqrt{-25}$ **52.** $\sqrt{-64}$

53. $\sqrt{-12}$ **54.** $\sqrt{-18}$ **55.** $\sqrt{-200}$ **56.** $\sqrt{-45}$

57. $\sqrt{(3 + 4i)(4i - 3)}$ **58.** $\sqrt{(4 + 3i)(3i - 4)}$

In Problems 59–78, solve each equation in the complex number system. Check your results using a graphing utility.

59. $x^2 + 4 = 0$ **60.** $x^2 - 4 = 0$ **61.** $x^2 - 16 = 0$ **62.** $x^2 + 25 = 0$

63. $x^2 - 6x + 13 = 0$ **64.** $x^2 + 4x + 8 = 0$ **65.** $x^2 - 6x + 10 = 0$ **66.** $x^2 - 2x + 5 = 0$

67. $8x^2 - 4x + 1 = 0$ **68.** $10x^2 + 6x + 1 = 0$ **69.** $5x^2 + 1 = 2x$ **70.** $13x^2 + 1 = 6x$

71. $x^2 + x + 1 = 0$ **72.** $x^2 - x + 1 = 0$ **73.** $x^3 - 8 = 0$ **74.** $x^3 + 27 = 0$

75. $x^4 = 16$ **76.** $x^4 = 1$ **77.** $x^4 + 13x^2 + 36 = 0$ **78.** $x^4 + 3x^2 - 4 = 0$

In Problems 79–84, without solving, determine the character of the solutions of each equation in the complex number system. Verify your answer using a graphing utility.

79. $3x^2 - 3x + 4 = 0$ **80.** $2x^2 - 4x + 1 = 0$ **81.** $2x^2 + 3x = 4$

82. $x^2 + 6 = 2x$

83. $9x^2 - 12x + 4 = 0$

84. $4x^2 + 12x + 9 = 0$

85. $2 + 3i$ is a solution of a quadratic equation with real coefficients. Find the other solution.

86. $4 - i$ is a solution of a quadratic equation with real coefficients. Find the other solution.

In Problems 87–90, $z = 3 - 4i$ and $w = 8 + 3i$. Write each expression in the standard form $a + bi$.

87. $z + \bar{z}$

88. $w - \bar{w}$

89. $z\bar{z}$

90. $\overline{z - w}$

Applications and Extensions

91. Electrical Circuits The impedance Z, in ohms, of a circuit element is defined as the ratio of the phasor voltage V, in volts, across the element to the phasor current I, in amperes, through the element. That is, $Z = \frac{V}{I}$. If the voltage across a circuit element is $18 + i$ volts and the current through the element is $3 - 4i$ amperes, determine the impedance.

92. Parallel Circuits In an ac circuit with two parallel pathways, the total impedance Z, in ohms, satisfies the formula $\frac{1}{Z} = \frac{1}{Z_1} + \frac{1}{Z_2}$, where Z_1 is the impedance of the first pathway and Z_2 is the impedance of the second pathway. Determine the total impedance if the impedances of the two pathways are $Z_1 = 2 + i$ ohms and $Z_2 = 4 - 3i$ ohms.

93. Use $z = a + bi$ to show that $z + \bar{z} = 2a$ and $z - \bar{z} = 2bi$.

94. Use $z = a + bi$ to show that $\bar{\bar{z}} = z$.

95. Use $z = a + bi$ and $w = c + di$ to show that $\overline{z + w} = \bar{z} + \bar{w}$.

96. Use $z = a + bi$ and $w = c + di$ to show that $\overline{z \cdot w} = \bar{z} \cdot \bar{w}$.

Explaining Concepts: Discussion and Writing

97. Explain to a friend how you would add two complex numbers and how you would multiply two complex numbers. Explain any differences between the two explanations.

98. Write a brief paragraph that compares the method used to rationalize denominators and the method used to write the quotient of two complex numbers in standard form.

99. Use an Internet search engine to investigate the origins of complex numbers. Write a paragraph describing what you find, and present it to the class.

100. Explain how the method of multiplying two complex numbers is related to multiplying two binomials.

101. What Went Wrong? A student multiplied $\sqrt{-9}$ and $\sqrt{-9}$ as follows:

$$\begin{aligned}\sqrt{-9} \cdot \sqrt{-9} &= \sqrt{(-9)(-9)} \\ &= \sqrt{81} \\ &= 9\end{aligned}$$

The instructor marked the problem incorrect. Why?

'Are You Prepared?' Answers

1. Integers: $\{-3, 0\}$; rational numbers: $\left\{-3, 0, \frac{6}{5}\right\}$

2. True

3. $3(2 - \sqrt{3})$

1.5 Radical Equations; Equations Quadratic in Form; Absolute Value Equations; Factorable Equations

PREPARING FOR THIS SECTION *Before getting started, review the following:*

- Absolute Value (Section R.2, p. 20)
- Square Roots (Section R.2, pp. 24–25)
- Factoring Polynomials (Section R.5, pp. 50–56)
- *n*th Roots; Rational Exponents (Section R.8, pp. 74–78)

Now Work the **'Are You Prepared?'** problems on page 135.

OBJECTIVES
1 Solve Radical Equations (p. 129)
2 Solve Equations Quadratic in Form (p. 131)
3 Solve Absolute Value Equations (p. 133)
4 Solve Equations by Factoring (p. 134)

1 Solve Radical Equations

When the variable in an equation occurs in a square root, cube root, and so on—that is, when it occurs in a radical—the equation is called a **radical equation**. Sometimes a suitable operation will change a radical equation to one that is linear or quadratic. A commonly used procedure is to isolate the most complicated radical on one side of the equation and then eliminate it by raising each side to a power equal to the index of the radical. Care must be taken, however, because apparent solutions that are not, in fact, solutions of the original equation may result. Recall that these are called **extraneous solutions**. In radical equations, extraneous solutions may occur when the index of the radical is even. Therefore, we need to check all answers when working with radical equations, and we check them in the *original* equation.

EXAMPLE 1 Solving a Radical Equation

Find the real solutions of the equation: $\sqrt[3]{2x-4} - 2 = 0$

Algebraic Solution

The equation contains a radical whose index is 3. Isolate it on the left side:

$$\sqrt[3]{2x-4} - 2 = 0$$

$$\sqrt[3]{2x-4} = 2 \quad \text{Add 2 to both sides.}$$

Because the index of the radical is 3, raise each side to the third power and solve.

$$(\sqrt[3]{2x-4})^3 = 2^3 \quad \text{Raise each side to the power 3.}$$

$$2x - 4 = 8 \quad \text{Simplify.}$$

$$2x = 12 \quad \text{Add 4 to both sides.}$$

$$x = 6 \quad \text{Divide both sides by 2.}$$

Graphing Solution

Figure 43 shows the graph of the equation $Y_1 = \sqrt[3]{2x-4} - 2$. From the graph, there is one x-intercept near 6. Using ZERO (or ROOT), find that the x-intercept is 6. The only solution is $x = 6$.

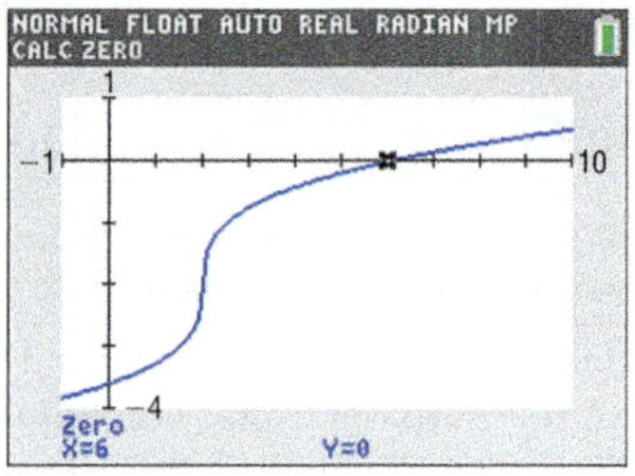

Figure 43 $Y_1 = \sqrt[3]{2x-4} - 2$

✓**Check:** $\sqrt[3]{2(6)-4} - 2 = \sqrt[3]{12-4} - 2 = \sqrt[3]{8} - 2 = 2 - 2 = 0$

The solution set is $\{6\}$.

Now Work PROBLEM 13

EXAMPLE 2 Solving a Radical Equation

Find the real solutions of the equation: $\sqrt{x-1} = x - 7$

Algebraic Solution

Square both sides since the index of a square root is 2.

$$\sqrt{x-1} = x - 7$$

$$(\sqrt{x-1})^2 = (x-7)^2 \quad \text{Square both sides.}$$

$$x - 1 = x^2 - 14x + 49 \quad \text{Remove parentheses.}$$

$$x^2 - 15x + 50 = 0 \quad \text{Put in standard form.}$$

$$(x-10)(x-5) = 0 \quad \text{Factor.}$$

$$x = 10 \quad \text{or} \quad x = 5 \quad \text{Apply the Zero-Product Property and solve.}$$

Graphing Solution

Graph $Y_1 = \sqrt{x-1}$ and $Y_2 = x - 7$. See Figure 44. From the graph, there is one point of intersection. Using INTERSECT, the point of intersection is $(10, 3)$, so the solution is $x = 10$.

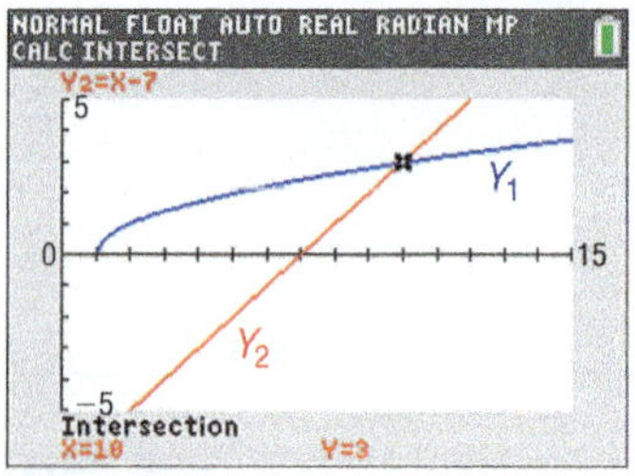

Figure 44 $Y_1 = \sqrt{x-1};\ Y_2 = x - 7$

There is a discrepancy between the algebraic solution and graphing solution. Let's check the results of our algebraic solution.

Check: $x = 10$: $\sqrt{x-1} = \sqrt{10-1} = \sqrt{9} = 3$ and $x - 7 = 10 - 7 = 3$

$x = 5$: $\sqrt{x-1} = \sqrt{5-1} = \sqrt{4} = 2$ and $x - 7 = 5 - 7 = -2$

The apparent algebraic solution $x = 5$ is extraneous; the only solution of the equation is $x = 10$. The solution set is $\{10\}$.

Now Work PROBLEM 25

Sometimes, it is necessary to raise each side to a power more than once in order to solve a radical equation algebraically.

EXAMPLE 3 Solving a Radical Equation

Find the real solutions of the equation: $\sqrt{2x+3} - \sqrt{x+2} = 2$

Algebraic Solution

First, isolate the more complicated radical expression (in this case, $\sqrt{2x+3}$) on the left side:

$$\sqrt{2x+3} = \sqrt{x+2} + 2$$

Now square both sides (the index of the radical is 2).

$$(\sqrt{2x+3})^2 = (\sqrt{x+2} + 2)^2 \quad \text{Square both sides.}$$

$$2x + 3 = (\sqrt{x+2})^2 + 4\sqrt{x+2} + 4 \quad \text{Remove parentheses.}$$

$$2x + 3 = x + 2 + 4\sqrt{x+2} + 4 \quad \text{Simplify.}$$

$$2x + 3 = x + 6 + 4\sqrt{x+2} \quad \text{Combine like terms.}$$

Because the equation still contains a radical, isolate the remaining radical on the right side and again square both sides.

$$x - 3 = 4\sqrt{x+2} \quad \text{Isolate the radical on the right side.}$$

$$(x-3)^2 = 16(x+2) \quad \text{Square both sides.}$$

$$x^2 - 6x + 9 = 16x + 32 \quad \text{Remove parentheses.}$$

$$x^2 - 22x - 23 = 0 \quad \text{Put in standard form.}$$

$$(x-23)(x+1) = 0 \quad \text{Factor.}$$

$$x = 23 \quad \text{or} \quad x = -1 \quad \text{Apply the Zero-Product Property and solve.}$$

Graphing Solution

Graph $Y_1 = \sqrt{2x+3} - \sqrt{x+2}$ and $Y_2 = 2$. See Figure 45. From the graph there is one point of intersection. Using INTERSECT, the point of intersection is $(23, 2)$, so the solution is $x = 23$.

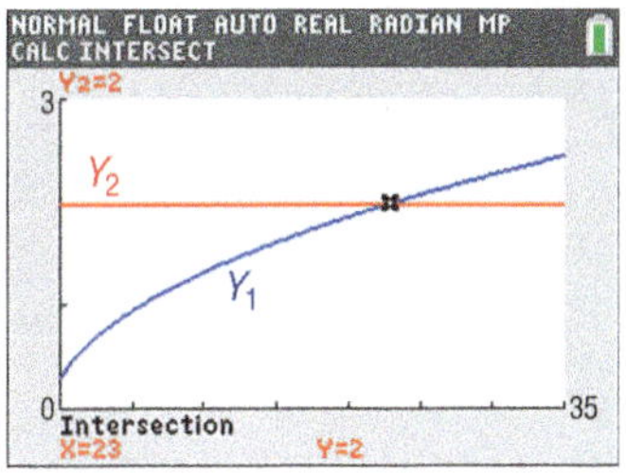

Figure 45 $Y_1 = \sqrt{2x+3} - \sqrt{x+2}$; $Y_2 = 2$

Check: $x = 23$: $\sqrt{2(23)+3} - \sqrt{23+2} = \sqrt{49} - \sqrt{25} = 7 - 5 = 2$

$x = -1$: $\sqrt{2(-1)+3} - \sqrt{-1+2} = \sqrt{1} - \sqrt{1} = 1 - 1 = 0$

The apparent solution $x = -1$ is extraneous; the only solution is $x = 23$. The solution set of the equation is $\{23\}$.

Now Work PROBLEM 35

2 Solve Equations Quadratic in Form

The equation $x^4 + x^2 - 12 = 0$ is not quadratic in x, but it is quadratic in x^2. That is, if we let $u = x^2$, we get $u^2 + u - 12 = 0$, a quadratic equation. This equation can be solved for u and, in turn, by using $u = x^2$, we can find the solutions x of the original equation.

In general, if an appropriate substitution u transforms an equation into one of the form

$$au^2 + bu + c = 0, \qquad a \neq 0$$

then the original equation is called an **equation of the quadratic type** or an **equation quadratic in form**.

The difficulty of solving such an equation lies in the determination that the equation is, in fact, quadratic in form. After you are told an equation is quadratic in form, it is easy enough to see it, but some practice is needed to enable you to recognize them on your own.

EXAMPLE 4 Solving an Equation That Is Quadratic in Form

Find the real solutions of the equation: $(x+2)^2 + 11(x+2) - 12 = 0$

Solution For this equation, let $u = x + 2$. Then $u^2 = (x+2)^2$, and the original equation,

$$(x+2)^2 + 11(x+2) - 12 = 0$$

becomes

WARNING Do not stop after finding values for u. Remember to finish solving for the original variable. ■

$$u^2 + 11u - 12 = 0 \qquad \text{Let } u = (x+2).$$
$$(u+12)(u-1) = 0 \qquad \text{Factor.}$$
$$u = -12 \quad \text{or} \quad u = 1 \qquad \text{Solve.}$$

But we want to solve for x. Because $u = x + 2$, we have

$$x + 2 = -12 \quad \text{or} \quad x + 2 = 1$$
$$x = -14 \qquad\qquad x = -1$$

✓**Check:** $x = -14$: $(-14+2)^2 + 11(-14+2) - 12$
$= (-12)^2 + 11(-12) - 12 = 144 - 132 - 12 = 0$

$x = -1$: $(-1+2)^2 + 11(-1+2) - 12 = 1 + 11 - 12 = 0$

The original equation has the solution set $\{-14, -1\}$. ■

✓**Check:** Verify the solution of Example 4 using a graphing utility.

EXAMPLE 5 Solving an Equation That Is Quadratic in Form

Find the real solutions of the equation: $x + 2\sqrt{x} - 3 = 0$

Solution For the equation $x + 2\sqrt{x} - 3 = 0$, let $u = \sqrt{x}$. Then $u^2 = x$, and the original equation,

$$x + 2\sqrt{x} - 3 = 0$$

becomes

$$u^2 + 2u - 3 = 0 \qquad \text{Let } u = \sqrt{x}. \text{ Then } u^2 = x.$$
$$(u+3)(u-1) = 0 \qquad \text{Factor.}$$
$$u = -3 \quad \text{or} \quad u = 1 \qquad \text{Solve.}$$

Since $u = \sqrt{x}$, we have $\sqrt{x} = -3$ or $\sqrt{x} = 1$. The first of these, $\sqrt{x} = -3$, has no real solution, since the square root of a real number is never negative. The second, $\sqrt{x} = 1$, has the solution $x = 1$.

✓**Check:** $1 + 2\sqrt{1} - 3 = 1 + 2 - 3 = 0$

The solution set of the original equation is $\{1\}$. ■

✓**Check:** Verify the solution to Example 5 using a graphing utility.

Another method for solving Example 5 would be to treat it as a radical equation. Solve it this way for practice.

The idea should now be clear. If an equation contains an expression and that same expression squared, make a substitution for the expression. You may get a quadratic equation.

Now Work PROBLEM 55

3 Solve Absolute Value Equations

Recall that on the real number line, the absolute value of a equals the distance from the origin to the point whose coordinate is a. For example, there are two points whose distance from the origin is 5 units, -5 and 5. Thus the equation $|x| = 5$ will have the solution set $\{-5, 5\}$.

Another way to obtain this result is to use the algebraic definition of absolute value, $|a| = a$ if $a \geq 0$, $|a| = -a$ if $a < 0$. The equation $|u| = a$ leads to two equations, depending on whether u is nonnegative (greater than or equal to zero) or negative.

$$|u| = a$$

If $u < 0$

$|u| = a$

$-u = a$ **$|u| = -u$ when $u < 0$**

$u = -a$ **Multiply both sides by -1.**

If $u \geq 0$

$|u| = a$

$u = a$ **$|u| = u$ when $u \geq 0$**

So we have the following result.

THEOREM

If a is a positive real number and if u is any algebraic expression, then

$$|u| = a \quad \text{is equivalent to} \quad u = a \quad \text{or} \quad u = -a \qquad \textbf{(1)}$$

EXAMPLE 6 Solving an Equation Involving Absolute Value

Solve the equation: $|2x - 3| + 2 = 7$

Algebraic Solution

$$|2x - 3| + 2 = 7$$

$|2x - 3| = 5$ **Subtract 2 from each side.**

$2x - 3 = 5$ or $2x - 3 = -5$ **Apply (1).**

$2x = 8$ or $2x = -2$ **Add 3 to both sides.**

$x = 4$ or $x = -1$ **Divide both sides by 2.**

The solution set is $\{-1, 4\}$.

Graphing Solution

For this equation, graph $Y_1 = |2x - 3| + 2$ and $Y_2 = 7$ on the same screen and find their point(s) of intersection, if any. See Figure 46. Using the INTERSECT command (twice), find the points of intersection to be $(-1, 7)$ and $(4, 7)$. The solution set is $\{-1, 4\}$.

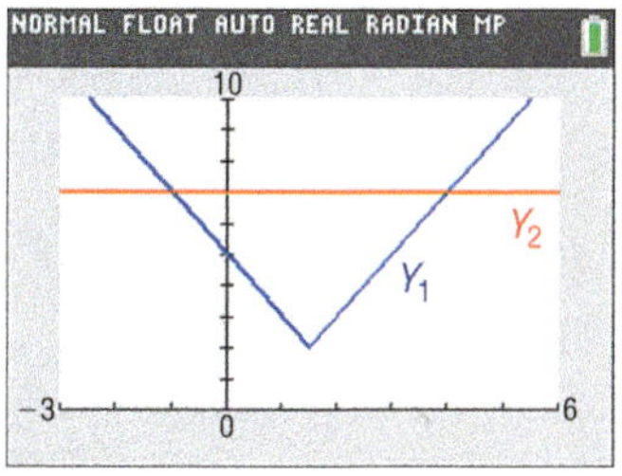

Figure 46 $Y_1 = |2x - 3| + 2$; $Y_2 = 7$

Now Work PROBLEM 71

4 Solve Equations by Factoring

We have already solved certain quadratic equations using factoring. Let's look at examples of other kinds of equations that can be solved by factoring.

EXAMPLE 7 Solving Equations by Factoring

Solve the equation: $x^4 = 4x^2$

Algebraic Solution

Begin by collecting all terms on one side. This results in 0 on one side and an expression to be factored on the other.

$$x^4 = 4x^2$$
$$x^4 - 4x^2 = 0$$
$$x^2(x^2 - 4) = 0 \quad \text{Factor.}$$
$$x^2 = 0 \quad \text{or} \quad x^2 - 4 = 0 \quad \text{Apply the Zero-Product Property.}$$
$$x^2 = 4$$
$$x = 0 \quad \text{or} \quad x = -2 \quad \text{or} \quad x = 2 \quad \text{Use the Square Root Method.}$$

Graphing Solution

Graph $Y_1 = x^4$ and $Y_2 = 4x^2$ on the same screen and find their point(s) of intersection, if any. See Figure 47. Using the INTERSECT command (three times), find the points of intersection to be $(-2, 16)$, $(0, 0)$, and $(2, 16)$. The solutions are $x = -2$, $x = 0$, and $x = 2$.

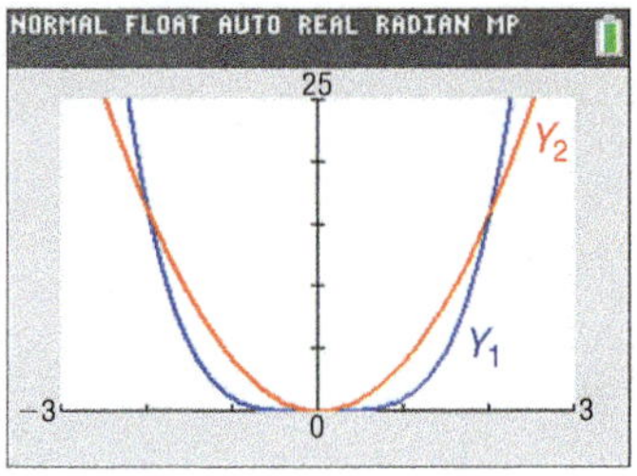

Figure 47 $Y_1 = x^4$; $Y_2 = 4x^2$

Check: $x = -2$: $(-2)^4 = 16$ and $4(-2)^2 = 16$ **−2 is a solution.**
$x = 0$: $0^4 = 0$ and $4 \cdot 0^2 = 0$ **0 is a solution.**
$x = 2$: $2^4 = 16$ and $4 \cdot 2^2 = 16$ **2 is a solution.**

The solution set is $\{-2, 0, 2\}$.

EXAMPLE 8 Solving Equations by Factoring

Solve the equation: $x^3 - x^2 - 4x + 4 = 0$

Algebraic Solution

Do you recall the method of factoring by grouping? (If not, review pp. 54–55.) Group the terms of $x^3 - x^2 - 4x + 4 = 0$ as follows:

$$(x^3 - x^2) - (4x - 4) = 0$$

Factor out x^2 from the first grouping and 4 from the second.

$$x^2(x - 1) - 4(x - 1) = 0$$

This reveals the common factor $(x - 1)$, so we have

$$(x^2 - 4)(x - 1) = 0$$
$$(x - 2)(x + 2)(x - 1) = 0 \quad \text{Factor again.}$$
$$x - 2 = 0 \quad \text{or} \quad x + 2 = 0 \quad \text{or} \quad x - 1 = 0 \quad \text{Set each factor equal to 0.}$$
$$x = 2 \qquad x = -2 \qquad x = 1 \quad \text{Solve.}$$

Graphing Solution

Graph $Y_1 = x^3 - x^2 - 4x + 4$. See Figure 48. Using ZERO (three times), the values of x for which $y = 0$ are -2, 1, and 2.

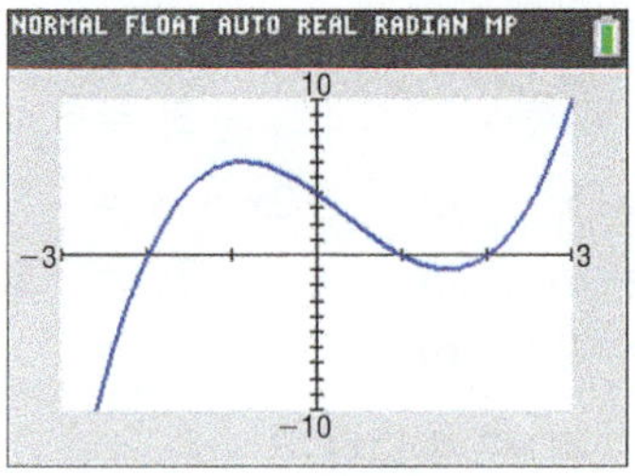

Figure 48 $Y_1 = x^3 - x^2 - 4x + 4$

Check:
$x = -2$: $(-2)^3 - (-2)^2 - 4(-2) + 4 = -8 - 4 + 8 + 4 = 0$ **−2 is a solution.**
$x = 1$: $1^3 - 1^2 - 4(1) + 4 = 1 - 1 - 4 + 4 = 0$ **1 is a solution.**
$x = 2$: $2^3 - 2^2 - 4(2) + 4 = 8 - 4 - 8 + 4 = 0$ **2 is a solution.**

The solution set is $\{-2, 1, 2\}$.

Now Work PROBLEM 93

1.5 Assess Your Understanding

'Are You Prepared?' *Answers are given at the end of these exercises. If you get a wrong answer, read the pages listed in red.*

1. ***True or False*** The principal square root of any nonnegative real number is always nonnegative. (p. 24)
2. $(\sqrt[3]{x})^3 =$ _____. (p. 75)
3. Factor $2x^2 - 7x - 4$. (pp. 55–56)
4. Factor $x^3 + 4x^2 - 9x - 36$. (pp. 54–55)
5. Use a real number line to describe why $|-4| = 4$. (p. 20)

Concepts and Vocabulary

6. ***True or False*** Factoring can be used to solve only quadratic equations or equations that are quadratic in form.
7. If u is an expression involving x, the equation $au^2 + bu + c = 0$, $a \neq 0$, is called an equation _______________.
8. ***True or False*** Radical equations sometimes have no real solution.
9. An apparent solution that does not satisfy the original equation is called a(n) _______ solution.
   (a) extraneous (b) imaginary
   (c) radical (d) conditional
10. Solving which equation is likely to require squaring each side more than once?
   (a) $\sqrt{x+2} = \sqrt{3x-5}$ (b) $x^4 - 3x^2 = 10$
   (c) $\sqrt{x+1} + \sqrt{x-4} = 8$ (d) $\sqrt{3x+1} = 5$

Skill Building

In Problems 11–44, find the real solutions of each equation. Verify your results using a graphing utility.

11. $\sqrt{y+3} = 5$
12. $\sqrt{t-3} = 7$
13. $\sqrt{2t-1} = 1$
14. $\sqrt{3t+4} = 2$
15. $\sqrt{3t+4} = -6$
16. $\sqrt{5t+3} = -2$
17. $\sqrt[3]{1-2x} - 3 = 0$
18. $\sqrt[3]{1-2x} - 1 = 0$
19. $\sqrt[4]{5x-4} = 2$
20. $\sqrt[5]{2x-3} = -1$
21. $\sqrt[5]{x^2+2x} = -1$
22. $\sqrt[4]{x^2+16} = \sqrt{5}$
23. $x = 8\sqrt{x}$
24. $x = 3\sqrt{x}$
25. $\sqrt{15-2x} = x$
26. $\sqrt{12-x} = x$
27. $\sqrt{3(x+10)} - 4 = x$
28. $\sqrt{1-x} - 3 = x + 2$
29. $\sqrt{x^2-x-4} = x + 2$
30. $\sqrt{x^2-x-8} = x + 5$
31. $3 + \sqrt{3x+1} = x$
32. $2 + \sqrt{12-2x} = x$
33. $\sqrt{2x+3} - \sqrt{x+1} = 1$
34. $\sqrt{3x+7} + \sqrt{x+2} = 1$
35. $\sqrt{3x+1} - \sqrt{x-1} = 2$
36. $\sqrt{3x-5} - \sqrt{x+7} = 2$
37. $\sqrt{3-2\sqrt{x}} = \sqrt{x}$
38. $\sqrt{10+3\sqrt{x}} = \sqrt{x}$
39. $(3x+1)^{1/2} = 4$
40. $(3x-5)^{1/2} = 2$
41. $(5x-2)^{1/3} = 2$
42. $(2x+1)^{1/3} = -1$
43. $(x^2+9)^{1/2} = 5$
44. $(x^2-16)^{1/2} = 9$

In Problems 45–70, find the real solutions of each equation. Verify your results using a graphing utility.

45. $t^4 - 16 = 0$
46. $y^4 - 4 = 0$
47. $x^4 - 5x^2 + 4 = 0$
48. $x^4 - 10x^2 + 24 = 0$
49. $3x^4 - 2x^2 - 1 = 0$
50. $2x^4 - 5x^2 - 12 = 0$
51. $x^6 + 7x^3 - 8 = 0$
52. $x^6 - 7x^3 - 8 = 0$
53. $(x+2)^2 + 7(x+2) + 12 = 0$
54. $(2x+5)^2 - (2x+5) - 6 = 0$
55. $2(s+1)^2 - 5(s+1) = 3$
56. $3(1-y)^2 + 5(1-y) + 2 = 0$
57. $x - 4\sqrt{x} = 0$
58. $x - 8\sqrt{x} = 0$
59. $x + \sqrt{x} = 20$
60. $x + \sqrt{x} = 6$
61. $t^{1/2} - 2t^{1/4} + 1 = 0$
62. $z^{1/2} - 4z^{1/4} + 4 = 0$
63. $4x^{1/2} - 9x^{1/4} + 4 = 0$
64. $x^{1/2} - 3x^{1/4} + 2 = 0$
65. $\dfrac{1}{(x+1)^2} = \dfrac{1}{x+1} + 2$
66. $\dfrac{1}{(x-1)^2} + \dfrac{1}{x-1} = 12$
67. $3x^{-2} - 7x^{-1} - 6 = 0$
68. $2x^{-2} - 3x^{-1} - 4 = 0$
69. $2x^{2/3} - 5x^{1/3} - 3 = 0$
70. $3x^{4/3} + 5x^{2/3} - 2 = 0$

In Problems 71–88, solve each equation. Verify your results using a graphing utility.

71. $|2x+3| = 5$
72. $|3x-1| = 2$
73. $|1-4t| + 8 = 13$

74. $|1 - 2z| + 6 = 9$
75. $|-2x| = 8$
76. $|-x| = 1$
77. $4 - |2x| = 3$
78. $5 - \left|\frac{1}{2}x\right| = 3$
79. $\frac{2}{3}|x| = 9$
80. $\frac{3}{4}|x| = 9$
81. $\left|\frac{x}{3} + \frac{2}{5}\right| = 2$
82. $\left|\frac{x}{2} - \frac{1}{3}\right| = 1$
83. $|u - 2| = -\frac{1}{2}$
84. $|2 - v| = -1$
85. $|x^2 - 9| = 0$
86. $|x^2 - 16| = 0$
87. $|x^2 - 2x| = 3$
88. $|x^2 + x| = 12$

In Problems 89–98, find the real solutions of each equation by factoring. Verify your results using a graphing utility.

89. $x^3 - 9x = 0$
90. $x^4 - 81x^2 = 0$
91. $x^3 + x^2 - 20x = 0$
92. $x^3 + 6x^2 - 7x = 0$
93. $x^3 + x^2 - x - 1 = 0$
94. $x^3 + 4x^2 - x - 4 = 0$
95. $x^3 - 3x^2 - 4x + 12 = 0$
96. $x^3 - 3x^2 - x + 3 = 0$
97. $2x^3 + 4 = x^2 + 8x$
98. $3x^3 + 4x^2 = 27x + 36$

In Problems 99–102, find the real solutions of each equation. Use a calculator to express any solutions rounded to two decimal places.

99. $x - 4x^{1/2} + 2 = 0$
100. $x^{2/3} + 4x^{1/3} + 2 = 0$
101. $x^4 + \sqrt{3}x^2 - 3 = 0$
102. $x^4 + \sqrt{2}x^2 - 2 = 0$

Mixed Practice

In Problems 103–122, find the real solutions of each equation. Verify your results using a graphing utility.

103. $3x^2 + 7x - 20 = 0$
104. $2x^2 - 13x + 21 = 0$
105. $5a^3 - 45a = -2a^2 + 18$
106. $3z^3 - 12z = -5z^2 + 20$
107. $-3|5x - 2| + 9 = 0$
108. $\frac{1}{4}|2x - 3| = \frac{3}{2}$
109. $4(w - 3) = w + 3$
110. $6(k + 3) - 2k = 12$
111. $\left(\frac{v}{v+1}\right)^2 + \frac{2v}{v+1} = 8$
112. $\left(\frac{y}{y-1}\right)^2 = \frac{6y}{y-1} + 7$
113. $|-3x + 2| = x + 10$
114. $|4x - 3| = x + 2$
115. $\sqrt{2x + 5} - x = 1$
116. $\sqrt{3x + 1} - 2x = -6$
117. $3m^2 + 6m = -1$
118. $4y^2 - 8y = 3$
119. $|x^2 + x - 1| = 1$
120. $|x^2 + 3x - 2| = 2$
121. $\sqrt[4]{5x^2 - 6} = x$
122. $\sqrt[4]{4 - 3x^2} = x$

In Problems 123–126, find all complex solutions of each equation.

123. $t^4 - 16 = 0$
124. $y^4 - 81 = 0$
125. $x^6 - 9x^3 + 8 = 0$
126. $z^6 + 28z^3 + 27 = 0$

Applications and Extensions

127. If $k = \frac{x+3}{x-3}$ and $k^2 - k = 12$, find x.

128. If $k = \frac{x+3}{x-4}$ and $k^2 - 3k = 28$, find x.

129. Find all points having an x-coordinate of 2 whose distance from the point $(-2, -1)$ is 5.

130. Find all points having a y-coordinate of -3 whose distance from the point $(1, 2)$ is 13.

131. Find all points on the x-axis that are 5 units from the point $(4, -3)$.

132. Find all points on the y-axis that are 5 units from the point $(4, 4)$.

133. Solve: $|8 - 3x| = |2x - 7|$

134. Solve: $|5x + 3| = |12 - 4x|$

135. **Physics: Using Sound to Measure Distance** The distance to the surface of the water in a well can sometimes be found by dropping an object into the well and measuring the time elapsed until a sound is heard. If t_1 is the time (measured in seconds) that it takes for the object to strike the water, then t_1 will obey the equation $s = 16t_1^2$, where s is the distance (measured in feet). It follows that $t_1 = \frac{\sqrt{s}}{4}$. Suppose that t_2 is the time that it takes for the sound of the impact to reach your ears. Because sound travels at a speed of approximately 1100 feet per second, the time t_2 for the sound to travel the distance s will be $t_2 = \frac{s}{1100}$. See the illustration. Now $t_1 + t_2$ is the total time that elapses from the moment that the object is dropped to the moment that a sound is heard. We have the equation

$$\text{Total time elapsed} = \frac{\sqrt{s}}{4} + \frac{s}{1100}$$

Find the distance to the water's surface if the total time elapsed from dropping a rock to hearing it hit water is 4 seconds.

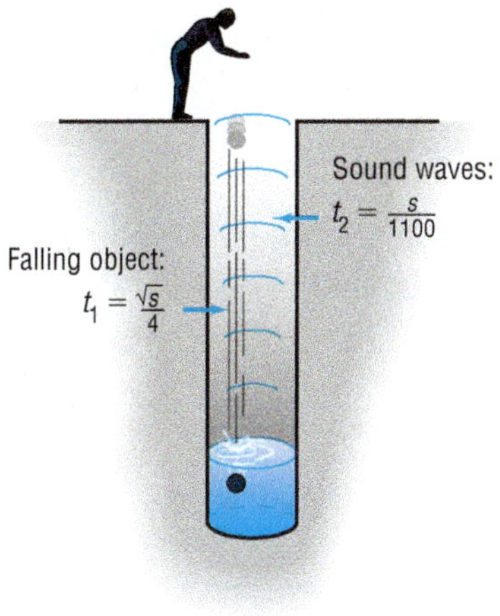

136. Crushing Load A civil engineer relates the thickness T, in inches, and height H, in feet, of a square wooden pillar to its crushing load L, in tons, using the model $T = \sqrt[4]{\frac{LH^2}{25}}$. If a square wooden pillar is 4 inches thick and 10 feet high, what is its crushing load?

137. Foucault's Pendulum The period of a pendulum is the time it takes the pendulum to make one full swing back and forth. The period T, in seconds, is given by the formula $T = 2\pi\sqrt{\frac{l}{32}}$, where l is the length, in feet, of the pendulum. In 1851, Jean-Bernard-Léon Foucault demonstrated the axial rotation of Earth using a large pendulum that he hung in the Panthéon in Paris. The period of Foucault's pendulum was approximately 16.5 seconds. What was its length?

Explaining Concepts: Discussion and Writing

138. Make up a radical equation that has no solution.

139. Make up a radical equation that has an extraneous solution.

140. Discuss the step in the solving process for radical equations that leads to the possibility of extraneous solutions. Why is there no such possibility for linear and quadratic equations?

141. The equation $|x| = -2$ has no real solution. Why?

142. What Went Wrong? On an exam, Jane solved the equation $\sqrt{2x+3} - x = 0$ and wrote that the solution set was $\{-1, 3\}$. Jane received 3 out of 5 points for the problem. Jane asks you why she received 3 out of 5 points. Provide an explanation.

'Are You Prepared?' Answers

1. True **2.** x **3.** $(2x + 1)(x - 4)$ **4.** $(x - 3)(x + 3)(x + 4)$

5. The distance from the origin to -4 on a real number line is 4 units.

1.6 Problem Solving: Interest, Mixture, Uniform Motion, Constant Rate Jobs

OBJECTIVES
1 Translate Verbal Descriptions into Mathematical Expressions (p. 138)
2 Solve Interest Problems (p. 139)
3 Solve Mixture Problems (p. 140)
4 Solve Uniform Motion Problems (p. 141)
5 Solve Constant Rate Job Problems (p. 143)

Applied (word) problems do not come in the form "Solve the equation. . . ." Instead, they supply information using words, a verbal description of the real problem. So, to solve applied problems, we must be able to translate the verbal description into the language of mathematics. This can be done by using variables to represent unknown quantities and then finding relationships (such as equations) that involve these variables. The process of doing all this is called **mathematical modeling**.

Any solution to the mathematical problem must be checked against the mathematical problem, the verbal description, and the real problem. See Figure 49 on page 138 for an illustration of the **modeling process**.

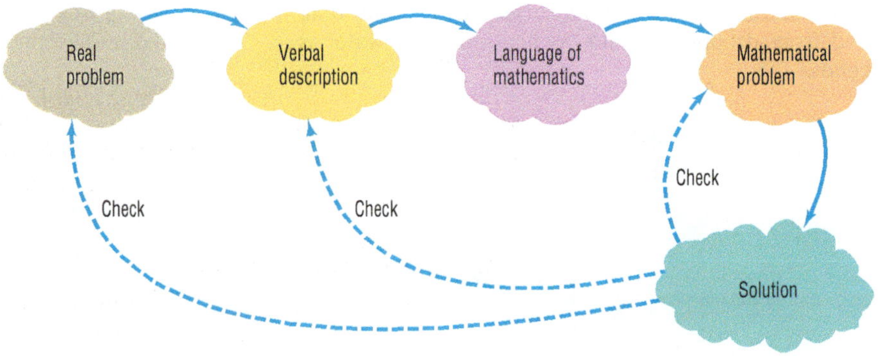

Figure 49 Modeling Process

1 Translate Verbal Descriptions into Mathematical Expressions

EXAMPLE 1 **Translating Verbal Descriptions into Mathematical Expressions**

(a) For uniform motion, the average speed of an object equals the distance traveled divided by the time required.

Translation: If r is the speed, d the distance, and t the time, then $r = \frac{d}{t}$.

(b) Let x denote a number.

The number 5 times as large as x is $5x$.

The number 3 less than x is $x - 3$.

The number that exceeds x by 4 is $x + 4$.

The number that, when added to x, gives 5 is $5 - x$. ■

Now Work PROBLEM 9

Always check the units used to measure the variables of an applied problem. In Example 1(a), if r is measured in miles per hour, then the distance d must be expressed in miles, and the time t must be expressed in hours. It is a good practice to check units to be sure that they are consistent and make sense.

The steps for solving applied problems, given earlier, are repeated next.

Steps for Solving Applied Problems

STEP 1: Read the problem carefully, perhaps two or three times. Pay particular attention to the question being asked in order to identify what you are looking for. Identify any relevant formulas you may need ($d = rt$, $A = \pi r^2$, etc.) If you can, determine realistic possibilities for the answer.

STEP 2: Assign a letter (variable) to represent what you are looking for, and, if necessary, express any remaining unknown quantities in terms of this variable.

STEP 3: Make a list of all the known facts, and translate them into mathematical expressions. These may take the form of an equation or an inequality involving the variable (the model). If possible, draw an appropriately labeled diagram to assist you. Sometimes a table or chart helps.

STEP 4: Solve the equation for the variable, and then answer the question using a complete sentence.

STEP 5: Check the answer with the facts in the problem. If it agrees, congratulations! If it does not agree, try again.

2 Solve Interest Problems

Interest is money paid for the use of money. The total amount borrowed (whether by an individual from a bank in the form of a loan or by a bank from an individual in the form of a savings account) is called the **principal**. The **rate of interest**, expressed as a percent, is the amount charged for the use of the principal for a given period of time, usually on a yearly (that is, per annum) basis.

Simple Interest Formula

If a principal of P dollars is borrowed for a period of t years at a per annum interest rate r, expressed as a decimal, the interest I charged is

$$I = Prt \qquad \textbf{(1)}$$

Interest charged according to formula (1) is called **simple interest**. When using formula (1), be sure to express r as a decimal. For example, if the rate of interest is 4%, then $r = 0.04$.

EXAMPLE 2 Finance: Computing Interest on a Loan

Suppose that Juanita borrows \$500 for 6 months at the simple interest rate of 9% per annum. What is the interest that Juanita will be charged on the loan? How much does Juanita owe after 6 months?

Solution The rate of interest is given per annum, so the actual time that the money is borrowed must be expressed in years. The interest charged would be the principal, \$500, times the rate of interest $(9\% = 0.09)$, times the time in years, $\frac{1}{2}$ (6 months $= \frac{1}{2}$ year):

$$\text{Interest charged} = I = Prt = (500)(0.09)\left(\frac{1}{2}\right) = \$22.50$$

After 6 months, Juanita will owe what she borrowed plus the interest:

$$\$500 + \$22.50 = \$522.50$$

EXAMPLE 3 Financial Planning

Candy has \$70,000 to invest and wants an annual return of \$2800, which requires an overall rate of return of 4%. She can invest in a safe, government-insured certificate of deposit, but it pays only 2%. To obtain 4%, she agrees to invest some of her money in noninsured corporate bonds paying 7%. How much should be placed in each investment to achieve her goal?

Solution

Step 1: The question is asking for two dollar amounts: the principal to invest in the corporate bonds and the principal to invest in the certificate of deposit.

Step 2: Let b represent the amount (in dollars) to be invested in the bonds. Then $70{,}000 - b$ is the amount that will be invested in the certificate. (Do you see why?)

Note: We could have also let c represent the amount invested in the certificate and $70{,}000 - c$ the amount invested in bonds.

Step 3: Set up a table:

	Principal (\$)	Rate	Time (yr)	Interest (\$)
Bonds	b	$7\% = 0.07$	1	$0.07b$
Certificate	$70{,}000 - b$	$2\% = 0.02$	1	$0.02(70{,}000 - b)$
Total	70,000	$4\% = 0.04$	1	$0.04(70{,}000) = 2800$

Since the combined interest from the investments is equal to the total interest, we have

$$\begin{aligned} \text{Bond interest} + \text{Certificate interest} &= \text{Total interest} \\ 0.07b + 0.02(70{,}000 - b) &= 2800 \end{aligned}$$

(Note that the units are consistent: the unit is dollars on each side.)

Step 4:
$$\begin{aligned} 0.07b + 1400 - 0.02b &= 2800 \\ 0.05b &= 1400 \\ b &= 28{,}000 \end{aligned}$$

Candy should place \$28,000 in the bonds and \$70,000 − \$28,000 = \$42,000 in the certificate.

Step 5: The interest on the bonds after 1 year is 0.07(\$28,000) = \$1960; the interest on the certificate after 1 year is 0.02(\$42,000) = \$840. The total annual interest is \$2800, the required amount. ■

Now Work PROBLEM 19

3 Solve Mixture Problems

Oil refineries sometimes produce gasoline that is a blend of two or more types of fuel; bakeries occasionally blend two or more types of flour for their bread. These problems are referred to as **mixture problems** because they combine two or more quantities to form a mixture.

EXAMPLE 4 Blending Coffees

The manager of a Starbucks store decides to experiment with a new blend of coffee. She will mix some B grade Colombian coffee that sells for \$5 per pound with some A grade Arabica coffee that sells for \$10 per pound to get 100 pounds of the new blend. The selling price of the new blend is to be \$7 per pound, and there is to be no difference in revenue from selling the new blend versus selling the other types. How many pounds of the B grade Colombian and A grade Arabica coffees are required?

Solution Let c represent the number of pounds of the B grade Colombian coffee. Then $100 - c$ equals the number of pounds of the A grade Arabica coffee. See Figure 50.

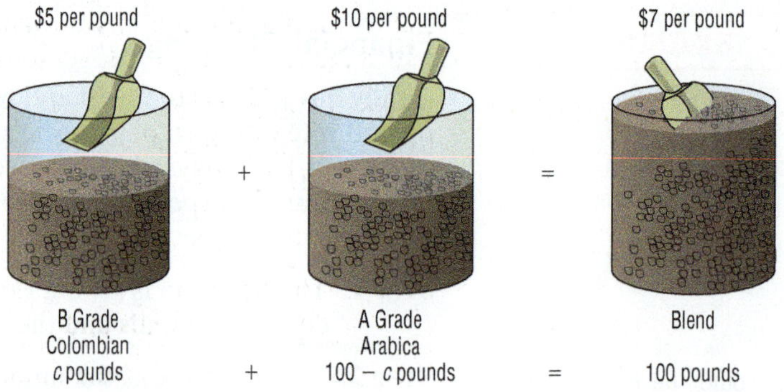

Figure 50 Mixing coffees

Since there is to be no difference in revenue between selling the A and B grades separately versus the blend, we have

$$\begin{array}{ccccc} \text{Revenue from B grade} & + & \text{Revenue from A grade} & = & \text{Revenue from blend} \\ \left\{\begin{matrix}\text{Price per pound}\\ \text{of B grade}\end{matrix}\right\}\left\{\begin{matrix}\text{Pounds of}\\ \text{B grade}\end{matrix}\right\} & + & \left\{\begin{matrix}\text{Price per pound}\\ \text{of A grade}\end{matrix}\right\}\left\{\begin{matrix}\text{Pounds of}\\ \text{A grade}\end{matrix}\right\} & = & \left\{\begin{matrix}\text{Price per pound}\\ \text{of blend}\end{matrix}\right\}\left\{\begin{matrix}\text{Pounds of}\\ \text{blend}\end{matrix}\right\} \\ \$5 \cdot c & + & \$10 \cdot (100 - c) & = & \$7 \cdot 100 \end{array}$$

Now solve the equation

$$\begin{aligned} 5c + 10(100 - c) &= 700 \\ 5c + 1000 - 10c &= 700 \\ -5c &= -300 \\ c &= 60 \end{aligned}$$

The manager should blend 60 pounds of B grade Colombian coffee with $100 - 60 = 40$ pounds of A grade Arabica coffee to get the desired blend.

Check: The 60 pounds of B grade coffee would sell for $(\$5)(60) = \$300$, and the 40 pounds of A grade coffee would sell for $(\$10)(40) = \$400$; the total revenue, \$700, equals the revenue obtained from selling the blend, as desired. ■

Now Work PROBLEM 23

4 Solve Uniform Motion Problems

Objects that move at a constant speed are said to be in **uniform motion**. When the average speed of an object is known, it can be interpreted as that object's constant speed. For example, a bicyclist traveling at an average speed of 25 miles per hour can be considered to be in uniform motion with a constant speed of 25 miles per hour.

Uniform Motion Formula

If an object moves at an average speed (rate) r, the distance d covered in time t is given by the formula

$$d = rt \quad \textbf{(2)}$$

That is, Distance = Rate · Time.

EXAMPLE 5 **Physics: Uniform Motion**

Tanya, who is a long-distance runner, runs at an average speed of 8 miles per hour (mi/h). Two hours after Tanya leaves your house, you leave in your Honda and follow the same route. If your average speed is 40 mi/h how long will it be before you catch up to Tanya? How far will each of you be from your home?

Solution Refer to Figure 51. We use t to represent the time (in hours) that it takes the Honda to catch up to Tanya. When this occurs, the total time elapsed for Tanya is $t + 2$ hours because she left 2 hours earlier.

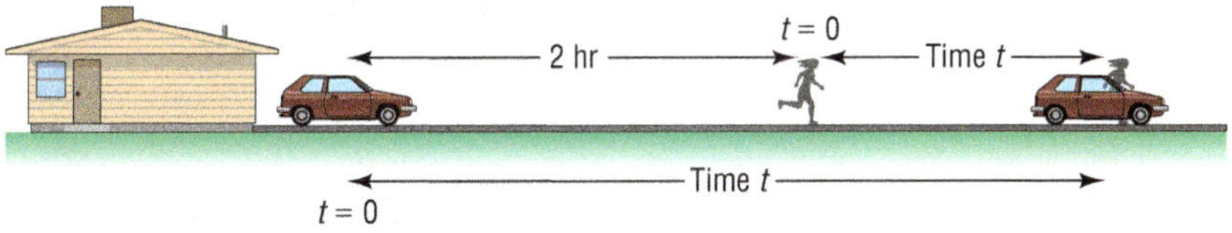

Figure 51

Set up the following table:

	Rate (mi/h)	Time (h)	Distance (mi)
Tanya	8	$t + 2$	$8(t + 2)$
Honda	40	t	$40t$

The distance traveled is the same for both, which leads to the equation

$$8(t+2) = 40t$$
$$8t + 16 = 40t$$
$$32t = 16$$
$$t = \frac{1}{2} \text{ hour}$$

It will take the Honda $\frac{1}{2}$ hour to catch up to Tanya. Each will have gone 20 miles.

Check: In 2.5 hours, Tanya travels a distance of $(2.5)(8) = 20$ miles. In $\frac{1}{2}$ hour, the Honda travels a distance of $\left(\frac{1}{2}\right)(40) = 20$ miles. ■

EXAMPLE 6

Physics: Uniform Motion

A motorboat heads upstream a distance of 24 miles on a river whose current is running at 3 miles per hour (mi/h). The trip up and back takes 6 hours. Assuming that the motorboat maintained a constant speed relative to the water, what was its speed?

Solution See Figure 52. Use r to represent the constant speed of the motorboat relative to the water. Then the true speed going upstream is $r - 3$ mi/h, and the true speed going downstream is $r + 3$ mi/h. Since Distance = Rate · Time, then Time $= \frac{\text{Distance}}{\text{Rate}}$. Set up a table.

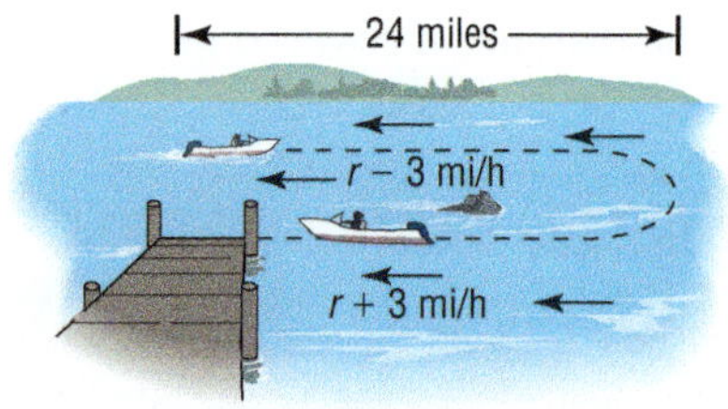

Figure 52

	Distance (mi)	Rate (mi/h)	Time (h) = Distance / Rate
Upstream	24	$r-3$	$\frac{24}{r-3}$
Downstream	24	$r+3$	$\frac{24}{r+3}$

The total time up and back is 6 hours, which gives the equation

$$\frac{24}{r-3} + \frac{24}{r+3} = 6$$

$$\frac{24(r+3) + 24(r-3)}{(r-3)(r+3)} = 6 \qquad \text{Add the quotients on the left.}$$

$$\frac{48r}{r^2-9} = 6 \qquad \text{Simplify.}$$

$$48r = 6(r^2-9) \qquad \text{Multiply both sides by } r^2-9.$$

$$6r^2 - 48r - 54 = 0 \qquad \text{Place in standard form.}$$

$$r^2 - 8r - 9 = 0 \qquad \text{Divide by 6.}$$

$$(r-9)(r+1) = 0 \qquad \text{Factor.}$$

$$r = 9 \quad \text{or} \quad r = -1 \qquad \text{Apply the Zero-Product Property and solve.}$$

Discard the solution $r = -1$ mi/h, and conclude that the speed of the motorboat relative to the water is 9 mi/h. ■

Now Work PROBLEM 29

5 Solve Constant Rate Job Problems

Here we look at jobs that are performed at a **constant rate**. The assumption is that if a job can be done in t units of time, then $\frac{1}{t}$ of the job is done in 1 unit of time. In other words, if a job takes 4 hours, then $\frac{1}{4}$ of the job is done in 1 hour.

EXAMPLE 7 Working Together to Do a Job

At 10 AM, Danny is asked by his father to weed the garden. From past experience, Danny knows that this will take him 4 hours, working alone. His older brother Mike, when it is his turn to do this job, requires 6 hours. Since Mike wants to go golfing with Danny and has a reservation for 1 PM, he agrees to help Danny. Assuming no gain or loss of efficiency, when will they finish if they work together? Can they make the golf date?

Solution Set up Table 4. In 1 hour, Danny does $\frac{1}{4}$ of the job, and in 1 hour, Mike does $\frac{1}{6}$ of the job. Let t be the time (in hours) that it takes them to do the job together. In 1 hour, then, $\frac{1}{t}$ of the job is completed. Reason as follows:

Table 4

	Hours to Do Job	Part of Job Done in 1 Hour
Danny	4	$\frac{1}{4}$
Mike	6	$\frac{1}{6}$
Together	t	$\frac{1}{t}$

$$\left(\begin{array}{c}\text{Part done by Danny}\\ \text{in 1 hour}\end{array}\right) + \left(\begin{array}{c}\text{Part done by Mike}\\ \text{in 1 hour}\end{array}\right) = \left(\begin{array}{c}\text{Part done together}\\ \text{in 1 hour}\end{array}\right)$$

From Table 2,

$$\frac{1}{4} + \frac{1}{6} = \frac{1}{t} \quad \text{The model.}$$

$$\frac{3}{12} + \frac{2}{12} = \frac{1}{t} \quad \text{LCD = 12 on the left.}$$

$$\frac{5}{12} = \frac{1}{t} \quad \text{Simplify.}$$

$$5t = 12 \quad \text{Multiply both sides by } 12t.$$

$$t = \frac{12}{5} \quad \text{Divide each side by 5.}$$

Working together, Mike and Danny can do the job in $\frac{12}{5}$ hours, or 2 hours, 24 minutes. They should make the golf date, since they will finish at 12:24 PM. ■

Now Work PROBLEM 35

1.6 Assess Your Understanding

Concepts and Vocabulary

1. The process of using variables to represent unknown quantities and then finding relationships that involve these variables is referred to as __________ __________.
2. The money paid for the use of money is ________.
3. Objects that move at a constant speed are said to be in ________ ________.
4. ***True or False*** The amount charged for the use of principal for a given period of time is called the rate of interest.
5. ***True or False*** If an object moves at an average speed r, the distance d covered in time t is given by the formula $d = rt$.
6. Suppose that you want to mix two coffees in order to obtain 100 pounds of a blend. If x represents the number of pounds of coffee A, write an algebraic expression that represents the number of pounds of coffee B.
   (a) $100 - x$ (b) $x - 100$ (c) $100x$ (d) $100 + x$
7. Which of the following is the simple interest formula?
   (a) $I = \frac{rt}{P}$ (b) $I = Prt$ (c) $I = \frac{P}{rt}$ (d) $I = P + rt$
8. If it takes 5 hours to complete a job, what fraction of the job is done in 1 hour?
   (a) $\frac{4}{5}$ (b) $\frac{5}{4}$ (c) $\frac{1}{5}$ (d) $\frac{1}{4}$

Applications and Extensions

In Problems 9–18, translate each sentence into a mathematical equation. Be sure to identify the meaning of all symbols.

9. **Geometry** The area of a circle is the product of the number π and the square of the radius.
10. **Geometry** The circumference of a circle is the product of the number π and twice the radius.
11. **Geometry** The area of a square is the square of the length of a side.
12. **Geometry** The perimeter of a square is four times the length of a side.
13. **Physics** Force equals the product of mass and acceleration.
14. **Physics** Pressure is force per unit area.
15. **Physics** Work equals force times distance.
16. **Physics** Kinetic energy is one-half the product of the mass and the square of the velocity.
17. **Business** The total variable cost of manufacturing x dishwashers is $150 per dishwasher times the number of dishwashers manufactured.
18. **Business** The total revenue derived from selling x dishwashers is $250 per dishwasher times the number of dishwashers sold.
19. **Financial Planning** Betsy, a recent retiree, requires $6000 per year in extra income. She has $50,000 to invest and can invest in B-rated bonds paying 15% per year or in a certificate of deposit (CD) paying 7% per year. How much money should Betsy invest in each to realize exactly $6000 in interest per year?
20. **Financial Planning** After 2 years, Betsy (see Problem 19) finds that she will now require $7000 per year. Assuming that the remaining information is the same, how should the money be reinvested?
21. **Banking** A bank loaned out $12,000, part of it at the rate of 8% per year and the rest at the rate of 18% per year. If the interest received totaled $1000, how much was loaned at 8%?
22. **Banking** Wendy, a loan officer at a bank, has $1,000,000 to lend and is required to obtain an average return of 18% per year. If she can lend at the rate of 19% or at the rate of 16%, how much can she lend at the 16% rate and still meet her requirement?
23. **Blending Teas** The manager of a store that specializes in selling tea decides to experiment with a new blend. She will mix some Earl Grey tea that sells for $5 per pound with some Orange Pekoe tea that sells for $3 per pound to get 100 pounds of the new blend. The selling price of the new blend is to be $4.50 per pound, and there is to be no difference in revenue between selling the new blend and selling the other types. How many pounds of the Earl Grey tea and of the Orange Pekoe tea are required?
24. **Business: Blending Coffee** A coffee manufacturer wants to market a new blend of coffee that sells for $3.90 per pound by mixing two coffees that sell for $2.75 and $5 per pound, respectively. What amounts of each coffee should be blended to obtain the desired mixture?

    [**Hint:** Assume that the total weight of the desired blend is 100 pounds.]
25. **Business: Mixing Nuts** A nut store normally sells cashews for $9.00 per pound and almonds for $3.50 per pound. But at the end of the month the almonds had not sold well, so, in order to sell 60 pounds of almonds, the manager decided to mix the 60 pounds of almonds with some cashews and sell the mixture for $7.50 per pound. How many pounds of cashews should be mixed with the almonds to ensure no change in the revenue?
26. **Business: Mixing Candy** A candy store sells boxes of candy containing caramels and cremes. Each box sells for $12.50 and holds 30 pieces of candy (all pieces are the same size). If the caramels cost $0.25 to produce and the cremes cost $0.45 to produce, how many of each should be in a box to yield a profit of $3?
27. **Physics: Uniform Motion** A motorboat can maintain a constant speed of 16 miles per hour relative to the water. The boat travels upstream to a certain point in 20 minutes; the return trip takes 15 minutes. What is the speed of the current? See the figure.

28. **Physics: Uniform Motion** A motorboat heads upstream on a river that has a current of 3 miles per hour. The trip upstream takes 5 hours, and the return trip takes 2.5 hours. What is the speed of the motorboat? (Assume that the boat maintains a constant speed relative to the water.)
29. **Physics: Uniform Motion** A motorboat maintained a constant speed of 15 miles per hour relative to the water in going 10 miles upstream and then returning. The total time for the trip was 1.5 hours. Use this information to find the speed of the current.
30. **Physics: Uniform Motion** Two cars enter the Florida Turnpike at Commercial Boulevard at 8:00 AM, each heading for Wildwood. One car's average speed is 10 miles per hour more than the other's. The faster car arrives at Wildwood at 11:00 AM, $\frac{1}{2}$ hour before the other car. What was the average speed of each car? How far did each travel?
31. **Moving Walkways** The speed of a moving walkway is typically about 2.5 feet per second. Walking on such a moving walkway, it takes Karen a total of 40 seconds to travel 50 feet with the movement of the walkway and then back again against the movement of the walkway. What is Karen's normal walking speed?

    Source: *Answers.com*
32. **High-Speed Walkways** Toronto's Pearson International Airport has a high-speed version of a moving walkway. If Liam walks while riding this moving walkway, he can travel 280 meters in 60 seconds less time than if he stands still on the moving walkway. If Liam walks at a normal rate of 1.5 meters per second, what is the speed of the walkway?

    Source: *Answers.com*
33. **Tennis** A regulation doubles tennis court has an area of 2808 square feet. If it is 6 feet longer than twice its width, determine the dimensions of the court.

    Source: *United States Tennis Association*

34. Laser Printers It takes an HP LaserJet M451dw laser printer 16 minutes longer to complete an 840-page print job by itself than it takes an HP LaserJet CP4025dn to complete the same job by itself. Together the two printers can complete the job in 15 minutes. How long does it take each printer to complete the print job alone? What is the speed of each printer?

Source: *Hewlett-Packard*

35. Working Together on a Job Trent can deliver his newspapers in 30 minutes. It takes Lois 20 minutes to do the same route. How long would it take them to deliver the newspapers if they worked together?

36. Working Together on a Job Patrice, by himself, can paint four rooms in 10 hours. If he hires April to help, they can do the same job together in 6 hours. If he lets April work alone, how long will it take her to paint four rooms?

37. Enclosing a Garden A gardener has 46 feet of fencing to be used to enclose a rectangular garden that has a border 2 feet wide surrounding it. See the figure.

(a) If the length of the garden is to be twice its width, what will be the dimensions of the garden?
(b) What is the area of the garden?
(c) If the length and width of the garden are to be the same, what will be the dimensions of the garden?
(d) What will be the area of the square garden?

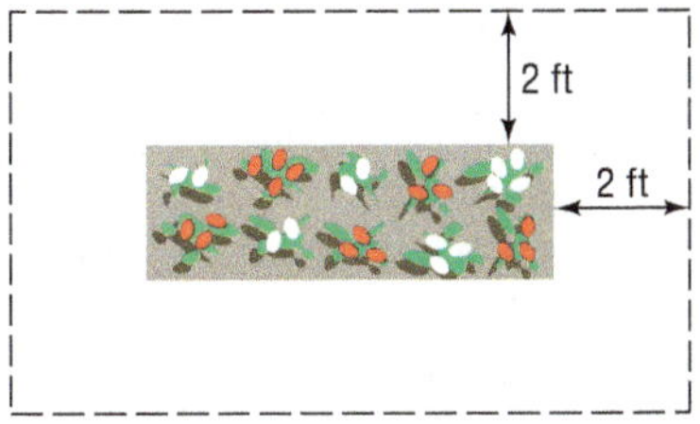

38. Construction A pond is enclosed by a wooden deck that is 3 feet wide. The fence surrounding the deck is 100 feet long.

(a) If the pond is square, what are its dimensions?
(b) If the pond is rectangular and the length of the pond is to be three times its width, what are its dimensions?
(c) If the pond is circular, what is its diameter?
(d) Which pond has the larger area?

39. Football A tight end can run the 100-yard dash in 12 seconds. A defensive back can do it in 10 seconds. The tight end catches a pass at his own 20-yard line with the defensive back at the 15-yard line. (See the figure.) If no other players are nearby, at what yard line will the defensive back catch up to the tight end?

[**Hint:** At time $t = 0$, the defensive back is 5 yards behind the tight end.]

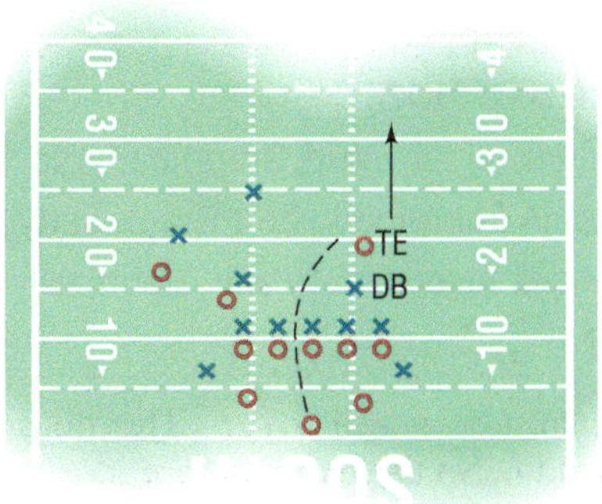

40. Computing Business Expense Therese, an outside salesperson, uses her car for both business and pleasure. Last year, she traveled 30,000 miles, using 900 gallons of gasoline. Her car gets 40 miles per gallon on the highway and 25 in the city. She can deduct all highway travel, but no city travel, on her taxes. How many miles should Therese deduct as a business expense?

41. Mixing Water and Antifreeze How much water should be added to 1 gallon of pure antifreeze to obtain a solution that is 60% antifreeze?

42. Mixing Water and Antifreeze The cooling system of a certain foreign-made car has a capacity of 15 liters. If the system is filled with a mixture that is 40% antifreeze, how much of this mixture should be drained and replaced by pure antifreeze so that the system is filled with a solution that is 60% antifreeze?

43. Chemistry: Salt Solutions How much water must be evaporated from 32 ounces of a 4% salt solution to make a 6% salt solution?

44. Chemistry: Salt Solutions How much water must be evaporated from 240 gallons of a 3% salt solution to produce a 5% salt solution?

45. Purity of Gold The purity of gold is measured in karats, with pure gold being 24 karats. Other purities of gold are expressed as proportional parts of pure gold. Thus, 18-karat gold is $\frac{18}{24}$, or 75% pure gold; 12-karat gold is $\frac{12}{24}$, or 50% pure gold; and so on. How much 12-karat gold should be mixed with pure gold to obtain 60 grams of 16-karat gold?

46. Chemistry: Sugar Molecules A sugar molecule has twice as many atoms of hydrogen as it does oxygen and one more atom of carbon than of oxygen. If a sugar molecule has a total of 45 atoms, how many are oxygen? How many are hydrogen?

47. Running a Race Mike can run the mile in 6 minutes, and Dan can run the mile in 9 minutes. If Mike gives Dan a head start of 1 minute, how far from the start will Mike pass Dan? How long does it take? See the figure.

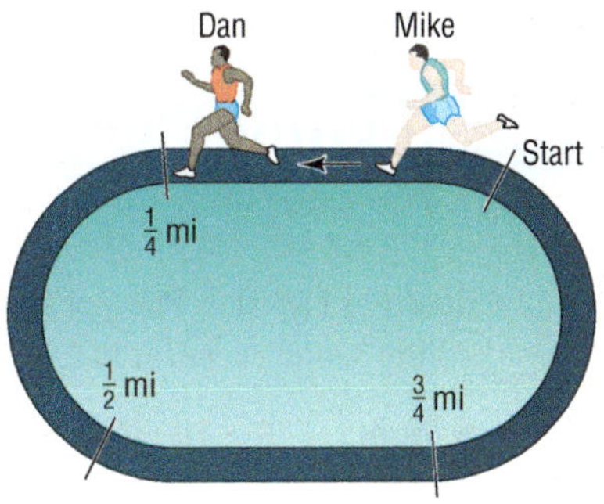

48. Range of an Airplane An air rescue plane averages 300 miles per hour in still air. It carries enough fuel for 5 hours of flying time. If, upon takeoff, it encounters a head wind of 30 mi/h, how far can it fly and return safely? (Assume that the wind remains constant.)

49. Emptying Oil Tankers An oil tanker can be emptied by the main pump in 4 hours. An auxiliary pump can empty the tanker in 9 hours. If the main pump is started at 9 AM, when should the auxiliary pump be started so that the tanker is emptied by noon?

50. Cement Mix A 20-pound bag of Economy brand cement mix contains 25% cement and 75% sand. How much pure cement must be added to produce a cement mix that is 40% cement?

51. Filling a Tub A bathroom tub will fill in 15 minutes with both faucets open and the stopper in place. With both faucets closed and the stopper removed, the tub will empty in 20 minutes. How long will it take for the tub to fill if both faucets are open and the stopper is removed?

52. Using Two Pumps A 5-horsepower (hp) pump can empty a pool in 5 hours. A smaller, 2-hp pump empties the same pool in 8 hours. The pumps are used together to begin emptying this pool. After two hours, the 2-hp pump breaks down. How long will it take the larger pump to finish emptying the pool?

53. A Biathlon Suppose that you have entered an 87-mile biathlon that consists of a run and a bicycle race. During your run, your average speed is 6 miles per hour, and during your bicycle race, your average speed is 25 miles per hour. You finish the race in 5 hours. What is the distance of the run? What is the distance of the bicycle race?

54. Cyclists Two cyclists leave a city at the same time, one going east and the other going west. The westbound cyclist bikes 5 mph faster than the eastbound cyclist. After 6 hours they are 246 miles apart. How fast is each cyclist riding?

55. Comparing Olympic Heroes In the 2012 Olympics, Usain Bolt of Jamaica won the gold medal in the 100-meter race with a time of 9.69 seconds. In the 1896 Olympics, Thomas Burke of the United States won the gold medal in the 100-meter race in 12.0 seconds. If they ran in the same race, repeating their respective times, by how many meters would Bolt beat Burke?

56. Constructing a Coffee Can A 39-ounce can of Hills Bros.® coffee requires 188.5 square inches of aluminum. If its height is 7 inches, what is its radius? **[Hint:** The surface area S of a right cylinder is $S = 2\pi r^2 + 2\pi rh$, where r is the radius and h is the height.]

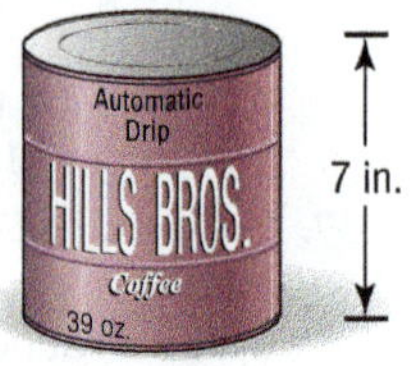

Explaining Concepts: Discussion and Writing

57. Critical Thinking You are the manager of a clothing store and have just purchased 100 dress shirts for \$20.00 each. After 1 month of selling the shirts at the regular price, you plan to have a sale giving 40% off the original selling price. However, you still want to make a profit of \$4 on each shirt at the sale price. What should you price the shirts at initially to ensure this? If, instead of 40% off at the sale, you give 50% off, by how much is your profit reduced?

58. Critical Thinking Make up a word problem that requires solving a linear equation as part of its solution. Exchange problems with a friend. Write a critique of your friend's problem.

59. Critical Thinking Without solving, explain what is wrong with the following mixture problem: How many liters of 25% ethanol should be added to 20 liters of 48% ethanol to obtain a solution of 58% ethanol? Now go through an algebraic solution. What happens?

60. Computing Average Speed In going from Chicago to Atlanta, a car averages 45 miles per hour, and in going from Atlanta to Miami, it averages 55 miles per hour. If Atlanta is halfway between Chicago and Miami, what is the average speed from Chicago to Miami? Discuss an intuitive solution. Write a paragraph defending your intuitive solution. Then solve the problem algebraically. Is your intuitive solution the same as the algebraic one? If not, find the flaw.

61. Speed of a Plane On a recent flight from Phoenix to Kansas City, a distance of 919 nautical miles, the plane arrived 20 minutes early. On leaving the aircraft, I asked the captain, "What was our tail wind?" He replied, "I don't know, but our ground speed was 550 knots." Has enough information been provided for you to find the tail wind? If possible, find the tail wind. (1 knot = 1 nautical mile per hour)

1.7 Solving Inequalities

PREPARING FOR THIS SECTION *Before getting started, review the following:*

- Inequalities (Section R.2, pp. 19–20)
- Absolute Value (Section R.2, p. 20)
- Union of Sets (Section R.1, pp. 2–3)

Now Work the 'Are You Prepared?' problems on page 154.

OBJECTIVES
1. Use Interval Notation (p. 147)
2. Use Properties of Inequalities (p. 148)
3. Solve Linear Inequalities Algebraically and Graphically (p. 150)
4. Solve Combined Inequalities Algebraically and Graphically (p. 151)
5. Solve Absolute Value Inequalities Algebraically and Graphically (p. 152)

Suppose that a and b are two real numbers and $a < b$. The notation $a < x < b$ means that x is a number *between* a and b. So the expression $a < x < b$ is equivalent to the two inequalities $a < x$ and $x < b$. Similarly, the expression $a \leq x \leq b$ is equivalent to the two inequalities $a \le x$ and $x \le b$. The remaining two possibilities, $a \le x < b$ and $a < x \le b$, are defined similarly.

Although it is acceptable to write $3 \ge x \ge 2$, it is preferable to reverse the inequality symbols and write instead $2 \le x \le 3$ so that the values go from smaller to larger, reading from left to right.

A statement such as $2 \le x \le 1$ is false because there is no number x for which $2 \le x$ and $x \le 1$. Finally, never mix inequality symbols, as in $2 \le x \ge 3$.

1 Use Interval Notation

Let a and b represent two real numbers with $a < b$:

In Words
The notation $[a, b]$ represents all real numbers between a and b, inclusive. The notation (a, b) represents all real numbers between a and b, not including either a or b.

A **closed interval**, denoted by $[a, b]$, consists of all real numbers x for which $a \le x \le b$.

An **open interval**, denoted by (a, b), consists of all real numbers x for which $a < x < b$.

The **half-open**, or **half-closed**, **intervals** are $(a, b]$, consisting of all real numbers x for which $a < x \le b$, and $[a, b)$, consisting of all real numbers x for which $a \le x < b$.

In each of these definitions, a is called the **left endpoint** and b the **right endpoint** of the interval.

The symbol ∞ (read as "infinity") is not a real number, but notation used to indicate unboundedness in the positive direction. The symbol $-\infty$ (read as "minus infinity" or "negative infinity") also is not a real number, but notation used to indicate unboundedness in the negative direction. The symbols ∞ and $-\infty$ are used to define five other kinds of intervals:

$[a, \infty)$	consists of all real numbers x for which $x \ge a$
(a, ∞)	consists of all real numbers x for which $x > a$
$(-\infty, a]$	consists of all real numbers x for which $x \le a$
$(-\infty, a)$	consists of all real numbers x for which $x < a$
$(-\infty, \infty)$	consists of all real numbers x

Note that ∞ and $-\infty$ are never included as endpoints since they are not real numbers.

Table 5 summarizes interval notation, corresponding inequality notation, and their graphs.

Table 5

Interval	Inequality	Graph
The open interval (a, b)	$a < x < b$	
The closed interval $[a, b]$	$a \le x \le b$	
The half-open interval $[a, b)$	$a \le x < b$	
The half-open interval $(a, b]$	$a < x \le b$	
The interval $[a, \infty)$	$x \ge a$	
The interval (a, ∞)	$x > a$	
The interval $(-\infty, a]$	$x \le a$	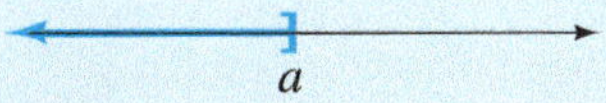
The interval $(-\infty, a)$	$x < a$	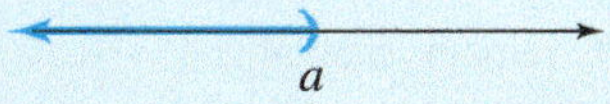
The interval $(-\infty, \infty)$	All real numbers	

EXAMPLE 1 **Writing Inequalities Using Interval Notation**

Write each inequality using interval notation.

(a) $1 \le x \le 3$
(b) $-4 < x < 0$
(c) $x > 5$
(d) $x \le 1$

Solution (a) $1 \le x \le 3$ represents all numbers x between 1 and 3, inclusive. In interval notation, we write $[1, 3]$.

(b) In interval notation, $-4 < x < 0$ is written $(-4, 0)$.

(c) $x > 5$ consists of all numbers x greater than 5. In interval notation, we write $(5, \infty)$.

(d) In interval notation, $x \le 1$ is written $(-\infty, 1]$. ■

EXAMPLE 2 **Writing Intervals Using Inequality Notation**

Write each interval as an inequality involving x.

(a) $[1, 4)$
(b) $(2, \infty)$
(c) $[2, 3]$
(d) $(-\infty, -3]$

Solution (a) $[1, 4)$ consists of all numbers x for which $1 \le x < 4$.

(b) $(2, \infty)$ consists of all numbers x for which $x > 2$.

(c) $[2, 3]$ consists of all numbers x for which $2 \le x \le 3$.

(d) $(-\infty, -3]$ consists of all numbers x for which $x \le -3$. ■

Now Work PROBLEMS 13, 23, AND 31

2 Use Properties of Inequalities

The product of two positive real numbers is positive, the product of two negative real numbers is positive, and the product of 0 and 0 is 0. For any real number a, the value of a^2 is 0 or positive; that is, a^2 is nonnegative. This is called the **nonnegative property**.

In Words
The square of a real number is never negative.

Nonnegative Property

For any real number a,

$$a^2 \ge 0 \qquad (1)$$

When the same number is added to both sides of an inequality, an equivalent inequality is obtained. For example, since $3 < 5$, then $3 + 4 < 5 + 4$ or $7 < 9$. This is called the **addition property** of inequalities.

In Words
The addition property states that the sense, or direction, of an inequality remains unchanged if the same number is added to each side.

Addition Property of Inequalities

For real numbers a, b, and c:

$$\text{If } a < b, \text{ then } a + c < b + c. \qquad \textbf{(2a)}$$

$$\text{If } a > b, \text{ then } a + c > b + c. \qquad \textbf{(2b)}$$

Figure 53 illustrates the addition property (2a). In Figure 53(a), we see that a lies to the left of b. If c is positive, then $a + c$ and $b + c$ lie c units to the right of a and c units to the right of b, respectively. Consequently, $a + c$ must lie to the left of $b + c$; that is, $a + c < b + c$. Figure 53(b) illustrates the situation if c is negative.

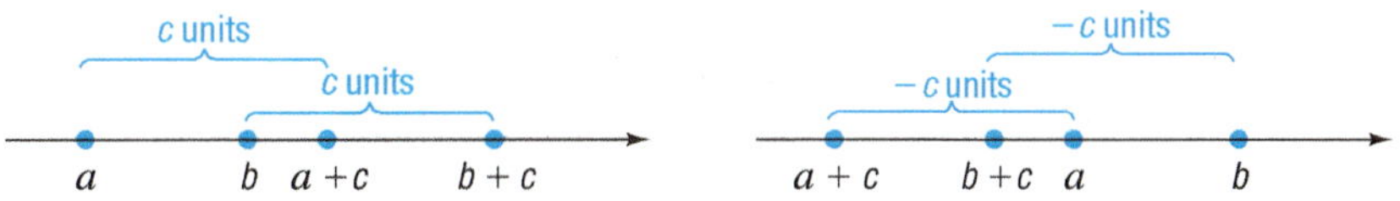

(a) If $a < b$ and $c > 0$, then $a + c < b + c$.

(b) If $a < b$ and $c < 0$, then $a + c < b + c$.

Figure 53 Addition property of inequalities

Draw an illustration similar to Figure 53 that illustrates the addition property (2b).

EXAMPLE 3 Addition Property of Inequalities

(a) If $x < -5$, then $x + 5 < -5 + 5$ or $x + 5 < 0$.

(b) If $x > 2$, then $x + (-2) > 2 + (-2)$ or $x - 2 > 0$. ■

Now Work PROBLEM 39

EXAMPLE 4 Multiplying an Inequality by a Positive Number

Express as an inequality the result of multiplying each side of the inequality $3 < 7$ by 2.

Solution Begin with

$$3 < 7$$

Multiplying each side by 2 yields the numbers 6 and 14, so we have

$$6 < 14$$ ■

EXAMPLE 5 Multiplying an Inequality by a Negative Number

Express as an inequality the result of multiplying each side of the inequality $9 > 2$ by -4.

Solution Begin with

$$9 > 2$$

Multiplying each side by -4 yields the numbers -36 and -8, so we have

$$-36 < -8$$ ■

Note that the effect of multiplying both sides of $9 > 2$ by the negative number -4 is that the direction of the inequality symbol is reversed.

Examples 4 and 5 illustrate the following general **multiplication properties** for inequalities:

In Words
The multiplication properties state that the sense, or direction, of an inequality *remains the same* if each side is multiplied by a *positive* real number, whereas the direction is *reversed* if each side is multiplied by a *negative* real number.

Multiplication Properties for Inequalities

For real numbers a, b, and c:

If $a < b$ and if $c > 0$, then $ac < bc$.
If $a < b$ and if $c < 0$, then $ac > bc$. **(3a)**

If $a > b$ and if $c > 0$, then $ac > bc$.
If $a > b$ and if $c < 0$, then $ac < bc$. **(3b)**

EXAMPLE 6 **Multiplication Property of Inequalities**

(a) If $2x < 6$, then $\frac{1}{2}(2x) < \frac{1}{2}(6)$ or $x < 3$.

(b) If $\frac{x}{-3} > 12$, then $-3\left(\frac{x}{-3}\right) < -3(12)$ or $x < -36$.

(c) If $-4x > -8$, then $\frac{-4x}{-4} < \frac{-8}{-4}$ or $x < 2$. ■

Now Work PROBLEM 45

3 Solve Linear Inequalities Algebraically and Graphically

An **inequality in one variable** is a statement involving two expressions, at least one containing the variable, separated by one of the inequality symbols, $<$, $\le$, $>$, or $\ge$. To **solve an inequality** means to find all values of the variable for which the statement is true. These values are called **solutions** of the inequality.

For example, the following are all inequalities involving one variable, x:

$$x + 5 < 8 \qquad 2x - 3 \ge 4 \qquad x^2 - 1 \le 3 \qquad \frac{x+1}{x-2} > 0$$

As with equations, one method for solving an inequality is to replace it by a series of equivalent inequalities until an inequality with an obvious solution, such as $x < 3$, is obtained. Equivalent inequalities are obtained by applying some of the same operations as those used to find equivalent equations. The addition property and the multiplication properties form the basis for the following procedures.

Procedures That Leave the Inequality Symbol Unchanged

1. Simplify both sides of the inequality by combining like terms and eliminating parentheses:

$$x + 2 + 6 > 2x + 5(x + 1)$$

is equivalent to $\quad x + 8 > 7x + 5$

2. Add or subtract the same expression on both sides of the inequality:

$$3x - 5 < 4$$

is equivalent to $\quad (3x - 5) + 5 < 4 + 5$

3. Multiply or divide both sides of the inequality by the same *positive* expression:

$$4x > 16 \quad \text{is equivalent to} \quad \frac{4x}{4} > \frac{16}{4}$$

Procedures That Reverse the Sense or Direction of the Inequality Symbol

1. Interchange the two sides of the inequality:

$$3 < x \quad \text{is equivalent to} \quad x > 3$$

2. Multiply or divide both sides of the inequality by the same *negative* expression:

$$-2x > 6 \quad \text{is equivalent to} \quad \frac{-2x}{-2} < \frac{6}{-2}$$

To solve an inequality using a graphing utility, follow these steps:

Steps for Solving Inequalities Graphically

STEP 1: Write the inequality in one of the following forms:

$$Y_1 < Y_2, \quad Y_1 > Y_2, \quad Y_1 \le Y_2, \quad Y_1 \ge Y_2$$

STEP 2: Graph Y_1 and Y_2 on the same screen.

STEP 3: If the inequality is of the form $Y_1 < Y_2$, determine on what interval Y_1 is below Y_2.

If the inequality is of the form $Y_1 > Y_2$, determine on what interval Y_1 is above Y_2.

If the inequality is not strict ($\le$ or $\ge$), include the x-coordinates of the points of intersection in the solution.

As the examples that follow illustrate, we solve linear inequalities using many of the same steps that we would use to solve linear equations. The goal is to get the variable on one side of the inequality and a constant on the other. In writing the solution of an inequality, either set notation or interval notation may be used, whichever is more convenient.

EXAMPLE 7 **Solving an Inequality Algebraically and Graphically**

Solve the inequality $4x + 7 \ge 2x - 3$, and graph the solution set.

Algebraic Solution

$4x + 7 \ge 2x - 3$	
$4x + 7 - 7 \ge 2x - 3 - 7$	Subtract 7 from both sides.
$4x \ge 2x - 10$	Simplify.
$4x - 2x \ge 2x - 10 - 2x$	Subtract $2x$ from both sides.
$2x \ge -10$	Simplify.
$\dfrac{2x}{2} \ge \dfrac{-10}{2}$	Divide both sides by 2. (The direction of the inequality symbol is unchanged.)
$x \ge -5$	Simplify.

The solution set is $\{x \mid x \ge -5\}$ or, using interval notation, all numbers in the interval $[-5, \infty)$.

Graphing Solution

Graph $Y_1 = 4x + 7$ and $Y_2 = 2x - 3$ on the same screen. See Figure 54. Using the INTERSECT command, we find that Y_1 and Y_2 intersect at $x = -5$. The graph of Y_1 is above that of Y_2, $Y_1 > Y_2$, to the right of the point of intersection. Since the inequality is not strict, the solution set is $\{x \mid x \ge -5\}$ or, using interval notation, $[-5, \infty)$.

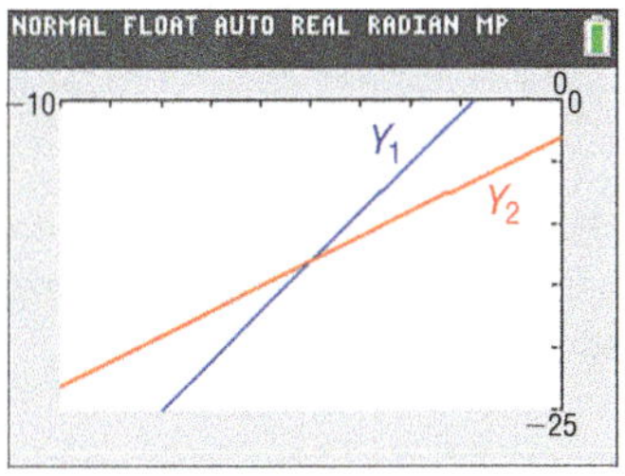

Figure 54 $Y_1 = 4x + 7$; $Y_2 = 2x - 3$

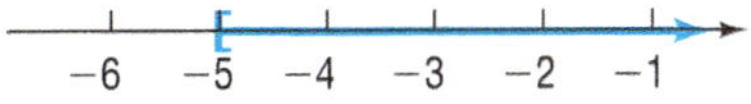

Figure 55 $\{x \mid x \ge -5\}$

See Figure 55 for the graph of the solution set. ■

Now Work PROBLEM 51

4 Solve Combined Inequalities Algebraically and Graphically

EXAMPLE 8 **Solving a Combined Inequality Algebraically and Graphically**

Solve the inequality: $-1 \le \dfrac{3 - 5x}{2} \le 9$

Algebraic Solution

$$-1 \le \frac{3-5x}{2} \le 9$$

$$2(-1) \le 2\left(\frac{3-5x}{2}\right) \le 2(9)$$ Multiply each part by 2 to remove the denominator.

$$-2 \le 3-5x \le 18$$ Simplify.

$$-2-3 \le 3-5x-3 \le 18-3$$ Subtract 3 from each part to isolate the term containing x.

$$-5 \le -5x \le 15$$ Simplify.

$$\frac{-5}{-5} \ge \frac{-5x}{-5} \ge \frac{15}{-5}$$ Divide each part by -5 (reverse the sense of each inequality symbol).

$$1 \ge x \ge -3$$ Simplify.

$$-3 \le x \le 1$$ Reverse the order so the numbers get larger as you read from left to right.

The solution set is $\{x \mid -3 \le x \le 1\}$. In interval notation, the solution is $[-3, 1]$.

Graphing Solution

To solve a combined inequality graphically, graph each part: $Y_1 = -1$, $Y_2 = \dfrac{3-5x}{2}$, $Y_3 = 9$. We seek the values of x for which the graph of Y_2 is between the graphs of Y_1 and Y_3. See Figure 56. The point of intersection of Y_1 and Y_2 is $(1, -1)$, and the point of intersection of Y_2 and Y_3 is $(-3, 9)$. The inequality is true for all values of x between these two intersection points. Since the inequality is nonstrict, the solution set is $\{x \mid -3 \le x \le 1\}$ or, using interval notation, $[-3, 1]$.

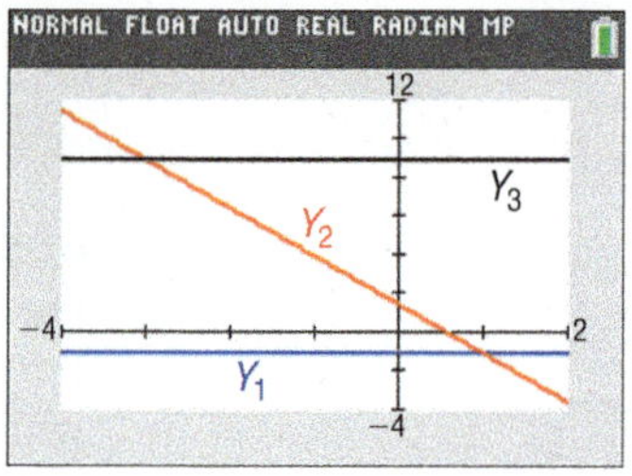

Figure 56 $Y_1 = -1$; $Y_2 = \dfrac{3-5x}{2}$; $Y_3 = 9$

See Figure 57 for the graph of the solution set.

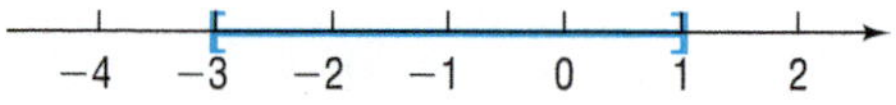

Figure 57 $\{x \mid -3 \le x \le 1\}$

Now Work PROBLEM 67

5 Solve Absolute Value Inequalities Algebraically and Graphically

EXAMPLE 9 **Solving an Inequality Involving Absolute Value**

Solve the inequality: $|x| < 4$

Algebraic Solution

We are looking for all points whose coordinate x is a distance less than 4 units from the origin. See Figure 58 for an illustration.

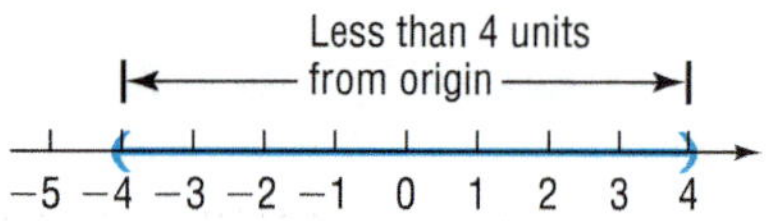

Figure 58 $|x| < 4$

Because any x between -4 and 4 satisfies the condition $|x| < 4$, the solution set consists of all numbers x for which $-4 < x < 4$, that is, all x in the interval $(-4, 4)$.

Graphing Solution

Graph $Y_1 = |x|$ and $Y_2 = 4$ on the same screen. See Figure 59. Using the INTERSECT command (twice), find that Y_1 and Y_2 intersect at $x = -4$ and at $x = 4$. The graph of Y_1 is below that of Y_2, $Y_1 < Y_2$, between the points of intersection. Because the inequality is strict, the solution set is $\{x \mid -4 < x < 4\}$ or, using interval notation, $(-4, 4)$.

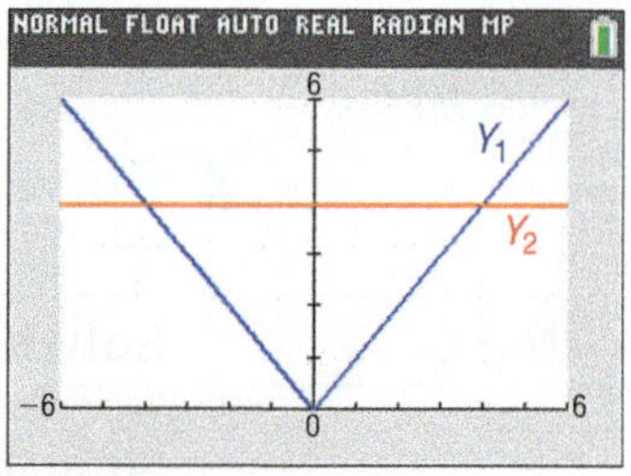

Figure 59 $Y_1 = |x|$; $Y_2 = 4$

We are led to the following results:

THEOREM

If a is any positive number and if u is any algebraic expression, then

$\lvert u \rvert < a$	is equivalent to	$-a < u < a$	**(4)**
$\lvert u \rvert \le a$	is equivalent to	$-a \le u \le a$	**(5)**

In other words, $\lvert u \rvert < a$ is equivalent to $-a < u$ and $u < a$.

EXAMPLE 10 Solving an Inequality Involving Absolute Value

Solve the inequality $\lvert 2x+4 \rvert \le 3$, and graph the solution set.

Algebraic Solution

$$\lvert 2x+4 \rvert \le 3$$ This follows the form of statement (5); the expression $u = 2x+4$ is inside the absolute value bars.

$$-3 \le 2x+4 \le 3$$ Apply statement (5).

$$-3-4 \le 2x+4-4 \le 3-4$$ Subtract 4 from each part.

$$-7 \le 2x \le -1$$ Simplify.

$$\frac{-7}{2} \le \frac{2x}{2} \le \frac{-1}{2}$$ Divide each part by 2.

$$-\frac{7}{2} \le x \le -\frac{1}{2}$$ Simplify.

The solution set is $\left\{x \middle| -\frac{7}{2} \le x \le -\frac{1}{2}\right\}$, that is, all x in the interval $\left[-\frac{7}{2}, -\frac{1}{2}\right]$.

Graphing Solution

Graph $Y_1 = \lvert 2x+4 \rvert$ and $Y_2 = 3$ on the same screen. See Figure 60. Using the INTERSECT command (twice), find that Y_1 and Y_2 intersect at $x = -3.5$ and at $x = -0.5$. The graph of Y_1 is below that of Y_2, $Y_1 < Y_2$, between the points of intersection. Because the inequality is not strict, the solution set is $\{x \mid -3.5 \le x \le -0.5\}$ or, using interval notation, $[-3.5, -0.5]$.

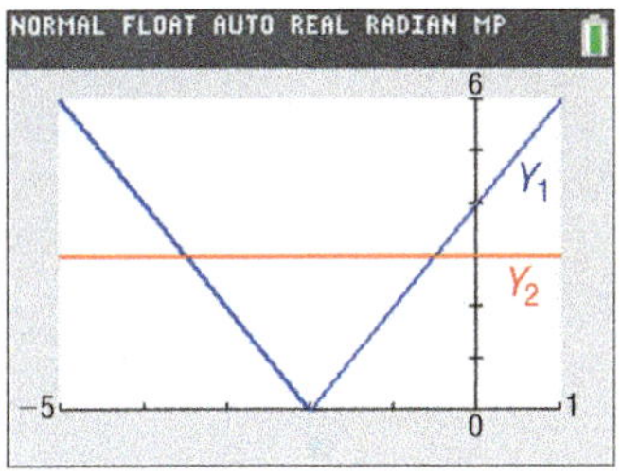

Figure 60 $Y_1 = \lvert 2x+4 \rvert$; $Y_2 = 3$

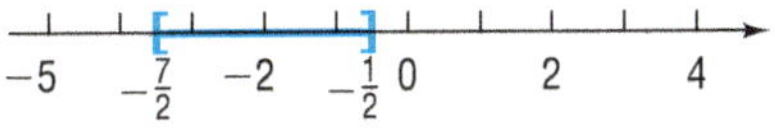

Figure 61 $\left\{x \middle| -\frac{7}{2} \le x \le -\frac{1}{2}\right\}$

See Figure 61 for a graph of the solution set. ■

Now Work PROBLEM 79

EXAMPLE 11 Solving an Inequality Involving Absolute Value

Solve the inequality $\lvert x \rvert > 3$.

Algebraic Solution

We are looking for all points whose coordinate x is a distance greater than 3 units from the origin. Figure 62 illustrates the situation.

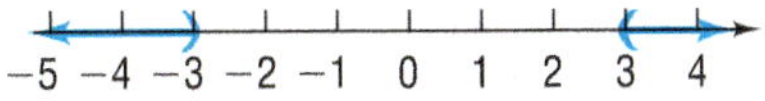

Figure 62 $\lvert x \rvert > 3$

Graphing Solution

Graph $Y_1 = \lvert x \rvert$ and $Y_2 = 3$ on the same screen. See Figure 63 on page 154. Using the INTERSECT command (twice), find that Y_1 and Y_2 intersect at $x = -3$ and at $x = 3$. The graph of Y_1 is above that of Y_2, $Y_1 > Y_2$, to the left of $x = -3$ and to the right of $x = 3$. Because the inequality is strict, the solution set is $\{x \mid x < -3 \text{ or } x > 3\}$. Using interval notation, the solution is $(-\infty, -3) \cup (3, \infty)$.*

*Recall that the symbol ∪ represents the union of two sets and means "or."

We conclude that any x less than -3 or greater than 3 satisfies the condition $|x| > 3$. Consequently, the solution set is $\{x \mid x < -3 \text{ or } x > 3\}$. Using interval notation, the solution is $(-\infty, -3) \cup (3, \infty)$.

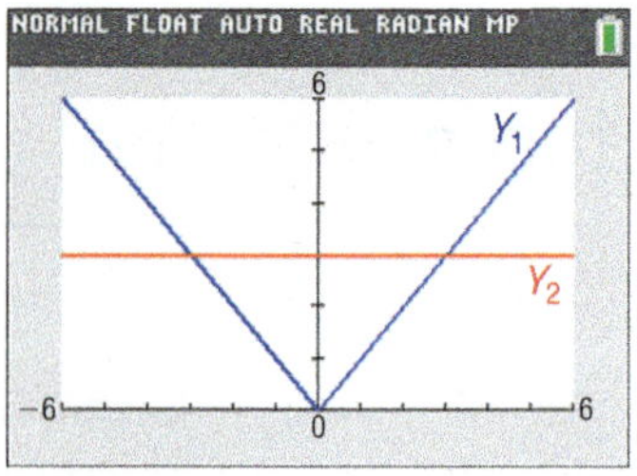

Figure 63 $Y_1 = |x|$; $Y_2 = 3$

THEOREM

If a is any positive number and u is any algebraic expression, then

$\lvert u\rvert > a$	is equivalent to	$u < -a$ or $u > a$	**(6)**
$\lvert u\rvert \geq a$	is equivalent to	$u \leq -a$ or $u \geq a$	**(7)**

WARNING A common error to be avoided is to attempt to write the solution $x < 1$ or $x > 4$ as $1 > x > 4$, which is incorrect, since there are no numbers x for which $1 > x$ and $x > 4$. Another common error is to "mix" the symbols and write $1 < x > 4$, which makes no sense.

EXAMPLE 12 Solving an Inequality Involving Absolute Value

Solve the inequality $|2x - 5| > 3$, and graph the solution set.

Algebraic Solution

$|2x - 5| > 3$ — This follows the form of statement (6); the expression $u = 2x - 5$ is inside the absolute value bars.

$2x - 5 < -3$	or	$2x - 5 > 3$	Apply statement (6).
$2x - 5 + 5 < -3 + 5$	or	$2x - 5 + 5 > 3 + 5$	Add 5 to each part.
$2x < 2$	or	$2x > 8$	Simplify.
$\frac{2x}{2} < \frac{2}{2}$	or	$\frac{2x}{2} > \frac{8}{2}$	Divide each part by 2.
$x < 1$	or	$x > 4$	Simplify.

The solution set is $\{x \mid x < 1 \text{ or } x > 4\}$. Using interval notation, the solution is $(-\infty, 1) \cup (4, \infty)$.

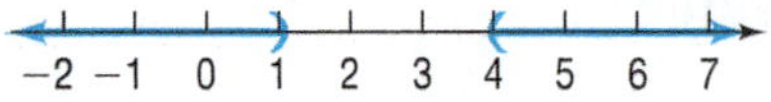

Figure 65 $\{x \mid x < 1 \text{ or } x > 4\}$

Graphing Solution

Graph $Y_1 = |2x - 5|$ and $Y_2 = 3$ on the same screen. See Figure 64. Using the INTERSECT command (twice), find that Y_1 and Y_2 intersect at $x = 1$ and at $x = 4$. The graph of Y_1 is above that of Y_2, $Y_1 > Y_2$, to the left of $x = 1$ and to the right of $x = 4$. Because the inequality is strict, the solution set is $\{x \mid x < 1 \text{ or } x > 4\}$. Using interval notation, the solution is $(-\infty, 1) \cup (4, \infty)$.

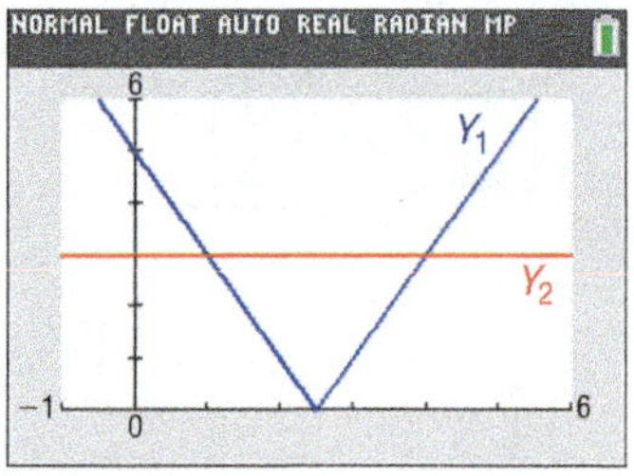

Figure 64 $Y_1 = |2x - 5|$; $Y_2 = 3$

See Figure 65 for a graph of the solution set.

Now Work PROBLEM 81

1.7 Assess Your Understanding

'Are You Prepared?' *Answers are given at the end of these exercises. If you get a wrong answer, read the pages listed in red.*

1. Graph the inequality: $x \geq -2$. (p. 20)
2. ***True or False*** The absolute value of a negative number is positive. (p. 20)
3. If $A = \{a, e, i, o, u\}$ and $B = \{a, b, c, d, e\}$, what is $A \cup B$? (pp. 2–3)

Concepts and Vocabulary

4. If each side of an inequality is multiplied by a(n) ________ number, then the sense of the inequality symbol is reversed.

5. A(n) ________ ________, denoted $[a, b]$, consists of all real numbers x for which $a \le x \le b$.

6. $|u| \le a$ is equivalent to ____________.

7. The interval ________ consists of all real numbers for which $x \le a$.

In Problems 8–10, assume that $a < b$ and $c < 0$.

8. ***True or False*** $a + c < b + c$

9. ***True or False*** $ac > bc$

10. ***True or False*** $\frac{a}{c} < \frac{b}{c}$

11. Which of the following has no solution?
(a) $|x| < -5$ (b) $|x| \le 0$ (c) $|x| > 0$ (d) $|x| \ge 0$

12. Which pair of inequalities is equivalent to $0 < x \le 3$?
(a) $x > 0$ and $x \ge 3$ (b) $x < 0$ and $x \ge 3$
(c) $x > 0$ and $x \le 3$ (d) $x < 0$ and $x \le 3$

Skill Building

In Problems 13–18, express the graph shown in color using interval notation. Also express each as an inequality involving x.

13. 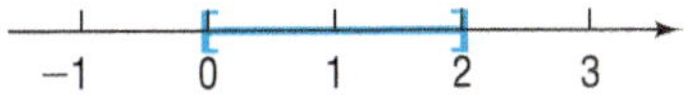

14. 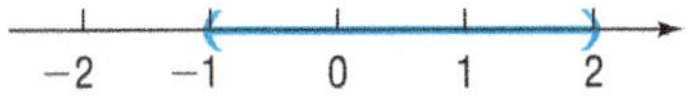

15. 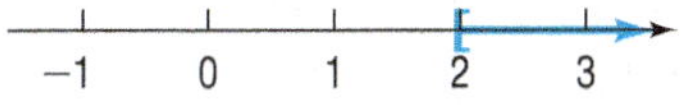

16. 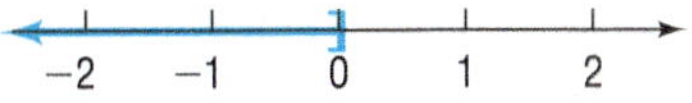

17. 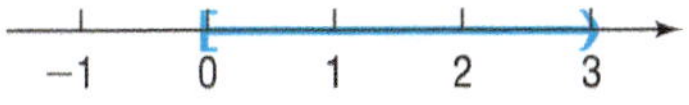

18. 

In Problems 19–22, an inequality is given. Write the inequality obtained by:

(a) Adding 3 to each side of the given inequality.
(b) Subtracting 5 from each side of the given inequality.
(c) Multiplying each side of the given inequality by 3.
(d) Multiplying each side of the given inequality by −2.

19. $3 < 5$ **20.** $2 > 1$ **21.** $2x + 1 < 2$ **22.** $1 - 2x > 5$

In Problems 23–30, write each inequality using interval notation, and illustrate each inequality using the real number line.

23. $0 \le x \le 4$ **24.** $-1 < x < 5$ **25.** $4 \le x < 6$ **26.** $-2 < x \le 0$

27. $x \ge 4$ **28.** $x \le 5$ **29.** $x < -4$ **30.** $x > 1$

In Problems 31–38, write each interval as an inequality involving x, and illustrate each inequality using the real number line.

31. $[2, 5]$ **32.** $(1, 2)$ **33.** $(-3, -2)$ **34.** $[0, 1)$

35. $[4, \infty)$ **36.** $(-\infty, 2]$ **37.** $(-\infty, -3)$ **38.** $(-8, \infty)$

In Problems 39–50, fill in the blank with the correct inequality symbol.

39. If $x < 5$, then $x - 5$ ________ 0.

40. If $x < -4$, then $x + 4$ ________ 0.

41. If $x > -4$, then $x + 4$ ________ 0.

42. If $x > 6$, then $x - 6$ ________ 0.

43. If $x \ge -4$, then $3x$ ________ -12.

44. If $x \le 3$, then $2x$ ________ 6.

45. If $x > 6$, then $-2x$ ________ -12.

46. If $x > -2$, then $-4x$ ________ 8.

47. If $2x < 6$, then x ________ 3.

48. If $3x \le 12$, then x ________ 4.

49. If $-\frac{1}{2}x \le 3$, then x ________ -6.

50. If $-\frac{1}{4}x > 1$, then x ________ -4.

In Problems 51–88, solve each inequality algebraically. Express your answer using set notation or interval notation. Graph the solution set. Verify your results using a graphing utility.

51. $3x - 7 > 2$ **52.** $2x + 5 > 1$ **53.** $1 - 2x \le 3$

54. $2 - 3x \le 5$ **55.** $3x - 1 \ge 3 + x$ **56.** $2x - 2 \ge 3 + x$

57. $-2(x + 3) < 8$ **58.** $-3(1 - x) < 12$ **59.** $4 - 3(1 - x) \le 3$

60. $8 - 4(2 - x) \le -2x$ **61.** $\frac{1}{2}(x - 4) > x + 8$ **62.** $3x + 4 > \frac{1}{3}(x - 2)$

63. $0 \le 2x - 6 \le 4$ **64.** $4 \le 2x + 2 \le 10$ **65.** $-5 \le 4 - 3x \le 2$

66. $-3 \le 3 - 2x \le 9$ **67.** $-3 < \frac{2x - 1}{4} < 0$ **68.** $0 < \frac{3x + 2}{2} < 4$

69. $1 < 1 - \frac{1}{2}x < 4$ **70.** $0 < 1 - \frac{1}{3}x < 1$ **71.** $(x + 2)(x - 3) > (x - 1)(x + 1)$

72. $(x - 1)(x + 1) > (x - 3)(x + 4)$ **73.** $|2x| < 8$ **74.** $|3x| < 15$

75. $|3x| > 12$

76. $|2x| > 6$

77. $|3t - 2| \leq 4$

78. $|2u + 5| \leq 7$

79. $|x - 2| + 2 < 3$

80. $|x + 4| + 3 < 5$

81. $|x - 3| \geq 2$

82. $|x + 4| \geq 2$

83. $|1 - 2x| > |-3|$

84. $|2 - 3x| > |-1|$

85. $|1 - 4x| - 7 < -2$

86. $|1 - 2x| - 4 < -1$

87. $|2x + 1| < -1$

88. $|3x - 4| \geq 0$

Mixed Practice

In Problems 89–108, solve each inequality algebraically. Express your answer using set notation or interval notation. Graph the solution set. Verify your results using a graphing utility.

89. $3 - 4x < 11$

90. $1 - 3x \leq 7$

91. $|2x + 1| - 5 \geq -1$

92. $|5x + 2| - 3 > 9$

93. $\frac{x}{2} \geq 1 - \frac{x}{4}$

94. $\frac{x}{3} \geq 2 + \frac{x}{6}$

95. $-\frac{1}{3} \leq \frac{x+1}{6} < \frac{4}{3}$

96. $-\frac{3}{2} < \frac{x-3}{4} \leq \frac{5}{4}$

97. $x(4x + 3) \leq (2x + 1)^2$

98. $x(9x - 5) \leq (3x - 1)^2$

99. $|(3x - 2) - 7| < \frac{1}{2}$

100. $|(4x - 1) - 11)| < \frac{1}{4}$

101. $-3 < 5 - 2x \leq 11$

102. $2 \leq 3 - 2(x + 1) < 8$

103. $7 - |x - 1| > 4$

104. $9 - |x + 3| \geq 5$

105. $-3 < x + 5 < 2x$

106. $2 < x - 3 < 2x$

107. $x + 2 < 2x - 1 < 5x$

108. $2x - 1 < 3x + 5 < 5x - 7$

Applications and Extensions

109. Express the fact that x differs from 2 by less than $\frac{1}{2}$ as an inequality involving an absolute value. Solve for x.

110. Express the fact that x differs from -1 by less than 1 as an inequality involving an absolute value. Solve for x.

111. Express the fact that x differs from -3 by more than 2 as an inequality involving an absolute value. Solve for x.

112. Express the fact that x differs from 2 by more than 3 as an inequality involving an absolute value. Solve for x.

113. A young adult may be defined as someone older than 21 but less than 30 years of age. Express this statement using inequalities.

114. Middle-aged may be defined as being 40 or more and less than 60. Express this statement using inequalities.

115. Body Temperature Normal human body temperature is 98.6°F. If a temperature x that differs from normal by at least 1.5°F is considered unhealthy, write the condition for an unhealthy temperature x as an inequality involving an absolute value, and solve for x.

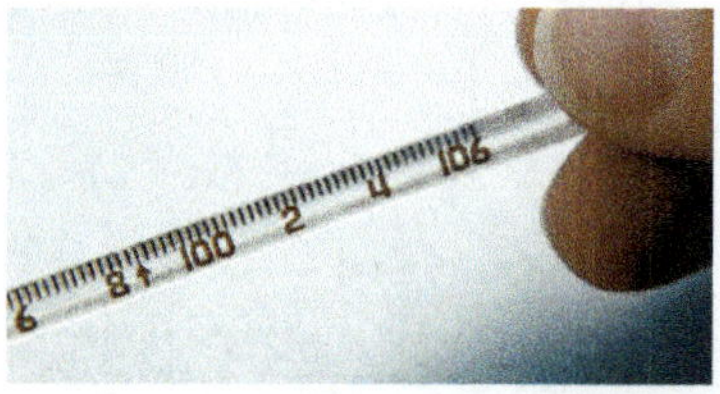

116. Household Voltage In the United States, normal household voltage is 115 volts. However, it is not uncommon for actual voltage to differ from normal voltage by at most 5 volts. Express this situation as an inequality involving an absolute value. Use x as the actual voltage and solve for x.

117. Life Expectancy The Social Security Administration determined that an average 30-year-old male in 2014 could expect to live at least 51.9 more years and an average 30-year-old female in 2014 could expect to live at least 55.6 more years.

(a) To what age can an average 30-year-old male expect to live? Express your answer as an inequality.

(b) To what age can an average 30-year-old female expect to live? Express your answer as an inequality.

(c) Who can expect to live longer, a male or a female? By how many years?

Source: *National Vital Statistics Report, 2014.*

118. General Chemistry For a certain ideal gas, the volume V (in cubic centimeters) equals 20 times the temperature T in kelvins (K). If the temperature varies from 353 K to 393 K, inclusive, what is the corresponding range of the volume of the gas?

119. Real Estate A real estate agent agrees to sell a large apartment complex according to the following commission schedule: \$45,000 plus 25% of the selling price in excess of \$900,000. Assuming that the complex will sell at some price between \$900,000 and \$1,100,000, inclusive, over what range does the agent's commission vary? How does the commission vary as a percent of selling price?

120. Sales Commission A used car salesperson is paid a commission of \$25 plus 40% of the selling price in excess of owner's cost. The owner claims that used cars typically sell for at least owner's cost plus \$70 and at most owner's cost plus \$300. For each sale made, over what range can the salesperson expect the commission to vary?

121. Federal Tax Withholding The percentage method of withholding for federal income tax (2014) states that a single person whose weekly wages, after subtracting withholding allowances, are over \$753, but not over \$1762, shall have \$97.75 plus 25% of the excess over \$753 withheld. Over what range does the amount withheld vary if the weekly wages vary from \$900 to \$1100, inclusive?

Source: *Employer's Tax Guide. Internal Revenue Service, 2014.*

122. Exercising Sue wants to lose weight. For healthy weight loss, the American College of Sports Medicine (ACSM) recommends 200 to 300 minutes of exercise per week. For the first six days of the week, Sue exercised 40, 45, 0, 50, 25, and 35 minutes. How long should Sue exercise on the seventh day in order to stay within the ACSM guidelines?

123. Electricity Rates Suppose Commonwealth Edison Company's charge for electricity in 2014 was 8.21¢ per kilowatt-hour. In addition, each monthly bill contained a customer charge of \$15.37. If that year's bills ranged from a low of \$72.84 to a high of \$237.04, over what range did usage vary (in kilowatt-hours)?

Source: *Commonwealth Edison Co., 2014.*

124. Water Bills The Village of Oak Lawn charges homeowners \$57.07 per quarter-year plus \$5.81 per 1000 gallons for water usage in excess of 10,000 gallons. In 2014, one homeowner's quarterly bill ranged from a high of \$150.03 to a low of \$97.74. Over what range did water usage vary?

Source: *Village of Oak Lawn, Illinois, 2014.*

125. Markup of a Used Car The markup over dealer's cost of a used car ranges from 12% to 18%. If the sticker price is \$8800, over what range will the dealer's cost vary?

126. IQ Tests A standard intelligence test has an average score of 100. According to statistical theory, of the people who take the test, the 2.5% with the highest scores will have scores of more than 1.96σ above the average, where σ (sigma, a number called the *standard deviation*) depends on the nature of the test. If $\sigma = 12$ for this test and there is (in principle) no upper limit to the score possible on the test, write the interval of possible test scores of the people in the top 2.5%.

127. Computing Grades In your Economics 101 class, you have scores of 68, 82, 87, and 89 on the first four of five tests. To get a grade of B, the average of the first five test scores must be greater than or equal to 80 and less than 90.

(a) Find the range of the score that you need on the last test to get a B.

(b) What score do you need if the fifth test counts double?

128. "Light" Foods For food products to be labeled "light," the U.S. Food and Drug Administration requires that the altered product must either contain at least one-third fewer calories than the regular product or contain at least one-half less fat than the regular product. If a serving of Miracle Whip® Light contains 20 calories and 1.5 grams of fat, then what must be true about either the number of calories or the grams of fat in a serving of regular Miracle Whip®?

129. Reading Books A HuffPost/YouGov poll found that in 2013 Americans read an average of 13.6 books per year. Suppose HuffPost/YouGov is 99% confident that the result from this poll is off by fewer than 1.8 books from the actual average x. Express this situation as an inequality involving absolute value, and solve the inequality for x to determine the interval in which the actual average is likely to fall. [**Note:** In statistics, this interval is called a 99% **confidence interval**.]

130. Speed of Sound According to data from the Hill Aerospace Museum (Hill Air Force Base, Utah), the speed of sound varies depending on altitude, barometric pressure, and temperature. For example, at 20,000 feet, 13.75 inches of mercury, and −12.3°F, the speed of sound is about 707 miles per hour, but the speed can vary from this result by as much as 55 miles per hour as conditions change.

(a) Express this situation as an inequality involving an absolute value.

(b) Using x for the speed of sound, solve for x to find an interval for the speed of sound.

131. Arithmetic Mean If $a < b$, show that $a < \frac{a+b}{2} < b$. The number $\frac{a+b}{2}$ is called the **arithmetic mean** of a and b.

132. Refer to Problem 131. Show that the arithmetic mean of a and b is equidistant from a and b.

133. Geometric Mean If $0 < a < b$, show that $a < \sqrt{ab} < b$. The number $\sqrt{ab}$ is called the **geometric mean** of a and b.

134. Refer to Problems 131 and 133. Show that the geometric mean of a and b is less than the arithmetic mean of a and b.

135. Harmonic Mean For $0 < a < b$, let h be defined by

$$\frac{1}{h} = \frac{1}{2}\left(\frac{1}{a} + \frac{1}{b}\right)$$

Show that $a < h < b$. The number h is called the **harmonic mean** of a and b.

136. Refer to Problems 131, 133, and 135. Show that the harmonic mean of a and b equals the geometric mean squared divided by the arithmetic mean.

Explaining Concepts: Discussion and Writing

137. Make up an inequality that has no solution. Make up one that has exactly one solution.

138. How would you explain to a fellow student the underlying reason for the multiplication properties for inequalities (page 149)? That is, the sense or direction of an inequality remains the same if each side is multiplied by a positive real number, while the direction is reversed if each side is multiplied by a negative real number.

139. The inequality $x^2 + 1 < -5$ has no solution. Explain why.

140. Do you prefer to use inequality notation or interval notation to express the solution to an inequality? Give your reasons. Are there particular circumstances when you prefer one to the other? Cite examples.

'Are You Prepared?' Answers

1. 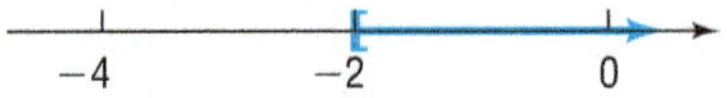

2. True

3. $A \cup B = \{a, b, c, d, e, i, o, u\}$

Chapter Review

Things to Know

Formulas

Distance formula (p. 85) $\quad d = \sqrt{(x_2 - x_1)^2 + (y_2 - y_1)^2}$

Midpoint formula (p. 88) $\quad M = (x, y) = \left(\frac{x_1 + x_2}{2}, \frac{y_1 + y_2}{2}\right)$

Quadratic equation and quadratic formula

The real solutions of the equation $ax^2 + bx + c = 0$, $a \neq 0$, are given by $x = \frac{-b \pm \sqrt{b^2 - 4ac}}{2a}$, provided $b^2 - 4ac \geq 0$.

If $b^2 - 4ac < 0$, there are no real solutions. (p. 114)

In the complex number system, the solutions of the equation $ax^2 + bx + c = 0$, $a \neq 0$, are given by $x = \frac{-b \pm \sqrt{b^2 - 4ac}}{2a}$. (p. 126)

Discriminant (pp. 114 and 127)

If $b^2 - 4ac > 0$, there are two unequal real solutions.

If $b^2 - 4ac = 0$, there is one repeated real solution—a double root, or a root of multiplicity 2.

If $b^2 - 4ac < 0$, there are no real solutions, but there are two distinct complex solutions that are not real; the complex solutions are conjugates of each other.

Interval notation (p. 147)

$[a, b]$	$\{x \mid a \leq x \leq b\}$	(a, b)	$\{x \mid a < x < b\}$	$(-\infty, a]$	$\{x \mid x \leq a\}$
$[a, b)$	$\{x \mid a \leq x < b\}$	$[a, \infty)$	$\{x \mid x \geq a\}$	$(-\infty, a)$	$\{x \mid x < a\}$
$(a, b]$	$\{x \mid a < x \leq b\}$	(a, ∞)	$\{x \mid x > a\}$	$(-\infty, \infty)$	All real numbers

Properties of inequalities

Addition property (p. 148)
- If $a < b$, then $a + c < b + c$.
- If $a > b$, then $a + c > b + c$.

Multiplication properties (p. 149)
- (a) If $a < b$ and if $c > 0$, then $ac < bc$.
  If $a < b$ and if $c < 0$, then $ac > bc$.
- (b) If $a > b$ and if $c > 0$, then $ac > bc$.
  If $a > b$ and if $c < 0$, then $ac < bc$.

Absolute value

If $|u| = a$, $a > 0$, then $u = -a$ or $u = a$. (p. 133)

If $|u| \leq a$, $a > 0$, then $-a \leq u \leq a$. (p. 153)

If $|u| \geq a$, $a > 0$, then $u \leq -a$ or $u \geq a$. (p. 154)

Objectives

Section		You should be able to	Example(s)	Review Exercises
1.1	1	Use the distance formula (p. 85)	1, 2, 3	45(a)–47(a), 53, 54
	2	Use the midpoint formula (p. 87)	4	45(b)–47(b)
	3	Graph equations by plotting points (p. 88)	5, 6, 7	50–52
	4	Graph equations using a graphing utility (p. 90)	8, 9	48, 50–52
	5	Use a graphing utility to create tables (p. 92)	10	48
	6	Find intercepts from a graph (p. 93)	11	49
	7	Use a graphing utility to approximate intercepts (p. 93)	12	50–52
1.2	1	Solve equations using a graphing utility (p. 100)	1, 2	27, 28
	2	Solve linear equations (p. 102)	3, 4	1–3
	3	Solve rational equations (p. 103)	5, 6	4, 25
	4	Solve problems that can be modeled by linear equations (p. 105)	7, 8	69
1.3	1	Solve quadratic equations by factoring (p. 110)	1, 2	5, 7, 18, 19
	2	Solve quadratic equations using the Square Root Method (p. 112)	3	26
	3	Solve quadratic equations by completing the square (p. 113)	4	5, 7, 8, 10, 11, 18, 19
	4	Solve quadratic equations using the quadratic formula (p. 113)	5, 6	5, 7, 8, 10, 11, 18, 19
	5	Solve problems that can be modeled by quadratic equations (p. 116)	7	63, 67, 71, 72
1.4	1	Add, subtract, multiply, and divide complex numbers (p. 122)	1–8	36–40
	2	Solve quadratic equations in the complex number system (p. 125)	10, 11	41–44
1.5	1	Solve radical equations (p. 129)	1–3	9, 13–15, 20
	2	Solve equations quadratic in form (p. 131)	4, 5	12, 17
	3	Solve absolute value equations (p. 133)	6	21, 22
	4	Solve equations by factoring (p. 134)	7, 8	23, 24
1.6	1	Translate verbal descriptions into mathematical expressions (p. 138)	1	55
	2	Solve interest problems (p. 139)	2, 3	56, 57
	3	Solve mixture problems (p. 140)	4	65, 66
	4	Solve uniform motion problems (p. 141)	5, 6	58, 60–62, 68, 73
	5	Solve constant rate job problems (p. 143)	7	64, 70
1.7	1	Use interval notation (p. 147)	1, 2	29–35
	2	Use properties of inequalities (p. 148)	3–6	29–35
	3	Solve linear inequalities algebraically and graphically (p. 150)	7	29
	4	Solve combined inequalities algebraically and graphically (p. 151)	8	30, 31, 59
	5	Solve absolute value inequalities algebraically and graphically (p. 152)	9–12	32–35

Review Exercises

In Problems 1–26, find all the real solutions, if any, of each equation. (Where they appear, a, b, m, and n are positive constants.) When possible, verify your results using a graphing utility.

1. $2 - \frac{x}{3} = 8$

2. $-2(5 - 3x) + 8 = 4 + 5x$

3. $\frac{3x}{4} - \frac{x}{3} = \frac{1}{12}$

4. $\frac{x}{x - 1} = \frac{6}{5}, \quad x \neq 1$

5. $x(1 - x) = 6$

6. $\frac{1 - 3x}{4} = \frac{x + 6}{3} + \frac{1}{2}$

7. $(x - 1)(2x + 3) = 3$

8. $2x + 3 = 4x^2$

9. $\sqrt[3]{x^2 - 1} = 2$

10. $x(x + 1) + 2 = 0$

11. $3x^2 - x + 1 = 0$

12. $x^4 - 5x^2 + 4 = 0$

13. $\sqrt{2x - 3} + x = 3$

14. $\sqrt[4]{2x + 3} = 2$

15. $\sqrt{x + 1} + \sqrt{x - 1} = \sqrt{2x + 1}$

16. $2x^{1/2} - 3 = 0$

17. $x^{-6} - 7x^{-3} - 8 = 0$

18. $x^2 + m^2 = 2mx + (nx)^2, \quad n \neq 1, n \neq -1$

19. $10a^2x^2 - 2abx - 36b^2 = 0$

20. $\sqrt{x^2 + 3x + 7} - \sqrt{x^2 - 3x + 9} + 2 = 0$

21. $|2x + 3| = 7$

22. $|2 - 3x| + 2 = 9$

23. $2x^3 = 3x^2$

24. $2x^3 + 5x^2 - 8x - 20 = 0$

25. $\dfrac{1}{x - 1} + \dfrac{3}{x + 2} = \dfrac{11}{x^2 + x - 2}$

26. $(x - 2)^2 = 9$

In Problems 27–28, use a graphing utility to approximate the solutions of each equation rounded to two decimal places. All solutions lie between −10 and 10.

27. $x^3 - 5x + 3 = 0$

28. $x^4 - 3 = 2x + 1$

In Problems 29–35, solve each inequality. Express your answer using set notation or interval notation. Graph the solution set. Verify your results using a graphing utility.

29. $\dfrac{2x - 3}{5} + 2 \leq \dfrac{x}{2}$

30. $-9 \leq \dfrac{2x + 3}{-4} \leq 7$

31. $2 < \dfrac{3 - 3x}{12} < 6$

32. $|3x + 4| < \dfrac{1}{2}$

33. $|2x - 5| \geq 9$

34. $2 + |2 - 3x| \leq 4$

35. $1 - |2 - 3x| < -4$

In Problems 36–40, use the complex number system and write each expression in the standard form $a + bi$.

36. $(6 + 3i) - (2 - 4i)$

37. $4(3 - i) + 3(-5 + 2i)$

38. $\dfrac{3}{3 + i}$

39. i^{50}

40. $(2 + 3i)^3$

In Problems 41–44, solve each equation in the complex number system.

41. $x^2 + x + 1 = 0$

42. $2x^2 + x - 2 = 0$

43. $x^2 + 3 = x$

44. $x(1 + x) = 2$

In Problems 45–47, find the following for each pair of points:
(a) *The distance between the points*
(b) *The midpoint of the line segment connecting the points*

45. $(0, 0); (4, 2)$

46. $(1, -1); (-2, 3)$

47. $(4, -4); (4, 8)$

48. Graph $y = -x^2 + 15$ using a graphing utility. Create a table of values to determine a good initial viewing window.

49. List the intercepts of the following graph.

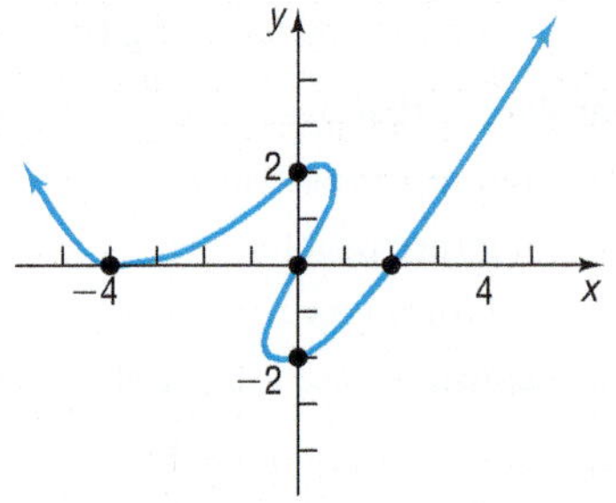

In Problems 50–52, graph each equation by plotting points. Verify your results using a graphing utility. Approximate the intercepts using a graphing utility, and label them on the graph.

50. $2x - 3y = 6$

51. $y = x^2 - 9$

52. $x^2 + 2y = 16$

---

53. Show that the points $A = (3, 4)$, $B = (1, 1)$, and $C = (-2, 3)$ are the vertices of an isosceles triangle.

54. Find two numbers y such that the distance from $(-3, 2)$ to $(5, y)$ is 10.

55. Translate the following statement into a mathematical expression: The perimeter p of a rectangle is the sum of two times the length l and two times the width w.

56. Banking A bank lends \$9000 at 7% simple interest. At the end of 1 year, how much interest is owed on the loan?

57. Financial Planning Steve, a recent retiree, requires \$5000 per year in extra income. He has \$70,000 to invest and can invest in A-rated bonds paying 8% per year or in a certificate of deposit (CD) paying 5% per year. How much money should be invested in each to realize exactly \$5000 in interest per year?

58. Lightning and Thunder A flash of lightning is seen, and the resulting thunderclap is heard 3 seconds later. If the speed of sound averages 1100 feet per second, how far away is the storm?

59. **Physics: Intensity of Light** The intensity I (in candlepower) of a certain light source obeys the equation $I = \frac{900}{x^2}$, where x is the distance (in meters) from the light. Over what range of distances can an object be placed from this light source so that the range of intensity of light is from 1600 to 3600 candlepower, inclusive?

60. **Extent of Search and Rescue** A search plane has a cruising speed of 250 miles per hour and carries enough fuel for at most 5 hours of flying. If there is a wind that averages 30 miles per hour and the direction of the search is with the wind one way and against it the other, how far can the search plane travel before it has to turn back?

61. **Rescue at Sea** A life raft, set adrift from a sinking ship 150 miles offshore, travels directly toward a Coast Guard station at the rate of 5 miles per hour. At the time that the raft is set adrift, a rescue helicopter is dispatched from the Coast Guard station. If the helicopter's average speed is 90 miles per hour, how long will it take the helicopter to reach the life raft?

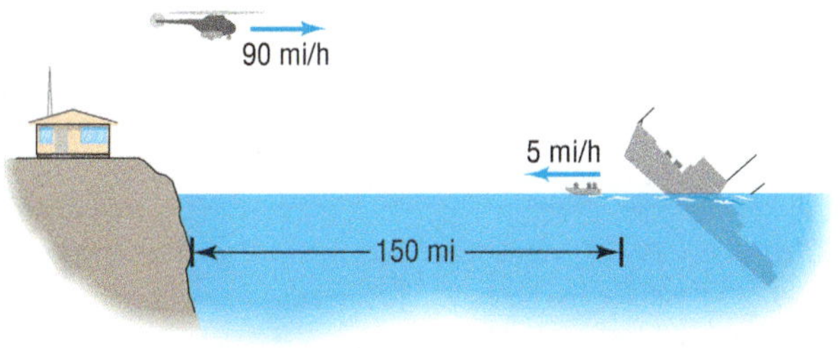

62. **Physics: Uniform Motion** Two bees leave two locations 150 meters apart and fly, without stopping, back and forth between these two locations at average speeds of 3 meters per second and 5 meters per second, respectively. How long is it until the bees meet for the first time? How long is it until they meet for the second time?

63. **Physics** An object is thrown down from the top of a building 1280 feet tall with an initial velocity of 32 feet per second. The distance s (in feet) of the object from the ground after t seconds is $s = 1280 - 32t - 16t^2$.

(a) When will the object strike ground?
(b) What is the height of the object after 4 seconds?

64. **Working Together to Get a Job Done** Clarissa and Shawna, working together, can paint the exterior of a house in 6 days. Clarissa by herself can complete this job in 5 days less than Shawna. How long will it take Clarissa to complete the job by herself?

65. **Business: Blending Coffee** A coffee house has 20 pounds of a coffee that sells for \$4 per pound. How many pounds of a coffee that sells for \$8 per pound should be mixed with the 20 pounds of \$4-per-pound coffee to obtain a blend that will sell for \$5 per pound? How much of the \$5-per-pound coffee is there to sell?

66. **Chemistry: Salt Solutions** How much water must be evaporated from 64 ounces of a 2% salt solution to make a 10% salt solution?

67. **Geometry** The hypotenuse of a right triangle measures 13 centimeters. Find the lengths of the legs if their sum is 17 centimeters.

68. **Physics: Uniform Motion** A man is walking at an average speed of 4 miles per hour alongside a railroad track. A freight train, going in the same direction at an average speed of 30 miles per hour, requires 5 seconds to pass the man. How long is the freight train? Give your answer in feet.

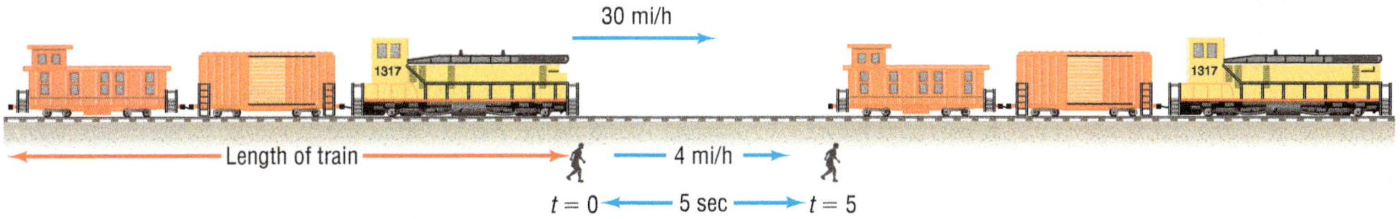

69. **Framing a Painting** An artist has 50 inches of oak trim to frame a painting. The frame is to have a border 3 inches wide surrounding the painting.

(a) If the painting is square, what are its dimensions? What are the dimensions of the frame?
(b) If the painting is rectangular with a length twice its width, what are the dimensions of the painting? What are the dimensions of the frame?

70. **Using Two Pumps** An 8-horsepower (hp) pump can fill a tank in 8 hours. A smaller, 3-hp pump fills the same tank in 12 hours. The pumps are used together to begin filling this tank. After 4 hours, the 8-hp pump breaks down. How long will it take the smaller pump to fill the tank?

71. **Pleasing Proportion** One formula stating the relationship between the length l and width w of a rectangle of "pleasing proportion" is $l^2 = w(l + w)$. How should a 4 foot by 8 foot sheet of plasterboard be cut so that the result is a rectangle of "pleasing proportion" with a width of 4 feet?

72. **Business: Determining the Cost of a Charter** A group of 20 senior citizens can charter a bus for a one-day excursion trip for \$15 per person. The charter company agrees to reduce the price of each ticket by 10¢ for each additional passenger in excess of 20 who goes on the trip, up to a maximum of 44 passengers (the capacity of the bus). If the final bill from the charter company was \$482.40, how many seniors went on the trip, and how much did each pay?

73. Evening Up a Race In a 100-meter race, Todd crosses the finish line 5 meters ahead of Scott. To even things up, Todd suggests to Scott that they race again, this time with Todd lining up 5 meters behind the start.

(a) Assuming that Todd and Scott run at the same pace as before, does the second race end in a tie?
(b) If not, who wins?
(c) By how many meters does he win?
(d) How far back should Todd start so that the race ends in a tie?

After running the race a second time, Scott, to even things up, suggests to Todd that he (Scott) line up 5 meters in front of the start.

(e) Assuming again that they run at the same pace as in the first race, does the third race result in a tie?
(f) If not, who wins?
(g) By how many meters?
(h) How far up should Scott start so that the race ends in a tie?

74. Explain the differences among the following three problems. Are there any similarities in their solution?

(a) Write the expression as a single quotient:

$$\frac{x}{x-2}+\frac{x}{x^2-4}$$

(b) Solve: $\frac{x}{x-2}+\frac{x}{x^2-4}=0$

(c) Solve: $\frac{x}{x-2}+\frac{x}{x^2-4}<0$

CHAPTER Test Prep VIDEOS

The Chapter Test Prep Videos are step-by-step solutions available in MyMathLab®, or on this text's YouTube™ Channel. Flip back to the Resources for Success page for a link to this text's YouTube channel.

Chapter Test

1. Suppose the points $(-2, -3)$ and $(4, 5)$ are the endpoints of a line segment.

(a) Find the distance between the two points.
(b) Find the midpoint of the line segment connecting the two points.

In Problems 2–9, solve each equation algebraically in the real number system. Answers should be exact.

2. $2x^2 + 6x = x - 3$

3. $x + 1 = \sqrt{x+7}$

4. $2 - \frac{3}{m} = \frac{2}{m+2}$

5. $5x - 8 = -4(x - 1) + 6$

6. $5|3 - 2b| - 7 = 8$

7. $x^4 + x^2 = 3x^2 + 8$

8. $x^2 - 4x + 2 = 0$

9. $2x^2 + x - 1 = x(x + 7) + 2$

In Problems 10 and 11, graph each equation by plotting points. Use a graphing utility to approximate the intercepts and label them on the graph.

10. $2x - 7y = 21$

11. $y = x^2 - 5$

In Problems 12–14, use a graphing utility to approximate the real solutions of each equation rounded to two decimal places. All solutions lie between −10 and 10.

12. $2x^3 - x^2 - 2x + 1 = 0$

13. $x^4 - 5x^2 - 8 = 0$

14. $-x^3 + 7x - 2 = x^2 + 3x - 3$

In Problems 15–18, solve each inequality. Express your answer in set-builder notation or interval notation. Graph the solution set.

15. $\frac{2x+3}{4} < -2$

16. $|2x + 3| - 4 \ge 3$

17. $-7 < 3 - 5x \le 8$

18. $|3x + 4| < 8$

In Problems 19–21, write each expression in the standard form $a + bi$.

19. $2(3 - 7i) - (4 + 11i)$

20. $(3 + 10i)(8 + i)$

21. $\frac{2+i}{5-3i}$

22. Solve the equation $4x^2 - 4x + 5 = 0$ in the complex number system.

23. Jamie is a cashier at a local supermarket. On average, she can check out a customer in 5 minutes. Scott, a cashier trainee, takes an average of 8.5 minutes to check out customers. How long would it take Jamie and Scott to check out 65 customers if they work together at different registers? Round to two decimal places if necessary.

24. A health food retailer sells a mixture of dried cherries, cranberries, and pecans for \$15.00 per pound and dried banana chips for \$2.25 per pound. The retailer decides to create a new mix by adding banana chips to the original mix. How many pounds of banana chips must be mixed with 40 pounds of the original mix to obtain a new mixture that sells for \$10.25 per pound with no loss in revenue? Round to two decimal places if necessary.

25. A Mountainsmith Auspex 4000 backpack originally sold for \$275.00 but is advertised at 42% off. What is the sale price of the backpack?

26. Glenn invests \$10,000 in a certificate of deposit (CD) that pays simple interest of 4% per annum. How much interest will Glenn earn after 3 months?

Chapter Projects

I. Financing a Purchase At some point in your life, you are likely to need to borrow money to finance a purchase. For example, most of us will finance the purchase of a car or a home. What is the mathematics behind financing a purchase? When you borrow money from a bank, the bank uses a rather complex equation (or formula) to determine how much you need to pay each month to repay the loan. There are a number of variables that determine the monthly payment. These variables include the amount borrowed, the interest rate, and the length of the loan. The interest rate is based on current economic conditions, the length of the loan, the type of item being purchased, and your credit history.

The following formula gives the monthly payment P required to pay off a loan amount L at an annual interest rate r, expressed as a decimal, but usually given as a percent. The time t, measured in months, is the length of the loan. For example, a 30-year loan requires $12 \times 30 = 360$ monthly payments.

$$P = L\left[\frac{\frac{r}{12}}{1-\left(1+\frac{r}{12}\right)^{-t}}\right]$$

P = monthly payment
L = loan amount
r = annual rate of interest expressed as a decimal
t = length of loan, in months

1. Interest rates change daily. Many websites post current interest rates on loans. Go to www.bankrate.com (or some other website that posts lenders' interest rates) and find the current best interest rate on a 60-month new-car purchase loan. Use this rate to determine the monthly payment on a $30,000 automobile loan.
2. Determine the total amount paid for the loan by multiplying the loan payment by the term of the loan.
3. Determine the total amount of interest paid by subtracting the loan amount from the total amount paid from question 2.
4. More often than not, we decide how much of a payment we can afford and use that information to determine the loan amount. Suppose you can afford a monthly payment of $500. Use the interest rate from question 1 to determine the maximum amount you can borrow. If you have $5000 to put down on the car, what is the maximum value of a car you can purchase?
5. Repeat questions 1 through 4 using a 72-month new-car purchase loan, a 60-month used-car purchase loan, and a 72-month used-car purchase loan.
6. We can use the power of a spreadsheet, such as Excel, to create a loan amortization schedule. A loan amortization schedule is a list of the monthly payments, a breakdown of interest and principal, along with a current loan balance. Create a loan amortization schedule for each of the four loan scenarios discussed above, using the following as a guide. You may want to use an Internet search engine to research specific keystrokes for creating an amortization schedule in a spreadsheet. We supply a sample spreadsheet with formulas included as a guide. Use the spreadsheet to verify your results from questions 1 through 5.

Loan Information		Payment Number	Payment Amount	Interest	Principal	Balance	Total Interest Paid
Loan Amount	$30,000.00	1	=PMT($B$3/12,$B$5,–$B$2,0)	=B2*$B$3/12	=E2–F2	=B2–G2	=B2*$B$3/12
Annual Interest Rate	0.045	2	=PMT($B$3/12,$B$5,–$B$2,0)	=H2*$B$3/12	=E3–F3	=H2–G3	=I2+F3
Length of Loan (years)	5	3	=PMT($B$3/12,$B$5,–$B$2,0)	=H3*$B$3/12	=E4–F4	=H3–G4	=I3+F4
Number of Payments	=B4*12	⋮	⋮	⋮	⋮	⋮	⋮

7. Go to an online automobile website such as www.cars.com, www.edmunds.com, or www.autobytel.com. Research the types of vehicles you can afford for a monthly payment of $500. Decide on a vehicle you would purchase based on your analysis in questions 1–6. Be sure to justify your decision, and include the impact the term of the loan has on your decision. You might consider other factors in your decision, such as expected maintenance costs and insurance costs.

The following project is also available on the Instructor's Resource Center (IRC):

II. Project at Motorola: How Many Cellular Phones Can I Make? An industrial engineer uses a model involving equations to be sure production levels meet customer demand.

2 Graphs

How to Value a House

Two things to consider in valuing a home are, first, how does it compare to similar homes that have sold recently? Is the asking price fair? And second, what value do you place on the advertised features and amenities? Yes, other people might value them highly, but do you?

Zestimate home valuation, RealestateABC.com, and Reply.com are among the many algorithmic (generated by a computer model) starting points in figuring out the value of a home. They show you how the home is priced relative to other homes in the area, but you need to add in all the things that only someone who has seen the house knows. You can do that using My Estimator, and then you create your own estimate and see how it stacks up against the asking price.

Looking at "Comps"

Knowing whether an asking price is fair will be important when you're ready to make an offer on a house. It will be even more important when your mortgage lender hires an appraiser to determine whether the house is worth the loan you want.

Check with your agent, Zillow.com, propertyshark.com, or other websites to see recent sales of homes in the area that are similar, or comparable, to what you're looking for. Print them out and keep these "comps" in a three-ring binder; you'll be referring to them quite a bit.

Note that "recent sales" usually means within the last six months. A sales price from a year ago may bear little or no relation to what is going on in your area right now. In fact, some lenders will not accept comps older than three months.

Market activity also determines how easy or difficult it is to find accurate comps. In a "hot" or busy market, with sales happening all the time, you're likely to have lots of comps to choose from. In a less active market, finding reasonable comps becomes harder. And if the home you're looking at has special design features, finding a comparable property is harder still. It's also necessary to know what's going on in a given sub-segment. Maybe large, high-end homes are selling like hotcakes, but owners of smaller houses are staying put, or vice versa.

Source: *http://allmyhome.blogspot.com/2008/07/how-to-value-house.html*

—See the Internet-based Chapter Project—

Outline

••• A Look Back

Chapter R reviewed algebra essentials and geometry essentials. Chapter 1 introduced the Cartesian plane to graph equations in two variables and examined equations in one variable.

A Look Ahead •••

Here we continue our discussion of the Cartesian plane by introducing additional techniques that can be used to obtain the graph of an equation with two variables. We also introduce two specific graphs: lines and circles. We conclude the chapter by looking at variation.

2.1 Intercepts; Symmetry; Graphing Key Equations

PREPARING FOR THIS SECTION *Before getting started, review the following:*

- Introduction to Graphing Equations (Section 1.1, p. 88–92)
- Solving Quadratic Equations (Section 1.3, pp. 110–116)
- Finding Intercepts Using a Graphing Utility (Section 1.1, pp. 93–94)
- Finding Intercepts from a Graph (Section 1.1, p. 93)

Now Work the 'Are You Prepared?' problems on page 170.

OBJECTIVES
1 Find Intercepts Algebraically from an Equation (p. 165)
2 Test an Equation for Symmetry (p. 166)
3 Know How to Graph Key Equations (p. 168)

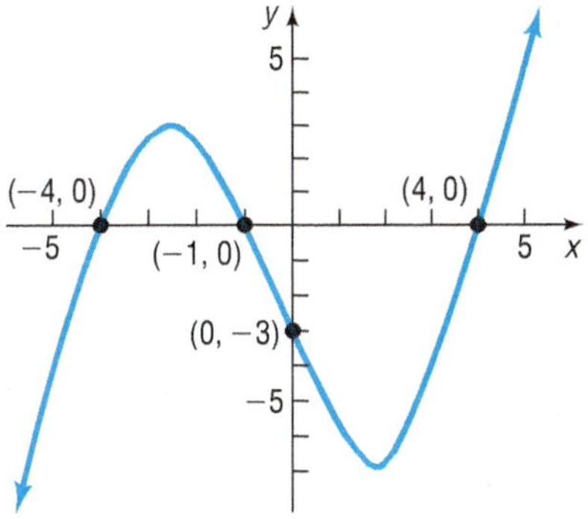

Figure 1

1 Find Intercepts Algebraically from an Equation

In Section 1.1, we discussed how to find intercepts from a graph and how to approximate intercepts from an equation using a graphing utility. Now we discuss how to find intercepts from an equation algebraically. To better understand the procedure, look at Figure 1. From the graph, note that the intercepts are $(-4, 0)$, $(-1, 0)$, $(4, 0)$, and $(0, -3)$. The x-intercepts are $-4, -1$, and 4. The y-intercept is -3. Notice that x-intercepts have y-coordinates that equal 0; y-intercepts have x-coordinates that equal 0. This leads to the following procedure for finding intercepts.

Procedure for Finding Intercepts

1. To find the x-intercept(s), if any, of the graph of an equation, let $y = 0$ in the equation and solve for x, where x is a real number.
2. To find the y-intercept(s), if any, of the graph of an equation, let $x = 0$ in the equation and solve for y, where y is a real number.

EXAMPLE 1 **Finding Intercepts from an Equation**

Find the x-intercept(s) and the y-intercept(s) of the graph of $y = x^2 - 4$. Then graph $y = x^2 - 4$ by plotting points.

Solution To find the x-intercept(s), let $y = 0$ and obtain the equation

$$x^2 - 4 = 0 \qquad y = x^2 - 4 \text{ with } y = 0$$
$$(x + 2)(x - 2) = 0 \qquad \text{Factor.}$$
$$x + 2 = 0 \text{ or } x - 2 = 0 \qquad \text{Zero-Product Property}$$
$$x = -2 \text{ or } x = 2 \qquad \text{Solve.}$$

The equation has two solutions, -2 and 2. The x-intercepts are -2 and 2.

To find the y-intercept(s), let $x = 0$ in the equation.

$$y = x^2 - 4$$
$$= 0^2 - 4 = -4$$

The y-intercept is -4.

Since $x^2 \geq 0$ for all x, we deduce from the equation $y = x^2 - 4$ that $y \geq -4$ for all x. This information, the intercepts, and the points from Table 1 on the next page enable us to graph $y = x^2 - 4$ by hand. See Figure 2.

Table 1

x	$y = x^2 - 4$	(x, y)
-3	$(-3)^2 - 4 = 5$	$(-3, 5)$
-1	-3	$(-1, -3)$
1	-3	$(1, -3)$
3	5	$(3, 5)$

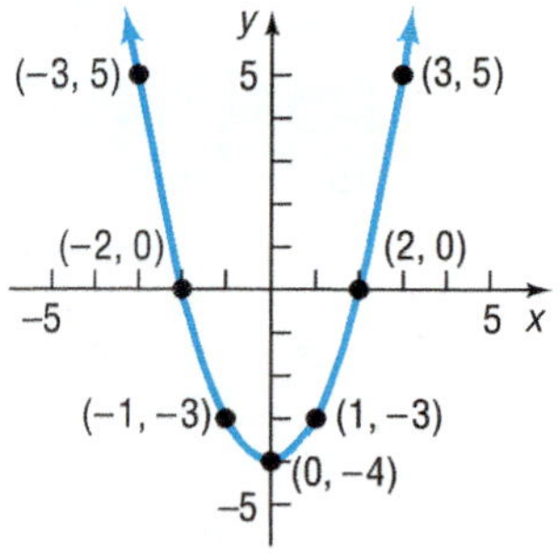

Figure 2 $y = x^2 - 4$

Now Work PROBLEM 15

2 Test an Equation for Symmetry

Another helpful tool for graphing equations by hand involves *symmetry*, particularly symmetry with respect to the x-axis, the y-axis, and the origin.

Symmetry often occurs in nature. Consider the picture of the butterfly. Do you see the symmetry?

DEFINITION

A graph is said to be **symmetric with respect to the x-axis** if, for every point (x, y) on the graph, the point $(x, -y)$ is also on the graph.

A graph is said to be **symmetric with respect to the y-axis** if, for every point (x, y) on the graph, the point $(-x, y)$ is also on the graph.

A graph is said to be **symmetric with respect to the origin** if, for every point (x, y) on the graph, the point $(-x, -y)$ is also on the graph.

Figure 3 illustrates the definition. Notice that, when a graph is symmetric with respect to the x-axis, the part of the graph above the x-axis is a reflection or mirror image of the part below it, and vice versa. When a graph is symmetric with respect to the y-axis, the part of the graph to the right of the y-axis is a reflection of the part to the left of it, and vice versa. Symmetry with respect to the origin may be viewed in two ways:

1. As a reflection about the y-axis, followed by a reflection about the x-axis
2. As a projection along a line through the origin so that the distances from the origin are equal

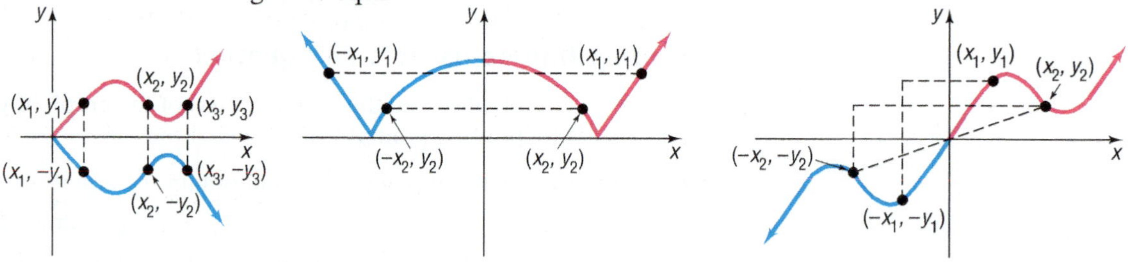

Symmetry with respect to the x-axis

Symmetry with respect to the y-axis

Symmetry with respect to the origin

Figure 3

EXAMPLE 2 **Symmetric Points**

(a) If a graph is symmetric with respect to the x-axis and the point $(4, 2)$ is on the graph, then the point $(4, -2)$ is also on the graph.

(b) If a graph is symmetric with respect to the y-axis and the point $(4, 2)$ is on the graph, then the point $(-4, 2)$ is also on the graph.

(c) If a graph is symmetric with respect to the origin and the point $(4, 2)$ is on the graph, then the point $(-4, -2)$ is also on the graph.

Now Work PROBLEM 23

When the graph of an equation is symmetric with respect to the x-axis, the y-axis, or the origin, the number of points that you need to plot in order to see the pattern is reduced. For example, if the graph of an equation is symmetric with respect to the y-axis, then once points to the right of the y-axis are plotted, an equal number of points on the graph can be obtained by reflecting them about the y-axis. Because of this, before we graph an equation, we first want to determine whether it has any symmetry. The following tests are used for this purpose.

Tests for Symmetry

To test the graph of an equation for symmetry with respect to the

x-Axis Replace y by $-y$ in the equation. If an equivalent equation results, the graph of the equation is symmetric with respect to the x-axis.

y-Axis Replace x by $-x$ in the equation. If an equivalent equation results, the graph of the equation is symmetric with respect to the y-axis.

Origin Replace x by $-x$ and y by $-y$ in the equation. If an equivalent equation results, the graph of the equation is symmetric with respect to the origin.

EXAMPLE 3 Finding Intercepts and Testing an Equation for Symmetry

For the equation $y = \dfrac{x^2 - 4}{x^2 + 1}$: (a) find the intercepts and (b) test for symmetry.

Solution (a) To obtain the x-intercept(s), let $y = 0$ in the equation and solve for x.

$$\frac{x^2 - 4}{x^2 + 1} = 0 \qquad \text{Let } y = 0.$$

$$x^2 - 4 = 0 \qquad \text{Multiply both sides by } x^2 + 1.$$

$$x = -2 \quad \text{or} \quad x = 2 \qquad \text{Factor and use the Zero-Product Property.}$$

To obtain the y-intercept(s), let $x = 0$ in the equation and solve for y.

$$y = \frac{x^2 - 4}{x^2 + 1} = \frac{0^2 - 4}{0^2 + 1} = \frac{-4}{1} = -4$$

The x-intercepts are -2 and 2; the y-intercept is -4.

(b) Now test the equation for symmetry with respect to the x-axis, the y-axis, and the origin.

x-Axis: To test for symmetry with respect to the x-axis, replace y by $-y$. Since

$$-y = \frac{x^2 - 4}{x^2 + 1} \text{ is not equivalent to } y = \frac{x^2 - 4}{x^2 + 1}$$

the graph of the equation is not symmetric with respect to the x-axis.

y-Axis: To test for symmetry with respect to the y-axis, replace x by $-x$. Since

$$y = \frac{(-x)^2 - 4}{(-x)^2 + 1} = \frac{x^2 - 4}{x^2 + 1} \text{ is equivalent to } y = \frac{x^2 - 4}{x^2 + 1}$$

the graph of the equation is symmetric with respect to the y-axis.

Origin: To test for symmetry with respect to the origin, replace x by $-x$ and y by $-y$.

$$-y = \frac{(-x)^2 - 4}{(-x)^2 + 1} \qquad \text{Replace } x \text{ by } -x \text{ and } y \text{ by } -y.$$

$$-y = \frac{x^2 - 4}{x^2 + 1} \qquad \text{Simplify.}$$

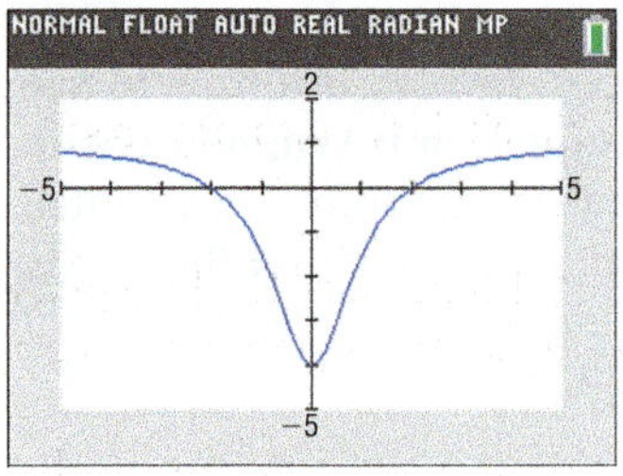

Figure 4 $y = \dfrac{x^2 - 4}{x^2 + 1}$

Since the result is not equivalent to the original equation, the graph of the equation $y = \dfrac{x^2 - 4}{x^2 + 1}$ is not symmetric with respect to the origin. ■

Seeing the Concept

Figure 4 shows the graph of $y = \dfrac{x^2 - 4}{x^2 + 1}$ using a TI-84 Plus C graphing calculator. Do you see the symmetry with respect to the y-axis? Also, did you notice that the point $(2, 0)$ is on the graph along with $(-2, 0)$? How could we have found the second x-intercept using symmetry? ■

Now Work PROBLEM 53

3 Know How to Graph Key Equations

There are certain equations whose graphs we should be able to easily visualize in our mind's eye. For example, you should know the graph of $y = x^2$ discussed in Example 7 in Section 1.1. The next three examples use intercepts, symmetry, and point plotting to obtain the graphs of additional key equations. It is important to know the graphs of these key equations because we use them later.

EXAMPLE 4 Graphing the Equation $y = x^3$ by Finding Intercepts and Checking for Symmetry

Graph the equation $y = x^3$ by hand by plotting points. Find any intercepts and check for symmetry first.

Solution First, find the intercepts of $y = x^3$. When $x = 0$, then $y = 0$; and when $y = 0$, then $x = 0$. The origin $(0, 0)$ is the only intercept. Now test $y = x^3$ for symmetry.

x-Axis: Replace y by $-y$. Since $-y = x^3$ is not equivalent to $y = x^3$, the graph is not symmetric with respect to the x-axis.

y-Axis: Replace x by $-x$. Since $y = (-x)^3 = -x^3$ is not equivalent to $y = x^3$, the graph is not symmetric with respect to the y-axis.

Origin: Replace x by $-x$ and y by $-y$. Since $-y = (-x)^3 = -x^3$ is equivalent to $y = x^3$ (multiply both sides by -1), the graph is symmetric with respect to the origin.

To graph by hand, use the equation to obtain several points on the graph. Because of the symmetry with respect to the origin, we only need to locate points on the graph for which $x \geq 0$. See Table 2. Points on the graph could also be obtained using the TABLE feature on a graphing utility. See Table 3. Do you see the symmetry with respect to the origin from the table? Figure 5 shows the graph.

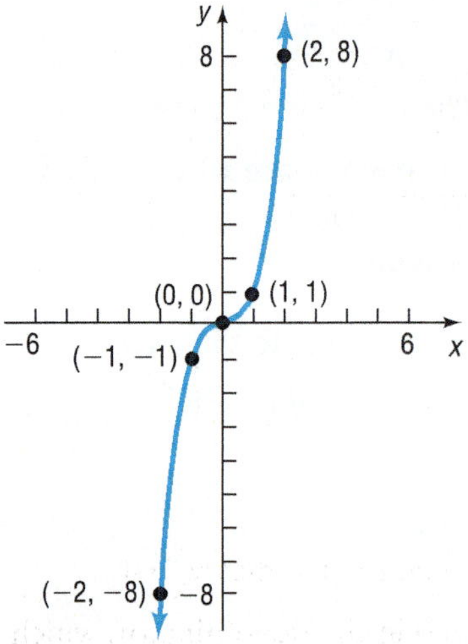

Figure 5 $y = x^3$

Table 2

x	$y = x^3$	(x, y)
0	0	(0, 0)
1	1	(1, 1)
2	8	(2, 8)
3	27	(3, 27)

Table 3

NORMAL FLOAT AUTO REAL RADIAN MP
PRESS ENTER TO EDIT

X	Y₁
-5	-125
-4	-64
-3	-27
-2	-8
-1	-1
0	0
1	1
2	8
3	27
4	64
5	125

$Y_1 = X^3$

■

EXAMPLE 5 Graphing the Equation $x = y^2$

(a) Graph the equation $x = y^2$. Find any intercepts and check for symmetry first.
(b) Graph $x = y^2$ where $y \geq 0$.

Solution (a) The lone intercept is $(0, 0)$. The graph is symmetric with respect to the x-axis since $x = (-y)^2$ is equivalent to $x = y^2$. The graph is not symmetric with respect to the y-axis or the origin.

To graph $x = y^2$ by hand, use the equation to obtain several points on the graph. Because the equation is solved for x, it is easier to assign values to y and use the equation to determine the corresponding values of x. Because of the symmetry, start by finding points whose y-coordinates are nonnegative. Then use the symmetry to find additional points on the graph. See Table 4. For example, since $(1, 1)$ is on the graph, so is $(1, -1)$. Since $(4, 2)$ is on the graph, so is $(4, -2)$, and so on. Plot these points and connect them with a smooth curve to obtain Figure 6.

Table 4

y	$x = y^2$	(x, y)
0	0	(0, 0)
1	1	(1, 1)
2	4	(4, 2)
3	9	(9, 3)

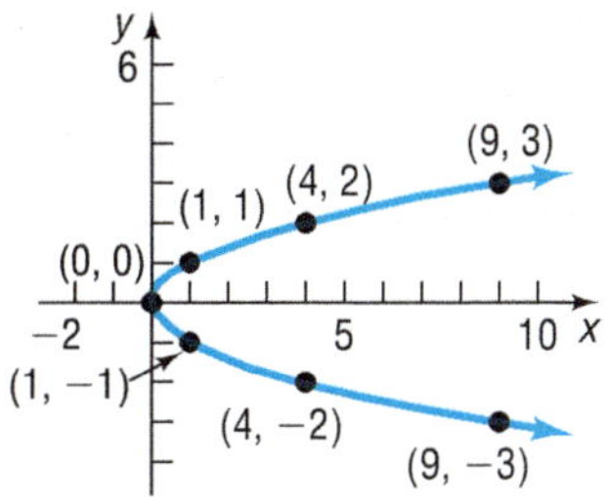

Figure 6 $x = y^2$

To graph the equation $x = y^2$ using a graphing utility, write the equation in the form $y = \{\text{expression in } x\}$. We proceed to solve for y.

$$x = y^2$$
$$y^2 = x$$
$$y = \pm\sqrt{x} \quad \textbf{Square Root Method}$$

To graph $x = y^2$, graph both $Y_1 = \sqrt{x}$ and $Y_2 = -\sqrt{x}$ on the same screen. Figure 7 shows the result. Table 5 shows various values of y for a given value of x when $Y_1 = \sqrt{x}$ and $Y_2 = -\sqrt{x}$. Notice that when $x < 0$ we get an error. Can you explain why?

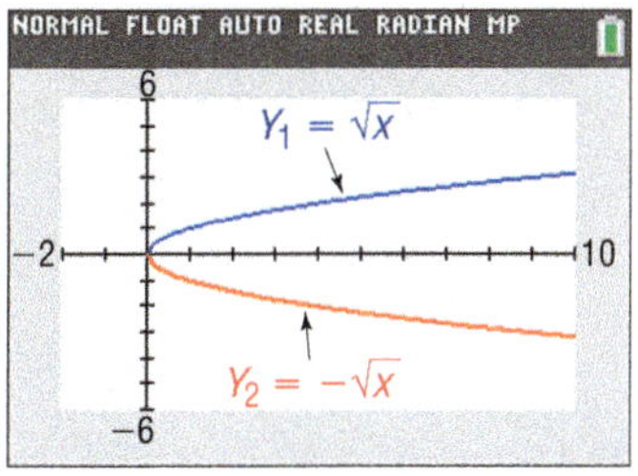

Figure 7

Table 5

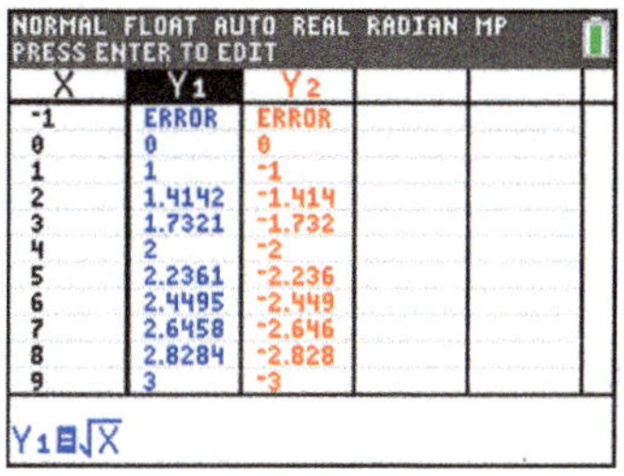
NORMAL FLOAT AUTO REAL RADIAN MP
PRESS ENTER TO EDIT

X	Y1	Y2
-1	ERROR	ERROR
0	0	0
1	1	-1
2	1.4142	-1.414
3	1.7321	-1.732
4	2	-2
5	2.2361	-2.236
6	2.4495	-2.449
7	2.6458	-2.646
8	2.8284	-2.828
9	3	-3

Y1=√X

(b) If we restrict y so that $y \geq 0$, the equation $x = y^2$, $y \geq 0$, may be written as $y = \sqrt{x}$. The portion of the graph of $x = y^2$ in quadrant I plus the origin is the graph of $y = \sqrt{x}$. See Figure 8.

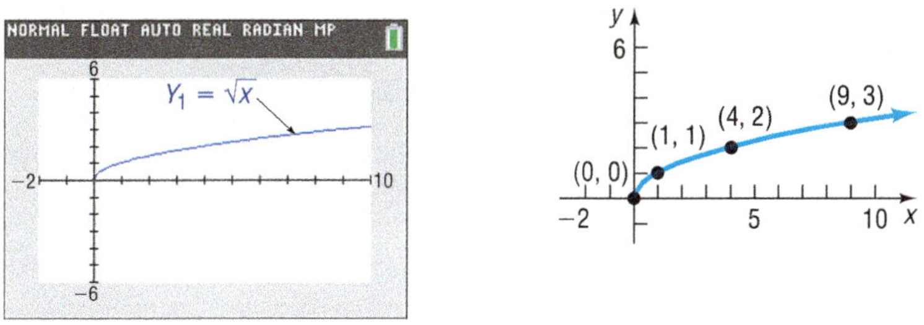

Figure 8 $y = \sqrt{x}$

■

EXAMPLE 6 Graphing the Equation $y = \dfrac{1}{x}$

Graph the equation $y = \dfrac{1}{x}$. Find any intercepts and check for symmetry first.

Solution Check for intercepts first. If we let $x = 0$, we obtain a 0 in the denominator, which is not defined. We conclude that there is no y-intercept. If we let $y = 0$, we get the

Table 6

x	$y = \frac{1}{x}$	(x, y)
$\frac{1}{10}$	10	$\left(\frac{1}{10}, 10\right)$
$\frac{1}{3}$	3	$\left(\frac{1}{3}, 3\right)$
$\frac{1}{2}$	2	$\left(\frac{1}{2}, 2\right)$
1	1	$(1, 1)$
2	$\frac{1}{2}$	$\left(2, \frac{1}{2}\right)$
3	$\frac{1}{3}$	$\left(3, \frac{1}{3}\right)$
10	$\frac{1}{10}$	$\left(10, \frac{1}{10}\right)$

equation $\frac{1}{x} = 0$, which has no solution. We conclude that there is no x-intercept. The graph of $y = \frac{1}{x}$ does not cross or touch the coordinate axes.

Next check for symmetry.

x-Axis: Replacing y by $-y$ yields $-y = \frac{1}{x}$, which is not equivalent to $y = \frac{1}{x}$.

y-Axis: Replacing x by $-x$ yields $y = \frac{1}{-x} = -\frac{1}{x}$, which is not equivalent to $y = \frac{1}{x}$.

Origin: Replacing x by $-x$ and y by $-y$ yields $-y = -\frac{1}{x}$, which is equivalent to $y = \frac{1}{x}$. The graph is symmetric only with respect to the origin.

Use the equation to form Table 6 and obtain some points on the graph. Because of symmetry, we only find points (x, y) for which x is positive. From Table 6 we infer that, if x is a large and positive number, then $y = \frac{1}{x}$ is a positive number close to 0. We also infer that if x is a positive number close to 0 then $y = \frac{1}{x}$ is a large and positive number. Armed with this information, we can graph the equation. Figure 9 illustrates some of these points and the graph of $y = \frac{1}{x}$. Observe how the absence of intercepts and the existence of symmetry with respect to the origin were utilized. Figure 10 confirms our algebraic analysis using a TI-84 Plus C.

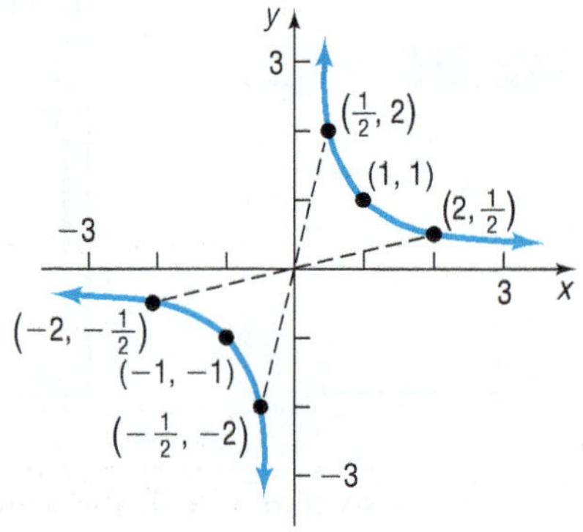

Figure 9 $y = \frac{1}{x}$

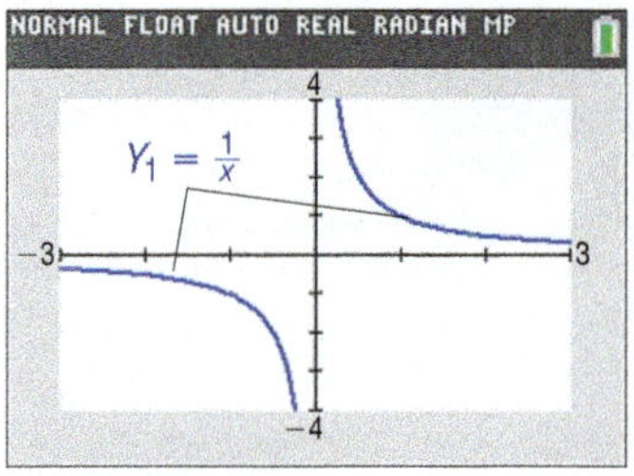

Figure 10 $y = \frac{1}{x}$

2.1 Assess Your Understanding

'Are You Prepared?' *Answers are given at the end of these exercises. If you get a wrong answer, read the pages listed in red.*

1. Graph $y = 2x - 4$ by plotting points. Based on the graph, determine the intercepts. (pp. 88–90, 93)

2. Solve: $x^2 - 4x - 12 = 0$ (pp. 110–116)

Concepts and Vocabulary

3. The points, if any, at which a graph crosses or touches the coordinate axes are called ________.

4. If for every point (x, y) on the graph of an equation the point $(-x, y)$ is also on the graph, then the graph is symmetric with respect to the ________.

5. If the graph of an equation is symmetric with respect to the y-axis and -4 is an x-intercept of this graph, then ________ is also an x-intercept.

6. If the graph of an equation is symmetric with respect to the origin and $(3, -4)$ is a point on the graph, then ________ is also a point on the graph.

7. ***True or False*** To find the y-intercepts of the graph of an equation, let $x = 0$ and solve for y.

8. ***True or False*** If a graph is symmetric with respect to the x-axis, then it cannot be symmetric with respect to the y-axis.

9. To find the x-intercept(s), if any, of the graph of an equation, let __________ in the equation and solve for x.
(a) $y = 0$ (b) $x = 0$
(c) $y = x$ (d) $x = -y$

10. To test whether the graph of an equation is symmetric with respect to the origin, replace __________ in the equation and simplify. If an equivalent equation results, then the graph is symmetric with respect to the origin.
(a) x by $-x$ (b) y by $-y$
(c) x by $-x$ and y by $-y$ (d) x by $-y$ and y by $-x$

Skill Building

In Problems 11–22, find the intercepts and graph each equation by plotting points. Be sure to label the intercepts.

11. $y = x + 2$ **12.** $y = x - 6$ **13.** $y = 2x + 8$ **14.** $y = 3x - 9$
15. $y = x^2 - 1$ **16.** $y = x^2 - 9$ **17.** $y = -x^2 + 4$ **18.** $y = -x^2 + 1$
19. $2x + 3y = 6$ **20.** $5x + 2y = 10$ **21.** $9x^2 + 4y = 36$ **22.** $4x^2 + y = 4$

In Problems 23–32, plot each point. Then plot the point that is symmetric to it with respect to (a) the x-axis; (b) the y-axis; (c) the origin.

23. $(3, 4)$ **24.** $(5, 3)$ **25.** $(-2, 1)$ **26.** $(4, -2)$ **27.** $(5, -2)$
28. $(-1, -1)$ **29.** $(-3, -4)$ **30.** $(4, 0)$ **31.** $(0, -3)$ **32.** $(-3, 0)$

In Problems 33–44, the graph of an equation is given. (a) Find the intercepts. (b) Indicate whether the graph is symmetric with respect to the x-axis, the y-axis, or the origin.

33.
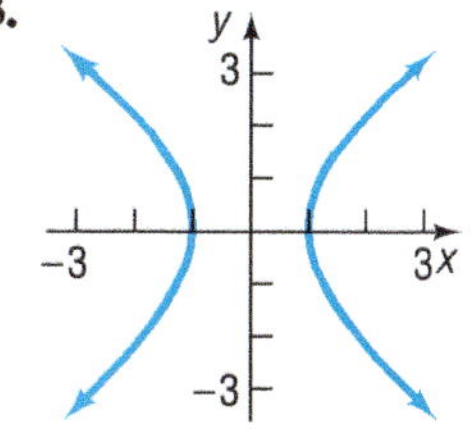

34.
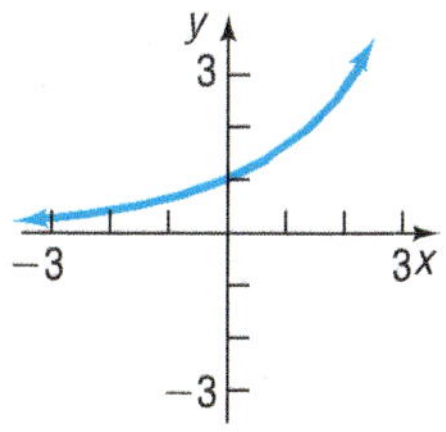

35.
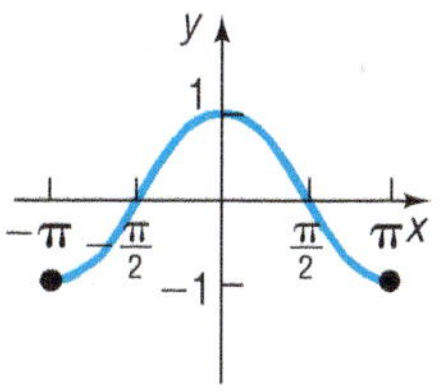

36.
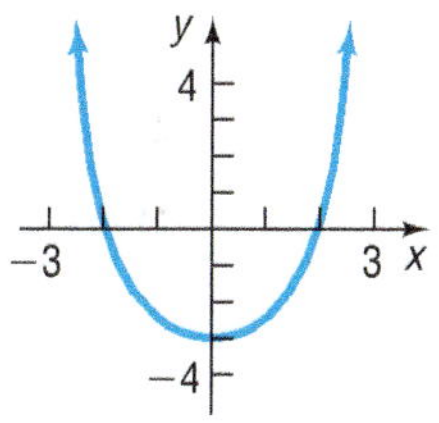

37.
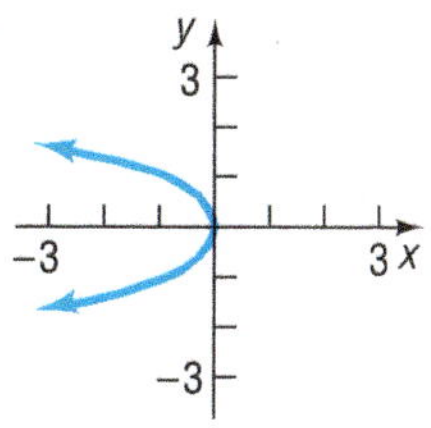

38.
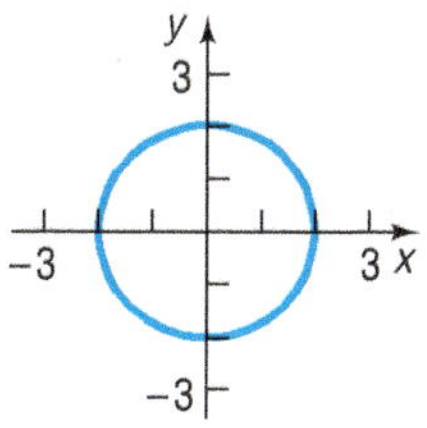

39.
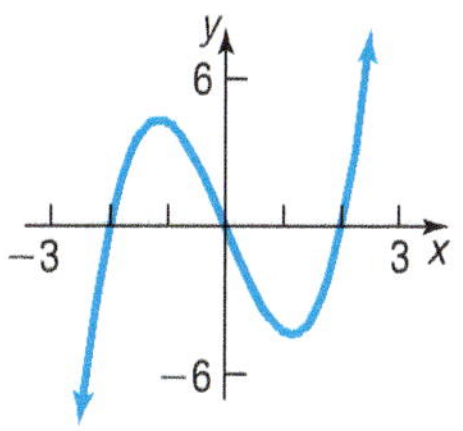

40.
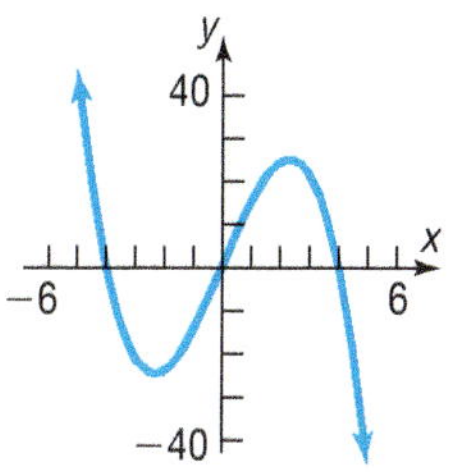

41.
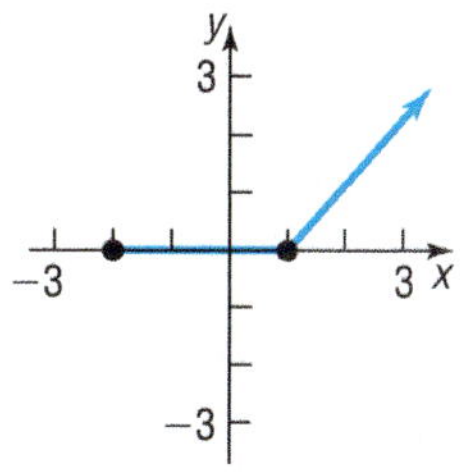

42.
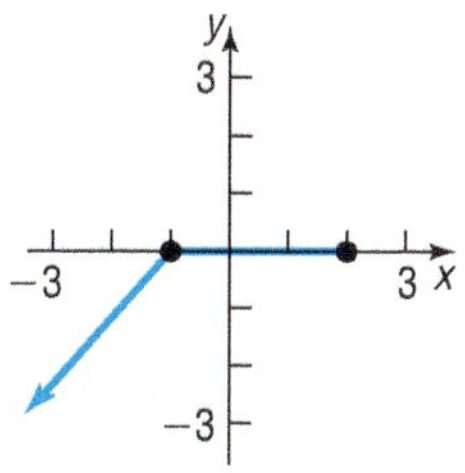

43.
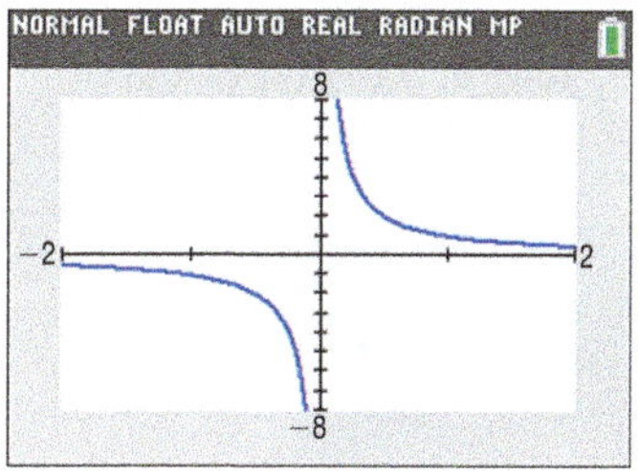

44.
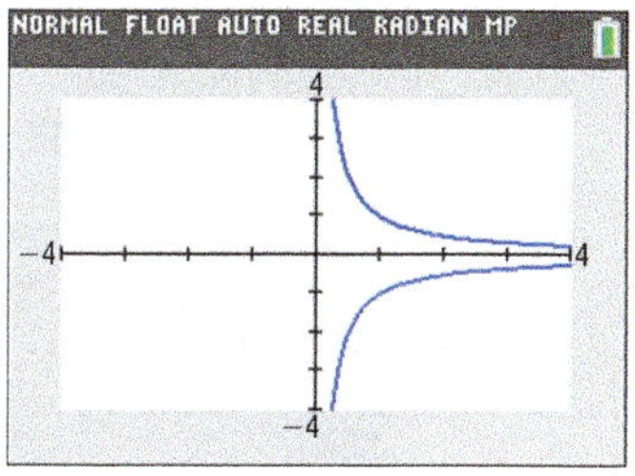

In Problems 45–48, draw a complete graph so that it has the type of symmetry indicated.

45. y-axis
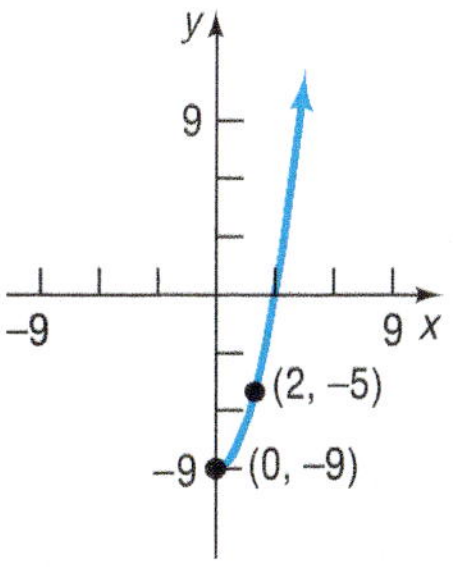

46. x-axis
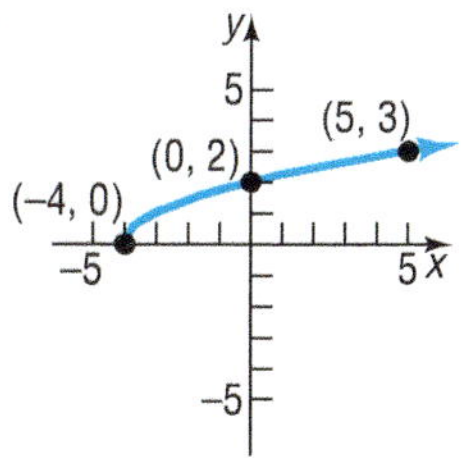

47. Origin
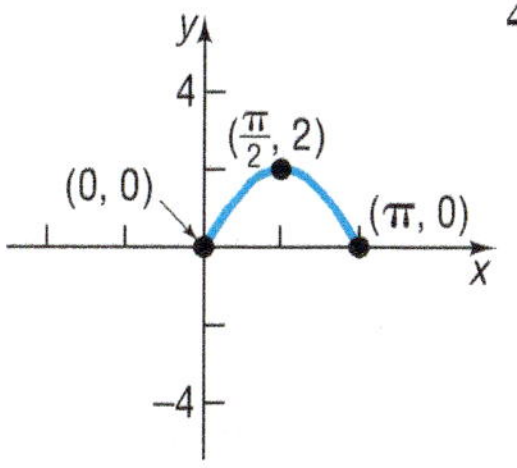

48. y-axis
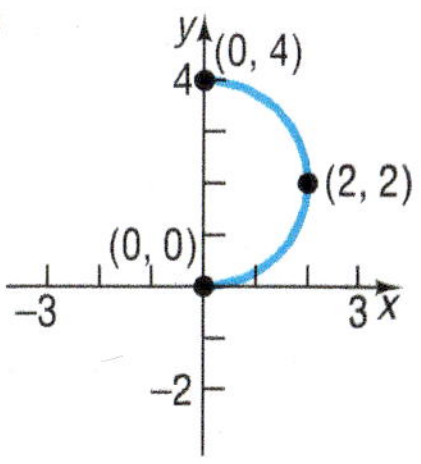

In Problems 49–64, list the intercepts and test for symmetry.

49. $y^2 = x + 4$

50. $y^2 = x + 9$

51. $y = \sqrt[3]{x}$

52. $y = \sqrt[5]{x}$

53. $y = x^4 - 8x^2 - 9$

54. $y = x^4 - 2x^2 - 8$

55. $9x^2 + 4y^2 = 36$

56. $4x^2 + y^2 = 4$

57. $y = x^3 - 27$

58. $y = x^4 - 1$

59. $y = x^2 - 3x - 4$

60. $y = x^2 + 4$

61. $y = \dfrac{3x}{x^2 + 9}$

62. $y = \dfrac{x^2 - 4}{2x}$

63. $y = \dfrac{-x^3}{x^2 - 9}$

64. $y = \dfrac{x^4 + 1}{2x^5}$

In Problems 65–68, draw a quick sketch of each equation.

65. $y = x^3$

66. $x = y^2$

67. $y = \sqrt{x}$

68. $y = \dfrac{1}{x}$

69. If $(3, b)$ is a point on the graph of $y = 4x + 1$, what is b?

70. If $(-2, b)$ is a point on the graph of $2x + 3y = 2$, what is b?

71. If $(a, 4)$ is a point on the graph of $y = x^2 + 3x$, what is a?

72. If $(a, -5)$ is a point on the graph of $y = x^2 + 6x$, what is a?

Mixed Practice

In Problems 73–80, (a) find the intercepts of each equation, (b) test each equation for symmetry with respect to the x-axis, the y-axis, and the origin, and (c) graph each equation by hand by plotting points. Be sure to label the intercepts on the graph and use any symmetry to assist in drawing the graph. Verify your results using a graphing utility.

73. $y = x^2 - 5$

74. $y = x^2 - 8$

75. $x - y^2 = -9$

76. $x + y^2 = 4$

77. $x^2 + y^2 = 9$

78. $x^2 + y^2 = 16$

79. $y = x^3 - 4x$

80. $y = x^3 - x$

Applications and Extensions

81. Given that the point (1, 2) is on the graph of an equation that is symmetric with respect to the origin, what other point is on the graph?

82. If the graph of an equation is symmetric with respect to the y-axis and 6 is an x-intercept of this graph, name another x-intercept.

83. If the graph of an equation is symmetric with respect to the origin and -4 is an x-intercept of this graph, name another x-intercept.

84. If the graph of an equation is symmetric with respect to the x-axis and 2 is a y-intercept, name another y-intercept.

85. Microphones In studios and on stages, cardioid microphones are often preferred for the richness they add to voices and for their ability to reduce the level of sound from the sides and rear of the microphone. Suppose one such cardioid pattern is given by the equation $(x^2 + y^2 - x)^2 = x^2 + y^2$.

(a) Find the intercepts of the graph of the equation.
(b) Test for symmetry with respect to the x-axis, y-axis, and origin.

Source: *www.notaviva.com*

86. Solar Energy The solar electric generating systems at Kramer Junction, California, use parabolic troughs to heat a heat-transfer fluid to a high temperature. This fluid is used to generate steam that drives a power conversion system to produce electricity. For troughs 7.5 feet wide, an equation for the cross-section is $16y^2 = 120x - 225$.

(a) Find the intercepts of the graph of the equation.
(b) Test for symmetry with respect to the x-axis, y-axis, and origin.

Source: *U.S. Department of Energy*

Explaining Concepts: Discussion and Writing

87. (a) Graph $y = \sqrt{x^2}$, $y = x$, $y = |x|$, and $y = (\sqrt{x})^2$, noting which graphs are the same.
(b) Explain why the graphs of $y = \sqrt{x^2}$ and $y = |x|$ are the same.
(c) Explain why the graphs of $y = x$ and $y = (\sqrt{x})^2$ are not the same.
(d) Explain why the graphs of $y = \sqrt{x^2}$ and $y = x$ are not the same.

88. Explain what is meant by a complete graph.

89. Draw a graph of an equation that contains two x-intercepts; at one the graph crosses the x-axis, and at the other the graph touches the x-axis.

90. Make up an equation with the intercepts $(2, 0)$, $(4, 0)$, and $(0, 1)$. Compare your equation with a friend's equation. Comment on any similarities.

91. Draw a graph that contains the points $(-2, -1)$, $(0, 1)$, $(1, 3)$, and $(3, 5)$. Compare your graph with those of other students. Are most of the graphs almost straight lines? How many are "curved"? Discuss the various ways that these points might be connected.

92. An equation is being tested for symmetry with respect to the x-axis, the y-axis, and the origin. Explain why, if two of these symmetries are present, the remaining one must also be present.

93. Draw a graph that contains the points $(-2, 5)$, $(-1, 3)$, and $(0, 2)$ that is symmetric with respect to the y-axis. Compare your graph with those of other students; comment on any similarities. Can a graph contain these points and be symmetric with respect to the x-axis? the origin? Why or why not?

Retain Your Knowledge

Problems 94–97 are based on material learned earlier in the course. The purpose of these problems is to keep the material fresh in your mind so that you are better prepared for the final exam.

94. Find the value of $\dfrac{x+y}{x-y}$ if $x = 6$ and $y = -2$.

95. Factor $3x^2 - 30x + 75$ completely.

96. Simplify: $\sqrt{-196}$

97. Solve $x^2 - 8x + 4 = 0$ by completing the square.

'Are You Prepared?' Answers

1.

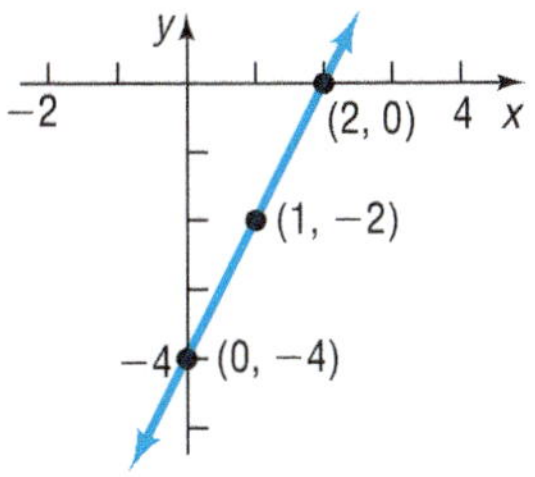

x-intercept: 2; y-intercept: -4

2. $\{-2, 6\}$

2.2 Lines

OBJECTIVES
1 Calculate and Interpret the Slope of a Line (p. 173)
2 Graph Lines Given a Point and the Slope (p. 176)
3 Find the Equation of a Vertical Line (p. 176)
4 Use the Point–Slope Form of a Line; Identify Horizontal Lines (p. 177)
5 Write the Equation of a Line in Slope–Intercept Form (p. 178)
6 Find the Equation of a Line Given Two Points (p. 179)
7 Graph Lines Written in General Form Using Intercepts (p. 180)
8 Find Equations of Parallel Lines (p. 181)
9 Find Equations of Perpendicular Lines (p. 182)

In this section we study a certain type of equation that contains two variables, called a *linear equation,* and its graph, a *line.*

1 Calculate and Interpret the Slope of a Line

Consider the staircase illustrated in Figure 11. Each step contains exactly the same horizontal **run** and the same vertical **rise**. The ratio of the rise to the run, called the *slope*, is a numerical measure of the steepness of the staircase. For example, if the run is increased and the rise remains the same, the staircase becomes less steep. If the run is kept the same but the rise is increased, the staircase becomes more steep. This important characteristic of a line is best defined using rectangular coordinates.

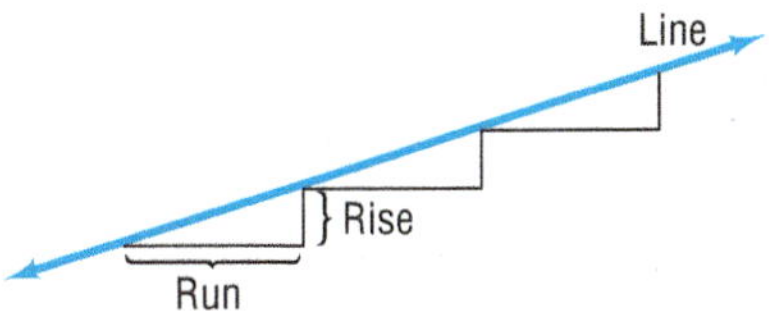

Figure 11

DEFINITION

Let $P = (x_1, y_1)$ and $Q = (x_2, y_2)$ be two distinct points. If $x_1 \neq x_2$, the **slope** m of the nonvertical line L containing P and Q is defined by the formula

$$m = \frac{y_2 - y_1}{x_2 - x_1} \qquad x_1 \neq x_2 \qquad \textbf{(1)}$$

If $x_1 = x_2$, L is a **vertical line** and the slope m of L is **undefined** (since this results in division by 0).

Figure 12(a) provides an illustration of the slope of a nonvertical line; Figure 12(b) illustrates a vertical line.

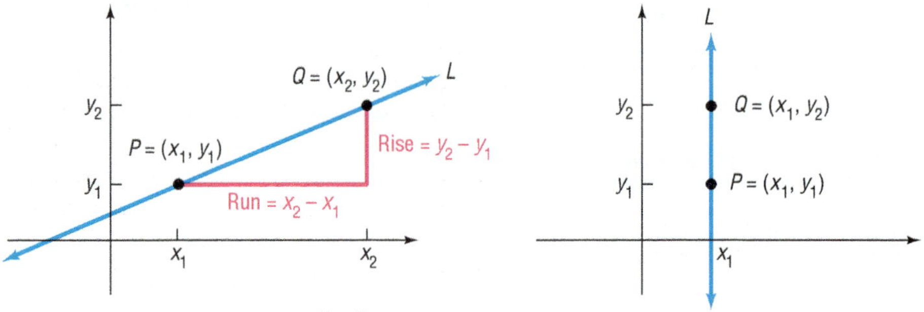

Figure 12 (a) Slope of L is $m = \frac{y_2 - y_1}{x_2 - x_1}$ (b) Slope is undefined; L is vertical

> **In Words**
> The symbol Δ is the Greek letter delta. In mathematics, Δ is read "change in," so $\frac{\Delta y}{\Delta x}$ is read "change in y divided by change in x."

As Figure 12(a) illustrates, the slope m of a nonvertical line may be viewed as

$$m = \frac{y_2 - y_1}{x_2 - x_1} = \frac{\text{Rise}}{\text{Run}} \quad \text{or} \quad m = \frac{y_2 - y_1}{x_2 - x_1} = \frac{\text{Change in } y}{\text{Change in } x} = \frac{\Delta y}{\Delta x}$$

That is, the slope m of a nonvertical line measures the amount y changes when x changes from x_1 to x_2. The expression $\frac{\Delta y}{\Delta x}$ is called the **average rate of change** of y with respect to x.

Two comments about computing the slope of a nonvertical line may prove helpful:

1. Any two distinct points on the line can be used to compute the slope of the line. (See Figure 13 for justification.) Since any two distinct points can be used to compute the slope of a line, the average rate of change of a line is always the same number.

Figure 13 Triangles ABC and PQR are similar (equal angles), so ratios of corresponding sides are equal. Then

Slope using P and $Q = \frac{y_2 - y_1}{x_2 - x_1} = \frac{d(B, C)}{d(A, C)} =$ Slope using A and B

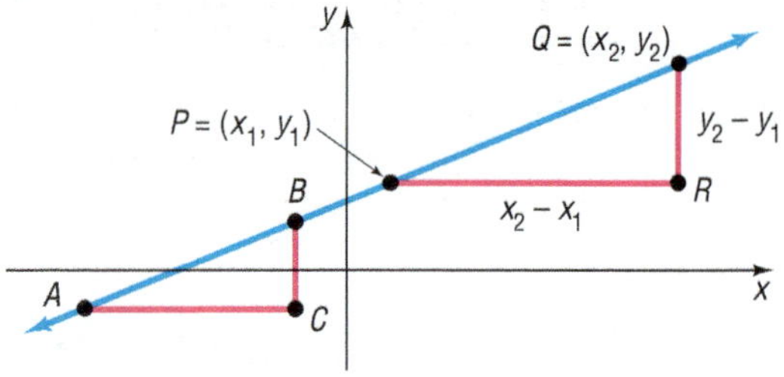

2. The slope of a line may be computed from $P = (x_1, y_1)$ to $Q = (x_2, y_2)$ or from Q to P because

$$\frac{y_2 - y_1}{x_2 - x_1} = \frac{y_1 - y_2}{x_1 - x_2}$$

EXAMPLE 1 Finding and Interpreting the Slope of a Line Given Two Points

The slope m of the line containing the points $(1, 2)$ and $(5, -3)$ may be computed as

$$m = \frac{-3 - 2}{5 - 1} = \frac{-5}{4} = -\frac{5}{4} \quad \text{or as} \quad m = \frac{2 - (-3)}{1 - 5} = \frac{5}{-4} = -\frac{5}{4}$$

For every 4-unit change in x, y will change by -5 units. That is, if x increases by 4 units, then y will decrease by 5 units. The average rate of change of y with respect to x is $-\frac{5}{4}$. ■

Now Work PROBLEMS 13 AND 19

Square Screens

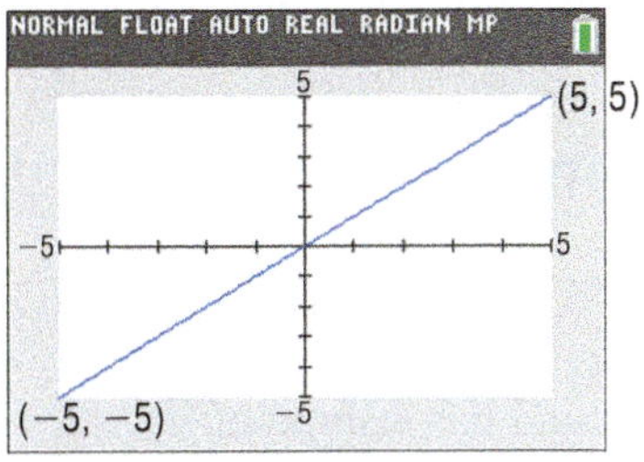

Figure 14 $y = x$

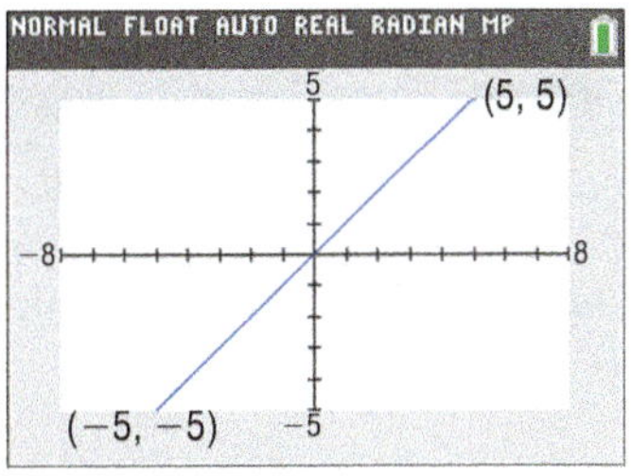

Figure 15 $y = x$

To get an undistorted view of slope, the same scale must be used on each axis. However, most graphing utilities have a rectangular screen. Because of this, using the same interval for both x and y will result in a distorted view. For example, Figure 14 shows the graph of the line $y = x$ connecting the points $(-5, -5)$ and $(5, 5)$. We expect the line to bisect the first and third quadrants, but it doesn't. We need to adjust the selections for Xmin, Xmax, Ymin, and Ymax so that a **square screen** results. On a TI-84 Plus C, this is accomplished by setting the ratio of x to y at 8 : 5.*

Figure 15 shows the graph of the line $y = x$ on a square screen using a TI-84 Plus C. Notice that the line now bisects the first and third quadrants. Compare this illustration to Figure 14.

To get a better idea of the meaning of the slope m of a line, consider the following.

Exploration On the same square screen, graph the following equations:

$Y_1 = 0$ Slope of the line is 0.

$Y_2 = \frac{1}{4}x$ Slope of the line is $\frac{1}{4}$.

$Y_3 = \frac{1}{2}x$ Slope of the line is $\frac{1}{2}$.

$Y_4 = x$ Slope of the line is 1.

$Y_5 = 2x$ Slope of the line is 2.

$Y_6 = 6x$ Slope of the line is 6.

See Figure 16. ■

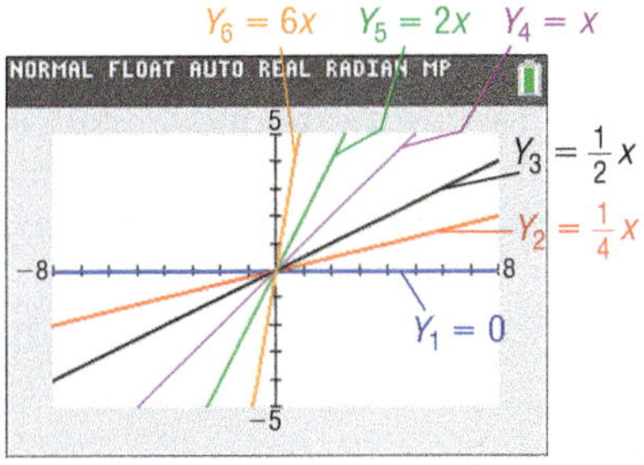

Figure 16 $y = mx$

Exploration On the same square screen, graph the following equations:

$Y_1 = 0$ Slope of the line is 0.

$Y_2 = -\frac{1}{4}x$ Slope of the line is $-\frac{1}{4}$.

$Y_3 = -\frac{1}{2}x$ Slope of the line is $-\frac{1}{2}$.

$Y_4 = -x$ Slope of the line is -1.

$Y_5 = -2x$ Slope of the line is -2.

$Y_6 = -6x$ Slope of the line is -6.

See Figure 17. ■

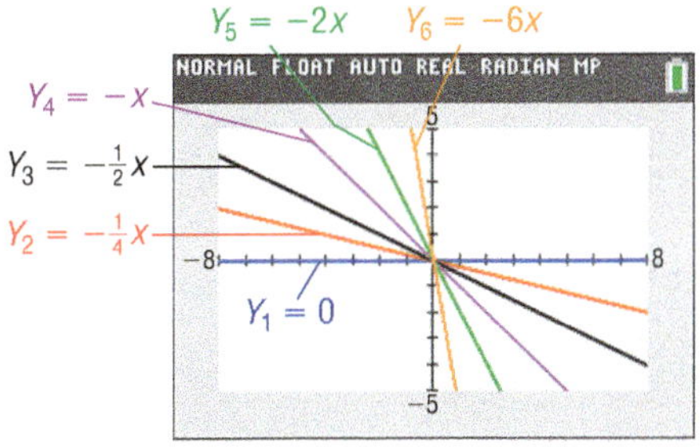

Figure 17 $y = mx$

*Most graphing utilities have a feature that automatically squares the viewing window. Consult your owner's manual for the appropriate keystrokes.

Figures 16 and 17 on the previous page illustrate the following facts:

1. When the slope of a line is positive, the line slants upward from left to right.
2. When the slope of a line is negative, the line slants downward from left to right.
3. When the slope is 0, the line is horizontal.

Figures 16 and 17 also illustrate that the closer the line is to the vertical position, the greater the magnitude of the slope. Thus, a line with slope 6 is steeper than a line whose slope is 3.

2 Graph Lines Given a Point and the Slope

EXAMPLE 2 **Graphing a Line Given a Point and a Slope**

Draw a graph of the line that contains the point $(3, 2)$ and has a slope of:

(a) $\frac{3}{4}$ (b) $-\frac{4}{5}$

Solution (a) Slope $= \frac{\text{Rise}}{\text{Run}}$. The slope $\frac{3}{4}$ means that for every horizontal movement (run) of 4 units to the right, there will be a vertical movement (rise) of 3 units. Start at the given point $(3, 2)$ and move 4 units to the right and 3 units up, arriving at the point $(7, 5)$. Drawing the line through this point and the point $(3, 2)$ gives the graph. See Figure 18.

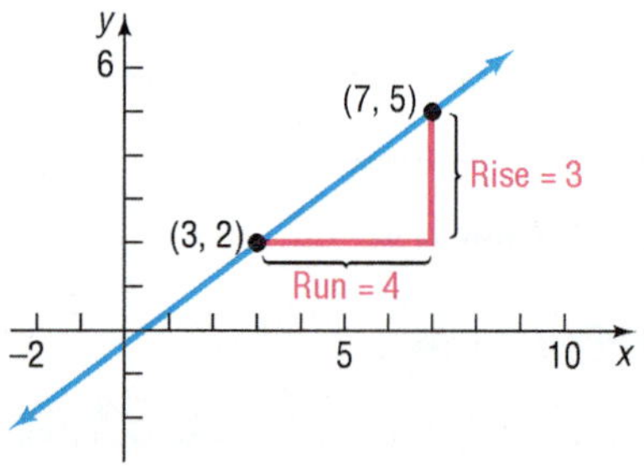

Figure 18

(b) The fact that the slope is

$$-\frac{4}{5} = \frac{-4}{5} = \frac{\text{Rise}}{\text{Run}}$$

means that for every horizontal movement of 5 units to the right there will be a corresponding vertical movement of -4 units (a downward movement of 4 units). Start at the given point $(3, 2)$ and move 5 units to the right and then 4 units down, arriving at the point $(8, -2)$. Drawing the line through these points gives the graph. See Figure 19.

Alternatively, consider that

$$-\frac{4}{5} = \frac{4}{-5} = \frac{\text{Rise}}{\text{Run}}$$

so for every horizontal movement of -5 units (a movement to the left of 5 units), there will be a corresponding vertical movement of 4 units (upward). This approach leads to the point $(-2, 6)$, which is also on the graph of the line in Figure 19. ■

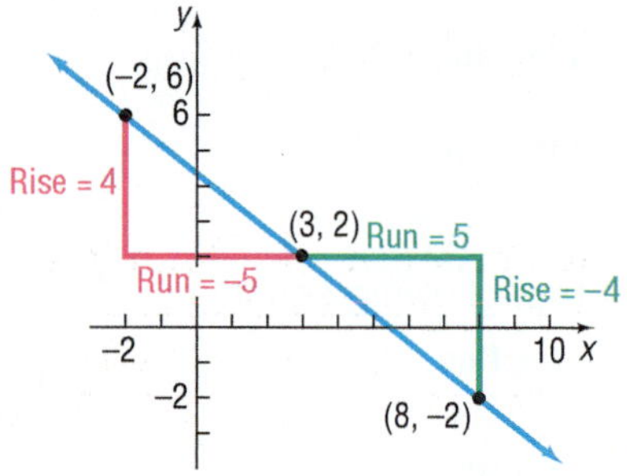

Figure 19

Now Work PROBLEM 25

3 Find the Equation of a Vertical Line

EXAMPLE 3 **Graphing a Line**

Graph the equation: $x = 3$

Solution To graph $x = 3$ by hand, find all points (x, y) in the plane for which $x = 3$. No matter what y-coordinate is used, the corresponding x-coordinate always equals 3. Consequently, the graph of the equation $x = 3$ is a vertical line with x-intercept 3 and undefined slope. See Figure 20(a).

In order for an equation to be graphed using a graphing utility, the equation must be expressed in the form $y = \{\text{expression in } x\}$. But $x = 3$ cannot be put into this form. To overcome this, most graphing utilities have special commands for drawing vertical lines. DRAW, LINE, PLOT, and VERT are among the more common ones. Consult your manual to determine the correct methodology for your graphing utility. Figure 20(b) shows the graph on a TI-84 Plus C.

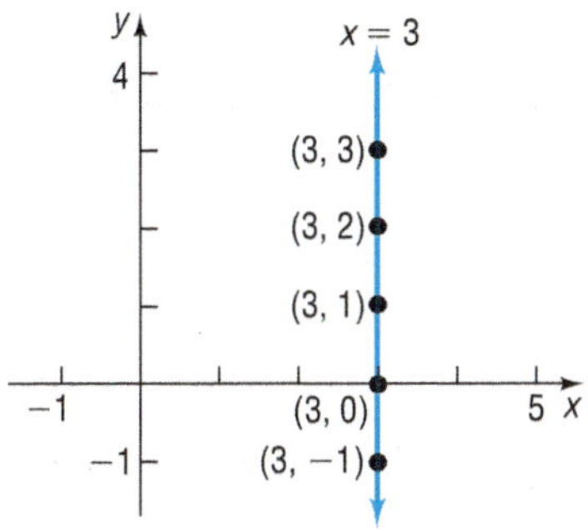

(a)

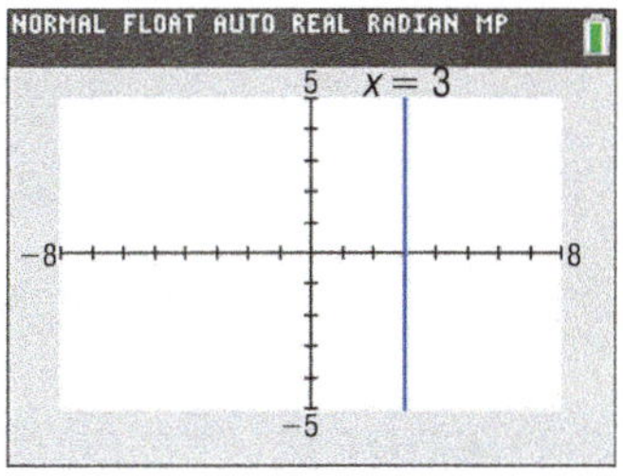

(b)

Figure 20

Example 3 suggests the following result:

THEOREM

Equation of a Vertical Line

A vertical line is given by an equation of the form

$$x = a$$

where a is the x-intercept.

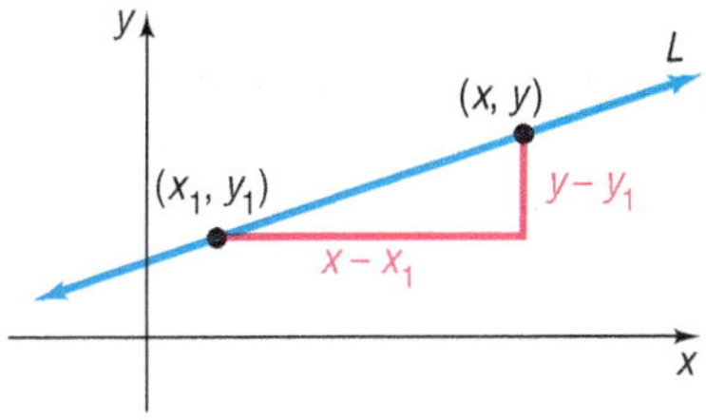

Figure 21

4 Use the Point–Slope Form of a Line; Identify Horizontal Lines

Let L be a nonvertical line with slope m that contains the point (x_1, y_1). See Figure 21. For any other point (x, y) on L, we have

$$m = \frac{y - y_1}{x - x_1} \quad \text{or} \quad y - y_1 = m(x - x_1)$$

THEOREM

Point–Slope Form of an Equation of a Line

An equation of a nonvertical line with slope m that contains the point (x_1, y_1) is

$$y - y_1 = m(x - x_1) \qquad \textbf{(2)}$$

EXAMPLE 4

Using the Point–Slope Form of a Line

An equation of the line with slope 4 and containing the point $(1, 2)$ can be found by using the point–slope form with $m = 4$, $x_1 = 1$, and $y_1 = 2$.

$$\begin{aligned} y - y_1 &= m(x - x_1) \\ y - 2 &= 4(x - 1) && m = 4, x_1 = 1, y_1 = 2 \\ y &= 4x - 2 && \text{Solve for } y. \end{aligned}$$

See Figure 22 for the graph.

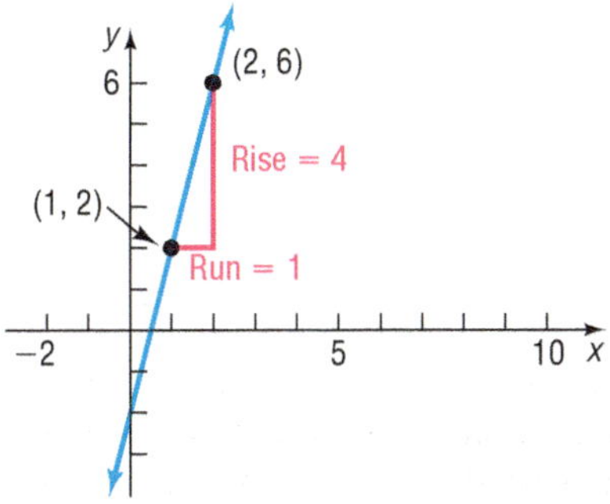

Figure 22 $y = 4x - 2$

Now Work PROBLEMS 39 AND 47

EXAMPLE 5 Finding the Equation of a Horizontal Line

Find an equation of the horizontal line containing the point $(3, 2)$.

Solution Because all the y-values are equal on a horizontal line, the slope of a horizontal line is 0. To get an equation, use the point–slope form with $m = 0, x_1 = 3$, and $y_1 = 2$.

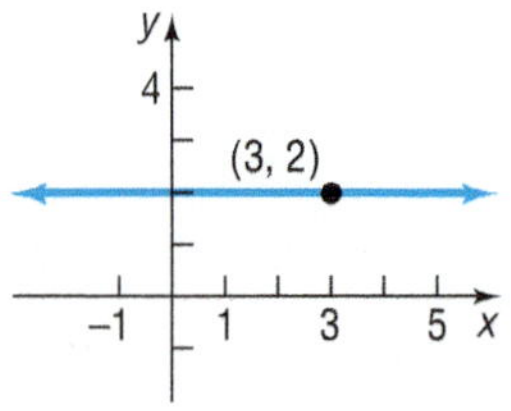

Figure 23 $y = 2$

$$y - y_1 = m(x - x_1)$$
$$y - 2 = 0 \cdot (x - 3) \quad m = 0, x_1 = 3, \text{ and } y_1 = 2$$
$$y - 2 = 0$$
$$y = 2$$

See Figure 23 for the graph.

Example 5 suggests the following result:

THEOREM

Equation of a Horizontal Line

A horizontal line is given by an equation of the form

$$y = b$$

where b is the y-intercept.

Now Work PROBLEM 59

5 Write the Equation of a Line in Slope–Intercept Form

Another useful equation of a line is obtained when the slope m and y-intercept b are known. In this event, both the slope m of the line and the point $(0, b)$ on the line are known; then use the point–slope form, equation (2), to obtain the following equation:

$$y - b = m(x - 0) \quad \text{or} \quad y = mx + b$$

THEOREM

Slope–Intercept Form of an Equation of a Line

An equation of a line with slope m and y-intercept b is

$$y = mx + b \qquad \textbf{(3)}$$

Now Work PROBLEM 51 (EXPRESS ANSWER IN SLOPE–INTERCEPT FORM)

Seeing the Concept

To see the role that the slope m plays, graph the following lines on the same screen.

$Y_1 = 2 \qquad m = 0$

$Y_2 = x + 2 \qquad m = 1$

$Y_3 = -x + 2 \qquad m = -1$

$Y_4 = 3x + 2 \qquad m = 3$

$Y_5 = -3x + 2 \qquad m = -3$

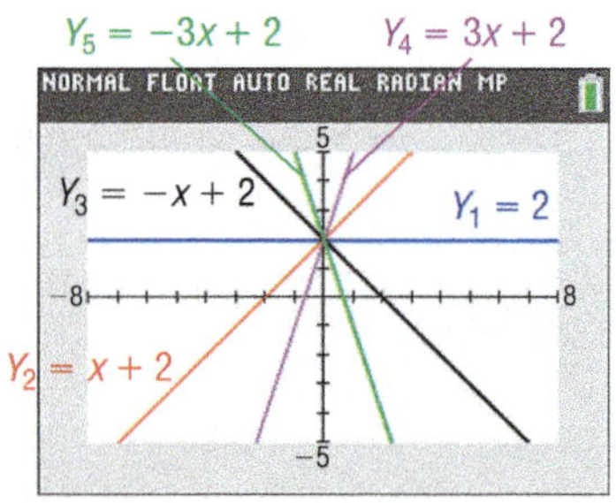

Figure 24 $y = mx + 2$

See Figure 24. What do you conclude about the lines $y = mx + 2$?

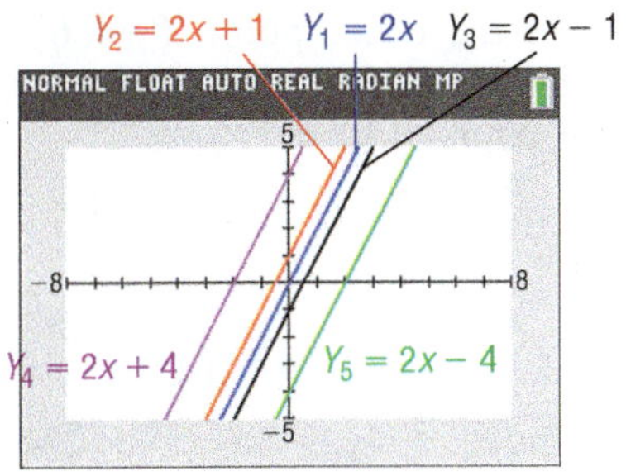

Figure 25 $y = 2x + b$

Seeing the Concept

To see the role of the y-intercept b, graph the following lines on the same screen.

$$Y_1 = 2x \qquad b = 0$$
$$Y_2 = 2x + 1 \qquad b = 1$$
$$Y_3 = 2x - 1 \qquad b = -1$$
$$Y_4 = 2x + 4 \qquad b = 4$$
$$Y_5 = 2x - 4 \qquad b = -4$$

See Figure 25. What do you conclude about the lines $y = 2x + b$? ■

When the equation of a line is written in slope–intercept form, it is easy to find the slope m and y-intercept b of the line. For example,

$$y = -2x + 7$$
$$\uparrow \quad \uparrow$$
$$y = mx + b$$

The slope of this line is -2 and its y-intercept is 7.

Now Work PROBLEM 73

EXAMPLE 6

Finding the Slope and y-Intercept

Find the slope m and y-intercept b of the equation $2x + 4y = 8$. Graph the equation.

Solution To obtain the slope and y-intercept, write the equation in slope–intercept form by solving for y.

$$2x + 4y = 8$$
$$4y = -2x + 8$$
$$y = -\frac{1}{2}x + 2 \qquad y = mx + b$$

The coefficient of x, $-\frac{1}{2}$, is the slope, and the constant, 2, is the y-intercept. Graph the line with the y-intercept 2 and the slope $-\frac{1}{2}$. Starting at the point $(0, 2)$, go to the right 2 units and then down 1 unit to the point $(2, 1)$. Draw the line through these points. See Figure 26. ■

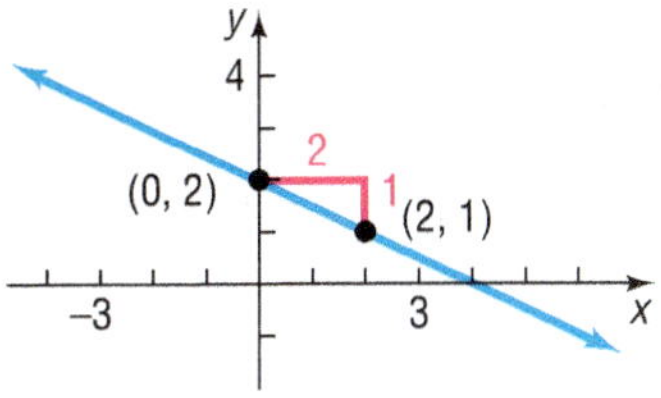

Figure 26 $y = -\frac{1}{2}x + 2$

Now Work PROBLEM 79

6 Find the Equation of a Line Given Two Points

EXAMPLE 7

Finding an Equation of a Line Given Two Points

Find an equation of the line containing the points $(2, 3)$ and $(-4, 6)$. Graph the line.

Solution First compute the slope of the line with $(x_1, y_1) = (2, 3)$ and $(x_2, y_2) = (-4, 6)$.

$$m = \frac{6 - 3}{-4 - 2} = \frac{3}{-6} = -\frac{1}{2} \qquad m = \frac{y_2 - y_1}{x_2 - x_1}$$

Use the point $(x_1, y_1) = (2, 3)$ and the slope $m = -\frac{1}{2}$ to get the point–slope form of the equation of the line.

$$y - 3 = -\frac{1}{2}(x - 2) \qquad y - y_1 = m(x - x_1)$$
$$y = -\frac{1}{2}x + 4 \qquad y = mx + b$$

See Figure 27 for the graph. ■

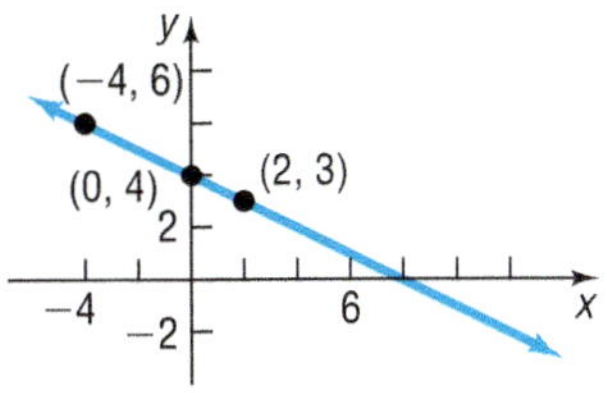

Figure 27 $y = -\frac{1}{2}x + 4$

In the solution to Example 7, we could have used the other point, $(-4, 6)$, instead of the point $(2, 3)$. The equation that results, when written in slope–intercept form, is the equation obtained in Example 7. (Try it for yourself.)

Now Work PROBLEM 53

7 Graph Lines Written in General Form Using Intercepts

Refer to Example 6. The form of the equation of the line $2x + 4y = 8$ is called the *general form.*

DEFINITION

The equation of a line is in **general form*** when it is written as

$$Ax + By = C \tag{4}$$

where A, B, and C are real numbers and A and B are not both 0.

If $B = 0$ in equation (4), then $A \neq 0$ and the graph of the equation is a vertical line: $x = \frac{C}{A}$. If $B \neq 0$ in equation (4), then we can solve the equation for y and write the equation in slope–intercept form as we did in Example 6.

Another approach to graphing equation (4) would be to find its intercepts. Remember, the intercepts of the graph of an equation are the points where the graph crosses or touches a coordinate axis.

EXAMPLE 8 Graphing an Equation in General Form Using Its Intercepts

Graph the equation $2x + 4y = 8$ by finding its intercepts.

Solution To obtain the x-intercept, let $y = 0$ in the equation and solve for x.

$$2x + 4y = 8$$
$$2x + 4(0) = 8 \quad \text{Let } y = 0.$$
$$2x = 8$$
$$x = 4 \quad \text{Divide both sides by 2.}$$

The x-intercept is 4 and the point $(4, 0)$ is on the graph of the equation.

To obtain the y-intercept, let $x = 0$ in the equation and solve for y.

$$2x + 4y = 8$$
$$2(0) + 4y = 8 \quad \text{Let } x = 0.$$
$$4y = 8$$
$$y = 2 \quad \text{Divide both sides by 4.}$$

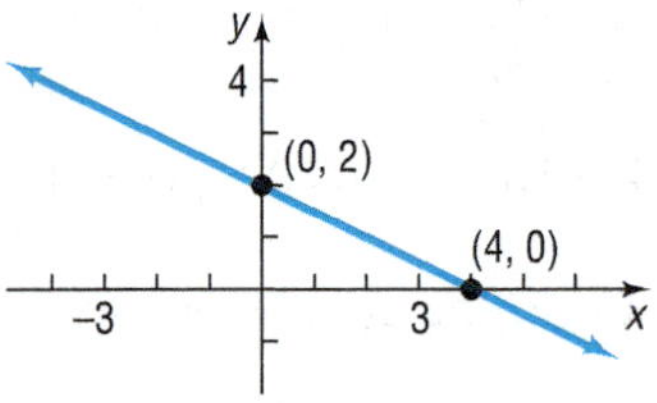

Figure 28 $2x + 4y = 8$

The y-intercept is 2 and the point $(0, 2)$ is on the graph of the equation.

Plot the points $(4, 0)$ and $(0, 2)$ and draw the line through the points. See Figure 28.

Now Work PROBLEM 93

*Some texts use the term **standard form**.

Every line has an equation that is equivalent to an equation written in general form. For example, a vertical line whose equation is

$$x = a$$

can be written in the general form

$$1 \cdot x + 0 \cdot y = a \quad A = 1, B = 0, C = a$$

A horizontal line whose equation is

$$y = b$$

can be written in the general form

$$0 \cdot x + 1 \cdot y = b \quad A = 0, B = 1, C = b$$

Lines that are neither vertical nor horizontal have general equations of the form

$$Ax + By = C \quad A \neq 0 \text{ and } B \neq 0$$

Because the equation of every line can be written in general form, any equation equivalent to equation (4) is called a **linear equation.**

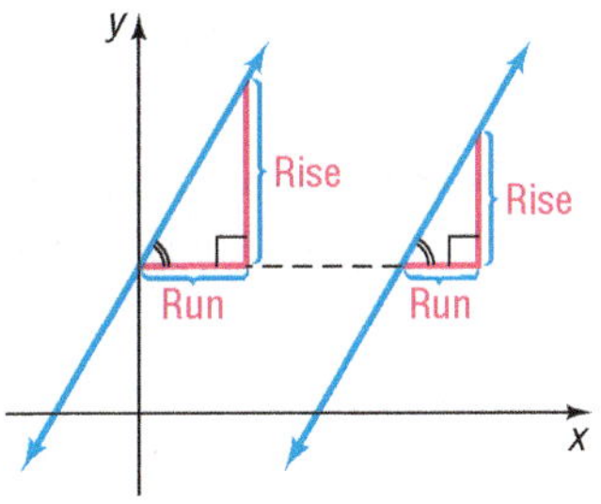

Figure 29 Parallel lines

8 Find Equations of Parallel Lines

When two lines (in the plane) do not intersect (that is, they have no points in common), they are **parallel**. Look at Figure 29. There we have drawn two parallel lines and have constructed two right triangles by drawing sides parallel to the coordinate axes. The right triangles are similar. (Do you see why? Two angles are equal.) Because the triangles are similar, the ratios of corresponding sides are equal.

THEOREM

Criteria for Parallel Lines

Two nonvertical lines are parallel if and only if their slopes are equal and they have different y-intercepts. ■

The use of the words "if and only if" in the preceding theorem means that actually two statements are being made, one the converse of the other.

If two nonvertical lines are parallel, then their slopes are equal and they have different y-intercepts.

If two nonvertical lines have equal slopes and they have different y-intercepts, then they are parallel.

EXAMPLE 9 Showing That Two Lines Are Parallel

Show that the lines given by the following equations are parallel.

$$L_1: \; 2x + 3y = 6 \qquad L_2: \; 4x + 6y = 0$$

Solution To determine whether these lines have equal slopes and different y-intercepts, write each equation in slope–intercept form.

$$L_1: \; 2x + 3y = 6 \qquad\qquad L_2: \; 4x + 6y = 0$$

$$3y = -2x + 6 \qquad\qquad 6y = -4x$$

$$y = -\frac{2}{3}x + 2 \qquad\qquad y = -\frac{2}{3}x$$

$$\text{Slope} = -\frac{2}{3};\; y\text{-intercept} = 2 \qquad \text{Slope} = -\frac{2}{3};\; y\text{-intercept} = 0$$

Because these lines have the same slope, $-\frac{2}{3}$, but different y-intercepts, the lines are parallel. See Figure 30.

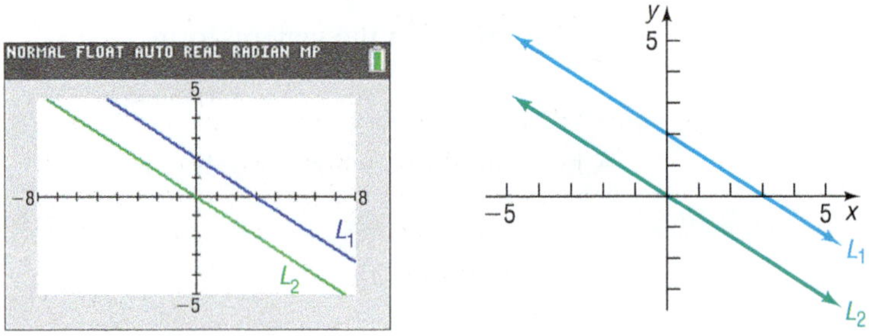

Figure 30 Parallel lines

EXAMPLE 10 Finding a Line That Is Parallel to a Given Line

Find an equation for the line that contains the point $(2, -3)$ and is parallel to the line $2x + y = 6$.

Solution Since the two lines are to be parallel, the slope of the line being sought equals the slope of the line $2x + y = 6$. Begin by writing the equation of the line $2x + y = 6$ in slope–intercept form.

$$2x + y = 6$$
$$y = -2x + 6 \quad \text{Place in the form } y = mx + b.$$

The slope is -2. Since the line being sought also has slope -2 and contains the point $(2, -3)$, use the point–slope form to obtain its equation.

$$y - y_1 = m(x - x_1) \quad \text{Point–slope form}$$
$$y - (-3) = -2(x - 2) \quad m = -2, x_1 = 2, y_1 = -3$$
$$y + 3 = -2x + 4 \quad \text{Simplify.}$$
$$y = -2x + 1 \quad \text{Slope–intercept form}$$
$$2x + y = 1 \quad \text{General form}$$

This line is parallel to the line $2x + y = 6$ and contains the point $(2, -3)$. See Figure 31.

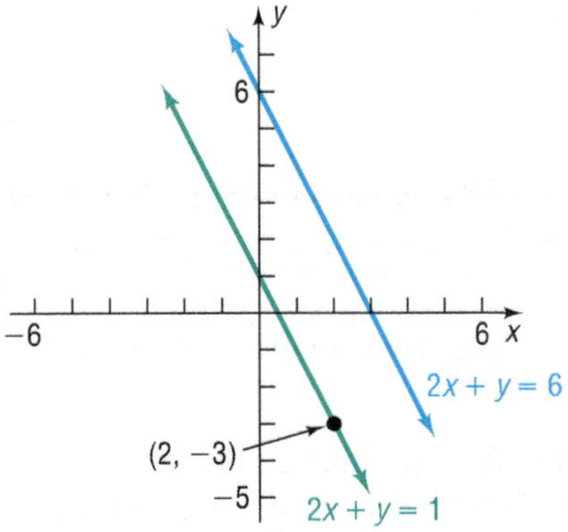

Figure 31

Now Work PROBLEM 61

9 Find Equations of Perpendicular Lines

When two lines intersect at a right angle (90°), they are **perpendicular**. See Figure 32.

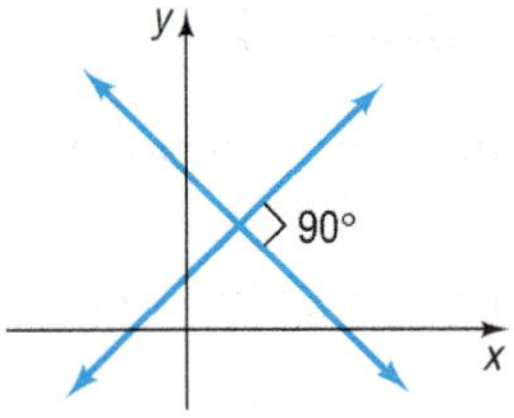

Figure 32 Perpendicular lines

The following result gives a condition, in terms of their slopes, for two lines to be perpendicular.

THEOREM **Criterion for Perpendicular Lines**

Two nonvertical lines are perpendicular if and only if the product of their slopes is −1.

Here we shall prove the "only if" part of the statement:

If two nonvertical lines are perpendicular, then the product of their slopes is −1.

In Problem 130 you are asked to prove the "if" part of the theorem:

If two nonvertical lines have slopes whose product is -1, then the lines are perpendicular.

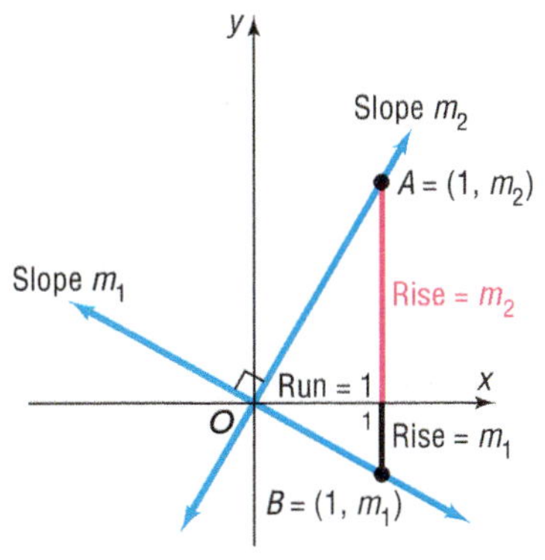

Figure 33

Proof Let m_1 and m_2 denote the slopes of the two lines. There is no loss in generality (that is, neither the angle nor the slopes are affected) if we situate the lines so that they meet at the origin. See Figure 33. The point $A = (1, m_2)$ is on the line having slope m_2, and the point $B = (1, m_1)$ is on the line having slope m_1. (Do you see why this must be true?)

Suppose that the lines are perpendicular. Then triangle OAB is a right triangle. As a result of the Pythagorean Theorem, it follows that

$$[d(O,A)]^2 + [d(O,B)]^2 = [d(A,B)]^2 \tag{5}$$

Using the distance formula, the squares of these distances are

$$[d(O,A)]^2 = (1-0)^2 + (m_2-0)^2 = 1 + m_2^2$$
$$[d(O,B)]^2 = (1-0)^2 + (m_1-0)^2 = 1 + m_1^2$$
$$[d(A,B)]^2 = (1-1)^2 + (m_2-m_1)^2 = m_2^2 - 2m_1m_2 + m_1^2$$

Using these facts in equation (5), we get

$$(1 + m_2^2) + (1 + m_1^2) = m_2^2 - 2m_1m_2 + m_1^2$$

which, upon simplification, can be written as

$$m_1m_2 = -1$$

If the lines are perpendicular, the product of their slopes is -1. ■

You may find it easier to remember the condition for two nonvertical lines to be perpendicular by observing that the equality $m_1m_2 = -1$ means that m_1 and m_2 are negative reciprocals of each other; that is, either $m_1 = -\frac{1}{m_2}$ or $m_2 = -\frac{1}{m_1}$.

EXAMPLE 11 Finding the Slope of a Line Perpendicular to Another Line

If a line has slope $\frac{3}{2}$, any line having slope $-\frac{2}{3}$ is perpendicular to it. ■

EXAMPLE 12 Finding the Equation of a Line Perpendicular to a Given Line

Find an equation of the line that contains the point $(1, -2)$ and is perpendicular to the line $x + 3y = 6$. Graph the two lines.

Solution First write the equation of the given line in slope–intercept form to find its slope.

$$x + 3y = 6$$
$$3y = -x + 6 \quad \text{Proceed to solve for } y.$$
$$y = -\frac{1}{3}x + 2 \quad \text{Place in the form } y = mx + b.$$

The given line has slope $-\frac{1}{3}$. Any line perpendicular to this line will have slope 3. Because the point $(1, -2)$ is on this line with slope 3, use the point–slope form of the equation of a line.

$$y - y_1 = m(x - x_1) \quad \text{Point–slope form}$$
$$y - (-2) = 3(x - 1) \quad m = 3, x_1 = 1, y_1 = -2$$

To obtain other forms of the equation, proceed as follows:

$$y + 2 = 3(x - 1)$$

$$y + 2 = 3x - 3 \quad \text{Simplify.}$$

$$y = 3x - 5 \quad \text{Slope–intercept form}$$

$$3x - y = 5 \quad \text{General form}$$

Figure 34 shows the graphs.

WARNING Be sure to use a square screen when you graph perpendicular lines. Otherwise, the angle between the two lines will appear distorted. ■

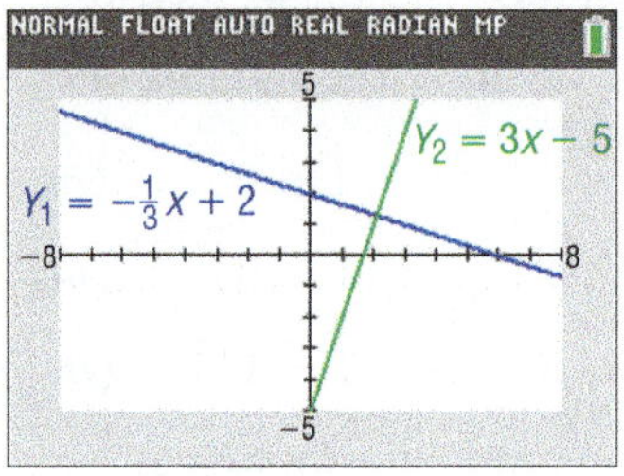

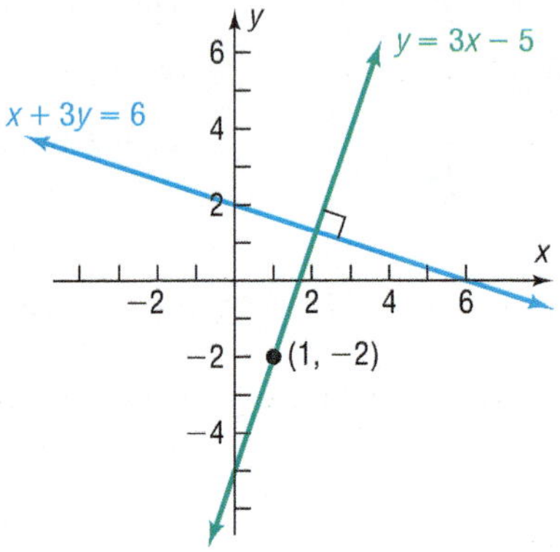

Figure 34

■

Now Work PROBLEM 67

2.2 Assess Your Understanding

Concepts and Vocabulary

1. The slope of a vertical line is __________; the slope of a horizontal line is _______.
2. For the line $2x + 3y = 6$, the x-intercept is _______ and the y-intercept is _______.
3. ***True or False*** The equation $3x + 4y = 6$ is written in general form.
4. ***True or False*** The slope of the line $2y = 3x + 5$ is 3.
5. ***True or False*** The point $(1, 2)$ is on the line $2x + y = 4$.
6. Two nonvertical lines have slopes m_1 and m_2, respectively. The lines are parallel if __________ and the __________ are unequal; the lines are perpendicular if __________.
7. The lines $y = 2x + 3$ and $y = ax + 5$ are parallel if $a =$ _______.
8. The lines $y = 2x - 1$ and $y = ax + 2$ are perpendicular if $a =$ _______.
9. ***True or False*** Perpendicular lines have slopes that are reciprocals of one another.
10. Choose the formula for finding the slope m of a nonvertical line that contains the two distinct points (x_1, y_1) and (x_2, y_2).
    (a) $m = \dfrac{y_2 - x_2}{y_1 - x_1} \quad x_1 \neq y_1$
    (b) $m = \dfrac{y_2 - x_1}{x_2 - y_1} \quad y_1 \neq x_2$
    (c) $m = \dfrac{x_2 - x_1}{y_2 - y_1} \quad y_1 \neq y_2$
    (d) $m = \dfrac{y_2 - y_1}{x_2 - x_1} \quad x_1 \neq x_2$
11. If a line slants downward from left to right, then which of the following describes its slope?
    (a) positive (b) zero (c) negative (d) undefined
12. Choose the correct statement about the graph of the line $y = -3$.
    (a) The graph is vertical with x-intercept -3.
    (b) The graph is horizontal with y-intercept -3.
    (c) The graph is vertical with y-intercept -3.
    (d) The graph is horizontal with x-intercept -3.

Skill Building

In Problems 13–16, (a) find the slope of the line and (b) interpret the slope.

13.

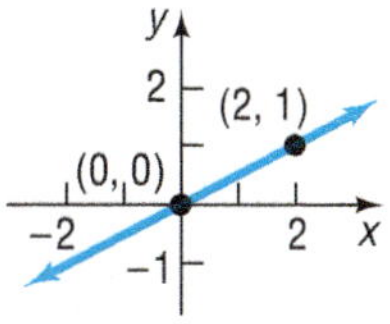

14.

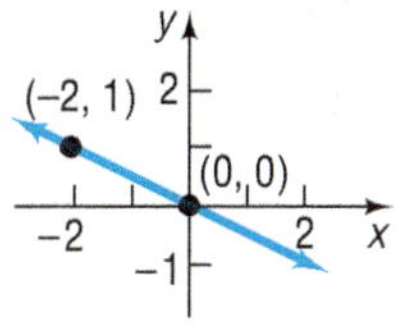

15.

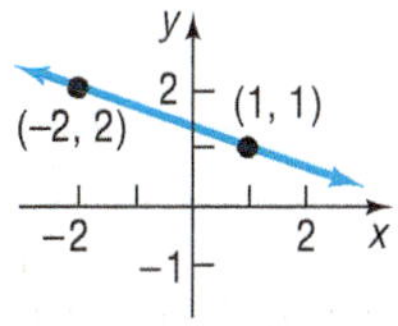

16.

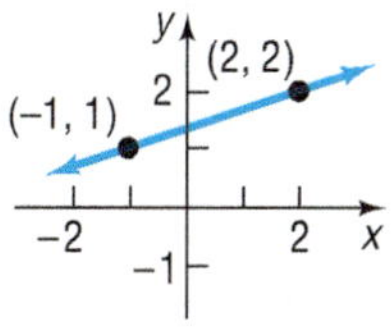

In Problems 17–24, plot each pair of points and determine the slope of the line containing them. Graph the line by hand.

17. $(2,3);\ (4,0)$
18. $(4,2);\ (3,4)$
19. $(-2,3);\ (2,1)$
20. $(-1,1);\ (2,3)$
21. $(-3,-1);\ (2,-1)$
22. $(4,2);\ (-5,2)$
23. $(-1,2);\ (-1,-2)$
24. $(2,0);\ (2,2)$

In Problems 25–32, graph the line containing the point P and having slope m.

25. $P=(1,2);\ m=3$
26. $P=(2,1);\ m=4$
27. $P=(2,4);\ m=-\frac{3}{4}$
28. $P=(1,3);\ m=-\frac{2}{5}$
29. $P=(-1,3);\ m=0$
30. $P=(2,-4);\ m=0$
31. $P=(0,3)$; slope undefined
32. $P=(-2,0)$; slope undefined

In Problems 33–38, the slope and a point on a line are given. Use this information to locate three additional points on the line. Answers may vary.
[**Hint:** It is not necessary to find the equation of the line. See Example 2.]

33. Slope 4; point $(1,2)$
34. Slope 2; point $(-2,3)$
35. Slope $-\frac{3}{2}$; point $(2,-4)$
36. Slope $\frac{4}{3}$; point $(-3,2)$
37. Slope -2; point $(-2,-3)$
38. Slope -1; point $(4,1)$

In Problems 39–46, find an equation of the line L.

39.

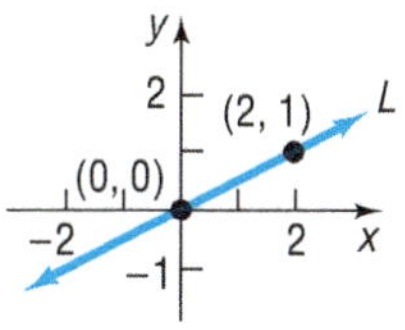

40.

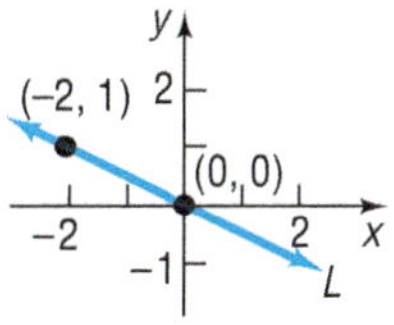

41.

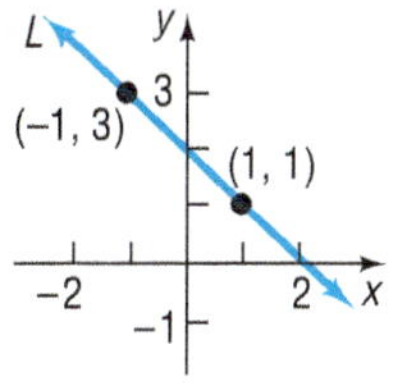

42.

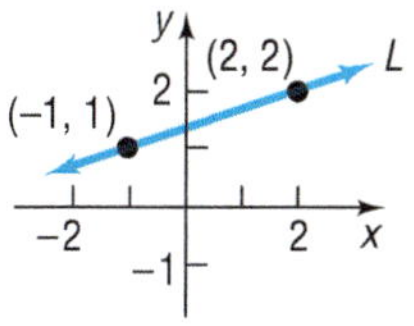

43.

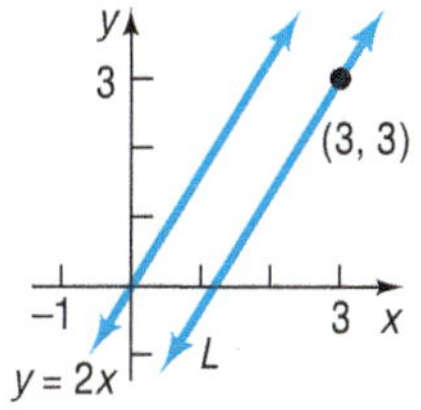

L is parallel to $y = 2x$

44.

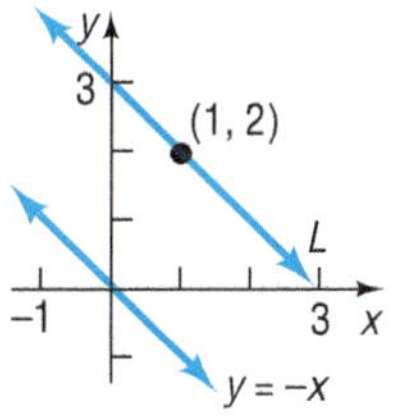

L is parallel to $y = -x$

45.

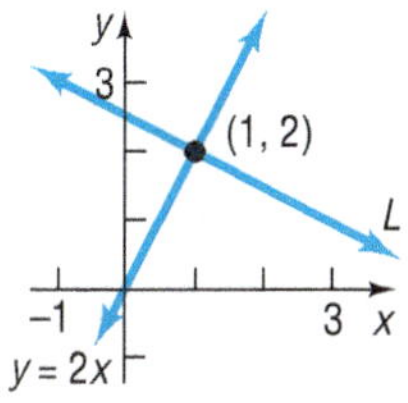

L is perpendicular to $y = 2x$

46.

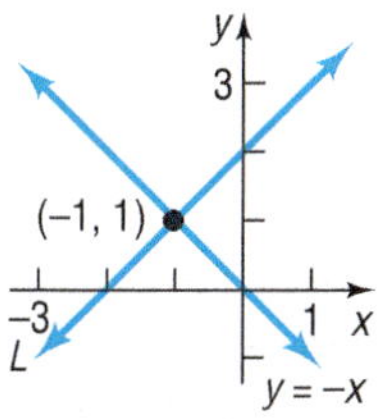

L is perpendicular to $y = -x$

In Problems 47–72, find an equation for the line with the given properties. Express your answer using either the general form or the slope–intercept form of the equation of a line, whichever you prefer.

47. Slope $= 3$; containing the point $(-2,3)$
48. Slope $= 2$; containing the point $(4,-3)$
49. Slope $= -\frac{2}{3}$; containing the point $(1,-1)$
50. Slope $= \frac{1}{2}$; containing the point $(3,1)$
51. Slope $= -3$; y-intercept $= 3$
52. Slope $= -5$; y-intercept $= -7$
53. Containing the points $(1,3)$ and $(-1,2)$
54. Containing the points $(-3,4)$ and $(2,5)$
55. x-intercept $= 2$; y-intercept $= -1$
56. x-intercept $= -4$; y-intercept $= 4$
57. Slope undefined; containing the point $(2,4)$
58. Slope undefined; containing the point $(3,8)$
59. Horizontal; containing the point $(-3,2)$
60. Vertical; containing the point $(4,-5)$
61. Parallel to the line $y = 4x$; containing the point $(-1,2)$
62. Parallel to the line $y = -3x$; containing the point $(-1,2)$
63. Parallel to the line $5x - y = -2$; containing the point $(0,0)$
64. Parallel to the line $x - 2y = -5$; containing the point $(0,0)$

65. Parallel to the line $x = 5$; containing the point $(4, 2)$

66. Parallel to the line $y = 5$; containing the point $(4, 2)$

67. Perpendicular to the line $y = \frac{1}{6}x + 4$; containing the point $(1, -6)$

68. Perpendicular to the line $y = 8x - 3$; containing the point $(10, -2)$

69. Perpendicular to the line $2x + 5y = 2$; containing the point $(-3, -6)$

70. Perpendicular to the line $x - 3y = -12$; containing the point $(0, 4)$

71. Perpendicular to the line $x = 8$; containing the point $(3, 4)$

72. Perpendicular to the line $y = 8$; containing the point $(3, 4)$

In Problems 73–92, find the slope and y-intercept of each line. Graph the line by hand. Verify your graph using a graphing utility.

73. $y = 2x + 3$
74. $y = -3x + 4$
75. $\frac{1}{4}y = x - 1$
76. $\frac{1}{3}x + y = 2$
77. $y = \frac{1}{2}x + 2$
78. $y = 2x + \frac{1}{2}$
79. $x + 4y = 4$
80. $-x + 3y = 6$
81. $2x - 3y = 6$
82. $3x + 2y = 6$
83. $x + y = 1$
84. $x - y = 2$
85. $x = -4$
86. $y = -1$
87. $y = 5$
88. $x = 2$
89. $y - x = 0$
90. $x + y = 0$
91. $2y - 3x = 0$
92. $3x + 2y = 0$

In Problems 93–102, (a) find the intercepts of the graph of each equation and (b) graph the equation.

93. $2x + 3y = 6$
94. $3x - 2y = 6$
95. $-4x + 5y = 40$
96. $6x - 4y = 24$
97. $7x + 2y = 21$
98. $5x + 3y = 18$
99. $\frac{1}{2}x + \frac{1}{3}y = 1$
100. $x - \frac{2}{3}y = 4$
101. $0.2x - 0.5y = 1$
102. $-0.3x + 0.4y = 1.2$

103. Find an equation of the x-axis.

104. Find an equation of the y-axis.

In Problems 105–108, the equations of two lines are given. Determine whether the lines are parallel, perpendicular, or neither.

105. $y = 2x - 3$; $y = 2x + 4$

106. $y = \frac{1}{2}x - 3$; $y = -2x + 4$

107. $y = 4x + 5$; $y = -4x + 2$

108. $y = -2x + 3$; $y = -\frac{1}{2}x + 2$

In Problems 109–112, write an equation of each line. Express your answer using either the general form or the slope–intercept form of the equation of a line, whichever you prefer.

109.

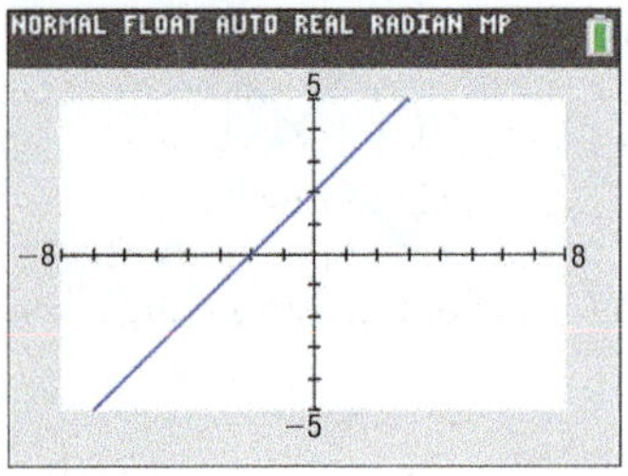

110.

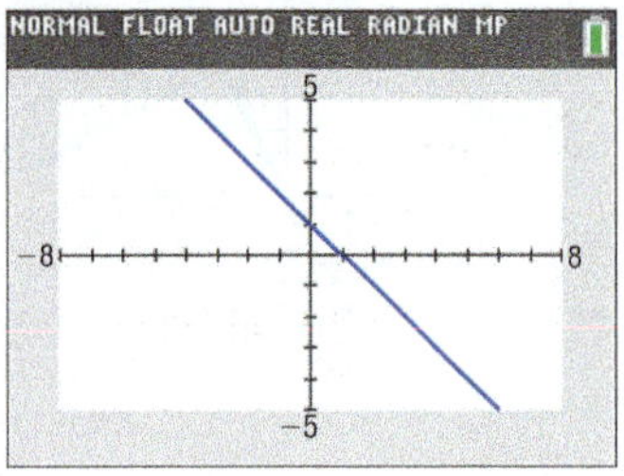

111.

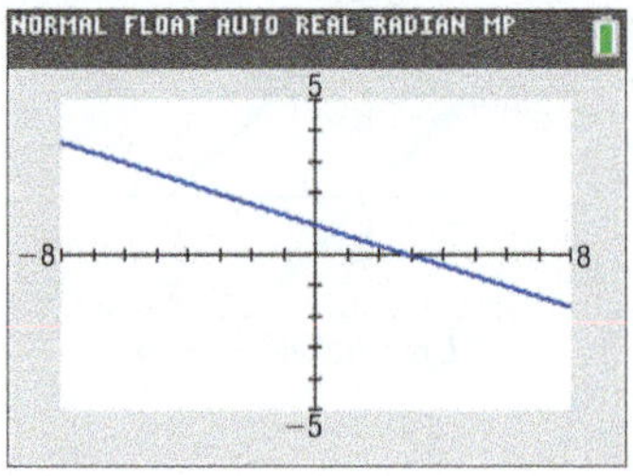

112.

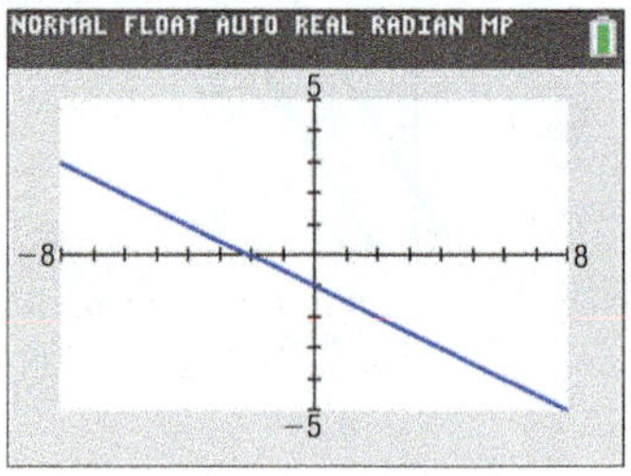

Applications and Extensions

113. **Geometry** Use slopes to show that the triangle whose vertices are $(-2, 5)$, $(1, 3)$, and $(-1, 0)$ is a right triangle.

114. **Geometry** Use slopes to show that the quadrilateral whose vertices are $(1, -1)$, $(4, 1)$, $(2, 2)$, and $(5, 4)$ is a parallelogram.

115. **Geometry** Use slopes to show that the quadrilateral whose vertices are $(-1, 0)$, $(2, 3)$, $(1, -2)$, and $(4, 1)$ is a rectangle.

116. **Geometry** Use slopes and the distance formula to show that the quadrilateral whose vertices are $(0, 0)$, $(1, 3)$, $(4, 2)$, and $(3, -1)$ is a square.

117. **Truck Rentals** A truck rental company rents a moving truck for one day by charging \$39 plus \$0.60 per mile. Write a linear equation that relates the cost C, in dollars, of renting the truck to the number x of miles driven. What is the cost of renting the truck if the truck is driven 110 miles? 230 miles?

118. **Cost Equation** The **fixed costs** of operating a business are the costs incurred regardless of the level of production. Fixed costs include rent, fixed salaries, and costs of leasing machinery. The **variable costs** of operating a business are the costs that change with the level of output. Variable costs include raw materials, hourly wages, and electricity. Suppose

that a manufacturer of jeans has fixed daily costs of \$500 and variable costs of \$8 for each pair of jeans manufactured. Write a linear equation that relates the daily cost C, in dollars, of manufacturing the jeans to the number x of jeans manufactured. What is the cost of manufacturing 400 pairs of jeans? 740 pairs?

119. **Cost of Driving a Car** The annual fixed costs for owning a small sedan are \$1461, assuming the car is completely paid for. The cost to drive the car is approximately \$0.16 per mile. Write a linear equation that relates the cost C and the number x of miles driven annually.
Source: AAA, 2014

120. **Wages of a Car Salesperson** Dan receives \$375 per week for selling new and used cars at a car dealership in Oak Lawn, Illinois. In addition, he receives 5% of the profit on any sales that he generates. Write a linear equation that represents Dan's weekly salary S when he has sales that generate a profit of x dollars.

121. **Electricity Rates in Illinois** Commonwealth Edison Company supplies electricity to residential customers for a monthly customer charge of \$15.14 plus 7.57 cents per kilowatt-hour for up to 800 kilowatt-hours (kW-h).

(a) Write a linear equation that relates the monthly charge C, in dollars, to the number x of kilowatt-hours used in a month, $0 \le x \le 800$.
(b) Graph this equation.
(c) What is the monthly charge for using 200 kilowatt-hours?
(d) What is the monthly charge for using 500 kilowatt-hours?
(e) Interpret the slope of the line.

Source: Commonwealth Edison Company, January 2015.

122. **Electricity Rates in Florida** Florida Power & Light Company supplies electricity to residential customers for a monthly customer charge of \$7.57 plus 9.01 cents per kilowatt-hour for up to 1000 kilowatt-hours.
(a) Write a linear equation that relates the monthly charge C, in dollars, to the number x of kilowatt-hours used in a month, $0 \le x \le 1000$.
(b) Graph this equation.
(c) What is the monthly charge for using 200 kilowatt-hours?
(d) What is the monthly charge for using 500 kilowatt-hours?
(e) Interpret the slope of the line.

Source: Florida Power & Light Company, March 2015.

123. **Measuring Temperature** The relationship between Celsius (°C) and Fahrenheit (°F) degrees of measuring temperature is linear. Find a linear equation relating °C and °F if 0°C corresponds to 32°F and 100°C corresponds to 212°F. Use the equation to find the Celsius measure of 70°F.

124. **Measuring Temperature** The Kelvin (K) scale for measuring temperature is obtained by adding 273 to the Celsius temperature.
(a) Write a linear equation relating K and °C.
(b) Write a linear equation relating K and °F (see Problem 123).

125. **Access Ramp** A wooden access ramp is being built to reach a platform that sits 30 inches above the floor. The ramp drops 2 inches for every 25-inch run.

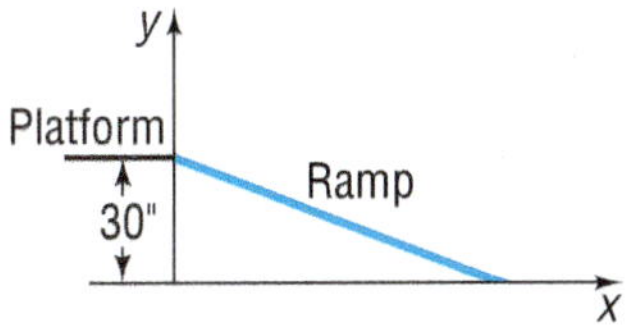

(a) Write a linear equation that relates the height y of the ramp above the floor to the horizontal distance x from the platform.
(b) Find and interpret the x-intercept of the graph of your equation.
(c) Design requirements stipulate that the maximum run be 30 feet and that the maximum slope be a drop of 1 inch for each 12 inches of run. Will this ramp meet the requirements? Explain.
(d) What slopes could be used to obtain the 30-inch rise and still meet design requirements?

Source: www.adaptiveaccess.com/wood_ramps.php

126. **Cigarette Use** A report in the Child Trends DataBase indicated that, in 2000, 20.6% of twelfth grade students reported daily use of cigarettes. In 2013, 8.5% of twelfth grade students reported daily use of cigarettes.
(a) Write a linear equation that relates the percent y of twelfth grade students who smoke cigarettes daily to the number x of years after 2000.
(b) Find the intercepts of the graph of your equation.
(c) Do the intercepts have any meaningful interpretation?
(d) Use your equation to predict the percent for the year 2025. Is this result reasonable?

Source: *www.childtrends.org*

127. **Product Promotion** A cereal company finds that the number of people who will buy one of its products in the first month that the product is introduced is linearly related to the amount of money it spends on advertising. If it spends \$40,000 on advertising, then 100,000 boxes of cereal will be sold, and if it spends \$60,000, then 200,000 boxes will be sold.
(a) Write a linear equation that relates the amount A spent on advertising to the number x of boxes the company aims to sell.
(b) How much expenditure on advertising is needed to sell 300,000 boxes of cereal?
(c) Interpret the slope.

128. Show that the line containing the points (a, b) and (b, a), $a \neq b$, is perpendicular to the line $y = x$. Also show that the midpoint of (a, b) and (b, a) lies on the line $y = x$.

129. The equation $2x - y = C$ defines a **family of lines**, one line for each value of C. On one set of coordinate axes, graph the members of the family when $C = -4$, $C = 0$, and $C = 2$. Can you draw a conclusion from the graph about each member of the family?

130. Prove that if two nonvertical lines have slopes whose product is -1, then the lines are perpendicular. **[Hint:** Refer to Figure 33 and use the converse of the Pythagorean Theorem.]

Explaining Concepts: Discussion and Writing

131. Which of the following equations might have the graph shown? (More than one answer is possible.)

(a) $2x + 3y = 6$
(b) $-2x + 3y = 6$
(c) $3x - 4y = -12$
(d) $x - y = 1$
(e) $x - y = -1$
(f) $y = 3x - 5$
(g) $y = 2x + 3$
(h) $y = -3x + 3$

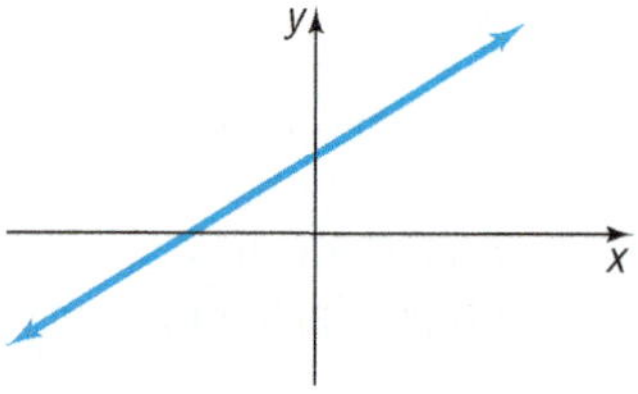

132. Which of the following equations might have the graph shown? (More than one answer is possible.)

(a) $2x + 3y = 6$
(b) $2x - 3y = 6$
(c) $3x + 4y = 12$
(d) $x - y = 1$
(e) $x - y = -1$
(f) $y = -2x - 1$
(g) $y = -\frac{1}{2}x + 10$
(h) $y = x + 4$

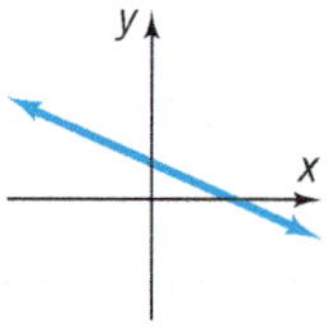

133. The figure shows the graph of two parallel lines. Which of the following pairs of equations might have such a graph?

(a) $x - 2y = 3$
 $x + 2y = 7$
(b) $x + y = 2$
 $x + y = -1$
(c) $x - y = -2$
 $x - y = 1$
(d) $x - y = -2$
 $2x - 2y = -4$
(e) $x + 2y = 2$
 $x + 2y = -1$

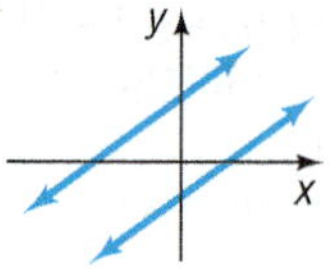

134. The figure shows the graph of two perpendicular lines. Which of the following pairs of equations might have such a graph?

(a) $y - 2x = 2$
 $y + 2x = -1$
(b) $y - 2x = 0$
 $2y + x = 0$
(c) $2y - x = 2$
 $2y + x = -2$
(d) $y - 2x = 2$
 $x + 2y = -1$
(e) $2x + y = -2$
 $2y + x = -2$

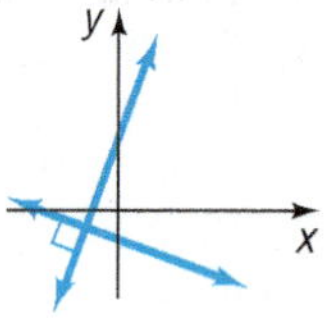

135. ***m* is for Slope** The accepted symbol used to denote the slope of a line is the letter m. Investigate the origin of this symbolism. Begin by consulting a French dictionary and looking up the French word *monter*. Write a brief essay on your findings.

136. **Grade of a Road** The term *grade* is used to describe the inclination of a road. How is this term related to the notion of slope of a line? Is a 4% grade very steep? Investigate the grades of some mountainous roads and determine their slopes. Write a brief essay on your findings.

137. **Carpentry** Carpenters use the term *pitch* to describe the steepness of staircases and roofs. How is pitch related to slope? Investigate typical pitches used for stairs and for roofs. Write a brief essay on your findings.

138. Can the equation of every line be written in slope–intercept form? Why?

139. Does every line have exactly one x-intercept and one y-intercept? Are there any lines that have no intercepts?

140. What can you say about two lines that have equal slopes and equal y-intercepts?

141. What can you say about two lines with the same x-intercept and the same y-intercept? Assume that the x-intercept is not 0.

142. If two distinct lines have the same slope but different x-intercepts, can they have the same y-intercept?

143. If two distinct lines have the same y-intercept but different slopes, can they have the same x-intercept?

144. Which form of the equation of a line do you prefer to use? Justify your position with an example that shows that your choice is better than another. Have reasons.

145. **What Went Wrong?** A student is asked to find the slope of the line joining $(-3, 2)$ and $(1, -4)$. He states that the slope is $\frac{3}{2}$. Is he correct? If not, what went wrong?

Retain Your Knowledge

Problems 146–149 are based on material learned earlier in the course. The purpose of these problems is to keep the material fresh in your mind so that you are better prepared for the final exam.

146. Simplify $\left(\dfrac{x^2y^{-3}}{x^4y^5}\right)^{-2}$. Assume $x \neq 0$ and $y \neq 0$. Express the answer so that all exponents are positive.

147. The lengths of the legs of a right triangle are $a = 8$ and $b = 15$. Find the hypotenuse.

148. Solve the equation: $(x - 3)^2 + 25 = 49$

149. Solve $|2x - 5| + 7 < 10$. Express the answer using set notation or interval notation. Graph the solution set.

2.3 Circles

PREPARING FOR THIS SECTION *Before getting started, review the following:*

- Completing the Square (Chapter R, Section R.5, p. 57)
- Square Root Method (Section 1.3, p. 112)

Now Work the **'Are You Prepared?'** problems on page 193.

OBJECTIVES
1 Write the Standard Form of the Equation of a Circle (p. 189)
2 Graph a Circle by Hand and by Using a Graphing Utility (p. 190)
3 Work with the General Form of the Equation of a Circle (p. 192)

1 Write the Standard Form of the Equation of a Circle

One advantage of a coordinate system is that it enables us to translate a geometric statement into an algebraic statement, and vice versa. Consider, for example, the following geometric statement that defines a circle.

DEFINITION

A **circle** is a set of points in the xy-plane that are a fixed distance r from a fixed point (h, k). The fixed distance r is called the **radius**, and the fixed point (h, k) is called the **center** of the circle.

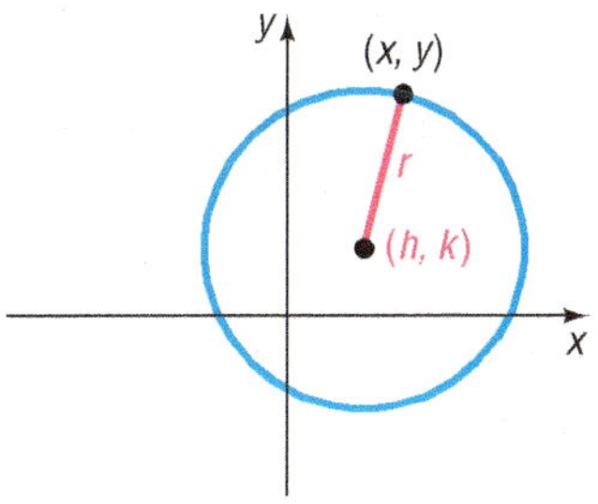

Figure 35
$(x - h)^2 + (y - k)^2 = r^2$

Figure 35 shows the graph of a circle. To find the equation, let (x, y) represent the coordinates of any point on a circle with radius r and center (h, k). Then the distance between the points (x, y) and (h, k) must always equal r. That is, by the distance formula,

$$\sqrt{(x - h)^2 + (y - k)^2} = r$$

or, equivalently,

$$(x - h)^2 + (y - k)^2 = r^2$$

DEFINITION

The **standard form of an equation of a circle** with radius r and center (h, k) is

$$(x - h)^2 + (y - k)^2 = r^2 \qquad \textbf{(1)}$$

THEOREM

The standard form of an equation of a circle of radius r with center at the origin $(0, 0)$ is

$$x^2 + y^2 = r^2$$

DEFINITION

If the radius $r = 1$, the circle whose center is at the origin is called the **unit circle** and has the equation

$$x^2 + y^2 = 1$$

See Figure 36. Notice that the graph of the unit circle is symmetric with respect to the x-axis, the y-axis, and the origin.

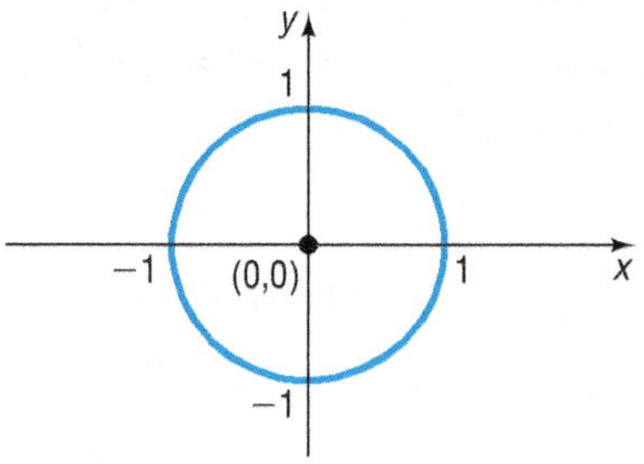

Figure 36
Unit circle $x^2 + y^2 = 1$

EXAMPLE 1 Writing the Standard Form of the Equation of a Circle

Write the standard form of the equation of the circle with radius 5 and center $(-3, 6)$.

Solution Substitute the values $r = 5$, $h = -3$, and $k = 6$ into equation (1).

$$(x - h)^2 + (y - k)^2 = r^2$$
$$(x + 3)^2 + (y - 6)^2 = 25$$

Now Work PROBLEM 9

2 Graph a Circle by Hand and by Using a Graphing Utility

EXAMPLE 2 Graphing a Circle by Hand and by Using a Graphing Utility

Graph the equation: $(x + 3)^2 + (y - 2)^2 = 16$

Solution Since the equation is in the form of equation (1), its graph is a circle. To graph the equation by hand, compare the given equation to the standard form of the equation of a circle. The comparison yields information about the circle.

$$(x + 3)^2 + (y - 2)^2 = 16$$
$$(x - (-3))^2 + (y - 2)^2 = 4^2$$
$$\uparrow \qquad \uparrow \qquad \uparrow$$
$$(x - h)^2 + (y - k)^2 = r^2$$

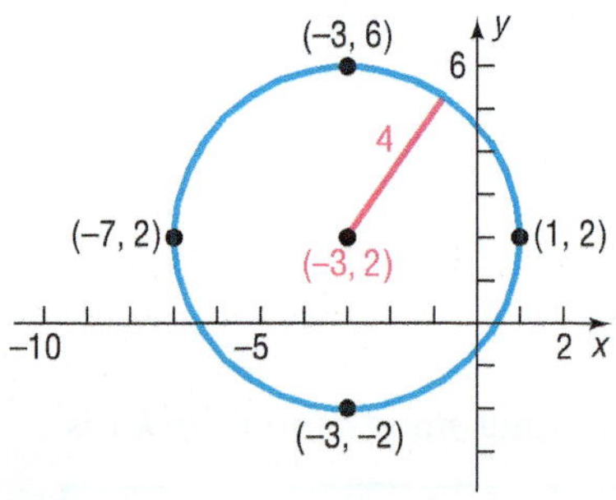

Figure 37
$(x + 3)^2 + (y - 2)^2 = 16$

We see that $h = -3$, $k = 2$, and $r = 4$. The circle has center $(-3, 2)$ and a radius of 4 units. To graph this circle, first plot the center $(-3, 2)$. Since the radius is 4, locate four points on the circle by plotting points 4 units to the left, to the right, up, and down from the center. These four points can then be used as guides to obtain the graph. See Figure 37.

> **In Words**
> The symbol $\pm$ is read "plus or minus." It means to add and subtract the quantity following the $\pm$ symbol. For example, 5 ± 2 means "$5 - 2 = 3$ or $5 + 2 = 7$."

To graph a circle on a graphing utility, we must write the equation in the form $y = \{\text{expression involving } x\}$.* We must solve for y in the equation

$$(x+3)^2 + (y-2)^2 = 16$$

$$(y-2)^2 = 16 - (x+3)^2 \qquad \text{Subtract } (x+3)^2 \text{ from both sides.}$$

$$y - 2 = \pm\sqrt{16-(x+3)^2} \qquad \text{Use the Square Root Method.}$$

$$y = 2 \pm \sqrt{16-(x+3)^2} \qquad \text{Add 2 to both sides.}$$

To graph the circle, graph the top half

$$Y_1 = 2 + \sqrt{16-(x+3)^2}$$

and the bottom half

$$Y_2 = 2 - \sqrt{16-(x+3)^2}$$

Also, be sure to use a square screen. Otherwise, the circle will appear distorted. Figure 38 shows the graph on a TI-84 Plus C. The graph is "disconnected" because of the resolution of the calculator. ■

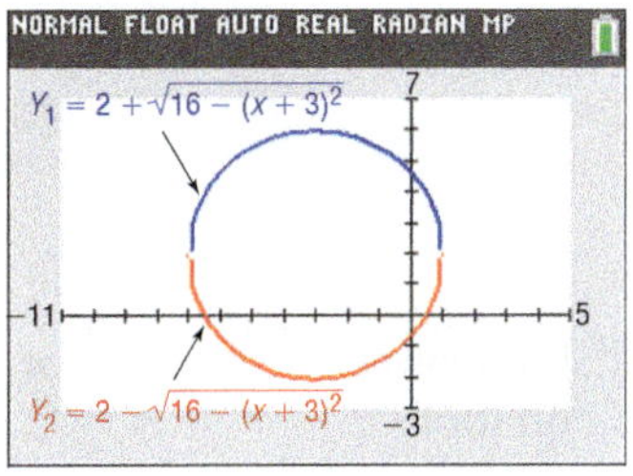

Figure 38
$(x+3)^2 + (y-2)^2 = 16$

Now Work PROBLEMS 25(a) AND (b)

EXAMPLE 3 Finding the Intercepts of a Circle

For the circle $(x+3)^2 + (y-2)^2 = 16$, find the intercepts, if any, of its graph.

Solution This is the equation discussed and graphed in Example 2. To find the x-intercepts, if any, let $y = 0$ and solve for x. Then

$$(x+3)^2 + (y-2)^2 = 16$$

$$(x+3)^2 + (0-2)^2 = 16 \qquad y = 0$$

$$(x+3)^2 + 4 = 16 \qquad \text{Simplify.}$$

$$(x+3)^2 = 12 \qquad \text{Subtract 4 from both sides.}$$

$$x + 3 = \pm\sqrt{12} \qquad \text{Use the Square Root Method.}$$

$$x = -3 \pm 2\sqrt{3} \qquad \text{Solve for } x.$$

The x-intercepts are $-3 - 2\sqrt{3} \approx -6.46$ and $-3 + 2\sqrt{3} \approx 0.46$.

To find the y-intercepts, if any, let $x = 0$ and solve for y. Then

$$(x+3)^2 + (y-2)^2 = 16$$

$$(0+3)^2 + (y-2)^2 = 16 \qquad x = 0$$

$$9 + (y-2)^2 = 16$$

$$(y-2)^2 = 7$$

$$y - 2 = \pm\sqrt{7} \qquad \text{Use the Square Root Method.}$$

$$y = 2 \pm \sqrt{7} \qquad \text{Solve for } y.$$

The y-intercepts are $2 - \sqrt{7} \approx -0.65$ and $2 + \sqrt{7} \approx 4.65$.

Look back at Figure 37 or 38 to verify the approximate locations of the intercepts. ■

Now Work PROBLEM 25 (c)

*Some graphing utilities (e.g., TI-83, TI-84, and TI-86) have a CIRCLE function that enables the user to enter only the coordinates of the center of the circle and its radius to graph the circle.

3 Work with the General Form of the Equation of a Circle

If we eliminate the parentheses from the standard form of the equation of the circle given in Example 3, we get

$$(x+3)^2 + (y-2)^2 = 16$$
$$x^2 + 6x + 9 + y^2 - 4y + 4 = 16$$

which simplifies to

$$x^2 + y^2 + 6x - 4y - 3 = 0$$

It can be shown that any equation of the form

$$x^2 + y^2 + ax + by + c = 0$$

has a graph that is a circle or a point, or has no graph at all. For example, the graph of the equation $x^2 + y^2 = 0$ is the single point $(0, 0)$. The equation $x^2 + y^2 + 5 = 0$, or $x^2 + y^2 = -5$, has no graph, because sums of squares of real numbers are never negative.

DEFINITION

When its graph is a circle, the equation

$$\boxed{x^2 + y^2 + ax + by + c = 0}$$

is the **general form of the equation of a circle.**

If an equation of a circle is in the general form, we use the method of completing the square to put the equation in standard form so that we can identify its center and radius.

EXAMPLE 4

Graphing a Circle Whose Equation Is in General Form

Graph the equation $x^2 + y^2 + 4x - 6y + 12 = 0$.

Solution Group the expression involving x, group the expression involving y, and put the constant on the right side of the equation. The result is

$$(x^2 + 4x) + (y^2 - 6y) = -12$$

Next, complete the square of each expression in parentheses. Remember that any number added on the left side of the equation must be added on the right.

$$(x^2 + 4x + 4) + (y^2 - 6y + 9) = -12 + 4 + 9$$

$$\left(\tfrac{4}{2}\right)^2 = 4 \qquad \left(\tfrac{-6}{2}\right)^2 = 9$$

$$(x+2)^2 + (y-3)^2 = 1 \quad \textbf{Factor.}$$

This equation is the standard form of the equation of a circle with radius 1 and center $(-2, 3)$. To graph the equation by hand, use the center $(-2, 3)$ and the radius 1. See Figure 39(a).

To graph the equation using a graphing utility, solve for y.

$$(y-3)^2 = 1 - (x+2)^2$$
$$y - 3 = \pm\sqrt{1 - (x+2)^2} \quad \textbf{Use the Square Root Method.}$$
$$y = 3 \pm \sqrt{1 - (x+2)^2} \quad \textbf{Add 3 to both sides.}$$

Figure 39(b) illustrates the graph on a TI-84 Plus C graphing calculator.

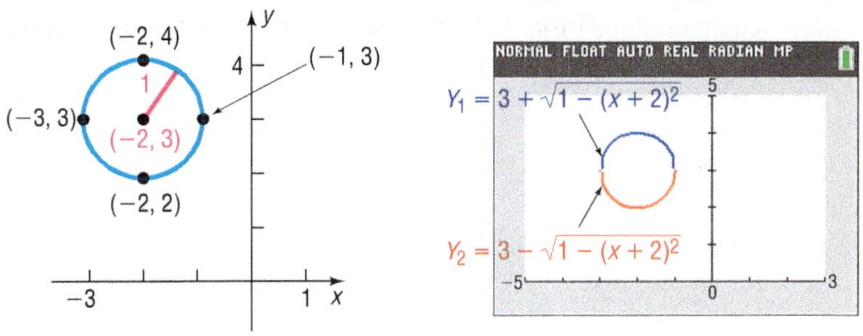

(a) (b)

Figure 39
$(x + 2)^2 + (y - 3)^2 = 1$

Now Work PROBLEM 29

EXAMPLE 5 **Finding the General Equation of a Circle**

Find the general equation of the circle whose center is $(1, -2)$ and whose graph contains the point $(4, -2)$.

Solution To find the equation of a circle, we need to know its center and its radius. Here, the center is $(1, -2)$. Since the point $(4, -2)$ is on the graph, the radius r will equal the distance from $(4, -2)$ to the center $(1, -2)$. See Figure 40. Thus,

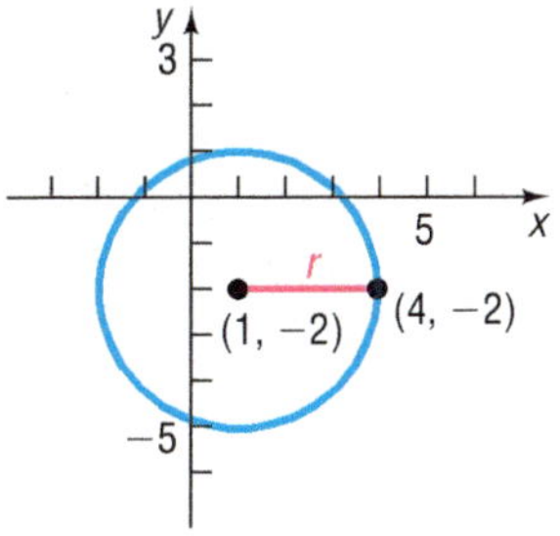

Figure 40
$(x - 1)^2 + (y + 2)^2 = 9$

$$r = \sqrt{(4 - 1)^2 + [-2 - (-2)]^2}$$
$$= \sqrt{9} = 3$$

The standard form of the equation of the circle is

$$(x - 1)^2 + (y + 2)^2 = 9$$

Eliminate the parentheses and rearrange the terms to get the general equation

$$x^2 + y^2 - 2x + 4y - 4 = 0$$

Now Work PROBLEM 15

Overview

The discussion in Sections 2.2 and 2.3 about lines and circles dealt with two main types of problems that can be generalized as follows:

1. Given an equation, classify it and graph it.
2. Given a graph, or information about a graph, find its equation.

This text deals with both types of problems. We shall study various equations, classify them, and graph them. The second type of problem is usually more difficult to solve than the first. In many instances a graphing utility can be used to solve problems when information about the problem (such as data) is given.

2.3 Assess Your Understanding

'Are You Prepared?' *Answers are given at the end of these exercises. If you get a wrong answer, read the pages listed in red.*

1. To complete the square of $x^2 + 10x$, you would ______ (*add/subtract*) the number ________. (p. 57)

2. Use the Square Root Method to solve the equation $(x - 2)^2 = 9$. (p. 112)

Concepts and Vocabulary

3. ***True or False*** Every equation of the form

$$x^2 + y^2 + ax + by + c = 0$$

has a circle as its graph.

4. For a circle, the ______ is the distance from the center to any point on the circle.

5. ***True or False*** The radius of the circle $x^2 + y^2 = 9$ is 3.

6. ***True or False*** The center of the circle

$$(x + 3)^2 + (y - 2)^2 = 13$$

is $(3, -2)$.

7. Choose the equation of a circle with radius 6 and center $(3, -5)$.
   (a) $(x - 3)^2 + (y + 5)^2 = 6$
   (b) $(x + 3)^2 + (y - 5)^2 = 36$
   (c) $(x + 3)^2 + (y - 5)^2 = 6$
   (d) $(x - 3)^2 + (y + 5)^2 = 36$

8. The equation of a circle can be changed from general form to standard form by doing which of the following?
   (a) completing the squares
   (b) solving for x
   (c) solving for y
   (d) squaring both sides

Skill Building

In Problems 9–12, find the center and radius of each circle. Write the standard form of the equation.

9.
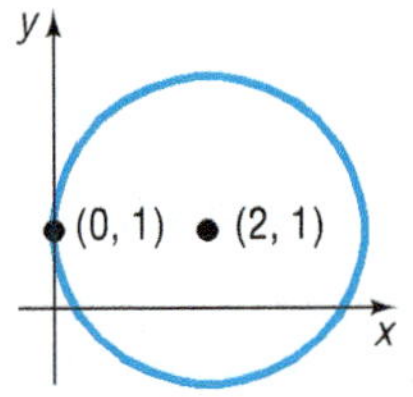

10.
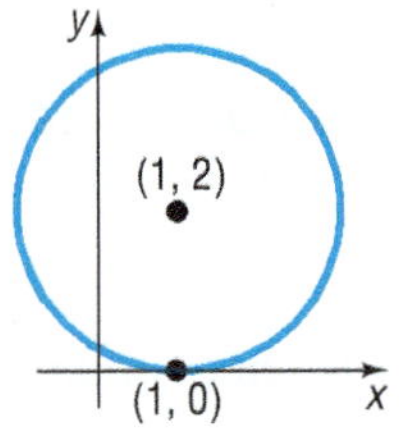

11.
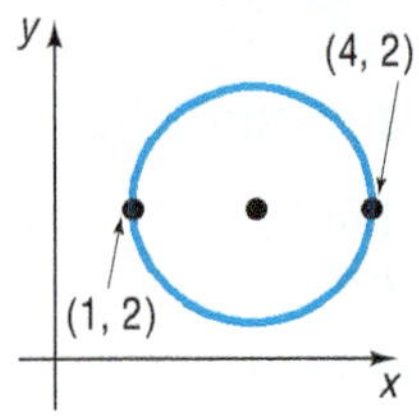

12.
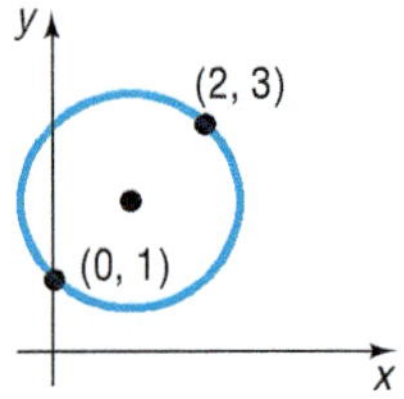

In Problems 13–22, write the standard form of the equation and the general form of the equation of each circle of radius r and center (h, k). Graph each circle.

13. $r = 2;\quad (h, k) = (0, 0)$
14. $r = 3;\quad (h, k) = (0, 0)$
15. $r = 2;\quad (h, k) = (0, 2)$
16. $r = 3;\quad (h, k) = (1, 0)$
17. $r = 5;\quad (h, k) = (4, -3)$
18. $r = 4;\quad (h, k) = (2, -3)$
19. $r = 4;\quad (h, k) = (-2, 1)$
20. $r = 7;\quad (h, k) = (-5, -2)$
21. $r = \frac{1}{2};\quad (h, k) = \left(\frac{1}{2}, 0\right)$
22. $r = \frac{1}{2};\quad (h, k) = \left(0, -\frac{1}{2}\right)$

In Problems 23–36, (a) find the center (h, k) and radius r of each circle; (b) graph each circle; (c) find the intercepts, if any.

23. $x^2 + y^2 = 4$
24. $x^2 + (y - 1)^2 = 1$
25. $2(x - 3)^2 + 2y^2 = 8$
26. $3(x + 1)^2 + 3(y - 1)^2 = 6$
27. $x^2 + y^2 - 2x - 4y - 4 = 0$
28. $x^2 + y^2 + 4x + 2y - 20 = 0$
29. $x^2 + y^2 + 4x - 4y - 1 = 0$
30. $x^2 + y^2 - 6x + 2y + 9 = 0$
31. $x^2 + y^2 - x + 2y + 1 = 0$
32. $x^2 + y^2 + x + y - \frac{1}{2} = 0$
33. $2x^2 + 2y^2 - 12x + 8y - 24 = 0$
34. $2x^2 + 2y^2 + 8x + 7 = 0$
35. $2x^2 + 8x + 2y^2 = 0$
36. $3x^2 + 3y^2 - 12y = 0$

In Problems 37–44, find the standard form of the equation of each circle.

37. Center at the origin and containing the point $(-2, 3)$
38. Center $(1, 0)$ and containing the point $(-3, 2)$
39. Center $(2, 3)$ and tangent to the x-axis
40. Center $(-3, 1)$ and tangent to the y-axis
41. With endpoints of a diameter at $(1, 4)$ and $(-3, 2)$
42. With endpoints of a diameter at $(4, 3)$ and $(0, 1)$
43. Center $(-1, 3)$ and tangent to the line $y = 2$
44. Center $(4, -2)$ and tangent to the line $x = 1$

In Problems 45–48, match each graph with the correct equation.

(a) $(x - 3)^2 + (y + 3)^2 = 9$ (b) $(x + 1)^2 + (y - 2)^2 = 4$ (c) $(x - 1)^2 + (y + 2)^2 = 4$ (d) $(x + 3)^2 + (y - 3)^2 = 9$

45.
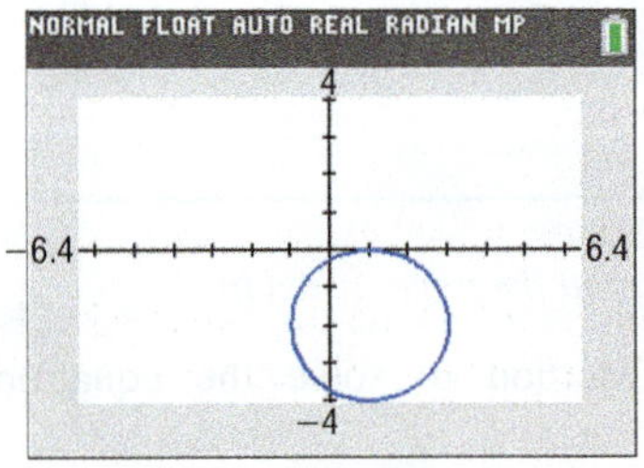

46.
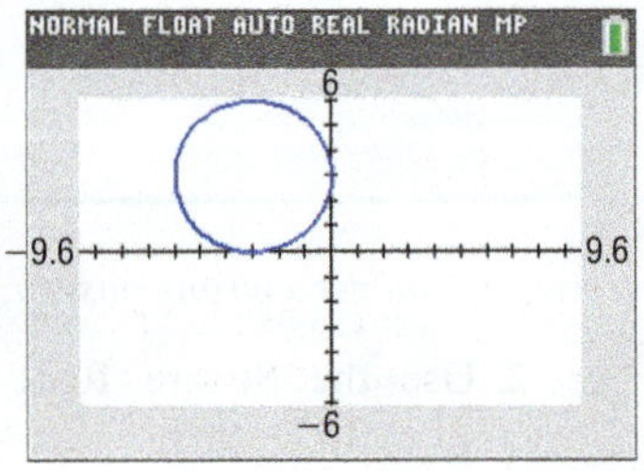

47.
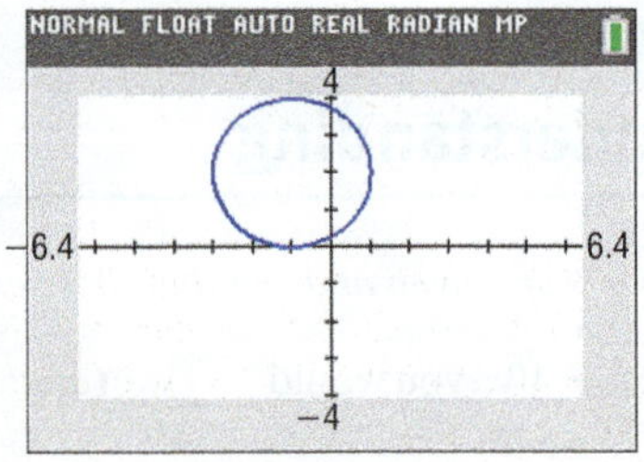

48.
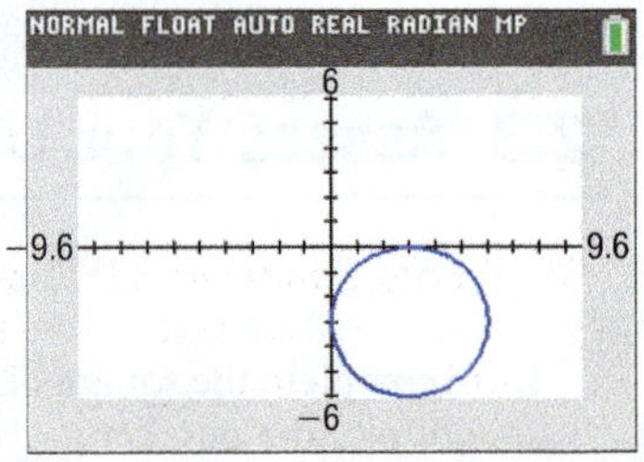

Applications and Extensions

49. Find the area of the square in the figure.

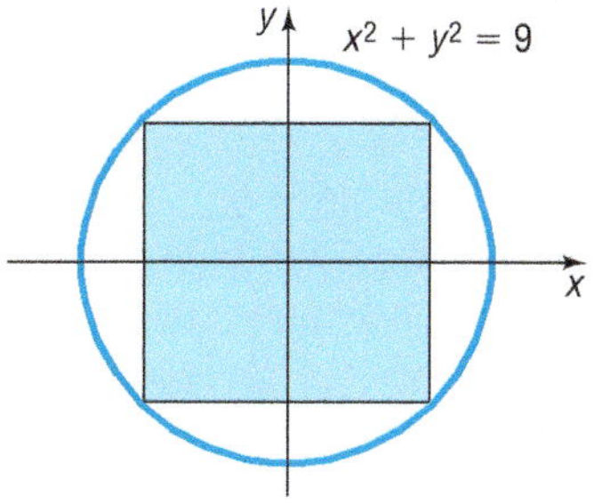

50. Find the area of the blue shaded region in the figure, assuming the quadrilateral inside the circle is a square.

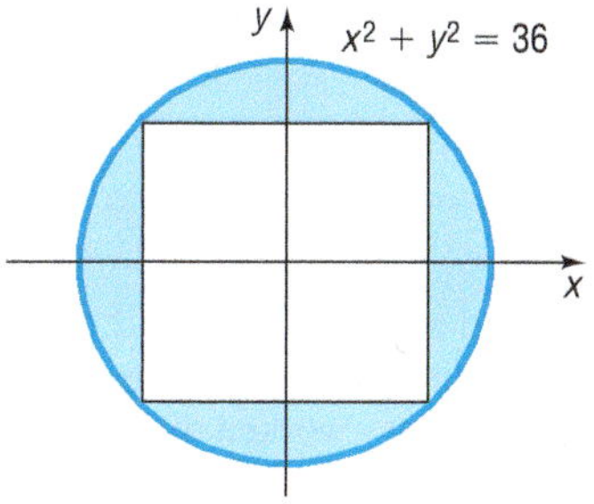

51. Ferris Wheel The original Ferris wheel was built in 1893 by Pittsburgh, Pennsylvania, bridge builder George W. Ferris. The Ferris wheel was originally built for the 1893 World's Fair in Chicago and was later reconstructed for the 1904 World's Fair in St. Louis. It had a maximum height of 264 feet and a wheel diameter of 250 feet. Find an equation for the wheel if the center of the wheel is on the y-axis.
Source: *guinnessworldrecords.com*

52. Ferris Wheel The High Roller observation wheel in Las Vegas has a maximum height of 550 feet and a diameter of 520 feet, with one full rotation taking approximately 30 minutes. Find an equation for the wheel if the center of the wheel is on the y-axis.
Source: *Las Vegas Review Journal*

53. Weather Satellites Earth is represented on a map of a portion of the solar system so that its surface is the circle with equation $x^2 + y^2 + 2x + 4y - 4091 = 0$. A weather satellite circles 0.6 unit above Earth with the center of its circular orbit at the center of Earth. Find the equation for the orbit of the satellite on this map.

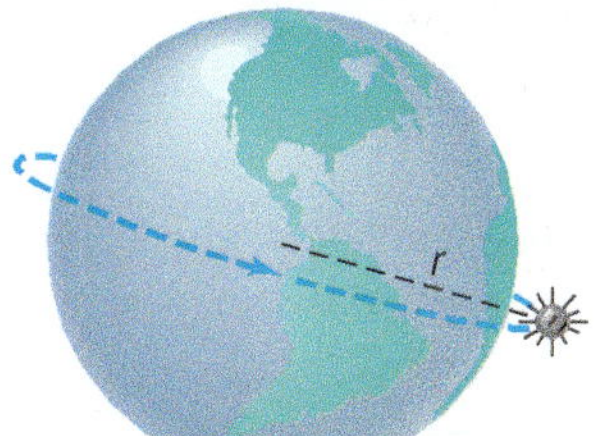

54. The **tangent line** to a circle may be defined as the line that intersects the circle in a single point, called the **point of tangency**. See the figure.

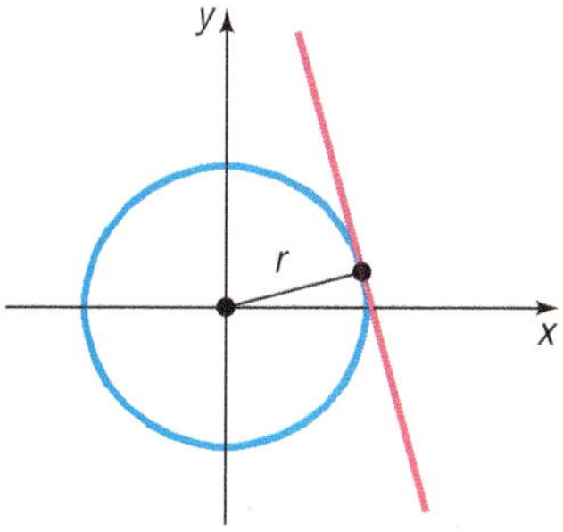

If the equation of the circle is $x^2 + y^2 = r^2$ and the equation of the tangent line is $y = mx + b$, show that:

(a) $r^2(1 + m^2) = b^2$

[**Hint:** The quadratic equation $x^2 + (mx + b)^2 = r^2$ has exactly one solution.]

(b) The point of tangency is $\left(\frac{-r^2m}{b}, \frac{r^2}{b}\right)$.

(c) The tangent line is perpendicular to the line containing the center of the circle and the point of tangency.

55. The Greek Method The Greek method for finding the equation of the tangent line to a circle uses the fact that at any point on a circle the lines containing the center and the tangent line are perpendicular (see Problem 54). Use this method to find an equation of the tangent line to the circle $x^2 + y^2 = 9$ at the point $(1, 2\sqrt{2})$.

56. Use the Greek method described in Problem 55 to find an equation of the tangent line to the circle $x^2 + y^2 - 4x + 6y + 4 = 0$ at the point $(3, 2\sqrt{2} - 3)$.

57. Refer to Problem 54. The line $x - 2y + 4 = 0$ is tangent to a circle at $(0, 2)$. The line $y = 2x - 7$ is tangent to the same circle at $(3, -1)$. Find the center of the circle.

58. Find an equation of the line containing the centers of the two circles

$$x^2 + y^2 - 4x + 6y + 4 = 0$$

and

$$x^2 + y^2 + 6x + 4y + 9 = 0$$

59. If a circle of radius 2 is made to roll along the x-axis, what is an equation for the path of the center of the circle?

60. If the circumference of a circle is 6π, what is its radius?

Explaining Concepts: Discussion and Writing

61. Which of the following equations might have the graph shown? (More than one answer is possible.)
(a) $(x-2)^2+(y+3)^2=13$
(b) $(x-2)^2+(y-2)^2=8$
(c) $(x-2)^2+(y-3)^2=13$
(d) $(x+2)^2+(y-2)^2=8$
(e) $x^2+y^2-4x-9y=0$
(f) $x^2+y^2+4x-2y=0$
(g) $x^2+y^2-9x-4y=0$
(h) $x^2+y^2-4x-4y=4$

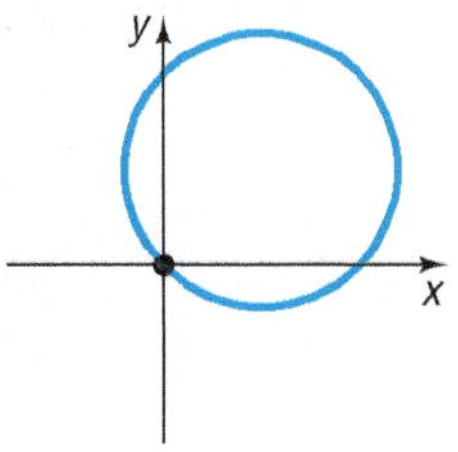

62. Which of the following equations might have the graph shown? (More than one answer is possible.)
(a) $(x-2)^2+y^2=3$
(b) $(x+2)^2+y^2=3$
(c) $x^2+(y-2)^2=3$
(d) $(x+2)^2+y^2=4$
(e) $x^2+y^2+10x+16=0$
(f) $x^2+y^2+10x-2y=1$
(g) $x^2+y^2+9x+10=0$
(h) $x^2+y^2-9x-10=0$

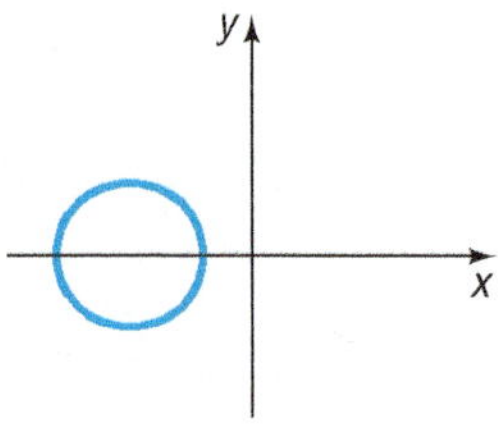

63. Explain how the center and radius of a circle can be used to graph the circle.

64. **What Went Wrong?** A student stated that the center and radius of the graph whose equation is $(x+3)^2+(y-2)^2=16$ are $(3,-2)$ and 4, respectively. Why is this incorrect?

Retain Your Knowledge

Problems 65–68 are based on material learned earlier in the course. The purpose of these problems is to keep the material fresh in your mind so that you are better prepared for the final exam.

65. Find the area and circumference of a circle of radius 13 cm.

66. Multiply $(3x-2)(x^2-2x+3)$. Express the answer as a polynomial in standard form.

67. Solve the equation: $\sqrt{2x^2+3x-1}=x+1$

68. Aaron can load a delivery van in 22 minutes. Elizabeth can load the same van in 28 minutes. How long would it take them to load the van if they worked together?

'Are You Prepared?' Answers

1. add; 25 **2.** $\{-1,5\}$

2.4 Variation

OBJECTIVES
1 Construct a Model Using Direct Variation (p. 197)
2 Construct a Model Using Inverse Variation (p. 197)
3 Construct a Model Using Joint Variation or Combined Variation (p. 198)

When a mathematical model is developed for a real-world problem, it often involves relationships between quantities that are expressed in terms of proportionality:

Force is proportional to acceleration.

When an ideal gas is held at a constant temperature, pressure and volume are inversely proportional.

The force of attraction between two heavenly bodies is inversely proportional to the square of the distance between them.

Revenue is directly proportional to sales.

Each of the preceding statements illustrates the idea of **variation**, or how one quantity varies in relation to another quantity. Quantities may vary *directly, inversely,* or *jointly*.

1 Construct a Model Using Direct Variation

DEFINITION

Let x and y denote two quantities. Then y **varies directly** with x, or y is **directly proportional to** x, if there is a nonzero number k such that

$$y = kx$$

The number k is called the **constant of proportionality**.

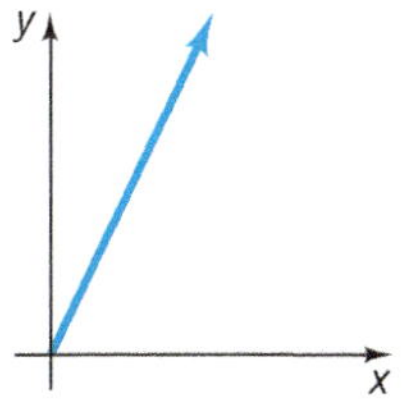

Figure 41
$y = kx;\ k > 0,\ x \geq 0$

The graph in Figure 41 illustrates the relationship between y and x if y varies directly with x and $k > 0$, $x \geq 0$. Note that the constant of proportionality is, in fact, the slope of the line.

If two quantities vary directly, then knowing the value of each quantity in one instance enables us to write a formula that is true in all cases.

EXAMPLE 1 **Mortgage Payments**

The monthly payment p on a mortgage varies directly with the amount borrowed B. If the monthly payment on a 30-year mortgage is \$6.65 for every \$1000 borrowed, find a formula that relates the monthly payment p to the amount borrowed B for a mortgage with these terms. Then find the monthly payment p when the amount borrowed B is \$120,000.

Solution Because p varies directly with B, we know that

$$p = kB$$

for some constant k. Because $p = 6.65$ when $B = 1000$, it follows that

$$6.65 = k(1000)$$

$$k = 0.00665 \quad \text{Solve for } k.$$

Since $p = kB$,

$$p = 0.00665B$$

In particular, when $B = \$120{,}000$,

$$p = 0.00665(\$120{,}000) = \$798$$

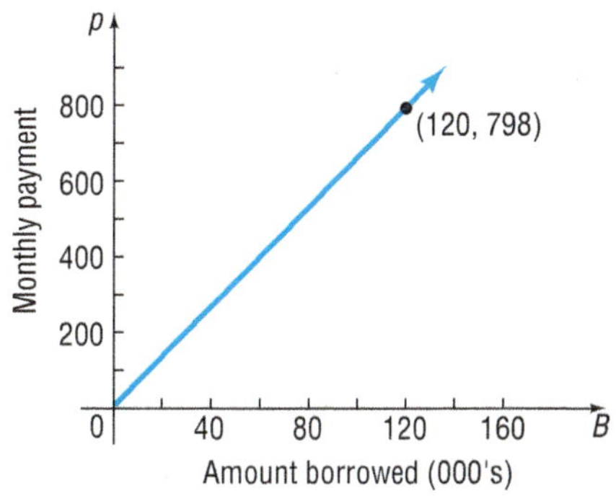

Figure 42

Figure 42 illustrates the relationship between the monthly payment p and the amount borrowed B.

Now Work PROBLEMS 5 AND 23

2 Construct a Model Using Inverse Variation

DEFINITION

Let x and y denote two quantities. Then y **varies inversely** with x, or y is **inversely proportional to** x, if there is a nonzero constant k such that

$$y = \frac{k}{x}$$

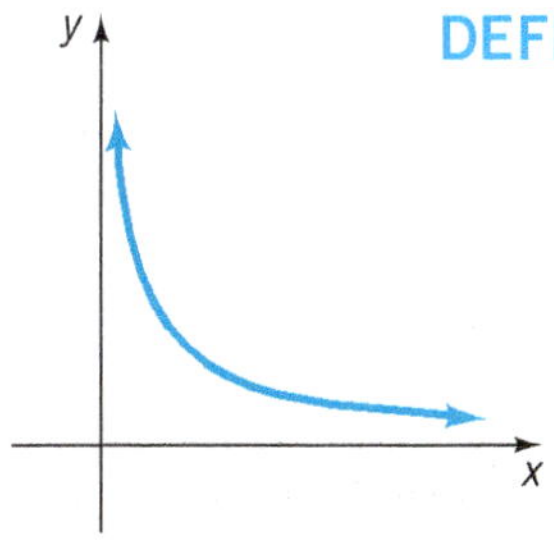

Figure 43 $y = \dfrac{k}{x};\ k > 0,\ x > 0$

The graph in Figure 43 illustrates the relationship between y and x if y varies inversely with x and $k > 0$, $x > 0$.

EXAMPLE 2 Maximum Weight That Can Be Supported by a Piece of Pine

See Figure 44. The maximum weight W that can be safely supported by a 2-inch by 4-inch piece of pine varies inversely with its length l. Experiments indicate that the maximum weight that a 10-foot-long 2-by-4 piece of pine can support is 500 pounds. Write a general formula relating the maximum weight W (in pounds) to length l (in feet). Find the maximum weight W that can be safely supported by a length of 25 feet.

Figure 44

Solution Because W varies inversely with l, we know that

$$W = \frac{k}{l}$$

for some constant k. Because $W = 500$ when $l = 10$, we have

$$500 = \frac{k}{10}$$

$$k = 5000$$

Since $W = \frac{k}{l}$,

$$W = \frac{5000}{l}$$

In particular, the maximum weight W that can be safely supported by a piece of pine 25 feet in length is

$$W = \frac{5000}{25} = 200 \text{ pounds}$$

Figure 45 illustrates the relationship between the weight W and the length l. ■

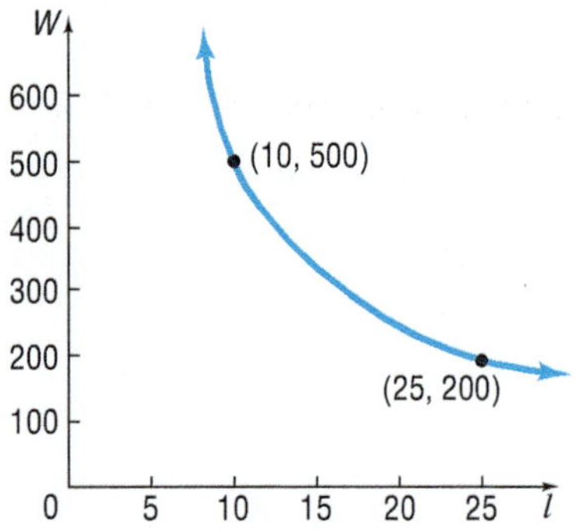

Figure 45 $W = \frac{5000}{l}$

Now Work PROBLEM 33

3 Construct a Model Using Joint Variation or Combined Variation

When a variable quantity Q is proportional to the product of two or more other variables, we say that Q **varies jointly** with these quantities. Finally, combinations of direct and/or inverse variation may occur. This is usually referred to as **combined variation**.

EXAMPLE 3 Loss of Heat through a Wall

The loss of heat through a wall varies jointly with the area of the wall and the difference between the inside and outside temperatures and varies inversely with the thickness of the wall. Write an equation that relates these quantities.

Solution Begin by assigning symbols to represent the quantities:

L = Heat loss T = Temperature difference
A = Area of wall d = Thickness of wall

Then

$$L = k\frac{AT}{d}$$

where k is the constant of proportionality. ■

In direct or inverse variation, the quantities that vary may be raised to powers. For example, in the early seventeenth century, Johannes Kepler (1571–1630) discovered that the square of the period of revolution T of a planet around the Sun varies directly with the cube of its mean distance a from the Sun. That is, $T^2 = ka^3$, where k is the constant of proportionality.

EXAMPLE 4 **Force of the Wind on a Window**

The force F of the wind on a flat surface positioned at a right angle to the direction of the wind varies jointly with the area A of the surface and the square of the speed v of the wind. A wind of 30 miles per hour blowing on a window measuring 4 feet by 5 feet has a force of 150 pounds. See Figure 46. What force does a wind of 50 miles per hour exert on a window measuring 3 feet by 4 feet?

Figure 46

Solution Since F varies jointly with A and v^2, we have

$$F = kAv^2$$

where k is the constant of proportionality. We are told that $F = 150$ when $A = 4 \cdot 5 = 20$ and $v = 30$. Then

$$150 = k(20)(900) \quad F = kAv^2, F = 150, A = 20, v = 30$$

$$k = \frac{1}{120}$$

Since $F = kAv^2$,

$$F = \frac{1}{120}Av^2$$

For a wind of 50 miles per hour blowing on a window whose area is $A = 3 \cdot 4 = 12$ square feet, the force F is

$$F = \frac{1}{120}(12)(2500) = 250 \text{ pounds}$$

Now Work PROBLEM 41

2.4 Assess Your Understanding

Concepts and Vocabulary

1. If x and y are two quantities, then y is directly proportional to x if there is a nonzero number k such that ________.

2. ***True or False*** If y varies directly with x, then $y = \frac{k}{x}$, where k is a constant.

3. Which equation represents a joint variation model?

(a) $y = 5x$ (b) $y = 5xzw$ (c) $y = \frac{5}{x}$ (d) $y = \frac{5xz}{w}$

4. Choose the best description for the model $y = \frac{kx}{z}$, if k is a nonzero constant.

(a) y varies jointly with x and z.
(b) y is inversely proportional to x and z.
(c) y varies directly with x and inversely with z.
(d) y is directly proportion to z and inversely proportional to x.

Skill Building

In Problems 5–16, write a general formula to describe each variation.

5. y varies directly with x; $y = 2$ when $x = 10$

6. v varies directly with t; $v = 16$ when $t = 2$

7. A varies directly with x^2; $A = 4\pi$ when $x = 2$

8. V varies directly with x^3; $V = 36\pi$ when $x = 3$

9. F varies inversely with d^2; $F = 10$ when $d = 5$

10. y varies inversely with $\sqrt{x}$; $y = 4$ when $x = 9$

11. z varies directly with the sum of the squares of x and y; $z = 5$ when $x = 3$ and $y = 4$

12. T varies jointly with the cube root of x and the square of d; $T = 18$ when $x = 8$ and $d = 3$

13. M varies directly with the square of d and inversely with the square root of x; $M = 24$ when $x = 9$ and $d = 4$

14. z varies directly with the sum of the cube of x and the square of y; $z = 1$ when $x = 2$ and $y = 3$

15. The square of T varies directly with the cube of a and inversely with the square of d; $T = 2$ when $a = 2$ and $d = 4$

16. The cube of z varies directly with the sum of the squares of x and y; $z = 2$ when $x = 9$ and $y = 4$

Applications and Extensions

In Problems 17–22, write an equation that relates the quantities.

17. **Geometry** The volume V of a sphere varies directly with the cube of its radius r. The constant of proportionality is $\frac{4\pi}{3}$.

18. **Geometry** The square of the length of the hypotenuse c of a right triangle varies jointly with the sum of the squares of the lengths of its legs a and b. The constant of proportionality is 1.

19. **Geometry** The area A of a triangle varies jointly with the lengths of the base b and the height h. The constant of proportionality is $\frac{1}{2}$.

20. **Geometry** The perimeter p of a rectangle varies jointly with the sum of the lengths of its sides l and w. The constant of proportionality is 2.

21. **Physics: Newton's Law** The force F (in newtons) of attraction between two bodies varies jointly with their masses m and M (in kilograms) and inversely with the square of the distance d (in meters) between them. The constant of proportionality is $G = 6.67 \times 10^{-11}$.

22. **Physics: Simple Pendulum** The **period** of a pendulum is the time required for one oscillation; the pendulum is usually referred to as **simple** when the angle made to the vertical is less than 5°. The period T of a simple pendulum (in seconds) varies directly with the square root of its length l (in feet). The constant of proportionality is $\frac{2\pi}{\sqrt{32}}$.

23. **Mortgage Payments** The monthly payment p on a mortgage varies directly with the amount borrowed B. If the monthly payment on a 30-year mortgage is \$6.49 for every \$1000 borrowed, find a linear equation that relates the monthly payment p to the amount borrowed B for a mortgage with the same terms. Then find the monthly payment p when the amount borrowed B is \$145,000.

24. **Mortgage Payments** The monthly payment p on a mortgage varies directly with the amount borrowed B. If the monthly payment on a 15-year mortgage is \$8.99 for every \$1000 borrowed, find a linear equation that relates the monthly payment p to the amount borrowed B for a mortgage with the same terms. Then find the monthly payment p when the amount borrowed B is \$175,000.

25. **Physics: Falling Objects** The distance s that an object falls is directly proportional to the square of the time t of the fall. If an object falls 16 feet in 1 second, how far will it fall in 3 seconds? How long will it take an object to fall 64 feet?

26. **Physics: Falling Objects** The velocity v of a falling object is directly proportional to the time t of the fall. If, after 2 seconds, the velocity of the object is 64 feet per second, what will its velocity be after 3 seconds?

27. **Physics: Stretching a Spring** The elongation E of a spring balance varies directly with the applied weight W (see the figure). If $E = 3$ when $W = 20$, find E when $W = 15$.

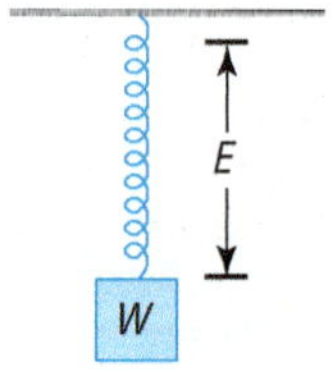

28. **Physics: Vibrating String** The rate of vibration of a string under constant tension varies inversely with the length of the string. If a string is 48 inches long and vibrates 256 times per second, what is the length of a string that vibrates 576 times per second?

29. **Revenue Equation** At the corner Shell station, the revenue R varies directly with the number g of gallons of gasoline sold. If the revenue is \$47.40 when the number of gallons sold is 12, find a linear equation that relates revenue R to the number g of gallons of gasoline. Then find the revenue R when the number of gallons of gasoline sold is 10.5.

30. **Cost Equation** The cost C of roasted almonds varies directly with the number A of pounds of almonds purchased. If the cost is \$23.75 when the number of pounds of roasted almonds purchased is 5, find a linear equation that relates the cost C to the number A of pounds of almonds purchased. Then find the cost C when the number of pounds of almonds purchased is 3.5.

31. **Demand** Suppose that the demand D for candy at the movie theater is inversely related to the price p.
    (a) When the price of candy is \$2.75 per bag, the theater sells 156 bags of candy. Express the demand for candy in terms of its price.
    (b) Determine the number of bags of candy that will be sold if the price is raised to \$3 a bag.

32. **Driving to School** The time t that it takes to get to school varies inversely with your average speed s.
    (a) Suppose that it takes you 40 minutes to get to school when your average speed is 30 miles per hour. Express the driving time to school in terms of average speed.
    (b) Suppose that your average speed to school is 40 miles per hour. How long will it take you to get to school?

33. **Pressure** The volume of a gas V held at a constant temperature in a closed container varies inversely with its pressure P. If the volume of a gas is 600 cubic centimeters (cm^3) when the pressure is 150 millimeters of mercury (mm Hg), find the volume when the pressure is 200 mm Hg.

34. Resistance The current i in a circuit is inversely proportional to its resistance Z measured in ohms. Suppose that when the current in a circuit is 30 amperes, the resistance is 8 ohms. Find the current in the same circuit when the resistance is 10 ohms.

35. Weight The weight of an object above the surface of Earth varies inversely with the square of the distance from the center of Earth. If Maria weighs 125 pounds when she is on the surface of Earth (3960 miles from the center), determine Maria's weight when she is at the top of Mount McKinley (3.8 miles from the surface of Earth).

36. Weight of a Body The weight of a body above the surface of Earth varies inversely with the square of the distance from the center of Earth. If a certain body weighs 55 pounds when it is 3960 miles from the center of Earth, how much will it weigh when it is 3965 miles from the center?

37. Geometry The volume V of a right circular cylinder varies jointly with the square of its radius r and its height h. The constant of proportionality is π. See the figure. Write an equation for V.

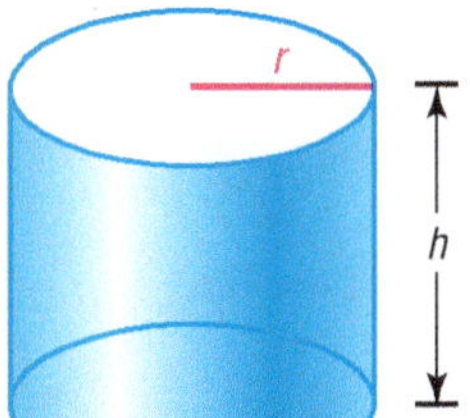

38. Geometry The volume V of a right circular cone varies jointly with the square of its radius r and its height h. The constant of proportionality is $\frac{\pi}{3}$. See the figure. Write an equation for V.

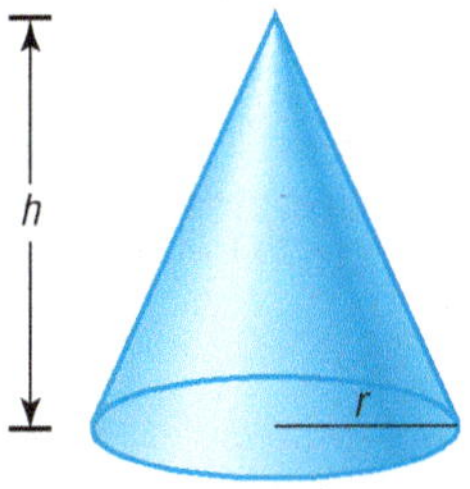

39. Intensity of Light The intensity I of light (measured in foot-candles) varies inversely with the square of the distance from the bulb. Suppose that the intensity of a 100-watt light bulb at a distance of 2 meters is 0.075 foot-candle. Determine the intensity of the bulb at a distance of 5 meters.

40. Force of the Wind on a Window The force exerted by the wind on a plane surface varies jointly with the area of the surface and the square of the velocity of the wind. If the force on an area of 20 square feet is 11 pounds when the wind velocity is 22 miles per hour, find the force on a surface area of 47.125 square feet when the wind velocity is 36.5 miles per hour.

41. Horsepower The horsepower (hp) that a shaft can safely transmit varies jointly with its speed (in revolutions per minute, rpm) and the cube of its diameter. If a shaft of a certain material 2 inches in diameter can transmit 36 hp at 75 rpm, what diameter must the shaft have in order to transmit 45 hp at 125 rpm?

42. Chemistry: Gas Laws The volume V of an ideal gas varies directly with the temperature T and inversely with the pressure P. Write an equation relating V, T, and P using k as the constant of proportionality. If a cylinder contains oxygen at a temperature of 300 K and a pressure of 15 atmospheres in a volume of 100 liters, what is the constant of proportionality k? If a piston is lowered into the cylinder, decreasing the volume occupied by the gas to 80 liters and raising the temperature to 310 K, what is the gas pressure?

43. Physics: Kinetic Energy The kinetic energy K of a moving object varies jointly with its mass m and the square of its velocity v. If an object weighing 25 kilograms and moving with a velocity of 10 meters per second has a kinetic energy of 1250 joules, find its kinetic energy when the velocity is 15 meters per second.

44. Electrical Resistance of a Wire The electrical resistance of a wire varies directly with the length of the wire and inversely with the square of the diameter of the wire. If a wire 432 feet long and 4 millimeters in diameter has a resistance of 1.24 ohms, find the length of a wire of the same material whose resistance is 1.44 ohms and whose diameter is 3 millimeters.

45. Measuring the Stress of Materials The stress in the material of a pipe subject to internal pressure varies jointly with the internal pressure and the internal diameter of the pipe and inversely with the thickness of the pipe. The stress is 100 pounds per square inch when the diameter is 5 inches, the thickness is 0.75 inch, and the internal pressure is 25 pounds per square inch. Find the stress when the internal pressure is 40 pounds per square inch if the diameter is 8 inches and the thickness is 0.50 inch.

46. Safe Load for a Beam The maximum safe load for a horizontal rectangular beam varies jointly with the width of the beam and the square of the thickness of the beam and inversely with its length. If an 8-foot beam will support up to 750 pounds when the beam is 4 inches wide and 2 inches thick, what is the maximum safe load in a similar beam 10 feet long, 6 inches wide, and 2 inches thick?

Explaining Concepts: Discussion and Writing

47. In the early 17th century, Johannes Kepler discovered that the square of the period T of the revolution of a planet around the Sun varies directly with the cube of its mean distance a from the Sun. Go to the library and research this law and Kepler's other two laws. Write a brief paper about these laws and Kepler's place in history.

48. Using a situation that has not been discussed in the text, write a real-world problem that you think involves two variables that vary directly. Exchange your problem with another student's to solve and critique.

49. Using a situation that has not been discussed in the text, write a real-world problem that you think involves two variables that vary inversely. Exchange your problem with another student's to solve and critique.

50. Using a situation that has not been discussed in the text, write a real-world problem that you think involves three variables that vary jointly. Exchange your problem with another student's to solve and critique.

Retain Your Knowledge

Problems 51–54 are based on material learned earlier in the course. The purpose of these problems is to keep the material fresh in your mind so that you are better prepared for the final exam.

51. Factor $3x^3 + 25x^2 - 12x - 100$ completely.

52. Add $\frac{5}{x+3} + \frac{x-2}{x^2+7x+12}$ and simplify the result.

53. Simplify: $\left(\frac{4}{25}\right)^{3/2}$

54. Rationalize the denominator of $\frac{3}{\sqrt{7}-2}$.

Chapter Review

Things to Know

Formulas

Slope (p. 174)	$m = \frac{y_2 - y_1}{x_2 - x_1}$ if $x_1 \neq x_2$; undefined if $x_1 = x_2$
Parallel lines (p. 181)	Equal slopes $(m_1 = m_2)$ and different y-intercepts $(b_1 \neq b_2)$
Perpendicular lines (p. 182)	Product of slopes is -1 $(m_1 \cdot m_2 = -1)$
Direct variation (p. 197)	$y = kx, k \neq 0$
Inverse variation (p. 197)	$y = \frac{k}{x}, k \neq 0$

Equations of Lines and Circles

Vertical line (p. 177)	$x = a$; a is the x-intercept
Point–slope form of the equation of a line (p. 177)	$y - y_1 = m(x - x_1)$; m is the slope of the line, (x_1, y_1) is a point on the line
Horizontal line (p. 178)	$y = b$; b is the y-intercept
Slope–intercept form of the equation of a line (p. 178)	$y = mx + b$; m is the slope of the line, b is the y-intercept
General form of the equation of a line (p. 180)	$Ax + By = C$; A, B not both 0
Standard form of the equation of a circle (p. 189)	$(x - h)^2 + (y - k)^2 = r^2$; r is the radius of the circle, (h, k) is the center of the circle
Equation of the unit circle (p. 190)	$x^2 + y^2 = 1$
General form of the equation of a circle (p. 192)	$x^2 + y^2 + ax + by + c = 0$, with restrictions on a, b, and c

Objectives

Section		You should be able to ...	Examples	Review Exercises
2.1	1	Find intercepts algebraically from an equation (p. 165)	1	5–9
	2	Test an equation for symmetry (p. 166)	2, 3	5–9
	3	Know how to graph key equations (p. 168)	4–6	27
2.2	1	Calculate and interpret the slope of a line (p. 173)	1	1–4, 29, 31
	2	Graph lines given a point and the slope (p. 176)	2	28
	3	Find the equation of a vertical line (p. 176)	3	17
	4	Use the point–slope form of a line; identify horizontal lines (p. 177)	4, 5	15, 16
	5	Write the equation of a line in slope–intercept form (p. 178)	6	15, 16, 18–24
	6	Find the equation of a line given two points (p. 179)	7	18–20
	7	Graph lines written in general form using intercepts (p. 180)	8	25, 26
	8	Find equations of parallel lines (p. 181)	9, 10	21
	9	Find equations of perpendicular lines (p. 182)	11, 12	22
2.3	1	Write the standard form of the equation of a circle (p. 189)	1	10, 11, 30
	2	Graph a circle by hand and by using a graphing utility (p. 190)	2, 3	12–14
	3	Work with the general form of the equation of a circle (p. 192)	4, 5	13, 14
2.4	1	Construct a model using direct variation (p. 197)	1	32
	2	Construct a model using inverse variation (p. 197)	2	33
	3	Construct a model using joint variation or combined variation (p. 198)	3, 4	34

Review Exercises

In Problems 1–4, find the following for each pair of points:
(a) The slope of the line containing the points
(b) Interpret the slope found in part (a)

1. $(0,0)$; $(4,2)$ **2.** $(1,-1)$; $(-2,3)$

3. $(4,-4)$; $(4,8)$ **4.** $(-2,-1)$; $(3,-1)$

In Problems 5–9, list the intercepts and test for symmetry with respect to the x-axis, the y-axis, and the origin.

5. $2x = 3y^2$

6. $x^2 + 4y^2 = 16$

7. $y = x^4 - 3x^2 - 4$

8. $y = x^3 - x$

9. $x^2 + x + y^2 + 2y = 0$

In Problems 10 and 11, find the standard form of the equation of the circle whose center and radius are given.

10. $(h,k) = (-2,3); r = 4$ **11.** $(h,k) = (-1,-2); r = 1$

In Problems 12–14, find the center and radius of each circle. Graph each circle by hand. Find the intercepts, if any, of each circle.

12. $x^2 + (y-1)^2 = 4$ **13.** $x^2 + y^2 - 2x + 4y - 4 = 0$

14. $3x^2 + 3y^2 - 6x + 12y = 0$

In Problems 15–22, find an equation of the line having the given characteristics. Express your answer using either the general form or the slope–intercept form of the equation of a line, whichever you prefer.

15. Slope $= -2$; containing the point $(3,-1)$

16. Slope $= 0$; containing the point $(-5,4)$

17. Vertical; containing the point $(-3,4)$

18. x-intercept $= 2$; containing the point $(4,-5)$

19. y-intercept $= -2$; containing the point $(5,-3)$

20. Containing the points $(3,-4)$ and $(2,1)$

21. Parallel to the line $2x - 3y = -4$; containing the point $(-5,3)$

22. Perpendicular to the line $3x - y = -4$; containing the point $(-2,4)$

In Problems 23 and 24, find the slope and y-intercept of each line. Graph the line, labeling any intercepts.

23. $4x - 5y = -20$ **24.** $\frac{1}{2}x - \frac{1}{3}y = -\frac{1}{6}$

In Problems 25 and 26, find the intercepts and graph each line.

25. $2x - 3y = 12$ **26.** $\frac{1}{2}x + \frac{1}{3}y = 2$

27. Sketch a graph of $y = x^3$.

28. Graph the line with slope $\frac{2}{3}$ containing the point $(1,2)$.

29. Show that the points $A = (-2,0), B = (-4,4)$, and $C = (8,5)$ are the vertices of a right triangle by using the slopes of the lines joining the vertices.

30. The endpoints of the diameter of a circle are $(-3,2)$ and $(5,-6)$. Find the center and radius of the circle. Write the standard equation of this circle.

31. Show that the points $A = (2,5)$, $B = (6,1)$, and $C = (8,-1)$ lie on a line by using slopes.

32. Mortgage Payments The monthly payment p on a mortgage varies directly with the amount borrowed B. If the monthly payment on a 30-year mortgage is \$854.00 when \$130,000 is borrowed, find an equation that relates the monthly payment p to the amount borrowed B for a mortgage with the same terms. Then find the monthly payment p when the amount borrowed B is \$165,000.

33. Weight of a Body The weight of a body varies inversely with the square of its distance from the center of Earth. Assuming that the radius of Earth is 3960 miles, how much would a man weigh at an altitude of 1 mile above Earth's surface if he weighs 200 pounds on Earth's surface?

34. Heat Loss The amount of heat transferred per hour through a glass window varies jointly with the surface area of the window and the difference in temperature between the areas separated by the glass. A window with a surface area of 7.5 square feet loses 135 Btu per hour when the temperature difference is 40°F. How much heat is lost per hour for a similar window with a surface are of 12 square feet when the temperature difference is 35°F?

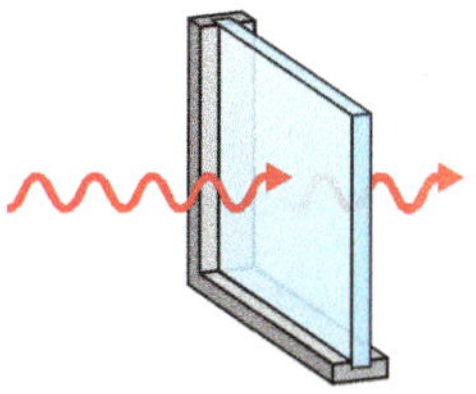

Chapter Test

CHAPTER **Test Prep** VIDEOS

The Chapter Test Prep Videos are step-by-step solutions available in MyMathLab®, or on this text's YouTube Channel. Flip back to the Resources for Success page for a link to this text's YouTube channel.

1. Use $P_1 = (-1, 3)$ and $P_2 = (5, -1)$.
   (a) Find the slope of the line containing P_1 and P_2.
   (b) Interpret this slope.
2. Graph $y = x^2 - 9$ by plotting points.
3. Sketch the graph of $y^2 = x$.
4. List the intercepts and test for symmetry: $x^2 + y = 9$.
5. Write the slope–intercept form of the line with slope -2 containing the point $(3, -4)$. Graph the line.
6. Find the slope and y-intercept: $2x + 3y = 9$.
7. Graph the line $3x - 4y = 24$ by finding the intercepts.
8. Write the general form of the circle with center $(4, -3)$ and radius 5.
9. Find the center and radius of the circle $x^2 + y^2 + 4x - 2y - 4 = 0$. Graph this circle.
10. For the line $2x + 3y = 6$, find a line parallel to it containing the point $(1, -1)$. Also find a line perpendicular to it containing the point $(0, 3)$.
11. **Resistance Due to a Conductor** The resistance (in ohms) of a circular conductor varies directly with the length of the conductor and inversely with the square of the radius of the conductor. If 50 feet of wire with a radius of 6×10^{-3} inch has a resistance of 10 ohms, what would be the resistance of 100 feet of the same wire if the radius is increased to 7×10^{-3} inch?

Cumulative Review

In Problems 1–8, find the real solution(s) of each equation.

1. $3x - 5 = 0$
2. $x^2 - x - 12 = 0$
3. $2x^2 - 5x - 3 = 0$
4. $x^2 - 2x - 2 = 0$
5. $x^2 + 2x + 5 = 0$
6. $\sqrt{2x + 1} = 3$
7. $|x - 2| = 1$
8. $\sqrt{x^2 + 4x} = 2$

In Problems 9 and 10, solve each equation in the complex number system.

9. $x^2 = -9$
10. $x^2 - 2x + 5 = 0$

In Problems 11–14, solve each inequality. Graph the solution set.

11. $2x - 3 \le 7$
12. $-1 < x + 4 < 5$
13. $|x - 2| \le 1$
14. $|2 + x| > 3$
15. Find the distance between the points $P = (-1, 3)$ and $Q = (4, -2)$. Find the midpoint of the line segment from P to Q.
16. Which of the following points are on the graph of $y = x^3 - 3x + 1$?
    (a) $(-2, -1)$ (b) $(2, 3)$ (c) $(3, 1)$
17. Sketch the graph of $y = x^3$.
18. Find the equation of the line containing the points $(-1, 4)$ and $(2, -2)$. Express your answer in slope–intercept form.
19. Find the equation of the line perpendicular to the line $y = 2x + 1$ and containing the point $(3, 5)$. Express your answer in slope–intercept form and graph the line.
20. Graph the equation $x^2 + y^2 - 4x + 8y - 5 = 0$.

Chapter Project

Determining the Selling Price of a Home Determining how much to pay for a home is one of the more difficult decisions that must be made when purchasing a home. There are many factors that play a role in a home's value. Location, size, number of bedrooms, number of bathrooms, lot size, and building materials are just a few. Fortunately, the website Zillow.com has developed its own formula for predicting the selling price of a home. This information is a great tool for predicting the actual sale price. For example, the data below show the "zestimate"—the selling price of a home as predicted by the folks at Zillow and the actual selling price of the home for homes in Oak Park, Illinois.

Zestimate (000s of dollars)	Sale Price (000s of dollars)
291.5	268
320	305
371.5	375
303.5	283
351.5	350
314	275
332.5	356
295	300
313	285
368	385

The graph below, called a scatter diagram, shows the points $(291.5, 268), (320, 305), \ldots, (368, 385)$ in a Cartesian plane. From the graph, it appears that the data follow a linear relation.

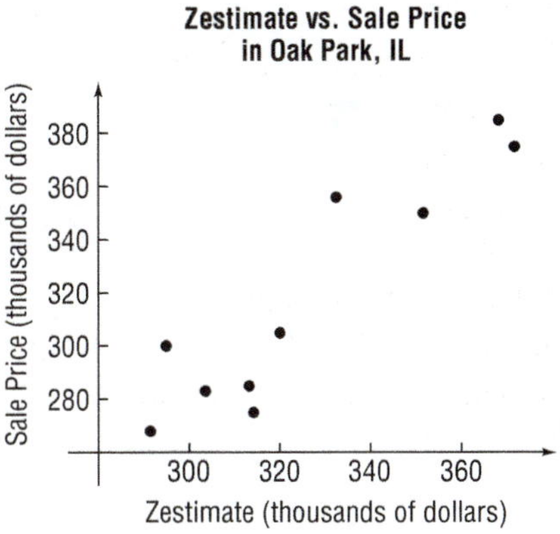

1. Imagine drawing a line through the data that appears to fit the data well. Do you believe the slope of the line would be positive, negative, or close to zero? Why?
2. Pick two points from the scatter diagram. Treat the zestimate as the value of x and treat the sale price as the corresponding value of y. Find the equation of the line through the two points you selected.
3. Interpret the slope of the line.
4. Use your equation to predict the selling price of a home whose zestimate is \$335,000.
5. Do you believe it would be a good idea to use the equation you found in part 2 if the zestimate is \$950,000? Why or why not?
6. Choose a location in which you would like to live. Go to www.zillow.com and randomly select at least ten homes that have recently sold. Record the Zestimate and sale price for each home.
   (a) Draw a scatter diagram of your data.
   (b) Select two points from the scatter diagram and find the equation of the line through the points.
   (c) Interpret the slope.
   (d) Find a home from the Zillow website that interests you under the "Make Me Move" option for which a zestimate is available. Use your equation to predict the sale price based on the estimate.

3 Functions and Their Graphs

Choosing a Wireless Data Plan

Most consumers choose a cellular provider first and then select an appropriate data plan from that provider. The choice as to the type of plan selected depends on your use of the device. For example, is online gaming important? Do you want to stream audio or video? The mathematics learned in this chapter can help you decide what plan is best suited to your particular needs.

—See the Internet-based Chapter Project—

Outline

••• A Look Back

So far, our discussion has focused on techniques for graphing equations containing two variables.

A Look Ahead •••

In this chapter, we look at a special type of equation involving two variables called a *function*. This chapter deals with what a function is, how to graph functions, properties of functions, and how functions are used in applications. The word *function* apparently was introduced by René Descartes in 1637. For him, a function was simply any positive integral power of a variable x. Gottfried Wilhelm Leibniz (1646–1716), who always emphasized the geometric side of mathematics, used the word *function* to denote any quantity associated with a curve, such as the coordinates of a point on the curve. Leonhard Euler (1707–1783) employed the word to mean any equation or formula involving variables and constants. His idea of a function is similar to the one most often seen in courses that precede calculus. Later, the use of functions in investigating heat flow equations led to a very broad definition that originated with Lejeune Dirichlet (1805–1859), which describes a function as a correspondence between two sets. That is the definition used in this text.

3.1 Functions

PREPARING FOR THIS SECTION *Before getting started, review the following:*

- Intervals (Section 1.7, pp. 147–148)
- Solving Inequalities (Section 1.7, pp. 150–151)
- Evaluating Algebraic Expressions, Domain of a Variable (Chapter R, Section R.2, pp. 21–22)
- Rationalizing Denominators (Chapter R, Section R.8, p. 76)

Now Work the **'Are You Prepared?'** problems on page 218.

OBJECTIVES
1. Determine Whether a Relation Represents a Function (p. 207)
2. Find the Value of a Function (p. 210)
3. Find the Difference Quotient of a Function (p. 213)
4. Find the Domain of a Function Defined by an Equation (p. 214)
5. Form the Sum, Difference, Product, and Quotient of Two Functions (p. 216)

1 Determine Whether a Relation Represents a Function

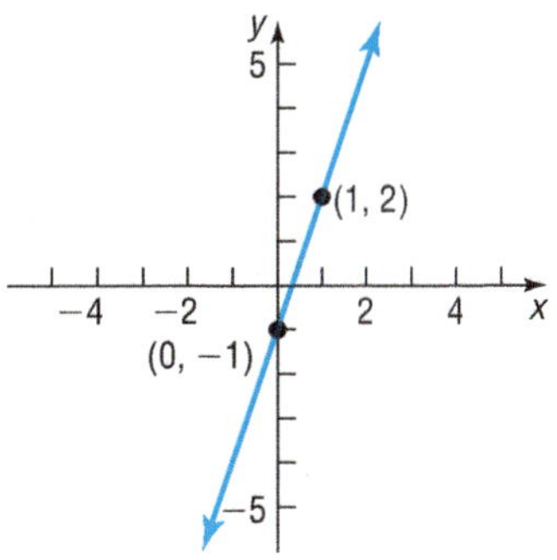

Figure 1 $y = 3x - 1$

Often there are situations where the value of one variable is somehow linked to the value of another variable. For example, an individual's level of education is linked to annual income. Engine size is linked to gas mileage. When the value of one variable is related to the value of a second variable, we have a *relation*. A **relation** is a correspondence between two sets. If x and y are two elements, one from each of these sets, and if a relation exists between x and y, then we say that x **corresponds** to y or that y **depends on** x, and we write $x \rightarrow y$.

There are a number of ways to express relations between two sets. For example, the equation $y = 3x - 1$ shows a relation between x and y. It says that if we take some number x, multiply it by 3, and then subtract 1, we obtain the corresponding value of y. In this sense, x serves as the **input** to the relation, and y is the **output** of the relation. This relation, expressed as a graph, is shown in Figure 1.

The set of all inputs for a relation is called the **domain** of the relation, and the set of all outputs is called the **range**.

In addition to being expressed as equations and graphs, relations can be expressed through a technique called *mapping*. A **map** illustrates a relation as a set of inputs with an arrow drawn from each element in the set of inputs to the corresponding element in the set of outputs. **Ordered pairs** can be used to represent $x \rightarrow y$ as (x, y).

EXAMPLE 1 **Maps and Ordered Pairs as Relations**

Figure 2 shows a relation between states and the number of representatives each state has in the House of Representatives. (***Source:*** *www.house.gov*). The relation might be named "number of representatives."

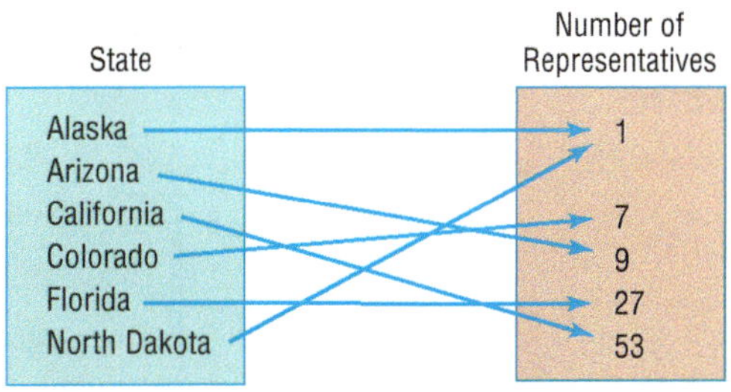

Figure 2 Number of representatives

In this relation, Alaska corresponds to 1, Arizona corresponds to 9, and so on. Using ordered pairs, this relation would be expressed as

$$\{(\text{Alaska}, 1), (\text{Arizona}, 9), (\text{California}, 53), (\text{Colorado}, 7), (\text{Florida}, 27), (\text{North Dakota}, 1)\}$$

The domain of the relation is {Alaska, Arizona, California, Colorado, Florida, North Dakota}, and the range is {1, 7, 9, 27, 53}. Since the range is a set, the output "1" is listed only once.

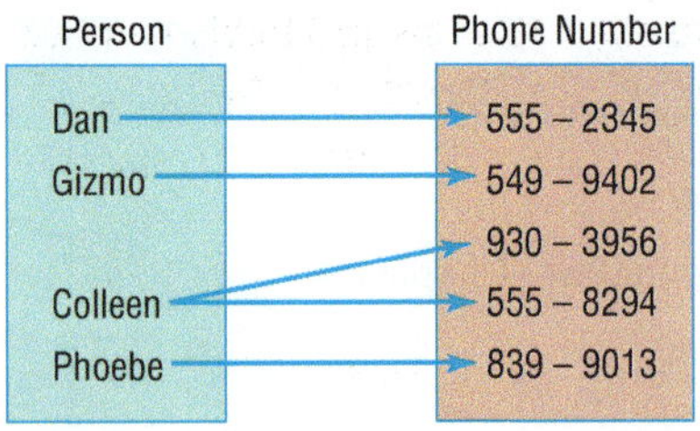

Figure 3 Phone numbers

One of the most important concepts in algebra is the *function*. A function is a special type of relation. To understand the idea behind a function, let's revisit the relation presented in Example 1. If we were to ask, "How many representatives does Alaska have?" you would respond "1." In fact, each input *state* corresponds to a single output *number of representatives*.

Let's consider a second relation, one that involves a correspondence between four people and their phone numbers. See Figure 3. Notice that Colleen has two telephone numbers. There is no single answer to the question "What is Colleen's phone number?"

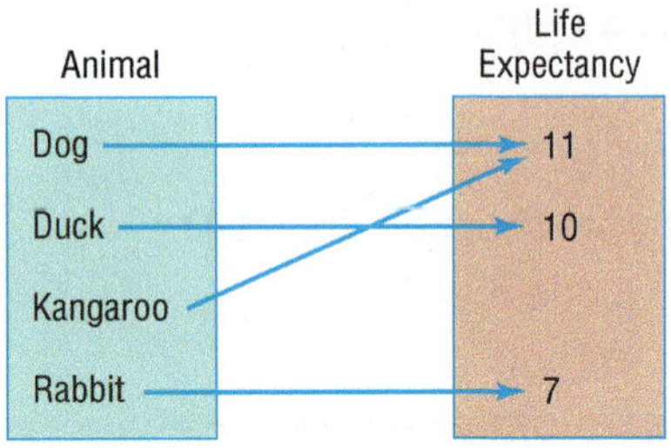

Figure 4 Animal life expectancy

Let's look at one more relation. Figure 4 is a relation that shows a correspondence between type of *animal* and *life expectancy*. If asked to determine the life expectancy of a dog, we would all respond, "11 years." If asked to determine the life expectancy of a rabbit, we would all respond, "7 years."

Notice that the relations presented in Figures 2 and 4 have something in common. What is it? In both of these relations, each input corresponds to exactly one output. This leads to the definition of a *function*.

DEFINITION

Let X and Y be two nonempty sets.* A **function** from X into Y is a relation that associates with each element of X exactly one element of Y.

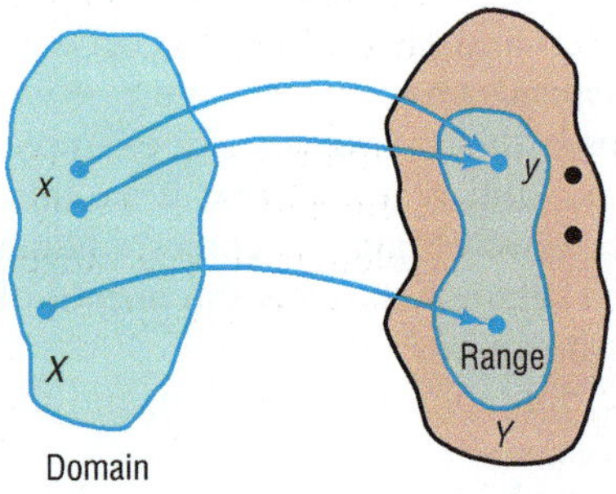

Figure 5

The set X is called the **domain** of the function. For each element x in X, the corresponding element y in Y is called the **value** of the function at x, or the **image** of x. The set of all images of the elements in the domain is called the **range** of the function. See Figure 5.

Since there may be some elements in Y that are not the image of some x in X, it follows that the range of a function may be a subset of Y, as shown in Figure 5. For example, consider the function $y = x^2$, where the domain is the set of all real numbers. Since $x^2 \geq 0$ for all real numbers x, the range of $y = x^2$ is $\{y \mid y \geq 0\}$, which is a subset of the set of all real numbers, Y.

Not all relations between two sets are functions. The next example shows how to determine whether a relation is a function.

EXAMPLE 2

Determining Whether a Relation Is a Function

For each relation in Figures 6, 7, and 8, state the domain and range. Then determine whether the relation is a function.

(a) See Figure 6. For this relation, the input is the number of calories in a fast-food sandwich, and the output is the fat content (in grams).

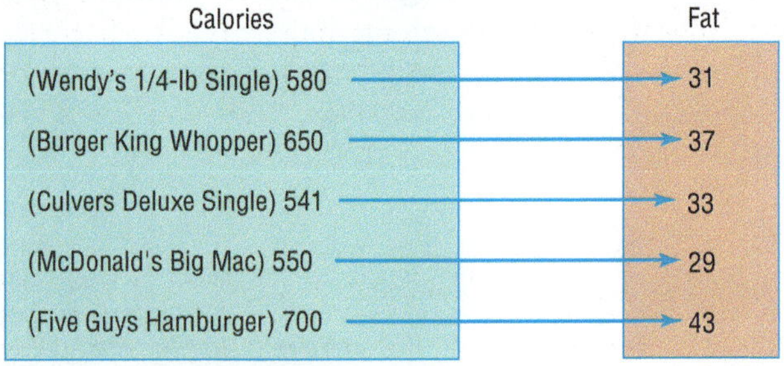

Figure 6 Fat content
***Source:** Each company's website*

*The sets X and Y will usually be sets of real numbers, in which case a (real) function results. The two sets can also be sets of complex numbers, and then we have defined a complex function. In the broad definition (proposed by Lejeune Dirichlet), X and Y can be any two sets.

(b) See Figure 7. For this relation, the inputs are gasoline stations in Harris County, Texas, and the outputs are the price per gallon of unleaded regular in March 2015.

(c) See Figure 8. For this relation, the inputs are the weight (in carats) of pear-cut diamonds and the outputs are the price (in dollars).

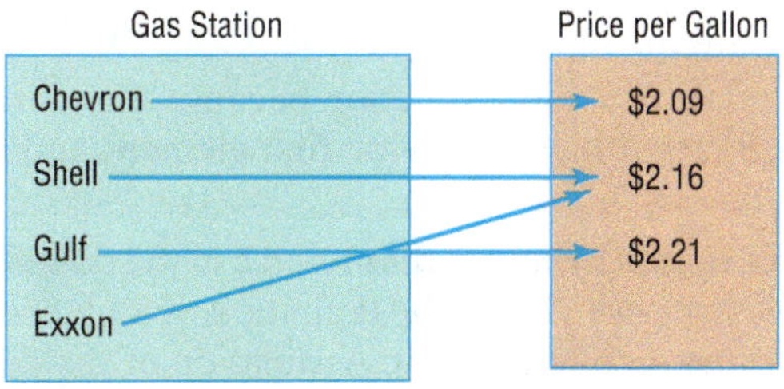

Figure 7 Unleaded price per gallon

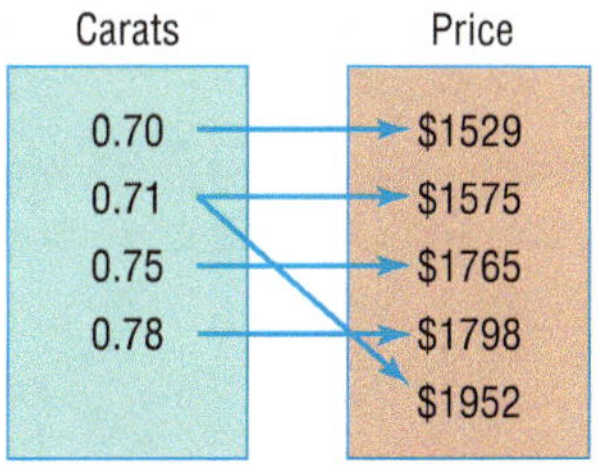

Figure 8 Diamond price
Source: *Used with permission of Diamonds.com*

Solution (a) The domain of the relation is {541, 550, 580, 650, 700}, and the range of the relation is {29, 31, 33, 37, 43}. The relation in Figure 6 is a function because each element in the domain corresponds to exactly one element in the range.

(b) The domain of the relation is {Chevron, Exxon, Gulf, Shell}. The range of the relation is {\$2.09, \$2.16, \$2.21}. The relation in Figure 7 is a function because each element in the domain corresponds to exactly one element in the range. Notice that it is okay for more than one element in the domain to correspond to the same element in the range (Shell and Exxon both sold gas for \$2.16 a gallon).

(c) The domain of the relation is {0.70, 0.71, 0.75, 0.78}, and the range is {\$1529, \$1575, \$1765, \$1798, \$1952}. The relation in Figure 8 is not a function because not every element in the domain corresponds to exactly one element in the range. If a 0.71-carat diamond is chosen from the domain, a single price cannot be assigned to it. ■

Now Work PROBLEM 19

The idea behind a function is its predictability. If the input is known, we can use the function to determine the output. With "nonfunctions," we don't have this predictability. Look back at Figure 6. If asked, "How many grams of fat are in a 580-calorie sandwich?" we could use the correspondence to answer, "31." Now consider Figure 8. If asked, "What is the price of a 0.71-carat diamond?" we could not give a single response because two outputs result from the single input "0.71." For this reason, the relation in Figure 8 is not a function.

> **In Words**
> For a function, no input has more than one output. The domain of a function is the set of all inputs; the range is the set of all outputs.

We may also think of a function as a set of ordered pairs (x, y) in which no ordered pairs have the same first element and different second elements. The set of all first elements x is the domain of the function, and the set of all second elements y is its range. Each element x in the domain corresponds to exactly one element y in the range.

EXAMPLE 3 Determining Whether a Relation Is a Function

For each relation, state the domain and range. Then determine whether the relation is a function.

(a) $\{(1,4),\ (2,5),\ (3,6),\ (4,7)\}$

(b) $\{(1,4),\ (2,4),\ (3,5),\ (6,10)\}$

(c) $\{(-3,9),\ (-2,4),\ (0,0),\ (1,1),\ (-3,8)\}$

Solution

(a) The domain of this relation is $\{1, 2, 3, 4\}$, and its range is $\{4, 5, 6, 7\}$. This relation is a function because there are no ordered pairs with the same first element and different second elements.

(b) The domain of this relation is $\{1, 2, 3, 6\}$, and its range is $\{4, 5, 10\}$. This relation is a function because there are no ordered pairs with the same first element and different second elements.

(c) The domain of this relation is $\{-3, -2, 0, 1\}$, and its range is $\{0, 1, 4, 8, 9\}$. This relation is not a function because there are two ordered pairs, $(-3, 9)$ and $(-3, 8)$, that have the same first element and different second elements. ■

In Example 3(b), notice that 1 and 2 in the domain both have the same image in the range. This does not violate the definition of a function; two different first elements can have the same second element. A violation of the definition occurs when two ordered pairs have the same first element and different second elements, as in Example 3(c).

Now Work PROBLEM 23

Up to now we have shown how to identify when a relation is a function for relations defined by mappings (Example 2) and ordered pairs (Example 3). But relations can also be expressed as equations. The circumstances under which equations are functions are discussed next.

To determine whether an equation, where y depends on x, is a function, it is often easiest to solve the equation for y. If any value of x in the domain corresponds to more than one y, the equation does not define a function; otherwise, it does define a function.

EXAMPLE 4 **Determining Whether an Equation Is a Function**

Determine whether the equation $y = 2x - 5$ defines y as a function of x.

Solution The equation tells us to take an input x, multiply it by 2, and then subtract 5. For any input x, these operations yield only one output y, so the equation is a function. For example, if $x = 1$, then $y = 2(1) - 5 = -3$. If $x = 3$, then $y = 2(3) - 5 = 1$. The graph of the equation $y = 2x - 5$ is a line with slope 2 and y-intercept -5. The function is called a *linear function*. ■

EXAMPLE 5 **Determining Whether an Equation Is a Function**

Determine whether the equation $x^2 + y^2 = 1$ defines y as a function of x.

Solution To determine whether the equation $x^2 + y^2 = 1$, which defines the unit circle, is a function, solve the equation for y.

$$x^2 + y^2 = 1$$
$$y^2 = 1 - x^2$$
$$y = \pm\sqrt{1 - x^2}$$

For values of x for which $-1 < x < 1$, two values of y result. For example, if $x = 0$, then $y = \pm 1$, so two different outputs result from the same input. This means that the equation $x^2 + y^2 = 1$ does not define a function. ■

Now Work PROBLEM 37

2 Find the Value of a Function

Functions are often denoted by letters such as f, F, g, G, and others. If f is a function, then for each number x in its domain, the corresponding image in the range is designated by the symbol $f(x)$, read as "f of x" or "f at x." We refer to $f(x)$ as the **value of f at the number x**; $f(x)$ is the number that results when x is given and the function f is applied; $f(x)$ is the output corresponding to x, or $f(x)$ is the image of x;

$f(x)$ does *not* mean "f times x." For example, the function given in Example 4 may be written as $y = f(x) = 2x - 5$. Then $f(1) = -3$ and $f(3) = 1$.

Figure 9 illustrates some other functions. Notice that in every function, for each x in the domain, there is one value in the range.

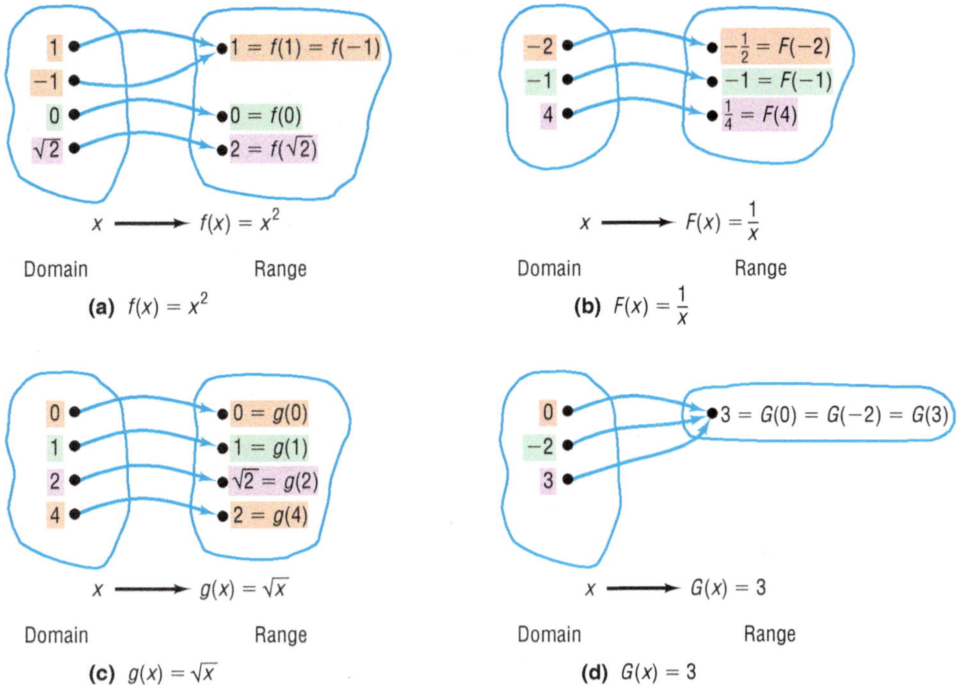

Figure 9

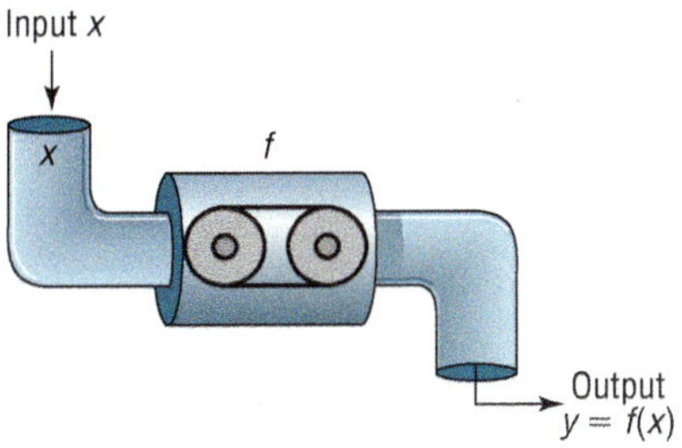

Figure 10 Input/output machine

Sometimes it is helpful to think of a function f as a machine that receives as input a number from the domain, manipulates it, and outputs a value. See Figure 10.

The restrictions on this input/output machine are as follows:

1. It accepts only numbers from the domain of the function.
2. For each input, there is exactly one output (which may be repeated for different inputs).

For a function $y = f(x)$, the variable x is called the **independent variable**, because it can be assigned any of the permissible numbers from the domain. The variable y is called the **dependent variable**, because its value depends on x.

Any symbols can be used to represent the independent and dependent variables. For example, if f is the *cube function*, then f can be given by $f(x) = x^3$ or $f(t) = t^3$ or $f(z) = z^3$. All three functions are the same. Each says to cube the independent variable to get the output. In practice, the symbols used for the independent and dependent variables are based on common usage, such as using C for cost in business.

The independent variable is also called the **argument** of the function. Thinking of the independent variable as an argument can sometimes make it easier to find the value of a function. For example, if f is the function defined by $f(x) = x^3$, then f tells us to cube the argument. Thus $f(2)$ means to cube 2, $f(a)$ means to cube the number a, and $f(x + h)$ means to cube the quantity $x + h$.

EXAMPLE 6 Finding Values of a Function

For the function f defined by $f(x) = 2x^2 - 3x$, evaluate

(a) $f(3)$ (b) $f(x) + f(3)$ (c) $3f(x)$ (d) $f(-x)$

(e) $-f(x)$ (f) $f(3x)$ (g) $f(x + 3)$

Solution (a) Substitute 3 for x in the equation for f, $f(x) = 2x^2 - 3x$, to get

$$f(3) = 2(3)^2 - 3(3) = 18 - 9 = 9$$

The image of 3 is 9.

(b) $f(x) + f(3) = (2x^2 - 3x) + (9) = 2x^2 - 3x + 9$

(c) Multiply the equation for f by 3.

$$3f(x) = 3(2x^2 - 3x) = 6x^2 - 9x$$

(d) Substitute $-x$ for x in the equation for f and simplify.

$$f(-x) = 2(-x)^2 - 3(-x) = 2x^2 + 3x$$ **Notice the use of parentheses here.**

(e) $-f(x) = -(2x^2 - 3x) = -2x^2 + 3x$

(f) Substitute $3x$ for x in the equation for f and simplify.

$$f(3x) = 2(3x)^2 - 3(3x) = 2(9x^2) - 9x = 18x^2 - 9x$$

(g) Substitute $x + 3$ for x in the equation for f and simplify.

$$\begin{aligned} f(x+3) &= 2(x+3)^2 - 3(x+3) \\ &= 2(x^2 + 6x + 9) - 3x - 9 \\ &= 2x^2 + 12x + 18 - 3x - 9 \\ &= 2x^2 + 9x + 9 \end{aligned}$$

Notice in this example that $f(x+3) \neq f(x) + f(3)$, $f(-x) \neq -f(x)$, and $3f(x) \neq f(3x)$.

Now Work PROBLEM 43

Most calculators have special keys that allow you to find the value of certain commonly used functions. For example, you should be able to find the square function $f(x) = x^2$, the square root function $f(x) = \sqrt{x}$, the reciprocal function $f(x) = \frac{1}{x} = x^{-1}$, and many others that will be discussed later in this text (such as $\ln x$ and $\log x$). Verify the results of the following example on your calculator.

EXAMPLE 7 Finding Values of a Function on a Calculator

(a) $f(x) = x^2$ $\quad f(1.234) = 1.234^2 = 1.522756$

(b) $F(x) = \frac{1}{x}$ $\quad F(1.234) = \frac{1}{1.234} \approx 0.8103727715$

(c) $g(x) = \sqrt{x}$ $\quad g(1.234) = \sqrt{1.234} \approx 1.110855526$

COMMENT Graphing calculators can be used to evaluate any function. Figure 11 shows the result obtained in Example 6(a) on a TI-84 Plus C graphing calculator with the function to be evaluated, $f(x) = 2x^2 - 3x$, in Y_1.

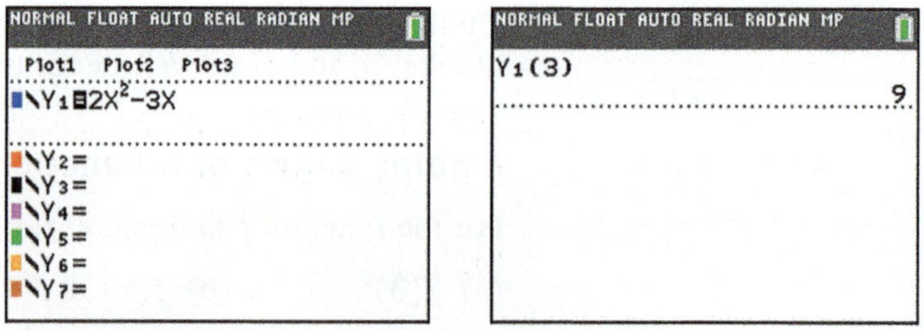

Figure 11 Evaluating $f(x) = 2x^2 - 3x$ for $x = 3$

Implicit Form of a Function

COMMENT The explicit form of a function is the form required by a graphing calculator. ■

In general, when a function f is defined by an equation in x and y, we say that the function f is given **implicitly**. If it is possible to solve the equation for y in terms of x, then we write $y = f(x)$ and say that the function is given **explicitly**. For example,

Implicit Form	**Explicit Form**
$3x + y = 5$	$y = f(x) = -3x + 5$
$x^2 - y = 6$	$y = f(x) = x^2 - 6$
$xy = 4$	$y = f(x) = \dfrac{4}{x}$

SUMMARY

Important Facts about Functions

(a) For each x in the domain of a function f, there is exactly one image $f(x)$ in the range; however, an element in the range can result from more than one x in the domain.

(b) f is the symbol that we use to denote the function. It is symbolic of the equation (rule) that we use to get from an x in the domain to $f(x)$ in the range.

(c) If $y = f(x)$, then x is called the independent variable or argument of f, and y is called the dependent variable or the value of f at x.

3 Find the Difference Quotient of a Function

An important concept in calculus involves looking at a certain quotient. For a given function $y = f(x)$, the inputs x and $x + h, h \neq 0$, result in the images $f(x)$ and $f(x + h)$. The quotient of their differences

$$\frac{f(x+h) - f(x)}{(x+h) - x} = \frac{f(x+h) - f(x)}{h}$$

with $h \neq 0$, is called the *difference quotient of f* at x.

DEFINITION

The **difference quotient** of a function f at x is given by

$$\frac{f(x+h) - f(x)}{h} \qquad h \neq 0 \qquad \textbf{(1)}$$

■

The difference quotient is used in calculus to define the derivative, which leads to applications such as the velocity of an object and optimization of resources.

When finding a difference quotient, it is necessary to simplify the expression in order to divide out the h in the denominator, as illustrated in the following example.

EXAMPLE 8 Finding the Difference Quotient of a Function

Find the difference quotient of each function.

(a) $f(x) = 2x^2 - 3x$

(b) $f(x) = \dfrac{4}{x}$

(c) $f(x) = \sqrt{x}$

Solution (a)

$$\frac{f(x+h)-f(x)}{h} = \frac{[2(x+h)^2 - 3(x+h)] - [2x^2 - 3x]}{h}$$

$\uparrow$ $f(x+h) = 2(x+h)^2 - 3(x+h)$

$$= \frac{2(x^2 + 2xh + h^2) - 3x - 3h - 2x^2 + 3x}{h} \quad \text{Simplify.}$$

$$= \frac{2x^2 + 4xh + 2h^2 - 3h - 2x^2}{h} \quad \text{Distribute and combine like terms.}$$

$$= \frac{4xh + 2h^2 - 3h}{h} \quad \text{Combine like terms.}$$

$$= \frac{h(4x + 2h - 3)}{h} \quad \text{Factor out } h.$$

$$= 4x + 2h - 3 \quad \text{Divide out the factor } h.$$

(b)

$$\frac{f(x+h)-f(x)}{h} = \frac{\dfrac{4}{x+h} - \dfrac{4}{x}}{h} \qquad f(x+h) = \frac{4}{x+h}$$

$$= \frac{\dfrac{4x - 4(x+h)}{x(x+h)}}{h} \quad \text{Subtract.}$$

$$= \frac{4x - 4x - 4h}{x(x+h)h} \quad \text{Divide and distribute.}$$

$$= \frac{-4h}{x(x+h)h} \quad \text{Simplify.}$$

$$= -\frac{4}{x(x+h)} \quad \text{Divide out the factor } h.$$

(c)

$$\frac{f(x+h)-f(x)}{h} = \frac{\sqrt{x+h} - \sqrt{x}}{h} \qquad f(x+h) = \sqrt{x+h}$$

$$= \frac{\sqrt{x+h} - \sqrt{x}}{h} \cdot \frac{\sqrt{x+h} + \sqrt{x}}{\sqrt{x+h} + \sqrt{x}} \quad \text{Rationalize the numerator.}$$

$$= \frac{(\sqrt{x+h})^2 - (\sqrt{x})^2}{h(\sqrt{x+h} + \sqrt{x})} \qquad (A - B)(A + B) = A^2 - B^2$$

$$= \frac{h}{h(\sqrt{x+h} + \sqrt{x})} \qquad (\sqrt{x+h})^2 - (\sqrt{x})^2 = x + h - x = h$$

$$= \frac{1}{\sqrt{x+h} + \sqrt{x}} \quad \text{Divide out the factor } h.$$

■

Now Work PROBLEM 79

4 Find the Domain of a Function Defined by an Equation

Often the domain of a function f is not specified; instead, only the equation defining the function is given. In such cases, we agree that the **domain of f** is the largest set of real numbers for which the value $f(x)$ is a real number. The domain of a function f is the same as the domain of the variable x in the expression $f(x)$.

EXAMPLE 9 Finding the Domain of a Function

Find the domain of each of the following functions.

(a) $f(x) = x^2 + 5x$ (b) $g(x) = \dfrac{3x}{x^2 - 4}$

(c) $h(t) = \sqrt{4 - 3t}$ (d) $F(x) = \dfrac{\sqrt{3x + 12}}{x - 5}$

Solution

(a) The function says to square a number and then add five times the number. Since these operations can be performed on any real number, the domain of f is the set of all real numbers.

(b) The function g says to divide $3x$ by $x^2 - 4$. Since division by 0 is not defined, the denominator $x^2 - 4$ can never be 0, so x can never equal -2 or 2. The domain of the function g is $\{x|x \neq -2, x \neq 2\}$.

(c) The function h says to take the square root of $4 - 3t$. But only nonnegative numbers have real square roots, so the expression under the square root (the radicand) must be nonnegative (greater than or equal to zero). This requires that

$$\begin{aligned} 4 - 3t &\geq 0 \\ -3t &\geq -4 \\ t &\leq \frac{4}{3} \end{aligned}$$

The domain of h is $\left\{t \middle| t \leq \frac{4}{3}\right\}$, or the interval $\left(-\infty, \frac{4}{3}\right]$.

(d) The function F says to take the square root of $3x + 12$ and divide this result by $x - 5$. This requires that $3x + 12 \geq 0$, so $x \geq -4$, and also that $x - 5 \neq 0$, so $x \neq 5$. Combining these two restrictions, the domain of F is

$$\{x|x \geq -4, \quad x \neq 5\}.$$

In Words

The domain of g found in Example 9(b) is $\{x|x \neq -2, x \neq 2\}$. This notation is read, "The domain of the function g is the set of all real numbers x such that x does not equal -2 and x does not equal 2."

The following steps may prove helpful for finding the domain of a function that is defined by an equation and whose domain is a subset of the real numbers.

Finding the Domain of a Function Defined by an Equation

1. Start with the domain as the set of all real numbers.
2. If the equation has a denominator, exclude any numbers that give a zero denominator.
3. If the equation has a radical of even index, exclude any numbers that cause the expression inside the radical (the radicand) to be negative.

Now Work PROBLEM 55

If x is in the domain of a function f, we shall say that **f is defined at x**, or **$f(x)$ exists**. If x is not in the domain of f, we say that **f is not defined at x**, or **$f(x)$ does not exist**. For example, if $f(x) = \frac{x}{x^2 - 1}$, then $f(0)$ exists, but $f(1)$ and $f(-1)$ do not exist. (Do you see why?)

We have not said much about finding the range of a function. We will say more about finding the range when we look at the graph of a function in the next section. When a function is defined by an equation, it can be difficult to find the range. Therefore, we shall usually be content to find just the domain of a function when the function is defined by an equation. We shall express the domain of a function using inequalities, interval notation, set notation, or words, whichever is most convenient.

When we use functions in applications, the domain may be restricted by physical or geometric considerations. For example, the domain of the function f defined by $f(x) = x^2$ is the set of all real numbers. However, if f is used to obtain the area of a square when the length x of a side is known, then we must restrict the domain of f to the positive real numbers, since the length of a side can never be 0 or negative.

EXAMPLE 10

Finding the Domain in an Application

Express the area of a circle as a function of its radius. Find the domain.

Solution See Figure 12. The formula for the area A of a circle of radius r is $A = \pi r^2$. Using r to represent the independent variable and A to represent the dependent variable, the function expressing this relationship is

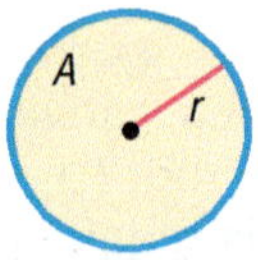

Figure 12 Circle of radius r

$$A(r) = \pi r^2$$

In this setting, the domain is $\{r|r > 0\}$. (Do you see why?)

Observe, in the solution to Example 10, that the symbol A is used in two ways: It is used to name the function, and it is used to symbolize the dependent variable. This double use is common in applications and should not cause any difficulty.

Now Work PROBLEM 97

5 Form the Sum, Difference, Product, and Quotient of Two Functions

Next we introduce some operations on functions. Functions, like numbers, can be added, subtracted, multiplied, and divided. For example, if $f(x) = x^2 + 9$ and $g(x) = 3x + 5$, then

$$f(x) + g(x) = (x^2 + 9) + (3x + 5) = x^2 + 3x + 14$$

The new function $y = x^2 + 3x + 14$ is called the *sum function* $f + g$. Similarly,

$$f(x) \cdot g(x) = (x^2 + 9)(3x + 5) = 3x^3 + 5x^2 + 27x + 45$$

The new function $y = 3x^3 + 5x^2 + 27x + 45$ is called the *product function* $f \cdot g$.

The general definitions are given next.

DEFINITION

If f and g are functions:
The **sum $f + g$** is the function defined by

$$(f + g)(x) = f(x) + g(x)$$

In Words
Remember, the symbol $\cap$ stands for intersection. It means you should find the elements that are common to two sets.

The domain of $f + g$ consists of the numbers x that are in the domains of both f and g. That is, domain of $f + g$ = domain of $f \cap$ domain of g.

DEFINITION

The **difference $f - g$** is the function defined by

$$(f - g)(x) = f(x) - g(x)$$

The domain of $f - g$ consists of the numbers x that are in the domains of both f and g. That is, domain of $f - g$ = domain of $f \cap$ domain of g.

DEFINITION

The **product $f \cdot g$** is the function defined by

$$(f \cdot g)(x) = f(x) \cdot g(x)$$

The domain of $f \cdot g$ consists of the numbers x that are in the domains of both f and g. That is, domain of $f \cdot g$ = domain of $f \cap$ domain of g.

DEFINITION

The **quotient** $\dfrac{f}{g}$ is the function defined by

$$\left(\frac{f}{g}\right)(x) = \frac{f(x)}{g(x)} \qquad g(x) \neq 0$$

The domain of $\dfrac{f}{g}$ consists of the numbers x for which $g(x) \neq 0$ and that are in the domains of both f and g. That is,

$$\text{domain of } \frac{f}{g} = \{x \mid g(x) \neq 0\} \cap \text{domain of } f \cap \text{domain of } g$$

EXAMPLE 11 **Operations on Functions**

Let f and g be two functions defined as

$$f(x) = \frac{1}{x+2} \quad \text{and} \quad g(x) = \frac{x}{x-1}$$

Find the following functions, and determine the domain in each case.

(a) $(f+g)(x)$ (b) $(f-g)(x)$ (c) $(f \cdot g)(x)$ (d) $\left(\dfrac{f}{g}\right)(x)$

Solution The domain of f is $\{x \mid x \neq -2\}$ and the domain of g is $\{x \mid x \neq 1\}$.

(a) $$(f+g)(x) = f(x) + g(x) = \frac{1}{x+2} + \frac{x}{x-1} = \frac{x-1}{(x+2)(x-1)} + \frac{x(x+2)}{(x+2)(x-1)} = \frac{x^2+3x-1}{(x+2)(x-1)}$$

The domain of $f+g$ consists of those numbers x that are in the domains of both f and g. Therefore, the domain of $f+g$ is $\{x \mid x \neq -2, x \neq 1\}$.

(b) $$(f-g)(x) = f(x) - g(x) = \frac{1}{x+2} - \frac{x}{x-1} = \frac{x-1}{(x+2)(x-1)} - \frac{x(x+2)}{(x+2)(x-1)} = \frac{-(x^2+x+1)}{(x+2)(x-1)}$$

The domain of $f-g$ consists of those numbers x that are in the domains of both f and g. Therefore, the domain of $f-g$ is $\{x \mid x \neq -2, x \neq 1\}$.

(c) $$(f \cdot g)(x) = f(x) \cdot g(x) = \frac{1}{x+2} \cdot \frac{x}{x-1} = \frac{x}{(x+2)(x-1)}$$

The domain of $f \cdot g$ consists of those numbers x that are in the domains of both f and g. Therefore, the domain of $f \cdot g$ is $\{x \mid x \neq -2, x \neq 1\}$.

(d) $$\left(\frac{f}{g}\right)(x) = \frac{f(x)}{g(x)} = \frac{\dfrac{1}{x+2}}{\dfrac{x}{x-1}} = \frac{1}{x+2} \cdot \frac{x-1}{x} = \frac{x-1}{x(x+2)}$$

The domain of $\dfrac{f}{g}$ consists of the numbers x for which $g(x) \neq 0$ and that are in the domains of both f and g. Since $g(x) = 0$ when $x = 0$, we exclude 0 as well as -2 and 1 from the domain. The domain of $\dfrac{f}{g}$ is $\{x \mid x \neq -2, x \neq 0, x \neq 1\}$. ■

Now Work PROBLEM 67

In calculus, it is sometimes helpful to view a complicated function as the sum, difference, product, or quotient of simpler functions. For example,

$$F(x) = x^2 + \sqrt{x} \text{ is the sum of } f(x) = x^2 \text{ and } g(x) = \sqrt{x}.$$

$$H(x) = \frac{x^2 - 1}{x^2 + 1} \text{ is the quotient of } f(x) = x^2 - 1 \text{ and } g(x) = x^2 + 1.$$

SUMMARY

Function	A relation between two sets of real numbers so that each number x in the first set, the domain, has corresponding to it exactly one number y in the second set, the range. A set of ordered pairs (x, y) or $(x, f(x))$ in which no first element is paired with two different second elements. The range is the set of y-values of the function that are the images of the x-values in the domain. A function f may be defined implicitly by an equation involving x and y or explicitly by writing $y = f(x)$.
Unspecified domain	If a function f is defined by an equation and no domain is specified, then the domain will be taken to be the largest set of real numbers for which the equation defines a real number.
Function notation	$y = f(x)$ f is a symbol for the function. x is the independent variable, or argument. y is the dependent variable. $f(x)$ is the value of the function at x, or the image of x.

3.1 Assess Your Understanding

'Are You Prepared?' *Answers are given at the end of these exercises. If you get a wrong answer, read the pages listed in red.*

1. The inequality $-1 < x < 3$ can be written in interval notation as ________. (pp. 147–148)
2. If $x = -2$, the value of the expression $3x^2 - 5x + \frac{1}{x}$ is ______. (pp. 21–22)
3. The domain of the variable in the expression $\frac{x - 3}{x + 4}$ is ____________. (p. 22)
4. Solve the inequality: $3 - 2x > 5$. Graph the solution set. (pp. 150–151)
5. To rationalize the denominator of $\frac{3}{\sqrt{5} - 2}$, multiply the numerator and denominator by ________ (p. 76)
6. A quotient is considered rationalized if its denominator contains no ________. (p. 76)

Concepts and Vocabulary

7. If f is a function defined by the equation $y = f(x)$, then x is called the ____________ variable, and y is the __________ variable.
8. If the domain of f is all real numbers in the interval $[0, 7]$, and the domain of g is all real numbers in the interval $[-2, 5]$, then the domain of $f + g$ is all real numbers in the interval _______.
9. The domain of $\frac{f}{g}$ consists of numbers x for which $g(x)$ ___ 0 that are in the domains of both ___ and ___.
10. If $f(x) = x + 1$ and $g(x) = x^3$, then ____________ $= x^3 - (x + 1)$.
11. ***True or False*** Every relation is a function.
12. ***True or False*** The domain of $(f \cdot g)(x)$ consists of the numbers x that are in the domains of both f and g.
13. ***True or False*** If no domain is specified for a function f, then the domain of f is taken to be the set of real numbers.
14. ***True or False*** The domain of the function $f(x) = \frac{x^2 - 4}{x}$ is $\{x \mid x \neq \pm 2\}$.
15. The set of all images of the elements in the domain of a function is called the ______.
    (a) range (b) domain (c) solution set (d) function
16. The independent variable is sometimes referred to as the ______ of the function.
    (a) range (b) value (c) argument (d) definition
17. The expression $\frac{f(x + h) - f(x)}{h}$ is called the ______ of f.
    (a) radicand (b) image
    (c) correspondence (d) difference quotient
18. When written as $y = f(x)$, a function is said to be defined ______.
    (a) explicitly (b) consistently
    (c) implicitly (d) rationally

Skill Building

In Problems 19–30, state the domain and range for each relation. Then determine whether each relation represents a function.

19.

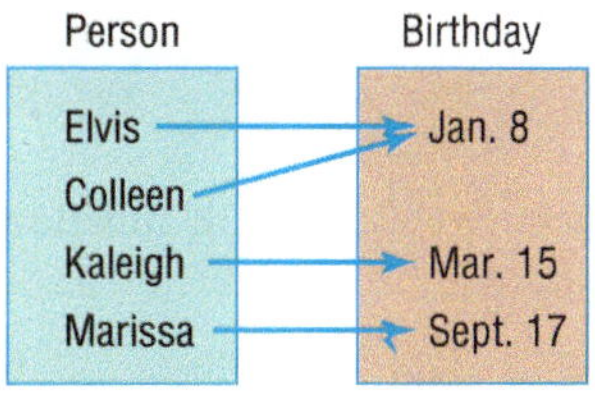

20.

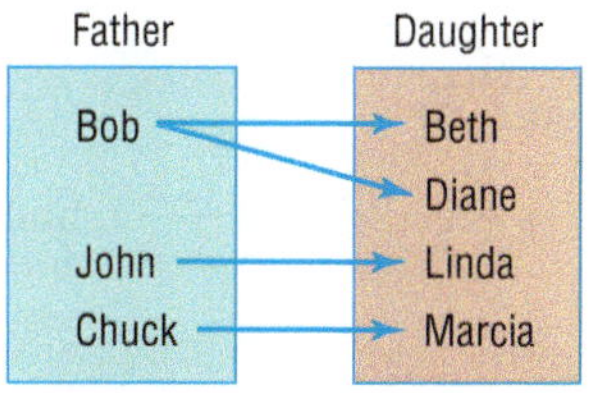

21.

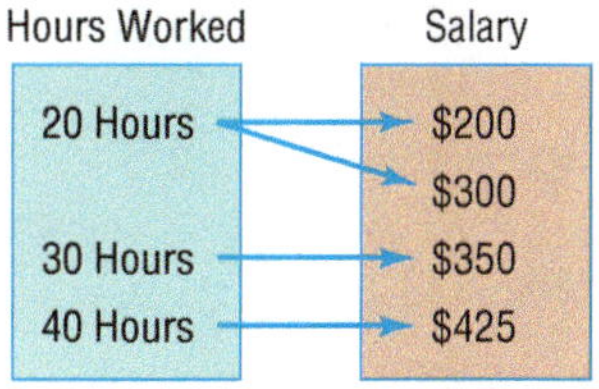

22.

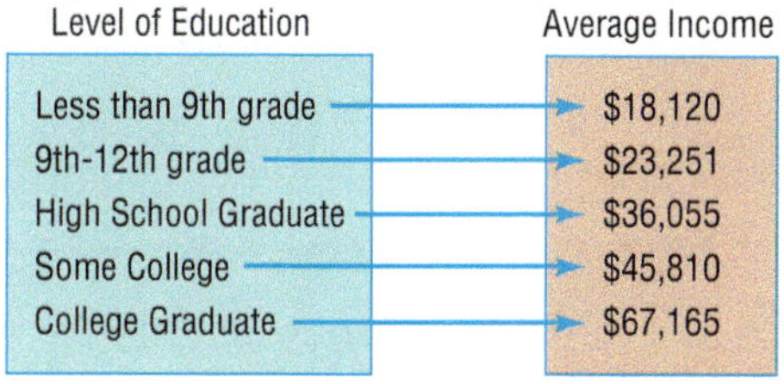

23. $\{(2,6), (-3,6), (4,9), (2,10)\}$ **24.** $\{(-2,5), (-1,3), (3,7), (4,12)\}$ **25.** $\{(1,3), (2,3), (3,3), (4,3)\}$

26. $\{(0,-2), (1,3), (2,3), (3,7)\}$ **27.** $\{(-2,4), (-2,6), (0,3), (3,7)\}$ **28.** $\{(-4,4), (-3,3), (-2,2), (-1,1), (-4,0)\}$

29. $\{(-2,4), (-1,1), (0,0), (1,1)\}$ **30.** $\{(-2,16), (-1,4), (0,3), (1,4)\}$

In Problems 31–42, determine whether the equation defines y as a function of x.

31. $y = 2x^2 - 3x + 4$ **32.** $y = x^3$ **33.** $y = \dfrac{1}{x}$ **34.** $y = |x|$

35. $y^2 = 4 - x^2$ **36.** $y = \pm\sqrt{1 - 2x}$ **37.** $x = y^2$ **38.** $x + y^2 = 1$

39. $y = \sqrt[3]{x}$ **40.** $y = \dfrac{3x - 1}{x + 2}$ **41.** $2x^2 + 3y^2 = 1$ **42.** $x^2 - 4y^2 = 1$

In Problems 43–50, find the following for each function:

(a) $f(0)$ *(b)* $f(1)$ *(c)* $f(-1)$ *(d)* $f(-x)$ *(e)* $-f(x)$ *(f)* $f(x+1)$ *(g)* $f(2x)$ *(h)* $f(x+h)$

43. $f(x) = 3x^2 + 2x - 4$ **44.** $f(x) = -2x^2 + x - 1$ **45.** $f(x) = \dfrac{x}{x^2 + 1}$ **46.** $f(x) = \dfrac{x^2 - 1}{x + 4}$

47. $f(x) = |x| + 4$ **48.** $f(x) = \sqrt{x^2 + x}$ **49.** $f(x) = \dfrac{2x + 1}{3x - 5}$ **50.** $f(x) = 1 - \dfrac{1}{(x+2)^2}$

In Problems 51–66, find the domain of each function.

51. $f(x) = -5x + 4$ **52.** $f(x) = x^2 + 2$ **53.** $f(x) = \dfrac{x}{x^2 + 1}$ **54.** $f(x) = \dfrac{x^2}{x^2 + 1}$

55. $g(x) = \dfrac{x}{x^2 - 16}$ **56.** $h(x) = \dfrac{2x}{x^2 - 4}$ **57.** $F(x) = \dfrac{x - 2}{x^3 + x}$ **58.** $G(x) = \dfrac{x + 4}{x^3 - 4x}$

59. $h(x) = \sqrt{3x - 12}$ **60.** $G(x) = \sqrt{1 - x}$ **61.** $p(x) = \sqrt{\dfrac{2}{x - 1}}$ **62.** $f(x) = \dfrac{4}{\sqrt{x - 9}}$

63. $f(x) = \dfrac{x}{\sqrt{x - 4}}$ **64.** $f(x) = \dfrac{-x}{\sqrt{-x - 2}}$ **65.** $P(t) = \dfrac{\sqrt{t - 4}}{3t - 21}$ **66.** $h(z) = \dfrac{\sqrt{z + 3}}{z - 2}$

In Problems 67–76, for the given functions f and g, find the following. For parts (a)–(d), also find the domain.

(a) $(f+g)(x)$ *(b)* $(f-g)(x)$ *(c)* $(f\cdot g)(x)$ *(d)* $\left(\dfrac{f}{g}\right)(x)$

(e) $(f+g)(3)$ *(f)* $(f-g)(4)$ *(g)* $(f\cdot g)(2)$ *(h)* $\left(\dfrac{f}{g}\right)(1)$

67. $f(x) = 3x + 4;\quad g(x) = 2x - 3$ **68.** $f(x) = 2x + 1;\quad g(x) = 3x - 2$

69. $f(x) = x - 1;\quad g(x) = 2x^2$ **70.** $f(x) = 2x^2 + 3;\quad g(x) = 4x^3 + 1$

71. $f(x) = \sqrt{x};\quad g(x) = 3x - 5$

72. $f(x) = |x|;\quad g(x) = x$

73. $f(x) = 1 + \frac{1}{x};\quad g(x) = \frac{1}{x}$

74. $f(x) = \sqrt{x-1};\quad g(x) = \sqrt{4-x}$

75. $f(x) = \frac{2x+3}{3x-2};\quad g(x) = \frac{4x}{3x-2}$

76. $f(x) = \sqrt{x+1};\quad g(x) = \frac{2}{x}$

77. Given $f(x) = 3x + 1$ and $(f + g)(x) = 6 - \frac{1}{2}x$, find the function g.

78. Given $f(x) = \frac{1}{x}$ and $\left(\frac{f}{g}\right)(x) = \frac{x+1}{x^2 - x}$, find the function g.

In Problems 79–90, find the difference quotient of f; that is, find $\frac{f(x+h) - f(x)}{h}$, $h \neq 0$, *for each function. Be sure to simplify.*

79. $f(x) = 4x + 3$

80. $f(x) = -3x + 1$

81. $f(x) = x^2 - 4$

82. $f(x) = 3x^2 + 2$

83. $f(x) = x^2 - x + 4$

84. $f(x) = 3x^2 - 2x + 6$

85. $f(x) = \frac{1}{x^2}$

86. $f(x) = \frac{1}{x+3}$

87. $f(x) = \frac{2x}{x+3}$

88. $f(x) = \frac{5x}{x-4}$

89. $f(x) = \sqrt{x-2}$
[**Hint:** Rationalize the numerator.]

90. $f(x) = \sqrt{x+1}$

Applications and Extensions

91. Given $f(x) = x^2 - 2x + 3$, find the value(s) for x such that $f(x) = 11$.

92. Given $f(x) = \frac{5}{6}x - \frac{3}{4}$, find the value(s) for x such that $f(x) = -\frac{7}{16}$.

93. If $f(x) = 2x^3 + Ax^2 + 4x - 5$ and $f(2) = 5$, what is the value of A?

94. If $f(x) = 3x^2 - Bx + 4$ and $f(-1) = 12$, what is the value of B?

95. If $f(x) = \frac{3x+8}{2x-A}$ and $f(0) = 2$, what is the value of A?

96. If $f(x) = \frac{2x-B}{3x+4}$ and $f(2) = \frac{1}{2}$, what is the value of B?

97. Geometry Express the area A of a rectangle as a function of the length x if the length of the rectangle is twice its width.

98. Geometry Express the area A of an isosceles right triangle as a function of the length x of one of the two equal sides.

99. Constructing Functions Express the gross salary G of a person who earns \$14 per hour as a function of the number x of hours worked.

100. Constructing Functions Tiffany, a commissioned salesperson, earns \$100 base pay plus \$10 per item sold. Express her gross salary G as a function of the number x of items sold.

101. Population as a Function of Age The function

$$P(a) = 0.014a^2 - 5.073a + 327.287$$

represents the population P (in millions) of Americans that are a years of age or older in 2012.

(a) Identify the dependent and independent variables.
(b) Evaluate $P(20)$. Provide a verbal explanation of the meaning of $P(20)$.
(c) Evaluate $P(0)$. Provide a verbal explanation of the meaning of $P(0)$.

Source: *U.S. Census Bureau*

102. Number of Rooms The function

$$N(r) = -1.35r^2 + 15.45r - 20.71$$

represents the number N of housing units (in millions) in 2012 that had r rooms, where r is an integer and $2 \leq r \leq 9$.

(a) Identify the dependent and independent variables.
(b) Evaluate $N(3)$. Provide a verbal explanation of the meaning of $N(3)$.

103. Effect of Gravity on Earth If a rock falls from a height of 20 meters on Earth, the height H (in meters) after x seconds is approximately

$$H(x) = 20 - 4.9x^2$$

(a) What is the height of the rock when $x = 1$ second? $x = 1.1$ seconds? $x = 1.2$ seconds? $x = 1.3$ seconds?
(b) When is the height of the rock 15 meters? When is it 10 meters? When is it 5 meters?
(c) When does the rock strike the ground?

104. Effect of Gravity on Jupiter If a rock falls from a height of 20 meters on the planet Jupiter, its height H (in meters) after x seconds is approximately

$$H(x) = 20 - 13x^2$$

(a) What is the height of the rock when $x = 1$ second? $x = 1.1$ seconds? $x = 1.2$ seconds?
(b) When is the height of the rock 15 meters? When is it 10 meters? When is it 5 meters?
(c) When does the rock strike the ground?

105. Cost of Transatlantic Travel A Boeing 747 crosses the Atlantic Ocean (3000 miles) with an airspeed of 500 miles per hour. The cost C (in dollars) per passenger is given by

$$C(x) = 100 + \frac{x}{10} + \frac{36{,}000}{x}$$

where x is the ground speed (airspeed $\pm$ wind).

(a) What is the cost per passenger for quiescent (no wind) conditions?

(b) What is the cost per passenger with a head wind of 50 miles per hour?

(c) What is the cost per passenger with a tail wind of 100 miles per hour?

(d) What is the cost per passenger with a head wind of 100 miles per hour?

106. Cross-sectional Area The cross-sectional area of a beam cut from a log with radius 1 foot is given by the function $A(x) = 4x\sqrt{1 - x^2}$, where x represents the length, in feet, of half the base of the beam. See the figure. Determine the cross-sectional area of the beam if the length of half the base of the beam is as follows:

(a) One-third of a foot

(b) One-half of a foot

(c) Two-thirds of a foot

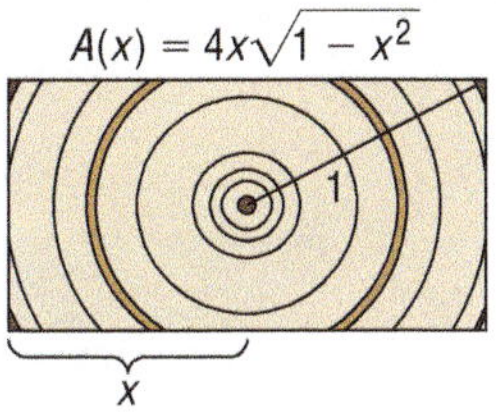

107. Economics The **participation rate** is the number of people in the labor force divided by the civilian population (excludes military). Let $L(x)$ represent the size of the labor force in year x, and $P(x)$ represent the civilian population in year x. Determine a function that represents the participation rate R as a function of x.

108. Crimes Suppose that $V(x)$ represents the number of violent crimes committed in year x and $P(x)$ represents the number of property crimes committed in year x. Determine a function T that represents the combined total of violent crimes and property crimes in year x.

109. Health Care Suppose that $P(x)$ represents the percentage of income spent on health care in year x and $I(x)$ represents income in year x. Determine a function H that represents total health care expenditures in year x.

110. Income Tax Suppose that $I(x)$ represents the income of an individual in year x before taxes and $T(x)$ represents the individual's tax bill in year x. Determine a function N that represents the individual's net income (income after taxes) in year x.

111. Profit Function Suppose that the revenue R, in dollars, from selling x cell phones, in hundreds, is $R(x) = -1.2x^2 + 220x$. The cost C, in dollars, of selling x cell phones, in hundreds, is $C(x) = 0.05x^3 - 2x^2 + 65x + 500$.

(a) Find the profit function, $P(x) = R(x) - C(x)$.

(b) Find the profit if $x = 15$ hundred cell phones are sold.

(c) Interpret $P(15)$.

112. Profit Function Suppose that the revenue R, in dollars, from selling x clocks is $R(x) = 30x$. The cost C, in dollars, of selling x clocks is $C(x) = 0.1x^2 + 7x + 400$.

(a) Find the profit function, $P(x) = R(x) - C(x)$.

(b) Find the profit if $x = 30$ clocks are sold.

(c) Interpret $P(30)$.

113. Stopping Distance When the driver of a vehicle observes an impediment, the total stopping distance involves both the reaction distance (the distance the vehicle travels while the driver moves his or her foot to the brake pedal) and the braking distance (the distance the vehicle travels once the brakes are applied). For a car traveling at a speed of v miles per hour, the reaction distance R, in feet, can be estimated by $R(v) = 2.2v$. Suppose that the braking distance B, in feet, for a car is given by $B(v) = 0.05v^2 + 0.4v - 15$.

(a) Find the stopping distance function

$$D(v) = R(v) + B(v).$$

(b) Find the stopping distance if the car is traveling at a speed of 60 mph.

(c) Interpret $D(60)$.

114. Some functions f have the property that $f(a + b) = f(a) + f(b)$ for all real numbers a and b. Which of the following functions have this property?

(a) $h(x) = 2x$ (b) $g(x) = x^2$

(c) $F(x) = 5x - 2$ (d) $G(x) = \dfrac{1}{x}$

Explaining Concepts: Discussion and Writing

115. Are the functions $f(x) = x - 1$ and $g(x) = \dfrac{x^2 - 1}{x + 1}$ the same? Explain.

116. Investigate when, historically, the use of the function notation $y = f(x)$ first appeared.

117. Find a function H that multiplies a number x by 3 and then subtracts the cube of x and divides the result by your age.

Retain Your Knowledge

Problems 118–121 are based on material learned earlier in the course. The purpose of these problems is to keep the material fresh in your mind so that you are better prepared for the final exam.

118. List the intercepts and test for symmetry: $(x + 12)^2 + y^2 = 16$

119. Determine which of the given points are on the graph of the equation $y = 3x^2 - 8\sqrt{x}$.

Points: $(-1, -5)$, $(4, 32)$, $(9, 171)$

120. How many pounds of lean hamburger that is 7% fat must be mixed with 12 pounds of ground chuck that is 20% fat to have a hamburger mixture that is 15% fat?

121. Solve $x^3 - 9x = 2x^2 - 18$.

'Are You Prepared?' Answers

1. $(-1, 3)$ **2.** 21.5 **3.** $\{x \mid x \neq -4\}$ **4.** $\{x \mid x < -1\}$ 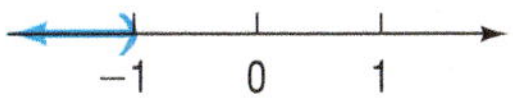

5. $\sqrt{5} + 2$ **6.** radicals

3.2 The Graph of a Function

PREPARING FOR THIS SECTION *Before getting started, review the following:*

- Graphs of Equations (Section 1.1, pp. 88–92)
- Intercepts (Section 2.1, pp. 165–166)

Now Work the **'Are You Prepared?'** problems on page 226.

OBJECTIVES 1 Identify the Graph of a Function (p. 222)
2 Obtain Information from or about the Graph of a Function (p. 223)

In applications, a graph often demonstrates more clearly the relationship between two variables than, say, an equation or table. For example, Table 1 shows the average price of gasoline in the United States for the years 1985–2014 (adjusted for inflation, based on 2014 dollars). If we plot these data and then connect the points, we obtain Figure 13.

Table 1

Year	Price	Year	Price	Year	Price
1985	2.55	1995	1.72	2005	2.74
1986	1.90	1996	1.80	2006	3.01
1987	1.89	1997	1.76	2007	3.19
1988	1.81	1998	1.49	2008	3.56
1989	1.87	1999	1.61	2009	2.58
1990	2.03	2000	2.03	2010	3.00
1991	1.91	2001	1.90	2011	3.69
1992	1.82	2002	1.76	2012	3.72
1993	1.74	2003	1.99	2013	3.54
1994	1.71	2004	2.31	2014	3.43

***Source:** U.S. Energy Information Administration*

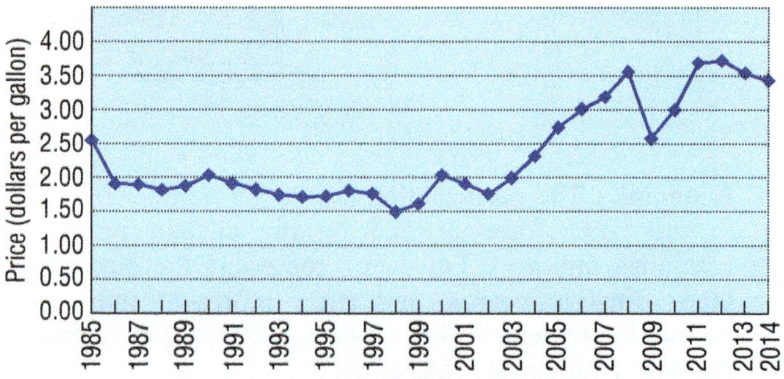

Figure 13 Average retail price of gasoline (2014 dollars)
***Source:** U.S. Energy Information Administration*

We can see from the graph that the price of gasoline (adjusted for inflation) stayed roughly the same from 1986 to 1991 and rose rapidly from 2002 to 2008. The graph also shows that the lowest price occurred in 1998. To learn information such as this from an equation requires that some calculations be made.

Look again at Figure 13. The graph shows that for each date on the horizontal axis, there is only one price on the vertical axis. The graph represents a function, although an exact rule for getting from date to price is not given.

When a function is defined by an equation in x and y, the **graph of the function** is the graph of the equation; that is, it is the set of points (x, y) in the xy-plane that satisfy the equation.

1 Identify the Graph of a Function

Not every collection of points in the xy-plane represents the graph of a function. Remember, for a function, each number x in the domain has exactly one image y in the range. This means that the graph of a function cannot contain two points with the same x-coordinate and different y-coordinates. Therefore, the graph of a function must satisfy the following **vertical-line test**.

In Words
If any vertical line intersects a graph at more than one point, the graph is not the graph of a function.

THEOREM

Vertical-Line Test

A set of points in the xy-plane is the graph of a function if and only if every vertical line intersects the graph in at most one point.

EXAMPLE 1 Identifying the Graph of a Function

Which of the graphs in Figure 14 are graphs of functions?

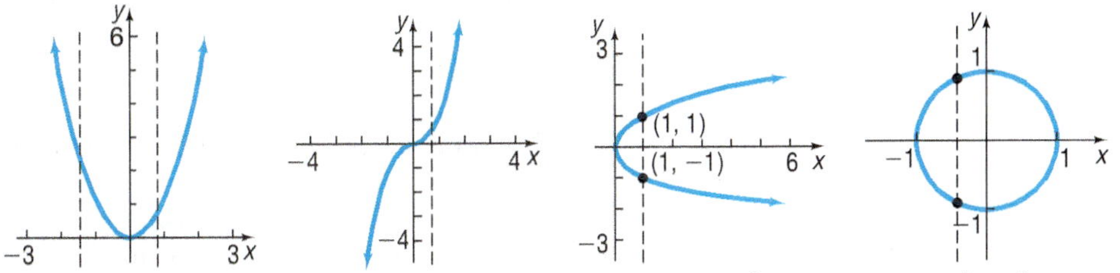

(a) $y = x^2$ (b) $y = x^3$ (c) $x = y^2$ (d) $x^2 + y^2 = 1$

Figure 14

Solution The graphs in Figures 14(a) and 14(b) are graphs of functions, because every vertical line intersects each graph in at most one point. The graphs in Figures 14(c) and 14(d) are not graphs of functions, because there is a vertical line that intersects each graph in more than one point. Notice in Figure 14(c) that the input 1 corresponds to two outputs, -1 and 1. This is why the graph does not represent a function. ■

Now Work PROBLEMS 15 AND 17

2 Obtain Information from or about the Graph of a Function

If (x, y) is a point on the graph of a function f, then y is the value of f at x; that is, $y = f(x)$. Also if $y = f(x)$, then (x, y) is a point on the graph of f. For example, if $(-2, 7)$ is on the graph of f, then $f(-2) = 7$, and if $f(5) = 8$, then the point $(5, 8)$ is on the graph of $y = f(x)$. The next example illustrates how to obtain information about a function if its graph is given.

EXAMPLE 2 Obtaining Information from the Graph of a Function

Let f be the function whose graph is given in Figure 15.

(a) What are $f(-5)$, $f(0)$, and $f(3)$?

(b) What is the domain of f?

(c) What is the range of f?

(d) List the intercepts. (Recall that these are the points, if any, where the graph crosses or touches the coordinate axes.)

(e) How many times does the line $y = 2$ intersect the graph?

(f) For what values of x does $f(x) = -8$?

(g) For what values of x is $f(x) > 0$?

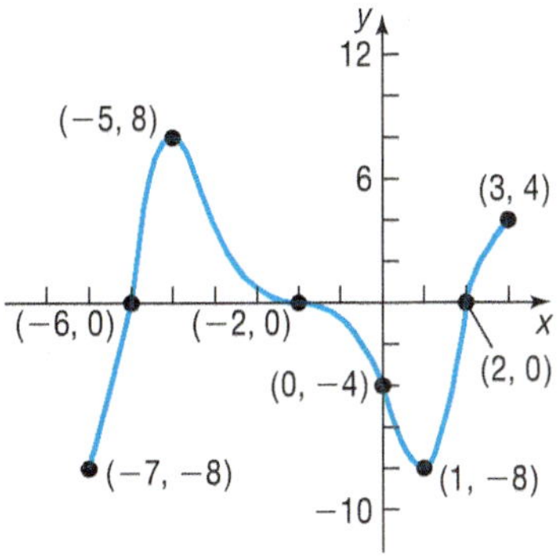

Figure 15

Solution

(a) Since $(-5, 8)$ is on the graph of f, the y-coordinate is the value of f at the x-coordinate -5; that is, $f(-5) = 8$. In a similar way, when $x = 0$, then $y = -4$, so $f(0) = -4$. When $x = 3$, then $y = 4$, so $f(3) = 4$.

(b) To determine the domain of f, notice that the points on the graph of f have x-coordinates between -7 and 3, inclusive; and for each number x between -7 and 3, there is a point $(x, f(x))$ on the graph. The domain of f is $\{x \mid -7 \le x \le 3\}$ or the interval $[-7, 3]$.

(c) The points on the graph all have y-coordinates between -8 and 8, inclusive; and for each number y, there is at least one number x in the domain. The range of f is $\{y \mid -8 \le y \le 8\}$ or the interval $[-8, 8]$.

(d) The intercepts are the points $(0, -4)$, $(-6, 0)$, $(-2, 0)$, and $(2, 0)$.

(e) Draw the horizontal line $y = 2$ on the graph in Figure 15. Notice that the line intersects the graph three times.

(f) Since $(-7, -8)$ and $(1, -8)$ are the only points on the graph for which $y = f(x) = -8$, we have $f(x) = -8$ when $x = -7$ and $x = 1$.

(g) To determine where $f(x) > 0$, look at Figure 15 and determine the x-values from -7 to 3 for which the y-coordinate is positive. This occurs on $(-6, -2) \cup (2, 3]$. Using inequality notation, $f(x) > 0$ for $-6 < x < -2$ or $2 < x \le 3$. ■

When the graph of a function is given, its domain may be viewed as the shadow created by the graph on the x-axis by vertical beams of light. Its range can be viewed as the shadow created by the graph on the y-axis by horizontal beams of light. Try this technique with the graph given in Figure 15.

Now Work PROBLEM 11

EXAMPLE 3 Obtaining Information about the Graph of a Function

Consider the function: $f(x) = \dfrac{x + 1}{x + 2}$

(a) Find the domain of f.

(b) Is the point $\left(1, \dfrac{1}{2}\right)$ on the graph of f?

(c) If $x = 2$, what is $f(x)$? What point is on the graph of f?

(d) If $f(x) = 2$, what is x? What point is on the graph of f?

(e) What are the x-intercepts of the graph of f (if any)? What point(s) are on the graph of f?

Solution

(a) The domain of f is $\{x \mid x \neq -2\}$, since $x = -2$ results in division by 0.

(b) When $x = 1$, then

$$f(1) = \frac{1 + 1}{1 + 2} = \frac{2}{3} \qquad f(x) = \frac{x + 1}{x + 2}$$

The point $\left(1, \dfrac{2}{3}\right)$ is on the graph of f; the point $\left(1, \dfrac{1}{2}\right)$ is not.

(c) If $x = 2$, then

$$f(2) = \frac{2 + 1}{2 + 2} = \frac{3}{4}$$

The point $\left(2, \dfrac{3}{4}\right)$ is on the graph of f.

(d) If $f(x) = 2$, then

$$\frac{x + 1}{x + 2} = 2 \qquad f(x) = 2$$

$$x + 1 = 2(x + 2) \qquad \text{Multiply both sides by } x + 2.$$

$$x + 1 = 2x + 4 \qquad \text{Distribute.}$$

$$x = -3 \qquad \text{Solve for } x.$$

If $f(x) = 2$, then $x = -3$. The point $(-3, 2)$ is on the graph of f.

(e) The x-intercepts of the graph of f are the real solutions of the equation $f(x) = 0$ that are in the domain of f.

$$\frac{x+1}{x+2} = 0$$

$$x + 1 = 0 \qquad \text{Multiply both sides by } x + 2.$$

$$x = -1 \qquad \text{Subtract 1 from both sides.}$$

The only real solution of the equation $f(x) = \dfrac{x+1}{x+2} = 0$ is $x = -1$, so -1 is the only x-intercept. Since $f(-1) = 0$, the point $(-1, 0)$ is on the graph of f. ■

Now Work PROBLEM 27

EXAMPLE 4 Average Cost Function

The average cost $\overline{C}$ per computer of manufacturing x computers per day is given by the function

$$\overline{C}(x) = 0.56x^2 - 34.39x + 1212.57 + \frac{20{,}000}{x}$$

Determine the average cost of manufacturing:

(a) 30 computers in a day
(b) 40 computers in a day
(c) 50 computers in a day
(d) Graph the function $\overline{C} = \overline{C}(x)$, $0 < x \le 80$.
(e) Create a TABLE with TblStart = 1 and ΔTbl = 1. Which value of x minimizes the average cost?

Solution (a) The average cost per computer of manufacturing $x = 30$ computers is

$$\overline{C}(30) = 0.56(30)^2 - 34.39(30) + 1212.57 + \frac{20{,}000}{30} = \$1351.54$$

(b) The average cost per computer of manufacturing $x = 40$ computers is

$$\overline{C}(40) = 0.56(40)^2 - 34.39(40) + 1212.57 + \frac{20{,}000}{40} = \$1232.97$$

(c) The average cost per computer of manufacturing $x = 50$ computers is

$$\overline{C}(50) = 0.56(50)^2 - 34.39(50) + 1212.57 + \frac{20{,}000}{50} = \$1293.07$$

(d) See Figure 16 for the graph of $\overline{C} = \overline{C}(x)$.
(e) With the function $\overline{C} = \overline{C}(x)$ in Y_1, we create Table 2. We scroll down until we find a value of x for which Y_1 is smallest. Table 3 shows that manufacturing $x = 41$ computers minimizes the average cost at \$1231.74 per computer.

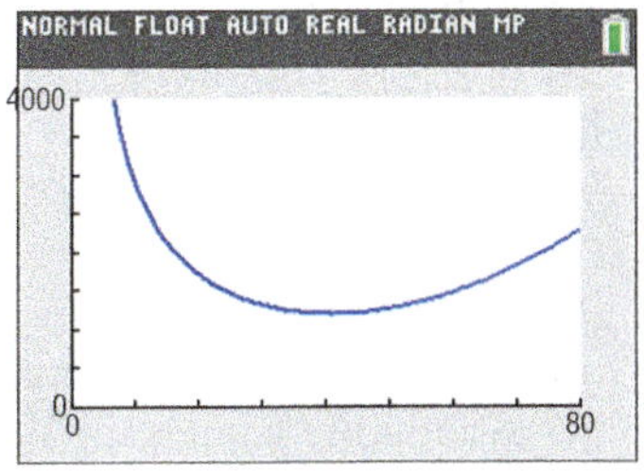

Figure 16 $\overline{C}(x) = 0.56x^2 - 34.39x + 1212.57 + \dfrac{20{,}000}{x}$

NORMAL FLOAT AUTO REAL RADIAN MP
PRESS ENTER TO EDIT

X	Y_1
1	21179
2	11146
3	7781.1
4	6084
5	5054.6
6	4359.7
7	3856.4
8	3473.3
9	3170.6
10	2924.7
11	2720.2

Y_1=0.56X²−34.39X+1212.57+2

Table 2

NORMAL FLOAT AUTO REAL RADIAN MP
PRESS ◆ TO EDIT FUNCTION

X	Y_1
38	1240.7
39	1235.9
40	1233
41	1231.7
42	1232.2
43	1234.4
44	1238.1
45	1243.5
46	1250.4
47	1258.8
48	1268.8

Y_1=1231.74487805

Table 3 ■

Now Work PROBLEM 35

SUMMARY

Graph of a Function	The collection of points (x, y) that satisfy the equation $y = f(x)$.
Vertical-Line Test	A collection of points is the graph of a function if and only if every vertical line intersects the graph in at most one point.

3.2 Assess Your Understanding

'Are You Prepared?' *Answers are given at the end of these exercises. If you get a wrong answer, read the pages listed in red.*

1. The intercepts of the equation $x^2 + 4y^2 = 16$ are ________. (pp. 165–166)

2. ***True or False*** The point $(-2, -6)$ is on the graph of the equation $x = 2y - 2$. (pp. 88–92)

Concepts and Vocabulary

3. A set of points in the xy-plane is the graph of a function if and only if every ________ line intersects the graph in at most one point.

4. If the point $(5, -3)$ is a point on the graph of f, then $f(\_\_\_) = \_\_\_\_$.

5. Find a so that the point $(-1, 2)$ is on the graph of $f(x) = ax^2 + 4$.

6. ***True or False*** Every graph represents a function.

7. ***True or False*** The graph of a function $y = f(x)$ always crosses the y-axis.

8. ***True or False*** The y-intercept of the graph of the function $y = f(x)$, whose domain is all real numbers, is $f(0)$.

9. If a function is defined by an equation in x and y, then the collection of points (x, y) in the xy-plane that satisfy the equation is called ________.
   (a) the domain of the function (b) the range of the function
   (c) the graph of the function (d) the relation of the function

10. The graph of a function $y = f(x)$ can have more than one of which type of intercept?
    (a) x-intercept (b) y-intercept (c) both (d) neither

Skill Building

11. Use the given graph of the function f to answer parts (a)–(n).

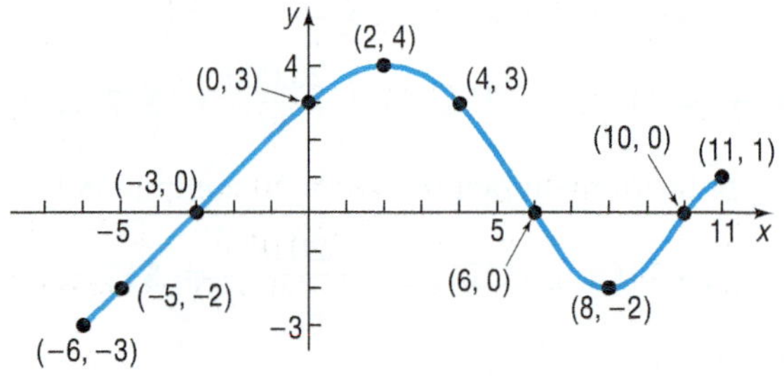

(a) Find $f(0)$ and $f(-6)$.
(b) Find $f(6)$ and $f(11)$.
(c) Is $f(3)$ positive or negative?
(d) Is $f(-4)$ positive or negative?
(e) For what values of x is $f(x) = 0$?
(f) For what values of x is $f(x) > 0$?
(g) What is the domain of f?
(h) What is the range of f?
(i) What are the x-intercepts?
(j) What is the y-intercept?
(k) How often does the line $y = \frac{1}{2}$ intersect the graph?
(l) How often does the line $x = 5$ intersect the graph?
(m) For what values of x does $f(x) = 3$?
(n) For what values of x does $f(x) = -2$?

12. Use the given graph of the function f to answer parts (a)–(n).

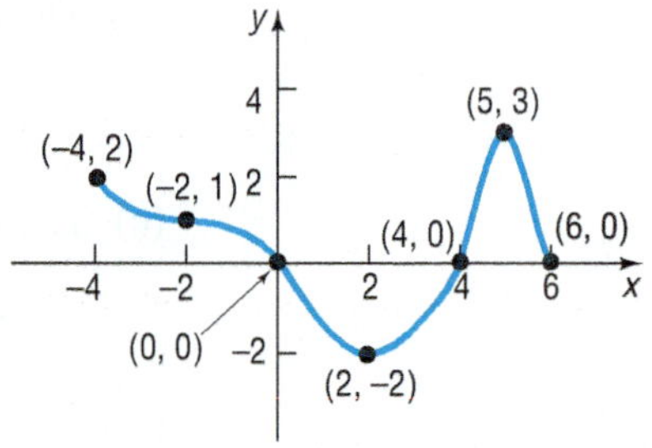

(a) Find $f(0)$ and $f(6)$.
(b) Find $f(2)$ and $f(-2)$.
(c) Is $f(3)$ positive or negative?
(d) Is $f(-1)$ positive or negative?
(e) For what values of x is $f(x) = 0$?
(f) For what values of x is $f(x) < 0$?
(g) What is the domain of f?
(h) What is the range of f?
(i) What are the x-intercepts?
(j) What is the y-intercept?
(k) How often does the line $y = -1$ intersect the graph?
(l) How often does the line $x = 1$ intersect the graph?
(m) For what value of x does $f(x) = 3$?
(n) For what value of x does $f(x) = -2$?

In Problems 13–24, determine whether the graph is that of a function by using the vertical-line test. If it is, use the graph to find:
(a) The domain and range *(b) The intercepts, if any* *(c) Any symmetry with respect to the x-axis, the y-axis, or the origin*

13.

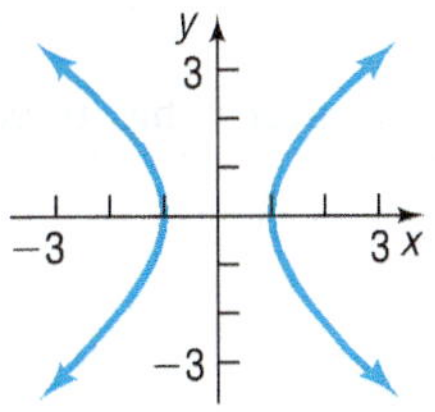

14.

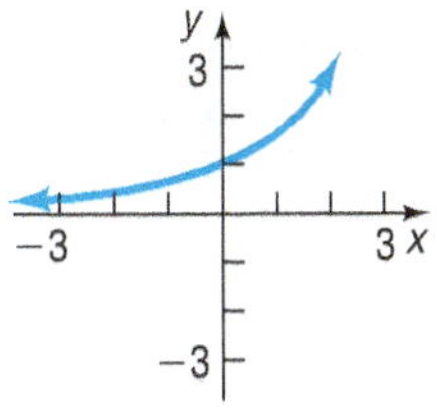

15.

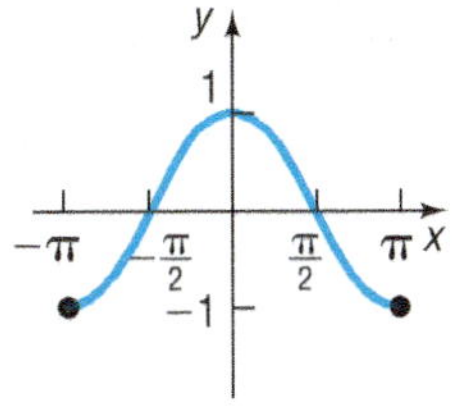

16.

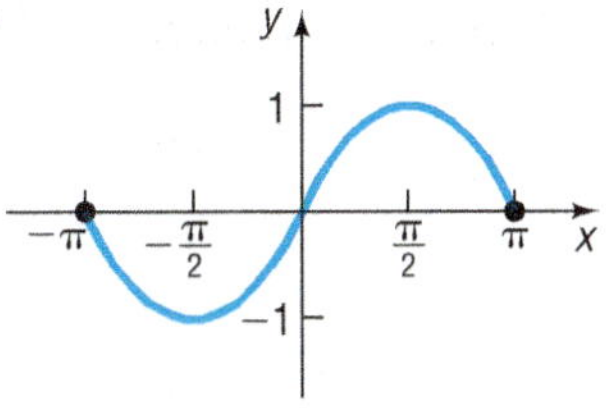

17.

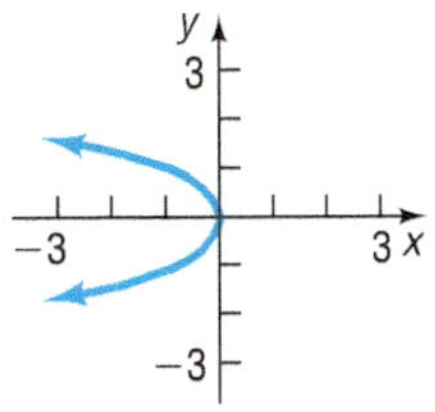

18.

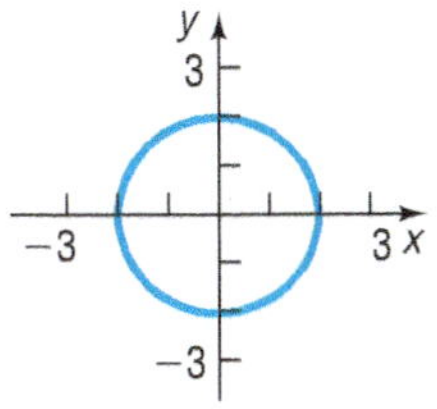

19.

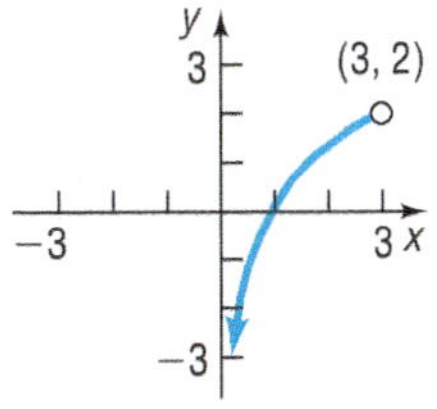

20.

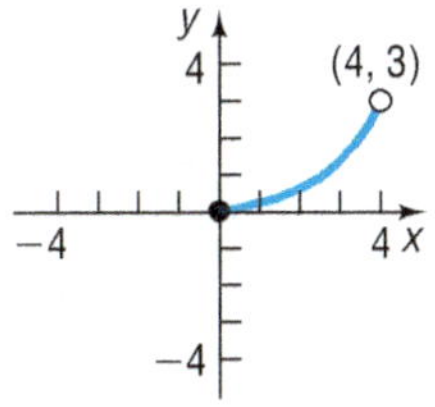

21.

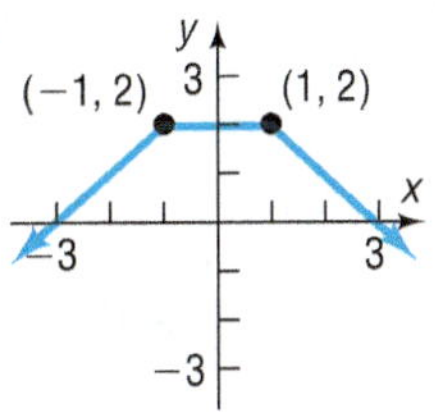

22.

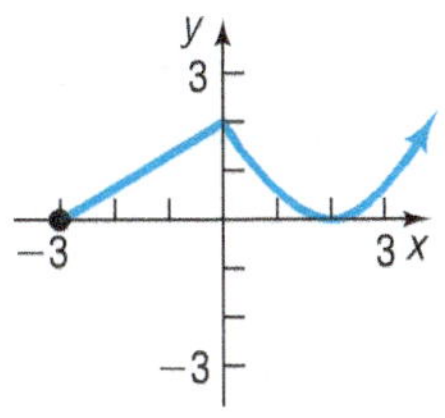

23.

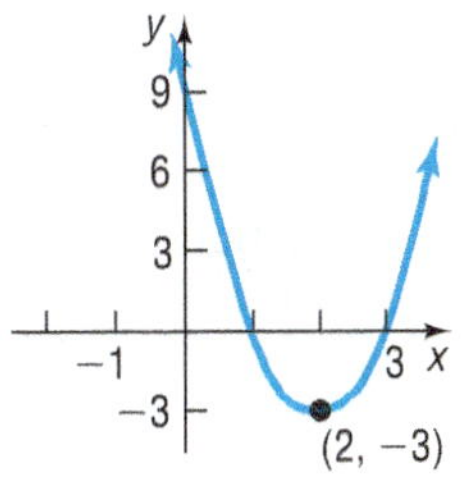

24.

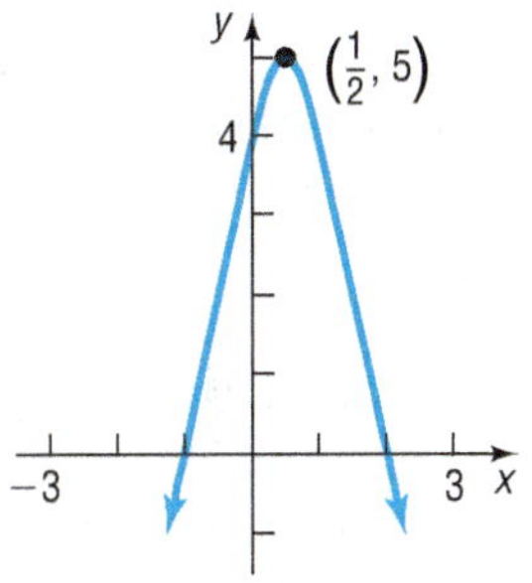

In Problems 25–30, answer the questions about the given function.

25. $f(x) = 2x^2 - x - 1$

(a) Is the point $(-1, 2)$ on the graph of f?
(b) If $x = -2$, what is $f(x)$? What point is on the graph of f?
(c) If $f(x) = -1$, what is x? What point(s) are on the graph of f?
(d) What is the domain of f?
(e) List the x-intercepts, if any, of the graph of f.
(f) List the y-intercept, if there is one, of the graph of f.

26. $f(x) = -3x^2 + 5x$

(a) Is the point $(-1, 2)$ on the graph of f?
(b) If $x = -2$, what is $f(x)$? What point is on the graph of f?
(c) If $f(x) = -2$, what is x? What point(s) are on the graph of f?
(d) What is the domain of f?
(e) List the x-intercepts, if any, of the graph of f.
(f) List the y-intercept, if there is one, of the graph of f.

27. $f(x) = \dfrac{x + 2}{x - 6}$

(a) Is the point $(3, 14)$ on the graph of f?
(b) If $x = 4$, what is $f(x)$? What point is on the graph of f?
(c) If $f(x) = 2$, what is x? What point(s) are on the graph of f?
(d) What is the domain of f?
(e) List the x-intercepts, if any, of the graph of f.
(f) List the y-intercept, if there is one, of the graph of f.

28. $f(x) = \dfrac{x^2 + 2}{x + 4}$

(a) Is the point $\left(1, \dfrac{3}{5}\right)$ on the graph of f?
(b) If $x = 0$, what is $f(x)$? What point is on the graph of f?
(c) If $f(x) = \dfrac{1}{2}$, what is x? What point(s) are on the graph of f?
(d) What is the domain of f?
(e) List the x-intercepts, if any, of the graph of f.
(f) List the y-intercept, if there is one, of the graph of f.

29. $f(x) = \dfrac{2x^2}{x^4 + 1}$

(a) Is the point $(-1, 1)$ on the graph of f?
(b) If $x = 2$, what is $f(x)$? What point is on the graph of f?
(c) If $f(x) = 1$, what is x? What point(s) are on the graph of f?
(d) What is the domain of f?
(e) List the x-intercepts, if any, of the graph of f.
(f) List the y-intercept, if there is one, of the graph of f.

30. $f(x) = \dfrac{2x}{x - 2}$

(a) Is the point $\left(\dfrac{1}{2}, -\dfrac{2}{3}\right)$ on the graph of f?
(b) If $x = 4$, what is $f(x)$? What point is on the graph of f?
(c) If $f(x) = 1$, what is x? What point(s) are on the graph of f?
(d) What is the domain of f?
(e) List the x-intercepts, if any, of the graph of f.
(f) List the y-intercept, if there is one, of the graph of f.

Applications and Extensions

31. The graphs of two functions, f and g, are illustrated. Use the graphs to answer parts (a)–(f).

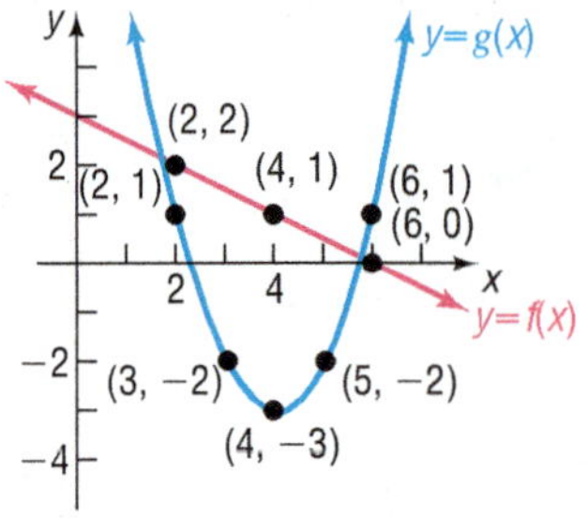

(a) $(f+g)(2)$ (b) $(f+g)(4)$

(c) $(f-g)(6)$ (d) $(g-f)(6)$

(e) $(f \cdot g)(2)$ (f) $\left(\frac{f}{g}\right)(4)$

32. Granny Shots The last player in the NBA to use an underhand foul shot (a "granny" shot) was Hall of Fame forward Rick Barry, who retired in 1980. Barry believes that current NBA players could increase their free-throw percentage if they were to use an underhand shot. Since underhand shots are released from a lower position, the angle of the shot must be increased. If a player shoots an underhand foul shot, releasing the ball at a 70-degree angle from a position 3.5 feet above the floor, then the path of the ball can be modeled by the function $h(x) = -\frac{136x^2}{v^2} + 2.7x + 3.5$, where h is the height of the ball above the floor, x is the forward distance of the ball in front of the foul line, and v is the initial velocity with which the ball is shot in feet per second.

(a) The center of the hoop is 10 feet above the floor and 15 feet in front of the foul line. Determine the initial velocity with which the ball must be shot in order for the ball to go through the hoop.

(b) Write the function for the path of the ball using the velocity found in part (a).

(c) Determine the height of the ball after it has traveled 9 feet in front of the foul line.

(d) Find additional points and graph the path of the basketball.

Source: *The Physics of Foul Shots,* Discover, *Vol. 21, No. 10, October 2000*

33. Free-throw Shots According to physicist Peter Brancazio, the key to a successful foul shot in basketball lies in the arc of the shot. Brancazio determined the optimal angle of the arc from the free-throw line to be 45 degrees. The arc also depends on the velocity with which the ball is shot. If a player shoots a foul shot, releasing the ball at a 45-degree angle from a position 6 feet above the floor, then the path of the ball can be modeled by the function

$$h(x) = -\frac{44x^2}{v^2} + x + 6$$

where h is the height of the ball above the floor, x is the forward distance of the ball in front of the foul line, and v is the initial velocity with which the ball is shot in feet per second. Suppose a player shoots a ball with an initial velocity of 28 feet per second.

(a) Determine the height of the ball after it has traveled 8 feet in front of the foul line.

(b) Determine the height of the ball after it has traveled 12 feet in front of the foul line.

(c) Find additional points and graph the path of the basketball.

(d) The center of the hoop is 10 feet above the floor and 15 feet in front of the foul line. Will the ball go through the hoop? Why or why not? If not, with what initial velocity must the ball be shot in order for the ball to go through the hoop?

Source: *The Physics of Foul Shots,* Discover, *Vol. 21, No. 10, October 2000*

34. Cross-sectional Area The cross-sectional area of a beam cut from a log with radius 1 foot is given by the function $A(x) = 4x\sqrt{1-x^2}$, where x represents the length, in feet, of half the base of the beam. See the figure.

(a) Find the domain of A.

(b) Use a graphing utility to graph the function $A = A(x)$.

(c) Create a TABLE with TblStart $= 0$ and ΔTbl $= 0.1$ for $0 \le x \le 1$. Which value of x maximizes the cross-sectional area? What should be the length of the base of the beam to maximize the cross-sectional area?

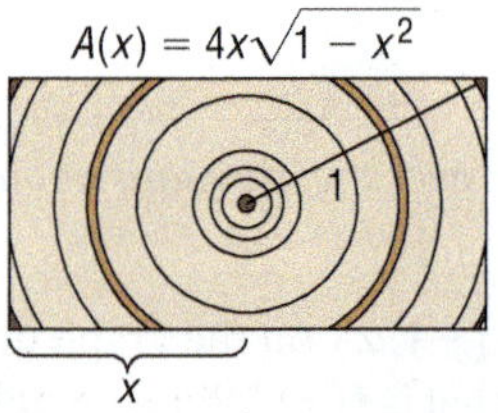

35. Motion of a Golf Ball A golf ball is hit with an initial velocity of 130 feet per second at an inclination of 45° to the horizontal. In physics, it is established that the height h of the golf ball is given by the function

$$h(x) = \frac{-32x^2}{130^2} + x$$

where x is the horizontal distance that the golf ball has traveled.

(a) Determine the height of the golf ball after it has traveled 100 feet.
(b) What is the height after it has traveled 300 feet?
(c) What is the height after it has traveled 500 feet?
(d) How far was the golf ball hit?
(e) Use a graphing utility to graph the function $h = h(x)$.
(f) Use a graphing utility to determine the distance that the ball has traveled when the height of the ball is 90 feet.
(g) Create a TABLE with TblStart = 0 and ΔTbl = 25. To the nearest 25 feet, how far does the ball travel before it reaches a maximum height? What is the maximum height?
(h) Adjust the value of ΔTbl until you determine the distance, to within 1 foot, that the ball travels before it reaches its maximum height.

36. Effect of Elevation on Weight If an object weighs m pounds at sea level, then its weight W (in pounds) at a height of h miles above sea level is given approximately by

$$W(h) = m\left(\frac{4000}{4000 + h}\right)^2$$

(a) If Amy weighs 120 pounds at sea level, how much will she weigh on Pikes Peak, which is 14,110 feet above sea level?
(b) Use a graphing utility to graph the function $W = W(h)$. Use $m = 120$ pounds.
(c) Create a TABLE with TblStart = 0 and ΔTbl = 0.5 to see how the weight W varies as h changes from 0 to 5 miles.
(d) At what height will Amy weigh 119.95 pounds?
(e) Does your answer to part (d) seem reasonable? Explain.

37. Cost of Transatlantic Travel A Boeing 747 crosses the Atlantic Ocean (3000 miles) with an airspeed of 500 miles per hour. The cost C (in dollars) per passenger is given by

$$C(x) = 100 + \frac{x}{10} + \frac{36{,}000}{x}$$

where x is the groundspeed (airspeed $\pm$ wind).

(a) What is the cost when the groundspeed is 480 miles per hour? 600 miles per hour?
(b) Find the domain of C.
(c) Use a graphing utility to graph the function $C = C(x)$.
(d) Create a TABLE with TblStart = 0 and ΔTbl = 50.
(e) To the nearest 50 miles per hour, what groundspeed minimizes the cost per passenger?

38. Reading and Interpreting Graphs Let C be the function whose graph is given in the next column. This graph represents the cost C of manufacturing q computers in a day.

(a) Determine $C(0)$. Interpret this value.
(b) Determine $C(10)$. Interpret this value.
(c) Determine $C(50)$. Interpret this value.
(d) What is the domain of C? What does this domain imply in terms of daily production?
(e) Describe the shape of the graph.
(f) The point $(30, 32\,000)$ is called an *inflection point.* Describe the behavior of the graph around the inflection point.

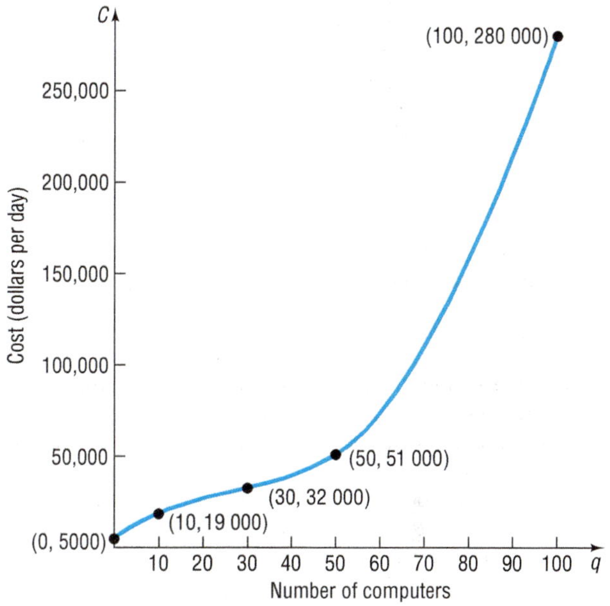

39. Reading and Interpreting Graphs Let C be the function whose graph is given below. This graph represents the cost C of using g gigabytes of data in a month for a shared-data family plan.

(a) Determine $C(0)$. Interpret this value.
(b) Determine $C(5)$. Interpret this value.
(c) Determine $C(15)$. Interpret this value.
(d) What is the domain of C? What does this domain imply in terms of the number of gigabytes?
(e) Describe the shape of the graph.

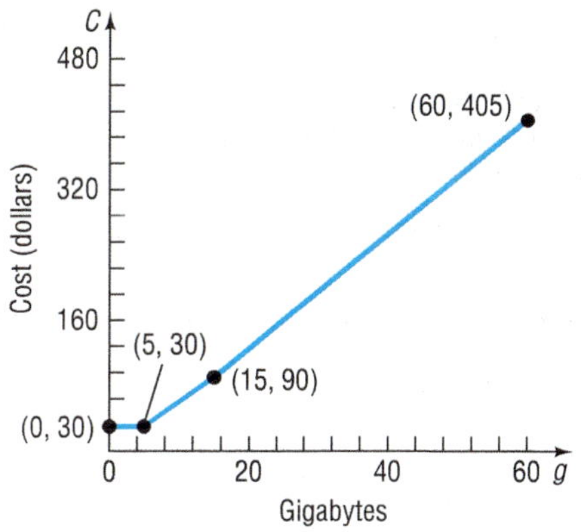

Explaining Concepts: Discussion and Writing

40. Describe how you would find the domain and range of a function if you were given its graph. How would your strategy change if you were given the equation defining the function instead of its graph?
41. How many x-intercepts can the graph of a function have? How many y-intercepts can the graph of a function have?
42. Is a graph that consists of a single point the graph of a function? Can you write the equation of such a function?

43. Match each of the following functions with the graph that best describes the situation.

(a) The cost of building a house as a function of its square footage
(b) The height of an egg dropped from a 300-foot building as a function of time
(c) The height of a human as a function of time
(d) The demand for Big Macs as a function of price
(e) The height of a child on a swing as a function of time

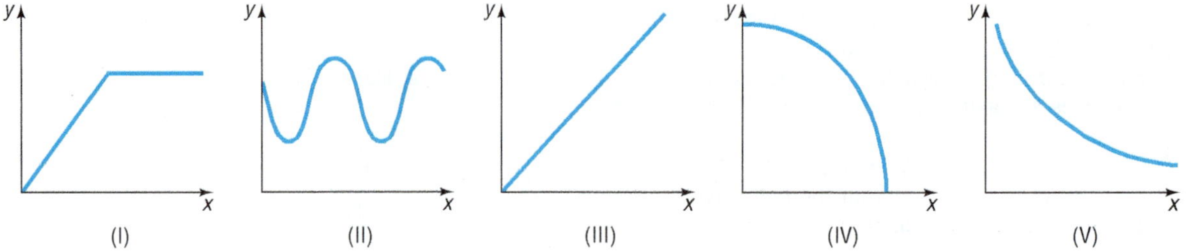

44. Match each of the following functions with the graph that best describes the situation.

(a) The temperature of a bowl of soup as a function of time
(b) The number of hours of daylight per day over a 2-year period
(c) The population of Florida as a function of time
(d) The distance traveled by a car going at a constant velocity as a function of time
(e) The height of a golf ball hit with a 7-iron as a function of time

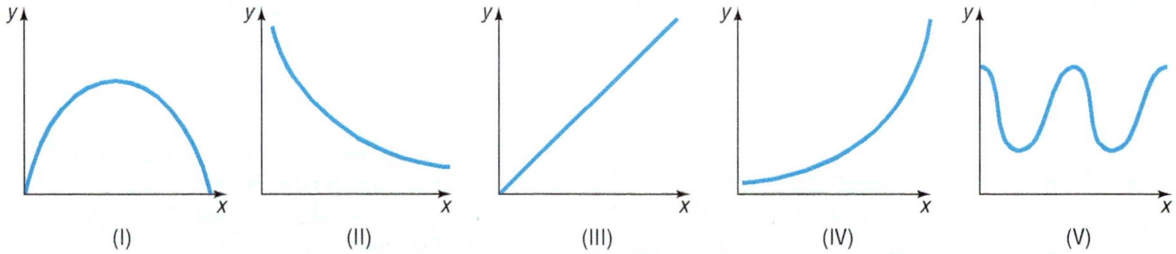

45. Consider the following scenario: Barbara decides to take a walk. She leaves home, walks 2 blocks in 5 minutes at a constant speed, and realizes that she forgot to lock the door. So Barbara runs home in 1 minute. While at her doorstep, it takes her 1 minute to find her keys and lock the door. Barbara walks 5 blocks in 15 minutes and then decides to jog home. It takes her 7 minutes to get home. Draw a graph of Barbara's distance from home (in blocks) as a function of time.

46. Consider the following scenario: Jayne enjoys riding her bicycle through the woods. At the forest preserve, she gets on her bicycle and rides up a 2000-foot incline in 10 minutes. She then travels down the incline in 3 minutes. The next 5000 feet is level terrain, and she covers the distance in 20 minutes. She rests for 15 minutes. Jayne then travels 10,000 feet in 30 minutes. Draw a graph of Jayne's distance traveled (in feet) as a function of time.

47. The following sketch represents the distance d (in miles) that Kevin was from home as a function of time t (in hours). Answer the questions by referring to the graph. In parts (a)–(g), how many hours elapsed and how far was Kevin from home during this time?

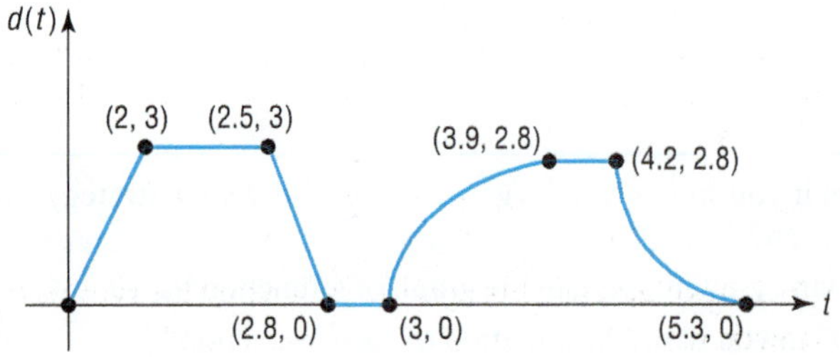

(a) From $t = 0$ to $t = 2$
(b) From $t = 2$ to $t = 2.5$
(c) From $t = 2.5$ to $t = 2.8$
(d) From $t = 2.8$ to $t = 3$
(e) From $t = 3$ to $t = 3.9$
(f) From $t = 3.9$ to $t = 4.2$
(g) From $t = 4.2$ to $t = 5.3$
(h) What is the farthest distance that Kevin was from home?
(i) How many times did Kevin return home?

48. The following sketch represents the speed v (in miles per hour) of Michael's car as a function of time t (in minutes).

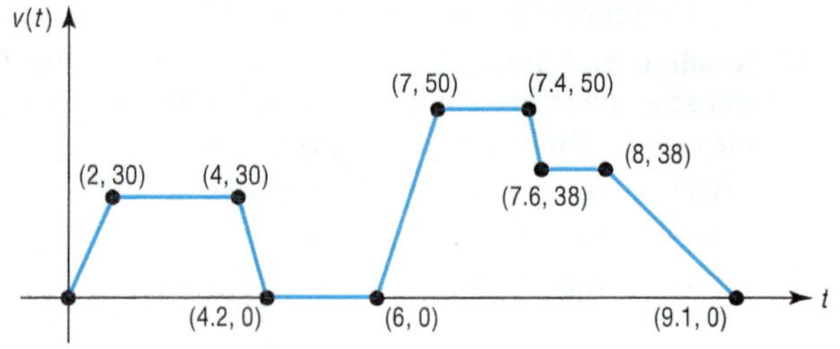

(a) Over what interval of time was Michael traveling fastest?
(b) Over what interval(s) of time was Michael's speed zero?
(c) What was Michael's speed between 0 and 2 minutes?
(d) What was Michael's speed between 4.2 and 6 minutes?
(e) What was Michael's speed between 7 and 7.4 minutes?
(f) When was Michael's speed constant?

49. Draw the graph of a function whose domain is $\{x \mid -3 \le x \le 8,\ x \ne 5\}$ and whose range is $\{y \mid -1 \le y \le 2,\ y \ne 0\}$. What point(s) in the rectangle $-3 \le x \le 8, -1 \le y \le 2$ cannot be on the graph? Compare your graph with those of other students. What differences do you see?

50. Is there a function whose graph is symmetric with respect to the x-axis? Explain.

51. Explain why the vertical-line test works.

Retain Your Knowledge

Problems 52–55 are based on material learned earlier in the course. The purpose of these problems is to keep the material fresh in your mind so that you are better prepared for the final exam.

52. Factor completely: $8x^2 + 24x + 18$

53. Find the distance between the points $(3, -6)$ and $(1, 0)$.

54. Write the equation of the line with slope $\frac{2}{3}$ that passes through the point $(-6, 4)$.

55. Subtract: $(4x^3 - 5x^2 + 2) - (3x^2 + 5x - 2)$

'Are You Prepared?' Answers

1. $(-4, 0), (4, 0), (0, -2), (0, 2)$

2. False

3.3 Properties of Functions

PREPARING FOR THIS SECTION *Before getting started, review the following:*

- Intervals (Section 1.7, pp. 147–148)
- Intercepts (Section 2.1, pp. 165–166)
- Slope of a Line (Section 2.2, pp. 173–176)
- Point–Slope Form of a Line (Section 2.2, p. 177)
- Symmetry (Section 2.1, pp. 166–168)

Now Work the **'Are You Prepared?'** problems on page 240.

OBJECTIVES

1 Determine Even and Odd Functions from a Graph (p. 231)
2 Identify Even and Odd Functions from an Equation (p. 233)
3 Use a Graph to Determine Where a Function Is Increasing, Decreasing, or Constant (p. 234)
4 Use a Graph to Locate Local Maxima and Local Minima (p. 235)
5 Use a Graph to Locate the Absolute Maximum and the Absolute Minimum (p. 236)
6 Use a Graphing Utility to Approximate Local Maxima and Local Minima and to Determine Where a Function Is Increasing or Decreasing (p. 237)
7 Find the Average Rate of Change of a Function (p. 238)

To obtain the graph of a function $y = f(x)$, it is often helpful to know certain properties that the function has and the impact of these properties on the way the graph will look.

1 Determine Even and Odd Functions from a Graph

The words *even* and *odd*, when applied to a function f, describe the symmetry that exists for the graph of the function.

A function f is even if and only if, whenever the point (x, y) is on the graph of f, the point $(-x, y)$ is also on the graph. Using function notation, we define an even function as follows:

DEFINITION

A function f is **even** if, for every number x in its domain, the number $-x$ is also in the domain and

$$f(-x) = f(x)$$

For the even function shown in Figure 17(a), notice that $f(x_1) = f(-x_1)$ and that $f(x_2) = f(-x_2)$.

A function f is odd if and only if, whenever the point (x, y) is on the graph of f, the point $(-x, -y)$ is also on the graph. Using function notation, we define an odd function as follows:

DEFINITION

A function f is **odd** if, for every number x in its domain, the number $-x$ is also in the domain and

$$f(-x) = -f(x)$$

For the odd function shown in Figure 17(b), notice that $f(x_1) = -f(-x_1)$, which is equivalent to $f(-x_1) = -f(x_1)$, and that $f(x_2) = -f(-x_2)$, which is equivalent to $f(-x_2) = -f(x_2)$.

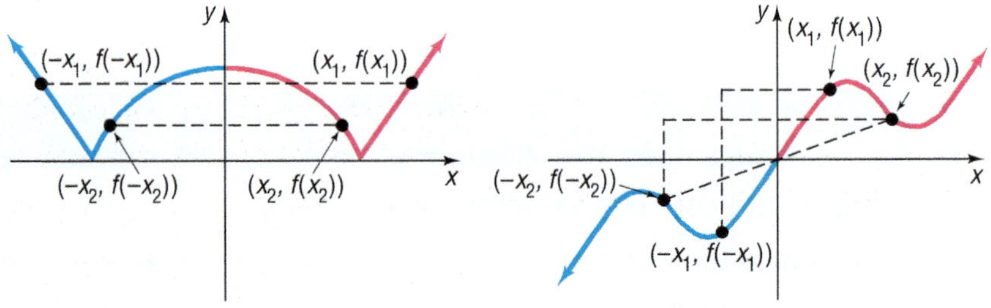

(a) Even function (b) Odd function

Figure 17

Refer to page 167, where the tests for symmetry are listed. The following results are then evident.

THEOREM

A function is even if and only if its graph is symmetric with respect to the y-axis. A function is odd if and only if its graph is symmetric with respect to the origin.

EXAMPLE 1

Determining Even and Odd Functions from the Graph

Determine whether each graph given in Figure 18 is the graph of an even function, an odd function, or a function that is neither even nor odd.

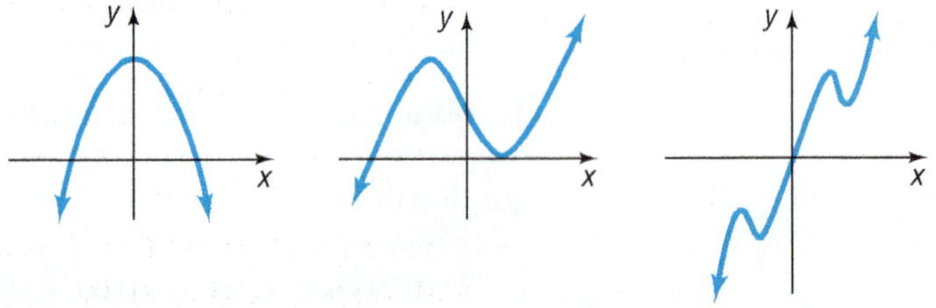

(a) (b) (c)

Figure 18

Solution (a) The graph in Figure 18(a) is that of an even function, because the graph is symmetric with respect to the y-axis.

(b) The function whose graph is given in Figure 18(b) is neither even nor odd, because the graph is neither symmetric with respect to the y-axis nor symmetric with respect to the origin.

(c) The function whose graph is given in Figure 18(c) is odd, because its graph is symmetric with respect to the origin. ■

Now Work PROBLEMS 25(a), (b), AND (d)

2 Identify Even and Odd Functions from an Equation

A graphing utility can be used to conjecture whether a function is even, odd, or neither. Remember that when the graph of an even function contains the point (x, y), it must also contain the point $(-x, y)$. Therefore, if the graph shows evidence of symmetry with respect to the y-axis, we would conjecture that the function is even. In addition, if the graph shows evidence of symmetry with respect to the origin, we would conjecture that the function is odd.

EXAMPLE 2 Identifying Even and Odd Functions

Use a graphing utility to conjecture whether each of the following functions is even, odd, or neither. Then algebraically determine whether the graph is symmetric with respect to the y-axis or with respect to the origin.

(a) $f(x) = x^2 - 5$ (b) $g(x) = x^3 - 1$ (c) $h(x) = 5x^3 - x$

Solution (a) Graph the function. See Figure 19. It appears that the graph is symmetric with respect to the y-axis. We conjecture that the function is even.

To algebraically verify the conjecture, replace x by $-x$ in $f(x) = x^2 - 5$. Then

$$f(-x) = (-x)^2 - 5 = x^2 - 5 = f(x)$$

Since $f(-x) = f(x)$, we conclude that f is an even function and that the graph is symmetric with respect to the y-axis.

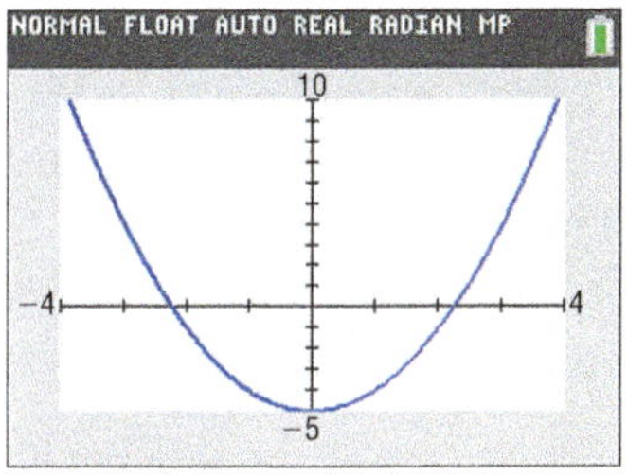

Figure 19

(b) Graph the function. See Figure 20. It appears that there is no symmetry. We conjecture that the function is neither even nor odd.

To algebraically verify that the function is not even, find $g(-x)$ and compare the result with $g(x)$.

$$g(-x) = (-x)^3 - 1 = -x^3 - 1; \quad g(x) = x^3 - 1$$

Since $g(-x) \neq g(x)$, the function is not even.

To algebraically verify that the function is not odd, find $-g(x)$ and compare the result with $g(-x)$.

$$-g(x) = -(x^3 - 1) = -x^3 + 1; \quad g(-x) = -x^3 - 1$$

Since $g(-x) \neq -g(x)$, the function is not odd. The graph is not symmetric with respect to the y-axis nor is it symmetric with respect to the origin.

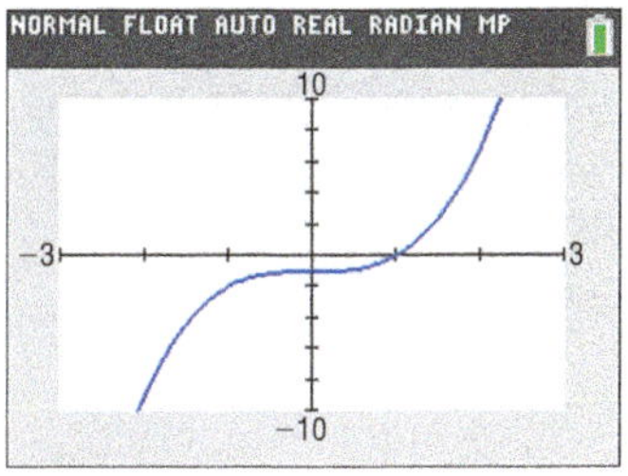

Figure 20

(c) Graph the function. See Figure 21. It appears that there is symmetry with respect to the origin. We conjecture that the function is odd.

To algebraically verify the conjecture, replace x by $-x$ in $h(x) = 5x^3 - x$. Then

$$h(-x) = 5(-x)^3 - (-x) = -5x^3 + x = -(5x^3 - x) = -h(x)$$

Since $h(-x) = -h(x)$, h is an odd function and the graph of h is symmetric with respect to the origin. ■

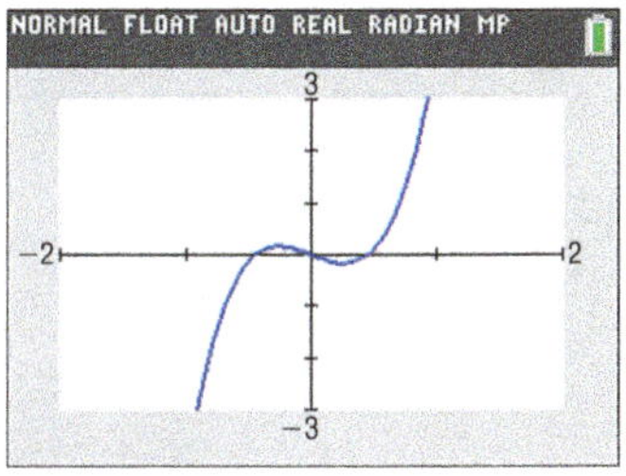

Figure 21

Now Work PROBLEM 37

3 Use a Graph to Determine Where a Function Is Increasing, Decreasing, or Constant

Consider the graph given in Figure 22. If you look from left to right along the graph of the function, you will notice that parts of the graph are going up, parts are going down, and parts are horizontal. In such cases, the function is described as *increasing, decreasing,* or *constant,* respectively.

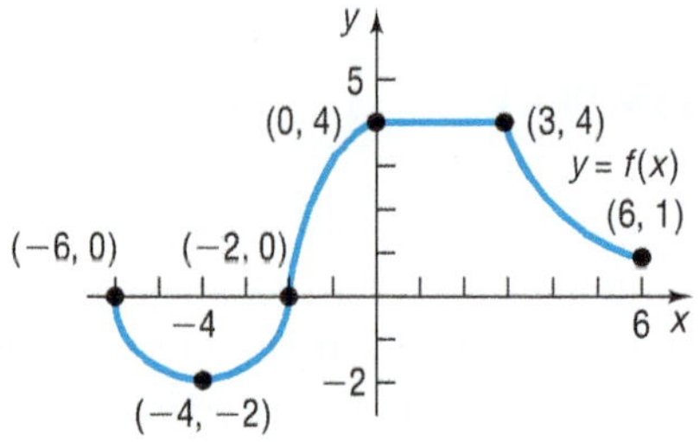

Figure 22

EXAMPLE 3 **Determining Where a Function Is Increasing, Decreasing, or Constant from Its Graph**

Determine the values of x for which the function in Figure 22 is increasing. Where is it decreasing? Where is it constant?

Solution When determining where a function is increasing, where it is decreasing, and where it is constant, we use inequalities involving the independent variable x, or we use intervals of x-coordinates. The function whose graph is given in Figure 22 is increasing on the interval $[-4, 0]$, or for $-4 \leq x \leq 0$. The function is decreasing on the intervals $[-6, -4]$ and $[3, 6]$, or for $-6 \leq x \leq -4$ and $3 \leq x \leq 6$. The function is constant on the closed interval $[0, 3]$, or for $0 \leq x \leq 3$. ■

WARNING Describe the behavior of a graph in terms of its x-values. Do not say the graph in Figure 22 is increasing from the point $(-4, -2)$ to the point $(0, 4)$. Rather, say it is increasing on the interval $[-4, 0]$. ■

More precise definitions follow:

In Words

If a function is decreasing, then as the values of x get bigger, the values of the function get smaller. If a function is increasing, then as the values of x get bigger, the values of the function also get bigger. If a function is constant, then as the values of x get bigger, the values of the function remain unchanged.

DEFINITIONS A function f is **increasing** on an interval I if, for any choice of x_1 and x_2 in I, with $x_1 < x_2$, we have $f(x_1) < f(x_2)$.

A function f is **decreasing** on an interval I if, for any choice of x_1 and x_2 in I, with $x_1 < x_2$, we have $f(x_1) > f(x_2)$.

A function f is **constant** on an interval I if, for all choices of x in I, the values $f(x)$ are equal. ■

Figure 23 illustrates the definitions. The graph of an increasing function goes up from left to right, the graph of a decreasing function goes down from left to right, and the graph of a constant function remains at a fixed height.

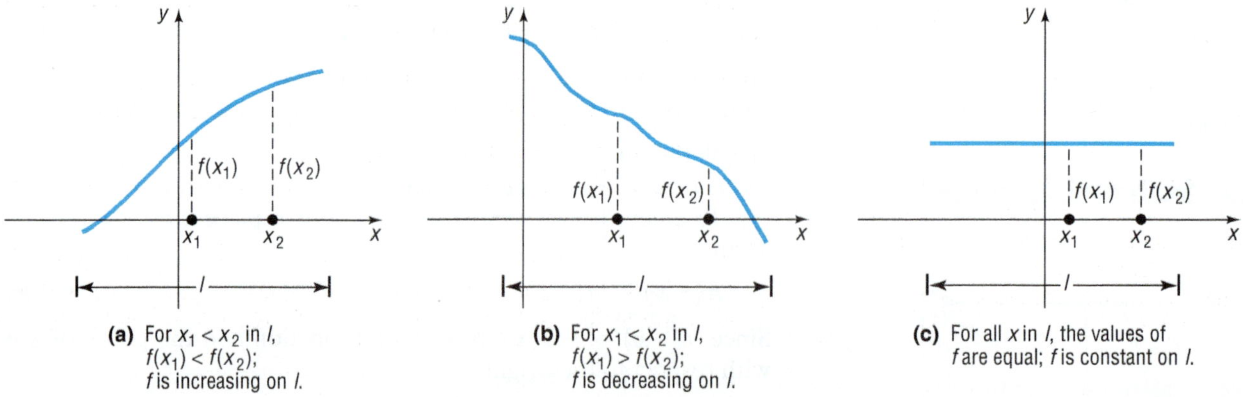

Figure 23

Now Work PROBLEMS 13, 15, 17, AND 25(c)

4 Use a Graph to Locate Local Maxima and Local Minima

Suppose f is a function defined on an open interval I containing c. If the value of f at c is greater than or equal to the values of f on I, then f has a *local maximum* at c.* See Figure 24(a).

If the value of f at c is less than or equal to the values of f on I, then f has a *local minimum* at c. See Figure 24(b).

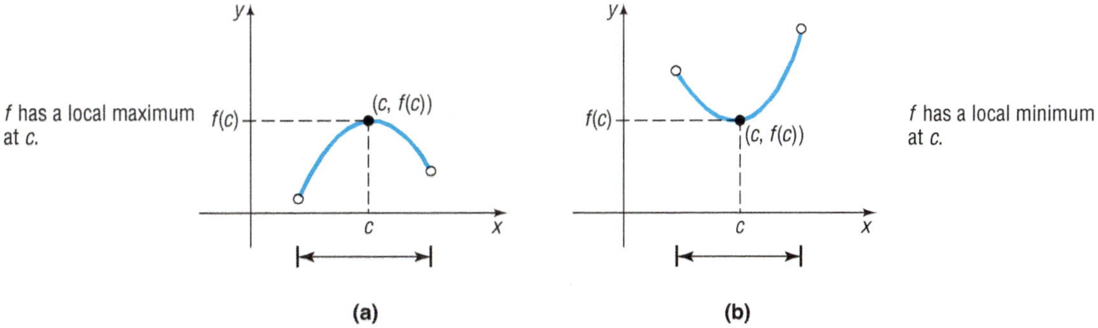

Figure 24 Local maximum and local minimum

DEFINITIONS

> Let f be a function defined on some interval I.
>
> A function f has a **local maximum** at c if there is an open interval in I containing c so that, for all x in this open interval, we have $f(x) \leq f(c)$. We call $f(c)$ a **local maximum value of f.**
>
> A function f has a **local minimum** at c if there is an open interval in I containing c so that, for all x in this open interval, we have $f(x) \geq f(c)$. We call $f(c)$ a **local minimum value of f.** ■

If f has a local maximum at c, then the value of f at c is greater than or equal to the values of f near c. If f has a local minimum at c, then the value of f at c is less than or equal to the values of f near c. The word *local* is used to suggest that it is only near c, not necessarily over the entire domain, that the value $f(c)$ has these properties.

EXAMPLE 4 Finding Local Maxima and Local Minima from the Graph of a Function and Determining Where the Function Is Increasing, Decreasing, or Constant

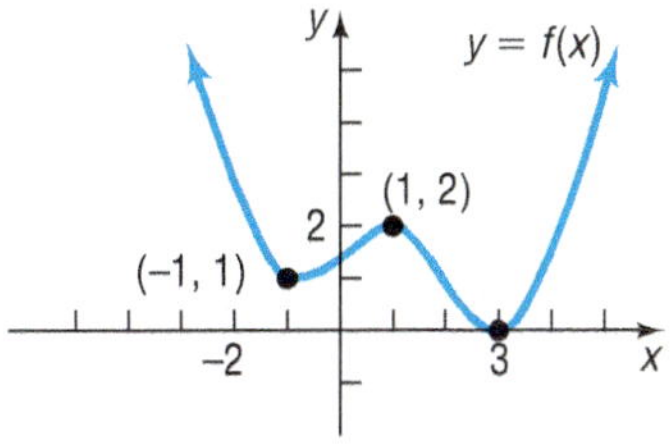

Figure 25

Figure 25 shows the graph of a function f.

(a) At what value(s) of x, if any, does f have a local maximum? List the local maximum value(s).

(b) At what value(s) of x, if any, does f have a local minimum? List the local minimum value(s).

(c) Find the intervals on which f is increasing. Find the intervals on which f is decreasing.

Solution The domain of f is the set of real numbers.

(a) f has a local maximum at 1, since for all x close to 1, we have $f(x) \leq f(1)$. The local maximum value is $f(1) = 2$.

(b) f has local minima at -1 and at 3. The local minimum values are $f(-1) = 1$ and $f(3) = 0$.

(c) The function whose graph is given in Figure 25 is increasing on the intervals $[-1, 1]$ and $[3, \infty)$, or for $-1 \leq x \leq 1$ and $x \geq 3$. The function is decreasing on the intervals $(-\infty, -1]$ and $[1, 3]$, or for $x \leq -1$ and $1 \leq x \leq 3$. ■

WARNING The y-value is the local maximum value or local minimum value, and it occurs at some x-value. For example, in Figure 25, we say f has a local maximum at 1 and the local maximum value is 2. ■

Now Work PROBLEMS 19 AND 21

*Some texts use the term *relative* instead of *local*.

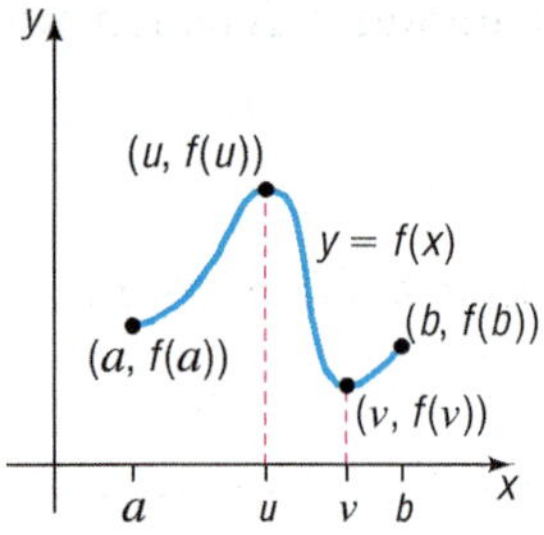

domain: $[a, b]$
for all x in $[a, b]$, $f(x) \le f(u)$
for all x in $[a, b]$, $f(x) \ge f(v)$
absolute maximum: $f(u)$
absolute minimum: $f(v)$

Figure 26

5 Use a Graph to Locate the Absolute Maximum and the Absolute Minimum

Look at the graph of the function f given in Figure 26. The domain of f is the closed interval $[a, b]$. Also, the largest value of f is $f(u)$ and the smallest value of f is $f(v)$. These are called, respectively, the *absolute maximum* and the *absolute minimum* of f on $[a, b]$.

DEFINITION Let f be a function defined on some interval I. If there is a number u in I for which $f(x) \le f(u)$ for all x in I, then f has an **absolute maximum at u**, and the number $f(u)$ is the **absolute maximum of f on I**.

If there is a number v in I for which $f(x) \ge f(v)$ for all x in I, then f has an **absolute minimum at v**, and the number $f(v)$ is the **absolute minimum of f on I**.

The absolute maximum and absolute minimum of a function f are sometimes called the **absolute extrema** or **extreme values** of f on I.

The absolute maximum or absolute minimum of a function f may not exist. Let's look at some examples.

EXAMPLE 5 Finding the Absolute Maximum and the Absolute Minimum from the Graph of a Function

For each graph of a function $y = f(x)$ in Figure 27, find the absolute maximum and the absolute minimum, if they exist. Also, find any local maxima or local minima.

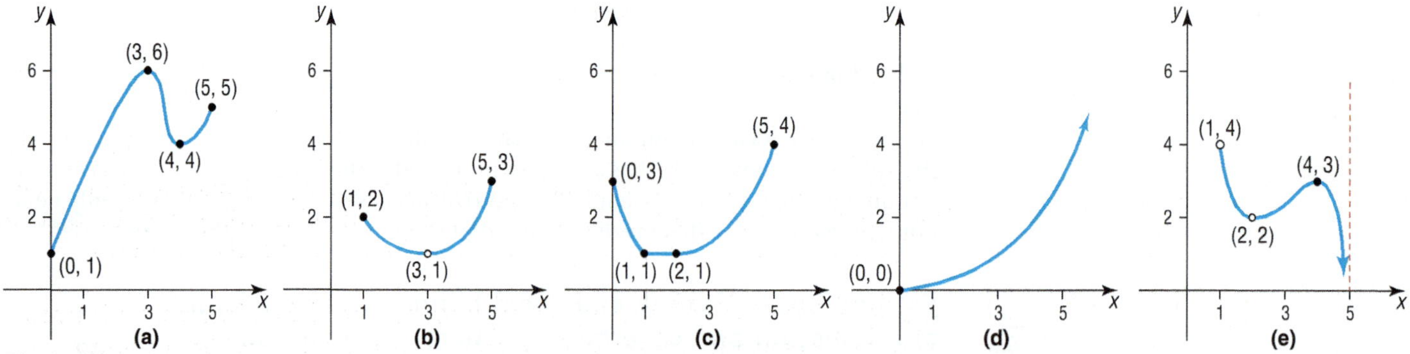

Figure 27

Solution

(a) The function f whose graph is given in Figure 27(a) has the closed interval $[0, 5]$ as its domain. The largest value of f is $f(3) = 6$, the absolute maximum. The smallest value of f is $f(0) = 1$, the absolute minimum. The function has a local maximum value of 6 at $x = 3$ and a local minimum value of 4 at $x = 4$.

(b) The function f whose graph is given in Figure 27(b) has the domain $\{x | 1 \le x \le 5, x \ne 3\}$. Note that we exclude 3 from the domain because of the "hole" at $(3, 1)$. The largest value of f on its domain is $f(5) = 3$, the absolute maximum. There is no absolute minimum. Do you see why? As you trace the graph, getting closer to the point $(3, 1)$, there is no single smallest value. [As soon as you claim a smallest value, we can trace closer to $(3, 1)$ and get a smaller value!] The function has no local maximum or local minimum.

(c) The function f whose graph is given in Figure 27(c) has the interval $[0, 5]$ as its domain. The absolute maximum of f is $f(5) = 4$. The absolute minimum is 1. Notice that the absolute minimum 1 occurs at any number in the interval $[1, 2]$. The function has a local minimum value of 1 at every x in the interval $[1, 2]$, but it has no local maximum value.

(d) The function f given in Figure 27(d) has the interval $[0, \infty)$ as its domain. The function has no absolute maximum; the absolute minimum is $f(0) = 0$. The function has no local maximum or local minimum.

(e) The function f in Figure 27(e) has the domain $\{x | 1 < x < 5, x \ne 2\}$. The function has no absolute maximum and no absolute minimum. Do you see why? The function has a local maximum value of 3 at $x = 4$, but no local minimum value.

WARNING A function may have an absolute maximum or an absolute minimum at an endpoint but not a local maximum or a local minimum. Why? Local maxima and local minima are found over some open interval I, and this interval cannot be created around an endpoint.

In calculus, there is a theorem with conditions that guarantee a function will have an absolute maximum and an absolute minimum.

THEOREM

Extreme Value Theorem

If f is a continuous function* whose domain is a closed interval $[a, b]$, then f has an absolute maximum and an absolute minimum on $[a, b]$.

The absolute maximum (minimum) can be found by selecting the largest (smallest) value of f from the following list:

1. The values of f at any local maxima or local minima of f in $[a, b]$.
2. The value of f at each endpoint of $[a, b]$—that is, $f(a)$ and $f(b)$.

For example, the graph of the function f given in Figure 27(a) is continuous on the closed interval $[0, 5]$. The Extreme Value Theorem guarantees that f has extreme values on $[0, 5]$. To find them, we list

1. The value of f at the local extrema: $f(3) = 6, f(4) = 4$
2. The value of f at the endpoints: $f(0) = 1, f(5) = 5$

The largest of these, 6, is the absolute maximum; the smallest of these, 1, is the absolute minimum.

Notice that absolute extrema may occur at the endpoints of a function defined on a closed interval. However, local extrema cannot occur at the endpoints because an open interval cannot be constructed around the endpoint. So, in Figure 27(b), for example, $f(1) = 2$ is not a local maximum.

Now Work PROBLEM 49

6 Use a Graphing Utility to Approximate Local Maxima and Local Minima and to Determine Where a Function Is Increasing or Decreasing

To locate the exact value at which a function f has a local maximum or a local minimum usually requires calculus. However, a graphing utility may be used to approximate these values using the MAXIMUM and MINIMUM features.

EXAMPLE 6 Using a Graphing Utility to Approximate Local Maxima and Minima and to Determine Where a Function Is Increasing or Decreasing

(a) Use a graphing utility to graph $f(x) = 6x^3 - 12x + 5$ for $-2 \le x \le 2$. Approximate where f has a local maximum and where f has a local minimum.

(b) Determine where f is increasing and where it is decreasing.

Solution (a) Graphing utilities have a feature that finds the maximum or minimum point of a graph within a given interval. Graph the function f for $-2 \le x \le 2$. The MAXIMUM and MINIMUM commands require us to first determine the open interval I. The graphing utility will then approximate the maximum or minimum value in the interval. Using MAXIMUM, we find that the local maximum value is 11.53 and that it occurs at $x = -0.82$, rounded to two decimal places. See Figure 28(a). Using MINIMUM, we find that the local minimum value is -1.53 and that it occurs at $x = 0.82$, rounded to two decimal places. See Figure 28(b).

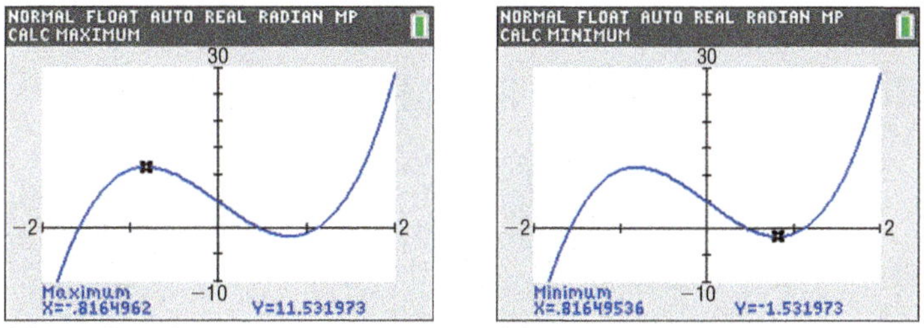

Figure 28 (a) Local maximum (b) Local minimum

*Although a precise definition requires calculus, we'll agree for now that a continuous function is one whose graph has no gaps or holes and can be traced without lifting the pencil from the paper.

(b) Looking at Figures 28(a) and (b), we see that the graph of f is increasing from $x = -2$ to $x = -0.82$ and from $x = 0.82$ to $x = 2$, so f is increasing on the intervals $[-2, -0.82]$ and $[0.82, 2]$, or for $-2 \le x \le -0.82$ and $0.82 \le x \le 2$. The graph is decreasing from $x = -0.82$ to $x = 0.82$, so f is decreasing on the interval $[-0.82, 0.82]$, or for $-0.82 \le x \le 0.82$. ■

Now Work PROBLEM 57

7 Find the Average Rate of Change of a Function

In Section 2.2, we said that the slope of a line can be interpreted as the average rate of change. To find the average rate of change of a function between any two points on its graph, calculate the slope of the line containing the two points.

DEFINITION

If a and b, $a \ne b$, are in the domain of a function $y = f(x)$, the **average rate of change of f** from a to b is defined as

$$\text{Average rate of change} = \frac{\Delta y}{\Delta x} = \frac{f(b) - f(a)}{b - a} \qquad a \ne b \qquad \textbf{(1)}$$

In Words
The symbol Δ is the Greek capital letter delta and is read "change in."

The symbol Δy in equation (1) is the "change in y," and Δx is the "change in x." The average rate of change of f is the change in y divided by the change in x.

EXAMPLE 7 **Finding the Average Rate of Change**

Find the average rate of change of $f(x) = 3x^2$:

(a) From 1 to 3 (b) From 1 to 5 (c) From 1 to 7

Solution (a) The average rate of change of $f(x) = 3x^2$ from 1 to 3 is

$$\frac{\Delta y}{\Delta x} = \frac{f(3) - f(1)}{3 - 1} = \frac{27 - 3}{3 - 1} = \frac{24}{2} = 12$$

(b) The average rate of change of $f(x) = 3x^2$ from 1 to 5 is

$$\frac{\Delta y}{\Delta x} = \frac{f(5) - f(1)}{5 - 1} = \frac{75 - 3}{5 - 1} = \frac{72}{4} = 18$$

(c) The average rate of change of $f(x) = 3x^2$ from 1 to 7 is

$$\frac{\Delta y}{\Delta x} = \frac{f(7) - f(1)}{7 - 1} = \frac{147 - 3}{7 - 1} = \frac{144}{6} = 24$$ ■

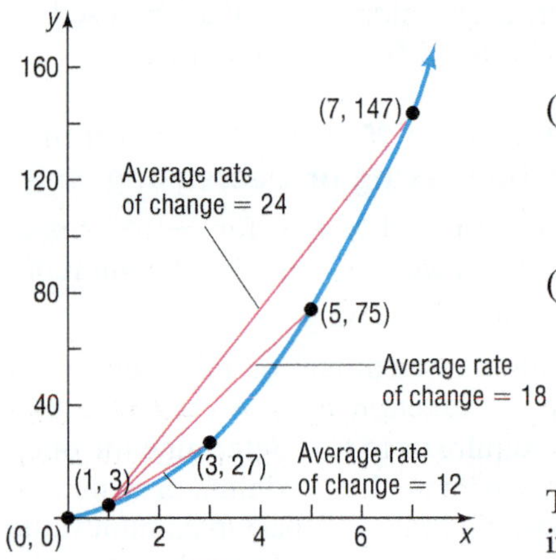

Figure 29 $f(x) = 3x^2$

See Figure 29 for a graph of $f(x) = 3x^2$. The function f is increasing for $x \ge 0$. The fact that the average rate of change is positive for any $x_1, x_2, x_1 \ne x_2$, in the interval $[1, 7]$ indicates that the graph is increasing on $1 \le x \le 7$. Further, the average rate of change is consistently getting larger for $1 \le x \le 7$, which indicates that the graph is increasing at an increasing rate.

Now Work PROBLEM 65

The Secant Line

The average rate of change of a function has an important geometric interpretation. Look at the graph of $y = f(x)$ in Figure 30. Two points are labeled on the graph: $(a, f(a))$ and $(b, f(b))$. The line containing these two points is called the **secant line**; its slope is

$$m_{\text{sec}} = \frac{f(b) - f(a)}{b - a} = \frac{f(a + h) - f(a)}{h}$$

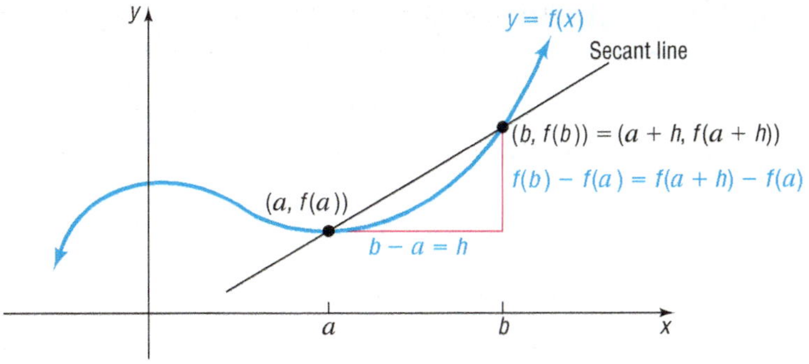

Figure 30 Secant line

THEOREM

Slope of the Secant Line

The average rate of change of a function from a to b equals the slope of the secant line containing the two points $(a, f(a))$ and $(b, f(b))$ on its graph. ■

EXAMPLE 8

Finding the Equation of a Secant Line

Suppose that $g(x) = 3x^2 - 2x + 3$.

(a) Find the average rate of change of g from -2 to 1.

(b) Find an equation of the secant line containing $(-2, g(-2))$ and $(1, g(1))$.

(c) Using a graphing utility, draw the graph of g and the secant line obtained in part (b) on the same screen.

Solution (a) The average rate of change of $g(x) = 3x^2 - 2x + 3$ from -2 to 1 is

$$\begin{aligned}\text{Average rate of change} &= \frac{g(1) - g(-2)}{1 - (-2)} \\ &= \frac{4 - 19}{3} \qquad g(1) = 3(1)^2 - 2(1) + 3 = 4 \\ & \qquad\qquad\quad g(-2) = 3(-2)^2 - 2(-2) + 3 = 19 \\ &= -\frac{15}{3} = -5\end{aligned}$$

(b) The slope of the secant line containing $(-2, g(-2)) = (-2, 19)$ and $(1, g(1)) = (1, 4)$ is $m_{\text{sec}} = -5$. Use the point–slope form to find an equation of the secant line.

$$\begin{aligned}y - y_1 &= m_{\text{sec}}(x - x_1) && \text{Point–slope form of the secant line} \\ y - 19 &= -5(x - (-2)) && x_1 = -2,\ y_1 = g(-2) = 19,\ m_{\text{sec}} = -5 \\ y - 19 &= -5x - 10 && \text{Distribute.} \\ y &= -5x + 9 && \text{Slope–intercept form of the secant line}\end{aligned}$$

(c) Figure 31 shows the graph of g along with the secant line $y = -5x + 9$. ■

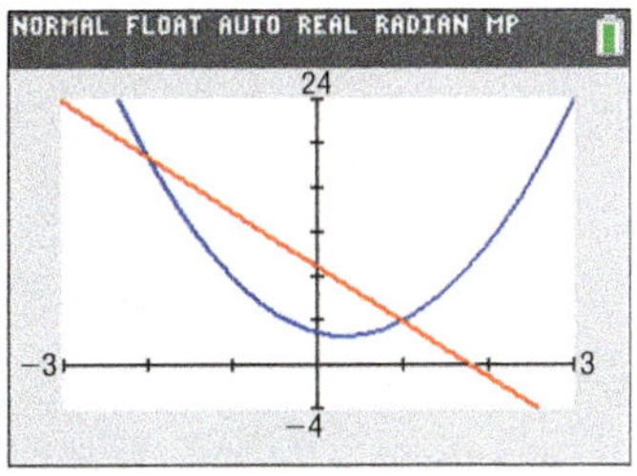

Figure 31 Graph of g and the secant line

Now Work PROBLEM 71

3.3 Assess Your Understanding

'Are You Prepared?' *Answers are given at the end of these exercises. If you get a wrong answer, read the pages listed in red.*

1. The interval $(2, 5)$ can be written as the inequality ______. (pp. 147–148)
2. The slope of the line containing the points $(-2, 3)$ and $(3, 8)$ is ___. (pp. 173–176)
3. Test the equation $y = 5x^2 - 1$ for symmetry with respect to the x-axis, the y-axis, and the origin. (pp. 166–168)
4. Write the point–slope form of the line with slope 5 containing the point $(3, -2)$. (p. 177)
5. The intercepts of the equation $y = x^2 - 9$ are ______. (pp. 165–166)

Concepts and Vocabulary

6. A function f is ________ on an interval I if, for any choice of x_1 and x_2 in I, with $x_1 < x_2$, we have $f(x_1) < f(x_2)$.
7. A(n) ____ function f is one for which $f(-x) = f(x)$ for every x in the domain of f; a(n) ____ function f is one for which $f(-x) = -f(x)$ for every x in the domain of f.
8. ***True or False*** A function f is decreasing on an interval I if, for any choice of x_1 and x_2 in I, with $x_1 < x_2$, we have $f(x_1) > f(x_2)$.
9. ***True or False*** A function f has a local maximum at c if there is an open interval I containing c such that for all x in I, $f(x) \le f(c)$.
10. ***True or False*** Even functions have graphs that are symmetric with respect to the origin.
11. An odd function is symmetric with respect to ______.
    (a) the x-axis (b) the y-axis
    (c) the origin (d) the line $y = x$
12. Which of the following intervals is required to guarantee a continuous function will have both an absolute maximum and an absolute minimum?
    (a) (a, b) (b) $(a, b]$
    (c) $[a, b)$ (d) $[a, b]$

Skill Building

In Problems 13–24, use the graph of the function f given.

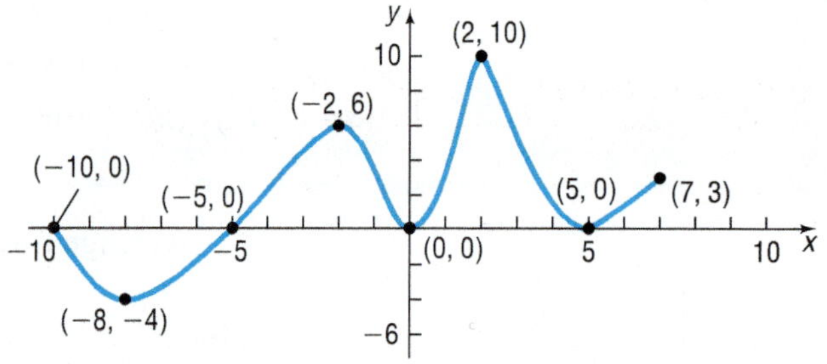

13. Is f increasing on the interval $[-8, -2]$?
14. Is f decreasing on the interval $[-8, -4]$?
15. Is f increasing on the interval $[-2, 6]$?
16. Is f decreasing on the interval $[2, 5]$?
17. List the interval(s) on which f is increasing.
18. List the interval(s) on which f is decreasing.
19. Is there a local maximum at 2? If yes, what is it?
20. Is there a local maximum at 5? If yes, what is it?
21. List the number(s) at which f has a local maximum. What are the local maximum values?
22. List the number(s) at which f has a local minimum. What are the local minimum values?
23. Find the absolute minimum of f on $[-10, 7]$.
24. Find the absolute maximum of f on $[-10, 7]$.

In Problems 25–32, the graph of a function is given. Use the graph to find:

(a) The intercepts, if any
(b) The domain and range
(c) The intervals on which the function is increasing, decreasing, or constant
(d) Whether the function is even, odd, or neither

25.
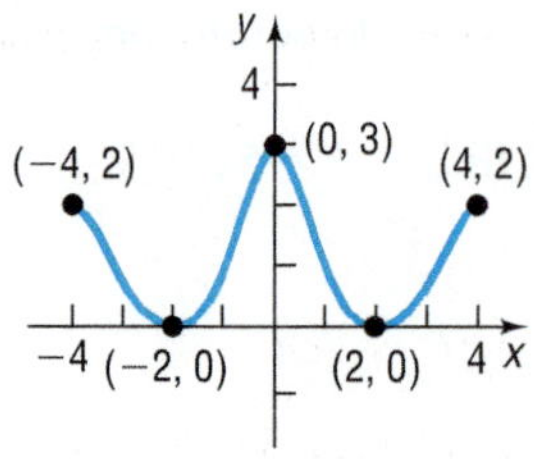

26.
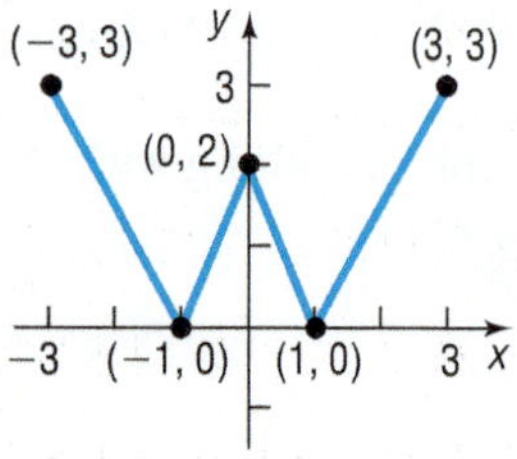

27.
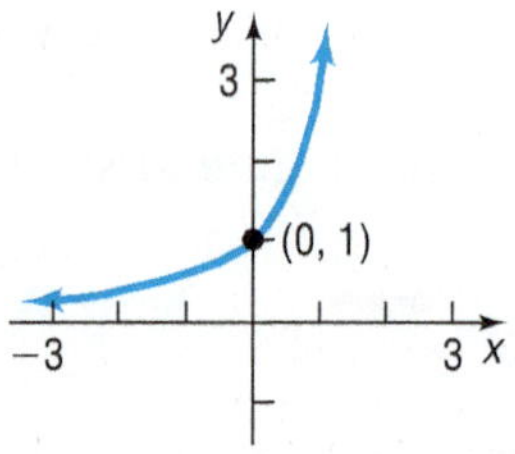

28.
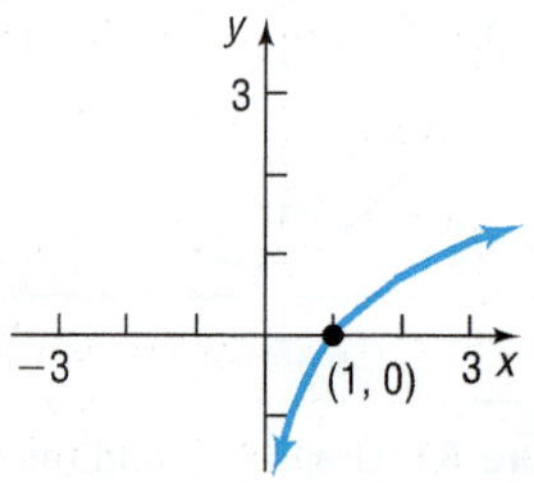

29.

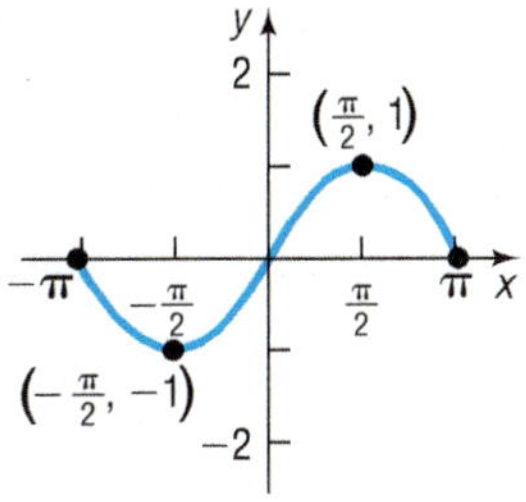

30.

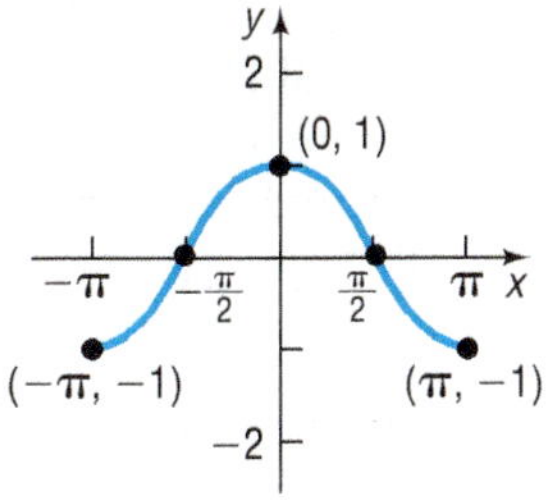

31.

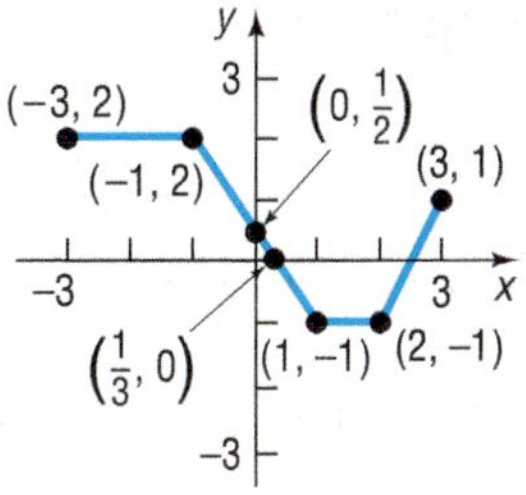

32.

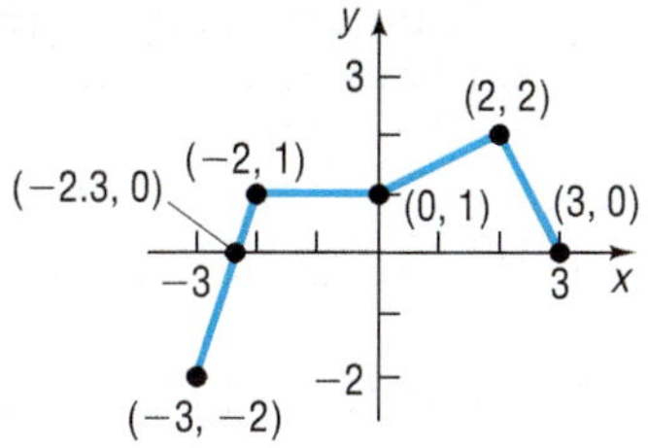

In Problems 33–36, the graph of a function f is given. Use the graph to find:

(a) The numbers, if any, at which f has a local maximum. What are the local maximum values?

(b) The numbers, if any, at which f has a local minimum. What are the local minimum values?

33.

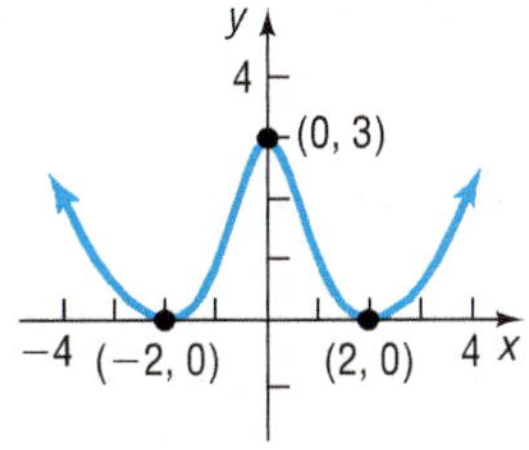

34.

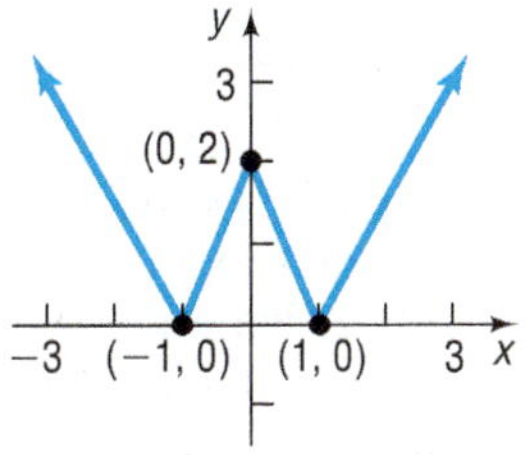

35.

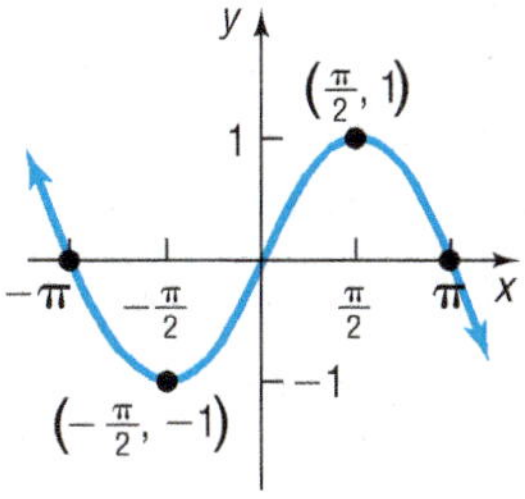

36.

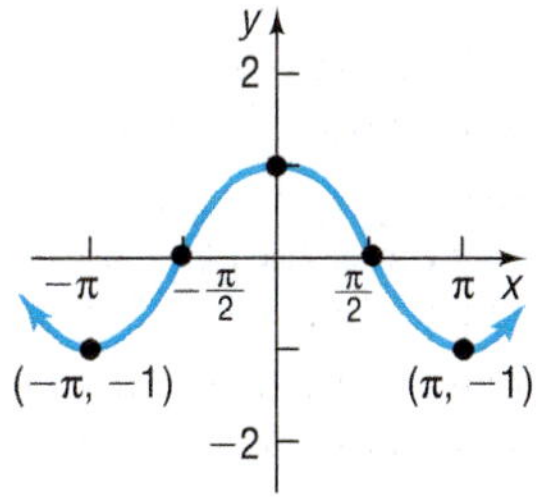

In Problems 37–48, determine algebraically whether each function is even, odd, or neither.

37. $f(x) = 4x^3$

38. $f(x) = 2x^4 - x^2$

39. $g(x) = -3x^2 - 5$

40. $h(x) = 3x^3 + 5$

41. $F(x) = \sqrt[3]{x}$

42. $G(x) = \sqrt{x}$

43. $f(x) = x + |x|$

44. $f(x) = \sqrt[3]{2x^2 + 1}$

45. $g(x) = \dfrac{1}{x^2}$

46. $h(x) = \dfrac{x}{x^2 - 1}$

47. $h(x) = \dfrac{-x^3}{3x^2 - 9}$

48. $F(x) = \dfrac{2x}{|x|}$

In Problems 49–56, for each graph of a function $y = f(x)$, find the absolute maximum and the absolute minimum, if they exist. Identify any local maximum values or local minimum values.

49.

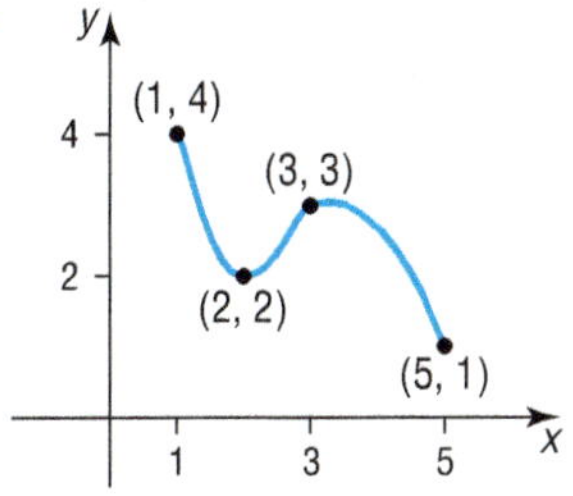

50.

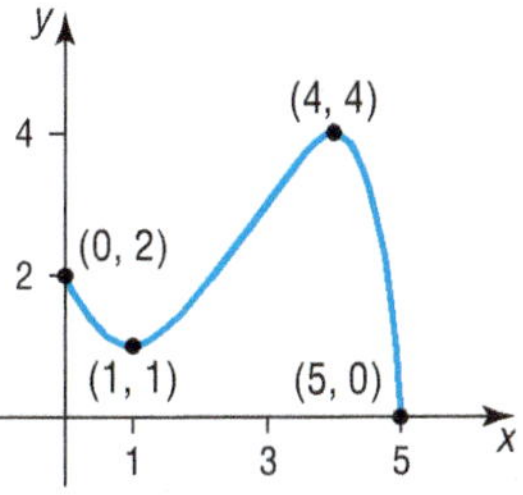

51.

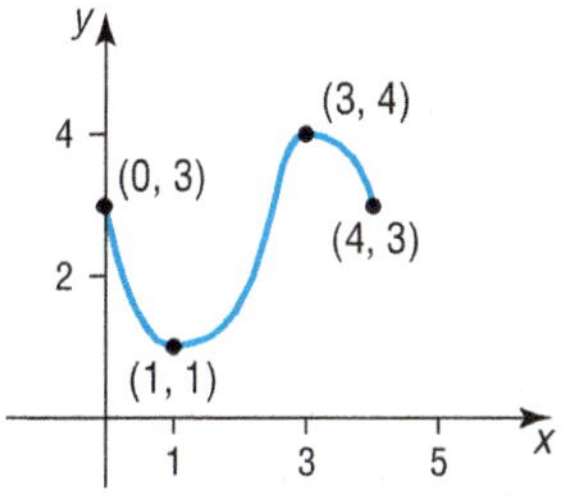

52.

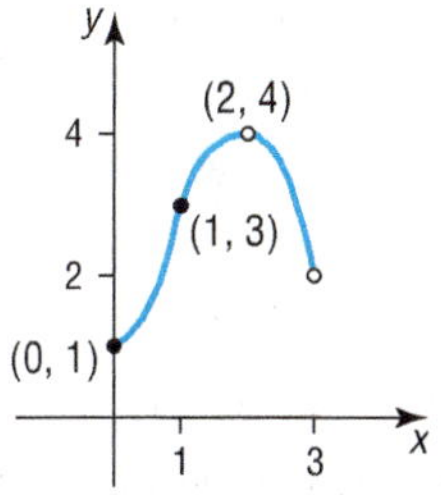

53.

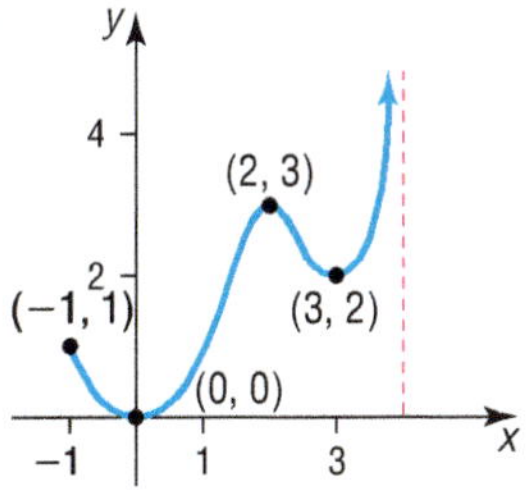

54.

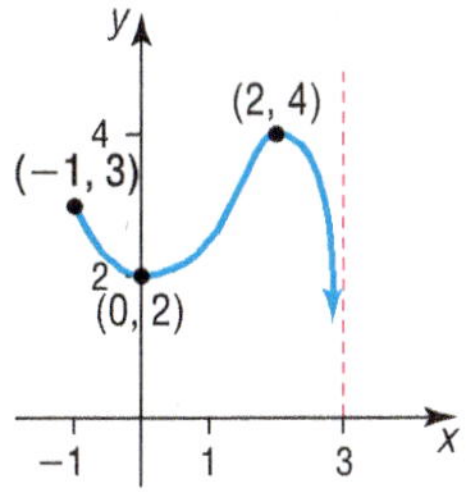

55.

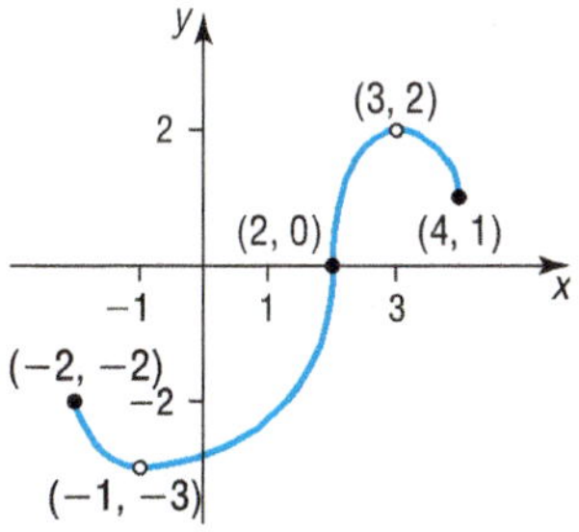

56.

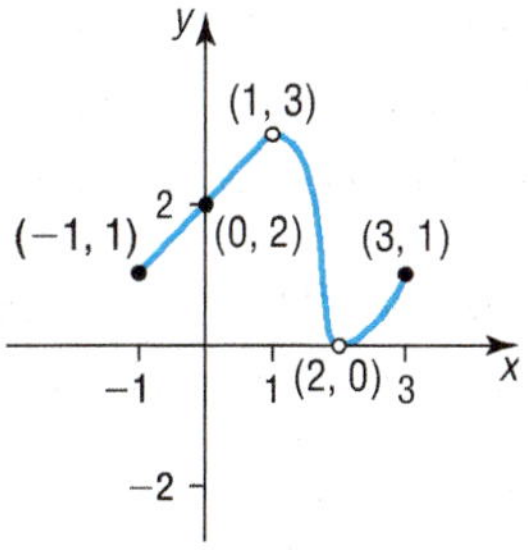

In Problems 57–64, use a graphing utility to graph each function over the indicated interval and approximate any local maximum values and local minimum values. Determine where the function is increasing and where it is decreasing. Round answers to two decimal places.

57. $f(x) = x^3 - 3x + 2 \quad [-2, 2]$

58. $f(x) = x^3 - 3x^2 + 5 \quad [-1, 3]$

59. $f(x) = x^5 - x^3 \quad [-2, 2]$

60. $f(x) = x^4 - x^2 \quad [-2, 2]$

61. $f(x) = -0.2x^3 - 0.6x^2 + 4x - 6 \quad [-6, 4]$

62. $f(x) = -0.4x^3 + 0.6x^2 + 3x - 2 \quad [-4, 5]$

63. $f(x) = 0.25x^4 + 0.3x^3 - 0.9x^2 + 3 \quad [-3, 2]$

64. $f(x) = -0.4x^4 - 0.5x^3 + 0.8x^2 - 2 \quad [-3, 2]$

65. Find the average rate of change of $f(x) = -2x^2 + 4$:
(a) From 0 to 2
(b) From 1 to 3
(c) From 1 to 4
(d) By hand, graph f and illustrate the average rate of change found in parts (a), (b), and (c).

66. Find the average rate of change of $f(x) = -x^3 + 1$:
(a) From 0 to 2
(b) From 1 to 3
(c) From -1 to 1
(d) By hand, graph f and illustrate the average rate of change found in parts (a), (b), and (c).

67. Find the average rate of change of $g(x) = x^3 - 2x + 1$:
(a) From -3 to -2
(b) From -1 to 1
(c) From 1 to 3

68. Find the average rate of change of $h(x) = x^2 - 2x + 3$:
(a) From -1 to 1
(b) From 0 to 2
(c) From 2 to 5

69. $f(x) = 5x - 2$
(a) Find the average rate of change from 1 to 3.
(b) Find an equation of the secant line containing $(1, f(1))$ and $(3, f(3))$.

70. $f(x) = -4x + 1$
(a) Find the average rate of change from 2 to 5.
(b) Find an equation of the secant line containing $(2, f(2))$ and $(5, f(5))$.

71. $g(x) = x^2 - 2$
(a) Find the average rate of change from -2 to 1.
(b) Find an equation of the secant line containing $(-2, g(-2))$ and $(1, g(1))$.
(c) Using a graphing utility, draw the graph of g and the secant line obtained in part (b) on the same screen.

72. $g(x) = x^2 + 1$
(a) Find the average rate of change from -1 to 2.
(b) Find an equation of the secant line containing $(-1, g(-1))$ and $(2, g(2))$.
(c) Using a graphing utility, draw the graph of g and the secant line obtained in part (b) on the same screen.

73. $h(x) = x^2 - 2x$
(a) Find the average rate of change from 2 to 4.
(b) Find an equation of the secant line containing $(2, h(2))$ and $(4, h(4))$.
(c) Using a graphing utility, draw the graph of h and the secant line obtained in part (b) on the same screen.

74. $h(x) = -2x^2 + x$
(a) Find the average rate of change from 0 to 3.
(b) Find an equation of the secant line containing $(0, h(0))$ and $(3, h(3))$.
(c) Using a graphing utility, draw the graph of h and the secant line obtained in part (b) on the same screen.

Mixed Practice

75. $g(x) = x^3 - 27x$
(a) Determine whether g is even, odd, or neither.
(b) There is a local minimum value of -54 at 3. Determine the local maximum value.

76. $f(x) = -x^3 + 12x$
(a) Determine whether f is even, odd, or neither.
(b) There is a local maximum value of 16 at 2. Determine the local minimum value.

77. $F(x) = -x^4 + 8x^2 + 9$
(a) Determine whether F is even, odd, or neither.
(b) There is a local maximum value of 25 at $x = 2$. Determine a second local maximum value.
(c) Suppose the area under the graph of F between $x = 0$ and $x = 3$ that is bounded from below by the x-axis is 50.4 square units. Using the result from part (a), determine the area under the graph of F between $x = -3$ and $x = 0$ that is bounded from below by the x-axis.

78. $G(x) = -x^4 + 32x^2 + 144$
(a) Determine whether G is even, odd, or neither.
(b) There is a local maximum value of 400 at $x = 4$. Determine a second local maximum value.
(c) Suppose the area under the graph of G between $x = 0$ and $x = 6$ that is bounded from below by the x-axis is 1612.8 square units. Using the result from part (a), determine the area under the graph of G between $x = -6$ and $x = 0$ that is bounded from below by the x-axis.

Applications and Extensions

79. **Minimum Average Cost** The average cost per hour in dollars, $\overline{C}$, of producing x riding lawn mowers can be modeled by the function

$$\overline{C}(x) = 0.3x^2 + 21x - 251 + \frac{2500}{x}$$

(a) Use a graphing utility to graph $\overline{C} = \overline{C}(x)$.
(b) Determine the number of riding lawn mowers to produce in order to minimize average cost.
(c) What is the minimum average cost?

80. **Medicine Concentration** The concentration C of a medication in the bloodstream t hours after being administered is modeled by the function

$$C(t) = -0.002x^4 + 0.039t^3 - 0.285t^2 + 0.766t + 0.085$$

(a) After how many hours will the concentration be highest?
(b) A woman nursing a child must wait until the concentration is below 0.5 before she can feed him. After taking the medication, how long must she wait before feeding her child?

81. **Data Plan Cost** The monthly cost C, in dollars, for wireless data plans with x gigabytes of data included is shown in the table on the top left column of the next page. Since each input value for x corresponds to exactly one output value for C, the plan cost is a function of the number of data gigabytes. Thus $C(x)$ represents the monthly cost for a wireless data plan with x gigabytes included.

GB	Cost ($)	GB	Cost ($)
4	70	20	150
6	80	30	225
10	100	40	300
15	130	50	375

(a) Plot the points $(4, 70)$, $(6, 80)$, $(10, 100)$, and so on in a Cartesian plane.
(b) Draw a line segment from the point $(10, 100)$ to $(30, 225)$. What does the slope of this line segment represent?
(c) Find the average rate of change of the monthly cost from 4 to 10 gigabytes.
(d) Find the average rate of change of the monthly cost from 10 to 30 gigabytes.
(e) Find the average rate of change of the monthly cost from 30 to 50 gigabytes.
(f) What is happening to the average rate of change as the gigabytes of data increase?

82. National Debt The size of the total debt owed by the United States federal government continues to grow. In fact, according to the Department of the Treasury, the debt per person living in the United States is approximately \$53,000 (or over \$140,000 per U.S. household). The following data represent the U.S. debt for the years 2001–2014. Since the debt D depends on the year y, and each input corresponds to exactly one output, the debt is a function of the year. So $D(y)$ represents the debt for each year y.

Year	Debt (billions of dollars)	Year	Debt (billions of dollars)
2001	5807	2008	10,025
2002	6228	2009	11,910
2003	6783	2010	13,562
2004	7379	2011	14,790
2005	7933	2012	16,066
2006	8507	2013	16,738
2007	9008	2014	17,824

Source: *www.treasurydirect.gov*

(a) Plot the points $(2001, 5807)$, $(2002, 6228)$, and so on in a Cartesian plane.
(b) Draw a line segment from the point $(2001, 5807)$ to $(2006, 8507)$. What does the slope of this line segment represent?
(c) Find the average rate of change of the debt from 2002 to 2004.
(d) Find the average rate of change of the debt from 2006 to 2008.
(e) Find the average rate of change of the debt from 2010 to 2012.
(f) What appears to be happening to the average rate of change as time passes?

83. *E. coli* Growth A strain of *E. coli* Beu 397-recA441 is placed into a nutrient broth at 30° Celsius and allowed to grow. The data shown in the table are collected. The population is measured in grams and the time in hours. Since population P depends on time t, and each input corresponds to exactly one output, we can say that population is a function of time. Thus $P(t)$ represents the population at time t.

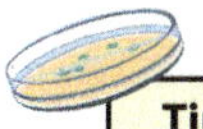

Time (hours), t	Population (grams), P
0	0.09
2.5	0.18
3.5	0.26
4.5	0.35
6	0.50

(a) Find the average rate of change of the population from 0 to 2.5 hours.
(b) Find the average rate of change of the population from 4.5 to 6 hours.
(c) What is happening to the average rate of change as time passes?

84. e-Filing Tax Returns The Internal Revenue Service Restructuring and Reform Act (RRA) was signed into law by President Bill Clinton in 1998. A major objective of the RRA was to promote electronic filing of tax returns. The data in the table that follows show the percentage of individual income tax returns filed electronically for filing years 2004–2013. Since the percentage P of returns filed electronically depends on the filing year y, and each input corresponds to exactly one output, the percentage of returns filed electronically is a function of the filing year; so $P(y)$ represents the percentage of returns filed electronically for filing year y.
(a) Find the average rate of change of the percentage of e-filed returns from 2004 to 2006.
(b) Find the average rate of change of the percentage of e-filed returns from 2007 to 2009.
(c) Find the average rate of change of the percentage of e-filed returns from 2010 to 2012.
(d) What is happening to the average rate of change as time passes?

Year	Percentage of returns e-filed
2004	46.5
2005	51.1
2006	53.8
2007	57.1
2008	58.5
2009	67.2
2010	69.8
2011	77.2
2012	82.7
2013	84.7

Source: *Internal Revenue Service*

85. For the function $f(x) = x^2$, compute the average rate of change:
(a) From 0 to 1
(b) From 0 to 0.5
(c) From 0 to 0.1
(d) From 0 to 0.01
(e) From 0 to 0.001
(f) Use a graphing utility to graph each of the secant lines along with f.
(g) What do you think is happening to the secant lines?
(h) What is happening to the slopes of the secant lines? Is there some number that they are getting closer to? What is that number?

86. For the function $f(x) = x^2$, compute the average rate of change:
(a) From 1 to 2
(b) From 1 to 1.5
(c) From 1 to 1.1
(d) From 1 to 1.01
(e) From 1 to 1.001
(f) Use a graphing utility to graph each of the secant lines along with f.
(g) What do you think is happening to the secant lines?
(h) What is happening to the slopes of the secant lines? Is there some number that they are getting closer to? What is that number?

Problems 87–94 require the following discussion of a secant line. The slope of the secant line containing the two points $(x, f(x))$ and $(x + h, f(x + h))$ on the graph of a function $y = f(x)$ may be given as

$$m_{sec} = \frac{f(x+h) - f(x)}{(x+h) - x} = \frac{f(x+h) - f(x)}{h}, \qquad h \neq 0$$

In calculus, this expression is called the **difference quotient of *f*.**
(a) Express the slope of the secant line of each function in terms of x and h. Be sure to simplify your answer.
(b) Find m_{sec} for $h = 0.5, 0.1$, and 0.01 at $x = 1$. What value does m_{sec} approach as h approaches 0?
(c) Find an equation for the secant line at $x = 1$ with $h = 0.01$.
(d) Use a graphing utility to graph f and the secant line found in part (c) in the same viewing window.

87. $f(x) = 2x + 5$

88. $f(x) = -3x + 2$

89. $f(x) = x^2 + 2x$

90. $f(x) = 2x^2 + x$

91. $f(x) = 2x^2 - 3x + 1$

92. $f(x) = -x^2 + 3x - 2$

93. $f(x) = \dfrac{1}{x}$

94. $f(x) = \dfrac{1}{x^2}$

Explaining Concepts: Discussion and Writing

95. Draw the graph of a function that has the following properties: domain: all real numbers; range: all real numbers; intercepts: $(0, -3)$ and $(3, 0)$; a local maximum value of -2 is at -1; a local minimum value of -6 is at 2. Compare your graph with those of others. Comment on any differences.

96. Redo Problem 95 with the following additional information: increasing on $(-\infty, -1], [2, \infty)$; decreasing on $[-1, 2]$. Again compare your graph with others and comment on any differences.

97. How many x-intercepts can a function defined on an interval have if it is increasing on that interval? Explain.

98. Suppose that a friend of yours does not understand the idea of increasing and decreasing functions. Provide an explanation, complete with graphs, that clarifies the idea.

99. Can a function be both even and odd? Explain.

100. Using a graphing utility, graph $y = 5$ on the interval $(-3, 3)$. Use MAXIMUM to find the local maximum values on $(-3, 3)$. Comment on the result provided by the calculator.

101. A function f has a positive average rate of change on the interval $[2, 5]$. Is f increasing on $[2, 5]$? Explain.

102. Show that a constant function $f(x) = b$ has an average rate of change of 0. Compute the average rate of change of $y = \sqrt{4 - x^2}$ on the interval $[-2, 2]$. Explain how this can happen.

Retain Your Knowledge

Problems 103–106 are based on material learned earlier in the course. The purpose of these problems is to keep the material fresh in your mind so that you are better prepared for the final exam.

103. Write each number in scientific notation.
(a) 0.00000701
(b) 2,305,000,000

104. Simplify: $\sqrt{540}$

105. Solve: $14 < 5 - 3x \leq 29$

106. The shelf life of a perishable commodity varies inversely with the storage temperature. If the shelf life at 10°C is 33 days, what is the shelf life at 40°C?

'Are You Prepared?' Answers

1. $2 < x < 5$ **2.** 1 **3.** symmetric with respect to the y-axis **4.** $y + 2 = 5(x - 3)$ **5.** $(-3, 0), (3, 0), (0, -9)$

3.4 Library of Functions; Piecewise-defined Functions

PREPARING FOR THIS SECTION *Before getting started, review the following:*

- Intercepts (Section 2.1, pp. 165–166)
- Graphs of Key Equations (Section 2.1: Example 4, p. 168; Example 5, p. 169; Example 6, pp. 169–170)

Now Work the **'Are You Prepared?'** problems on page 252.

OBJECTIVES
1 Graph the Functions Listed in the Library of Functions (p. 245)
2 Graph Piecewise-defined Functions (p. 250)

1 Graph the Functions Listed in the Library of Functions

First we introduce a few more functions, beginning with the *square root function*.

On page 169, we graphed the equation $y = \sqrt{x}$. Figure 32 shows a graph of the function $f(x) = \sqrt{x}$. Based on the graph, we have the following properties:

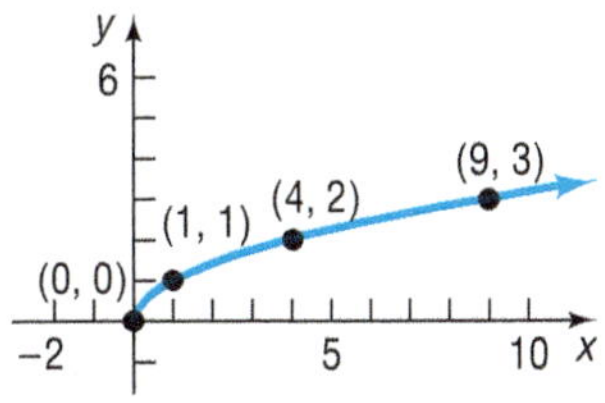

Figure 32 Square root function

Properties of $f(x) = \sqrt{x}$

1. The domain and the range are the set of nonnegative real numbers.
2. The x-intercept of the graph of $f(x) = \sqrt{x}$ is 0. The y-intercept of the graph of $f(x) = \sqrt{x}$ is also 0.
3. The function is neither even nor odd.
4. The function is increasing on the interval $[0, \infty)$.
5. The function has an absolute minimum of 0 at $x = 0$.

EXAMPLE 1 **Graphing the Cube Root Function**

(a) Determine whether $f(x) = \sqrt[3]{x}$ is even, odd, or neither. State whether the graph of f is symmetric with respect to the y-axis or symmetric with respect to the origin.

(b) Determine the intercepts, if any, of the graph of $f(x) = \sqrt[3]{x}$.

(c) Graph $f(x) = \sqrt[3]{x}$.

Solution (a) Because

$$f(-x) = \sqrt[3]{-x} = -\sqrt[3]{x} = -f(x)$$

the function is odd. The graph of f is symmetric with respect to the origin.

(b) The y-intercept is $f(0) = \sqrt[3]{0} = 0$. The x-intercept is found by solving the equation $f(x) = 0$.

$$f(x) = 0$$

$$\sqrt[3]{x} = 0 \quad f(x) = \sqrt[3]{x}$$

$$x = 0 \quad \text{Cube both sides of the equation.}$$

The x-intercept is also 0.

(c) Use the function to form Table 4 (on page 246) and obtain some points on the graph. Because of the symmetry with respect to the origin, we find only points (x, y) for which $x \geq 0$. Figure 33 shows the graph of $f(x) = \sqrt[3]{x}$.

Table 4

x	$y = f(x) = \sqrt[3]{x}$	(x, y)
0	0	(0, 0)
$\frac{1}{8}$	$\frac{1}{2}$	$\left(\frac{1}{8}, \frac{1}{2}\right)$
1	1	(1, 1)
2	$\sqrt[3]{2} \approx 1.26$	$(2, \sqrt[3]{2})$
8	2	(8, 2)

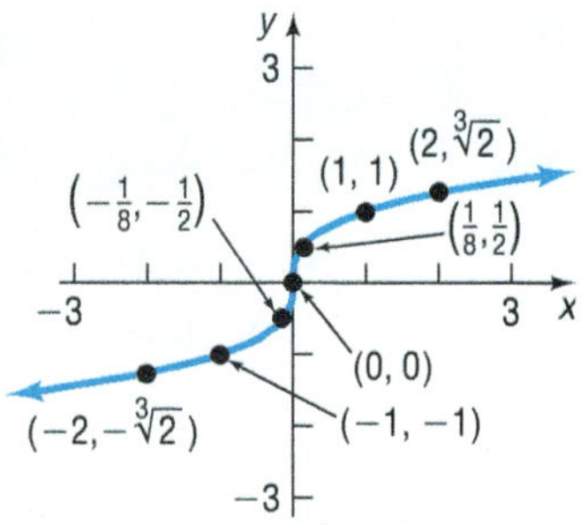

Figure 33 Cube root function

From the results of Example 1 and Figure 33, we have the following properties of the cube root function.

Properties of $f(x) = \sqrt[3]{x}$

1. The domain and the range are the set of all real numbers.
2. The x-intercept of the graph of $f(x) = \sqrt[3]{x}$ is 0. The y-intercept of the graph of $f(x) = \sqrt[3]{x}$ is also 0.
3. The function is odd. The graph is symmetric with respect to the origin.
4. The function is increasing on the interval $(-\infty, \infty)$.
5. The function does not have any local minima or any local maxima.

EXAMPLE 2 Graphing the Absolute Value Function

(a) Determine whether $f(x) = |x|$ is even, odd, or neither. State whether the graph of f is symmetric with respect to the y-axis, symmetric with respect to the origin, or neither.

(b) Determine the intercepts, if any, of the graph of $f(x) = |x|$.

(c) Graph $f(x) = |x|$.

Solution (a) Because

$$\begin{aligned} f(-x) &= |-x| \\ &= |x| = f(x) \end{aligned}$$

the function is even. The graph of f is symmetric with respect to the y-axis.

(b) The y-intercept is $f(0) = |0| = 0$. The x-intercept is found by solving the equation $f(x) = 0$, or $|x| = 0$. The x-intercept is 0.

(c) Use the function to form Table 5 and obtain some points on the graph. Because of the symmetry with respect to the y-axis, we only need to find points (x, y) for which $x \geq 0$. Figure 34 shows the graph of $f(x) = |x|$.

Table 5

x	$y = f(x) = \|x\|$	(x, y)
0	0	(0, 0)
1	1	(1, 1)
2	2	(2, 2)
3	3	(3, 3)

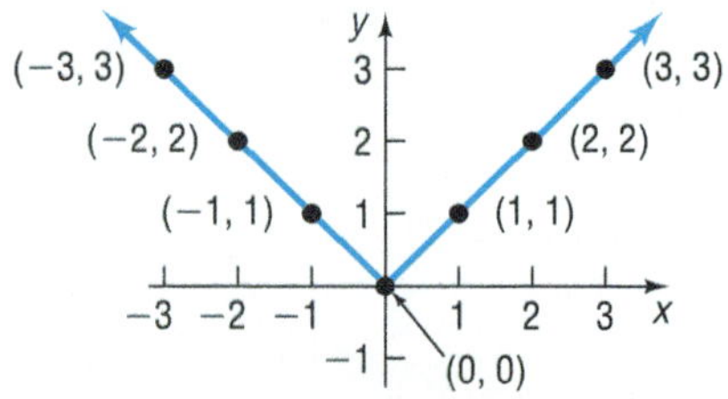

Figure 34 Absolute value function

From the results of Example 2 and Figure 34, we have the following properties of the absolute value function.

Properties of $f(x) = |x|$

1. The domain is the set of all real numbers. The range of f is $\{y | y \geq 0\}$.
2. The x-intercept of the graph of $f(x) = |x|$ is 0. The y-intercept of the graph of $f(x) = |x|$ is also 0.
3. The function is even. The graph is symmetric with respect to the y-axis.
4. The function is decreasing on the interval $(-\infty, 0]$. It is increasing on the interval $[0, \infty)$.
5. The function has an absolute minimum of 0 at $x = 0$.

Seeing the Concept

Graph $y = |x|$ on a square screen and compare what you see with Figure 34. Note that some graphing calculators use abs(x) for absolute value. ■

Below is a list of the key functions that we have discussed. In going through this list, pay special attention to the properties of each function, particularly to the shape of each graph. Knowing these graphs, along with key points on each graph, will lay the foundation for further graphing techniques.

Constant Function

$$f(x) = b \qquad b \text{ is a real number}$$

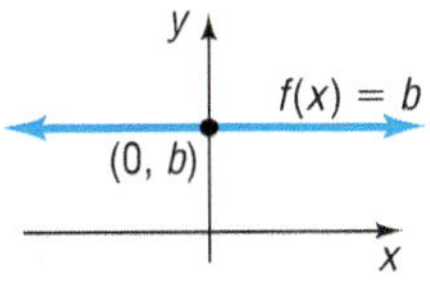

Figure 35 Constant function

See Figure 35.

The domain of a **constant function** is the set of all real numbers; its range is the set consisting of a single number b. Its graph is a horizontal line whose y-intercept is b. The constant function is an even function.

Identity Function

$$f(x) = x$$

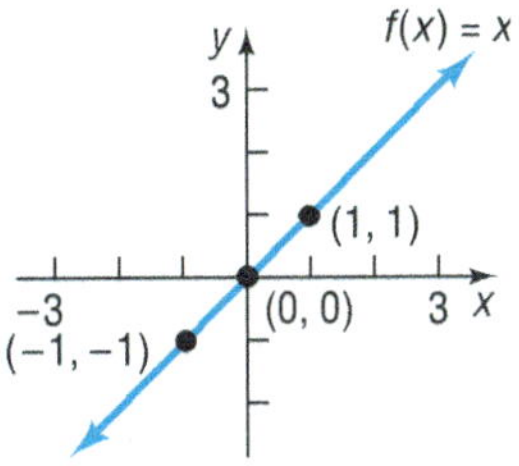

Figure 36 Identity function

See Figure 36.

The domain and the range of the **identity function** are the set of all real numbers. Its graph is a line whose slope is 1 and whose y-intercept is 0. The line consists of all points for which the x-coordinate equals the y-coordinate. The identity function is an odd function that is increasing over its domain. Note that the graph bisects quadrants I and III.

Square Function

$$f(x) = x^2$$

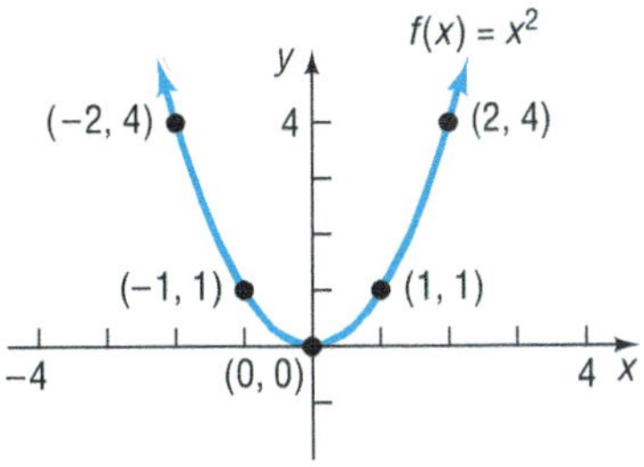

Figure 37 Square function

See Figure 37.

The domain of the **square function** is the set of all real numbers; its range is the set of nonnegative real numbers. The graph of this function is a parabola whose intercept is at $(0, 0)$. The square function is an even function that is decreasing on the interval $(-\infty, 0]$ and increasing on the interval $[0, \infty)$.

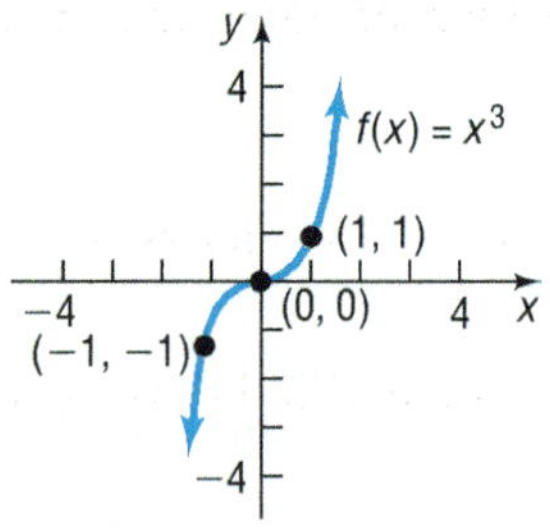

Figure 38 Cube function

Cube Function

$$f(x) = x^3$$

See Figure 38.

The domain and the range of the **cube function** are the set of all real numbers. The intercept of the graph is at $(0, 0)$. The cube function is odd and is increasing on the interval $(-\infty, \infty)$.

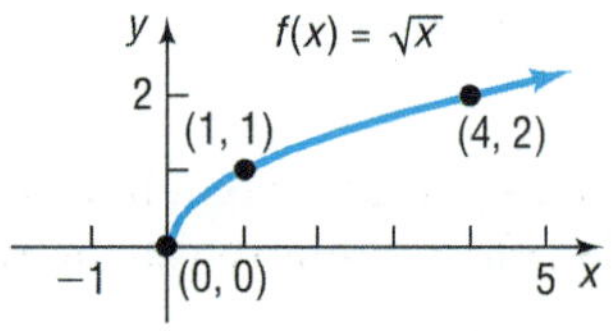

Figure 39 Square root function

Square Root Function

$$f(x) = \sqrt{x}$$

See Figure 39.

The domain and the range of the **square root function** are the set of nonnegative real numbers. The intercept of the graph is at $(0, 0)$. The square root function is neither even nor odd and is increasing on the interval $[0, \infty)$.

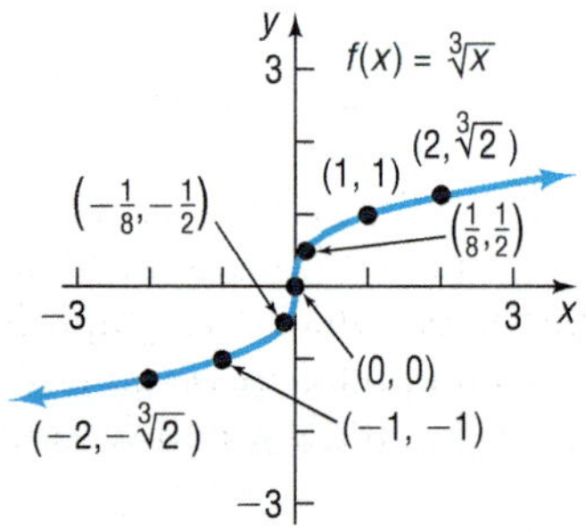

Figure 40 Cube root function

Cube Root Function

$$f(x) = \sqrt[3]{x}$$

See Figure 40.

The domain and the range of the **cube root function** are the set of all real numbers. The intercept of the graph is at $(0, 0)$. The cube root function is an odd function that is increasing on the interval $(-\infty, \infty)$.

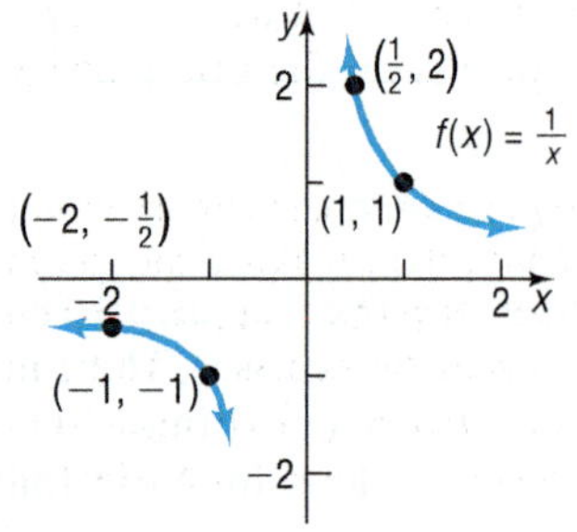

Figure 41 Reciprocal function

Reciprocal Function

$$f(x) = \frac{1}{x}$$

Refer to Example 6, page 169, for a discussion of the equation $y = \frac{1}{x}$. See Figure 41.

The domain and the range of the **reciprocal function** are the set of all nonzero real numbers. The graph has no intercepts. The reciprocal function is decreasing on the intervals $(-\infty, 0)$ and $(0, \infty)$ and is an odd function.

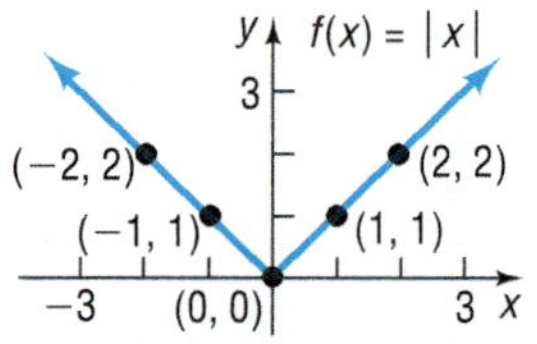

Figure 42 Absolute value function

Absolute Value Function

$$f(x) = |x|$$

See Figure 42.

The domain of the **absolute value function** is the set of all real numbers; its range is the set of nonnegative real numbers. The intercept of the graph is at $(0, 0)$. If $x \ge 0$, then $f(x) = x$, and the graph of f is part of the line $y = x$; if $x < 0$, then $f(x) = -x$, and the graph of f is part of the line $y = -x$. The absolute value function is an even function; it is decreasing on the interval $(-\infty, 0]$ and increasing on the interval $[0, \infty)$.

The notation $\text{int}(x)$ stands for the largest integer less than or equal to x. For example,

$$\text{int}(1) = 1, \quad \text{int}(2.5) = 2, \quad \text{int}\left(\frac{1}{2}\right) = 0, \quad \text{int}\left(-\frac{3}{4}\right) = -1, \quad \text{int}(\pi) = 3$$

Table 6

x	$y = f(x) = \text{int}(x)$	(x, y)
-1	-1	$(-1, -1)$
$-\frac{1}{2}$	-1	$\left(-\frac{1}{2}, -1\right)$
$-\frac{1}{4}$	-1	$\left(-\frac{1}{4}, -1\right)$
0	0	$(0, 0)$
$\frac{1}{4}$	0	$\left(\frac{1}{4}, 0\right)$
$\frac{1}{2}$	0	$\left(\frac{1}{2}, 0\right)$
$\frac{3}{4}$	0	$\left(\frac{3}{4}, 0\right)$

This type of correspondence occurs frequently enough in mathematics that we give it a name.

DEFINITION Greatest Integer Function

$$f(x) = \text{int}(x)^* = \text{greatest integer less than or equal to } x$$

We obtain the graph of $f(x) = \text{int}(x)$ by plotting several points. See Table 6. For values of x, $-1 \le x < 0$, the value of $f(x) = \text{int}(x)$ is -1; for values of x, $0 \le x < 1$, the value of f is 0. See Figure 43 for the graph.

The domain of the **greatest integer function** is the set of all real numbers; its range is the set of integers. The y-intercept of the graph is 0. The x-intercepts lie in the interval $[0, 1)$. The greatest integer function is neither even nor odd. It is constant on every interval of the form $[k, k + 1)$, for k an integer. In Figure 43, a solid dot is used to indicate, for example, that at $x = 1$ the value of f is $f(1) = 1$; an open circle is used to illustrate that the function does not assume the value of 0 at $x = 1$.

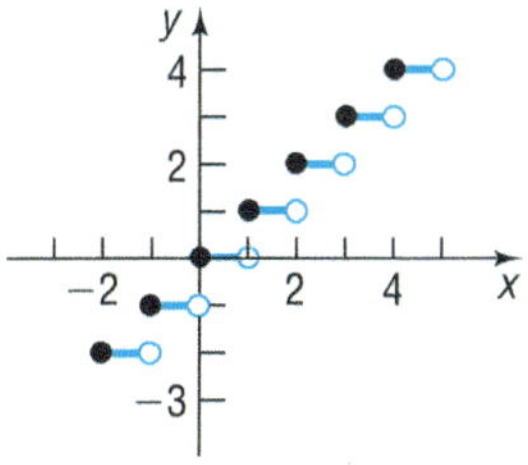

Figure 43 Greatest integer function

Although a precise definition requires the idea of a limit (discussed in calculus), in a rough sense, a function is said to be *continuous* if its graph has no gaps or holes and can be drawn without lifting a pencil from the paper on which the graph is drawn. We contrast this with a *discontinuous* function. A function is discontinuous if its graph has gaps or holes and so cannot be drawn without lifting a pencil from the paper.

From the graph of the greatest integer function, we can see why it is also called a **step function**. At $x = 0$, $x = \pm 1$, $x = \pm 2$, and so on, this function is discontinuous because, at integer values, the graph suddenly "steps" from one value to another without taking on any of the intermediate values. For example, to the immediate left of $x = 3$, the y-coordinates of the points on the graph are 2, and at $x = 3$ and to the immediate right of $x = 3$, the y-coordinates of the points on the graph are 3. Consequently, the graph has gaps in it.

COMMENT When graphing a function using a graphing utility, typically you can choose either **connected mode**, in which points plotted on the screen are connected, making the graph appear without any breaks, or **dot mode**, in which only the points plotted appear. When graphing the greatest integer function with a graphing utility, it may be necessary to be in **dot mode**. This is to prevent the utility from "connecting the dots" when $f(x)$ changes from one integer value to the next. However, some utilities will display the gaps even when in "connected" mode. See Figure 44.

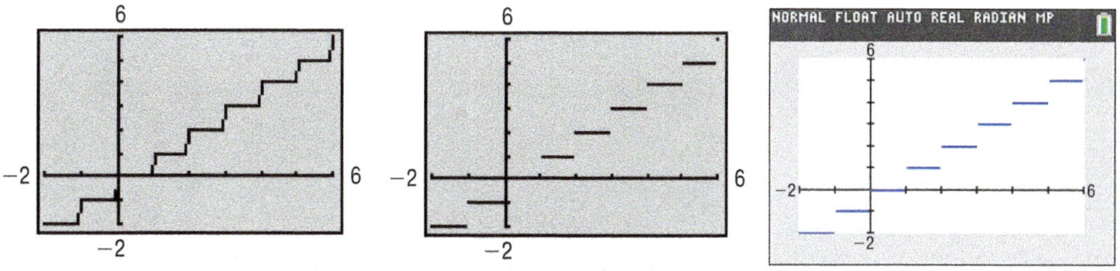

Figure 44 $f(x) = \text{int}(x)$ **(a)** TI-83 Plus, connected mode **(b)** TI-83 Plus, dot mode **(c)** TI-84 Plus C

* Some texts use the notation $f(x) = [x]$ instead of $\text{int}(x)$.

The functions discussed so far are basic. Whenever you encounter one of them, you should see a mental picture of its graph. For example, if you encounter the function $f(x) = x^2$, you should see in your mind's eye a picture like Figure 37.

Now Work PROBLEMS 11 THROUGH 18

2 Graph Piecewise-defined Functions

Sometimes a function is defined using different equations on different parts of its domain. For example, the absolute value function $f(x) = |x|$ is actually defined by two equations: $f(x) = x$ if $x \geq 0$ and $f(x) = -x$ if $x < 0$. For convenience, these equations are generally combined into one expression as

$$f(x) = |x| = \begin{cases} x & \text{if } x \geq 0 \\ -x & \text{if } x < 0 \end{cases}$$

When a function is defined by different equations on different parts of its domain, it is called a **piecewise-defined** function.

EXAMPLE 3 Graphing a Piecewise-defined Function

The function f is defined as

$$f(x) = \begin{cases} 2x + 3 & \text{if } x < 0 \\ -\frac{1}{2}x + 3 & \text{if } x \geq 0 \end{cases}$$

(a) Find $f(-2)$, $f(0)$, and $f(2)$.
(b) Graph f.

Solution (a) To find $f(-2)$, observe that when $x = -2$, the equation for f is given by $f(x) = 2x + 3$, so

$$f(-2) = 2(-2) + 3 = -1$$

When $x = 0$, the equation for f is $f(x) = -\frac{1}{2}x + 3$, so

$$f(0) = -\frac{1}{2}(0) + 3 = 3$$

When $x = 2$, the equation for f is $f(x) = -\frac{1}{2}x + 3$, so

$$f(2) = -\frac{1}{2}(2) + 3 = 2$$

(b) To graph f, graph each "piece." First graph the line $y = 2x + 3$ and keep only the part for which $x < 0$. Next, graph the line $y = -\frac{1}{2}x + 3$ and keep only the part for which $x \geq 0$. See Figure 45. ■

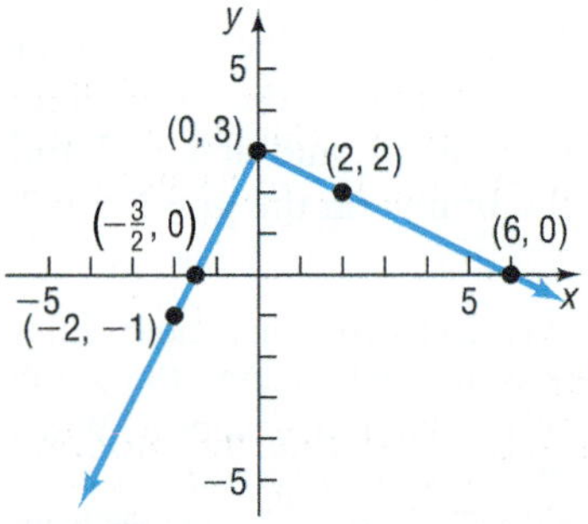

Figure 45

EXAMPLE 4 Analyzing a Piecewise-defined Function

The function f is defined as

$$f(x) = \begin{cases} -2x + 1 & \text{if } -3 \leq x < 1 \\ 2 & \text{if } x = 1 \\ x^2 & \text{if } x > 1 \end{cases}$$

(a) Determine the domain of f.
(b) Locate any intercepts.
(c) Graph f.
(d) Use the graph to find the range of f.

Solution

(a) To find the domain of f, look at its definition. Since f is defined for all x greater than or equal to -3, the domain of f is $\{x \mid x \geq -3\}$, or the interval $[-3, \infty)$.

(b) The y-intercept of the graph of the function is $f(0)$. Because the equation for f when $x = 0$ is $f(x) = -2x + 1$, the y-intercept is $f(0) = -2(0) + 1 = 1$. The x-intercepts of the graph of a function f are the real solutions to the equation $f(x) = 0$. To find the x-intercepts of f, solve $f(x) = 0$ for each "piece" of the function, and then determine what values of x, if any, satisfy the condition that defines the piece.

$$\begin{aligned} f(x) &= 0 \\ -2x + 1 &= 0 \quad -3 \leq x < 1 \\ -2x &= -1 \\ x &= \frac{1}{2} \end{aligned} \qquad \begin{aligned} f(x) &= 0 \\ 2 &= 0 \quad x = 1 \\ &\text{No solution} \end{aligned} \qquad \begin{aligned} f(x) &= 0 \\ x^2 &= 0 \quad x > 1 \\ x &= 0 \end{aligned}$$

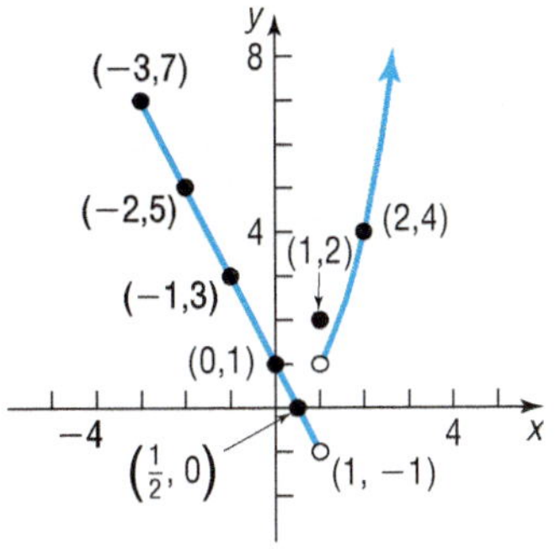

Figure 46

The first potential x-intercept, $x = \frac{1}{2}$, satisfies the condition $-3 \leq x < 1$, so $x = \frac{1}{2}$ is an x-intercept. The second potential x-intercept, $x = 0$, does not satisfy the condition $x > 1$, so $x = 0$ is not an x-intercept. The only x-intercept is $\frac{1}{2}$. The intercepts are $(0, 1)$ and $\left(\frac{1}{2}, 0\right)$.

(c) To graph f, graph each "piece." First graph the line $y = -2x + 1$ and keep only the part for which $-3 \leq x < 1$. Then plot the point $(1, 2)$ because, when $x = 1$, $f(x) = 2$. Finally, graph the parabola $y = x^2$ and keep only the part for which $x > 1$. See Figure 46.

(d) From the graph, we conclude that the range of f is $\{y \mid y > -1\}$, or the interval $(-1, \infty)$. ■

Now Work PROBLEM 31

EXAMPLE 5 Cost of Electricity

In the spring of 2015, Duke Energy Progress supplied electricity to residences in South Carolina for a monthly customer charge of \$6.50 plus 9.997¢ per kilowatt-hour (kWh) for the first 800 kWh supplied in the month and 8.997¢ per kWh for all usage over 800 kWh in the month.

(a) What is the charge for using 300 kWh in a month?
(b) What is the charge for using 1500 kWh in a month?
(c) If C is the monthly charge for x kWh, develop a model relating the monthly charge and kilowatt-hours used. That is, express C as a function of x.

Source: *Duke Energy Progress, 2015*

Solution

(a) For 300 kWh, the charge is \$6.50 plus (9.997¢ = \$0.09997) per kWh. That is,

$$\text{Charge} = \$6.50 + \$0.09997(300) = \$36.49$$

(b) For 1500 kWh, the charge is \$6.50 plus 9.997¢ per kWh for the first 800 kWh plus 8.997¢ per kWh for the 700 in excess of 800. That is,

$$\text{Charge} = \$6.50 + \$0.09997(800) + \$0.08997(700) = \$149.46$$

(c) Let x represent the number of kilowatt-hours used. If $0 \leq x \leq 800$, then the monthly charge C (in dollars) can be found by multiplying x times \$0.09997 and adding the monthly customer charge of \$6.50. So if $0 \leq x \leq 800$, then

$$C(x) = 0.09997x + 6.50$$

For $x > 800$, the charge is $0.09997(800) + 6.50 + 0.08997(x - 800)$, since $(x - 800)$ equals the usage in excess of 800 kWh, which costs \$0.08997 per kWh. That is, if $x > 800$, then

$$\begin{aligned} C(x) &= 0.09997(800) + 6.50 + 0.08997(x - 800) \\ &= 79.976 + 6.50 + 0.08997x - 71.976 \\ &= 0.08997x + 14.50 \end{aligned}$$

The rule for computing C follows two equations:

$$C(x) = \begin{cases} 0.09997x + 6.50 & \text{if } 0 \le x \le 800 \\ 0.08997x + 14.50 & \text{if } x > 800 \end{cases}$$ **The Model**

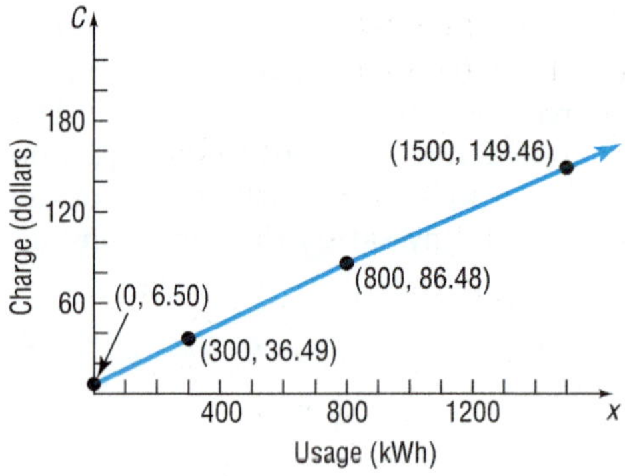

Figure 47

See Figure 47 for the graph. Note that the two "pieces" are linear, but they have different slopes (rates), and meet at the point $(800, 86.48)$.

3.4 Assess Your Understanding

'Are You Prepared?' *Answers are given at the end of these exercises. If you get a wrong answer, read the pages listed in red.*

1. Sketch the graph of $y = \sqrt{x}$. (p. 169)
2. Sketch the graph of $y = \dfrac{1}{x}$. (p. 169)
3. List the intercepts of the equation $y = x^3 - 8$. (pp. 165–166)

Concepts and Vocabulary

4. The function $f(x) = x^2$ is decreasing on the interval ________.
5. When functions are defined by more than one equation, they are called ______________ functions.
6. ***True or False*** The cube function is odd and is increasing on the interval $(-\infty, \infty)$.
7. ***True or False*** The cube root function is odd and is decreasing on the interval $(-\infty, \infty)$.
8. ***True or False*** The domain and the range of the reciprocal function are the set of all real numbers.
9. Which of the following functions has a graph that is symmetric about the y-axis?
 (a) $y = \sqrt{x}$ (b) $y = |x|$ (c) $y = x^3$ (d) $y = \dfrac{1}{x}$
10. Consider the following function.

$$f(x) = \begin{cases} 3x - 2 & \text{if} \quad x < 2 \\ x^2 + 5 & \text{if} \quad 2 \le x < 10 \\ 3 & \text{if} \quad x \ge 10 \end{cases}$$

 Which "piece(s)" should be used to find the y-intercept?
 (a) $3x - 2$ (b) $x^2 + 5$ (c) 3 (d) all three

Skill Building

In Problems 11–18, match each graph to its function.

A. Constant function *B. Identity function* *C. Square function* *D. Cube function*
E. Square root function *F. Reciprocal function* *G. Absolute value function* *H. Cube root function*

11. 

12. 

13. 

14. 

15. 

16. 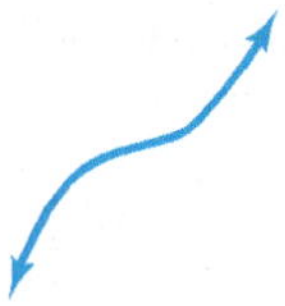

17. 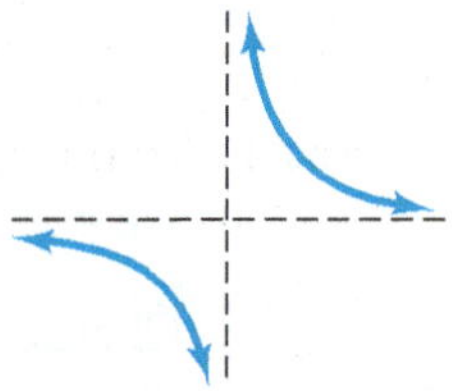

18. 

In Problems 19–26, sketch the graph of each function. Be sure to label three points on the graph.

19. $f(x) = x$
20. $f(x) = x^2$
21. $f(x) = x^3$
22. $f(x) = \sqrt{x}$
23. $f(x) = \dfrac{1}{x}$
24. $f(x) = |x|$
25. $f(x) = \sqrt[3]{x}$
26. $f(x) = 3$

27. If $f(x) = \begin{cases} x^2 & \text{if } x < 0 \\ 2 & \text{if } x = 0 \\ 2x + 1 & \text{if } x > 0 \end{cases}$

find: (a) $f(-2)$ (b) $f(0)$ (c) $f(2)$

28. If $f(x) = \begin{cases} -3x & \text{if } x < -1 \\ 0 & \text{if } x = -1 \\ 2x^2 + 1 & \text{if } x > -1 \end{cases}$

find: (a) $f(-2)$ (b) $f(-1)$ (c) $f(0)$

29. If $f(x) = \begin{cases} 2x - 4 & \text{if } -1 \le x \le 2 \\ x^3 - 2 & \text{if } 2 < x \le 3 \end{cases}$

find: (a) $f(0)$ (b) $f(1)$ (c) $f(2)$ (d) $f(3)$

30. If $f(x) = \begin{cases} x^3 & \text{if } -2 \le x < 1 \\ 3x + 2 & \text{if } 1 \le x \le 4 \end{cases}$

find: (a) $f(-1)$ (b) $f(0)$ (c) $f(1)$ (d) $f(3)$

In Problems 31–42:

(a) Find the domain of each function.
(b) Locate any intercepts.
(c) Graph each function.
(d) Based on the graph, find the range.

31. $f(x) = \begin{cases} 2x & \text{if } x \ne 0 \\ 1 & \text{if } x = 0 \end{cases}$

32. $f(x) = \begin{cases} 3x & \text{if } x \ne 0 \\ 4 & \text{if } x = 0 \end{cases}$

33. $f(x) = \begin{cases} -2x + 3 & \text{if } x < 1 \\ 3x - 2 & \text{if } x \ge 1 \end{cases}$

34. $f(x) = \begin{cases} x + 3 & \text{if } x < -2 \\ -2x - 3 & \text{if } x \ge -2 \end{cases}$

35. $f(x) = \begin{cases} x + 3 & \text{if } -2 \le x < 1 \\ 5 & \text{if } x = 1 \\ -x + 2 & \text{if } x > 1 \end{cases}$

36. $f(x) = \begin{cases} 2x + 5 & \text{if } -3 \le x < 0 \\ -3 & \text{if } x = 0 \\ -5x & \text{if } x > 0 \end{cases}$

37. $f(x) = \begin{cases} 1 + x & \text{if } x < 0 \\ x^2 & \text{if } x \ge 0 \end{cases}$

38. $f(x) = \begin{cases} \dfrac{1}{x} & \text{if } x < 0 \\ \sqrt[3]{x} & \text{if } x \ge 0 \end{cases}$

39. $f(x) = \begin{cases} |x| & \text{if } -2 \le x < 0 \\ x^3 & \text{if } x > 0 \end{cases}$

40. $f(x) = \begin{cases} 2 - x & \text{if } -3 \le x < 1 \\ \sqrt{x} & \text{if } x > 1 \end{cases}$

41. $f(x) = 2\,\text{int}(x)$

42. $f(x) = \text{int}(2x)$

In Problems 43–46, the graph of a piecewise-defined function is given. Write a definition for each function.

43.

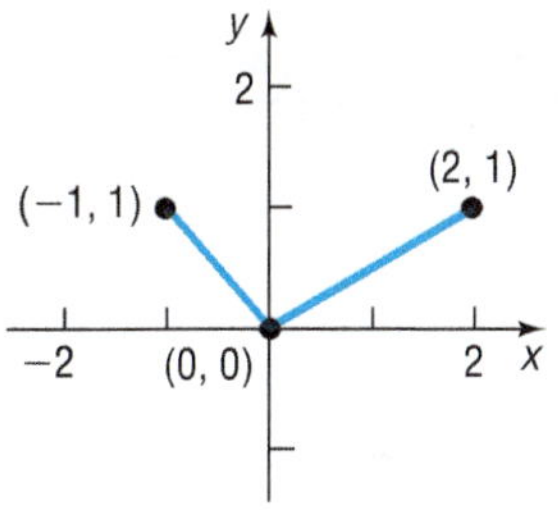

44.

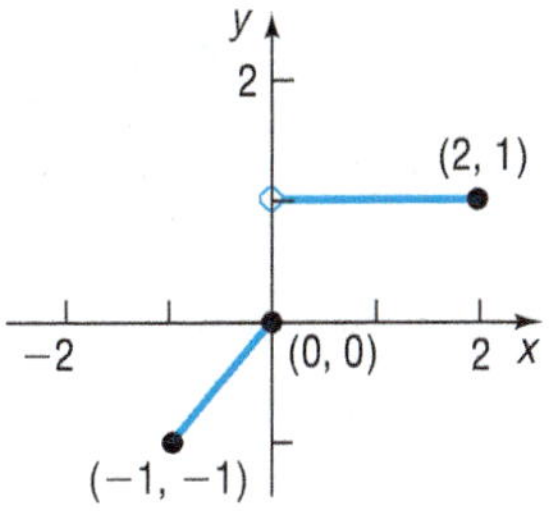

45.

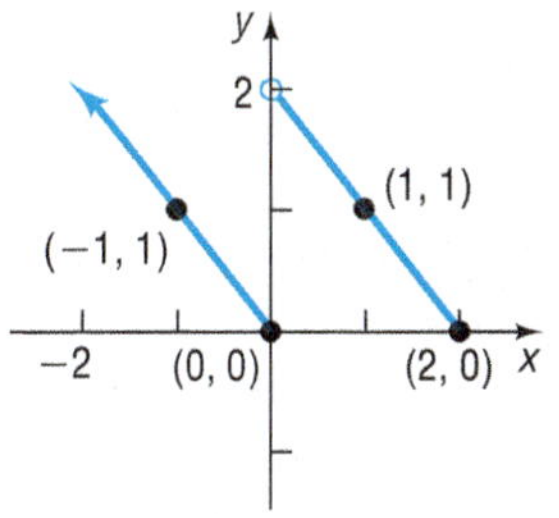

46.

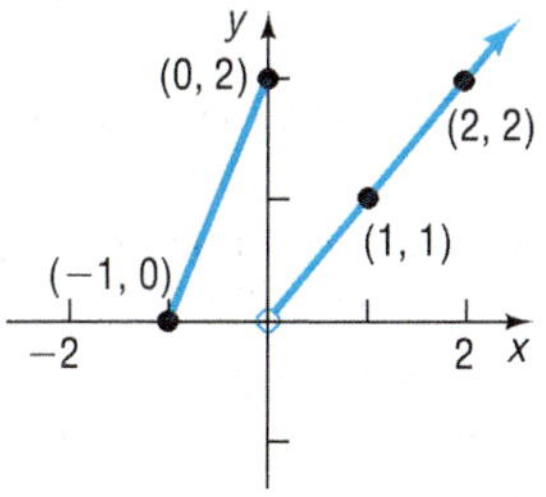

47. If $f(x) = \text{int}(2x)$, find

(a) $f(1.2)$ (b) $f(1.6)$ (c) $f(-1.8)$

48. If $f(x) = \text{int}\left(\dfrac{x}{2}\right)$, find

(a) $f(1.2)$ (b) $f(1.6)$ (c) $f(-1.8)$

49. (a) Graph $f(x) = \begin{cases} (x-1)^2 & \text{if } 0 \le x < 2 \\ -2x + 10 & \text{if } 2 \le x \le 6 \end{cases}$

(b) Find the domain of f.
(c) Find the absolute maximum and the absolute minimum, if they exist.
(d) Find the local maximum and the local minimum values, if they exist.

50. (a) Graph $f(x) = \begin{cases} -x + 1 & \text{if } -2 \le x < 0 \\ 2 & \text{if } \quad x = 0 \\ x + 1 & \text{if } 0 < x \le 2 \end{cases}$

(b) Find the domain of f.
(c) Find the absolute maximum and the absolute minimum, if they exist.
(d) Find the local maximum and the local minimum values, if they exist.

Applications and Extensions

51. Tablet Service A monthly tablet plan costs \$34.99. It includes 3 gigabytes of data and charges \$15 per gigabyte for additional gigabytes. The following function is used to compute the monthly cost for a subscriber.

$$C(x) = \begin{cases} 34.99 & \text{if } 0 \le x \le 3 \\ 15x - 10.01 & \text{if } x > 3 \end{cases}$$

Compute the monthly cost for each of the following gigabytes of use.

(a) 2 (b) 5 (c) 13

52. Parking at O'Hare International Airport The short-term (no more than 24 hours) parking fee F (in dollars) for parking x hours on a weekday at O'Hare International Airport's main parking garage can be modeled by the function

$$F(x) = \begin{cases} 4 & \text{if } \ 0 < x \le 3 \\ 2\,\text{int}(x+1) + 3 & \text{if } \ 3 < x \le 11 \\ 35 & \text{if } 11 < x \le 24 \end{cases}$$

Determine the fee for parking in the short-term parking garage for

(a) 2 hours (b) 7 hours
(c) 15 hours (d) 8 hours and 24 minutes

Source: *O'Hare International Airport*

53. **Cost of Natural Gas** In March 2015, Laclede Gas had the following rate schedule for natural gas usage in single-family residences.

Monthly service charge	\$19.50
Delivery charge	
First 30 therms	\$0.91686/therm
Over 30 therms	\$0
Natural gas cost	
First 30 therms	\$0.348/therm
Over 30 therms	\$0.5922/therm

(a) What is the charge for using 20 therms in a month?
(b) What is the charge for using 150 therms in a month?
(c) Develop a function that models the monthly charge C for x therms of gas.
(d) Graph the function found in part (c).

Source: *Laclede Gas*

54. **Cost of Natural Gas** In April 2015, Nicor Gas had the following rate schedule for natural gas usage in small businesses.

Monthly customer charge	\$72.60
Distribution charge	
1st 150 therms	\$0.1201/therm
Next 4850 therms	\$0.0549/therm
Over 5000 therms	\$0.0482/therm
Gas supply charge	\$0.35/therm

(a) What is the charge for using 1000 therms in a month?
(b) What is the charge for using 6000 therms in a month?
(c) Develop a function that models the monthly charge C for x therms of gas.
(d) Graph the function found in part (c).

Source: *Nicor Gas, 2015*

55. **Federal Income Tax** Two 2015 Tax Rate Schedules are given in the accompanying table. If x equals taxable income and y equals the tax due, construct a function $y = f(x)$ for Schedule X.

2015 Tax Rate Schedules

Schedule X—Single						Schedule Y-1—Married Filing Jointly or Qualified Widow(er)					
If Taxable Income is Over	But Not Over	The Tax is This Amount		Plus This %	Of the Excess Over	If Taxable Income is Over	But Not Over	The Tax is This Amount		Plus This %	Of the Excess Over
\$0	\$9,225	\$0	+	10%	\$0	\$0	\$18,450	\$0	+	10%	\$0
9,225	37,450	922.50	+	15%	9,225	18,450	74,900	1,845	+	15%	18,450
37,450	90,750	5,156.25	+	25%	37,450	74,900	151,200	10,312.50	+	25%	74,900
90,750	189,300	18,481.25	+	28%	90,750	151,200	230,450	29,387.50	+	28%	151,200
189,300	411,500	46,075.25	+	33%	189,300	230,450	411,500	51,577.50	+	33%	230,450
411,500	413,200	119,401.25	+	35%	411,500	411,500	464,850	111,324.00	+	35%	411,500
413,200	–	119,996.25	+	39.6%	413,200	464,850	–	129,996.50	+	39.6%	464,850

56. **Federal Income Tax** Refer to the 2015 tax rate schedules. If x equals taxable income and y equals the tax due, construct a function $y = f(x)$ for Schedule Y-1.

57. **Cost of Transporting Goods** A trucking company transports goods between Chicago and New York, a distance of 960 miles. The company's policy is to charge, for each pound, \$0.50 per mile for the first 100 miles, \$0.40 per mile for the next 300 miles, \$0.25 per mile for the next 400 miles, and no charge for the remaining 160 miles.

(a) Graph the relationship between the cost of transportation in dollars and mileage over the entire 960-mile route.
(b) Find the cost as a function of mileage for hauls between 100 and 400 miles from Chicago.
(c) Find the cost as a function of mileage for hauls between 400 and 800 miles from Chicago.

58. **Car Rental Costs** An economy car rented in Florida from Enterprise® on a weekly basis costs \$185 per week. Extra days cost \$37 per day until the day rate exceeds the weekly rate, in which case the weekly rate applies. Also, any part of a day used counts as a full day. Find the cost C of renting an economy car as a function of the number x of days used, where $7 \le x \le 14$. Graph this function.

59. **Mortgage Fees** Fannie Mae charges a loan-level price adjustment (LLPA) on all mortgages, which represents a fee homebuyers seeking a loan must pay. The rate paid depends on the credit score of the borrower, the amount borrowed, and the loan-to-value (LTV) ratio. The LTV ratio is the ratio of amount borrowed to appraised value of the home. For example, a homebuyer who wishes to borrow \$250,000 with a credit score of 730 and an LTV ratio of 80% will pay 0.5% (0.005) of \$250,000, or \$1250. The table shows the LLPA for various credit scores and an LTV ratio of 80%.

Credit Score	Loan-Level Price Adjustment Rate
≤659	3.00%
660–679	2.50%
680–699	1.75%
700–719	1%
720–739	0.5%
≥740	0.25%

Source: *Fannie Mae*

(a) Construct a function $C = C(s)$, where C is the loan-level price adjustment (LLPA) and s is the credit score of an individual who wishes to borrow \$300,000 with an 80% LTV ratio.

(b) What is the LLPA on a \$300,000 loan with an 80% LTV ratio for a borrower whose credit score is 725?

(c) What is the LLPA on a \$300,000 loan with an 80% LTV ratio for a borrower whose credit score is 670?

60. Minimum Payments for Credit Cards Holders of credit cards issued by banks, department stores, oil companies, and so on receive bills each month that state minimum amounts that must be paid by a certain due date. The minimum due depends on the total amount owed. One such credit card company uses the following rules: For a bill of less than \$10, the entire amount is due. For a bill of at least \$10 but less than \$500, the minimum due is \$10. A minimum of \$30 is due on a bill of at least \$500 but less than \$1000, a minimum of \$50 is due on a bill of at least \$1000 but less than \$1500, and a minimum of \$70 is due on bills of \$1500 or more. Find the function f that describes the minimum payment due on a bill of x dollars. Graph f.

61. Wind Chill The wind chill factor represents the air temperature at a standard wind speed that would produce the same heat loss as the given temperature and wind speed. One formula for computing the equivalent temperature is

$$W = \begin{cases} t & 0 \le v < 1.79 \\ 33 - \dfrac{(10.45 + 10\sqrt{v} - v)(33 - t)}{22.04} & 1.79 \le v \le 20 \\ 33 - 1.5958(33 - t) & v > 20 \end{cases}$$

where v represents the wind speed (in meters per second) and t represents the air temperature (°C). Compute the wind chill for the following:

(a) An air temperature of 10°C and a wind speed of 1 meter per second (m/sec)

(b) An air temperature of 10°C and a wind speed of 5 m/sec

(c) An air temperature of 10°C and a wind speed of 15 m/sec

(d) An air temperature of 10°C and a wind speed of 25 m/sec

(e) Explain the physical meaning of the equation corresponding to $0 \le v < 1.79$.

(f) Explain the physical meaning of the equation corresponding to $v > 20$.

62. Wind Chill Redo Problem 61(a)–(d) for an air temperature of −10°C.

63. First-class Mail In 2015 the U.S. Postal Service charged \$0.98 postage for first-class mail retail flats (such as an 8.5" by 11" envelope) weighing up to 1 ounce, plus \$0.22 for each additional ounce up to 13 ounces. First-class rates do not apply to flats weighing more than 13 ounces. Develop a model that relates C, the first-class postage charged, for a flat weighing x ounces. Graph the function.

Source: *United States Postal Service*

Explaining Concepts: Discussion and Writing

In Problems 64–71, use a graphing utility.

64. Exploration Graph $y = x^2$. Then on the same screen graph $y = x^2 + 2$, followed by $y = x^2 + 4$, followed by $y = x^2 - 2$. What pattern do you observe? Can you predict the graph of $y = x^2 - 4$? Of $y = x^2 + 5$?

65. Exploration Graph $y = x^2$. Then on the same screen graph $y = (x - 2)^2$, followed by $y = (x - 4)^2$, followed by $y = (x + 2)^2$. What pattern do you observe? Can you predict the graph of $y = (x + 4)^2$? Of $y = (x - 5)^2$?

66. Exploration Graph $y = |x|$. Then on the same screen graph $y = 2|x|$, followed by $y = 4|x|$, followed by $y = \frac{1}{2}|x|$. What pattern do you observe? Can you predict the graph of $y = \frac{1}{4}|x|$? Of $y = 5|x|$?

67. Exploration Graph $y = x^2$. Then on the same screen graph $y = -x^2$. Now try $y = |x|$ and $y = -|x|$. What do you conclude?

68. Exploration Graph $y = \sqrt{x}$. Then on the same screen graph $y = \sqrt{-x}$. Now try $y = 2x + 1$ and $y = 2(-x) + 1$. What do you conclude?

69. Exploration Graph $y = x^3$. Then on the same screen graph $y = (x - 1)^3 + 2$. Could you have predicted the result?

70. Exploration Graph $y = x^2$, $y = x^4$, and $y = x^6$ on the same screen. What do you notice is the same about each graph? What do you notice is different?

71. Exploration Graph $y = x^3$, $y = x^5$, and $y = x^7$ on the same screen. What do you notice is the same about each graph? What do you notice is different?

72. Consider the equation

$$y = \begin{cases} 1 & \text{if } x \text{ is rational} \\ 0 & \text{if } x \text{ is irrational} \end{cases}$$

Is this a function? What is its domain? What is its range? What is its y-intercept, if any? What are its x-intercepts, if any? Is it even, odd, or neither? How would you describe its graph?

73. Define some functions that pass through $(0, 0)$ and $(1, 1)$ and are increasing for $x \ge 0$. Begin your list with $y = \sqrt{x}$, $y = x$, and $y = x^2$. Can you propose a general result about such functions?

Retain Your Knowledge

Problems 74–77 are based on material learned earlier in the course. The purpose of these problems is to keep the material fresh in your mind so that you are better prepared for the final exam.

74. Simplify: $(3 + 2i)(4 - 5i)$

75. Find the center and radius of the circle $x^2 + y^2 = 6y + 16$.

76. Solve: $4x - 5(2x - 1) = 4 - 7(x + 1)$

77. Ethan has \$60,000 to invest. He puts part of the money in a CD that earns 3% simple interest per year and the rest in a mutual fund that earns 8% simple interest per year. How much did he invest in each if his earned interest the first year was \$3700?

'Are You Prepared?' Answers

1. 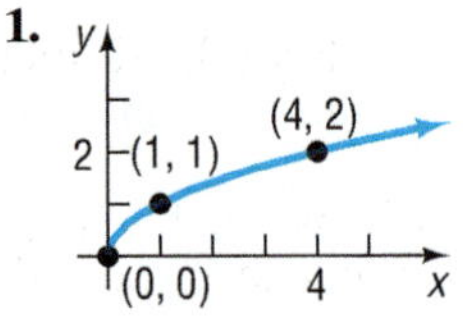

2. 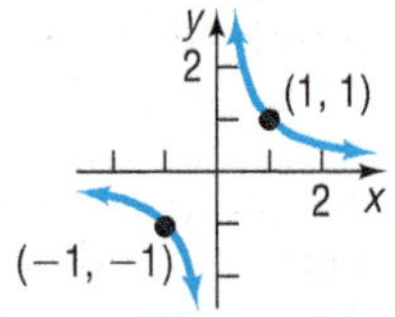

3. $(0, -8)$, $(2, 0)$

3.5 Graphing Techniques: Transformations

OBJECTIVES

At this stage, if you were asked to graph any of the functions defined by $y = x$, $y = x^2$, $y = x^3$, $y = \sqrt{x}$, $y = \sqrt[3]{x}$, $y = |x|$, or $y = \frac{1}{x}$, your response should be, "Yes, I recognize these functions and know the general shapes of their graphs." (If this is not your answer, review the previous section, Figures 36 through 42.)

Sometimes we are asked to graph a function that is "almost" like one that we already know how to graph. In this section, we develop techniques for graphing such functions. Collectively, these techniques are referred to as **transformations**. We introduce the method of transformations because it is a more efficient method of graphing than point-plotting.

1 Graph Functions Using Vertical and Horizontal Shifts

Exploration On the same screen, graph each of the following functions:

$$Y_1 = x^2, \; Y_2 = x^2 + 2, \; Y_3 = x^2 - 2$$

What do you observe? Now create a table of values for Y_1, Y_2, and Y_3. What do you observe?

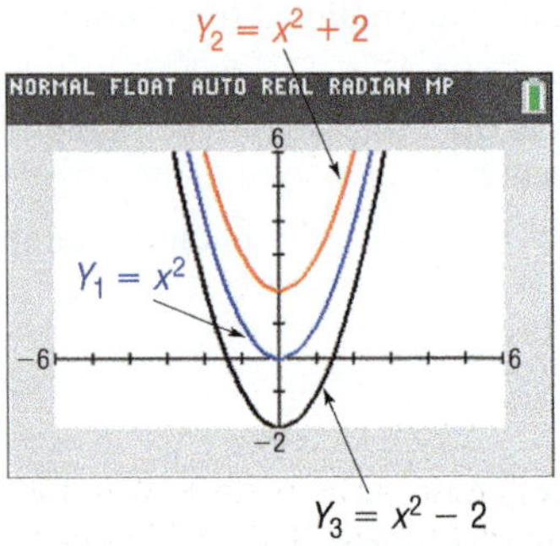

Figure 48 Vertical shift

Result Figure 48 illustrates the graphs. You should have observed a general pattern. With $Y_1 = x^2$ on the screen, the graph of $Y_2 = x^2 + 2$ is identical to that of $Y_1 = x^2$, except that it is shifted vertically up 2 units. The graph of $Y_3 = x^2 - 2$ is identical to that of $Y_1 = x^2$, except that it is shifted vertically down 2 units. From Table 7(a), we see that the y-coordinates on $Y_2 = x^2 + 2$ are 2 units larger than the y-coordinates on $Y_1 = x^2$ for any given x-coordinate. From Table 7(b), we see that the y-coordinates on $Y_3 = x^2 - 2$ are 2 units smaller than the y-coordinates on $Y_1 = x^2$ for any given x-coordinate.

Notice a vertical shift only affects the range of a function, not the domain. For example, the range of Y_1 is $[0, \infty)$ while the range of Y_2 is $[2, \infty)$. The domain of both functions is all real numbers.

Table 7

NORMAL FLOAT AUTO REAL RADIAN MP
PRESS ENTER TO EDIT

X	Y1	Y2
-5	25	27
-4	16	18
-3	9	11
-2	4	6
-1	1	3
0	0	2
1	1	3
2	4	6
3	9	11
4	16	18
5	25	27

Y2=X²+2

(a)

NORMAL FLOAT AUTO REAL RADIAN MP
PRESS ENTER TO EDIT

X	Y1	Y3
-5	25	23
-4	16	14
-3	9	7
-2	4	2
-1	1	-1
0	0	-2
1	1	-1
2	4	2
3	9	7
4	16	14
5	25	23

Y3=X²−2

(b)

■

In Words
For $y = f(x) + k$, $k > 0$, add k to each y-coordinate on the graph of $y = f(x)$ to shift the graph up k units.
For $y = f(x) - k$, $k > 0$, subtract k from each y-coordinate to shift the graph down k units.

We are led to the following conclusions:

If a positive real number k is added to the outputs of a function $y = f(x)$, the graph of the new function $y = f(x) + k$ is the graph of f **shifted vertically up** k units.

If a positive real number k is subtracted from the outputs of a function $y = f(x)$, the graph of the new function $y = f(x) - k$ is the graph of f **shifted vertically down** k units.

EXAMPLE 1 Vertical Shift Down

Use the graph of $f(x) = x^2$ to obtain the graph of $h(x) = x^2 - 4$. Find the domain and range of h.

Solution Table 8 lists some points on the graphs of $Y_1 = f(x) = x^2$ and $Y_2 = h(x) = f(x) - 4 = x^2 - 4$. Notice that each y-coordinate of h is 4 units less than the corresponding y-coordinate of f.

To obtain the graph of h from the graph of f, subtract 4 from each y-coordinate on the graph of f. The graph of h is identical to that of f, except that it is shifted down 4 units. See Figure 49.

Table 8

NORMAL FLOAT AUTO REAL RADIAN MP
PRESS ENTER TO EDIT

X	Y1	Y2
-5	25	21
-4	16	12
-3	9	5
-2	4	0
-1	1	-3
0	0	-4
1	1	-3
2	4	0
3	9	5
4	16	12
5	25	21

Y2=X²-4

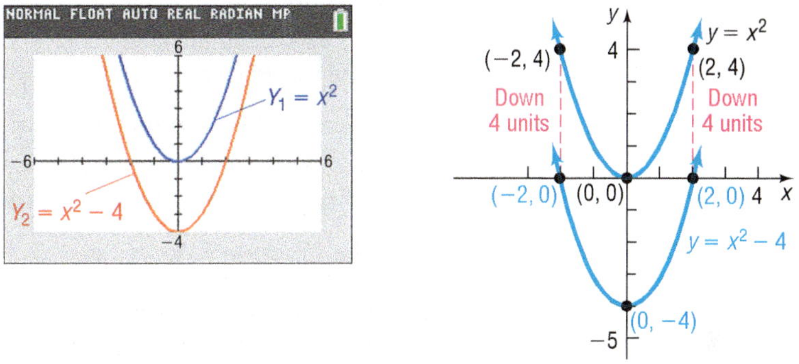

Figure 49

The domain of h is the set of all real numbers. The range of h is $[-4, \infty)$. ■

Now Work PROBLEM 39

Exploration

On the same screen, graph each of the following functions:

$$Y_1 = x^2$$
$$Y_2 = (x - 3)^2$$
$$Y_3 = (x + 2)^2$$

What do you observe?

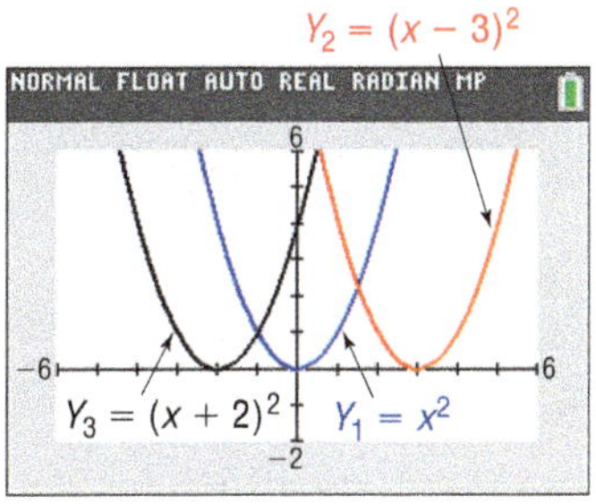

Figure 50 Horizontal shift

Result Figure 50 illustrates the graphs. You should have observed the following pattern. With the graph of $Y_1 = x^2$ on the screen, the graph of $Y_2 = (x - 3)^2$ is identical to that of $Y_1 = x^2$, except it is shifted horizontally to the right 3 units. The graph of $Y_3 = (x + 2)^2$ is identical to that of $Y_1 = x^2$, except it is shifted horizontally to the left 2 units.

From Table 9(a), we see the x-coordinates on $Y_2 = (x - 3)^2$ are 3 units larger than they are for $Y_1 = x^2$ for any given y-coordinate. For example, when $Y_1 = 0$, then $x = 0$, and when $Y_2 = 0$, then $x = 3$. Also, when $Y_1 = 1$, then $x = -1$ or 1, and when $Y_2 = 1$, then $x = 2$ or 4. From Table 9(b), we see the x-coordinates on $Y_3 = (x + 2)^2$ are 2 units smaller than they are for $Y_1 = x^2$ for any given y-coordinate. For example, when $Y_1 = 0$, then $x = 0$, and when $Y_3 = 0$, then $x = -2$. Also, when $Y_1 = 4$, then $x = -2$ or 2, and when $Y_3 = 4$, then $x = -4$ or 0.

Table 9

(a)

NORMAL FLOAT AUTO REAL RADIAN MP
PRESS ENTER TO EDIT

X	Y1	Y2
-2	4	25
-1	1	16
0	0	9
1	1	4
2	4	1
3	9	0
4	16	1
5	25	4
6	36	9
7	49	16
8	64	25

Y2=(X-3)²

(b)

NORMAL FLOAT AUTO REAL RADIAN MP
PRESS ENTER TO EDIT

X	Y1	Y3
-6	36	16
-5	25	9
-4	16	4
-3	9	1
-2	4	0
-1	1	1
0	0	4
1	1	9
2	4	16
3	9	25
4	16	36

Y3=(X+2)²

■

In Words

For $y = f(x - h)$, $h > 0$, add h to each x-coordinate on the graph of $y = f(x)$ to shift the graph right h units. For $y = f(x + h)$, $h > 0$, subtract h from each x-coordinate on the graph of $y = f(x)$ to shift the graph left h units.

We are led to the following conclusions:

If the argument x of a function f is replaced by $x - h$, $h > 0$, the graph of the new function $y = f(x - h)$ is the graph of f **shifted horizontally right** h units.

If the argument x of a function f is replaced by $x + h$, $h > 0$, the graph of the new function $y = f(x + h)$ is the graph of f **shifted horizontally left** h units.

Now Work PROBLEM 43

Note: Vertical shifts result when adding or subtracting a real number k after performing the operation suggested by the basic function, while horizontal shifts result when adding or subtracting a real number h to or from x before performing the operation suggested by the basic function. For example, the graph of $f(x) = \sqrt{x} + 3$ is obtained by shifting the graph of $y = \sqrt{x}$ up 3 units, because we evaluate the square root function first and then add 3. The graph of $g(x) = \sqrt{x + 3}$ is obtained by shifting the graph of $y = \sqrt{x}$ left 3 units, because we first add 3 to x before we evaluate the square root function. ■

Vertical and horizontal shifts are sometimes combined.

EXAMPLE 2 **Combining Vertical and Horizontal Shifts**

Graph the function $f(x) = |x + 3| - 5$. Find the domain and range of f.

Solution We graph f in steps. First, note that the rule for f is basically an absolute value function, so begin with the graph of $y = |x|$ as shown in Figure 51(a). Next, to get the graph of $y = |x + 3|$, shift the graph of $y = |x|$ horizontally 3 units to the left. See Figure 51(b). Finally, to get the graph of $y = |x + 3| - 5$, shift the graph of $y = |x + 3|$ vertically down 5 units. See Figure 51(c). Note the points plotted on each graph. Using key points can be helpful in keeping track of the transformation that has taken place.

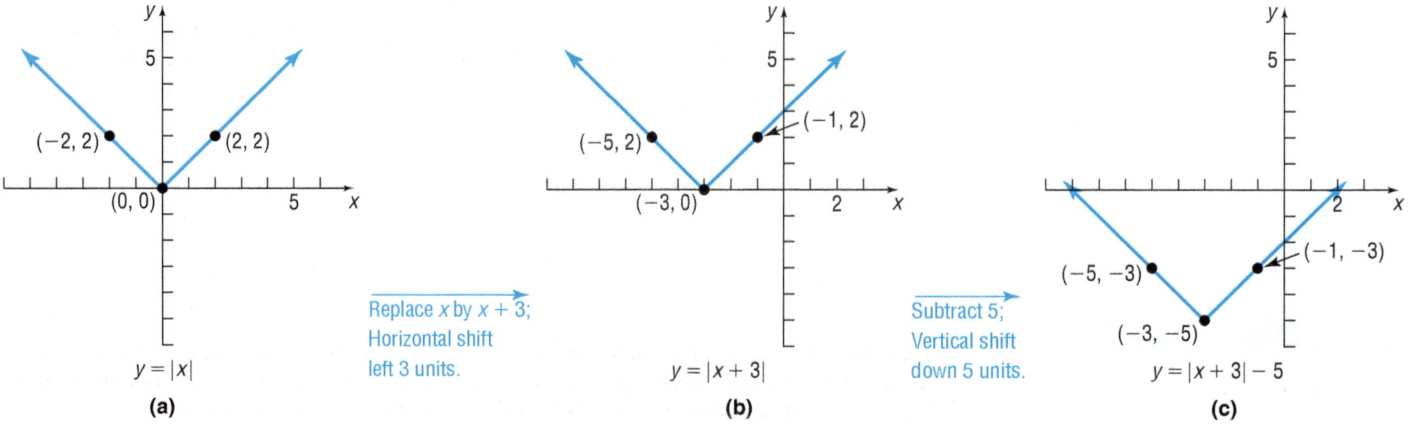

Figure 51

The domain of f is all real numbers, or $(-\infty, \infty)$. The range of f is $[-5, \infty)$. ■

Check: Graph $Y_1 = f(x) = |x + 3| - 5$ and compare the graph to Figure 51(c).

In Example 2, if the vertical shift had been done first, followed by the horizontal shift, the final graph would have been the same. (Try it for yourself.)

Now Work PROBLEMS 45 AND 71

2 Graph Functions Using Compressions and Stretches

Exploration On the same screen, graph each of the following functions:

$$Y_1 = |x|$$
$$Y_2 = 2|x|$$
$$Y_3 = \frac{1}{2}|x|$$

Then create a table of values and compare the y-coordinates for any given x-coordinate.

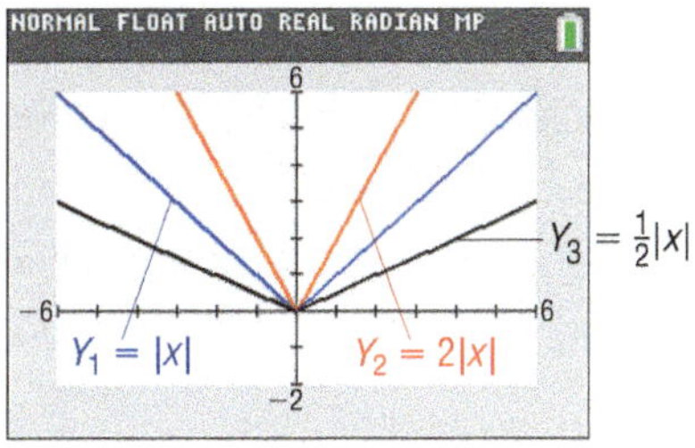

Figure 52 Vertical stretch or compression

Result Figure 52 illustrates the graphs.

Now look at Table 10, where $Y_1 = |x|$ and $Y_2 = 2|x|$. Notice that the values for Y_2 are two times the values of Y_1 for each x-value. This means that the graph of $Y_2 = 2|x|$ can be obtained from the graph of $Y_1 = |x|$ by multiplying each y-coordinate of $Y_1 = |x|$ by 2. Therefore, the graph of Y_2 will be the graph of Y_1 vertically *stretched* by a factor of 2.

Look at Table 11, where $Y_1 = |x|$ and $Y_3 = \frac{1}{2}|x|$. The values of Y_3 are half the values of Y_1 for each x-value. So, the graph of $Y_3 = \frac{1}{2}|x|$ can be obtained from the graph of $Y_1 = |x|$ by multiplying each y-coordinate by $\frac{1}{2}$. Therefore, the graph of Y_3 will be the graph of Y_1 vertically *compressed* by a factor of $\frac{1}{2}$.

Table 10

NORMAL FLOAT AUTO REAL RADIAN MP
PRESS ENTER TO EDIT

X	Y1	Y2
-3	3	6
-2	2	4
-1	1	2
0	0	0
1	1	2
2	2	4
3	3	6
4	4	8
5	5	10
6	6	12
7	7	14

Y2=2|X|

Table 11

NORMAL FLOAT AUTO REAL RADIAN MP
PRESS ENTER TO EDIT

X	Y1	Y3
-2	2	1
-1	1	$\frac{1}{2}$
0	0	0
1	1	$\frac{1}{2}$
2	2	1
3	3	$\frac{3}{2}$
4	4	2

Y3=$\frac{1}{2}$|X|

■

> **In Words**
> For $y = af(x)$, $a > 0$, the factor a is "outside" the function, so it affects the y-coordinates. Multiply each y-coordinate on the graph of $y = f(x)$ by a.

Based on the Exploration, we have the following result:

When the right side of a function $y = f(x)$ is multiplied by a positive number a, the graph of the new function $y = af(x)$ is obtained by multiplying each y-coordinate on the graph of $y = f(x)$ by a. The new graph is a **vertically compressed** (if $0 < a < 1$) or a **vertically stretched** (if $a > 1$) version of the graph of $y = f(x)$.

Now Work PROBLEM 47

What happens if the argument x of a function $y = f(x)$ is multiplied by a positive number a, creating a new function $y = f(ax)$? To find the answer, look at the following Exploration.

Exploration

On the same screen, graph each of the following functions:

$$Y_1 = f(x) = \sqrt{x} \qquad Y_2 = f(2x) = \sqrt{2x} \qquad Y_3 = f\left(\frac{1}{2}x\right) = \sqrt{\frac{1}{2}x} = \sqrt{\frac{x}{2}}$$

Create a table of values to explore the relation between the x- and y-coordinates of each function.

Result You should have obtained the graphs in Figure 53. Look at Table 12(a). Note that (1, 1), (4, 2), and (9, 3) are points on the graph of $Y_1 = \sqrt{x}$. Also, (0.5, 1), (2, 2), and (4.5, 3) are points on the graph of $Y_2 = \sqrt{2x}$. For a given y-coordinate, the x-coordinate on the graph of Y_2 is $\frac{1}{2}$ of the x-coordinate on Y_1.

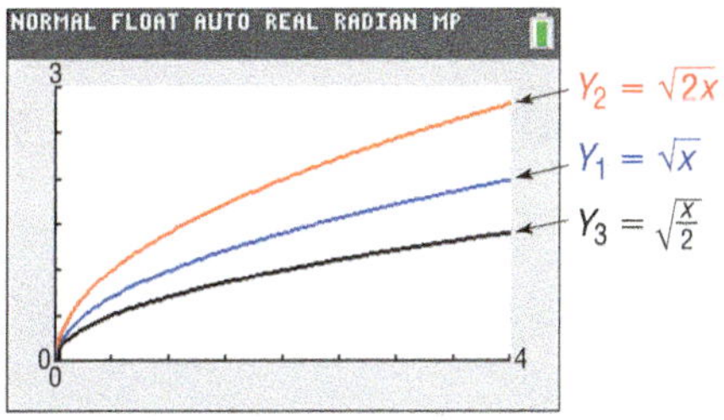

Figure 53 Horizontal stretch or compression

Table 12

(a)

NORMAL FLOAT AUTO REAL RADIAN MP
PRESS ENTER TO EDIT

X	Y1	Y2
0	0	0
.5	.70711	1
1	1	1.4142
2	1.4142	2
4	2	2.8284
4.5	2.1213	3
8	2.8284	4
9	3	4.2426
16	4	5.6569
12.5	3.5355	5
25	5	7.0711

Y2=$\sqrt{2X}$

(b)

NORMAL FLOAT AUTO REAL RADIAN MP
PRESS ENTER TO EDIT

X	Y1	Y3
0	0	0
1	1	.70711
2	1.4142	1
4	2	1.4142
8	2.8284	2
9	3	2.1213
16	4	2.8284
18	4.2426	3
25	5	3.5355
32	5.6569	4
50	7.0711	5

Y3=$\sqrt{X/2}$

We conclude that the graph of $Y_2 = \sqrt{2x}$ is obtained by multiplying the x-coordinate of each point on the graph of $Y_1 = \sqrt{x}$ by $\frac{1}{2}$. The graph of $Y_2 = \sqrt{2x}$ is the graph of $Y_1 = \sqrt{x}$ *compressed* horizontally.

Look at Table 12(b). Notice that (1, 1), (4, 2), and (9, 3) are points on the graph of $Y_1 = \sqrt{x}$. Also notice that (2, 1), (8, 2), and (18, 3) are points on the graph of $Y_3 = \sqrt{\frac{x}{2}}$. For a given y-coordinate, the x-coordinate on the graph of Y_3 is 2 times the x-coordinate on Y_1. We conclude that the graph of $Y_3 = \sqrt{\frac{x}{2}}$ is obtained by multiplying the x-coordinate of each point on the graph of $Y_1 = \sqrt{x}$ by 2. The graph of $Y_3 = \sqrt{\frac{x}{2}}$ is the graph of $Y_1 = \sqrt{x}$ *stretched* horizontally.

Based on the Exploration, we have the following result:

> **In Words**
> For $y = f(ax)$, $a > 0$, the factor a is "inside" the function, so it affects the x-coordinates. Multiply each x-coordinate on the graph of $y = f(x)$ by $\frac{1}{a}$.

If the argument x of a function $y = f(x)$ is multiplied by a positive number a, then the graph of the new function $y = f(ax)$ is obtained by multiplying each x-coordinate of $y = f(x)$ by $\frac{1}{a}$. A **horizontal compression** results if $a > 1$, and a **horizontal stretch** results if $0 < a < 1$.

Let's look at an example.

EXAMPLE 3 **Graphing Using Stretches and Compressions**

The graph of $y = f(x)$ is given in Figure 54. Use this graph to find the graphs of

(a) $y = 2f(x)$ (b) $y = f(3x)$

Solution (a) The graph of $y = 2f(x)$ is obtained by multiplying each y-coordinate of $y = f(x)$ by 2. See Figure 55.

(b) The graph of $y = f(3x)$ is obtained from the graph of $y = f(x)$ by multiplying each x-coordinate of $y = f(x)$ by $\frac{1}{3}$. See Figure 56.

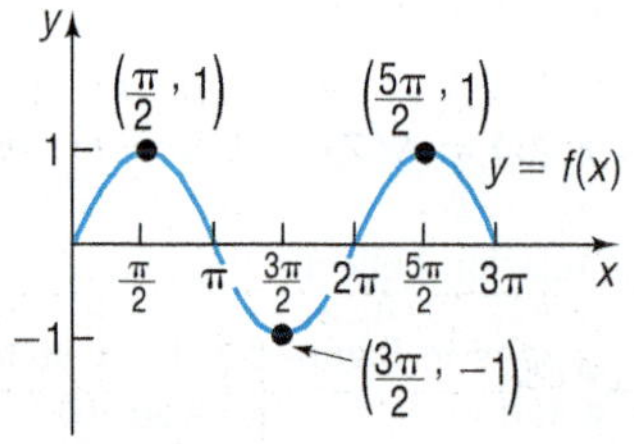

Figure 54 $y = f(x)$

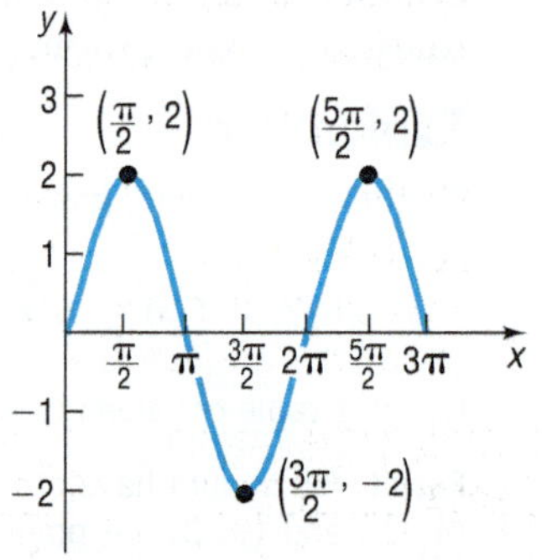

Figure 55 $y = 2f(x)$

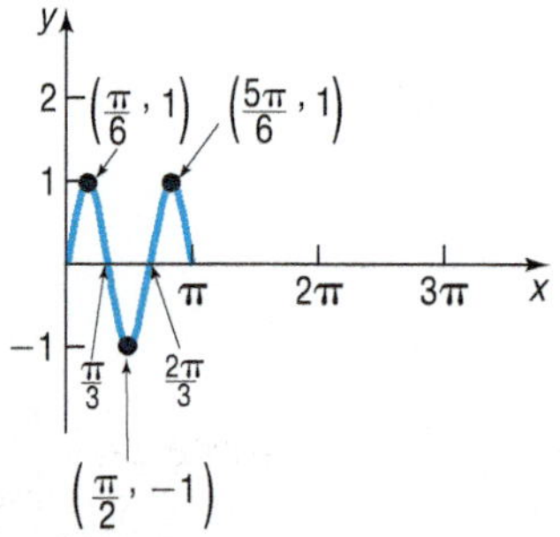

Figure 56 $y = f(3x)$

Now Work PROBLEMS 63(e) AND (g)

3 Graph Functions Using Reflections about the x-Axis or y-Axis

Exploration

Reflection about the x-axis:

(a) Graph and create a table of $Y_1 = x^2$ and $Y_2 = -x^2$.

(b) Graph and create a table of $Y_1 = |x|$ and $Y_2 = -|x|$.

(c) Graph and create a table of $Y_1 = x^2 - 4$ and $Y_2 = -(x^2 - 4) = -x^2 + 4$.

Result See Tables 13(a), (b), and (c) and Figures 57(a), (b), and (c). For each point (x, y) on the graph of Y_1, the point $(x, -y)$ is on the graph of Y_2. Put another way, Y_2 is the reflection about the x-axis of Y_1.

Table 13

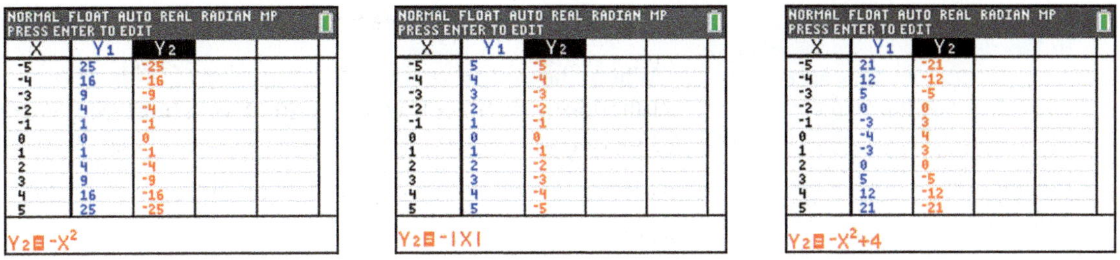

NORMAL FLOAT AUTO REAL RADIAN MP
PRESS ENTER TO EDIT

X	Y1	Y2
-5	25	-25
-4	16	-16
-3	9	-9
-2	4	-4
-1	1	-1
0	0	0
1	1	-1
2	4	-4
3	9	-9
4	16	-16
5	25	-25

Y2=-X²

(a)

NORMAL FLOAT AUTO REAL RADIAN MP
PRESS ENTER TO EDIT

X	Y1	Y2
-5	5	-5
-4	4	-4
-3	3	-3
-2	2	-2
-1	1	-1
0	0	0
1	1	-1
2	2	-2
3	3	-3
4	4	-4
5	5	-5

Y2=-|X|

(b)

NORMAL FLOAT AUTO REAL RADIAN MP
PRESS ENTER TO EDIT

X	Y1	Y2
-5	21	-21
-4	12	-12
-3	5	-5
-2	0	0
-1	-3	3
0	-4	4
1	-3	3
2	0	0
3	5	-5
4	12	-12
5	21	-21

Y2=-X²+4

(c)

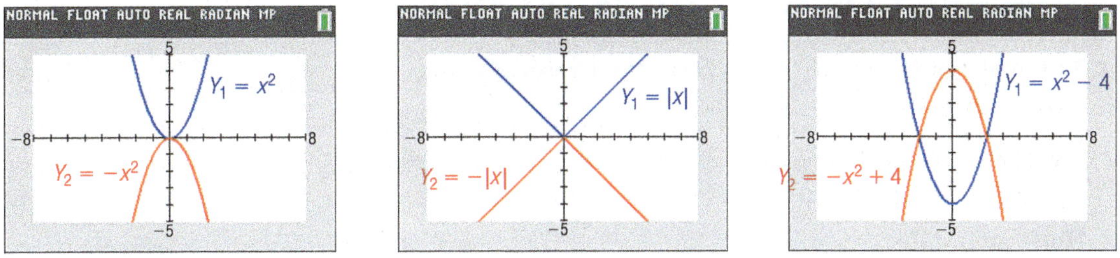

(a) (b) (c)

Figure 57 Reflection about the *x*-axis ■

Based on the previous Exploration, we have the following result:

When the right side of the function $y = f(x)$ is multiplied by -1, the graph of the new function $y = -f(x)$ is the **reflection about the *x*-axis** of the graph of the function $y = f(x)$.

Now Work PROBLEM 49

Exploration

Reflection about the *y*-axis:

(a) Graph $Y_1 = \sqrt{x}$, followed by $Y_2 = \sqrt{-x}$.

(b) Graph $Y_1 = x + 1$, followed by $Y_2 = -x + 1$.

(c) Graph $Y_1 = x^4 + x$, followed by $Y_2 = (-x)^4 + (-x) = x^4 - x$.

Result See Tables 14(a), (b), and (c) and Figures 58(a), (b), and (c). For each point (x, y) on the graph of Y_1, the point $(-x, y)$ is on the graph of Y_2. Put another way, Y_2 is the reflection about the *y*-axis of Y_1.

Table 14

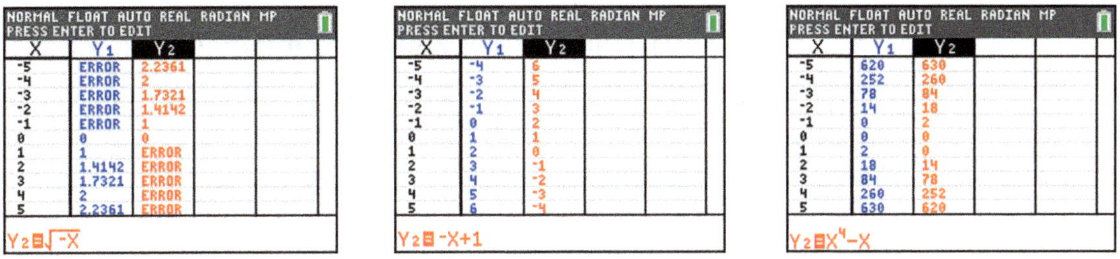

NORMAL FLOAT AUTO REAL RADIAN MP
PRESS ENTER TO EDIT

X	Y1	Y2
-5	ERROR	2.2361
-4	ERROR	2
-3	ERROR	1.7321
-2	ERROR	1.4142
-1	ERROR	1
0	0	0
1	1	ERROR
2	1.4142	ERROR
3	1.7321	ERROR
4	2	ERROR
5	2.2361	ERROR

Y2=√(-X)

(a)

NORMAL FLOAT AUTO REAL RADIAN MP
PRESS ENTER TO EDIT

X	Y1	Y2
-5	-4	6
-4	-3	5
-3	-2	4
-2	-1	3
-1	0	2
0	1	1
1	2	0
2	3	-1
3	4	-2
4	5	-3
5	6	-4

Y2=-X+1

(b)

NORMAL FLOAT AUTO REAL RADIAN MP
PRESS ENTER TO EDIT

X	Y1	Y2
-5	620	630
-4	252	260
-3	78	84
-2	14	18
-1	0	2
0	0	0
1	2	0
2	18	14
3	84	78
4	260	252
5	630	620

Y2=X⁴-X

(c)

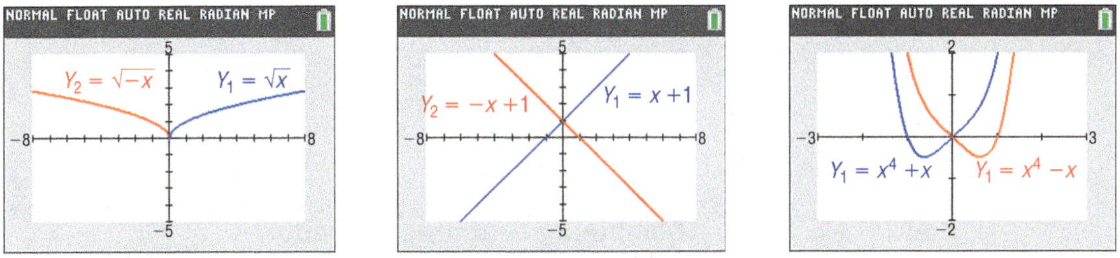

(a) (b) (c)

Figure 58 Reflection about the *y*-axis ■

In Words

For $y = -f(x)$, multiply each y-coordinate on the graph of $y = f(x)$ by -1.

For $y = f(-x)$, multiply each x-coordinate by -1.

Based on the previous Exploration, we have the following result:

When the graph of the function $y = f(x)$ is known, the graph of the new function $y = f(-x)$ is the **reflection about the y-axis** of the graph of the function $y = f(x)$.

SUMMARY OF GRAPHING TECHNIQUES

To Graph:	Draw the Graph of f and:	Functional Change to $f(x)$
Vertical shifts		
$y = f(x) + k,\quad k > 0$	Raise the graph of f by k units.	Add k to $f(x)$.
$y = f(x) - k,\quad k > 0$	Lower the graph of f by k units.	Subtract k from $f(x)$.
Horizontal shifts		
$y = f(x + h),\quad h > 0$	Shift the graph of f to the left h units.	Replace x by $x + h$.
$y = f(x - h),\quad h > 0$	Shift the graph of f to the right h units.	Replace x by $x - h$.
Compressing or stretching		
$y = af(x),\quad a > 0$	Multiply each y-coordinate of $y = f(x)$ by a. Stretch the graph of f vertically if $a > 1$. Compress the graph of f vertically if $0 < a < 1$.	Multiply $f(x)$ by a.
$y = f(ax),\quad a > 0$	Multiply each x-coordinate of $y = f(x)$ by $\frac{1}{a}$. Stretch the graph of f horizontally if $0 < a < 1$. Compress the graph of f horizontally if $a > 1$.	Replace x by ax.
Reflection about the x-axis		
$y = -f(x)$	Reflect the graph of f about the x-axis.	Multiply $f(x)$ by -1.
Reflection about the y-axis		
$y = f(-x)$	Reflect the graph of f about the y-axis.	Replace x by $-x$.

EXAMPLE 4 Determining the Function Obtained from a Series of Transformations

Find the function that is finally graphed after the following three transformations are applied to the graph of $y = |x|$.

1. Shift left 2 units.
2. Shift up 3 units.
3. Reflect about the y-axis.

Solution

1. Shift left 2 units:	Replace x by $x + 2$.	$y = \lvert x + 2\rvert$
2. Shift up 3 units:	Add 3.	$y = \lvert x + 2\rvert + 3$
3. Reflect about the y-axis:	Replace x by $-x$.	$y = \lvert -x + 2\rvert + 3$

Now Work PROBLEM 27

EXAMPLE 5 Combining Graphing Procedures

Graph the function $f(x) = \dfrac{3}{x - 2} + 1$. Find the domain and range of f.

Solution It is helpful to write f as $f(x) = 3\left(\frac{1}{x-2}\right) + 1$. Now use the following steps to obtain the graph of f.

Step 1: $y = \frac{1}{x}$ **Reciprocal function**

Step 2: $y = 3 \cdot \left(\frac{1}{x}\right) = \frac{3}{x}$ **Multiply by 3; vertical stretch by a factor of 3.**

Step 3: $y = \frac{3}{x-2}$ **Replace x by $x - 2$; horizontal shift to the right 2 units.**

Step 4: $y = \frac{3}{x-2} + 1$ **Add 1; vertical shift up 1 unit.**

See Figure 59.

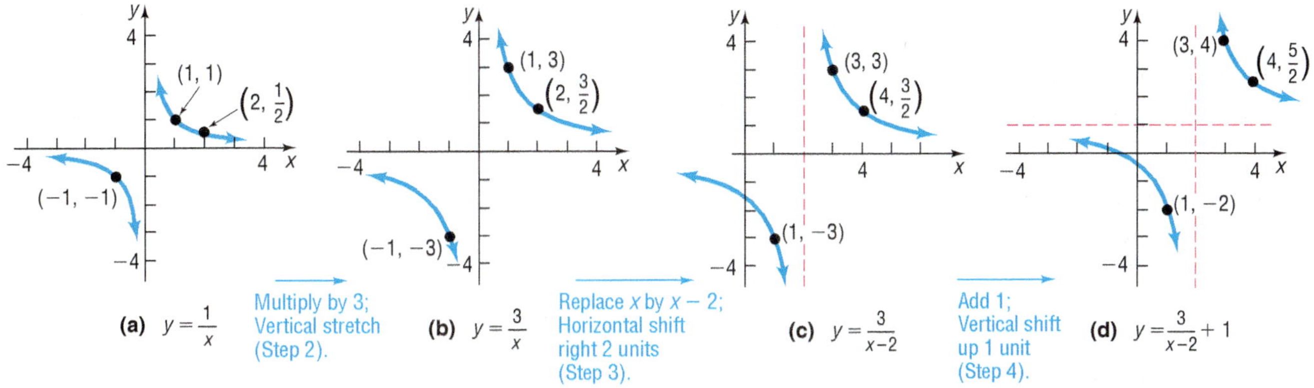

Figure 59

The domain of $y = \frac{1}{x}$ is $\{x | x \neq 0\}$ and its range is $\{y | y \neq 0\}$. Because we shifted right 2 units and up 1 unit to obtain f, the domain of f is $\{x | x \neq 2\}$ and its range is $\{y | y \neq 1\}$.

Other orderings of the steps shown in Example 5 would also result in the graph of f. For example, try this one:

Hint: Although the order in which transformations are performed can be altered, you may consider using the following order for consistency:

1. Reflections
2. Compressions and stretches
3. Shifts

Step 1: $y = \frac{1}{x}$ **Reciprocal function**

Step 2: $y = \frac{1}{x-2}$ **Replace x by $x - 2$; horizontal shift to the right 2 units.**

Step 3: $y = \frac{3}{x-2}$ **Multiply by 3; vertical stretch of the graph of $y = \frac{1}{x-2}$ by a factor of 3.**

Step 4: $y = \frac{3}{x-2} + 1$ **Add 1; vertical shift up 1 unit.**

EXAMPLE 6 Combining Graphing Procedures

Graph the function $f(x) = \sqrt{1-x} + 2$. Find the domain and range of f.

Solution Because horizontal shifts require the form $x - h$, begin by rewriting $f(x)$ as $f(x) = \sqrt{1-x} + 2 = \sqrt{-(x-1)} + 2$. Now use the following steps:

Step 1: $y = \sqrt{x}$ **Square root function**

Step 2: $y = \sqrt{-x}$ **Replace x by $-x$; reflect about the y-axis.**

Step 3: $y = \sqrt{-(x-1)} = \sqrt{1-x}$ **Replace x by $x - 1$; horizontal shift to the right 1 unit.**

Step 4: $y = \sqrt{1-x} + 2$ **Add 2; vertical shift up 2 units.**

See Figure 60.

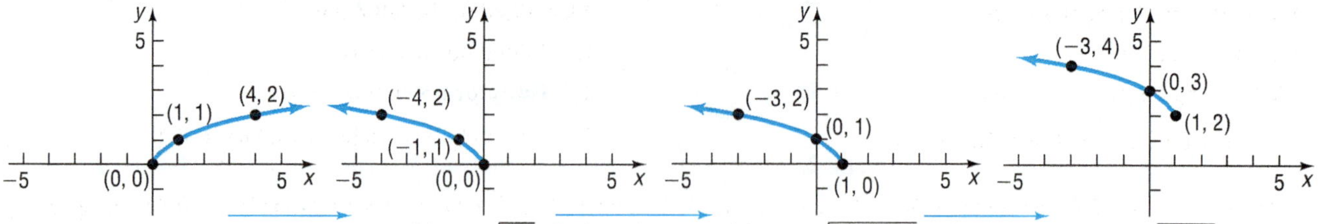

(a) $y = \sqrt{x}$ Replace x by $-x$; Reflect about y-axis (Step 2). (b) $y = \sqrt{-x}$ Replace x by $x - 1$; Horizontal shift right 1 unit (Step 3). (c) $y = \sqrt{-(x-1)} = \sqrt{-x+1} = \sqrt{1-x}$ Add 2; Vertical shift up 2 units (Step 4). (d) $y = \sqrt{1-x} + 2$

Figure 60

The domain of f is $(-\infty, 1]$ and the range is $[2, \infty)$.

Now Work PROBLEM 55

3.5 Assess Your Understanding

Concepts and Vocabulary

1. Suppose that the graph of a function f is known. Then the graph of $y = f(x-2)$ may be obtained by a(n) ___________ shift of the graph of f to the ______ a distance of 2 units.

2. Suppose that the graph of a function f is known. Then the graph of $y = f(-x)$ may be obtained by a reflection about the ___-axis of the graph of the function $y = f(x)$.

3. ***True or False*** The graph of $y = \frac{1}{3}g(x)$ is the graph of $y = g(x)$ stretched by a factor of 3.

4. ***True or False*** The graph of $y = -f(x)$ is the reflection about the x-axis of the graph of $y = f(x)$.

5. Which of the following functions has a graph that is the graph of $y = \sqrt{x}$ shifted down 3 units?
   (a) $y = \sqrt{x+3}$ (b) $y = \sqrt{x-3}$
   (c) $y = \sqrt{x} + 3$ (d) $y = \sqrt{x} - 3$

6. Which of the following functions has a graph that is the graph of $y = f(x)$ compressed horizontally by a factor of 4?
   (a) $y = f(4x)$ (b) $y = f\left(\frac{1}{4}x\right)$
   (c) $y = 4f(x)$ (d) $y = \frac{1}{4}f(x)$

Skill Building

In Problems 7–18, match each graph to one of the following functions:

A. $y = x^2 + 2$ *B.* $y = -x^2 + 2$ *C.* $y = |x| + 2$ *D.* $y = -|x| + 2$
E. $y = (x-2)^2$ *F.* $y = -(x+2)^2$ *G.* $y = |x-2|$ *H.* $y = -|x+2|$
I. $y = 2x^2$ *J.* $y = -2x^2$ *K.* $y = 2|x|$ *L.* $y = -2|x|$

7.
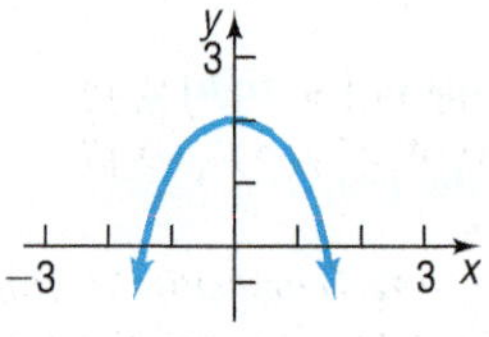

8.
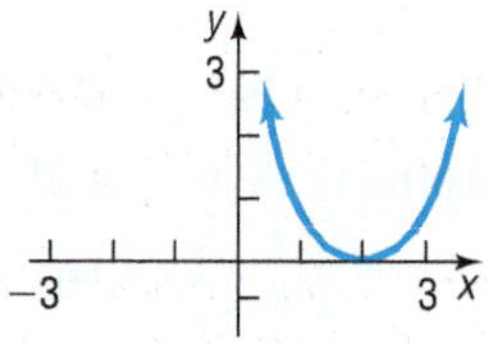

9.
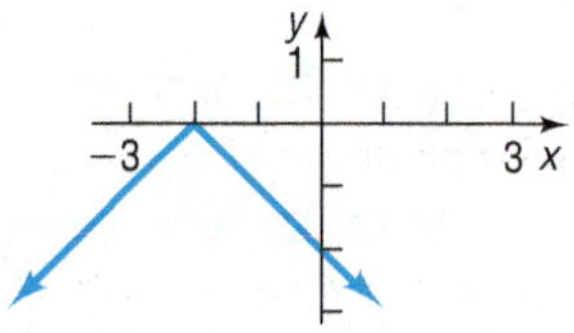

10.
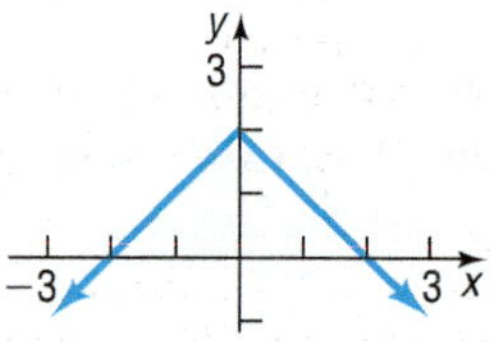

11.
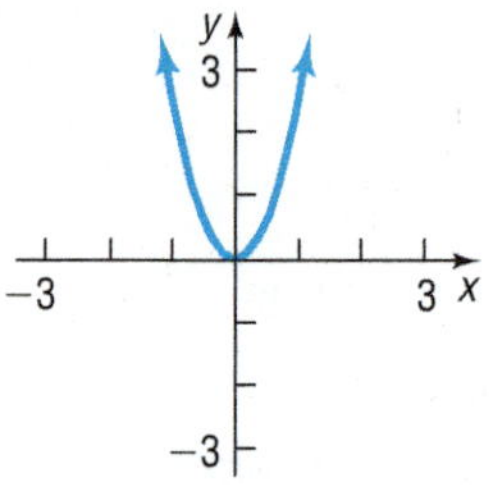

12.
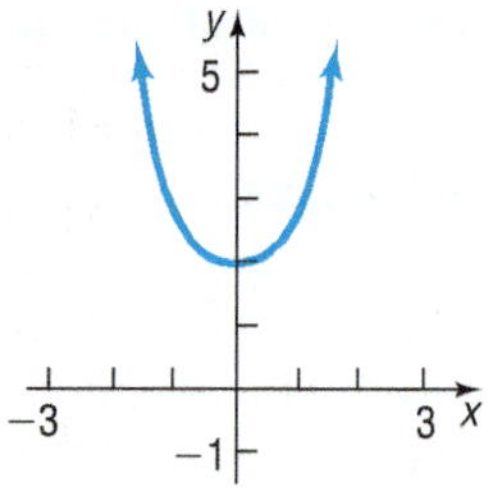

13.
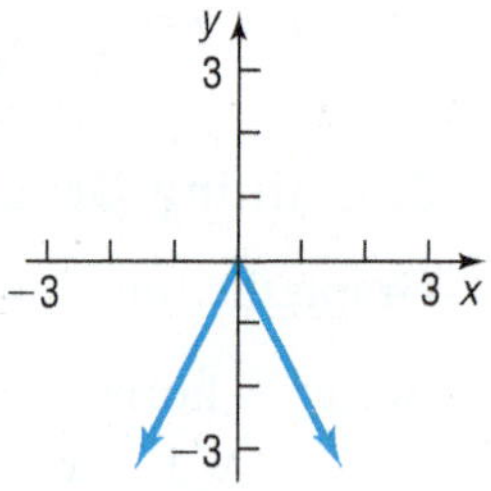

14.
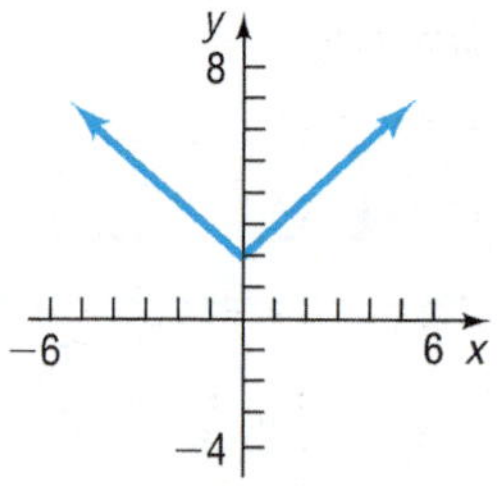

15.
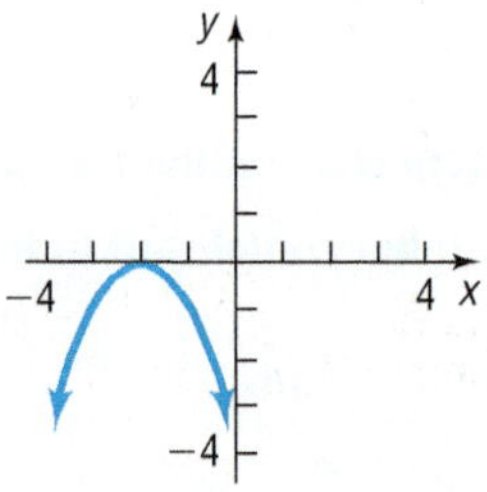

16.
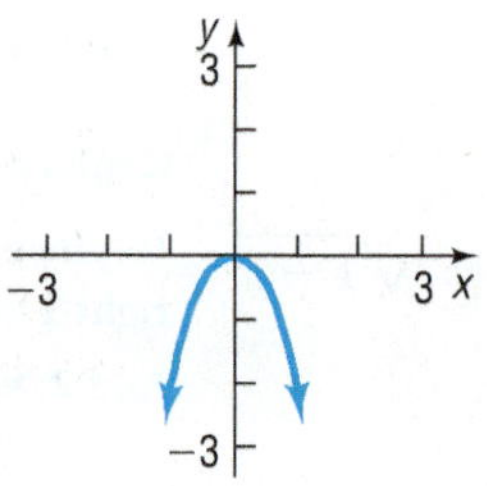

17.
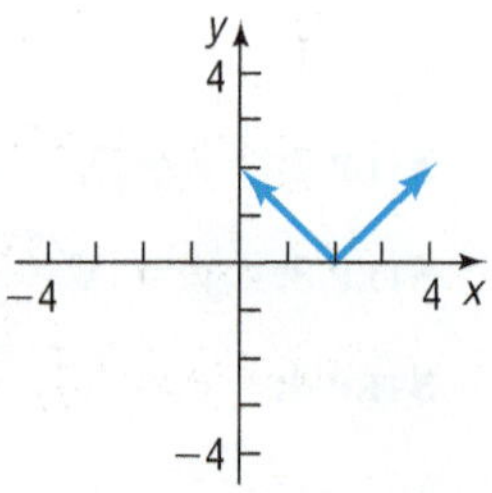

18.
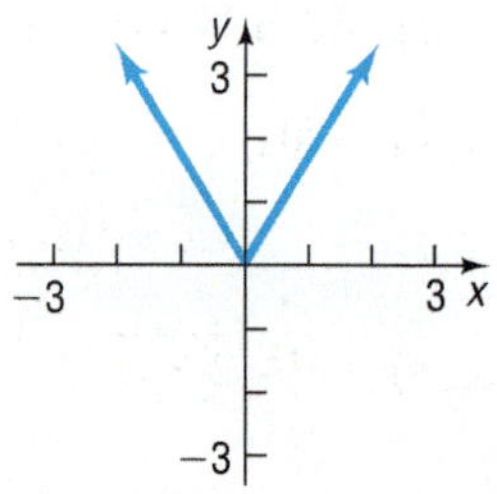

In Problems 19–26, write the function whose graph is the graph of $y = x^3$, but is:

19. Shifted to the right 4 units

20. Shifted to the left 4 units

21. Shifted up 4 units

22. Shifted down 4 units

23. Reflected about the y-axis

24. Reflected about the x-axis

25. Vertically stretched by a factor of 4

26. Horizontally stretched by a factor of 4

In Problems 27–30, find the function that is finally graphed after each of the following transformations is applied to the graph of $y = \sqrt{x}$ in the order stated.

27. (1) Shift up 2 units
(2) Reflect about the x-axis
(3) Reflect about the y-axis

28. (1) Reflect about the x-axis
(2) Shift right 3 units
(3) Shift down 2 units

29. (1) Reflect about the x-axis
(2) Shift up 2 units
(3) Shift left 3 units

30. (1) Shift up 2 units
(2) Reflect about the y-axis
(3) Shift left 3 units

31. If $(3, 6)$ is a point on the graph of $y = f(x)$, which of the following points must be on the graph of $y = -f(x)$?

(a) $(6, 3)$ (b) $(6, -3)$
(c) $(3, -6)$ (d) $(-3, 6)$

32. If $(3, 6)$ is a point on the graph of $y = f(x)$, which of the following points must be on the graph of $y = f(-x)$?

(a) $(6, 3)$ (b) $(6, -3)$
(c) $(3, -6)$ (d) $(-3, 6)$

33. If $(1, 3)$ is a point on the graph of $y = f(x)$, which of the following points must be on the graph of $y = 2f(x)$?

(a) $\left(1, \frac{3}{2}\right)$ (b) $(2, 3)$
(c) $(1, 6)$ (d) $\left(\frac{1}{2}, 3\right)$

34. If $(4, 2)$ is a point on the graph of $y = f(x)$, which of the following points must be on the graph of $y = f(2x)$?

(a) $(4, 1)$ (b) $(8, 2)$
(c) $(2, 2)$ (d) $(4, 4)$

35. Suppose that the x-intercepts of the graph of $y = f(x)$ are -5 and 3.

(a) What are the x-intercepts of the graph of $y = f(x + 2)$?
(b) What are the x-intercepts of the graph of $y = f(x - 2)$?
(c) What are the x-intercepts of the graph of $y = 4f(x)$?
(d) What are the x-intercepts of the graph of $y = f(-x)$?

36. Suppose that the x-intercepts of the graph of $y = f(x)$ are -8 and 1.

(a) What are the x-intercepts of the graph of $y = f(x + 4)$?
(b) What are the x-intercepts of the graph of $y = f(x - 3)$?
(c) What are the x-intercepts of the graph of $y = 2f(x)$?
(d) What are the x-intercepts of the graph of $y = f(-x)$?

37. Suppose that the function $y = f(x)$ is increasing on the interval $[-1, 5]$.

(a) Over what interval is the graph of $y = f(x + 2)$ increasing?
(b) Over what interval is the graph of $y = f(x - 5)$ increasing?
(c) What can be said about the graph of $y = -f(x)$?
(d) What can be said about the graph of $y = f(-x)$?

38. Suppose that the function $y = f(x)$ is decreasing on the interval $[-2, 7]$.

(a) Over what interval is the graph of $y = f(x + 2)$ decreasing?
(b) Over what interval is the graph of $y = f(x - 5)$ decreasing?
(c) What can be said about the graph of $y = -f(x)$?
(d) What can be said about the graph of $y = f(-x)$?

In Problems 39–62, graph each function using the techniques of shifting, compressing, stretching, and/or reflecting. Start with the graph of the basic function (for example, $y = x^2$) and show all stages. Be sure to show at least three key points. Find the domain and the range of each function.

39. $f(x) = x^2 - 1$

40. $f(x) = x^2 + 4$

41. $g(x) = x^3 + 1$

42. $g(x) = x^3 - 1$

43. $h(x) = \sqrt{x + 2}$

44. $h(x) = \sqrt{x} + 1$

45. $f(x) = (x - 1)^3 + 2$

46. $f(x) = (x + 2)^3 - 3$

47. $g(x) = 4\sqrt{x}$

48. $g(x) = \frac{1}{2}\sqrt{x}$

49. $f(x) = -\sqrt[3]{x}$

50. $f(x) = -\sqrt{x}$

51. $f(x) = 2(x + 1)^2 - 3$

52. $f(x) = 3(x - 2)^2 + 1$

53. $g(x) = 2\sqrt{x - 2} + 1$

54. $g(x) = 3|x + 1| - 3$

55. $h(x) = \sqrt{-x} - 2$

56. $h(x) = \frac{4}{x} + 2$

57. $f(x) = -(x + 1)^3 - 1$

58. $f(x) = -4\sqrt{x - 1}$

59. $g(x) = 2|1 - x|$

60. $g(x) = 4\sqrt{2 - x}$

61. $h(x) = \frac{1}{2x}$

62. $h(x) = \sqrt[3]{x - 1} + 3$

In Problems 63–66, the graph of a function f is illustrated. Use the graph of f as the first step toward graphing each of the following functions:

(a) $F(x) = f(x) + 3$ (b) $G(x) = f(x + 2)$ (c) $P(x) = -f(x)$ (d) $H(x) = f(x + 1) - 2$

(e) $Q(x) = \frac{1}{2}f(x)$ (f) $g(x) = f(-x)$ (g) $h(x) = f(2x)$

63.

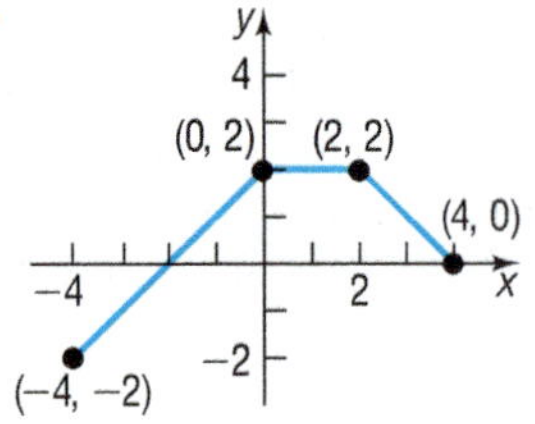

64.

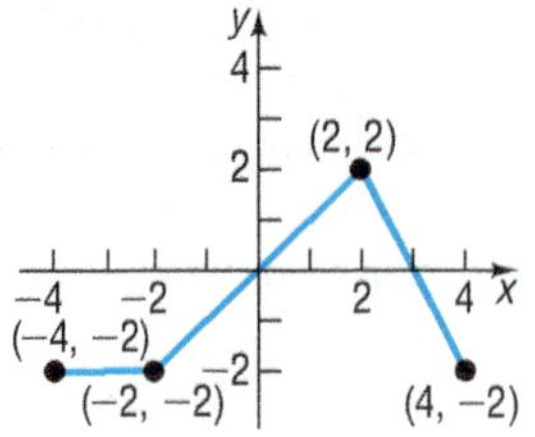

65.

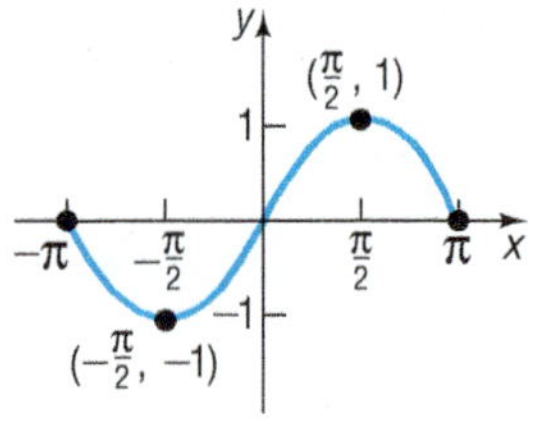

66.

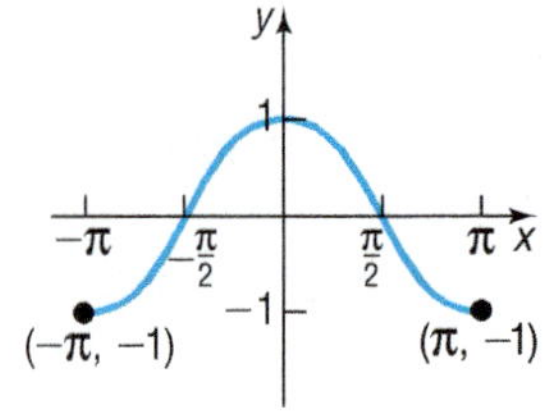

Mixed Practice

67. (a) Using a graphing utility, graph $f(x) = x^3 - 9x$ for $-4 \le x \le 4$.
(b) Find the x-intercepts of the graph of f.
(c) Approximate any local maxima and local minima.
(d) Determine where f is increasing and where it is decreasing.
(e) Without using a graphing utility, repeat parts (b)–(d) for $y = f(x + 2)$.
(f) Without using a graphing utility, repeat parts (b)–(d) for $y = 2f(x)$.
(g) Without using a graphing utility, repeat parts (b)–(d) for $y = f(-x)$.

68. (a) Using a graphing utility, graph $f(x) = x^3 - 4x$ for $-3 \le x \le 3$.
(b) Find the x-intercepts of the graph of f.
(c) Approximate any local maxima and local minima.
(d) Determine where f is increasing and where it is decreasing.
(e) Without using a graphing utility, repeat parts (b)–(d) for $y = f(x - 4)$.
(f) Without using a graphing utility, repeat parts (b)–(d) for $y = f(2x)$.
(g) Without using a graphing utility, repeat parts (b)–(d) for $y = -f(x)$.

In Problems 69–76, complete the square of each quadratic expression. Then graph each function using the technique of shifting. (If necessary, refer to Chapter R, Section R.5 to review completing the square.)

69. $f(x) = x^2 + 2x$
70. $f(x) = x^2 - 6x$
71. $f(x) = x^2 - 8x + 1$
72. $f(x) = x^2 + 4x + 2$
73. $f(x) = 2x^2 - 12x + 19$
74. $f(x) = 3x^2 + 6x + 1$
75. $f(x) = -3x^2 - 12x - 17$
76. $f(x) = -2x^2 - 12x - 13$

Applications and Extensions

77. The graph of a function f is illustrated in the figure.
(a) Draw the graph of $y = |f(x)|$.
(b) Draw the graph of $y = f(|x|)$.

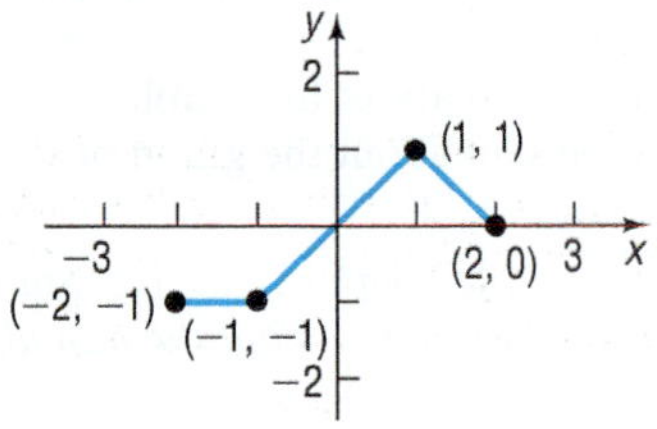

78. The graph of a function f is illustrated in the figure.
(a) Draw the graph of $y = |f(x)|$.
(b) Draw the graph of $y = f(|x|)$.

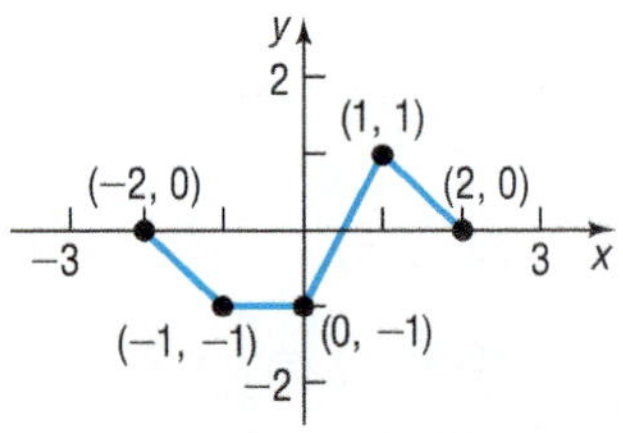

79. Suppose $(1, 3)$ is a point on the graph of $y = f(x)$.
(a) What point is on the graph of $y = f(x + 3) - 5$?
(b) What point is on the graph of $y = -2f(x - 2) + 1$?
(c) What point is on the graph of $y = f(2x + 3)$?

80. Suppose $(-3, 5)$ is a point on the graph of $y = g(x)$.
(a) What point is on the graph of $y = g(x + 1) - 3$?
(b) What point is on the graph of $y = -3g(x - 4) + 3$?
(c) What point is on the graph of $y = g(3x + 9)$?

81. Graph the following functions using transformations.
(a) $f(x) = \text{int}(-x)$ (b) $g(x) = -\text{int}(x)$

82. Graph the following functions using transformations.
(a) $f(x) = \text{int}(x - 1)$ (b) $g(x) = \text{int}(1 - x)$

83. (a) Graph $f(x) = |x - 3| - 3$ using transformations.
(b) Find the area of the region that is bounded by f and the x-axis and lies below the x-axis.

84. (a) Graph $f(x) = -2|x - 4| + 4$ using transformations.
(b) Find the area of the region that is bounded by f and the x-axis and lies above the x-axis.

85. Thermostat Control Energy conservation experts estimate that homeowners can save 5% to 10% on winter heating bills by programming their thermostats 5 to 10 degrees lower while sleeping. In the graph (next page), the temperature T (in degrees Fahrenheit) of a home is given as a function of time t (in hours after midnight) over a 24-hour period.
(a) At what temperature is the thermostat set during daytime hours? At what temperature is the thermostat set overnight?

(b) The homeowner reprograms the thermostat to $y = T(t) - 2$. Explain how this affects the temperature in the house. Graph this new function.

(c) The homeowner reprograms the thermostat to $y = T(t + 1)$. Explain how this affects the temperature in the house. Graph this new function.

Source: Roger Albright, *547 Ways to Be Fuel Smart,* 2000

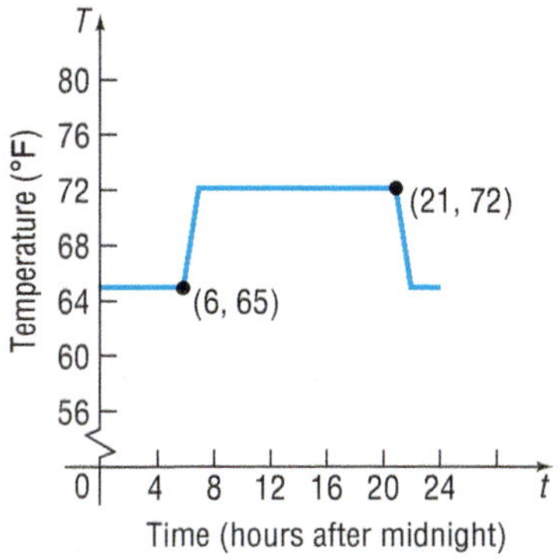

86. Digital Music Revenues The total projected worldwide digital music revenues R, in millions of dollars, for the years 2012 through 2017 can be estimated by the function

$$R(x) = 28.6x^2 + 300x + 4843$$

where x is the number of years after 2012.

(a) Find $R(0), R(3)$, and $R(5)$ and explain what each value represents.

(b) Find $r(x) = R(x - 2)$.

(c) Find $r(2), r(5)$, and $r(7)$ and explain what each value represents.

(d) In the model $r = r(x)$, what does x represent?

(e) Would there be an advantage in using the model r when estimating the projected revenues for a given year instead of the model R?

Source: *IFPI Digital Music Report*

87. Temperature Measurements The relationship between the Celsius (°C) and Fahrenheit (°F) scales for measuring temperature is given by the equation

$$F = \frac{9}{5}C + 32$$

The relationship between the Celsius (°C) and Kelvin (K) scales is $K = C + 273$. Graph the equation $F = \frac{9}{5}C + 32$ using degrees Fahrenheit on the y-axis and degrees Celsius on the x-axis. Use the techniques introduced in this section to obtain the graph showing the relationship between Kelvin and Fahrenheit temperatures.

88. Period of a Pendulum The period T (in seconds) of a simple pendulum is a function of its length l (in feet) defined by the equation

$$T = 2\pi\sqrt{\frac{l}{g}}$$

where $g \approx 32.2$ feet per second per second is the acceleration due to gravity.

(a) Use a graphing utility to graph the function $T = T(l)$.

(b) Now graph the functions $T = T(l + 1), T = T(l + 2)$, and $T = T(l + 3)$.

(c) Discuss how adding to the length l changes the period T.

(d) Now graph the functions $T = T(2l), T = T(3l)$, and $T = T(4l)$.

(e) Discuss how multiplying the length l by factors of 2, 3, and 4 changes the period T.

89. The equation $y = (x - c)^2$ defines a *family of parabolas,* one parabola for each value of c. On one set of coordinate axes, graph the members of the family for $c = 0, c = 3$, and $c = -2$.

90. Repeat Problem 89 for the family of parabolas $y = x^2 + c$.

Explaining Concepts: Discussion and Writing

91. Suppose that the graph of a function f is known. Explain how the graph of $y = 4f(x)$ differs from the graph of $y = f(4x)$.

92. Suppose that the graph of a function f is known. Explain how the graph of $y = f(x) - 2$ differs from the graph of $y = f(x - 2)$.

93. The area under the curve $y = \sqrt{x}$ bounded from below by the x-axis and on the right by $x = 4$ is $\frac{16}{3}$ square units. Using the ideas presented in this section, what do you think is the area under the curve of $y = \sqrt{-x}$ bounded from below by the x-axis and on the left by $x = -4$? Justify your answer.

94. Explain how the range of the function $f(x) = x^2$ compares to the range of $g(x) = f(x) + k$.

95. Explain how the domain of $g(x) = \sqrt{x}$ compares to the domain of $g(x - k)$, where $k \geq 0$.

Retain Your Knowledge

Problems 96–99 are based on material learned earlier in the course. The purpose of these problems is to keep the material fresh in your mind so that you are better prepared for the final exam.

96. Determine the slope and y-intercept of the graph of $3x - 5y = 30$.

97. Simplify $\frac{(x^{-2}y^3)^4}{(x^2y^{-5})^{-2}}$.

98. The amount of water used when taking a shower varies directly with the number of minutes the shower is run. If a 4-minute shower uses 7 gallons of water, how much water is used in a 9-minute shower?

99. List the intercepts and test for symmetry: $y^2 = x + 4$

3.6 Mathematical Models: Building Functions

OBJECTIVE 1 Build and Analyze Functions (p. 268)

1 Build and Analyze Functions

Real-world problems often result in mathematical models that involve functions. These functions need to be constructed or built based on the information given. In building functions, we must be able to translate the verbal description into the language of mathematics. This is done by assigning symbols to represent the independent and dependent variables and then by finding the function or rule that relates these variables.

EXAMPLE 1 Finding the Distance from the Origin to a Point on a Graph

Let $P = (x, y)$ be a point on the graph of $y = x^2 - 1$.

(a) Express the distance d from P to the origin O as a function of x.

(b) What is d if $x = 0$?

(c) What is d if $x = 1$?

(d) What is d if $x = \dfrac{\sqrt{2}}{2}$?

(e) Use a graphing utility to graph the function $d = d(x)$, $x \geq 0$. Rounding to two decimal places, find the value(s) of x at which d has a local minimum. [This gives the point(s) on the graph of $y = x^2 - 1$ closest to the origin.]

Solution

(a) Figure 61 illustrates the graph of $y = x^2 - 1$. The distance d from P to O is

$$d = \sqrt{(x - 0)^2 + (y - 0)^2} = \sqrt{x^2 + y^2}$$

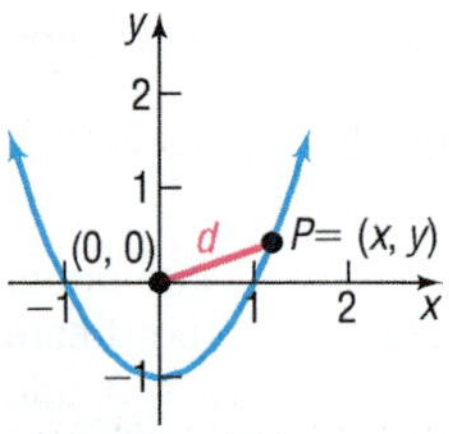

Figure 61 $y = x^2 - 1$

Since P is a point on the graph of $y = x^2 - 1$, substitute $x^2 - 1$ for y. Then

$$d(x) = \sqrt{x^2 + (x^2 - 1)^2} = \sqrt{x^4 - x^2 + 1}$$

The distance d is expressed as a function of x.

(b) If $x = 0$, the distance d is

$$d(0) = \sqrt{0^4 - 0^2 + 1} = \sqrt{1} = 1$$

(c) If $x = 1$, the distance d is

$$d(1) = \sqrt{1^4 - 1^2 + 1} = 1$$

(d) If $x = \dfrac{\sqrt{2}}{2}$, the distance d is

$$d\left(\frac{\sqrt{2}}{2}\right) = \sqrt{\left(\frac{\sqrt{2}}{2}\right)^4 - \left(\frac{\sqrt{2}}{2}\right)^2 + 1} = \sqrt{\frac{1}{4} - \frac{1}{2} + 1} = \frac{\sqrt{3}}{2}$$

(e) Figure 62 shows the graph of $Y_1 = \sqrt{x^4 - x^2 + 1}$. Using the MINIMUM feature on a graphing utility, we find that when $x \approx 0.71$ the value of d is smallest. The local minimum value is $d \approx 0.87$ rounded to two decimal places. Since $d(x)$ is even, it follows by symmetry that when $x \approx -0.71$, the value of d is the same local minimum value. Since $(\pm 0.71)^2 - 1 \approx -0.50$, the points $(-0.71, -0.50)$ and $(0.71, -0.50)$ on the graph of $y = x^2 - 1$ are closest to the origin. ■

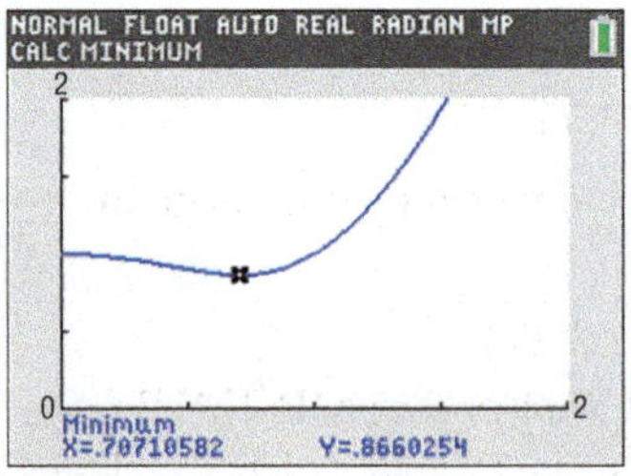

Figure 62 $d(x) = \sqrt{x^4 - x^2 + 1}$

Now Work PROBLEM 1

EXAMPLE 2

Area of a Rectangle

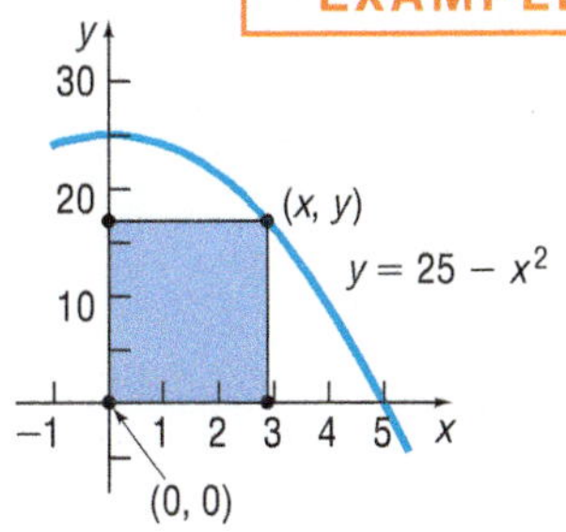

Figure 63

A rectangle has one corner in quadrant I on the graph of $y = 25 - x^2$, another at the origin, a third on the positive y-axis, and the fourth on the positive x-axis. See Figure 63.

(a) Express the area A of the rectangle as a function of x.
(b) What is the domain of A?
(c) Graph $A = A(x)$.
(d) For what value of x is the area largest?

Solution

(a) The area A of the rectangle is $A = xy$, where $y = 25 - x^2$. Substituting this expression for y, we obtain $A(x) = x(25 - x^2) = 25x - x^3$.
(b) Since (x, y) is in quadrant I, we have $x > 0$. Also, $y = 25 - x^2 > 0$, which implies that $x^2 < 25$, so $-5 < x < 5$. Combining these restrictions, we have the domain of A as $\{x | 0 < x < 5\}$, or $(0, 5)$ using interval notation.
(c) See Figure 64 for the graph of $A = A(x)$.
(d) Using MAXIMUM, we find that the maximum area is 48.11 square units at $x = 2.89$ units, each rounded to two decimal places. See Figure 65.

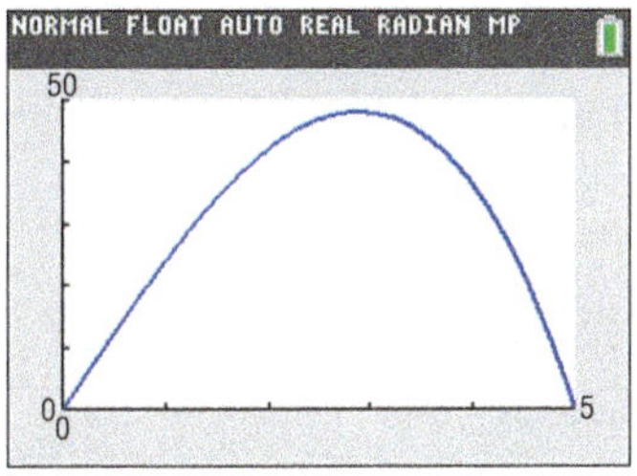

Figure 64 $A(x) = 25x - x^3$

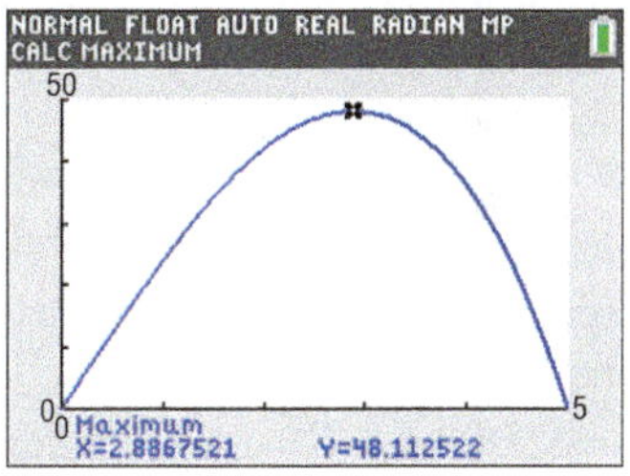

Figure 65

Now Work PROBLEM 7

EXAMPLE 3

Close Call?

Suppose two planes flying at the same altitude are headed toward each other. One plane is flying due south at a groundspeed of 400 miles per hour and is 600 miles from the potential intersection point of the planes. The other plane is flying due west with a groundspeed of 250 miles per hour and is 400 miles from the potential intersection point of the planes. See Figure 66.

(a) Build a model that expresses the distance d between the planes as a function of time t.
(b) Use a graphing utility to graph $d = d(t)$. How close do the planes come to each other? At what time are the planes closest?

Solution

(a) Refer to Figure 66. The distance d between the two planes is the hypotenuse of a right triangle. At any time t, the length of the north/south leg of the triangle is $600 - 400t$. At any time t, the length of the east/west leg of the triangle is $400 - 250t$. Use the Pythagorean Theorem to find that the square of the distance between the two planes is

$$d^2 = (600 - 400t)^2 + (400 - 250t)^2$$

Therefore, the distance between the two planes as a function of time is given by the model

$$d(t) = \sqrt{(600 - 400t)^2 + (400 - 250t)^2}$$

(b) Figure 67(a) on the next page shows the graph of $d = d(t)$. Using MINIMUM, the minimum distance between the planes is 21.20 miles, and the time at which the planes are closest is after 1.53 hours, each rounded to two decimal places. See Figure 67(b).

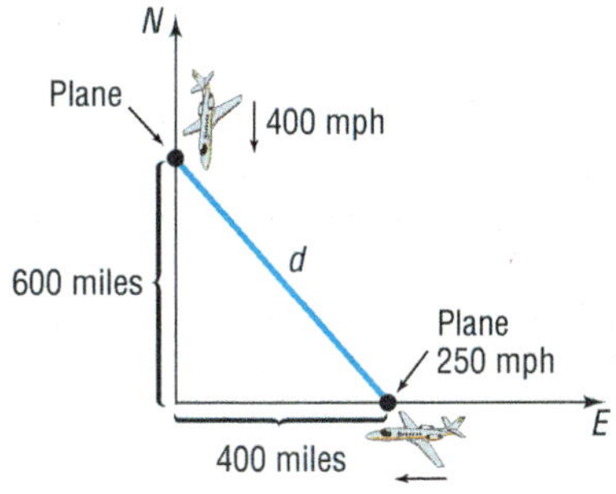

Figure 66

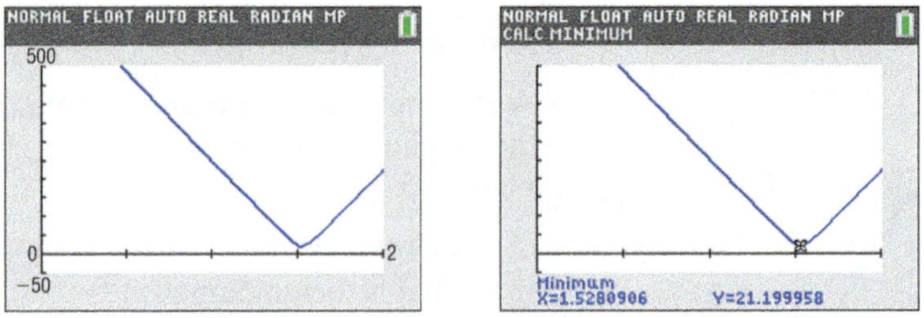

(a) (b)

Figure 67

Now Work PROBLEM 19

3.6 Assess Your Understanding

Applications and Extensions

1. Let $P = (x, y)$ be a point on the graph of $y = x^2 - 8$.
   (a) Express the distance d from P to the origin as a function of x.
   (b) What is d if $x = 0$?
   (c) What is d if $x = 1$?
   (d) Use a graphing utility to graph $d = d(x)$.
   (e) For what values of x is d smallest?
2. Let $P = (x, y)$ be a point on the graph of $y = x^2 - 8$.
   (a) Express the distance d from P to the point $(0, -1)$ as a function of x.
   (b) What is d if $x = 0$?
   (c) What is d if $x = -1$?
   (d) Use a graphing utility to graph $d = d(x)$.
   (e) For what values of x is d smallest?
3. Let $P = (x, y)$ be a point on the graph of $y = \sqrt{x}$.
   (a) Express the distance d from P to the point $(1, 0)$ as a function of x.
   (b) Use a graphing utility to graph $d = d(x)$.
   (c) For what values of x is d smallest?
4. Let $P = (x, y)$ be a point on the graph of $y = \dfrac{1}{x}$.
   (a) Express the distance d from P to the origin as a function of x.
   (b) Use a graphing utility to graph $d = d(x)$.
   (c) For what values of x is d smallest?
5. A right triangle has one vertex on the graph of $y = x^3, x > 0$, at (x, y), another at the origin, and the third on the positive y-axis at $(0, y)$, as shown in the figure. Express the area A of the triangle as a function of x.

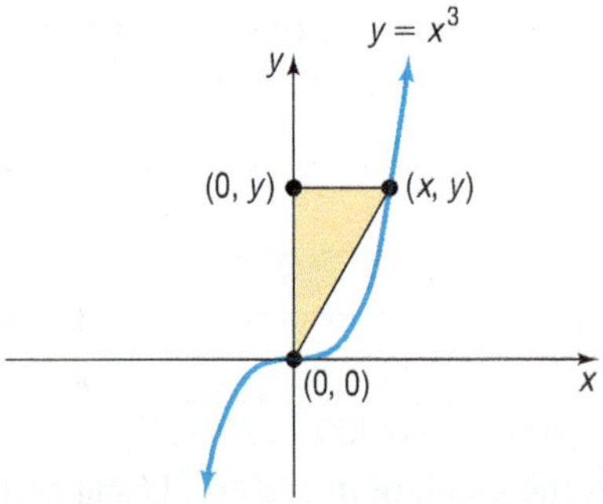

6. A right triangle has one vertex on the graph of $y = 9 - x^2, x > 0$, at (x, y), another at the origin, and the third on the positive x-axis at $(x, 0)$. Express the area A of the triangle as a function of x.
7. A rectangle has one corner in quadrant I on the graph of $y = 16 - x^2$, another at the origin, a third on the positive y-axis, and the fourth on the positive x-axis. See the figure below.

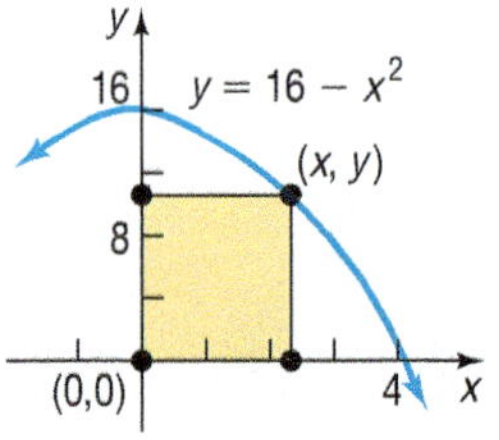

   (a) Express the area A of the rectangle as a function of x.
   (b) What is the domain of A?
   (c) Graph $A = A(x)$. For what value of x is A largest?
8. A rectangle is inscribed in a semicircle of radius 2. See the figure. Let $P = (x, y)$ be the point in quadrant I that is a vertex of the rectangle and is on the circle.

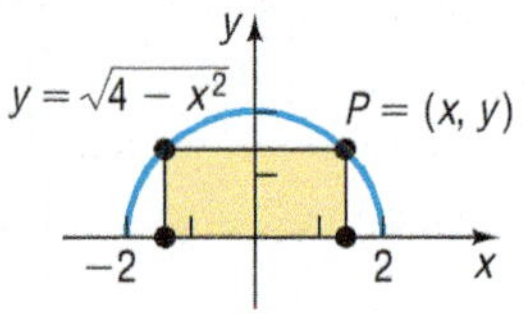

   (a) Express the area A of the rectangle as a function of x.
   (b) Express the perimeter p of the rectangle as a function of x.
   (c) Graph $A = A(x)$. For what value of x is A largest?
   (d) Graph $p = p(x)$. For what value of x is p largest?
9. A rectangle is inscribed in a circle of radius 2. See the figure. Let $P = (x, y)$ be the point in quadrant I that is a vertex of the rectangle and is on the circle.

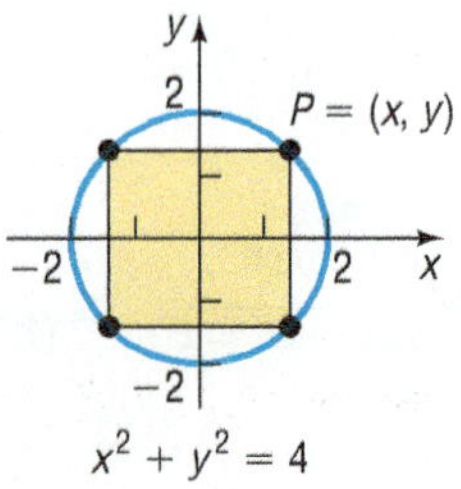

(a) Express the area A of the rectangle as a function of x.
(b) Express the perimeter p of the rectangle as a function of x.
(c) Graph $A = A(x)$. For what value of x is A largest?
(d) Graph $p = p(x)$. For what value of x is p largest?

10. A circle of radius r is inscribed in a square. See the figure.

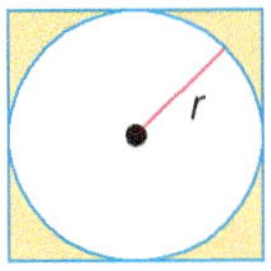

(a) Express the area A of the square as a function of the radius r of the circle.
(b) Express the perimeter p of the square as a function of r.

11. Geometry A wire 10 meters long is to be cut into two pieces. One piece will be shaped as a square, and the other piece will be shaped as a circle. See the figure.

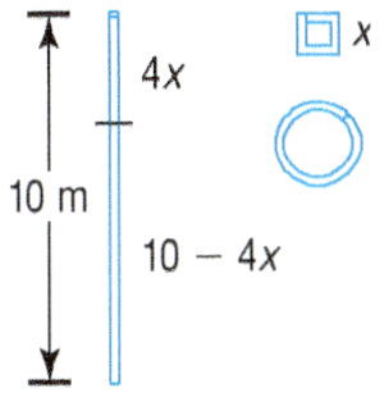

(a) Express the total area A enclosed by the pieces of wire as a function of the length x of a side of the square.
(b) What is the domain of A?
(c) Graph $A = A(x)$. For what value of x is A smallest?

12. Geometry A wire 10 meters long is to be cut into two pieces. One piece will be shaped as an equilateral triangle, and the other piece will be shaped as a circle.

(a) Express the total area A enclosed by the pieces of wire as a function of the length x of a side of the equilateral triangle.
(b) What is the domain of A ?
(c) Graph $A = A(x)$. For what value of x is A smallest?

13. Geometry A wire of length x is bent into the shape of a circle.

(a) Express the circumference C of the circle as a function of x.
(b) Express the area A of the circle as a function of x.

14. Geometry A wire of length x is bent into the shape of a square.

(a) Express the perimeter p of the square as a function of x.
(b) Express the area A of the square as a function of x.

15. Geometry A semicircle of radius r is inscribed in a rectangle so that the diameter of the semicircle is the length of the rectangle. See the figure.

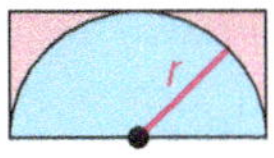

(a) Express the area A of the rectangle as a function of the radius r of the semicircle.
(b) Express the perimeter p of the rectangle as a function of r.

16. Geometry An equilateral triangle is inscribed in a circle of radius r. See the figure. Express the circumference C of the circle as a function of the length x of a side of the triangle.

[**Hint:** First show that $r^2 = \frac{x^2}{3}$.]

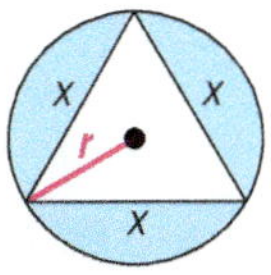

17. Geometry An equilateral triangle is inscribed in a circle of radius r. See the figure in Problem 16. Express the area A within the circle, but outside the triangle, as a function of the length x of a side of the triangle.

18. Uniform Motion Two cars leave an intersection at the same time. One is headed south at a constant speed of 30 miles per hour, and the other is headed west at a constant speed of 40 miles per hour (see the figure). Build a model that expresses the distance d between the cars as a function of the time t.

[**Hint:** At $t = 0$, the cars leave the intersection.]

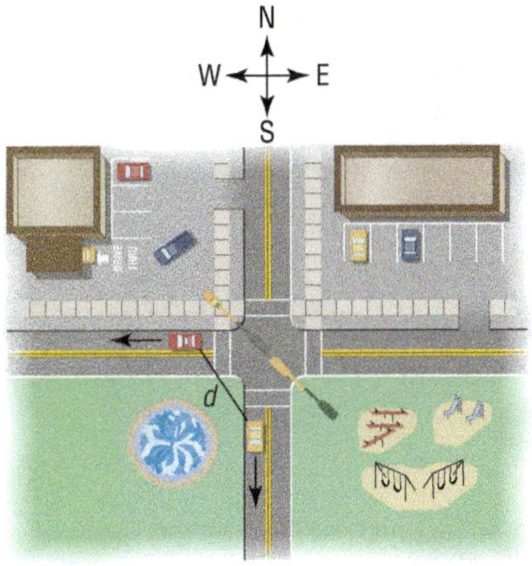

19. Uniform Motion Two cars are approaching an intersection. One is 2 miles south of the intersection and is moving at a constant speed of 30 miles per hour. At the same time, the other car is 3 miles east of the intersection and is moving at a constant speed of 40 miles per hour.

(a) Build a model that expresses the distance d between the cars as a function of time t.
[**Hint:** At $t = 0$, the cars are 2 miles south and 3 miles east of the intersection, respectively.]
(b) Use a graphing utility to graph $d = d(t)$. For what value of t is d smallest?

20. Inscribing a Cylinder in a Sphere Inscribe a right circular cylinder of height h and radius r in a sphere of fixed radius R. See the illustration. Express the volume V of the cylinder as a function of h.

[**Hint:** $V = \pi r^2 h$. Note also the right triangle.]

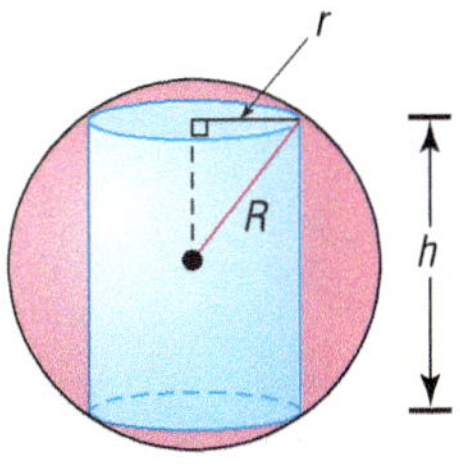

Sphere

21. Inscribing a Cylinder in a Cone Inscribe a right circular cylinder of height h and radius r in a cone of fixed radius R and fixed height H. See the illustration. Express the volume V of the cylinder as a function of r.

[**Hint:** $V = \pi r^2 h$. Note also the similar triangles.]

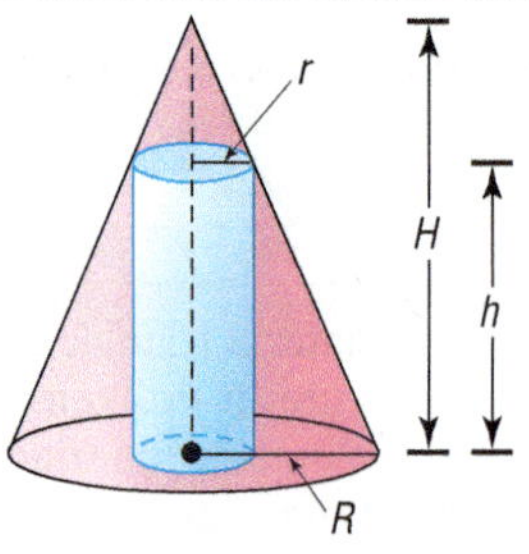

Cone

22. Installing Cable TV MetroMedia Cable is asked to provide service to a customer whose house is located 2 miles from the road along which the cable is buried. The nearest connection box for the cable is located 5 miles down the road. See the figure.

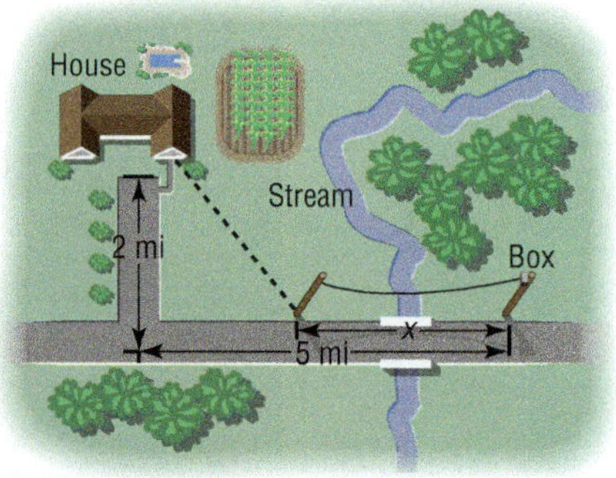

(a) If the installation cost is \$500 per mile along the road and \$700 per mile off the road, build a model that expresses the total cost C of installation as a function of the distance x (in miles) from the connection box to the point where the cable installation turns off the road. Find the domain of $C = C(x)$.
(b) Compute the cost if $x = 1$ mile.
(c) Compute the cost if $x = 3$ miles.
(d) Graph the function $C = C(x)$. Use TRACE to see how the cost C varies as x changes from 0 to 5.
(e) What value of x results in the least cost?

23. Time Required to Go from an Island to a Town An island is 2 miles from the nearest point P on a straight shoreline. A town is 12 miles down the shore from P. See the illustration.

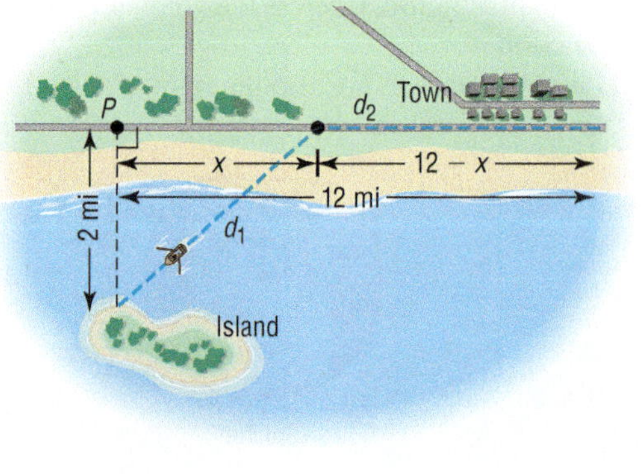

(a) If a person can row a boat at an average speed of 3 miles per hour and the same person can walk 5 miles per hour, build a model that expresses the time T that it takes to go from the island to town as a function of the distance x from P to where the person lands the boat.
(b) What is the domain of T?
(c) How long will it take to travel from the island to town if the person lands the boat 4 miles from P?
(d) How long will it take if the person lands the boat 8 miles from P?

24. Filling a Conical Tank Water is poured into a container in the shape of a right circular cone with radius 4 feet and height 16 feet. See the figure. Express the volume V of the water in the cone as a function of the height h of the water.

[**Hint:** The volume V of a cone of radius r and height h is $V = \frac{1}{3}\pi r^2 h$.]

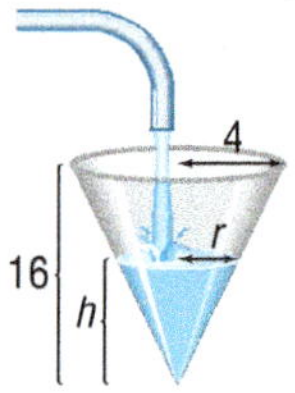

25. Constructing an Open Box An open box with a square base is to be made from a square piece of cardboard 24 inches on a side by cutting out a square from each corner and turning up the sides. See the figure.

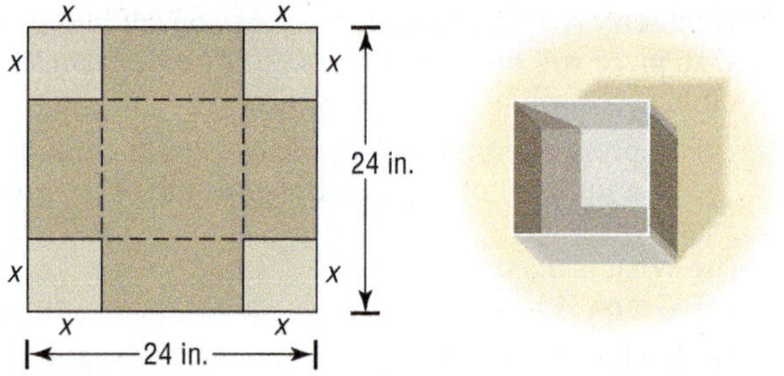

(a) Express the volume V of the box as a function of the length x of the side of the square cut from each corner.
(b) What is the volume if a 3-inch square is cut out?
(c) What is the volume if a 10-inch square is cut out?
(d) Graph $V = V(x)$. For what value of x is V largest?

26. Constructing an Open Box An open box with a square base is required to have a volume of 10 cubic feet.

(a) Express the amount A of material used to make such a box as a function of the length x of a side of the square base.
(b) How much material is required for a base 1 foot by 1 foot?
(c) How much material is required for a base 2 feet by 2 feet?
(d) Use a graphing utility to graph $A = A(x)$. For what value of x is A smallest?

Retain Your Knowledge

Problems 27–30 are based on material learned earlier in the course. The purpose of these problems is to keep the material fresh in your mind so that you are better prepared for the final exam.

27. Solve: $|2x - 3| - 5 = -2$

28. A 16-foot long Ford Fusion wants to pass a 50-foot truck traveling at 55 mi/h. How fast must the car travel to completely pass the truck in 5 seconds?

29. Find the slope of the line containing the points $(3, -2)$ and $(1, 6)$.

30. Find the missing length x for the given pair of similar triangles.

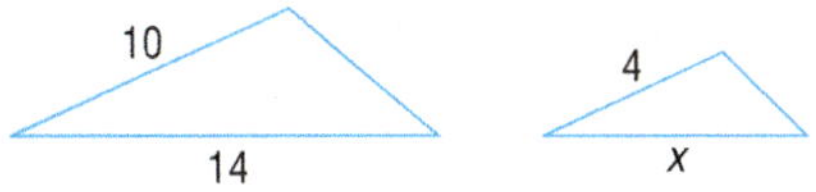

Chapter Review

Library of Functions

Constant function (p. 247)

$f(x) = b$

The graph is a horizontal line with y-intercept b.

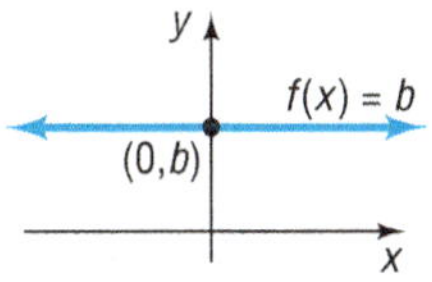

Identity function (p. 247)

$f(x) = x$

The graph is a line with slope 1 and y-intercept 0.

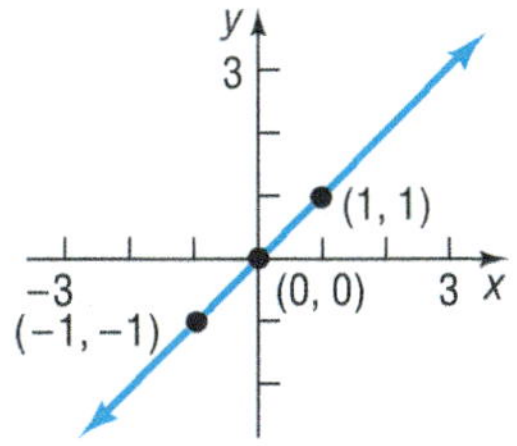

Square function (p. 247)

$f(x) = x^2$

The graph is a parabola with intercept at $(0, 0)$.

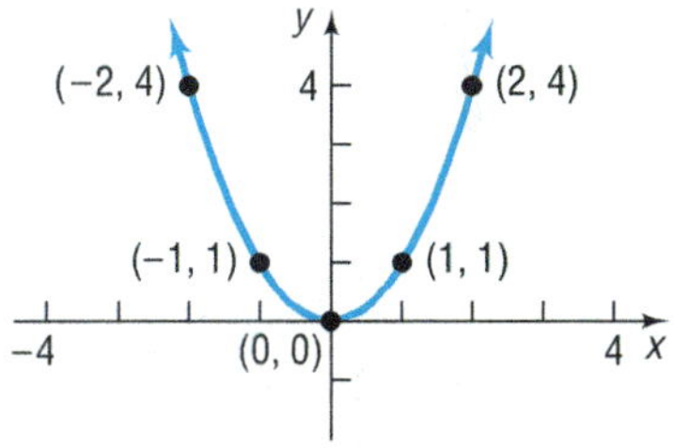

Cube function (p. 248)

$f(x) = x^3$

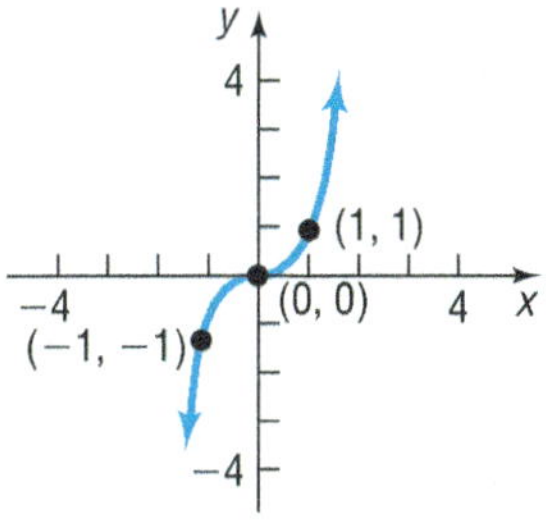

Square root function (p. 248)

$f(x) = \sqrt{x}$

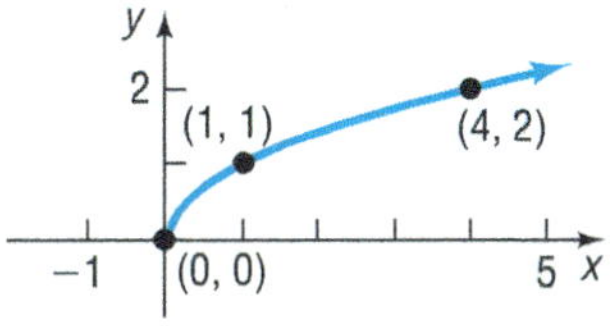

Cube root function (p. 248)

$f(x) = \sqrt[3]{x}$

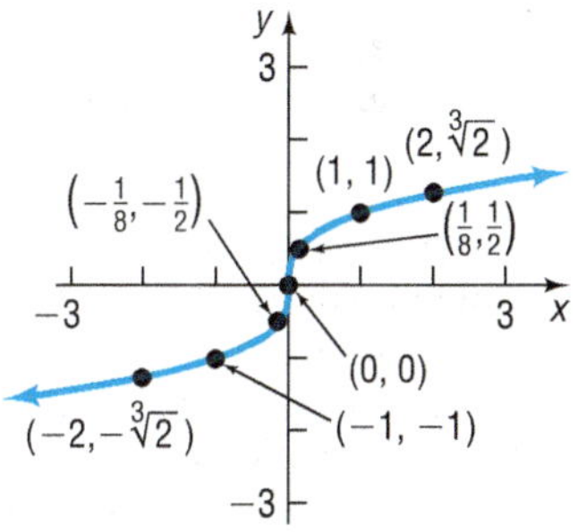

Reciprocal function (p. 248)

$f(x) = \dfrac{1}{x}$

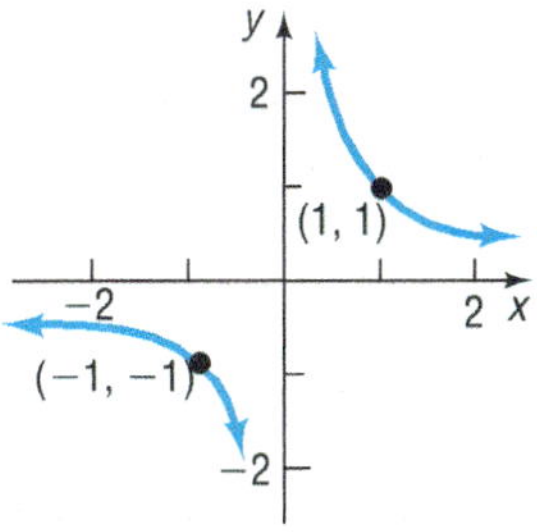

Absolute value function (p. 248)

$f(x) = |x|$

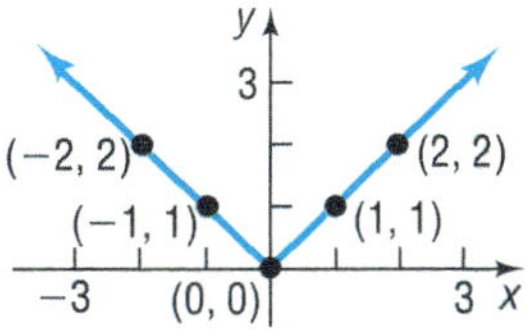

Greatest integer function (p. 249)

$f(x) = \text{int}(x)$

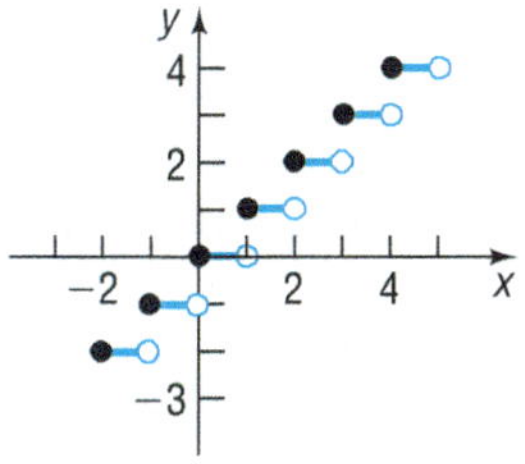

Things to Know

Function (pp. 207–210)
A relation between two sets so that each element x in the first set, the domain, has corresponding to it exactly one element y in the second set, the range. The range is the set of y-values of the function for the x-values in the domain.

A function can also be described as a set of ordered pairs (x, y) in which no first element is paired with two different second elements.

Function notation (pp. 210–213)
$y = f(x)$

f is a symbol for the function.

x is the argument, or independent variable.

y is the dependent variable.

$f(x)$ is the value of the function at x, or the image of x.

A function f may be defined implicitly by an equation involving x and y or explicitly by writing $y = f(x)$.

Difference quotient of f (p. 213)
$$\frac{f(x+h) - f(x)}{h} \quad h \neq 0$$

Domain (pp. 214–216)
If unspecified, the domain of a function f defined by an equation is the largest set of real numbers for which $f(x)$ is a real number.

Vertical-line test (p. 222)
A set of points in the xy-plane is the graph of a function if and only if every vertical line intersects the graph in at most one point.

Even function f (p. 232)
$f(-x) = f(x)$ for every x in the domain ($-x$ must also be in the domain).

Odd function f (p. 232)
$f(-x) = -f(x)$ for every x in the domain ($-x$ must also be in the domain).

Increasing function (p. 234)
A function f is increasing on an interval I if, for any choice of x_1 and x_2 in I, with $x_1 < x_2$, we have $f(x_1) < f(x_2)$.

Decreasing function (p. 234)
A function f is decreasing on an interval I if, for any choice of x_1 and x_2 in I, with $x_1 < x_2$, we have $f(x_1) > f(x_2)$.

Constant function (p. 234)
A function f is constant on an interval I if, for all choices of x in I, the values of $f(x)$ are equal.

Local maximum (p. 235)
A function f, defined on some interval I, has a local maximum at c if there is an open interval in I containing c such that, for all x in this open interval, $f(x) \leq f(c)$.

Local minimum (p. 235)
A function f, defined on some interval I, has a local minimum at c if there is an open interval in I containing c such that, for all x in this open interval, $f(x) \geq f(c)$.

Absolute maximum and Absolute minimum (p. 236)
Let f denote a function defined on some interval I.
If there is a number u in I for which $f(x) \leq f(u)$ for all x in I, then f has an absolute maximum at u, and the number $f(u)$ is the absolute maximum of f on I.

If there is a number v in I for which $f(x) \geq f(v)$, for all x in I, then f has an absolute minimum at v, and the number $f(v)$ is the absolute minimum of f on I.

Average rate of change of a function (p. 238)
The average rate of change of f from a to b is
$$\frac{\Delta y}{\Delta x} = \frac{f(b) - f(a)}{b - a} \quad a \neq b$$

Objectives

Section		You should be able to . . .	Examples	Review Exercises
3.1	1	Determine whether a relation represents a function (p. 207)	1–5	1, 2
	2	Find the value of a function (p. 210)	6, 7	3–5, 39
	3	Find the difference quotient of a function (p. 213)	8	15
	4	Find the domain of a function defined by an equation (p. 214)	9, 10	6–11
	5	Form the sum, difference, product, and quotient of two functions (p. 216)	11	12–14
3.2	1	Identify the graph of a function (p. 222)	1	27, 28
	2	Obtain information from or about the graph of a function (p. 223)	2–4	16(a)–(e), 17(a), 17(e), 17(g)

Section		You should be able to . . .	Examples	Review Exercises
3.3	1	Determine even and odd functions from a graph (p. 231)	1	17(f)
	2	Identify even and odd functions from an equation (p. 233)	2	18–21
	3	Use a graph to determine where a function is increasing, decreasing, or constant (p. 234)	3	17(b)
	4	Use a graph to locate local maxima and local minima (p. 235)	4	17(c)
	5	Use a graph to locate the absolute maximum and the absolute minimum (p. 236)	5	17(d)
	6	Use a graphing utility to approximate local maxima and local minima and to determine where a function is increasing or decreasing (p. 237)	6	22, 23, 40(d), 41(b)
	7	Find the average rate of change of a function (p. 238)	7, 8	24–26
3.4	1	Graph the functions listed in the library of functions (p. 245)	1, 2	29, 30
	2	Graph piecewise-defined functions (p. 250)	3, 4, 5	37, 38
3.5	1	Graph functions using vertical and horizontal shifts (p. 256)	1, 2, 4–6	16(f), 31, 33–36
	2	Graph functions using compressions and stretches (p. 258)	3, 5	16(g), 32, 36
	3	Graph functions using reflections about the x-axis or y-axis (p. 260)	4, 6	16(h), 32, 34, 36
3.6	1	Build and analyze functions (p. 268)	1–3	40, 41

Review Exercises

In Problems 1 and 2, determine whether each relation represents a function. For each function, state the domain and range.

1. $\{(-1, 0), (2, 3), (4, 0)\}$

2. $\{(4, -1), (2, 1), (4, 2)\}$

In Problems 3–5, find the following for each function:

(a) $f(2)$ (b) $f(-2)$ (c) $f(-x)$ (d) $-f(x)$ (e) $f(x-2)$ (f) $f(2x)$

3. $f(x) = \dfrac{3x}{x^2 - 1}$

4. $f(x) = \sqrt{x^2 - 4}$

5. $f(x) = \dfrac{x^2 - 4}{x^2}$

In Problems 6–11, find the domain of each function.

6. $f(x) = \dfrac{x}{x^2 - 9}$

7. $f(x) = \sqrt{2 - x}$

8. $g(x) = \dfrac{|x|}{x}$

9. $f(x) = \dfrac{x}{x^2 + 2x - 3}$

10. $f(x) = \dfrac{\sqrt{x+1}}{x^2 - 4}$

11. $g(x) = \dfrac{x}{\sqrt{x+8}}$

In Problems 12–14, find $f + g$, $f - g$, $f \cdot g$, and $\dfrac{f}{g}$ for each pair of functions. State the domain of each of these functions.

12. $f(x) = 2 - x$; $g(x) = 3x + 1$

13. $f(x) = 3x^2 + x + 1$; $g(x) = 3x$

14. $f(x) = \dfrac{x+1}{x-1}$; $g(x) = \dfrac{1}{x}$

15. Find the difference quotient of $f(x) = -2x^2 + x + 1$; that is, find $\dfrac{f(x+h) - f(x)}{h}$, $h \neq 0$.

16. Consider the graph of the function f on the right.

(a) Find the domain and the range of f.
(b) List the intercepts.
(c) Find $f(-2)$.
(d) For what value of x does $f(x) = -3$?
(e) Solve $f(x) > 0$.
(f) Graph $y = f(x - 3)$.
(g) Graph $y = f\left(\dfrac{1}{2}x\right)$.
(h) Graph $y = -f(x)$.

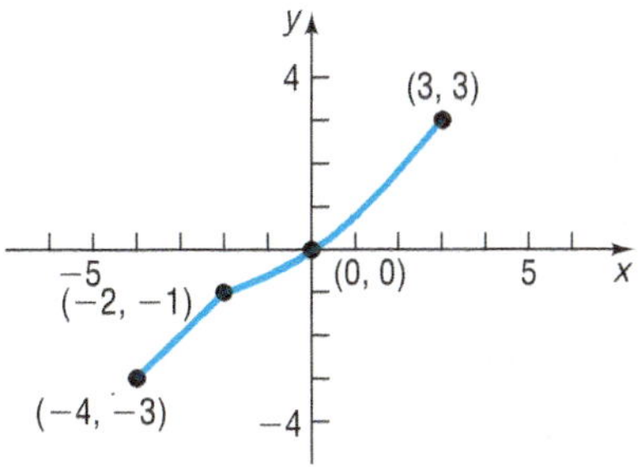

17. Use the graph of the function f shown to find:

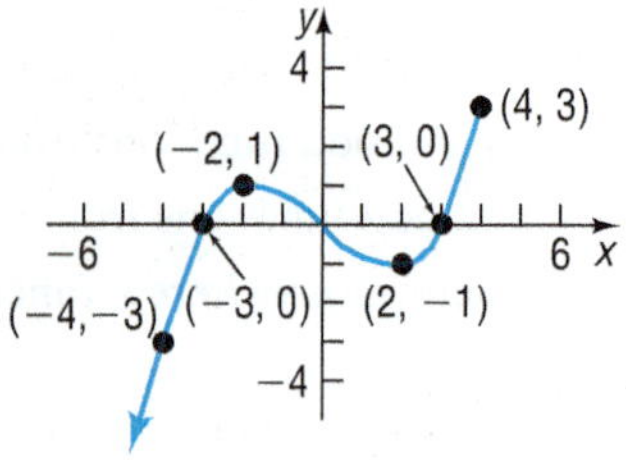

(a) The domain and the range of f.
(b) The intervals on which f is increasing, decreasing, or constant.
(c) The local minimum values and local maximum values.
(d) The absolute maximum and absolute minimum.
(e) Whether the graph is symmetric with respect to the x-axis, the y-axis, or the origin.
(f) Whether the function is even, odd, or neither.
(g) The intercepts, if any.

In Problems 18–21, determine (algebraically) whether the given function is even, odd, or neither.

18. $f(x) = x^3 - 4x$

19. $g(x) = \dfrac{4 + x^2}{1 + x^4}$

20. $G(x) = 1 - x + x^3$

21. $f(x) = \dfrac{x}{1 + x^2}$

In Problems 22 and 23, use a graphing utility to graph each function over the indicated interval. Approximate any local maximum values and local minimum values. Determine where the function is increasing and where it is decreasing.

22. $f(x) = 2x^3 - 5x + 1 \quad [-3, 3]$

23. $f(x) = 2x^4 - 5x^3 + 2x + 1 \quad [-2, 3]$

24. Find the average rate of change of $f(x) = 8x^2 - x$:

(a) From 1 to 2 (b) From 0 to 1 (c) From 2 to 4

In Problems 25 and 26, find the average rate of change from 2 to 3 for each function f. Be sure to simplify.

25. $f(x) = 2 - 5x$

26. $f(x) = 3x - 4x^2$

In Problems 27 and 28, is the graph shown the graph of a function?

27.

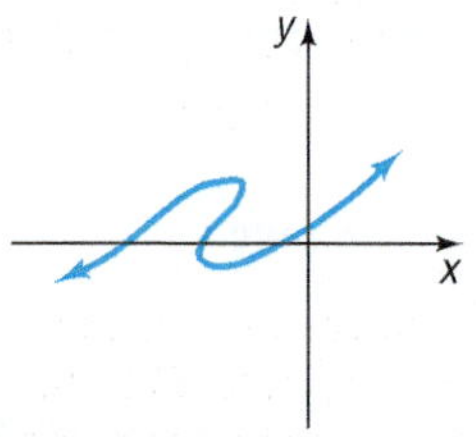

28.

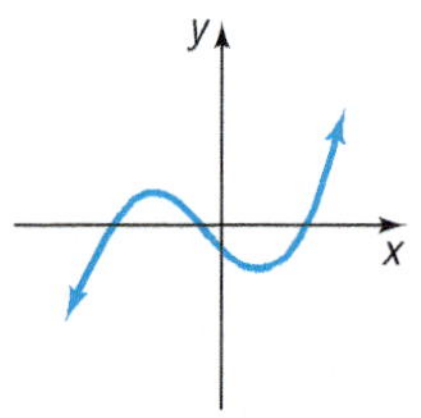

In Problems 29 and 30, graph each function. Be sure to label at least three points.

29. $f(x) = |x|$

30. $f(x) = \sqrt{x}$

In Problems 31–36, graph each function using the techniques of shifting, compressing or stretching, and reflections. Identify any intercepts of the graph. State the domain and, based on the graph, find the range.

31. $F(x) = |x| - 4$

32. $g(x) = -2|x|$

33. $h(x) = \sqrt{x - 1}$

34. $f(x) = \sqrt{1 - x}$

35. $h(x) = (x - 1)^2 + 2$

36. $g(x) = -2(x + 2)^3 - 8$

In Problems 37 and 38:

(a) Find the domain of each function.
(b) Locate any intercepts.
(c) Graph each function.
(d) Based on the graph, find the range.

37. $f(x) = \begin{cases} 3x & \text{if } -2 < x \le 1 \\ x + 1 & \text{if } x > 1 \end{cases}$

38. $f(x) = \begin{cases} x & \text{if } -4 \le x < 0 \\ 1 & \text{if } x = 0 \\ 3x & \text{if } x > 0 \end{cases}$

39. A function f is defined by

$$f(x) = \frac{Ax + 5}{6x - 2}$$

If $f(1) = 4$, find A.

40. Constructing a Closed Box A closed box with a square base is required to have a volume of 10 cubic feet.

(a) Build a model that expresses the amount A of material used to make such a box as a function of the length x of a side of the square base.
(b) How much material is required for a base 1 foot by 1 foot?
(c) How much material is required for a base 2 feet by 2 feet?
(d) Graph $A = A(x)$. For what value of x is A smallest?

41. Area of a Rectangle A rectangle has one vertex in quadrant I on the graph of $y = 10 - x^2$, another at the origin, one on the positive x-axis, and one on the positive y-axis.

(a) Express the area A of the rectangle as a function of x.
(b) Find the largest area A that can be enclosed by the rectangle.

Chapter Test

CHAPTER **Test Prep** VIDEOS | The Chapter Test Prep Videos are step-by-step solutions available in MyMathLab®, or on this text's YouTube™ Channel. Flip back to the Resources for Success page for a link to this text's YouTube channel.

1. Determine whether each relation represents a function. For each function, state the domain and the range.

(a) $\{(2,5), (4,6), (6,7), (8,8)\}$

(b) $\{(1,3), (4,-2), (-3,5), (1,7)\}$

(c)

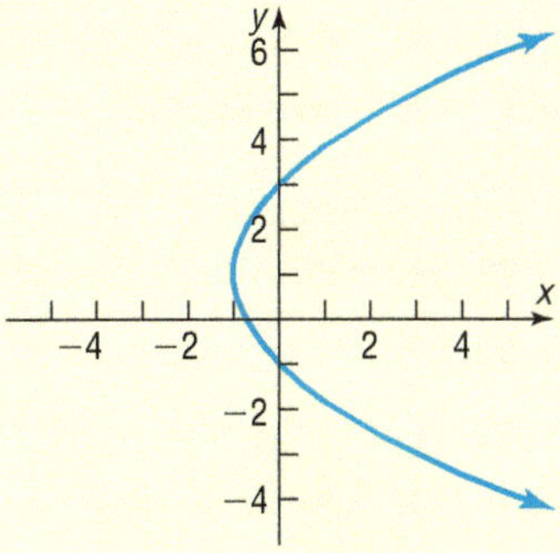

(d)

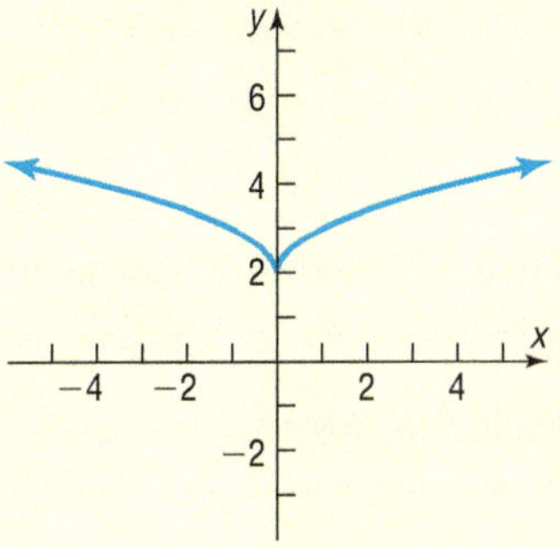

In Problems 2–4, find the domain of each function and evaluate each function at $x = -1$.

2. $f(x) = \sqrt{4 - 5x}$

3. $g(x) = \dfrac{x+2}{|x+2|}$

4. $h(x) = \dfrac{x-4}{x^2 + 5x - 36}$

5. Consider the graph of the function f below.

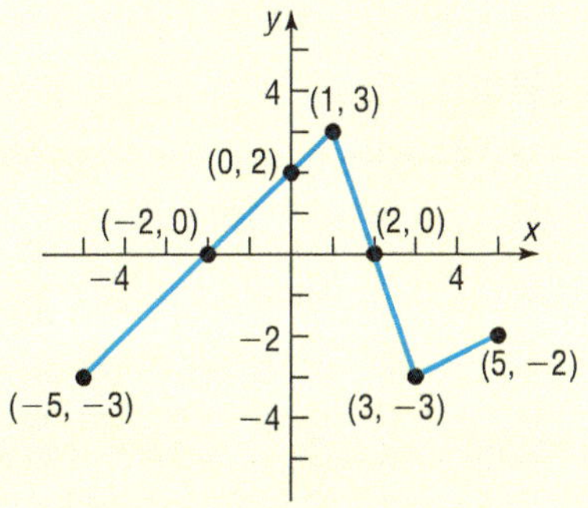

(a) Find the domain and the range of f.

(b) List the intercepts.

(c) Find $f(1)$.

(d) For what value(s) of x does $f(x) = -3$?

(e) Solve $f(x) < 0$.

6. Use a graphing utility to graph the function $f(x) = -x^4 + 2x^3 + 4x^2 - 2$ on the interval $[-5, 5]$. Then approximate any local maximum values and local minimum values rounded to two decimal places. Determine where the function is increasing and where it is decreasing.

7. Consider the function $g(x) = \begin{cases} 2x + 1 & \text{if } x < -1 \\ x - 4 & \text{if } x \geq -1 \end{cases}$

(a) Graph the function.

(b) List the intercepts.

(c) Find $g(-5)$.

(d) Find $g(2)$.

8. For the function $f(x) = 3x^2 - 2x + 4$, find the average rate of change of f from 3 to 4.

9. For the functions $f(x) = 2x^2 + 1$ and $g(x) = 3x - 2$, find the following and simplify.

(a) $(f - g)(x)$

(b) $(f \cdot g)(x)$

(c) $f(x + h) - f(x)$

10. Graph each function using the techniques of shifting, compressing or stretching, and reflecting. Start with the graph of the basic function and show all stages.

(a) $h(x) = -2(x + 1)^3 + 3$

(b) $g(x) = |x + 4| + 2$

11. The variable interest rate on a student loan changes each July 1 based on the bank prime loan rate. For the years 1992–2007, this rate can be approximated by the model

$$r(x) = -0.115x^2 + 1.183x + 5.623$$

where x is the number of years since 1992 and r is the interest rate as a percent.

(a) Use a graphing utility to estimate the highest rate during this time period. During which year was the interest rate the highest?

(b) Use the model to estimate the rate in 2010. Does this value seem reasonable?

Source: *U.S. Federal Reserve*

12. A community skating rink is in the shape of a rectangle with semicircles attached at the ends. The length of the rectangle is 20 feet less than twice the width. The thickness of the ice is 0.75 inch.

(a) Build a model that expresses the ice volume, V, as a function of the width, x.

(b) How much ice is in the rink if the width is 90 feet?

Cumulative Review

In Problems 1–6, find the real solutions of each equation.

1. $3x - 8 = 10$
2. $3x^2 - x = 0$
3. $x^2 - 8x - 9 = 0$
4. $6x^2 - 5x + 1 = 0$
5. $|2x + 3| = 4$
6. $\sqrt{2x + 3} = 2$

In Problems 7–9, solve each inequality. Graph the solution set.

7. $2 - 3x > 6$
8. $|2x - 5| < 3$
9. $|4x + 1| \geq 7$

10. (a) Find the distance from $P_1 = (-2, -3)$ to $P_2 = (3, -5)$.
    (b) What is the midpoint of the line segment from P_1 to P_2?
    (c) What is the slope of the line containing the points P_1 and P_2?

In Problems 11–14, graph each equation.

11. $3x - 2y = 12$
12. $x = y^2$
13. $x^2 + (y - 3)^2 = 16$
14. $y = \sqrt{x}$

15. For the equation $3x^2 - 4y = 12$, find the intercepts and check for symmetry.
16. Find the slope–intercept form of the equation of the line containing the points $(-2, 4)$ and $(6, 8)$.

In Problems 17–19, graph each function.

17. $f(x) = (x + 2)^2 - 3$
18. $f(x) = \dfrac{1}{x}$
19. $f(x) = \begin{cases} 2 - x & \text{if } x \leq 2 \\ |x| & \text{if } x > 2 \end{cases}$

Chapter Projects

I. Choosing a Wireless Data Plan Collect information from your family, friends, or consumer agencies such as Consumer Reports. Then decide on a cellular provider, choosing the company that you feel offers the best service. Once you have selected a service provider, research the various types of individual plans offered by the company by visiting the provider's website. Many providers offer family plans that include unlimited talk and text. The monthly cost is primarily determined by the amount of data used and the number of devices.

1. Suppose you expect to use 10 gigabytes of data for a single smartphone. What would be the monthly cost of each plan you are considering?
2. Suppose you expect to use 30 gigabytes of data and want a personal hotspot, but you still have only a single smartphone. What would be the monthly cost of each plan you are considering?
3. Suppose you expect to use 20 gigabytes of data with three smartphones sharing the data. What would be the monthly cost of each plan you are considering?
4. Suppose you expect to use 20 gigabytes of data with a single smartphone and a personal hotspot. What would be the monthly cost of each plan you are considering?
5. Build a model that describes the monthly cost C, in dollars, as a function of the number of data gigabytes used, g, assuming a single smartphone and a personal hotspot for each plan you are considering.
6. Graph each function from Problem 5.
7. Based on your particular usage, which plan is best for you?
8. Now, develop an Excel spreadsheet to analyze the various plans you are considering. Suppose you want a family plan with unlimited talk and text that offers 10 gigabytes of shared data and costs \$100 per month. Additional gigabytes of data cost \$15 per gigabyte, extra phones can be added to the plan for \$15 each per month, and each hotspot costs \$20 per month. Because wireless

data plans have a cost structure based on piecewise-defined functions, we need an "if/then" statement within Excel to analyze the cost of the plan. Use the accompanying Excel spreadsheet as a guide in developing your spreadsheet. Enter into your spreadsheet a variety of possible amounts of data and various numbers of additional phones and hotspots.

	A	B	C	D	
1					
2	Monthly fee	$100			
3	Allotted data per month (GB)	10			
4	Data used (GB)	12			
5	Cost per additional GB of data	$15			
6					
7	Monthly cost of hotspot	$20			
8	Number of hotspots	1			
9	Monthly cost of additional phone	$15			
10	Number of additional phones	2			
11					
12	Cost of data	=IF(B4<B3,B2,B2+B5*(B4-B3))			
13	Cost of additional devices/hotspots	=B8*B7+B10*B9			
14					
15	Total Cost	=B12+B13			
16					

9. Write a paragraph supporting the choice in plans that best meets your needs.

10. How are "if/then" loops similar to a piecewise-defined function?

The following projects are available on the Instructor's Resource Center (IRC).

II. **Project at Motorola:** ***Wireless Internet Service*** Use functions and their graphs to analyze the total cost of various wireless Internet service plans.

III. **Cost of Cable** When government regulations and customer preference influence the path of a new cable line, the Pythagorean Theorem can be used to assess the cost of installation.

IV. **Oil Spill** Functions are used to analyze the size and spread of an oil spill from a leaking tanker.

4 Linear and Quadratic Functions

The Beta of a Stock

Investing in the stock market can be rewarding and fun, but how does one go about selecting which stocks to purchase? Financial investment firms hire thousands of analysts who track individual stocks (equities) and assess the value of the underlying company. One measure the analysts consider is the *beta* of the stock. **Beta** measures the relative risk of an individual company's equity to that of a market basket of stocks, such as the Standard & Poor's 500. But how is beta computed?

—See the Internet-based Chapter Project I—

Outline

••• A Look Back

Up to now, our discussion has focused on graphs of equations and functions. We learned how to graph equations using the point-plotting method, intercepts, and the tests for symmetry. In addition, we learned what a function is and how to identify whether a relation represents a function. We also discussed properties of functions, such as domain/range, increasing/decreasing, even/odd, and average rate of change.

A Look Ahead •••

Going forward, we will look at classes of functions. This chapter focuses on linear and quadratic functions, their properties, and their applications.

4.1 Properties of Linear Functions and Linear Models

PREPARING FOR THIS SECTION *Before getting started, review the following:*

- Lines (Section 2.2, pp. 173–184)
- Graphs of Equations in Two Variables; Intercepts; Symmetry (Section 2.1, pp. 165–170)
- Linear Equations (Section 1.2, pp. 102–103)
- Functions (Section 3.1, pp. 207–218)
- The Graph of a Function (Section 3.2, pp. 222–226)
- Properties of Functions (Section 3.3, pp. 231–239)

Now Work the 'Are You Prepared?' problems on page 287.

OBJECTIVES
1 Graph Linear Functions (p. 281)
2 Use Average Rate of Change to Identify Linear Functions (p. 281)
3 Determine Whether a Linear Function Is Increasing, Decreasing, or Constant (p. 284)
4 Build Linear Models from Verbal Descriptions (p. 285)

1 Graph Linear Functions

In Section 2.2 we discussed lines. In particular, for nonvertical lines we developed the slope–intercept form of the equation of a line $y = mx + b$. When the slope–intercept form of a line is written using function notation, the result is a *linear function*.

DEFINITION

A **linear function** is a function of the form

$$f(x) = mx + b$$

The graph of a linear function is a line with slope m and y-intercept b. Its domain is the set of all real numbers.

Functions that are not linear are said to be **nonlinear**.

EXAMPLE 1 **Graphing a Linear Function**

Graph the linear function $f(x) = -3x + 7$. What are the domain and the range of f?

Solution This is a linear function with slope $m = -3$ and y-intercept $b = 7$. To graph this function, plot the point $(0, 7)$, the y-intercept, and use the slope to find an additional point by moving right 1 unit and down 3 units. See Figure 1. The domain and range of f are each the set of all real numbers.

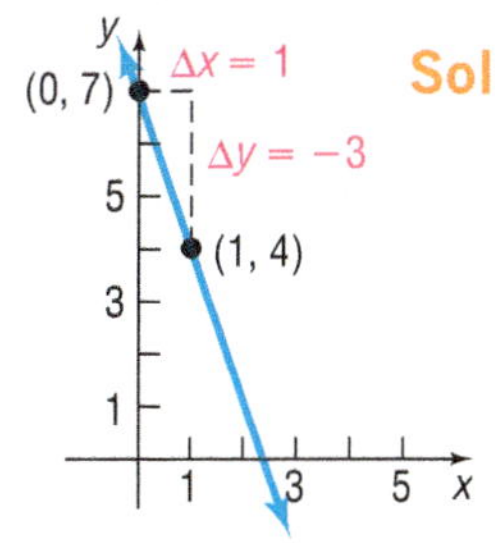

Figure 1 $f(x) = -3x + 7$

Alternatively, an additional point could have been found by evaluating the function at some $x \neq 0$. For $x = 1, f(1) = -3(1) + 7 = 4$ and the point $(1, 4)$ lies on the graph.

Now Work PROBLEMS 13(a) AND (b)

2 Use Average Rate of Change to Identify Linear Functions

Look at Table 1 on the next page, which shows certain values of the independent variable x and corresponding values of the dependent variable y for the function $f(x) = -3x + 7$. Notice that as the value of the independent variable, x, increases by 1, the value of the dependent variable y decreases by 3. That is, the average rate of change of y with respect to x is a constant, -3.

Table 1

x	$y = f(x) = -3x + 7$	Average Rate of Change $= \dfrac{\Delta y}{\Delta x}$
−2	13	
		$\dfrac{10 - 13}{-1 - (-2)} = \dfrac{-3}{1} = -3$
−1	10	
		$\dfrac{7 - 10}{0 - (-1)} = \dfrac{-3}{1} = -3$
0	7	
		−3
1	4	
		−3
2	1	
		−3
3	−2	

It is not a coincidence that the average rate of change of the linear function $f(x) = -3x + 7$ is the slope of the linear function. That is, $\dfrac{\Delta y}{\Delta x} = m = -3$. The following theorem states this fact.

THEOREM

Average Rate of Change of a Linear Function

Linear functions have a constant average rate of change. That is, the average rate of change of a linear function $f(x) = mx + b$ is

$$\frac{\Delta y}{\Delta x} = m$$

Proof The average rate of change of $f(x) = mx + b$ from x_1 to x_2, $x_1 \neq x_2$, is

$$\frac{\Delta y}{\Delta x} = \frac{f(x_2) - f(x_1)}{x_2 - x_1} = \frac{(mx_2 + b) - (mx_1 + b)}{x_2 - x_1}$$
$$= \frac{mx_2 - mx_1}{x_2 - x_1} = \frac{m(x_2 - x_1)}{x_2 - x_1} = m$$

Based on the theorem just proved, the average rate of change of the function $g(x) = -\frac{2}{5}x + 5$ is $-\frac{2}{5}$.

Now Work PROBLEM 13(c)

As it turns out, only linear functions have a constant average rate of change. Because of this, the average rate of change can be used to determine whether a function is linear. This is especially useful if the function is defined by a data set.

EXAMPLE 2 **Using the Average Rate of Change to Identify Linear Functions**

(a) A strain of *E. coli* known as Beu 397-recA441 is placed into a Petri dish at 30° Celsius and allowed to grow. The data shown in Table 2 are collected. The population is measured in grams and the time in hours. Plot the ordered pairs (x, y) in the Cartesian plane, and use the average rate of change to determine whether the function is linear.

(b) The data in Table 3 represent the maximum number of heartbeats that healthy individuals of different ages should have during a 15-second interval of time while exercising. Plot the ordered pairs (x, y) in the Cartesian plane, and use the average rate of change to determine whether the function is linear.

Table 2

Time (hours), x	Population (grams), y	(x, y)
0	0.09	(0, 0.09)
1	0.12	(1, 0.12)
2	0.16	(2, 0.16)
3	0.22	(3, 0.22)
4	0.29	(4, 0.29)
5	0.39	(5, 0.39)

Table 3

Age, x	Maximum Number of Heartbeats, y	(x, y)
20	50	(20, 50)
30	47.5	(30, 47.5)
40	45	(40, 45)
50	42.5	(50, 42.5)
60	40	(60, 40)
70	37.5	(70, 37.5)

Source: American Heart Association

Solution Compute the average rate of change of each function. If the average rate of change is constant, the function is linear. If the average rate of change is not constant, the function is nonlinear.

(a) Figure 2 shows the points listed in Table 2 plotted in the Cartesian plane. Note that it is impossible to draw a straight line that contains all the points. Table 4 displays the average rate of change of the population.

Table 4

Time (hours), x	Population (grams), y	Average Rate of Change $= \dfrac{\Delta y}{\Delta x}$
0	0.09	
		$\dfrac{0.12 - 0.09}{1 - 0} = 0.03$
1	0.12	
		0.04
2	0.16	
		0.06
3	0.22	
		0.07
4	0.29	
		0.10
5	0.39	

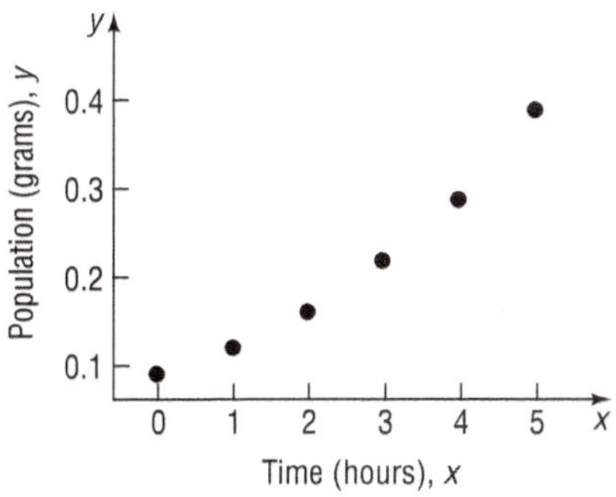

Figure 2

Because the average rate of change is not constant, the function is not linear. In fact, because the average rate of change is increasing as the value of the independent variable increases, the function is increasing at an increasing rate. So not only is the population increasing over time, but it is also growing more rapidly as time passes.

(b) Figure 3 on the next page shows the points listed in Table 3 plotted in the Cartesian plane. Note that the data in Figure 3 lie on a straight line. Table 5 on the next page contains the average rate of change of the maximum number of heartbeats. The average rate of change of the heartbeat data is constant, -0.25 beat per year, so the function is linear. To find the linear function, use the point-slope formula with $x_1 = 20$, $y_1 = 50$, and $m = -0.25$.

$$y - 50 = -0.25(x - 20) \quad y - y_1 = m(x - x_1)$$
$$y - 50 = -0.25x + 5$$
$$y = -0.25x + 55$$

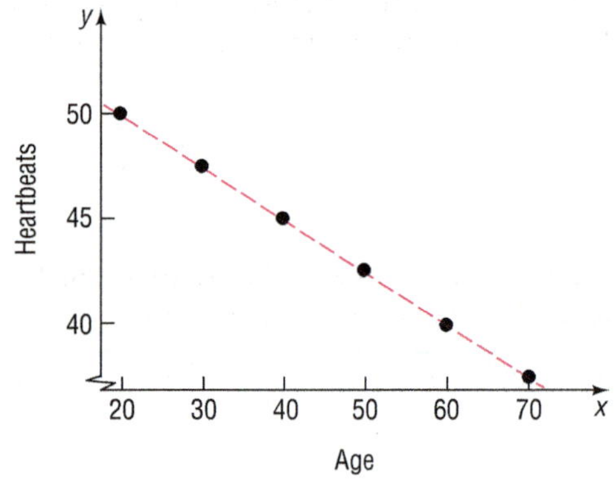

Figure 3

Table 5

Age, x	Maximum Number of Heartbeats, y	Average Rate of Change $= \dfrac{\Delta y}{\Delta x}$
20	50	
		$\dfrac{47.5 - 50}{30 - 20} = -0.25$
30	47.5	
		-0.25
40	45	
		-0.25
50	42.5	
		-0.25
60	40	
		-0.25
70	37.5	

Now Work PROBLEM 21

3 Determine Whether a Linear Function Is Increasing, Decreasing, or Constant

Look back at the Seeing the Concept on page 178. When the slope m of a linear function is positive $(m > 0)$, the line slants upward from left to right. When the slope m of a linear function is negative $(m < 0)$, the line slants downward from left to right. When the slope m of a linear function is zero $(m = 0)$, the line is horizontal.

THEOREM

Increasing, Decreasing, and Constant Linear Functions

A linear function $f(x) = mx + b$ is increasing over its domain if its slope, m, is positive. It is decreasing over its domain if its slope, m, is negative. It is constant over its domain if its slope, m, is zero.

EXAMPLE 3 **Determining Whether a Linear Function Is Increasing, Decreasing, or Constant**

Determine whether the following linear functions are increasing, decreasing, or constant.

(a) $f(x) = 5x - 2$

(b) $g(x) = -2x + 8$

(c) $s(t) = \dfrac{3}{4}t - 4$

(d) $h(z) = 7$

Solution

(a) For the linear function $f(x) = 5x - 2$, the slope is 5, which is positive. The function f is increasing on the interval $(-\infty, \infty)$.

(b) For the linear function $g(x) = -2x + 8$, the slope is -2, which is negative. The function g is decreasing on the interval $(-\infty, \infty)$.

(c) For the linear function $s(t) = \dfrac{3}{4}t - 4$, the slope is $\dfrac{3}{4}$, which is positive. The function s is increasing on the interval $(-\infty, \infty)$.

(d) The linear function h can be written as $h(z) = 0z + 7$. Because the slope is 0, the function h is constant on the interval $(-\infty, \infty)$. ■

Now Work PROBLEM 13(d)

4 Build Linear Models from Verbal Descriptions

When the average rate of change of a function is constant, a linear function can model the relation between the two variables. For example, if a recycling company pays \$0.52 per pound for aluminum cans, then the relation between the price paid p and the pounds recycled x can be modeled as the linear function $p(x) = 0.52x$, with slope $m = \dfrac{0.52 \text{ dollar}}{1 \text{ pound}}$.

Modeling with a Linear Function

If the average rate of change of a function is a constant m, a linear function f can be used to model the relation between the two variables as follows:

$$f(x) = mx + b$$

where b is the value of f at 0; that is, $b = f(0)$.

EXAMPLE 4 **Straight-line Depreciation**

Book value is the value of an asset that a company uses to create its balance sheet. Some companies depreciate their assets using straight-line depreciation so that the value of the asset declines by a fixed amount each year. The amount of the decline depends on the useful life that the company assigns to the asset. Suppose that a company just purchased a fleet of new cars for its sales force at a cost of \$31,500 per car. The company chooses to depreciate each vehicle using the straight-line method over 7 years. This means that each car will depreciate by $\dfrac{\$31{,}500}{7} = \$4500$ per year.

(a) Write a linear function that expresses the book value V of each car as a function of its age, x.

(b) Graph the linear function.

(c) What is the book value of each car after 3 years?

(d) Interpret the slope.

(e) When will the book value of each car be \$9000?
[**Hint:** Solve the equation $V(x) = 9000$.]

Solution

(a) If we let $V(x)$ represent the value of each car after x years, then $V(0)$ represents the original value of each car, so $V(0) = \$31{,}500$. The y-intercept of the linear function is \$31,500. Because each car depreciates by \$4500 per year, the slope of the linear function is -4500. The linear function that represents the book value V of each car after x years is

$$V(x) = -4500x + 31{,}500$$

(b) Figure 4 shows the graph of V.

(c) The book value of each car after 3 years is

$$\begin{aligned} V(3) &= -4500(3) + 31{,}500 \\ &= \$18{,}000 \end{aligned}$$

(d) Since the slope of $V(x) = -4500x + 31{,}500$ is -4500, the average rate of change of the book value is –\$4500/year. So for each additional year that passes, the book value of the car decreases by \$4500.

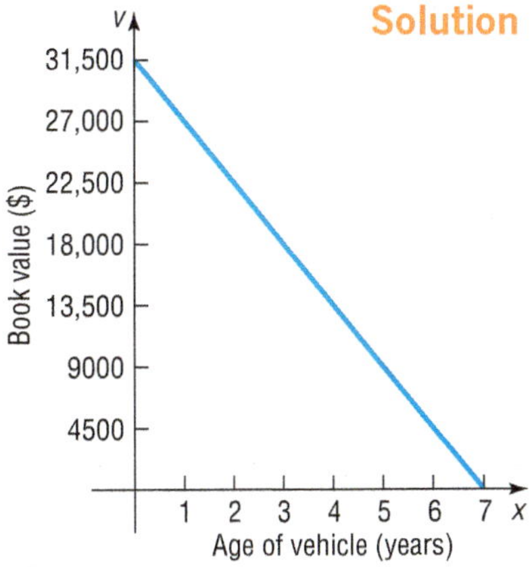

Figure 4
$V(x) = -4500x + 31{,}500$

(e) To find when the book value will be \$9000, solve the equation

$$
\begin{aligned}
V(x) &= 9000 \\
-4500x + 31{,}500 &= 9000 \\
-4500x &= -22{,}500 && \text{Subtract 31,500 from each side.} \\
x &= \frac{-22{,}500}{-4500} = 5 && \text{Divide by } -4500.
\end{aligned}
$$

Each car will have a book value of \$9000 when it is 5 years old. ■

Now Work PROBLEM 45

EXAMPLE 5 Supply and Demand

The **quantity supplied** of a good is the amount of a product that a company is willing to make available for sale at a given price. The **quantity demanded** of a good is the amount of a product that consumers are willing to purchase at a given price. Suppose that the quantity supplied, S, and quantity demanded, D, of cell phones each month are given by the following functions:

$$
\begin{aligned}
S(p) &= 30p - 900 \\
D(p) &= -7.5p + 2850
\end{aligned}
$$

where p is the price (in dollars) of the cell phone.

(a) The **equilibrium price** of a product is defined as the price at which quantity supplied equals quantity demanded. That is, the equilibrium price is the price at which $S(p) = D(p)$. Find the equilibrium price of cell phones. What is the **equilibrium quantity**, the amount demanded (or supplied) at the equilibrium price?

(b) Determine the prices for which quantity supplied is greater than quantity demanded. That is, solve the inequality $S(p) > D(p)$.

(c) Graph $S = S(p)$, $D = D(p)$ and label the equilibrium point.

Solution (a) To find the equilibrium price, solve the equation $S(p) = D(p)$.

$$
\begin{aligned}
30p - 900 &= -7.5p + 2850 && S(p) = 30p - 900;\ D(p) = -7.5p + 2850 \\
30p &= -7.5p + 3750 && \text{Add 900 to each side.} \\
37.5p &= 3750 && \text{Add } 7.5p \text{ to each side.} \\
p &= 100 && \text{Divide each side by 37.5.}
\end{aligned}
$$

The equilibrium price is \$100 per cell phone. To find the equilibrium quantity, evaluate either $S(p)$ or $D(p)$ at $p = 100$.

$$S(100) = 30(100) - 900 = 2100$$

The equilibrium quantity is 2100 cell phones. At a price of \$100 per phone, the company will produce and sell 2100 phones each month and have no shortages or excess inventory.

(b) The inequality $S(p) > D(p)$ is

$$
\begin{aligned}
30p - 900 &> -7.5p + 2850 && S(p) > D(p) \\
30p &> -7.5p + 3750 && \text{Add 900 to each side.} \\
37.5p &> 3750 && \text{Add } 7.5p \text{ to each side.} \\
p &> 100 && \text{Divide each side by 37.5.}
\end{aligned}
$$

If the company charges more than \$100 per phone, quantity supplied will exceed quantity demanded. In this case the company will have excess phones in inventory.

(c) Figure 5 shows the graphs of $S = S(p)$ and $D = D(p)$ with the equilibrium point labeled.

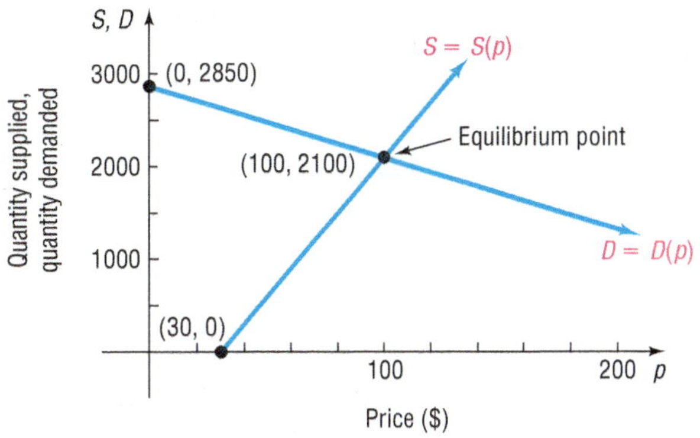

Figure 5 Supply and demand functions

Now Work PROBLEM 39

4.1 Assess Your Understanding

'Are You Prepared?' *Answers are given at the end of these exercises. If you get a wrong answer, read the pages listed in red.*

1. Graph $y = 2x - 3$. (pp. 176–179)
2. Find the slope of the line joining the points $(2, 5)$ and $(-1, 3)$. (pp. 173–175)
3. Find the average rate of change of $f(x) = 3x^2 - 2$, from 2 to 4. (pp. 238–239)
4. Solve: $60x - 900 = -15x + 2850$. (pp. 102–103)
5. If $f(x) = x^2 - 4$, find $f(-2)$. (pp. 210–212)
6. ***True or False*** The graph of the function $f(x) = x^2$ is increasing on the interval $[0, \infty)$. (p. 234)

Concepts and Vocabulary

7. For the graph of the linear function $f(x) = mx + b$, m is the _____ and b is the __________.
8. If the slope m of the graph of a linear function is _______, the function is increasing over its domain.
9. ***True or False*** The slope of a nonvertical line is the average rate of change of the linear function.
10. ***True or False*** The average rate of change of $f(x) = 2x + 8$ is 8.
11. What is the only type of function that has a constant average rate of change?
    (a) linear function (b) quadratic function
    (c) step function (d) absolute value function
12. A car has 12,500 miles on its odometer. Say the car is driven an average of 40 miles per day. Choose the model that expresses the number of miles N that will be on its odometer after x days.
    (a) $N(x) = -40x + 12{,}500$ (b) $N(x) = 40x - 12{,}500$
    (c) $N(x) = 12{,}500x + 40$ (d) $N(x) = 40x + 12{,}500$

Skill Building

In Problems 13–20, a linear function is given.
(a) Determine the slope and y-intercept of each function.
(b) Use the slope and y-intercept to graph the linear function.
(c) Determine the average rate of change of each function.
(d) Determine whether the linear function is increasing, decreasing, or constant.

13. $f(x) = 2x + 3$
14. $g(x) = 5x - 4$
15. $h(x) = -3x + 4$
16. $p(x) = -x + 6$
17. $f(x) = \frac{1}{4}x - 3$
18. $h(x) = -\frac{2}{3}x + 4$
19. $F(x) = 4$
20. $G(x) = -2$

In Problems 21–28, determine whether the given function is linear or nonlinear. If it is linear, determine the equation of the line.

21.

x	y = f(x)
−2	4
−1	1
0	−2
1	−5
2	−8

22.

x	y = f(x)
−2	1/4
−1	1/2
0	1
1	2
2	4

23.

x	y = f(x)
−2	−8
−1	−3
0	0
1	1
2	0

24.

x	y = f(x)
−2	−4
−1	0
0	4
1	8
2	12

25.

x	$y = f(x)$
-2	-26
-1	-4
0	2
1	-2
2	-10

26.

x	$y = f(x)$
-2	-4
-1	-3.5
0	-3
1	-2.5
2	-2

27.

x	$y = f(x)$
-2	8
-1	8
0	8
1	8
2	8

28.

x	$y = f(x)$
-2	0
-1	1
0	4
1	9
2	16

Applications and Extensions

29. Suppose that $f(x) = 4x - 1$ and $g(x) = -2x + 5$.
(a) Solve $f(x) = 0$. (b) Solve $f(x) > 0$.
(c) Solve $f(x) = g(x)$. (d) Solve $f(x) \leq g(x)$.
(e) Graph $y = f(x)$ and $y = g(x)$ and label the point that represents the solution to the equation $f(x) = g(x)$.

30. Suppose that $f(x) = 3x + 5$ and $g(x) = -2x + 15$.
(a) Solve $f(x) = 0$. (b) Solve $f(x) < 0$.
(c) Solve $f(x) = g(x)$. (d) Solve $f(x) \geq g(x)$.
(e) Graph $y = f(x)$ and $y = g(x)$ and label the point that represents the solution to the equation $f(x) = g(x)$.

31. In parts (a)–(f), use the following figure.

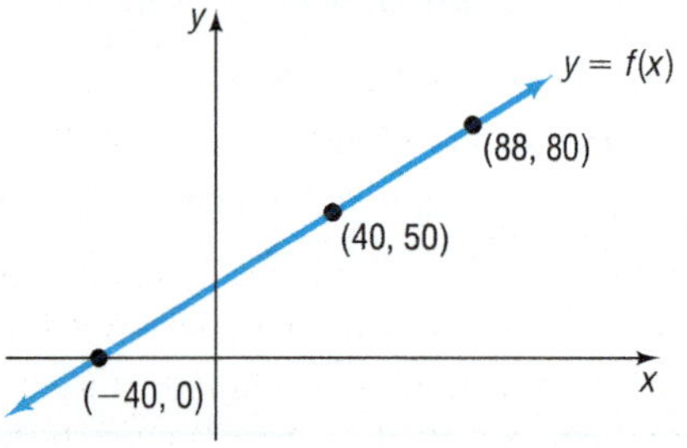

(a) Solve $f(x) = 50$. (b) Solve $f(x) = 80$.
(c) Solve $f(x) = 0$. (d) Solve $f(x) > 50$.
(e) Solve $f(x) \leq 80$. (f) Solve $0 < f(x) < 80$.

32. In parts (a)–(f), use the following figure.

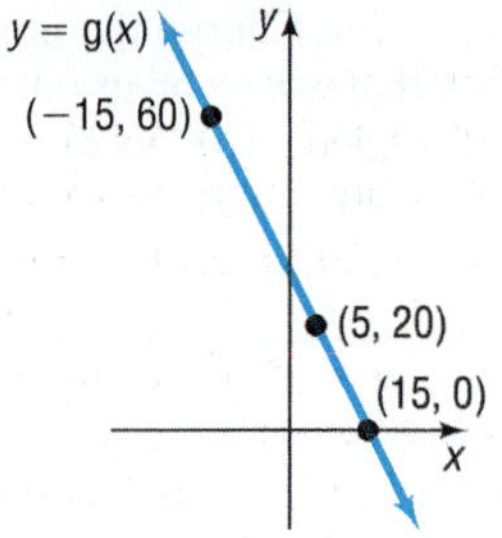

(a) Solve $g(x) = 20$. (b) Solve $g(x) = 60$.
(c) Solve $g(x) = 0$. (d) Solve $g(x) > 20$.
(e) Solve $g(x) \leq 60$. (f) Solve $0 < g(x) < 60$.

33. In parts (a) and (b), use the following figure.

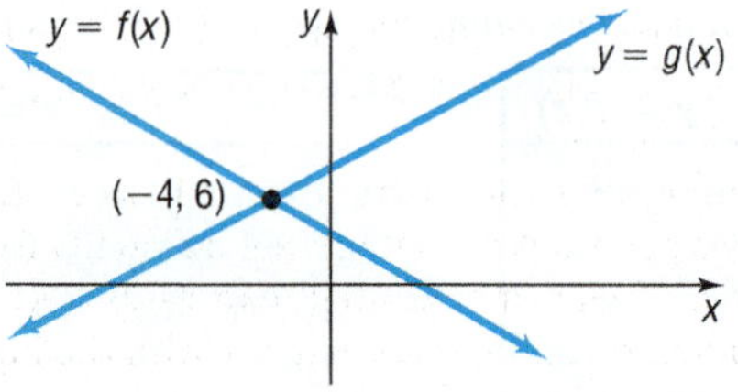

(a) Solve the equation: $f(x) = g(x)$.
(b) Solve the inequality: $f(x) > g(x)$.

34. In parts (a) and (b), use the following figure.

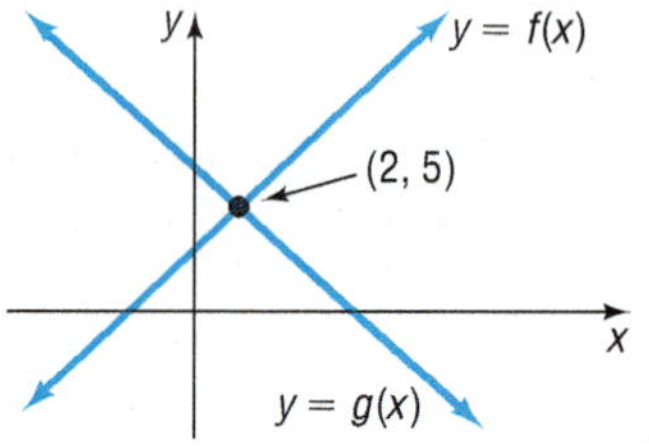

(a) Solve the equation: $f(x) = g(x)$.
(b) Solve the inequality: $f(x) \leq g(x)$.

35. In parts (a) and (b), use the following figure.

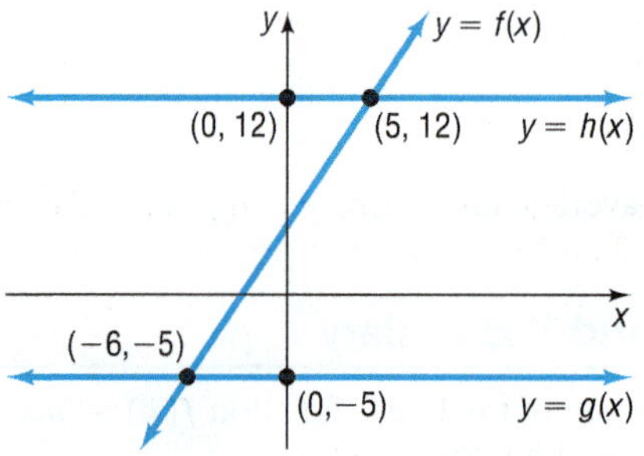

(a) Solve the equation: $f(x) = g(x)$.
(b) Solve the inequality: $g(x) \leq f(x) < h(x)$.

36. In parts (a) and (b), use the following figure.

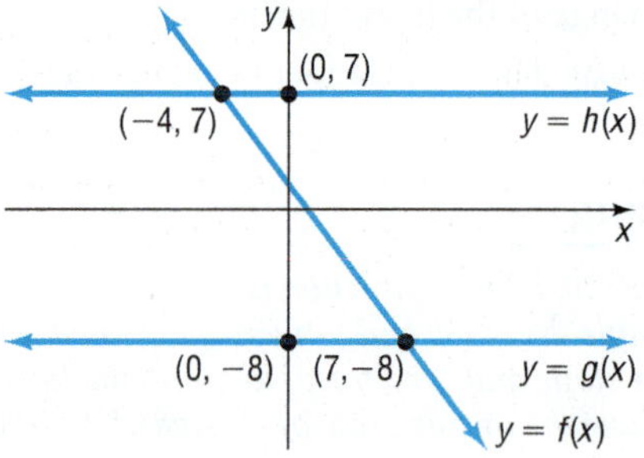

(a) Solve the equation: $f(x) = g(x)$.
(b) Solve the inequality: $g(x) < f(x) \leq h(x)$.

37. Car Rentals The cost C, in dollars, of a one-day car rental is modeled by the function $C(x) = 0.35x + 45$, where x is the number of miles driven.
(a) What is the cost if you drive $x = 40$ miles?
(b) If the cost of renting the car is \$108, how many miles did you drive?
(c) Suppose that you want the cost to be no more than \$150. What is the maximum number of miles that you can drive?
(d) What is the implied domain of C?
(e) Interpret the slope.
(f) Interpret the y-intercept.

38. Phone Charges The monthly cost C, in dollars, for calls from the United States to Japan on a certain phone plan is modeled by the function $C(x) = 0.24x + 5$, where x is the number of minutes used.
(a) What is the cost if you talk on the phone for $x = 50$ minutes?
(b) Suppose that your monthly bill is \$20.36. How many minutes did you use the phone?
(c) Suppose that you budget yourself \$40 per month for the phone. What is the maximum number of minutes that you can talk?
(d) What is the implied domain of C if there are 30 days in the month?
(e) Interpret the slope.
(f) Interpret the y-intercept.

39. Supply and Demand Suppose that the quantity supplied S and the quantity demanded D of T-shirts at a concert are given by the following functions:

$$S(p) = -600 + 50p$$
$$D(p) = 1200 - 25p$$

where p is the price of a T-shirt.
(a) Find the equilibrium price for T-shirts at this concert. What is the equilibrium quantity?
(b) Determine the prices for which quantity demanded is greater than quantity supplied.
(c) What do you think will eventually happen to the price of T-shirts if quantity demanded is greater than quantity supplied?

40. Supply and Demand Suppose that the quantity supplied S and the quantity demanded D of hot dogs at a baseball game are given by the following functions:

$$S(p) = -2000 + 3000p$$
$$D(p) = 10{,}000 - 1000p$$

where p is the price of a hot dog.
(a) Find the equilibrium price for hot dogs at the baseball game. What is the equilibrium quantity?
(b) Determine the prices for which quantity demanded is less than quantity supplied.
(c) What do you think will eventually happen to the price of hot dogs if quantity demanded is less than quantity supplied?

41. Taxes The function $T(x) = 0.15(x - 9225) + 922.50$ represents the tax bill T of a single person whose adjusted gross income is x dollars for income between \$9225 and \$37,450, inclusive, in 2015.
Source: *Internal Revenue Service*
(a) What is the domain of this linear function?
(b) What is a single filer's tax bill if adjusted gross income is \$20,000?
(c) Which variable is independent and which is dependent?
(d) Graph the linear function over the domain specified in part (a).
(e) What is a single filer's adjusted gross income if the tax bill is \$3663.75?
(f) Interpret the slope.

42. Competitive Balance Tax In 2011, major league baseball signed a labor agreement with the players. In this agreement, any team whose payroll exceeded \$189 million in 2015 had to pay a competitive balance tax of 50% (for four or more consecutive offenses). The linear function $T(p) = 0.50(p - 189)$ describes the competitive balance tax T of a team whose payroll was p (in millions of dollars).
Source: *Major League Baseball*
(a) What is the implied domain of this linear function?
(b) What was the competitive balance tax for the New York Yankees whose 2015 payroll was \$214.2 million?
(c) Graph the linear function.
(d) What was the payroll of a team that paid a competitive balance tax of \$15.7 million?
(e) Interpret the slope.

The point at which a company's profits equal zero is called the company's ***break-even point.*** *For Problems 43 and 44, let R represent a company's revenue, let C represent the company's costs, and let x represent the number of units produced and sold each day.*
(a) *Find the firm's break-even point; that is, find x so that $R = C$.*
(b) *Find the values of x such that $R(x) > C(x)$. This represents the number of units that the company must sell to earn a profit.*

43. $R(x) = 8x$
$C(x) = 4.5x + 17{,}500$

44. $R(x) = 12x$
$C(x) = 10x + 15{,}000$

45. Straight-line Depreciation Suppose that a company has just purchased a new computer for \$3000. The company chooses to depreciate the computer using the straight-line method over 3 years.
(a) Write a linear model that expresses the book value V of the computer as a function of its age x.
(b) What is the implied domain of the function found in part (a)?
(c) Graph the linear function.
(d) What is the book value of the computer after 2 years?
(e) When will the computer have a book value of \$2000?

46. Straight-line Depreciation Suppose that a company has just purchased a new machine for its manufacturing facility for \$120,000. The company chooses to depreciate the machine using the straight-line method over 10 years.
(a) Write a linear model that expresses the book value V of the machine as a function of its age x.
(b) What is the implied domain of the function found in part (a)?
(c) Graph the linear function.
(d) What is the book value of the machine after 4 years?
(e) When will the machine have a book value of \$72,000?

47. Cost Function The simplest cost function is the linear cost function, $C(x) = mx + b$, where the y-intercept b represents the fixed costs of operating a business and the slope m represents the cost of each item produced. Suppose that a small bicycle manufacturer has daily fixed costs of \$1800, and each bicycle costs \$90 to manufacture.
(a) Write a linear model that expresses the cost C of manufacturing x bicycles in a day.
(b) Graph the model.
(c) What is the cost of manufacturing 14 bicycles in a day?
(d) How many bicycles could be manufactured for \$3780?

48. Cost Function Refer to Problem 47. Suppose that the landlord of the building increases the bicycle manufacturer's rent by \$100 per month.
(a) Assuming that the manufacturer is open for business 20 days per month, what are the new daily fixed costs?
(b) Write a linear model that expresses the cost C of manufacturing x bicycles in a day with the higher rent.
(c) Graph the model.
(d) What is the cost of manufacturing 14 bicycles in a day?
(e) How many bicycles can be manufactured for \$3780?

49. Truck Rentals A truck rental company rents a truck for one day by charging \$39.95 plus \$0.89 per mile.

(a) Write a linear model that relates the cost C, in dollars, of renting the truck to the number x of miles driven.

(b) What is the cost of renting the truck if the truck is driven 110 miles? 230 miles?

50. International Calling A cell phone company offers an international plan by charging \$30 for the first 80 minutes, plus \$0.50 for each minute over 80.

(a) Write a linear model that relates the cost C, in dollars, of talking x minutes, assuming $x \geq 80$.

(b) What is the cost of talking 105 minutes? 120 minutes?

Mixed Practice

51. Developing a Linear Model from Data How many songs can an iPod hold? The following data represent the memory m and the number of songs n.

Memory, m (gigabytes)	Number of songs, n
8	1750
16	3500
32	7000
64	14,000

(a) Plot the ordered pairs (m, n) in a Cartesian plane.

(b) Show that the number of songs n is a linear function of memory m.

(c) Determine the linear function that describes the relation between m and n.

(d) What is the implied domain of the linear function?

(e) Graph the linear function in the Cartesian plane drawn in part (a).

(f) Interpret the slope.

52. Developing a Linear Model from Data The following data represent the various combinations of soda and hot dogs that Yolanda can buy at a baseball game with \$60.

Soda, s	Hot Dogs, h
20	0
15	3
10	6
5	9

(a) Plot the ordered pairs (s, h) in a Cartesian plane.

(b) Show that the number of hot dogs purchased h is a linear function of the number of sodas purchased s.

(c) Determine the linear function that describes the relation between s and h.

(d) What is the implied domain of the linear function?

(e) Graph the linear function in the Cartesian plane drawn in part (a).

(f) Interpret the slope.

(g) Interpret the values of the intercepts.

Explaining Concepts: Discussion and Writing

53. Which of the following functions might have the graph shown? (More than one answer is possible.)

(a) $f(x) = 2x - 7$

(b) $g(x) = -3x + 4$

(c) $H(x) = 5$

(d) $F(x) = 3x + 4$

(e) $G(x) = \frac{1}{2}x + 2$

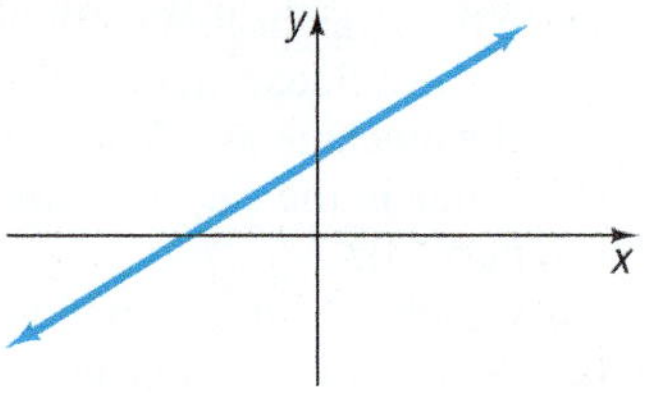

54. Which of the following functions might have the graph shown? (More than one answer is possible.)

(a) $f(x) = 3x + 1$

(b) $g(x) = -2x + 3$

(c) $H(x) = 3$

(d) $F(x) = -4x - 1$

(e) $G(x) = -\frac{2}{3}x + 3$

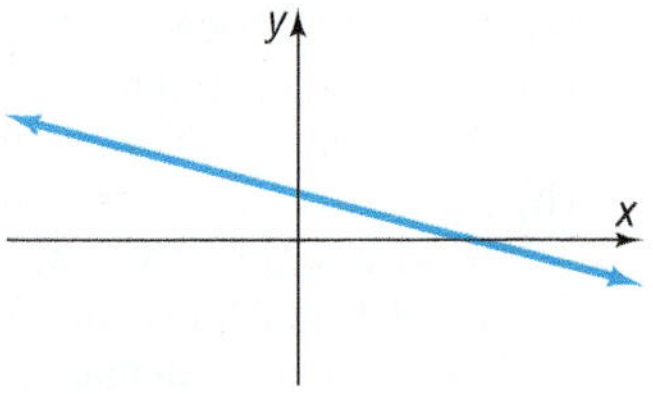

55. Under what circumstances is a linear function $f(x) = mx + b$ odd? Can a linear function ever be even?

56. Explain how the graph of $f(x) = mx + b$ can be used to solve $mx + b > 0$.

Retain Your Knowledge

Problems 57–60 are based on material learned earlier in the course. The purpose of these problems is to keep the material fresh in your mind so that you are better prepared for the final exam.

57. Graph $x^2 - 4x + y^2 + 10y - 7 = 0$.

58. If $f(x) = \dfrac{2x + B}{x - 3}$ and $f(5) = 8$, what is the value of B?

59. Find the average rate of change of $f(x) = 3x^2 - 5x$ from 1 to 3.

60. Graph $g(x) = \begin{cases} x^2 & x \leq 0 \\ \sqrt{x} + 1 & x > 0 \end{cases}$

'Are You Prepared?' Answers

1.

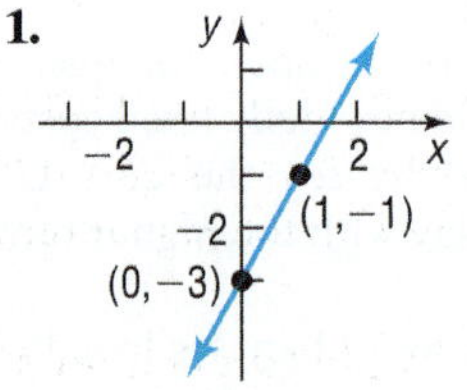

2. $\frac{2}{3}$ **3.** 18 **4.** $\{50\}$ **5.** 0 **6.** True

4.2 Building Linear Models from Data

PREPARING FOR THIS SECTION *Before getting started, review the following:*

- Rectangular Coordinates (Section 1.1, pp. 83–85)
- Lines (Section 2.2, pp. 173–184)
- Functions (Section 3.1, pp. 270–218)

Now Work the 'Are You Prepared?' problems on page 294.

OBJECTIVES
1 Draw and Interpret Scatter Diagrams (p. 291)
2 Distinguish between Linear and Nonlinear Relations (p. 292)
3 Use a Graphing Utility to Find the Line of Best Fit (p. 293)

1 Draw and Interpret Scatter Diagrams

In Section 4.1, we built linear models from verbal descriptions. Linear models can also be constructed by fitting a linear function to data. The first step is to plot the ordered pairs using rectangular coordinates. The resulting graph is a **scatter diagram.**

EXAMPLE 1 Drawing and Interpreting a Scatter Diagram

In baseball, the on-base percentage for a team represents the percentage of time that the players safely reach base. The data given in Table 6 represent the number of runs scored y and the on-base percentage x for teams in the National League during the 2014 baseball season.

Table 6

Team	On-base Percentage, x	Runs Scored, y	(x, y)
Arizona	30.2	615	(30.2, 615)
Atlanta	30.5	573	(30.5, 573)
Chicago Cubs	30.0	614	(30.0, 614)
Cincinnati	29.6	595	(29.6, 595)
Colorado	32.7	755	(32.7, 755)
LA Dodgers	33.3	718	(33.3, 718)
Miami	31.7	645	(31.7, 645)
Milwaukee	31.1	650	(31.1, 650)
NY Mets	30.8	629	(30.8, 629)
Philadelphia	30.2	619	(30.2, 619)
Pittsburgh	33.0	682	(33.0, 682)
San Diego	29.2	535	(29.2, 535)
San Francisco	31.1	665	(31.1, 665)
St. Louis	32.0	619	(32.0, 619)
Washington	32.1	686	(32.1, 686)

Source: *espn.go.com*

(a) Draw a scatter diagram of the data, treating on-base percentage as the independent variable.
(b) Use a graphing utility to draw a scatter diagram.
(c) Describe what happens to runs scored as the on-base percentage increases.

Solution
(a) To draw a scatter diagram, plot the ordered pairs listed in Table 6, with the on-base percentage as the x-coordinate and the runs scored as the y-coordinate. See Figure 6(a) on the next page. Notice that the points in the scatter diagram are not connected.
(b) Figure 6(b) shows a scatter diagram using a TI-84 Plus C graphing calculator.

(c) The scatter diagrams show that as the on-base percentage increases, the number of runs scored also increases.

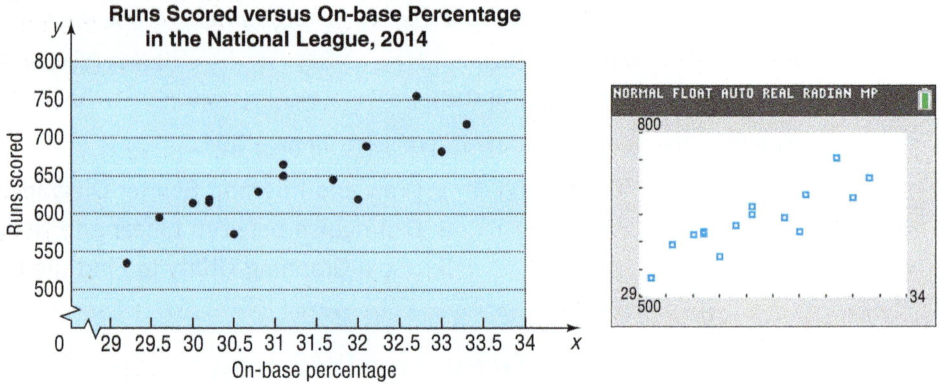

Figure 6 On-base percentage

(a) (b)

Now Work PROBLEM 11(a)

2 Distinguish between Linear and Nonlinear Relations

Notice that the points in Figure 6 do not follow a perfect linear relation (as they do in Figure 3 in Section 4.1). However, the data do exhibit a linear pattern. There are numerous possible explanations why the data are not perfectly linear, but one easy explanation is the fact that other variables besides on-base percentage (such as number of home runs hit) play a role in determining runs scored.

Scatter diagrams are used to help us see the type of relation that exists between two variables. In this text, we will discuss a variety of different relations that may exist between two variables. For now, we concentrate on distinguishing between linear and nonlinear relations. See Figure 7.

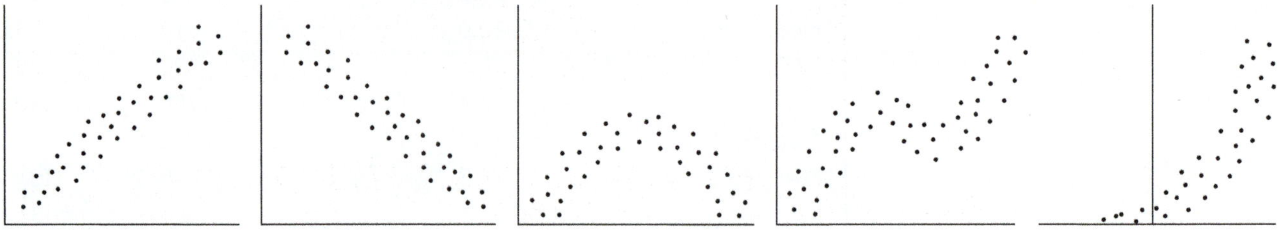

Figure 7 **(a)** Linear $y = mx + b, m > 0$ **(b)** Linear $y = mx + b, m < 0$ **(c)** Nonlinear **(d)** Nonlinear **(e)** Nonlinear

EXAMPLE 2 Distinguishing between Linear and Nonlinear Relations

Determine whether the relation between the two variables in each scatter diagram in Figure 8 is linear or nonlinear.

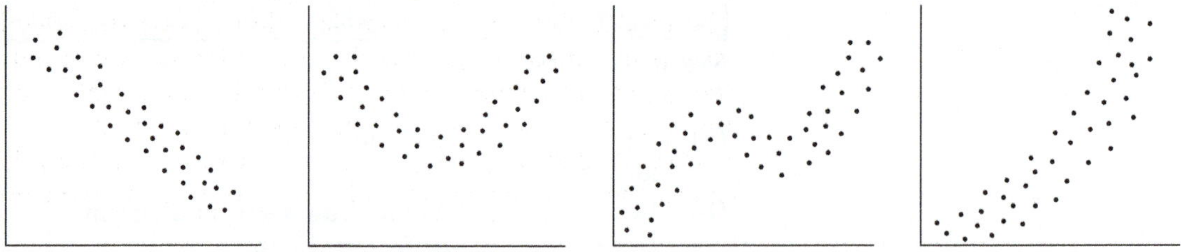

Figure 8 (a) (b) (c) (d)

Solution (a) Linear (b) Nonlinear (c) Nonlinear (d) Nonlinear

Now Work PROBLEM 5

This section considers data whose scatter diagrams suggest that a linear relation exists between the two variables.

Suppose that the scatter diagram of a set of data indicates a linear relationship, as in Figure 7(a) or (b). We might want to model the data by finding an equation of a line that relates the two variables. One way to obtain a model for such data is to draw a line through two points on the scatter diagram and determine the equation of the line.

EXAMPLE 3 Finding a Model for Linearly Related Data

Use the data in Table 6 from Example 1.

(a) Select two points and find an equation of the line containing the points.

(b) Graph the line on the scatter diagram obtained in Example 1(a).

Solution (a) Select two points, say $(30.8, 629)$ and $(32.1, 686)$. The slope of the line joining the points $(30.8, 629)$ and $(32.1, 686)$ is

$$m = \frac{686 - 629}{32.1 - 30.8} = \frac{57}{1.3} \approx 43.85$$

The equation of the line with slope 43.85 and passing through $(30.8, 629)$ is found using the point–slope form with $m = 43.85$, $x_1 = 30.8$, and $y_1 = 629$.

$$y - y_1 = m(x - x_1) \quad \text{Point–slope form of a line}$$
$$y - 629 = 43.85(x - 30.8) \quad x_1 = 30.8, y_1 = 629, m = 43.85$$
$$y - 629 = 43.85x - 1350.58$$
$$y = 43.85x - 721.58$$

(b) Figure 9 shows the scatter diagram with the graph of the line found in part (a).

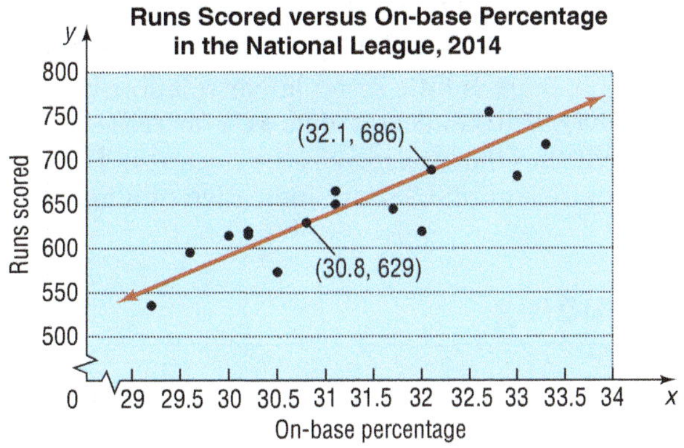

Figure 9

Select two other points and complete the solution. Graph the line on the scatter diagram obtained in Figure 6.

Now Work PROBLEM 11(b) AND (c)

3 Use a Graphing Utility to Find the Line of Best Fit

The model obtained in Example 3 depends on the selection of points, which will vary from person to person. So the model that we found might be different from the model you found. Although the model in Example 3 appears to fit the data well, there may be a model that "fits it better." Do you think your model fits the data better? Is there a *line of best fit*? As it turns out, there is a method for finding a model that best fits linearly related data (called the **line of best fit**).*

EXAMPLE 4 Finding a Model for Linearly Related Data

Use the data in Table 6 from Example 1.

(a) Use a graphing utility to find the line of best fit that models the relation between on-base percentage and runs scored.

*We shall not discuss the underlying mathematics of lines of best fit in this text.

(b) Graph the line of best fit on the scatter diagram obtained in Example 1(b).
(c) Interpret the slope.
(d) Use the line of best fit to predict the number of runs a team will score if their on-base percentage is 31.5.

Solution (a) Graphing utilities contain built-in programs that find the line of best fit for a collection of points in a scatter diagram. Executing the LINear REGression program provides the results shown in Figure 10. This output shows the equation $y = ax + b$, where a is the slope of the line and b is the y-intercept. The line of best fit that relates on-base percentage to runs scored may be expressed as the line

$$y = 38.02x - 544.86 \quad \textbf{The model}$$

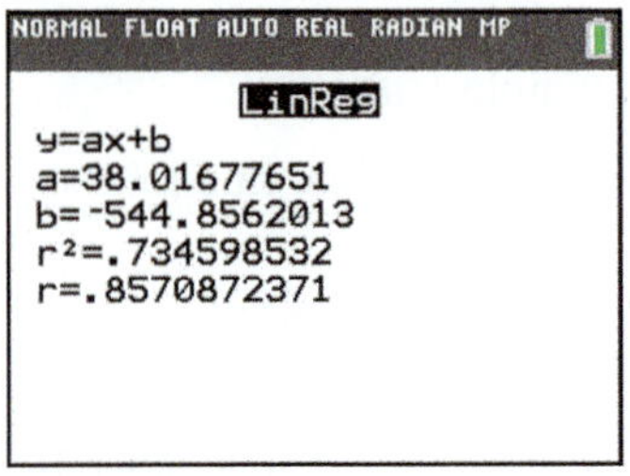

Figure 10

(b) Figure 11 shows the graph of the line of best fit, along with the scatter diagram.
(c) The slope of the line of best fit is 38.02, which means that for every 1 percent increase in the on-base percentage, runs scored increase 38.02, on average.
(d) Letting $x = 31.5$ in the equation of the line of best fit, we obtain $y = 38.02(31.5) - 544.86 \approx 653$ runs. ■

Now Work PROBLEM 11 (d) AND (e)

Does the line of best fit appear to be a good fit? In other words, does the line appear to accurately describe the relation between on-base percentage and runs scored?

And just how "good" is this line of best fit? Look again at Figure 10. The last line of output is $r = 0.857$. This number, called the **correlation coefficient**, $r, -1 \le r \le 1$, is a measure of the strength of the linear relation that exists between two variables. The closer $|r|$ is to 1, the more nearly perfect the linear relationship is. If r is close to 0, there is little or no linear relationship between the variables. A negative value of $r, r < 0$, indicates that as x increases, y decreases; a positive value of $r, r > 0$, indicates that as x increases, y does also. The data given in Table 6, having a correlation coefficient of 0.857, are indicative of a linear relationship with positive slope.

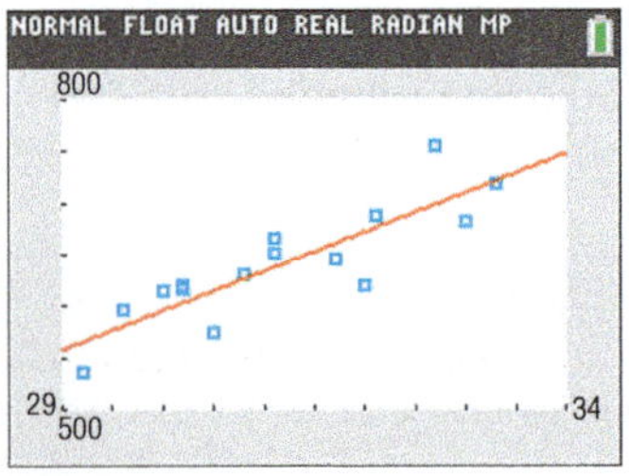

Figure 11

4.2 Assess Your Understanding

'Are You Prepared?' *Answers are given at the end of these exercises. If you get a wrong answer, read the pages listed in red.*

1. Plot the points $(1, 5)$, $(2, 6)$, $(3, 9)$, $(1, 12)$ in the Cartesian plane. Is the relation $\{(1, 5), (2, 6), (3, 9), (1, 12)\}$ a function? Why? (pp. 83 and 207–210)

2. Find an equation of the line containing the points $(1, 4)$ and $(3, 8)$. (pp. 179–180)

Concepts and Vocabulary

3. A ______________ is used to help us to see what type of relation, if any, may exist between two variables.

4. If the independent variable in a line of best fit $y = -0.008x + 14$ is credit score, and the dependent variable is the interest rate on a used-car loan, then the slope is interpreted as follows: "If credit score increases by 1 point, the interest rate will __________ (increase/decrease) by _____ percent, on average."

Skill Building

In Problems 5–10, examine the scatter diagram and determine whether the type of relation is linear or nonlinear.

5.

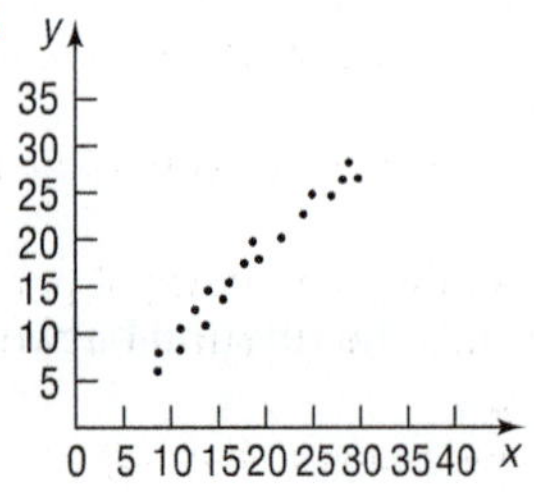

6.

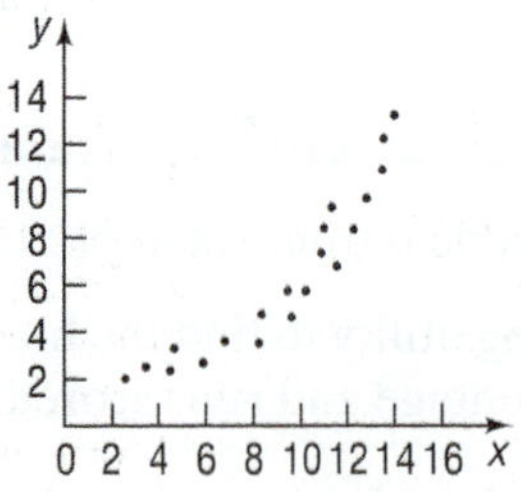

7.

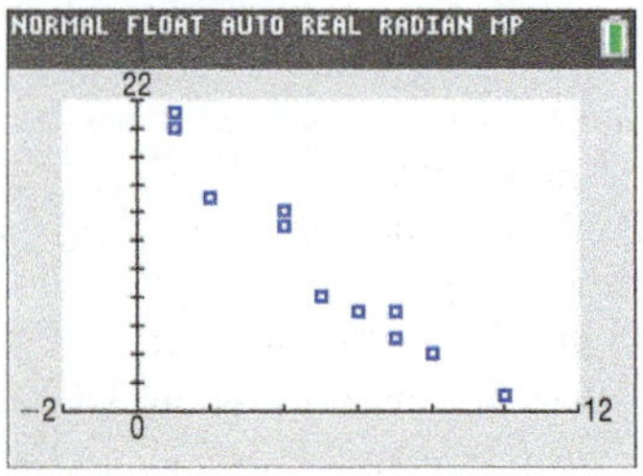

8. 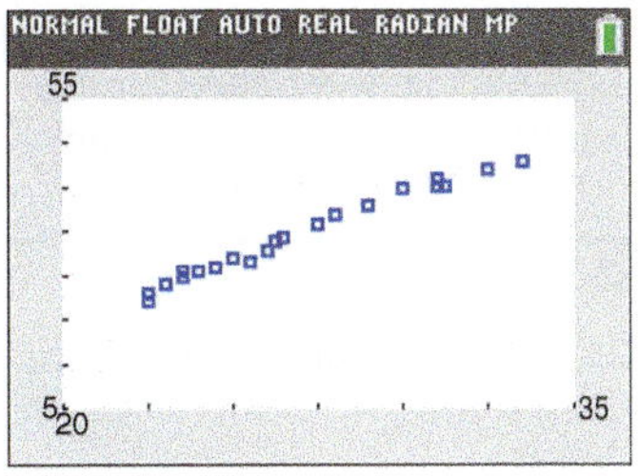

9. 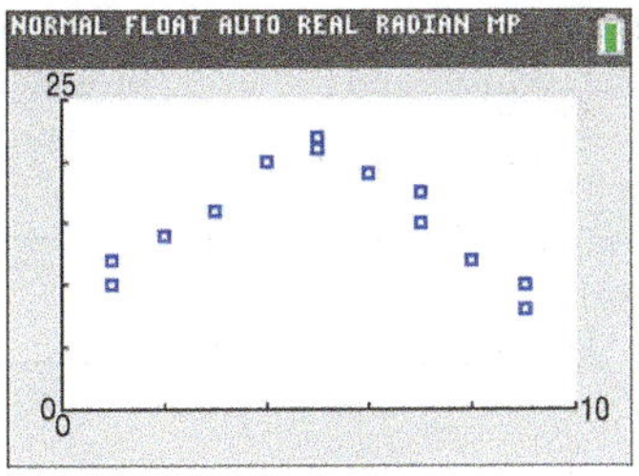

10. 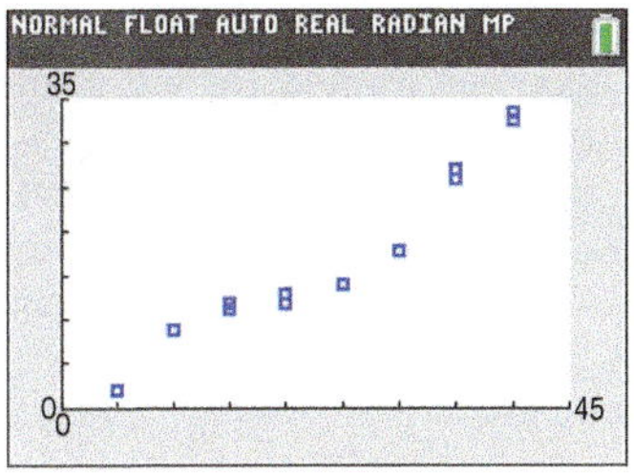

In Problems 11–16,

(a) Draw a scatter diagram.

(b) Select two points from the scatter diagram and find the equation of the line containing the points selected.

(c) Graph the line found in part (b) on the scatter diagram.

(d) Use a graphing utility to find the line of best fit.

(e) Use a graphing utility to draw the scatter diagram and graph the line of best fit on it.

11.

x	3	4	5	6	7	8	9
y	4	6	7	10	12	14	16

12.

x	3	5	7	9	11	13
y	0	2	3	6	9	11

13.

x	−2	−1	0	1	2
y	−4	0	1	4	5

14.

x	−2	−1	0	1	2
y	7	6	3	2	0

15.

x	−20	−17	−15	−14	−10
y	100	120	118	130	140

16.

x	−30	−27	−25	−20	−14
y	10	12	13	13	18

Applications and Extensions

17. Candy The following data represent the weight (in grams) of various candy bars and the corresponding number of calories.

Candy Bar	Weight, x	Calories, y
Hershey's Milk Chocolate®	44.28	230
Nestle's Crunch®	44.84	230
Butterfinger®	61.30	270
Baby Ruth®	66.45	280
Almond Joy®	47.33	220
Twix® (with caramel)	58.00	280
Snickers®	61.12	280
Heath®	39.52	210

Source: Megan Pocius, student at Joliet Junior College

(a) Draw a scatter diagram of the data, treating weight as the independent variable.

(b) What type of relation appears to exist between the weight of a candy bar and the number of calories?

(c) Select two points and find a linear model that contains the points.

(d) Graph the line on the scatter diagram drawn in part (a).

(e) Use the linear model to predict the number of calories in a candy bar that weighs 62.3 grams.

(f) Interpret the slope of the line found in part (c).

18. Tornadoes The following data represent the width (in yards) and length (in miles) of tornadoes.

Width (yards), w	Length (miles), L
200	2.5
350	4.8
180	2.0
300	2.5
500	5.8
400	4.5
500	8.0
800	8.0
100	3.4
50	0.5
700	9.0
600	5.7

Source: NOAA

(a) Draw a scatter diagram of the data, treating width as the independent variable.

(b) What type of relation appears to exist between the width and the length of tornadoes?

(c) Select two points and find a linear model that contains the points.

(d) Graph the line on the scatter diagram drawn in part (b).

(e) Use the linear model to predict the length of a tornado that has a width of 450 yards.

(f) Interpret the slope of the line found in part (c).

19. Video Games and Grade-Point Average Professor Grant Alexander wanted to find a linear model that relates the number of hours a student plays video games each week, h, to the cumulative grade-point average, G, of the student. He obtained a random sample of 10 full-time students at his college and asked each student to disclose the number of hours spent playing video games and the student's cumulative grade-point average.

Hours of Video Games per Week, h	Grade-Point Average, G
0	3.49
0	3.05
2	3.24
3	2.82
3	3.19
5	2.78
8	2.31
8	2.54
10	2.03
12	2.51

(a) Explain why the number of hours spent playing video games is the independent variable and cumulative grade-point average is the dependent variable.
(b) Use a graphing utility to draw a scatter diagram.
(c) Use a graphing utility to find the line of best fit that models the relation between number of hours of video game playing each week and grade-point average. Express the model using function notation.
(d) Interpret the slope.
(e) Predict the grade-point average of a student who plays video games for 8 hours each week.
(f) How many hours of video game playing do you think a student plays whose grade-point average is 2.40?

20. Hurricanes The following data represent the atmospheric pressure p (in millibars) and the wind speed w (in knots) measured during various tropical systems in the Atlantic Ocean.

Atmospheric Pressure (millibars), p	Wind Speed (knots), w
993	50
994	60
997	45
1003	45
1004	40
1000	55
994	55
942	105
1006	40
942	120
986	50
983	70
940	120
966	100
982	55

Source: National Hurricane Center

(a) Use a graphing utility to draw a scatter diagram of the data, treating atmospheric pressure as the independent variable.
(b) Use a graphing utility to find the line of best fit that models the relation between atmospheric pressure and wind speed. Express the model using function notation.
(c) Interpret the slope.
(d) Predict the wind speed of a tropical storm if the atmospheric pressure measures 990 millibars.
(e) What is the atmospheric pressure of a hurricane if the wind speed is 85 knots?

Mixed Practice

21. Homeruns A baseball analyst wishes to find a function that relates the distance, d, of a homerun and the speed, s, of the ball off the bat. Consider the data shown to the right.

(a) Does the relation defined by the set of orders pairs (s, d) represent a function?
(b) Draw a scatter diagram of the data, treating speed off bat as the independent variable.
(c) Using a graphing utility, find the line of best fit that models the relation between the speed off the bat and the distance of the homerun?
(d) Interpret the slope.
(e) Express the relationship found in part (c), using function notation.
(f) What is the domain of the function?
(g) Predict the homerun distance if the speed of the ball off the bat is 103 miles per hour.

Speed Off Bat (mph), s	Homerun Distance (feet), d
98	369
99	381
100	380
100	397
101	400
102	383
102	408
104	392
104	406
104	421
105	411
107	396
107	429
109	404
109	418

Source: Major League Baseball, 2014

22. Demand for Jeans The marketing manager at Levi-Strauss wishes to find a function that relates the demand D for men's jeans and p, the price of the jeans. The following data were obtained based on a price history of the jeans.

Price ($/Pair), p	Demand (Pairs of Jeans Sold per Day), D
20	60
22	57
23	56
23	53
27	52
29	49
30	44

(a) Does the relation defined by the set of ordered pairs (p, D) represent a function?
(b) Draw a scatter diagram of the data.
(c) Using a graphing utility, find the line of best fit that models the relation between price and quantity demanded.
(d) Interpret the slope.
(e) Express the relationship found in part (c) using function notation.
(f) What is the domain of the function?
(g) How many jeans will be demanded if the price is $28 a pair?

Explaining Concepts: Discussion and Writing

23. Maternal Age versus Down Syndrome A biologist would like to know how the age of the mother affects the incidence of Down syndrome. The data to the right represent the age of the mother and the incidence of Down syndrome per 1000 pregnancies. Draw a scatter diagram treating age of the mother as the independent variable. Would it make sense to find the line of best fit for these data? Why or why not?

Age of Mother, x	Incidence of Down Syndrome, y
33	2.4
34	3.1
35	4
36	5
37	6.7
38	8.3
39	10
40	13.3
41	16.7
42	22.2
43	28.6
44	33.3
45	50

Source: Hook, E.B., *Journal of the American Medical Association*, 249, 2034-2038, 1983.

24. Find the line of best fit for the ordered pairs $(1, 5)$ and $(3, 8)$. What is the correlation coefficient for these data? Why is this result reasonable?

25. What does a correlation coefficient of 0 imply?

26. Explain why it does not make sense to interpret the y-intercept in Problem 17.

27. Refer to Problem 19. Solve $G(h) = 0$. Provide an interpretation of this result. Find $G(0)$. Provide an interpretation of this result.

Retain Your Knowledge

Problems 28–31 are based on material learned earlier in the course. The purpose of these problems is to keep the material fresh in your mind so that you are better prepared for the final exam.

28. Find an equation for the line containing the points $(-1, 5)$ and $(3, -3)$. Express your answer using either the general form or the slope-intercept form of the equation of a line, whichever you prefer.

29. Find the domain of $f(x) = \dfrac{x-1}{x^2 - 25}$.

30. For $f(x) = 5x - 8$ and $g(x) = x^2 - 3x + 4$, find $(g - f)(x)$.

31. Write the function whose graph is the graph of $y = x^2$, but shifted to the left 3 units and shifted down 4 units.

'Are You Prepared?' Answers

1.

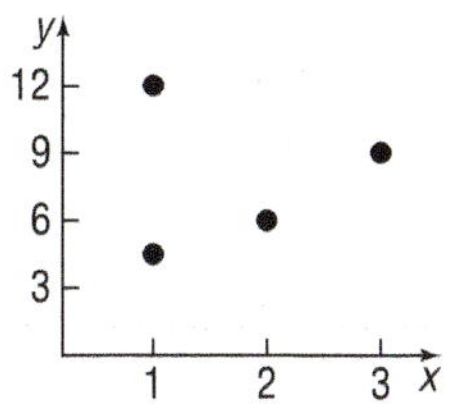

No, because the input, 1, corresponds to two different outputs.

2. $y = 2x + 2$

4.3 Quadratic Functions and Their Properties

PREPARING FOR THIS SECTION *Before getting started, review the following:*

- Intercepts (Section 2.1, pp. 165–166)
- Graphing Techniques: Transformations (Section 3.5, pp. 256–264)
- Completing the Square (Section R.5, p. 57)
- Quadratic Equations (Section 1.3, pp. 110–117)

Now Work the 'Are You Prepared?' problems on page 306.

OBJECTIVES
1. Graph a Quadratic Function Using Transformations (p. 299)
2. Identify the Vertex and Axis of Symmetry of a Quadratic Function (p. 301)
3. Graph a Quadratic Function Using Its Vertex, Axis, and Intercepts (p. 301)
4. Find a Quadratic Function Given Its Vertex and One Other Point (p. 304)
5. Find the Maximum or Minimum Value of a Quadratic Function (p. 305)

Quadratic Functions

DEFINITION

A **quadratic function** is a function of the form

$$f(x) = ax^2 + bx + c$$

where a, b, and c are real numbers and $a \neq 0$. The domain of a quadratic function is the set of all real numbers.

In Words
A quadratic function is a function defined by a second-degree polynomial in one variable.

Some examples of quadratic functions are

$$F(x) = 3x^2 - 5x + 1 \quad g(x) = -6x^2 + 1 \quad H(x) = \frac{1}{2}x^2 + \frac{2}{3}x$$

Many applications require a knowledge of quadratic functions. For example, suppose that Texas Instruments collects the data shown in Table 7, which relate the number of calculators sold to the price p (in dollars) per calculator. Since the price of a product determines the quantity that will be purchased, we treat price as the independent variable. The relationship between the number x of calculators sold and the price p per calculator is given by the linear equation

$$x = 21{,}000 - 150p$$

Table 7

Price per Calculator, p (dollars)	Number of Calculators, x
60	12,000
65	11,250
70	10,500
75	9,750
80	9,000
85	8,250
90	7,500

Then the revenue R derived from selling x calculators at the price p per calculator is equal to the unit selling price p of the calculator times the number x of units actually sold. That is,

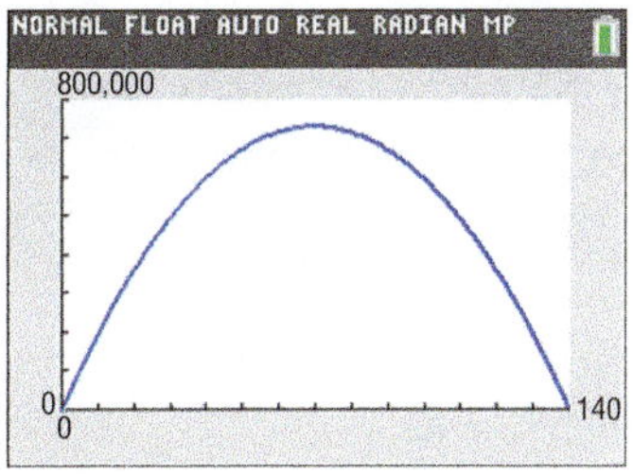

Figure 12
$R(p) = -150p^2 + 21{,}000p$

$$R = xp$$
$$R(p) = (21{,}000 - 150p)p \quad x = 21{,}000 - 150p$$
$$= -150p^2 + 21{,}000p$$

So the revenue R is a quadratic function of the price p. Figure 12 illustrates the graph of this revenue function, whose domain is $0 \le p \le 140$, since both x and p must be nonnegative.

A second situation in which a quadratic function appears involves the motion of a projectile. Based on Newton's Second Law of Motion (force equals mass times acceleration, $F = ma$), it can be shown that, ignoring air resistance, the path of a projectile propelled upward at an inclination to the horizontal is the graph of a quadratic function. See Figure 13 for an illustration.

Figure 13
Path of a cannonball

1 Graph a Quadratic Function Using Transformations

We know how to graph the square function $f(x) = x^2$. Figure 14 shows the graph of three functions of the form $f(x) = ax^2, a > 0$, for $a = 1, a = \frac{1}{2}$, and $a = 3$. Notice that the larger the value of a, the "narrower" the graph is, and the smaller the value of a, the "wider" the graph is.

Figure 15 shows the graphs of $f(x) = ax^2$ for $a < 0$. Notice that these graphs are reflections about the x-axis of the graphs in Figure 14. Based on the results of these two figures, general conclusions can be drawn about the graph of $f(x) = ax^2$. First, as $|a|$ increases, the graph is stretched vertically (becomes "taller"), and as $|a|$ gets closer to zero, the graph is compressed vertically (becomes "shorter"). Second, if a is positive, the graph opens "up," and if a is negative, the graph opens "down."

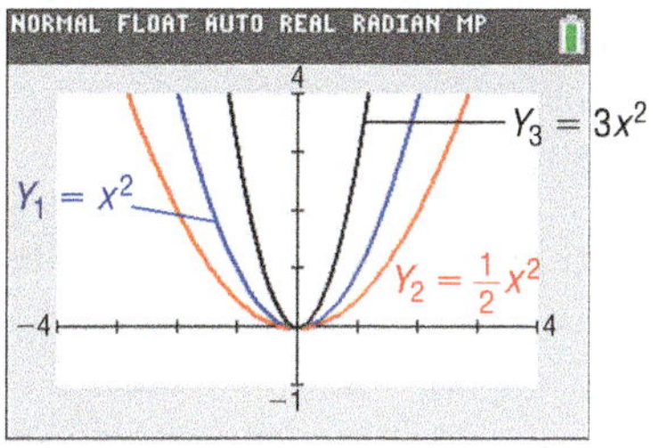

Figure 14

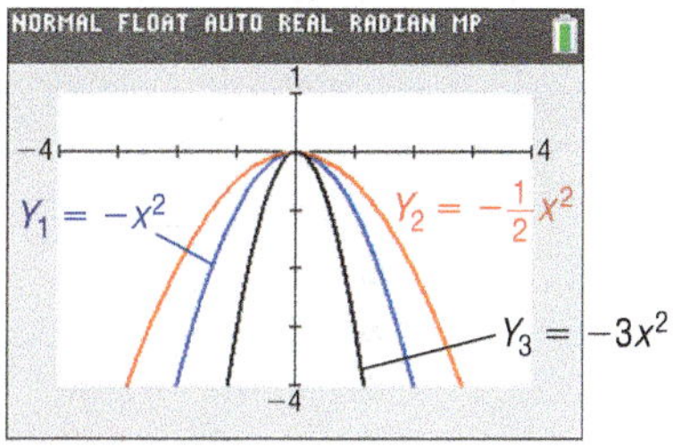

Figure 15

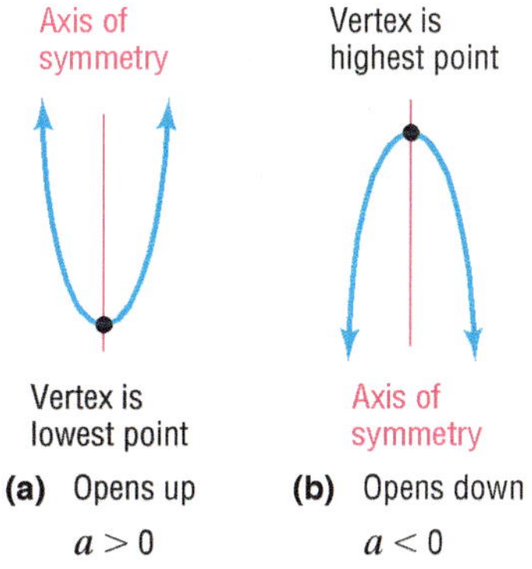

Figure 16
Graphs of a quadratic function, $f(x) = ax^2 + bx + c, a \ne 0$

The graphs in Figures 14 and 15 are typical of the graphs of all quadratic functions, which are called **parabolas**.* Refer to Figure 16, where two parabolas are pictured. The one on the left **opens up** and has a lowest point; the one on the right **opens down** and has a highest point. The lowest or highest point of a parabola is called the **vertex**. The vertical line passing through the vertex in each parabola in Figure 16 is called the **axis of symmetry** (usually abbreviated to **axis**) of the parabola. Because the parabola is symmetric about its axis, the axis of symmetry of a parabola can be used to find additional points on the parabola.

The parabolas shown in Figure 16 are the graphs of a quadratic function $f(x) = ax^2 + bx + c, a \ne 0$. Notice that the coordinate axes are not included in the figure. Depending on the values of a, b, and c, the axes could be placed anywhere. The important fact is that the shape of the graph of a quadratic function will look like one of the parabolas in Figure 16.

In the following example, techniques from Section 3.5 are used to graph a quadratic function $f(x) = ax^2 + bx + c, a \ne 0$. The method of completing the square is used to write the function f in the form $f(x) = a(x - h)^2 + k$.

*Parabolas will be studied using a geometric definition later in this text.

EXAMPLE 1 Graphing a Quadratic Function Using Transformations

Graph the function $f(x) = 2x^2 + 8x + 5$. Find the vertex and axis of symmetry.

Solution Begin by completing the square on the right side.

$$\begin{aligned} f(x) &= 2x^2 + 8x + 5 \\ &= 2(x^2 + 4x) + 5 && \text{Factor out the 2 from } 2x^2 + 8x. \\ &= 2(x^2 + 4x + 4) + 5 - 8 && \text{Complete the square of } x^2 + 4x \text{ by adding 4.} \\ &= 2(x + 2)^2 - 3 && \text{Notice that the factor of 2 requires that 8 be subtracted.} \end{aligned}$$

The graph of f can be obtained from the graph of $y = x^2$ in three stages, as shown in Figure 17. Now compare this graph to the graph in Figure 16(a). The graph of $f(x) = 2x^2 + 8x + 5$ is a parabola that opens up and has its vertex (lowest point) at $(-2, -3)$. Its axis of symmetry is the line $x = -2$.

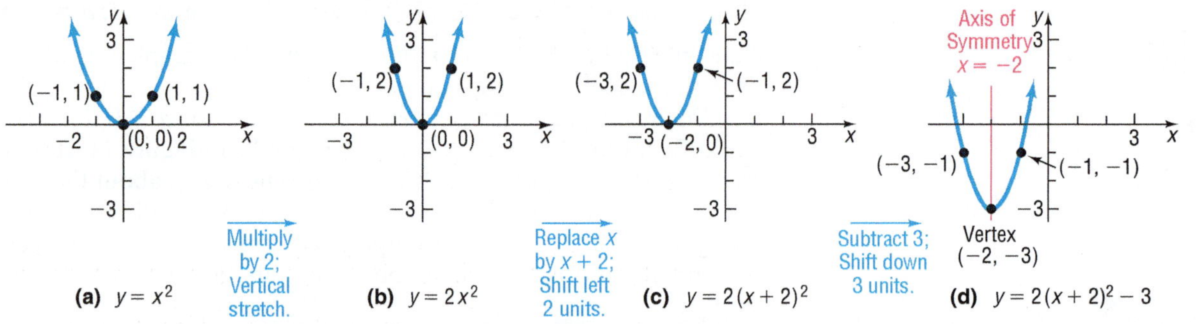

Figure 17 (a) $y = x^2$ (b) $y = 2x^2$ (c) $y = 2(x+2)^2$ (d) $y = 2(x+2)^2 - 3$

Check: Use a graphing utility to graph $Y_1 = f(x) = 2x^2 + 8x + 5$ and use the MINIMUM command to locate its vertex.

Now Work PROBLEM 25

The method used in Example 1 can be used to graph any quadratic function $f(x) = ax^2 + bx + c$, $a \neq 0$, as follows:

$$\begin{aligned} f(x) &= ax^2 + bx + c \\ &= a\left(x^2 + \frac{b}{a}x\right) + c && \text{Factor out } a \text{ from } ax^2 + bx. \\ &= a\left(x^2 + \frac{b}{a}x + \frac{b^2}{4a^2}\right) + c - a\left(\frac{b^2}{4a^2}\right) && \text{Complete the square by adding } \frac{b^2}{4a^2}. \text{ Look closely at this step!} \\ &= a\left(x + \frac{b}{2a}\right)^2 + c - \frac{b^2}{4a} && \text{Factor; simplify.} \\ &= a\left(x + \frac{b}{2a}\right)^2 + \frac{4ac - b^2}{4a} && c - \frac{b^2}{4a} = c \cdot \frac{4a}{4a} - \frac{b^2}{4a} = \frac{4ac - b^2}{4a} \end{aligned}$$

These results lead to the following conclusion:

If $h = -\dfrac{b}{2a}$ and $k = \dfrac{4ac - b^2}{4a}$, then

$$f(x) = ax^2 + bx + c = a(x - h)^2 + k \qquad (1)$$

The graph of $f(x) = a(x - h)^2 + k$ is the parabola $y = ax^2$ shifted horizontally h units (replace x by $x - h$) and vertically k units (add k). As a result, the vertex is at (h, k), and the graph opens up if $a > 0$ and down if $a < 0$. The axis of symmetry is the vertical line $x = h$.

For example, compare equation (1) with the solution given in Example 1.

$$\begin{aligned} f(x) &= 2(x+2)^2 - 3 \\ &= 2(x-(-2))^2 + (-3) \\ &= a(x-h)^2 + k \end{aligned}$$

Because $a = 2$, the graph opens up. Also, because $h = -2$ and $k = -3$, its vertex is at $(-2, -3)$.

2 Identify the Vertex and Axis of Symmetry of a Quadratic Function

We do not need to complete the square to obtain the vertex. In almost every case, it is easier to obtain the vertex of a quadratic function f by remembering that its x-coordinate is $h = -\frac{b}{2a}$. The y-coordinate k can then be found by evaluating f at $-\frac{b}{2a}$. That is, $k = f\left(-\frac{b}{2a}\right)$.

Properties of the Graph of a Quadratic Function

$$f(x) = ax^2 + bx + c \qquad a \neq 0$$

$$\text{Vertex} = \left(-\frac{b}{2a}, f\left(-\frac{b}{2a}\right)\right) \quad \text{Axis of symmetry: the vertical line } x = -\frac{b}{2a} \qquad \textbf{(2)}$$

Parabola opens up if $a > 0$; the vertex is a minimum point.
Parabola opens down if $a < 0$; the vertex is a maximum point.

EXAMPLE 2 **Locating the Vertex without Graphing**

Without graphing, locate the vertex and axis of symmetry of the parabola defined by $f(x) = -3x^2 + 6x + 1$. Does it open up or down?

Solution For this quadratic function, $a = -3$, $b = 6$, and $c = 1$. The x-coordinate of the vertex is

$$h = -\frac{b}{2a} = -\frac{6}{2(-3)} = 1$$

The y-coordinate of the vertex is

$$k = f\left(-\frac{b}{2a}\right) = f(1) = -3(1)^2 + 6(1) + 1 = 4$$

The vertex is located at the point $(1, 4)$. The axis of symmetry is the line $x = 1$. Because $a = -3 < 0$, the parabola opens down. ■

3 Graph a Quadratic Function Using Its Vertex, Axis, and Intercepts

The location of the vertex and intercepts, along with knowledge of whether the graph opens up or down, usually provides enough information to graph $f(x) = ax^2 + bx + c$, $a \neq 0$.

The y-intercept is the value of f at $x = 0$; that is, the y-intercept is $f(0) = c$. The x-intercepts, if there are any, are found by solving the quadratic equation

$$ax^2 + bx + c = 0$$

This equation has two, one, or no real solutions, depending on whether the discriminant $b^2 - 4ac$ is positive, 0, or negative. Depending on the value of the discriminant, the graph of f has x-intercepts, as follows:

The x-Intercepts of a Quadratic Function

1. If the discriminant $b^2 - 4ac > 0$, the graph of $f(x) = ax^2 + bx + c$ has two distinct x-intercepts so it crosses the x-axis in two places.
2. If the discriminant $b^2 - 4ac = 0$, the graph of $f(x) = ax^2 + bx + c$ has one x-intercept so it touches the x-axis at its vertex.
3. If the discriminant $b^2 - 4ac < 0$, the graph of $f(x) = ax^2 + bx + c$ has no x-intercepts so it does not cross or touch the x-axis.

Figure 18 illustrates these possibilities for parabolas that open up.

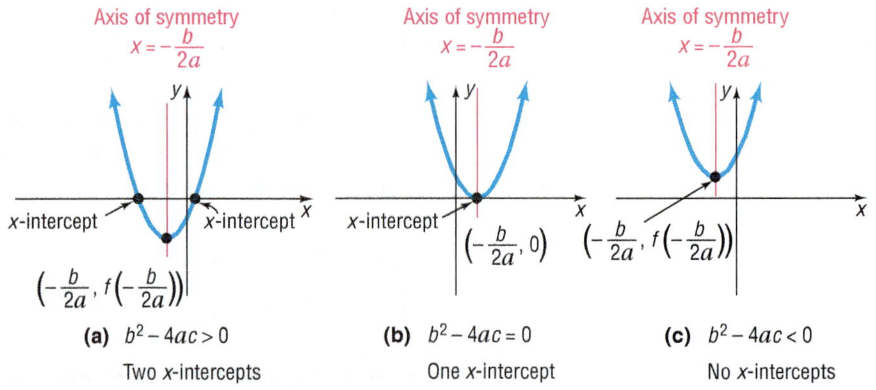

Figure 18 $f(x) = ax^2 + bx + c, a > 0$

EXAMPLE 3

How to Graph a Quadratic Function by Hand Using Its Properties

Graph $f(x) = -3x^2 + 6x + 1$ using its properties. Determine the domain and the range of f. Determine where f is increasing and where it is decreasing.

Step-by-Step Solution

Step 1: Determine whether the graph of f opens up or down.

In Example 2, it was determined that the graph of $f(x) = -3x^2 + 6x + 1$ opens down because $a = -3 < 0$.

Step 2: Determine the vertex and axis of symmetry of the graph of f.

In Example 2, the vertex was found to be at the point whose coordinates are $(1, 4)$. The axis of symmetry is the line $x = 1$.

Step 3: Determine the intercepts of the graph of f.

The y-intercept is found by letting $x = 0$. The y-intercept is $f(0) = 1$. The x-intercepts are found by solving the equation $f(x) = 0$.

$$f(x) = 0$$

$$-3x^2 + 6x + 1 = 0 \quad a = -3, b = 6, c = 1$$

The discriminant $b^2 - 4ac = (6)^2 - 4(-3)(1) = 36 + 12 = 48 > 0$, so the equation has two real solutions and the graph has two x-intercepts. Use the quadratic formula to find that

$$x = \frac{-b + \sqrt{b^2 - 4ac}}{2a} = \frac{-6 + \sqrt{48}}{2(-3)} = \frac{-6 + 4\sqrt{3}}{-6} \approx -0.15$$

and

$$x = \frac{-b - \sqrt{b^2 - 4ac}}{2a} = \frac{-6 - \sqrt{48}}{2(-3)} = \frac{-6 - 4\sqrt{3}}{-6} \approx 2.15$$

The x-intercepts are approximately -0.15 and 2.15.

Step 4: Use the information in Steps 1 through 3 to graph f.

The graph is illustrated in Figure 19. Note how the y-intercept and the axis of symmetry, $x = 1$, are used to obtain the additional point $(2, 1)$ on the graph.

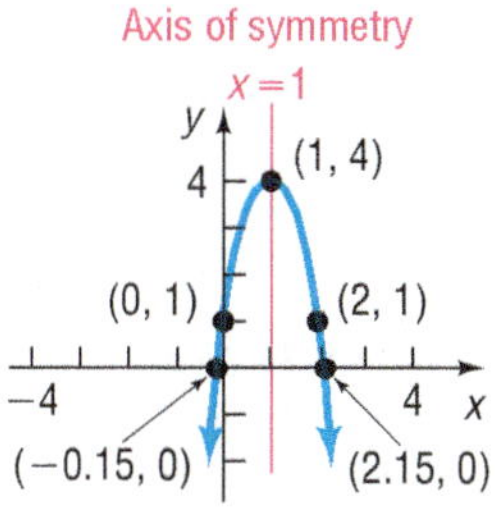

Figure 19 $f(x) = -3x^2 + 6x + 1$

The domain of f is the set of all real numbers. Based on the graph, the range of f is the interval $(-\infty, 4]$. The function f is increasing on the interval $(-\infty, 1]$ and decreasing on the interval $[1, \infty)$.

Graph the function in Example 3 by completing the square and using transformations. Which method do you prefer?

Now Work PROBLEM 33

If the graph of a quadratic function has only one x-intercept or no x-intercepts, it is usually necessary to plot an additional point to obtain the graph.

EXAMPLE 4

Graphing a Quadratic Function Using Its Vertex, Axis, and Intercepts

(a) Graph $f(x) = x^2 - 6x + 9$ by determining whether the graph opens up or down and by finding its vertex, axis of symmetry, y-intercept, and x-intercepts, if any.

(b) Determine the domain and the range of f.

(c) Determine where f is increasing and where it is decreasing.

Solution

(a) **STEP 1:** For $f(x) = x^2 - 6x + 9$, note that $a = 1$, $b = -6$, and $c = 9$. Because $a = 1 > 0$, the parabola opens up.

STEP 2: The x-coordinate of the vertex is

$$h = -\frac{b}{2a} = -\frac{-6}{2(1)} = 3$$

The y-coordinate of the vertex is

$$k = f(3) = (3)^2 - 6(3) + 9 = 0$$

The vertex is at $(3, 0)$. The axis of symmetry is the line $x = 3$.

STEP 3: The y-intercept is $f(0) = 9$. Since the vertex $(3, 0)$ lies on the x-axis, the graph touches the x-axis at the x-intercept.

STEP 4: By using the axis of symmetry and the y-intercept at $(0, 9)$, we can locate the additional point $(6, 9)$ on the graph. See Figure 20.

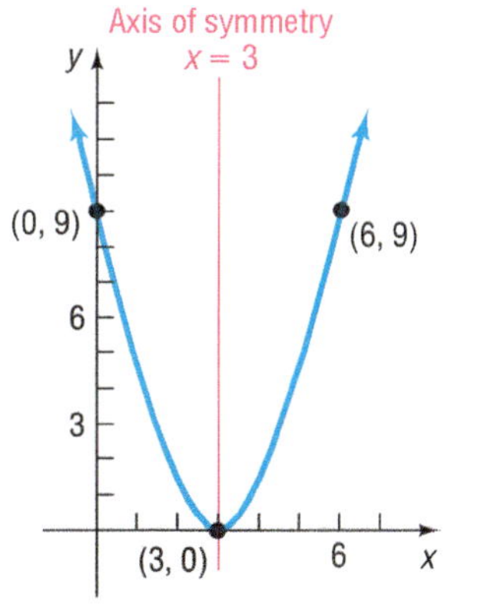

Figure 20 $f(x) = x^2 - 6x + 9$

(b) The domain of f is the set of all real numbers. Based on the graph, the range of f is the interval $[0, \infty)$.

(c) The function f is decreasing on the interval $(-\infty, 3]$ and increasing on the interval $[3, \infty)$.

Now Work PROBLEM 39

EXAMPLE 5 Graphing a Quadratic Function Using Its Vertex, Axis, and Intercepts

(a) Graph $f(x) = 2x^2 + x + 1$ by determining whether the graph opens up or down and by finding its vertex, axis of symmetry, y-intercept, and x-intercepts, if any.

(b) Determine the domain and the range of f.

(c) Determine where f is increasing and where it is decreasing.

Solution

(a) **STEP 1:** For $f(x) = 2x^2 + x + 1$, we have $a = 2$, $b = 1$, and $c = 1$. Because $a = 2 > 0$, the parabola opens up.

STEP 2: The x-coordinate of the vertex is

$$h = -\frac{b}{2a} = -\frac{1}{4}$$

The y-coordinate of the vertex is

$$k = f\left(-\frac{1}{4}\right) = 2\left(\frac{1}{16}\right) + \left(-\frac{1}{4}\right) + 1 = \frac{7}{8}$$

The vertex is at $\left(-\frac{1}{4}, \frac{7}{8}\right)$. The axis of symmetry is the line $x = -\frac{1}{4}$.

STEP 3: The y-intercept is $f(0) = 1$. The x-intercept(s), if any, obey the equation $2x^2 + x + 1 = 0$. The discriminant $b^2 - 4ac = (1)^2 - 4(2)(1) = -7 < 0$. This equation has no real solutions, which means the graph has no x-intercepts.

STEP 4: Use the point $(0, 1)$ and the axis of symmetry $x = -\frac{1}{4}$ to locate the additional point $\left(-\frac{1}{2}, 1\right)$ on the graph. See Figure 21.

(b) The domain of f is the set of all real numbers. Based on the graph, the range of f is the interval $\left[\frac{7}{8}, \infty\right)$.

(c) The function f is decreasing on the interval $\left(-\infty, -\frac{1}{4}\right]$ and is increasing on the interval $\left[-\frac{1}{4}, \infty\right)$. ■

Note: In Example 5, since the vertex is above the x-axis and the parabola opens up, we can conclude that the graph of the quadratic function will have no x-intercepts. ■

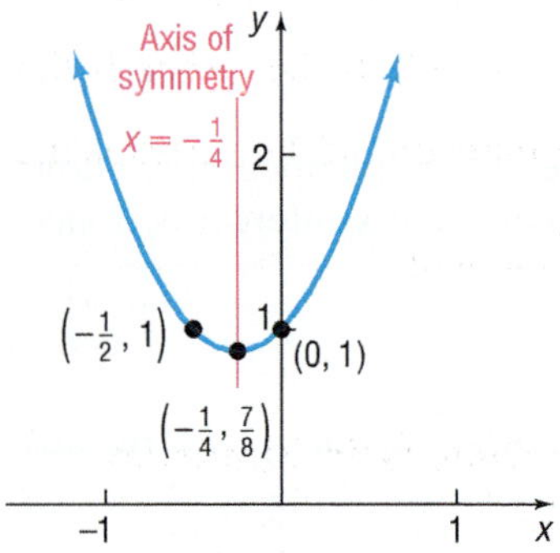

Figure 21 $f(x) = 2x^2 + x + 1$

Now Work PROBLEM 43

4 Find a Quadratic Function Given Its Vertex and One Other Point

If the vertex (h, k) and one additional point on the graph of a quadratic function $f(x) = ax^2 + bx + c$, $a \neq 0$, are known, then

$$f(x) = a(x - h)^2 + k \tag{3}$$

can be used to obtain the quadratic function.

EXAMPLE 6 Finding the Quadratic Function Given Its Vertex and One Other Point

Determine the quadratic function whose vertex is $(1, -5)$ and whose y-intercept is -3. The graph of the parabola is shown in Figure 22.

Solution The vertex is $(1, -5)$, so $h = 1$ and $k = -5$. Substitute these values into equation (3).

$$f(x) = a(x - h)^2 + k \quad \textbf{Equation (3)}$$
$$f(x) = a(x - 1)^2 - 5 \quad h = 1, k = -5$$

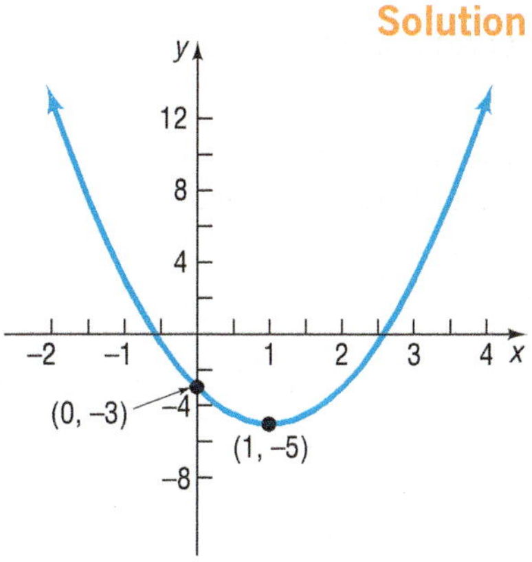

Figure 22 $f(x) = 2x^2 - 4x - 3$

To determine the value of a, use the fact that $f(0) = -3$ (the y-intercept).

$$f(x) = a(x - 1)^2 - 5$$
$$-3 = a(0 - 1)^2 - 5 \quad x = 0, y = f(0) = -3$$
$$-3 = a - 5$$
$$a = 2$$

The quadratic function whose graph is shown in Figure 22 is

$$f(x) = a(x - h)^2 + k = 2(x - 1)^2 - 5 = 2x^2 - 4x - 3$$

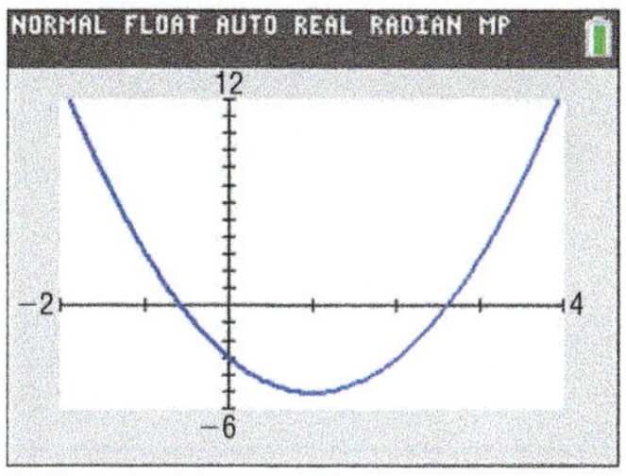

Figure 23 $Y_1 = 2x^2 - 4x - 3$

✓**Check:** Figure 23 shows the graph of $Y_1 = 2x^2 - 4x - 3$ using a TI-84 Plus C. ■

Now Work PROBLEM 49

5 Find the Maximum or Minimum Value of a Quadratic Function

The graph of a quadratic function

$$f(x) = ax^2 + bx + c \qquad a \neq 0$$

is a parabola with vertex at $\left(-\frac{b}{2a}, f\left(-\frac{b}{2a}\right)\right)$. This vertex is the highest point on the graph if $a < 0$ and the lowest point on the graph if $a > 0$. If the vertex is the highest point $(a < 0)$, then $f\left(-\frac{b}{2a}\right)$ is the **maximum value** of f. If the vertex is the lowest point $(a > 0)$, then $f\left(-\frac{b}{2a}\right)$ is the **minimum value** of f.

EXAMPLE 7 Finding the Maximum or Minimum Value of a Quadratic Function

Determine whether the quadratic function

$$f(x) = x^2 - 4x - 5$$

has a maximum or a minimum value. Then find the maximum or minimum value.

Solution Compare $f(x) = x^2 - 4x - 5$ to $f(x) = ax^2 + bx + c$. Then $a = 1, b = -4$, and $c = -5$. Because $a > 0$, the graph of f opens up, which means the vertex is a minimum point. The minimum value occurs at

$$x = -\frac{b}{2a} \underset{\uparrow}{=} -\frac{-4}{2(1)} = \frac{4}{2} = 2$$
$$a = 1, b = -4$$

The minimum value is

$$f\left(-\frac{b}{2a}\right) = f(2) = 2^2 - 4(2) - 5 = 4 - 8 - 5 = -9$$

✓**Check:** Graph $Y_1 = f(x) = x^2 - 4x - 5$. Use MINIMUM to verify the vertex is $(2, -9)$. ■

Now Work PROBLEM 57

SUMMARY

Steps for Graphing a Quadratic Function $f(x) = ax^2 + bx + c, a \neq 0$

Option 1

STEP 1: Complete the square in x to write the quadratic function in the form $f(x) = a(x - h)^2 + k$.

STEP 2: Graph the function in stages using transformations.

Option 2

STEP 1: Determine whether the parabola opens up $(a > 0)$ or down $(a < 0)$.

STEP 2: Determine the vertex $\left(-\dfrac{b}{2a}, f\left(-\dfrac{b}{2a}\right)\right)$.

STEP 3: Determine the axis of symmetry, $x = -\dfrac{b}{2a}$.

STEP 4: Determine the y-intercept, $f(0)$, and the x-intercepts, if any.

(a) If $b^2 - 4ac > 0$, the graph of the quadratic function has two x-intercepts, which are found by solving the equation $ax^2 + bx + c = 0$.

(b) If $b^2 - 4ac = 0$, the vertex is the x-intercept.

(c) If $b^2 - 4ac < 0$, there are no x-intercepts.

STEP 5: Determine an additional point by using the y-intercept and the axis of symmetry.

STEP 6: Plot the points and draw the graph.

4.3 Assess Your Understanding

'Are You Prepared?' *Answers are given at the end of these exercises. If you get a wrong answer, read the pages listed in red.*

1. List the intercepts of the equation $y = x^2 - 9$. (pp. 165–166)
2. Find the real solutions of the equation $2x^2 + 7x - 4 = 0$. (pp. 110–117)
3. To complete the square of $x^2 - 5x$, you add the number ________. (p. 57)
4. To graph $y = (x - 4)^2$, you shift the graph of $y = x^2$ to the ______ a distance of ______ units. (pp. 256–258)

Concepts and Vocabulary

5. The graph of a quadratic function is called a(n) ________.
6. The vertical line passing through the vertex of a parabola is called the ________________.
7. The x-coordinate of the vertex of $f(x) = ax^2 + bx + c$, $a \neq 0$, is ________.
8. ***True or False*** The graph of $f(x) = 2x^2 + 3x - 4$ opens up.
9. ***True or False*** The y-coordinate of the vertex of $f(x) = -x^2 + 4x + 5$ is $f(2)$.
10. ***True or False*** If the discriminant $b^2 - 4ac = 0$, the graph of $f(x) = ax^2 + bx + c, a \neq 0$, will touch the x-axis at its vertex.
11. If $b^2 - 4ac > 0$, which of the following conclusions can be made about the graph of $f(x) = ax^2 + bx + c$, $a \neq 0$?
    (a) The graph has two distinct x-intercepts.
    (b) The graph has no x-intercepts.
    (c) The graph has three distinct x-intercepts.
    (d) The graph has one x-intercept.
12. If the graph of $f(x) = ax^2 + bx + c$, $a \neq 0$, has a maximum value at its vertex, which of the following conditions must be true?
    (a) $-\dfrac{b}{2a} > 0$ (b) $-\dfrac{b}{2a} < 0$ (c) $a > 0$ (d) $a < 0$

Skill Building

In Problems 13–20, match each graph to one the following functions.

13. $f(x) = x^2 - 1$
14. $f(x) = -x^2 - 1$
15. $f(x) = x^2 - 2x + 1$
16. $f(x) = x^2 + 2x + 1$
17. $f(x) = x^2 - 2x + 2$
18. $f(x) = x^2 + 2x$
19. $f(x) = x^2 - 2x$
20. $f(x) = x^2 + 2x + 2$

A

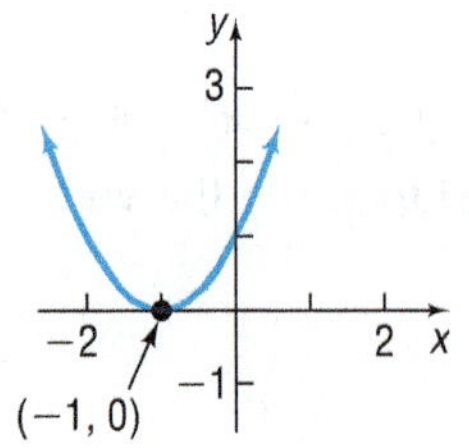

B

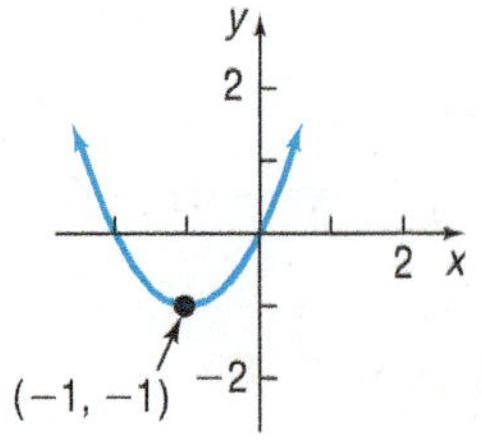

C

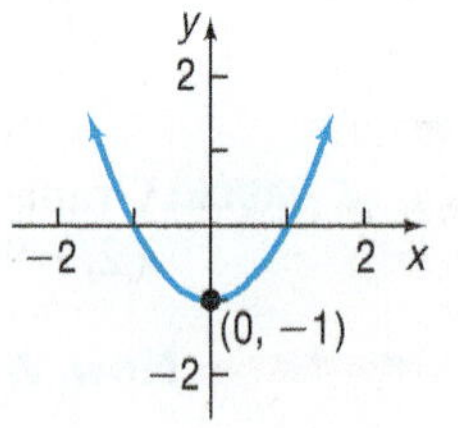

D

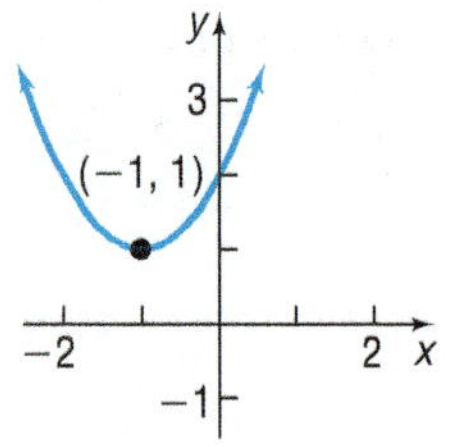

E

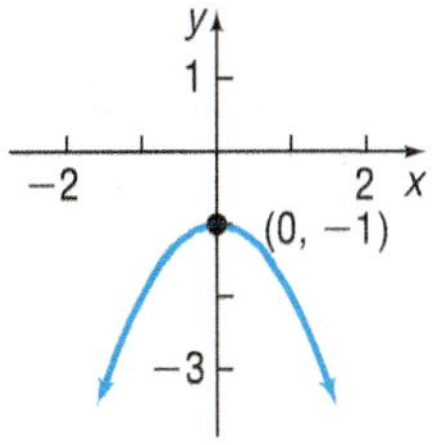

F

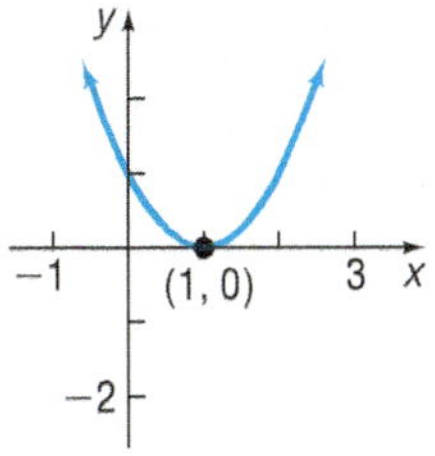

G

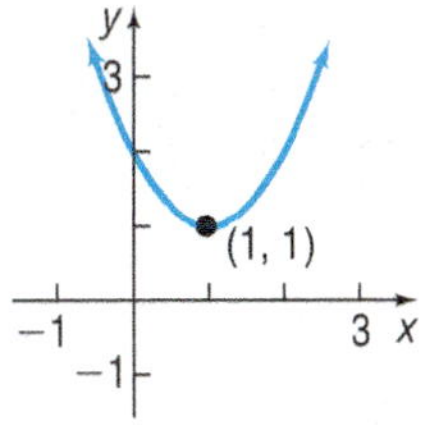

H

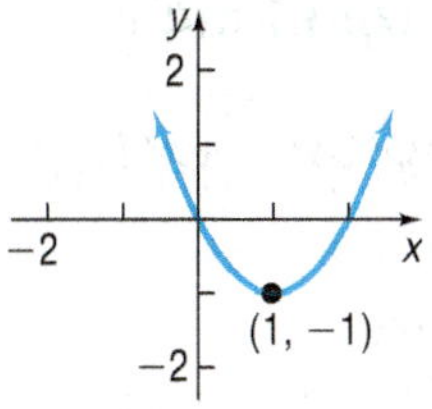

In Problems 21–32, graph the function f by starting with the graph of $y = x^2$ and using transformations (shifting, compressing, stretching, and/or reflection). Verify your results using a graphing utility.
[**Hint:** If necessary, write f in the form $f(x) = a(x - h)^2 + k$.]

21. $f(x) = \frac{1}{4}x^2$ **22.** $f(x) = 2x^2 + 4$ **23.** $f(x) = (x + 2)^2 - 2$ **24.** $f(x) = (x - 3)^2 - 10$

25. $f(x) = x^2 + 4x + 2$ **26.** $f(x) = x^2 - 6x - 1$ **27.** $f(x) = 2x^2 - 4x + 1$ **28.** $f(x) = 3x^2 + 6x$

29. $f(x) = -x^2 - 2x$ **30.** $f(x) = -2x^2 + 6x + 2$ **31.** $f(x) = \frac{1}{2}x^2 + x - 1$ **32.** $f(x) = \frac{2}{3}x^2 + \frac{4}{3}x - 1$

In Problems 33–48, (a) graph each quadratic function by determining whether its graph opens up or down and by finding its vertex, axis of symmetry, y-intercept, and x-intercepts, if any. (b) Determine the domain and the range of the function. (c) Determine where the function is increasing and where it is decreasing. Verify your results using a graphing utility.

33. $f(x) = x^2 + 2x$ **34.** $f(x) = x^2 - 4x$ **35.** $f(x) = -x^2 - 6x$ **36.** $f(x) = -x^2 + 4x$

37. $f(x) = x^2 + 2x - 8$ **38.** $f(x) = x^2 - 2x - 3$ **39.** $f(x) = x^2 + 2x + 1$ **40.** $f(x) = x^2 + 6x + 9$

41. $f(x) = 2x^2 - x + 2$ **42.** $f(x) = 4x^2 - 2x + 1$ **43.** $f(x) = -2x^2 + 2x - 3$ **44.** $f(x) = -3x^2 + 3x - 2$

45. $f(x) = 3x^2 + 6x + 2$ **46.** $f(x) = 2x^2 + 5x + 3$ **47.** $f(x) = -4x^2 - 6x + 2$ **48.** $f(x) = 3x^2 - 8x + 2$

In Problems 49–54, determine the quadratic function whose graph is given.

49.

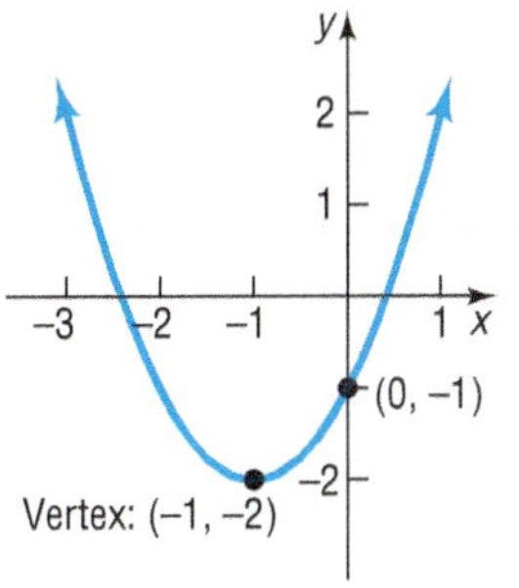

50.

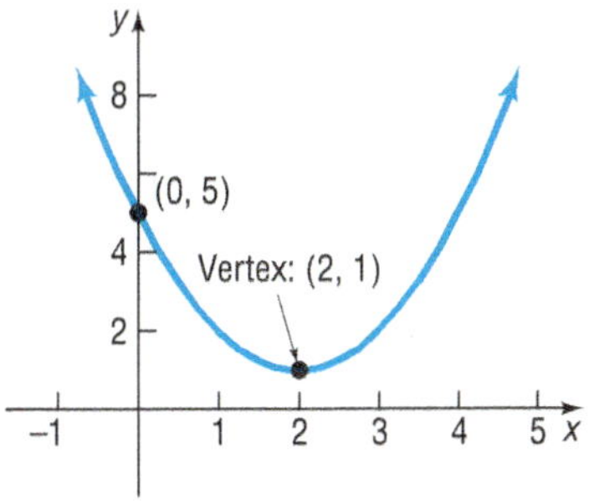

51.

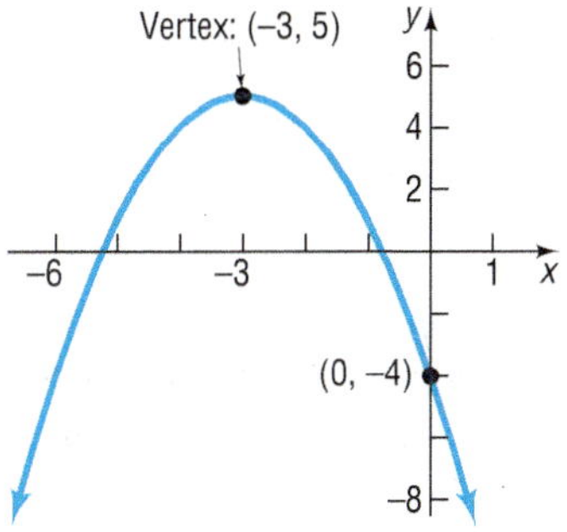

52.

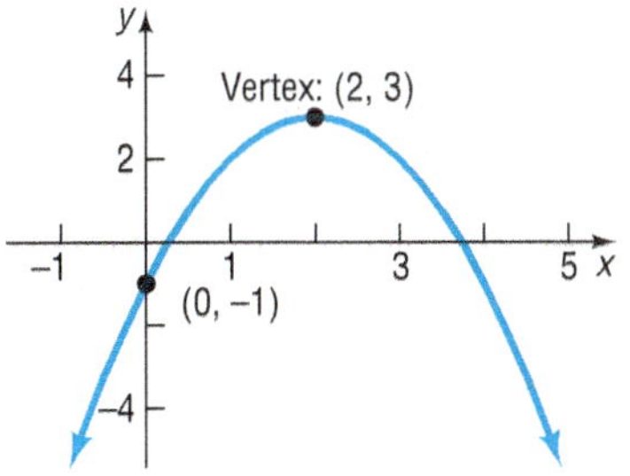

53.

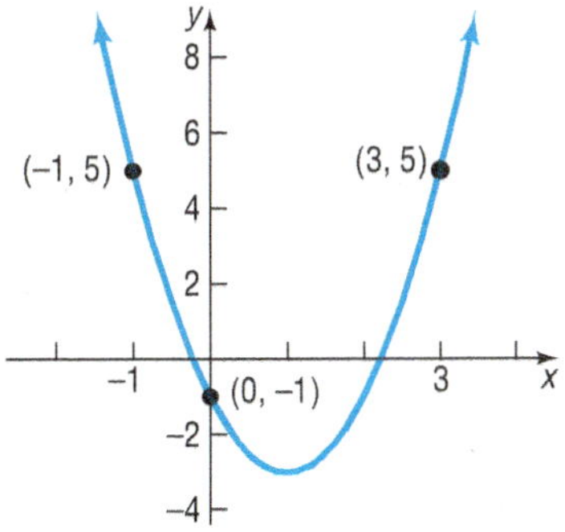

54.

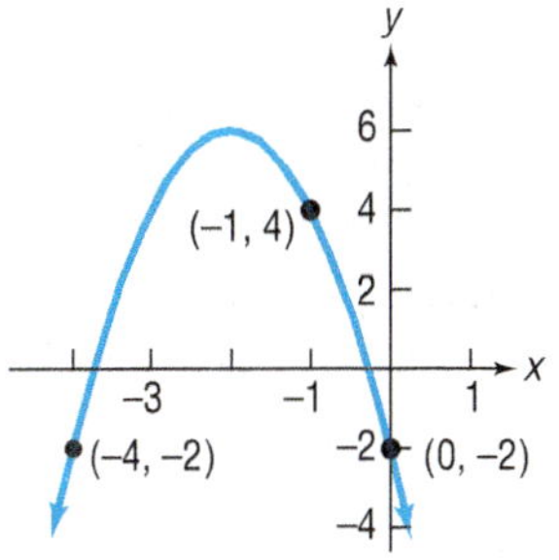

In Problems 55–62, determine, without graphing, whether the given quadratic function has a maximum value or a minimum value and then find the value.

55. $f(x) = 2x^2 + 12x$ **56.** $f(x) = -2x^2 + 12x$ **57.** $f(x) = 2x^2 + 12x - 3$ **58.** $f(x) = 4x^2 - 8x + 3$

59. $f(x) = -x^2 + 10x - 4$ **60.** $f(x) = -2x^2 + 8x + 3$ **61.** $f(x) = -3x^2 + 12x + 1$ **62.** $f(x) = 4x^2 - 4x$

Mixed Practice

In Problems 63–74, (a) graph each function, (b) determine the domain and the range of the function, and (c) determine where the function is increasing and where it is decreasing.

63. $f(x) = x^2 - 2x - 15$

64. $g(x) = x^2 - 2x - 8$

65. $F(x) = 2x - 5$

66. $f(x) = \frac{3}{2}x - 2$

67. $g(x) = -2(x - 3)^2 + 2$

68. $h(x) = -3(x + 1)^2 + 4$

69. $f(x) = 2x^2 + x + 1$

70. $G(x) = 3x^2 + 2x + 5$

71. $h(x) = -\frac{2}{5}x + 4$

72. $f(x) = -3x + 2$

73. $H(x) = -4x^2 - 4x - 1$

74. $F(x) = -4x^2 + 20x - 25$

Applications and Extensions

75. The graph of the function $f(x) = ax^2 + bx + c$ has vertex at $(0, 2)$ and passes through the point $(1, 8)$. Find $a, b,$ and c.

76. The graph of the function $f(x) = ax^2 + bx + c$ has vertex at $(1, 4)$ and passes through the point $(-1, -8)$. Find $a, b,$ and c.

In Problems 77–82, for the given functions f and g,

(a) Graph f and g on the same Cartesian plane.
(b) Solve $f(x) = g(x)$.
(c) Use the result of part (b) to label the points of intersection of the graphs of f and g.
(d) Shade the region for which $f(x) > g(x)$, that is, the region below f and above g.

77. $f(x) = 2x - 1;\ \ g(x) = x^2 - 4$

78. $f(x) = -2x - 1;\ \ g(x) = x^2 - 9$

79. $f(x) = -x^2 + 4;\ \ g(x) = -2x + 1$

80. $f(x) = -x^2 + 9;\ \ g(x) = 2x + 1$

81. $f(x) = -x^2 + 5x;\ \ g(x) = x^2 + 3x - 4$

82. $f(x) = -x^2 + 7x - 6;\ \ g(x) = x^2 + x - 6$

Answer Problems 83 and 84 using the following: A quadratic function of the form $f(x) = ax^2 + bx + c$ with $b^2 - 4ac > 0$ may also be written in the form $f(x) = a(x - r_1)(x - r_2)$, where r_1 and r_2 are the x-intercepts of the graph of the quadratic function.

83. (a) Find a quadratic function whose x-intercepts are -3 and 1 with $a = 1; a = 2; a = -2; a = 5$.
(b) How does the value of a affect the intercepts?
(c) How does the value of a affect the axis of symmetry?
(d) How does the value of a affect the vertex?
(e) Compare the x-coordinate of the vertex with the midpoint of the x-intercepts. What might you conclude?

84. (a) Find a quadratic function whose x-intercepts are -5 and 3 with $a = 1; a = 2; a = -2; a = 5$.
(b) How does the value of a affect the intercepts?
(c) How does the value of a affect the axis of symmetry?
(d) How does the value of a affect the vertex?
(e) Compare the x-coordinate of the vertex with the midpoint of the x-intercepts. What might you conclude?

---

85. Suppose that $f(x) = x^2 + 4x - 21$.
(a) What is the vertex of f?
(b) What are the x-intercepts of the graph of f?
(c) Solve $f(x) = -21$ for x. What points are on the graph of f?
(d) Use the information obtained in parts (a)–(c) to graph $f(x) = x^2 + 4x - 21$.

86. Suppose that $f(x) = x^2 + 2x - 8$.
(a) What is the vertex of f?
(b) What are the x-intercepts of the graph of f?
(c) Solve $f(x) = -8$ for x. What points are on the graph of f?
(d) Use the information obtained in parts (a)–(c) to graph $f(x) = x^2 + 2x - 8$.

87. Find the point on the line $y = x$ that is closest to the point $(3, 1)$.
[**Hint:** Express the distance d from the point to the line as a function of x, and then find the minimum value of $[d(x)]^2$.]

88. Find the point on the line $y = x + 1$ that is closest to the point $(4, 1)$.

89. Maximizing Revenue Suppose that the manufacturer of a gas clothes dryer has found that, when the unit price is p dollars, the revenue R (in dollars) is

$$R(p) = -4p^2 + 4000p$$

What unit price should be established for the dryer to maximize revenue? What is the maximum revenue?

90. Maximizing Revenue The John Deere company has found that the revenue, in dollars, from sales of riding mowers is a function of the unit price p, in dollars, that it charges. If the revenue R is

$$R(p) = -\frac{1}{2}p^2 + 1900p$$

what unit price p should be charged to maximize revenue? What is the maximum revenue?

91. Minimizing Marginal Cost The **marginal cost** of a product can be thought of as the cost of producing one additional unit of output. For example, if the marginal cost of producing the 50th product is \$6.20, it cost \$6.20 to increase production from 49 to 50 units of output. Suppose the marginal cost C

(in dollars) to produce x thousand digital music players is given by the function

$$C(x) = x^2 - 140x + 7400$$

(a) How many players should be produced to minimize the marginal cost?
(b) What is the minimum marginal cost?

92. Minimizing Marginal Cost (See Problem 91.) The marginal cost C (in dollars) of manufacturing x cell phones (in thousands) is given by

$$C(x) = 5x^2 - 200x + 4000$$

(a) How many cell phones should be manufactured to minimize the marginal cost?
(b) What is the minimum marginal cost?

93. Business The monthly revenue R achieved by selling x wristwatches is figured to be $R(x) = 75x - 0.2x^2$. The monthly cost C of selling x wristwatches is $C(x) = 32x + 1750$.
(a) How many wristwatches must the firm sell to maximize revenue? What is the maximum revenue?
(b) Profit is given as $P(x) = R(x) - C(x)$. What is the profit function?
(c) How many wristwatches must the firm sell to maximize profit? What is the maximum profit?
(d) Provide a reasonable explanation as to why the answers found in parts (a) and (c) differ. Explain why a quadratic function is a reasonable model for revenue.

94. Business The daily revenue R achieved by selling x boxes of candy is figured to be $R(x) = 9.5x - 0.04x^2$. The daily cost C of selling x boxes of candy is $C(x) = 1.25x + 250$.
(a) How many boxes of candy must the firm sell to maximize revenue? What is the maximum revenue?
(b) Profit is given as $P(x) = R(x) - C(x)$. What is the profit function?
(c) How many boxes of candy must the firm sell to maximize profit? What is the maximum profit?
(d) Provide a reasonable explanation as to why the answers found in parts (a) and (c) differ. Explain why a quadratic function is a reasonable model for revenue.

95. Stopping Distance An accepted relationship between stopping distance, d (in feet), and the speed of a car, v (in mph), is $d = 1.1v + 0.06v^2$ on dry, level concrete.
(a) How many feet will it take a car traveling 45 mph to stop on dry, level concrete?
(b) If an accident occurs 200 feet ahead of you, what is the maximum speed you can be traveling to avoid being involved?
(c) What might the term $1.1v$ represent?

96. Birthrate for Unmarried Women In the United States, the birthrate B for unmarried women (births per 1000 unmarried women) whose age is a is modeled by the function $B(a) = -0.30a^2 + 16.26a - 158.90$.
(a) What is the age of unmarried women with the highest birthrate?
(b) What is the highest birthrate of unmarried women?
(c) Evaluate and interpret $B(40)$.

Source: *National Vital Statistics System, 2013*

97. Let $f(x) = ax^2 + bx + c$, where $a, b,$ and c are odd integers. If x is an integer, show that $f(x)$ must be an odd integer.
[**Hint:** x is either an even integer or an odd integer.]

Explaining Concepts: Discussion and Writing

98. Make up a quadratic function that opens down and has only one x-intercept. Compare yours with others in the class. What are the similarities? What are the differences?

99. On one set of coordinate axes, graph the family of parabolas $f(x) = x^2 + 2x + c$ for $c = -3$, $c = 0$, and $c = 1$. Describe the characteristics of a member of this family.

100. On one set of coordinate axes, graph the family of parabolas $f(x) = x^2 + bx + 1$ for $b = -4$, $b = 0$, and $b = 4$. Describe the general characteristics of this family.

101. State the circumstances that cause the graph of a quadratic function $f(x) = ax^2 + bx + c$ to have no x-intercepts.

102. Why does the graph of a quadratic function open up if $a > 0$ and down if $a < 0$?

103. Can a quadratic function have a range of $(-\infty, \infty)$? Justify your answer.

104. What are the possibilities for the number of times the graphs of two different quadratic functions intersect?

Retain Your Knowledge

Problems 105–108 are based on material learned earlier in the course. The purpose of these problems is to keep the material fresh in your mind so that you are better prepared for the final exam.

105. Determine whether $x^2 + 4y^2 = 16$ is symmetric respect to the x-axis, the y-axis, and/or the origin.

106. Solve the inequality $27 - x \geq 5x + 3$. Write the solution in both set notation and interval notation.

107. Find the center and radius of the circle $x^2 + y^2 - 10x + 4y + 20 = 0$.

108. Write the function whose graph is the graph of $y = \sqrt{x}$, but reflected about the y-axis.

'Are You Prepared?' Answers

1. $(0, -9), (-3, 0), (3, 0)$ **2.** $\left\{-4, \frac{1}{2}\right\}$ **3.** $\frac{25}{4}$ **4.** right; 4

4.4 Build Quadratic Models from Verbal Descriptions and from Data

PREPARING FOR THIS SECTION *Before getting started, review the following:*

- Problem Solving (Section 1.6, pp. 137–143)
- Linear Models: Building Linear Functions from Data (Section 4.2, pp. 291–294)

Now Work the 'Are You Prepared?' problems on page 315.

OBJECTIVES 1 Build Quadratic Models from Verbal Descriptions (p. 310)
2 Build Quadratic Models from Data (p. 314)

In this section we will first discuss models in the form of a quadratic function when a verbal description of the problem is given. We end the section by fitting a quadratic function to data, which is another form of modeling.

When a mathematical model is in the form of a quadratic function, the properties of the graph of the function can provide important information about the model. In particular, we can use the quadratic function to determine the maximum or minimum value of the function. The fact that the graph of a quadratic function has a maximum or minimum value enables us to answer questions involving **optimization**—that is, finding the maximum or minimum values in models.

1 Build Quadratic Models from Verbal Descriptions

In economics, revenue R, in dollars, is defined as the amount of money received from the sale of an item and is equal to the unit selling price p, in dollars, of the item times the number x of units actually sold. That is,

$$R = xp$$

The Law of Demand states that p and x are related: As one increases, the other decreases. The equation that relates p and x is called the **demand equation**. When the demand equation is linear, the revenue model is a quadratic function.

EXAMPLE 1 Maximizing Revenue

The marketing department at Texas Instruments has found that when certain calculators are sold at a price of p dollars per unit, the number x of calculators sold is given by the demand equation

$$x = 21{,}000 - 150p$$

(a) Find a model that expresses the revenue R as a function of the price p.
(b) What is the domain of R?
(c) What unit price should be used to maximize revenue?
(d) If this price is charged, what is the maximum revenue?
(e) How many units are sold at this price?
(f) Graph R.
(g) What price should Texas Instruments charge to collect at least \$675,000 in revenue?

Solution (a) The revenue R is $R = xp$, where $x = 21{,}000 - 150p$.

$$R = xp = (21{,}000 - 150p)p = -150p^2 + 21{,}000p$$

(b) Because x represents the number of calculators sold, we have $x \geq 0$, so $21{,}000 - 150p \geq 0$. Solving this linear inequality gives $p \leq 140$. In addition, Texas Instruments will charge only a positive price for the calculator, so $p > 0$. Combining these inequalities gives the domain of R, which is $\{p \mid 0 < p \leq 140\}$.

(c) The function R is a quadratic function with $a = -150$, $b = 21{,}000$, and $c = 0$. Because $a < 0$, the vertex is the highest point on the parabola. The revenue R is a maximum when the price p is

$$p = -\frac{b}{2a} = -\frac{21{,}000}{2(-150)} = \$70.00$$

$a = -150,\ b = 21{,}000$

(d) The maximum revenue R is

$$R(70) = -150(70)^2 + 21{,}000(70) = \$735{,}000$$

(e) The number of calculators sold is given by the demand equation $x = 21{,}000 - 150p$. At a price of $p = \$70$,

$$x = 21{,}000 - 150(70) = 10{,}500$$

calculators are sold.

(f) To graph R, plot the intercept $(140, 0)$ and the vertex $(70, 735\,000)$. See Figure 24 for the graph.

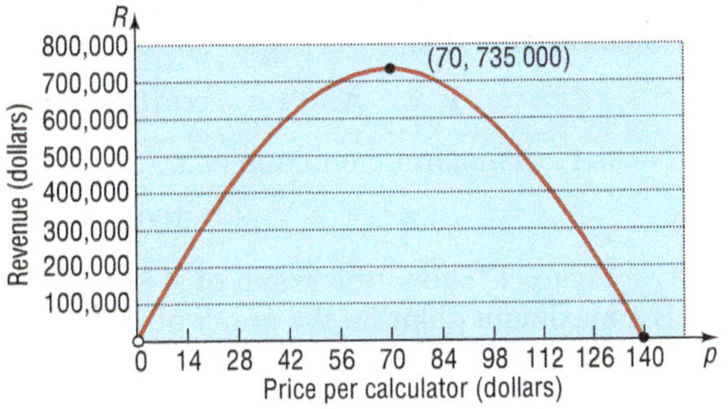

Figure 24

(g) Graph $R = 675{,}000$ and $R(p) = -150p^2 + 21{,}000p$ on the same Cartesian plane. See Figure 25. We find where the graphs intersect by solving

$$\begin{aligned} 675{,}000 &= -150p^2 + 21{,}000p \\ 150p^2 - 21{,}000p + 675{,}000 &= 0 \quad \text{Add } 150p^2 - 21{,}000p \text{ to both sides.} \\ p^2 - 140p + 4500 &= 0 \quad \text{Divide both sides by 150.} \\ (p - 50)(p - 90) &= 0 \quad \text{Factor.} \\ p = 50 \text{ or } p &= 90 \quad \text{Use the Zero-Product Property.} \end{aligned}$$

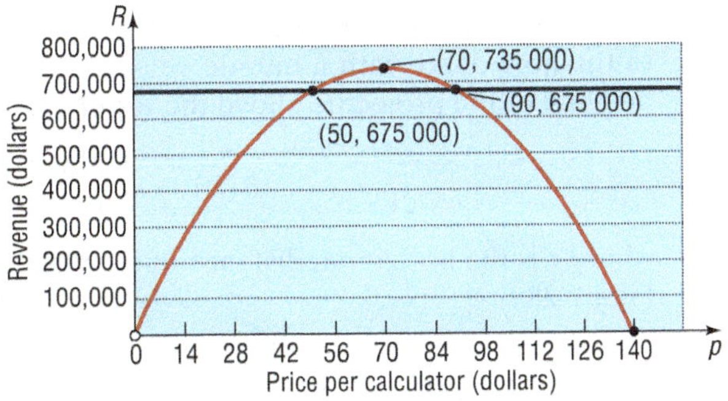

Figure 25

The graphs intersect at $(50, 675\,000)$ and $(90, 675\,000)$. Based on the graph in Figure 25, Texas Instruments should charge between \$50 and \$90 to earn at least \$675,000 in revenue.

Now Work PROBLEM 3

EXAMPLE 2 Maximizing the Area Enclosed by a Fence

A farmer has 2000 yards of fence to enclose a rectangular field. What are the dimensions of the rectangle that encloses the most area?

Solution Figure 26 illustrates the situation. The available fence represents the perimeter of the rectangle. If x is the length and w is the width, then

$$2x + 2w = 2000 \qquad \textbf{(1)}$$

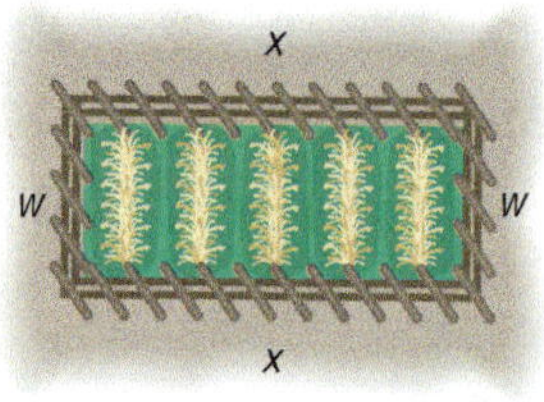

Figure 26

The area A of the rectangle is

$$A = xw$$

To express A in terms of a single variable, solve equation (1) for w and substitute the result in $A = xw$. Then A involves only the variable x. [You could also solve equation (1) for x and express A in terms of w alone. Try it!]

$$\begin{aligned} 2x + 2w &= 2000 \\ 2w &= 2000 - 2x \\ w &= \frac{2000 - 2x}{2} = 1000 - x \end{aligned}$$

Then the area A is

$$A = xw = x(1000 - x) = -x^2 + 1000x$$

Now, A is a quadratic function of x.

$$A(x) = -x^2 + 1000x \qquad a = -1, b = 1000, c = 0$$

Figure 27 shows the graph of $A(x) = -x^2 + 1000x$. Because $a < 0$, the vertex is a maximum point on the graph of A. The maximum value occurs at

$$x = -\frac{b}{2a} = -\frac{1000}{2(-1)} = 500$$

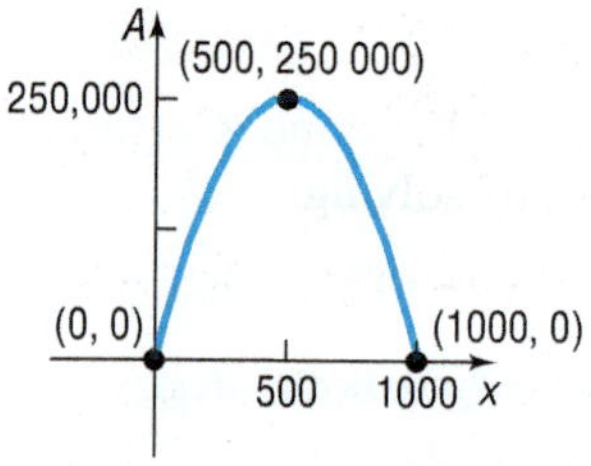

Figure 27 $A(x) = -x^2 + 1000x$

The maximum value of A is

$$A\left(-\frac{b}{2a}\right) = A(500) = -500^2 + 1000(500) = -250{,}000 + 500{,}000 = 250{,}000$$

The largest rectangle that can be enclosed by 2000 yards of fence has an area of 250,000 square yards. Its dimensions are 500 yards by 500 yards. ■

Now Work PROBLEM 7

EXAMPLE 3 Analyzing the Motion of a Projectile

A projectile is fired from a cliff 500 feet above the water at an inclination of 45° to the horizontal, with a muzzle velocity of 400 feet per second. From physics, the height h of the projectile above the water can be modeled by

$$h(x) = \frac{-32x^2}{(400)^2} + x + 500$$

where x is the horizontal distance of the projectile from the base of the cliff. See Figure 28.

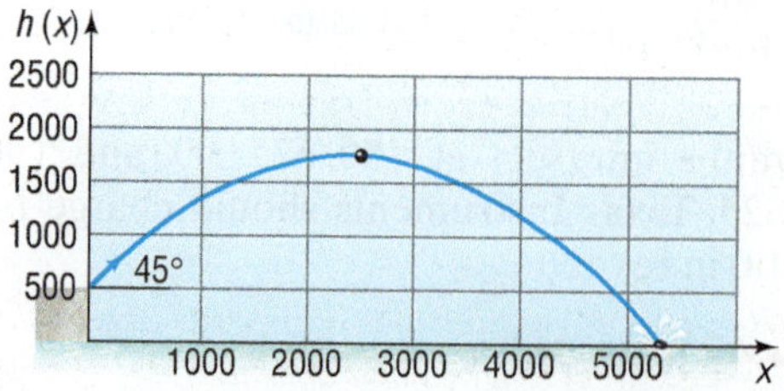

Figure 28

(a) Find the maximum height of the projectile.
(b) How far from the base of the cliff will the projectile strike the water?

Solution (a) The height of the projectile is given by a quadratic function.

$$h(x) = \frac{-32x^2}{(400)^2} + x + 500 = \frac{-1}{5000}x^2 + x + 500$$

We are looking for the maximum value of h. Because $a < 0$, the maximum value occurs at the vertex, whose x-coordinate is

$$x = -\frac{b}{2a} = -\frac{1}{2\left(-\frac{1}{5000}\right)} = \frac{5000}{2} = 2500$$

The maximum height of the projectile is

$$h(2500) = \frac{-1}{5000}(2500)^2 + 2500 + 500 = -1250 + 2500 + 500 = 1750 \text{ ft}$$

(b) The projectile will strike the water when the height is zero. To find the distance x traveled, solve the equation

$$h(x) = \frac{-1}{5000}x^2 + x + 500 = 0$$

The discriminant of this quadratic equation is

$$b^2 - 4ac = 1^2 - 4\left(\frac{-1}{5000}\right)(500) = 1.4$$

Then

$$x = \frac{-b \pm \sqrt{b^2 - 4ac}}{2a} = \frac{-1 \pm \sqrt{1.4}}{2\left(-\frac{1}{5000}\right)} \approx \begin{cases} -458 \\ 5458 \end{cases}$$

Discard the negative solution. The projectile will strike the water at a distance of about 5458 feet from the base of the cliff. ■

Seeing the Concept

Graph

$$h(x) = \frac{-1}{5000}x^2 + x + 500$$
$$0 \le x \le 5500$$

Use MAXIMUM to find the maximum height of the projectile, and use ROOT or ZERO to find the distance from the base of the cliff to where it strikes the water. Compare your results with those obtained in Example 3. ■

Now Work PROBLEM 11

EXAMPLE 4 The Golden Gate Bridge

The Golden Gate Bridge, a suspension bridge, spans the entrance to San Francisco Bay. Its 746-foot-tall towers are 4200 feet apart. The bridge is suspended from two huge cables more than 3 feet in diameter; the 90-foot-wide roadway is 220 feet above the water. The cables are parabolic in shape* and touch the road surface at the center of the bridge. Find the height of the cable above the road at a distance of 1000 feet from the center.

Solution See Figure 29 on the next page. Begin by choosing the placement of the coordinate axes so that the x-axis coincides with the road surface and the origin coincides with the center of the bridge. As a result, the twin towers will be vertical (height $746 - 220 = 526$ feet above the road) and located 2100 feet from the center. Also, the cable, which has the shape of a parabola, will extend from the towers, open up, and have its vertex at $(0, 0)$. This choice of placement of the axes enables the equation of the parabola to have the form $y = ax^2$, $a > 0$. Note that the points $(-2100, 526)$ and $(2100, 526)$ are on the graph.

*A cable suspended from two towers is in the shape of a **catenary**, but when a horizontal roadway is suspended from the cable, the cable takes the shape of a parabola.

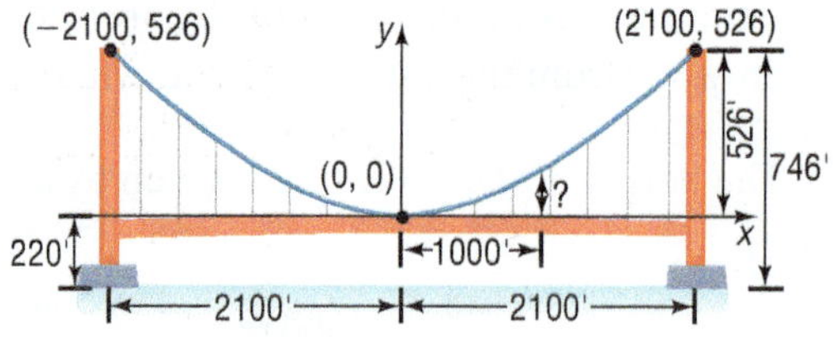

Figure 29

Use these facts to find the value of a in $y = ax^2$.

$$y = ax^2$$

$$526 = a(2100)^2 \qquad x = 2100,\ y = 526$$

$$a = \frac{526}{(2100)^2}$$

The equation of the parabola is

$$y = \frac{526}{(2100)^2}x^2$$

When $x = 1000$, the height of the cable is

$$y = \frac{526}{(2100)^2}(1000)^2 \approx 119.3 \text{ feet}$$

The cable is 119.3 feet above the road at a distance of 1000 feet from the center of the bridge. ■

Now Work PROBLEM 13

2 Build Quadratic Models from Data

In Section 4.2, we found the line of best fit for data that appeared to be linearly related. It was noted that data may also follow a nonlinear relation. Figures 30(a) and (b) show scatter diagrams of data that follow a quadratic relation.

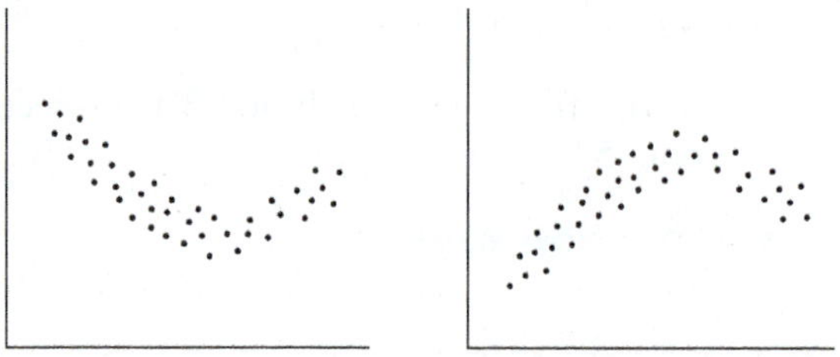

$y = ax^2 + bx + c,\ a > 0$ (a) $y = ax^2 + bx + c,\ a < 0$ (b)

Figure 30

EXAMPLE 5 Fitting a Quadratic Function to Data

The data in Table 8 represent the percentage D of the population that is divorced for various ages x.

(a) Draw a scatter diagram of the data treating age as the independent variable. Comment on the type of relation that may exist between age and percentage of the population divorced.

(b) Use a graphing utility to find the quadratic function of best fit that models the relation between age and percentage of the population divorced.

(c) Use the model found in part (b) to approximate the age at which the percentage of the population divorced is greatest.

(d) Use the model found in part (b) to approximate the highest percentage of the population that is divorced.

(e) Use a graphing utility to draw the quadratic function of best fit on the scatter diagram.

Table 8

Age, x	Percentage Divorced, D
22	0.9
27	3.6
32	7.4
37	10.4
42	12.7
50	15.7
60	16.2
70	13.1
80	6.5

Source: United States Statistical Abstract, 2012

Solution

(a) Figure 31 shows the scatter diagram, from which it appears the data follow a quadratic relation, with $a < 0$.

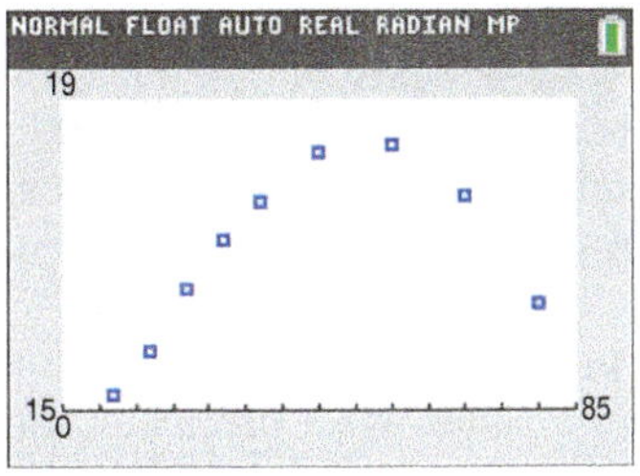

Figure 31

(b) Execute the QUADratic REGression program to obtain the results shown in Figure 32. The output shows the equation $y = ax^2 + bx + c$. The quadratic function of best fit that models the relation between age and percentage divorced is

$$D(x) = -0.0143x^2 + 1.5861x - 28.1886 \quad \text{The model}$$

where x represents age and D represents the percentage divorced.

Figure 32

(c) Based on the quadratic function of best fit, the age with the greatest percentage divorced is

$$-\frac{b}{2a} = -\frac{1.5861}{2(-0.0143)} \approx 55 \text{ years}$$

(d) Evaluate the function $D(x)$ at $x = 55$.

$$D(55) = -0.0143(55)^2 + 1.5861(55) - 28.1886 \approx 15.8 \text{ percent}$$

According to the model, 55-year-olds have the highest percentage divorced at 15.8 percent.

(e) Figure 33 shows the graph of the quadratic function found in part (b) drawn on the scatter diagram. ■

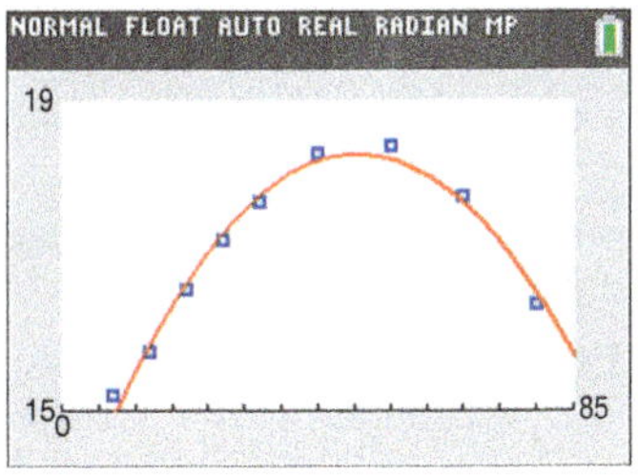

Figure 33

Look again at Figure 32. Notice that the output given by the graphing calculator does not include r, the correlation coefficient. Recall that the correlation coefficient is a measure of the strength of a linear relation that exists between two variables. The graphing calculator does not provide an indication of how well the function fits the data in terms of r, since a quadratic function cannot be expressed as a linear function.

Now Work PROBLEM 25

4.4 Assess Your Understanding

'Are You Prepared?' *Answers are given at the end of these exercises. If you get a wrong answer, read the pages listed in red.*

1. Translate the following sentence into a mathematical equation: The total revenue R from selling x hot dogs is \$3 times the number of hot dogs sold. (p. 138)

2. Use a graphing utility to find the line of best fit for the following data: (pp. 293–294)

x	3	5	5	6	7	8
y	10	13	12	15	16	19

Applications and Extensions

3. **Maximizing Revenue** The price p (in dollars) and the quantity x sold of a certain product obey the demand equation

$$p = -\frac{1}{6}x + 100$$

(a) Find a model that expresses the revenue R as a function of x. (Remember, $R = xp$.)
(b) What is the domain of R?
(c) What is the revenue if 200 units are sold?
(d) What quantity x maximizes revenue? What is the maximum revenue?
(e) What price should the company charge to maximize revenue?

4. **Maximizing Revenue** The price p (in dollars) and the quantity x sold of a certain product obey the demand equation

$$p = -\frac{1}{3}x + 100$$

(a) Find a model that expresses the revenue R as a function of x.
(b) What is the domain of R?
(c) What is the revenue if 100 units are sold?
(d) What quantity x maximizes revenue? What is the maximum revenue?
(e) What price should the company charge to maximize revenue?

5. **Maximizing Revenue** The price p (in dollars) and the quantity x sold of a certain product obey the demand equation

$$x = -5p + 100 \qquad 0 < p \le 20$$

(a) Express the revenue R as a function of x.
(b) What is the revenue if 15 units are sold?
(c) What quantity x maximizes revenue? What is the maximum revenue?
(d) What price should the company charge to maximize revenue?
(e) What price should the company charge to earn at least \$480 in revenue?

6. **Maximizing Revenue** The price p (in dollars) and the quantity x sold of a certain product obey the demand equation

$$x = -20p + 500 \qquad 0 < p \le 25$$

(a) Express the revenue R as a function of x.
(b) What is the revenue if 20 units are sold?
(c) What quantity x maximizes revenue? What is the maximum revenue?
(d) What price should the company charge to maximize revenue?
(e) What price should the company charge to earn at least \$3000 in revenue?

7. **Enclosing a Rectangular Field** David has 400 yards of fencing and wishes to enclose a rectangular area.
(a) Express the area A of the rectangle as a function of the width w of the rectangle.
(b) For what value of w is the area largest?
(c) What is the maximum area?

8. **Enclosing a Rectangular Field** Beth has 3000 feet of fencing available to enclose a rectangular field.
(a) Express the area A of the rectangle as a function of x, where x is the length of the rectangle.
(b) For what value of x is the area largest?
(c) What is the maximum area?

9. **Enclosing the Most Area with a Fence** A farmer with 4000 meters of fencing wants to enclose a rectangular plot that borders on a river. If the farmer does not fence the side along the river, what is the largest area that can be enclosed? (See the figure.)

10. **Enclosing the Most Area with a Fence** A farmer with 2000 meters of fencing wants to enclose a rectangular plot that borders on a straight highway. If the farmer does not fence the side along the highway, what is the largest area that can be enclosed?

11. **Analyzing the Motion of a Projectile** A projectile is fired from a cliff 200 feet above the water at an inclination of 45° to the horizontal, with a muzzle velocity of 50 feet per second. The height h of the projectile above the water is modeled by

$$h(x) = \frac{-32x^2}{(50)^2} + x + 200$$

where x is the horizontal distance of the projectile from the face of the cliff.
(a) At what horizontal distance from the face of the cliff is the height of the projectile a maximum?
(b) Find the maximum height of the projectile.
(c) At what horizontal distance from the face of the cliff will the projectile strike the water?
(d) Using a graphing utility, graph the function h, $0 \le x \le 200$.
(e) Use a graphing utility to verify the solutions found in parts (b) and (c).
(f) When the height of the projectile is 100 feet above the water, how far is it from the cliff?

12. **Analyzing the Motion of a Projectile** A projectile is fired at an inclination of 45° to the horizontal, with a muzzle velocity of 100 feet per second. The height h of the projectile is modeled by

$$h(x) = \frac{-32x^2}{(100)^2} + x$$

where x is the horizontal distance of the projectile from the firing point.
(a) At what horizontal distance from the firing point is the height of the projectile a maximum?
(b) Find the maximum height of the projectile.
(c) At what horizontal distance from the firing point will the projectile strike the ground?
(d) Using a graphing utility, graph the function h, $0 \le x \le 350$.

(e) Use a graphing utility to verify the results obtained in parts (b) and (c).

(f) When the height of the projectile is 50 feet above the ground, how far has it traveled horizontally?

13. Suspension Bridge A suspension bridge with weight uniformly distributed along its length has twin towers that extend 75 meters above the road surface and are 400 meters apart. The cables are parabolic in shape and are suspended from the tops of the towers. The cables touch the road surface at the center of the bridge. Find the height of the cables at a point 100 meters from the center. (Assume that the road is level.)

14. Architecture A parabolic arch has a span of 120 feet and a maximum height of 25 feet. Choose suitable rectangular coordinate axes and find the equation of the parabola. Then calculate the height of the arch at points 10 feet, 20 feet, and 40 feet from the center.

15. Constructing Rain Gutters A rain gutter is to be made of aluminum sheets that are 12 inches wide by turning up the edges 90°. See the illustration.

(a) What depth will provide maximum cross-sectional area and hence allow the most water to flow?

(b) What depths will allow at least 16 square inches of water to flow?

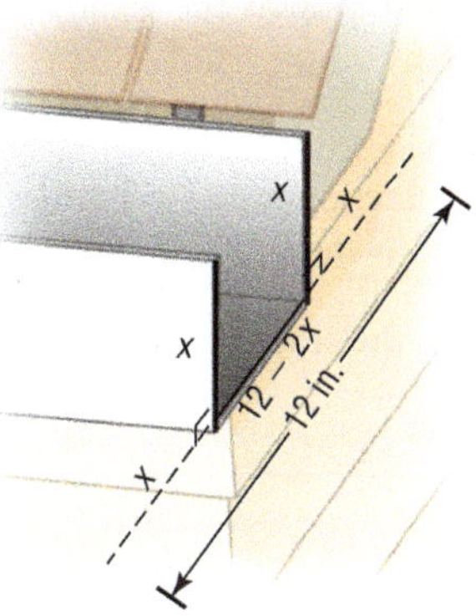

16. Norman Windows A **Norman window** has the shape of a rectangle surmounted by a semicircle of diameter equal to the width of the rectangle. See the picture. If the perimeter of the window is 20 feet, what dimensions will admit the most light (maximize the area)?

[**Hint:** Circumference of a circle $= 2\pi r$; area of a circle $= \pi r^2$, where r is the radius of the circle.]

17. Constructing a Stadium A track-and-field playing area is in the shape of a rectangle with semicircles at each end. See the figure (top, right). The inside perimeter of the track is to be 1500 meters. What should the dimensions of the rectangle be so that the area of the rectangle is a maximum?

18. Architecture A special window has the shape of a rectangle surmounted by an equilateral triangle. See the figure. If the perimeter of the window is 16 feet, what dimensions will admit the most light?

[**Hint:** Area of an equilateral triangle $= \left(\frac{\sqrt{3}}{4}\right)x^2$, where x is the length of a side of the triangle.]

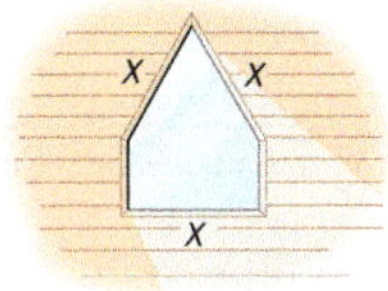

19. Chemical Reactions A self-catalytic chemical reaction results in the formation of a compound that causes the formation ratio to increase. If the reaction rate V is modeled by

$$V(x) = kx(a - x), \qquad 0 \le x \le a$$

where k is a positive constant, a is the initial amount of the compound, and x is the variable amount of the compound, for what value of x is the reaction rate a maximum?

20. Calculus: Simpson's Rule The figure shows the graph of $y = ax^2 + bx + c$. Suppose that the points $(-h, y_0)$, $(0, y_1)$, and (h, y_2) are on the graph. It can be shown that the area enclosed by the parabola, the x-axis, and the lines $x = -h$ and $x = h$ is

$$\text{Area} = \frac{h}{3}(2ah^2 + 6c)$$

Show that this area may also be given by

$$\text{Area} = \frac{h}{3}(y_0 + 4y_1 + y_2)$$

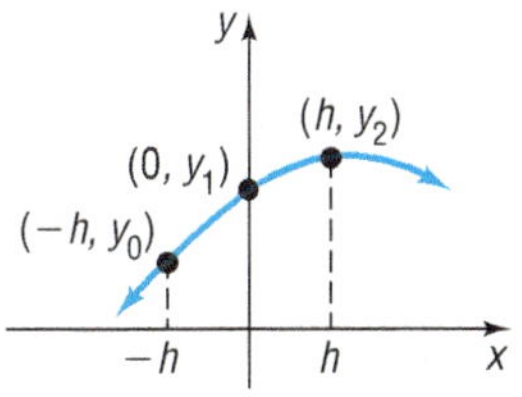

21. Use the result obtained in Problem 20 to find the area enclosed by $f(x) = -5x^2 + 8$, the x-axis, and the lines $x = -1$ and $x = 1$.

22. Use the result obtained in Problem 20 to find the area enclosed by $f(x) = 2x^2 + 8$, the x-axis, and the lines $x = -2$ and $x = 2$.

23. Use the result obtained in Problem 20 to find the area enclosed by $f(x) = x^2 + 3x + 5$, the x-axis, and the lines $x = -4$ and $x = 4$.

24. Use the result obtained in Problem 20 to find the area enclosed by $f(x) = -x^2 + x + 4$, the x-axis, and the lines $x = -1$ and $x = 1$.

25. **Life Cycle Hypothesis** An individual's income varies with his or her age. The following table shows the median income I of males of different age groups within the United States for 2012. For each age group, let the class midpoint represent the independent variable, x. For the class "65 years and older," we will assume that the class midpoint is 69.5.

Age	Class Midpoint, x	Median Income, I
15–24 years	19.5	\$10,869
25–34 years	29.5	\$34,113
35–44 years	39.5	\$45,225
45–54 years	49.5	\$46,466
55–64 years	59.5	\$42,176
65 years and older	69.5	\$27,612

Source: U.S. Census Bureau

(a) Use a graphing utility to draw a scatter diagram of the data. Comment on the type of relation that may exist between the two variables.
(b) Use a graphing utility to find the quadratic function of best fit that models the relation between age and median income.
(c) Use the function found in part (b) to determine the age at which an individual can expect to earn the most income.
(d) Use the function found in part (b) to predict the peak income earned.
(e) With a graphing utility, graph the quadratic function of best fit on the scatter diagram.

26. **Height of a Ball** A shot-putter throws a ball at an inclination of 45° to the horizontal. The following data represent the height of the ball h, in feet, at the instant that it has traveled x feet horizontally.

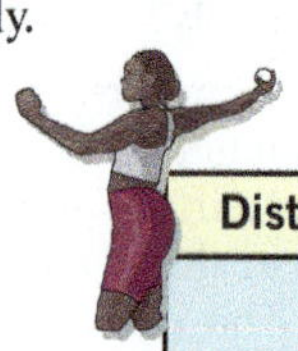

Distance, x	Height, h
20	25
40	40
60	55
80	65
100	71
120	77
140	77
160	75
180	71
200	64

(a) Use a graphing utility to draw a scatter diagram of the data. Comment on the type of relation that may exist between the two variables.
(b) Use a graphing utility to find the quadratic function of best fit that models the relation between distance and height.
(c) Use the function found in part (b) to determine how far the ball will travel before it reaches its maximum height.
(d) Use the function found in part (b) to find the maximum height of the ball.
(e) With a graphing utility, graph the quadratic function of best fit on the scatter diagram.

Mixed Practice

27. **Which Model?** The following data represent the square footage and rents (dollars per month) for apartments in the La Jolla area of San Diego, California.

Square Footage, x	Rent per Month, R
520	\$1525
621	\$1750
718	\$1785
753	\$1850
850	\$1900
968	\$2130
1020	\$2180

Source: apartments.com, 2014

(a) Using a graphing utility, draw a scatter diagram of the data treating square footage as the independent variable. What type of relation appears to exist between square footage and rent?
(b) Based on your response to part (a), find either a linear or a quadratic model that describes the relation between square footage and rent.
(c) Use your model to predict the rent for an apartment in San Diego that is 875 square feet.

28. **Which Model?** An engineer collects the following data showing the speed s of a Toyota Camry and its average miles per gallon, M.

Speed, s	Miles per Gallon, M
30	18
35	20
40	23
40	25
45	25
50	28
55	30
60	29
65	26
65	25
70	25

(a) Using a graphing utility, draw a scatter diagram of the data, treating speed as the independent variable. What type of relation appears to exist between speed and miles per gallon?

(b) Based on your response to part (a), find either a linear model or a quadratic model that describes the relation between speed and miles per gallon.

(c) Use your model to predict the miles per gallon for a Camry that is traveling 63 miles per hour.

29. **Which Model?** The following data represent the birth rate (births per 1000 population) for women whose age is a, in 2012.

Age, a	Birth Rate, B
16	14.1
19	51.4
22	83.1
27	106.5
32	97.3
37	48.3
42	10.4

Source: National Vital Statistics System, 2013

(a) Using a graphing utility, draw a scatter diagram of the data, treating age as the independent variable. What type of relation appears to exist between age and birth rate?

(b) Based on your response to part (a), find either a linear or a quadratic model that describes the relation between age and birth rate.

(c) Use your model to predict the birth rate for 35-year-old women.

30. **Which Model?** A cricket makes a chirping noise by sliding its wings together rapidly. Perhaps you have noticed that the rapidity of chirps seems to increase with the temperature. The following data list the temperature (in degrees Fahrenheit) and the number of chirps per second for the striped ground cricket.

Temperature (°F), x	Chirps per Second, C
88.6	20.0
93.3	19.8
80.6	17.1
69.7	14.7
69.4	15.4
79.6	15.0
80.6	16.0
76.3	14.4
75.2	15.5

Source: Pierce, George W. *The Songs of Insects*. Cambridge, MA Harvard University Press, 1949, pp. 12 – 21

(a) Using a graphing utility, draw a scatter diagram of the data, treating temperature as the independent variable. What type of relation appears to exist between temperature and chirps per second?

(b) Based on your response to part (a), find either a linear or a quadratic model that best describes the relation between temperature and chirps per second.

(c) Use your model to predict the chirps per second if the temperature is 80°F.

Explaining Concepts: Discussion and Writing

31. Refer to Example 1 in this section. Notice that if the price charged for the calculators is \$0 or \$140, then the revenue is \$0. It is easy to explain why revenue would be \$0 if the price charged were \$0, but how can revenue be \$0 if the price charged is \$140?

Retain Your Knowledge

Problems 32–35 are based on material learned earlier in the course. The purpose of these problems is to keep the material fresh in your mind so that you are better prepared for the final exam.

32. Express as a complex number: $\sqrt{-225}$

33. Find the distance between the points $P_1 = (4, -7)$ and $P_2 = (-1, 5)$.

34. Find the equation of the circle with center $(-6, 0)$ and radius $r = \sqrt{7}$.

35. Solve: $5x^2 + 8x - 3 = 0$

'Are You Prepared?' Answers

1. $R = 3x$ **2.** $y = 1.7826x + 4.0652$

4.5 Inequalities Involving Quadratic Functions

PREPARING FOR THIS SECTION *Before getting started, review the following:*

- Solve Inequalities (Section 1.7, pp. 150–154)
- Use Interval Notation (Section 1.7, pp. 147–148)

Now Work the 'Are You Prepared?' problems on page 322.

OBJECTIVE 1 Solve Inequalities Involving a Quadratic Function (p. 320)

1 Solve Inequalities Involving a Quadratic Function

In this section we solve inequalities that involve quadratic functions. We will accomplish this by using their graphs. For example, to solve the inequality

$$ax^2 + bx + c > 0 \qquad a \neq 0$$

graph the function $f(x) = ax^2 + bx + c$ and, from the graph, determine where it is above the x-axis—that is, where $f(x) > 0$. To solve the inequality $ax^2 + bx + c < 0, a \neq 0$, graph the function $f(x) = ax^2 + bx + c$ and determine where the graph is below the x-axis. If the inequality is not strict, include the x-intercepts in the solution.

EXAMPLE 1 **Solving an Inequality**

Solve the inequality $x^2 - 4x - 12 \leq 0$ and graph the solution set.

By Hand Solution

Graph the function $f(x) = x^2 - 4x - 12$.

y-intercept: $f(0) = -12$ **Evaluate f at 0.**

x-intercepts (if any): $x^2 - 4x - 12 = 0$ **Solve $f(x) = 0$.**

$(x - 6)(x + 2) = 0$ **Factor.**

$x - 6 = 0$ or $x + 2 = 0$ **Apply the Zero Product Property.**

$x = 6$ or $x = -2$

The y-intercept is -12; the x-intercepts are -2 and 6.

The vertex is at $x = -\dfrac{b}{2a} = -\dfrac{-4}{2(1)} = 2$. Since $f(2) = -16$, the vertex is $(2, -16)$. See Figure 34 for the graph.

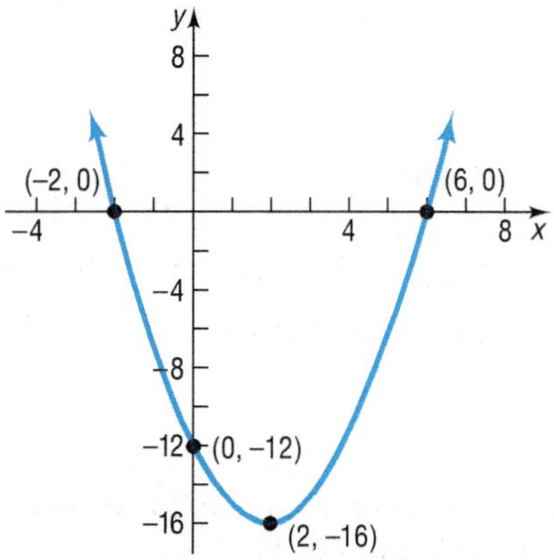

Figure 34 $f(x) = x^2 - 4x - 12$

Because we are solving $f(x) \leq 0$, we must find the x-values for which the graph is below the x-axis, which is between -2 and 6. Since the original inequality is not strict, include the x-intercepts. The solution set is $\{x | -2 \leq x \leq 6\}$ or, using interval notation, $[-2, 6]$.

Graphing Utility Solution

Graph $Y_1 = x^2 - 4x - 12$. See Figure 35. Use the ZERO command to find that the x-intercepts of Y_1 are -2 and 6. Because we are solving $f(x) \leq 0$, we must find the x-values for which the graph is below the x-axis, which is between -2 and 6. Since the inequality is not strict, the solution set is $\{x | -2 \leq x \leq 6\}$ or, using interval notation, $[-2, 6]$.

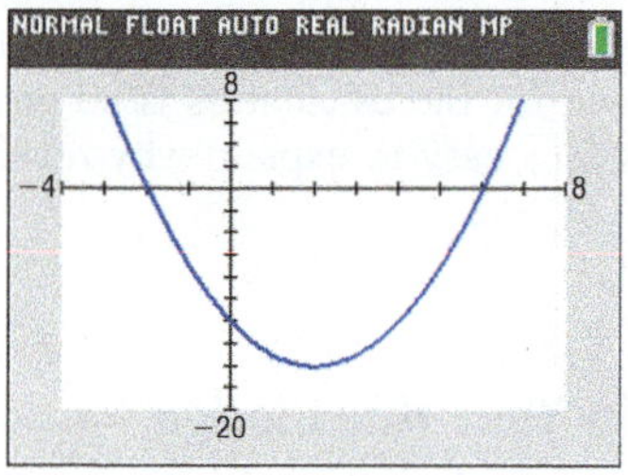

Figure 35 $Y_1 = x^2 - 4x - 12$

See Figure 36 for the graph of the solution set.

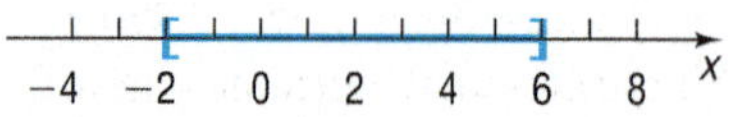

Figure 36

Now Work PROBLEM 9

EXAMPLE 2 Solving an Inequality

Solve the inequality $2x^2 < x + 10$ and graph the solution set.

Solution

Option 1 Rearrange the inequality so that 0 is on the right side.

$$2x^2 < x + 10$$

$$2x^2 - x - 10 < 0 \qquad \text{Subtract } x + 10 \text{ from both sides.}$$

This inequality is equivalent to the original inequality.

Next graph the function $f(x) = 2x^2 - x - 10$ to find where $f(x) < 0$.

y-intercept: $\quad f(0) = -10 \qquad$ **Evaluate f at 0.**

x-intercepts (if any): $\quad 2x^2 - x - 10 = 0 \qquad$ **Solve $f(x) = 0$.**

$$(2x - 5)(x + 2) = 0 \qquad \text{Factor.}$$

$$2x - 5 = 0 \quad \text{or} \quad x + 2 = 0 \qquad \text{Apply the Zero-Product Property.}$$

$$x = \frac{5}{2} \quad \text{or} \quad x = -2$$

The y-intercept is -10; the x-intercepts are -2 and $\frac{5}{2}$.

The vertex is at $x = -\frac{b}{2a} = -\frac{-1}{2(2)} = \frac{1}{4}$. Since $f\left(\frac{1}{4}\right) = -10.125$, the vertex is $\left(\frac{1}{4}, -10.125\right)$. See Figure 37 for the graph.

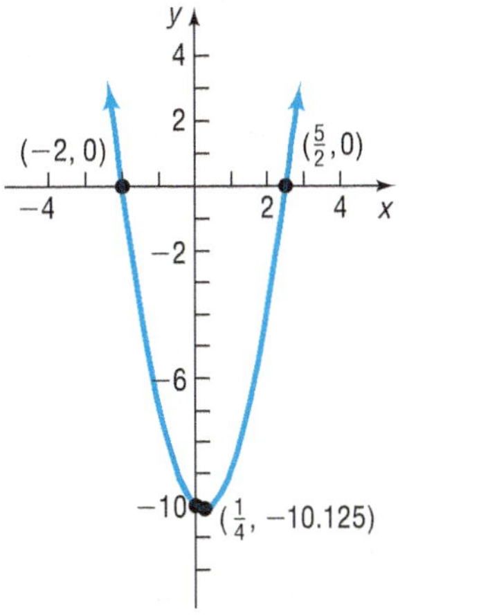

Figure 37 $f(x) = 2x^2 - x - 10$

Because we are solving $f(x) < 0$, we must find where the graph is below the x-axis. This occurs between $x = -2$ and $x = \frac{5}{2}$. Since the inequality is strict, the solution set is $\left\{x \middle| -2 < x < \frac{5}{2}\right\}$ or, using interval notation, $\left(-2, \frac{5}{2}\right)$.

✓**Check:** Verify the result of Option 1 by graphing $Y_1 = 2x^2 - x - 10$ and using the ZERO feature.

Option 2 If $f(x) = 2x^2$ and $g(x) = x + 10$, then the inequality to be solved is $f(x) < g(x)$. Graph the functions $f(x) = 2x^2$ and $g(x) = x + 10$. See Figure 38. The graphs intersect where $f(x) = g(x)$. Then

$$2x^2 = x + 10 \qquad f(x) = g(x)$$

$$2x^2 - x - 10 = 0$$

$$(2x - 5)(x + 2) = 0 \qquad \text{Factor.}$$

$$2x - 5 = 0 \quad \text{or} \quad x + 2 = 0 \qquad \text{Apply the Zero-Product Property}$$

$$x = \frac{5}{2} \quad \text{or} \quad x = -2$$

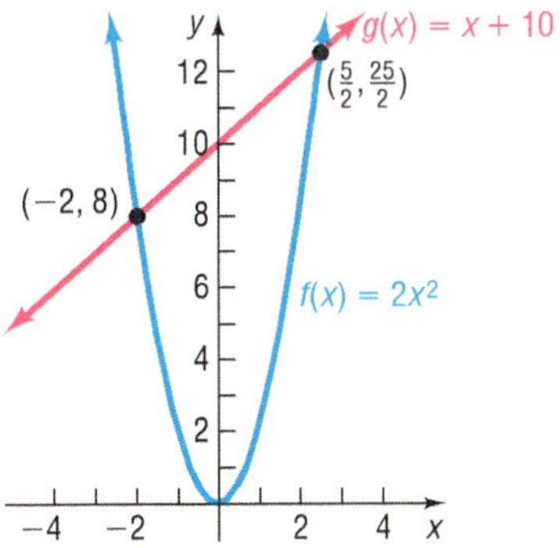

Figure 38

The graphs intersect at the points $(-2, 8)$ and $\left(\frac{5}{2}, \frac{25}{2}\right)$. To solve $f(x) < g(x)$, find where the graph of f is below that of g. This happens between the points of intersection. Since the inequality is strict, the solution set is $\left\{x \middle| -2 < x < \frac{5}{2}\right\}$ or, using interval notation, $\left(-2, \frac{5}{2}\right)$.

✓**Check:** Verify the result of Option 2 by graphing $Y_1 = 2x^2$ and $Y_2 = x + 10$ on the same screen. Use INTERSECT to find the x-coordinates of the points of intersection. Determine where Y_1 is below Y_2 to solve $Y_1 < Y_2$. ■

See Figure 39 for the graph of the solution set.

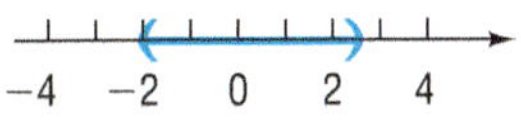

Figure 39

Now Work PROBLEMS 5 AND 13

EXAMPLE 3 **Solving an Inequality**

Solve the inequality $x^2 + x + 1 > 0$ and graph the solution set.

Solution Graph the function $f(x) = x^2 + x + 1$. The y-intercept is 1; there are no x-intercepts (Do you see why? Check the discriminant). The vertex is at $x = -\frac{b}{2a} = -\frac{1}{2}$. Since $f\left(-\frac{1}{2}\right) = \frac{3}{4}$, the vertex is at $\left(-\frac{1}{2}, \frac{3}{4}\right)$. The points $(1, 3)$ and $(-1, 1)$ are also on the graph. See Figure 40.

Because we are solving $f(x) > 0$, we must find where the graph is above the x-axis. The graph of f lies above the x-axis for all x. The solution set is the set of all real numbers, or $(-\infty, \infty)$. See Figure 41.

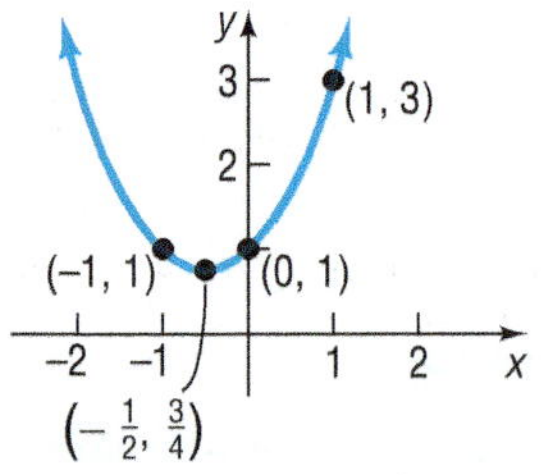

Figure 40 $f(x) = x^2 + x + 1$

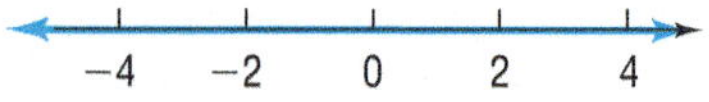

Figure 41

Now Work PROBLEM 17

4.5 Assess Your Understanding

'Are You Prepared?' *Answers are given at the end of these exercises. If you get a wrong answer, read the pages listed in red.*

1. Solve the inequality $-3x - 2 < 7$. (pp. 150–151)

2. Write $(-2, 7]$ using inequality notation. (pp. 147–148)

Skill Building

In Problems 3–6, use the figure to solve each inequality.

3.

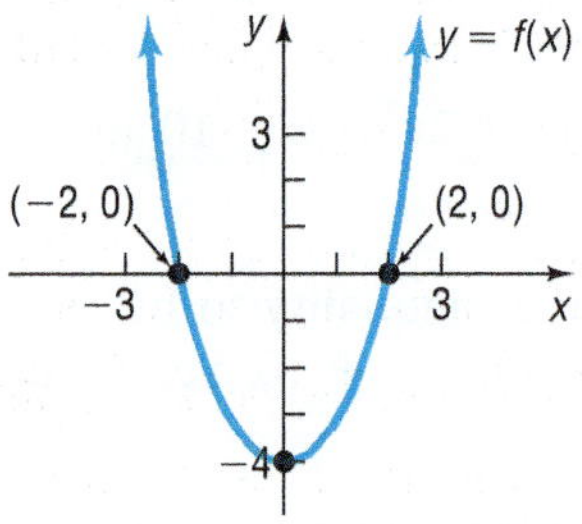

(a) $f(x) > 0$
(b) $f(x) \le 0$

4.

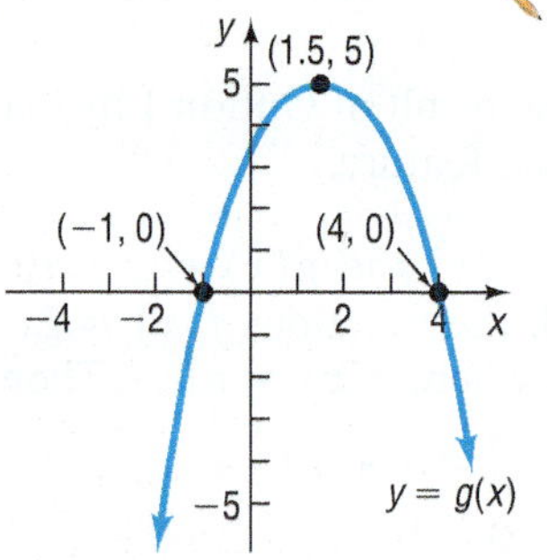

(a) $g(x) < 0$
(b) $g(x) \ge 0$

5.

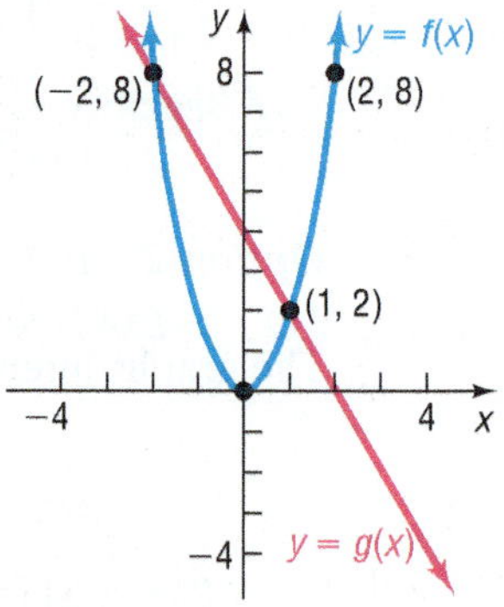

(a) $g(x) \ge f(x)$
(b) $f(x) > g(x)$

6.

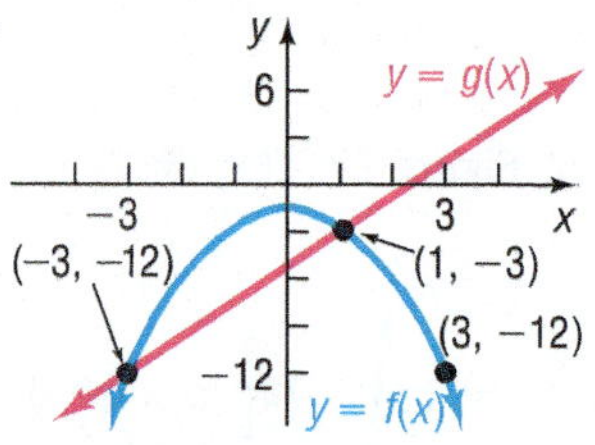

(a) $f(x) < g(x)$
(b) $f(x) \ge g(x)$

In Problems 7–22, solve each inequality.

7. $x^2 - 3x - 10 < 0$

8. $x^2 + 3x - 10 > 0$

9. $x^2 - 4x > 0$

10. $x^2 + 8x > 0$

11. $x^2 - 9 < 0$

12. $x^2 - 1 < 0$

13. $x^2 + x > 12$

14. $x^2 + 7x < -12$

15. $2x^2 < 5x + 3$

16. $6x^2 < 6 + 5x$

17. $x^2 - x + 1 \le 0$

18. $x^2 + 2x + 4 > 0$

19. $4x^2 + 9 < 6x$

20. $25x^2 + 16 < 40x$

21. $6(x^2 - 1) > 5x$

22. $2(2x^2 - 3x) > -9$

Mixed Practice

23. What is the domain of the function $f(x) = \sqrt{x^2 - 16}$?

24. What is the domain of the function $f(x) = \sqrt{x - 3x^2}$?

In Problems 25–32, use the given functions f and g.

(a) Solve $f(x) = 0$. *(b) Solve $g(x) = 0$.* *(c) Solve $f(x) = g(x)$.* *(d) Solve $f(x) > 0$.*
(e) Solve $g(x) \le 0$. *(f) Solve $f(x) > g(x)$.* *(g) Solve $f(x) \ge 1$.*

25. $f(x) = x^2 - 1$
$g(x) = 3x + 3$

26. $f(x) = -x^2 + 3$
$g(x) = -3x + 3$

27. $f(x) = -x^2 + 1$
$g(x) = 4x + 1$

28. $f(x) = -x^2 + 4$
$g(x) = -x - 2$

29. $f(x) = x^2 - 4$
$g(x) = -x^2 + 4$

30. $f(x) = x^2 - 2x + 1$
$g(x) = -x^2 + 1$

31. $f(x) = x^2 - x - 2$
$g(x) = x^2 + x - 2$

32. $f(x) = -x^2 - x + 1$
$g(x) = -x^2 + x + 6$

Applications and Extensions

33. **Physics** A ball is thrown vertically upward with an initial velocity of 80 feet per second. The distance s (in feet) of the ball from the ground after t seconds is $s(t) = 80t - 16t^2$.

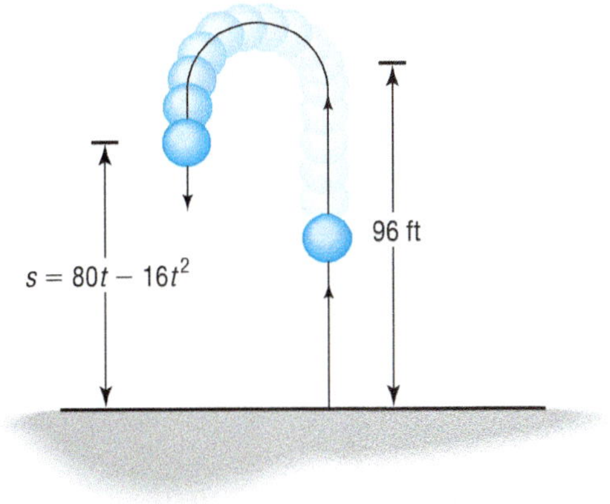

(a) At what time t will the ball strike the ground?
(b) For what time t is the ball more than 96 feet above the ground?

34. **Physics** A ball is thrown vertically upward with an initial velocity of 96 feet per second. The distance s (in feet) of the ball from the ground after t seconds is $s(t) = 96t - 16t^2$.
(a) At what time t will the ball strike the ground?
(b) For what time t is the ball more than 128 feet above the ground?

35. **Revenue** Suppose that the manufacturer of a gas clothes dryer has found that when the unit price is p dollars, the revenue R (in dollars) is

$$R(p) = -4p^2 + 4000p$$

(a) At what prices p is revenue zero?
(b) For what range of prices will revenue exceed \$800,000?

36. **Revenue** The John Deere company has found that the revenue from sales of heavy-duty tractors is a function of the unit price p, in dollars, that it charges. If the revenue R, in dollars, is

$$R(p) = -\frac{1}{2}p^2 + 1900p$$

(a) At what prices p is revenue zero?
(b) For what range of prices will revenue exceed \$1,200,000?

37. **Artillery** A projectile fired from the point $(0, 0)$ at an angle to the positive x-axis has a trajectory given by

$$y = cx - (1 + c^2)\left(\frac{g}{2}\right)\left(\frac{x}{v}\right)^2$$

where
x = horizontal distance in meters
y = height in meters
v = initial muzzle velocity in meters per second (m/sec)
g = acceleration due to gravity = 9.81 meters per second squared (m/sec^2)
$c > 0$ is a constant determined by the angle of elevation.

A howitzer fires an artillery round with a muzzle velocity of 897 m/sec.
(a) If the round must clear a hill 200 meters high at a distance of 2000 meters in front of the howitzer, what c values are permitted in the trajectory equation?
(b) If the goal in part (a) is to hit a target on the ground 75 kilometers away, is it possible to do so? If so, for what values of c? If not, what is the maximum distance the round will travel?

Source: *www.answers.com*

38. **Runaway Car** Using Hooke's Law, we can show that the *work* done in compressing a spring a distance of x feet from its at-rest position is $W = \frac{1}{2}kx^2$, where k is a stiffness constant depending on the spring. It can also be shown that the work done by a body in motion before it comes to rest is given by $\widetilde{W} = \frac{w}{2g}v^2$, where w = weight of the object (lb), g = acceleration due to gravity (32.2 ft/sec^2), and v = object's velocity (in ft/sec). A parking garage has a spring shock absorber at the end of a ramp to stop runaway cars. The spring has a stiffness constant $k = 9450$ lb/ft and must be able to stop a 4000-lb car traveling at 25 mph. What is the least compression required of the spring? Express your answer using feet to the nearest tenth.
[**Hint:** Solve $W > \widetilde{W}, x \geq 0$].

Source: *www.sciforums.com*

Explaining Concepts: Discussion and Writing

39. Show that the inequality $(x - 4)^2 \leq 0$ has exactly one solution.

40. Show that the inequality $(x - 2)^2 > 0$ has one real number that is not a solution.

41. Explain why the inequality $x^2 + x + 1 > 0$ has all real numbers as the solution set.

42. Explain why the inequality $x^2 - x + 1 < 0$ has the empty set as the solution set.

43. Explain the circumstances under which the x-intercepts of the graph of a quadratic function are included in the solution set of a quadratic inequality.

Retain Your Knowledge

Problems 44–47 are based on material learned earlier in the course. The purpose of these problems is to keep the material fresh in your mind so that you are better prepared for the final exam.

44. Determine the domain of $f(x) = \sqrt{10 - 2x}$.

45. Consider the linear function $f(x) = \frac{2}{3}x - 6$.
(a) Find the intercepts of the graph of f.
(b) Graph f.

46. Determine algebraically whether $f(x) = \frac{-x}{x^2 + 9}$ is even, odd, or neither.

47. Multiply $(6 - 5i)(4 - i)$. Write the answer in the form $a + bi$.

'Are You Prepared?' Answers

1. $\{x | x > -3\}$ or $(-3, \infty)$ **2.** $-2 < x \le 7$

Chapter Review

Things to Know

Linear function (p. 281)

$f(x) = mx + b$ Average rate of change $= m$

The graph is a line with slope m and y-intercept b.

Quadratic function (pp. 298–302)

$f(x) = ax^2 + bx + c, a \neq 0$ The graph is a parabola that opens up if $a > 0$ and opens down if $a < 0$.

Vertex: $\left(-\frac{b}{2a}, f\left(-\frac{b}{2a}\right)\right)$

Axis of symmetry: $x = -\frac{b}{2a}$

y-intercept: $f(0) = c$

x-intercept(s): If any, found by finding the real solutions of the equation $ax^2 + bx + c = 0$

Objectives

Section		You should be able to ...	Examples	Review Exercises
4.1	1	Graph linear functions (p. 281)	1	1(a)–3(a), 1(c)–3(c)
	2	Use average rate of change to identify linear functions (p. 281)	2	1(b)–3(b), 4, 5
	3	Determine whether a linear function is increasing, decreasing, or constant (p. 284)	3	1(d)–3(d)
	4	Build linear models from verbal descriptions (p. 285)	4, 5	22
4.2	1	Draw and interpret scatter diagrams (p. 291)	1	28(a), 29(a)
	2	Distinguish between linear and nonlinear relations (p. 292)	2	28(b), 29(a)
	3	Use a graphing utility to find the line of best fit (p. 293)	4	28(c)
4.3	1	Graph a quadratic function using transformations (p. 299)	1	6–8
	2	Identify the vertex and axis of symmetry of a quadratic function (p. 301)	2	9–14
	3	Graph a quadratic function using its vertex, axis, and intercepts (p. 301)	3–5	9–14
	4	Find a quadratic function given its vertex and one other point (p. 304)	6	20, 21
	5	Find the maximum or minimum value of a quadratic function (p. 305)	7	15–17, 23–26
4.4	1	Build quadratic models from verbal descriptions (p. 310)	1–4	23–27
	2	Build quadratic models from data (p. 314)	5	29
4.5	1	Solve inequalities involving a quadratic function (p. 320)	1–3	18, 19

Review Exercises

In Problems 1–3:

(a) Determine the slope and y-intercept of each linear function.
(b) Find the average rate of change of each function.
(c) Graph each function. Label the intercepts.
(d) Determine whether the function is increasing, decreasing, or constant.

1. $f(x) = 2x - 5$ **2.** $F(x) = -\frac{1}{3}x + 1$ **3.** $G(x) = 4$

In Problems 4 and 5, determine whether the function is linear or nonlinear. If the function is linear, find the equation of the line.

4.

x	$y = f(x)$
−1	−2
0	3
1	8
2	13
3	18

5.

x	$y = g(x)$
−1	−3
0	4
1	7
2	6
3	1

In Problems 6–8, graph each quadratic function using transformations (shifting, compressing, stretching, and/or reflecting).

6. $f(x) = (x - 2)^2 + 2$ **7.** $f(x) = -(x - 4)^2$ **8.** $f(x) = 2(x + 1)^2 + 4$

In Problems 9–14, (a) graph each quadratic function by determining whether its graph opens up or down and by finding its vertex, axis of symmetry, y-intercept, and x-intercepts, if any. (b) Determine the domain and the range of the function. (c) Determine where the function is increasing and where it is decreasing.

9. $f(x) = x^2 - 4x + 6$ **10.** $f(x) = -\frac{1}{2}x^2 + 2$ **11.** $f(x) = -4x^2 + 4x$

12. $f(x) = 9x^2 + 6x + 1$ **13.** $f(x) = -x^2 + x + \frac{1}{2}$ **14.** $f(x) = 3x^2 + 4x - 1$

In Problems 15–17, determine whether the given quadratic function has a maximum value or a minimum value, and then find the value.

15. $f(x) = 3x^2 - 6x + 4$ **16.** $f(x) = -x^2 + 8x - 4$ **17.** $f(x) = -2x^2 + 4$

In Problems 18–19, solve each quadratic inequality.

18. $x^2 + 6x - 16 < 0$ **19.** $3x^2 \geq 14x + 5$

In Problems 20 and 21, find the quadratic function for which:

20. Vertex is $(-1, 2)$; contains the point $(1, 6)$ **21.** Contains the points $(0, 5)$, $(1, 2)$, and $(3, 2)$

---

22. Sales Commissions Bill has just been offered a sales position for a computer company. His salary would be \$25,000 per year plus 1% of his total annual sales.
(a) Find a linear function that relates Bill's annual salary, S, to his total annual sales, x.
(b) If Bill's total annual sales were \$1,000,000, what would be Bill's salary?
(c) What would Bill have to sell to earn \$100,000?
(d) Determine the sales required of Bill for his salary to exceed \$150,000.

23. Demand Equation The price p (in dollars) and the quantity x sold of a certain product obey the demand equation

$$p = -\frac{1}{10}x + 150 \quad 0 \leq x \leq 1500$$

(a) Express the revenue R as a function of x.
(b) What is the revenue if 100 units are sold?
(c) What quantity x maximizes revenue? What is the maximum revenue?
(d) What price should the company charge to maximize revenue?

24. Enclosing the Most Area with a Fence A farmer with 10,000 meters of fencing wants to enclose a rectangular field and then divide it into two plots with a fence parallel to one of the sides. See the figure. What is the largest area that can be enclosed?

25. Minimizing Marginal Cost Callaway Golf Company has determined that the marginal cost C of manufacturing x Big Bertha golf clubs may be expressed by the quadratic function

$$C(x) = 4.9x^2 - 617.4x + 19{,}600$$

(a) How many clubs should be manufactured to minimize the marginal cost?
(b) At this level of production, what is the marginal cost?

26. Maximizing Area A rectangle has one vertex on the line $y = 10 - x, x > 0$, another at the origin, one on the positive x-axis, and one on the positive y-axis. Express the area A of the rectangle as a function of x. Find the largest area A that can be enclosed by the rectangle.

27. Parabolic Arch Bridge A horizontal bridge is in the shape of a parabolic arch. Given the information shown in the figure, what is the height h of the arch 2 feet from shore?

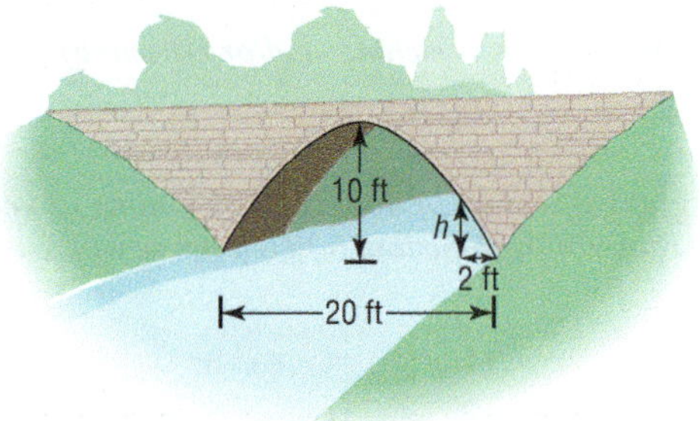

28. Bone Length Research performed at NASA, led by Dr. Emily R. Morey-Holton, measured the lengths of the right humerus and right tibia in 11 rats that were sent to space on Spacelab Life Sciences 2. The following data were collected.

Right Humerus (mm), x	Right Tibia (mm), y
24.80	36.05
24.59	35.57
24.59	35.57
24.29	34.58
23.81	34.20
24.87	34.73
25.90	37.38
26.11	37.96
26.63	37.46
26.31	37.75
26.84	38.50

Source: NASA Life Sciences Data Archive

(a) Draw a scatter diagram of the data treating length of the right humerus as the independent variable.
(b) Based on the scatter diagram, do you think that there is a linear relation between the length of the right humerus and the length of the right tibia?
(c) Use a graphing utility to find the line of best fit relating length of the right humerus and length of the right tibia.
(d) Predict the length of the right tibia on a rat whose right humerus is 26.5 millimeters (mm).

29. Advertising A small manufacturing firm collected the following data on advertising expenditures A (in thousands of dollars) and total revenue R (in thousands of dollars).

Advertising	Total Revenue
20	$6101
22	$6222
25	$6350
25	$6378
27	$6453
28	$6423
29	$6360
31	$6231

(a) Draw a scatter diagram of the data. Comment on the type of relation that may exist between the two variables.
(b) The quadratic function of best fit to these data is

$$R(A) = -7.76A^2 + 411.88A + 942.72$$

Use this function to determine the optimal level of advertising.
(c) Use the function to predict the total revenue when the optimal level of advertising is spent.
(d) Use a graphing utility to verify that the function given in part (b) is the quadratic function of best fit.
(e) Use a graphing utility to draw a scatter diagram of the data and then graph the quadratic function of best fit on the scatter diagram.

Chapter Test

CHAPTER **Test Prep** VIDEOS — The Chapter Test Prep Videos are step-by-step solutions available in MyMathLab®, or on this text's YouTube™ Channel. Flip back to the Resources for Success page for a link to this text's YouTube channel.

1. For the linear function $f(x) = -4x + 3$,
   (a) Find the slope and y-intercept.
   (b) Determine whether f is increasing, decreasing, or constant.
   (c) Graph f.

2. Determine whether the given function is linear or nonlinear. If it is linear, determine the equation of the line.

x	y
−2	12
−1	7
0	2
1	−3
2	−8

3. Graph $f(x) = (x - 3)^2 - 2$ using transformations.

In Problems 4 and 5,
(a) Determine whether the graph opens up or down.
(b) Determine the vertex of the graph of the quadratic function.
(c) Determine the axis of symmetry of the graph of the quadratic function.
(d) Determine the intercepts of the graph of the quadratic function.
(e) Use the information in parts (a)–(d) to graph the quadratic function.
(f) Based on the graph, determine the domain and the range of the quadratic function.
(g) Based on the graph, determine where the function is increasing and where it is decreasing.

4. $f(x) = 3x^2 - 12x + 4$
5. $g(x) = -2x^2 + 4x - 5$
6. Determine the quadratic function for the given graph.

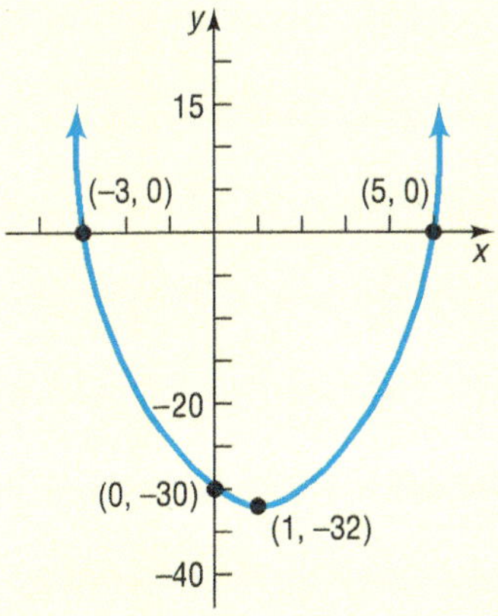

7. Determine whether $f(x) = -2x^2 + 12x + 3$ has a maximum or minimum value. Then find the maximum or minimum value.
8. Solve $x^2 - 10x + 24 \geq 0$.
9. The weekly rental cost of a 20-foot recreational vehicle is \$129.50 plus \$0.15 per mile.
   (a) Find a linear function that expresses the cost C as a function of miles driven m.
   (b) What is the rental cost if 860 miles are driven?
   (c) How many miles were driven if the rental cost is \$213.80?
10. The price p (in dollars) and the quantity x sold of a certain product obey the demand equation $p = -\frac{1}{10}x + 1000$.
   (a) Find a model that expresses the revenue R as a function of x.
   (b) What is the revenue if 400 units are sold?
   (c) What quantity x maximizes revenue? What is the maximum revenue?
   (d) What price should the company charge to maximize revenue?
11. Consider these two data sets:

Set A

x	−2	−1	0	0	1	2	2
y	5	2	1	−3	−8	−12	−10

Set B

x	−2	−1	0	0	1	2	2
y	10	4	2	3	5	10	12

One data set follows a linear pattern and one data set follows a quadratic relation.
   (a) Draw a scatter diagram of each data set. Determine which is linear and which is quadratic. For the linear data, indicate whether the relation shows a positive or negative slope. For the quadratic relation, indicate whether the quadratic function of best fit will open up or down.
   (b) For the linear data set, find the line of best fit.
   (c) For the quadratic data set, find the quadratic function of best fit.

Cumulative Review

1. Find the distance between the points $P = (-1, 3)$ and $Q = (4, -2)$. Find the midpoint of the line segment from P to Q.
2. Which of the following points are on the graph of $y = x^3 - 3x + 1$?
   (a) $(-2, -1)$ (b) $(2, 3)$ (c) $(3, 1)$
3. Solve the inequality $5x + 3 \geq 0$ and graph the solution set.
4. Find the equation of the line containing the points $(-1, 4)$ and $(2, -2)$. Express your answer in slope–intercept form and graph the line.
5. Find the equation of the line perpendicular to the line $y = 2x + 1$ and containing the point $(3, 5)$. Express your answer in slope–intercept form and graph the line.
6. Graph the equation $x^2 + y^2 - 4x + 8y - 5 = 0$.
7. Does the following relation represent a function? $\{(-3, 8), (1, 3), (2, 5), (3, 8)\}$.
8. For the function f defined by $f(x) = x^2 - 4x + 1$, find:
   (a) $f(2)$
   (b) $f(x) + f(2)$
   (c) $f(-x)$
   (d) $-f(x)$
   (e) $f(x + 2)$
   (f) $\dfrac{f(x+h) - f(x)}{h} \quad h \neq 0$
9. Find the domain of $h(z) = \dfrac{3z - 1}{6z - 7}$.
10. Is the following graph the graph of a function?

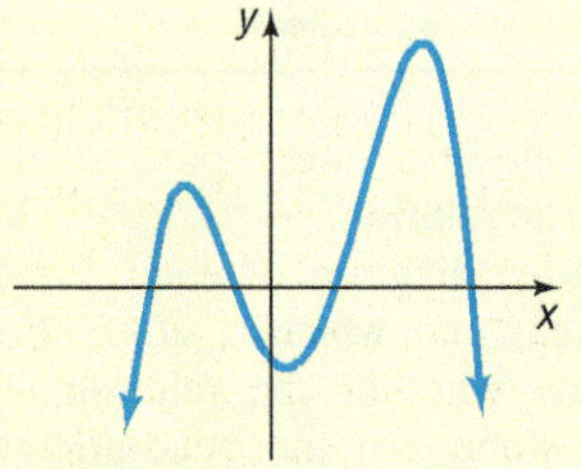

11. Consider the function $f(x) = \dfrac{x}{x + 4}$.
    (a) Is the point $\left(1, \dfrac{1}{4}\right)$ on the graph of f?
    (b) If $x = -2$, what is $f(x)$? What point is on the graph of f?
    (c) If $f(x) = 2$, what is x? What point is on the graph of f?
12. Is the function $f(x) = \dfrac{x^2}{2x + 1}$ even, odd, or neither?
13. Approximate the local maximum values and local minimum values of $f(x) = x^3 - 5x + 1$ on $[-4, 4]$. Determine where the function is increasing and where it is decreasing.
14. If $f(x) = 3x + 5$ and $g(x) = 2x + 1$,
    (a) Solve $f(x) = g(x)$.
    (b) Solve $f(x) > g(x)$.
15. For the graph of the function f,

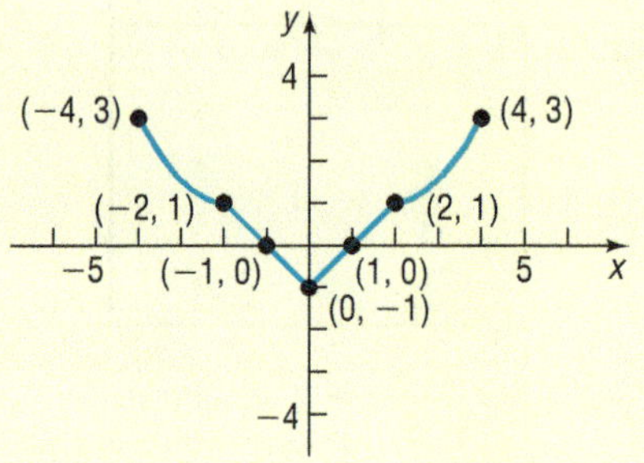

    (a) Find the domain and the range of f.
    (b) Find the intercepts.
    (c) Is the graph of f symmetric with respect to the x-axis, the y-axis, or the origin?
    (d) Find $f(2)$.
    (e) For what value(s) of x is $f(x) = 3$?
    (f) Solve $f(x) < 0$.
    (g) Graph $y = f(x) + 2$.
    (h) Graph $y = f(-x)$.
    (i) Graph $y = 2f(x)$.
    (j) Is f even, odd, or neither?
    (k) Find the interval(s) on which f is increasing.

Chapter Projects

I. The Beta of a Stock You want to invest in the stock market but are not sure which stock to purchase. Information is the key to making an informed investment decision. One piece of information that many stock analysts use is the beta of the stock. Go to Wikipedia (*http://en.wikipedia.org/wiki/Beta_%28finance%29*) and research what beta measures and what it represents.

1. **Approximating the beta of a stock.** Choose a well-known company such as Google or Coca-Cola. Go to a website such as Yahoo! Finance (*http://finance.yahoo.com/*) and find the weekly closing price of the company's stock for the past year. Then find the closing price of the Standard & Poor's 500 (S&P500) for the same time period.

   To get the historical prices in Yahoo! Finance, select Historical Prices from the left menu. Choose the appropriate time period. Select Weekly and Get Prices. Finally, select Download to Spreadsheet. Repeat this for the S&P500, and copy the data into the same spreadsheet. Finally, rearrange the data in chronological order. Be sure to expand the selection to sort all the data. Now, using the adjusted close price, compute the percentage change in price for each week, using the formula % change $= \dfrac{P_1 - P_o}{P_o}$. For example, if week 1 price is in cell D1 and week 2 price is in cell D2, then % change $= \dfrac{D2 - D1}{D1}$. Repeat this for the S&P500 data.

2. **Using Excel to draw a scatter diagram.** Treat the percentage change in the S&P500 as the independent variable and the percentage change in the stock you chose as the dependent variable. The easiest way to draw a scatter diagram in Excel is to place the two columns of data next to each other (for example, have the percentage change in the S&P500 in column F and the percentage change in the stock you chose in column G). Then highlight the data and select the Scatter Diagram icon under Insert. Comment on the type of relation that appears to exist between the two variables.

3. **Finding beta.** To find beta requires that we find the line of best fit using least-squares regression. The easiest approach is to click inside the scatter diagram. Select the Chart Elements icon (+). Check the box for Trendline, select the arrow to the right, and choose More Options. Select Linear and check the box for Display Equation on chart. The line of best fit appears on the scatter diagram. See below.

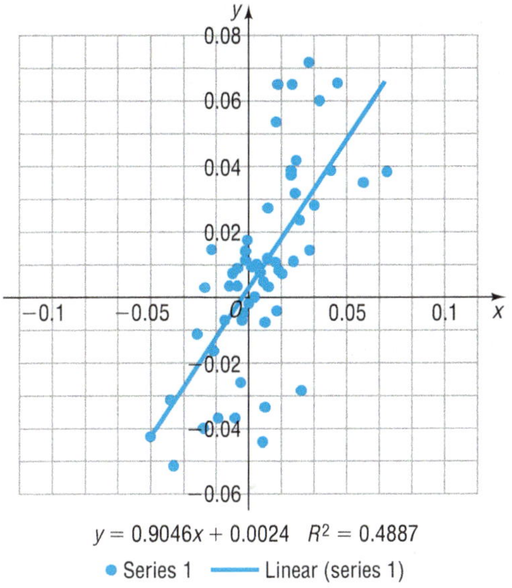

The line of best fit for this data is $y = 0.9046x + 0.0024$. You may click on Chart Title or either axis title and insert the appropriate names. The beta is the slope of the line of best fit, 0.9046. We interpret this by saying, "If the S&P500 increases by 1%, then this stock will increase by 0.9%, on average." Find the beta of your stock and provide an interpretation. NOTE: Another way to use Excel to find the line of best fit requires using the Data Analysis Tool Pack under add-ins.

The following projects are available on the Instructor's Resource Center (IRC):

II. Cannons A battery commander uses the weight of a missile, its initial velocity, and the position of its gun to determine where the missile will travel.

III. First and Second Differences Finite differences provide a numerical method that is used to estimate the graph of an unknown function.

IV. CBL Experiment Computer simulation is used to study the physical properties of a bouncing ball.

5 Polynomial and Rational Functions

Day Length

Day length is the length of time each day from the moment the upper limb of the sun's disk appears above the horizon during sunrise to the moment when the upper limb disappears below the horizon during sunset. The length of a day depends on the day of the year as well as on the latitude of the location. Latitude gives the location of a point on Earth north or south of the equator. In the Internet Project at the end of this chapter, we use information from the chapter to investigate the relation between day length and latitude for a specific day of the year.

—See the Internet-based Chapter Project I—

Outline

••• A Look Back

In Chapter 3, we began our discussion of functions. We defined domain, range, and independent and dependent variables, found the value of a function, and graphed functions. We continued our study of functions by listing the properties that a function might have, such as being even or odd, and created a library of functions, naming key functions and listing their properties, including their graphs.

In Chapter 4, we discussed linear functions and quadratic functions, which belong to the class of *polynomial functions*.

A Look Ahead •••

In this chapter, we look at two general classes of functions, polynomial functions and rational functions, and examine their properties. Polynomial functions are arguably the simplest expressions in algebra. For this reason, they are often used to approximate other, more complicated functions. Rational functions are ratios of polynomial functions.

5.1 Polynomial Functions and Models

PREPARING FOR THIS SECTION *Before getting started, review the following:*

- Polynomials (Chapter R, Section R.4, pp. 41–42)
- Using a Graphing Utility to Approximate Local Maxima and Local Minima (Section 3.3, pp. 237–238)
- Intercepts of a Function (Section 3.2, pp. 223–225)
- Graphing Techniques: Transformations (Section 3.5, pp. 256–264)
- Intercepts (Section 2.1, pp. 165–166)

Now Work the 'Are You Prepared?' problems on page 346.

OBJECTIVES
1. Identify Polynomial Functions and Their Degree (p. 331)
2. Graph Polynomial Functions Using Transformations (p. 334)
3. Identify the Real Zeros of a Polynomial Function and Their Multiplicity (p. 335)
4. Analyze the Graph of a Polynomial Function (p. 342)
5. Build Cubic Models from Data (p. 345)

1 Identify Polynomial Functions and Their Degree

In Chapter 4, we studied the linear function $f(x) = mx + b$, which can be written as

$$f(x) = a_1x + a_0$$

and the quadratic function $f(x) = ax^2 + bx + c, a \neq 0$, which can be written as

$$f(x) = a_2x^2 + a_1x + a_0 \qquad a_2 \neq 0$$

Each of these functions is an example of a *polynomial function.*

DEFINITION

A **polynomial function** in one variable is a function of the form

$$f(x) = a_nx^n + a_{n-1}x^{n-1} + \cdots + a_1x + a_0 \qquad (1)$$

where $a_n, a_{n-1}, \ldots, a_1, a_0$ are constants, called the **coefficients** of the polynomial, $n \geq 0$ is an integer, and x is the variable. If $a_n \neq 0$, it is called the **leading coefficient**, and n is the **degree** of the polynomial.

The domain of a polynomial function is the set of all real numbers.

In Words
A polynomial function is a sum of monomials.

The monomials that make up a polynomial are called its **terms**. If $a_n \neq 0$, a_nx^n is called the **leading term**; a_0 is called the **constant term**. If all of the coefficients are 0, the polynomial is called the **zero polynomial**, which has no degree.

Polynomials are usually written in **standard form**, beginning with the nonzero term of highest degree and continuing with terms in descending order according to degree. If a power of x is missing, it is because its coefficient is zero.

Polynomial functions are among the simplest expressions in algebra. They are easy to evaluate: only addition and repeated multiplication are required. Because of this, they are often used to approximate other, more complicated functions. In this section, we investigate properties of this important class of functions.

EXAMPLE 1 **Identifying Polynomial Functions**

Determine which of the following are polynomial functions. For those that are, state the degree; for those that are not, tell why not. Write each polynomial in standard form, and then identify the leading term and the constant term.

(a) $p(x) = 5x^3 - \frac{1}{4}x^2 - 9$ (b) $f(x) = x + 2 - 3x^4$ (c) $g(x) = \sqrt{x}$

(d) $h(x) = \dfrac{x^2 - 2}{x^3 - 1}$ (e) $G(x) = 8$ (f) $H(x) = -2x^3(x - 1)^2$

Solution

(a) p is a polynomial function of degree 3, and it is already in standard form. The leading term is $5x^3$, and the constant term is -9.

(b) f is a polynomial function of degree 4. Its standard form is $f(x) = -3x^4 + x + 2$. The leading term is $-3x^4$, and the constant term is 2.

(c) g is not a polynomial function because $g(x) = \sqrt{x} = x^{\frac{1}{2}}$, so the variable x is raised to the $\frac{1}{2}$ power, which is not a nonnegative integer.

(d) h is not a polynomial function. It is the ratio of two distinct polynomials, and the polynomial in the denominator is of positive degree.

(e) G is a nonzero constant polynomial function so it is of degree 0. The polynomial is in standard form. The leading term and constant term are both 8.

(f) $H(x) = -2x^3(x-1)^2 = -2x^3(x^2 - 2x + 1) = -2x^5 + 4x^4 - 2x^3$. So, H is a polynomial function of degree 5. Because $H(x) = -2x^5 + 4x^4 - 2x^3$, the leading term is $-2x^5$. Since no constant term is shown, the constant term is 0. Do you see a way to find the degree of H without multiplying it out? ■

Now Work PROBLEMS 17 AND 21

We have already discussed in detail polynomial functions of degrees 0, 1, and 2. See Table 1 for a summary of the properties of the graphs of these polynomial functions.

Table 1

Degree	Form	Name	Graph
No degree	$f(x) = 0$	Zero function	The x-axis
0	$f(x) = a_0, \quad a_0 \neq 0$	Constant function	Horizontal line with y-intercept a_0
1	$f(x) = a_1x + a_0, \quad a_1 \neq 0$	Linear function	Nonvertical, nonhorizontal line with slope a_1 and y-intercept a_0
2	$f(x) = a_2x^2 + a_1x + a_0, \quad a_2 \neq 0$	Quadratic function	Parabola: graph opens up if $a_2 > 0$; graph opens down if $a_2 < 0$

One objective of this section is to analyze the graph of a polynomial function. If you take a course in calculus, you will learn that the graph of every polynomial function is both smooth and continuous. By **smooth**, we mean that the graph contains no sharp corners or cusps; by **continuous**, we mean that the graph has no gaps or holes and can be drawn without lifting your pencil from the paper. See Figures 1(a) and 1(b).

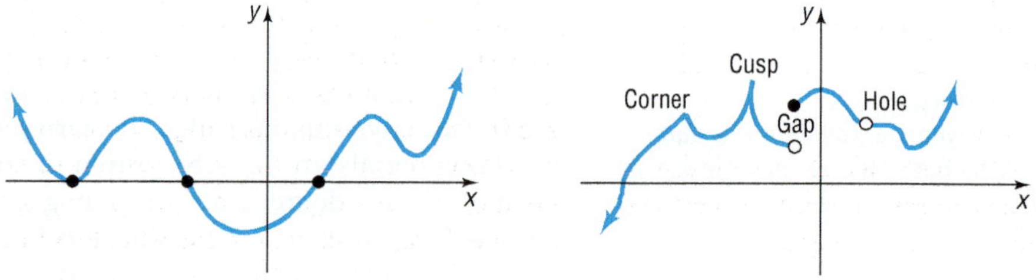

(a) Graph of a polynomial function: smooth, continuous

(b) Cannot be the graph of a polynomial function

Figure 1

Power Functions

We begin the analysis of the graph of a polynomial function by discussing *power functions,* a special kind of polynomial function.

DEFINITION

A **power function of degree n** is a monomial function of the form

$$f(x) = ax^n \tag{2}$$

where a is a real number, $a \neq 0$, and $n > 0$ is an integer. ■

In Words
A power function is defined by a single monomial.

Examples of power functions are

$f(x) = 3x$	$f(x) = -5x^2$	$f(x) = 8x^3$	$f(x) = -5x^4$
degree 1	**degree 2**	**degree 3**	**degree 4**

The graph of a power function of degree 1, $f(x) = ax$, is a straight line, with slope a, that passes through the origin. The graph of a power function of degree 2, $f(x) = ax^2$, is a parabola, with vertex at the origin, that opens up if $a > 0$ and opens down if $a < 0$.

If we know how to graph a power function of the form $f(x) = x^n$, a compression or stretch and, perhaps, a reflection about the x-axis will enable us to obtain the graph of $g(x) = ax^n$. Consequently, we shall concentrate on graphing power functions of the form $f(x) = x^n$.

We begin with power functions of even degree of the form $f(x) = x^n$, $n \geq 2$ and n even.

Exploration

Using your graphing utility and the viewing window $-2 \leq x \leq 2$, $-4 \leq y \leq 16$, graph the function $Y_1 = f(x) = x^4$. On the same screen, graph $Y_2 = g(x) = x^8$. Now, also on the same screen, graph $Y_3 = h(x) = x^{12}$. What do you notice about the graphs as the magnitude of the exponent increases? Repeat this procedure for the viewing window $-1 \leq x \leq 1$, $0 \leq y \leq 1$. What do you notice?

Result See Figures 2(a) and 2(b).

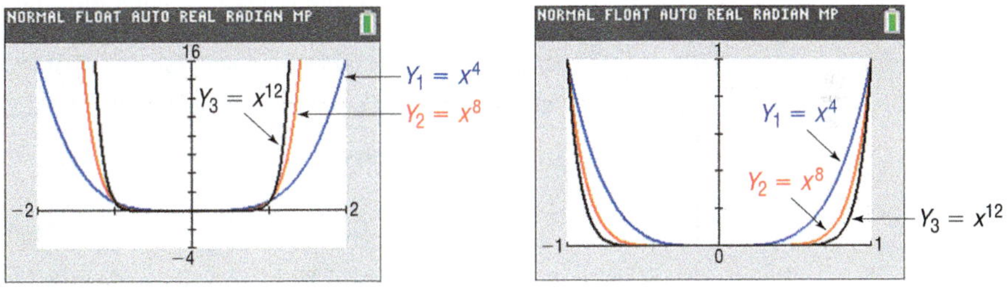

(a) (b)

Figure 2 $Y_1 = x^4$; $Y_2 = x^8$; $Y_3 = x^{12}$

Table 2

NORMAL FLOAT AUTO REAL RADIAN MP
PRESS ENTER TO EDIT

X	Y_1	Y_2	Y_3
-1	1	1	1
-.5	.0625	.00391	2.4E-4
-.1	1E-4	1E-8	1E-12
-.01	1E-8	1E-16	1E-24
-.001	1E-12	1E-24	1E-36
0	0	0	0
.001	1E-12	1E-24	1E-36
.01	1E-8	1E-16	1E-24
.1	1E-4	1E-8	1E-12
.5	.0625	.00391	2.4E-4
1	1	1	1

$Y_3 = X^{12}$

Note: Don't forget how graphing calculators express scientific notation. In Table 2, 1E −8 means 1×10^{-8}. ■

The domain of $f(x) = x^n$, $n \geq 2$ and n even, is the set of all real numbers, and the range is the set of nonnegative real numbers. Such a power function is an even function (do you see why?), so its graph is symmetric with respect to the y-axis. Its graph always contains the origin $(0, 0)$ and the points $(-1, 1)$ and $(1, 1)$.

If $n = 2$, the graph is the parabola $y = x^2$ that opens up, with vertex at the origin. For large n, it appears that the graph coincides with the x-axis near the origin, but it does not; the graph actually touches the x-axis only at the origin. See Table 2, where $Y_1 = x^4$, $Y_2 = x^8$, and $Y_3 = x^{12}$. For x close to 0, the values of y are positive and close to 0. Also, for large n, it may appear that for $x < -1$ or for $x > 1$ the graph is vertical, but it is not; it is only increasing very rapidly. That is, as the values of x approach negative infinity $(-\infty)$, the values of $f(x)$ approach ∞. In calculus, we denote this as $\lim_{x \to -\infty} f(x) = \infty$. Using this notation, we also see $\lim_{x \to \infty} f(x) = \infty$. These limits describe the end behavior of the graph of f, as we will see later in this section. If you TRACE along one of the graphs, these distinctions will be clear. ■

To summarize:

Properties of Power Functions, $f(x) = x^n$, n Is an Even Integer

1. f is an even function, so its graph is symmetric with respect to the y-axis.
2. The domain is the set of all real numbers. The range is the set of nonnegative real numbers.
3. The graph always contains the points $(-1, 1)$, $(0, 0)$, and $(1, 1)$.
4. As the exponent n increases in magnitude, the graph is steeper when $x < -1$ or $x > 1$; but for x near the origin, the graph tends to flatten out and lie closer to the x-axis.

Now we consider power functions of odd degree of the form $f(x) = x^n$, n odd.

Exploration

Using your graphing utility and the viewing window $-2 \le x \le 2$, $-16 \le y \le 16$, graph the function $Y_1 = f(x) = x^3$. On the same screen, graph $Y_2 = g(x) = x^7$ and $Y_3 = h(x) = x^{11}$. What do you notice about the graphs as the magnitude of the exponent increases? Repeat this procedure for the viewing window $-1 \le x \le 1$, $-1 \le y \le 1$. What do you notice?

Result The graphs on your screen should look like Figures 3(a) and 3(b).

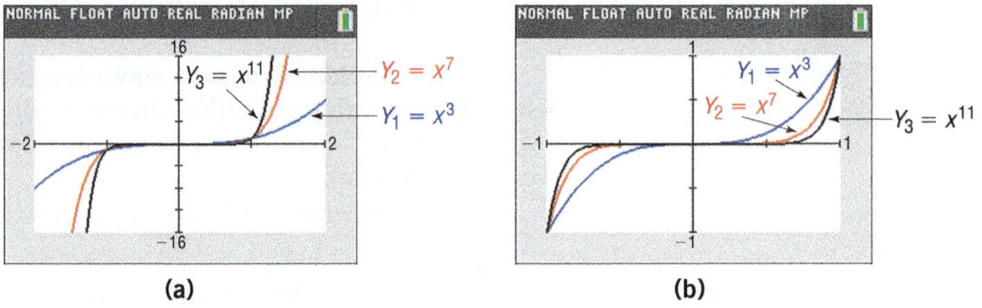

Figure 3 $Y_1 = x^3$; $Y_2 = x^7$; $Y_3 = x^{11}$

The domain and the range of $f(x) = x^n$, $n \ge 3$ and n odd, are the set of real numbers. Such a power function is an odd function (do you see why?), so its graph is symmetric with respect to the origin. Its graph always contains the origin $(0, 0)$ and the points $(-1, -1)$ and $(1, 1)$.

It appears that the graph coincides with the x-axis near the origin, but it does not; the graph actually crosses the x-axis only at the origin. Also, it appears that as x increases the graph is vertical, but it is not; it is increasing very rapidly. That is, $\lim_{x \to -\infty} f(x) = -\infty$ and $\lim_{x \to \infty} f(x) = \infty$. These limits describe the end behavior of the graph of f. TRACE along the graphs to verify these distinctions. ■

To summarize:

Properties of Power Functions, $f(x) = x^n$, n Is an Odd Integer

1. f is an odd function, so its graph is symmetric with respect to the origin.
2. The domain and the range are the set of all real numbers.
3. The graph always contains the points $(-1, -1)$, $(0, 0)$, and $(1, 1)$.
4. As the exponent n increases in magnitude, the graph is steeper when $x < -1$ or $x > 1$, but for x near the origin, the graph tends to flatten out and lie closer to the x-axis.

2 Graph Polynomial Functions Using Transformations

The methods of shifting, compression, stretching, and reflection studied in Section 3.5, when used with the facts just presented, enable us to graph polynomial functions that are transformations of power functions.

EXAMPLE 2 Graphing a Polynomial Function Using Transformations

Graph: $f(x) = 1 - x^5$

Solution It is helpful to rewrite f as $f(x) = -x^5 + 1$. Figure 4 shows the required stages.

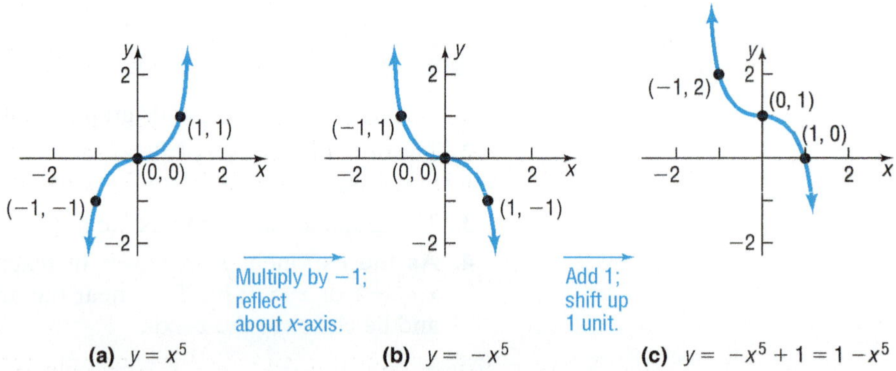

Figure 4 (a) $y = x^5$ (b) $y = -x^5$ (c) $y = -x^5 + 1 = 1 - x^5$

Check: Verify the graph of f by graphing $Y_1 = 1 - x^5$ on a graphing utility. ■

EXAMPLE 3 **Graphing a Polynomial Function Using Transformations**

Graph: $f(x) = \frac{1}{2}(x - 1)^4$

Solution Figure 5 shows the required stages.

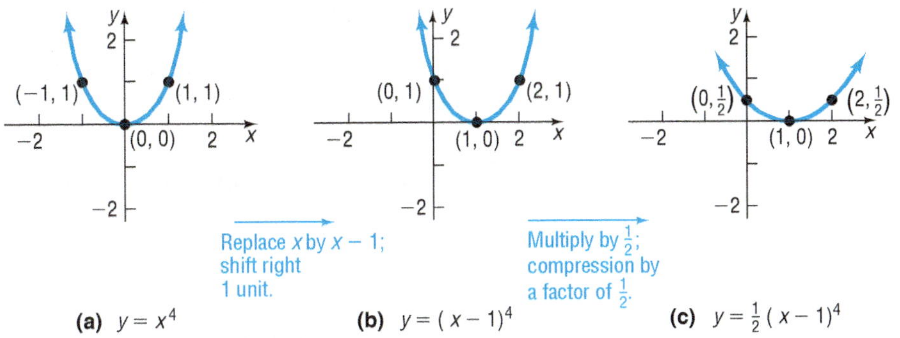

Figure 5

Check: Verify the graph of f by graphing $Y_1 = \frac{1}{2}(x - 1)^4$ on a graphing utility. ■

Now Work PROBLEMS 29 AND 35

3 Identify the Real Zeros of a Polynomial Function and Their Multiplicity

Figure 6 shows the graph of a polynomial function with four x-intercepts. Notice that at the x-intercepts, the graph must either cross the x-axis or touch the x-axis. Consequently, between consecutive x-intercepts the graph is either above the x-axis or below the x-axis.

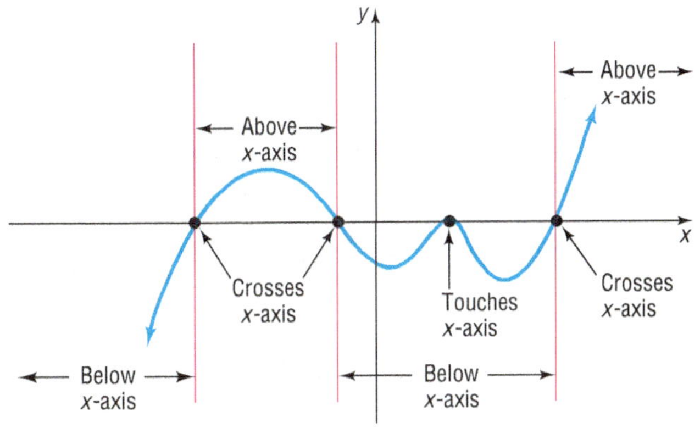

Figure 6 Graph of a polynomial function

If a polynomial function f is factored completely, it is easy to locate the x-intercepts of the graph by solving the equation $f(x) = 0$ using the Zero-Product Property. For example, if $f(x) = (x - 1)^2(x + 3)$ then the solutions of the equation

$$f(x) = (x - 1)^2(x + 3) = 0$$

are identified as 1 and -3. That is, $f(1) = 0$ and $f(-3) = 0$.

DEFINITION If f is a function and r is a real number for which $f(r) = 0$, then r is called a **real zero** of f. ■

As a consequence of this definition, the following statements are equivalent.

- r is a real zero of a polynomial function f.
- r is an x-intercept of the graph of f.
- $x - r$ is a factor of f.
- r is a real solution to the equation $f(x) = 0$.

So the real zeros of a polynomial function are the x-intercepts of its graph, and they are found by solving the equation $f(x) = 0$.

EXAMPLE 4 Finding a Polynomial Function from Its Zeros

(a) Find a polynomial of degree 3 whose zeros are -3, 2, and 5.

(b) Use a graphing utility to graph the polynomial found in part (a) to verify your result.

Solution

(a) If r is a real zero of a polynomial function f, then $x - r$ is a factor of f. This means that $x - (-3) = x + 3$, $x - 2$, and $x - 5$ are factors of f. As a result, any polynomial function of the form

$$f(x) = a(x + 3)(x - 2)(x - 5)$$

where a is a nonzero real number, qualifies. The value of a causes a stretch, compression, or reflection, but it does not affect the x-intercepts of the graph. Do you know why?

(b) We choose to graph f with $a = 1$. Then

$$f(x) = (x + 3)(x - 2)(x - 5) = x^3 - 4x^2 - 11x + 30$$

Figure 7 shows the graph of f. Notice that the x-intercepts are -3, 2, and 5. ■

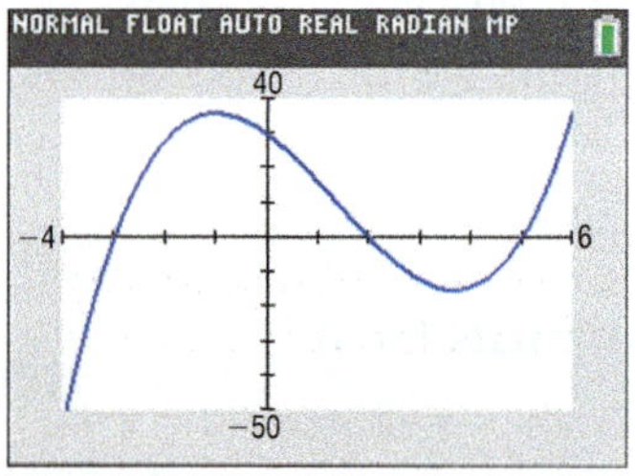

Figure 7 $f(x) = x^3 - 4x^2 - 11x + 30$

Seeing the Concept

Graph the function found in Example 4 for $a = 2$ and $a = -1$. Does the value of a affect the zeros of f? How does the value of a affect the graph of f? ■

Now Work PROBLEM 43

If the same factor $x - r$ occurs more than once, r is called a **repeated**, or **multiple**, **zero of** f. More precisely, we have the following definition.

DEFINITION

If $(x - r)^m$ is a factor of a polynomial f and $(x - r)^{m+1}$ is not a factor of f, then r is called a **zero of multiplicity m of f**.* ■

EXAMPLE 5 Identifying Zeros and Their Multiplicities

For the polynomial function

$$f(x) = 5x^2(x + 2)\left(x - \frac{1}{2}\right)^4$$

0 is a zero of multiplicity 2 because the exponent on the factor $x = x - 0$ is 2.

-2 is a zero of multiplicity 1 because the exponent on the factor $x + 2$ is 1.

$\frac{1}{2}$ is a zero of multiplicity 4 because the exponent on the factor $x - \frac{1}{2}$ is 4. ■

In Words
The multiplicity of a zero is the number of times its corresponding factor occurs.

Now Work PROBLEM 57(a)

*Some texts use the terms **multiple root** and **root of multiplicity** m.

In Example 5, notice that if you add the multiplicities $(1 + 2 + 4 = 7)$, you obtain the degree of the polynomial function.

Suppose that it is possible to factor completely a polynomial function and, as a result, locate all the x-intercepts of its graph (the real zeros of the function). The following example illustrates the role that the multiplicity of an x-intercept plays.

EXAMPLE 6 Investigating the Role of Multiplicity

For the polynomial function $f(x) = (x+1)^2(x-2)$:

(a) Find the x- and y-intercepts of the graph of f.

(b) Using a graphing utility, graph the polynomial function.

(c) For each x-intercept, determine whether it is of odd or even multiplicity.

Solution

(a) The y-intercept is $f(0) = (0+1)^2(0-2) = -2$. The x-intercepts satisfy the equation

$$f(x) = (x+1)^2(x-2) = 0$$

from which we find that

$$\begin{aligned} (x+1)^2 &= 0 \quad \text{or} \quad x - 2 = 0 \\ x &= -1 \quad \text{or} \quad\quad\; x = 2 \end{aligned}$$

The x-intercepts are -1 and 2.

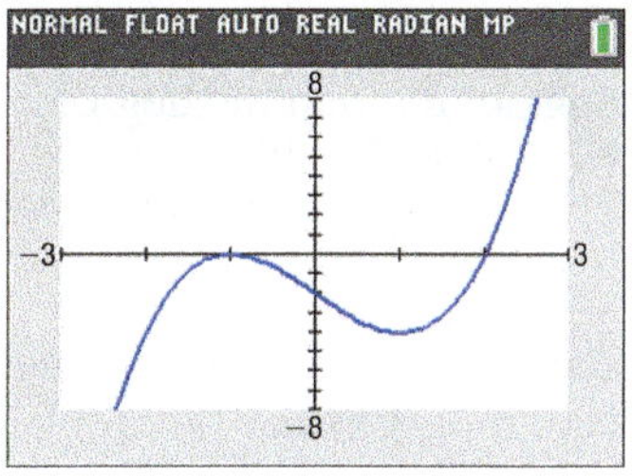

Figure 8 $Y_1 = (x+1)^2(x-2)$

(b) See Figure 8 for the graph of f.

(c) We can see from the factored form of f that -1 is a zero or root of multiplicity 2, and 2 is a zero or root of multiplicity 1; so -1 is of even multiplicity and 2 is of odd multiplicity. ■

Table 3

NORMAL FLOAT AUTO REAL RADIAN MP
PRESS ENTER TO EDIT

X	Y1
-5	-112
-4	-54
-3	-20
-2	-4
-1	0
0	-2
1	-4
2	0
3	16
4	50
5	108

$Y_1 = (X+1)^2(X-2)$

We can use a TABLE to further analyze the graph. See Table 3. The sign of $f(x)$ is the same on each side of $x = -1$, and the graph of f just *touches* the x-axis at $x = -1$ (a zero of *even* multiplicity). The sign of $f(x)$ changes from one side of $x = 2$ to the other, and the graph of f *crosses* the x-axis at $x = 2$ (a zero of *odd* multiplicity). These observations suggest the following result:

If r Is a Zero of Even Multiplicity

Numerically: The sign of $f(x)$ does not change from one side to the other side of r.

Graphically: The graph of f **touches** the x-axis at r.

If r Is a Zero of Odd Multiplicity

Numerically: The sign of $f(x)$ changes from one side to the other side of r.

Graphically: The graph of f **crosses** the x-axis at r.

Now Work PROBLEM 57(b)

Turning Points

Points on the graph where the graph changes from an increasing function to a decreasing function, or vice versa, are called **turning points.*** Each turning point yields a local maximum or a local minimum (see Section 3.3).

*Graphing utilities can be used to approximate turning points. Calculus is needed to find the exact location of turning points for most polynomial functions.

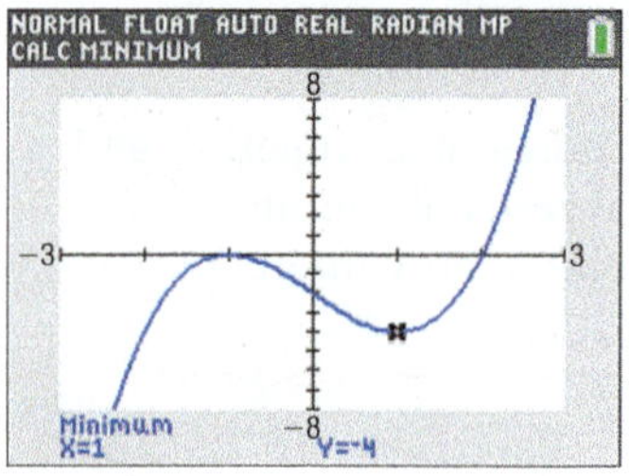

Figure 9 Local minimum of $f(x) = (x + 1)^2(x - 2)$

Look at Figure 9. The graph of $f(x) = (x + 1)^2(x - 2) = x^3 - 3x - 2$ has a turning point at $(-1, 0)$. After utilizing MINIMUM, we find that the graph also has a turning point at $(1, -4)$.

Exploration

Graph $Y_1 = x^3$, $Y_2 = x^3 - x$, and $Y_3 = x^3 + 3x^2 + 4$. How many turning points do you see? Graph $Y_1 = x^4$, $Y_2 = x^4 - \frac{4}{3}x^3$, and $Y_3 = x^4 - 2x^2$. How many turning points do you see? How does the number of turning points compare to the degree? ■

The following theorem from calculus supplies the answer to the question posed in the Exploration.

THEOREM

Turning Points

If f is a polynomial function of degree n, then f has at most $n - 1$ turning points.

If the graph of a polynomial function f has $n - 1$ turning points, then the degree of f is at least n. ■

Based on the first part of the theorem, a polynomial function of degree 5 will have at most $5 - 1 = 4$ turning points. Based on the second part of the theorem, if a polynomial function has 3 turning points, then its degree must be at least 4.

EXAMPLE 7 Identifying the Graph of a Polynomial Function

Which of the graphs in Figure 10 could be the graph of a polynomial function? For those that could, list the real zeros and state the least degree the polynomial function can have. For those that could not, say why not.

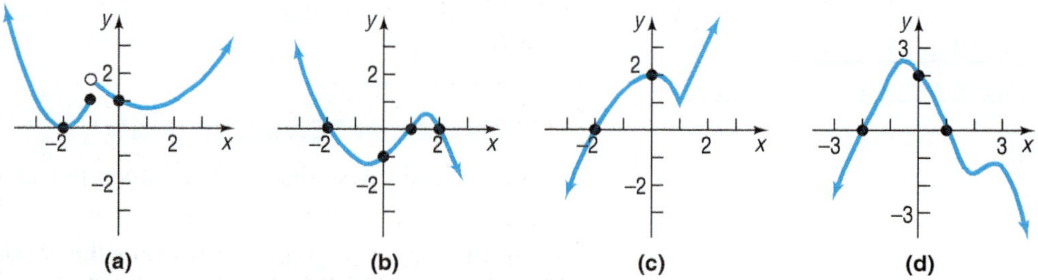

Figure 10

Solution

(a) The graph in Figure 10(a) cannot be the graph of a polynomial function because of the gap that occurs at $x = -1$. Remember, the graph of a polynomial function is continuous—no gaps or holes. (See Figure 1.)

(b) The graph in Figure 10(b) could be the graph of a polynomial function because the graph is smooth and continuous. It has three real zeros, at -2, 1, and 2. Since the graph has two turning points, the degree of the polynomial function must be at least 3.

(c) The graph in Figure 10(c) cannot be the graph of a polynomial function because of the cusp at $x = 1$. Remember, the graph of a polynomial function is smooth.

(d) The graph in Figure 10(d) could be the graph of a polynomial function. It has two real zeros, at -2 and 1. Since the graph has three turning points, the degree of the polynomial function is at least 4. ■

Now Work PROBLEMS 57(c) AND 69

End Behavior

One last remark about Figure 8. For very large values of x, either positive or negative, the graph of $f(x) = (x + 1)^2(x - 2)$ looks like the graph of $y = x^3$. To see why, write f in the form

$$f(x) = (x + 1)^2(x - 2) = x^3 - 3x - 2 = x^3\left(1 - \frac{3}{x^2} - \frac{2}{x^3}\right)$$

For large values of x, either positive or negative, the terms $\frac{3}{x^2}$ and $\frac{2}{x^3}$ are close to 0. Do you see why? Evaluate $\frac{3}{x^2}$ and $\frac{2}{x^3}$ for $x = 10, 100, 1000$ and for $x = -10, -100, -1000$. What happens to the value of each expression for large values of $|x|$? So, for large values of $|x|$,

$$f(x) = x^3 - 3x - 2 = x^3\left(1 - \frac{3}{x^2} - \frac{2}{x^3}\right) \approx x^3$$

The behavior of the graph of a function for large values of x, either positive or negative, is referred to as its **end behavior**.

THEOREM

End Behavior

For large values of x, either positive or negative, the graph of the polynomial function

$$f(x) = a_n x^n + a_{n-1}x^{n-1} + \cdots + a_1 x + a_0 \quad a_n \neq 0$$

resembles the graph of the power function

$$y = a_n x^n$$

In Words

The end behavior of a polynomial function resembles that of its leading term.

For example, if $f(x) = -2x^3 + 5x^2 + x - 4$, then the graph of f will behave like the graph of $y = -2x^3$ for very large values of x, either positive or negative. We can see that the graphs of f and $y = -2x^3$ "behave" the same by considering Table 4 and Figure 11.

Table 4

x	$f(x)$	$y = -2x^3$
10	−1494	−2000
100	−1,949,904	−2,000,000
500	−248,749,504	−250,000,000
1000	−1,994,999,004	−2,000,000,000

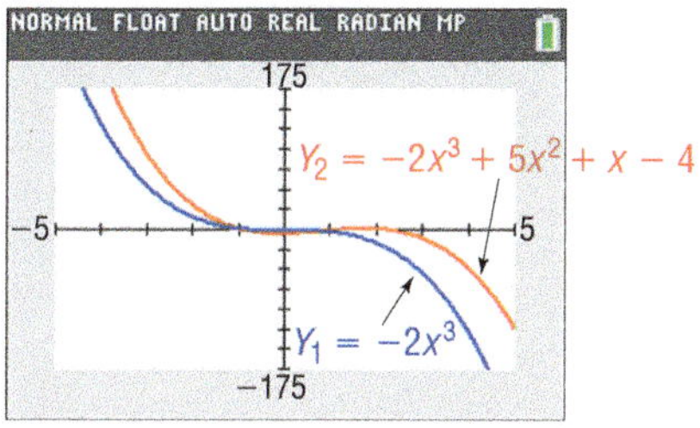

Figure 11

Notice that, as x becomes a larger and larger positive number, the values of f become larger and larger negative numbers. When this happens, we say that f is **unbounded in the negative direction**. Rather than using words to describe the behavior of the graph of the function, we explain its behavior using notation. We can symbolize "the value of f becomes a larger and larger negative number as x becomes a larger and larger positive number" by writing $f(x) \to -\infty$ as $x \to \infty$ (read "the values of f approach negative infinity as x approaches infinity"). In calculus, **limits** are used to convey these ideas. There we use the symbolism $\lim_{x\to\infty} f(x) = -\infty$, read "the limit of $f(x)$ as x approaches infinity equals negative infinity," to mean that $f(x) \to -\infty$ as $x \to \infty$.

When the value of a limit equals infinity (or negative infinity), we mean that the values of the function are unbounded in the positive (or negative) direction and call the limit an **infinite limit**. When we discuss limits as x becomes unbounded in the negative direction or unbounded in the positive direction, we are discussing **limits at infinity**.

Note: Infinity (∞) and negative infinity ($-\infty$) are not numbers. Rather, they are symbols that represent unboundedness.

Look back at Figures 2 and 3. Based on the preceding theorem and the previous discussion on power functions, the end behavior of a polynomial function can be of only four types. See Figure 12.

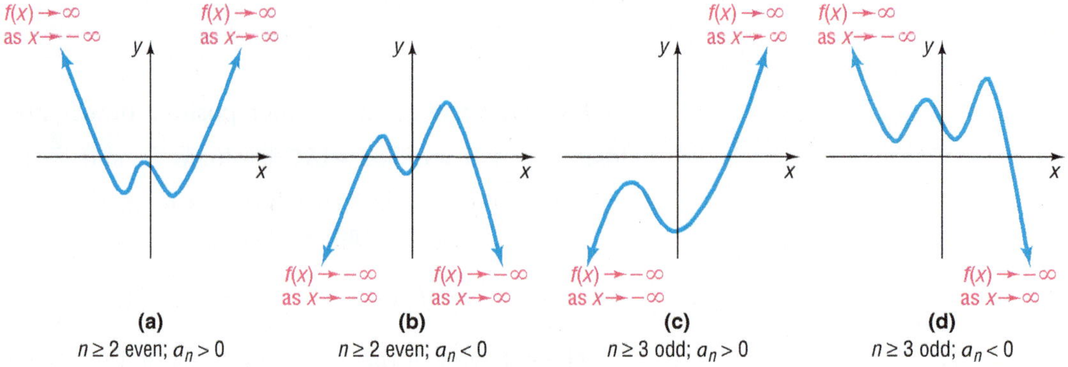

Figure 12 End behavior of $f(x) = a_nx^n + a_{n-1}x^{n-1} + \cdots + a_1x + a_0$

For example, if $f(x) = -2x^4 + x^3 + 4x^2 - 7x + 1$, the graph of f will resemble the graph of the power function $y = -2x^4$ for large $|x|$. The graph of f will behave like Figure 12(b) for large $|x|$.

Now Work PROBLEM 57(d)

EXAMPLE 8

Identifying the Graph of a Polynomial Function

Which of the graphs in Figure 13 could be the graph of

$$f(x) = x^4 + ax^3 + bx^2 - 5x - 6$$

where $a > 0, b > 0$?

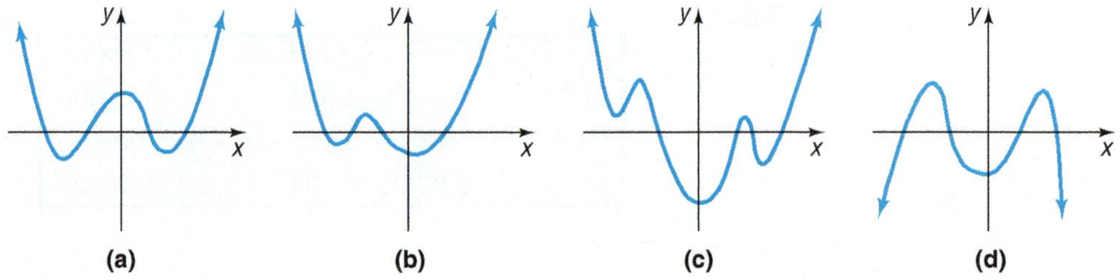

Figure 13

Solution The y-intercept of f is $f(0) = -6$. We can eliminate the graph in Figure 13(a), whose y-intercept is positive.

We are not able to solve $f(x) = 0$ to find the x-intercepts of f, so we move on to investigate the turning points of each graph. Since f is of degree 4, the graph of f has at most 3 turning points. Eliminate the graph in Figure 13(c) because that graph has 5 turning points.

Now look at end behavior. For large values of x, the graph of f will behave like the graph of $y = x^4$. This eliminates the graph in Figure 13(d), whose end behavior is like the graph of $y = -x^4$.

Only the graph in Figure 13(b) could be the graph of

$$f(x) = x^4 + ax^3 + bx^2 - 5x - 6$$

where $a > 0, b > 0$. ■

EXAMPLE 9 **Writing a Polynomial Function from Its Graph**

Write a polynomial function whose graph is shown in Figure 14 (use the smallest degree possible).

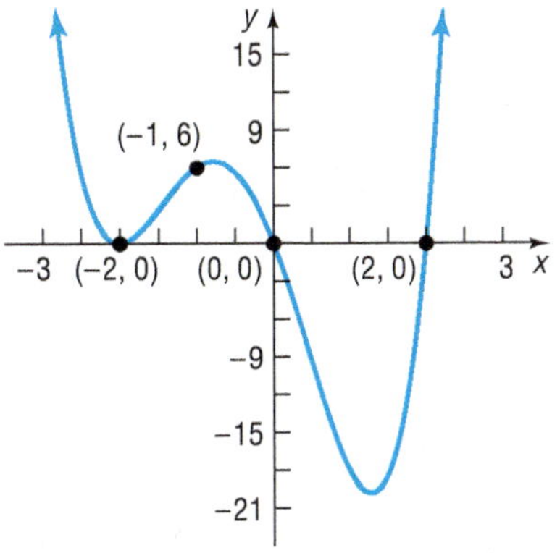

Figure 14

Solution The x-intercepts are -2, 0, and 2. Therefore, the polynomial must have the factors $(x+2)$, x, and $(x-2)$, respectively. There are three turning points, so the degree of the polynomial must be at least 4. The graph touches the x-axis at $x=-2$, so -2 must have an even multiplicity. The graph crosses the x-axis at $x=0$ and $x=2$, so 0 and 2 must have odd multiplicities. Using the smallest degree possible (1 for odd multiplicity and 2 for even multiplicity), we can write

$$f(x) = ax(x+2)^2(x-2)$$

All that remains is to find the leading coefficient, a. From Figure 14, the point $(-1, 6)$ must lie on the graph.

$$6 = a(-1)(-1+2)^2(-1-2) \quad f(-1) = 6$$
$$6 = 3a$$
$$2 = a$$

The polynomial function $f(x) = 2x(x+2)^2(x-2)$ would have the graph in Figure 14.

✓**Check:** Graph $Y_1 = 2x(x+2)^2(x-2)$ using a graphing utility to verify this result.

Now Work PROBLEMS 73 AND 77

SUMMARY

Graph of a Polynomial Function $f(x) = a_n x^n + a_{n-1}x^{n-1} + \cdots + a_1 x + a_0 \quad a_n \neq 0$

Degree of the polynomial function f: n
y-intercept: $f(0) = a_0$.
Graph is smooth and continuous.
Maximum number of turning points: $n - 1$
At a zero of even multiplicity: The graph of f touches the x-axis.
At a zero of odd multiplicity: The graph of f crosses the x-axis.
Between zeros, the graph of f is either above or below the x-axis.
End behavior: For large $|x|$, the graph of f behaves like the graph of $y = a_n x^n$.

4 Analyze the Graph of a Polynomial Function

EXAMPLE 10 **How to Analyze the Graph of a Polynomial Function**

Analyze the graph of the polynomial function $f(x) = (2x + 1)(x - 3)^2$.

Step-by-Step Solution

Step 1: Determine the end behavior of the graph of the function.

$$\begin{aligned} f(x) &= (2x + 1)(x - 3)^2 \\ &= (2x + 1)(x^2 - 6x + 9) && \text{Square the binomial difference.} \\ &= 2x^3 - 12x^2 + 18x + x^2 - 6x + 9 && \text{Multiply.} \\ &= 2x^3 - 11x^2 + 12x + 9 && \text{Combine like terms.} \end{aligned}$$

The polynomial function f is of degree 3. The graph of f behaves like $y = 2x^3$ for large values of $|x|$.

Step 2: Find the x- and y-intercepts of the graph of the function.

The y-intercept is $f(0) = 9$. To find the x-intercepts, solve $f(x) = 0$.

$$f(x) = 0$$

$$(2x + 1)(x - 3)^2 = 0$$

$$2x + 1 = 0 \quad \text{or} \quad (x - 3)^2 = 0$$

$$x = -\frac{1}{2} \quad \text{or} \quad x - 3 = 0$$

$$x = 3$$

The x-intercepts are $-\frac{1}{2}$ and 3.

Step 3: Determine the zeros of the function and their multiplicity. Use this information to determine whether the graph crosses or touches the x-axis at each x-intercept.

The zeros of f are $-\frac{1}{2}$ and 3. The zero $-\frac{1}{2}$ is a zero of multiplicity 1, so the graph of f crosses the x-axis at $x = -\frac{1}{2}$. The zero 3 is a zero of multiplicity 2, so the graph of f touches the x-axis at $x = 3$.

Step 4: Use a graphing utility to graph the function.

See Figure 15 for the graph of f.

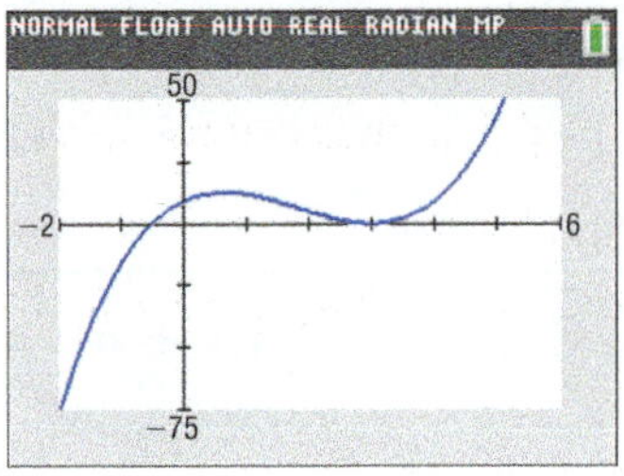

Figure 15 $Y_1 = (2x + 1)(x - 3)^2$

Step 5: Approximate the turning points of the graph.

From the graph of f shown in Figure 15, we see that f has two turning points. Using MAXIMUM, one turning point is at $(0.67, 12.70)$, rounded to two decimal places. Using MINIMUM, the other turning point is at $(3, 0)$.

Step 6: Use the information in Steps 1 to 5 to draw a complete graph of the function by hand.

Figure 16 shows a graph of f using the information in Steps 1 through 5.

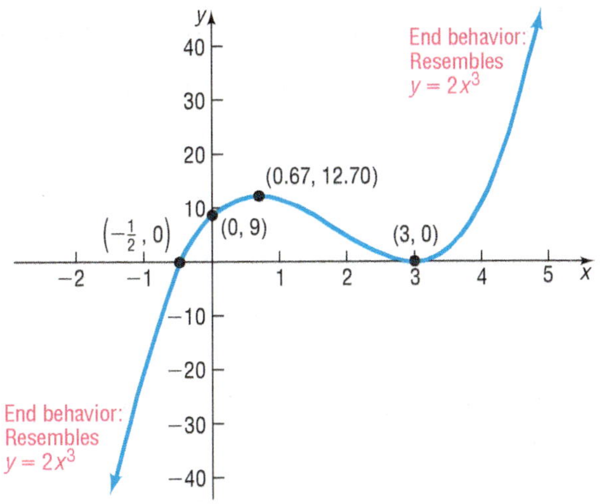

Figure 16 $f(x) = (2x + 1)(x - 3)^2$

Step 7: Find the domain and the range of the function.

The domain and the range of f is the set of all real numbers.

Step 8: Use the graph to determine where the function is increasing and where it is decreasing.

Based on the graph, f is increasing on the intervals $(-\infty, 0.67]$ and $[3, \infty)$. Also, f is decreasing on the interval $[0.67, 3]$. ■

SUMMARY

Analyzing the Graph of a Polynomial Function

STEP 1: Determine the end behavior of the graph of the function.

STEP 2: Find the x- and y-intercepts of the graph of the function.

STEP 3: Determine the zeros of the function and their multiplicity. Use this information to determine whether the graph crosses or touches the x-axis at each x-intercept.

STEP 4: Use a graphing utility to graph the function.

STEP 5: Approximate the turning points of the graph.

STEP 6: Use the information in Steps 1 through 5 to draw a complete graph of the function by hand.

STEP 7: Find the domain and the range of the function.

STEP 8: Use the graph to determine where the function is increasing and where it is decreasing.

Now Work PROBLEM 81

For polynomial functions that have noninteger coefficients and for polynomials that are not easily factored, we utilize the graphing utility early in the analysis. This is because the amount of information that can be obtained from algebraic analysis is limited.

EXAMPLE 11 How to Use a Graphing Utility to Analyze the Graph of a Polynomial Function

Analyze the graph of the polynomial function

$$f(x) = x^3 + 2.48x^2 - 4.3155x + 2.484406$$

Step-by-Step Solution

Step 1: Determine the end behavior of the graph of the function.

The polynomial function f is of degree 3. The graph of f behaves like $y = x^3$ for large values of $|x|$.

Step 2: Graph the function using a graphing utility.

See Figure 17 for the graph of f.

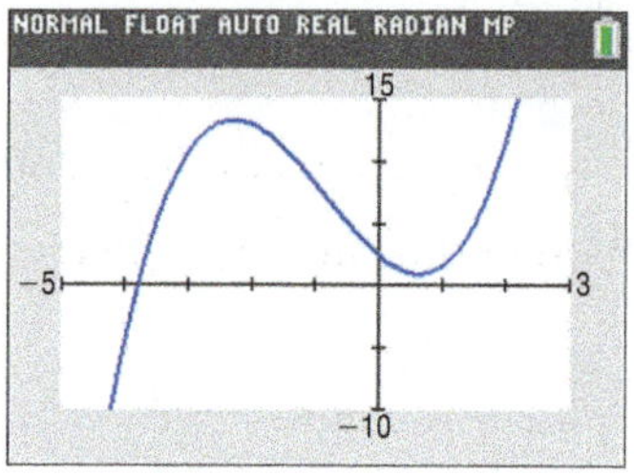

Figure 17 $Y_1 = x^3 + 2.48x^2 - 4.3155x + 2.484406$

Step 3: Use a graphing utility to approximate the x- and y-intercepts of the graph.

The y-intercept is $f(0) = 2.484406$. In Example 10 the polynomial function was factored, so it was easy to find the x-intercepts algebraically. However, it is not readily apparent how to factor f in this example. Therefore, use a graphing utility's ZERO (or ROOT or SOLVE) feature to find the lone x-intercept, -3.79, rounded to two decimal places.

Step 4: Use a graphing utility to create a TABLE to find points on the graph around each x-intercept.

Table 5 shows values of x around the x-intercept. The points $(-4, -4.57)$ and $(-2, 13.04)$ are on the graph.

Table 5

NORMAL FLOAT AUTO REAL RADIAN MP
PRESS ENTER TO EDIT

X	Y1
-4	-4.574
-2	13.035

Y1=X³+2.48X²−4.3155X+2.4844

Step 5: Approximate the turning points of the graph.

From the graph of f shown in Figure 17, we see that f has two turning points. Using MAXIMUM, one turning point is at $(-2.28, 13.36)$, rounded to two decimal places. Using MINIMUM, the other turning point is at $(0.63, 1)$, rounded to two decimal places.

Step 6: Use the information in Steps 1 through 5 to draw a complete graph of the function by hand.

Figure 18 shows a graph of f using the information in Steps 1 to 5.

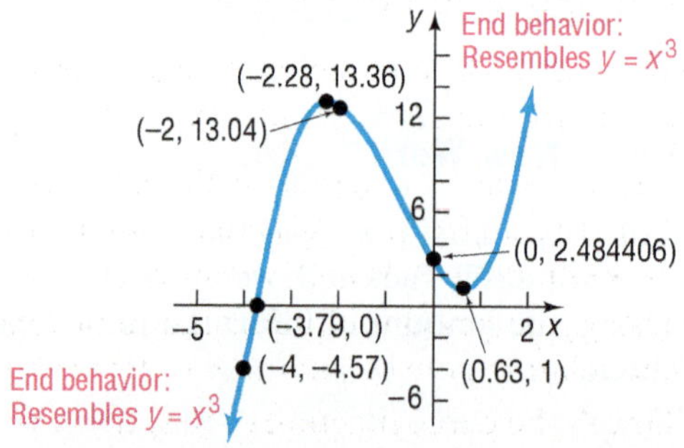

Figure 18 $f(x) = x^3 + 2.48x^2 - 4.3155x + 2.484406$

Step 7: Find the domain and the range of the function.

The domain and the range of f are the set of all real numbers.

Step 8: Use the graph to determine where the function is increasing and where it is decreasing.

Based on the graph, f is increasing on the intervals $(-\infty, -2.28]$ and $[0.63, \infty)$. Also, f is decreasing on the interval $[-2.28, 0.63]$.

SUMMARY

Using a Graphing Utility to Analyze the Graph of a Polynomial Function

STEP 1: Determine the end behavior of the graph of the function.

STEP 2: Graph the function using a graphing utility.

STEP 3: Use a graphing utility to approximate the x- and y-intercepts of the graph.

STEP 4: Use a graphing utility to create a TABLE to find points on the graph around each x-intercept.

STEP 5: Approximate the turning points of the graph.

STEP 6: Use the information in Steps 1 through 5 to draw a complete graph of the function by hand.

STEP 7: Find the domain and the range of the function.

STEP 8: Use the graph to determine where the function is increasing and where it is decreasing.

Now Work PROBLEM 99

5 Build Cubic Models from Data

In Section 4.2 we found the line of best fit from data, and in Section 4.4 we found the quadratic function of best fit. It is also possible to find polynomial functions of best fit. However, most statisticians do not recommend finding polynomial functions of best fit of degree higher than 3.

Data that follow a cubic relation should look like Figure 19(a) or 19(b).

$y = ax^3 + bx^2 + cx + d, a > 0$
(a)

$y = ax^3 + bx^2 + cx + d, a < 0$
(b)

Figure 19 Cubic relation

EXAMPLE 12 A Cubic Function of Best Fit

The data in Table 6 on the next page represent the weekly cost C (in thousands of dollars) of printing x thousand textbooks.

(a) Draw a scatter diagram of the data using x as the independent variable and C as the dependent variable. Comment on the type of relation that may exist between the two variables x and C.

(b) Using a graphing utility, find the cubic function of best fit $C = C(x)$ that models the relation between number of texts and cost.

(c) Graph the cubic function of best fit on your scatter diagram.

(d) Use the function found in part (b) to predict the cost of printing 22 thousand texts per week.

Table 6

Number of Textbooks, x (thousands)	Cost, C ($1000s)
0	100
5	128.1
10	144
13	153.5
17	161.2
18	162.6
20	166.3
23	178.9
25	190.2
27	221.8

Solution

(a) Figure 20 shows the scatter diagram. A cubic relation may exist between the two variables.

(b) Upon executing the CUBIC REGression program, we obtain the results shown in Figure 21. The output that the utility provides shows the equation $y = ax^3 + bx^2 + cx + d$. The cubic function of best fit to the data is $C(x) = 0.0155x^3 - 0.5951x^2 + 9.1502x + 98.4327$.

(c) Figure 22 shows the graph of the cubic function of best fit on the scatter diagram. The function fits the data reasonably well.

(d) Evaluate the function $C(x)$ at $x = 22$.

$$C(22) = 0.0155(22)^3 - 0.5951(22)^2 + 9.1502(22) + 98.4327 \approx 176.8$$

The model predicts that the cost of printing 22 thousand textbooks in a week will be 176.8 thousand dollars—that is, $176,800.

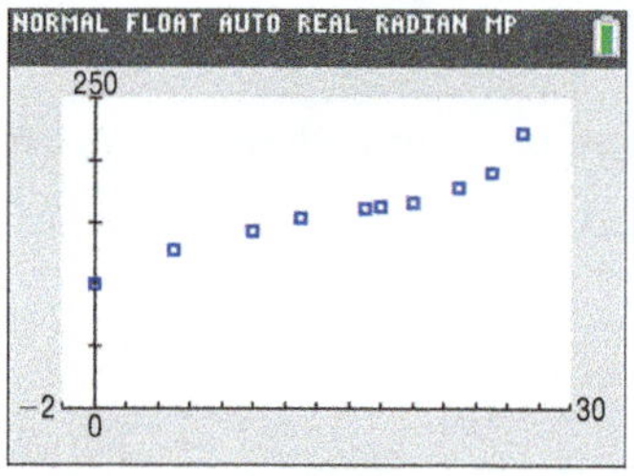

Figure 20

Figure 21

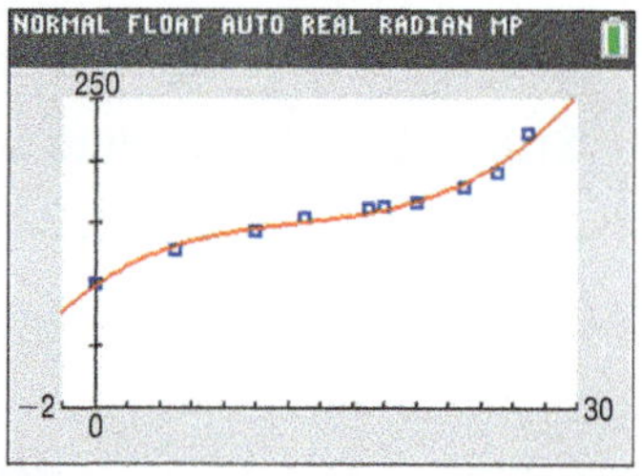

Figure 22

In Example 12, notice that the cubic function of best fit suggests that as the number of textbooks printed increases, cost also increases. That is, $\lim_{x \to \infty} C(x) = \infty$.

5.1 Assess Your Understanding

'Are You Prepared?' *Answers are given at the end of these exercises. If you get a wrong answer, read the pages listed in red.*

1. The intercepts of the equation $9x^2 + 4y = 36$ are _____. (pp. 165–166)

2. Is the expression $4x^3 - 3.6x^2 - \sqrt{2}$ a polynomial? If so, what is its degree? (pp. 41–42)

3. To graph $y = x^2 - 4$, you would shift the graph of $y = x^2$ _______ a distance of ___ units. (pp. 256–258)

4. Use a graphing utility to approximate (rounded to two decimal places) the local maximum value and local minimum value of $f(x) = x^3 - 2x^2 - 4x + 5$, for $-3 \le x \le 3$. (pp. 237–238)

5. ***True or False*** The x-intercepts of the graph of a function $y = f(x)$ are the real solutions of the equation $f(x) = 0$. (pp. 223–225)

6. If $g(5) = 0$, what point is on the graph of g? What is the corresponding x-intercept of the graph of g? (pp. 223–225)

Concepts and Vocabulary

7. The graph of every polynomial function is both ________ and ____________.

8. If r is a real zero of even multiplicity of a polynomial function f, then the graph of f _________ (crosses/touches) the x-axis at r.

9. The graphs of power functions of the form $f(x) = x^n$, where n is an even integer, always contain the points ________, ________, and ________.

10. If r is a solution to the equation $f(x) = 0$, name three additional statements that can be made about f and r, assuming f is a polynomial function.

11. The points at which a graph changes direction (from increasing to decreasing or decreasing to increasing) are called ______________.

12. For the function $f(x) = 3x^4$, $\lim_{x\to-\infty} f(x) =$ ____ and $\lim_{x\to\infty} f(x) =$ ____.

13. If $f(x) = -2x^5 + x^3 - 5x^2 + 7$, then $\lim_{x\to-\infty} f(x) =$ ____ and $\lim_{x\to\infty} f(x) =$ ____.

14. Explain what the notation $\lim_{x\to\infty} f(x) = -\infty$ means.

15. The ________ of a zero is the number of times its corresponding factor occurs.
(a) degree (b) multiplicity (c) turning point (d) limit

16. The graph of $y = 5x^6 - 3x^4 + 2x - 9$ has at most how many turning points?
(a) -9 (b) 14 (c) 6 (d) 5

Skill Building

In Problems 17–28, determine which functions are polynomial functions. For those that are, state the degree. For those that are not, tell why not. Write each polynomial in standard form. Then identify the leading term and the constant term.

17. $f(x) = 4x + x^3$

18. $f(x) = 5x^2 + 4x^4$

19. $g(x) = \dfrac{1 - x^2}{2}$

20. $h(x) = 3 - \dfrac{1}{2}x$

21. $f(x) = 1 - \dfrac{1}{x}$

22. $f(x) = x(x - 1)$

23. $g(x) = x^{3/2} - x^2 + 2$

24. $h(x) = \sqrt{x}(\sqrt{x} - 1)$

25. $F(x) = 5x^4 - \pi x^3 + \dfrac{1}{2}$

26. $F(x) = \dfrac{x^2 - 5}{x^3}$

27. $G(x) = 2(x - 1)^2(x^2 + 1)$

28. $G(x) = -3x^2(x + 2)^3$

In Problems 29–42, use transformations of the graph of $y = x^4$ or $y = x^5$ to graph each function.

29. $f(x) = (x + 1)^4$

30. $f(x) = (x - 2)^5$

31. $f(x) = x^5 - 3$

32. $f(x) = x^4 + 2$

33. $f(x) = \dfrac{1}{2}x^4$

34. $f(x) = 3x^5$

35. $f(x) = -x^5$

36. $f(x) = -x^4$

37. $f(x) = (x - 1)^5 + 2$

38. $f(x) = (x + 2)^4 - 3$

39. $f(x) = 2(x + 1)^4 + 1$

40. $f(x) = \dfrac{1}{2}(x - 1)^5 - 2$

41. $f(x) = 4 - (x - 2)^5$

42. $f(x) = 3 - (x + 2)^4$

In Problems 43–50, form a polynomial function whose real zeros and degree are given. Answers will vary depending on the choice of a leading coefficient.

43. Zeros: $-1, 1, 3$; degree 3

44. Zeros: $-2, 2, 3$; degree 3

45. Zeros: $-3, 0, 4$; degree 3

46. Zeros: $-4, 0, 2$; degree 3

47. Zeros: $-4, -1, 2, 3$; degree 4

48. Zeros: $-3, -1, 2, 5$; degree 4

49. Zeros: -1, multiplicity 1; 3, multiplicity 2; degree 3

50. Zeros: -2, multiplicity 2; 4, multiplicity 1; degree 3

In Problems 51–56, find the polynomial function with the given zeros whose graph passes through the given point.

51. Zeros: $-3, 1, 4$
Point: $(6, 180)$

52. Zeros: $-2, 0, 2$
Point: $(-4, 16)$

53. Zeros: $-1, 0, 2, 4$
Point: $\left(\dfrac{1}{2}, 63\right)$

54. Zeros: $-5, -1, 2, 6$
Point: $\left(\dfrac{5}{2}, 15\right)$

55. Zeros: -1 (multiplicity 2), 1 (multiplicity 2)
Point: $(-2, 45)$

56. Zeros: 0 (multiplicity 1), $-1, 3$ (multiplicity 2)
Point: $(1, -48)$

In Problems 57–68, for each polynomial function:
(a) List each real zero and its multiplicity.
(b) Determine whether the graph crosses or touches the x-axis at each x-intercept.
(c) Determine the maximum number of turning points on the graph.
(d) Determine the end behavior; that is, find the power function that the graph of f resembles for large values of $|x|$.

57. $f(x) = 3(x - 7)(x + 3)^2$

58. $f(x) = 4(x + 4)(x + 3)^3$

59. $f(x) = 4(x^2 + 1)(x - 2)^3$

60. $f(x) = 2(x - 3)(x^2 + 4)^3$

61. $f(x) = -2\left(x + \dfrac{1}{2}\right)^2(x + 4)^3$

62. $f(x) = \left(x - \dfrac{1}{3}\right)^2(x - 1)^3$

63. $f(x) = (x - 5)^3(x + 4)^2$

64. $f(x) = (x + \sqrt{3})^2(x - 2)^4$

65. $f(x) = 3(x^2 + 8)(x^2 + 9)^2$

66. $f(x) = -2(x^2 + 3)^3$

67. $f(x) = -2x^2(x^2 - 2)$

68. $f(x) = 4x(x^2 - 3)$

In Problems 69–72, identify which of the graphs could be the graph of a polynomial function. For those that could, list the real zeros and state the least degree the polynomial can have. For those that could not, say why not.

69.

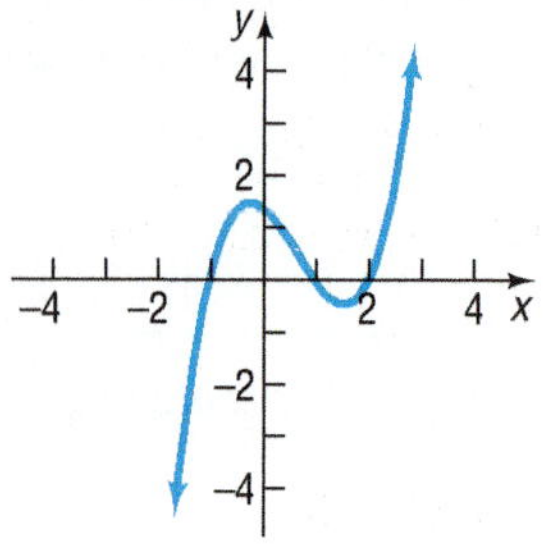

70.

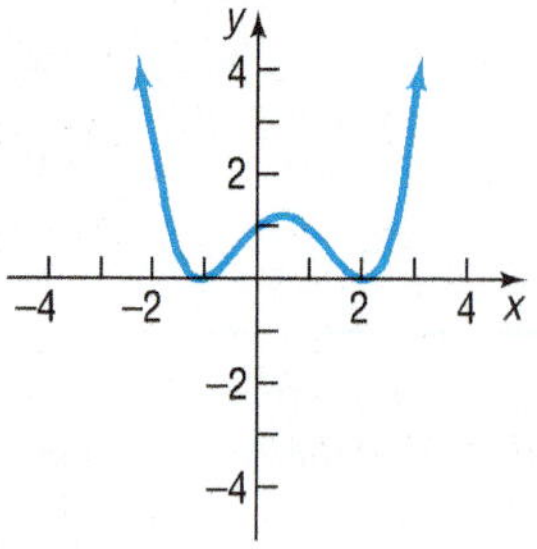

71.

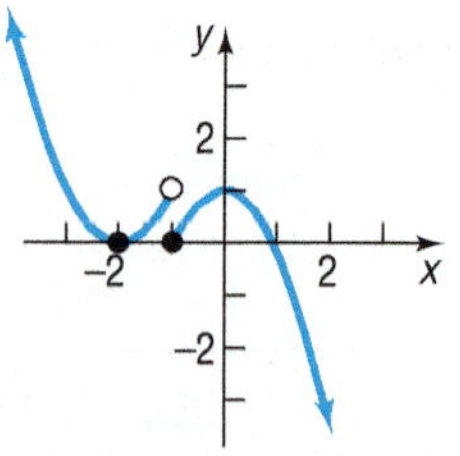

72.

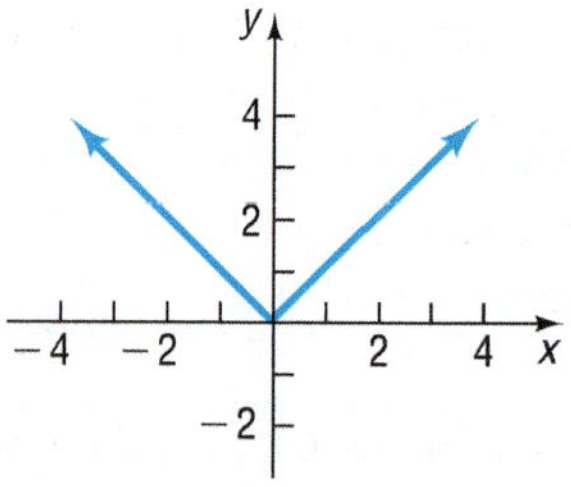

In Problems 73–76, construct a polynomial function that might have the given graph. (More than one answer may be possible.)

73.

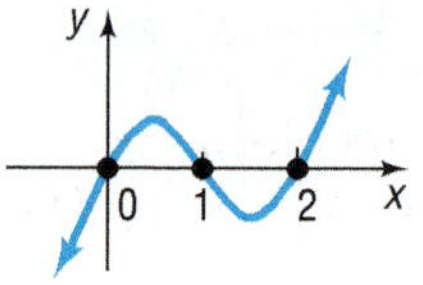

74.

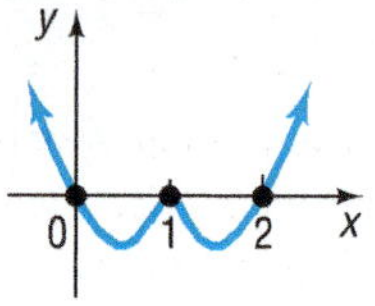

75.

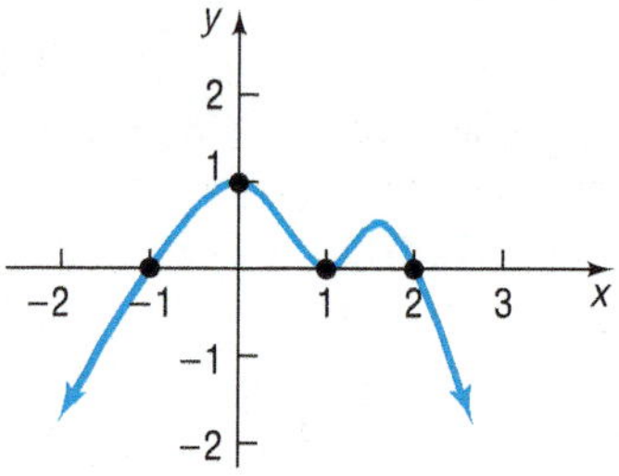

76.

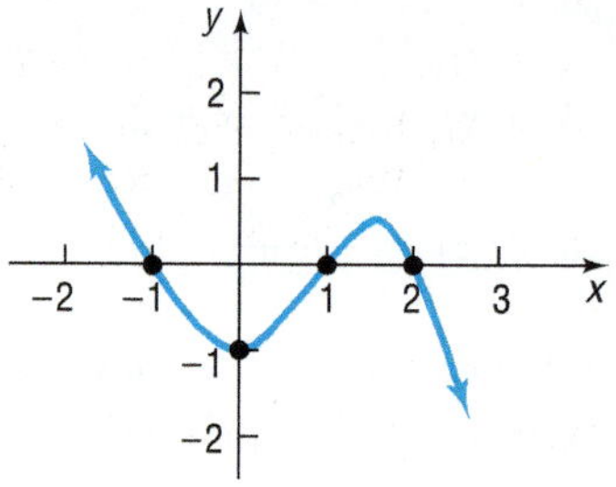

In Problems 77–80, write a polynomial function whose graph is shown (use the smallest degree possible).

77.

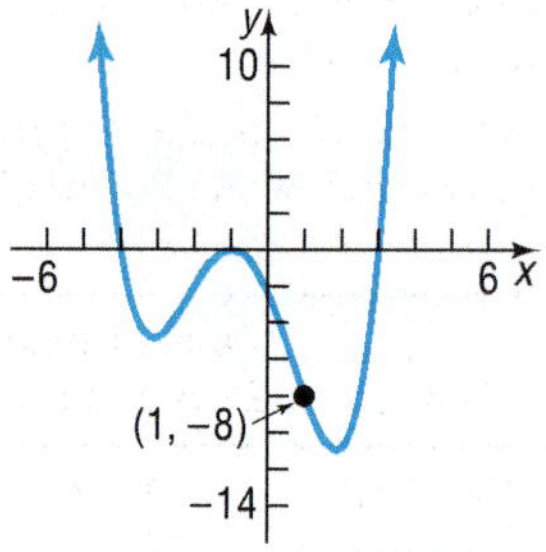

78.

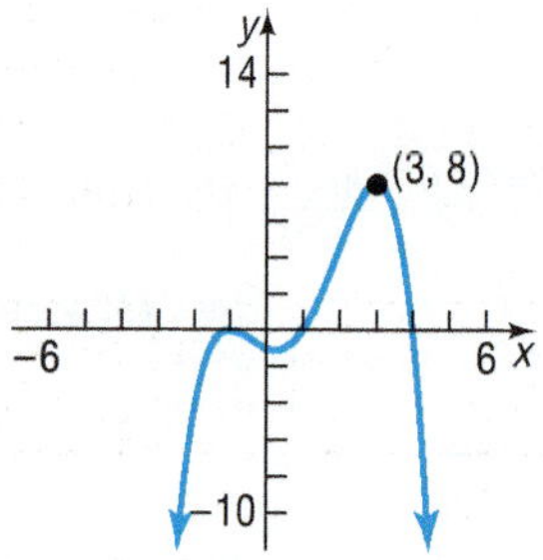

79.

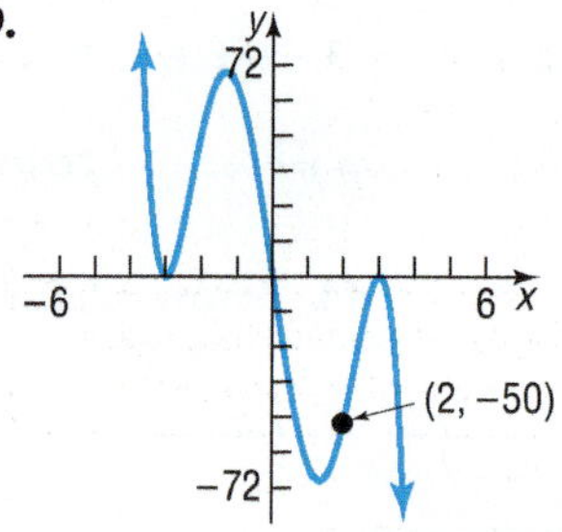

80.

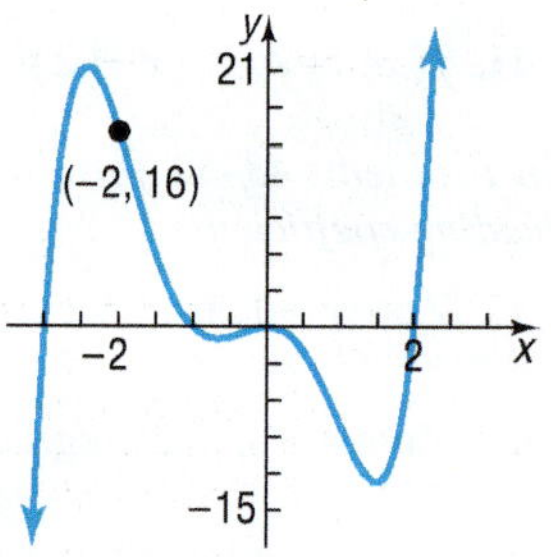

In Problems 81–98, analyze each polynomial function by following Steps 1 through 8 on page 343.

81. $f(x) = x^2(x-3)$

82. $f(x) = x(x+2)^2$

83. $f(x) = (x+4)^2(1-x)$

84. $f(x) = (x-1)(x+3)^2$

85. $f(x) = -2(x+2)(x-2)^3$

86. $f(x) = -\frac{1}{2}(x+4)(x-1)^3$

87. $f(x) = (x+1)(x-2)(x+4)$

88. $f(x) = (x-1)(x+4)(x-3)$

89. $f(x) = x^2(x-2)(x+2)$

90. $f(x) = x^2(x-3)(x+4)$

91. $f(x) = (x+1)^2(x-2)^2$

92. $f(x) = (x-4)^2(x+2)^2$

93. $f(x) = x^2(x+3)(x+1)$

94. $f(x) = x^2(x-3)(x-1)$

95. $f(x) = 5x(x^2-4)(x+3)$

96. $f(x) = (x-2)^2(x+2)(x+4)$

97. $f(x) = x^2(x-2)(x^2+3)$

98. $f(x) = x^2(x^2+1)(x+4)$

In Problems 99–106, analyze each polynomial function f by following Steps 1 through 8 on page 345.

99. $f(x) = x^3 + 0.2x^2 - 1.5876x - 0.31752$

100. $f(x) = x^3 - 0.8x^2 - 4.6656x + 3.73248$

101. $f(x) = x^3 + 2.56x^2 - 3.31x + 0.89$

102. $f(x) = x^3 - 2.91x^2 - 7.668x - 3.8151$

103. $f(x) = x^4 - 2.5x^2 + 0.5625$

104. $f(x) = x^4 - 18.5x^2 + 50.2619$

105. $f(x) = 2x^4 - \pi x^3 + \sqrt{5}x - 4$

106. $f(x) = -1.2x^4 + 0.5x^2 - \sqrt{3}x + 2$

Mixed Practice

In Problems 107–114, analyze each polynomial function by following Steps 1 through 8 on page 343.

[**Hint:** You will need to first factor the polynomial].

107. $f(x) = 4x - x^3$

108. $f(x) = x - x^3$

109. $f(x) = x^3 + x^2 - 12x$

110. $f(x) = x^3 + 2x^2 - 8x$

111. $f(x) = 2x^4 + 12x^3 - 8x^2 - 48x$

112. $f(x) = 4x^3 + 10x^2 - 4x - 10$

113. $f(x) = -x^5 - x^4 + x^3 + x^2$

114. $f(x) = -x^5 + 5x^4 + 4x^3 - 20x^2$

In Problems 115–118, construct a polynomial function f with the given characteristics.

115. Zeros: $-3, 1, 4$; degree 3; y-intercept: 36

116. Zeros: $-4, -1, 2$; degree 3; y-intercept: 16

117. Zeros: -5(multiplicity 2); 2 (multiplicity 1); 4 (multiplicity 1); degree 4; contains the point $(3, 128)$

118. Zeros: -4 (multiplicity 1); 0 (multiplicity 3); 2 (multiplicity 1); degree 5; contains the point $(-2, 64)$

119. $G(x) = (x + 3)^2(x - 2)$

(a) Identify the x-intercepts of the graph of G.

(b) What are the x-intercepts of the graph of $y = G(x + 3)$?

120. $h(x) = (x + 2)(x - 4)^3$

(a) Identify the x-intercepts of the graph of h.

(b) What are the x-intercepts of the graph of $y = h(x - 2)$?

Applications and Extensions

121. Hurricanes In 2012, Hurricane Sandy struck the East Coast of the United States, killing 147 people and causing an estimated \$75 billion in damage. With a gale diameter of about 1000 miles, it was the largest ever to form over the Atlantic Basin. The accompanying data represent the number of major hurricane strikes in the Atlantic Basin (category 3, 4, or 5) each decade from 1921 to 2010.

Decade, x	Major Hurricanes Striking Atlantic Basin, H
1921–1930, 1	17
1931–1940, 2	16
1941–1950, 3	29
1951–1960, 4	33
1961–1970, 5	27
1971–1980, 6	16
1981–1990, 7	16
1991–2000, 8	27
2001–2010, 9	33

Source: *National Oceanic & Atmospheric Administration*

(a) Draw a scatter diagram of the data. Comment on the type of relation that may exist between the two variables.

(b) Use a graphing utility to find the cubic function of best fit that models the relation between decade and number of major hurricanes.

(c) Use the model found in part (b) to predict the number of major hurricanes that struck the Atlantic Basin between 1961 and 1970.

(d) With a graphing utility, draw a scatter diagram of the data and then graph the cubic function of best fit on the scatter diagram.

(e) Concern has risen about the increase in the number and intensity of hurricanes, but some scientists believe this is just a natural fluctuation that could last another decade or two. Use your model to predict the number of major hurricanes that will strike the Atlantic Basin between 2011 and 2020. Is your result reasonable? How does this result suggest using end behavior of models to make long-term predictions is dangerous?

122. Poverty Rates The following data represent the percentage of people in the United States living below the poverty level.

Year, t	Percent below Poverty Level, p	Year, t	Percent below Poverty Level, p
1990, 1	13.5	2001, 12	11.7
1991, 2	14.2	2002, 13	12.1
1992, 3	14.8	2003, 14	12.5
1993, 4	15.1	2004, 15	12.7
1994, 5	14.5	2005, 16	12.6
1995, 6	13.8	2006, 17	12.3
1996, 7	13.7	2007, 18	12.5
1997, 8	13.3	2008, 19	13.2
1998, 9	12.7	2009, 20	14.3
1999, 10	11.9	2010, 21	15.1
2000, 11	11.3	2011, 22	15.0

Source: *U.S. Census Bureau*

(a) With a graphing utility, draw a scatter diagram of the data. Comment on the type of relation that appears to exist between the two variables.

(b) Decide on a function of best fit to these data (linear, quadratic, or cubic), and use this function to predict the percentage of people in the United States who were living below the poverty level in 2013 $(t = 24)$. Compare your prediction to the actual value of 14.5.

(c) Draw the function of best fit on the scatter diagram drawn in part (a).

123. Temperature The following data represent the temperature T (°Fahrenheit) in Kansas City, Missouri, x hours after midnight on March 15, 2015.

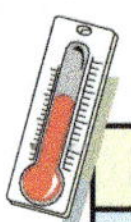

Hours after Midnight, x	Temperature (°F), T
3	43.0
6	39.0
9	44.1
12	62.1
15	71.1
18	71.6
21	60.1
24	59.0

Source: *The Weather Underground*

(a) Draw a scatter diagram of the data. Comment on the type of relation that may exist between the two variables.
(b) Find the average rate of change in temperature from 9 AM to 12 noon.
(c) What is the average rate of change in temperature from 3 PM to 6 PM?
(d) Decide on a function of best fit to these data (linear, quadratic, or cubic) and use this function to predict the temperature at 5 PM.
(e) With a graphing utility, draw a scatter diagram of the data and then graph the function of best fit on the scatter diagram.
(f) Interpret the y-intercept.

124. Future Value of Money Suppose that you make deposits of \$500 at the beginning of every year into an Individual Retirement Account (IRA) earning interest r (expressed as a decimal). At the beginning of the first year, the value of the account will be \$500; at the beginning of the second year, the value of the account, will be

$$\underbrace{\$500 + \$500r}_{\text{Value of 1st deposit}} + \underbrace{\$500}_{\text{Value of 2nd deposit}} = \$500(1+r) + \$500 = 500r + 1000$$

(a) Verify that the value of the account at the beginning of the third year is $T(r) = 500r^2 + 1500r + 1500$.
(b) The account value at the beginning of the fourth year is $F(r) = 500r^3 + 2000r^2 + 3000r + 2000$. If the annual rate of interest is $5\% = 0.05$, what will be the value of the account at the beginning of the fourth year?

125. A Geometric Series In calculus, you will learn that certain functions can be approximated by polynomial functions. We will explore one such function now.
(a) Using a graphing utility, create a table of values with

$$Y_1 = f(x) = \frac{1}{1-x} \text{ and } Y_2 = g_2(x) = 1 + x + x^2 + x^3$$

for $-1 < x < 1$ with $\Delta\text{Tbl} = 0.1$.
(b) Using a graphing utility, create a table of values with

$$Y_1 = f(x) = \frac{1}{1-x} \text{ and}$$
$$Y_2 = g_3(x) = 1 + x + x^2 + x^3 + x^4$$

for $-1 < x < 1$ with $\Delta\text{Tbl} = 0.1$.
(c) Using a graphing utility, create a table of values with

$$Y_1 = f(x) = \frac{1}{1-x} \text{ and}$$
$$Y_2 = g_4(x) = 1 + x + x^2 + x^3 + x^4 + x^5$$

for $-1 < x < 1$ with $\Delta\text{Tbl} = 0.1$.
(d) What do you notice about the values of the function as more terms are added to the polynomial? Are there some values of x for which the approximations are better?

126. If $f(x) = x^3$, graph $f(2x)$.

Explaining Concepts: Discussion and Writing

127. Write a few paragraphs that provide a general strategy for graphing a polynomial function. Be sure to mention the following: degree, intercepts, end behavior, and turning points.

128. Make up a polynomial that has the following characteristics: crosses the x-axis at -1 and 4, touches the x-axis at 0 and 2, and is above the x-axis between 0 and 2. Give your polynomial to a fellow classmate and ask for a written critique.

129. Make up two polynomials, not of the same degree, with the following characteristics: crosses the x-axis at -2, touches the x-axis at 1, and is above the x-axis between -2 and 1. Give your polynomials to a fellow classmate and ask for a written critique.

130. The graph of a polynomial function is always smooth and continuous. Name a function studied earlier that is smooth but not continuous. Name one that is continuous but not smooth.

131. Which of the following statements are true regarding the graph of the cubic polynomial $f(x) = x^3 + bx^2 + cx + d$? (Give reasons for your conclusions.)
(a) It intersects the y-axis in one and only one point.
(b) It intersects the x-axis in at most three points.
(c) It intersects the x-axis at least once.
(d) For $|x|$ very large, it behaves like the graph of $y = x^3$.
(e) It is symmetric with respect to the origin.
(f) It passes through the origin.

132. The illustration shows the graph of a polynomial function.

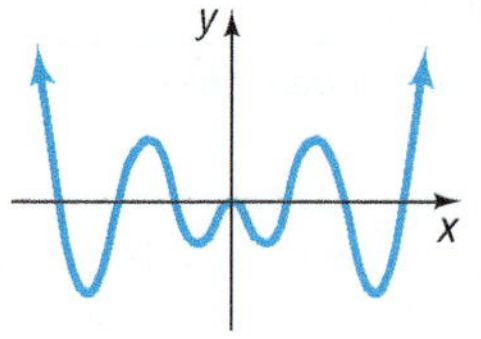

(a) Is the degree of the polynomial even or odd?
(b) Is the leading coefficient positive or negative?
(c) Is the function even, odd, or neither?
(d) Why is x^2 necessarily a factor of the polynomial?
(e) What is the minimum degree of the polynomial?
(f) Formulate five different polynomials whose graphs could look like the one shown. Compare yours to those of other students. What similarities do you see? What differences?

133. Design a polynomial function with the following characteristics: degree 6; four distinct real zeros, one of multiplicity 3; y-intercept 3; behaves like $y = -5x^6$ for large values of $|x|$. Is this polynomial unique? Compare your polynomial with those of other students. What terms will be the same as everyone else's? Add some more characteristics, such as symmetry or naming the real zeros. How does this modify the polynomial?

134. Can the graph of a polynomial function have no y-intercept? Can it have no x-intercepts? Explain.

Retain Your Knowledge

Problems 135–138 are based on material learned earlier in the course. The purpose of these problems is to keep the material fresh in your mind so that you are better prepared for the final exam.

135. Find the equation of the line that contains the point $(2, -3)$ and is perpendicular to the line $5x - 2y = 6$.

136. Find the domain of the function $h(x) = \dfrac{x-3}{x+5}$.

137. Use the quadratic formula to find the zeros of the function $f(x) = 4x^2 + 8x - 3$.

138. Solve: $|5x - 3| = 7$.

'Are You Prepared?' Answers

1. $(-2, 0), (2, 0), (0, 9)$ **2.** Yes; 3 **3.** Down; 4 **4.** Local maximum value 6.48 at $x = -0.67$; local minimum value -3 at $x = 2$

5. True **6.** $(5, 0); 5$

5.2 The Real Zeros of a Polynomial Function

PREPARING FOR THIS SECTION *Before getting started, review the following:*

- Evaluating Functions (Section 3.1 pp. 210–213)
- Factoring Polynomials (Chapter R, Section R.5, pp. 50–56)
- Synthetic Division (Chapter R, Section R.6, pp. 59–62)
- Polynomial Division (Chapter R, Section R.4, pp. 45–48)
- Solve a Quadratic Equation (Section 1.3, pp. 110–116)

Now Work the 'Are You Prepared?' problems on page 363.

OBJECTIVES
1 Use the Remainder and Factor Theorems (p. 352)
2 Use Descartes' Rule of Signs to Determine the Number of Positive and the Number of Negative Real Zeros of a Polynomial Function (p. 354)
3 Use the Rational Zeros Theorem to List the Potential Rational Zeros of a Polynomial Function (p. 355)
4 Find the Real Zeros of a Polynomial Function (p. 356)
5 Solve Polynomial Equations (p. 359)
6 Use the Theorem for Bounds on Zeros (p. 359)
7 Use the Intermediate Value Theorem (p. 362)

In Section 5.1, we were able to identify the real zeros of a polynomial function because either the polynomial function was in factored form or it could be easily factored. But how do we find the real zeros of a polynomial function if it is not factored or cannot be easily factored?

Recall that if r is a real zero of a polynomial function f then $f(r) = 0$, r is an x-intercept of the graph of f, $x - r$ is a factor of f, and r is a solution of the equation $f(x) = 0$. For example, if $x - 4$ is a factor of f, then 4 is a real zero of f and 4 is a solution to the equation $f(x) = 0$. For polynomial functions, we have seen the importance of the real zeros for graphing. In most cases, however, the real zeros of a polynomial function are difficult to find using algebraic methods. No nice formulas like the quadratic formula are available to help us find zeros for polynomial functions of degree 3 or higher. Formulas do exist for solving any third- or fourth-degree polynomial equation, but they are somewhat complicated. No general formulas exist for polynomial equations of degree 5 or higher. Refer to the Historical Feature at the end of this section for more information.

1 Use the Remainder and Factor Theorems

When one polynomial (the dividend) is divided by another (the divisor), a quotient polynomial and a remainder are obtained, the remainder being either the zero polynomial or a polynomial whose degree is less than the degree of the divisor. To check, verify that

$$(\text{Quotient})(\text{Divisor}) + \text{Remainder} = \text{Dividend}$$

This checking routine is the basis for a famous theorem called the **division algorithm* for polynomials**, which we now state without proof.

THEOREM

Division Algorithm for Polynomials

If $f(x)$ and $g(x)$ denote polynomial functions and if $g(x)$ is a polynomial function whose degree is greater than zero, then there are unique polynomial functions $q(x)$ and $r(x)$ such that

$$\frac{f(x)}{g(x)} = q(x) + \frac{r(x)}{g(x)} \quad \text{or} \quad \underset{\text{dividend}}{f(x)} = \underset{\text{quotient}}{q(x)}\underset{\text{divisor}}{g(x)} + \underset{\text{remainder}}{r(x)} \tag{1}$$

where $r(x)$ is either the zero polynomial or a polynomial function of degree less than that of $g(x)$.

In equation (1), $f(x)$ is the **dividend**, $g(x)$ is the **divisor**, $q(x)$ is the **quotient**, and $r(x)$ is the **remainder**.

If the divisor $g(x)$ is a first-degree polynomial function of the form

$$g(x) = x - c \quad c \text{ a real number}$$

then the remainder $r(x)$ is either the zero polynomial or a polynomial function of degree 0. As a result, for such divisors, the remainder is some number, say R, and

$$f(x) = (x - c)q(x) + R \tag{2}$$

This equation is an identity in x and is true for all real numbers x. Suppose that $x = c$. Then equation (2) becomes

$$f(c) = (c - c)q(c) + R$$
$$f(c) = R$$

Substitute $f(c)$ for R in equation (2) to obtain

$$f(x) = (x - c)q(x) + f(c) \tag{3}$$

which proves the **Remainder Theorem**.

REMAINDER THEOREM

Let f be a polynomial function. If $f(x)$ is divided by $x - c$, then the remainder is $f(c)$.

*A systematic process in which certain steps are repeated a finite number of times is called an **algorithm**. For example, long division is an algorithm.

EXAMPLE 1 Using the Remainder Theorem

Find the remainder if $f(x) = x^3 - 4x^2 - 5$ is divided by

(a) $x - 3$ (b) $x + 2$

Solution (a) We could use long division or synthetic division, but it is easier to use the Remainder Theorem, which says that the remainder is $f(3)$.

$$f(3) = (3)^3 - 4(3)^2 - 5 = 27 - 36 - 5 = -14$$

The remainder is -14.

(b) To find the remainder when $f(x)$ is divided by $x + 2 = x - (-2)$, evaluate $f(-2)$.

$$f(-2) = (-2)^3 - 4(-2)^2 - 5 = -8 - 16 - 5 = -29$$

The remainder is -29. ■

Compare the method used in Example 1(a) with the method used in Example 1 of Chapter R, Section R.6. Which method do you prefer? Give reasons.

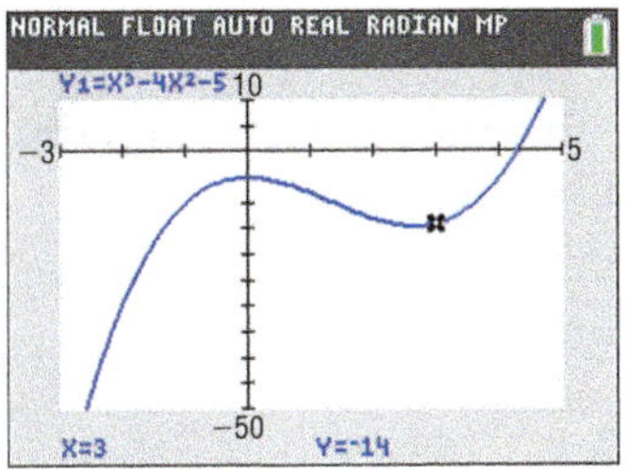

Figure 23 Value of a function

COMMENT A graphing utility provides another way to find the value of a function using the eVALUEate feature. Consult your manual for details. See Figure 23 for the result of Example 1(a). ■

An important and useful consequence of the Remainder Theorem is the **Factor Theorem**.

FACTOR THEOREM

Let f be a polynomial function. Then $x - c$ is a factor of $f(x)$ if and only if $f(c) = 0$. ■

The Factor Theorem actually consists of two separate statements:

1. If $f(c) = 0$, then $x - c$ is a factor of $f(x)$.
2. If $x - c$ is a factor of $f(x)$, then $f(c) = 0$.

The proof requires two parts.

Proof

1. Suppose that $f(c) = 0$. Then, by equation (3), we have

$$f(x) = (x - c)q(x)$$

for some polynomial function $q(x)$. That is, $x - c$ is a factor of $f(x)$.

2. Suppose that $x - c$ is a factor of $f(x)$. Then there is a polynomial function q such that

$$f(x) = (x - c)q(x)$$

Replacing x by c, we find that

$$f(c) = (c - c)q(c) = 0 \cdot q(c) = 0$$

This completes the proof. ■

One use of the Factor Theorem is to determine whether a polynomial has a particular factor.

EXAMPLE 2 Using the Factor Theorem

Use the Factor Theorem to determine whether the function

$$f(x) = 2x^3 - x^2 + 2x - 3$$

has the factor

(a) $x - 1$ (b) $x + 2$

Solution The Factor Theorem states that if $f(c) = 0$ then $x - c$ is a factor.

(a) Because $x - 1$ is of the form $x - c$ with $c = 1$, find the value of $f(1)$. We choose to use substitution.

$$f(1) = 2(1)^3 - (1)^2 + 2(1) - 3 = 2 - 1 + 2 - 3 = 0$$

See also Figure 24(a). By the Factor Theorem, $x - 1$ is a factor of $f(x)$.

(b) To test the factor $x + 2$, first write it in the form $x - c$. Since $x + 2 = x - (-2)$, find the value of $f(-2)$. See Figure 24(b). Because $f(-2) = -27 \neq 0$, conclude from the Factor Theorem that $x - (-2) = x + 2$ is not a factor of $f(x)$. ■

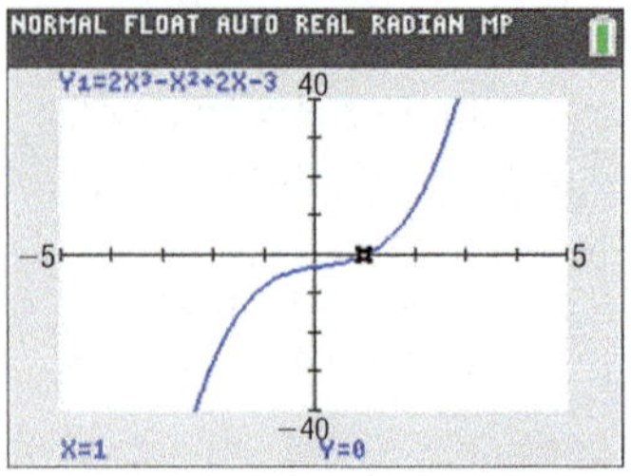

(a)

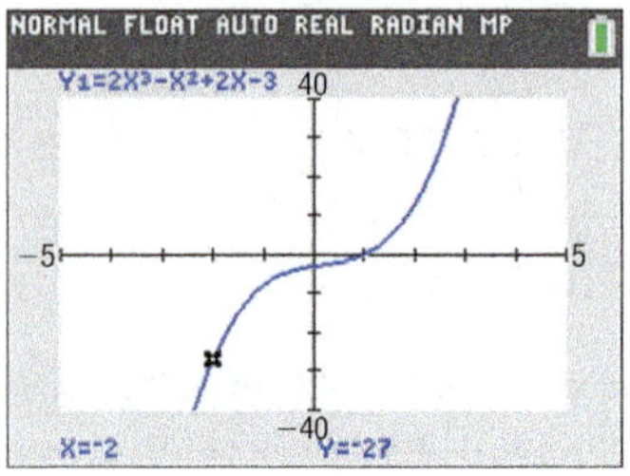

(b)

Figure 24

Now Work PROBLEM 11

In Example 2(a), $x - 1$ was found to be a factor of f. To write f in factored form, use long division or synthetic division. Using synthetic division,

$$\begin{array}{r|rrrr} 1 & 2 & -1 & 2 & -3 \\ & & 2 & 1 & 3 \\ \hline & 2 & 1 & 3 & 0 \end{array}$$

The quotient is $q(x) = 2x^2 + x + 3$ with a remainder of 0, as expected. Write f in factored form as

$$f(x) = 2x^3 - x^2 + 2x - 3 = (x - 1)(2x^2 + x + 3)$$

WARNING Remember that in order for synthetic division to be used, the divisor must be of the form $x - c$. ■

The next theorem concerns the number of real zeros that a polynomial function may have. In counting the zeros of a polynomial function, count each zero as many times as its multiplicity.

THEOREM

Number of Real Zeros

A polynomial function cannot have more real zeros than its degree. ■

Proof The proof is based on the Factor Theorem. If r is a real zero of a polynomial function f, then $f(r) = 0$ and, hence, $x - r$ is a factor of $f(x)$. Each real zero corresponds to a factor of degree 1. Because f cannot have more first-degree factors than its degree, the result follows. ■

2 Use Descartes' Rule of Signs to Determine the Number of Positive and the Number of Negative Real Zeros of a Polynomial Function

Descartes' Rule of Signs provides information about the number and location of the real zeros of a polynomial function written in standard form (omitting terms with a 0 coefficient). It utilizes the number of variations in the sign of the coefficients of $f(x)$ and $f(-x)$.

For example, the following polynomial function has two variations in the signs of the coefficients.

$$f(x) = -3x^7 + 4x^4 + 3x^2 - 2x - 1$$

$-$ to $+$ $+$ to $-$

Replacing x by $-x$ gives

$$\begin{aligned} f(-x) &= -3(-x)^7 + 4(-x)^4 + 3(-x)^2 - 2(-x) - 1 \\ &= 3x^7 + 4x^4 + 3x^2 + 2x - 1 \end{aligned}$$

$+$ to $-$

which has one variation in sign.

THEOREM

Descartes' Rule of Signs

Let f denote a polynomial function written in standard form.

- The number of positive real zeros of f either equals the number of variations in the sign of the nonzero coefficients of $f(x)$ or else equals that number less an even integer.
- The number of negative real zeros of f either equals the number of variations in the sign of the nonzero coefficients of $f(-x)$ or else equals that number less an even integer.

We shall not prove Descartes' Rule of Signs. Let's see how it is used.

EXAMPLE 3

Using the Number of Real Zeros Theorem and Descartes' Rule of Signs

Discuss the real zeros of $f(x) = 3x^7 - 4x^4 + 3x^3 + 2x^2 - x - 3$.

Solution Because the polynomial is of degree 7, by the Number of Real Zeros Theorem there are at most seven real zeros. Since there are three variations in the sign of the nonzero coefficients of $f(x)$, by Descartes' Rule of Signs we expect either three positive real zeros or one positive real zero. To continue, look at $f(-x)$.

$$f(-x) = -3x^7 - 4x^4 - 3x^3 + 2x^2 + x - 3$$

There are two variations in sign, so we expect either two negative real zeros or no negative real zeros. Equivalently, we now know that the graph of f has either three positive x-intercepts or one positive x-intercept and two negative x intercepts or no negative x-intercepts.

Now Work PROBLEM 21

3 Use the Rational Zeros Theorem to List the Potential Rational Zeros of a Polynomial Function

The next result, called the **Rational Zeros Theorem**, provides information about the rational zeros of a polynomial function *with integer coefficients*.

THEOREM

Rational Zeros Theorem

Let f be a polynomial function of degree 1 or higher of the form

$$f(x) = a_nx^n + a_{n-1}x^{n-1} + \cdots + a_1x + a_0 \qquad a_n \neq 0, \quad a_0 \neq 0$$

where each coefficient is an integer. If $\frac{p}{q}$, in lowest terms, is a rational zero of f, then p must be a factor of a_0 and q must be a factor of a_n.

EXAMPLE 4

Listing Potential Rational Zeros

List the potential rational zeros of

$$f(x) = 2x^3 + 11x^2 - 7x - 6$$

Solution Because f has integer coefficients, we may use the Rational Zeros Theorem. First, list all the integers p that are factors of the constant term $a_0 = -6$ and all the integers q that are factors of the leading coefficient $a_3 = 2$.

$$p: \quad \pm 1, \pm 2, \pm 3, \pm 6 \qquad \text{Factors of } -6$$

$$q: \quad \pm 1, \pm 2 \qquad \text{Factors of } 2$$

Now form all possible ratios $\frac{p}{q}$.

$$\frac{p}{q}: \pm 1, \pm 2, \pm 3, \pm 6, \pm\frac{1}{2}, \pm\frac{3}{2}$$

If f has a rational zero, it will be found in this list, which contains 12 possibilities. ■

Now Work PROBLEM 33

In Words

For the polynomial function $f(x) = 2x^3 + 11x^2 - 7x - 6$, we know 5 is not a zero, because 5 is not in the list of potential rational zeros. However, -1 may or may not be a zero.

Be sure that you understand what the Rational Zeros Theorem says: For a polynomial function with integer coefficients, *if* there is a rational zero, it is one of those listed. It may be the case that the function does not have any rational zeros.

The Rational Zeros Theorem provides a list of potential rational zeros of a function f. If we graph f, we can get a better sense of the location of the x-intercepts and test to see if they are rational. We can also use the potential rational zeros to select our initial viewing window to graph f and then adjust the window based on the results. The graphs shown throughout the text will be those obtained after setting the final viewing window.

4 Find the Real Zeros of a Polynomial Function

EXAMPLE 5 How to Find the Real Zeros of a Polynomial Function

Find the real zeros of the polynomial function $f(x) = 2x^3 + 11x^2 - 7x - 6$. Write f in factored form.

Step-by-Step Solution

Step 1: Determine the maximum number of zeros. Also determine the number of positive and negative real zeros.

Since f is a polynomial function of degree 3, there are at most three real zeros. From Descartes' Rule of Signs, there is one positive real zero. Also, since $f(-x) = -2x^3 + 11x^2 + 7x - 6$, there are two negative real zeros or no negative real zeros.

Step 2: If the polynomial function has integer coefficients, use the Rational Zeros Theorem to identify those rational numbers that potentially can be zeros.

List the potential rational zeros obtained in Example 4:

$$\pm 1, \pm 2, \pm 3, \pm 6, \pm\frac{1}{2}, \pm\frac{3}{2}$$

Step 3: Using a graphing utility, graph the polynomial function.

Figure 25 shows the graph of f. We see that f has three zeros: one near -6, one between -1 and 0, and one near 1.

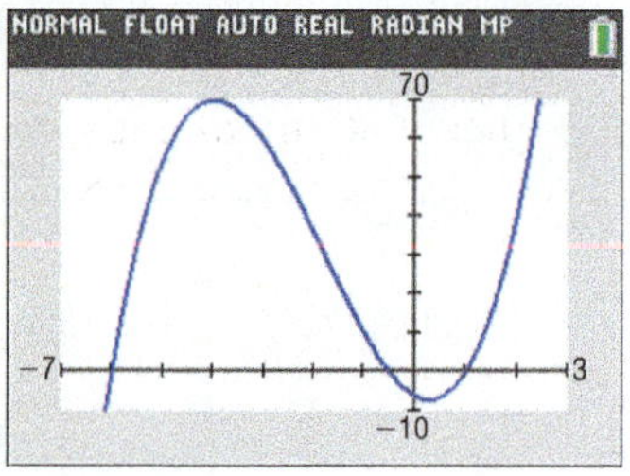

Figure 25 $Y_1 = 2x^3 + 11x^2 - 7x - 6$

Step 4: Use the Factor Theorem to determine if the potential rational zero is a zero. If it is, use synthetic division or long division to factor the polynomial function. Repeat Step 4 until all the zeros of the polynomial function have been identified and the polynomial function is completely factored.

From our list of potential rational zeros, we test -6 to determine if it is a zero of f. Because

$$\begin{aligned} f(-6) &= 2(-6)^3 + 11(-6)^2 - 7(-6) - 6 \\ &= 2(-216) + 11(36) + 42 - 6 \\ &= -432 + 396 + 36 \\ &= 0 \end{aligned}$$

we know that -6 is a zero and $x - (-6) = x + 6$ is a factor of f. Use long division or synthetic division to factor f. (We will not show the division here, but you are encouraged to verify the results shown.) After dividing f by $x + 6$, the quotient is $2x^2 - x - 1$, so

$$\begin{aligned} f(x) &= 2x^3 + 11x^2 - 7x - 6 \\ &= (x + 6)(2x^2 - x - 1) \end{aligned}$$

Now any solution of the equation $2x^2 - x - 1 = 0$ will be a zero of f. Because of this, the equation $2x^2 - x - 1 = 0$ is called a **depressed equation** of f. Because any solution to the equation $2x^2 - x - 1 = 0$ is a zero of f, work with the depressed equation to find the remaining zeros of f.

The depressed equation $2x^2 - x - 1 = 0$ is a quadratic equation with discriminant $b^2 - 4ac = (-1)^2 - 4(2)(-1) = 9 > 0$. The equation has two real solutions, which can be found by factoring.

$$2x^2 - x - 1 = (2x + 1)(x - 1) = 0$$

$$2x + 1 = 0 \quad \text{or} \quad x - 1 = 0$$

$$x = -\frac{1}{2} \quad \text{or} \quad x = 1$$

The zeros of f are -6, $-\frac{1}{2}$, and 1.

Factor f completely as follows:

$$f(x) = 2x^3 + 11x^2 - 7x - 6 = (x + 6)(2x^2 - x - 1) = (x + 6)(2x + 1)(x - 1)$$

Notice that the three zeros of f are in the list of potential rational zeros in Step 2, and confirm what was expected from Descartes' Rule of Signs. ■

SUMMARY

Steps for Finding the Real Zeros of a Polynomial Function

STEP 1: Use the degree of the polynomial function to determine the maximum number of zeros. Use Descartes' Rule of Signs to determine the number of positive and negative real zeros.

STEP 2: If the polynomial function has integer coefficients, use the Rational Zeros Theorem to identify those rational numbers that potentially can be zeros.

STEP 3: Graph the polynomial function using a graphing utility to find the best choice of potential rational zeros to test.

STEP 4: Use the Factor Theorem to determine if the potential rational zero is a zero. If it is, use synthetic division or long division to factor the polynomial function. Each time that a zero (and thus a factor) is found, repeat Step 4 on the depressed equation. In attempting to find the zeros, remember to use (if possible) the factoring techniques that you already know (special products, factoring by grouping, and so on).

EXAMPLE 6 Finding the Real Zeros of a Polynomial Function

Find the real zeros of $f(x) = x^6 + 4x^5 - 16x^3 - 37x^2 - 84x - 84$. Write f in factored form.

Solution

STEP 1: There are at most six real zeros. There is one positive real zero and there are five, three, or one negative real zeros.

STEP 2: To obtain the list of potential rational zeros, write the factors p of $a_0 = -84$ and the factors q of the leading coefficient $a_6 = 1$.

p: $\pm1, \pm2, \pm3, \pm4, \pm6, \pm7, \pm12, \pm14, \pm21, \pm28, \pm42, \pm84$

q: ±1

The potential rational zeros consist of all possible quotients $\frac{p}{q}$:

$\frac{p}{q}$: $\pm1, \pm2, \pm3, \pm4, \pm6, \pm7, \pm12, \pm14, \pm21, \pm28, \pm42, \pm84$

Step 3: Figure 26 shows the graph of f. The graph has the characteristics expected of the given polynomial function of degree 6: no more than five turning points, y-intercept -84, and it behaves like $y = x^6$ for large $|x|$.

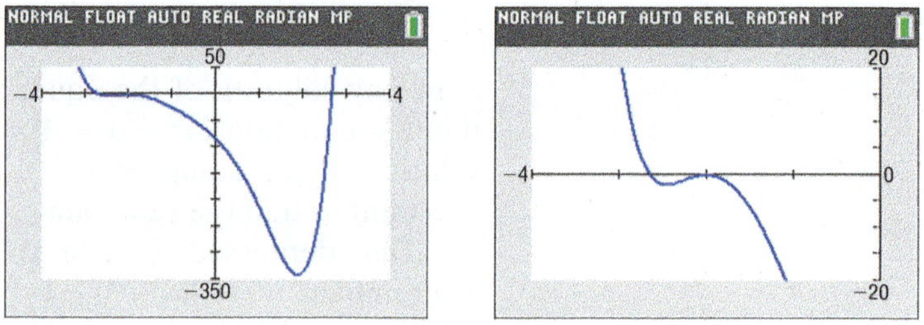

(a) **(b)**

Figure 26 $Y_1 = x^6 + 4x^5 - 16x^3 - 37x^2 - 84x - 84$

Step 4: From Figure 26(b), -2 appears to be a zero. Additionally, -2 is a potential rational zero. Evaluate $f(-2)$ and find that $f(-2) = 0$. By the Factor Theorem, $x + 2$ is a factor of f. We use synthetic division to factor f.

$$\begin{array}{r|rrrrrrr} -2 & 1 & 4 & 0 & -16 & -37 & -84 & -84 \\ & & -2 & -4 & 8 & 16 & 42 & 84 \\ \hline & 1 & 2 & -4 & -8 & -21 & -42 & 0 \end{array}$$

Factor f as

$$f(x) = (x + 2)(x^5 + 2x^4 - 4x^3 - 8x^2 - 21x - 42)$$

Now work with the first depressed equation:

$$q_1(x) = x^5 + 2x^4 - 4x^3 - 8x^2 - 21x - 42 = 0$$

Repeat Step 4: In looking back at Figure 26(b), it appears that -2 might be a zero of even multiplicity. Check the potential rational zero -2 again using synthetic division.

$$\begin{array}{r|rrrrrr} -2 & 1 & 2 & -4 & -8 & -21 & -42 \\ & & -2 & 0 & 8 & 0 & 42 \\ \hline & 1 & 0 & -4 & 0 & -21 & 0 \end{array}$$

Since $q_1(-2) = 0$, then $x + 2$ is a factor and

$$f(x) = (x + 2)(x + 2)(x^4 - 4x^2 - 21)$$

Repeat Step 4: The depressed equation $q_2(x) = x^4 - 4x^2 - 21 = 0$ can be factored.

$$x^4 - 4x^2 - 21 = (x^2 - 7)(x^2 + 3) = 0$$
$$x^2 - 7 = 0 \quad \text{or} \quad x^2 + 3 = 0$$
$$x^2 = 7$$
$$x = \pm\sqrt{7}$$

Since $x^2 + 3 = 0$ has no real solutions, the real zeros of f are $-\sqrt{7}, \sqrt{7}$, and -2, with -2 being a zero of multiplicity 2. The factored form of f is

$$\begin{aligned} f(x) &= x^6 + 4x^5 - 16x^3 - 37x^2 - 84x - 84 \\ &= (x + 2)^2(x + \sqrt{7})(x - \sqrt{7})(x^2 + 3) \end{aligned}$$

Note that one positive real zero and three negative real zeros agrees with our result from Step 1. ■

Now Work PROBLEM 51

5 Solve Polynomial Equations

EXAMPLE 7 **Solving a Polynomial Equation**

Solve the equation: $x^6 + 4x^5 - 16x^3 - 37x^2 - 84x - 84 = 0$

Solution The solutions of this equation are the zeros of the polynomial function

$$f(x) = x^6 + 4x^5 - 16x^3 - 37x^2 - 84x - 84$$

Using the result of Example 6, the real zeros are $-2, -\sqrt{7}$, and $\sqrt{7}$. These are the real solutions of the equation $x^6 + 4x^5 - 16x^3 - 37x^2 - 84x - 84 = 0$.

Now Work PROBLEM 75

In Example 6, the quadratic factor $x^2 + 3$ that appears in the factored form of $f(x)$ is called *irreducible,* because the polynomial $x^2 + 3$ cannot be factored over the real numbers. In general, a quadratic factor $ax^2 + bx + c$ is **irreducible** if it cannot be factored over the real numbers, that is, if it is prime over the real numbers.

Refer again to Examples 5 and 6. The polynomial function of Example 5 has three real zeros, and its factored form contains three linear factors. The polynomial function of Example 6 has three distinct real zeros, and its factored form contains three distinct linear factors and one irreducible quadratic factor.

THEOREM

Every polynomial function (with real coefficients) can be uniquely factored into a product of linear factors and/or irreducible quadratic factors.

We prove this result in Section 5.3, and, in fact, shall draw several additional conclusions about the zeros of a polynomial function. One conclusion is worth noting now. If a polynomial function (with real coefficients) is of odd degree, then it must contain at least one linear factor. (Do you see why? Consider the end behavior of polynomial functions of odd degree.) This means that it must have at least one real zero.

COROLLARY

A polynomial function (with real coefficients) of odd degree has at least one real zero.

6 Use the Theorem for Bounds on Zeros

One challenge in using a graphing utility is to set the viewing window so that a complete graph is obtained. The next theorem is a tool that can be used to find bounds on the zeros. This will assure that the function does not have any zeros outside these bounds. Then using these bounds to set *X*min and *X*max assures that all the x-intercepts appear in the viewing window.

A number M is an **upper bound** to the zeros of a polynomial f if no zero of f is greater than M. The number m is a **lower bound** if no zero of f is less than m. Accordingly, if m is a lower bound and M is an upper bound to the zeros of a polynomial function f, then

$$m \le \text{any zero of } f \le M$$

COMMENT Knowing the values of a lower bound m and an upper bound M may enable you to eliminate some potential rational zeros—that is, any potential zeros outside of the interval $[m, M]$.

THEOREM

Bounds on Zeros

Let f denote a polynomial function whose leading coefficient is positive.

- If $M > 0$ is a real number and if the third row in the process of synthetic division of f by $x - M$ contains only numbers that are positive or zero, then M is an upper bound to the zeros of f.
- If $m < 0$ is a real number and if the third row in the process of synthetic division of f by $x - m$ contains numbers that alternate positive (or 0) and negative (or 0), then m is a lower bound to the zeros of f.

Note: When finding a lower bound, remember that a 0 can be treated as either positive or negative, but not both. For example, 3, 0, 5 would be considered to alternate sign, whereas 3, 0, −5 would not. ■

Proof (Outline) We give only an outline of the proof of the first part of the theorem. Suppose that M is a positive real number, and the third row in the process of synthetic division of the polynomial f by $x - M$ contains only numbers that are positive or 0. Then there are a quotient q and a remainder R such that

$$f(x) = (x - M)q(x) + R$$

where the coefficients of $q(x)$ are positive or 0 and the remainder $R \geq 0$. Then, for any $x > M$, we must have $x - M > 0$, $q(x) > 0$, and $R \geq 0$, so that $f(x) > 0$. That is, there is no zero of f larger than M. The proof of the second part follows similar reasoning. ■

In finding bounds, it is preferable to find the smallest upper bound and largest lower bound. This will require repeated synthetic division until a desired pattern is observed. For simplicity, we will consider only potential rational zeros that are integers. If a bound is not found using these values, continue checking positive and/or negative integers until you find both an upper and a lower bound.

EXAMPLE 8 Finding Upper and Lower Bounds of Zeros

For the polynomial function $f(x) = 2x^3 + 11x^2 - 7x - 6$, use the Bounds on Zeros Theorem to find integer upper and lower bounds to the zeros of f.

Solution From Example 4, the potential rational zeros of f are $\pm 1, \pm 2, \pm 3, \pm 6, \pm \frac{1}{2}, \pm \frac{3}{2}$.

To find an upper bound, start with the smallest positive integer that is a potential rational zero, which is 1. Continue checking 2, 3, and 6 (and then subsequent positive integers), if necessary, until an upper bound is found. To find a lower bound, start with the largest negative integer that is a potential rational zero, which is −1. Continue checking −2, −3, and −6 (and then subsequent negative integers), if necessary, until a lower bound is found. Table 7 summarizes the results of doing repeated synthetic divisions by showing only the third row of each division. For example, the first row of the table shows the result of dividing $f(x)$ by $x - 1$.

$$\begin{array}{r|rrrr} 1 & 2 & 11 & -7 & -6 \\ & & 2 & 13 & 6 \\ \hline & 2 & 13 & 6 & 0 \end{array}$$

Table 7 Synthetic Division Summary

	r	Coefficients of $q(x)$			Remainder	
Upper bound →	(1)	2	13	6	0	All nonnegative
	−1	2	9	−16	10	
	−2	2	7	−21	36	
	−3	2	5	−22	60	
	−6	2	−1	−1	0	
Lower bound →	(−7)	2	−3	14	−104	Alternating Signs

Note: Keep track of any zeros that are found when looking for bounds. ■

For $r = 1$, the third row of synthetic division contains only numbers that are positive or 0, so we know there are no zeros greater than 1. Since the third row of synthetic division for $r = -7$ results in alternating positive (or 0) and negative (or 0) values, we know that −7 is a lower bound. There are no zeros less than −7. Notice that in looking for bounds, two zeros were discovered. These zeros are 1 and −6. ■

EXAMPLE 9 Obtaining Graphs Using Bounds on Zeros

Obtain a graph for the polynomial function.

$$f(x) = 2x^3 + 11x^2 - 7x - 6$$

Solution Based on Example 8, every zero lies between −7 and 1. Using Xmin = −7 and Xmax = 1, we graph $Y_1 = f(x) = 2x^3 + 11x^2 - 7x - 6$. Figure 27 shows the graph obtained using ZOOM-FIT.

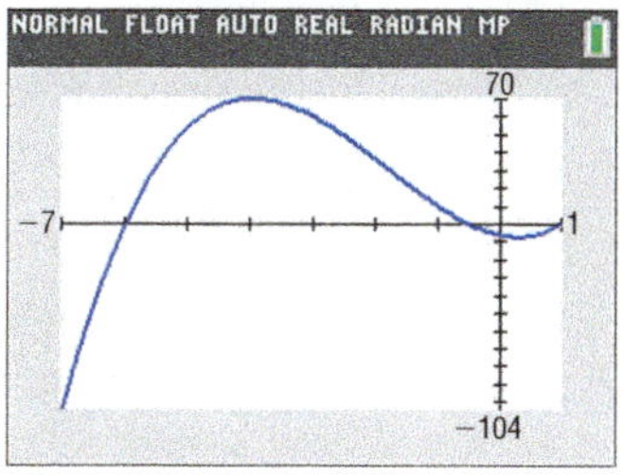

Figure 27

Now Work PROBLEM 45

The next example shows how to proceed when some of the coefficients of the polynomial are not integers.

EXAMPLE 10 Finding the Real Zeros of a Polynomial Function

Find all the real zeros of the polynomial function

$$f(x) = x^5 - 1.8x^4 - 17.78x^3 + 31.61x^2 + 37.9x - 8.7$$

Round answers to two decimal places.

Solution

Step 1: There are at most five real zeros. There are three positive zeros or one positive zero, and there are two negative zeros or no negative zeros.

Step 2: Since there are noninteger coefficients, the Rational Zeros Theorem does not apply.

Step 3: Determine the bounds on the zeros of f. Using synthetic division with successive integers, beginning with ±1, a lower bound is −5 and an upper bound is 6.

Every real zero of f lies between −5 and 6. Figure 28(a) shows the graph of f with Xmin = −5 and Xmax = 6. Figure 28(b) shows a graph of f after adjusting the viewing window to improve the graph.

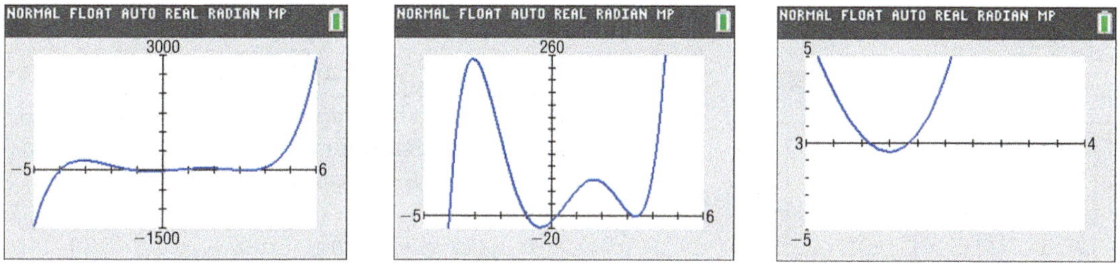

(a) (b) (c)

Figure 28

Step 4: From Figure 28(b), we see that f appears to have four x-intercepts: one near −4, one near −1, one between 0 and 1, and one near 3. The x-intercept near 3 might be a zero of even multiplicity since the graph seems to touch the x-axis at that point.

Using ZERO (or ROOT), the zero between 0 and 1 is found to be 0.20 (rounded to two decimal places), and the zeros −4 and −1 are confirmed. Zooming in shows that there are in fact two distinct zeros near 3. See Figure 28(c). The two remaining zeros are 3.23 and 3.37, each of which is rounded to two decimal places.

Now Work PROBLEM 69

7 Use the Intermediate Value Theorem

The Intermediate Value Theorem requires that the function be *continuous*. Although calculus is needed to explain the meaning precisely, we have already said that, very basically, a function f is continuous when its graph can be drawn without lifting pencil from paper, that is, when the graph contains no holes or jumps or gaps. Every polynomial function is continuous.

INTERMEDIATE VALUE THEOREM

Let f denote a continuous function. If $a < b$ and if $f(a)$ and $f(b)$ are of opposite sign, then f has at least one zero between a and b.

Although the proof of this result requires advanced methods in calculus, it is easy to "see" why the result is true. Look at Figure 29.

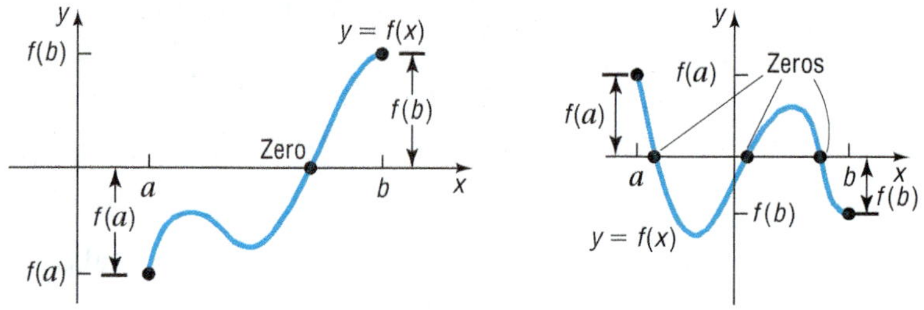

(a) (b)

Figure 29 If $f(a)$ and $f(b)$ are of opposite sign and if f is continuous, there is at least one zero between a and b.

The Intermediate Value Theorem together with the TABLE feature of a graphing utility provides a basis for finding zeros.

EXAMPLE 11 Using the Intermediate Value Theorem and a Graphing Utility to Locate Zeros

Find the positive zero of $f(x) = x^5 - x^3 - 1$ correct to two decimal places.

Solution Because $f(1) = -1 < 0$ and $f(2) = 23 > 0$, we know from the Intermediate Value Theorem that the zero lies between 1 and 2. Divide the interval $[1, 2]$ into 10 equal subintervals, and use the TABLE feature of a graphing utility to evaluate f at the endpoints of the intervals. See Table 8.

We can conclude that the zero is between 1.2 and 1.3 since $f(1.2) < 0$ and $f(1.3) > 0$. Now divide the interval $[1.2, 1.3]$ into 10 equal subintervals and evaluate f at each endpoint. See Table 9. The zero lies between 1.23 and 1.24, and so, correct to two decimal places, the zero is 1.23. ■

Table 8

NORMAL FLOAT AUTO REAL RADIAN MP
PRESS ENTER TO EDIT

X	Y_1
1	-1
1.1	-.7205
1.2	-.2397
1.3	.51593
1.4	1.6342
1.5	3.2188
1.6	5.3898
1.7	8.2856
1.8	12.064
1.9	16.902
2	23

$Y_1 = X^5 - X^3 - 1$

Table 9

NORMAL FLOAT AUTO REAL RADIAN MP
PRESS ENTER TO EDIT

X	Y_1
1.2	-.2397
1.21	-.1778
1.22	-.1131
1.23	-.0456
1.24	.025
1.25	.09863
1.26	.17542
1.27	.25545
1.28	.33882
1.29	.42562
1.3	.51593

$Y_1 = X^5 - X^3 - 1$

Now Work PROBLEM 87

There are many other numerical techniques for approximating the zeros of a polynomial function. The one outlined in Example 11 (a variation of the *bisection method*) has the advantages that it will always work, it can be programmed rather easily on a computer, and each time it is used another decimal place of accuracy is achieved. See Problem 111 for the bisection method, which places the zero in a succession of intervals, with each new interval being half the length of the preceding one.

Historical Feature

Formulas for the solution of third- and fourth-degree polynomial equations exist, and, while not very practical, they do have an interesting history.

In the 1500s in Italy, mathematical contests were a popular pastime, and people who possessed methods for solving problems kept them secret. (Solutions that were published were already common knowledge.) Niccolo of Brescia (1499–1557), commonly referred to as Tartaglia ("the stammerer"), had the secret for solving cubic (third-degree) equations, which gave him a decided advantage in the contests. Girolamo Cardano (1501–1576) found out that Tartaglia had the secret, and, being interested in cubics, he requested it from Tartaglia. The reluctant Tartaglia hesitated for some time, but finally, swearing Cardano to secrecy with midnight oaths by candlelight, told him the secret. Cardano then published the solution in his book *Ars Magna* (1545), giving Tartaglia the credit but rather compromising the secrecy. Tartaglia exploded into bitter recriminations, and each wrote pamphlets that reflected on the other's mathematics, moral character, and ancestry.

The quartic (fourth-degree) equation was solved by Cardano's student Lodovico Ferrari, and this solution also was included, with credit and this time with permission, in the *Ars Magna*.

Attempts were made to solve the fifth-degree equation in similar ways, all of which failed. In the early 1800s, P. Ruffini, Niels Abel, and Evariste Galois all found ways to show that it is not possible to solve fifth-degree equations by formula, but the proofs required the introduction of new methods. Galois's methods eventually developed into a large part of modern algebra.

Historical Problems

Problems 1–8 develop the Tartaglia–Cardano solution of the cubic equation, and show why it is not altogether practical.

1. Show that the general cubic equation $y^3 + by^2 + cy + d = 0$ can be transformed into an equation of the form $x^3 + px + q = 0$ by using the substitution $y = x - \frac{b}{3}$.
2. In the equation $x^3 + px + q = 0$, replace x by $H + K$. Let $3HK = -p$, and show that $H^3 + K^3 = -q$.
3. Based on Problem 2, we have the two equations
$$3HK = -p \quad \text{and} \quad H^3 + K^3 = -q$$
Solve for K in $3HK = -p$ and substitute into $H^3 + K^3 = -q$. Then show that
$$H = \sqrt[3]{\frac{-q}{2} + \sqrt{\frac{q^2}{4} + \frac{p^3}{27}}}$$
[**Hint:** Look for an equation that is quadratic in form.]
4. Use the solution for H from Problem 3 and the equation $H^3 + K^3 = -q$ to show that
$$K = \sqrt[3]{\frac{-q}{2} - \sqrt{\frac{q^2}{4} + \frac{p^3}{27}}}$$
5. Use the results from Problems 2 to 4 to show that the solution of $x^3 + px + q = 0$ is
$$x = \sqrt[3]{\frac{-q}{2} + \sqrt{\frac{q^2}{4} + \frac{p^3}{27}}} + \sqrt[3]{\frac{-q}{2} - \sqrt{\frac{q^2}{4} + \frac{p^3}{27}}}$$
6. Use the result of Problem 5 to solve the equation $x^3 - 6x - 9 = 0$.
7. Use a calculator and the result of Problem 5 to solve the equation $x^3 + 3x - 14 = 0$.
8. Use the methods of this section to solve the equation $x^3 + 3x - 14 = 0$.

5.2 Assess Your Understanding

'Are You Prepared?' *Answers are given at the end of these exercises. If you get a wrong answer, read the pages listed in red.*

1. Find $f(-1)$ if $f(x) = 2x^2 - x$. (pp. 210–212)
2. Factor the expression $6x^2 + x - 2$. (pp. 55–56)
3. Find the quotient and remainder if $3x^4 - 5x^3 + 7x - 4$ is divided by $x - 3$. (pp. 45–48 or 59–62)
4. Find the zeros of $f(x) = x^2 + x - 3$. (pp. 113–115)

Concepts and Vocabulary

5. If $f(x) = q(x)g(x) + r(x)$, the function $r(x)$ is called the ________.
   (a) remainder (b) dividend (c) quotient (d) divisor
6. When a polynomial function f is divided by $x - c$, the remainder is ______.
7. Given $f(x) = 3x^4 - 2x^3 + 7x - 2$, how many sign changes are there in the coefficients of $f(-x)$?
   (a) 0 (b) 1 (c) 2 (d) 3
8. ***True or False*** Every polynomial function of degree 3 with real coefficients has exactly three real zeros.
9. If f is a polynomial function and $x - 4$ is a factor of f, then $f(4) =$ ____.
10. ***True or False*** If f is a polynomial function of degree 4 and if $f(2) = 5$, then
$$\frac{f(x)}{x-2} = p(x) + \frac{5}{x-2}$$
where $p(x)$ is a polynomial function of degree 3.

Skill Building

In Problems 11–20, use the Remainder Theorem to find the remainder when $f(x)$ is divided by $x - c$. Then use the Factor Theorem to determine whether $x - c$ is a factor of $f(x)$.

11. $f(x) = 4x^3 - 3x^2 - 8x + 4;\ x - 2$

12. $f(x) = -4x^3 + 5x^2 + 8;\ x + 3$

13. $f(x) = 3x^4 - 6x^3 - 5x + 10;\ x - 2$

14. $f(x) = 4x^4 - 15x^2 - 4;\ x - 2$

15. $f(x) = 3x^6 + 82x^3 + 27;\ x + 3$

16. $f(x) = 2x^6 - 18x^4 + x^2 - 9;\ x + 3$

17. $f(x) = 4x^6 - 64x^4 + x^2 - 15;\ x + 4$

18. $f(x) = x^6 - 16x^4 + x^2 - 16;\ x + 4$

19. $f(x) = 2x^4 - x^3 + 2x - 1;\ x - \frac{1}{2}$

20. $f(x) = 3x^4 + x^3 - 3x + 1;\ x + \frac{1}{3}$

In Problems 21–32, use Descartes' Rule of Signs to determine how many positive and how many negative zeros each polynomial function may have. Do not attempt to find the zeros.

21. $f(x) = -4x^7 + x^3 - x^2 + 2$

22. $f(x) = 5x^4 + 2x^2 - 6x - 5$

23. $f(x) = 2x^6 - 3x^2 - x + 1$

24. $f(x) = -3x^5 + 4x^4 + 2$

25. $f(x) = 3x^3 - 2x^2 + x + 2$

26. $f(x) = -x^3 - x^2 + x + 1$

27. $f(x) = -x^4 + x^2 - 1$

28. $f(x) = x^4 + 5x^3 - 2$

29. $f(x) = x^5 + x^4 + x^2 + x + 1$

30. $f(x) = x^5 - x^4 + x^3 - x^2 + x - 1$

31. $f(x) = x^6 - 1$

32. $f(x) = x^6 + 1$

In Problems 33–44, determine the maximum number of real zeros that each polynomial function may have. Then list the potential rational zeros of each polynomial function. Do not attempt to find the zeros.

33. $f(x) = 3x^4 - 3x^3 + x^2 - x + 1$

34. $f(x) = x^5 - x^4 + 2x^2 + 3$

35. $f(x) = x^5 - 6x^2 + 9x - 3$

36. $f(x) = 2x^5 - x^4 - x^2 + 1$

37. $f(x) = -4x^3 - x^2 + x + 2$

38. $f(x) = 6x^4 - x^2 + 2$

39. $f(x) = 6x^4 - x^2 + 9$

40. $f(x) = -4x^3 + x^2 + x + 6$

41. $f(x) = 2x^5 - x^3 + 2x^2 + 12$

42. $f(x) = 3x^5 - x^2 + 2x + 18$

43. $f(x) = 6x^4 + 2x^3 - x^2 + 20$

44. $f(x) = -6x^3 - x^2 + x + 10$

In Problems 45–50, find the bounds to the zeros of each polynomial function. Use the bounds to obtain a complete graph of f.

45. $f(x) = 2x^3 + x^2 - 1$

46. $f(x) = 3x^3 - 2x^2 + x + 4$

47. $f(x) = x^3 - 5x^2 - 11x + 11$

48. $f(x) = 2x^3 - x^2 - 11x - 6$

49. $f(x) = x^4 + 3x^3 - 5x^2 + 9$

50. $f(x) = 4x^4 - 12x^3 + 27x^2 - 54x + 81$

In Problems 51–68, find the real zeros of f. Use the real zeros to factor f.

51. $f(x) = x^3 + 2x^2 - 5x - 6$

52. $f(x) = x^3 + 8x^2 + 11x - 20$

53. $f(x) = 2x^3 - 13x^2 + 24x - 9$

54. $f(x) = 2x^3 - 5x^2 - 4x + 12$

55. $f(x) = 3x^3 + 4x^2 + 4x + 1$

56. $f(x) = 3x^3 - 7x^2 + 12x - 28$

57. $f(x) = x^3 - 10x^2 + 28x - 16$

58. $f(x) = x^3 + 6x^2 + 6x - 4$

59. $f(x) = x^4 + x^3 - 3x^2 - x + 2$

60. $f(x) = x^4 - x^3 - 6x^2 + 4x + 8$

61. $f(x) = 21x^4 + 22x^3 - 99x^2 - 72x + 28$

62. $f(x) = 54x^4 - 57x^3 - 323x^2 + 20x + 12$

63. $f(x) = x^3 - 8x^2 + 17x - 6$

64. $f(x) = 2x^4 + 11x^3 - 5x^2 - 43x + 35$

65. $f(x) = 4x^4 + 7x^2 - 2$

66. $f(x) = 4x^4 + 15x^2 - 4$

67. $f(x) = 4x^5 - 8x^4 - x + 2$

68. $f(x) = 4x^5 + 12x^4 - x - 3$

In Problems 69–74, find the real zeros of f. If necessary, round to two decimal places.

69. $f(x) = x^3 + 3.2x^2 - 16.83x - 5.31$

70. $f(x) = x^3 + 3.2x^2 - 7.25x - 6.3$

71. $f(x) = x^4 - 1.4x^3 - 33.71x^2 + 23.94x + 292.41$

72. $f(x) = x^4 + 1.2x^3 - 7.46x^2 - 4.692x + 15.2881$

73. $f(x) = x^3 + 19.5x^2 - 1021x + 1000.5$

74. $f(x) = x^3 + 42.2x^2 - 664.8x + 1490.4$

In Problems 75–84, find the real solutions of each equation.

75. $x^4 - x^3 + 2x^2 - 4x - 8 = 0$

76. $2x^3 + 3x^2 + 2x + 3 = 0$

77. $3x^3 + 4x^2 - 7x + 2 = 0$

78. $2x^3 - 3x^2 - 3x - 5 = 0$

79. $3x^3 - x^2 - 15x + 5 = 0$

80. $2x^3 - 11x^2 + 10x + 8 = 0$

81. $x^4 + 4x^3 + 2x^2 - x + 6 = 0$

82. $x^4 - 2x^3 + 10x^2 - 18x + 9 = 0$

83. $x^3 - \frac{2}{3}x^2 + \frac{8}{3}x + 1 = 0$

84. $x^3 - \frac{2}{3}x^2 + 3x - 2 = 0$

In Problems 85–90, use the Intermediate Value Theorem to show that each function has a zero in the given interval. Approximate the zero correct to two decimal places.

85. $f(x) = 8x^4 - 2x^2 + 5x - 1;\ [0, 1]$

86. $f(x) = x^4 + 8x^3 - x^2 + 2;\ [-1, 0]$

87. $f(x) = 2x^3 + 6x^2 - 8x + 2;\ [-5, -4]$

88. $f(x) = 3x^3 - 10x + 9;\ [-3, -2]$

89. $f(x) = x^5 - x^4 + 7x^3 - 7x^2 - 18x + 18;\ [1.4, 1.5]$

90. $f(x) = x^5 - 3x^4 - 2x^3 + 6x^2 + x + 2;\ [1.7, 1.8]$

Mixed Practice

In Problems 91–98, analyze each polynomial function using Steps 1 through 8 on page 343 in Section 5.1.

91. $f(x) = x^3 + 2x^2 - 5x - 6$
[**Hint:** See Problem 51.]

92. $f(x) = x^3 + 8x^2 + 11x - 20$
[**Hint:** See Problem 52.]

93. $f(x) = x^4 + x^3 - 3x^2 - x + 2$
[**Hint:** See Problem 59.]

94. $f(x) = x^4 - x^3 - 6x^2 + 4x + 8$
[**Hint:** See Problem 60.]

95. $f(x) = 4x^5 - 8x^4 - x + 2$
[**Hint:** See Problem 67.]

96. $f(x) = 4x^5 + 12x^4 - x - 3$
[**Hint:** See Problem 68.]

97. $f(x) = 6x^4 - 37x^3 + 58x^2 + 3x - 18$

98. $f(x) = 20x^4 + 73x^3 + 46x^2 - 52x - 24$

Applications and Extensions

99. Find k such that $f(x) = x^3 - kx^2 + kx + 2$ has the factor $x - 2$.

100. Find k such that $f(x) = x^4 - kx^3 + kx^2 + 1$ has the factor $x + 2$.

101. What is the remainder when $f(x) = 2x^{20} - 8x^{10} + x - 2$ is divided by $x - 1$?

102. What is the remainder when $f(x) = -3x^{17} + x^9 - x^5 + 2x$ is divided by $x + 1$?

103. Use the Factor Theorem to prove that $x - c$ is a factor of $x^n - c^n$ for any positive integer n.

104. Use the Factor Theorem to prove that $x + c$ is a factor of $x^n + c^n$ if $n \geq 1$ is an odd integer.

105. One solution of the equation $x^3 - 8x^2 + 16x - 3 = 0$ is 3. Find the sum of the remaining solutions.

106. One solution of the equation $x^3 + 5x^2 + 5x - 2 = 0$ is -2. Find the sum of the remaining solutions.

107. Geometry What is the length of the edge of a cube if, after a slice 1-inch thick is cut from one side, the volume remaining is 294 cubic inches?

108. Geometry What is the length of the edge of a cube if its volume could be doubled by an increase of 6 centimeters in one edge, an increase of 12 centimeters in a second edge, and a decrease of 4 centimeters in the third edge?

109. Let $f(x)$ be a polynomial function whose coefficients are integers. Suppose that r is a real zero of f and that the leading coefficient of f is 1. Use the Rational Zeros Theorem to show that r is either an integer or an irrational number.

110. Prove the Rational Zeros Theorem.

[**Hint:** Let $\frac{p}{q}$, where p and q have no common factors except 1 and -1, be a zero of the polynomial function

$$f(x) = a_n x^n + a_{n-1}x^{n-1} + \cdots + a_1 x + a_0$$

whose coefficients are all integers. Show that

$$a_n p^n + a_{n-1}p^{n-1}q + \cdots + a_1 pq^{n-1} + a_0 q^n = 0$$

Now, because p is a factor of the first n terms of this equation, p must also be a factor of the term $a_0 q^n$. Since p is not a factor of q (why?), p must be a factor of a_0. Similarly, q must be a factor of a_n.]

111. Bisection Method for Approximating Zeros of a Function f We begin with two consecutive integers, a and $a + 1$, such that $f(a)$ and $f(a + 1)$ are of opposite sign. Evaluate f at the midpoint m_1 of a and $a + 1$. If $f(m_1) = 0$, then m_1 is the zero of f, and we are finished. Otherwise, $f(m_1)$ is of opposite sign to either $f(a)$ or $f(a + 1)$. Suppose that it is

$f(a)$ and $f(m_1)$ that are of opposite sign. Now evaluate f at the midpoint m_2 of a and m_1. Repeat this process until the desired degree of accuracy is obtained. Note that each iteration places the zero in an interval whose length is half that of the previous interval. Use the bisection method to approximate the zero of $f(x) = 8x^4 - 2x^2 + 5x - 1$ in the interval $[0, 1]$ correct to three decimal places. Verify your result using a graphing utility.

[**Hint:** The process ends when both endpoints agree to the desired number of decimal places.]

Discussion and Writing

112. Is $\frac{1}{3}$ a zero of $f(x) = 2x^3 + 3x^2 - 6x + 7$? Explain.

113. Is $\frac{1}{3}$ a zero of $f(x) = 4x^3 - 5x^2 - 3x + 1$? Explain.

114. Is $\frac{3}{5}$ a zero of $f(x) = 2x^6 - 5x^4 + x^3 - x + 1$? Explain.

115. Is $\frac{2}{3}$ a zero of $f(x) = x^7 + 6x^5 - x^4 + x + 2$? Explain.

Retain Your Knowledge

Problems 116–119 are based on material learned earlier in the course. The purpose of these problems is to keep the material fresh in your mind so that you are better prepared for the final exam.

116. Solve $2x - 5y = 3$ for y.

117. Express the inequality $3 \le x < 8$ using interval notation.

118. Find the intercepts of the graph of the equation $3x + y^2 = 12$.

119. Use the figure to determine the interval(s) on which the function is increasing.

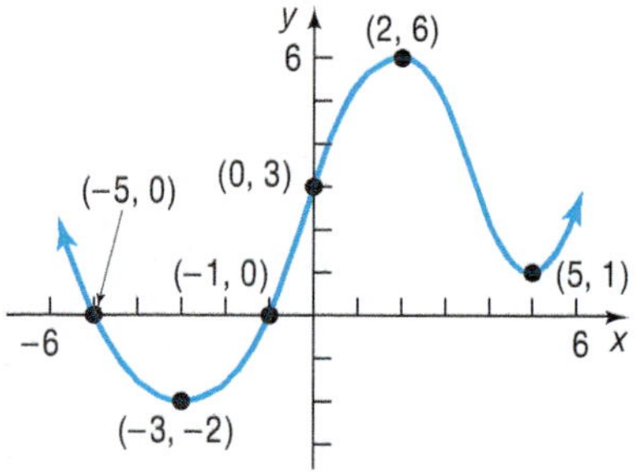

'Are You Prepared?' Answers

1. 3 **2.** $(3x + 2)(2x - 1)$ **3.** Quotient: $3x^3 + 4x^2 + 12x + 43$; remainder: 125 **4.** $\left\{\frac{-1 - \sqrt{13}}{2}, \frac{-1 + \sqrt{13}}{2}\right\}$

5.3 Complex Zeros; Fundamental Theorem of Algebra

PREPARING FOR THIS SECTION *Before getting started, review the following:*

- Complex Numbers (Section 1.4, pp. 121–125)
- Complex Solutions of a Quadratic Equation (Section 1.4, pp. 125–127)

Now Work the **'Are You Prepared?'** problems on page 370.

OBJECTIVES
1 Use the Conjugate Pairs Theorem (p. 367)
2 Find a Polynomial Function with Specified Zeros (p. 368)
3 Find the Complex Zeros of a Polynomial Function (p. 369)

In Section 1.3, we found the real solutions of a quadratic equation. That is, we found the real zeros of a polynomial function of degree 2. Then, in Section 1.4 we found the complex solutions of a quadratic equation. That is, we found the complex zeros of a polynomial function of degree 2.

In Section 5.2, we found the real zeros of polynomial functions of degree 3 or higher. In this section we will find the *complex zeros* of polynomial functions of degree 3 or higher.

DEFINITION

A variable in the complex number system is referred to as a **complex variable**. A **complex polynomial function** f of degree n is a function of the form

$$f(x) = a_n x^n + a_{n-1}x^{n-1} + \cdots + a_1 x + a_0 \qquad (1)$$

where $a_n, a_{n-1}, \ldots, a_1, a_0$ are complex numbers, $a_n \neq 0$, n is a nonnegative integer, and x is a complex variable. As before, a_n is called the **leading coefficient** of f. A complex number r is called a **complex zero** of f if $f(r) = 0$.

In most of our work the coefficients in (1) will be real numbers.

We have learned that some quadratic equations have no real solutions, but that in the complex number system every quadratic equation has a solution, either real or complex. The next result, proved by Karl Friedrich Gauss (1777–1855) when he was 22 years old,* extends this idea to polynomial equations of degree 3 or higher. In fact, this result is so important and useful that it has become known as the **Fundamental Theorem of Algebra**.

FUNDAMENTAL THEOREM OF ALGEBRA

Every complex polynomial function $f(x)$ of degree $n \geq 1$ has at least one complex zero.

We shall not prove this result, as the proof is beyond the scope of this book. However, using the Fundamental Theorem of Algebra and the Factor Theorem, we can prove the following result:

THEOREM

Every complex polynomial function $f(x)$ of degree $n \geq 1$ can be factored into n linear factors (not necessarily distinct) of the form

$$f(x) = a_n(x - r_1)(x - r_2) \cdot \cdots \cdot (x - r_n) \qquad (2)$$

where $a_n, r_1, r_2, \ldots, r_n$ are complex numbers. That is, every complex polynomial function of degree $n \geq 1$ has exactly n complex zeros, some of which may repeat.

Proof Let

$$f(x) = a_n x^n + a_{n-1}x^{n-1} + \cdots + a_1 x + a_0$$

By the Fundamental Theorem of Algebra, f has at least one zero, say r_1. Then, by the Factor Theorem, $x - r_1$ is a factor, and

$$f(x) = (x - r_1)q_1(x)$$

where $q_1(x)$ is a complex polynomial function of degree $n - 1$ whose leading coefficient is a_n. Repeating this argument n times, we arrive at

$$f(x) = (x - r_1)(x - r_2) \cdot \cdots \cdot (x - r_n)q_n(x)$$

where $q_n(x)$ is a complex polynomial function of degree $n - n = 0$ whose leading coefficient is a_n. That is, $q_n(x) = a_n x^0 = a_n$, and so

$$f(x) = a_n(x - r_1)(x - r_2) \cdot \cdots \cdot (x - r_n)$$

We conclude that every complex polynomial function $f(x)$ of degree $n \geq 1$ has exactly n (not necessarily distinct) zeros. ■

1 Use the Conjugate Pairs Theorem

The Fundamental Theorem of Algebra can be used to obtain valuable information about the complex zeros of polynomial functions whose coefficients are real numbers.

*In all, Gauss gave four different proofs of this theorem, the first one in 1799 being the subject of his doctoral dissertation.

CONJUGATE PAIRS THEOREM

Let $f(x)$ be a polynomial function whose coefficients are real numbers. If $r = a + bi$ is a zero of f, the complex conjugate $\bar{r} = a - bi$ is also a zero of f.

In other words, for polynomial functions whose coefficients are real numbers, the complex zeros occur in conjugate pairs. This result should not be all that surprising since the complex zeros of a quadratic function occurred in conjugate pairs.

Proof Let

$$f(x) = a_n x^n + a_{n-1}x^{n-1} + \cdots + a_1 x + a_0$$

where $a_n, a_{n-1}, \ldots, a_1, a_0$ are real numbers and $a_n \neq 0$. If $r = a + bi$ is a zero of f, then $f(r) = f(a + bi) = 0$, so

$$a_n r^n + a_{n-1}r^{n-1} + \cdots + a_1 r + a_0 = 0$$

Take the conjugate of both sides to get

$$\overline{a_n r^n + a_{n-1}r^{n-1} + \cdots + a_1 r + a_0} = \bar{0}$$

$$\overline{a_n r^n} + \overline{a_{n-1}r^{n-1}} + \cdots + \overline{a_1 r} + \bar{a}_0 = \bar{0}$$ The conjugate of a sum equals the sum of the conjugates (see Section 1.4).

$$\bar{a}_n (\bar{r})^n + \overline{a_{n-1}} (\bar{r})^{n-1} + \cdots + \bar{a}_1 \bar{r} + \bar{a}_0 = \bar{0}$$ The conjugate of a product equals the product of the conjugates.

$$a_n (\bar{r})^n + a_{n-1} (\bar{r})^{n-1} + \cdots + a_1 \bar{r} + a_0 = 0$$ The conjugate of a real number equals the real number.

This last equation states that $f(\bar{r}) = 0$; that is, $\bar{r} = a - bi$ is a zero of f. ■

The importance of this result should be clear. Once we know that, say, $3 + 4i$ is a zero of a polynomial function with real coefficients, then we know that $3 - 4i$ is also a zero. This result has an important corollary.

COROLLARY

A polynomial function f of odd degree with real coefficients has at least one real zero.

Proof Because complex zeros occur as conjugate pairs in a polynomial function with real coefficients, there will always be an even number of zeros that are not real numbers. Consequently, since f is of odd degree, one of its zeros has to be a real number. ■

For example, the polynomial function $f(x) = x^5 - 3x^4 + 4x^3 - 5$ has at least one zero that is a real number, since f is of degree 5 (odd) and has real coefficients.

EXAMPLE 1 **Using the Conjugate Pairs Theorem**

A polynomial function f of degree 5 whose coefficients are real numbers has the zeros $1, 5i$, and $1 + i$. Find the remaining two zeros.

Solution Since f has coefficients that are real numbers, complex zeros appear as conjugate pairs. It follows that $-5i$, the conjugate of $5i$, and $1 - i$, the conjugate of $1 + i$, are the two remaining zeros.

Now Work PROBLEM 7

2 Find a Polynomial Function with Specified Zeros

EXAMPLE 2 **Finding a Polynomial Function Whose Zeros Are Given**

(a) Find a polynomial function f of degree 4 whose coefficients are real numbers and that has the zeros $1, 1$, and $-4 + i$.

(b) Graph the function found in part (a) to verify your result.

Solution (a) Since $-4 + i$ is a zero, by the Conjugate Pairs Theorem, $-4 - i$ must also be a zero of f. Because of the Factor Theorem, if $f(c) = 0$, then $x - c$ is a factor of $f(x)$. So f can now be written as

$$f(x) = a(x - 1)(x - 1)[x - (-4 + i)][x - (-4 - i)]$$

where a is any real number. Letting $a = 1$, we obtain

$$\begin{aligned} f(x) &= (x - 1)(x - 1)[x - (-4 + i)][x - (-4 - i)] \\ &= (x^2 - 2x + 1)[(x + 4) - i][(x + 4) + i] \\ &= (x^2 - 2x + 1)((x + 4)^2 - i^2) \\ &= (x^2 - 2x + 1)(x^2 + 8x + 16 - (-1)) \\ &= (x^2 - 2x + 1)(x^2 + 8x + 17) \\ &= x^4 + 8x^3 + 17x^2 - 2x^3 - 16x^2 - 34x + x^2 + 8x + 17 \\ &= x^4 + 6x^3 + 2x^2 - 26x + 17 \end{aligned}$$

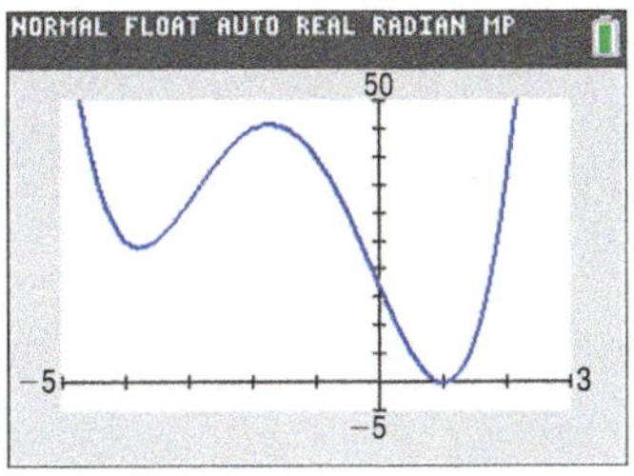

Figure 30
$f(x) = x^4 + 6x^3 + 2x^2 - 26x + 17$

(b) A quick analysis of the polynomial function f tells us what to expect:

At most three turning points

For large $|x|$, the graph behaves like $y = x^4$.

A repeated real zero at 1 so the graph touches the x-axis at 1

The only x-intercept is at 1. The y-intercept is 17.

Figure 30 shows the complete graph. (Do you see why? The graph has exactly three turning points and the degree of the polynomial function is 4.) ■

Now Work PROBLEM 17

Now we can prove the theorem we conjectured earlier in Section 5.2.

THEOREM

Every polynomial function with real coefficients can be uniquely factored over the real numbers into a product of linear factors and/or irreducible quadratic factors. ■

Exploration

Graph the function found in Example 2 for $a = 2$ and $a = -1$. Does the value of a affect the zeros of f? How does the value of a affect the graph of f? ■

Proof Every complex polynomial function f of degree n has exactly n zeros and can be factored into a product of n linear factors. If its coefficients are real, then those zeros that are complex numbers always occur as conjugate pairs. As a result, if $r = a + bi$ is a complex zero, then so is $\bar{r} = a - bi$. Consequently, when the linear factors $x - r$ and $x - \bar{r}$ of f are multiplied, we have

$$(x - r)(x - \bar{r}) = x^2 - (r + \bar{r})x + r\bar{r} = x^2 - 2ax + a^2 + b^2$$

This second-degree polynomial has real coefficients and is irreducible (over the real numbers). Thus, the factors of f are either linear or irreducible quadratic factors. ■

3 Find the Complex Zeros of a Polynomial Function

The steps for finding the complex zeros of a polynomial function are the same as those for finding the real zeros.

EXAMPLE 3 **Finding the Complex Zeros of a Polynomial Function**

Find the complex zeros of the polynomial function

$$f(x) = 3x^4 + 5x^3 + 25x^2 + 45x - 18$$

Write f in factored form.

Solution **Step 1:** The degree of f is 4. So f has four complex zeros. From Descartes' Rule of Signs, there is one positive real zero. Also, since $f(-x) = 3x^4 - 5x^3 + 25x^2 - 45x - 18$, there are three negative real zeros or one negative real zero.

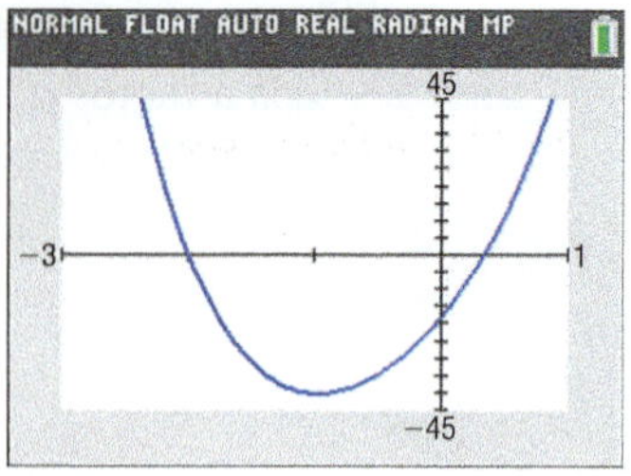

Figure 31

$f(x) = 3x^4 + 5x^3 + 25x^2 + 45x - 18$

STEP 2: The Rational Zeros Theorem provides information about the potential rational zeros of polynomial functions with integer coefficients. For this polynomial function (which has integer coefficients), the potential rational zeros are

$$\pm\frac{1}{3},\ \pm\frac{2}{3},\ \pm 1,\ \pm 2,\ \pm 3,\ \pm 6,\ \pm 9,\ \pm 18$$

STEP 3: Figure 31 shows the graph of f. The graph has the characteristics expected of this polynomial function of degree 4: It behaves like $y = 3x^4$ for large $|x|$ and has y-intercept -18. There are x-intercepts near -2 and between 0 and 1.

STEP 4: Because $f(-2) = 0$, we know that -2 is a zero of f and $x - (-2) = x + 2$ is a factor of f. Use long division or synthetic division to factor f. Using synthetic division,

$$\begin{array}{r|rrrrr} -2 & 3 & 5 & 25 & 45 & -18 \\ & & -6 & 2 & -54 & 18 \\ \hline & 3 & -1 & 27 & -9 & 0 \end{array}$$

So $f(x) = (x + 2)(3x^3 - x^2 + 27x - 9)$. The depressed equation is

$$q_1(x) = 3x^3 - x^2 + 27x - 9 = 0$$

Repeat Step 4: The depressed equation $3x^3 - x^2 + 27x - 9 = 0$ can be factored by grouping.

$$3x^3 - x^2 + 27x - 9 = 0$$

$$x^2(3x - 1) + 9(3x - 1) = 0 \quad \text{Factor } x^2 \text{ from } 3x^3 - x^2 \text{ and 9 from } 27x - 9.$$

$$(x^2 + 9)(3x - 1) = 0 \quad \text{Factor out the common factor } 3x - 1.$$

$$x^2 + 9 = 0 \quad \text{or} \quad 3x - 1 = 0 \quad \text{Apply the Zero-Product Property.}$$

$$x^2 = -9 \quad \text{or} \quad 3x = 1$$

$$x = -3i,\ x = 3i \quad \text{or} \quad x = \frac{1}{3}$$

The four complex zeros of f are $-3i, 3i, -2$, and $\frac{1}{3}$.

The factored form of f is

$$\begin{aligned} f(x) &= 3x^4 + 5x^3 + 25x^2 + 45x - 18 \\ &= (x + 3i)(x - 3i)(x + 2)(3x - 1) \\ &= 3(x + 3i)(x - 3i)(x + 2)\left(x - \frac{1}{3}\right) \end{aligned}$$

Now Work PROBLEM 33

5.3 Assess Your Understanding

'Are You Prepared?' *Answers are given at the end of these exercises. If you get a wrong answer, read the pages listed in red.*

1. Find the sum and the product of the complex numbers $3 - 2i$ and $-3 + 5i$. (pp. 122–123)

2. In the complex number system, find the complex solutions of the equation $x^2 + 2x + 2 = 0$. (pp. 125–127)

Concepts and Vocabulary

3. Every polynomial function of odd degree with real coefficients has at least _____ real zero(s).

4. If $3 + 4i$ is a zero of a polynomial function of degree 5 with real coefficients, then so is ________.

5. ***True or False*** A polynomial function of degree n with real coefficients has exactly n complex zeros. At most n of them are real zeros.

6. ***True or False*** A polynomial function of degree 4 with real coefficients could have $-3, 2 + i, 2 - i$, and $-3 + 5i$ as its zeros.

Skill Building

In Problems 7–16, information is given about a polynomial function $f(x)$ whose coefficients are real numbers. Find the remaining zeros of f.

7. Degree 3; zeros: $3, 4 - i$
8. Degree 3; zeros: $4, 3 + i$
9. Degree 4; zeros: $i, 1 + i$
10. Degree 4; zeros: $1, 2, 2 + i$
11. Degree 5; zeros: $1, i, 2i$
12. Degree 5; zeros: $0, 1, 2, i$
13. Degree 4; zeros: $i, 2, -2$
14. Degree 4; zeros: $2 - i, -i$
15. Degree 6; zeros: $2, 2 + i, -3 - i, 0$
16. Degree 6; zeros: $i, 3 - 2i, -2 + i$

In Problems 17–22, form a polynomial function f(x) with real coefficients having the given degree and zeros. Answers will vary depending on the choice of the leading coefficient. Use a graphing utility to graph the function and verify the result.

17. Degree 4; zeros: $3 + 2i$; 4, multiplicity 2
18. Degree 4; zeros: i; $1 + 2i$
19. Degree 5; zeros: 2; $-i$; $1 + i$
20. Degree 6; zeros: i; $4 - i$; $2 + i$
21. Degree 4; zeros: 3, multiplicity 2; $-i$
22. Degree 5; zeros: 1, multiplicity 3; $1 + i$

In Problems 23–30, use the given zero to find the remaining zeros of each function.

23. $f(x) = x^3 - 4x^2 + 4x - 16$; zero: $2i$
24. $g(x) = x^3 + 3x^2 + 25x + 75$; zero: $-5i$
25. $f(x) = 2x^4 + 5x^3 + 5x^2 + 20x - 12$; zero: $-2i$
26. $h(x) = 3x^4 + 5x^3 + 25x^2 + 45x - 18$; zero: $3i$
27. $h(x) = x^4 - 9x^3 + 21x^2 + 21x - 130$; zero: $3 - 2i$
28. $f(x) = x^4 - 7x^3 + 14x^2 - 38x - 60$; zero: $1 + 3i$
29. $h(x) = 3x^5 + 2x^4 + 15x^3 + 10x^2 - 528x - 352$; zero: $-4i$
30. $g(x) = 2x^5 - 3x^4 - 5x^3 - 15x^2 - 207x + 108$; zero: $3i$

In Problems 31–40, find the complex zeros of each polynomial function. Write f in factored form.

31. $f(x) = x^3 - 1$
32. $f(x) = x^4 - 1$
33. $f(x) = x^3 - 8x^2 + 25x - 26$
34. $f(x) = x^3 + 13x^2 + 57x + 85$
35. $f(x) = x^4 + 5x^2 + 4$
36. $f(x) = x^4 + 13x^2 + 36$
37. $f(x) = x^4 + 2x^3 + 22x^2 + 50x - 75$
38. $f(x) = x^4 + 3x^3 - 19x^2 + 27x - 252$
39. $f(x) = 3x^4 - x^3 - 9x^2 + 159x - 52$
40. $f(x) = 2x^4 + x^3 - 35x^2 - 113x + 65$

Mixed Practice

41. Given $f(x) = 2x^3 - 14x^2 + bx - 3$ with $f(2) = 0$, $g(x) = x^3 + cx^2 - 8x + 30$, with the zero $x = 3 - i$, and b and c real numbers, find $(f \cdot g)(1)$.†
42. Let f be the polynomial function of degree 4 with real coefficients, leading coefficient 1, and zeros $x = 3 + i, 2, -2$. Let g be the polynomial function of degree 4 with intercept $(0, -4)$ and zeros $x = i, 2i$. Find $(f + g)(1)$.†
43. **The complex zeros of $f(x) = x^4 + 1$** For the function $f(x) = x^4 + 1$:
    (a) Factor f into the product of two irreducible quadratics. (***Hint:*** Complete the square by adding and subtracting $2x^2$.)
    (b) Find the zeros of f by finding the zeros of each irreducible quadratic.

† Courtesy of the Joliet Junior College Mathematics Department

Discussion and Writing

In Problems 44 and 45, explain why the facts given are contradictory.

44. f is a polynomial function of degree 3 whose coefficients are real numbers; its zeros are $2, i,$ and $3 + i$.
45. f is a polynomial function of degree 3 whose coefficients are real numbers; its zeros are $4 + i, 4 - i,$ and $2 + i$.
46. f is a polynomial function of degree 4 whose coefficients are real numbers; two of its zeros are -3 and $4 - i$. Explain why one of the remaining zeros must be a real number. Write down one of the missing zeros.

47. f is a polynomial function of degree 4 whose coefficients are real numbers; three of its zeros are $2, 1 + 2i$, and $1 - 2i$. Explain why the remaining zero must be a real number.

48. For the polynomial function $f(x) = x^2 + 2ix - 10$:
(a) Verify that $3 - i$ is a zero of f.
(b) Verify that $3 + i$ is not a zero of f.
(c) Explain why these results do not contradict the Conjugate Pairs Theorem.

Retain Your Knowledge

Problems 49–52 are based on material learned earlier in the course. The purpose of these problems is to keep the material fresh in your mind so that you are better prepared for the final exam.

49. Draw a scatter diagram for the given data.

x	−1	1	2	5	8	10
y	−4	0	3	1	5	7

50. Solve: $\sqrt{3 - x} = 5$

51. Multiply: $(2x - 5)(3x^2 + x - 4)$

52. Find the area and circumference of a circle with a diameter of 6 feet.

'Are You Prepared?' Answers

1. Sum: $3i$; product: $1 + 21i$ **2.** $-1 - i, \quad -1 + i$

5.4 Properties of Rational Functions

PREPARING FOR THIS SECTION *Before getting started, review the following:*

- Rational Expressions (Chapter R, Section R.7, pp. 63–71)
- Polynomial Division (Chapter R, Section R.4, pp. 45–48)
- Graph of $f(x) = \dfrac{1}{x}$ (Section 2.1, Example 6, pp. 169–170)
- Graphing Techniques: Transformations (Section 3.5, pp. 256–264)

Now Work the **'Are You Prepared?'** problems on page 379.

OBJECTIVES
1 Find the Domain of a Rational Function (p. 372)
2 Find the Vertical Asymptotes of a Rational Function (p. 376)
3 Find the Horizontal or Oblique Asymptote of a Rational Function (p. 377)

Ratios of integers are called *rational numbers*. Similarly, ratios of polynomial functions are called *rational functions*. Examples of rational functions are

$$R(x) = \frac{x^2 - 4}{x^2 + x + 1} \qquad F(x) = \frac{x^3}{x^2 - 4} \qquad G(x) = \frac{3x^2}{x^4 - 1}$$

DEFINITION

A **rational function** is a function of the form

$$R(x) = \frac{p(x)}{q(x)}$$

where p and q are polynomial functions and q is not the zero polynomial. The domain of a rational function is the set of all real numbers except those for which the denominator q is 0.

1 Find the Domain of a Rational Function

EXAMPLE 1 **Finding the Domain of a Rational Function**

(a) The domain of $R(x) = \dfrac{2x^2 - 4}{x + 5}$ is the set of all real numbers x except -5; that is, the domain is $\{x \mid x \neq -5\}$.

(b) The domain of $R(x) = \dfrac{1}{x^2 - 4}$ is the set of all real numbers x except -2 and 2; that is, the domain is $\{x \mid x \neq -2, x \neq 2\}$.

(c) The domain of $R(x) = \dfrac{x^3}{x^2 + 1}$ is the set of all real numbers.

(d) The domain of $R(x) = \dfrac{x^2 - 1}{x - 1}$ is the set of all real numbers x except 1; that is, the domain is $\{x \mid x \neq 1\}$. ■

Table 10

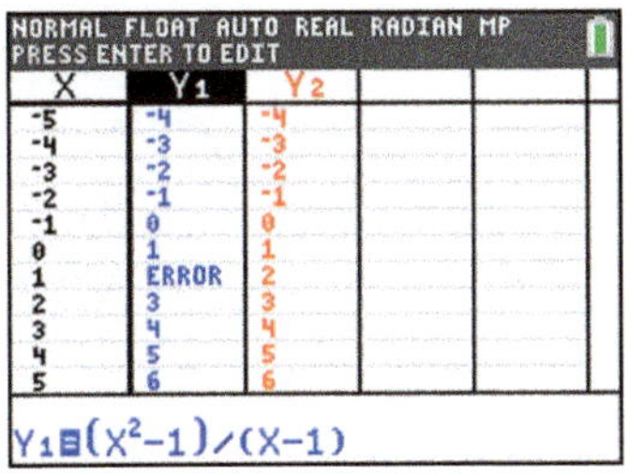

NORMAL FLOAT AUTO REAL RADIAN MP
PRESS ENTER TO EDIT

X	Y1	Y2
-5	-4	-4
-4	-3	-3
-3	-2	-2
-2	-1	-1
-1	0	0
0	1	1
1	ERROR	2
2	3	3
3	4	4
4	5	5
5	6	6

Y1=(X²-1)/(X-1)

Although $\dfrac{x^2 - 1}{x - 1}$ simplifies to $x + 1$, it is important to observe that the functions

$$R(x) = \frac{x^2 - 1}{x - 1} \quad \text{and} \quad f(x) = x + 1$$

are not equal, since the domain of R is $\{x \mid x \neq 1\}$ and the domain of f is the set of all real numbers. Notice in Table 10 there is an error message for $Y_1 = R(x) = \dfrac{x^2 - 1}{x - 1}$, but there is no error message for $Y_2 = f(x) = x + 1$.

Now Work PROBLEM 17

If $R(x) = \dfrac{p(x)}{q(x)}$ is a rational function and if p and q have no common factors, then the rational function R is said to be in **lowest terms**. For a rational function $R(x) = \dfrac{p(x)}{q(x)}$ in lowest terms, the real zeros, if any, of the numerator in the domain of R are the x-intercepts of the graph of R and so play a major role in the graph of R. The real zeros of the denominator of R [that is, the numbers x, if any, for which $q(x) = 0$], although not in the domain of R, also play a major role in the graph of R.

We have already discussed the properties of the rational function $y = \dfrac{1}{x}$. (Refer to Example 6, page 169.) The next rational function that we take up is $H(x) = \dfrac{1}{x^2}$.

EXAMPLE 2 Graphing $y = \dfrac{1}{x^2}$

Analyze the graph of $H(x) = \dfrac{1}{x^2}$.

Solution The domain of $H(x) = \dfrac{1}{x^2}$ consists of all real numbers x except 0. The graph has no y-intercept, since x can never equal 0. The graph has no x-intercept because the equation $H(x) = 0$ has no solution. Therefore, the graph of H will not cross either coordinate axis.

Because

$$H(-x) = \frac{1}{(-x)^2} = \frac{1}{x^2} = H(x)$$

H is an even function, so its graph is symmetric with respect to the y-axis.

See Figure 32. Notice that the graph confirms the conclusions just reached. But what happens to the graph as the values of x get closer and closer to 0? We use a TABLE to answer the question. See Table 11 on the following page. The first four rows show that as the values of x approach (get closer to) 0, the values of $H(x)$ become larger and larger positive numbers. When this happens, we say that $H(x)$ is **unbounded in the positive direction**. We symbolize this by writing $H(x) \to \infty$ [read as "$H(x)$ **approaches infinity**"]. In calculus, we use limit notation, $\lim_{x \to 0} H(x) = \infty$, read as "the limit of $H(x)$ as x approaches 0 is infinity," to mean that $H(x) \to \infty$ as $x \to 0$.

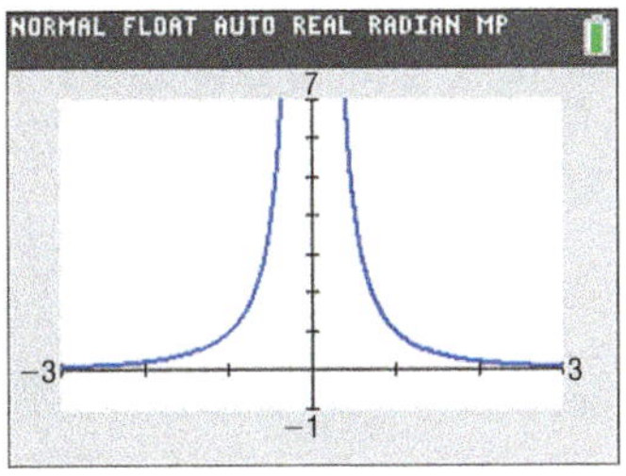

Figure 32 $Y_1 = \dfrac{1}{x^2}$

Table 11

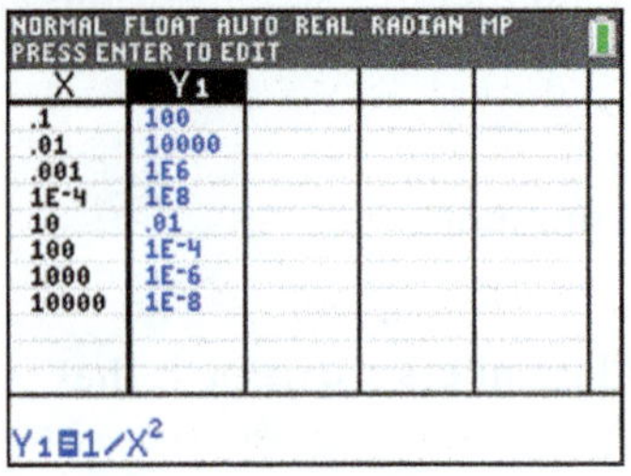
NORMAL FLOAT AUTO REAL RADIAN MP
PRESS ENTER TO EDIT

X	Y_1
.1	100
.01	10000
.001	1E6
1E-4	1E8
10	.01
100	1E-4
1000	1E-6
10000	1E-8

$Y_1 = 1/X^2$

Look at the last four rows of Table 11. As $x \to \infty$, the values of $H(x)$ approach 0 (the end behavior of the graph). This is expressed in calculus by writing $\lim_{x \to \infty} H(x) = 0$. Remember, on the calculator 1E−4 means 1×10^{-4} or 0.0001.

Figure 33 shows the graph of $H(x) = \frac{1}{x^2}$ drawn by hand. Notice the use of red dashed lines to convey the ideas discussed.

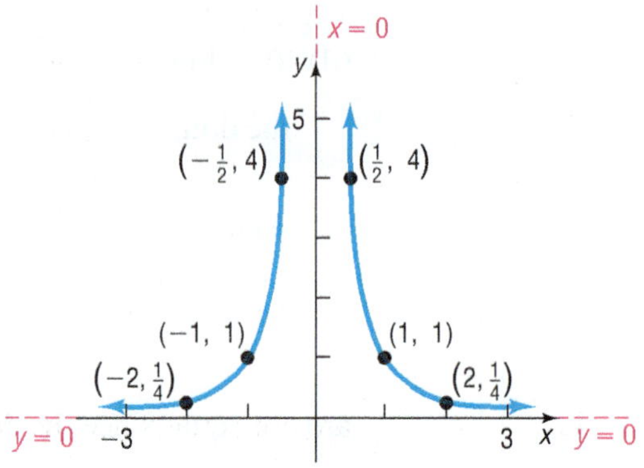

Figure 33 $H(x) = \frac{1}{x^2}$

EXAMPLE 3 Using Transformations to Graph a Rational Function

Graph the rational function: $R(x) = \frac{1}{(x-2)^2} + 1$

Solution The domain of R is the set of all real numbers except $x = 2$. To graph R, start with the graph of $y = \frac{1}{x^2}$. See Figure 34 for the stages.

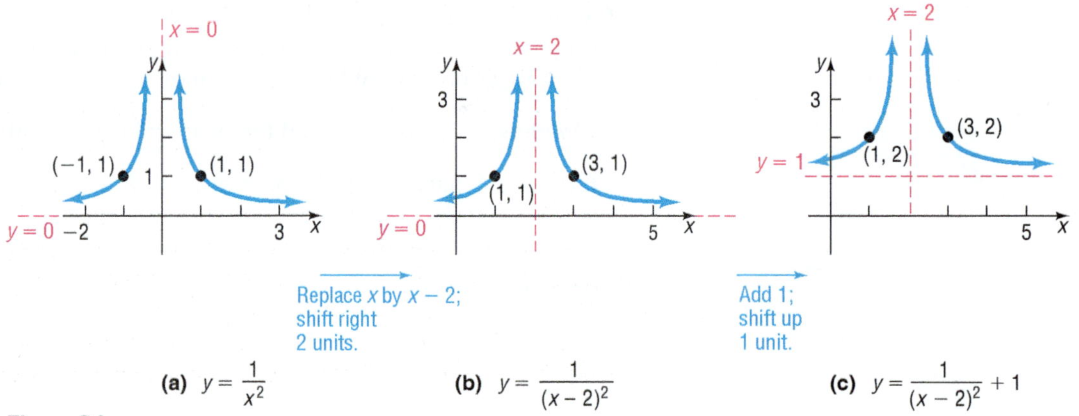

Figure 34

Check: Graph $Y_1 = \frac{1}{(x-2)^2} + 1$ using a graphing utility to verify the graph obtained in Figure 34(c).

Now Work PROBLEM 35

Asymptotes

Notice that the y-axis in Figure 34(a) is transformed into the vertical line $x = 2$ in Figure 34(c), and the x-axis in Figure 34(a) is transformed into the horizontal line $y = 1$ in Figure 34(c). The **Exploration** that follows will help us analyze the role of these lines.

Exploration

(a) Using a graphing utility and the TABLE feature, evaluate the function $H(x) = \frac{1}{(x-2)^2} + 1$ at $x = 10$, 100, 1000, and 10,000. What happens to the values of H as x becomes unbounded in the positive direction, expressed as $\lim_{x \to \infty} H(x)$?

(b) Evaluate H at $x = -10, -100, -1000$, and $-10{,}000$. What happens to the values of H as x becomes unbounded in the negative direction, expressed as $\lim_{x \to -\infty} H(x)$?

(c) Evaluate H at $x = 1.5$, 1.9, 1.99, 1.999, and 1.9999. What happens to the values of H as x approaches 2, $x < 2$, expressed as $\lim_{x \to 2^-} H(x)$?

(d) Evaluate H at $x = 2.5$, 2.1, 2.01, 2.001, and 2.0001. What happens to the values of H as x approaches 2, $x > 2$, expressed as $\lim_{x \to 2^+} H(x)$?

Result

(a) Table 12 shows the values of $Y_1 = H(x)$ as x approaches ∞. Notice that the values of H are approaching 1, so $\lim_{x \to \infty} H(x) = 1$.

(b) Table 13 shows the values of $Y_1 = H(x)$ as x approaches $-\infty$. Again the values of H are approaching 1, so $\lim_{x \to -\infty} H(x) = 1$.

(c) From Table 14 we see that, as x approaches 2, $x < 2$, the values of H are increasing without bound, so $\lim_{x \to 2^-} H(x) = \infty$.

(d) Finally, Table 15 reveals that, as x approaches 2, $x > 2$, the values of H are increasing without bound, so $\lim_{x \to 2^+} H(x) = \infty$.

Table 12

NORMAL FLOAT AUTO REAL RADIAN MP
PRESS ENTER TO EDIT

X	Y1
10	1.0156
100	1.0001
1000	1
10000	1

Y1=1/(X−2)²+1

Table 13

NORMAL FLOAT AUTO REAL RADIAN MP
PRESS ENTER TO EDIT

X	Y1
-10	1.0069
-100	1.0001
-1000	1
-10000	1

Y1=1/(X−2)²+1

Table 14

NORMAL FLOAT AUTO REAL RADIAN MP
PRESS ENTER TO EDIT

X	Y1
1.5	5
1.9	101
1.99	10001
1.999	1E6
1.9999	1E8

Y1=1/(X−2)²+1

Table 15

NORMAL FLOAT AUTO REAL RADIAN MP
PRESS ENTER TO EDIT

X	Y1
2.5	5
2.1	101
2.01	10001
2.001	1E6
2.0001	1E8

Y1=1/(X−2)²+1

The results of the Exploration reveal an important property of rational functions. The vertical line $x = 2$ and the horizontal line $y = 1$ in Figure 34(c) are called *asymptotes* of the graph of H.

DEFINITION

Let R denote a function:

If, as $x \to -\infty$ or as $x \to \infty$, the values of $R(x)$ approach some fixed number L, then the line $y = L$ is a **horizontal asymptote** of the graph of R. [Refer to Figures 35(a) and (b).]

If, as x approaches some number c, the values $|R(x)| \to \infty$, then the line $x = c$ is a **vertical asymptote** of the graph of R. The graph of R never intersects a vertical asymptote. [Refer to Figures 35(c) and (d).]

A horizontal asymptote, when it occurs, describes the end behavior of the graph as $x \to \infty$ or as $x \to -\infty$. **The graph of a function may intersect a horizontal asymptote.**

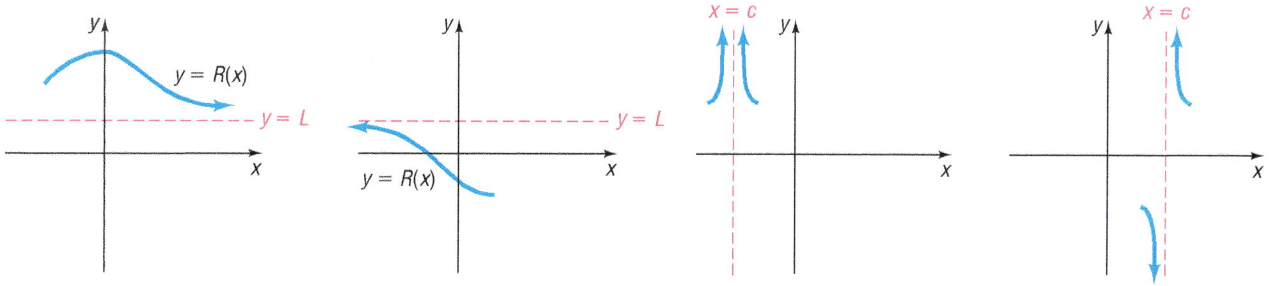

(a) End behavior: As $x \to \infty$, the values of $R(x)$ approach L [expressed as $\lim_{x \to \infty} R(x) = L$]. That is, the points on the graph of R are getting closer to the line $y = L$; $y = L$ is a horizontal asymptote.

(b) End behavior: As $x \to -\infty$, the values of $R(x)$ approach L [expressed as $\lim_{x \to -\infty} R(x) = L$]. That is, the points on the graph of R are getting closer to the line $y = L$; $y = L$ is a horizontal asymptote.

(c) As x approaches c, the values of $R(x) \to \infty$ [for $x < c$, this is expressed as $\lim_{x \to c^-} R(x) = \infty$; for $x > c$, this is expressed as $\lim_{x \to c^+} R(x) = \infty$]. That is, the points on the graph of R are getting closer to the line $x = c$; $x = c$ is a vertical asymptote.

(d) As x approaches c, the values of $|R(x)| \to \infty$ [for $x < c$, this is expressed as $\lim_{x \to c^-} R(x) = -\infty$; for $x > c$, this is expressed as $\lim_{x \to c^+} R(x) = \infty$]. That is, the points on the graph of R are getting closer to the line $x = c$; $x = c$ is a vertical asymptote.

Figure 35

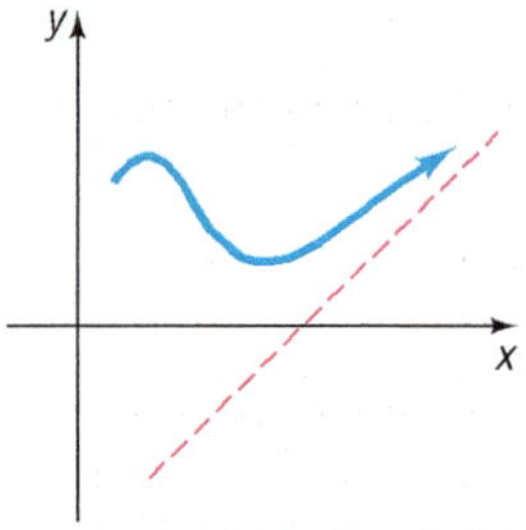

Figure 36 Oblique asymptote

A vertical asymptote, when it occurs, describes the behavior of the graph when x is close to some number c. **The graph of a function will never intersect a vertical asymptote.**

There is a third possibility. If, as $x \to -\infty$ or as $x \to \infty$, the value of a rational function $R(x)$ approaches a linear expression $ax + b$, $a \neq 0$, then the line $y = ax + b$, $a \neq 0$, is an **oblique (or slant) asymptote** of R. Figure 36 shows an oblique asymptote. An oblique asymptote, when it occurs, describes the end behavior of the graph. **The graph of a function may intersect an oblique asymptote.**

Now Work PROBLEM 27

2 Find the Vertical Asymptotes of a Rational Function

The vertical asymptotes of a rational function $R(x) = \dfrac{p(x)}{q(x)}$, in lowest terms, are located at the real zeros of the denominator, $q(x)$. Suppose that r is a real zero of q, so $x - r$ is a factor of q. As x approaches r, symbolized as $x \to r$, the values of $x - r$ approach 0, causing the ratio to become unbounded; that is, $|R(x)| \to \infty$. Based on the definition, we conclude that the line $x = r$ is a vertical asymptote.

THEOREM

Locating Vertical Asymptotes

A rational function $R(x) = \dfrac{p(x)}{q(x)}$, *in lowest terms,* will have a vertical asymptote $x = r$ if r is a real zero of the denominator q. That is, if $x - r$ is a factor of the denominator q of a rational function $R(x) = \dfrac{p(x)}{q(x)}$, in lowest terms, R will have the vertical asymptote $x = r$.

WARNING If a rational function is not in lowest terms, an application of this theorem may result in an incorrect listing of vertical asymptotes.

EXAMPLE 4 **Finding Vertical Asymptotes**

Find the vertical asymptotes, if any, of the graph of each rational function.

(a) $F(x) = \dfrac{x + 3}{x - 1}$ (b) $R(x) = \dfrac{x}{x^2 - 4}$

(c) $H(x) = \dfrac{x^2}{x^2 + 1}$ (d) $G(x) = \dfrac{x^2 - 9}{x^2 + 4x - 21}$

Solution

(a) F is in lowest terms, and the only zero of the denominator is 1. The line $x = 1$ is the vertical asymptote of the graph of F.

(b) R is in lowest terms, and the zeros of the denominator $x^2 - 4$ are -2 and 2. The lines $x = -2$ and $x = 2$ are the vertical asymptotes of the graph of R.

(c) H is in lowest terms, and the denominator has no real zeros, because the equation $x^2 + 1$ has no real solutions. The graph of H has no vertical asymptotes.

(d) Factor $G(x)$ to determine whether it is in lowest terms.

$$G(x) = \frac{x^2 - 9}{x^2 + 4x - 21} = \frac{(x + 3)(x - 3)}{(x + 7)(x - 3)} = \frac{x + 3}{x + 7} \qquad x \neq 3$$

The only zero of the denominator of $G(x)$ in lowest terms is -7. The line $x = -7$ is the only vertical asymptote of the graph of G.

WARNING In Example 4(b), the vertical asymptotes are $x = -2$ and $x = 2$. Do not say that the vertical asymptotes are -2 and 2.

As Example 4 points out, rational functions can have no vertical asymptotes, one vertical asymptote, or more than one vertical asymptote.

For polynomial functions, we saw how the multiplicities of zeros could be used to determine the behavior of the graph around each x-intercept. In a similar way, the multiplicities of the zeros in the denominator of a rational function, in lowest terms, can be used to determine the behavior of the graph around each vertical asymptote. Consider the following Exploration.

Exploration

Graph each of the following rational functions using a graphing utility and describe the behavior of the graph near the vertical asymptote $x = 1$.

$$R_1(x) = \frac{1}{x-1} \quad R_2(x) = \frac{1}{(x-1)^2} \quad R_3(x) = -\frac{1}{(x-1)^3} \quad R_4(x) = -\frac{1}{(x-1)^4}$$

Result For R_1, as x approaches 1 (odd multiplicity) from the left side of the vertical asymptote, the values of $R_1(x)$ approach negative infinity. That is, $\lim_{x \to 1^-} R_1(x) = -\infty$. As x approaches 1 from the right side of the vertical asymptote, the values of $R_1(x)$ approach infinity. That is, $\lim_{x \to 1^+} R_1(x) = \infty$. See Figure 37(a). Below, we summarize the behavior near $x = 1$ for the remaining rational functions:

- For R_2: $\lim_{x \to 1^-} R_2(x) = \infty$ and $\lim_{x \to 1^+} R_2(x) = \infty$ [See Figure 37(b)]
- For R_3: $\lim_{x \to 1^-} R_3(x) = \infty$ and $\lim_{x \to 1^+} R_3(x) = -\infty$ [See Figure 37(c)]
- For R_4: $\lim_{x \to 1^-} R_4(x) = -\infty$ and $\lim_{x \to 1^+} R_4(x) = -\infty$ [See Figure 37(d)] ■

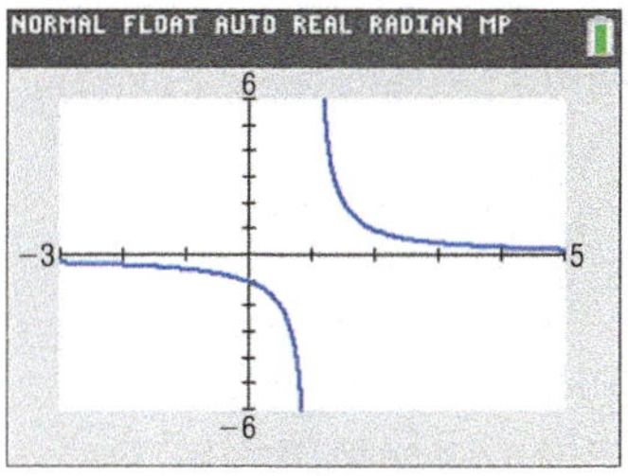

(a) $R_1(x)$; odd multiplicity

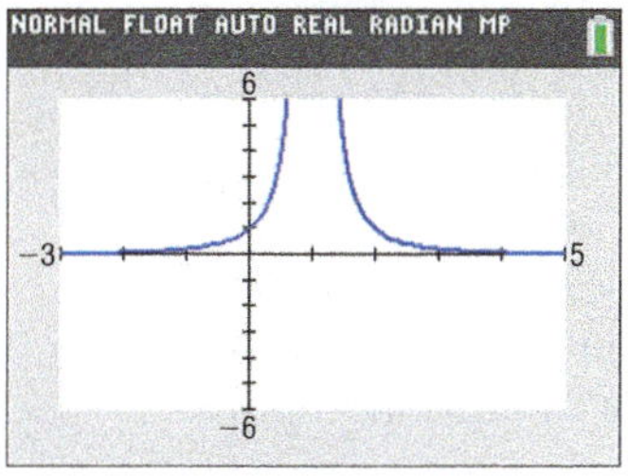

(b) $R_2(x)$; even multiplicity

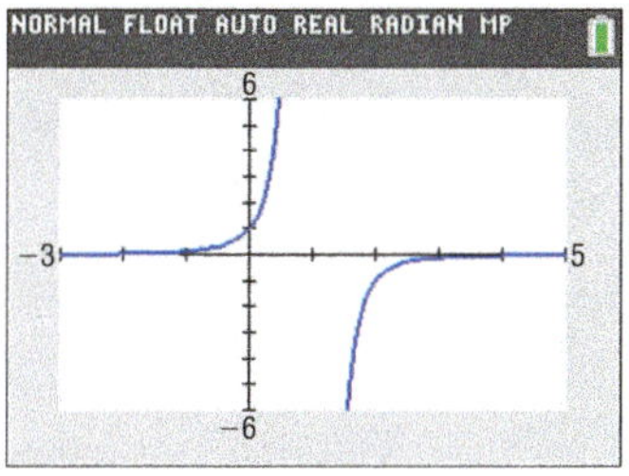

(c) $R_3(x)$; odd multiplicity

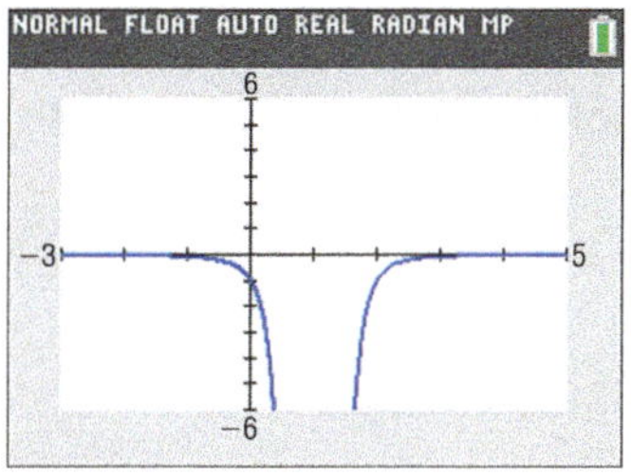

(d) $R_4(x)$; even multiplicity

Figure 37

The results of the Exploration lead to the following conclusion.

Multiplicity and Vertical Asymptotes

- If the multiplicity of any zero of the denominator of a rational function in lowest terms that gives rise to a vertical asymptote is odd, the graph approaches ∞ on one side of the vertical asymptote and approaches $-\infty$ on the other side.
- If the multiplicity of any zero of the denominator of a rational function in lowest terms that gives rise to a vertical asymptote is even, the graph approaches either ∞ or $-\infty$ on both sides of the vertical asymptote.

Now Work PROBLEMS 45, 47, AND 49 (FIND THE VERTICAL ASYMPTOTES, IF ANY)

3 Find the Horizontal or Oblique Asymptote of a Rational Function

To find horizontal or oblique asymptotes, we need to know how the value of the function behaves as $x \to -\infty$ or as $x \to \infty$. That is, we need to determine the end behavior of the function. This can be done by examining the degrees of the numerator and denominator, and the respective power functions that each resembles. For example, consider the rational function

$$R(x) = \frac{3x-2}{5x^2-7x+1}$$

The degree of the numerator, 1, is less than the degree of the denominator, 2. When $|x|$ is very large, the numerator of R can be approximated by the power function $y = 3x$, and the denominator can be approximated by the power function $y = 5x^2$. This means

$$R(x) = \frac{3x-2}{5x^2-7x+1} \underset{\text{For } |x| \text{ very large}}{\approx} \frac{3x}{5x^2} = \frac{3}{5x} \underset{\text{As } x \to -\infty \text{ or } x \to \infty}{\to} 0$$

Table 16

NORMAL FLOAT AUTO REAL RADIAN MP
PRESS ENTER TO EDIT

X	Y_1
-10	-.056
-100	-.006
-1000	-6E-4
-10000	-6E-5
-1E5	-6E-6
10	.06497
100	.00604
1000	6E-4
10000	6E-5
100000	6E-6

$Y_1 = (3X-2)/(5X^2-7X+1)$

which shows that the line $y = 0$ is a horizontal asymptote. We verify this in Table 16.

This result is true for all rational functions that are **proper** (that is, the degree of the numerator is less than the degree of the denominator). If a rational function is **improper** (that is, if the degree of the numerator is greater than or equal to the degree of the denominator), there could be a horizontal asymptote, an oblique asymptote, or neither. The summary on the next page details how to find horizontal or oblique asymptotes.

Finding a Horizontal or Oblique Asymptote of a Rational Function

Consider the rational function

$$R(x) = \frac{p(x)}{q(x)} = \frac{a_n x^n + a_{n-1}x^{n-1} + \cdots + a_1 x + a_0}{b_m x^m + b_{m-1}x^{m-1} + \cdots + b_1 x + b_0}$$

in which the degree of the numerator is n and the degree of the denominator is m.

1. If $n < m$ (the degree of the numerator is less than the degree of the denominator), the line $y = 0$ is a horizontal asymptote.
2. If $n = m$ (the degree of the numerator equals the degree of the denominator), the line $y = \frac{a_n}{b_m}$ is a horizontal asymptote. (That is, the horizontal asymptote equals the ratio of the leading coefficients.)
3. If $n = m + 1$ (the degree of the numerator is one more than the degree of the denominator), the line $y = ax + b$ is an oblique asymptote, which is the quotient found using long division. However, if R in lowest terms is a linear function, then we agree that R has no horizontal or oblique asymptote.
4. If $n \geq m + 2$ (the degree of the numerator is two or more greater than the degree of the denominator), there are no horizontal or oblique asymptotes. The end behavior of the graph will resemble the power function $y = \frac{a_n}{b_m}x^{n-m}$.

Note: A rational function will never have both a horizontal asymptote and an oblique asymptote. A rational function may have neither a horizontal nor an oblique asymptote. ■

We illustrate each of the possibilities in Examples 5 through 8.

EXAMPLE 5 Finding a Horizontal Asymptote

Find the horizontal asymptote, if one exists, of the graph of

$$R(x) = \frac{4x^3 - 5x + 2}{7x^5 + 2x^4 - 3x}$$

Solution Since the degree of the numerator, 3, is less than the degree of the denominator, 5, the rational function R is proper. The line $y = 0$ is a horizontal asymptote of the graph of R. ■

EXAMPLE 6 Finding a Horizontal or Oblique Asymptote

Find the horizontal or oblique asymptote, if one exists, of the graph of

$$H(x) = \frac{3x^4 - x^2}{x^3 - x^2 + 1}$$

Solution Since the degree of the numerator, 4, is exactly one greater than the degree of the denominator, 3, the rational function H has an oblique asymptote. Find the asymptote by using long division.

$$\begin{array}{r} 3x + 3 \\ x^3 - x^2 + 1 \overline{\big) 3x^4 - x^2 } \\ \underline{3x^4 - 3x^3 + 3x } \\ 3x^3 - x^2 - 3x \\ \underline{3x^3 - 3x^2 + 3} \\ 2x^2 - 3x - 3 \end{array}$$

Table 17

X	Y_1	Y_2
-10	-27.21	-27
-100	-297	-297
-1000	-2997	-2997
-10000	-29997	-29997
-1E5	-3E5	-3E5
10	33.185	33
100	303.02	303
1000	3003	3003
10000	30003	30003
100000	300003	300003

NORMAL FLOAT AUTO REAL RADIAN MP
PRESS ENTER TO EDIT
Y1=(3X^4-X^2)/(X^3-X^2+1)

As a result,

$$H(x) = \frac{3x^4 - x^2}{x^3 - x^2 + 1} = 3x + 3 + \frac{2x^2 - 3x - 3}{x^3 - x^2 + 1}$$

As $x \to -\infty$ or as $x \to \infty$,

$$\frac{2x^2 - 3x - 3}{x^3 - x^2 + 1} \approx \frac{2x^2}{x^3} = \frac{2}{x} \to 0$$

So, as $x \to -\infty$ or as $x \to \infty$, we have $H(x) \to 3x + 3$. See Table 17 with $Y_1 = H(x)$ and $Y_2 = 3x + 3$. As $x \to \infty$ or $x \to -\infty$, the difference in values between Y_1 and Y_2 becomes indistinguishable. Put another way, as $x \to \pm\infty$, the graph of H will behave like the graph of $y = 3x + 3$. The graph of the rational function H has an oblique asymptote $y = 3x + 3$. ■

EXAMPLE 7 **Finding a Horizontal or Oblique Asymptote**

Find the horizontal or oblique asymptote, if one exists, of the graph of

$$R(x) = \frac{8x^2 - x + 2}{4x^2 - 1}$$

Solution Since the degree of the numerator, 2, equals the degree of the denominator, 2, the rational function R has a horizontal asymptote equal to the ratio of the leading coefficients.

$$y = \frac{a_n}{b_m} = \frac{8}{4} = 2$$

To see why the horizontal asymptote equals the ratio of the leading coefficients, investigate the behavior of R as $x \to -\infty$ or as $x \to \infty$. When $|x|$ is very large, the numerator of R can be approximated by the power function $y = 8x^2$, and the denominator can be approximated by the power function $y = 4x^2$. This means that as $x \to -\infty$ or as $x \to \infty$,

$$R(x) = \frac{8x^2 - x + 2}{4x^2 - 1} \approx \frac{8x^2}{4x^2} = \frac{8}{4} = 2$$

The graph of the rational function R has a horizontal asymptote $y = 2$. The graph of R will behave like $y = 2$ as $x \to \pm\infty$.

✓**Check:** Verify the results by creating a TABLE with $Y_1 = R(x)$ and $Y_2 = 2$. ■

EXAMPLE 8 **Finding a Horizontal or Oblique Asymptote**

Find the horizontal or oblique asymptote, if one exists, of the graph of

$$G(x) = \frac{2x^5 - x^3 + 2}{x^3 - 1}$$

Solution Since the degree of the numerator, 5, is greater than the degree of the denominator, 3, by more than one, the rational function G has no horizontal or oblique asymptote. The end behavior of the graph will resemble the power function $y = 2x^{5-3} = 2x^2$.

To see why this is the case, investigate the behavior of G as $x \to -\infty$ or as $x \to \infty$. When $|x|$ is very large, the numerator of G can be approximated by the power function $y = 2x^5$, and the denominator can be approximated by the power function $y = x^3$. This means as $x \to -\infty$ or as $x \to \infty$,

$$G(x) = \frac{2x^5 - x^3 + 2}{x^3 - 1} \approx \frac{2x^5}{x^3} = 2x^{5-3} = 2x^2$$

Since this is not linear, the graph of G has no horizontal or oblique asymptote. The graph of G will behave like $y = 2x^2$ as $x \to \pm\infty$. ■

Now Work PROBLEMS 45, 47, AND 49 (FIND THE HORIZONTAL OR OBLIQUE ASYMPTOTE, IF ONE EXISTS.)

5.4 Assess Your Understanding

'Are You Prepared?' *Answers are given at the end of these exercises. If you get a wrong answer, read the pages listed in red.*

1. ***True or False*** The quotient of two polynomial expressions is a rational expression. (p. 63)
2. What are the quotient and remainder when $3x^4 - x^2$ is divided by $x^3 - x^2 + 1$. (pp. 45–48)
3. Graph $y = \frac{1}{x}$. (pp. 169–170)
4. Graph $y = 2(x + 1)^2 - 3$ using transformations. (pp. 256–264)

Concepts and Vocabulary

5. ***True or False*** The domain of every rational function is the set of all real numbers.

6. If, as $x \to -\infty$ or as $x \to \infty$, the values of $R(x)$ approach some fixed number L, then the line $y = L$ is a __________ __________ of the graph of R.

7. If, as x approaches some number c, the values of $|R(x)| \to \infty$, then the line $x = c$ is a __________ __________ of the graph of R.

8. For a rational function R, if the degree of the numerator is less than the degree of the denominator, then R is ________.

9. ***True or False*** The graph of a rational function may intersect a horizontal asymptote.

10. ***True or False*** The graph of a rational function may intersect a vertical asymptote.

11. If a rational function is proper, then ________ is a horizontal asymptote.

12. ***True or False*** If the degree of the numerator of a rational function equals the degree of the denominator, then the ratio of the leading coefficients gives rise to the horizontal asymptote.

13. If $R(x) = \dfrac{p(x)}{q(x)}$ is a rational function and if p and q have no common factors, then R is ________.
(a) improper (b) proper
(c) undefined (d) in lowest terms

14. Which type of asymptote, when it occurs, describes the behavior of a graph when x is close to some number?
(a) vertical (b) horizontal
(c) oblique (d) all of these

Skill Building

In Problems 15–26, find the domain of each rational function.

15. $R(x) = \dfrac{4x}{x-3}$

16. $R(x) = \dfrac{5x^2}{3+x}$

17. $H(x) = \dfrac{-4x^2}{(x-2)(x+4)}$

18. $G(x) = \dfrac{6}{(x+3)(4-x)}$

19. $F(x) = \dfrac{3x(x-1)}{2x^2-5x-3}$

20. $Q(x) = \dfrac{-x(1-x)}{3x^2+5x-2}$

21. $R(x) = \dfrac{x}{x^3-8}$

22. $R(x) = \dfrac{x}{x^4-1}$

23. $H(x) = \dfrac{3x^2+x}{x^2+4}$

24. $G(x) = \dfrac{x-3}{x^4+1}$

25. $R(x) = \dfrac{3(x^2-x-6)}{4(x^2-9)}$

26. $F(x) = \dfrac{-2(x^2-4)}{3(x^2+4x+4)}$

In Problems 27–32, use the graph shown to find
(a) The domain and range of each function
(b) The intercepts, if any
(c) Horizontal asymptotes, if any
(d) Vertical asymptotes, if any
(e) Oblique asymptotes, if any

27.

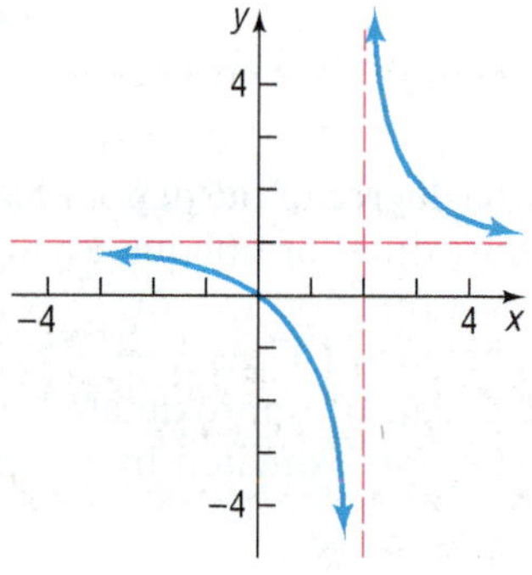

28.

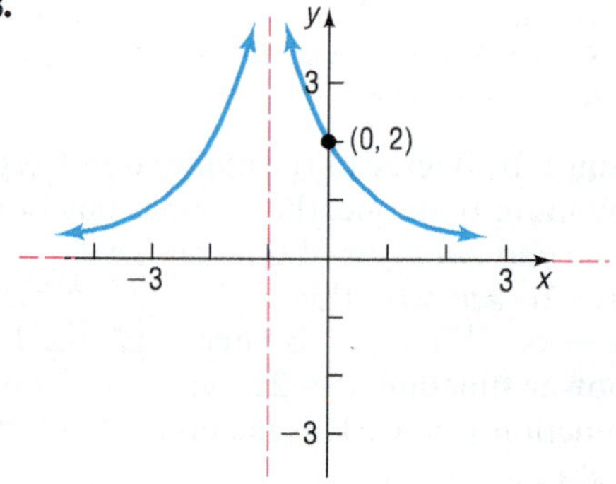

29.

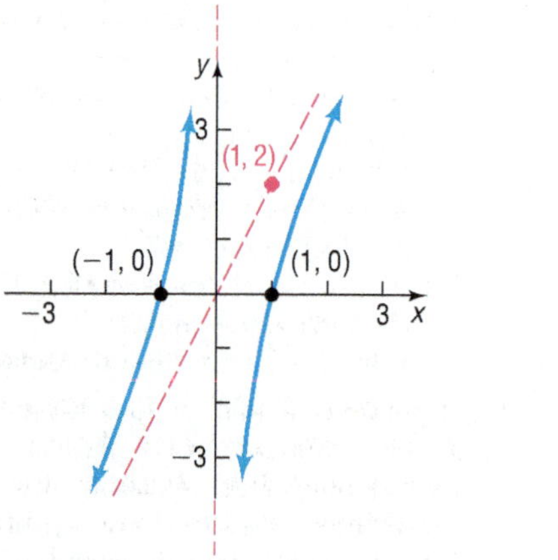

30.

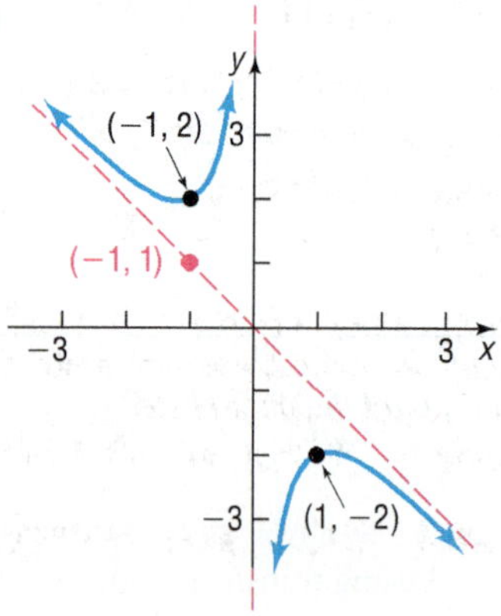

31.

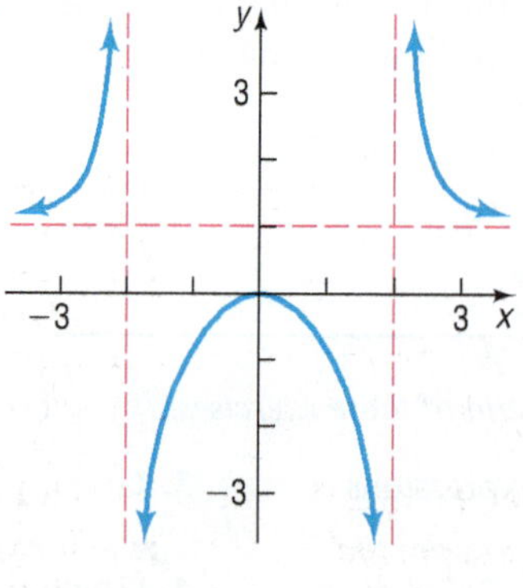

32.

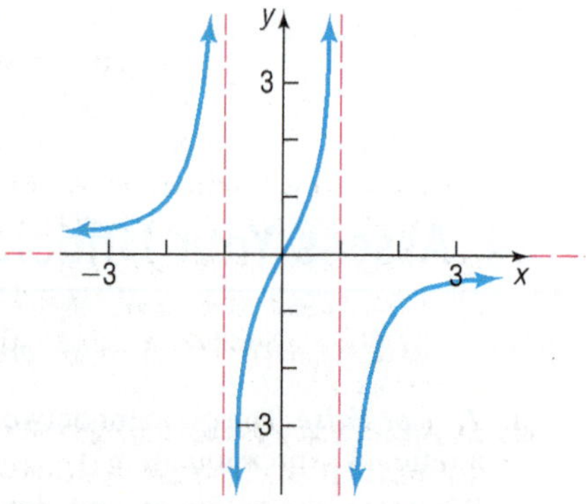

In Problems 33–44, (a) graph the rational function using transformations, (b) use the final graph to find the domain and range, and (c) use the final graph to list any vertical, horizontal, or oblique asymptotes.

33. $F(x) = 2 + \frac{1}{x}$

34. $Q(x) = 3 + \frac{1}{x^2}$

35. $R(x) = \frac{1}{(x-1)^2}$

36. $R(x) = \frac{3}{x}$

37. $H(x) = \frac{-2}{x+1}$

38. $G(x) = \frac{2}{(x+2)^2}$

39. $R(x) = \frac{-1}{x^2 + 4x + 4}$

40. $R(x) = \frac{1}{x-1} + 1$

41. $G(x) = 1 + \frac{2}{(x-3)^2}$

42. $F(x) = 2 - \frac{1}{x+1}$

43. $R(x) = \frac{x^2 - 4}{x^2}$

44. $R(x) = \frac{x-4}{x}$

In Problems 45–56, find the vertical, horizontal, and oblique asymptotes, if any, of each rational function.

45. $R(x) = \frac{3x}{x+4}$

46. $R(x) = \frac{3x+5}{x-6}$

47. $H(x) = \frac{x^3 - 8}{x^2 - 5x + 6}$

48. $G(x) = \frac{x^3 + 1}{x^2 - 5x - 14}$

49. $T(x) = \frac{x^3}{x^4 - 1}$

50. $P(x) = \frac{4x^2}{x^3 - 1}$

51. $Q(x) = \frac{2x^2 - 5x - 12}{3x^2 - 11x - 4}$

52. $F(x) = \frac{x^2 + 6x + 5}{2x^2 + 7x + 5}$

53. $R(x) = \frac{6x^2 + 7x - 5}{3x + 5}$

54. $R(x) = \frac{8x^2 + 26x - 7}{4x - 1}$

55. $G(x) = \frac{x^4 - 1}{x^2 - x}$

56. $F(x) = \frac{x^4 - 16}{x^2 - 2x}$

Applications and Extensions

57. Gravity In physics, it is established that the acceleration due to gravity, g (in meters/sec^2), at a height h meters above sea level is given by

$$g(h) = \frac{3.99 \times 10^{14}}{(6.374 \times 10^6 + h)^2}$$

where 6.374×10^6 is the radius of Earth in meters.

(a) What is the acceleration due to gravity at sea level?

(b) The Willis Tower in Chicago, Illinois, is 443 meters tall. What is the acceleration due to gravity at the top of the Willis Tower?

(c) The peak of Mount Everest is 8848 meters above sea level. What is the acceleration due to gravity on the peak of Mount Everest?

(d) Find the end behavior of g. That is, find $\lim_{h \to \infty} g(h)$. What does the result suggest?

(e) Solve $g(h) = 0$. How do you interpret your answer?

58. Population Model A rare species of insect was discovered in the Amazon Rain Forest. To protect the species, environmentalists declared the insect endangered and transplanted the insect into a protected area. The population P of the insect t months after being transplanted is

$$P(t) = \frac{50(1 + 0.5t)}{2 + 0.01t}$$

(a) How many insects were discovered? In other words, what was the population when $t = 0$?

(b) What will the population be after 5 years?

(c) Determine the end behavior of P. What is the largest population that the protected area can sustain?

59. Resistance in Parallel Circuits From Ohm's Law for circuits, it follows that the total resistance R_{tot} of two components hooked in parallel is given by the equation

$$R_{tot} = \frac{R_1 R_2}{R_1 + R_2}$$

where R_1 and R_2 are the individual resistances.

(a) Let $R_1 = 10$ ohms, and graph R_{tot} as a function of R_2.

(b) Find and interpret any asymptotes of the graph obtained in part (a).

(c) If $R_2 = 2\sqrt{R_1}$, what value of R_1 will yield an R_{tot} of 17 ohms?

60. Newton's Method In calculus you will learn that if

$$p(x) = a_n x^n + a_{n-1}x^{n-1} + \cdots + a_1 x + a_0$$

is a polynomial function, then the *derivative* of $p(x)$ is

$$p'(x) = na_n x^{n-1} + (n-1)a_{n-1}x^{n-2} + \cdots + 2a_2 x + a_1$$

Newton's Method is an efficient method for approximating the x-intercepts (or real zeros) of a function, such as $p(x)$. The following steps outline Newton's Method.

STEP 1: Select an initial value x_0 that is somewhat close to the x-intercept being sought.

STEP 2: Find values for x using the relation

$$x_{n+1} = x_n - \frac{p(x_n)}{p'(x_n)} \qquad n = 1, 2, \ldots$$

until you get two consecutive values x_n and x_{n+1} that agree to whatever decimal place accuracy you desire.

STEP 3: The approximate zero will be x_{n+1}.

Consider the polynomial $p(x) = x^3 - 7x - 40$.

(a) Evaluate $p(5)$ and $p(-3)$.

(b) What might we conclude about a zero of p? Explain.

(c) Use Newton's Method to approximate an x-intercept, r, $-3 < r < 5$, of $p(x)$ to four decimal places.

(d) Use a graphing utility to graph $p(x)$ and verify your answer in part (c).

(e) Using a graphing utility, evaluate $p(r)$ to verify your result.

61. Exploration The standard form of the rational function $R(x) = \frac{mx + b}{cx + d}$, where $c \neq 0$,

is $R(x) = a\left(\frac{1}{x - h}\right) + k$. To write a rational function in standard form requires long division.

(a) Write the rational function $R(x) = \frac{2x + 3}{x - 1}$ in standard form by writing R in the form

$$\text{Quotient} + \frac{\text{remainder}}{\text{divisor}}$$

(b) Graph R using transformations.

(c) Determine the vertical asymptote and the horizontal asymptote of R.

62. Exploration Repeat Problem 61 for the rational function $R(x) = \frac{-6x + 16}{2x - 7}$.

Explaining Concepts: Discussion and Writing

63. If the graph of a rational function R has the vertical asymptote $x = 4$, the factor $x - 4$ must be present in the denominator of R. Explain why.

64. If the graph of a rational function R has the horizontal asymptote $y = 2$, the degree of the numerator of R equals the degree of the denominator of R. Explain why.

65. The graph of a rational function cannot have both a horizontal and an oblique asymptote? Explain why.

66. Make up a rational function that has $y = 2x + 1$ as an oblique asymptote. Explain the methodology that you used.

Retain Your Knowledge

Problems 67–70 are based on material learned earlier in the course. The purpose of these problems is to keep the material fresh in your mind so that you are better prepared for the final exam.

67. Find the equation of a vertical line passing through the point $(5, -3)$.

68. Solve: $\frac{2}{5}(3x - 7) + 1 = \frac{x}{4} - 2$

69. Determine whether the graph of the equation $2x^3 - xy^2 = 4$ is symmetric with respect to the x-axis, the y-axis, the origin, or none of these.

70. What are the points of intersection of the graphs of the functions $f(x) = -3x + 2$ and $g(x) = x^2 - 2x - 4$?

'Are You Prepared?' Answers

1. True **2.** Quotient: $3x + 3$; remainder: $2x^2 - 3x - 3$

3.

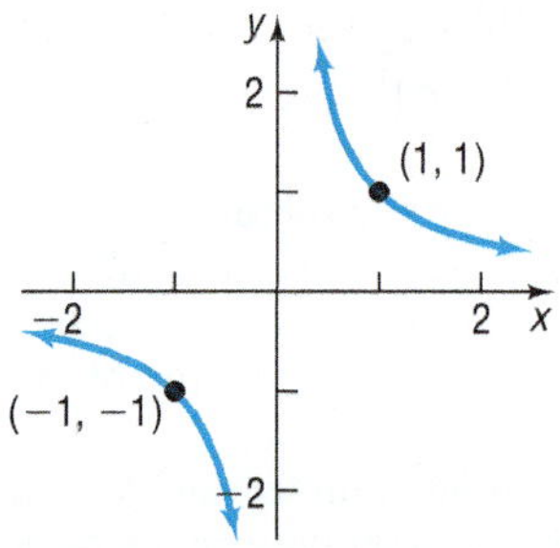

4.

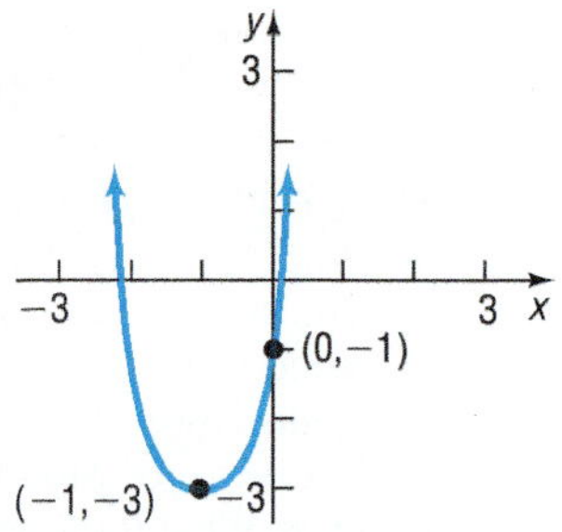

5.5 The Graph of a Rational Function

PREPARING FOR THIS SECTION *Before getting started, review the following:*

- Finding Intercepts (Section 2.1, pp. 165–166)

Now Work the **'Are You Prepared?'** problem on page 390.

OBJECTIVES 1 Analyze the Graph of a Rational Function (p. 382)
2 Solve Applied Problems Involving Rational Functions (p. 389)

1 Analyze the Graph of a Rational Function

Graphing utilities make the task of graphing rational functions less time consuming. However, the results of algebraic analysis must be taken into account before drawing conclusions based on the graph provided by the utility. In the next example we illustrate how to use the information collected in the last section in conjunction with the graphing utility to analyze the graph of a rational function $R(x) = \frac{p(x)}{q(x)}$.

EXAMPLE 1 How to Analyze the Graph of a Rational Function

Analyze the graph of the rational function: $R(x) = \dfrac{x-1}{x^2-4}$

Step-by-Step Solution

Step 1: Factor the numerator and denominator of R. Find the domain of the rational function.

$$R(x) = \frac{x-1}{x^2-4} = \frac{x-1}{(x+2)(x-2)}$$

The domain of R is $\{x \mid x \neq -2, x \neq 2\}$.

Step 2: Write R in lowest terms.

Because there are no common factors between the numerator and denominator, R is in lowest terms.

Step 3: Locate the intercepts of the graph. Use multiplicity to determine the bahavior of the graph of R at each x-intercept.

Since 0 is in the domain of R, the y-intercept is $R(0) = \frac{1}{4}$. The x-intercepts are found by determining the real zeros of the numerator of R written in lowest terms. By solving $x - 1 = 0$, the only real zero of the numerator is 1. So the only x-intercept of the graph of R is 1. The multiplicity of 1 is odd, so the graph will cross the x-axis at $x = 1$.

Step 4: Locate the vertical asymptotes. Determine the behavior of the graph on either side of each vertical asymptote.

To locate the vertical asymptotes, find the zeros of the denominator with the rational function in lowest terms. With R written in lowest terms, we find that the graph of R has two vertical asymptotes: the lines $x = -2$ and $x = 2$. The multiplicities of the zeros that give rise to the vertical asymptotes are both odd. Therefore, the graph will approach ∞ on one side of each vertical asymptote, and it will approach $-\infty$ on the other side.

Step 5: Locate the horizontal or oblique asymptote. Determine points, if any, at which the graph of R intersects this asymptote.

Because the degree of the numerator is less than the degree of the denominator, R is proper and the line $y = 0$ (the x-axis) is a horizontal asymptote of the graph. To determine if the graph of R intersects the horizontal asymptote, solve the equation $R(x) = 0$:

$$\frac{x-1}{x^2-4} = 0$$
$$x - 1 = 0$$
$$x = 1$$

The only solution is $x = 1$, so the graph of R intersects the horizontal asymptote at $(1, 0)$.

Step 6: Graph R using a graphing utility.

The analysis in Steps 1 through 5 helps us to determine an appropriate viewing window to obtain a complete graph. Figure 38 shows the graph of $R(x) = \dfrac{x-1}{x^2-4}$. You should confirm that all the algebraic conclusions that we came to in Steps 1 through 5 are part of the graph. For example, the graph has a horizontal asymptote at $y = 0$ and vertical asymptotes at $x = -2$ and $x = 2$. The y-intercept is $\frac{1}{4}$ and the x-intercept is 1.

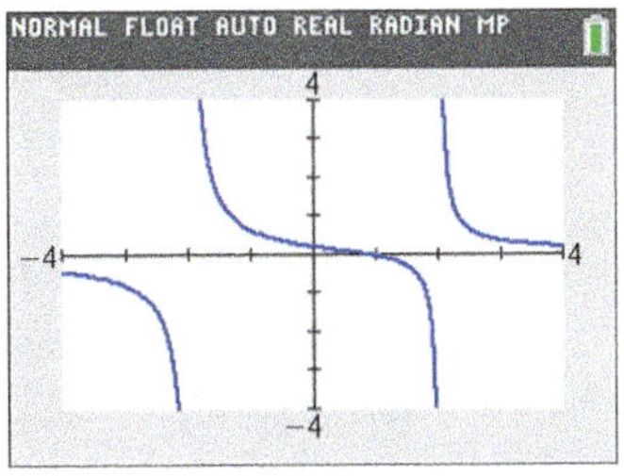

Figure 38 $Y_1 = (x-1)/(x^2-4)$

Step 7: Use the results obtained in Steps 1 through 6 to graph R by hand.

Figure 39 shows the graph of R.

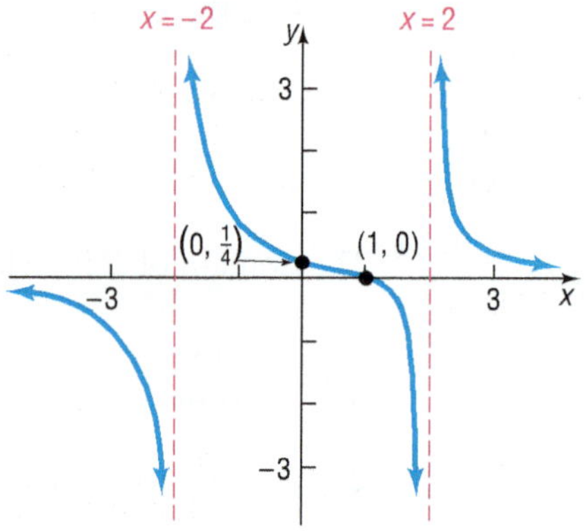

Figure 39 $R(x) = \dfrac{x - 1}{x^2 - 4}$

SUMMARY

Analyzing the Graph of a Rational Function

Step 1: Factor the numerator and denominator of R. Find the domain of the rational function.

Step 2: Write R in lowest terms.

Step 3: Locate the intercepts of the graph. The y-intercept, if there is one, is $R(0)$. The x-intercepts, if any, of $R(x) = \dfrac{p(x)}{q(x)}$ in lowest terms satisfy the equation $p(x) = 0$. Use multiplicity to determine the behavior of the graph of R at each x-intercept.

Step 4: Locate the vertical asymptotes. The vertical asymptotes, if any, of $R(x) = \dfrac{p(x)}{q(x)}$ in lowest terms are found by identifying the real zeros of $q(x)$. Each zero of the denominator gives rise to a vertical asymptote. Use multiplicity to determine the behavior of the graph of R on either side of each vertical asymptote.

Step 5: Locate the horizontal or oblique asymptote, if one exists, using the procedure given in Section 5.4. Determine points, if any, at which the graph of R intersects this asymptote.

Step 6: Graph R using a graphing utility.

Step 7: Use the results obtained in Steps 1 through 6 to graph R by hand.

Now Work PROBLEM 7

EXAMPLE 2 Analyzing the Graph of a Rational Function

Analyze the graph of the rational function: $R(x) = \dfrac{x^2 - 1}{x}$

Solution

Step 1: $R(x) = \dfrac{x^2 - 1}{x} = \dfrac{(x + 1)(x - 1)}{x}$

The domain of R is $\{x | x \neq 0\}$.

Step 2: R is in lowest terms.

Step 3: The graph has two x-intercepts, -1 and 1, each with odd multiplicity. The graph will cross the x-axis at each point. There is no y-intercept, since x cannot equal 0.

Note: Because the denominator of the rational function is a monomial, we can also find the oblique asymptote as follows:

$$\frac{x^2-1}{x} = \frac{x^2}{x} - \frac{1}{x} = x - \frac{1}{x}$$

Since $\frac{1}{x} \to 0$ as $x \to \infty$, $y = x$ is an oblique asymptote. ■

STEP 4: The real zero of the denominator with R in lowest terms is 0, so the graph of $R(x)$ has the line $x = 0$ (the y-axis) as a vertical asymptote. The multiplicity is odd, so the graph will approach ∞ on one side of the asymptote and $-\infty$ on the other.

STEP 5: Since the degree of the numerator, 2, is one more than the degree of the denominator, 1, the rational function R is improper and will have an oblique asymptote. To find the oblique asymptote, use long division.

$$\begin{array}{r} x \\ x\overline{)\,x^2 - 1} \\ \underline{x^2} \\ -1 \end{array}$$

The quotient is x, so the line $y = x$ is an oblique asymptote of the graph. To determine whether the graph of R intersects the asymptote $y = x$, solve the equation $R(x) = x$.

$$\begin{aligned} R(x) = \frac{x^2-1}{x} &= x \\ x^2 - 1 &= x^2 \\ -1 &= 0 \quad \text{Impossible} \end{aligned}$$

The equation $\frac{x^2-1}{x} = x$ has no solution, so the graph of $R(x)$ does not intersect the line $y = x$.

STEP 6: See Figure 40. We see from the graph that there is no y-intercept and there are two x-intercepts, -1 and 1. We can also see that there is a vertical asymptote at $x = 0$.

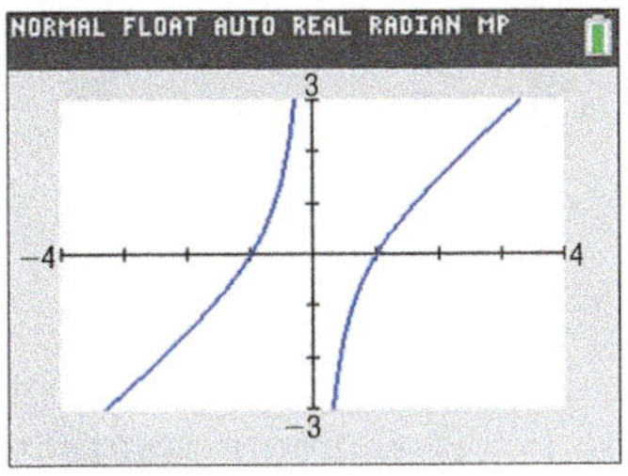

Figure 40 $Y_1 = \frac{x^2-1}{x}$

STEP 7: Using the information gathered in Steps 1 through 6, we obtain the graph of R shown in Figure 41. Notice how the oblique asymptote is used as a guide in graphing the rational function by hand. ■

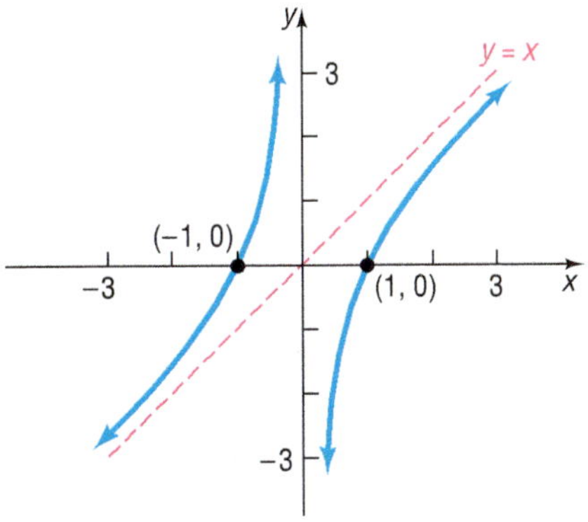

Figure 41 $R(x) = \frac{x^2-1}{x}$

Now Work PROBLEM 15

EXAMPLE 3 Analyzing the Graph of a Rational Function

Analyze the graph of the rational function: $R(x) = \dfrac{x^4+1}{x^2}$

Solution

STEP 1: R is completely factored. The domain of R is $\{x \mid x \neq 0\}$.

STEP 2: R is in lowest terms.

STEP 3: There is no y-intercept. Since $x^4 + 1 = 0$ has no real solutions, there are no x-intercepts.

STEP 4: R is in lowest terms, so $x = 0$ (the y-axis) is a vertical asymptote of R. The multiplicity of 0 is even, so the graph will approach either ∞ or $-\infty$ on both sides of the asymptote.

STEP 5: Since the degree of the numerator, 4, is two more than the degree of the denominator, 2, the rational function will not have a horizontal or oblique asymptote. Find the end behavior of R. As $|x| \to \infty$,

$$R(x) = \frac{x^4+1}{x^2} \approx \frac{x^4}{x^2} = x^2$$

The graph of R will approach the graph of $y = x^2$ as $x \to -\infty$ and as $x \to \infty$. The graph of R does not intersect $y = x^2$. Do you know why?

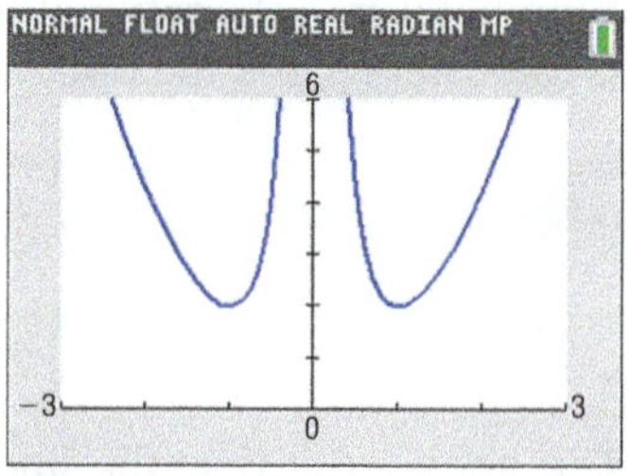

Figure 42 $Y_1 = \frac{x^4 + 1}{x^2}$

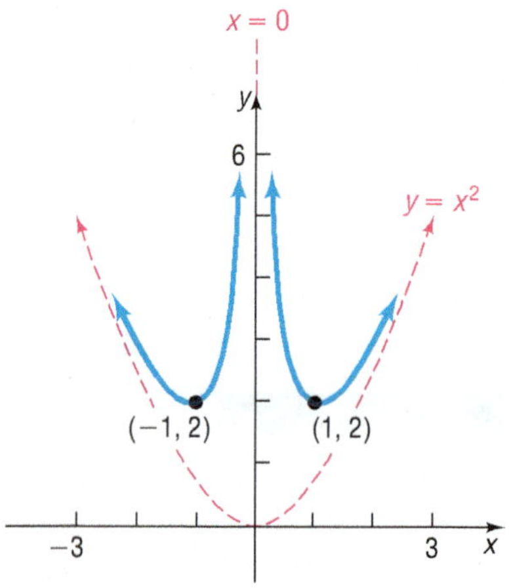

Figure 43

Step 6: See Figure 42. Notice the vertical asymptote at $x = 0$, and note that there are no intercepts.

Note: Notice that R in Example 3 is an even function. Do you see the symmetry about the y-axis in the graph of R?

Step 7: Since the graph of R is above the x-axis, and the multiplicity of the zero that gives rise to the vertical asymptote, $x = 0$, is even, the graph will approach the vertical asymptote $x = 0$ at the top to the left of $x = 0$ and at the top to the right of $x = 0$. See Figure 43.

EXAMPLE 4 Analyzing the Graph of a Rational Function

Analyze the graph of the rational function: $R(x) = \frac{3x^2 - 3x}{x^2 + x - 12}$

Solution

Step 1: Factor R to get

$$R(x) = \frac{3x(x-1)}{(x+4)(x-3)}$$

The domain of R is $\{x \mid x \neq -4, x \neq 3\}$.

Step 2: R is in lowest terms.

Step 3: The graph has two x-intercepts, 0 and 1, each with odd multiplicity. The graph will cross the x-axis at both points. The y-intercept is $R(0) = 0$.

Step 4: The real zeros of the denominator of R with R in lowest terms are -4 and 3. So the graph of R has two vertical asymptotes: $x = -4$ and $x = 3$. The multiplicities of the values that give rise to the asymptotes are both odd, so the graph will approach ∞ on one side of each asymptote and $-\infty$ on the other side.

Step 5: Since the degree of the numerator equals the degree of the denominator, the graph has a horizontal asymptote. To find it, form the quotient of the leading coefficient of the numerator, 3, and the leading coefficient of the denominator, 1. The graph of R has the horizontal asymptote $y = 3$. To find out whether the graph of R intersects the asymptote, solve the equation $R(x) = 3$.

$$R(x) = \frac{3x^2 - 3x}{x^2 + x - 12} = 3$$

$$3x^2 - 3x = 3x^2 + 3x - 36$$

$$-6x = -36$$

$$x = 6$$

The graph intersects the line $y = 3$ at $x = 6$, and $(6, 3)$ is a point on the graph of R.

Step 6: Figure 44(a) shows the graph of R.

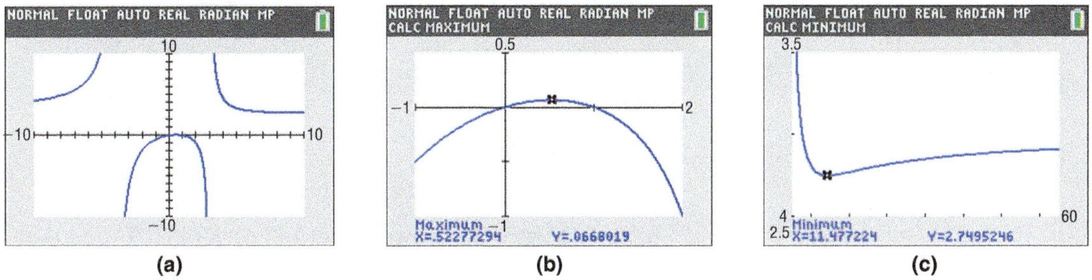

Figure 44 $Y_1 = \dfrac{3x^2 - 3x}{x^2 + x - 12}$

Step 7: Figure 44(a) does not clearly show the graph between the two x-intercepts, 0 and 1. Because the zeros in the numerator, 0 and 1, are of odd multiplicity (both are multiplicity 1), we know that the graph of R crosses the x-axis at 0 and 1. Therefore, the graph of R is above the x-axis for $0 < x < 1$. To see this part better, we graph R for $-1 \le x \le 2$ in Figure 44(b). Using MAXIMUM, we approximate the turning point to be $(0.52, 0.07)$, rounded to two decimal places.

Figure 44(a) also does not display the graph of R crossing the horizontal asymptote at $(6, 3)$. To see this part better, we graph R for $4 \le x \le 60$ in Figure 44(c). Using MINIMUM, we approximate the turning point to be $(11.48, 2.75)$, rounded to two decimal places.

Using this information along with the information gathered in Steps 1 through 6, we obtain the graph of R shown in Figure 45.

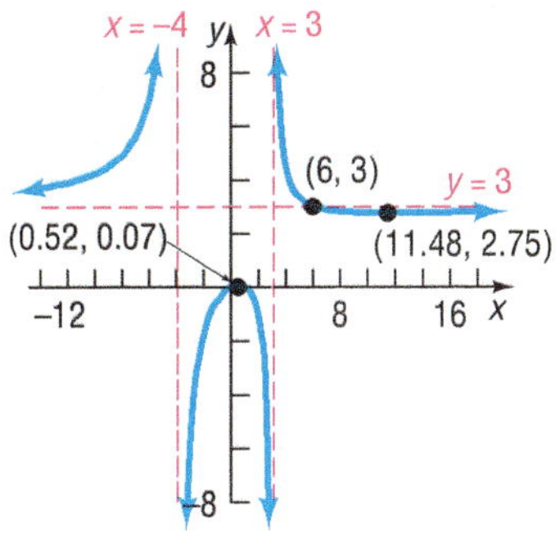

Figure 45 $R(x) = \dfrac{3x^2 - 3x}{x^2 + x - 12}$

EXAMPLE 5 Analyzing the Graph of a Rational Function with a Hole

Analyze the graph of the rational function: $R(x) = \dfrac{2x^2 - 5x + 2}{x^2 - 4}$

Solution

Step 1: Factor R and obtain

$$R(x) = \frac{(2x - 1)(x - 2)}{(x + 2)(x - 2)}$$

The domain of R is $\{x \mid x \ne -2, x \ne 2\}$.

Step 2: In lowest terms,

$$R(x) = \frac{2x - 1}{x + 2} \qquad x \ne -2, x \ne 2$$

Step 3: The graph has one x-intercept, 0.5, with odd multiplicity. The graph will cross the x-axis at $x = \frac{1}{2}$. The y-intercept is $R(0) = -0.5$.

Step 4: Since $x + 2$ is the only factor of the denominator of $R(x)$ *in lowest terms*, the graph has one vertical asymptote, $x = -2$. However, the rational function is undefined at both $x = 2$ and $x = -2$. The multiplicity of -2 is odd, so the graph will approach ∞ on one side of the asymptote and $-\infty$ on the other side.

Step 5: Since the degree of the numerator equals the degree of the denominator, the graph has a horizontal asymptote. To find it, form the quotient of the leading coefficient of the numerator, 2, and the leading coefficient of the denominator, 1. The graph of R has the horizontal asymptote $y = 2$. To find whether the graph of R intersects the asymptote, solve the equation $R(x) = 2$.

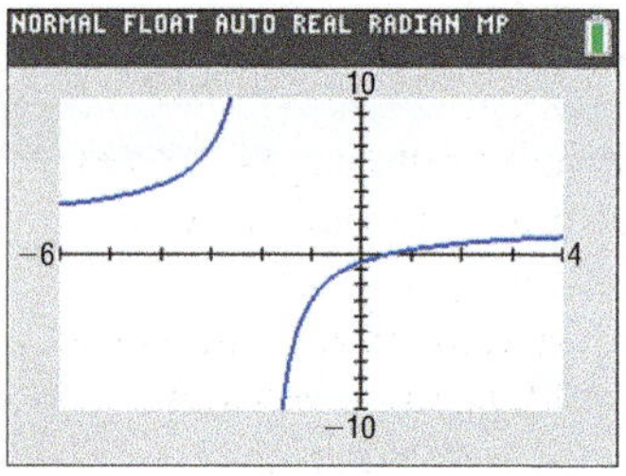

Figure 46 $Y_1 = \dfrac{2x^2 - 5x + 2}{x^2 - 4}$

$$R(x) = \frac{2x - 1}{x + 2} = 2$$
$$2x - 1 = 2(x + 2)$$
$$2x - 1 = 2x + 4$$
$$-1 = 4 \qquad \text{Impossible}$$

The graph does not intersect the line $y = 2$.

Step 6: Figure 46 shows the graph of $R(x)$. Notice that the graph has one vertical asymptote at $x = -2$. Also, the function appears to be continuous at $x = 2$.

Step 7: The analysis presented thus far does not explain the behavior of the graph at $x = 2$. We use the TABLE feature of our graphing utility to determine the behavior of the graph of R as x approaches 2. See Table 18. From the table, we conclude that the value of R approaches 0.75 as x approaches 2. This result is further verified by evaluating R in lowest terms at $x = 2$. We conclude that there is a hole in the graph at $(2, 0.75)$. Using the information gathered in Steps 1 through 6, we obtain the graph of R shown in Figure 47.

Table 18

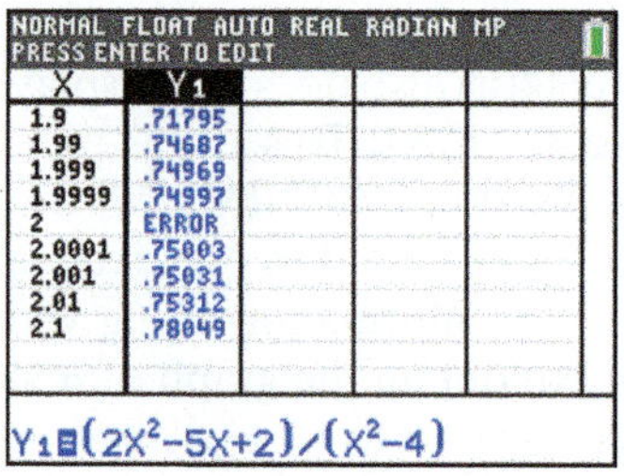

NORMAL FLOAT AUTO REAL RADIAN MP
PRESS ENTER TO EDIT

X	Y1
1.9	.71795
1.99	.74687
1.999	.74969
1.9999	.74997
2	ERROR
2.0001	.75003
2.001	.75031
2.01	.75312
2.1	.78049

Y1=(2X²−5X+2)/(X²−4)

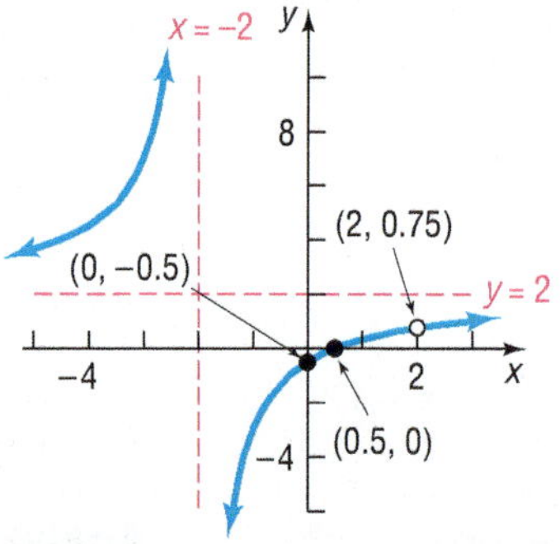

Figure 47 $R(x) = \dfrac{2x^2 - 5x + 2}{x^2 - 4}$

As Example 5 shows, **the values excluded from the domain of a rational function give rise to either vertical asymptotes or holes.**

Now Work PROBLEM 33

EXAMPLE 6 Constructing a Rational Function from Its Graph

Make up a rational function that might have the graph shown in Figure 48.

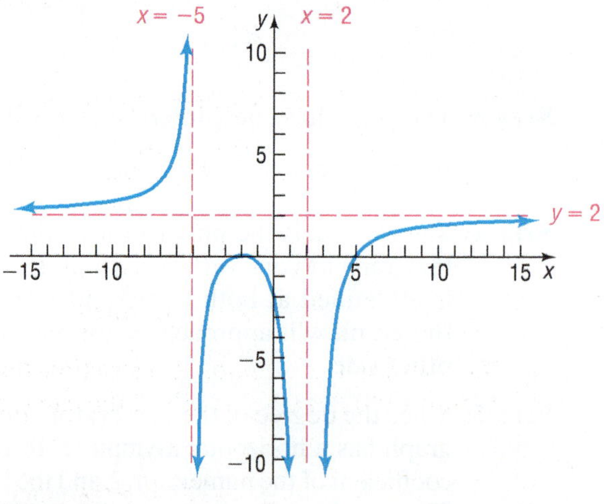

Figure 48

Solution The numerator of a rational function $R(x) = \dfrac{p(x)}{q(x)}$ in lowest terms determines the x-intercepts of its graph. The graph shown in Figure 48 has x-intercepts -2 (even multiplicity; graph touches the x-axis) and 5 (odd multiplicity; graph crosses the x-axis). So one possibility for the numerator is $p(x) = (x+2)^2(x-5)$.

The denominator of a rational function in lowest terms determines the vertical asymptotes of its graph. The vertical asymptotes of the graph are $x = -5$ and $x = 2$. Since $R(x)$ approaches ∞ from the left of $x = -5$ and $R(x)$ approaches $-\infty$ from the right of $x = -5$, we know that $(x+5)$ is a factor of odd multiplicity in $q(x)$. Also, $R(x)$ approaches $-\infty$ from both sides of $x = 2$, so $(x-2)$ is a factor of even multiplicity in $q(x)$. A possibility for the denominator is $q(x) = (x+5)(x-2)^2$.

So far we have $R(x) = \dfrac{(x+2)^2(x-5)}{(x+5)(x-2)^2}$. The horizontal asymptote of the graph given in Figure 48 is $y = 2$, so we know that the degree of the numerator must equal the degree of the denominator, and the quotient of leading coefficients must be $\dfrac{2}{1}$. This leads to $R(x) = \dfrac{2(x+2)^2(x-5)}{(x+5)(x-2)^2}$. Figure 49 shows the graph of R drawn on a graphing utility. Since Figure 49 looks similar to Figure 48, we have found a rational function R for the graph in Figure 48. ■

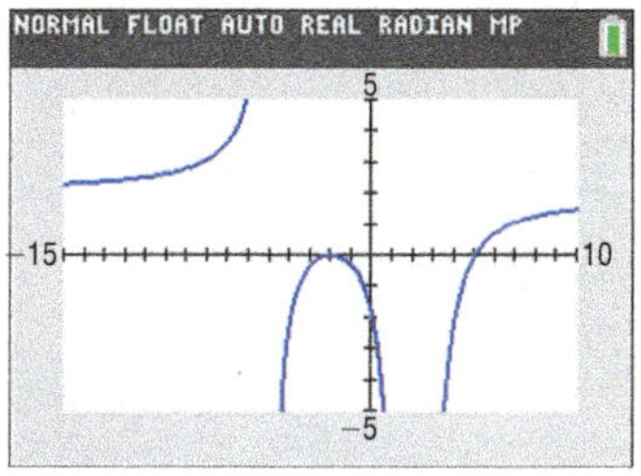

Figure 49 $Y_1 = \dfrac{2(x+2)^2(x-5)}{(x+5)(x-2)^2}$

Now Work PROBLEM 51

2 Solve Applied Problems Involving Rational Functions

EXAMPLE 7 Finding the Least Cost of a Can

Reynolds Metal Company manufactures aluminum cans in the shape of a cylinder with a capacity of 500 cubic centimeters (cm^3), or $\frac{1}{2}$ liter. The top and bottom of the can are made of a special aluminum alloy that costs 0.05¢ per square centimeter (cm^2). The sides of the can are made of material that costs 0.02¢ per square centimeter.

(a) Express the cost of material for the can as a function of the radius r of the can.
(b) Use a graphing utility to graph the function $C = C(r)$.
(c) What value of r will result in the least cost?
(d) What is this least cost?

Solution (a) Figure 50 illustrates the situation. Notice that the material required to produce a cylindrical can of height h and radius r consists of a rectangle of area $2\pi rh$ and two circles, each of area πr^2. The total cost C (in cents) of manufacturing the can is therefore

$$C = \text{Cost of the top and bottom} + \text{Cost of the side}$$

$$= \underbrace{2(\pi r^2)}_{\substack{\text{Total area}\\ \text{of top and}\\ \text{bottom}}}\ \underbrace{(0.05)}_{\substack{\text{Cost/unit}\\ \text{area}}} + \underbrace{(2\pi rh)}_{\substack{\text{Total}\\ \text{area of}\\ \text{side}}}\ \underbrace{(0.02)}_{\substack{\text{Cost/unit}\\ \text{area}}}$$

$$= 0.10\pi r^2 + 0.04\pi rh$$

But there is the additional restriction that the height h and radius r must be chosen so that the volume V of the can is 500 cm³. Since $V = \pi r^2 h$,

$$500 = \pi r^2 h \quad \text{or} \quad h = \frac{500}{\pi r^2}$$

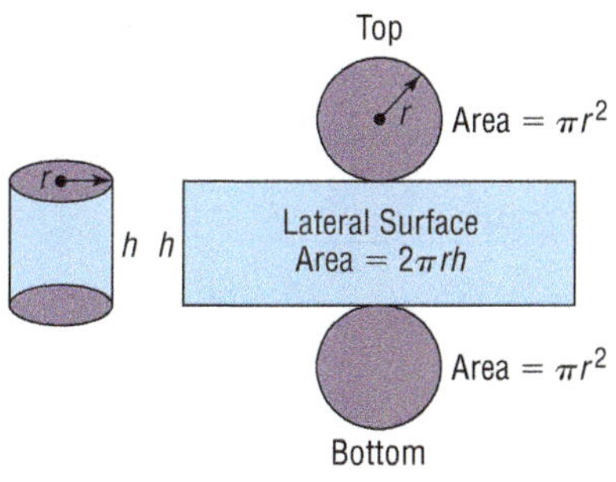

Figure 50 Surface of a cylinder

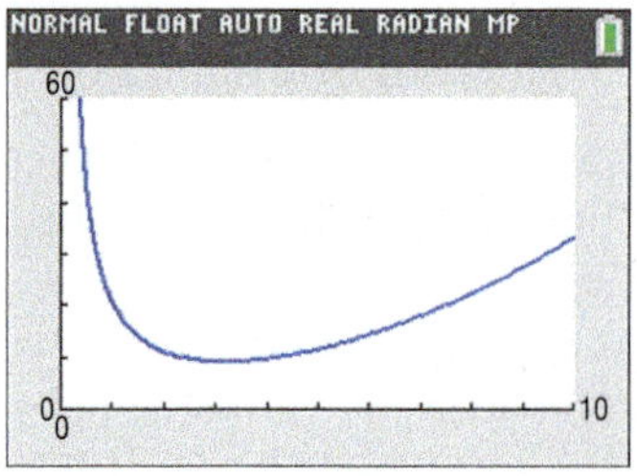

Figure 51 $Y_1 = \dfrac{0.10\pi x^3 + 20}{x}$

Substituting this expression for h, the cost C, in cents, as a function of the radius r, is

$$C(r) = 0.10\pi r^2 + 0.04\pi r \frac{500}{\pi r^2} = 0.10\pi r^2 + \frac{20}{r} = \frac{0.10\pi r^3 + 20}{r}$$

(b) See Figure 51 for the graph of $C = C(r)$.

(c) Using the MINIMUM command, the cost is least for a radius of about 3.17 cm.

(d) The least cost is $C(3.17) \approx 9.47$¢.

Now Work PROBLEM 63

5.5 Assess Your Understanding

'Are You Prepared?' *The answer is given at the end of these exercises. If you get the wrong answer, read the pages listed in red.*

1. Find the intercepts of the graph of the equation $y = \dfrac{x^2 - 1}{x^2 - 4}$. (pp. 165–166)

Concepts and Vocabulary

2. ***True or False*** Every rational function has at least one asymptote.

3. Which type of asymptote will never intersect the graph of a rational function?
(a) horizontal (b) oblique (c) vertical (d) all of these

4. Identify the y-intercept of the graph of $R(x) = \dfrac{6(x-1)}{(x+1)(x+2)}$.
(a) -3 (b) -2 (c) -1 (d) 1

5. $R(x) = \dfrac{x(x-2)^2}{x-2}$
(a) Find the domain of R.
(b) Find the x-intercepts of R.

6. ***True or False*** The graph of a rational function sometimes has a hole.

Skill Building

In Problems 7–50, follow Steps 1 through 7 on page 384 to analyze the graph of each function.

7. $R(x) = \dfrac{x+1}{x(x+4)}$

8. $R(x) = \dfrac{x}{(x-1)(x+2)}$

9. $R(x) = \dfrac{3x+3}{2x+4}$

10. $R(x) = \dfrac{2x+4}{x-1}$

11. $R(x) = \dfrac{3}{x^2-4}$

12. $R(x) = \dfrac{6}{x^2-x-6}$

13. $P(x) = \dfrac{x^4+x^2+1}{x^2-1}$

14. $Q(x) = \dfrac{x^4-1}{x^2-4}$

15. $H(x) = \dfrac{x^3-1}{x^2-9}$

16. $G(x) = \dfrac{x^3+1}{x^2+2x}$

17. $R(x) = \dfrac{x^2}{x^2+x-6}$

18. $R(x) = \dfrac{x^2+x-12}{x^2-4}$

19. $G(x) = \dfrac{x}{x^2-4}$

20. $G(x) = \dfrac{3x}{x^2-1}$

21. $R(x) = \dfrac{3}{(x-1)(x^2-4)}$

22. $R(x) = \dfrac{-4}{(x+1)(x^2-9)}$

23. $H(x) = \dfrac{x^2-1}{x^4-16}$

24. $H(x) = \dfrac{x^2+4}{x^4-1}$

25. $F(x) = \dfrac{x^2-3x-4}{x+2}$

26. $F(x) = \dfrac{x^2+3x+2}{x-1}$

27. $R(x) = \dfrac{x^2+x-12}{x-4}$

28. $R(x) = \dfrac{x^2-x-12}{x+5}$

29. $F(x) = \dfrac{x^2+x-12}{x+2}$

30. $G(x) = \dfrac{x^2-x-12}{x+1}$

31. $R(x) = \dfrac{x(x-1)^2}{(x+3)^3}$

32. $R(x) = \dfrac{(x-1)(x+2)(x-3)}{x(x-4)^2}$

33. $R(x) = \dfrac{x^2+x-12}{x^2-x-6}$

34. $R(x) = \dfrac{x^2+3x-10}{x^2+8x+15}$

35. $R(x) = \dfrac{6x^2-7x-3}{2x^2-7x+6}$

36. $R(x) = \dfrac{8x^2+26x+15}{2x^2-x-15}$

37. $R(x) = \dfrac{x^2+5x+6}{x+3}$

38. $R(x) = \dfrac{x^2+x-30}{x+6}$

39. $H(x) = \dfrac{3x-6}{4-x^2}$

40. $H(x) = \dfrac{2-2x}{x^2-1}$

41. $F(x) = \dfrac{x^2-5x+4}{x^2-2x+1}$

42. $F(x) = \dfrac{x^2-2x-15}{x^2+6x+9}$

43. $G(x) = \dfrac{x}{(x+2)^2}$

44. $G(x) = \dfrac{2-x}{(x-1)^2}$

45. $f(x) = x + \dfrac{1}{x}$

46. $f(x) = 2x + \dfrac{9}{x}$

47. $f(x) = x^2 + \dfrac{1}{x}$

48. $f(x) = 2x^2 + \dfrac{16}{x}$

49. $f(x) = x + \dfrac{1}{x^3}$

50. $f(x) = 2x + \dfrac{9}{x^3}$

In Problems 51–54, find a rational function that might have the given graph. (More than one answer might be possible.)

51.

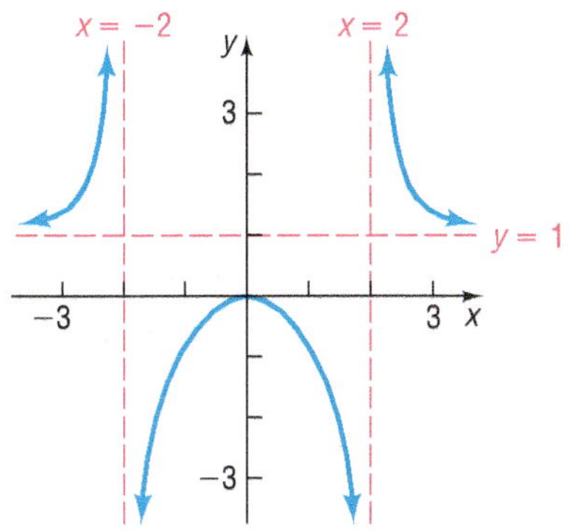

52.

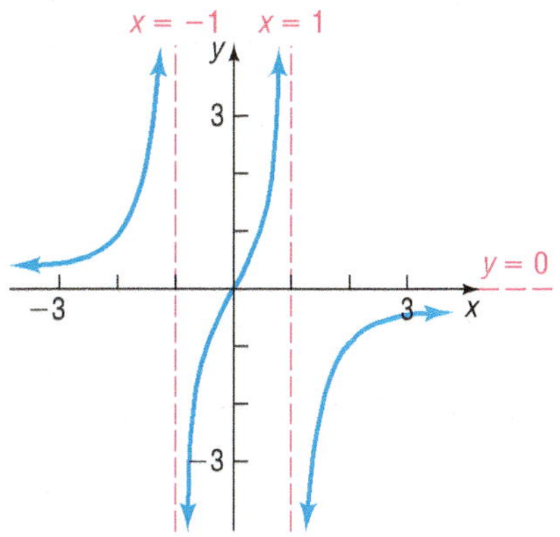

53.

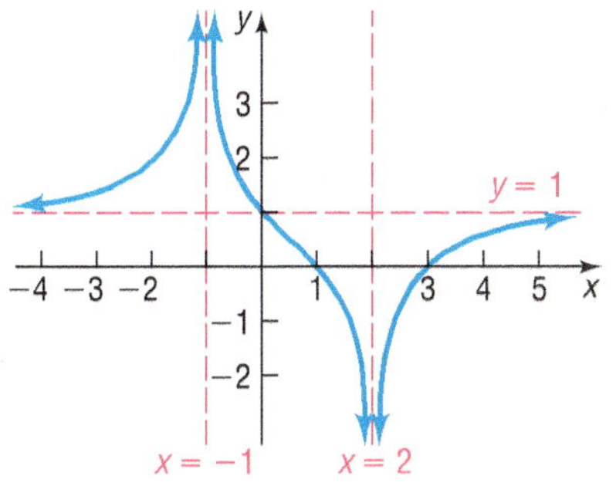

54.

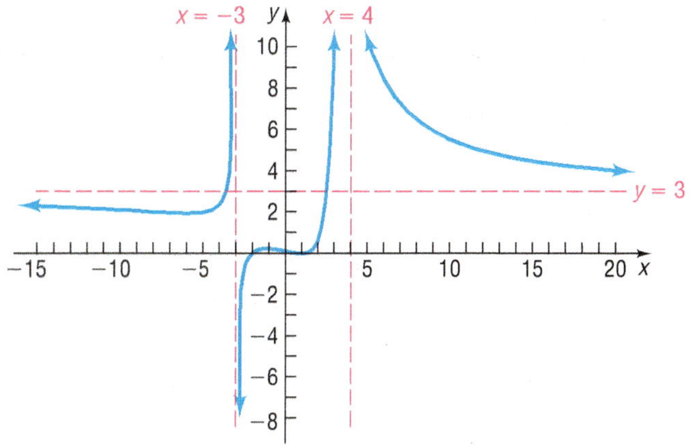

Applications and Extensions

55. **Probability** Suppose you attend a fundraiser where each person in attendance is given a ball, each with a different number. The balls are numbered 1 through x. Each person in attendance places his or her ball in an urn. After dinner, a ball is chosen at random from the urn. The probability that your ball is selected is $\frac{1}{x}$. Therefore, the probability that your ball is not chosen is $1 - \frac{1}{x}$. Graph $P(x) = 1 - \frac{1}{x}$ using transformations. Comment on the likelihood of your ball *not* being chosen as x increases.

56. **Waiting in Line** Suppose that two employees at a fast-food restaurant can serve customers at the rate of 6 customers per minute. Further suppose that customers are arriving at the restaurant at the rate of x customers per minute. Then the average waiting time T, in minutes, spent waiting in line and having your order taken and filled is given by the function $T(x) = -\dfrac{1}{x-6}$, where $0 < x < 6$. Graph this function using transformations.

57. **Drug Concentration** The concentration C of a certain drug in a patient's bloodstream t hours after injection is given by

$$C(t) = \frac{t}{2t^2 + 1}$$

(a) Find the horizontal asymptote of $C(t)$. What happens to the concentration of the drug as t increases?
(b) Using your graphing utility, graph $C = C(t)$.
(c) Determine the time at which the concentration is highest.

58. **Drug Concentration** The concentration C of a certain drug in a patient's bloodstream t minutes after injection is given by

$$C(t) = \frac{50t}{t^2 + 25}$$

(a) Find the horizontal asymptote of $C(t)$. What happens to the concentration of the drug as t increases?
(b) Using your graphing utility, graph $C = C(t)$.
(c) Determine the time at which the concentration is highest.

59. Minimum Cost A rectangular area adjacent to a river is to be fenced in; no fence is needed on the river side. The enclosed area is to be 1000 square feet. Fencing for the side parallel to the river is \$5 per linear foot, and fencing for the other two sides is \$8 per linear foot; the four corner posts are \$25 apiece. Let x be the length of one of the sides perpendicular to the river.
(a) Write a function $C(x)$ that describes the cost of the project.
(b) What is the domain of C?
(c) Use a graphing utility to graph $C = C(x)$.
(d) Find the dimensions of the cheapest enclosure.

Source: *www.uncwil.edu/courses/math111hb/PandR/rational/rational.html*

60. Doppler Effect The Doppler effect (named after Christian Doppler) is the change in the pitch (frequency) of the sound from a source (s) as heard by an observer (o) when one or both are in motion. If we assume both the source and the observer are moving in the same direction, the relationship is

$$f' = f_a\left(\frac{v - v_o}{v - v_s}\right)$$

where f' = perceived pitch by the observer
f_a = actual pitch of the source
v = speed of sound in air (assume 772.4 mph)
v_o = speed of the observer
v_s = speed of the source

Suppose that you are traveling down the road at 45 mph and you hear an ambulance (with siren) coming toward you from the rear. The actual pitch of the siren is 600 hertz (Hz).
(a) Write a function $f'(v_s)$ that describes this scenario.
(b) If $f' = 620$ Hz, find the speed of the ambulance.
(c) Use a graphing utility to graph the function.
(d) Verify your answer from part (b).

Source: *www.kettering.edu/~drussell/*

61. Minimizing Surface Area United Parcel Service has contracted you to design a closed box with a square base that has a volume of 10,000 cubic inches. See the illustration.

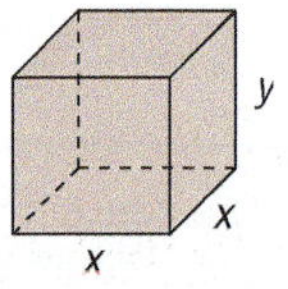

(a) Express the surface area S of the box as a function of x.
(b) Using a graphing utility, graph the function found in part (a).
(c) What is the minimum amount of cardboard that can be used to construct the box?
(d) What are the dimensions of the box that minimize the surface area?
(e) Why might UPS be interested in designing a box that minimizes the surface area?

62. Minimizing Surface Area United Parcel Service has contracted you to design an open box with a square base that has a volume of 5000 cubic inches. See the illustration.

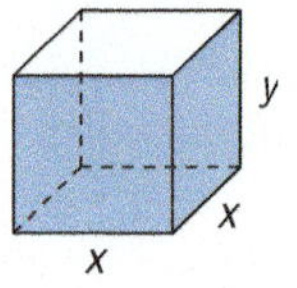

(a) Express the surface area S of the box as a function of x.
(b) Using a graphing utility, graph the function found in part (a).
(c) What is the minimum amount of cardboard that can be used to construct the box?
(d) What are the dimensions of the box that minimize the surface area?
(e) Why might UPS be interested in designing a box that minimizes the surface area?

63. Cost of a Can A can in the shape of a right circular cylinder is required to have a volume of 500 cubic centimeters. The top and bottom are made of material that costs 0.06¢ per square centimeter, while the sides are made of material that costs 0.04¢ per square centimeter.
(a) Express the total cost C of the material as a function of the radius r of the cylinder. (Refer to Figure 50.)
(b) Graph $C = C(r)$. For what value of r is the cost C a minimum?

64. Material Needed to Make a Drum A steel drum in the shape of a right circular cylinder is required to have a volume of 100 cubic feet.

(a) Express the amount A of material required to make the drum as a function of the radius r of the cylinder.
(b) How much material is required if the drum's radius is 3 feet?
(c) How much material is required if the drum's radius is 4 feet?
(d) How much material is required if the drum's radius is 5 feet?
(e) Graph $A = A(r)$. For what value of r is A smallest?

Explaining Concepts: Discussion and Writing

65. Graph each of the following functions:

$$y = \frac{x^2 - 1}{x - 1} \qquad y = \frac{x^3 - 1}{x - 1}$$

$$y = \frac{x^4 - 1}{x - 1} \qquad y = \frac{x^5 - 1}{x - 1}$$

Is $x = 1$ a vertical asymptote? Why not? What is happening for $x = 1$? What do you conjecture about $y = \frac{x^n - 1}{x - 1}$, $n \geq 1$ an integer, for $x = 1$?

66. Graph each of the following functions:

$$y = \frac{x^2}{x-1} \quad y = \frac{x^4}{x-1} \quad y = \frac{x^6}{x-1} \quad y = \frac{x^8}{x-1}$$

What similarities do you see? What differences?

67. Write a few paragraphs that provide a general strategy for graphing a rational function. Be sure to mention the following: proper, improper, intercepts, and asymptotes.

68. Create a rational function that has the following characteristics: crosses the x-axis at 2; touches the x-axis at -1; one vertical asymptote at $x = -5$ and another at $x = 6$; and one horizontal asymptote, $y = 3$. Compare your function to a fellow classmate's. How do they differ? What are their similarities?

69. Create a rational function that has the following characteristics: crosses the x-axis at 3; touches the x-axis at -2; one vertical asymptote, $x = 1$; and one horizontal asymptote, $y = 2$. Give your rational function to a fellow classmate and ask for a written critique of your rational function.

70. Create a rational function with the following characteristics: three real zeros, one of multiplicity 2; y-intercept 1; vertical asymptotes, $x = -2$ and $x = 3$; oblique asymptote, $y = 2x + 1$. Is this rational function unique? Compare your function with those of other students. What will be the same as everyone else's? Add some more characteristics, such as symmetry or naming the real zeros. How does this modify the rational function?

71. Explain the circumstances under which the graph of a rational function will have a hole.

Retain Your Knowledge

Problems 72–75 are based on material learned earlier in the course. The purpose of these problems is to keep the material fresh in your mind so that you are better prepared for the final exam.

72. Subtract: $(4x^3 - 7x - 1) - (5x^2 - 9x - 3)$

73. Solve: $\dfrac{3x}{3x+1} = \dfrac{x-2}{x+5}$

74. Find the maximum value of $f(x) = -\dfrac{2}{3}x^2 + 6x - 5$.

75. Approximate $\dfrac{\sqrt{5}-3}{\sqrt{7}+2}$. Round your answer to three decimal places.

'Are You Prepared?' Answer

1. $\left(0, \frac{1}{4}\right), (1, 0), (-1, 0)$

5.6 Polynomial and Rational Inequalities

PREPARING FOR THIS SECTION *Before getting started, review the following:*

- Solving Linear Inequalities (Section 1.7, pp. 150–151)
- Solving Quadratic Inequalities (Section 4.5, pp. 320–322)

Now Work the **'Are You Prepared?'** problems on page 398.

OBJECTIVES
1 Solve Polynomial Inequalities Algebraically and Graphically (p. 393)
2 Solve Rational Inequalities Algebraically and Graphically (p. 395)

1 Solve Polynomial Inequalities Algebraically and Graphically

In this section we solve inequalities that involve polynomials of degree 3 and higher, along with inequalities that involve rational functions. To help understand the algebraic procedure for solving such inequalities, we use the information obtained in Sections 5.1, 5.2, and 5.5 about the graphs of polynomial and rational functions. The approach follows the same logic used to solve inequalities involving quadratic functions.

EXAMPLE 1 Solving a Polynomial Inequality Using Its Graph

Solve $(x + 3)(x - 1)^2 > 0$ by graphing $f(x) = (x + 3)(x - 1)^2$.

By Hand Graphical Solution

Graph $f(x) = (x + 3)(x - 1)^2$ and determine the intervals of x for which the graph is above the x-axis. Do you see why? These values of x result in $f(x)$ being positive. Using Steps 1 through 8 on page 343, we obtain the graph shown in Figure 52.

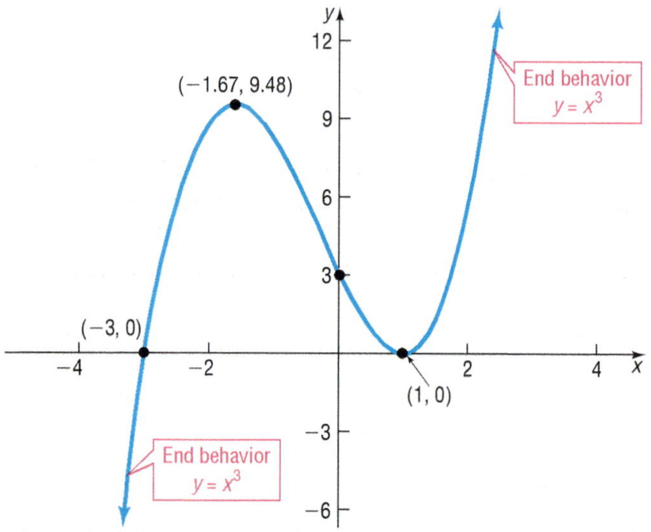

Figure 52 $f(x) = (x + 3)(x - 1)^2$

From the graph, we can see that $f(x) > 0$ for $-3 < x < 1$ or $x > 1$. The solution set is $\{x \mid -3 < x < 1 \text{ or } x > 1\}$ or, using interval notation, $(-3, 1) \cup (1, \infty)$.

Graphing Utility Solution

Graph $Y_1 = (x + 3)(x - 1)^2$. See Figure 53. Using the ZERO command, find that the x-intercepts of the graph of Y_1 are -3 and 1. The graph of Y_1 is above the x-axis (and therefore f is positive) for $-3 < x < 1$ or $x > 1$. Therefore, the solution set is $\{x \mid -3 < x < 1 \text{ or } x > 1\}$ or, using interval notation, $(-3, 1) \cup (1, \infty)$.

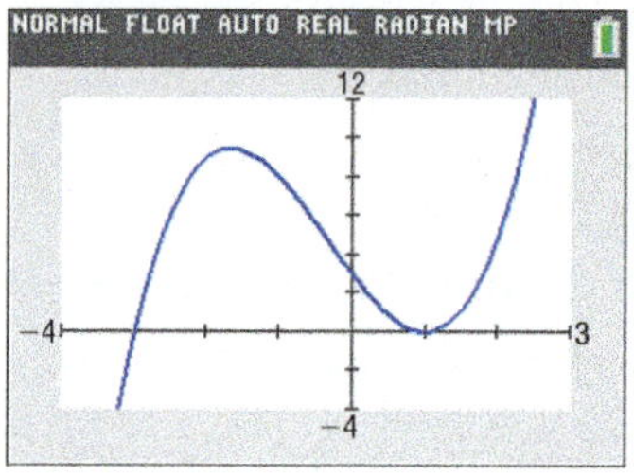

Figure 53 $Y_1 = (x + 3)(x - 1)^2$

Now Work PROBLEM 9

The results of Example 1 lead to the following approach to solving polynomial inequalities algebraically. Suppose that the polynomial inequality is in one of the forms

$$f(x) < 0 \qquad f(x) > 0 \qquad f(x) \le 0 \qquad f(x) \ge 0$$

Locate the zeros (x-intercepts of the graph) of the polynomial function f. We know that the sign of f can change on either side of an x-intercept, so we use these zeros to divide the real number line into intervals. Then we know that on each interval the graph of f is either above $[f(x) > 0]$ or below $[f(x) < 0]$ the x-axis.

EXAMPLE 2 How to Solve a Polynomial Inequality Algebraically

Solve the inequality $x^4 > x$ algebraically, and graph the solution set.

Step-by-Step Solution

Step 1: Write the inequality so that a polynomial expression f is on the left side and zero is on the right side.

Rearrange the inequality so that 0 is on the right side.

$$x^4 > x$$

$$x^4 - x > 0 \quad \text{Subtract } x \text{ from both sides of the inequality.}$$

This inequality is equivalent to the one we wish to solve.

Step 2: Determine the real zeros (x-intercepts of the graph) of f.

Find the zeros of $f(x) = x^4 - x$ by solving $x^4 - x = 0$.

$$x^4 - x = 0$$

$$x(x^3 - 1) = 0 \quad \text{Factor out } x.$$

$$x(x-1)(x^2 + x + 1) = 0 \quad \text{Factor the difference of two cubes.}$$

$$x = 0 \quad \text{or} \quad x - 1 = 0 \quad \text{or} \quad x^2 + x + 1 = 0 \quad \text{Set each factor equal to zero and solve.}$$

$$x = 0 \quad \text{or} \quad x = 1$$

The equation $x^2 + x + 1 = 0$ has no real solutions. Do you see why?

Step 3: Use the zeros found in Step 2 to divide the real number line into intervals.

Use the zeros to separate the real number line into three intervals:

$$(-\infty, 0) \qquad (0, 1) \qquad (1, \infty)$$

Step 4: Select a number in each interval, evaluate f at the number, and determine whether f is positive or negative. If f is positive, all values of f in the interval are positive. If f is negative, all values of f in the interval are negative.

Select a test number in each interval found in Step 3 and evaluate $f(x) = x^4 - x$ at each number to determine if $f(x)$ is positive or negative. See Table 19.

Table 19

	0	1	x
Interval	$(-\infty, 0)$	$(0, 1)$	$(1, \infty)$
Number chosen	-1	$\frac{1}{2}$	2
Value of f	$f(-1) = 2$	$f\left(\frac{1}{2}\right) = -\frac{7}{16}$	$f(2) = 14$
Conclusion	Positive	Negative	Positive

Note: If the inequality is not strict ($\leq$ or $\geq$), include the solutions of $f(x) = 0$ in the solution set. ■

Since we want to know where $f(x)$ is positive, conclude that $f(x) > 0$ for all numbers x for which $x < 0$ or $x > 1$. Because the original inequality is strict, numbers x that satisfy the equation $x^4 = x$ are not solutions. The solution set of the inequality $x^4 > x$ is $\{x \mid x < 0 \text{ or } x > 1\}$ or, using interval notation, $(-\infty, 0) \cup (1, \infty)$.

Figure 54 shows the graph of the solution set.

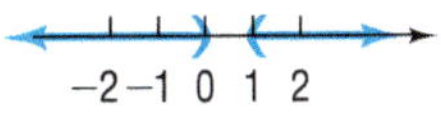

Figure 54

✓**Check:** Graph $Y_1 = x^4$ and $Y_2 = x$ on the same screen. Use INTERSECT to find where $Y_1 = Y_2$. Then determine where $Y_1 > Y_2$. ■

The Role of Multiplicity in Solving Polynomial Inequalities

In Example 2, we used the number -1 and found that f is positive for all $x < 0$. Because the "cut point" of 0 is the result of a zero of odd multiplicity (x is a factor to the first power), we know that the sign of f will change on either side of 0, so for $0 < x < 1$, f will be negative. Similarly, we know that f will be positive for $x > 1$, since the multiplicity of the zero 1 is odd. Therefore, the solution set of $x^4 > x$ is $\{x \mid x < 0 \text{ or } x > 1\}$ or, using interval notation, $(-\infty, 0) \cup (1, \infty)$.

Now Work PROBLEM 21

2 Solve Rational Inequalities Algebraically and Graphically

Just as we presented a graphical approach to help understand the algebraic procedure for solving inequalities involving polynomials, we present a graphical approach to help understand the algebraic procedure for solving inequalities involving rational expressions.

EXAMPLE 3 Solving a Rational Inequality Using Its Graph

Solve $\frac{x-1}{x^2-4} \geq 0$ by graphing $R(x) = \frac{x-1}{x^2-4}$.

By Hand Graphical Solution

Graph $R(x) = \frac{x-1}{x^2-4}$ and determine the intervals of x such that the graph is above or on the x-axis. Do you see why? These values of x result in $R(x)$ being positive or zero. We graphed $R(x) = \frac{x-1}{x^2-4}$ in Example 1 from Section 5.5 (p. 383). We reproduce the graph in Figure 55.

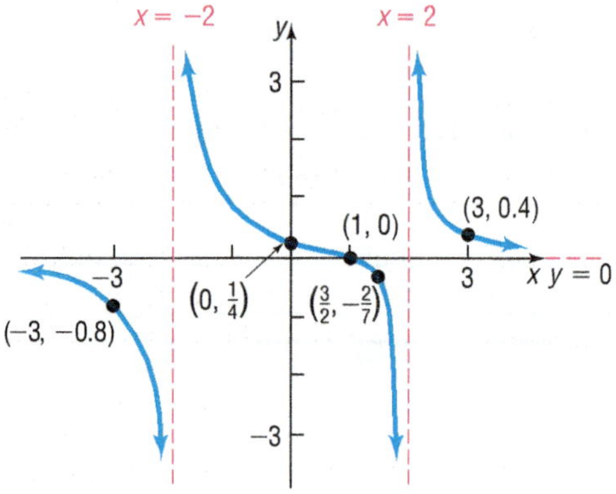

Figure 55 $R(x) = \frac{x-1}{x^2-4}$

From the graph, we can see that $R(x) \geq 0$ for $-2 < x \leq 1$ or $x > 2$. The solution set is $\{x \mid -2 < x \leq 1 \text{ or } x > 2\}$ or, using interval notation, $(-2, 1] \cup (2, \infty)$.

Graphing Utility Solution

Graph $Y_1 = \frac{x-1}{x^2-4}$. See Figure 56. Using the ZERO command, find that the x-intercept of the graph of Y_1 is 1. The graph of Y_1 is above the x-axis (and, therefore, R is positive) for $-2 < x < 1$ or $x > 2$. Since the inequality is not strict, include 1 in the solution set. Therefore, the solution set is $\{x \mid -2 < x \leq 1 \text{ or } x > 2\}$ or, using interval notation, $(-2, 1] \cup (2, \infty)$. Do you see why we do not include 2?

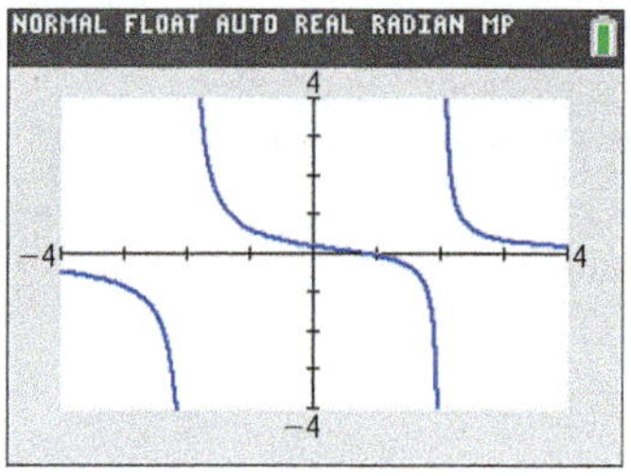

Figure 56 $Y_1 = \frac{x-1}{x^2-4}$

Now Work PROBLEMS 15 AND 33

To solve a rational inequality algebraically, we follow the same approach that we used to solve a polynomial inequality algebraically. However, we must also identify the zeros of the denominator of the rational function, because the sign of a rational function may change on either side of a vertical asymptote. Convince yourself of this by looking at Figure 55. Notice the function values are negative for $x < -2$ and are positive for $x > -2$ (but less than 1).

EXAMPLE 4 How to Solve a Rational Inequality Algebraically

Solve the inequality $\frac{3x^2 + 13x + 9}{(x+2)^2} \leq 3$ algebraically, and graph the solution set.

Step-by-Step Solution

Step 1: Write the inequality so that a rational expression f is on the left side and zero is on the right side.

Rearrange the inequality so that 0 is on the right side.

$$\frac{3x^2 + 13x + 9}{(x+2)^2} \leq 3$$

$$\frac{3x^2 + 13x + 9}{x^2 + 4x + 4} - 3 \leq 0$$

Subtract 3 from both sides of the inequality; Expand $(x+2)^2$.

$$\frac{3x^2 + 13x + 9}{x^2 + 4x + 4} - 3 \cdot \frac{x^2 + 4x + 4}{x^2 + 4x + 4} \leq 0$$

Multiply 3 by $\frac{x^2 + 4x + 4}{x^2 + 4x + 4}$.

$$\frac{3x^2 + 13x + 9 - 3x^2 - 12x - 12}{x^2 + 4x + 4} \leq 0$$

Write as a single quotient.

$$\frac{x-3}{(x+2)^2} \leq 0$$

Combine like terms.

Step 2: Determine the real zeros (x-intercepts of the graph) of f and the real numbers for which f is undefined.

The zero of $f(x) = \dfrac{x-3}{(x+2)^2}$ is 3. Also, f is undefined for $x = -2$.

Step 3: Use the zeros and undefined values found in Step 2 to divide the real number line into intervals.

Use the zero and the undefined value to separate the real number line into three intervals:

$$(-\infty, -2) \quad (-2, 3) \quad (3, \infty)$$

Step 4: Select a number in each interval, evaluate f at the number, and determine whether $f(x)$ is positive or negative. If $f(x)$ is positive, all values of f in the interval are positive. If $f(x)$ is negative, all values of f in the interval are negative.

Note: If the inequality is not strict ($\le$ or $\ge$), include the solutions of $f(x) = 0$ in the solution set. ■

Select a test number in each interval from Step 3, and evaluate f at each number to determine whether $f(x)$ is positive or negative. See Table 20.

Table 20

	−2	3	x
Interval	$(-\infty, -2)$	$(-2, 3)$	$(3, \infty)$
Number chosen	−3	0	4
Value of *f*	$f(-3) = -6$	$f(0) = -\frac{3}{4}$	$f(4) = \frac{1}{36}$
Conclusion	Negative	Negative	Positive

Since we want to know where $f(x)$ is negative or zero, we conclude that $f(x) \le 0$ for all numbers for which $x < -2$ or $-2 < x \le 3$. Notice that we do not include -2 in the solution because -2 is not in the domain of f. The solution set of the inequality $\dfrac{3x^2 + 13x + 9}{(x+2)^2} \le 3$ is $\{x \mid x < -2 \text{ or } -2 < x \le 3\}$ or, using interval notation, $(-\infty, -2) \cup (-2, 3]$. Figure 57 shows the graph of the solution set.

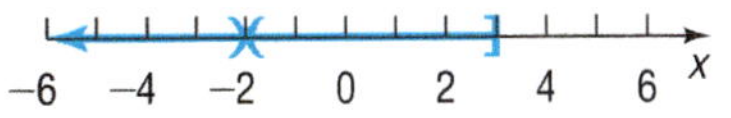

Figure 57

✓**Check:** Graph $Y_1 = \dfrac{3x^2 + 13x + 9}{(x+2)^2}$ and $Y_2 = 3$ on the same screen. Use INTERSECT to find where $Y_1 = Y_2$. Then determine where $Y_1 \le Y_2$. ■

The Role of Multiplicity in Solving Rational Inequalities

In Example 4, we used the number -3 and found that R is negative for all $x < -2$. Because the "cut point" of -2 is the result of a zero of even multiplicity, we know the sign of R will not change on either side of -2, so for $-2 < x < 3$, R will also be negative. Because the "cut point" of 3 is the result of a zero of odd multiplicity, the sign of R will change on either side of 3, so for $x > 3$, R will be positive. Therefore, the solution set of $\dfrac{3x^2 + 13x + 9}{(x+2)^2} \le 3$ is $\{x \mid x < -2 \text{ or } -2 < x \le 3\}$ or, using interval notation, $(-\infty, -2) \cup (-2, 3]$.

Now Work PROBLEM 39

SUMMARY

Steps for Solving Polynomial and Rational Inequalities Algebraically

STEP 1: Write the inequality so that a polynomial or rational expression f is on the left side and zero is on the right side in one of the following forms:

$$f(x) > 0 \qquad f(x) \ge 0 \qquad f(x) < 0 \qquad f(x) \le 0$$

For rational expressions, be sure that the left side is written as a single quotient, and find the domain of f.

STEP 2: Determine the real numbers at which the expression f equals zero and, if the expression is rational, the real numbers at which the expression f is undefined.

STEP 3: Use the numbers found in Step 2 to separate the real number line into intervals.

STEP 4: Select a number in each interval and evaluate f at the number.
(a) If the value of f is positive, then $f(x) > 0$ for all numbers x in the interval.
(b) If the value of f is negative, then $f(x) < 0$ for all numbers x in the interval.
If the inequality is not strict ($\ge$ or $\le$), include the solutions of $f(x) = 0$ that are in the domain of f in the solution set. Be careful to exclude values of x where f is undefined.

5.6 Assess Your Understanding

'Are You Prepared?' *Answers are given at the end of these exercises. If you get a wrong answer, read the pages listed in red.*

1. Solve the inequality $3 - 4x > 5$. Graph the solution set. (pp. 150–151)

2. Solve the inequality $x^2 - 5x \le 24$. Graph the solution set. (pp. 320–322)

Concepts and Vocabulary

3. Which of the following could be a test number for the interval $-2 < x < 5$?

(a) -3 (b) -2 (c) 4 (d) 7

4. ***True or False*** The graph of $f(x) = \dfrac{x}{x-3}$ is above the x-axis for $x < 0$ or $x > 3$, so the solution set of the inequality $\dfrac{x}{x-3} \ge 0$ is $\{x \mid x \le 0 \text{ or } x \ge 3\}$.

Skill Building

In Problems 5–8, use the graph of the function f to solve the inequality.

5. (a) $f(x) > 0$
(b) $f(x) \le 0$

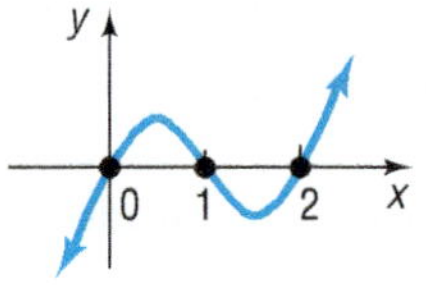

6. (a) $f(x) < 0$
(b) $f(x) \ge 0$

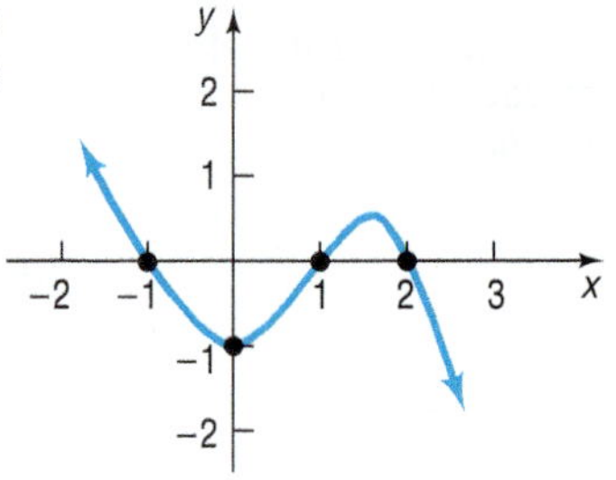

7. (a) $f(x) < 0$
(b) $f(x) \ge 0$

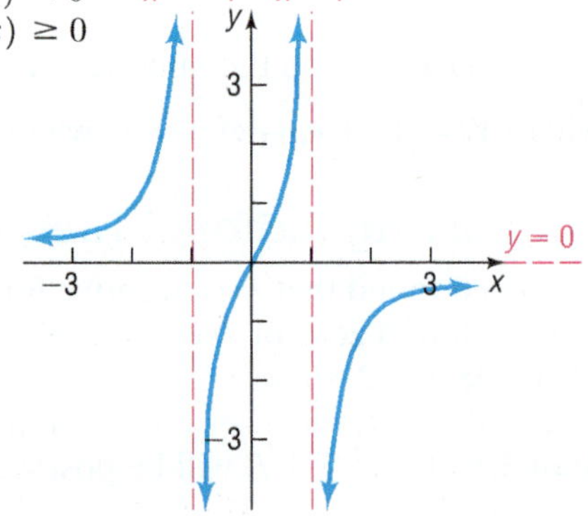

8. (a) $f(x) > 0$
(b) $f(x) \le 0$

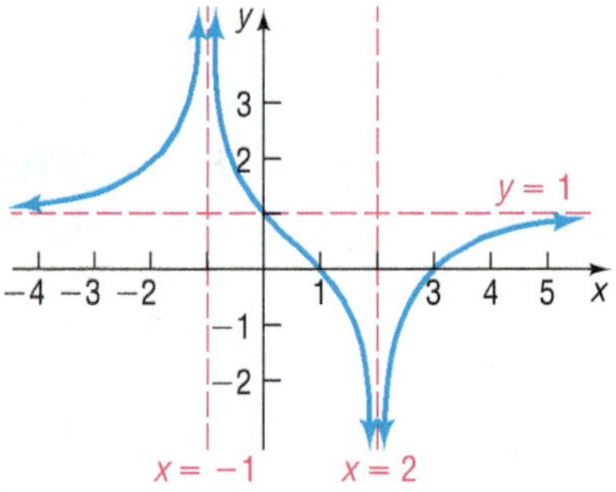

In Problems 9–14, solve the inequality by using the graph of the function.
[**Hint:** The graphs were drawn in Problems 81–86 of Section 5.1.]

9. Solve $f(x) < 0$, where $f(x) = x^2(x - 3)$.

10. Solve $f(x) \le 0$, where $f(x) = x(x + 2)^2$.

11. Solve $f(x) \ge 0$, where $f(x) = (x + 4)^2(1 - x)$.

12. Solve $f(x) > 0$, where $f(x) = (x - 1)(x + 3)^2$.

13. Solve $f(x) \le 0$, where $f(x) = -2(x + 2)(x - 2)^3$.

14. Solve $f(x) < 0$, where $f(x) = -\dfrac{1}{2}(x + 4)(x - 1)^3$.

In Problems 15–18, solve the inequality by using the graph of the function.
[**Hint:** The graphs were drawn in Problems 7–10 of Section 5.5.]

15. Solve $R(x) > 0$, where $R(x) = \dfrac{x + 1}{x(x + 4)}$.

16. Solve $R(x) < 0$, where $R(x) = \dfrac{x}{(x - 1)(x + 2)}$.

17. Solve $R(x) \le 0$, where $R(x) = \dfrac{3x + 3}{2x + 4}$.

18. Solve $R(x) \ge 0$, where $R(x) = \dfrac{2x + 4}{x - 1}$.

In Problems 19–48, solve each inequality algebraically.

19. $(x - 5)^2(x + 2) < 0$

20. $(x - 5)(x + 2)^2 > 0$

21. $x^3 - 4x^2 > 0$

22. $x^3 + 8x^2 < 0$

23. $2x^3 > -8x^2$

24. $3x^3 < -15x^2$

25. $(x - 1)(x - 2)(x - 3) \leq 0$

26. $(x + 1)(x + 2)(x + 3) \leq 0$

27. $x^3 - 2x^2 - 3x > 0$

28. $x^3 + 2x^2 - 3x > 0$

29. $x^4 > x^2$

30. $x^4 < 9x^2$

31. $x^4 > 1$

32. $x^3 > 1$

33. $\dfrac{x + 1}{x - 1} > 0$

34. $\dfrac{x - 3}{x + 1} > 0$

35. $\dfrac{(x - 1)(x + 1)}{x} \leq 0$

36. $\dfrac{(x - 3)(x + 2)}{x - 1} \leq 0$

37. $\dfrac{(x - 2)^2}{x^2 - 1} \geq 0$

38. $\dfrac{(x + 5)^2}{x^2 - 4} \geq 0$

39. $\dfrac{x + 4}{x - 2} \leq 1$

40. $\dfrac{x + 2}{x - 4} \geq 1$

41. $\dfrac{3x - 5}{x + 2} \leq 2$

42. $\dfrac{x - 4}{2x + 4} \geq 1$

43. $\dfrac{1}{x - 2} < \dfrac{2}{3x - 9}$

44. $\dfrac{5}{x - 3} > \dfrac{3}{x + 1}$

45. $\dfrac{x^2(3 + x)(x + 4)}{(x + 5)(x - 1)} \geq 0$

46. $\dfrac{x(x^2 + 1)(x - 2)}{(x - 1)(x + 1)} \geq 0$

47. $\dfrac{(3 - x)^3(2x + 1)}{x^3 - 1} < 0$

48. $\dfrac{(2 - x)^3(3x - 2)}{x^3 + 1} < 0$

Mixed Practice

In Problems 49–60, solve each inequality algebraically.

49. $(x + 1)(x - 3)(x - 5) > 0$

50. $(2x - 1)(x + 2)(x + 5) < 0$

51. $7x - 4 \geq -2x^2$

52. $x^2 + 3x \geq 10$

53. $\dfrac{x + 1}{x - 3} \leq 2$

54. $\dfrac{x - 1}{x + 2} \geq -2$

55. $3(x^2 - 2) < 2(x - 1)^2 + x^2$

56. $(x - 3)(x + 2) < x^2 + 3x + 5$

57. $6x - 5 < \dfrac{6}{x}$

58. $x + \dfrac{12}{x} < 7$

59. $x^3 - 9x \leq 0$

60. $x^3 - x \geq 0$

In Problems 61 and 62, (a) find the zeros of each function, (b) factor each function over the real numbers, (c) graph each function by hand, and (d) solve $f(x) < 0$.

61. $f(x) = 2x^4 + 11x^3 - 11x^2 - 104x - 48$

62. $f(x) = 4x^5 - 19x^4 + 32x^3 - 31x^2 + 28x - 12$

In Problems 63–66, (a) graph each function by hand, and (b) solve $f(x) \geq 0$.

63. $f(x) = \dfrac{x^2 + 5x - 6}{x^2 - 4x + 4}$

64. $f(x) = \dfrac{2x^2 + 9x + 9}{x^2 - 4}$

65. $f(x) = \dfrac{x^3 + 2x^2 - 11x - 12}{x^2 - x - 6}$

66. $f(x) = \dfrac{x^3 - 6x^2 + 9x - 4}{x^2 + x - 20}$

Applications and Extensions

67. For what positive numbers will the cube of a number exceed four times its square?

68. For what positive numbers will the cube of a number be less than the number?

69. What is the domain of the function $f(x) = \sqrt{x^4 - 16}$?

70. What is the domain of the function $f(x) = \sqrt{x^3 - 3x^2}$?

71. What is the domain of the function $f(x) = \sqrt{\dfrac{x - 2}{x + 4}}$?

72. What is the domain of the function $f(x) = \sqrt{\dfrac{x - 1}{x + 4}}$?

In Problems 73–76, determine where the graph of f is below the graph of g by solving the inequality $f(x) \leq g(x)$. Graph f and g together.

73. $f(x) = x^4 - 1$
$g(x) = -2x^2 + 2$

74. $f(x) = x^4 - 1$
$g(x) = x - 1$

75. $f(x) = x^4 - 4$
$g(x) = 3x^2$

76. $f(x) = x^4$
$g(x) = 2 - x^2$

77. Average Cost Suppose that the daily cost C of manufacturing bicycles is given by $C(x) = 80x + 5000$. Then the average daily cost $\overline{C}$ is given by $\overline{C}(x) = \dfrac{80x + 5000}{x}$. How many bicycles must be produced each day for the average cost to be no more than $100?

78. Average Cost See Problem 77. Suppose that the government imposes a $1000-per-day tax on the bicycle manufacturer so that the daily cost C of manufacturing x bicycles is now given by $C(x) = 80x + 6000$. Now the average daily cost $\overline{C}$ is given by $\overline{C}(x) = \dfrac{80x + 6000}{x}$. How many bicycles must be produced each day for the average cost to be no more than $100?

79. Bungee Jumping Originating on Pentecost Island in the Pacific, the practice of a person jumping from a high place harnessed to a flexible attachment was introduced to Western culture in 1979 by the Oxford University Dangerous Sport Club. One important parameter to know before attempting a bungee jump is the amount the cord will stretch at the bottom of the fall. The stiffness of the cord is related to the amount of stretch by the equation

$$K = \frac{2W(S+L)}{S^2}$$

where W = weight of the jumper (pounds)
K = cord's stiffness (pounds per foot)
L = free length of the cord (feet)
S = stretch (feet)

(a) A 150-pound person plans to jump off a ledge attached to a cord of length 42 feet. If the stiffness of the cord is no less than 16 pounds per foot, how much will the cord stretch?
(b) If safety requirements will not permit the jumper to get any closer than 3 feet to the ground, what is the minimum height required for the ledge in part (a)?

Source: *American Institute of Physics, Physics News Update, No. 150, November 5, 1993.*

80. Gravitational Force According to Newton's Law of Universal Gravitation, the attractive force F between two bodies is given by

$$F = G\frac{m_1 m_2}{r^2}$$

where m_1, m_2 = the masses of the two bodies
r = distance between the two bodies
G = gravitational constant $= 6.6742 \times 10^{-11}$ newtons · meter^2· kilogram^{-2}

Suppose an object is traveling directly from Earth to the moon. The mass of Earth is 5.9742×10^{24} kilograms, the mass of the moon is 7.349×10^{22} kilograms, and the mean distance from Earth to the moon is 384,400 kilometers. For an object between Earth and the moon, how far from Earth is the force on the object due to the moon greater than the force on the object due to Earth?

Source: *www.solarviews.com;en.wikipedia.org*

81. Field Trip Mrs. West has decided to take her fifth grade class to a play. The manager of the theater agreed to discount the regular \$40 price of the ticket by \$0.20 for each ticket sold. The cost of the bus, \$500, will be split equally among the students. How many students must attend to keep the cost per student at or below \$40?

Explaining Concepts: Discussion and Writing

82. Make up an inequality that has no solution. Make up one that has exactly one solution.

83. The inequality $x^4 + 1 < -5$ has no solution. Explain why.

84. A student attempted to solve the inequality $\frac{x+4}{x-3} \le 0$ by multiplying both sides of the inequality by $x - 3$ to get $x + 4 \le 0$. This led to a solution of $\{x | x \le -4\}$. Is the student correct? Explain.

85. Write a rational inequality whose solution set is $\{x | -3 < x \le 5\}$.

Retain Your Knowledge

Problems 86–89 are based on material learned earlier in the course. The purpose of these problems is to keep the material fresh in your mind so that you are better prepared for the final exam.

86. Solve: $9 - 2x \le 4x + 1$

87. Given $f(x) = x^2 + 3x - 2$, find $f(x-2)$.

88. Factor completely: $6x^4y^4 + 3x^3y^5 - 18x^2y^6$

89. Suppose y varies directly with $\sqrt{x}$. Write a general formula to describe the variation if $y = 2$ when $x = 9$.

'Are You Prepared?' Answers

1. $\left\{x \middle| x < -\frac{1}{2}\right\}$ or $\left(-\infty, -\frac{1}{2}\right)$

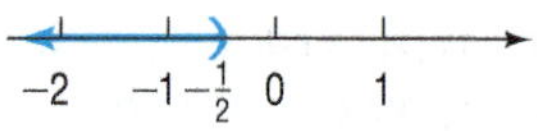

2. $\{x | -3 \le x \le 8\}$ or $[-3, 8]$

Chapter Review

Things to Know

Power function (pp. 332–334)

$f(x) = x^n$, $n \ge 2$ even

Domain: all real numbers Range: nonnegative real numbers
Passes through $(-1, 1)$, $(0, 0)$, $(1, 1)$
Even function
Decreasing on $(-\infty, 0]$, increasing on $[0, \infty)$

$f(x) = x^n$, $n \geq 3$ odd	Domain: all real numbers Range: all real numbers Passes through $(-1, -1)$, $(0, 0)$, $(1, 1)$ Odd function Increasing on $(-\infty, \infty)$
Polynomial function (pp. 331–341) $f(x) = a_n x^n + a_{n-1}x^{n-1} + \cdots + a_1 x + a_0$, $a_n \neq 0$	Domain: all real numbers At most $n - 1$ turning points End behavior: Behaves like $y = a_n x^n$ for large $\lvert x \rvert$
Real zeros of a polynomial function f (p. 335)	Real numbers for which $f(x) = 0$; the real zeros of f are the x-intercepts of the graph of f.
Remainder Theorem (p. 352)	If a polynomial function $f(x)$ is divided by $x - c$, then the remainder is $f(c)$.
Factor Theorem (p. 353)	$x - c$ is a factor of a polynomial function $f(x)$ if and only if $f(c) = 0$.
Descartes' Rule of Signs (p. 355)	Let f denote a polynomial function written in standard form. • The number of positive real zeros of f either equals the number of variations in the sign of the nonzero coefficients of $f(x)$ or else equals that number less an even integer. • The number of negative real zeros of f either equals the number of variations in the sign of the nonzero coefficients of $f(-x)$ or else equals that number less an even integer.
Rational Zeros Theorem (p. 355)	Let f be a polynomial function of degree 1 or higher of the form $f(x) = a_n x^n + a_{n-1}x^{n-1} + \cdots + a_1 x + a_0 \quad a_n \neq 0, a_0 \neq 0$ where each coefficient is an integer. If $\frac{p}{q}$, in lowest terms, is a rational zero of f, then p must be a factor of a_0, and q must be a factor of a_n.
Intermediate Value Theorem (p. 362)	Let f be a continuous function. If $a < b$ and $f(a)$ and $f(b)$ are of opposite sign, then there is at least one real zero of f between a and b.
Fundamental Theorem of Algebra (p. 367)	Every complex polynomial function $f(x)$ of degree $n \geq 1$ has at least one complex zero.
Conjugate Pairs Theorem (p. 368)	Let $f(x)$ be a polynomial function whose coefficients are real numbers. If $r = a + bi$ is a zero of f, then its complex conjugate $\bar{r} = a - bi$ is also a zero of f.
Rational function (pp. 372–379) $R(x) = \dfrac{p(x)}{q(x)}$	Domain: $\{x \mid q(x) \neq 0\}$
p, q are polynomial functions and q is not the zero polynomial.	Vertical asymptotes: With $R(x)$ in lowest terms, if $q(r) = 0$ for some real number, then $x = r$ is a vertical asymptote. Horizontal or oblique asymptote: See the summary on pages 377–378.

Objectives

Section		You should be able to . . .	Examples	Review Exercises
5.1	1	Identify polynomial functions and their degree (p. 331)	1	1–4
	2	Graph polynomial functions using transformations (p. 334)	2, 3	5–7
	3	Identify the real zeros of a polynomial function and their multiplicity (p. 335)	4–9	8–11
	4	Analyze the graph of a polynomial function (p. 342)	10, 11	8–11
	5	Build cubic models from data (p. 345)	12	51
5.2	1	Use the Remainder and Factor Theorems (p. 352)	1, 2	12–14
	2	Use Descartes' Rule of Signs to determine the number of positive and the number of negative real zeros of a polynomial function. (p. 354)	3	15, 16
	3	Use the Rational Zeros Theorem to list the potential rational zeros of a polynomial function (p. 355)	4	17–20
	4	Find the real zeros of a polynomial function (p. 356)	5, 6	18–20
	5	Solve polynomial equations (p. 359)	7	21, 22
	6	Use the Theorem for Bounds on Zeros (p. 359)	8, 9, 10	23, 24
	7	Use the Intermediate Value Theorem (p. 362)	11	25–28
5.3	1	Use the Conjugate Pairs Theorem (p. 367)	1	29, 30
	2	Find a polynomial function with specified zeros (p.368)	2	29, 30
	3	Find the complex zeros of a polynomial function (p. 369)	3	31–34

Section		You should be able to . . .	Examples	Review Exercises
5.4	1	Find the domain of a rational function (p. 372)	1–3	35, 36
	2	Find the vertical asymptotes of a rational function (p. 376)	4	35, 36, 44
	3	Find the horizontal or oblique asymptote of a rational function (p. 377)	5–8	35, 36, 44
5.5	1	Analyze the graph of a rational function (p. 382)	1–6	37–42
	2	Solve applied problems involving rational functions (p. 389)	7	50
5.6	1	Solve polynomial inequalities algebraically and graphically (p. 393)	1, 2	43, 45, 46
	2	Solve rational inequalities algebraically and graphically (p. 395)	3, 4	44, 47–49

Review Exercises

In Problems 1–4, determine whether the function is a polynomial function, rational function, or neither. For those that are polynomial functions, state the degree. For those that are not polynomial functions, tell why not.

1. $f(x) = 4x^5 - 3x^2 + 5x - 2$ **2.** $f(x) = \dfrac{3x^5}{2x + 1}$ **3.** $f(x) = 3x^2 + 5x^{1/2} - 1$ **4.** $f(x) = 3$

In Problems 5–7, graph each function using transformations (shifting, compressing, stretching, and reflection). Show all the stages.

5. $f(x) = (x + 2)^3$ **6.** $f(x) = -(x - 1)^4$ **7.** $f(x) = (x - 1)^4 + 2$

In Problems 8–11, analyze each polynomial function by following Steps 1 through 8 on page 343.

8. $f(x) = x(x + 2)(x + 4)$

9. $f(x) = (x - 2)^2(x + 4)$

10. $f(x) = -2x^3 + 4x^2$

11. $f(x) = (x - 1)^2(x + 3)(x + 1)$

In Problems 12 and 13, find the remainder R when $f(x)$ is divided by $g(x)$. Is g a factor of f?

12. $f(x) = 8x^3 - 3x^2 + x + 4;\quad g(x) = x - 1$

13. $f(x) = x^4 - 2x^3 + 15x - 2;\quad g(x) = x + 2$

14. Find the value of $f(x) = 12x^6 - 8x^4 + 1$ at $x = 4$.

In Problems 15 and 16, use Descartes' Rule of Signs to determine how many positive and how many negative zeros each polynomial function may have. Do not attempt to find the zeros.

15. $f(x) = 12x^8 - x^7 + 8x^4 - 2x^3 + x + 3$

16. $f(x) = -6x^5 + x^4 + 5x^3 + x + 1$

17. List all the potential rational zeros of $f(x) = 12x^8 - x^7 + 6x^4 - x^3 + x - 3$.

In Problems 18–20, use the Rational Zeros Theorem to find all the real zeros of each polynomial function. Use the zeros to factor f over the real numbers.

18. $f(x) = x^3 - 3x^2 - 6x + 8$ **19.** $f(x) = 4x^3 + 4x^2 - 7x + 2$ **20.** $f(x) = x^4 - 4x^3 + 9x^2 - 20x + 20$

In Problems 21 and 22, solve each equation in the real number system.

21. $2x^4 + 2x^3 - 11x^2 + x - 6 = 0$

22. $2x^4 + 7x^3 + x^2 - 7x - 3 = 0$

In Problems 23 and 24, find bounds to the real zeros of each polynomial function. Obtain a complete graph of f using a graphing utility.

23. $f(x) = x^3 - x^2 - 4x + 2$

24. $f(x) = 2x^3 - 7x^2 - 10x + 35$

In Problems 25 and 26, use the Intermediate Value Theorem to show that each polynomial function has a zero in the given interval.

25. $f(x) = 3x^3 - x - 1;\ [0, 1]$

26. $f(x) = 8x^4 - 4x^3 - 2x - 1;\ [0, 1]$

In Problems 27 and 28, each polynomial function has exactly one positive zero. Approximate the zero correct to two decimal places.

27. $f(x) = x^3 - x - 2$

28. $f(x) = 8x^4 - 4x^3 - 2x - 1$

In Problems 29 and 30, information is given about a complex polynomial function $f(x)$ whose coefficients are real numbers. Find the remaining zeros of f. Then find a polynomial function with real coefficients that has the zeros.

29. Degree 3; zeros: $4 + i$, 6

30. Degree 4; zeros: i, $1 + i$

In Problems 31–34, find the complex zeros of each polynomial function $f(x)$. Write f in factored form.

31. $f(x) = x^3 - 3x^2 - 6x + 8$

32. $f(x) = 4x^3 + 4x^2 - 7x + 2$

33. $f(x) = x^4 - 4x^3 + 9x^2 - 20x + 20$

34. $f(x) = 2x^4 + 2x^3 - 11x^2 + x - 6$

In Problems 35 and 36, find the domain of each rational function. Find any horizontal, vertical, or oblique asymptotes.

35. $R(x) = \dfrac{x+2}{x^2-9}$

36. $R(x) = \dfrac{x^2+3x+2}{(x+2)^2}$

In Problems 37–42, analyze each rational function following Steps 1–7 given on page 384.

37. $R(x) = \dfrac{2x-6}{x}$

38. $H(x) = \dfrac{x+2}{x(x-2)}$

39. $R(x) = \dfrac{x^2+x-6}{x^2-x-6}$

40. $F(x) = \dfrac{x^3}{x^2-4}$

41. $R(x) = \dfrac{2x^4}{(x-1)^2}$

42. $G(x) = \dfrac{x^2-4}{x^2-x-2}$

43. Use the graph below of a polynomial function $y = f(x)$ to solve (a) $f(x) = 0$, (b) $f(x) > 0$, (c) $f(x) \le 0$, and (d) determine f.

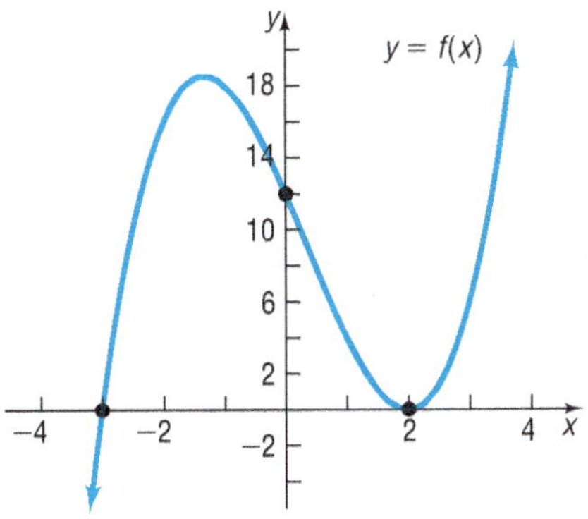

44. Use the graph below of a rational function $y = R(x)$ to (a) identify the horizontal asymptote of R, (b) identify the vertical asymptotes of R, (c) solve $R(x) < 0$, (d) solve $R(x) \ge 0$, and (e) determine R.

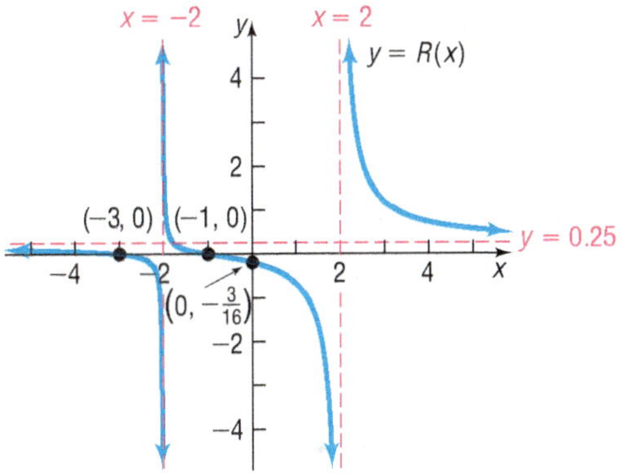

In Problems 45–49, solve each inequality. Graph the solution set.

45. $x^3 + x^2 < 4x + 4$

46. $x^3 + 4x^2 \ge x + 4$

47. $\dfrac{2x-6}{1-x} < 2$

48. $\dfrac{(x-2)(x-1)}{x-3} \ge 0$

49. $\dfrac{x^2-8x+12}{x^2-16} > 0$

50. Making a Can A can in the shape of a right circular cylinder is required to have a volume of 250 cubic centimeters.

(a) Express the amount A of material to make the can as a function of the radius r of the cylinder.

(b) How much material is required if the can is of radius 3 centimeters?

(c) How much material is required if the can is of radius 5 centimeters?

(d) Graph $A = A(r)$. For what value of r is A smallest?

51. Housing Prices The data in the table represent the January median new-home prices in the United States for the years shown.

(a) With a graphing utility, draw a scatter diagram of the data. Comment on the type of relation that appears to exist between the two variables.

(b) Decide on the function of best fit to these data (linear, quadratic, or cubic), and use this function to predict the median new-home price in the United States for January 2020 ($t = 9$).

(c) Draw the function of best fit on the scatter diagram obtained in part (a).

Year, t	Median Price, P ($1000s)
2004, 1	209.5
2006, 2	244.9
2008, 3	232.4
2010, 4	218.2
2012, 5	221.7
2014, 6	262.7

52. The illustration shows the graph of a polynomial function.

(a) Is the degree of the polynomial even or odd?

(b) Is the leading coefficient positive or negative?

(c) Is the function even, odd, or neither?

(d) Why is x^2 necessarily a factor of the polynomial?

(e) What is the minimum degree of the polynomial?

(f) Formulate five different polynomial functions whose graphs could look like the one shown. Compare yours to those of other students. What similarities do you see? What differences?

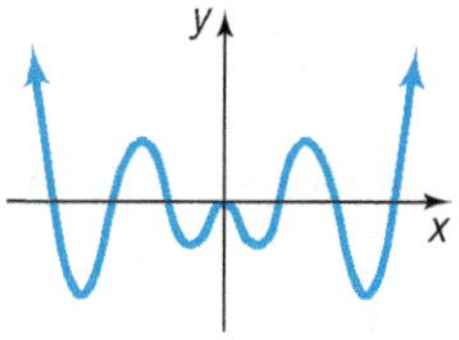

Chapter Test

CHAPTER Test Prep VIDEOS

The Chapter Test Prep Videos are step-by-step solutions available in MyMathLab®, or on this text's YouTube Channel. Flip back to the Resources for Success page for a link to this text's YouTube channel.

1. Graph $f(x) = (x-3)^4 - 2$ using transformations.

2. For the polynomial function $g(x) = 2x^3 + 5x^2 - 28x - 15$,

(a) Determine the maximum number of real zeros that the function may have.

(b) List the potential rational zeros.

(c) Determine the real zeros of g. Factor g over the reals.

(d) Find the x- and y-intercepts of the graph of g.

(e) Determine whether the graph crosses or touches the x-axis at each x-intercept.

(f) Find the power function that the graph of g resembles for large values of $|x|$.

(g) Approximate the turning points on the graph of g.

(h) Put all the information together to obtain the graph of g.

3. Find the complex zeros of $f(x) = x^3 - 4x^2 + 25x - 100$.

4. Solve $3x^3 + 2x - 1 = 8x^2 - 4$ in the complex number system.

In Problems 5 and 6, find the domain of each function. Find any horizontal, vertical, or oblique asymptotes.

5. $g(x) = \dfrac{2x^2 - 14x + 24}{x^2 + 6x - 40}$ **6.** $r(x) = \dfrac{x^2 + 2x - 3}{x + 1}$

7. Sketch the graph of the function in Problem 6. Label all intercepts, vertical asymptotes, horizontal asymptotes, and oblique asymptotes.

In Problems 8 and 9, write a function that meets the given conditions.

8. Fourth-degree polynomial function with real coefficients; zeros: $-2, 0, 3 + i$.

9. Rational function; asymptotes: $y = 2, x = 4$; domain: $\{x \mid x \neq 4, x \neq 9\}$.

10. Use the Intermediate Value Theorem to show that the function $f(x) = -2x^2 - 3x + 8$ has at least one real zero on the interval $[0, 4]$.

11. Solve: $\dfrac{x+2}{x-3} < 2$

Cumulative Review

1. Find the distance between the points $P = (1, 3)$ and $Q = (-4, 2)$.

2. Solve the inequality $x^2 \geq x$ and graph the solution set.

3. Solve the inequality $x^2 - 3x < 4$ and graph the solution set.

4. Find a linear function with slope -3 that contains the point $(-1, 4)$. Graph the function.

5. Find the equation of the line parallel to the line $y = 2x + 1$ and containing the point $(3, 5)$. Express your answer in slope–intercept form and graph the line.

6. Graph the equation $y = x^3$.

7. Does the relation $\{(3, 6), (1, 3), (2, 5), (3, 8)\}$ represent a function? Why or why not?

8. Solve the equation $x^3 - 6x^2 + 8x = 0$.

9. Solve the inequality $3x + 2 \leq 5x - 1$ and graph the solution set.

10. Find the center and the radius of the circle described by the equation $x^2 + 4x + y^2 - 2y - 4 = 0$. Graph the circle.

11. For the equation $y = x^3 - 9x$, determine the intercepts and test for symmetry.

12. Find an equation of the line perpendicular to $3x - 2y = 7$ that contains the point $(1, 5)$.

13. Is the following the graph of a function? Why or why not?

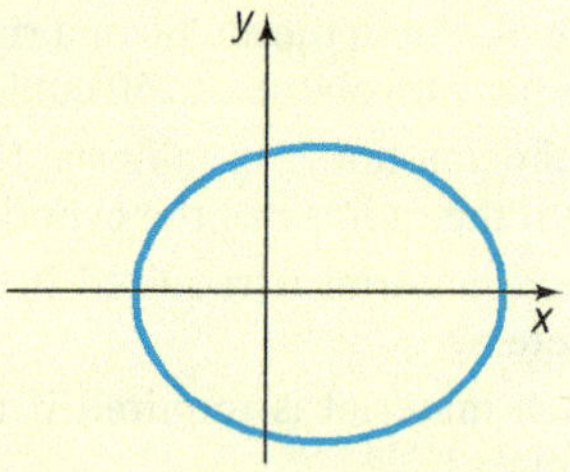

14. For the function $f(x) = x^2 + 5x - 2$, find

(a) $f(3)$ (b) $f(-x)$

(c) $-f(x)$ (d) $f(3x)$

(e) $\dfrac{f(x+h) - f(x)}{h}$ $h \neq 0$

15. Answer the following questions regarding the function

$$f(x) = \frac{x+5}{x-1}$$

(a) What is the domain of f?

(b) Is the point $(2, 6)$ on the graph of f?

(c) If $x = 3$, what is $f(x)$? What point is on the graph of f?

(d) If $f(x) = 9$, what is x? What point is on the graph of f?

(e) Is f a polynomial or rational function?

16. Graph the function $f(x) = -3x + 7$.

17. Graph $f(x) = 2x^2 - 4x + 1$ by determining whether its graph opens up or down and by finding its vertex, axis of symmetry, y-intercept, and x-intercepts, if any.

18. Find the average rate of change of $f(x) = x^2 + 3x + 1$ from 1 to 2. Use this result to find the equation of the secant line containing $(1, f(1))$ and $(2, f(2))$.

19. In parts (a) to (f), use the following graph.

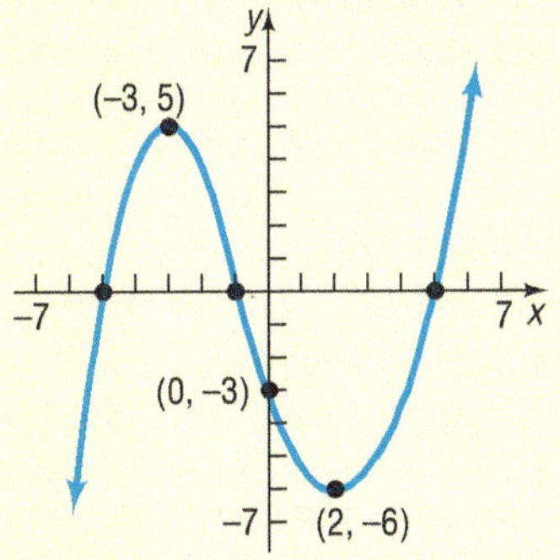

(a) Determine the intercepts.

(b) Based on the graph, tell whether the graph is symmetric with respect to the x-axis, the y-axis, and/or the origin.

(c) Based on the graph, tell whether the function is even, odd, or neither.

(d) List the intervals on which f is increasing. List the intervals on which f is decreasing.

(e) List the numbers, if any, at which f has a local maximum value. What are these local maximum values?

(f) List the numbers, if any, at which f has a local minimum value. What are these local minimum values?

20. Determine algebraically whether the function

$$f(x) = \frac{5x}{x^2 - 9}$$

is even, odd, or neither.

21. For the function $f(x) = \begin{cases} 2x + 1 & \text{if} \quad -3 < x < 2 \\ -3x + 4 & \text{if} \quad x \geq 2 \end{cases}$

(a) Find the domain of f.

(b) Locate any intercepts.

(c) Graph the function.

(d) Based on the graph, find the range.

22. Graph the function $f(x) = -3(x + 1)^2 + 5$ using transformations.

23. Suppose that $f(x) = x^2 - 5x + 1$ and $g(x) = -4x - 7$.

(a) Find $f + g$ and state its domain.

(b) Find $\dfrac{f}{g}$ and state its domain.

24. Demand Equation The price p (in dollars) and the quantity x sold of a certain product obey the demand equation

$$p = -\frac{1}{10}x + 150, \qquad 0 \leq x \leq 1500$$

(a) Express the revenue R as a function of x.

(b) What is the revenue if 100 units are sold?

(c) What quantity x maximizes revenue? What is the maximum revenue?

(d) What price should the company charge to maximize revenue?

Chapter Projects

I. Length of Day Go to *http://en.wikipedia.org/wiki/Latitude* and read about latitude through the subhead "Effect of Latitude." Now go to *http://www.orchidculture.com/COD/daylength.html#0N*.

1. For a particular day of the year, record in a table the length of day for the equator (0°N), 5°N, 10°N, . . . , 60°N. Enter the data into an Excel spreadsheet, TI graphing calculator, or some other spreadsheet capable of finding linear, quadratic, and cubic functions of best fit.

2. Draw a scatter diagram of the data with latitude as the independent variable and length of day as the dependent variable using Excel, a TI graphing calculator, or some other spreadsheet. The Chapter 4 project describes how to draw a scatter diagram in Excel.

3. Determine the linear function of best fit. Graph the linear function of best fit on the scatter diagram. To do this in Excel, right click on any data point in the scatter diagram. Now click the Add Chart Element menu, select Trendline, and then select More Trendline Options. Select the Linear radio button and select Display Equation on Chart. See Figure 58. Move the Trendline Options window off to the side and you will see the linear function of best fit displayed on the scatter diagram. Do you think the function accurately describes the relation between latitude and length of day?

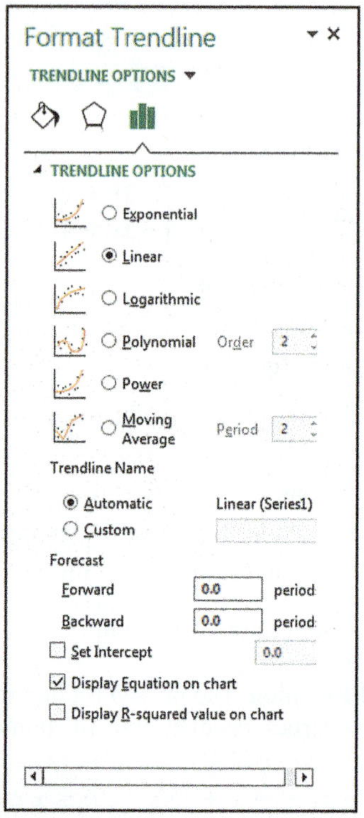

Figure 58

4. Determine the quadratic function of best fit. Graph the quadratic function of best fit on the scatter diagram. To do this in Excel, click on any data point in the scatter diagram. Now click the Add Chart Element menu, select Trendline, and then select More Trendline Options. Select the Polynomial radio button with Order set to 2. Select Display Equation on chart. Move the Trendline Options window off to the side and you will see the quadratic function of best fit displayed on the scatter diagram. Do you think the function accurately describes the relation between latitude and length of day?
5. Determine the cubic function of best fit. Graph the cubic function of best fit on the scatter diagram. To do this in Excel, click on any data point in the scatter diagram. Now click the Add Chart Element menu, select Trendline, and then select More Trendline Options. Select the Polynomial radio button with Order set to 3. Select Display Equation on chart. Move the Trendline Options window off to the side and you will see the cubic function of best fit displayed on the scatter diagram. Do you think the function accurately describes the relation between latitude and length of day?
6. Which of the three models seems to fit the data best? Explain your reasoning.
7. Use your model to predict the hours of daylight on the day you selected for Chicago (41.85 degrees north latitude). Go to the Old Farmer's Almanac or another website to determine the hours of daylight in Chicago for the day you selected. How do the two compare?

The following project is available at the Instructor's Resource Center (IRC):

II. **Theory of Equations** The coefficients of a polynomial function can be found if its zeros are known, an advantage of using polynomials in modeling.

6 Exponential and Logarithmic Functions

Depreciation of Cars

You are ready to buy that first new car. You know that cars lose value over time due to depreciation and that different cars have different rates of depreciation. So you will research the depreciation rates for the cars you are thinking of buying. After all, for cars that sell for about the same price, the lower the depreciation rate, the more the car will be worth each year.

—See the Internet-based Chapter Project I—

••• A Look Back

Until now, our study of functions has concentrated on polynomial and rational functions. These functions belong to the class of **algebraic functions**—that is, functions that can be expressed in terms of sums, differences, products, quotients, powers, or roots of polynomials. Functions that are not algebraic are termed **transcendental** (they transcend, or go beyond, algebraic functions).

A Look Ahead •••

In this chapter, we study two transcendental functions: the exponential function and the logarithmic function. These functions occur frequently in a wide variety of applications, such as biology, chemistry, economics, and psychology.

The chapter begins with a discussion of composite, one-to-one, and inverse functions—concepts that are needed to explain the relationship between exponential and logarithmic functions.

Outline

6.1 Composite Functions

PREPARING FOR THIS SECTION *Before getting started, review the following:*

- Find the Value of a Function (Section 3.1, pp. 210–212)
- Domain of a Function (Section 3.1, pp. 214–216)

Now Work the 'Are You Prepared?' problems on page 413.

OBJECTIVES 1 Form a Composite Function (p. 408)
2 Find the Domain of a Composite Function (p. 409)

1 Form a Composite Function

Suppose that an oil tanker is leaking oil and you want to determine the area of the circular oil patch around the ship. See Figure 1. It is determined that the oil is leaking from the tanker in such a way that the radius of the circular patch of oil around the ship is increasing at a rate of 3 feet per minute. Therefore, the radius r of the oil patch at any time t, in minutes, is given by $r(t) = 3t$. So after 20 minutes, the radius of the oil patch is $r(20) = 3(20) = 60$ feet.

The area A of a circle as a function of the radius r is given by $A(r) = \pi r^2$. The area of the circular patch of oil after 20 minutes is $A(60) = \pi(60)^2 = 3600\pi$ square feet. Note that $60 = r(20)$, so $A(60) = A(r(20))$. The argument of the function A is the output of the function r!

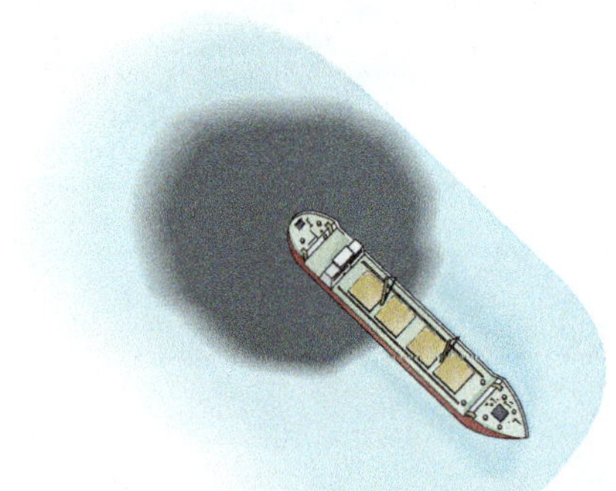

Figure 1

In general, the area of the oil patch can be expressed as a function of time t by evaluating $A(r(t))$ and obtaining $A(r(t)) = A(3t) = \pi(3t)^2 = 9\pi t^2$. The function $A(r(t))$ is a special type of function called a *composite function.*

As another example, consider the function $y = (2x + 3)^2$. Let $y = f(u) = u^2$ and $u = g(x) = 2x + 3$. Then by a substitution process, the original function is obtained as follows: $y = f(u) = f(g(x)) = (2x + 3)^2$.

In general, suppose that f and g are two functions and that x is a number in the domain of g. Evaluating g at x yields $g(x)$. If $g(x)$ is in the domain of f, then evaluating f at $g(x)$ yields the expression $f(g(x))$. The correspondence from x to $f(g(x))$ is called a *composite function* $f \circ g$.

DEFINITION

Given two functions f and g, the **composite function**, denoted by $f \circ g$ (read as "f composed with g"), is defined by

$$(f \circ g)(x) = f(g(x))$$

The domain of $f \circ g$ is the set of all numbers x in the domain of g such that $g(x)$ is in the domain of f.

Look carefully at Figure 2. Only those values of x in the domain of g for which $g(x)$ is in the domain of f can be in the domain of $f \circ g$. The reason is that if $g(x)$ is not in the domain of f, then $f(g(x))$ is not defined. Because of this, the domain of $f \circ g$ is a subset of the domain of g; the range of $f \circ g$ is a subset of the range of f.

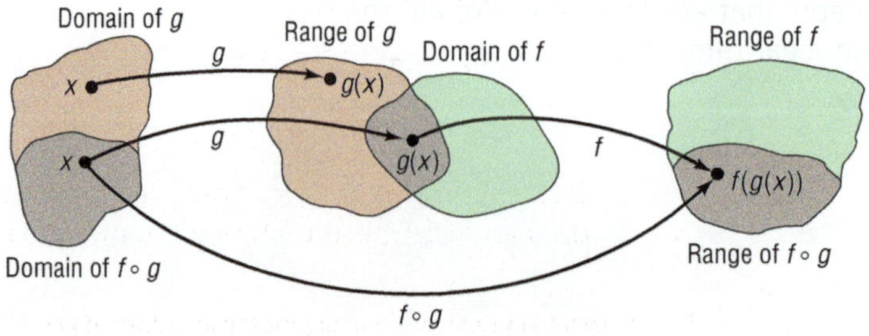

Figure 2

Figure 3 provides a second illustration of the definition. Here x is the input to the function g, yielding $g(x)$. Then $g(x)$ is the input to the function f, yielding $f(g(x))$. Note that the "inside" function g in $f(g(x))$ is "processed" first.

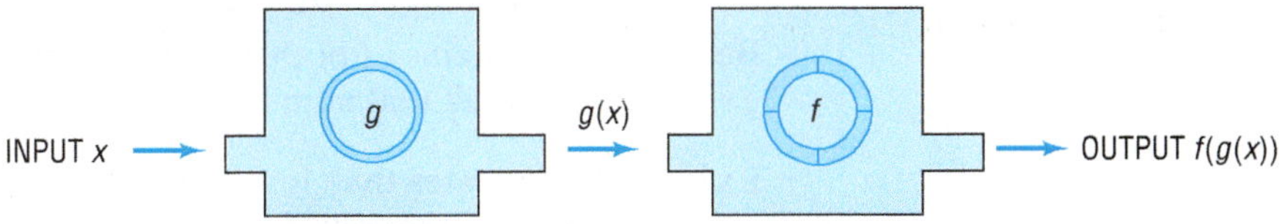

Figure 3

EXAMPLE 1 Evaluating a Composite Function

Suppose that $f(x) = 2x^2 - 3$ and $g(x) = 4x$. Find:

(a) $(f \circ g)(1)$ (b) $(g \circ f)(1)$ (c) $(f \circ f)(-2)$ (d) $(g \circ g)(-1)$

Solution

(a) $(f \circ g)(1) = f(g(1)) = f(4) = 2 \cdot 4^2 - 3 = 29$

$g(x) = 4x$, $g(1) = 4$; $f(x) = 2x^2 - 3$

(b) $(g \circ f)(1) = g(f(1)) = g(-1) = 4 \cdot (-1) = -4$

$f(x) = 2x^2 - 3$, $f(1) = -1$; $g(x) = 4x$

(c) $(f \circ f)(-2) = f(f(-2)) = f(5) = 2 \cdot 5^2 - 3 = 47$

$f(-2) = 2(-2)^2 - 3 = 5$

(d) $(g \circ g)(-1) = g(g(-1)) = g(-4) = 4 \cdot (-4) = -16$

$g(-1) = -4$

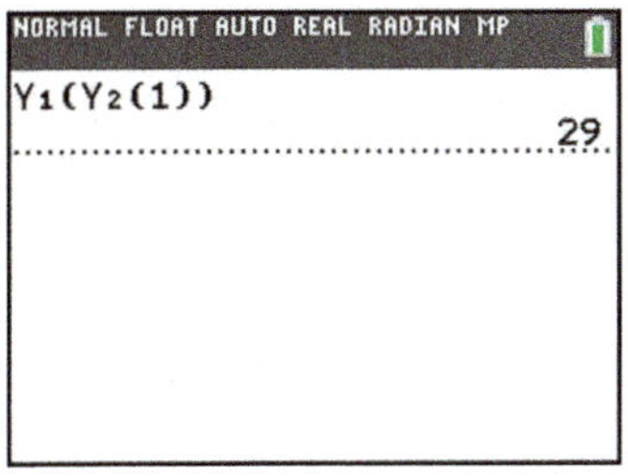

Figure 4

COMMENT Graphing calculators can be used to evaluate composite functions.*
Let $Y_1 = f(x) = 2x^2 - 3$ and $Y_2 = g(x) = 4x$. Then, using a TI-84 Plus C graphing calculator, find $(f \circ g)(1)$ as shown in Figure 4. Note that this is the result obtained in Example 1(a).

Now Work PROBLEM 13

2 Find the Domain of a Composite Function

EXAMPLE 2 Finding a Composite Function and Its Domain

Suppose that $f(x) = x^2 + 3x - 1$ and $g(x) = 2x + 3$.

Find: (a) $f \circ g$ (b) $g \circ f$

Then find the domain of each composite function.

Solution The domain of f and the domain of g are the set of all real numbers.

(a) $(f \circ g)(x) = f(g(x)) = f(2x + 3) = (2x + 3)^2 + 3(2x + 3) - 1$

$f(x) = x^2 + 3x - 1$

$= 4x^2 + 12x + 9 + 6x + 9 - 1 = 4x^2 + 18x + 17$

Because the domains of both f and g are the set of all real numbers, the domain of $f \circ g$ is the set of all real numbers.

*Consult your owner's manual for the appropriate keystrokes.

(b) $(g \circ f)(x) = g(f(x)) = g(x^2 + 3x - 1) = 2(x^2 + 3x - 1) + 3$

$\uparrow$ $g(x) = 2x + 3$

$= 2x^2 + 6x - 2 + 3 = 2x^2 + 6x + 1$

Because the domains of both f and g are the set of all real numbers, the domain of $g \circ f$ is the set of all real numbers.

Example 2 illustrates that, in general, $f \circ g \neq g \circ f$. Sometimes $f \circ g$ does equal $g \circ f$, as we shall see in Example 5.

Look back at Figure 2 on page 408. In determining the domain of the composite function $(f \circ g)(x) = f(g(x))$, keep the following two thoughts in mind about the input x.

1. Any x not in the domain of g must be excluded.
2. Any x for which $g(x)$ is not in the domain of f must be excluded.

EXAMPLE 3 Finding the Domain of $f \circ g$

Find the domain of $f \circ g$ if $f(x) = \dfrac{1}{x + 2}$ and $g(x) = \dfrac{4}{x - 1}$.

Solution For $(f \circ g)(x) = f(g(x))$, first note that the domain of g is $\{x | x \neq 1\}$, so 1 is excluded from the domain of $f \circ g$. Next note that the domain of f is $\{x | x \neq -2\}$, which means that $g(x)$ cannot equal -2. Solve the equation $g(x) = -2$ to determine what additional value(s) of x to exclude.

$$\frac{4}{x-1} = -2 \qquad g(x) = -2$$

$$4 = -2(x - 1) \qquad \text{Multiply both sides by } x - 1.$$

$$4 = -2x + 2 \qquad \text{Apply the Distributive Property.}$$

$$2x = -2 \qquad \text{Add } 2x \text{ to both sides. Subtract 4 from both sides.}$$

$$x = -1 \qquad \text{Divide both sides by 2.}$$

Also exclude -1 from the domain of $f \circ g$.

The domain of $f \circ g$ is $\{x | x \neq -1, x \neq 1\}$.

✓**Check:** For $x = 1$, $g(x) = \dfrac{4}{x - 1}$ is not defined, so $(f \circ g)(x) = f(g(x))$ is not defined.

For $x = -1$, $g(-1) = -2$, and $(f \circ g)(-1) = f(g(-1)) = f(-2)$ is not defined.

EXAMPLE 4 Finding a Composite Function and Its Domain

Suppose that $f(x) = \dfrac{1}{x + 2}$ and $g(x) = \dfrac{4}{x - 1}$.

Find: (a) $f \circ g$ (b) $f \circ f$

Then find the domain of each composite function.

Solution The domain of f is $\{x | x \neq -2\}$ and the domain of g is $\{x | x \neq 1\}$.

(a) $$(f \circ g)(x) = f(g(x)) = f\left(\frac{4}{x-1}\right) = \frac{1}{\frac{4}{x-1} + 2} = \frac{x-1}{4 + 2(x-1)} = \frac{x-1}{2x+2} = \frac{x-1}{2(x+1)}$$

$\uparrow$ $f(x) = \dfrac{1}{x+2}$ $\qquad$ $\uparrow$ Multiply by $\dfrac{x-1}{x-1}$.

In Example 3, the domain of $f \circ g$ was found to be $\{x | x \neq -1, x \neq 1\}$.

The domain of $f \circ g$ also can be found by first looking at the domain of g: $\{x | x \neq 1\}$. Exclude 1 from the domain of $f \circ g$ as a result. Then look at $f \circ g$ and note that x cannot equal -1, because $x = -1$ results in division by 0. So exclude -1 from the domain of $f \circ g$. Therefore, the domain of $f \circ g$ is $\{x | x \neq -1, x \neq 1\}$.

(b) $$(f \circ f)(x) = f(f(x)) = f\left(\frac{1}{x+2}\right) = \frac{1}{\frac{1}{x+2} + 2} = \frac{x+2}{1 + 2(x+2)} = \frac{x+2}{2x+5}$$

$f(x) = \frac{1}{x+2}$ Multiply by $\frac{x+2}{x+2}$.

The domain of $f \circ f$ consists of all values of x in the domain of f, $\{x | x \neq -2\}$, for which

$$f(x) = \frac{1}{x+2} \neq -2 \qquad \frac{1}{x+2} = -2$$
$$1 = -2(x+2)$$
$$1 = -2x - 4$$
$$2x = -5$$
$$x = -\frac{5}{2}$$

or, equivalently,

$$x \neq -\frac{5}{2}$$

The domain of $f \circ f$ is $\left\{x \middle| x \neq -\frac{5}{2}, x \neq -2\right\}$.

The domain of $f \circ f$ also can be found by recognizing that -2 is not in the domain of f and so should be excluded from the domain of $f \circ f$. Then, looking at $f \circ f$, note that x cannot equal $-\frac{5}{2}$. Do you see why? Therefore, the domain of $f \circ f$ is $\left\{x \middle| x \neq -\frac{5}{2}, x \neq -2\right\}$. ■

Now Work PROBLEMS 27 AND 29

EXAMPLE 5 Showing That Two Composite Functions Are Equal

If $f(x) = 3x - 4$ and $g(x) = \frac{1}{3}(x+4)$, show that

$$(f \circ g)(x) = (g \circ f)(x) = x$$

for every x in the domain of $f \circ g$ and $g \circ f$.

Solution

$$(f \circ g)(x) = f(g(x))$$
$$= f\left(\frac{x+4}{3}\right) \qquad g(x) = \frac{1}{3}(x+4) = \frac{x+4}{3}$$
$$= 3\left(\frac{x+4}{3}\right) - 4 \qquad f(x) = 3x - 4$$
$$= x + 4 - 4 = x$$

Seeing the Concept

Using a graphing calculator, let

$Y_1 = f(x) = 3x - 4$

$Y_2 = g(x) = \frac{1}{3}(x + 4)$

$Y_3 = f \circ g, \; Y_4 = g \circ f$

Using the viewing window $-3 \leq x \leq 3$, $-2 \leq y \leq 2$, graph only Y_3 and Y_4. What do you see? TRACE to verify that $Y_3 = Y_4$.

$$\begin{aligned}(g \circ f)(x) &= g(f(x)) \\ &= g(3x - 4) && f(x) = 3x - 4 \\ &= \frac{1}{3}[(3x - 4) + 4] && g(x) = \frac{1}{3}(x + 4) \\ &= \frac{1}{3}(3x) = x\end{aligned}$$

We conclude that $(f \circ g)(x) = (g \circ f)(x) = x$.

In Section 6.2, we shall see that there is an important relationship between functions f and g for which $(f \circ g)(x) = (g \circ f)(x) = x$.

Now Work PROBLEM 39

Calculus Application

Some techniques in calculus require the ability to determine the components of a composite function. For example, the function $H(x) = \sqrt{x + 1}$ is the composition of the functions f and g, where $f(x) = \sqrt{x}$ and $g(x) = x + 1$, because $H(x) = (f \circ g)(x) = f(g(x)) = f(x + 1) = \sqrt{x + 1}$.

EXAMPLE 6 Finding the Components of a Composite Function

Find functions f and g such that $f \circ g = H$ if $H(x) = (x^2 + 1)^{50}$.

Solution The function H takes $x^2 + 1$ and raises it to the power 50. A natural way to decompose H is to raise the function $g(x) = x^2 + 1$ to the power 50. Let $f(x) = x^{50}$ and $g(x) = x^2 + 1$. Then

$$\begin{aligned}(f \circ g)(x) &= f(g(x)) \\ &= f(x^2 + 1) \\ &= (x^2 + 1)^{50} = H(x)\end{aligned}$$

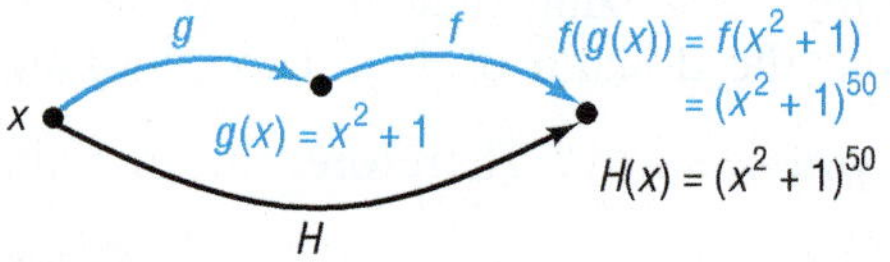

Figure 5

See Figure 5.

Other functions f and g may be found for which $f \circ g = H$ in Example 6. For instance, if $f(x) = x^2$ and $g(x) = (x^2 + 1)^{25}$, then

$$(f \circ g)(x) = f(g(x)) = f((x^2 + 1)^{25}) = [(x^2 + 1)^{25}]^2 = (x^2 + 1)^{50}$$

Although the functions f and g found as a solution to Example 6 are not unique, there is usually a "natural" selection for f and g that comes to mind first.

EXAMPLE 7 Finding the Components of a Composite Function

Find functions f and g such that $f \circ g = H$ if $H(x) = \dfrac{1}{x + 1}$.

Solution Here H is the reciprocal of $g(x) = x + 1$. Let $f(x) = \dfrac{1}{x}$ and $g(x) = x + 1$. Then

$$(f \circ g)(x) = f(g(x)) = f(x + 1) = \frac{1}{x + 1} = H(x)$$

Now Work PROBLEM 47

6.1 Assess Your Understanding

'Are You Prepared?' *Answers are given at the end of these exercises. If you get a wrong answer, read the pages listed in red.*

1. Find $f(3)$ if $f(x) = -4x^2 + 5x$. (pp. 210–212)

2. Find $f(3x)$ if $f(x) = 4 - 2x^2$. (pp. 210–212)

3. Find the domain of the function $f(x) = \dfrac{x^2 - 1}{x^2 - 25}$. (pp. 214–216)

Concepts and Vocabulary

4. Given two functions f and g, the __________ __________, denoted $f \circ g$, is defined by $(f \circ g)(x) =$ ________.

5. ***True or False*** If $f(x) = x^2$ and $g(x) = \sqrt{x + 9}$, then $(f \circ g)(4) = 5$.

6. If $f(x) = \sqrt{x + 2}$ and $g(x) = \dfrac{3}{x}$, which of the following does $(f \circ g)(x)$ equal?

(a) $\dfrac{3}{\sqrt{x + 2}}$ (b) $\dfrac{3}{\sqrt{x}} + 2$ (c) $\sqrt{\dfrac{3}{x} + 2}$ (d) $\sqrt{\dfrac{3}{x + 2}}$

7. If $H = f \circ g$ and $H(x) = \sqrt{25 - 4x^2}$, which of the following cannot be the component functions f and g?

(a) $f(x) = \sqrt{25 - x^2}$; $g(x) = 4x$

(b) $f(x) = \sqrt{x}$; $g(x) = 25 - 4x^2$

(c) $f(x) = \sqrt{25 - x}$; $g(x) = 4x^2$

(d) $f(x) = \sqrt{25 - 4x}$; $g(x) = x^2$

8. ***True or False*** The domain of the composite function $(f \circ g)(x)$ is the same as the domain of $g(x)$.

Skill Building

In Problems 9 and 10, evaluate each expression using the values given in the table.

9.

x	−3	−2	−1	0	1	2	3
$f(x)$	−7	−5	−3	−1	3	5	7
$g(x)$	8	3	0	−1	0	3	8

(a) $(f \circ g)(1)$ (b) $(f \circ g)(-1)$
(c) $(g \circ f)(-1)$ (d) $(g \circ f)(0)$
(e) $(g \circ g)(-2)$ (f) $(f \circ f)(-1)$

10.

x	−3	−2	−1	0	1	2	3
$f(x)$	11	9	7	5	3	1	−1
$g(x)$	−8	−3	0	1	0	−3	−8

(a) $(f \circ g)(1)$ (b) $(f \circ g)(2)$
(c) $(g \circ f)(2)$ (d) $(g \circ f)(3)$
(e) $(g \circ g)(1)$ (f) $(f \circ f)(3)$

In Problems 11 and 12, evaluate each expression using the graphs of $y = f(x)$ and $y = g(x)$ shown in the figure.

11. (a) $(g \circ f)(-1)$ (b) $(g \circ f)(0)$
(c) $(f \circ g)(-1)$ (d) $(f \circ g)(4)$

12. (a) $(g \circ f)(1)$ (b) $(g \circ f)(5)$
(c) $(f \circ g)(0)$ (d) $(f \circ g)(2)$

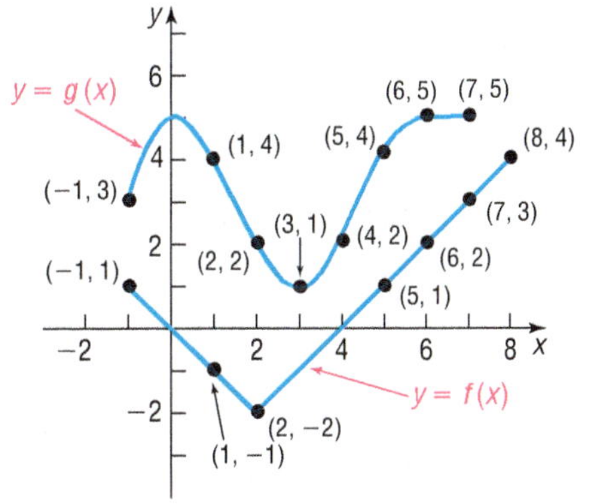

In Problems 13–22, for the given functions f and g, find:
(a) $(f \circ g)(4)$ *(b)* $(g \circ f)(2)$ *(c)* $(f \circ f)(1)$ *(d)* $(g \circ g)(0)$

13. $f(x) = 2x$; $g(x) = 3x^2 + 1$

14. $f(x) = 3x + 2$; $g(x) = 2x^2 - 1$

15. $f(x) = 4x^2 - 3$; $g(x) = 3 - \frac{1}{2}x^2$

16. $f(x) = 2x^2$; $g(x) = 1 - 3x^2$

17. $f(x) = \sqrt{x}$; $g(x) = 2x$

18. $f(x) = \sqrt{x + 1}$; $g(x) = 3x$

19. $f(x) = |x|$; $g(x) = \dfrac{1}{x^2 + 1}$

20. $f(x) = |x - 2|$; $g(x) = \dfrac{3}{x^2 + 2}$

21. $f(x) = \dfrac{3}{x + 1}$; $g(x) = \sqrt[3]{x}$

22. $f(x) = x^{3/2}$; $g(x) = \dfrac{2}{x + 1}$

In Problems 23–38, for the given functions f and g, find:
(a) $f \circ g$ (b) $g \circ f$ (c) $f \circ f$ (d) $g \circ g$
State the domain of each composite function.

23. $f(x) = 2x + 3;\quad g(x) = 3x$

24. $f(x) = -x;\quad g(x) = 2x - 4$

25. $f(x) = 3x + 1;\quad g(x) = x^2$

26. $f(x) = x + 1;\quad g(x) = x^2 + 4$

27. $f(x) = x^2;\quad g(x) = x^2 + 4$

28. $f(x) = x^2 + 1;\quad g(x) = 2x^2 + 3$

29. $f(x) = \dfrac{3}{x-1};\quad g(x) = \dfrac{2}{x}$

30. $f(x) = \dfrac{1}{x+3};\quad g(x) = -\dfrac{2}{x}$

31. $f(x) = \dfrac{x}{x-1};\quad g(x) = -\dfrac{4}{x}$

32. $f(x) = \dfrac{x}{x+3};\quad g(x) = \dfrac{2}{x}$

33. $f(x) = \sqrt{x};\quad g(x) = 2x + 3$

34. $f(x) = \sqrt{x-2};\quad g(x) = 1 - 2x$

35. $f(x) = x^2 + 1;\quad g(x) = \sqrt{x-1}$

36. $f(x) = x^2 + 4;\quad g(x) = \sqrt{x-2}$

37. $f(x) = \dfrac{x-5}{x+1};\quad g(x) = \dfrac{x+2}{x-3}$

38. $f(x) = \dfrac{2x-1}{x-2};\quad g(x) = \dfrac{x+4}{2x-5}$

In Problems 39–46, show that $(f \circ g)(x) = (g \circ f)(x) = x$.

39. $f(x) = 2x;\quad g(x) = \dfrac{1}{2}x$

40. $f(x) = 4x;\quad g(x) = \dfrac{1}{4}x$

41. $f(x) = x^3;\quad g(x) = \sqrt[3]{x}$

42. $f(x) = x + 5;\quad g(x) = x - 5$

43. $f(x) = 2x - 6;\quad g(x) = \dfrac{1}{2}(x + 6)$

44. $f(x) = 4 - 3x;\quad g(x) = \dfrac{1}{3}(4 - x)$

45. $f(x) = ax + b;\quad g(x) = \dfrac{1}{a}(x - b)\quad a \neq 0$

46. $f(x) = \dfrac{1}{x};\quad g(x) = \dfrac{1}{x}$

In Problems 47–52, find functions f and g so that $f \circ g = H$.

47. $H(x) = (2x + 3)^4$

48. $H(x) = (1 + x^2)^3$

49. $H(x) = \sqrt{x^2 + 1}$

50. $H(x) = \sqrt{1 - x^2}$

51. $H(x) = |2x + 1|$

52. $H(x) = |2x^2 + 3|$

Applications and Extensions

53. If $f(x) = 2x^3 - 3x^2 + 4x - 1$ and $g(x) = 2$, find $(f \circ g)(x)$ and $(g \circ f)(x)$.

54. If $f(x) = \dfrac{x+1}{x-1}$, find $(f \circ f)(x)$.

55. If $f(x) = 2x^2 + 5$ and $g(x) = 3x + a$, find a so that the graph of $f \circ g$ crosses the y-axis at 23.

56. If $f(x) = 3x^2 - 7$ and $g(x) = 2x + a$, find a so that the graph of $f \circ g$ crosses the y-axis at 68.

In Problems 57 and 58, use the functions f and g to find:
(a) $f \circ g$ (b) $g \circ f$
(c) the domain of $f \circ g$ and of $g \circ f$
(d) the conditions for which $f \circ g = g \circ f$

57. $f(x) = ax + b\quad g(x) = cx + d$

58. $f(x) = \dfrac{ax+b}{cx+d}\quad g(x) = mx$

59. Surface Area of a Balloon The surface area S (in square meters) of a hot-air balloon is given by

$$S(r) = 4\pi r^2$$

where r is the radius of the balloon (in meters). If the radius r is increasing with time t (in seconds) according to the formula $r(t) = \dfrac{2}{3}t^3, t \geq 0$, find the surface area S of the balloon as a function of the time t.

60. Volume of a Balloon The volume V (in cubic meters) of the hot-air balloon described in Problem 59 is given by $V(r) = \dfrac{4}{3}\pi r^3$. If the radius r is the same function of t as in Problem 59, find the volume V as a function of the time t.

61. Automobile Production The number N of cars produced at a certain factory in one day after t hours of operation is given by $N(t) = 100t - 5t^2, 0 \leq t \leq 10$. If the cost C (in dollars) of producing N cars is $C(N) = 15{,}000 + 8000N$, find the cost C as a function of the time t of operation of the factory.

62. Environmental Concerns The spread of oil leaking from a tanker is in the shape of a circle. If the radius r (in feet) of the spread after t hours is $r(t) = 200\sqrt{t}$, find the area A of the oil slick as a function of the time t.

63. Production Cost The price p, in dollars, of a certain product and the quantity x sold obey the demand equation

$$p = -\frac{1}{4}x + 100\quad 0 \leq x \leq 400$$

Suppose that the cost C, in dollars, of producing x units is

$$C = \frac{\sqrt{x}}{25} + 600$$

Assuming that all items produced are sold, find the cost C as a function of the price p.

[**Hint:** Solve for x in the demand equation and then form the composite function.]

64. Cost of a Commodity The price p, in dollars, of a certain commodity and the quantity x sold obey the demand equation

$$p = -\frac{1}{5}x + 200 \quad 0 \le x \le 1000$$

Suppose that the cost C, in dollars, of producing x units is

$$C = \frac{\sqrt{x}}{10} + 400$$

Assuming that all items produced are sold, find the cost C as a function of the price p.

65. Volume of a Cylinder The volume V of a right circular cylinder of height h and radius r is $V = \pi r^2 h$. If the height is twice the radius, express the volume V as a function of r.

66. Volume of a Cone The volume V of a right circular cone is $V = \frac{1}{3}\pi r^2 h$. If the height is twice the radius, express the volume V as a function of r.

67. Foreign Exchange Traders often buy foreign currency in the hope of making money when the currency's value changes. For example, on April 15, 2015, one U.S. dollar could purchase 0.9428 euro, and one euro could purchase 126.457 yen. Let $f(x)$ represent the number of euros you can buy with x dollars, and let $g(x)$ represent the number of yen you can buy with x euros.

(a) Find a function that relates dollars to euros.

(b) Find a function that relates euros to yen.

(c) Use the results of parts (a) and (b) to find a function that relates dollars to yen. That is, find $(g \circ f)(x) = g(f(x))$.

(d) What is $g(f(1000))$?

68. Temperature Conversion The function $C(F) = \frac{5}{9}(F - 32)$ converts a temperature in degrees Fahrenheit, F, to a temperature in degrees Celsius, C. The function $K(C) = C + 273$, converts a temperature in degrees Celsius to a temperature in kelvins, K.

(a) Find a function that converts a temperature in degrees Fahrenheit to a temperature in kelvins.

(b) Determine 80 degrees Fahrenheit in kelvins.

69. Discounts The manufacturer of a computer is offering two discounts on last year's model computer. The first discount is a \$200 rebate and the second discount is 20% off the regular price, p.

(a) Write a function f that represents the sale price if only the rebate applies.

(b) Write a function g that represents the sale price if only the 20% discount applies.

(c) Find $f \circ g$ and $g \circ f$. What does each of these functions represent? Which combination of discounts represents a better deal for the consumer? Why?

70. Taxes Suppose that you work for \$15 per hour. Write a function that represents gross salary G as a function of hours worked h. Your employer is required to withhold taxes (federal income tax, Social Security, Medicare) from your paycheck. Suppose your employer withholds 20% of your income for taxes. Write a function that represents net salary N as a function of gross salary G. Find and interpret $N \circ G$.

71. Let $f(x) = ax + b$ and $g(x) = bx + a$, where a and b are integers. If $f(1) = 8$ and $f(g(20)) - g(f(20)) = -14$, find the product of a and b.*

72. If f and g are odd functions, show that the composite function $f \circ g$ is also odd.

73. If f is an odd function and g is an even function, show that the composite functions $f \circ g$ and $g \circ f$ are both even.

*Courtesy of the Joliet Junior College Mathematics Department

Retain Your Knowledge

Problems 74–77 are based on material learned earlier in the course. The purpose of these problems is to keep the material fresh in your mind so that you are better prepared for the final exam.

74. Given $f(x) = 3x + 8$ and $g(x) = x - 5$, find $(f + g)(x)$, $(f - g)(x)$, $(f \cdot g)(x)$, and $\left(\frac{f}{g}\right)(x)$. State the domain of each.

75. Find the real zeros of $f(x) = 2x - 5\sqrt{x} + 2$.

76. Use a graphing utility to graph $f(x) = -x^3 + 4x - 2$ over the interval $[-3, 3]$. Approximate any local maxima and local minima. Determine where the function is increasing and where it is decreasing.

77. Find the domain of $R(x) = \dfrac{x^2 + 6x + 5}{x - 3}$. Find any horizontal, vertical, or oblique asymptotes.

'Are You Prepared?' Answers

1. -21 **2.** $4 - 18x^2$ **3.** $\{x \mid x \neq -5, x \neq 5\}$

6.2 One-to-One Functions; Inverse Functions

PREPARING FOR THIS SECTION *Before getting started, review the following:*

- Functions (Section 3.1, pp. 207–218)
- Increasing/Decreasing Functions (Section 3.3, p. 234)
- Rational Expressions (Chapter R, Section R.7, pp. 63–71)

Now Work the 'Are You Prepared?' problems on page 424.

OBJECTIVES
1 Determine Whether a Function Is One-to-One (p. 416)
2 Determine the Inverse of a Function Defined by a Map or a Set of Ordered Pairs (p. 418)
3 Obtain the Graph of the Inverse Function from the Graph of the Function (p. 421)
4 Find the Inverse of a Function Defined by an Equation (p. 422)

1 Determine Whether a Function Is One-to-One

Section 3.1 presented four different ways to represent a function: (1) a map, (2) a set of ordered pairs, (3) a graph, and (4) an equation. For example, Figures 6 and 7 illustrate two different functions represented as mappings. The function in Figure 6 shows the correspondence between states and their populations (in millions). The function in Figure 7 shows a correspondence between animals and life expectancies (in years).

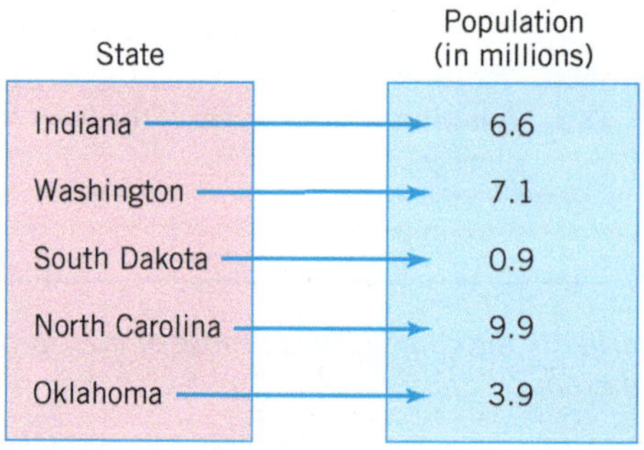

Figure 6

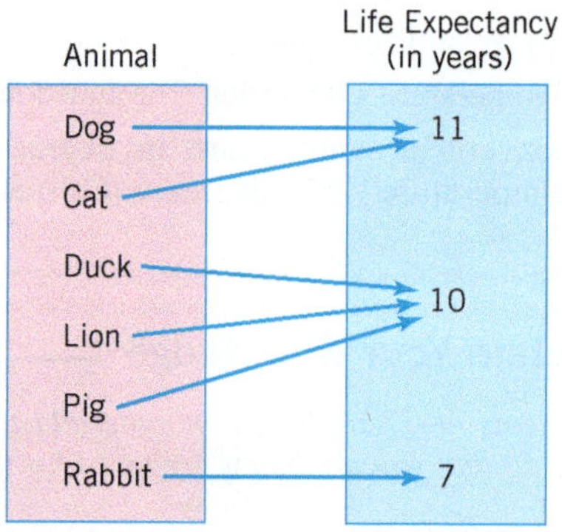

Figure 7

Suppose several people are asked to name a state that has a population of 0.9 million based on the function in Figure 6. Everyone will respond "South Dakota." Now, if the same people are asked to name an animal whose life expectancy is 11 years based on the function in Figure 7, some may respond "dog," while others may respond "cat." What is the difference between the functions in Figures 6 and 7? In Figure 6, no two elements in the domain correspond to the same element in the range. In Figure 7, this is not the case: Different elements in the domain correspond to the same element in the range. Functions such as the one in Figure 6 are given a special name.

DEFINITION

A function is **one-to-one** if any two different inputs in the domain correspond to two different outputs in the range. That is, if x_1 and x_2 are two different inputs of a function f, then f is one-to-one if $f(x_1) \neq f(x_2)$.

In Words
A function is not one-to-one if two different inputs correspond to the same output.

Put another way, a function f is one-to-one if no y in the range is the image of more than one x in the domain. A function is not one-to-one if any two (or more) different elements in the domain correspond to the same element in the range. So the function in Figure 7 is not one-to-one because two different elements in the domain, *dog* and *cat*, both correspond to 11 (and also because three different elements in the domain correspond to 10). Figure 8 illustrates the distinction among one-to-one functions, functions that are not one-to-one, and relations that are not functions.

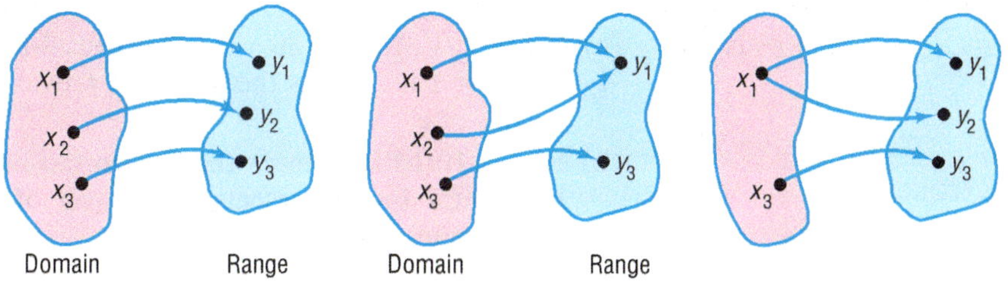

Figure 8

(a) One-to-one function: Each x in the domain has one and only one image in the range.

(b) Not a one-to-one function: y_1 is the image of both x_1 and x_2.

(c) Not a function: x_1 has two images, y_1 and y_2.

EXAMPLE 1

Determining Whether a Function Is One-to-One

Determine whether the following functions are one-to-one.

(a) For the following function, the domain represents the ages of five males, and the range represents their HDL (good) cholesterol scores (mg/dL).

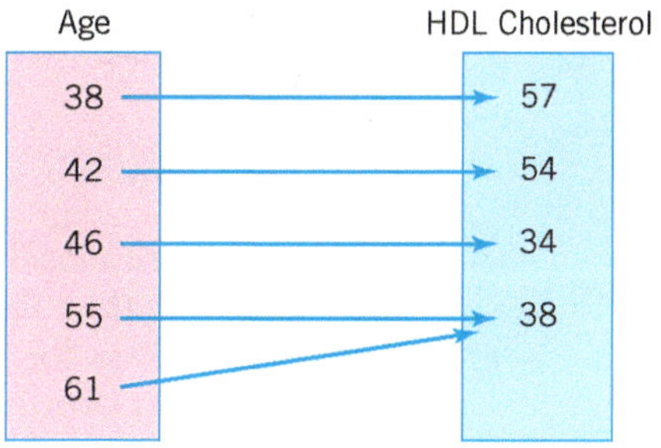

(b) $\{(-2, 6), (-1, 3), (0, 2), (1, 5), (2, 8)\}$

Solution

(a) The function is not one-to-one because there are two different inputs, 55 and 61, that correspond to the same output, 38.

(b) The function is one-to-one because no two distinct inputs correspond to the same output. ■

Now Work PROBLEMS 13 AND 17

For functions defined by an equation $y = f(x)$ and for which the graph of f is known, there is a simple test, called the **horizontal-line test**, to determine whether f is one-to-one.

THEOREM

Horizontal-line Test

If every horizontal line intersects the graph of a function f in at most one point, then f is one-to-one. ■

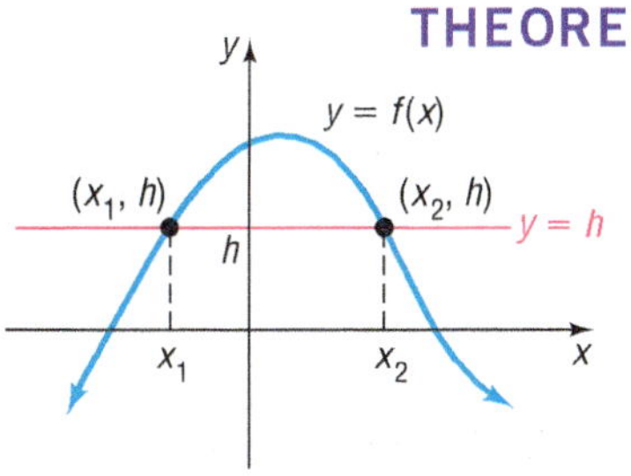

Figure 9
$f(x_1) = f(x_2) = h$ and $x_1 \neq x_2$; f is not a one-to-one function.

The reason why this test works can be seen in Figure 9, where the horizontal line $y = h$ intersects the graph at two distinct points, (x_1, h) and (x_2, h). Since h is the image of both x_1 and x_2 and $x_1 \neq x_2$, f is not one-to-one. Based on Figure 9, the horizontal-line test can be stated in another way: If the graph of any horizontal line intersects the graph of a function f at more than one point, then f is not one-to-one.

EXAMPLE 2 Using the Horizontal-line Test

For each function, use its graph to determine whether the function is one-to-one.

(a) $f(x) = x^2$ (b) $g(x) = x^3$

Solution

(a) Figure 10(a) illustrates the horizontal-line test for $f(x) = x^2$. The horizontal line $y = 1$ intersects the graph of f twice, at $(1, 1)$ and at $(-1, 1)$, so f is not one-to-one.

(b) Figure 10(b) illustrates the horizontal-line test for $g(x) = x^3$. Because every horizontal line intersects the graph of g exactly once, it follows that g is one-to-one.

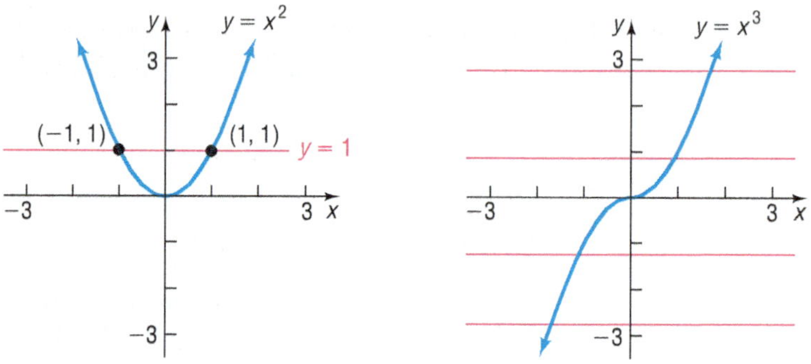

Figure 10 **(a)** A horizontal line intersects the graph twice; f is not one-to-one. **(b)** Every horizontal line intersects the graph exactly once; g is one-to-one.

Now Work PROBLEM 21

Look more closely at the one-to-one function $g(x) = x^3$. This function is an increasing function. Because an increasing (or decreasing) function will always have different y-values for unequal x-values, it follows that a function that is increasing (or decreasing) over its domain is also a one-to-one function.

THEOREM

A function that is increasing on an interval I is a one-to-one function on I.
A function that is decreasing on an interval I is a one-to-one function on I.

2 Determine the Inverse of a Function Defined by a Map or a Set of Ordered Pairs

DEFINITION

Suppose that f is a one-to-one function. Then, corresponding to each x in the domain of f, there is exactly one y in the range (because f is a function); and corresponding to each y in the range of f, there is exactly one x in the domain (because f is one-to-one). The correspondence from the range of f back to the domain of f is called the **inverse function of f**. The symbol f^{-1} is used to denote the inverse function of f.

In Words

Suppose that f is a one-to-one function so that the input 5 corresponds to the output 10. In the inverse function f^{-1}, the input 10 will correspond to the output 5.

We will discuss how to find inverses for all four representations of functions: (1) maps, (2) sets of ordered pairs, (3) equations, and (4) graphs. We begin with finding inverses of functions represented by maps or sets of ordered pairs.

EXAMPLE 3 Finding the Inverse of a Function Defined by a Map

Find the inverse of the function defined by the map. Let the domain of the function represent certain states, and let the range represent the states' populations (in millions). Find the domain and the range of the inverse function.

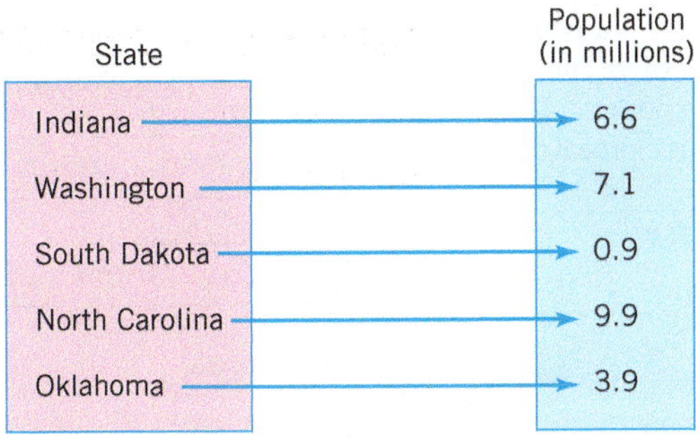

Solution The function is one-to-one. To find the inverse function, interchange the elements in the domain with the elements in the range. For example, the function receives as input Indiana and outputs 6.6 million. So the inverse receives as input 6.6 million and outputs Indiana. The inverse function is shown next.

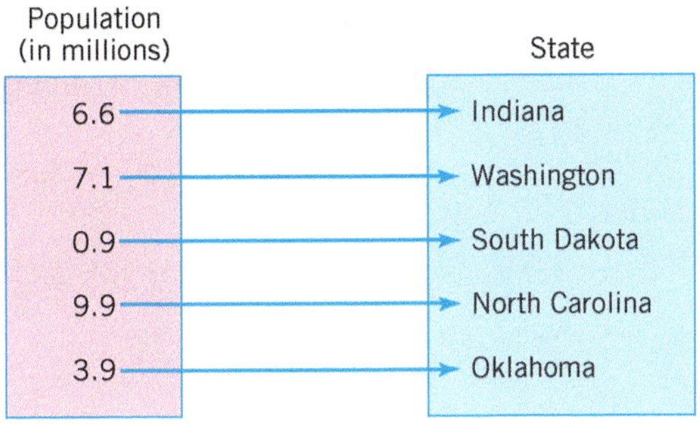

The domain of the inverse function is $\{6.6, 7.1, 0.9, 9.9, 3.9\}$. The range of the inverse function is {Indiana, Washington, South Dakota, North Carolina, Oklahoma}.

If the function f is a set of ordered pairs (x, y), then the inverse function of f, denoted f^{-1}, is the set of ordered pairs (y, x).

EXAMPLE 4 Finding the Inverse of a Function Defined by a Set of Ordered Pairs

Find the inverse of the following one-to-one function:

$$\{(-3, -27), (-2, -8), (-1, -1), (0, 0), (1, 1), (2, 8), (3, 27)\}$$

State the domain and the range of the function and its inverse.

Solution The inverse of the given function is found by interchanging the entries in each ordered pair and so is given by

$$\{(-27, -3), (-8, -2), (-1, -1), (0, 0), (1, 1), (8, 2), (27, 3)\}$$

The domain of the function is $\{-3, -2, -1, 0, 1, 2, 3\}$. The range of the function is $\{-27, -8, -1, 0, 1, 8, 27\}$. The domain of the inverse function is $\{-27, -8, -1, 0, 1, 8, 27\}$. The range of the inverse function is $\{-3, -2, -1, 0, 1, 2, 3\}$.

Now Work PROBLEMS 27 AND 31

Remember, if f is a one-to-one function, it has an inverse function, f^{-1}. See Figure 11.

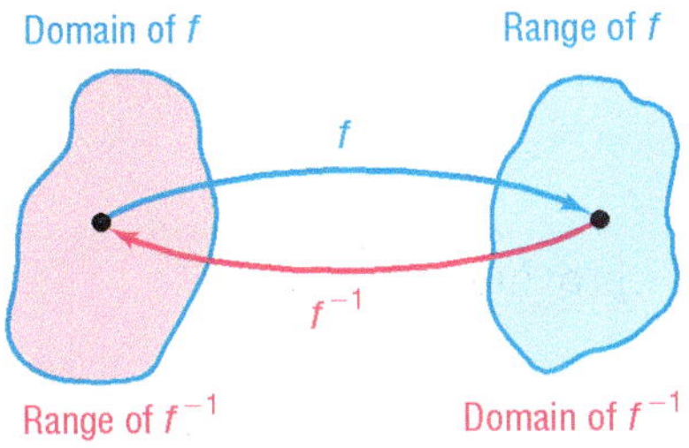

Figure 11

Based on the results of Example 4 and Figure 11, two facts are now apparent about a one-to-one function f and its inverse f^{-1}.

Domain of f = Range of f^{-1} Range of f = Domain of f^{-1}

Look again at Figure 11 to visualize the relationship. Starting with x, applying f, and then applying f^{-1} gets x back again. Starting with x, applying f^{-1}, and then applying f

WARNING Be careful! f^{-1} is a symbol for the inverse function of f. The -1 used in f^{-1} is not an exponent. That is, f^{-1} does *not* mean the reciprocal of f; $f^{-1}(x)$ is not equal to $\dfrac{1}{f(x)}$. ■

gets the number x back again. To put it simply, what f does, f^{-1} undoes, and vice versa. See the illustration that follows.

Input x from domain of f $\xrightarrow{\text{Apply } f}$ $f(x)$ $\xrightarrow{\text{Apply } f^{-1}}$ $f^{-1}(f(x)) = x$

Input x from domain of f^{-1} $\xrightarrow{\text{Apply } f^{-1}}$ $f^{-1}(x)$ $\xrightarrow{\text{Apply } f}$ $f(f^{-1}(x)) = x$

In other words,

$$f^{-1}(f(x)) = x \text{ where } x \text{ is in the domain of } f$$
$$f(f^{-1}(x)) = x \text{ where } x \text{ is in the domain of } f^{-1}$$

Consider the function $f(x) = 2x$, which multiplies the argument x by 2. The inverse function f^{-1} undoes whatever f does. So the inverse function of f is $f^{-1}(x) = \frac{1}{2}x$, which divides the argument by 2. For example, $f(3) = 2(3) = 6$ and $f^{-1}(6) = \frac{1}{2}(6) = 3$, so f^{-1} undoes what f did. This is verified by showing that

$$f^{-1}(f(x)) = f^{-1}(2x) = \frac{1}{2}(2x) = x \quad \text{and} \quad f(f^{-1}(x)) = f\left(\frac{1}{2}x\right) = 2\left(\frac{1}{2}x\right) = x$$

See Figure 12.

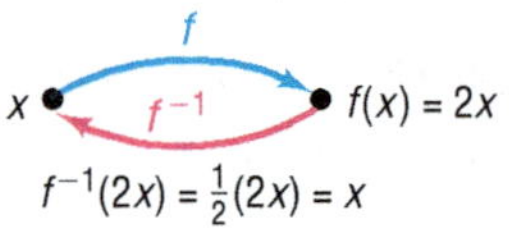

Figure 12

EXAMPLE 5 **Verifying Inverse Functions**

(a) Verify that the inverse of $g(x) = x^3$ is $g^{-1}(x) = \sqrt[3]{x}$.

(b) Verify that the inverse of $f(x) = 2x + 3$ is $f^{-1}(x) = \frac{1}{2}(x - 3)$.

Solution (a) $g^{-1}(g(x)) = g^{-1}(x^3) = \sqrt[3]{x^3} = x$ for all x in the domain of g

$g(g^{-1}(x)) = g(\sqrt[3]{x}) = (\sqrt[3]{x})^3 = x$ for all x in the domain of g^{-1}

(b) $f^{-1}(f(x)) = f^{-1}(2x + 3) = \frac{1}{2}[(2x + 3) - 3] = \frac{1}{2}(2x) = x$ for all x in the domain of f

$f(f^{-1}(x)) = f\left(\frac{1}{2}(x - 3)\right) = 2\left[\frac{1}{2}(x - 3)\right] + 3 = (x - 3) + 3 = x$ for all x in the domain of f^{-1} ■

EXAMPLE 6 **Verifying Inverse Functions**

Verify that the inverse of $f(x) = \dfrac{1}{x - 1}$ is $f^{-1}(x) = \dfrac{1}{x} + 1$. For what values of x is $f^{-1}(f(x)) = x$? For what values of x is $f(f^{-1}(x)) = x$?

Solution The domain of f is $\{x \mid x \neq 1\}$ and the domain of f^{-1} is $\{x \mid x \neq 0\}$. Now

$$f^{-1}(f(x)) = f^{-1}\left(\frac{1}{x - 1}\right) = \frac{1}{\frac{1}{x - 1}} + 1 = x - 1 + 1 = x \quad \text{provided } x \neq 1$$

$$f(f^{-1}(x)) = f\left(\frac{1}{x} + 1\right) = \frac{1}{\left(\frac{1}{x} + 1\right) - 1} = \frac{1}{\frac{1}{x}} = x \quad \text{provided } x \neq 0$$

■

Now Work PROBLEMS 35 AND 39

3 Obtain the Graph of the Inverse Function from the Graph of the Function

For the functions in Example 5(b), we list points on the graph of $f = Y_1$ and on the graph of $f^{-1} = Y_2$ in Table 1. Note that whenever (a, b) is on the graph of f then (b, a) is on the graph of f^{-1}. Figure 13 shows these points plotted. Also shown is the graph of $y = x$, which you should observe is a line of symmetry of the points.

Table 1

NORMAL FLOAT AUTO REAL RADIAN MP
PRESS ENTER TO EDIT

X	Y_1
-6	-9
-5	-7
-4	-5
-3	-3
-2	-1
-1	1
0	3
1	5
2	7
3	9
4	11

$Y_1 = 2X+3$

NORMAL FLOAT AUTO REAL RADIAN MP
PRESS ENTER TO EDIT

X	Y_2
-9	-6
-7	-5
-5	-4
-3	-3
-1	-2
1	-1
3	0
5	1
7	2
9	3
11	4

$Y_2 = \frac{1}{2}(X-3)$

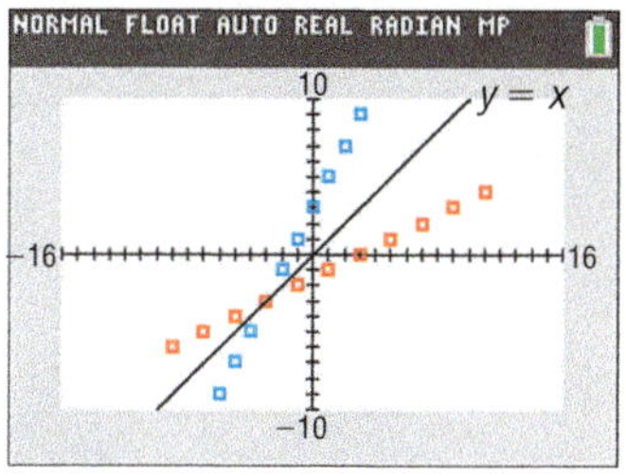

Figure 13

Exploration

Simultaneously graph $Y_1 = x$, $Y_2 = x^3$, and $Y_3 = \sqrt[3]{x}$ on a square screen with $-3 \le x \le 3$. What do you observe about the graphs of $Y_2 = x^3$, its inverse $Y_3 = \sqrt[3]{x}$, and the line $Y_1 = x$?

Repeat this experiment by simultaneously graphing $Y_1 = x$, $Y_2 = 2x + 3$, and $Y_3 = \dfrac{1}{2}(x - 3)$ on a square screen with $-6 \le x \le 3$. Do you see the symmetry of the graph of Y_2 and its inverse Y_3 with respect to the line $Y_1 = x$? ■

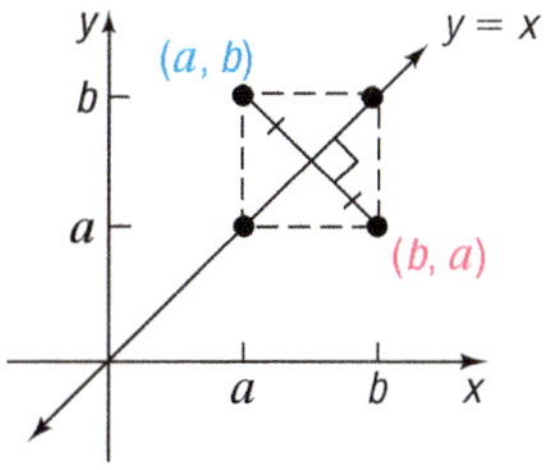

Figure 14

Suppose that (a, b) is a point on the graph of a one-to-one function f defined by $y = f(x)$. Then $b = f(a)$. This means that $a = f^{-1}(b)$, so (b, a) is a point on the graph of the inverse function f^{-1}. The relationship between the point (a, b) on f and the point (b, a) on f^{-1} is shown in Figure 14. The line segment with endpoints (a, b) and (b, a) is perpendicular to the line $y = x$ and is bisected by the line $y = x$. (Do you see why?) It follows that the point (b, a) on f^{-1} is the reflection about the line $y = x$ of the point (a, b) on f.

THEOREM

The graph of a one-to-one function f and the graph of its inverse function f^{-1} are symmetric with respect to the line $y = x$. ■

Figure 15 illustrates this result. Once the graph of f is known, the graph of f^{-1} may be obtained by reflecting the graph of f about the line $y = x$.

EXAMPLE 7 Graphing the Inverse Function

The graph in Figure 16(a) is that of a one-to-one function $y = f(x)$. Draw the graph of its inverse.

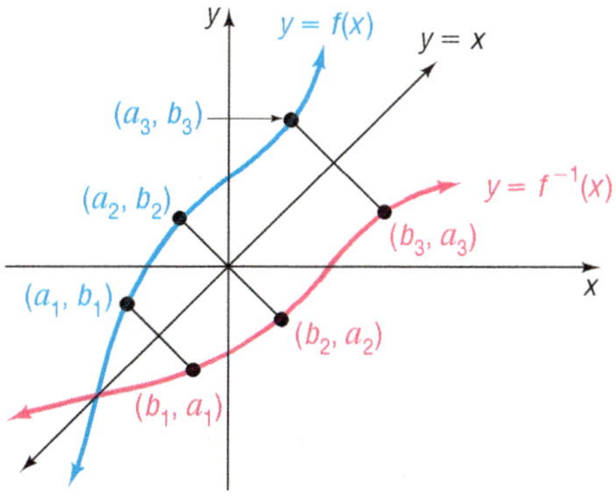

Figure 15

Solution Begin by adding the graph of $y = x$ to Figure 16(a). Since the points $(-2, -1)$, $(-1, 0)$, and $(2, 1)$ are on the graph of f, the points $(-1, -2)$, $(0, -1)$, and $(1, 2)$ must be on the graph of f^{-1}. Keeping in mind that the graph of f^{-1} is the reflection about the line $y = x$ of the graph of f, draw the graph of f^{-1}. See Figure 16(b).

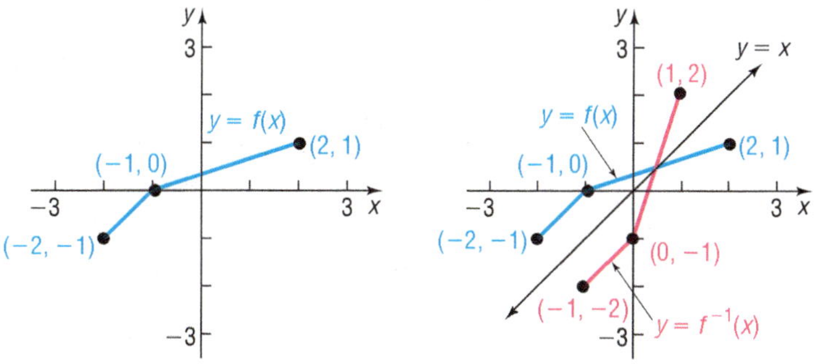

Figure 16 (a) (b) ■

Now Work PROBLEM 45

4 Find the Inverse of a Function Defined by an Equation

The fact that the graphs of a one-to-one function f and its inverse function f^{-1} are symmetric with respect to the line $y = x$ tells us more. It says that we can obtain f^{-1} by interchanging the roles of x and y in f. Look again at Figure 15. If f is defined by the equation

$$y = f(x)$$

then f^{-1} is defined by the equation

$$x = f(y)$$

The equation $x = f(y)$ defines f^{-1} *implicitly*. If we can solve this equation for y, we will have the *explicit* form of f^{-1}, that is,

$$y = f^{-1}(x)$$

Let's use this procedure to find the inverse of $f(x) = 2x + 3$. (Because f is a linear function and is increasing, f is one-to-one and so has an inverse function.)

EXAMPLE 8 **How to Find the Inverse Function**

Find the inverse of $f(x) = 2x + 3$. Graph f and f^{-1} on the same coordinate axes.

Step-by-Step Solution

Step 1: Replace $f(x)$ with y. In $y = f(x)$, interchange the variables x and y to obtain $x = f(y)$. This equation defines the inverse function f^{-1} implicitly.

Replace $f(x)$ with y in $f(x) = 2x + 3$ and obtain $y = 2x + 3$. Now interchange the variables x and y to obtain

$$x = 2y + 3$$

This equation defines the inverse function f^{-1} implicitly.

Step 2: If possible, solve the implicit equation for y in terms of x to obtain the explicit form of f^{-1}, $y = f^{-1}(x)$.

To find the explicit form of the inverse, solve $x = 2y + 3$ for y.

$$x = 2y + 3$$

$$2y + 3 = x \quad \text{Reflexive Property; If } a = b, \text{ then } b = a.$$

$$2y = x - 3 \quad \text{Subtract 3 from both sides.}$$

$$y = \frac{1}{2}(x - 3) \quad \text{Multiply both sides by } \frac{1}{2}.$$

The explicit form of the inverse function f^{-1} is

$$f^{-1}(x) = \frac{1}{2}(x - 3)$$

Step 3: Check the result by showing that $f^{-1}(f(x)) = x$ and $f(f^{-1}(x)) = x$.

We verified that f and f^{-1} are inverses in Example 5(b).

The graphs of $f(x) = 2x + 3$ and its inverse $f^{-1}(x) = \frac{1}{2}(x - 3)$ are shown in Figure 17. Note the symmetry of the graphs with respect to the line $y = x$. ■

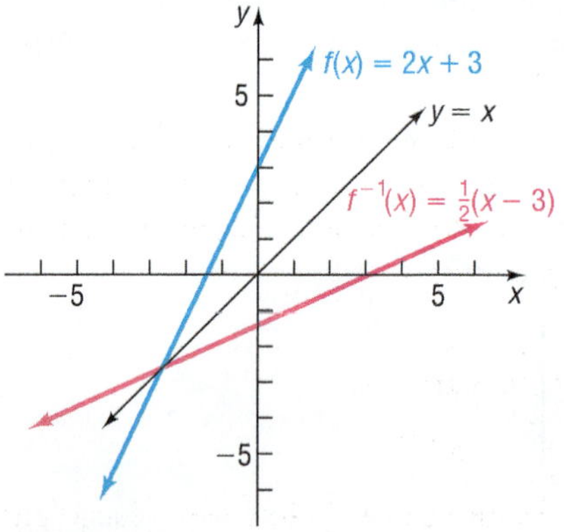

Figure 17

Procedure for Finding the Inverse of a One-to-One Function

STEP 1: In $y = f(x)$, interchange the variables x and y to obtain

$$x = f(y)$$

This equation defines the inverse function f^{-1} implicitly.

STEP 2: If possible, solve the implicit equation for y in terms of x to obtain the explicit form of f^{-1}:

$$y = f^{-1}(x)$$

STEP 3: Check the result by showing that

$$f^{-1}(f(x)) = x \quad \text{and} \quad f(f^{-1}(x)) = x$$

EXAMPLE 9

Finding the Inverse Function

The function

$$f(x) = \frac{2x+1}{x-1} \qquad x \neq 1$$

is one-to-one. Find its inverse function and check the result.

Solution

STEP 1: Replace $f(x)$ with y and interchange the variables x and y in

$$y = \frac{2x+1}{x-1}$$

to obtain

$$x = \frac{2y+1}{y-1}$$

STEP 2: Solve for y.

$$x = \frac{2y+1}{y-1}$$

$$x(y-1) = 2y+1 \qquad \text{Multiply both sides by } y-1.$$

$$xy - x = 2y+1 \qquad \text{Apply the Distributive Property.}$$

$$xy - 2y = x+1 \qquad \text{Subtract } 2y \text{ from both sides; add } x \text{ to both sides.}$$

$$(x-2)y = x+1 \qquad \text{Factor.}$$

$$y = \frac{x+1}{x-2} \qquad \text{Divide by } x-2.$$

The inverse function is

$$f^{-1}(x) = \frac{x+1}{x-2} \qquad x \neq 2 \qquad \text{Replace } y \text{ by } f^{-1}(x).$$

STEP 3: ✓Check:

$$f^{-1}(f(x)) = f^{-1}\left(\frac{2x+1}{x-1}\right) = \frac{\frac{2x+1}{x-1}+1}{\frac{2x+1}{x-1}-2} = \frac{2x+1+x-1}{2x+1-2(x-1)} = \frac{3x}{3} = x, \quad x \neq 1$$

$$f(f^{-1}(x)) = f\left(\frac{x+1}{x-2}\right) = \frac{2\left(\frac{x+1}{x-2}\right)+1}{\frac{x+1}{x-2}-1} = \frac{2(x+1)+x-2}{x+1-(x-2)} = \frac{3x}{3} = x, \quad x \neq 2$$

■

Exploration

In Example 9, we found that if $f(x) = \frac{2x+1}{x-1}$, then $f^{-1}(x) = \frac{x+1}{x-2}$. Compare the vertical and horizontal asymptotes of f and f^{-1}.

Result The vertical asymptote of f is $x = 1$, and the horizontal asymptote is $y = 2$. The vertical asymptote of f^{-1} is $x = 2$, and the horizontal asymptote is $y = 1$. ■

Now Work PROBLEMS 53 AND 67

If a function is not one-to-one, it has no inverse function. Sometimes, though, an appropriate restriction on the domain of such a function will yield a new function that *is* one-to-one. Then the function defined on the restricted domain has an inverse function. Let's look at an example of this common practice.

EXAMPLE 10

Finding the Inverse of a Domain-restricted Function

Find the inverse of $y = f(x) = x^2$ if $x \geq 0$. Graph f and f^{-1}.

Solution

The function $y = x^2$ is not one-to-one. [Refer to Example 2(a).] However, restricting the domain of this function to $x \geq 0$, as indicated, results in a new function that is increasing and therefore is one-to-one. Consequently, the function defined by $y = f(x) = x^2, x \geq 0$, has an inverse function, f^{-1}.

Follow the steps given previously to find f^{-1}.

STEP 1: In the equation $y = x^2, x \geq 0$, interchange the variables x and y. The result is

$$x = y^2 \qquad y \geq 0$$

This equation defines the inverse function implicitly.

STEP 2: Solve for y to get the explicit form of the inverse. Because $y \geq 0$, only one solution for y is obtained: $y = \sqrt{x}$. So $f^{-1}(x) = \sqrt{x}$.

STEP 3: **Check:** $f^{-1}(f(x)) = f^{-1}(x^2) = \sqrt{x^2} = |x| = x$ because $x \geq 0$

$$f(f^{-1}(x)) = f(\sqrt{x}) = (\sqrt{x})^2 = x$$

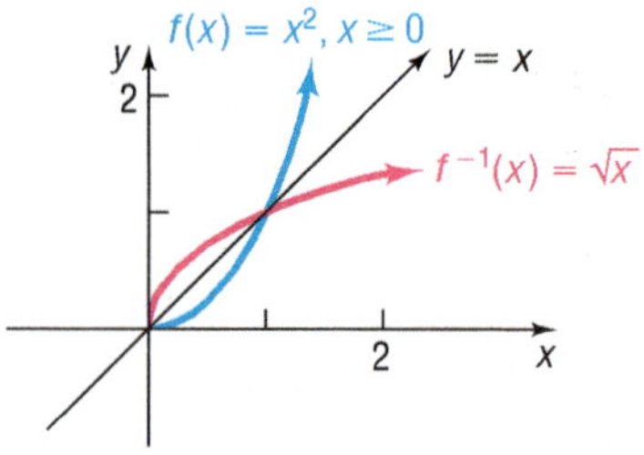

Figure 18

Figure 18 illustrates the graphs of $f(x) = x^2, x \geq 0$, and $f^{-1}(x) = \sqrt{x}$. Note that the domain of f = range of $f^{-1} = [0, \infty)$, and the domain of f^{-1} = range of $f = [0, \infty)$.

SUMMARY

1. If a function f is one-to-one, then it has an inverse function f^{-1}.
2. Domain of f = Range of f^{-1}; Range of f = Domain of f^{-1}.
3. To verify that f^{-1} is the inverse of f, show that $f^{-1}(f(x)) = x$ for every x in the domain of f and that $f(f^{-1}(x)) = x$ for every x in the domain of f^{-1}.
4. The graphs of f and f^{-1} are symmetric with respect to the line $y = x$.

6.2 Assess Your Understanding

'Are You Prepared?' *Answers are given at the end of these exercises. If you get a wrong answer, read the pages listed in red.*

1. Is the set of ordered pairs $\{(1, 3), (2, 3), (-1, 2)\}$ a function? Why or why not? (pp. 207–210)
2. Where is the function $f(x) = x^2$ increasing? Where is it decreasing? (p. 234)
3. What is the domain of $f(x) = \dfrac{x + 5}{x^2 + 3x - 18}$? (pp. 214–216)
4. Simplify: $\dfrac{\frac{1}{x} + 1}{\frac{1}{x^2} - 1}$ (pp. 69–70)

Concepts and Vocabulary

5. If x_1 and x_2 are two different inputs of a function f, then f is one-to-one if ________________.
6. If every horizontal line intersects the graph of a function f at no more than one point, then f is a(n) ______________ function.
7. If f is a one-to-one function and $f(3) = 8$, then $f^{-1}(8) =$ _________.
8. If f^{-1} denotes the inverse of a function f, then the graphs of f and f^{-1} are symmetric with respect to the line _________.
9. If the domain of a one-to-one function f is $[4, \infty)$, then the range of its inverse function f^{-1} is _______________.
10. ***True or False*** If f and g are inverse functions, then the domain of f is the same as the range of g.
11. If $(-2, 3)$ is a point on the graph of a one-to-one function f, which of the following points is on the graph of f^{-1}?
(a) $(3, -2)$ (b) $(2, -3)$ (c) $(-3, 2)$ (d) $(-2, -3)$
12. Suppose f is a one-to-one function with a domain of $\{x \mid x \neq 3\}$ and a range of $\left\{x \,\middle|\, x \neq \frac{2}{3}\right\}$. Which of the following is the domain of f^{-1}?
(a) $\{x \mid x \neq 3\}$ (b) All real numbers
(c) $\left\{x \,\middle|\, x \neq \frac{2}{3}, x \neq 3\right\}$ (d) $\left\{x \,\middle|\, x \neq \frac{2}{3}\right\}$

Skill Building

In Problems 13–20, determine whether the function is one-to-one.

13. 

14. 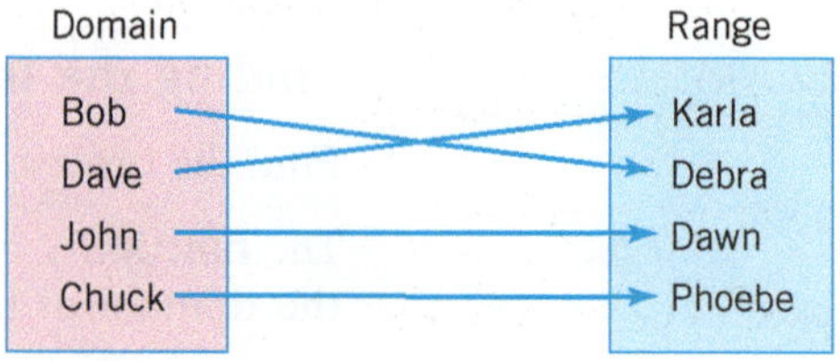

15.

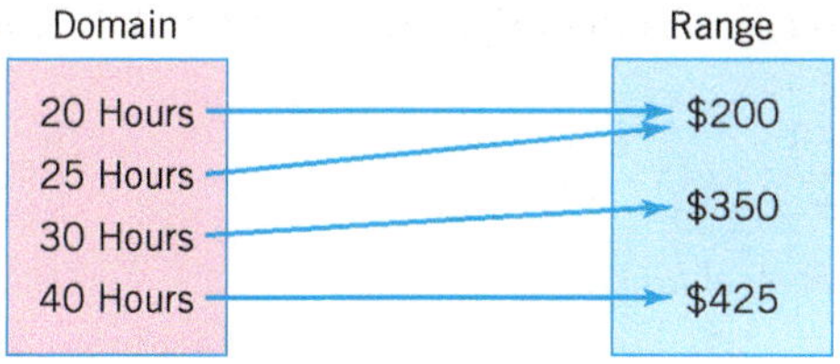

16.

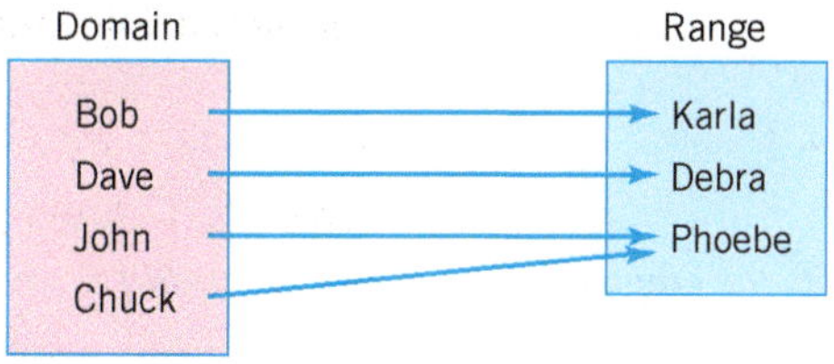

17. $\{(2, 6), (-3, 6), (4, 9), (1, 10)\}$

18. $\{(-2, 5), (-1, 3), (3, 7), (4, 12)\}$

19. $\{(0, 0), (1, 1), (2, 16), (3, 81)\}$

20. $\{(1, 2), (2, 8), (3, 18), (4, 32)\}$

In Problems 21–26, the graph of a function f is given. Use the horizontal-line test to determine whether f is one-to-one.

21.

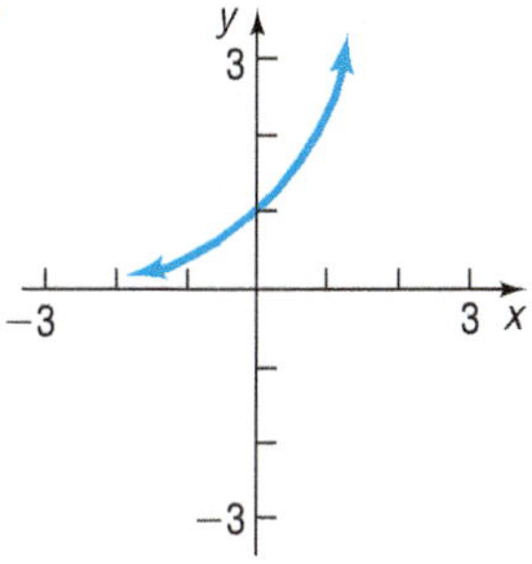

22.

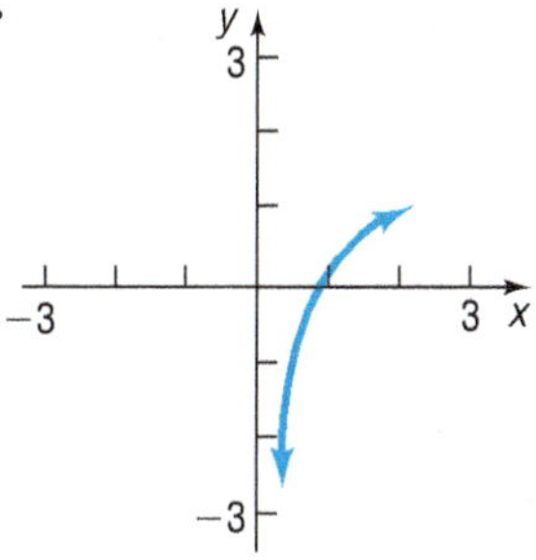

23.

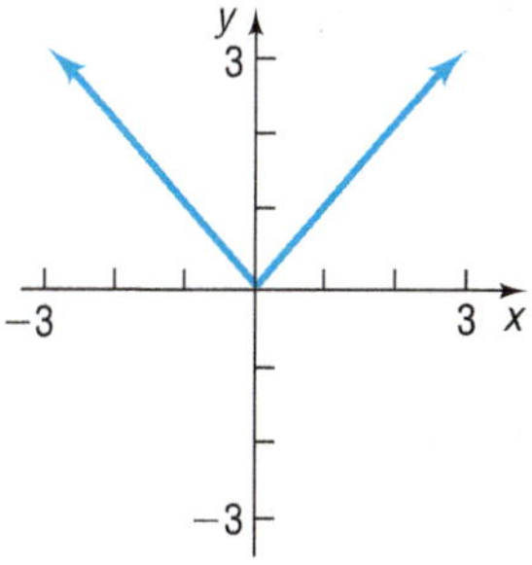

24.

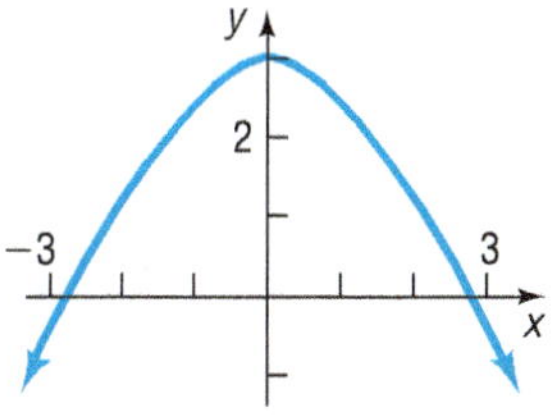

25.

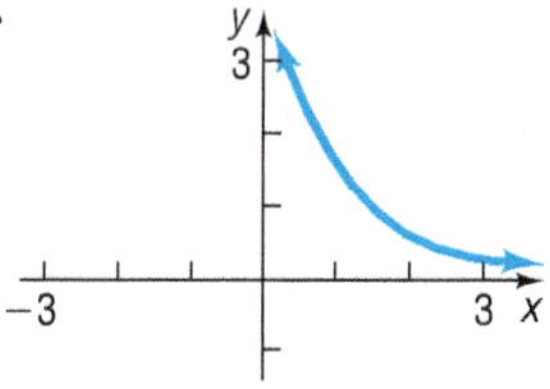

26.

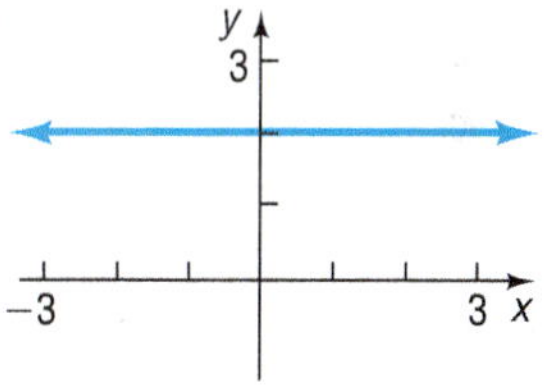

In Problems 27–34, find the inverse of each one-to-one function. State the domain and the range of each inverse function.

27.

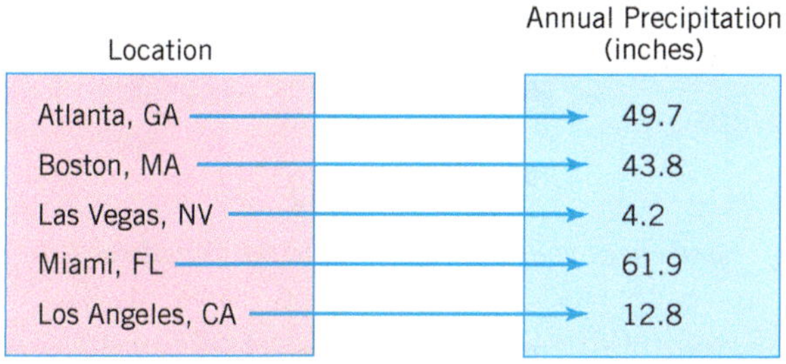

Source: *currentresults.com*

28.

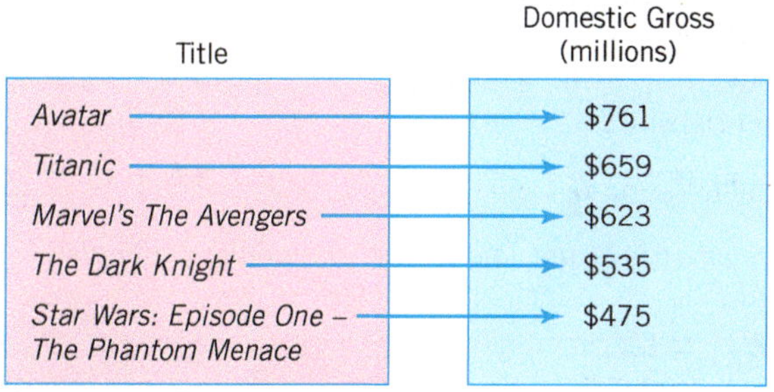

Source: *boxofficemojo.com*

29.

Source: *tiaa-cref.org*

30.

Source: *United States Bureau of Labor Statistics, March 2015*

31. $\{(-3, 5), (-2, 9), (-1, 2), (0, 11), (1, -5)\}$

32. $\{(-2, 2), (-1, 6), (0, 8), (1, -3), (2, 9)\}$

33. $\{(-2, 1), (-3, 2), (-10, 0), (1, 9), (2, 4)\}$

34. $\{(-2, -8), (-1, -1), (0, 0), (1, 1), (2, 8)\}$

In Problems 35–44, verify that the functions f and g are inverses of each other by showing that $f(g(x)) = x$ and $g(f(x)) = x$. Give any values of x that need to be excluded from the domain of f and the domain of g.

35. $f(x) = 3x + 4; \quad g(x) = \frac{1}{3}(x - 4)$

36. $f(x) = 3 - 2x; \quad g(x) = -\frac{1}{2}(x - 3)$

37. $f(x) = 4x - 8; \quad g(x) = \frac{x}{4} + 2$

38. $f(x) = 2x + 6; \quad g(x) = \frac{1}{2}x - 3$

39. $f(x) = x^3 - 8; \quad g(x) = \sqrt[3]{x + 8}$

40. $f(x) = (x - 2)^2, x \geq 2; \quad g(x) = \sqrt{x} + 2$

41. $f(x) = \frac{1}{x};\quad g(x) = \frac{1}{x}$

42. $f(x) = x;\quad g(x) = x$

43. $f(x) = \frac{2x+3}{x+4};\quad g(x) = \frac{4x-3}{2-x}$

44. $f(x) = \frac{x-5}{2x+3};\quad g(x) = \frac{3x+5}{1-2x}$

In Problems 45–50, the graph of a one-to-one function f is given. Draw the graph of the inverse function f^{-1}.

45.

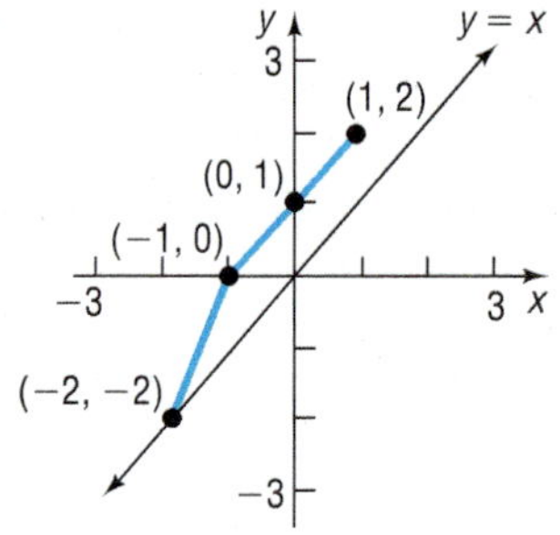

46.

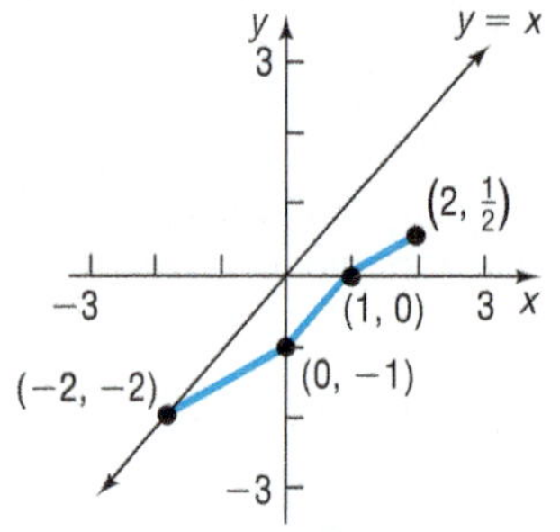

47.

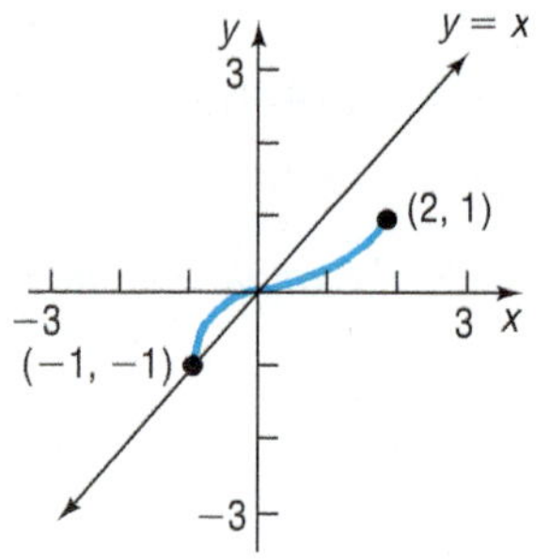

48.

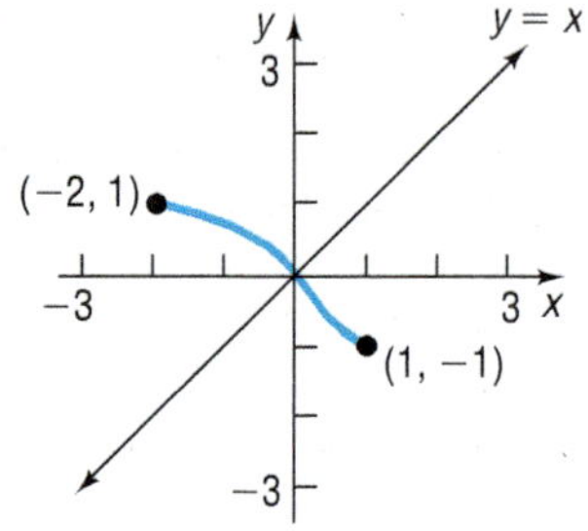

49.

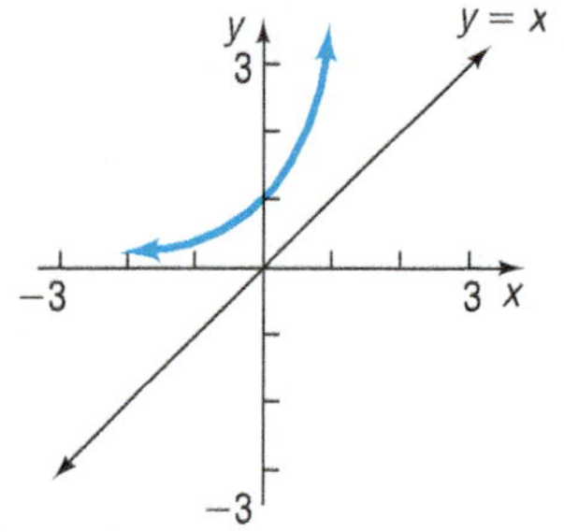

50.

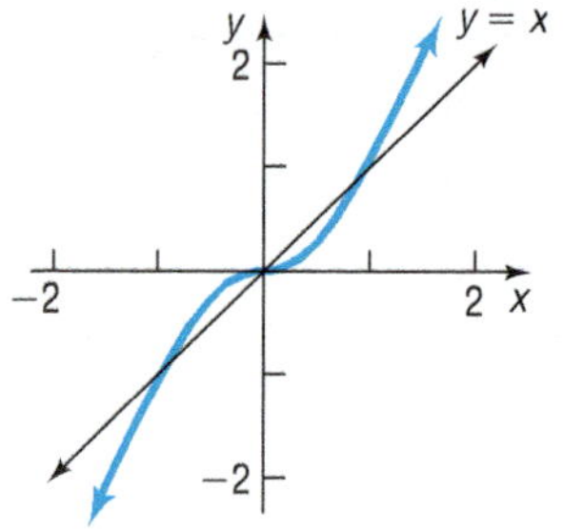

In Problems 51–62, the function f is one-to-one. (a) Find its inverse function f^{-1} and check your answer. (b) Find the domain and the range of f and f^{-1}. (c) Graph f, f^{-1}, and $y = x$ on the same coordinate axes.

51. $f(x) = 3x$

52. $f(x) = -4x$

53. $f(x) = 4x + 2$

54. $f(x) = 1 - 3x$

55. $f(x) = x^3 - 1$

56. $f(x) = x^3 + 1$

57. $f(x) = x^2 + 4,\quad x \ge 0$

58. $f(x) = x^2 + 9,\quad x \ge 0$

59. $f(x) = \frac{4}{x}$

60. $f(x) = -\frac{3}{x}$

61. $f(x) = \frac{1}{x-2}$

62. $f(x) = \frac{4}{x+2}$

In Problems 63–74, the function f is one-to-one. (a) Find its inverse function f^{-1} and check your answer. (b) Find the domain and the range of f and f^{-1}.

63. $f(x) = \frac{2}{3+x}$

64. $f(x) = \frac{4}{2-x}$

65. $f(x) = \frac{3x}{x+2}$

66. $f(x) = -\frac{2x}{x-1}$

67. $f(x) = \frac{2x}{3x-1}$

68. $f(x) = -\frac{3x+1}{x}$

69. $f(x) = \frac{3x+4}{2x-3}$

70. $f(x) = \frac{2x-3}{x+4}$

71. $f(x) = \frac{2x+3}{x+2}$

72. $f(x) = \frac{-3x-4}{x-2}$

73. $f(x) = \frac{x^2-4}{2x^2},\quad x > 0$

74. $f(x) = \frac{x^2+3}{3x^2}\quad x > 0$

Applications and Extensions

75. Use the graph of $y = f(x)$ given in Problem 45 to evaluate the following:
(a) $f(-1)$ (b) $f(1)$ (c) $f^{-1}(1)$ (d) $f^{-1}(2)$

76. Use the graph of $y = f(x)$ given in Problem 46 to evaluate the following:
(a) $f(2)$ (b) $f(1)$ (c) $f^{-1}(0)$ (d) $f^{-1}(-1)$

77. If $f(7) = 13$ and f is one-to-one, what is $f^{-1}(13)$?

78. If $g(-5) = 3$ and g is one-to-one, what is $g^{-1}(3)$?

79. The domain of a one-to-one function f is $[5, \infty)$, and its range is $[-2, \infty)$. State the domain and the range of f^{-1}.

80. The domain of a one-to-one function f is $[0, \infty)$, and its range is $[5, \infty)$. State the domain and the range of f^{-1}.

81. The domain of a one-to-one function g is $(-\infty, 0]$, and its range is $[0, \infty)$. State the domain and the range of g^{-1}.

82. The domain of a one-to-one function g is $[0, 15]$, and its range is $(0, 8)$. State the domain and the range of g^{-1}.

83. A function $y = f(x)$ is increasing on the interval $[0, 5]$. What conclusions can you draw about the graph of $y = f^{-1}(x)$?

84. A function $y = f(x)$ is decreasing on the interval $[0, 5]$. What conclusions can you draw about the graph of $y = f^{-1}(x)$?

85. Find the inverse of the linear function

$$f(x) = mx + b, \quad m \neq 0$$

86. Find the inverse of the function

$$f(x) = \sqrt{r^2 - x^2}, \quad 0 \leq x \leq r$$

87. A function f has an inverse function f^{-1}. If the graph of f lies in quadrant I, in which quadrant does the graph of f^{-1} lie?

88. A function f has an inverse function f^{-1}. If the graph of f lies in quadrant II, in which quadrant does the graph of f^{-1} lie?

89. The function $f(x) = |x|$ is not one-to-one. Find a suitable restriction on the domain of f so that the new function that results is one-to-one. Then find the inverse of the new function.

90. The function $f(x) = x^4$ is not one-to-one. Find a suitable restriction on the domain of f so that the new function that results is one-to-one. Then find the inverse of the new function.

In applications, the symbols used for the independent and dependent variables are often based on common usage. So, rather than using $y = f(x)$ to represent a function, an applied problem might use $C = C(q)$ to represent the cost C of manufacturing q units of a good. Because of this, the inverse notation f^{-1} used in a pure mathematics problem is not used when finding inverses of applied problems. Rather, the inverse of a function such as $C = C(q)$ will be $q = q(C)$. So $C = C(q)$ is a function that represents the cost C as a function of the number q of units manufactured, and $q = q(C)$ is a function that represents the number q as a function of the cost C. Problems 91–94 illustrate this idea.

91. Vehicle Stopping Distance Taking into account reaction time, the distance d (in feet) that a car requires to come to a complete stop while traveling r miles per hour is given by the function

$$d(r) = 6.97r - 90.39$$

(a) Express the speed r at which the car is traveling as a function of the distance d required to come to a complete stop.
(b) Verify that $r = r(d)$ is the inverse of $d = d(r)$ by showing that $r(d(r)) = r$ and $d(r(d)) = d$.
(c) Predict the speed that a car was traveling if the distance required to stop was 300 feet.

92. Height and Head Circumference The head circumference C of a child is related to the height H of the child (both in inches) through the function

$$H(C) = 2.15C - 10.53$$

(a) Express the head circumference C as a function of height H.
(b) Verify that $C = C(H)$ is the inverse of $H = H(C)$ by showing that $H(C(H)) = H$ and $C(H(C)) = C$.
(c) Predict the head circumference of a child who is 26 inches tall.

93. Ideal Body Weight One model for the ideal body weight W for men (in kilograms) as a function of height h (in inches) is given by the function

$$W(h) = 50 + 2.3(h - 60)$$

(a) What is the ideal weight of a 6-foot male?
(b) Express the height h as a function of weight W.
(c) Verify that $h = h(W)$ is the inverse of $W = W(h)$ by showing that $h(W(h)) = h$ and $W(h(W)) = W$.
(d) What is the height of a male who is at his ideal weight of 80 kilograms?
[**Note:** The ideal body weight W for women (in kilograms) as a function of height h (in inches) is given by $W(h) = 45.5 + 2.3(h - 60)$.]

94. Temperature Conversion The function $F(C) = \frac{9}{5}C + 32$ converts a temperature from C degrees Celsius to F degrees Fahrenheit.
(a) Express the temperature in degrees Celsius C as a function of the temperature in degrees Fahrenheit F.
(b) Verify that $C = C(F)$ is the inverse of $F = F(C)$ by showing that $C(F(C)) = C$ and $F(C(F)) = F$.
(c) What is the temperature in degrees Celsius if it is 70 degrees Fahrenheit?

95. Income Taxes The function

$$T(g) = 5156.25 + 0.25(g - 37{,}450)$$

represents the 2015 federal income tax T (in dollars) due for a "single" filer whose modified adjusted gross income is g dollars, where $37{,}450 \leq g \leq 90{,}750$.
(a) What is the domain of the function T?
(b) Given that the tax due T is an increasing linear function of modified adjusted gross income g, find the range of the function T.
(c) Find adjusted gross income g as a function of federal income tax T. What are the domain and the range of this function?

96. Income Taxes The function

$$T(g) = 1845 + 0.15(g - 18{,}450)$$

represents the 2015 federal income tax T (in dollars) due for a "married filing jointly" filer whose modified adjusted gross income is g dollars, where $18{,}450 \leq g \leq 74{,}900$.
(a) What is the domain of the function T?
(b) Given that the tax due T is an increasing linear function of modified adjusted gross income g, find the range of the function T.
(c) Find adjusted gross income g as a function of federal income tax T. What are the domain and the range of this function?

97. Gravity on Earth If a rock falls from a height of 100 meters on Earth, the height H (in meters) after t seconds is approximately

$$H(t) = 100 - 4.9t^2$$

(a) In general, quadratic functions are not one-to-one. However, the function H is one-to-one. Why?
(b) Find the inverse of H and verify your result.
(c) How long will it take a rock to fall 80 meters?

98. Period of a Pendulum The period T (in seconds) of a simple pendulum as a function of its length l (in feet) is given by

$$T(l) = 2\pi\sqrt{\frac{l}{32.2}}$$

(a) Express the length l as a function of the period T.
(a) How long is a pendulum whose period is 3 seconds?

99. Given

$$f(x) = \frac{ax + b}{cx + d}$$

find $f^{-1}(x)$. If $c \neq 0$, under what conditions on $a, b, c,$ and d is $f = f^{-1}$?

Explaining Concepts: Discussion and Writing

100. Can a one-to-one function and its inverse be equal? What must be true about the graph of f for this to happen? Give some examples to support your conclusion.

101. Draw the graph of a one-to-one function that contains the points $(-2, -3)$, $(0, 0)$, and $(1, 5)$. Now draw the graph of its inverse. Compare your graph to those of other students. Discuss any similarities. What differences do you see?

102. Give an example of a function whose domain is the set of real numbers and that is neither increasing nor decreasing on its domain, but is one-to-one.
[**Hint:** Use a piecewise-defined function.]

103. Is every odd function one-to-one? Explain.

104. Suppose that $C(g)$ represents the cost C, in dollars, of manufacturing g cars. Explain what $C^{-1}(800{,}000)$ represents.

105. Explain why the horizontal-line test can be used to identify one-to-one functions from a graph.

106. Explain why a function must be one-to-one in order to have an inverse that is a function. Use the function $y = x^2$ to support your explanation.

Retain Your Knowledge

Problems 107–110 are based on material learned earlier in the course. The purpose of these problems is to keep the material fresh in your mind so that you are better prepared for the final exam.

107. Use the techniques of shifting, compressing or stretching, and reflections to graph $f(x) = -|x + 2| + 3$.

108. Find the zeros of the quadratic function $f(x) = 3x^2 + 5x + 1$. What are the x-intercepts, if any, of the graph of the function?

109. Find the domain of $R(x) = \dfrac{6x^2 - 11x - 2}{2x^2 - x - 6}$. Find any horizontal, vertical, or oblique asymptotes.

110. If $f(x) = 3x^2 - 7x$, find $f(x + h) - f(x)$.

'Are You Prepared?' Answers

1. Yes; for each input x there is one output y.

2. Increasing on $[0, \infty)$; decreasing on $(-\infty, 0]$

3. $\{x \mid x \neq -6, x \neq 3\}$

4. $\dfrac{x}{1 - x}$, $x \neq 0, x \neq -1$

6.3 Exponential Functions

PREPARING FOR THIS SECTION *Before getting started, review the following:*

- Exponents (Chapter R, Section R.2, pp. 22–24, and Section R.8, pp. 77–78)
- Graphing Techniques: Transformations (Section 3.5, pp. 256–264)
- Solving Linear and Quadratic Equations (Section 1.2, pp. 102–103, and Section 1.3, pp. 110–115)
- Average Rate of Change (Section 3.3, pp. 238–239)
- Quadratic Functions (Section 4.3, pp. 298–306)
- Linear Functions (Section 4.1, pp. 281–287)
- Horizontal Asymptotes (Section 5.4, pp. 377–379)

Now Work the 'Are You Prepared?' problems on page 439.

OBJECTIVES
1 Evaluate Exponential Functions (p. 428)
2 Graph Exponential Functions (p. 432)
3 Define the Number e (p. 436)
4 Solve Exponential Equations (p. 437)

1 Evaluate Exponential Functions

Chapter R, Section R.8, gives a definition for raising a real number a to a rational power. That discussion provides meaning to expressions of the form

$$a^r$$

where the base a is a positive real number and the exponent r is a rational number.

But what is the meaning of a^x, where the base a is a positive real number and the exponent x is an irrational number? Although a rigorous definition requires methods discussed in calculus, the basis for the definition is easy to follow: Select a rational number r that is formed by truncating (removing) all but a finite number of digits from the irrational number x. Then it is reasonable to expect that

$$a^x \approx a^r$$

For example, take the irrational number $\pi = 3.14159\ldots$. Then an approximation to a^π is

$$a^\pi \approx a^{3.14}$$

where the digits after the hundredths position have been removed from the value for π. A better approximation would be

$$a^\pi \approx a^{3.14159}$$

where the digits after the hundred-thousandths position have been removed. Continuing in this way, we can obtain approximations to a^π to any desired degree of accuracy.

Most calculators have an $\boxed{x^y}$ key or a caret key $\boxed{\wedge}$ for working with exponents. To evaluate expressions of the form a^x, enter the base a, then press the $\boxed{x^y}$ key (or the $\boxed{\wedge}$ key), enter the exponent x, and press $\boxed{=}$ (or $\boxed{\text{ENTER}}$).

EXAMPLE 1

Using a Calculator to Evaluate Powers of 2

Using a calculator, evaluate:

(a) $2^{1.4}$ (b) $2^{1.41}$ (c) $2^{1.414}$ (d) $2^{1.4142}$ (e) $2^{\sqrt{2}}$

Solution Figure 19 shows the solution to parts (a) and (e) using a TI-84 Plus C graphing calculator.

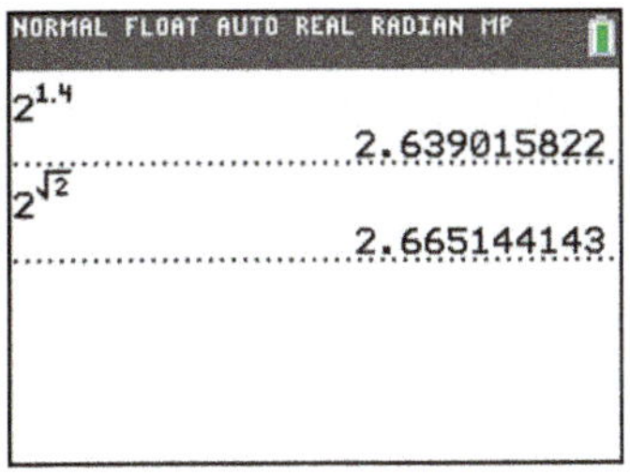

Figure 19

(a) $2^{1.4} \approx 2.639015822$ (b) $2^{1.41} \approx 2.657371628$

(c) $2^{1.414} \approx 2.66474965$ (d) $2^{1.4142} \approx 2.665119089$

(e) $2^{\sqrt{2}} \approx 2.665144143$

Now Work PROBLEM 15

It can be shown that the familiar laws for rational exponents hold for real exponents.

THEOREM

Laws of Exponents

If s, t, a, and b are real numbers with $a > 0$ and $b > 0$, then

$$a^s \cdot a^t = a^{s+t} \qquad (a^s)^t = a^{st} \qquad (ab)^s = a^s \cdot b^s$$
$$1^s = 1 \qquad a^{-s} = \frac{1}{a^s} = \left(\frac{1}{a}\right)^s \qquad a^0 = 1 \qquad \textbf{(1)}$$

Introduction to Exponential Growth

Suppose a function f has the following two properties:

1. The value of f doubles with every 1-unit increase in the independent variable x.
2. The value of f at $x = 0$ is 5, so $f(0) = 5$.

Table 2 shows values of the function f for $x = 0, 1, 2, 3$, and 4.

Table 2

x	$f(x)$
0	5
1	10
2	20
3	40
4	80

Let's find an equation $y = f(x)$ that describes this function f. The key fact is that the value of f doubles for every 1-unit increase in x.

$f(0) = 5$

$f(1) = 2f(0) = 2 \cdot 5 = 5 \cdot 2^1$ **Double the value of f at 0 to get the value at 1.**

$f(2) = 2f(1) = 2(5 \cdot 2) = 5 \cdot 2^2$ **Double the value of f at 1 to get the value at 2.**

$$f(3) = 2f(2) = 2(5 \cdot 2^2) = 5 \cdot 2^3$$
$$f(4) = 2f(3) = 2(5 \cdot 2^3) = 5 \cdot 2^4$$

The pattern leads to

$$f(x) = 2f(x-1) = 2(5 \cdot 2^{x-1}) = 5 \cdot 2^x$$

DEFINITION

An **exponential function** is a function of the form

$$f(x) = Ca^x$$

where a is a positive real number $(a > 0)$, $a \neq 1$, and $C \neq 0$ is a real number. The domain of f is the set of all real numbers. The base a is the **growth factor**, and because $f(0) = Ca^0 = C$, C is called the **initial value**.

WARNING It is important to distinguish a power function, $g(x) = ax^n$, $n \geq 2$, an integer, from an exponential function, $f(x) = C \cdot a^x$, $a \neq 1$, $a > 0$. In a power function, the base is a variable and the exponent is a constant. In an exponential function, the base is a constant and the exponent is a variable.

In the definition of an exponential function, the base $a = 1$ is excluded because this function is simply the constant function $f(x) = C \cdot 1^x = C$. Bases that are negative are also excluded, otherwise, many values of x would have to be excluded from the domain, such as $x = \frac{1}{2}$ and $x = \frac{3}{4}$. [Recall that $(-2)^{1/2} = \sqrt{-2}$, $(-3)^{3/4} = \sqrt[4]{(-3)^3} = \sqrt[4]{-27}$, and so on, are not defined in the set of real numbers.]

Transformations (vertical shifts, horizontal shifts, reflections, and so on) of a function of the form $f(x) = Ca^x$ also represent exponential functions. Some examples of exponential functions are

$$f(x) = 2^x \qquad F(x) = \left(\frac{1}{3}\right)^x + 5 \qquad G(x) = 2 \cdot 3^{x-3}$$

For each function, note that the base of the exponential expression is a constant and the exponent contains a variable.

In the function $f(x) = 5 \cdot 2^x$, notice that the ratio of consecutive outputs is constant for 1-unit increases in the input. This ratio equals the constant 2, the base of the exponential function. In other words,

$$\frac{f(1)}{f(0)} = \frac{5 \cdot 2^1}{5} = 2 \qquad \frac{f(2)}{f(1)} = \frac{5 \cdot 2^2}{5 \cdot 2^1} = 2 \qquad \frac{f(3)}{f(2)} = \frac{5 \cdot 2^3}{5 \cdot 2^2} = 2 \quad \text{and so on}$$

This leads to the following result.

THEOREM

For an exponential function $f(x) = Ca^x$, where $a > 0$ and $a \neq 1$, if x is any real number, then

$$\frac{f(x+1)}{f(x)} = a \quad \text{or} \quad f(x+1) = af(x)$$

In Words
For 1-unit changes in the input x of an exponential function $f(x) = C \cdot a^x$, the ratio of consecutive outputs is the constant a.

Proof

$$\frac{f(x+1)}{f(x)} = \frac{Ca^{x+1}}{Ca^x} = a^{x+1-x} = a^1 = a$$

EXAMPLE 2 Identifying Linear or Exponential Functions

Determine whether the given function is linear, exponential, or neither. For those that are linear, find a linear function that models the data. For those that are exponential, find an exponential function that models the data.

(a)

x	y
−1	5
0	2
1	−1
2	−4
3	−7

(b)

x	y
−1	32
0	16
1	8
2	4
3	2

(c)

x	y
−1	2
0	4
1	7
2	11
3	16

Solution For each function, compute the average rate of change of y with respect to x and the ratio of consecutive outputs. If the average rate of change is constant, then the function is linear. If the ratio of consecutive outputs is constant, then the function is exponential.

Table 3 (a)

x	y	Average Rate of Change	Ratio of Consecutive Outputs
−1	5		
		$\frac{\Delta y}{\Delta x} = \frac{2-5}{0-(-1)} = -3$	$\frac{2}{5}$
0	2		
		$\frac{-1-2}{1-0} = -3$	$\frac{-1}{2} = -\frac{1}{2}$
1	−1		
		$\frac{-4-(-1)}{2-1} = -3$	$\frac{-4}{-1} = 4$
2	−4		
		$\frac{-7-(-4)}{3-2} = -3$	$\frac{-7}{-4} = \frac{7}{4}$
3	−7		

(b)

x	y	Average Rate of Change	Ratio of Consecutive Outputs
−1	32		
		$\frac{\Delta y}{\Delta x} = \frac{16-32}{0-(-1)} = -16$	$\frac{16}{32} = \frac{1}{2}$
0	16		
		−8	$\frac{8}{16} = \frac{1}{2}$
1	8		
		−4	$\frac{4}{8} = \frac{1}{2}$
2	4		
		−2	$\frac{2}{4} = \frac{1}{2}$
3	2		

(c)

x	y	Average Rate of Change	Ratio of Consecutive Outputs
−1	2		
		$\frac{\Delta y}{\Delta x} = \frac{4-2}{0-(-1)} = 2$	2
0	4		
		3	$\frac{7}{4}$
1	7		
		4	$\frac{11}{7}$
2	11		
		5	$\frac{16}{11}$
3	16		

(a) See Table 3(a) on the previous page. The average rate of change for every 1-unit increase in x is -3. Therefore, the function is a linear function. In a linear function the average rate of change is the slope m, so $m = -3$. The y-intercept b is the value of the function at $x = 0$, so $b = 2$. The linear function that models the data is $f(x) = mx + b = -3x + 2$.

(b) See Table 3(b) on the previous page. For this function, the average rate of change from -1 to 0 is -16, and the average rate of change from 0 to 1 is -8. Because the average rate of change is not constant, the function is not a linear function. The ratio of consecutive outputs for a 1-unit increase in the inputs is a constant, $\frac{1}{2}$. Because the ratio of consecutive outputs is constant, the function is an exponential function with growth factor $a = \frac{1}{2}$. The initial value of the exponential function is $C = 16$. Therefore, the exponential function that models the data is $g(x) = Ca^x = 16 \cdot \left(\frac{1}{2}\right)^x$.

(c) See Table 3(c) on the previous page. For this function, the average rate of change from -1 to 0 is 2, and the average rate of change from 0 to 1 is 3. Because the average rate of change is not constant, the function is not a linear function. The ratio of consecutive outputs from -1 to 0 is 2, and the ratio of consecutive outputs from 0 to 1 is $\frac{7}{4}$. Because the ratio of consecutive outputs is not a constant, the function is not an exponential function. ■

Now Work PROBLEM 27

2 Graph Exponential Functions

If we know how to graph an exponential function of the form $f(x) = a^x$, then we could use transformations (shifting, stretching, and so on) to obtain the graph of any exponential function.

First, let's graph the exponential function $f(x) = 2^x$.

EXAMPLE 3 Graphing an Exponential Function

Graph the exponential function: $f(x) = 2^x$

Solution

The domain of $f(x) = 2^x$ is the set of all real numbers. Begin by locating some points on the graph of $f(x) = 2^x$, as listed in Table 4.

Because $2^x > 0$ for all x, the range of f is $(0, \infty)$. Therefore, the graph has no x-intercepts, and in fact the graph will lie above the x-axis for all x. As Table 4 indicates, the y-intercept is 1. Table 4 also indicates that as $x \to -\infty$, the values of $f(x) = 2^x$ get closer and closer to 0. Therefore, the x-axis ($y = 0$) is a horizontal asymptote to the graph as $x \to -\infty$. This provides us the end behavior for x large and negative.

To determine the end behavior for x large and positive, look again at Table 4. As $x \to \infty$, $f(x) = 2^x$ grows very quickly, causing the graph of $f(x) = 2^x$ to rise very rapidly. It is apparent that f is an increasing function and so is one-to-one.

Using all this information, plot some of the points from Table 4 and connect them with a smooth, continuous curve, as shown in Figure 20.

Table 4

NORMAL FLOAT AUTO REAL RADIAN MP
PRESS ENTER TO EDIT

X	Y1
-10	9.8E-4
-3	.125
-2	.25
-1	.5
0	1
1	2
2	4
3	8
10	1024

Y1=2^X

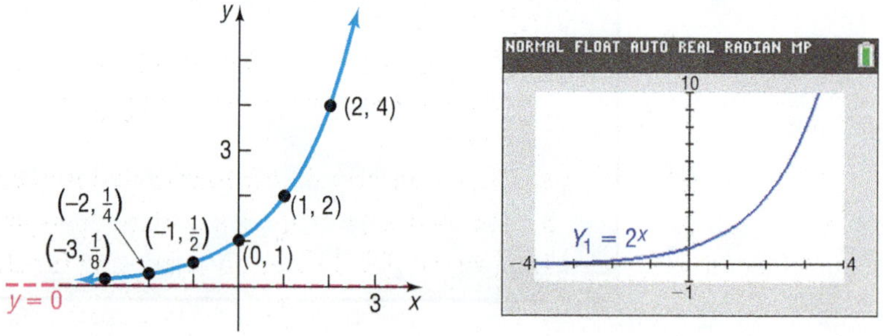

Figure 20 $f(x) = 2^x$

■

Graphs that look like the one shown in Figure 20 occur very frequently in a variety of situations. For example, the graph in Figure 21 shows the total monthly data used globally by mobile devices (uploads and downloads) from the first quarter of 2010 through the fourth quarter of 2014. One might conclude from this graph that global mobile data usage is growing *exponentially*.

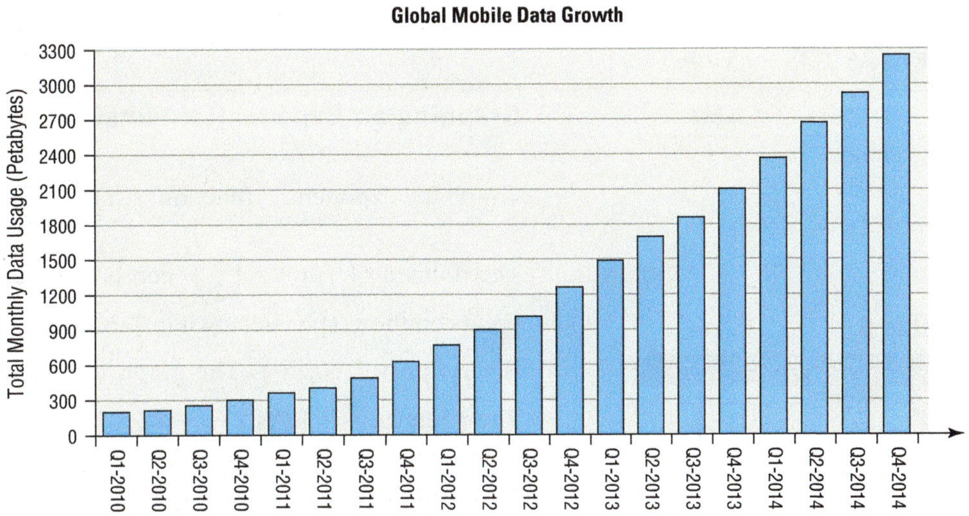

Figure 21 *Source: Ericsson Mobility Report, February 2015*

Later in this chapter, more will be said about situations that lead to exponential growth. For now, let's continue to explore the properties of exponential functions.

The graph of $f(x) = 2^x$ in Figure 20 is typical of all exponential functions of the form $f(x) = a^x$ with $a > 1$. Such functions are increasing functions and, hence, are one-to-one. Their graphs lie above the x-axis, pass through the point $(0, 1)$, and thereafter rise rapidly as $x \rightarrow \infty$. As $x \rightarrow -\infty$, the x-axis $(y = 0)$ is a horizontal asymptote. There are no vertical asymptotes. Finally, the graphs are smooth and continuous with no corners or gaps.

Figures 22 and 23 illustrate the graphs of two more exponential functions whose bases are larger than 1. Notice that the larger the base, the steeper the graph is when $x > 0$, and when $x < 0$, the larger the base, the closer the graph of the equation is to the x-axis.

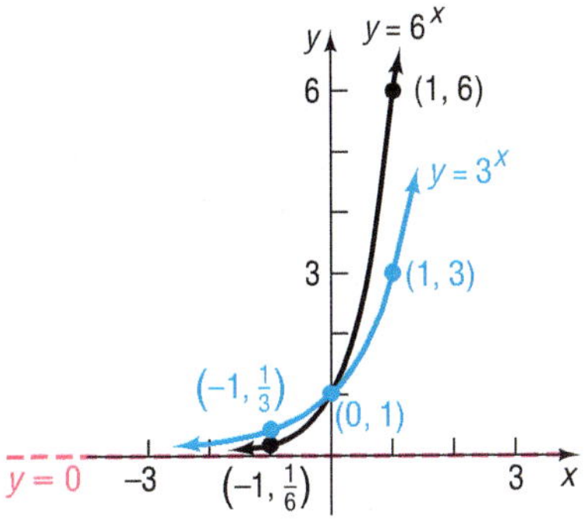

Figure 22

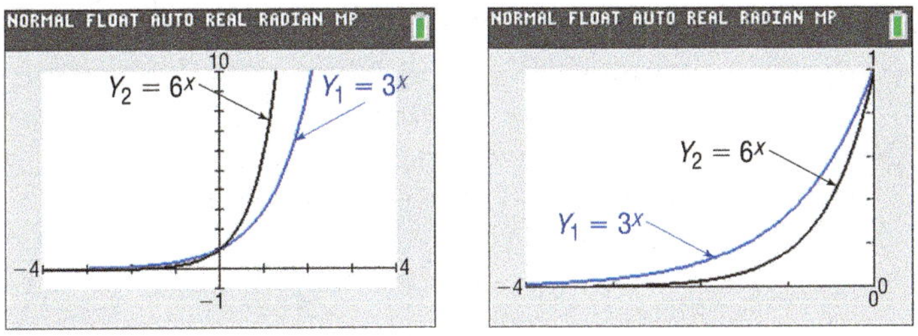

Figure 23 (a) (b)

Properties of the Exponential Function $f(x) = a^x$, $a > 1$

1. The domain is the set of all real numbers, or $(-\infty, \infty)$ using interval notation; the range is the set of positive real numbers, or $(0, \infty)$ using interval notation.
2. There are no x-intercepts; the y-intercept is 1.
3. The x-axis $(y = 0)$ is a horizontal asymptote as $x \rightarrow -\infty$ $\left[\lim_{x \to -\infty} a^x = 0\right]$
4. $f(x) = a^x$, $a > 1$, is an increasing function and is one-to-one.

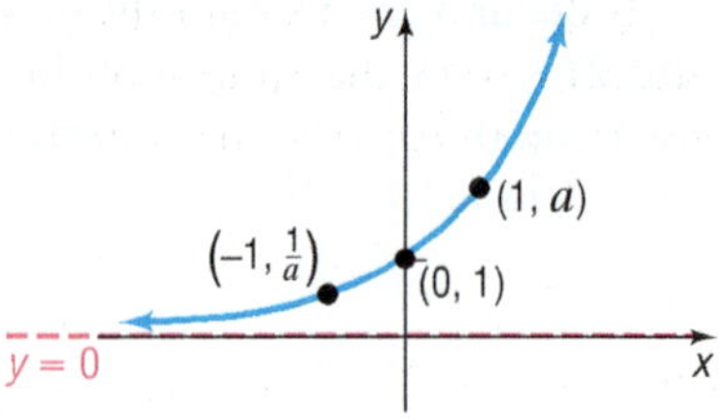

Figure 24 $f(x) = a^x$, $a > 1$

5. The graph of f contains the points $(0, 1)$, $(1, a)$, and $\left(-1, \frac{1}{a}\right)$.
6. The graph of f is smooth and continuous, with no corners or gaps. See Figure 24.

Now consider $f(x) = a^x$ when $0 < a < 1$.

EXAMPLE 4 Graphing an Exponential Function

Graph the exponential function: $f(x) = \left(\frac{1}{2}\right)^x$

Solution The domain of $f(x) = \left(\frac{1}{2}\right)^x$ consists of all real numbers. As before, locate some points on the graph, as shown in Table 5. Because $\left(\frac{1}{2}\right)^x > 0$ for all x, the range of f is the interval $(0, \infty)$. The graph lies above the x-axis and has no x-intercepts. The y-intercept is 1. As $x \to -\infty$, $f(x) = \left(\frac{1}{2}\right)^x$ grows very quickly. As $x \to \infty$, the values of $f(x)$ approach 0. The x-axis $(y = 0)$ is a horizontal asymptote as $x \to \infty$. It is apparent that f is a decreasing function and so is one-to-one. Figure 25 illustrates the graph.

Table 5

NORMAL FLOAT AUTO REAL RADIAN MP
PRESS ENTER TO EDIT

X	Y1
-10	1024
-3	8
-2	4
-1	2
0	1
1	.5
2	.25
3	.125
10	9.8E-4

Y1=(1/2)^X

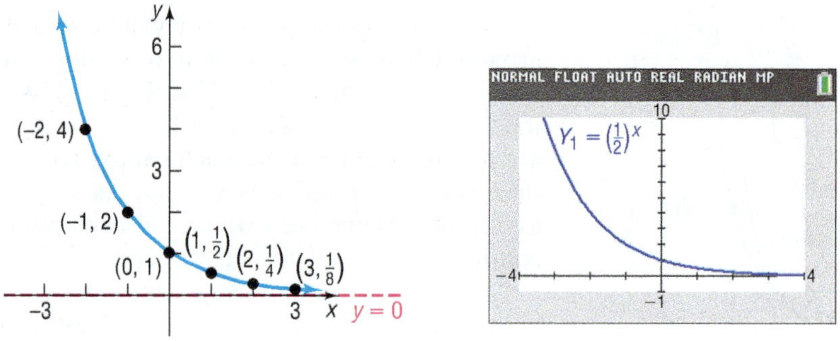

Figure 25 $f(x) = \left(\frac{1}{2}\right)^x$

The graph of $y = \left(\frac{1}{2}\right)^x$ also can be obtained from the graph of $y = 2^x$ using transformations. The graph of $y = \left(\frac{1}{2}\right)^x = 2^{-x}$ is a reflection about the y-axis of the graph of $y = 2^x$ (replace x by $-x$). See Figures 26(a) and 26(b).

Seeing the Concept

Using a graphing utility, simultaneously graph:

(a) $Y_1 = 3^x$, $Y_2 = \left(\frac{1}{3}\right)^x$

(b) $Y_1 = 6^x$, $Y_2 = \left(\frac{1}{6}\right)^x$

Conclude that the graph of $Y_2 = \left(\frac{1}{a}\right)^x$, for $a > 0$, is the reflection about the y-axis of the graph of $Y_1 = a^x$.

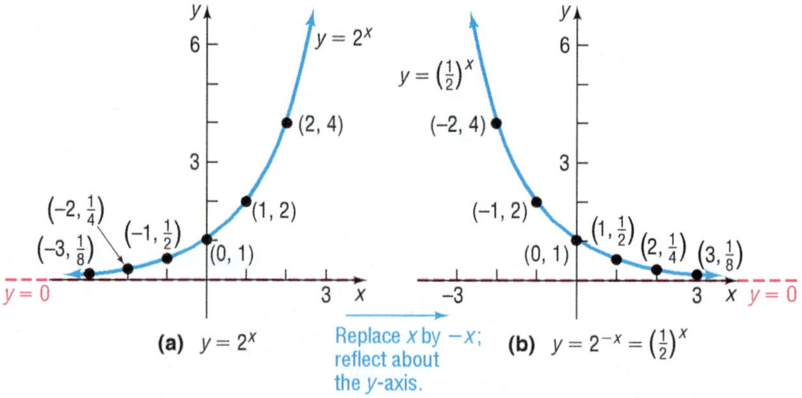

Figure 26 **(a)** $y = 2^x$ **(b)** $y = 2^{-x} = \left(\frac{1}{2}\right)^x$

The graph of $f(x) = \left(\frac{1}{2}\right)^x$ in Figure 25 is typical of all exponential functions of the form $f(x) = a^x$ with $0 < a < 1$. Such functions are decreasing and one-to-one.

Their graphs lie above the x-axis and pass through the point $(0, 1)$. The graphs rise rapidly as $x \to -\infty$. As $x \to \infty$, the x-axis $(y = 0)$ is a horizontal asymptote. There are no vertical asymptotes. Finally, the graphs are smooth and continuous, with no corners or gaps.

Figures 27 and 28 illustrate the graphs of two more exponential functions whose bases are between 0 and 1. Notice that the smaller base results in a graph that is steeper when $x < 0$. When $x > 0$, the graph of the equation with the smaller base is closer to the x-axis.

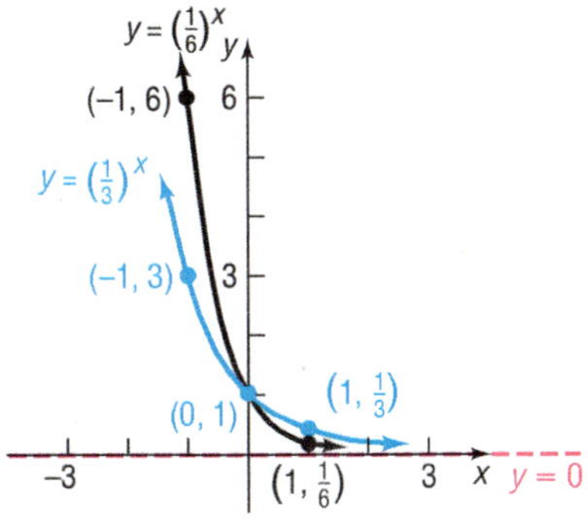

Figure 27

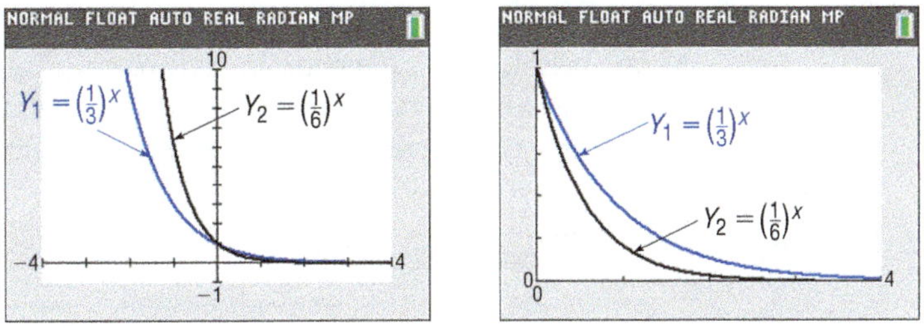

Figure 28 (a) (b)

Properties of the Exponential Function $f(x) = a^x$, $0 < a < 1$

1. The domain is the set of all real numbers, or $(-\infty, \infty)$ using interval notation; the range is the set of positive real numbers, or $(0, \infty)$ using interval notation.
2. There are no x-intercepts; the y-intercept is 1.
3. The x-axis $(y = 0)$ is a horizontal asymptote as $x \to \infty$ $\left[\lim_{x\to\infty} a^x = 0\right]$.
4. $f(x) = a^x$, $0 < a < 1$, is a decreasing function and is one-to-one.
5. The graph of f contains the points $\left(-1, \frac{1}{a}\right)$, $(0, 1)$, and $(1, a)$.
6. The graph of f is smooth and continuous, with no corners or gaps. See Figure 29.

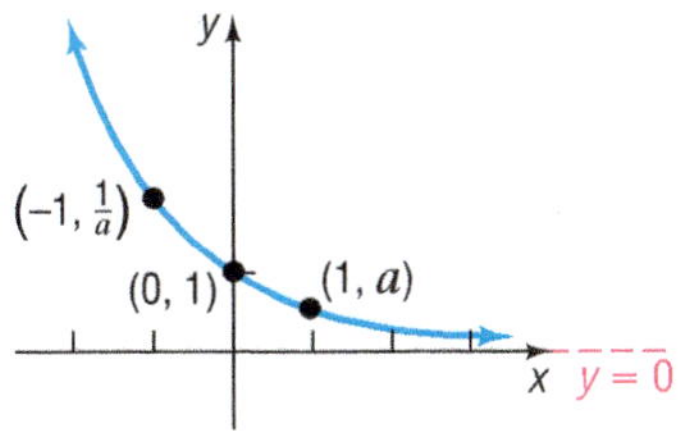

Figure 29 $f(x) = a^x, 0 < a < 1$

EXAMPLE 5 Graphing Exponential Functions Using Transformations

Graph $f(x) = 2^{-x} - 3$, and determine the domain, range, and horizontal asymptote of f.

Solution Begin with the graph of $y = 2^x$. Figure 30 shows the stages.

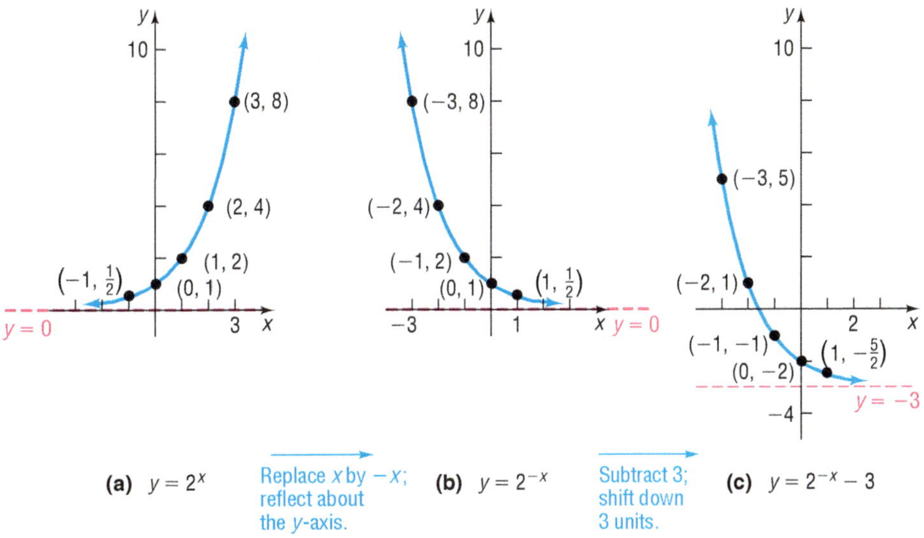

Figure 30 (a) $y = 2^x$ (b) $y = 2^{-x}$ (c) $y = 2^{-x} - 3$

Historical Feature

The number e is named in honor of the Swiss mathematician Leonhard Euler (1707–1783).

As Figure 30(c) on the previous page illustrates, the domain of $f(x) = 2^{-x} - 3$ is the interval $(-\infty, \infty)$ and the range is the interval $(-3, \infty)$. The horizontal asymptote of f is the line $y = -3$.

Check: Graph $Y_1 = 2^{-x} - 3$ to verify the graph obtained in Figure 30(c).

Now Work PROBLEM 43

3 Define the Number e

Many problems that occur in nature require the use of an exponential function whose base is a certain irrational number, symbolized by the letter e.

One way of arriving at this important number e is given next.

DEFINITION

The **number e** is defined as the number that the expression

$$\left(1 + \frac{1}{n}\right)^n \tag{2}$$

approaches as $n \to \infty$. In calculus, this is expressed using limit notation as

$$e = \lim_{n\to\infty}\left(1 + \frac{1}{n}\right)^n$$

Table 6 illustrates what happens to the defining expression (2) as n takes on increasingly large values. The last number in the right column in the table approximates e correct to nine decimal places. That is, $e = 2.718281828\ldots$. Remember, the three dots indicate that the decimal places continue. Because these decimal places continue but do not repeat, e is an irrational number. The number e is often expressed as a decimal rounded to a specific number of places. For example, $e \approx 2.71828$ is rounded to five decimal places.

The exponential function $f(x) = e^x$, whose base is the number e, occurs with such frequency in applications that it is usually referred to as *the* exponential function. Indeed, most calculators have the key $\boxed{e^x}$ or $\boxed{\exp(x)}$, which may be used to evaluate the exponential function for a given value of x.

Table 6

n	$\frac{1}{n}$	$1 + \frac{1}{n}$	$\left(1 + \frac{1}{n}\right)^n$
1	1	2	2
2	0.5	1.5	2.25
5	0.2	1.2	2.48832
10	0.1	1.1	2.59374246
100	0.01	1.01	2.704813829
1,000	0.001	1.001	2.716923932
10,000	0.0001	1.0001	2.718145927
100,000	0.00001	1.00001	2.718268237
1,000,000	0.000001	1.000001	2.718280469
10,000,000,000	10^{-10}	$1 + 10^{-10}$	2.718281828

Table 7

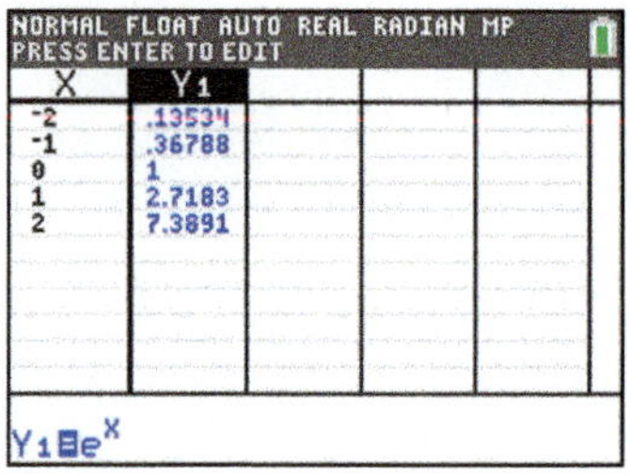

Now use your calculator to approximate e^x for $x = -2, x = -1, x = 0, x = 1$, and $x = 2$. See Table 7. The graph of the exponential function $f(x) = e^x$ is given in Figures 31(a) and (b). Since $2 < e < 3$, the graph of $y = e^x$ lies between the graphs of $y = 2^x$ and $y = 3^x$. See Figure 31(c).

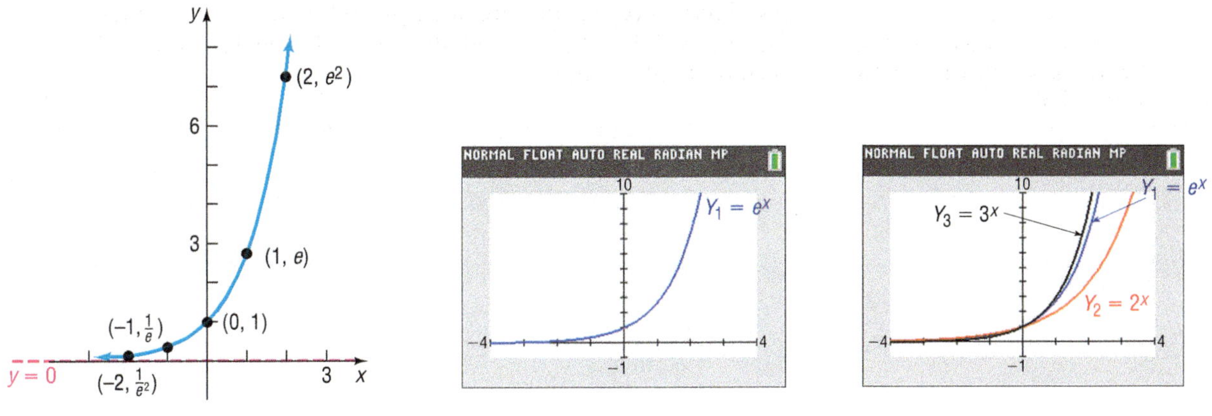

Figure 31 $y = e^x$ (a) (b) (c)

EXAMPLE 6 Graphing Exponential Functions Using Transformations

Graph $f(x) = -e^{x-3}$ and determine the domain, range, and horizontal asymptote of f.

Solution Begin with the graph of $y = e^x$. Figure 32 shows the stages.

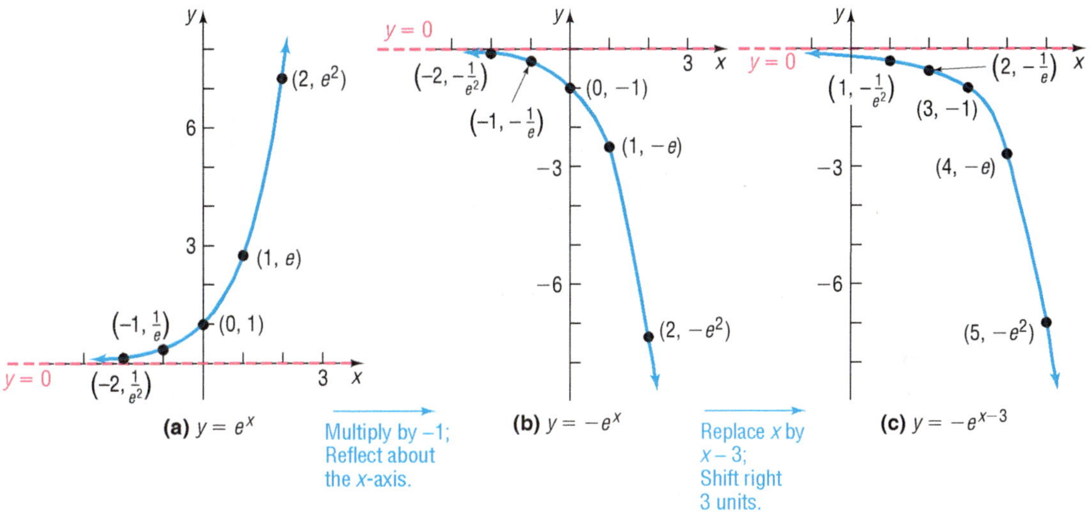

Figure 32

As Figure 32(c) illustrates, the domain of $f(x) = -e^{x-3}$ is the interval $(-\infty, \infty)$, and the range is the interval $(-\infty, 0)$. The horizontal asymptote is the line $y = 0$.

✓**Check:** Graph $Y_1 = -e^{x-3}$ to verify the graph obtained in Figure 32(c). ■

Now Work PROBLEM 55

4 Solve Exponential Equations

Equations that involve terms of the form a^x, where $a > 0$ and $a \neq 1$, are referred to as **exponential equations**. Such equations can sometimes be solved by appropriately applying the Laws of Exponents and property (3):

In Words
When two exponential expressions with the same base are equal, then their exponents are equal.

$$\text{If } a^u = a^v, \text{ then } u = v. \qquad (3)$$

Property (3) is a consequence of the fact that exponential functions are one-to-one. To use property (3), each side of the equality must be written with the same base.

EXAMPLE 7 Solving an Exponential Equation

Solve: $4^{2x-1} = 8^x$

Algebraic Solution

Write each exponential expression so each has the same base.

$$4^{2x-1} = 8^x$$
$$(2^2)^{(2x-1)} = (2^3)^x \qquad 4 = 2^2;\ 8 = 2^3$$
$$2^{2(2x-1)} = 2^{3x} \qquad (a^r)^s = a^{rs}$$
$$2(2x-1) = 3x \qquad \text{If } a^u = a^v, \text{ then } u = v.$$
$$4x - 2 = 3x$$
$$x = 2$$

The solution set is $\{2\}$. ■

Graphing Solution

Graph $Y_1 = 4^{2x-1}$ and $Y_2 = 8^x$. Use INTERSECT to determine the point of intersection. See Figure 33.

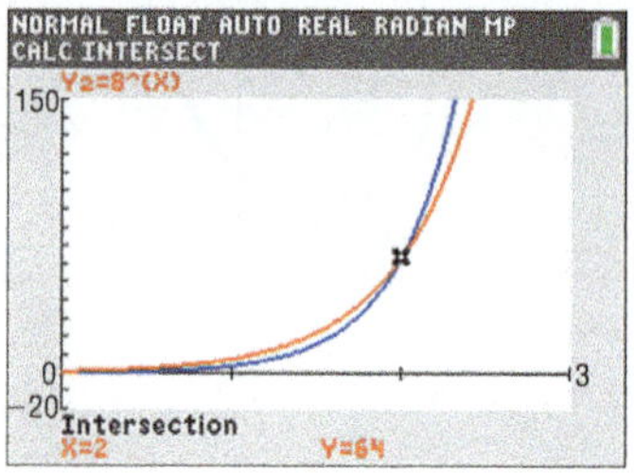

Figure 33

The graphs intersect at $(2, 64)$, so the solution set is $\{2\}$. ■

Now Work PROBLEMS 65 AND 75

EXAMPLE 8 Solving an Exponential Equation

Solve: $e^{-x^2} = (e^x)^2 \cdot \dfrac{1}{e^3}$

Solution Use the Laws of Exponents first to get a single expression with the base e on the right side.

$$(e^x)^2 \cdot \frac{1}{e^3} = e^{2x} \cdot e^{-3} = e^{2x-3}$$

As a result,

$$e^{-x^2} = e^{2x-3}$$
$$-x^2 = 2x - 3 \qquad \text{Apply property (3).}$$
$$x^2 + 2x - 3 = 0 \qquad \text{Place the quadratic equation in standard form.}$$
$$(x+3)(x-1) = 0 \qquad \text{Factor.}$$
$$x = -3 \quad \text{or} \quad x = 1 \qquad \text{Use the Zero-Product Property.}$$

The solution set is $\{-3, 1\}$. ■

Now Work PROBLEM 81

EXAMPLE 9 Exponential Probability

Between 9:00 PM and 10:00 PM, cars arrive at Burger King's drive-thru at the rate of 12 cars per hour (0.2 car per minute). The following formula from statistics can be used to determine the probability that a car will arrive within t minutes of 9:00 PM.

$$F(t) = 1 - e^{-0.2t}$$

(a) Determine the probability that a car will arrive within 5 minutes of 9 PM (that is, before 9:05 PM).

(b) Determine the probability that a car will arrive within 30 minutes of 9 PM (before 9:30 PM).

(c) Graph F using your graphing utility.

(d) What value does F approach as t increases without bound in the positive direction?

Solution (a) The probability that a car will arrive within 5 minutes is found by evaluating $F(t)$ at $t = 5$.

$$F(5) = 1 - e^{-0.2(5)} \approx 0.6321$$

↑ Use a calculator.

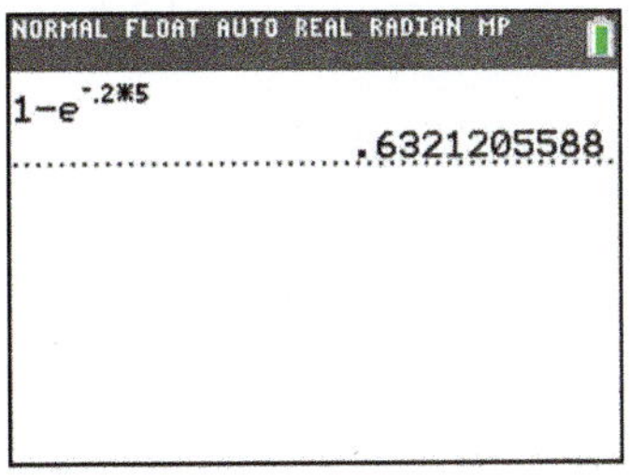

Figure 34 $F(5)$

See Figure 34. There is a 63.21% probability that a car will arrive within 5 minutes.

(b) The probability that a car will arrive within 30 minutes is found by evaluating $F(t)$ at $t = 30$.

$$F(30) = 1 - e^{-0.2(30)} \approx 0.9975$$

↑ Use a calculator.

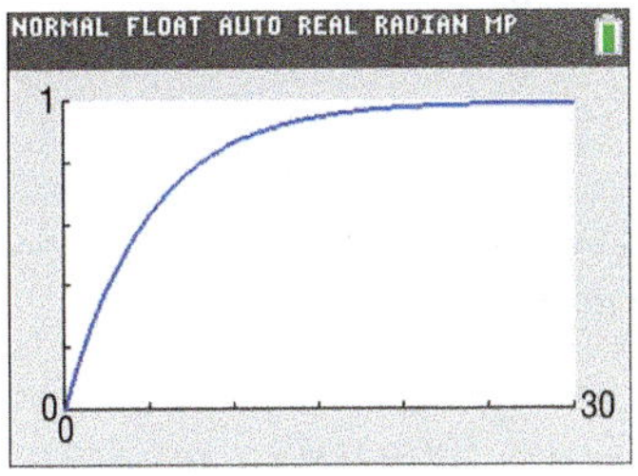

Figure 35 $F(t) = 1 - e^{-0.2t}$

There is a 99.75% probability that a car will arrive within 30 minutes.

(c) See Figure 35 for the graph of F.

(d) As time passes, the probability that a car will arrive increases. The value that F approaches can be found by letting $t \to \infty$. Since $e^{-0.2t} = \frac{1}{e^{0.2t}}$, it follows that $e^{-0.2t} \to 0$ as $t \to \infty$. Therefore, F approaches 1 as t gets large. The algebraic analysis is confirmed by Figure 35. ■

Now Work PROBLEM 113

SUMMARY

Properties of the Exponential Function

$f(x) = a^x, \quad a > 1$	Domain: the interval $(-\infty, \infty)$; range: the interval $(0, \infty)$ x-intercepts: none; y-intercept: 1 Horizontal asymptote: x-axis $(y = 0)$ as $x \to -\infty$ Increasing; one-to-one; smooth; continuous See Figure 24 for a typical graph.
$f(x) = a^x, \quad 0 < a < 1$	Domain: the interval $(-\infty, \infty)$; range: the interval $(0, \infty)$ x-intercepts: none; y-intercept: 1 Horizontal asymptote: x-axis $(y = 0)$ as $x \to \infty$ Decreasing; one-to-one; smooth; continuous See Figure 29 for a typical graph.

If $a^u = a^v$, then $u = v$.

6.3 Assess Your Understanding

'Are You Prepared?' *Answers are given at the end of these exercises. If you get a wrong answer, read the pages listed in red.*

1. $4^3 =$ ____; $8^{2/3} =$ ____; $3^{-2} =$ ____. (pp. 22–24 and pp. 77–78)

2. Solve: $x^2 + 3x = 4$ (pp. 110–115)

3. ***True or False*** To graph $y = (x - 2)^3$, shift the graph of $y = x^3$ to the left 2 units. (pp. 256–264)

4. Find the average rate of change of $f(x) = 3x - 5$ from $x = 0$ to $x = 4$. (pp. 238–239; 281–284)

5. ***True or False*** The function $f(x) = \frac{2x}{x - 3}$ has $y = 2$ as a horizontal asymptote. (pp. 377–379)

Concepts and Vocabulary

6. A(n) __________ ______ is a function of the form $f(x) = Ca^x$, where $a > 0$, $a \neq 1$, and $C \neq 0$ are real numbers. The base a is the ______ _____ and C is the _____ _____.

7. For an exponential function $f(x) = Ca^x$, $\dfrac{f(x+1)}{f(x)} =$ __.

8. ***True or False*** The domain of the exponential function $f(x) = a^x$, where $a > 0$ and $a \neq 1$, is the set of all real numbers.

9. ***True or False*** The graph of the exponential function $f(x) = a^x$, where $a > 0$ and $a \neq 1$, has no x-intercept.

10. The graph of every exponential function $f(x) = a^x$, where $a > 0$ and $a \neq 1$, passes through three points: _____, _____, and _____.

11. If $3^x = 3^4$, then $x =$ __.

12. ***True or False*** The graphs of $y = 3^x$ and $y = \left(\dfrac{1}{3}\right)^x$ are identical.

13. Which of the following exponential functions is an increasing function?
(a) $f(x) = 0.5^x$ (b) $f(x) = \left(\dfrac{5}{2}\right)^x$
(c) $f(x) = \left(\dfrac{2}{3}\right)^x$ (d) $f(x) = 0.9^x$

14. Which of the following is the range of the exponential function $f(x) = a^x$, $a > 0$ and $a \neq 1$?
(a) $(-\infty, \infty)$ (b) $(-\infty, 0)$
(c) $(0, \infty)$ (d) $(-\infty, 0) \cup (0, \infty)$

Skill Building

In Problems 15–26, approximate each number using a calculator. Express your answer rounded to three decimal places.

15. (a) $2^{3.14}$ (b) $2^{3.141}$ (c) $2^{3.1415}$ (d) 2^{π}

16. (a) $2^{2.7}$ (b) $2^{2.71}$ (c) $2^{2.718}$ (d) 2^{e}

17. (a) $3.1^{2.7}$ (b) $3.14^{2.71}$ (c) $3.141^{2.718}$ (d) π^{e}

18. (a) $2.7^{3.1}$ (b) $2.71^{3.14}$ (c) $2.718^{3.141}$ (d) e^{π}

19. $(1 + 0.04)^6$

20. $\left(1 + \dfrac{0.09}{12}\right)^{24}$

21. $8.4\left(\dfrac{1}{3}\right)^{2.9}$

22. $158\left(\dfrac{5}{6}\right)^{8.63}$

23. $e^{1.2}$

24. $e^{-1.3}$

25. $125e^{0.026(7)}$

26. $83.6e^{-0.157(9.5)}$

In Problems 27–34, determine whether the given function is linear, exponential, or neither. For those that are linear functions, find a linear function that models the data; for those that are exponential, find an exponential function that models the data.

27.

x	$f(x)$
−1	3
0	6
1	12
2	18
3	30

28.

x	$g(x)$
−1	2
0	5
1	8
2	11
3	14

29.

x	$H(x)$
−1	$\frac{1}{4}$
0	1
1	4
2	16
3	64

30.

x	$F(x)$
−1	$\frac{2}{3}$
0	1
1	$\frac{3}{2}$
2	$\frac{9}{4}$
3	$\frac{27}{8}$

31.

x	$f(x)$
−1	$\frac{3}{2}$
0	3
1	6
2	12
3	24

32.

x	$g(x)$
−1	6
0	1
1	0
2	3
3	10

33.

x	$H(x)$
−1	2
0	4
1	6
2	8
3	10

34.

x	$F(x)$
−1	$\frac{1}{2}$
0	$\frac{1}{4}$
1	$\frac{1}{8}$
2	$\frac{1}{16}$
3	$\frac{1}{32}$

In Problems 35–42, the graph of an exponential function is given. Match each graph to one of the following functions:

(A) $y = 3^x$ (B) $y = 3^{-x}$ (C) $y = -3^x$ (D) $y = -3^{-x}$
(E) $y = 3^x - 1$ (F) $y = 3^{x-1}$ (G) $y = 3^{1-x}$ (H) $y = 1 - 3^x$

35.

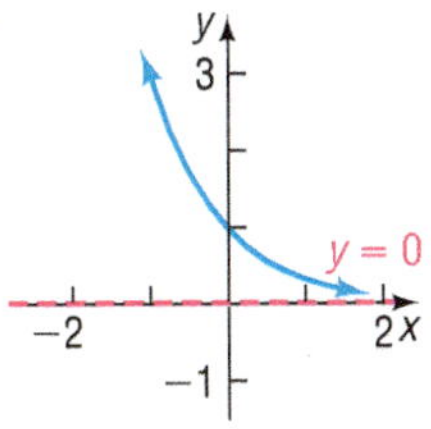

36.

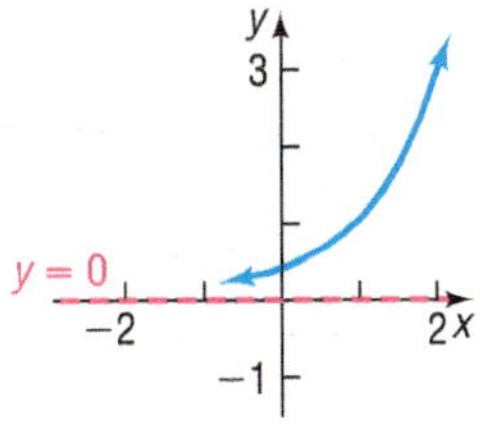

37.

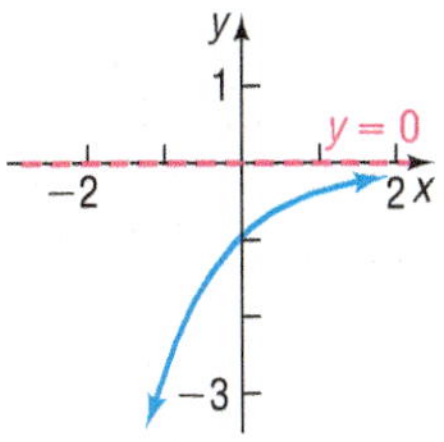

38.

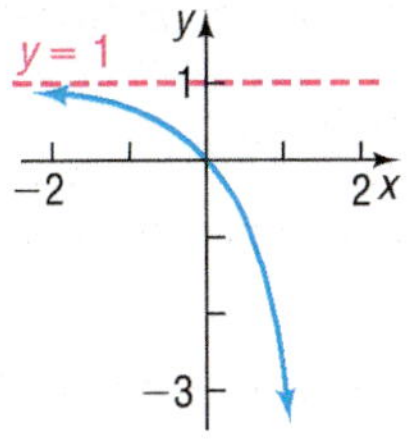

39.

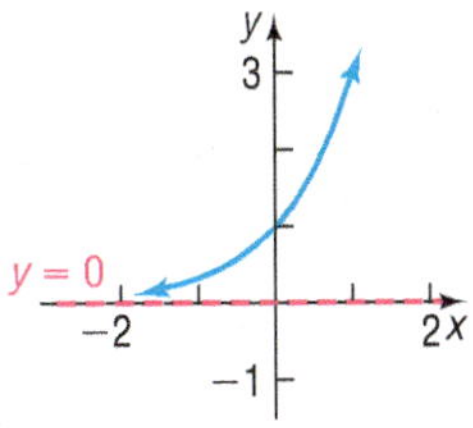

40.

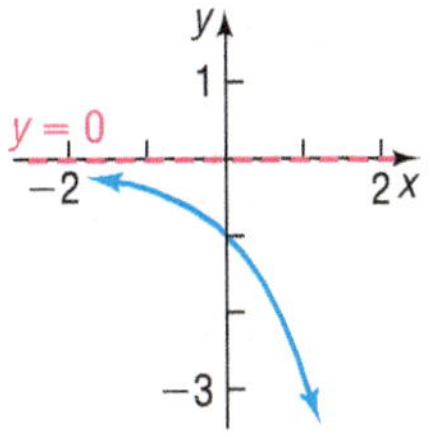

41.

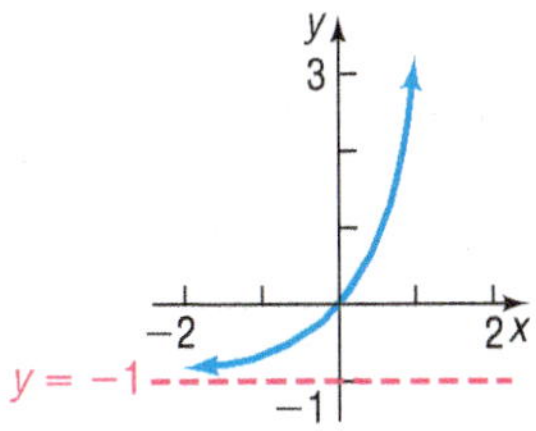

42.

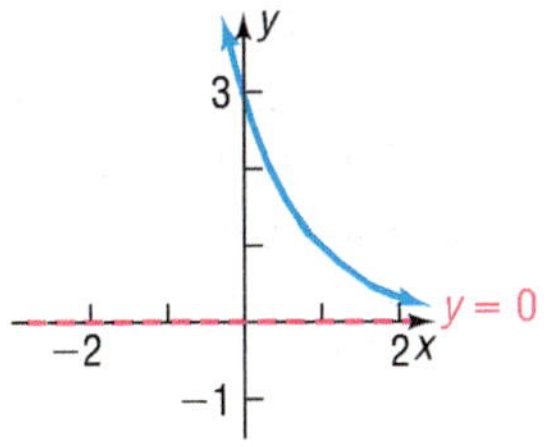

In Problems 43–54, use transformations to graph each function. Determine the domain, range, and horizontal asymptote of each function.

43. $f(x) = 2^x + 1$
44. $f(x) = 3^x - 2$
45. $f(x) = 3^{x-1}$
46. $f(x) = 2^{x+2}$
47. $f(x) = 3 \cdot \left(\frac{1}{2}\right)^x$
48. $f(x) = 4 \cdot \left(\frac{1}{3}\right)^x$
49. $f(x) = 3^{-x} - 2$
50. $f(x) = -3^x + 1$
51. $f(x) = 2 + 4^{x-1}$
52. $f(x) = 1 - 2^{x+3}$
53. $f(x) = 2 + 3^{x/2}$
54. $f(x) = 1 - 2^{-x/3}$

In Problems 55–62, begin with the graph of $y = e^x$ (Figure 31) and use transformations to graph each function. Determine the domain, range, and horizontal asymptote of each function.

55. $f(x) = e^{-x}$
56. $f(x) = -e^x$
57. $f(x) = e^{x+2}$
58. $f(x) = e^x - 1$
59. $f(x) = 5 - e^{-x}$
60. $f(x) = 9 - 3e^{-x}$
61. $f(x) = 2 - e^{-x/2}$
62. $f(x) = 7 - 3e^{2x}$

In Problems 63–82, solve each equation.

63. $7^x = 7^3$
64. $5^x = 5^{-6}$
65. $2^{-x} = 16$
66. $3^{-x} = 81$
67. $\left(\frac{1}{5}\right)^x = \frac{1}{25}$
68. $\left(\frac{1}{4}\right)^x = \frac{1}{64}$
69. $2^{2x-1} = 4$
70. $5^{x+3} = \frac{1}{5}$
71. $3^{x^3} = 9^x$
72. $4^{x^2} = 2^x$
73. $8^{-x+14} = 16^x$
74. $9^{-x+15} = 27^x$
75. $3^{x^2-7} = 27^{2x}$
76. $5^{x^2+8} = 125^{2x}$
77. $4^x \cdot 2^{x^2} = 16^2$
78. $9^{2x} \cdot 27^{x^2} = 3^{-1}$
79. $e^x = e^{3x+8}$
80. $e^{3x} = e^{2-x}$
81. $e^{x^2} = e^{3x} \cdot \frac{1}{e^2}$
82. $(e^4)^x \cdot e^{x^2} = e^{12}$

83. If $4^x = 7$, what does 4^{-2x} equal?
84. If $2^x = 3$, what does 4^{-x} equal?
85. If $3^{-x} = 2$, what does 3^{2x} equal?
86. If $5^{-x} = 3$, what does 5^{3x} equal?
87. If $9^x = 25$, what does 3^x equal?
88. If $2^{-3x} = \frac{1}{1000}$, what does 2^x equal?

In Problems 89–92, determine the exponential function whose graph is given.

89.

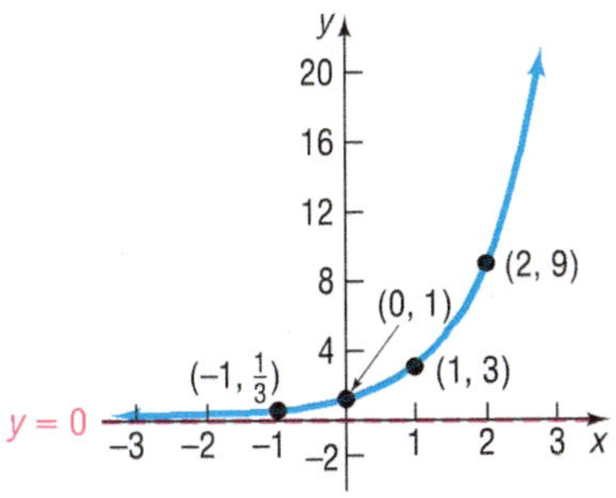

90.

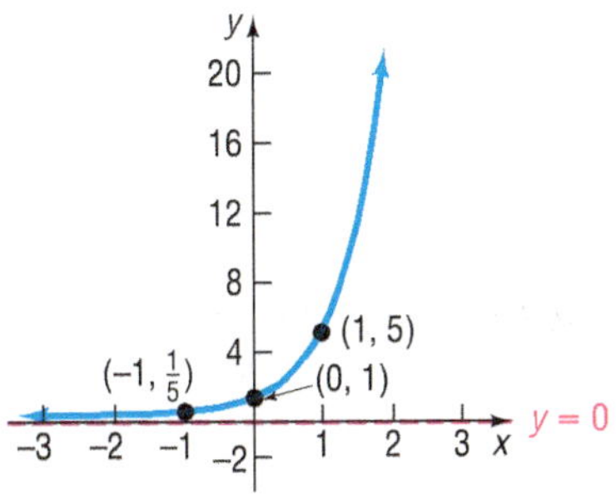

91.

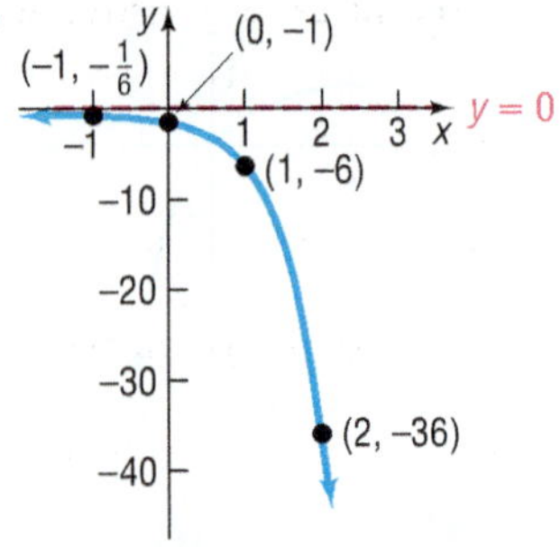

92.

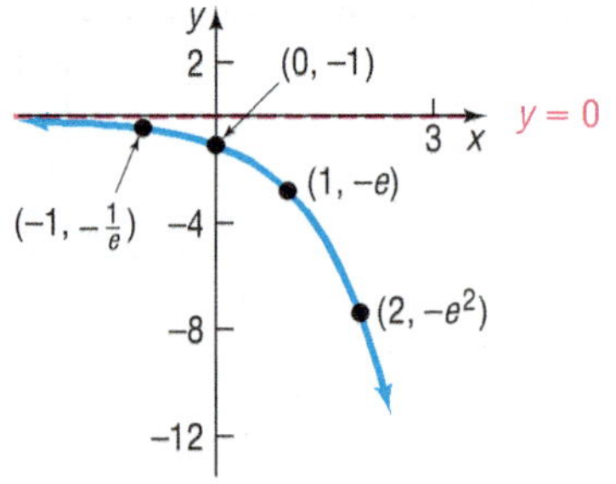

93. Find an exponential function with horizontal asymptote $y = 2$ whose graph contains the points $(0, 3)$ and $(1, 5)$.

94. Find an exponential function with horizontal asymptote $y = -3$ whose graph contains the points $(0, -2)$ and $(-2, 1)$.

Mixed Practice

95. Suppose that $f(x) = 2^x$.
(a) What is $f(4)$? What point is on the graph of f?
(b) If $f(x) = \frac{1}{16}$, what is x? What point is on the graph of f?

96. Suppose that $f(x) = 3^x$.
(a) What is $f(4)$? What point is on the graph of f?
(b) If $f(x) = \frac{1}{9}$, what is x? What point is on the graph of f?

97. Suppose that $g(x) = 4^x + 2$.
(a) What is $g(-1)$? What point is on the graph of g?
(b) If $g(x) = 66$, what is x? What point is on the graph of g?

98. Suppose that $g(x) = 5^x - 3$.
(a) What is $g(-1)$? What point is on the graph of g?
(b) If $g(x) = 122$, what is x? What point is on the graph of g?

99. Suppose that $H(x) = \left(\frac{1}{2}\right)^x - 4$.
(a) What is $H(-6)$? What point is on the graph of H?
(b) If $H(x) = 12$, what is x? What point is on the graph of H?
(c) Find the zero of H.

100. Suppose that $F(x) = \left(\frac{1}{3}\right)^x - 3$.
(a) What is $F(-5)$? What point is on the graph of F?
(b) If $F(x) = 24$, what is x? What point is on the graph of F?
(c) Find the zero of F.

In Problems 101–104, graph each function. Based on the graph, state the domain and the range, and find any intercepts.

101. $f(x) = \begin{cases} e^{-x} & \text{if } x < 0 \\ e^x & \text{if } x \geq 0 \end{cases}$

102. $f(x) = \begin{cases} e^{x} & \text{if } x < 0 \\ e^{-x} & \text{if } x \geq 0 \end{cases}$

103. $f(x) = \begin{cases} -e^{x} & \text{if } x < 0 \\ -e^{-x} & \text{if } x \geq 0 \end{cases}$

104. $f(x) = \begin{cases} -e^{-x} & \text{if } x < 0 \\ -e^{x} & \text{if } x \geq 0 \end{cases}$

Applications and Extensions

105. Optics If a single pane of glass obliterates 3% of the light passing through it, the percent p of light that passes through n successive panes is given approximately by the function

$$p(n) = 100(0.97)^n$$

(a) What percent of light will pass through 10 panes?
(b) What percent of light will pass through 25 panes?
(c) Explain the meaning of the base 0.97 in this problem.

106. Atmospheric Pressure The atmospheric pressure p on a balloon or airplane decreases with increasing height. This pressure, measured in millimeters of mercury, is related to the height h (in kilometers) above sea level by the function

$$p(h) = 760e^{-0.145h}$$

(a) Find the atmospheric pressure at a height of 2 km (over a mile).
(b) What is it at a height of 10 kilometers (over 30,000 feet)?

107. Depreciation The price p, in dollars, of a Honda Civic EX-L sedan that is x years old is modeled by

$$p(x) = 22{,}265(0.90)^x$$

(a) How much should a 3-year-old Civic EX-L sedan cost?
(b) How much should a 9-year-old Civic EX-L sedan cost?
(c) Explain the meaning of the base 0.90 in this problem.

108. Healing of Wounds The normal healing of wounds can be modeled by an exponential function. If A_0 represents the original area of the wound and if A equals the area of the wound, then the function

$$A(n) = A_0e^{-0.35n}$$

describes the area of a wound after n days following an injury when no infection is present to retard the healing. Suppose that a wound initially had an area of 100 square millimeters.
(a) If healing is taking place, how large will the area of the wound be after 3 days?
(b) How large will it be after 10 days?

109. Advanced-Stage Pancreatic Cancer The percentage of patients P who have survived t years after initial diagnosis of advanced-stage pancreatic cancer is modeled by the function

$$P(t) = 100(0.3)^t$$

Source: *Cancer Treatment Centers of America*

(a) According to the model, what percent of patients survive 1 year after initial diagnosis?
(b) What percent of patients survive 2 years after initial diagnosis?
(c) Explain the meaning of the base 0.3 in the context of this problem.

110. Endangered Species In a protected environment, the population P of a certain endangered species recovers over time t (in years) according to the model

$$P(t) = 30(1.149)^t$$

(a) What is the size of the initial population of the species?
(b) According to the model, what will be the population of the species in 5 years?
(c) According to the model, what will be the population of the species in 10 years?
(d) According to the model, what will be the population of the species in 15 years?
(e) What is happening to the population every 5 years?

111. Drug Medication The function

$$D(h) = 5e^{-0.4h}$$

can be used to find the number of milligrams D of a certain drug that is in a patient's bloodstream h hours after the drug has been administered. How many milligrams will be present after 1 hour? After 6 hours?

112. Spreading of Rumors A model for the number N of people in a college community who have heard a certain rumor is

$$N = P(1 - e^{-0.15d})$$

where P is the total population of the community and d is the number of days that have elapsed since the rumor began. In a community of 1000 students, how many students will have heard the rumor after 3 days?

113. Exponential Probability Between 12:00 PM and 1:00 PM, cars arrive at Citibank's drive-thru at the rate of 6 cars per hour (0.1 car per minute). The following formula from probability can be used to determine the probability that a car will arrive within t minutes of 12:00 PM.

$$F(t) = 1 - e^{-0.1t}$$

(a) Determine the probability that a car will arrive within 10 minutes of 12:00 PM (that is, before 12:10 PM).
(b) Determine the probability that a car will arrive within 40 minutes of 12:00 PM (before 12:40 PM).
(c) What value does F approach as t becomes unbounded in the positive direction?
(d) Graph F using a graphing utility.
(e) Using INTERSECT, determine how many minutes are needed for the probability to reach 50%.

114. Exponential Probability Between 5:00 PM and 6:00 PM, cars arrive at Jiffy Lube at the rate of 9 cars per hour (0.15 car per minute). This formula from probability can be used to determine the probability that a car will arrive within t minutes of 5:00 PM:

$$F(t) = 1 - e^{-0.15t}$$

(a) Determine the probability that a car will arrive within 15 minutes of 5:00 PM (that is, before 5:15 PM).
(b) Determine the probability that a car will arrive within 30 minutes of 5:00 PM (before 5:30 PM).
(c) What value does F approach as t becomes unbounded in the positive direction?
(d) Graph F using a graphing utility.
(e) Using INTERSECT, determine how many minutes are needed for the probability to reach 60%.

115. Poisson Probability Between 5:00 PM and 6:00 PM, cars arrive at a McDonald's drive-thru at the rate of 20 cars per hour. The following formula from probability can be used to determine the probability that x cars will arrive between 5:00 PM and 6:00 PM.

$$P(x) = \frac{20^x e^{-20}}{x!}$$

where

$$x! = x \cdot (x-1) \cdot (x-2) \cdot \cdots \cdot 3 \cdot 2 \cdot 1$$

(a) Determine the probability that $x = 15$ cars will arrive between 5:00 PM and 6:00 PM.
(b) Determine the probability that $x = 20$ cars will arrive between 5:00 PM and 6:00 PM.

116. Poisson Probability People enter a line for the *Demon Roller Coaster* at the rate of 4 per minute. The following formula from probability can be used to determine the probability that x people will arrive within the next minute.

$$P(x) = \frac{4^x e^{-4}}{x!}$$

where

$$x! = x \cdot (x-1) \cdot (x-2) \cdot \cdots \cdot 3 \cdot 2 \cdot 1$$

(a) Determine the probability that $x = 5$ people will arrive within the next minute.
(b) Determine the probability that $x = 8$ people will arrive within the next minute.

117. Relative Humidity The relative humidity is the ratio (expressed as a percent) of the amount of water vapor in the air to the maximum amount that the air can hold at a specific temperature. The relative humidity, R, is found using the following formula:

$$R = 10^{\left(\frac{4221}{T+459.4} - \frac{4221}{D+459.4} + 2\right)}$$

where T is the air temperature (in °F) and D is the dew point temperature (in °F).

(a) Determine the relative humidity if the air temperature is 50° Fahrenheit and the dew point temperature is 41° Fahrenheit.
(b) Determine the relative humidity if the air temperature is 68° Fahrenheit and the dew point temperature is 59° Fahrenheit.
(c) What is the relative humidity if the air temperature and the dew point temperature are the same?

118. Learning Curve Suppose that a student has 500 vocabulary words to learn. If the student learns 15 words after 5 minutes, the function

$$L(t) = 500(1 - e^{-0.0061t})$$

approximates the number of words L that the student will have learned after t minutes.

(a) How many words will the student have learned after 30 minutes?

(b) How many words will the student have learned after 60 minutes?

119. Current in an RL Circuit The equation governing the amount of current I (in amperes) after time t (in seconds) in a single RL circuit consisting of a resistance R (in ohms), an inductance L (in henrys), and an electromotive force E (in volts) is

$$I = \frac{E}{R}[1 - e^{-(R/L)t}]$$

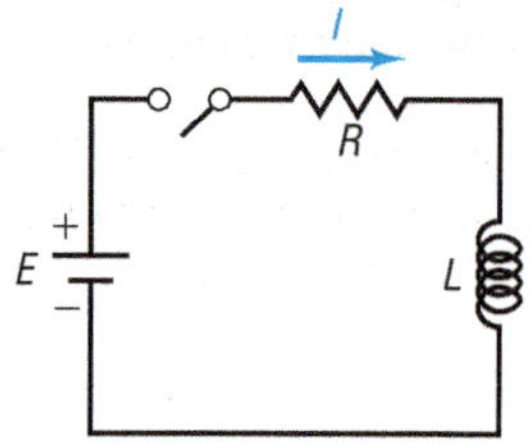

(a) If $E = 120$ volts, $R = 10$ ohms, and $L = 5$ henrys, how much current I_1 is flowing after 0.3 second? After 0.5 second? After 1 second?

(b) What is the maximum current?

(c) Graph this function $I = I_1(t)$, measuring I along the y-axis and t along the x-axis.

(d) If $E = 120$ volts, $R = 5$ ohms, and $L = 10$ henrys, how much current I_2 is flowing after 0.3 second? After 0.5 second? After 1 second?

(e) What is the maximum current?

(f) Graph the function $I = I_2(t)$ on the same coordinate axes as $I_1(t)$.

120. Current in an RC Circuit The equation governing the amount of current I (in amperes) after time t (in microseconds) in a single RC circuit consisting of a resistance R (in ohms), a capacitance C (in microfarads), and an electromotive force E (in volts) is

$$I = \frac{E}{R}e^{-t/(RC)}$$

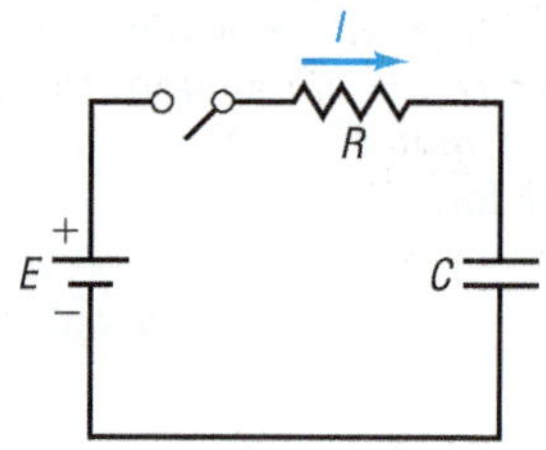

(a) If $E = 120$ volts, $R = 2000$ ohms, and $C = 1.0$ microfarad, how much current I_1 is flowing initially ($t = 0$)? After 1000 microseconds? After 3000 microseconds?

(b) What is the maximum current?

(c) Graph the function $I = I_1(t)$, measuring I along the y-axis and t along the x-axis.

(d) If $E = 120$ volts, $R = 1000$ ohms, and $C = 2.0$ microfarads, how much current I_2 is flowing initially? After 1000 microseconds? After 3000 microseconds?

(e) What is the maximum current?

(f) Graph the function $I = I_2(t)$ on the same coordinate axes as $I_1(t)$.

121. If f is an exponential function of the form $f(x) = Ca^x$ with growth factor 3, and if $f(6) = 12$, what is $f(7)$?

122. Another Formula for *e* Use a calculator to compute the values of

$$2 + \frac{1}{2!} + \frac{1}{3!} + \cdots + \frac{1}{n!}$$

for $n = 4, 6, 8$, and 10. Compare each result with e.

[**Hint:** $1! = 1, 2! = 2 \cdot 1, 3! = 3 \cdot 2 \cdot 1,$
$n! = n(n-1) \cdot \cdots \cdot (3)(2)(1)$.]

123. Another Formula for *e* Use a calculator to compute the various values of the expression. Compare the values to e.

$$2 + \cfrac{1}{1 + \cfrac{1}{2 + \cfrac{2}{3 + \cfrac{3}{4 + \cfrac{4}{\text{etc.}}}}}}$$

124. Difference Quotient If $f(x) = a^x$, show that

$$\frac{f(x+h) - f(x)}{h} = a^x \cdot \frac{a^h - 1}{h} \qquad h \neq 0$$

125. If $f(x) = a^x$, show that $f(A + B) = f(A) \cdot f(B)$.

126. If $f(x) = a^x$, show that $f(-x) = \dfrac{1}{f(x)}$.

127. If $f(x) = a^x$, show that $f(\alpha x) = [f(x)]^\alpha$.

Problems 128 and 129 provide definitions for two other transcendental functions.

128. The **hyperbolic sine function**, designated by $\sinh x$, is defined as

$$\sinh x = \frac{1}{2}(e^x - e^{-x})$$

(a) Show that $f(x) = \sinh x$ is an odd function.

(b) Graph $f(x) = \sinh x$ using a graphing utility.

129. The **hyperbolic cosine function**, designated by $\cosh x$, is defined as

$$\cosh x = \frac{1}{2}(e^x + e^{-x})$$

(a) Show that $f(x) = \cosh x$ is an even function.

(b) Graph $f(x) = \cosh x$ using a graphing utility.

(c) Refer to Problem 128. Show that, for every x,

$$(\cosh x)^2 - (\sinh x)^2 = 1$$

130. Historical Problem Pierre de Fermat (1601–1665) conjectured that the function

$$f(x) = 2^{(2^x)} + 1$$

for $x = 1, 2, 3, \ldots$, would always have a value equal to a prime number. But Leonhard Euler (1707–1783) showed that this formula fails for $x = 5$. Use a calculator to determine the prime numbers produced by f for $x = 1, 2, 3, 4$. Then show that $f(5) = 641 \times 6{,}700{,}417$, which is not prime.

Explaining Concepts: Discussion and Writing

131. The bacteria in a 4-liter container double every minute. After 60 minutes the container is full. How long did it take to fill half the container?

132. Explain in your own words what the number *e* is. Provide at least two applications that use this number.

133. Do you think that there is a power function that increases more rapidly than an exponential function whose base is greater than 1? Explain.

134. As the base a of an exponential function $f(x) = a^x$, where $a > 1$, increases, what happens to the behavior of its graph for $x > 0$? What happens to the behavior of its graph for $x < 0$?

135. The graphs of $y = a^{-x}$ and $y = \left(\frac{1}{a}\right)^x$ are identical. Why?

Retain Your Knowledge

Problems 136–139 are based on material learned earlier in the course. The purpose of these problems is to keep the material fresh in your mind so that you are better prepared for the final exam.

136. Solve the inequality: $x^3 + 5x^2 \leq 4x + 20$.

137. Solve the inequality: $\dfrac{x+1}{x-2} \geq 1$.

138. Find the equation of the quadratic function f that has its vertex at $(3, 5)$ and contains the point $(2, 3)$.

139. Consider the quadratic function $f(x) = x^2 + 2x - 3$.

(a) Graph f by determining whether its graph opens up or down and by finding its vertex, axis of symmetry, y-intercept, and x-intercepts, if any.

(b) Determine the domain and range of f.

(c) Determine where f is increasing and where it is decreasing.

'Are You Prepared?' Answers

1. $64; 4; \frac{1}{9}$ **2.** $\{-4, 1\}$ **3.** False **4.** 3 **5.** True

6.4 Logarithmic Functions

PREPARING FOR THIS SECTION *Before getting started, review the following:*

- Solve Linear Inequalities (Section 1.7, pp. 150–151)
- Solve Quadratic Inequalities (Section 4.5, pp. 320–322)
- Polynomial and Rational Inequalities (Section 5.6, pp. 393–397)
- Solve Linear Equations (Section 1.2, pp. 102–103)

Now Work the **'Are You Prepared?'** problems on page 454.

OBJECTIVES
1. Change Exponential Statements to Logarithmic Statements and Logarithmic Statements to Exponential Statements (p. 446)
2. Evaluate Logarithmic Expressions (p. 446)
3. Determine the Domain of a Logarithmic Function (p. 447)
4. Graph Logarithmic Functions (p. 448)
5. Solve Logarithmic Equations (p. 452)

Recall that a one-to-one function $y = f(x)$ has an inverse function that is defined (implicitly) by the equation $x = f(y)$. In particular, the exponential function $y = f(x) = a^x$, where $a > 0$ and $a \neq 1$, is one-to-one and, hence, has an inverse function that is defined implicitly by the equation

$$x = a^y, \quad a > 0, \quad a \neq 1$$

This inverse function is so important that it is given a name, the *logarithmic function*.

DEFINITION

The **logarithmic function with base a**, where $a > 0$ and $a \neq 1$, is denoted by $y = \log_a x$ (read as "y is the logarithm with base a of x") and is defined by

$$y = \log_a x \quad \text{if and only if} \quad x = a^y$$

The domain of the logarithmic function $y = \log_a x$ is $x > 0$.

In Words

When you read $\log_a x$, think to yourself "a raised to what power gives me x."

As this definition illustrates, **a logarithm is a name for a certain exponent.** So $\log_a x$ represents the exponent to which a must be raised to obtain x.

EXAMPLE 1 Relating Logarithms to Exponents

(a) If $y = \log_3 x$, then $x = 3^y$. For example, the logarithmic statement $4 = \log_3 81$ is equivalent to the exponential statement $81 = 3^4$.

(b) If $y = \log_5 x$, then $x = 5^y$. For example, $-1 = \log_5\left(\frac{1}{5}\right)$ is equivalent to $\frac{1}{5} = 5^{-1}$.

1 Change Exponential Statements to Logarithmic Statements and Logarithmic Statements to Exponential Statements

The definition of a logarithm can be used to convert from exponential form to logarithmic form, and vice versa, as the following two examples illustrate.

EXAMPLE 2 Changing Exponential Statements to Logarithmic Statements

Change each exponential statement to an equivalent statement involving a logarithm.

(a) $1.2^3 = m$ (b) $e^b = 9$ (c) $a^4 = 24$

Solution Use the fact that $y = \log_a x$ and $x = a^y$, where $a > 0$ and $a \neq 1$, are equivalent.

(a) If $1.2^3 = m$, then $3 = \log_{1.2} m$. (b) If $e^b = 9$, then $b = \log_e 9$.
(c) If $a^4 = 24$, then $4 = \log_a 24$.

Now Work PROBLEM 11

EXAMPLE 3 Changing Logarithmic Statements to Exponential Statements

Change each logarithmic statement to an equivalent statement involving an exponent.

(a) $\log_a 4 = 5$ (b) $\log_e b = -3$ (c) $\log_3 5 = c$

Solution (a) If $\log_a 4 = 5$, then $a^5 = 4$. (b) If $\log_e b = -3$, then $e^{-3} = b$.
(c) If $\log_3 5 = c$, then $3^c = 5$.

Now Work PROBLEM 19

2 Evaluate Logarithmic Expressions

To find the exact value of a logarithm, write the logarithm in exponential notation using the fact that $y = \log_a x$ is equivalent to $a^y = x$, and use the fact that if $a^u = a^v$, then $u = v$.

EXAMPLE 4 Finding the Exact Value of a Logarithmic Expression

Find the exact value of:

(a) $\log_2 16$ (b) $\log_3 \frac{1}{27}$

Solution

(a) To evaluate $\log_2 16$, think "2 raised to what power yields 16?" So,

$y = \log_2 16$

$2^y = 16$ **Change to exponential form.**

$2^y = 2^4$ **$16 = 2^4$**

$y = 4$ **Equate exponents.**

Therefore, $\log_2 16 = 4$.

(b) To evaluate $\log_3 \frac{1}{27}$, think "3 raised to what power yields $\frac{1}{27}$?" So,

$y = \log_3 \frac{1}{27}$

$3^y = \frac{1}{27}$ **Change to exponential form.**

$3^y = 3^{-3}$ **$\frac{1}{27} = \frac{1}{3^3} = 3^{-3}$**

$y = -3$ **Equate exponents.**

Therefore, $\log_3 \frac{1}{27} = -3$. ■

Now Work PROBLEM 27

3 Determine the Domain of a Logarithmic Function

The logarithmic function $y = \log_a x$ has been defined as the inverse of the exponential function $y = a^x$. That is, if $f(x) = a^x$, then $f^{-1}(x) = \log_a x$. Based on the discussion in Section 6.2 on inverse functions, for a function f and its inverse f^{-1},

$$\text{Domain of } f^{-1} = \text{Range of } f \quad \text{and} \quad \text{Range of } f^{-1} = \text{Domain of } f$$

Consequently, it follows that

> Domain of the logarithmic function = Range of the exponential function = $(0, \infty)$
>
> Range of the logarithmic function = Domain of the exponential function = $(-\infty, \infty)$

The next box summarizes some properties of the logarithmic function.

> $y = \log_a x$ (defining equation: $x = a^y$)
>
> Domain: $(0, \infty)$ Range: $(-\infty, \infty)$

The domain of a logarithmic function consists of the *positive* real numbers, so the argument of a logarithmic function must be greater than zero.

EXAMPLE 5 **Finding the Domain of a Logarithmic Function**

Find the domain of each logarithmic function.

(a) $F(x) = \log_2(x + 3)$ (b) $g(x) = \log_5\left(\frac{1 + x}{1 - x}\right)$ (c) $h(x) = \log_{1/2}|x|$

Solution

(a) The domain of F consists of all x for which $x + 3 > 0$, that is, $x > -3$. Using interval notation, the domain of F is $(-3, \infty)$.

(b) The domain of g is restricted to

$$\frac{1 + x}{1 - x} > 0$$

Solve this inequality to find that the domain of g consists of all x between -1 and 1, that is, $-1 < x < 1$ or, using interval notation, $(-1, 1)$.

(c) Since $|x| > 0$, provided that $x \neq 0$, the domain of h consists of all real numbers except zero or, using interval notation, $(-\infty, 0) \cup (0, \infty)$. ■

Now Work PROBLEMS 41 AND 47

4 Graph Logarithmic Functions

Because exponential functions and logarithmic functions are inverses of each other, the graph of the logarithmic function $y = \log_a x$ is the reflection about the line $y = x$ of the graph of the exponential function $y = a^x$, as shown in Figure 36.

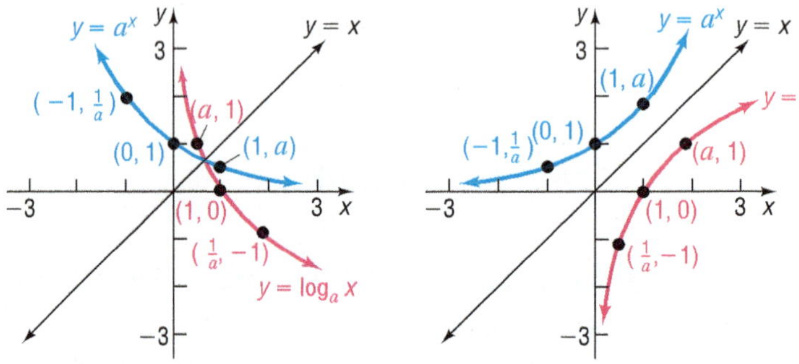

Figure 36 (a) $0 < a < 1$ (b) $a > 1$

For example, to graph $y = \log_2 x$, graph $y = 2^x$ and reflect it about the line $y = x$. See Figure 37. To graph $y = \log_{1/3} x$, graph $y = \left(\frac{1}{3}\right)^x$ and reflect it about the line $y = x$. See Figure 38.

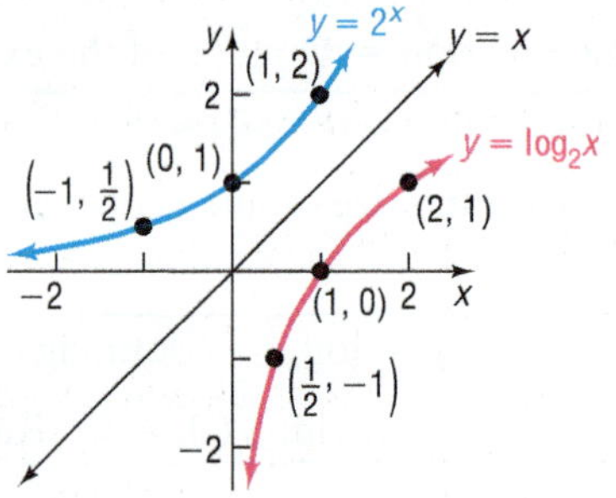

Figure 37

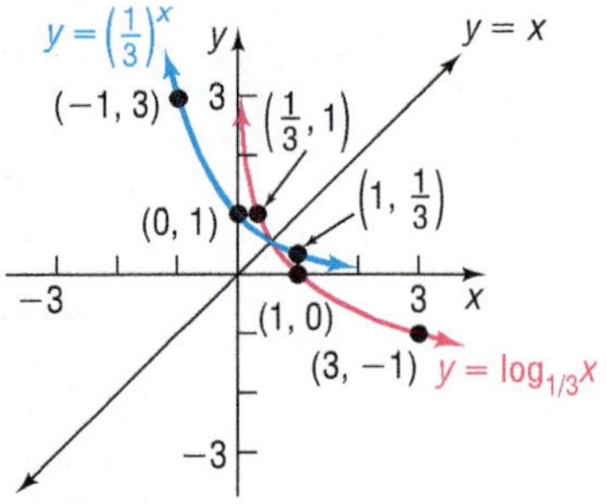

Figure 38

Now Work PROBLEM 61

The graphs of $y = \log_a x$ in Figures 36(a) and (b) lead to the following properties.

Properties of the Logarithmic Function $f(x) = \log_a x$; $a > 0$, $a \neq 1$

1. The domain is the set of positive real numbers, or $(0, \infty)$ using interval notation; the range is the set of all real numbers, or $(-\infty, \infty)$ using interval notation.
2. The x-intercept of the graph is 1. There is no y-intercept.
3. The y-axis $(x = 0)$ is a vertical asymptote of the graph.
4. A logarithmic function is decreasing if $0 < a < 1$ and is increasing if $a > 1$.
5. The graph of f contains the points $(1, 0)$, $(a, 1)$, and $\left(\frac{1}{a}, -1\right)$.
6. The graph is smooth and continuous, with no corners or gaps.

In Words
$y = \log_e x$ is written $y = \ln x$.

If the base of a logarithmic function is the number e, the result is the **natural logarithm function**. This function occurs so frequently in applications that it is given a special symbol, **ln** (from the Latin, *logarithmus naturalis*). That is,

$$y = \ln x \quad \text{if and only if} \quad x = e^y \tag{1}$$

Because $y = \ln x$ and the exponential function $y = e^x$ are inverse functions, the graph of $y = \ln x$ can be obtained by reflecting the graph of $y = e^x$ about the line $y = x$. See Figure 39.

Using a calculator with an ln key, we can obtain other points on the graph of $f(x) = \ln x$. See Table 8.

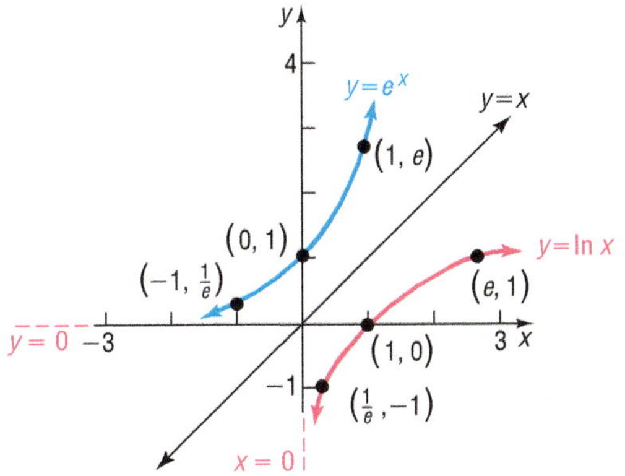

Figure 39

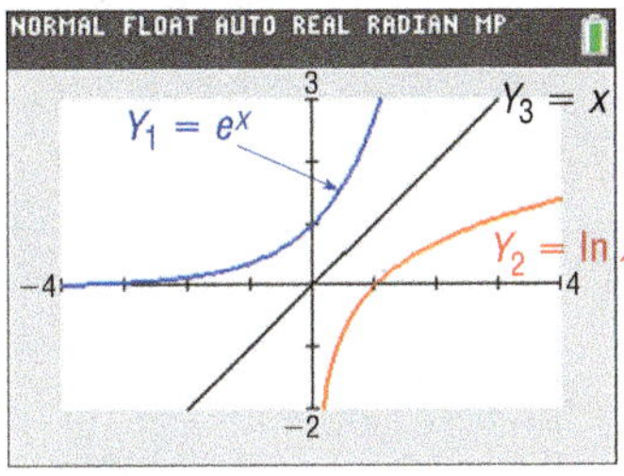

Table 8

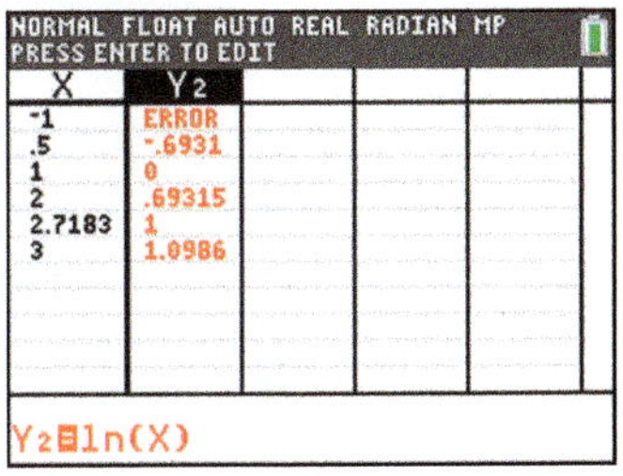

NORMAL FLOAT AUTO REAL RADIAN MP
PRESS ENTER TO EDIT

X	Y2
-1	ERROR
.5	-.6931
1	0
2	.69315
2.7183	1
3	1.0986

Y2=ln(X)

EXAMPLE 6 **Graphing a Logarithmic Function and Its Inverse**

(a) Find the domain of the logarithmic function $f(x) = -\ln(x - 2)$.
(b) Graph f.
(c) From the graph, determine the range and vertical asymptote of f.
(d) Find f^{-1}, the inverse of f.
(e) Find the domain and the range of f^{-1}.
(f) Graph f^{-1}.

Solution

(a) The domain of f consists of all x for which $x - 2 > 0$, or equivalently, $x > 2$. The domain of f is $\{x \mid x > 2\}$, or $(2, \infty)$ in interval notation.

(b) To obtain the graph of $y = -\ln(x - 2)$, begin with the graph of $y = \ln x$ and use transformations. See Figure 40.

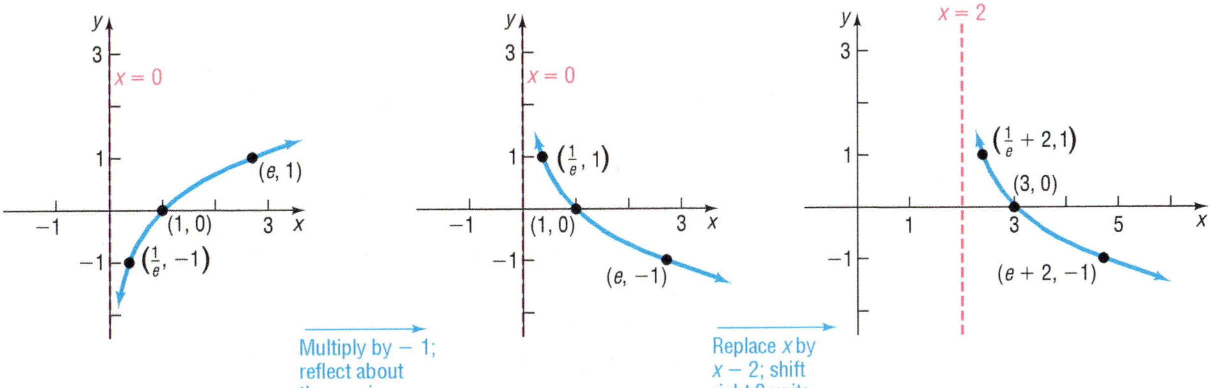

Figure 40 (a) $y = \ln x$ (b) $y = -\ln x$ (c) $y = -\ln(x - 2)$

(c) The range of $f(x) = -\ln(x - 2)$ is the set of all real numbers. The vertical asymptote is $x = 2$. [Do you see why? The original asymptote $(x = 0)$ is shifted to the right 2 units.]

(d) To find f^{-1}, begin with $y = -\ln(x-2)$. The inverse function is defined (implicitly) by the equation

$$x = -\ln(y-2)$$

Now solve for y.

$$-x = \ln(y-2) \quad \text{Isolate the logarithm.}$$
$$e^{-x} = y - 2 \quad \text{Change to exponential form.}$$
$$y = e^{-x} + 2 \quad \text{Solve for } y.$$

The inverse of f is $f^{-1}(x) = e^{-x} + 2$.

(e) The domain of f^{-1} equals the range of f, which is the set of all real numbers, from part (c). The range of f^{-1} is the domain of f, which is $(2, \infty)$ in interval notation.

(f) To graph f^{-1}, use the graph of f in Figure 40(c) and reflect it about the line $y = x$. See Figure 41. We could also graph $f^{-1}(x) = e^{-x} + 2$ using transformations.

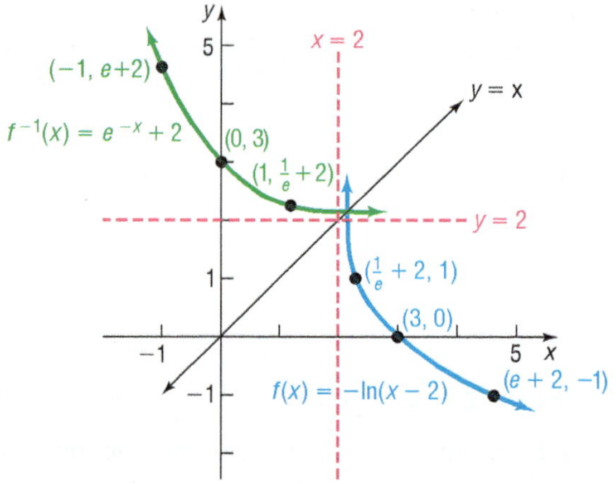

Figure 41

Now Work PROBLEM 73

If the base of a logarithmic function is the number 10, the result is the **common logarithm function**. If the base a of the logarithmic function is not indicated, it is understood to be 10. That is,

$$y = \log x \quad \text{if and only if} \quad x = 10^y$$

Because $y = \log x$ and the exponential function $y = 10^x$ are inverse functions, the graph of $y = \log x$ can be obtained by reflecting the graph of $y = 10^x$ about the line $y = x$. See Figure 42.

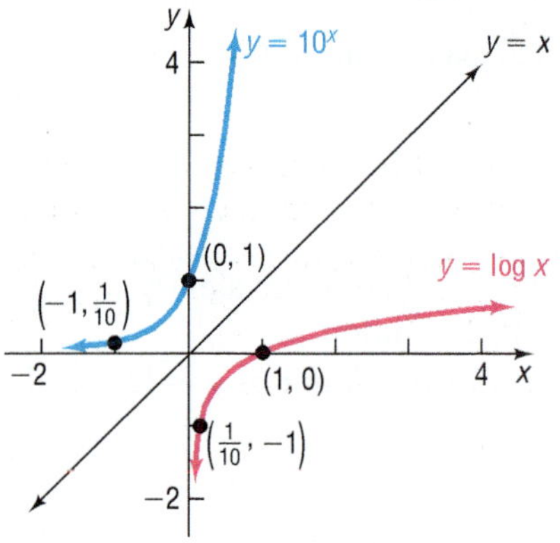

Figure 42

EXAMPLE 7 Graphing a Logarithmic Function and Its Inverse

(a) Find the domain of the logarithmic function $f(x) = 3\log(x-1)$.
(b) Graph f.
(c) From the graph, determine the range and vertical asymptote of f.
(d) Find f^{-1}, the inverse of f.
(e) Find the domain and the range of f^{-1}.
(f) Graph f^{-1}.

Solution

(a) The domain of f consists of all x for which $x - 1 > 0$, or equivalently, $x > 1$. The domain of f is $\{x \mid x > 1\}$, or $(1, \infty)$ in interval notation.

(b) To obtain the graph of $y = 3\log(x-1)$, begin with the graph of $y = \log x$ and use transformations. See Figure 43.

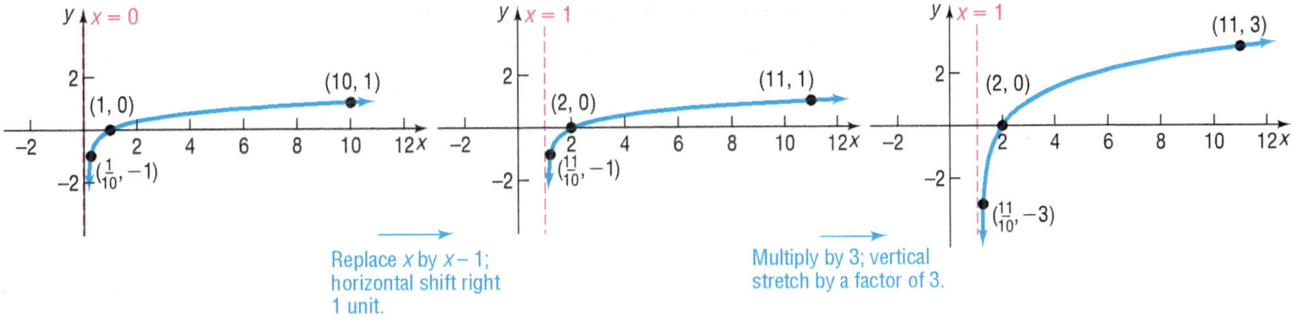

Figure 43 **(a)** $y = \log x$ **(b)** $y = \log(x - 1)$ **(c)** $y = 3\log(x - 1)$

(c) The range of $f(x) = 3\log(x - 1)$ is the set of all real numbers. The vertical asymptote is $x = 1$.

(d) Begin with $y = 3\log(x - 1)$. The inverse function is defined implicitly by the equation

$$x = 3\log(y - 1)$$

Proceed to solve for y.

$$\frac{x}{3} = \log(y - 1) \quad \text{Isolate the logarithm.}$$

$$10^{x/3} = y - 1 \quad \text{Change to exponential form.}$$

$$y = 10^{x/3} + 1 \quad \text{Solve for } y.$$

The inverse of f is $f^{-1}(x) = 10^{x/3} + 1$.

(e) The domain of f^{-1} is the range of f, which is the set of all real numbers, from part (c). The range of f^{-1} is the domain of f, which is $(1, \infty)$ in interval notation.

(f) To graph f^{-1}, we use the graph of f in Figure 43(c) and reflect it about the line $y = x$. See Figure 44. We could also graph $f^{-1}(x) = 10^{x/3} + 1$ using transformations.

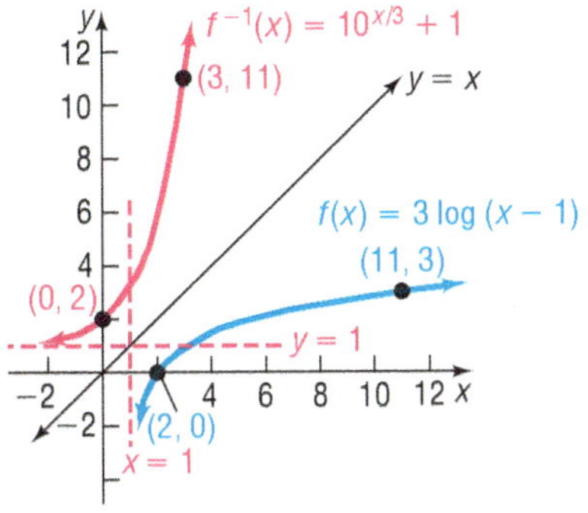

Figure 44

Now Work PROBLEM 81

5 Solve Logarithmic Equations

Equations that contain logarithms are called **logarithmic equations**. Care must be taken when solving logarithmic equations algebraically. In the expression $\log_a M$, remember that a and M are positive and $a \neq 1$. Be sure to check each apparent solution in the original equation and discard any that are extraneous.

Some logarithmic equations can be solved by changing the logarithmic equation to exponential form using the fact that $y = \log_a x$ means $a^y = x$.

EXAMPLE 8 Solving Logarithmic Equations

Solve:

(a) $\log_3(4x - 7) = 2$ (b) $\log_x 64 = 2$

Solution (a) To solve, change the logarithmic equation to exponential form.

$$\log_3(4x - 7) = 2$$

$$4x - 7 = 3^2 \quad \text{Change to exponential form.}$$

$$4x - 7 = 9$$

$$4x = 16$$

$$x = 4$$

Check: $\log_3(4x - 7) = \log_3(4 \cdot 4 - 7) = \log_3 9 = 2$ $\quad 3^2 = 9$

The solution set is $\{4\}$.

(b) To solve, change the logarithmic equation to exponential form.

$$\log_x 64 = 2$$

$$x^2 = 64 \quad \text{Change to exponential form.}$$

$$x = \pm\sqrt{64} = \pm 8 \quad \text{Square Root Method}$$

Because the base of a logarithm must be positive, discard -8. Check the potential solution 8.

Check: $\log_8 64 = 2$ $\quad 8^2 = 64$

The solution set is $\{8\}$. ■

EXAMPLE 9 Using Logarithms to Solve an Exponential Equation

Solve: $e^{2x} = 5$

Solution To solve, change the exponential equation to logarithmic form.

$$e^{2x} = 5$$

$$\ln 5 = 2x \quad \text{Change to logarithmic form.}$$

$$x = \frac{\ln 5}{2} \quad \text{Exact solution}$$

$$\approx 0.805 \quad \text{Approximate solution}$$

The solution set is $\left\{\frac{\ln 5}{2}\right\}$. ■

Now Work PROBLEMS 89 AND 101

EXAMPLE 10 Alcohol and Driving

Blood alcohol concentration (BAC) is a measure of the amount of alcohol in a person's bloodstream. A BAC of 0.04% means that a person has 4 parts alcohol per 10,000 parts blood in the body. Relative risk is defined as the likelihood of one event occurring divided by the likelihood of a second event occurring. For example, if an individual with a BAC of 0.02% is 1.4 times as likely to have a car accident as an individual who has not been drinking, the relative risk of an accident with a BAC of 0.02% is 1.4. Recent medical research suggests that the relative risk R of having an accident while driving a car can be modeled by an equation of the form

$$R = e^{kx}$$

where x is the percent of concentration of alcohol in the bloodstream and k is a constant.

(a) Research indicates that the relative risk of a person having an accident with a BAC of 0.02% is 1.4. Find the constant k in the equation.
(b) Using this value of k, what is the relative risk if the concentration is 0.17%?
(c) Using this same value of k, what BAC corresponds to a relative risk of 100?
(d) If the law asserts that anyone with a relative risk of 4 or more should not have driving privileges, at what concentration of alcohol in the bloodstream should a driver be arrested and charged with DUI (driving under the influence)?

Solution

(a) For a concentration of alcohol in the blood of 0.02% and a relative risk of 1.4, let $x = 0.02$ and $R = 1.4$ in the equation and solve for k.

$$\begin{aligned} R &= e^{kx} & \\ 1.4 &= e^{k(0.02)} & \quad R = 1.4;\ x = 0.02 \\ 0.02k &= \ln 1.4 & \quad \text{Change to logarithmic form.} \\ k &= \frac{\ln 1.4}{0.02} \approx 16.82 & \quad \text{Solve for } k. \end{aligned}$$

(b) A concentration of 0.17% means $x = 0.17$. Use $k = 16.82$ in the equation to find the relative risk R:

$$R = e^{kx} = e^{(16.82)(0.17)} \approx 17.5$$

For a concentration of alcohol in the blood of 0.17%, the relative risk of an accident is about 17.5. That is, a person with a BAC of 0.17% is 17.5 times as likely to have a car accident as a person with no alcohol in the bloodstream.

(c) A relative risk of 100 means $R = 100$. Use $k = 16.82$ in the equation $R = e^{kx}$. The concentration x of alcohol in the blood obeys

$$\begin{aligned} 100 &= e^{16.82x} & \quad R = e^{kx};\ R = 100;\ k = 16.82 \\ 16.82x &= \ln 100 & \quad \text{Change to logarithmic form.} \\ x &= \frac{\ln 100}{16.82} \approx 0.27 & \quad \text{Solve for } x. \end{aligned}$$

For a concentration of alcohol in the blood of 0.27%, the relative risk of an accident is 100.

Note: A BAC of 0.30% results in a loss of consciousness in most people. ■

(d) A relative risk of 4 means $R = 4$. Use $k = 16.82$ in the equation $R = e^{kx}$. The concentration x of alcohol in the bloodstream obeys

$$\begin{aligned} 4 &= e^{16.82x} \\ 16.82x &= \ln 4 \\ x &= \frac{\ln 4}{16.82} \approx 0.082 \end{aligned}$$

A driver with a BAC of 0.082% or more should be arrested and charged with DUI. ■

Note: The blood alcohol content at which a DUI citation is given is 0.08%. ■

SUMMARY

Properties of the Logarithmic Function

$f(x) = \log_a x, \quad a > 1$ $(y = \log_a x$ means $x = a^y)$	Domain: the interval $(0, \infty)$; Range: the interval $(-\infty, \infty)$ x-intercept: 1; y-intercept: none; vertical asymptote: $x = 0$ (y-axis); increasing; one-to-one See Figure 45(a) for a typical graph.
$f(x) = \log_a x, \quad 0 < a < 1$ $(y = \log_a x$ means $x = a^y)$	Domain: the interval $(0, \infty)$; Range: the interval $(-\infty, \infty)$ x-intercept: 1; y-intercept: none; vertical asymptote: $x = 0$ (y-axis); decreasing; one-to-one See Figure 45(b) for a typical graph.

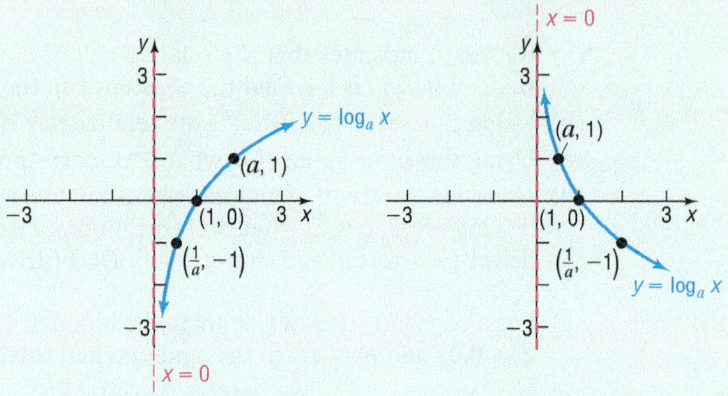

Figure 45 (a) $a > 1$ (b) $0 < a < 1$

6.4 Assess Your Understanding

'Are You Prepared?' *Answers are given at the end of these exercises. If you get a wrong answer, read the pages listed in red.*

1. Solve each inequality:
   (a) $3x - 7 \le 8 - 2x$ (pp. 150–151)
   (b) $x^2 - x - 6 > 0$ (pp. 320–322)
2. Solve the inequality: $\dfrac{x-1}{x+4} > 0$ (pp. 393–397)
3. Solve: $2x + 3 = 9$ (pp. 102–103)

Concepts and Vocabulary

4. The domain of the logarithmic function $f(x) = \log_a x$ is ________.
5. The graph of every logarithmic function $f(x) = \log_a x$, where $a > 0$ and $a \ne 1$, passes through three points: ________, ________, and ________.
6. If the graph of a logarithmic function $f(x) = \log_a x$, where $a > 0$ and $a \ne 1$, is increasing, then its base must be larger than ______.
7. ***True or False*** If $y = \log_a x$, then $y = a^x$.
8. ***True or False*** The graph of $f(x) = \log_a x$, where $a > 0$ and $a \ne 1$, has an x-intercept equal to 1 and no y-intercept.
9. Select the answer that completes the statement: $y = \ln x$ if and only if ________.
   (a) $x = e^y$ (b) $y = e^x$ (c) $x = 10^y$ (d) $y = 10^x$
10. Choose the domain of $f(x) = \log_3(x+2)$.
   (a) $(-\infty, \infty)$ (b) $(2, \infty)$ (c) $(-2, \infty)$ (d) $(0, \infty)$

Skill Building

In Problems 11–18, change each exponential statement to an equivalent statement involving a logarithm.

11. $9 = 3^2$
12. $16 = 4^2$
13. $a^2 = 1.6$
14. $a^3 = 2.1$
15. $2^x = 7.2$
16. $3^x = 4.6$
17. $e^x = 8$
18. $e^{2.2} = M$

In Problems 19–26, change each logarithmic statement to an equivalent statement involving an exponent.

19. $\log_2 8 = 3$
20. $\log_3\left(\dfrac{1}{9}\right) = -2$
21. $\log_a 3 = 6$
22. $\log_b 4 = 2$
23. $\log_3 2 = x$
24. $\log_2 6 = x$
25. $\ln 4 = x$
26. $\ln x = 4$

In Problems 27–38, find the exact value of each logarithm without using a calculator.

27. $\log_2 1$ **28.** $\log_8 8$ **29.** $\log_5 25$ **30.** $\log_3\left(\dfrac{1}{9}\right)$

31. $\log_{1/2} 16$ **32.** $\log_{1/3} 9$ **33.** $\log_{10} \sqrt{10}$ **34.** $\log_5 \sqrt[3]{25}$

35. $\log_{\sqrt{2}} 4$ **36.** $\log_{\sqrt{3}} 9$ **37.** $\ln \sqrt{e}$ **38.** $\ln e^3$

In Problems 39–50, find the domain of each function.

39. $f(x) = \ln(x - 3)$ **40.** $g(x) = \ln(x - 1)$ **41.** $F(x) = \log_2 x^2$

42. $H(x) = \log_5 x^3$ **43.** $f(x) = 3 - 2\log_4\left(\dfrac{x}{2} - 5\right)$ **44.** $g(x) = 8 + 5\ln(2x + 3)$

45. $f(x) = \ln\left(\dfrac{1}{x + 1}\right)$ **46.** $g(x) = \ln\left(\dfrac{1}{x - 5}\right)$ **47.** $g(x) = \log_5\left(\dfrac{x + 1}{x}\right)$

48. $h(x) = \log_3\left(\dfrac{x}{x - 1}\right)$ **49.** $f(x) = \sqrt{\ln x}$ **50.** $g(x) = \dfrac{1}{\ln x}$

In Problems 51–58, use a calculator to evaluate each expression. Round your answer to three decimal places.

51. $\ln\dfrac{5}{3}$ **52.** $\dfrac{\ln 5}{3}$ **53.** $\dfrac{\ln\frac{10}{3}}{0.04}$ **54.** $\dfrac{\ln\frac{2}{3}}{-0.1}$

55. $\dfrac{\ln 4 + \ln 2}{\log 4 + \log 2}$ **56.** $\dfrac{\log 15 + \log 20}{\ln 15 + \ln 20}$ **57.** $\dfrac{2\ln 5 + \log 50}{\log 4 - \ln 2}$ **58.** $\dfrac{3\log 80 - \ln 5}{\log 5 + \ln 20}$

59. Find a so that the graph of $f(x) = \log_a x$ contains the point $(2, 2)$.

60. Find a so that the graph of $f(x) = \log_a x$ contains the point $\left(\dfrac{1}{2}, -4\right)$.

In Problems 61–64, graph each function and its inverse on the same set of axes.

61. $f(x) = 3^x; f^{-1}(x) = \log_3 x$ **62.** $f(x) = 4^x; f^{-1}(x) = \log_4 x$

63. $f(x) = \left(\dfrac{1}{2}\right)^x; f^{-1}(x) = \log_{1/2} x$ **64.** $f(x) = \left(\dfrac{1}{3}\right)^x; f^{-1}(x) = \log_{1/3} x$

In Problems 65–72, the graph of a logarithmic function is given. Match each graph to one of the following functions:

(A) $y = \log_3 x$ (B) $y = \log_3(-x)$ (C) $y = -\log_3 x$ (D) $y = -\log_3(-x)$
(E) $y = \log_3 x - 1$ (F) $y = \log_3(x - 1)$ (G) $y = \log_3(1 - x)$ (H) $y = 1 - \log_3 x$

65.
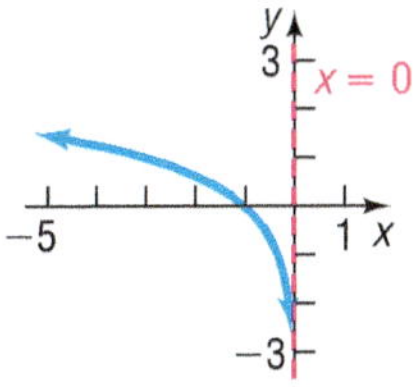

66.
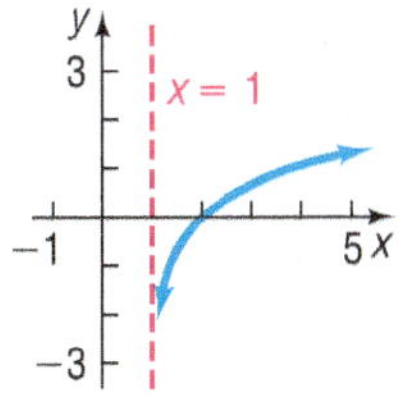

67.
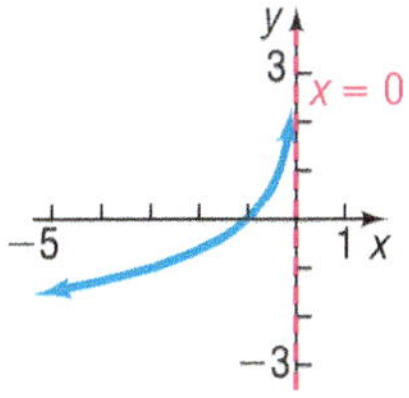

68.
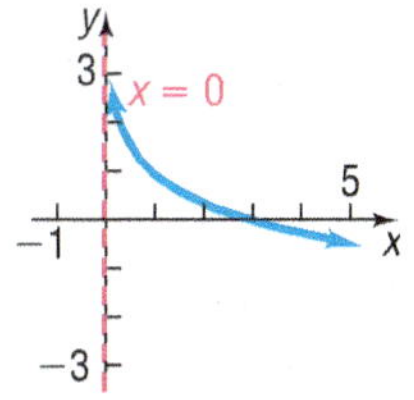

69.
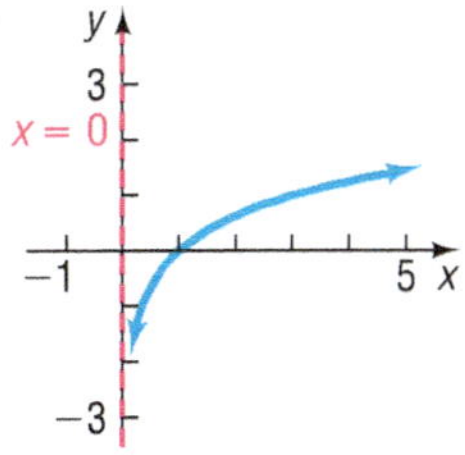

70.
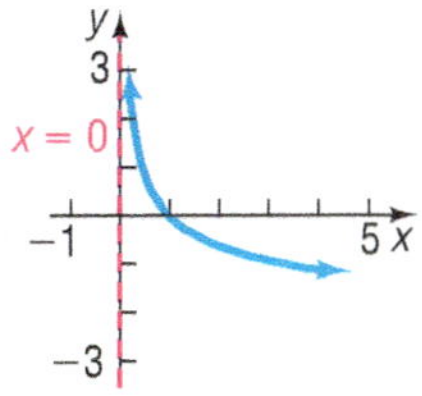

71.
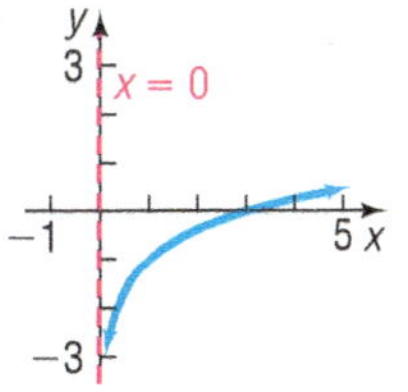

72.
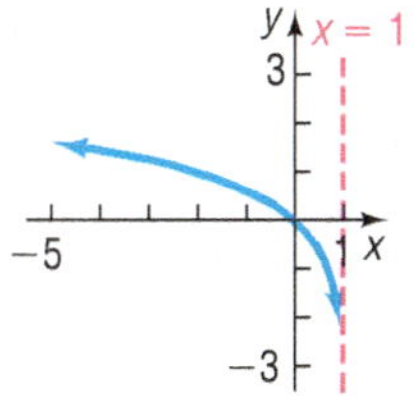

In Problems 73–88, use the given function f.
(a) Find the domain of f. (b) Graph f. (c) From the graph, determine the range and any asymptotes of f.
(d) Find f^{-1}, the inverse of f. (e) Find the domain and the range of f^{-1}. (f) Graph f^{-1}.

73. $f(x) = \ln(x + 4)$ **74.** $f(x) = \ln(x - 3)$ **75.** $f(x) = 2 + \ln x$ **76.** $f(x) = -\ln(-x)$

77. $f(x) = \ln(2x) - 3$ **78.** $f(x) = -2\ln(x + 1)$ **79.** $f(x) = \log(x - 4) + 2$ **80.** $f(x) = \dfrac{1}{2}\log x - 5$

81. $f(x) = \dfrac{1}{2}\log(2x)$ **82.** $f(x) = \log(-2x)$ **83.** $f(x) = 3 + \log_3(x + 2)$ **84.** $f(x) = 2 - \log_3(x + 1)$

85. $f(x) = e^{x+2} - 3$ **86.** $f(x) = 3e^x + 2$ **87.** $f(x) = 2^{x/3} + 4$ **88.** $f(x) = -3^{x+1}$

In Problems 89–112, solve each equation.

89. $\log_3 x = 2$
90. $\log_5 x = 3$
91. $\log_2(2x + 1) = 3$
92. $\log_3(3x - 2) = 2$
93. $\log_x 4 = 2$
94. $\log_x\left(\frac{1}{8}\right) = 3$
95. $\ln e^x = 5$
96. $\ln e^{-2x} = 8$
97. $\log_4 64 = x$
98. $\log_5 625 = x$
99. $\log_3 243 = 2x + 1$
100. $\log_6 36 = 5x + 3$
101. $e^{3x} = 10$
102. $e^{-2x} = \frac{1}{3}$
103. $e^{2x+5} = 8$
104. $e^{-2x+1} = 13$
105. $\log_3(x^2 + 1) = 2$
106. $\log_5(x^2 + x + 4) = 2$
107. $\log_2 8^x = -3$
108. $\log_3 3^x = -1$
109. $5e^{0.2x} = 7$
110. $8 \cdot 10^{2x-7} = 3$
111. $2 \cdot 10^{2-x} = 5$
112. $4e^{x+1} = 5$

Mixed Practice

113. Suppose that $G(x) = \log_3(2x + 1) - 2$.
(a) What is the domain of G?
(b) What is $G(40)$? What point is on the graph of G?
(c) If $G(x) = 3$, what is x? What point is on the graph of G?
(d) What is the zero of G?

114. Suppose that $F(x) = \log_2(x + 1) - 3$.
(a) What is the domain of F?
(b) What is $F(7)$? What point is on the graph of F?
(c) If $F(x) = -1$, what is x? What point is on the graph of F?
(d) What is the zero of F?

In Problems 115–118, graph each function. Based on the graph, state the domain and the range, and find any intercepts.

115. $f(x) = \begin{cases} \ln(-x) & \text{if } x < 0 \\ \ln x & \text{if } x > 0 \end{cases}$

116. $f(x) = \begin{cases} \ln(-x) & \text{if } x \le -1 \\ -\ln(-x) & \text{if } -1 < x < 0 \end{cases}$

117. $f(x) = \begin{cases} -\ln x & \text{if } 0 < x < 1 \\ \ln x & \text{if } x \ge 1 \end{cases}$

118. $f(x) = \begin{cases} \ln x & \text{if } 0 < x < 1 \\ -\ln x & \text{if } x \ge 1 \end{cases}$

Applications and Extensions

119. Chemistry The pH of a chemical solution is given by the formula

$$\text{pH} = -\log_{10}[H^+]$$

where $[H^+]$ is the concentration of hydrogen ions in moles per liter. Values of pH range from 0 (acidic) to 14 (alkaline).
(a) What is the pH of a solution for which $[H^+]$ is 0.1?
(b) What is the pH of a solution for which $[H^+]$ is 0.01?
(c) What is the pH of a solution for which $[H^+]$ is 0.001?
(d) What happens to pH as the hydrogen ion concentration decreases?
(e) Determine the hydrogen ion concentration of an orange ($\text{pH} = 3.5$).
(f) Determine the hydrogen ion concentration of human blood ($\text{pH} = 7.4$).

120. Diversity Index Shannon's diversity index is a measure of the diversity of a population. The diversity index is given by the formula

$$H = -(p_1 \log p_1 + p_2 \log p_2 + \cdots + p_n \log p_n)$$

where p_1 is the proportion of the population that is species 1, p_2 is the proportion of the population that is species 2, and so on. In this problem, the population is people in the United States and the species is race.
(a) According to the U.S. Census Bureau, the distribution of race in the United States in 2015 was as follows:

Race	Proportion
White	0.617
Black or African American	0.124
American Indian and Alaskan Native	0.007
Asian	0.053
Native Hawaiian and Other Pacific Islander	0.002
Hispanic	0.177
Two or More Races	0.020

Source: *U.S. Census Bureau*

Compute the diversity index of the United States in 2015.
(b) The largest value of the diversity index is given by $H_{max} = \log(S)$, where S is the number of categories of race. Compute H_{max}.
(c) The **evenness ratio** is given by $E_H = \frac{H}{H_{max}}$, where $0 \le E_H \le 1$. If $E_H = 1$, there is complete evenness. Compute the evenness ratio for the United States.
(d) Obtain the distribution of race for the United States in 2010 from the Census Bureau. Compute Shannon's diversity index. Is the United States becoming more diverse? Why?

121. Atmospheric Pressure The atmospheric pressure p on an object decreases with increasing height. This pressure, measured in millimeters of mercury, is related to the height h (in kilometers) above sea level by the function

$$p(h) = 760e^{-0.145h}$$

(a) Find the height of an aircraft if the atmospheric pressure is 320 millimeters of mercury.
(b) Find the height of a mountain if the atmospheric pressure is 667 millimeters of mercury.

122. Healing of Wounds The normal healing of wounds can be modeled by an exponential function. If A_0 represents the original area of the wound, and if A equals the area of the wound, then the function

$$A(n) = A_0 e^{-0.35n}$$

describes the area of a wound after n days following an injury when no infection is present to retard the healing. Suppose that a wound initially had an area of 100 square millimeters.
(a) If healing is taking place, after how many days will the wound be one-half its original size?
(b) How long before the wound is 10% of its original size?

123. Exponential Probability Between 12:00 PM and 1:00 PM, cars arrive at Citibank's drive-thru at the rate of 6 cars per hour (0.1 car per minute). The following formula from statistics can be used to determine the probability that a car will arrive within t minutes of 12:00 PM.

$$F(t) = 1 - e^{-0.1t}$$

(a) Determine how many minutes are needed for the probability to reach 50%.
(b) Determine how many minutes are needed for the probability to reach 80%.
(c) Is it possible for the probability to equal 100%? Explain.

124. Exponential Probability Between 5:00 PM and 6:00 PM, cars arrive at Jiffy Lube at the rate of 9 cars per hour (0.15 car per minute). The following formula from statistics can be used to determine the probability that a car will arrive within t minutes of 5:00 PM.

$$F(t) = 1 - e^{-0.15t}$$

(a) Determine how many minutes are needed for the probability to reach 50%.
(b) Determine how many minutes are needed for the probability to reach 80%.

125. Drug Medication The formula

$$D = 5e^{-0.4h}$$

can be used to find the number of milligrams D of a certain drug that is in a patient's bloodstream h hours after the drug was administered. When the number of milligrams reaches 2, the drug is to be administered again. What is the time between injections?

126. Spreading of Rumors A model for the number N of people in a college community who have heard a certain rumor is

$$N = P(1 - e^{-0.15d})$$

where P is the total population of the community and d is the number of days that have elapsed since the rumor began. In a community of 1000 students, how many days will elapse before 450 students have heard the rumor?

127. Current in an RL Circuit The equation governing the amount of current I (in amperes) after time t (in seconds) in a simple RL circuit consisting of a resistance R (in ohms), an inductance L (in henrys), and an electromotive force E (in volts) is

$$I = \frac{E}{R}[1 - e^{-(R/L)t}]$$

If $E = 12$ volts, $R = 10$ ohms, and $L = 5$ henrys, how long does it take to obtain a current of 0.5 ampere? Of 1.0 ampere? Graph the equation.

128. Learning Curve Psychologists sometimes use the function

$$L(t) = A(1 - e^{-kt})$$

to measure the amount L learned at time t. Here A represents the amount to be learned, and the number k measures the rate of learning. Suppose that a student has an amount A of 200 vocabulary words to learn. A psychologist determines that the student has learned 20 vocabulary words after 5 minutes.
(a) Determine the rate of learning k.
(b) Approximately how many words will the student have learned after 10 minutes?
(c) After 15 minutes?
(d) How long does it take for the student to learn 180 words?

Loudness of Sound *Problems 129–132 use the following discussion: The* ***loudness*** $L(x)$*, measured in decibels (dB), of a sound of intensity x, measured in watts per square meter, is defined as* $L(x) = 10\log\frac{x}{I_0}$*, where* $I_0 = 10^{-12}$ *watt per square meter is the least intense sound that a human ear can detect. Determine the loudness, in decibels, of each of the following sounds.*

129. Normal conversation: intensity of $x = 10^{-7}$ watt per square meter.

130. Amplified rock music: intensity of 10^{-1} watt per square meter.

131. Heavy city traffic: intensity of $x = 10^{-3}$ watt per square meter.

132. Diesel truck traveling 40 miles per hour 50 feet away: intensity 10 times that of a passenger car traveling 50 miles per hour 50 feet away, whose loudness is 70 decibels.

The Richter Scale *Problems 133 and 134 on the next page use the following discussion: The* ***Richter scale*** *is one way of converting seismographic readings into numbers that provide an easy reference for measuring the magnitude M of an earthquake. All earthquakes are compared to a* ***zero-level earthquake*** *whose seismographic reading measures 0.001 millimeter at a distance of 100 kilometers from the epicenter. An earthquake whose seismographic reading measures x millimeters has* ***magnitude*** $M(x)$*, given by*

$$M(x) = \log\left(\frac{x}{x_0}\right)$$

where $x_0 = 10^{-3}$ *is the reading of a zero-level earthquake the same distance from its epicenter. In Problems 133 and 134, determine the magnitude of each earthquake.*

133. Magnitude of an Earthquake Mexico City in 1985: seismographic reading of 125,892 millimeters 100 kilometers from the center

134. Magnitude of an Earthquake San Francisco in 1906: seismographic reading of 50,119 millimeters 100 kilometers from the center

135. Alcohol and Driving The concentration of alcohol in a person's bloodstream is measurable. Suppose that the relative risk R of having an accident while driving a car can be modeled by an equation of the form

$$R = e^{kx}$$

where x is the percent concentration of alcohol in the bloodstream and k is a constant.

(a) Suppose that a concentration of alcohol in the bloodstream of 0.03 percent results in a relative risk of an accident of 1.4. Find the constant k in the equation.

(b) Using this value of k, what is the relative risk if the concentration is 0.17 percent?

(c) Using the same value of k, what concentration of alcohol corresponds to a relative risk of 100?

(d) If the law asserts that anyone with a relative risk of having an accident of 5 or more should not have driving privileges, at what concentration of alcohol in the bloodstream should a driver be arrested and charged with a DUI?

(e) Compare this situation with that of Example 10. If you were a lawmaker, which situation would you support? Give your reasons.

Explaining Concepts: Discussion and Writing

136. Is there any function of the form $y = x^{\alpha}$, $0 < \alpha < 1$, that increases more slowly than a logarithmic function whose base is greater than 1? Explain.

137. In the definition of the logarithmic function, the base a is not allowed to equal 1. Why?

138. **Critical Thinking** In buying a new car, one consideration might be how well the price of the car holds up over time. Different makes of cars have different depreciation rates. One way to compute a depreciation rate for a car is given here. Suppose that the current prices of a certain automobile are as shown in the table.

Age in Years					
New	1	2	3	4	5
\$38,000	\$36,600	\$32,400	\$28,750	\$25,400	\$21,200

Use the formula $\text{New} = \text{Old}(e^{Rt})$ to find R, the annual depreciation rate, for a specific time t. When might be the best time to trade in the car? Consult the NADA ("blue") book and compare two like models that you are interested in. Which has the better depreciation rate?

Retain Your Knowledge

Problems 139–142 are based on material learned earlier in the course. The purpose of these problems is to keep the material fresh in your mind so that you are better prepared for the final exam.

139. Find the real zeros of $g(x) = 4x^4 - 37x^2 + 9$. What are the x-intercepts of the graph of g?

140. Find the average rate of change of $f(x) = 9^x$ from $\frac{1}{2}$ to 1.

141. Use the Intermediate Value Theorem to show that the function $f(x) = 4x^3 - 2x^2 - 7$ has a real zero in the interval $[1, 2]$.

142. A complex polynomial function f of degree 4 with real coefficients has the zeros $-1, 2$, and $3 - i$. Find the remaining zero(s) of f. Then find a polynomial function that has the zeros.

'Are You Prepared?' Answers

1. (a) $x \le 3$ (b) $x < -2$ or $x > 3$ **2.** $x < -4$ or $x > 1$ **3.** $\{3\}$

6.5 Properties of Logarithms

OBJECTIVES
1 Work with the Properties of Logarithms (p. 458)
2 Write a Logarithmic Expression as a Sum or Difference of Logarithms (p. 460)
3 Write a Logarithmic Expression as a Single Logarithm (p. 461)
4 Evaluate a Logarithm Whose Base Is Neither 10 Nor e (p. 462)
5 Graph a Logarithmic Function Whose Base Is Neither 10 Nor e (p. 464)

1 Work with the Properties of Logarithms

Logarithms have some very useful properties that can be derived directly from the definition and the laws of exponents.

EXAMPLE 1

Establishing Properties of Logarithms

(a) Show that $\log_a 1 = 0$. (b) Show that $\log_a a = 1$.

Solution (a) This fact was established when we graphed $y = \log_a x$ (see Figure 36 on page 448). To show the result algebraically, let $y = \log_a 1$. Then

$$\begin{aligned} y &= \log_a 1 \\ a^y &= 1 && \text{Change to exponential form.} \\ a^y &= a^0 && a^0 = 1 \text{ since } a > 0, a \neq 1 \\ y &= 0 && \text{Solve for } y. \\ \log_a 1 &= 0 && y = \log_a 1 \end{aligned}$$

(b) Let $y = \log_a a$. Then

$$\begin{aligned} y &= \log_a a \\ a^y &= a && \text{Change to exponential form.} \\ a^y &= a^1 && a = a^1 \\ y &= 1 && \text{Solve for } y. \\ \log_a a &= 1 && y = \log_a a \end{aligned}$$

To summarize:

$$\log_a 1 = 0 \qquad \log_a a = 1$$

THEOREM

Properties of Logarithms

In the properties given next, M and a are positive real numbers, $a \neq 1$, and r is any real number.

The number $\log_a M$ is the exponent to which a must be raised to obtain M. That is,

$$a^{\log_a M} = M \qquad \textbf{(1)}$$

The logarithm with base a of a raised to a power equals that power. That is,

$$\log_a a^r = r \qquad \textbf{(2)}$$

The proof uses the fact that $y = a^x$ and $y = \log_a x$ are inverses.

Proof of Property (1) For inverse functions,

$$f(f^{-1}(x)) = x \quad \text{for all } x \text{ in the domain of } f^{-1}$$

Use $f(x) = a^x$ and $f^{-1}(x) = \log_a x$ to find

$$f(f^{-1}(x)) = a^{\log_a x} = x \quad \text{for } x > 0$$

Now let $x = M$ to obtain $a^{\log_a M} = M$, where $M > 0$. ■

Proof of Property (2) For inverse functions,

$$f^{-1}(f(x)) = x \quad \text{for all } x \text{ in the domain of } f$$

Use $f(x) = a^x$ and $f^{-1}(x) = \log_a x$ to find

$$f^{-1}(f(x)) = \log_a a^x = x \quad \text{for all real numbers } x$$

Now let $x = r$ to obtain $\log_a a^r = r$, where r is any real number. ■

EXAMPLE 2

Using Properties (1) and (2)

(a) $2^{\log_2 \pi} = \pi$ (b) $\log_{0.2} 0.2^{-\sqrt{2}} = -\sqrt{2}$ (c) $\ln e^{kt} = kt$

Now Work PROBLEM 15

Other useful properties of logarithms are given next.

THEOREM

Properties of Logarithms

In the following properties, M, N, and a are positive real numbers, $a \neq 1$, and r is any real number.

The Log of a Product Equals the Sum of the Logs

$$\log_a(MN) = \log_a M + \log_a N \tag{3}$$

The Log of a Quotient Equals the Difference of the Logs

$$\log_a\left(\frac{M}{N}\right) = \log_a M - \log_a N \tag{4}$$

The Log of a Power Equals the Product of the Power and the Log

$$\log_a M^r = r \log_a M \tag{5}$$

$$a^r = e^{r \ln a} \tag{6}$$

We shall derive properties (3), (5), and (6) and leave the derivation of property (4) as an exercise (see Problem 109).

Proof of Property (3) Let $A = \log_a M$ and let $B = \log_a N$. These expressions are equivalent to the exponential expressions

$$a^A = M \quad \text{and} \quad a^B = N$$

Now

$$\begin{aligned} \log_a(MN) = \log_a(a^A a^B) &= \log_a a^{A+B} && \text{Law of Exponents} \\ &= A + B && \text{Property (2) of logarithms} \\ &= \log_a M + \log_a N \end{aligned}$$

■

Proof of Property (5) Let $A = \log_a M$. This expression is equivalent to

$$a^A = M$$

Now

$$\begin{aligned} \log_a M^r = \log_a(a^A)^r &= \log_a a^{rA} && \text{Law of Exponents} \\ &= rA && \text{Property (2) of logarithms} \\ &= r \log_a M \end{aligned}$$

■

Proof of Property (6) Property (1), with $a = e$, gives

$$e^{\ln M} = M$$

Now let $M = a^r$ and apply property (5).

$$e^{\ln a^r} = e^{r \ln a} = (e^{\ln a})^r = a^r$$

■

Now Work PROBLEM 19

2 Write a Logarithmic Expression as a Sum or Difference of Logarithms

Logarithms can be used to transform products into sums, quotients into differences, and powers into factors. Such transformations prove useful in certain types of calculus problems.

EXAMPLE 3 **Writing a Logarithmic Expression as a Sum of Logarithms**

Write $\log_a(x\sqrt{x^2+1}), x > 0$, as a sum of logarithms. Express all powers as factors.

Solution

$$\log_a(x\sqrt{x^2+1}) = \log_a x + \log_a\sqrt{x^2+1} \qquad \log_a(M\cdot N) = \log_a M + \log_a N$$
$$= \log_a x + \log_a(x^2+1)^{1/2}$$
$$= \log_a x + \frac{1}{2}\log_a(x^2+1) \qquad \log_a M^r = r\log_a M$$

EXAMPLE 4 Writing a Logarithmic Expression as a Difference of Logarithms

Write

$$\ln\frac{x^2}{(x-1)^3} \qquad x>1$$

as a difference of logarithms. Express all powers as factors.

Solution

$$\ln\frac{x^2}{(x-1)^3} = \ln x^2 - \ln(x-1)^3 = 2\ln x - 3\ln(x-1)$$

$\log_a\left(\frac{M}{N}\right) = \log_a M - \log_a N$ $\quad$ $\log_a M^r = r\log_a M$

EXAMPLE 5 Writing a Logarithmic Expression as a Sum and Difference of Logarithms

Write

$$\log_a\frac{\sqrt{x^2+1}}{x^3(x+1)^4} \qquad x>0$$

as a sum and difference of logarithms. Express all powers as factors.

Solution

$$\log_a\frac{\sqrt{x^2+1}}{x^3(x+1)^4} = \log_a\sqrt{x^2+1} - \log_a[x^3(x+1)^4] \qquad \text{Property (4)}$$
$$= \log_a\sqrt{x^2+1} - [\log_a x^3 + \log_a(x+1)^4] \qquad \text{Property (3)}$$
$$= \log_a(x^2+1)^{1/2} - \log_a x^3 - \log_a(x+1)^4$$
$$= \frac{1}{2}\log_a(x^2+1) - 3\log_a x - 4\log_a(x+1) \qquad \text{Property (5)}$$

WARNING In using properties (3) through (5), be careful about the values that the variable may assume. For example, the domain of the variable for $\log_a x$ is $x>0$ and for $\log_a(x-1)$ is $x>1$. If these functions are added, the domain is $x>1$. That is, the equality $\log_a x + \log_a(x-1) = \log_a[x(x-1)]$ is true only for $x>1$.

Now Work PROBLEM 51

3 Write a Logarithmic Expression as a Single Logarithm

Another use of properties (3) through (5) is to write sums and/or differences of logarithms with the same base as a single logarithm. This skill will be needed to solve certain logarithmic equations discussed in the next section.

EXAMPLE 6 Writing Expressions as a Single Logarithm

Write each of the following as a single logarithm.

(a) $\log_a 7 + 4\log_a 3$ $\qquad$ (b) $\frac{2}{3}\ln 8 - \ln(5^2-1)$

(c) $\log_a x + \log_a 9 + \log_a(x^2+1) - \log_a 5$

Solution

(a)
$$\log_a 7 + 4\log_a 3 = \log_a 7 + \log_a 3^4 \qquad r\log_a M = \log_a M^r$$
$$= \log_a 7 + \log_a 81$$
$$= \log_a(7\cdot 81) \qquad \log_a M + \log_a N = \log_a(M\cdot N)$$
$$= \log_a 567$$

(b) $\frac{2}{3}\ln 8 - \ln(5^2 - 1) = \ln 8^{2/3} - \ln(25 - 1)$ $\quad r\log_a M = \log_a M^r$

$= \ln 4 - \ln 24$ $\quad 8^{2/3} = (\sqrt[3]{8})^2 = 2^2 = 4$

$= \ln\left(\frac{4}{24}\right)$ $\quad \log_a M - \log_a N = \log_a\left(\frac{M}{N}\right)$

$= \ln\left(\frac{1}{6}\right)$

$= \ln 1 - \ln 6$

$= -\ln 6$ $\quad \ln 1 = 0$

(c) $\log_a x + \log_a 9 + \log_a(x^2 + 1) - \log_a 5 = \log_a(9x) + \log_a(x^2 + 1) - \log_a 5$

$= \log_a[9x(x^2 + 1)] - \log_a 5$

$= \log_a\left[\frac{9x(x^2 + 1)}{5}\right]$ ■

WARNING A common error made by some students is to express the logarithm of a sum as the sum of logarithms.

$\log_a(M + N)$ is not equal to $\log_a M + \log_a N$

Correct statement $\log_a(MN) = \log_a M + \log_a N$ **Property (3)**

Another common error is to express the difference of logarithms as the quotient of logarithms.

$\log_a M - \log_a N$ is not equal to $\frac{\log_a M}{\log_a N}$

Correct statement $\log_a M - \log_a N = \log_a\left(\frac{M}{N}\right)$ **Property (4)**

A third common error is to express a logarithm raised to a power as the product of the power times the logarithm.

$(\log_a M)^r$ is not equal to $r \log_a M$

Correct statement $\log_a M^r = r \log_a M$ **Property (5)** ■

Now Work PROBLEMS 57 AND 63

Two other important properties of logarithms are consequences of the fact that the logarithmic function $y = \log_a x$ is a one-to-one function.

THEOREM

Properties of Logarithms

In the following properties, M, N, and a are positive real numbers, $a \neq 1$.

If $M = N$, then $\log_a M = \log_a N$. **(7)**

If $\log_a M = \log_a N$, then $M = N$. **(8)**

■

Property (7) is used as follows: Starting with the equation $M = N$, "take the logarithm of both sides" to obtain $\log_a M = \log_a N$.

Properties (7) and (8) are useful for solving *exponential and logarithmic equations*, a topic discussed in the next section.

4 Evaluate a Logarithm Whose Base Is Neither 10 Nor *e*

Logarithms with base 10—common logarithms—were used to facilitate arithmetic computations before the widespread use of calculators. (See the Historical Feature at the end of this section.) Natural logarithms—that is, logarithms whose base is the number *e*—remain very important because they arise frequently in the study of natural phenomena.

Common logarithms are usually abbreviated by writing **log**, with the base understood to be 10, just as natural logarithms are abbreviated by **ln**, with the base understood to be *e*.

Most calculators have both [log] and [ln] keys to calculate the common logarithm and the natural logarithm of a number, respectively. Let's look at an example to see how to approximate logarithms having a base other than 10 or *e*.

EXAMPLE 7 Approximating a Logarithm Whose Base Is Neither 10 Nor e

Approximate $\log_2 7$. Round the answer to four decimal places.

Solution Remember, $\log_2 7$ means "2 raised to what exponent equals 7." Let $y = \log_2 7$. Then $2^y = 7$. Because $2^2 = 4$ and $2^3 = 8$, the value of $\log_2 7$ is between 2 and 3.

$$\begin{aligned} 2^y &= 7 \\ \ln 2^y &= \ln 7 && \text{Property (7)} \\ y \ln 2 &= \ln 7 && \text{Property (5)} \\ y &= \frac{\ln 7}{\ln 2} && \text{Exact value} \\ y &\approx 2.8074 && \text{Approximate value rounded to four decimal places} \end{aligned}$$

Example 7 shows how to approximate a logarithm whose base is 2 by changing to logarithms involving the base e. In general, the **Change-of-Base Formula** is used.

THEOREM

Change-of-Base Formula

If $a \neq 1$, $b \neq 1$, and M are positive real numbers, then

$$\log_a M = \frac{\log_b M}{\log_b a} \tag{9}$$

Proof Let $y = \log_a M$. Then

$$\begin{aligned} a^y &= M \\ \log_b a^y &= \log_b M && \text{Property (7)} \\ y \log_b a &= \log_b M && \text{Property (5)} \\ y &= \frac{\log_b M}{\log_b a} && \text{Solve for } y. \\ \log_a M &= \frac{\log_b M}{\log_b a} && y = \log_a M \end{aligned}$$

Technology Note

Some calculators have features for evaluating logarithms with bases other than 10 or e. For example, the TI-84 Plus C has the logBASE function (under Math > Math > A: logBASE). Consult the user's manual for your calculator.

Because most calculators have keys only for [log] and [ln], in practice, the Change-of-Base Formula uses either $b = 10$ or $b = e$. That is,

$$\log_a M = \frac{\log M}{\log a} \quad \text{and} \quad \log_a M = \frac{\ln M}{\ln a} \tag{10}$$

EXAMPLE 8 Using the Change-of-Base Formula

Approximate:

(a) $\log_5 89$ (b) $\log_{\sqrt{2}} \sqrt{5}$

Round answers to four decimal places.

Solution (a) $\log_5 89 = \dfrac{\log 89}{\log 5} \approx \dfrac{1.949390007}{0.6989700043} \approx 2.7889$ or $\log_5 89 = \dfrac{\ln 89}{\ln 5} \approx \dfrac{4.48863637}{1.609437912} \approx 2.7889$

(b) $\log_{\sqrt{2}}\sqrt{5} = \dfrac{\log\sqrt{5}}{\log\sqrt{2}} = \dfrac{\frac{1}{2}\log 5}{\frac{1}{2}\log 2} = \dfrac{\log 5}{\log 2} \approx 2.3219$ or $\log_{\sqrt{2}}\sqrt{5} = \dfrac{\ln\sqrt{5}}{\ln\sqrt{2}} = \dfrac{\frac{1}{2}\ln 5}{\frac{1}{2}\ln 2} = \dfrac{\ln 5}{\ln 2} \approx 2.3219$ ■

Now Work PROBLEMS 23 AND 71

5 Graph a Logarithmic Function Whose Base Is Neither 10 Nor *e*

The Change-of-Base Formula also can be used to graph logarithmic functions whose bases are neither 10 nor *e*.

EXAMPLE 9 **Graphing a Logarithmic Function Whose Base Is Neither 10 Nor *e***

Use a graphing utility to graph $y = \log_2 x$.

Solution Let's use the Change-of-Base Formula to express $y = \log_2 x$ in terms of logarithms with base 10 or base *e*. Graph either $y = \dfrac{\ln x}{\ln 2}$ or $y = \dfrac{\log x}{\log 2}$ to obtain the graph of $y = \log_2 x$. See Figure 46.

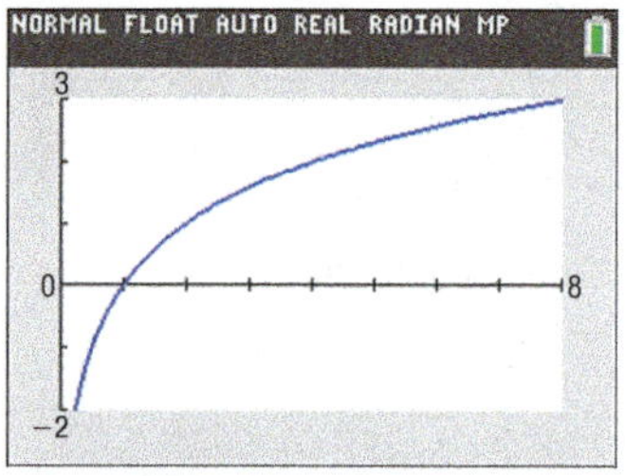

Figure 46 $y = \log_2 x$

✓**Check:** Verify that $y = \dfrac{\ln x}{\ln 2}$ and $y = \dfrac{\log x}{\log 2}$ result in the same graph by graphing each on the same screen. ■

Now Work PROBLEM 79

SUMMARY

Properties of Logarithms

In the list that follows, a, b, M, N, and r are real numbers. Also, $a > 0, a \neq 1, b > 0, b \neq 1, M > 0$, and $N > 0$.

Definition	$y = \log_a x$ means $x = a^y$	
Properties of logarithms	$\log_a 1 = 0;\ \log_a a = 1$	$\log_a M^r = r\log_a M$
	$a^{\log_a M} = M;\ \log_a a^r = r$	$a^r = e^{r\ln a}$
	$\log_a(MN) = \log_a M + \log_a N$	If $M = N$, then $\log_a M = \log_a N$.
	$\log_a\left(\dfrac{M}{N}\right) = \log_a M - \log_a N$	If $\log_a M = \log_a N$, then $M = N$.
Change-of-Base Formula	$\log_a M = \dfrac{\log_b M}{\log_b a}$	

Historical Feature

John Napier (1550–1617)

Logarithms were invented about 1590 by John Napier (1550–1617) and Joost Bürgi (1552–1632), working independently. Napier, whose work had the greater influence, was a Scottish lord, a secretive man whose neighbors were inclined to believe him to be in league with the devil. His approach to logarithms was very different from ours; it was based on the relationship between arithmetic and geometric sequences—discussed in a later chapter—and not on the inverse function relationship of logarithms to exponential functions (described in Section 6.4). Napier's tables, published in 1614, listed what would now be called *natural logarithms* of sines and were rather difficult to use. A London professor, Henry Briggs, became interested in the tables and visited Napier. In their conversations, they developed the idea of common logarithms, which were published in 1617. Their importance for calculation was immediately recognized, and by 1650 they were being printed as far away as China. They remained an important calculation tool until the advent of the inexpensive handheld calculator about 1972, which has decreased their calculation—but not their theoretical—importance.

A side effect of the invention of logarithms was the popularization of the decimal system of notation for real numbers.

6.5 Assess Your Understanding

Concepts and Vocabulary

1. $\log_a 1 =$ ____
2. $a^{\log_a M} =$ ____
3. $\log_a a^r =$ ____
4. $\log_a (MN) =$ ____ + ____
5. $\log_a\left(\frac{M}{N}\right) =$ ____ − ____
6. $\log_a M^r =$ ____
7. If $\log_8 M = \frac{\log_5 7}{\log_5 8}$, then $M =$ ____.
8. ***True or False*** $\ln(x+3) - \ln(2x) = \frac{\ln(x+3)}{\ln(2x)}$
9. ***True or False*** $\log_2(3x^4) = 4\log_2(3x)$
10. ***True or False*** $\log\left(\frac{2}{3}\right) = \frac{\log 2}{\log 3}$
11. Choose the expression equivalent to 2^x.
    (a) e^{2x} (b) $e^{x\ln 2}$ (c) $e^{\log_2 x}$ (d) $e^{2\ln x}$
12. Writing $\log_a x - \log_a y + 2\log_a z$ as a single logarithm results in which of the following?
    (a) $\log_a(x - y + 2z)$ (b) $\log_a\left(\frac{xz^2}{y}\right)$
    (c) $\log_a\left(\frac{2xz}{y}\right)$ (d) $\log_a\left(\frac{x}{yz^2}\right)$

Skill Building

In Problems 13–28, use properties of logarithms to find the exact value of each expression. Do not use a calculator.

13. $\log_3 3^{71}$
14. $\log_2 2^{-13}$
15. $\ln e^{-4}$
16. $\ln e^{\sqrt{2}}$
17. $2^{\log_2 7}$
18. $e^{\ln 8}$
19. $\log_8 2 + \log_8 4$
20. $\log_6 9 + \log_6 4$
21. $\log_6 18 - \log_6 3$
22. $\log_8 16 - \log_8 2$
23. $\log_2 6 \cdot \log_6 8$
24. $\log_3 8 \cdot \log_8 9$
25. $3^{\log_3 5 - \log_3 4}$
26. $5^{\log_5 6 + \log_5 7}$
27. $e^{\log_{e^2} 16}$
28. $e^{\log_{e^2} 9}$

In Problems 29–36, suppose that $\ln 2 = a$ *and* $\ln 3 = b$. *Use properties of logarithms to write each logarithm in terms of a and b.*

29. $\ln 6$
30. $\ln \frac{2}{3}$
31. $\ln 1.5$
32. $\ln 0.5$
33. $\ln 8$
34. $\ln 27$
35. $\ln \sqrt[5]{6}$
36. $\ln \sqrt[4]{\frac{2}{3}}$

In Problems 37–56, write each expression as a sum and/or difference of logarithms. Express powers as factors.

37. $\log_5(25x)$
38. $\log_3 \frac{x}{9}$
39. $\log_2 z^3$
40. $\log_7 x^5$
41. $\ln(ex)$
42. $\ln \frac{e}{x}$
43. $\ln \frac{x}{e^x}$
44. $\ln(xe^x)$
45. $\log_a(u^2v^3) \quad u > 0, v > 0$
46. $\log_2\left(\frac{a}{b^2}\right) \quad a > 0, b > 0$
47. $\ln(x^2\sqrt{1-x}) \quad 0 < x < 1$
48. $\ln(x\sqrt{1+x^2}) \quad x > 0$
49. $\log_2\left(\frac{x^3}{x-3}\right) \quad x > 3$
50. $\log_5\left(\frac{\sqrt[3]{x^2+1}}{x^2-1}\right) \quad x > 1$
51. $\log\left[\frac{x(x+2)}{(x+3)^2}\right] \quad x > 0$
52. $\log\left[\frac{x^3\sqrt{x+1}}{(x-2)^2}\right] \quad x > 2$
53. $\ln\left[\frac{x^2-x-2}{(x+4)^2}\right]^{1/3} \quad x > 2$
54. $\ln\left[\frac{(x-4)^2}{x^2-1}\right]^{2/3} \quad x > 4$
55. $\ln \frac{5x\sqrt{1+3x}}{(x-4)^3} \quad x > 4$
56. $\ln\left[\frac{5x^2\sqrt[3]{1-x}}{4(x+1)^2}\right] \quad 0 < x < 1$

In Problems 57–70, write each expression as a single logarithm.

57. $3\log_5 u + 4\log_5 v$
58. $2\log_3 u - \log_3 v$
59. $\log_3 \sqrt{x} - \log_3 x^3$
60. $\log_2\left(\frac{1}{x}\right) + \log_2\left(\frac{1}{x^2}\right)$
61. $\log_4(x^2-1) - 5\log_4(x+1)$
62. $\log(x^2+3x+2) - 2\log(x+1)$
63. $\ln\left(\frac{x}{x-1}\right) + \ln\left(\frac{x+1}{x}\right) - \ln(x^2-1)$
64. $\log\left(\frac{x^2+2x-3}{x^2-4}\right) - \log\left(\frac{x^2+7x+6}{x+2}\right)$
65. $8\log_2\sqrt{3x-2} - \log_2\left(\frac{4}{x}\right) + \log_2 4$
66. $21\log_3\sqrt[3]{x} + \log_3(9x^2) - \log_3 9$
67. $2\log_a(5x^3) - \frac{1}{2}\log_a(2x+3)$
68. $\frac{1}{3}\log(x^3+1) + \frac{1}{2}\log(x^2+1)$
69. $2\log_2(x+1) - \log_2(x+3) - \log_2(x-1)$
70. $3\log_5(3x+1) - 2\log_5(2x-1) - \log_5 x$

In Problems 71–78, use the Change-of-Base Formula and a calculator to evaluate each logarithm. Round your answer to three decimal places.

71. $\log_3 21$ **72.** $\log_5 18$ **73.** $\log_{1/3} 71$ **74.** $\log_{1/2} 15$

75. $\log_{\sqrt{2}} 7$ **76.** $\log_{\sqrt{5}} 8$ **77.** $\log_\pi e$ **78.** $\log_\pi \sqrt{2}$

In Problems 79–84, graph each function using a graphing utility and the Change-of-Base Formula.

79. $y = \log_4 x$ **80.** $y = \log_5 x$ **81.** $y = \log_2(x + 2)$

82. $y = \log_4(x - 3)$ **83.** $y = \log_{x-1}(x + 1)$ **84.** $y = \log_{x+2}(x - 2)$

Mixed Practice

85. If $f(x) = \ln x$, $g(x) = e^x$, and $h(x) = x^2$, find:
(a) $(f \circ g)(x)$. What is the domain of $f \circ g$?
(b) $(g \circ f)(x)$. What is the domain of $g \circ f$?
(c) $(f \circ g)(5)$
(d) $(f \circ h)(x)$. What is the domain of $f \circ h$?
(e) $(f \circ h)(e)$

86. If $f(x) = \log_2 x$, $g(x) = 2^x$, and $h(x) = 4x$, find:
(a) $(f \circ g)(x)$. What is the domain of $f \circ g$?
(b) $(g \circ f)(x)$. What is the domain of $g \circ f$?
(c) $(f \circ g)(3)$
(d) $(f \circ h)(x)$. What is the domain of $f \circ h$?
(e) $(f \circ h)(8)$

Applications and Extensions

In Problems 87–96, express y as a function of x. The constant C is a positive number.

87. $\ln y = \ln x + \ln C$

88. $\ln y = \ln(x + C)$

89. $\ln y = \ln x + \ln(x + 1) + \ln C$

90. $\ln y = 2\ln x - \ln(x + 1) + \ln C$

91. $\ln y = 3x + \ln C$

92. $\ln y = -2x + \ln C$

93. $\ln(y - 3) = -4x + \ln C$

94. $\ln(y + 4) = 5x + \ln C$

95. $3\ln y = \frac{1}{2}\ln(2x + 1) - \frac{1}{3}\ln(x + 4) + \ln C$

96. $2\ln y = -\frac{1}{2}\ln x + \frac{1}{3}\ln(x^2 + 1) + \ln C$

97. Find the value of $\log_2 3 \cdot \log_3 4 \cdot \log_4 5 \cdot \log_5 6 \cdot \log_6 7 \cdot \log_7 8$.

98. Find the value of $\log_2 4 \cdot \log_4 6 \cdot \log_6 8$.

99. Find the value of $\log_2 3 \cdot \log_3 4 \cdot \cdots \cdot \log_n(n + 1) \cdot \log_{n+1} 2$.

100. Find the value of $\log_2 2 \cdot \log_2 4 \cdot \cdots \cdot \log_2 2^n$.

101. Show that $\log_a(x + \sqrt{x^2 - 1}) + \log_a(x - \sqrt{x^2 - 1}) = 0$.

102. Show that $\log_a(\sqrt{x} + \sqrt{x - 1}) + \log_a(\sqrt{x} - \sqrt{x - 1}) = 0$.

103. Show that $\ln(1 + e^{2x}) = 2x + \ln(1 + e^{-2x})$.

104. Difference Quotient If $f(x) = \log_a x$, show that $\frac{f(x + h) - f(x)}{h} = \log_a\left(1 + \frac{h}{x}\right)^{1/h}$, $h \neq 0$.

105. If $f(x) = \log_a x$, show that $-f(x) = \log_{1/a} x$.

106. If $f(x) = \log_a x$, show that $f(AB) = f(A) + f(B)$.

107. If $f(x) = \log_a x$, show that $f\left(\frac{1}{x}\right) = -f(x)$.

108. If $f(x) = \log_a x$, show that $f(x^\alpha) = \alpha f(x)$.

109. Show that $\log_a\left(\frac{M}{N}\right) = \log_a M - \log_a N$, where a, M, and N are positive real numbers and $a \neq 1$.

110. Show that $\log_a\left(\frac{1}{N}\right) = -\log_a N$, where a and N are positive real numbers and $a \neq 1$.

Explaining Concepts: Discussion and Writing

111. Graph $Y_1 = \log(x^2)$ and $Y_2 = 2\log(x)$ using a graphing utility. Are they equivalent? What might account for any differences in the two functions?

112. Write an example that illustrates why $(\log_a x)^r \neq r\log_a x$.

113. Write an example that illustrates why

$$\log_2(x + y) \neq \log_2 x + \log_2 y.$$

114. Does $3^{\log_3(-5)} = -5$? Why or why not?

Retain Your Knowledge

Problems 115–118 are based on material learned earlier in the course. The purpose of these problems is to keep the material fresh in your mind so that you are better prepared for the final exam.

115. Use a graphing utility to solve $x^3 - 3x^2 - 4x + 8 = 0$. Round answers to two decimal places.

116. Without solving, determine the character of the solution of the quadratic equation $4x^2 - 28x + 49 = 0$ in the complex number system.

117. Find the real zeros of

$$f(x) = 5x^5 + 44x^4 + 116x^3 + 95x^2 - 4x - 4$$

118. Graph $f(x) = \sqrt{2 - x}$ using the techniques of shifting, compressing or stretching, and reflecting. State the domain and the range of f.

6.6 Logarithmic and Exponential Equations

PREPARING FOR THIS SECTION *Before getting started, review the following:*

- Solve Equations Using a Graphing Utility (Section 1.2, pp. 100–101)
- Solve Quadratic Equations (Section 1.3, pp. 110–115)
- Solve Equations Quadratic in Form (Section 1.5, pp. 131–133)

Now Work the 'Are You Prepared?' problems on page 472.

OBJECTIVES
1 Solve Logarithmic Equations (p. 467)
2 Solve Exponential Equations (p. 469)
3 Solve Logarithmic and Exponential Equations Using a Graphing Utility (p. 471)

1 Solve Logarithmic Equations

In Section 6.4 we solved logarithmic equations by changing a logarithmic equation to an exponential equation. That is, we used the definition of a logarithm:

$$y = \log_a x \quad \text{is equivalent to} \quad x = a^y \qquad a > 0, a \neq 1$$

For example, to solve the equation $\log_2(1 - 2x) = 3$, use the equivalent exponential equation $1 - 2x = 2^3$ and solve for x.

$$\log_2(1 - 2x) = 3$$

$$1 - 2x = 2^3 \qquad \text{Change to exponential form.}$$

$$-2x = 7 \qquad \text{Simplify}$$

$$x = -\frac{7}{2} \qquad \text{Divide both sides by } -2.$$

You should check this solution for yourself.

For most logarithmic equations, some manipulation of the equation (usually using properties of logarithms) is required to obtain a solution. Also, to avoid extraneous solutions with logarithmic equations, determine the domain of the variable first.

Our practice will be to solve equations, whenever possible, by finding exact solutions using algebraic methods and exact or approximate solutions using a graphing utility. When algebraic methods cannot be used, approximate solutions will be obtained using a graphing utility. The reader is encouraged to pay particular attention to the form of equations for which exact solutions are possible.

Let's begin with an example of a logarithmic equation that requires using the fact that a logarithmic function is a one-to-one function.

$$\text{If } \log_a M = \log_a N, \text{ then } M = N \qquad M, N, \text{ and } a \text{ are positive and } a \neq 1$$

EXAMPLE 1 **Solving a Logarithmic Equation**

Solve: $2\log_5 x = \log_5 9$

Algebraic Solution

The domain of the variable in this equation is $x > 0$. Note that each logarithm is to the same base, 5. Find the exact solution as follows:

$$2\log_5 x = \log_5 9$$

$$\log_5 x^2 = \log_5 9 \qquad \log_a M^r = r\log_a M$$

$$x^2 = 9 \qquad \text{If } \log_a M = \log_a N, \text{ then } M = N.$$

$$x = 3 \quad \text{or} \quad \cancel{x = -3}$$

The domain of the variable is $x > 0$. Therefore, -3 is extraneous and must be discarded.

Check:

$$2\log_5 3 \stackrel{?}{=} \log_5 9$$

$$\log_5 3^2 \stackrel{?}{=} \log_5 9 \qquad r\log_a M = \log_a M^r$$

$$\log_5 9 = \log_5 9$$

The solution set is $\{3\}$. ■

Graphing Solution

To solve the equation using a graphing utility, graph $Y_1 = 2\log_5 x = \dfrac{2\log x}{\log 5}$ and $Y_2 = \log_5 9 = \dfrac{\log 9}{\log 5}$, and determine the point of intersection. See Figure 47. The point of intersection is $(3, 1.3652124)$; so $x = 3$ is the only solution. The solution set is $\{3\}$.

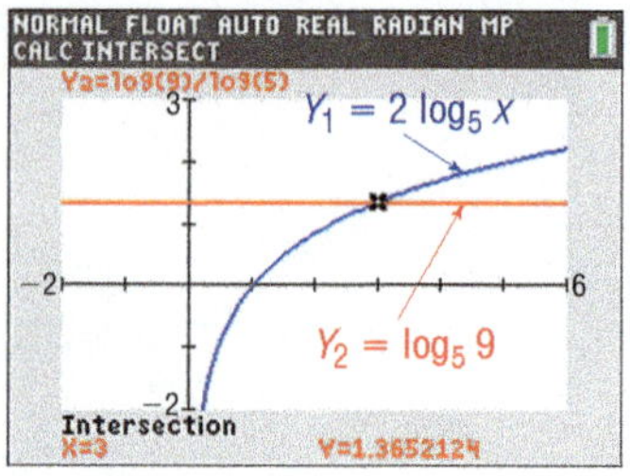

Figure 47 ■

Now Work PROBLEM 13

EXAMPLE 2 Solving a Logarithmic Equation

Solve: $\log_5(x + 6) + \log_5(x + 2) = 1$

Algebraic Solution

The domain of the variable requires that $x + 6 > 0$ and $x + 2 > 0$, so $x > -6$ and $x > -2$. This means any solution must satisfy $x > -2$. To obtain an exact solution, first express the left side as a single logarithm. Then change the equation to exponential form.

$$\log_5(x + 6) + \log_5(x + 2) = 1$$

$$\log_5[(x + 6)(x + 2)] = 1 \qquad \log_a M + \log_a N = \log_a(MN)$$

$$(x + 6)(x + 2) = 5^1 = 5 \qquad \text{Change to exponential form.}$$

$$x^2 + 8x + 12 = 5 \qquad \text{Multiply out.}$$

$$x^2 + 8x + 7 = 0 \qquad \text{Place the quadratic equation in standard form.}$$

$$(x + 7)(x + 1) = 0 \qquad \text{Factor.}$$

$$x = -7 \quad \text{or} \quad x = -1 \qquad \text{Zero-Product Property}$$

Only $x = -1$ satisfies the restriction that $x > -2$, so $x = -7$ is extraneous. The solution set is $\{-1\}$, which you should check. ■

Graphing Solution

Graph $Y_1 = \log_5(x + 6) + \log_5(x + 2) = \dfrac{\log(x + 6)}{\log 5} + \dfrac{\log(x + 2)}{\log 5}$ and $Y_2 = 1$, and determine the point(s) of intersection. See Figure 48. The point of intersection is $(-1, 1)$, so $x = -1$ is the only solution. The solution set is $\{-1\}$.

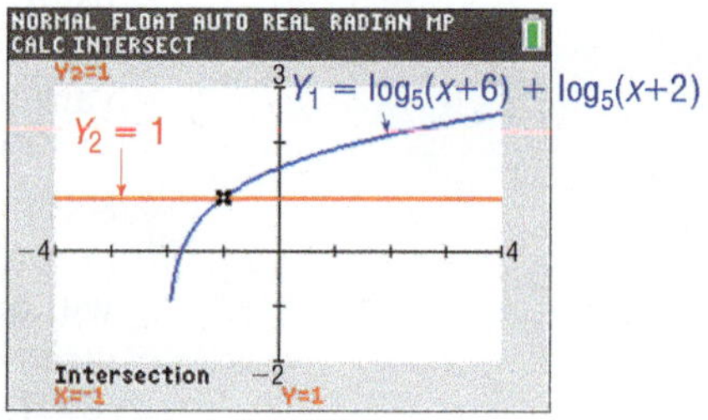

Figure 48 ■

WARNING A negative solution is not automatically extraneous. You must determine whether the potential solution causes the argument of any logarithmic expression in the equation to be negative or 0. ■

Now Work PROBLEM 21

EXAMPLE 3 Solving a Logarithmic Equation

Solve: $\ln x + \ln(x - 4) = \ln(x + 6)$

Algebraic Solution

The domain of the variable requires that $x > 0$, $x - 4 > 0$, and $x + 6 > 0$. As a result, the domain of the variable is $x > 4$. Begin the solution using the log of a product property.

$$\ln x + \ln(x - 4) = \ln(x + 6)$$

$$\ln[x(x - 4)] = \ln(x + 6) \qquad \ln M + \ln N = \ln(MN)$$

$$x(x - 4) = x + 6 \qquad \text{If } \ln M = \ln N, \text{ then } M = N.$$

$$x^2 - 4x = x + 6 \qquad \text{Multiply out.}$$

$$x^2 - 5x - 6 = 0 \qquad \text{Place the quadratic equation in standard form.}$$

$$(x - 6)(x + 1) = 0 \qquad \text{Factor.}$$

$$x = 6 \quad \text{or} \quad x = -1 \qquad \text{Zero-Product Property}$$

Because the domain of the variable is $x > 4$, discard -1 as extraneous. The solution set is $\{6\}$, which you should check. ■

Graphing Solution

Graph $Y_1 = \ln x + \ln(x - 4)$ and $Y_2 = \ln(x + 6)$, and determine the point(s) of intersection. See Figure 49. The x-coordinate of the point of intersection is 6, so the solution set is $\{6\}$.

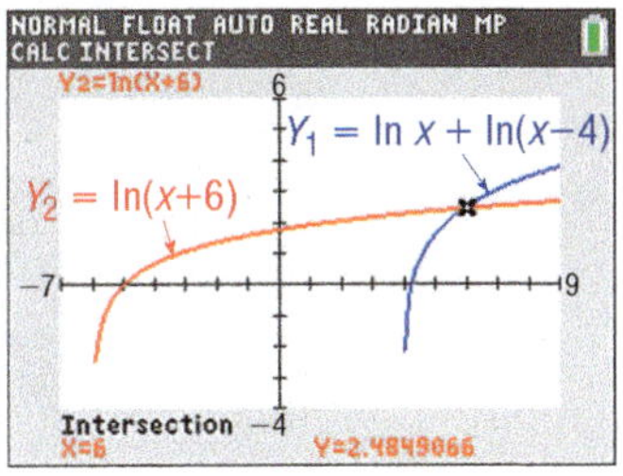

Figure 49 ■

WARNING In using properties of logarithms to solve logarithmic equations, avoid using the property $\log_a x^r = r \log_a x$, when r is even. The reason can be seen in this example:

Solve: $\log_3 x^2 = 4$

Solution: The domain of the variable x is all real numbers except 0.

(a) $\log_3 x^2 = 4$

$$x^2 = 3^4 = 81 \qquad \text{Change to exponential form.}$$

$$x = -9 \text{ or } x = 9$$

(b) $\log_3 x^2 = 4 \qquad \log_a x^r = r \log_a x$

$$2 \log_3 x = 4 \qquad \text{Domain of variable is } x > 0.$$

$$\log_3 x = 2$$

$$x = 9$$

Both -9 and 9 are solutions of $\log_3 x^2 = 4$ (as you can verify). The process in part (b) does not find the solution -9 because the domain of the variable was further restricted to $x > 0$ due to the application of the property $\log_a x^r = r \log_a x$. ■

Now Work PROBLEM 31

2 Solve Exponential Equations

In Sections 6.3 and 6.4, we solved exponential equations algebraically by expressing each side of the equation using the same base. That is, we used the one-to-one property of the exponential function:

$$\text{If } a^u = a^v, \text{ then } u = v \qquad a > 0, a \neq 1$$

For example, to solve the exponential equation $4^{2x+1} = 16$, notice that $16 = 4^2$ and apply the property above to obtain the equation $2x + 1 = 2$, from which we find $x = \frac{1}{2}$.

Not all exponential equations can be readily expressed so that each side of the equation has the same base. For such equations, algebraic techniques often can be used to obtain exact solutions. When algebraic techniques cannot be used, a graphing utility can be used to obtain approximate solutions. You should pay particular attention to the form of equations for which exact solutions are obtained.

EXAMPLE 4 Solving an Exponential Equation

Solve: $2^x = 5$

Algebraic Solution

Because 5 cannot be written as an integer power of 2 ($2^2 = 4$ and $2^3 = 8$), write the exponential equation as the equivalent logarithmic equation.

$$2^x = 5$$

$$x = \log_2 5 = \frac{\ln 5}{\ln 2}$$

↑ **Change-of-Base Formula (10), Section 6.5**

Alternatively, the equation $2^x = 5$ can be solved by taking the natural logarithm (or common logarithm) of each side.

$$2^x = 5$$

$$\ln 2^x = \ln 5 \quad \textbf{If } M = N, \textbf{ then } \ln M = \ln N.$$

$$x \ln 2 = \ln 5 \quad \ln M^r = r \ln M$$

$$x = \frac{\ln 5}{\ln 2} \quad \textbf{Exact solution}$$

$$\approx 2.322 \quad \textbf{Approximate solution}$$

The solution set is $\left\{\frac{\ln 5}{\ln 2}\right\}$. ■

Graphing Solution

Graph $Y_1 = 2^x$ and $Y_2 = 5$, and determine the x-coordinate of the point of intersection. See Figure 50.

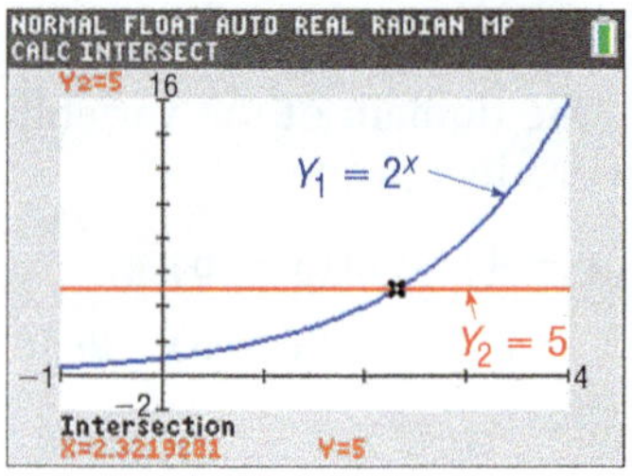

Figure 50

The approximate solution, rounded to three decimal places, is 2.322. ■

Now Work PROBLEM 43

EXAMPLE 5 Solving an Exponential Equation

Solve: $8 \cdot 3^x = 5$

Algebraic Solution

Isolate the exponential expression and then rewrite the statement as an equivalent logarithm.

$$8 \cdot 3^x = 5$$

$$3^x = \frac{5}{8} \quad \textbf{Solve for } 3^x.$$

$$x = \log_3\left(\frac{5}{8}\right) = \frac{\ln\left(\frac{5}{8}\right)}{\ln 3} \quad \textbf{Exact solution}$$

$$\approx -0.428 \quad \textbf{Approximate solution}$$

The solution set is $\left\{\log_3\left(\frac{5}{8}\right)\right\}$. ■

Graphing Solution

Graph $Y_1 = 8 \cdot 3^x$ and $Y_2 = 5$, and determine the x-coordinate of the point of intersection. See Figure 51.

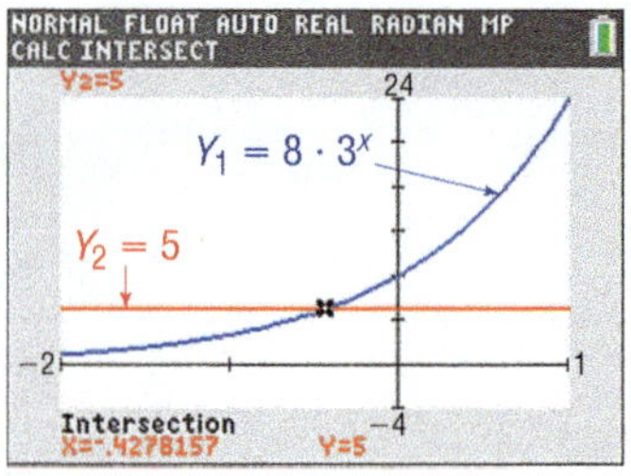

Figure 51

The approximate solution, rounded to three decimal places, is -0.428. ■

EXAMPLE 6 Solving an Exponential Equation

Solve: $5^{x-2} = 3^{3x+2}$

Algebraic Solution

Because the bases are different, first apply property (7), Section 6.5 (take the natural logarithm of each side), and then use appropriate properties of logarithms. The result is a linear equation in x that can be solved.

Graphing Solution

Graph $Y_1 = 5^{x-2}$ and $Y_2 = 3^{3x+2}$, and determine the x-coordinate of the point of intersection. See Figure 52.

$$5^{x-2} = 3^{3x+2}$$

$$\ln 5^{x-2} = \ln 3^{3x+2} \qquad \text{If } M = N, \ln M = \ln N.$$

$$(x-2)\ln 5 = (3x+2)\ln 3 \qquad \ln M^r = r \ln M$$

$$(\ln 5)x - 2\ln 5 = (3\ln 3)x + 2\ln 3 \qquad \text{Distribute. The equation is now linear in } x.$$

$$(\ln 5)x - (3\ln 3)x = 2\ln 3 + 2\ln 5 \qquad \text{Place terms involving } x \text{ on the left.}$$

$$(\ln 5 - 3\ln 3)x = 2(\ln 3 + \ln 5) \qquad \text{Factor.}$$

$$x = \frac{2(\ln 3 + \ln 5)}{\ln 5 - 3\ln 3} \qquad \text{Exact solution}$$

$$\approx -3.212 \qquad \text{Approximate solution}$$

The solution set is $\left\{\dfrac{2(\ln 3 + \ln 5)}{\ln 5 - 3\ln 3}\right\}$. ■

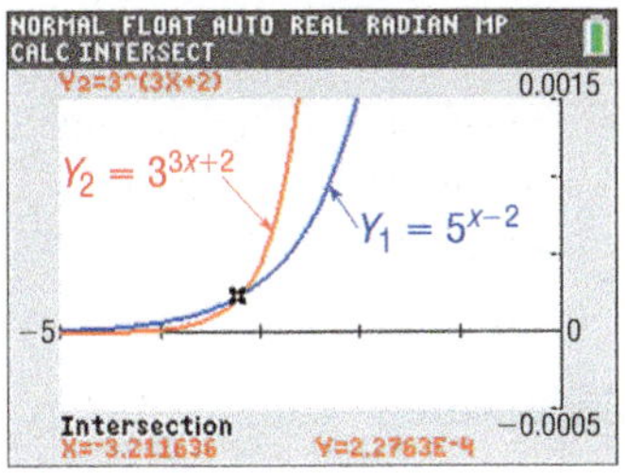

Figure 52

The approximate solution, rounded to three decimal places, is −3.212. Note that the y-coordinate, 2.2763E–4, is in scientific notation and means $2.2763 \times 10^{-4} = 0.0002763$. ■

Now Work PROBLEM 53

Note: Because of the properties of logarithms, exact solutions involving logarithms often can be expressed in multiple ways. For example, the solution to $5^{x-2} = 3^{3x+2}$ from Example 6 can be expressed equivalently as $\dfrac{2\ln 15}{\ln 5 - \ln 27}$ or as $\dfrac{\ln 225}{\ln(5/27)}$, among others. Do you see why? ■

The next example deals with an exponential equation that is quadratic in form.

EXAMPLE 7 Solving an Exponential Equation That Is Quadratic in Form

Solve: $4^x - 2^x - 12 = 0$

Algebraic Solution

Note that $4^x = (2^2)^x = 2^{2x} = (2^x)^2$, so the equation is quadratic in form and can be written as

$$(2^x)^2 - 2^x - 12 = 0 \qquad \text{Let } u = 2^x\text{; then } u^2 - u - 12 = 0.$$

Now we can factor as usual.

$$(2^x - 4)(2^x + 3) = 0 \qquad (u-4)(u+3) = 0$$

$$2^x - 4 = 0 \quad \text{or} \quad 2^x + 3 = 0 \qquad u - 4 = 0 \quad \text{or} \quad u + 3 = 0$$

$$2^x = 4 \qquad 2^x = -3 \qquad u = 2^x = 4 \quad \text{or} \quad u = 2^x = -3$$

The equation on the left has the solution $x = 2$, since $2^x = 4 = 2^2$; the equation on the right has no solution, since $2^x > 0$ for all x. The only solution is 2. The solution set is $\{2\}$. ■

Graphing Solution

Graph $Y_1 = 4^x - 2^x - 12$, and determine the x-intercept. See Figure 53. The x-intercept is 2, so the solution set is $\{2\}$.

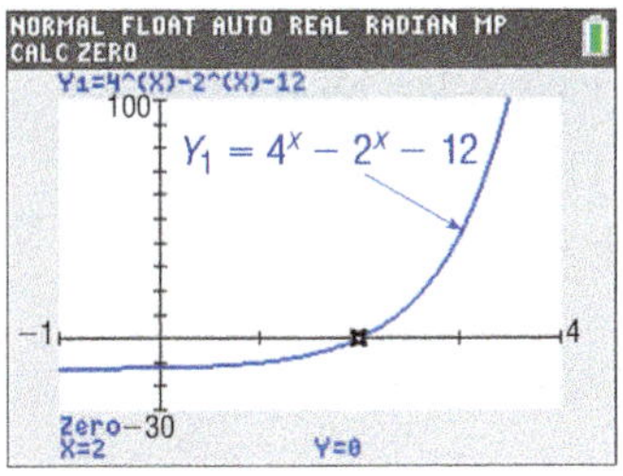

Figure 53 ■

Now Work PROBLEM 61

3 Solve Logarithmic and Exponential Equations Using a Graphing Utility

The algebraic techniques introduced in this section to obtain exact solutions apply only to certain types of logarithmic and exponential equations. Solutions for other types are usually studied in calculus, using numerical methods. For such types, we can use a graphing utility to approximate the solution.

EXAMPLE 8 **Solving Equations Using a Graphing Utility**

Solve: $x + e^x = 2$

Express the solution(s) rounded to two decimal places.

Solution The solution is found by graphing $Y_1 = x + e^x$ and $Y_2 = 2$. Since Y_1 is an increasing function (do you know why?), there is only one point of intersection for Y_1 and Y_2. Figure 54 shows the graphs of Y_1 and Y_2. Using the INTERSECT command reveals that the solution is 0.44, rounded to two decimal places. ■

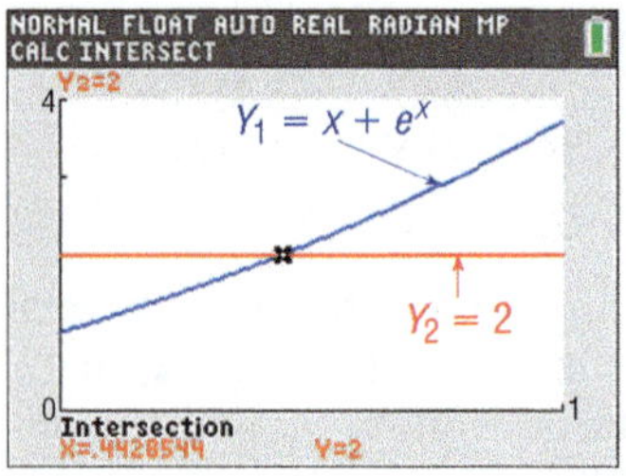

Figure 54

Now Work PROBLEM 71

6.6 Assess Your Understanding

'Are You Prepared?' *Answers are given at the end of these exercises. If you get a wrong answer, read the pages listed in red.*

1. Solve $x^2 - 7x - 30 = 0$. (pp. 110–115)
2. Solve $(x + 3)^2 - 4(x + 3) + 3 = 0$. (pp. 131–133)
3. Approximate the solution(s) to $x^3 = x^2 - 5$ using a graphing utility. (pp. 100–101)
4. Approximate the solution(s) to $x^3 - 2x + 2 = 0$ using a graphing utility. (pp. 100–101)

Skill Building

In Problems 5–40, solve each logarithmic equation. Express irrational solutions in exact form and as a decimal rounded to three decimal places.

5. $\log_4 x = 2$
6. $\log(x + 6) = 1$
7. $\log_2(5x) = 4$
8. $\log_3(3x - 1) = 2$
9. $\log_4(x + 2) = \log_4 8$
10. $\log_5(2x + 3) = \log_5 3$
11. $\frac{1}{2}\log_3 x = 2\log_3 2$
12. $-2\log_4 x = \log_4 9$
13. $3\log_2 x = -\log_2 27$
14. $2\log_5 x = 3\log_5 4$
15. $3\log_2(x - 1) + \log_2 4 = 5$
16. $2\log_3(x + 4) - \log_3 9 = 2$
17. $\log x + \log(x + 15) = 2$
18. $\log x + \log(x - 21) = 2$
19. $\log(2x + 1) = 1 + \log(x - 2)$
20. $\log(2x) - \log(x - 3) = 1$
21. $\log_2(x + 7) + \log_2(x + 8) = 1$
22. $\log_6(x + 4) + \log_6(x + 3) = 1$
23. $\log_8(x + 6) = 1 - \log_8(x + 4)$
24. $\log_5(x + 3) = 1 - \log_5(x - 1)$
25. $\ln x + \ln(x + 2) = 4$
26. $\ln(x + 1) - \ln x = 2$
27. $\log_3(x + 1) + \log_3(x + 4) = 2$
28. $\log_2(x + 1) + \log_2(x + 7) = 3$
29. $\log_{1/3}(x^2 + x) - \log_{1/3}(x^2 - x) = -1$
30. $\log_4(x^2 - 9) - \log_4(x + 3) = 3$
31. $\log_a(x - 1) - \log_a(x + 6) = \log_a(x - 2) - \log_a(x + 3)$
32. $\log_a x + \log_a(x - 2) = \log_a(x + 4)$
33. $2\log_5(x - 3) - \log_5 8 = \log_5 2$
34. $\log_3 x - 2\log_3 5 = \log_3(x + 1) - 2\log_3 10$
35. $2\log_6(x + 2) = 3\log_6 2 + \log_6 4$
36. $3(\log_7 x - \log_7 2) = 2\log_7 4$
37. $2\log_{13}(x + 2) = \log_{13}(4x + 7)$
38. $\log(x - 1) = \frac{1}{3}\log 2$
39. $(\log_3 x)^2 - 5(\log_3 x) = 6$
40. $\ln x - 3\sqrt{\ln x} + 2 = 0$

In Problems 41–68, solve each exponential equation. Express irrational solutions in exact form and as a decimal rounded to three decimal places.

41. $2^{x-5} = 8$
42. $5^{-x} = 25$
43. $2^x = 10$
44. $3^x = 14$
45. $8^{-x} = 1.2$
46. $2^{-x} = 1.5$
47. $5(2^{3x}) = 8$
48. $0.3(4^{0.2x}) = 0.2$
49. $3^{1-2x} = 4^x$
50. $2^{x+1} = 5^{1-2x}$
51. $\left(\frac{3}{5}\right)^x = 7^{1-x}$
52. $\left(\frac{4}{3}\right)^{1-x} = 5^x$
53. $1.2^x = (0.5)^{-x}$
54. $0.3^{1+x} = 1.7^{2x-1}$
55. $\pi^{1-x} = e^x$
56. $e^{x+3} = \pi^x$

57. $2^{2x} + 2^x - 12 = 0$

58. $3^{2x} + 3^x - 2 = 0$

59. $3^{2x} + 3^{x+1} - 4 = 0$

60. $2^{2x} + 2^{x+2} - 12 = 0$

61. $16^x + 4^{x+1} - 3 = 0$

62. $9^x - 3^{x-1} + 1 = 0$

63. $25^x - 8 \cdot 5^x = -16$

64. $36^x - 6 \cdot 6^x = -9$

65. $3 \cdot 4^x + 4 \cdot 2^x + 8 = 0$

66. $2 \cdot 49^x + 11 \cdot 7^x + 5 = 0$

67. $4^x - 10 \cdot 4^{-x} = 3$

68. $3^x - 14 \cdot 3^{-x} = 5$

In Problems 69–82, use a graphing utility to solve each equation. Express your answer rounded to two decimal places.

69. $\log_5(x + 1) - \log_4(x - 2) = 1$

70. $\log_2(x - 1) - \log_6(x + 2) = 2$

71. $e^x = -x$

72. $e^{2x} = x + 2$

73. $e^x = x^2$

74. $e^x = x^3$

75. $\ln x = -x$

76. $\ln(2x) = -x + 2$

77. $\ln x = x^3 - 1$

78. $\ln x = -x^2$

79. $e^x + \ln x = 4$

80. $e^x - \ln x = 4$

81. $e^{-x} = \ln x$

82. $e^{-x} = -\ln x$

Mixed Practice

In Problems 83–94, solve each equation. Express irrational solutions in exact form and as a decimal rounded to three decimal places.

83. $\log_2(x + 1) - \log_4 x = 1$
[**Hint:** Change $\log_4 x$ to base 2.]

84. $\log_2(3x + 2) - \log_4 x = 3$

85. $\log_{16} x + \log_4 x + \log_2 x = 7$

86. $\log_9 x + 3 \log_3 x = 14$

87. $(\sqrt[3]{2})^{2-x} = 2^{x^2}$

88. $\log_2 x^{\log_2 x} = 4$

89. $\dfrac{e^x + e^{-x}}{2} = 1$
[**Hint:** Multiply each side by e^x.]

90. $\dfrac{e^x + e^{-x}}{2} = 3$

91. $\dfrac{e^x - e^{-x}}{2} = 2$

92. $\dfrac{e^x - e^{-x}}{2} = -2$

93. $\log_5 x + \log_3 x = 1$
[**Hint:** Use the Change-of-Base Formula.]

94. $\log_2 x - \log_6 x = 3$

95. $f(x) = \log_2(x + 3)$ and $g(x) = \log_2(3x + 1)$.
(a) Solve $f(x) = 3$. What point is on the graph of f?
(b) Solve $g(x) = 4$. What point is on the graph of g?
(c) Solve $f(x) = g(x)$. Do the graphs of f and g intersect? If so, where?
(d) Solve $(f + g)(x) = 7$.
(e) Solve $(f - g)(x) = 2$.

96. $f(x) = \log_3(x + 5)$ and $g(x) = \log_3(x - 1)$.
(a) Solve $f(x) = 2$. What point is on the graph of f?
(b) Solve $g(x) = 3$. What point is on the graph of g?
(c) Solve $f(x) = g(x)$. Do the graphs of f and g intersect? If so, where?
(d) Solve $(f + g)(x) = 3$.
(e) Solve $(f - g)(x) = 2$.

97. (a) If $f(x) = 3^{x+1}$ and $g(x) = 2^{x+2}$, graph f and g on the same Cartesian plane.
(b) Find the point(s) of intersection of the graphs of f and g by solving $f(x) = g(x)$. Round answers to three decimal places. Label any intersection points on the graph drawn in part (a).
(c) Based on the graph, solve $f(x) > g(x)$.

98. (a) If $f(x) = 5^{x-1}$ and $g(x) = 2^{x+1}$, graph f and g on the same Cartesian plane.
(b) Find the point(s) of intersection of the graphs of f and g by solving $f(x) = g(x)$. Label any intersection points on the graph drawn in part (a).
(c) Based on the graph, solve $f(x) > g(x)$.

99. (a) Graph $f(x) = 3^x$ and $g(x) = 10$ on the same Cartesian plane.
(b) Shade the region bounded by the y-axis, $f(x) = 3^x$, and $g(x) = 10$ on the graph drawn in part (a).
(c) Solve $f(x) = g(x)$ and label the point of intersection on the graph drawn in part (a).

100. (a) Graph $f(x) = 2^x$ and $g(x) = 12$ on the same Cartesian plane.
(b) Shade the region bounded by the y-axis, $f(x) = 2^x$, and $g(x) = 12$ on the graph drawn in part (a).
(c) Solve $f(x) = g(x)$ and label the point of intersection on the graph drawn in part (a).

101. (a) Graph $f(x) = 2^{x+1}$ and $g(x) = 2^{-x+2}$ on the same Cartesian plane.
(b) Shade the region bounded by the y-axis, $f(x) = 2^{x+1}$, and $g(x) = 2^{-x+2}$ on the graph drawn in part (a).
(c) Solve $f(x) = g(x)$ and label the point of intersection on the graph drawn in part (a).

102. (a) Graph $f(x) = 3^{-x+1}$ and $g(x) = 3^{x-2}$ on the same Cartesian plane.
(b) Shade the region bounded by the y-axis, $f(x) = 3^{-x+1}$, and $g(x) = 3^{x-2}$ on the graph drawn in part (a).
(c) Solve $f(x) = g(x)$ and label the point of intersection on the graph drawn in part (a).

103. (a) Graph $f(x) = 2^x - 4$.
(b) Find the zero of f.
(c) Based on the graph, solve $f(x) < 0$.

104. (a) Graph $g(x) = 3^x - 9$.
(b) Find the zero of g.
(c) Based on the graph, solve $g(x) > 0$.

Applications and Extensions

105. A Population Model The resident population of the United States in 2015 was 320 million people and was growing at a rate of 0.7% per year. Assuming that this growth rate continues, the model $P(t) = 320(1.007)^{t-2015}$ represents the population P (in millions of people) in year t.

(a) According to this model, when will the population of the United States be 400 million people?

(b) According to this model, when will the population of the United States be 435 million people?

Source: *U.S. Census Bureau*

106. A Population Model The population of the world in 2015 was 7.21 billion people and was growing at a rate of 1.1% per year. Assuming that this growth rate continues, the model $P(t) = 7.21(1.011)^{t-2015}$ represents the population P (in billions of people) in year t.

(a) According to this model, when will the population of the world be 9 billion people?

(b) According to this model, when will the population of the world be 12.5 billion people?

Source: *U.S. Census Bureau*

107. Depreciation The value V of a Chevy Cruze LS that is t years old can be modeled by $V(t) = 18{,}700(0.84)^t$.

(a) According to the model, when will the car be worth \$9000?

(b) According to the model, when will the car be worth \$6000?

(c) According to the model, when will the car be worth \$2000?

Source: *Kelley Blue Book*

108. Depreciation The value V of a Honda Civic SE that is t years old can be modeled by $V(t) = 18{,}955(0.905)^t$.

(a) According to the model, when will the car be worth \$16,000?

(b) According to the model, when will the car be worth \$10,000?

(c) According to the model, when will the car be worth \$7500?

Source: *Kelley Blue Book*

Explaining Concepts: Discussion and Writing

109. Fill in the reason for each step in the following two solutions.

Solve: $\log_3(x-1)^2 = 2$

Solution A

$\log_3(x-1)^2 = 2$

$(x-1)^2 = 3^2 = 9$______

$(x-1) = \pm 3$______

$x - 1 = -3$ or $x - 1 = 3$______

$x = -2$ or $x = 4$______

Solution B

$\log_3(x-1)^2 = 2$

$2\log_3(x-1) = 2$______

$\log_3(x-1) = 1$______

$x - 1 = 3^1 = 3$______

$x = 4$______

Both solutions given in Solution A check. Explain what caused the solution $x = -2$ to be lost in Solution B.

Retain Your Knowledge

Problems 110–113 are based on material learned earlier in the course. The purpose of these problems is to keep the material fresh in your mind so that you are better prepared for the final exam.

110. Solve: $4x^3 + 3x^2 - 25x + 6 = 0$

111. Determine whether the function $\{(0,-4), (2,-2), (4,0), (6,2)\}$ is one-to-one.

112. For $f(x) = \dfrac{x}{x-2}$ and $g(x) = \dfrac{x+5}{x-3}$, find $f \circ g$. Then find the domain of $f \circ g$.

113. Find the domain of $f(x) = \sqrt{x+3} + \sqrt{x-1}$.

'Are You Prepared?' Answers

1. $\{-3, 10\}$ **2.** $\{-2, 0\}$ **3.** $\{-1.43\}$ **4.** $\{-1.77\}$

6.7 Financial Models

PREPARING FOR THIS SECTION *Before getting started, review the following:*

- Simple Interest (Section 1.6, pp. 139–140)

Now Work the 'Are You Prepared?' problems on page 481.

OBJECTIVES
1 Determine the Future Value of a Lump Sum of Money (p. 475)
2 Calculate Effective Rates of Return (p. 478)
3 Determine the Present Value of a Lump Sum of Money (p. 479)
4 Determine the Rate of Interest or the Time Required to Double a Lump Sum of Money (p. 480)

1 Determine the Future Value of a Lump Sum of Money

Interest is money paid for the use of money. The total amount borrowed (whether by an individual from a bank in the form of a loan or by a bank from an individual in the form of a savings account) is called the **principal**. The **rate of interest**, expressed as a percent, is the amount charged for the use of the principal for a given period of time, usually on a yearly (that is, per annum) basis.

THEOREM

Simple Interest Formula

If a principal of P dollars is borrowed for a period of t years at a per annum interest rate r, expressed as a decimal, the interest I charged is

$$I = Prt \qquad \textbf{(1)}$$

Interest charged according to formula (1) is called **simple interest**.

In problems involving interest, the term **payment period** is defined as follows.

Annually:	Once per year	**Monthly:**	12 times per year
Semiannually:	Twice per year	**Daily:**	365 times per year*
Quarterly:	Four times per year		

When the interest due at the end of a payment period is added to the principal so that the interest computed at the end of the next payment period is based on this new principal amount (old principal + interest), the interest is said to have been **compounded**. **Compound interest** is interest paid on the principal and on previously earned interest.

EXAMPLE 1 **Computing Compound Interest**

A credit union pays interest of 2% per annum compounded quarterly on a certain savings plan. If \$1000 is deposited in such a plan and the interest is left to accumulate, how much is in the account after 1 year?

Solution Use the simple interest formula, $I = Prt$. The principal P is \$1000 and the rate of interest is $2\% = 0.02$. After the first quarter of a year, the time t is $\frac{1}{4}$ year, so the interest earned is

$$I = Prt = (\$1000)(0.02)\left(\frac{1}{4}\right) = \$5$$

The new principal is $P + I = \$1000 + \$5 = \$1005$. At the end of the second quarter, the interest on this principal is

$$I = (\$1005)(0.02)\left(\frac{1}{4}\right) = \$5.03$$

*Most banks use a 360-day "year." Why do you think they do?

At the end of the third quarter, the interest on the new principal of $1005 + $5.03 = $1010.03 is

$$I = (\$1010.03)(0.02)\left(\frac{1}{4}\right) = \$5.05$$

Finally, after the fourth quarter, the interest is

$$I = (\$1015.08)(0.02)\left(\frac{1}{4}\right) = \$5.08$$

After 1 year the account contains $1015.08 + $5.08 = $1020.16. ■

The pattern of the calculations performed in Example 1 leads to a general formula for compound interest. For this purpose, let P represent the principal to be invested at a per annum interest rate r that is compounded n times per year, so the time of each compounding period is $\frac{1}{n}$ years. (For computing purposes, r is expressed as a decimal.) The interest earned after each compounding period is given by formula (1).

$$\text{Interest} = \text{principal} \times \text{rate} \times \text{time} = P \cdot r \cdot \frac{1}{n} = P \cdot \left(\frac{r}{n}\right)$$

The amount A after one compounding period is

$$A = P + P \cdot \left(\frac{r}{n}\right) = P \cdot \left(1 + \frac{r}{n}\right)$$

After two compounding periods, the amount A, based on the new principal $P \cdot \left(1 + \frac{r}{n}\right)$, is

$$A = \underbrace{P\cdot\left(1 + \frac{r}{n}\right)}_{\text{New principal}} + \underbrace{P\cdot\left(1 + \frac{r}{n}\right)\left(\frac{r}{n}\right)}_{\text{Interest on new principal}} \underset{\uparrow}{=} P\cdot\left(1 + \frac{r}{n}\right)\left(1 + \frac{r}{n}\right) = P\cdot\left(1 + \frac{r}{n}\right)^2$$

Factor out $P \cdot \left(1 + \frac{r}{n}\right)$

After three compounding periods, the amount A is

$$A = P \cdot \left(1 + \frac{r}{n}\right)^2 + P \cdot \left(1 + \frac{r}{n}\right)^2\left(\frac{r}{n}\right) = P \cdot \left(1 + \frac{r}{n}\right)^2 \cdot \left(1 + \frac{r}{n}\right) = P \cdot \left(1 + \frac{r}{n}\right)^3$$

Continuing this way, after n compounding periods (1 year), the amount A is

$$A = P \cdot \left(1 + \frac{r}{n}\right)^n$$

Because t years will contain $n \cdot t$ compounding periods, the amount after t years is

$$A = P \cdot \left(1 + \frac{r}{n}\right)^{nt}$$

THEOREM

Compound Interest Formula

The amount A after t years due to a principal P invested at an annual interest rate r, expressed as a decimal, compounded n times per year is

$$A = P \cdot \left(1 + \frac{r}{n}\right)^{nt} \qquad \textbf{(2)}$$

■

Exploration

To see the effects of compounding interest monthly on an initial deposit of $1, graph $Y_1 = \left(1 + \frac{r}{12}\right)^{12x}$ with $r = 0.06$ and $r = 0.12$ for $0 \leq x \leq 30$. What is the future value of $1 in 30 years when the interest rate per annum is $r = 0.06$ (6%)? What is the future value of $1 in 30 years when the interest rate per annum is $r = 0.12$ (12%)? Does doubling the interest rate double the future value? ■

For example, to rework Example 1, use $P = \$1000$, $r = 0.02$, $n = 4$ (quarterly compounding), and $t = 1$ year to obtain

$$A = P\cdot\left(1 + \frac{r}{n}\right)^{nt} = 1000\left(1 + \frac{0.02}{4}\right)^{4\cdot 1} = \$1020.15^{*}$$

In equation (2), the amount A is typically referred to as the **future value** of the account, while P is called the **present value**.

Now Work PROBLEM 7

*The result shown here differs from Example 1 due to rounding.

EXAMPLE 2 **Comparing Investments Using Different Compounding Periods**

Investing \$1000 at an annual rate of 10% compounded annually, semiannually, quarterly, monthly, and daily will yield the following amounts after 1 year:

Annual compounding $(n = 1)$:
$$A = P\cdot(1 + r) = (\$1000)(1 + 0.10) = \$1100.00$$

Semiannual compounding $(n = 2)$:
$$A = P\cdot\left(1 + \frac{r}{2}\right)^2 = (\$1000)(1 + 0.05)^2 = \$1102.50$$

Quarterly compounding $(n = 4)$:
$$A = P\cdot\left(1 + \frac{r}{4}\right)^4 = (\$1000)(1 + 0.025)^4 = \$1103.81$$

Monthly compounding $(n = 12)$:
$$A = P\cdot\left(1 + \frac{r}{12}\right)^{12} = (\$1000)\left(1 + \frac{0.10}{12}\right)^{12} = \$1104.71$$

Daily compounding $(n = 365)$:
$$A = P\cdot\left(1 + \frac{r}{365}\right)^{365} = (\$1000)\left(1 + \frac{0.10}{365}\right)^{365} = \$1105.16$$

From Example 2, note that the effect of compounding more frequently is that the amount after 1 year is higher: \$1000 compounded 4 times a year at 10% results in \$1103.81, \$1000 compounded 12 times a year at 10% results in \$1104.71, and \$1000 compounded 365 times a year at 10% results in \$1105.16. This leads to the following question: What would happen to the amount after 1 year if the number of times that the interest is compounded were increased without bound?

Let's find the answer. Suppose that P is the principal, r is the per annum interest rate, and n is the number of times that the interest is compounded each year. The amount after 1 year is

$$A = P\cdot\left(1 + \frac{r}{n}\right)^n$$

Rewrite this expression as follows:

$$A = P\cdot\left(1 + \frac{r}{n}\right)^n = P\cdot\left(1 + \frac{1}{\frac{n}{r}}\right)^n = P\cdot\left[\left(1 + \frac{1}{\frac{n}{r}}\right)^{n/r}\right]^r \underset{h = \frac{n}{r}}{=} P\cdot\left[\left(1 + \frac{1}{h}\right)^h\right]^r \quad \textbf{(3)}$$

Now suppose that the number n of times that the interest is compounded per year gets larger and larger; that is, suppose that $n \to \infty$. Then $h = \frac{n}{r} \to \infty$, and the expression in brackets in equation (3) equals e. That is, $A \to Pe^r$.

Table 9 compares $\left(1 + \frac{r}{n}\right)^n$, for large values of n, to e^r for $r = 0.05$, $r = 0.10$, $r = 0.15$, and $r = 1$. The larger that n gets, the closer $\left(1 + \frac{r}{n}\right)^n$ gets to e^r. No matter how frequent the compounding, the amount after 1 year has the definite ceiling Pe^r.

Table 9

	$(1+\frac{r}{n})^n$			
	$n = 100$	$n = 1000$	$n = 10{,}000$	e^r
$r = 0.05$	1.0512580	1.0512698	1.051271	1.0512711
$r = 0.10$	1.1051157	1.1051654	1.1051704	1.1051709
$r = 0.15$	1.1617037	1.1618212	1.1618329	1.1618342
$r = 1$	2.7048138	2.7169239	2.7181459	2.7182818

When interest is compounded so that the amount after 1 year is Pe^r, the interest is said to be **compounded continuously**.

THEOREM

Continuous Compounding

The amount A after t years due to a principal P invested at an annual interest rate r compounded continuously is

$$A = Pe^{rt} \quad (4)$$

EXAMPLE 3

Using Continuous Compounding

The amount A that results from investing a principal P of \$1000 at an annual rate r of 10% compounded continuously for a time t of 1 year is

$$A = \$1000e^{0.10} = (\$1000)(1.10517) = \$1105.17$$

Now Work PROBLEM 13

2 Calculate Effective Rates of Return

Suppose that you have \$1000 and a bank offers to pay you 3% annual interest on a savings account with interest compounded monthly. What annual interest rate must be earned for you to have the same amount at the end of the year as you would have if the interest had been compounded annually (once per year)? To answer this question, first determine the value of the \$1000 in the account that earns 3% compounded monthly.

$$A = \$1000\left(1 + \frac{0.03}{12}\right)^{12} \quad \text{Use } A = P\left(1+\frac{r}{n}\right)^n \text{ with } P = \$1000, r = 0.03, n = 12.$$

$$= \$1030.42$$

So the interest earned is \$30.42. Using $I = Prt$ with $t = 1$, $I = \$30.42$, and $P = \$1000$, the annual simple interest rate is $0.03042 = 3.042\%$. This interest rate is known as the *effective rate of interest.*

The **effective rate of interest** is the annual simple interest rate that would yield the same amount as compounding n times per year, or continuously, after 1 year.

THEOREM

Effective Rate of Interest

The effective rate of interest r_e of an investment earning an annual interest rate r is given by

$$\text{Compounding } n \text{ times per year: } r_e = \left(1 + \frac{r}{n}\right)^n - 1$$

$$\text{Continuous compounding: } r_e = e^r - 1$$

EXAMPLE 4

Computing the Effective Rate of Interest—Which Is the Best Deal?

Suppose you want to buy a 5-year certificate of deposit (CD). You visit three banks to determine their CD rates. American Express offers you 2.15% annual interest compounded monthly, and First Internet Bank offers you 2.20% compounded quarterly. Discover offers 2.12% compounded daily. Determine which bank is offering the best deal.

Solution The bank that offers the best deal is the one with the highest effective interest rate.

American Express	First Internet Bank	Discover
$r_e = \left(1 + \frac{0.0215}{12}\right)^{12} - 1$	$r_e = \left(1 + \frac{0.022}{4}\right)^{4} - 1$	$r_e = \left(1 + \frac{0.0212}{365}\right)^{365} - 1$
$\approx 1.02171 - 1$	$\approx 1.02218 - 1$	$\approx 1.02143 - 1$
$= 0.02171$	$= 0.02218$	$= 0.02143$
$= 2.171\%$	$= 2.218\%$	$= 2.143\%$

The effective rate of interest is highest for First Internet Bank, so First Internet Bank is offering the best deal. ■

Now Work PROBLEM 23

3 Determine the Present Value of a Lump Sum of Money

When people in finance speak of the "time value of money," they are usually referring to the *present value* of money. The **present value** of A dollars to be received at a future date is the principal that you would need to invest now so that it will grow to A dollars in the specified time period. The present value of money to be received at a future date is always less than the amount to be received, since the amount to be received will equal the present value (money invested now) *plus* the interest accrued over the time period.

The compound interest formula (2) is used to develop a formula for present value. If P is the present value of A dollars to be received after t years at a per annum interest rate r compounded n times per year, then, by formula (2),

$$A = P \cdot \left(1 + \frac{r}{n}\right)^{nt}$$

To solve for P, divide both sides by $\left(1 + \frac{r}{n}\right)^{nt}$. The result is

$$\frac{A}{\left(1 + \frac{r}{n}\right)^{nt}} = P \quad \text{or} \quad P = A \cdot \left(1 + \frac{r}{n}\right)^{-nt}$$

THEOREM

Present Value Formulas

The present value P of A dollars to be received after t years, assuming a per annum interest rate r compounded n times per year, is

$$P = A \cdot \left(1 + \frac{r}{n}\right)^{-nt} \tag{5}$$

If the interest is compounded continuously, then

$$P = Ae^{-rt} \tag{6}$$

To derive formula (6), solve formula (4) for P.

EXAMPLE 5

Computing the Value of a Zero-Coupon Bond

A zero-coupon (noninterest-bearing) bond can be redeemed in 10 years for \$1000. How much should you be willing to pay for it now if you want a return of

(a) 8% compounded monthly? (b) 7% compounded continuously?

Solution

(a) To find the present value of \$1000, use formula (5) with $A = \$1000$, $n = 12$, $r = 0.08$, and $t = 10$.

$$P = A \cdot \left(1 + \frac{r}{n}\right)^{-nt} = \$1000\left(1 + \frac{0.08}{12}\right)^{-12(10)} = \$450.52$$

For a return of 8% compounded monthly, pay \$450.52 for the bond.

(b) Here use formula (6) with $A = \$1000$, $r = 0.07$, and $t = 10$.

$$P = Ae^{-rt} = \$1000e^{-(0.07)(10)} = \$496.59$$

For a return of 7% compounded continuously, pay \$496.59 for the bond. ■

Now Work PROBLEM 15

4 Determine the Rate of Interest or the Time Required to Double a Lump Sum of Money

EXAMPLE 6 **Rate of Interest Required to Double an Investment**

What annual rate of interest compounded annually is needed in order to double an investment in 5 years?

Solution If P is the principal and P is to double, then the amount A will be $2P$. Use the compound interest formula with $n = 1$ and $t = 5$ to find r.

$$A = P \cdot \left(1 + \frac{r}{n}\right)^{nt}$$

$$2P = P \cdot (1 + r)^5 \qquad A = 2P, n = 1, t = 5$$

$$2 = (1 + r)^5 \qquad \text{Divide both sides by } P.$$

$$1 + r = \sqrt[5]{2} \qquad \text{Take the fifth root of each side.}$$

$$r = \sqrt[5]{2} - 1 \approx 1.148698 - 1 = 0.148698$$

The annual rate of interest needed to double the principal in 5 years is 14.87%. ■

Now Work PROBLEM 31

EXAMPLE 7 **Time Required to Double or Triple an Investment**

(a) How long will it take for an investment to double in value if it earns 5% compounded continuously?
(b) How long will it take to triple at this rate?

Solution (a) If P is the initial investment and P is to double, then the amount A will be $2P$. Use formula (4) for continuously compounded interest with $r = 0.05$. Then

$$A = Pe^{rt}$$

$$2P = Pe^{0.05t} \qquad A = 2P, r = 0.05$$

$$2 = e^{0.05t} \qquad \text{Divide out the } P\text{'s.}$$

$$0.05t = \ln 2 \qquad \text{Rewrite as a logarithm.}$$

$$t = \frac{\ln 2}{0.05} \approx 13.86 \qquad \text{Solve for } t.$$

It will take about 14 years to double the investment.

(b) To triple the investment, let $A = 3P$ in formula (4).

$$A = Pe^{rt}$$

$$3P = Pe^{0.05t} \qquad A = 3P, r = 0.05$$

$$3 = e^{0.05t} \qquad \text{Divide out the } P\text{'s.}$$

$$0.05t = \ln 3 \qquad \text{Rewrite as a logarithm.}$$

$$t = \frac{\ln 3}{0.05} \approx 21.97 \qquad \text{Solve for } t.$$

It will take about 22 years to triple the investment. ■

Now Work PROBLEM 35

6.7 Assess Your Understanding

'Are You Prepared?' *Answers are given at the end of these exercises. If you get a wrong answer, read the page listed in red.*

1. What is the interest due if $500 is borrowed for 6 months at a simple interest rate of 6% per annum? (pp. 139–140)

2. If you borrow $5000 and, after 9 months, pay off the loan in the amount of $5500, what per annum rate of interest was charged? (pp. 139–140)

Concepts and Vocabulary

3. The total amount borrowed (whether by an individual from a bank in the form of a loan or by a bank from an individual in the form of a savings account) is called the ________.

4. If a principal of P dollars is borrowed for a period of t years at a per annum interest rate r, expressed as a decimal, the interest I charged is _____ = _______. Interest charged according to this formula is called ______ _______.

5. In working problems involving interest, if the payment period of the interest is quarterly, then interest is paid _____ times per year.

6. The ________ _____ ___ _______ is the equivalent annual simple interest rate that would yield the same amount as compounding n times per year, or continuously, after 1 year.

Skill Building

In Problems 7–14, find the amount that results from each investment.

7. $100 invested at 4% compounded quarterly after a period of 2 years

8. $50 invested at 6% compounded monthly after a period of 3 years

9. $500 invested at 8% compounded quarterly after a period of $2\frac{1}{2}$ years

10. $300 invested at 12% compounded monthly after a period of $1\frac{1}{2}$ years

11. $600 invested at 5% compounded daily after a period of 3 years

12. $700 invested at 6% compounded daily after a period of 2 years

13. $1000 invested at 11% compounded continuously after a period of 2 years

14. $400 invested at 7% compounded continuously after a period of 3 years

In Problems 15–22, find the principal needed now to get each amount; that is, find the present value.

15. To get $100 after 2 years at 6% compounded monthly

16. To get $75 after 3 years at 8% compounded quarterly

17. To get $1000 after $2\frac{1}{2}$ years at 6% compounded daily

18. To get $800 after $3\frac{1}{2}$ years at 7% compounded monthly

19. To get $600 after 2 years at 4% compounded quarterly

20. To get $300 after 4 years at 3% compounded daily

21. To get $80 after $3\frac{1}{4}$ years at 9% compounded continuously

22. To get $800 after $2\frac{1}{2}$ years at 8% compounded continuously

In Problems 23–26, find the effective rate of interest.

23. For 5% compounded quarterly

24. For 6% compounded monthly

25. For 5% compounded continuously

26. For 6% compounded continuously

In Problems 27–30, determine the rate that represents the better deal.

27. 6% compounded quarterly or $6\frac{1}{4}$% compounded annually

28. 9% compounded quarterly or $9\frac{1}{4}$% compounded annually

29. 9% compounded monthly or 8.8% compounded daily

30. 8% compounded semiannually or 7.9% compounded daily

31. What rate of interest compounded annually is required to double an investment in 3 years?

32. What rate of interest compounded annually is required to double an investment in 6 years?

33. What rate of interest compounded annually is required to triple an investment in 5 years?

34. What rate of interest compounded annually is required to triple an investment in 10 years?

35. (a) How long does it take for an investment to double in value if it is invested at 8% compounded monthly?
(b) How long does it take if the interest is compounded continuously?

36. (a) How long does it take for an investment to triple in value if it is invested at 6% compounded monthly?
(b) How long does it take if the interest is compounded continuously?

37. What rate of interest compounded quarterly will yield an effective interest rate of 7%?

38. What rate of interest compounded continuously will yield an effective interest rate of 6%?

Applications and Extensions

39. **Time Required to Reach a Goal** If Tanisha has \$100 to invest at 4% per annum compounded monthly, how long will it be before she has \$150? If the compounding is continuous, how long will it be?

40. **Time Required to Reach a Goal** If Angela has \$100 to invest at 2.5% per annum compounded monthly, how long will it be before she has \$175? If the compounding is continuous, how long will it be?

41. **Time Required to Reach a Goal** How many years will it take for an initial investment of \$10,000 to grow to \$25,000? Assume a rate of interest of 6% compounded continuously.

42. **Time Required to Reach a Goal** How many years will it take for an initial investment of \$25,000 to grow to \$80,000? Assume a rate of interest of 7% compounded continuously.

43. **Price Appreciation of Homes** What will a \$90,000 condominium cost 5 years from now if the price appreciation for condos over that period averages 3% compounded annually?

44. **Credit Card Interest** A department store charges 1.25% per month on the unpaid balance for customers with charge accounts (interest is compounded monthly). A customer charges \$200 and does not pay her bill for 6 months. What is the bill at that time?

45. **Saving for a Car** Jerome will be buying a used car for \$15,000 in 3 years. How much money should he ask his parents for now so that, if he invests it at 5% compounded continuously, he will have enough to buy the car?

46. **Paying off a Loan** John requires \$3000 in 6 months to pay off a loan that has no prepayment privileges. If he has the \$3000 now, how much of it should he save in an account paying 3% compounded monthly so that in 6 months he will have exactly \$3000?

47. **Return on a Stock** George contemplates the purchase of 100 shares of a stock selling for \$15 per share. The stock pays no dividends. The history of the stock indicates that it should grow at an annual rate of 15% per year. How much should the 100 shares of stock be worth in 5 years?

48. **Return on an Investment** A business purchased for \$650,000 in 2012 is sold in 2015 for \$850,000. What is the annual rate of return for this investment?

49. **Comparing Savings Plans** Jim places \$1000 in a bank account that pays 5.6% compounded continuously. After 1 year, will he have enough money to buy a computer system that costs \$1060? If another bank will pay Jim 5.9% compounded monthly, is this a better deal?

50. **Savings Plans** On January 1, Kim places \$1000 in a certificate of deposit that pays 6.8% compounded continuously and matures in 3 months. Then Kim places the \$1000 and the interest in a passbook account that pays 5.25% compounded monthly. How much does Kim have in the passbook account on May 1?

51. **Comparing IRA Investments** Will invests \$2000 in his IRA in a bond trust that pays 9% interest compounded semiannually. His friend Henry invests \$2000 in his IRA in a certificate of deposit that pays $8\frac{1}{2}\%$ compounded continuously. Who has more money after 20 years, Will or Henry?

52. **Comparing Two Alternatives** Suppose that April has access to an investment that will pay 10% interest compounded continuously. Which is better: to be given \$1000 now so that she can take advantage of this investment opportunity or to be given \$1325 after 3 years?

53. **College Costs** The average annual cost of college at 4-year private colleges was \$31,231 in the 2014–2015 academic year. This was a 3.7% increase from the previous year.
Source: The College Board
(a) If the cost of college increases by 3.7% each year, what will be the average cost of college at a 4-year private college for the 2034–2035 academic year?
(b) College savings plans, such as a 529 plan, allow individuals to put money aside now to help pay for college later. If one such plan offers a rate of 2% compounded continuously, how much should be put in a college savings plan in 2016 to pay for 1 year of the cost of college at a 4-year private college for an incoming freshman in 2034?

54. **Analyzing Interest Rates on a Mortgage** Colleen and Bill have just purchased a house for \$650,000, with the seller holding a second mortgage of \$100,000. They promise to pay the seller \$100,000 plus all accrued interest 5 years from now. The seller offers them three interest options on the second mortgage:
(a) Simple interest at 6% per annum
(b) 5.5% interest compounded monthly
(c) 5.25% interest compounded continuously
Which option is best? That is, which results in paying the least interest on the loan?

55. **2009 Federal Stimulus Package** In February 2009, President Obama signed into law a \$787 billion federal stimulus package. At that time, 20-year Series EE bonds had a fixed rate of 1.3% compounded semiannually. If the federal government financed the stimulus through EE bonds, how much would it have to pay back in 2029? How much interest was paid to finance the stimulus?
Source: U.S. Treasury Department

56. **Per Capita Federal Debt** In 2015, the federal debt was about \$18 trillion. In 2015, the U.S. population was about 320 million. Assuming that the federal debt is increasing about 4.5% per year and the U.S. population is increasing about 0.7% per year, determine the per capita debt (total debt divided by population) in 2030.

Inflation *Problems 57–62 require the following discussion.* ***Inflation*** *is a term used to describe the erosion of the purchasing power of money. For example, if the annual inflation rate is 3%, then \$1000 worth of purchasing power now will have only \$970 worth of purchasing power in 1 year because 3% of the original \$1000 (0.03 × 1000 = 30) has been eroded due to inflation. In general, if the rate of inflation averages r per annum over n years, the amount A that \$P will purchase after n years is*

$$A = P \cdot (1 - r)^n$$

where r is expressed as a decimal.

57. **Inflation** If the inflation rate averages 3%, how much will \$1000 purchase in 2 years?

58. **Inflation** If the inflation rate averages 2%, how much will \$1000 purchase in 3 years?

59. **Inflation** If the amount that \$1000 will purchase is only \$950 after 2 years, what was the average inflation rate?

60. **Inflation** If the amount that \$1000 will purchase is only \$930 after 2 years, what was the average inflation rate?

61. **Inflation** If the average inflation rate is 2%, how long is it until purchasing power is cut in half?

62. **Inflation** If the average inflation rate is 4%, how long is it until purchasing power is cut in half?

Problems 63–66 involve zero-coupon bonds. A ***zero-coupon bond*** *is a bond that is sold now at a discount and will pay its face value at the time when it matures; no interest payments are made.*

63. **Zero-Coupon Bonds** A zero-coupon bond can be redeemed in 20 years for \$10,000. How much should you be willing to pay for it now if you want a return of:
    (a) 5% compounded monthly?
    (b) 5% compounded continuously?

64. **Zero-Coupon Bonds** A child's grandparents are considering buying a \$80,000 face-value, zero-coupon bond at birth so that she will have money for her college education 17 years later. If they want a rate of return of 6% compounded annually, what should they pay for the bond?

65. **Zero-Coupon Bonds** How much should a \$10,000 face-value, zero-coupon bond, maturing in 10 years, be sold for now if its rate of return is to be 4.5% compounded annually?

66. **Zero-Coupon Bonds** If Pat pays \$15,334.65 for a \$25,000 face-value, zero-coupon bond that matures in 8 years, what is his annual rate of return?

67. **Time to Double or Triple an Investment** The formula

$$t = \frac{\ln m}{n \ln\left(1 + \frac{r}{n}\right)}$$

can be used to find the number of years t required to multiply an investment m times when r is the per annum interest rate compounded n times a year.
    (a) How many years will it take to double the value of an IRA that compounds annually at the rate of 6%?
    (b) How many years will it take to triple the value of a savings account that compounds quarterly at an annual rate of 5%?
    (c) Give a derivation of this formula.

68. **Time to Reach an Investment Goal** The formula

$$t = \frac{\ln A - \ln P}{r}$$

can be used to find the number of years t required for an investment P to grow to a value A when compounded continuously at an annual rate r.
    (a) How long will it take to increase an initial investment of \$1000 to \$4500 at an annual rate of 5.75%?
    (b) What annual rate is required to increase the value of a \$2000 IRA to \$30,000 in 35 years?
    (c) Give a derivation of this formula.

Problems 69–72 require the following discussion. The ***consumer price index (CPI)*** *indicates the relative change in price over time for a fixed basket of goods and services. It is a cost of living index that helps measure the effect of inflation on the cost of goods and services. The CPI uses the base period 1982–1984 for comparison (the CPI for this period is 100). The CPI for March 2015 was 236.12. This means that \$100 in the period 1982–1984 had the same purchasing power as \$236.12 in March 2015. In general, if the rate of inflation averages r percent per annum over n years, then the CPI index after n years is*

$$\text{CPI} = \text{CPI}_0\left(1 + \frac{r}{100}\right)^n$$

where CPI_0 is the CPI index at the beginning of the n-year period.
Source: *U.S. Bureau of Labor Statistics*

69. **Consumer Price Index**
    (a) The CPI was 214.5 for 2009 and 236.7 for 2014. Assuming that annual inflation remained constant for this time period, determine the average annual inflation rate.
    (b) Using the inflation rate from part (a), in what year will the CPI reach 300?

70. **Consumer Price Index** If the current CPI is 234.2 and the average annual inflation rate is 2.8%, what will be the CPI in 5 years?

71. **Consumer Price Index** If the average annual inflation rate is 3.1%, how long will it take for the CPI index to double? (A doubling of the CPI index means purchasing power is cut in half.)

72. **Consumer Price Index** The base period for the CPI changed in 1998. Under the previous weight and item structure, the CPI for 1995 was 456.5. If the average annual inflation rate was 5.57%, what year was used as the base period for the CPI?

Explaining Concepts: Discussion and Writing

73. Explain in your own words what the term *compound interest* means. What does *continuous compounding* mean?

74. Explain in your own words the meaning of *present value*.

75. **Critical Thinking** You have just contracted to buy a house and will seek financing in the amount of \$100,000. You go to several banks. Bank 1 will lend you \$100,000 at the rate of 4.125% amortized over 30 years with a loan origination fee of 0.45%. Bank 2 will lend you \$100,000 at the rate of 3.375% amortized over 15 years with a loan origination fee of 0.95%. Bank 3 will lend you \$100,000 at the rate of 4.25% amortized over 30 years with no loan origination fee. Bank 4 will lend you \$100,000 at the rate of 3.625% amortized over

15 years with no loan origination fee. Which loan would you take? Why? Be sure to have sound reasons for your choice. Use the information in the table to assist you. If the amount of the monthly payment does not matter to you, which loan would you take? Again, have sound reasons for your choice. Compare your final decision with others in the class. Discuss.

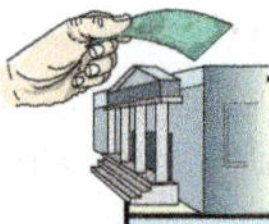

	Monthly Payment	Loan Origination Fee
Bank 1	\$485.00	\$450.00
Bank 2	\$709.00	\$950.00
Bank 3	\$492.00	\$0.00
Bank 4	\$721.00	\$0.00

Retain Your Knowledge

Problems 76–79 are based on material learned earlier in the course. The purpose of these problems is to keep the material fresh in your mind so that you are better prepared for the final exam.

76. Find the remainder R when $f(x) = 6x^3 + 3x^2 + 2x - 11$ is divided by $g(x) = x - 1$. Is g a factor of f?

77. The function $f(x) = \dfrac{x}{x-2}$ is one-to-one. Find f^{-1}.

78. Find the real zeros of
$$f(x) = x^5 - x^4 - 15x^3 - 21x^2 - 16x - 20.$$
Then write f in factored form.

79. Solve: $\log_2(x + 3) = 2\log_2(x - 3)$

'Are You Prepared?' Answers

1. \$15 **2.** $13\frac{1}{3}\%$

6.8 Exponential Growth and Decay Models; Newton's Law; Logistic Growth and Decay Models

OBJECTIVES
1 Find Equations of Populations That Obey the Law of Uninhibited Growth (p. 484)
2 Find Equations of Populations That Obey the Law of Decay (p. 487)
3 Use Newton's Law of Cooling (p. 488)
4 Use Logistic Models (p. 489)

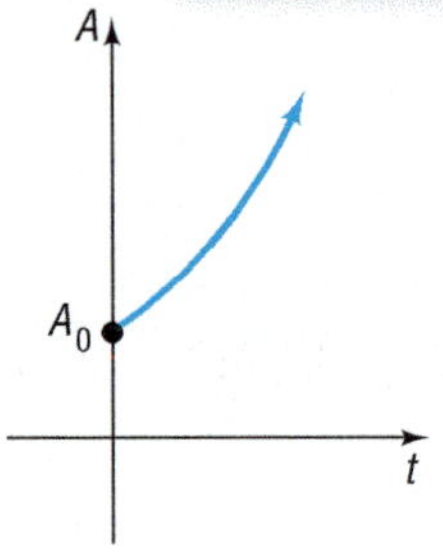

(a) $A(t) = A_0e^{kt}, k > 0$
Exponential growth

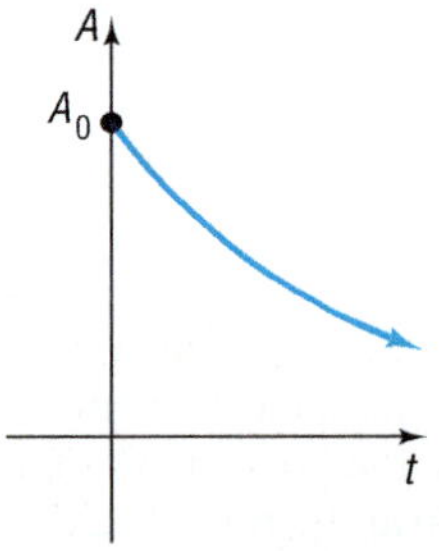

(b) $A(t) = A_0e^{kt}, k < 0$
Exponential decay

Figure 55

1 Find Equations of Populations That Obey the Law of Uninhibited Growth

Many natural phenomena have been found to follow the law that an amount A varies with time t according to the function

$$A(t) = A_0e^{kt} \qquad (1)$$

Here A_0 is the original amount $(t = 0)$ and $k \neq 0$ is a constant.

If $k > 0$, then equation (1) states that the amount A is increasing over time; if $k < 0$, the amount A is decreasing over time. In either case, when an amount A varies over time according to equation (1), it is said to follow the **exponential law** or the **law of uninhibited growth** $(k > 0)$ **or decay** $(k < 0)$. See Figure 55.

For example, in Section 6.7, continuously compounded interest was shown to follow the law of uninhibited growth. In this section we shall look at some additional phenomena that follow the exponential law.

Cell division is the growth process of many living organisms, such as amoebas, plants, and human skin cells. Based on an ideal situation in which no cells die and no by-products are produced, the number of cells present at a given time follows the law of uninhibited growth. Actually, however, after enough time has passed, growth at an exponential rate will cease due to the influence of factors such as lack of living space and dwindling food supply. The law of uninhibited growth accurately models only the early stages of the cell division process.

The cell division process begins with a culture containing N_0 cells. Each cell in the culture grows for a certain period of time and then divides into two identical cells. Assume that the time needed for each cell to divide in two is constant and does not change as the number of cells increases. These new cells then grow, and eventually each divides in two, and so on.

Uninhibited Growth of Cells

A model that gives the number N of cells in a culture after a time t has passed (in the early stages of growth) is

$$N(t) = N_0 e^{kt} \qquad k > 0 \qquad \textbf{(2)}$$

where N_0 is the initial number of cells and k is a positive constant that represents the growth rate of the cells.

Using formula (2) to model the growth of cells employs a function that yields positive real numbers, even though the number of cells being counted must be an integer. This is a common practice in many applications.

EXAMPLE 1 Bacterial Growth

A colony of bacteria that grows according to the law of uninhibited growth is modeled by the function $N(t) = 100e^{0.045t}$, where N is measured in grams and t is measured in days.

(a) Determine the initial amount of bacteria.
(b) What is the growth rate of the bacteria?
(c) Graph the function using a graphing utility.
(d) What is the population after 5 days?
(e) How long will it take for the population to reach 140 grams?
(f) What is the doubling time for the population?

Solution

(a) The initial amount of bacteria, N_0, is obtained when $t = 0$, so

$$N_0 = N(0) = 100e^{0.045(0)} = 100 \text{ grams}$$

(b) Compare $N(t) = 100e^{0.045t}$ to $N(t) = N_0 e^{kt}$. The value of k, 0.045, indicates a growth rate of 4.5%.

(c) Figure 56 shows the graph of $N(t) = 100e^{0.045t}$.

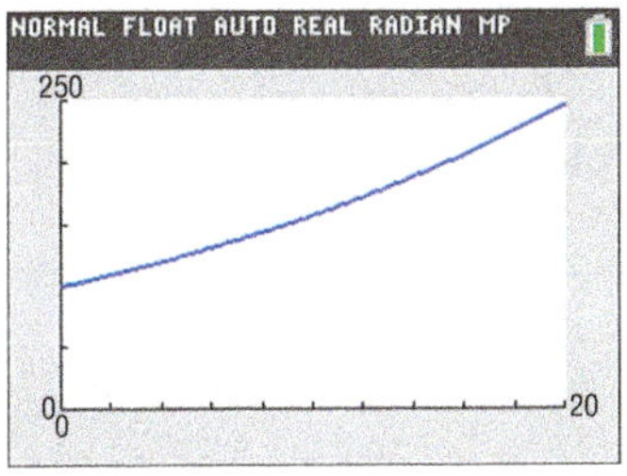

Figure 56 $Y_1 = 100e^{0.045x}$

(d) The population after 5 days is $N(5) = 100e^{0.045(5)} \approx 125.2$ grams.

(e) To find how long it takes for the population to reach 140 grams, solve the equation $N(t) = 140$.

$$100e^{0.045t} = 140$$

$$e^{0.045t} = 1.4 \qquad \text{Divide both sides of the equation by 100.}$$

$$0.045t = \ln 1.4 \qquad \text{Rewrite as a logarithm.}$$

$$t = \frac{\ln 1.4}{0.045} \qquad \text{Divide both sides of the equation by 0.045.}$$

$$\approx 7.5 \text{ days}$$

The population reaches 140 grams in about 7.5 days.

(f) The population doubles when $N(t) = 200$ grams, so the doubling time is found by solving the equation $200 = 100e^{0.045t}$ for t.

$$200 = 100e^{0.045t}$$

$$2 = e^{0.045t} \quad \text{Divide both sides of the equation by 100.}$$

$$\ln 2 = 0.045t \quad \text{Rewrite as a logarithm.}$$

$$t = \frac{\ln 2}{0.045} \quad \text{Divide both sides of the equation by 0.045.}$$

$$\approx 15.4 \text{ days}$$

The population doubles approximately every 15.4 days.

Now Work PROBLEM 1

EXAMPLE 2 Bacterial Growth

A colony of bacteria increases according to the law of uninhibited growth.

(a) If N is the number of cells and t is the time in hours, express N as a function of t.

(b) If the number of bacteria doubles in 3 hours, find the function that gives the number of cells in the culture.

(c) How long will it take for the size of the colony to triple?

(d) How long will it take for the population to double a second time (that is, increase four times)?

Solution (a) Using formula (2), the number N of cells at time t is

$$N(t) = N_0e^{kt}$$

where N_0 is the initial number of bacteria present and k is a positive number.

(b) To find the growth rate k, note that the number of cells doubles in 3 hours, so

$$N(3) = 2N_0$$

But $N(3) = N_0e^{k(3)}$, so

$$N_0e^{k(3)} = 2N_0$$

$$e^{3k} = 2 \quad \text{Divide both sides by } N_0.$$

$$3k = \ln 2 \quad \text{Rewrite as a logarithm.}$$

$$k = \frac{1}{3}\ln 2 \approx 0.23105$$

The function that models this growth process is, therefore,

$$N(t) = N_0e^{0.23105t}$$

(c) The time t needed for the size of the colony to triple requires that $N = 3N_0$. Substitute $3N_0$ for N to get

$$3N_0 = N_0e^{0.23105t}$$

$$3 = e^{0.23105t} \quad \text{Divide both sides by } N_0.$$

$$0.23105t = \ln 3 \quad \text{Rewrite as a logarithm.}$$

$$t = \frac{\ln 3}{0.23105} \approx 4.755 \text{ hours}$$

It will take about 4.755 hours, or 4 hours and 45 minutes, for the size of the colony to triple.

(d) If a population doubles in 3 hours, it will double a second time in 3 more hours, for a total time of 6 hours.

2 Find Equations of Populations That Obey the Law of Decay

Radioactive materials follow the law of uninhibited decay.

Uninhibited Radioactive Decay

The amount A of a radioactive material present at time t is given by

$$A(t) = A_0 e^{kt} \qquad k < 0 \qquad \textbf{(3)}$$

where A_0 is the original amount of radioactive material and k is a negative number that represents the rate of decay.

All radioactive substances have a specific **half-life**, which is the time required for half of the radioactive substance to decay. **Carbon dating** uses the fact that all living organisms contain two kinds of carbon, carbon-12 (a stable carbon) and carbon-14 (a radioactive carbon with a half-life of 5730 years). While an organism is living, the ratio of carbon-12 to carbon-14 is constant. But when an organism dies, the original amount of carbon-12 present remains unchanged, whereas the amount of carbon-14 begins to decrease. This change in the amount of carbon-14 present relative to the amount of carbon-12 present makes it possible to calculate when the organism died.

EXAMPLE 3 Estimating the Age of Ancient Tools

Traces of burned wood along with ancient stone tools in an archeological dig in Chile were found to contain approximately 1.67% of the original amount of carbon-14.

(a) If the half-life of carbon-14 is 5730 years, approximately when was the tree cut and burned?

(b) Using a graphing utility, graph the relation between the percentage of carbon-14 remaining and time.

(c) Use a graphing utility to determine the time that elapses until half of the carbon-14 remains. This answer should equal the half-life of carbon-14.

(d) Use a graphing utility to verify the answer found in part (a).

Solution (a) Using formula (3), the amount A of carbon-14 present at time t is

$$A(t) = A_0 e^{kt}$$

where A_0 is the original amount of carbon-14 present and k is a negative number. We first seek the number k. To find it, we use the fact that after 5730 years half of the original amount of carbon-14 remains, so $A(5730) = \frac{1}{2}A_0$. Then

$$\frac{1}{2}A_0 = A_0 e^{k(5730)}$$

$$\frac{1}{2} = e^{5730k} \qquad \text{Divide both sides of the equation by } A_0.$$

$$5730k = \ln\frac{1}{2} \qquad \text{Rewrite as a logarithm.}$$

$$k = \frac{1}{5730}\ln\frac{1}{2} \approx -0.000120968$$

Formula (3), therefore, becomes

$$A(t) = A_0 e^{-0.000120968t}$$

If the amount A of carbon-14 now present is 1.67% of the original amount, it follows that

$$0.0167A_0 = A_0 e^{-0.000120968t}$$

$$0.0167 = e^{-0.000120968t} \qquad \text{Divide both sides of the equation by } A_0.$$

$$-0.000120968t = \ln 0.0167 \qquad \text{Rewrite as a logarithm.}$$

$$t = \frac{\ln 0.0167}{-0.000120968} \approx 33{,}830 \text{ years}$$

The tree was cut and burned about 33,830 years ago. Some archeologists use this conclusion to argue that humans lived in the Americas nearly 34,000 years ago, much earlier than is generally accepted.

(b) Figure 57 shows the graph of $y = e^{-0.000120968x}$, where y is the fraction of carbon-14 present and x is the time.

(c) Graph $Y_1 = e^{-0.000120968x}$ and $Y_2 = 0.5$, where x is time. Use INTERSECT to find that it takes 5730 years until half the carbon-14 remains. The half-life of carbon-14 is 5730 years.

(d) Graph $Y_1 = e^{-0.000120968x}$ and $Y_2 = 0.0167$ where x is time. Use INTERSECT to find that it takes 33,830 years until 1.67% of the carbon-14 remains.

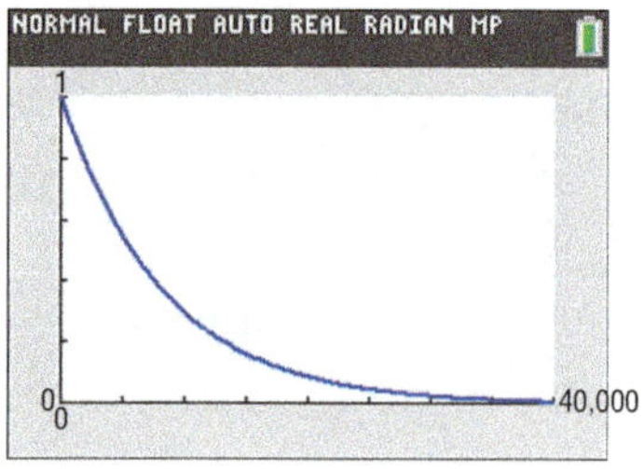

Figure 57 $Y_1 = e^{-0.000120968x}$

Now Work PROBLEM 3

3 Use Newton's Law of Cooling

Newton's Law of Cooling* states that the temperature of a heated object decreases exponentially over time toward the temperature of the surrounding medium.

Newton's Law of Cooling

The temperature u of a heated object at a given time t can be modeled by the following function:

$$u(t) = T + (u_0 - T)e^{kt} \qquad k < 0 \qquad \textbf{(4)}$$

where T is the constant temperature of the surrounding medium, u_0 is the initial temperature of the heated object, and k is a negative constant.

EXAMPLE 4 Using Newton's Law of Cooling

An object is heated to 100°C (degrees Celsius) and is then allowed to cool in a room whose air temperature is 30°C.

(a) If the temperature of the object is 80°C after 5 minutes, when will its temperature be 50°C?

(b) Using a graphing utility, graph the relation found between the temperature and time.

(c) Using a graphing utility, verify the results from part (a).

(d) Using a graphing utility, determine the elapsed time before the object is 35°C.

(e) What do you notice about the temperature as time passes?

Solution (a) Using formula (4) with $T = 30$ and $u_0 = 100$, the temperature (in degrees Celsius) of the object at time t (in minutes) is

$$u(t) = 30 + (100 - 30)e^{kt} = 30 + 70e^{kt}$$

where k is a negative constant. To find k, use the fact that $u = 80$ when $t = 5$ [that is, $u(5) = 80$]. Then

$$80 = 30 + 70e^{k(5)} \qquad u(5) = 80$$

$$50 = 70e^{5k} \qquad \text{Simplify.}$$

$$e^{5k} = \frac{50}{70} \qquad \text{Solve for } e^{5k}.$$

$$5k = \ln\frac{5}{7} \qquad \text{Rewrite as a logarithm.}$$

$$k = \frac{1}{5}\ln\frac{5}{7} \approx -0.0673 \qquad \text{Solve for } k.$$

Formula (4), therefore, becomes

$$u(t) = 30 + 70e^{-0.0673t}$$

*Named after Sir Isaac Newton (1642–1727), one of the cofounders of calculus.

To find t when $u = 50°C$, solve the equation

$$50 = 30 + 70e^{-0.0673t}$$

$$20 = 70e^{-0.0673t} \qquad \text{Subtract 30 from both sides.}$$

$$e^{-0.0673t} = \frac{20}{70} \qquad \text{Solve for } e^{-0.0673t}$$

$$-0.0673t = \ln\frac{2}{7} \qquad \text{Rewrite as a logarithm.}$$

$$t = \frac{\ln\frac{2}{7}}{-0.0673} \approx 18.6 \text{ minutes} \qquad \text{Solve for } t.$$

The temperature of the object will be 50°C after about 18.6 minutes, or 18 minutes and 37 seconds.

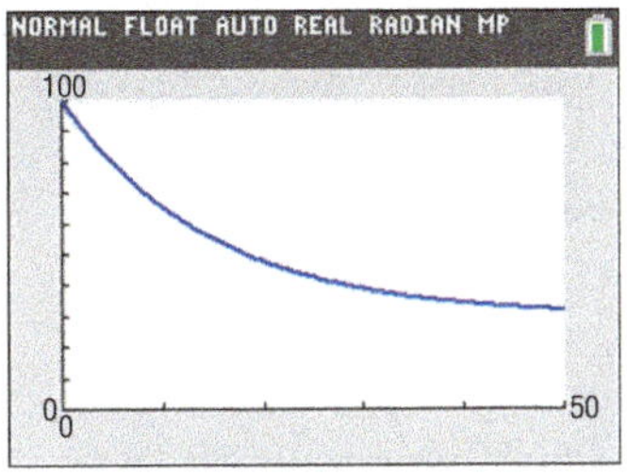

Figure 58 $Y_1 = 30 + 70e^{-0.0673x}$

(b) Figure 58 shows the graph of $y = 30 + 70e^{-0.0673x}$, where y is the temperature and x is the time.

(c) Graph $Y_1 = 30 + 70e^{-0.0673x}$ and $Y_2 = 50$, where x is time. Use INTERSECT to find that it takes $x = 18.6$ minutes (18 minutes, 37 seconds) for the temperature to cool to 50°C.

(d) Graph $Y_1 = 30 + 70e^{-0.0673x}$ and $Y_2 = 35$, where x is time. Use INTERSECT to find that it takes $x = 39.21$ minutes (39 minutes, 13 seconds) for the temperature to cool to 35°C.

(e) As t increases, the value of $e^{-0.0673t}$ approaches zero, so the value of u, the temperature of the object, approaches 30°C, the air temperature of the room. ■

Now Work PROBLEM 13

4 Use Logistic Models

The exponential growth model $A(t) = A_0e^{kt}$, $k > 0$, assumes uninhibited growth, meaning that the value of the function grows without limit. Recall that cell division can be modeled using this function, assuming that no cells die and no by-products are produced. However, cell division eventually is limited by factors such as living space and food supply. The **logistic model**, given next, can describe situations where the growth or decay of the dependent variable is limited.

Logistic Model

In a logistic model, the population P after time t is given by the function

$$P(t) = \frac{c}{1 + ae^{-bt}} \qquad \textbf{(5)}$$

where a, b, and c are constants with $a > 0$ and $c > 0$. The model is a growth model if $b > 0$; the model is a decay model if $b < 0$.

The number c is called the **carrying capacity** (for growth models) because the value $P(t)$ approaches c as t approaches infinity; that is, $\lim_{t \to \infty} P(t) = c$. The number $|b|$ is the growth rate for $b > 0$ and the decay rate for $b < 0$. Figure 59(a) on the next page shows the graph of a typical logistic growth function, and Figure 59(b) shows the graph of a typical logistic decay function.

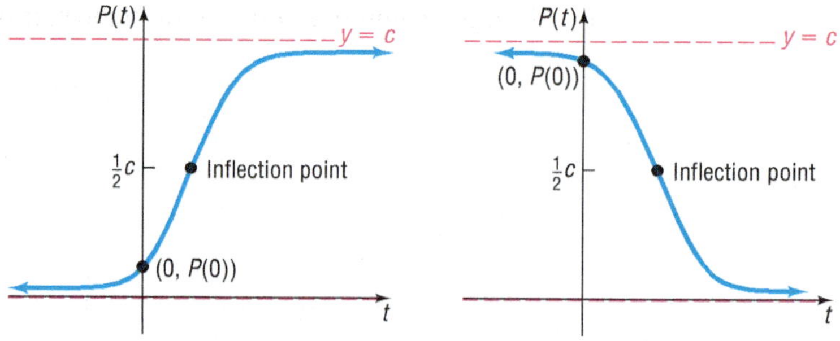

Figure 59

(a) $P(t) = \dfrac{c}{1 + ae^{-bt}}, b > 0$

Logistic growth

(b) $P(t) = \dfrac{c}{1 + ae^{-bt}}, b < 0$

Logistic decay

Based on the figures, the following properties of logistic functions emerge.

Properties of the Logistic Model, Equation (5)

1. The domain is the set of all real numbers. The range is the interval $(0, c)$, where c is the carrying capacity.
2. There are no x-intercepts; the y-intercept is $P(0)$.
3. There are two horizontal asymptotes: $y = 0$ and $y = c$.
4. $P(t)$ is an increasing function if $b > 0$ and a decreasing function if $b < 0$.
5. There is an **inflection point** where $P(t)$ equals $\frac{1}{2}$ of the carrying capacity. The inflection point is the point on the graph where the graph changes from being curved upward to being curved downward for growth functions, and the point where the graph changes from being curved downward to being curved upward for decay functions.
6. The graph is smooth and continuous, with no corners or gaps.

EXAMPLE 5 **Fruit Fly Population**

Fruit flies are placed in a half-pint milk bottle with a banana (for food) and yeast plants (for food and to provide a stimulus to lay eggs). Suppose that the fruit fly population after t days is given by

$$P(t) = \frac{230}{1 + 56.5e^{-0.37t}}$$

(a) State the carrying capacity and the growth rate.
(b) Determine the initial population.
(c) What is the population after 5 days?
(d) How long does it take for the population to reach 180?
(e) Use a graphing utility to determine how long it takes for the population to reach one-half of the carrying capacity.

Solution (a) As $t \to \infty$, $e^{-0.37t} \to 0$ and $P(t) \to \frac{230}{1}$. The carrying capacity of the half-pint bottle is 230 fruit flies. The growth rate is $|b| = |0.37| = 37\%$ per day.

(b) To find the initial number of fruit flies in the half-pint bottle, evaluate $P(0)$.

$$P(0) = \frac{230}{1 + 56.5e^{-0.37(0)}} = \frac{230}{1 + 56.5} = 4$$

So, initially, there were 4 fruit flies in the half-pint bottle.

(c) To find the number of fruit flies in the half-pint bottle after 5 days, evaluate $P(5)$.

$$P(5) = \frac{230}{1 + 56.5e^{-0.37(5)}} \approx 23 \text{ fruit flies}$$

After 5 days, there are approximately 23 fruit flies in the bottle.

(d) To determine when the population of fruit flies will be 180, solve the equation $P(t) = 180$.

$$\frac{230}{1 + 56.5e^{-0.37t}} = 180$$

$$230 = 180(1 + 56.5e^{-0.37t})$$

$$1.2778 = 1 + 56.5e^{-0.37t} \qquad \text{Divide both sides by 180.}$$

$$0.2778 = 56.5e^{-0.37t} \qquad \text{Subtract 1 from both sides.}$$

$$0.0049 = e^{-0.37t} \qquad \text{Divide both sides by 56.5.}$$

$$\ln(0.0049) = -0.37t \qquad \text{Rewrite as a logarithmic expression.}$$

$$t \approx 14.4 \text{ days} \qquad \text{Divide both sides by } -0.37.$$

It will take approximately 14.4 days (14 days, 10 hours) for the population to reach 180 fruit flies.

(e) One-half of the carrying capacity is 115 fruit flies. Solve $P(t) = 115$ by graphing $Y_1 = \frac{230}{1 + 56.5e^{-0.37x}}$ and $Y_2 = 115$ and using INTERSECT. See Figure 60. The population will reach one-half of the carrying capacity in about 10.9 days (10 days, 22 hours). ■

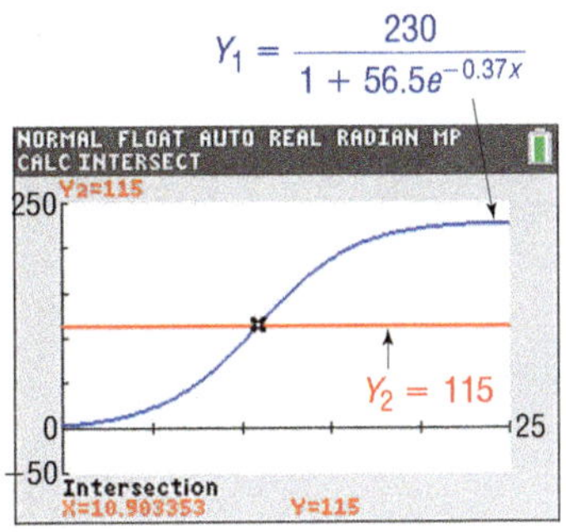

Figure 60

Look at Figure 60. Notice the point where the graph reaches 115 fruit flies (one-half of the carrying capacity): The graph changes from being curved upward to being curved downward. Using the language of calculus, we say the graph changes from increasing at an increasing rate to increasing at a decreasing rate. For any logistic growth function, when the population reaches one-half the carrying capacity, the population growth starts to slow down.

Exploration

On the same viewing rectangle, graph

$Y_1 = \frac{500}{1 + 24e^{-0.03x}}$ and $Y_2 = \frac{500}{1 + 24e^{-0.08x}}$

What effect does the growth rate $|b|$ have on the logistic growth function? ■

Now Work PROBLEM 23

EXAMPLE 6 **Wood Products**

The EFISCEN wood product model classifies wood products according to their life-span. There are four classifications: short (1 year), medium short (4 years), medium long (16 years), and long (50 years). Based on data obtained from the European Forest Institute, the percentage of remaining wood products after t years for wood products with long life-spans (such as those used in the building industry) is given by

$$P(t) = \frac{100.3952}{1 + 0.0316e^{0.0581t}}$$

(a) What is the decay rate?
(b) What is the percentage of remaining wood products after 10 years?
(c) How long does it take for the percentage of remaining wood products to reach 50%?
(d) Explain why the numerator given in the model is reasonable.

Solution (a) The decay rate is $|b| = |-0.0581| = 5.81\%$.

(b) Evaluate $P(10)$.

$$P(10) = \frac{100.3952}{1 + 0.0316e^{0.0581(10)}} \approx 95.0$$

So 95% of long-life-span wood products remain after 10 years.

(c) Solve the equation $P(t) = 50$.

$$\frac{100.3952}{1 + 0.0316e^{0.0581t}} = 50$$

$$100.3952 = 50(1 + 0.0316e^{0.0581t})$$

$$2.0079 = 1 + 0.0316e^{0.0581t} \quad \text{Divide both sides by 50.}$$

$$1.0079 = 0.0316e^{0.0581t} \quad \text{Subtract 1 from both sides.}$$

$$31.8956 = e^{0.0581t} \quad \text{Divide both sides by 0.0316.}$$

$$\ln(31.8956) = 0.0581t \quad \text{Rewrite as a logarithmic expression.}$$

$$t \approx 59.6 \text{ years} \quad \text{Divide both sides by 0.0581.}$$

It will take approximately 59.6 years for the percentage of long-life-span wood products remaining to reach 50%.

(d) The numerator of 100.3952 is reasonable because the maximum percentage of wood products remaining that is possible is 100%. ■

Now Work PROBLEM 27

6.8 Assess Your Understanding

Applications and Extensions

1. **Growth of an Insect Population** The size P of a certain insect population at time t (in days) obeys the model $P(t) = 500e^{0.02t}$.
   (a) Determine the number of insects at $t = 0$ days.
   (b) What is the growth rate of the insect population?
   (c) Graph the function using a graphing utility.
   (d) What is the population after 10 days?
   (e) When will the insect population reach 800?
   (f) When will the insect population double?

2. **Growth of Bacteria** The number N of bacteria present in a culture at time t (in hours) obeys the model $N(t) = 1000e^{0.01t}$.
   (a) Determine the number of bacteria at $t = 0$ hours.
   (b) What is the growth rate of the bacteria?
   (c) Graph the function using a graphing utility.
   (d) What is the population after 4 hours?
   (e) When will the number of bacteria reach 1700?
   (f) When will the number of bacteria double?

3. **Radioactive Decay** Strontium-90 is a radioactive material that decays according to the function $A(t) = A_0e^{-0.0244t}$, where A_0 is the initial amount present and A is the amount present at time t (in years). Assume that a scientist has a sample of 500 grams of strontium-90.
   (a) What is the decay rate of strontium-90?
   (b) Graph the function using a graphing utility.
   (c) How much strontium-90 is left after 10 years?
   (d) When will 400 grams of strontium-90 be left?
   (e) What is the half-life of strontium-90?

4. **Radioactive Decay** Iodine-131 is a radioactive material that decays according to the function $A(t) = A_0e^{-0.087t}$, where A_0 is the initial amount present and A is the amount present at time t (in days). Assume that a scientist has a sample of 100 grams of iodine-131.
   (a) What is the decay rate of iodine-131?
   (b) Graph the function using a graphing utility.
   (c) How much iodine-131 is left after 9 days?
   (d) When will 70 grams of iodine-131 be left?
   (e) What is the half-life of iodine-131?

5. **Growth of a Colony of Mosquitoes** The population of a colony of mosquitoes obeys the law of uninhibited growth.
   (a) If N is the population of the colony and t is the time in days, express N as a function of t.
   (b) If there are 1000 mosquitoes initially and there are 1800 after 1 day, what is the size of the colony after 3 days?
   (c) How long is it until there are 10,000 mosquitoes?

6. **Bacterial Growth** A culture of bacteria obeys the law of uninhibited growth.
   (a) If N is the number of bacteria in the culture and t is the time in hours, express N as a function of t.
   (b) If 500 bacteria are present initially and there are 800 after 1 hour, how many will be present in the culture after 5 hours?
   (c) How long is it until there are 20,000 bacteria?

7. **Population Growth** The population of a southern city follows the exponential law.
   (a) If N is the population of the city and t is the time in years, express N as a function of t.
   (b) If the population doubled in size over an 18-month period and the current population is 10,000, what will the population be 2 years from now?

8. **Population Decline** The population of a midwestern city follows the exponential law.
   (a) If N is the population of the city and t is the time in years, express N as a function of t.

(b) If the population decreased from 900,000 to 800,000 from 2013 to 2015, what will the population be in 2017?

9. Radioactive Decay The half-life of radium is 1690 years. If 10 grams is present now, how much will be present in 50 years?

10. Radioactive Decay The half-life of radioactive potassium is 1.3 billion years. If 10 grams is present now, how much will be present in 100 years? In 1000 years?

11. Estimating the Age of a Tree A piece of charcoal is found to contain 30% of the carbon-14 that it originally had.
(a) When did the tree from which the charcoal came die? Use 5730 years as the half-life of carbon-14.
(b) Using a graphing utility, graph the relation between the percentage of carbon-14 remaining and time.
(c) Using INTERSECT, determine the time that elapses until half of the carbon-14 remains.
(d) Verify the answer found in part (a).

12. Estimating the Age of a Fossil A fossilized leaf contains 70% of its normal amount of carbon-14.
(a) How old is the fossil? Use 5730 years as the half-life of carbon-14.
(b) Using a graphing utility, graph the relation between the percentage of carbon-14 remaining and time.
(c) Using INTERSECT, determine the time that elapses until one-fourth of the carbon-14 remains.
(d) Verify the answer found in part (a).

13. Cooling Time of a Pizza A pizza baked at 450°F is removed from the oven at 5:00 PM and placed in a room that is a constant 70°F. After 5 minutes, the pizza is at 300°F.
(a) At what time can you begin eating the pizza if you want its temperature to be 135°F?
(b) Using a graphing utility, graph the relation between temperature and time.
(c) Using INTERSECT, determine the time that needs to elapse before the pizza is 160°F.
(d) TRACE the function for large values of time. What do you notice about y, the temperature?

14. Newton's Law of Cooling A thermometer reading 72°F is placed in a refrigerator where the temperature is a constant 38°F.
(a) If the thermometer reads 60°F after 2 minutes, what will it read after 7 minutes?
(b) How long will it take before the thermometer reads 39°F?
(c) Using a graphing utility, graph the relation between temperature and time.
(d) Using INTERSECT, determine the time that must elapse before the thermometer reads 45°F.
(e) TRACE the function for large values of time. What do you notice about y, the temperature?

15. Newton's Law of Heating A thermometer reading 8°C is brought into a room with a constant temperature of 35°C.
(a) If the thermometer reads 15°C after 3 minutes, what will it read after being in the room for 5 minutes? For 10 minutes?
(b) Graph the relation between temperature and time. TRACE to verify that your answers are correct.

[**Hint:** You need to construct a formula similar to equation (4).]

16. Warming Time of a Beer Stein A beer stein has a temperature of 28°F. It is placed in a room with a constant temperature of 70°F. After 10 minutes, the temperature of the stein has risen to 35°F. What will the temperature of the stein be after 30 minutes? How long will it take the stein to reach a temperature of 45°F? (See the hint given for Problem 15.)

17. Decomposition of Chlorine in a Pool Under certain water conditions, the free chlorine (hypochlorous acid, HOCl) in a swimming pool decomposes according to the law of uninhibited decay. After shocking his pool, Ben tested the water and found the amount of free chlorine to be 2.5 parts per million (ppm). Twenty-four hours later, Ben tested the water again and found the amount of free chlorine to be 2.2 ppm. What will be the reading after 3 days (that is, 72 hours)? When the chlorine level reaches 1.0 ppm, Ben must shock the pool again. How long can Ben go before he must shock the pool again?

18. Decomposition of Dinitrogen Pentoxide At 45°C, dinitrogen pentoxide (N_2O_5) decomposes into nitrous dioxide (NO_2) and oxygen (O_2) according to the law of uninhibited decay. An initial amount of 0.25 mole of dinitrogen pentoxide decomposes to 0.15 mole in 17 minutes. How much dinitrogen pentoxide will remain after 30 minutes? How long will it take until 0.01 mole of dinitrogen pentoxide remains?

19. Decomposition of Sucrose Reacting with water in an acidic solution at 35°C, sucrose ($C_{12}H_{22}O_{11}$) decomposes into glucose ($C_6H_{12}O_6$) and fructose ($C_6H_{12}O_6$)* according to the law of uninhibited decay. An initial amount of 0.40 mole of sucrose decomposes to 0.36 mole in 30 minutes. How much sucrose will remain after 2 hours? How long will it take until 0.10 mole of sucrose remains?

20. Decomposition of Salt in Water Salt (NaCl) decomposes in water into sodium (Na^-) and chloride (Cl^-) ions according to the law of uninhibited decay. If the initial amount of salt is 25 kilograms and, after 10 hours, 15 kilograms of salt is left, how much salt is left after 1 day? How long does it take until $\frac{1}{2}$ kilogram of salt is left?

21. Radioactivity from Chernobyl After the release of radioactive material into the atmosphere from a nuclear power plant at Chernobyl (Ukraine) in 1986, the hay in Austria was contaminated by iodine-131 (half-life 8 days). If it is safe to feed the hay to cows when 10% of the iodine-131 remains, how long did the farmers need to wait to use this hay?

22. Pig Roasts The hotel Bora-Bora is having a pig roast. At noon, the chef put the pig in a large earthen oven. The pig's original temperature was 75°F. At 2:00 PM the chef checked

*Author's Note: Surprisingly, the chemical formulas for glucose and fructose are the same: This is not a typo.

the pig's temperature and was upset because it had reached only 100°F. If the oven's temperature remains a constant 325°F, at what time may the hotel serve its guests, assuming that pork is done when it reaches 175°F?

23. Population of a Bacteria Culture The logistic growth model

$$P(t) = \frac{1000}{1 + 32.33e^{-0.439t}}$$

represents the population (in grams) of a bacterium after t hours.

(a) Determine the carrying capacity of the environment.
(b) What is the growth rate of the bacteria?
(c) Determine the initial population size.
(d) Use a graphing utility to graph $P = P(t)$.
(e) What is the population after 9 hours?
(f) When will the population be 700 grams?
(g) How long does it take for the population to reach one-half the carrying capacity?

24. Population of an Endangered Species Environmentalists often capture an endangered species and transport the species to a controlled environment where the species can produce offspring and regenerate its population. Suppose that six American bald eagles are captured, transported to Montana, and set free. Based on experience, the environmentalists expect the population to grow according to the model

$$P(t) = \frac{500}{1 + 82.33e^{-0.162t}}$$

where t is measured in years.

(a) Determine the carrying capacity of the environment.
(b) What is the growth rate of the bald eagle?
(c) Use a graphing utility to graph $P = P(t)$.
(d) What is the population after 3 years?
(e) When will the population be 300 eagles?
(f) How long does it take for the population to reach one-half of the carrying capacity?

25. Invasive Species A habitat can be altered by invasive species that crowd out or replace native species. The logistic model

$$P(t) = \frac{431}{1 + 7.91e^{-0.017t}}$$

represents the number of invasive species present in the Great Lakes t years after 1900.

(a) Evaluate and interpret $P(0)$.
(b) What is the growth rate of invasive species?
(c) Use a graphing utility to graph $P = P(t)$.
(d) How many invasive species were present in the Great Lakes in 2000?
(e) In what year was the number of invasive species 175?

Source: *NOAA*

26. Word Users According to a survey by Olsten Staffing Services, the percentage of companies reporting usage of Microsoft Word t years since 1984 is given by

$$P(t) = \frac{99.744}{1 + 3.014e^{-0.799t}}$$

(a) What is the growth rate in the percentage of Microsoft Word users?
(b) Use a graphing utility to graph $P = P(t)$.
(c) What was the percentage of Microsoft Word users in 1990?
(d) During what year did the percentage of Microsoft Word users reach 90%?
(e) Explain why the numerator given in the model is reasonable. What does it imply?

27. Home Computers The logistic model

$$P(t) = \frac{95.4993}{1 + 0.0405e^{0.1968t}}$$

represents the percentage of households that do not own a personal computer t years since 1984.

(a) Evaluate and interpret $P(0)$.
(b) Use a graphing utility to graph $P = P(t)$.
(c) What percentage of households did not own a personal computer in 1995?
(d) In what year did the percentage of households that do not own a personal computer reach 10%?

Source: *U.S. Department of Commerce*

28. Farmers The logistic model

$$W(t) = \frac{14{,}656{,}248}{1 + 0.059e^{0.057t}}$$

represents the number of farm workers in the United States t years after 1910.

(a) Evaluate and interpret $W(0)$.
(b) Use a graphing utility to graph $W = W(t)$.
(c) How many farm workers were there in the United States in 2010?
(d) When did the number of farm workers in the United States reach 10,000,000?
(e) According to this model, what happens to the number of farm workers in the United States as t approaches ∞? Based on this result, do you think that it is reasonable to use this model to predict the number of farm workers in the United States in 2060? Why?

Source: *U.S. Department of Agriculture*

29. Birthdays The logistic model

$$P(n) = \frac{113.3198}{1 + 0.115e^{0.0912n}}$$

models the probability that, in a room of n people, no two people share the same birthday.

(a) Use a graphing utility to graph $P = P(n)$.
(b) In a room of $n = 15$ people, what is the probability that no two share the same birthday?

(c) How many people must be in a room before the probability that no two people share the same birthday falls below 10%?

(d) What happens to the probability as n increases? Explain what this result means.

30. Social Networking The logistic model

$$P(t) = \frac{30.3}{1 + 5.31e^{-0.703t}}$$

gives the percentage of Americans who report using social network websites "several times per day," where t represents the number of years after 2008.

(a) Evaluate and interpret $P(0)$.

(b) What is the growth rate?

(c) Use a graphing utility to graph $P = P(t)$.

(d) During 2012, what percentage of Americans visited social network websites several times per day?

(e) In what year did 28.2% of Americans visit social network websites several times per day?

Source: *Edison Research, 2014*

Problems 31 and 32 refer to the following discussion: Uninhibited growth can be modeled by exponential functions other than $A(t) = A_0e^{kt}$. For example, if an initial population P_0 requires n units of time to double, then the function $P(t) = P_0 \cdot 2^{t/n}$ models the size of the population at time t. Likewise, a population requiring n units of time to triple can be modeled by $P(t) = P_0 \cdot 3^{t/n}$.

31. Growth of a Human Population The population of a town is growing exponentially.

(a) If its population doubled in size over an 8-year period and the current population is 25,000, write an exponential function of the form $P(t) = P_0 \cdot 2^{t/n}$ that models the population.

(b) Graph the function using a graphing utility.

(c) What will the population be in 3 years?

(d) When will the population reach 80,000?

(e) Express the model from part (a) in the form $A(t) = A_0e^{kt}$.

32. Growth of an Insect Population An insect population grows exponentially.

(a) If the population triples in 20 days, and 50 insects are present initially, write an exponential function of the form $P(t) = P_0 \cdot 3^{t/n}$ that models the population.

(b) Graph the function using a graphing utility.

(c) What will the population be in 47 days?

(d) When will the population reach 700?

(e) Express the model from part (a) in the form $A(t) = A_0e^{kt}$.

Retain Your Knowledge

Problems 33–36 are based on material learned earlier in the course. The purpose of these problems is to keep the material fresh in your mind so that you are better prepared for the final exam.

33. Find the equation of the linear function f that passes through the points $(4, 1)$ and $(8, -5)$.

34. Determine whether the graphs of the linear functions $f(x) = 5x - 1$ and $g(x) = \frac{1}{5}x + 1$ are parallel, perpendicular, or neither.

35. Write the logarithmic expression $\ln\left(\frac{x^2\sqrt{y}}{z}\right)$ as the sum and/or difference of logarithms. Express powers as factors.

36. Rationalize the denominator of $\frac{10}{\sqrt[3]{25}}$.

6.9 Building Exponential, Logarithmic, and Logistic Models from Data

PREPARING FOR THIS SECTION *Before getting started, review the following:*

- Building Linear Models from Data (Section 4.2, pp. 291–294)
- Building Cubic Models from Data (Section 5.1, pp. 345–346)
- Building Quadratic Models from Data (Section 4.4, pp. 314–315)

OBJECTIVES
1 Build an Exponential Model from Data (p. 496)
2 Build a Logarithmic Model from Data (p. 498)
3 Build a Logistic Model from Data (p. 498)

Finding the linear function of best fit $(y = ax + b)$ for a set of data was discussed in Section 4.2. Likewise, finding the quadratic function of best fit $(y = ax^2 + bx + c)$ and finding the cubic function of best fit $(y = ax^3 + bx^2 + cx + d)$ were discussed in Sections 4.4 and 5.1, respectively.

In this section we discuss how to use a graphing utility to find equations of best fit that describe the relation between two variables when the relation is thought to be exponential ($y = ab^x$), logarithmic ($y = a + b \ln x$), or logistic $\left(y = \dfrac{c}{1 + ae^{-bx}}\right)$. As before, a scatter diagram of the data is drawn to help to determine the appropriate model to use.

Figure 61 shows scatter diagrams that will typically be observed for the three models. Below each scatter diagram are any restrictions on the values of the parameters.

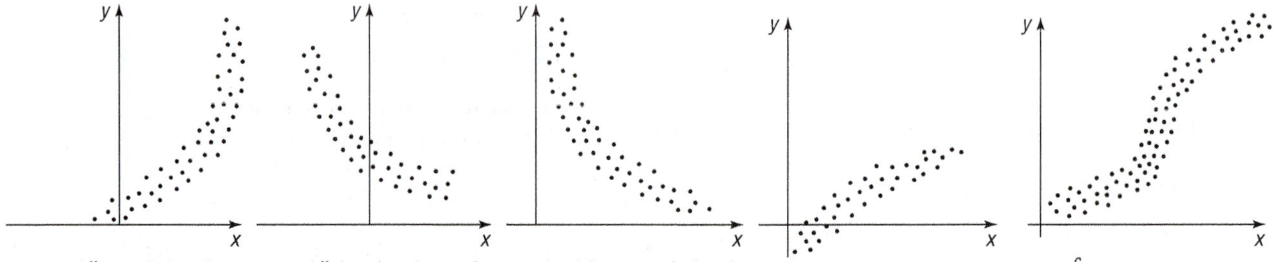

$y = ab^x, a > 0, b > 1$ — Exponential

$y = ab^x, 0 < b < 1, a > 0$ — Exponential

$y = a + b \ln x, a > 0, b < 0$ — Logarithmic

$y = a + b \ln x, a > 0, b > 0$ — Logarithmic

$y = \dfrac{c}{1 + ae^{-bx}}, a > 0, b > 0, c > 0$ — Logistic

Figure 61

Most graphing utilities have REGression options that fit data to a specific type of curve. Once the data have been entered and a scatter diagram obtained, the type of curve that you want to fit to the data is selected. Then that REGression option is used to obtain the curve of *best fit* of the type selected.

The correlation coefficient r will appear only if the model can be written as a linear expression. As it turns out, r will appear for the linear, power, exponential, and logarithmic models, since these models can be written as a linear expression. Remember, the closer $|r|$ is to 1, the better the fit.

1 Build an Exponential Model from Data

We saw in Section 6.7 that the future value of money behaves exponentially, and we saw in Section 6.8 that growth and decay models also behave exponentially. The next example shows how data can lead to an exponential model.

EXAMPLE 1 Fitting an Exponential Function to Data

Mariah deposited \$20,000 in a well-diversified mutual fund 6 years ago. The data in Table 10 represent the value of the account at the beginning of each year for the last 7 years.

Table 10

Year, x	Account Value, y
0	20,000
1	21,516
2	23,355
3	24,885
4	27,484
5	30,053
6	32,622

(a) Using a graphing utility, draw a scatter diagram with year as the independent variable.

(b) Using a graphing utility, build an exponential model from the data.

(c) Express the function found in part (b) in the form $A = A_0 e^{kt}$.

(d) Graph the exponential function found in part (b) or (c) on the scatter diagram.

(e) Using the solution to part (b) or (c), predict the value of the account after 10 years.

(f) Interpret the value of k found in part (c).

Solution

(a) Enter the data into the graphing utility and draw the scatter diagram as shown in Figure 62.

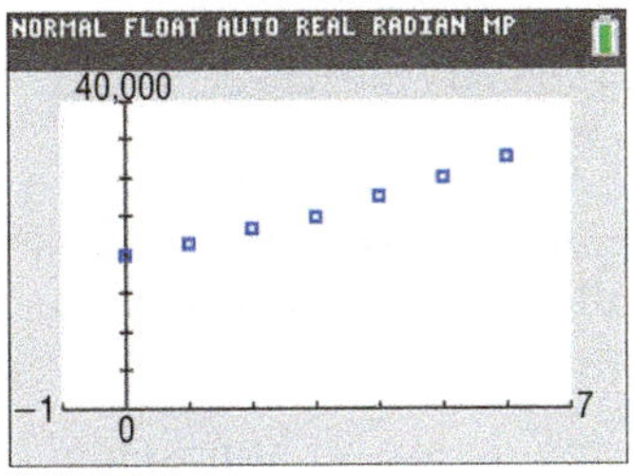

Figure 62

(b) A graphing utility fits the data in Table 10 to an exponential model of the form $y = ab^x$ using the EXPonential REGression option. Figure 63 shows that $y = ab^x = 19{,}820.43(1.085568)^x$. Notice that $|r| = 0.999$, which is close to 1, indicating a good fit.

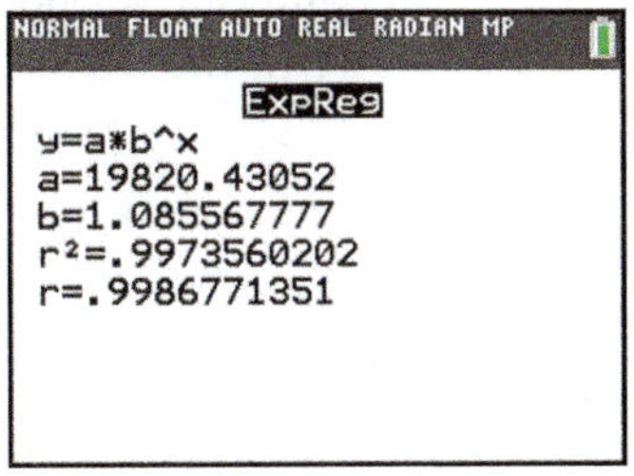

Figure 63

(c) To express $y = ab^x$ in the form $A = A_0e^{kt}$, where $x = t$ and $y = A$, proceed as follows:

$$ab^x = A_0e^{kt}$$

If $x = t = 0$, then $a = A_0$. This leads to

$$b^x = e^{kt}$$

$$b^x = (e^k)^t$$

$$b = e^k \qquad x = t$$

Because $y = ab^x = 19{,}820.43(1.085568)^x$, this means that $a = 19{,}820.43$ and $b = 1.085568$.

$$a = A_0 = 19{,}820.43 \quad \text{and} \quad b = e^k = 1.085568$$

To find k, rewrite $e^k = 1.085568$ as a logarithm to obtain

$$k = \ln(1.085568) \approx 0.08210$$

As a result, $A = A_0e^{kt} = 19{,}820.43e^{0.08210t}$.

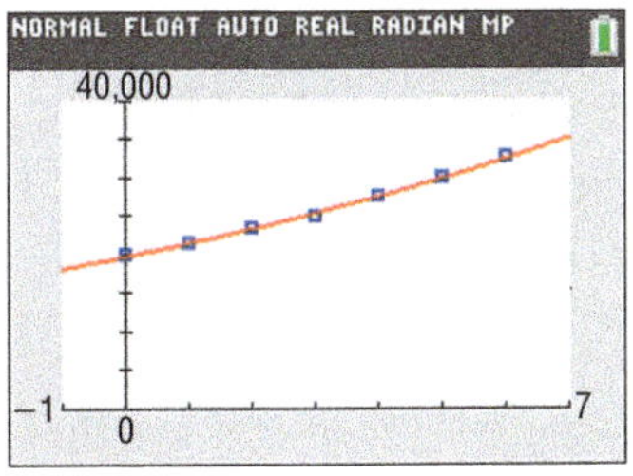

Figure 64

(d) See Figure 64 for the graph of the exponential function of best fit.

(e) Let $t = 10$ in the function found in part (c). The predicted value of the account after 10 years is

$$A = A_0e^{kt} = 19{,}820.43e^{0.08210(10)} \approx \$45{,}047$$

(f) The value of $k = 0.08210 = 8.210\%$ represents the annual growth rate of the account. It represents the rate of interest earned, assuming the account is growing continuously. ■

Now Work PROBLEM 1

2 Build a Logarithmic Model from Data

Some relations between variables follow a logarithmic model.

EXAMPLE 2 Fitting a Logarithmic Function to Data

Jodi, a meteorologist, is interested in finding a function that explains the relation between the height of a weather balloon (in kilometers) and the atmospheric pressure (measured in millimeters of mercury) on the balloon. She collects the data shown in Table 11.

Table 11

Atmospheric Pressure, p	Height, h
760	0
740	0.184
725	0.328
700	0.565
650	1.079
630	1.291
600	1.634
580	1.862
550	2.235

(a) Using a graphing utility, draw a scatter diagram of the data with atmospheric pressure as the independent variable.

(b) It is known that the relation between atmospheric pressure and height follows a logarithmic model. Using a graphing utility, build a logarithmic model from the data.

(c) Draw the logarithmic function found in part (b) on the scatter diagram.

(d) Use the function found in part (b) to predict the height of the weather balloon if the atmospheric pressure is 560 millimeters of mercury.

Solution

(a) Enter the data into the graphing utility, and draw the scatter diagram. See Figure 65.

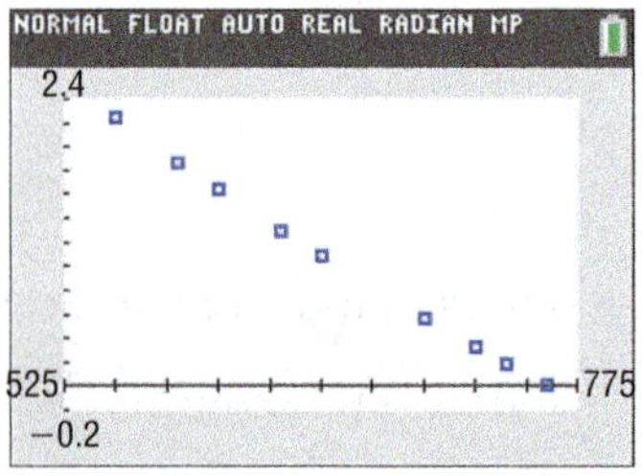

Figure 65

(b) A graphing utility fits the data in Table 11 to a logarithmic function of the form $y = a + b \ln x$ by using the LOGarithm REGression option. See Figure 66. The logarithmic model from the data is

$$h(p) = 45.7863 - 6.9025 \ln p$$

where h is the height of the weather balloon and p is the atmospheric pressure. Notice that $|r|$ is close to 1, indicating a good fit.

(c) Figure 67 shows the graph of $h(p) = 45.7863 - 6.9025 \ln p$ on the scatter diagram.

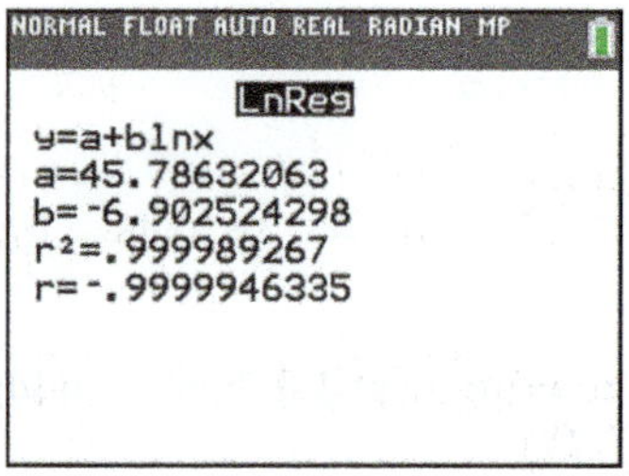

Figure 66

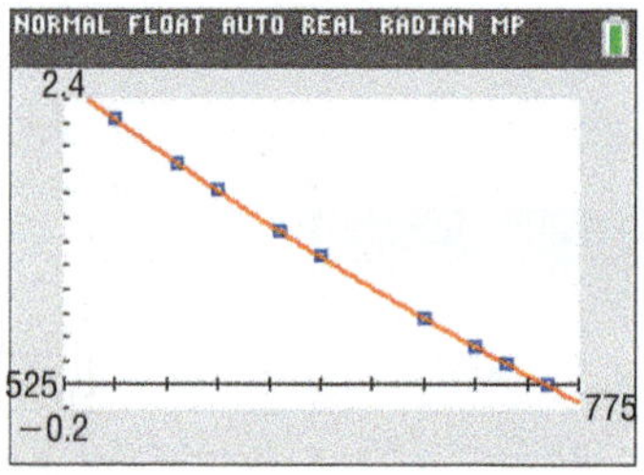

Figure 67

(d) Using the function found in part (b), Jodi predicts the height of the weather balloon when the atmospheric pressure is 560 to be

$$h(560) = 45.7863 - 6.9025 \ln 560$$
$$\approx 2.108 \text{ kilometers}$$

Now Work PROBLEM 5

3 Build a Logistic Model from Data

Logistic growth models can be used to model situations for which the value of the dependent variable is limited. Many real-world situations conform to this scenario. For example, the population of the human race is limited by the availability of natural resources such as food and shelter. When the value of the dependent variable is limited, a logistic growth model is often appropriate.

EXAMPLE 3

Fitting a Logistic Function to Data

The data in Table 12 represent the amount of yeast biomass in a culture after t hours.

Table 12

Time (hours)	Yeast Biomass	Time (hours)	Yeast Biomass	Time (hours)	Yeast Biomass
0	9.6	7	257.3	14	640.8
1	18.3	8	350.7	15	651.1
2	29.0	9	441.0	16	655.9
3	47.2	10	513.3	17	659.6
4	71.1	11	559.7	18	661.8
5	119.1	12	594.8		
6	174.6	13	629.4		

Source: *Tor Carlson (Über Geschwindigkeit and Grösse der Hefevermehrung in Würze, Biochemische Zeitschrift, Bd. 57, pp. 313–334, 1913)*

(a) Using a graphing utility, draw a scatter diagram of the data with time as the independent variable.

(b) Using a graphing utility, build a logistic model from the data.

(c) Using a graphing utility, graph the function found in part (b) on the scatter diagram.

(d) What is the predicted carrying capacity of the culture?

(e) Use the function found in part (b) to predict the population of the culture at $t = 19$ hours.

Solution

(a) See Figure 68 for a scatter diagram of the data.

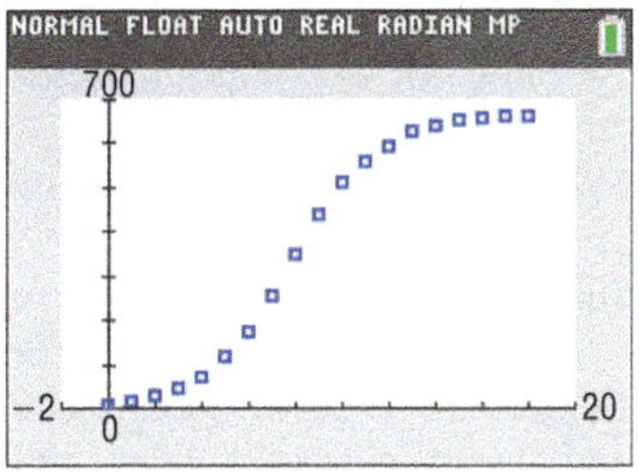

Figure 68

(b) A graphing utility fits the data in Table 12 to a logistic growth model of the form $y = \frac{c}{1 + ae^{-bx}}$ by using the LOGISTIC regression option. See Figure 69. The logistic model from the data is

$$y = \frac{663.0}{1 + 71.6e^{-0.5470x}}$$

where y is the amount of yeast biomass in the culture and x is the time.

(c) See Figure 70 for the graph of the logistic model.

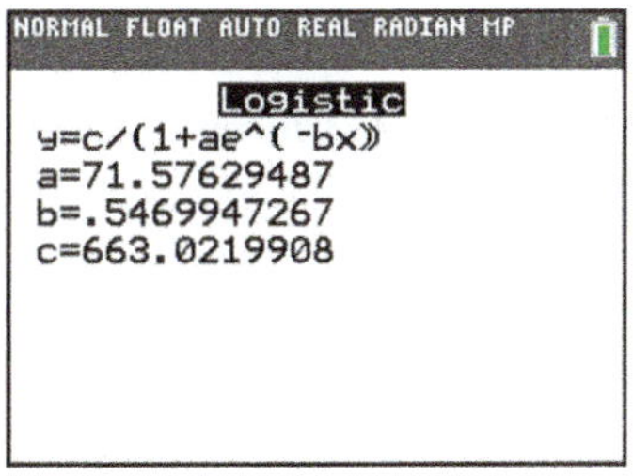

Figure 69

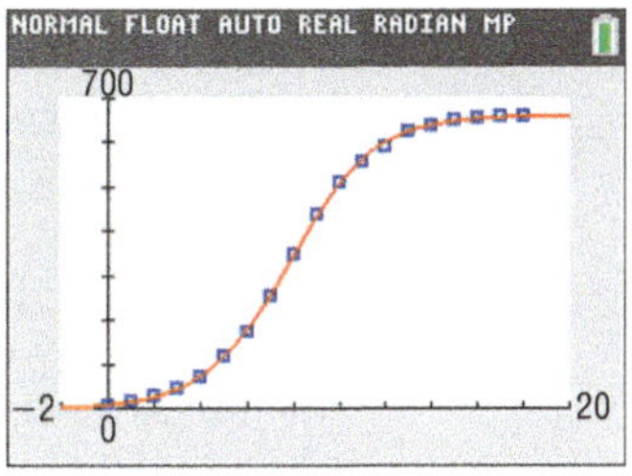

Figure 70

(d) Based on the logistic growth model found in part (b), the carrying capacity of the culture is 663.

(e) Using the logistic growth model found in part (b), the predicted amount of yeast biomass at $t = 19$ hours is

$$y = \frac{663.0}{1 + 71.6e^{-0.5470(19)}} \approx 661.5$$

Now Work PROBLEM 7

6.9 Assess Your Understanding

Applications and Extensions

1. **Biology** A strain of *E. coli*, Beu 397-recA441, is placed into a nutrient broth at 30° Celsius and allowed to grow. The following data are collected. Theory states that the number of bacteria in the petri dish will initially grow according to the law of uninhibited growth. The population is measured using an optical device in which the amount of light that passes through the petri dish is measured.

Time (hours), x	Population, y
0	0.09
2.5	0.18
3.5	0.26
4.5	0.35
6	0.50

Source: *Dr. Polly Lavery, Joliet Junior College*

(a) Draw a scatter diagram treating time as the independent variable.
(b) Using a graphing utility, build an exponential model from the data.
(c) Express the function found in part (b) in the form $N(t) = N_0e^{kt}$.
(d) Graph the exponential function found in part (b) or (c) on the scatter diagram.
(e) Use the exponential function from part (b) or (c) to predict the population at $x = 7$ hours.
(f) Use the exponential function from part (b) or (c) to predict when the population will reach 0.75.

2. **Ethanol Production** The data in the table below represent ethanol production (in billions of gallons) in the United States from 2000 to 2014.

Year	Ethanol Produced (billion gallons)	Year	Ethanol Produced (billion gallons)
2000 ($x = 0$)	1.6	2008 ($x = 8$)	9.2
2001 ($x = 1$)	1.8	2009 ($x = 9$)	10.8
2002 ($x = 2$)	2.1	2010 ($x = 10$)	13.2
2003 ($x = 3$)	2.8	2011 ($x = 11$)	13.9
2004 ($x = 4$)	3.4	2012 ($x = 12$)	13.3
2005 ($x = 5$)	3.9	2013 ($x = 13$)	13.3
2006 ($x = 6$)	4.9	2014 ($x = 14$)	14.3
2007 ($x = 7$)	6.5		

Source: *Renewable Fuels Association, 2015*

(a) Using a graphing utility, draw a scatter diagram of the data using 0 for 2000, 1 for 2001, and so on, as the independent variable.
(b) Using a graphing utility, build an exponential model from the data.
(c) Express the function found in part (b) in the form $A(t) = A_0e^{kt}$.
(d) Graph the exponential function found in part (b) or (c) on the scatter diagram.
(e) Use the model to predict the amount of ethanol that will be produced in 2016.
(f) Interpret the meaning of k in the function found in part (c).

3. **Advanced-Stage Breast Cancer** The data in the table below represent the percentage of patients who have survived after diagnosis of advanced-stage breast cancer at 6-month intervals of time.

Time after Diagnosis (years)	Percentage Surviving
0.5	95.7
1	83.6
1.5	74.0
2	58.6
2.5	47.4
3	41.9
3.5	33.6

Source: *Cancer Treatment Centers of America*

(a) Using a graphing utility, draw a scatter diagram of the data with time after diagnosis as the independent variable.
(b) Using a graphing utility, build an exponential model from the data.
(c) Express the function found in part (b) in the form $A(t) = A_0e^{kt}$.
(d) Graph the exponential function found in part (b) or (c) on the scatter diagram.
(e) What percentage of patients diagnosed with advanced-stage cancer are expected to survive for 4 years after initial diagnosis?
(f) Interpret the meaning of k in the function found in part (c).

4. **Chemistry** A chemist has a 100-gram sample of a radioactive material. He records the amount of radioactive material every week for 7 weeks and obtains the following data:

Week	Weight (grams)
0	100.0
1	88.3
2	75.9
3	69.4
4	59.1
5	51.8
6	45.5

(a) Using a graphing utility, draw a scatter diagram with week as the independent variable.
(b) Using a graphing utility, build an exponential model from the data.
(c) Express the function found in part (b) in the form $A(t) = A_0 e^{kt}$.
(d) Graph the exponential function found in part (b) or (c) on the scatter diagram.
(e) From the result found in part (b), determine the half-life of the radioactive material.
(f) How much radioactive material will be left after 50 weeks?
(g) When will there be 20 grams of radioactive material?

5. **Milk Production** The data in the table below represent the number of dairy farms (in thousands) and the amount of milk produced (in billions of pounds) in the United States for various years.

Year	Dairy Farms (thousands)	Milk Produced (billion pounds)
1980	334	128
1985	269	143
1990	193	148
1995	140	155
2000	105	167
2005	78	177
2010	63	193

Source: *Statistical Abstract of the United States, 2012*

(a) Using a graphing utility, draw a scatter diagram of the data with the number of dairy farms as the independent variable.
(b) Using a graphing utility, build a logarithmic model from the data.
(c) Graph the logarithmic function found in part (b) on the scatter diagram.
(d) In 2008, there were 67 thousand dairy farms in the United States. Use the function in part (b) to predict the amount of milk produced in 2008.
(e) The actual amount of milk produced in 2008 was 190 billion pounds. How does your prediction in part (d) compare to this?

6. **Social Networking** The data in the table below represent the percent of U.S. citizens aged 12 and older who have a profile on at least one social network.

Year	Percent on a Social Networking Site
2008 ($x = 8$)	24
2009 ($x = 9$)	34
2010 ($x = 10$)	48
2011 ($x = 11$)	52
2012 ($x = 12$)	53
2013 ($x = 13$)	62
2014 ($x = 14$)	67

Source: *Edison Research, 2014*

(a) Using a graphing utility, draw a scatter diagram of the data using 8 for 2008, 9 for 2009, and so on, as the independent variable, and percent on social networking site as the dependent variable.
(b) Using a graphing utility, build a logarithmic model from the data.
(c) Graph the logarithmic function found in part (b) on the scatter diagram.
(d) Use the model to predict the percent of U.S. citizens on social networking sites in 2015.
(e) Use the model to predict the year in which 90% of U.S. citizens will be on social networking sites.

7. **Population Model** The following data represent the population of the United States. An ecologist is interested in building a model that describes the population of the United States.

Year	Population
1900	76,212,168
1910	92,228,496
1920	106,021,537
1930	123,202,624
1940	132,164,569
1950	151,325,798
1960	179,323,175
1970	203,302,031
1980	226,542,203
1990	248,709,873
2000	281,421,906
2010	308,745,538

Source: *U.S. Census Bureau*

(a) Using a graphing utility, draw a scatter diagram of the data using years since 1900 as the independent variable and population as the dependent variable.
(b) Using a graphing utility, build a logistic model from the data.
(c) Using a graphing utility, draw the function found in part (b) on the scatter diagram.
(d) Based on the function found in part (b), what is the carrying capacity of the United States?
(e) Use the function found in part (b) to predict the population of the United States in 2012.
(f) When will the United States population be 350,000,000?
(g) Compare actual U.S. Census figures to the predictions found in parts (e) and (f). Discuss any differences.

8. Population Model The data on the right represent the world population. An ecologist is interested in building a model that describes the world population.

(a) Using a graphing utility, draw a scatter diagram of the data using years since 2000 as the independent variable and population as the dependent variable.
(b) Using a graphing utility, build a logistic model from the data.
(c) Using a graphing utility, draw the function found in part (b) on the scatter diagram.
(d) Based on the function found in part (b), what is the carrying capacity of the world?
(e) Use the function found in part (b) to predict the population of the world in 2021.
(f) When will world population be 10 billion?

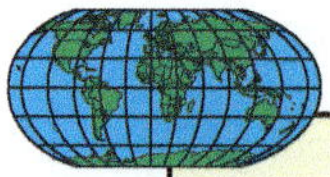

Year	Population (billions)	Year	Population (billions)
2001	6.17	2008	6.71
2002	6.24	2009	6.79
2003	6.32	2010	6.87
2004	6.40	2011	6.94
2005	6.47	2012	7.02
2006	6.55	2013	7.10
2007	6.63	2014	7.18

Source: *U.S. Census Bureau*

9. Cell Phone Towers The following data represent the number of cell sites in service in the United States from 1985 to 2013 at the end of each year.

Year	Cell Sites (thousands)	Year	Cell Sites (thousands)	Year	Cell Sites (thousands)
1985 ($x = 1$)	0.9	1995 ($x = 11$)	22.7	2005 ($x = 21$)	183.7
1986 ($x = 2$)	1.5	1996 ($x = 12$)	30.0	2006 ($x = 22$)	195.6
1987 ($x = 3$)	2.3	1997 ($x = 13$)	51.6	2007 ($x = 23$)	213.3
1988 ($x = 4$)	3.2	1998 ($x = 14$)	65.9	2008 ($x = 24$)	242.1
1989 ($x = 5$)	4.2	1999 ($x = 15$)	81.7	2009 ($x = 25$)	247.1
1990 ($x = 6$)	5.6	2000 ($x = 16$)	104.3	2010 ($x = 26$)	253.1
1991 ($x = 7$)	7.8	2001 ($x = 17$)	127.5	2011 ($x = 27$)	283.4
1992 ($x = 8$)	10.3	2002 ($x = 18$)	139.3	2012 ($x = 28$)	301.8
1993 ($x = 9$)	12.8	2003 ($x = 19$)	163.0	2013 ($x = 29$)	304.4
1994 ($x = 10$)	17.9	2004 ($x = 20$)	175.7		

Source: *©2014 CTIA-The Wireless Association®. All Rights Reserved.*

(a) Using a graphing utility, draw a scatter diagram of the data using 1 for 1985, 2 for 1986, and so on as the independent variable, and number of cell sites as the dependent variable.
(b) Using a graphing utility, build a logistic model from the data.
(c) Graph the logistic function found in part (b) on the scatter diagram.
(d) What is the predicted carrying capacity for cell sites in the United States?
(e) Use the model to predict the number of cell sites in the United States at the end of 2019.

10. Cable Rates The data on the right represent the average monthly rate charged for expanded basic cable television in the United States from 1995 to 2014. A market researcher believes that external factors, such as the growth of satellite television and internet programming, have affected the cost of basic cable. She is interested in building a model that will describe the average monthly cost of basic cable.

(a) Using a graphing utility, draw a scatter diagram of the data using 0 for 1995, 1 for 1996, and so on, as the independent variable and average monthly rate as the dependent variable.
(b) Using a graphing utility, build a logistic model from the data.
(c) Graph the logistic function found in part (b) on the scatter diagram.
(d) Based on the model found in part (b), what is the maximum possible average monthly rate for basic cable?
(e) Use the model to predict the average rate for basic cable in 2018.

Year	Average Monthly Rate (dollars)	Year	Average Monthly Rate (dollars)
1995 ($x = 0$)	22.35	2005 ($x = 10$)	43.04
1996 ($x = 1$)	24.28	2006 ($x = 11$)	45.26
1997 ($x = 2$)	26.31	2007 ($x = 12$)	47.27
1998 ($x = 3$)	27.88	2008 ($x = 13$)	49.65
1999 ($x = 4$)	28.94	2009 ($x = 14$)	52.37
2000 ($x = 5$)	31.22	2010 ($x = 15$)	54.44
2001 ($x = 6$)	33.75	2011 ($x = 16$)	57.46
2002 ($x = 7$)	36.47	2012 ($x = 17$)	61.63
2003 ($x = 8$)	38.95	2013 ($x = 18$)	64.41
2004 ($x = 9$)	41.04	2014 ($x = 19$)	66.61

Source: *Federal Communications Commission, 2014*

Mixed Practice

11. **Online Advertising Revenue** The data in the table below represent the U.S. online advertising revenues for the years 2005–2014.

Year	U.S. Online Advertising Revenue ($ billions)
2005 ($x = 0$)	12.5
2006 ($x = 1$)	16.9
2007 ($x = 2$)	21.2
2008 ($x = 3$)	23.4
2009 ($x = 4$)	22.7
2010 ($x = 5$)	26.0
2011 ($x = 6$)	31.7
2012 ($x = 7$)	36.6
2013 ($x = 8$)	42.8
2014 ($x = 9$)	49.5

Source: *marketingcharts.com*

(a) Using a graphing utility, draw a scatter diagram of the data using 0 for 2005, 1 for 2006, and so on as the independent variable, and online advertising revenue as the dependent variable.
(b) Based on the scatter diagram drawn in part (a), decide what model (linear, quadratic, cubic, exponential, logarithmic, or logistic) that you think best describes the relation between year and revenue.
(c) Using a graphing utitlity, find the model of best fit.
(d) Using a graphing utility, draw the model of best fit on the scatter diagram drawn in part (a).
(e) Use your model to predict the online advertising revenue in 2016.

12. **Age versus Total Cholesterol** The following data represent the age and average total cholesterol for adult males at various ages.

Age	Total Cholesterol
27	189
40	205
50	215
60	210
70	210
80	194

(a) Using a graphing utility, draw a scatter diagram of the data using age, x, as the independent variable and total cholesterol, y, as the dependent variable.
(b) Based on the scatter diagram drawn in part (a), decide on a model (linear, quadratic, cubic, exponential, logarithmic, or logistic) that you think best describes the relation between age and total cholesterol. Be sure to justify your choice of model.
(c) Using a graphing utility, find the model of best fit.
(d) Using a graphing utility, draw the model of best fit on the scatter diagram drawn in part (a).
(e) Use your model to predict the total cholesterol of a 35-year-old male.

13. **Golfing** The data below represent the expected percentage of putts that will be made by professional golfers on the PGA Tour depending on distance. For example, it is expected that 99.3% of 2-foot putts will be made.

Distance (feet)	Expected Percentage	Distance (feet)	Expected Percentage
2	99.3	14	25.0
3	94.8	15	22.0
4	85.8	16	20.0
5	74.7	17	19.0
6	64.7	18	17.0
7	55.6	19	16.0
8	48.5	20	14.0
9	43.4	21	13.0
10	38.3	22	12.0
11	34.2	23	11.0
12	30.1	24	11.0
13	27.0	25	10.0

Source: *TheSandTrap.com*

(a) Using a graphing utility, draw a scatter diagram of the data with distance as the independent variable.
(b) Based on the scatter diagram drawn in part (a), decide on a model (linear, quadratic, cubic, exponential, logarithmic, or logistic) that you think best describes the relation between distance and expected percentage. Be sure to justify your choice of model.
(c) Using a graphing utility, find the model of best fit.
(d) Graph the function found in part (c) on the scatter diagram.
(e) Use the function found in part (c) to predict what percentage of 30-foot putts will be made.

Retain Your Knowledge

Problems 14–17 are based on material learned earlier in the course. The purpose of these problems is to keep the material fresh in your mind so that you are better prepared for the final exam.

14. Construct a polynomial function that might have the graph shown. (More than one answer is possible.)

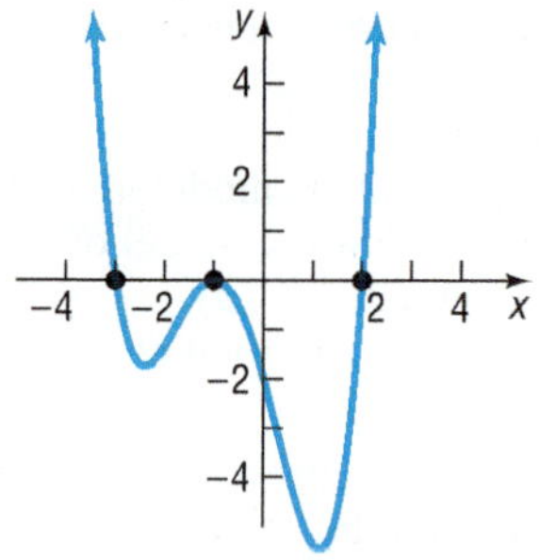

15. Rationalize the denominator of $\frac{3}{\sqrt{2}}$.

16. Use the Pythagorean Theorem to find the exact length of the unlabeled side in the given right triangle.

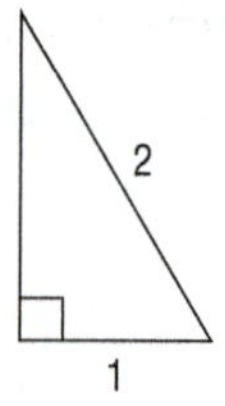

17. Graph the equation $(x-3)^2 + y^2 = 25$.

Chapter Review

Things to Know

Composite function (p. 408)
$(f \circ g)(x) = f(g(x))$ The domain of $f \circ g$ is the set of all numbers x in the domain of g for which $g(x)$ is in the domain of f.

One-to-one function (p. 416)
A function for which any two different inputs in the domain correspond to two different outputs in the range

For any choice of elements x_1, x_2 in the domain of f, if $x_1 \neq x_2$, then $f(x_1) \neq f(x_2)$.

Horizontal-line test (p. 417)
If every horizontal line intersects the graph of a function f in at most one point, f is one-to-one.

Inverse function f^{-1} of f (pp. 418–420, 421)
Domain of f = range of f^{-1}; range of f = domain of f^{-1}

$f^{-1}(f(x)) = x$ for all x in the domain of f

$f(f^{-1}(x)) = x$ for all x in the domain of f^{-1}

The graphs of f and f^{-1} are symmetric with respect to the line $y = x$.

Properties of the exponential function (pp. 430, 433–434, 435)

$f(x) = Ca^x, \quad a > 1, C > 0$
- Domain: the interval $(-\infty, \infty)$
- Range: the interval $(0, \infty)$
- x-intercepts: none; y-intercept: C
- Horizontal asymptote: x-axis $(y = 0)$ as $x \to -\infty$
- Increasing; one-to-one; smooth; continuous
- See Figure 24 for a typical graph.

$f(x) = Ca^x, \quad 0 < a < 1, C > 0$
- Domain: the interval $(-\infty, \infty)$
- Range: the interval $(0, \infty)$
- x-intercepts: none; y-intercept: C
- Horizontal asymptote: x-axis $(y = 0)$ as $x \to \infty$
- Decreasing; one-to-one; smooth; continuous
- See Figure 29 for a typical graph.

Number e (p. 436)
Value approached by the expression $\left(1 + \frac{1}{n}\right)^n$ as $n \to \infty$; that is, $\lim_{n\to\infty}\left(1 + \frac{1}{n}\right)^n = e$.

Property of exponents (p. 437)
If $a^u = a^v$, then $u = v$.

Properties of the logarithmic function (pp. 445–446, 447, 448)

$f(x) = \log_a x, \quad a > 1$

$(y = \log_a x$ means $x = a^y)$
- Domain: the interval $(0, \infty)$
- Range: the interval $(-\infty, \infty)$
- x-intercept: 1; y-intercept: none
- Vertical asymptote: $x = 0$ (y-axis)
- Increasing; one-to-one; smooth; continuous
- See Figure 45(a) for a typical graph.

	$f(x) = \log_a x, \quad 0 < a < 1$ $(y = \log_a x \text{ means } x = a^y)$	Domain: the interval $(0, \infty)$ Range: the interval $(-\infty, \infty)$ x-intercept: 1; y-intercept: none Vertical asymptote: $x = 0$ (y-axis) Decreasing; one-to-one; smooth; continuous See Figure 45(b) for a typical graph.

Natural logarithm (p. 449) $y = \ln x$ means $x = e^y$.

Properties of logarithms (pp. 459–460, 462)

$\log_a 1 = 0 \qquad \log_a a = 1 \qquad a^{\log_a M} = M \qquad \log_a a^r = r$

$\log_a(MN) = \log_a M + \log_a N \qquad \log_a\left(\dfrac{M}{N}\right) = \log_a M - \log_a N$

$\log_a M^r = r \log_a M \qquad a^r = e^{r \ln a}$

If $M = N$, then $\log_a M = \log_a N$.

If $\log_a M = \log_a N$, then $M = N$.

Formulas

Change-of-Base Formula (p. 463) $\log_a M = \dfrac{\log_b M}{\log_b a}$

Compound Interest Formula (p. 476) $A = P \cdot \left(1 + \dfrac{r}{n}\right)^{nt}$

Continuous compounding (p. 478) $A = Pe^{rt}$

Effective rate of interest (p. 478) Compounding n times per year: $r_e = \left(1 + \dfrac{r}{n}\right)^n - 1$

Continuous compounding: $r_e = e^r - 1$

Present Value Formulas (p. 479) $P = A \cdot \left(1 + \dfrac{r}{n}\right)^{-nt}$ or $P = Ae^{-rt}$

Growth and decay (p. 485, 487) $A(t) = A_0 e^{kt}$

Newton's Law of Cooling (p. 488) $u(t) = T + (u_0 - T)e^{kt} \quad k < 0$

Logistic model (p. 489) $P(t) = \dfrac{c}{1 + ae^{-bt}}$

Objectives

Section		You should be able to ...	Examples	Review Exercises
6.1	1	Form a composite function (p. 408)	1, 2, 4, 5	1–7
	2	Find the domain of a composite function (p. 409)	2–4	5–7
6.2	1	Determine whether a function is one-to-one (p. 416)	1, 2	8(a), 9
	2	Determine the inverse of a function defined by a map or a set of ordered pairs (p. 418)	3, 4	8(b)
	3	Obtain the graph of the inverse function from the graph of the function (p. 421)	7	9
	4	Find the inverse of a function defined by an equation (p. 422)	8, 9, 10	10–13
6.3	1	Evaluate exponential functions (p. 428)	1	14(a), (c), 47(a)
	2	Graph exponential functions (p. 432)	3–6	31–33
	3	Define the number e (p. 436)	pg. 436	33
	4	Solve exponential equations (p. 437)	7, 8	35, 36, 39, 41
6.4	1	Change exponential statements to logarithmic statements and logarithmic statements to exponential statements (p. 446)	2, 3	15, 16
	2	Evaluate logarithmic expressions (p. 446)	4	14(b), (d), 19, 46(b), 48(a), 49
	3	Determine the domain of a logarithmic function (p. 447)	5	17, 18, 34(a)
	4	Graph logarithmic functions (p. 448)	6, 7	34(b), 46(a)
	5	Solve logarithmic equations (p. 452)	8, 9	37, 40, 46(c), 48(b)

Section	You should be able to ...	Examples	Review Exercises
6.5	1 Work with the properties of logarithms (p. 458)	1, 2	20, 21
	2 Write a logarithmic expression as a sum or difference of logarithms (p. 460)	3–5	22–25
	3 Write a logarithmic expression as a single logarithm (p. 461)	6	26–28
	4 Evaluate a logarithm whose base is neither 10 nor e (p. 462)	7, 8	29
	5 Graph a logarithmic function whose base is neither 10 nor e (p. 464)	9	30
6.6	1 Solve logarithmic equations (p. 467)	1–3	37, 43
	2 Solve exponential equations (p. 469)	4–7	38, 42, 44, 45
	3 Solve logarithmic and exponential equations using a graphing utility (p. 471)	8	35–45
6.7	1 Determine the future value of a lump sum of money (p. 475)	1–3	50
	2 Calculate effective rates of return (p. 478)	4	50
	3 Determine the present value of a lump sum of money (p. 479)	5	51
	4 Determine the rate of interest or the time required to double a lump sum of money (p. 480)	6, 7	50
6.8	1 Find equations of populations that obey the law of uninhibited growth (p. 484)	1, 2	54
	2 Find equations of populations that obey the law of decay (p. 487)	3	52, 55
	3 Use Newton's Law of Cooling (p. 488)	4	53
	4 Use logistic models (p. 489)	5, 6	56
6.9	1 Build an exponential model from data (p. 496)	1	57
	2 Build a logarithmic model from data (p. 498)	2	58
	3 Build a logistic model from data (p. 498)	3	59

Review Exercises

1. Evaluate each expression using the graphs of $y = f(x)$ and $y = g(x)$ shown in the figure.

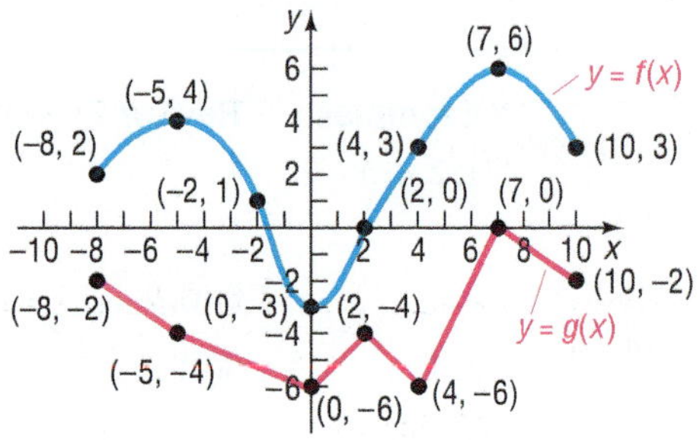

(a) $(g \circ f)(-8)$
(b) $(f \circ g)(-8)$
(c) $(g \circ g)(7)$
(d) $(g \circ f)(-5)$

In Problems 2–4, for the given functions f and g find:
(a) $(f \circ g)(2)$
(b) $(g \circ f)(-2)$
(c) $(f \circ f)(4)$
(d) $(g \circ g)(-1)$

2. $f(x) = 3x - 5;\quad g(x) = 1 - 2x^2$

3. $f(x) = \sqrt{x+2}:\quad g(x) = 2x^2 + 1$

4. $f(x) = e^x;\quad g(x) = 3x - 2$

In Problems 5–7, find $f \circ g$, $g \circ f$, $f \circ f$, and $g \circ g$ for each pair of functions. State the domain of each composite function.

5. $f(x) = 2 - x;\quad g(x) = 3x + 1$

6. $f(x) = \sqrt{3x};\quad g(x) = 1 + x + x^2$

7. $f(x) = \dfrac{x+1}{x-1};\quad g(x) = \dfrac{1}{x}$

In Problem 8, (a) verify that the function is one-to-one, and (b) find the inverse of the given function.

8. $\{(1, 2), (3, 5), (5, 8), (6, 10)\}$

In Problem 9, state why the graph of the function is one-to-one. Then draw the graph of the inverse function f^{-1}. For convenience (and as a hint), the graph of $y = x$ is also given.

9.

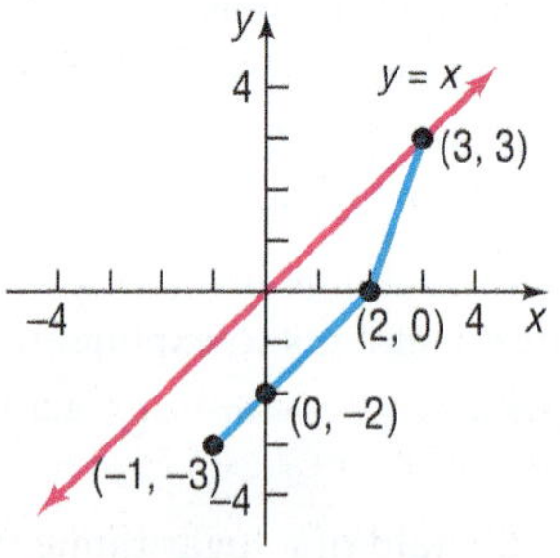

In Problems 10–13, the function f is one-to-one. Find the inverse of each function and check your answer. State the domain and the range of f and f^{-1}.

10. $f(x) = \dfrac{2x+3}{5x-2}$

11. $f(x) = \dfrac{1}{x-1}$

12. $f(x) = \sqrt{x-2}$

13. $f(x) = x^{1/3} + 1$

In Problem 14, $f(x) = 3^x$ and $g(x) = \log_3 x$.

14. Evaluate: (a) $f(4)$ (b) $g(9)$ (c) $f(-2)$ (d) $g\left(\dfrac{1}{27}\right)$

15. Convert $5^2 = z$ to an equivalent statement involving a logarithm.

16. Convert $\log_5 u = 13$ to an equivalent statement involving an exponent.

In Problems 17 and 18, find the domain of each logarithmic function.

17. $f(x) = \log(3x-2)$

18. $H(x) = \log_2(x^2 - 3x + 2)$

In Problems 19–21, evaluate each expression. Do not use a calculator.

19. $\log_2\left(\dfrac{1}{8}\right)$

20. $\ln e^{\sqrt{2}}$

21. $2^{\log_2 0.4}$

In Problems 22–25, write each expression as the sum and/or difference of logarithms. Express powers as factors.

22. $\log_3\left(\dfrac{uv^2}{w}\right)$, $u > 0, v > 0, w > 0$

23. $\log_2(a^2\sqrt{b})^4$, $a > 0, b > 0$

24. $\log(x^2\sqrt{x^3+1})$, $x > 0$

25. $\ln\left(\dfrac{2x+3}{x^2-3x+2}\right)^2$, $x > 2$

In Problems 26–28, write each expression as a single logarithm.

26. $3\log_4 x^2 + \dfrac{1}{2}\log_4\sqrt{x}$

27. $\ln\left(\dfrac{x-1}{x}\right) + \ln\left(\dfrac{x}{x+1}\right) - \ln(x^2-1)$

28. $\dfrac{1}{2}\ln(x^2+1) - 4\ln\dfrac{1}{2} - \dfrac{1}{2}[\ln(x-4) + \ln x]$

29. Use the Change-of-Base Formula and a calculator to evaluate $\log_4 19$. Round your answer to three decimal places.

30. Graph $y = \log_3 x$ using a graphing utility and the Change-of-Base Formula.

In Problems 31–34, use the given function f to:

(a) Find the domain of f. *(b) Graph f.* *(c) From the graph, determine the range and any asymptotes of f.*
(d) Find f^{-1}, the inverse of f. *(e) Find the domain and the range of f^{-1}.* *(f) Graph f^{-1}.*

31. $f(x) = 2^{x-3}$

32. $f(x) = 1 + 3^{-x}$

33. $f(x) = 3e^{x-2}$

34. $f(x) = \dfrac{1}{2}\ln(x+3)$

In Problems 35–45, solve each equation. Express irrational solutions in exact form and as a decimal rounded to 3 decimal places. Verify your results using a graphing utility.

35. $8^{6+3x} = 4$

36. $3^{x^2+x} = \sqrt{3}$

37. $\log_x 64 = -3$

38. $5^x = 3^{x+2}$

39. $25^{2x} = 5^{x^2-12}$

40. $\log_3\sqrt{x-2} = 2$

41. $8 = 4^{x^2} \cdot 2^{5x}$

42. $2^x \cdot 5 = 10^x$

43. $\log_6(x+3) + \log_6(x+4) = 1$

44. $e^{1-x} = 5$

45. $9^x + 4 \cdot 3^x - 3 = 0$

---

46. Suppose that $f(x) = \log_2(x-2) + 1$.

(a) Graph f.

(b) What is $f(6)$? What point is on the graph of f?

(c) Solve $f(x) = 4$. What point is on the graph of f?

(d) Based on the graph drawn in part (a), solve $f(x) > 0$.

(e) Find $f^{-1}(x)$. Graph f^{-1} on the same Cartesian plane as f.

47. Amplifying Sound An amplifier's power output P (in watts) is related to its decibel voltage gain d by the formula

$$P = 25e^{0.1d}$$

(a) Find the power output for a decibel voltage gain of 4 decibels.

(b) For a power output of 50 watts, what is the decibel voltage gain?

48. Limiting Magnitude of a Telescope A telescope is limited in its usefulness by the brightness of the star that it is aimed at and by the diameter of its lens. One measure of a star's brightness is its *magnitude;* the dimmer the star, the larger its magnitude. A formula for the limiting magnitude L of a telescope, that is, the magnitude of the dimmest star that it can be used to view, is given by

$$L = 9 + 5.1 \log d$$

where d is the diameter (in inches) of the lens.
(a) What is the limiting magnitude of a 3.5-inch telescope?
(b) What diameter is required to view a star of magnitude 14?

49. Salvage Value The number of years n for a piece of machinery to depreciate to a known salvage value can be found using the formula

$$n = \frac{\log s - \log i}{\log(1 - d)}$$

where s is the salvage value of the machinery, i is its initial value, and d is the annual rate of depreciation.
(a) How many years will it take for a piece of machinery to decline in value from \$90,000 to \$10,000 if the annual rate of depreciation is 0.20 (20%)?
(b) How many years will it take for a piece of machinery to lose half of its value if the annual rate of depreciation is 15%?

50. Funding a College Education A child's grandparents purchase a \$10,000 bond fund that matures in 18 years to be used for her college education. The bond fund pays 4% interest compounded semiannually. How much will the bond fund be worth at maturity? What is the effective rate of interest? How long will it take the bond to double in value under these terms?

51. Funding a College Education A child's grandparents wish to purchase a bond that matures in 18 years to be used for her college education. The bond pays 4% interest compounded semiannually. How much should they pay so that the bond will be worth \$85,000 at maturity?

52. Estimating the Date That a Prehistoric Man Died The bones of a prehistoric man found in the desert of New Mexico contain approximately 5% of the original amount of carbon 14. If the half-life of carbon 14 is 5730 years, approximately how long ago did the man die?

53. Temperature of a Skillet A skillet is removed from an oven whose temperature is 450°F and placed in a room whose temperature is 70°F. After 5 minutes, the temperature of the skillet is 400°F. How long will it be until its temperature is 150°F?

54. World Population The annual growth rate of the world's population in 2015 was $k = 1.08\% = 0.0108$. The population of the world in 2015 was 7,214,958,996. Letting $t = 0$ represent 2015, use the uninhibited growth model to predict the world's population in the year 2020.
Source: *U.S. Census Bureau*

55. Radioactive Decay The half-life of cobalt is 5.27 years. If 100 grams of radioactive cobalt is present now, how much will be present in 20 years? In 40 years?

56. Logistic Growth The logistic growth model

$$P(t) = \frac{0.8}{1 + 1.67e^{-0.16t}}$$

represents the proportion of new cars with a global positioning system (GPS). Let $t = 0$ represent 2006, $t = 1$ represent 2007, and so on.
(a) What proportion of new cars in 2006 had a GPS?
(b) Determine the maximum proportion of new cars that have a GPS.
(c) Using a graphing utility, graph $P = P(t)$.
(d) When will 75% of new cars have a GPS?

57. Rising Tuition The following data represent the average in-state tuition and fees (in 2013 dollars) at public four-year colleges and universities in the United States from the academic year 1983–84 to the academic year 2013–14.

Academic Year	Tuition and Fees (2013 dollars)
1983–84 ($x = 0$)	2684
1988–89 ($x = 5$)	3111
1993–94 ($x = 10$)	4101
1998–99 ($x = 15$)	4648
2003–04 ($x = 20$)	5900
2008–09 ($x = 25$)	7008
2013–14 ($x = 30$)	8893

Source: *The College Board*

(a) Using a graphing utility, draw a scatter diagram with academic year as the independent variable.
(b) Using a graphing utility, build an exponential model from the data.
(c) Express the function found in part (b) in the form $A(t) = A_0 e^{kt}$.
(d) Graph the exponential function found in part (b) or (c) on the scatter diagram.
(e) Predict the academic year when the average tuition will reach \$12,000.

58. Wind Chill Factor The following data represent the wind speed (mph) and wind chill factor at an air temperature of 15°F.

Wind Speed (mph)	Wind Chill Factor (°F)
5	7
10	3
15	0
20	−2
25	−4
30	−5
35	−7

Source: *U.S. National Weather Service*

(a) Using a graphing utility, draw a scatter diagram with wind speed as the independent variable.
(b) Using a graphing utility, build a logarithmic model from the data.
(c) Using a graphing utility, draw the logarithmic function found in part (b) on the scatter diagram.
(d) Use the function found in part (b) to predict the wind chill factor if the air temperature is 15°F and the wind speed is 23 mph.

59. **Spreading of a Disease** Jack and Diane live in a small town of 50 people. Unfortunately, both Jack and Diane have a cold. Those who come in contact with someone who has this cold will themselves catch the cold. The following data represent the number of people in the small town who have caught the cold after t days.

Days, t	Number of People with Cold, C
0	2
1	4
2	8
3	14
4	22
5	30
6	37
7	42
8	44

(a) Using a graphing utility, draw a scatter diagram of the data. Comment on the type of relation that appears to exist between the days and number of people with a cold.

(b) Using a graphing utility, build a logistic model from the data.

(c) Graph the function found in part (b) on the scatter diagram.

(d) According to the function found in part (b), what is the maximum number of people who will catch the cold? In reality, what is the maximum number of people who could catch the cold?

(e) Sometime between the second and third day, 10 people in the town had a cold. According to the model found in part (b), when did 10 people have a cold?

(f) How long will it take for 46 people to catch the cold?

Chapter Test

CHAPTER **Test Prep** VIDEOS

The Chapter Test Prep Videos are step-by-step solutions available in MyMathLab®, or on this text's YouTube™ Channel. Flip back to the Resources for Success page for a link to this text's YouTube channel.

1. Given $f(x) = \dfrac{x+2}{x-2}$ and $g(x) = 2x + 5$, find:
   (a) $f \circ g$ and state its domain (b) $(g \circ f)(-2)$
   (c) $(f \circ g)(-2)$

2. Determine whether the function is one-to-one.
   (a) $y = 4x^2 + 3$
   (b) $y = \sqrt{x+3} - 5$

3. Find the inverse of $f(x) = \dfrac{2}{3x-5}$ and check your answer. State the domain and the range of f and f^{-1}.

4. If the point $(3, -5)$ is on the graph of a one-to-one function f, what point must be on the graph of f^{-1}?

In Problems 5–7, solve each equation.

5. $3^x = 243$ 6. $\log_b 16 = 2$ 7. $\log_5 x = 4$

In Problems 8–11, use a calculator to evaluate each expression. Round your answer to three decimal places.

8. $e^3 + 2$ 9. $\log 20$

10. $\log_3 21$ 11. $\ln 133$

In Problems 12 and 13, use the given function f to:

(a) Find the domain of f.
(b) Graph f.
(c) From the graph, determine the range and any asymptotes of f.
(d) Find f^{-1}, the inverse of f.
(e) Find the domain and the range of f^{-1}.
(f) Graph f^{-1}.

12. $f(x) = 4^{x+1} - 2$ 13. $f(x) = 1 - \log_5(x-2)$

In Problems 14–19, solve each equation.

14. $5^{x+2} = 125$ 15. $\log(x+9) = 2$

16. $8 - 2e^{-x} = 4$ 17. $\log(x^2 + 3) = \log(x + 6)$

18. $7^{x+3} = e^x$ 19. $\log_2(x-4) + \log_2(x+4) = 3$

20. Write $\log_2\left(\dfrac{4x^3}{x^2 - 3x - 18}\right)$ as the sum and/or difference of logarithms. Express powers as factors.

21. A 50-mg sample of a radioactive substance decays to 34 mg after 30 days. How long will it take for there to be 2 mg remaining?

22. (a) If \$1000 is invested at 5% compounded monthly, how much is there after 8 months?
    (b) If you want to have \$1000 in 9 months, how much do you need to place in a savings account now that pays 5% compounded quarterly?
    (c) How long does it take to double your money if you can invest it at 6% compounded annually?

23. The decibel level, D, of sound is given by the equation $D = 10\log\left(\dfrac{I}{I_0}\right)$, where I is the intensity of the sound and $I_0 = 10^{-12}$ watt per square meter.
    (a) If the shout of a single person measures 80 decibels, how loud will the sound be if two people shout at the same time? That is, how loud would the sound be if the intensity doubled?
    (b) The pain threshold for sound is 125 decibels. If the Athens Olympic Stadium 2004 (Olympiako Stadio Athinas 'Spyros Louis') can seat 74,400 people, how many people in the crowd need to shout at the same time for the resulting sound level to meet or exceed the pain threshold? (Ignore any possible sound dampening.)

Cumulative Review

1. (a) Is the following graph the graph of a function? If it is, is the function one-to-one?

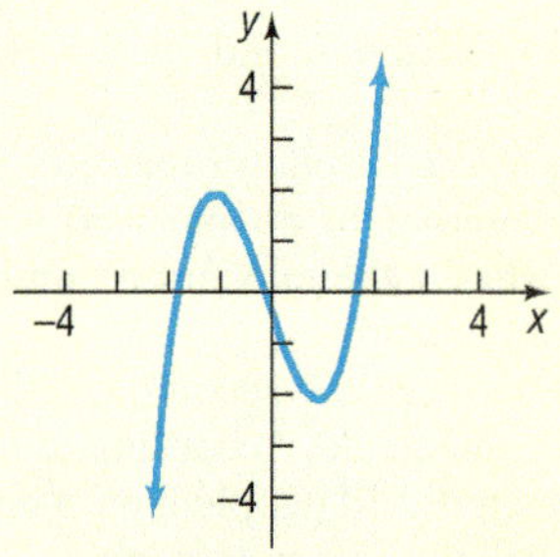

(b) Assuming the graph is a function, what type of function might it be (polynomial, exponential, and so on)? Why?

2. For the function $f(x) = 2x^2 - 3x + 1$, find the following:
(a) $f(3)$ (b) $f(-x)$ (c) $f(x+h)$

3. Determine which of the following points are on the graph of $x^2 + y^2 = 1$.
(a) $\left(\frac{1}{2}, \frac{1}{2}\right)$ (b) $\left(\frac{1}{2}, \frac{\sqrt{3}}{2}\right)$

4. Solve the equation $3(x - 2) = 4(x + 5)$.

5. Graph the line $2x - 4y = 16$.

6. (a) Graph the quadratic function $f(x) = -x^2 + 2x - 3$ by determining whether its graph opens up or down and by finding its vertex, axis of symmetry, y-intercept, and x-intercept(s), if any.
(b) Solve $f(x) \le 0$.

7. Determine the quadratic function whose graph is given in the figure.

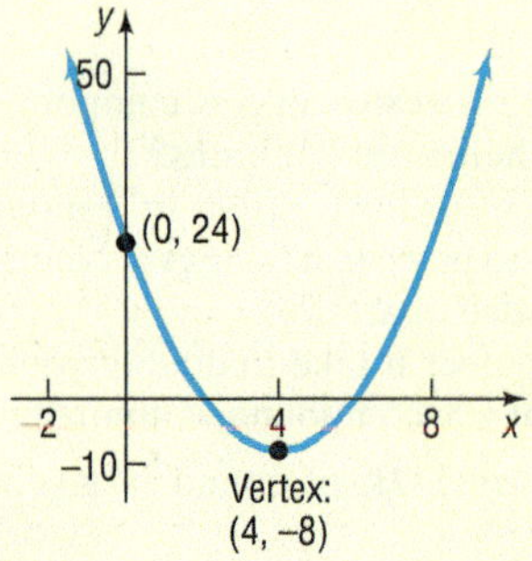

8. Is the graph of the following function a polynomial, exponential, or logarithmic function? Determine the function whose graph is given. Exponential;

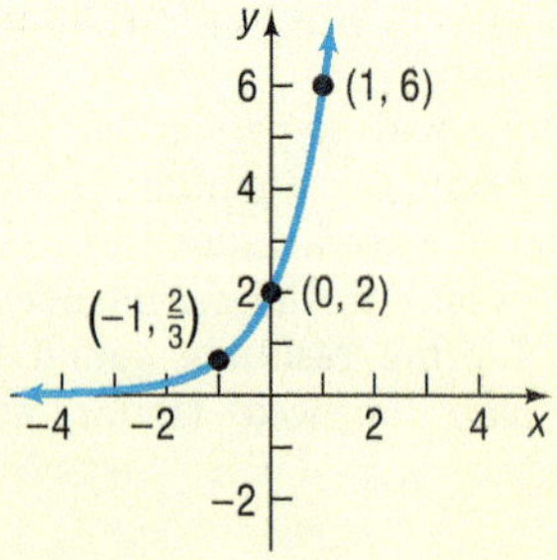

9. Graph $f(x) = 3(x + 1)^3 - 2$ using transformations.

10. (a) Given that $f(x) = x^2 + 2$ and $g(x) = \dfrac{2}{x - 3}$, find $f(g(x))$ and state its domain. What is $f(g(5))$?
(b) If $f(x) = x + 2$ and $g(x) = \log_2 x$, find $(f(g(x))$ and state its domain. What is $f(g(14))$?

11. For the polynomial function $f(x) = 4x^3 + 9x^2 - 30x - 8$:
(a) Find the real zeros of f.
(b) Determine the intercepts of the graph of f.
(c) Use a graphing utility to approximate the local maxima and local minima.
(d) Draw a complete graph of f. Be sure to label the intercepts and turning points.

12. For the function $g(x) = 3^x + 2$:
(a) Graph g using transformations. State the domain, range, and horizontal asymptote of g.
(b) Determine the inverse of g. State the domain, range, and vertical asymptote of g^{-1}.
(c) On the same graph as g, graph g^{-1}.

13. Solve the equation $4^{x-3} = 8^{2x}$.

14. Solve the equation: $\log_3(x + 1) + \log_3(2x - 3) = \log_9 9$

15. Suppose that $f(x) = \log_3(x + 2)$. Solve:
(a) $f(x) = 0$ (b) $f(x) > 0$
(c) $f(x) = 3$

16. Data Analysis The following data represent the percent of all drivers by age that have been stopped by the police for any reason within the past year. The median age represents the midpoint of the upper and lower limit for the age range.

Age Range	Median Age, x	Percentage Stopped, y
16–19	17.5	18.2
20–29	24.5	16.8
30–39	34.5	11.3
40–49	44.5	9.4
50–59	54.5	7.7
≥60	69.5	3.8

(a) Using your graphing utility, draw a scatter diagram of the data treating median age, x, as the independent variable.
(b) Determine a model that you feel best describes the relation between median age and percentage stopped. You may choose from among linear, quadratic, cubic, exponential, logarithmic, or logistic models.
(c) Provide a justification for the model that you selected in part (b).

Chapter Projects

I. Depreciation of Cars Kelley Blue Book is a guide that provides the current retail price of cars. You can access the Kelley Blue Book at your library or online at *www.kbb.com*.

1. Identify three cars that you are considering purchasing, and find the Kelley Blue Book value of the cars for 0 (brand new), 1, 2, 3, 4, and 5 years of age. Online, the value of the car can be found by selecting What should I pay for a used car? Enter the year, make, and model of the car you are selecting. To be consistent, we will assume the cars will be driven 12,000 miles per year, so a 1-year-old car will have 12,000 miles, a 2-year-old car will have 24,000 miles, and so on. Choose the same options for each year, and finally determine the suggested retail price for cars that are in Excellent, Good, and Fair shape. You should have a total of 16 observations (1 for a brand new car, 3 for a 1-year-old car, 3 for a 2-year-old car, and so on).
2. Draw a scatter diagram of the data with age as the independent variable and value as the dependent variable using Excel, a TI-graphing calculator, or some other spreadsheet. The Chapter 4 project describes how to draw a scatter diagram in Excel.
3. Determine the exponential function of best fit. Graph the exponential function of best fit on the scatter diagram. To do this in Excel, click on any data point in the scatter diagram. Now select the Chart Element icon (+). Check the box for Trendline, select the arrow to the right, and choose More Options. Select the Exponential radio button and select Display Equation on Chart. See Figure 71. Move the Trendline Options window off to the side, and you will see the exponential function of best fit displayed on the scatter diagram. Do you think the function accurately describes the relation between age of the car and suggested retail price?

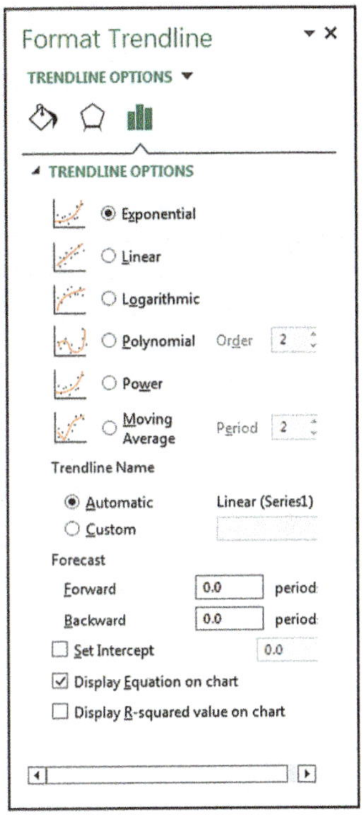

Figure 71

4. The exponential function of best fit is of the form $y = Ce^{rx}$, where y is the suggested retail value of the car and x is the age of the car (in years). What does the value of C represent? What does the value of r represent? What is the depreciation rate for each car that you are considering?
5. Write a report detailing which car you would purchase based on the depreciation rate you found for each car.

The following projects are available on the Instructor's Resource Center (IRC):

II. Hot Coffee A fast-food restaurant wants a special container to hold coffee. The restaurant wishes the container to quickly cool the coffee from 200° to 130°F and keep the liquid between 110° and 130°F as long as possible. The restaurant has three containers to select from. Which one should be purchased?

III. Project at Motorola ***Thermal Fatigue of Solder Connections*** Product reliability is a major concern of a manufacturer. Here a logarithmic transformation is used to simplify the analysis of a cell phone's ability to withstand temperature change.

7 Trigonometric Functions

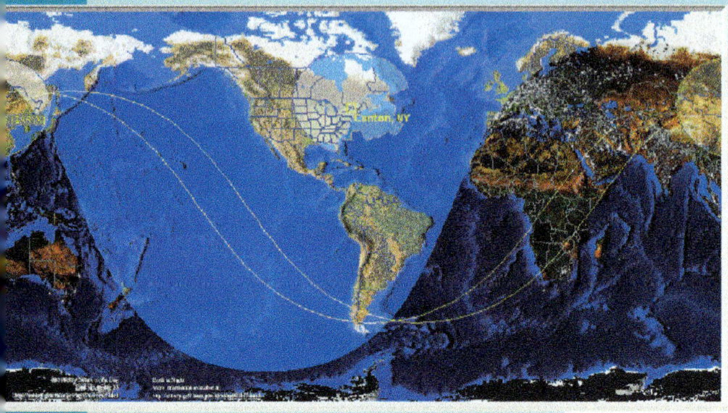

Length of Day Revisited

The length of a day depends on the day of the year as well as on the latitude of the location. Latitude gives the location of a point on Earth north or south of the equator. In Chapter 5 we found a model that describes the relation between the length of day and latitude for a specific day of the year. In the Internet Project at the end of this chapter, we will find a model that describes the relation between the length of day and day of the year for a specific latitude.

—See the Internet-based Chapter Project I—

Outline

••• A Look Back

In Chapter 3, we began our discussion of functions. We defined domain and range and independent and dependent variables; we found the value of a function and graphed functions. We continued our study of functions by listing properties that a function might have, such as being even or odd, and we created a library of functions, naming key functions and listing their properties, including the graph.

A Look Ahead •••

In this chapter we define the trigonometric functions, six functions that have wide application. We shall talk about their domain and range, see how to find values for them, graph them, and develop a list of their properties.

There are two widely accepted approaches to the development of the trigonometric functions: one uses right triangles; the other uses circles, especially the unit circle. In this text, we develop the trigonometric functions using right triangles. In Section 7.5, we introduce trigonometric functions using the unit circle and show that this approach leads to the definition using right triangles.

7.1 Angles and Their Measure

PREPARING FOR THIS SECTION *Before getting started, review the following:*

- Circumference and Area of a Circle (Chapter R, Review, Section R.3, p. 33)
- Uniform Motion (Chapter 1, Section 1.6, pp. 141–142)

Now Work the 'Are You Prepared?' problems on page 522.

OBJECTIVES
1. Convert between Decimal and Degree, Minute, Second Measures for Angles (p. 515)
2. Find the Length of an Arc of a Circle (p. 516)
3. Convert from Degrees to Radians and from Radians to Degrees (p. 517)
4. Find the Area of a Sector of a Circle (p. 520)
5. Find the Linear Speed of an Object Traveling in Circular Motion (p. 520)

Figure 1 A ray or half-line

A **ray**, or **half-line**, is that portion of a line that starts at a point V on the line and extends indefinitely in one direction. The starting point V of a ray is called its **vertex**. See Figure 1.

When two rays are drawn with a common vertex, they form an **angle**. We call one ray of an angle the **initial side** and the other the **terminal side**. The angle formed is identified by showing the direction and amount of rotation from the initial side to the terminal side. If the rotation is in the counterclockwise direction, the angle is **positive**; if the rotation is clockwise, the angle is **negative**. See Figure 2.

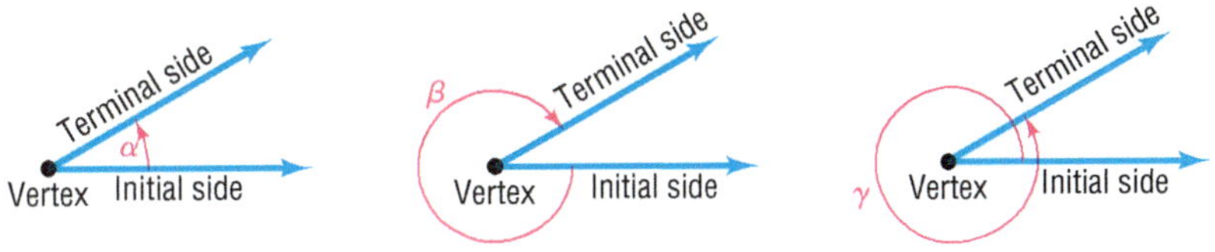

Figure 2 (a) Counterclockwise rotation Positive angle (b) Clockwise rotation Negative angle (c) Counterclockwise rotation Positive angle

Lowercase Greek letters, such as α (alpha), β (beta), γ (gamma), and θ (theta), are often used to denote angles. Notice in Figure 2(a) that the angle α is positive because the direction of the rotation from the initial side to the terminal side is counterclockwise. The angle β in Figure 2(b) is negative because the rotation is clockwise. The angle γ in Figure 2(c) is positive. Notice that the angle α in Figure 2(a) and the angle γ in Figure 2(c) have the same initial side and the same terminal side. However, α and γ are unequal, because the amount of rotation required to go from the initial side to the terminal side is greater for angle γ than for angle α.

An angle θ is said to be in **standard position** if its vertex is at the origin of a rectangular coordinate system and its initial side coincides with the positive x-axis. See Figure 3.

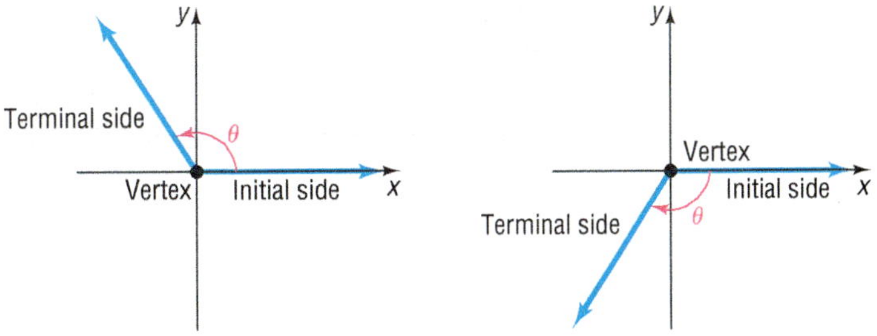

Figure 3 Standard position of an angle (a) θ is in standard position; θ is positive (b) θ is in standard position; θ is negative

When an angle θ is in standard position, either the terminal side will lie in a quadrant, in which case we say that θ **lies in that quadrant**, or the terminal side will lie on the x-axis or the y-axis, in which case we say that θ is a **quadrantal angle**. For example, the angle θ in Figure 4(a) lies in quadrant II, the angle θ in Figure 4(b) lies in quadrant IV, and the angle θ in Figure 4(c) is a quadrantal angle.

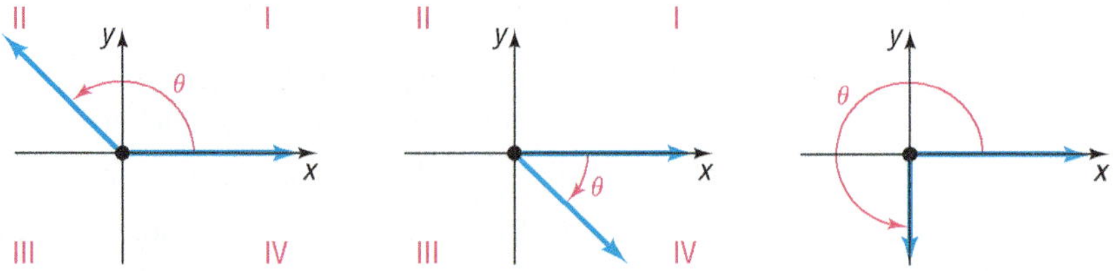

Figure 4 **(a)** θ lies in quadrant II **(b)** θ lies in quadrant IV **(c)** θ is a quadrantal angle

Angles are measured by determining the amount of rotation needed for the initial side to become coincident with the terminal side. The two commonly used measures for angles are *degrees* and *radians*.

Degrees

Historical Note:

One counterclockwise rotation was said to measure 360° because the Babylonian year had 360 days. ■

The angle formed by rotating the initial side exactly once in the counterclockwise direction until it coincides with itself (1 revolution) is said to measure 360 degrees, abbreviated 360°. **One degree, 1°**, is $\frac{1}{360}$ revolution. A **right angle** is an angle that measures 90°, or $\frac{1}{4}$ revolution; a **straight angle** is an angle that measures 180°, or $\frac{1}{2}$ revolution. See Figure 5. As Figure 5(b) shows, it is customary to indicate a right angle by using the symbol ⊾.

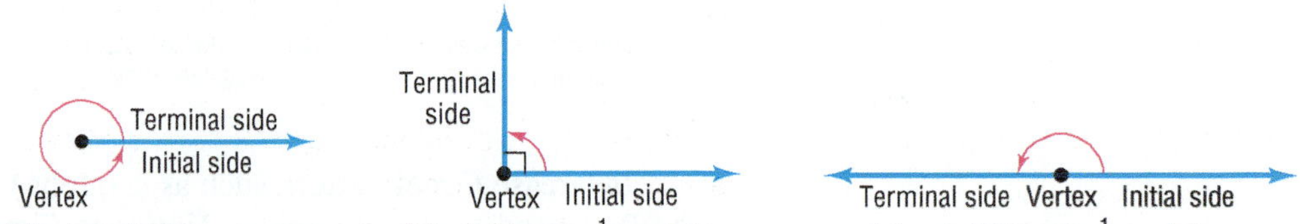

(a) 1 revolution counterclockwise, 360° **(b)** right angle, $\frac{1}{4}$ revolution counterclockwise, 90° **(c)** straight angle, $\frac{1}{2}$ revolution counterclockwise, 180°

Figure 5

It is also customary to refer to an angle that measures θ degrees as an angle of θ degrees.

EXAMPLE 1

Drawing an Angle

Draw each angle in standard position.

(a) 45° (b) −90° (c) 225° (d) 405°

Solution

(a) An angle of 45° is $\frac{1}{2}$ of a right angle. See Figure 6.

(b) An angle of −90° is $\frac{1}{4}$ revolution in the clockwise direction. See Figure 7.

(c) An angle of 225° consists of a rotation through 180° followed by a rotation through 45°. See Figure 8.

(d) An angle of 405° consists of 1 revolution (360°) followed by a rotation through 45°. See Figure 9.

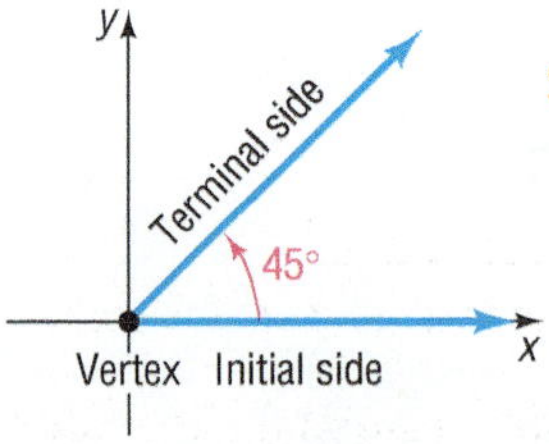

Figure 6 45° angle

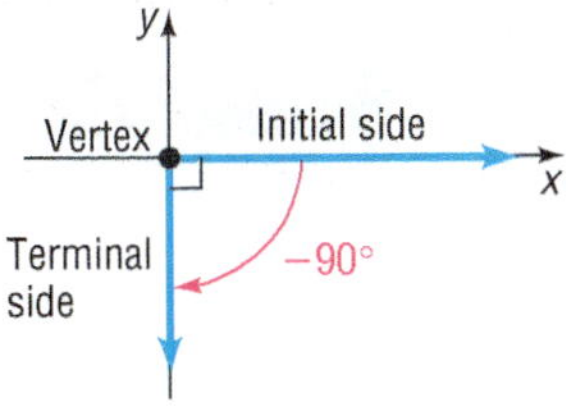

Figure 7 −90° angle

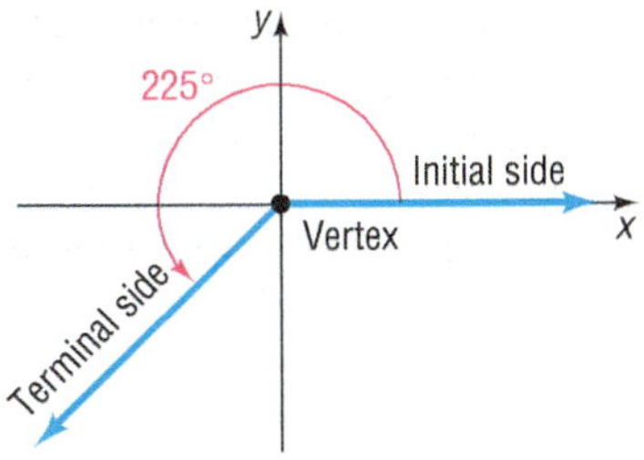

Figure 8 225° angle

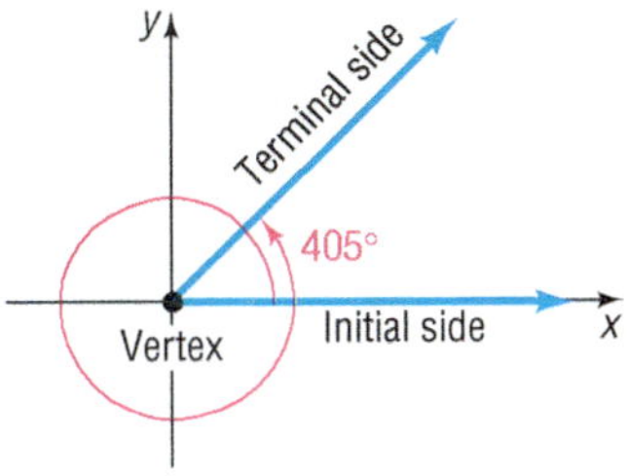

Figure 9 405° angle

Now Work PROBLEM 11

1 Convert between Decimal and Degree, Minute, Second Measures for Angles

Although subdivisions of a degree may be obtained by using decimals, the notion of *minutes* and *seconds* may also be used. **One minute**, denoted by **1′**, is defined as $\frac{1}{60}$ degree. **One second**, denoted by **1″**, is defined as $\frac{1}{60}$ minute or, equivalently, $\frac{1}{3600}$ degree. An angle of, say, 30 degrees, 40 minutes, 10 seconds is written compactly as 30°40′10″. To summarize:

$$1 \text{ counterclockwise revolution} = 360^\circ$$
$$1^\circ = 60' \qquad 1' = 60'' \qquad \textbf{(1)}$$

It is sometimes necessary to convert from the degree, minute, second notation (D°M′S″) to a decimal form, and vice versa. Check your calculator; it should be capable of doing the conversion for you.

Before using your calculator, though, you must set the mode to degrees because there are two common ways to measure angles: degree mode and radian mode. (We will define radians shortly.) Usually, a menu is used to change from one mode to another. Check your owner's manual to find out how your particular calculator works.

EXAMPLE 2 Converting between Degrees, Minutes, Seconds, and Decimal Forms

(a) Convert 50°6′21″ to a decimal in degrees. Round the answer to four decimal places.

(b) Convert 21.256° to the degree, minute, second notation. Round the answer to the nearest second.

Algebraic Solution

(a) Because $1' = \left(\frac{1}{60}\right)^\circ$ and $1'' = \left(\frac{1}{60}\right)' = \left(\frac{1}{60}\cdot\frac{1}{60}\right)^\circ$, convert as follows:

$$50^\circ 6' 21'' = 50^\circ + 6' + 21'' = 50^\circ + 6 \cdot 1' + 21 \cdot 1''$$
$$= 50^\circ + 6 \cdot \left(\frac{1}{60}\right)^\circ + 21 \cdot \left(\frac{1}{60}\cdot\frac{1}{60}\right)^\circ \quad \text{Convert minutes and seconds to degrees.}$$
$$\approx 50^\circ + 0.1^\circ + 0.0058^\circ$$
$$= 50.1058^\circ$$

Graphing Solution

(a) Figure 10 shows the solution using a TI-84 Plus C graphing calculator.

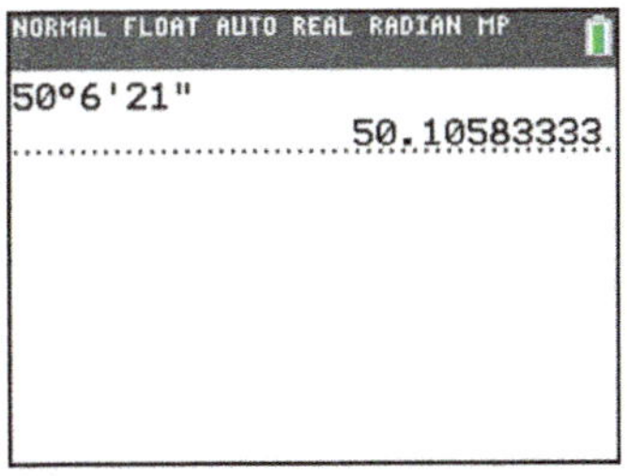

Figure 10 Converting D°M′S″ to a decimal

So 50°6′21″ = 51.1058° rounded to four decimal places.

(b) Because $1° = 60'$ and $1' = 60''$, proceed as follows:

$$
\begin{aligned}
21.256° &= 21° + 0.256° \\
&= 21° + (0.256) \cdot 1° \\
&= 21° + (0.256)(60') && \text{Convert fraction of degree to minutes; } 1° = 60'. \\
&= 21° + 15.36' \\
&= 21° + 15' + 0.36' \\
&= 21° + 15' + (0.36) \cdot 1' \\
&= 21° + 15' + (0.36)(60'') && \text{Convert fraction of minute to seconds; } 1' = 60''. \\
&= 21° + 15' + 21.6'' \\
&\approx 21°15'22'' && \text{Round to the nearest second.}
\end{aligned}
$$

(b) Figure 11 shows the solution using a T1-84 Plus C graphing calculator.

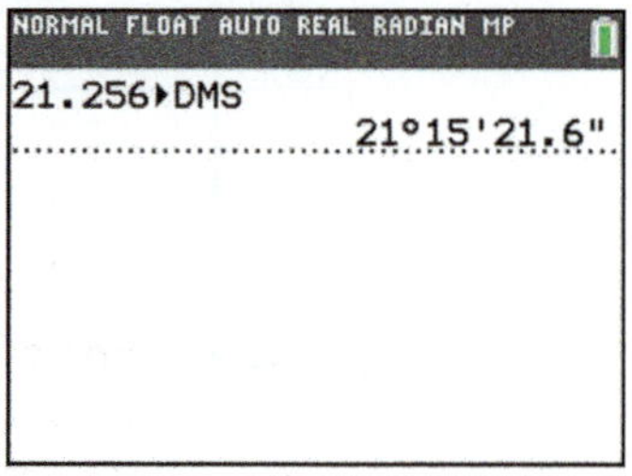

Figure 11 Converting a decimal to D°M′S″

So $21.256° = 21°15'22''$ rounded to the nearest second.

Now Work PROBLEMS 59 AND 65

In many applications, such as describing the exact location of a star or the precise position of a ship at sea, angles measured in degrees, minutes, and even seconds are used. For calculation purposes, these are transformed to decimal form. In other applications, especially those in calculus, angles are measured using *radians*.

Radians

A **central angle** is a positive angle whose vertex is at the center of a circle. The rays of a central angle subtend (intersect) an arc on the circle. If the radius of the circle is r and the length of the arc subtended by the central angle is also r, then the measure of the angle is **1 radian**. See Figure 12(a).

For a circle of radius 1, the rays of a central angle with measure 1 radian subtend an arc of length 1. For a circle of radius 3, the rays of a central angle with measure 1 radian subtend an arc of length 3. See Figure 12(b).

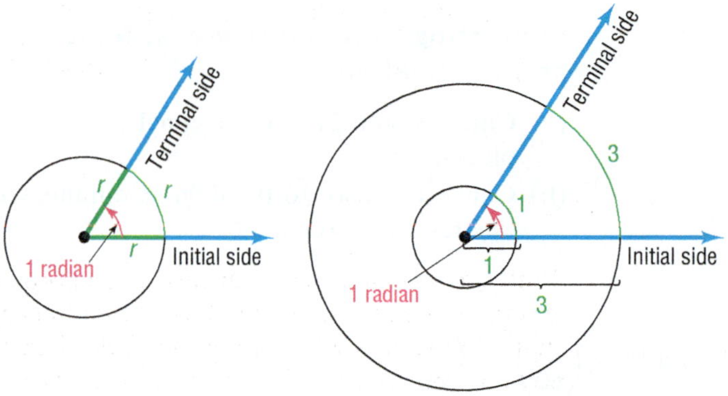

Figure 12 (a) (b)

2 Find the Length of an Arc of a Circle

Now consider a circle of radius r and two central angles, θ and θ_1, measured in radians. Suppose that these central angles subtend arcs of lengths s and s_1, respectively, as shown in Figure 13. From geometry, the ratio of the measures of the angles equals the ratio of the corresponding lengths of the arcs subtended by these angles; that is,

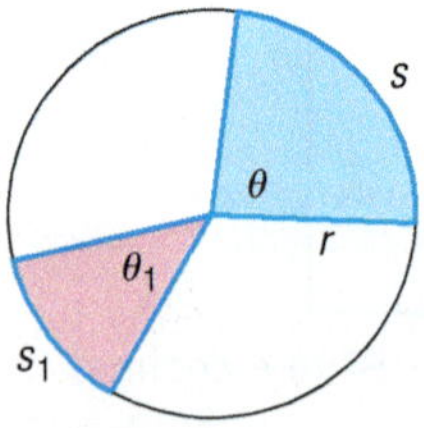

Figure 13 $\dfrac{\theta}{\theta_1} = \dfrac{s}{s_1}$

$$\frac{\theta}{\theta_1} = \frac{s}{s_1} \tag{2}$$

Suppose that $\theta_1 = 1$ radian. Refer again to Figure 12(a). The length s_1 of the arc subtended by the central angle $\theta_1 = 1$ radian equals the radius r of the circle. Then $s_1 = r$, so equation (2) reduces to

$$\frac{\theta}{1} = \frac{s}{r} \quad \text{or} \quad s = r\theta \tag{3}$$

THEOREM

Arc Length

For a circle of radius r, a central angle of θ radians subtends an arc whose length s is

$$s = r\theta \tag{4}$$

Note: Formulas must be consistent with regard to the units used. In equation (4), we write

$$s = r\theta$$

To see the units, however, we must go back to equation (3) and write

$$\frac{\theta \text{ radians}}{1 \text{ radian}} = \frac{s \text{ length units}}{r \text{ length units}}$$

$$s \text{ length units} = r \text{ length units} \frac{\theta \text{ radians}}{1 \text{ radian}}$$

The radians divide out, leaving

$$s \text{ length units} = (r \text{ length units})\theta \qquad s = r\theta$$

where θ appears to be "dimensionless" but, in fact, is measured in radians. So, in using the formula $s = r\theta$, the dimension for θ is radians, and any convenient unit of length (such as inches or meters) may be used for s and r.

EXAMPLE 3 **Finding the Length of an Arc of a Circle**

Find the length of the arc of a circle of radius 2 meters subtended by a central angle of 0.25 radian.

Solution Use equation (4) with $r = 2$ meters and $\theta = 0.25$. The length s of the arc is

$$s = r\theta = 2(0.25) = 0.5 \text{ meter}$$

Now Work PROBLEM 71

3 Convert from Degrees to Radians and from Radians to Degrees

With two ways to measure angles, it is important to be able to convert from one to the other. Consider a circle of radius r. A central angle of 1 revolution will subtend an arc equal to the circumference of the circle (Figure 14). Because the circumference of a circle of radius r equals $2\pi r$, substitute $2\pi r$ for s in equation (4) to find that, for an angle θ of 1 revolution,

$$s = r\theta$$

$$2\pi r = r\theta \qquad \theta = 1 \text{ revolution}; s = 2\pi r$$

$$\theta = 2\pi \text{ radians} \qquad \text{Solve for } \theta.$$

From this, we have

$$1 \text{ revolution} = 2\pi \text{ radians} \tag{5}$$

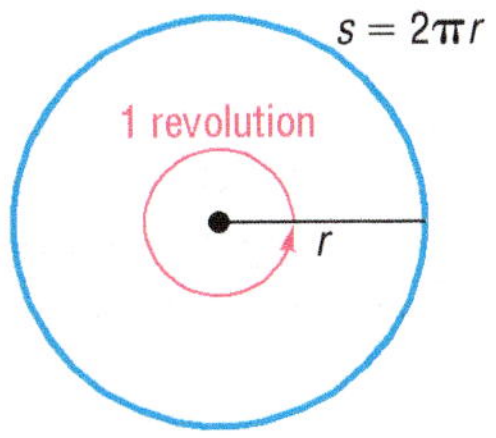

Figure 14 1 revolution = 2π radians

Since 1 revolution = 360°, we have

$$360° = 2\pi \text{ radians}$$

Dividing both sides by 2 yields

$$180° = \pi \text{ radians} \quad \textbf{(6)}$$

Divide both sides of equation (6) by 180. Then

$$1 \text{ degree} = \frac{\pi}{180} \text{ radian}$$

Divide both sides of (6) by π. Then

$$\frac{180}{\pi} \text{ degrees} = 1 \text{ radian}$$

We have the following two conversion formulas:*

$$1 \text{ degree} = \frac{\pi}{180} \text{ radian} \qquad 1 \text{ radian} = \frac{180}{\pi} \text{ degrees} \quad \textbf{(7)}$$

EXAMPLE 4

Converting from Degrees to Radians

Convert each angle in degrees to radians.

(a) 60° (b) 150° (c) −45° (d) 90° (e) 107°

Solution

(a) $60° = 60 \cdot 1 \text{ degree} = 60 \cdot \frac{\pi}{180} \text{ radian} = \frac{\pi}{3} \text{ radians}$

(b) $150° = 150 \cdot 1° = 150 \cdot \frac{\pi}{180} \text{ radian} = \frac{5\pi}{6} \text{ radians}$

(c) $-45° = -45 \cdot \frac{\pi}{180} \text{ radian} = -\frac{\pi}{4} \text{ radian}$

(d) $90° = 90 \cdot \frac{\pi}{180} \text{ radian} = \frac{\pi}{2} \text{ radians}$

(e) $107° = 107 \cdot \frac{\pi}{180} \text{ radian} \approx 1.868 \text{ radians}$ ■

Example 4, parts (a)–(d), illustrates that angles that are "nice" fractions of a revolution are expressed in radian measure as fractional multiples of π, rather than as decimals. For example, a right angle, as in Example 4(d), is left in the form $\frac{\pi}{2}$ radians, which is exact, rather than using the approximation $\frac{\pi}{2} \approx \frac{3.1416}{2} = 1.5708$ radians. When the fractions are not "nice," use the decimal approximation of the angle, as in Example 4(e).

Now Work PROBLEMS 23 AND 49

EXAMPLE 5

Converting from Radians to Degrees

Convert each angle in radians to degrees.

(a) $\frac{\pi}{6}$ radian (b) $\frac{3\pi}{2}$ radians (c) $-\frac{3\pi}{4}$ radians

(d) $\frac{7\pi}{3}$ radians (e) 3 radians

*Some students prefer instead to use the proportion $\frac{\text{Degree}}{180°} = \frac{\text{Radian}}{\pi}$. Then they substitute for what is given and solve for the measurement sought.

Solution

(a) $\frac{\pi}{6}$ radian $= \frac{\pi}{6} \cdot 1$ radian $= \frac{\pi}{6} \cdot \frac{180}{\pi}$ degrees $= 30°$

(b) $\frac{3\pi}{2}$ radians $= \frac{3\pi}{2} \cdot \frac{180}{\pi}$ degrees $= 270°$

(c) $-\frac{3\pi}{4}$ radians $= -\frac{3\pi}{4} \cdot \frac{180}{\pi}$ degrees $= -135°$

(d) $\frac{7\pi}{3}$ radians $= \frac{7\pi}{3} \cdot \frac{180}{\pi}$ degrees $= 420°$

(e) 3 radians $= 3 \cdot \frac{180}{\pi}$ degrees $\approx 171.89°$ ■

Now Work PROBLEM 35

Table 1 lists the degree and radian measures of some commonly encountered angles. You should learn to feel equally comfortable using degree measure and using radian measure for these angles.

Table 1

Degrees	0°	30°	45°	60°	90°	120°	135°	150°	180°
Radians	0	$\frac{\pi}{6}$	$\frac{\pi}{4}$	$\frac{\pi}{3}$	$\frac{\pi}{2}$	$\frac{2\pi}{3}$	$\frac{3\pi}{4}$	$\frac{5\pi}{6}$	π
Degrees		210°	225°	240°	270°	300°	315°	330°	360°
Radians		$\frac{7\pi}{6}$	$\frac{5\pi}{4}$	$\frac{4\pi}{3}$	$\frac{3\pi}{2}$	$\frac{5\pi}{3}$	$\frac{7\pi}{4}$	$\frac{11\pi}{6}$	2π

EXAMPLE 6 Finding the Distance between Two Cities

The latitude of a location L is the measure of the angle formed by a ray drawn from the center of Earth to the equator and a ray drawn from the center of Earth to L. See Figure 15(a). Sioux Falls, South Dakota, is due north of Dallas, Texas. Find the distance between Sioux Falls (43° 33′ north latitude) and Dallas (32° 46′ north latitude). See Figure 15(b). Assume that the radius of Earth is 3960 miles.

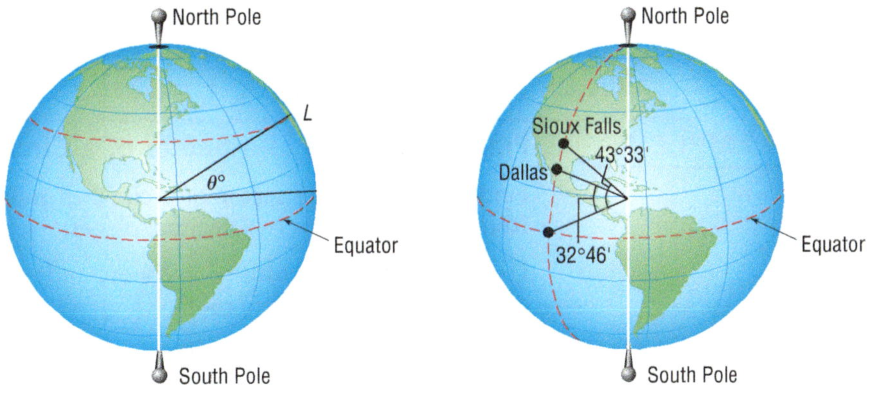

Figure 15 (a) (b)

Solution The measure of the central angle between the two cities is 43° 33′ − 32° 46′ = 10° 47′. Use equation (4), $s = r\theta$. But remember to first convert the angle of 10° 47′ to radians.

$$\theta = 10°\,47' \approx 10.7833° = 10.7833 \cdot \frac{\pi}{180} \text{ radian} \approx 0.188 \text{ radian}$$

$$\uparrow$$

$$47' = 47\left(\frac{1}{60}\right)^{\circ}$$

Use $\theta = 0.188$ radian and $r = 3960$ miles in equation (4). The distance between the two cities is

$$s = r\theta = 3960 \cdot 0.188 \approx 744 \text{ miles}$$

Note: If the measure of an angle is given as 5, it is understood to mean 5 radians; if the measure of an angle is given as 5°, it means 5 degrees.

When an angle is measured in degrees, the degree symbol will always be shown. However, when an angle is measured in radians, the usual practice is to omit the word *radians*. So if the measure of an angle is given as $\frac{\pi}{6}$, it is understood to mean $\frac{\pi}{6}$ radian.

Now Work PROBLEM 107

4 Find the Area of a Sector of a Circle

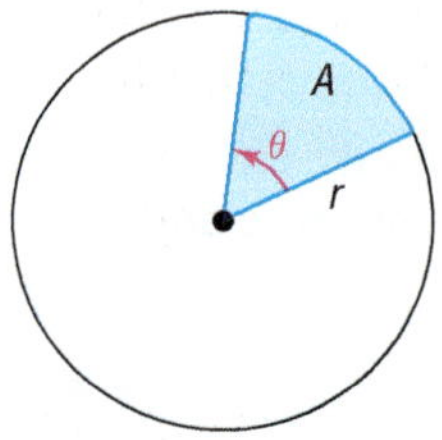

Figure 16 Sector of a Circle

Consider a circle of radius r. Suppose that θ, measured in radians, is a central angle of this circle. See Figure 16. We seek a formula for the area A of the sector (shown in blue) formed by the angle θ.

Now consider a circle of radius r and two central angles θ and θ_1, both measured in radians. See Figure 17. From geometry, the ratio of the measures of the angles equals the ratio of the corresponding areas of the sectors formed by these angles. That is,

$$\frac{\theta}{\theta_1} = \frac{A}{A_1}$$

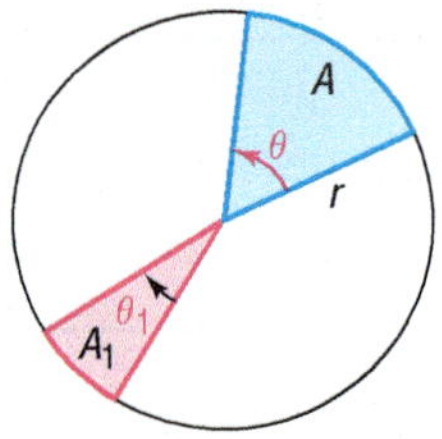

Figure 17 $\frac{\theta}{\theta_1} = \frac{A}{A_1}$

Suppose that $\theta_1 = 2\pi$ radians. Then A_1 = area of the circle $= \pi r^2$. Solving for A, we find

$$A = A_1 \frac{\theta}{\theta_1} = \pi r^2 \frac{\theta}{2\pi} = \frac{1}{2} r^2 \theta$$

$$A_1 = \pi r^2$$
$$\theta_1 = 2\pi$$

THEOREM

Area of a Sector

The area A of the sector of a circle of radius r formed by a central angle of θ radians is

$$A = \frac{1}{2} r^2 \theta \qquad \textbf{(8)}$$

EXAMPLE 7 **Finding the Area of a Sector of a Circle**

Find the area of the sector of a circle of radius 2 feet formed by an angle of 30°. Round the answer to two decimal places.

Solution Use equation (8) with $r = 2$ feet and $\theta = 30° = \frac{\pi}{6}$ radian. [Remember, in equation (8), θ must be in radians.]

$$A = \frac{1}{2} r^2 \theta = \frac{1}{2}(2)^2 \frac{\pi}{6} = \frac{\pi}{3} \approx 1.05$$

The area A of the sector is 1.05 square feet, rounded to two decimal places.

Now Work PROBLEM 79

5 Find the Linear Speed of an Object Traveling in Circular Motion

In Chapter 1, Section 1.6, we defined the average speed of an object as the distance traveled divided by the elapsed time. For motion along a circle, we distinguish between *linear speed* and *angular speed*.

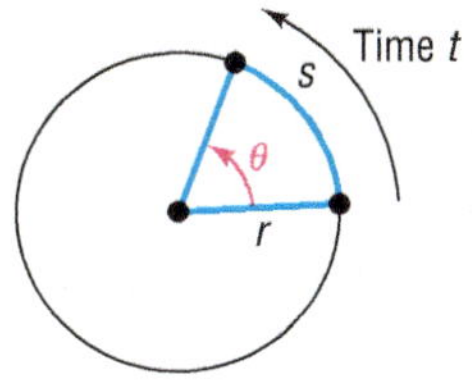

Figure 18 $v = \frac{s}{t}$

DEFINITION

Suppose that an object moves on a circle of radius r at a constant speed. If s is the distance traveled in time t on this circle, then the **linear speed** v of the object is defined as

$$v = \frac{s}{t} \quad \textbf{(9)}$$

As this object travels on the circle, suppose that θ (measured in radians) is the central angle swept out in time t. See Figure 18.

DEFINITION

The **angular speed** ω (the Greek lowercase letter omega) of this object is the angle θ (measured in radians) swept out, divided by the elapsed time t; that is,

$$\omega = \frac{\theta}{t} \quad \textbf{(10)}$$

Angular speed is the way the turning rate of an engine is described. For example, an engine idling at 900 rpm (revolutions per minute) is one that rotates at an angular speed of

$$900\,\frac{\text{revolutions}}{\text{minute}} = 900\,\frac{\cancel{\text{revolutions}}}{\text{minute}} \cdot 2\pi\,\frac{\text{radians}}{\cancel{\text{revolution}}} = 1800\pi\,\frac{\text{radians}}{\text{minute}}$$

There is an important relationship between linear speed and angular speed:

$$\text{linear speed} = v = \underset{\substack{\uparrow \\ (9)}}{\frac{s}{t}} = \underset{\substack{\uparrow \\ s = r\theta}}{\frac{r\theta}{t}} = r\left(\frac{\theta}{t}\right) = \underset{\substack{\uparrow \\ (10)}}{r \cdot \omega}$$

$$v = r\omega \quad \textbf{(11)}$$

where ω is measured in radians per unit time.

When using equation (11), remember that $v = \frac{s}{t}$ (the linear speed) has the dimensions of length per unit of time (such as feet per second or miles per hour), r (the radius of the circular motion) has the same length dimension as s, and ω (the angular speed) has the dimension of radians per unit of time. If the angular speed is given in terms of *revolutions* per unit of time (as is often the case), be sure to convert it to *radians* per unit of time, using the fact that 1 revolution $= 2\pi$ radians, before attempting to use equation (11).

EXAMPLE 8 Finding Linear Speed

A child is spinning a rock at the end of a 2-foot rope at the rate of 180 revolutions per minute (rpm). Find the linear speed of the rock when it is released.

Solution Look at Figure 19. The rock is moving around a circle of radius $r = 2$ feet. The angular speed ω of the rock is

$$\omega = 180\,\frac{\text{revolutions}}{\text{minute}} = 180\,\frac{\cancel{\text{revolutions}}}{\text{minute}} \cdot 2\pi\,\frac{\text{radians}}{\cancel{\text{revolution}}} = 360\pi\,\frac{\text{radians}}{\text{minute}}$$

From equation (11), the linear speed v of the rock is

$$v = r\omega = 2\text{ feet} \cdot 360\pi\,\frac{\text{radians}}{\text{minute}} = 720\pi\,\frac{\text{feet}}{\text{minute}} \approx 2262\,\frac{\text{feet}}{\text{minute}}$$

The linear speed of the rock when it is released is 2262 ft/min $\approx$ 25.7 mi/h. ■

Figure 19

Now Work PROBLEM 99

Historical Feature

Trigonometry was developed by Greek astronomers, who regarded the sky as the inside of a sphere, so it was natural that triangles on a sphere were investigated early (by Menelaus of Alexandria about AD 100) and that triangles in the plane were studied much later. The first book containing a systematic treatment of plane and spherical trigonometry was written by the Persian astronomer Nasir Eddin (about AD 1250).

Regiomontanus (1436–1476) is the person most responsible for moving trigonometry from astronomy into mathematics. His work was improved by Copernicus (1473–1543) and Copernicus' student Rhaeticus (1514–1576). Rhaeticus's book was the first to define the six trigonometric functions as ratios of sides of triangles, although he did not give the functions their present names. Credit for this is due to Thomas Finck (1583), but Finck's notation was by no means universally accepted at the time. The notation was finally stabilized by the textbooks of Leonhard Euler (1707–1783).

Trigonometry has since evolved from its use by surveyors, navigators, and engineers to present applications involving ocean tides, the rise and fall of food supplies in certain ecologies, brain wave patterns, and many other phenomena.

7.1 Assess Your Understanding

'Are You Prepared?' *Answers are given at the end of these exercises. If you get a wrong answer, read the pages listed in red.*

1. What is the formula for the circumference C of a circle of radius r? What is the formula for the area A of a circle of radius r? (p. 33)

2. If an object has a speed of r feet per second and travels a distance d (in feet) in time t (in seconds), then $d =$ ______. (pp. 141–142)

Concepts and Vocabulary

3. An angle θ is in __________ __________ if its vertex is at the origin of a rectangular coordinate system and its initial side coincides with the positive x-axis.

4. A __________ __________ is a positive angle whose vertex is at the center of a circle.

5. If the radius of a circle is r and the length of the arc subtended by a central angle is also r, then the measure of the angle is 1 ______.
(a) degree (b) minute (c) second (d) radian

6. On a circle of radius r, a central angle of θ radians subtends an arc of length $s =$ ___; the area of the sector formed by this angle θ is $A =$ ______.

7. $180° =$ ______ radians
(a) $\frac{\pi}{2}$ (b) π (c) $\frac{3\pi}{2}$ (d) 2π

8. An object travels on a circle of radius r with constant speed. If s is the distance traveled in time t on the circle and θ is the central angle (in radians) swept out in time t, then the linear speed of the object is $v =$ ______ and the angular speed of the object is $\omega =$ ______.

9. ***True or False*** The angular speed ω of an object traveling on a circle of radius r is the angle θ (measured in radians) swept out, divided by the elapsed time t.

10. ***True or False*** For circular motion on a circle of radius r, linear speed equals angular speed divided by r.

Skill Building

In Problems 11–22, draw each angle in standard position.

11. $30°$	**12.** $60°$	**13.** $135°$	**14.** $-120°$	**15.** $450°$	**16.** $540°$
17. $\frac{3\pi}{4}$	**18.** $\frac{4\pi}{3}$	**19.** $-\frac{\pi}{6}$	**20.** $-\frac{2\pi}{3}$	**21.** $\frac{16\pi}{3}$	**22.** $\frac{21\pi}{4}$

In Problems 23–34, convert each angle in degrees to radians. Express your answer as a multiple of π.

23. $30°$	**24.** $120°$	**25.** $240°$	**26.** $330°$	**27.** $-60°$	**28.** $-30°$
29. $180°$	**30.** $270°$	**31.** $-135°$	**32.** $-225°$	**33.** $-90°$	**34.** $-180°$

In Problems 35–46, convert each angle in radians to degrees.

35. $\frac{\pi}{3}$	**36.** $\frac{5\pi}{6}$	**37.** $-\frac{5\pi}{4}$	**38.** $-\frac{2\pi}{3}$	**39.** $\frac{\pi}{2}$	**40.** 4π
41. $\frac{\pi}{12}$	**42.** $\frac{5\pi}{12}$	**43.** $-\frac{\pi}{2}$	**44.** $-\pi$	**45.** $-\frac{\pi}{6}$	**46.** $-\frac{3\pi}{4}$

In Problems 47–52, convert each angle in degrees to radians. Express your answer in decimal form, rounded to two decimal places.

47. $17°$	**48.** $73°$	**49.** $-40°$	**50.** $-51°$	**51.** $125°$	**52.** $350°$

In Problems 53–58, convert each angle in radians to degrees. Express your answer in decimal form, rounded to two decimal places.

53. 3.14	**54.** 0.75	**55.** 2	**56.** 3	**57.** 6.32	**58.** $\sqrt{2}$

In Problems 59–64, convert each angle to a decimal in degrees. Round your answer to two decimal places.

59. $40°10'25''$ **60.** $61°42'21''$ **61.** $50°14'20''$ **62.** $73°40'40''$ **63.** $9°9'9''$ **64.** $98°22'45''$

In Problems 65–70, convert each angle to D°M′S″ form. Round your answer to the nearest second.

65. $40.32°$ **66.** $61.24°$ **67.** $18.255°$ **68.** $29.411°$ **69.** $19.99°$ **70.** $44.01°$

In Problems 71–78, s denotes the length of the arc of a circle of radius r subtended by the central angle θ. Find the missing quantity. Round answers to three decimal places.

71. $r = 10$ meters, $\theta = \frac{1}{2}$ radian, $s = ?$

72. $r = 6$ feet, $\theta = 2$ radians, $s = ?$

73. $\theta = \frac{1}{3}$ radian, $s = 2$ feet, $r = ?$

74. $\theta = \frac{1}{4}$ radian, $s = 6$ centimeters, $r = ?$

75. $r = 5$ miles, $s = 3$ miles, $\theta = ?$

76. $r = 6$ meters, $s = 8$ meters, $\theta = ?$

77. $r = 2$ inches, $\theta = 30°$, $s = ?$

78. $r = 3$ meters, $\theta = 120°$, $s = ?$

In Problems 79–86, A denotes the area of the sector of a circle of radius r formed by the central angle θ. Find the missing quantity. Round answers to three decimal places.

79. $r = 10$ meters, $\theta = \frac{1}{2}$ radian, $A = ?$

80. $r = 6$ feet, $\theta = 2$ radians, $A = ?$

81. $\theta = \frac{1}{3}$ radian, $A = 2$ square feet, $r = ?$

82. $\theta = \frac{1}{4}$ radian, $A = 6$ square centimeters, $r = ?$

83. $r = 5$ miles, $A = 3$ square miles, $\theta = ?$

84. $r = 6$ meters, $A = 8$ square meters, $\theta = ?$

85. $r = 2$ inches, $\theta = 30°$, $A = ?$

86. $r = 3$ meters, $\theta = 120°$, $A = ?$

In Problems 87–90, find the length s and the area A. Round answers to three decimal places.

87.

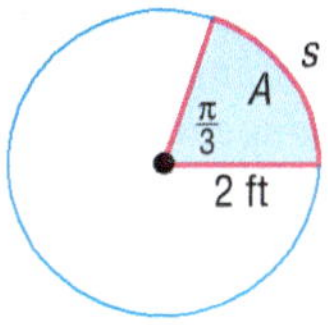

88.

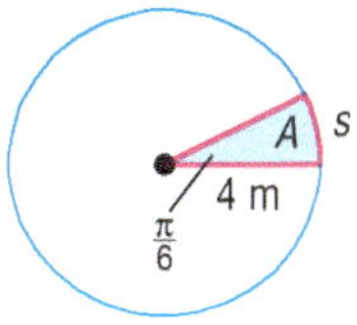

89.

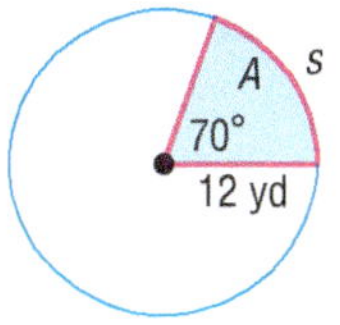

90.

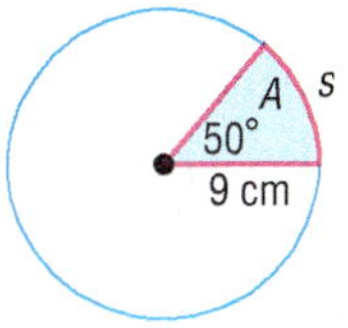

Applications and Extensions

91. Movement of a Minute Hand The minute hand of a clock is 6 inches long. How far does the tip of the minute hand move in 15 minutes? How far does it move in 25 minutes? Round answers to two decimal places.

92. Movement of a Pendulum A pendulum swings through an angle of 20° each second. If the pendulum is 40 inches long, how far does its tip move each second? Round answers to two decimal places.

93. Area of a Sector Find the area of the sector of a circle of radius 4 meters formed by an angle of 45°. Round the answer to two decimal places.

94. Area of a Sector Find the area of the sector of a circle of radius 3 centimeters formed by an angle of 60°. Round the answer to two decimal places.

95. Watering a Lawn A water sprinkler sprays water over a distance of 30 feet while rotating through an angle of 135°. What area of lawn receives water?

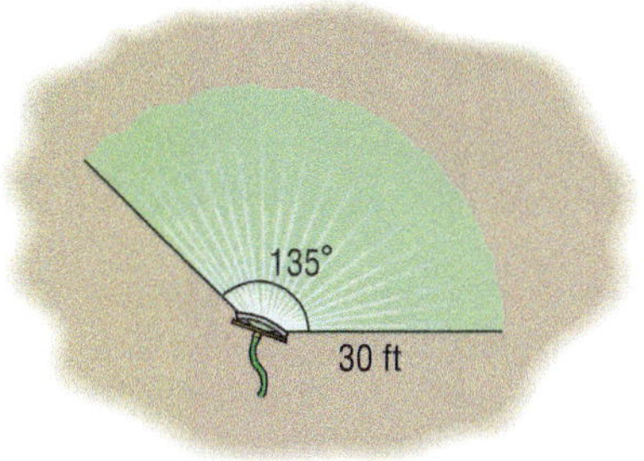

96. Designing a Water Sprinkler An engineer is asked to design a water sprinkler that will cover a field of 100 square yards that is in the shape of a sector of a circle of radius 15 yards. Through what angle should the sprinkler rotate?

97. Windshield Wiper The arm and blade of a windshield wiper have a total length of 34 inches. If the blade is 25 inches long and the wiper sweeps out an angle of 120°, how much window area can the blade clean?

98. Windshield Wiper The arm and blade of a windshield wiper have a total length of 30 inches. If the blade is 24 inches long and the wiper sweeps out an angle of 125°, how much window area can the blade clean?

99. Motion on a Circle An object is traveling on a circle with a radius of 5 centimeters. If in 20 seconds a central angle of $\frac{1}{3}$ radian is swept out, what is the angular speed of the object? What is its linear speed?

100. Motion on a Circle An object is traveling on a circle with a radius of 2 meters. If in 20 seconds the object travels 5 meters, what is its angular speed? What is its linear speed?

101. Amusement Park Ride A gondola on an amusement park ride, similar to the Spin Cycle at Silverwood Theme Park, spins at a speed of 13 revolutions per minute. If the gondola is 25 feet from the ride's center, what is the linear speed of the gondola in miles per hour?

102. Amusement Park Ride A centrifugal force ride, similar to the Gravitron, spins at a speed of 22 revolutions per minute. If the diameter of the ride is 13 meters, what is the linear speed of the passengers in kilometers per hour?

103. Blu-ray Drive A Blu-ray drive has a maximum speed of 10,000 revolutions per minute. If a Blu-ray disc has a diameter of 12 cm, what is the linear speed, in km/h, of a point 4 cm from the center if the disc is spinning at a rate of 8000 revolutions per minute?

104. DVD Drive A DVD drive has a maximum speed of 7200 revolutions per minute. If a DVD has a diameter of 12 cm, what is the linear speed, in km/h, of a point 5 cm from the disc's center if it is spinning at a rate of 5400 revolutions per minute?

105. Bicycle Wheels The diameter of each wheel of a bicycle is 26 inches. If you are traveling at a speed of 35 miles per hour on this bicycle, through how many revolutions per minute are the wheels turning?

106. Car Wheels The radius of each wheel of a car is 15 inches. If the wheels are turning at the rate of 3 revolutions per second, how fast is the car moving? Express your answer in inches per second and in miles per hour.

In Problems 107–110, the latitude of a location L is the angle formed by a ray drawn from the center of Earth to the equator and a ray drawn from the center of Earth to L. See the figure.

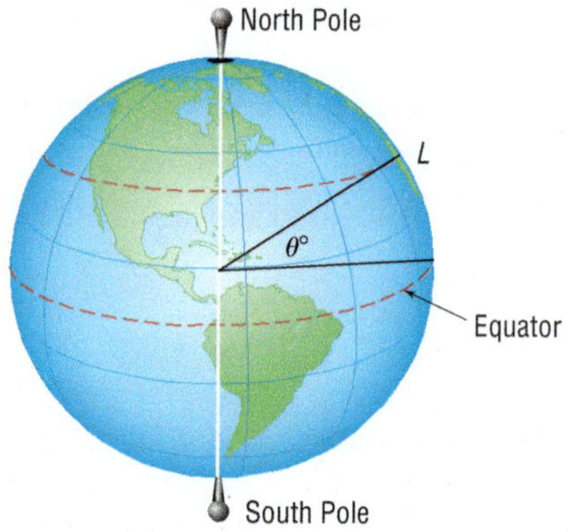

107. Distance between Cities Memphis, Tennessee, is due north of New Orleans, Louisiana. Find the distance between Memphis (35°9′ north latitude) and New Orleans (29°57′ north latitude). Assume that the radius of Earth is 3960 miles.

108. Distance between Cities Charleston, West Virginia, is due north of Jacksonville, Florida. Find the distance between Charleston (38°21′ north latitude) and Jacksonville (30°20′ north latitude). Assume that the radius of Earth is 3960 miles.

109. Linear Speed on Earth Earth rotates on an axis through its poles. The distance from the axis to a location on Earth at 30° north latitude is about 3429.5 miles. Therefore, a location on Earth at 30° north latitude is spinning on a circle of radius 3429.5 miles. Compute the linear speed on the surface of Earth at 30° north latitude.

110. Linear Speed on Earth Earth rotates on an axis through its poles. The distance from the axis to a location on Earth at 40° north latitude is about 3033.5 miles. Therefore, a location on Earth at 40° north latitude is spinning on a circle of radius 3033.5 miles. Compute the linear speed on the surface of Earth at 40° north latitude.

111. Speed of the Moon The mean distance of the moon from Earth is 2.39×10^5 miles. Assuming that the orbit of the moon around Earth is circular and that 1 revolution takes 27.3 days, find the linear speed of the moon. Express your answer in miles per hour.

112. Speed of Earth The mean distance of Earth from the Sun is 9.29×10^7 miles. Assuming that the orbit of Earth around the Sun is circular and that 1 revolution takes 365 days, find the linear speed of Earth. Express your answer in miles per hour.

113. Pulleys Two pulleys, one with radius 2 inches and the other with radius 8 inches, are connected by a belt. (See figure on the next page.) If the 2-inch pulley is caused to rotate at 3 revolutions per minute, determine the revolutions per minute of the 8-inch pulley.

[**Hint:** The linear speeds of the pulleys are the same; both equal the speed of the belt.]

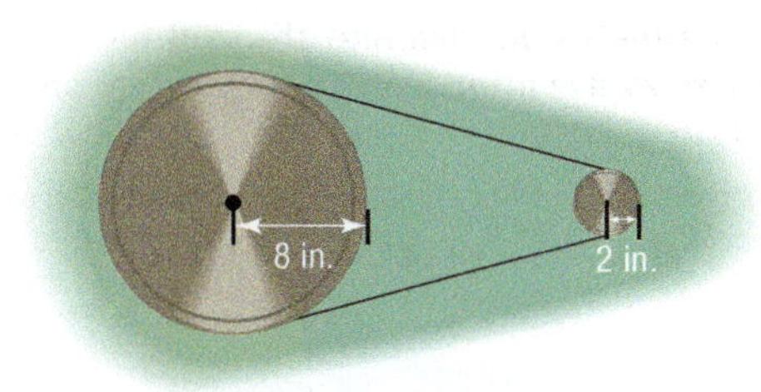

114. Ferris Wheels A neighborhood carnival has a Ferris wheel with a radius of 30 feet. You measure the time it takes for one revolution to be 70 seconds. What is the linear speed (in feet per second) of this Ferris wheel? What is the angular speed in radians per second?

115. Computing the Speed of a River Current To approximate the speed of the current of a river, a circular paddle wheel with radius 4 feet is lowered into the water. If the current causes the wheel to rotate at a speed of 10 revolutions per minute, what is the speed of the current? Express your answer in miles per hour.

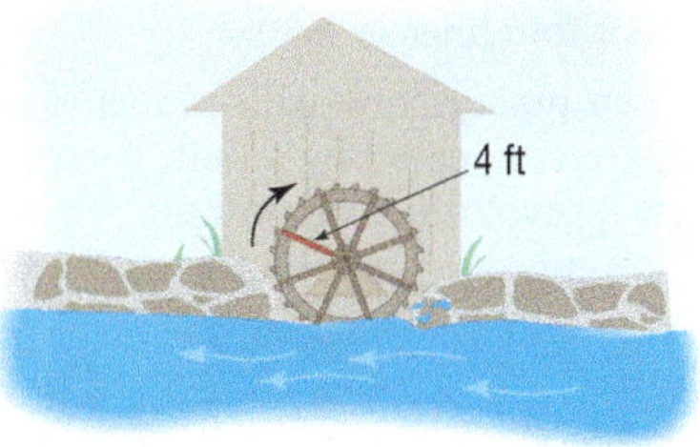

116. Spin Balancing Tires A spin balancer rotates the wheel of a car at 480 revolutions per minute. If the diameter of the wheel is 26 inches, what road speed is being tested? Express your answer in miles per hour. At how many revolutions per minute should the balancer be set to test a road speed of 80 miles per hour?

117. The Cable Cars of San Francisco At the Cable Car Museum you can see the four cable lines that are used to pull cable cars up and down the hills of San Francisco. Each cable travels at a speed of 9.55 miles per hour, driven by a rotating wheel whose diameter is 8.5 feet. How fast is the wheel rotating? Express your answer in revolutions per minute.

118. Difference in Time of Sunrise Naples, Florida, is about 90 miles due west of Ft. Lauderdale. How much sooner would a person in Ft. Lauderdale first see the rising Sun than a person in Naples? See the hint at the top of the next column.

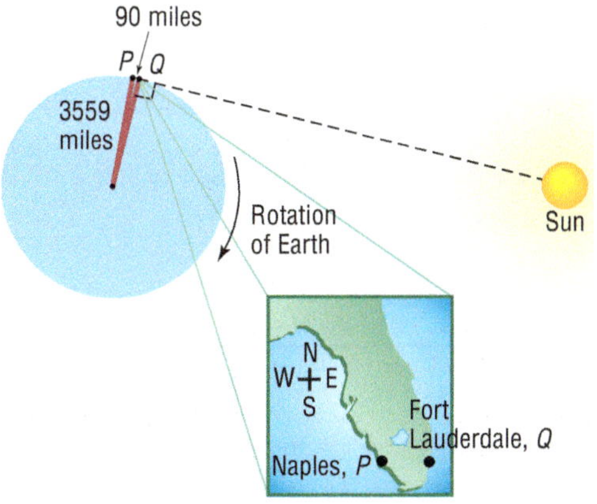

[**Hint:** Consult the figure. When a person at Q sees the first rays of the Sun, a person at P is still in the dark. The person at P sees the first rays after Earth has rotated until P is at the location Q. Now use the fact that at the latitude of Ft. Lauderdale, in 24 hours an arc of length $2\pi(3559)$ miles is subtended.]

119. Let the Dog Roam A dog is attached to a 9-foot rope fastened to the outside corner of a fenced-in garden that measures 6 feet by 10 feet. Assuming that the dog cannot enter the garden, compute the exact area that the dog can wander. Write the exact area in square feet.*

120. Area of a Region The measure of arc $\widehat{BE}$ is 2π. Find the exact area of the portion of the rectangle $ABCD$ that falls outside of the circle whose center is at A.*

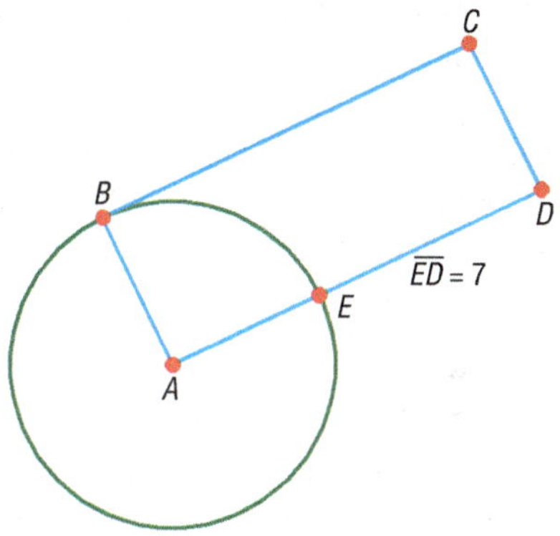

121. Keeping Up with the Sun How fast would you have to travel on the surface of Earth at the equator to keep up with the Sun (that is, so that the Sun would appear to remain in the same position in the sky)?

122. Nautical Miles A **nautical mile** equals the length of the arc subtended by a central angle of 1 minute on a great circle† on the surface of Earth. See the figure below. If the radius of Earth is taken as 3960 miles, express 1 nautical mile in terms of ordinary, or **statute**, miles.

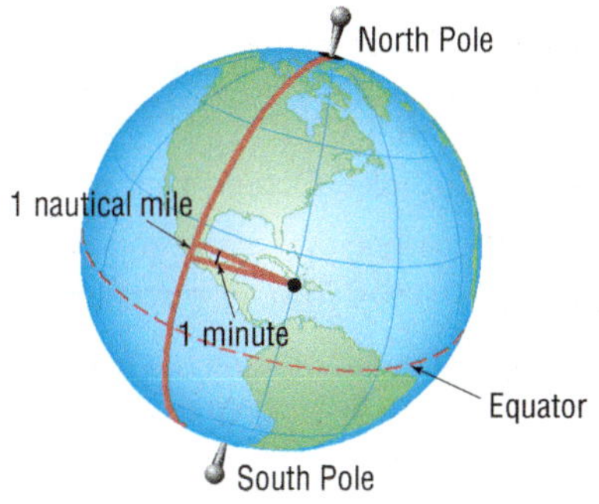

123. Approximating the Circumference of Earth Eratosthenes of Cyrene (276–195 BC) was a Greek scholar who lived and worked in Cyrene and Alexandria. One day while visiting in Syene, he noticed that the Sun's rays shone directly down a well. On this date 1 year later, in Alexandria,

*Courtesy of the Joliet Junior College Mathematics Department

†Any circle drawn on the surface of Earth that divides Earth into two equal hemispheres.

which is 500 miles due north of Syene, he measured the angle of the Sun to be about 7.2 degrees. See the figure. Use this information to approximate the radius and circumference of Earth.

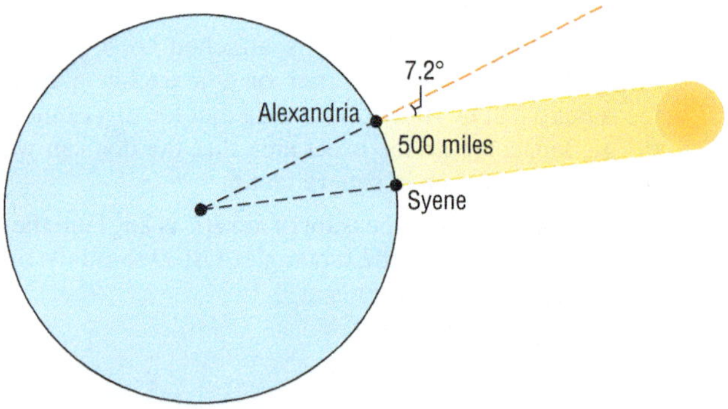

124. Designing a Little League Field For a 60-foot Little League Baseball field, the distance from home base to the nearest fence (or other obstruction) in fair territory should be a minimum of 200 feet. The commissioner of parks and recreation is making plans for a new 60-foot field. Because of limited ground availability, he will use the minimum required distance to the outfield fence. To increase safety, however, he plans to include a 10-foot-wide warning track on the inside of the fence. To further increase safety, the fence and warning track will extend both directions into foul territory. In total, the arc formed by the outfield fence (including the extensions into the foul territories) will be subtended by a central angle at home plate measuring 96°, as illustrated.

(a) Determine the length of the outfield fence.

(b) Determine the area of the warning track.

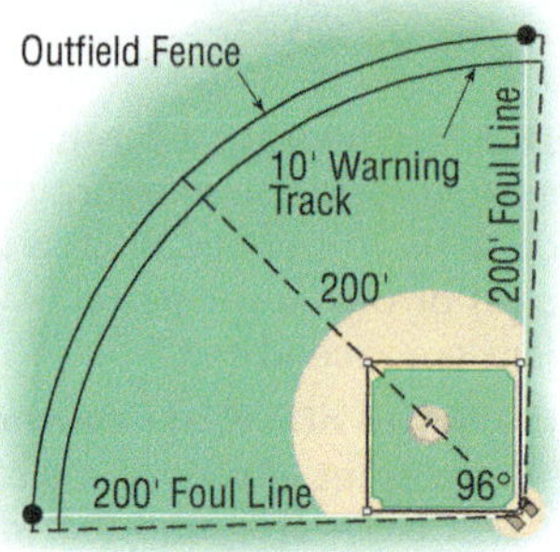

Source: *www.littleleague.org*

[**Note:** There is a 90° angle between the two foul lines. Then there are two 3° angles between the foul lines and the dashed lines shown. The angle between the two dashed lines outside the 200-foot foul lines is 96°.]

125. Pulleys Two pulleys, one with radius r_1 and the other with radius r_2, are connected by a belt. The pulley with radius r_1 rotates at ω_1 revolutions per minute, whereas the pulley with radius r_2 rotates at ω_2 revolutions per minute. Show that

$$\frac{r_1}{r_2} = \frac{\omega_2}{\omega_1}.$$

Explaining Concepts: Discussion and Writing

126. Do you prefer to measure angles using degrees or radians? Provide justification and a rationale for your choice.

127. What is 1 radian? What is 1 degree?

128. Which angle has the larger measure: 1 degree or 1 radian? Or are they equal?

129. Explain the difference between linear speed and angular speed.

130. For a circle of radius r, a central angle of θ degrees subtends an arc whose length s is $s = \frac{\pi}{180} r\theta$. Discuss whether this statement is true or false. Defend your position.

131. Discuss why ships and airplanes use nautical miles to measure distance. Explain the difference between a nautical mile and a statute mile.

132. Investigate the way that speed bicycles work. In particular, explain the differences and similarities between 5-speed and 9-speed derailleurs. Be sure to include a discussion of linear speed and angular speed.

133. In Example 6, we found that the distance between Dallas, Texas, and Sioux Falls, South Dakota, is approximately 744 miles. According to *mapquest.com*, the distance is approximately 850 miles. What might account for the difference?

Retain Your Knowledge

Problems 134–137 are based on material learned earlier in the course. The purpose of these problems is to keep the material fresh in your mind so that you are better prepared for the final exam.

134. Find the zero of $f(x) = 3x + 7$.

135. Solve: $5x^2 + 2 = 5 - 14x$

136. Write the function that is finally graphed if all of the following transformations are applied to the graph of $y = |x|$.
(a) Shift left 3 units. (b) Reflect about the x-axis. (c) Shift down 4 units.

137. Find the horizontal and vertical asymptotes of $R(x) = \frac{3x^2 - 12}{x^2 - 5x - 14}$.

'Are You Prepared?' Answers

1. $C = 2\pi r$; $A = \pi r^2$ **2.** $r \cdot t$

7.2 Right Triangle Trigonometry

PREPARING FOR THIS SECTION *Before getting started, review the following:*

- Geometry Essentials (Chapter R, Review, Section R.3, pp. 31–36)
- Functions (Section 3.1, pp. 207–218)

Now Work the 'Are You Prepared?' problems on page 535.

OBJECTIVES
1. Find the Values of Trigonometric Functions of Acute Angles (p. 527)
2. Use Fundamental Identities (p. 528)
3. Find the Values of the Remaining Trigonometric Functions, Given the Value of One of Them (p. 531)
4. Use the Complementary Angle Theorem (p. 533)

1 Find the Values of Trigonometric Functions of Acute Angles

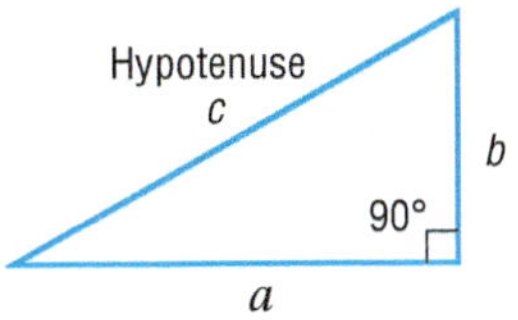

Figure 20 Right triangle

A triangle in which one angle is a right angle (90°) is called a **right triangle**. Recall that the side opposite the right angle is called the **hypotenuse**, and the remaining two sides are called the **legs** of the triangle. In Figure 20 we have labeled the hypotenuse as c to indicate that its length is c units, and in a like manner, we have labeled the legs as a and b. Because the triangle is a right triangle, the Pythagorean Theorem tells us that

$$c^2 = a^2 + b^2$$

Now, suppose that θ is an **acute angle**; that is, $0° < \theta < 90°$ (if θ is measured in degrees) and $0 < \theta < \frac{\pi}{2}$ (if θ is measured in radians). See Figure 21(a). Using this acute angle θ, we can form a right triangle, like the one illustrated in Figure 21(b), with hypotenuse of length c and legs of lengths a and b. Using the three sides of this triangle, exactly six ratios can be formed:

$$\frac{b}{c} \quad \frac{a}{c} \quad \frac{b}{a} \quad \frac{c}{b} \quad \frac{c}{a} \quad \frac{a}{b}$$

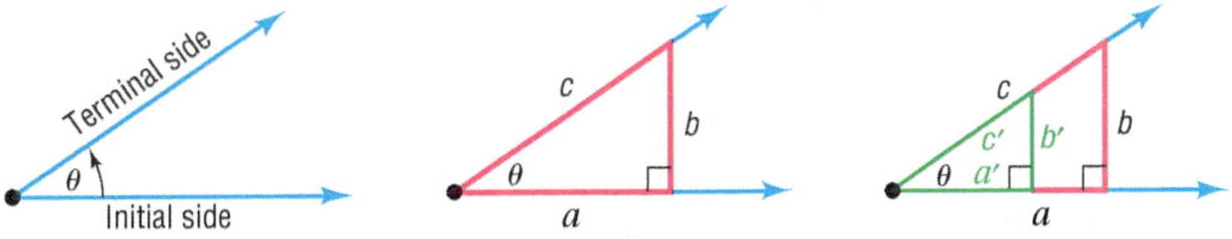

Figure 21 **(a)** Acute angle **(b)** Right triangle **(c)** Similar triangles

In fact, these ratios depend only on the size of the angle θ and not on the triangle formed. To see why, look at Figure 21(c). Any two right triangles formed using the angle θ will be similar, so corresponding ratios will be equal. As a result,

$$\frac{b}{c} = \frac{b'}{c'} \quad \frac{a}{c} = \frac{a'}{c'} \quad \frac{b}{a} = \frac{b'}{a'} \quad \frac{c}{b} = \frac{c'}{b'} \quad \frac{c}{a} = \frac{c'}{a'} \quad \frac{a}{b} = \frac{a'}{b'}$$

Because the ratios depend only on the angle θ and not on the triangle itself, each ratio is given a name that involves θ: sine of θ, cosine of θ, tangent of θ, cosecant of θ, secant of θ, and cotangent of θ.

DEFINITION

The six ratios of the lengths of the sides of a right triangle are called **trigonometric functions of acute angles** and are defined as follows:

Function Name	Abbreviation	Value	Function Name	Abbreviation	Value
sine of θ	$\sin\theta$	$\frac{b}{c}$	cosecant of θ	$\csc\theta$	$\frac{c}{b}$
cosine of θ	$\cos\theta$	$\frac{a}{c}$	secant of θ	$\sec\theta$	$\frac{c}{a}$
tangent of θ	$\tan\theta$	$\frac{b}{a}$	cotangent of θ	$\cot\theta$	$\frac{a}{b}$

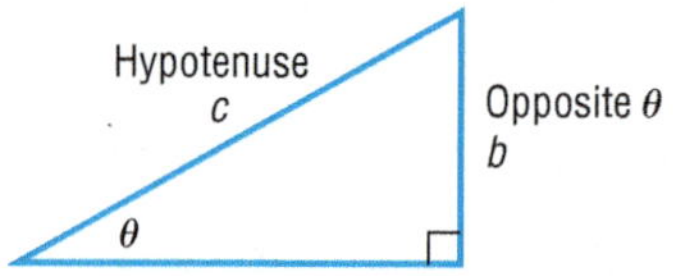

Figure 22

As an aid to remembering these definitions, it may be helpful to refer to the lengths of the sides of the triangle by the names *hypotenuse* (c), *opposite* (b), and *adjacent* (a). See Figure 22. In terms of these names, we have the following ratios:

$$\sin\theta = \frac{\text{opposite}}{\text{hypotenuse}} = \frac{b}{c} \quad \cos\theta = \frac{\text{adjacent}}{\text{hypotenuse}} = \frac{a}{c} \quad \tan\theta = \frac{\text{opposite}}{\text{adjacent}} = \frac{b}{a}$$
$$\csc\theta = \frac{\text{hypotenuse}}{\text{opposite}} = \frac{c}{b} \quad \sec\theta = \frac{\text{hypotenuse}}{\text{adjacent}} = \frac{c}{a} \quad \cot\theta = \frac{\text{adjacent}}{\text{opposite}} = \frac{a}{b} \tag{1}$$

Since a, b, and c are positive, the value of each of the trigonometric functions of an acute angle θ is positive.

EXAMPLE 1

Finding the Values of Trigonometric Functions

Find the value of each of the six trigonometric functions of the angle θ in Figure 23.

Solution See Figure 23. The two given sides of the triangle are

$$c = \text{hypotenuse} = 5 \qquad a = \text{adjacent} = 3$$

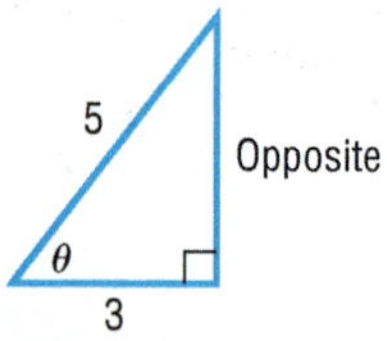

Figure 23

To find the length of the opposite side, use the Pythagorean Theorem.

$$(\text{adjacent})^2 + (\text{opposite})^2 = (\text{hypotenuse})^2$$
$$3^2 + (\text{opposite})^2 = 5^2$$
$$(\text{opposite})^2 = 25 - 9 = 16$$
$$\text{opposite} = 4$$

Now that the lengths of the three sides are known, use the ratios in (1) to find the value of each of the six trigonometric functions:

$$\sin\theta = \frac{\text{opposite}}{\text{hypotenuse}} = \frac{4}{5} \quad \cos\theta = \frac{\text{adjacent}}{\text{hypotenuse}} = \frac{3}{5} \quad \tan\theta = \frac{\text{opposite}}{\text{adjacent}} = \frac{4}{3}$$
$$\csc\theta = \frac{\text{hypotenuse}}{\text{opposite}} = \frac{5}{4} \quad \sec\theta = \frac{\text{hypotenuse}}{\text{adjacent}} = \frac{5}{3} \quad \cot\theta = \frac{\text{adjacent}}{\text{opposite}} = \frac{3}{4}$$

WARNING When writing the values of the trigonometric functions, do not forget the argument of the function

$\sin\theta = \frac{4}{5}$ **correct**

$\sin = \frac{4}{5}$ **incorrect**

Now Work PROBLEM 11

2 Use Fundamental Identities

You may have observed some relationships that exist among the six trigonometric functions of acute angles. For example, $\csc\theta = \frac{c}{b} = \frac{1}{\frac{b}{c}} = \frac{1}{\sin\theta}$. This identity is one of several known as the **reciprocal identities.**

Reciprocal Identities

$$\csc\theta = \frac{1}{\sin\theta} \qquad \sec\theta = \frac{1}{\cos\theta} \qquad \cot\theta = \frac{1}{\tan\theta} \qquad \textbf{(2)}$$

Two other fundamental identities are the **quotient identities**.

Quotient Identities

$$\tan\theta = \frac{\sin\theta}{\cos\theta} \qquad \cot\theta = \frac{\cos\theta}{\sin\theta} \qquad \textbf{(3)}$$

If $\sin\theta$ and $\cos\theta$ are known, the reciprocal and quotient identities in (2) and (3) make it easy to find the values of the remaining trigonometric functions.

EXAMPLE 2

Finding the Values of the Remaining Trigonometric Functions, Given $\sin\theta$ and $\cos\theta$

Given $\sin\theta = \frac{\sqrt{5}}{5}$ and $\cos\theta = \frac{2\sqrt{5}}{5}$, find the value of each of the four remaining trigonometric functions of θ.

Solution Based on the quotient identity in formula (3),

$$\tan\theta = \frac{\sin\theta}{\cos\theta} = \frac{\frac{\sqrt{5}}{5}}{\frac{2\sqrt{5}}{5}} = \frac{1}{2}$$

Then use the reciprocal identities from formula (2) to get

$$\csc\theta = \frac{1}{\sin\theta} = \frac{1}{\frac{\sqrt{5}}{5}} = \frac{5}{\sqrt{5}} = \sqrt{5} \qquad \sec\theta = \frac{1}{\cos\theta} = \frac{1}{\frac{2\sqrt{5}}{5}} = \frac{5}{2\sqrt{5}} = \frac{\sqrt{5}}{2} \qquad \cot\theta = \frac{1}{\tan\theta} = \frac{1}{\frac{1}{2}} = 2$$

■

Now Work PROBLEM 21

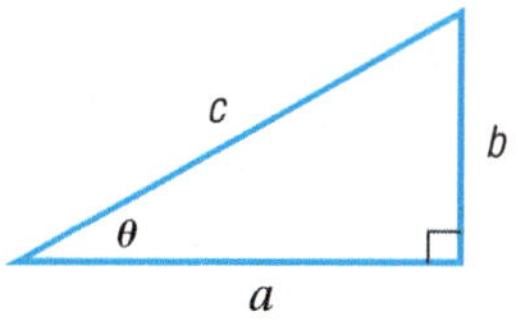

Figure 24 Right triangle

Refer now to the right triangle in Figure 24. The Pythagorean Theorem states that $a^2 + b^2 = c^2$, which can be rewritten as

$$b^2 + a^2 = c^2$$

Dividing each side by c^2, we get

$$\frac{b^2}{c^2} + \frac{a^2}{c^2} = 1 \quad \text{or} \quad \left(\frac{b}{c}\right)^2 + \left(\frac{a}{c}\right)^2 = 1$$

In terms of trigonometric functions of the angle θ, this equation states that

$$(\sin\theta)^2 + (\cos\theta)^2 = 1 \qquad \textbf{(4)}$$

Equation (4) is an identity, since the equation is true for any acute angle θ.

It is customary to write $\sin^2\theta$ instead of $(\sin\theta)^2$, $\cos^2\theta$ instead of $(\cos\theta)^2$, and so on. With this notation, equation (4) can be rewritten as

$$\sin^2\theta + \cos^2\theta = 1 \qquad \textbf{(5)}$$

Another identity can be obtained from equation (5) by dividing each side by $\cos^2\theta$.

$$\frac{\sin^2\theta}{\cos^2\theta} + \frac{\cos^2\theta}{\cos^2\theta} = \frac{1}{\cos^2\theta}$$

$$\left(\frac{\sin\theta}{\cos\theta}\right)^2 + 1 = \left(\frac{1}{\cos\theta}\right)^2$$

Now use identities (2) and (3) to get

$$\tan^2\theta + 1 = \sec^2\theta \qquad \textbf{(6)}$$

Similarly, dividing each side of equation (5) by $\sin^2\theta$, and using identities (2) and (3), gives $1 + \cot^2\theta = \csc^2\theta$, which can be rewritten as

$$\cot^2\theta + 1 = \csc^2\theta \qquad \textbf{(7)}$$

Collectively, the identities in equations (5), (6), and (7) are referred to as the **Pythagorean Identities**.

Fundamental Identities

$$\tan\theta = \frac{\sin\theta}{\cos\theta} \qquad \cot\theta = \frac{\cos\theta}{\sin\theta}$$

$$\csc\theta = \frac{1}{\sin\theta} \qquad \sec\theta = \frac{1}{\cos\theta} \qquad \cot\theta = \frac{1}{\tan\theta}$$

$$\sin^2\theta + \cos^2\theta = 1 \qquad \tan^2\theta + 1 = \sec^2\theta \qquad \cot^2\theta + 1 = \csc^2\theta$$

EXAMPLE 3 Finding the Exact Value of a Trigonometric Expression Using Identities

Find the exact value of each expression. Do not use a calculator.

(a) $\tan 20° - \dfrac{\sin 20°}{\cos 20°}$

(b) $\sin^2\dfrac{\pi}{12} + \dfrac{1}{\sec^2\dfrac{\pi}{12}}$

Solution (a) $\tan 20° - \dfrac{\sin 20°}{\cos 20°} = \tan 20° - \tan 20° = 0$

$\uparrow$ $\dfrac{\sin\theta}{\cos\theta} = \tan\theta$

(b) $\sin^2\dfrac{\pi}{12} + \dfrac{1}{\sec^2\dfrac{\pi}{12}} = \sin^2\dfrac{\pi}{12} + \cos^2\dfrac{\pi}{12} = 1$

$\uparrow$ $\cos\theta = \dfrac{1}{\sec\theta}$ $\qquad$ $\uparrow$ $\sin^2\theta + \cos^2\theta = 1$ ■

Now Work PROBLEM 39

3 Find the Values of the Remaining Trigonometric Functions, Given the Value of One of Them

EXAMPLE 4 **Finding the Values of the Remaining Trigonometric Functions, Given $\sin\theta$, θ Acute**

Given that $\sin\theta = \frac{1}{3}$ and θ is an acute angle, find the exact value of each of the remaining five trigonometric functions of θ.

Solution We solve this problem in two ways: The first uses the definition of the trigonometric functions, and the second uses fundamental identities.

Option 1 Using the Definition

Because $\sin\theta = \frac{1}{3} = \frac{b}{c}$, draw a right triangle with acute angle θ, opposite side of length $b = 1$, and hypotenuse of length $c = 3$. See Figure 25. The adjacent side a can be found by using the Pythagorean Theorem.

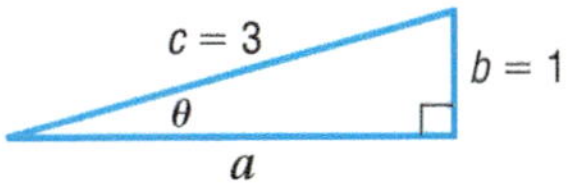

Figure 25 $\sin\theta = \frac{1}{3}$

$$a^2 + 1^2 = 3^2 \quad a^2 + b^2 = c^2;\ b = 1, c = 3$$
$$a^2 + 1 = 9$$
$$a^2 = 8$$
$$a = 2\sqrt{2}$$

Now the definitions given in equation (1) can be used to find the value of each of the remaining five trigonometric functions. Using $a = 2\sqrt{2}$, $b = 1$, and $c = 3$, we have

$$\cos\theta = \frac{a}{c} = \frac{2\sqrt{2}}{3} \qquad \tan\theta = \frac{b}{a} = \frac{1}{2\sqrt{2}} = \frac{\sqrt{2}}{4}$$

$$\csc\theta = \frac{c}{b} = \frac{3}{1} = 3 \quad \sec\theta = \frac{c}{a} = \frac{3}{2\sqrt{2}} = \frac{3\sqrt{2}}{4} \quad \cot\theta = \frac{a}{b} = \frac{2\sqrt{2}}{1} = 2\sqrt{2}$$

Option 2 Using Identities

Begin by finding $\cos\theta$ using the Pythagorean Identity from equation (5).

$$\sin^2\theta + \cos^2\theta = 1 \quad \text{Equation (5)}$$
$$\frac{1}{9} + \cos^2\theta = 1 \quad \sin\theta = \frac{1}{3}$$
$$\cos^2\theta = 1 - \frac{1}{9} = \frac{8}{9}$$

Recall that the trigonometric functions of an acute angle are positive. In particular, $\cos\theta > 0$ for an acute angle θ, so

$$\cos\theta = \sqrt{\frac{8}{9}} = \frac{2\sqrt{2}}{3}$$

Now, use $\sin\theta = \frac{1}{3}$ and $\cos\theta = \frac{2\sqrt{2}}{3}$ and fundamental identities.

$$\tan\theta = \frac{\sin\theta}{\cos\theta} = \frac{\frac{1}{3}}{\frac{2\sqrt{2}}{3}} = \frac{1}{2\sqrt{2}} = \frac{\sqrt{2}}{4} \qquad \cot\theta = \frac{1}{\tan\theta} = \frac{1}{\frac{\sqrt{2}}{4}} = \frac{4}{\sqrt{2}} = 2\sqrt{2}$$

$$\sec\theta = \frac{1}{\cos\theta} = \frac{1}{\frac{2\sqrt{2}}{3}} = \frac{3}{2\sqrt{2}} = \frac{3\sqrt{2}}{4} \qquad \csc\theta = \frac{1}{\sin\theta} = \frac{1}{\frac{1}{3}} = 3$$

■

Finding the Values of the Trigonometric Functions When One Is Known

Given the value of one trigonometric function of an acute angle θ, the exact value of each of the remaining five trigonometric functions of θ can be found in either of two ways.

Option 1 Using the Definition

STEP 1: Draw a right triangle showing the acute angle θ.
STEP 2: Two of the sides can then be assigned values based on the value of the given trigonometric function.
STEP 3: Find the length of the third side by using the Pythagorean Theorem.
STEP 4: Use the definitions in equation (1) to find the value of each of the remaining trigonometric functions.

Option 2 Using Identities

Use appropriately selected identities to find the value of each of the remaining trigonometric functions.

EXAMPLE 5 Given One Value of a Trigonometric Function, Find the Values of the Remaining Trigonometric Functions

Given $\tan\theta = \frac{1}{2}$, θ an acute angle, find the exact value of each of the remaining five trigonometric functions of θ.

Solution We solve this problem two ways.

Option 1 Using the Definition Figure 26 shows a right triangle with acute angle θ, where

$$\tan\theta = \frac{1}{2} = \frac{\text{opposite}}{\text{adjacent}} = \frac{b}{a}$$

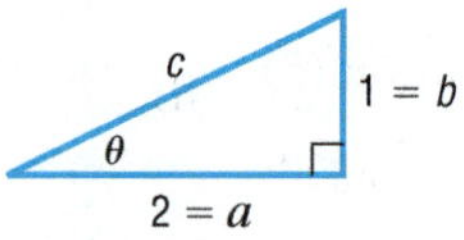

Figure 26 $\tan\theta = \frac{1}{2}$

With $b = 1$ and $a = 2$, the hypotenuse c can be found by using the Pythagorean Theorem.

$$c^2 = a^2 + b^2 = 2^2 + 1^2 = 5$$
$$c = \sqrt{5}$$

Now apply the definitions using $a = 2$, $b = 1$, and $c = \sqrt{5}$.

$$\sin\theta = \frac{b}{c} = \frac{1}{\sqrt{5}} = \frac{\sqrt{5}}{5} \qquad \cos\theta = \frac{a}{c} = \frac{2}{\sqrt{5}} = \frac{2\sqrt{5}}{5}$$

$$\csc\theta = \frac{c}{b} = \frac{\sqrt{5}}{1} = \sqrt{5} \qquad \sec\theta = \frac{c}{a} = \frac{\sqrt{5}}{2} \qquad \cot\theta = \frac{a}{b} = \frac{2}{1} = 2$$

Option 2 Using Identities Because the value of $\tan\theta$ is known, use the Pythagorean Identity that involves $\tan\theta$:

$$\tan^2\theta + 1 = \sec^2\theta \qquad \text{Equation (6)}$$

$$\left(\frac{1}{2}\right)^2 + 1 = \sec^2\theta \qquad \tan\theta = \frac{1}{2}$$

$$\sec^2\theta = \frac{1}{4} + 1 = \frac{5}{4} \qquad \text{Proceed to solve for } \sec\theta.$$

$$\sec\theta = \frac{\sqrt{5}}{2} \qquad \sec\theta > 0 \text{ since } \theta \text{ is acute.}$$

We now know $\tan\theta = \frac{1}{2}$ and $\sec\theta = \frac{\sqrt{5}}{2}$. Using reciprocal identities,

$$\cos\theta = \frac{1}{\sec\theta} = \frac{1}{\frac{\sqrt{5}}{2}} = \frac{2}{\sqrt{5}} = \frac{2\sqrt{5}}{5}$$

$$\cot\theta = \frac{1}{\tan\theta} = \frac{1}{\frac{1}{2}} = 2$$

To find $\sin\theta$, use the following reasoning:

$$\tan\theta = \frac{\sin\theta}{\cos\theta}, \quad \text{so} \quad \sin\theta = (\tan\theta)(\cos\theta) = \frac{1}{2}\cdot\frac{2\sqrt{5}}{5} = \frac{\sqrt{5}}{5}$$

Then

$$\csc\theta = \frac{1}{\sin\theta} = \frac{1}{\frac{\sqrt{5}}{5}} = \sqrt{5}$$

■

Now Work PROBLEM 25

4 Use the Complementary Angle Theorem

Two acute angles are called **complementary** if their sum is a right angle, or 90°. Because the sum of the angles of any triangle is 180°, it follows that for a right triangle, the sum of the acute angles is 90°, so the two acute angles are complementary.

Refer now to Figure 27; we have labeled the angle opposite side b as B and the angle opposite side a as A. Notice that side a is adjacent to angle B and is opposite angle A. Similarly, side b is opposite angle B and is adjacent to angle A. As a result,

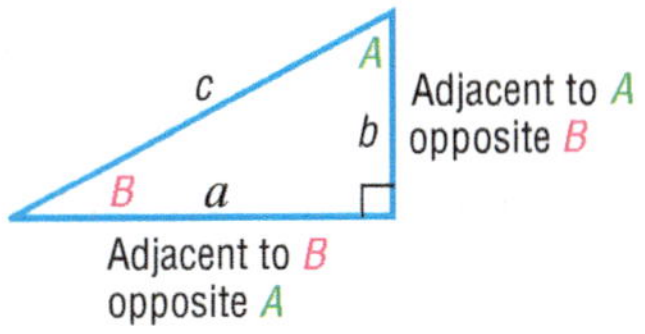

Figure 27

$$\sin B = \frac{b}{c} = \cos A \qquad \cos B = \frac{a}{c} = \sin A \qquad \tan B = \frac{b}{a} = \cot A$$
$$\csc B = \frac{c}{b} = \sec A \qquad \sec B = \frac{c}{a} = \csc A \qquad \cot B = \frac{a}{b} = \tan A \qquad \textbf{(8)}$$

Because of these relationships, the functions sine and cosine, tangent and cotangent, and secant and cosecant are called **cofunctions** of each other. The identities (8) may be expressed in words as follows:

THEOREM

Complementary Angle Theorem

Cofunctions of complementary angles are equal. ■

Here are examples of this theorem.

Complementary angles	Complementary angles	Complementary angles
$\sin 30° = \cos 60°$	$\tan 40° = \cot 50°$	$\sec 80° = \csc 10°$
Cofunctions	Cofunctions	Cofunctions

If an angle θ is measured in degrees, we will use the degree symbol when writing a trigonometric function of θ, as, for example, in $\sin 30°$ and $\tan 45°$. If an angle θ is measured in radians, then no symbol is used when writing a trigonometric function of θ, as, for example, in $\cos\pi$ and $\sec\frac{\pi}{3}$.

If θ is an acute angle measured in degrees, the angle $90° - \theta$ (or $\frac{\pi}{2} - \theta$, if θ is in radians) is the angle complementary to θ. Table 2 restates the preceding theorem on cofunctions.

Table 2

θ (Degrees)	θ (Radians)
$\sin\theta = \cos(90° - \theta)$	$\sin\theta = \cos\left(\frac{\pi}{2} - \theta\right)$
$\cos\theta = \sin(90° - \theta)$	$\cos\theta = \sin\left(\frac{\pi}{2} - \theta\right)$
$\tan\theta = \cot(90° - \theta)$	$\tan\theta = \cot\left(\frac{\pi}{2} - \theta\right)$
$\csc\theta = \sec(90° - \theta)$	$\csc\theta = \sec\left(\frac{\pi}{2} - \theta\right)$
$\sec\theta = \csc(90° - \theta)$	$\sec\theta = \csc\left(\frac{\pi}{2} - \theta\right)$
$\cot\theta = \tan(90° - \theta)$	$\cot\theta = \tan\left(\frac{\pi}{2} - \theta\right)$

The angle θ in Table 2 is acute. We will see later (Section 8.5) that these results are valid for any angle θ.

EXAMPLE 6 **Using the Complementary Angle Theorem**

(a) $\sin 62° = \cos(90° - 62°) = \cos 28°$

(b) $\tan\frac{\pi}{12} = \cot\left(\frac{\pi}{2} - \frac{\pi}{12}\right) = \cot\frac{5\pi}{12}$

(c) $\cos\frac{\pi}{4} = \sin\left(\frac{\pi}{2} - \frac{\pi}{4}\right) = \sin\frac{\pi}{4}$

(d) $\csc\frac{\pi}{6} = \sec\left(\frac{\pi}{2} - \frac{\pi}{6}\right) = \sec\frac{\pi}{3}$ ■

EXAMPLE 7 **Using the Complementary Angle Theorem**

Find the exact value of each expression. Do not use a calculator.

(a) $\sec 28° - \csc 62°$ (b) $\dfrac{\sin 35°}{\cos 55°}$

Solution (a) $\sec 28° - \csc 62° = \csc(90° - 28°) - \csc 62°$

$$= \csc 62° - \csc 62° = 0$$

(b) $\dfrac{\sin 35°}{\cos 55°} = \dfrac{\cos(90° - 35°)}{\cos 55°} = \dfrac{\cos 55°}{\cos 55°} = 1$ ■

Now Work PROBLEM 43

Historical Feature

The name *sine* for the sine function arose from a medieval confusion. The name comes from the Sanskrit word *jiva*, (meaning "chord"), first used in India by Araybhata the Elder (AD 510). He really meant half-chord, but abbreviated it. This was brought into Arabic as *jiba*, which was meaningless. Because the proper Arabic word *jaib* would be written the same way (short vowels are not written out in Arabic), *jiba*, was pronounced as *jaib*, which meant "bosom" or "hollow," and *jaib* remains the Arabic word for *sine* to this day. Scholars translating the Arabic works into Latin found that the word *sinus* also meant "bosom" or "hollow," and from *sinus* we get our *sine*.

The name *tangent*, introduced by Thomas Finck (1583), can be understood by looking at Figure 28. The line segment $\overline{DC}$ is tangent to the circle at C. If $d(O, B) = d(O, C) = 1$, then the length of the line segment $\overline{DC}$ is

$$d(D, C) = \frac{d(D, C)}{1} = \frac{d(D, C)}{d(O, C)} = \tan\alpha$$

The old name for the tangent is *umbra versa* (meaning turned shadow), referring to the use of the tangent in solving height problems with shadows.

The names of the cofunctions came about as follows. If α and β are complementary angles, then $\cos\alpha = \sin\beta$. Because β is the complement of α, it was natural to write the cosine of α as *sin co* α. Probably for reasons involving ease of pronunciation, the *co* migrated to the front, and then cosine received a three-letter abbreviation to match sin, sec, and tan. The two other cofunctions were similarly treated, except that the long forms *cotan* and *cosec* survive to this day in some countries.

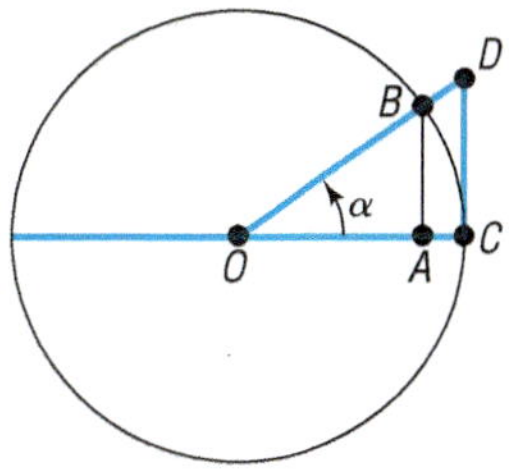

Figure 28

7.2 Assess Your Understanding

'Are You Prepared?'

Answers are given at the end of these exercises. If you get a wrong answer, read the pages listed in red.

1. In a right triangle with legs $a = 6$ and $b = 10$, the Pythagorean Theorem tells us that the hypotenuse is $c =$ ________. (pp. 31–32)

2. The value of the function $f(x) = 3x - 7$ at 5 is _____. (pp. 210–213)

Concepts and Vocabulary

3. Two acute angles whose sum is a right angle are called ____________ angles.

4. The sine and ______ functions are cofunctions.

5. $\sec 28° = \csc$ ____.

6. Which of the following ratios is used to find the value of $\cot\theta$?

(a) $\dfrac{\text{opposite}}{\text{hypotenuse}}$ (b) $\dfrac{\text{opposite}}{\text{adjacent}}$

(c) $\dfrac{\text{adjacent}}{\text{opposite}}$ (d) $\dfrac{\text{hypotenuse}}{\text{adjacent}}$

7. ***True or False*** $\tan\theta = \dfrac{\sin\theta}{\cos\theta}$.

8. Which of the following expressions does *not* equal 1?

(a) $\cos^2\theta - \sin^2\theta$ (b) $\sec^2\theta - \tan^2\theta$

(c) $\csc^2\theta - \cot^2\theta$ (d) $\sin^2\theta + \cos^2\theta$

9. ***True or False*** If θ is an acute angle and $\sec\theta = 3$, then $\cos\theta = \dfrac{1}{3}$.

10. ***True or False*** $\tan\dfrac{\pi}{5} = \cot\dfrac{4\pi}{5}$.

Skill Building

In Problems 11–20, find the value of the six trigonometric functions of the angle θ in each figure.

11.

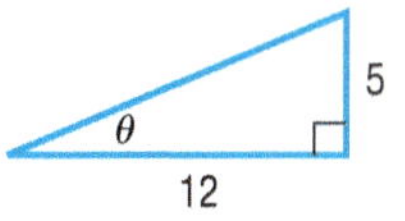

12.

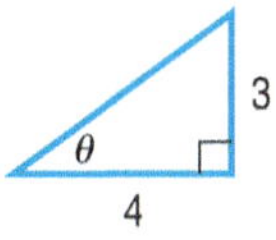

13.

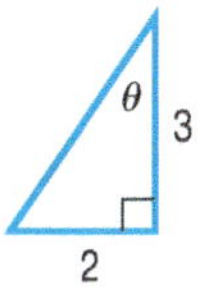

14.

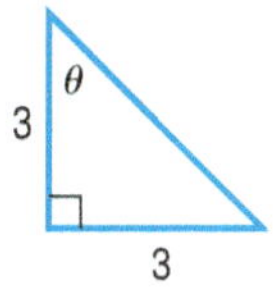

15.

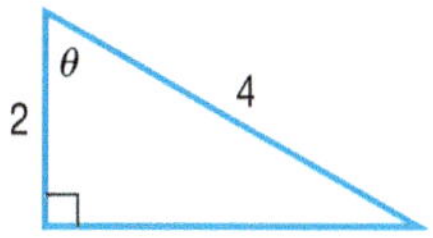

16.

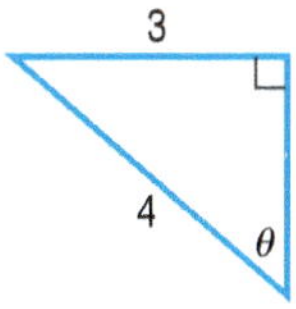

17.

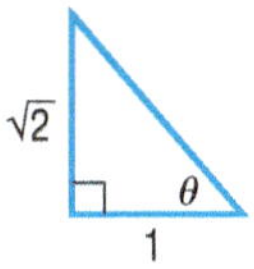

18.

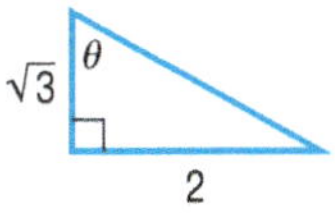

19.

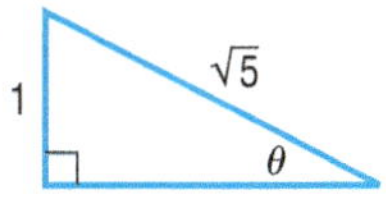

20.

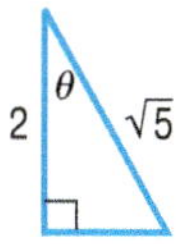

In Problems 21–24, use identities to find the exact value of each of the four remaining trigonometric functions of the acute angle θ.

21. $\sin\theta = \dfrac{1}{2}$, $\cos\theta = \dfrac{\sqrt{3}}{2}$

22. $\sin\theta = \dfrac{\sqrt{3}}{2}$, $\cos\theta = \dfrac{1}{2}$

23. $\sin\theta = \dfrac{2}{3}$, $\cos\theta = \dfrac{\sqrt{5}}{3}$

24. $\sin\theta = \dfrac{1}{3}$, $\cos\theta = \dfrac{2\sqrt{2}}{3}$

In Problems 25–36, use the definition or identities to find the exact value of each of the remaining five trigonometric functions of the acute angle θ.

25. $\sin\theta = \dfrac{\sqrt{2}}{2}$

26. $\cos\theta = \dfrac{\sqrt{2}}{2}$

27. $\cos\theta = \dfrac{1}{3}$

28. $\sin\theta = \dfrac{\sqrt{3}}{4}$

29. $\tan\theta = \dfrac{1}{2}$

30. $\cot\theta = \dfrac{1}{2}$

31. $\sec\theta = 3$

32. $\csc\theta = 5$

33. $\tan\theta = \sqrt{2}$

34. $\sec\theta = \dfrac{5}{3}$

35. $\csc\theta = 2$

36. $\cot\theta = 2$

In Problems 37–54, use Fundamental Identities and/or the Complementary Angle Theorem to find the exact value of each expression. Do not use a calculator.

37. $\sin^2 20° + \cos^2 20°$

38. $\sec^2 28° - \tan^2 28°$

39. $\sin 80° \csc 80°$

40. $\tan 10° \cot 10°$

41. $\tan 50° - \dfrac{\sin 50°}{\cos 50°}$

42. $\cot 25° - \dfrac{\cos 25°}{\sin 25°}$

43. $\sin 38° - \cos 52°$

44. $\tan 12° - \cot 78°$

45. $\dfrac{\cos 10°}{\sin 80°}$

46. $\dfrac{\cos 40°}{\sin 50°}$

47. $1 - \cos^2 20° - \cos^2 70°$

48. $1 + \tan^2 5° - \csc^2 85°$

49. $\tan 20° - \dfrac{\cos 70°}{\cos 20°}$

50. $\cot 40° - \dfrac{\sin 50°}{\sin 40°}$

51. $\tan 35° \cdot \sec 55° \cdot \cos 35°$

52. $\cot 25° \cdot \csc 65° \cdot \sin 25°$

53. $\cos 35° \sin 55° + \cos 55° \sin 35°$

54. $\sec 35° \csc 55° - \tan 35° \cot 55°$

---

55. Given $\sin 30° = \dfrac{1}{2}$, use trigonometric identities to find the exact value of

(a) $\cos 60°$ (b) $\cos^2 30°$

(c) $\csc \dfrac{\pi}{6}$ (d) $\sec \dfrac{\pi}{3}$

56. Given $\sin 60° = \dfrac{\sqrt{3}}{2}$, use trigonometric identities to find the exact value of

(a) $\cos 30°$ (b) $\cos^2 60°$

(c) $\sec \dfrac{\pi}{6}$ (d) $\csc \dfrac{\pi}{3}$

57. Given $\tan \theta = 4$, use trigonometric identities to find the exact value of

(a) $\sec^2 \theta$ (b) $\cot \theta$

(c) $\cot\left(\dfrac{\pi}{2} - \theta\right)$ (d) $\csc^2 \theta$

58. Given $\sec \theta = 3$, use trigonometric identities to find the exact value of

(a) $\cos \theta$ (b) $\tan^2 \theta$ (c) $\csc(90° - \theta)$ (d) $\sin^2 \theta$

59. Given $\csc \theta = 4$, use trigonometric identities to find the exact value of

(a) $\sin \theta$ (b) $\cot^2 \theta$

(c) $\sec(90° - \theta)$ (d) $\sec^2 \theta$

60. Given $\cot \theta = 2$, use trigonometric identities to find the exact value of

(a) $\tan \theta$ (b) $\csc^2 \theta$

(c) $\tan\left(\dfrac{\pi}{2} - \theta\right)$ (d) $\sec^2 \theta$

61. Given the approximation $\sin 38° \approx 0.62$, use trigonometric identities to find the approximate value of

(a) $\cos 38°$ (b) $\tan 38°$

(c) $\cot 38°$ (d) $\sec 38°$

(e) $\csc 38°$ (f) $\sin 52°$

(g) $\cos 52°$ (h) $\tan 52°$

62. Given the approximation $\cos 21° \approx 0.93$, use trigonometric identities to find the approximate value of

(a) $\sin 21°$ (b) $\tan 21°$

(c) $\cot 21°$ (d) $\sec 21°$

(e) $\csc 21°$ (f) $\sin 69°$

(g) $\cos 69°$ (h) $\tan 69°$

63. If $\sin \theta = 0.3$, find the exact value of $\sin \theta + \cos\left(\dfrac{\pi}{2} - \theta\right)$.

64. If $\tan \theta = 4$, find the exact value of $\tan \theta + \tan\left(\dfrac{\pi}{2} - \theta\right)$.

65. Find an acute angle θ that satisfies the equation

$$\sin \theta = \cos(2\theta + 30°)$$

66. Find an acute angle θ that satisfies the equation

$$\tan \theta = \cot(\theta + 45°)$$

Applications and Extensions

In Problems 67–72, simplify each expression for the given value of x. Assume the angle θ is acute.

67. $\sqrt{25 - x^2}$ if $x = 5 \sin \theta$

68. $\sqrt{16 - 9x^2}$ if $x = \dfrac{4}{3} \sin \theta$

69. $\sqrt{49 + x^2}$ if $x = 7 \tan \theta$

70. $\sqrt{4 + x^2}$ if $x = -2 \tan \theta$

71. $\sqrt{x^2 - 36}$ if $x = -6 \sec \theta$

72. $\sqrt{100x^2 - 81}$ if $x = \dfrac{9}{10} \sec \theta$

---

73. Calculating the Time of a Trip From a parking lot you want to walk to a house on the ocean. The house is located 1500 feet down a paved path that parallels the beach, which is 500 feet wide. Along the path you can walk 300 feet per minute, but in the sand on the beach you can walk only 100 feet per minute. See the illustration.

(a) Calculate the time T if you walk 1500 feet along the paved path and then 500 feet in the sand to the house.

(b) Calculate the time T if you walk in the sand directly toward the ocean for 500 feet and then turn left and walk along the beach for 1500 feet to the house.

(c) Express the time T to get from the parking lot to the beachhouse as a function of the angle θ shown in the illustration.

(d) Calculate the time T if you walk directly from the parking lot to the house.

$\left[\textbf{Hint: } \tan\theta = \frac{500}{1500}\right]$

(e) Calculate the time T if you walk 1000 feet along the paved path and then walk directly to the house.

(f) Graph $T = T(\theta)$. For what angle θ is T least? What is x for this angle? What is the minimum time?

(g) Explain why $\tan\theta = \frac{1}{3}$ gives the smallest angle θ that is possible.

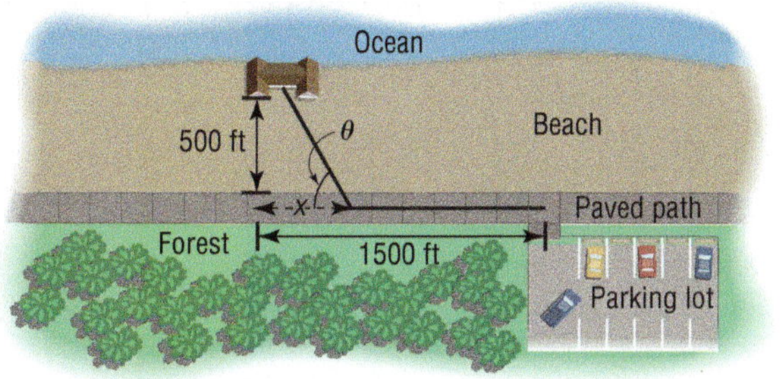

74. Carrying a Ladder around a Corner Two hallways, one of width 3 feet, the other of width 4 feet, meet at a right angle. See the illustration below.

(a) Express the length L of the line segment shown as a function of the angle θ.

(b) Discuss why the length of the longest ladder that can be carried around the corner is equal to the smallest value of L.

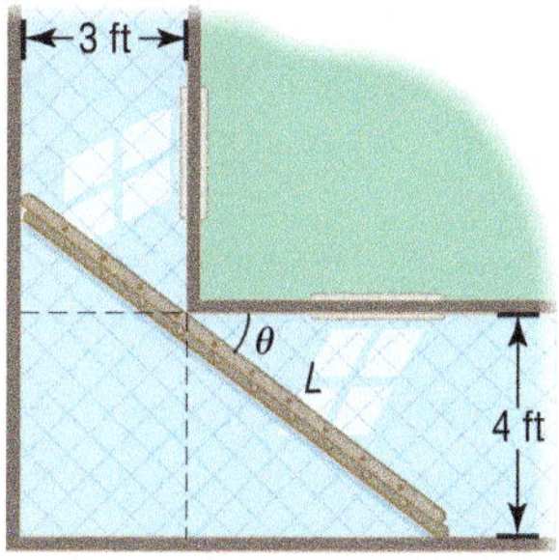

75. Electrical Engineering A resistor and an inductor connected in a series network impede the flow of an alternating current. This impedance Z is determined by the reactance X of the inductor and the resistance R of the resistor. The three quantities, all measured in ohms, can be represented by the sides of a right triangle as illustrated, so $Z^2 = X^2 + R^2$. The angle ϕ is called the **phase angle**. Suppose a series network has an inductive reactance of $X = 400$ ohms and a resistance of $R = 600$ ohms.

(a) Find the impedance Z.

(b) Find the values of the six trigonometric functions of the phase angle ϕ.

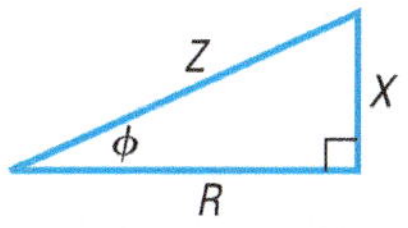

76. Electrical Engineering Refer to Problem 75. A series network has a resistance of $R = 588$ ohms. The phase angle ϕ is such that $\tan\phi = \frac{5}{12}$.

(a) Determine the inductive reactance X and the impedance Z.

(b) Determine the values of the remaining five trigonometric functions of the phase angle ϕ.

77. Geometry Suppose that the angle θ is a central angle of a circle of radius 1 (see the figure). Show that

(a) Angle $OAC = \frac{\theta}{2}$

(b) $|CD| = \sin\theta$ and $|OD| = \cos\theta$

(c) $\tan\frac{\theta}{2} = \frac{\sin\theta}{1+\cos\theta}$

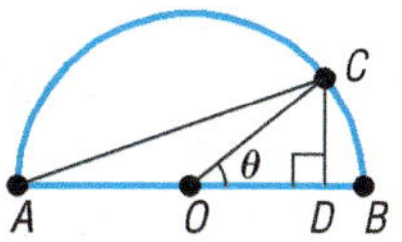

78. Geometry Show that the area A of an isosceles triangle is $A = a^2 \sin\theta\cos\theta$, where a is the length of one of the two equal sides and θ is the measure of one of the two equal angles (see the figure).

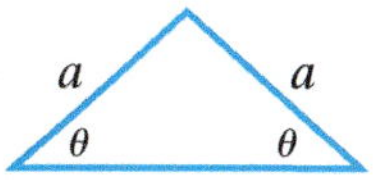

79. Geometry Let $n \geq 1$ be any real number, and let θ be any angle for which $0 < n\theta < \frac{\pi}{2}$. Then we can draw a triangle with the angles θ and $n\theta$ and included side of length 1 (do you see why?) and place it on the unit circle as illustrated. Now, drop the perpendicular from C to $D = (x, 0)$ and show that

$$x = \frac{\tan(n\theta)}{\tan\theta + \tan(n\theta)}$$

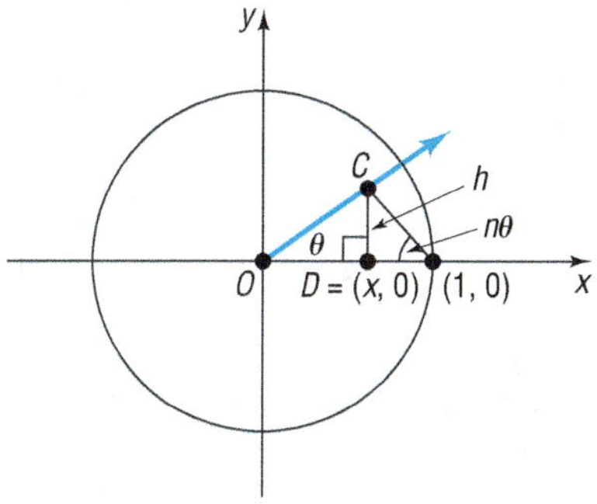

80. Geometry Refer to the figure below. The smaller circle, whose radius is a, is tangent to the larger circle, whose radius is b. The ray $\overrightarrow{OA}$ contains a diameter of each circle, and the ray $\overrightarrow{OB}$ is tangent to each circle. Show that

$$\cos\theta = \frac{\sqrt{ab}}{\frac{a+b}{2}}$$

(This shows that $\cos\theta$ equals the ratio of the geometric mean of a and b to the arithmetic mean of a and b.)

[**Hint:** First show that $\sin\theta = (b-a)/(b+a)$.]

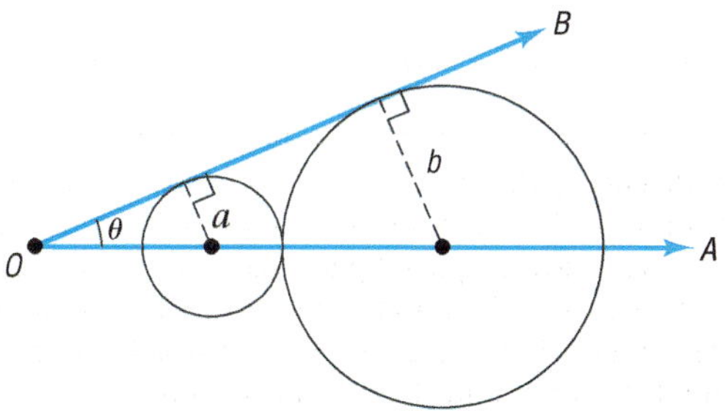

81. Geometry Refer to the figure. If $|OA| = 1$, show that

(a) Area $\Delta OAC = \frac{1}{2}\sin\alpha\cos\alpha$

(b) Area $\Delta OCB = \frac{1}{2}|OB|^2\sin\beta\cos\beta$

(c) Area $\Delta OAB = \frac{1}{2}|OB|\sin(\alpha+\beta)$

(d) $|OB| = \frac{\cos\alpha}{\cos\beta}$

(e) $\sin(\alpha+\beta) = \sin\alpha\cos\beta + \cos\alpha\sin\beta$

[**Hint:** Area ΔOAB = Area ΔOAC + Area ΔOCB]

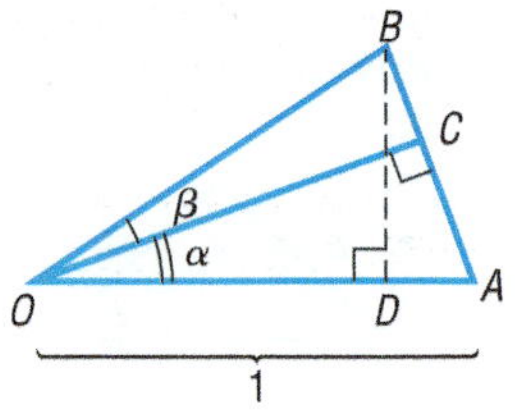

82. Geometry Refer to the figure, where a unit circle is drawn. The line segment $\overline{DB}$ is tangent to the circle.

(a) Express the area of ΔOBC in terms of $\sin\theta$ and $\cos\theta$. [**Hint:** Use the altitude from C to the base $\overline{OB} = 1$.]

(b) Express the area of ΔOBD in terms of $\sin\theta$ and $\cos\theta$.

(c) The area of the sector OBC of the circle is $\frac{1}{2}\theta$, where θ is measured in radians. Use the results of parts (a) and (b) and the fact that

Area $\Delta OBC <$ Area of sector $OBC <$ Area ΔOBD

to show that

$$1 < \frac{\theta}{\sin\theta} < \frac{1}{\cos\theta}$$

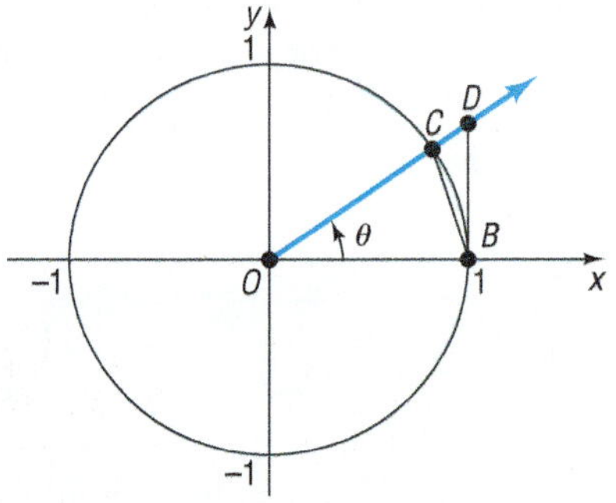

83. If $\cos\alpha = \tan\beta$ and $\cos\beta = \tan\alpha$, where α and β are acute angles, show that

$$\sin\alpha = \sin\beta = \sqrt{\frac{3-\sqrt{5}}{2}}$$

84. If θ is an acute angle and $\tan\theta = x$, express the remaining five trigonometric functions in terms of x.

Explaining Concepts: Discussion and Writing

85. If θ is an acute angle, explain why $\sec\theta > 1$.

86. If θ is an acute angle, explain why $0 < \sin\theta < 1$.

87. How would you explain the meaning of the sine function to a fellow student who has just completed College Algebra?

88. Look back at Example 5. Which of the two solutions do you prefer? Explain your reasoning.

Retain Your Knowledge

Problems 89–92 are based on material learned earlier in the course. The purpose of these problems is to keep the material fresh in your mind so that you are better prepared for the final exam.

89. State the domain of $f(x) = \ln(5x+2)$.

90. Given that the polynomial function $P(x) = x^4 - 5x^3 - 9x^2 + 155x - 250$ has zeros of $4 + 3i$ and 2, find the remaining zeros of the function.

91. Find the remainder when $P(x) = 8x^4 - 2x^3 + x - 8$ is divided by $x + 2$.

92. Sidewalk Area A sidewalk with a uniform width of 3 feet is to be placed around a circular garden with a diameter of 24 feet. Find the exact area of the sidewalk.

'Are You Prepared?' Answers

1. $2\sqrt{34}$ **2.** 8

7.3 Computing the Values of Trigonometric Functions of Acute Angles

OBJECTIVES
1. Find the Exact Values of the Trigonometric Functions of $\frac{\pi}{4} = 45°$ (p. 539)
2. Find the Exact Values of the Trigonometric Functions of $\frac{\pi}{6} = 30°$ and $\frac{\pi}{3} = 60°$ (p. 540)
3. Use a Calculator to Approximate the Values of the Trigonometric Functions of Acute Angles (p. 541)
4. Model and Solve Applied Problems Involving Right Triangles (p. 542)

In the previous section, we developed ways to find the value of each trigonometric function of an acute angle when the value of one of the functions is known. In this section, we discuss the problem of finding the value of each trigonometric function of an acute angle when the angle is given.

For three special acute angles, we can use some results from plane geometry to find the exact value of each of the six trigonometric functions.

1 Find the Exact Values of the Trigonometric Functions of $\frac{\pi}{4} = 45°$

EXAMPLE 1 **Finding the Exact Values of the Trigonometric Functions of $\frac{\pi}{4} = 45°$**

Find the exact values of the six trigonometric functions of $\frac{\pi}{4} = 45°$.

Solution Using the right triangle in Figure 29(a), in which one of the angles is $\frac{\pi}{4} = 45°$, it follows that the other acute angle is also $\frac{\pi}{4} = 45°$, so the triangle is isosceles. As a result, side a and side b are equal in length. Since the values of the trigonometric functions of an angle depend only on the angle and not on the size of the triangle, we may assign any values to a and b for which $a = b > 0$. For simplicity, we use the triangle for which

$$a = b = 1$$

Then, by the Pythagorean Theorem,

$$c^2 = a^2 + b^2 = 1 + 1 = 2$$
$$c = \sqrt{2}$$

As a result, we have the triangle in Figure 29(b), from which we find

$$\sin\frac{\pi}{4} = \sin 45° = \frac{b}{c} = \frac{1}{\sqrt{2}} = \frac{\sqrt{2}}{2} \qquad \cos\frac{\pi}{4} = \cos 45° = \frac{a}{c} = \frac{1}{\sqrt{2}} = \frac{\sqrt{2}}{2}$$

Using Quotient and Reciprocal Identities,

$$\tan\frac{\pi}{4} = \tan 45° = \frac{\sin 45°}{\cos 45°} = \frac{\frac{\sqrt{2}}{2}}{\frac{\sqrt{2}}{2}} = 1 \qquad \cot\frac{\pi}{4} = \cot 45° = \frac{1}{\tan 45°} = \frac{1}{1} = 1$$

$$\sec\frac{\pi}{4} = \sec 45° = \frac{1}{\cos 45°} = \frac{1}{\frac{1}{\sqrt{2}}} = \sqrt{2} \qquad \csc\frac{\pi}{4} = \csc 45° = \frac{1}{\sin 45°} = \frac{1}{\frac{1}{\sqrt{2}}} = \sqrt{2}$$

■

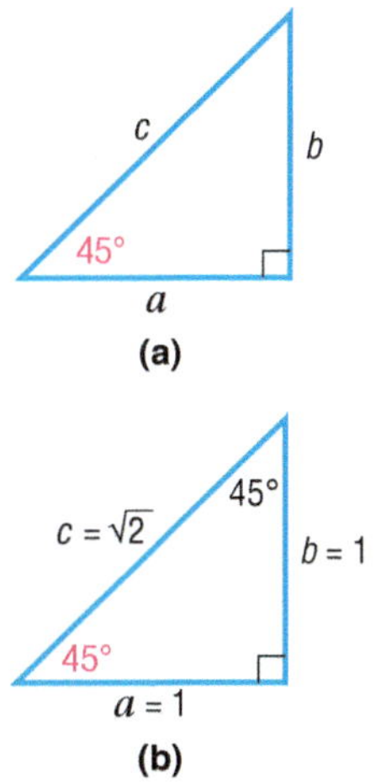

Figure 29 Isosceles right triangle

EXAMPLE 2 Finding the Exact Value of a Trigonometric Expression

Find the exact value of each expression.

(a) $(\sin 45°)(\tan 45°)$ (b) $\left(\sec \dfrac{\pi}{4}\right)\left(\cot \dfrac{\pi}{4}\right)$

Solution Use the results obtained in Example 1.

(a) $(\sin 45°)(\tan 45°) = \dfrac{\sqrt{2}}{2} \cdot 1 = \dfrac{\sqrt{2}}{2}$ (b) $\left(\sec \dfrac{\pi}{4}\right)\left(\cot \dfrac{\pi}{4}\right) = \sqrt{2} \cdot 1 = \sqrt{2}$

Now Work PROBLEMS 7 AND 19

2 Find the Exact Values of the Trigonometric Functions of $\dfrac{\pi}{6} = 30°$ and $\dfrac{\pi}{3} = 60°$

EXAMPLE 3 Finding the Exact Values of the Trigonometric Functions of $\dfrac{\pi}{6} = 30°$ and $\dfrac{\pi}{3} = 60°$

Find the exact values of the six trigonometric functions of $\dfrac{\pi}{6} = 30°$ and $\dfrac{\pi}{3} = 60°$.

Solution Form a right triangle in which one of the angles is $\dfrac{\pi}{6} = 30°$. It then follows that the third angle is $\dfrac{\pi}{3} = 60°$. Figure 30(a) illustrates such a triangle with hypotenuse of length 2. Our problem is to determine a and b.

Begin by placing next to the triangle in Figure 30(a) another triangle congruent to the first, as shown in Figure 30(b). Notice that you now have a triangle whose angles are each 60°. This triangle is therefore equilateral, so each side is of length 2. In particular, the base is $2a = 2$, so $a = 1$. By the Pythagorean Theorem, b satisfies the equation $a^2 + b^2 = c^2$, so

$$\begin{aligned} a^2 + b^2 &= c^2 \\ 1^2 + b^2 &= 2^2 \quad a = 1, c = 2 \\ b^2 &= 4 - 1 = 3 \\ b &= \sqrt{3} \end{aligned}$$

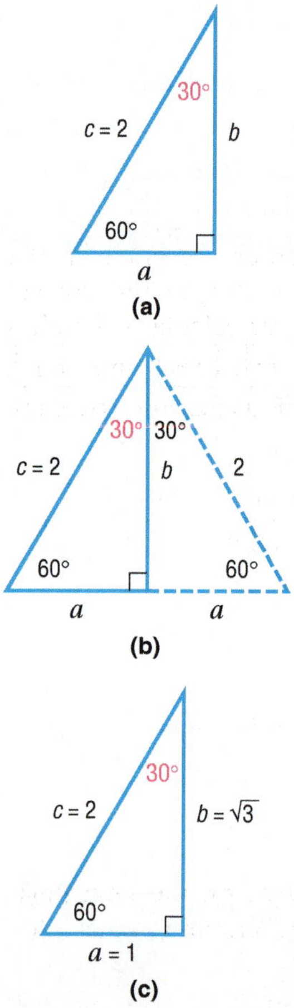

Figure 30 30°–60°–90° triangle

Using the triangle in Figure 30(c) and the fact that $\dfrac{\pi}{6} = 30°$ and $\dfrac{\pi}{3} = 60°$ are complementary angles, we find

$$\sin \frac{\pi}{6} = \sin 30° = \frac{\text{opposite}}{\text{hypotenuse}} = \frac{1}{2} \qquad \cos \frac{\pi}{3} = \cos 60° = \frac{1}{2}$$

$$\cos \frac{\pi}{6} = \cos 30° = \frac{\text{adjacent}}{\text{hypotenuse}} = \frac{\sqrt{3}}{2} \qquad \sin \frac{\pi}{3} = \sin 60° = \frac{\sqrt{3}}{2}$$

$$\tan \frac{\pi}{6} = \tan 30° = \frac{\sin 30°}{\cos 30°} = \frac{\frac{1}{2}}{\frac{\sqrt{3}}{2}} = \frac{1}{\sqrt{3}} = \frac{\sqrt{3}}{3} \qquad \cot \frac{\pi}{3} = \cot 60° = \frac{\sqrt{3}}{3}$$

$$\csc \frac{\pi}{6} = \csc 30° = \frac{1}{\sin 30°} = \frac{1}{\frac{1}{2}} = 2 \qquad \sec \frac{\pi}{3} = \sec 60° = 2$$

$$\sec \frac{\pi}{6} = \sec 30° = \frac{1}{\cos 30°} = \frac{1}{\frac{\sqrt{3}}{2}} = \frac{2}{\sqrt{3}} = \frac{2\sqrt{3}}{3} \qquad \csc \frac{\pi}{3} = \csc 60° = \frac{2\sqrt{3}}{3}$$

$$\cot \frac{\pi}{6} = \cot 30° = \frac{1}{\tan 30°} = \frac{1}{\frac{\sqrt{3}}{3}} = \frac{3}{\sqrt{3}} = \sqrt{3} \qquad \tan \frac{\pi}{3} = \tan 60° = \sqrt{3}$$

Table 3 summarizes the information just derived for the angles $\frac{\pi}{6} = 30°, \frac{\pi}{4} = 45°$, and $\frac{\pi}{3} = 60°$. Rather than memorize the entries in Table 3, you can draw the appropriate triangle to determine the values given in the table.

Table 3

θ (Radians)	θ (Degrees)	$\sin\theta$	$\cos\theta$	$\tan\theta$	$\csc\theta$	$\sec\theta$	$\cot\theta$
$\frac{\pi}{6}$	30°	$\frac{1}{2}$	$\frac{\sqrt{3}}{2}$	$\frac{\sqrt{3}}{3}$	2	$\frac{2\sqrt{3}}{3}$	$\sqrt{3}$
$\frac{\pi}{4}$	45°	$\frac{\sqrt{2}}{2}$	$\frac{\sqrt{2}}{2}$	1	$\sqrt{2}$	$\sqrt{2}$	1
$\frac{\pi}{3}$	60°	$\frac{\sqrt{3}}{2}$	$\frac{1}{2}$	$\sqrt{3}$	$\frac{2\sqrt{3}}{3}$	2	$\frac{\sqrt{3}}{3}$

EXAMPLE 4 **Finding the Exact Value of a Trigonometric Expression**

Find the exact value of each expression.

(a) $\sin 45° \cos 30°$ (b) $\tan\frac{\pi}{4} - \sin\frac{\pi}{3}$ (c) $\tan^2\frac{\pi}{6} + \sin^2\frac{\pi}{4}$

Solution

(a) $\sin 45° \cos 30° = \frac{\sqrt{2}}{2}\cdot\frac{\sqrt{3}}{2} = \frac{\sqrt{6}}{4}$

(b) $\tan\frac{\pi}{4} - \sin\frac{\pi}{3} = 1 - \frac{\sqrt{3}}{2} = \frac{2-\sqrt{3}}{2}$

(c) $\tan^2\frac{\pi}{6} + \sin^2\frac{\pi}{4} = \left(\frac{\sqrt{3}}{3}\right)^2 + \left(\frac{\sqrt{2}}{2}\right)^2 = \frac{1}{3} + \frac{1}{2} = \frac{5}{6}$ ■

Now Work PROBLEMS 11 AND 21

3 Use a Calculator to Approximate the Values of the Trigonometric Functions of Acute Angles

Before getting started, you must first decide whether to enter the angle in the calculator using radians or degrees and then set the calculator to the correct MODE. (Check your instruction manual to find out how your calculator handles degrees and radians.) Your calculator has the keys marked [sin], [cos], and [tan]. To find the values of the remaining three trigonometric functions (secant, cosecant, and cotangent), use the reciprocal identities.

WARNING On your calculator the second functions $\sin^{-1}$, $\cos^{-1}$, and $\tan^{-1}$ do not represent the reciprocal of sin, cos, and tan. ■

$$\sec\theta = \frac{1}{\cos\theta} \qquad \csc\theta = \frac{1}{\sin\theta} \qquad \cot\theta = \frac{1}{\tan\theta}$$

EXAMPLE 5 **Using a Calculator to Approximate the Values of Trigonometric Functions**

Use a calculator to find the approximate value of each of the following:

(a) $\cos 48°$ (b) $\csc 21°$ (c) $\tan\frac{\pi}{12}$

Express your answer rounded to two decimal places.

Solution

(a) First, set the MODE to receive degrees. See Figure 31(a) on the next page. Figure 31(b) shows the solution using a TI-84 Plus C graphing calculator. Rounded to two decimal places,

$$\cos 48° = 0.67$$

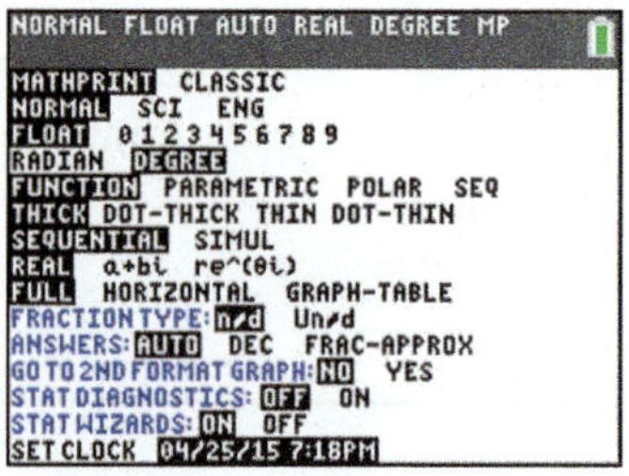

(a)

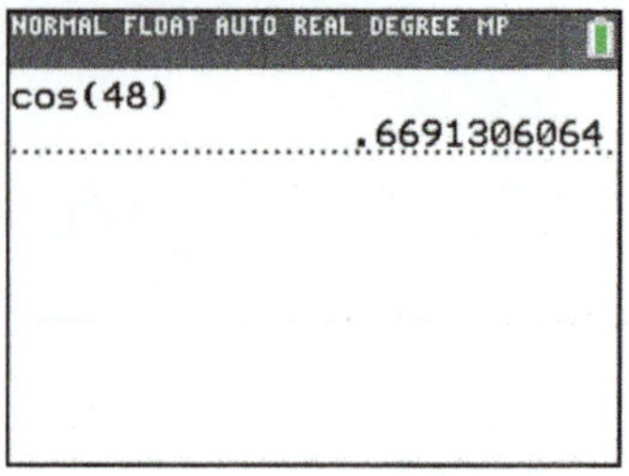

(b)

Figure 31 cos 48°

(b) Most calculators do not have a csc key. The manufacturers assume that the user knows some trigonometry. To find the value of csc 21°, use the fact that $\csc 21° = \dfrac{1}{\sin 21°}$. Figure 32 shows the solution using a TI-84 Plus C graphing calculator. Rounded to two decimal places,

$$\csc 21° = 2.79$$

(c) Set the MODE to receive radians. Figure 33 shows the solution using a TI-84 Plus C graphing calculator. Rounded to two decimal places,

$$\tan \frac{\pi}{12} = 0.27$$

Degree mode

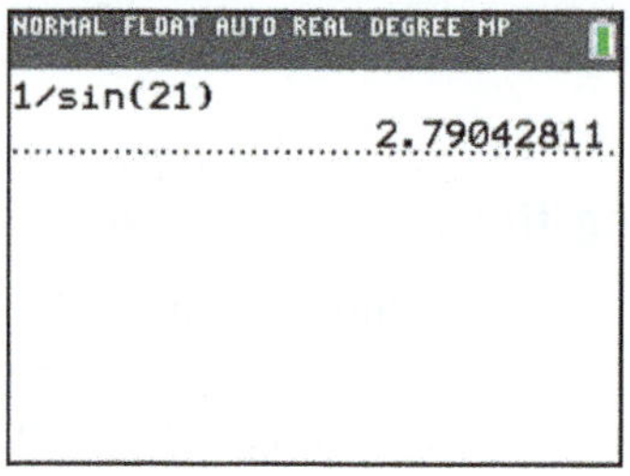

Figure 32 csc 21°

Radian mode

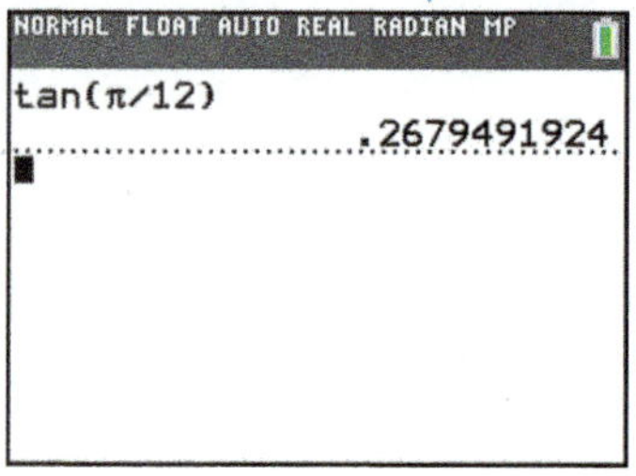

Figure 33 $\tan \dfrac{\pi}{12}$

Now Work PROBLEM 31

4 Model and Solve Applied Problems Involving Right Triangles

Right triangles can be used to model many types of situations, such as the optimal design of a rain gutter.*

EXAMPLE 6 **Constructing a Rain Gutter**

A rain gutter is to be constructed of aluminum sheets 12 inches wide. After marking off a length of 4 inches from each edge, the sides are bent up at an angle θ. See Figure 34.

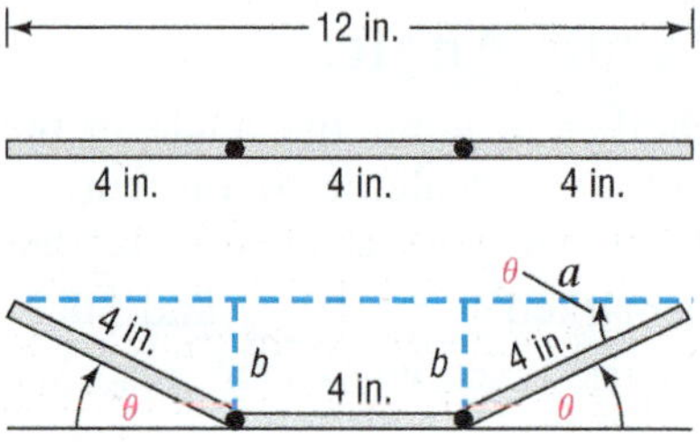

Figure 34

(a) Express the area A of the opening as a function of θ.
[**Hint:** Let b denote the vertical height of the bend.]

(b) Find the area A of the opening for $\theta = 30°$, $\theta = 45°$, $\theta = 60°$, and $\theta = 75°$.

(c) Graph $A = A(\theta)$. Find the angle θ that makes A largest. (This bend will allow the most water to flow through the gutter.)

Solution (a) Look again at Figure 34. The area A of the opening is the sum of the areas of two congruent right triangles and one rectangle. Look at Figure 35, which shows the triangle on the right in Figure 34 redrawn. Note that

$$\cos\theta = \frac{a}{4}, \quad \text{so} \quad a = 4\cos\theta \qquad \sin\theta = \frac{b}{4}, \quad \text{so} \quad b = 4\sin\theta$$

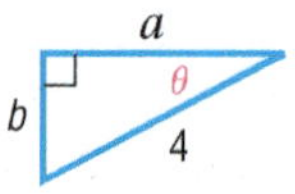

Figure 35

The area of the triangle is

$$\text{area of triangle} = \frac{1}{2}(\text{base})(\text{height}) = \frac{1}{2}ab = \frac{1}{2}(4\cos\theta)(4\sin\theta) = 8\sin\theta\cos\theta$$

*In applied problems, it is important that answers be reported with both justifiable accuracy and appropriate significant figures. We shall assume that angles are measured to the nearest tenth and sides are measured to the nearest hundredth. This results in sides being rounded to two decimal places and angles being rounded to one decimal place.

So the area of the two congruent triangles is $16 \sin \theta \cos \theta$.

The rectangle has length 4 and height b, so its area is

$$\text{area of rectangle} = 4b = 4(4 \sin \theta) = 16 \sin \theta$$

The area A of the opening is

$$A = \text{area of the two triangles} + \text{area of the rectangle}$$
$$A(\theta) = 16 \sin \theta \cos \theta + 16 \sin \theta = 16 \sin \theta(\cos \theta + 1)$$

(b) For $\theta = 30°$: $\quad A(30°) = 16 \sin 30°(\cos 30° + 1)$

$$= 16\left(\frac{1}{2}\right)\left(\frac{\sqrt{3}}{2} + 1\right) = 4\sqrt{3} + 8 \approx 14.93$$

The area of the opening for $\theta = 30°$ is about 14.93 square inches.

For $\theta = 45°$: $\quad A(45°) = 16 \sin 45°(\cos 45° + 1)$

$$= 16\left(\frac{\sqrt{2}}{2}\right)\left(\frac{\sqrt{2}}{2} + 1\right) = 8 + 8\sqrt{2} \approx 19.31$$

The area of the opening for $\theta = 45°$ is about 19.31 square inches.

For $\theta = 60°$: $\quad A(60°) = 16 \sin 60°(\cos 60° + 1)$

$$= 16\left(\frac{\sqrt{3}}{2}\right)\left(\frac{1}{2} + 1\right) = 12\sqrt{3} \approx 20.78$$

The area of the opening for $\theta = 60°$ is about 20.78 square inches.

For $\theta = 75°$: $\quad A(75°) = 16 \sin 75°(\cos 75° + 1) \approx 19.45$

The area of the opening for $\theta = 75°$ is about 19.45 square inches.

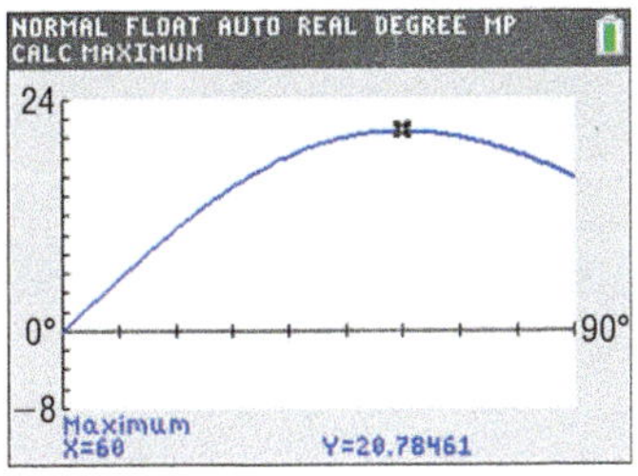

Figure 36

(c) Figure 36 shows the graph of $A = A(\theta)$. Using MAXIMUM, the angle θ that makes A largest is 60°. The maximum area is about 20.78 square inches. ■

Now Work PROBLEM 67

In addition to developing models using right triangles, right triangle trigonometry can be used to measure heights and distances that are either awkward or impossible to measure by ordinary means. When using right triangles to solve these problems, pay attention to the known measures. This will indicate which trigonometric function to use. For example, if we know the measure of an angle and the length of the side adjacent to the angle, and we wish to find the length of the opposite side, we would use the tangent function. Do you know why?

EXAMPLE 7 Finding the Width of a River

A surveyor can measure the width of a river by setting up a transit at a point C on one side of the river and taking a sighting of a point A on the other side. Refer to Figure 37 on page 544. After turning through an angle of 90° at C, the surveyor walks a distance of 200 meters to point B. Using the transit at B, the angle θ is measured and found to be 20°. What is the width of the river rounded to the nearest meter?

Solution We seek the length of side b. We know a and θ, so we use the fact that b is opposite θ and a is adjacent to θ and write

$$\tan \theta = \frac{b}{a}$$

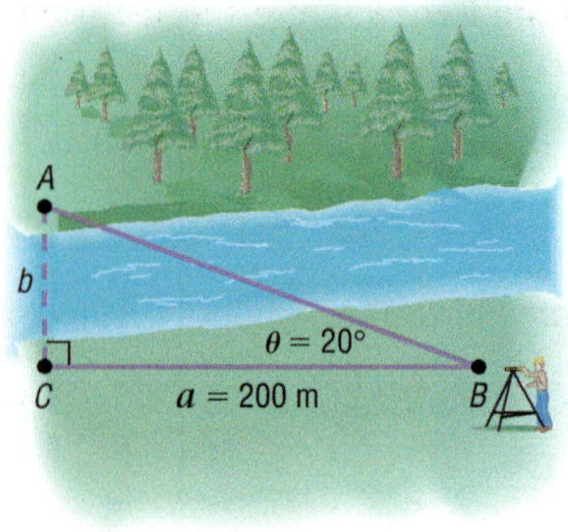

Figure 37

which leads to

$$\tan 20° = \frac{b}{200}$$

$$b = 200 \tan 20° \approx 72.79 \text{ meters}$$

The width of the river is 73 meters, rounded to the nearest meter. ■

Now Work PROBLEM 73

Vertical heights can sometimes be measured using either the *angle of elevation* or the *angle of depression*. If a person is looking up at an object, the acute angle measured from the horizontal to a line of sight to the object is called the **angle of elevation**. See Figure 38(a).

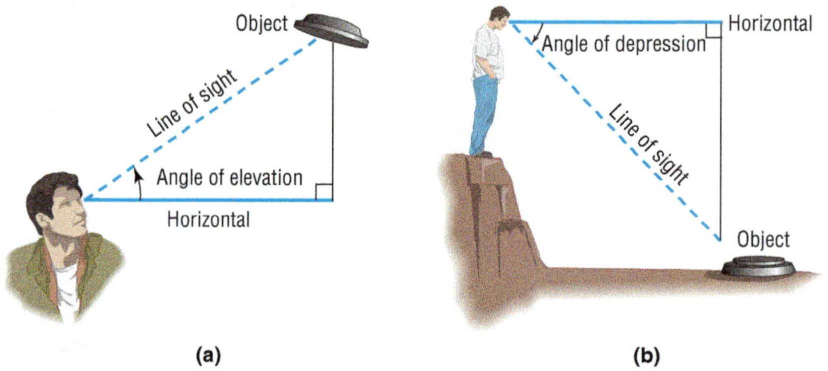

Figure 38

If a person is looking down at an object, the acute angle made by the line of sight to the object and the horizontal is called the **angle of depression**. See Figure 38(b).

EXAMPLE 8 Finding the Height of a Cloud

Meteorologists find the height of a cloud using an instrument called a **ceilometer**. A ceilometer consists of a **light projector** that directs a vertical light beam up to the cloud base and a **light detector** that scans the cloud to detect the light beam. See Figure 39(a). At Midway Airport in Chicago, a ceilometer was employed to find the height of the cloud cover. It was set up with its light detector 300 feet from its light projector. If the angle of elevation from the light detector to the base of the cloud was 75°, what was the height of the cloud cover?

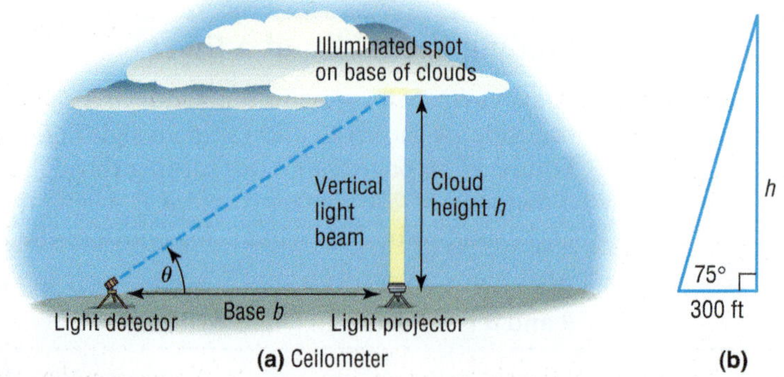

Figure 39

Solution Figure 39(b) illustrates the situation. To determine the height h, use the fact that $\tan 75° = \frac{h}{300}$, so

$$h = 300 \tan 75° \approx 1120 \text{ feet}$$

The ceiling (height to the base of the cloud cover) was approximately 1120 feet. ■

Now Work PROBLEM 75

The idea behind Example 8 can also be used to find the height of an object with a base that is not accessible to the horizontal.

EXAMPLE 9 **Finding the Height of a Statue on a Building**

Adorning the top of the Board of Trade building in Chicago is a statue of Ceres, the Roman goddess of wheat. From street level, two observations are taken 400 feet from the center of the building. The angle of elevation to the base of the statue is found to be 55.1°, and the angle of elevation to the top of the statue is 56.5°. See Figure 40(a). What is the height of the statue?

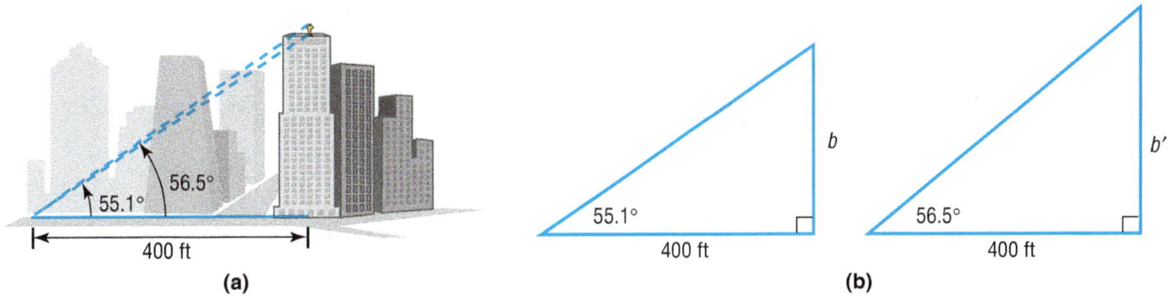

Figure 40

Solution Figure 40(b) shows two triangles that replicate Figure 40(a). The height of the statue of Ceres will be $b' - b$. To find b and b', refer to Figure 40(b).

$$\tan 55.1° = \frac{b}{400} \qquad \tan 56.5° = \frac{b'}{400}$$

$$b = 400 \tan 55.1° \qquad b' = 400 \tan 56.5°$$

$$\approx 573.39 \qquad \approx 604.33$$

The height of the statue is approximately $604.33 - 573.39 = 30.94$ feet ≈ 31 feet. ■

Now Work PROBLEM 81

7.3 Assess Your Understanding

Concepts and Vocabulary

1. $\tan 45° + \sin 30° =$ ____.

2. Using a calculator, $\sin 1 =$ ______, rounded to two decimal places.

3. ***True or False*** Exact values can be found for the trigonometric functions of 60°.

4. ***True or False*** Exact values can be found for the sine of any angle.

5. Find the exact value of $\csc \frac{\pi}{6}$.
   (a) 1 (b) 2 (c) $\sqrt{2}$ (d) $\frac{2\sqrt{3}}{3}$

6. Use a calculator to approximate $\cot 35°$.
   (a) 0.47 (b) 0.70 (c) 1.43 (d) 2.11

Skill Building

7. Write down the exact value of each of the six trigonometric functions of 45°.

8. Write down the exact value of each of the six trigonometric functions of 30° and of 60°.

In Problems 9–18, $f(\theta) = \sin\theta$ and $g(\theta) = \cos\theta$. Find the exact value of each expression if $\theta = 60°$. Do not use a calculator.

9. $f(\theta)$ **10.** $g(\theta)$ **11.** $f\left(\frac{\theta}{2}\right)$ **12.** $g\left(\frac{\theta}{2}\right)$ **13.** $[f(\theta)]^2$

14. $[g(\theta)]^2$ **15.** $2f(\theta)$ **16.** $2g(\theta)$ **17.** $\frac{f(\theta)}{2}$ **18.** $\frac{g(\theta)}{2}$

In Problems 19–30, find the exact value of each expression. Do not use a calculator.

19. $4\cos 45° - 2\sin 45°$ **20.** $2\sin 45° + 4\cos 30°$ **21.** $6\tan 45° - 8\cos 60°$

22. $\sin 30° \cdot \tan 60°$ **23.** $\sec\frac{\pi}{4} + 2\csc\frac{\pi}{3}$ **24.** $\tan\frac{\pi}{4} + \cot\frac{\pi}{4}$

25. $\sec^2\frac{\pi}{6} - 4$ **26.** $4 + \tan^2\frac{\pi}{3}$ **27.** $\sin^2 30° + \cos^2 60°$

28. $\sec^2 60° - \tan^2 45°$ **29.** $1 - \cos^2 30° - \cos^2 60°$ **30.** $1 + \tan^2 30° - \csc^2 45°$

In Problems 31–48, use a calculator to find the approximate value of each expression. Round the answer to two decimal places.

31. $\sin 28°$ **32.** $\cos 14°$ **33.** $\tan 21°$ **34.** $\cot 70°$ **35.** $\sec 41°$ **36.** $\csc 55°$

37. $\sin\frac{\pi}{10}$ **38.** $\cos\frac{\pi}{8}$ **39.** $\tan\frac{5\pi}{12}$ **40.** $\cot\frac{\pi}{18}$ **41.** $\sec\frac{\pi}{12}$ **42.** $\csc\frac{5\pi}{13}$

43. $\sin 1$ **44.** $\tan 1$ **45.** $\sin 1°$ **46.** $\tan 1°$ **47.** $\tan 0.3$ **48.** $\tan 0.1$

Mixed Practice

In Problems 49–58, $f(x) = \sin x$, $g(x) = \cos x$, $h(x) = 2x$, and $p(x) = \frac{x}{2}$. Find the value of each of the following:

49. $(f+g)(30°)$ **50.** $(f-g)(60°)$ **51.** $(f\cdot g)\left(\frac{\pi}{4}\right)$ **52.** $(f\cdot g)\left(\frac{\pi}{3}\right)$

53. $(f\circ h)\left(\frac{\pi}{6}\right)$ **54.** $(g\circ p)(60°)$ **55.** $(p\circ g)(45°)$ **56.** $(h\circ f)\left(\frac{\pi}{6}\right)$

57. (a) Find $f\left(\frac{\pi}{4}\right)$. What point is on the graph of f?

(b) Assuming $0 < x < \frac{\pi}{2}$, f is one-to-one. Use the result of part (a) to find a point on the graph of f^{-1}.

(c) What point is on the graph of $y = f\left(x + \frac{\pi}{4}\right) - 3$ if $x = \frac{\pi}{4}$?

58. (a) Find $g\left(\frac{\pi}{6}\right)$. What point is on the graph of g?

(b) Assuming $0 < x < \frac{\pi}{2}$, g is one-to-one. Use the result of part (a) to find a point on the graph of g^{-1}.

(c) What point is on the graph of $y = 2g\left(x - \frac{\pi}{6}\right)$ if $x = \frac{\pi}{6}$?

Applications and Extensions

59. Geometry A right triangle has a hypotenuse of length 8 inches. If one angle is 35°, find the length of each leg.

60. Geometry A right triangle has a hypotenuse of length 10 centimeters. If one angle is 40°, find the length of each leg.

61. Geometry A right triangle contains a 25° angle.

(a) If one leg is of length 5 inches, what is the length of the hypotenuse?

(b) There are two answers. How is this possible?

62. Geometry A right triangle contains an angle of $\frac{\pi}{8}$ radian.

(a) If one leg is of length 3 meters, what is the length of the hypotenuse?

(b) There are two answers. How is this possible?

Problems 63–66 require the following discussion.

Projectile Motion The path of a projectile fired at an inclination θ to the horizontal with initial speed v_0 is a parabola. See the figure. The **range *R*** of the projectile—that is, the horizontal distance that the projectile travels—is found by using the function

$$R(\theta) = \frac{2v_0^2 \sin\theta\cos\theta}{g}$$

where $g \approx 32.2$ feet per second per second ≈ 9.8 meters per second per second is the acceleration due to gravity. The maximum height H of the projectile is given by the function

$$H(\theta) = \frac{v_0^2 \sin^2 \theta}{2g}$$

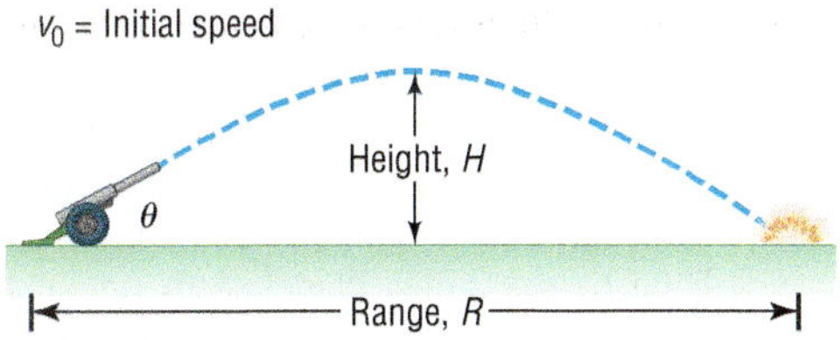

In Problems 63–66, find the range R and maximum height H of the projectile. Round answers to two decimal places.

63. The projectile is fired at an angle of 45° to the horizontal with an initial speed of 100 feet per second.

64. The projectile is fired at an angle of 30° to the horizontal with an initial speed of 150 meters per second.

65. The projectile is fired at an angle of 25° to the horizontal with an initial speed of 500 meters per second.

66. The projectile is fired at an angle of 50° to the horizontal with an initial speed of 200 feet per second.

67. Inclined Plane See the illustration. If friction is ignored, the time t (in seconds) required for a block to slide down an inclined plane is modeled by the function

$$t(\theta) = \sqrt{\frac{2a}{g \sin \theta \cos \theta}}$$

where a is the length (in feet) of the base, and $g \approx 32$ feet per second per second is the acceleration due to gravity. How long does it take a block to slide down an inclined plane with base $a = 10$ feet when

(a) $\theta = 30°$?

(b) $\theta = 45°$?

(c) $\theta = 60°$?

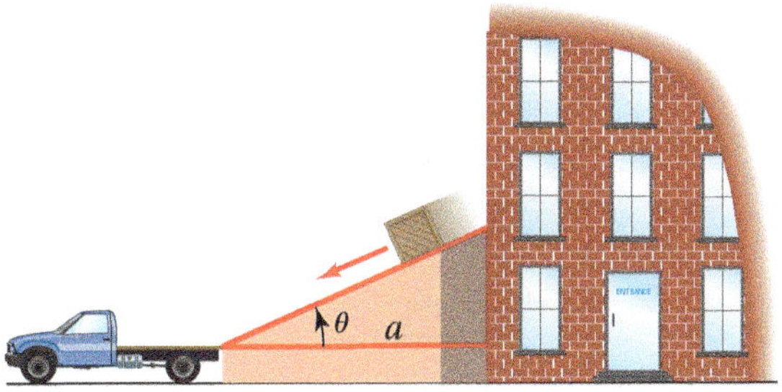

68. Piston Engines See the illustration. In a certain piston engine, the distance x (in inches) from the center of the drive shaft to the head of the piston is modeled by the function

$$x(\theta) = \cos \theta + \sqrt{16 + 0.5(2\cos^2 \theta - 1)}$$

where θ is the angle between the crank and the path of the piston head. Find x when $\theta = 30°$ and when $\theta = 45°$.

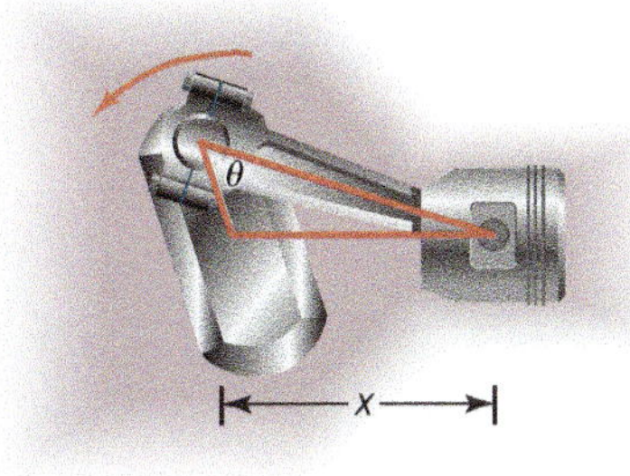

69. Find the exact value of $\tan 1° \cdot \tan 2° \cdot \tan 3° \cdot \cdots \cdot \tan 89°$.

70. Find the exact value of $\cot 1° \cdot \cot 2° \cdot \cot 3° \cdot \cdots \cdot \cot 89°$.

71. Find the exact value of
$\cos 1° \cdot \cos 2° \cdot \cdots \cdot \cos 45° \cdot \csc 46° \cdot \cdots \cdot \csc 89°$.

72. Find the exact value of
$\sin 1° \cdot \sin 2° \cdot \cdots \cdot \sin 45° \cdot \sec 46° \cdot \cdots \cdot \sec 89°$.

73. Finding the Width of a Gorge Find the distance from A to C across the gorge illustrated in the figure.

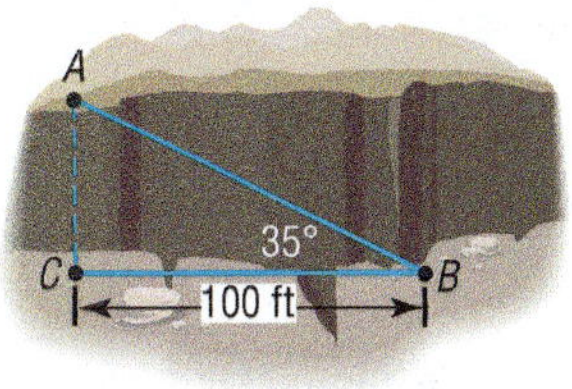

74. Finding the Distance across a Pond Find the distance from A to C across the pond illustrated in the figure.

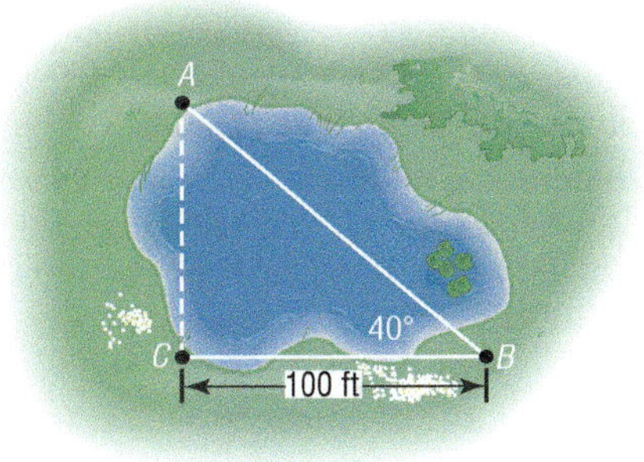

75. The Eiffel Tower The tallest tower built before the era of television masts, the Eiffel Tower was completed on March 31, 1889. Find the height of the Eiffel Tower (before a television mast was added to the top) using the information given in the illustration.

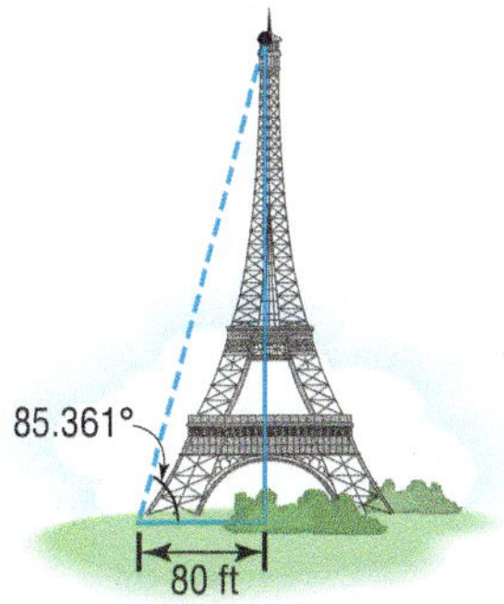

76. Finding the Distance of a Ship from Shore A person in a small boat, offshore from a vertical cliff known to be 100 feet in height, takes a sighting of the top of the cliff. If the angle of elevation is found to be 30°, how far offshore is the boat?

77. Finding the Distance to a Plateau Suppose that you are headed toward a plateau 50 meters high. If the angle of elevation to the top of the plateau is 60°, how far are you from the base of the plateau?

78. Finding the Reach of a Ladder A 22-foot extension ladder leaning against a building makes a 70° angle with the ground. How far up the building does the ladder touch?

79. Finding the Distance between Two Objects A blimp flying at an altitude of 500 feet is directly over a line from Soldier Field to the Adler Planetarium on Lake Michigan (see the figure). If the angle of depression from the blimp to the stadium is 32° and that from the blimp to the planetarium is 23°, find the distance between Soldier Field and the Adler Planetarium.

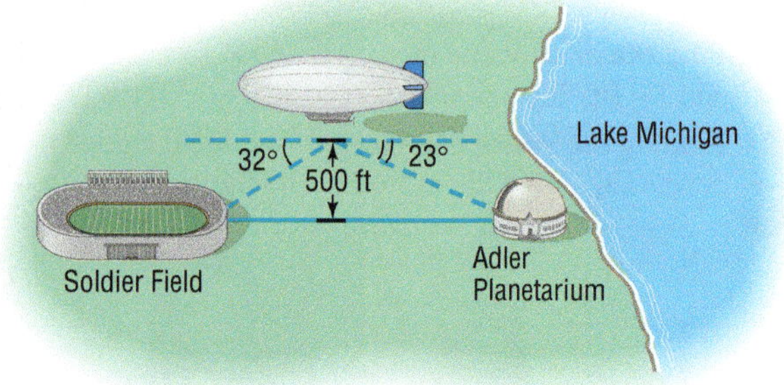

80. Hot-air Balloon While taking a ride in a hot-air balloon in Napa Valley, Francisco wonders how high he is. To find out, he chooses a landmark that is to the east of the balloon and measures the angle of depression to be 54°. A few minutes later, after traveling 100 feet east, the angle of depression to the same landmark is determined to be 61°. Use this information to determine the height of the balloon.

81. Mt. Rushmore To measure the height of Lincoln's caricature on Mt. Rushmore, two sightings 800 feet from the base of the mountain are taken. If the angle of elevation to the bottom of Lincoln's face is 32° and the angle of elevation to the top is 35°, what is the height of Lincoln's face?

82. The CN Tower The CN Tower, located in Toronto, Canada, is the tallest structure in the Americas. While visiting Toronto, a tourist wondered what the height of the tower above the top of the Sky Pod is. While standing 4000 feet from the tower, she measured the angle to the top of the Sky Pod to be 20.1°. At this same distance, the angle of elevation to the top of the tower was found to be 24.4°. Use this information to determine the height of the tower above the Sky Pod.

83. Finding the Length of a Guy Wire A radio transmission tower is 200 feet high. How long should a guy wire be if it is to be attached to the tower 10 feet from the top and is to make an angle of 45° with the ground?

84. Washington Monument The angle of elevation of the Sun is 35.1° at the instant the shadow cast by the Washington Monument is 789 feet long. Use this information to calculate the height of the monument.

85. Constructing a Highway A highway is being constructed along the coast of Florida. Near Naples, a bay obstructs the straight path of the road. Since the cost of a bridge is prohibitive, engineers decide to go around the bay. The illustration shows the path that they decide on and the measurements taken. What is the length of highway needed to go around the bay?

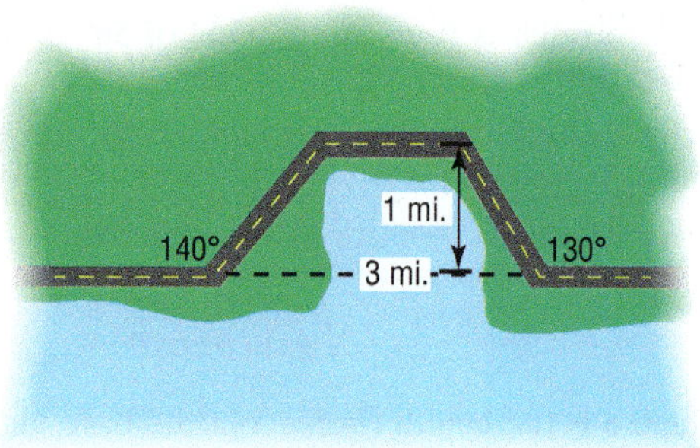

86. Photography A camera is mounted on a tripod 4 feet high at a distance of 10 feet from George, who is 6 feet tall. See the illustration. If the camera lens has angles of depression and elevation of 20°, will George's feet and head be seen by the lens? If not, how far back will the camera need to be moved to include George's feet and head?

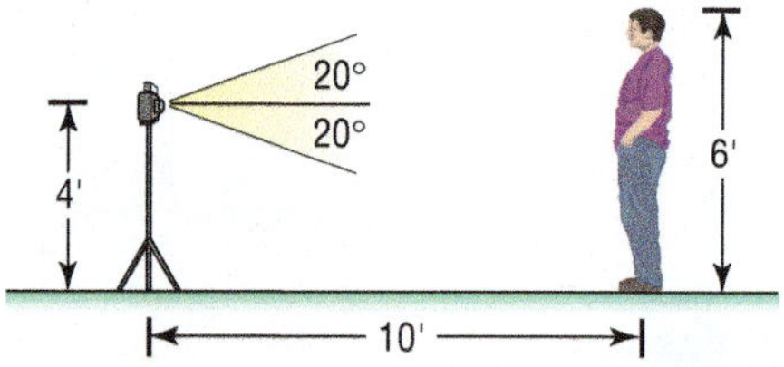

87. Calculating the Time of a Trip Two oceanfront homes are located 8 miles apart on a straight stretch of beach, each a distance of 1 mile from a paved path that parallels the ocean. Sally can jog 8 miles per hour along the paved path, but only 3 miles per hour in the sand on the beach. Because a river flows between the two houses, it is necessary to jog on the sand to the path, continue on the path, and then jog on the sand to get from one house to the other. See the illustration.

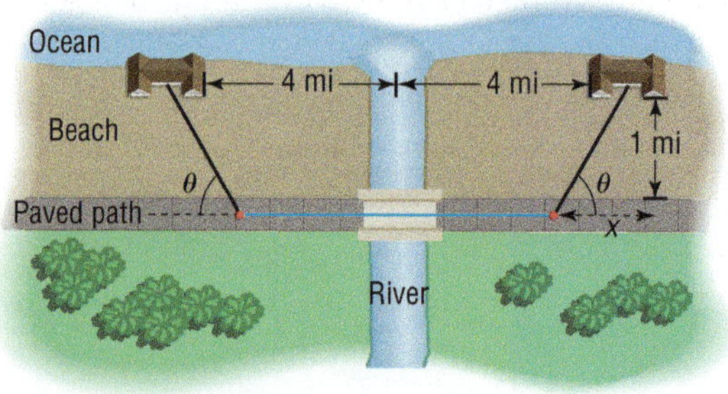

(a) Express the time T to get from one house to the other as a function of the angle θ shown in the illustration.

(b) Calculate the time T for $\theta = 30°$. How long is Sally on the paved path?

(c) Calculate the time T for $\theta = 45°$. How long is Sally on the paved path?

(d) Calculate the time T for $\theta = 60°$. How long is Sally on the paved path?

(e) Calculate the time T for $\theta = 90°$. Describe the route taken.

(f) Calculate the time T for $\tan\theta = \frac{1}{4}$. Describe the route taken. Explain why θ must be larger than 14°.

(g) Graph $T = T(\theta)$. What angle θ results in the least time? What is the least time? How long is Sally on the paved path?

88. Designing Fine Decorative Pieces A designer of decorative art plans to market solid gold spheres encased in clear crystal cones. Each sphere is of fixed radius R and will be enclosed in a cone of height h and radius r. See the illustration below. Many cones can be used to enclose the sphere, each having a different slant angle θ.

(a) Express the volume V of the cone as a function of the slant angle θ of the cone.
[**Hint:** The volume V of a cone of height h and radius r is

$$V = \frac{1}{3}\pi r^2 h.]$$

(b) What volume V is necessary to enclose a sphere of radius 2 centimeters in a cone whose slant angle θ is 30°? 45°? 60°?

(c) What slant angle θ should be used for the volume V of the cone to be a minimum? (This choice minimizes the amount of crystal required and gives maximum emphasis to the gold sphere.)

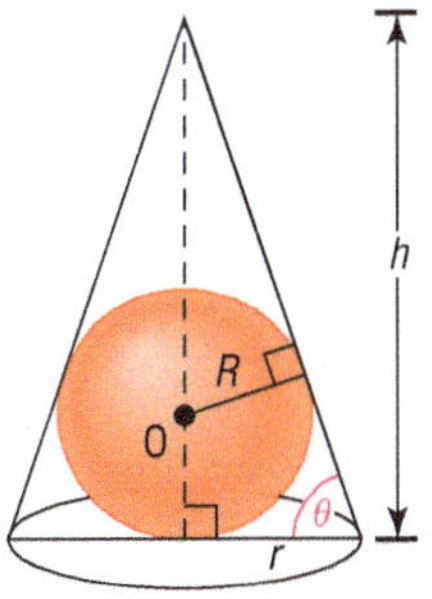

89. Calculating Pool Shots A pool player located at X wants to shoot the white ball off the top cushion and hit the red ball dead center. He knows from physics that the white ball will come off a cushion at the same angle as that at which it hit the cushion. Where on the top cushion should he hit the white ball?

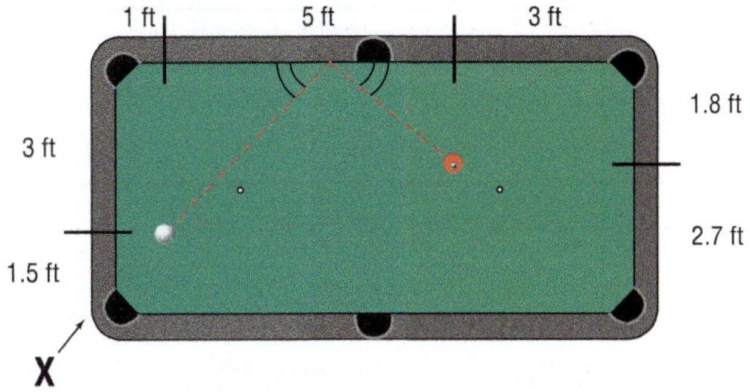

90. One World Trade Center One World Trade Center is the centerpiece of the rebuilding of the World Trade Center in New York City. The tower is 1776 feet tall (including its spire). The angle of elevation from the base of an office building to the tip of the spire is 34°. The angle of elevation from the helipad on the roof of the office building to the tip of the spire is 20°.

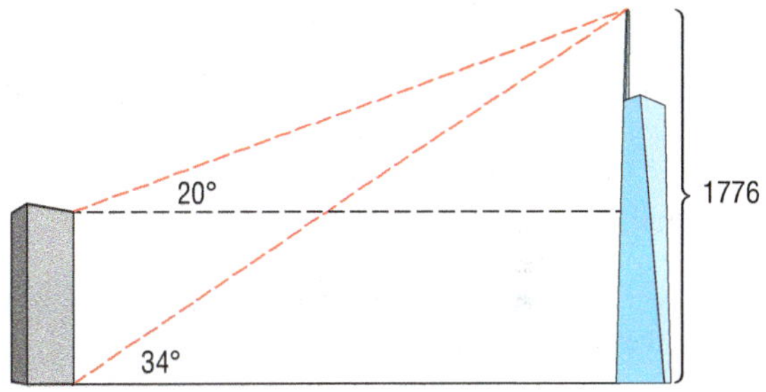

(a) How far away is the office building from One World Trade Center? Assume the side of the tower is vertical. Round to the nearest foot.

(b) How tall is the office building? Round to the nearest foot.

91. Guy Wires Two poles, 10 feet and 15 feet in height, are located 30 feet apart. They are to be supported by a single guy wire attached to their tops and anchored to the ground at a point on the line joining the two poles. See the illustration. Suppose that θ is the angle between the guy wire and the taller pole and that L is the length of the guy wire.

(a) Express $L\ (= L_1 + L_2)$ as a function of θ.

(b) What is the domain of L?

(c) Find L if $\theta = 45°$. What is the corresponding value of x?

(d) Graph L and find the angle θ that minimizes L.

(e) What is the minimum length for the guy wire? Where is it anchored to the ground?

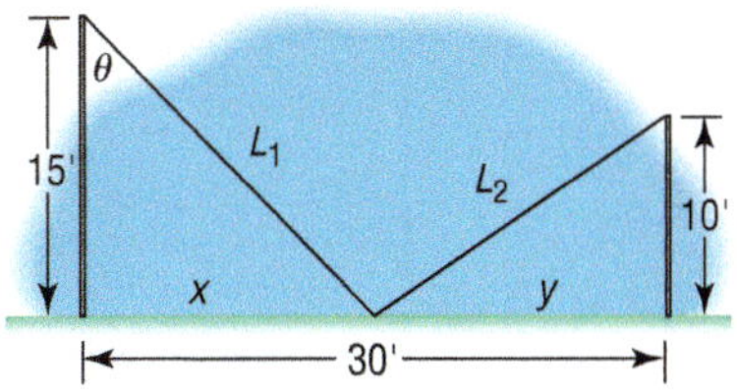

92. Use a calculator set in radian mode to complete the following table. What can you conclude about the value of $g(\theta) = \dfrac{\cos\theta - 1}{\theta}$ as θ approaches 0?

θ	0.5	0.4	0.2	0.1	0.01	0.001	0.0001	0.00001
$g(\theta) = \dfrac{\cos\theta - 1}{\theta}$								

93. Use a calculator set in radian mode to complete the following table. What can you conclude about the value of $f(\theta) = \dfrac{\sin\theta}{\theta}$ as θ approaches 0?

θ	0.5	0.4	0.2	0.1	0.01	0.001	0.0001	0.00001
$f(\theta) = \dfrac{\sin\theta}{\theta}$								

Explaining Concepts: Discussion and Writing

94. Write a brief paragraph that explains how to quickly compute the trigonometric functions of 30°, 45°, and 60°.

95. Explain how you would measure the width of the Grand Canyon from a point on its ridge.

96. Explain how you would measure the height of a TV tower that is on the roof of a tall building.

Retain Your Knowledge

Problems 97–100 are based on material learned earlier in the course. The purpose of these problems is to keep the material fresh in your mind so that you are better prepared for the final exam.

97. Solve: $x^2 - 5 = (x + 3)(x - 3) + 4$

98. Find the vertex and the axis of symmetry of $f(x) = -2x^2 + 12x - 13$.

99. Solve exactly: $e^{x-4} = 6$

100. Find the real zeros of $f(x) = x^3 - 9x^2 + 3x - 27$.

7.4 Trigonometric Functions of Any Angle

OBJECTIVES

1. Find the Exact Values of the Trigonometric Functions for Any Angle (p. 550)
2. Use Coterminal Angles to Find the Exact Value of a Trigonometric Function (p. 553)
3. Determine the Signs of the Trigonometric Functions of an Angle in a Given Quadrant (p. 554)
4. Find the Reference Angle of an Angle (p. 555)
5. Use a Reference Angle to Find the Exact Value of a Trigonometric Function (p. 556)
6. Find the Exact Values of the Trigonometric Functions of an Angle, Given Information about the Functions (p. 557)

1 Find the Exact Values of the Trigonometric Functions for Any Angle

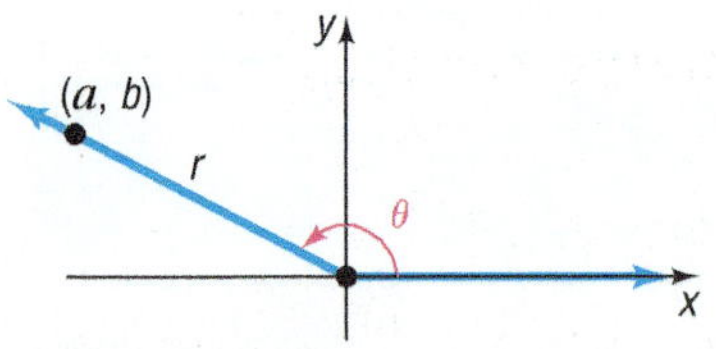

Figure 41

To extend the definition of the trigonometric functions to include angles that are not acute, we employ a rectangular coordinate system and place the angle in standard position so that its vertex is at the origin and its initial side is along the positive x-axis. See Figure 41.

DEFINITION

Let θ be any angle in standard position, and let (a, b) denote the coordinates of any point, except the origin $(0, 0)$, on the terminal side of θ. If $r = \sqrt{a^2 + b^2}$ denotes the distance from $(0, 0)$ to (a, b), then the **six trigonometric functions of θ** are defined as the ratios

$$\sin\theta = \frac{b}{r} \qquad \cos\theta = \frac{a}{r} \qquad \tan\theta = \frac{b}{a}$$

$$\csc\theta = \frac{r}{b} \qquad \sec\theta = \frac{r}{a} \qquad \cot\theta = \frac{a}{b}$$

provided that no denominator equals 0. If a denominator equals 0, that trigonometric function of the angle θ is not defined.

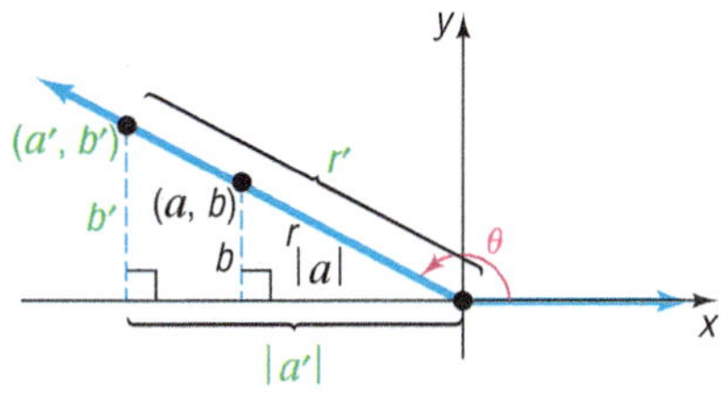

Figure 42

Notice in the preceding definitions that if $a = 0$, that is, if the point (a, b) is on the y-axis, then the tangent function and the secant function are undefined. Also, if $b = 0$, that is, if the point (a, b) is on the x-axis, then the cosecant function and the cotangent function are undefined.

By constructing similar triangles, you should be convinced that the values of the six trigonometric functions of an angle θ do not depend on the selection of the point (a, b) on the terminal side of θ, but rather depend only on the angle θ itself. See Figure 42 for an illustration of this when θ lies in quadrant II. Since the triangles are similar, the ratio $\frac{b}{r}$ equals the ratio $\frac{b'}{r'}$, which equals $\sin\theta$. Also, the ratio $\frac{|a|}{r}$ equals the ratio $\frac{|a'|}{r'}$, so $\frac{a}{r} = \frac{a'}{r'}$, which equals $\cos\theta$. And so on.

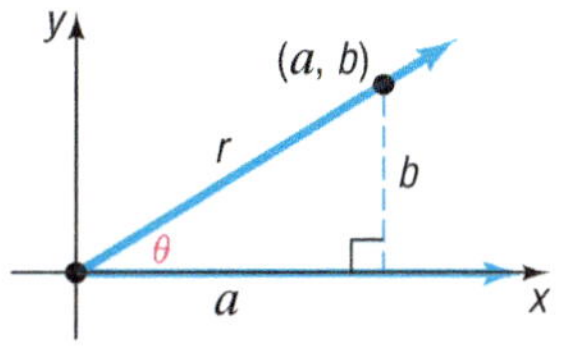

Figure 43

Also, in Figure 43 observe that if θ is acute these definitions reduce to the right triangle definitions given in Section 7.2.

Finally, from the definition of the six trigonometric functions of any angle, we see that the Quotient and Reciprocal Identities hold. Also, using $r^2 = a^2 + b^2$ and dividing each side by r^2, we can derive the Pythagorean Identities for any angle.

EXAMPLE 1 Finding the Exact Values of the Six Trigonometric Functions of θ, Given a Point on the Terminal Side

Find the exact value of each of the six trigonometric functions of a positive angle θ if $(4, -3)$ is a point on its terminal side.

Solution Figure 44 illustrates the situation. For the point $(a, b) = (4, -3)$, we have $a = 4$ and $b = -3$. Then $r = \sqrt{a^2 + b^2} = \sqrt{16 + 9} = 5$, so

$$\sin\theta = \frac{b}{r} = -\frac{3}{5} \qquad \cos\theta = \frac{a}{r} = \frac{4}{5} \qquad \tan\theta = \frac{b}{a} = -\frac{3}{4}$$

$$\csc\theta = \frac{r}{b} = -\frac{5}{3} \qquad \sec\theta = \frac{r}{a} = \frac{5}{4} \qquad \cot\theta = \frac{a}{b} = -\frac{4}{3}$$

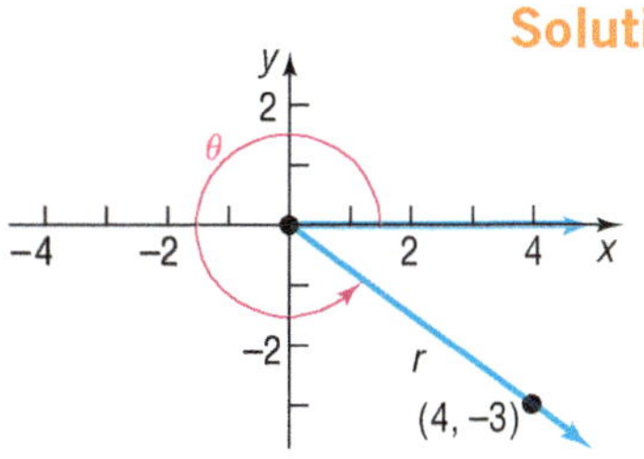

Figure 44

Now Work PROBLEM 11

In the next example, we find the exact value of each of the six trigonometric functions at the quadrantal angles $0, \frac{\pi}{2}, \pi$, and $\frac{3\pi}{2}$.

EXAMPLE 2 Finding the Exact Values of the Six Trigonometric Functions of Quadrantal Angles

Find the exact value of each of the six trigonometric functions of

(a) $\theta = 0 = 0°$ (b) $\theta = \frac{\pi}{2} = 90°$ (c) $\theta = \pi = 180°$ (d) $\theta = \frac{3\pi}{2} = 270°$

Solution

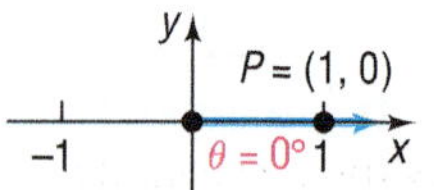

Figure 45 $\theta = 0 = 0°$

(a) We can choose any point on the terminal side of $\theta = 0 = 0°$. For convenience, we choose the point $P = (1, 0) = (a, b)$, which is a distance of $r = 1$ unit from the origin. See Figure 45. Then

$$\sin 0 = \sin 0° = \frac{b}{r} = \frac{0}{1} = 0 \qquad \cos 0 = \cos 0° = \frac{a}{r} = \frac{1}{1} = 1$$

$$\tan 0 = \tan 0° = \frac{b}{a} = \frac{0}{1} = 0 \qquad \sec 0 = \sec 0° = \frac{r}{a} = \frac{1}{1} = 1$$

Since the y-coordinate of P is 0, csc 0 and cot 0 are not defined.

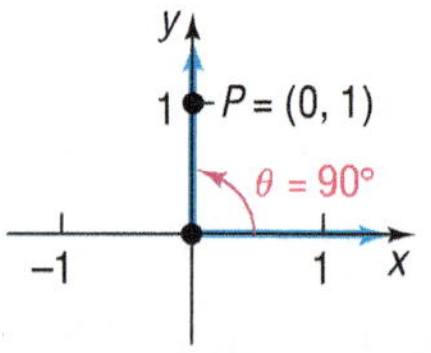

Figure 46 $\theta = \frac{\pi}{2} = 90°$

(b) The point $P = (0, 1) = (a, b)$ is on the terminal side of $\theta = \frac{\pi}{2} = 90°$ and is a distance of $r = 1$ unit from the origin. See Figure 46. Then

$$\sin\frac{\pi}{2} = \sin 90° = \frac{b}{r} = \frac{1}{1} = 1 \qquad \cos\frac{\pi}{2} = \cos 90° = \frac{a}{r} = \frac{0}{1} = 0$$

$$\csc\frac{\pi}{2} = \csc 90° = \frac{r}{b} = \frac{1}{1} = 1 \qquad \cot\frac{\pi}{2} = \cot 90° = \frac{a}{b} = \frac{0}{1} = 0$$

Since the x-coordinate of P is 0, $\tan\frac{\pi}{2}$ and $\sec\frac{\pi}{2}$ are not defined.

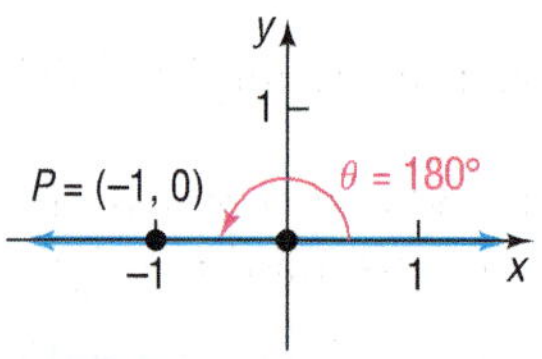

Figure 47 $\theta = \pi = 180°$

(c) The point $P = (-1, 0)$ is on the terminal side of $\theta = \pi = 180°$ and is a distance of $r = 1$ unit from the origin. See Figure 47. Then

$$\sin\pi = \sin 180° = \frac{0}{1} = 0 \qquad \cos\pi = \cos 180° = \frac{-1}{1} = -1$$

$$\tan\pi = \tan 180° = \frac{0}{-1} = 0 \qquad \sec\pi = \sec 180° = \frac{1}{-1} = -1$$

Since the y-coordinate of P is 0, $\csc\pi$ and $\cot\pi$ are not defined.

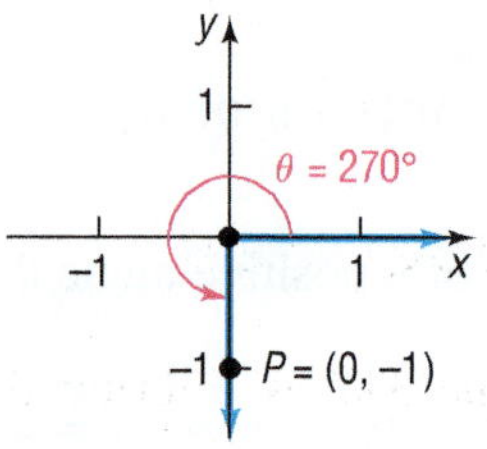

Figure 48 $\theta = \frac{3\pi}{2} = 270°$

(d) The point $P = (0, -1)$ is on the terminal side of $\theta = \frac{3\pi}{2} = 270°$ and is a distance of $r = 1$ unit from the origin. See Figure 48. Then

$$\sin\frac{3\pi}{2} = \sin 270° = \frac{-1}{1} = -1 \qquad \cos\frac{3\pi}{2} = \cos 270° = \frac{0}{1} = 0$$

$$\csc\frac{3\pi}{2} = \csc 270° = \frac{1}{-1} = -1 \qquad \cot\frac{3\pi}{2} = \cot 270° = \frac{0}{-1} = 0$$

Since the x-coordinate of P is 0, $\tan\frac{3\pi}{2}$ and $\sec\frac{3\pi}{2}$ are not defined. ■

Table 4 summarizes the values of the trigonometric functions found in Example 2.

Table 4

θ (Radians)	θ (Degrees)	$\sin\theta$	$\cos\theta$	$\tan\theta$	$\csc\theta$	$\sec\theta$	$\cot\theta$
0	0°	0	1	0	Not defined	1	Not defined
$\frac{\pi}{2}$	90°	1	0	Not defined	1	Not defined	0
π	180°	0	−1	0	Not defined	−1	Not defined
$\frac{3\pi}{2}$	270°	−1	0	Not defined	−1	Not defined	0

There is no need to memorize Table 4. To find the value of a trigonometric function of a quadrantal angle, draw the angle and apply the definition, as done in Example 2.

2 Use Coterminal Angles to Find the Exact Value of a Trigonometric Function

DEFINITION

Two angles in standard position are said to be **coterminal** if they have the same terminal side.

See Figure 49.

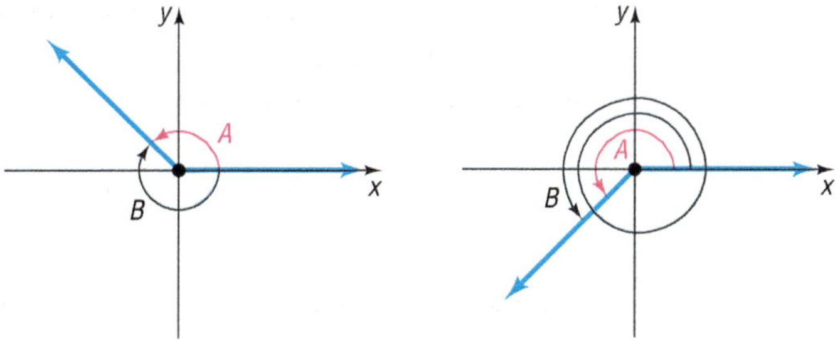

(a) *A* and *B* are coterminal (b) *A* and *B* are coterminal

Figure 49 Coterminal angles

For example, the angles 60° and 420° are coterminal, as are the angles −40° and 320°.

In general, if θ is an angle measured in degrees, then $\theta + 360°k$, where k is any integer, is an angle coterminal with θ. If θ is measured in radians, then $\theta + 2\pi k$, where k is any integer, is an angle coterminal with θ.

Because coterminal angles have the same terminal side, it follows that the values of the trigonometric functions of coterminal angles are equal.

EXAMPLE 3 Using a Coterminal Angle to Find the Exact Value of a Trigonometric Function

Find the exact value of each of the following:

(a) $\sin 390°$ (b) $\cos 420°$ (c) $\tan \dfrac{9\pi}{4}$ (d) $\sec\left(-\dfrac{7\pi}{4}\right)$ (e) $\csc(-270°)$

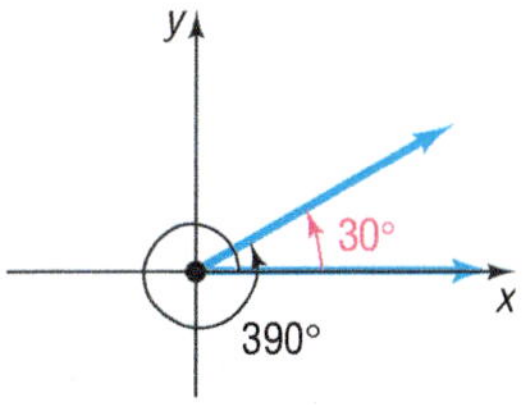

Figure 50

Solution

(a) It is best to sketch the angle first. See Figure 50. The angle 390° is coterminal with 30°.

$$\sin 390° = \sin(30° + 360°)$$

$$= \sin 30° = \frac{1}{2}$$

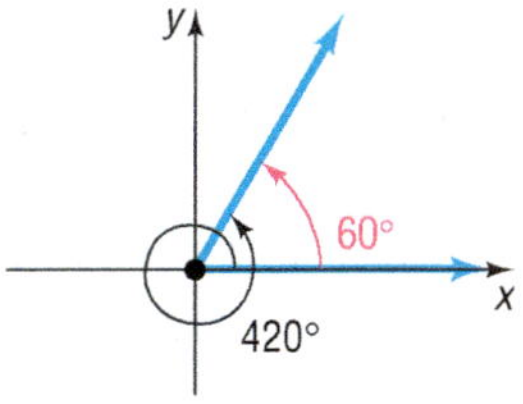

Figure 51

(b) See Figure 51. The angle 420° is coterminal with 60°.

$$\cos 420° = \cos(60° + 360°)$$

$$= \cos 60° = \frac{1}{2}$$

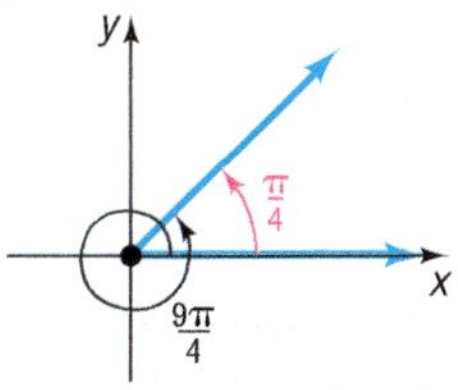

Figure 52

(c) See Figure 52. The angle $\dfrac{9\pi}{4}$ is coterminal with $\dfrac{\pi}{4}$.

$$\tan\frac{9\pi}{4} = \tan\left(\frac{\pi}{4} + 2\pi\right)$$

$$= \tan\frac{\pi}{4} = 1$$

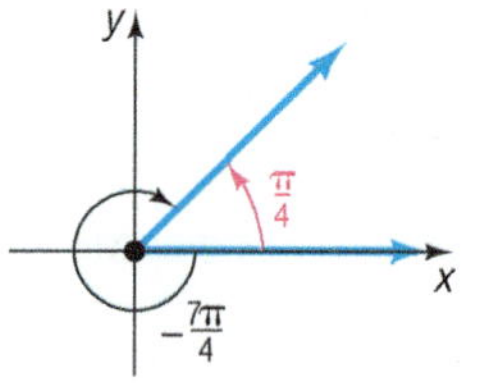

Figure 53

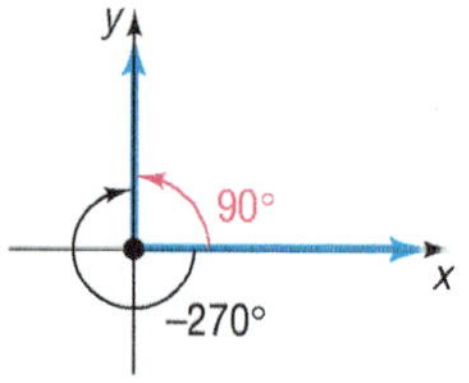

Figure 54

(d) See Figure 53. The angle $-\frac{7\pi}{4}$ is coterminal with $\frac{\pi}{4}$.

$$\sec\left(-\frac{7\pi}{4}\right) = \sec\left(\frac{\pi}{4} + 2\pi(-1)\right) = \sec\frac{\pi}{4} = \sqrt{2}$$

(e) See Figure 54. The angle $-270°$ is coterminal with $90°$.

$$\csc(-270°) = \csc(90° + 360°(-1))$$
$$= \csc 90° = 1$$

As Example 3 illustrates, the value of a trigonometric function of any angle is equal to the value of the same trigonometric function of an angle θ coterminal to the given angle, where $0° \le \theta < 360°$ (or $0 \le \theta < 2\pi$). Because the angles θ and $\theta + 360°k$ (or $\theta + 2\pi k$), where k is any integer, are coterminal, and because the values of the trigonometric functions are equal for coterminal angles, it follows that

θ degrees	**θ radians**	
$\sin(\theta + 360°k) = \sin\theta$	$\sin(\theta + 2\pi k) = \sin\theta$	
$\cos(\theta + 360°k) = \cos\theta$	$\cos(\theta + 2\pi k) = \cos\theta$	
$\tan(\theta + 360°k) = \tan\theta$	$\tan(\theta + 2\pi k) = \tan\theta$	**(1)**
$\csc(\theta + 360°k) = \csc\theta$	$\csc(\theta + 2\pi k) = \csc\theta$	
$\sec(\theta + 360°k) = \sec\theta$	$\sec(\theta + 2\pi k) = \sec\theta$	
$\cot(\theta + 360°k) = \cot\theta$	$\cot(\theta + 2\pi k) = \cot\theta$	

where k is any integer.

These formulas show that the values of the trigonometric functions repeat themselves every 360° (or 2π radians).

Now Work PROBLEM 21

3 Determine the Signs of the Trigonometric Functions of an Angle in a Given Quadrant

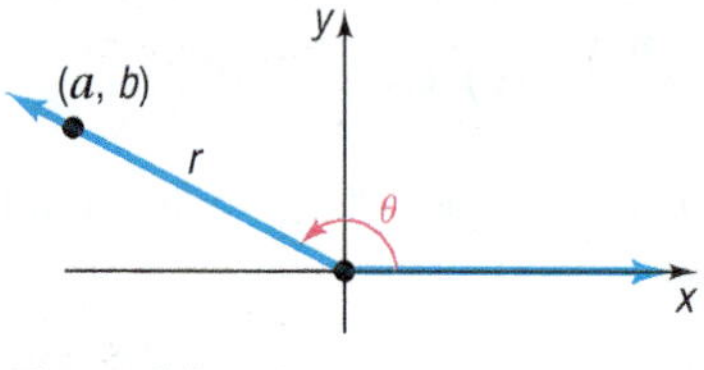

Figure 55 θ in quadrant II, $a < 0$, $b > 0$, $r > 0$

If θ is not a quadrantal angle, then it will lie in a particular quadrant. In such a case, the signs of the x-coordinate and y-coordinate of a point (a, b) on the terminal side of θ are known. Because $r = \sqrt{a^2 + b^2} > 0$, it follows that we can find the signs of the trigonometric functions of an angle θ if we know in which quadrant θ lies.

For example, if θ lies in quadrant II, as shown in Figure 55, then a point (a, b) on the terminal side of θ has a negative x-coordinate and a positive y-coordinate. Then

$$\sin\theta = \frac{b}{r} > 0 \qquad \cos\theta = \frac{a}{r} < 0 \qquad \tan\theta = \frac{b}{a} < 0$$

$$\csc\theta = \frac{r}{b} > 0 \qquad \sec\theta = \frac{r}{a} < 0 \qquad \cot\theta = \frac{a}{b} < 0$$

Table 5 lists the signs of the six trigonometric functions for each quadrant. Figure 56 provides two illustrations.

Table 5

Quadrant of θ	$\sin\theta$, $\csc\theta$	$\cos\theta$, $\sec\theta$	$\tan\theta$, $\cot\theta$
I	Positive	Positive	Positive
II	Positive	Negative	Negative
III	Negative	Negative	Positive
IV	Negative	Positive	Negative

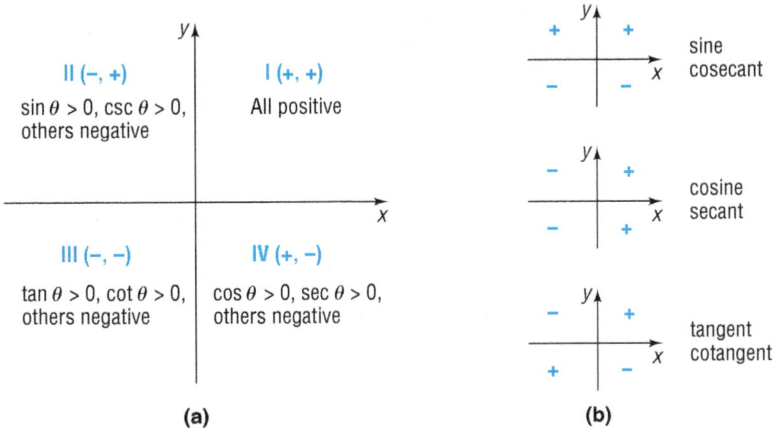

Figure 56 Signs of the trigonometric functions

EXAMPLE 4 **Finding the Quadrant in Which an Angle Lies**

If $\sin\theta < 0$ and $\cos\theta < 0$, name the quadrant in which the angle θ lies.

Solution If $\sin\theta < 0$, then θ lies in quadrant III or IV. If $\cos\theta < 0$, then θ lies in quadrant II or III. Therefore, θ lies in quadrant III. ■

Now Work PROBLEM 33

4 Find the Reference Angle of an Angle

Once we know in which quadrant an angle lies, we know the sign of each trigonometric function of that angle. This information, along with the *reference angle,* will enable us to evaluate the trigonometric functions of such an angle.

DEFINITION Let θ denote an angle that lies in a quadrant. The acute angle formed by the terminal side of θ and the x-axis is called the **reference angle** for θ. ■

Figure 57 illustrates the reference angle for some general angles θ. Note that a reference angle is always an acute angle. That is, a reference angle has a measure between 0° and 90°.

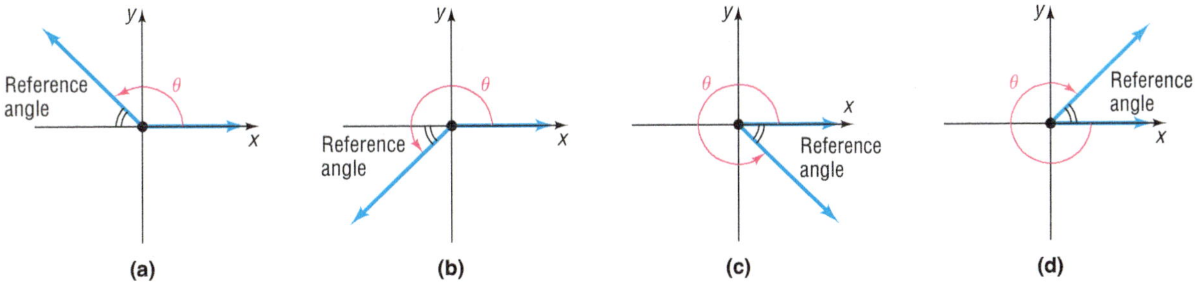

Figure 57 Reference angles

Although formulas can be given for calculating reference angles, usually it is easier to find the reference angle for a given angle by making a quick sketch of the angle.

EXAMPLE 5 **Finding Reference Angles**

Find the reference angle for each of the following angles:

(a) 150° (b) −45° (c) $\frac{9\pi}{4}$ (d) $-\frac{5\pi}{6}$

Solution (a) Refer to Figure 58. The reference angle for 150° is 30°.

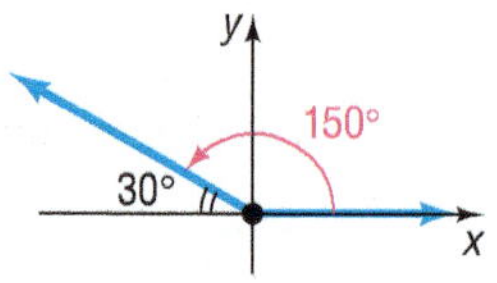

Figure 58

(b) Refer to Figure 59. The reference angle for −45° is 45°.

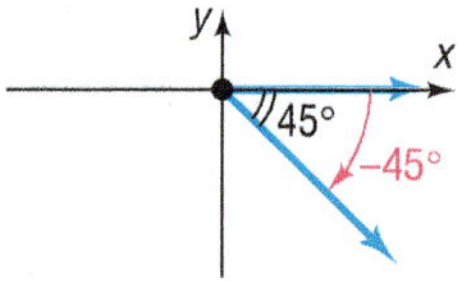

Figure 59

(c) Refer to Figure 60. The reference angle for $\frac{9\pi}{4}$ is $\frac{\pi}{4}$.

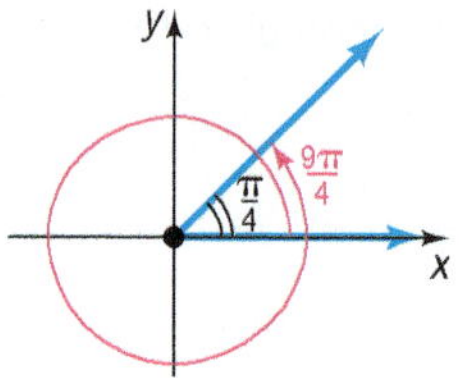

Figure 60

(d) Refer to Figure 61. The reference angle for $-\frac{5\pi}{6}$ is $\frac{\pi}{6}$.

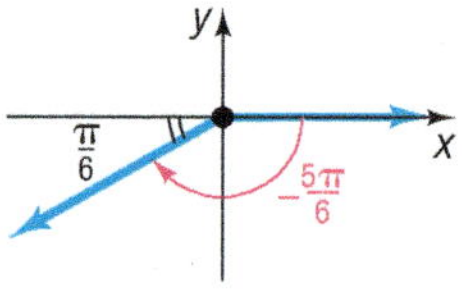

Figure 61

Now Work PROBLEM 41

5 Use a Reference Angle to Find the Exact Value of a Trigonometric Function

The advantage of using reference angles is that, except for the correct sign, the values of the trigonometric functions of any angle θ equal the values of the trigonometric functions of its reference angle.

THEOREM

Reference Angles

If θ is an angle that lies in a quadrant and if α is its reference angle, then

$\sin\theta = \pm\sin\alpha$	$\cos\theta = \pm\cos\alpha$	$\tan\theta = \pm\tan\alpha$	**(2)**
$\csc\theta = \pm\csc\alpha$	$\sec\theta = \pm\sec\alpha$	$\cot\theta = \pm\cot\alpha$	

where the + or − sign depends on the quadrant in which θ lies.

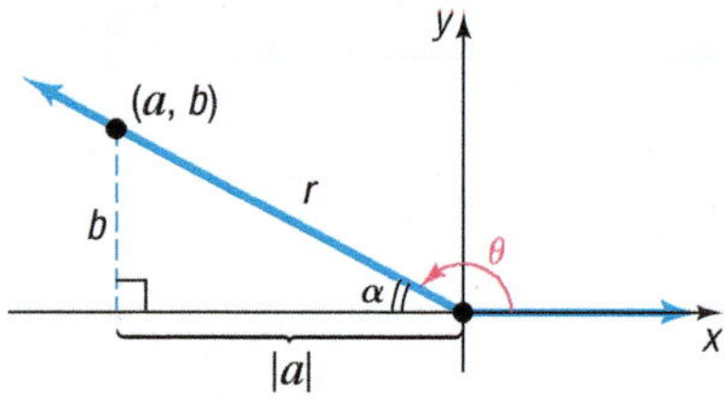

Figure 62

For example, suppose that θ lies in quadrant II and α is its reference angle. See Figure 62. If (a, b) is a point on the terminal side of θ and if $r = \sqrt{a^2 + b^2}$, we have

$$\sin\theta = \frac{b}{r} = \sin\alpha \qquad \cos\theta = \frac{a}{r} = \frac{-|a|}{r} = -\cos\alpha$$

and so on.

EXAMPLE 6 **Using the Reference Angle to Find the Exact Value of a Trigonometric Function**

Find the exact value of each of the following trigonometric functions using reference angles.

(a) $\sin 135°$ (b) $\cos 600°$ (c) $\cos\frac{17\pi}{6}$ (d) $\tan\left(-\frac{\pi}{3}\right)$

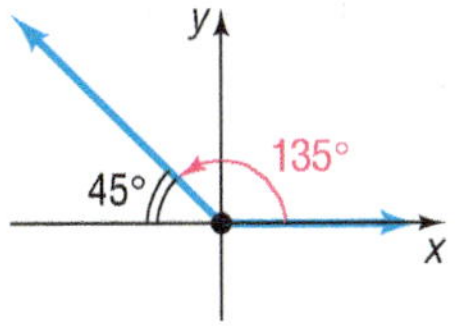

Figure 63

Solution (a) Refer to Figure 63. The reference angle for 135° is 45°, and $\sin 45° = \frac{\sqrt{2}}{2}$. The angle 135° is in quadrant II, where the sine function is positive, so

$$\sin 135° = \sin 45° = \frac{\sqrt{2}}{2}$$

(b) Refer to Figure 64. The reference angle for 600° is 60°, and $\cos 60° = \frac{1}{2}$. The angle 600° is in quadrant III, where the cosine function is negative, so

$$\cos 600° = -\cos 60° = -\frac{1}{2}$$

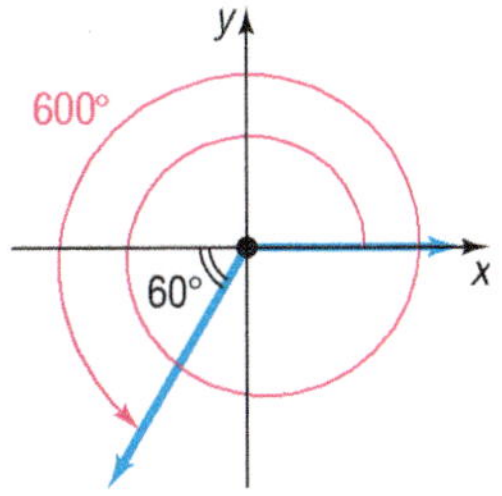

Figure 64

(c) Refer to Figure 65. The reference angle for $\frac{17\pi}{6}$ is $\frac{\pi}{6}$, and $\cos \frac{\pi}{6} = \frac{\sqrt{3}}{2}$. The angle $\frac{17\pi}{6}$ is in quadrant II, where the cosine function is negative, so

$$\cos \frac{17\pi}{6} = -\cos \frac{\pi}{6} = -\frac{\sqrt{3}}{2}$$

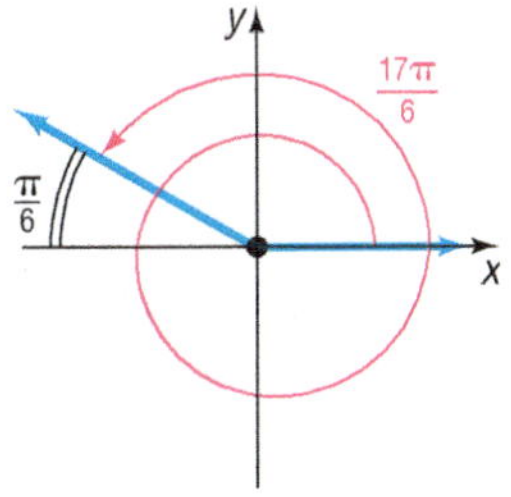

Figure 65

(d) Refer to Figure 66. The reference angle for $-\frac{\pi}{3}$ is $\frac{\pi}{3}$, and $\tan \frac{\pi}{3} = \sqrt{3}$. The angle $-\frac{\pi}{3}$ is in quadrant IV, where the tangent function is negative, so

$$\tan\left(-\frac{\pi}{3}\right) = -\tan \frac{\pi}{3} = -\sqrt{3}$$

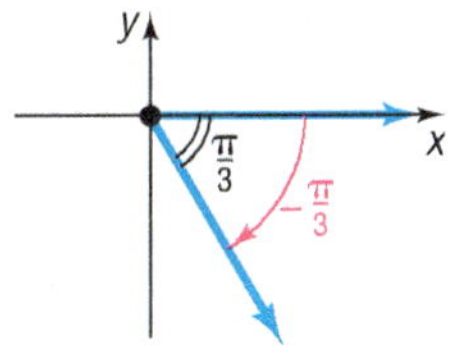

Figure 66

Finding the Values of the Trigonometric Functions of Any Angle

- If the angle θ is a quadrantal angle, draw the angle, pick a point on its terminal side, and apply the definition of the trigonometric functions.
- If the angle θ lies in a quadrant:
  1. Find the reference angle α of θ.
  2. Find the value of the trigonometric function at α.
  3. Adjust the sign (+ or −) of the value of the trigonometric function based on the quadrant in which θ lies.

Now Work PROBLEMS 59 AND 61

6 Find the Exact Values of the Trigonometric Functions of an Angle, Given Information about the Functions

EXAMPLE 7 **Finding the Exact Values of Trigonometric Functions**

Given that $\cos \theta = -\frac{2}{3}, \frac{\pi}{2} < \theta < \pi$, find the exact value of each of the remaining trigonometric functions.

Solution The angle θ lies in quadrant II, so $\sin \theta$ and $\csc \theta$ are positive and the other four trigonometric functions are negative. If α is the reference angle for θ, then $\cos \alpha = \frac{2}{3} = \frac{\text{adjacent}}{\text{hypotenuse}}$. The values of the remaining trigonometric functions of the reference angle α can be found by drawing the appropriate triangle and using

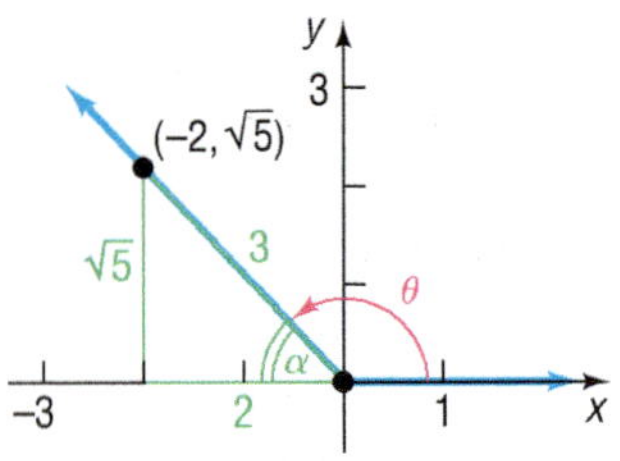

Figure 67 $\cos \alpha = \frac{2}{3}$

the Pythagorean Theorem to determine that the side opposite α is $\sqrt{5}$. Use Figure 67 to obtain

$$\sin \alpha = \frac{\sqrt{5}}{3} \qquad \cos \alpha = \frac{2}{3} \qquad \tan \alpha = \frac{\sqrt{5}}{2}$$

$$\csc \alpha = \frac{3}{\sqrt{5}} = \frac{3\sqrt{5}}{5} \qquad \sec \alpha = \frac{3}{2} \qquad \cot \alpha = \frac{2}{\sqrt{5}} = \frac{2\sqrt{5}}{5}$$

Now assign the appropriate signs to each of these values to find the values of the trigonometric functions of θ.

$$\sin \theta = \frac{\sqrt{5}}{3} \qquad \cos \theta = -\frac{2}{3} \qquad \tan \theta = -\frac{\sqrt{5}}{2}$$

$$\csc \theta = \frac{3\sqrt{5}}{5} \qquad \sec \theta = -\frac{3}{2} \qquad \cot \theta = -\frac{2\sqrt{5}}{5}$$

Now Work PROBLEM 83

EXAMPLE 8

Finding the Exact Values of Trigonometric Functions

If $\tan \theta = -4$ and $\sin \theta < 0$, find the exact value of each of the remaining trigonometric functions of θ.

Solution Since $\tan \theta = -4 < 0$ and $\sin \theta < 0$, it follows that θ lies in quadrant IV. If α is the reference angle for θ, then $\tan \alpha = 4 = \frac{4}{1} = \frac{b}{a}$. With $a = 1$ and $b = 4$, we find $r = \sqrt{1^2 + 4^2} = \sqrt{17}$. See Figure 68. Then

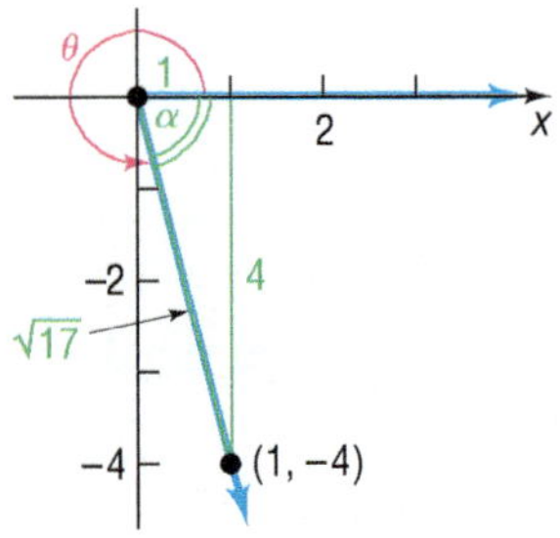

Figure 68 $\tan \alpha = 4$

$$\sin \alpha = \frac{4}{\sqrt{17}} = \frac{4\sqrt{17}}{17} \qquad \cos \alpha = \frac{1}{\sqrt{17}} = \frac{\sqrt{17}}{17} \qquad \tan \alpha = \frac{4}{1} = 4$$

$$\csc \alpha = \frac{\sqrt{17}}{4} \qquad \sec \alpha = \frac{\sqrt{17}}{1} = \sqrt{17} \qquad \cot \alpha = \frac{1}{4}$$

Assign the appropriate sign to each of these to obtain the values of the trigonometric functions of θ.

$$\sin \theta = -\frac{4\sqrt{17}}{17} \qquad \cos \theta = \frac{\sqrt{17}}{17} \qquad \tan \theta = -4$$

$$\csc \theta = -\frac{\sqrt{17}}{4} \qquad \sec \theta = \sqrt{17} \qquad \cot \theta = -\frac{1}{4}$$

Now Work PROBLEM 93

7.4 Assess Your Understanding

Concepts and Vocabulary

1. For an angle θ that lies in quadrant III, the trigonometric functions ________ and ________ are positive.
2. Two angles in standard position that have the same terminal side are ________.
3. The reference angle of 240° is _____.
4. ***True or False*** $\sin 182° = \cos 2°$.
5. ***True or False*** $\tan \frac{\pi}{2}$ is not defined.
6. ***True or False*** The reference angle of an angle is always an acute angle.
7. What is the reference angle of 600°?
8. Which function is positive in quadrant IV?
(a) $\sin \theta$ (b) $\cos \theta$ (c) $\tan \theta$ (d) $\csc \theta$
9. If $0 \le \theta < 2\pi$, for what angles θ, if any, is $\tan \theta$ undefined?
10. What is the reference angle of $\frac{13\pi}{3}$?
(a) $-\frac{2\pi}{3}$ (b) $-\frac{\pi}{3}$ (c) $\frac{2\pi}{3}$ (d) $\frac{\pi}{3}$

Skill Building

In Problems 11–20, a point on the terminal side of an angle θ in standard position is given. Find the exact value of each of the six trigonometric functions of θ.

11. $(-3, 4)$
12. $(5, -12)$
13. $(2, -3)$
14. $(-1, -2)$
15. $(-3, -3)$
16. $(2, -2)$
17. $\left(\frac{\sqrt{3}}{2}, \frac{1}{2}\right)$
18. $\left(-\frac{1}{2}, \frac{\sqrt{3}}{2}\right)$
19. $\left(\frac{\sqrt{2}}{2}, -\frac{\sqrt{2}}{2}\right)$
20. $\left(-\frac{\sqrt{2}}{2}, -\frac{\sqrt{2}}{2}\right)$

In Problems 21–32, use a coterminal angle to find the exact value of each expression. Do not use a calculator.

21. $\sin 405°$
22. $\cos 420°$
23. $\tan 405°$
24. $\sin 390°$
25. $\csc 450°$
26. $\sec 540°$
27. $\cot 390°$
28. $\sec 420°$
29. $\cos \frac{33\pi}{4}$
30. $\sin \frac{9\pi}{4}$
31. $\tan(21\pi)$
32. $\csc \frac{9\pi}{2}$

In Problems 33–40, name the quadrant in which the angle θ lies.

33. $\sin\theta > 0, \quad \cos\theta < 0$
34. $\sin\theta < 0, \quad \cos\theta > 0$
35. $\sin\theta < 0, \quad \tan\theta < 0$
36. $\cos\theta > 0, \quad \tan\theta > 0$
37. $\cos\theta > 0, \quad \cot\theta < 0$
38. $\sin\theta < 0, \quad \cot\theta > 0$
39. $\sec\theta < 0, \quad \tan\theta > 0$
40. $\csc\theta > 0, \quad \cot\theta < 0$

In Problems 41–58, find the reference angle of each angle.

41. $-30°$
42. $-60°$
43. $120°$
44. $210°$
45. $300°$
46. $330°$
47. $\frac{5\pi}{4}$
48. $\frac{5\pi}{6}$
49. $\frac{8\pi}{3}$
50. $\frac{7\pi}{4}$
51. $-135°$
52. $-240°$
53. $-\frac{2\pi}{3}$
54. $-\frac{7\pi}{6}$
55. $440°$
56. $490°$
57. $\frac{15\pi}{4}$
58. $\frac{19\pi}{6}$

In Problems 59–82, use the reference angle to find the exact value of each expression. Do not use a calculator.

59. $\sin 150°$
60. $\cos 210°$
61. $\sin 510°$
62. $\cos 600°$
63. $\cos(-45°)$
64. $\sin(-240°)$
65. $\sec 240°$
66. $\csc 300°$
67. $\cot 330°$
68. $\tan 225°$
69. $\sin \frac{3\pi}{4}$
70. $\cos \frac{2\pi}{3}$
71. $\cos \frac{13\pi}{4}$
72. $\tan \frac{8\pi}{3}$
73. $\sin\left(-\frac{2\pi}{3}\right)$
74. $\cot\left(-\frac{\pi}{6}\right)$
75. $\tan \frac{14\pi}{3}$
76. $\sec \frac{11\pi}{4}$
77. $\sin(8\pi)$
78. $\cos(-2\pi)$
79. $\tan(7\pi)$
80. $\cot(5\pi)$
81. $\sec(-3\pi)$
82. $\csc\left(-\frac{5\pi}{2}\right)$

In Problems 83–100, find the exact value of each of the remaining trigonometric functions of θ.

83. $\sin\theta = \frac{12}{13}, \quad \theta$ in quadrant II
84. $\cos\theta = \frac{3}{5}, \quad \theta$ in quadrant IV
85. $\cos\theta = -\frac{4}{5}, \quad \theta$ in quadrant III
86. $\sin\theta = -\frac{5}{13}, \quad \theta$ in quadrant III
87. $\sin\theta = \frac{5}{13}, \quad 90° < \theta < 180°$
88. $\cos\theta = \frac{4}{5}, \quad 270° < \theta < 360°$
89. $\cos\theta = -\frac{1}{3}, \quad 180° < \theta < 270°$
90. $\sin\theta = -\frac{2}{3}, \quad 180° < \theta < 270°$
91. $\sin\theta = \frac{2}{3}, \quad \tan\theta < 0$
92. $\cos\theta = -\frac{1}{4}, \quad \tan\theta > 0$
93. $\sec\theta = 2, \quad \sin\theta < 0$
94. $\csc\theta = 3, \quad \cot\theta < 0$
95. $\tan\theta = \frac{3}{4}, \quad \sin\theta < 0$
96. $\cot\theta = \frac{4}{3}, \quad \cos\theta < 0$
97. $\tan\theta = -\frac{1}{3}, \quad \sin\theta > 0$
98. $\sec\theta = -2, \quad \tan\theta > 0$
99. $\csc\theta = -2, \quad \tan\theta > 0$
100. $\cot\theta = -2, \quad \sec\theta > 0$

101. Find the exact value of $\sin 40° + \sin 130° + \sin 220° + \sin 310°$.

102. Find the exact value of $\tan 40° + \tan 140°$.

Mixed Practice

In Problems 103–106, $f(x) = \sin x$, $g(x) = \cos x$, $h(x) = \tan x$, $F(x) = \csc x$, $G(x) = \sec x$, *and* $H(x) = \cot x$.

103. (a) Find $f(315°)$. What point is on the graph of f?
(b) Find $G(315°)$. What point is on the graph of G?
(c) Find $h(315°)$. What point is on the graph of h?

104. (a) Find $g(120°)$. What point is on the graph of g?
(b) Find $F(120°)$. What point is on the graph of F?
(c) Find $H(120°)$. What point is on the graph of H?

105. (a) Find $g\left(\frac{7\pi}{6}\right)$. What point is on the graph of g?
(b) Find $F\left(\frac{7\pi}{6}\right)$. What point is on the graph of F?
(c) Find $H(-315°)$. What point is on the graph of H?

106. (a) Find $f\left(\frac{7\pi}{4}\right)$. What point is on the graph of f?
(b) Find $G\left(\frac{7\pi}{4}\right)$. What point is on the graph of G?
(c) Find $F(-225°)$. What point is on the graph of F?

Applications and Extensions

107. If $f(\theta) = \sin\theta = 0.2$, find $f(\theta + \pi)$.

108. If $g(\theta) = \cos\theta = 0.4$, find $g(\theta + \pi)$.

109. If $F(\theta) = \tan\theta = 3$, find $F(\theta + \pi)$.

110. If $G(\theta) = \cot\theta = -2$, find $G(\theta + \pi)$.

111. If $\sin\theta = \frac{1}{5}$, find $\csc(\theta + \pi)$.

112. If $\cos\theta = \frac{2}{3}$, find $\sec(\theta + \pi)$.

113. Find the exact value of
$\sin 1° + \sin 2° + \sin 3° + \cdots + \sin 358° + \sin 359°$

114. Find the exact value of
$\cos 1° + \cos 2° + \cos 3° + \cdots + \cos 358° + \cos 359°$

115. Projectile Motion An object is propelled upward at an angle θ, $45° < \theta < 90°$, to the horizontal with an initial velocity of v_0 feet per second from the base of a plane that makes an angle of 45° with the horizontal. See the illustration. If air resistance is ignored, the distance R that it travels up the inclined plane is given by the function

$$R(\theta) = \frac{v_0^2\sqrt{2}}{32}[\sin(2\theta) - \cos(2\theta) - 1]$$

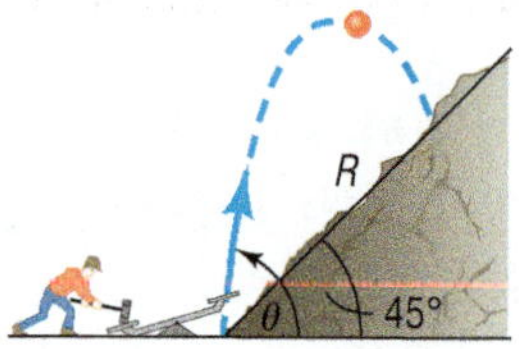

(a) Find the distance R that the object travels along the inclined plane if the initial velocity is 32 feet per second and $\theta = 60°$.
(b) Graph $R = R(\theta)$ if the initial velocity is 32 feet per second.
(c) What value of θ makes R largest?

116. Projectile Distance An object is fired at an angle θ to the horizontal with an initial speed of v_0 feet per second. Ignoring air resistance, the length of the projectile's path is given by

$$L(\theta) = \frac{v_0^2}{32}\left[\sin\theta - \cos^2\theta\left(\ln\left[\tan\left(\frac{\pi - 2\theta}{4}\right)\right]\right)\right]$$

where $0 < \theta < \frac{\pi}{2}$.

(a) Find the length of the object's path for angles $\theta = \frac{\pi}{6}, \frac{\pi}{4}$, and $\frac{\pi}{3}$ if the initial velocity is 128 feet per second.
(b) Using a graphing utility, determine the angle required for the object to have a path length of 550 feet if the initial velocity is 128 feet per second.
(c) What angle will result in the longest path? How does this angle compare to the angles that result in the longest range? (See Problems 63–66 in Section 7.3)

117. Beehive Cell Surface Area A beehive cell is in the shape of a regular hexagonal prism that is open on one end and capped at the other with three rhombuses that form a trihedral angle θ. See the diagram.

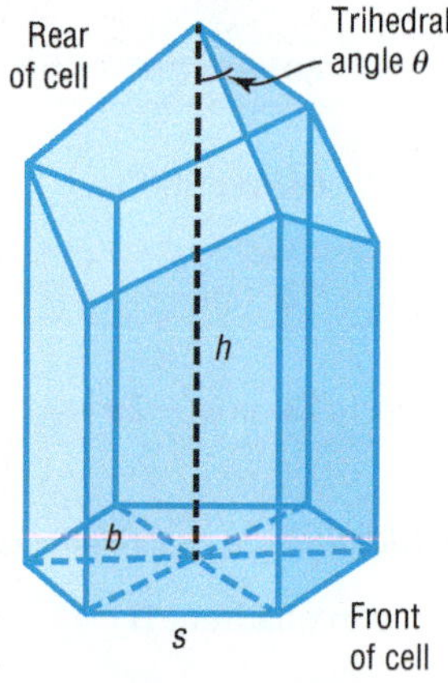

The surface area of the cell is given by

$$S(\theta) = 6sh + \frac{3s^2}{2}\left(\frac{\sqrt{3} - \cos\theta}{\sin\theta}\right) \qquad 0° < \theta < 90°$$

where s is the length of each side of the hexagonal opening and h is the height of the cell. Assuming s and h are constants, determine the trihedral angle that minimizes surface area.

Explaining Concepts: Discussion and Writing

118. Give three examples that demonstrate how to use the theorem on reference angles.

119. Write a brief paragraph that explains how to quickly compute the value of the trigonometric functions of 0°, 90°, 180°, and 270°.

120. Explain what a reference angle is. What role does it play in finding the value of a trigonometric function?

Retain Your Knowledge

Problems 121–124 are based on material learned earlier in the course. The purpose of these problems is to keep the material fresh in your mind so that you are better prepared for the final exam.

121. Solve: $25^{x+3} = 5^{x-4}$

122. The demand function for a certain product is $D = 3500 - x^2$, $0 \le x \le 59.16$, where x is the price per unit in dollars and D is in thousands of units demanded. Determine the number of units demanded if the price per unit is \$32.

123. A function f has an average rate of change of $\frac{3}{8}$ over the interval $[0, 12]$. If $f(0) = \frac{1}{2}$, find $f(12)$.

124. Given that $(f \circ g)(x) = x^2 - 8x + 19$ and $f(x) = x^2 + 3$, find $g(x)$.

7.5 Unit Circle Approach; Properties of the Trigonometric Functions

PREPARING FOR THIS SECTION *Before getting started, review the following:*

- Unit Circle (Section 2.3, p. 190)
- Functions (Section 3.1, pp. 207–218)
- Even and Odd Functions (Section 3.3, pp. 231–233)

Now Work the 'Are You Prepared?' problems on page 569.

OBJECTIVES
1. Find the Exact Values of the Trigonometric Functions Using the Unit Circle (p. 561)
2. Know the Domain and Range of the Trigonometric Functions (p. 565)
3. Use Periodic Properties to Find the Exact Values of the Trigonometric Functions (p. 567)
4. Use Even-Odd Properties to Find the Exact Values of the Trigonometric Functions (p. 568)

In this section, we develop important properties of the trigonometric functions. We begin by introducing the trigonometric functions using the unit circle. This approach will lead to the definition given earlier of the trigonometric functions of any angle (page 551).

1 Find the Exact Values of the Trigonometric Functions Using the Unit Circle

Recall that the unit circle is a circle whose radius is 1 and whose center is at the origin of a rectangular coordinate system. Also recall that any circle of radius r has circumference of length $2\pi r$. Therefore, the unit circle $(\text{radius} = 1)$ has a circumference of length 2π. So, for 1 revolution around the unit circle, the length of the arc is 2π units.

The following discussion sets the stage for defining the trigonometric functions using the unit circle.

Draw a vertical real number line with the origin of the number line at $(1, 0)$ in the Cartesian plane. Let $t \ge 0$ be any real number and let s be the distance from the origin to t on the real number line. See the red portion of Figure 69(a) on the next page. Now look at the unit circle in Figure 69(a). Beginning at the point $(1, 0)$ on the unit circle, travel $s = t$ units in the counterclockwise direction along the circle to arrive at the point $P = (a, b)$. In this sense, the length $s = t$ units is being **wrapped** around the unit circle.

If $t < 0$, we begin at the point $(1, 0)$ on the unit circle and travel $s = |t|$ units in the clockwise direction to arrive at the point $P = (a, b)$. See Figure 69(b).

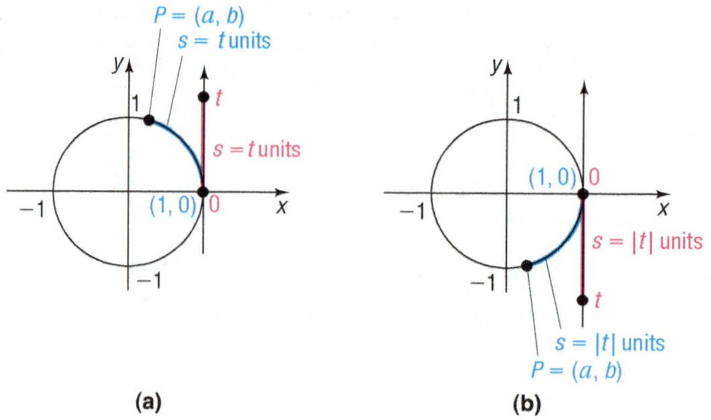

Figure 69

If $t > 2\pi$ or if $t < -2\pi$, it will be necessary to travel around the unit circle more than once before arriving at point P. Do you see why?

Let's describe this process another way. Picture a string of length $s = |t|$ units being wrapped around a circle of radius 1 unit. Start wrapping the string around the circle at the point $(1, 0)$. If $t \geq 0$, wrap the string in the counterclockwise direction; if $t < 0$, wrap the string in the clockwise direction. The point $P = (a, b)$ is the point where the string ends.

This discussion tells us that for any real number t, we can locate a unique point $P = (a, b)$ on the unit circle. We call this point **the point P on the unit circle that corresponds to t.** This is the important idea here. No matter what real number t is chosen, there is a unique point P on the unit circle corresponding to it. The coordinates of the point $P = (a, b)$ on the unit circle corresponding to the real number t are used to define the **six trigonometric functions of t.** Be sure to consult Figures 69(a) and (b) as you read these definitions.

DEFINITION

Let t be a real number, and let $P = (a, b)$ be the point on the unit circle that corresponds to t.

The **sine function** associates with t the y-coordinate of P and is denoted by

$$\sin t = b$$

The **cosine function** associates with t the x-coordinate of P and is denoted by

$$\cos t = a$$

If $a \neq 0$, the **tangent function** associates with t the ratio of the y-coordinate to the x-coordinate of P and is denoted by

$$\tan t = \frac{b}{a}$$

If $b \neq 0$, the **cosecant function** is defined as

$$\csc t = \frac{1}{b}$$

If $a \neq 0$, the **secant function** is defined as

$$\sec t = \frac{1}{a}$$

If $b \neq 0$, the **cotangent function** is defined as

$$\cot t = \frac{a}{b}$$

In Words

The point $P = (a, b)$ on the unit circle corresponding to a real number t is $(\cos t, \sin t)$.

Once again, notice in these definitions that if $a = 0$ (that is, if the point P is on the y-axis), the tangent function and the secant function are undefined. Also, if $b = 0$ (that is, if the point P is on the x-axis), the cosecant function and the cotangent function are undefined.

Because the unit circle is used in these definitions of the trigonometric functions, they are also sometimes referred to as **circular functions**.

EXAMPLE 1 Finding the Values of the Trigonometric Functions Using a Point on the Unit Circle

Find the values of $\sin t$, $\cos t$, $\tan t$, $\csc t$, $\sec t$, and $\cot t$ if $P = \left(-\frac{1}{2}, \frac{\sqrt{3}}{2}\right)$ is the point on the unit circle that corresponds to the real number t.

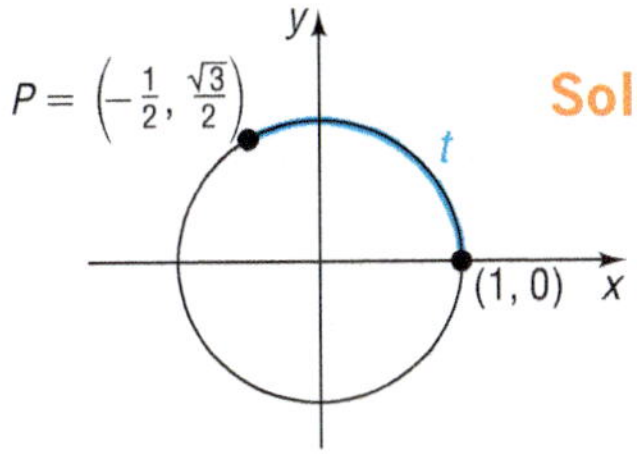

Figure 70

Solution See Figure 70. Follow the definition of the six trigonometric functions using $P = \left(-\frac{1}{2}, \frac{\sqrt{3}}{2}\right) = (a, b)$. Then, with $a = -\frac{1}{2}$ and $b = \frac{\sqrt{3}}{2}$, we have

$$\sin t = b = \frac{\sqrt{3}}{2} \qquad \cos t = a = -\frac{1}{2} \qquad \tan t = \frac{b}{a} = \frac{\frac{\sqrt{3}}{2}}{-\frac{1}{2}} = -\sqrt{3}$$

$$\csc t = \frac{1}{b} = \frac{1}{\frac{\sqrt{3}}{2}} = \frac{2}{\sqrt{3}} = \frac{2\sqrt{3}}{3} \qquad \sec t = \frac{1}{a} = \frac{1}{-\frac{1}{2}} = -2 \qquad \cot t = \frac{a}{b} = \frac{-\frac{1}{2}}{\frac{\sqrt{3}}{2}} = -\frac{1}{\sqrt{3}} = -\frac{\sqrt{3}}{3}$$

■

Now Work PROBLEM 11

Trigonometric Functions of Angles

Let $P = (a, b)$ be the point on the unit circle corresponding to the real number t. See Figure 71(a). Let θ be the angle in standard position, measured in radians, whose terminal side is the ray from the origin through P and whose arc length is $|t|$. See Figure 71(b). Since the unit circle has radius 1 unit, if $s = |t|$ units, then from the arc length formula $s = r|\theta|$, we have $\theta = t$ radians. See Figures 71(c) and (d).

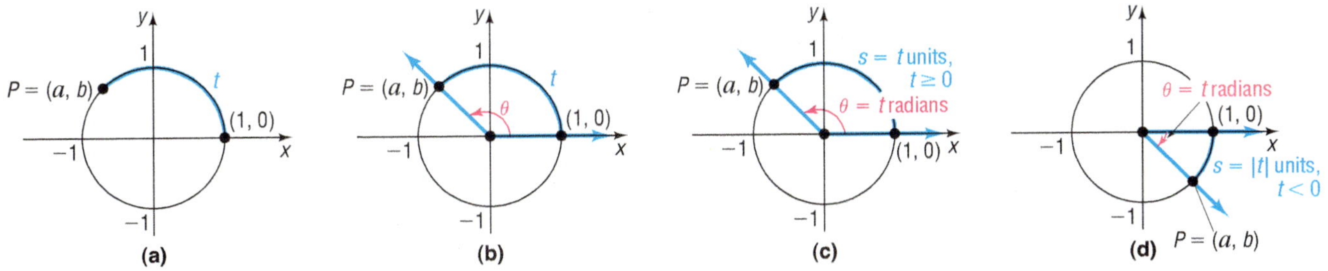

Figure 71

The point $P = (a, b)$ on the unit circle that corresponds to the real number t is also the point P on the terminal side of the angle $\theta = t$ radians. As a result, we can say that

$$\underset{\text{Real number}}{\sin t} = \underset{\theta\,=\,t\text{ radians}}{\sin \theta}$$

and so on. We can now define the trigonometric functions of the angle θ.

DEFINITION

If $\theta = t$ radians, the six **trigonometric functions of the angle θ** are defined as

$$\sin\theta = \sin t \qquad \cos\theta = \cos t \qquad \tan\theta = \tan t$$
$$\csc\theta = \csc t \qquad \sec\theta = \sec t \qquad \cot\theta = \cot t$$

Even though the trigonometric functions can be viewed both as functions of real numbers and as functions of angles, it is customary to refer to trigonometric functions of real numbers and trigonometric functions of angles collectively as *the trigonometric functions.* We will follow this practice from now on.

Since the values of the trigonometric functions of an angle θ are determined by the coordinates of the point $P = (a, b)$ on the unit circle corresponding to θ, the units used to measure the angle θ are irrelevant. For example, it does not matter whether we write $\theta = \frac{\pi}{2}$ radians or $\theta = 90°$. In either case, the point on the unit circle corresponding to this angle is $P = (0, 1)$. As a result,

$$\sin\frac{\pi}{2} = \sin 90° = 1 \quad \text{and} \quad \cos\frac{\pi}{2} = \cos 90° = 0$$

The discussion based on Figure 71 implies the following: To find the exact value of a trigonometric function of an angle θ requires that we locate the corresponding point $P^* = (a^*, b^*)$ on the unit circle. In fact, though, any circle whose center is at the origin can be used.

Let θ be any nonquadrantal angle placed in standard position. Let $P = (a, b)$ be the point on the circle $x^2 + y^2 = r^2$ that corresponds to θ, and let $P^* = (a^*, b^*)$ be the point on the unit circle that corresponds to θ. See Figure 72, where θ is shown in quadrant II.

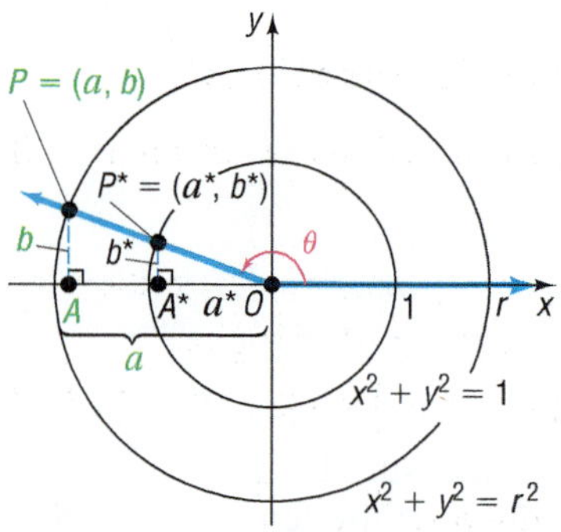

Figure 72

Note that the triangles OA^*P^* and OAP are similar, so ratios of corresponding sides are equal.

$$\frac{b^*}{1} = \frac{b}{r} \qquad \frac{a^*}{1} = \frac{a}{r} \qquad \frac{b^*}{a^*} = \frac{b}{a}$$
$$\frac{1}{b^*} = \frac{r}{b} \qquad \frac{1}{a^*} = \frac{r}{a} \qquad \frac{a^*}{b^*} = \frac{a}{b}$$

These results lead us to formulate the following theorem:

THEOREM

For an angle θ in standard position, let $P = (a, b)$ be any point on the terminal side of θ that is also on the circle $x^2 + y^2 = r^2$. Then

$$\sin\theta = \frac{b}{r} \qquad \cos\theta = \frac{a}{r} \qquad \tan\theta = \frac{b}{a} \quad a \neq 0$$
$$\csc\theta = \frac{r}{b} \quad b \neq 0 \qquad \sec\theta = \frac{r}{a} \quad a \neq 0 \qquad \cot\theta = \frac{a}{b} \quad b \neq 0$$

This result agrees with the definition given in Section 7.4 for the six trigonometric functions of any angle θ (page 551).

Consider Figure 73, which shows a circle of radius $\sqrt{2}$. Inside the circle we have drawn a right triangle whose acute angles both measure 45°. The lengths of the legs are then 1 unit each. This is the right triangle in Figure 29(b) on page 539. The point on the circle that corresponds to an angle of 45° is $(1, 1) = (a, b)$. From this, $\sin 45° = \frac{b}{r} = \frac{1}{\sqrt{2}} = \frac{\sqrt{2}}{2}$, $\cos 45° = \frac{a}{r} = \frac{1}{\sqrt{2}} = \frac{\sqrt{2}}{2}$, and so on.

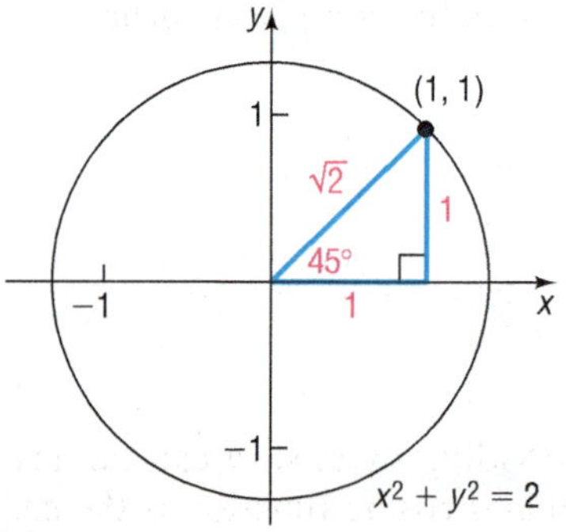

Figure 73

Now consider Figure 74(a), which shows a circle of radius 2. Inside the circle we have drawn a right triangle whose acute angles measure 30° and 60°. Refer to Figure 30(c) on page 540. If we place the 30° angle at the origin, we find that the point on the circle that corresponds to an angle of 30° is $(\sqrt{3}, 1)$. From this, $\sin 30° = \frac{b}{r} = \frac{1}{2}$, $\cos 30° = \frac{a}{r} = \frac{\sqrt{3}}{2}$, and so on. Figure 74(b) also shows a circle of radius 2, but the angle at the origin is 60°. Now we can see that $\sin 60° = \frac{b}{r} = \frac{\sqrt{3}}{2}$, and so on.

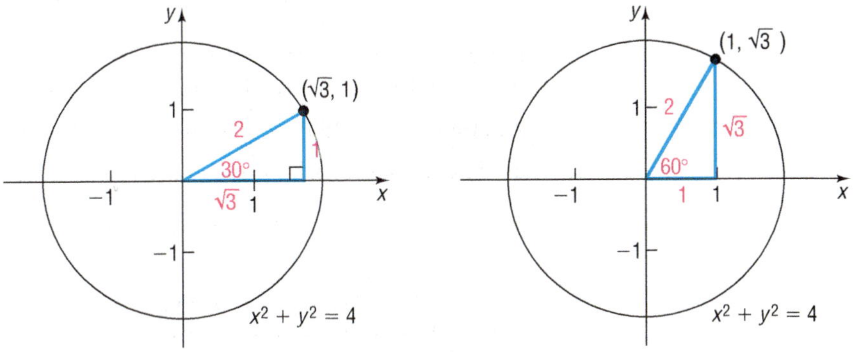

Figure 74 (a) (b)

EXAMPLE 2 **Finding the Exact Values of the Six Trigonometric Functions**

Find the exact values of each of the six trigonometric functions of an angle θ if $(4, -3)$ is a point on its terminal side.

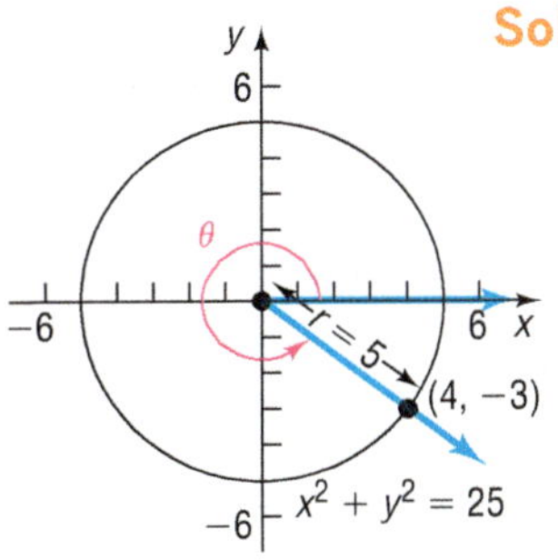

Figure 75

Solution See Figure 75. The point $(4, -3)$ is on a circle that has a radius of $r = \sqrt{4^2 + (-3)^2} = \sqrt{16 + 9} = \sqrt{25} = 5$ with the center at the origin.

For the point $(a, b) = (4, -3)$, we have $a = 4$ and $b = -3$. Since $r = 5$, we find

$$\sin\theta = \frac{b}{r} = -\frac{3}{5} \qquad \cos\theta = \frac{a}{r} = \frac{4}{5} \qquad \tan\theta = \frac{b}{a} = -\frac{3}{4}$$

$$\csc\theta = \frac{r}{b} = -\frac{5}{3} \qquad \sec\theta = \frac{r}{a} = \frac{5}{4} \qquad \cot\theta = \frac{a}{b} = -\frac{4}{3}$$

Now Work PROBLEM 17

2 Know the Domain and Range of the Trigonometric Functions

Let θ be an angle in standard position, and let $P = (a, b)$ be the point on the unit circle that corresponds to θ. See Figure 76. Then, by the definition given earlier,

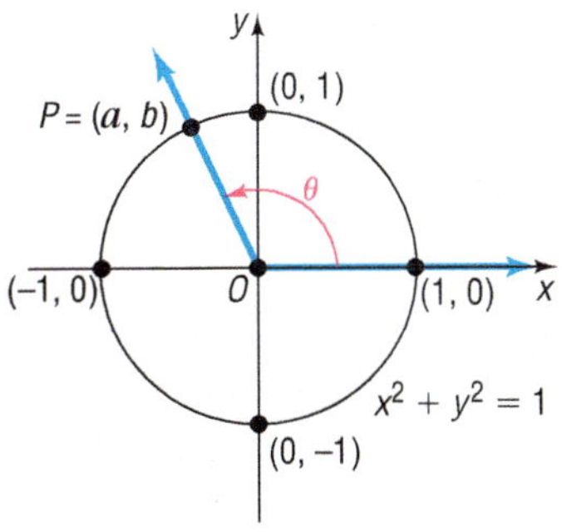

Figure 76

$$\sin\theta = b \qquad \cos\theta = a \qquad \tan\theta = \frac{b}{a} \quad a \neq 0$$

$$\csc\theta = \frac{1}{b} \quad b \neq 0 \qquad \sec\theta = \frac{1}{a} \quad a \neq 0 \qquad \cot\theta = \frac{a}{b} \quad b \neq 0$$

For $\sin\theta$ and $\cos\theta$, there is no concern about dividing by 0, so θ can be any angle. It follows that the domain of the sine function and cosine function is the set of all real numbers.

The domain of the sine function is the set of all real numbers.

The domain of the cosine function is the set of all real numbers.

For the tangent function and secant function, the x-coordinate of $P = (a, b)$ cannot be 0 since this results in division by 0. See Figure 76. On the unit circle, there are two such points, $(0, 1)$ and $(0, -1)$. These two points correspond to the angles $\frac{\pi}{2}$ $(90°)$ and $\frac{3\pi}{2}$ $(270°)$ or, more generally, to any angle that is an odd integer multiple of $\frac{\pi}{2}$ $(90°)$, such as $\pm\frac{\pi}{2}$ $(\pm 90°)$, $\pm\frac{3\pi}{2}$ $(\pm 270°)$, and $\pm\frac{5\pi}{2}$ $(\pm 450°)$. Such angles must be excluded from the domain of the tangent function and secant function.

The domain of the tangent function is the set of all real numbers, except odd integer multiples of $\frac{\pi}{2}$ $(90°)$.

The domain of the secant function is the set of all real numbers, except odd integer multiples of $\frac{\pi}{2}$ $(90°)$.

For the cotangent function and cosecant function, the y-coordinate of $P = (a, b)$ cannot be 0 since this results in division by 0. See Figure 76. On the unit circle, there are two such points, $(1, 0)$ and $(-1, 0)$. These two points correspond to the angles 0 $(0°)$ and π $(180°)$ or, more generally, to any angle that is an integer multiple of π $(180°)$, such as $0(0°)$, $\pm\pi(\pm 180°)$, $\pm 2\pi(\pm 360°)$, and $\pm 3\pi(\pm 540°)$. Such angles must be excluded from the domain of the cotangent function and cosecant function.

The domain of the cotangent function is the set of all real numbers, except integer multiples of π $(180°)$.

The domain of the cosecant function is the set of all real numbers, except integer multiples of π $(180°)$.

Next, we determine the range of each of the six trigonometric functions. Refer again to Figure 76. Let $P = (a, b)$ be the point on the unit circle that corresponds to the angle θ. It follows that $-1 \le a \le 1$ and $-1 \le b \le 1$. Since $\sin\theta = b$ and $\cos\theta = a$, we have

$$-1 \le \sin\theta \le 1 \quad \text{and} \quad -1 \le \cos\theta \le 1$$

The range of both the sine function and the cosine function consists of all real numbers between -1 and 1, inclusive. Using absolute value notation, we have $|\sin\theta| \le 1$ and $|\cos\theta| \le 1$.

If θ is not an integer multiple of π $(180°)$, then $\csc\theta = \frac{1}{b}$. Since $b = \sin\theta$ and $|b| = |\sin\theta| \le 1$, it follows that $|\csc\theta| = \frac{1}{|\sin\theta|} = \frac{1}{|b|} \ge 1 \left(\frac{1}{b} \le -1 \text{ or } \frac{1}{b} \ge 1\right)$. Since $\csc\theta = \frac{1}{b}$, the range of the cosecant function consists of all real numbers less than or equal to -1 or greater than or equal to 1. That is,

$$\csc\theta \le -1 \quad \text{or} \quad \csc\theta \ge 1$$

If θ is not an odd integer multiple of $\frac{\pi}{2}$ $(90°)$, then $\sec\theta = \frac{1}{a}$. Since $a = \cos\theta$ and $|a| = |\cos\theta| \le 1$, it follows that $|\sec\theta| = \frac{1}{|\cos\theta|} = \frac{1}{|a|} \ge 1 \left(\frac{1}{a} \le -1 \text{ or } \frac{1}{a} \ge 1\right)$.

Since $\sec\theta = \frac{1}{a}$, the range of the secant function consists of all real numbers less than or equal to -1 or greater than or equal to 1. That is,

$$\sec\theta \le -1 \quad \text{or} \quad \sec\theta \ge 1$$

The range of both the tangent function and the cotangent function consists of all real numbers. That is,

$$-\infty < \tan\theta < \infty \quad \text{and} \quad -\infty < \cot\theta < \infty$$

You are asked to prove this in Problems 99 and 100.

Table 6 summarizes these results.

Table 6

Function	Symbol	Domain	Range
sine	$f(\theta) = \sin\theta$	All real numbers	All real numbers from -1 to 1, inclusive
cosine	$f(\theta) = \cos\theta$	All real numbers	All real numbers from -1 to 1, inclusive
tangent	$f(\theta) = \tan\theta$	All real numbers, except odd integer multiples of $\frac{\pi}{2}$ (90°)	All real numbers
cosecant	$f(\theta) = \csc\theta$	All real numbers, except integer multiples of π (180°)	All real numbers greater than or equal to 1 or less than or equal to -1
secant	$f(\theta) = \sec\theta$	All real numbers, except odd integer multiples of $\frac{\pi}{2}$ (90°)	All real numbers greater than or equal to 1 or less than or equal to -1
cotangent	$f(\theta) = \cot\theta$	All real numbers, except integer multiples of π (180°)	All real numbers

Now Work PROBLEMS 63 AND 67

3 Use Periodic Properties to Find the Exact Values of the Trigonometric Functions

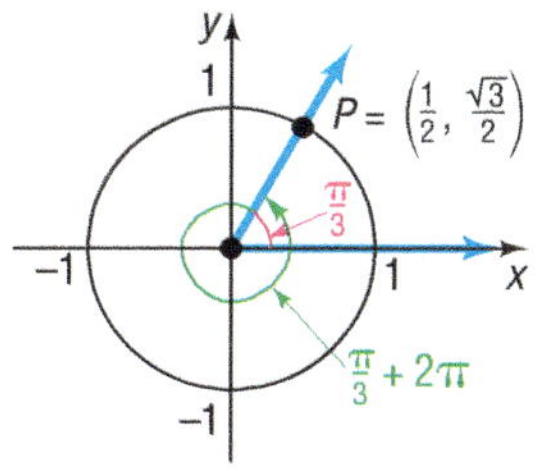

Figure 77

Look at Figure 77, which shows that for an angle of $\frac{\pi}{3}$ radians, the corresponding point P on the unit circle is $\left(\frac{1}{2}, \frac{\sqrt{3}}{2}\right)$. Notice that for an angle of $\frac{\pi}{3} + 2\pi$ radians, the corresponding point P on the unit circle is also $\left(\frac{1}{2}, \frac{\sqrt{3}}{2}\right)$. As a result,

$$\sin\frac{\pi}{3} = \frac{\sqrt{3}}{2} \quad \text{and} \quad \sin\left(\frac{\pi}{3} + 2\pi\right) = \frac{\sqrt{3}}{2}$$

$$\cos\frac{\pi}{3} = \frac{1}{2} \quad \text{and} \quad \cos\left(\frac{\pi}{3} + 2\pi\right) = \frac{1}{2}$$

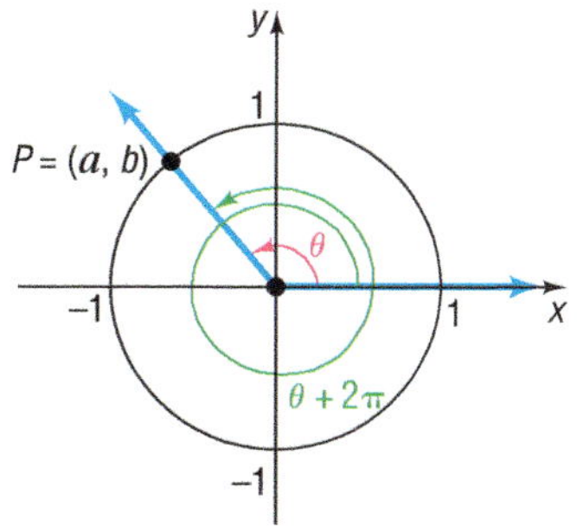

Figure 78

This example illustrates a more general situation. For a given angle θ, measured in radians, suppose that we know the corresponding point $P = (a, b)$ on the unit circle. Now add 2π to θ. The point on the unit circle corresponding to $\theta + 2\pi$ is identical to the point P on the unit circle corresponding to θ. See Figure 78. The values of the trigonometric functions of $\theta + 2\pi$ are equal to the values of the corresponding trigonometric functions of θ.

If we add (or subtract) integer multiples of 2π to θ, the values of the sine and cosine function remain unchanged. That is, for all θ,

$$\sin(\theta + 2\pi k) = \sin\theta \qquad \cos(\theta + 2\pi k) = \cos\theta \tag{1}$$

where k is any integer

Functions that exhibit this kind of behavior are called *periodic functions.*

DEFINITION

A function f is called **periodic** if there is a positive number p such that whenever θ is in the domain of f, so is $\theta + p$, and

$$f(\theta + p) = f(\theta)$$

If there is a smallest such number p, this smallest value is called the **(fundamental) period** of f.

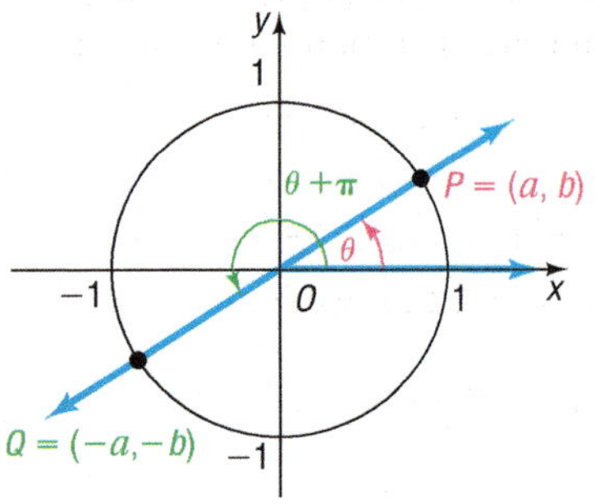

Figure 79

$\tan\theta = \dfrac{b}{a} = \dfrac{-b}{-a} = \tan(\theta + \pi)$

Based on equation (1), the sine and cosine functions are periodic. In fact, the sine, cosine, secant, and cosecant functions have period 2π. You are asked to prove this in Problems 93 through 96.

The tangent and cotangent functions are periodic with period π. See Figure 79 for a partial justification. You are asked to prove this statement in Problems 97 and 98.

In Words
Tangent and cotangent have period π; the others have period 2π.

Periodic Properties

$\sin(\theta + 2\pi) = \sin\theta$	$\cos(\theta + 2\pi) = \cos\theta$	$\tan(\theta + \pi) = \tan\theta$
$\csc(\theta + 2\pi) = \csc\theta$	$\sec(\theta + 2\pi) = \sec\theta$	$\cot(\theta + \pi) = \cot\theta$

Because the sine, cosine, secant, and cosecant functions have period 2π, once their values for $0 \le \theta < 2\pi$ are known, all their values are known. Similarly, since the tangent and cotangent functions have period π, once their values for $0 \le \theta < \pi$ are known, all their values are known.

EXAMPLE 3 Using Periodic Properties to Find Exact Values

Find the exact value of each of the following:

(a) $\sin 420°$ (b) $\tan \dfrac{5\pi}{4}$ (c) $\cos \dfrac{11\pi}{4}$

Solution

(a) $\sin 420° = \sin(60° + 360°) = \sin 60° = \dfrac{\sqrt{3}}{2}$

(b) $\tan\dfrac{5\pi}{4} = \tan\left(\dfrac{\pi}{4} + \pi\right) = \tan\dfrac{\pi}{4} = 1$

(c) $\cos\dfrac{11\pi}{4} = \cos\left(\dfrac{3\pi}{4} + \dfrac{8\pi}{4}\right) = \cos\left(\dfrac{3\pi}{4} + 2\pi\right) = \cos\dfrac{3\pi}{4} = -\dfrac{\sqrt{2}}{2}$

The periodic properties of the trigonometric functions will be very helpful when we study their graphs later in this chapter.

Now Work PROBLEMS 23 AND 81

4 Use Even-Odd Properties to Find the Exact Values of the Trigonometric Functions

Recall that a function f is even if $f(-\theta) = f(\theta)$ for all θ in the domain of f; a function f is odd if $f(-\theta) = -f(\theta)$ for all θ in the domain of f. The trigonometric functions sine, tangent, cotangent, and cosecant are odd functions and the functions cosine and secant are even functions.

THEOREM

Even-Odd Properties

$\sin(-\theta) = -\sin\theta$	$\cos(-\theta) = \cos\theta$	$\tan(-\theta) = -\tan\theta$
$\csc(-\theta) = -\csc\theta$	$\sec(-\theta) = \sec\theta$	$\cot(-\theta) = -\cot\theta$

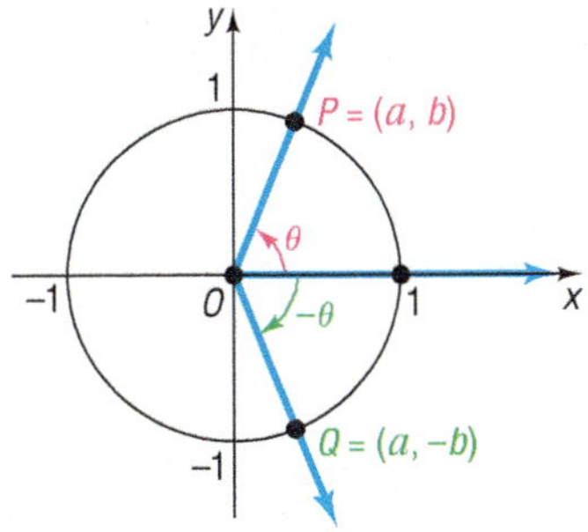

Figure 80

Proof Let $P = (a, b)$ be the point on the unit circle that corresponds to the angle θ. See Figure 80. The point Q on the unit circle that corresponds to the angle $-\theta$ will have coordinates $(a, -b)$. Using the definition for the trigonometric functions, we have

$$\sin\theta = b \qquad \sin(-\theta) = -b \qquad \cos\theta = a \qquad \cos(-\theta) = a$$

so

$$\sin(-\theta) = -\sin\theta \qquad\qquad \cos(-\theta) = \cos\theta$$

Now, using these results and some of the Fundamental Identities, we have

$$\tan(-\theta) = \frac{\sin(-\theta)}{\cos(-\theta)} = \frac{-\sin\theta}{\cos\theta} = -\tan\theta$$

$$\cot(-\theta) = \frac{1}{\tan(-\theta)} = \frac{1}{-\tan\theta} = -\cot\theta$$

$$\sec(-\theta) = \frac{1}{\cos(-\theta)} = \frac{1}{\cos\theta} = \sec\theta$$

$$\csc(-\theta) = \frac{1}{\sin(-\theta)} = \frac{1}{-\sin\theta} = -\csc\theta$$

■

In Words

Cosine and secant are even functions; the others are odd functions.

EXAMPLE 4

Finding Exact Values Using Even-Odd Properties

Find the exact value of each of the following:

(a) $\sin(-45°)$ (b) $\cos(-\pi)$ (c) $\cot\left(-\frac{7\pi}{4}\right)$ (d) $\tan\left(-\frac{37\pi}{4}\right)$

Solution

(a) $\sin(-45°) = -\sin 45° = -\frac{\sqrt{2}}{2}$
↑ Odd function

(b) $\cos(-\pi) = \cos\pi = -1$
↑ Even function

(c) $\cot\left(-\frac{7\pi}{4}\right) = -\cot\frac{7\pi}{4} = 1$
↑ Odd function

(d) $\tan\left(-\frac{37\pi}{4}\right) = -\tan\frac{37\pi}{4} = -\tan\left(\frac{\pi}{4} + 9\pi\right) = -\tan\frac{\pi}{4} = -1$
↑ Odd function ↑ Period is .

■

Now Work PROBLEMS 39 AND 75

7.5 Assess Your Understanding

'Are You Prepared?' *Answers are given at the end of these exercises. If you get a wrong answer, read the pages listed in red.*

1. What is the equation of the unit circle? (p. 190)

2. The domain of the function $f(x) = \frac{3x - 6}{x - 4}$ is __________. (pp. 207–218)

3. A function for which $f(x) = f(-x)$ for all x in the domain of f is called a(n) ______ function. (pp. 231–233)

Concepts and Vocabulary

4. The sine, cosine, cosecant, and secant functions have period _____; the tangent and cotangent functions have period ____.

5. Let t be a real number and let $P = (a, b)$ be the point on the unit circle that corresponds to t. Then $\sin t =$ ______ and $\cos t =$ ______.

6. For any angle θ in standard position, let $P = (a, b)$ be any point on the terminal side of θ that is also on the circle $x^2 + y^2 = r^2$. Then $\sin\theta =$ ______ and $\cos\theta =$ ______.

7. If $\sin\theta = 0.2$, then $\sin(-\theta) =$ ________ and $\sin(\theta + 2\pi) =$ ______.

8. ***True or False*** The only trigonometric functions with a domain of all real numbers are the sine and cosine functions.

9. Which of the following is not in the range of the sine function?
(a) $\frac{\pi}{4}$ (b) $\frac{3}{2}$ (c) -0.37 (d) -1

10. Which of the following functions is even?
(a) cosine (b) sine (c) tangent (d) cosecant

Skill Building

In Problems 11–16, the point P on the unit circle that corresponds to a real number t is given. Find sin t, cos t, tan t, csc t, sec t, and cot t.

11. $\left(\frac{\sqrt{3}}{2}, -\frac{1}{2}\right)$ 12. $\left(-\frac{\sqrt{3}}{2}, -\frac{1}{2}\right)$ 13. $\left(-\frac{\sqrt{2}}{2}, -\frac{\sqrt{2}}{2}\right)$ 14. $\left(\frac{\sqrt{2}}{2}, -\frac{\sqrt{2}}{2}\right)$ 15. $\left(\frac{\sqrt{5}}{3}, \frac{2}{3}\right)$ 16. $\left(-\frac{\sqrt{5}}{5}, \frac{2\sqrt{5}}{5}\right)$

In Problems 17–22, the point P on the circle $x^2 + y^2 = r^2$ that is also on the terminal side of an angle θ in standard position is given. Find $\sin\theta$, $\cos\theta$, $\tan\theta$, $\csc\theta$, $\sec\theta$, and $\cot\theta$.

17. $(3, -4)$ 18. $(5, -12)$ 19. $(-2, 3)$ 20. $(2, -4)$ 21. $(-1, -1)$ 22. $(-3, 1)$

In Problems 23–38, use the fact that the trigonometric functions are periodic to find the exact value of each expression. Do not use a calculator.

23. $\sin 405°$ 24. $\cos 420°$ 25. $\tan 405°$ 26. $\sin 390°$

27. $\csc 450°$ 28. $\sec 540°$ 29. $\cot 390°$ 30. $\sec 420°$

31. $\cos\frac{33\pi}{4}$ 32. $\sin\frac{9\pi}{4}$ 33. $\tan(21\pi)$ 34. $\csc\frac{9\pi}{2}$

35. $\sec\frac{17\pi}{4}$ 36. $\cot\frac{17\pi}{4}$ 37. $\tan\frac{19\pi}{6}$ 38. $\sec\frac{25\pi}{6}$

In Problems 39–56, use the even–odd properties to find the exact value of each expression. Do not use a calculator.

39. $\sin(-60°)$ 40. $\cos(-30°)$ 41. $\tan(-30°)$ 42. $\sin(-135°)$ 43. $\sec(-60°)$

44. $\csc(-30°)$ 45. $\sin(-90°)$ 46. $\cos(-270°)$ 47. $\tan\left(-\frac{\pi}{4}\right)$ 48. $\sin(-\pi)$

49. $\cos\left(-\frac{\pi}{4}\right)$ 50. $\sin\left(-\frac{\pi}{3}\right)$ 51. $\tan(-\pi)$ 52. $\sin\left(-\frac{3\pi}{2}\right)$

53. $\csc\left(-\frac{\pi}{4}\right)$ 54. $\sec(-\pi)$ 55. $\sec\left(-\frac{\pi}{6}\right)$ 56. $\csc\left(-\frac{\pi}{3}\right)$

In Problems 57–62, find the exact value of each expression. Do not use a calculator.

57. $\sin(-\pi) + \cos(5\pi)$ 58. $\tan\left(-\frac{5\pi}{6}\right) - \cot\frac{7\pi}{2}$ 59. $\sec(-\pi) + \csc\left(-\frac{\pi}{2}\right)$

60. $\tan(-6\pi) + \cos\frac{9\pi}{4}$ 61. $\sin\left(-\frac{9\pi}{4}\right) - \tan\left(-\frac{9\pi}{4}\right)$ 62. $\cos\left(-\frac{17\pi}{4}\right) - \sin\left(-\frac{3\pi}{2}\right)$

63. What is the domain of the sine function?

64. What is the domain of the cosine function?

65. For what numbers θ is $f(\theta) = \tan\theta$ not defined?

66. For what numbers θ is $f(\theta) = \cot\theta$ not defined?

67. For what numbers θ is $f(\theta) = \sec\theta$ not defined?

68. For what numbers θ is $f(\theta) = \csc\theta$ not defined?

69. What is the range of the sine function?

70. What is the range of the cosine function?

71. What is the range of the tangent function?

72. What is the range of the cotangent function?

73. What is the range of the secant function?

74. What is the range of the cosecant function?

75. Is the sine function even, odd, or neither? Is its graph symmetric? With respect to what?

76. Is the cosine function even, odd, or neither? Is its graph symmetric? With respect to what?

77. Is the tangent function even, odd, or neither? Is its graph symmetric? With respect to what?

78. Is the cotangent function even, odd, or neither? Is its graph symmetric? With respect to what?

79. Is the secant function even, odd, or neither? Is its graph symmetric? With respect to what?

80. Is the cosecant function even, odd, or neither? Is its graph symmetric? With respect to what?

81. If $\sin\theta = 0.3$, find the value of

$$\sin\theta + \sin(\theta + 2\pi) + \sin(\theta + 4\pi)$$

82. If $\cos\theta = 0.2$, find the value of

$$\cos\theta + \cos(\theta + 2\pi) + \cos(\theta + 4\pi)$$

83. If $\tan\theta = 3$, find the value of

$$\tan\theta + \tan(\theta + \pi) + \tan(\theta + 2\pi)$$

84. If $\cot\theta = -2$, find the value of

$$\cot\theta + \cot(\theta - \pi) + \cot(\theta - 2\pi)$$

In Problems 85–90, use the periodic and even–odd properties.

85. If $f(x) = \sin x$ and $f(a) = \frac{1}{3}$, find the exact value of:

(a) $f(-a)$ (b) $f(a) + f(a + 2\pi) + f(a + 4\pi)$

86. If $f(x) = \cos x$ and $f(a) = \frac{1}{4}$, find the exact value of:

(a) $f(-a)$ (b) $f(a) + f(a + 2\pi) + f(a - 2\pi)$

87. If $f(x) = \tan x$ and $f(a) = 2$, find the exact value of:

(a) $f(-a)$ (b) $f(a) + f(a + \pi) + f(a + 2\pi)$

88. If $f(x) = \cot x$ and $f(a) = -3$, find the exact value of:

(a) $f(-a)$ (b) $f(a) + f(a + \pi) + f(a + 4\pi)$

89. If $f(x) = \sec x$ and $f(a) = -4$, find the exact value of:

(a) $f(-a)$ (b) $f(a) + f(a + 2\pi) + f(a + 4\pi)$

90. If $f(x) = \csc x$ and $f(a) = 2$, find the exact value of:

(a) $f(-a)$ (b) $f(a) + f(a + 2\pi) + f(a + 4\pi)$

Applications and Extensions

In Problems 91 and 92, use the figure to approximate the value of the six trigonometric functions at t to the nearest tenth. Then use a calculator to approximate each of the six trigonometric functions at t.

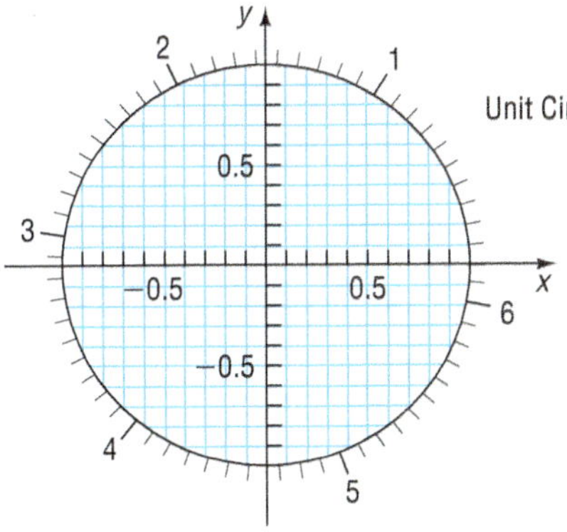

91. (a) $t = 1$ (b) $t = 5.1$

92. (a) $t = 2$ (b) $t = 4$

93. Show that the period of $f(\theta) = \sin\theta$ is 2π.
[**Hint:** Assume that $0 < p < 2\pi$ exists so that $\sin(\theta + p) = \sin\theta$ for all θ. Let $\theta = 0$ to find p. Then let $\theta = \frac{\pi}{2}$ to obtain a contradiction.]

94. Show that the period of $f(\theta) = \cos\theta$ is 2π.

95. Show that the period of $f(\theta) = \sec\theta$ is 2π.

96. Show that the period of $f(\theta) = \csc\theta$ is 2π.

97. Show that the period of $f(\theta) = \tan\theta$ is π.

98. Show that the period of $f(\theta) = \cot\theta$ is π.

99. Show that the range of the tangent function is the set of all real numbers.

100. Show that the range of the cotangent function is the set of all real numbers.

101. If θ, $0 < \theta < \pi$, is the angle between a horizontal ray directed to the right (say, the positive x-axis) and a nonhorizontal, nonvertical line L, show that the slope m of L equals $\tan\theta$. The angle θ is called the **inclination** of L.
[**Hint:** See the illustration, where we have drawn the line M parallel to L and passing through the origin. Use the fact that M intersects the unit circle at the point $(\cos\theta, \sin\theta)$.]

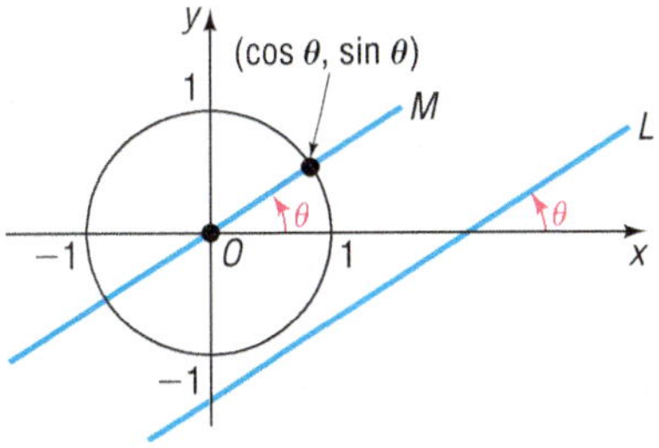

Explaining Concepts: Discussion and Writing

102. Explain how you would find the value of $\sin 390°$ using periodic properties.

103. Explain how you would find the value of $\cos(-45°)$ using even–odd properties.

104. Write down five properties of the tangent function. Explain the meaning of each.

105. Describe your understanding of the meaning of a periodic function.

Retain Your Knowledge

Problems 106–109 are based on material learned earlier in the course. The purpose of these problems is to keep the material fresh in your mind so that you are better prepared for the final exam.

106. Find the real zeros of $f(x) = 3x^2 - 10x + 5$.

107. Find an equation of the line that contains the point $(-3, 7)$ and is perpendicular to the line $y = -4x - 5$.

108. Determine the interest rate required for an investment of \$1500 to be worth \$1800 after 3 years if interest is compounded quarterly. Round your answer to two decimal places.

109. Find any vertical or horizontal asymptotes for the graph of $f(x) = \dfrac{5x - 2}{x + 3}$.

'Are You Prepared?' Answers

1. $x^2 + y^2 = 1$ **2.** $\{x \mid x \neq 4\}$ **3.** even

7.6 Graphs of the Sine and Cosine Functions*

PREPARING FOR THIS SECTION *Before getting started, review the following:*

- Graphing Techniques: Transformations (Section 3.5, pp. 256–264)

Now Work the 'Are You Prepared?' problems on page 582.

OBJECTIVES
1. Graph Functions of the Form $y = A\sin(\omega x)$ Using Transformations (p. 573)
2. Graph Functions of the Form $y = A\cos(\omega x)$ Using Transformations (p. 575)
3. Determine the Amplitude and Period of Sinusoidal Functions (p. 576)
4. Graph Sinusoidal Functions Using Key Points (p. 578)
5. Find an Equation for a Sinusoidal Graph (p. 581)

Since we want to graph the trigonometric functions in the xy-plane, we shall use the traditional symbols x for the independent variable (or argument) and y for the dependent variable (or value at x) for each function. So the six trigonometric functions can be written as

$$y = f(x) = \sin x \qquad y = f(x) = \cos x \qquad y = f(x) = \tan x$$
$$y = f(x) = \csc x \qquad y = f(x) = \sec x \qquad y = f(x) = \cot x$$

Here the independent variable x represents an angle, measured in radians. In calculus, x will usually be treated as a real number. As noted earlier, these are equivalent ways of viewing x.

The Graph of the Sine Function $y = \sin x$

Because the sine function has period 2π, it is only necessary to graph $y = \sin x$ on the interval $[0, 2\pi]$. The remainder of the graph will consist of repetitions of this portion of the graph.

To begin, consider Table 7, which lists some points on the graph of $y = \sin x$, $0 \le x \le 2\pi$. As the table shows, the graph of $y = \sin x, 0 \le x \le 2\pi$, begins at the origin. As x increases from 0 to $\frac{\pi}{2}$, the value of $y = \sin x$ increases from 0 to 1; as x increases from $\frac{\pi}{2}$ to π to $\frac{3\pi}{2}$, the value of y decreases from 1 to 0 to -1;

*For those who wish to include phase shifts here, Section 7.8 can be covered immediately after Section 7.6 without loss of continuity.

Table 7

x	$y = \sin x$	(x, y)
0	0	$(0, 0)$
$\frac{\pi}{6}$	$\frac{1}{2}$	$\left(\frac{\pi}{6}, \frac{1}{2}\right)$
$\frac{\pi}{2}$	1	$\left(\frac{\pi}{2}, 1\right)$
$\frac{5\pi}{6}$	$\frac{1}{2}$	$\left(\frac{5\pi}{6}, \frac{1}{2}\right)$
π	0	$(\pi, 0)$
$\frac{7\pi}{6}$	$-\frac{1}{2}$	$\left(\frac{7\pi}{6}, -\frac{1}{2}\right)$
$\frac{3\pi}{2}$	-1	$\left(\frac{3\pi}{2}, -1\right)$
$\frac{11\pi}{6}$	$-\frac{1}{2}$	$\left(\frac{11\pi}{6}, -\frac{1}{2}\right)$
2π	0	$(2\pi, 0)$

as x increases from $\frac{3\pi}{2}$ to 2π, the value of y increases from -1 to 0. Plotting the points listed in Table 7 and connecting them with a smooth curve yields the graph shown in Figure 81.

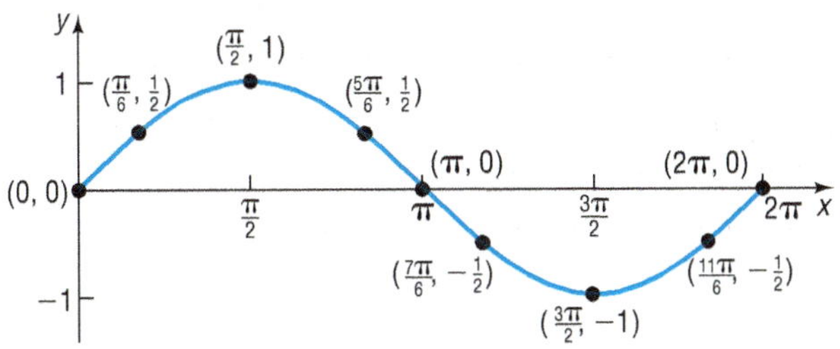

Figure 81 $y = \sin x,\ 0 \le x \le 2\pi$

The graph in Figure 81 is one period, or **cycle**, of the graph of $y = \sin x$. To obtain a more complete graph of $y = \sin x$, continue the graph in each direction, as shown in Figure 82(a). Figure 82(b) shows the graph on a TI-84 Plus C graphing calculator.

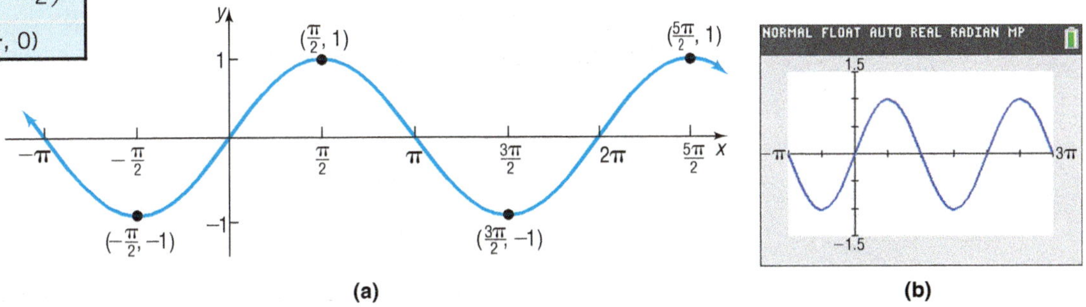

Figure 82 $y = \sin x,\ -\infty < x < \infty$

The graph of $y = \sin x$ illustrates some of the facts already discussed about the sine function.

Properties of the Sine Function $y = \sin x$

1. The domain is the set of all real numbers.
2. The range consists of all real numbers from -1 to 1, inclusive.
3. The sine function is an odd function, as the symmetry of the graph with respect to the origin indicates.
4. The sine function is periodic, with period 2π.
5. The x-intercepts are $\dots, -2\pi, -\pi, 0, \pi, 2\pi, 3\pi, \dots$; the y-intercept is 0.
6. The maximum value is 1 and occurs at $x = \dots, -\frac{3\pi}{2}, \frac{\pi}{2}, \frac{5\pi}{2}, \frac{9\pi}{2}, \dots$; the minimum value is -1 and occurs at $x = \dots, -\frac{\pi}{2}, \frac{3\pi}{2}, \frac{7\pi}{2}, \frac{11\pi}{2}, \dots$.

Now Work PROBLEM 11

1 Graph Functions of the Form $y = A\sin(\omega x)$ Using Transformations

EXAMPLE 1 **Graphing Functions of the Form $y = A\sin(\omega x)$ Using Transformations**

Graph $y = 3\sin x$ using transformations. Use the graph to determine the domain and the range of the function.

Solution Figure 83 illustrates the steps.

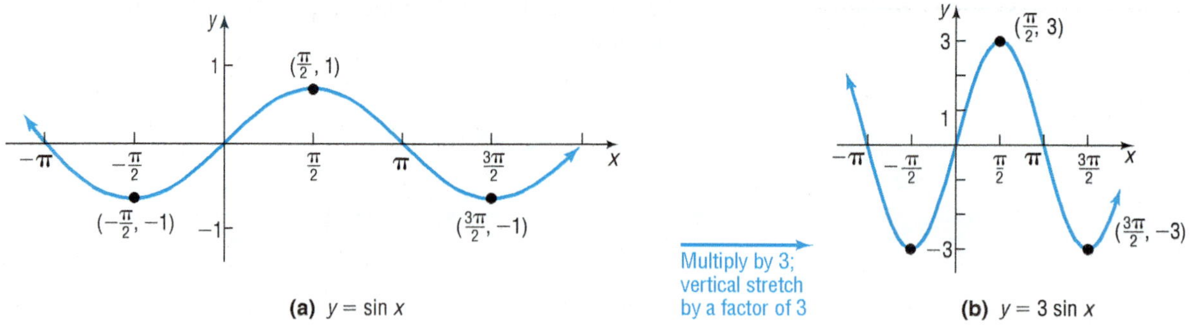

Figure 83

The domain of $y = 3 \sin x$ is the set of all real numbers, or $(-\infty, \infty)$. The range is $\{y \mid -3 \le y \le 3\}$, or $[-3, 3]$.

EXAMPLE 2 **Graphing Functions of the Form $y = A \sin(\omega x)$ Using Transformations**

Graph $y = -\sin(2x)$ using transformations. Use the graph to determine the domain and the range of the function.

Solution Figure 84 illustrates the steps.

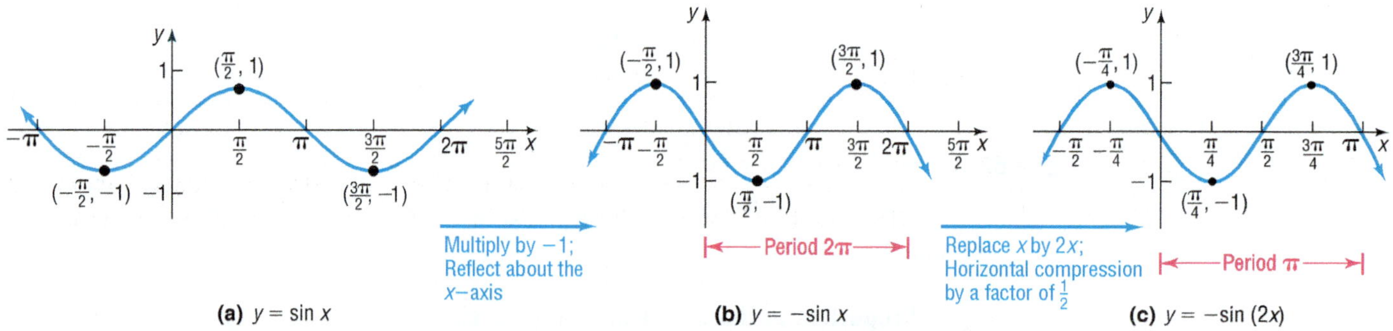

Figure 84

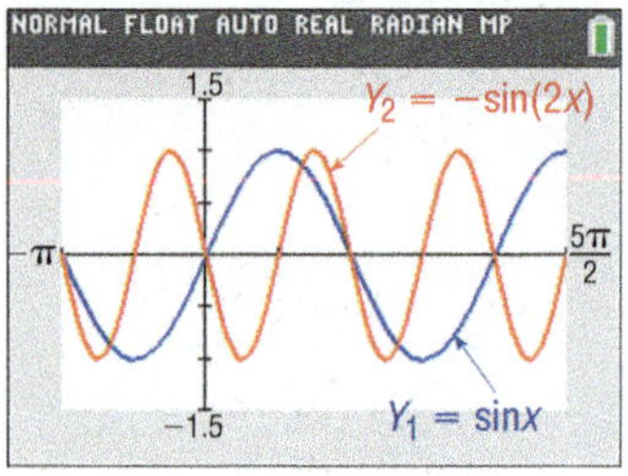

Figure 85 $Y_1 = \sin x$; $Y_2 = -\sin(2x)$

The domain of $y = -\sin(2x)$ is the set of all real numbers, or $(-\infty, \infty)$. The range is $\{y \mid -1 \le y \le 1\}$, or $[-1, 1]$.

Note in Figure 84(c) that the period of the function $y = -\sin(2x)$ is π because of the horizontal compression of the original period 2π by a factor of $\frac{1}{2}$. Figure 85 shows the graph on a TI-84 Plus C graphing calculator, along with the graph of $y = \sin x$.

Now Work PROBLEM 35 USING TRANSFORMATIONS

The Graph of the Cosine Function $y = \cos x$

The cosine function also has period 2π. Proceed as with the sine function by constructing Table 8, which lists some points on the graph of $y = \cos x, 0 \le x \le 2\pi$. As the table shows, the graph of $y = \cos x, 0 \le x \le 2\pi$, begins at the point $(0, 1)$. As x increases from 0 to $\frac{\pi}{2}$ to π, the value of y decreases from 1 to 0 to -1; as x increases from π to $\frac{3\pi}{2}$ to 2π, the value of y increases from -1 to 0 to 1. As before, plot the points in Table 8 to get one period or cycle of the graph. See Figure 86.

Table 8

x	$y = \cos x$	(x, y)
0	1	$(0, 1)$
$\frac{\pi}{3}$	$\frac{1}{2}$	$\left(\frac{\pi}{3}, \frac{1}{2}\right)$
$\frac{\pi}{2}$	0	$\left(\frac{\pi}{2}, 0\right)$
$\frac{2\pi}{3}$	$-\frac{1}{2}$	$\left(\frac{2\pi}{3}, -\frac{1}{2}\right)$
π	-1	$(\pi, -1)$
$\frac{4\pi}{3}$	$-\frac{1}{2}$	$\left(\frac{4\pi}{3}, -\frac{1}{2}\right)$
$\frac{3\pi}{2}$	0	$\left(\frac{3\pi}{2}, 0\right)$
$\frac{5\pi}{3}$	$\frac{1}{2}$	$\left(\frac{5\pi}{3}, \frac{1}{2}\right)$
2π	1	$(2\pi, 1)$

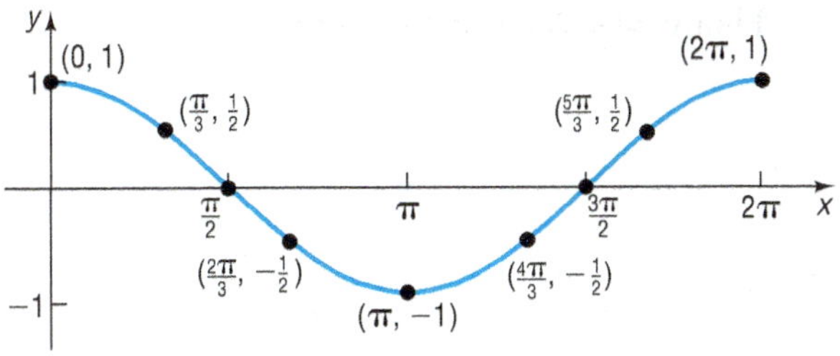

Figure 86 $y = \cos x$, $0 \le x \le 2\pi$

A more complete graph of $y = \cos x$ is obtained by continuing the graph in each direction, as shown in Figure 87(a). Figure 87(b) shows the graph on a TI-84 Plus C graphing calculator.

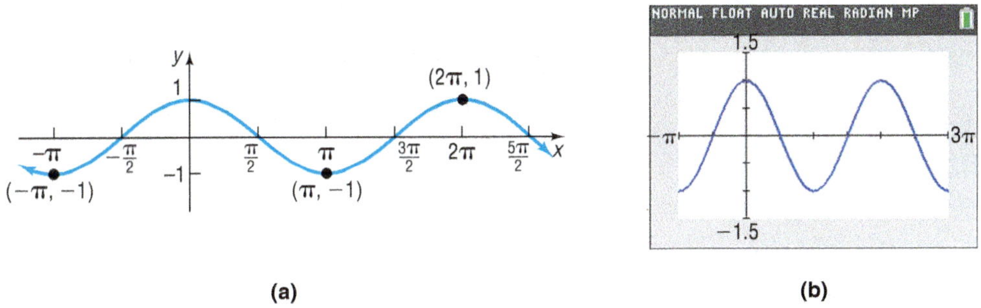

Figure 87 $y = \cos x$, $-\infty < x < \infty$

The graph of $y = \cos x$ illustrates some of the facts already discussed about the cosine function.

Properties of the Cosine Function

1. The domain is the set of all real numbers.
2. The range consists of all real numbers from -1 to 1, inclusive.
3. The cosine function is an even function, as the symmetry of the graph with respect to the y-axis indicates.
4. The cosine function is periodic, with period 2π.
5. The x-intercepts are $\ldots, -\frac{3\pi}{2}, -\frac{\pi}{2}, \frac{\pi}{2}, \frac{3\pi}{2}, \frac{5\pi}{2}, \ldots$; the y-intercept is 1.
6. The maximum value is 1 and occurs at $x = \ldots, -2\pi, 0, 2\pi, 4\pi, 6\pi, \ldots$; the minimum value is -1 and occurs at $x = \ldots, -\pi, \pi, 3\pi, 5\pi, \ldots$.

2 Graph Functions of the Form $y = A\cos(\omega x)$ Using Transformations

EXAMPLE 3

Graphing Functions of the Form $y = A\cos(\omega x)$ Using Transformations

Graph $y = 2\cos(3x)$ using transformations. Use the graph to determine the domain and the range of the function.

Solution Figure 88 shows the steps.

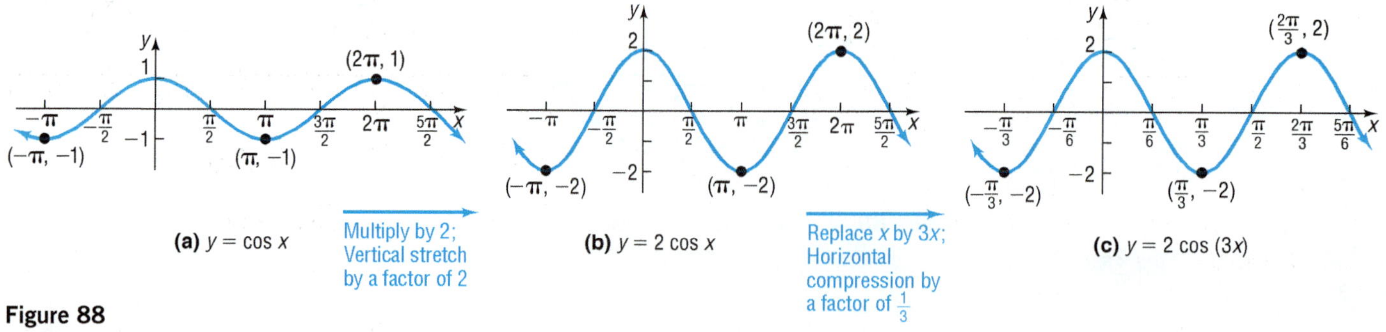

Figure 88

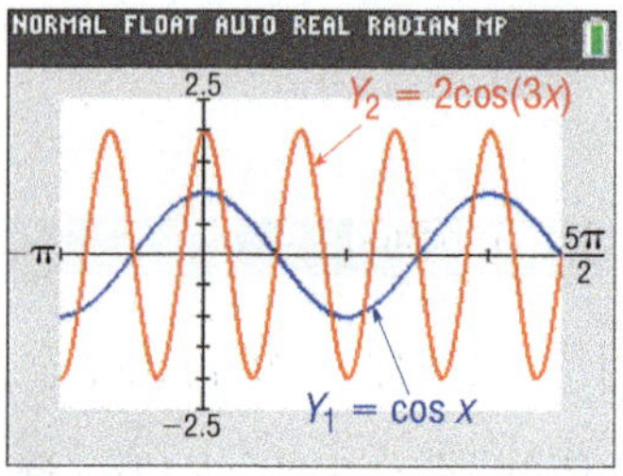

Figure 89

The domain of $y = 2\cos(3x)$ is the set of all real numbers, or $(-\infty, \infty)$. The range is $\{y \mid -2 \le y \le 2\}$, or $[-2, 2]$. ■

Notice in Figure 88(c) that the period of the function $y = 2\cos(3x)$ is $\frac{2\pi}{3}$ because of the compression of the original period 2π by a factor of $\frac{1}{3}$. Figure 89 shows the graph on a TI-84 Plus C graphing calculator, along with the graph of $y = \cos x$.

Now Work PROBLEM 43 USING TRANSFORMATIONS

Sinusoidal Graphs

Shift the graph of $y = \cos x$ to the right $\frac{\pi}{2}$ units to obtain the graph of $y = \cos\left(x - \frac{\pi}{2}\right)$. See Figure 90(a). Now look at the graph of $y = \sin x$ in Figure 90(b). Notice that the graph of $y = \sin x$ is the same as the graph of $y = \cos\left(x - \frac{\pi}{2}\right)$.

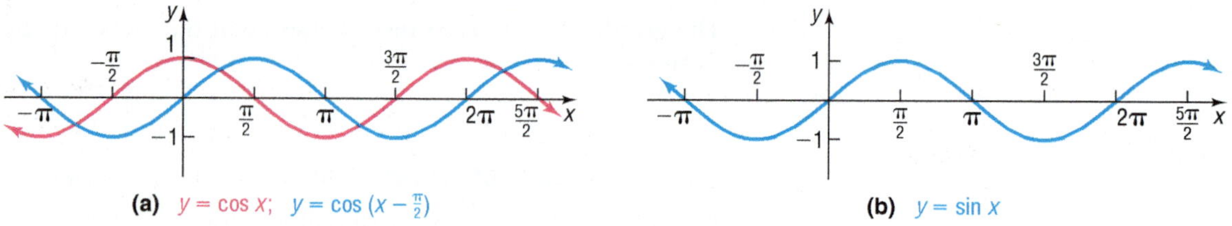

Figure 90

Based on Figure 90, we conjecture that

$$\sin x = \cos\left(x - \frac{\pi}{2}\right)$$

Seeing the Concept

Graph $Y_1 = \sin x$ and $Y_2 = \cos\left(x - \frac{\pi}{2}\right)$. How many graphs do you see? ■

(We shall prove this fact in Chapter 8.) Because of this relationship, the graphs of functions of the form $y = A\sin(\omega x)$ or $y = A\cos(\omega x)$ are referred to as **sinusoidal graphs**.

3 Determine the Amplitude and Period of Sinusoidal Functions

Figure 91 uses transformations to obtain the graph of $y = 2\cos x$. Note that the values of $y = 2\cos x$ lie between -2 and 2, inclusive.

In general, the values of the functions $y = A\sin x$ and $y = A\cos x$, where $A \ne 0$, will always satisfy the inequalities

$$-|A| \le A\sin x \le |A| \quad \text{and} \quad -|A| \le A\cos x \le |A|$$

respectively. The number $|A|$ is called the **amplitude** of $y = A\sin x$ or $y = A\cos x$. See Figure 92.

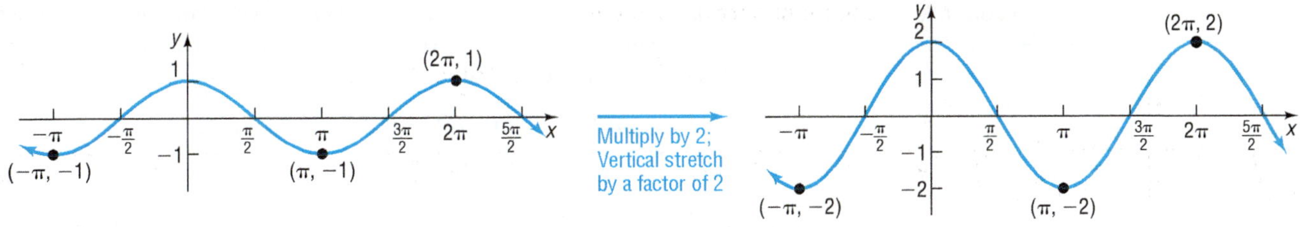

(a) $y = \cos x$

(b) $y = 2\cos x$

Figure 91

Figure 92 **(a)** $y = A \sin x$, $A > 0$; period $= 2\pi$

(b) $y = A \cos x$, $A > 0$; period $= 2\pi$

Figure 93 uses transformations to obtain the graph of $y = \cos(3x)$. Note that the period of this function is $\frac{2\pi}{3}$, because of the horizontal compression of the original period 2π by a factor of $\frac{1}{3}$.

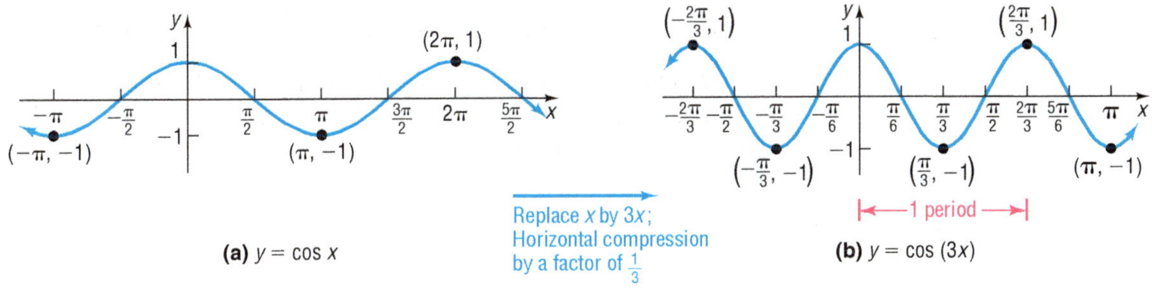

(a) $y = \cos x$

(b) $y = \cos(3x)$

Figure 93

If $\omega > 0$, the functions $y = \sin(\omega x)$ and $y = \cos(\omega x)$ will have period $T = \frac{2\pi}{\omega}$. To see why, recall that the graph of $y = \sin(\omega x)$ is obtained from the graph of $y = \sin x$ by performing a horizontal compression or stretch by a factor $\frac{1}{\omega}$. This horizontal compression replaces the interval $[0, 2\pi]$, which contains one period of the graph of $y = \sin x$, by the interval $\left[0, \frac{2\pi}{\omega}\right]$, which contains one period of the graph of $y = \sin(\omega x)$. So the function $y = \cos(3x)$, graphed in Figure 93(b) with $\omega = 3$, has period $\frac{2\pi}{\omega} = \frac{2\pi}{3}$.

One period of the graph of $y = \sin(\omega x)$ or $y = \cos(\omega x)$ is called a **cycle**. Figure 94 illustrates the general situation. The blue portion of the graph is one cycle.

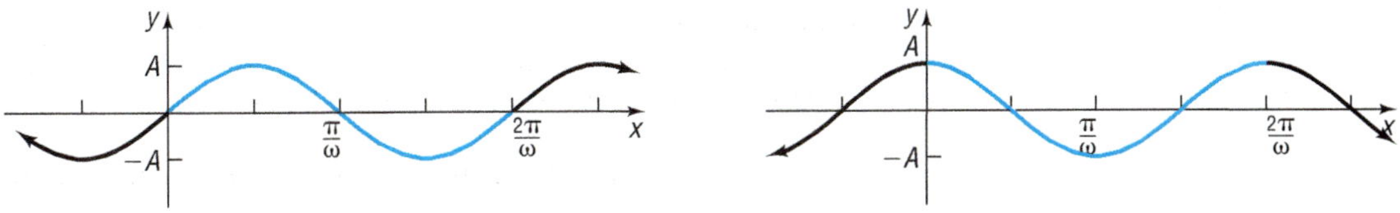

Figure 94 **(a)** $y = A \sin(\omega x)$, $A > 0$, $\omega > 0$; period $= \frac{2\pi}{\omega}$

(b) $y = A \cos(\omega x)$, $A > 0$, $\omega > 0$; period $= \frac{2\pi}{\omega}$

Note: Recall that a function f is even if $f(-x) = f(x)$; a function f is odd if $f(-x) = -f(x)$. Since the sine function is odd, $\sin(-x) = -\sin x$; since the cosine function is even, $\cos(-x) = \cos x$. ■

When graphing $y = \sin(\omega x)$ or $y = \cos(\omega x)$, we want ω to be positive. To graph either $y = \sin(-\omega x), \omega > 0$, or $y = \cos(-\omega x), \omega > 0$, use the even–odd properties of the sine and cosine functions as follows:

$$\sin(-\omega x) = -\sin(\omega x) \quad \text{and} \quad \cos(-\omega x) = \cos(\omega x)$$

This provides an equivalent form in which the coefficient of x in the argument is positive. For example,

$$\sin(-2x) = -\sin(2x) \quad \text{and} \quad \cos(-\pi x) = \cos(\pi x)$$

Because of this, we can assume that $\omega > 0$.

THEOREM

If $\omega > 0$, the amplitude and period of $y = A\sin(\omega x)$ and $y = A\cos(\omega x)$ are given by

$$\text{Amplitude} = |A| \qquad \text{Period} = T = \frac{2\pi}{\omega} \tag{1}$$

■

EXAMPLE 4

Finding the Amplitude and Period of a Sinusoidal Function

Determine the amplitude and period of $y = 3\sin(4x)$.

Solution

Comparing $y = 3\sin(4x)$ to $y = A\sin(\omega x)$, note that $A = 3$ and $\omega = 4$. From equation (1),

$$\text{Amplitude} = |A| = 3 \qquad \text{Period} = T = \frac{2\pi}{\omega} = \frac{2\pi}{4} = \frac{\pi}{2}$$

■

Now Work PROBLEM 17

4 Graph Sinusoidal Functions Using Key Points

So far, we have graphed functions of the form $y = A\sin(\omega x)$ or $y = A\cos(\omega x)$ using transformations. We now introduce another method that can be used to graph these functions.

Figure 95 shows one cycle of the graphs of $y = \sin x$ and $y = \cos x$ on the interval $[0, 2\pi]$. Notice that each graph consists of four parts corresponding to the four subintervals:

$$\left[0, \frac{\pi}{2}\right] \quad \left[\frac{\pi}{2}, \pi\right] \quad \left[\pi, \frac{3\pi}{2}\right] \quad \left[\frac{3\pi}{2}, 2\pi\right]$$

Each subinterval is of length $\frac{\pi}{2}$ (the period 2π divided by 4, the number of parts), and the endpoints of these intervals $x = 0, x = \frac{\pi}{2}, x = \pi, x = \frac{3\pi}{2}, x = 2\pi$ give rise to five key points on each graph:

$$\text{For } y = \sin x: \quad (0, 0), \left(\frac{\pi}{2}, 1\right), (\pi, 0), \left(\frac{3\pi}{2}, -1\right), (2\pi, 0)$$

$$\text{For } y = \cos x: \quad (0, 1), \left(\frac{\pi}{2}, 0\right), (\pi, -1), \left(\frac{3\pi}{2}, 0\right), (2\pi, 1)$$

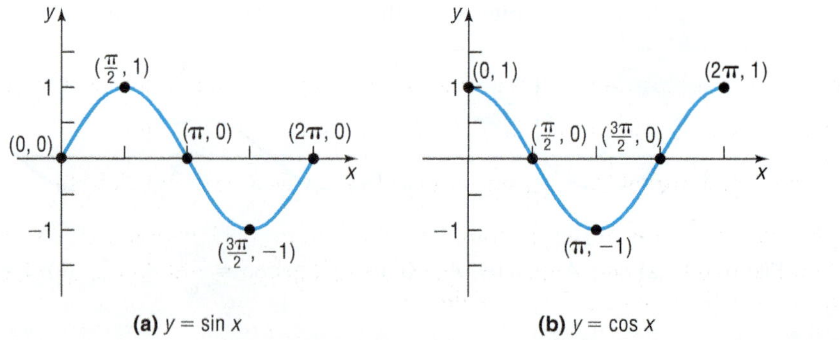

Figure 95

EXAMPLE 5 Graphing a Sinusoidal Function Using Key Points

Graph $y = 3\sin(4x)$ using key points.

Step-by-Step Solution

Step 1: Determine the amplitude and period of the sinusoidal function.

Comparing $y = 3\sin(4x)$ to $y = A\sin(\omega x)$, note that $A = 3$ and $\omega = 4$, so the amplitude is $|A| = 3$ and the period is $\frac{2\pi}{\omega} = \frac{2\pi}{4} = \frac{\pi}{2}$. Because the amplitude is 3, the graph of $y = 3\sin(4x)$ will lie between -3 and 3 on the y-axis. Because the period is $\frac{\pi}{2}$, one cycle will begin at $x = 0$ and end at $x = \frac{\pi}{2}$.

Step 2: Divide the interval $\left[0, \frac{2\pi}{\omega}\right]$ into four subintervals of the same length.

Divide the interval $\left[0, \frac{\pi}{2}\right]$ into four subintervals, each of length $\frac{\pi}{2} \div 4 = \frac{\pi}{8}$, as follows:

$$\left[0, \frac{\pi}{8}\right]\left[\frac{\pi}{8}, \frac{\pi}{8} + \frac{\pi}{8}\right] = \left[\frac{\pi}{8}, \frac{\pi}{4}\right]\left[\frac{\pi}{4}, \frac{\pi}{4} + \frac{\pi}{8}\right] = \left[\frac{\pi}{4}, \frac{3\pi}{8}\right]\left[\frac{3\pi}{8}, \frac{3\pi}{8} + \frac{\pi}{8}\right] = \left[\frac{3\pi}{8}, \frac{\pi}{2}\right]$$

The endpoints of the subintervals are $0, \frac{\pi}{8}, \frac{\pi}{4}, \frac{3\pi}{8}, \frac{\pi}{2}$. These values represent the x-coordinates of the five key points on the graph.

Step 3: Use the endpoints of the subintervals from Step 2 to obtain five key points on the graph.

Note: The five key points could also be obtained by evaluating $y = 3\sin(4x)$ at each endpoint. ■

To obtain the y-coordinates of the five key points of $y = 3\sin(4x)$, multiply the y-coordinates of the five key points for $y = \sin x$ in Figure 95(a) by $A = 3$. The five key points are

$$(0, 0) \quad \left(\frac{\pi}{8}, 3\right) \quad \left(\frac{\pi}{4}, 0\right) \quad \left(\frac{3\pi}{8}, -3\right) \quad \left(\frac{\pi}{2}, 0\right)$$

Step 4: Plot the five key points and draw a sinusoidal graph to obtain the graph of one cycle. Extend the graph in each direction to make it complete.

Plot the five key points obtained in Step 3, and fill in the graph of the sine curve as shown in Figure 96(a). Extend the graph in each direction to obtain the complete graph shown in Figure 96(b). Notice that additional key points appear every $\frac{\pi}{8}$ radian.

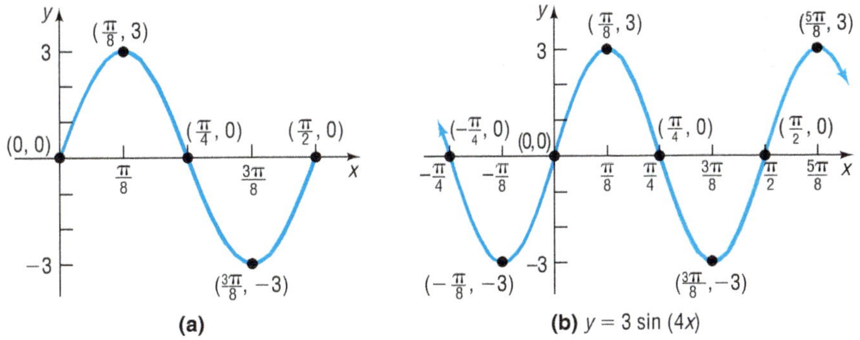

Figure 96

✓**Check:** Graph $y = 3\sin(4x)$ using transformations. Which graphing method do you prefer? ■

Now Work PROBLEM 35 USING KEY POINTS

SUMMARY

Steps for Graphing a Sinusoidal Function of the Form $y = A\sin(\omega x)$ or $y = A\cos(\omega x)$ Using Key Points

STEP 1: Determine the amplitude and period of the sinusoidal function.

STEP 2: Divide the interval $\left[0, \frac{2\pi}{\omega}\right]$ into four subintervals of the same length.

STEP 3: Use the endpoints of these subintervals to obtain five key points on the graph.

STEP 4: Plot the five key points, and draw a sinusoidal graph to obtain the graph of one cycle. Extend the graph in each direction to make it complete.

EXAMPLE 6 Graphing a Sinusoidal Function Using Key Points

Graph $y = 2\sin\left(-\frac{\pi}{2}x\right)$ using key points.

Solution Since the sine function is odd, use the equivalent form:

$$y = -2\sin\left(\frac{\pi}{2}x\right)$$

Step 1: Comparing $y = -2\sin\left(\frac{\pi}{2}x\right)$ to $y = A\sin(\omega x)$, note that $A = -2$ and $\omega = \frac{\pi}{2}$. The amplitude is $|A| = |-2| = 2$, and the period is $T = \frac{2\pi}{\omega} = \frac{2\pi}{\frac{\pi}{2}} = 4$. The graph of $y = -2\sin\left(\frac{\pi}{2}x\right)$ lies between -2 and 2 on the y-axis. One cycle will begin at $x = 0$ and end at $x = 4$.

Step 2: Divide the interval $[0, 4]$ into four subintervals, each of length $4 \div 4 = 1$. The x-coordinates of the five key points are

0	$0 + 1 = 1$	$1 + 1 = 2$	$2 + 1 = 3$	$3 + 1 = 4$
1st x-coordinate	2nd x-coordinate	3rd x-coordinate	4th x-coordinate	5th x-coordinate

Step 3: Since $y = -2\sin\left(\frac{\pi}{2}x\right)$, multiply the y-coordinates of the five key points in Figure 95(a) by $A = -2$. The five key points on the graph are

$$(0, 0) \quad (1, -2) \quad (2, 0) \quad (3, 2) \quad (4, 0)$$

Step 4: Plot these five points, and fill in the graph of the sine function as shown in Figure 97(a). Extend the graph in each direction to obtain Figure 97(b). Figure 97(c) shows the graph of one period using a graphing utility.

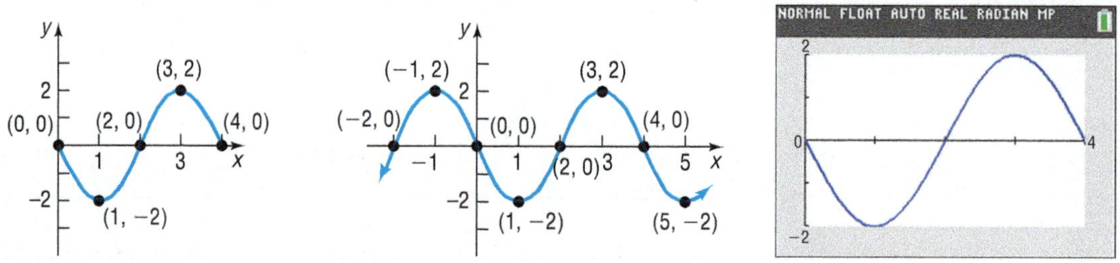

(a) (b) $y = 2\sin\left(-\frac{\pi}{2}x\right)$ (c)

Figure 97

COMMENT To graph a sinusoidal function of the form $y = A\sin(\omega x)$ or $y = A\cos(\omega x)$ using a graphing utility, use the amplitude to set Y_{min} and Y_{max}, and use the period to set X_{min} and X_{max}. ■

✓**Check:** Graph $y = 2\sin\left(-\frac{\pi}{2}x\right)$ using transformations. Which graphing method do you prefer? ■

Now Work PROBLEM 39 USING KEY POINTS

If the function to be graphed is of the form $y = A\sin(\omega x) + B$ [or $y = A\cos(\omega x) + B$], first graph $y = A\sin(\omega x)$ [or $y = A\cos(\omega x)$] and then use a vertical shift.

EXAMPLE 7 Graphing a Sinusoidal Function Using Key Points

Graph $y = -4\cos(\pi x) - 2$ using key points. Use the graph to determine the domain and the range of $y = -4\cos(\pi x) - 2$.

Solution Begin by graphing the function $y = -4\cos(\pi x)$. Comparing $y = -4\cos(\pi x)$ with $y = A\cos(\omega x)$, note that $A = -4$ and $\omega = \pi$. The amplitude is $|A| = |-4| = 4$, and the period is $T = \frac{2\pi}{\omega} = \frac{2\pi}{\pi} = 2$.

The graph of $y = -4\cos(\pi x)$ will lie between -4 and 4 on the y-axis. One cycle will begin at $x = 0$ and end at $x = 2$.

Divide the interval $[0, 2]$ into four subintervals, each of length $2 \div 4 = \frac{1}{2}$. The x-coordinates of the five key points are

0	$0 + \frac{1}{2} = \frac{1}{2}$	$\frac{1}{2} + \frac{1}{2} = 1$	$1 + \frac{1}{2} = \frac{3}{2}$	$\frac{3}{2} + \frac{1}{2} = 2$
1st x-coordinate	2nd x-coordinate	3rd x-coordinate	4th x-coordinate	5th x-coordinate

Since $y = -4\cos(\pi x)$, multiply the y-coordinates of the five key points of $y = \cos x$ shown in Figure 95(b) by $A = -4$ to obtain the five key points on the graph of $y = -4\cos(\pi x)$:

$$(0, -4) \quad \left(\frac{1}{2}, 0\right) \quad (1, 4) \quad \left(\frac{3}{2}, 0\right) \quad (2, -4)$$

Plot these five points, and fill in the graph of the cosine function as shown in Figure 98(a). Extend the graph in each direction to obtain Figure 98(b), the graph of $y = -4\cos(\pi x)$.

A vertical shift down 2 units gives the graph of $y = -4\cos(\pi x) - 2$, as shown in Figure 98(c).

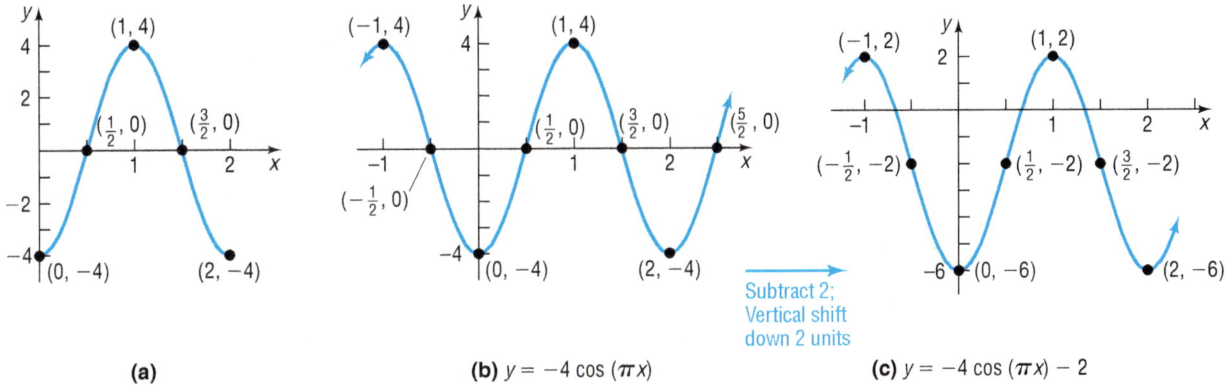

(a) (b) $y = -4\cos(\pi x)$ (c) $y = -4\cos(\pi x) - 2$

Figure 98

The domain of $y = -4\cos(\pi x) - 2$ is the set of all real numbers, or $(-\infty, \infty)$. The range of $y = -4\cos(\pi x) - 2$ is $\{y \mid -6 \le y \le 2\}$, or $[-6, 2]$. ■

Now Work PROBLEM 49

5 Find an Equation for a Sinusoidal Graph

EXAMPLE 8 Finding an Equation for a Sinusoidal Graph

Find an equation for the graph shown in Figure 99.

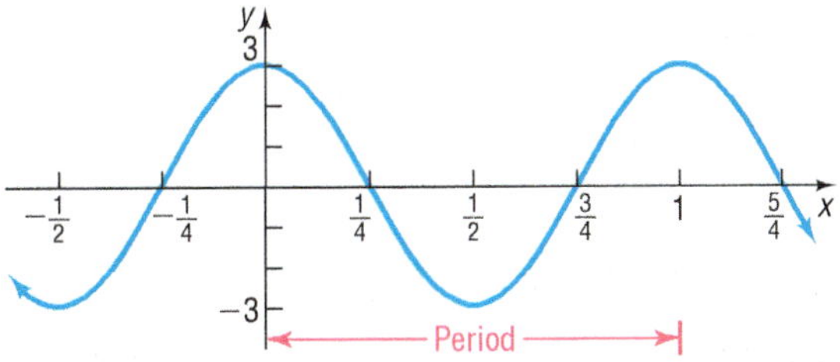

Figure 99

Solution The graph has the characteristics of a cosine function. Do you see why? The maximum value, 3, occurs at $x = 0$. So the equation can be viewed as a cosine function $y = A\cos(\omega x)$ with $A = 3$ and period $T = 1$. Then $\frac{2\pi}{\omega} = 1$, so $\omega = 2\pi$. The cosine function whose graph is given in Figure 99 is

$$y = A\cos(\omega x) = 3\cos(2\pi x)$$

✓**Check:** Graph $Y_1 = 3\cos(2\pi x)$ and compare the result with Figure 99. ■

EXAMPLE 9 **Finding an Equation for a Sinusoidal Graph**

Find an equation for the graph shown in Figure 100.

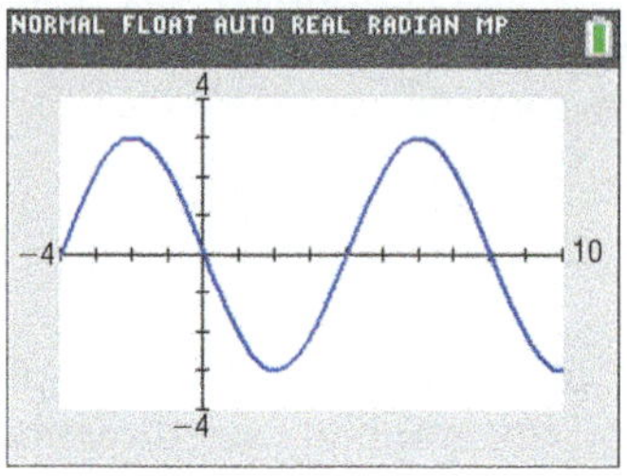

Figure 100

Solution The graph is sinusoidal, with amplitude $|A| = 3$. The period is 8, so $\frac{2\pi}{\omega} = 8$, or $\omega = \frac{\pi}{4}$. Since the graph passes through the origin, it is easier to view the equation as a sine function,[†] but note that the graph is actually the reflection of a sine function about the x-axis (since the graph is decreasing near the origin). This requires that $A = -3$. The sine function whose graph is given in Figure 100 is

$$y = A\sin(\omega x) = -3\sin\left(\frac{\pi}{4}x\right)$$

✓**Check:** Graph $Y_1 = -3\sin\left(\frac{\pi}{4}x\right)$ and compare the result with Figure 100. ■

Now Work PROBLEMS 57 AND 61

7.6 Assess Your Understanding

'Are You Prepared?' *Answers are given at the end of these exercises. If you get a wrong answer, read the pages listed in red.*

1. Use transformations to graph $y = 3x^2$. (pp. 256–264)
2. Use transformations to graph $y = \sqrt{2x}$. (pp. 256–264)

Concepts and Vocabulary

3. The maximum value of $y = \sin x$, $0 \le x \le 2\pi$, is _____ and occurs at $x =$ ______.
4. The function $y = A\sin(\omega x)$, $A > 0$, has amplitude 3 and period 2; then $A =$ _____ and $\omega =$ _____.
5. The function $y = 3\cos(6x)$ has amplitude _____ and period ____ .
6. ***True or False*** The graphs of $y = \sin x$ and $y = \cos x$ are identical except for a horizontal shift.
7. ***True or False*** For $y = 2\sin(\pi x)$, the amplitude is 2 and the period is $\frac{\pi}{2}$.
8. ***True or False*** The graph of the sine function has infinitely many x-intercepts.
9. One period of the graph of $y = \sin(\omega x)$ or $y = \cos(\omega x)$ is called a(n) ______.
   (a) amplitude (b) phase shift
   (c) transformation (d) cycle
10. To graph $y = 3\sin(-2x)$ using key points, the equivalent form ______ could be graphed instead.
   (a) $y = -3\sin(-2x)$ (b) $y = -2\sin(3x)$
   (c) $y = 3\sin(2x)$ (d) $y = -3\sin(2x)$

[†]The equation could also be viewed as a cosine function with a horizontal shift, but viewing it as a sine function is easier.

Skill Building

11. $f(x) = \sin x$

(a) What is the y-intercept of the graph of f?

(b) For what numbers x, $-\pi \le x \le \pi$, is the graph of f increasing?

(c) What is the absolute maximum of f?

(d) For what numbers x, $0 \le x \le 2\pi$, does $f(x) = 0$?

(e) For what numbers x, $-2\pi \le x \le 2\pi$, does $f(x) = 1$? Where does $f(x) = -1$?

(f) For what numbers x, $-2\pi \le x \le 2\pi$, does $f(x) = -\frac{1}{2}$?

(g) What are the x-intercepts of f?

12. $g(x) = \cos x$

(a) What is the y-intercept of the graph of g?

(b) For what numbers x, $-\pi \le x \le \pi$, is the graph of g decreasing?

(c) What is the absolute minimum of g?

(d) For what numbers x, $0 \le x \le 2\pi$, does $g(x) = 0$?

(e) For what numbers x, $-2\pi \le x \le 2\pi$, does $g(x) = 1$? Where does $g(x) = -1$?

(f) For what numbers x, $-2\pi \le x \le 2\pi$, does $g(x) = \frac{\sqrt{3}}{2}$?

(g) What are the x-intercepts of g?

In Problems 13–22, determine the amplitude and period of each function without graphing.

13. $y = 2\sin x$

14. $y = 3\cos x$

15. $y = -4\cos(2x)$

16. $y = -\sin\left(\frac{1}{2}x\right)$

17. $y = 6\sin(\pi x)$

18. $y = -3\cos(3x)$

19. $y = -\frac{1}{2}\cos\left(\frac{3}{2}x\right)$

20. $y = \frac{4}{3}\sin\left(\frac{2}{3}x\right)$

21. $y = \frac{5}{3}\sin\left(-\frac{2\pi}{3}x\right)$

22. $y = \frac{9}{5}\cos\left(-\frac{3\pi}{2}x\right)$

In Problems 23–32, match the given function to one of the graphs (A)–(J).

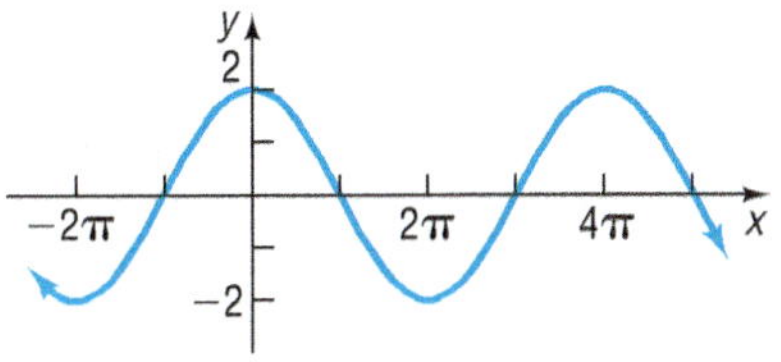

(A)

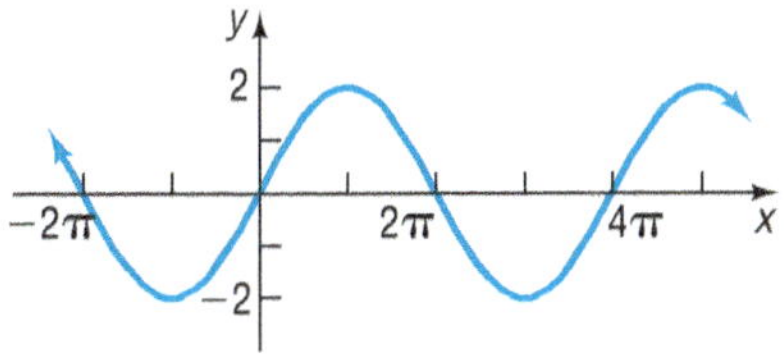

(B)

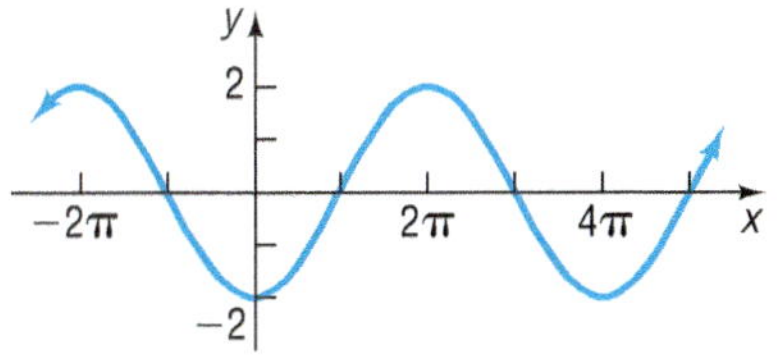

(C)

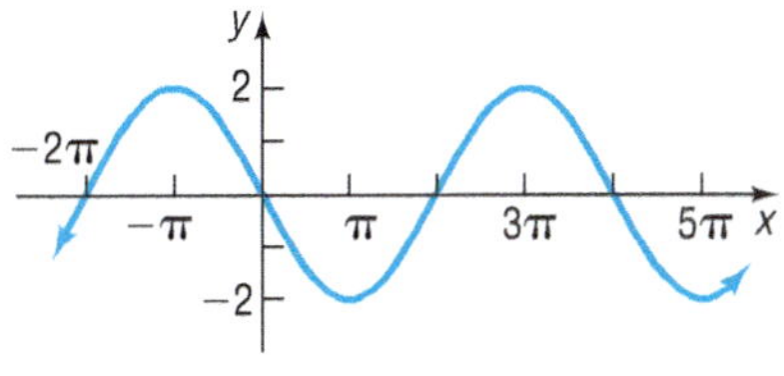

(D)

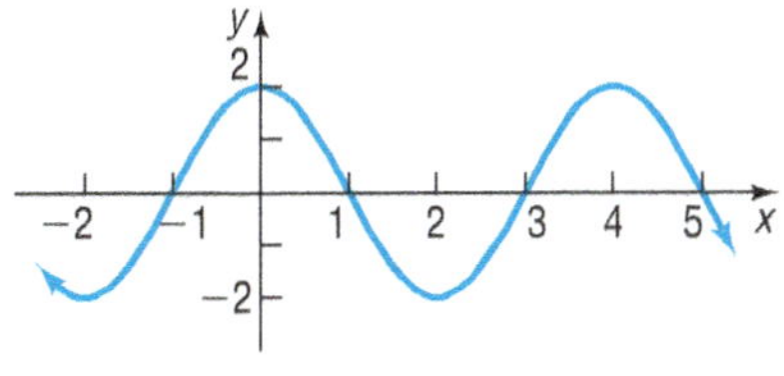

(E)

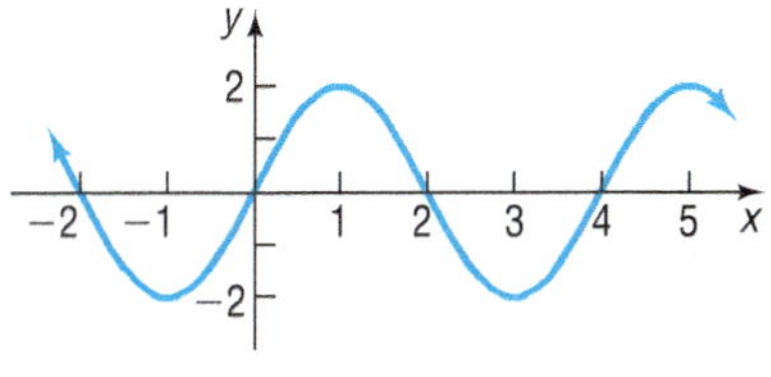

(F)

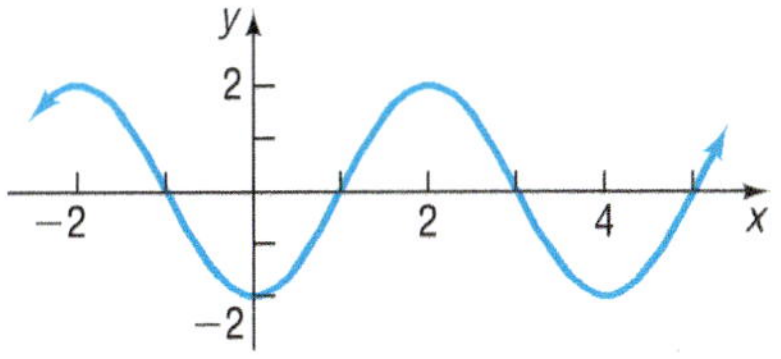

(G)

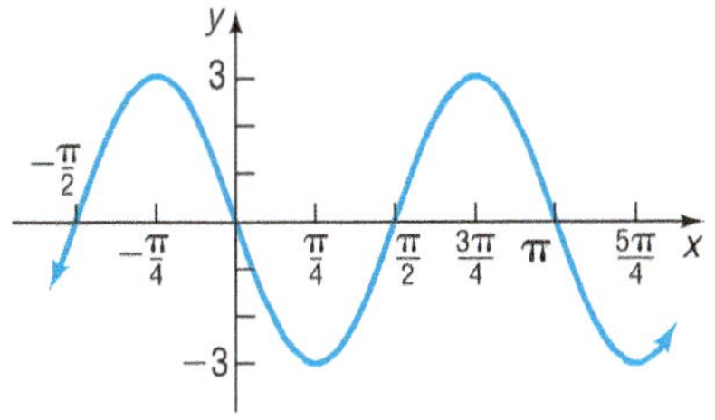

(H)

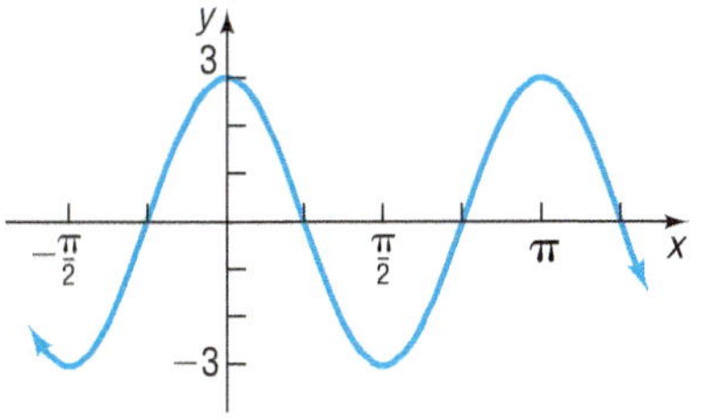

(I)

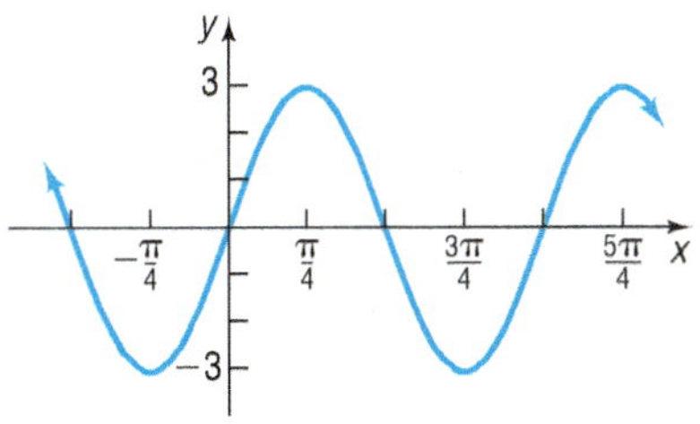

(J)

23. $y = 2\sin\left(\frac{\pi}{2}x\right)$

24. $y = 2\cos\left(\frac{\pi}{2}x\right)$

25. $y = 2\cos\left(\frac{1}{2}x\right)$

26. $y = 3\cos(2x)$

27. $y = -3\sin(2x)$

28. $y = 2\sin\left(\frac{1}{2}x\right)$

29. $y = -2\cos\left(\frac{1}{2}x\right)$

30. $y = -2\cos\left(\frac{\pi}{2}x\right)$

31. $y = 3\sin(2x)$

32. $y = -2\sin\left(\frac{1}{2}x\right)$

In Problems 33–56, graph each function using transformations or the method of key points. Be sure to label key points and show at least two cycles. Use the graph to determine the domain and the range of each function.

33. $y = 4\cos x$
34. $y = 3\sin x$
35. $y = -4\sin x$
36. $y = -3\cos x$
37. $y = \cos(4x)$
38. $y = \sin(3x)$
39. $y = \sin(-2x)$
40. $y = \cos(-2x)$
41. $y = 2\sin\left(\frac{1}{2}x\right)$
42. $y = 2\cos\left(\frac{1}{4}x\right)$
43. $y = -\frac{1}{2}\cos(2x)$
44. $y = -4\sin\left(\frac{1}{8}x\right)$
45. $y = 2\sin x + 3$
46. $y = 3\cos x + 2$
47. $y = 5\cos(\pi x) - 3$
48. $y = 4\sin\left(\frac{\pi}{2}x\right) - 2$
49. $y = -6\sin\left(\frac{\pi}{3}x\right) + 4$
50. $y = -3\cos\left(\frac{\pi}{4}x\right) + 2$
51. $y = 5 - 3\sin(2x)$
52. $y = 2 - 4\cos(3x)$
53. $y = \frac{5}{3}\sin\left(-\frac{2\pi}{3}x\right)$
54. $y = \frac{9}{5}\cos\left(-\frac{3\pi}{2}x\right)$
55. $y = -\frac{3}{2}\cos\left(\frac{\pi}{4}x\right) + \frac{1}{2}$
56. $y = -\frac{1}{2}\sin\left(\frac{\pi}{8}x\right) + \frac{3}{2}$

In Problems 57–60, write the equation of a sine function that has the given characteristics.

57. Amplitude: 3
Period: π

58. Amplitude: 2
Period: 4π

59. Amplitude: 3
Period: 2

60. Amplitude: 4
Period: 1

In Problems 61–74, find an equation for each graph.

61.

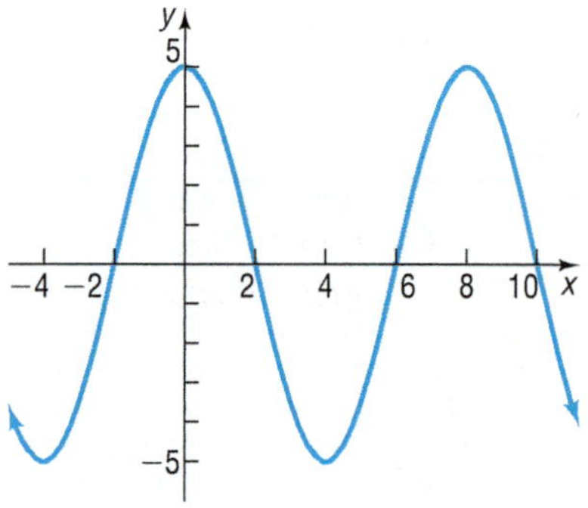

62.

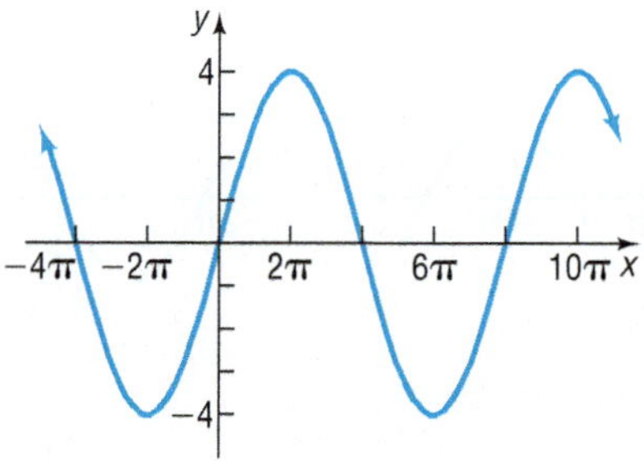

63.

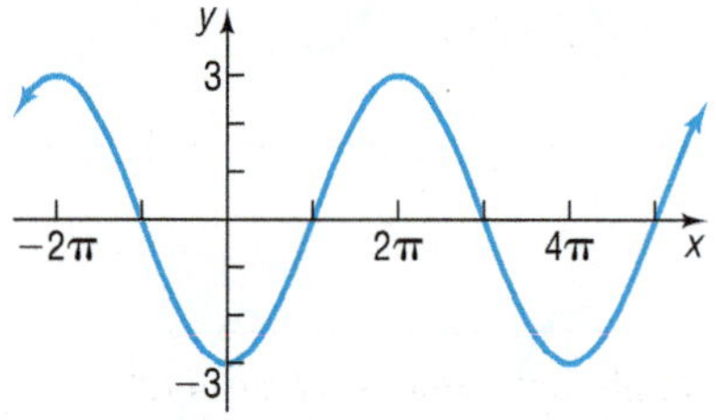

64.

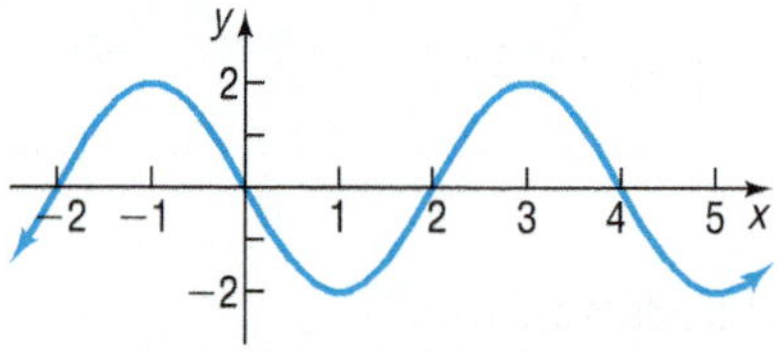

65.

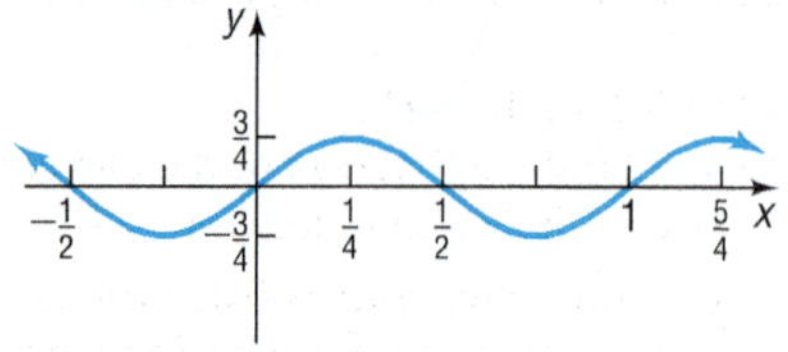

66.

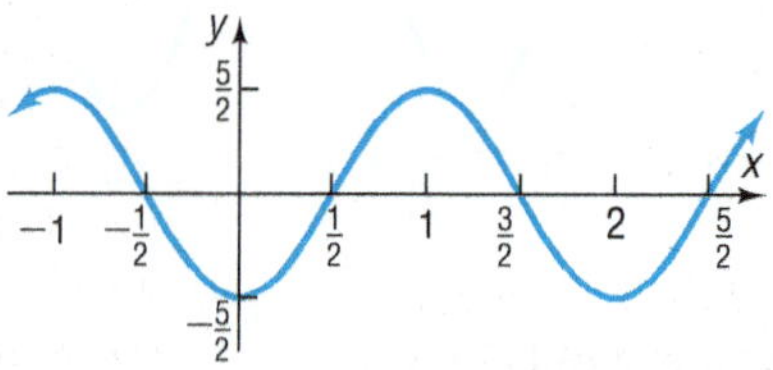

67.

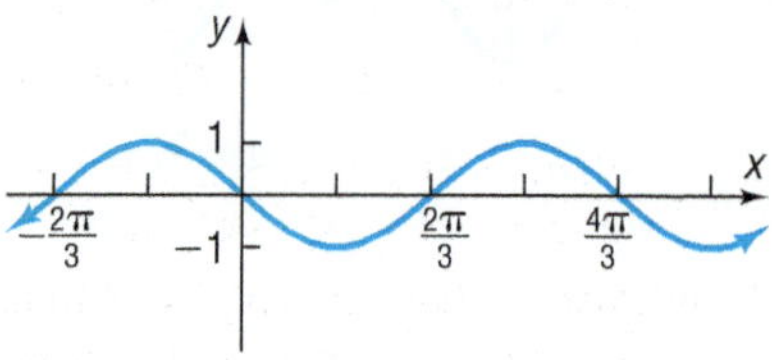

68.

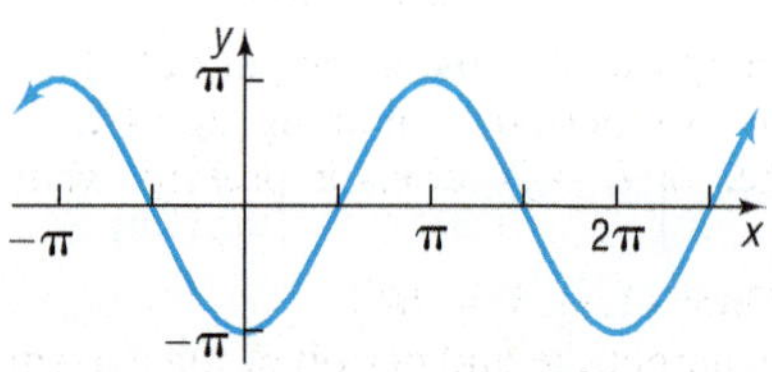

69.

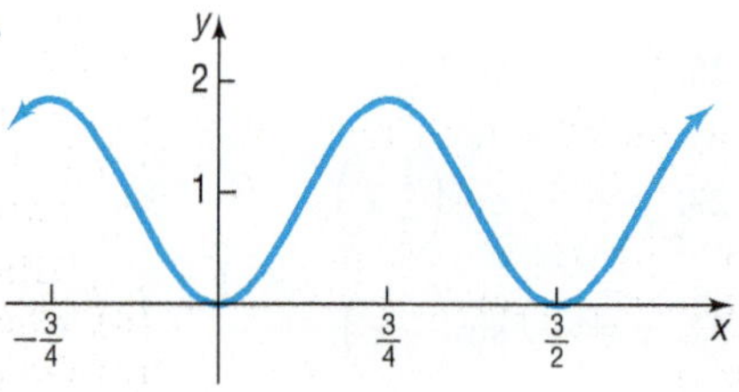

70.

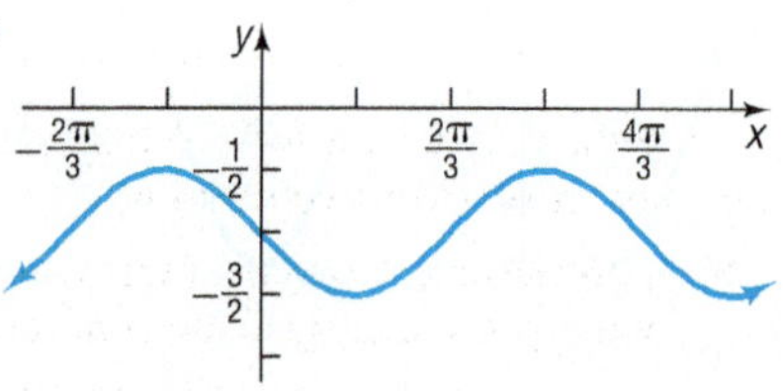

71.
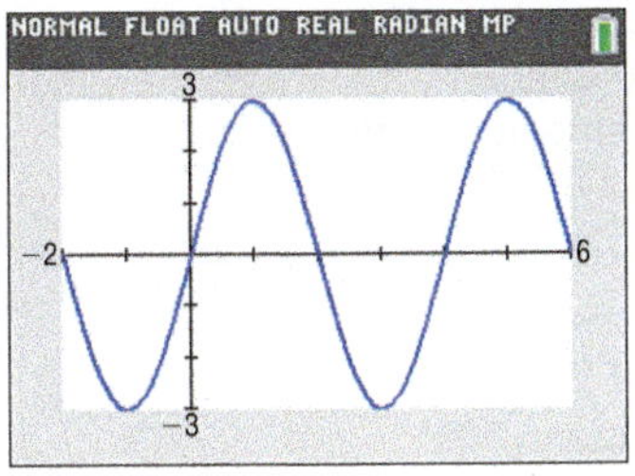

72.
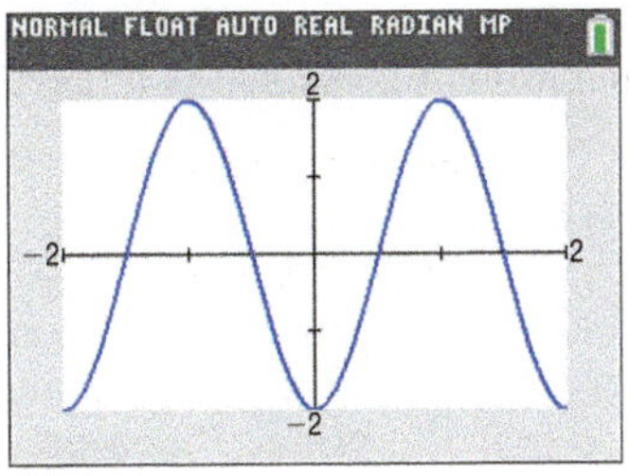

73.
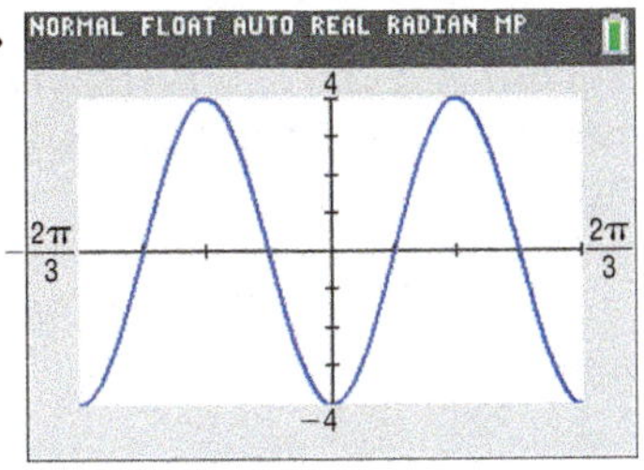

74.
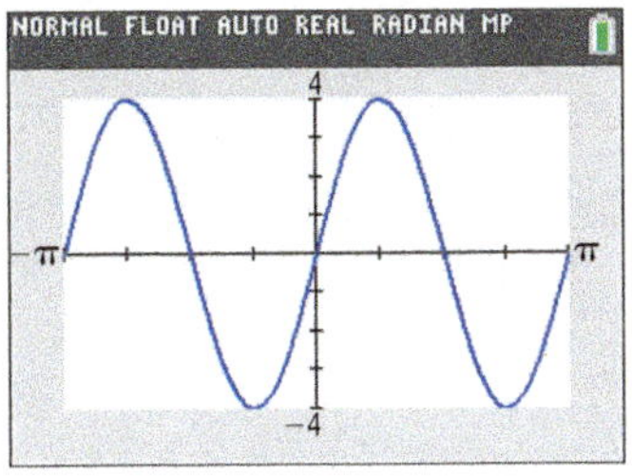

Mixed Practice

In Problems 75–78, find the average rate of change of f from 0 to $\frac{\pi}{2}$.

75. $f(x) = \sin x$ **76.** $f(x) = \cos x$ **77.** $f(x) = \sin \frac{x}{2}$ **78.** $f(x) = \cos(2x)$

In Problems 79–82, find $(f \circ g)(x)$ *and* $(g \circ f)(x)$*, and graph each of these functions.*

79. $f(x) = \sin x$, $g(x) = 4x$

80. $f(x) = \cos x$, $g(x) = \frac{1}{2}x$

81. $f(x) = -2x$, $g(x) = \cos x$

82. $f(x) = -3x$, $g(x) = \sin x$

In Problems 83 and 84, graph each function.

83. $f(x) = \begin{cases} \sin x & 0 \le x < \frac{5\pi}{4} \\ \cos x & \frac{5\pi}{4} \le x \le 2\pi \end{cases}$

84. $g(x) = \begin{cases} 2\sin x & 0 \le x \le \pi \\ \cos x + 1 & \pi < x \le 2\pi \end{cases}$

Applications and Extensions

85. Alternating Current (ac) Circuits The current I, in amperes, flowing through an ac (alternating current) circuit at time t, in seconds, is

$$I(t) = 220 \sin(60\pi t) \qquad t \ge 0$$

What is the period? What is the amplitude? Graph this function over two periods.

86. Alternating Current (ac) Circuits The current I, in amperes, flowing through an ac (alternating current) circuit at time t, in seconds, is

$$I(t) = 120 \sin(30\pi t) \qquad t \ge 0$$

What is the period? What is the amplitude? Graph this function over two periods.

87. Alternating Current (ac) Generators The voltage V, in volts, produced by an ac generator at time t, in seconds, is

$$V(t) = 220 \sin(120\pi t)$$

(a) What is the amplitude? What is the period?
(b) Graph V over two periods, beginning at $t = 0$.
(c) If a resistance of $R = 10$ ohms is present, what is the current I?
[**Hint:** Use Ohm's Law, $V = IR$.]
(d) What are the amplitude and period of the current I?
(e) Graph I over two periods, beginning at $t = 0$.

88. Alternating Current (ac) Generators The voltage V, in volts, produced by an ac generator at time t, in seconds, is

$$V(t) = 120 \sin(120\pi t)$$

(a) What is the amplitude? What is the period?
(b) Graph V over two periods, beginning at $t = 0$.
(c) If a resistance of $R = 20$ ohms is present, what is the current I?
[**Hint:** Use Ohm's Law, $V = IR$.]
(d) What are the amplitude and period of the current I?
(e) Graph I over two periods, beginning at $t = 0$.

89. Alternating Current (ac) Generators The voltage V produced by an ac generator is sinusoidal. As a function of time, the voltage V is

$$V(t) = V_0 \sin(2\pi f t)$$

where f is the **frequency**, the number of complete oscillations (cycles) per second. [In the United States and Canada, f is 60 hertz (Hz).] The **power** P delivered to a resistance R at any time t is defined as

$$P(t) = \frac{[V(t)]^2}{R}$$

(a) Show that $P(t) = \frac{V_0^2}{R}\sin^2(2\pi f t)$.

(b) The graph of P is shown in the figure. Express P as a sinusoidal function.

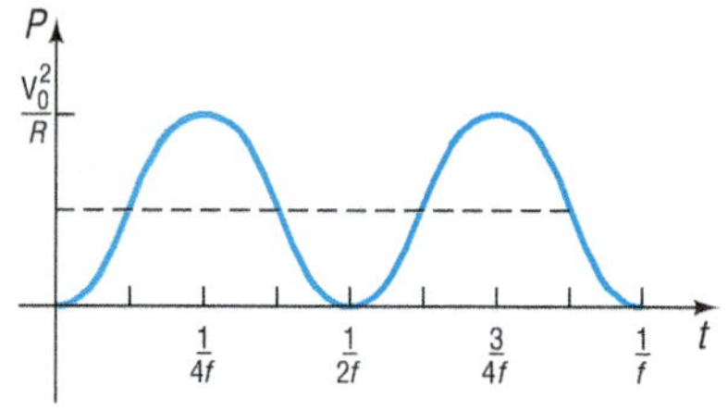

Power in an ac generator

(c) Deduce that

$$\sin^2(2\pi f t) = \frac{1}{2}[1 - \cos(4\pi f t)]$$

90. Bridge Clearance A one-lane highway runs through a tunnel in the shape of one-half a sine curve cycle. The opening is 28 feet wide at road level and is 15 feet tall at its highest point.

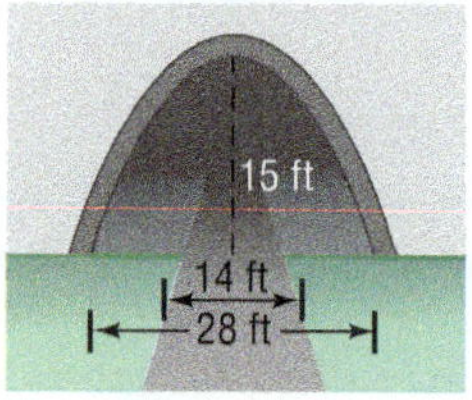

(a) Find an equation for the sine curve that fits the opening. Place the origin at the left end of the opening.

(b) If the road is 14 feet wide with 7-foot shoulders on each side, what is the height of the tunnel at the edge of the road?

Source: *en.wikipedia.org/wiki/Interstate_Highway_standards and Ohio Revised Code*

91. Blood Pressure Blood pressure is a way of measuring the amount of force exerted on the walls of blood vessels. It is measured using two numbers: systolic (as the heart beats) blood pressure and diastolic (as the heart rests) blood pressure. Typical blood pressure is 120/80, which means the systolic blood pressure is 120 mmHg and the diastolic blood pressure is 80 mmHg. Assuming that a person's heart beats 70 times per minute, the blood pressure P of an individual after t seconds can be modeled by the function

$$P(t) = 20\sin\left(\frac{7}{3}\pi t\right) + 100$$

Graph this function for two periods.

92. Ferris Wheel The function

$$h(t) = 100\cos\left(\frac{\pi}{15}t\right) + 105$$

represents the height h, in feet, of a seat on a Ferris wheel as a function of time t, where t is measured in seconds. Graph the function for one period.

93. Hours of Daylight For a certain town in Alaska, the function

$$y = 6.97\cos\left(\frac{2\pi}{365}x - 3\right) + 12.45$$

models the number of hours of daylight y, in hours, on the xth day of the year. Graph this function using transformations.

94. Holding Pattern The function

$$d(t) = 50\cos\left(\frac{\pi}{10}t\right) + 60$$

represents the distance d, in miles, of an airplane asked to fly in a holding pattern at time t, in minutes. Graph the function for one period.

95. Biorhythms In the theory of biorhythms, a sine function of the form

$$P(t) = 50\sin(\omega t) + 50$$

is used to measure the percent P of a person's potential at time t, where t is measured in days and $t = 0$ is the day the person is born. Three characteristics are commonly measured:

Physical potential: period of 23 days
Emotional potential: period of 28 days
Intellectual potential: period of 33 days

(a) Find ω for each characteristic.

(b) Using a graphing utility, graph all three functions on the same screen.

(c) Is there a time t when all three characteristics have 100% potential? When is it?

(d) Suppose that you are 20 years old today ($t = 7305$ days). Describe your physical, emotional, and intellectual potential for the next 30 days.

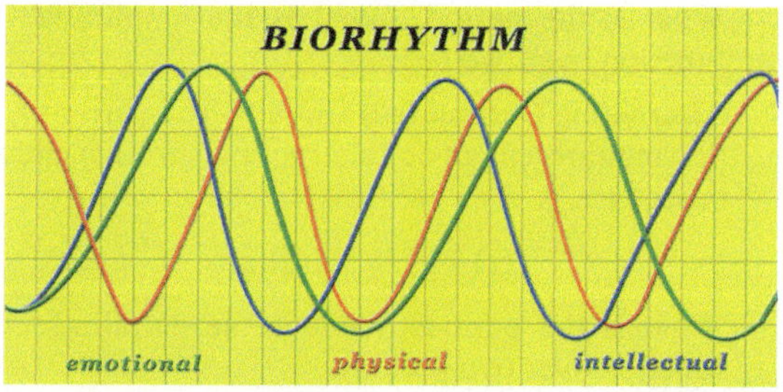

96. Graph $y = |\cos x|, -2\pi \le x \le 2\pi$.

97. Graph $y = |\sin x|, -2\pi \le x \le 2\pi$.

In Problems 98–101, the graphs of the given pairs of functions intersect infinitely many times. In each problem, find four of these points of intersection.

98. $y = \sin x$
$y = \frac{1}{2}$

99. $y = \cos x$
$y = \frac{1}{2}$

100. $y = 2\sin x$
$y = -2$

101. $y = \tan x$
$y = 1$

Explaining Concepts: Discussion and Writing

102. Explain how you would scale the x-axis and y-axis before graphing $y = 3\cos(\pi x)$.

103. Explain the term *amplitude* as it relates to the graph of a sinusoidal function.

104. Explain the term *period* as it relates to the graph of a sinusoidal function.

105. Explain how the amplitude and period of a sinusoidal graph are used to establish the scale on each coordinate axis.

106. Find an application in your major field that leads to a sinusoidal graph. Write a summary of your findings.

Retain Your Knowledge

Problems 107–110 are based on material learned earlier in the course. The purpose of these problems is to keep the material fresh in your mind so that you are better prepared for the final exam.

107. If $f(x) = x^2 - 5x + 1$, find $\dfrac{f(x+h) - f(x)}{h}$.

108. Find the vertex of the graph of $g(x) = -3x^2 + 12x - 7$.

109. Find the intercepts of the graph of $h(x) = 3|x + 2| - 1$.

110. Solve: $3x - 2(5x + 16) = -3x + 4(8 + x)$

'Are You Prepared?' Answers

1. Vertical stretch by a factor of 3

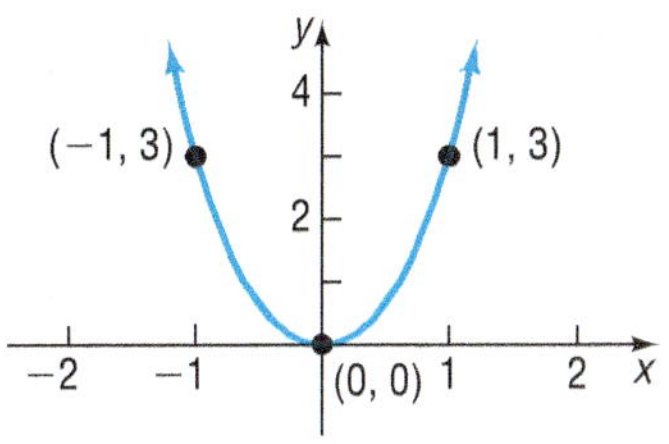

2. Horizontal compression by a factor of $\frac{1}{2}$

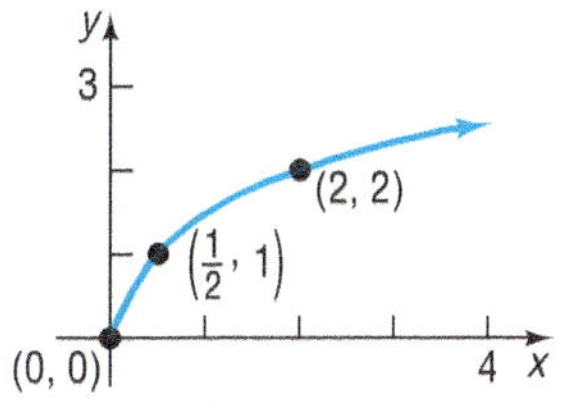

7.7 Graphs of the Tangent, Cotangent, Cosecant, and Secant Functions

PREPARING FOR THIS SECTION *Before getting started, review the following:*

- Vertical Asymptotes (Section 5.4, pp. 374–377)

Now Work the 'Are You Prepared?' problems on page 593.

OBJECTIVES
1 Graph Functions of the Form $y = A\tan(\omega x) + B$ and $y = A\cot(\omega x) + B$ (p. 589)
2 Graph Functions of the Form $y = A\csc(\omega x) + B$ and $y = A\sec(\omega x) + B$ (p. 592)

The Graph of the Tangent Function $y = \tan x$

Because the tangent function has period π, we only need to determine the graph over some interval of length π. The rest of the graph will consist of repetitions of that graph. Because the tangent function is not defined at $\ldots, -\frac{3\pi}{2}, -\frac{\pi}{2}, \frac{\pi}{2}, \frac{3\pi}{2}, \ldots$, we will concentrate on the interval $\left(-\frac{\pi}{2}, \frac{\pi}{2}\right)$, of length π, and construct Table 9, which lists some points on the graph of $y = \tan x$, $-\frac{\pi}{2} < x < \frac{\pi}{2}$. We plot the points in the table and connect them with a smooth curve. See Figure 101 for a partial graph of $y = \tan x$, where $-\frac{\pi}{3} \le x \le \frac{\pi}{3}$.

Table 9

x	$y = \tan x$	(x, y)
$-\frac{\pi}{3}$	$-\sqrt{3} \approx -1.73$	$\left(-\frac{\pi}{3}, -\sqrt{3}\right)$
$-\frac{\pi}{4}$	-1	$\left(-\frac{\pi}{4}, -1\right)$
$-\frac{\pi}{6}$	$-\frac{\sqrt{3}}{3} \approx -0.58$	$\left(-\frac{\pi}{6}, -\frac{\sqrt{3}}{3}\right)$
0	0	(0, 0)
$\frac{\pi}{6}$	$\frac{\sqrt{3}}{3} \approx 0.58$	$\left(\frac{\pi}{6}, \frac{\sqrt{3}}{3}\right)$
$\frac{\pi}{4}$	1	$\left(\frac{\pi}{4}, 1\right)$
$\frac{\pi}{3}$	$\sqrt{3} \approx 1.73$	$\left(\frac{\pi}{3}, \sqrt{3}\right)$

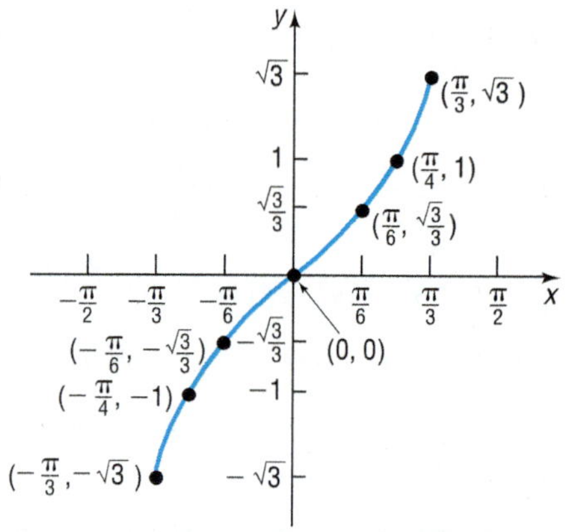

Figure 101 $y = \tan x,\ -\frac{\pi}{3} \le x \le \frac{\pi}{3}$

To complete one period of the graph of $y = \tan x$, we need to investigate the behavior of the function as x approaches $-\frac{\pi}{2}$ and $\frac{\pi}{2}$. We must be careful, though, because $y = \tan x$ is not defined at these numbers. To determine this behavior, we use the identity

$$\tan x = \frac{\sin x}{\cos x}$$

See Table 10. If x is close to $\frac{\pi}{2} \approx 1.5708$ but remains less than $\frac{\pi}{2}$, then $\sin x$ will be close to 1, and $\cos x$ will be positive and close to 0. (To see this, refer to the graphs of the sine function and the cosine function.) So the ratio $\frac{\sin x}{\cos x}$ will be positive and large. In fact, the closer x gets to $\frac{\pi}{2}$, the closer $\sin x$ gets to 1 and $\cos x$ gets to 0, so $\tan x$ approaches ∞ $\left(\lim_{x \to \frac{\pi}{2}^-} \tan x = \infty\right)$. In other words, the vertical line $x = \frac{\pi}{2}$ is a vertical asymptote to the graph of $y = \tan x$.

Table 10

x	$\sin x$	$\cos x$	$y = \tan x$
$\frac{\pi}{3} \approx 1.05$	$\frac{\sqrt{3}}{2}$	$\frac{1}{2}$	$\sqrt{3} \approx 1.73$
1.5	0.9975	0.0707	14.1
1.57	0.9999	7.96×10^{-4}	1255.8
1.5707	0.9999	9.6×10^{-5}	10,381
$\frac{\pi}{2} \approx 1.5708$	1	0	Undefined

If x is close to $-\frac{\pi}{2}$ but remains greater than $-\frac{\pi}{2}$, then $\sin x$ is close to -1, and $\cos x$ is positive and close to 0. The ratio $\frac{\sin x}{\cos x}$ approaches $-\infty$ $\left(\lim_{x \to -\frac{\pi}{2}^+} \tan x = -\infty\right)$. In other words, the vertical line $x = -\frac{\pi}{2}$ is also a vertical asymptote to the graph.

With these observations, one period of the graph can be completed. Obtain the complete graph of $y = \tan x$, $-\infty < x < \infty$, by repeating this period, as shown in Figure 102. Figure 103 shows the graph using a graphing utility.

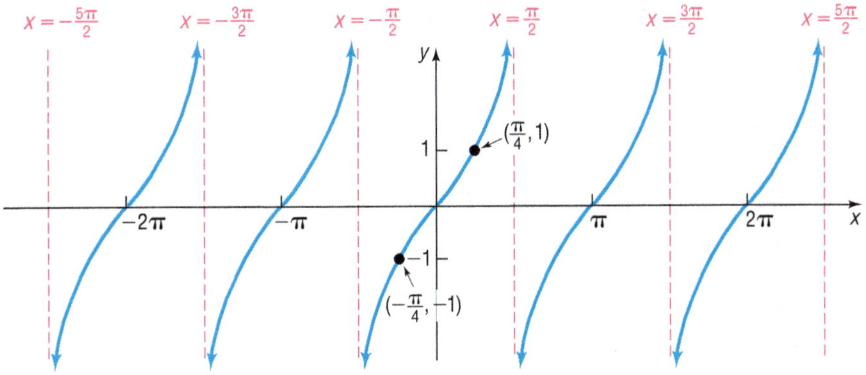

Figure 102 $y = \tan x$, $-\infty < x < \infty$, x not equal to odd multiples of $\frac{\pi}{2}$, $-\infty < y < \infty$

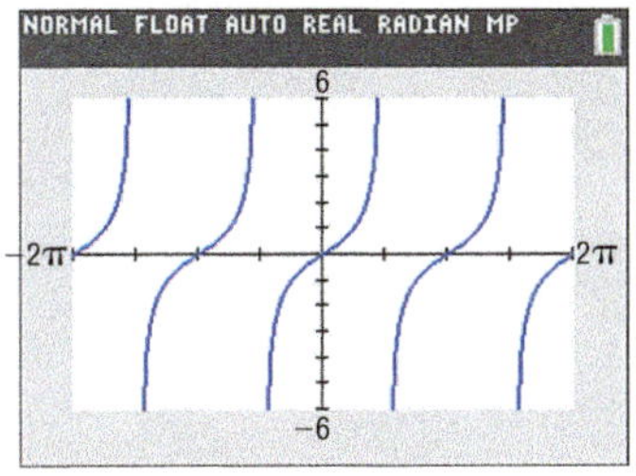

Figure 103

The graph of $y = \tan x$ in Figure 102 illustrates the following properties.

Properties of the Tangent Function

1. The domain is the set of all real numbers, except odd multiples of $\frac{\pi}{2}$.
2. The range is the set of all real numbers.
3. The tangent function is an odd function, as the symmetry of the graph with respect to the origin indicates.
4. The tangent function is periodic, with period π.
5. The x-intercepts are $\ldots, -2\pi, -\pi, 0, \pi, 2\pi, 3\pi, \ldots$; the y-intercept is 0.
6. Vertical asymptotes occur at $x = \ldots, -\frac{3\pi}{2}, -\frac{\pi}{2}, \frac{\pi}{2}, \frac{3\pi}{2}, \ldots$.

Now Work PROBLEMS 7 AND 15

1 Graph Functions of the Form $y = A\tan(\omega x) + B$ and $y = A\cot(\omega x) + B$

For tangent functions, there is no concept of amplitude since the range of the tangent function is $(-\infty, \infty)$. The role of A in $y = A\tan(\omega x) + B$ is to provide the magnitude of the vertical stretch. The period of $y = \tan x$ is π, so the period of $y = A\tan(\omega x) + B$ is $\frac{\pi}{\omega}$, caused by the horizontal compression (or stretch) of the graph by a factor of $\frac{1}{\omega}$. Finally, the presence of B indicates that a vertical shift is required.

EXAMPLE 1 **Graphing Functions of the Form $y = A\tan(\omega x) + B$**

Graph $y = 2\tan x - 1$. Use the graph to determine the domain and the range of the function $y = 2\tan x - 1$.

Solution Figure 104 shows the steps using transformations.

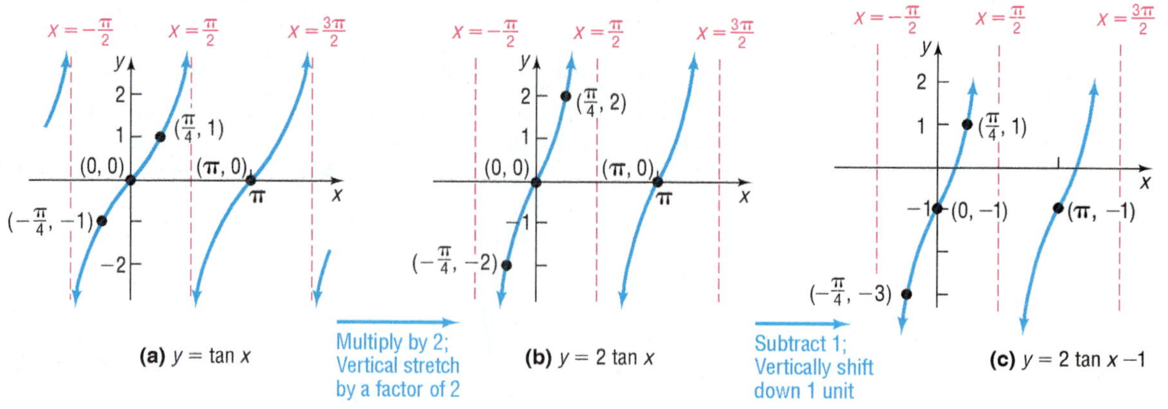

Figure 104

Check: Graph $Y_1 = 2\tan x - 1$ to verify the graph shown in Figure 104(c).

The domain of $y = 2\tan x - 1$ is $\left\{x \middle| x \neq \frac{k\pi}{2}, k \text{ is an odd integer}\right\}$, and the range is the set of all real numbers, or $(-\infty, \infty)$.

EXAMPLE 2 **Graphing Functions of the Form $y = A\tan(\omega x) + B$**

Graph $y = 3\tan(2x)$. Use the graph to determine the domain and the range of $y = 3\tan(2x)$.

Solution Figure 105 shows the steps using transformations.

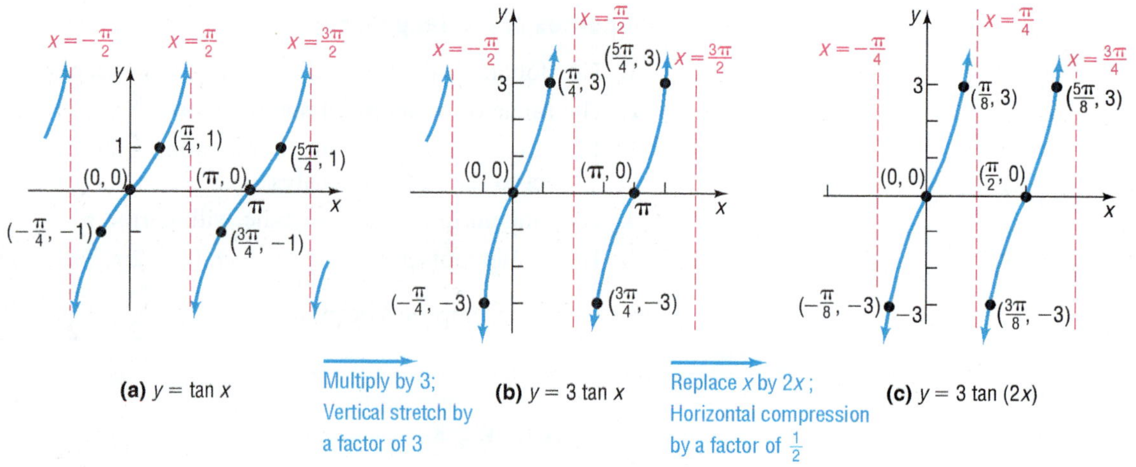

Figure 105

The domain of $y = 3\tan(2x)$ is $\left\{x \middle| x \neq \frac{k\pi}{4}, k \text{ is an odd integer}\right\}$, and the range is the set of all real numbers, or $(-\infty, \infty)$.

Check: Graph $Y_1 = 3\tan(2x)$ to verify the graph in Figure 105(c).

Note in Figure 105(c) that the period of $y = 3\tan(2x)$ is $\frac{\pi}{2}$ because of the compression of the original period π by a factor of $\frac{1}{2}$. Notice that the asymptotes are $x = -\frac{\pi}{4}, x = \frac{\pi}{4}, x = \frac{3\pi}{4}$, and so on, also because of the compression.

Now Work PROBLEM 21

The Graph of the Cotangent Function $y = \cot x$

Table 11

x	$y = \cot x$	(x, y)
$\frac{\pi}{6}$	$\sqrt{3}$	$\left(\frac{\pi}{6}, \sqrt{3}\right)$
$\frac{\pi}{4}$	1	$\left(\frac{\pi}{4}, 1\right)$
$\frac{\pi}{3}$	$\frac{\sqrt{3}}{3}$	$\left(\frac{\pi}{3}, \frac{\sqrt{3}}{3}\right)$
$\frac{\pi}{2}$	0	$\left(\frac{\pi}{2}, 0\right)$
$\frac{2\pi}{3}$	$-\frac{\sqrt{3}}{3}$	$\left(\frac{2\pi}{3}, -\frac{\sqrt{3}}{3}\right)$
$\frac{3\pi}{4}$	-1	$\left(\frac{3\pi}{4}, -1\right)$
$\frac{5\pi}{6}$	$-\sqrt{3}$	$\left(\frac{5\pi}{6}, -\sqrt{3}\right)$

The graph of $y = \cot x$ can be obtained in the same manner as the graph of $y = \tan x$. The period of $y = \cot x$ is π. Because the cotangent function is not defined for integer multiples of π, concentrate on the interval $(0, \pi)$. Table 11 lists some points on the graph of $y = \cot x$, $0 < x < \pi$. As x approaches 0 but remains greater than 0, the value of $\cos x$ will be close to 1, and the value of $\sin x$ will be positive and close to 0. The ratio $\frac{\cos x}{\sin x} = \cot x$ will be positive and large; so as x approaches 0, with $x > 0$, then $\cot x$ approaches ∞ $\left(\lim_{x \to 0^+} \cot x = \infty\right)$. Similarly, as x approaches π but remains less than π, the value of $\cos x$ will be close to -1, and the value of $\sin x$ will be positive and close to 0. So the ratio $\frac{\cos x}{\sin x} = \cot x$ will be negative and will approach $-\infty$ as x approaches π $\left(\lim_{x \to \pi^-} \cot x = -\infty\right)$. Figure 106 shows the graph of $y = \cot x$. Figure 107 shows the graph using a graphing utility.

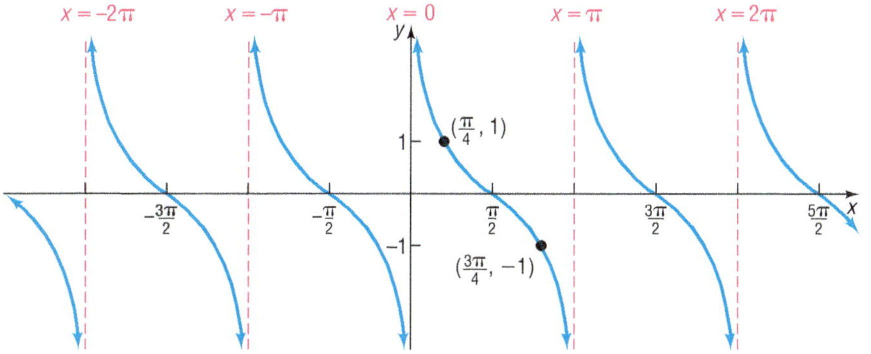

Figure 106 $y = \cot x$, $-\infty < x < \infty$, x not equal to integer multiples of π, $-\infty < y < \infty$

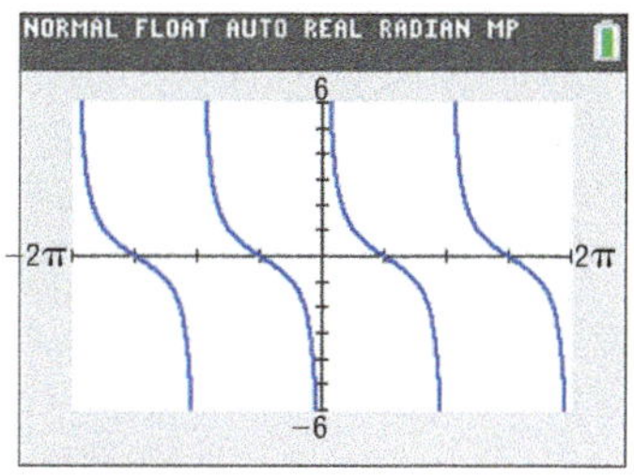

Figure 107

The graph of $y = A\cot(\omega x) + B$ has characteristics similar to those of the tangent function. The cotangent function $y = A\cot(\omega x) + B$ has period $\frac{\pi}{\omega}$. The cotangent function has no amplitude. The role of A is to provide the magnitude of the vertical stretch; the presence of B indicates that a vertical shift is required.

Now Work PROBLEM 23

The Graphs of the Cosecant Function and the Secant Function

The cosecant and secant functions, sometimes referred to as **reciprocal functions**, are graphed by making use of the reciprocal identities

$$\csc x = \frac{1}{\sin x} \quad \text{and} \quad \sec x = \frac{1}{\cos x}$$

For example, the value of the cosecant function $y = \csc x$ at a given number x equals the reciprocal of the corresponding value of the sine function, provided that the value of the sine function is not 0. If the value of $\sin x$ is 0, then x is an integer multiple of π. At such numbers, the cosecant function is not defined. In fact, the graph of the cosecant function has vertical asymptotes at integer multiples of π. Figure 108 on the next page shows the graph.

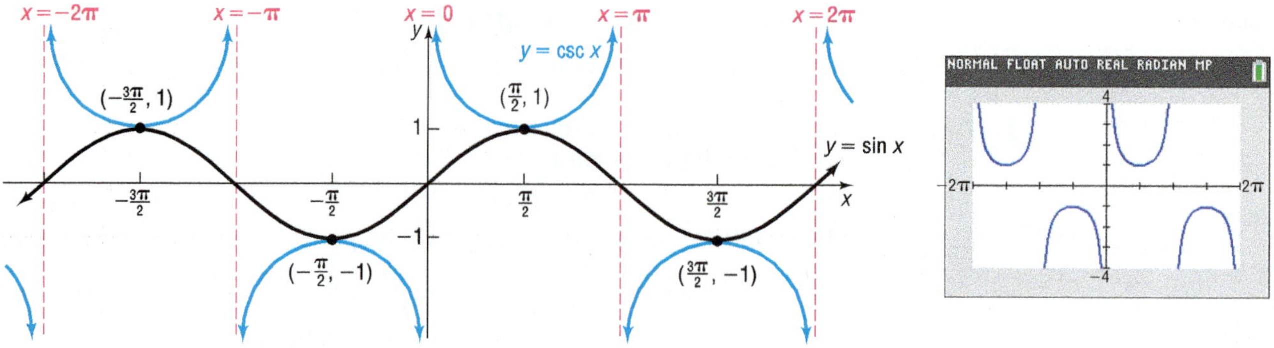

Figure 108 $y = \csc x$, $-\infty < x < \infty$, x not equal to integer multiples of π, $|y| \geq 1$

Using the idea of reciprocals, the graph of $y = \sec x$ can be obtained in a similar manner. See Figure 109.

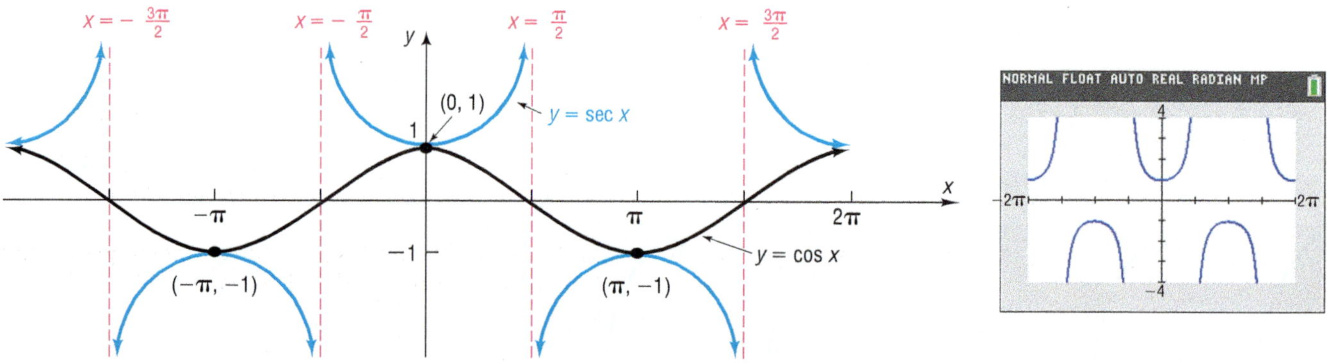

Figure 109 $y = \sec x$, $-\infty < x < \infty$, x not equal to odd multiples of $\frac{\pi}{2}$, $|y| \geq 1$

2 Graph Functions of the Form $y = A\csc(\omega x) + B$ and $y = A\sec(\omega x) + B$

The role of A in these functions is to set the range. The range of $y = \csc x$ is $\{y \mid y \leq -1 \text{ or } y \geq 1\}$ or $\{y \mid |y| \geq 1\}$; the range of $y = A\csc x$ is $\{y \mid |y| \geq |A|\}$ because of the vertical stretch of the graph by a factor of $|A|$. Just as with the sine and cosine functions, the period of $y = \csc(\omega x)$ and $y = \sec(\omega x)$ becomes $\frac{2\pi}{\omega}$ because of the horizontal compression of the graph by a factor of $\frac{1}{\omega}$. The presence of B indicates that a vertical shift is required.

EXAMPLE 3 **Graphing Functions of the Form $y = A\csc(\omega x) + B$**

Graph $y = 2\csc x - 1$. Use the graph to determine the domain and the range of $y = 2\csc x - 1$.

Solution We use transformations. Figure 110 shows the required steps.

The domain of $y = 2\csc x - 1$ is $\{x \mid x \neq k\pi, k \text{ is an integer}\}$ and the range is $\{y \mid y \leq -3 \text{ or } y \geq 1\}$, or, using interval notation, $(-\infty, -3] \cup [1, \infty)$.

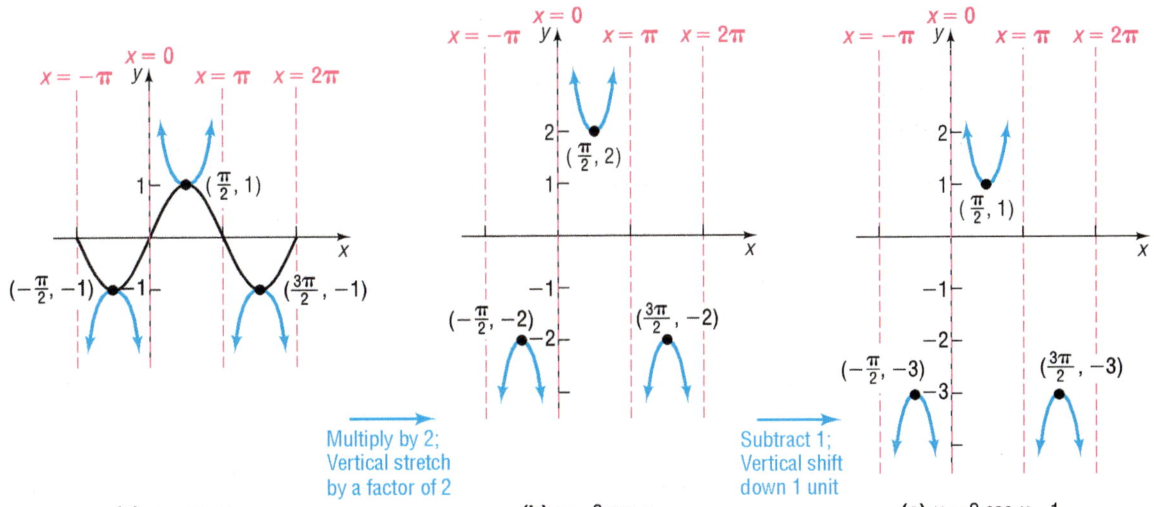

(a) $y = \csc x$ **(b)** $y = 2 \csc x$ **(c)** $y = 2 \csc x - 1$

Figure 110

Check: Graph $Y_1 = 2 \csc x - 1$ to verify the graph shown in Figure 110(c). ■

Now Work PROBLEM 29

7.7 Assess Your Understanding

'Are You Prepared?' *Answers are given at the end of these exercises. If you get a wrong answer, read the pages listed in red.*

1. The graph of $y = \dfrac{3x - 6}{x - 4}$ has a vertical asymptote. What is it? (pp. 374–377)

2. ***True or False*** If $x = 3$ is a vertical asymptote of a rational function R, then $\lim_{x \to 3} |R(x)| = \infty$. (pp. 374–377)

Concepts and Vocabulary

3. The graph of $y = \tan x$ is symmetric with respect to the ________ and has vertical asymptotes at ________________.

4. The graph of $y = \sec x$ is symmetric with respect to the ________ and has vertical asymptotes at ________________.

5. It is easiest to graph $y = \sec x$ by first sketching the graph of ______.
(a) $y = \sin x$ (b) $y = \cos x$ (c) $y = \tan x$ (d) $y = \csc x$

6. ***True or False*** The graphs of $y = \tan x$, $y = \cot x$, $y = \sec x$, and $y = \csc x$ each have infinitely many vertical asymptotes.

Skill Building

In Problems 7–16, if necessary, refer to the graphs of the functions to answer each question.

7. What is the y-intercept of $y = \tan x$?

8. What is the y-intercept of $y = \cot x$?

9. What is the y-intercept of $y = \sec x$?

10. What is the y-intercept of $y = \csc x$?

11. For what numbers x, $-2\pi \le x \le 2\pi$, does $\sec x = 1$? For what numbers x does $\sec x = -1$?

12. For what numbers x, $-2\pi \le x \le 2\pi$, does $\csc x = 1$? For what numbers x does $\csc x = -1$?

13. For what numbers x, $-2\pi \le x \le 2\pi$, does the graph of $y = \sec x$ have vertical asymptotes?

14. For what numbers x, $-2\pi \le x \le 2\pi$, does the graph of $y = \csc x$ have vertical asymptotes?

15. For what numbers x, $-2\pi \le x \le 2\pi$, does the graph of $y = \tan x$ have vertical asymptotes?

16. For what numbers x, $-2\pi \le x \le 2\pi$, does the graph of $y = \cot x$ have vertical asymptotes?

In Problems 17–40, graph each function. Be sure to label key points and show at least two cycles. Use the graph to determine the domain and the range of each function.

17. $y = 3 \tan x$

18. $y = -2 \tan x$

19. $y = 4 \cot x$

20. $y = -3 \cot x$

21. $y = \tan\left(\dfrac{\pi}{2}x\right)$

22. $y = \tan\left(\dfrac{1}{2}x\right)$

23. $y = \cot\left(\frac{1}{4}x\right)$

24. $y = \cot\left(\frac{\pi}{4}x\right)$

25. $y = 2 \sec x$

26. $y = \frac{1}{2} \csc x$

27. $y = -3 \csc x$

28. $y = -4 \sec x$

29. $y = 4 \sec\left(\frac{1}{2}x\right)$

30. $y = \frac{1}{2} \csc(2x)$

31. $y = -2 \csc(\pi x)$

32. $y = -3 \sec\left(\frac{\pi}{2}x\right)$

33. $y = \tan\left(\frac{1}{4}x\right) + 1$

34. $y = 2 \cot x - 1$

35. $y = \sec\left(\frac{2\pi}{3}x\right) + 2$

36. $y = \csc\left(\frac{3\pi}{2}x\right)$

37. $y = \frac{1}{2} \tan\left(\frac{1}{4}x\right) - 2$

38. $y = 3 \cot\left(\frac{1}{2}x\right) - 2$

39. $y = 2 \csc\left(\frac{1}{3}x\right) - 1$

40. $y = 3 \sec\left(\frac{1}{4}x\right) + 1$

Mixed Practice

In Problems 41–44, find the average rate of change of f from 0 to $\frac{\pi}{6}$.

41. $f(x) = \tan x$

42. $f(x) = \sec x$

43. $f(x) = \tan(2x)$

44. $f(x) = \sec(2x)$

In Problems 45–48, find $(f \circ g)(x)$ and $(g \circ f)(x)$, and graph each of these functions.

45. $f(x) = \tan x$
$g(x) = 4x$

46. $f(x) = 2 \sec x$
$g(x) = \frac{1}{2}x$

47. $f(x) = -2x$
$g(x) = \cot x$

48. $f(x) = \frac{1}{2}x$
$g(x) = 2 \csc x$

In Problems 49 and 50, graph each function.

49. $f(x) = \begin{cases} \tan x & 0 \le x < \frac{\pi}{2} \\ 0 & x = \frac{\pi}{2} \\ \sec x & \frac{\pi}{2} < x \le \pi \end{cases}$

50. $g(x) = \begin{cases} \csc x & 0 < x < \pi \\ 0 & x = \pi \\ \cot x & \pi < x < 2\pi \end{cases}$

Applications and Extensions

51. **Carrying a Ladder around a Corner** Two hallways, one of width 3 feet, the other of width 4 feet, meet at a right angle. See the illustration.

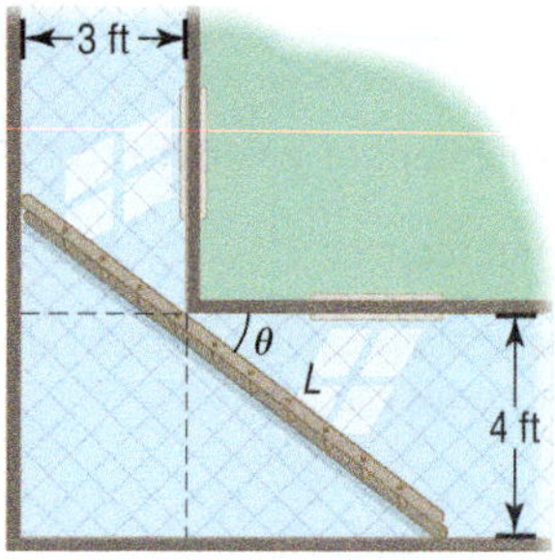

(a) Show that the length L of the ladder shown as a function of the angle θ is

$$L(\theta) = 3 \sec \theta + 4 \csc \theta$$

(b) Graph $L = L(\theta), 0 < \theta < \frac{\pi}{2}$.

(c) For what value of θ is L the least?

(d) What is the length of the longest ladder that can be carried around the corner? Why is this also the least value of L?

52. **A Rotating Beacon** Suppose that a fire truck is parked in front of a building as shown in the figure.

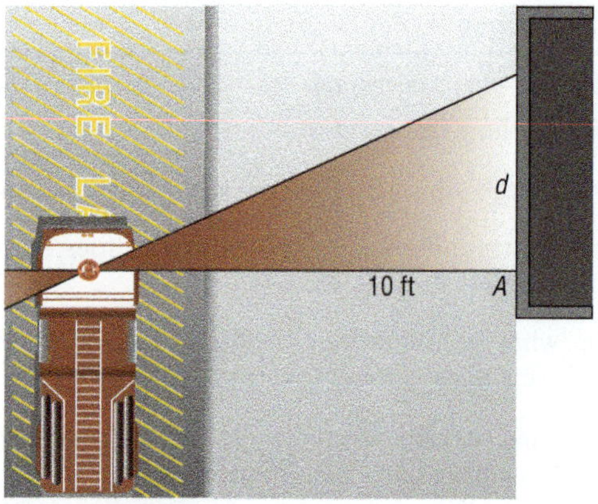

The beacon light on top of the fire truck is located 10 feet from the wall and has a light on each side. If the beacon light rotates 1 revolution every 2 seconds, then a model for determining the distance d, in feet, that the beacon of light is from point A on the wall after t seconds is given by

$$d(t) = |10 \tan(\pi t)|$$

(a) Graph $d(t) = |10 \tan(\pi t)|$ for $0 \le t \le 2$.

(b) For what values of t is the function undefined? Explain what this means in terms of the beam of light on the wall.

(c) Fill in the following table.

t	0	0.1	0.2	0.3	0.4
$d(t) = 10\tan(\pi t)$					

(d) Compute $\dfrac{d(0.1) - d(0)}{0.1 - 0}$, $\dfrac{d(0.2) - d(0.1)}{0.2 - 0.1}$, and so on, for each consecutive value of t. These are called **first differences**.

(e) Interpret the first differences found in part (d). What is happening to the speed of the beam of light as d increases?

53. Exploration Graph

$$y = \tan x \quad \text{and} \quad y = -\cot\left(x + \frac{\pi}{2}\right)$$

Do you think that $\tan x = -\cot\left(x + \dfrac{\pi}{2}\right)$?

Retain Your Knowledge

Problems 54–57 are based on material learned earlier in the course. The purpose of these problems is to keep the material fresh in your mind so that you are better prepared for the final exam.

54. Factor: $125p^3 - 8q^6$

55. Painting a Room Hazel can paint a room in 2 hours less time than her friend Gwyneth. Working together, they can paint the room in 2.4 hours. How long does it take each woman to paint the room by herself?

56. Solve: $9^{x-1} = 3^{x^2-5}$

57. Use the slope and the y-intercept to graph the linear function $f(x) = \frac{1}{4}x - 3$.

'Are You Prepared?' Answers

1. $x = 4$ **2.** True

7.8 Phase Shift; Sinusoidal Curve Fitting

OBJECTIVES 1 Graph Sinusoidal Functions of the Form $y = A\sin(\omega x - \phi) + B$ (p. 595)
2 Build Sinusoidal Models from Data (p. 599)

1 Graph Sinusoidal Functions of the Form $y = A\sin(\omega x - \phi) + B$

We have seen that the graph of $y = A\sin(\omega x)$, $\omega > 0$, has amplitude $|A|$ and period $T = \dfrac{2\pi}{\omega}$. One cycle can be drawn as x varies from 0 to $\dfrac{2\pi}{\omega}$, or, equivalently, as ωx varies from 0 to 2π. See Figure 111.

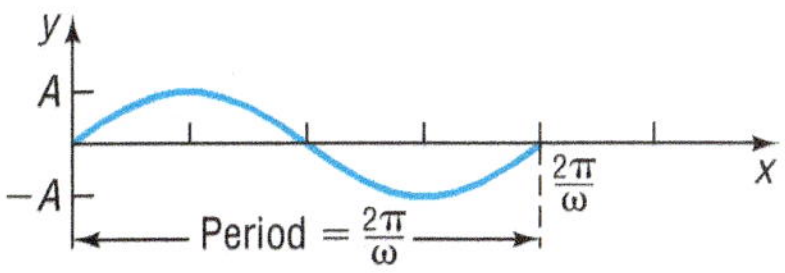

Figure 111 One cycle of $y = A\sin(\omega x)$, $A > 0$, $\omega > 0$

Now consider the graph of

$$y = A\sin(\omega x - \phi)$$

which may also be written as

$$y = A\sin\left[\omega\left(x - \frac{\phi}{\omega}\right)\right]$$

where $\omega > 0$ and ϕ (the Greek letter phi) are real numbers. The graph is a sine curve with amplitude $|A|$. As $\omega x - \phi$ varies from 0 to 2π, one period will be traced out. This period begins when

$$\omega x - \phi = 0 \quad \text{or} \quad x = \frac{\phi}{\omega}$$

Note: The beginning and end of the period can also be found by solving the inequality:

$$0 \le \omega x - \phi \le 2\pi$$
$$\phi \le \omega x \le 2\pi + \phi$$
$$\frac{\phi}{\omega} \le x \le \frac{2\pi}{\omega} + \frac{\phi}{\omega}$$ ■

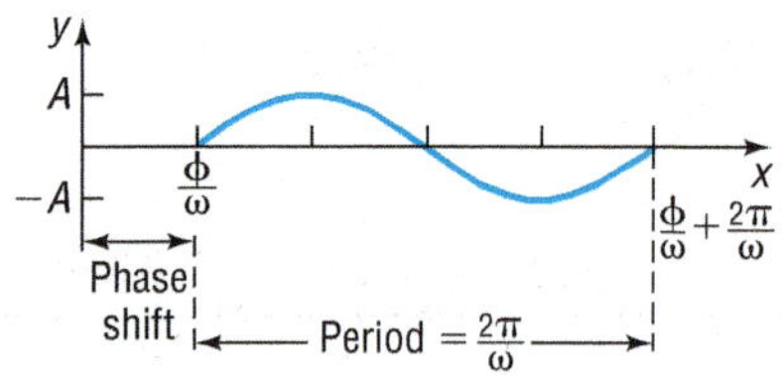

Figure 112 One cycle of $y = A\sin(\omega x - \phi)$, $A > 0$, $\omega > 0$, $\phi > 0$

and ends when

$$\omega x - \phi = 2\pi \quad \text{or} \quad x = \frac{\phi}{\omega} + \frac{2\pi}{\omega}$$

See Figure 112.

Notice that the graph of $y = A\sin(\omega x - \phi) = A\sin\left[\omega\left(x - \frac{\phi}{\omega}\right)\right]$ is the same as the graph of $y = A\sin(\omega x)$, except that it has been shifted $\left|\frac{\phi}{\omega}\right|$ units (to the right if $\phi > 0$ and to the left if $\phi < 0$). This number $\frac{\phi}{\omega}$ is called the **phase shift** of the graph of $y = A\sin(\omega x - \phi)$.

For the graphs of $y = A\sin(\omega x - \phi)$ or $y = A\cos(\omega x - \phi)$, $\omega > 0$,

$$\text{Amplitude} = |A| \qquad \text{Period} = T = \frac{2\pi}{\omega} \qquad \text{Phase shift} = \frac{\phi}{\omega}$$

The phase shift is to the left if $\phi < 0$ and to the right if $\phi > 0$.

EXAMPLE 1 Finding the Amplitude, Period, and Phase Shift of a Sinusoidal Function and Graphing It

Find the amplitude, period, and phase shift of $y = 3\sin(2x - \pi)$, and graph the function.

Solution Use the same four steps used to graph sinusoidal functions of the form $y = A\sin(\omega x)$ or $y = A\cos(\omega x)$ given on page 579.

Step 1: Comparing

$$y = 3\sin(2x - \pi) = 3\sin\left[2\left(x - \frac{\pi}{2}\right)\right]$$

to

$$y = A\sin(\omega x - \phi) = A\sin\left[\omega\left(x - \frac{\phi}{\omega}\right)\right]$$

note that $A = 3$, $\omega = 2$, and $\phi = \pi$. The graph is a sine curve with amplitude $|A| = 3$, period $T = \frac{2\pi}{\omega} = \frac{2\pi}{2} = \pi$, and phase shift $= \frac{\phi}{\omega} = \frac{\pi}{2}$.

Step 2: The graph of $y = 3\sin(2x - \pi)$ will lie between -3 and 3 on the y-axis. One cycle will begin at $x = \frac{\phi}{\omega} = \frac{\pi}{2}$ and end at $x = \frac{\phi}{\omega} + \frac{2\pi}{\omega} = \frac{\pi}{2} + \pi = \frac{3\pi}{2}$. To find the five key points, divide the interval $\left[\frac{\pi}{2}, \frac{3\pi}{2}\right]$ into four subintervals, each of length $\pi \div 4 = \frac{\pi}{4}$, by finding the following values of x:

$\frac{\pi}{2}$	$\frac{\pi}{2} + \frac{\pi}{4} = \frac{3\pi}{4}$	$\frac{3\pi}{4} + \frac{\pi}{4} = \pi$	$\pi + \frac{\pi}{4} = \frac{5\pi}{4}$	$\frac{5\pi}{4} + \frac{\pi}{4} = \frac{3\pi}{2}$
1st x-coordinate	2nd x-coordinate	3rd x-coordinate	4th x-coordinate	5th x-coordinate

Note: The interval defining one cycle can also be found by solving the inequality

$$0 \le 2x - \pi \le 2\pi$$

Then

$$\pi \le 2x \le 3\pi$$
$$\frac{\pi}{2} \le x \le \frac{3\pi}{2}$$

Step 3: Use these values of x to determine the five key points on the graph:

$$\left(\frac{\pi}{2}, 0\right) \quad \left(\frac{3\pi}{4}, 3\right) \quad (\pi, 0) \quad \left(\frac{5\pi}{4}, -3\right) \quad \left(\frac{3\pi}{2}, 0\right)$$

Step 4: Plot these five points and fill in the graph of the sine function as shown in Figure 113(a). Extend the graph in each direction to obtain Figure 113(b).

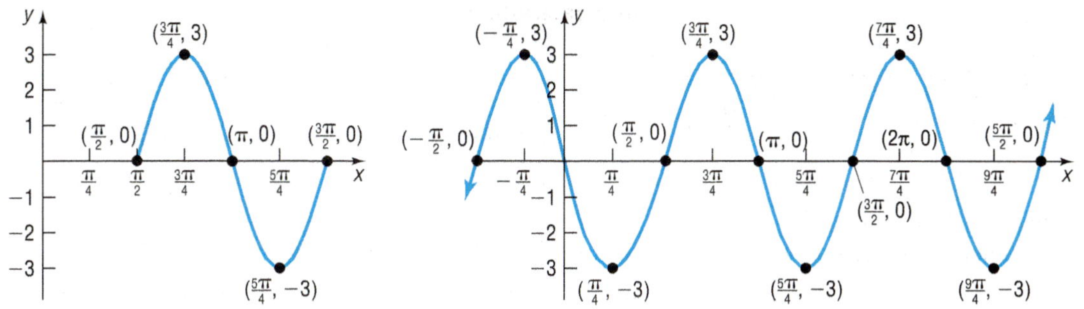

Figure 113 (a) (b) $y = 3\sin(2x - \pi)$

The graph of $y = 3\sin(2x - \pi) = 3\sin\left[2\left(x - \frac{\pi}{2}\right)\right]$ may also be obtained using transformations. See Figure 114.

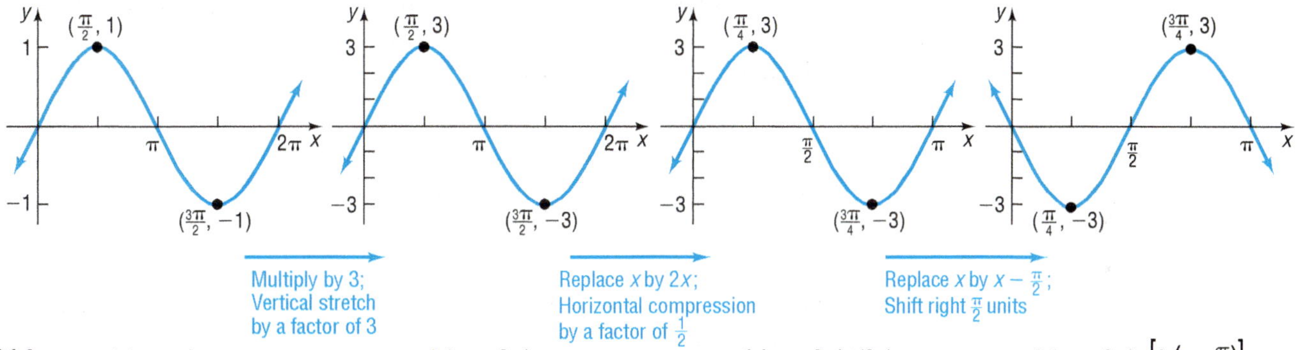

Figure 114 (a) $y = \sin x$ (b) $y = 3\sin x$ (c) $y = 3\sin(2x)$ (d) $y = 3\sin\left[2\left(x - \frac{\pi}{2}\right)\right] = 3\sin(2x - \pi)$

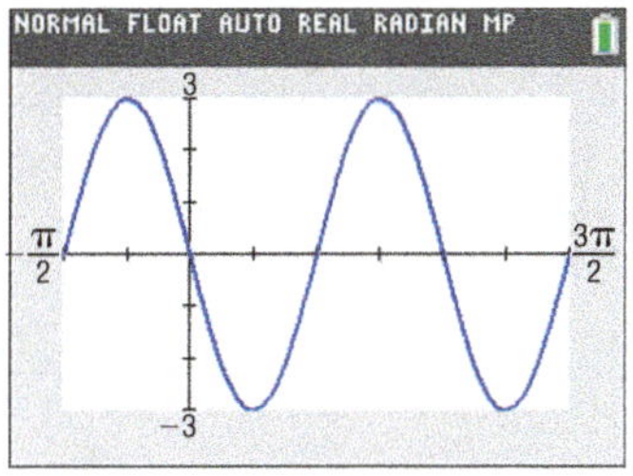

Figure 115

✓**Check:** Figure 115 shows the graph of $Y_1 = 3\sin(2x - \pi)$ using a graphing utility. ■

To graph a sinusoidal function of the form $y = A\sin(\omega x - \phi) + B$, first graph the function $y = A\sin(\omega x - \phi)$ and then apply a vertical shift.

EXAMPLE 2 Finding the Amplitude, Period, and Phase Shift of a Sinusoidal Function and Graphing It

Find the amplitude, period, and phase shift of $y = 2\cos(4x + 3\pi) + 1$, and graph the function.

Solution **Step 1:** Begin by comparing

$$y = 2\cos(4x + 3\pi) = 2\cos\left[4\left(x + \frac{3\pi}{4}\right)\right]$$

to

$$y = A\cos(\omega x - \phi) = A\cos\left[\omega\left(x - \frac{\phi}{\omega}\right)\right]$$

Note that $A = 2$, $\omega = 4$, and $\phi = -3\pi$. The graph is a cosine curve with amplitude $|A| = 2$, period $T = \frac{2\pi}{\omega} = \frac{2\pi}{4} = \frac{\pi}{2}$, and phase shift $= \frac{\phi}{\omega} = -\frac{3\pi}{4}$.

Note: The interval defining one cycle can also be found by solving the inequality

$$0 \le 4x + 3\pi \le 2\pi$$

Then

$$-3\pi \le 4x \le -\pi$$

$$-\frac{3\pi}{4} \le x \le -\frac{\pi}{4}$$

■

STEP 2: The graph of $y = 2\cos(4x + 3\pi)$ will lie between -2 and 2 on the y-axis. One cycle begins at $x = \frac{\phi}{\omega} = -\frac{3\pi}{4}$ and ends at $x = \frac{\phi}{\omega} + \frac{2\pi}{\omega} = -\frac{3\pi}{4} + \frac{\pi}{2} = -\frac{\pi}{4}$. To find the five key points, divide the interval $\left[-\frac{3\pi}{4}, -\frac{\pi}{4}\right]$ into four subintervals, each of the length $\frac{\pi}{2} \div 4 = \frac{\pi}{8}$, by finding the following values.

$-\frac{3\pi}{4}$	$-\frac{3\pi}{4} + \frac{\pi}{8} = -\frac{5\pi}{8}$	$-\frac{5\pi}{8} + \frac{\pi}{8} = -\frac{\pi}{2}$	$-\frac{\pi}{2} + \frac{\pi}{8} = -\frac{3\pi}{8}$	$-\frac{3\pi}{8} + \frac{\pi}{8} = -\frac{\pi}{4}$
1st x-coordinate	2nd x-coordinate	3rd x-coordinate	4th x-coordinate	5th x-coordinate

STEP 3: The five key points on the graph of $y = 2\cos(4x + 3\pi)$ are

$$\left(-\frac{3\pi}{4}, 2\right) \quad \left(-\frac{5\pi}{8}, 0\right) \quad \left(-\frac{\pi}{2}, -2\right) \quad \left(-\frac{3\pi}{8}, 0\right) \quad \left(-\frac{\pi}{4}, 2\right)$$

STEP 4: Plot these five points and fill in the graph of the cosine function as shown in Figure 116(a). Extend the graph in each direction to obtain Figure 116(b), the graph of $y = 2\cos(4x + 3\pi)$.

STEP 5: A vertical shift up 1 unit gives the final graph. See Figure 116(c). The graph of $y = 2\cos(4x + 3\pi) + 1 = 2\cos\left[4\left(x + \frac{3\pi}{4}\right)\right] + 1$ may also be obtained using transformations. See Figure 117.

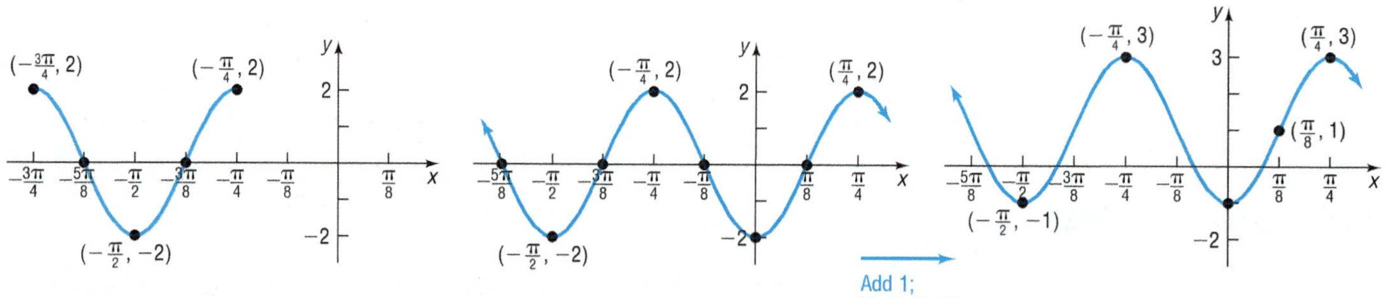

Figure 116 (a) (b) $y = 2\cos(4x + 3\pi)$ (c) $y = 2\cos(4x + 3\pi) + 1$

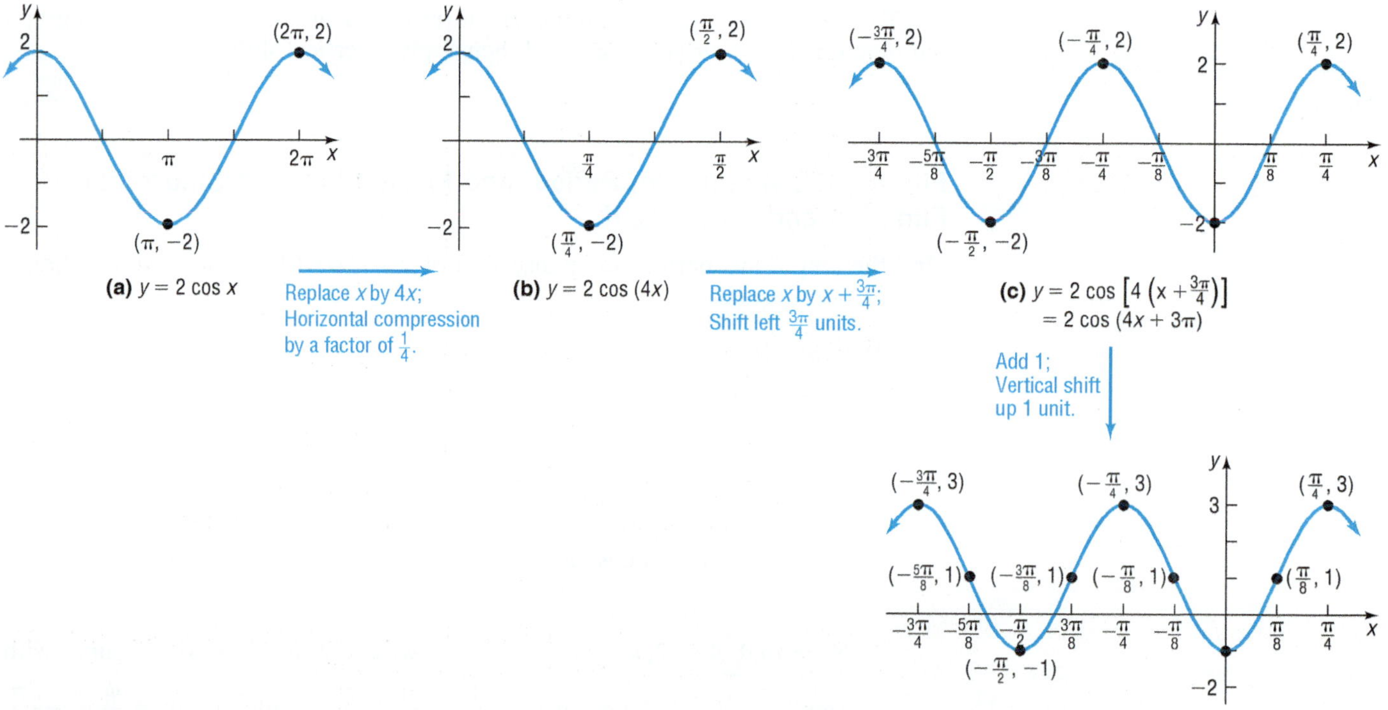

Figure 117 (a) $y = 2\cos x$ (b) $y = 2\cos(4x)$ (c) $y = 2\cos\left[4\left(x + \frac{3\pi}{4}\right)\right] = 2\cos(4x + 3\pi)$ (d) $y = 2\cos(4x + 3\pi) + 1$

■

Now Work PROBLEM 3

SUMMARY

Steps for Graphing Sinusoidal Functions $y = A\sin(\omega x - \phi) + B$ or $y = A\cos(\omega x - \phi) + B$

STEP 1: Determine the amplitude $|A|$, period $T = \dfrac{2\pi}{\omega}$, and phase shift $\dfrac{\phi}{\omega}$.

STEP 2: Determine the starting point of one cycle of the graph, $\dfrac{\phi}{\omega}$. Determine the ending point of one cycle of the graph, $\dfrac{\phi}{\omega} + \dfrac{2\pi}{\omega}$. Divide the interval $\left[\dfrac{\phi}{\omega}, \dfrac{\phi}{\omega} + \dfrac{2\pi}{\omega}\right]$ into four subintervals, each of length $\dfrac{2\pi}{\omega} \div 4$.

STEP 3: Use the endpoints of the subintervals to find the five key points on the graph.

STEP 4: Plot the five key points, and connect them with a sinusoidal graph to obtain one cycle of the graph. Extend the graph in each direction to make it complete.

STEP 5: If $B \neq 0$, apply a vertical shift.

2 Build Sinusoidal Models from Data

Scatter diagrams of data sometimes resemble the graph of a sinusoidal function. Let's look at an example.

The data given in Table 12 represent the average monthly temperatures in Denver, Colorado. Since the data represent *average* monthly temperatures collected over many years, the data will not vary much from year to year and so will essentially repeat each year. In other words, the data are periodic. Figure 118 shows the scatter diagram of these data, where $x = 1$ represents January, $x = 2$ represents February, and so on.

Table 12

Month, x	Average Monthly Temperature, °F
January, 1	30.7
February, 2	32.5
March, 3	40.4
April, 4	47.4
May, 5	57.1
June, 6	67.4
July, 7	74.2
August, 8	72.5
September, 9	62.4
October, 10	50.9
November, 11	38.3
December, 12	30.0

Source: *U.S. National Oceanic and Atmospheric Administration*

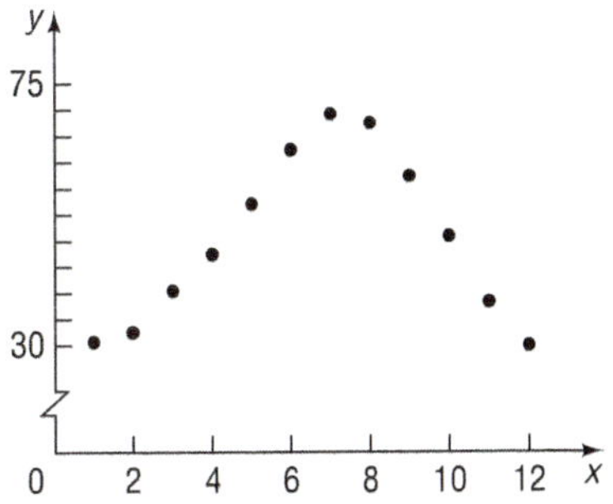

Figure 118 Denver average monthly temperature

Notice that the scatter diagram looks like the graph of a sinusoidal function. We choose to fit the data to a sine function of the form

$$y = A\sin(\omega x - \phi) + B$$

where A, B, ω, and ϕ are constants.

EXAMPLE 3 Finding a Sinusoidal Function from Temperature Data

Fit a sine function to the data in Table 12.

Solution Begin with a scatter diagram of the data for one year. See Figure 119. The data will be fitted to a sine function of the form

$$y = A\sin(\omega x - \phi) + B$$

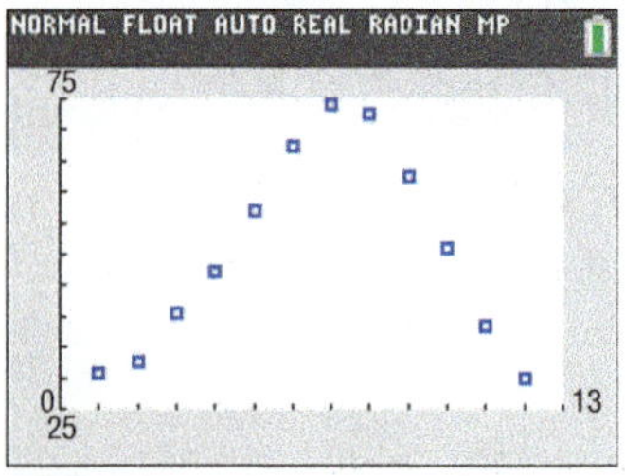

Figure 119

STEP 1: To find the amplitude A, compute

$$\text{Amplitude} = \frac{\text{largest data value} - \text{smallest data value}}{2}$$

$$= \frac{74.2 - 30.0}{2} = 22.1$$

To see the remaining steps in this process, superimpose the graph of the function $y = 22.1\sin x$, where x represents months, on the scatter diagram.

Figure 120 shows the two graphs. To fit the data, the graph needs to be shifted vertically, shifted horizontally, and stretched horizontally.

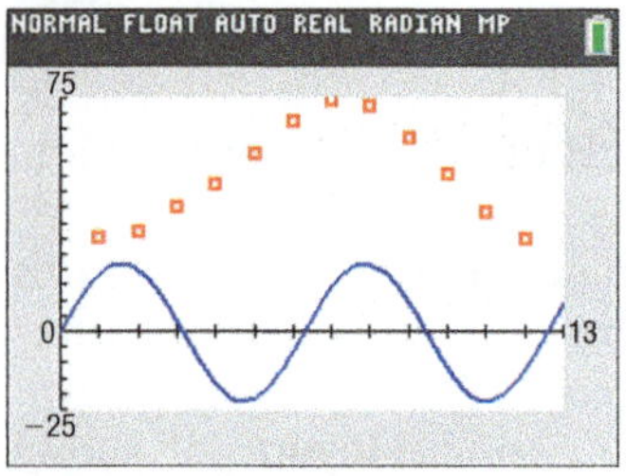

Figure 120

STEP 2: Determine the vertical shift by finding the average of the highest and lowest data values.

$$\text{Vertical shift} = \frac{74.2 + 30.0}{2} = 52.1$$

Now superimpose the graph of $y = 22.1\sin x + 52.1$ on the scatter diagram. See Figure 121.

We see that the graph needs to be shifted horizontally and stretched horizontally.

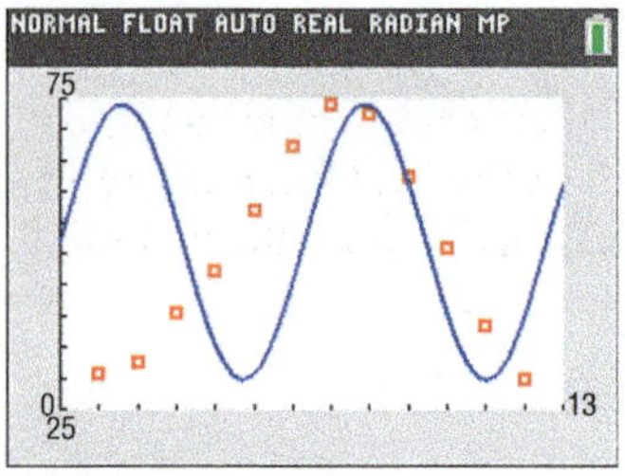

Figure 121

STEP 3: It is easier to find the horizontal stretch factor first. Since the temperatures repeat every 12 months, the period of the function is $T = 12$. Because $T = \frac{2\pi}{\omega} = 12$, this means

$$\omega = \frac{2\pi}{12} = \frac{\pi}{6}$$

Now superimpose the graph of $y = 22.1\sin\left(\frac{\pi}{6}x\right) + 52.1$ on the scatter diagram. See Figure 122, where it is clear that the graph still needs to be shifted horizontally.

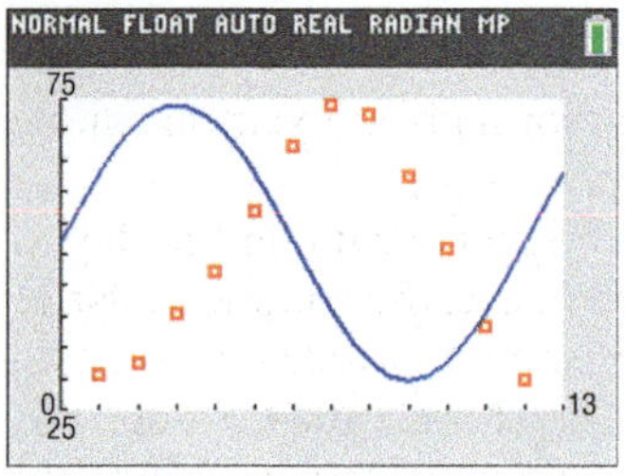

Figure 122

STEP 4: To determine the horizontal shift, use the period $T = 12$ and divide the interval $[0, 12]$ into four subintervals of length $12 \div 4 = 3$:

$$[0, 3] \quad [3, 6] \quad [6, 9] \quad [9, 12]$$

The sine curve is increasing on the interval $[0, 3]$ and is decreasing on the interval $[3, 9]$, so a local maximum occurs at $x = 3$. The data indicate that a maximum occurs at $x = 7$ (corresponding to July's temperature), so the graph of the function must be shifted 4 units to the right by replacing x by $x - 4$. Doing this yields

$$y = 22.1\sin\left(\frac{\pi}{6}(x - 4)\right) + 52.1$$

Multiplying out reveals that a sine function of the form $y = A\sin(\omega x - \phi) + B$ that fits the data is

$$y = 22.1\sin\left(\frac{\pi}{6}x - \frac{2\pi}{3}\right) + 52.1$$

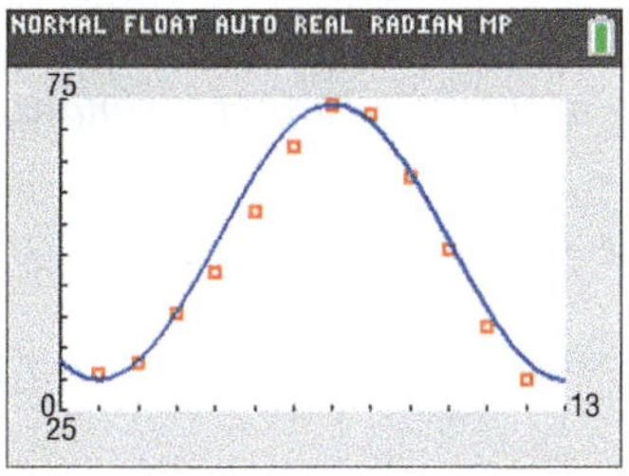

Figure 123

The graph of $y = 22.1 \sin\left(\frac{\pi}{6}x - \frac{2\pi}{3}\right) + 52.1$ and the scatter diagram of the data are shown in Figure 123. ■

The steps to fit a sine function

$$y = A \sin(\omega x - \phi) + B$$

to sinusoidal data follow.

Steps for Fitting a Sine Function $y = A \sin(\omega x - \phi) + B$ to Data

Step 1: Determine A, the amplitude of the function.

$$\text{Amplitude} = \frac{\text{largest data value} - \text{smallest data value}}{2}$$

Step 2: Determine B, the vertical shift of the function.

$$\text{Vertical shift} = \frac{\text{largest data value} + \text{smallest data value}}{2}$$

Step 3: Determine ω. Since the period T, the time it takes for the data to repeat, is $T = \frac{2\pi}{\omega}$, we have

$$\omega = \frac{2\pi}{T}$$

Step 4: Determine the horizontal shift of the function by using the period of the data. Divide the period into four subintervals of equal length. Determine the x-coordinate for the maximum of the sine function and the x-coordinate for the maximum value of the data. Use this information to determine the value of the phase shift, $\frac{\phi}{\omega}$.

Now Work PROBLEMS 29(a)–(c)

Let's look at another example. Since the number of hours of sunlight in a day cycles annually, the number of hours of sunlight in a day for a given location can be modeled by a sinusoidal function.

The longest day of the year (in terms of hours of sunlight) occurs on the day of the summer solstice. For locations in the Northern Hemisphere, the summer solstice is the time when the Sun is farthest north. In 2015, the summer solstice occurred on June 21 (the 172nd day of the year) at 12:38 PM EDT. The shortest day of the year occurs on the day of the winter solstice, the time when the Sun is farthest south (for locations in the Northern Hemisphere). In 2015, the winter solstice occurred on December 21 (the 355th day of the year) at 11:48 PM (EST).

EXAMPLE 4 Finding a Sinusoidal Function for Hours of Daylight

According to the *Old Farmer's Almanac*, the number of hours of sunlight in Boston on the day of the summer solstice is 15.28, and the number of hours of sunlight on the day of the winter solstice is 9.08.

(a) Find a sinusoidal function of the form $y = A \sin(\omega x - \phi) + B$ that fits the data.

(b) Use the function found in part (a) to predict the number of hours of sunlight in Boston on April 1, the 91st day of the year.

(c) Graph the function found in part (a).

(d) Look up the number of hours of sunlight for April 1 in the *Old Farmer's Almanac* and compare it to the results found in part (b).

Source: *The Old Farmer's Almanac, www.almanac.com/rise*

Solution (a) **STEP 1:** Amplitude $= \dfrac{\text{largest data value} - \text{smallest data value}}{2}$

$$= \frac{15.28 - 9.08}{2} = 3.10$$

STEP 2: Vertical shift $= \dfrac{\text{largest data value} + \text{smallest data value}}{2}$

$$= \frac{15.28 + 9.08}{2} = 12.18$$

STEP 3: The data repeat every 365 days. Since $T = \dfrac{2\pi}{\omega} = 365$, we find

$$\omega = \frac{2\pi}{365}$$

So far, we have $y = 3.10 \sin\left(\dfrac{2\pi}{365}x - \phi\right) + 12.18$.

STEP 4: To determine the horizontal shift, use the period $T = 365$ and divide the interval $[0, 365]$ into four subintervals of length $365 \div 4 = 91.25$:

$$[0, 91.25] \quad [91.25, 182.5] \quad [182.5, 273.75] \quad [273.75, 365]$$

The sine curve is increasing on the interval $[0, 91.25]$ and is decreasing on the interval $[91.25, 273.75]$, so a local maximum occurs at $x = 91.25$. Since the maximum occurs on the summer solstice at $x = 172$, we must shift the graph of the function $172 - 91.25 = 80.75$ units to the right by replacing x by $x - 80.75$. Doing this yields

$$y = 3.10 \sin\left(\frac{2\pi}{365}(x - 80.75)\right) + 12.18$$

Next, multiply out to obtain a sine function of the form $y = A \sin(\omega x - \phi) + B$ that fits the data.

$$y = 3.10 \sin\left(\frac{2\pi}{365}x - \frac{323\pi}{730}\right) + 12.18$$

(b) To predict the number of hours of daylight on April 1, let $x = 91$ in the function found in part (a) and obtain

$$y = 3.10 \sin\left(\frac{2\pi}{365} \cdot 91 - \frac{323}{730}\pi\right) + 12.18$$

$$\approx 12.72$$

The prediction is that there will be about 12.72 hours = 12 hours 43 minutes of sunlight on April 1 in Boston.

(c) The graph of the function found in part (a) is given in Figure 124.

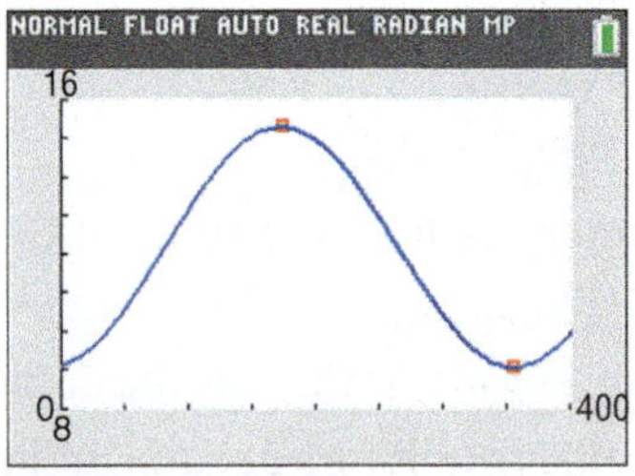

Figure 124

(d) According to the *Old Farmer's Almanac*, there will be 12 hours 45 minutes of sunlight on April 1 in Boston. ■

Now Work PROBLEM 35

Certain graphing utilities (such as the TI-84 Plus, TI-84 Plus C, and TI-89) have the capability of finding the sine function of best fit for sinusoidal data. At least four data points are required for this process.

EXAMPLE 5 **Finding the Sine Function of Best Fit**

Use a graphing utility to find the sine function of best fit for the data in Table 12. Graph this function with the scatter diagram of the data.

Solution Enter the data from Table 12 and execute the SINe REGression program. The result is shown in Figure 125.

The output that the utility provides shows the equation

$$y = a\sin(bx + c) + d$$

The sinusoidal function of best fit is

$$y = 21.43\sin(0.56x - 2.44) + 51.71$$

where x represents the month and y represents the average monthly temperature.

Figure 126 shows the graph of the sinusoidal function of best fit on the scatter diagram.

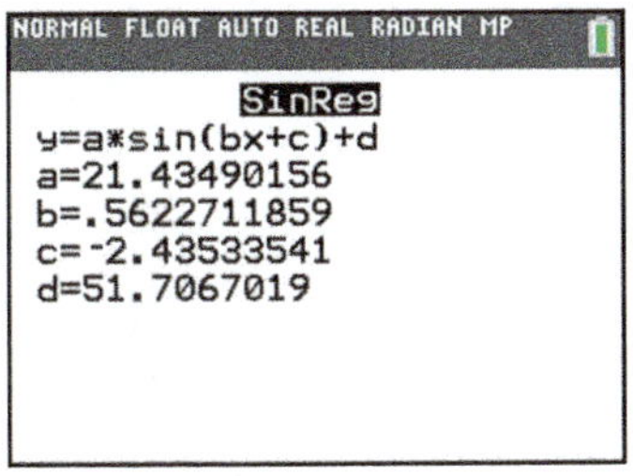

Figure 125

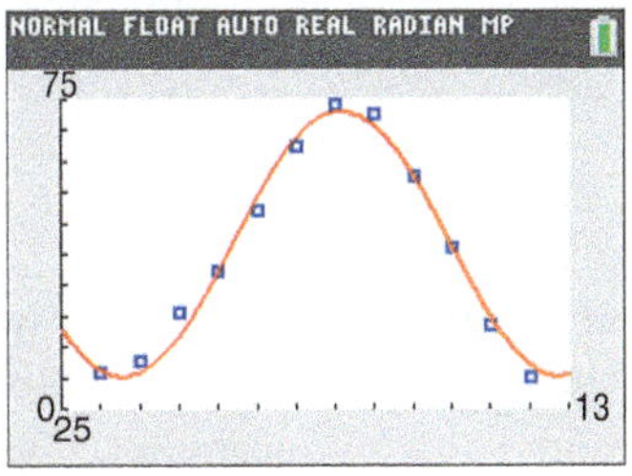

Figure 126

Now Work PROBLEMS 29(d) AND (e)

7.8 Assess Your Understanding

Concepts and Vocabulary

1. For the graph of $y = A\sin(\omega x - \phi)$, the number $\dfrac{\phi}{\omega}$ is called the ________ ________.

2. ***True or False*** A graphing utility requires only two data points to find the sine function of best fit.

Skill Building

In Problems 3–14, find the amplitude, period, and phase shift of each function. Graph each function. Be sure to label key points. Show at least two periods.

3. $y = 4\sin(2x - \pi)$

4. $y = 3\sin(3x - \pi)$

5. $y = 2\cos\left(3x + \dfrac{\pi}{2}\right)$

6. $y = 3\cos(2x + \pi)$

7. $y = -3\sin\left(2x + \dfrac{\pi}{2}\right)$

8. $y = -2\cos\left(2x - \dfrac{\pi}{2}\right)$

9. $y = 4\sin(\pi x + 2) - 5$

10. $y = 2\cos(2\pi x + 4) + 4$

11. $y = 3\cos(\pi x - 2) + 5$

12. $y = 2\cos(2\pi x - 4) - 1$

13. $y = -3\sin\left(-2x + \dfrac{\pi}{2}\right)$

14. $y = -3\cos\left(-2x + \dfrac{\pi}{2}\right)$

In Problems 15–18, write the equation of a sine function that has the given characteristics.

15. Amplitude: 2
Period: π
Phase shift: $\dfrac{1}{2}$

16. Amplitude: 3
Period: $\dfrac{\pi}{2}$
Phase shift: 2

17. Amplitude: 3
Period: 3π
Phase shift: $-\dfrac{1}{3}$

18. Amplitude: 2
Period: π
Phase shift: -2

Mixed Practice

In Problems 19–26, apply the methods of this and the previous section to graph each function. Be sure to label key points and show at least two periods.

19. $y = 2\tan(4x - \pi)$

20. $y = \frac{1}{2}\cot(2x - \pi)$

21. $y = 3\csc\left(2x - \frac{\pi}{4}\right)$

22. $y = \frac{1}{2}\sec(3x - \pi)$

23. $y = -\cot\left(2x + \frac{\pi}{2}\right)$

24. $y = -\tan\left(3x + \frac{\pi}{2}\right)$

25. $y = -\sec(2\pi x + \pi)$

26. $y = -\csc\left(-\frac{1}{2}\pi x + \frac{\pi}{4}\right)$

Applications and Extensions

27. Alternating Current (ac) Circuits The current I, in amperes, flowing through an ac (alternating current) circuit at time t, in seconds, is

$$I(t) = 120\sin\left(30\pi t - \frac{\pi}{3}\right) \qquad t \geq 0$$

What is the period? What is the amplitude? What is the phase shift? Graph this function over two periods.

28. Alternating Current (ac) Circuits The current I, in amperes, flowing through an ac (alternating current) circuit at time t, in seconds, is

$$I(t) = 220\sin\left(60\pi t - \frac{\pi}{6}\right) \qquad t \geq 0$$

What is the period? What is the amplitude? What is the phase shift? Graph this function over two periods.

29. Hurricanes Hurricanes are categorized using the Saffir-Simpson Hurricane Scale, with winds 111–130 miles per hour (mph) corresponding to a category 3 hurricane, winds 131–155 mph corresponding to a category 4 hurricane, and winds in excess of 155 mph corresponding to a category 5 hurricane. The following data represent the number of major hurricanes in the Atlantic Basin (category 3, 4, or 5) each decade from 1921 to 2010.

(a) Draw a scatter diagram of the data.

(b) Find a sinusoidal function of the form $y = A\sin(\omega x - \phi) + B$ that models the data.

(c) Draw the sinusoidal function found in part (b) on the scatter diagram.

(d) Use a graphing utility to find the sinusoidal function of best fit.

(e) Graph the sinusoidal function of best fit on a scatter diagram of the data.

Decade, x	Major Hurricanes, H
1921–1930, 1	17
1931–1940, 2	16
1941–1950, 3	29
1951–1960, 4	33
1961–1970, 5	27
1971–1980, 6	16
1981–1990, 7	16
1991–2000, 8	27
2001–2010, 9	33

Source: *U.S. National Oceanic and Atmospheric Administration*

30. Monthly Temperature The data below represent the average monthly temperatures for Washington, D.C.

(a) Draw a scatter diagram of the data for one period.

(b) Find a sinusoidal function of the form $y = A\sin(\omega x - \phi) + B$ that models the data.

(c) Draw the sinusoidal function found in part (b) on the scatter diagram.

(d) Use a graphing utility to find the sinusoidal function of best fit.

(e) Graph the sinusoidal function of best fit on a scatter diagram of the data.

Month, x	Average Monthly Temperature, °F
January, 1	36.0
February, 2	39.0
March, 3	46.8
April, 4	56.8
May, 5	66.0
June, 6	75.2
July, 7	79.8
August, 8	78.1
September, 9	71.0
October, 10	59.5
November, 11	49.6
December, 12	39.7

Source: *U.S. National Oceanic and Atmospheric Administration*

31. Monthly Temperature The given data represent the average monthly temperatures for Indianapolis, Indiana.

(a) Draw a scatter diagram of the data for one period.

(b) Find a sinusoidal function of the form

$$y = A\sin(\omega x - \phi) + B$$

that models the data.

(c) Draw the sinusoidal function found in part (b) on the scatter diagram.

(d) Use a graphing utility to find the sinusoidal function of best fit.

(e) Graph the sinusoidal function of best fit on a scatter diagram of the data.

Month, x	Average Monthly Temperature, °F
January, 1	28.1
February, 2	32.1
March, 3	42.2
April, 4	53.0
May, 5	62.7
June, 6	72.0
July, 7	75.4
August, 8	74.2
September, 9	66.9
October, 10	55.0
November, 11	43.6
December, 12	31.6

Source: *U.S. National Oceanic and Atmospheric Administration*

32. Monthly Temperature The following data represent the average monthly temperatures for Baltimore, Maryland.
(a) Draw a scatter diagram of the data for one period.
(b) Find a sinusoidal function of the form $y = A\sin(\omega x - \phi) + B$ that models the data.
(c) Draw the sinusoidal function found in part (b) on the scatter diagram.
(d) Use a graphing utility to find the sinusoidal function of best fit.
(e) Graph the sinusoidal function of best fit on a scatter diagram of the data.

Month, x	Average Monthly Temperature, °F
January, 1	32.9
February, 2	35.8
March, 3	43.6
April, 4	53.7
May, 5	62.9
June, 6	72.4
July, 7	77.0
August, 8	75.1
September, 9	67.8
October, 10	56.1
November, 11	46.5
December, 12	36.7

Source: *U.S. National Oceanic and Atmospheric Administration*

33. Tides The length of time between consecutive high tides is 12 hours and 25 minutes. According to the National Oceanic and Atmospheric Administration, on Saturday, March 28, 2015, in Charleston, South Carolina, high tide occurred at 2:12 AM (2.2 hours) and low tide occurred at 8:18 AM (8.3 hours). Water heights are measured as the amounts above or below the mean lower low water. The height of the water at high tide was 5.27 feet, and the height of the water at low tide was 0.87 foot.
(a) Approximately when will the next high tide occur?
(b) Find a sinusoidal function of the form

$$y = A\sin(\omega x - \phi) + B$$

that models the data.
(c) Use the function found in part (b) to predict the height of the water at 11 AM on March 28, 2015.

34. Tides The length of time between consecutive high tides is 12 hours and 25 minutes. According to the National Oceanic and Atmospheric Administration, on Saturday, March 28, 2015, in Sitka, Alaska, high tide occurred at 8:06 AM (8.10 hours) and low tide occurred at 2:48 PM (14.8 hours). Water heights are measured as the amounts above or below the mean lower low water. The height of the water at high tide was 8.13 feet, and the height of the water at low tide was 1.95 feet.
(a) Approximately when will the next high tide occur?
(b) Find a sinusoidal function of the form

$$y = A\sin(\omega x - \phi) + B$$

that models the data.
(c) Use the function found in part (b) to predict the height of the water at 6 PM on March 28, 2015.

35. Hours of Daylight According to the *Old Farmer's Almanac*, in Miami, Florida, the number of hours of sunlight on the summer solstice of 2015 was 13.75, and the number of hours of sunlight on the winter solstice was 10.52.
(a) Find a sinusoidal function of the form

$$y = A\sin(\omega x - \phi) + B$$

that models the data.
(b) Use the function found in part (a) to predict the number of hours of sunlight on April 1, the 91st day of the year.
(c) Draw a graph of the function found in part (a).
(d) Look up the number of hours of sunlight for April 1 in the *Old Farmer's Almanac*, and compare the actual hours of daylight to the results found in part (b).

36. Hours of Daylight According to the *Old Farmer's Almanac*, in Detroit, Michigan, the number of hours of sunlight on the summer solstice of 2015 was 15.27, and the number of hours of sunlight on the winter solstice was 9.07.
(a) Find a sinusoidal function of the form

$$y = A\sin(\omega x - \phi) + B$$

that models the data.
(b) Use the function found in part (a) to predict the number of hours of sunlight on April 1, the 91st day of the year.
(c) Draw a graph of the function found in part (a).
(d) Look up the number of hours of sunlight for April 1 in the *Old Farmer's Almanac*, and compare the actual hours of daylight to the results found in part (b).

37. Hours of Daylight According to the *Old Farmer's Almanac*, in Anchorage, Alaska, the number of hours of sunlight on the summer solstice of 2015 was 19.37, and the number of hours of sunlight on the winter solstice was 5.45.
(a) Find a sinusoidal function of the form

$$y = A\sin(\omega x - \phi) + B$$

that models the data.

(continued on next page)

(b) Use the function found in part (a) to predict the number of hours of sunlight on April 1, the 91st day of the year.

(c) Draw a graph of the function found in part (a).

(d) Look up the number of hours of sunlight for April 1 in the *Old Farmer's Almanac*, and compare the actual hours of daylight to the results found in part (b).

38. Hours of Daylight According to the *Old Farmer's Almanac*, in Honolulu, Hawaii, the number of hours of sunlight on the summer solstice of 2015 was 13.42, and the number of hours of sunlight on the winter solstice was 10.83.

(a) Find a sinusoidal function of the form

$$y = A\sin(\omega x - \phi) + B$$

that models the data.

(b) Use the function found in part (a) to predict the number of hours of sunlight on April 1, the 91st day of the year.

(c) Draw a graph of the function found in part (a).

(d) Look up the number of hours of sunlight for April 1 in the *Old Farmer's Almanac*, and compare the actual hours of daylight to the results found in part (b).

Discussion and Writing

39. Explain how the amplitude and period of a sinusoidal graph are used to establish the scale on each coordinate axis.

40. Find an application in your major field that leads to a sinusoidal graph. Write an account of your findings.

Retain Your Knowledge

Problems 41–44 are based on material learned earlier in the course. The purpose of these problems is to keep the material fresh in your mind so that you are better prepared for the final exam.

41. Given $f(x) = \dfrac{4x + 9}{2}$, find $f^{-1}(x)$.

42. Solve: $0.25(0.4x + 0.8) = 3.7 - 1.4x$

43. Multiply: $(8x + 15y)^2$

44. Find the exact distance between the points $(4, -1)$ and $(10, 3)$.

Chapter Review

Things to Know

Definition

Angle in standard position (p. 513)	Vertex is at the origin; initial side is along the positive x-axis.
1 Degree (1°) (p. 514)	$1° = \dfrac{1}{360}$ revolution
1 Radian (p. 516)	The measure of a central angle of a circle whose rays subtend an arc whose length is equal to the radius of the circle
Acute angle (p. 527)	An angle θ whose measure is $0° < \theta < 90°$ or $0 < \theta < \dfrac{\pi}{2}$, θ in radians
Complementary angles (p. 533)	Two acute angles whose sum is $90°\left(\dfrac{\pi}{2}\text{ radians}\right)$
Cofunction (p. 533)	The following pairs of functions are cofunctions of each other: sine and cosine; tangent and cotangent; secant and cosecant.
Trigonometric functions of any angle (p. 551)	$P = (a, b)$ is a point, not the origin, on the terminal side of an angle θ in standard position a distance r from the origin: $\sin\theta = \dfrac{b}{r}$ $\quad\cos\theta = \dfrac{a}{r}$ $\quad\tan\theta = \dfrac{b}{a},\ a \neq 0$ $\csc\theta = \dfrac{r}{b},\ b \neq 0$ $\quad\sec\theta = \dfrac{r}{a},\ a \neq 0$ $\quad\cot\theta = \dfrac{a}{b},\ b \neq 0$
Reference angle of θ (p. 555)	The acute angle formed by the terminal side of θ and either the positive or negative x-axis
Periodic function (p. 568)	$f(\theta + p) = f(\theta)$, for all θ, $p > 0$, where the smallest such p is the fundamental period

Formulas

1 counterclockwise revolution = 360° (p. 515) = 2π radians (p. 517)	$1° = \frac{\pi}{180}$ radian (p. 518); 1 radian $= \frac{180}{\pi}$ degrees (p. 518)
Arc length: $s = r\,\theta$ (p. 517)	θ is measured in radians; s is the length of the arc subtended by the central angle θ of the circle of radius r.
Area of a sector: $A = \frac{1}{2}r^2\theta$ (p. 520)	A is the area of the sector of a circle of radius r formed by a central angle of θ radians.
Linear speed: $v = \frac{s}{t}$ (p. 521)	v is the linear speed along the circle of radius r; ω is the angular speed (measured in radians per unit time).
Angular speed: $\omega = \frac{\theta}{t}$ (p. 521) $v = r\omega$ (p. 521)	

Table of Values

θ (Radians)	θ (Degrees)	sin θ	cos θ	tan θ	csc θ	sec θ	cot θ
0	0°	0	1	0	Not defined	1	Not defined
$\frac{\pi}{6}$	30°	$\frac{1}{2}$	$\frac{\sqrt{3}}{2}$	$\frac{\sqrt{3}}{3}$	2	$\frac{2\sqrt{3}}{3}$	$\sqrt{3}$
$\frac{\pi}{4}$	45°	$\frac{\sqrt{2}}{2}$	$\frac{\sqrt{2}}{2}$	1	$\sqrt{2}$	$\sqrt{2}$	1
$\frac{\pi}{3}$	60°	$\frac{\sqrt{3}}{2}$	$\frac{1}{2}$	$\sqrt{3}$	$\frac{2\sqrt{3}}{3}$	2	$\frac{\sqrt{3}}{3}$
$\frac{\pi}{2}$	90°	1	0	Not defined	1	Not defined	0
π	180°	0	−1	0	Not defined	−1	Not defined
$\frac{3\pi}{2}$	270°	−1	0	Not defined	−1	Not defined	0

Fundamental identities (p. 530)

Quotient identities: $\tan\theta = \frac{\sin\theta}{\cos\theta}$ $\quad \cot\theta = \frac{\cos\theta}{\sin\theta}$ $\quad$ Reciprocal identities: $\csc\theta = \frac{1}{\sin\theta}$ $\quad \sec\theta = \frac{1}{\cos\theta}$ $\quad \cot\theta = \frac{1}{\tan\theta}$

Pythagorean identities: $\sin^2\theta + \cos^2\theta = 1$ $\quad \tan^2\theta + 1 = \sec^2\theta$ $\quad \cot^2\theta + 1 = \csc^2\theta$

Properties of the trigonometric functions

$y = \sin x$ (p. 573)

Domain: $-\infty < x < \infty$
Range: $-1 \le y \le 1$
Periodic: period $= 2\pi\,(360°)$
Odd function

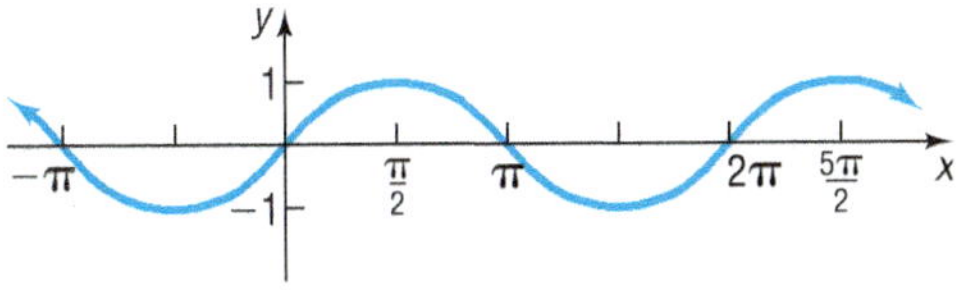

$y = \cos x$ (p. 575)

Domain: $-\infty < x < \infty$
Range: $-1 \le y \le 1$
Periodic: period $= 2\pi\,(360°)$
Even function

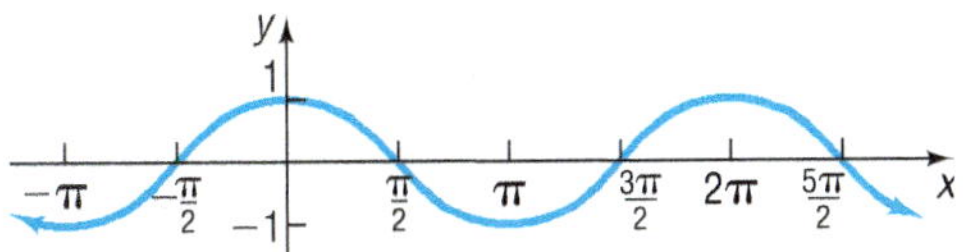

$y = \tan x$ (p. 589)

Domain: $-\infty < x < \infty$, except odd integer multiples of $\frac{\pi}{2}$ (90°)
Range: $-\infty < y < \infty$
Periodic: period $= \pi\,(180°)$
Odd function
Vertical asymptotes at odd integer multiples of $\frac{\pi}{2}$

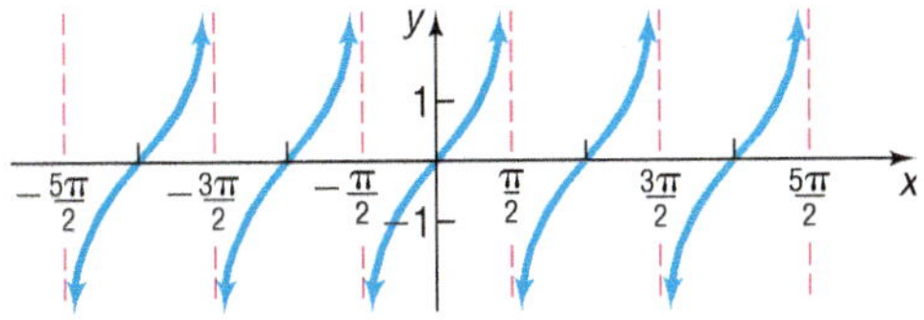

$y = \cot x$ (p. 591)

Domain: $-\infty < x < \infty$, except integer multiples of π (180°)

Range: $-\infty < y < \infty$

Periodic: period $= \pi$ (180°)

Odd function

Vertical asymptotes at integer multiples of π

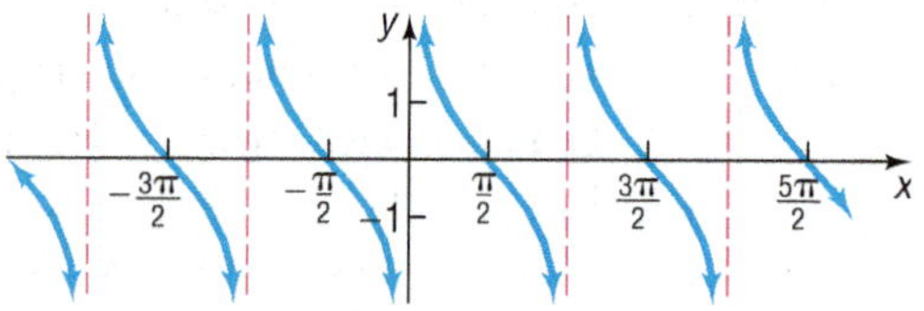

$y = \csc x$ (pp. 591–592)

Domain: $-\infty < x < \infty$, except integer multiples of π (180°)

Range: $|y| \geq 1$

Periodic: period $= 2\pi$ (360°)

Odd function

Vertical asymptotes at integer multiples of π

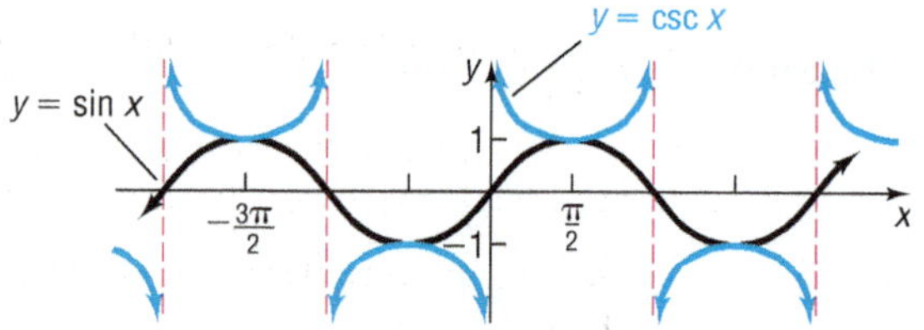

$y = \sec x$ (pp. 591–592)

Domain: $-\infty < x < \infty$, except odd integer multiples of $\frac{\pi}{2}$ (90°)

Range: $|y| \geq 1$

Periodic: period $= 2\pi$ (360°)

Even function

Vertical asymptotes at odd integer multiples of $\frac{\pi}{2}$

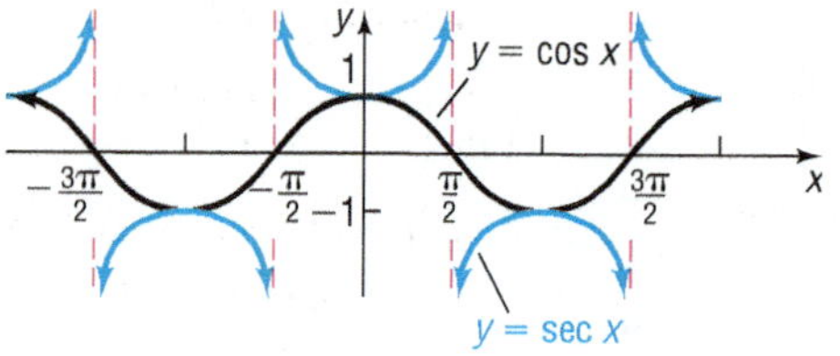

Sinusoidal graphs

$y = A \sin(\omega x) + B, \quad \omega > 0$ Period $= \dfrac{2\pi}{\omega}$ (pp. 578, 596)

$y = A \cos(\omega x) + B, \quad \omega > 0$ Amplitude $= |A|$ (pp. 578, 596)

$y = A \sin(\omega x - \phi) + B = A \sin\left[\omega\left(x - \dfrac{\phi}{\omega}\right)\right] + B$ Phase shift $= \dfrac{\phi}{\omega}$ (p. 596)

$y = A \cos(\omega x - \phi) + B = A \cos\left[\omega\left(x - \dfrac{\phi}{\omega}\right)\right] + B$

Objectives

Section		You should be able to . . .	Examples	Review Exercises
7.1	1	Convert between decimal and degree, minute, second measures for angles (p. 515)	2	48
	2	Find the length of an arc of a circle (p. 516)	3	49, 50
	3	Convert from degrees to radians and from radians to degrees (p. 517)	4–6	1–4
	4	Find the area of a sector of a circle (p. 520)	7	49
	5	Find the linear speed of an object traveling in circular motion (p. 520)	8	51, 52
7.2	1	Find the values of trigonometric functions of acute angles (p. 527)	1	41
	2	Use Fundamental Identities (p. 528)	2, 3	11, 12
	3	Find the values of the remaining trigonometric functions, given the value of one of them (p. 531)	4, 5	16
	4	Use the Complementary Angle Theorem (p. 533)	6, 7	13–15
7.3	1	Find the exact values of the trigonometric functions of $\frac{\pi}{4} = 45°$ (p. 539)	1, 2	5, 6
	2	Find the exact values of the trigonometric functions of $\frac{\pi}{6} = 30°$ and $\frac{\pi}{3} = 60°$ (p. 540)	3, 4	5, 6
	3	Use a calculator to approximate the values of the trigonometric functions of acute angles (p. 541)	5	42
	4	Model and solve applied problems involving right triangles (p. 542)	6–9	53–55

Section		You should be able to . . .	Examples	Review Exercises
7.4	1	Find the exact values of the trigonometric functions for any angle (p. 550)	1, 2	9, 46
	2	Use coterminal angles to find the exact value of a trigonometric function (p. 553)	3	10
	3	Determine the signs of the trigonometric functions of an angle in a given quadrant (p. 554)	4	43
	4	Find the reference angle of an angle (p. 555)	5	44
	5	Use a reference angle to find the exact value of a trigonometric function (p. 556)	6	7, 8, 10
	6	Find the exact values of the trigonometric functions of an angle, given information about the functions (p. 557)	7, 8	16–23
7.5	1	Find the exact values of the trigonometric functions using the unit circle (p. 561)	1, 2	45, 46
	2	Know the domain and range of the trigonometric functions (p. 565)		47
	3	Use periodic properties to find the exact values of the trigonometric functions (p. 567)	3	10
	4	Use even–odd properties to find the exact values of the trigonometric functions (p. 568)	4	7, 8, 10, 14, 15
7.6	1	Graph functions of the form $y = A \sin(\omega x)$ using transformations (p. 573)	1, 2	24
	2	Graph functions of the form $y = A \cos(\omega x)$ using transformations (p. 575)	3	25
	3	Determine the amplitude and period of sinusoidal functions (p. 576)	4	33, 34
	4	Graph sinusoidal functions using key points (p. 578)	5–7	24, 25, 35
	5	Find an equation for a sinusoidal graph (p. 581)	8, 9	39, 40
7.7	1	Graph functions of the form $y = A \tan(\omega x) + B$ and $y = A \cot(\omega x) + B$ (p. 589)	1, 2	26–28, 32
	2	Graph functions of the form $y = A \csc(\omega x) + B$ and $y = A \sec(\omega x) + B$ (p. 592)	3	29, 30
7.8	1	Graph sinusoidal functions of the form $y = A \sin(\omega x - \phi) + B$ (p. 595)	1, 2	31, 36–38, 56
	2	Build sinusoidal models from data (p. 599)	3–5	57

Review Exercises

In Problems 1 and 2, convert each angle in degrees to radians. Express your answer as a multiple of π.

1. $135°$

2. $15°$

In Problems 3 and 4, convert each angle in radians to degrees.

3. $\dfrac{2\pi}{3}$

4. $-\dfrac{5\pi}{2}$

In Problems 5–15, find the exact value of each expression. Do not use a calculator.

5. $\tan\dfrac{\pi}{4} - \sin\dfrac{\pi}{6}$

6. $3\sin 45° - 4\tan\dfrac{\pi}{6}$

7. $6\cos\dfrac{3\pi}{4} + 2\tan\left(-\dfrac{\pi}{3}\right)$

8. $\sec\left(-\dfrac{\pi}{3}\right) - \cot\left(-\dfrac{5\pi}{4}\right)$

9. $\tan\pi + \sin\pi$

10. $\cos 540° - \tan(-405°)$

11. $\sin^2 20° + \dfrac{1}{\sec^2 20°}$

12. $\sec 50° \cos 50°$

13. $\dfrac{\sin 50°}{\cos 40°}$

14. $\dfrac{\sin(-40°)}{\cos 50°}$

15. $\sin 400° \sec(-50°)$

In Problems 16–23, find the exact value of each of the remaining trigonometric functions.

16. $\sin\theta = \frac{4}{5}$, θ is acute

17. $\tan\theta = \frac{12}{5}$, $\sin\theta < 0$

18. $\sec\theta = -\frac{5}{4}$, $\tan\theta < 0$

19. $\cos\theta = -\frac{3}{5}$, θ in quadrant III

20. $\sin\theta = -\frac{5}{13}$, $\frac{3\pi}{2} < \theta < 2\pi$

21. $\tan\theta = \frac{1}{3}$, $180° < \theta < 270°$

22. $\csc\theta = -4$, $\pi < \theta < \frac{3\pi}{2}$

23. $\cot\theta = -2$, $\frac{\pi}{2} < \theta < \pi$

In Problems 24–32, graph each function. Each graph should contain at least two periods. Use the graph to determine the domain and the range of each function.

24. $y = 2\sin(4x)$

25. $y = -3\cos(2x)$

26. $y = \tan(x + \pi)$

27. $y = -2\tan(3x)$

28. $y = \cot\left(x + \frac{\pi}{4}\right)$

29. $y = 4\sec(2x)$

30. $y = \csc\left(x + \frac{\pi}{4}\right)$

31. $y = 4\sin(2x + 4) - 2$

32. $y = 5\cot\left(\frac{x}{3} - \frac{\pi}{4}\right)$

In Problems 33 and 34, determine the amplitude and the period of each function without graphing.

33. $y = \sin(2x)$

34. $y = -2\cos(3\pi x)$

In Problems 35–38, find the amplitude, period, and phase shift of each function. Graph each function. Show at least two periods.

35. $y = 4\sin(3x)$

36. $y = -\cos\left(\frac{1}{2}x + \frac{\pi}{2}\right)$

37. $y = \frac{1}{2}\sin\left(\frac{3}{2}x - \pi\right)$

38. $y = -\frac{2}{3}\cos(\pi x - 6)$

In Problems 39 and 40, find a function whose graph is given.

39.

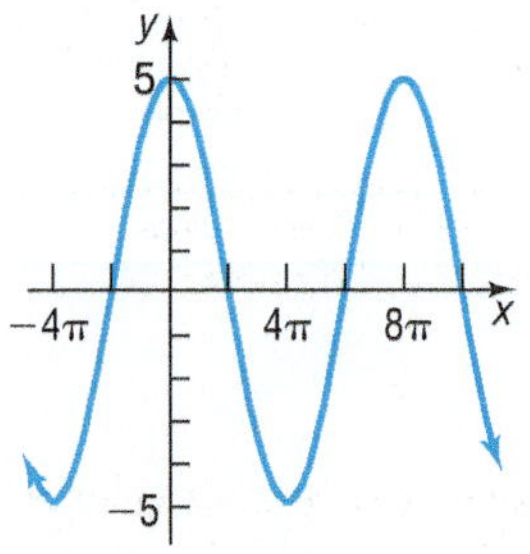

40.

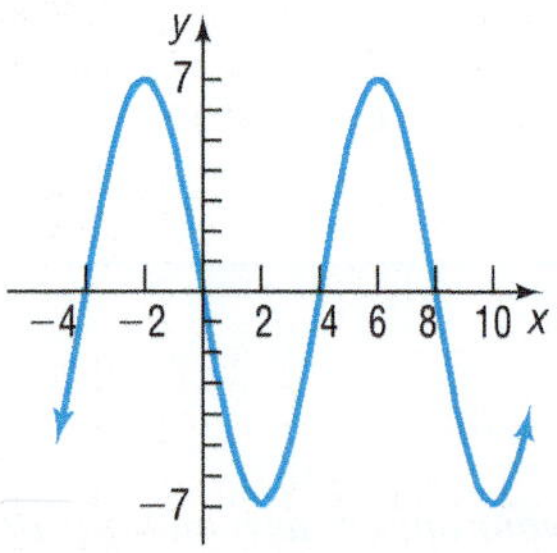

41. Find the value of each of the six trigonometric functions of the acute angle θ in a right triangle where the length of the side opposite θ is 3 and the length of the hypotenuse is 7.

42. Use a calculator to approximate $\sin\frac{\pi}{8}$. Use a calculator to approximate sec 10°. Round answers to two decimal places.

43. Determine the signs of the six trigonometric functions of an angle θ whose terminal side is in quadrant III.

44. Find the reference angle of 750°.

45. Find the exact values of the six trigonometric functions of t if $P = \left(-\frac{1}{3}, \frac{2\sqrt{2}}{3}\right)$ is the point on the unit circle that corresponds to t.

46. Find the exact values of $\sin t$, $\cos t$, and $\tan t$ if $P = (-2, 5)$ is the point on a circle that corresponds to t.

47. What are the domain and the range of the secant function? What is the period?

48. (a) Convert the angle 32°20′35″ to a decimal in degrees. Round the answer to two decimal places.
(b) Convert the angle 63.18° to D°M′S″ form. Express the answer to the nearest second.

49. Find the length of the arc subtended by a central angle of 30° on a circle of radius 2 feet. What is the area of the sector?

50. The minute hand of a clock is 8 inches long. How far does the tip of the minute hand move in 30 minutes? How far does it move in 20 minutes?

51. Angular Speed of a Race Car A race car is driven around a circular track at a constant speed of 180 miles per hour. If the diameter of the track is $\frac{1}{2}$ mile, what is the angular speed of the car? Express your answer in revolutions per hour (which is equivalent to laps per hour).

52. Lighthouse Beacons The Montauk Point Lighthouse on Long Island has dual beams (two light sources opposite each other). Ships at sea observe a blinking light every 5 seconds. What rotation speed is required to do this?

53. Measuring the Length of a Lake From a stationary hot-air balloon 500 feet above the ground, two sightings of a lake are made (see the figure). How long is the lake?

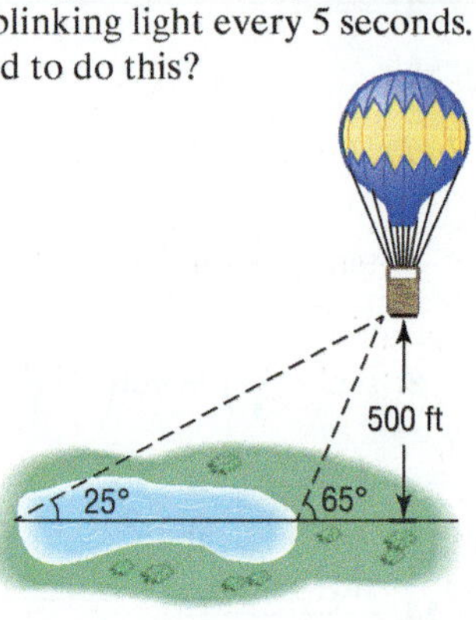

54. Finding the Width of a River Find the distance from A to C across the river illustrated in the figure.

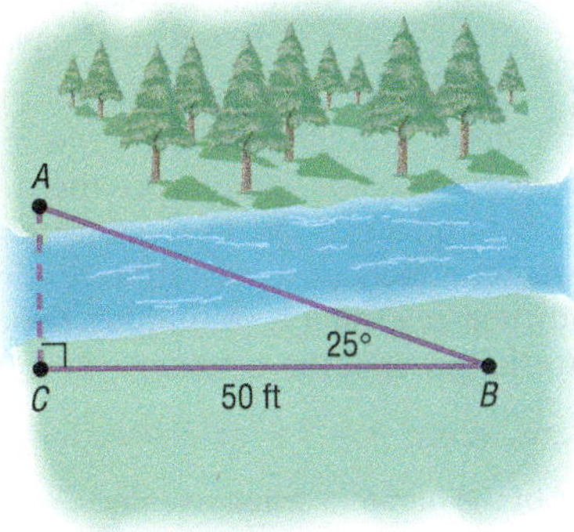

55. Finding the Distance to Shore The Willis Tower in Chicago is 1454 feet tall and is situated about 1 mile inland from the shore of Lake Michigan, as indicated in the figure. An observer in a pleasure boat on the lake directly in front of the Willis Tower looks at the top of the tower and measures the angle of elevation as 5°. How far offshore is the boat?

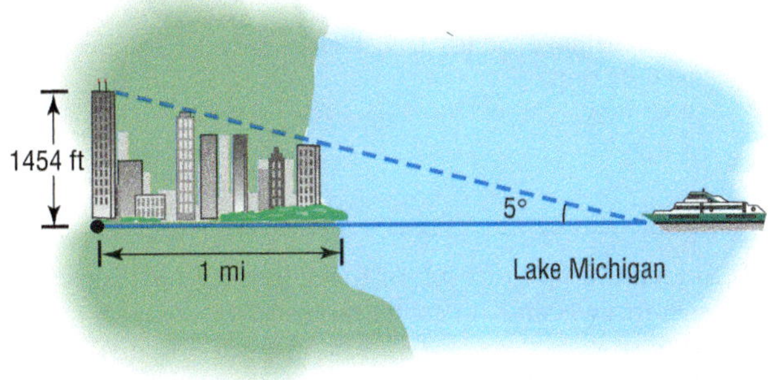

56. Alternating Current The current I, in amperes, flowing through an ac (alternating current) circuit at time t is

$$I = 220\sin\left(30\pi t + \frac{\pi}{6}\right) \qquad t \geq 0$$

(a) What is the period?
(b) What is the amplitude?
(c) What is the phase shift?
(d) Graph this function over two periods.

57. Monthly Temperature The following data represent the average monthly temperatures for Phoenix, Arizona.
(a) Draw a scatter diagram of the data for one period.
(b) Find a sinusoidal function of the form

$$y = A\sin(\omega x - \phi) + B$$

that models the data.
(c) Draw the sinusoidal function found in part (b) on the scatter diagram.
(d) Use a graphing utility to find the sinusoidal function of best fit.
(e) Graph the sinusoidal function of best fit on the scatter diagram.

Month, x	Average Monthly Temperature, °F
January, 1	56
February, 2	60
March, 3	65
April, 4	73
May, 5	82
June, 6	91
July, 7	95
August, 8	94
September, 9	88
October, 10	77
November, 11	64
December, 12	55

Source: *U.S. National Oceanic and Atmospheric Administration*

Chapter Test

CHAPTER **Test Prep** VIDEOS

The Chapter Test Prep Videos are step-by-step solutions available in MyMathLab®, or on this text's YouTube™ Channel. Flip back to the Resources for Success page for a link to this text's YouTube channel.

In Problems 1–3, convert each angle in degrees to radians. Express your answer as a multiple of π.

1. 260° **2.** −400° **3.** 13°

In Problems 4–6, convert each angle in radians to degrees.

4. $-\frac{\pi}{8}$ **5.** $\frac{9\pi}{2}$ **6.** $\frac{3\pi}{4}$

In Problems 7–12, find the exact value of each expression.

7. $\sin\frac{\pi}{6}$ **8.** $\cos\left(-\frac{5\pi}{4}\right) - \cos\frac{3\pi}{4}$

9. $\cos(-120°)$ **10.** $\tan 330°$

11. $\sin\frac{\pi}{2} - \tan\frac{19\pi}{4}$ **12.** $2\sin^2 60° - 3\cos 45°$

In Problems 13–16, use a calculator to evaluate each expression. Round your answer to three decimal places.

13. $\sin 17°$ **14.** $\cos\frac{2\pi}{5}$ **15.** $\sec 229°$ **16.** $\cot\frac{28\pi}{9}$

17. Fill in each table entry with the sign of each function.

	$\sin\theta$	$\cos\theta$	$\tan\theta$	$\sec\theta$	$\csc\theta$	$\cot\theta$
θ in QI						
θ in QII						
θ in QIII						
θ in QIV						

18. If $f(x) = \sin x$ and $f(a) = \frac{3}{5}$, find $f(-a)$.

In Problems 19–21, find the value of the remaining five trigonometric functions of θ.

19. $\sin\theta = \frac{5}{7}$, θ in quadrant II **20.** $\cos\theta = \frac{2}{3}$, $\frac{3\pi}{2} < \theta < 2\pi$

21. $\tan\theta = -\frac{12}{5}$, $\frac{\pi}{2} < \theta < \pi$

In Problems 22–24, the point (x, y) is on the terminal side of angle θ in standard position. Find the exact value of the given trigonometric function.

22. $(2, 7)$, $\sin\theta$ **23.** $(-5, 11)$, $\cos\theta$

24. $(6, -3)$, $\tan\theta$

In Problems 25 and 26, graph the function.

25. $y = 2\sin\left(\frac{x}{3} - \frac{\pi}{6}\right)$ **26.** $y = \tan\left(-x + \frac{\pi}{4}\right) + 2$

27. Write an equation for a sinusoidal graph with the following properties:

$$A = -3 \qquad \text{period} = \frac{2\pi}{3} \qquad \text{phase shift} = -\frac{\pi}{4}$$

28. Logan has a garden in the shape of a sector of a circle; the outer rim of the garden is 25 feet long and the central angle of the sector is 50°. She wants to add a 3-foot-wide walk to the outer rim. How many square feet of paving blocks will she need to build the walk?

29. Hungarian Adrian Annus won the gold medal for the hammer throw at the 2004 Olympics in Athens with a winning distance of 83.19 meters.* The event consists of swinging a 16-pound weight attached to a wire 190 centimeters long in a circle and then releasing it. Assuming his release is at a 45° angle to the ground, the hammer will travel a distance of $\frac{v_0^2}{g}$ meters, where $g = 9.8$ meters/second^2 and v_0 is the linear speed of the hammer when released. At what rate (rpm) was he swinging the hammer upon release?

30. A ship is just offshore of New York City. A sighting is taken of the Statue of Liberty, which is about 305 feet tall. If the angle of elevation to the top of the statue is 20°, how far is the ship from the base of the statue?

31. To measure the height of a building, two sightings are taken a distance of 50 feet apart. If the first angle of elevation is 40° and the second is 32°, what is the height of the building?

*Annus was stripped of his medal after refusing to cooperate with postmedal drug testing.

Cumulative Review

1. Find the real solutions, if any, of the equation $2x^2 + x - 1 = 0$.
2. Find an equation for the line with slope -3 containing the point $(-2, 5)$.
3. Find an equation for a circle of radius 4 and center at the point $(0, -2)$.
4. Discuss the equation $2x - 3y = 12$. Graph it.
5. Discuss the equation $x^2 + y^2 - 2x + 4y - 4 = 0$. Graph it.
6. Use transformations to graph the function $y = (x - 3)^2 + 2$.
7. Sketch a graph of each of the following functions. Label at least three points on each graph.
   (a) $y = x^2$ (b) $y = x^3$ (c) $y = e^x$
   (d) $y = \ln x$ (e) $y = \sin x$ (f) $y = \tan x$
8. Find the inverse function of $f(x) = 3x - 2$.
9. Find the exact value of $(\sin 14°)^2 + (\cos 14°)^2 - 3$.
10. Graph $y = 3\sin(2x)$.
11. Find the exact value of $\tan\frac{\pi}{4} - 3\cos\frac{\pi}{6} + \csc\frac{\pi}{6}$.

12. Find an exponential function for the following graph. Express your answer in the form $y = Ab^x$.

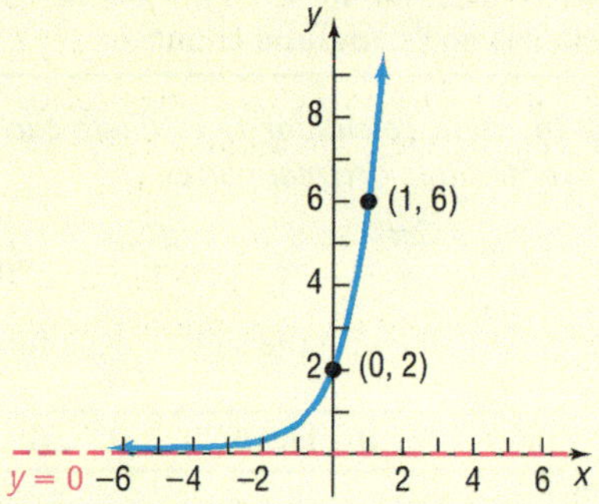

13. Find a sinusoidal function for the following graph.

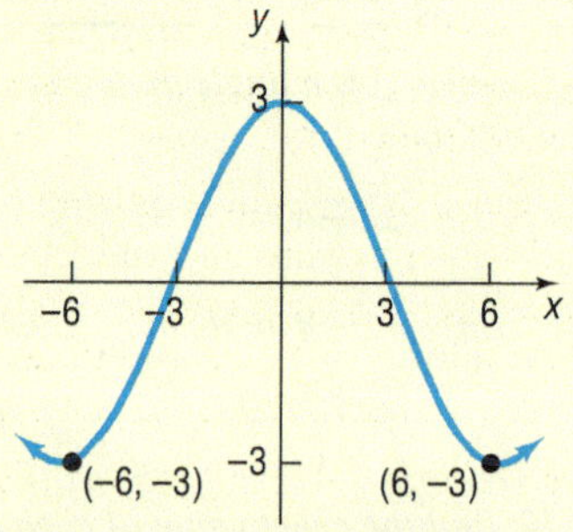

14. (a) Find a linear function that contains the points $(-2, 3)$ and $(1, -6)$. What is the slope? What are the intercepts of the function? Graph the function. Be sure to label the intercepts.
(b) Find a quadratic function with vertex $(1, -6)$ that contains the point $(-2, 3)$. What are the intercepts of the function? Graph the function.
(c) Show that there is no exponential function of the form $f(x) = ae^x$ that contains the points $(-2, 3)$ and $(1, -6)$.

15. (a) Find a polynomial function of degree 3 whose y-intercept is 5 and whose x-intercepts are -2, 3, and 5. Graph the function.
(b) Find a rational function whose y-intercept is 5 and whose x-intercepts are -2, 3, and 5 that has the line $x = 2$ as a vertical asymptote. Graph the function.

Chapter Projects

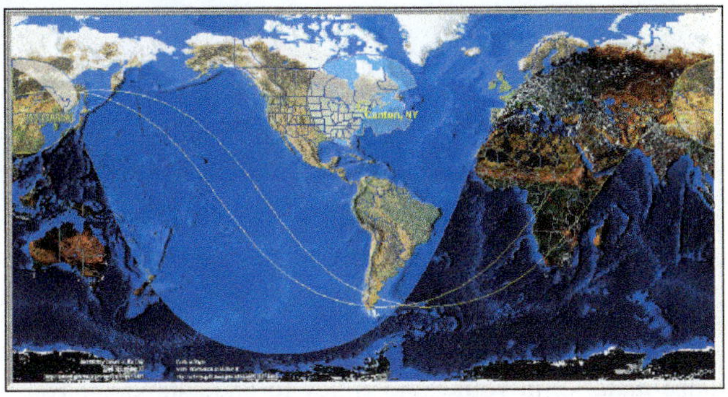

I. Length of Day Revisited Go to *http://en.wikipedia.org/wiki/latitude* and read about latitude. Then go to *http://www.orchidculture.com/COD/daylength.html.*

1. For a particular latitude, record in a table the length of day for the various days of the year. For January 1, use 1 as the day, for January 16, use 16 as the day, for February 1, use 32 as the day, and so on. Enter the data into an Excel spreadsheet using column B for the day of the year and column C for the length of day.
2. Draw a scatter diagram of the data with day of the year as the independent variable and length of day as the dependent variable using Excel. (The Chapter 4 project describes how to draw a scatter diagram in Excel.)
3. Determine the sinusoidal function of best fit, $y = A\sin(Bx + C) + D$, as follows:
   (a) Enter initial guesses for the values of A, B, C, and D into column A with the value of A in cell A1, B in cell A2, C in cell A3, and D in cell A4.
   (b) Enter "=A$1*sin(A$2*B1+A$3)+A$4" into cell D1. Copy this cell entry into the cells below D1 for as many rows as there are data. For example, if column C goes to row 23, then column D should also go to row 23.
   (c) Enter "= (D1−C1)^2" into cell E1. Copy this entry below as described in part 3(b).
   (d) The idea behind curve fitting is to make the sum of the squared differences between what is predicted and actual observations as small as possible. Enter "=sum(E1..E#)" into cell A6, where # represents the row number of the last data point. For example, if you have 23 rows of data, enter "=sum(E1..E23)" in cell A6.
   (e) Now, install the Solver feature of Excel. To do this, click the File tab, and then select Options. Select Add-Ins. In the drop-down menu entitled "Manage," choose Excel Add-ins, and then click Go . . . . Check the box entitled "Solver Add-in" and click OK. The Solver add-in is now available in the Data tab. Choose Solver. Fill in the screen as shown below:

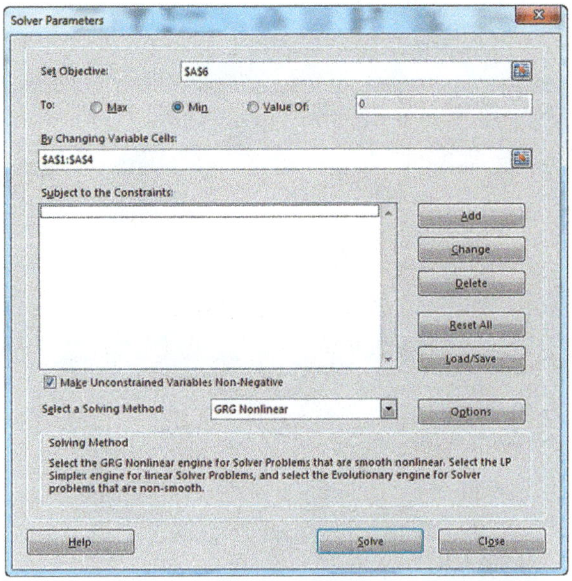

Citation: Excel © 2013 Microsoft Corporation. Used with permission from Microsoft.

Click Solve. The values for A, B, C, and D are located in cells A1–A4. What is the sinusoidal function of best fit?

4. Determine the longest day of the year according to your model. What is the day length on the longest day of the year? Determine the shortest day of the year according to your model. What is the day length on the shortest day of the year?
5. On which days is the day length exactly 12 hours according to your model?
6. Look up the day on which the vernal equinox and autumnal equinox occur. How do they match up with the results obtained in part 5?
7. Do you think your model accurately describes the relation between day of the year and length of day?
8. Use your model to predict the hours of daylight for the latitude you selected for various days of the year. Go to the *Old Farmer's Almanac* or other website (such as *http://astro.unl.edu/classaction/animations/coordsmotion/daylighthoursexplorer.html*) to determine the hours of daylight for the latitude you selected. How do the two compare?

The following projects are available on the Instructor's Resource Center (IRC):

II. Tides Data from a tide table are used to build a sine function that models tides.

III. Project at Motorola ***Digital Transmission over the Air*** Learn how Motorola Corporation transmits digital sequences by modulating the phase of the carrier waves.

IV. Identifying Mountain Peaks in Hawaii The visibility of a mountain is affected by its altitude, its distance from the viewer, and the curvature of Earth's surface. Trigonometry can be used to determine whether a distant object can be seen.

V. CBL Experiment Technology is used to model and study the effects of damping on sound waves.

8 Analytic Trigonometry

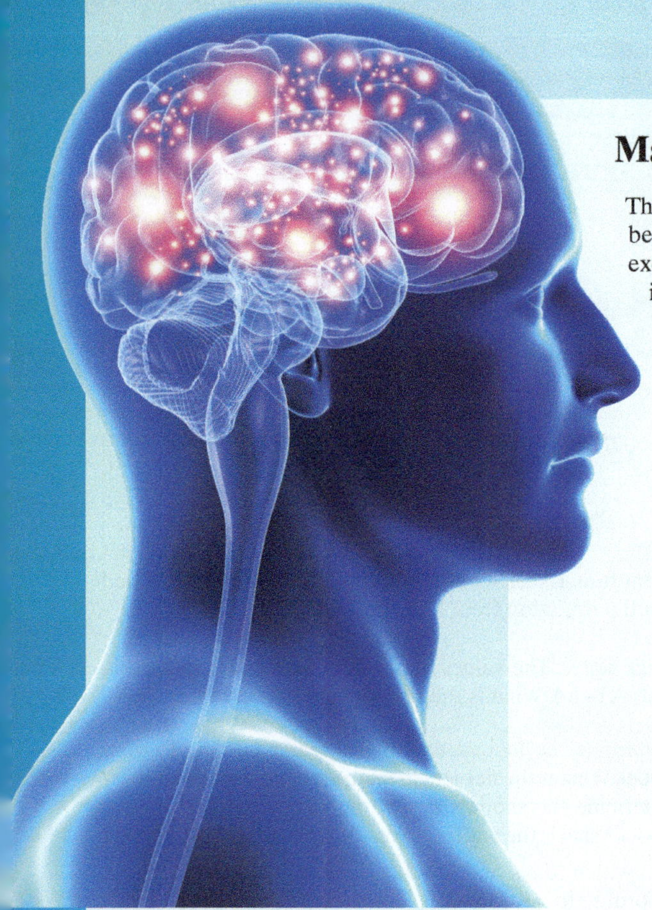

Mapping Your Mind

The ability to organize material in your mind is key to understanding. You have been exposed to a lot of concepts at this point in the course, and it is a worthwhile exercise to organize the material. In the past, we might organize material using index cards or an outline. But in today's digital world, we can use interesting software that enables us to digitally organize the material that is in our mind and share it with anyone on the Web.

—See the Internet-based Chapter Project I—

Outline

••• A Look Back

Chapter 6 introduced inverse functions and developed their properties, particularly the relationship between the domain and range of a function and those of its inverse. We learned that the graphs of a function and its inverse are symmetric with respect to the line $y = x$.

Chapter 6 continued by defining two transcendental functions: the exponential function and the inverse of the exponential function, the logarithmic function. Chapter 7 defined six more transcendental functions, the trigonometric functions, and discussed their properties.

A Look Ahead •••

The first two sections of this chapter define the six inverse trigonometric functions and investigate their properties. Section 8.3 discusses equations that contain trigonometric functions. Sections 8.4 through 8.7 continue the derivation of identities. These identities play an important role in calculus, the physical and life sciences, and economics, where they are used to simplify complicated expressions.

8.1 The Inverse Sine, Cosine, and Tangent Functions

PREPARING FOR THIS SECTION *Before getting started, review the following:*

- Inverse Functions (Section 6.2, pp. 418–424)
- Values of the Trigonometric Functions (Section 7.3, pp. 539–541; Section 7.4, pp. 550–558)
- Properties of the Sine, Cosine, and Tangent Functions (Section 7.5, pp. 561–569)
- Graphs of the Sine, Cosine, and Tangent Functions (Section 7.6, pp. 572–582; Section 7.7, pp. 587–590)

Now Work the 'Are You Prepared?' problems on page 625.

OBJECTIVES
1 Find the Exact Value of an Inverse Sine Function (p. 616)
2 Find an Approximate Value of an Inverse Sine Function (p. 617)
3 Use Properties of Inverse Functions to Find Exact Values of Certain Composite Functions (p. 618)
4 Find the Inverse Function of a Trigonometric Function (p. 624)
5 Solve Equations Involving Inverse Trigonometric Functions (p. 625)

In Section 6.2 we discussed inverse functions, and we concluded that if a function is one-to-one, it has an inverse function. We also observed that if a function is not one-to-one, it may be possible to restrict its domain in some suitable manner so that the restricted function is one-to-one. For example, the function $y = x^2$ is not one-to-one; however, if the domain is restricted to $x \geq 0$, the function is one-to-one.

Other properties of a one-to-one function f and its inverse function f^{-1} that were discussed in Section 6.2 are summarized next.

1. $f^{-1}(f(x)) = x$ for every x in the domain of f, and $f(f^{-1}(x)) = x$ for every x in the domain of f^{-1}.
2. The domain of $f =$ the range of f^{-1}, and the range of $f =$ the domain of f^{-1}.
3. The graph of f and the graph of f^{-1} are reflections of one another about the line $y = x$.
4. If a function $y = f(x)$ has an inverse function, the implicit equation of the inverse function is $x = f(y)$. Solving this equation for y yields the explicit equation $y = f^{-1}(x)$.

The Inverse Sine Function

Figure 1 shows the graph of $y = \sin x$. Because every horizontal line $y = b$, where b is between -1 and 1, inclusive, intersects the graph of $y = \sin x$ infinitely many times, it follows from the horizontal-line test that the function $y = \sin x$ is not one-to-one.

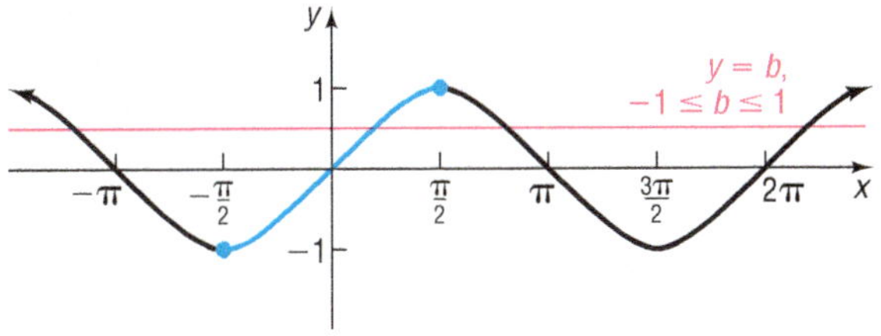

Figure 1 $y = \sin x,\ -\infty < x < \infty,\ -1 \leq y \leq 1$

However, if the domain of $y = \sin x$ is restricted to the interval $\left[-\frac{\pi}{2}, \frac{\pi}{2}\right]$, the restricted function

$$y = \sin x \qquad -\frac{\pi}{2} \leq x \leq \frac{\pi}{2}$$

is one-to-one and has an inverse function.* See Figure 2.

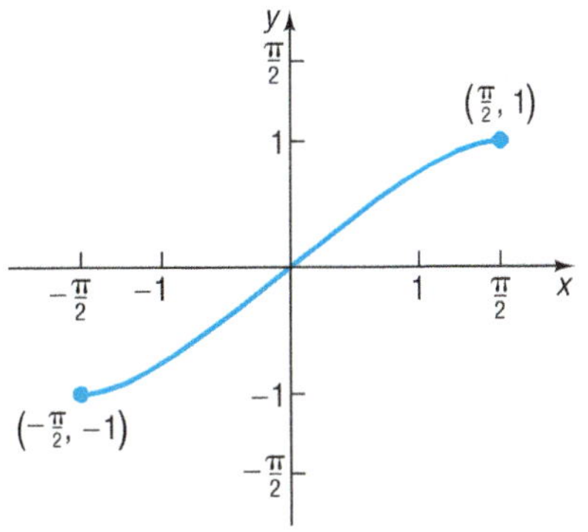

Figure 2 $y = \sin x,\ -\frac{\pi}{2} \leq x \leq \frac{\pi}{2},\ -1 \leq y \leq 1$

*Although there are many other ways to restrict the domain and obtain a one-to-one function, mathematicians have agreed to use the interval $\left[-\frac{\pi}{2}, \frac{\pi}{2}\right]$ to define the inverse of $y = \sin x$.

An equation for the inverse of $y = f(x) = \sin x$ is obtained by interchanging x and y. The implicit form of the inverse function is $x = \sin y$, $-\frac{\pi}{2} \le y \le \frac{\pi}{2}$. The explicit form is called the **inverse sine** of x and is symbolized by $y = f^{-1}(x) = \sin^{-1} x$.

DEFINITION

$$y = \sin^{-1} x \quad \text{means} \quad x = \sin y$$
$$\text{where} \quad -1 \le x \le 1 \quad \text{and} \quad -\frac{\pi}{2} \le y \le \frac{\pi}{2} \tag{1}$$

Because $y = \sin^{-1} x$ means $x = \sin y$, $y = \sin^{-1} x$ is read as "y is the angle or real number whose sine equals x," or, alternatively, is read "y is the inverse sine of x." Be careful about the notation used. The superscript -1 that appears in $y = \sin^{-1} x$ is not an exponent but the symbol used to denote the inverse function f^{-1} of f. (To avoid confusion, some texts use the notation $y = \text{Arcsin}\, x$ instead of $y = \sin^{-1} x$.)

The inverse of a function f receives as input an element from the range of f and returns as output an element in the domain of f. The restricted sine function, $y = f(x) = \sin x$, receives as input an angle or real number x in the interval $\left[-\frac{\pi}{2}, \frac{\pi}{2}\right]$ and outputs a real number in the interval $[-1, 1]$. Therefore, the inverse sine function, $y = \sin^{-1} x$, receives as input a real number in the interval $[-1, 1]$ or $-1 \le x \le 1$, its domain, and outputs an angle or real number in the interval $\left[-\frac{\pi}{2}, \frac{\pi}{2}\right]$ or $-\frac{\pi}{2} \le y \le \frac{\pi}{2}$, its range.

Note: Remember, the domain of a function f equals the range of its inverse function f^{-1}, and the range of a function f equals the domain of its inverse function f^{-1}. Because the restricted domain of the sine function is $\left[-\frac{\pi}{2}, \frac{\pi}{2}\right]$, the range of the inverse sine function is $\left[-\frac{\pi}{2}, \frac{\pi}{2}\right]$, and because the range of the sine function is $[-1, 1]$, the domain of the inverse sine function is $[-1, 1]$. ■

The graph of the inverse sine function can be obtained by reflecting the restricted portion of the graph of $y = f(x) = \sin x$ about the line $y = x$, as shown in Figure 3(a). Figure 3(b) shows the graph using a TI-84 Plus C graphing calculator.

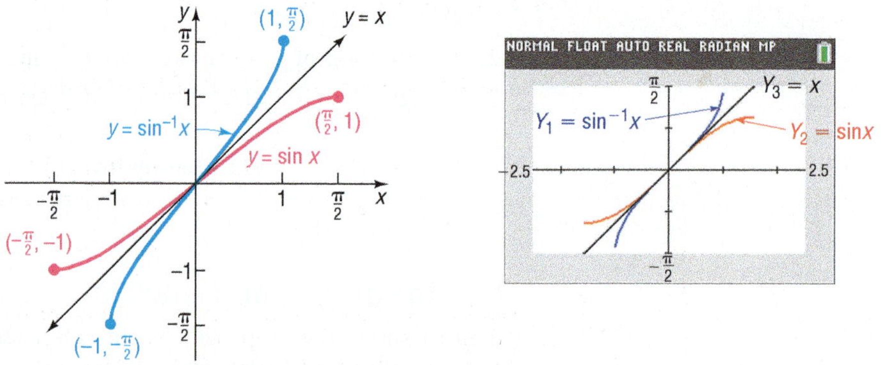

Figure 3 (a) (b)

$y = \sin^{-1} x$, $-1 \le x \le 1$, $-\frac{\pi}{2} \le y \le \frac{\pi}{2}$

1 Find the Exact Value of an Inverse Sine Function

For some numbers x, it is possible to find the exact value of $y = \sin^{-1} x$.

EXAMPLE 1 **Finding the Exact Value of an Inverse Sine Function**

Find the exact value of: $\sin^{-1} 1$

Solution Let $\theta = \sin^{-1} 1$. Then θ is the angle, $-\frac{\pi}{2} \le \theta \le \frac{\pi}{2}$, whose sine equals 1.

$$\theta = \sin^{-1} 1 \qquad -\frac{\pi}{2} \le \theta \le \frac{\pi}{2}$$

$$\sin \theta = 1 \qquad -\frac{\pi}{2} \le \theta \le \frac{\pi}{2} \qquad \textbf{By definition of } y = \sin^{-1} x$$

Now look at Table 1 and Figure 4.

Table 1

θ	$-\frac{\pi}{2}$	$-\frac{\pi}{3}$	$-\frac{\pi}{4}$	$-\frac{\pi}{6}$	0	$\frac{\pi}{6}$	$\frac{\pi}{4}$	$\frac{\pi}{3}$	$\frac{\pi}{2}$
$\sin\theta$	-1	$-\frac{\sqrt{3}}{2}$	$-\frac{\sqrt{2}}{2}$	$-\frac{1}{2}$	0	$\frac{1}{2}$	$\frac{\sqrt{2}}{2}$	$\frac{\sqrt{3}}{2}$	1

The only angle θ in the interval $\left[-\frac{\pi}{2}, \frac{\pi}{2}\right]$ whose sine is 1 is $\frac{\pi}{2}$. (Note that $\sin\frac{5\pi}{2}$ also equals 1, but $\frac{5\pi}{2}$ lies outside the interval $\left[-\frac{\pi}{2}, \frac{\pi}{2}\right]$, which is not allowed.) Therefore,

$$\sin^{-1} 1 = \frac{\pi}{2}$$

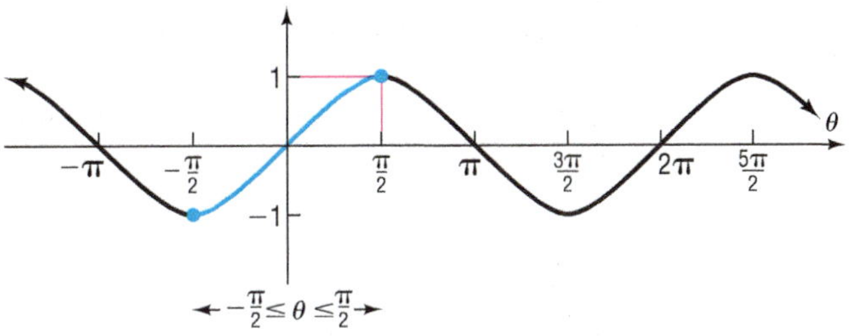

Figure 4

Now Work PROBLEM 15

EXAMPLE 2 Finding the Exact Value of an Inverse Sine Function

Find the exact value of: $\sin^{-1}\left(-\frac{1}{2}\right)$

Solution Let $\theta = \sin^{-1}\left(-\frac{1}{2}\right)$. Then θ is the angle, $-\frac{\pi}{2} \le \theta \le \frac{\pi}{2}$, whose sine equals $-\frac{1}{2}$.

$$\theta = \sin^{-1}\left(-\frac{1}{2}\right) \qquad -\frac{\pi}{2} \le \theta \le \frac{\pi}{2}$$

$$\sin\theta = -\frac{1}{2} \qquad -\frac{\pi}{2} \le \theta \le \frac{\pi}{2}$$

(Refer to Table 1 and Figure 4, if necessary.) The only angle within the interval $\left[-\frac{\pi}{2}, \frac{\pi}{2}\right]$ whose sine is $-\frac{1}{2}$ is $-\frac{\pi}{6}$. So,

$$\sin^{-1}\left(-\frac{1}{2}\right) = -\frac{\pi}{6}$$

Now Work PROBLEM 21

2 Find an Approximate Value of an Inverse Sine Function

For most numbers x, the value $y = \sin^{-1} x$ must be approximated.

EXAMPLE 3 Finding an Approximate Value of an Inverse Sine Function

Find an approximate value of each of the following functions:

(a) $\sin^{-1}\frac{1}{3}$ (b) $\sin^{-1}\left(-\frac{1}{4}\right)$

Express the answer in radians rounded to two decimal places.

Solution Because the angle is to be measured in radians, first set the mode of the calculator to radians.

(a) Figure 5(a) shows the solution using a TI-84 Plus C graphing calculator.

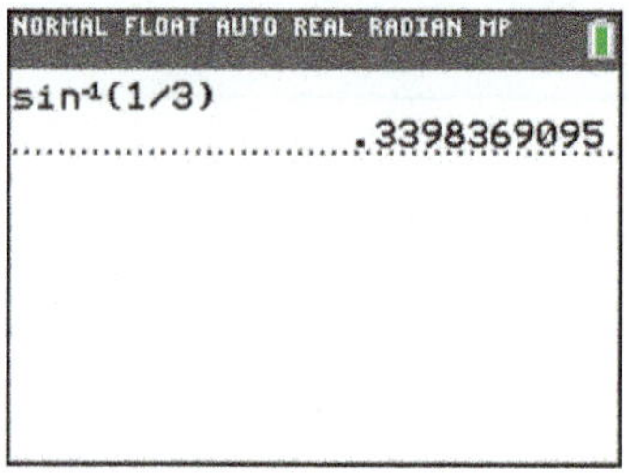

Figure 5(a)

Rounded to two decimal places, $\sin^{-1}\frac{1}{3} = 0.34$.

(b) Figure 5(b) shows the solution using a TI-84 Plus C graphing calculator.

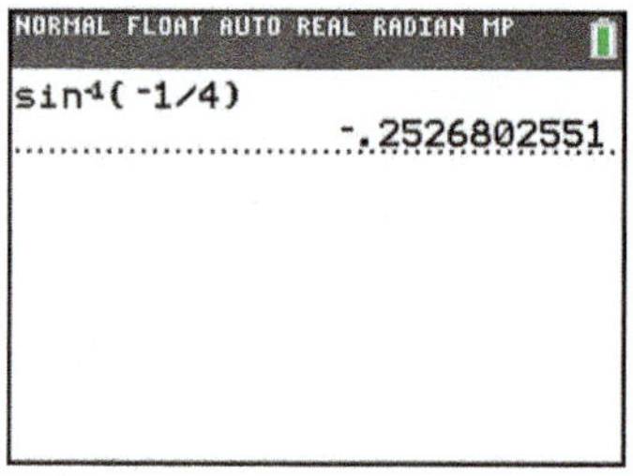

Figure 5(b)

Rounded to two decimal places, $\sin^{-1}\left(-\frac{1}{4}\right) = -0.25$. ■

Now Work PROBLEM 27

3 Use Properties of Inverse Functions to Find Exact Values of Certain Composite Functions

Recall from the discussion of functions and their inverses in Section 6.2, that $f^{-1}(f(x)) = x$ for all x in the domain of f and that $f(f^{-1}(x)) = x$ for all x in the domain of f^{-1}. In terms of the sine function and its inverse, these properties are of the form

$$f^{-1}(f(x)) = \sin^{-1}(\sin x) = x \quad \text{where } -\frac{\pi}{2} \le x \le \frac{\pi}{2} \quad \textbf{(2a)}$$

$$f(f^{-1}(x)) = \sin(\sin^{-1} x) = x \quad \text{where } -1 \le x \le 1 \quad \textbf{(2b)}$$

EXAMPLE 4 **Finding the Exact Value of Certain Composite Functions**

Find the exact value of each of the following composite functions:

(a) $\sin^{-1}\left(\sin\frac{\pi}{8}\right)$ (b) $\sin^{-1}\left(\sin\frac{5\pi}{8}\right)$

Solution (a) The composite function $\sin^{-1}\left(\sin\frac{\pi}{8}\right)$ follows the form of property (2a). Because $\frac{\pi}{8}$ is in the interval $\left[-\frac{\pi}{2}, \frac{\pi}{2}\right]$, property (2a) can be used. So

$$\sin^{-1}\left(\sin\frac{\pi}{8}\right) = \frac{\pi}{8}$$

Figure 6 verifies the calculation using a TI-84 Plus C graphing calculator.

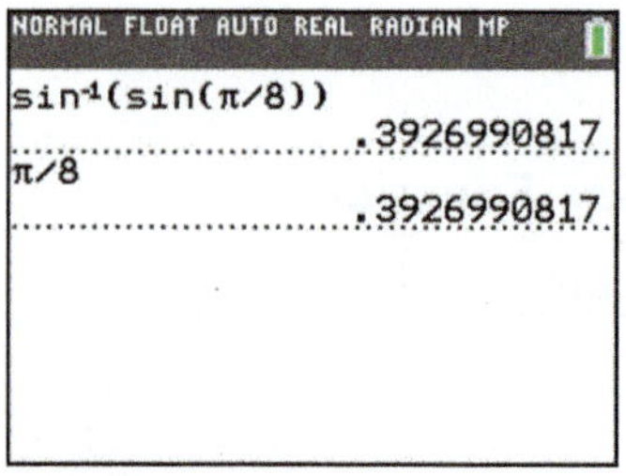

Figure 6

(b) The composite function $\sin^{-1}\left(\sin\frac{5\pi}{8}\right)$ follows the form of property (2a), but $\frac{5\pi}{8}$ is not in the interval $\left[-\frac{\pi}{2}, \frac{\pi}{2}\right]$. To use (2a), first find an angle θ in the interval $\left[-\frac{\pi}{2}, \frac{\pi}{2}\right]$ for which $\sin\theta = \sin\frac{5\pi}{8}$. Then, using (2a),

$$\sin^{-1}\left(\sin\frac{5\pi}{8}\right) = \sin^{-1}(\sin\theta) = \theta$$

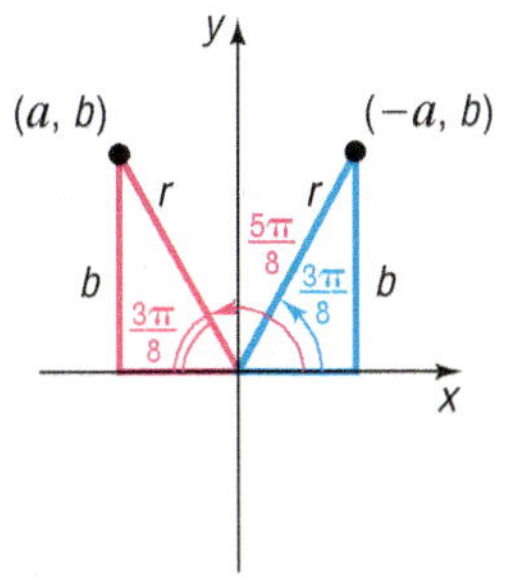

Figure 7

Look at Figure 7. The angle $\frac{5\pi}{8}$ is in quadrant II. The reference angle of $\frac{5\pi}{8}$ is $\frac{3\pi}{8}$ and $\sin \frac{5\pi}{8} = \frac{b}{r} = \sin \frac{3\pi}{8}$. Because $\frac{3\pi}{8}$ is in the interval $\left[-\frac{\pi}{2}, \frac{\pi}{2}\right]$, this means

$$\sin^{-1}\left(\sin \frac{5\pi}{8}\right) = \sin^{-1}\left(\sin \frac{3\pi}{8}\right) = \frac{3\pi}{8}$$

Use property (2a).

These results can be verified with a graphing calculator. Figure 8(a) shows that $\sin^{-1}\left(\sin \frac{5\pi}{8}\right) \neq \frac{5\pi}{8}$. Figure 8(b) confirms that

$$\sin^{-1}\left(\sin \frac{5\pi}{8}\right) = \sin^{-1}\left(\sin \frac{3\pi}{8}\right) = \frac{3\pi}{8}$$

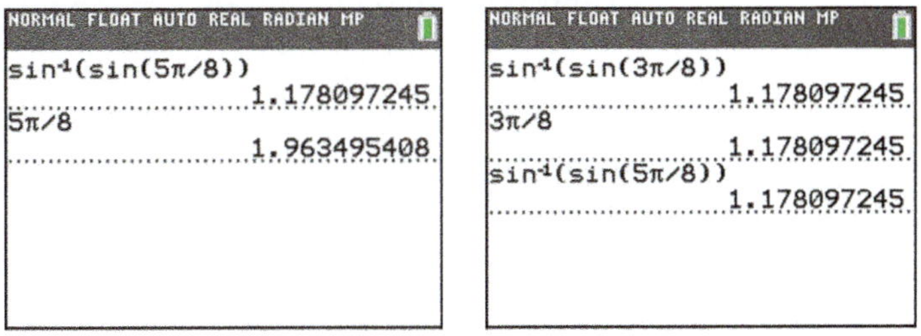

Figure 8 (a) (b)

Now Work PROBLEM 43

EXAMPLE 5 Finding the Exact Value of Certain Composite Functions

Find the exact value, if any, of each composite function.

(a) $\sin(\sin^{-1} 0.8)$ (b) $\sin(\sin^{-1} 1.3)$

Solution

(a) The composite function $\sin(\sin^{-1} 0.8)$ follows the form of property (2b), and 0.8 is in the interval $[-1, 1]$. Using (2b) reveals that

$$\sin(\sin^{-1} 0.8) = 0.8$$

Figure 9 verifies the result.

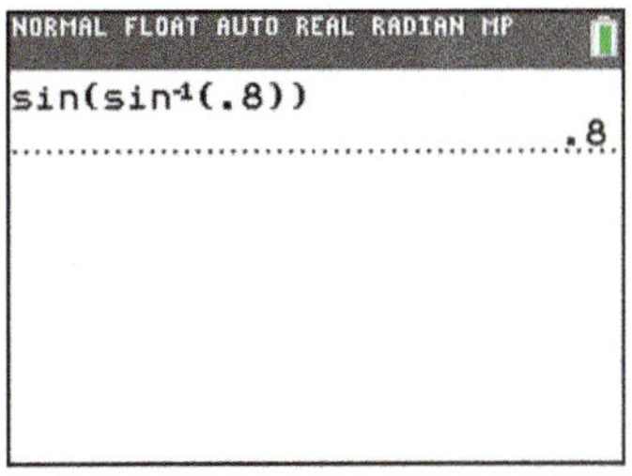

Figure 9

(b) The composite function $\sin(\sin^{-1} 1.3)$ follows the form of property (2b). Since the domain of the inverse sine function is $[-1, 1]$, $\sin^{-1} 1.3$ is not defined. Therefore, $\sin(\sin^{-1} 1.3)$ is also not defined. See Figure 10.

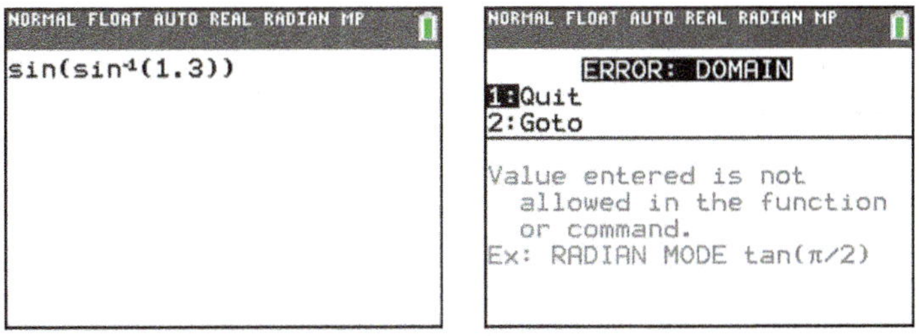

Figure 10

Now Work PROBLEM 55

The Inverse Cosine Function

Figure 11 shows the graph of $y = \cos x$. Because every horizontal line $y = b$, where b is between -1 and 1, inclusive, intersects the graph of $y = \cos x$ infinitely many times, it follows that the cosine function is not one-to-one.

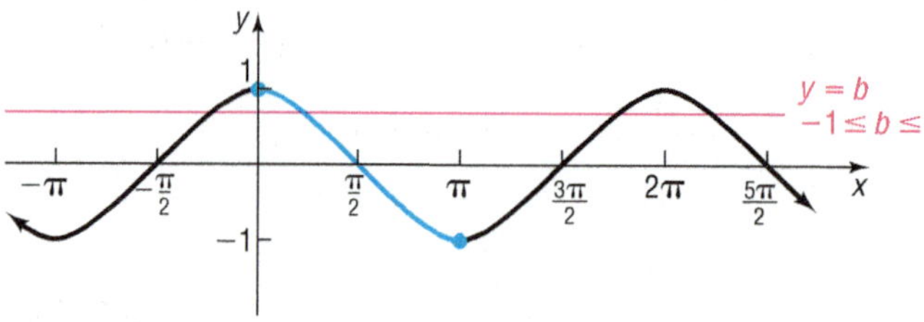

Figure 11 $y = \cos x, -\infty < x < \infty, -1 \le y \le 1$

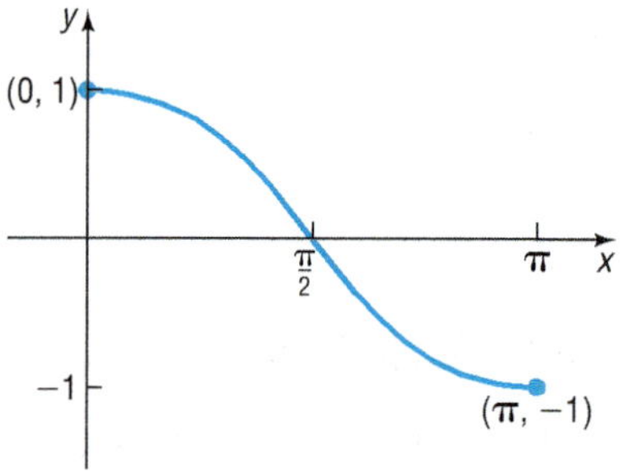

Figure 12
$y = \cos x, 0 \le x \le \pi, -1 \le y \le 1$

However, if the domain of $y = \cos x$ is restricted to the interval $[0, \pi]$, the restricted function

$$y = \cos x \qquad 0 \le x \le \pi$$

is one-to-one and has an inverse function.* See Figure 12.

An equation for the inverse of $y = f(x) = \cos x$ is obtained by interchanging x and y. The implicit form of the inverse function is $x = \cos y$, $0 \le y \le \pi$. The explicit form is called the **inverse cosine** of x and is symbolized by $y = f^{-1}(x) = \cos^{-1} x$ (or by $y = \text{Arccos } x$).

DEFINITION

$$y = \cos^{-1} x \quad \text{means} \quad x = \cos y$$
$$\text{where} \quad -1 \le x \le 1 \quad \text{and} \quad 0 \le y \le \pi \qquad (3)$$

Here y is the angle whose cosine is x. Because the range of the cosine function, $y = \cos x$, is $-1 \le y \le 1$, the domain of the inverse function $y = \cos^{-1} x$ is $-1 \le x \le 1$. Because the restricted domain of the cosine function, $y = \cos x$, is $0 \le x \le \pi$, the range of the inverse function $y = \cos^{-1} x$ is $0 \le y \le \pi$.

The graph of $y = \cos^{-1} x$ can be obtained by reflecting the restricted portion of the graph of $y = \cos x$ about the line $y = x$, as shown in Figure 13.

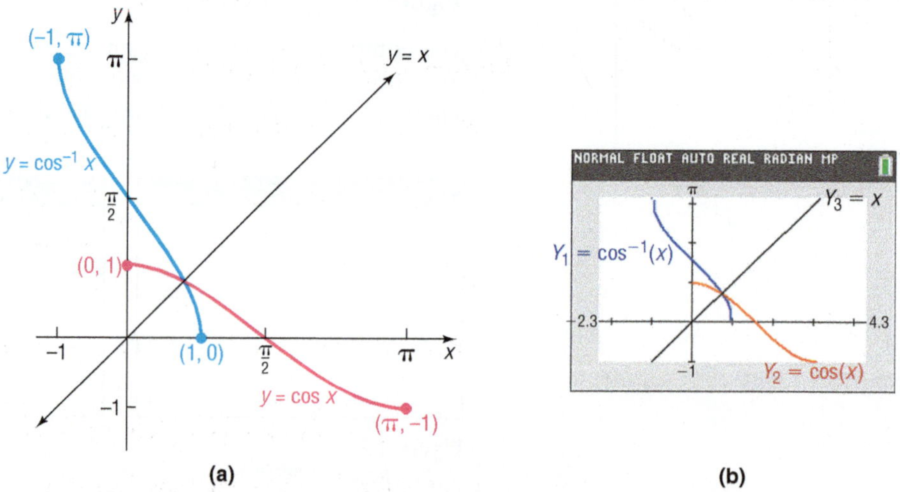

Figure 13
$y = \cos^{-1} x, -1 \le x \le 1, 0 \le y \le \pi$

*This is the generally accepted restriction to define the inverse cosine function.

EXAMPLE 6

Finding the Exact Value of an Inverse Cosine Function

Find the exact value of: $\cos^{-1} 0$

Solution Let $\theta = \cos^{-1} 0$. Then θ is the angle, $0 \le \theta \le \pi$, whose cosine equals 0.

$$\theta = \cos^{-1} 0 \qquad 0 \le \theta \le \pi$$
$$\cos \theta = 0 \qquad 0 \le \theta \le \pi$$

Table 2

θ	$\cos \theta$
0	1
$\frac{\pi}{6}$	$\frac{\sqrt{3}}{2}$
$\frac{\pi}{4}$	$\frac{\sqrt{2}}{2}$
$\frac{\pi}{3}$	$\frac{1}{2}$
$\frac{\pi}{2}$	0
$\frac{2\pi}{3}$	$-\frac{1}{2}$
$\frac{3\pi}{4}$	$-\frac{\sqrt{2}}{2}$
$\frac{5\pi}{6}$	$-\frac{\sqrt{3}}{2}$
π	-1

Look at Table 2 and Figure 14.

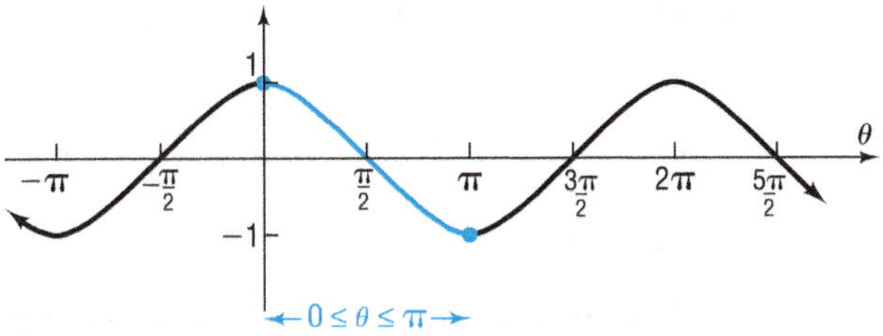

Figure 14

The only angle θ within the interval $[0, \pi]$ whose cosine is 0 is $\frac{\pi}{2}$. [Note that $\cos \frac{3\pi}{2}$ and $\cos\left(-\frac{\pi}{2}\right)$ also equal 0, but they lie outside the interval $[0, \pi]$, so these values are not allowed.] Therefore,

$$\cos^{-1} 0 = \frac{\pi}{2}$$

EXAMPLE 7

Finding the Exact Value of an Inverse Cosine Function

Find the exact value of: $\cos^{-1}\left(-\frac{\sqrt{2}}{2}\right)$

Solution Let $\theta = \cos^{-1}\left(-\frac{\sqrt{2}}{2}\right)$. Then θ is the angle, $0 \le \theta \le \pi$, whose cosine equals $-\frac{\sqrt{2}}{2}$.

$$\theta = \cos^{-1}\left(-\frac{\sqrt{2}}{2}\right) \qquad 0 \le \theta \le \pi$$

$$\cos \theta = -\frac{\sqrt{2}}{2} \qquad 0 \le \theta \le \pi$$

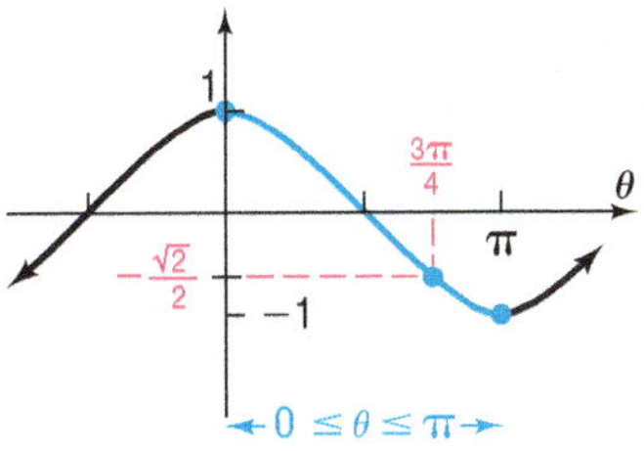

Figure 15

Look at Table 2 and Figure 15.

The only angle θ within the interval $[0, \pi]$ whose cosine is $-\frac{\sqrt{2}}{2}$ is $\frac{3\pi}{4}$, so

$$\cos^{-1}\left(-\frac{\sqrt{2}}{2}\right) = \frac{3\pi}{4}$$

Now Work PROBLEM 25

For the cosine function and its inverse, the following properties hold.

$$f^{-1}(f(x)) = \cos^{-1}(\cos x) = x \quad \text{where } 0 \le x \le \pi \qquad \textbf{(4a)}$$

$$f(f^{-1}(x)) = \cos(\cos^{-1} x) = x \quad \text{where } -1 \le x \le 1 \qquad \textbf{(4b)}$$

EXAMPLE 8 **Using Properties of Inverse Functions to Find the Exact Value of Certain Composite Functions**

Find the exact value of:

(a) $\cos^{-1}\left(\cos\dfrac{\pi}{12}\right)$ (b) $\cos\left[\cos^{-1}(-0.4)\right]$

(c) $\cos^{-1}\left[\cos\left(-\dfrac{2\pi}{3}\right)\right]$ (d) $\cos\left(\cos^{-1}\pi\right)$

Solution (a) $\cos^{-1}\left(\cos\dfrac{\pi}{12}\right)=\dfrac{\pi}{12}$ $\quad$ **$\dfrac{\pi}{12}$ is in the interval $[0,\pi]$; use Property (4a).**

(b) $\cos\left[\cos^{-1}(-0.4)\right]=-0.4$ $\quad$ **-0.4 is in the interval $[-1,1]$; use Property (4b).**

(c) The angle $-\dfrac{2\pi}{3}$ is not in the interval $[0,\pi]$ so property (4a) cannot be used. However, because the cosine function is even, $\cos\left(-\dfrac{2\pi}{3}\right)=\cos\dfrac{2\pi}{3}$. Because $\dfrac{2\pi}{3}$ is in the interval $[0,\pi]$, property (4a) can now be used, and

$$\cos^{-1}\left[\cos\left(-\frac{2\pi}{3}\right)\right]=\cos^{-1}\left(\cos\frac{2\pi}{3}\right)=\frac{2\pi}{3}$$

(d) Because π is not in the interval $[-1,1]$, the domain of the inverse cosine function, $\cos^{-1}\pi$ is not defined. This means the composite function $\cos(\cos^{-1}\pi)$ is also not defined. ■

Now Work PROBLEMS 39 AND 59

The Inverse Tangent Function

Figure 16 shows the graph of $y=\tan x$. Because every horizontal line intersects the graph infinitely many times, it follows that the tangent function is not one-to-one.

However, if the domain of $y=\tan x$ is restricted to the interval $\left(-\dfrac{\pi}{2},\dfrac{\pi}{2}\right)$, the restricted function

$$y=\tan x \qquad -\frac{\pi}{2}<x<\frac{\pi}{2}$$

is one-to-one and has an inverse function.* See Figure 17.

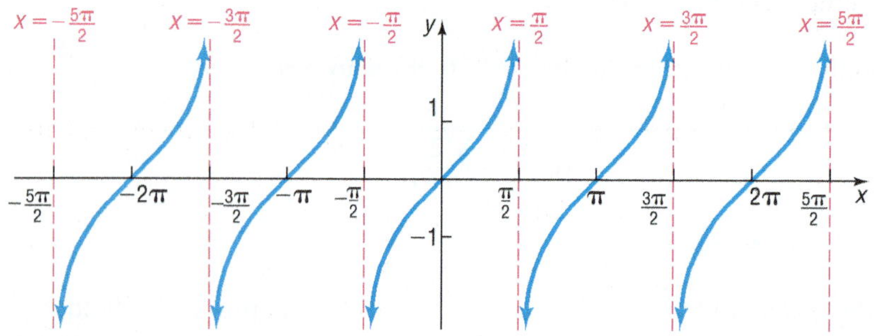

Figure 16
$y=\tan x$, $-\infty<x<\infty$, x not equal to odd multiples of $\dfrac{\pi}{2}$, $-\infty<y<\infty$

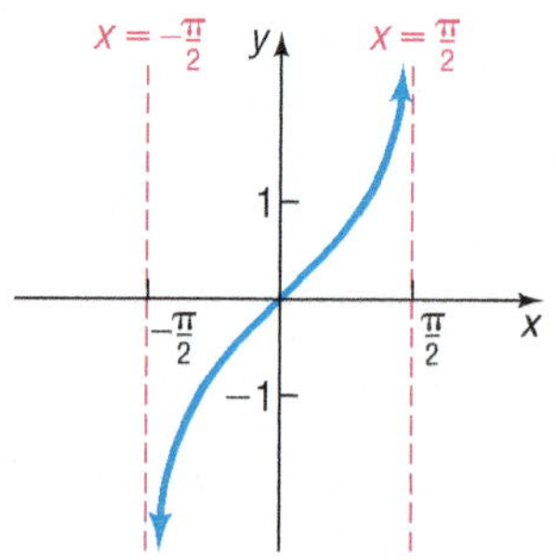

Figure 17
$y=\tan x$, $-\dfrac{\pi}{2}<x<\dfrac{\pi}{2}$, $-\infty<y<\infty$

*This is the generally accepted restriction.

An equation for the inverse of $y = f(x) = \tan x$ is obtained by interchanging x and y. The implicit form of the inverse function is $x = \tan y$, $-\frac{\pi}{2} < y < \frac{\pi}{2}$. The explicit form is called the **inverse tangent** of x and is symbolized by $y = f^{-1}(x) = \tan^{-1} x$ (or by $y = \text{Arctan}\, x$).

DEFINITION

$$y = \tan^{-1} x \quad \text{means} \quad x = \tan y$$
$$\text{where} \quad -\infty < x < \infty \quad \text{and} \quad -\frac{\pi}{2} < y < \frac{\pi}{2} \qquad \textbf{(5)}$$

Here y is the angle whose tangent is x. The domain of the function $y = \tan^{-1} x$ is $-\infty < x < \infty$, and its range is $-\frac{\pi}{2} < y < \frac{\pi}{2}$. The graph of $y = \tan^{-1} x$ can be obtained by reflecting the restricted portion of the graph of $y = \tan x$ about the line $y = x$, as shown in Figure 18.

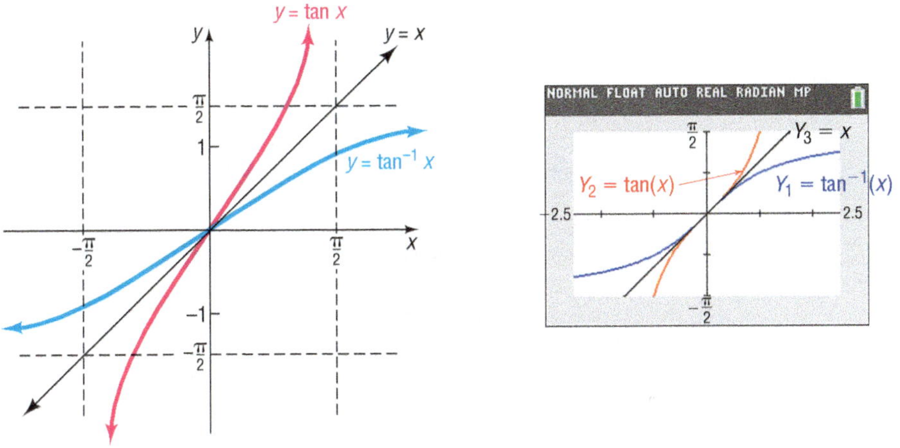

(a) (b)

Figure 18 $y = \tan^{-1} x$, $\infty < x < \infty$, $-\frac{\pi}{2} < y < \frac{\pi}{2}$

EXAMPLE 9 **Finding the Exact Value of an Inverse Tangent Function**

Find the exact value of:

(a) $\tan^{-1} 1$ (b) $\tan^{-1}(-\sqrt{3})$

Solution (a) Let $\theta = \tan^{-1} 1$. Then θ is the angle, $-\frac{\pi}{2} < \theta < \frac{\pi}{2}$, whose tangent equals 1.

$$\theta = \tan^{-1} 1 \qquad -\frac{\pi}{2} < \theta < \frac{\pi}{2}$$

$$\tan \theta = 1 \qquad -\frac{\pi}{2} < \theta < \frac{\pi}{2}$$

Look at Table 3 on the next page. The only angle θ within the interval $\left(-\frac{\pi}{2}, \frac{\pi}{2}\right)$ whose tangent is 1 is $\frac{\pi}{4}$, so

$$\tan^{-1} 1 = \frac{\pi}{4}$$

Table 3

θ	$\tan\theta$
$-\frac{\pi}{2}$	Undefined
$-\frac{\pi}{3}$	$-\sqrt{3}$
$-\frac{\pi}{4}$	-1
$-\frac{\pi}{6}$	$-\frac{\sqrt{3}}{3}$
0	0
$\frac{\pi}{6}$	$\frac{\sqrt{3}}{3}$
$\frac{\pi}{4}$	1
$\frac{\pi}{3}$	$\sqrt{3}$
$\frac{\pi}{2}$	Undefined

(b) Let $\theta = \tan^{-1}(-\sqrt{3})$. Then θ is the angle, $-\frac{\pi}{2} < \theta < \frac{\pi}{2}$, whose tangent equals $-\sqrt{3}$.

$$\theta = \tan^{-1}(-\sqrt{3}) \qquad -\frac{\pi}{2} < \theta < \frac{\pi}{2}$$

$$\tan\theta = -\sqrt{3} \qquad -\frac{\pi}{2} < \theta < \frac{\pi}{2}$$

Look at Table 3. The only angle θ within the interval $\left(-\frac{\pi}{2}, \frac{\pi}{2}\right)$ whose tangent is $-\sqrt{3}$ is $-\frac{\pi}{3}$, so

$$\tan^{-1}(-\sqrt{3}) = -\frac{\pi}{3}$$

■

Now Work PROBLEM 19

For the tangent function and its inverse, the following properties hold.

$$f^{-1}(f(x)) = \tan^{-1}(\tan x) = x \quad \text{where } -\frac{\pi}{2} < x < \frac{\pi}{2}$$

$$f(f^{-1}(x)) = \tan(\tan^{-1} x) = x \quad \text{where } -\infty < x < \infty$$

Now Work PROBLEM 47

4 Find the Inverse Function of a Trigonometric Function

EXAMPLE 10 **Finding the Inverse Function of a Trigonometric Function**

(a) Find the inverse function f^{-1} of $f(x) = 2\sin x - 1$, $-\frac{\pi}{2} \le x \le \frac{\pi}{2}$.

(b) Find the range of f and the domain and range of f^{-1}.

Solution (a) The function f is one-to-one and so has an inverse function. Follow the steps on page 422 for finding the inverse function.

$$y = 2\sin x - 1$$

$$x = 2\sin y - 1 \quad \text{Interchange } x \text{ and } y.$$

$$x + 1 = 2\sin y \quad \text{Proceed to solve for } y.$$

$$\sin y = \frac{x+1}{2}$$

$$y = \sin^{-1}\frac{x+1}{2} \quad \text{Apply the definition (1).}$$

The inverse function is $f^{-1}(x) = \sin^{-1}\frac{x+1}{2}$.

(b) To find the range of f, solve $y = 2\sin x - 1$ for $\sin x$, and use the fact that $-1 \le \sin x \le 1$.

$$y = 2\sin x - 1$$

$$\sin x = \frac{y+1}{2}$$

$$-1 \le \frac{y+1}{2} \le 1$$

$$-2 \le y + 1 \le 2$$

$$-3 \le y \le 1$$

Note: The range of f also can be found using transformations. The range of $y = \sin x$ is $[-1, 1]$. The range of $y = 2\sin x$ is $[-2, 2]$ due to the vertical stretch by a factor of 2. The range of $f(x) = 2\sin x - 1$ is $[-3, 1]$ due to the shift down of 1 unit. ■

The range of f is $\{y \mid -3 \le y \le 1\}$, or $[-3, 1]$ using interval notation.

The domain of f^{-1} equals the range of f, $[-3, 1]$.

The range of f^{-1} equals the domain of f, $\left[-\frac{\pi}{2}, \frac{\pi}{2}\right]$.

Now Work PROBLEM 65

5 Solve Equations Involving Inverse Trigonometric Functions

Equations that contain inverse trigonometric functions are called **inverse trigonometric equations.**

EXAMPLE 11 Solving an Equation Involving an Inverse Trigonometric Function

Solve the equation: $3 \sin^{-1} x = \pi$

Solution To solve an equation involving a single inverse trigonometric function, first isolate the inverse trigonometric function.

$$3 \sin^{-1} x = \pi$$

$$\sin^{-1} x = \frac{\pi}{3} \qquad \text{Divide both sides by 3.}$$

$$x = \sin \frac{\pi}{3} \qquad y = \sin^{-1} x \text{ means } x = \sin y.$$

$$x = \frac{\sqrt{3}}{2}$$

The solution set is $\left\{\frac{\sqrt{3}}{2}\right\}$.

Now Work PROBLEM 71

8.1 Assess Your Understanding

'Are You Prepared?' *Answers are given at the end of these exercises. If you get a wrong answer, read the pages listed in red.*

1. What is the domain and the range of $y = \sin x$? (p. 566)
2. A suitable restriction on the domain of the function $f(x) = (x-1)^2$ to make it one-to-one would be ______. (pp. 423–424)
3. If the domain of a one-to-one function is $[3, \infty)$, the range of its inverse is ______. (pp. 419–420)
4. ***True or False*** The graph of $y = \cos x$ is decreasing on the interval $[0, \pi]$. (pp. 574–575)
5. $\tan \frac{\pi}{4} =$ ______; $\sin \frac{\pi}{3} =$ ______ (pp. 539–541)
6. $\sin\left(-\frac{\pi}{6}\right) =$ ______; $\cos \pi =$ ______ (pp. 550–558)

Concepts and Vocabulary

7. $y = \sin^{-1} x$ means ______, where $-1 \le x \le 1$ and $-\frac{\pi}{2} \le y \le \frac{\pi}{2}$.
8. $\cos^{-1}(\cos x) = x$, where ______.
9. $\tan(\tan^{-1} x) = x$, where ______.
10. ***True or False*** The domain of $y = \sin^{-1} x$ is $-\frac{\pi}{2} \le x \le \frac{\pi}{2}$.
11. ***True or False*** $\sin(\sin^{-1} 0) = 0$ and $\cos(\cos^{-1} 0) = 0$.
12. ***True or False*** $y = \tan^{-1} x$ means $x = \tan y$, where $-\infty < x < \infty$ and $-\frac{\pi}{2} < y < \frac{\pi}{2}$.

13. Which of the following inequalities describes where $\sin^{-1}(\sin x) = x$?
(a) $-\infty < x < \infty$ (b) $0 \le x \le \pi$
(c) $-1 \le x \le 1$ (d) $-\frac{\pi}{2} \le x \le \frac{\pi}{2}$

14. Choose the inverse function f^{-1} of
$$f(x) = \frac{1}{2}\tan x, -\frac{\pi}{2} < x < \frac{\pi}{2}$$
(a) $f^{-1}(x) = \tan^{-1}(2x); -\infty < x < \infty$
(b) $f^{-1}(x) = 2\tan^{-1} x; -\infty < x < \infty$
(c) $f^{-1}(x) = \frac{1}{2}\tan^{-1} x; -\infty < x < \infty$
(d) $f^{-1}(x) = \tan^{-1}\left(\frac{1}{2}x\right); -\infty < x < \infty$

Skill Building

In Problems 15–26, find the exact value of each expression.

15. $\sin^{-1} 0$ **16.** $\cos^{-1} 1$ **17.** $\sin^{-1}(-1)$ **18.** $\cos^{-1}(-1)$

19. $\tan^{-1} 0$ **20.** $\tan^{-1}(-1)$ **21.** $\sin^{-1}\frac{\sqrt{2}}{2}$ **22.** $\tan^{-1}\frac{\sqrt{3}}{3}$

23. $\tan^{-1}\sqrt{3}$ **24.** $\sin^{-1}\left(-\frac{\sqrt{3}}{2}\right)$ **25.** $\cos^{-1}\left(-\frac{\sqrt{3}}{2}\right)$ **26.** $\sin^{-1}\left(-\frac{\sqrt{2}}{2}\right)$

In Problems 27–38, use a calculator to find the value of each expression rounded to two decimal places.

27. $\sin^{-1} 0.1$ **28.** $\cos^{-1} 0.6$ **29.** $\tan^{-1} 5$ **30.** $\tan^{-1} 0.2$

31. $\cos^{-1}\frac{7}{8}$ **32.** $\sin^{-1}\frac{1}{8}$ **33.** $\tan^{-1}(-0.4)$ **34.** $\tan^{-1}(-3)$

35. $\sin^{-1}(-0.12)$ **36.** $\cos^{-1}(-0.44)$ **37.** $\cos^{-1}\frac{\sqrt{2}}{3}$ **38.** $\sin^{-1}\frac{\sqrt{3}}{5}$

In Problems 39–62, find the exact value, if any, of each composite function. If there is no value, say it is "not defined." Do not use a calculator.

39. $\cos^{-1}\left(\cos\frac{4\pi}{5}\right)$ **40.** $\sin^{-1}\left[\sin\left(-\frac{\pi}{10}\right)\right]$ **41.** $\tan^{-1}\left[\tan\left(-\frac{3\pi}{8}\right)\right]$ **42.** $\sin^{-1}\left[\sin\left(-\frac{3\pi}{7}\right)\right]$

43. $\sin^{-1}\left(\sin\frac{9\pi}{8}\right)$ **44.** $\sin^{-1}\left(\sin\frac{11\pi}{4}\right)$ **45.** $\cos^{-1}\left[\cos\left(-\frac{5\pi}{3}\right)\right]$ **46.** $\cos^{-1}\left(\cos\frac{7\pi}{6}\right)$

47. $\tan^{-1}\left(\tan\frac{4\pi}{5}\right)$ **48.** $\tan^{-1}\left[\tan\left(-\frac{10\pi}{9}\right)\right]$ **49.** $\tan^{-1}\left[\tan\left(-\frac{2\pi}{3}\right)\right]$ **50.** $\cos^{-1}\left(\cos\frac{4\pi}{3}\right)$

51. $\cos^{-1}\left[\cos\left(-\frac{\pi}{4}\right)\right]$ **52.** $\sin^{-1}\left[\sin\left(-\frac{3\pi}{4}\right)\right]$ **53.** $\tan^{-1}\left[\tan\left(\frac{\pi}{2}\right)\right]$ **54.** $\tan^{-1}\left[\tan\left(-\frac{3\pi}{2}\right)\right]$

55. $\sin\left(\sin^{-1}\frac{1}{4}\right)$ **56.** $\cos\left[\cos^{-1}\left(-\frac{2}{3}\right)\right]$ **57.** $\tan(\tan^{-1}4)$ **58.** $\tan[\tan^{-1}(-2)]$

59. $\cos(\cos^{-1}1.2)$ **60.** $\sin[\sin^{-1}(-2)]$ **61.** $\tan(\tan^{-1}\pi)$ **62.** $\sin[\sin^{-1}(-1.5)]$

In Problems 63–70, find the inverse function f^{-1} of each function f. Find the range of f and the domain and range of f^{-1}.

63. $f(x) = 5\sin x + 2; -\frac{\pi}{2} \le x \le \frac{\pi}{2}$ **64.** $f(x) = 2\tan x - 3; -\frac{\pi}{2} < x < \frac{\pi}{2}$

65. $f(x) = -2\cos(3x); 0 \le x \le \frac{\pi}{3}$ **66.** $f(x) = 3\sin(2x); -\frac{\pi}{4} \le x \le \frac{\pi}{4}$

67. $f(x) = -\tan(x+1) - 3; -1 - \frac{\pi}{2} < x < \frac{\pi}{2} - 1$ **68.** $f(x) = \cos(x+2) + 1; -2 \le x \le \pi - 2$

69. $f(x) = 3\sin(2x+1); -\frac{1}{2} - \frac{\pi}{4} \le x \le -\frac{1}{2} + \frac{\pi}{4}$ **70.** $f(x) = 2\cos(3x+2); -\frac{2}{3} \le x \le -\frac{2}{3} + \frac{\pi}{3}$

In Problems 71–78, find the exact solution of each equation.

71. $4\sin^{-1}x = \pi$ **72.** $2\cos^{-1}x = \pi$ **73.** $3\cos^{-1}(2x) = 2\pi$

74. $-6\sin^{-1}(3x) = \pi$ **75.** $3\tan^{-1}x = \pi$ **76.** $-4\tan^{-1}x = \pi$

77. $4\cos^{-1}x - 2\pi = 2\cos^{-1}x$ **78.** $5\sin^{-1}x - 2\pi = 2\sin^{-1}x - 3\pi$

Applications and Extensions

In Problems 79–84, use the following discussion. The formula

$$D = 24\left[1 - \frac{\cos^{-1}(\tan i \tan \theta)}{\pi}\right]$$

can be used to approximate the number of hours of daylight D when the declination of the Sun is $i°$ at a location $\theta°$ north latitude for any date between the vernal equinox and autumnal equinox. The declination of the Sun is defined as the angle i between the equatorial plane and any ray of light from the Sun. The latitude of a location is the angle θ between the Equator and the location on the surface of Earth, with the vertex of the angle located at the center of Earth. See the figure. To use the formula, $\cos^{-1}(\tan i \tan \theta)$ *must be expressed in radians.*

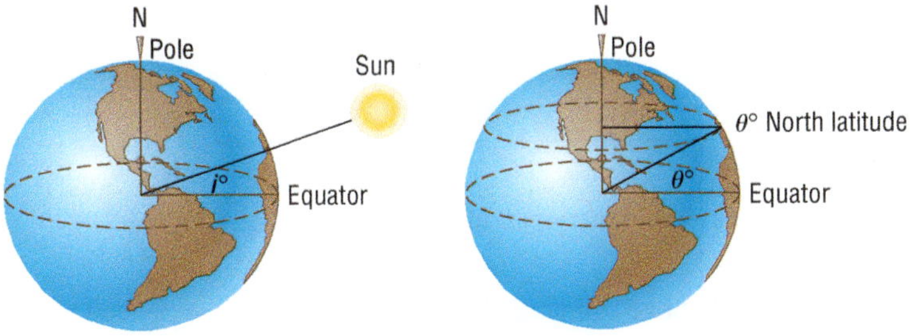

79. Approximate the number of hours of daylight in Houston, Texas (29°45′ north latitude), for the following dates:
(a) Summer solstice $(i = 23.5°)$
(b) Vernal equinox $(i = 0°)$
(c) July 4 $(i = 22°48')$

80. Approximate the number of hours of daylight in New York, New York (40°45′ north latitude), for the following dates:
(a) Summer solstice $(i = 23.5°)$
(b) Vernal equinox $(i = 0°)$
(c) July 4 $(i = 22°48')$

81. Approximate the number of hours of daylight in Honolulu, Hawaii (21°18′ north latitude), for the following dates:
(a) Summer solstice $(i = 23.5°)$
(b) Vernal equinox $(i = 0°)$
(c) July 4 $(i = 22°48')$

82. Approximate the number of hours of daylight in Anchorage, Alaska (61°10′ north latitude), for the following dates:
(a) Summer solstice $(i = 23.5°)$
(b) Vernal equinox $(i = 0°)$
(c) July 4 $(i = 22°48')$

83. Approximate the number of hours of daylight at the Equator (0° north latitude) for the following dates:
(a) Summer solstice $(i = 23.5°)$
(b) Vernal equinox $(i = 0°)$
(c) July 4 $(i = 22°48')$
(d) What do you conclude about the number of hours of daylight throughout the year for a location at the Equator?

84. Approximate the number of hours of daylight for any location that is 66°30′ north latitude for the following dates:
(a) Summer solstice $(i = 23.5°)$
(b) Vernal equinox $(i = 0°)$
(c) July 4 $(i = 22°48')$
(d) Thanks to the symmetry of the orbital path of Earth around the Sun, the number of hours of daylight on the winter solstice may be found by computing the number of hours of daylight on the summer solstice and subtracting this result from 24 hours. Compute the number of hours of daylight for this location on the winter solstice. What do you conclude about daylight for a location at 66°30′ north latitude?

85. Being the First to See the Rising Sun Cadillac Mountain, elevation 1530 feet, is located in Acadia National Park, Maine, and is the highest peak on the east coast of the United States. It is said that a person standing on the summit will be the first person in the United States to see the rays of the rising Sun. How much sooner would a person atop Cadillac Mountain see the first rays than a person standing below, at sea level?

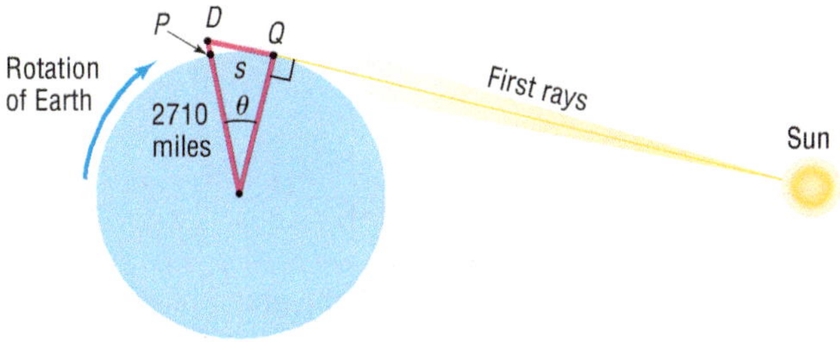

[**Hint:** Consult the figure. When the person at D sees the first rays of the Sun, the person at P does not. The person at P sees the first rays of the Sun only after Earth has rotated so that P is at location Q. Compute the length of the arc subtended by the central angle θ. Then use the fact that at the latitude of Cadillac Mountain, in 24 hours a length of $2\pi(2710) \approx 17{,}027.4$ miles is subtended.]

86. Movie Theater Screens Suppose that a movie theater has a screen that is 28 feet tall. When you sit down, the bottom of the screen is 6 feet above your eye level. The angle formed by drawing a line from your eye to the bottom of the screen and another line from your eye to the top of the screen is called the **viewing angle**. In the figure, θ is the viewing angle. Suppose that you sit x feet from the screen. The viewing angle θ is given by the function

$$\theta(x) = \tan^{-1}\left(\frac{34}{x}\right) - \tan^{-1}\left(\frac{6}{x}\right)$$

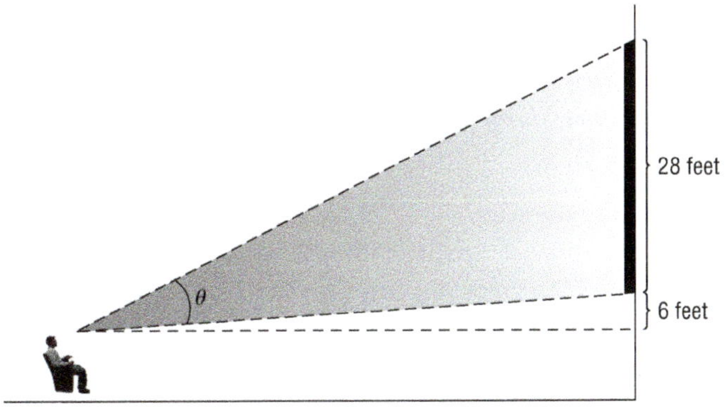

(a) What is your viewing angle if you sit 10 feet from the screen? 15 feet? 20 feet?
(b) If there are 5 feet between the screen and the first row of seats and there are 3 feet between each row and the row behind it, which row results in the largest viewing angle?

(c) Using a graphing utility, graph

$$\theta(x) = \tan^{-1}\left(\frac{34}{x}\right) - \tan^{-1}\left(\frac{6}{x}\right)$$

What value of x results in the largest viewing angle?

87. Area under a Curve The area under the graph of $y = \frac{1}{1+x^2}$ and above the x-axis between $x = a$ and $x = b$ is given by

$$\tan^{-1} b - \tan^{-1} a$$

See the figure.

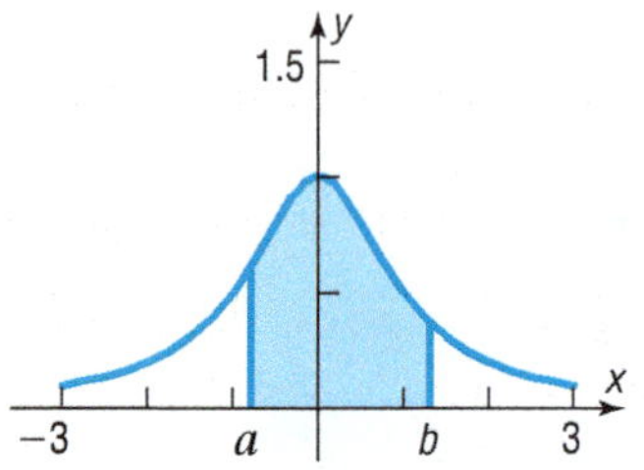

(a) Find the exact area under the graph of $y = \frac{1}{1+x^2}$ and above the x-axis between $x = 0$ and $x = \sqrt{3}$.

(b) Find the exact area under the graph of $y = \frac{1}{1+x^2}$ and above the x-axis between $x = -\frac{\sqrt{3}}{3}$ and $x = 1$.

88. Area under a Curve The area under the graph of $y = \frac{1}{\sqrt{1-x^2}}$ and above the x-axis between $x = a$ and $x = b$ is given by

$$\sin^{-1} b - \sin^{-1} a$$

See the figure.

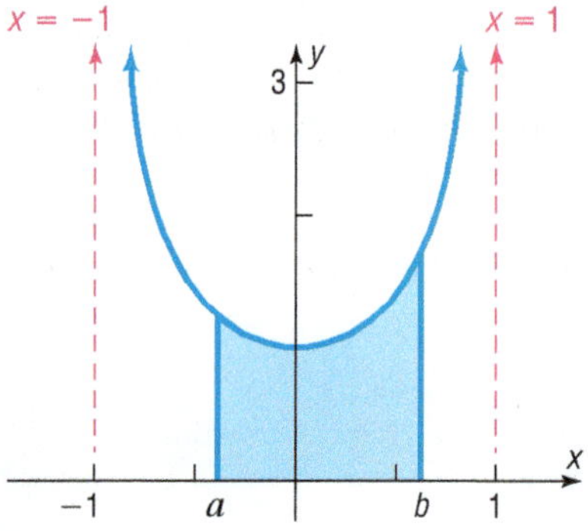

(a) Find the exact area under the graph of $y = \frac{1}{\sqrt{1-x^2}}$ and above the x-axis between $x = 0$ and $x = \frac{\sqrt{3}}{2}$.

(b) Find the exact area under the graph of $y = \frac{1}{\sqrt{1-x^2}}$ and above the x-axis between $x = -\frac{1}{2}$ and $x = \frac{1}{2}$.

Problems 89 and 90 require the following discussion:
The shortest distance between two points on Earth's surface can be determined from the latitude and longitude of the two locations. For example, if location 1 has (lat, lon) $= (\alpha_1, \beta_1)$ *and location 2 has* (lat, lon) $= (\alpha_2, \beta_2)$, *the shortest distance between the two locations is approximately* $d = r\cos^{-1}[(\cos\alpha_1\cos\beta_1\cos\alpha_2\cos\beta_2) + (\cos\alpha_1\sin\beta_1\cos\alpha_2\sin\beta_2) + (\sin\alpha_1\sin\alpha_2)]$, *where* r = *radius of Earth* $\approx$ 3960 *miles and the inverse cosine function is expressed in radians. Also, N latitude and E longitude are positive angles, and S latitude and W longitude are negative angles.*

City	Latitude	Longitude
Chicago, IL	41°50′N	87°37′W
Honolulu, HI	21°18′N	157°50′W
Melbourne, Australia	37°47′S	144°58′E

Source: *www.infoplease.com*

89. Shortest Distance from Chicago to Honolulu Find the shortest distance from Chicago, latitude 41°50′N, longitude 87°37′W, to Honolulu, latitude 21°18′N, longitude 157°50′W. Round your answer to the nearest mile.

90. Shortest Distance from Honolulu to Melbourne, Australia Find the shortest distance from Honolulu to Melbourne, Australia, latitude 37°47′S, longitude 144°58′E. Round your answer to the nearest mile.

Retain Your Knowledge

Problems 91–94 are based on material learned earlier in the course. The purpose of these problems is to keep the material fresh in your mind so that you are better prepared for the final exam.

91. Solve: $|3x - 2| + 5 \le 9$

92. State why the graph of the function f shown to the right is one-to-one. Then draw the graph of the inverse function f^{-1}.
[**Hint:** The graph of $y = x$ is given.]

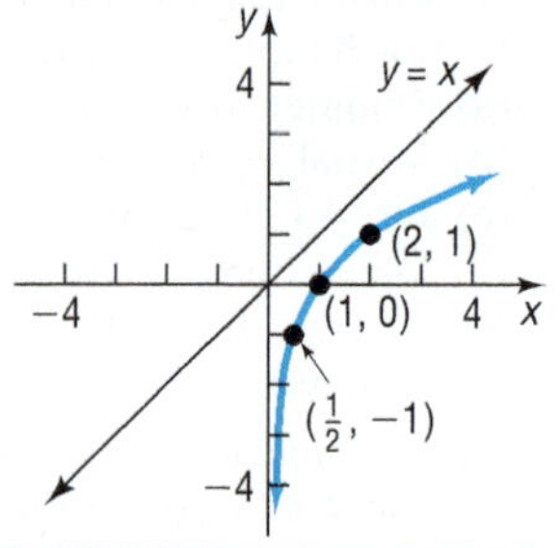

93. The exponential function $f(x) = 1 + 2^x$ is one-to-one. Find f^{-1}.

94. Factor: $(2x+1)^{-\frac{1}{2}}(x^2+3)^{-\frac{1}{2}} - (x^2+3)^{-\frac{3}{2}}x(2x+1)^{\frac{1}{2}}$

'Are You Prepared?' Answers

1. domain: the set of all real numbers; range: $-1 \le y \le 1$ **2.** Two answers are possible: $x \le 1$ or $x \ge 1$

3. $[3, \infty)$ **4.** True **5.** $1; \frac{\sqrt{3}}{2}$ **6.** $-\frac{1}{2}; -1$

8.2 The Inverse Trigonometric Functions (Continued)

PREPARING FOR THIS SECTION *Before getting started, review the following:*

- Finding Exact Values of the Trigonometric Functions, Given the Value of a Trigonometric Function and the Quadrant of the Angle (Section 7.4, pp. 557–558)
- Graphs of the Secant, Cosecant, and Cotangent Functions (Section 7.7, pp. 589–593)
- Domain and Range of the Secant, Cosecant, and Cotangent Functions (Section 7.5, pp. 565–567)

Now Work the 'Are You Prepared?' problems on page 632.

OBJECTIVES
1. Find the Exact Value of Expressions Involving the Inverse Sine, Cosine, and Tangent Functions (p. 629)
2. Define the Inverse Secant, Cosecant, and Cotangent Functions (p. 630)
3. Use a Calculator to Evaluate $\sec^{-1} x$, $\csc^{-1} x$, and $\cot^{-1} x$ (p. 631)
4. Write a Trigonometric Expression as an Algebraic Expression (p. 632)

1 Find the Exact Value of Expressions Involving the Inverse Sine, Cosine, and Tangent Functions

EXAMPLE 1 Finding the Exact Value of an Expression Involving Inverse Trigonometric Functions

Find the exact value of: $\sin\left(\tan^{-1}\frac{1}{2}\right)$

Solution Let $\theta = \tan^{-1}\frac{1}{2}$. Then $\tan\theta = \frac{1}{2}$, where $-\frac{\pi}{2} < \theta < \frac{\pi}{2}$. We seek $\sin\theta$. Because $\tan\theta > 0$, it follows that $0 < \theta < \frac{\pi}{2}$, so θ lies in quadrant I. Figure 19 shows a triangle in quadrant I. Because $\tan\theta = \frac{b}{a} = \frac{1}{2}$, the side opposite θ is $b = 1$ and the side adjacent to θ is $a = 2$. The hypotenuse of this triangle is found using $a^2 + b^2 = c^2$ or $2^2 + 1^2 = c^2$, so $c = \sqrt{5}$. Then $\sin\theta = \frac{b}{c} = \frac{1}{\sqrt{5}}$, and

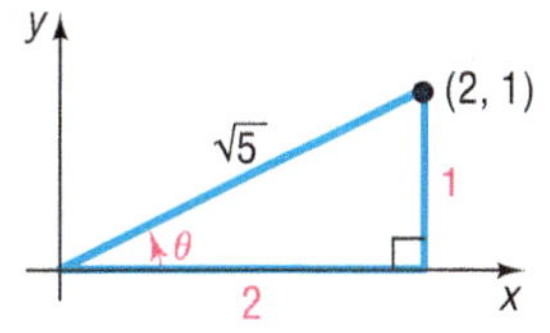

Figure 19 $\tan\theta = \frac{1}{2}$

$$\sin\left(\tan^{-1}\frac{1}{2}\right) = \sin\theta = \frac{1}{\sqrt{5}} = \frac{\sqrt{5}}{5}$$

EXAMPLE 2 Finding the Exact Value of an Expression Involving Inverse Trigonometric Functions

Find the exact value of: $\cos\left[\sin^{-1}\left(-\frac{1}{3}\right)\right]$

Solution Let $\theta = \sin^{-1}\left(-\frac{1}{3}\right)$. Then $\sin\theta = -\frac{1}{3}$ and $-\frac{\pi}{2} \le \theta \le \frac{\pi}{2}$. We seek $\cos\theta$. Because $\sin\theta < 0$, it follows that $-\frac{\pi}{2} \le \theta < 0$, so θ lies in quadrant IV. Figure 20 illustrates $\sin\theta = -\frac{1}{3}$ for θ in quadrant IV. Then

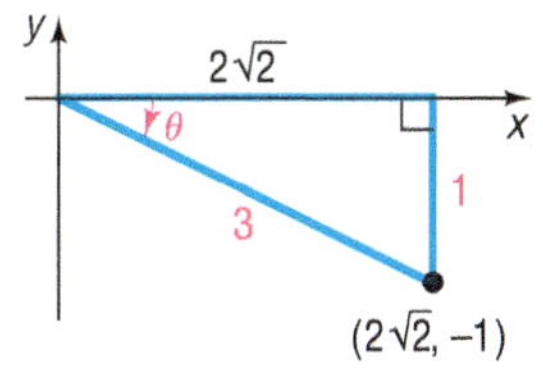

Figure 20 $\sin\theta = -\frac{1}{3}$

$$\cos\left[\sin^{-1}\left(-\frac{1}{3}\right)\right] = \cos\theta = \frac{2\sqrt{2}}{3}$$

$\uparrow$ $\cos\theta = \frac{a}{r}$

EXAMPLE 3 **Finding the Exact Value of an Expression Involving Inverse Trigonometric Functions**

Find the exact value of: $\tan\left[\cos^{-1}\left(-\frac{1}{3}\right)\right]$

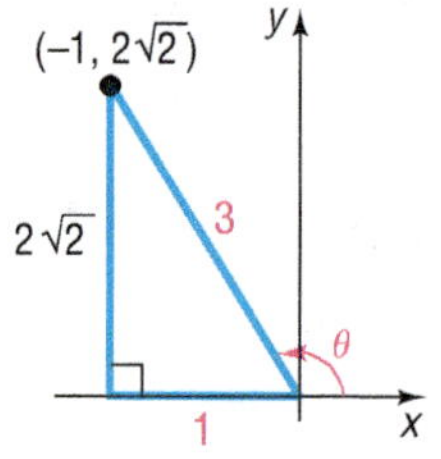

Figure 21 $\cos\theta = -\frac{1}{3}$

Solution Let $\theta = \cos^{-1}\left(-\frac{1}{3}\right)$. Then $\cos\theta = -\frac{1}{3}$ and $0 \le \theta \le \pi$. We seek $\tan\theta$. Because $\cos\theta < 0$, it follows that $\frac{\pi}{2} < \theta \le \pi$, so θ lies in quadrant II. Figure 21 illustrates $\cos\theta = -\frac{1}{3}$ for θ in quadrant II. Then

$$\tan\left[\cos^{-1}\left(-\frac{1}{3}\right)\right] = \tan\theta = \frac{2\sqrt{2}}{-1} = -2\sqrt{2}$$

Now Work PROBLEMS 9 AND 27

2 Define the Inverse Secant, Cosecant, and Cotangent Functions

The inverse secant, inverse cosecant, and inverse cotangent functions are defined as follows:

DEFINITION

$$y = \sec^{-1} x \quad \text{means} \quad x = \sec y \qquad \textbf{(1)}$$
$$\text{where} \quad |x| \ge 1 \quad \text{and} \quad 0 \le y \le \pi, \quad y \ne \frac{\pi}{2}{}^{*}$$

$$y = \csc^{-1} x \quad \text{means} \quad x = \csc y \qquad \textbf{(2)}$$
$$\text{where} \quad |x| \ge 1 \quad \text{and} \quad -\frac{\pi}{2} \le y \le \frac{\pi}{2}, \quad y \ne 0^{\dagger}$$

$$y = \cot^{-1} x \quad \text{means} \quad x = \cot y \qquad \textbf{(3)}$$
$$\text{where} \quad -\infty < x < \infty \quad \text{and} \quad 0 < y < \pi$$

Take time to review the graphs of the cotangent, cosecant, and secant functions in Figures 106, 108, and 109 in Section 7.7 to see the basis for these definitions.

EXAMPLE 4 **Finding the Exact Value of an Inverse Cosecant Function**

Find the exact value of: $\csc^{-1} 2$

Solution Let $\theta = \csc^{-1} 2$. Then θ is the angle, $-\frac{\pi}{2} \le \theta \le \frac{\pi}{2}, \theta \ne 0$, whose cosecant equals 2 $\left(\text{or, equivalently, whose sine equals } \frac{1}{2}\right)$.

$$\theta = \csc^{-1} 2 \qquad -\frac{\pi}{2} \le \theta \le \frac{\pi}{2}, \quad \theta \ne 0$$
$$\csc\theta = 2 \qquad -\frac{\pi}{2} \le \theta \le \frac{\pi}{2}, \quad \theta \ne 0 \qquad \sin\theta = \frac{1}{2}$$

The only angle θ in the interval $-\frac{\pi}{2} \le \theta \le \frac{\pi}{2}, \theta \ne 0$, whose cosecant is 2 $\left[\sin\theta = \frac{1}{2}\right]$ is $\frac{\pi}{6}$, so $\csc^{-1} 2 = \frac{\pi}{6}$.

Now Work PROBLEM 39

*Most texts use this definition. A few use the restriction $0 \le y < \frac{\pi}{2}, \pi \le y < \frac{3\pi}{2}$.

†Most texts use this definition. A few use the restriction $-\pi < y \le -\frac{\pi}{2}, 0 < y \le \frac{\pi}{2}$.

3 Use a Calculator to Evaluate $\sec^{-1} x$, $\csc^{-1} x$, and $\cot^{-1} x$

Note: Remember that the range of $y = \sin^{-1} x$ is $\left[-\frac{\pi}{2}, \frac{\pi}{2}\right]$ and that the range of $y = \cos^{-1} x$ is $[0, \pi]$. ■

Most calculators do not have keys for evaluating the inverse cotangent, cosecant, and secant functions. The easiest way to evaluate them is to convert each to an inverse trigonometric function whose range is the same as the one to be evaluated. In this regard, notice that $y = \cot^{-1} x$ and $y = \sec^{-1} x$, except where undefined, have the same range as $y = \cos^{-1} x$ and that $y = \csc^{-1} x$, except where undefined, has the same range as $y = \sin^{-1} x$.

EXAMPLE 5 **Approximating the Value of Inverse Trigonometric Functions**

Use a calculator to approximate each expression in radians rounded to two decimal places.

(a) $\sec^{-1} 3$ (b) $\csc^{-1}(-4)$ (c) $\cot^{-1} \frac{1}{2}$ (d) $\cot^{-1}(-2)$

Solution First, set your calculator to radian mode.

(a) Let $\theta = \sec^{-1} 3$. Then $\sec\theta = 3$ and $0 \le \theta \le \pi, \theta \ne \frac{\pi}{2}$. Now find $\cos\theta$ because $y = \cos^{-1} x$ has the same range as $y = \sec^{-1} x$, except where undefined. Because $\sec\theta = \frac{1}{\cos\theta} = 3$, this means $\cos\theta = \frac{1}{3}$. Then $\theta = \cos^{-1}\frac{1}{3}$ and

$$\sec^{-1} 3 = \theta = \cos^{-1}\frac{1}{3} \approx 1.23$$

↑ Use a calculator.

(b) Let $\theta = \csc^{-1}(-4)$. Then $\csc\theta = -4$, $-\frac{\pi}{2} \le \theta \le \frac{\pi}{2}, \theta \ne 0$. Now find $\sin\theta$ because $y = \sin^{-1} x$ has the same range as $y = \csc^{-1} x$, except where undefined. Because $\csc\theta = \frac{1}{\sin\theta} = -4$, this means $\sin\theta = -\frac{1}{4}$. Then $\theta = \sin^{-1}\left(-\frac{1}{4}\right)$, and

$$\csc^{-1}(-4) = \theta = \sin^{-1}\left(-\frac{1}{4}\right) \approx -0.25$$

(c) Let $\theta = \cot^{-1}\frac{1}{2}$. Then $\cot\theta = \frac{1}{2}$, $0 < \theta < \pi$. These facts indicate that θ lies in quadrant I. Now find $\cos\theta$ because $y = \cos^{-1} x$ has the same range as $y = \cot^{-1} x$, except where undefined. Use Figure 22 to find that $\cos\theta = \frac{1}{\sqrt{5}}$, $0 < \theta < \frac{\pi}{2}$. So, $\theta = \cos^{-1}\left(\frac{1}{\sqrt{5}}\right)$, and

$$\cot^{-1}\frac{1}{2} = \theta = \cos^{-1}\left(\frac{1}{\sqrt{5}}\right) \approx 1.11$$

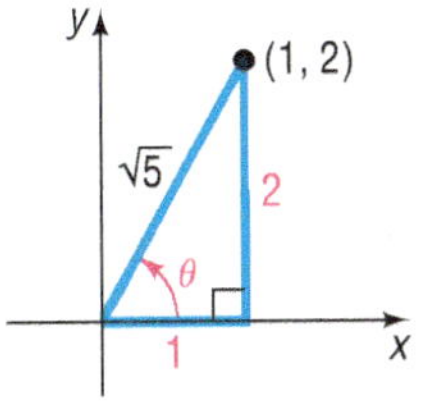

Figure 22
$\cot\theta = \frac{1}{2}$, $0 < \theta < \pi$

(d) Let $\theta = \cot^{-1}(-2)$. Then $\cot\theta = -2$, $0 < \theta < \pi$. These facts indicate that θ lies in quadrant II. Now find $\cos\theta$. Use Figure 23 to find that $\cos\theta = -\frac{2}{\sqrt{5}}$, $\frac{\pi}{2} < \theta < \pi$. This means $\theta = \cos^{-1}\left(-\frac{2}{\sqrt{5}}\right)$, and

$$\cot^{-1}(-2) = \theta = \cos^{-1}\left(-\frac{2}{\sqrt{5}}\right) \approx 2.68$$ ■

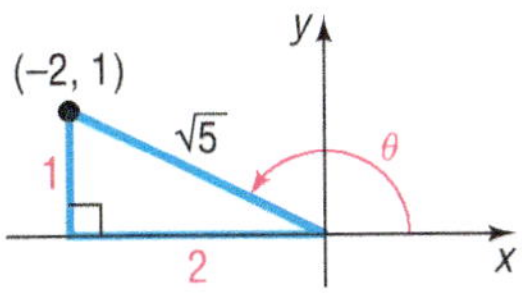

Figure 23
$\cot\theta = -2$, $0 < \theta < \pi$

Now Work PROBLEM 45

4 Write a Trigonometric Expression as an Algebraic Expression

EXAMPLE 6 **Writing a Trigonometric Expression as an Algebraic Expression**

Write $\sin(\tan^{-1}u)$ as an algebraic expression containing u.

Solution Let $\theta = \tan^{-1}u$ so that $\tan\theta = u$, $-\frac{\pi}{2} < \theta < \frac{\pi}{2}$, $-\infty < u < \infty$. This means $\sec\theta > 0$. Then

$$\sin(\tan^{-1}u) = \sin\theta = \sin\theta \cdot \frac{\cos\theta}{\cos\theta} = \tan\theta\cos\theta = \frac{\tan\theta}{\sec\theta} = \frac{\tan\theta}{\sqrt{1+\tan^2\theta}} = \frac{u}{\sqrt{1+u^2}}$$

Multiply by 1: $\frac{\cos\theta}{\cos\theta}$; $\frac{\sin\theta}{\cos\theta} = \tan\theta$; $\sec^2\theta = 1 + \tan^2\theta$, $\sec\theta > 0$ ■

Now Work PROBLEM 57

8.2 Assess Your Understanding

'Are You Prepared?' *Answers are given at the end of these exercises. If you get a wrong answer, read the pages listed in red.*

1. What are the domain and the range of $y = \sec x$? (pp. 565–567)
2. ***True or False*** The graph of $y = \sec x$ is one-to-one on the interval $\left[0, \frac{\pi}{2}\right)$ and on the interval $\left(\frac{\pi}{2}, \pi\right]$. (pp. 591–592)
3. If $\tan\theta = \frac{1}{2}$, $-\frac{\pi}{2} < \theta < \frac{\pi}{2}$, then $\sin\theta =$ __________. (pp. 557–558)

Concepts and Vocabulary

4. $y = \sec^{-1}x$ means ________, where $|x|$ ______ and ______ $\le y \le$ ___, $y \ne \frac{\pi}{2}$.
5. To find the inverse secant of a real number x such that $|x| \ge 1$, convert the inverse secant to an inverse ________.
6. ***True or False*** It is impossible to obtain exact values for the inverse secant function.
7. ***True or False*** $\csc^{-1} 0.5$ is not defined.
8. ***True or False*** The domain of the inverse cotangent function is the set of real numbers.

Skill Building

In Problems 9–36, find the exact value of each expression.

9. $\cos\left(\sin^{-1}\frac{\sqrt{2}}{2}\right)$
10. $\sin\left(\cos^{-1}\frac{1}{2}\right)$
11. $\tan\left[\cos^{-1}\left(-\frac{\sqrt{3}}{2}\right)\right]$
12. $\tan\left[\sin^{-1}\left(-\frac{1}{2}\right)\right]$
13. $\sec\left(\cos^{-1}\frac{1}{2}\right)$
14. $\cot\left[\sin^{-1}\left(-\frac{1}{2}\right)\right]$
15. $\csc(\tan^{-1}1)$
16. $\sec(\tan^{-1}\sqrt{3})$
17. $\sin[\tan^{-1}(-1)]$
18. $\cos\left[\sin^{-1}\left(-\frac{\sqrt{3}}{2}\right)\right]$
19. $\sec\left[\sin^{-1}\left(-\frac{1}{2}\right)\right]$
20. $\csc\left[\cos^{-1}\left(-\frac{\sqrt{3}}{2}\right)\right]$
21. $\cos^{-1}\left(\sin\frac{5\pi}{4}\right)$
22. $\tan^{-1}\left(\cot\frac{2\pi}{3}\right)$
23. $\sin^{-1}\left[\cos\left(-\frac{7\pi}{6}\right)\right]$
24. $\cos^{-1}\left[\tan\left(-\frac{\pi}{4}\right)\right]$
25. $\tan\left(\sin^{-1}\frac{1}{3}\right)$
26. $\tan\left(\cos^{-1}\frac{1}{3}\right)$
27. $\sec\left(\tan^{-1}\frac{1}{2}\right)$
28. $\cos\left(\sin^{-1}\frac{\sqrt{2}}{3}\right)$
29. $\cot\left[\sin^{-1}\left(-\frac{\sqrt{2}}{3}\right)\right]$
30. $\csc[\tan^{-1}(-2)]$
31. $\sin[\tan^{-1}(-3)]$
32. $\cot\left[\cos^{-1}\left(-\frac{\sqrt{3}}{3}\right)\right]$
33. $\sec\left(\sin^{-1}\frac{2\sqrt{5}}{5}\right)$
34. $\csc\left(\tan^{-1}\frac{1}{2}\right)$
35. $\sin^{-1}\left(\cos\frac{3\pi}{4}\right)$
36. $\cos^{-1}\left(\sin\frac{7\pi}{6}\right)$

In Problems 37–44, find the exact value of each expression.

37. $\cot^{-1}\sqrt{3}$
38. $\cot^{-1}1$
39. $\csc^{-1}(-1)$
40. $\csc^{-1}\sqrt{2}$
41. $\sec^{-1}\frac{2\sqrt{3}}{3}$
42. $\sec^{-1}(-2)$
43. $\cot^{-1}\left(-\frac{\sqrt{3}}{3}\right)$
44. $\csc^{-1}\left(-\frac{2\sqrt{3}}{3}\right)$

In Problems 45–56, use a calculator to find the value of each expression rounded to two decimal places.

45. $\sec^{-1} 4$ **46.** $\csc^{-1} 5$ **47.** $\cot^{-1} 2$ **48.** $\sec^{-1}(-3)$

49. $\csc^{-1}(-3)$ **50.** $\cot^{-1}\left(-\frac{1}{2}\right)$ **51.** $\cot^{-1}(-\sqrt{5})$ **52.** $\cot^{-1}(-8.1)$

53. $\csc^{-1}\left(-\frac{3}{2}\right)$ **54.** $\sec^{-1}\left(-\frac{4}{3}\right)$ **55.** $\cot^{-1}\left(-\frac{3}{2}\right)$ **56.** $\cot^{-1}(-\sqrt{10})$

In Problems 57–66, write each trigonometric expression as an algebraic expression in u.

57. $\cos(\tan^{-1} u)$ **58.** $\sin(\cos^{-1} u)$ **59.** $\tan(\sin^{-1} u)$ **60.** $\tan(\cos^{-1} u)$ **61.** $\sin(\sec^{-1} u)$

62. $\sin(\cot^{-1} u)$ **63.** $\cos(\csc^{-1} u)$ **64.** $\cos(\sec^{-1} u)$ **65.** $\tan(\cot^{-1} u)$ **66.** $\tan(\sec^{-1} u)$

Mixed Practice

In Problems 67–78, $f(x) = \sin x, -\frac{\pi}{2} \le x \le \frac{\pi}{2}$; $g(x) = \cos x, 0 \le x \le \pi$; *and* $h(x) = \tan x, -\frac{\pi}{2} < x < \frac{\pi}{2}$. *Find the exact value of each composite function.*

67. $g\left(f^{-1}\left(\frac{12}{13}\right)\right)$ **68.** $f\left(g^{-1}\left(\frac{5}{13}\right)\right)$ **69.** $g^{-1}\left(f\left(-\frac{\pi}{4}\right)\right)$ **70.** $f^{-1}\left(g\left(\frac{5\pi}{6}\right)\right)$

71. $h\left(f^{-1}\left(-\frac{3}{5}\right)\right)$ **72.** $h\left(g^{-1}\left(-\frac{4}{5}\right)\right)$ **73.** $g\left(h^{-1}\left(\frac{12}{5}\right)\right)$ **74.** $f\left(h^{-1}\left(\frac{5}{12}\right)\right)$

75. $g^{-1}\left(f\left(-\frac{\pi}{3}\right)\right)$ **76.** $g^{-1}\left(f\left(-\frac{\pi}{6}\right)\right)$ **77.** $h\left(g^{-1}\left(-\frac{1}{4}\right)\right)$ **78.** $h\left(f^{-1}\left(-\frac{2}{5}\right)\right)$

Applications and Extensions

Problems 79 and 80 require the following discussion: When granular materials are allowed to fall freely, they form conical (cone-shaped) piles. The naturally occurring angle of slope, measured from the horizontal, at which the loose material comes to rest is called the ***angle of repose*** *and varies for different materials. The angle of repose θ is related to the height h and the base radius r of the conical pile by the equation* $\theta = \cot^{-1}\frac{r}{h}$. *See the illustration.*

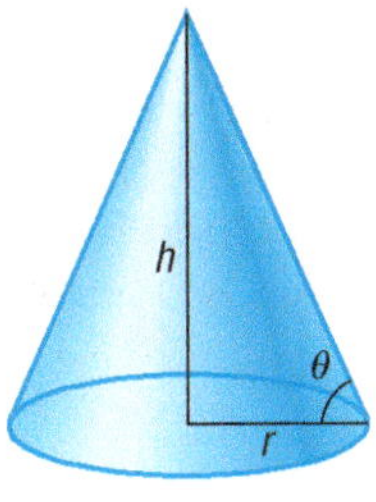

79. Angle of Repose: De-icing Salt Due to potential transportation issues (for example, frozen waterways) de-icing salt used by highway departments in the Midwest must be ordered early and stored for future use. When de-icing salt is stored in a pile 14 feet high, the diameter of the base of the pile is 45 feet.
(a) Find the angle of repose for de-icing salt.
(b) What is the base diameter of a pile that is 17 feet high?
(c) What is the height of a pile that has a base diameter of approximately 122 feet?
Source: Salt Institute, *The Salt Storage Handbook*, 2013

80. Angle of Repose: Bunker Sand The steepness of sand bunkers on a golf course is affected by the angle of repose of the sand (a larger angle of repose allows for steeper bunkers). A freestanding pile of loose sand from a United States Golf Association (USGA) bunker had a height of 4 feet and a base diameter of approximately 6.68 feet.
(a) Find the angle of repose for USGA bunker sand.
(b) What is the height of such a pile if the diameter of the base is 8 feet?
(c) A 6-foot-high pile of loose Tour Grade 50/50 sand has a base diameter of approximately 8.44 feet. Which type of sand (USGA or Tour Grade 50/50) would be better suited for steep bunkers?
Source: *Purdue University Turfgrass Science Program*

81. Artillery A projectile fired into the first quadrant from the origin of a coordinate system will pass through the point (x, y) at time t according to the relationship $\cot\theta = \frac{2x}{2y + gt^2}$, where $\theta =$ the angle of elevation of the launcher and $g =$ the acceleration due to gravity $= 32.2$ feet/second^2. An artilleryman is firing at an enemy bunker located 2450 feet up the side of a hill that is 6175 feet away. He fires a round, and exactly 2.27 seconds later he scores a direct hit.
(a) What angle of elevation did he use?
(b) If the angle of elevation is also given by $\sec\theta = \frac{v_0 t}{x}$, where v_0 is the muzzle velocity of the weapon, find the muzzle velocity of the artillery piece he used.
Source: *www.egwald.com/geometry/projectile3d.php*

82. Using a graphing utility, graph $y = \cot^{-1} x$.

83. Using a graphing utility, graph $y = \sec^{-1} x$.

84. Using a graphing utility, graph $y = \csc^{-1} x$.

Explaining Concepts: Discussion and Writing

85. Explain in your own words how you would use your calculator to find the value of $\cot^{-1} 10$.

86. Consult three books on calculus and write down the definition in each of $y = \sec^{-1} x$ and $y = \csc^{-1} x$. Compare these with the definitions given in this book.

Retain Your Knowledge

Problems 87–90 are based on material learned earlier in the course. The purpose of these problems is to keep the material fresh in your mind so that you are better prepared for the final exam.

87. Find the complex zeros of $f(x) = x^4 + 21x^2 - 100$.

88. Determine algebraically whether $f(x) = x^3 + x^2 - x$ is even, odd, or neither.

89. Convert 315° to radians.

90. Find the length of the arc subtended by a central angle of 75° on a circle of radius 6 inches. Give both the exact length and an approximation rounded to two decimal places.

'Are You Prepared?' Answers

1. Domain: $\left\{x \middle| x \neq \text{odd integer multiples of } \frac{\pi}{2}\right\}$; range: $\{y \leq -1 \text{ or } y \geq 1\}$ **2.** True **3.** $\frac{\sqrt{5}}{5}$

8.3 Trigonometric Equations

PREPARING FOR THIS SECTION *Before getting started, review the following:*

- Linear Equations (Section 1.2, pp. 102–103)
- Values of the Trigonometric Functions (Section 7.3, pp. 539–541; Section 7.4, pp. 550–558)
- Quadratic Equations (Section 1.3, pp. 110–116)
- Equations Quadratic in Form (Section 1.5, pp. 131–133)
- Using a Graphing Utility to Solve Equations (Section 1.2, pp. 100–101)

Now Work the **'Are You Prepared?'** problems on page 639.

OBJECTIVES
1. Solve Equations Involving a Single Trigonometric Function (p. 634)
2. Solve Trigonometric Equations Using a Calculator (p. 637)
3. Solve Trigonometric Equations Quadratic in Form (p. 638)
4. Solve Trigonometric Equations Using Fundamental Identities (p. 638)
5. Solve Trigonometric Equations Using a Graphing Utility (p. 639)

1 Solve Equations Involving a Single Trigonometric Function

In this section, we discuss **trigonometric equations**—that is, equations involving trigonometric functions that are satisfied only by some values of the variable (or, possibly, are not satisfied by any values of the variable). The values that satisfy the equation are called **solutions** of the equation.

EXAMPLE 1 **Checking Whether a Given Number Is a Solution of a Trigonometric Equation**

Determine whether $\theta = \frac{\pi}{4}$ is a solution of the equation $2\sin\theta - 1 = 0$. Is $\theta = \frac{\pi}{6}$ a solution?

Solution Replace θ by $\frac{\pi}{4}$ in the given equation. The result is

$$2\sin\frac{\pi}{4} - 1 = 2\cdot\frac{\sqrt{2}}{2} - 1 = \sqrt{2} - 1 \neq 0$$

Therefore, $\frac{\pi}{4}$ is not a solution.

Next replace θ by $\frac{\pi}{6}$ in the equation. The result is

$$2\sin\frac{\pi}{6} - 1 = 2\cdot\frac{1}{2} - 1 = 0$$

Therefore, $\frac{\pi}{6}$ is a solution of the given equation. ■

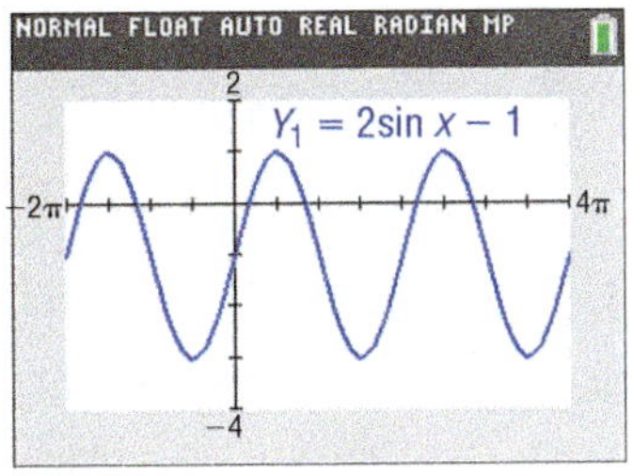

Figure 24 $y = 2\sin x - 1$

The equation given in Example 1 has other solutions besides $\theta = \frac{\pi}{6}$. For example, $\theta = \frac{5\pi}{6}$ is also a solution, as is $\theta = \frac{13\pi}{6}$. (Check this for yourself.) In fact, the equation has an infinite number of solutions because of the periodicity of the sine function, as can be seen in Figure 24, which shows the graph of $y = 2\sin x - 1$. Each x-intercept of the graph represents a solution to the equation $2\sin x - 1 = 0$.

Unless the domain of the variable is restricted, we need to find *all* the solutions of a trigonometric equation. As the next example illustrates, finding all the solutions can be accomplished by first finding solutions over an interval whose length equals the period of the function and then adding multiples of that period to the solutions found.

EXAMPLE 2 Finding All the Solutions of a Trigonometric Equation

Solve the equation: $\cos\theta = \frac{1}{2}$

Give a general formula for all the solutions. List eight of the solutions.

In Words

To solve the equation $\cos\theta = \frac{1}{2}$ means to find all the angles θ whose cosine is $\frac{1}{2}$.

Solution The period of the cosine function is 2π. In the interval $[0, 2\pi)$, there are two angles θ for which $\cos\theta = \frac{1}{2}$: $\theta = \frac{\pi}{3}$ and $\theta = \frac{5\pi}{3}$. See Figure 25. Because the cosine function has period 2π, all the solutions of $\cos\theta = \frac{1}{2}$ may be given by the general formula

$$\theta = \frac{\pi}{3} + 2k\pi \quad \text{or} \quad \theta = \frac{5\pi}{3} + 2k\pi \qquad k \text{ any integer}$$

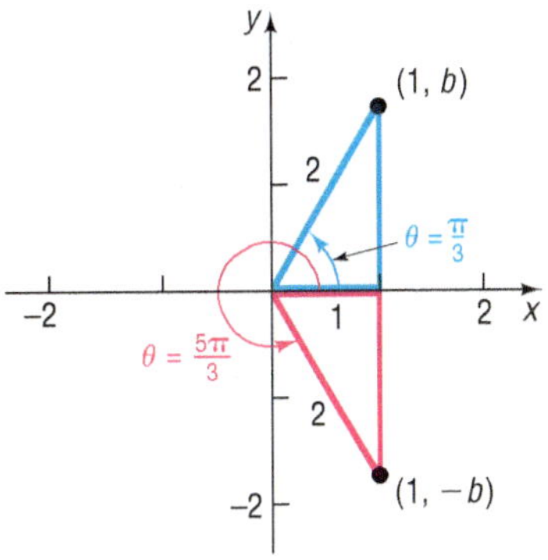

Figure 25

Eight of the solutions are

$$\underbrace{-\frac{5\pi}{3}, -\frac{\pi}{3}}_{k=-1},\ \underbrace{\frac{\pi}{3}, \frac{5\pi}{3}}_{k=0},\ \underbrace{\frac{7\pi}{3}, \frac{11\pi}{3}}_{k=1},\ \underbrace{\frac{13\pi}{3}, \frac{17\pi}{3}}_{k=2}$$

Check: To verify the solutions, graph $Y_1 = \cos x$ and $Y_2 = \frac{1}{2}$ and determine where the graphs intersect. (Be sure to graph in radian mode.) See Figure 26. The graph of Y_1 intersects the graph of Y_2 at $x = 1.05\left(\approx\frac{\pi}{3}\right)$, $5.24\left(\approx\frac{5\pi}{3}\right)$, $7.33\left(\approx\frac{7\pi}{3}\right)$, and $11.52\left(\approx\frac{11\pi}{3}\right)$, rounded to two decimal places. ■

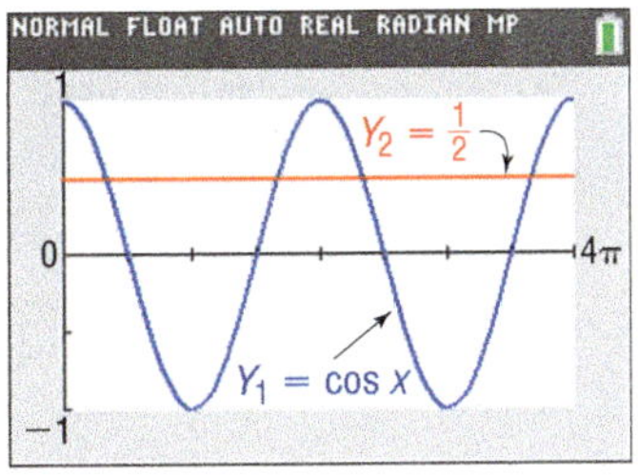

Figure 26

Now Work PROBLEM 37

In most of the work we do, we shall be interested only in finding solutions of trigonometric equations for $0 \le \theta < 2\pi$.

EXAMPLE 3 Solving a Linear Trigonometric Equation

Solve the equation: $2\sin\theta + \sqrt{3} = 0, \quad 0 \le \theta < 2\pi$

Solution First solve the equation for $\sin\theta$.

$$2\sin\theta + \sqrt{3} = 0$$

$$2\sin\theta = -\sqrt{3} \qquad \text{Subtract } \sqrt{3} \text{ from both sides.}$$

$$\sin\theta = -\frac{\sqrt{3}}{2} \qquad \text{Divide both sides by 2.}$$

In the interval $[0, 2\pi)$, there are two angles θ for which $\sin\theta = -\frac{\sqrt{3}}{2}$: $\theta = \frac{4\pi}{3}$ and $\theta = \frac{5\pi}{3}$. The solution set is $\left\{\frac{4\pi}{3}, \frac{5\pi}{3}\right\}$. ■

WARNING For Example 3, if a calculator is used to find $\sin^{-1}\left(-\frac{\sqrt{3}}{2}\right)$, the result given is ≈ -1.047, which is not in the interval $[0, 2\pi)$. This occurs because, as we noted in Section 8.1, the range of $y = \sin^{-1} x$ is $\left[-\frac{\pi}{2}, \frac{\pi}{2}\right]$, so the calculator provides output in this range. This limitation of the calculator highlights the importance of knowing the common angles and the values of the six trigonometric functions of these angles. ■

Now Work PROBLEM 13

When the argument of the trigonometric function in an equation is a multiple of θ, the general formula must be used to solve the equation.

EXAMPLE 4 Solving a Trigonometric Equation

Solve the equation: $\sin(2\theta) = \frac{1}{2}$, $0 \le \theta < 2\pi$

Solution In the interval $[0, 2\pi)$, the sine function equals $\frac{1}{2}$ at $\frac{\pi}{6}$ and $\frac{5\pi}{6}$. See Figure 27(a). So 2θ must equal $\frac{\pi}{6}$ and $\frac{5\pi}{6}$. Here's the problem, however. The period of $y = \sin(2\theta)$ is $\frac{2\pi}{2} = \pi$. So in the interval $[0, 2\pi)$, the graph of $y = \sin(2\theta)$ completes two cycles, and the graph of $y = \sin(2\theta)$ intersects the graph of $y = \frac{1}{2}$ four times. See Figure 27(b). For this reason, there are four solutions to the equation $\sin(2\theta) = \frac{1}{2}$ in $[0, 2\pi)$. To find these solutions, write the general formula that gives all the solutions.

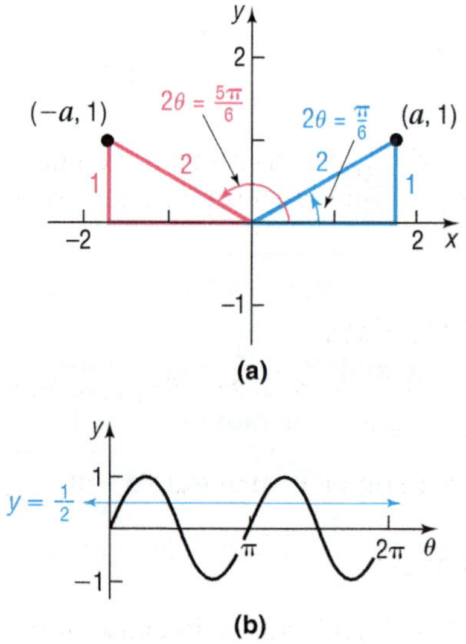

Figure 27

$$2\theta = \frac{\pi}{6} + 2k\pi \quad\text{or}\quad 2\theta = \frac{5\pi}{6} + 2k\pi \qquad k \text{ any integer}$$

$$\theta = \frac{\pi}{12} + k\pi \quad\text{or}\quad \theta = \frac{5\pi}{12} + k\pi \qquad \text{Divide by 2.}$$

Then

$$\theta = \frac{\pi}{12} + (-1)\pi = \frac{-11\pi}{12} \qquad k = -1 \qquad \theta = \frac{5\pi}{12} + (-1)\pi = \frac{-7\pi}{12}$$

$$\theta = \frac{\pi}{12} + (0)\pi = \frac{\pi}{12} \qquad k = 0 \qquad \theta = \frac{5\pi}{12} + (0)\pi = \frac{5\pi}{12}$$

$$\theta = \frac{\pi}{12} + (1)\pi = \frac{13\pi}{12} \qquad k = 1 \qquad \theta = \frac{5\pi}{12} + (1)\pi = \frac{17\pi}{12}$$

$$\theta = \frac{\pi}{12} + (2)\pi = \frac{25\pi}{12} \qquad k = 2 \qquad \theta = \frac{5\pi}{12} + (2)\pi = \frac{29\pi}{12}$$

In the interval $[0, 2\pi)$, the solutions of $\sin(2\theta) = \frac{1}{2}$ are $\theta = \frac{\pi}{12}$, $\theta = \frac{5\pi}{12}$, $\theta = \frac{13\pi}{12}$, and $\theta = \frac{17\pi}{12}$. The solution set is $\left\{\frac{\pi}{12}, \frac{5\pi}{12}, \frac{13\pi}{12}, \frac{17\pi}{12}\right\}$. This means the graph of $y = \sin(2\theta)$ intersects $y = \frac{1}{2}$ at $\left(\frac{\pi}{12}, \frac{1}{2}\right)$, $\left(\frac{5\pi}{12}, \frac{1}{2}\right)$, $\left(\frac{13\pi}{12}, \frac{1}{2}\right)$, and $\left(\frac{17\pi}{12}, \frac{1}{2}\right)$ in the interval $[0, 2\pi)$.

✓**Check:** Verify these solutions by graphing $Y_1 = \sin(2x)$ and $Y_2 = \frac{1}{2}$ for $0 \le x \le 2\pi$. ■

WARNING In solving a trigonometric equation for θ, $0 \le \theta < 2\pi$, in which the argument is not θ (as in Example 4), you must write down all the solutions first and then list those that are in the interval $[0, 2\pi)$. Otherwise, solutions may be lost. For example, in solving $\sin(2\theta) = \frac{1}{2}$, if you write only the solutions $2\theta = \frac{\pi}{6}$ and $2\theta = \frac{5\pi}{6}$, you will find only $\theta = \frac{\pi}{12}$ and $\theta = \frac{5\pi}{12}$ and miss the other solutions. ■

EXAMPLE 5 Solving a Trigonometric Equation

Solve the equation: $\tan\left(\theta - \frac{\pi}{2}\right) = 1, \quad 0 \le \theta < 2\pi$

Solution The period of the tangent function is π. In the interval $[0, \pi)$, the tangent function has the value 1 when the argument is $\frac{\pi}{4}$. Because the argument is $\theta - \frac{\pi}{2}$ in the given equation, write the general formula that gives all the solutions.

$$\theta - \frac{\pi}{2} = \frac{\pi}{4} + k\pi \qquad k \text{ any integer}$$

$$\theta = \frac{3\pi}{4} + k\pi$$

In the interval $[0, 2\pi)$, $\theta = \frac{3\pi}{4}$ and $\theta = \frac{3\pi}{4} + \pi = \frac{7\pi}{4}$ are the only solutions. The solution set is $\left\{\frac{3\pi}{4}, \frac{7\pi}{4}\right\}$. ■

Now Work PROBLEM 27

2 Solve Trigonometric Equations Using a Calculator

The next example illustrates how to solve trigonometric equations using a calculator. Remember that the function keys on a calculator give only values consistent with the definition of the function.

EXAMPLE 6 Solving a Trigonometric Equation with a Calculator

Use a calculator to solve the equation $\tan\theta = -2, 0 \le \theta < 2\pi$. Express any solutions in radians, rounded to two decimal places.

Solution To solve $\tan\theta = -2$ on a calculator, first set the mode to radians. Then use the $\boxed{\tan^{-1}}$ key to obtain

$$\theta = \tan^{-1}(-2) \approx -1.1071487$$

Rounded to two decimal places, $\theta = \tan^{-1}(-2) = -1.11$ radian. Because of the definition of $y = \tan^{-1} x$, the angle θ that is obtained is the angle $-\frac{\pi}{2} < \theta < \frac{\pi}{2}$ for which $\tan\theta = -2$. Because we seek solutions for which $0 \le \theta < 2\pi$, we express the angle as $2\pi - 1.11$.

See Figure 28. Another angle for which $\tan\theta = -2$ is $\pi - 1.11$. The angle $\pi - 1.11$ is the angle in quadrant II, where $\tan\theta = -2$. The solutions for $\tan\theta = -2, 0 \le \theta < 2\pi$, are

$$\theta = \pi - 1.11 \approx 2.03 \text{ radians} \quad \text{and} \quad \theta = 2\pi - 1.11 \approx 5.17 \text{ radians}$$

The solution set is {2.03, 5.17}. ■

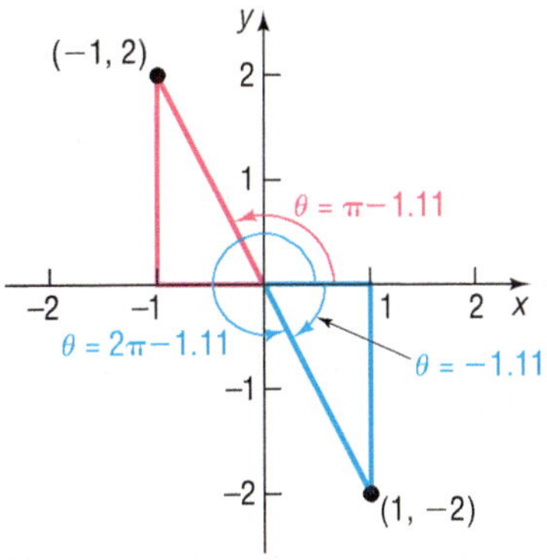

Figure 28

WARNING Example 6 illustrates that caution must be exercised when solving trigonometric equations on a calculator. Remember that the calculator supplies only an angle within the restrictions of the definition of the inverse trigonometric function. To find the remaining solutions, you must identify other quadrants, if any, in which a solution may be located. Using the definition of the inverse trigonometric functions to find one of the angles for the general formula will work. However, we choose to directly find those angles in the interval $[0, 2\pi)$. ■

Now Work PROBLEM 47

3 Solve Trigonometric Equations Quadratic in Form

Many trigonometric equations can be solved by applying techniques that we already know, such as applying the quadratic formula (if the equation is a second-degree polynomial) or factoring.

EXAMPLE 7 Solving a Trigonometric Equation Quadratic in Form

Solve the equation: $2\sin^2\theta - 3\sin\theta + 1 = 0, \quad 0 \le \theta < 2\pi$

Solution This equation is a quadratic equation (in $\sin\theta$) that can be factored.

$$2\sin^2\theta - 3\sin\theta + 1 = 0 \qquad 2u^2 - 3u + 1 = 0,\ u = \sin\theta$$

$$(2\sin\theta - 1)(\sin\theta - 1) = 0 \qquad (2u - 1)(u - 1) = 0$$

$$2\sin\theta - 1 = 0 \quad \text{or} \quad \sin\theta - 1 = 0 \qquad \text{Use the Zero-Product Property.}$$

$$\sin\theta = \frac{1}{2} \quad \text{or} \quad \sin\theta = 1$$

Solving each equation in the interval $[0, 2\pi)$ yields

$$\theta = \frac{\pi}{6}, \qquad \theta = \frac{5\pi}{6}, \qquad \theta = \frac{\pi}{2}$$

The solution set is $\left\{\frac{\pi}{6}, \frac{5\pi}{6}, \frac{\pi}{2}\right\}$.

Now Work PROBLEM 61

4 Solve Trigonometric Equations Using Fundamental Identities

When a trigonometric equation contains more than one trigonometric function, identities sometimes can be used to obtain an equivalent equation that contains only one trigonometric function.

EXAMPLE 8 Solving a Trigonometric Equation Using Identities

Solve the equation: $3\cos\theta + 3 = 2\sin^2\theta, \quad 0 \le \theta < 2\pi$

Solution The equation in its present form contains a sine and a cosine. However, a form of the Pythagorean Identity, $\sin^2\theta + \cos^2\theta = 1$, can be used to transform the equation into an equivalent one containing only cosines.

$$3\cos\theta + 3 = 2\sin^2\theta$$

$$3\cos\theta + 3 = 2(1 - \cos^2\theta) \qquad \sin^2\theta = 1 - \cos^2\theta$$

$$3\cos\theta + 3 = 2 - 2\cos^2\theta$$

$$2\cos^2\theta + 3\cos\theta + 1 = 0 \qquad \text{Quadratic in } \cos\theta$$

$$(2\cos\theta + 1)(\cos\theta + 1) = 0 \qquad \text{Factor.}$$

$$2\cos\theta + 1 = 0 \quad \text{or} \quad \cos\theta + 1 = 0 \qquad \text{Use the Zero-Product Property.}$$

$$\cos\theta = -\frac{1}{2} \quad \text{or} \quad \cos\theta = -1$$

Solving each equation in the interval $[0, 2\pi)$ yields

$$\theta = \frac{2\pi}{3}, \qquad \theta = \frac{4\pi}{3}, \qquad \theta = \pi$$

The solution set is $\left\{\frac{2\pi}{3}, \pi, \frac{4\pi}{3}\right\}$.

Check: Graph $Y_1 = 3\cos x + 3$ and $Y_2 = 2\sin^2 x$, $0 \le x \le 2\pi$, and find the points of intersection. How close are your approximate solutions to the exact solutions?

EXAMPLE 9 Solving a Trigonometric Equation Using Identities

Solve the equation: $\cos^2\theta + \sin\theta = 2$, $0 \le \theta < 2\pi$

Solution

This equation involves two trigonometric functions: sine and cosine. By using a Pythagorean Identity, we can express the equation in terms of just sine functions.

$$\cos^2\theta + \sin\theta = 2$$
$$(1 - \sin^2\theta) + \sin\theta = 2 \qquad \cos^2\theta = 1 - \sin^2\theta$$
$$\sin^2\theta - \sin\theta + 1 = 0$$

This is a quadratic equation in $\sin\theta$. The discriminant is $b^2 - 4ac = 1 - 4 = -3 < 0$. Therefore, the equation has no real solution. The solution set is the empty set, $\varnothing$.

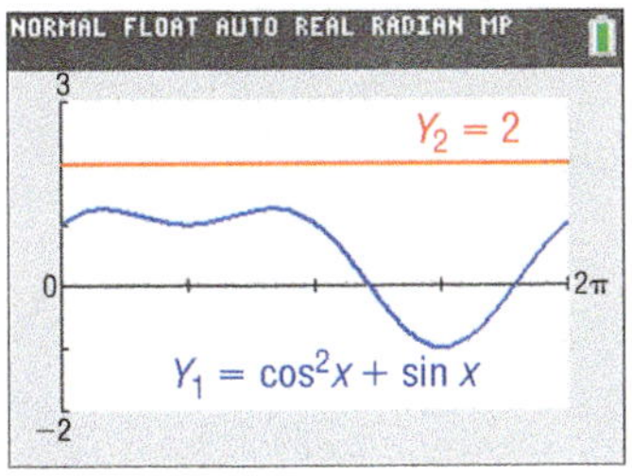

Figure 29

Check: Graph $Y_1 = \cos^2 x + \sin x$ and $Y_2 = 2$ to see that the two graphs never intersect, so the equation $Y_1 = Y_2$ has no real solution. See Figure 29.

5 Solve Trigonometric Equations Using a Graphing Utility

The techniques introduced in this section apply only to certain types of trigonometric equations. Solutions for other types are usually studied in calculus, using numerical methods.

EXAMPLE 10 Solving a Trigonometric Equation Using a Graphing Utility

Solve: $5\sin x + x = 3$

Express the solution(s) rounded to two decimal places.

Solution

This type of trigonometric equation cannot be solved by previous methods. A graphing utility, though, can be used here. Each solution of this equation is the x-coordinate of a point of intersection of the graphs of $Y_1 = 5\sin x + x$ and $Y_2 = 3$. See Figure 30.

There are three points of intersection; the x-coordinates are the solutions that we seek. Using INTERSECT, we find

$$x = 0.52, \qquad x = 3.18, \qquad x = 5.71$$

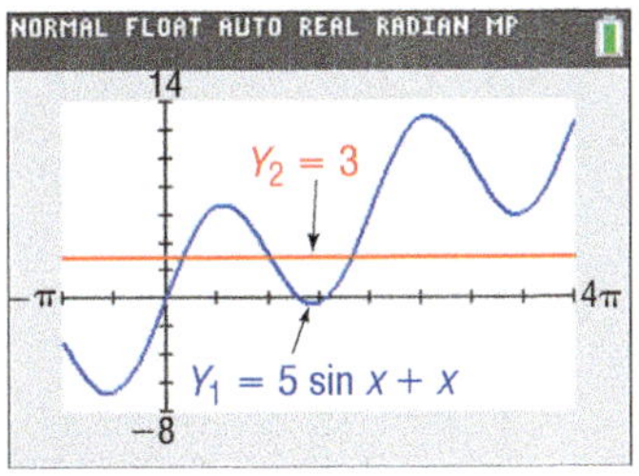

Figure 30

The solution set is $\{0.52, 3.18, 5.71\}$.

Now Work PROBLEM 83

8.3 Assess Your Understanding

'Are You Prepared?' *Answers are given at the end of these exercises. If you get a wrong answer, read the pages listed in red.*

1. Solve: $3x - 5 = -x + 1$ (pp. 102–103)
2. $\sin\left(\frac{\pi}{4}\right) =$ ____; $\cos\left(\frac{8\pi}{3}\right) =$ ____. (pp. 539 and 553–554)
3. Find the real solutions of $4x^2 - x - 5 = 0$. (pp. 110–111)
4. Find the real solutions of $x^2 - x - 1 = 0$. (pp. 113–116)
5. Find the real solutions of $(2x - 1)^2 - 3(2x - 1) - 4 = 0$. (pp. 131–133)
6. Use a graphing utility to solve $5x^3 - 2 = x - x^2$. Round answers to two decimal places. (pp. 100–101)

Concepts and Vocabulary

7. ***True or False*** Most trigonometric equations have unique solutions.

8. ***True or False*** Two solutions of the equation $\sin\theta = \frac{1}{2}$ are $\frac{\pi}{6}$ and $\frac{5\pi}{6}$.

9. ***True or False*** The set of all solutions of the equation $\tan\theta = 1$ is given by $\left\{\theta \middle| \theta = \frac{\pi}{4} + k\pi, k \text{ is any integer}\right\}$.

10. ***True or False*** The equation $\sin\theta = 2$ has a real solution that can be found using a calculator.

11. If all solutions of a trigonometric equation are given by the general formula $\theta = \frac{\pi}{6} + 2k\pi$ or $\theta = \frac{11\pi}{6} + 2k\pi$, where k is any integer, then which of the following is *not* a solution of the equation?

(a) $\frac{35\pi}{6}$ (b) $\frac{23\pi}{6}$ (c) $\frac{13\pi}{6}$ (d) $\frac{7\pi}{6}$

12. Suppose $\theta = \frac{\pi}{2}$ is the only solution of a trigonometric equation in the interval $0 \le \theta < 2\pi$. Assuming a period of 2π, which of the following formulas gives all solutions of the equation, where k is any integer?

(a) $\theta = \frac{\pi}{2} + 2k\pi$ (b) $\theta = \frac{\pi}{2} + k\pi$

(c) $\theta = \frac{k\pi}{2}$ (d) $\theta = \frac{\pi + k\pi}{2}$

Skill Building

In Problems 13–36, solve each equation on the interval $0 \le \theta < 2\pi$.

13. $2\sin\theta + 3 = 2$
14. $1 - \cos\theta = \frac{1}{2}$
15. $2\sin\theta + 1 = 0$
16. $\cos\theta + 1 = 0$
17. $\tan\theta + 1 = 0$
18. $\sqrt{3}\cot\theta + 1 = 0$
19. $4\sec\theta + 6 = -2$
20. $5\csc\theta - 3 = 2$
21. $3\sqrt{2}\cos\theta + 2 = -1$
22. $4\sin\theta + 3\sqrt{3} = \sqrt{3}$
23. $4\cos^2\theta = 1$
24. $\tan^2\theta = \frac{1}{3}$
25. $2\sin^2\theta - 1 = 0$
26. $4\cos^2\theta - 3 = 0$
27. $\sin(3\theta) = -1$
28. $\tan\frac{\theta}{2} = \sqrt{3}$
29. $\cos(2\theta) = -\frac{1}{2}$
30. $\tan(2\theta) = -1$
31. $\sec\frac{3\theta}{2} = -2$
32. $\cot\frac{2\theta}{3} = -\sqrt{3}$
33. $\cos\left(2\theta - \frac{\pi}{2}\right) = -1$
34. $\sin\left(3\theta + \frac{\pi}{18}\right) = 1$
35. $\tan\left(\frac{\theta}{2} + \frac{\pi}{3}\right) = 1$
36. $\cos\left(\frac{\theta}{3} - \frac{\pi}{4}\right) = \frac{1}{2}$

In Problems 37–46, solve each equation. Give a general formula for all the solutions. List six solutions.

37. $\sin\theta = \frac{1}{2}$
38. $\tan\theta = 1$
39. $\tan\theta = -\frac{\sqrt{3}}{3}$
40. $\cos\theta = -\frac{\sqrt{3}}{2}$
41. $\cos\theta = 0$
42. $\sin\theta = \frac{\sqrt{2}}{2}$
43. $\cos(2\theta) = -\frac{1}{2}$
44. $\sin(2\theta) = -1$
45. $\sin\frac{\theta}{2} = -\frac{\sqrt{3}}{2}$
46. $\tan\frac{\theta}{2} = -1$

In Problems 47–58, use a calculator to solve each equation on the interval $0 \le \theta < 2\pi$. Round answers to two decimal places.

47. $\sin\theta = 0.4$
48. $\cos\theta = 0.6$
49. $\tan\theta = 5$
50. $\cot\theta = 2$
51. $\cos\theta = -0.9$
52. $\sin\theta = -0.2$
53. $\sec\theta = -4$
54. $\csc\theta = -3$
55. $5\tan\theta + 9 = 0$
56. $4\cot\theta = -5$
57. $3\sin\theta - 2 = 0$
58. $4\cos\theta + 3 = 0$

In Problems 59–82, solve each equation on the interval $0 \le \theta < 2\pi$.

59. $2\cos^2\theta + \cos\theta = 0$

60. $\sin^2\theta - 1 = 0$

61. $2\sin^2\theta - \sin\theta - 1 = 0$

62. $2\cos^2\theta + \cos\theta - 1 = 0$

63. $(\tan\theta - 1)(\sec\theta - 1) = 0$

64. $(\cot\theta + 1)\left(\csc\theta - \frac{1}{2}\right) = 0$

65. $\sin^2\theta - \cos^2\theta = 1 + \cos\theta$

66. $\cos^2\theta - \sin^2\theta + \sin\theta = 0$

67. $\sin^2\theta = 6(\cos(-\theta) + 1)$

68. $2\sin^2\theta = 3(1 - \cos(-\theta))$

69. $\cos\theta = -\sin(-\theta)$

70. $\cos\theta - \sin(-\theta) = 0$

71. $\tan\theta = 2\sin\theta$

72. $\tan\theta = \cot\theta$

73. $1 + \sin\theta = 2\cos^2\theta$

74. $\sin^2\theta = 2\cos\theta + 2$

75. $2\sin^2\theta - 5\sin\theta + 3 = 0$

76. $2\cos^2\theta - 7\cos\theta - 4 = 0$

77. $3(1 - \cos\theta) = \sin^2\theta$

78. $4(1 + \sin\theta) = \cos^2\theta$

79. $\tan^2\theta = \frac{3}{2}\sec\theta$

80. $\csc^2\theta = \cot\theta + 1$

81. $\sec^2\theta + \tan\theta = 0$

82. $\sec\theta = \tan\theta + \cot\theta$

In Problems 83–94, use a graphing utility to solve each equation. Express the solution(s) rounded to two decimal places.

83. $x + 5\cos x = 0$

84. $x - 4\sin x = 0$

85. $22x - 17\sin x = 3$

86. $19x + 8\cos x = 2$

87. $\sin x + \cos x = x$

88. $\sin x - \cos x = x$

89. $x^2 - 2\cos x = 0$

90. $x^2 + 3\sin x = 0$

91. $x^2 - 2\sin(2x) = 3x$

92. $x^2 = x + 3\cos(2x)$

93. $6\sin x - e^x = 2, \quad x > 0$

94. $4\cos(3x) - e^x = 1, \quad x > 0$

Mixed Practice

95. What are the zeros of $f(x) = 4\sin^2 x - 3$ on the interval $[0, 2\pi]$?

96. What are the zeros of $f(x) = 2\cos(3x) + 1$ on the interval $[0, \pi]$?

97. $f(x) = 3\sin x$
(a) Find the zeros of f on the interval $[-2\pi, 4\pi]$.
(b) Graph $f(x) = 3\sin x$ on the interval $[-2\pi, 4\pi]$.
(c) Solve $f(x) = \frac{3}{2}$ on the interval $[-2\pi, 4\pi]$. What points are on the graph of f? Label these points on the graph drawn in part (b).
(d) Use the graph drawn in part (b) along with the results of part (c) to determine the values of x such that $f(x) > \frac{3}{2}$ on the interval $[-2\pi, 4\pi]$.

98. $f(x) = 2\cos x$
(a) Find the zeros of f on the interval $[-2\pi, 4\pi]$.
(b) Graph $f(x) = 2\cos x$ on the interval $[-2\pi, 4\pi]$.
(c) Solve $f(x) = -\sqrt{3}$ on the interval $[-2\pi, 4\pi]$. What points are on the graph of f? Label these points on the graph drawn in part (b).
(d) Use the graph drawn in part (b) along with the results of part (c) to determine the values of x such that $f(x) < -\sqrt{3}$ on the interval $[-2\pi, 4\pi]$.

99. $f(x) = 4\tan x$
(a) Solve $f(x) = -4$.
(b) For what values of x is $f(x) < -4$ on the interval $\left(-\frac{\pi}{2}, \frac{\pi}{2}\right)$?

100. $f(x) = \cot x$
(a) Solve $f(x) = -\sqrt{3}$.
(b) For what values of x is $f(x) > -\sqrt{3}$ on the interval $(0, \pi)$?

101. (a) Graph $f(x) = 3\sin(2x) + 2$ and $g(x) = \frac{7}{2}$ on the same Cartesian plane for the interval $[0, \pi]$.
(b) Solve $f(x) = g(x)$ on the interval $[0, \pi]$, and label the points of intersection on the graph drawn in part (a).
(c) Solve $f(x) > g(x)$ on the interval $[0, \pi]$.
(d) Shade the region bounded by $f(x) = 3\sin(2x) + 2$ and $g(x) = \frac{7}{2}$ between the two points found in part (b) on the graph drawn in part (a).

102. (a) Graph $f(x) = 2\cos\frac{x}{2} + 3$ and $g(x) = 4$ on the same Cartesian plane for the interval $[0, 4\pi]$.
(b) Solve $f(x) = g(x)$ on the interval $[0, 4\pi]$, and label the points of intersection on the graph drawn in part (a).
(c) Solve $f(x) < g(x)$ on the interval $[0, 4\pi]$.
(d) Shade the region bounded by $f(x) = 2\cos\frac{x}{2} + 3$ and $g(x) = 4$ between the two points found in part (b) on the graph drawn in part (a).

103. (a) Graph $f(x) = -4\cos x$ and $g(x) = 2\cos x + 3$ on the same Cartesian plane for the interval $[0, 2\pi]$.
(b) Solve $f(x) = g(x)$ on the interval $[0, 2\pi]$, and label the points of intersection on the graph drawn in part (a).
(c) Solve $f(x) > g(x)$ on the interval $[0, 2\pi]$.
(d) Shade the region bounded by $f(x) = -4\cos x$ and $g(x) = 2\cos x + 3$ between the two points found in part (b) on the graph drawn in part (a).

104. (a) Graph $f(x) = 2\sin x$ and $g(x) = -2\sin x + 2$ on the same Cartesian plane for the interval $[0, 2\pi]$.
(b) Solve $f(x) = g(x)$ on the interval $[0, 2\pi]$, and label the points of intersection on the graph drawn in part (a).
(c) Solve $f(x) > g(x)$ on the interval $[0, 2\pi]$.
(d) Shade the region bounded by $f(x) = 2\sin x$ and $g(x) = -2\sin x + 2$ between the two points found in part (b) on the graph drawn in part (a).

Applications and Extensions

105. Blood Pressure Blood pressure is a way of measuring the amount of force exerted on the walls of blood vessels. It is measured using two numbers: systolic (as the heart beats) blood pressure and diastolic (as the heart rests) blood pressure. Blood pressures vary substantially from person to person, but a typical blood pressure is 120/80, which means the systolic blood pressure is 120 mmHg and the diastolic blood pressure is 80 mmHg. Assuming that a person's heart beats 70 times per minute, the blood pressure P of an individual after t seconds can be modeled by the function

$$P(t) = 100 + 20\sin\left(\frac{7\pi}{3}t\right)$$

(a) In the interval $[0, 1]$, determine the times at which the blood pressure is 100 mmHg.
(b) In the interval $[0, 1]$, determine the times at which the blood pressure is 120 mmHg.
(c) In the interval $[0, 1]$, determine the times at which the blood pressure is between 100 and 105 mmHg.

106. The Ferris Wheel In 1893, George Ferris engineered the Ferris wheel. It was 250 feet in diameter. If a Ferris wheel makes 1 revolution every 40 seconds, then the function

$$h(t) = 125\sin\left(0.157t - \frac{\pi}{2}\right) + 125$$

represents the height h, in feet, of a seat on the wheel as a function of time t, where t is measured in seconds. The ride begins when $t = 0$.
(a) During the first 40 seconds of the ride, at what time t is an individual on the Ferris wheel exactly 125 feet above the ground?
(b) During the first 80 seconds of the ride, at what time t is an individual on the Ferris wheel exactly 250 feet above the ground?
(c) During the first 40 seconds of the ride, over what interval of time t is an individual on the Ferris wheel more than 125 feet above the ground?

107. Holding Pattern An airplane is asked to stay within a holding pattern near Chicago's O'Hare International Airport. The function $d(x) = 70\sin(0.65x) + 150$ represents the distance d, in miles, of the airplane from the airport at time x, in minutes.
(a) When the plane enters the holding pattern, $x = 0$, how far is it from O'Hare?
(b) During the first 20 minutes after the plane enters the holding pattern, at what time x is the plane exactly 100 miles from the airport?
(c) During the first 20 minutes after the plane enters the holding pattern, at what time x is the plane more than 100 miles from the airport?
(d) While the plane is in the holding pattern, will it ever be within 70 miles of the airport? Why?

108. Projectile Motion A golfer hits a golf ball with an initial velocity of 100 miles per hour. The range R of the ball as a function of the angle θ to the horizontal is given by $R(\theta) = 672\sin(2\theta)$, where R is measured in feet.
(a) At what angle θ should the ball be hit if the golfer wants the ball to travel 450 feet (150 yards)?
(b) At what angle θ should the ball be hit if the golfer wants the ball to travel 540 feet (180 yards)?
(c) At what angle θ should the ball be hit if the golfer wants the ball to travel at least 480 feet (160 yards)?
(d) Can the golfer hit the ball 720 feet (240 yards)?

109. Heat Transfer In the study of heat transfer, the equation $x + \tan x = 0$ occurs. Graph $Y_1 = -x$ and $Y_2 = \tan x$ for $x \geq 0$. Conclude that there are an infinite number of points of intersection of these two graphs. Now find the first two positive solutions of $x + \tan x = 0$ rounded to two decimal places.

110. Carrying a Ladder around a Corner Two hallways, one of width 3 feet, the other of width 4 feet, meet at a right angle. See the illustration. It can be shown that the length L of the ladder as a function of θ is $L(\theta) = 4\csc\theta + 3\sec\theta$.
(a) In calculus, you will be asked to find the length of the longest ladder that can turn the corner by solving the equation

$$3\sec\theta\tan\theta - 4\csc\theta\cot\theta = 0 \quad 0° < \theta < 90°$$

Solve this equation for θ.

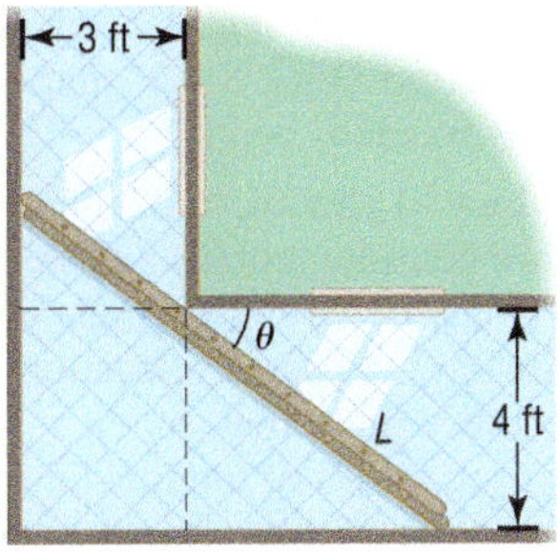

(b) What is the length of the longest ladder that can be carried around the corner?
(c) Graph $L = L(\theta), 0° \le \theta \le 90°$, and find the angle θ that minimizes the length L.
(d) Compare the result with the one found in part (a). Explain why the two answers are the same.

111. Projectile Motion The horizontal distance that a projectile will travel in the air (ignoring air resistance) is given by the equation

$$R(\theta) = \frac{v_0^2 \sin(2\theta)}{g}$$

where v_0 is the initial velocity of the projectile, θ is the angle of elevation, and g is acceleration due to gravity (9.8 meters per second squared).
(a) If you can throw a baseball with an initial speed of 34.8 meters per second, at what angle of elevation θ should you direct the throw so that the ball travels a distance of 107 meters before striking the ground?
(b) Determine the maximum distance that you can throw the ball.
(c) Graph $R = R(\theta)$, with $v_0 = 34.8$ meters per second.
(d) Verify the results obtained in parts (a) and (b) using a graphing utility.

112. Projectile Motion Refer to Problem 111.
(a) If you can throw a baseball with an initial speed of 40 meters per second, at what angle of elevation θ should you direct the throw so that the ball travels a distance of 110 meters before striking the ground?
(b) Determine the maximum distance that you can throw the ball.
(c) Graph $R = R(\theta)$, with $v_0 = 40$ meters per second.
(d) Verify the results obtained in parts (a) and (b) using a graphing utility.

The following discussion of Snell's Law of Refraction (named after Willebrord Snell, 1580–1626) is needed for Problems 113–120. Light, sound, and other waves travel at different speeds, depending on the medium (air, water, wood, and so on) through which they pass. Suppose that light travels from a point A in one medium, where its speed is v_1, to a point B in another medium, where its speed is v_2. Refer to the figure, where the angle θ_1 is called the angle of incidence and the angle θ_2 is the angle of refraction. Snell's Law, which can be proved using calculus, states that*

$$\frac{\sin\theta_1}{\sin\theta_2} = \frac{v_1}{v_2}$$

The ratio $\frac{v_1}{v_2}$ is called the index of refraction. Some values are given in the table shown to the right.

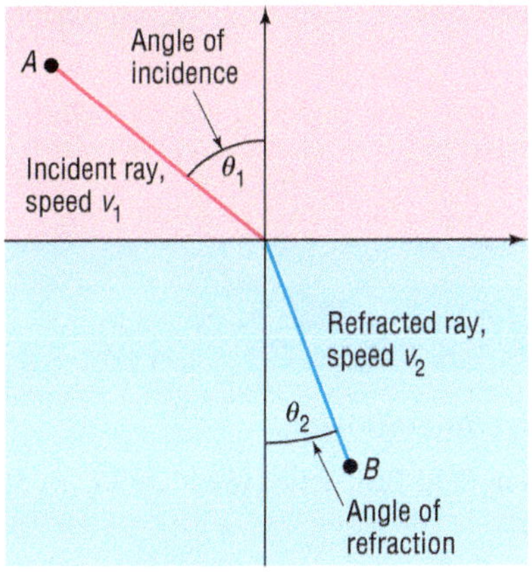

Some Indexes of Refraction	
Medium	**Index of Refraction**[†]
Water	1.33
Ethyl alcohol (20°C)	1.36
Carbon disulfide	1.63
Air (1 atm and 0°C)	1.00029
Diamond	2.42
Fused quartz	1.46
Glass, crown	1.52
Glass, dense flint	1.66
Sodium chloride	1.54

113. The index of refraction of light in passing from a vacuum into water is 1.33. If the angle of incidence is 40°, determine the angle of refraction.

114. The index of refraction of light in passing from a vacuum into dense flint glass is 1.66. If the angle of incidence is 50°, determine the angle of refraction.

115. Ptolemy, who lived in the city of Alexandria in Egypt during the second century AD, gave the measured values in the following table for the angle of incidence θ_1 and the angle of refraction θ_2 for a light beam passing from air into water. Do these values agree with Snell's Law? If so, what index of refraction results? (These data are of interest as the oldest recorded physical measurements.)

θ_1	θ_2	θ_1	θ_2
10°	8°	50°	35°0′
20°	15°30′	60°	40°30′
30°	22°30′	70°	45°30′
40°	29°0′	80°	50°0′

*Because this law was also deduced by René Descartes in France, it is also known as Descartes' Law.

†For light of wavelength 589 nanometers, measured with respect to a vacuum. The index with respect to air is negligibly different in most cases.

116. Bending Light The speed of yellow sodium light (wavelength, 589 nanometers) in a certain liquid is measured to be 1.92×10^8 meters per second. What is the index of refraction of this liquid, with respect to air, for sodium light?*

[**Hint:** The speed of light in air is approximately 2.998×10^8 meters per second.]

117. Bending Light A beam of light with a wavelength of 589 nanometers traveling in air makes an angle of incidence of 40° on a slab of transparent material, and the refracted beam makes an angle of refraction of 26°. Find the index of refraction of the material.*

118. Bending Light A light ray with a wavelength of 589 nanometers (produced by a sodium lamp) traveling through air makes an angle of incidence of 30° on a smooth, flat slab of crown glass. Find the angle of refraction.*

119. A light beam passes through a thick slab of material whose index of refraction is n_2. Show that the emerging beam is parallel to the incident beam.*

120. Brewster's Law If the angle of incidence and the angle of refraction are complementary angles, the angle of incidence is referred to as the Brewster angle θ_B. The Brewster angle is related to the indices of refraction of the two media, n_1 and n_2, by the equation $n_1 \sin \theta_B = n_2 \cos \theta_B$, where n_1 is the index of refraction of the incident medium and n_2 is the index of refraction of the refractive medium. Determine the Brewster angle for a light beam traveling through water (at 20°C) that makes an angle of incidence with a smooth, flat slab of crown glass.

*Adapted from Halliday, Resnick, and Walker, *Fundamentals of Physics*, 10th ed., 2014, John Wiley & Sons.

Explaining Concepts: Discussion and Writing

121. Explain in your own words how you would use your calculator to solve the equation $\cos x = -0.6, 0 \le x < 2\pi$. How would you modify your approach to solve the equation $\cot x = 5, 0 < x < 2\pi$?

122. Explain why no further points of intersection (and therefore no further solutions) exist in Figure 30 for $x < -\pi$ or $x > 4\pi$.

Retain Your Knowledge

Problems 123–126 are based on material learned earlier in the course. The purpose of these problems is to keep the material fresh in your mind so that you are better prepared for the final exam.

123. Convert $6^x = y$ to an equivalent statement involving a logarithm.

124. Find the zeros of $f(x) = 2x^2 - 9x + 8$.

125. Given $\sin \theta = -\dfrac{\sqrt{10}}{10}$ and $\cos \theta = \dfrac{3\sqrt{10}}{10}$, find the exact value of each of the four remaining trigonometric functions.

126. Determine the amplitude, period, and phase shift of the function $y = 2 \sin (2x - \pi)$. Graph the function. Show at least two periods.

'Are You Prepared?' Answers

1. $\left\{\dfrac{3}{2}\right\}$ **2.** $\dfrac{\sqrt{2}}{2}; -\dfrac{1}{2}$ **3.** $\left\{-1, \dfrac{5}{4}\right\}$ **4.** $\left\{\dfrac{1-\sqrt{5}}{2}, \dfrac{1+\sqrt{5}}{2}\right\}$ **5.** $\left\{0, \dfrac{5}{2}\right\}$ **6.** $\{0.76\}$

8.4 Trigonometric Identities

PREPARING FOR THIS SECTION *Before getting started, review the following:*

- Fundamental Identities (Section 7.2, pp. 529–530)
- Even–Odd Properties (Section 7.5, pp. 568–569)

Now Work the 'Are You Prepared?' problems on page 649.

OBJECTIVES 1 Use Algebra to Simplify Trigonometric Expressions (p. 646)
2 Establish Identities (p. 646)

This section establishes some additional identities involving trigonometric functions. First, let's review the definition of an *identity*.

DEFINITION

Two functions f and g are **identically equal** if

$$f(x) = g(x)$$

for every value of x for which both functions are defined. Such an equation is referred to as an **identity**. An equation that is not an identity is called a **conditional equation**.

For example, the following are identities:

$$(x+1)^2 = x^2 + 2x + 1 \qquad \sin^2 x + \cos^2 x = 1 \qquad \csc x = \frac{1}{\sin x}$$

The following are conditional equations:

$2x + 5 = 0$ **True only if** $x = -\frac{5}{2}$

$\sin x = 0$ **True only if** $x = k\pi$, k **an integer**

$\sin x = \cos x$ **True only if** $x = \frac{\pi}{4} + 2k\pi$ **or** $x = \frac{5\pi}{4} + 2k\pi$, k **an integer**

The following summarizes the trigonometric identities that have been established thus far.

Quotient Identities

$$\tan\theta = \frac{\sin\theta}{\cos\theta} \qquad \cot\theta = \frac{\cos\theta}{\sin\theta}$$

Reciprocal Identities

$$\csc\theta = \frac{1}{\sin\theta} \qquad \sec\theta = \frac{1}{\cos\theta} \qquad \cot\theta = \frac{1}{\tan\theta}$$

Pythagorean Identities

$$\sin^2\theta + \cos^2\theta = 1 \qquad \tan^2\theta + 1 = \sec^2\theta$$
$$\cot^2\theta + 1 = \csc^2\theta$$

Even–Odd Identities

$$\sin(-\theta) = -\sin\theta \qquad \cos(-\theta) = \cos\theta \qquad \tan(-\theta) = -\tan\theta$$
$$\csc(-\theta) = -\csc\theta \qquad \sec(-\theta) = \sec\theta \qquad \cot(-\theta) = -\cot\theta$$

This list comprises what shall be referred to as the **basic trigonometric identities**. These identities should not merely be memorized but should be *known* (just as you know your name rather than have it memorized). In fact, minor variations of a basic identity are often used. For example,

$$\sin^2\theta = 1 - \cos^2\theta \quad \text{or} \quad \cos^2\theta = 1 - \sin^2\theta$$

might be used instead of $\sin^2\theta + \cos^2\theta = 1$. For this reason, among others, it is very important to know these relationships and be comfortable with variations of them.

1 Use Algebra to Simplify Trigonometric Expressions

The ability to use algebra to manipulate trigonometric expressions is a key skill that one must have to establish identities. Four basic algebraic techniques are used to establish identities:

1. Rewriting a trigonometric expression in terms of sine and cosine only
2. Multiplying the numerator and denominator of a ratio by a "well-chosen 1"
3. Writing sums of trigonometric ratios as a single ratio
4. Factoring

EXAMPLE 1 Using Algebraic Techniques to Simplify Trigonometric Expressions

(a) Simplify $\dfrac{\cot\theta}{\csc\theta}$ by rewriting each trigonometric function in terms of sine and cosine functions.

(b) Show that $\dfrac{\cos\theta}{1+\sin\theta}=\dfrac{1-\sin\theta}{\cos\theta}$ by multiplying the numerator and denominator by $1-\sin\theta$.

(c) Simplify $\dfrac{1+\sin u}{\sin u}+\dfrac{\cot u-\cos u}{\cos u}$ by rewriting the expression as a single ratio.

(d) Simplify $\dfrac{\sin^2 v-1}{\tan v\sin v-\tan v}$ by factoring.

Solution

(a) $$\frac{\cot\theta}{\csc\theta}=\frac{\dfrac{\cos\theta}{\sin\theta}}{\dfrac{1}{\sin\theta}}=\frac{\cos\theta}{\sin\theta}\cdot\frac{\sin\theta}{1}=\cos\theta$$

(b) $$\frac{\cos\theta}{1+\sin\theta}=\frac{\cos\theta}{1+\sin\theta}\cdot\frac{1-\sin\theta}{1-\sin\theta}=\frac{\cos\theta(1-\sin\theta)}{1-\sin^2\theta}$$

↑ Multiply by a well-chosen 1: $\dfrac{1-\sin\theta}{1-\sin\theta}$.

$$=\frac{\cos\theta(1-\sin\theta)}{\cos^2\theta}=\frac{1-\sin\theta}{\cos\theta}$$

(c) $$\frac{1+\sin u}{\sin u}+\frac{\cot u-\cos u}{\cos u}=\frac{1+\sin u}{\sin u}\cdot\frac{\cos u}{\cos u}+\frac{\cot u-\cos u}{\cos u}\cdot\frac{\sin u}{\sin u}$$

$$=\frac{\cos u+\sin u\cos u+\cot u\sin u-\cos u\sin u}{\sin u\cos u}=\frac{\cos u+\cot u\sin u}{\sin u\cos u}$$

$$=\frac{\cos u+\dfrac{\cos u}{\sin u}\cdot\sin u}{\sin u\cos u}=\frac{\cos u+\cos u}{\sin u\cos u}=\frac{2\cos u}{\sin u\cos u}=\frac{2}{\sin u}$$

↑ $\cot u=\dfrac{\cos u}{\sin u}$

(d) $$\frac{\sin^2 v-1}{\tan v\sin v-\tan v}=\frac{(\sin v+1)(\sin v-1)}{\tan v(\sin v-1)}=\frac{\sin v+1}{\tan v}$$ ■

Now Work PROBLEMS 11, 13, AND 15

2 Establish Identities

In the examples that follow, the directions read "Establish the identity...." This is accomplished by starting with one side of the given equation (usually the side containing the more complicated expression) and, using appropriate basic identities and algebraic manipulations, arriving at the other side. The selection of appropriate basic identities to obtain the desired result is learned only through experience and lots of practice.

EXAMPLE 2 **Establishing an Identity**

Establish the identity: $\csc\theta \cdot \tan\theta = \sec\theta$

Solution Start with the left side, because it contains the more complicated expression, and apply a quotient identity and a reciprocal identity.

$$\csc\theta \cdot \tan\theta = \frac{1}{\sin\theta} \cdot \frac{\sin\theta}{\cos\theta} = \frac{1}{\cos\theta} = \sec\theta$$

The right side has been reached, so the identity is established. ■

Note: A graphing utility can be used to provide evidence of an identity. For example, if we graph $Y_1 = \csc\theta \cdot \tan\theta$ and $Y_2 = \sec\theta$, the graphs appear to be the same. This provides evidence that $Y_1 = Y_2$. However, it does not prove their equality. A graphing utility *cannot be used to establish an identity*—identities must be established algebraically. ■

Now Work PROBLEM 21

EXAMPLE 3 **Establishing an Identity**

Establish the identity: $\sin^2(-\theta) + \cos^2(-\theta) = 1$

Solution Begin with the left side and, because the arguments are $-\theta$, apply Even–Odd Identities.

$$\begin{aligned}\sin^2(-\theta) + \cos^2(-\theta) &= [\sin(-\theta)]^2 + [\cos(-\theta)]^2 \\ &= (-\sin\theta)^2 + (\cos\theta)^2 && \text{Even–Odd Identities} \\ &= (\sin\theta)^2 + (\cos\theta)^2 \\ &= 1 && \text{Pythagorean Identity}\end{aligned}$$

■

EXAMPLE 4 **Establishing an Identity**

Establish the identity: $\dfrac{\sin^2(-\theta) - \cos^2(-\theta)}{\sin(-\theta) - \cos(-\theta)} = \cos\theta - \sin\theta$

Solution Observe that the left side contains the more complicated expression. Also, the left side contains expressions with the argument $-\theta$, whereas the right side contains expressions with the argument θ. So start with the left side and apply Even–Odd Identities.

$$\begin{aligned}\frac{\sin^2(-\theta) - \cos^2(-\theta)}{\sin(-\theta) - \cos(-\theta)} &= \frac{[\sin(-\theta)]^2 - [\cos(-\theta)]^2}{\sin(-\theta) - \cos(-\theta)} \\ &= \frac{(-\sin\theta)^2 - (\cos\theta)^2}{-\sin\theta - \cos\theta} && \text{Even–Odd Identities} \\ &= \frac{(\sin\theta)^2 - (\cos\theta)^2}{-\sin\theta - \cos\theta} && \text{Simplify.} \\ &= \frac{(\sin\theta - \cos\theta)(\sin\theta + \cos\theta)}{-(\sin\theta + \cos\theta)} && \text{Factor} \\ &= \cos\theta - \sin\theta && \text{Divide out and simplify.}\end{aligned}$$

■

EXAMPLE 5 **Establishing an Identity**

Establish the identity: $\dfrac{1 + \tan u}{1 + \cot u} = \tan u$

Solution

$$\frac{1 + \tan u}{1 + \cot u} = \frac{1 + \tan u}{1 + \dfrac{1}{\tan u}} = \frac{1 + \tan u}{\dfrac{\tan u + 1}{\tan u}} = \frac{\tan u(1 + \tan u)}{\tan u + 1} = \tan u$$

■

Now Work PROBLEMS 25 AND 29

When sums or differences of quotients appear, it is usually best to rewrite them as a single quotient, especially if the other side of the identity consists of only one term.

EXAMPLE 6 Establishing an Identity

Establish the identity: $\dfrac{\sin\theta}{1+\cos\theta}+\dfrac{1+\cos\theta}{\sin\theta}=2\csc\theta$

Solution The left side is more complicated. Start with it and proceed to add.

$$\begin{aligned}\frac{\sin\theta}{1+\cos\theta}+\frac{1+\cos\theta}{\sin\theta} &= \frac{\sin^2\theta+(1+\cos\theta)^2}{(1+\cos\theta)(\sin\theta)} && \text{Add the quotients.}\\ &= \frac{\sin^2\theta+1+2\cos\theta+\cos^2\theta}{(1+\cos\theta)(\sin\theta)} && \text{Remove parentheses in the numerator.}\\ &= \frac{(\sin^2\theta+\cos^2\theta)+1+2\cos\theta}{(1+\cos\theta)(\sin\theta)} && \text{Regroup}\\ &= \frac{2+2\cos\theta}{(1+\cos\theta)(\sin\theta)} && \text{Pythagorean Identity}\\ &= \frac{2\cancel{(1+\cos\theta)}}{\cancel{(1+\cos\theta)}(\sin\theta)} && \text{Factor.}\\ &= \frac{2}{\sin\theta} && \text{Divide out } 1+\cos\theta.\\ &= 2\csc\theta && \text{Reciprocal Identity}\end{aligned}$$

■

Now Work PROBLEM 51

Sometimes it helps to write one side in terms of sine and cosine functions only.

EXAMPLE 7 Establishing an Identity

Establish the identity: $\dfrac{\tan v+\cot v}{\sec v\csc v}=1$

Solution

$$\frac{\tan v+\cot v}{\sec v\csc v}=\frac{\dfrac{\sin v}{\cos v}+\dfrac{\cos v}{\sin v}}{\dfrac{1}{\cos v}\cdot\dfrac{1}{\sin v}}=\frac{\dfrac{\sin^2 v+\cos^2 v}{\cos v\sin v}}{\dfrac{1}{\cos v\sin v}}$$

Change to sines and cosines. Add the quotients in the numerator.

$$=\frac{1}{\cos v\sin v}\cdot\frac{\cos v\sin v}{1}=1$$

Divide the quotients; $\sin^2 v+\cos^2 v=1$.

■

Now Work PROBLEM 71

Sometimes, multiplying the numerator and the denominator by an appropriate factor simplifies an expression.

EXAMPLE 8 **Establishing an Identity**

Establish the identity: $\frac{1 - \sin\theta}{\cos\theta} = \frac{\cos\theta}{1 + \sin\theta}$

Solution Start with the left side and multiply the numerator and the denominator by $1 + \sin\theta$. (Alternatively, we could multiply the numerator and the denominator of the right side by $1 - \sin\theta$.)

$$\frac{1 - \sin\theta}{\cos\theta} = \frac{1 - \sin\theta}{\cos\theta} \cdot \frac{1 + \sin\theta}{1 + \sin\theta}$$ Multiply the numerator and the denominator by $1 + \sin\theta$.

$$= \frac{1 - \sin^2\theta}{\cos\theta(1 + \sin\theta)}$$

$$= \frac{\cos^2\theta}{\cos\theta(1 + \sin\theta)}$$ $1 - \sin^2\theta = \cos^2\theta$

$$= \frac{\cos\theta}{1 + \sin\theta}$$ Divide out $\cos\theta$. ■

Now Work PROBLEM 55

Although a lot of practice is the only real way to learn how to establish identities, the following guidelines should prove helpful.

WARNING Be careful not to handle identities to be established as if they were conditional equations. You *cannot* establish an identity by such methods as adding the same expression to each side and obtaining a true statement. This practice is not allowed, because the original statement is precisely the one that you are trying to establish. You do not know until it has been established that it is, in fact, true. ■

Guidelines for Establishing Identities

1. It is almost always preferable to start with the side containing the more complicated expression.
2. Rewrite sums or differences of quotients as a single quotient.
3. Sometimes it will help to rewrite one side in terms of sine and cosine functions only.
4. Always keep the goal in mind. As you manipulate one side of the expression, keep in mind the form of the expression on the other side.

8.4 Assess Your Understanding

'Are You Prepared?' *Answers are given at the end of these exercises. If you get a wrong answer, read the pages listed in red.*

1. ***True or False*** $\sin^2\theta = 1 - \cos^2\theta$. (pp. 529–530)

2. ***True or False*** $\sin(-\theta) + \cos(-\theta) = \cos\theta - \sin\theta$. (pp. 568–569)

Concepts and Vocabulary

3. Suppose that f and g are two functions with the same domain. If $f(x) = g(x)$ for every x in the domain, the equation is called a(n) ________. Otherwise, it is called a(n) ________ equation.

4. $\tan^2\theta - \sec^2\theta =$ ______.

5. $\cos(-\theta) - \cos\theta =$ ______.

6. ***True or False*** $\sin(-\theta) + \sin\theta = 0$ for any value of θ.

7. ***True or False*** In establishing an identity, it is often easiest to just multiply both sides by a well-chosen nonzero expression involving the variable.

8. ***True or False*** $\tan\theta \cdot \cos\theta = \sin\theta$ for any $\theta \neq (2k + 1)\frac{\pi}{2}$.

9. Which of the following equations is *not* an identity?
(a) $\cot^2\theta + 1 = \csc^2\theta$ (b) $\tan(-\theta) = -\tan\theta$
(c) $\tan\theta = \frac{\cos\theta}{\sin\theta}$ (d) $\csc\theta = \frac{1}{\sin\theta}$

10. The expression $\frac{1}{1 - \sin\theta} + \frac{1}{1 + \sin\theta}$ simplifies to which of the following?
(a) $2\cos^2\theta$ (b) $2\sec^2\theta$ (c) $2\sin^2\theta$ (d) $2\csc^2\theta$

Skill Building

In Problems 11–20, simplify each trigonometric expression by following the indicated direction.

11. Rewrite in terms of sine and cosine functions:

$$\tan\theta\cdot\csc\theta$$

12. Rewrite in terms of sine and cosine functions:

$$\cot\theta\cdot\sec\theta$$

13. Multiply $\dfrac{\cos\theta}{1-\sin\theta}$ by $\dfrac{1+\sin\theta}{1+\sin\theta}$.

14. Multiply $\dfrac{\sin\theta}{1+\cos\theta}$ by $\dfrac{1-\cos\theta}{1-\cos\theta}$.

15. Rewrite over a common denominator:

$$\frac{\sin\theta+\cos\theta}{\cos\theta}+\frac{\cos\theta-\sin\theta}{\sin\theta}$$

16. Rewrite over a common denominator:

$$\frac{1}{1-\cos v}+\frac{1}{1+\cos v}$$

17. Multiply and simplify: $\dfrac{(\sin\theta+\cos\theta)(\sin\theta+\cos\theta)-1}{\sin\theta\cos\theta}$

18. Multiply and simplify: $\dfrac{(\tan\theta+1)(\tan\theta+1)-\sec^2\theta}{\tan\theta}$

19. Factor and simplify: $\dfrac{3\sin^2\theta+4\sin\theta+1}{\sin^2\theta+2\sin\theta+1}$

20. Factor and simplify: $\dfrac{\cos^2\theta-1}{\cos^2\theta-\cos\theta}$

In Problems 21–100, establish each identity.

21. $\csc\theta\cdot\cos\theta=\cot\theta$

22. $\sec\theta\cdot\sin\theta=\tan\theta$

23. $1+\tan^2(-\theta)=\sec^2\theta$

24. $1+\cot^2(-\theta)=\csc^2\theta$

25. $\cos\theta(\tan\theta+\cot\theta)=\csc\theta$

26. $\sin\theta(\cot\theta+\tan\theta)=\sec\theta$

27. $\tan u\cot u-\cos^2 u=\sin^2 u$

28. $\sin u\csc u-\cos^2 u=\sin^2 u$

29. $(\sec\theta-1)(\sec\theta+1)=\tan^2\theta$

30. $(\csc\theta-1)(\csc\theta+1)=\cot^2\theta$

31. $(\sec\theta+\tan\theta)(\sec\theta-\tan\theta)=1$

32. $(\csc\theta+\cot\theta)(\csc\theta-\cot\theta)=1$

33. $\cos^2\theta(1+\tan^2\theta)=1$

34. $(1-\cos^2\theta)(1+\cot^2\theta)=1$

35. $(\sin\theta+\cos\theta)^2+(\sin\theta-\cos\theta)^2=2$

36. $\tan^2\theta\cos^2\theta+\cot^2\theta\sin^2\theta=1$

37. $\sec^4\theta-\sec^2\theta=\tan^4\theta+\tan^2\theta$

38. $\csc^4\theta-\csc^2\theta=\cot^4\theta+\cot^2\theta$

39. $\sec u-\tan u=\dfrac{\cos u}{1+\sin u}$

40. $\csc u-\cot u=\dfrac{\sin u}{1+\cos u}$

41. $3\sin^2\theta+4\cos^2\theta=3+\cos^2\theta$

42. $9\sec^2\theta-5\tan^2\theta=5+4\sec^2\theta$

43. $1-\dfrac{\cos^2\theta}{1+\sin\theta}=\sin\theta$

44. $1-\dfrac{\sin^2\theta}{1-\cos\theta}=-\cos\theta$

45. $\dfrac{1+\tan v}{1-\tan v}=\dfrac{\cot v+1}{\cot v-1}$

46. $\dfrac{\csc v-1}{\csc v+1}=\dfrac{1-\sin v}{1+\sin v}$

47. $\dfrac{\sec\theta}{\csc\theta}+\dfrac{\sin\theta}{\cos\theta}=2\tan\theta$

48. $\dfrac{\csc\theta-1}{\cot\theta}=\dfrac{\cot\theta}{\csc\theta+1}$

49. $\dfrac{1+\sin\theta}{1-\sin\theta}=\dfrac{\csc\theta+1}{\csc\theta-1}$

50. $\dfrac{\cos\theta+1}{\cos\theta-1}=\dfrac{1+\sec\theta}{1-\sec\theta}$

51. $\dfrac{1-\sin v}{\cos v}+\dfrac{\cos v}{1-\sin v}=2\sec v$

52. $\dfrac{\cos v}{1+\sin v}+\dfrac{1+\sin v}{\cos v}=2\sec v$

53. $\dfrac{\sin\theta}{\sin\theta-\cos\theta}=\dfrac{1}{1-\cot\theta}$

54. $1-\dfrac{\sin^2\theta}{1+\cos\theta}=\cos\theta$

55. $\dfrac{1-\sin\theta}{1+\sin\theta}=(\sec\theta-\tan\theta)^2$

56. $\dfrac{1-\cos\theta}{1+\cos\theta}=(\csc\theta-\cot\theta)^2$

57. $\dfrac{\cos\theta}{1-\tan\theta}+\dfrac{\sin\theta}{1-\cot\theta}=\sin\theta+\cos\theta$

58. $\dfrac{\cot\theta}{1-\tan\theta}+\dfrac{\tan\theta}{1-\cot\theta}=1+\tan\theta+\cot\theta$

59. $\tan\theta+\dfrac{\cos\theta}{1+\sin\theta}=\sec\theta$

60. $\dfrac{\sin\theta\cos\theta}{\cos^2\theta-\sin^2\theta}=\dfrac{\tan\theta}{1-\tan^2\theta}$

61. $\dfrac{\tan\theta+\sec\theta-1}{\tan\theta-\sec\theta+1}=\tan\theta+\sec\theta$

62. $\dfrac{\sin\theta-\cos\theta+1}{\sin\theta+\cos\theta-1}=\dfrac{\sin\theta+1}{\cos\theta}$

63. $\dfrac{\tan\theta-\cot\theta}{\tan\theta+\cot\theta}=\sin^2\theta-\cos^2\theta$

64. $\dfrac{\sec\theta-\cos\theta}{\sec\theta+\cos\theta}=\dfrac{\sin^2\theta}{1+\cos^2\theta}$

65. $\dfrac{\tan u-\cot u}{\tan u+\cot u}+1=2\sin^2 u$

66. $\dfrac{\tan u-\cot u}{\tan u+\cot u}+2\cos^2 u=1$

67. $\dfrac{\sec\theta+\tan\theta}{\cot\theta+\cos\theta}=\tan\theta\sec\theta$

68. $\dfrac{\sec\theta}{1+\sec\theta}=\dfrac{1-\cos\theta}{\sin^2\theta}$

69. $\dfrac{1-\tan^2\theta}{1+\tan^2\theta}+1=2\cos^2\theta$

70. $\dfrac{1-\cot^2\theta}{1+\cot^2\theta}+2\cos^2\theta=1$

71. $\dfrac{\sec\theta-\csc\theta}{\sec\theta\csc\theta}=\sin\theta-\cos\theta$

72. $\dfrac{\sin^2\theta-\tan\theta}{\cos^2\theta-\cot\theta}=\tan^2\theta$

73. $\sec\theta-\cos\theta=\sin\theta\tan\theta$

74. $\tan\theta+\cot\theta=\sec\theta\csc\theta$

75. $\dfrac{1}{1-\sin\theta}+\dfrac{1}{1+\sin\theta}=2\sec^2\theta$

76. $\dfrac{1+\sin\theta}{1-\sin\theta}-\dfrac{1-\sin\theta}{1+\sin\theta}=4\tan\theta\sec\theta$

77. $\dfrac{\sec\theta}{1-\sin\theta}=\dfrac{1+\sin\theta}{\cos^3\theta}$

78. $\dfrac{1+\sin\theta}{1-\sin\theta}=(\sec\theta+\tan\theta)^2$

79. $\dfrac{(\sec v-\tan v)^2+1}{\csc v(\sec v-\tan v)}=2\tan v$

80. $\dfrac{\sec^2 v-\tan^2 v+\tan v}{\sec v}=\sin v+\cos v$

81. $\dfrac{\sin\theta+\cos\theta}{\cos\theta}-\dfrac{\sin\theta-\cos\theta}{\sin\theta}=\sec\theta\csc\theta$

82. $\dfrac{\sin\theta+\cos\theta}{\sin\theta}-\dfrac{\cos\theta-\sin\theta}{\cos\theta}=\sec\theta\csc\theta$

83. $\dfrac{\sin^3\theta+\cos^3\theta}{\sin\theta+\cos\theta}=1-\sin\theta\cos\theta$

84. $\dfrac{\sin^3\theta+\cos^3\theta}{1-2\cos^2\theta}=\dfrac{\sec\theta-\sin\theta}{\tan\theta-1}$

85. $\dfrac{\cos^2\theta-\sin^2\theta}{1-\tan^2\theta}=\cos^2\theta$

86. $\dfrac{\cos\theta+\sin\theta-\sin^3\theta}{\sin\theta}=\cot\theta+\cos^2\theta$

87. $\dfrac{(2\cos^2\theta-1)^2}{\cos^4\theta-\sin^4\theta}=1-2\sin^2\theta$

88. $\dfrac{1-2\cos^2\theta}{\sin\theta\cos\theta}=\tan\theta-\cot\theta$

89. $\dfrac{1+\sin\theta+\cos\theta}{1+\sin\theta-\cos\theta}=\dfrac{1+\cos\theta}{\sin\theta}$

90. $\dfrac{1+\cos\theta+\sin\theta}{1+\cos\theta-\sin\theta}=\sec\theta+\tan\theta$

91. $(a\sin\theta+b\cos\theta)^2+(a\cos\theta-b\sin\theta)^2=a^2+b^2$

92. $(2a\sin\theta\cos\theta)^2+a^2(\cos^2\theta-\sin^2\theta)^2=a^2$

93. $\dfrac{\tan\alpha+\tan\beta}{\cot\alpha+\cot\beta}=\tan\alpha\tan\beta$

94. $(\tan\alpha+\tan\beta)(1-\cot\alpha\cot\beta)+(\cot\alpha+\cot\beta)(1-\tan\alpha\tan\beta)=0$

95. $(\sin\alpha+\cos\beta)^2+(\cos\beta+\sin\alpha)(\cos\beta-\sin\alpha)=2\cos\beta(\sin\alpha+\cos\beta)$

96. $(\sin\alpha-\cos\beta)^2+(\cos\beta+\sin\alpha)(\cos\beta-\sin\alpha)=-2\cos\beta(\sin\alpha-\cos\beta)$

97. $\ln|\sec\theta|=-\ln|\cos\theta|$

98. $\ln|\tan\theta|=\ln|\sin\theta|-\ln|\cos\theta|$

99. $\ln|1+\cos\theta|+\ln|1-\cos\theta|=2\ln|\sin\theta|$

100. $\ln|\sec\theta+\tan\theta|+\ln|\sec\theta-\tan\theta|=0$

In Problems 101–104, show that the functions f and g are identically equal.

101. $f(x)=\sin x\cdot\tan x \quad g(x)=\sec x-\cos x$

102. $f(x)=\cos x\cdot\cot x \quad g(x)=\csc x-\sin x$

103. $f(\theta)=\dfrac{1-\sin\theta}{\cos\theta}-\dfrac{\cos\theta}{1+\sin\theta} \quad g(\theta)=0$

104. $f(\theta)=\tan\theta+\sec\theta \quad g(\theta)=\dfrac{\cos\theta}{1-\sin\theta}$

105. Show that $\sqrt{16+16\tan^2\theta}=4\sec\theta$ if $-\dfrac{\pi}{2}<\theta<\dfrac{\pi}{2}$.

106. Show that $\sqrt{9\sec^2\theta-9}=3\tan\theta$ if $\pi\le\theta<\dfrac{3\pi}{2}$.

Applications and Extensions

107. Searchlights A searchlight at the grand opening of a new car dealership casts a spot of light on a wall located 75 meters from the searchlight. The acceleration $\ddot{r}$ of the spot of light is found to be $\ddot{r}=1200\sec\theta(2\sec^2\theta-1)$. Show that this is equivalent to $\ddot{r}=1200\left(\dfrac{1+\sin^2\theta}{\cos^3\theta}\right)$.

Source: Adapted from Hibbeler, *Engineering Mechanics: Dynamics,* 13th ed., Pearson © 2013.

108. Optical Measurement Optical methods of measurement often rely on the interference of two light waves. If two light waves, identical except for a phase lag, are mixed together, the resulting intensity, or irradiance, is given by

$$I_t=4A^2\,\frac{(\csc\theta-1)(\sec\theta+\tan\theta)}{\csc\theta\sec\theta}.$$

Show that this is equivalent to $I_t=(2A\cos\theta)^2$.

Source: *Experimental Techniques,* July/August 2002

Explaining Concepts: Discussion and Writing

109. Write a few paragraphs outlining your strategy for establishing identities.

110. Write down the three Pythagorean Identities.

111. Why do you think it is usually preferable to start with the side containing the more complicated expression when establishing an identity?

112. Make up an identity that is not a basic identity.

Retain Your Knowledge

Problems 113–116 are based on material learned earlier in the course. The purpose of these problems is to keep the material fresh in your mind so that you are better prepared for the final exam.

113. Determine whether $f(x)=-3x^2+120x+50$ has a maximum or a minimum value, and then find the value.

114. Given $f(x)=\dfrac{x+1}{x-2}$ and $g(x)=3x-4$, find $f\circ g$.

115. Find the exact values of the six trigonometric functions of an angle θ in standard position if $(-12, 5)$ is a point on its terminal side.

116. Find the average rate of change of $f(x)=\cos x$ from 0 to $\dfrac{\pi}{2}$.

'Are You Prepared?' Answers

1. True **2.** True

8.5 Sum and Difference Formulas

PREPARING FOR THIS SECTION *Before getting started, review the following:*

- Distance Formula (Section 1.1, pp. 85–87)
- Values of the Trigonometric Functions (Section 7.3, pp. 539–541; Section 7.4, pp. 550–558)
- Finding Exact Values Given the Value of a Trigonometric Function and the Quadrant of the Angle (Section 7.4, pp. 557–558)

Now Work the 'Are You Prepared?' problems on page 661.

OBJECTIVES
1. Use Sum and Difference Formulas to Find Exact Values (p. 653)
2. Use Sum and Difference Formulas to Establish Identities (p. 654)
3. Use Sum and Difference Formulas Involving Inverse Trigonometric Functions (p. 658)
4. Solve Trigonometric Equations Linear in Sine and Cosine (p. 659)

This section continues the derivation of trigonometric identities by obtaining formulas that involve the sum or the difference of two angles, such as $\cos(\alpha + \beta)$, $\cos(\alpha - \beta)$, and $\sin(\alpha + \beta)$. These formulas are referred to as the **sum and difference formulas**. We begin with the formulas for $\cos(\alpha + \beta)$ and $\cos(\alpha - \beta)$.

THEOREM

Sum and Difference Formulas for the Cosine Function

$$\cos(\alpha + \beta) = \cos\alpha\cos\beta - \sin\alpha\sin\beta \qquad (1)$$

$$\cos(\alpha - \beta) = \cos\alpha\cos\beta + \sin\alpha\sin\beta \qquad (2)$$

In Words

Formula (1) states that the cosine of the sum of two angles equals the cosine of the first angle times the cosine of the second angle minus the sine of the first angle times the sine of the second angle.

Proof We will prove formula (2) first. Although this formula is true for all numbers α and β, we shall assume in our proof that $0 < \beta < \alpha < 2\pi$. Begin with the unit circle and place the angles α and β in standard position, as shown in Figure 31(a). The point P_1 lies on the terminal side of β, so its coordinates are $(\cos\beta, \sin\beta)$; and the point P_2 lies on the terminal side of α, so its coordinates are $(\cos\alpha, \sin\alpha)$.

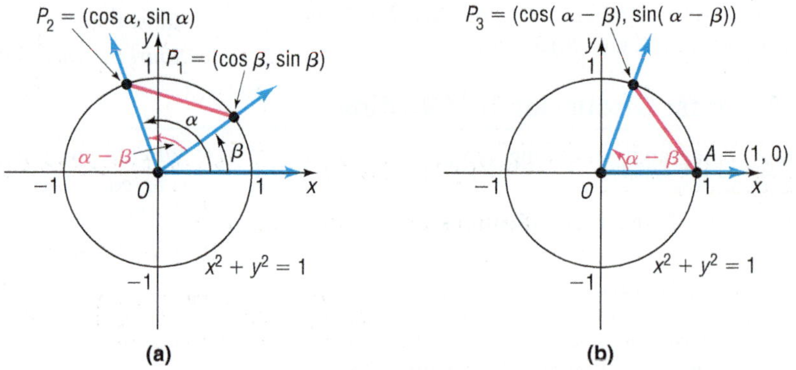

Figure 31

Now place the angle $\alpha - \beta$ in standard position, as shown in Figure 31(b). The point A has coordinates $(1, 0)$, and the point P_3 is on the terminal side of the angle $\alpha - \beta$, so its coordinates are $(\cos(\alpha - \beta), \sin(\alpha - \beta))$.

Looking at triangle OP_1P_2 in Figure 31(a) and triangle OAP_3 in Figure 31(b), note that these triangles are congruent. (Do you see why? SAS: two sides and the included angle, $\alpha - \beta$, are equal.) As a result, the unknown side of triangle OP_1P_2 and the unknown side of triangle OAP_3 must be equal; that is,

$$d(A, P_3) = d(P_1, P_2)$$

Now use the distance formula to obtain

$$\begin{aligned}
\sqrt{[\cos(\alpha-\beta) - 1]^2 + [\sin(\alpha-\beta) - 0]^2} &= \sqrt{(\cos\alpha - \cos\beta)^2 + (\sin\alpha - \sin\beta)^2} && d(A, P_3) = d(P_1, P_2)\\
[\cos(\alpha-\beta) - 1]^2 + \sin^2(\alpha-\beta) &= (\cos\alpha - \cos\beta)^2 + (\sin\alpha - \sin\beta)^2 && \text{Square both sides.}\\
\cos^2(\alpha-\beta) - 2\cos(\alpha-\beta) + 1 + \sin^2(\alpha-\beta) &= \cos^2\alpha - 2\cos\alpha\cos\beta + \cos^2\beta && \text{Multiply out the squared terms.}\\
&\quad + \sin^2\alpha - 2\sin\alpha\sin\beta + \sin^2\beta\\
2 - 2\cos(\alpha-\beta) &= 2 - 2\cos\alpha\cos\beta - 2\sin\alpha\sin\beta && \text{Apply a Pythagorean Identity (3 times).}\\
-2\cos(\alpha-\beta) &= -2\cos\alpha\cos\beta - 2\sin\alpha\sin\beta && \text{Subtract 2 from each side.}\\
\cos(\alpha-\beta) &= \cos\alpha\cos\beta + \sin\alpha\sin\beta && \text{Divide each side by } -2.
\end{aligned}$$

This is formula (2). ■

The proof of formula (1) follows from formula (2) and the Even–Odd Identities. Because $\alpha + \beta = \alpha - (-\beta)$, it follows that

$$\begin{aligned}
\cos(\alpha+\beta) &= \cos[\alpha - (-\beta)]\\
&= \cos\alpha\cos(-\beta) + \sin\alpha\sin(-\beta) && \text{Use formula (2).}\\
&= \cos\alpha\cos\beta - \sin\alpha\sin\beta && \text{Even–Odd Identities}
\end{aligned}$$

1 Use Sum and Difference Formulas to Find Exact Values

One use of formulas (1) and (2) is to obtain the exact value of the cosine of an angle that can be expressed as the sum or difference of angles whose sine and cosine are known exactly.

EXAMPLE 1 Using the Sum Formula to Find an Exact Value

Find the exact value of $\cos 75°$.

Solution Because $75° = 45° + 30°$, use formula (1) to obtain

$$\cos 75° = \cos(45° + 30°) \underset{\text{Formula (1)}}{\uparrow}= \cos 45° \cos 30° - \sin 45° \sin 30°$$

$$= \frac{\sqrt{2}}{2} \cdot \frac{\sqrt{3}}{2} - \frac{\sqrt{2}}{2} \cdot \frac{1}{2} = \frac{1}{4}(\sqrt{6} - \sqrt{2})$$ ■

EXAMPLE 2 Using the Difference Formula to Find an Exact Value

Find the exact value of $\cos \frac{\pi}{12}$.

Solution

$$\begin{aligned}
\cos\frac{\pi}{12} &= \cos\left(\frac{3\pi}{12} - \frac{2\pi}{12}\right) = \cos\left(\frac{\pi}{4} - \frac{\pi}{6}\right)\\
&= \cos\frac{\pi}{4}\cos\frac{\pi}{6} + \sin\frac{\pi}{4}\sin\frac{\pi}{6} && \text{Use formula (2).}\\
&= \frac{\sqrt{2}}{2} \cdot \frac{\sqrt{3}}{2} + \frac{\sqrt{2}}{2} \cdot \frac{1}{2} = \frac{1}{4}(\sqrt{6} + \sqrt{2})
\end{aligned}$$ ■

Now Work PROBLEMS 13 AND 19

2 Use Sum and Difference Formulas to Establish Identities

Another use of formulas (1) and (2) is to establish other identities. Two important identities conjectured earlier, in Section 7.2, are given next.

Seeing the Concept

Graph $Y_1 = \cos\left(\frac{\pi}{2} - x\right)$ and $Y_2 = \sin x$ on the same screen. Does doing this demonstrate the formula (3a)? How would you demonstrate the formula (3b)? ■

$$\cos\left(\frac{\pi}{2} - \theta\right) = \sin\theta \qquad \textbf{(3a)}$$

$$\sin\left(\frac{\pi}{2} - \theta\right) = \cos\theta \qquad \textbf{(3b)}$$

Proof To prove formula (3a), use the formula for $\cos(\alpha - \beta)$ with $\alpha = \frac{\pi}{2}$ and $\beta = \theta$.

$$\begin{aligned}\cos\left(\frac{\pi}{2} - \theta\right) &= \cos\frac{\pi}{2}\cos\theta + \sin\frac{\pi}{2}\sin\theta\\ &= 0\cdot\cos\theta + 1\cdot\sin\theta\\ &= \sin\theta\end{aligned}$$

To prove formula (3b), make use of the identity (3a) just established.

$$\sin\left(\frac{\pi}{2} - \theta\right) = \cos\left[\frac{\pi}{2} - \left(\frac{\pi}{2} - \theta\right)\right] = \cos\theta$$

Use formula (3a). ■

Also, because

$$\cos\left(\frac{\pi}{2} - \theta\right) = \cos\left[-\left(\theta - \frac{\pi}{2}\right)\right] = \cos\left(\theta - \frac{\pi}{2}\right)$$

Even Property of Cosine

and because

$$\cos\left(\frac{\pi}{2} - \theta\right) = \sin\theta$$

Formula (3a)

it follows that $\cos\left(\theta - \frac{\pi}{2}\right) = \sin\theta$. The graphs of $y = \cos\left(\theta - \frac{\pi}{2}\right)$ and $y = \sin\theta$ are identical.

Having established the identities in formulas (3a) and (3b), we now can derive the sum and difference formulas for $\sin(\alpha + \beta)$ and $\sin(\alpha - \beta)$.

Proof

$$\begin{aligned}\sin(\alpha + \beta) &= \cos\left[\frac{\pi}{2} - (\alpha + \beta)\right] && \text{Formula (3a)}\\ &= \cos\left[\left(\frac{\pi}{2} - \alpha\right) - \beta\right]\\ &= \cos\left(\frac{\pi}{2} - \alpha\right)\cos\beta + \sin\left(\frac{\pi}{2} - \alpha\right)\sin\beta && \text{Formula (2)}\\ &= \sin\alpha\cos\beta + \cos\alpha\sin\beta && \text{Formulas (3a) and (3b)}\end{aligned}$$

$$\begin{aligned}\sin(\alpha - \beta) &= \sin[\alpha + (-\beta)]\\ &= \sin\alpha\cos(-\beta) + \cos\alpha\sin(-\beta) && \text{Use the sum formula for sine just obtained.}\\ &= \sin\alpha\cos\beta + \cos\alpha(-\sin\beta) && \text{Even–Odd Identities}\\ &= \sin\alpha\cos\beta - \cos\alpha\sin\beta\end{aligned}$$

■

In Words
Formula (4) states that the sine of the sum of two angles equals the sine of the first angle times the cosine of the second angle plus the cosine of the first angle times the sine of the second angle.

THEOREM

Sum and Difference Formulas for the Sine Function

$$\sin(\alpha + \beta) = \sin\alpha\cos\beta + \cos\alpha\sin\beta \quad (4)$$

$$\sin(\alpha - \beta) = \sin\alpha\cos\beta - \cos\alpha\sin\beta \quad (5)$$

EXAMPLE 3 Using the Sum Formula to Find an Exact Value

Find the exact value of $\sin\frac{7\pi}{12}$.

Solution

$$\sin\frac{7\pi}{12} = \sin\left(\frac{3\pi}{12} + \frac{4\pi}{12}\right) = \sin\left(\frac{\pi}{4} + \frac{\pi}{3}\right)$$

$$= \sin\frac{\pi}{4}\cos\frac{\pi}{3} + \cos\frac{\pi}{4}\sin\frac{\pi}{3} \qquad \text{Formula (4)}$$

$$= \frac{\sqrt{2}}{2}\cdot\frac{1}{2} + \frac{\sqrt{2}}{2}\cdot\frac{\sqrt{3}}{2} = \frac{1}{4}(\sqrt{2} + \sqrt{6})$$

Now Work PROBLEM 21

EXAMPLE 4 Using the Difference Formula to Find an Exact Value

Find the exact value of $\sin 80^\circ \cos 20^\circ - \cos 80^\circ \sin 20^\circ$.

Solution The form of the expression $\sin 80^\circ \cos 20^\circ - \cos 80^\circ \sin 20^\circ$ is that of the right side of formula (5) for $\sin(\alpha - \beta)$ with $\alpha = 80^\circ$ and $\beta = 20^\circ$. That is,

$$\sin 80^\circ \cos 20^\circ - \cos 80^\circ \sin 20^\circ = \sin(80^\circ - 20^\circ) = \sin 60^\circ = \frac{\sqrt{3}}{2}$$

Now Work PROBLEMS 27 AND 31

EXAMPLE 5 Finding Exact Values

Say it is known that $\sin\alpha = \frac{4}{5}$, $\frac{\pi}{2} < \alpha < \pi$, and that $\sin\beta = -\frac{2}{\sqrt{5}} = -\frac{2\sqrt{5}}{5}$, $\pi < \beta < \frac{3\pi}{2}$. Find the exact value of each of the following.

(a) $\cos\alpha$ (b) $\cos\beta$ (c) $\cos(\alpha + \beta)$ (d) $\sin(\alpha + \beta)$

Solution (a) Because $\sin\alpha = \frac{4}{5} = \frac{b}{r}$ and $\frac{\pi}{2} < \alpha < \pi$, let $b = 4$ and $r = 5$, and place α in quadrant II. See Figure 32. The point $P = (a, b) = (a, 4)$ is in quadrant II, so $a < 0$. The distance from $(a, 4)$ to $(0, 0)$ is 5, so

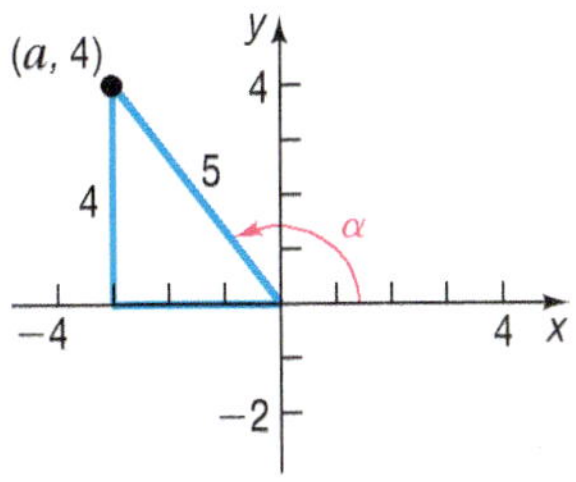

Figure 32 $\sin\alpha = \frac{4}{5}, \frac{\pi}{2} < \alpha < \pi$

$$a^2 + b^2 = 25$$

$$a^2 + 16 = 25 \qquad b = 4$$

$$a^2 = 25 - 16 = 9$$

$$a = -3 \qquad a < 0$$

Then

$$\cos\alpha = \frac{a}{r} = -\frac{3}{5}$$

Alternatively, $\cos\alpha$ can be found using identities, as follows:

$$\cos\alpha = -\sqrt{1-\sin^2\alpha} = -\sqrt{1-\frac{16}{25}} = -\sqrt{\frac{9}{25}} = -\frac{3}{5}$$

↑ α in quadrant II, $\cos\alpha < 0$

(b) Because $\sin\beta = \frac{-2}{\sqrt{5}} = \frac{b}{r}$, and $\pi < \beta < \frac{3\pi}{2}$, let $b = -2$ and $r = \sqrt{5}$, and place β in quadrant III. See Figure 33. Because the point $P = (a, b) = (a, -2)$ is in quadrant III, $a < 0$. The distance from $(a, -2)$ to $(0, 0)$ is $\sqrt{5}$, so

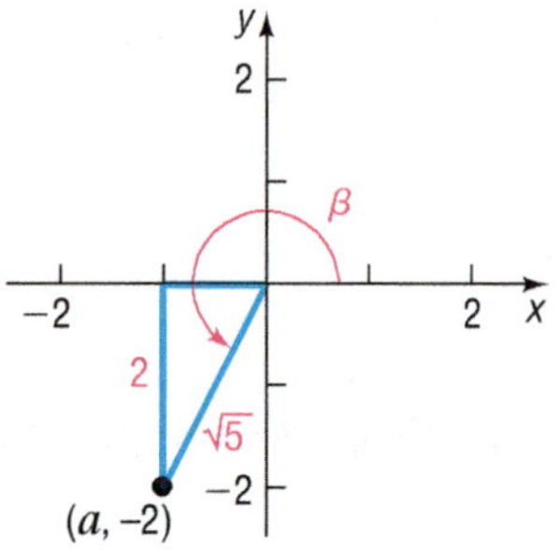

Figure 33 $\sin\beta = \frac{-2}{\sqrt{5}},\ \pi < \beta < \frac{3\pi}{2}$

$$a^2 + b^2 = 5$$
$$a^2 + 4 = 5 \qquad b = -2$$
$$a^2 = 1$$
$$a = -1 \qquad a < 0$$

Then

$$\cos\beta = \frac{a}{r} = \frac{-1}{\sqrt{5}} = -\frac{\sqrt{5}}{5}$$

Alternatively, $\cos\beta$ can be found using identities, as follows:

$$\cos\beta = -\sqrt{1-\sin^2\beta} = -\sqrt{1-\frac{4}{5}} = -\sqrt{\frac{1}{5}} = -\frac{\sqrt{5}}{5}$$

(c) Use the results found in parts (a) and (b) and formula (1) to obtain

$$\cos(\alpha+\beta) = \cos\alpha\cos\beta - \sin\alpha\sin\beta$$
$$= -\frac{3}{5}\left(-\frac{\sqrt{5}}{5}\right) - \frac{4}{5}\left(-\frac{2\sqrt{5}}{5}\right) = \frac{11\sqrt{5}}{25}$$

(d) $\sin(\alpha+\beta) = \sin\alpha\cos\beta + \cos\alpha\sin\beta$

$$= \frac{4}{5}\left(-\frac{\sqrt{5}}{5}\right) + \left(-\frac{3}{5}\right)\left(-\frac{2\sqrt{5}}{5}\right) = \frac{2\sqrt{5}}{25}$$

■

Now Work PROBLEMS 35(a), (b), AND (c)

EXAMPLE 6 Establishing an Identity

Establish the identity: $\frac{\cos(\alpha-\beta)}{\sin\alpha\sin\beta} = \cot\alpha\cot\beta + 1$

Solution

$$\frac{\cos(\alpha-\beta)}{\sin\alpha\sin\beta} = \frac{\cos\alpha\cos\beta + \sin\alpha\sin\beta}{\sin\alpha\sin\beta}$$
$$= \frac{\cos\alpha\cos\beta}{\sin\alpha\sin\beta} + \frac{\sin\alpha\sin\beta}{\sin\alpha\sin\beta}$$
$$= \frac{\cos\alpha}{\sin\alpha}\cdot\frac{\cos\beta}{\sin\beta} + 1$$
$$= \cot\alpha\cot\beta + 1$$

■

Now Work PROBLEMS 49 AND 61

Use the identity $\tan\theta = \dfrac{\sin\theta}{\cos\theta}$ and the sum formulas for $\sin(\alpha+\beta)$ and $\cos(\alpha+\beta)$ to derive a formula for $\tan(\alpha+\beta)$.

Proof $\tan(\alpha+\beta) = \dfrac{\sin(\alpha+\beta)}{\cos(\alpha+\beta)} = \dfrac{\sin\alpha\cos\beta + \cos\alpha\sin\beta}{\cos\alpha\cos\beta - \sin\alpha\sin\beta}$

Now divide the numerator and the denominator by $\cos\alpha\cos\beta$.

$$\tan(\alpha+\beta) = \frac{\dfrac{\sin\alpha\cos\beta + \cos\alpha\sin\beta}{\cos\alpha\cos\beta}}{\dfrac{\cos\alpha\cos\beta - \sin\alpha\sin\beta}{\cos\alpha\cos\beta}} = \frac{\dfrac{\sin\alpha\cancel{\cos\beta}}{\cos\alpha\cancel{\cos\beta}} + \dfrac{\cancel{\cos\alpha}\sin\beta}{\cancel{\cos\alpha}\cos\beta}}{\dfrac{\cancel{\cos\alpha}\cancel{\cos\beta}}{\cancel{\cos\alpha}\cancel{\cos\beta}} - \dfrac{\sin\alpha\sin\beta}{\cos\alpha\cos\beta}}$$

$$= \frac{\dfrac{\sin\alpha}{\cos\alpha} + \dfrac{\sin\beta}{\cos\beta}}{1 - \dfrac{\sin\alpha}{\cos\alpha}\cdot\dfrac{\sin\beta}{\cos\beta}} = \frac{\tan\alpha + \tan\beta}{1 - \tan\alpha\tan\beta}$$ ■

Proof Use the sum formula for $\tan(\alpha+\beta)$ and Even–Odd Properties to get the difference formula.

$$\tan(\alpha-\beta) = \tan[\alpha + (-\beta)] = \frac{\tan\alpha + \tan(-\beta)}{1 - \tan\alpha\tan(-\beta)} = \frac{\tan\alpha - \tan\beta}{1 + \tan\alpha\tan\beta}$$ ■

We have proved the following results:

THEOREM

Sum and Difference Formulas for the Tangent Function

$$\tan(\alpha+\beta) = \frac{\tan\alpha + \tan\beta}{1 - \tan\alpha\tan\beta} \quad (6)$$

$$\tan(\alpha-\beta) = \frac{\tan\alpha - \tan\beta}{1 + \tan\alpha\tan\beta} \quad (7)$$

In Words

Formula (6) states that the tangent of the sum of two angles equals the tangent of the first angle plus the tangent of the second angle, all divided by 1 minus their product.

Now Work PROBLEMS 15 AND 35(d)

EXAMPLE 7 Establishing an Identity

Prove the identity: $\tan(\theta+\pi) = \tan\theta$

Solution

$$\tan(\theta+\pi) = \frac{\tan\theta + \tan\pi}{1 - \tan\theta\tan\pi} = \frac{\tan\theta + 0}{1 - \tan\theta\cdot 0} = \tan\theta$$ ■

Example 7 verifies that the tangent function is periodic with period π, a fact that was discussed earlier.

EXAMPLE 8 Establishing an Identity

Prove the identity: $\tan\left(\theta + \dfrac{\pi}{2}\right) = -\cot\theta$

Solution Formula (6) cannot be used because $\tan\dfrac{\pi}{2}$ is not defined. Instead, proceed as follows:

$$\tan\left(\theta+\frac{\pi}{2}\right) = \frac{\sin\left(\theta+\dfrac{\pi}{2}\right)}{\cos\left(\theta+\dfrac{\pi}{2}\right)} = \frac{\sin\theta\cos\dfrac{\pi}{2} + \cos\theta\sin\dfrac{\pi}{2}}{\cos\theta\cos\dfrac{\pi}{2} - \sin\theta\sin\dfrac{\pi}{2}}$$

$$= \frac{(\sin\theta)(0) + (\cos\theta)(1)}{(\cos\theta)(0) - (\sin\theta)(1)} = \frac{\cos\theta}{-\sin\theta} = -\cot\theta$$ ■

WARNING Be careful when using formulas (6) and (7). These formulas can be used only for angles α and β for which $\tan\alpha$ and $\tan\beta$ are defined. That is, they can be used for all angles except odd integer multiples of $\dfrac{\pi}{2}$. ■

3 Use Sum and Difference Formulas Involving Inverse Trigonometric Functions

EXAMPLE 9

Finding the Exact Value of an Expression Involving Inverse Trigonometric Functions

Find the exact value of: $\sin\left(\cos^{-1}\frac{1}{2} + \sin^{-1}\frac{3}{5}\right)$

Solution We seek the sine of the sum of two angles, $\alpha = \cos^{-1}\frac{1}{2}$ and $\beta = \sin^{-1}\frac{3}{5}$. Then

$$\cos\alpha = \frac{1}{2} \quad 0 \le \alpha \le \pi \quad \text{and} \quad \sin\beta = \frac{3}{5} \quad -\frac{\pi}{2} \le \beta \le \frac{\pi}{2}$$

Use Pythagorean Identities to obtain $\sin\alpha$ and $\cos\beta$. Because $\sin\alpha \ge 0$ (since $0 \le \alpha \le \pi$) and $\cos\beta \ge 0$ $\left(\text{since } -\frac{\pi}{2} \le \beta \le \frac{\pi}{2}\right)$, this means that

$$\sin\alpha = \sqrt{1-\cos^2\alpha} = \sqrt{1-\frac{1}{4}} = \sqrt{\frac{3}{4}} = \frac{\sqrt{3}}{2}$$

$$\cos\beta = \sqrt{1-\sin^2\beta} = \sqrt{1-\frac{9}{25}} = \sqrt{\frac{16}{25}} = \frac{4}{5}$$

Note: In Example 9, $\sin\alpha$ also can be found by using $\cos\alpha = \frac{1}{2} = \frac{a}{r}$, so $a = 1$ and $r = 2$. Then $b = \sqrt{3}$ and $\sin\alpha = \frac{b}{r} = \frac{\sqrt{3}}{2}$. Also $\cos\beta$ can be found in a similar fashion. ■

As a result,

$$\sin\left(\cos^{-1}\frac{1}{2} + \sin^{-1}\frac{3}{5}\right) = \sin(\alpha+\beta) = \sin\alpha\cos\beta + \cos\alpha\sin\beta$$

$$= \frac{\sqrt{3}}{2}\cdot\frac{4}{5} + \frac{1}{2}\cdot\frac{3}{5} = \frac{4\sqrt{3}+3}{10}$$ ■

Now Work PROBLEM 77

EXAMPLE 10

Writing a Trigonometric Expression as an Algebraic Expression

Write $\sin(\sin^{-1} u + \cos^{-1} v)$ as an algebraic expression containing u and v (that is, without any trigonometric functions). Give the restrictions on u and v.

Solution First, for $\sin^{-1} u$, the restriction on u is $-1 \le u \le 1$, and for $\cos^{-1} v$, the restriction on v is $-1 \le v \le 1$. Now let $\alpha = \sin^{-1} u$ and $\beta = \cos^{-1} v$. Then

$$\sin\alpha = u \quad -\frac{\pi}{2} \le \alpha \le \frac{\pi}{2} \quad -1 \le u \le 1$$

$$\cos\beta = v \quad 0 \le \beta \le \pi \quad -1 \le v \le 1$$

Because $-\frac{\pi}{2} \le \alpha \le \frac{\pi}{2}$, this means that $\cos\alpha \ge 0$. As a result,

$$\cos\alpha = \sqrt{1-\sin^2\alpha} = \sqrt{1-u^2}$$

Similarly, because $0 \le \beta \le \pi$, this means that $\sin\beta \ge 0$. Then

$$\sin\beta = \sqrt{1-\cos^2\beta} = \sqrt{1-v^2}$$

As a result,

$$\sin(\sin^{-1} u + \cos^{-1} v) = \sin(\alpha+\beta) = \sin\alpha\cos\beta + \cos\alpha\sin\beta$$

$$= uv + \sqrt{1-u^2}\sqrt{1-v^2}$$ ■

Now Work PROBLEM 87

4 Solve Trigonometric Equations Linear in Sine and Cosine

Sometimes it is necessary to square both sides of an equation to obtain expressions that allow the use of identities. Remember, squaring both sides of an equation may introduce extraneous solutions. As a result, apparent solutions must be checked.

EXAMPLE 11 Solving a Trigonometric Equation Linear in Sine and Cosine

Solve the equation: $\sin\theta + \cos\theta = 1, \quad 0 \le \theta < 2\pi$

Option 1 Attempts to use available identities do not lead to equations that are easy to solve. (Try it yourself.) So, given the form of this equation, square each side.

$$\sin\theta + \cos\theta = 1$$

$$(\sin\theta + \cos\theta)^2 = 1 \qquad \text{Square each side.}$$

$$\sin^2\theta + 2\sin\theta\cos\theta + \cos^2\theta = 1 \qquad \text{Remove parentheses.}$$

$$2\sin\theta\cos\theta = 0 \qquad \sin^2\theta + \cos^2\theta = 1$$

$$\sin\theta\cos\theta = 0$$

Setting each factor equal to zero leads to

$$\sin\theta = 0 \quad \text{or} \quad \cos\theta = 0$$

The apparent solutions are

$$\theta = 0, \qquad \theta = \pi, \qquad \theta = \frac{\pi}{2}, \qquad \theta = \frac{3\pi}{2}$$

Because both sides of the original equation were squared, these apparent solutions must be checked to see whether any are extraneous.

$$\theta = 0: \quad \sin 0 + \cos 0 = 0 + 1 = 1 \qquad \text{A solution}$$

$$\theta = \pi: \quad \sin\pi + \cos\pi = 0 + (-1) = -1 \qquad \text{Not a solution}$$

$$\theta = \frac{\pi}{2}: \quad \sin\frac{\pi}{2} + \cos\frac{\pi}{2} = 1 + 0 = 1 \qquad \text{A solution}$$

$$\theta = \frac{3\pi}{2}: \quad \sin\frac{3\pi}{2} + \cos\frac{3\pi}{2} = -1 + 0 = -1 \qquad \text{Not a solution}$$

The values $\theta = \pi$ and $\theta = \frac{3\pi}{2}$ are extraneous. The solution set is $\left\{0, \frac{\pi}{2}\right\}$.

Option 2 Start with the equation

$$\sin\theta + \cos\theta = 1$$

and divide each side by $\sqrt{2}$. (The reason for this choice will become apparent shortly.) Then

$$\frac{1}{\sqrt{2}}\sin\theta + \frac{1}{\sqrt{2}}\cos\theta = \frac{1}{\sqrt{2}}$$

The left side now resembles the formula for the sine of the sum of two angles, one of which is θ. The other angle is unknown (call it ϕ). Then

$$\sin(\theta + \phi) = \sin\theta\cos\phi + \cos\theta\sin\phi = \frac{1}{\sqrt{2}} = \frac{\sqrt{2}}{2} \qquad \textbf{(8)}$$

where

$$\cos\phi = \frac{1}{\sqrt{2}} = \frac{\sqrt{2}}{2} \qquad \sin\phi = \frac{1}{\sqrt{2}} = \frac{\sqrt{2}}{2} \qquad 0 \le \phi < 2\pi$$

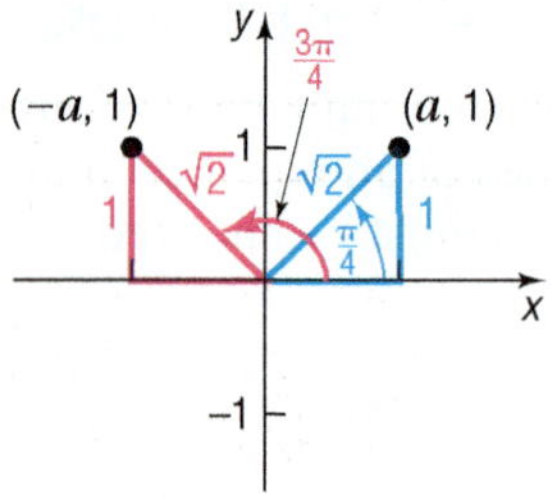

Figure 34

The angle ϕ is therefore $\frac{\pi}{4}$. As a result, equation (8) becomes

$$\sin\left(\theta + \frac{\pi}{4}\right) = \frac{\sqrt{2}}{2}$$

In the interval $[0, 2\pi)$, there are two angles whose sine is $\frac{\sqrt{2}}{2}$: $\frac{\pi}{4}$ and $\frac{3\pi}{4}$. See Figure 34. As a result,

$$\theta + \frac{\pi}{4} = \frac{\pi}{4} \quad \text{or} \quad \theta + \frac{\pi}{4} = \frac{3\pi}{4}$$

$$\theta = 0 \quad \text{or} \quad \theta = \frac{\pi}{2}$$

The solution set is $\left\{0, \frac{\pi}{2}\right\}$. ■

This second method of solution can be used to solve any linear equation in the variables $\sin\theta$ and $\cos\theta$.

EXAMPLE 12 Solving a Trigonometric Equation Linear in $\sin\theta$ and $\cos\theta$

Solve:

$$a\sin\theta + b\cos\theta = c \tag{9}$$

where a, b, and c are constants and either $a \neq 0$ or $b \neq 0$.

Solution Divide each side of equation (9) by $\sqrt{a^2 + b^2}$. Then

$$\frac{a}{\sqrt{a^2 + b^2}}\sin\theta + \frac{b}{\sqrt{a^2 + b^2}}\cos\theta = \frac{c}{\sqrt{a^2 + b^2}} \tag{10}$$

There is a unique angle ϕ, $0 \leq \phi < 2\pi$, for which

$$\cos\phi = \frac{a}{\sqrt{a^2 + b^2}} \quad \text{and} \quad \sin\phi = \frac{b}{\sqrt{a^2 + b^2}} \tag{11}$$

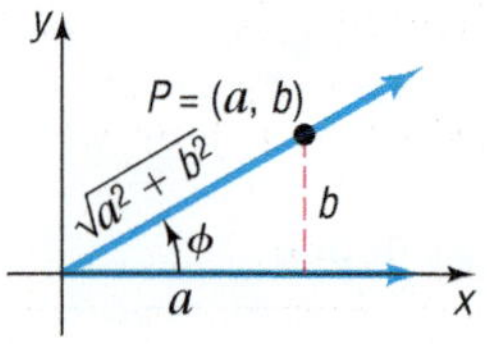

Figure 35

Figure 35 shows the situation for $a > 0$ and $b > 0$. Equation (10) may be written as

$$\sin\theta\cos\phi + \cos\theta\sin\phi = \frac{c}{\sqrt{a^2 + b^2}}$$

or, equivalently,

$$\sin(\theta + \phi) = \frac{c}{\sqrt{a^2 + b^2}} \tag{12}$$

where ϕ satisfies equation (11).

If $|c| > \sqrt{a^2 + b^2}$, then $\sin(\theta + \phi) > 1$ or $\sin(\theta + \phi) < -1$, and equation (12) has no solution.

If $|c| \leq \sqrt{a^2 + b^2}$, then the solutions of equation (12) are

$$\theta + \phi = \sin^{-1}\frac{c}{\sqrt{a^2 + b^2}} \quad \text{or} \quad \theta + \phi = \pi - \sin^{-1}\frac{c}{\sqrt{a^2 + b^2}}$$

Because the angle ϕ is determined by equations (11), these give the solutions to equation (9). ■

Now Work PROBLEM 95

SUMMARY

Sum and Difference Formulas

$\cos(\alpha+\beta) = \cos\alpha\cos\beta - \sin\alpha\sin\beta$ $\qquad$ $\cos(\alpha-\beta) = \cos\alpha\cos\beta + \sin\alpha\sin\beta$

$\sin(\alpha+\beta) = \sin\alpha\cos\beta + \cos\alpha\sin\beta$ $\qquad$ $\sin(\alpha-\beta) = \sin\alpha\cos\beta - \cos\alpha\sin\beta$

$\tan(\alpha+\beta) = \dfrac{\tan\alpha + \tan\beta}{1 - \tan\alpha\tan\beta}$ $\qquad$ $\tan(\alpha-\beta) = \dfrac{\tan\alpha - \tan\beta}{1 + \tan\alpha\tan\beta}$

8.5 Assess Your Understanding

'Are You Prepared?' *Answers are given at the end of these exercises. If you get a wrong answer, read the pages listed in red.*

1. The distance d from the point $(2, -3)$ to the point $(5, 1)$ is _____. (pp. 85–87)

2. If $\sin\theta = \frac{4}{5}$ and θ is in quadrant II, then $\cos\theta =$ _____. (pp. 557–558)

3. (a) $\sin\frac{\pi}{4}\cdot\cos\frac{\pi}{3} =$ _____. (pp. 539–541)
   (b) $\tan\frac{\pi}{4} - \sin\frac{\pi}{6} =$ _____. (pp. 539–541)

4. If $\sin\alpha = -\frac{4}{5}$, $\pi < \alpha < \frac{3\pi}{2}$, then $\cos\alpha =$ _____. (p. 557–558)

Concepts and Vocabulary

5. $\cos(\alpha+\beta) = \cos\alpha\cos\beta$ ____ $\sin\alpha\sin\beta$

6. $\sin(\alpha-\beta) = \sin\alpha\cos\beta$ ____ $\cos\alpha\sin\beta$

7. ***True or False*** $\sin(\alpha+\beta) = \sin\alpha + \sin\beta + 2\sin\alpha\sin\beta$

8. ***True or False*** $\tan 75° = \tan 30° + \tan 45°$

9. ***True or False*** $\cos\left(\frac{\pi}{2} - \theta\right) = \cos\theta$

10. ***True or False*** If $f(x) = \sin x$ and $g(x) = \cos x$, then $g(\alpha+\beta) = g(\alpha)g(\beta) - f(\alpha)f(\beta)$

11. Choose the expression that completes the sum formula for tangent functions: $\tan(\alpha+\beta) =$ ________
    (a) $\dfrac{\tan\alpha + \tan\beta}{1 - \tan\alpha\tan\beta}$ (b) $\dfrac{\tan\alpha - \tan\beta}{1 + \tan\alpha\tan\beta}$
    (c) $\dfrac{\tan\alpha + \tan\beta}{1 + \tan\alpha\tan\beta}$ (d) $\dfrac{\tan\alpha - \tan\beta}{1 - \tan\alpha\tan\beta}$

12. Choose the expression that is equivalent to $\sin 60°\cos 20° + \cos 60°\sin 20°$.
    (a) $\cos 40°$ (b) $\sin 40°$ (c) $\cos 80°$ (d) $\sin 80°$

Skill Building

In Problems 13–24, find the exact value of each expression.

13. $\cos 165°$
14. $\sin 105°$
15. $\tan 15°$
16. $\tan 195°$
17. $\sin\frac{5\pi}{12}$
18. $\sin\frac{\pi}{12}$
19. $\cos\frac{7\pi}{12}$
20. $\tan\frac{7\pi}{12}$
21. $\sin\frac{17\pi}{12}$
22. $\tan\frac{19\pi}{12}$
23. $\sec\left(-\frac{\pi}{12}\right)$
24. $\cot\left(-\frac{5\pi}{12}\right)$

In Problems 25–34, find the exact value of each expression.

25. $\sin 20°\cos 10° + \cos 20°\sin 10°$
26. $\sin 20°\cos 80° - \cos 20°\sin 80°$
27. $\cos 70°\cos 20° - \sin 70°\sin 20°$
28. $\cos 40°\cos 10° + \sin 40°\sin 10°$
29. $\dfrac{\tan 20° + \tan 25°}{1 - \tan 20°\tan 25°}$
30. $\dfrac{\tan 40° - \tan 10°}{1 + \tan 40°\tan 10°}$
31. $\sin\frac{\pi}{12}\cos\frac{7\pi}{12} - \cos\frac{\pi}{12}\sin\frac{7\pi}{12}$
32. $\cos\frac{5\pi}{12}\cos\frac{7\pi}{12} - \sin\frac{5\pi}{12}\sin\frac{7\pi}{12}$
33. $\cos\frac{\pi}{12}\cos\frac{5\pi}{12} + \sin\frac{5\pi}{12}\sin\frac{\pi}{12}$
34. $\sin\frac{\pi}{18}\cos\frac{5\pi}{18} + \cos\frac{\pi}{18}\sin\frac{5\pi}{18}$

In Problems 35–40, find the exact value of each of the following under the given conditions:
(a) $\sin(\alpha+\beta)$ *(b)* $\cos(\alpha+\beta)$ *(c)* $\sin(\alpha-\beta)$ *(d)* $\tan(\alpha-\beta)$

35. $\sin\alpha = \frac{3}{5}$, $0 < \alpha < \frac{\pi}{2}$; $\cos\beta = \frac{2\sqrt{5}}{5}$, $-\frac{\pi}{2} < \beta < 0$
36. $\cos\alpha = \frac{\sqrt{5}}{5}$, $0 < \alpha < \frac{\pi}{2}$; $\sin\beta = -\frac{4}{5}$, $-\frac{\pi}{2} < \beta < 0$

37. $\tan\alpha = -\frac{4}{3}, \frac{\pi}{2} < \alpha < \pi;\quad \cos\beta = \frac{1}{2}, 0 < \beta < \frac{\pi}{2}$

38. $\tan\alpha = \frac{5}{12}, \pi < \alpha < \frac{3\pi}{2};\quad \sin\beta = -\frac{1}{2}, \pi < \beta < \frac{3\pi}{2}$

39. $\sin\alpha = \frac{5}{13}, -\frac{3\pi}{2} < \alpha < -\pi;\quad \tan\beta = -\sqrt{3}, \frac{\pi}{2} < \beta < \pi$

40. $\cos\alpha = \frac{1}{2}, -\frac{\pi}{2} < \alpha < 0;\quad \sin\beta = \frac{1}{3}, 0 < \beta < \frac{\pi}{2}$

41. If $\sin\theta = \frac{1}{3}$, θ in quadrant II, find the exact value of:

(a) $\cos\theta$ (b) $\sin\left(\theta + \frac{\pi}{6}\right)$

(c) $\cos\left(\theta - \frac{\pi}{3}\right)$ (d) $\tan\left(\theta + \frac{\pi}{4}\right)$

42. If $\cos\theta = \frac{1}{4}$, θ in quadrant IV, find the exact value of:

(a) $\sin\theta$ (b) $\sin\left(\theta - \frac{\pi}{6}\right)$

(c) $\cos\left(\theta + \frac{\pi}{3}\right)$ (d) $\tan\left(\theta - \frac{\pi}{4}\right)$

In Problems 43–48, use the figures to evaluate each function if $f(x) = \sin x$, $g(x) = \cos x$, and $h(x) = \tan x$.

43. $f(\alpha + \beta)$ **44.** $g(\alpha + \beta)$

45. $g(\alpha - \beta)$ **46.** $f(\alpha - \beta)$

47. $h(\alpha + \beta)$ **48.** $h(\alpha - \beta)$

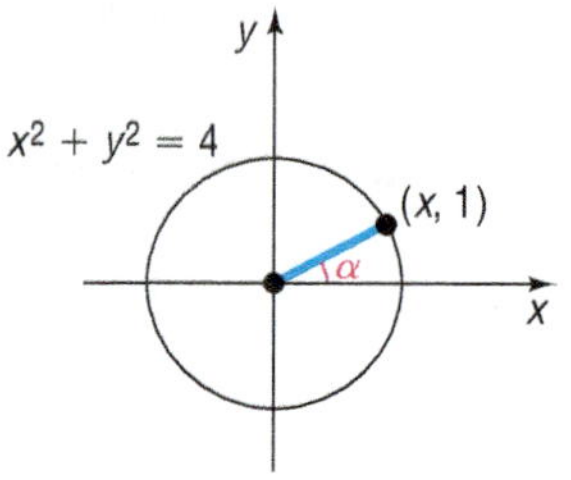

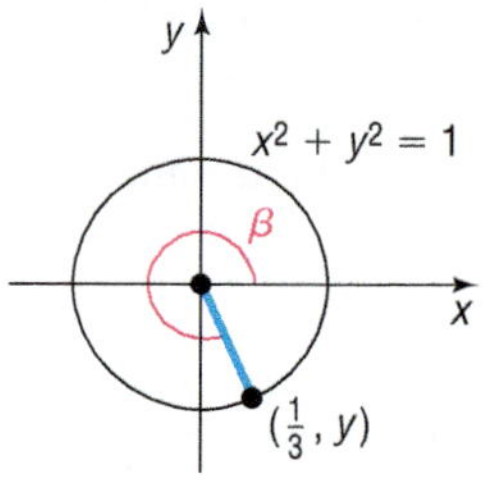

In Problems 49–74, establish each identity.

49. $\sin\left(\frac{\pi}{2} + \theta\right) = \cos\theta$

50. $\cos\left(\frac{\pi}{2} + \theta\right) = -\sin\theta$

51. $\sin(\pi - \theta) = \sin\theta$

52. $\cos(\pi - \theta) = -\cos\theta$

53. $\sin(\pi + \theta) = -\sin\theta$

54. $\cos(\pi + \theta) = -\cos\theta$

55. $\tan(\pi - \theta) = -\tan\theta$

56. $\tan(2\pi - \theta) = -\tan\theta$

57. $\sin\left(\frac{3\pi}{2} + \theta\right) = -\cos\theta$

58. $\cos\left(\frac{3\pi}{2} + \theta\right) = \sin\theta$

59. $\sin(\alpha + \beta) + \sin(\alpha - \beta) = 2\sin\alpha\cos\beta$

60. $\cos(\alpha + \beta) + \cos(\alpha - \beta) = 2\cos\alpha\cos\beta$

61. $\frac{\sin(\alpha + \beta)}{\sin\alpha\cos\beta} = 1 + \cot\alpha\tan\beta$

62. $\frac{\sin(\alpha + \beta)}{\cos\alpha\cos\beta} = \tan\alpha + \tan\beta$

63. $\frac{\cos(\alpha + \beta)}{\cos\alpha\cos\beta} = 1 - \tan\alpha\tan\beta$

64. $\frac{\cos(\alpha - \beta)}{\sin\alpha\cos\beta} = \cot\alpha + \tan\beta$

65. $\frac{\sin(\alpha + \beta)}{\sin(\alpha - \beta)} = \frac{\tan\alpha + \tan\beta}{\tan\alpha - \tan\beta}$

66. $\frac{\cos(\alpha + \beta)}{\cos(\alpha - \beta)} = \frac{1 - \tan\alpha\tan\beta}{1 + \tan\alpha\tan\beta}$

67. $\cot(\alpha + \beta) = \frac{\cot\alpha\cot\beta - 1}{\cot\beta + \cot\alpha}$

68. $\cot(\alpha - \beta) = \frac{\cot\alpha\cot\beta + 1}{\cot\beta - \cot\alpha}$

69. $\sec(\alpha + \beta) = \frac{\csc\alpha\csc\beta}{\cot\alpha\cot\beta - 1}$

70. $\sec(\alpha - \beta) = \frac{\sec\alpha\sec\beta}{1 + \tan\alpha\tan\beta}$

71. $\sin(\alpha - \beta)\sin(\alpha + \beta) = \sin^2\alpha - \sin^2\beta$

72. $\cos(\alpha - \beta)\cos(\alpha + \beta) = \cos^2\alpha - \sin^2\beta$

73. $\sin(\theta + k\pi) = (-1)^k\sin\theta$, k any integer

74. $\cos(\theta + k\pi) = (-1)^k\cos\theta$, k any integer

In Problems 75–86, find the exact value of each expression.

75. $\sin\left(\sin^{-1}\frac{1}{2} + \cos^{-1}0\right)$

76. $\sin\left(\sin^{-1}\frac{\sqrt{3}}{2} + \cos^{-1}1\right)$

77. $\sin\left[\sin^{-1}\frac{3}{5} - \cos^{-1}\left(-\frac{4}{5}\right)\right]$

78. $\sin\left[\sin^{-1}\left(-\frac{4}{5}\right) - \tan^{-1}\frac{3}{4}\right]$

79. $\cos\left(\tan^{-1}\frac{4}{3} + \cos^{-1}\frac{5}{13}\right)$

80. $\cos\left[\tan^{-1}\frac{5}{12} - \sin^{-1}\left(-\frac{3}{5}\right)\right]$

81. $\cos\left(\sin^{-1}\frac{5}{13} - \tan^{-1}\frac{3}{4}\right)$

82. $\cos\left(\tan^{-1}\frac{4}{3} + \cos^{-1}\frac{12}{13}\right)$

83. $\tan\left(\sin^{-1}\frac{3}{5} + \frac{\pi}{6}\right)$

84. $\tan\left(\frac{\pi}{4} - \cos^{-1}\frac{3}{5}\right)$

85. $\tan\left(\sin^{-1}\frac{4}{5} + \cos^{-1} 1\right)$

86. $\tan\left(\cos^{-1}\frac{4}{5} + \sin^{-1} 1\right)$

In Problems 87–92, write each trigonometric expression as an algebraic expression containing u and v. Give the restrictions required on u and v.

87. $\cos(\cos^{-1} u + \sin^{-1} v)$

88. $\sin(\sin^{-1} u - \cos^{-1} v)$

89. $\sin(\tan^{-1} u - \sin^{-1} v)$

90. $\cos(\tan^{-1} u + \tan^{-1} v)$

91. $\tan(\sin^{-1} u - \cos^{-1} v)$

92. $\sec(\tan^{-1} u + \cos^{-1} v)$

In Problems 93–98, solve each equation on the interval $0 \le \theta < 2\pi$.

93. $\sin\theta - \sqrt{3}\cos\theta = 1$

94. $\sqrt{3}\sin\theta + \cos\theta = 1$

95. $\sin\theta + \cos\theta = \sqrt{2}$

96. $\sin\theta - \cos\theta = -\sqrt{2}$

97. $\tan\theta + \sqrt{3} = \sec\theta$

98. $\cot\theta + \csc\theta = -\sqrt{3}$

Applications and Extensions

99. Show that $\sin^{-1} v + \cos^{-1} v = \frac{\pi}{2}$.

100. Show that $\tan^{-1} v + \cot^{-1} v = \frac{\pi}{2}$.

101. Show that $\tan^{-1}\left(\frac{1}{v}\right) = \frac{\pi}{2} - \tan^{-1} v$, if $v > 0$.

102. Show that $\cot^{-1} e^{v} = \tan^{-1} e^{-v}$.

103. Show that $\sin(\sin^{-1} v + \cos^{-1} v) = 1$.

104. Show that $\cos(\sin^{-1} v + \cos^{-1} v) = 0$.

---

105. Calculus Show that the difference quotient for $f(x) = \sin x$ is given by

$$\frac{f(x+h) - f(x)}{h} = \frac{\sin(x+h) - \sin x}{h} = \cos x \cdot \frac{\sin h}{h} - \sin x \cdot \frac{1 - \cos h}{h}$$

106. Calculus Show that the difference quotient for $f(x) = \cos x$ is given by

$$\frac{f(x+h) - f(x)}{h} = \frac{\cos(x+h) - \cos x}{h} = -\sin x \cdot \frac{\sin h}{h} - \cos x \cdot \frac{1 - \cos h}{h}$$

107. One, Two, Three
(a) Show that $\tan(\tan^{-1} 1 + \tan^{-1} 2 + \tan^{-1} 3) = 0$.
(b) Conclude from part (a) that

$$\tan^{-1} 1 + \tan^{-1} 2 + \tan^{-1} 3 = \pi$$

Source: *College Mathematics Journal,* Vol. 37, No. 3, May 2006

108. Electric Power In an alternating current (ac) circuit, the instantaneous power p at time t is given by

$$p(t) = V_m I_m \cos\phi \sin^2(\omega t) - V_m I_m \sin\phi \sin(\omega t)\cos(\omega t)$$

Show that this is equivalent to

$$p(t) = V_m I_m \sin(\omega t)\sin(\omega t - \phi)$$

Source: *HyperPhysics, hosted by Georgia State University*

109. Geometry: Angle between Two Lines Let L_1 and L_2 denote two nonvertical intersecting lines, and let θ denote the acute angle between L_1 and L_2 (see the figure). Show that

$$\tan\theta = \frac{m_2 - m_1}{1 + m_1 m_2}$$

where m_1 and m_2 are the slopes of L_1 and L_2, respectively. [**Hint:** Use the facts that $\tan\theta_1 = m_1$ and $\tan\theta_2 = m_2$.]

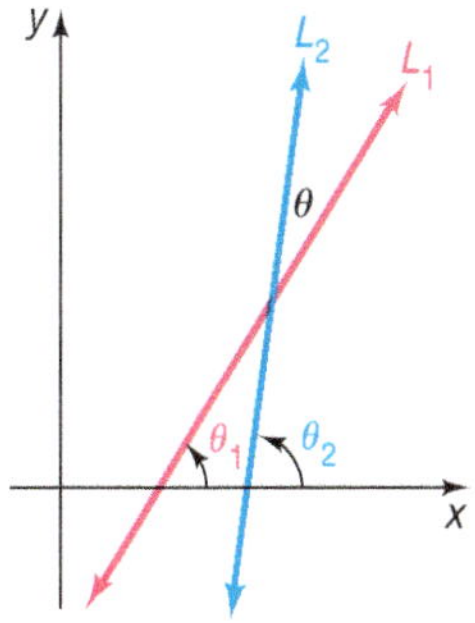

110. If $\alpha + \beta + \gamma = 180°$ and

$$\cot\theta = \cot\alpha + \cot\beta + \cot\gamma \quad 0 < \theta < 90°$$

show that

$$\sin^3\theta = \sin(\alpha - \theta)\sin(\beta - \theta)\sin(\gamma - \theta)$$

111. If $\tan\alpha = x + 1$ and $\tan\beta = x - 1$, show that

$$2\cot(\alpha - \beta) = x^2$$

Explaining Concepts: Discussion and Writing

112. Discuss the following derivation:

$$\tan\left(\theta + \frac{\pi}{2}\right) = \frac{\tan\theta + \tan\frac{\pi}{2}}{1 - \tan\theta\tan\frac{\pi}{2}} = \frac{\dfrac{\tan\theta}{\tan\frac{\pi}{2}} + 1}{\dfrac{1}{\tan\frac{\pi}{2}} - \tan\theta} = \frac{0+1}{0-\tan\theta} = \frac{1}{-\tan\theta} = -\cot\theta$$

Can you justify each step?

113. Explain why formula (7) cannot be used to show that

$$\tan\left(\frac{\pi}{2} - \theta\right) = \cot\theta$$

Establish this identity by using formulas (3a) and (3b).

Retain Your Knowledge

Problems 114–117 are based on material learned earlier in the course. The purpose of these problems is to keep the material fresh in your mind so that you are better prepared for the final exam.

114. Determine the points of intersection of the graphs of $f(x) = x^2 + 5x + 1$ and $g(x) = -2x^2 - 11x - 4$ by solving $f(x) = g(x)$.

115. Convert $\frac{17\pi}{6}$ to degrees.

116. Find the area of the sector of a circle of radius 6 centimeters formed by an angle of 45°. Give both the exact area and an approximation rounded to two decimal places.

117. Given $\tan\theta = -2$, $270° < \theta < 360°$, find the exact value of the remaining five trigonometric functions.

'Are You Prepared' Answers

1. 5 **2.** $-\frac{3}{5}$ **3.** (a) $\frac{\sqrt{2}}{4}$ (b) $\frac{1}{2}$ **4.** $-\frac{3}{5}$

8.6 Double-angle and Half-angle Formulas

OBJECTIVES
1 Use Double-angle Formulas to Find Exact Values (p. 665)
2 Use Double-angle Formulas to Establish Identities (p. 665)
3 Use Half-angle Formulas to Find Exact Values (p. 668)

In this section, formulas for $\sin(2\theta)$, $\cos(2\theta)$, $\sin\left(\frac{1}{2}\theta\right)$, and $\cos\left(\frac{1}{2}\theta\right)$ are established in terms of $\sin\theta$ and $\cos\theta$. They are derived using the sum formulas.

$$\sin(\alpha + \beta) = \sin\alpha\cos\beta + \cos\alpha\sin\beta$$

$$\sin(\theta + \theta) = \sin\theta\cos\theta + \cos\theta\sin\theta \qquad \text{Let } \alpha = \beta = \theta.$$

$$\sin(2\theta) = 2\sin\theta\cos\theta$$

and

$$\cos(\alpha + \beta) = \cos\alpha\cos\beta - \sin\alpha\sin\beta$$

$$\cos(\theta + \theta) = \cos\theta\cos\theta - \sin\theta\sin\theta \qquad \text{Let } \alpha = \beta = \theta.$$

$$\cos(2\theta) = \cos^2\theta - \sin^2\theta$$

An application of the Pythagorean Identity $\sin^2\theta + \cos^2\theta = 1$ results in two other ways to express $\cos(2\theta)$.

$$\cos(2\theta) = \cos^2\theta - \sin^2\theta = (1 - \sin^2\theta) - \sin^2\theta = 1 - 2\sin^2\theta$$

and

$$\cos(2\theta) = \cos^2\theta - \sin^2\theta = \cos^2\theta - (1 - \cos^2\theta) = 2\cos^2\theta - 1$$

The following theorem summarizes the **Double-angle Formulas.**

THEOREM

Double-angle Formulas

$$\sin(2\theta) = 2\sin\theta\cos\theta \quad \textbf{(1)}$$
$$\cos(2\theta) = \cos^2\theta - \sin^2\theta \quad \textbf{(2)}$$
$$\cos(2\theta) = 1 - 2\sin^2\theta \quad \textbf{(3)}$$
$$\cos(2\theta) = 2\cos^2\theta - 1 \quad \textbf{(4)}$$

1 Use Double-angle Formulas to Find Exact Values

EXAMPLE 1 **Finding Exact Values Using the Double-angle Formulas**

If $\sin\theta = \frac{3}{5}$, $\frac{\pi}{2} < \theta < \pi$, find the exact value of:

(a) $\sin(2\theta)$ (b) $\cos(2\theta)$

Solution (a) Because $\sin(2\theta) = 2\sin\theta\cos\theta$, and because $\sin\theta = \frac{3}{5}$ is known, begin by finding $\cos\theta$. Since $\sin\theta = \frac{3}{5} = \frac{b}{r}$, $\frac{\pi}{2} < \theta < \pi$, let $b = 3$ and $r = 5$, and place θ in quadrant II. See Figure 36. The point $P = (a, b) = (a, 3)$ is in quadrant II, so $a < 0$. The distance from $(a, 3)$ to $(0, 0)$ is 5, so

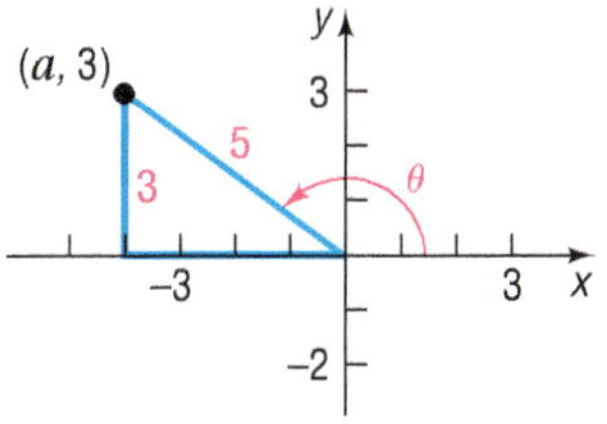

Figure 36

$$a^2 + 3^2 = 5^2 \qquad a^2 + b^2 = r^2$$
$$a^2 = 25 - 9 = 16$$
$$a = -4 \qquad a < 0$$

This means that $\cos\theta = \frac{a}{r} = \frac{-4}{5} = -\frac{4}{5}$. Now use formula (1) to obtain

$$\sin(2\theta) = 2\sin\theta\cos\theta = 2\left(\frac{3}{5}\right)\left(-\frac{4}{5}\right) = -\frac{24}{25}$$

(b) Because $\sin\theta = \frac{3}{5}$ is given, it is easiest to use formula (3) to get $\cos(2\theta)$.

$$\cos(2\theta) = 1 - 2\sin^2\theta = 1 - 2\left(\frac{9}{25}\right) = 1 - \frac{18}{25} = \frac{7}{25}$$

WARNING In finding $\cos(2\theta)$ in Example 1(b), a version of the Double-angle Formula, formula (3), was used. Note that it is not possible to use the Pythagorean Identity $\cos(2\theta) = \pm\sqrt{1 - \sin^2(2\theta)}$, with $\sin(2\theta) = -\frac{24}{25}$, because there is no way of knowing which sign to choose.

Now Work PROBLEMS 9(a) AND (b)

2 Use Double-angle Formulas to Establish Identities

EXAMPLE 2 **Establishing Identities**

(a) Develop a formula for $\tan(2\theta)$ in terms of $\tan\theta$.
(b) Develop a formula for $\sin(3\theta)$ in terms of $\sin\theta$ and $\cos\theta$.

Solution (a) In the sum formula for $\tan(\alpha + \beta)$, let $\alpha = \beta = \theta$. Then

$$\tan(\alpha + \beta) = \frac{\tan \alpha + \tan \beta}{1 - \tan \alpha \tan \beta}$$

$$\tan(\theta + \theta) = \frac{\tan \theta + \tan \theta}{1 - \tan \theta \tan \theta}$$

$$\tan(2\theta) = \frac{2 \tan \theta}{1 - \tan^2 \theta} \qquad \textbf{(5)}$$

(b) To get a formula for $\sin(3\theta)$, write 3θ as $2\theta + \theta$ and use the sum formula.

$$\sin(3\theta) = \sin(2\theta + \theta) = \sin(2\theta)\cos\theta + \cos(2\theta)\sin\theta$$

Now use the Double-angle Formulas to get

$$\begin{aligned}\sin(3\theta) &= (2\sin\theta\cos\theta)(\cos\theta) + (\cos^2\theta - \sin^2\theta)(\sin\theta)\\ &= 2\sin\theta\cos^2\theta + \sin\theta\cos^2\theta - \sin^3\theta\\ &= 3\sin\theta\cos^2\theta - \sin^3\theta\end{aligned}$$

The formula obtained in Example 2(b) also can be written as

$$\begin{aligned}\sin(3\theta) = 3\sin\theta\cos^2\theta - \sin^3\theta &= 3\sin\theta(1 - \sin^2\theta) - \sin^3\theta\\ &= 3\sin\theta - 4\sin^3\theta\end{aligned}$$

That is, $\sin(3\theta)$ is a third-degree polynomial in the variable $\sin\theta$. In fact, $\sin(n\theta)$, n a positive odd integer, can always be written as a polynomial of degree n in the variable $\sin\theta$.*

Now Work PROBLEMS 9(e) AND 67

Rearranging the Double-angle Formulas (3) and (4) leads to other formulas that will be used later in this section.

Begin with formula (3) and proceed to solve for $\sin^2\theta$.

$$\cos(2\theta) = 1 - 2\sin^2\theta$$

$$2\sin^2\theta = 1 - \cos(2\theta)$$

$$\sin^2\theta = \frac{1 - \cos(2\theta)}{2} \qquad \textbf{(6)}$$

Similarly, using formula (4), solve for $\cos^2\theta$.

$$\cos(2\theta) = 2\cos^2\theta - 1$$

$$2\cos^2\theta = 1 + \cos(2\theta)$$

$$\cos^2\theta = \frac{1 + \cos(2\theta)}{2} \qquad \textbf{(7)}$$

Formulas (6) and (7) can be used to develop a formula for $\tan^2\theta$.

$$\tan^2\theta = \frac{\sin^2\theta}{\cos^2\theta} = \frac{\dfrac{1 - \cos(2\theta)}{2}}{\dfrac{1 + \cos(2\theta)}{2}}$$

*Because of the work done by P. L. Chebyshëv, these polynomials are sometimes called *Chebyshëv polynomials*.

$$\tan^2\theta = \frac{1-\cos(2\theta)}{1+\cos(2\theta)} \quad \textbf{(8)}$$

Formulas (6) through (8) do not have to be memorized since their derivations are so straightforward.

Formulas (6) and (7) are important in calculus. The next example illustrates a problem that arises in calculus requiring the use of formula (7).

EXAMPLE 3 Establishing an Identity

Write an equivalent expression for $\cos^4\theta$ that does not involve any powers of sine or cosine greater than 1.

Solution The idea here is to apply formula (7) twice.

$$\begin{aligned}
\cos^4\theta = (\cos^2\theta)^2 &= \left(\frac{1+\cos(2\theta)}{2}\right)^2 && \text{Formula (7)}\\
&= \frac{1}{4}[1 + 2\cos(2\theta) + \cos^2(2\theta)]\\
&= \frac{1}{4} + \frac{1}{2}\cos(2\theta) + \frac{1}{4}\cos^2(2\theta)\\
&= \frac{1}{4} + \frac{1}{2}\cos(2\theta) + \frac{1}{4}\left\{\frac{1+\cos[2(2\theta)]}{2}\right\} && \text{Formula (7)}\\
&= \frac{1}{4} + \frac{1}{2}\cos(2\theta) + \frac{1}{8}[1+\cos(4\theta)]\\
&= \frac{3}{8} + \frac{1}{2}\cos(2\theta) + \frac{1}{8}\cos(4\theta)
\end{aligned}$$

Now Work PROBLEM 43

EXAMPLE 4 Solving a Trigonometric Equation Using Identities

Solve the equation: $\sin\theta\cos\theta = -\frac{1}{2}, \quad 0 \le \theta < 2\pi$

Solution The left side of the given equation is in the form of the Double-angle Formula $2\sin\theta\cos\theta = \sin(2\theta)$, except for a factor of 2. Multiply each side by 2.

$$\begin{aligned}
\sin\theta\cos\theta &= -\frac{1}{2}\\
2\sin\theta\cos\theta &= -1 && \text{Multiply each side by 2.}\\
\sin(2\theta) &= -1 && \text{Double-angle Formula}
\end{aligned}$$

The argument here is 2θ. Write the general formula that gives all the solutions of this equation, and then list those that are in the interval $[0, 2\pi)$. Because $\sin\left(\frac{3\pi}{2} + 2\pi k\right) = -1$, for any integer k, this means that

$$2\theta = \frac{3\pi}{2} + 2k\pi$$

$$\theta = \frac{3\pi}{4} + k\pi$$

$$\underset{k=-1}{\theta = \frac{3\pi}{4} + (-1)\pi = -\frac{\pi}{4}}, \quad \underset{k=0}{\theta = \frac{3\pi}{4} + (0)\pi = \frac{3\pi}{4}}, \quad \underset{k=1}{\theta = \frac{3\pi}{4} + (1)\pi = \frac{7\pi}{4}}, \quad \underset{k=2}{\theta = \frac{3\pi}{4} + (2)\pi = \frac{11\pi}{4}}$$

The solutions in the interval $[0, 2\pi)$ are

$$\theta = \frac{3\pi}{4}, \qquad \theta = \frac{7\pi}{4}$$

The solution set is $\left\{\frac{3\pi}{4}, \frac{7\pi}{4}\right\}$.

Now Work PROBLEM 71

EXAMPLE 5 Projectile Motion

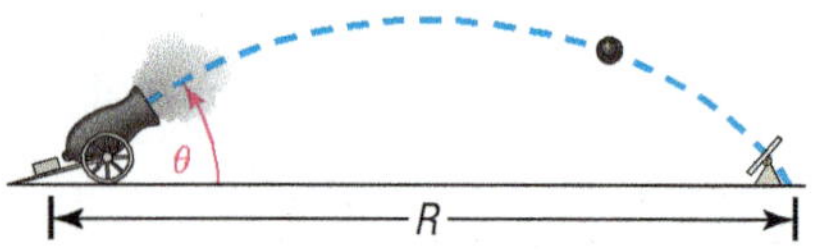

Figure 37

An object is propelled upward at an angle θ to the horizontal with an initial velocity of v_0 feet per second. See Figure 37. If air resistance is ignored, the **range** R—the horizontal distance that the object travels—is given by the function

$$R(\theta) = \frac{1}{16} v_0^2 \sin\theta \cos\theta$$

(a) Show that $R(\theta) = \frac{1}{32} v_0^2 \sin(2\theta)$.

(b) Find the angle θ for which R is a maximum.

Solution (a) Rewrite the given expression for the range using the Double-angle Formula $\sin(2\theta) = 2\sin\theta\cos\theta$. Then

$$R(\theta) = \frac{1}{16} v_0^2 \sin\theta\cos\theta = \frac{1}{16} v_0^2 \frac{2\sin\theta\cos\theta}{2} = \frac{1}{32} v_0^2 \sin(2\theta)$$

(b) In this form, the largest value for the range R can be found. For a fixed initial speed v_0, the angle θ of inclination to the horizontal determines the value of R. The largest value of a sine function is 1, which occurs when the argument 2θ is $90°$. For maximum R it follows that

$$2\theta = 90°$$
$$\theta = 45°$$

An inclination to the horizontal of $45°$ results in the maximum range.

3 Use Half-angle Formulas to Find Exact Values

Another important use of formulas (6) through (8) is to prove the *Half-angle Formulas*. In formulas (6) through (8), let $\theta = \frac{\alpha}{2}$. Then

$$\sin^2\frac{\alpha}{2} = \frac{1-\cos\alpha}{2} \qquad \cos^2\frac{\alpha}{2} = \frac{1+\cos\alpha}{2} \qquad \tan^2\frac{\alpha}{2} = \frac{1-\cos\alpha}{1+\cos\alpha} \qquad \textbf{(9)}$$

The identities in box (9) will prove useful in integral calculus.

Solving for the trigonometric functions on the left sides of equations (9) gives the Half-angle Formulas.

THEOREM **Half-angle Formulas**

$$\sin\frac{\alpha}{2} = \pm\sqrt{\frac{1-\cos\alpha}{2}} \qquad \textbf{(10)}$$

$$\cos\frac{\alpha}{2} = \pm\sqrt{\frac{1+\cos\alpha}{2}} \qquad \textbf{(11)}$$

$$\tan\frac{\alpha}{2} = \pm\sqrt{\frac{1-\cos\alpha}{1+\cos\alpha}} \qquad \textbf{(12)}$$

where the + or − sign is determined by the quadrant of the angle $\frac{\alpha}{2}$.

EXAMPLE 6 Finding Exact Values Using Half-angle Formulas

Use a Half-angle Formula to find the exact value of:

(a) $\cos 15°$ (b) $\sin(-15°)$

Solution (a) Because $15° = \frac{30°}{2}$, use the Half-angle Formula for $\cos \frac{\alpha}{2}$ with $\alpha = 30°$. Also, because $15°$ is in quadrant I, $\cos 15° > 0$, so choose the + sign when using formula (11):

$$\cos 15° = \cos \frac{30°}{2} = \sqrt{\frac{1 + \cos 30°}{2}}$$

$$= \sqrt{\frac{1 + \frac{\sqrt{3}}{2}}{2}} = \sqrt{\frac{2 + \sqrt{3}}{4}} = \frac{\sqrt{2 + \sqrt{3}}}{2}$$

(b) Use the fact that $\sin(-15°) = -\sin 15°$, and then apply formula (10).

$$\sin(-15°) = -\sin \frac{30°}{2} = -\sqrt{\frac{1 - \cos 30°}{2}}$$

$$= -\sqrt{\frac{1 - \frac{\sqrt{3}}{2}}{2}} = -\sqrt{\frac{2 - \sqrt{3}}{4}} = -\frac{\sqrt{2 - \sqrt{3}}}{2}$$ ■

It is interesting to compare the answer found in Example 6(a) with the answer to Example 2 of Section 8.5. There it was calculated that

$$\cos \frac{\pi}{12} = \cos 15° = \frac{1}{4}(\sqrt{6} + \sqrt{2})$$

Based on this and the result of Example 6(a),

$$\frac{1}{4}(\sqrt{6} + \sqrt{2}) \quad \text{and} \quad \frac{\sqrt{2 + \sqrt{3}}}{2}$$

are equal. (Since each expression is positive, you can verify this equality by squaring each expression.) Two very different looking, yet correct, answers can be obtained, depending on the approach taken to solve a problem.

Now Work PROBLEM 21

EXAMPLE 7 Finding Exact Values Using Half-angle Formulas

If $\cos \alpha = -\frac{3}{5}$, $\pi < \alpha < \frac{3\pi}{2}$, find the exact value of:

(a) $\sin \frac{\alpha}{2}$ (b) $\cos \frac{\alpha}{2}$ (c) $\tan \frac{\alpha}{2}$

Solution First, observe that if $\pi < \alpha < \frac{3\pi}{2}$, then $\frac{\pi}{2} < \frac{\alpha}{2} < \frac{3\pi}{4}$. As a result, $\frac{\alpha}{2}$ lies in quadrant II.

(a) Because $\frac{\alpha}{2}$ lies in quadrant II, $\sin \frac{\alpha}{2} > 0$, so use the + sign in formula (10) to get

$$\sin \frac{\alpha}{2} = \sqrt{\frac{1 - \cos \alpha}{2}} = \sqrt{\frac{1 - \left(-\frac{3}{5}\right)}{2}}$$

$$= \sqrt{\frac{\frac{8}{5}}{2}} = \sqrt{\frac{4}{5}} = \frac{2}{\sqrt{5}} = \frac{2\sqrt{5}}{5}$$

(b) Because $\frac{\alpha}{2}$ lies in quadrant II, $\cos\frac{\alpha}{2} < 0$, so use the $-$ sign in formula (11) to get

$$\cos\frac{\alpha}{2} = -\sqrt{\frac{1+\cos\alpha}{2}} = -\sqrt{\frac{1+\left(-\frac{3}{5}\right)}{2}}$$

$$= -\sqrt{\frac{\frac{2}{5}}{2}} = -\frac{1}{\sqrt{5}} = -\frac{\sqrt{5}}{5}$$

(c) Because $\frac{\alpha}{2}$ lies in quadrant II, $\tan\frac{\alpha}{2} < 0$, so use the $-$ sign in formula (12) to get

$$\tan\frac{\alpha}{2} = -\sqrt{\frac{1-\cos\alpha}{1+\cos\alpha}} = -\sqrt{\frac{1-\left(-\frac{3}{5}\right)}{1+\left(-\frac{3}{5}\right)}} = -\sqrt{\frac{\frac{8}{5}}{\frac{2}{5}}} = -2$$

Another way to solve Example 7(c) is to use the results of parts (a) and (b).

$$\tan\frac{\alpha}{2} = \frac{\sin\frac{\alpha}{2}}{\cos\frac{\alpha}{2}} = \frac{\frac{2\sqrt{5}}{5}}{-\frac{\sqrt{5}}{5}} = -2$$

Now Work PROBLEM 9(c), (d), AND (f)

There is a formula for $\tan\frac{\alpha}{2}$ that does not contain $+$ and $-$ signs, making it more useful than formula (12). To derive it, use the formulas

$$1-\cos\alpha = 2\sin^2\frac{\alpha}{2} \qquad \text{Formula (9)}$$

and

$$\sin\alpha = \sin\left[2\left(\frac{\alpha}{2}\right)\right] = 2\sin\frac{\alpha}{2}\cos\frac{\alpha}{2} \qquad \text{Double-angle Formula (1)}$$

Then

$$\frac{1-\cos\alpha}{\sin\alpha} = \frac{2\sin^2\frac{\alpha}{2}}{2\sin\frac{\alpha}{2}\cos\frac{\alpha}{2}} = \frac{\sin\frac{\alpha}{2}}{\cos\frac{\alpha}{2}} = \tan\frac{\alpha}{2}$$

Because it also can be shown that

$$\frac{1-\cos\alpha}{\sin\alpha} = \frac{\sin\alpha}{1+\cos\alpha}$$

this results in the following two Half-angle Formulas:

Half-angle Formulas for $\tan\frac{\alpha}{2}$

$$\tan\frac{\alpha}{2} = \frac{1-\cos\alpha}{\sin\alpha} = \frac{\sin\alpha}{1+\cos\alpha} \qquad \textbf{(13)}$$

With this formula, the solution to Example 7(c) can be obtained as follows:

$$\cos\alpha = -\frac{3}{5} \qquad \pi < \alpha < \frac{3\pi}{2}$$

$$\sin\alpha = -\sqrt{1-\cos^2\alpha} = -\sqrt{1-\frac{9}{25}} = -\sqrt{\frac{16}{25}} = -\frac{4}{5}$$

Then, by equation (13),

$$\tan\frac{\alpha}{2} = \frac{1-\cos\alpha}{\sin\alpha} = \frac{1-\left(-\frac{3}{5}\right)}{-\frac{4}{5}} = \frac{\frac{8}{5}}{-\frac{4}{5}} = -2$$

8.6 Assess Your Understanding

Concepts and Vocabulary

1. $\cos(2\theta) = \cos^2\theta -$ _____ $=$ _______ $- 1 = 1 -$ _______.

2. $\sin^2\frac{\theta}{2} = \frac{\_\_\_\_\_\_}{2}$.

3. $\tan\frac{\theta}{2} = \frac{1-\cos\theta}{\_\_\_\_\_}$.

4. ***True or False*** $\tan(2\theta) = \dfrac{2\tan\theta}{1-\tan^2\theta}$

5. ***True or False*** $\sin(2\theta)$ has two equivalent forms:
   $2\sin\theta\cos\theta$ and $\sin^2\theta - \cos^2\theta$

6. ***True or False*** $\tan(2\theta) + \tan(2\theta) = \tan(4\theta)$

7. Choose the expression that completes the Half-angle Formula for cosine functions: $\cos\frac{\alpha}{2} =$ ______
   (a) $\pm\sqrt{\dfrac{1-\cos\alpha}{2}}$ (b) $\pm\sqrt{\dfrac{1+\cos\alpha}{2}}$
   (c) $\pm\sqrt{\dfrac{\cos\alpha-\sin\alpha}{2}}$ (d) $\pm\sqrt{\dfrac{1-\cos\alpha}{1+\cos\alpha}}$

8. If $\sin\alpha = \pm\sqrt{\dfrac{1-\cos\theta}{2}}$, then which of the following describes how the value of θ is related to the value of α?
   (a) $\theta = \alpha$ (b) $\theta = \frac{\alpha}{2}$ (c) $\theta = 2\alpha$ (d) $\theta = \alpha^2$

Skill Building

In Problems 9–20, use the information given about the angle θ, $0 \le \theta < 2\pi$, to find the exact value of:

(a) $\sin(2\theta)$ (b) $\cos(2\theta)$ (c) $\sin\frac{\theta}{2}$ (d) $\cos\frac{\theta}{2}$ (e) $\tan(2\theta)$ (f) $\tan\frac{\theta}{2}$

9. $\sin\theta = \frac{3}{5}$, $0 < \theta < \frac{\pi}{2}$

10. $\cos\theta = \frac{3}{5}$, $0 < \theta < \frac{\pi}{2}$

11. $\tan\theta = \frac{4}{3}$, $\pi < \theta < \frac{3\pi}{2}$

12. $\tan\theta = \frac{1}{2}$, $\pi < \theta < \frac{3\pi}{2}$

13. $\cos\theta = -\frac{\sqrt{6}}{3}$, $\frac{\pi}{2} < \theta < \pi$

14. $\sin\theta = -\frac{\sqrt{3}}{3}$, $\frac{3\pi}{2} < \theta < 2\pi$

15. $\sec\theta = 3$, $\sin\theta > 0$

16. $\csc\theta = -\sqrt{5}$, $\cos\theta < 0$

17. $\cot\theta = -2$, $\sec\theta < 0$

18. $\sec\theta = 2$, $\csc\theta < 0$

19. $\tan\theta = -3$, $\sin\theta < 0$

20. $\cot\theta = 3$, $\cos\theta < 0$

In Problems 21–30, use the Half-angle Formulas to find the exact value of each expression.

21. $\sin 22.5°$

22. $\cos 22.5°$

23. $\tan\frac{7\pi}{8}$

24. $\tan\frac{9\pi}{8}$

25. $\cos 165°$

26. $\sin 195°$

27. $\sec\frac{15\pi}{8}$

28. $\csc\frac{7\pi}{8}$

29. $\sin\left(-\frac{\pi}{8}\right)$

30. $\cos\left(-\frac{3\pi}{8}\right)$

In Problems 31–42, use the figures to evaluate each function, given that $f(x) = \sin x$, $g(x) = \cos x$, and $h(x) = \tan x$.

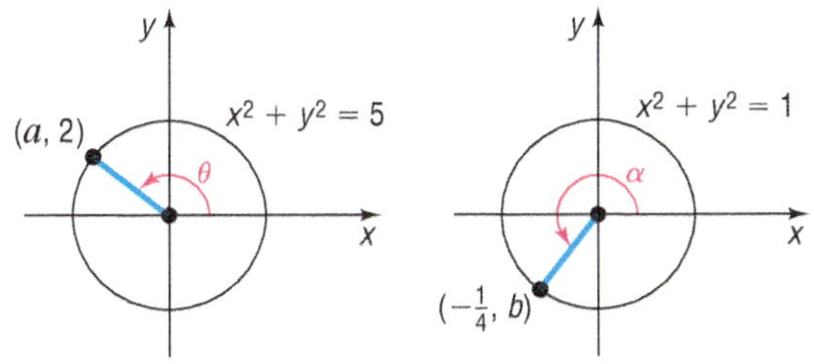

31. $f(2\theta)$

32. $g(2\theta)$

33. $g\left(\frac{\theta}{2}\right)$

34. $f\left(\frac{\theta}{2}\right)$

35. $h(2\theta)$

36. $h\left(\frac{\theta}{2}\right)$

37. $g(2\alpha)$

38. $f(2\alpha)$

39. $f\left(\frac{\alpha}{2}\right)$ **40.** $g\left(\frac{\alpha}{2}\right)$ **41.** $h\left(\frac{\alpha}{2}\right)$ **42.** $h(2\alpha)$

43. Show that $\sin^4\theta = \frac{3}{8} - \frac{1}{2}\cos(2\theta) + \frac{1}{8}\cos(4\theta)$.

44. Show that $\sin(4\theta) = (\cos\theta)(4\sin\theta - 8\sin^3\theta)$.

45. Develop a formula for $\cos(3\theta)$ as a third-degree polynomial in the variable $\cos\theta$.

46. Develop a formula for $\cos(4\theta)$ as a fourth-degree polynomial in the variable $\cos\theta$.

47. Find an expression for $\sin(5\theta)$ as a fifth-degree polynomial in the variable $\sin\theta$.

48. Find an expression for $\cos(5\theta)$ as a fifth-degree polynomial in the variable $\cos\theta$.

In Problems 49–70, establish each identity.

49. $\cos^4\theta - \sin^4\theta = \cos(2\theta)$

50. $\dfrac{\cot\theta - \tan\theta}{\cot\theta + \tan\theta} = \cos(2\theta)$

51. $\cot(2\theta) = \dfrac{\cot^2\theta - 1}{2\cot\theta}$

52. $\cot(2\theta) = \frac{1}{2}(\cot\theta - \tan\theta)$

53. $\sec(2\theta) = \dfrac{\sec^2\theta}{2 - \sec^2\theta}$

54. $\csc(2\theta) = \frac{1}{2}\sec\theta\csc\theta$

55. $\cos^2(2u) - \sin^2(2u) = \cos(4u)$

56. $(4\sin u\cos u)(1 - 2\sin^2 u) = \sin(4u)$

57. $\dfrac{\cos(2\theta)}{1 + \sin(2\theta)} = \dfrac{\cot\theta - 1}{\cot\theta + 1}$

58. $\sin^2\theta\cos^2\theta = \frac{1}{8}[1 - \cos(4\theta)]$

59. $\sec^2\frac{\theta}{2} = \dfrac{2}{1 + \cos\theta}$

60. $\csc^2\frac{\theta}{2} = \dfrac{2}{1 - \cos\theta}$

61. $\cot^2\frac{v}{2} = \dfrac{\sec v + 1}{\sec v - 1}$

62. $\tan\frac{v}{2} = \csc v - \cot v$

63. $\cos\theta = \dfrac{1 - \tan^2\frac{\theta}{2}}{1 + \tan^2\frac{\theta}{2}}$

64. $1 - \frac{1}{2}\sin(2\theta) = \dfrac{\sin^3\theta + \cos^3\theta}{\sin\theta + \cos\theta}$

65. $\dfrac{\sin(3\theta)}{\sin\theta} - \dfrac{\cos(3\theta)}{\cos\theta} = 2$

66. $\dfrac{\cos\theta + \sin\theta}{\cos\theta - \sin\theta} - \dfrac{\cos\theta - \sin\theta}{\cos\theta + \sin\theta} = 2\tan(2\theta)$

67. $\tan(3\theta) = \dfrac{3\tan\theta - \tan^3\theta}{1 - 3\tan^2\theta}$

68. $\tan\theta + \tan(\theta + 120°) + \tan(\theta + 240°) = 3\tan(3\theta)$

69. $\ln|\sin\theta| = \frac{1}{2}(\ln|1 - \cos(2\theta)| - \ln 2)$

70. $\ln|\cos\theta| = \frac{1}{2}(\ln|1 + \cos(2\theta)| - \ln 2)$

In Problems 71–80, solve each equation on the interval $0 \le \theta < 2\pi$.

71. $\cos(2\theta) + 6\sin^2\theta = 4$

72. $\cos(2\theta) = 2 - 2\sin^2\theta$

73. $\cos(2\theta) = \cos\theta$

74. $\sin(2\theta) = \cos\theta$

75. $\sin(2\theta) + \sin(4\theta) = 0$

76. $\cos(2\theta) + \cos(4\theta) = 0$

77. $3 - \sin\theta = \cos(2\theta)$

78. $\cos(2\theta) + 5\cos\theta + 3 = 0$

79. $\tan(2\theta) + 2\sin\theta = 0$

80. $\tan(2\theta) + 2\cos\theta = 0$

Mixed Practice

In Problems 81–92, find the exact value of each expression.

81. $\sin\left(2\sin^{-1}\frac{1}{2}\right)$

82. $\sin\left[2\sin^{-1}\frac{\sqrt{3}}{2}\right]$

83. $\cos\left(2\sin^{-1}\frac{3}{5}\right)$

84. $\cos\left(2\cos^{-1}\frac{4}{5}\right)$

85. $\tan\left[2\cos^{-1}\left(-\frac{3}{5}\right)\right]$

86. $\tan\left(2\tan^{-1}\frac{3}{4}\right)$

87. $\sin\left(2\cos^{-1}\frac{4}{5}\right)$

88. $\cos\left[2\tan^{-1}\left(-\frac{4}{3}\right)\right]$

89. $\sin^2\left(\frac{1}{2}\cos^{-1}\frac{3}{5}\right)$

90. $\cos^2\left(\frac{1}{2}\sin^{-1}\frac{3}{5}\right)$

91. $\sec\left(2\tan^{-1}\frac{3}{4}\right)$

92. $\csc\left[2\sin^{-1}\left(-\frac{3}{5}\right)\right]$

In Problems 93–95, find the real zeros of each trigonometric function on the interval $0 \le \theta < 2\pi$.

93. $f(x) = \sin(2x) - \sin x$

94. $f(x) = \cos(2x) + \cos x$

95. $f(x) = \cos(2x) + \sin^2 x$

Applications and Extensions

96. Constructing a Rain Gutter A rain gutter is to be constructed of aluminum sheets 12 inches wide. After marking off a length of 4 inches from each edge, the builder bends this length up at an angle θ. See the illustration. The area A of the opening as a function of θ is given by

$$A(\theta) = 16 \sin\theta(\cos\theta + 1) \quad 0° < \theta < 90°$$

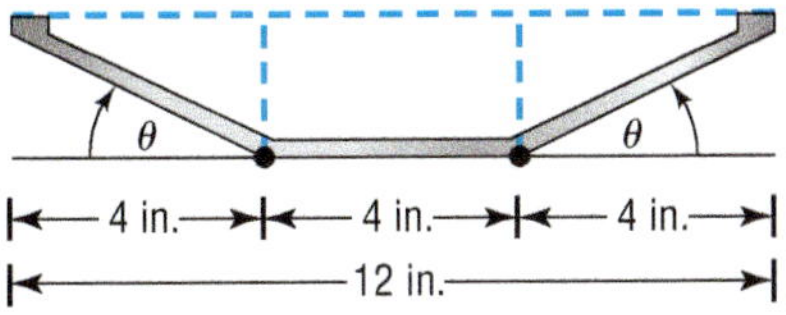

(a) In calculus, you will be asked to find the angle θ that maximizes A by solving the equation

$$\cos(2\theta) + \cos\theta = 0, \quad 0° < \theta < 90°$$

Solve this equation for θ.

(b) What is the maximum area A of the opening?

(c) Graph $A = A(\theta), 0° \le \theta \le 90°$, and find the angle θ that maximizes the area A. Also find the maximum area.

97. Laser Projection In a laser projection system, the **optical angle** or **scanning angle** θ is related to the throw distance D from the scanner to the screen and the projected image width W by the equation

$$D = \frac{\frac{1}{2}W}{\csc\theta - \cot\theta}$$

(a) Show that the projected image width is given by

$$W = 2D\tan\frac{\theta}{2}$$

(b) Find the optical angle if the throw distance is 15 feet and the projected image width is 6.5 feet.

Source: *Pangolin Laser Systems, Inc.*

98. Product of Inertia The **product of inertia** for an area about inclined axes is given by the formula

$$I_{uv} = I_x \sin\theta\cos\theta - I_y \sin\theta\cos\theta + I_{xy}(\cos^2\theta - \sin^2\theta)$$

Show that this is equivalent to

$$I_{uv} = \frac{I_x - I_y}{2}\sin(2\theta) + I_{xy}\cos(2\theta)$$

Source: Adapted from Hibbeler, *Engineering Mechanics: Statics,* 13th ed., Pearson © 2013.

99. Projectile Motion An object is propelled upward at an angle θ, $45° < \theta < 90°$, to the horizontal with an initial velocity of v_0 feet per second from the base of a plane that makes an angle of 45° with the horizontal. See the illustration. If air resistance is ignored, the distance R that it travels up the inclined plane is given by the function

$$R(\theta) = \frac{v_0^2\sqrt{2}}{16}\cos\theta(\sin\theta - \cos\theta)$$

(a) Show that

$$R(\theta) = \frac{v_0^2\sqrt{2}}{32}[\sin(2\theta) - \cos(2\theta) - 1]$$

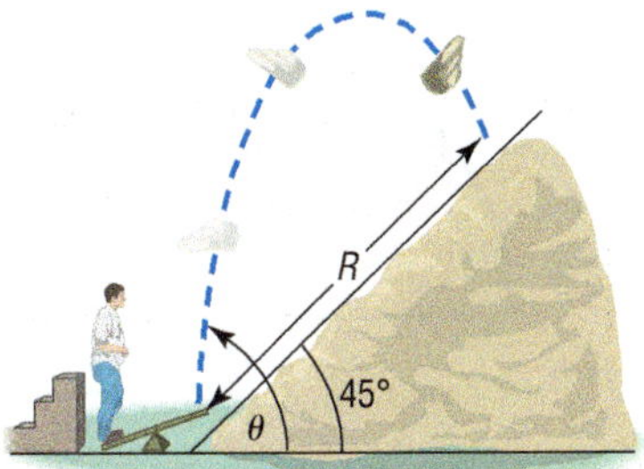

(b) In calculus, you will be asked to find the angle θ that maximizes R by solving the equation

$$\sin(2\theta) + \cos(2\theta) = 0$$

Solve this equation for θ.

(c) What is the maximum distance R if $v_0 = 32$ feet per second?

(d) Graph $R = R(\theta), 45° \le \theta \le 90°$, and find the angle θ that maximizes the distance R. Also find the maximum distance. Use $v_0 = 32$ feet per second. Compare the results with the answers found in parts (b) and (c).

100. Sawtooth Curve An oscilloscope often displays a sawtooth curve. This curve can be approximated by sinusoidal curves of varying periods and amplitudes. A first approximation to the sawtooth curve is given by

$$y = \frac{1}{2}\sin(2\pi x) + \frac{1}{4}\sin(4\pi x)$$

Show that $y = \sin(2\pi x)\cos^2(\pi x)$.

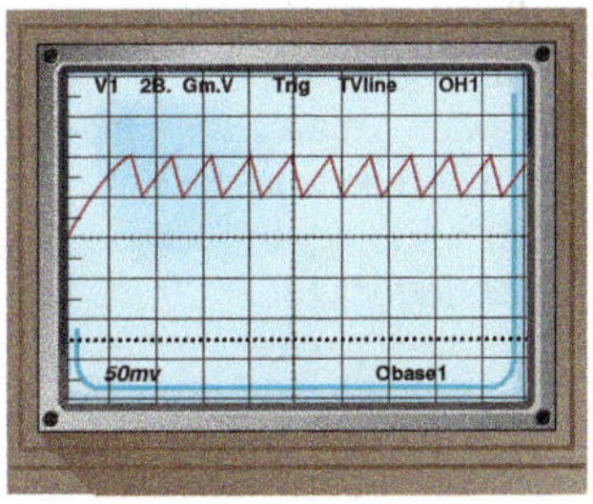

101. Area of an Isosceles Triangle Show that the area A of an isosceles triangle whose equal sides are of length s, and where θ is the angle between them, is

$$A = \frac{1}{2}s^2\sin\theta$$

[**Hint:** See the illustration. The height h bisects the angle θ and is the perpendicular bisector of the base.]

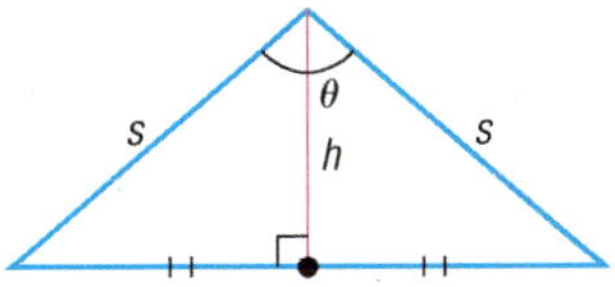

102. Geometry A rectangle is inscribed in a semicircle of radius 1. See the illustration.

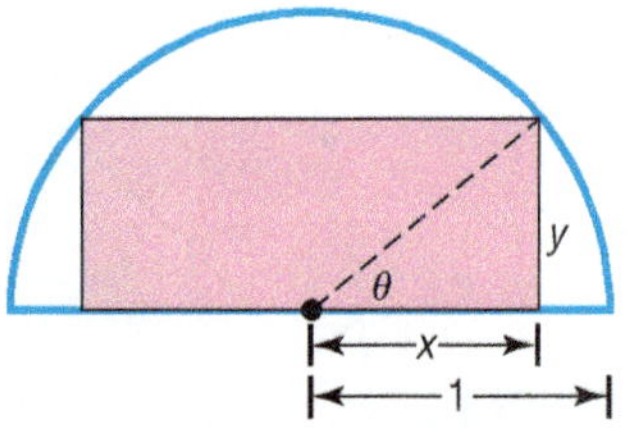

(a) Express the area A of the rectangle as a function of the angle θ shown in the illustration.
(b) Show that $A(\theta) = \sin(2\theta)$.
(c) Find the angle θ that results in the largest area A.
(d) Find the dimensions of this largest rectangle.

103. Geometry A regular dodecagon is a regular polygon with sides of equal length where all interior angles are 150°. See the figure.

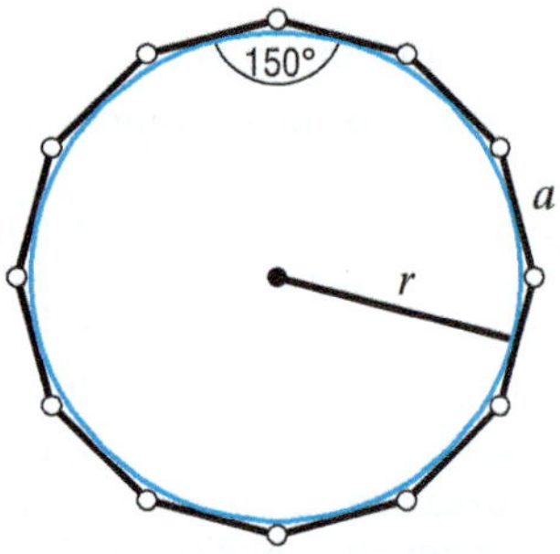

The area A of a regular dodecagon with side a is $A = 3\cot\left(\frac{\pi}{12}\right)a^2$. If r is the radius of the inscribed circle, then $A = 12\tan\left(\frac{\pi}{12}\right)r^2$.

(a) Find the exact area of a regular dodecagon with side $a = 5$ cm.
(b) Find the radius of the inscribed circle for the regular dodecagon from part (a).
(c) Find the exact area of the circle inscribed in a regular dodecagon with side $a = 5$ cm.
(d) Find the exact area of the region between the circle and the regular dodecagon.

104. (a) If $x = 2\tan\theta$, express $\sin(2\theta)$ as a function of x.
(b) If $x = 2\tan\theta$, express $\cos(2\theta)$ as a function of x.

105. Find the value of the number C:

$$\frac{1}{2}\sin^2 x + C = -\frac{1}{4}\cos(2x)$$

106. Find the value of the number C:

$$\frac{1}{2}\cos^2 x + C = \frac{1}{4}\cos(2x)$$

107. If $z = \tan\frac{\alpha}{2}$, show that $\sin\alpha = \frac{2z}{1+z^2}$.

108. If $z = \tan\frac{\alpha}{2}$, show that $\cos\alpha = \frac{1-z^2}{1+z^2}$.

109. Graph $f(x) = \sin^2 x = \frac{1-\cos(2x)}{2}$ for $0 \le x \le 2\pi$ by using transformations.

110. Repeat Problem 109 for $g(x) = \cos^2 x$.

111. Use the fact that

$$\cos\frac{\pi}{12} = \frac{1}{4}(\sqrt{6}+\sqrt{2})$$

to find $\sin\frac{\pi}{24}$ and $\cos\frac{\pi}{24}$.

112. Show that

$$\cos\frac{\pi}{8} = \frac{\sqrt{2+\sqrt{2}}}{2}$$

and use it to find $\sin\frac{\pi}{16}$ and $\cos\frac{\pi}{16}$.

113. Show that

$$\sin^3\theta + \sin^3(\theta + 120°) + \sin^3(\theta + 240°) = -\frac{3}{4}\sin(3\theta)$$

114. If $\tan\theta = a\tan\frac{\theta}{3}$, express $\tan\frac{\theta}{3}$ in terms of a.

115. For $\cos(2x) + (2m-1)\sin x + m - 1 = 0$, find m such that there is exactly one real solution for x, $-\frac{\pi}{2} \le x \le \frac{\pi}{2}$.†

†Courtesy of Joliet Junior College Mathematics Department

Explaining Concepts: Discussion and Writing

116. Go to the library and research Chebyshëv polynomials. Write a report on your findings.

Retain Your Knowledge

Problems 117–120 are based on material learned earlier in the course. The purpose of these problems is to keep the material fresh in your mind so that you are better prepared for the final exam.

117. Find an equation of the line that contains the point $(2, -3)$ and is perpendicular to the line $y = -2x + 9$.

118. Graph $f(x) = -x^2 + 6x + 7$. Label the vertex and any intercepts.

119. Find the exact value of $\sin\left(\frac{2\pi}{3}\right) - \cos\left(\frac{4\pi}{3}\right)$.

120. Graph $y = -2\cos\left(\frac{\pi}{2}x\right)$. Show at least two periods.

8.7 Product-to-Sum and Sum-to-Product Formulas

OBJECTIVES 1 Express Products as Sums (p. 675)
2 Express Sums as Products (p. 676)

1 Express Products as Sums

Sum and difference formulas can be used to derive formulas for writing the products of sines and/or cosines as sums or differences. These identities are usually called the **Product-to-Sum Formulas.**

THEOREM

Product-to-Sum Formulas

$$\sin\alpha\sin\beta = \frac{1}{2}[\cos(\alpha-\beta) - \cos(\alpha+\beta)] \quad \textbf{(1)}$$

$$\cos\alpha\cos\beta = \frac{1}{2}[\cos(\alpha-\beta) + \cos(\alpha+\beta)] \quad \textbf{(2)}$$

$$\sin\alpha\cos\beta = \frac{1}{2}[\sin(\alpha+\beta) + \sin(\alpha-\beta)] \quad \textbf{(3)}$$

These formulas do not have to be memorized. Instead, remember how they are derived. Then, when you want to use them, either look them up or derive them, as needed.

To derive formulas (1) and (2), write down the sum and difference formulas for the cosine:

$$\cos(\alpha-\beta) = \cos\alpha\cos\beta + \sin\alpha\sin\beta \quad \textbf{(4)}$$

$$\cos(\alpha+\beta) = \cos\alpha\cos\beta - \sin\alpha\sin\beta \quad \textbf{(5)}$$

To derive formula (1), subtract equation (5) from equation (4) to get

$$\cos(\alpha-\beta) - \cos(\alpha+\beta) = 2\sin\alpha\sin\beta$$

from which

$$\sin\alpha\sin\beta = \frac{1}{2}[\cos(\alpha-\beta) - \cos(\alpha+\beta)]$$

To derive formula (2), add equations (4) and (5) to get

$$\cos(\alpha-\beta) + \cos(\alpha+\beta) = 2\cos\alpha\cos\beta$$

from which

$$\cos\alpha\cos\beta = \frac{1}{2}[\cos(\alpha-\beta) + \cos(\alpha+\beta)]$$

To derive Product-to-Sum Formula (3), use the sum and difference formulas for sine in a similar way. (You are asked to do this in Problem 53.)

EXAMPLE 1 **Expressing Products as Sums**

Express each of the following products as a sum containing only sines or only cosines.

(a) $\sin(6\theta)\sin(4\theta)$ (b) $\cos(3\theta)\cos\theta$ (c) $\sin(3\theta)\cos(5\theta)$

Solution (a) Use formula (1) to get

$$\sin(6\theta)\sin(4\theta) = \frac{1}{2}[\cos(6\theta - 4\theta) - \cos(6\theta + 4\theta)]$$
$$= \frac{1}{2}[\cos(2\theta) - \cos(10\theta)]$$

(b) Use formula (2) to get

$$\cos(3\theta)\cos\theta = \frac{1}{2}[\cos(3\theta - \theta) + \cos(3\theta + \theta)]$$
$$= \frac{1}{2}[\cos(2\theta) + \cos(4\theta)]$$

(c) Use formula (3) to get

$$\sin(3\theta)\cos(5\theta) = \frac{1}{2}[\sin(3\theta + 5\theta) + \sin(3\theta - 5\theta)]$$
$$= \frac{1}{2}[\sin(8\theta) + \sin(-2\theta)] = \frac{1}{2}[\sin(8\theta) - \sin(2\theta)]$$ ■

Now Work PROBLEM 7

2 Express Sums as Products

The **Sum-to-Product Formulas** are given next.

THEOREM

Sum-to-Product Formulas

$$\sin\alpha + \sin\beta = 2\sin\frac{\alpha+\beta}{2}\cos\frac{\alpha-\beta}{2} \quad \textbf{(6)}$$

$$\sin\alpha - \sin\beta = 2\sin\frac{\alpha-\beta}{2}\cos\frac{\alpha+\beta}{2} \quad \textbf{(7)}$$

$$\cos\alpha + \cos\beta = 2\cos\frac{\alpha+\beta}{2}\cos\frac{\alpha-\beta}{2} \quad \textbf{(8)}$$

$$\cos\alpha - \cos\beta = -2\sin\frac{\alpha+\beta}{2}\sin\frac{\alpha-\beta}{2} \quad \textbf{(9)}$$

Formula (6) is derived here. The derivations of formulas (7) through (9) are left as exercises (see Problems 54 through 56).

Proof

$$2\sin\frac{\alpha+\beta}{2}\cos\frac{\alpha-\beta}{2} = 2\cdot\frac{1}{2}\left[\sin\left(\frac{\alpha+\beta}{2}+\frac{\alpha-\beta}{2}\right) + \sin\left(\frac{\alpha+\beta}{2}-\frac{\alpha-\beta}{2}\right)\right]$$

↑ Product-to-Sum Formula (3)

$$= \sin\frac{2\alpha}{2} + \sin\frac{2\beta}{2} = \sin\alpha + \sin\beta$$ ■

EXAMPLE 2

Expressing Sums (or Differences) as Products

Express each sum or difference as a product of sines and/or cosines.

(a) $\sin(5\theta) - \sin(3\theta)$ (b) $\cos(3\theta) + \cos(2\theta)$

Solution (a) Use formula (7) to get

$$\sin(5\theta) - \sin(3\theta) = 2\sin\frac{5\theta - 3\theta}{2}\cos\frac{5\theta + 3\theta}{2}$$
$$= 2\sin\theta\cos(4\theta)$$

(b) $$\cos(3\theta) + \cos(2\theta) = 2\cos\frac{3\theta + 2\theta}{2}\cos\frac{3\theta - 2\theta}{2} \quad \text{Formula (8)}$$
$$= 2\cos\frac{5\theta}{2}\cos\frac{\theta}{2}$$

Now Work PROBLEM 17

8.7 Assess Your Understanding

Skill Building

In Problems 1–6, find the exact value of each expression.

1. $\sin 195° \cdot \cos 75°$
2. $\cos 285° \cdot \cos 195°$
3. $\sin 285° \cdot \sin 75°$
4. $\sin 75° + \sin 15°$
5. $\cos 255° - \cos 195°$
6. $\sin 255° - \sin 15°$

In Problems 7–16, express each product as a sum containing only sines or only cosines.

7. $\sin(4\theta)\sin(2\theta)$
8. $\cos(4\theta)\cos(2\theta)$
9. $\sin(4\theta)\cos(2\theta)$
10. $\sin(3\theta)\sin(5\theta)$
11. $\cos(3\theta)\cos(5\theta)$
12. $\sin(4\theta)\cos(6\theta)$
13. $\sin\theta\sin(2\theta)$
14. $\cos(3\theta)\cos(4\theta)$
15. $\sin\frac{3\theta}{2}\cos\frac{\theta}{2}$
16. $\sin\frac{\theta}{2}\cos\frac{5\theta}{2}$

In Problems 17–24, express each sum or difference as a product of sines and/or cosines.

17. $\sin(4\theta) - \sin(2\theta)$
18. $\sin(4\theta) + \sin(2\theta)$
19. $\cos(2\theta) + \cos(4\theta)$
20. $\cos(5\theta) - \cos(3\theta)$
21. $\sin\theta + \sin(3\theta)$
22. $\cos\theta + \cos(3\theta)$
23. $\cos\frac{\theta}{2} - \cos\frac{3\theta}{2}$
24. $\sin\frac{\theta}{2} - \sin\frac{3\theta}{2}$

In Problems 25–42, establish each identity.

25. $\dfrac{\sin\theta + \sin(3\theta)}{2\sin(2\theta)} = \cos\theta$
26. $\dfrac{\cos\theta + \cos(3\theta)}{2\cos(2\theta)} = \cos\theta$
27. $\dfrac{\sin(4\theta) + \sin(2\theta)}{\cos(4\theta) + \cos(2\theta)} = \tan(3\theta)$
28. $\dfrac{\cos\theta - \cos(3\theta)}{\sin(3\theta) - \sin\theta} = \tan(2\theta)$
29. $\dfrac{\cos\theta - \cos(3\theta)}{\sin\theta + \sin(3\theta)} = \tan\theta$
30. $\dfrac{\cos\theta - \cos(5\theta)}{\sin\theta + \sin(5\theta)} = \tan(2\theta)$
31. $\sin\theta[\sin\theta + \sin(3\theta)] = \cos\theta[\cos\theta - \cos(3\theta)]$
32. $\sin\theta[\sin(3\theta) + \sin(5\theta)] = \cos\theta[\cos(3\theta) - \cos(5\theta)]$
33. $\dfrac{\sin(4\theta) + \sin(8\theta)}{\cos(4\theta) + \cos(8\theta)} = \tan(6\theta)$
34. $\dfrac{\sin(4\theta) - \sin(8\theta)}{\cos(4\theta) - \cos(8\theta)} = -\cot(6\theta)$
35. $\dfrac{\sin(4\theta) + \sin(8\theta)}{\sin(4\theta) - \sin(8\theta)} = -\dfrac{\tan(6\theta)}{\tan(2\theta)}$
36. $\dfrac{\cos(4\theta) - \cos(8\theta)}{\cos(4\theta) + \cos(8\theta)} = \tan(2\theta)\tan(6\theta)$

37. $\dfrac{\sin\alpha + \sin\beta}{\sin\alpha - \sin\beta} = \tan\dfrac{\alpha+\beta}{2}\cot\dfrac{\alpha-\beta}{2}$

38. $\dfrac{\cos\alpha + \cos\beta}{\cos\alpha - \cos\beta} = -\cot\dfrac{\alpha+\beta}{2}\cot\dfrac{\alpha-\beta}{2}$

39. $\dfrac{\sin\alpha + \sin\beta}{\cos\alpha + \cos\beta} = \tan\dfrac{\alpha+\beta}{2}$

40. $\dfrac{\sin\alpha - \sin\beta}{\cos\alpha - \cos\beta} = -\cot\dfrac{\alpha+\beta}{2}$

41. $1 + \cos(2\theta) + \cos(4\theta) + \cos(6\theta) = 4\cos\theta\cos(2\theta)\cos(3\theta)$

42. $1 - \cos(2\theta) + \cos(4\theta) - \cos(6\theta) = 4\sin\theta\cos(2\theta)\sin(3\theta)$

In Problems 43–46, solve each equation on the interval $0 \le \theta < 2\pi$.

43. $\sin(2\theta) + \sin(4\theta) = 0$

44. $\cos(2\theta) + \cos(4\theta) = 0$

45. $\cos(4\theta) - \cos(6\theta) = 0$

46. $\sin(4\theta) - \sin(6\theta) = 0$

Applications and Extensions

47. Touch-Tone Phones On a Touch-Tone phone, each button produces a unique sound. The sound produced is the sum of two tones, given by

$$y = \sin(2\pi l t) \quad \text{and} \quad y = \sin(2\pi h t)$$

where l and h are the low and high frequencies (cycles per second) shown on the illustration. For example, if you touch 7, the low frequency is $l = 852$ cycles per second and the high frequency is $h = 1209$ cycles per second. The sound emitted when you touch 7 is

$$y = \sin[2\pi(852)t] + \sin[2\pi(1209)t]$$

Touch-Tone phone

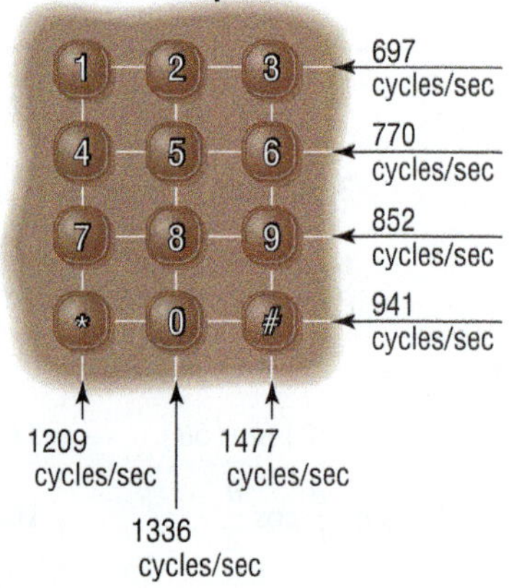

(a) Write this sound as a product of sines and/or cosines.
(b) Determine the maximum value of y.
(c) Graph the sound emitted when 7 is touched.

48. Touch-Tone Phones
(a) Write, as a product of sines and/or cosines, the sound emitted when the # key is touched.
(b) Determine the maximum value of y.
(c) Graph the sound emitted when the # key is touched.

49. Moment of Inertia The moment of inertia I of an object is a measure of how easy it is to rotate the object about some fixed point. In engineering mechanics, it is sometimes necessary to compute moments of inertia with respect to a set of rotated axes. These moments are given by the equations

$$I_u = I_x\cos^2\theta + I_y\sin^2\theta - 2I_{xy}\sin\theta\cos\theta$$
$$I_v = I_x\sin^2\theta + I_y\cos^2\theta + 2I_{xy}\sin\theta\cos\theta$$

Use Product-to-Sum Formulas to show that

$$I_u = \frac{I_x + I_y}{2} + \frac{I_x - I_y}{2}\cos(2\theta) - I_{xy}\sin(2\theta)$$

and

$$I_v = \frac{I_x + I_y}{2} - \frac{I_x - I_y}{2}\cos(2\theta) + I_{xy}\sin(2\theta)$$

Source: Adapted from Hibbeler, *Engineering Mechanics: Statics,* 13th ed., Pearson © 2013.

50. Projectile Motion The range R of a projectile propelled downward from the top of an inclined plane at an angle θ to the inclined plane is given by

$$R(\theta) = \frac{2v_0^2\sin\theta\cos(\theta - \phi)}{g\cos^2\phi}$$

where v_0 is the initial velocity of the projectile, ϕ is the angle the plane makes with respect to the horizontal, and g is acceleration due to gravity.

(a) Show that for fixed v_0 and ϕ, the maximum range down the incline is given by $R_{\max} = \dfrac{v_0^2}{g(1 - \sin\phi)}$.
(b) Determine the maximum range if the projectile has an initial velocity of 50 meters/second, the angle of the plane is $\phi = 35°$, and $g = 9.8$ meters/second^2.

51. If $\alpha + \beta + \gamma = \pi$, show that

$$\sin(2\alpha) + \sin(2\beta) + \sin(2\gamma) = 4\sin\alpha\sin\beta\sin\gamma$$

52. If $\alpha + \beta + \gamma = \pi$, show that

$$\tan\alpha + \tan\beta + \tan\gamma = \tan\alpha\tan\beta\tan\gamma$$

53. Derive formula (3).

54. Derive formula (7).

55. Derive formula (8).

56. Derive formula (9).

Retain Your Knowledge

Problems 57–60 are based on material learned earlier in the course. The purpose of these problems is to keep the material fresh in your mind so that you are better prepared for the final exam.

57. Solve: $27^{x-1} = 9^{x+5}$

58. For $y = 5\cos(4x - \pi)$, find the amplitude, the period, and the phase shift.

59. Find the exact value of $\cos\left(\csc^{-1}\dfrac{7}{5}\right)$.

60. Find the inverse function f^{-1} of $f(x) = 3\sin x - 5$, $-\dfrac{\pi}{2} \le x \le \dfrac{\pi}{2}$. Find the range of f and the domain and range of f^{-1}.

Chapter Review

Things to Know

Definitions of the six inverse trigonometric functions

$y = \sin^{-1} x$ means $x = \sin y$ where $-1 \le x \le 1,\ -\frac{\pi}{2} \le y \le \frac{\pi}{2}$ (p. 616)

$y = \cos^{-1} x$ means $x = \cos y$ where $-1 \le x \le 1,\ 0 \le y \le \pi$ (p. 620)

$y = \tan^{-1} x$ means $x = \tan y$ where $-\infty < x < \infty,\ -\frac{\pi}{2} < y < \frac{\pi}{2}$ (p. 623)

$y = \sec^{-1} x$ means $x = \sec y$ where $|x| \ge 1,\ 0 \le y \le \pi,\ y \ne \frac{\pi}{2}$ (p. 630)

$y = \csc^{-1} x$ means $x = \csc y$ where $|x| \ge 1,\ -\frac{\pi}{2} \le y \le \frac{\pi}{2},\ y \ne 0$ (p. 630)

$y = \cot^{-1} x$ means $x = \cot y$ where $-\infty < x < \infty,\ 0 < y < \pi$ (p. 630)

Sum and Difference Formulas (pp. 652, 655, and 657)

$\cos(\alpha + \beta) = \cos\alpha\cos\beta - \sin\alpha\sin\beta$ $\quad$ $\cos(\alpha - \beta) = \cos\alpha\cos\beta + \sin\alpha\sin\beta$

$\sin(\alpha + \beta) = \sin\alpha\cos\beta + \cos\alpha\sin\beta$ $\quad$ $\sin(\alpha - \beta) = \sin\alpha\cos\beta - \cos\alpha\sin\beta$

$\tan(\alpha + \beta) = \dfrac{\tan\alpha + \tan\beta}{1 - \tan\alpha\tan\beta}$ $\quad$ $\tan(\alpha - \beta) = \dfrac{\tan\alpha - \tan\beta}{1 + \tan\alpha\tan\beta}$

Double-angle Formulas (pp. 665 and 666)

$\sin(2\theta) = 2\sin\theta\cos\theta$ $\quad$ $\cos(2\theta) = \cos^2\theta - \sin^2\theta$ $\quad$ $\tan(2\theta) = \dfrac{2\tan\theta}{1 - \tan^2\theta}$

$\cos(2\theta) = 2\cos^2\theta - 1$ $\quad$ $\cos(2\theta) = 1 - 2\sin^2\theta$

Half-angle Formulas (pp. 668 and 670)

$\sin^2\dfrac{\alpha}{2} = \dfrac{1 - \cos\alpha}{2}$ $\quad$ $\cos^2\dfrac{\alpha}{2} = \dfrac{1 + \cos\alpha}{2}$ $\quad$ $\tan^2\dfrac{\alpha}{2} = \dfrac{1 - \cos\alpha}{1 + \cos\alpha}$

$\sin\dfrac{\alpha}{2} = \pm\sqrt{\dfrac{1 - \cos\alpha}{2}}$ $\quad$ $\cos\dfrac{\alpha}{2} = \pm\sqrt{\dfrac{1 + \cos\alpha}{2}}$ $\quad$ $\tan\dfrac{\alpha}{2} = \pm\sqrt{\dfrac{1 - \cos\alpha}{1 + \cos\alpha}} = \dfrac{1 - \cos\alpha}{\sin\alpha} = \dfrac{\sin\alpha}{1 + \cos\alpha}$

where the $+$ or $-$ is determined by the quadrant of $\frac{\alpha}{2}$.

Product-to-Sum Formulas (p. 675)

$\sin\alpha\sin\beta = \frac{1}{2}[\cos(\alpha - \beta) - \cos(\alpha + \beta)]$

$\cos\alpha\cos\beta = \frac{1}{2}[\cos(\alpha - \beta) + \cos(\alpha + \beta)]$

$\sin\alpha\cos\beta = \frac{1}{2}[\sin(\alpha + \beta) + \sin(\alpha - \beta)]$

Sum-to-Product Formulas (p. 676)

$\sin\alpha + \sin\beta = 2\sin\dfrac{\alpha + \beta}{2}\cos\dfrac{\alpha - \beta}{2}$ $\quad$ $\sin\alpha - \sin\beta = 2\sin\dfrac{\alpha - \beta}{2}\cos\dfrac{\alpha + \beta}{2}$

$\cos\alpha + \cos\beta = 2\cos\dfrac{\alpha + \beta}{2}\cos\dfrac{\alpha - \beta}{2}$ $\quad$ $\cos\alpha - \cos\beta = -2\sin\dfrac{\alpha + \beta}{2}\sin\dfrac{\alpha - \beta}{2}$

Objectives

Section		You should be able to ...	Examples	Review Exercises
8.1	1	Find the exact value of an inverse sine function (p. 616)	1, 2, 6, 7, 9	1–6
	2	Find an approximate value of an inverse sine function (p. 617)	3	76–78
	3	Use properties of inverse functions to find exact values of certain composite functions (p. 618)	4, 5, 8	9–17
	4	Find the inverse function of a trigonometric function (p. 624)	10	24, 25
	5	Solve equations involving inverse trigonometric functions (p. 625)	11	84, 85
8.2	1	Find the exact value of expressions involving the inverse sine, cosine, and tangent functions (p. 629)	1–3	18–21, 23
	2	Define the inverse secant, cosecant, and cotangent functions (p. 630)	4	7, 8, 22

Section	You should be able to ...	Examples	Review Exercises
	3 Use a calculator to evaluate $\sec^{-1} x$, $\csc^{-1} x$, and $\cot^{-1} x$ (p. 631)	5	79, 80
	4 Write a trigonometric expression as an algebraic expression (p. 632)	6	26, 27
8.3	1 Solve equations involving a single trigonometric function (p. 634)	1–5	64–68
	2 Solve trigonometric equations using a calculator (p. 637)	6	69
	3 Solve trigonometric equations quadratic in form (p. 638)	7	72
	4 Solve trigonometric equations using fundamental identities (p. 638)	8, 9	70, 71, 73, 74
	5 Solve trigonometric equations using a graphing utility (p. 639)	10	81–83
8.4	1 Use algebra to simplify trigonometric expressions (p. 646)	1	28–44
	2 Establish identities (p. 646)	2–8	28–36
8.5	1 Use sum and difference formulas to find exact values (p. 653)	1–5	45–50, 53–57(a) – (d), 86
	2 Use sum and difference formulas to establish identities (p. 654)	6–8	37, 38, 42
	3 Use sum and difference formulas involving inverse trigonometric functions (p. 658)	9, 10	58–61
	4 Solve trigonometric equations linear in sine and cosine (p. 659)	11, 12	75
8.6	1 Use Double-angle Formulas to find exact values (p. 665)	1	53–57(e), (f), 62, 63, 87
	2 Use Double-angle Formulas to establish identities (p. 665)	2–5	40, 41
	3 Use Half-angle Formulas to find exact values (p. 668)	6, 7	51, 52, 53 – 57(g), (h), 86
8.7	1 Express products as sums (p. 675)	1	42
	2 Express sums as products (p. 676)	2	43, 44

Review Exercises

In Problems 1–8, find the exact value of each expression. Do not use a calculator.

1. $\sin^{-1} 1$ **2.** $\cos^{-1} 0$ **3.** $\tan^{-1} 1$ **4.** $\sin^{-1}\left(-\frac{1}{2}\right)$

5. $\cos^{-1}\left(-\frac{\sqrt{3}}{2}\right)$ **6.** $\tan^{-1}(-\sqrt{3})$ **7.** $\sec^{-1}\sqrt{2}$ **8.** $\cot^{-1}(-1)$

In Problems 9–23, find the exact value, if any, of each composite function. If there is no value, say it is "not defined." Do not use a calculator.

9. $\sin^{-1}\left(\sin \frac{3\pi}{8}\right)$ **10.** $\cos^{-1}\left(\cos \frac{3\pi}{4}\right)$ **11.** $\tan^{-1}\left(\tan \frac{2\pi}{3}\right)$ **12.** $\cos^{-1}\left(\cos \frac{15\pi}{7}\right)$

13. $\sin^{-1}\left[\sin\left(-\frac{8\pi}{9}\right)\right]$ **14.** $\sin(\sin^{-1} 0.9)$ **15.** $\cos(\cos^{-1} 0.6)$ **16.** $\tan[\tan^{-1} 5]$

17. $\cos[\cos^{-1}(-1.6)]$ **18.** $\sin^{-1}\left(\cos \frac{2\pi}{3}\right)$ **19.** $\cos^{-1}\left(\tan \frac{3\pi}{4}\right)$ **20.** $\tan\left[\sin^{-1}\left(-\frac{\sqrt{3}}{2}\right)\right]$

21. $\sec\left(\tan^{-1} \frac{\sqrt{3}}{3}\right)$ **22.** $\sin\left(\cot^{-1} \frac{3}{4}\right)$ **23.** $\tan\left[\sin^{-1}\left(-\frac{4}{5}\right)\right]$

In Problems 24 and 25, find the inverse function f^{-1} of each function f. Find the range of f and the domain and range of f^{-1}.

24. $f(x) = 2\sin(3x)$, $-\frac{\pi}{6} \le x \le \frac{\pi}{6}$ **25.** $f(x) = -\cos x + 3$, $0 \le x \le \pi$

In Problems 26 and 27, write each trigonometric expression as an algebraic expression in u.

26. $\cos(\sin^{-1} u)$ **27.** $\tan(\csc^{-1} u)$

In Problems 28 – 44, establish each identity.

28. $\tan\theta\cot\theta - \sin^2\theta = \cos^2\theta$ **29.** $\sin^2\theta(1 + \cot^2\theta) = 1$ **30.** $5\cos^2\theta + 3\sin^2\theta = 3 + 2\cos^2\theta$

31. $\frac{1-\cos\theta}{\sin\theta} + \frac{\sin\theta}{1-\cos\theta} = 2\csc\theta$ **32.** $\frac{\cos\theta}{\cos\theta - \sin\theta} = \frac{1}{1-\tan\theta}$ **33.** $\frac{\csc\theta}{1+\csc\theta} = \frac{1-\sin\theta}{\cos^2\theta}$

34. $\csc\theta - \sin\theta = \cos\theta\cot\theta$

35. $\dfrac{1-\sin\theta}{\sec\theta} = \dfrac{\cos^3\theta}{1+\sin\theta}$

36. $\dfrac{1-2\sin^2\theta}{\sin\theta\cos\theta} = \cot\theta - \tan\theta$

37. $\dfrac{\cos(\alpha+\beta)}{\cos\alpha\sin\beta} = \cot\beta - \tan\alpha$

38. $\dfrac{\cos(\alpha-\beta)}{\cos\alpha\cos\beta} = 1 + \tan\alpha\tan\beta$

39. $(1+\cos\theta)\tan\dfrac{\theta}{2} = \sin\theta$

40. $2\cot\theta\cot(2\theta) = \cot^2\theta - 1$

41. $1 - 8\sin^2\theta\cos^2\theta = \cos(4\theta)$

42. $\dfrac{\sin(3\theta)\cos\theta - \sin\theta\cos(3\theta)}{\sin(2\theta)} = 1$

43. $\dfrac{\sin(2\theta)+\sin(4\theta)}{\cos(2\theta)+\cos(4\theta)} = \tan(3\theta)$

44. $\dfrac{\cos(2\theta)-\cos(4\theta)}{\cos(2\theta)+\cos(4\theta)} - \tan\theta\tan(3\theta) = 0$

In Problems 45–52, find the exact value of each expression.

45. $\sin 165°$

46. $\tan 105°$

47. $\cos\dfrac{5\pi}{12}$

48. $\sin\left(-\dfrac{\pi}{12}\right)$

49. $\cos 80°\cos 20° + \sin 80°\sin 20°$

50. $\sin 70°\cos 40° - \cos 70°\sin 40°$

51. $\tan\dfrac{\pi}{8}$

52. $\sin\dfrac{5\pi}{8}$

In Problems 53–57, use the information given about the angles α and β to find the exact value of:

(a) $\sin(\alpha+\beta)$ (b) $\cos(\alpha+\beta)$ (c) $\sin(\alpha-\beta)$ (d) $\tan(\alpha+\beta)$

(e) $\sin(2\alpha)$ (f) $\cos(2\beta)$ (g) $\sin\dfrac{\beta}{2}$ (h) $\cos\dfrac{\alpha}{2}$

53. $\sin\alpha = \dfrac{4}{5}, 0 < \alpha < \dfrac{\pi}{2}; \sin\beta = \dfrac{5}{13}, \dfrac{\pi}{2} < \beta < \pi$

54. $\sin\alpha = -\dfrac{3}{5}, \pi < \alpha < \dfrac{3\pi}{2}; \cos\beta = \dfrac{12}{13}, \dfrac{3\pi}{2} < \beta < 2\pi$

55. $\tan\alpha = \dfrac{3}{4}, \pi < \alpha < \dfrac{3\pi}{2}; \tan\beta = \dfrac{12}{5}, 0 < \beta < \dfrac{\pi}{2}$

56. $\sec\alpha = 2, -\dfrac{\pi}{2} < \alpha < 0; \sec\beta = 3, \dfrac{3\pi}{2} < \beta < 2\pi$

57. $\sin\alpha = -\dfrac{2}{3}, \pi < \alpha < \dfrac{3\pi}{2}; \cos\beta = -\dfrac{2}{3}, \pi < \beta < \dfrac{3\pi}{2}$

In Problems 58–63, find the exact value of each expression.

58. $\cos\left(\sin^{-1}\dfrac{3}{5} - \cos^{-1}\dfrac{1}{2}\right)$

59. $\sin\left(\cos^{-1}\dfrac{5}{13} - \cos^{-1}\dfrac{4}{5}\right)$

60. $\tan\left[\sin^{-1}\left(-\dfrac{1}{2}\right) - \tan^{-1}\dfrac{3}{4}\right]$

61. $\cos\left[\tan^{-1}(-1) + \cos^{-1}\left(-\dfrac{4}{5}\right)\right]$

62. $\sin\left[2\cos^{-1}\left(-\dfrac{3}{5}\right)\right]$

63. $\cos\left(2\tan^{-1}\dfrac{4}{3}\right)$

In Problems 64–75, solve each equation on the interval $0 \le \theta < 2\pi$.

64. $\cos\theta = \dfrac{1}{2}$

65. $\tan\theta + \sqrt{3} = 0$

66. $\sin(2\theta) + 1 = 0$

67. $\tan(2\theta) = 0$

68. $\sec^2\theta = 4$

69. $0.2\sin\theta = 0.05$

70. $\sin\theta + \sin(2\theta) = 0$

71. $\sin(2\theta) - \cos\theta - 2\sin\theta + 1 = 0$

72. $2\sin^2\theta - 3\sin\theta + 1 = 0$

73. $4\sin^2\theta = 1 + 4\cos\theta$

74. $\sin(2\theta) = \sqrt{2}\cos\theta$

75. $\sin\theta - \cos\theta = 1$

In Problems 76 – 80, use a calculator to find an approximate value for each expression, rounded to two decimal places.

76. $\sin^{-1} 0.7$

77. $\tan^{-1}(-2)$

78. $\cos^{-1}(-0.2)$

79. $\sec^{-1} 3$

80. $\cot^{-1}(-4)$

In Problems 81– 83, use a graphing utility to solve each equation on the interval $0 \le x \le 2\pi$. Approximate any solutions rounded to two decimal places.

81. $2x = 5\cos x$

82. $2\sin x + 3\cos x = 4x$

83. $\sin x = \ln x$

In Problems 84 and 85, find the exact solution of each equation.

84. $-3\sin^{-1} x = \pi$

85. $2\cos^{-1} x + \pi = 4\cos^{-1} x$

86. Use a Half-angle Formula to find the exact value of $\sin 15°$. Then use a difference formula to find the exact value of $\sin 15°$. Show that the answers found are the same.

87. If you are given the value of $\cos\theta$ and want the exact value of $\cos(2\theta)$, what form of the Double-angle Formula for $\cos(2\theta)$ is most efficient to use?

Chapter Test

CHAPTER Test Prep VIDEOS

The Chapter Test Prep Videos are step-by-step solutions available in MyMathLab®, or on this text's YouTube Channel. Flip back to the Resources for Success page for a link to this text's YouTube channel.

In Problems 1–6, find the exact value of each expression. Express angles in radians.

1. $\sec^{-1}\left(\frac{2}{\sqrt{3}}\right)$

2. $\sin^{-1}\left(-\frac{\sqrt{2}}{2}\right)$

3. $\sin^{-1}\left(\sin\frac{11\pi}{5}\right)$

4. $\tan\left(\tan^{-1}\frac{7}{3}\right)$

5. $\cot(\csc^{-1}\sqrt{10})$

6. $\sec\left(\cos^{-1}\left(-\frac{3}{4}\right)\right)$

In Problems 7–10, use a calculator to evaluate each expression. Express angles in radians rounded to two decimal places.

7. $\sin^{-1} 0.382$

8. $\sec^{-1} 1.4$

9. $\tan^{-1} 3$

10. $\cot^{-1} 5$

In Problems 11–16 establish each identity.

11. $\dfrac{\csc\theta + \cot\theta}{\sec\theta + \tan\theta} = \dfrac{\sec\theta - \tan\theta}{\csc\theta - \cot\theta}$

12. $\sin\theta\tan\theta + \cos\theta = \sec\theta$

13. $\tan\theta + \cot\theta = 2\csc(2\theta)$

14. $\dfrac{\sin(\alpha + \beta)}{\tan\alpha + \tan\beta} = \cos\alpha\cos\beta$

15. $\sin(3\theta) = 3\sin\theta - 4\sin^3\theta$

16. $\dfrac{\tan\theta - \cot\theta}{\tan\theta + \cot\theta} = 1 - 2\cos^2\theta$

In Problems 17–24, use sum, difference, product, or Half-angle Formulas to find the exact value of each expression.

17. $\cos 15°$

18. $\tan 75°$

19. $\sin\left(\frac{1}{2}\cos^{-1}\frac{3}{5}\right)$

20. $\tan\left(2\sin^{-1}\frac{6}{11}\right)$

21. $\cos\left(\sin^{-1}\frac{2}{3} + \tan^{-1}\frac{3}{2}\right)$

22. $\sin 75°\cos 15°$

23. $\sin 75° + \sin 15°$

24. $\cos 65°\cos 20° + \sin 65°\sin 20°$

In Problems 25–29, solve each equation on $0 \le \theta < 2\pi$.

25. $4\sin^2\theta - 3 = 0$

26. $-3\cos\left(\frac{\pi}{2} - \theta\right) = \tan\theta$

27. $\cos^2\theta + 2\sin\theta\cos\theta - \sin^2\theta = 0$

28. $\sin(\theta + 1) = \cos\theta$

29. $4\sin^2\theta + 7\sin\theta = 2$

Cumulative Review

1. Find the real solutions, if any, of the equation $3x^2 + x - 1 = 0$.

2. Find an equation for the line containing the points $(-2, 5)$ and $(4, -1)$. What is the distance between these points? What is their midpoint?

3. Test the equation $3x + y^2 = 9$ for symmetry with respect to the x-axis, y-axis, and origin. List the intercepts.

4. Use transformations to graph the equation $y = |x - 3| + 2$.

5. Use transformations to graph the equation $y = 3e^x - 2$.

6. Use transformations to graph the equation

$$y = \cos\left(x - \frac{\pi}{2}\right) - 1.$$

7. Sketch a graph of each of the following functions. Label at least three points on each graph. Name the inverse function of each and show its graph.

(a) $y = x^3$

(b) $y = e^x$

(c) $y = \sin x,\ -\frac{\pi}{2} \le x \le \frac{\pi}{2}$

(d) $y = \cos x,\quad 0 \le x \le \pi$

8. If $\sin\theta = -\frac{1}{3}$ and $\pi < \theta < \frac{3\pi}{2}$, find the exact value of:

(a) $\cos\theta$

(b) $\tan\theta$

(c) $\sin(2\theta)$

(d) $\cos(2\theta)$

(e) $\sin\left(\frac{1}{2}\theta\right)$

(f) $\cos\left(\frac{1}{2}\theta\right)$

9. Find the exact value of $\cos(\tan^{-1} 2)$.

10. If $\sin\alpha = \frac{1}{3}, \frac{\pi}{2} < \alpha < \pi$, and $\cos\beta = -\frac{1}{3}, \pi < \beta < \frac{3\pi}{2}$, find the exact value of:

(a) $\cos\alpha$ (b) $\sin\beta$ (c) $\cos(2\alpha)$

(d) $\cos(\alpha + \beta)$ (e) $\sin\frac{\beta}{2}$

11. For the function

$$f(x) = 2x^5 - x^4 - 4x^3 + 2x^2 + 2x - 1:$$

(a) Find the real zeros and their multiplicity.
(b) Find the intercepts.
(c) Find the power function that the graph of f resembles for large $|x|$.
(d) Graph f using a graphing utility.
(e) Approximate the turning points, if any exist.
(f) Use the information obtained in parts (a)–(e) to sketch a graph of f by hand.
(g) Identify the intervals on which f is increasing, decreasing, or constant.

12. If $f(x) = 2x^2 + 3x + 1$ and $g(x) = x^2 + 3x + 2$, solve:

(a) $f(x) = 0$ (b) $f(x) = g(x)$

(c) $f(x) > 0$ (d) $f(x) \geq g(x)$

Chapter Projects

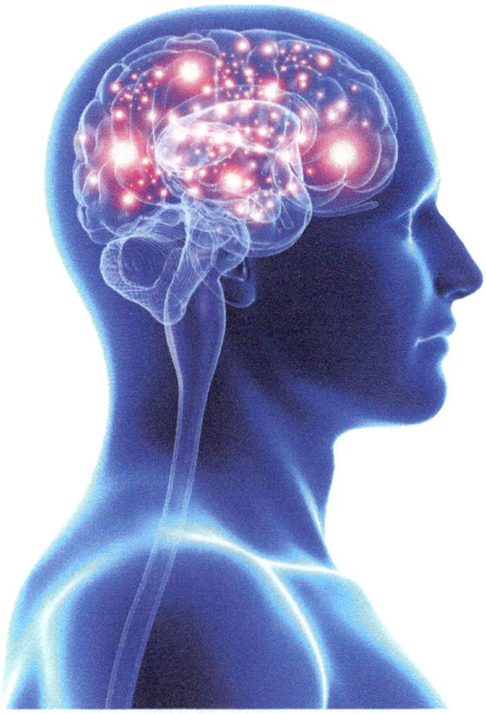

I. Mapping Your Mind The goal of this project is to organize in your mind the material you have learned recently. To do this, we will use mind-mapping software called Mindomo. Mindomo is free software that enables you to organize your thoughts digitally and share these thoughts with anyone on the Web. By organizing your thoughts, you can see the big picture and then communicate this big picture to others. You are also able to see how various concepts are related to each other.

1. Go to *http://www.mindomo.com* and register. Learn how to use Mindomo. A video on using Mindomo can be found at *http://www.screencast.com/t/ZPwJQDs4*
2. Use an Internet search engine to research Mind Mapping. Write a few paragraphs that explain the history and benefit of mind mapping.
3. Create a MindMap that explains the following:
   (a) The six trigonometric functions and their properties (including the inverses of these functions)
   (b) The fundamental trigonometric identities
   When creating your map, be creative! Perhaps you can share ideas about when a particular identity might be used, or when a particular identity cannot be used.
4. Share the MindMap so that students in your class can view it.

The following projects are available on the Instructor's Resource Center (IRC):

II. Waves Wave motion is described by a sinusoidal equation. The Principle of Superposition of two waves is discussed.

III. Project at Motorola ***Sending Pictures Wirelessly*** The electronic transmission of pictures is made practical by image compression, mathematical methods that greatly reduce the number of bits of data used to compose the picture.

IV. Calculus of Differences Finding consecutive difference quotients is called finding finite differences and is used to analyze the graph of an unknown function.

9 Applications of Trigonometric Functions

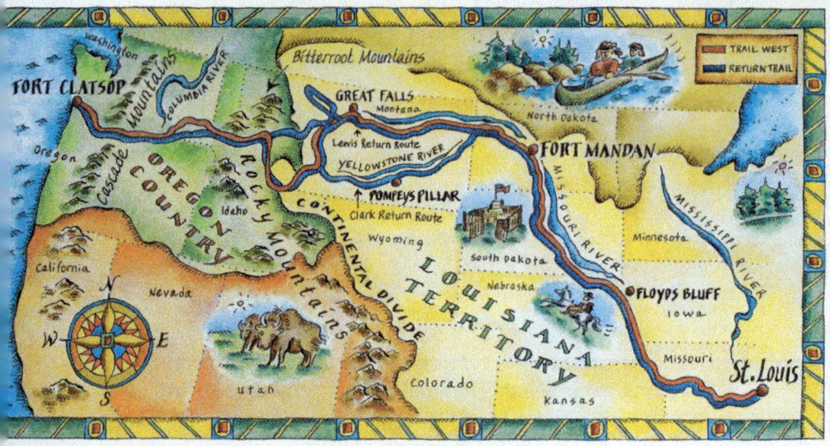

The Lewis and Clark Expedition

In today's world of GPS and smart phone apps that can precisely track one's whereabouts, it is difficult to fathom the magnitude of the challenge that confronted Meriwether Lewis and William Clark in 1804.

But Lewis and Clark managed. Commissioned by President Thomas Jefferson to explore the newly purchased Louisiana Territory, the co-captains led their expedition—the Corps of Discovery—on a journey that took nearly two and a half years and carried them more than 7000 miles. Starting at St. Louis, Missouri, they traveled up the Missouri River, across the Great Plains, over the Rocky Mountains, down the Columbia River to the Pacific Ocean, and then back. Along the way, using limited tools such as a compass and octant, they created more than 130 maps of the area with remarkable detail and accuracy.

—See Chapter Project II—

Outline

••• A Look Back

In Chapter 7, we defined the six trigonometric functions using right triangles and then extended this definition to include any angle. In particular, we learned to evaluate the trigonometric functions. We also learned how to graph sinusoidal functions. In Chapter 8, we defined the inverse trigonometric functions and solved equations involving the trigonometric functions.

A Look Ahead •••

In this chapter, we use the trigonometric functions to solve applied problems. The first four sections deal with applications involving right triangles and *oblique triangles*, triangles that do not have a right angle. To solve problems involving oblique triangles, we will develop the Law of Sines and the Law of Cosines. We will also develop formulas for finding the area of a triangle.

The final section deals with applications of sinusoidal functions involving simple harmonic motion and damped motion.

9.1 Applications Involving Right Triangles

PREPARING FOR THIS SECTION *Before getting started, review the following:*

- Pythagorean Theorem (Chapter R, Review, Section R.3, p. 31)
- Trigonometric Equations (Section 8.3, pp. 634–639)
- Complementary Angles (Section 7.2, pp. 533–534)

Now Work the 'Are You Prepared?' problems on page 688.

OBJECTIVES 1 Solve Right Triangles (p. 685)
2 Solve Applied Problems (p. 686)

1 Solve Right Triangles

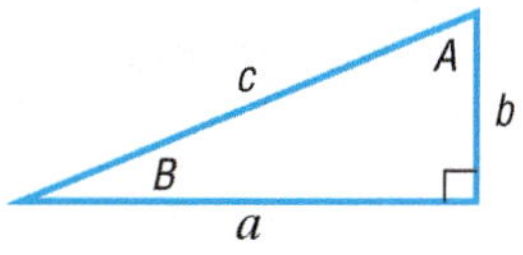

Figure 1 Right triangle

In the discussion that follows, a right triangle is always labeled so that side a is opposite angle A, side b is opposite angle B, and side c is the hypotenuse, as shown in Figure 1. **To solve a right triangle** means to find the missing lengths of its sides and the measurements of its angles. We shall follow the practice of expressing the lengths of the sides rounded to two decimal places and expressing angles in degrees rounded to one decimal place. (Be sure that your calculator is in degree mode.)

To solve a right triangle we need to know one of the acute angles A or B and a side, or else two sides, (in which case the Pythagorean Theorem can be used). Also, because the sum of the angles of a triangle is 180°, the sum of the angles A and B in a right triangle must be 90°.

THEOREM

For the right triangle shown in Figure 1, we have

$$c^2 = a^2 + b^2 \qquad A + B = 90°$$

EXAMPLE 1 **Solving a Right Triangle**

Use Figure 2. If $b = 2$ and $A = 40°$, find a, c, and B.

Solution Because $A = 40°$ and $A + B = 90°$, it follows that $B = 50°$. To find the sides a and c, use the facts that

$$\tan 40° = \frac{a}{2} \quad \text{and} \quad \cos 40° = \frac{2}{c}$$

Now solve for a and c.

$$a = 2 \tan 40° \approx 1.68 \quad \text{and} \quad c = \frac{2}{\cos 40°} \approx 2.61$$

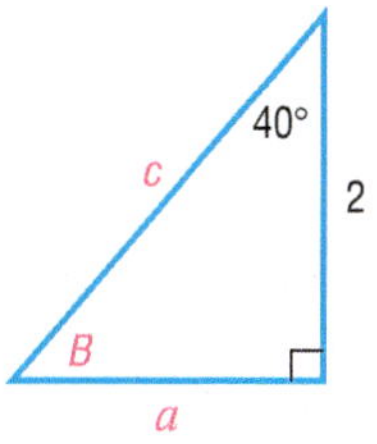

Figure 2

Now Work PROBLEM 9

EXAMPLE 2 **Solving a Right Triangle**

Use Figure 3. If $a = 3$ and $b = 2$, find c, A, and B.

Solution Since $a = 3$ and $b = 2$, then, by the Pythagorean Theorem,

$$c^2 = a^2 + b^2 = 3^2 + 2^2 = 9 + 4 = 13$$
$$c = \sqrt{13} \approx 3.61$$

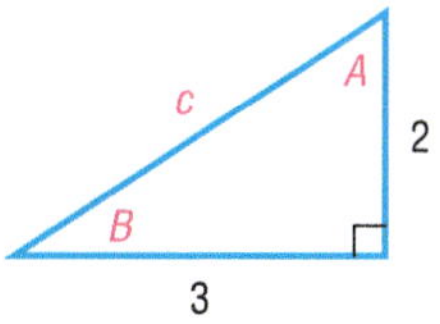

Figure 3

Note: To avoid round-off errors when using a calculator, store unrounded values in memory for use in subsequent calculations. ■

To find angle A, use the fact that

$$\tan A = \frac{3}{2} \quad \text{so} \quad A = \tan^{-1}\frac{3}{2}$$

Use a calculator with the mode set to degrees to find that $A = 56.3°$ rounded to one decimal place. Because $A + B = 90°$, this means that $B = 33.7°$. ■

Now Work PROBLEM 19

2 Solve Applied Problems

In Section 7.3, we used right triangle trigonometry to find the lengths of unknown sides of a right triangle, given the measure of an angle and the length of a side. Drawing on our understanding of the concept of inverse trigonometric functions and how to solve trigonometric equations, we can solve applied problems that require finding the measure of an angle given the lengths of two sides in a right triangle.

EXAMPLE 3 Finding the Inclination of a Mountain Trail

A straight trail leads from the Alpine Hotel, elevation 8000 feet, to a scenic overlook, elevation 11,100 feet. The length of the trail is 14,100 feet. What is the inclination (grade) of the trail? That is, what is the angle B in Figure 4?

Solution Figure 4 shows that the length of the side opposite angle B is $11{,}100 - 8000 = 3100$ feet, and the length of the hypotenuse is 14,100 feet. So, angle B obeys the equation

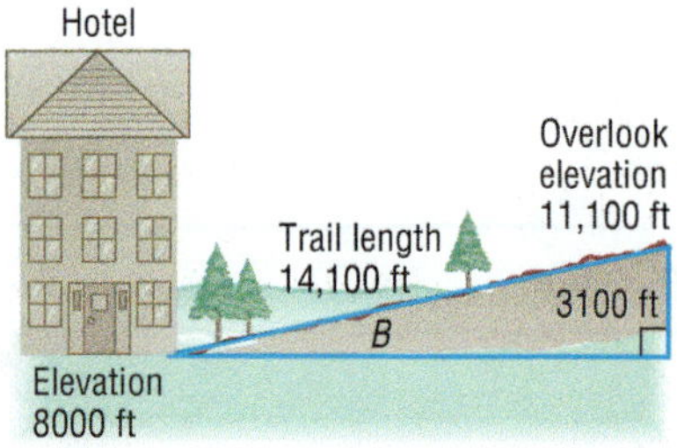

Figure 4

$$\sin B = \frac{3100}{14{,}100}$$

Using a calculator* reveals that

$$B = \sin^{-1}\frac{3100}{14{,}100} \approx 12.7°$$

The inclination (grade) of the trail is approximately 12.7°. ■

Now Work PROBLEM 27

EXAMPLE 4 The Gibb's Hill Lighthouse, Southampton, Bermuda

In operation since 1846, the Gibb's Hill Lighthouse stands 117 feet high on a hill 245 feet high, so its beam of light is 362 feet above sea level. A brochure states that the light can be seen on the horizon about 26 miles distant. Verify the accuracy of this statement.

Solution Figure 5 illustrates the situation. The central angle θ, positioned at the center of Earth, radius 3960 miles, obeys the equation

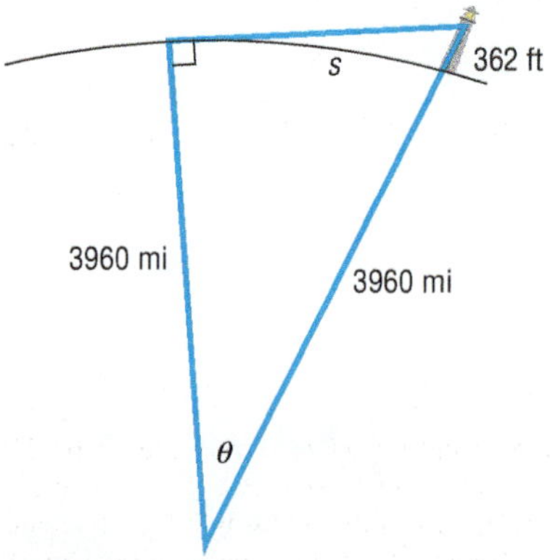

Figure 5

$$\cos\theta = \frac{3960}{3960 + \dfrac{362}{5280}} \approx 0.999982687 \qquad \textbf{1 mile = 5280 feet}$$

Solving for θ yields

$$\theta = \cos^{-1} 0.999982687 \approx 0.33715° \approx 20.23'$$

The brochure does not indicate whether the distance is measured in nautical miles or statute miles. Let's calculate both distances.

The distance d in nautical miles (refer to Problem 122, p. 525) is the measure of the angle θ in minutes, so $d \approx 20.23$ nautical miles.

*Because the angles are acute, the $\sin^{-1}$ key will return the correct value.

The distance s in statute miles is given by the formula $s = r\theta$, where θ is measured in radians. Then, because

$$\theta \approx 20.23' \approx 0.33715° \approx 0.00588 \text{ radian}$$

$$1' = \left(\frac{1}{60}\right)^{\circ} \qquad 1° = \frac{\pi}{180} \text{ radian}$$

this means that

$$s = r\theta \approx (3960)(0.00588) \approx 23.3 \text{ miles}$$

In either case, it would seem that the brochure overstated the distance somewhat.

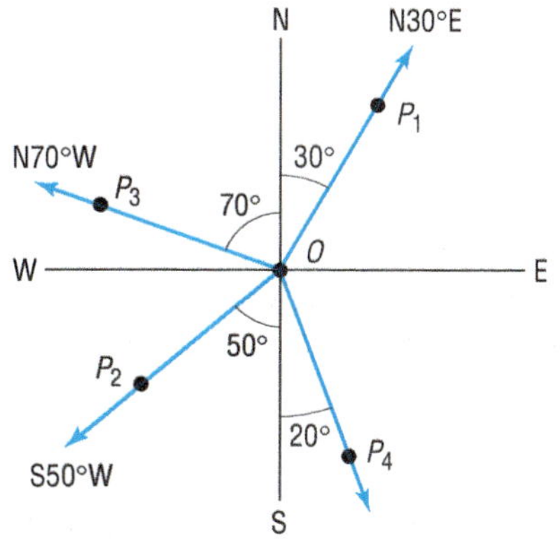

Figure 6

In navigation and surveying, the **direction** or **bearing** from a point O to a point P equals the acute angle θ between the ray OP and the vertical line through O, the north–south line.

Figure 6 illustrates some bearings. Notice that the bearing from O to P_1 is denoted N30°E, indicating that the bearing is 30° east of north. In writing the bearing from O to P, the direction north or south always appears first, followed by an acute angle, followed by east or west. In Figure 6, the bearing from O to P_2 is S50°W, and from O to P_3 it is N70°W.

EXAMPLE 5 Finding the Bearing of an Object

In Figure 6, what is the bearing from O to an object at P_4?

Solution The acute angle between the ray OP_4 and the north–south line through O is 20°. The bearing from O to P_4 is S20°E.

EXAMPLE 6 Finding the Bearing of an Airplane

A Boeing 777 aircraft takes off from Nashville International Airport on runway 2 LEFT, which has a bearing of N20°E.* After flying for 1 mile, the pilot of the aircraft requests permission to turn 90° and head toward the northwest. The request is granted. After the plane goes 2 miles in this direction, what bearing should the control tower use to locate the aircraft?

Solution Figure 7 illustrates the situation. After flying 1 mile from the airport O (the control tower), the aircraft is at P. After turning 90° toward the northwest and flying 2 miles, the aircraft is at the point Q. In triangle OPQ, the angle θ obeys the equation

$$\tan\theta = \frac{2}{1} = 2 \quad \text{so} \quad \theta = \tan^{-1} 2 \approx 63.4°$$

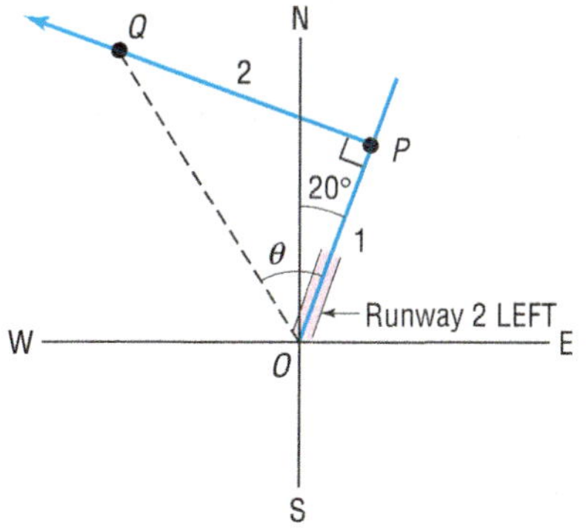

Figure 7

The acute angle between north and the ray OQ is $63.4° - 20° = 43.4°$. The bearing of the aircraft from O to Q is N43.4°W.

Now Work PROBLEM 33

*In air navigation, the term **azimuth** denotes the positive angle measured clockwise from the north (N) to a ray OP. In Figure 6, the azimuth from O to P_1 is 30°; the azimuth from O to P_2 is 230°; the azimuth from O to P_3 is 290°. In naming runways, the units digit is left off the azimuth. Runway 2 LEFT means the left runway with a direction of azimuth 20° (bearing N20° E). Runway 23 is the runway with azimuth 230° and bearing S50° W.

9.1 Assess Your Understanding

'Are You Prepared?'

Answers are given at the end of these exercises. If you get a wrong answer, read the pages listed in red.

1. In a right triangle, if the length of the hypotenuse is 5 and the length of one of the other sides is 3, what is the length of the third side? (p. 31)

2. ***True or False*** $\sin 52° = \cos 48°$. (pp. 533–534)

3. If θ is an acute angle, solve the equation $\tan\theta = \frac{1}{2}$. Express your answer in degrees, rounded to one decimal place. (pp. 634–639)

4. If θ is an acute angle, solve the equation $\sin\theta = \frac{1}{2}$. (pp. 634–639)

Concepts and Vocabulary

5. The sum of the measures of the two acute angles in a right triangle is ________.

(a) 45° (b) 90° (c) 180° (d) 360°

6. In navigation or surveying, the ________ or ________ from a point O to a point P equals the acute angle θ between ray OP and the vertical line through O, the north–south line.

7. ***True or False*** In a right triangle, if two sides are known, we can solve the triangle.

8. ***True or False*** In a right triangle, if we know the two acute angles, we can solve the triangle.

Skill Building

In Problems 9–22, use the right triangle shown below. Then, using the given information, solve the triangle.

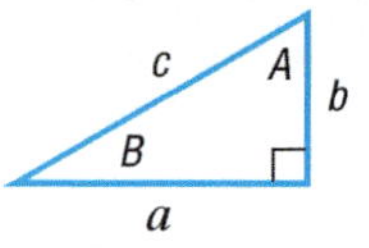

9. $b = 5$, $B = 20°$; find $a, c,$ and A

10. $b = 4$, $B = 10°$; find $a, c,$ and A

11. $a = 6$, $B = 40°$; find $b, c,$ and A

12. $a = 7$, $B = 50°$; find $b, c,$ and A

13. $b = 4$, $A = 10°$; find $a, c,$ and B

14. $b = 6$, $A = 20°$; find $a, c,$ and B

15. $a = 5$, $A = 25°$; find $b, c,$ and B

16. $a = 6$, $A = 40°$; find $b, c,$ and B

17. $c = 9$, $B = 20°$; find $b, a,$ and A

18. $c = 10$, $A = 40°$; find $b, a,$ and B

19. $a = 5$, $b = 3$; find $c, A,$ and B

20. $a = 2$, $b = 8$; find $c, A,$ and B

21. $a = 2$, $c = 5$; find $b, A,$ and B

22. $b = 4$, $c = 6$; find $a, A,$ and B

Applications and Extensions

23. Geometry The hypotenuse of a right triangle is 5 inches. If one leg is 2 inches, find the degree measure of each angle.

24. Geometry The hypotenuse of a right triangle is 3 feet. If one leg is 1 foot, find the degree measure of each angle.

25. Geometry A right triangle has a hypotenuse of length 8 inches. If one angle is 35°, find the length of each leg.

26. Geometry A right triangle has a hypotenuse of length 10 centimeters. If one angle is 40°, find the length of each leg.

27. Finding the Angle of Elevation of the Sun At 10 AM on April 26, 2015, a building 300 feet high cast a shadow 50 feet long. What was the angle of elevation of the Sun?

28. Directing a Laser Beam A laser beam is to be directed through a small hole in the center of a circle of radius 10 feet. The origin of the beam is 35 feet from the circle (see the figure). At what angle of elevation should the beam be aimed to ensure that it goes through the hole?

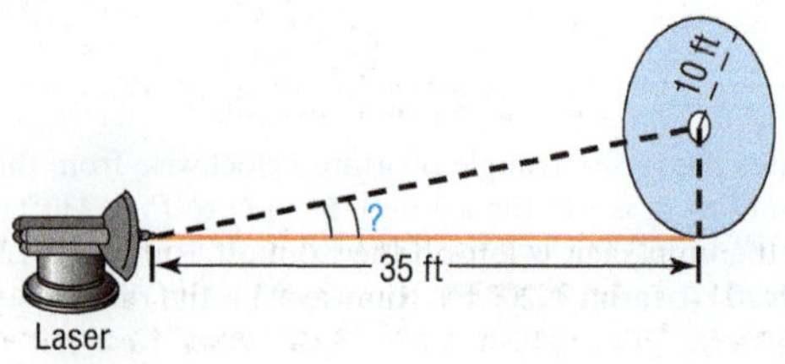

29. Finding the Speed of a Truck A state trooper is hidden 30 feet from a highway. One second after a truck passes, the angle θ between the highway and the line of observation from the patrol car to the truck is measured. See the illustration.

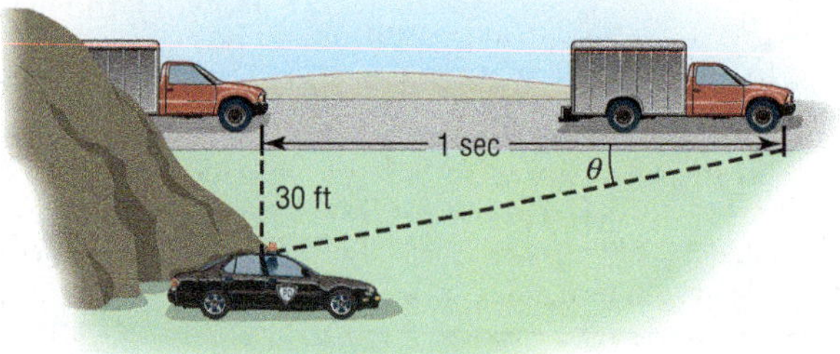

(a) If the angle measures 15°, how fast is the truck traveling? Express the answer in feet per second and in miles per hour.

(b) If the angle measures 20°, how fast is the truck traveling? Express the answer in feet per second and in miles per hour.

(c) If the speed limit is 55 miles per hour and a speeding ticket is issued for speeds of 5 miles per hour or more over the limit, for what angles should the trooper issue a ticket?

30. Security A security camera in a neighborhood bank is mounted on a wall 9 feet above the floor. What angle of depression should be used if the camera is to be directed to a spot 6 feet above the floor and 12 feet from the wall?

31. Parallax One method of measuring the distance from Earth to a star is the parallax method. The idea behind computing this distance is to measure the angle formed between Earth and the star at two different points in time. Typically, the measurements are taken so that the side opposite the angle is as large as possible. Therefore, the optimal approach is to measure the angle when Earth is on opposite sides of the Sun, as shown in the figure.

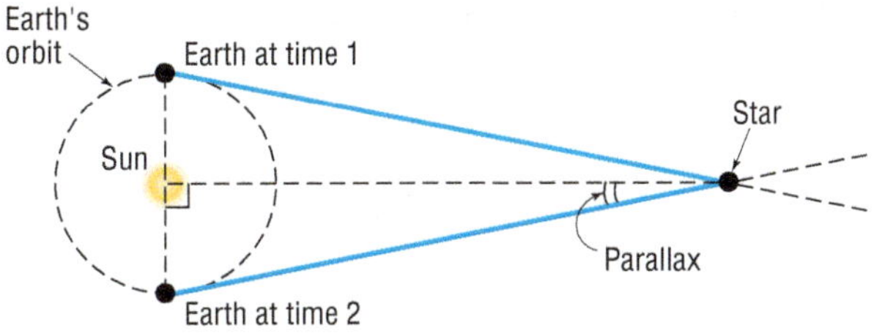

(a) Proxima Centauri is 4.22 light-years from Earth. If 1 light-year is about 5.9 trillion miles, how many miles is Proxima Centauri from Earth?

(b) The mean distance from Earth to the Sun is 93,000,000 miles. What is the parallax of Proxima Centauri?

32. Parallax See Problem 31. The star 61 Cygni, sometimes called Bessel's Star (after Friedrich Bessel, who measured the distance from Earth to the star in 1838), is a star in the constellation Cygnus.

(a) If 61 Cygni is 11.14 light-years from Earth and 1 light-year is about 5.9 trillion miles, how many miles is 61 Cygni from Earth?

(b) The mean distance from Earth to the Sun is 93,000,000 miles. What is the parallax of 61 Cygni?

33. Finding the Bearing of an Aircraft A DC-9 aircraft leaves Midway Airport from runway 4 RIGHT, whose bearing is N40°E. After flying for $\frac{1}{2}$ mile, the pilot requests permission to turn 90° and head toward the southeast. The permission is granted. After the airplane goes 1 mile in this direction, what bearing should the control tower use to locate the aircraft?

34. Finding the Bearing of a Ship A ship leaves the port of Miami with a bearing of S80°E and a speed of 15 knots. After 1 hour, the ship turns 90° toward the south. After 2 hours, maintaining the same speed, what is the bearing to the ship from the port?

35. Niagara Falls Incline Railway Situated between Portage Road and the Niagara Parkway directly across from the Canadian Horseshoe Falls, the Falls Incline Railway is a funicular that carries passengers up an embankment to Table Rock Observation Point. If the length of the track is 51.8 meters and the angle of inclination is 36°2′, determine the height of the embankment.

Source: *www.niagaraparks.com*

36. Willis Tower The Willis Tower in Chicago is the 2nd tallest building in the United States and is topped by a high antenna. A surveyor on the ground makes the following measurements:

- The angle of elevation from her position to the top of the building is 34°.
- The distance from her position to the top of the building is 2595 feet.
- The distance from her position to the top of the antenna is 2760 feet.

(a) How far away from the base of the building is the surveyor located?

(b) How tall is the building?

(c) What is the angle of elevation from the surveyor to the top of the antenna?

(d) How tall is the antenna?

Source: *Council on Tall Buildings and Urban Habitat*

37. Chicago Skyscrapers The angle of inclination from the base of the John Hancock Center to the top of the main structure of the Willis Tower is approximately 10.3°. If the main structure of the Willis Tower is 1451 feet tall, how far apart are the two skyscrapers? Assume the bases of the two buildings are at the same elevation.

Source: *www.emporis.com*

38. Estimating the Width of the Mississippi River A tourist at the top of the Gateway Arch (height 630 feet) in St. Louis, Missouri, observes a boat moored on the Illinois side of the Mississippi River 2070 feet directly across from the Arch. She also observes a boat moored on the Missouri side directly across from the first boat (see the figure). Given that $B = \cot^{-1}\frac{67}{55}$, estimate the width of the Mississippi River at the St. Louis riverfront.

Source: *U.S. Army Corps of Engineers*

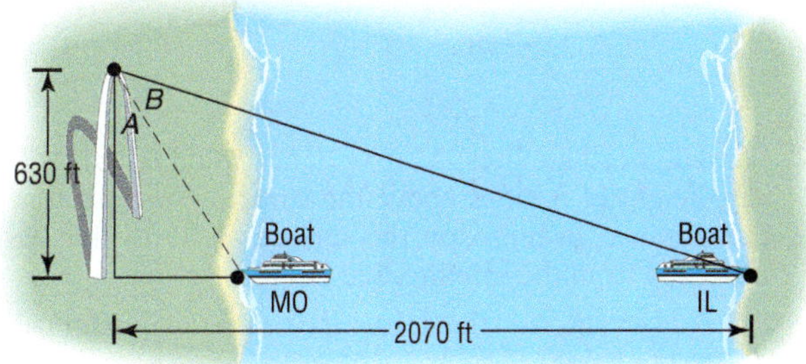

39. Finding the Pitch of a Roof A carpenter is preparing to put a roof on a garage that is 20 feet by 40 feet by 20 feet. A steel support beam 46 feet in length is positioned in the center of the garage. To support the roof, another beam will be attached to the top of the center beam (see the figure). At what angle of elevation is the new beam? In other words, what is the pitch of the roof?

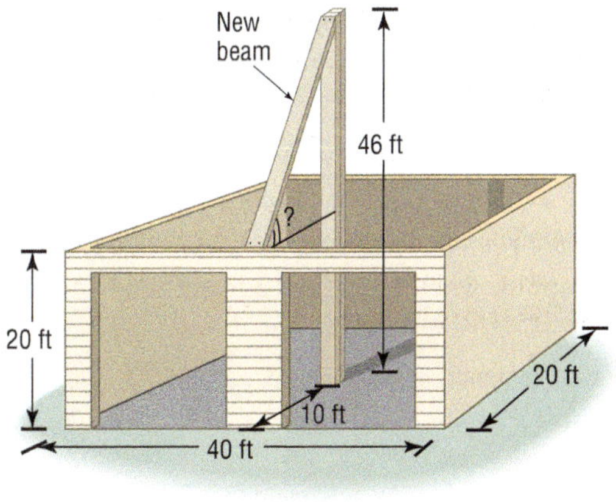

40. Shooting Free Throws in Basketball The eyes of a basketball player are 6 feet above the floor. The player is at the free-throw line, which is 15 feet from the center of the

basket rim (see the figure). What is the angle of elevation from the player's eyes to the center of the rim?
[**Hint:** The rim is 10 feet above the floor.]

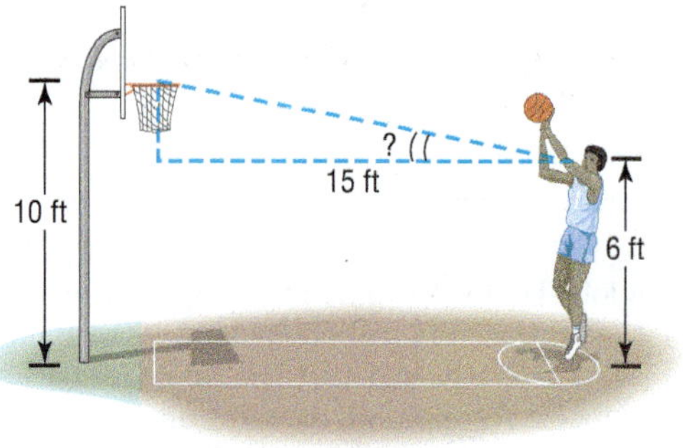

41. Geometry Find the value of the angle θ (see the figure) in degrees rounded to the nearest tenth of a degree.

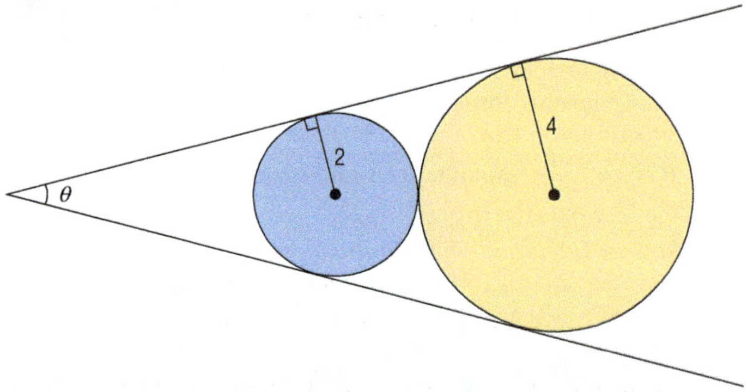

42. Surveillance Satellites A surveillance satellite circles Earth at a height of h miles above the surface. Suppose that d is the distance, in miles, on the surface of Earth that can be observed from the satellite. See the illustration.

(a) Find an equation that relates the central angle θ (in radians) to the height h.
(b) Find an equation that relates the observable distance d and θ.
(c) Find an equation that relates d and h.
(d) If d is to be 2500 miles, how high must the satellite orbit above Earth?
(e) If the satellite orbits at a height of 300 miles, what distance d on the surface can be observed?

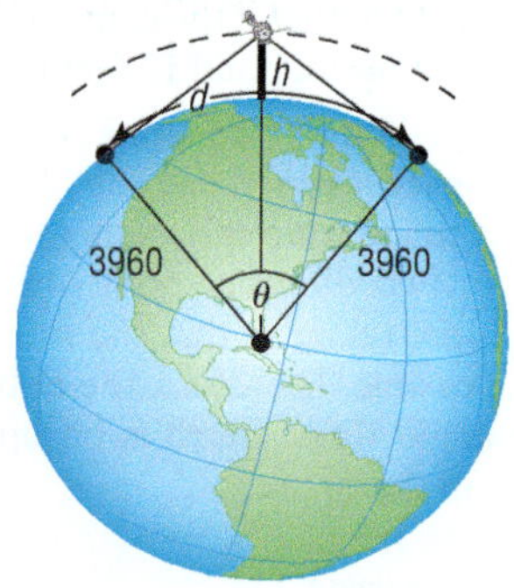

43. Drive Wheel of an Engine The drive wheel of an engine is 13 inches in diameter, and the pulley on the rotary pump is 5 inches in diameter. If the shafts of the drive wheel and the pulley are 2 feet apart, what length of belt is required to join them as shown in the figure?

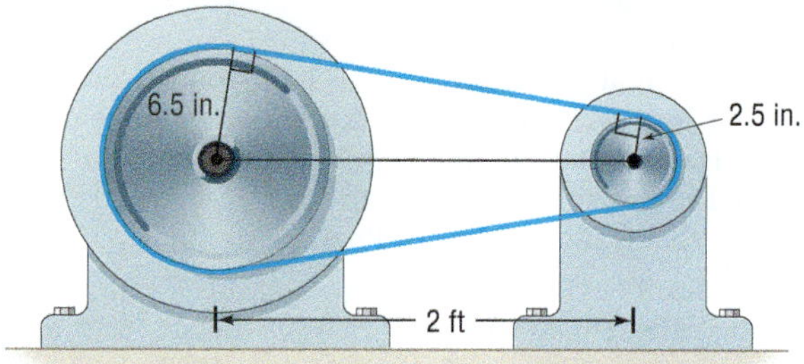

44. Drive Wheel of an Engine Rework Problem 43 if the belt is crossed, as shown in the figure.

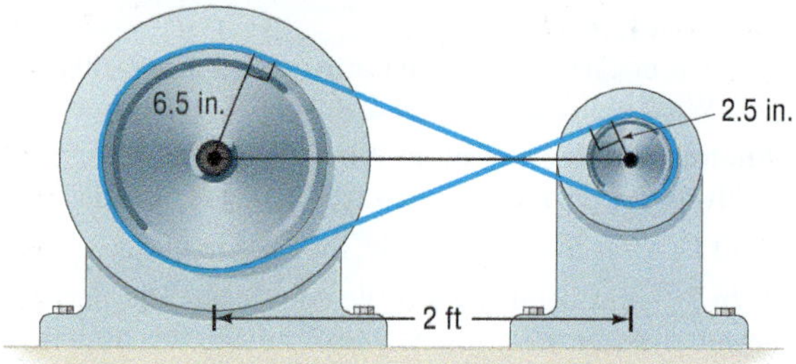

Explaining Concepts: Discussion and Writing

45. The Gibb's Hill Lighthouse, Southampton, Bermuda In operation since 1846, the Gibb's Hill Lighthouse stands 117 feet high on a hill 245 feet high, so its beam of light is 362 feet above sea level. A brochure states that ships 40 miles away can see the light and planes flying at 10,000 feet can see it 120 miles away. Verify the accuracy of these statements. What assumption did the brochure make about the height of the ship?

Retain Your Knowledge

Problems 46–49 are based on material learned earlier in the course. The purpose of these problems is to keep the material fresh in your mind so that you are better prepared for the final exam.

46. Determine whether $x - 3$ is a factor of $x^4 + 2x^3 - 21x^2 + 19x - 3$.

47. Find the exact value of $\sin 15°$.
Hint: $15° = 45° - 30°$

48. Evaluate $\dfrac{f(x) - f(4)}{x - 4}$, where $f(x) = \sqrt{x}$ for $x = 5, 4.5,$ and 4.1. Round results to three decimal places.

49. Solve $2\sin^2\theta - \sin\theta - 1 = 0$ for $0 \le \theta < 2\pi$.

'Are You Prepared?' Answers

1. 4 **2.** False **3.** 26.6° **4.** 30°

9.2 The Law of Sines

PREPARING FOR THIS SECTION *Before getting started, review the following:*

- Trigonometric Equations (Section 8.3, pp. 634–639)
- Difference Formula for the Sine Function (Section 8.5, p. 655)
- Geometry Essentials (Chapter R, Review, Section R.3, pp. 31–36)

Now Work the **'Are You Prepared?'** problems on page 697.

OBJECTIVES
1 Solve SAA or ASA Triangles (p. 692)
2 Solve SSA Triangles (p. 693)
3 Solve Applied Problems (p. 695)

If none of the angles of a triangle is a right angle, the triangle is called **oblique**. An oblique triangle will have either three acute angles or two acute angles and one obtuse angle (an angle between 90° and 180°). See Figure 8.

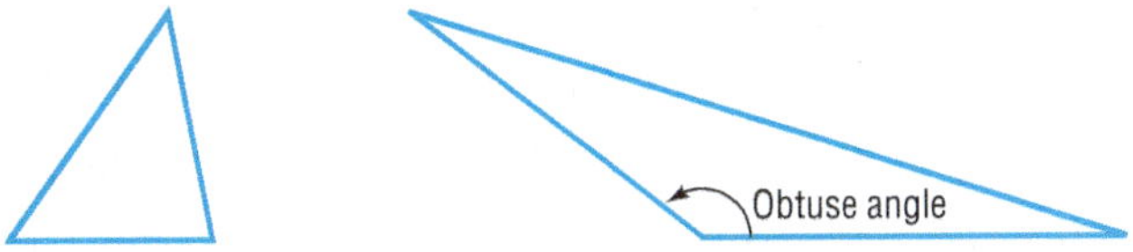

Figure 8 **(a)** All angles are acute **(b)** Two acute angles and one obtuse angle

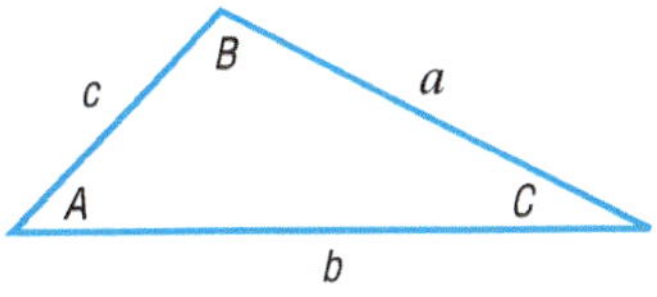

Figure 9 Oblique triangle

In the discussion that follows, an oblique triangle is always labeled so that side a is opposite angle A, side b is opposite angle B, and side c is opposite angle C, as shown in Figure 9.

To **solve an oblique triangle** means to find the lengths of its sides and the measurements of its angles. To do this, we need to know the length of one side,* along with (i) two angles, (ii) one angle and one other side, or (iii) the other two sides. There are four possibilities to consider.

CASE 1: One side and two angles are known (ASA or SAA).
CASE 2: Two sides and the angle opposite one of them are known (SSA).
CASE 3: Two sides and the included angle are known (SAS).
CASE 4: Three sides are known (SSS).

Figure 10 illustrates the four cases, where the known measurements are shown in blue.

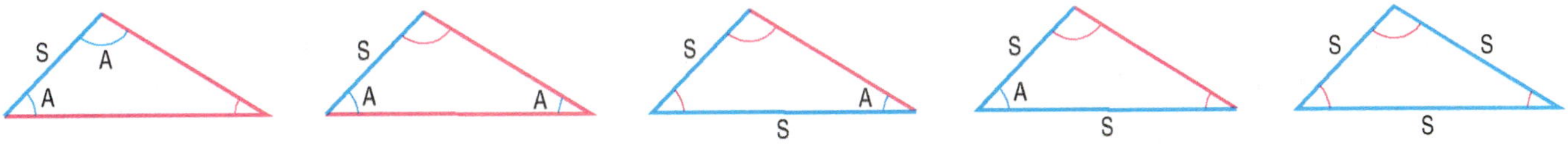

Case 1: ASA Case 1: SAA Case 2: SSA Case 3: SAS Case 4: SSS

Figure 10

The **Law of Sines** is used to solve triangles for which Case 1 or 2 holds. Cases 3 and 4 are considered when we study the Law of Cosines in the next section.

THEOREM

WARNING Oblique triangles cannot be solved using the methods of Section 9.1. Do you know why? ■

Law of Sines

For a triangle with sides a, b, c and opposite angles A, B, C, respectively,

$$\frac{\sin A}{a} = \frac{\sin B}{b} = \frac{\sin C}{c} \qquad \textbf{(1)}$$

■

*The length of one side must be known, because knowing only the angles will reveal only a family of *similar triangles*.

A proof of the Law of Sines is given at the end of this section. The Law of Sines actually consists of three equalities:

$$\frac{\sin A}{a} = \frac{\sin B}{b} \qquad \frac{\sin A}{a} = \frac{\sin C}{c} \qquad \frac{\sin B}{b} = \frac{\sin C}{c}$$

Formula (1) is a compact way to write these three equations.

Typically, applying the Law of Sines to solve triangles uses the fact that the sum of the angles of any triangle equals 180°; that is,

$$A + B + C = 180° \quad \textbf{(2)}$$

1 Solve SAA or ASA Triangles

The first two examples show how to solve a triangle when one side and two angles are known (Case 1: SAA or ASA).

EXAMPLE 1 Using the Law of Sines to Solve an SAA Triangle

Solve the triangle: $A = 40°, B = 60°, a = 4$

Solution Figure 11 shows the triangle to be solved. The third angle C is found using equation (2).

$$A + B + C = 180°$$
$$40° + 60° + C = 180°$$
$$C = 80°$$

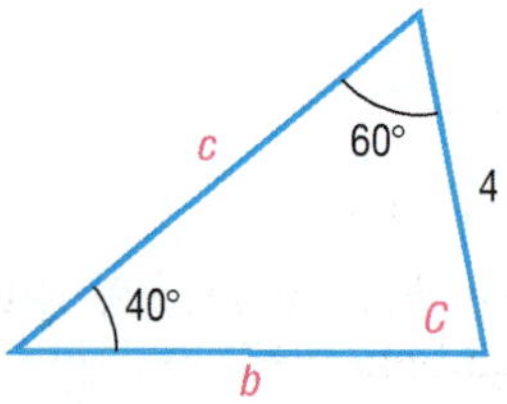

Figure 11

Now use the Law of Sines (twice) to find the unknown sides b and c.

$$\frac{\sin A}{a} = \frac{\sin B}{b} \qquad \frac{\sin A}{a} = \frac{\sin C}{c}$$

Because $a = 4, A = 40°, B = 60°$, and $C = 80°$, we have

$$\frac{\sin 40°}{4} = \frac{\sin 60°}{b} \qquad \frac{\sin 40°}{4} = \frac{\sin 80°}{c}$$

Solving for b and c yields

$$b = \frac{4 \sin 60°}{\sin 40°} \approx 5.39 \qquad c = \frac{4 \sin 80°}{\sin 40°} \approx 6.13$$

Note: Although it is not a check, the reasonableness of answers can be verified by determining whether the longest side is opposite the largest angle and the shortest side is opposite the smallest angle.

Notice in Example 1 that b and c are found by working with the given side a. This is better than finding b first and working with a rounded value of b to find c.

Now Work PROBLEM 9

EXAMPLE 2 Using the Law of Sines to Solve an ASA Triangle

Solve the triangle: $A = 35°, B = 15°, c = 5$

Solution Figure 12 illustrates the triangle to be solved. Two angles are known ($A = 35°$ and $B = 15°$). Find the third angle using equation (2):

$$A + B + C = 180°$$
$$35° + 15° + C = 180°$$
$$C = 130°$$

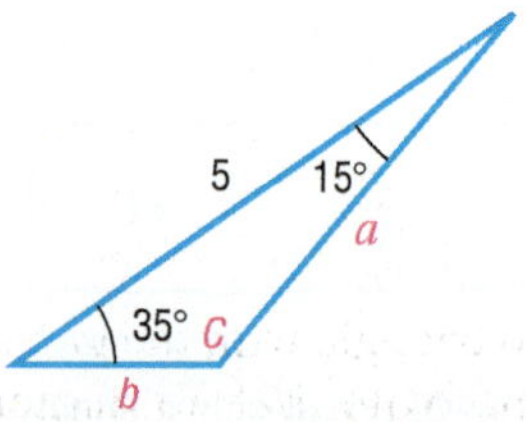

Figure 12

Now the three angles and one side ($c = 5$) of the triangle are known. To find the remaining two sides a and b, use the Law of Sines (twice).

$$\frac{\sin A}{a} = \frac{\sin C}{c} \qquad \frac{\sin B}{b} = \frac{\sin C}{c}$$

$$\frac{\sin 35°}{a} = \frac{\sin 130°}{5} \qquad \frac{\sin 15°}{b} = \frac{\sin 130°}{5}$$

$$a = \frac{5 \sin 35°}{\sin 130°} \approx 3.74 \qquad b = \frac{5 \sin 15°}{\sin 130°} \approx 1.69$$

Now Work PROBLEM 23

2 Solve SSA Triangles

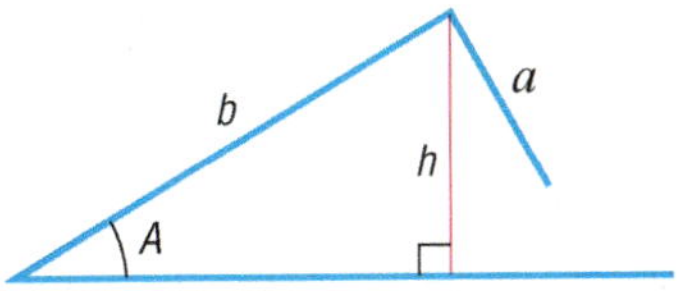

Figure 13 $\sin A = \frac{h}{b}$

Case 2 (SSA), which applies to triangles for which two sides and the angle opposite one of them are known, is referred to as the **ambiguous case**, because the known information may result in one triangle, two triangles, or no triangle at all. Suppose that sides a and b and angle A are given, as illustrated in Figure 13. The key to determining how many triangles, if any, may be formed from the given information lies primarily with the relative size of side a, the height h, and the fact that $h = b \sin A$.

No Triangle If $a < h = b \sin A$, then side a is not sufficiently long to form a triangle. See Figure 14.

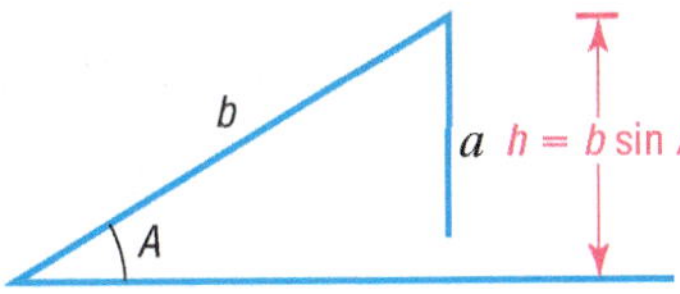

Figure 14 $a < h = b \sin A$

One Right Triangle If $a = h = b \sin A$, then side a is just long enough to form a right triangle. See Figure 15.

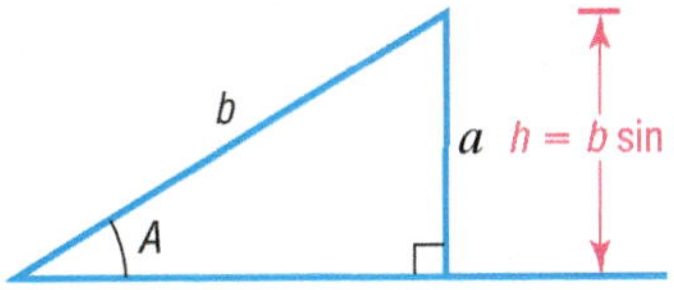

Figure 15 $a = h = b \sin A$

Two Triangles If $h = b \sin A < a$ and $a < b$, then two distinct triangles can be formed from the given information. See Figure 16.

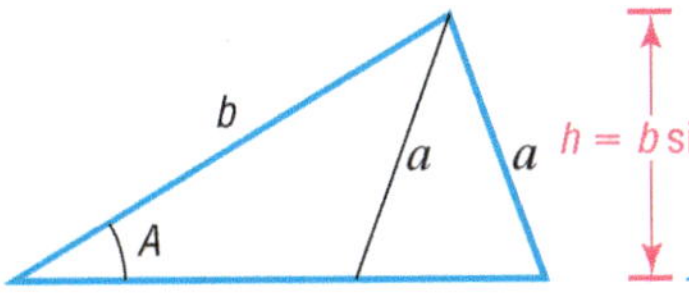

Figure 16 $b \sin A < a$ and $a < b$

One Triangle If $a \geq b$, only one triangle can be formed. See Figure 17.

Figure 17 $a \geq b$

Fortunately, it is not necessary to rely on an illustration or on complicated relationships to draw the correct conclusion in the ambiguous case. The Law of Sines will lead us to the correct determination. Let's see how.

EXAMPLE 3 Using the Law of Sines to Solve an SSA Triangle (No Solution)

Solve the triangle: $a = 2, c = 1, C = 50°$

Solution Because $a = 2$, $c = 1$, and $C = 50°$ are known, use the Law of Sines to find the angle A.

$$\frac{\sin A}{a} = \frac{\sin C}{c}$$

$$\frac{\sin A}{2} = \frac{\sin 50°}{1}$$

$$\sin A = 2 \sin 50° \approx 1.53$$

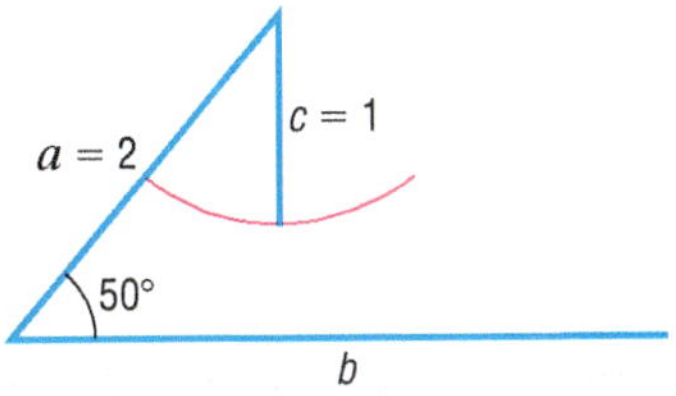

Figure 18

Since there is no angle A for which $\sin A > 1$, there can be no triangle with the given measurements. Figure 18 illustrates the measurements given. Note that no matter how side c is positioned, it will never touch side b to form a triangle.

EXAMPLE 4 Using the Law of Sines to Solve an SSA Triangle (One Solution)

Solve the triangle: $a = 3, b = 2, A = 40°$

Solution See Figure 19(a). Because $a = 3$, $b = 2$, and $A = 40°$ are known, use the Law of Sines to find the angle B.

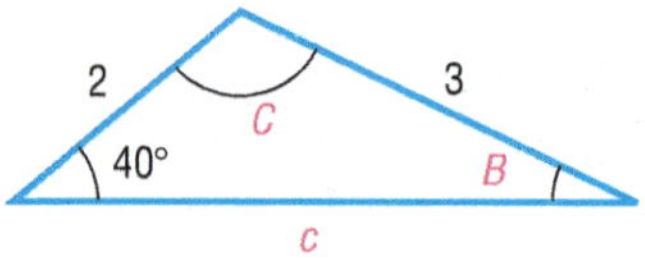

Figure 19(a)

$$\frac{\sin A}{a} = \frac{\sin B}{b}$$

Then

$$\frac{\sin 40°}{3} = \frac{\sin B}{2}$$

$$\sin B = \frac{2 \sin 40°}{3} \approx 0.43$$

Note: The angle B_1 was determined by finding the value of $\sin^{-1}\left(\frac{2 \sin 40°}{3}\right)$. Using the rounded value and evaluating $\sin^{-1}(0.43)$ will yield a slightly different result. ■

There are two angles B, $0° < B < 180°$, for which $\sin B \approx 0.43$.

$$B_1 \approx 25.4° \quad \text{and} \quad B_2 \approx 180° - 25.4° = 154.6°$$

The second possibility, $B_2 \approx 154.6°$, is ruled out, because $A = 40°$ makes $A + B_2 \approx 194.6° > 180°$. Now use $B_1 \approx 25.4°$ to find that

$$C = 180° - A - B_1 \approx 180° - 40° - 25.4° = 114.6°$$

The third side c may now be determined using the Law of Sines.

$$\frac{\sin A}{a} = \frac{\sin C}{c}$$

$$\frac{\sin 40°}{3} = \frac{\sin 114.6°}{c}$$

$$c = \frac{3 \sin 114.6°}{\sin 40°} \approx 4.24$$

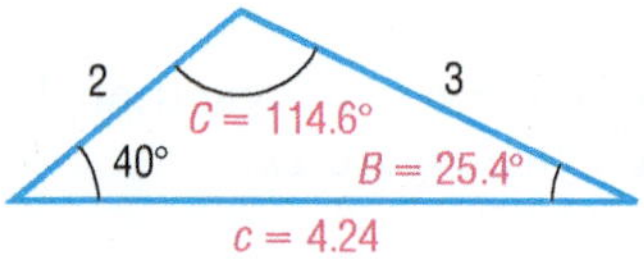

Figure 19(b)

Figure 19(b) illustrates the solved triangle. ■

EXAMPLE 5 Using the Law of Sines to Solve an SSA Triangle (Two Solutions)

Solve the triangle: $a = 6, b = 8, A = 35°$

Solution Because $a = 6$, $b = 8$, and $A = 35°$ are known, use the Law of Sines to find the angle B.

$$\frac{\sin A}{a} = \frac{\sin B}{b}$$

Then

$$\frac{\sin 35°}{6} = \frac{\sin B}{8}$$

$$\sin B = \frac{8 \sin 35°}{6} \approx 0.76$$

$$B_1 \approx 49.9° \quad \text{or} \quad B_2 \approx 180° - 49.9° = 130.1°$$

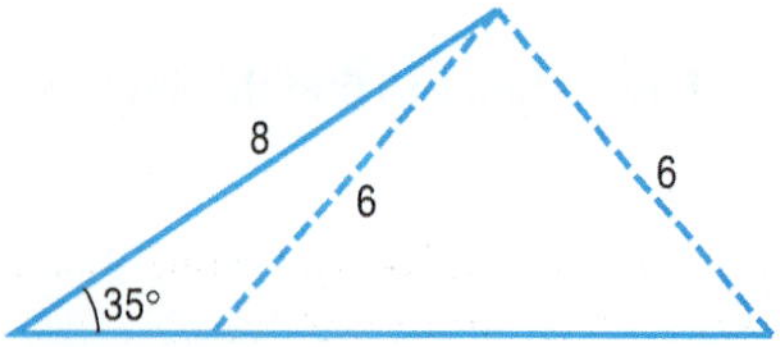

Figure 20(a)

Both choices of B result in $A + B < 180°$. There are two triangles, one containing the angle $B_1 \approx 49.9°$ and the other containing the angle $B_2 \approx 130.1°$. See Figure 20(a).

The third angle C is either

$$C_1 = 180° - A - B_1 \approx 95.1° \quad \text{or} \quad C_2 = 180° - A - B_2 \approx 14.9°$$

($A = 35°$, $B_1 = 49.9°$) ($A = 35°$, $B_2 = 130.1°$)

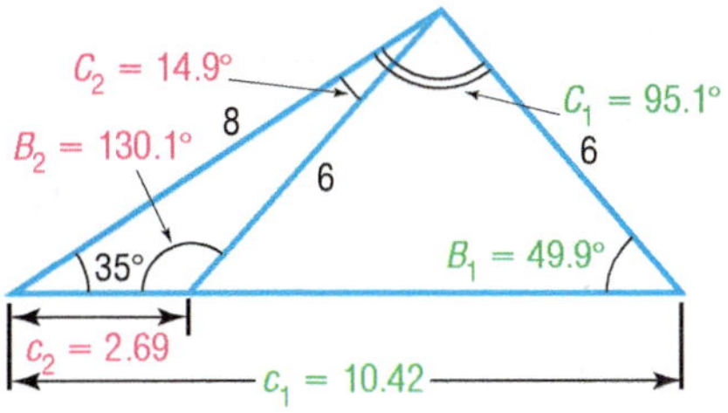

Figure 20(b)

The third side c obeys the Law of Sines, so

$$\frac{\sin A}{a} = \frac{\sin C_1}{c_1} \qquad \frac{\sin A}{a} = \frac{\sin C_2}{c_2}$$

$$\frac{\sin 35°}{6} = \frac{\sin 95.1°}{c_1} \qquad \frac{\sin 35°}{6} = \frac{\sin 14.9°}{c_2}$$

$$c_1 = \frac{6 \sin 95.1°}{\sin 35°} \approx 10.42 \qquad c_2 = \frac{6 \sin 14.9°}{\sin 35°} \approx 2.69$$

The two solved triangles are illustrated in Figure 20(b).

Now Work PROBLEMS 25, 29, AND 31

3 Solve Applied Problems

EXAMPLE 6 Finding the Height of a Mountain

To measure the height of a mountain, a surveyor takes two sightings of the peak at a distance 900 meters apart on a direct line to the mountain.* See Figure 21(a). The first observation results in an angle of elevation of 47°, and the second results in an angle of elevation of 35°. If the transit is 2 meters high, what is the height h of the mountain?

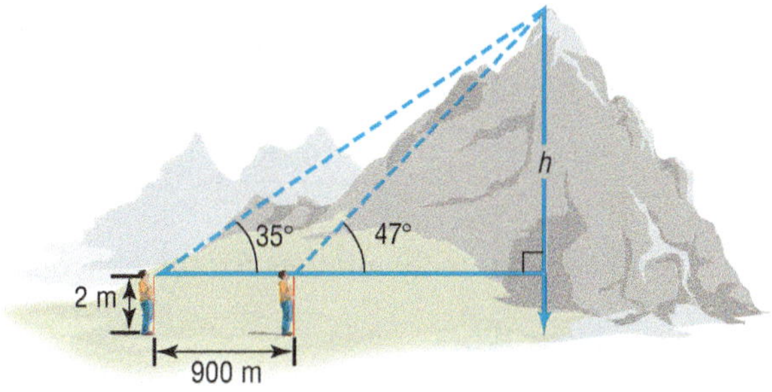

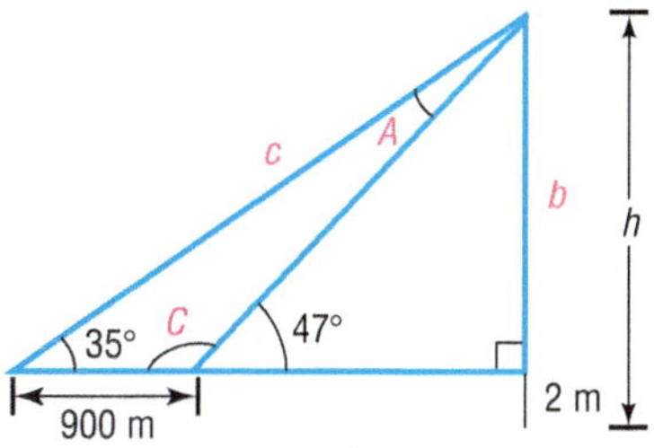

Figure 21 (a) (b)

Solution Figure 21(b) shows the triangles that replicate the illustration in Figure 21(a). Since $C + 47° = 180°$, this means that $C = 133°$. Also, since $A + C + 35° = 180°$, this means that $A = 180° - 35° - C = 145° - 133° = 12°$. Use the Law of Sines to find c.

$$\frac{\sin A}{a} = \frac{\sin C}{c} \qquad A = 12°, C = 133°, a = 900$$

$$c = \frac{900 \sin 133°}{\sin 12°} \approx 3165.86$$

Using the larger right triangle gives

$$\sin 35° = \frac{b}{c}$$

$$b = 3165.86 \sin 35° \approx 1815.86 \approx 1816 \text{ meters}$$

The height of the peak from ground level is approximately $1816 + 2 = 1818$ meters.

Now Work PROBLEM 37

EXAMPLE 7 Rescue at Sea

Coast Guard Station Zulu is located 120 miles due west of Station X-ray. A ship at sea sends an SOS call that is received by each station. The call to Station Zulu indicates that the bearing of the ship from Zulu is N40°E (40° east of north). The call to Station X-ray indicates that the bearing of the ship from X-ray is N30°W (30° west of north).

(a) How far is each station from the ship?

(b) If a helicopter capable of flying 200 miles per hour is dispatched from the nearest station to the ship, how long will it take to reach the ship?

*For simplicity, assume that these sightings are at the same level.

Solution

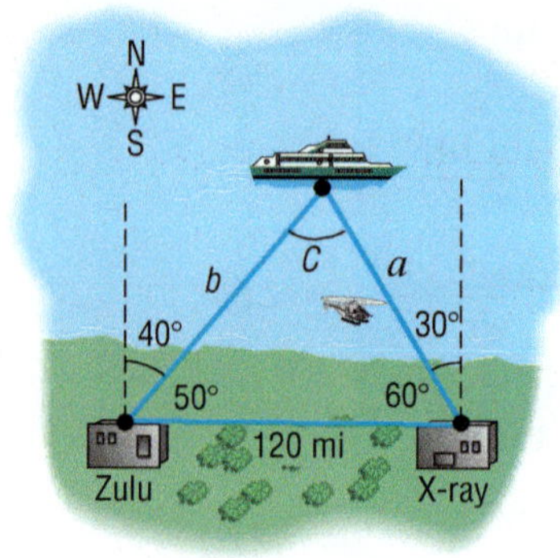

Figure 22

(a) Figure 22 illustrates the situation. The angle C is found to be

$$C = 180° - 50° - 60° = 70°$$

The Law of Sines can now be used to find the two distances a and b that are needed.

$$\frac{\sin 50°}{a} = \frac{\sin 70°}{120}$$

$$a = \frac{120 \sin 50°}{\sin 70°} \approx 97.82 \text{ miles}$$

$$\frac{\sin 60°}{b} = \frac{\sin 70°}{120}$$

$$b = \frac{120 \sin 60°}{\sin 70°} \approx 110.59 \text{ miles}$$

Station Zulu is about 111 miles from the ship, and Station X-ray is about 98 miles from the ship.

(b) The time t needed for the helicopter to reach the ship from Station X-ray is found by using the formula

$$(\text{Rate}, r)\,(\text{Time}, t) = \text{Distance}, a$$

Then

$$t = \frac{a}{r} = \frac{97.82}{200} \approx 0.49 \text{ hour} \approx 29 \text{ minutes}$$

It will take about 29 minutes for the helicopter to reach the ship. ■

Now Work PROBLEM 47

Proof of the Law of Sines To prove the Law of Sines, construct an altitude of length h from one of the vertices of a triangle. Figure 23(a) shows h for a triangle with three acute angles, and Figure 23(b) shows h for a triangle with an obtuse angle. In each case, the altitude is drawn from the vertex at B. Using either illustration

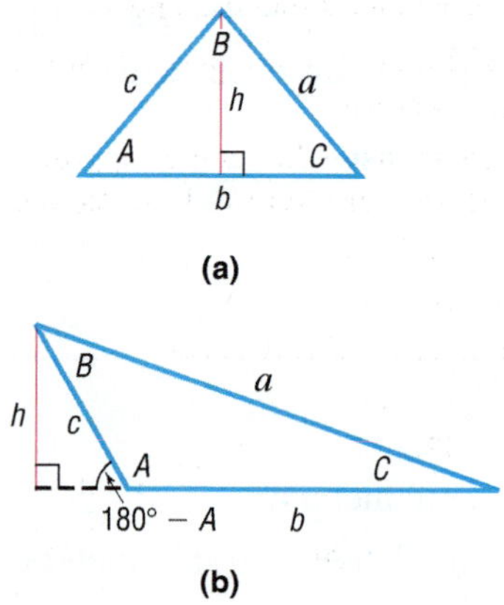

Figure 23

$$\sin C = \frac{h}{a}$$

from which

$$h = a \sin C \tag{3}$$

From Figure 23(a), it also follows that

$$\sin A = \frac{h}{c}$$

from which

$$h = c \sin A \tag{4}$$

From Figure 23(b), it follows that

$$\sin(180° - A) = \sin A = \frac{h}{c}$$

↑ $\sin(180° - A) = \sin 180° \cos A - \cos 180° \sin A = \sin A$

which again gives

$$h = c \sin A$$

Thus, whether the triangle has three acute angles or has two acute angles and one obtuse angle, equations (3) and (4) hold. As a result, the expressions for h in equations (3) and (4) are equal. That is,

$$a \sin C = c \sin A$$

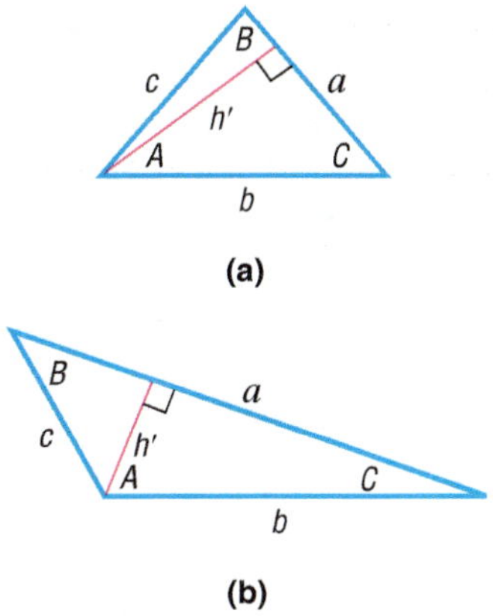

Figure 24

from which

$$\frac{\sin A}{a} = \frac{\sin C}{c} \tag{5}$$

In a similar manner, constructing the altitude h' from the vertex of angle A, as shown in Figure 24, reveals that

$$\sin B = \frac{h'}{c} \quad \text{and} \quad \sin C = \frac{h'}{b}$$

Equating the expressions for h' gives

$$h' = c \sin B = b \sin C$$

from which

$$\frac{\sin B}{b} = \frac{\sin C}{c} \tag{6}$$

When equations (5) and (6) are combined, the result is equation (1), the Law of Sines. ■

9.2 Assess Your Understanding

'Are You Prepared?' *Answers are given at the end of these exercises. If you get a wrong answer, read the pages listed in red.*

1. The difference formula for the sine function is $\sin(A - B) =$ _______. (p. 655)

2. If θ is an acute angle, solve the equation $\cos\theta = \dfrac{\sqrt{3}}{2}$. (pp. 634–639)

3. The two triangles shown are similar. Find the missing length. (pp. 31–36)

Concepts and Vocabulary

4. If none of the angles of a triangle is a right angle, the triangle is called _______.

(a) oblique (b) isosceles (c) right (d) scalene

5. For a triangle with sides a, b, c and opposite angles A, B, C, the Law of Sines states that _________________.

6. ***True or False*** An oblique triangle in which two sides and an angle are given always results in at least one triangle.

7. ***True or False*** The Law of Sines can be used to solve triangles where three sides are known.

8. Triangles for which two sides and the angle opposite one of them are known (SSA) are referred to as the _________ ____.

Skill Building

In Problems 9–16, solve each triangle.

9.

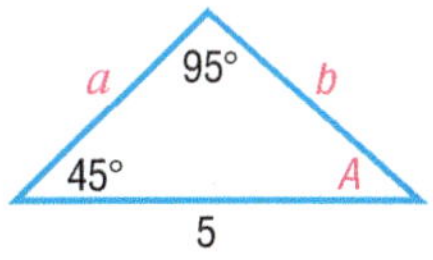

10.

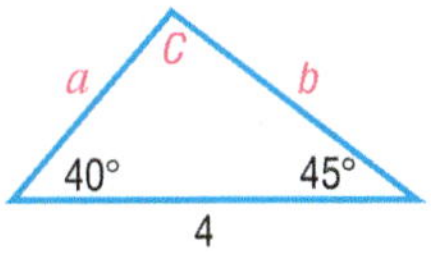

11.

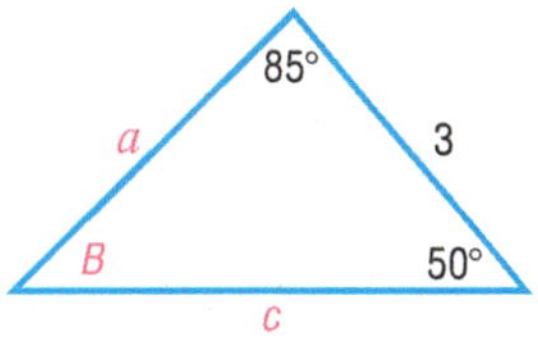

12.

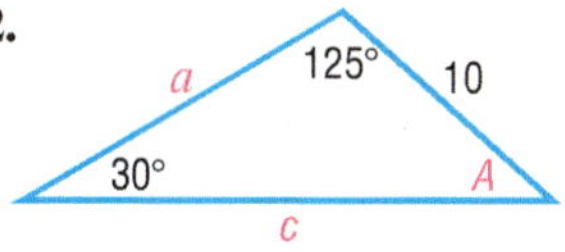

13.

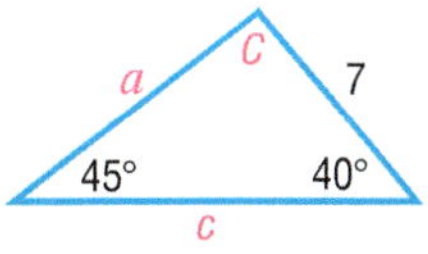

14.

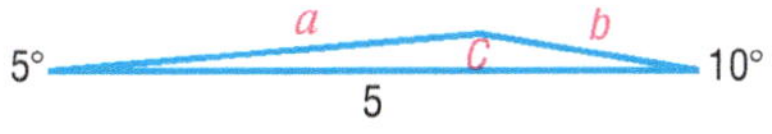

15.

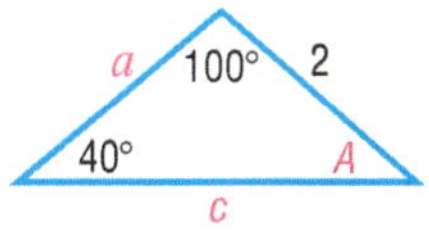

16.

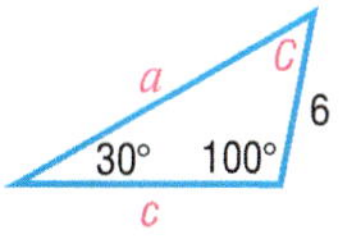

In Problems 17–24, solve each triangle.

17. $A = 40°$, $B = 20°$, $a = 2$

18. $A = 50°$, $C = 20°$, $a = 3$

19. $B = 70°$, $C = 10°$, $b = 5$

20. $A = 70°$, $B = 60°$, $c = 4$

21. $A = 110°$, $C = 30°$, $c = 3$

22. $B = 10°$, $C = 100°$, $b = 2$

23. $A = 40°$, $B = 40°$, $c = 2$

24. $B = 20°$, $C = 70°$, $a = 1$

In Problems 25–36, two sides and an angle are given. Determine whether the given information results in one triangle, two triangles, or no triangle at all. Solve any resulting triangle(s).

25. $a = 3, \quad b = 2, \quad A = 50°$

26. $b = 4, \quad c = 3, \quad B = 40°$

27. $b = 5, \quad c = 3, \quad B = 100°$

28. $a = 2, \quad c = 1, \quad A = 120°$

29. $a = 4, \quad b = 5, \quad A = 60°$

30. $b = 2, \quad c = 3, \quad B = 40°$

31. $b = 4, \quad c = 6, \quad B = 20°$

32. $a = 3, \quad b = 7, \quad A = 70°$

33. $a = 2, \quad c = 1, \quad C = 100°$

34. $b = 4, \quad c = 5, \quad B = 95°$

35. $a = 2, \quad c = 1, \quad C = 25°$

36. $b = 4, \quad c = 5, \quad B = 40°$

Applications and Extensions

37. Finding the Length of a Ski Lift Consult the figure. To find the length of the span of a proposed ski lift from P to Q, a surveyor measures $\angle DPQ$ to be 25° and then walks back a distance of 1000 feet to R and measures $\angle PRQ$ to be 15°. What is the distance from P to Q?

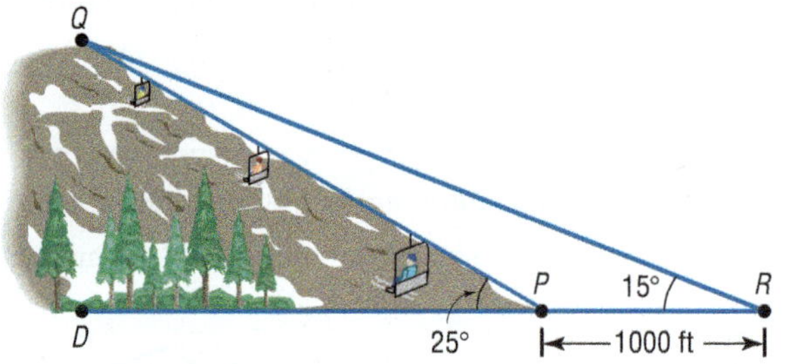

38. Finding the Height of a Mountain Use the illustration in Problem 37 to find the height QD of the mountain.

39. Finding the Height of an Airplane An aircraft is spotted by two observers who are 1000 feet apart. As the airplane passes over the line joining them, each observer takes a sighting of the angle of elevation to the plane, as indicated in the figure. How high is the airplane?

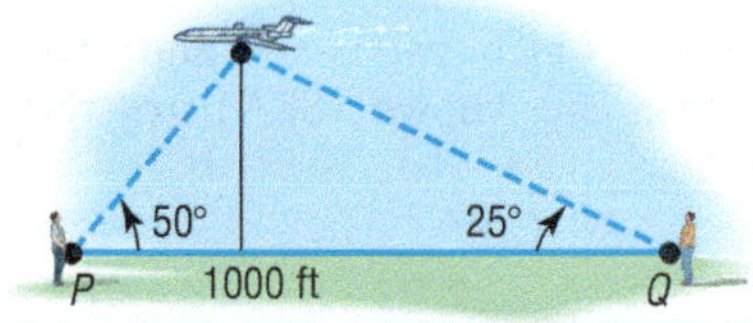

40. Finding the Height of the Bridge over the Royal Gorge The highest bridge in the world is the bridge over the Royal Gorge of the Arkansas River in Colorado. Sightings to the same point at water level directly under the bridge are taken from each side of the 880-foot-long bridge, as indicated in the figure. How high is the bridge?

Source: *Guinness Book of World Records*

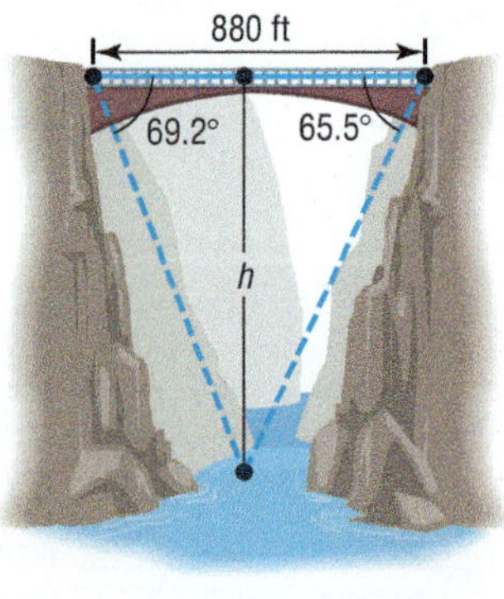

41. Land Dimensions A triangular plot of land has one side along a straight road measuring 200 feet. A second side makes a 50° angle with the road, and the third side makes a 43° angle with the road. How long are the other two sides?

42. Distance between Runners Two runners in a marathon determine that the angles of elevation of a news helicopter covering the race are 38° and 45°. If the helicopter is 1700 feet directly above the finish line, how far apart are the runners?

43. Landscaping Pat needs to determine the height of a tree before cutting it down to be sure that it will not fall on a nearby fence. The angle of elevation of the tree from one position on a flat path from the tree is 30°, and from a second position 40 feet farther along this path it is 20°. What is the height of the tree?

44. Construction A loading ramp 10 feet long that makes an angle of 18° with the horizontal is to be replaced by one that makes an angle of 12° with the horizontal. How long is the new ramp?

45. Commercial Navigation Adam must fly home to St. Louis from a business meeting in Oklahoma City. One flight option flies directly to St. Louis, a distance of about 461.1 miles. A second flight option flies first to Kansas City and then connects to St. Louis. The bearing from Oklahoma City to Kansas City is N29.6°E, and the bearing from Oklahoma City to St. Louis is N57.7°E. The bearing from St. Louis to Oklahoma City is S57.7°W, and the bearing from St. Louis to Kansas City is N79.4°W. How many more frequent flyer miles will Adam receive if he takes the connecting flight rather than the direct flight?

Source: *www.landings.com*

46. Time Lost to a Navigation Error In attempting to fly from city P to city Q, an aircraft followed a course that was 10° in error, as indicated in the figure. After flying a distance of 50 miles, the pilot corrected the course by turning at point R and flying 300 miles farther. If the constant speed of the aircraft was 250 miles per hour, how much time was lost due to the error?

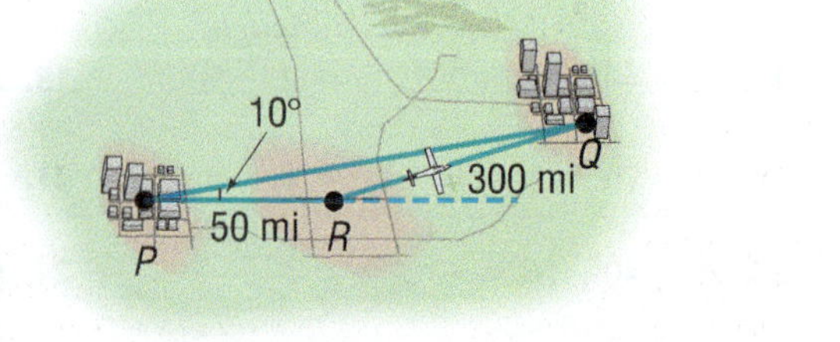

47. **Rescue at Sea** Coast Guard Station Able is located 150 miles due south of Station Baker. A ship at sea sends an SOS call that is received by each station. The call to Station Able indicates that the ship is located N55°E; the call to Station Baker indicates that the ship is located S60°E.
   (a) How far is each station from the ship?
   (b) If a helicopter capable of flying 200 miles per hour is dispatched from the station nearest the ship, how long will it take to reach the ship?

48. **Distance to the Moon** At exactly the same time, Tom and Alice measured the angle of elevation to the moon while standing exactly 300 km apart. The angle of elevation to the moon for Tom was 49.8974°, and the angle of elevation to the moon for Alice was 49.9312°. See the figure. To the nearest 1000 km, how far was the moon from Earth when the measurement was obtained?

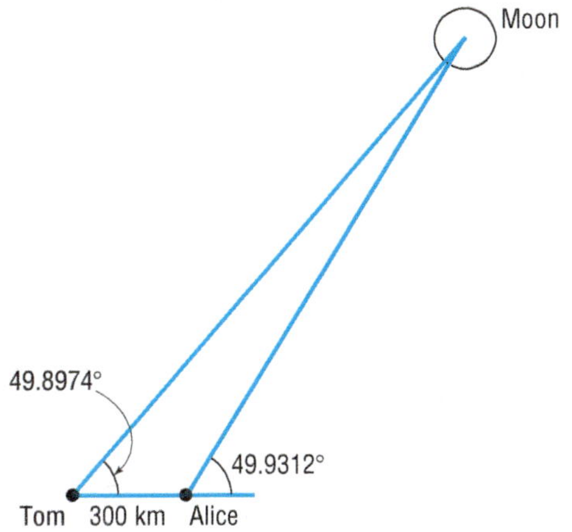

49. **Finding the Lean of the Leaning Tower of Pisa** The famous Leaning Tower of Pisa was originally 184.5 feet high.* At a distance of 123 feet from the base of the tower, the angle of elevation to the top of the tower is found to be 60°. Find $\angle RPQ$ indicated in the figure. Also, find the perpendicular distance from R to PQ.

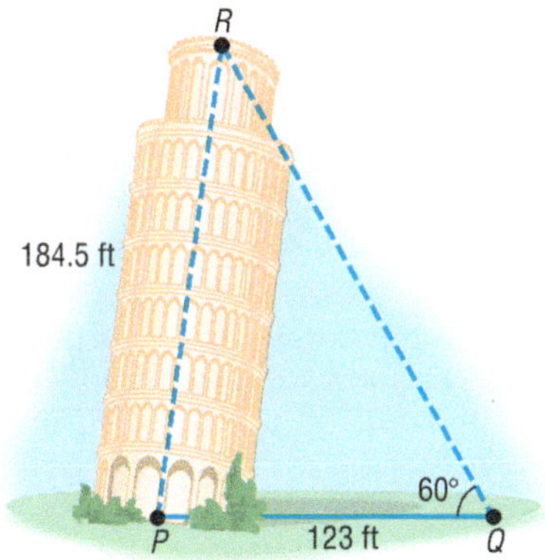

50. **Crankshafts on Cars** On a certain automobile, the crankshaft is 3 inches long and the connecting rod is 9 inches long (see the figure, top, right). At the time when $\angle OPQ$ is 15°, how far is the piston (P) from the center (O) of the crankshaft?

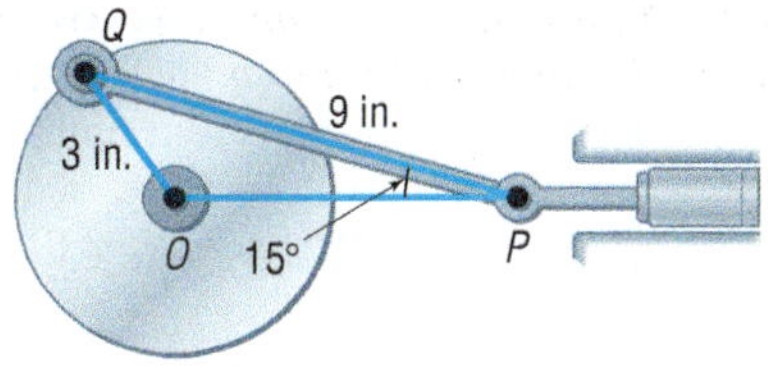

51. **Constructing a Highway** A highway whose primary directions are north–south, is being constructed along the west coast of Florida. Near Naples, a bay obstructs the straight path of the road. Since the cost of a bridge is prohibitive, engineers decide to go around the bay. The illustration shows the path that they decide on and the measurements taken. What is the length of highway needed to go around the bay?

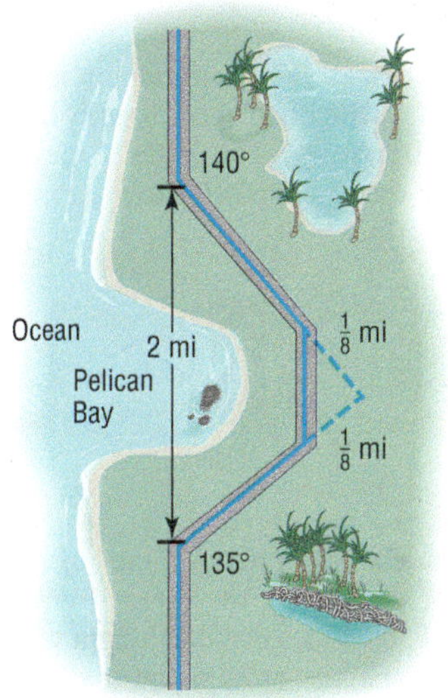

52. **Calculating Distances at Sea** The navigator of a ship at sea spots two lighthouses that she knows to be 3 miles apart along a straight seashore. She determines that the angles formed between two line-of-sight observations of the lighthouses and the line from the ship directly to shore are 15° and 35°. See the illustration.
   (a) How far is the ship from lighthouse P?
   (b) How far is the ship from lighthouse Q?
   (c) How far is the ship from shore?

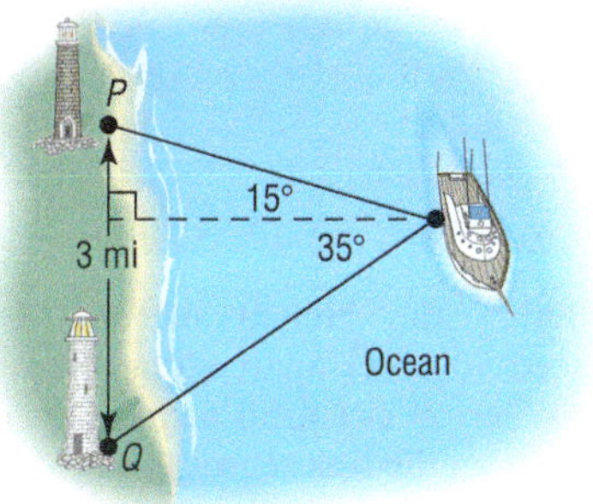

*On February 27, 1964, the government of Italy requested aid in preventing the tower from toppling. A multinational task force of engineers, mathematicians, and historians was assigned to discuss stabilization methods. After over two decades of work on the subject, the tower was closed to the public in January 1990. During the time that the tower was closed, the bells were removed to relieve it of some weight, and cables were cinched around the third level and anchored several hundred meters away. After a decade of corrective reconstruction and stabilization efforts, the tower was reopened to the public in December, 2001. Many methods were proposed to stabilize the tower, including the addition of 800 metric tons of lead counterweights to the raised end of the base. The final solution was to remove 38 cubic meters of soil from underneath the raised end. The tower has been declared stable for at least another 300 years.
Source: *http://www.history.com/this-day-in-history/leaning-tower-needs-help*

53. Designing an Awning An awning that covers a sliding glass door that is 88 inches tall forms an angle of 50° with the wall. The purpose of the awning is to prevent sunlight from entering the house when the angle of elevation of the Sun is more than 65°. See the figure. Find the length L of the awning.

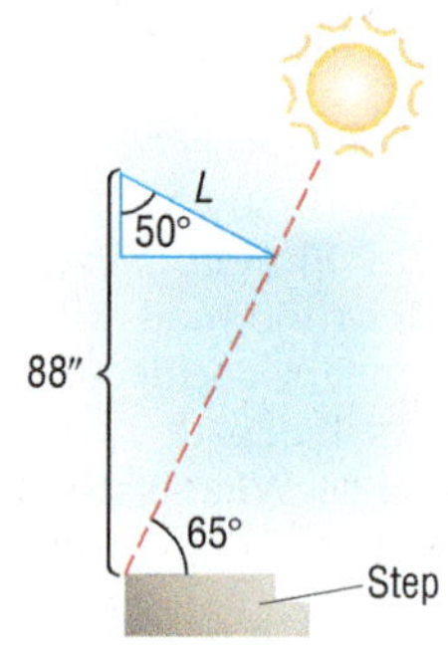

54. Finding Distances A forest ranger is walking on a path inclined at 5° to the horizontal directly toward a 100-foot-tall fire observation tower. The angle of elevation from the path to the top of the tower is 40°. How far is the ranger from the tower at this time?

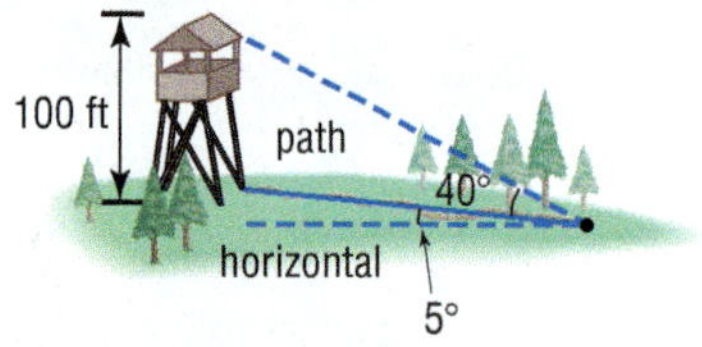

55. Great Pyramid of Cheops One of the original Seven Wonders of the World, the Great Pyramid of Cheops was built about 2580 BC. Its original height was 480 feet 11 inches, but owing to the loss of its topmost stones, it is now shorter. Find the current height of the Great Pyramid using the information given in the illustration.

Source: *Guinness Book of World Records*

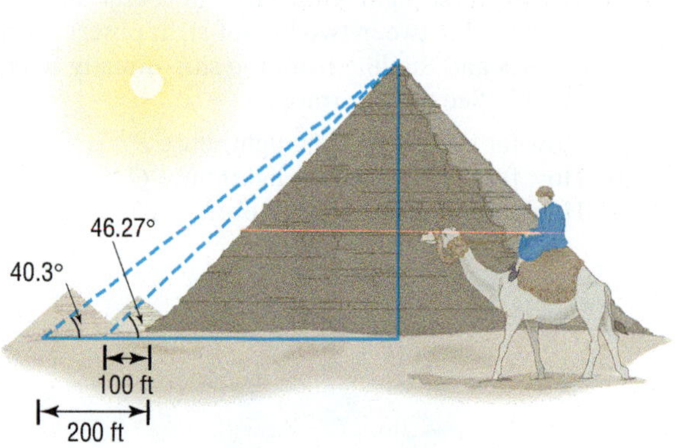

56. Determining the Height of an Aircraft Two sensors are spaced 700 feet apart along the approach to a small airport. When an aircraft is nearing the airport, the angle of elevation from the first sensor to the aircraft is 20°, and from the second sensor to the aircraft it is 15°. Determine how high the aircraft is at this time.

57. Mercury The distance from the Sun to Earth is approximately 149,600,000 kilometers (km). The distance from the Sun to Mercury is approximately 57,910,000 km. The **elongation angle** α is the angle formed between the line of sight from Earth to the Sun and the line of sight from Earth to Mercury. See the figure. Suppose that the elongation angle for Mercury is 15°. Use this information to find the possible distances between Earth and Mercury.

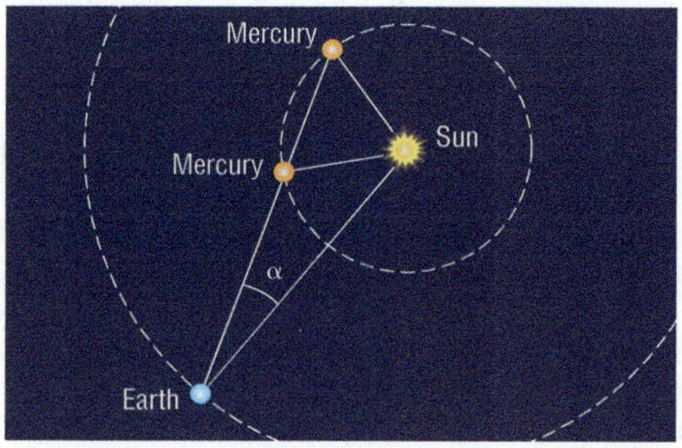

58. Venus The distance from the Sun to Earth is approximately 149,600,000 km. The distance from the Sun to Venus is approximately 108,200,000 km. The elongation angle α is the angle formed between the line of sight from Earth to the Sun and the line of sight from Earth to Venus. Suppose that the elongation angle for Venus is 10°. Use this information to find the possible distances between Earth and Venus.

59. The Original Ferris Wheel George Washington Gale Ferris, Jr., designed the original Ferris wheel for the 1893 World's Columbian Exposition in Chicago, Illinois. The wheel had 36 equally spaced cars each the size of a school bus. The distance between adjacent cars was approximately 22 feet. Determine the diameter of the wheel to the nearest foot.

Source: *Carnegie Library of Pittsburgh, www.clpgh.org*

60. Mollweide's Formula For any triangle, **Mollweide's Formula** (named after Karl Mollweide, 1774–1825) states that

$$\frac{a+b}{c} = \frac{\cos\left[\frac{1}{2}(A-B)\right]}{\sin\left(\frac{1}{2}C\right)}$$

Derive it.

[**Hint:** Use the Law of Sines and then a Sum-to-Product Formula. Notice that this formula involves all six parts of a triangle. As a result, it is sometimes used to check the solution of a triangle.]

61. Mollweide's Formula Another form of Mollweide's Formula is

$$\frac{a-b}{c} = \frac{\sin\left[\frac{1}{2}(A-B)\right]}{\cos\left(\frac{1}{2}C\right)}$$

Derive it.

62. For any triangle, derive the formula

$$a = b\cos C + c\cos B$$

[**Hint:** Use the fact that $\sin A = \sin(180° - B - C)$.]

63. Law of Tangents For any triangle, derive the **Law of Tangents:**

$$\frac{a-b}{a+b} = \frac{\tan\left[\frac{1}{2}(A-B)\right]}{\tan\left[\frac{1}{2}(A+B)\right]}$$

[**Hint:** Use Mollweide's Formula.]

64. Circumscribing a Triangle Show that

$$\frac{\sin A}{a} = \frac{\sin B}{b} = \frac{\sin C}{c} = \frac{1}{2r}$$

where r is the radius of the circle circumscribing the triangle PQR whose sides are a, b, and c, as shown in the figure.

[**Hint:** Draw the diameter PP'. Then $B = \angle PQR = \angle PP'R$, and angle $\angle PRP' = 90°$.]

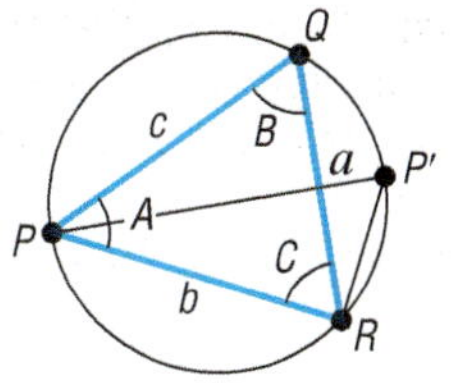

Explaining Concepts: Discussion and Writing

65. Make up three problems involving oblique triangles. One should result in one triangle, the second in two triangles, and the third in no triangle.

66. What do you do first if you are asked to solve a triangle and are given one side and two angles?

67. What do you do first if you are asked to solve a triangle and are given two sides and the angle opposite one of them?

68. Solve Example 6 using right-triangle geometry. Comment on which solution, using the Law of Sines or using right triangles, you prefer. Give reasons.

Retain Your Knowledge

Problems 69–72 are based on material learned earlier in the course. The purpose of these problems is to keep the material fresh in your mind so that you are better prepared for the final exam.

69. Solve: $3x^3 + 4x^2 - 27x - 36 = 0$

70. Find the exact distance between $P_1 = (-1, -7)$ and $P_2 = (2, -1)$. Then approximate the distance to two decimal places.

71. Find the exact value of $\tan\left[\cos^{-1}\left(-\frac{7}{8}\right)\right]$.

72. Graph $y = 4\sin\left(\frac{1}{2}x\right)$. Show at least two periods.

'Are You Prepared?' Answers

1. $\sin A\cos B - \cos A\sin B$ **2.** $30°$ or $\frac{\pi}{6}$ **3.** $\frac{15}{2}$

9.3 The Law of Cosines

PREPARING FOR THIS SECTION *Before getting started, review the following:*

- Trigonometric Equations (Section 8.3, pp. 634–639)
- Distance Formula (Section 1.1, p. 85)

Now Work the **'Are You Prepared?'** problems on page 704.

OBJECTIVES
1 Solve SAS Triangles (p. 702)
2 Solve SSS Triangles (p. 703)
3 Solve Applied Problems (p. 703)

In the previous section, the Law of Sines was used to solve Case 1 (SAA or ASA) and Case 2 (SSA) of an oblique triangle. In this section, the Law of Cosines is derived and used to solve Cases 3 and 4.

CASE 3: Two sides and the included angle are known (SAS).
CASE 4: Three sides are known (SSS).

THEOREM

Law of Cosines

For a triangle with sides a, b, c and opposite angles A, B, C, respectively,

$$c^2 = a^2 + b^2 - 2ab\cos C \quad (1)$$

$$b^2 = a^2 + c^2 - 2ac\cos B \quad (2)$$

$$a^2 = b^2 + c^2 - 2bc\cos A \quad (3)$$

Proof Only formula (1) is proved here. Formulas (2) and (3) may be proved using the same argument.

Begin by strategically placing a triangle on a rectangular coordinate system so that the vertex of angle C is at the origin and side b lies along the positive x-axis. Regardless of whether C is acute, as in Figure 25(a), or obtuse, as in Figure 25(b), the vertex of angle B has coordinates $(a\cos C, a\sin C)$. The vertex of angle A has coordinates $(b, 0)$.

Use the distance formula to compute c^2.

$$\begin{aligned} c^2 &= (b - a\cos C)^2 + (0 - a\sin C)^2 \\ &= b^2 - 2ab\cos C + a^2\cos^2 C + a^2\sin^2 C \\ &= b^2 - 2ab\cos C + a^2(\cos^2 C + \sin^2 C) \\ &= a^2 + b^2 - 2ab\cos C \end{aligned}$$

■

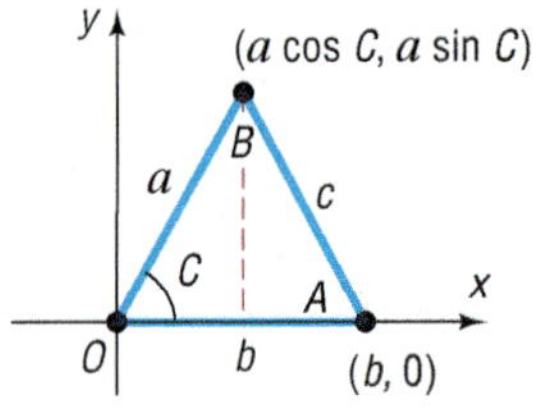

(a) Angle C is acute.

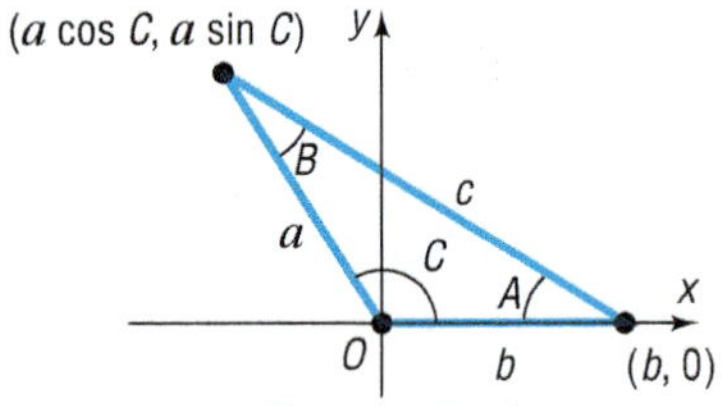

(b) Angle C is obtuse.

Figure 25

Each of formulas (1), (2), and (3) may be stated in words as follows:

THEOREM

Law of Cosines

The square of one side of a triangle equals the sum of the squares of the other two sides, minus twice their product times the cosine of their included angle. ■

Observe that if the triangle is a right triangle (so that, say, $C = 90°$), formula (1) becomes the familiar Pythagorean Theorem: $c^2 = a^2 + b^2$. The Pythagorean Theorem is a special case of the Law of Cosines!

1 Solve SAS Triangles

The Law of Cosines is used to solve Case 3 (SAS), which applies to triangles for which two sides and the included angle are known.

EXAMPLE 1 **Using the Law of Cosines to Solve an SAS Triangle**

Solve the triangle: $a = 2, \quad b = 3, \quad C = 60°$

Solution See Figure 26. Because two sides, a and b, and the included angle, $C = 60°$, are known, the Law of Cosines makes it easy to find the third side, c.

$$\begin{aligned} c^2 &= a^2 + b^2 - 2ab\cos C \\ &= 2^2 + 3^2 - 2\cdot 2\cdot 3\cdot \cos 60° \quad \color{blue}{a = 2, b = 3, C = 60°} \\ &= 13 - \left(12\cdot\frac{1}{2}\right) = 7 \\ c &= \sqrt{7} \end{aligned}$$

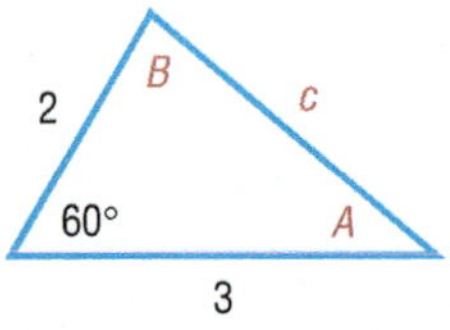

Figure 26

Side c is of length $\sqrt{7}$. To find the angles A and B, either the Law of Sines or the Law of Cosines may be used. It is preferable to use the Law of Cosines because it will lead to an equation with one solution. Using the Law of Sines would lead to an equation with two solutions that would need to be checked to determine which solution fits the given data.* We choose to use formulas (2) and (3) of the Law of Cosines to find A and B.

For A:

$$\begin{aligned} a^2 &= b^2 + c^2 - 2bc\cos A \\ 2bc\cos A &= b^2 + c^2 - a^2 \\ \cos A &= \frac{b^2 + c^2 - a^2}{2bc} = \frac{9 + 7 - 4}{2\cdot 3\sqrt{7}} = \frac{12}{6\sqrt{7}} = \frac{2\sqrt{7}}{7} \\ A &= \cos^{-1}\frac{2\sqrt{7}}{7} \approx 40.9° \end{aligned}$$

*The Law of Sines can be used if the angle sought is opposite the smaller side, thus ensuring that it is acute. (In Figure 26, use the Law of Sines to find A, the angle opposite the smaller side.)

For B:

$$b^2 = a^2 + c^2 - 2ac\cos B$$

$$\cos B = \frac{a^2 + c^2 - b^2}{2ac} = \frac{4 + 7 - 9}{4\sqrt{7}} = \frac{2}{4\sqrt{7}} = \frac{\sqrt{7}}{14}$$

$$B = \cos^{-1}\frac{\sqrt{7}}{14} \approx 79.1°$$

Note: The angle B could also have been found by using the fact that the sum $A + B + C = 180°$, so $B = 180° - 40.9° - 60° = 79.1°$. However, using the Law of Cosines twice allows for a check. ■

Notice that $A + B + C = 40.9° + 79.1° + 60° = 180°$, as required. ■

Now Work PROBLEM 9

2 Solve SSS Triangles

The next example illustrates how the Law of Cosines is used when three sides of a triangle are known, Case 4 (SSS).

EXAMPLE 2 Using the Law of Cosines to Solve an SSS Triangle

Solve the triangle: $a = 4, b = 3, c = 6$

Solution See Figure 27. To find the angles A, B, and C, proceed as in the solution to Example 1.

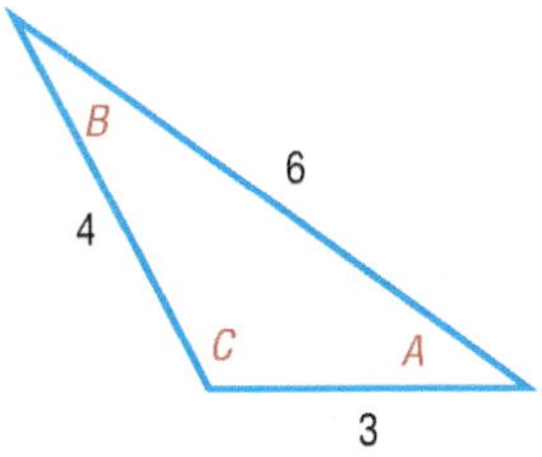

Figure 27

For A:

$$\cos A = \frac{b^2 + c^2 - a^2}{2bc} = \frac{9 + 36 - 16}{2 \cdot 3 \cdot 6} = \frac{29}{36}$$

$$A = \cos^{-1}\frac{29}{36} \approx 36.3°$$

For B:

$$\cos B = \frac{a^2 + c^2 - b^2}{2ac} = \frac{16 + 36 - 9}{2 \cdot 4 \cdot 6} = \frac{43}{48}$$

$$B = \cos^{-1}\frac{43}{48} \approx 26.4°$$

Now use A and B to find C:

$$C = 180° - A - B \approx 180° - 36.3° - 26.4° = 117.3°$$ ■

Now Work PROBLEM 15

3 Solve Applied Problems

EXAMPLE 3 Correcting a Navigational Error

A motorized sailboat leaves Naples, Florida, bound for Key West, 150 miles away. Maintaining a constant speed of 15 miles per hour, but encountering heavy crosswinds and strong currents, the crew finds, after 4 hours, that the sailboat is off course by 20°.

(a) How far is the sailboat from Key West at this time?
(b) Through what angle should the sailboat turn to correct its course?
(c) How much time has been added to the trip because of this? (Assume that the speed remains at 15 miles per hour.)

Solution See Figure 28. With a speed of 15 miles per hour, the sailboat has gone 60 miles after 4 hours. The distance x of the sailboat from Key West is to be found, along with the angle θ that the sailboat should turn through to correct its course.

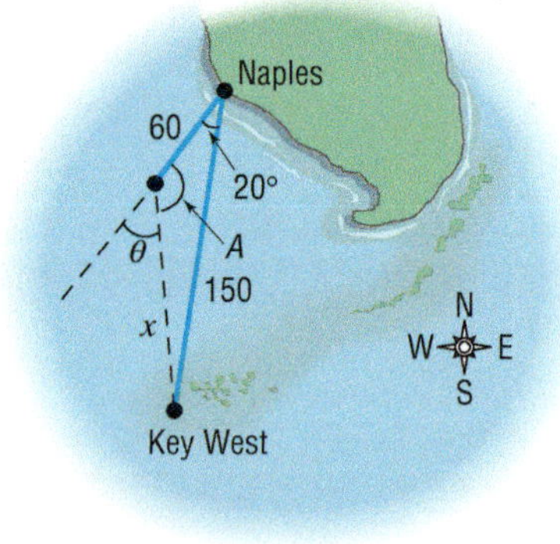

Figure 28

(a) To find x, use the Law of Cosines, because two sides and the included angle are known.

$$x^2 = 150^2 + 60^2 - 2(150)(60)\cos 20° \approx 9185.53$$
$$x \approx 95.8$$

The sailboat is about 96 miles from Key West.

(b) With all three sides of the triangle now known, use the Law of Cosines again to find the angle A opposite the side of length 150 miles.

$$150^2 = 96^2 + 60^2 - 2(96)(60)\cos A$$
$$9684 = -11{,}520\cos A$$
$$\cos A \approx -0.8406$$
$$A \approx 147.2°$$

So,

$$\theta = 180° - A \approx 180° - 147.2° = 32.8°$$

The sailboat should turn through an angle of about 33° to correct its course.

(c) The total length of the trip is now $60 + 96 = 156$ miles. The extra 6 miles will only require about 0.4 hour, or 24 minutes, more if the speed of 15 miles per hour is maintained.

Now Work PROBLEM 45

Historical Feature

The Law of Sines was known vaguely long before it was explicitly stated by Nasir Eddin (about AD 1250). Ptolemy (about AD 150) was aware of it in a form using a chord function instead of the sine function. But it was first clearly stated in Europe by Regiomontanus, writing in 1464.

The Law of Cosines appears first in Euclid's *Elements* (Book II), but in a well-disguised form in which squares built on the sides of triangles are added and a rectangle representing the cosine term is subtracted. It was thus known to all mathematicians because of their familiarity with Euclid's work. An early modern form of the Law of Cosines, that for finding the angle when the sides are known, was stated by François Viète (in 1593).

The Law of Tangents (see Problem 63 in Section 9.2) has become obsolete. In the past it was used in place of the Law of Cosines, because the Law of Cosines was very inconvenient for calculation with logarithms or slide rules. Mixing of addition and multiplication is now very easy on a calculator, however, and the Law of Tangents has been shelved along with the slide rule.

9.3 Assess Your Understanding

'Are You Prepared?' *Answers are given at the end of these exercises. If you get a wrong answer, read the pages listed in red.*

1. Write the formula for the distance d from $P_1 = (x_1, y_1)$ to $P_2 = (x_2, y_2)$. (p. 85)

2. If θ is an acute angle, solve the equation $\cos\theta = \frac{\sqrt{2}}{2}$. (pp. 634–639)

Concepts and Vocabulary

3. If three sides of a triangle are given, the Law of ________ is used to solve the triangle.

4. If one side and two angles of a triangle are given, which law can be used to solve the triangle?
(a) Law of Sines (b) Law of Cosines
(c) Either a or b (d) The triangle cannot be solved.

5. If two sides and the included angle of a triangle are given, which law can be used to solve the triangle?
(a) Law of Sines (b) Law of Cosines
(c) Either a or b (d) The triangle cannot be solved.

6. ***True or False*** Given only the three sides of a triangle, there is insufficient information to solve the triangle.

7. ***True or False*** The Law of Cosines states that the square of one side of a triangle equals the sum of the squares of the other two sides, minus twice their product.

8. ***True or False*** A special case of the Law of Cosines is the Pythagorean Theorem.

Skill Building

In Problems 9–16, solve each triangle.

9.

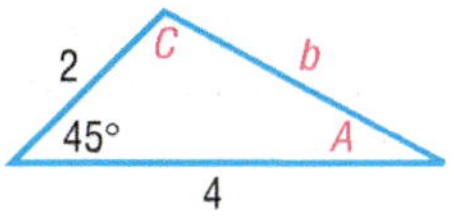

10.

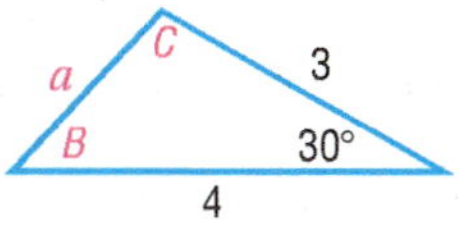

11.

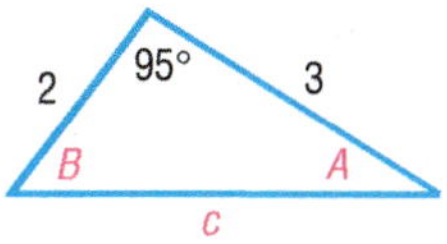

12.

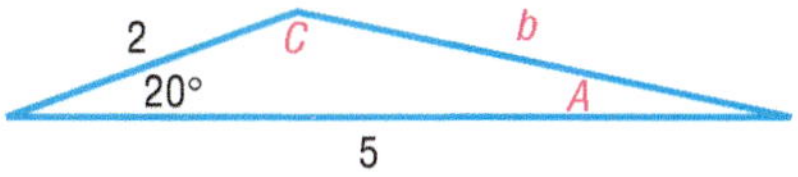

13.

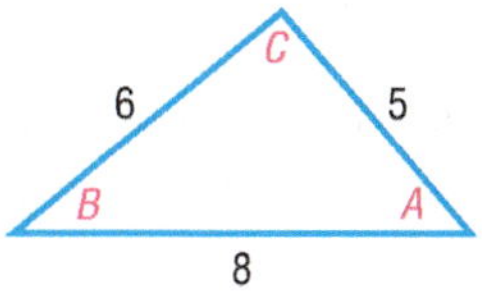

14.

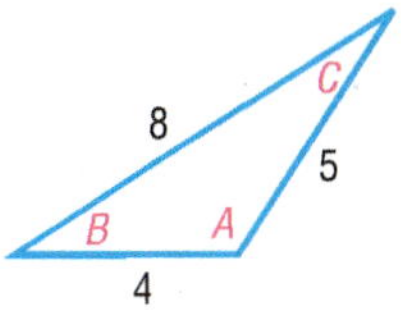

15.

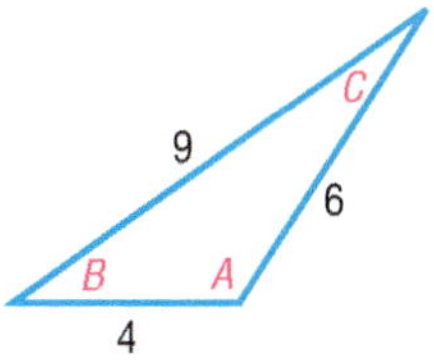

16.

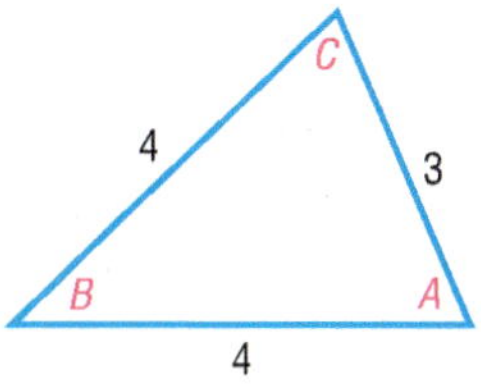

In Problems 17–32, solve each triangle.

17. $a = 3,\quad b = 4,\quad C = 40°$

18. $a = 2,\quad c = 1,\quad B = 10°$

19. $b = 1,\quad c = 3,\quad A = 80°$

20. $a = 6,\quad b = 4,\quad C = 60°$

21. $a = 3,\quad c = 2,\quad B = 110°$

22. $b = 4,\quad c = 1,\quad A = 120°$

23. $a = 2,\quad b = 2,\quad C = 50°$

24. $a = 3,\quad c = 2,\quad B = 90°$

25. $a = 12,\quad b = 13,\quad c = 5$

26. $a = 4,\quad b = 5,\quad c = 3$

27. $a = 2,\quad b = 2,\quad c = 2$

28. $a = 3,\quad b = 3,\quad c = 2$

29. $a = 5,\quad b = 8,\quad c = 9$

30. $a = 4,\quad b = 3,\quad c = 6$

31. $a = 10,\quad b = 8,\quad c = 5$

32. $a = 9,\quad b = 7,\quad c = 10$

Mixed Practice

In Problems 33–42, solve each triangle using either the Law of Sines or the Law of Cosines.

33. $B = 20°, C = 75°, b = 5$

34. $A = 50°, B = 55°, c = 9$

35. $a = 6, b = 8, c = 9$

36. $a = 14, b = 7, A = 85°$

37. $B = 35°, C = 65°, a = 15$

38. $a = 4, c = 5, B = 55°$

39. $A = 10°, a = 3, b = 10$

40. $A = 65°, B = 72°, b = 7$

41. $b = 5, c = 12, A = 60°$

42. $a = 10, b = 10, c = 15$

Applications and Extensions

43. Distance to the Green A golfer hits an errant tee shot that lands in the rough. A marker in the center of the fairway is 150 yards from the center of the green. While standing on the marker and facing the green, the golfer turns 110° toward his ball. He then paces off 35 yards to his ball. See the figure. How far is the ball from the center of the green?

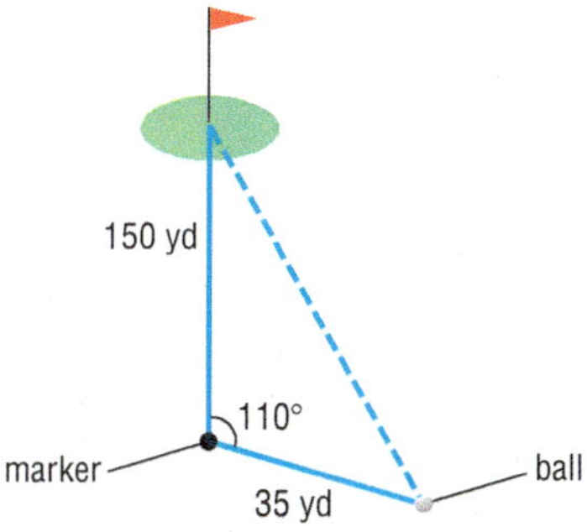

44. Navigation An airplane flies due north from Ft. Myers to Sarasota, a distance of 150 miles, and then turns through an angle of 50° and flies to Orlando, a distance of 100 miles. See the figure.

(a) How far is it directly from Ft. Myers to Orlando?

(b) What bearing should the pilot use to fly directly from Ft. Myers to Orlando?

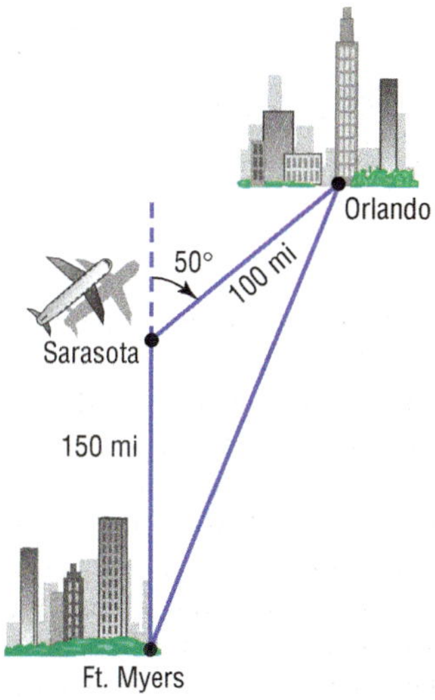

45. **Avoiding a Tropical Storm** A cruise ship maintains an average speed of 15 knots in going from San Juan, Puerto Rico, to Barbados, West Indies, a distance of 600 nautical miles. To avoid a tropical storm, the captain heads out of San Juan in a direction of 20° off a direct heading to Barbados. The captain maintains the 15-knot speed for 10 hours, after which time the path to Barbados becomes clear of storms.
(a) Through what angle should the captain turn to head directly to Barbados?
(b) Once the turn is made, how long will it be before the ship reaches Barbados if the same 15-knot speed is maintained?

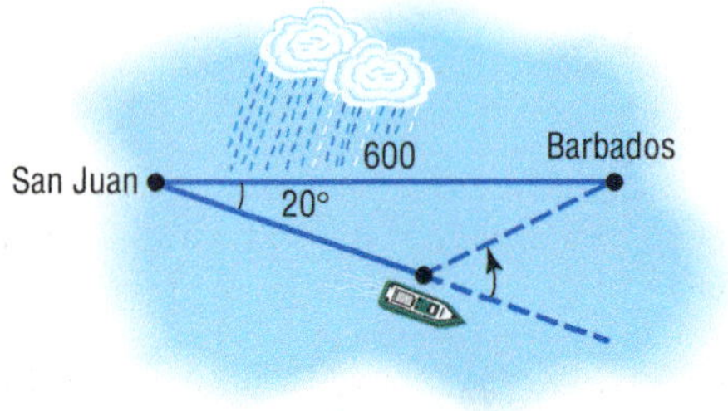

46. **Revising a Flight Plan** In attempting to fly from Chicago to Louisville, a distance of 330 miles, a pilot inadvertently took a course that was 10° in error, as indicated in the figure.
(a) If the aircraft maintains an average speed of 220 miles per hour, and if the error in direction is discovered after 15 minutes, through what angle should the pilot turn to head toward Louisville?
(b) What new average speed should the pilot maintain so that the total time of the trip is 90 minutes?

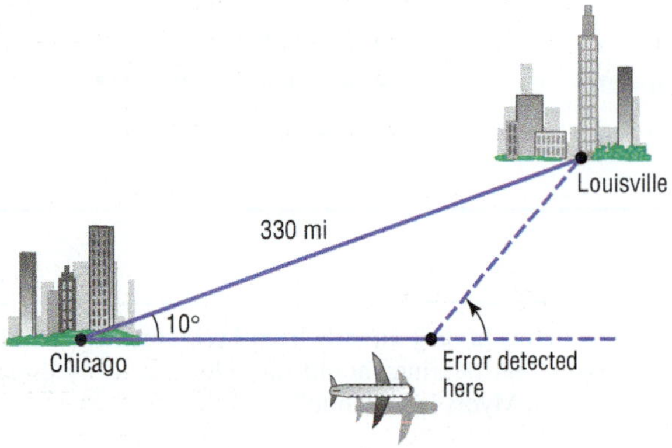

47. **Major League Baseball Field** A major league baseball diamond is actually a square 90 feet on a side. The pitching rubber is located 60.5 feet from home plate on a line joining home plate and second base.
(a) How far is it from the pitching rubber to first base?
(b) How far is it from the pitching rubber to second base?
(c) If a pitcher faces home plate, through what angle does he need to turn to face first base?

48. **Little League Baseball Field** According to Little League baseball official regulations, the diamond is a square 60 feet on a side. The pitching rubber is located 46 feet from home plate on a line joining home plate and second base.
(a) How far is it from the pitching rubber to first base?
(b) How far is it from the pitching rubber to second base?
(c) If a pitcher faces home plate, through what angle does he need to turn to face first base?

49. **Finding the Length of a Guy Wire** The height of a radio tower is 500 feet, and the ground on one side of the tower slopes upward at an angle of 10° (see the figure).
(a) How long should a guy wire be if it is to connect to the top of the tower and be secured at a point on the sloped side 100 feet from the base of the tower?
(b) How long should a second guy wire be if it is to connect to the middle of the tower and be secured at a position 100 feet from the base on the flat side?

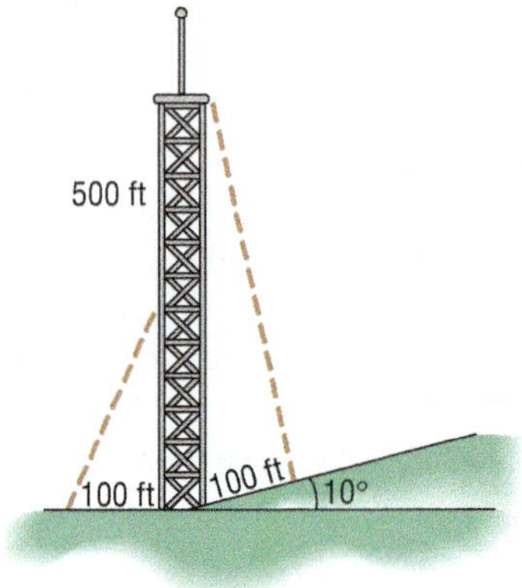

50. **Finding the Length of a Guy Wire** A radio tower 500 feet high is located on the side of a hill with an inclination to the horizontal of 5°. See the figure. How long should two guy wires be if they are to connect to the top of the tower and be secured at two points 100 feet directly above and directly below the base of the tower?

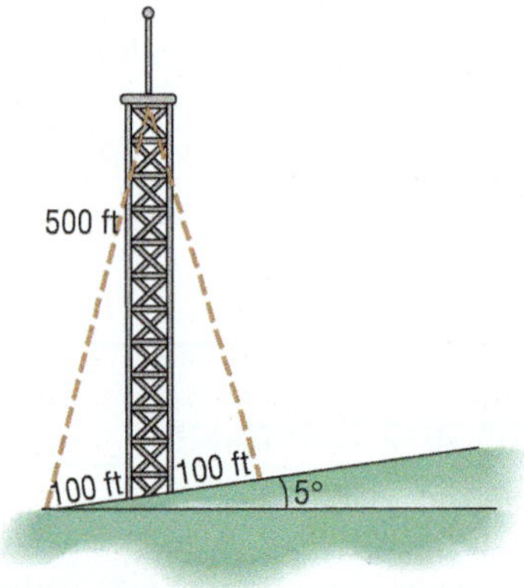

51. **Identifying Remains** The Purkait triangle, located at the proximal end of the femur, has been used to identify the gender of fragmented skeletal remains. See the figure.
(a) Given $\overline{AB} = 30.1$ mm, $\overline{AC} = 51.4$ mm, and $A = 89.2°$, find the length of $\overline{BC}$.
(b) If the average length of $\overline{BC}$ is 59.4 mm for males and 53.3 mm for females, which gender would be identified for the measurements in part (a)?

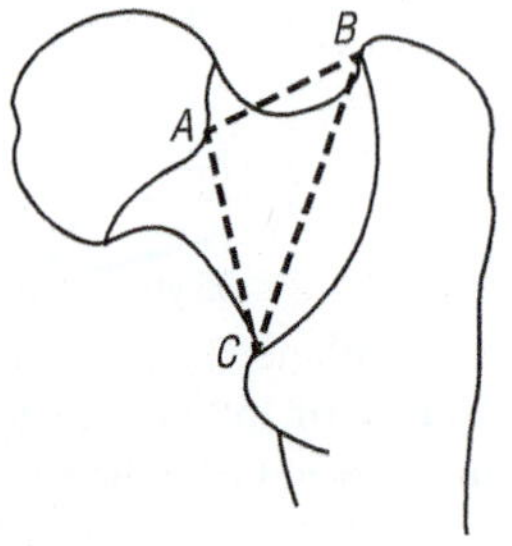

52. Identifying Remains Like the Purkait triangle in Problem 51, the metric triangle is located at the proximal end of the femur and has been used to identify the gender of fragmented skeletal remains. See the figure.

(a) If $\overline{AC} = 48.8$ mm, $\overline{BC} = 62.2$ mm, and $C = 89°$, find the length of $\overline{AB}$.

(b) If $\overline{AB} < 80$ mm typically indicates a female and $\overline{AB} > 100$ mm typically indicates a male, which gender, if any, would be identified from the measurements in part (a)?

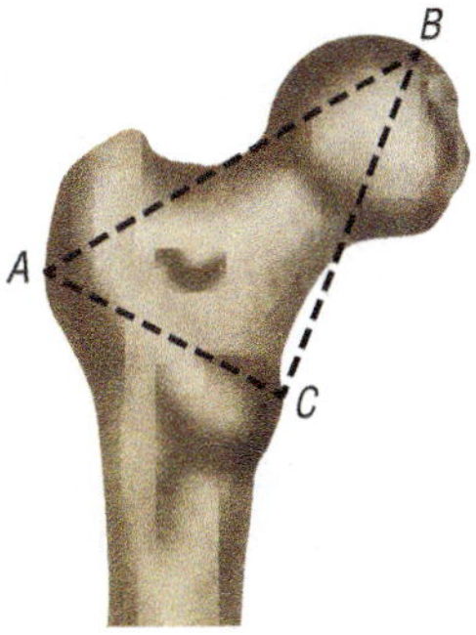

53. Soccer Angles A soccer goal is 8 yards wide. Suppose a goalie is standing on her line in the center of her goal as a striker from the opposing team moves the ball towards her. The near post angle, α, is formed by rays extending from the ball to the near post and the goalie. Similarly, the far post angle, β, is formed by rays extending from the ball to the far post and the goalie. See the figure.

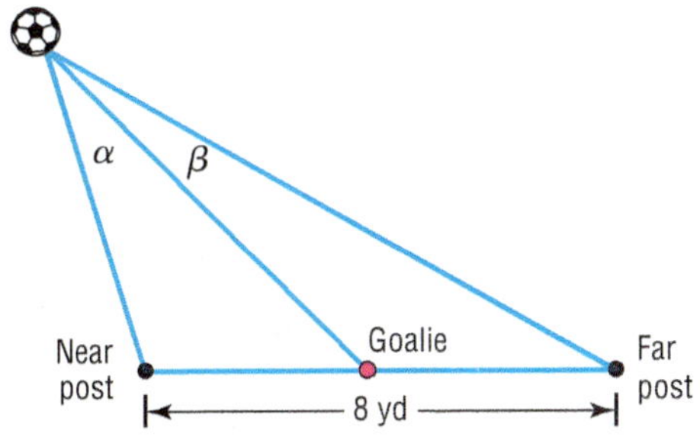

(a) Determine the near post angle and the far post angle when the striker is 20 yards from the near post and 24 yards from the far post.

(b) How far is the goalie from the ball?

(c) To cover the near post, the goalie moves toward the near post in order to make the near post angle and the far post angle equal. How far toward her near post does the goalie need to move?

54. Covering the Angles In soccer, a defending goalkeeper wants to take up a position that bisects the angle that needs to be covered. The keeper stands square to the ball—that is, perpendicular to the line of bisection—at a point such that the area covered (shaded) lies completely outside the goal. How far will the goalkeeper be from the center of the goal line if an attacking striker is 24 yards from the near post and 30 yards from the far post?

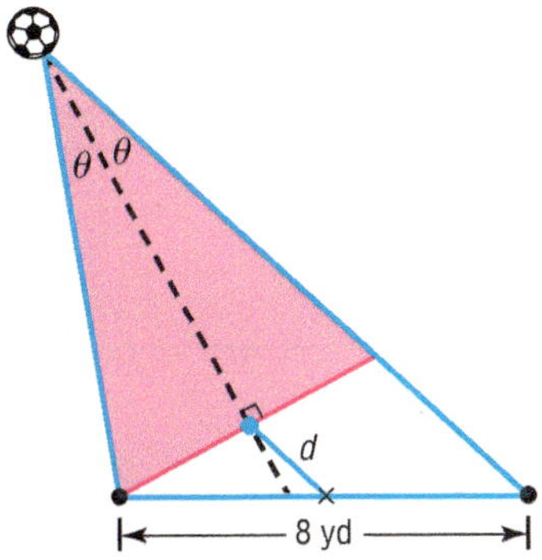

55. Wrigley Field, Home of the Chicago Cubs The distance from home plate to the fence in dead center in Wrigley Field is 400 feet (see the figure). How far is it from the fence in dead center to third base?

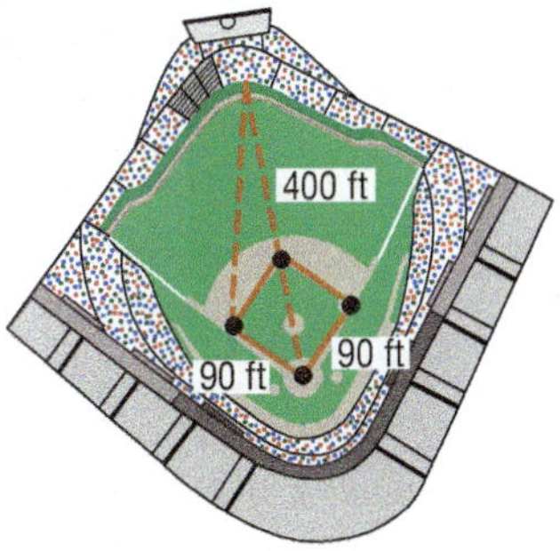

56. Little League Baseball The distance from home plate to the fence in dead center at the Oak Lawn Little League field is 280 feet. How far is it from the fence in dead center to third base?
[**Hint:** The distance between the bases in Little League is 60 feet.]

57. Building a Swing Set Clint is building a wooden swing set for his children. Each supporting end of the swing set is to be an A-frame constructed with two 10-foot-long 4 by 4's joined at a 45° angle. To prevent the swing set from tipping over, Clint wants to secure the base of each A-frame to concrete footings. How far apart should the footings for each A-frame be?

58. Rods and Pistons Rod OA rotates about the fixed point O so that point A travels on a circle of radius r. Connected to point A is another rod AB of length $L > 2r$, and point B is connected to a piston. See the figure. Show that the distance x between point O and point B is given by

$$x = r\cos\theta + \sqrt{r^2\cos^2\theta + L^2 - r^2}$$

where θ is the angle of rotation of rod OA.

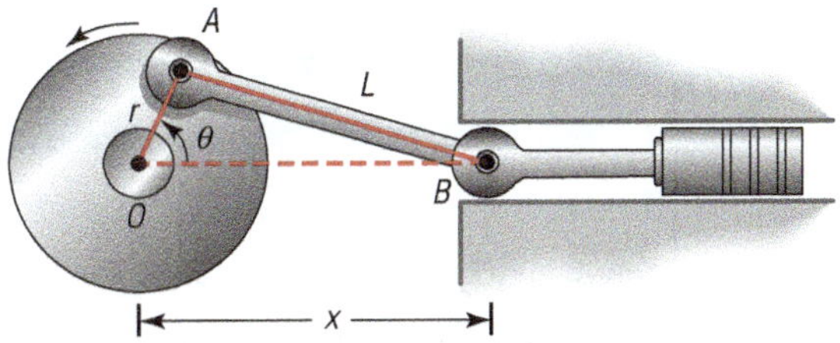

59. Geometry Show that the length d of a chord of a circle of radius r is given by the formula

$$d = 2r\sin\frac{\theta}{2}$$

where θ is the central angle formed by the radii to the ends of the chord. See the figure. Use this result to derive the fact that $\sin\theta < \theta$, where $\theta > 0$ is measured in radians.

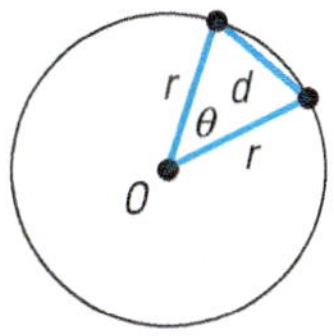

60. For any triangle, show that

$$\cos\frac{C}{2} = \sqrt{\frac{s(s-c)}{ab}}$$

where $s = \frac{1}{2}(a + b + c)$.

[**Hint:** Use a Half-angle Formula and the Law of Cosines.]

61. For any triangle, show that

$$\sin\frac{C}{2} = \sqrt{\frac{(s-a)(s-b)}{ab}}$$

where $s = \frac{1}{2}(a + b + c)$.

62. Use the Law of Cosines to prove the identity

$$\frac{\cos A}{a} + \frac{\cos B}{b} + \frac{\cos C}{c} = \frac{a^2 + b^2 + c^2}{2abc}$$

Explaining Concepts: Discussion and Writing

63. What do you do first if you are asked to solve a triangle and are given two sides and the included angle?

64. What do you do first if you are asked to solve a triangle and are given three sides?

65. Make up an applied problem that requires using the Law of Cosines.

66. Write down your strategy for solving an oblique triangle.

67. State the Law of Cosines in words.

Retain Your Knowledge

Problems 68–71 are based on material learned earlier in the course. The purpose of these problems is to keep the material fresh in your mind so that you are better prepared for the final exam.

68. Graph: $R(x) = \frac{2x + 1}{x - 3}$

69. Solve $4^x = 3^{x+1}$. If the solution is irrational, express it both in exact form and as a decimal rounded to three places.

70. Given $\tan\theta = -\frac{2\sqrt{6}}{5}$ and $\cos\theta = -\frac{5}{7}$, find the exact value of each of the four remaining trigonometric functions.

71. Find an equation for the graph.

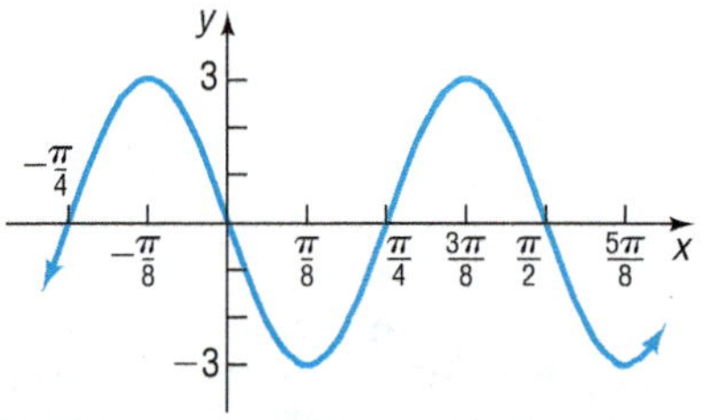

'Are You Prepared?' Answers

1. $d = \sqrt{(x_2 - x_1)^2 + (y_2 - y_1)^2}$ **2.** $\theta = 45°$ or $\frac{\pi}{4}$

9.4 Area of a Triangle

PREPARING FOR THIS SECTION *Before getting started, review the following:*

- Geometry Essentials (Chapter R, Review, Section R.3, pp. 31–36)

Now Work the 'Are You Prepared?' problem on page 711.

OBJECTIVES 1 Find the Area of SAS Triangles (p. 709)
2 Find the Area of SSS Triangles (p. 710)

In this section, several formulas for calculating the area of a triangle are derived. The most familiar of these follows.

THEOREM

The area K of a triangle is

$$K = \frac{1}{2}bh \quad (1)$$

where b is the base and h is an altitude drawn to that base.

Note: Typically, A is used for area. However, because A is also used as the measure of an angle, K is used here for area to avoid confusion.

Proof Look at the triangle in Figure 29. Around the triangle construct a rectangle of height h and base b, as shown in Figure 30.

Triangles 1 and 2 in Figure 30 are equal in area, as are triangles 3 and 4. Consequently, the area of the triangle with base b and altitude h is exactly half the area of the rectangle, which is bh.

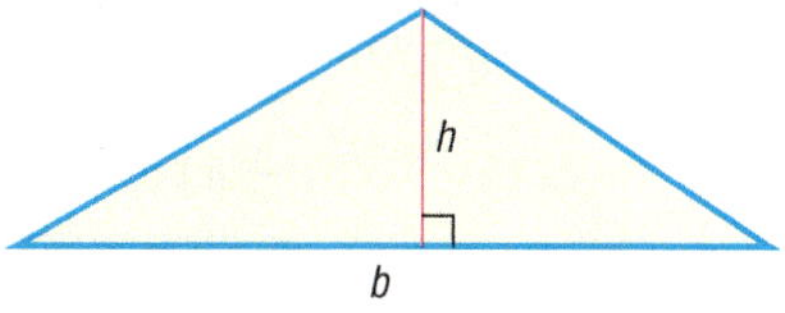

Figure 29

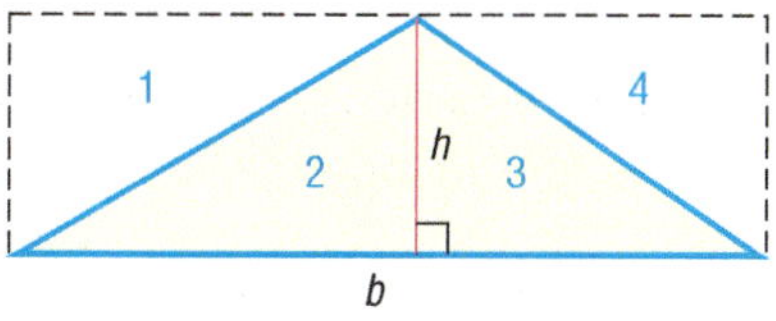

Figure 30 ■

1 Find the Area of SAS Triangles

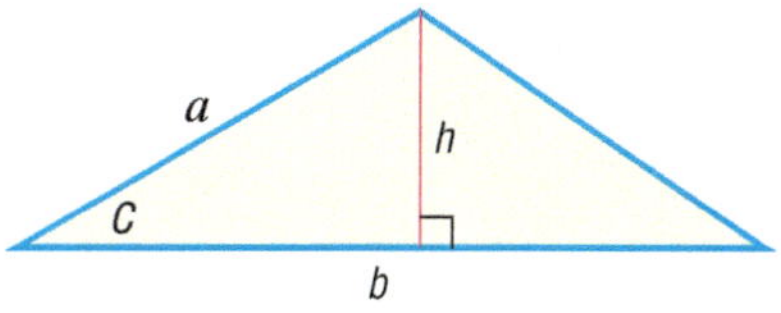

Figure 31

If the base b and the altitude h to that base are known, then the area of such a triangle can be found using formula (1). Usually, though, the information required to use formula (1) is not given. Suppose, for example, that two sides a and b and the included angle C are known. See Figure 31. Then the altitude h can be found by noting that

$$\frac{h}{a} = \sin C$$

so

$$h = a \sin C$$

Using this fact in formula (1) produces

$$K = \frac{1}{2}bh = \frac{1}{2}b(a \sin C) = \frac{1}{2}ab \sin C$$

The area K of the triangle is given by the formula

$$K = \frac{1}{2}ab \sin C \qquad \textbf{(2)}$$

Dropping altitudes from the other two vertices of the triangle leads to the following corresponding formulas:

$$K = \frac{1}{2}bc \sin A \qquad \textbf{(3)}$$

$$K = \frac{1}{2}ac \sin B \qquad \textbf{(4)}$$

It is easiest to remember these formulas by using the following wording:

THEOREM

The area K of a triangle equals one-half the product of two of its sides times the sine of their included angle. ■

EXAMPLE 1 **Finding the Area of an SAS Triangle**

Find the area K of the triangle for which $a = 8$, $b = 6$, and $C = 30°$.

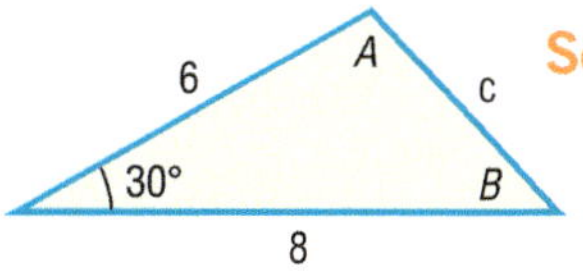

Figure 32

Solution See Figure 32. Use formula (2) to get

$$K = \frac{1}{2}ab \sin C = \frac{1}{2} \cdot 8 \cdot 6 \cdot \sin 30° = 12 \text{ square units}$$ ■

Now Work PROBLEM 7

2 Find the Area of SSS Triangles

If the three sides of a triangle are known, another formula, called **Heron's Formula** (named after Heron of Alexandria), can be used to find the area of a triangle.

THEOREM

Heron's Formula

The area K of a triangle with sides a, b, and c is

$$K = \sqrt{s(s-a)(s-b)(s-c)} \tag{5}$$

where $s = \frac{1}{2}(a+b+c)$.

EXAMPLE 2 **Finding the Area of an SSS Triangle**

Find the area of a triangle whose sides are 4, 5, and 7.

Solution Let $a = 4$, $b = 5$, and $c = 7$. Then

$$s = \frac{1}{2}(a+b+c) = \frac{1}{2}(4+5+7) = 8$$

Heron's Formula gives the area K as

$$K = \sqrt{s(s-a)(s-b)(s-c)} = \sqrt{8\cdot 4\cdot 3\cdot 1} = \sqrt{96} = 4\sqrt{6} \text{ square units}$$

Now Work PROBLEM 13

Proof of Heron's Formula The proof given here uses the Law of Cosines and is quite different from the proof given by Heron.

From the Law of Cosines,

$$c^2 = a^2 + b^2 - 2ab\cos C$$

and the Half-angle Formula,

$$\cos^2\frac{C}{2} = \frac{1+\cos C}{2}$$

it follows that

$$\cos^2\frac{C}{2} = \frac{1+\cos C}{2} = \frac{1+\dfrac{a^2+b^2-c^2}{2ab}}{2}$$

$$= \frac{a^2+2ab+b^2-c^2}{4ab} = \frac{(a+b)^2-c^2}{4ab}$$

$$= \frac{(a+b-c)(a+b+c)}{4ab} = \frac{2(s-c)\cdot 2s}{4ab} = \frac{s(s-c)}{ab} \tag{6}$$

Factor. ($a + b - c = a + b + c - 2c = 2s - 2c = 2(s - c)$)

Similarly, using $\sin^2\frac{C}{2} = \frac{1-\cos C}{2}$, it follows that

$$\sin^2\frac{C}{2} = \frac{(s-a)(s-b)}{ab} \tag{7}$$

Now use formula (2) for the area.

$$K = \frac{1}{2}ab \sin C$$

$$= \frac{1}{2}ab \cdot 2 \sin \frac{C}{2} \cos \frac{C}{2} \qquad \sin C = \sin\left[2\left(\frac{C}{2}\right)\right] = 2 \sin \frac{C}{2} \cos \frac{C}{2}$$

$$= ab\sqrt{\frac{(s-a)(s-b)}{ab}}\sqrt{\frac{s(s-c)}{ab}} \qquad \text{Use equations (6) and (7).}$$

$$= \sqrt{s(s-a)(s-b)(s-c)}$$

■

Historical Feature

Heron's Formula (also known as *Hero's Formula*) was first expressed by Heron of Alexandria (first century AD), who had, besides his mathematical talents, a good deal of engineering skills. In various temples his mechanical devices produced effects that seemed supernatural, and visitors presumably were thus moved to generosity. Heron's book *Metrica*, on making such devices, has survived and was discovered in 1896 in the city of Constantinople.

Heron's Formulas for the area of a triangle caused some mild discomfort in Greek mathematics, because a product with two factors was an area and one with three factors was a volume, but four factors seemed contradictory in Heron's time.

9.4 Assess Your Understanding

'Are You Prepared?' *The answer is given at the end of these exercises. If you get the wrong answer, read the pages listed in red.*

1. The area K of a triangle whose base is b and whose height is h is ________. (pp. 32–33)

Concepts and Vocabulary

2. If two sides a and b and the included angle C are known in a triangle, then the area K is found using the formula $K =$ ________.

3. The area K of a triangle with sides a, b, and c is $K =$ ________________, where $s =$ ________.

4. ***True or False*** The area of a triangle equals one-half the product of the lengths of two of its sides times the sine of their included angle.

5. Given two sides of a triangle, b and c, and the included angle A, the altitude h from angle B to side b is given by ________.

(a) $\frac{1}{2}ab \sin A$ (b) $b \sin A$ (c) $c \sin A$ (d) $\frac{1}{2}bc \sin A$

6. Heron's Formula is used to find the area of ________ triangles.
(a) ASA (b) SAS (c) SSS (d) AAS

Skill Building

In Problems 7–14, find the area of each triangle. Round answers to two decimal places.

7. 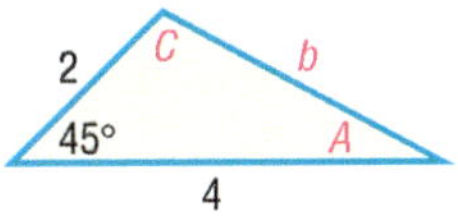

8. 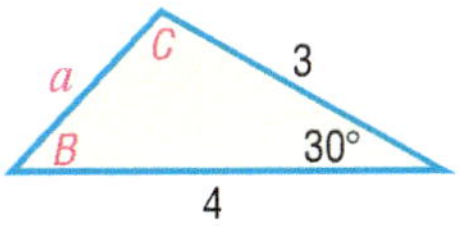

9. 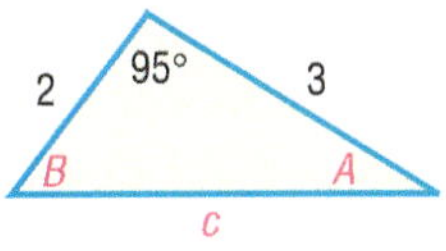

10. 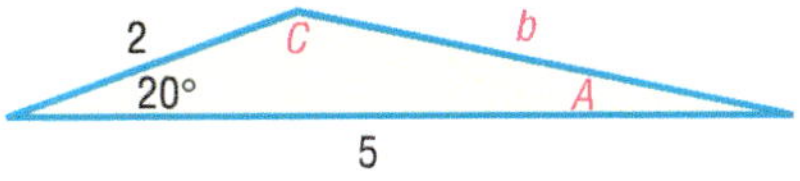

11. 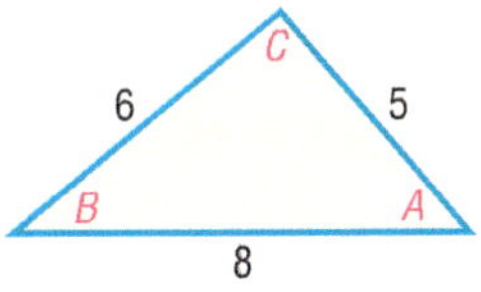

12. 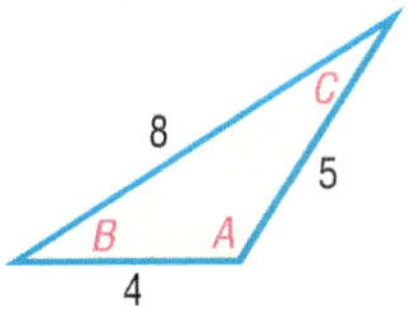

13. 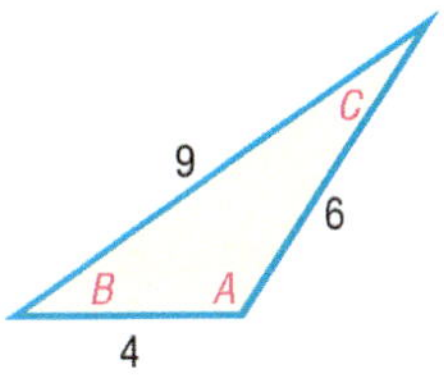

14. 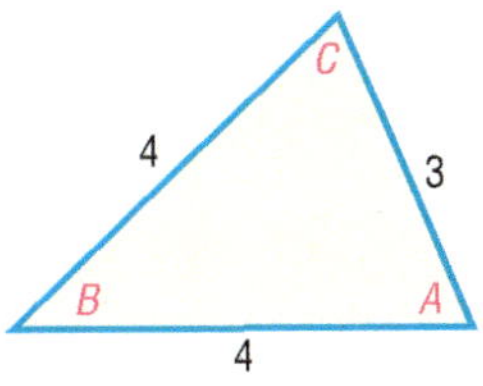

In Problems 15–26, find the area of each triangle. Round answers to two decimal places.

15. $a = 3, \quad b = 4, \quad C = 40°$

16. $a = 2, \quad c = 1, \quad B = 10°$

17. $b = 1, \quad c = 3, \quad A = 80°$

18. $a = 6, \quad b = 4, \quad C = 60°$

19. $a = 3, \quad c = 2, \quad B = 110°$

20. $b = 4, \quad c = 1, \quad A = 120°$

21. $a = 12, \quad b = 13, \quad c = 5$

22. $a = 4, \quad b = 5, \quad c = 3$

23. $a = 2, \quad b = 2, \quad c = 2$

24. $a = 3, \quad b = 3, \quad c = 2$

25. $a = 5, \quad b = 8, \quad c = 9$

26. $a = 4, \quad b = 3, \quad c = 6$

Applications and Extensions

27. Area of an ASA Triangle If two angles and the included side are given, the third angle is easy to find. Use the Law of Sines to show that the area K of a triangle with side a and angles A, B, and C is

$$K = \frac{a^2 \sin B \sin C}{2 \sin A}$$

28. Area of a Triangle Prove the two other forms of the formula given in Problem 27.

$$K = \frac{b^2 \sin A \sin C}{2 \sin B} \quad \text{and} \quad K = \frac{c^2 \sin A \sin B}{2 \sin C}$$

In Problems 29–34, use the results of Problem 27 or 28 to find the area of each triangle. Round answers to two decimal places.

29. $A = 40°, \quad B = 20°, \quad a = 2$

30. $A = 50°, \quad C = 20°, \quad a = 3$

31. $B = 70°, \quad C = 10°, \quad b = 5$

32. $A = 70°, \quad B = 60°, \quad c = 4$

33. $A = 110°, \quad C = 30°, \quad c = 3$

34. $B = 10°, \quad C = 100°, \quad b = 2$

35. Area of a Segment Find the area of the segment (shaded in blue in the figure) of a circle whose radius is 8 feet, formed by a central angle of 70°.
[**Hint:** Subtract the area of the triangle from the area of the sector to obtain the area of the segment.]

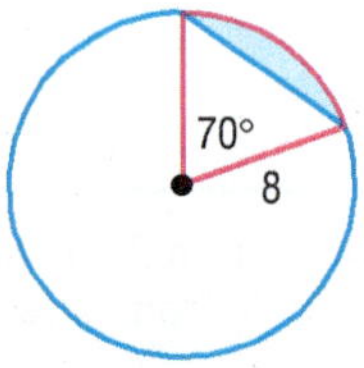

36. Area of a Segment Find the area of the segment of a circle whose radius is 5 inches, formed by a central angle of 40°.

37. Cost of a Triangular Lot The dimensions of a triangular lot are 100 feet by 50 feet by 75 feet. If the price of such land is \$3 per square foot, how much does the lot cost?

38. Amount of Material to Make a Tent A cone-shaped tent is made from a circular piece of canvas 24 feet in diameter by removing a sector with central angle 100° and connecting the ends. What is the surface area of the tent?

39. Property Area A lot for sale in a subdivision has the shape of a quadrilateral as shown in the figure. Find the area of the lot to the nearest square foot.

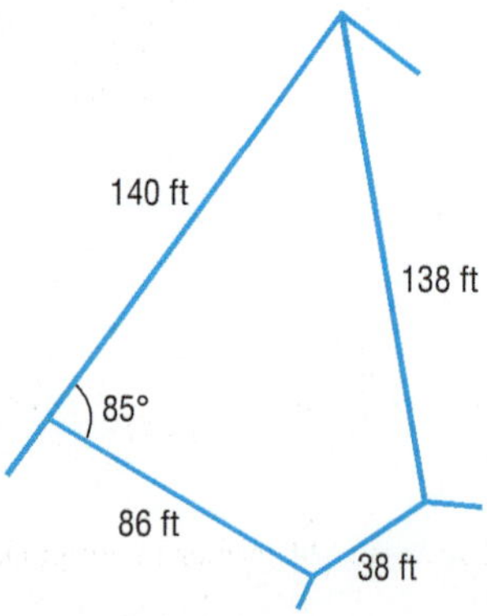

40. Dimensions of Home Plate The dimensions of home plate at any major league baseball stadium are shown. Find the area of home plate.

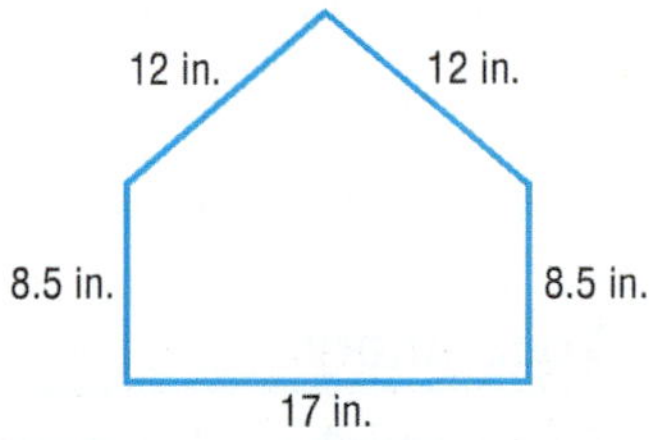

41. Computing Areas See the figure. Find the area of the shaded region enclosed in a semicircle of diameter 10 inches. The length of the chord PQ is 8 inches.
[**Hint:** Triangle PQR is a right triangle.]

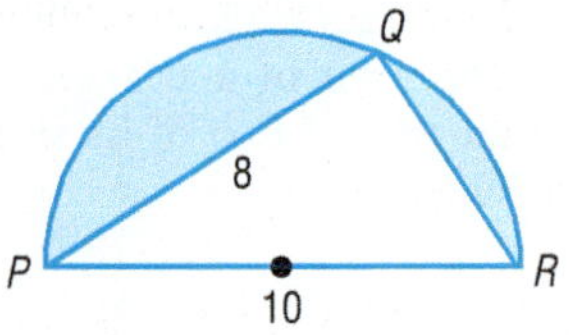

42. Geometry See the figure, which shows a circle of radius r with center at O. Find the area K of the shaded region as a function of the central angle θ.

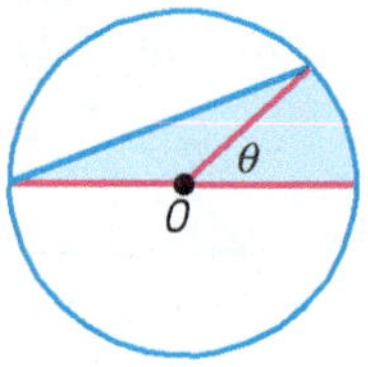

43. Approximating the Area of a Lake To approximate the area of a lake, a surveyor walks around the perimeter of the lake, taking the measurements shown in the illustration. Using this technique, what is the approximate area of the lake?
[**Hint:** Use the Law of Cosines on the three triangles shown, and then find the sum of their areas.]

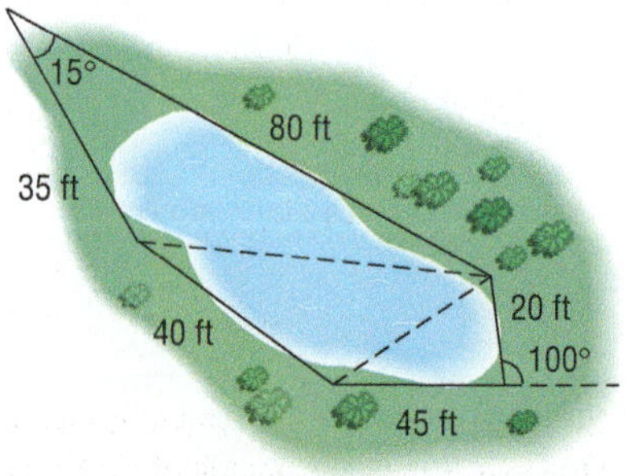

44. Bermuda Triangle The Bermuda Triangle is roughly defined by Hamilton, Bermuda; San Juan, Puerto Rico; and Fort Lauderdale, Florida. The distances from Hamilton to Fort Lauderdale, Fort Lauderdale to San Juan, and San Juan to Hamilton are approximately 1028, 1046, and 965 miles, respectively. Ignoring the curvature of Earth, approximate the area of the Bermuda Triangle.

Source: *www.worldatlas.com*

45. The Flatiron Building Completed in 1902 in New York City, the Flatiron Building is triangular shaped and bounded by 22nd Street, Broadway, and 5th Avenue. The building measures approximately 87 feet on the 22nd Street side, 190 feet on the Broadway side, and 173 feet on the 5th Avenue side. Approximate the ground area covered by the building.

Source: Sarah Bradford Landau and Carl W. Condit, *Rise of the New York Skyscraper: 1865–1913.* New Haven, CT: Yale University Press, 1996

46. Area of a Quadrilateral Bretschneider's Formula is Heron-type formula that can be used to find the area of a general quadrilateral.

$$K = \sqrt{(s-a)(s-b)(s-c)(s-d) - abcd\cos^2\theta}$$

where a, b, c, and d are the side lengths, θ is half the sum of two opposite angles, and s is half the perimeter.

(a) Show that if a triangle is considered a quadrilateral with one side equal to 0, Bretschneider's Formula reduces to Heron's Formula.

(b) Use Bretschneiders's Formula to find the area of the lot for sale in Problem 39, rounded to the nearest square foot. Do the results agree?

47. Geometry Refer to the figure. If $|OA| = 1$, show that:

(a) Area $\Delta OAC = \frac{1}{2}\sin\alpha\cos\alpha$

(b) Area $\Delta OCB = \frac{1}{2}|OB|^2\sin\beta\cos\beta$

(c) Area $\Delta OAB = \frac{1}{2}|OB|\sin(\alpha+\beta)$

(d) $|OB| = \dfrac{\cos\alpha}{\cos\beta}$

(e) $\sin(\alpha+\beta) = \sin\alpha\cos\beta + \cos\alpha\sin\beta$

[**Hint:** Area ΔOAB = Area ΔOAC + Area ΔOCB]

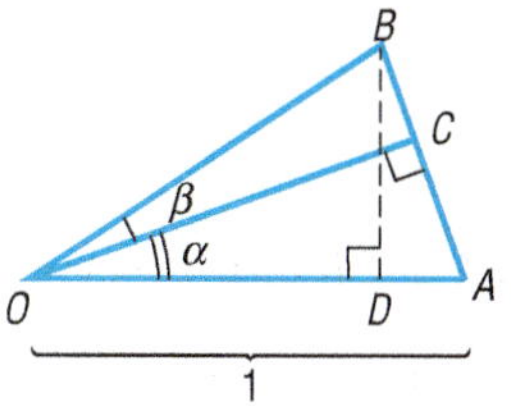

48. Geometry Refer to the figure, in which a unit circle is drawn. The line segment DB is tangent to the circle, and θ is acute.

(a) Express the area of ΔOBC in terms of $\sin\theta$ and $\cos\theta$.

(b) Express the area of ΔOBD in terms of $\sin\theta$ and $\cos\theta$.

(c) The area of the sector $\widehat{OBC}$ of the circle is $\frac{1}{2}\theta$, where θ is measured in radians. Use the results of parts (a) and (b) and the fact that

$$\text{Area } \Delta OBC < \text{Area } \widehat{OBC} < \text{Area } \Delta OBD$$

to show that

$$1 < \frac{\theta}{\sin\theta} < \frac{1}{\cos\theta}$$

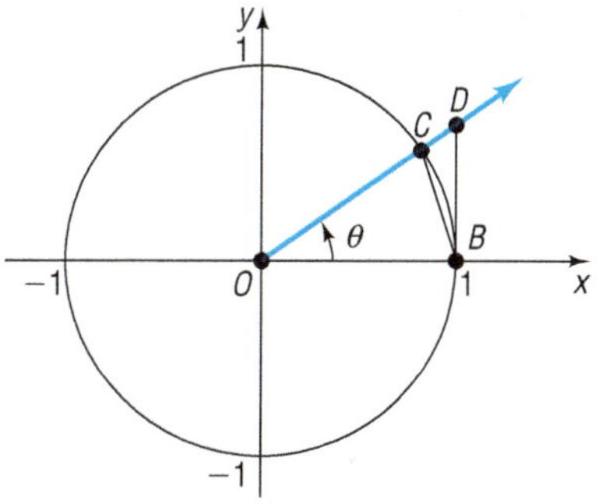

49. The Cow Problem* A cow is tethered to one corner of a square barn, 10 feet by 10 feet, with a rope 100 feet long. What is the maximum grazing area for the cow? See the illustration.

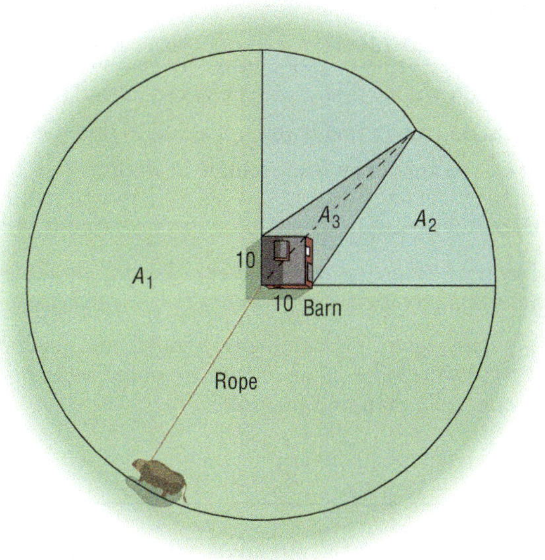

*Suggested by Professor Teddy Koukounas of Suffolk Community College, who learned of it from an old farmer in Virginia. Solution provided by Professor Kathleen Miranda of SUNY at Old Westbury.

50. **Another Cow Problem** If the barn in Problem 49 is rectangular, 10 feet by 20 feet, what is the maximum grazing area for the cow?

51. **Perfect Triangles** A *perfect triangle* is one having integers for sides for which the area is numerically equal to the perimeter. Show that the triangles with the given side lengths are perfect.

(a) 9, 10, 17 (b) 6, 25, 29

Source: M.V. Bonsangue, G. E. Gannon, E. Buchman, and N. Gross, "In Search of Perfect Triangles," *Mathematics Teacher,* Vol. 92, No. 1, 1999: 56–61

52. If h_1, h_2, and h_3 are the altitudes dropped from P, Q, and R, respectively, in a triangle (see the figure), show that

$$\frac{1}{h_1}+\frac{1}{h_2}+\frac{1}{h_3}=\frac{s}{K}$$

where K is the area of the triangle and $s=\frac{1}{2}(a+b+c)$.

[**Hint:** $h_1=\frac{2K}{a}$.]

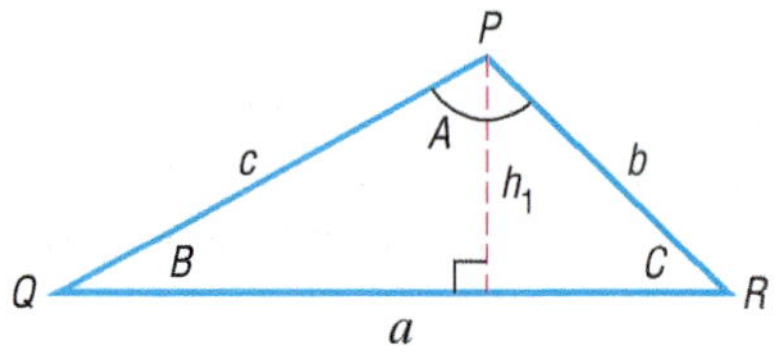

53. Show that a formula for the altitude h from a vertex to the opposite side a of a triangle is

$$h=\frac{a\sin B\sin C}{\sin A}$$

54. Refer to Problem 103 in Section 8.6.

(a) Show that the area of a regular dodecagon is given by $K=3\cot\left(\frac{\pi}{12}\right)a^2$ or $K=12\tan\left(\frac{\pi}{12}\right)r^2$, where a is the length of one of the sides and r is the radius of the inscribed circle.

(b) Given that each interior angle of a regular n-sided polygon $(n\geq 3)$ measures $\frac{(n-2)\cdot 180^\circ}{n}$, generalize these formulas for any such polygon.

Inscribed Circle *For Problems 55–58, the lines that bisect each angle of a triangle meet in a single point O, and the perpendicular distance r from O to each side of the triangle is the same. The circle with center at O and radius r is called the inscribed circle of the triangle (see the figure).*

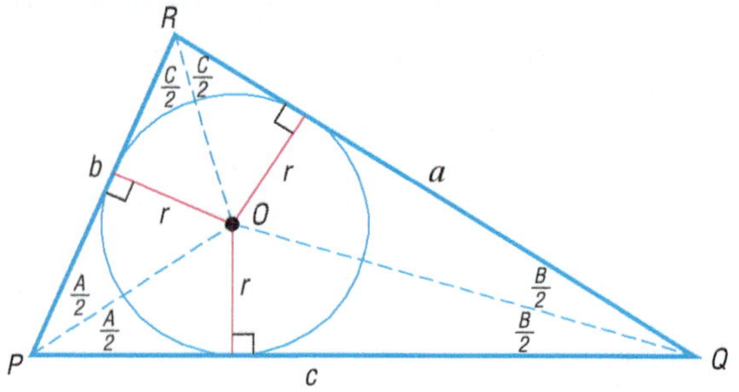

55. Apply the formula from Problem 53 to triangle OPQ to show that

$$r=\frac{c\sin\frac{A}{2}\sin\frac{B}{2}}{\cos\frac{C}{2}}$$

56. Use the result of Problem 55 and the results of Problem 60 and 61 in Section 9.3 to show that

$$\cot\frac{C}{2}=\frac{s-c}{r}$$

where $s=\frac{1}{2}(a+b+c)$.

57. Show that $\cot\frac{A}{2}+\cot\frac{B}{2}+\cot\frac{C}{2}=\frac{s}{r}$.

58. Show that the area K of triangle PQR is $K=rs$, where $s=\frac{1}{2}(a+b+c)$. Then show that

$$r=\sqrt{\frac{(s-a)(s-b)(s-c)}{s}}$$

59. A triangle has vertices $A(0,0)$, $B(1,0)$, and C, where C is the point on the unit circle corresponding to an angle of 105° when it is drawn in standard position. Find the area of the triangle. State the answer in complete simplified form with a rationalized denominator.

Explaining Concepts: Discussion and Writing

60. What do you do first if you are asked to find the area of a triangle and are given two sides and the included angle?

61. What do you do first if you are asked to find the area of a triangle and are given three sides?

62. State the area of an SAS triangle in words.

Retain Your Knowledge

Problems 63–66 are based on material learned earlier in the course. The purpose of these problems is to keep the material fresh in your mind so that you are better prepared for the final exam.

63. Without graphing, determine whether the quadratic function $f(x)=-3x^2+12x+5$ has a maximum value or a minimum value, and then find the value.

64. Solve the inequality: $\frac{x+1}{x^2-9}\leq 0$

65. $P=\left(-\frac{\sqrt{7}}{3},\frac{\sqrt{2}}{3}\right)$ is the point on the unit circle that corresponds to a real number t. Find the exact values of the six trigonometric functions of t.

66. Establish the identity: $\csc\theta-\sin\theta=\cos\theta\cot\theta$

'Are You Prepared?' Answer

1. $K=\frac{1}{2}bh$

9.5 Simple Harmonic Motion; Damped Motion; Combining Waves

PREPARING FOR THIS SECTION *Before getting started, review the following:*

- Sinusoidal Graphs (Section 7.6, pp. 576–582)

Now Work the **'Are You Prepared?'** problem on page 721.

OBJECTIVES
1. Build a Model for an Object in Simple Harmonic Motion (p. 715)
2. Analyze Simple Harmonic Motion (p. 717)
3. Analyze an Object in Damped Motion (p. 718)
4. Graph the Sum of Two Functions (p. 719)

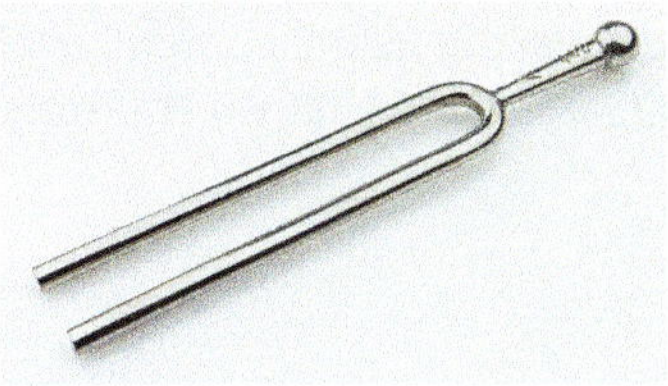

Tuning fork

1 Build a Model for an Object in Simple Harmonic Motion

Many physical phenomena can be described as simple harmonic motion. Radio and television waves, light waves, sound waves, and water waves exhibit motion that is simple harmonic.

The swinging of a pendulum, the vibrations of a tuning fork, and the bobbing of a weight attached to a coiled spring are examples of vibrational motion. In this type of motion, an object swings back and forth over the same path. In Figure 33, the point B is the **equilibrium (rest) position** of the vibrating object. The **amplitude** is the distance from the object's rest position to its point of greatest displacement (either point A or point C in Figure 33). The **period** is the time required to complete one vibration—that is, the time it takes to go from, say, point A through B to C and back to A.

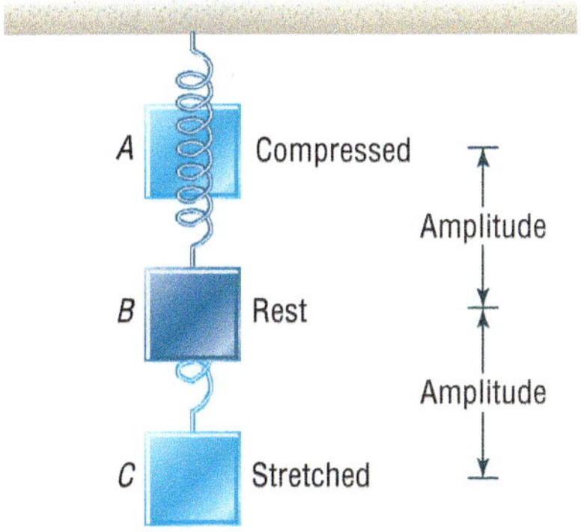

Figure 33 Coiled spring

Simple harmonic motion is a special kind of vibrational motion in which the acceleration a of the object is directly proportional to the negative of its displacement d from its rest position. That is, $a = -kd$, $k > 0$.

For example, when the mass hanging from the spring in Figure 33 is pulled down from its rest position B to the point C, the force of the spring tries to restore the mass to its rest position. Assuming that there is no frictional force* to retard the motion, the amplitude will remain constant. The force increases in direct proportion to the distance that the mass is pulled from its rest position. Since the force increases directly, the acceleration of the mass of the object must do likewise, because (by Newton's Second Law of Motion) force is directly proportional to acceleration. As a result, the acceleration of the object varies directly with its displacement, and the motion is an example of simple harmonic motion.

Simple harmonic motion is related to circular motion. To see this relationship, consider a circle of radius a, with center at $(0,0)$. See Figure 34. Suppose that an object initially placed at $(a,0)$ moves counterclockwise around the circle at a constant angular speed ω. Suppose further that after time t has elapsed the object is at the point $P = (x, y)$ on the circle. The angle θ, in radians, swept out by the ray $\overrightarrow{OP}$ in this time t is

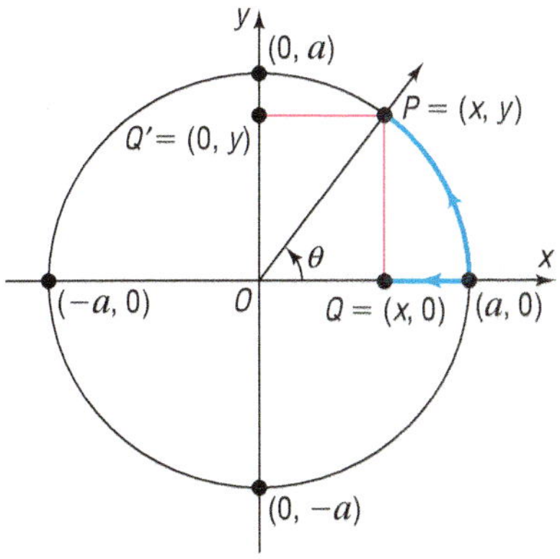

Figure 34

$$\theta = \omega t \qquad \omega = \frac{\theta}{t}$$

*If friction is present, the amplitude will decrease with time to 0. This type of motion is an example of **damped motion**, which is discussed later in this section.

The coordinates of the point P at time t are

$$x = a\cos\theta = a\cos(\omega t)$$
$$y = a\sin\theta = a\sin(\omega t)$$

Corresponding to each position $P = (x, y)$ of the object moving about the circle, there is the point $Q = (x, 0)$, called the **projection of P on the x-axis.** As P moves around the circle at a constant rate, the point Q moves back and forth between the points $(a, 0)$ and $(-a, 0)$ along the x-axis with a motion that is simple harmonic. Similarly, for each point P there is a point $Q' = (0, y)$, called the **projection of P on the y-axis.** As P moves around the circle, the point Q' moves back and forth between the points $(0, a)$ and $(0, -a)$ on the y-axis with a motion that is simple harmonic. Simple harmonic motion can be described as the projection of constant circular motion on a coordinate axis.

To put it another way, again consider a mass hanging from a spring where the mass is pulled down from its rest position to the point C and then released. See Figure 35(a). The graph shown in Figure 35(b) describes the displacement d of the object from its rest position as a function of time t, assuming that no frictional force is present.

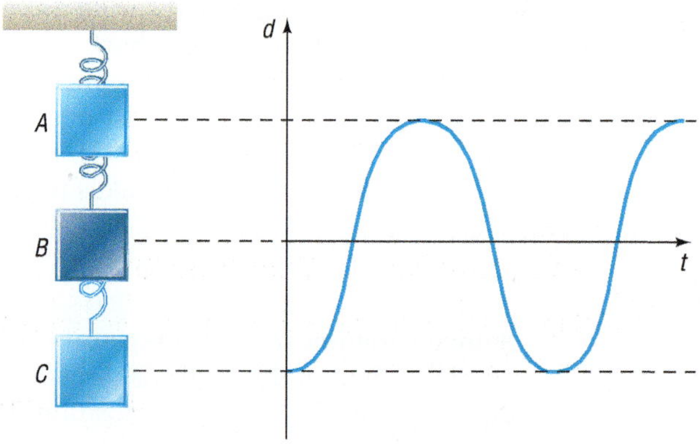

Figure 35 (a) (b)

THEOREM

Simple Harmonic Motion

An object that moves on a coordinate axis so that the displacement d from its rest position at time t is given by either

$$d = a\cos(\omega t) \quad \text{or} \quad d = a\sin(\omega t)$$

where a and $\omega > 0$ are constants, moves with simple harmonic motion. The motion has amplitude $|a|$ and period $\frac{2\pi}{\omega}$.

The **frequency** f of an object in simple harmonic motion is the number of oscillations per unit time. Since the period is the time required for one oscillation, it follows that the frequency is the reciprocal of the period; that is,

$$f = \frac{\omega}{2\pi} \qquad \omega > 0$$

EXAMPLE 1 Build a Model for an Object in Harmonic Motion

Suppose that an object attached to a coiled spring is pulled down a distance of 5 inches from its rest position and then released. If the time for one oscillation is 3 seconds, develop a model that relates the displacement d of the object from its rest position after time t (in seconds). Assume no friction.

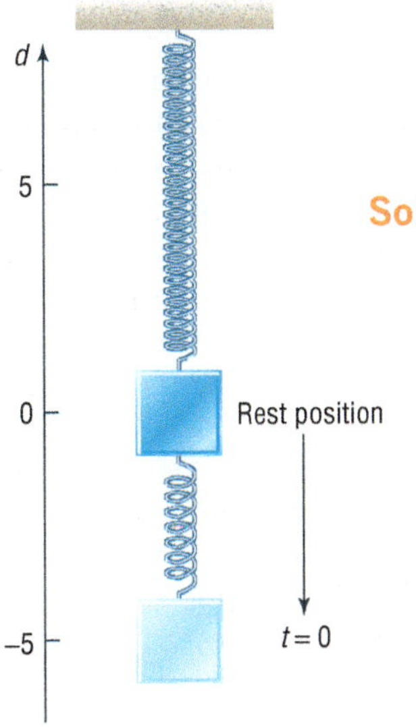

Figure 36

Note: In the solution to Example 1, $a = -5$ because the object is initially pulled down. (If the initial direction is up, then use $a = 5$.) ■

Solution The motion of the object is simple harmonic. See Figure 36. When the object is released $(t = 0)$, the displacement of the object from the rest position is -5 units (since the object was pulled down). Because $d = -5$ when $t = 0$, it is easier to use the cosine function*

$$d = a\cos(\omega t)$$

to describe the motion. Now the amplitude is $|-5| = 5$ and the period is 3, so

$$a = -5 \quad \text{and} \quad \frac{2\pi}{\omega} = \text{period} = 3, \quad \text{so } \omega = \frac{2\pi}{3}$$

An equation that models the motion of the object is

$$d = -5\cos\left[\frac{2\pi}{3}t\right]$$

■

Now Work PROBLEM 5

2 Analyze Simple Harmonic Motion

EXAMPLE 2 Analyzing the Motion of an Object

Suppose that the displacement d (in meters) of an object at time t (in seconds) satisfies the equation

$$d = 10\sin(5t)$$

(a) Describe the motion of the object.
(b) What is the maximum displacement from its resting position?
(c) What is the time required for one oscillation?
(d) What is the frequency?

Solution Observe that the given equation is of the form

$$d = a\sin(\omega t) \qquad d = 10\sin(5t)$$

where $a = 10$ and $\omega = 5$.

(a) The motion is simple harmonic.
(b) The maximum displacement of the object from its resting position is the amplitude: $|a| = 10$ meters.
(c) The time required for one oscillation is the period:

$$\text{Period} = \frac{2\pi}{\omega} = \frac{2\pi}{5} \text{ seconds}$$

(d) The frequency is the reciprocal of the period.

$$\text{Frequency} = f = \frac{5}{2\pi} \text{ oscillation per second}$$

■

Now Work PROBLEM 13

*No phase shift is required when a cosine function is used.

3 Analyze an Object in Damped Motion

Most physical phenomena are affected by friction or other resistive forces. These forces remove energy from a moving system and thereby damp its motion. For example, when a mass hanging from a spring is pulled down a distance a and released, the friction in the spring causes the distance that the mass moves from its at-rest position to decrease over time. As a result, the amplitude of any real oscillating spring or swinging pendulum decreases with time due to air resistance, friction, and so forth. See Figure 37.

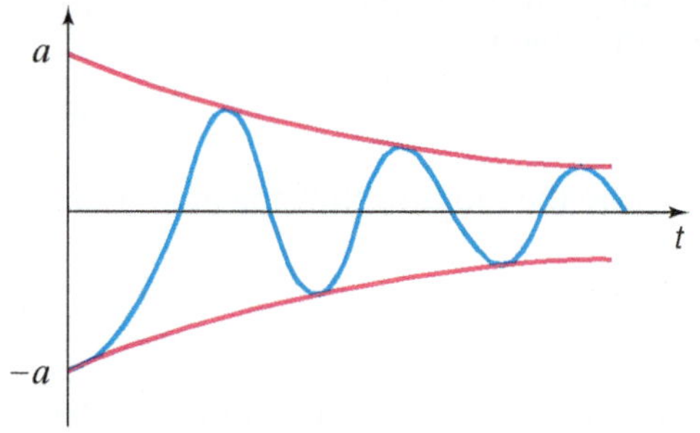

Figure 37

A model that describes this phenomenon maintains a sinusoidal component, but the amplitude of this component will decrease with time to account for the damping effect. In addition, the period of the oscillating component will be affected by the damping. The next result, from physics, describes damped motion.

THEOREM

Damped Motion

The displacement d of an oscillating object from its at-rest position at time t is given by

$$d(t) = ae^{-bt/(2m)} \cos\left(\sqrt{\omega^2 - \frac{b^2}{4m^2}}\, t\right)$$

where b is the **damping factor** or **damping coefficient** and m is the mass of the oscillating object. Here $|a|$ is the displacement at $t = 0$, and $\dfrac{2\pi}{\omega}$ is the period under simple harmonic motion (no damping). ■

Notice that for $b = 0$ (zero damping), we have the formula for simple harmonic motion with amplitude $|a|$ and period $\dfrac{2\pi}{\omega}$.

EXAMPLE 3

Analyzing a Damped Vibration Curve

Analyze the damped vibration curve

$$d(t) = e^{-t/\pi} \cos t \quad t \geq 0$$

Solution The displacement d is the product of $y = e^{-t/\pi}$ and $y = \cos t$. Using properties of absolute value and the fact that $|\cos t| \leq 1$, it follows that

$$|d(t)| = |e^{-t/\pi} \cos t| = |e^{-t/\pi}|\,|\cos t| \leq |e^{-t/\pi}| \underset{\substack{\uparrow \\ e^{-t/\pi} > 0}}{=} e^{-t/\pi}$$

As a result,

$$-e^{-t/\pi} \leq d(t) \leq e^{-t/\pi}$$

This means that the graph of d will lie between the graphs of $y = e^{-t/\pi}$ and $y = -e^{-t/\pi}$, called the **bounding curves** of d.

Also, the graph of d will touch these graphs when $|\cos t| = 1$; that is, when $t = 0, \pi, 2\pi$, and so on. The x-intercepts of the graph of d occur when $\cos t = 0$, that is, at $\frac{\pi}{2}, \frac{3\pi}{2}, \frac{5\pi}{2}$, and so on. See Table 1.

Table 1

t	0	$\frac{\pi}{2}$	π	$\frac{3\pi}{2}$	2π
$e^{-t/\pi}$	1	$e^{-1/2}$	e^{-1}	$e^{-3/2}$	e^{-2}
$\cos t$	1	0	-1	0	1
$d(t) = e^{-t/\pi}\cos t$	1	0	$-e^{-1}$	0	e^{-2}
Point on graph of d	$(0, 1)$	$\left(\frac{\pi}{2}, 0\right)$	$(\pi, -e^{-1})$	$\left(\frac{3\pi}{2}, 0\right)$	$(2\pi, e^{-2})$

The graphs of $y = \cos t$, $y = e^{-t/\pi}$, $y = -e^{-t/\pi}$, and $d(t) = e^{-t/\pi}\cos t$ are shown in in Figure 38.

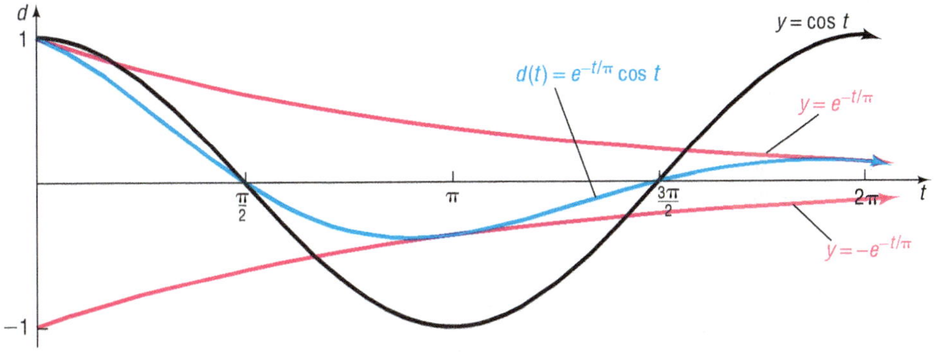

Figure 38

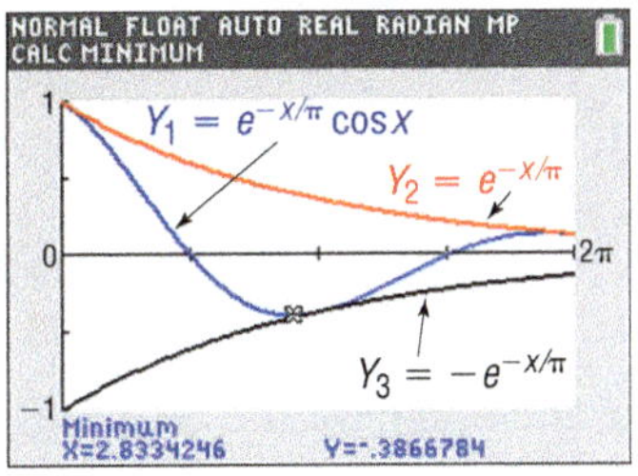

Figure 39

Exploration

Graph $Y_1 = e^{-x/\pi}\cos x$, along with $Y_2 = e^{-x/\pi}$ and $Y_3 = -e^{-x/\pi}$, for $0 \le x \le 2\pi$. Determine where Y_1 has its first turning point (local minimum). Compare this to where Y_1 intersects Y_3.

Result Figure 39 shows the graphs of $Y_1 = e^{-x/\pi}\cos x$, $Y_2 = e^{-x/\pi}$, and $Y_3 = -e^{-x/\pi}$. Using MINIMUM, the first turning point occurs at $x \approx 2.83$; Y_1 INTERSECTS Y_3 at $x = \pi \approx 3.14$.

Now Work PROBLEM 21

4 Graph the Sum of Two Functions

Many physical and biological applications require the graph of the sum of two functions, such as

$$f(x) = x + \sin x \quad \text{or} \quad g(x) = \sin x + \cos(2x)$$

For example, if two tones are emitted, the sound produced is the sum of the waves produced by the two tones. See Problem 57 for an explanation of Touch-Tone phones.

To graph the sum of two (or more) functions, use the method of adding y-coordinates described next.

EXAMPLE 4

Graphing the Sum of Two Functions

Use the method of adding y-coordinates to graph $f(x) = x + \sin x$.

Solution First, graph the component functions,

$$y = f_1(x) = x \qquad y = f_2(x) = \sin x$$

on the same coordinate system. See Figure 40(a). Now, select several values of x say $x = 0$, $x = \frac{\pi}{2}$, $x = \pi$, $x = \frac{3\pi}{2}$, and $x = 2\pi$, and use them to compute $f(x) = f_1(x) + f_2(x)$. Table 2 shows the computations. Plot these points and connect them to get the graph, as shown in Figure 40(b).

Table 2

x	0	$\frac{\pi}{2}$	π	$\frac{3\pi}{2}$	2π
$y = f_1(x) = x$	0	$\frac{\pi}{2}$	π	$\frac{3\pi}{2}$	2π
$y = f_2(x) = \sin x$	0	1	0	-1	0
$f(x) = x + \sin x$	0	$\frac{\pi}{2} + 1 \approx 2.57$	π	$\frac{3\pi}{2} - 1 \approx 3.71$	2π
Point on graph of f	$(0, 0)$	$\left(\frac{\pi}{2}, 2.57\right)$	(π, π)	$\left(\frac{3\pi}{2}, 3.71\right)$	$(2\pi, 2\pi)$

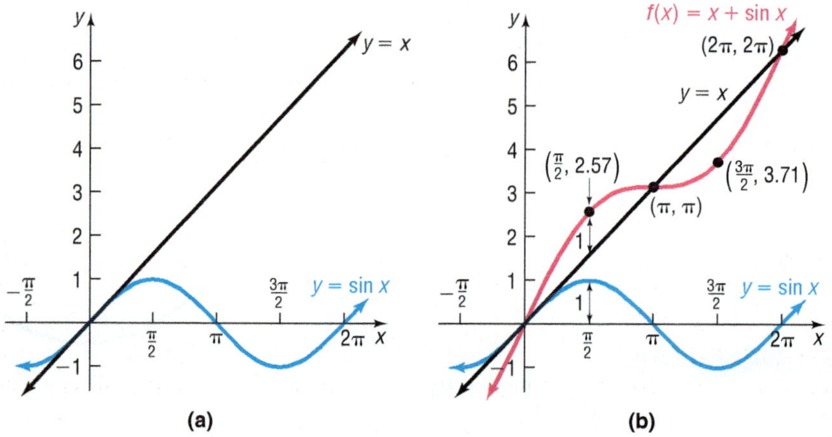

Figure 40

In Figure 40(b), notice that the graph of $f(x) = x + \sin x$ intersects the line $y = x$ whenever $\sin x = 0$. Also, notice that the graph of f is not periodic. ■

✓**Check:** Graph $Y_1 = x$, $Y_2 = \sin x$, and $Y_3 = x + \sin x$ and compare the result with Figure 40(b). Use INTERSECT to verify that the graphs of Y_1 and Y_3 intersect when $\sin x = 0$.

The next example shows a periodic graph.

EXAMPLE 5 Graphing the Sum of Two Sinusoidal Functions

Use the method of adding y-coordinates to graph

$$f(x) = \sin x + \cos(2x)$$

Solution Table 3 shows the steps for computing several points on the graph of f. Figure 41 illustrates the graphs of the component functions, $y = f_1(x) = \sin x$ and $y = f_2(x) = \cos(2x)$, and the graph of $f(x) = \sin x + \cos(2x)$, which is shown in red.

Table 3

x	$-\frac{\pi}{2}$	0	$\frac{\pi}{2}$	π	$\frac{3\pi}{2}$	2π
$y = f_1(x) = \sin x$	-1	0	1	0	-1	0
$y = f_2(x) = \cos(2x)$	-1	1	-1	1	-1	1
$f(x) = \sin x + \cos(2x)$	-2	1	0	1	-2	1
Point on graph of f	$\left(-\frac{\pi}{2}, -2\right)$	$(0, 1)$	$\left(\frac{\pi}{2}, 0\right)$	$(\pi, 1)$	$\left(\frac{3\pi}{2}, -2\right)$	$(2\pi, 1)$

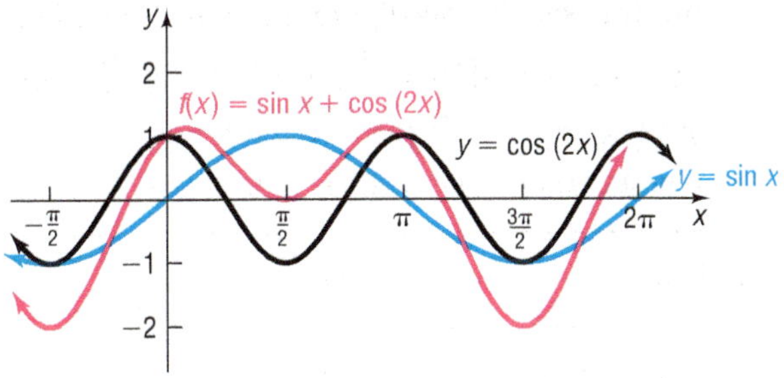

Figure 41

Notice that f is periodic, with period 2π.

✓Check: Graph $Y_1 = \sin x$, $Y_2 = \cos(2x)$, and $Y_3 = \sin x + \cos(2x)$ and compare the result with Figure 41.

Now Work PROBLEM 25

9.5 Assess Your Understanding

'Are You Prepared?' *The answers are given at the end of these exercises. If you get a wrong answer, read the pages listed in red.*

1. The amplitude A and period T of $f(x) = 5\sin(4x)$ are __ and __. (pp. 576–582)

Concepts and Vocabulary

2. The motion of an object obeys the equation $d = 4\cos(6t)$. Such motion is described as _______ _________. The number 4 is called the _________.

3. When a mass hanging from a spring is pulled down and then released, the motion is called _______ _________ if there is no frictional force to retard the motion, and the motion is called ________ if there is such friction.

4. ***True or False*** If the distance d of an object from its rest position at time t is given by a sinusoidal graph, the motion of the object is simple harmonic motion.

Skill Building

In Problems 5–8, an object attached to a coiled spring is pulled down a distance a from its rest position and then released. Assuming that the motion is simple harmonic with period T, write an equation that relates the displacement d of the object from its rest position after t seconds. Also assume that the positive direction of the motion is up.

5. $a = 5$; $T = 2$ seconds

6. $a = 10$; $T = 3$ seconds

7. $a = 6$; $T = \pi$ seconds

8. $a = 4$; $T = \frac{\pi}{2}$ seconds

9. Rework Problem 5 under the same conditions, except that at time $t = 0$, the object is at its resting position and moving down.

10. Rework Problem 6 under the same conditions, except that at time $t = 0$, the object is at its resting position and moving down.

11. Rework Problem 7 under the same conditions, except that at time $t = 0$, the object is at its resting position and moving down.

12. Rework Problem 8 under the same conditions, except that at time $t = 0$, the object is at its resting position and moving down.

In Problems 13–20, the displacement d (in meters) of an object at time t (in seconds) is given.
(a) Describe the motion of the object.
(b) What is the maximum displacement from its resting position?
(c) What is the time required for one oscillation?
(d) What is the frequency?

13. $d = 5\sin(3t)$

14. $d = 4\sin(2t)$

15. $d = 6\cos(\pi t)$

16. $d = 5\cos\left(\frac{\pi}{2}t\right)$

17. $d = -3\sin\left(\frac{1}{2}t\right)$

18. $d = -2\cos(2t)$

19. $d = 6 + 2\cos(2\pi t)$

20. $d = 4 + 3\sin(\pi t)$

In Problems 21–24, graph each damped vibration curve for $0 \le t \le 2\pi$.

21. $d(t) = e^{-t/\pi}\cos(2t)$
22. $d(t) = e^{-t/2\pi}\cos(2t)$
23. $d(t) = e^{-t/2\pi}\cos t$
24. $d(t) = e^{-t/4\pi}\cos t$

In Problems 25–32, use the method of adding y-coordinates to graph each function.

25. $f(x) = x + \cos x$
26. $f(x) = x + \cos(2x)$
27. $f(x) = x - \sin x$
28. $f(x) = x - \cos x$
29. $f(x) = \sin x + \cos x$
30. $f(x) = \sin(2x) + \cos x$
31. $g(x) = \sin x + \sin(2x)$
32. $g(x) = \cos(2x) + \cos x$

Mixed Practice

In Problems 33–38, (a) use the Product-to-Sum Formulas to express each product as a sum, and (b) use the method of adding y-coordinates to graph each function on the interval $[0, 2\pi]$.

33. $f(x) = \sin(2x)\sin x$
34. $F(x) = \sin(3x)\sin x$
35. $G(x) = \cos(4x)\cos(2x)$
36. $h(x) = \cos(2x)\cos(x)$
37. $H(x) = 2\sin(3x)\cos(x)$
38. $g(x) = 2\sin x\cos(3x)$

Applications and Extensions

In Problems 39–44, an object of mass m (in grams) attached to a coiled spring with damping factor b (in grams per second) is pulled down a distance a (in centimeters) from its rest position and then released. Assume that the positive direction of the motion is up and the period is T (in seconds) under simple harmonic motion.

(a) Write an equation that relates the displacement d of the object from its rest position after t seconds.
(b) Graph the equation found in part (a) for 5 oscillations using a graphing utility.

39. $m = 25, \quad a = 10, \quad b = 0.7, \quad T = 5$
40. $m = 20, \quad a = 15, \quad b = 0.75, \quad T = 6$
41. $m = 30, \quad a = 18, \quad b = 0.6, \quad T = 4$
42. $m = 15, \quad a = 16, \quad b = 0.65, \quad T = 5$
43. $m = 10, \quad a = 5, \quad b = 0.8, \quad T = 3$
44. $m = 10, \quad a = 5, \quad b = 0.7, \quad T = 3$

In Problems 45–50, the distance d (in meters) of the bob of a pendulum of mass m (in kilograms) from its rest position at time t (in seconds) is given. The bob is released from the left of its rest position and represents a negative direction.

(a) Describe the motion of the object. Be sure to give the mass and damping factor.
(b) What is the initial displacement of the bob? That is, what is the displacement at $t = 0$?
(c) Graph the motion using a graphing utility.
(d) What is the displacement of the bob at the start of the second oscillation?
(e) What happens to the displacement of the bob as time increases without bound?

45. $d = -20e^{-0.7t/40}\cos\left(\sqrt{\left(\frac{2\pi}{5}\right)^2 - \frac{0.49}{1600}}\,t\right)$
46. $d = -20e^{-0.8t/40}\cos\left(\sqrt{\left(\frac{2\pi}{5}\right)^2 - \frac{0.64}{1600}}\,t\right)$
47. $d = -30e^{-0.6t/80}\cos\left(\sqrt{\left(\frac{2\pi}{7}\right)^2 - \frac{0.36}{6400}}\,t\right)$
48. $d = -30e^{-0.5t/70}\cos\left(\sqrt{\left(\frac{\pi}{2}\right)^2 - \frac{0.25}{4900}}\,t\right)$
49. $d = -15e^{-0.9t/30}\cos\left(\sqrt{\left(\frac{\pi}{3}\right)^2 - \frac{0.81}{900}}\,t\right)$
50. $d = -10e^{-0.8t/50}\cos\left(\sqrt{\left(\frac{2\pi}{3}\right)^2 - \frac{0.64}{2500}}\,t\right)$

51. **Loudspeaker** A loudspeaker diaphragm is oscillating in simple harmonic motion described by the equation $d = a\cos(\omega t)$ with a frequency of 520 hertz (cycles per second) and a maximum displacement of 0.80 millimeter. Find ω and then determine the equation that describes the movement of the diaphragm.

52. **Colossus** Added to Six Flags St. Louis in 1986, the Colossus is a giant Ferris wheel. Its diameter is 165 feet, it rotates at a rate of about 1.6 revolutions per minute, and the bottom of the wheel is 15 feet above the ground. Determine an equation that relates a rider's height above the ground at time t. Assume the passenger begins the ride at the bottom of the wheel.

    Source: *Six Flags Theme Parks, Inc.*

53. **Tuning Fork** The end of a tuning fork moves in simple harmonic motion described by the equation $d = a\sin(\omega t)$. If a tuning fork for the note A above middle C on an even-tempered scale (A_4, the tone by which an orchestra tunes itself) has a frequency of 440 hertz (cycles per second), find ω. If the maximum displacement of the end of the tuning fork is 0.01 millimeter, determine the equation that describes the movement of the tuning fork.

    Source: David Lapp. *Physics of Music and Musical Instruments.* Medford, MA: Tufts University, 2003

54. **Tuning Fork** The end of a tuning fork moves in simple harmonic motion described by the equation $d = a\sin(\omega t)$. If a tuning fork for the note E above middle C on an even-tempered scale (E_4) has a frequency of approximately 329.63 hertz (cycles per second), find ω. If the maximum displacement of the end of the tuning fork is 0.025 millimeter, determine the equation that describes the movement of the tuning fork.

    Source: David Lapp. *Physics of Music and Musical Instruments.* Medford, MA: Tufts University, 2003

55. Charging a Capacitor See the illustration. If a charged capacitor is connected to a coil by closing a switch, energy is transferred to the coil and then back to the capacitor in an oscillatory motion. The voltage V (in volts) across the capacitor will gradually diminish to 0 with time t (in seconds).

(a) Graph the function relating V and t:

$$V(t) = e^{-t/3}\cos(\pi t) \qquad 0 \le t \le 3$$

(b) At what times t will the graph of V touch the graph of $y = e^{-t/3}$? When does the graph of V touch the graph of $y = -e^{-t/3}$?

(c) When will the voltage V be between -0.4 and 0.4 volt?

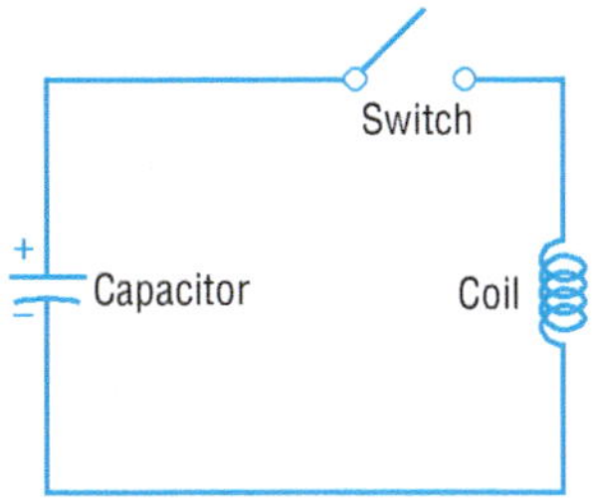

56. The Sawtooth Curve An oscilloscope often displays a *sawtooth curve*. This curve can be approximated by sinusoidal curves of varying periods and amplitudes.

(a) Use a graphing utility to graph the following function, which can be used to approximate the sawtooth curve.

$$f(x) = \frac{1}{2}\sin(2\pi x) + \frac{1}{4}\sin(4\pi x) \qquad 0 \le x \le 4$$

(b) A better approximation to the sawtooth curve is given by

$$f(x) = \frac{1}{2}\sin(2\pi x) + \frac{1}{4}\sin(4\pi x) + \frac{1}{8}\sin(8\pi x)$$

Use a graphing utility to graph this function for $0 \le x \le 4$ and compare the result to the graph obtained in part (a).

(c) A third and even better approximation to the sawtooth curve is given by

$$f(x) = \frac{1}{2}\sin(2\pi x) + \frac{1}{4}\sin(4\pi x) + \frac{1}{8}\sin(8\pi x) + \frac{1}{16}\sin(16\pi x)$$

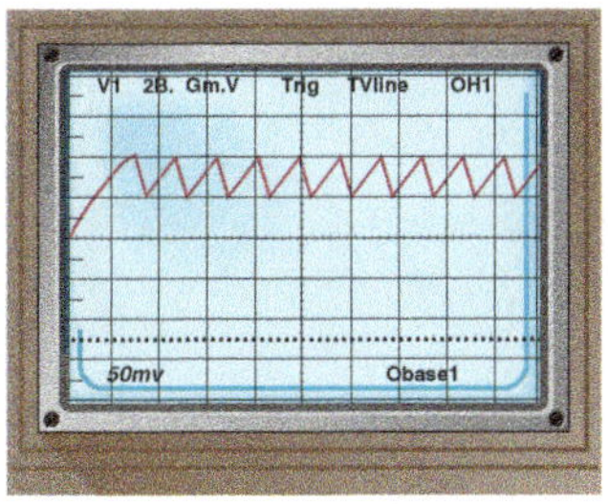

Use a graphing utility to graph this function for $0 \le x \le 4$ and compare the result to the graphs obtained in parts (a) and (b).

(d) What do you think the next approximation to the sawtooth curve is?

57. Touch-Tone Phones On a Touch-Tone phone, each button produces a unique sound. The sound produced is the sum of two tones, given by

$$y = \sin(2\pi l t) \quad \text{and} \quad y = \sin(2\pi h t)$$

where l and h are the low and high frequencies (cycles per second) shown in the illustration. For example, if you touch 7, the low frequency is $l = 852$ cycles per second and the high frequency is $h = 1209$ cycles per second. The sound emitted when someone touches 7 is

$$y = \sin[2\pi(852)t] + \sin[2\pi(1209)t]$$

Use a graphing utility to graph the sound emitted when 7 is touched.

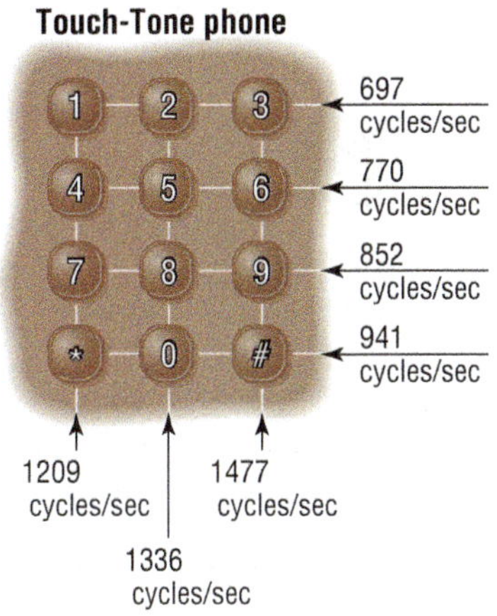

58. Use a graphing utility to graph the sound emitted by the * key on a Touch-Tone phone. See Problem 57.

59. CBL Experiment Pendulum motion is analyzed to estimate simple harmonic motion. A plot is generated with the position of the pendulum over time. The graph is used to find a sinusoidal curve of the form $y = A\cos[B(x - C)] + D$. Determine the amplitude, period, and frequency. (Activity 16, Real-World Math with the CBL System.)

60. CBL Experiment The sound from a tuning fork is collected over time. A model of the form $y = A\cos[B(x - C)]$ is fitted to the data. Determine the amplitude, frequency, and period of the graph.
(Activity 23, Real-World Math with the CBL System.)

Explaining Concepts: Discussion and Writing

61. Use a graphing utility to graph the function $f(x) = \dfrac{\sin x}{x}$, $x > 0$. Based on the graph, what do you conjecture about the value of $\dfrac{\sin x}{x}$ for x close to 0?

62. Use a graphing utility to graph $y = x\sin x$, $y = x^2\sin x$, and $y = x^3\sin x$ for $x > 0$. What patterns do you observe?

63. Use a graphing utility to graph $y = \dfrac{1}{x}\sin x$, $y = \dfrac{1}{x^2}\sin x$, and $y = \dfrac{1}{x^3}\sin x$ for $x > 0$. What patterns do you observe?

64. How would you explain to a friend what simple harmonic motion is? How would you explain damped motion?

Retain Your Knowledge

Problems 65–68 are based on material learned earlier in the course. The purpose of these problems is to keep the material fresh in your mind so that you are better prepared for the final exam.

65. The function $f(x) = \dfrac{x-3}{x-4}$, $x \neq 4$, is one-to-one. Find its inverse function.

66. Write as a single logarithm: $\log_7 x + 3\log_7 y - \log_7(x+y)$

67. Solve: $\log(x+1) + \log(x-2) = 1$

68. Given $\cos\alpha = \dfrac{4}{5}$, $0 < \alpha < \dfrac{\pi}{2}$, find the exact value of:

(a) $\cos\dfrac{\alpha}{2}$ (b) $\sin\dfrac{\alpha}{2}$ (c) $\tan\dfrac{\alpha}{2}$

'Are You Prepared?' Answer

1. $A = 5$; $T = \dfrac{\pi}{2}$

Chapter Review

Things to Know

Formulas

Law of Sines (p. 691)

$$\frac{\sin A}{a} = \frac{\sin B}{b} = \frac{\sin C}{c}$$

Law of Cosines (p. 701)

$$c^2 = a^2 + b^2 - 2ab\cos C$$
$$b^2 = a^2 + c^2 - 2ac\cos B$$
$$a^2 = b^2 + c^2 - 2bc\cos A$$

Area of a triangle (pp. 708–711)

$$K = \frac{1}{2}bh \quad K = \frac{1}{2}ab\sin C \quad K = \frac{1}{2}bc\sin A \quad K = \frac{1}{2}ac\sin B$$

$$K = \sqrt{s(s-a)(s-b)(s-c)}, \text{ where } s = \frac{1}{2}(a+b+c)$$

Objectives

Section		You should be able to . . .	Examples	Review Exercises
9.1	1	Solve right triangles (p. 685)	1, 2	1, 2
	2	Solve applied problems (p. 686)	3–6	21, 28
9.2	1	Solve SAA or ASA triangles (p. 692)	1, 2	3, 14
	2	Solve SSA triangles (p. 693)	3–5	4, 5, 7, 11, 13
	3	Solve applied problems (p. 695)	6, 7	22, 23
9.3	1	Solve SAS triangles (p. 702)	1	6, 10, 15
	2	Solve SSS triangles (p. 703)	2	8, 9, 12
	3	Solve applied problems (p. 703)	3	24
9.4	1	Find the area of SAS triangles (p. 709)	1	16, 17, 26, 27
	2	Find the area of SSS triangles (p. 710)	2	18, 19, 25
9.5	1	Build a model for an object in simple harmonic motion (p. 715)	1	29
	2	Analyze simple harmonic motion (p. 717)	2	30, 31
	3	Analyze an object in damped motion (p. 718)	3	32, 33
	4	Graph the sum of two functions (p. 719)	4, 5	34

Review Exercises

In Problems 1 and 2, solve each triangle.

1.
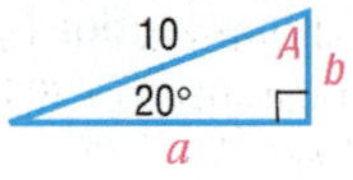

2.
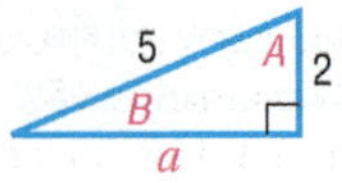

In Problems 3–15, find the remaining angle(s) and side(s) of each triangle that exists. If no triangle exists, say "No triangle."

3. $A = 50°, \quad B = 30°, \quad a = 1$

4. $A = 100°, \quad a = 5, \quad c = 2$

5. $a = 3, \quad c = 1, \quad C = 110°$

6. $a = 3, \quad c = 1, \quad B = 100°$

7. $a = 3, \quad b = 5, \quad B = 80°$

8. $a = 2, \quad b = 3, \quad c = 1$

9. $a = 10, \quad b = 7, \quad c = 8$

10. $a = 1, \quad b = 3, \quad C = 40°$

11. $a = 5, \quad b = 3, \quad A = 80°$

12. $a = 1, \quad b = \frac{1}{2}, \quad c = \frac{4}{3}$

13. $a = 3, \quad A = 10°, \quad b = 4$

14. $a = 4, \quad A = 20°, \quad B = 100°$

15. $c = 5, \quad b = 4, \quad A = 70°$

In Problems 16–20, find the area of each triangle.

16. $a = 2, \quad b = 3, \quad C = 40°$

17. $b = 4, \quad c = 10, \quad A = 70°$

18. $a = 4, \quad b = 3, \quad c = 5$

19. $a = 4, \quad b = 2, \quad c = 5$

20. $A = 50°, \quad B = 30°, \quad a = 1$

---

21. Finding the Grade of a Mountain Trail A straight trail with a uniform inclination leads from a hotel, elevation 5000 feet, to a lake in a valley, elevation 4100 feet. The length of the trail is 4100 feet. What is the inclination (grade) of the trail?

22. Finding the Height of a Helicopter Two observers simultaneously measure the angle of elevation of a helicopter. One angle is measured as 25°, the other as 40° (see the figure). If the observers are 100 feet apart and the helicopter lies over the line joining them, how high is the helicopter?

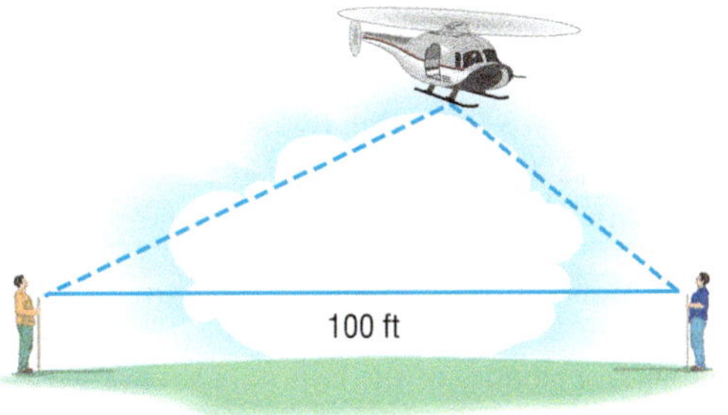

23. Constructing a Highway A highway whose primary directions are north–south is being constructed along the west coast of Florida. Near Naples, a bay obstructs the straight path of the road. Because the cost of a bridge is prohibitive, engineers decide to go around the bay. The illustration shows the path that they decide on and the measurements taken. What is the length of highway needed to go around the bay?

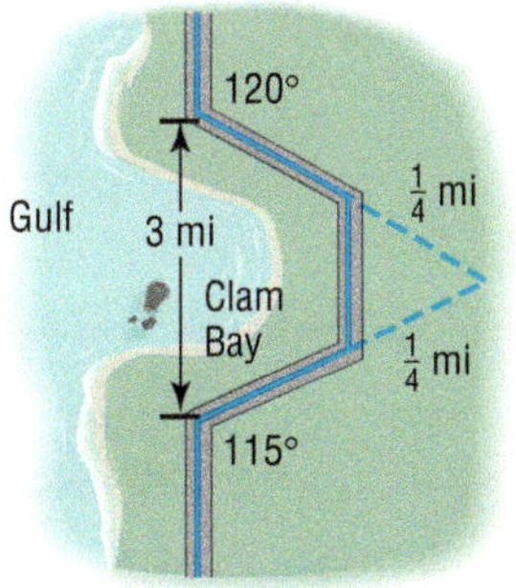

24. Correcting a Navigational Error A sailboat leaves St. Thomas bound for an island in the British West Indies, 200 miles away. Maintaining a constant speed of 18 miles per hour, but encountering heavy crosswinds and strong currents, the crew finds after 4 hours that the sailboat is off course by 15°.

(a) How far is the sailboat from the island at this time?

(b) Through what angle should the sailboat turn to correct its course?

(c) How much time has been added to the trip because of this? (Assume that the speed remains at 18 miles per hour.)

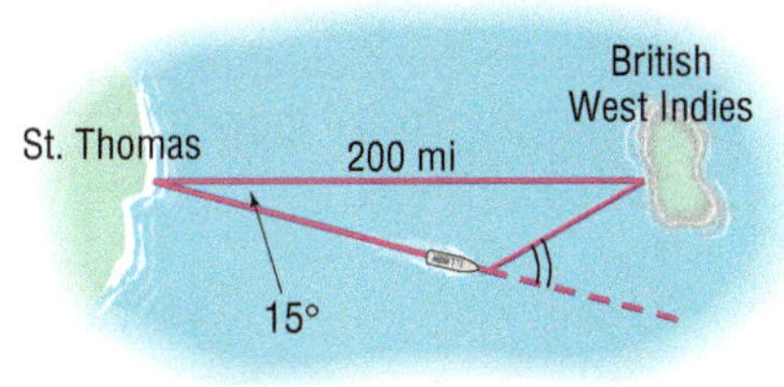

25. Approximating the Area of a Lake To approximate the area of a lake, Cindy walks around the perimeter of the lake, taking the measurements shown in the illustration. Using this technique, what approximate area does she find? [**Hint:** Use the Law of Cosines on the three triangles shown, and then find the sum of their areas.]

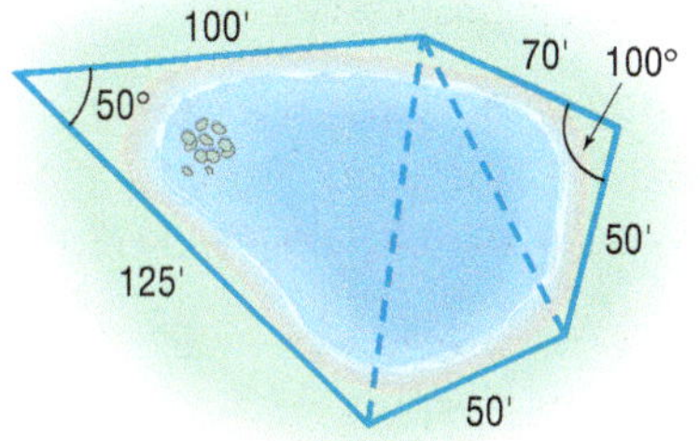

26. Calculating the Cost of Land The irregular parcel of land shown in the figure is being sold for \$100 per square foot. What is the cost of this parcel?

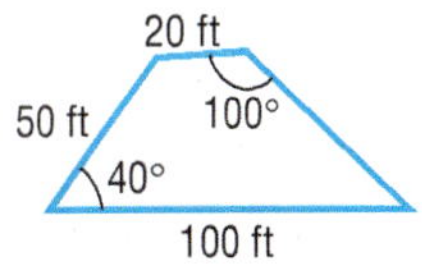

27. Area of a Segment Find the area of the segment of a circle whose radius is 6 inches formed by a central angle of 50°.

28. Finding the Bearing of a Ship The *Majesty* leaves the Port at Boston for Bermuda with a bearing of S80°E at an average speed of 10 knots. After 1 hour, the ship turns 90° toward the southwest. After 2 hours at an average speed of 20 knots, what is the bearing of the ship from Boston?

29. An object attached to a coiled spring is pulled down a distance $a = 3$ from its rest position and then released. Assuming that the motion is simple harmonic with period $T = 4$ seconds, develop a model that relates the displacement d of the object from its rest position after t seconds. Also assume that the positive direction of the motion is up.

In Problems 30 and 31, the displacement d (in feet) of an object at time t (in seconds) is given.

(a) *Describe the motion of the object.*
(b) *What is the maximum displacement from its rest position?*
(c) *What is the time required for one oscillation?*
(d) *What is the frequency?*

30. $d = 6\sin(2t)$ **31.** $d = -2\cos(\pi t)$

32. An object of mass $m = 40$ grams attached to a coiled spring with damping factor $b = 0.75$ gram/second is pulled down a distance $a = 15$ cm from its rest position and then released. Assume that the positive direction of the motion is up and the period is $T = 5$ seconds under simple harmonic motion.
(a) Develop a model that relates the displacement d of the object from its rest position after t seconds.
(b) Graph the equation found in part (a) for 5 oscillations.

33. The displacement d (in meters) of the bob of a pendulum of mass 20 kilograms from its rest position at time t (in seconds) is given as

$$d = -15e^{-0.6t/40}\cos\left(\sqrt{\left(\frac{2\pi}{5}\right)^2 - \frac{0.36}{1600}}\,t\right)$$

(a) Describe the motion of the object.
(b) What is the initial displacement of the bob? That is, what is the displacement at $t = 0$?
(c) Graph the motion using a graphing utility.
(d) What is the displacement of the bob at the start of the second oscillation?
(e) What happens to the displacement of the bob as time increases without bound?

34. Use the method of adding y-coordinates to graph

$$y = 2\sin x + \cos(2x).$$

Chapter Test

CHAPTER **Test Prep** VIDEOS

The Chapter Test Prep Videos are step-by-step solutions available in MyMathLab®, or on this text's YouTube™ Channel. Flip back to the Resources for Success page for a link to this text's YouTube channel.

1. A 12-foot ladder leans against a building. The top of the ladder leans against the wall 10.5 feet from the ground. What is the angle formed by the ground and the ladder?

2. A hot-air balloon is flying at a height of 600 feet and is directly above the Marshall Space Flight Center in Huntsville, Alabama. The pilot of the balloon looks down at the airport, which is known to be 5 miles from the Marshall Space Flight Center. What is the angle of depression from the balloon to the airport?

In Problems 3–5, use the given information to determine the three remaining parts of each triangle.

3.

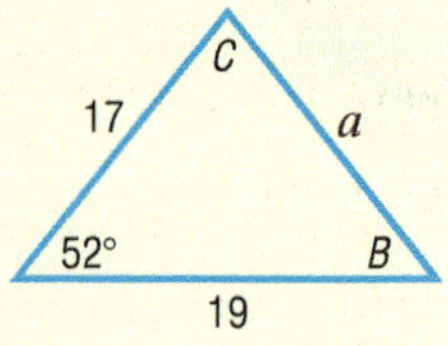

4.

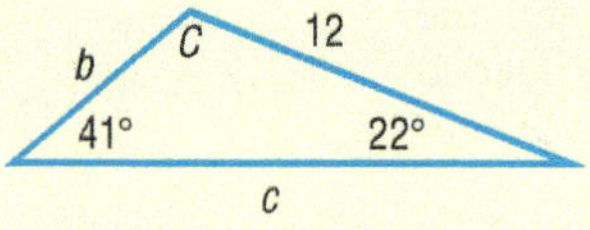

5.

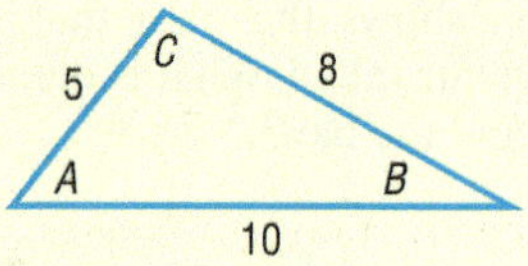

In Problems 6–8, solve each triangle.

6. $A = 55°,\ C = 20°,\ a = 4$ **7.** $a = 3,\ b = 7,\ A = 40°$ **8.** $a = 8,\ b = 4,\ C = 70°$

9. Find the area of the triangle described in Problem 8.

10. Find the area of the triangle described in Problem 5.

11. Find the area of the shaded region enclosed in a semicircle of diameter 8 centimeters. The length of the chord AB is 6 centimeters.

[**Hint:** Triangle ABC is a right triangle.]

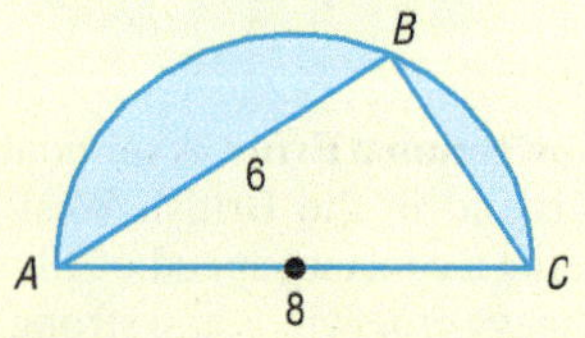

12. Find the area of the quadrilateral shown.

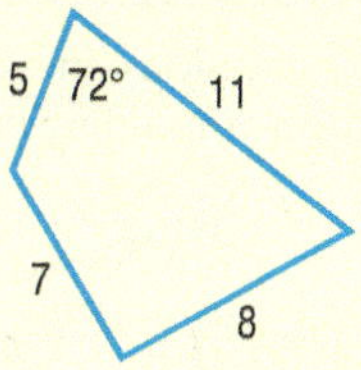

13. Madison wants to swim across Lake William from the fishing lodge (point A) to the boat ramp (point B), but she wants to know the distance first. Highway 20 goes right past the boat ramp, and County Road 3 goes to the lodge. The two roads intersect at point C, 4.2 miles from the ramp and 3.5 miles from the lodge. Madison uses a transit to measure the angle of intersection of the two roads to be 32°. How far will she need to swim?

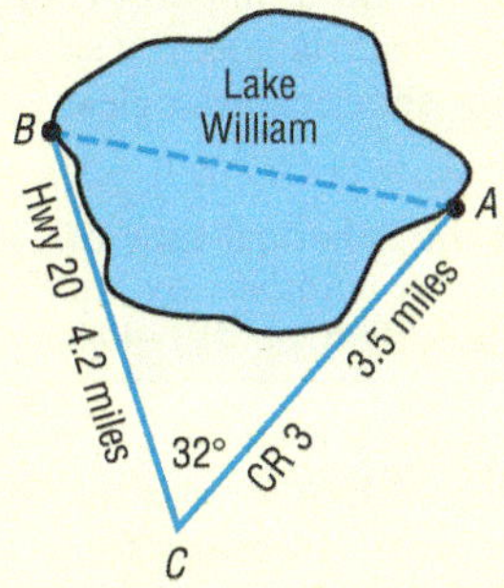

14. Given that $\triangle OAB$ is an isosceles triangle and the shaded sector is a semicircle, find the area of the entire region. Express your answer as a decimal rounded to two places.

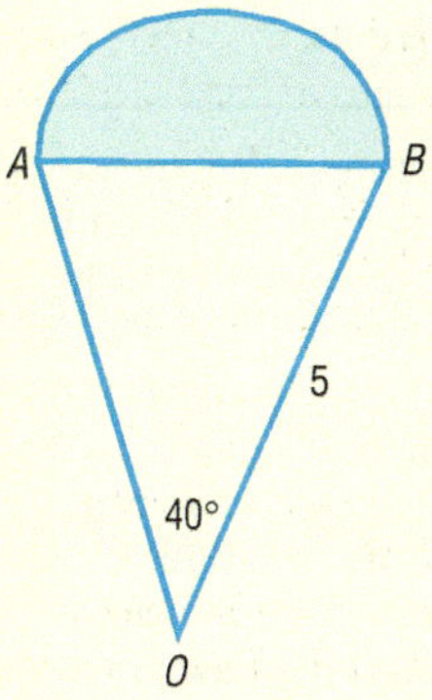

15. The area of the triangle shown below is $54\sqrt{6}$ square units; find the lengths of the sides.

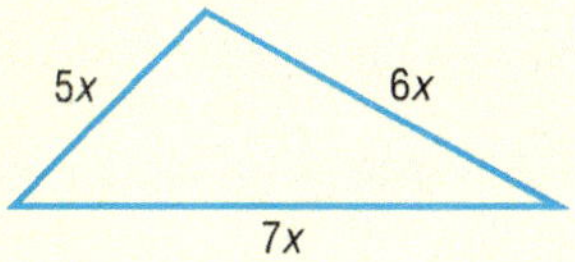

16. Logan is playing on her swing. One full swing (front to back to front) takes 6 seconds, and at the peak of her swing she is at an angle of 42° with the vertical. If her swing is 5 feet long, and we ignore all resistive forces, build a model that relates her horizontal displacement (from the rest position) after time t.

Cumulative Review

1. Find the real solutions, if any, of the equation $3x^2 + 1 = 4x$.
2. Find an equation for the circle with center at the point $(-5, 1)$ and radius 3. Graph this circle.
3. Determine the domain of the function
$$f(x) = \sqrt{x^2 - 3x - 4}$$
4. Graph the function: $y = 3\sin(\pi x)$
5. Graph the function: $y = -2\cos(2x - \pi)$
6. If $\tan\theta = -2$ and $\frac{3\pi}{2} < \theta < 2\pi$, find the exact value of:
   (a) $\sin\theta$ (b) $\cos\theta$ (c) $\sin(2\theta)$
   (d) $\cos(2\theta)$ (e) $\sin\left(\frac{1}{2}\theta\right)$ (f) $\cos\left(\frac{1}{2}\theta\right)$
7. Graph each of the following functions on the interval $[0, 4]$:
   (a) $y = e^x$ (b) $y = \sin x$
   (c) $y = e^x \sin x$ (d) $y = 2x + \sin x$
8. Graph each the following functions:
   (a) $y = x$ (b) $y = x^2$
   (c) $y = \sqrt{x}$ (d) $y = x^3$
   (e) $y = e^x$ (f) $y = \ln x$
   (g) $y = \sin x$ (h) $y = \cos x$
   (i) $y = \tan x$
9. Solve the triangle for which side a is 20, side c is 15, and angle C is 40°.
10. In the complex number system, solve the equation
$$3x^5 - 10x^4 + 21x^3 - 42x^2 + 36x - 8 = 0$$
11. Analyze the graph of the rational function
$$R(x) = \frac{2x^2 - 7x - 4}{x^2 + 2x - 15}$$
12. Solve $3^x = 12$. Round your answer to two decimal places.
13. Solve: $\log_3(x + 8) + \log_3 x = 2$
14. Suppose that $f(x) = 4x + 5$ and $g(x) = x^2 + 5x - 24$.
   (a) Solve $f(x) = 0$. (b) Solve $f(x) = 13$.
   (c) Solve $f(x) = g(x)$. (d) Solve $f(x) > 0$.
   (e) Solve $g(x) \le 0$. (f) Graph $y = f(x)$.
   (g) Graph $y = g(x)$.

Chapter Projects

I. **Spherical Trigonometry** When the distance between two locations on the surface of Earth is small, we can compute the distance in statutory miles. Using this assumption, we can use the Law of Sines and the Law of Cosines to approximate distances and angles. However, if you look at a globe, you notice that Earth is a sphere, so as the distance between two points on its surface increases, the linear distance is less accurate because of curvature. Under this circumstance, we need to take into account the curvature of Earth when using the Law of Sines and the Law of Cosines.

1. See the figure. The points A, B and C are the vertices of a spherical triangle with sides a, b, and c, a three-sided figure drawn on the surface of a sphere with center at the point O. Then connect each vertex by a radius to the center O of the sphere. Now draw tangent lines to the sides a and b of the triangle that go through C. Extend the lines OA and OB to intersect the tangent lines at P and Q, respectively. List the plane right triangles. Determine the measures of the central angles.
2. Apply the Law of Cosines to triangles OPQ and CPQ to find two expressions for the length of PQ.
3. Subtract the expressions in part (2) from each other. Solve for the term containing $\cos c$.

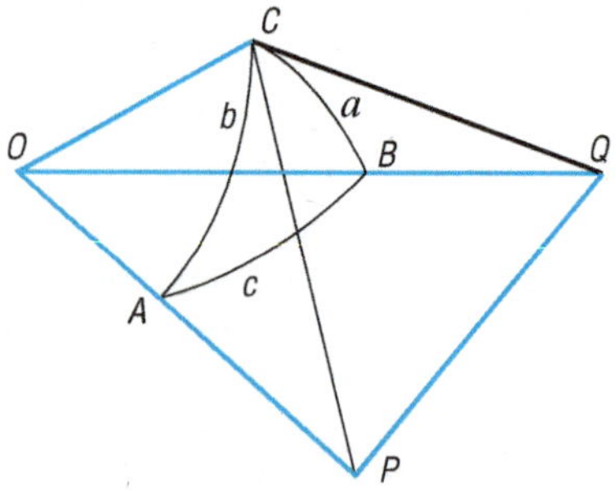

4. Use the Pythagorean Theorem to find another value for $OQ^2 - CQ^2$ and $OP^2 - CP^2$. Now solve for $\cos c$.
5. Replacing the ratios in part (4) by the cosines of the sides of the spherical triangle, you should now have the Law of Cosines for spherical triangles:

$$\cos c = \cos a \cos b + \sin a \sin b \cos C$$

Source: For the spherical Law of Cosines, see *Mathematics from the Birth of Numbers* by Jan Gullberg. W. W. Norton & Co., Publishers, 1996, pp. 491–494.

II. **The Lewis and Clark Expedition** Lewis and Clark followed several rivers in their trek from what is now Great Falls, Montana, to the Pacific coast. First, they went down the Missouri and Jefferson rivers from Great Falls to Lemhi, Idaho. Because the two cities are at different longitudes and different latitudes, we must account for the curvature of Earth when computing the distance that they traveled. Assume that the radius of Earth is 3960 miles.

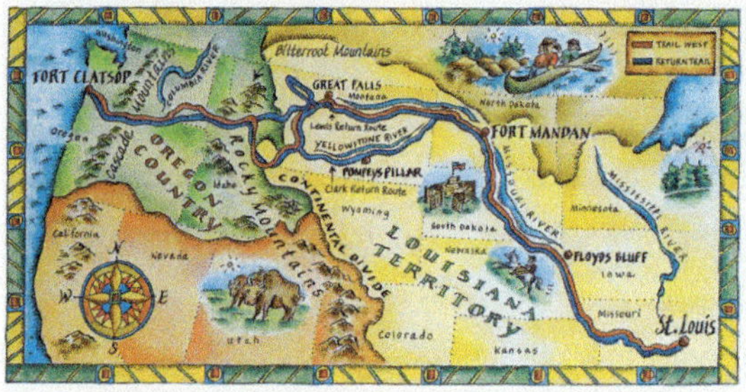

1. Great Falls is at approximately 47.5°N and 111.3°W. Lemhi is at approximately 45.5°N and 113.5°W. (We will assume that the rivers flow straight from Great Falls to Lemhi on the surface of Earth.) This line is called a geodesic line. Apply the Law of Cosines for a spherical triangle to find the angle between Great Falls and Lemhi. (The central angles are found by using the differences in the latitudes and longitudes of the towns. See the diagram.) Then find the length of the arc joining the two towns. (Recall $s = r\theta$.)

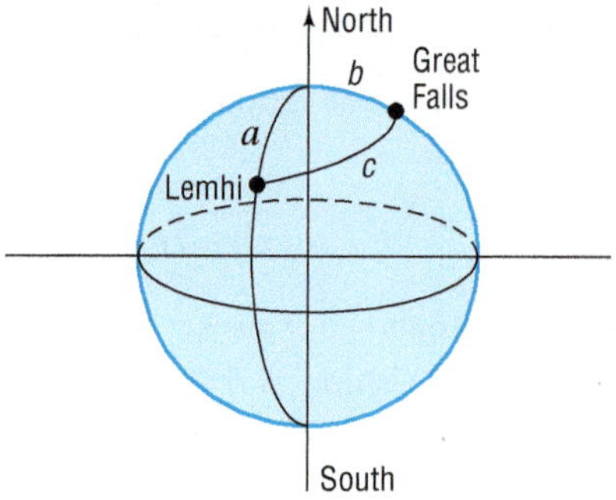

2. From Lemhi, they went up the Bitteroot River and the Snake River to what is now Lewiston and Clarkston on the border of Idaho and Washington. Although this is not really a side to a triangle, we will make a side that goes from Lemhi to Lewiston and Clarkston. If Lewiston and Clarkston are at about 46.5°N 117.0°W, find the distance from Lemhi using the Law of Cosines for a spherical triangle and the arc length.
3. How far did the explorers travel just to get that far?
4. Draw a plane triangle connecting the three towns. If the distance from Lewiston to Great Falls is 282 miles and the angle at Great Falls is 42° and the angle at Lewiston is 48.5°, find the distance from Great Falls to Lemhi and from Lemhi to Lewiston. How do these distances compare with the ones computed in parts (1) and (2)?

Source: For Lewis and Clark Expedition: *American Journey: The Quest for Liberty to 1877, Texas Edition.* Prentice Hall, 1992, p. 345.

Source: For map coordinates: *National Geographic Atlas of the World,* published by National Geographic Society, 1981, pp. 74–75.

Citation: Used with permission of *Technology Review*, from W. Roush, "From Lewis and Clark to Landsat: David Rumsey's Digital Maps Marry Past and Present," 108, no. 7, © 2005; permission conveyed through Copyright Clearance Center, Inc.

The following projects are available at the Instructor's Resource Center (IRC).

III. Project at Motorola ***How Can You Build or Analyze a Vibration Profile?*** Fourier functions not only are important to analyze vibrations but also are what a mathematician would call interesting. Complete the project to see why.

IV. Leaning Tower of Pisa Trigonometry is used to analyze the apparent height and tilt of the Leaning Tower of Pisa.

V. Locating Lost Treasure Clever treasure seekers who know the Law of Sines are able to find a buried treasure efficiently.

VI. Jacob's Field Angles of elevation and the Law of Sines are used to determine the height of the stadium wall and the distance from home plate to the top of the wall.

10 Polar Coordinates; Vectors

How Airplanes Fly

Four aerodynamic forces act on an airplane in flight: **lift**, **drag**, **thrust**, and **weight** (gravity). Drag is the resistance of air molecules hitting the airplane (the *backward* force), thrust is the power of the airplane's engine (the *forward* force), lift is the *upward* force, and weight is the *downward* force. So for airplanes to fly and stay airborne, the thrust must be greater than the drag, and the lift must be greater than the weight (*so, drag opposes thrust, and lift opposes weight*). This is certainly the case when an airplane takes off or climbs. However, when it is in straight and level flight, the opposing forces of lift, and weight are balanced. During a descent, weight exceeds lift, and to slow the airplane, drag has to overcome thrust.

Thrust is generated by the airplane's engine (propeller or jet), weight is created by the natural force of gravity acting on the airplane, and drag comes from friction as the plane moves through air molecules. Drag is also a *reaction* to lift, and this lift must be generated by the airplane in flight. This is done by the **wings** of the airplane.

A cross section of a typical airplane wing shows the top surface to be more curved than the bottom surface. This shaped profile is called an **airfoil** (or aerofoil), and the shape is used because an airfoil generates significantly more lift than opposing drag. In other words, it is very **efficient** at generating lift.

During flight, air naturally flows over and beneath the wing and is deflected upward over the top surface and downward beneath the lower surface. Any difference in deflection causes a difference in air pressure (pressure gradient), and because of the airfoil shape, the pressure of the deflected air is lower above the airfoil than below it. As a result the wing is "pushed" upward by the higher pressure beneath, or, you can argue, it is "sucked" upward by the lower pressure above.

Source: Adapted from Pete Carpenter. How Airplanes Fly—The Basic Principles of Flight http://www.rc-airplane-world.com/how-airplanes-fly.html, accessed May 2015. © rc-airplane-world.com

—See Chapter Project I—

Outline

••• A Look Back, A Look Ahead •••

This chapter is in two parts: Polar Coordinates (Sections 10.1–10.3) and Vectors (Sections 10.4–10.5). They are independent of each other and may be covered in either order.

Sections 10.1–10.3: In Chapter 2 we introduced rectangular coordinates (the *xy*-plane) and discussed the graph of an equation in two variables involving *x* and *y*. In Sections 10.1 and 10.2, we introduce polar coordinates, an alternative to rectangular coordinates, and discuss graphing equations that involve polar coordinates. In Section 6.3, we discussed raising a real number to a real power. In Section 10.3, we extend this idea by raising a complex number to a real power. As it turns out, polar coordinates are useful for the discussion.

Sections 10.4–10.5: We have seen in many chapters that we are often required to solve an equation to obtain a solution to applied problems. In the last two sections of this chapter, we develop the notion of a vector and show how it can be used to model applied problems in physics and engineering.

10.1 Polar Coordinates

PREPARING FOR THIS SECTION *Before getting started, review the following:*

- Rectangular Coordinates (Section 1.1, pp. 83–84)
- Definition of the Trigonometric Functions (Section 7.4, pp. 550–552)
- Inverse Tangent Function (Section 8.1, pp. 622–624)
- Completing the Square (Chapter R, Review, Section R.5, p. 57)

Now Work the 'Are You Prepared?' problems on page 738.

OBJECTIVES
1 Plot Points Using Polar Coordinates (p. 731)
2 Convert from Polar Coordinates to Rectangular Coordinates (p. 733)
3 Convert from Rectangular Coordinates to Polar Coordinates (p. 735)
4 Transform Equations between Polar and Rectangular Forms (p. 737)

So far, we have always used a system of rectangular coordinates to plot points in the plane. Now we are ready to describe another system, called *polar coordinates*. In many instances, polar coordinates offer certain advantages over rectangular coordinates.

In a rectangular coordinate system, you will recall, a point in the plane is represented by an ordered pair of numbers (x, y), where x and y equal the signed distances of the point from the y-axis and the x-axis, respectively. In a polar coordinate system, we select a point, called the **pole**, and then a ray with vertex at the pole, called the **polar axis**. See Figure 1. Comparing the rectangular and polar coordinate systems, note that the origin in rectangular coordinates coincides with the pole in polar coordinates, and the positive x-axis in rectangular coordinates coincides with the polar axis in polar coordinates.

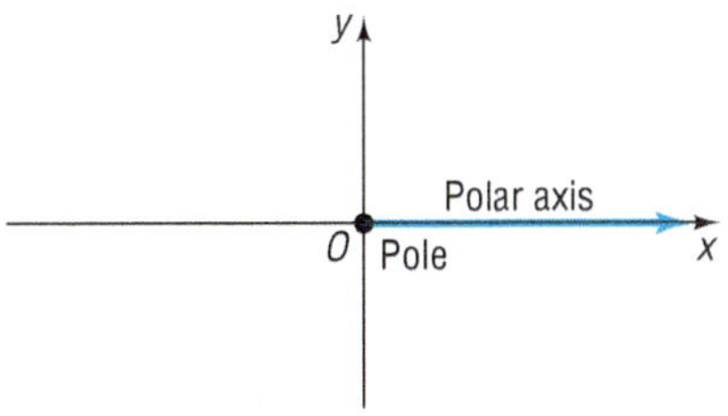

Figure 1

1 Plot Points Using Polar Coordinates

A point P in a polar coordinate system is represented by an ordered pair of numbers (r, θ). If $r > 0$, then r is the distance of the point from the pole; θ is an angle (in degrees or radians) formed by the polar axis and a ray from the pole through the point. We call the ordered pair (r, θ) the **polar coordinates** of the point. See Figure 2.

As an example, suppose that a point P has polar coordinates $\left(2, \frac{\pi}{4}\right)$. Locate P by first drawing an angle of $\frac{\pi}{4}$ radian, placing its vertex at the pole and its initial side along the polar axis. Then go out a distance of 2 units along the terminal side of the angle to reach the point P. See Figure 3.

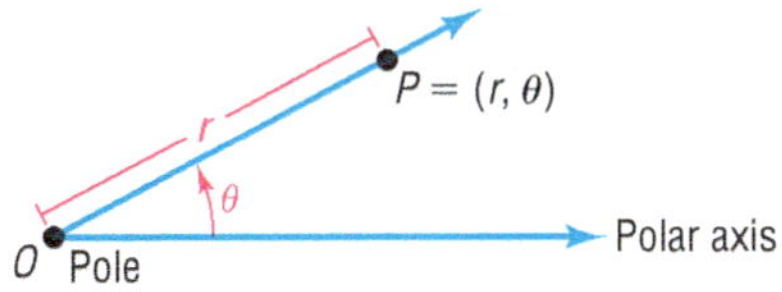

Figure 2

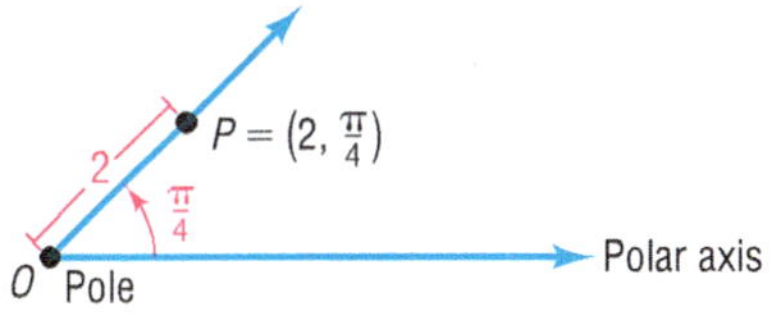

Figure 3

In using polar coordinates (r, θ), it is possible for r to be negative. When this happens, instead of the point being on the terminal side of θ, it is on the ray from the pole extending in the direction *opposite* the terminal side of θ at a distance $|r|$ units from the pole. See Figure 4 on the next page for an illustration.

For example, to plot the point $\left(-3, \frac{2\pi}{3}\right)$, use the ray in the opposite direction of $\frac{2\pi}{3}$ and go out $|-3| = 3$ units along that ray. See Figure 5 on the next page.

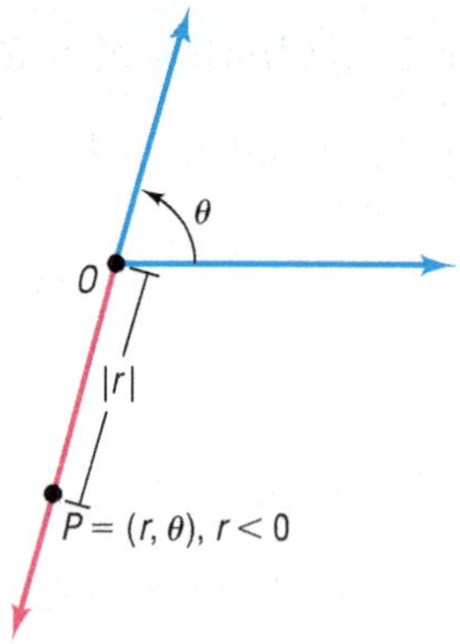

Figure 4

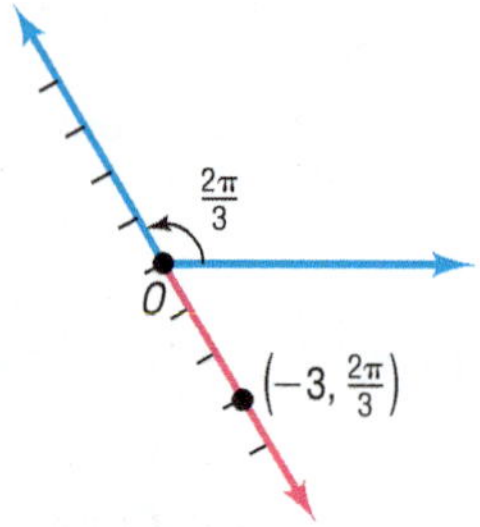

Figure 5

EXAMPLE 1 Plotting Points Using Polar Coordinates

Plot the points with the following polar coordinates:

(a) $\left(3, \dfrac{5\pi}{3}\right)$ (b) $\left(2, -\dfrac{\pi}{4}\right)$ (c) $(3, 0)$ (d) $\left(-2, \dfrac{\pi}{4}\right)$

Solution Figure 6 shows the points.

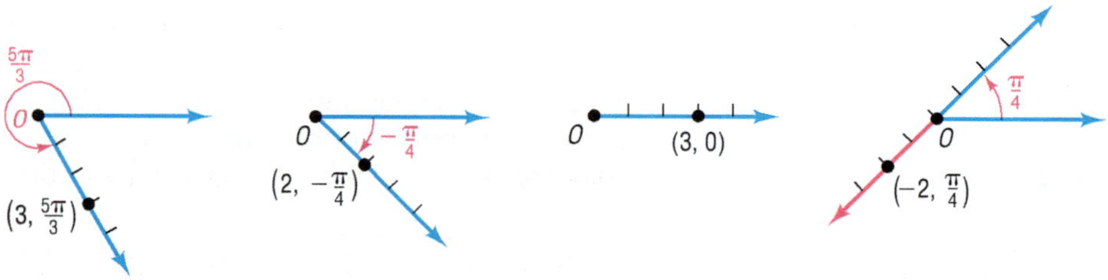

Figure 6 (a) (b) (c) (d)

Now Work PROBLEMS 11, 19, AND 29

Recall that an angle measured counterclockwise is positive and an angle measured clockwise is negative. This convention has some interesting consequences related to polar coordinates.

EXAMPLE 2 Finding Several Polar Coordinates of a Single Point

Consider again the point P with polar coordinates $\left(2, \dfrac{\pi}{4}\right)$, as shown in Figure 7(a). Because $\dfrac{\pi}{4}, \dfrac{9\pi}{4}$, and $-\dfrac{7\pi}{4}$ all have the same terminal side, this point P also can be located by using the polar coordinates $\left(2, \dfrac{9\pi}{4}\right)$ or the polar coordinates $\left(2, -\dfrac{7\pi}{4}\right)$, as shown in Figures 7(b) and (c). The point $\left(2, \dfrac{\pi}{4}\right)$ can also be represented by the polar coordinates $\left(-2, \dfrac{5\pi}{4}\right)$. See Figure 7(d).

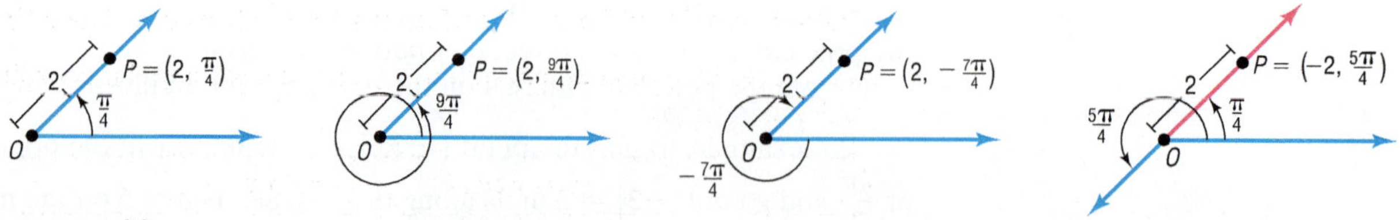

Figure 7 (a) (b) (c) (d)

EXAMPLE 3 Finding Other Polar Coordinates of a Given Point

Plot the point P with polar coordinates $\left(3, \frac{\pi}{6}\right)$, and find other polar coordinates (r, θ) of this same point for which:

(a) $r > 0, \quad 2\pi \le \theta < 4\pi$ (b) $r < 0, \quad 0 \le \theta < 2\pi$
(c) $r > 0, \quad -2\pi \le \theta < 0$

Solution The point $\left(3, \frac{\pi}{6}\right)$ is plotted in Figure 8.

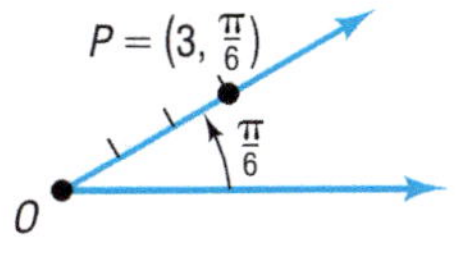

Figure 8

(a) Add 1 revolution (2π radians) to the angle $\frac{\pi}{6}$ to get $P = \left(3, \frac{\pi}{6} + 2\pi\right) = \left(3, \frac{13\pi}{6}\right)$. See Figure 9.

(b) Add $\frac{1}{2}$ revolution (π radians) to the angle $\frac{\pi}{6}$, and replace 3 by -3 to get $P = \left(-3, \frac{\pi}{6} + \pi\right) = \left(-3, \frac{7\pi}{6}\right)$. See Figure 10.

(c) Subtract 2π from the angle $\frac{\pi}{6}$ to get $P = \left(3, \frac{\pi}{6} - 2\pi\right) = \left(3, -\frac{11\pi}{6}\right)$. See Figure 11.

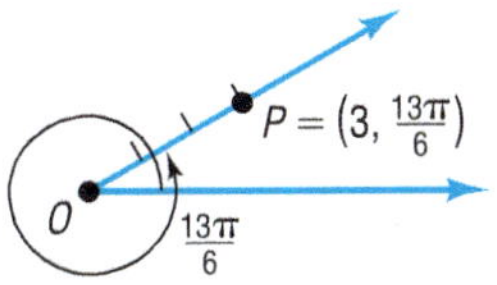

Figure 9

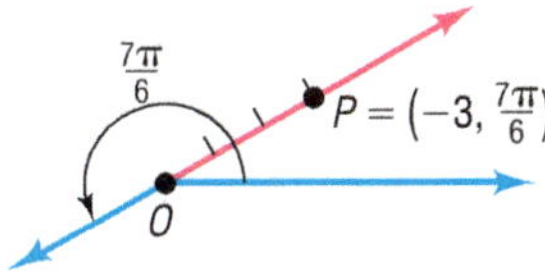

Figure 10

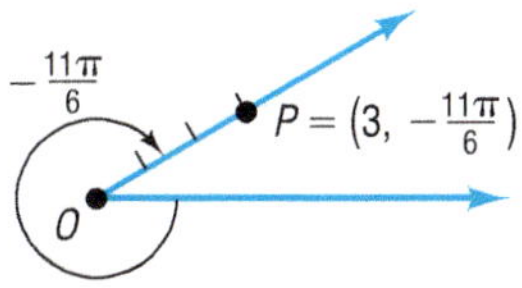

Figure 11

These examples show a major difference between rectangular coordinates and polar coordinates. A point has exactly one pair of rectangular coordinates; however, a point has infinitely many pairs of polar coordinates.

SUMMARY

A point with polar coordinates (r, θ), θ in radians, can also be represented by either of the following:

$$(r, \theta + 2\pi k) \quad \text{or} \quad (-r, \theta + \pi + 2\pi k) \qquad k \text{ any integer}$$

The polar coordinates of the pole are $(0, \theta)$, where θ can be any angle.

Now Work PROBLEM 33

2 Convert from Polar Coordinates to Rectangular Coordinates

Sometimes it is necessary to convert coordinates or equations in rectangular form to polar form, or vice versa. To do this, recall that the origin in rectangular coordinates is the pole in polar coordinates and that the positive x-axis in rectangular coordinates is the polar axis in polar coordinates.

THEOREM

Conversion from Polar Coordinates to Rectangular Coordinates

If P is a point with polar coordinates (r, θ), the rectangular coordinates (x, y) of P are given by

$$x = r\cos\theta \qquad y = r\sin\theta \tag{1}$$

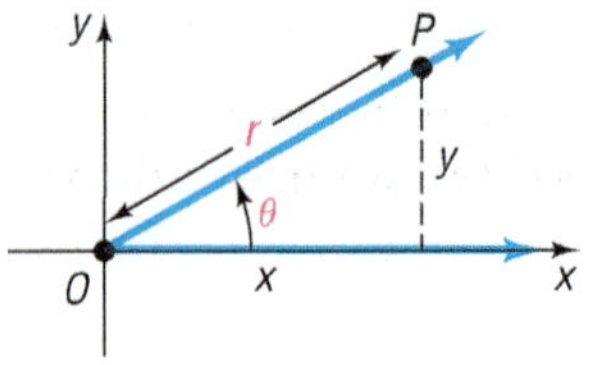

Figure 12

Proof Suppose that P has the polar coordinates (r, θ). We seek the rectangular coordinates (x, y) of P. Refer to Figure 12.

If $r = 0$, then, regardless of θ, the point P is the pole, for which the rectangular coordinates are $(0, 0)$. Formula (1) is valid for $r = 0$.

If $r > 0$, the point P is on the terminal side of θ, and $r = d(O, P) = \sqrt{x^2 + y^2}$. Because

$$\cos\theta = \frac{x}{r} \qquad \sin\theta = \frac{y}{r}$$

this means

$$x = r\cos\theta \qquad y = r\sin\theta$$

If $r < 0$ and θ is in radians, the point $P = (r, \theta)$ can be represented as $(-r, \pi + \theta)$, where $-r > 0$. Because

$$\cos(\pi + \theta) = -\cos\theta = \frac{x}{-r} \qquad \sin(\pi + \theta) = -\sin\theta = \frac{y}{-r}$$

this means

$$x = r\cos\theta \qquad y = r\sin\theta$$

■

EXAMPLE 4 Converting from Polar Coordinates to Rectangular Coordinates

Find the rectangular coordinates of the points with the following polar coordinates:

(a) $\left(6, \frac{\pi}{6}\right)$ (b) $\left(-4, -\frac{\pi}{4}\right)$

Solution Use formula (1): $x = r\cos\theta$ and $y = r\sin\theta$.

(a) Figure 13(a) shows $\left(6, \frac{\pi}{6}\right)$ plotted. Notice that $\left(6, \frac{\pi}{6}\right)$ lies in quadrant I of the rectangular coordinate system. So both the x-coordinate and the y-coordinate will be positive. Substituting $r = 6$ and $\theta = \frac{\pi}{6}$ gives

$$x = r\cos\theta = 6\cos\frac{\pi}{6} = 6\cdot\frac{\sqrt{3}}{2} = 3\sqrt{3}$$

$$y = r\sin\theta = 6\sin\frac{\pi}{6} = 6\cdot\frac{1}{2} = 3$$

The rectangular coordinates of the point $\left(6, \frac{\pi}{6}\right)$ are $(3\sqrt{3}, 3)$, which lies in quadrant I, as expected.

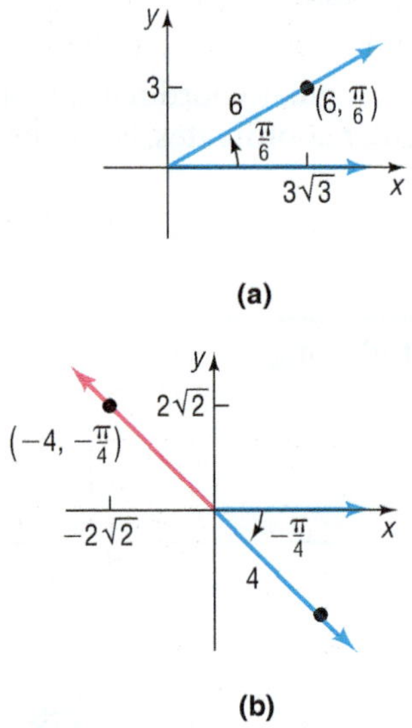

Figure 13

(b) Figure 13(b) shows $\left(-4, -\frac{\pi}{4}\right)$ plotted. Notice that $\left(-4, -\frac{\pi}{4}\right)$ lies in quadrant II of the rectangular coordinate system. Substituting $r = -4$ and $\theta = -\frac{\pi}{4}$ gives

$$x = r\cos\theta = -4\cos\left(-\frac{\pi}{4}\right) = -4\cdot\frac{\sqrt{2}}{2} = -2\sqrt{2}$$

$$y = r\sin\theta = -4\sin\left(-\frac{\pi}{4}\right) = -4\left(-\frac{\sqrt{2}}{2}\right) = 2\sqrt{2}$$

The rectangular coordinates of the point $\left(-4, -\frac{\pi}{4}\right)$ are $(-2\sqrt{2}, 2\sqrt{2})$, which lies in quadrant II, as expected. ■

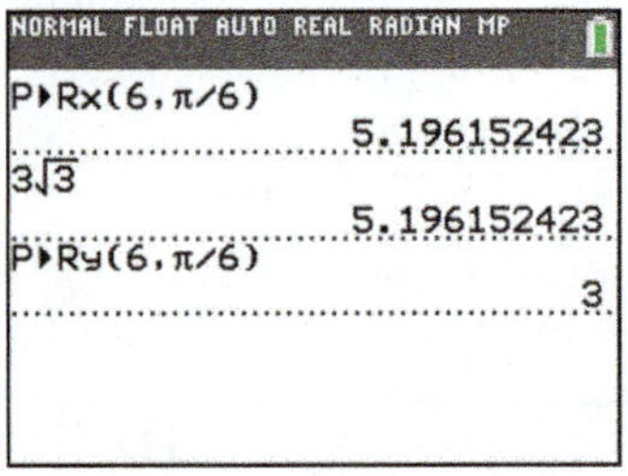

Figure 14

Most calculators have the capability of converting from polar coordinates to rectangular coordinates. Consult your owner's manual for the proper keystrokes. In most cases this procedure is tedious, so you will probably find that using formula (1) is faster.

Figure 14 verifies the result obtained in Example 4(a) using a TI-84 Plus C. Note that the calculator is in radian mode.

Now Work PROBLEMS 41 AND 53

3 Convert from Rectangular Coordinates to Polar Coordinates

Converting from rectangular coordinates (x, y) to polar coordinates (r, θ) is a little more complicated. Notice that each solution begins by plotting the given rectangular coordinates.

EXAMPLE 5 How to Convert from Rectangular Coordinates to Polar Coordinates with the Point on a Coordinate Axis

Find polar coordinates of a point whose rectangular coordinates are $(0, 3)$.

Step-by-Step Solution

Step 1: Plot the point (x, y) and note the quadrant the point lies in or the coordinate axis the point lies on.

Plot the point $(0, 3)$ in a rectangular coordinate system. See Figure 15. The point lies on the positive y-axis.

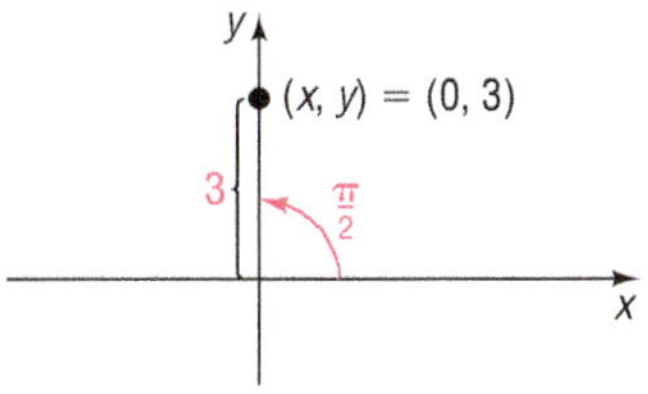

Figure 15

Step 2: Determine the distance r from the origin to the point.

The point $(0, 3)$ lies on the y-axis a distance of 3 units from the origin (pole), so $r = 3$.

Step 3: Determine θ.

A ray with vertex at the pole through $(0, 3)$ forms an angle $\theta = \dfrac{\pi}{2}$ with the polar axis.

Polar coordinates for this point can be given by $\left(3, \dfrac{\pi}{2}\right)$. Other possible representations include $\left(-3, -\dfrac{\pi}{2}\right)$ and $\left(3, \dfrac{5\pi}{2}\right)$. ■

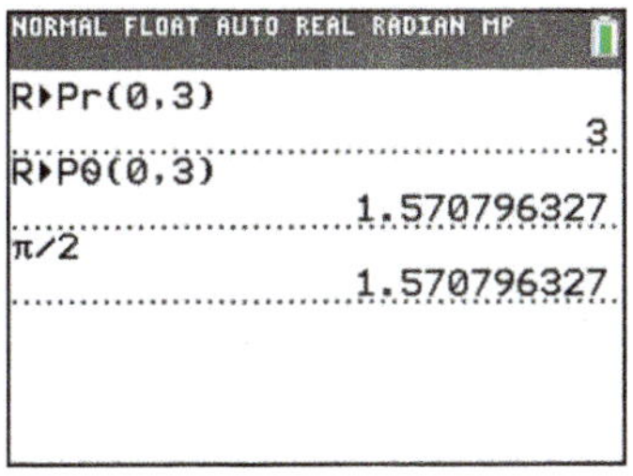

Figure 16

Most graphing calculators have the capability of converting from rectangular coordinates to polar coordinates. Consult your owner's manual for the proper keystrokes. Figure 16 verifies the results obtained in Example 5 using a TI-84 Plus C. Note that the calculator is in radian mode.

Figure 17 shows polar coordinates of points that lie on either the x-axis or the y-axis. In each illustration, $a > 0$.

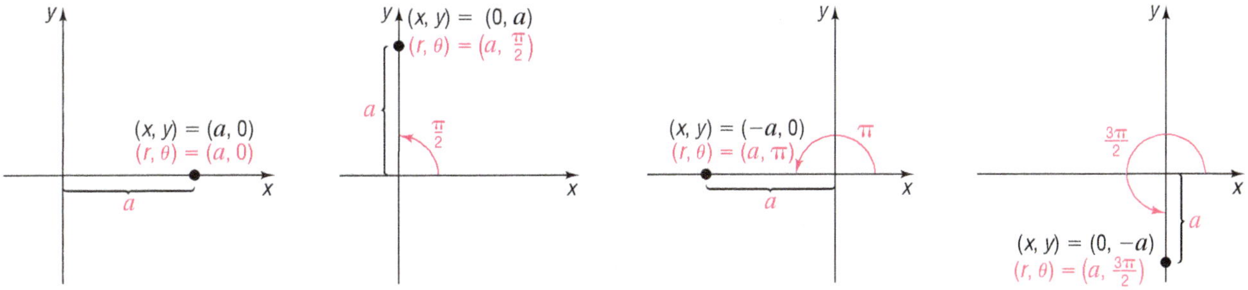

Figure 17 **(a)** $(x, y) = (a, 0), a > 0$ **(b)** $(x, y) = (0, a), a > 0$ **(c)** $(x, y) = (-a, 0), a > 0$ **(d)** $(x, y) = (0, -a), a > 0$

Now Work PROBLEM 57

EXAMPLE 6 How to Convert from Rectangular Coordinates to Polar Coordinates with the Point in a Quadrant

Find the polar coordinates of a point whose rectangular coordinates are $(2, -2)$.

Step-by-Step Solution

Step 1: Plot the point (x, y) and note the quadrant the point lies in or the coordinate axis the point lies on.

Plot the point $(2, -2)$ in a rectangular coordinate system. See Figure 18. The point lies in quadrant IV.

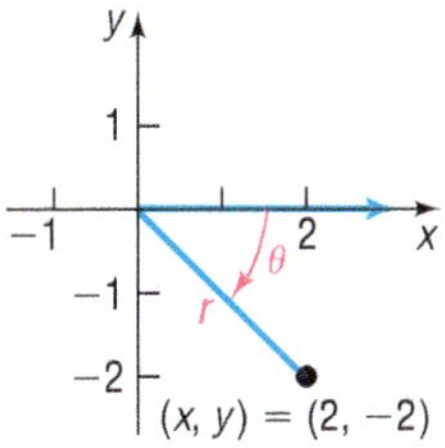

Figure 18

Step 2: Determine the distance r from the origin to the point using $r = \sqrt{x^2 + y^2}$.

$$r = \sqrt{x^2 + y^2} = \sqrt{(2)^2 + (-2)^2} = \sqrt{8} = 2\sqrt{2}$$

Step 3: Determine θ.

Find θ by recalling that $\tan\theta = \frac{y}{x}$, so $\theta = \tan^{-1}\frac{y}{x}$, $-\frac{\pi}{2} < \theta < \frac{\pi}{2}$. Because $(2, -2)$ lies in quadrant IV, this means that $-\frac{\pi}{2} < \theta < 0$. As a result,

$$\theta = \tan^{-1}\frac{y}{x} = \tan^{-1}\left(\frac{-2}{2}\right) = \tan^{-1}(-1) = -\frac{\pi}{4}$$

A set of polar coordinates for the point $(2, -2)$ is $\left(2\sqrt{2}, -\frac{\pi}{4}\right)$. Other possible representations include $\left(2\sqrt{2}, \frac{7\pi}{4}\right)$ and $\left(-2\sqrt{2}, \frac{3\pi}{4}\right)$. ■

EXAMPLE 7 Converting from Rectangular Coordinates to Polar Coordinates

Find polar coordinates of a point whose rectangular coordinates are $(-1, -\sqrt{3})$.

Solution

Step 1: See Figure 19. The point lies in quadrant III.

Step 2: The distance r from the origin to the point $(-1, -\sqrt{3})$ is

$$r = \sqrt{(-1)^2 + (-\sqrt{3})^2} = \sqrt{4} = 2$$

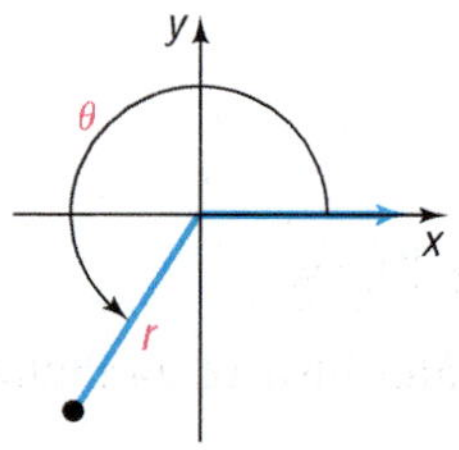

$(x, y) = (-1, -\sqrt{3})$

Figure 19

Step 3: To find θ, use $\alpha = \tan^{-1}\frac{y}{x} = \tan^{-1}\frac{-\sqrt{3}}{-1} = \tan^{-1}\sqrt{3}$, $-\frac{\pi}{2} < \alpha < \frac{\pi}{2}$. Since the point $(-1, -\sqrt{3})$ lies in quadrant III and the inverse tangent function gives an angle in quadrant I, add π to the result to obtain an angle in quadrant III.

$$\theta = \pi + \tan^{-1}\left(\frac{-\sqrt{3}}{-1}\right) = \pi + \tan^{-1}\sqrt{3} = \pi + \frac{\pi}{3} = \frac{4\pi}{3}$$

A set of polar coordinates for this point is $\left(2, \frac{4\pi}{3}\right)$. Other possible representations include $\left(-2, \frac{\pi}{3}\right)$ and $\left(2, -\frac{2\pi}{3}\right)$. ■

Figure 20 shows how to find polar coordinates of a point that lies in a quadrant when its rectangular coordinates (x, y) are given.

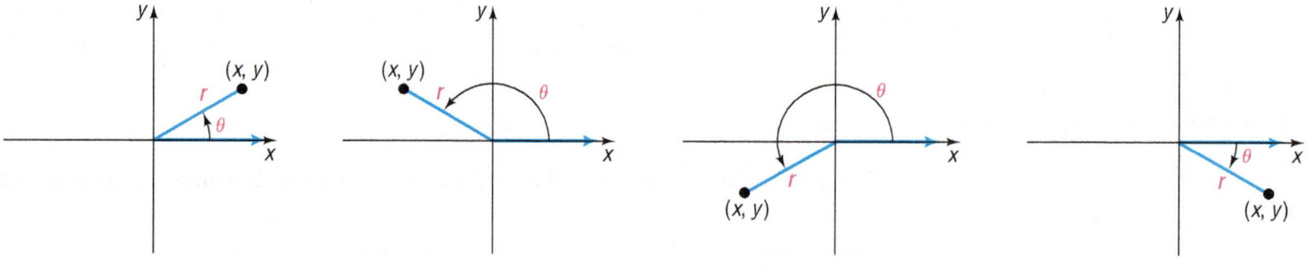

Figure 20 **(a)** $r = \sqrt{x^2 + y^2}$, $\theta = \tan^{-1}\frac{y}{x}$ **(b)** $r = \sqrt{x^2 + y^2}$, $\theta = \pi + \tan^{-1}\frac{y}{x}$ **(c)** $r = \sqrt{x^2 + y^2}$, $\theta = \pi + \tan^{-1}\frac{y}{x}$ **(d)** $r = \sqrt{x^2 + y^2}$, $\theta = \tan^{-1}\frac{y}{x}$

The preceding discussion provides the formulas

$$r^2 = x^2 + y^2 \qquad \tan\theta = \frac{y}{x} \qquad \text{if } x \neq 0 \tag{2}$$

To use formula (2) effectively, follow these steps:

Steps for Converting from Rectangular to Polar Coordinates

STEP 1: Always plot the point (x, y) first, as shown in Examples 5, 6, and 7. Note the quadrant the point lies in or the coordinate axis the point lies on.

STEP 2: If $x = 0$ or $y = 0$, use your illustration to find r. If $x \neq 0$ and $y \neq 0$, then $r = \sqrt{x^2 + y^2}$.

STEP 3: Find θ. If $x = 0$ or $y = 0$, use your illustration to find θ. If $x \neq 0$ and $y \neq 0$, note the quadrant in which the point lies.

$$\text{Quadrant I or IV:} \quad \theta = \tan^{-1}\frac{y}{x}$$

$$\text{Quadrant II or III:} \quad \theta = \pi + \tan^{-1}\frac{y}{x}$$

Now Work PROBLEM 61

4 Transform Equations between Polar and Rectangular Forms

Formulas (1) and (2) may also be used to transform equations from polar form to rectangular form, and vice versa. Two common techniques for transforming an equation from polar form to rectangular form are

1. Multiplying both sides of the equation by r
2. Squaring both sides of the equation

EXAMPLE 8 **Transforming an Equation from Polar to Rectangular Form**

Transform the equation $r = 6\cos\theta$ from polar coordinates to rectangular coordinates, and identify the graph.

Solution Multiplying each side by r makes it easier to apply formulas (1) and (2).

$$\begin{aligned} r &= 6\cos\theta & \\ r^2 &= 6r\cos\theta & \text{Multiply each side by } r. \\ x^2 + y^2 &= 6x & r^2 = x^2 + y^2;\ x = r\cos\theta \end{aligned}$$

This is the equation of a circle. Proceed to complete the square to obtain the standard form of the equation.

$$\begin{aligned} x^2 + y^2 &= 6x & \\ (x^2 - 6x) + y^2 &= 0 & \text{General form} \\ (x^2 - 6x + 9) + y^2 &= 9 & \text{Complete the square in } x. \\ (x - 3)^2 + y^2 &= 9 & \text{Factor.} \end{aligned}$$

This is the standard form of the equation of a circle with center $(3, 0)$ and radius 3. ■

Now Work PROBLEM 77

EXAMPLE 9 **Transforming an Equation from Rectangular to Polar Form**

Transform the equation $4xy = 9$ from rectangular coordinates to polar coordinates.

Solution Use formula (1): $x = r\cos\theta$ and $y = r\sin\theta$.

$$\begin{aligned} 4xy &= 9 & \\ 4(r\cos\theta)(r\sin\theta) &= 9 & x = r\cos\theta,\ y = r\sin\theta \\ 4r^2\cos\theta\sin\theta &= 9 & \end{aligned}$$

This is the polar form of the equation. It can be simplified as follows:

$$2r^2(2\sin\theta\cos\theta) = 9 \quad \text{Factor out } 2r^2.$$

$$2r^2\sin(2\theta) = 9 \quad \text{Double-angle Formula}$$

Now Work PROBLEM 71

10.1 Assess Your Understanding

'Are You Prepared?' *Answers are given at the end of these exercises. If you get a wrong answer, read the pages listed in red.*

1. Plot the point whose rectangular coordinates are $(3, -1)$. What quadrant does the point lie in? (pp. 83–84)

2. To complete the square of $x^2 + 6x$, add _____. (p. 57)

3. If $P = (a, b)$ is a point on the terminal side of the angle θ at a distance r from the origin, then $\tan\theta =$ _________. (pp. 550–552)

4. $\tan^{-1}(-1) =$ _____. (pp. 622–624)

Concepts and Vocabulary

5. The origin in rectangular coordinates coincides with the _____ in polar coordinates; the positive x-axis in rectangular coordinates coincides with the _____ _____ in polar coordinates.

6. If P is a point with polar coordinates (r, θ), the rectangular coordinates (x, y) of P are given by $x =$ ______ and $y =$ ______.

7. For the point with polar coordinates $\left(1, -\frac{\pi}{2}\right)$, which of the following best describes the location of the point in a rectangular coordinate system?
(a) in quadrant IV (b) on the y-axis
(c) in quadrant II (d) on the x-axis

8. The point $\left(5, \frac{\pi}{6}\right)$ can also be represented by which of the following polar coordinates?
(a) $\left(5, -\frac{\pi}{6}\right)$ (b) $\left(-5, \frac{13\pi}{6}\right)$
(c) $\left(5, -\frac{5\pi}{6}\right)$ (d) $\left(-5, \frac{7\pi}{6}\right)$

9. ***True or False*** In the polar coordinates (r, θ), r can be negative.

10. ***True or False*** The polar coordinates of a point are unique.

Skill Building

In Problems 11–18, match each point in polar coordinates with either A, B, C, or D on the graph.

11. $\left(2, -\frac{11\pi}{6}\right)$ **12.** $\left(-2, -\frac{\pi}{6}\right)$ **13.** $\left(-2, \frac{\pi}{6}\right)$ **14.** $\left(2, \frac{7\pi}{6}\right)$

15. $\left(2, \frac{5\pi}{6}\right)$ **16.** $\left(-2, \frac{5\pi}{6}\right)$ **17.** $\left(-2, \frac{7\pi}{6}\right)$ **18.** $\left(2, \frac{11\pi}{6}\right)$

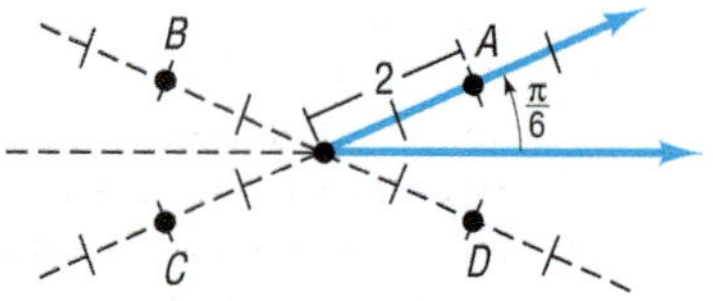

In Problems 19–32, plot each point given in polar coordinates.

19. $(3, 90°)$ **20.** $(4, 270°)$ **21.** $(-2, 0)$ **22.** $(-3, \pi)$ **23.** $\left(6, \frac{\pi}{6}\right)$

24. $\left(5, \frac{5\pi}{3}\right)$ **25.** $(-2, 135°)$ **26.** $(-3, 120°)$ **27.** $\left(4, -\frac{2\pi}{3}\right)$ **28.** $\left(2, -\frac{5\pi}{4}\right)$

29. $\left(-1, -\frac{\pi}{3}\right)$ **30.** $\left(-3, -\frac{3\pi}{4}\right)$ **31.** $(-2, -\pi)$ **32.** $\left(-3, -\frac{\pi}{2}\right)$

In Problems 33–40, plot each point given in polar coordinates, and find other polar coordinates (r, θ) of the point for which:
(a) $r > 0,\ -2\pi \le \theta < 0$ (b) $r < 0,\ 0 \le \theta < 2\pi$ (c) $r > 0,\ 2\pi \le \theta < 4\pi$

33. $\left(5, \frac{2\pi}{3}\right)$ **34.** $\left(4, \frac{3\pi}{4}\right)$ **35.** $(-2, 3\pi)$ **36.** $(-3, 4\pi)$

37. $\left(1, \frac{\pi}{2}\right)$ **38.** $(2, \pi)$ **39.** $\left(-3, -\frac{\pi}{4}\right)$ **40.** $\left(-2, -\frac{2\pi}{3}\right)$

In Problems 41–56, polar coordinates of a point are given. Find the rectangular coordinates of each point.

41. $\left(3, \frac{\pi}{2}\right)$
42. $\left(4, \frac{3\pi}{2}\right)$
43. $(-2, 0)$
44. $(-3, \pi)$
45. $(6, 150°)$
46. $(5, 300°)$
47. $\left(-2, \frac{3\pi}{4}\right)$
48. $\left(-2, \frac{2\pi}{3}\right)$
49. $\left(-1, -\frac{\pi}{3}\right)$
50. $\left(-3, -\frac{3\pi}{4}\right)$
51. $(-2, -180°)$
52. $(-3, -90°)$
53. $(7.5, 110°)$
54. $(-3.1, 182°)$
55. $(6.3, 3.8)$
56. $(8.1, 5.2)$

In Problems 57–68, the rectangular coordinates of a point are given. Find polar coordinates for each point.

57. $(3, 0)$
58. $(0, 2)$
59. $(-1, 0)$
60. $(0, -2)$
61. $(1, -1)$
62. $(-3, 3)$
63. $(\sqrt{3}, 1)$
64. $(-2, -2\sqrt{3})$
65. $(1.3, -2.1)$
66. $(-0.8, -2.1)$
67. $(8.3, 4.2)$
68. $(-2.3, 0.2)$

In Problems 69–76, the letters x and y represent rectangular coordinates. Write each equation using polar coordinates (r, θ).

69. $2x^2 + 2y^2 = 3$
70. $x^2 + y^2 = x$
71. $x^2 = 4y$
72. $y^2 = 2x$
73. $2xy = 1$
74. $4x^2y = 1$
75. $x = 4$
76. $y = -3$

In Problems 77–84, the letters r and θ represent polar coordinates. Write each equation using rectangular coordinates (x, y).

77. $r = \cos\theta$
78. $r = \sin\theta + 1$
79. $r^2 = \cos\theta$
80. $r = \sin\theta - \cos\theta$
81. $r = 2$
82. $r = 4$
83. $r = \dfrac{4}{1 - \cos\theta}$
84. $r = \dfrac{3}{3 - \cos\theta}$

Applications and Extensions

85. **Chicago** In Chicago, the road system is set up like a Cartesian plane, where streets are indicated by the number of blocks they are from Madison Street and State Street. For example, Wrigley Field in Chicago is located at 1060 West Addison, which is 10 blocks west of State Street and 36 blocks north of Madison Street. Treat the intersection of Madison Street and State Street as the origin of a coordinate system, with east being the positive x-axis.
(a) Write the location of Wrigley Field using rectangular coordinates.
(b) Write the location of Wrigley Field using polar coordinates. Use the east direction for the polar axis. Express θ in degrees.
(c) U.S. Cellular Field, home of the White Sox, is located at 35th and Princeton, which is 3 blocks west of State Street and 35 blocks south of Madison. Write the location of U.S. Cellular Field using rectangular coordinates.
(d) Write the location of U.S. Cellular Field using polar coordinates. Use the east direction for the polar axis. Express θ in degrees.

City of Chicago, Illinois

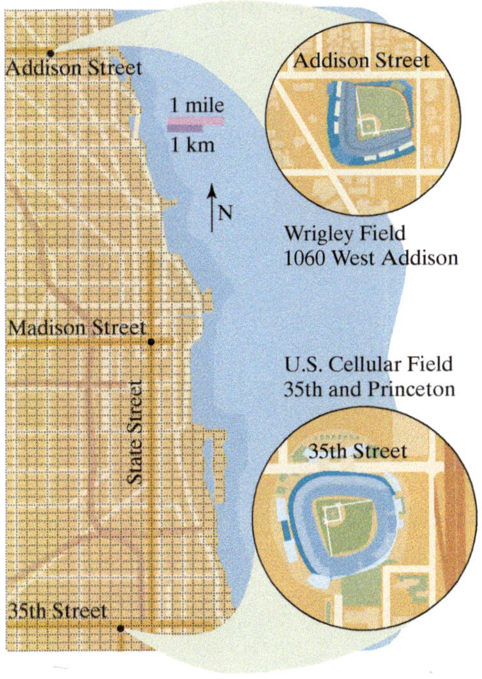

86. Show that the formula for the distance d between two points $P_1 = (r_1, \theta_1)$ and $P_2 = (r_2, \theta_2)$ is

$$d = \sqrt{r_1^2 + r_2^2 - 2r_1 r_2 \cos(\theta_2 - \theta_1)}$$

Explaining Concepts: Discussion and Writing

87. In converting from polar coordinates to rectangular coordinates, what formulas will you use?

88. Explain how to convert from rectangular coordinates to polar coordinates.

89. Is the street system in your town based on a rectangular coordinate system, a polar coordinate system, or some other system? Explain.

Retain Your Knowledge

Problems 90–93 are based on material learned earlier in the course. The purpose of these problems is to keep the material fresh in your mind so that you are better prepared for the final exam.

90. Solve: $\log_4(x+3) - \log_4(x-1) = 2$

91. Use Descartes' Rule of Signs to determine the possible number of positive or negative real zeros for the function $f(x) = -2x^3 + 6x^2 - 7x - 8$.

92. Find the midpoint of the line segment connecting the points $(-3, 7)$ and $\left(\frac{1}{2}, 2\right)$.

93. Given that the point (3, 8) is on the graph of $y = f(x)$, what is the corresponding point on the graph of $y = -2f(x+3) + 5$?

'Are You Prepared?' Answers

1. 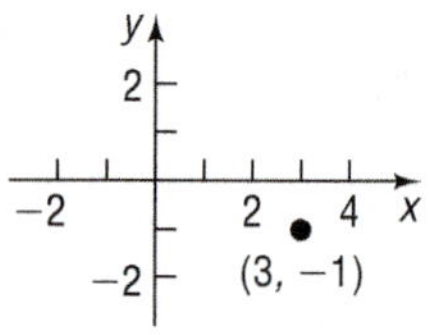

; quadrant IV **2.** 9 **3.** $\frac{b}{a}$ **4.** $-\frac{\pi}{4}$

10.2 Polar Equations and Graphs

PREPARING FOR THIS SECTION *Before getting started, review the following:*

- Symmetry (Section 2.1, pp. 166–168)
- Circles (Section 2.3, pp. 189–193)
- Even–Odd Properties of Trigonometric Functions (Section 7.5, pp. 568–569)
- Difference Formulas for Sine and Cosine (Section 8.5, pp. 652 and 655)
- Values of the Sine and Cosine Functions at Certain Angles (Section 7.3, pp. 539–541; Section 7.4, pp. 550–558)

Now Work the **'Are You Prepared?'** problems on page 753.

OBJECTIVES
1 Identify and Graph Polar Equations by Converting to Rectangular Equations (p. 741)
2 Graph Polar Equations Using a Graphing Utility (p. 742)
3 Test Polar Equations for Symmetry (p. 745)
4 Graph Polar Equations by Plotting Points (p. 746)

Just as a rectangular grid can be used to plot points given by rectangular coordinates, such as the points $(-3, 1)$ and $(1, 2)$ shown in Figure 21(a), a grid consisting of concentric circles (with centers at the pole) and rays (with vertices at the pole) can be used to plot points given by polar coordinates, such as the points $\left(4, \frac{5\pi}{4}\right)$ and $\left(2, \frac{\pi}{4}\right)$ shown in Figure 21(b). Such **polar grids** are used to graph *polar equations*.

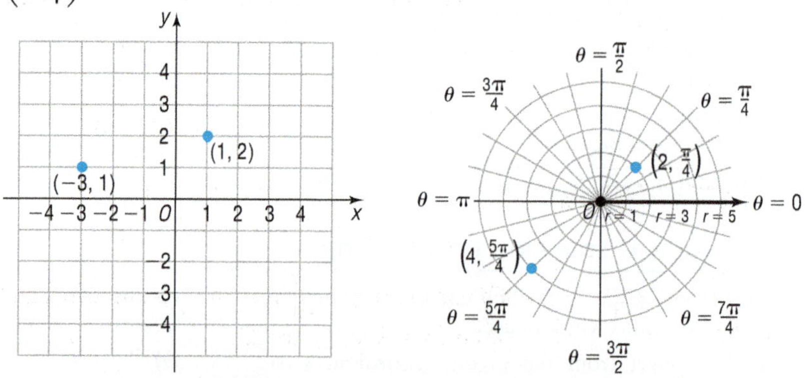

Figure 21 **(a)** Rectangular grid **(b)** Polar grid

DEFINITION

> An equation whose variables are polar coordinates is called a **polar equation**. The **graph of a polar equation** consists of all points whose polar coordinates satisfy the equation.

1 Identify and Graph Polar Equations by Converting to Rectangular Equations

One method that can be used to graph a polar equation is to convert the equation to rectangular coordinates. In the following discussion, (x, y) represents the rectangular coordinates of a point P, and (r, θ) represents polar coordinates of the point P.

EXAMPLE 1 Identifying and Graphing a Polar Equation (Circle)

Identify and graph the equation: $r = 3$

Solution Convert the polar equation to a rectangular equation.

$$r = 3$$

$$r^2 = 9 \quad \text{Square both sides.}$$

$$x^2 + y^2 = 9 \quad r^2 = x^2 + y^2$$

The graph of $r = 3$ is a circle, with center at the pole and radius 3. See Figure 22.

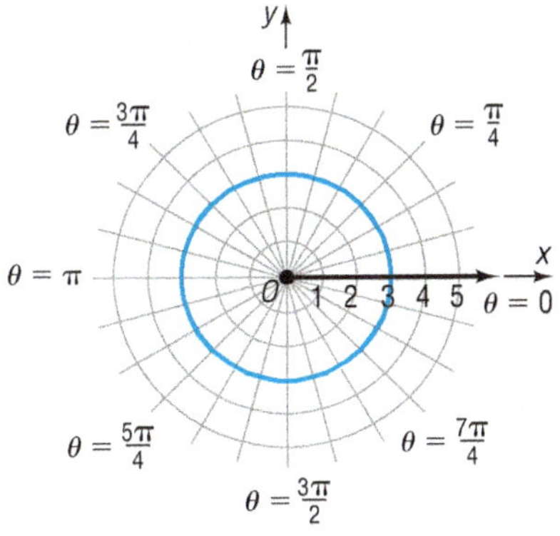

Figure 22 $r = 3$, or $x^2 + y^2 = 9$

Now Work PROBLEM 15

EXAMPLE 2 Identifying and Graphing a Polar Equation (Line)

Identify and graph the equation: $\theta = \dfrac{\pi}{4}$

Solution Convert the polar equation to a rectangular equation.

$$\theta = \frac{\pi}{4}$$

$$\tan\theta = \tan\frac{\pi}{4} \quad \text{Take the tangent of both sides.}$$

$$\frac{y}{x} = 1 \quad \tan\theta = \frac{y}{x};\ \tan\frac{\pi}{4} = 1$$

$$y = x$$

The graph of $\theta = \dfrac{\pi}{4}$ is a line passing through the pole making an angle of $\dfrac{\pi}{4}$ with the polar axis. See Figure 23.

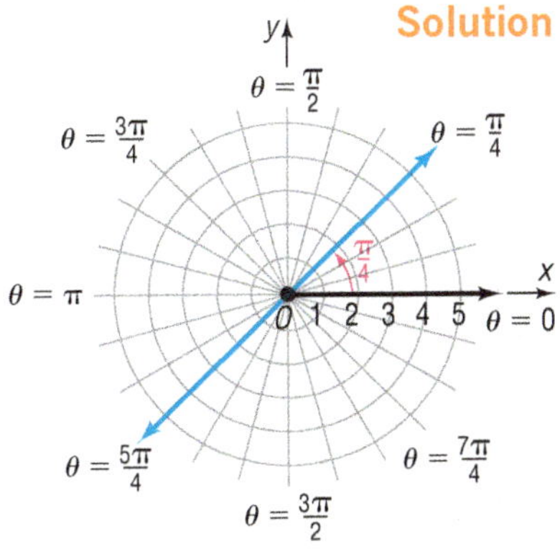

Figure 23 $\theta = \dfrac{\pi}{4}$, or $y = x$

Now Work PROBLEM 17

EXAMPLE 3

Identifying and Graphing a Polar Equation (Horizontal Line)

Identify and graph the equation: $r \sin \theta = 2$

Solution Because $y = r \sin \theta$, we can write the equation as

$$y = 2$$

Therefore, the graph of $r \sin \theta = 2$ is a horizontal line 2 units above the pole. See Figure 24.

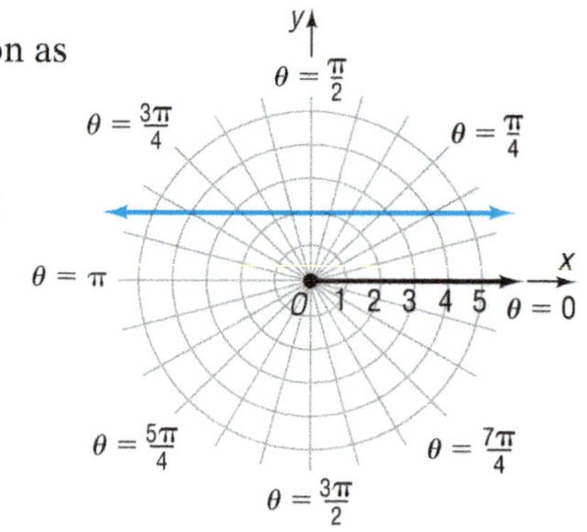

Figure 24 $r \sin \theta = 2$, or $y = 2$

2 Graph Polar Equations Using a Graphing Utility

A second method for graphing polar equations is to use a graphing utility.

Graphing a Polar Equation Using a Graphing Utility

STEP 1: Solve the equation for r in terms of θ.

STEP 2: Select the viewing window in POLar mode. In addition to setting Xmin, Xmax, Xscl, and so forth, the viewing window in polar mode requires setting minimum and maximum values for θ and an increment setting for θ(θstep). Finally, a square screen and radian measure should be used.

STEP 3: Enter the expression involving θ that you found in Step 1. (Consult your manual for the correct way to enter the expression.)

STEP 4: Press graph.

EXAMPLE 4

Graphing a Polar Equation Using a Graphing Utility

Use a graphing utility to graph the polar equation $r \sin \theta = 2$.

Solution **STEP 1:** Solve the equation for r in terms of θ.

$$r \sin \theta = 2$$
$$r = \frac{2}{\sin \theta}$$

STEP 2: From the polar mode, select a square viewing window. We use the one given next.

$$\theta\text{min} = 0 \qquad X\text{min} = -8 \qquad Y\text{min} = -5$$
$$\theta\text{max} = 2\pi \qquad X\text{max} = 8 \qquad Y\text{max} = 5$$
$$\theta\text{step} = \frac{\pi}{24} \qquad X\text{scl} = 1 \qquad Y\text{scl} = 1$$

θstep determines the number of points the graphing utility plots. For example, if θstep is $\frac{\pi}{24}$, then the graphing utility evaluates r at $\theta = 0$ (θmin), $\frac{\pi}{24}, \frac{2\pi}{24}, \frac{3\pi}{24}$, and so forth, up to 2π (θmax).

The smaller θstep is the more points the graphing utility plots. Experiment with different values for θmin, θmax, and θstep to see how the graph is affected.

STEP 3: Enter the expression $\frac{2}{\sin \theta}$ after the prompt $r =$.

STEP 4: Graph.

The graph is shown in Figure 25.

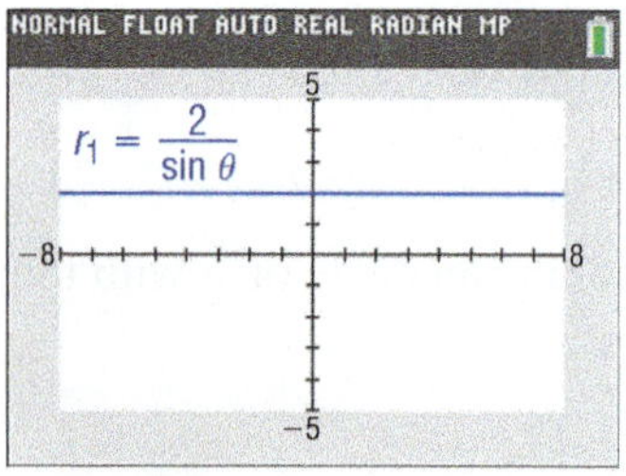

Figure 25 $r \sin \theta = 2$

EXAMPLE 5 Identifying and Graphing a Polar Equation (Vertical Line)

Identify and graph the equation: $r \cos \theta = -3$

Solution Since $x = r \cos \theta$, we can write the equation as

$$x = -3$$

Therefore, the graph of $r \cos \theta = -3$ is a vertical line 3 units to the left of the pole. Figure 26(a) shows the graph drawn by hand. Figure 26(b) shows the graph using a TI-84 Plus C with θmin $= 0$, θmax $= 2\pi$, and θstep $= \frac{\pi}{24}$.

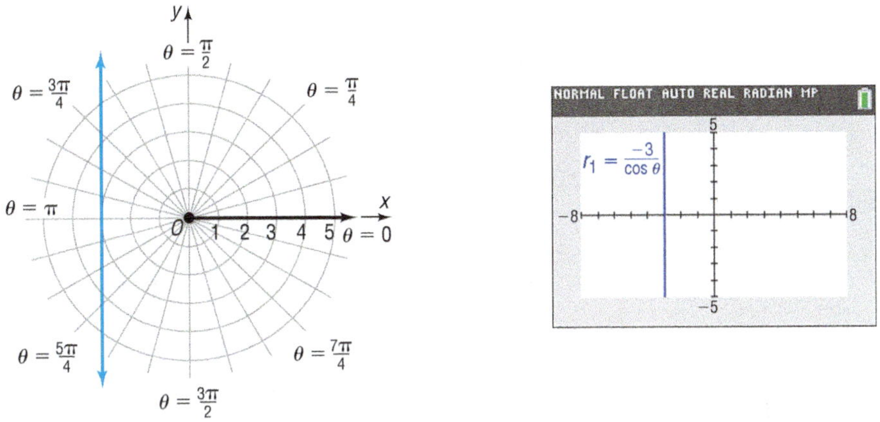

Figure 26 $r \cos \theta = -3$, or $x = -3$

(a) (b)

Examples 3, 4, and 5 lead to the following results. (The proofs are left as exercises. See Problems 83 and 84.)

THEOREM

Let a be a nonzero real number. Then the graph of the equation

$$r \sin \theta = a$$

is a horizontal line. It lies a units above the pole if $a > 0$ and lies $|a|$ units below the pole if $a < 0$.

The graph of the equation

$$r \cos \theta = a$$

is a vertical line. It lies a units to the right of the pole if $a > 0$ and lies $|a|$ units to the left of the pole if $a < 0$.

Now Work PROBLEM 21

EXAMPLE 6 Identifying and Graphing a Polar Equation (Circle)

Identify and graph the equation: $r = 4 \sin \theta$

Solution To transform the equation to rectangular coordinates, multiply each side by r.

$$r^2 = 4r \sin \theta$$

Now use the facts that $r^2 = x^2 + y^2$ and $y = r \sin \theta$. Then

$$x^2 + y^2 = 4y$$

$$x^2 + (y^2 - 4y) = 0$$

$$x^2 + (y^2 - 4y + 4) = 4 \quad \text{Complete the square in } y.$$

$$x^2 + (y - 2)^2 = 4 \quad \text{Factor.}$$

This is the standard equation of a circle with center at $(0, 2)$ in rectangular coordinates and radius 2. Figure 27(a) shows the graph drawn by hand. Figure 27(b) shows the graph using a TI-84 Plus C with θmin $= 0$, θmax $= 2\pi$, and θstep $= \frac{\pi}{24}$.

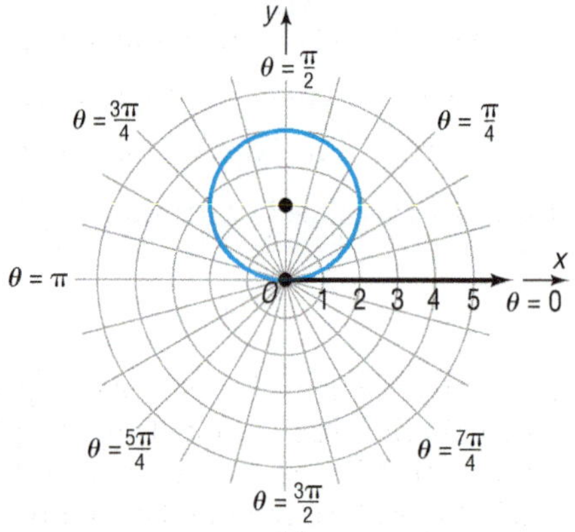

(a)

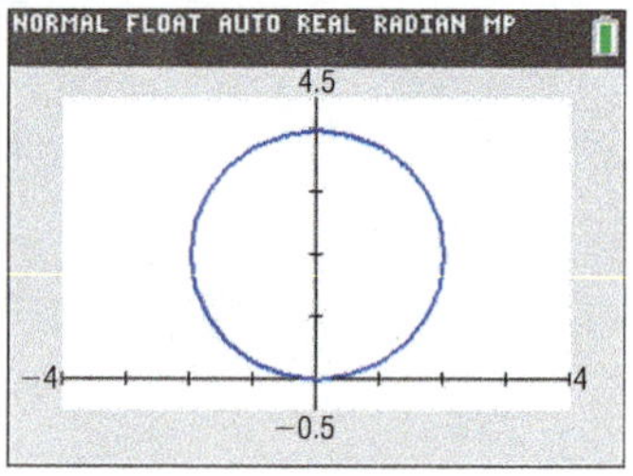

(b)

Figure 27
$r = 4\sin\theta$, or $x^2 + (y-2)^2 = 4$

EXAMPLE 7 Identifying and Graphing a Polar Equation (Circle)

Identify and graph the equation: $r = -2\cos\theta$

Solution To transform the equation to rectangular coordinates, multiply each side by r.

$$r^2 = -2r\cos\theta \qquad \text{Multiply both sides by } r.$$
$$x^2 + y^2 = -2x \qquad r^2 = x^2 + y^2; \quad x = r\cos\theta$$
$$x^2 + 2x + y^2 = 0$$
$$(x^2 + 2x + 1) + y^2 = 1 \qquad \text{Complete the square in } x.$$
$$(x+1)^2 + y^2 = 1 \qquad \text{Factor.}$$

This is the standard equation of a circle with center at $(-1, 0)$ in rectangular coordinates and radius 1. Figure 28(a) shows the graph drawn by hand. Figure 28(b) shows the graph using a TI-84 Plus C with θmin $= 0$, θmax $= 2\pi$, and θstep $= \frac{\pi}{24}$.

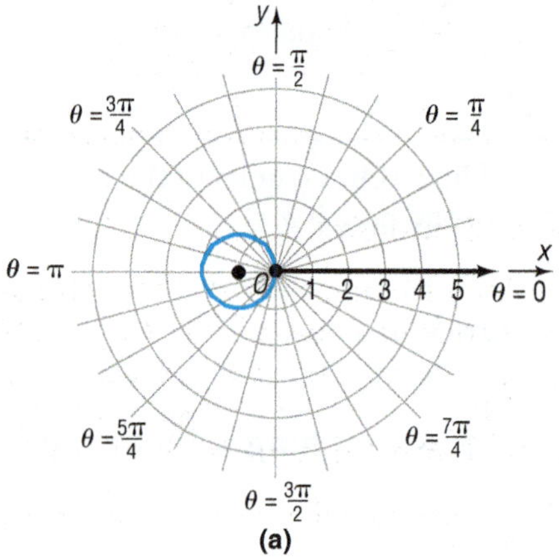

(a)

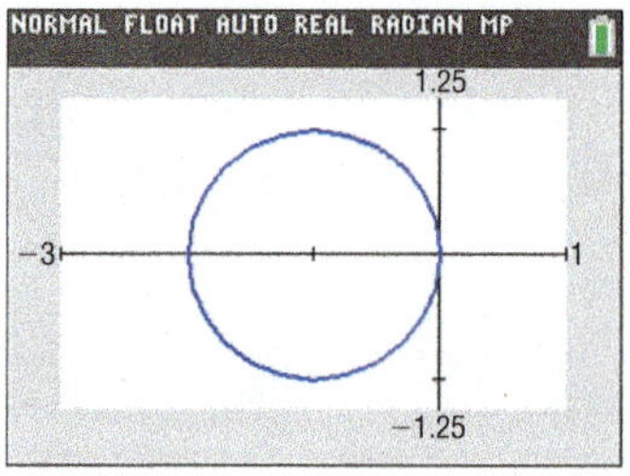

(b)

Figure 28
$r = -2\cos\theta$, or $(x+1)^2 + y^2 = 1$

Exploration

Using a square screen, graph $r_1 = \sin\theta$, $r_2 = 2\sin\theta$, and $r_3 = 3\sin\theta$. Do you see the pattern? Clear the screen and graph $r_1 = -\sin\theta$, $r_2 = -2\sin\theta$, and $r_3 = -3\sin\theta$. Do you see the pattern? Clear the screen and graph $r_1 = \cos\theta$, $r_2 = 2\cos\theta$, and $r_3 = 3\cos\theta$. Do you see the pattern? Clear the screen and graph $r_1 = -\cos\theta$, $r_2 = -2\cos\theta$, and $r_3 = -3\cos\theta$. Do you see the pattern?

Examples 6 and 7 and the preceding Exploration lead to the following results. (The proofs are left as exercises. See Problems 85–88.)

THEOREM

Let a be a positive real number. Then

Equation	**Description**
(a) $r = 2a \sin \theta$	Circle: radius a; center at $(0, a)$ in rectangular coordinates
(b) $r = -2a \sin \theta$	Circle: radius a; center at $(0, -a)$ in rectangular coordinates
(c) $r = 2a \cos \theta$	Circle: radius a; center at $(a, 0)$ in rectangular coordinates
(d) $r = -2a \cos \theta$	Circle: radius a; center at $(-a, 0)$ in rectangular coordinates

Each circle passes through the pole.

Now Work PROBLEM 23

The method of converting a polar equation to an identifiable rectangular equation to obtain the graph is not always helpful, nor is it always necessary. Usually, a table is created that lists several points on the graph. By checking for symmetry, it may be possible to reduce the number of points needed to draw the graph.

3 Test Polar Equations for Symmetry

In polar coordinates, the points (r, θ) and $(r, -\theta)$ are symmetric with respect to the polar axis (and to the x-axis). See Figure 29(a). The points (r, θ) and $(r, \pi - \theta)$ are symmetric with respect to the line $\theta = \frac{\pi}{2}$ (the y-axis). See Figure 29(b). The points (r, θ) and $(-r, \theta)$ are symmetric with respect to the pole (the origin). See Figure 29(c).

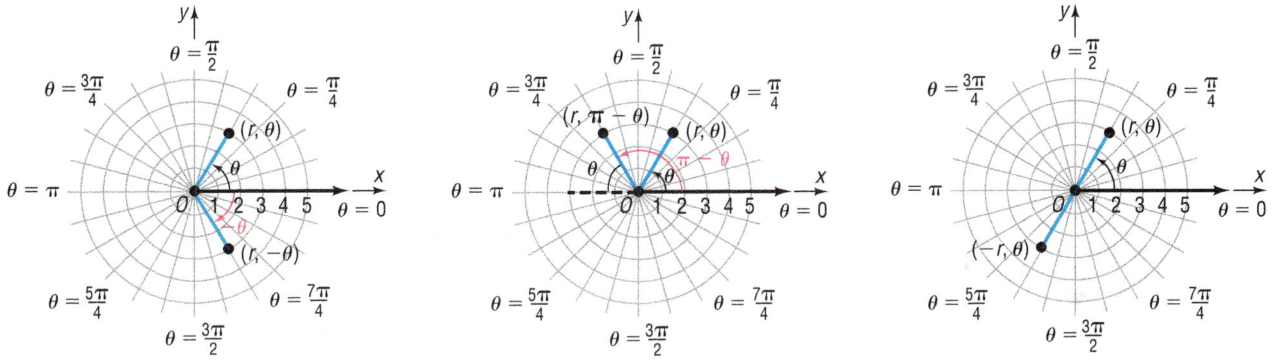

Figure 29 **(a)** Points symmetric with respect to the polar axis **(b)** Points symmetric with respect to the line $\theta = \frac{\pi}{2}$ **(c)** Points symmetric with respect to the pole

The following tests are a consequence of these observations.

THEOREM

Tests for Symmetry

Symmetry with Respect to the Polar Axis (x-Axis)

In a polar equation, replace θ by $-\theta$. If an equivalent equation results, the graph is symmetric with respect to the polar axis.

Symmetry with Respect to the Line $\theta = \frac{\pi}{2}$ (y-Axis)

In a polar equation, replace θ by $\pi - \theta$. If an equivalent equation results, the graph is symmetric with respect to the line $\theta = \frac{\pi}{2}$.

Symmetry with Respect to the Pole (Origin)

In a polar equation, replace r by $-r$ or θ by $\theta + \pi$. If an equivalent equation results, the graph is symmetric with respect to the pole.

The three tests for symmetry given here are *sufficient* conditions for symmetry, but they are not *necessary* conditions. That is, an equation may fail these tests and still have a graph that is symmetric with respect to the polar axis, the line $\theta = \frac{\pi}{2}$, or the pole. For example, the graph of $r = \sin(2\theta)$ turns out to be symmetric with respect to the polar axis, the line $\theta = \frac{\pi}{2}$, and the pole, but only the test for symmetry with respect to the pole (replace θ by $\theta + \pi$) works. See also Problems 89–91.

4 Graph Polar Equations by Plotting Points

EXAMPLE 8 Graphing a Polar Equation (Cardioid)

Graph the equation: $r = 1 - \sin\theta$

Solution Check for symmetry first.

Polar Axis: Replace θ by $-\theta$. The result is

$$r = 1 - \sin(-\theta) = 1 + \sin\theta \quad \sin(-\theta) = -\sin\theta$$

The test fails, so the graph may or may not be symmetric with respect to the polar axis.

The Line $\theta = \frac{\pi}{2}$: Replace θ by $\pi - \theta$. The result is

$$\begin{aligned} r = 1 - \sin(\pi - \theta) &= 1 - (\sin\pi\cos\theta - \cos\pi\sin\theta) \\ &= 1 - [0\cdot\cos\theta - (-1)\sin\theta] = 1 - \sin\theta \end{aligned}$$

The test is satisfied, so the graph is symmetric with respect to the line $\theta = \frac{\pi}{2}$.

The Pole: Replace r by $-r$. Then the result is $-r = 1 - \sin\theta$, so $r = -1 + \sin\theta$. The test fails. Replace θ by $\theta + \pi$. The result is

$$\begin{aligned} r &= 1 - \sin(\theta + \pi) \\ &= 1 - [\sin\theta\cos\pi + \cos\theta\sin\pi] \\ &= 1 - [\sin\theta\cdot(-1) + \cos\theta\cdot 0] \\ &= 1 + \sin\theta \end{aligned}$$

This test also fails, so the graph may or may not be symmetric with respect to the pole.

Next, identify points on the graph by assigning values to the angle θ and calculating the corresponding values of r. Due to the periodicity of the sine function and the symmetry with respect to the line $\theta = \frac{\pi}{2}$, just assign values to θ from $-\frac{\pi}{2}$ to $\frac{\pi}{2}$, as given in Table 1.

Table 1

θ	$r = 1 - \sin\theta$
$-\frac{\pi}{2}$	$1 - (-1) = 2$
$-\frac{\pi}{3}$	$1 - \left(-\frac{\sqrt{3}}{2}\right) \approx 1.87$
$-\frac{\pi}{6}$	$1 - \left(-\frac{1}{2}\right) = \frac{3}{2}$
0	$1 - 0 = 1$
$\frac{\pi}{6}$	$1 - \frac{1}{2} = \frac{1}{2}$
$\frac{\pi}{3}$	$1 - \frac{\sqrt{3}}{2} \approx 0.13$
$\frac{\pi}{2}$	$1 - 1 = 0$

Now plot the points (r, θ) from Table 1 and trace out the graph, beginning at the point $\left(2, -\frac{\pi}{2}\right)$ and ending at the point $\left(0, \frac{\pi}{2}\right)$. Then reflect this portion of the graph about the line $\theta = \frac{\pi}{2}$ (the y-axis) to obtain the complete graph. Figure 30(a) shows the graph drawn by hand. Figure 30(b) shows the graph using a TI-84 Plus C with $\theta\text{min} = 0$, $\theta\text{max} = 2\pi$, and $\theta\text{step} = \frac{\pi}{24}$.

Exploration

Graph $r_1 = 1 + \sin\theta$. Clear the screen and graph $r_1 = 1 - \cos\theta$. Clear the screen and graph $r_1 = 1 + \cos\theta$. Do you see a pattern?

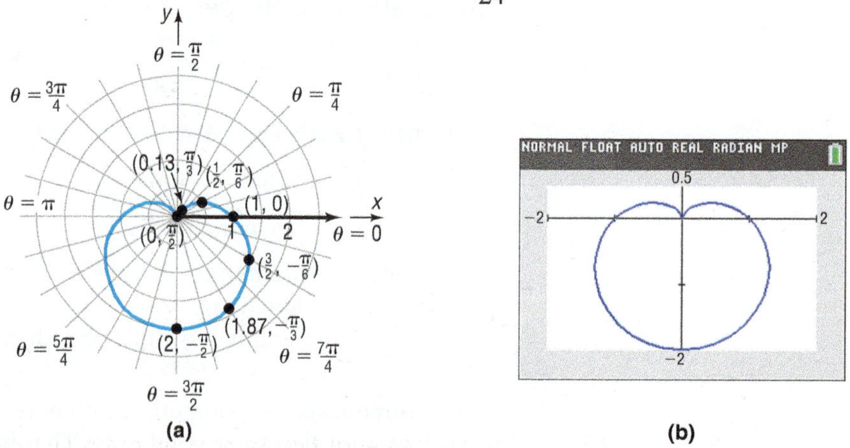

Figure 30 $r = 1 - \sin\theta$

The curve in Figure 30 is an example of a *cardioid* (a heart-shaped curve).

DEFINITION

Cardioids are characterized by equations of the form

$$r = a(1 + \cos\theta) \qquad r = a(1 + \sin\theta)$$
$$r = a(1 - \cos\theta) \qquad r = a(1 - \sin\theta)$$

where $a > 0$. The graph of a cardioid passes through the pole.

Now Work PROBLEM 39

EXAMPLE 9 Graphing a Polar Equation (Limaçon without Inner Loop)

Graph the equation: $r = 3 + 2\cos\theta$

Solution Check for symmetry first.

Polar Axis: Replace θ by $-\theta$. The result is

$$r = 3 + 2\cos(-\theta) = 3 + 2\cos\theta \quad \cos(-\theta) = \cos\theta$$

The test is satisfied, so the graph is symmetric with respect to the polar axis.

The Line $\theta = \frac{\pi}{2}$: Replace θ by $\pi - \theta$. The result is

$$r = 3 + 2\cos(\pi - \theta) = 3 + 2(\cos\pi\cos\theta + \sin\pi\sin\theta)$$
$$= 3 - 2\cos\theta$$

The test fails, so the graph may or may not be symmetric with respect to the line $\theta = \frac{\pi}{2}$.

The Pole: Replace r by $-r$. The test fails. Replace θ by $\theta + \pi$. This test also fails, so the graph may or may not be symmetric with respect to the pole.

Next, identify points on the graph by assigning values to the angle θ and calculating the corresponding values of r. Due to the periodicity of the cosine function and the symmetry with respect to the polar axis, just assign values to θ from 0 to π, as given in Table 2.

Now plot the points (r, θ) from Table 2 and trace out the graph, beginning at the point $(5, 0)$ and ending at the point $(1, \pi)$. Then reflect this portion of the graph about the polar axis (the x-axis) to obtain the complete graph. Figure 31(a) shows the graph drawn by hand. Figure 31(b) shows the graph using a TI-84 Plus C with θmin $= 0$, θmax $= 2\pi$, and θstep $= \frac{\pi}{24}$.

Table 2

θ	$r = 3 + 2\cos\theta$
0	$3 + 2(1) = 5$
$\frac{\pi}{6}$	$3 + 2\left(\frac{\sqrt{3}}{2}\right) \approx 4.73$
$\frac{\pi}{3}$	$3 + 2\left(\frac{1}{2}\right) = 4$
$\frac{\pi}{2}$	$3 + 2(0) = 3$
$\frac{2\pi}{3}$	$3 + 2\left(-\frac{1}{2}\right) = 2$
$\frac{5\pi}{6}$	$3 + 2\left(-\frac{\sqrt{3}}{2}\right) \approx 1.27$
π	$3 + 2(-1) = 1$

Exploration

Graph $r_1 = 3 - 2\cos\theta$. Clear the screen and graph $r_1 = 3 + 2\sin\theta$. Clear the screen and graph $r_1 = 3 - 2\sin\theta$. Do you see a pattern?

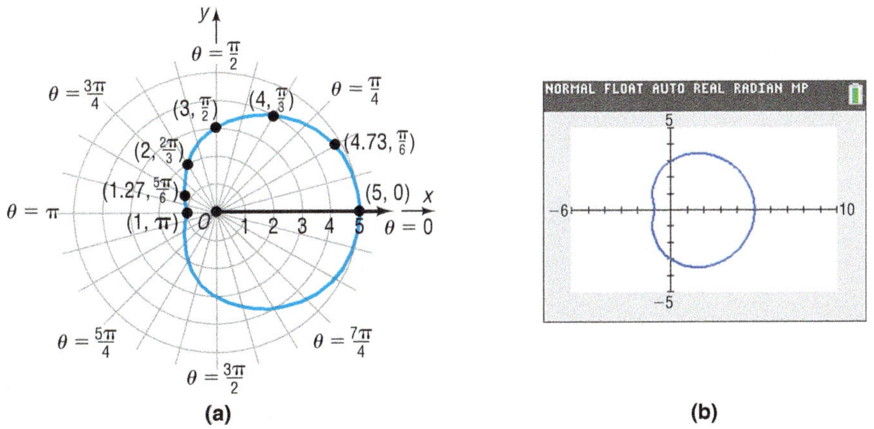

Figure 31 $r = 3 + 2\cos\theta$

The curve in Figure 31 is an example of a *limaçon* (the French word for *snail*) *without an inner loop*.

DEFINITION

Limaçons without an inner loop are characterized by equations of the form

$$r = a + b\cos\theta \qquad r = a + b\sin\theta$$
$$r = a - b\cos\theta \qquad r = a - b\sin\theta$$

where $a > 0$, $b > 0$, and $a > b$. The graph of a limaçon without an inner loop does not pass through the pole.

Now Work PROBLEM 45

EXAMPLE 10 Graphing a Polar Equation (Limaçon with Inner Loop)

Graph the equation: $r = 1 + 2\cos\theta$

Solution First, check for symmetry.

Polar Axis: Replace θ by $-\theta$. The result is

$$r = 1 + 2\cos(-\theta) = 1 + 2\cos\theta$$

The test is satisfied, so the graph is symmetric with respect to the polar axis.

The Line $\theta = \frac{\pi}{2}$: Replace θ by $\pi - \theta$. The result is

$$r = 1 + 2\cos(\pi - \theta) = 1 + 2(\cos\pi\cos\theta + \sin\pi\sin\theta)$$
$$= 1 - 2\cos\theta$$

The test fails, so the graph may or may not be symmetric with respect to the line $\theta = \frac{\pi}{2}$.

The Pole: Replace r by $-r$. The test fails. Replace θ by $\theta + \pi$. This test also fails, so the graph may or may not be symmetric with respect to the pole.

Next, identify points on the graph of $r = 1 + 2\cos\theta$ by assigning values to the angle θ and calculating the corresponding values of r. Due to the periodicity of the cosine function and the symmetry with respect to the polar axis, just assign values to θ from 0 to π, as given in Table 3.

Now plot the points (r, θ) from Table 3, beginning at $(3, 0)$ and ending at $(-1, \pi)$. See Figure 32(a). Finally, reflect this portion of the graph about the polar axis (the x-axis) to obtain the complete graph. See Figure 32(b). Figure 32(c) shows the graph using a TI-84 Plus C with $\theta\text{min} = 0$, $\theta\text{max} = 2\pi$, and $\theta\text{step} = \frac{\pi}{24}$.

Table 3

θ	$r = 1 + 2\cos\theta$
0	$1 + 2(1) = 3$
$\frac{\pi}{6}$	$1 + 2\left(\frac{\sqrt{3}}{2}\right) \approx 2.73$
$\frac{\pi}{3}$	$1 + 2\left(\frac{1}{2}\right) = 2$
$\frac{\pi}{2}$	$1 + 2(0) = 1$
$\frac{2\pi}{3}$	$1 + 2\left(-\frac{1}{2}\right) = 0$
$\frac{5\pi}{6}$	$1 + 2\left(-\frac{\sqrt{3}}{2}\right) \approx -0.73$
π	$1 + 2(-1) = -1$

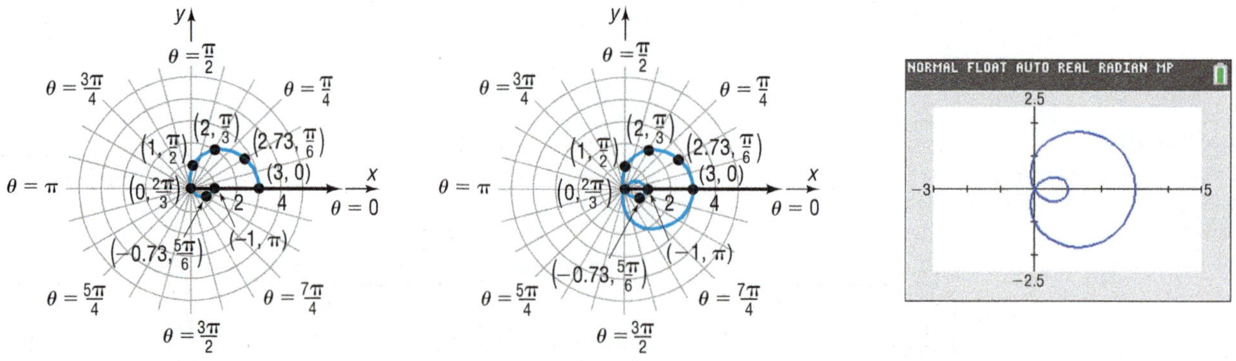

Figure 32 (a) (b) $r = 1 + 2\cos\theta$ (c)

The curve in Figure 32(b) or Figure 32(c) is an example of a *limaçon with an inner loop*.

DEFINITION

Limaçons with an inner loop are characterized by equations of the form

$$r = a + b\cos\theta \qquad r = a + b\sin\theta$$
$$r = a - b\cos\theta \qquad r = a - b\sin\theta$$

where $a > 0, b > 0$, and $a < b$. The graph of a limaçon with an inner loop passes through the pole twice.

Exploration

Graph $r_1 = 1 - 2\cos\theta$. Clear the screen and graph $r_1 = 1 + 2\sin\theta$. Clear the screen and graph $r_1 = 1 - 2\sin\theta$. Do you see a pattern?

Now Work PROBLEM 47

EXAMPLE 11 Graphing a Polar Equation (Rose)

Graph the equation: $r = 2\cos(2\theta)$

Solution Check for symmetry.

Polar Axis: Replace θ by $-\theta$. The result is

$$r = 2\cos[2(-\theta)] = 2\cos(2\theta)$$

The test is satisfied, so the graph is symmetric with respect to the polar axis.

The Line $\theta = \frac{\pi}{2}$: Replace θ by $\pi - \theta$. The result is

$$r = 2\cos[2(\pi - \theta)] = 2\cos(2\pi - 2\theta) = 2\cos(2\theta)$$

The test is satisfied, so the graph is symmetric with respect to the line $\theta = \frac{\pi}{2}$.

The Pole: Since the graph is symmetric with respect to both the polar axis and the line $\theta = \frac{\pi}{2}$, it must be symmetric with respect to the pole.

Next, construct Table 4. Because of the periodicity of the cosine function and the symmetry with respect to the polar axis, the line $\theta = \frac{\pi}{2}$, and the pole, consider only values of θ from 0 to $\frac{\pi}{2}$.

Plot and connect these points as shown in Figure 33(a). Finally, because of symmetry, reflect this portion of the graph first about the polar axis (the x-axis) and then about the line $\theta = \frac{\pi}{2}$ (the y-axis) to obtain the complete graph. See Figure 33(b). Figure 33(c) shows the graph using a TI-84 Plus C with $\theta\text{min} = 0$, $\theta\text{max} = 2\pi$, and $\theta\text{step} = \frac{\pi}{24}$.

Table 4

θ	$r = 2\cos(2\theta)$
0	$2(1) = 2$
$\frac{\pi}{6}$	$2\left(\frac{1}{2}\right) = 1$
$\frac{\pi}{4}$	$2(0) = 0$
$\frac{\pi}{3}$	$2\left(-\frac{1}{2}\right) = -1$
$\frac{\pi}{2}$	$2(-1) = -2$

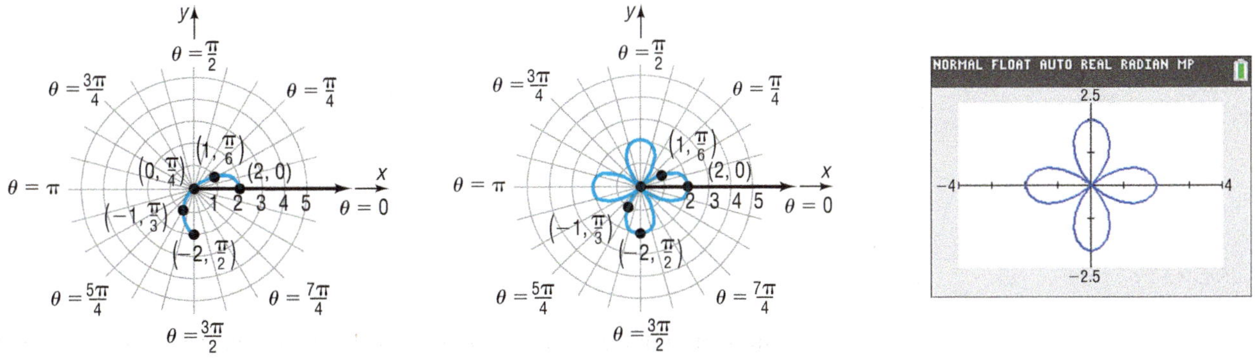

Figure 33 (a) (b) $r = 2\cos(2\theta)$ (c)

Exploration

Graph $r_1 = 2\cos(4\theta)$; clear the screen and graph $r_1 = 2\cos(6\theta)$. How many petals did each of these graphs have?

Clear the screen and graph, in order, each on a clear screen, $r_1 = 2\cos(3\theta)$, $r_1 = 2\cos(5\theta)$, and $r_1 = 2\cos(7\theta)$. What do you notice about the number of petals?

The curve in Figure 33(b) or 33(c) is called a *rose* with four petals.

DEFINITION

Rose curves are characterized by equations of the form

$$r = a\cos(n\theta), \qquad r = a\sin(n\theta), \qquad a \neq 0$$

and have graphs that are rose shaped. If $n \neq 0$ is even, the rose has $2n$ petals; if $n \neq \pm 1$ is odd, the rose has n petals.

Now Work PROBLEM 51

EXAMPLE 12 Graphing a Polar Equation (Lemniscate)

Graph the equation: $r^2 = 4\sin(2\theta)$

Solution

We leave it to you to verify that the graph is symmetric with respect to the pole. Because of the symmetry with respect to the pole, consider only those values of θ between $\theta = 0$ and $\theta = \pi$. Note that there are no points on the graph for $\frac{\pi}{2} < \theta < \pi$ (quadrant II), because $\sin(2\theta) < 0$ for such values. Table 5 lists points on the graph for values of $\theta = 0$ through $\theta = \frac{\pi}{2}$. The points from Table 5 where $r \geq 0$ are plotted in Figure 34(a). The remaining points on the graph may be obtained by using symmetry. Figure 34(b) shows the final graph drawn by hand. Figure 34(c) shows the graph using a TI-84 Plus C with θmin $= 0$, θmax $= 2\pi$, and θstep $= \frac{\pi}{360}$.

Table 5

θ	$r^2 = 4\sin(2\theta)$	r
0	$4(0) = 0$	0
$\frac{\pi}{6}$	$4\left(\frac{\sqrt{3}}{2}\right) = 2\sqrt{3}$	± 1.9
$\frac{\pi}{4}$	$4(1) = 4$	± 2
$\frac{\pi}{3}$	$4\left(\frac{\sqrt{3}}{2}\right) = 2\sqrt{3}$	± 1.9
$\frac{\pi}{2}$	$4(0) = 0$	0

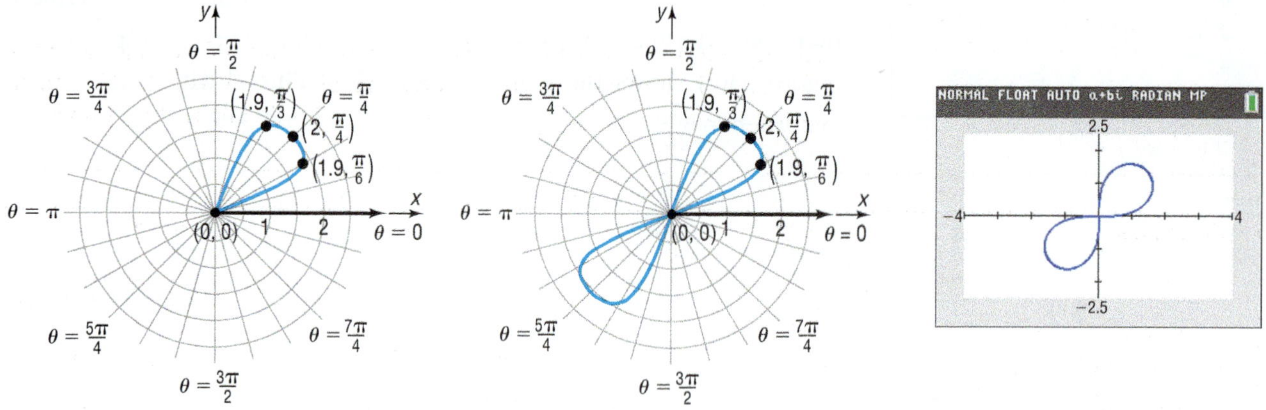

Figure 34 (a) (b) $r^2 = 4\sin(2\theta)$ (c)

The curve in Figure 34(b) or 34(c) is an example of a *lemniscate* (from the Greek word for *ribbon*).

DEFINITION

Lemniscates are characterized by equations of the form

$$r^2 = a^2 \sin(2\theta) \qquad r^2 = a^2 \cos(2\theta)$$

where $a \neq 0$ and have graphs that are propeller shaped.

Now Work PROBLEM 55

EXAMPLE 13 Graphing a Polar Equation (Spiral)

Graph the equation: $r = e^{\theta/5}$

Solution The tests for symmetry with respect to the pole, the polar axis, and the line $\theta = \frac{\pi}{2}$ fail. Furthermore, there is no number θ for which $r = 0$, so the graph does not pass through the pole. Observe that r is positive for all θ, r increases as θ increases, $r \to 0$ as $\theta \to -\infty$, and $r \to \infty$ as $\theta \to \infty$. With the help of a calculator, the values in Table 6 can be obtained. See Figure 35(a) for the graph drawn by hand. Figure 35(b) shows the graph using a TI-84 Plus C with θmin $= -6\pi$, θmax $= 3\pi$, and θstep $= \frac{\pi}{24}$.

Table 6

θ	$r = e^{\theta/5}$
$-\frac{3\pi}{2}$	0.39
$-\pi$	0.53
$-\frac{\pi}{2}$	0.73
$-\frac{\pi}{4}$	0.85
0	1
$\frac{\pi}{4}$	1.17
$\frac{\pi}{2}$	1.37
π	1.87
$\frac{3\pi}{2}$	2.57
2π	3.51

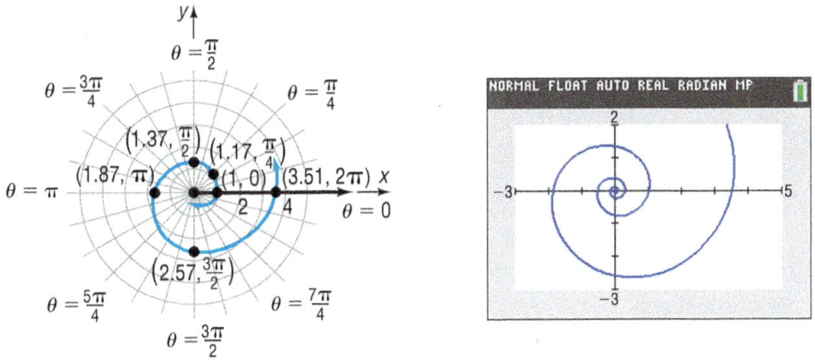

Figure 35 $r = e^{\theta/5}$

(a) (b)

The curve in Figure 35 is called a **logarithmic spiral**, since its equation may be written as $\theta = 5 \ln r$ and it spirals infinitely both toward the pole and away from it.

Classification of Polar Equations

The equations of some lines and circles in polar coordinates and their corresponding equations in rectangular coordinates are given in Table 7. Also included are the names and graphs of a few of the more frequently encountered polar equations.

Table 7

Lines			
Description	Line passing through the pole making an angle α with the polar axis	Vertical line	Horizontal line
Rectangular equation	$y = (\tan \alpha)x$	$x = a$	$y = b$
Polar equation	$\theta = \alpha$	$r \cos \theta = a$	$r \sin \theta = b$
Typical graph	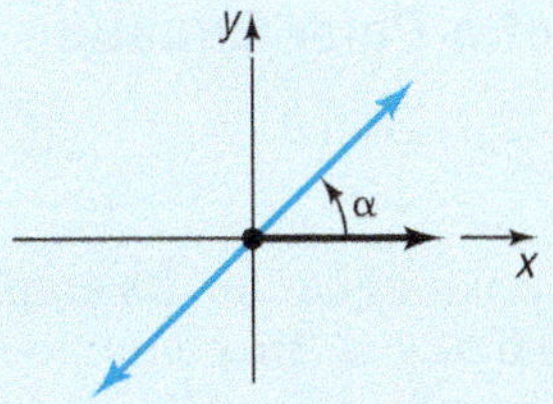	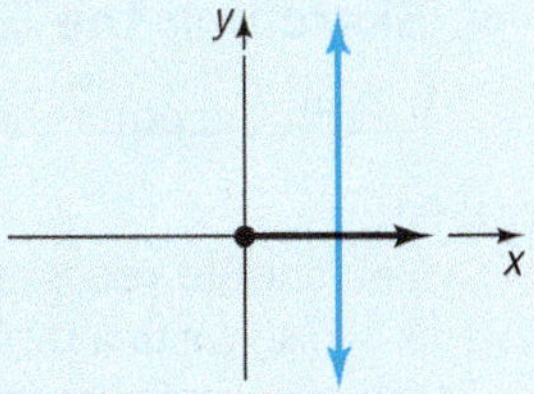	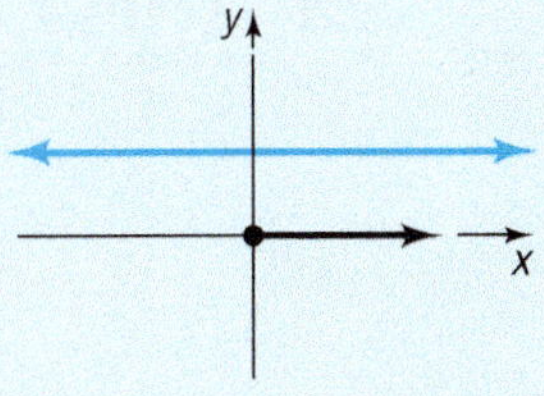

(continued)

Table 7 (Continued)

	Circles		
Description	Center at the pole, radius a	Passing through the pole, tangent to the line $\theta = \frac{\pi}{2}$, center on the polar axis, radius a	Passing through the pole, tangent to the polar axis, center on the line $\theta = \frac{\pi}{2}$, radius a
Rectangular equation	$x^2 + y^2 = a^2, \quad a > 0$	$x^2 + y^2 = \pm 2ax, \quad a > 0$	$x^2 + y^2 = \pm 2ay, \quad a > 0$
Polar equation	$r = a, \quad a > 0$	$r = \pm 2a\cos\theta, \quad a > 0$	$r = \pm 2a\sin\theta, \quad a > 0$
Typical graph	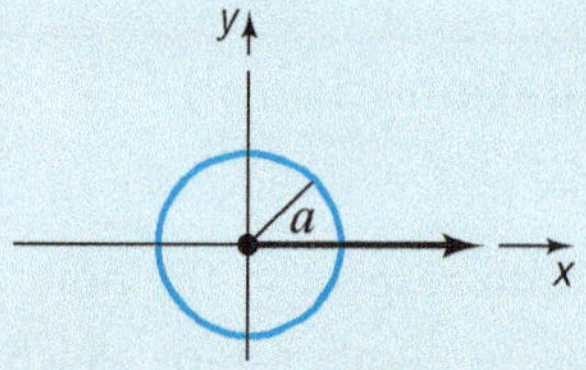	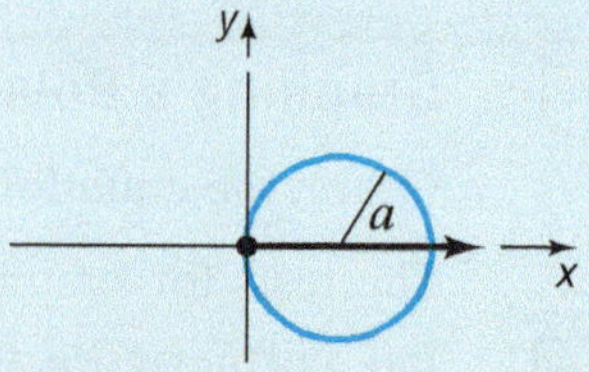	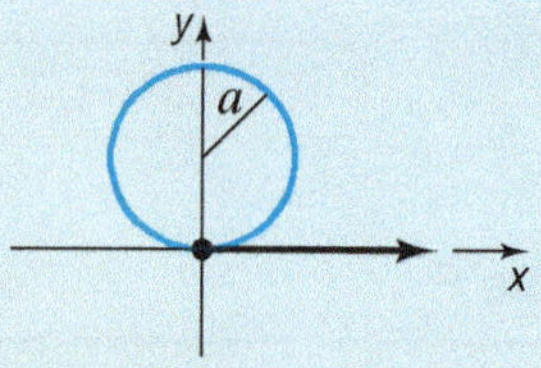
	Other Equations		
Name	Cardioid	Limaçon without inner loop	Limaçon with inner loop
Polar equations	$r = a \pm a\cos\theta, \quad a > 0$	$r = a \pm b\cos\theta, \quad 0 < b < a$	$r = a \pm b\cos\theta, \quad 0 < a < b$
	$r = a \pm a\sin\theta, \quad a > 0$	$r = a \pm b\sin\theta, \quad 0 < b < a$	$r = a \pm b\sin\theta, \quad 0 < a < b$
Typical graph	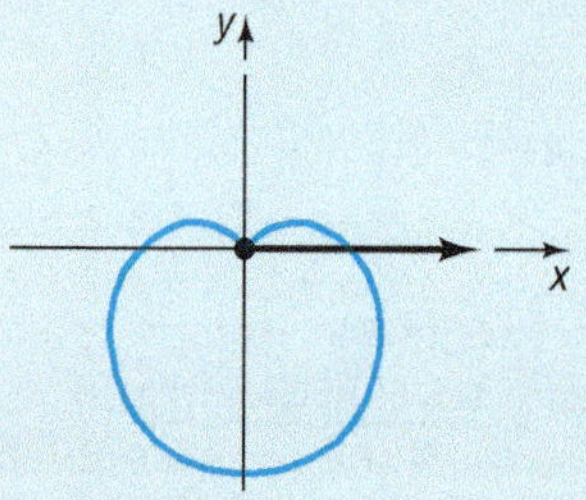	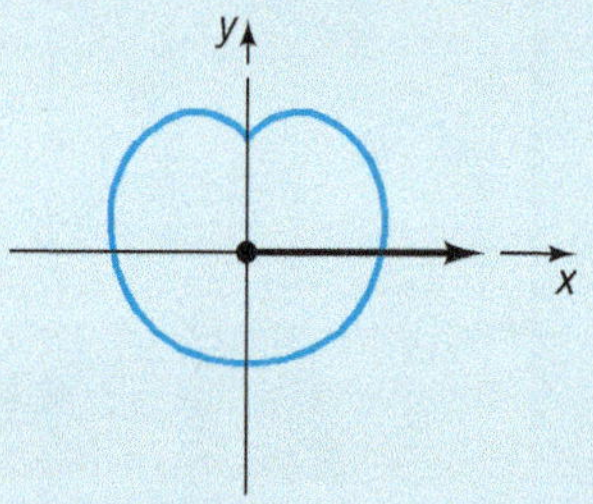	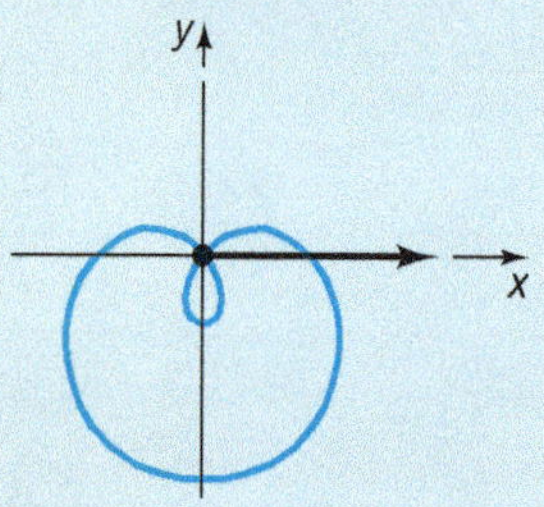
Name	Lemniscate	Rose with three petals	Rose with four petals
Polar equations	$r^2 = a^2\cos(2\theta), \quad a \neq 0$	$r = a\sin(3\theta), \quad a > 0$	$r = a\sin(2\theta), \quad a > 0$
	$r^2 = a^2\sin(2\theta), \quad a \neq 0$	$r = a\cos(3\theta), \quad a > 0$	$r = a\cos(2\theta), \quad a > 0$
Typical graph	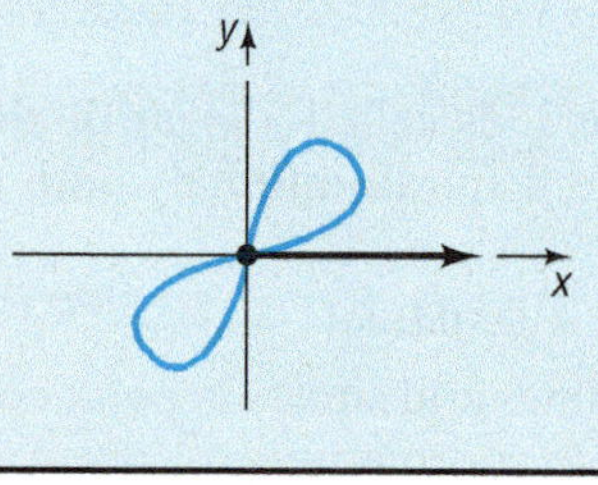	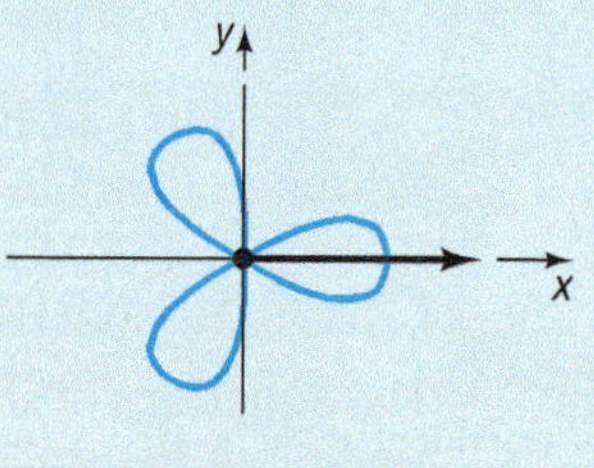	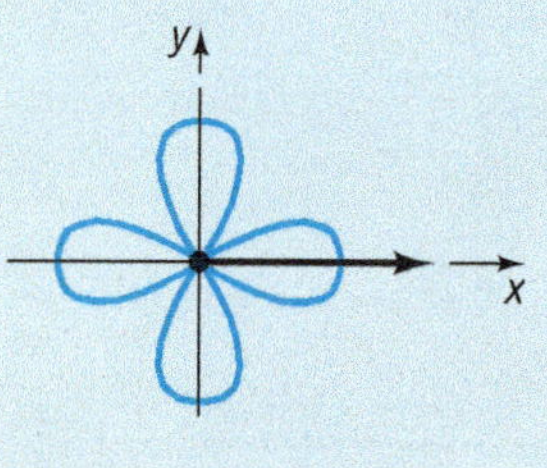

Sketching Quickly

If a polar equation involves only a sine (or cosine) function, you can quickly obtain a sketch of its graph by making use of Table 7, periodicity, and a short table.

EXAMPLE 14 **Sketching the Graph of a Polar Equation Quickly**

Graph the equation: $r = 2 + 2\sin\theta$

Solution You should recognize the polar equation: Its graph is a cardioid. The period of $\sin\theta$ is 2π, so form a table using $0 \le \theta \le 2\pi$, compute r, plot the points (r, θ), and sketch the graph of a cardioid as θ varies from 0 to 2π. See Table 8 and Figure 36.

Table 8

θ	$r = 2 + 2\sin\theta$
0	$2 + 2(0) = 2$
$\frac{\pi}{2}$	$2 + 2(1) = 4$
π	$2 + 2(0) = 2$
$\frac{3\pi}{2}$	$2 + 2(-1) = 0$
2π	$2 + 2(0) = 2$

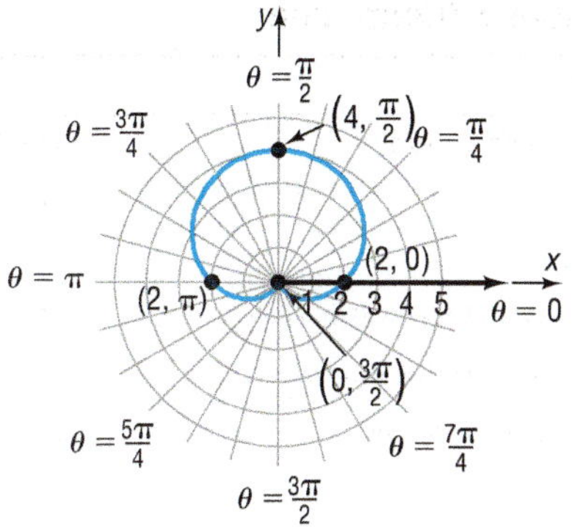

Figure 36 $r = 2 + 2\sin\theta$

Calculus Comment For those of you who are planning to study calculus, a comment about one important role of polar equations is in order.

In rectangular coordinates, the equation $x^2 + y^2 = 1$, whose graph is the unit circle, is not the graph of a function. In fact, it requires two functions to obtain the graph of the unit circle:

$$y_1 = \sqrt{1 - x^2} \quad \textbf{Upper semicircle} \qquad y_2 = -\sqrt{1 - x^2} \quad \textbf{Lower semicircle}$$

In polar coordinates, the equation $r = 1$, whose graph is also the unit circle, does define a function. For each choice of θ, there is only one corresponding value of r, that is, $r = 1$. Since many problems in calculus require the use of functions, the opportunity to express nonfunctions in rectangular coordinates as functions in polar coordinates becomes extremely useful.

Note also that the vertical-line test for functions is valid only for equations in rectangular coordinates.

Historical Feature

Jakob Bernoulli (1654–1705)

Polar coordinates seem to have been invented by Jakob Bernoulli (1654–1705) in about 1691, although, as with most such ideas, earlier traces of the notion exist. Early users of calculus remained committed to rectangular coordinates, and polar coordinates did not become widely used until the early 1800s. Even then, it was mostly geometers who used them for describing odd curves. Finally, about the mid-1800s, applied mathematicians realized the tremendous simplification that polar coordinates make possible in the description of objects with circular or cylindrical symmetry. From then on, their use became widespread.

10.2 Assess Your Understanding

'Are You Prepared?' *Answers are given at the end of these exercises. If you get a wrong answer, read the pages listed in red.*

1. If the rectangular coordinates of a point are $(4, -6)$, the point symmetric to it with respect to the origin is ________. (pp. 166–168)

2. The difference formula for cosine is $\cos(A - B) =$ ________. (p. 652)

3. The standard equation of a circle with center at $(-2, 5)$ and radius 3 is ________. (pp. 189–193)

4. Is the sine function even, odd, or neither? (pp. 568–569)

5. $\sin\frac{5\pi}{4} =$ ______. (pp. 550–558)

6. $\cos\frac{2\pi}{3} =$ ______. (pp. 550–558)

Concepts and Vocabulary

7. An equation whose variables are polar coordinates is called a(n) _____ ________.

8. ***True or False*** The tests for symmetry in polar coordinates are always conclusive.

9. To test whether the graph of a polar equation may be symmetric with respect to the polar axis, replace θ by ____.

10. To test whether the graph of a polar equation may be symmetric with respect to the line $\theta = \frac{\pi}{2}$, replace θ by ______.

11. ***True or False*** A cardioid passes through the pole.

12. Rose curves are characterized by equations of the form $r = a\cos(n\theta)$ or $r = a\sin(n\theta)$, $a \neq 0$. If $n \neq 0$ is even, the rose has ___ petals; if $n \neq \pm 1$ is odd, the rose has ____ petals.

13. For a positive real number a, the graph of which of the following polar equations is a circle with radius a and center at $(a, 0)$ in rectangular coordinates?
(a) $r = 2a\sin\theta$ (c) $r = 2a\cos\theta$
(b) $r = -2a\sin\theta$ (d) $r = -2a\cos\theta$

14. In polar coordinates, the points (r, θ) and $(-r, \theta)$ are symmetric with respect to which of the following?
(a) the polar axis (or x-axis) (b) the pole (or origin)
(c) the line $\theta = \frac{\pi}{2}$ (or y-axis) (d) the line $\theta = \frac{\pi}{4}$ (or $y = x$)

Skill Building

In Problems 15–30, transform each polar equation to an equation in rectangular coordinates. Then identify and graph the equation.

15. $r = 4$ **16.** $r = 2$ **17.** $\theta = \frac{\pi}{3}$ **18.** $\theta = -\frac{\pi}{4}$

19. $r\sin\theta = 4$ **20.** $r\cos\theta = 4$ **21.** $r\cos\theta = -2$ **22.** $r\sin\theta = -2$

23. $r = 2\cos\theta$ **24.** $r = 2\sin\theta$ **25.** $r = -4\sin\theta$ **26.** $r = -4\cos\theta$

27. $r\sec\theta = 4$ **28.** $r\csc\theta = 8$ **29.** $r\csc\theta = -2$ **30.** $r\sec\theta = -4$

In Problems 31–38, match each of the graphs (A) through (H) to one of the following polar equations.

31. $r = 2$ **32.** $\theta = \frac{\pi}{4}$ **33.** $r = 2\cos\theta$ **34.** $r\cos\theta = 2$

35. $r = 1 + \cos\theta$ **36.** $r = 2\sin\theta$ **37.** $\theta = \frac{3\pi}{4}$ **38.** $r\sin\theta = 2$

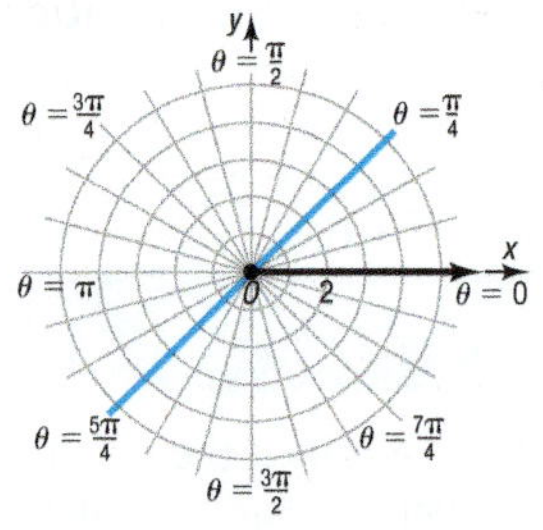

(A)

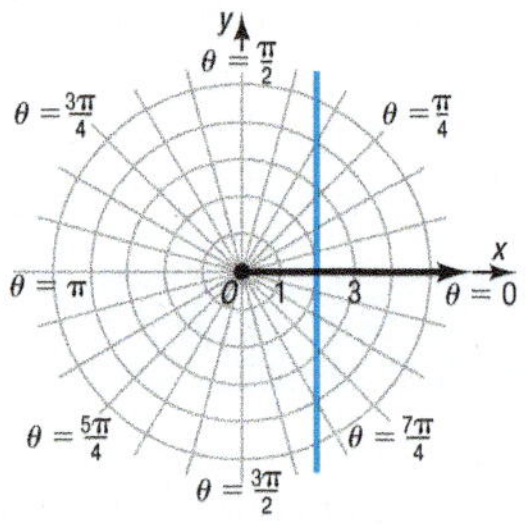

(B)

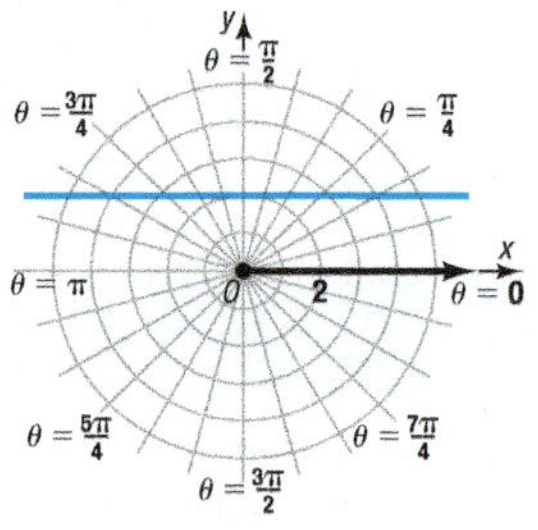

(C)

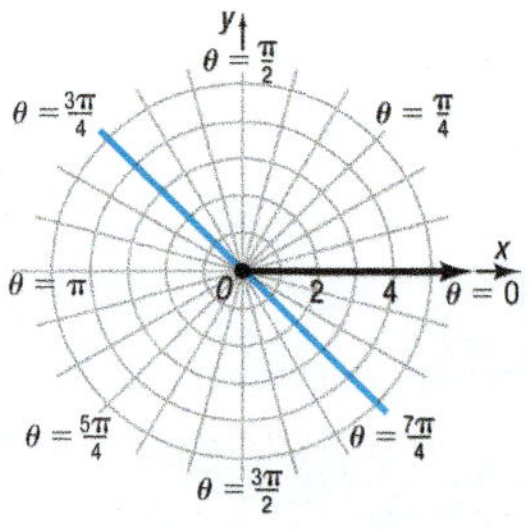

(D)

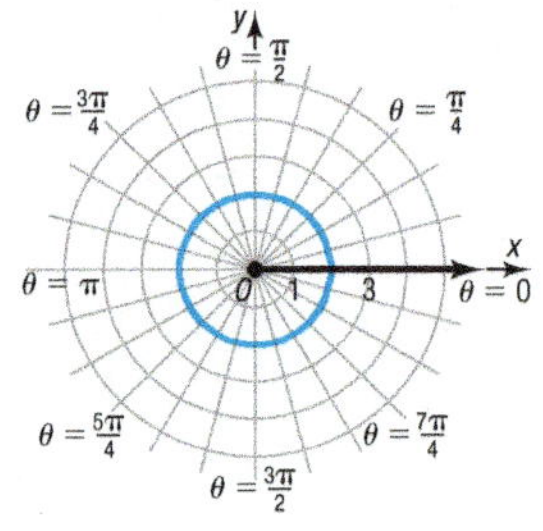

(E)

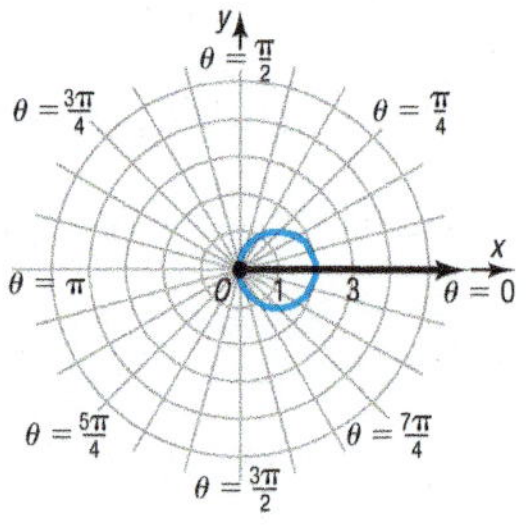

(F)

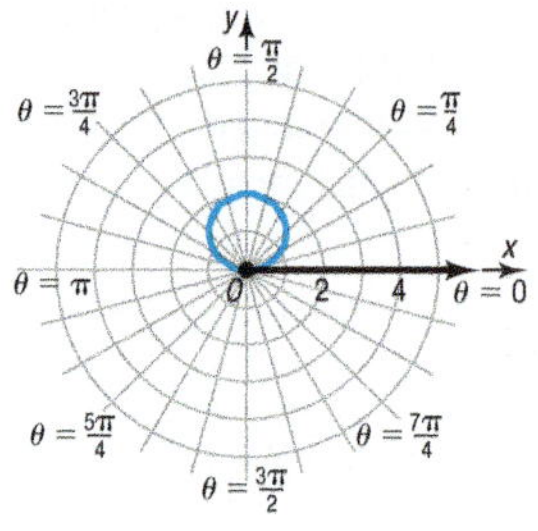

(G)

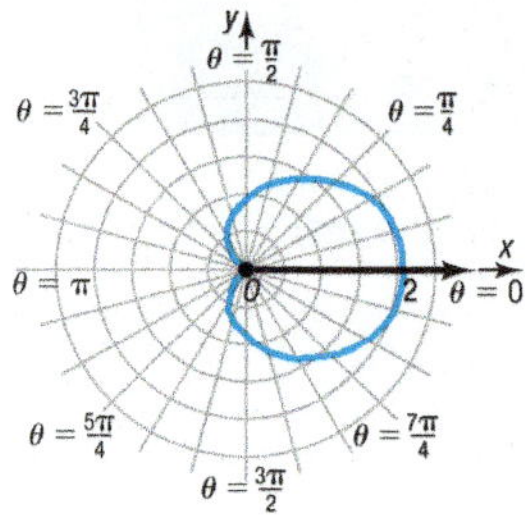

(H)

In Problems 39–62, identify and graph each polar equation.

39. $r = 2 + 2\cos\theta$ **40.** $r = 1 + \sin\theta$ **41.** $r = 3 - 3\sin\theta$ **42.** $r = 2 - 2\cos\theta$

43. $r = 2 + \sin\theta$ **44.** $r = 2 - \cos\theta$ **45.** $r = 4 - 2\cos\theta$ **46.** $r = 4 + 2\sin\theta$

47. $r = 1 + 2\sin\theta$ **48.** $r = 1 - 2\sin\theta$ **49.** $r = 2 - 3\cos\theta$ **50.** $r = 2 + 4\cos\theta$

51. $r = 3\cos(2\theta)$ **52.** $r = 2\sin(3\theta)$ **53.** $r = 4\sin(5\theta)$ **54.** $r = 3\cos(4\theta)$

55. $r^2 = 9\cos(2\theta)$ **56.** $r^2 = \sin(2\theta)$ **57.** $r = 2^\theta$ **58.** $r = 3^\theta$

59. $r = 1 - \cos\theta$ **60.** $r = 3 + \cos\theta$ **61.** $r = 1 - 3\cos\theta$ **62.** $r = 4\cos(3\theta)$

Mixed Practice

In Problems 63–68, graph each pair of polar equations on the same polar grid. Find the polar coordinates of the point(s) of intersection and label the point(s) on the graph.

63. $r = 8\cos\theta;\ r = 2\sec\theta$

64. $r = 8\sin\theta;\ r = 4\csc\theta$

65. $r = \sin\theta;\ r = 1 + \cos\theta$

66. $r = 3;\ r = 2 + 2\cos\theta$

67. $r = 1 + \sin\theta;\ r = 1 + \cos\theta$

68. $r = 1 + \cos\theta;\ r = 3\cos\theta$

Applications and Extensions

In Problems 69–72, the polar equation for each graph is either $r = a + b\cos\theta$ or $r = a + b\sin\theta$, $a > 0$. Select the correct equation and find the values of a and b.

69.

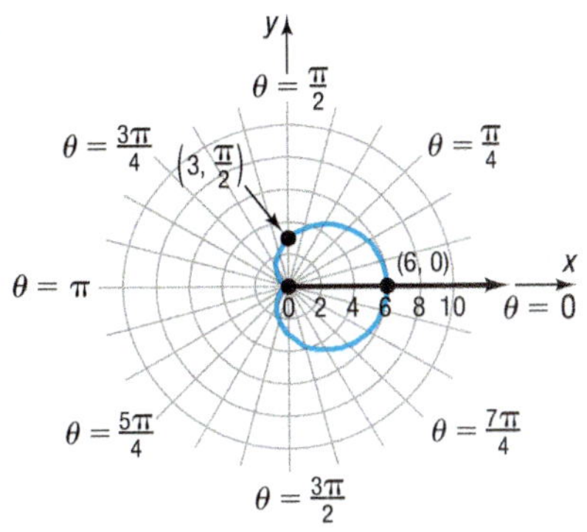

70.

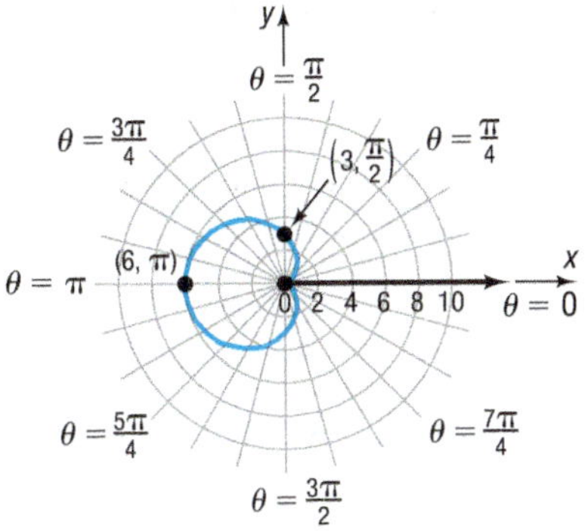

71.

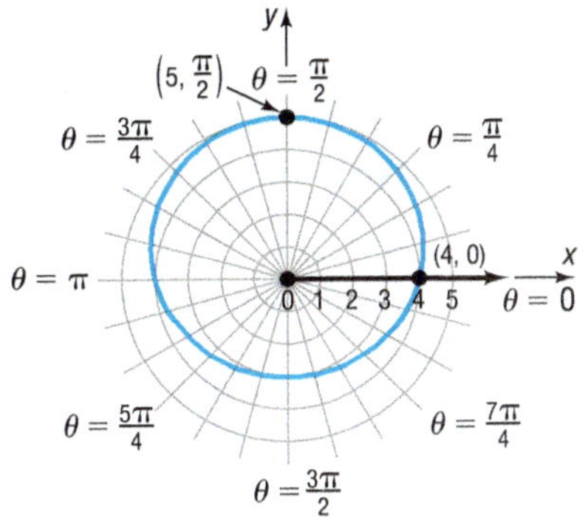

72.

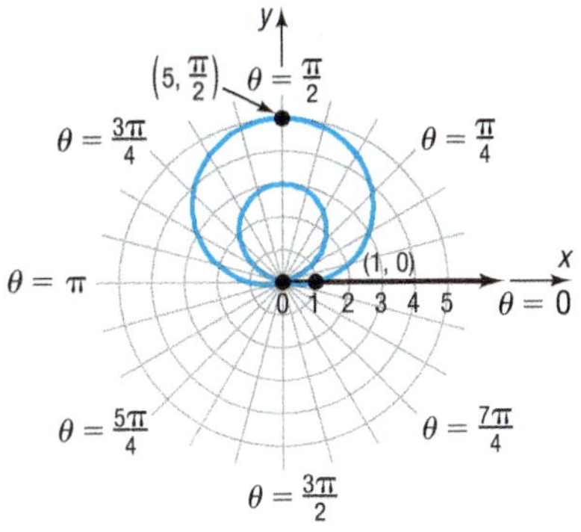

In Problems 73–82, graph each polar equation.

73. $r = \dfrac{2}{1 - \cos\theta}$ *(parabola)*

74. $r = \dfrac{2}{1 - 2\cos\theta}$ *(hyperbola)*

75. $r = \dfrac{1}{3 - 2\cos\theta}$ *(ellipse)*

76. $r = \dfrac{1}{1 - \cos\theta}$ *(parabola)*

77. $r = \theta,\ \ \theta \ge 0$ *(spiral of Archimedes)*

78. $r = \dfrac{3}{\theta}$ *(reciprocal spiral)*

79. $r = \csc\theta - 2,\ \ 0 < \theta < \pi$ *(conchoid)*

80. $r = \sin\theta\tan\theta$ *(cissoid)*

81. $r = \tan\theta,\ \ -\dfrac{\pi}{2} < \theta < \dfrac{\pi}{2}$ *(kappa curve)*

82. $r = \cos\dfrac{\theta}{2}$

83. Show that the graph of the equation $r\sin\theta = a$ is a horizontal line a units above the pole if $a \ge 0$ and $|a|$ units below the pole if $a < 0$.

84. Show that the graph of the equation $r\cos\theta = a$ is a vertical line a units to the right of the pole if $a \ge 0$ and $|a|$ units to the left of the pole if $a < 0$.

85. Show that the graph of the equation $r = 2a\sin\theta$, $a > 0$, is a circle of radius a with center at $(0, a)$ in rectangular coordinates.

86. Show that the graph of the equation $r = -2a\sin\theta$, $a > 0$, is a circle of radius a with center at $(0, -a)$ in rectangular coordinates.

87. Show that the graph of the equation $r = 2a\cos\theta$, $a > 0$, is a circle of radius a with center at $(a, 0)$ in rectangular coordinates.

88. Show that the graph of the equation $r = -2a\cos\theta$, $a > 0$, is a circle of radius a with center at $(-a, 0)$ in rectangular coordinates.

Explaining Concepts: Discussion and Writing

89. Explain why the following test for symmetry is valid: Replace r by $-r$ and θ by $-\theta$ in a polar equation. If an equivalent equation results, the graph is symmetric with respect to the line $\theta = \frac{\pi}{2}$ (y-axis).
   (a) Show that the test on page 745 fails for $r^2 = \cos\theta$, yet this new test works.
   (b) Show that the test on page 745 works for $r^2 = \sin\theta$, yet this new test fails.

90. Write down two different tests for symmetry with respect to the polar axis. Find examples in which one test works and the other fails. Which test do you prefer to use? Justify your answer.

91. The tests for symmetry given on page 745 are sufficient, but not necessary. Explain what this means.

92. Explain why the vertical-line test used to identify functions in rectangular coordinates does not work for equations expressed in polar coordinates.

Retain Your Knowledge

Problems 93–96 are based on material learned earlier in the course. The purpose of these problems is to keep the material fresh in your mind so that you are better prepared for the final exam.

93. Solve: $\dfrac{5}{x-3} \geq 1$

94. Convert $\dfrac{7\pi}{3}$ radians to degrees.

95. Determine the amplitude and period of $y = -2\sin(5x)$ without graphing.

96. Find any asymptotes for the graph of $R(x) = \dfrac{x+3}{x^2 - x - 12}$.

'Are You Prepared?' Answers

1. $(-4, 6)$ **2.** $\cos A \cos B + \sin A \sin B$ **3.** $(x+2)^2 + (y-5)^2 = 9$ **4.** Odd **5.** $-\dfrac{\sqrt{2}}{2}$ **6.** $-\dfrac{1}{2}$

10.3 The Complex Plane; De Moivre's Theorem

PREPARING FOR THIS SECTION *Before getting started, review the following:*

- Complex Numbers (Section 1.4, pp. 121–125)
- Value of the Sine and Cosine Functions at Certain Angles (Section 7.3, pp. 539–541; Section 7.4, pp. 550–558)
- Sum and Difference Formulas for Sine and Cosine (Section 8.5, pp. 652 and 655)

Now Work the **'Are You Prepared?'** problems on page 762.

OBJECTIVES
1. Plot Points in the Complex Plane (p. 756)
2. Convert a Complex Number between Rectangular Form and Polar Form (p. 757)
3. Find Products and Quotients of Complex Numbers in Polar Form (p. 758)
4. Use De Moivre's Theorem (p. 759)
5. Find Complex Roots (p. 760)

1 Plot Points in the Complex Plane

Complex numbers are discussed in Chapter 1, Section 1.4. In that discussion, we were not prepared to give a geometric interpretation of a complex number. Now we are ready.

A complex number $z = x + yi$ can be interpreted geometrically as the point (x, y) in the xy-plane. Each point in the plane corresponds to a complex number and, conversely, each complex number corresponds to a point in the plane. The collection of such points is referred to as the **complex plane**. The x-axis is referred to as the **real axis**, because any point that lies on the real axis is of the form $z = x + 0i = x$, a real number. The y-axis is called the **imaginary axis**, because any point that lies on it is of the form $z = 0 + yi = yi$, a pure imaginary number. See Figure 37.

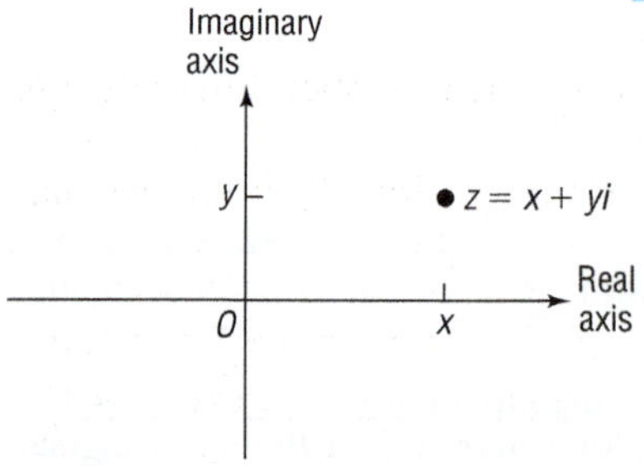

Figure 37 Complex plane

EXAMPLE 1 **Plotting a Point in the Complex Plane**

Plot the point corresponding to $z = \sqrt{3} - i$ in the complex plane.

Solution The point corresponding to $z = \sqrt{3} - i$ has the rectangular coordinates $(\sqrt{3}, -1)$. This point, located in quadrant IV, is plotted in Figure 38.

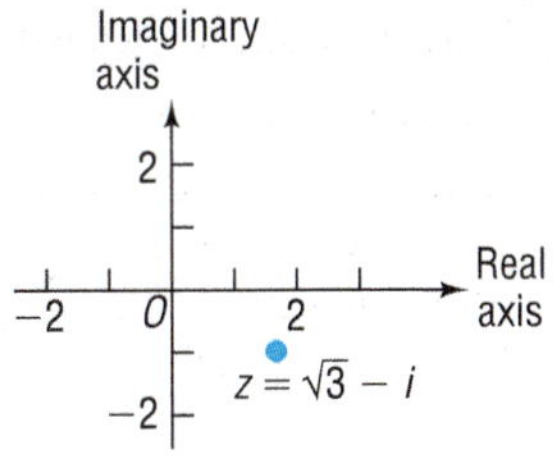

Figure 38

DEFINITION

Let $z = x + yi$ be a complex number. The **magnitude** or **modulus** of z, denoted by $|z|$, is defined as the distance from the origin to the point (x, y). That is,

$$|z| = \sqrt{x^2 + y^2} \quad \textbf{(1)}$$

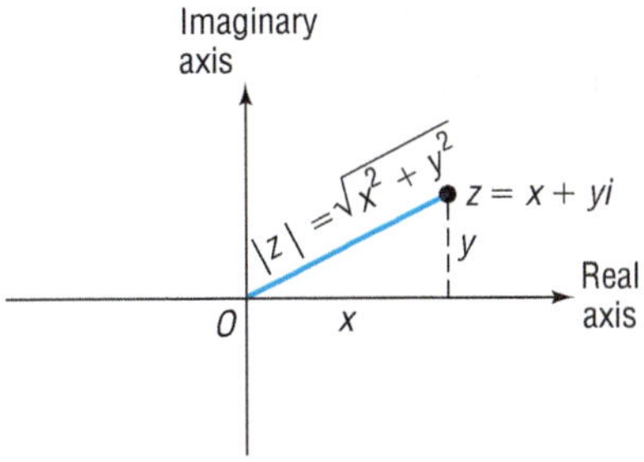

Figure 39

See Figure 39 for an illustration.

This definition for $|z|$ is consistent with the definition for the absolute value of a real number: If $z = x + yi$ is real, then $z = x + 0i$ and

$$|z| = \sqrt{x^2 + 0^2} = \sqrt{x^2} = |x|$$

For this reason, the magnitude of z is sometimes called the **absolute value of** z.

Recall that if $z = x + yi$, then its **conjugate**, denoted by $\bar{z}$, is $\bar{z} = x - yi$. Because $z\bar{z} = x^2 + y^2$, which is a nonnegative real number, it follows from equation (1) that the magnitude of z can be written as

$$|z| = \sqrt{z\bar{z}} \quad \textbf{(2)}$$

2 Convert a Complex Number between Rectangular Form and Polar Form

When a complex number is written in the standard form $z = x + yi$, it is in **rectangular**, or **Cartesian**, **form**, because (x, y) are the rectangular coordinates of the corresponding point in the complex plane. Suppose that (r, θ) are the polar coordinates of this point. Then

$$x = r\cos\theta, \qquad y = r\sin\theta \quad \textbf{(3)}$$

DEFINITION

If $r \geq 0$ and $0 \leq \theta < 2\pi$, the complex number $z = x + yi$ may be written in **polar form** as

$$z = x + yi = (r\cos\theta) + (r\sin\theta)i = r(\cos\theta + i\sin\theta) \quad \textbf{(4)}$$

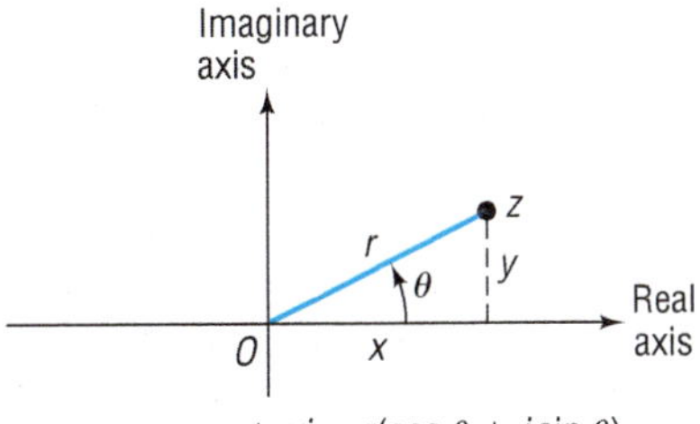

$z = x + yi = r(\cos\theta + i\sin\theta)$, $r \geq 0$, $0 \leq \theta < 2\pi$

Figure 40

See Figure 40.

If $z = r(\cos\theta + i\sin\theta)$ is the polar form of a complex number,* the angle θ, $0 \leq \theta < 2\pi$, is called the **argument of** z.

Also, because $r \geq 0$, we have $r = \sqrt{x^2 + y^2}$. From equation (1), it follows that the magnitude of $z = r(\cos\theta + i\sin\theta)$ is

$$|z| = r$$

*Some texts abbreviate the polar form using $z = r(\cos\theta + i\sin\theta) = r\,\text{cis}\,\theta$.

EXAMPLE 2 **Writing a Complex Number in Polar Form**

Write an expression for $z = \sqrt{3} - i$ in polar form.

Solution The point, located in quadrant IV, is plotted in Figure 38 on the previous page. Because $x = \sqrt{3}$ and $y = -1$, it follows that

$$r = \sqrt{x^2 + y^2} = \sqrt{(\sqrt{3})^2 + (-1)^2} = \sqrt{4} = 2$$

so

$$\sin\theta = \frac{y}{r} = \frac{-1}{2}, \qquad \cos\theta = \frac{x}{r} = \frac{\sqrt{3}}{2}, \qquad 0 \le \theta < 2\pi$$

The angle θ, $0 \le \theta < 2\pi$, that satisfies both equations is $\theta = \frac{11\pi}{6}$. With $\theta = \frac{11\pi}{6}$ and $r = 2$, the polar form of $z = \sqrt{3} - i$ is

$$z = r(\cos\theta + i\sin\theta) = 2\left(\cos\frac{11\pi}{6} + i\sin\frac{11\pi}{6}\right)$$

Now Work PROBLEM 13

EXAMPLE 3 **Plotting a Point in the Complex Plane and Converting from Polar to Rectangular Form**

Plot the point corresponding to $z = 2(\cos 30° + i\sin 30°)$ in the complex plane, and write an expression for z in rectangular form.

Solution To plot the complex number $z = 2(\cos 30° + i\sin 30°)$, plot the point whose polar coordinates are $(r, \theta) = (2, 30°)$, as shown in Figure 41. In rectangular form,

$$z = 2(\cos 30° + i\sin 30°) = 2\left(\frac{\sqrt{3}}{2} + \frac{1}{2}i\right) = \sqrt{3} + i$$

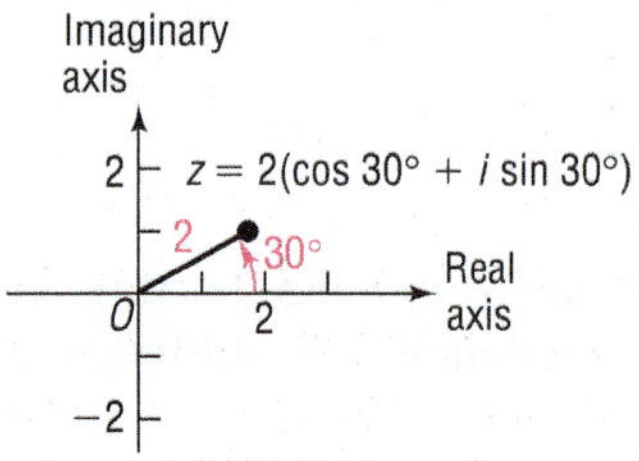

Figure 41

Now Work PROBLEM 25

3 Find Products and Quotients of Complex Numbers in Polar Form

The polar form of a complex number provides an alternative method for finding products and quotients of complex numbers.

THEOREM

Let $z_1 = r_1(\cos\theta_1 + i\sin\theta_1)$ and $z_2 = r_2(\cos\theta_2 + i\sin\theta_2)$ be two complex numbers. Then

$$z_1 z_2 = r_1 r_2[\cos(\theta_1 + \theta_2) + i\sin(\theta_1 + \theta_2)] \quad \textbf{(5)}$$

If $z_2 \ne 0$, then

$$\frac{z_1}{z_2} = \frac{r_1}{r_2}[\cos(\theta_1 - \theta_2) + i\sin(\theta_1 - \theta_2)] \quad \textbf{(6)}$$

In Words
The magnitude of a complex number z is r, and its argument is θ, so when
$z = r(\cos\theta + i\sin\theta)$
the magnitude of the product (quotient) of two complex numbers equals the product (quotient) of their magnitudes; the argument of the product (quotient) of two complex numbers is determined by the sum (difference) of their arguments.

Proof We will prove formula (5). The proof of formula (6) is left as an exercise (see Problem 68).

$$\begin{aligned} z_1 z_2 &= [r_1(\cos\theta_1 + i\sin\theta_1)][r_2(\cos\theta_2 + i\sin\theta_2)] \\ &= r_1 r_2[(\cos\theta_1 + i\sin\theta_1)(\cos\theta_2 + i\sin\theta_2)] \end{aligned}$$

$$= r_1 r_2[(\cos\theta_1\cos\theta_2 - \sin\theta_1\sin\theta_2) + i(\sin\theta_1\cos\theta_2 + \cos\theta_1\sin\theta_2)]$$
$$= r_1 r_2[\cos(\theta_1 + \theta_2) + i\sin(\theta_1 + \theta_2)]$$ ■

Let's look at an example of how this theorem can be used.

EXAMPLE 4 **Finding Products and Quotients of Complex Numbers in Polar Form**

If $z = 3(\cos 20° + i\sin 20°)$ and $w = 5(\cos 100° + i\sin 100°)$, find the following (leave your answers in polar form).

(a) zw (b) $\dfrac{z}{w}$

Solution (a) $zw = [3(\cos 20° + i\sin 20°)][5(\cos 100° + i\sin 100°)]$

$$= (3\cdot 5)[\cos(20° + 100°) + i\sin(20° + 100°)]$$ Apply equation (5).

$$= 15(\cos 120° + i\sin 120°)$$

(b) $$\frac{z}{w} = \frac{3(\cos 20° + i\sin 20°)}{5(\cos 100° + i\sin 100°)}$$

$$= \frac{3}{5}[\cos(20° - 100°) + i\sin(20° - 100°)]$$ Apply equation (6).

$$= \frac{3}{5}[\cos(-80°) + i\sin(-80°)]$$

$$= \frac{3}{5}(\cos 280° + i\sin 280°)$$ The argument must lie between 0° and 360°. ■

Now Work PROBLEM 35

4 Use De Moivre's Theorem

De Moivre's Theorem, stated by Abraham De Moivre (1667–1754) in 1730, but already known to many people by 1710, is important for the following reason: The fundamental processes of algebra are the four operations of addition, subtraction, multiplication, and division, together with powers and the extraction of roots. De Moivre's Theorem allows these latter fundamental algebraic operations to be applied to complex numbers.

De Moivre's Theorem, in its most basic form, is a formula for raising a complex number z to the power n, where $n \geq 1$ is a positive integer. Let's try to conjecture the form of the result.

Let $z = r(\cos\theta + i\sin\theta)$ be a complex number. Then equation (5) yields

$n = 2$: $z^2 = r^2[\cos(2\theta) + i\sin(2\theta)]$ Equation (5)

$n = 3$: $z^3 = z^2\cdot z$

$$= \{r^2[\cos(2\theta) + i\sin(2\theta)]\}[r(\cos\theta + i\sin\theta)]$$
$$= r^3[\cos(3\theta) + i\sin(3\theta)]$$ Equation (5)

$n = 4$: $z^4 = z^3\cdot z$

$$= \{r^3[\cos(3\theta) + i\sin(3\theta)]\}[r(\cos\theta + i\sin\theta)]$$
$$= r^4[\cos(4\theta) + i\sin(4\theta)]$$ Equation (5)

Do you see the pattern?

THEOREM

De Moivre's Theorem

If $z = r(\cos\theta + i\sin\theta)$ is a complex number, then

$$z^n = r^n[\cos(n\theta) + i\sin(n\theta)] \quad \textbf{(7)}$$

where $n \geq 1$ is a positive integer.

The proof of De Moivre's Theorem requires mathematical induction (which is not discussed until Section 13.4), so it is omitted here. The theorem is actually true for all integers n. You are asked to prove this in Problem 69.

EXAMPLE 5 Using De Moivre's Theorem

Write $[2(\cos 20° + i\sin 20°)]^3$ in the standard form $a + bi$.

Solution

$$[2(\cos 20° + i\sin 20°)]^3 = 2^3[\cos(3\cdot 20°) + i\sin(3\cdot 20°)] \quad \text{Apply De Moivre's Theorem.}$$

$$= 8(\cos 60° + i\sin 60°)$$

$$= 8\left(\frac{1}{2} + \frac{\sqrt{3}}{2}i\right) = 4 + 4\sqrt{3}i$$

Now Work PROBLEM 43

EXAMPLE 6 Using De Moivre's Theorem

Write $(1 + i)^5$ in the standard form $a + bi$.

Algebraic Solution

To apply De Moivre's Theorem, first write the complex number in polar form. Since the magnitude of $1 + i$ is $\sqrt{1^2 + 1^2} = \sqrt{2}$, begin by writing

$$1 + i = \sqrt{2}\left(\frac{1}{\sqrt{2}} + \frac{1}{\sqrt{2}}i\right) = \sqrt{2}\left(\cos\frac{\pi}{4} + i\sin\frac{\pi}{4}\right)$$

Now

$$(1 + i)^5 = \left[\sqrt{2}\left(\cos\frac{\pi}{4} + i\sin\frac{\pi}{4}\right)\right]^5$$

$$= (\sqrt{2})^5\left[\cos\left(5\cdot\frac{\pi}{4}\right) + i\sin\left(5\cdot\frac{\pi}{4}\right)\right]$$

$$= 4\sqrt{2}\left(\cos\frac{5\pi}{4} + i\sin\frac{5\pi}{4}\right)$$

$$= 4\sqrt{2}\left[-\frac{1}{\sqrt{2}} + \left(-\frac{1}{\sqrt{2}}\right)i\right] = -4 - 4i$$

Graphing Solution

A TI-84 Plus C graphing calculator provides the solution shown in Figure 42.

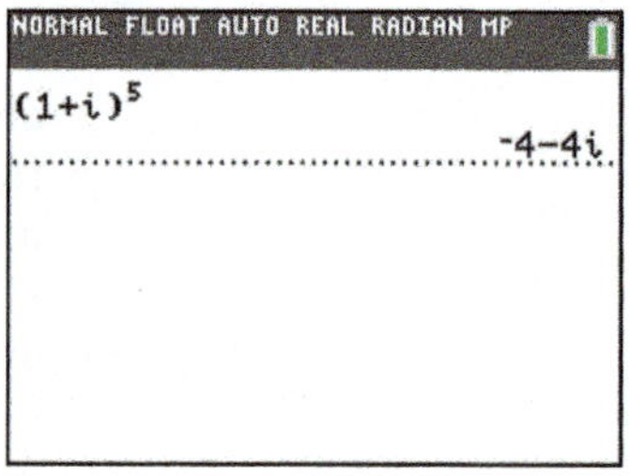

Figure 42

5 Find Complex Roots

Let w be a given complex number, and let $n \geq 2$ denote a positive integer. Any complex number z that satisfies the equation

$$z^n = w$$

is a **complex nth root** of w. In keeping with previous usage, if $n = 2$, the solutions of the equation $z^2 = w$ are called **complex square roots** of w, and if $n = 3$, the solutions of the equation $z^3 = w$ are called **complex cube roots** of w.

Note: In the algebraic solution of Example 6, the approach used in Example 2 could also be used to write $1 + i$ in polar form.

THEOREM

Finding Complex Roots

Let $w = r(\cos\theta_0 + i\sin\theta_0)$ be a complex number, and let $n \geq 2$ be an integer. If $w \neq 0$, there are n distinct complex nth roots of w, given by the formula

$$z_k = \sqrt[n]{r}\left[\cos\left(\frac{\theta_0}{n} + \frac{2k\pi}{n}\right) + i\sin\left(\frac{\theta_0}{n} + \frac{2k\pi}{n}\right)\right] \tag{8}$$

where $k = 0, 1, 2, \ldots, n-1$.

Proof (Outline) We will not prove this result in its entirety. Instead, we shall show only that each z_k in equation (8) satisfies the equation $z_k^n = w$, proving that each z_k is a complex nth root of w.

$$\begin{aligned} z_k^n &= \left\{\sqrt[n]{r}\left[\cos\left(\frac{\theta_0}{n} + \frac{2k\pi}{n}\right) + i\sin\left(\frac{\theta_0}{n} + \frac{2k\pi}{n}\right)\right]\right\}^n \\ &= (\sqrt[n]{r})^n\left\{\cos\left[n\left(\frac{\theta_0}{n} + \frac{2k\pi}{n}\right)\right] + i\sin\left[n\left(\frac{\theta_0}{n} + \frac{2k\pi}{n}\right)\right]\right\} && \text{Apply De Moivre's Theorem.} \\ &= r[\cos(\theta_0 + 2k\pi) + i\sin(\theta_0 + 2k\pi)] && \text{Simplify.} \\ &= r(\cos\theta_0 + i\sin\theta_0) = w && \text{The Periodic Property} \end{aligned}$$

So each z_k, $k = 0, 1, \ldots, n-1$, is a complex nth root of w. To complete the proof, we would need to show that each z_k, $k = 0, 1, \ldots, n-1$, is, in fact, distinct and that there are no complex nth roots of w other than those given by equation (8).

EXAMPLE 7

Finding Complex Cube Roots

Find the complex cube roots of $-1 + \sqrt{3}i$. Leave your answers in polar form, with the argument in degrees.

Solution First, express $-1 + \sqrt{3}i$ in polar form using degrees.

$$-1 + \sqrt{3}i = 2\left(-\frac{1}{2} + \frac{\sqrt{3}}{2}i\right) = 2(\cos 120° + i\sin 120°)$$

The three complex cube roots of $-1 + \sqrt{3}i = 2(\cos 120° + i\sin 120°)$ are

$$\begin{aligned} z_k &= \sqrt[3]{2}\left[\cos\left(\frac{120°}{3} + \frac{360°k}{3}\right) + i\sin\left(\frac{120°}{3} + \frac{360°k}{3}\right)\right] \\ &= \sqrt[3]{2}[\cos(40° + 120°k) + i\sin(40° + 120°k)] \quad k = 0, 1, 2 \end{aligned}$$

so

$$\begin{aligned} z_0 &= \sqrt[3]{2}[\cos(40° + 120°\cdot 0) + i\sin(40° + 120°\cdot 0)] = \sqrt[3]{2}(\cos 40° + i\sin 40°) \\ z_1 &= \sqrt[3]{2}[\cos(40° + 120°\cdot 1) + i\sin(40° + 120°\cdot 1)] = \sqrt[3]{2}(\cos 160° + i\sin 160°) \\ z_2 &= \sqrt[3]{2}[\cos(40° + 120°\cdot 2) + i\sin(40° + 120°\cdot 2)] = \sqrt[3]{2}(\cos 280° + i\sin 280°) \end{aligned}$$

WARNING Most graphing utilities will provide only the answer z_0 to the calculation $(-1 + \sqrt{3}i)$ ^ $(1/3)$. The paragraph following Example 7 explains how to obtain z_1 and z_2 from z_0.

Notice that all of the three complex roots of $-1 + \sqrt{3}i$ have the same magnitude, $\sqrt[3]{2}$. This means that the points corresponding to each cube root lie the same distance from the origin; that is, the three points lie on a circle with center at the origin and radius $\sqrt[3]{2}$. Furthermore, the arguments of these cube roots are 40°, 160°, and 280°, the difference of consecutive pairs being $120° = \frac{360°}{3}$. This means that the three points are equally spaced on the circle, as shown in Figure 43 on the next page. These results are not coincidental. In fact, you are asked to show that these results hold for complex nth roots in Problems 65 through 67.

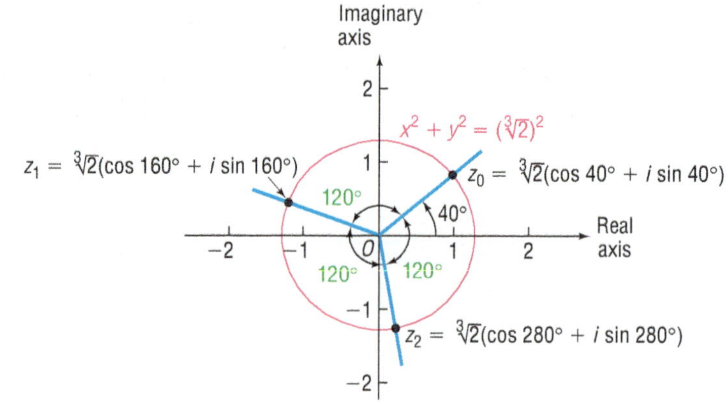

Figure 43

Now Work PROBLEM 55

Historical Feature

John Wallis

The Babylonians, Greeks, and Arabs considered square roots of negative quantities to be impossible and equations with complex solutions to be unsolvable. The first hint that there was some connection between real solutions of equations and complex numbers came when Girolamo Cardano (1501–1576) and Tartaglia (1499–1557) found *real* roots of cubic equations by taking cube roots of *complex* quantities. For centuries thereafter, mathematicians worked with complex numbers without much belief in their actual existence. In 1673, John Wallis appears to have been the first to suggest the graphical representation of complex numbers, a truly significant idea that was not pursued further until about 1800. Several people, including Karl Friedrich Gauss (1777–1855), then rediscovered the idea, and graphical representation helped to establish complex numbers as equal members of the number family. In practical applications, complex numbers have found their greatest uses in the study of alternating current, where they are a commonplace tool, and in the field of subatomic physics.

Historical Problems

1. The quadratic formula works perfectly well if the coefficients are complex numbers. Solve the following. [**Hint:** The answers are "nice."]
   (a) $z^2 - (2 + 5i)z - 3 + 5i = 0$ (b) $z^2 - (1 + i)z - 2 - i = 0$

10.3 Assess Your Understanding

'Are You Prepared?' *Answers are given at the end of these exercises. If you get a wrong answer, read the pages listed in red.*

1. The conjugate of $-4 - 3i$ is ________. (pp. 121–125)
2. The sum formula for the sine function is $\sin(A + B) =$ ______. (p. 655)
3. The sum formula for the cosine function is $\cos(A + B) =$ ______. (p. 652)
4. $\sin 120° =$ ______; $\cos 240° =$ ______. (pp. 550–558)

Concepts and Vocabulary

5. In the complex plane, the x-axis is referred to as the _____ axis, and the y-axis is called the _________ axis.
6. When a complex number z is written in the polar form $z = r(\cos\theta + i\sin\theta)$, the nonnegative number r is the _________ or _________ of z, and the angle θ, $0 \le \theta < 2\pi$, is the _________ of z.
7. Let $z_1 = r_1(\cos\theta_1 + i\sin\theta_1)$ and $z_2 = r_2(\cos\theta_2 + i\sin\theta_2)$ be two complex numbers. Then $z_1 z_2 =$ ___ [cos (______) + i sin (______)].
8. If $z = r(\cos\theta + i\sin\theta)$ is a complex number, then $z^n =$ ___ [cos(__) + i sin(__)].
9. Every nonzero complex number will have exactly _____ distinct complex cube roots.

10. ***True or False*** The polar form of a nonzero complex number is unique.

11. If $z = x + yi$ is a complex number, then $|z|$ equals which of the following?

(a) $x^2 + y^2$ (b) $|x| + |y|$
(c) $\sqrt{x^2 + y^2}$ (d) $\sqrt{|x| + |y|}$

12. If $z_1 = r_1(\cos\theta_1 + i\sin\theta_1)$ and $z_2 = r_2(\cos\theta_2 + i\sin\theta_2)$ are complex numbers, then $\frac{z_1}{z_2}$, $z_2 \neq 0$, equals which of the following?

(a) $\frac{r_1}{r_2}[\cos(\theta_1 - \theta_2) + i\sin(\theta_1 - \theta_2)]$

(b) $\frac{r_1}{r_2}\left[\cos\left(\frac{\theta_1}{\theta_2}\right) + i\sin\left(\frac{\theta_1}{\theta_2}\right)\right]$

(c) $\frac{r_1}{r_2}[\cos(\theta_1 + \theta_2) - i\sin(\theta_1 + \theta_2)]$

(d) $\frac{r_1}{r_2}\left[\cos\left(\frac{\theta_1}{\theta_2}\right) - i\sin\left(\frac{\theta_1}{\theta_2}\right)\right]$

Skill Building

In Problems 13–24, plot each complex number in the complex plane and write it in polar form. Express the argument in degrees.

13. $1 + i$ **14.** $-1 + i$ **15.** $\sqrt{3} - i$ **16.** $1 - \sqrt{3}i$ **17.** $-3i$ **18.** -2

19. $4 - 4i$ **20.** $9\sqrt{3} + 9i$ **21.** $3 - 4i$ **22.** $2 + \sqrt{3}i$ **23.** $-2 + 3i$ **24.** $\sqrt{5} - i$

In Problems 25–34, write each complex number in rectangular form.

25. $2(\cos 120° + i\sin 120°)$ **26.** $3(\cos 210° + i\sin 210°)$ **27.** $4\left(\cos\frac{7\pi}{4} + i\sin\frac{7\pi}{4}\right)$

28. $2\left(\cos\frac{5\pi}{6} + i\sin\frac{5\pi}{6}\right)$ **29.** $3\left(\cos\frac{3\pi}{2} + i\sin\frac{3\pi}{2}\right)$ **30.** $4\left(\cos\frac{\pi}{2} + i\sin\frac{\pi}{2}\right)$

31. $0.2(\cos 100° + i\sin 100°)$ **32.** $0.4(\cos 200° + i\sin 200°)$

33. $2\left(\cos\frac{\pi}{18} + i\sin\frac{\pi}{18}\right)$ **34.** $3\left(\cos\frac{\pi}{10} + i\sin\frac{\pi}{10}\right)$

In Problems 35–42, find zw and $\frac{z}{w}$. Leave your answers in polar form.

35. $z = 2(\cos 40° + i\sin 40°)$
$w = 4(\cos 20° + i\sin 20°)$

36. $z = \cos 120° + i\sin 120°$
$w = \cos 100° + i\sin 100°$

37. $z = 3(\cos 130° + i\sin 130°)$
$w = 4(\cos 270° + i\sin 270°)$

38. $z = 2(\cos 80° + i\sin 80°)$
$w = 6(\cos 200° + i\sin 200°)$

39. $z = 2\left(\cos\frac{\pi}{8} + i\sin\frac{\pi}{8}\right)$
$w = 2\left(\cos\frac{\pi}{10} + i\sin\frac{\pi}{10}\right)$

40. $z = 4\left(\cos\frac{3\pi}{8} + i\sin\frac{3\pi}{8}\right)$
$w = 2\left(\cos\frac{9\pi}{16} + i\sin\frac{9\pi}{16}\right)$

41. $z = 2 + 2i$
$w = \sqrt{3} - i$

42. $z = 1 - i$
$w = 1 - \sqrt{3}i$

In Problems 43–54, write each expression in the standard form $a + bi$.

43. $[4(\cos 40° + i\sin 40°)]^3$ **44.** $[3(\cos 80° + i\sin 80°)]^3$ **45.** $\left[2\left(\cos\frac{\pi}{10} + i\sin\frac{\pi}{10}\right)\right]^5$

46. $\left[\sqrt{2}\left(\cos\frac{5\pi}{16} + i\sin\frac{5\pi}{16}\right)\right]^4$ **47.** $[\sqrt{3}(\cos 10° + i\sin 10°)]^6$ **48.** $\left[\frac{1}{2}(\cos 72° + i\sin 72°)\right]^5$

49. $\left[\sqrt{5}\left(\cos\frac{3\pi}{16} + i\sin\frac{3\pi}{16}\right)\right]^4$ **50.** $\left[\sqrt{3}\left(\cos\frac{5\pi}{18} + i\sin\frac{5\pi}{18}\right)\right]^6$ **51.** $(1 - i)^5$

52. $(\sqrt{3} - i)^6$ **53.** $(\sqrt{2} - i)^6$ **54.** $(1 - \sqrt{5}i)^8$

In Problems 55–62, find all the complex roots. Leave your answers in polar form with the argument in degrees.

55. The complex cube roots of $1 + i$

56. The complex fourth roots of $\sqrt{3} - i$

57. The complex fourth roots of $4 - 4\sqrt{3}i$

58. The complex cube roots of $-8 - 8i$

59. The complex fourth roots of $-16i$

60. The complex cube roots of -8

61. The complex fifth roots of i

62. The complex fifth roots of $-i$

Applications and Extensions

63. Find the four complex fourth roots of unity (1) and plot them.

64. Find the six complex sixth roots of unity (1) and plot them.

65. Show that each complex nth root of a nonzero complex number w has the same magnitude.

66. Use the result of Problem 65 to draw the conclusion that each complex nth root lies on a circle with center at the origin. What is the radius of this circle?

67. Refer to Problem 66. Show that the complex nth roots of a nonzero complex number w are equally spaced on the circle.

68. Prove formula (6).

69. Prove that De Moivre's Theorem is true for *all* integers n by assuming it is true for integers $n \geq 1$ and then showing it is true for 0 and for negative integers.

Hint: Multiply the numerator and the denominator by the conjugate of the denominator, and use even-odd properties.

70. Mandelbrot Sets

(a) Consider the expression $a_n = (a_{n-1})^2 + z$, where z is some complex number (called the **seed**) and $a_0 = z$. Compute $a_1(=a_0^2+z), a_2(=a_1^2+z), a_3(=a_2^2+z), a_4, a_5$, and a_6 for the following seeds: $z_1 = 0.1 - 0.4i$, $z_2 = 0.5 + 0.8i$, $z_3 = -0.9 + 0.7i$, $z_4 = -1.1 + 0.1i$, $z_5 = 0 - 1.3i$, and $z_6 = 1 + 1i$.

(b) The dark portion of the graph represents the set of all values $z = x + yi$ that are in the Mandelbrot set. Determine which complex numbers in part (a) are in this set by plotting them on the graph. Do the complex numbers that are not in the Mandelbrot set have any common characteristics regarding the values of a_6 found in part (a)?

(c) Compute $|z| = \sqrt{x^2 + y^2}$ for each of the complex numbers in part (a). Now compute $|a_6|$ for each of the complex numbers in part (a). For which complex numbers is $|a_6| \leq |z|$ and $|z| \leq 2$? Conclude that the criterion for a complex number to be in the Mandelbrot set is that $|a_n| \leq |z|$ and $|z| \leq 2$.

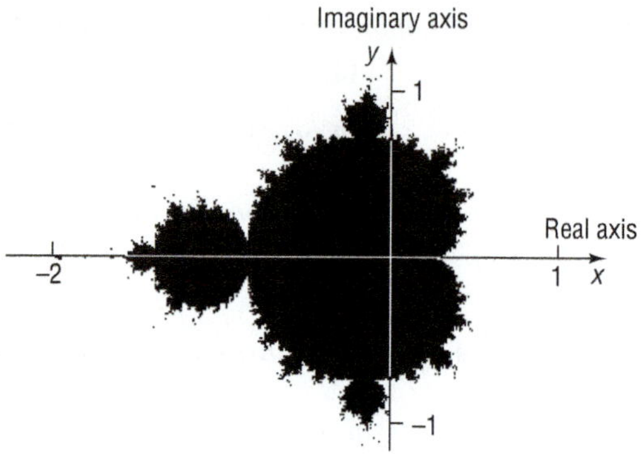

Retain Your Knowledge

Problems 71–74 are based on material learned earlier in the course. The purpose of these problems is to keep the material fresh in your mind so that you are better prepared for the final exam.

71. Find the area of the triangle with $a = 8$, $b = 11$, and $C = 113°$.

72. Convert 240° to radians. Express your answer as a multiple of π.

73. Simplify: $\sqrt[3]{24x^2y^5}$

74. Determine whether $f(x) = 5x^2 - 12x + 4$ has a maximum value or a minimum value, and then find the value.

'Are You Prepared?' Answers

1. $-4 + 3i$ **2.** $\sin A \cos B + \cos A \sin B$ **3.** $\cos A \cos B - \sin A \sin B$ **4.** $\frac{\sqrt{3}}{2}; -\frac{1}{2}$

10.4 Vectors

OBJECTIVES
1 Graph Vectors (p. 767)
2 Find a Position Vector (p. 767)
3 Add and Subtract Vectors Algebraically (p. 769)
4 Find a Scalar Multiple and the Magnitude of a Vector (p. 770)
5 Find a Unit Vector (p. 770)
6 Find a Vector from Its Direction and Magnitude (p. 771)
7 Model with Vectors (p. 772)

In simple terms, a **vector** (derived from the Latin *vehere,* meaning "to carry") is a quantity that has both magnitude and direction. It is customary to represent a vector by using an arrow. The length of the arrow represents the **magnitude** of the vector, and the arrowhead indicates the **direction** of the vector.

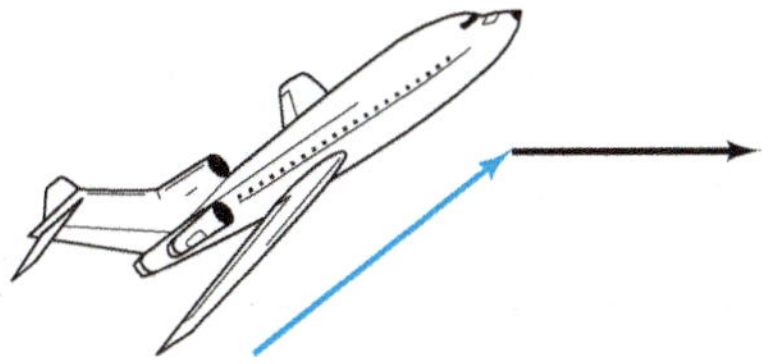

Figure 44

Many quantities in physics can be represented by vectors. For example, the velocity of an aircraft can be represented by an arrow that points in the direction of movement; the length of the arrow represents the speed. If the aircraft speeds up, we lengthen the arrow; if the aircraft changes direction, we introduce an arrow in the new direction. See Figure 44. Based on this representation, it is not surprising that vectors and *directed line segments* are somehow related.

Geometric Vectors

If P and Q are two distinct points in the xy-plane, there is exactly one line containing both P and Q [Figure 45(a)]. The points on that part of the line that joins P to Q, including P and Q, form what is called the **line segment** $\overline{PQ}$ [Figure 45(b)]. Ordering the points so that they proceed from P to Q results in a **directed line segment** from P to Q, or a **geometric vector**, which is denoted by $\overrightarrow{PQ}$. In a directed line segment $\overrightarrow{PQ}$, P is called the **initial point** and Q the **terminal point**, as indicated in Figure 45(c).

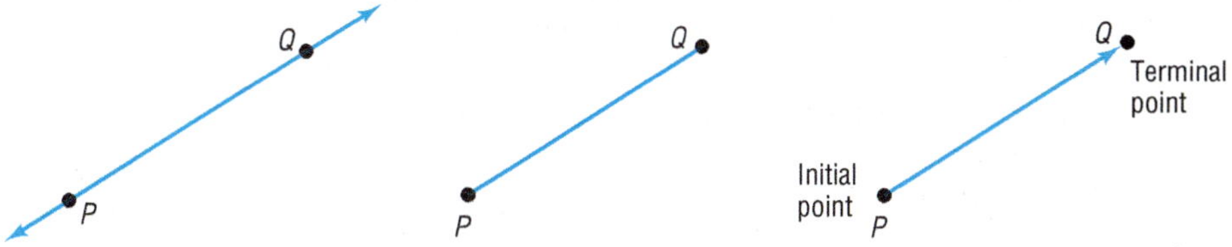

Figure 45 **(a)** Line containing P and Q **(b)** Line segment $\overline{PQ}$ **(c)** Directed line segment $\overrightarrow{PQ}$

The magnitude of the directed line segment $\overrightarrow{PQ}$ is the distance from the point P to the point Q; that is, it is the length of the line segment. The direction of $\overrightarrow{PQ}$ is from P to Q. If a vector $\mathbf{v}$* has the same magnitude and the same direction as the directed line segment $\overrightarrow{PQ}$, write

$$\mathbf{v} = \overrightarrow{PQ}$$

The vector $\mathbf{v}$ whose magnitude is 0 is called the **zero vector, 0.** The zero vector is assigned no direction.

Two vectors $\mathbf{v}$ and $\mathbf{w}$ are **equal**, written

$$\mathbf{v} = \mathbf{w}$$

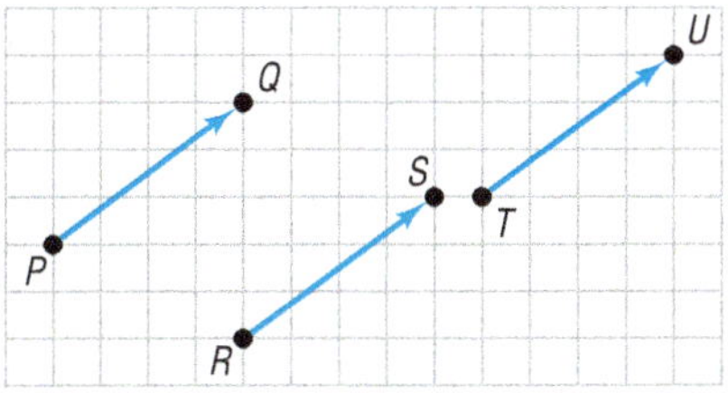

Figure 46 Equal vectors

if they have the same magnitude and the same direction.

For example, the three vectors shown in Figure 46 have the same magnitude and the same direction, so they are equal, even though they have different initial points and different terminal points. As a result, it is useful to think of a vector simply as an arrow, keeping in mind that two arrows (vectors) are equal if they have the same direction and the same magnitude (length).

Adding Vectors Geometrically

The **sum** $\mathbf{v} + \mathbf{w}$ of two vectors is defined as follows: Position the vectors $\mathbf{v}$ and $\mathbf{w}$ so that the terminal point of $\mathbf{v}$ coincides with the initial point of $\mathbf{w}$, as shown in Figure 47. The vector $\mathbf{v} + \mathbf{w}$ is then the unique vector whose initial point coincides with the initial point of $\mathbf{v}$ and whose terminal point coincides with the terminal point of $\mathbf{w}$.

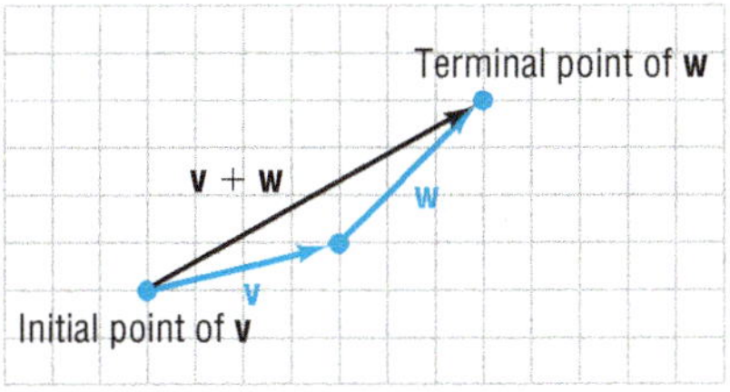

Figure 47 Adding vectors

Vector addition is **commutative**. That is, if $\mathbf{v}$ and $\mathbf{w}$ are any two vectors, then

$$\mathbf{v} + \mathbf{w} = \mathbf{w} + \mathbf{v}$$

*Boldface letters will be used to denote vectors, to distinguish them from numbers. For handwritten work, an arrow is placed over the letter to signify a vector. For example, write a vector by hand as $\vec{v}$.

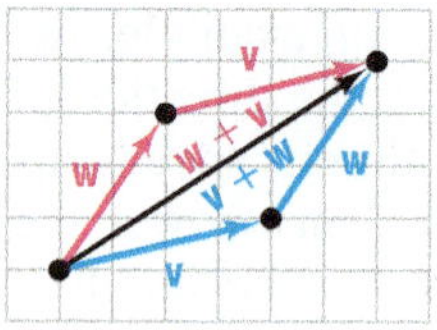

Figure 48 $\mathbf{v} + \mathbf{w} = \mathbf{w} + \mathbf{v}$

Figure 48 illustrates the commutative property for vector addition. (Observe that the commutative property is another way of saying that opposite sides of a parallelogram are equal and parallel.)

Vector addition is also **associative.** That is, if **u**, **v**, and **w** are vectors, then

$$\mathbf{u} + (\mathbf{v} + \mathbf{w}) = (\mathbf{u} + \mathbf{v}) + \mathbf{w}$$

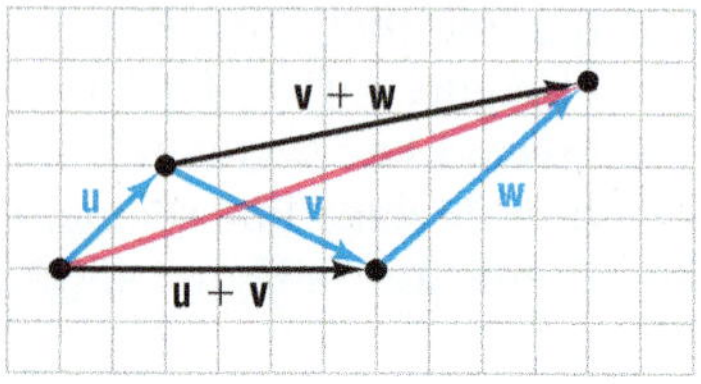

Figure 49
$(\mathbf{u} + \mathbf{v}) + \mathbf{w} = \mathbf{u} + (\mathbf{v} + \mathbf{w})$

Figure 49 illustrates the associative property for vector addition.

The zero vector **0** has the property that

$$\mathbf{v} + \mathbf{0} = \mathbf{0} + \mathbf{v} = \mathbf{v}$$

for any vector **v**.

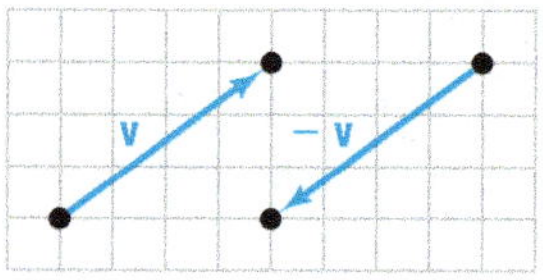

Figure 50 Opposite vectors

If **v** is a vector, then $-\mathbf{v}$ is the vector that has the same magnitude as **v**, but whose direction is opposite to **v**, as shown in Figure 50.

Furthermore,

$$\mathbf{v} + (-\mathbf{v}) = \mathbf{0}$$

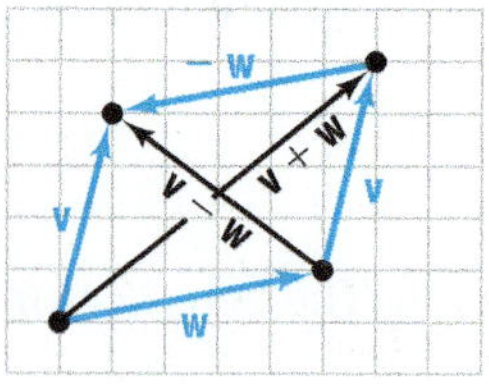

Figure 51

If **v** and **w** are two vectors, then the **difference** $\mathbf{v} - \mathbf{w}$ is defined as

$$\mathbf{v} - \mathbf{w} = \mathbf{v} + (-\mathbf{w})$$

Figure 51 illustrates the relationships among **v**, **w**, $\mathbf{v} + \mathbf{w}$, and $\mathbf{v} - \mathbf{w}$.

Multiplying Vectors by Numbers Geometrically

When dealing with vectors, real numbers are referred to as **scalars**. Scalars are quantities that have only magnitude. Examples of scalar quantities from physics are temperature, speed, and time. We now define how to multiply a vector by a scalar.

DEFINITION

If α is a scalar and **v** is a vector, the **scalar multiple** $\alpha\mathbf{v}$ is defined as follows:

1. If $\alpha > 0$, $\alpha\mathbf{v}$ is the vector whose magnitude is α times the magnitude of **v** and whose direction is the same as that of **v**.
2. If $\alpha < 0$, $\alpha\mathbf{v}$ is the vector whose magnitude is $|\alpha|$ times the magnitude of **v** and whose direction is opposite that of **v**.
3. If $\alpha = 0$ or if $\mathbf{v} = \mathbf{0}$, then $\alpha\mathbf{v} = \mathbf{0}$.

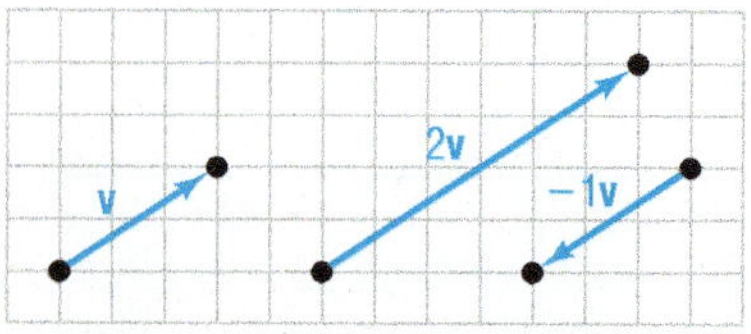

Figure 52 Scalar multiples

See Figure 52 for some illustrations.

For example, if **a** is the acceleration of an object of mass m due to a force **F** being exerted on it, then, by Newton's second law of motion, $\mathbf{F} = m\mathbf{a}$. Here, $m\mathbf{a}$ is the product of the scalar m and the vector **a**.

Scalar multiples have the following properties:

$$0\mathbf{v} = \mathbf{0} \qquad 1\mathbf{v} = \mathbf{v} \qquad -1\mathbf{v} = -\mathbf{v}$$
$$(\alpha + \beta)\mathbf{v} = \alpha\mathbf{v} + \beta\mathbf{v} \qquad \alpha(\mathbf{v} + \mathbf{w}) = \alpha\mathbf{v} + \alpha\mathbf{w}$$
$$\alpha(\beta\mathbf{v}) = (\alpha\beta)\mathbf{v}$$

1 Graph Vectors

EXAMPLE 1 **Graphing Vectors**

Use the vectors illustrated in Figure 53 to graph each of the following vectors:

(a) $\mathbf{v} - \mathbf{w}$ (b) $2\mathbf{v} + 3\mathbf{w}$ (c) $2\mathbf{v} - \mathbf{w} + \mathbf{u}$

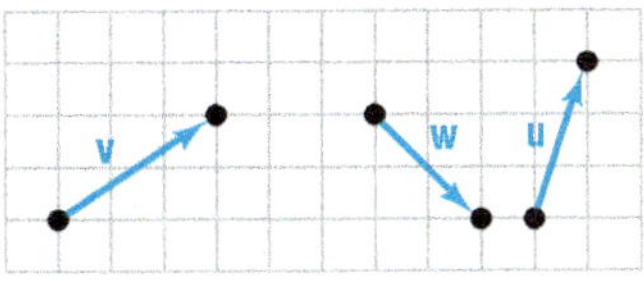

Figure 53

Solution Figure 54 shows each graph.

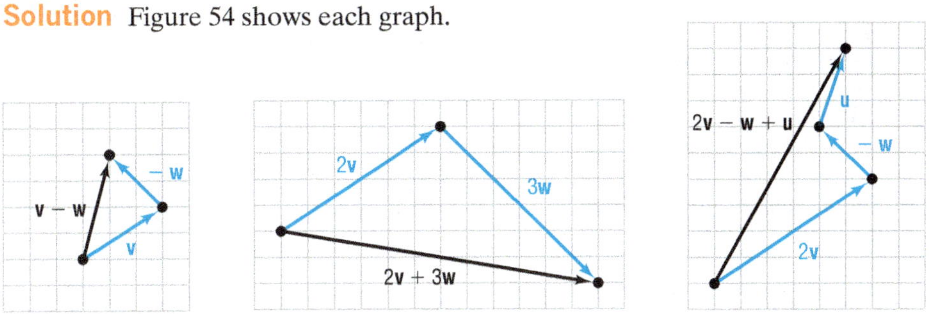

Figure 54 (a) $\mathbf{v} - \mathbf{w}$ (b) $2\mathbf{v} + 3\mathbf{w}$ (c) $2\mathbf{v} - \mathbf{w} + \mathbf{u}$

Now Work PROBLEMS 11 AND 13

Magnitude of Vectors

The symbol $\|\mathbf{v}\|$ represents the **magnitude** of a vector $\mathbf{v}$. Since $\|\mathbf{v}\|$ equals the length of a directed line segment, it follows that $\|\mathbf{v}\|$ has the following properties:

THEOREM

Properties of $\|\mathbf{v}\|$

If $\mathbf{v}$ is a vector and if α is a scalar, then

(a) $\|\mathbf{v}\| \geq 0$ (b) $\|\mathbf{v}\| = 0$ if and only if $\mathbf{v} = \mathbf{0}$

(c) $\|-\mathbf{v}\| = \|\mathbf{v}\|$ (d) $\|\alpha\mathbf{v}\| = |\alpha|\,\|\mathbf{v}\|$

Property (a) is a consequence of the fact that distance is a nonnegative number. Property (b) follows because the length of the directed line segment $\overrightarrow{PQ}$ is positive unless P and Q are the same point, in which case the length is 0. Property (c) follows because the length of the line segment $\overline{PQ}$ equals the length of the line segment $\overline{QP}$. Property (d) is a direct consequence of the definition of a scalar multiple.

DEFINITION

A vector $\mathbf{u}$ for which $\|\mathbf{u}\| = 1$ is called a **unit vector**.

2 Find a Position Vector

To compute the magnitude and direction of a vector, an algebraic way of representing vectors is needed.

DEFINITION

An **algebraic vector v** is represented as

$$\mathbf{v} = \langle a, b\rangle$$

where a and b are real numbers (scalars) called the **components** of the vector $\mathbf{v}$.

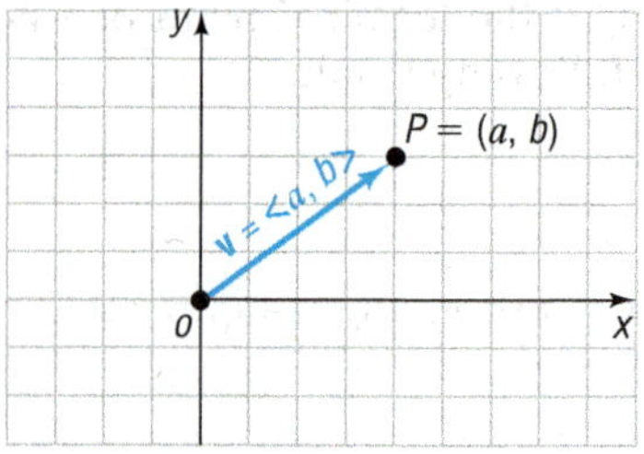

Figure 55 Position vector **v**

A rectangular coordinate system is used to represent algebraic vectors in the plane. If $\mathbf{v} = \langle a, b \rangle$ is an algebraic vector whose initial point is at the origin, then **v** is called a **position vector**. See Figure 55. Notice that the terminal point of the position vector $\mathbf{v} = \langle a, b \rangle$ is $P = (a, b)$.

The next result states that any vector whose initial point is not at the origin is equal to a unique position vector.

THEOREM

Suppose that **v** is a vector with initial point $P_1 = (x_1, y_1)$, not necessarily the origin, and terminal point $P_2 = (x_2, y_2)$. If $\mathbf{v} = \overrightarrow{P_1 P_2}$, then **v** is equal to the position vector

$$\mathbf{v} = \langle x_2 - x_1, y_2 - y_1 \rangle \qquad (1)$$

In Words

An algebraic vector represents "driving directions" to get from the initial point to the terminal point of a vector. So if $\mathbf{v} = \langle 5, 4 \rangle$, travel 5 units right and 4 units up from the initial point to arrive at the terminal point.

To see why this is true, look at Figure 56.

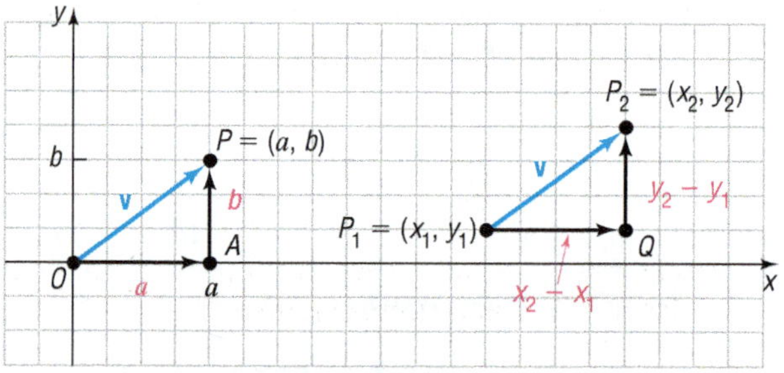

Figure 56 $\mathbf{v} = \langle a, b \rangle = \langle x_2 - x_1, y_2 - y_1 \rangle$

Triangle OPA and triangle P_1P_2Q are congruent. [Do you see why? The line segments have the same magnitude, so $d(O, P) = d(P_1, P_2)$; and they have the same direction, so $\angle POA = \angle P_2P_1Q$. Since the triangles are right triangles, we have angle–side–angle.] It follows that corresponding sides are equal. As a result, $x_2 - x_1 = a$ and $y_2 - y_1 = b$, so **v** may be written as

$$\mathbf{v} = \langle a, b \rangle = \langle x_2 - x_1, y_2 - y_1 \rangle$$

Because of this result, any algebraic vector can be replaced by a unique position vector, and vice versa. This flexibility is one of the main reasons for the wide use of vectors.

EXAMPLE 2 Finding a Position Vector

Find the position vector of the vector $\mathbf{v} = \overrightarrow{P_1P_2}$ if $P_1 = (-1, 2)$ and $P_2 = (4, 6)$.

Solution By equation (1), the position vector equal to **v** is

$$\mathbf{v} = \langle 4 - (-1), 6 - 2 \rangle = \langle 5, 4 \rangle$$

See Figure 57.

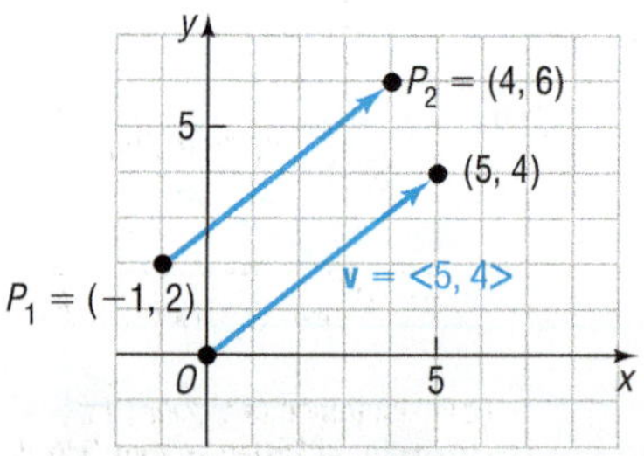

Figure 57

Two position vectors **v** and **w** are equal if and only if the terminal point of **v** is the same as the terminal point of **w**. This leads to the following result:

THEOREM

Equality of Vectors

Two vectors **v** and **w** are equal if and only if their corresponding components are equal. That is,

$$\text{If } \mathbf{v} = \langle a_1, b_1 \rangle \quad \text{and} \quad \mathbf{w} = \langle a_2, b_2 \rangle$$
$$\text{then} \quad \mathbf{v} = \mathbf{w} \quad \text{if and only if} \quad a_1 = a_2 \quad \text{and} \quad b_1 = b_2.$$

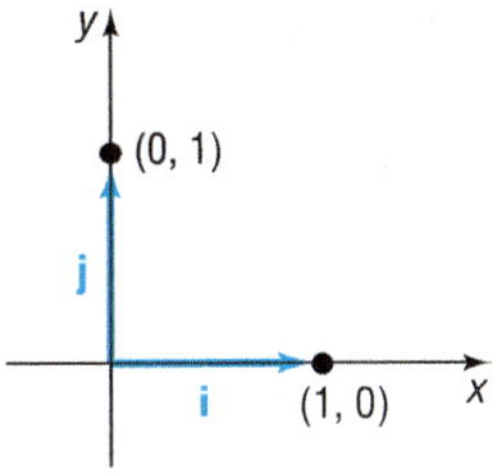

Figure 58 Unit vectors **i** and **j**

We now present an alternative representation of a vector in the plane that is common in the physical sciences. Let **i** denote the unit vector whose direction is along the positive x-axis; let **j** denote the unit vector whose direction is along the positive y-axis. Then $\mathbf{i} = \langle 1, 0 \rangle$ and $\mathbf{j} = \langle 0, 1 \rangle$, as shown in Figure 58. Any vector $\mathbf{v} = \langle a, b \rangle$ can be written using the unit vectors **i** and **j** as follows:

$$\mathbf{v} = \langle a, b \rangle = a\langle 1, 0 \rangle + b\langle 0, 1 \rangle = a\mathbf{i} + b\mathbf{j}$$

The quantities a and b are called the **horizontal** and **vertical components** of **v**, respectively. For example, if $\mathbf{v} = \langle 5, 4 \rangle = 5\mathbf{i} + 4\mathbf{j}$, then 5 is the horizontal component and 4 is the vertical component.

Now Work PROBLEM 31

3 Add and Subtract Vectors Algebraically

The sum, difference, scalar multiple, and magnitude of algebraic vectors are defined in terms of their components.

DEFINITION

> **In Words**
> To add two vectors, add corresponding components. To subtract two vectors, subtract corresponding components.

Let $\mathbf{v} = a_1\mathbf{i} + b_1\mathbf{j} = \langle a_1, b_1 \rangle$ and $\mathbf{w} = a_2\mathbf{i} + b_2\mathbf{j} = \langle a_2, b_2 \rangle$ be two vectors, and let α be a scalar. Then

$$\mathbf{v} + \mathbf{w} = (a_1 + a_2)\mathbf{i} + (b_1 + b_2)\mathbf{j} = \langle a_1 + a_2, b_1 + b_2 \rangle \quad \textbf{(2)}$$
$$\mathbf{v} - \mathbf{w} = (a_1 - a_2)\mathbf{i} + (b_1 - b_2)\mathbf{j} = \langle a_1 - a_2, b_1 - b_2 \rangle \quad \textbf{(3)}$$
$$\alpha\mathbf{v} = (\alpha a_1)\mathbf{i} + (\alpha b_1)\mathbf{j} = \langle \alpha a_1, \alpha b_1 \rangle \quad \textbf{(4)}$$
$$\|\mathbf{v}\| = \sqrt{a_1^2 + b_1^2} \quad \textbf{(5)}$$

These definitions are compatible with the geometric definitions given earlier in this section. See Figure 59.

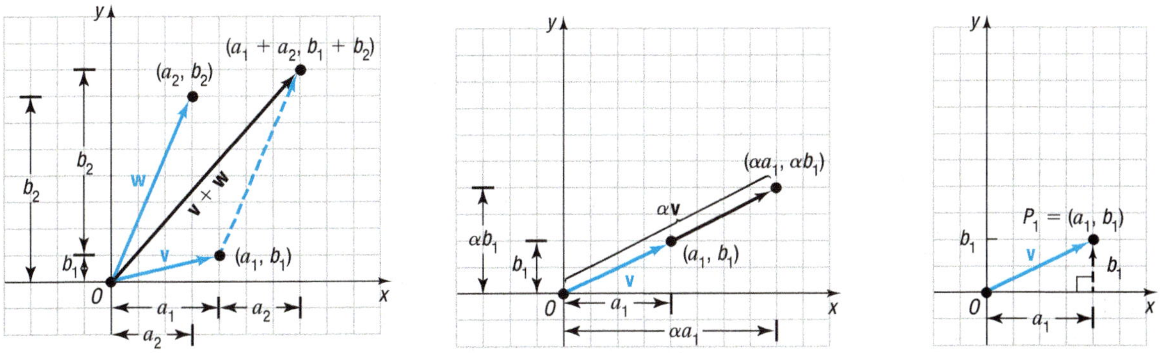

Figure 59 **(a)** Illustration of property (2) **(b)** Illustration of property (4), $\alpha > 0$ **(c)** Illustration of property (5): $\|\mathbf{v}\|$ = Distance from O to P_1; $\|\mathbf{v}\| = \sqrt{a_1^2 + b_1^2}$

EXAMPLE 3 **Adding and Subtracting Vectors**

If $\mathbf{v} = 2\mathbf{i} + 3\mathbf{j} = \langle 2, 3\rangle$ and $\mathbf{w} = 3\mathbf{i} - 4\mathbf{j} = \langle 3, -4\rangle$, find:

(a) $\mathbf{v} + \mathbf{w}$ (b) $\mathbf{v} - \mathbf{w}$

Solution (a) $\mathbf{v} + \mathbf{w} = (2\mathbf{i} + 3\mathbf{j}) + (3\mathbf{i} - 4\mathbf{j}) = (2 + 3)\mathbf{i} + (3 - 4)\mathbf{j} = 5\mathbf{i} - \mathbf{j}$

or

$\mathbf{v} + \mathbf{w} = \langle 2, 3\rangle + \langle 3, -4\rangle = \langle 2 + 3, 3 + (-4)\rangle = \langle 5, -1\rangle$

(b) $\mathbf{v} - \mathbf{w} = (2\mathbf{i} + 3\mathbf{j}) - (3\mathbf{i} - 4\mathbf{j}) = (2 - 3)\mathbf{i} + [3 - (-4)]\mathbf{j} = -\mathbf{i} + 7\mathbf{j}$

or

$\mathbf{v} - \mathbf{w} = \langle 2, 3\rangle - \langle 3, -4\rangle = \langle 2 - 3, 3 - (-4)\rangle = \langle -1, 7\rangle$ ■

4 Find a Scalar Multiple and the Magnitude of a Vector

EXAMPLE 4 **Finding Scalar Multiples and Magnitudes of Vectors**

If $\mathbf{v} = 2\mathbf{i} + 3\mathbf{j} = \langle 2, 3\rangle$ and $\mathbf{w} = 3\mathbf{i} - 4\mathbf{j} = \langle 3, -4\rangle$, find:

(a) $3\mathbf{v}$ (b) $2\mathbf{v} - 3\mathbf{w}$ (c) $\|\mathbf{v}\|$

Solution (a) $3\mathbf{v} = 3(2\mathbf{i} + 3\mathbf{j}) = 6\mathbf{i} + 9\mathbf{j}$

or

$3\mathbf{v} = 3\langle 2, 3\rangle = \langle 6, 9\rangle$

(b) $2\mathbf{v} - 3\mathbf{w} = 2(2\mathbf{i} + 3\mathbf{j}) - 3(3\mathbf{i} - 4\mathbf{j}) = 4\mathbf{i} + 6\mathbf{j} - 9\mathbf{i} + 12\mathbf{j}$

$= -5\mathbf{i} + 18\mathbf{j}$

or

$2\mathbf{v} - 3\mathbf{w} = 2\langle 2, 3\rangle - 3\langle 3, -4\rangle = \langle 4, 6\rangle - \langle 9, -12\rangle$

$= \langle 4 - 9, 6 - (-12)\rangle = \langle -5, 18\rangle$

(c) $\|\mathbf{v}\| = \|2\mathbf{i} + 3\mathbf{j}\| = \sqrt{2^2 + 3^2} = \sqrt{13}$ ■

Now Work PROBLEMS 37 AND 43

For the remainder of the section, we will express a vector $\mathbf{v}$ in the form $a\mathbf{i} + b\mathbf{j}$.

5 Find a Unit Vector

Recall that a unit vector $\mathbf{u}$ is a vector for which $\|\mathbf{u}\| = 1$. In many applications, it is useful to be able to find a unit vector $\mathbf{u}$ that has the same direction as a given vector $\mathbf{v}$.

THEOREM **Unit Vector in the Direction of v**

For any nonzero vector $\mathbf{v}$, the vector

$$\mathbf{u} = \frac{\mathbf{v}}{\|\mathbf{v}\|} \tag{6}$$

is a unit vector that has the same direction as $\mathbf{v}$. ■

Proof Let $\mathbf{v} = a\mathbf{i} + b\mathbf{j}$. Then $\|\mathbf{v}\| = \sqrt{a^2 + b^2}$ and

$$\mathbf{u} = \frac{\mathbf{v}}{\|\mathbf{v}\|} = \frac{a\mathbf{i} + b\mathbf{j}}{\sqrt{a^2 + b^2}} = \frac{a}{\sqrt{a^2 + b^2}}\mathbf{i} + \frac{b}{\sqrt{a^2 + b^2}}\mathbf{j}$$

The vector **u** is in the same direction as **v**, since $\|\mathbf{v}\| > 0$. Furthermore,

$$\|\mathbf{u}\| = \sqrt{\frac{a^2}{a^2+b^2} + \frac{b^2}{a^2+b^2}} = \sqrt{\frac{a^2+b^2}{a^2+b^2}} = 1$$

That is, **u** is a unit vector in the direction of **v**. ■

As a consequence of this theorem, if **u** is a unit vector in the same direction as a vector **v**, then **v** may be expressed as

$$\mathbf{v} = \|\mathbf{v}\|\mathbf{u} \qquad (7)$$

This way of expressing a vector is useful in many applications.

EXAMPLE 5

Finding a Unit Vector

Find a unit vector in the same direction as $\mathbf{v} = 4\mathbf{i} - 3\mathbf{j}$.

Solution Find $\|\mathbf{v}\|$ first.

$$\|\mathbf{v}\| = \|4\mathbf{i} - 3\mathbf{j}\| = \sqrt{16+9} = 5$$

Now multiply **v** by the scalar $\frac{1}{\|\mathbf{v}\|} = \frac{1}{5}$. A unit vector in the same direction as **v** is

$$\frac{\mathbf{v}}{\|\mathbf{v}\|} = \frac{4\mathbf{i} - 3\mathbf{j}}{5} = \frac{4}{5}\mathbf{i} - \frac{3}{5}\mathbf{j}$$

Check: This vector is indeed a unit vector because

$$\left\|\frac{\mathbf{v}}{\|\mathbf{v}\|}\right\| = \sqrt{\left(\frac{4}{5}\right)^2 + \left(-\frac{3}{5}\right)^2} = \sqrt{\frac{16}{25} + \frac{9}{25}} = \sqrt{\frac{25}{25}} = 1$$ ■

Now Work PROBLEM 53

6 Find a Vector from Its Direction and Magnitude

If a vector represents the speed and direction of an object, it is called a **velocity vector**. If a vector represents the direction and amount of a force acting on an object, it is called a **force vector**. In many applications, a vector is described in terms of its magnitude and direction, rather than in terms of its components. For example, a ball thrown with an initial speed of 25 miles per hour at an angle of 30° to the horizontal is a velocity vector.

Suppose that we are given the magnitude $\|\mathbf{v}\|$ of a nonzero vector **v** and the **direction angle** α, $0° \le \alpha < 360°$, between **v** and **i**. To express **v** in terms of $\|\mathbf{v}\|$ and α, first find the unit vector **u** having the same direction as **v**.

Look at Figure 60. The coordinates of the terminal point of **u** are $(\cos\alpha, \sin\alpha)$. Then $\mathbf{u} = \cos\alpha\,\mathbf{i} + \sin\alpha\,\mathbf{j}$ and, from equation (7),

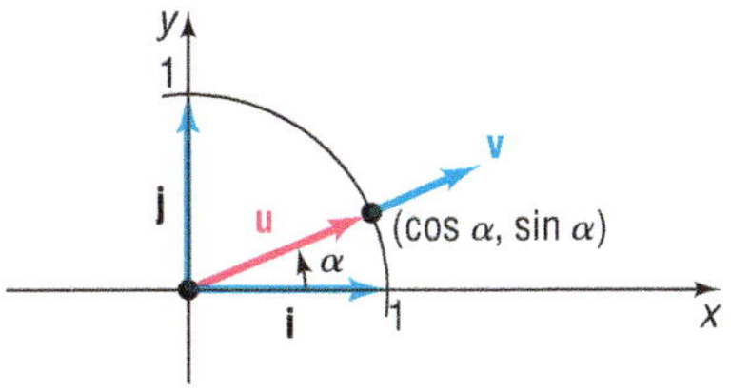

Figure 60 $\mathbf{v} = \|\mathbf{v}\|\,(\cos\alpha\,\mathbf{i} + \sin\alpha\,\mathbf{j})$

$$\mathbf{v} = \|\mathbf{v}\|(\cos\alpha\,\mathbf{i} + \sin\alpha\,\mathbf{j}) \qquad (8)$$

where α is the direction angle between **v** and **i**.

EXAMPLE 6 Finding a Vector When Its Magnitude and Direction Are Given

A ball is thrown with an initial speed of 25 miles per hour in a direction that makes an angle of 30° with the positive x-axis. Express the velocity vector $\mathbf{v}$ in terms of $\mathbf{i}$ and $\mathbf{j}$. What is the initial speed in the horizontal direction? What is the initial speed in the vertical direction?

Solution The magnitude of $\mathbf{v}$ is $\|\mathbf{v}\| = 25$ miles per hour, and the angle between the direction of $\mathbf{v}$ and $\mathbf{i}$, the positive x-axis, is $\alpha = 30°$. By equation (8),

$$\mathbf{v} = \|\mathbf{v}\|(\cos\alpha\mathbf{i} + \sin\alpha\mathbf{j}) = 25(\cos 30°\mathbf{i} + \sin 30°\mathbf{j})$$
$$= 25\left(\frac{\sqrt{3}}{2}\mathbf{i} + \frac{1}{2}\mathbf{j}\right) = \frac{25\sqrt{3}}{2}\mathbf{i} + \frac{25}{2}\mathbf{j}$$

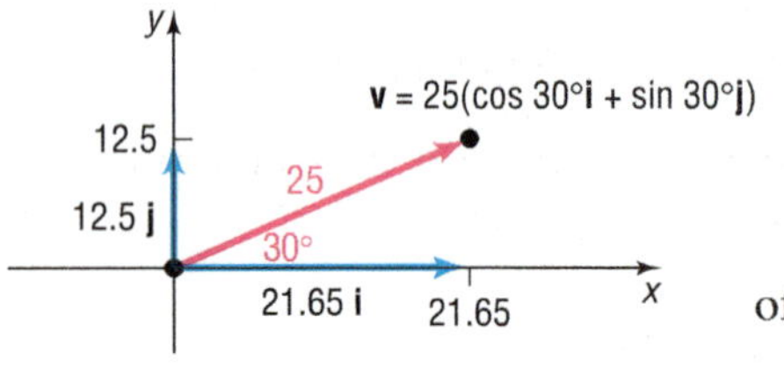

Figure 61

The initial speed of the ball in the horizontal direction is the horizontal component of $\mathbf{v}$, $\frac{25\sqrt{3}}{2} \approx 21.65$ miles per hour. The initial speed in the vertical direction is the vertical component of $\mathbf{v}$, $\frac{25}{2} = 12.5$ miles per hour. See Figure 61. ■

Now Work PROBLEM 59

EXAMPLE 7 Finding the Direction Angle of a Vector

Find the direction angle α of $\mathbf{v} = 4\mathbf{i} - 4\mathbf{j}$.

Solution See Figure 62. The direction angle α of $\mathbf{v} = 4\mathbf{i} - 4\mathbf{j}$ can be found by solving

$$\tan\alpha = \frac{-4}{4} = -1$$

Because $0° \le \alpha < 360°$, the direction angle is $\alpha = 315°$. ■

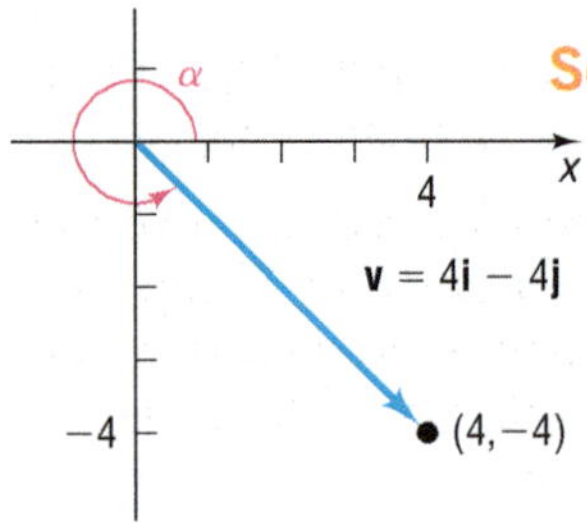

Figure 62

Now Work PROBLEM 65

7 Model with Vectors

Because forces can be represented by vectors, two forces "combine" the way that vectors "add." If $\mathbf{F}_1$ and $\mathbf{F}_2$ are two forces simultaneously acting on an object, the vector sum $\mathbf{F}_1 + \mathbf{F}_2$ is the **resultant force**. The resultant force produces the same effect on the object as that obtained when the two forces $\mathbf{F}_1$ and $\mathbf{F}_2$ act on the object. See Figure 63.

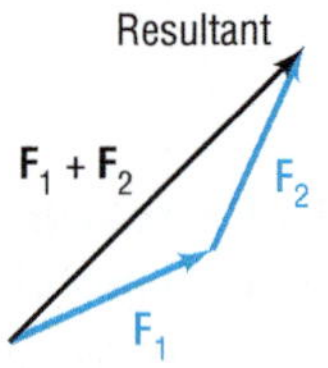

Figure 63 Resultant force

EXAMPLE 8 Finding the Actual Speed and Direction of an Aircraft

A Boeing 737 aircraft maintains a constant airspeed of 500 miles per hour headed due south. The jet stream is 80 miles per hour in the northeasterly direction.

(a) Express the velocity $\mathbf{v}_a$ of the 737 relative to the air and the velocity $\mathbf{v}_w$ of the jet stream in terms of $\mathbf{i}$ and $\mathbf{j}$.

(b) Find the velocity of the 737 relative to the ground.

(c) Find the actual speed and direction of the 737 relative to the ground.

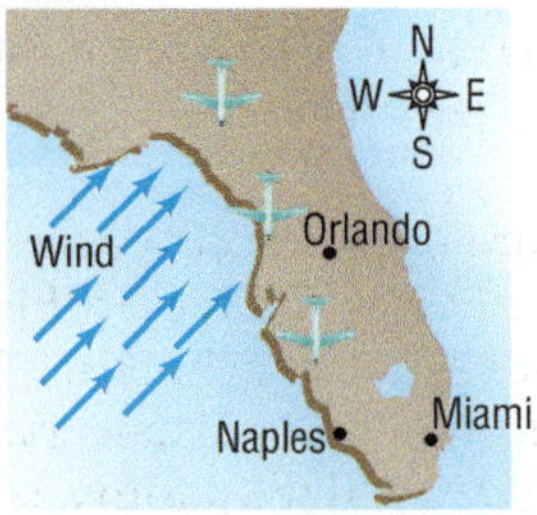

Solution

(a) Set up a coordinate system in which north (N) is along the positive y-axis. See Figure 64. The velocity of the 737 relative to the air is $\mathbf{v}_a = -500\mathbf{j}$. The velocity of the jet stream $\mathbf{v}_w$ has magnitude 80 and direction NE (northeast), so the angle between $\mathbf{v}_w$ and $\mathbf{i}$ is 45°. Express $\mathbf{v}_w$ in terms of $\mathbf{i}$ and $\mathbf{j}$ as

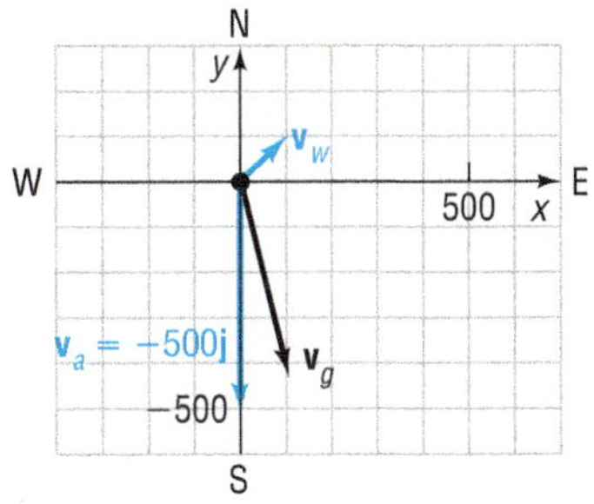

Figure 64

$$\mathbf{v}_w = 80(\cos 45°\,\mathbf{i} + \sin 45°\,\mathbf{j}) = 80\left(\frac{\sqrt{2}}{2}\mathbf{i} + \frac{\sqrt{2}}{2}\mathbf{j}\right) = 40\sqrt{2}\,(\mathbf{i} + \mathbf{j})$$

(b) The velocity of the 737 relative to the ground $\mathbf{v}_g$ is

$$\mathbf{v}_g = \mathbf{v}_a + \mathbf{v}_w = -500\mathbf{j} + 40\sqrt{2}(\mathbf{i} + \mathbf{j}) = 40\sqrt{2}\,\mathbf{i} + (40\sqrt{2} - 500)\mathbf{j}$$

(c) The actual speed of the 737 is

$$\|\mathbf{v}_g\| = \sqrt{(40\sqrt{2})^2 + (40\sqrt{2} - 500)^2} \approx 447 \text{ miles per hour}$$

To find the actual direction of the 737 relative to the ground, determine the direction angle of $\mathbf{v}_g$. The direction angle is found by solving

$$\tan \alpha = \frac{40\sqrt{2} - 500}{40\sqrt{2}}$$

Then $\alpha \approx -82.7°$. The 737 is traveling S 7.3° E. ■

Now Work PROBLEM 77

EXAMPLE 9 Finding the Weight of a Piano

Two movers require a magnitude of force of 300 pounds to push a piano up a ramp inclined at an angle 20° from the horizontal. How much does the piano weigh?

Solution

Let $\mathbf{F}_1$ represent the force of gravity, $\mathbf{F}_2$ represent the force required to move the piano up the ramp, and $\mathbf{F}_3$ represent the force of the piano against the ramp. See Figure 65. The angle between the ground and the ramp is the same as the angle between $\mathbf{F}_1$ and $\mathbf{F}_3$ because triangles ABC and BDE are similar, so $\angle BAC = \angle DBE = 20°$. To find the magnitude of $\mathbf{F}_1$ (the weight of the piano), calculate

$$\sin 20° = \frac{\|\mathbf{F}_2\|}{\|\mathbf{F}_1\|} = \frac{300}{\|\mathbf{F}_1\|}$$

$$\|\mathbf{F}_1\| = \frac{300 \text{ lb}}{\sin 20°} \approx 877 \text{ lb}$$

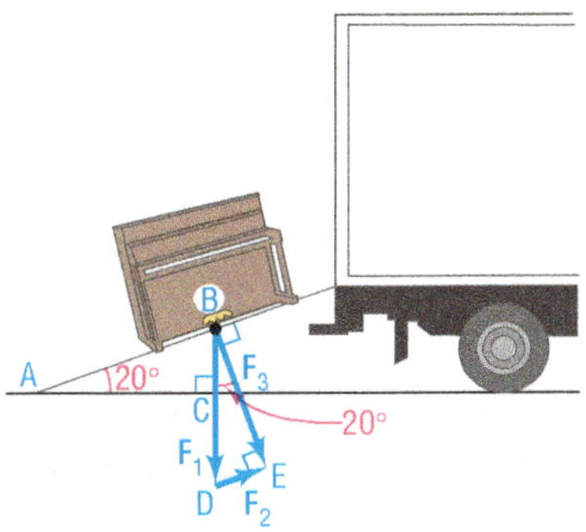

Figure 65

The piano weighs approximately 877 pounds. ■

An object is said to be in **static equilibrium** if the object is at rest and the sum of all forces acting on the object is zero—that is, if the resultant force is **0**.

EXAMPLE 10 Analyzing an Object in Static Equilibrium

A box of supplies that weighs 1200 pounds is suspended by two cables attached to the ceiling, as shown in Figure 66 on the next page. What are the tensions in the two cables?

Solution

Draw a force diagram using the vectors as shown in Figure 67 on the next page. The tensions in the cables are the magnitudes $\|\mathbf{F}_1\|$ and $\|\mathbf{F}_2\|$ of the force vectors $\mathbf{F}_1$ and $\mathbf{F}_2$. The magnitude of the force vector $\mathbf{F}_3$ equals 1200 pounds, the weight of the box. Now write each force vector in terms of the unit vectors $\mathbf{i}$ and $\mathbf{j}$. For $\mathbf{F}_1$ and $\mathbf{F}_2$, use equation (8). Remember that α is the angle between the vector and the positive x-axis.

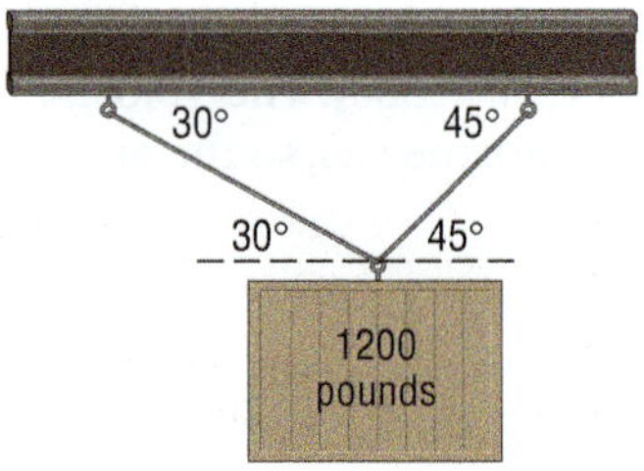

Figure 66

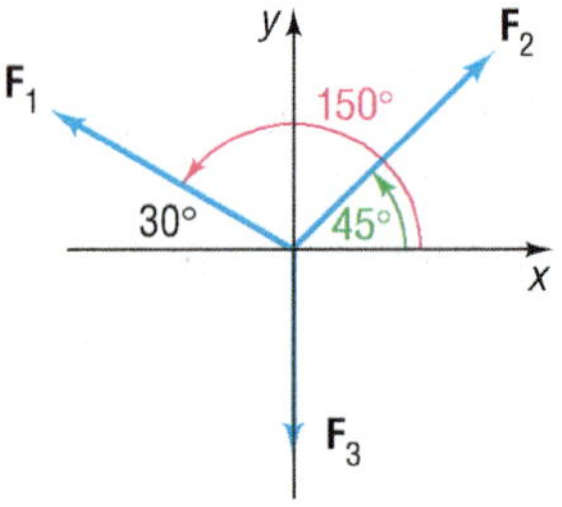

Figure 67 Force diagram

$$\mathbf{F}_1 = \|\mathbf{F}_1\|(\cos 150°\mathbf{i} + \sin 150°\mathbf{j}) = \|\mathbf{F}_1\|\left(-\frac{\sqrt{3}}{2}\mathbf{i} + \frac{1}{2}\mathbf{j}\right) = -\frac{\sqrt{3}}{2}\|\mathbf{F}_1\|\mathbf{i} + \frac{1}{2}\|\mathbf{F}_1\|\mathbf{j}$$

$$\mathbf{F}_2 = \|\mathbf{F}_2\|(\cos 45°\mathbf{i} + \sin 45°\mathbf{j}) = \|\mathbf{F}_2\|\left(\frac{\sqrt{2}}{2}\mathbf{i} + \frac{\sqrt{2}}{2}\mathbf{j}\right) = \frac{\sqrt{2}}{2}\|\mathbf{F}_2\|\mathbf{i} + \frac{\sqrt{2}}{2}\|\mathbf{F}_2\|\mathbf{j}$$

$$\mathbf{F}_3 = -1200\mathbf{j}$$

For static equilibrium, the sum of the force vectors must equal zero.

$$\mathbf{F}_1 + \mathbf{F}_2 + \mathbf{F}_3 = -\frac{\sqrt{3}}{2}\|\mathbf{F}_1\|\mathbf{i} + \frac{1}{2}\|\mathbf{F}_1\|\mathbf{j} + \frac{\sqrt{2}}{2}\|\mathbf{F}_2\|\mathbf{i} + \frac{\sqrt{2}}{2}\|\mathbf{F}_2\|\mathbf{j} - 1200\mathbf{j} = \mathbf{0}$$

The **i** component and **j** component will each equal zero. This results in the two equations

$$-\frac{\sqrt{3}}{2}\|\mathbf{F}_1\| + \frac{\sqrt{2}}{2}\|\mathbf{F}_2\| = 0 \qquad \textbf{(9)}$$

$$\frac{1}{2}\|\mathbf{F}_1\| + \frac{\sqrt{2}}{2}\|\mathbf{F}_2\| - 1200 = 0 \qquad \textbf{(10)}$$

Solve equation (9) for $\|\mathbf{F}_2\|$ to obtain

$$\|\mathbf{F}_2\| = \frac{\sqrt{3}}{\sqrt{2}}\|\mathbf{F}_1\| \qquad \textbf{(11)}$$

Substituting into equation (10) and solving for $\|\mathbf{F}_1\|$ yields

$$\frac{1}{2}\|\mathbf{F}_1\| + \frac{\sqrt{2}}{2}\left(\frac{\sqrt{3}}{\sqrt{2}}\|\mathbf{F}_1\|\right) - 1200 = 0$$

$$\frac{1}{2}\|\mathbf{F}_1\| + \frac{\sqrt{3}}{2}\|\mathbf{F}_1\| - 1200 = 0$$

$$\frac{1+\sqrt{3}}{2}\|\mathbf{F}_1\| = 1200$$

$$\|\mathbf{F}_1\| = \frac{2400}{1+\sqrt{3}} \approx 878.5 \text{ pounds}$$

Substituting this value into equation (11) gives $\|\mathbf{F}_2\|$.

$$\|\mathbf{F}_2\| = \frac{\sqrt{3}}{\sqrt{2}}\|\mathbf{F}_1\| = \frac{\sqrt{3}}{\sqrt{2}} \cdot \frac{2400}{1+\sqrt{3}} \approx 1075.9 \text{ pounds}$$

The left cable has tension of approximately 878.5 pounds, and the right cable has tension of approximately 1075.9 pounds.

Now Work PROBLEM 87

Historical Feature

Josiah Gibbs (1839–1903)

The history of vectors is surprisingly complicated for such a natural concept. In the *xy*-plane, complex numbers do a good job of imitating vectors. About 1840, mathematicians became interested in finding a system that would do for three dimensions what the complex numbers do for two dimensions. Hermann Grassmann (1809–1877), in Germany, and William Rowan Hamilton (1805–1865), in Ireland, both attempted to find solutions.

Hamilton's system was the *quaternions*, which are best thought of as a real number plus a vector; they do for four dimensions what complex numbers do for two dimensions. In this system the order of multiplication matters; that is, ab ≠ ba. Also, two products of vectors emerged, the scalar product (or dot product) and the vector product (or cross product).

Grassmann's abstract style, although easily read today, was almost impenetrable during the nineteenth century, and only a few of his ideas were appreciated. Among those few were the same scalar and vector products that Hamilton had found.

About 1880, the American physicist Josiah Willard Gibbs (1839–1903) worked out an algebra involving only the simplest concepts: the vectors and the two products. He then added some calculus, and the resulting system was simple, flexible, and well adapted to expressing a large number of physical laws. This system remains in use essentially unchanged. Hamilton's and Grassmann's more extensive systems each gave birth to much interesting mathematics, but little of it is seen at elementary levels.

10.4 Assess Your Understanding

Concepts and Vocabulary

1. A ______ is a quantity that has both magnitude and direction.
2. If **v** is a vector, then $\mathbf{v} + (-\mathbf{v}) =$ ___.
3. A vector **u** for which $\|\mathbf{u}\| = 1$ is called a(n) _____ vector.
4. If $\mathbf{v} = \langle a, b \rangle$ is an algebraic vector whose initial point is the origin, then **v** is called a(n) ________ vector.
5. If $\mathbf{v} = a\mathbf{i} + b\mathbf{j}$, then a is called the __________ component of **v** and b is called the ________ component of **v**.
6. If $\mathbf{F}_1$ and $\mathbf{F}_2$ are two forces simultaneously acting on an object, the vector sum $\mathbf{F}_1 + \mathbf{F}_2$ is called the ________ force.
7. ***True or False*** Force is an example of a vector.
8. ***True or False*** Mass is an example of a vector.
9. If **v** is a vector with initial point (x_1, y_1) and terminal point (x_2, y_2), then which of the following is the position vector that equals **v**?
   (a) $\langle x_2 - x_1, y_2 - y_1 \rangle$ (b) $\langle x_1 - x_2, y_1 - y_2 \rangle$
   (c) $\left\langle \frac{x_2 - x_1}{2}, \frac{y_2 - y_1}{2} \right\rangle$ (d) $\left\langle \frac{x_1 + x_2}{2}, \frac{y_1 + y_2}{2} \right\rangle$
10. If **v** is a nonzero vector with direction angle α, $0° \le \alpha < 360°$, between **v** and **i**, then **v** equals which of the following?
   (a) $\|\mathbf{v}\|(\cos\alpha\mathbf{i} - \sin\alpha\mathbf{j})$ (b) $\|\mathbf{v}\|(\cos\alpha\mathbf{i} + \sin\alpha\mathbf{j})$
   (c) $\|\mathbf{v}\|(\sin\alpha\mathbf{i} - \cos\alpha\mathbf{j})$ (d) $\|\mathbf{v}\|(\sin\alpha\mathbf{i} + \cos\alpha\mathbf{j})$

Skill Building

In Problems 11–18, use the vectors in the figure at the right to graph each of the following vectors.

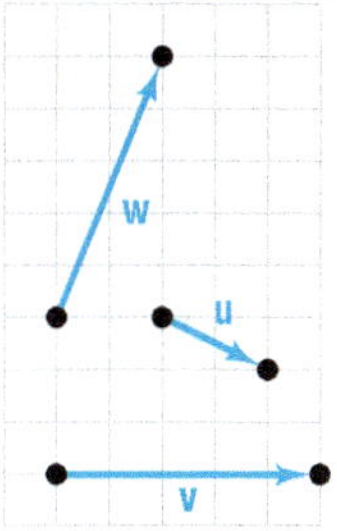

11. $\mathbf{v} + \mathbf{w}$
12. $\mathbf{u} + \mathbf{v}$
13. $3\mathbf{v}$
14. $2\mathbf{w}$
15. $\mathbf{v} - \mathbf{w}$
16. $\mathbf{u} - \mathbf{v}$
17. $3\mathbf{v} + \mathbf{u} - 2\mathbf{w}$
18. $2\mathbf{u} - 3\mathbf{v} + \mathbf{w}$

In Problems 19–26, use the figure at the right. Determine whether each statement given is true or false.

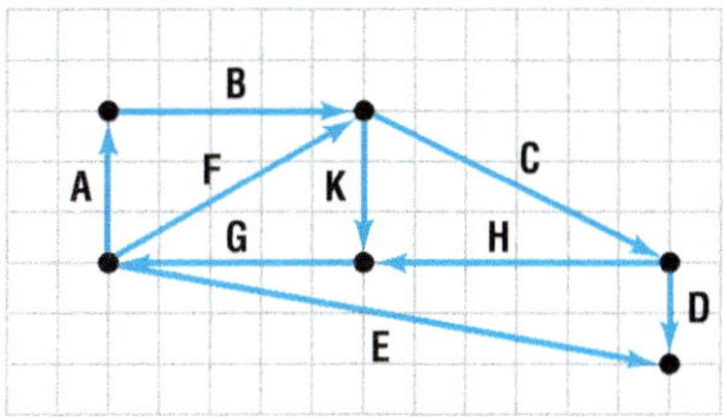

19. $\mathbf{A} + \mathbf{B} = \mathbf{F}$
20. $\mathbf{K} + \mathbf{G} = \mathbf{F}$
21. $\mathbf{C} = \mathbf{D} - \mathbf{E} + \mathbf{F}$
22. $\mathbf{G} + \mathbf{H} + \mathbf{E} = \mathbf{D}$
23. $\mathbf{E} + \mathbf{D} = \mathbf{G} + \mathbf{H}$
24. $\mathbf{H} - \mathbf{C} = \mathbf{G} - \mathbf{F}$
25. $\mathbf{A} + \mathbf{B} + \mathbf{K} + \mathbf{G} = \mathbf{0}$
26. $\mathbf{A} + \mathbf{B} + \mathbf{C} + \mathbf{H} + \mathbf{G} = \mathbf{0}$
27. If $\|\mathbf{v}\| = 4$, what is $\|3\mathbf{v}\|$?
28. If $\|\mathbf{v}\| = 2$, what is $\|-4\mathbf{v}\|$?

In Problems 29–36, the vector **v** *has initial point P and terminal point Q. Write* **v** *in the form* $a\mathbf{i} + b\mathbf{j}$; *that is, find its position vector.*

29. $P = (0, 0)$; $Q = (3, 4)$
30. $P = (0, 0)$; $Q = (-3, -5)$
31. $P = (3, 2)$; $Q = (5, 6)$
32. $P = (-3, 2)$; $Q = (6, 5)$
33. $P = (-2, -1)$; $Q = (6, -2)$
34. $P = (-1, 4)$; $Q = (6, 2)$
35. $P = (1, 0)$; $Q = (0, 1)$
36. $P = (1, 1)$; $Q = (2, 2)$

In Problems 37–42, find $\|\mathbf{v}\|$.

37. $\mathbf{v} = 3\mathbf{i} - 4\mathbf{j}$
38. $\mathbf{v} = -5\mathbf{i} + 12\mathbf{j}$
39. $\mathbf{v} = \mathbf{i} - \mathbf{j}$
40. $\mathbf{v} = -\mathbf{i} - \mathbf{j}$
41. $\mathbf{v} = -2\mathbf{i} + 3\mathbf{j}$
42. $\mathbf{v} = 6\mathbf{i} + 2\mathbf{j}$

In Problems 43–48, find each quantity if $\mathbf{v} = 3\mathbf{i} - 5\mathbf{j}$ *and* $\mathbf{w} = -2\mathbf{i} + 3\mathbf{j}$.

43. $2\mathbf{v} + 3\mathbf{w}$
44. $3\mathbf{v} - 2\mathbf{w}$
45. $\|\mathbf{v} - \mathbf{w}\|$
46. $\|\mathbf{v} + \mathbf{w}\|$
47. $\|\mathbf{v}\| - \|\mathbf{w}\|$
48. $\|\mathbf{v}\| + \|\mathbf{w}\|$

In Problems 49–54, find the unit vector in the same direction as **v**.

49. $\mathbf{v} = 5\mathbf{i}$
50. $\mathbf{v} = -3\mathbf{j}$
51. $\mathbf{v} = 3\mathbf{i} - 4\mathbf{j}$
52. $\mathbf{v} = -5\mathbf{i} + 12\mathbf{j}$
53. $\mathbf{v} = \mathbf{i} - \mathbf{j}$
54. $\mathbf{v} = 2\mathbf{i} - \mathbf{j}$

55. Find a vector **v** whose magnitude is 4 and whose component in the **i** direction is twice the component in the **j** direction.

56. Find a vector **v** whose magnitude is 3 and whose component in the **i** direction is equal to the component in the **j** direction.

57. If $\mathbf{v} = 2\mathbf{i} - \mathbf{j}$ and $\mathbf{w} = x\mathbf{i} + 3\mathbf{j}$, find all numbers x for which $\|\mathbf{v} + \mathbf{w}\| = 5$.

58. If $P = (-3, 1)$ and $Q = (x, 4)$, find all numbers x such that the vector represented by $\overrightarrow{PQ}$ has length 5.

In Problems 59–64, write the vector **v** *in the form* $a\mathbf{i} + b\mathbf{j}$*, given its magnitude* $\|\mathbf{v}\|$ *and the angle* α *it makes with the positive x-axis.*

59. $\|\mathbf{v}\| = 5, \quad \alpha = 60°$

60. $\|\mathbf{v}\| = 8, \quad \alpha = 45°$

61. $\|\mathbf{v}\| = 14, \quad \alpha = 120°$

62. $\|\mathbf{v}\| = 3, \quad \alpha = 240°$

63. $\|\mathbf{v}\| = 25, \quad \alpha = 330°$

64. $\|\mathbf{v}\| = 15, \quad \alpha = 315°$

In Problems 65–72, find the direction angle of **v**.

65. $\mathbf{v} = 3\mathbf{i} + 3\mathbf{j}$

66. $\mathbf{v} = \mathbf{i} + \sqrt{3}\mathbf{j}$

67. $\mathbf{v} = -3\sqrt{3}\mathbf{i} + 3\mathbf{j}$

68. $\mathbf{v} = -5\mathbf{i} - 5\mathbf{j}$

69. $\mathbf{v} = 4\mathbf{i} - 2\mathbf{j}$

70. $\mathbf{v} = 6\mathbf{i} - 4\mathbf{j}$

71. $\mathbf{v} = -\mathbf{i} - 5\mathbf{j}$

72. $\mathbf{v} = -\mathbf{i} + 3\mathbf{j}$

Applications and Extensions

73. Force Vectors A child pulls a wagon with a force of 40 pounds. The handle of the wagon makes an angle of 30° with the ground. Express the force vector **F** in terms of **i** and **j**.

74. Force Vectors A man pushes a wheelbarrow up an incline of 20° with a force of 100 pounds. Express the force vector **F** in terms of **i** and **j**.

75. Resultant Force Two forces of magnitude 40 newtons (N) and 60 N act on an object at angles of 30° and −45° with the positive x-axis, as shown in the figure. Find the direction and magnitude of the resultant force; that is, find $\mathbf{F}_1 + \mathbf{F}_2$.

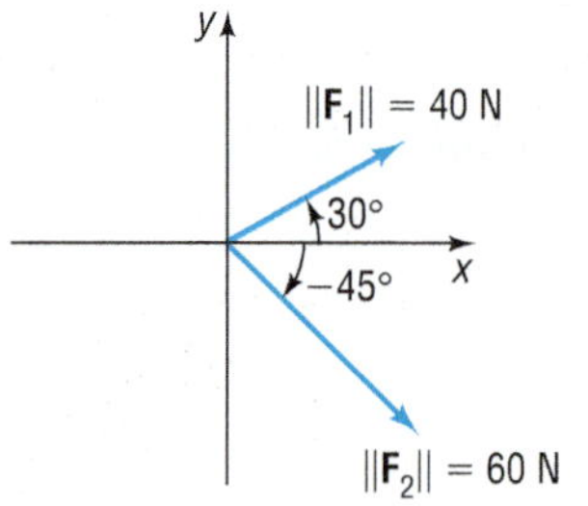

76. Resultant Force Two forces of magnitude 30 newtons (N) and 70 N act on an object at angles of 45° and 120° with the positive x-axis, as shown in the figure. Find the direction and magnitude of the resultant force; that is, find $\mathbf{F}_1 + \mathbf{F}_2$.

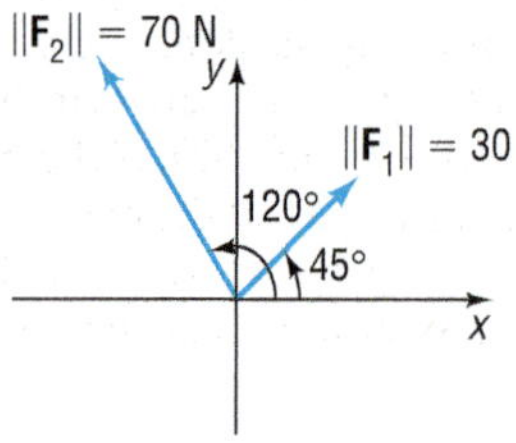

77. Finding the Actual Speed and Direction of an Aircraft A Boeing 747 jumbo jet maintains a constant airspeed of 550 miles per hour (mph) headed due north. The jet stream is 100 mph in the northeasterly direction.

(a) Express the velocity $\mathbf{v}_a$ of the 747 relative to the air and the velocity $\mathbf{v}_w$ of the jet stream in terms of **i** and **j**.
(b) Find the velocity of the 747 relative to the ground.
(c) Find the actual speed and direction of the 747 relative to the ground.

78. Finding the Actual Speed and Direction of an Aircraft An Airbus A320 jet maintains a constant airspeed of 500 mph headed due west. The jet stream is 100 mph in the southeasterly direction.

(a) Express the velocity $\mathbf{v}_a$ of the A320 relative to the air and the velocity $\mathbf{v}_w$ of the jet stream in terms of **i** and **j**.
(b) Find the velocity of the A320 relative to the ground.
(c) Find the actual speed and direction of the A320 relative to the ground.

79. Ground Speed and Direction of an Airplane An airplane has an airspeed of 500 kilometers per hour (km/h) bearing N45°E. The wind velocity is 60 km/h in the direction N30°W. Find the resultant vector representing the path of the plane relative to the ground. What is the groundspeed of the plane? What is its direction?

80. Ground Speed and Direction of an Airplane An airplane has an airspeed of 600 km/h bearing S30°E. The wind velocity is 40 km/h in the direction S45°E. Find the resultant vector representing the path of the plane relative to the ground. What is the groundspeed of the plane? What is its direction?

81. Weight of a Boat A magnitude of 700 pounds of force is required to hold a boat and its trailer in place on a ramp whose incline is 10° to the horizontal. What is the combined weight of the boat and its trailer?

82. Weight of a Car A magnitude of 1200 pounds of force is required to prevent a car from rolling down a hill whose incline is 15° to the horizontal. What is the weight of the car?

83. Correct Direction for Crossing a River A river has a constant current of 3 km/h. At what angle to a boat dock should a motorboat capable of maintaining a constant speed of 20 km/h be headed in order to reach a point directly opposite the dock? If the river is $\frac{1}{2}$ kilometer wide, how long will it take to cross?

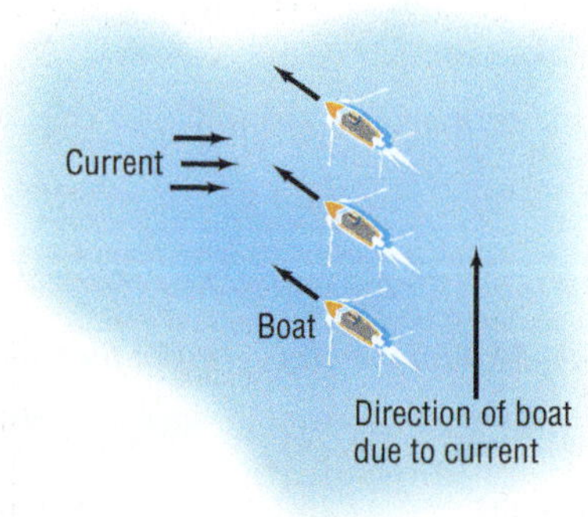

84. Finding the Correct Compass Heading The pilot of an aircraft wishes to head directly east but is faced with a wind speed of 40 mph from the northwest. If the pilot maintains an airspeed of 250 mph, what compass heading should be maintained to head directly east? What is the actual speed of the aircraft?

85. Charting a Course A helicopter pilot needs to travel to a regional airport 25 miles away. She flies at an actual heading of N16.26°E with an airspeed of 120 mph, and there is a wind blowing directly east at 20 mph.

(a) Determine the compass heading that the pilot needs to reach her destination.

(b) How long will it take her to reach her destination? Round to the nearest minute.

86. Crossing a River A captain needs to pilot a boat across a river that is 2 km wide. The current of the river is 2 km/h and the speed of the boat in still water is 10 km/h. The desired landing point on the other side is 1 km upstream.

(a) Determine the direction in which the captain should aim the boat.

(b) How long will the trip take?

87. Static Equilibrium A weight of 1000 pounds is suspended from two cables, as shown in the figure. What are the tensions in the two cables?

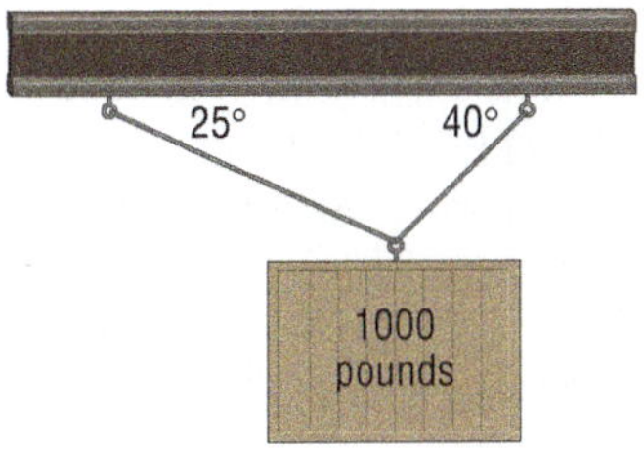

88. Static Equilibrium A weight of 800 pounds is suspended from two cables, as shown in the figure. What are the tensions in the two cables?

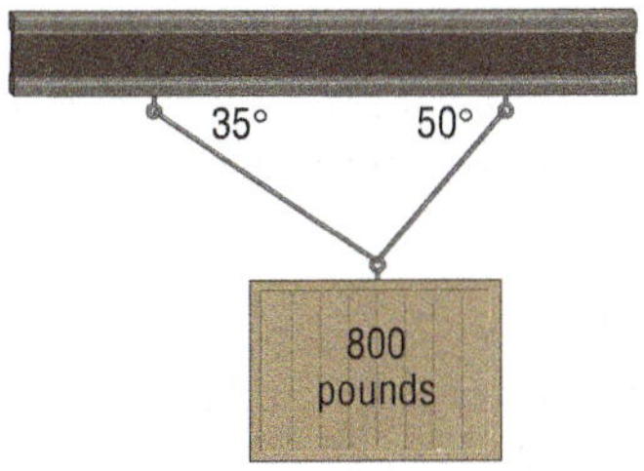

89. Static Equilibrium A tightrope walker located at a certain point deflects the rope as indicated in the figure. If the weight of the tightrope walker is 150 pounds, how much tension is in each part of the rope?

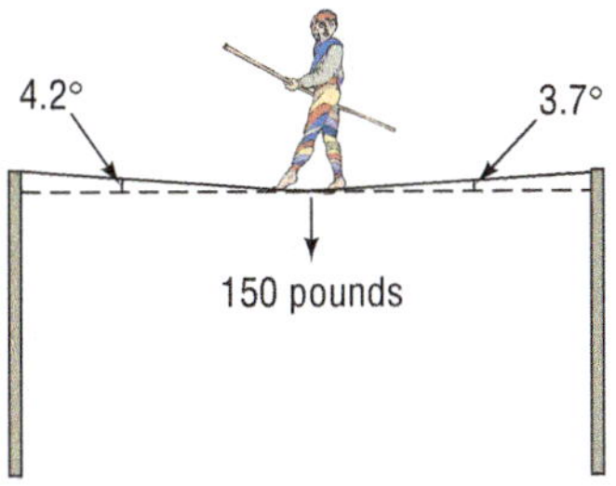

90. Static Equilibrium Repeat Problem 89 if the angle on the left is 3.8°, the angle on the right is 2.6°, and the weight of the tightrope walker is 135 pounds.

91. Static Friction A 20-pound box sits at rest on a horizontal surface, and there is friction between the box and the surface. One side of the surface is raised slowly to create a ramp. The friction force **f** opposes the direction of motion and is proportional to the normal force $\mathbf{F}_N$ exerted by the surface on the box. The proportionality constant is called the **coefficient of friction**, μ. When the angle of the ramp, θ, reaches 20°, the box begins to slide. Find the value of μ to two decimal places.

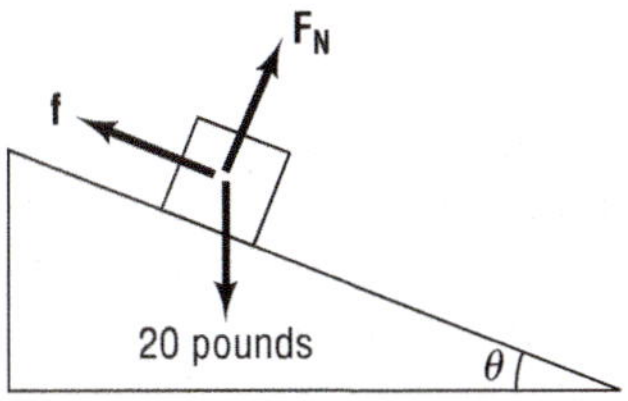

92. Inclined Ramp A 2-pound weight is attached to a 3-pound weight by a rope that passes over an ideal pulley. The smaller weight hangs vertically, while the larger weight sits on a frictionless inclined ramp with angle θ. The rope exerts a tension force **T** on both weights along the direction of the rope. Find the angle measure for θ that is needed to keep the larger weight from sliding down the ramp. Round your answer to the nearest tenth of a degree.

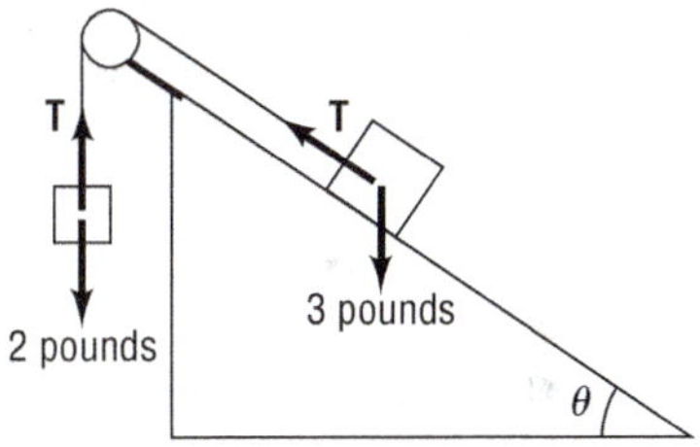

93. Inclined Ramp A box sitting on a horizontal surface is attached to a second box sitting on an inclined ramp by a rope that passes over an ideal pulley. The rope exerts a tension force **T** on both weights along the direction of the rope, and the coefficient of friction between the surface and boxes is 0.6 (see Problems 91 and 92). If the box on the right weighs 100 pounds and the angle of the ramp is 35°, how much must the box on the left weigh for the system to be in static equilibrium? Round your answer to two decimal places.

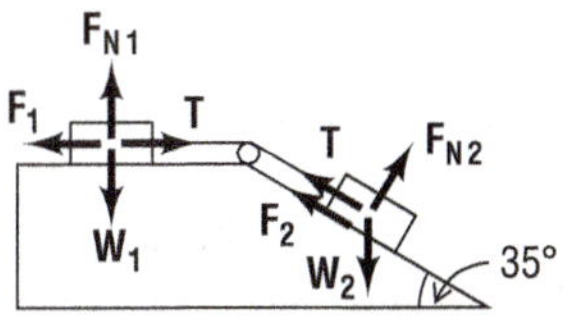

94. Muscle Force Two muscles exert force on a bone at the same point. The first muscle exerts a force of 800 N at a 10° angle with the bone. The second muscle exerts a force of 710 N at a 35° angle with the bone. What are the direction and magnitude of the resulting force on the bone?

95. **Truck Pull** At a county fair truck pull, two pickup trucks are attached to the back end of a monster truck as illustrated in the figure. One of the pickups pulls with a force of 2000 pounds, and the other pulls with a force of 3000 pounds. There is an angle of 45° between them. With how much force must the monster truck pull in order to remain unmoved? [**Hint:** Find the resultant force of the two trucks.]

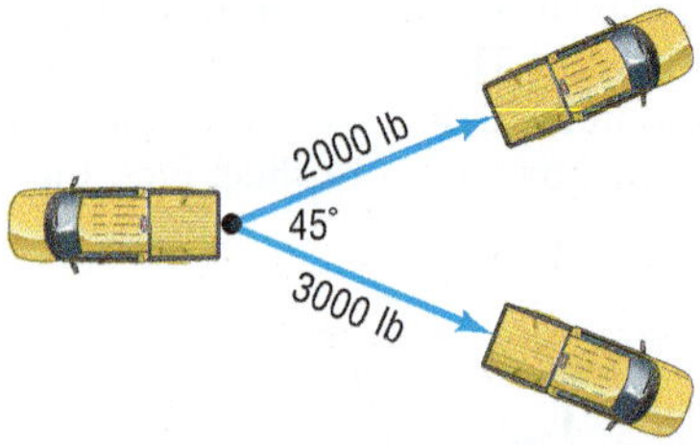

96. **Removing a Stump** A farmer wishes to remove a stump from a field by pulling it out with his tractor. Having removed many stumps before, he estimates that he will need 6 tons (12,000 pounds) of force to remove the stump. However, his tractor is only capable of pulling with a force of 7000 pounds, so he asks his neighbor to help. His neighbor's tractor can pull with a force of 5500 pounds. They attach the two tractors to the stump with a 40° angle between the forces, as shown in the figure.
   (a) Assuming the farmer's estimate of a needed 6-ton force is correct, will the farmer be successful in removing the stump? Explain.
   (b) Had the farmer arranged the tractors with a 25° angle between the forces, would he have been successful in removing the stump? Explain.

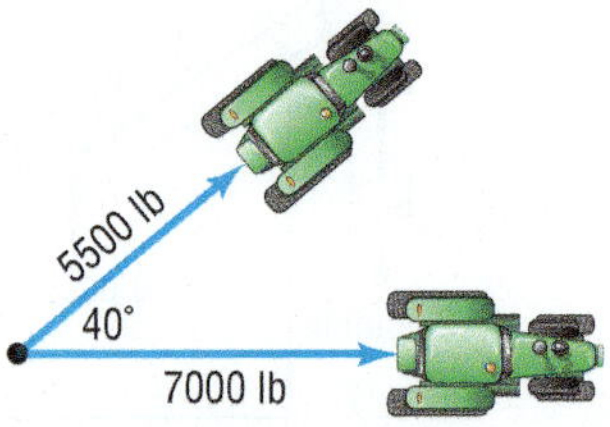

97. **Computer Graphics** The field of computer graphics utilizes vectors to compute translations of points. For example, if the point $(-3, 2)$ is to be translated by $\mathbf{v} = \langle 5, 2 \rangle$, then the new location will be $\mathbf{u}' = \mathbf{u} + \mathbf{v} = \langle -3, 2 \rangle + \langle 5, 2 \rangle = \langle 2, 4 \rangle$. As illustrated in the figure, the point $(-3, 2)$ is translated to $(2, 4)$ by $\mathbf{v}$.
   (a) Determine the new coordinates of $(3, -1)$ if it is translated by $\mathbf{v} = \langle -4, 5 \rangle$.
   (b) Illustrate this translation graphically.

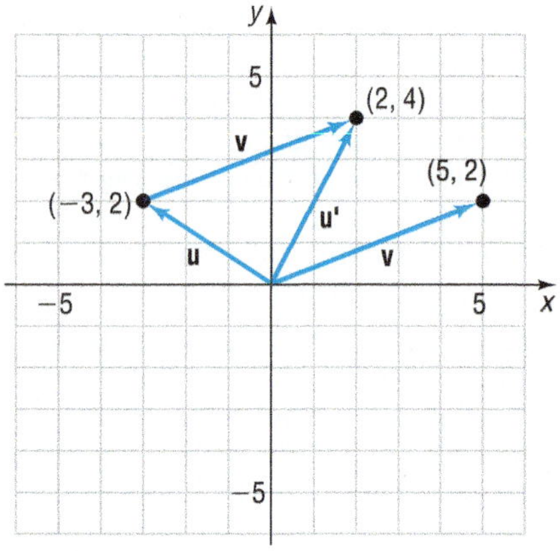

Source: *Phil Dadd. Vectors and Matrices: A Primer. www.gamedev.net/reference/articles/article1832.asp*

98. **Computer Graphics** Refer to Problem 97. The points $(-3, 0)$, $(-1, -2)$, $(3, 1)$, and $(1, 3)$ are the vertices of a parallelogram $ABCD$.
   (a) Find the new vertices of a parallelogram $A'B'C'D'$ if it is translated by $\mathbf{v} = \langle 3, -2 \rangle$.
   (b) Find the new vertices of a parallelogram $A'B'C'D'$ if it is translated by $-\frac{1}{2}\mathbf{v}$.

99. **Static Equilibrium** Show on the following graph the force needed for the object at P to be in static equilibrium.

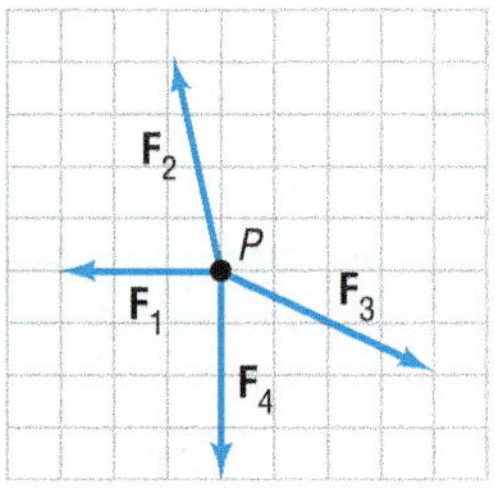

Explaining Concepts: Discussion and Writing

100. Explain in your own words what a vector is. Give an example of a vector.

101. Write a brief paragraph comparing the algebra of complex numbers and the algebra of vectors.

102. Explain the difference between an algebraic vector and a position vector.

Retain Your Knowledge

Problems 103–106 are based on material learned earlier in the course. The purpose of these problems is to keep the material fresh in your mind so that you are better prepared for the final exam.

103. Solve: $\sqrt[3]{x-2} = 3$.

104. Factor $-3x^3 + 12x^2 + 36x$ completely.

105. Find the exact value of $\tan\left[\cos^{-1}\left(\frac{1}{2}\right)\right]$.

106. Find the amplitude, period, and phase shift of $y = \frac{3}{2}\cos(6x + 3\pi)$. Graph the function, showing at least two periods.

10.5 The Dot Product

PREPARING FOR THIS SECTION *Before getting started, review the following:*

- Law of Cosines (Section 9.3, p. 701)

Now Work the **'Are You Prepared?'** problem on page 784.

OBJECTIVES
1. Find the Dot Product of Two Vectors (p. 779)
2. Find the Angle between Two Vectors (p. 780)
3. Determine Whether Two Vectors Are Parallel (p. 781)
4. Determine Whether Two Vectors Are Orthogonal (p. 781)
5. Decompose a Vector into Two Orthogonal Vectors (p. 781)
6. Compute Work (p. 783)

1 Find the Dot Product of Two Vectors

The definition for a product of two vectors is somewhat unexpected. However, such a product has meaning in many geometric and physical applications.

DEFINITION

If $\mathbf{v} = a_1\mathbf{i} + b_1\mathbf{j}$ and $\mathbf{w} = a_2\mathbf{i} + b_2\mathbf{j}$ are two vectors, the **dot product** $\mathbf{v} \cdot \mathbf{w}$ is defined as

$$\mathbf{v} \cdot \mathbf{w} = a_1 a_2 + b_1 b_2 \qquad (1)$$

EXAMPLE 1 **Finding Dot Products**

If $\mathbf{v} = 2\mathbf{i} - 3\mathbf{j}$ and $\mathbf{w} = 5\mathbf{i} + 3\mathbf{j}$, find:

(a) $\mathbf{v} \cdot \mathbf{w}$ (b) $\mathbf{w} \cdot \mathbf{v}$ (c) $\mathbf{v} \cdot \mathbf{v}$ (d) $\mathbf{w} \cdot \mathbf{w}$ (e) $\|\mathbf{v}\|$ (f) $\|\mathbf{w}\|$

Solution

(a) $\mathbf{v} \cdot \mathbf{w} = 2(5) + (-3)3 = 1$ (b) $\mathbf{w} \cdot \mathbf{v} = 5(2) + 3(-3) = 1$

(c) $\mathbf{v} \cdot \mathbf{v} = 2(2) + (-3)(-3) = 13$ (d) $\mathbf{w} \cdot \mathbf{w} = 5(5) + 3(3) = 34$

(e) $\|\mathbf{v}\| = \sqrt{2^2 + (-3)^2} = \sqrt{13}$ (f) $\|\mathbf{w}\| = \sqrt{5^2 + 3^2} = \sqrt{34}$

COMMENT A scalar multiple $\alpha\mathbf{v}$ is a vector. A dot product $\mathbf{u} \cdot \mathbf{v}$ is a scalar (real number).

Since the dot product $\mathbf{v} \cdot \mathbf{w}$ of two vectors $\mathbf{v}$ and $\mathbf{w}$ is a real number (a scalar), it is sometimes referred to as the **scalar product**.

The results obtained in Exmple 1 suggest some general properties of the dot product.

THEOREM

Properties of the Dot Product

If $\mathbf{u}$, $\mathbf{v}$, and $\mathbf{w}$ are vectors, then

Commutative Property

$$\mathbf{u} \cdot \mathbf{v} = \mathbf{v} \cdot \mathbf{u} \qquad (2)$$

Distributive Property

$$\mathbf{u} \cdot (\mathbf{v} + \mathbf{w}) = \mathbf{u} \cdot \mathbf{v} + \mathbf{u} \cdot \mathbf{w} \qquad (3)$$

$$\mathbf{v} \cdot \mathbf{v} = \|\mathbf{v}\|^2 \qquad (4)$$

$$\mathbf{0} \cdot \mathbf{v} = 0 \qquad (5)$$

Proof We prove properties (2) and (4) here and leave properties (3) and (5) as exercises (see Problems 38 and 39).

To prove property (2), let $\mathbf{u} = a_1\mathbf{i} + b_1\mathbf{j}$ and $\mathbf{v} = a_2\mathbf{i} + b_2\mathbf{j}$. Then

$$\mathbf{u}\cdot\mathbf{v} = a_1a_2 + b_1b_2 = a_2a_1 + b_2b_1 = \mathbf{v}\cdot\mathbf{u}$$

To prove property (4), let $\mathbf{v} = a\mathbf{i} + b\mathbf{j}$. Then

$$\mathbf{v}\cdot\mathbf{v} = a^2 + b^2 = \|\mathbf{v}\|^2$$

■

2 Find the Angle between Two Vectors

One use of the dot product is to calculate the angle between two vectors.

Let $\mathbf{u}$ and $\mathbf{v}$ be two vectors with the same initial point A. Then the vectors $\mathbf{u}$, $\mathbf{v}$, and $\mathbf{u}-\mathbf{v}$ form a triangle. The angle θ at vertex A of the triangle is the angle between the vectors $\mathbf{u}$ and $\mathbf{v}$. See Figure 68. We wish to find a formula for calculating the angle θ.

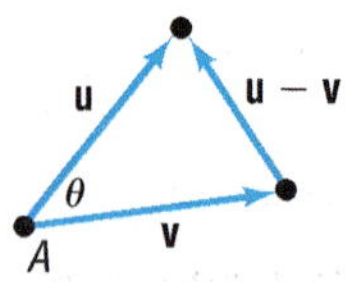

Figure 68

The sides of the triangle have lengths $\|\mathbf{v}\|$, $\|\mathbf{u}\|$, and $\|\mathbf{u}-\mathbf{v}\|$, and θ is the included angle between the sides of length $\|\mathbf{v}\|$ and $\|\mathbf{u}\|$. The Law of Cosines (Section 9.3) can be used to find the cosine of the included angle.

$$\|\mathbf{u}-\mathbf{v}\|^2 = \|\mathbf{u}\|^2 + \|\mathbf{v}\|^2 - 2\|\mathbf{u}\|\,\|\mathbf{v}\|\cos\theta$$

Now use property (4) to rewrite this equation in terms of dot products.

$$(\mathbf{u}-\mathbf{v})\cdot(\mathbf{u}-\mathbf{v}) = \mathbf{u}\cdot\mathbf{u} + \mathbf{v}\cdot\mathbf{v} - 2\|\mathbf{u}\|\,\|\mathbf{v}\|\cos\theta \tag{6}$$

Then apply the distributive property (3) twice on the left side of (6) to obtain

$$\begin{aligned}(\mathbf{u}-\mathbf{v})\cdot(\mathbf{u}-\mathbf{v}) &= \mathbf{u}\cdot(\mathbf{u}-\mathbf{v}) - \mathbf{v}\cdot(\mathbf{u}-\mathbf{v})\\ &= \mathbf{u}\cdot\mathbf{u} - \mathbf{u}\cdot\mathbf{v} - \mathbf{v}\cdot\mathbf{u} + \mathbf{v}\cdot\mathbf{v}\\ &= \mathbf{u}\cdot\mathbf{u} + \mathbf{v}\cdot\mathbf{v} - 2\,\mathbf{u}\cdot\mathbf{v}\end{aligned} \tag{7}$$

↑ Property (2)

Combining equations (6) and (7) gives

$$\begin{aligned}\mathbf{u}\cdot\mathbf{u} + \mathbf{v}\cdot\mathbf{v} - 2\,\mathbf{u}\cdot\mathbf{v} &= \mathbf{u}\cdot\mathbf{u} + \mathbf{v}\cdot\mathbf{v} - 2\|\mathbf{u}\|\,\|\mathbf{v}\|\cos\theta\\ \mathbf{u}\cdot\mathbf{v} &= \|\mathbf{u}\|\,\|\mathbf{v}\|\cos\theta\end{aligned}$$

THEOREM

Angle between Vectors

If $\mathbf{u}$ and $\mathbf{v}$ are two nonzero vectors, the angle θ, $0 \le \theta \le \pi$, between $\mathbf{u}$ and $\mathbf{v}$ is determined by the formula

$$\cos\theta = \frac{\mathbf{u}\cdot\mathbf{v}}{\|\mathbf{u}\|\,\|\mathbf{v}\|} \tag{8}$$

■

EXAMPLE 2 **Finding the Angle θ between Two Vectors**

Find the angle θ between $\mathbf{u} = 4\mathbf{i} - 3\mathbf{j}$ and $\mathbf{v} = 2\mathbf{i} + 5\mathbf{j}$.

Solution Find $\mathbf{u}\cdot\mathbf{v}$, $\|\mathbf{u}\|$, and $\|\mathbf{v}\|$.

$$\begin{aligned}\mathbf{u}\cdot\mathbf{v} &= 4(2) + (-3)(5) = -7\\ \|\mathbf{u}\| &= \sqrt{4^2 + (-3)^2} = 5\\ \|\mathbf{v}\| &= \sqrt{2^2 + 5^2} = \sqrt{29}\end{aligned}$$

By formula (8), if θ is the angle between $\mathbf{u}$ and $\mathbf{v}$, then

$$\cos\theta = \frac{\mathbf{u}\cdot\mathbf{v}}{\|\mathbf{u}\|\,\|\mathbf{v}\|} = \frac{-7}{5\sqrt{29}} \approx -0.26$$

Therefore, $\theta \approx \cos^{-1}(-0.26) \approx 105°$. See Figure 69. ■

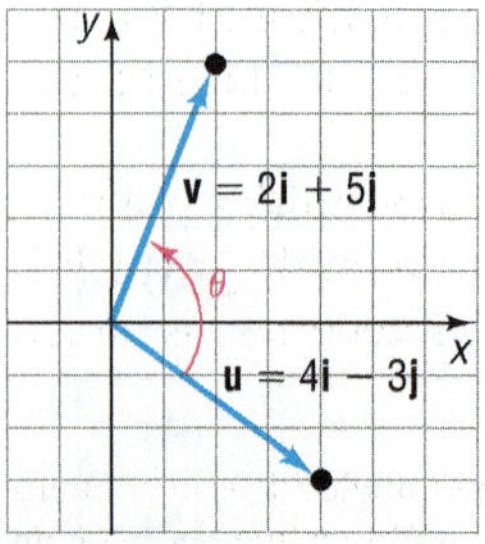

Figure 69

Now Work PROBLEMS 9(a) AND (b)

3 Determine Whether Two Vectors Are Parallel

Two vectors $\mathbf{v}$ and $\mathbf{w}$ are said to be **parallel** if there is a nonzero scalar α so that $\mathbf{v} = \alpha\mathbf{w}$. In this case, the angle θ between $\mathbf{v}$ and $\mathbf{w}$ is 0 or π.

EXAMPLE 3 Determining Whether Two Vectors Are Parallel

The vectors $\mathbf{v} = 3\mathbf{i} - \mathbf{j}$ and $\mathbf{w} = 6\mathbf{i} - 2\mathbf{j}$ are parallel, since $\mathbf{v} = \frac{1}{2}\mathbf{w}$. Furthermore, since

$$\cos\theta = \frac{\mathbf{v}\cdot\mathbf{w}}{\|\mathbf{v}\|\,\|\mathbf{w}\|} = \frac{18+2}{\sqrt{10}\,\sqrt{40}} = \frac{20}{\sqrt{400}} = 1$$

the angle θ between $\mathbf{v}$ and $\mathbf{w}$ is 0. ■

4 Determine Whether Two Vectors Are Orthogonal

If the angle θ between two nonzero vectors $\mathbf{v}$ and $\mathbf{w}$ is $\frac{\pi}{2}$, the vectors $\mathbf{v}$ and $\mathbf{w}$ are called **orthogonal**.* See Figure 70.

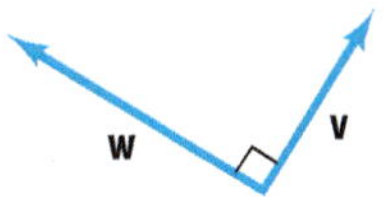

Figure 70 $\mathbf{v}$ is orthogonal to $\mathbf{w}$.

Since $\cos\frac{\pi}{2} = 0$, it follows from formula (8) that if the vectors $\mathbf{v}$ and $\mathbf{w}$ are orthogonal, then $\mathbf{v}\cdot\mathbf{w} = 0$.

On the other hand, if $\mathbf{v}\cdot\mathbf{w} = 0$, then $\mathbf{v} = \mathbf{0}$ or $\mathbf{w} = \mathbf{0}$ or $\cos\theta = 0$. If $\cos\theta = 0$, then $\theta = \frac{\pi}{2}$, and $\mathbf{v}$ and $\mathbf{w}$ are orthogonal. If $\mathbf{v}$ or $\mathbf{w}$ is the zero vector, then, since the zero vector has no specific direction, we adopt the convention that the zero vector is orthogonal to every vector.

THEOREM

Two vectors $\mathbf{v}$ and $\mathbf{w}$ are orthogonal if and only if

$$\mathbf{v}\cdot\mathbf{w} = 0$$

EXAMPLE 4 Determining Whether Two Vectors Are Orthogonal

The vectors

$$\mathbf{v} = 2\mathbf{i} - \mathbf{j} \quad \text{and} \quad \mathbf{w} = 3\mathbf{i} + 6\mathbf{j}$$

are orthogonal, since

$$\mathbf{v}\cdot\mathbf{w} = 6 - 6 = 0$$

See Figure 71. ■

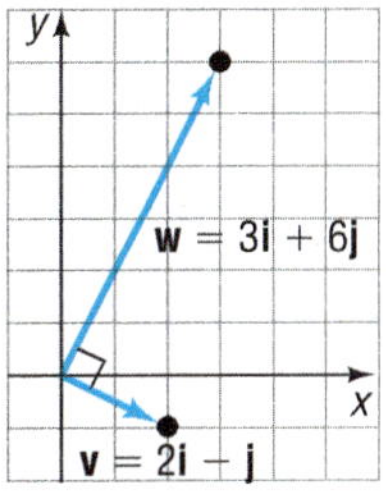

Figure 71

Now Work PROBLEM 9(c)

5 Decompose a Vector into Two Orthogonal Vectors

The last section discussed how to add two vectors to find the resultant vector. Now, the reverse problem is discussed, namely that of decomposing a vector into the sum of two components.

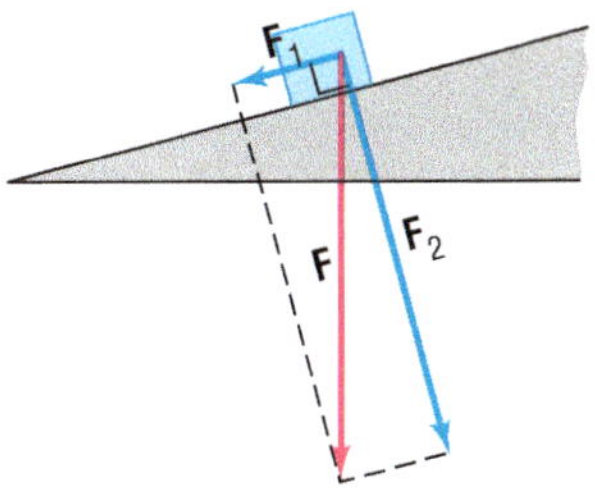

Figure 72

In many physical applications, it is necessary to find "how much" of a vector is applied in a given direction. Look at Figure 72. The force $\mathbf{F}$ due to gravity is pulling straight down (toward the center of Earth) on the block. To study the effect of gravity on the block, it is necessary to determine how much of $\mathbf{F}$ is actually pushing the block down the incline ($\mathbf{F}_1$) and how much is pressing the block against the incline ($\mathbf{F}_2$), at a right angle to the incline. Knowing the **decomposition** of $\mathbf{F}$ often enables us to determine when friction (the force holding the block in place on the incline) is overcome and the block will slide down the incline.

Suppose that $\mathbf{v}$ and $\mathbf{w}$ are two nonzero vectors with the same initial point P. We seek to decompose $\mathbf{v}$ into two vectors: $\mathbf{v}_1$, which is parallel to $\mathbf{w}$, and $\mathbf{v}_2$, which is orthogonal to $\mathbf{w}$. See Figure 73(a) and (b) on the next page. The vector $\mathbf{v}_1$ is called the **vector projection of v onto w**.

*_Orthogonal_, _perpendicular_, and _normal_ are all terms that mean "meet at a right angle." It is customary to refer to two vectors as being _orthogonal_, to two lines as being _perpendicular_, and to a line and a plane or a vector and a plane as being _normal_.

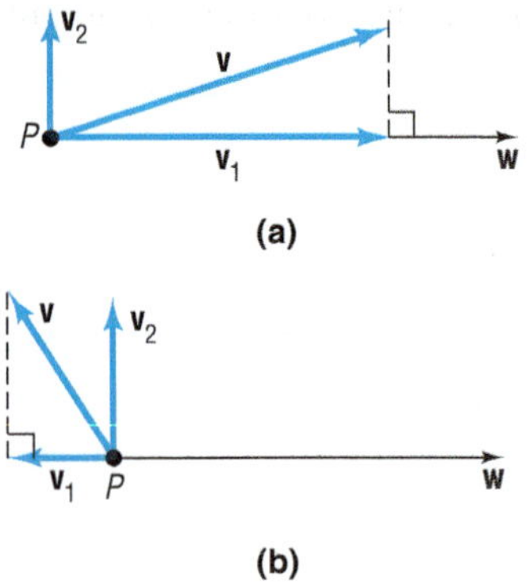

Figure 73

The vector $\mathbf{v}_1$ is obtained as follows: From the terminal point of $\mathbf{v}$, drop a perpendicular to the line containing $\mathbf{w}$. The vector $\mathbf{v}_1$ is the vector from P to the foot of this perpendicular. The vector $\mathbf{v}_2$ is given by $\mathbf{v}_2 = \mathbf{v} - \mathbf{v}_1$. Note that $\mathbf{v} = \mathbf{v}_1 + \mathbf{v}_2$, the vector $\mathbf{v}_1$ is parallel to $\mathbf{w}$, and the vector $\mathbf{v}_2$ is orthogonal to $\mathbf{w}$. This is the decomposition of $\mathbf{v}$ that was sought.

Now we seek a formula for $\mathbf{v}_1$ that is based on a knowledge of the vectors $\mathbf{v}$ and $\mathbf{w}$. Since $\mathbf{v} = \mathbf{v}_1 + \mathbf{v}_2$, we have

$$\mathbf{v} \cdot \mathbf{w} = (\mathbf{v}_1 + \mathbf{v}_2) \cdot \mathbf{w} = \mathbf{v}_1 \cdot \mathbf{w} + \mathbf{v}_2 \cdot \mathbf{w} \qquad (9)$$

Since $\mathbf{v}_2$ is orthogonal to $\mathbf{w}$, we have $\mathbf{v}_2 \cdot \mathbf{w} = 0$. Since $\mathbf{v}_1$ is parallel to $\mathbf{w}$, we have $\mathbf{v}_1 = \alpha\mathbf{w}$ for some scalar α. Equation (9) can be written as

$$\mathbf{v} \cdot \mathbf{w} = \alpha\mathbf{w} \cdot \mathbf{w} = \alpha\|\mathbf{w}\|^2 \qquad \mathbf{v}_1 = \alpha\mathbf{w}; \mathbf{v}_2 \cdot \mathbf{w} = 0$$

$$\alpha = \frac{\mathbf{v} \cdot \mathbf{w}}{\|\mathbf{w}\|^2}$$

Then

$$\mathbf{v}_1 = \alpha\mathbf{w} = \frac{\mathbf{v} \cdot \mathbf{w}}{\|\mathbf{w}\|^2}\mathbf{w}$$

THEOREM

If $\mathbf{v}$ and $\mathbf{w}$ are two nonzero vectors, the vector projection of $\mathbf{v}$ onto $\mathbf{w}$ is

$$\mathbf{v}_1 = \frac{\mathbf{v} \cdot \mathbf{w}}{\|\mathbf{w}\|^2}\mathbf{w} \qquad (10)$$

The decomposition of $\mathbf{v}$ into $\mathbf{v}_1$ and $\mathbf{v}_2$, where $\mathbf{v}_1$ is parallel to $\mathbf{w}$, and $\mathbf{v}_2$ is orthogonal to $\mathbf{w}$, is

$$\mathbf{v}_1 = \frac{\mathbf{v} \cdot \mathbf{w}}{\|\mathbf{w}\|^2}\mathbf{w} \qquad \mathbf{v}_2 = \mathbf{v} - \mathbf{v}_1 \qquad (11)$$

EXAMPLE 5 Decomposing a Vector into Two Orthogonal Vectors

Find the vector projection of $\mathbf{v} = \mathbf{i} + 3\mathbf{j}$ onto $\mathbf{w} = \mathbf{i} + \mathbf{j}$. Decompose $\mathbf{v}$ into two vectors, $\mathbf{v}_1$ and $\mathbf{v}_2$, where $\mathbf{v}_1$ is parallel to $\mathbf{w}$, and $\mathbf{v}_2$ is orthogonal to $\mathbf{w}$.

Solution Use formulas (10) and (11).

$$\mathbf{v}_1 = \frac{\mathbf{v} \cdot \mathbf{w}}{\|\mathbf{w}\|^2}\mathbf{w} = \frac{1 + 3}{(\sqrt{2})^2}\mathbf{w} = 2\mathbf{w} = 2(\mathbf{i} + \mathbf{j})$$

$$\mathbf{v}_2 = \mathbf{v} - \mathbf{v}_1 = (\mathbf{i} + 3\mathbf{j}) - 2(\mathbf{i} + \mathbf{j}) = -\mathbf{i} + \mathbf{j}$$

See Figure 74.

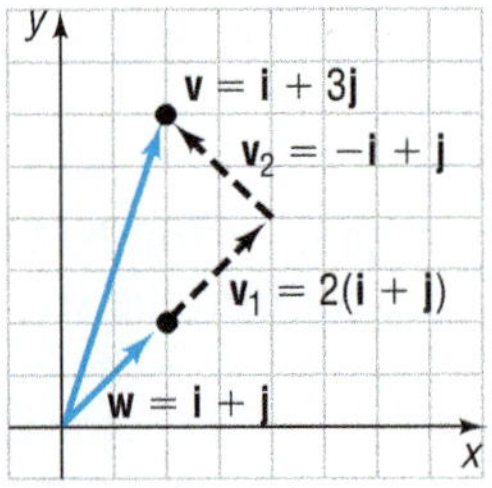

Figure 74

Now Work PROBLEM 21

EXAMPLE 6 Finding the Force Required to Hold a Wagon on a Hill

A wagon with two small children as occupants weighs 100 pounds and is on a hill with a grade of 20°. What is the magnitude of the force that is required to keep the wagon from rolling down the hill?

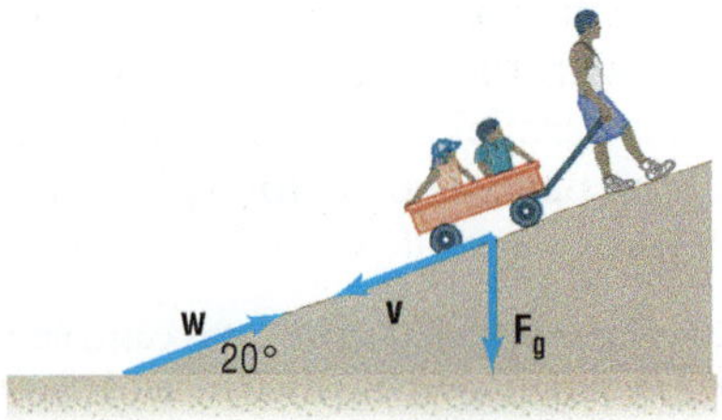

Figure 75

Solution See Figure 75. We wish to find the magnitude of the force $\mathbf{v}$ that is acting to cause the wagon to roll down the hill. A force with the same magnitude in the opposite direction of $\mathbf{v}$ will keep the wagon from rolling down the hill. The force of gravity is orthogonal to the level ground, so the force of the wagon due to gravity can be represented by the vector

$$\mathbf{F}_g = -100\mathbf{j}$$

Determine the vector projection of $\mathbf{F}_g$ onto $\mathbf{w}$, which is the force parallel to the hill. The vector $\mathbf{w}$ is given by

$$\mathbf{w} = \cos 20°\mathbf{i} + \sin 20°\mathbf{j}$$

The vector projection of $\mathbf{F}_g$ onto $\mathbf{w}$ is

$$\begin{aligned}\mathbf{v} &= \frac{\mathbf{F}_g \cdot \mathbf{w}}{\|\mathbf{w}\|^2}\mathbf{w}\\ &= \frac{-100(\sin 20°)}{\left(\sqrt{\cos^2 20° + \sin^2 20°}\right)^2}(\cos 20°\mathbf{i} + \sin 20°\mathbf{j})\\ &= -34.2(\cos 20°\mathbf{i} + \sin 20°\mathbf{j})\end{aligned}$$

The magnitude of $\mathbf{v}$ is 34.2 pounds, so the magnitude of the force required to keep the wagon from rolling down the hill is 34.2 pounds. ■

6 Compute Work

In elementary physics, the **work** W done by a constant force $\mathbf{F}$ in moving an object from a point A to a point B is defined as

$$W = (\text{magnitude of force})(\text{distance}) = \|\mathbf{F}\|\,\|\overrightarrow{AB}\|$$

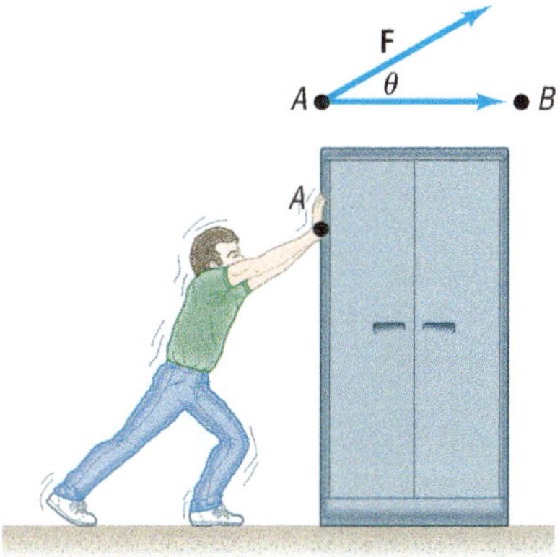

Figure 76

Work is commonly measured in foot-pounds or in newton-meters (joules).

In this definition, it is assumed that the force $\mathbf{F}$ is applied along the line of motion. If the constant force $\mathbf{F}$ is not along the line of motion, but instead is at an angle θ to the direction of the motion, as illustrated in Figure 76, then the **work** W **done by F** in moving an object from A to B is defined as

$$W = \mathbf{F} \cdot \overrightarrow{AB} \tag{12}$$

This definition is compatible with the force-times-distance definition, since

$$\begin{aligned}W &= (\text{amount of force in the direction of } \overrightarrow{AB})(\text{distance})\\ &= \|\text{projection of } \mathbf{F} \text{ on } AB\|\,\|\overrightarrow{AB}\| = \frac{\mathbf{F}\cdot\overrightarrow{AB}}{\|\overrightarrow{AB}\|^2}\|\overrightarrow{AB}\|\,\|\overrightarrow{AB}\| = \mathbf{F}\cdot\overrightarrow{AB}\end{aligned}$$

↑ Use formula (10)

EXAMPLE 7 **Computing Work**

A girl is pulling a wagon with a force of 50 pounds. How much work is done in moving the wagon 100 feet if the handle makes an angle of 30° with the ground? See Figure 77(a).

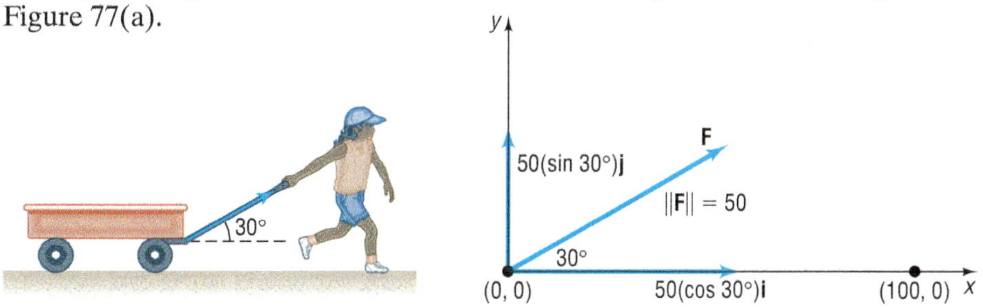

Figure 77 (a) (b)

Solution Position the vectors in a coordinate system in such a way that the wagon is moved from $(0, 0)$ to $(100, 0)$. The motion is from $A = (0, 0)$ to $B = (100, 0)$, so $\overrightarrow{AB} = 100\mathbf{i}$. The force vector $\mathbf{F}$, as shown in Figure 77(b), is

$$\mathbf{F} = 50(\cos 30°\mathbf{i} + \sin 30°\mathbf{j}) = 50\left(\frac{\sqrt{3}}{2}\mathbf{i} + \frac{1}{2}\mathbf{j}\right) = 25(\sqrt{3}\,\mathbf{i} + \mathbf{j})$$

By formula (12), the work done is

$$W = \mathbf{F}\cdot\overrightarrow{AB} = 25(\sqrt{3}\,\mathbf{i} + \mathbf{j}) \cdot 100\,\mathbf{i} = 2500\sqrt{3} \text{ foot-pounds}$$ ■

Now Work PROBLEM 29

Historical Feature

We stated in the Historical Feature in Section 10.4 that complex numbers were used as vectors in the plane before the general notion of a vector was clarified. Suppose that we make the correspondence

$$\text{Vector} \leftrightarrow \text{Complex number}$$
$$a\mathbf{i} + b\mathbf{j} \leftrightarrow a + bi$$
$$c\mathbf{i} + d\mathbf{j} \leftrightarrow c + di$$

Show that

$$(a\mathbf{i} + b\mathbf{j}) \cdot (c\mathbf{i} + d\mathbf{j}) = \text{real part } [(\overline{a + bi})(c + di)]$$

This is how the dot product was found originally. The imaginary part is also interesting. It is a determinant (see Section 12.3) and represents the area of the parallelogram whose edges are the vectors. This is close to some of Hermann Grassmann's ideas and is also connected with the scalar triple product of three-dimensional vectors.

10.5 Assess Your Understanding

'Are You Prepared?' *The answer is given at the end of these exercises. If you get the wrong answer, read the page listed in red.*

1. In a triangle with sides a, b, c and angles A, B, C, the Law of Cosines states that ______. (p. 701)

Concepts and Vocabulary

2. If $\mathbf{v} = a_1\mathbf{i} + b_1\mathbf{j}$ and $\mathbf{w} = a_2\mathbf{i} + b_2\mathbf{j}$ are two vectors, then the ____ ________ is defined as $\mathbf{v} \cdot \mathbf{w} = a_1a_2 + b_1b_2$.
3. If $\mathbf{v} \cdot \mathbf{w} = 0$, then the two vectors $\mathbf{v}$ and $\mathbf{w}$ are __________.
4. If $\mathbf{v} = 3\mathbf{w}$, then the two vectors $\mathbf{v}$ and $\mathbf{w}$ are ________.
5. ***True or False*** Given two nonzero vectors $\mathbf{v}$ and $\mathbf{w}$, it is always possible to decompose $\mathbf{v}$ into two vectors, one parallel to $\mathbf{w}$ and the other orthogonal to $\mathbf{w}$.
6. ***True or False*** Work is a physical example of a vector.
7. The angle θ, $0 \le \theta \le \pi$, between two nonzero vectors $\mathbf{u}$ and $\mathbf{v}$ can be found using which of the following formulas?
   (a) $\sin\theta = \dfrac{\|\mathbf{u}\|}{\|\mathbf{v}\|}$ (b) $\cos\theta = \dfrac{\|\mathbf{u}\|}{\|\mathbf{v}\|}$
   (b) $\sin\theta = \dfrac{\mathbf{u} \cdot \mathbf{v}}{\|\mathbf{u}\|\,\|\mathbf{v}\|}$ (d) $\cos\theta = \dfrac{\mathbf{u} \cdot \mathbf{v}}{\|\mathbf{u}\|\,\|\mathbf{v}\|}$
8. If two nonzero vectors $\mathbf{v}$ and $\mathbf{w}$ are orthogonal, then the angle between them has which of the following measures?
   (a) π (b) $\dfrac{\pi}{2}$ (c) $\dfrac{3\pi}{2}$ (d) 2π

Skill Building

In Problems 9–18, (a) find the dot product $\mathbf{v} \cdot \mathbf{w}$; *(b) find the angle between* $\mathbf{v}$ *and* $\mathbf{w}$; *(c) state whether the vectors are parallel, orthogonal, or neither.*

9. $\mathbf{v} = \mathbf{i} - \mathbf{j}, \quad \mathbf{w} = \mathbf{i} + \mathbf{j}$
10. $\mathbf{v} = \mathbf{i} + \mathbf{j}, \quad \mathbf{w} = -\mathbf{i} + \mathbf{j}$
11. $\mathbf{v} = 2\mathbf{i} + \mathbf{j}, \quad \mathbf{w} = \mathbf{i} - 2\mathbf{j}$
12. $\mathbf{v} = 2\mathbf{i} + 2\mathbf{j}, \quad \mathbf{w} = \mathbf{i} + 2\mathbf{j}$
13. $\mathbf{v} = \sqrt{3}\mathbf{i} - \mathbf{j}, \quad \mathbf{w} = \mathbf{i} + \mathbf{j}$
14. $\mathbf{v} = \mathbf{i} + \sqrt{3}\mathbf{j}, \quad \mathbf{w} = \mathbf{i} - \mathbf{j}$
15. $\mathbf{v} = 3\mathbf{i} + 4\mathbf{j}, \quad \mathbf{w} = -6\mathbf{i} - 8\mathbf{j}$
16. $\mathbf{v} = 3\mathbf{i} - 4\mathbf{j}, \quad \mathbf{w} = 9\mathbf{i} - 12\mathbf{j}$
17. $\mathbf{v} = 4\mathbf{i}, \quad \mathbf{w} = \mathbf{j}$
18. $\mathbf{v} = \mathbf{i}, \quad \mathbf{w} = -3\mathbf{j}$
19. Find a so that the vectors $\mathbf{v} = \mathbf{i} - a\mathbf{j}$ and $\mathbf{w} = 2\mathbf{i} + 3\mathbf{j}$ are orthogonal.
20. Find b so that the vectors $\mathbf{v} = \mathbf{i} + \mathbf{j}$ and $\mathbf{w} = \mathbf{i} + b\mathbf{j}$ are orthogonal.

In Problems 21–26, decompose $\mathbf{v}$ *into two vectors* $\mathbf{v}_1$ *and* $\mathbf{v}_2$, *where* $\mathbf{v}_1$ *is parallel to* $\mathbf{w}$, *and* $\mathbf{v}_2$ *is orthogonal to* $\mathbf{w}$.

21. $\mathbf{v} = 2\mathbf{i} - 3\mathbf{j}, \quad \mathbf{w} = \mathbf{i} - \mathbf{j}$
22. $\mathbf{v} = -3\mathbf{i} + 2\mathbf{j}, \quad \mathbf{w} = 2\mathbf{i} + \mathbf{j}$
23. $\mathbf{v} = \mathbf{i} - \mathbf{j}, \quad \mathbf{w} = -\mathbf{i} - 2\mathbf{j}$
24. $\mathbf{v} = 2\mathbf{i} - \mathbf{j}, \quad \mathbf{w} = \mathbf{i} - 2\mathbf{j}$
25. $\mathbf{v} = 3\mathbf{i} + \mathbf{j}, \quad \mathbf{w} = -2\mathbf{i} - \mathbf{j}$
26. $\mathbf{v} = \mathbf{i} - 3\mathbf{j}, \quad \mathbf{w} = 4\mathbf{i} - \mathbf{j}$

Applications and Extensions

27. Given vectors $\mathbf{u} = \mathbf{i} + 5\mathbf{j}$ and $\mathbf{v} = 4\mathbf{i} + y\mathbf{j}$, find y so that the angle between the vectors is 60°.†
28. Given vectors $\mathbf{u} = x\mathbf{i} + 2\mathbf{j}$ and $\mathbf{v} = 7\mathbf{i} - 3\mathbf{j}$, find x so that the angle between the vectors is 30°.
29. **Computing Work** Find the work done by a force of 3 pounds acting in the direction 60° to the horizontal in moving an object 6 feet from $(0, 0)$ to $(6, 0)$.
30. **Computing Work** A wagon is pulled horizontally by exerting a force of 20 pounds on the handle at an angle of 30° with the horizontal. How much work is done in moving the wagon 100 feet?
31. **Solar Energy** The amount of energy collected by a solar panel depends on the intensity of the sun's rays and the area of the panel. Let the vector **I** represent the intensity, in watts

†Courtesy of the Joliet Junior College Mathematics Department

per square centimeter, having the direction of the sun's rays. Let the vector $\mathbf{A}$ represent the area, in square centimeters, whose direction is the orientation of a solar panel. See the figure. The total number of watts collected by the panel is given by $W = |\mathbf{I} \cdot \mathbf{A}|$.

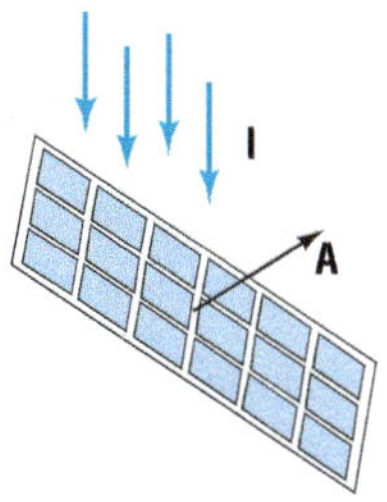

Suppose that $\mathbf{I} = \langle -0.02, -0.01 \rangle$ and $\mathbf{A} = \langle 300, 400 \rangle$.

(a) Find $\|\mathbf{I}\|$ and $\|\mathbf{A}\|$, and interpret the meaning of each.

(b) Compute W and interpret its meaning.

(c) If the solar panel is to collect the maximum number of watts, what must be true about $\mathbf{I}$ and $\mathbf{A}$?

32. Rainfall Measurement Let the vector $\mathbf{R}$ represent the amount of rainfall, in inches, whose direction is the inclination of the rain to a rain gauge. Let the vector $\mathbf{A}$ represent the area, in square inches, whose direction is the orientation of the opening of the rain gauge. See the figure. The volume of rain collected in the gauge, in cubic inches, is given by $V = |\mathbf{R} \cdot \mathbf{A}|$, even when the rain falls in a slanted direction or the gauge is not perfectly vertical.

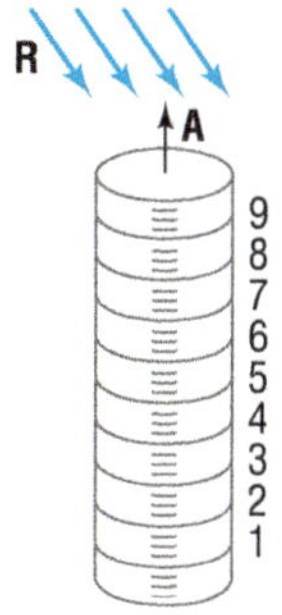

Suppose that $\mathbf{R} = \langle 0.75, -1.75 \rangle$ and $\mathbf{A} = \langle 0.3, 1 \rangle$.

(a) Find $\|\mathbf{R}\|$ and $\|\mathbf{A}\|$, and interpret the meaning of each.

(b) Compute V and interpret its meaning.

(c) If the gauge is to collect the maximum volume of rain, what must be true about $\mathbf{R}$ and $\mathbf{A}$?

33. Braking Load A Toyota Sienna with a gross weight of 5300 pounds is parked on a street with an 8° grade. See the figure. Find the magnitude of the force required to keep the Sienna from rolling down the hill. What is the magnitude of the force perpendicular to the hill?

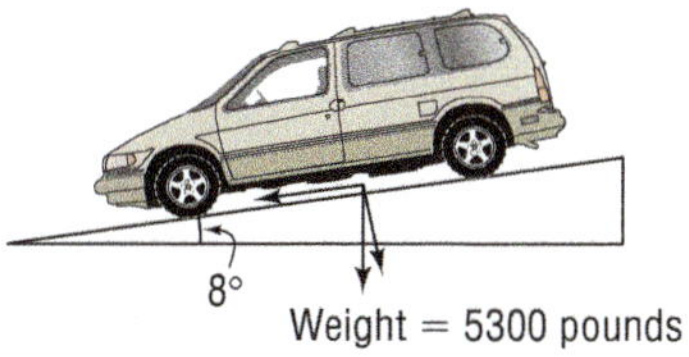

34. Braking Load A Chevrolet Silverado with a gross weight of 4500 pounds is parked on a street with a 10° grade. Find the magnitude of the force required to keep the Silverado from rolling down the hill. What is the magnitude of the force perpendicular to the hill?

35. Ramp Angle Billy and Timmy are using a ramp to load furniture into a truck. While rolling a 250-pound piano up the ramp, they discover that the truck is too full of other furniture for the piano to fit. Timmy holds the piano in place on the ramp while Billy repositions other items to make room for it in the truck. If the angle of inclination of the ramp is 20°, how many pounds of force must Timmy exert to hold the piano in position?

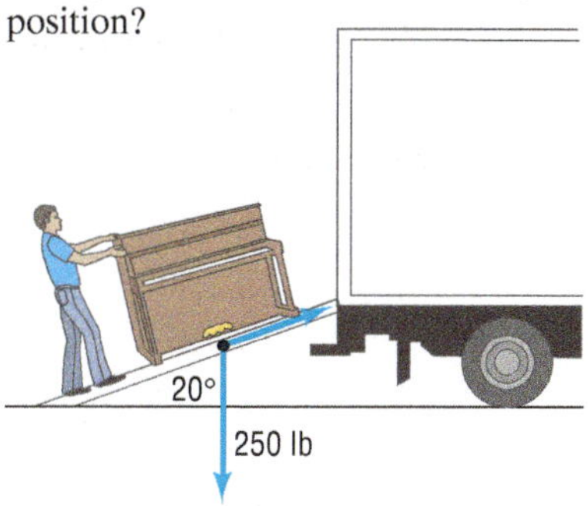

36. Incline Angle A bulldozer exerts 1000 pounds of force to prevent a 5000-pound boulder from rolling down a hill. Determine the angle of inclination of the hill.

37. Find the acute angle that a constant unit force vector makes with the positive x-axis if the work done by the force in moving a particle from $(0, 0)$ to $(4, 0)$ equals 2.

38. Prove the distributive property:

$$\mathbf{u} \cdot (\mathbf{v} + \mathbf{w}) = \mathbf{u} \cdot \mathbf{v} + \mathbf{u} \cdot \mathbf{w}$$

39. Prove property (5): $\mathbf{0} \cdot \mathbf{v} = 0$.

40. If $\mathbf{v}$ is a unit vector and the angle between $\mathbf{v}$ and $\mathbf{i}$ is α, show that $\mathbf{v} = \cos \alpha \mathbf{i} + \sin \alpha \mathbf{j}$.

41. Suppose that $\mathbf{v}$ and $\mathbf{w}$ are unit vectors. If the angle between $\mathbf{v}$ and $\mathbf{i}$ is α and the angle between $\mathbf{w}$ and $\mathbf{i}$ is β, use the idea of the dot product $\mathbf{v} \cdot \mathbf{w}$ to prove that

$$\cos(\alpha - \beta) = \cos \alpha \cos \beta + \sin \alpha \sin \beta$$

42. Show that the projection of $\mathbf{v}$ onto $\mathbf{i}$ is $(\mathbf{v} \cdot \mathbf{i})\mathbf{i}$. Then show that we can always write a vector $\mathbf{v}$ as

$$\mathbf{v} = (\mathbf{v} \cdot \mathbf{i})\mathbf{i} + (\mathbf{v} \cdot \mathbf{j})\mathbf{j}$$

43. (a) If $\mathbf{u}$ and $\mathbf{v}$ have the same magnitude, show that $\mathbf{u} + \mathbf{v}$ and $\mathbf{u} - \mathbf{v}$ are orthogonal.

(b) Use this to prove that an angle inscribed in a semicircle is a right angle (see the figure).

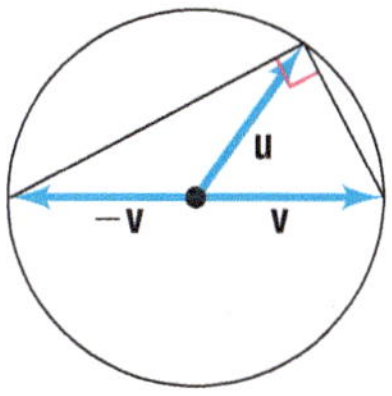

44. Let $\mathbf{v}$ and $\mathbf{w}$ denote two nonzero vectors. Show that the vector $\mathbf{v} - \alpha\mathbf{w}$ is orthogonal to $\mathbf{w}$ if $\alpha = \dfrac{\mathbf{v} \cdot \mathbf{w}}{\|\mathbf{w}\|^2}$.

45. Let $\mathbf{v}$ and $\mathbf{w}$ denote two nonzero vectors. Show that the vectors $\|\mathbf{w}\|\mathbf{v} + \|\mathbf{v}\|\mathbf{w}$ and $\|\mathbf{w}\|\mathbf{v} - \|\mathbf{v}\|\mathbf{w}$ are orthogonal.

46. In the definition of work given in this section, what is the work done if $\mathbf{F}$ is orthogonal to $\overrightarrow{AB}$?

47. Prove the **polarization identity**,

$$\|\mathbf{u} + \mathbf{v}\|^2 - \|\mathbf{u} - \mathbf{v}\|^2 = 4(\mathbf{u} \cdot \mathbf{v})$$

Explaining Concepts: Discussion and Writing

48. Create an application (different from any found in the text) that requires a dot product.

Retain Your Knowledge

Problems 49–52 are based on material learned earlier in the course. The purpose of these problems is to keep the material fresh in your mind so that you are better prepared for the final exam.

49. Find the average rate of change of $f(x) = x^3 - 5x^2 + 27$ from -3 to 2.

50. Find the exact value of $5\cos 60° + 2\tan\frac{\pi}{4}$. Do not use a calculator.

51. Establish the identity: $(1 - \sin^2\theta)(1 + \tan^2\theta) = 1$

52. **Volume of a Box** An open-top box is made from a sheet of metal by cutting squares from each corner and folding up the sides. The sheet has a length of 19 inches and a width of 13 inches. If x is the length of one side of the square to be cut out, write a function, $V(x)$, for the volume of the box in terms of x.

'Are You Prepared?' Answer

1. $c^2 = a^2 + b^2 - 2ab\cos C$

Chapter Review

Things to Know

Polar Coordinates

Relationship between polar coordinates (r, θ) and rectangular coordinates (x, y) (pp. 733 and 736)	$x = r\cos\theta,\ y = r\sin\theta$ $r^2 = x^2 + y^2,\ \tan\theta = \frac{y}{x},\quad x \neq 0$
Polar form of a complex number (p. 757)	If $z = x + yi$, then $z = r(\cos\theta + i\sin\theta)$, where $r = \lvert z\rvert = \sqrt{x^2 + y^2},\quad \sin\theta = \frac{y}{r},\quad \cos\theta = \frac{x}{r},\quad 0 \leq \theta < 2\pi$.
De Moivre's Theorem (p. 760)	If $z = r(\cos\theta + i\sin\theta)$, then $z^n = r^n[\cos(n\theta) + i\sin(n\theta)]$, where $n \geq 1$ is a positive integer.
nth root of a complex number $w = r(\cos\theta_0 + i\sin\theta_0)$ (p. 761)	$z_k = \sqrt[n]{r}\left[\cos\left(\frac{\theta_0}{n} + \frac{2k\pi}{n}\right) + i\sin\left(\frac{\theta_0}{n} + \frac{2k\pi}{n}\right)\right],\quad k = 0, \ldots, n-1,$ where $n \geq 2$ is an integer

Vectors

Vectors	Quantity having magnitude and direction; equivalent to a directed line segment $\overrightarrow{PQ}$
Position vector (p. 768)	Vector whose initial point is at the origin
Unit vector (pp. 767 and 770)	Vector whose magnitude is 1
Direction angle of a vector $\mathbf{v}$ (p. 771)	The angle α, $0° \leq \alpha < 360°$ between $\mathbf{i}$ and $\mathbf{v}$
Dot product (p. 779)	If $\mathbf{v} = a_1\mathbf{i} + b_1\mathbf{j}$ and $\mathbf{w} = a_2\mathbf{i} + b_2\mathbf{j}$, then $\mathbf{v}\cdot\mathbf{w} = a_1a_2 + b_1b_2$.
Angle θ between two nonzero vectors $\mathbf{u}$ and $\mathbf{v}$ (p. 780)	$\cos\theta = \frac{\mathbf{u}\cdot\mathbf{v}}{\lVert\mathbf{u}\rVert\,\lVert\mathbf{v}\rVert}$

Objectives

Section		You should be able to ...	Examples	Review Exercises
10.1	1	Plot points using polar coordinates (p. 731)	1–3	1–3
	2	Convert from polar coordinates to rectangular coordinates (p. 733)	4	1–3
	3	Convert from rectangular coordinates to polar coordinates (p. 735)	5–7	4–6
	4	Transform equations between polar and rectangular forms (p. 737)	8, 9	7(a)–10(a)

Section		You should be able to ...	Examples	Review Exercises
10.2	1	Identify and graph polar equations by converting to rectangular equations (p. 741)	1–3, 5–7	7(b)–10(b)
	2	Graph polar equations using a graphing utility (p. 742)	4–7	
	3	Test polar equations for symmetry (p. 745)	8–11	11–13
	4	Graph polar equations by plotting points (p. 746)	8–13	11–13
10.3	1	Plot points in the complex plane (p. 756)	1	16–18
	2	Convert a complex number between rectangular form and polar form (p. 757)	2, 3	14–18
	3	Find products and quotients of complex numbers in polar form (p. 758)	4	19–21
	4	Use De Moivre's Theorem (p. 759)	5, 6	22–25
	5	Find complex roots (p. 760)	7	26
10.4	1	Graph vectors (p. 767)	1	27, 28
	2	Find a position vector (p. 767)	2	29, 30
	3	Add and subtract vectors algebraically (p. 769)	3	31
	4	Find a scalar multiple and the magnitude of a vector (p. 770)	4	29, 30, 32–34
	5	Find a unit vector (p. 770)	5	35
	6	Find a vector from its direction and magnitude (p. 771)	6	36, 37
	7	Model with vectors (p. 772)	8–10	45, 46
10.5	1	Find the dot product of two vectors (p. 779)	1	38, 39
	2	Find the angle between two vectors (p. 780)	2	38, 39
	3	Determine whether two vectors are parallel (p. 781)	3	40–42
	4	Determine whether two vectors are orthogonal (p. 781)	4	40–42
	5	Decompose a vector into two orthogonal vectors (p. 781)	5, 6	43, 44, 48
	6	Compute work (p. 783)	7	47

Review Exercises

In Problems 1–3, plot each point given in polar coordinates, and find its rectangular coordinates.

1. $\left(3, \frac{\pi}{6}\right)$ **2.** $\left(-2, \frac{4\pi}{3}\right)$ **3.** $\left(-3, -\frac{\pi}{2}\right)$

In Problems 4–6, the rectangular coordinates of a point are given. Find two pairs of polar coordinates (r, θ) for each point, one with $r > 0$ and the other with $r < 0$. Express θ in radians.

4. $(-3, 3)$ **5.** $(0, -2)$ **6.** $(3, 4)$

In Problems 7–10, the variables r and θ represent polar coordinates. (a) Write each polar equation as an equation in rectangular coordinates (x, y). (b) Identify the equation and graph it.

7. $r = 2 \sin \theta$ **8.** $r = 5$ **9.** $\theta = \frac{\pi}{4}$ **10.** $r^2 + 4r \sin \theta - 8r \cos \theta = 5$

In Problems 11–13, sketch the graph of each polar equation. Be sure to test for symmetry.

11. $r = 4 \cos \theta$ **12.** $r = 3 - 3 \sin \theta$ **13.** $r = 4 - \cos \theta$

In Problems 14 and 15, write each complex number in polar form. Express each argument in degrees.

14. $-1 - i$ **15.** $4 - 3i$

In Problems 16–18, write each complex number in the standard form $a + bi$ and plot each in the complex plane.

16. $2(\cos 150° + i \sin 150°)$ **17.** $3\left(\cos \frac{2\pi}{3} + i \sin \frac{2\pi}{3}\right)$ **18.** $0.1(\cos 350° + i \sin 350°)$

In Problems 19–21, find zw and $\frac{z}{w}$. Leave your answers in polar form.

19. $z = \cos 80° + i \sin 80°$
$w = \cos 50° + i \sin 50°$

20. $z = 3\left(\cos \frac{9\pi}{5} + i \sin \frac{9\pi}{5}\right)$
$w = 2\left(\cos \frac{\pi}{5} + i \sin \frac{\pi}{5}\right)$

21. $z = 5(\cos 10° + i \sin 10°)$
$w = \cos 355° + i \sin 355°$

In Problems 22–26, write each expression in the standard form $a + bi$.

22. $[3(\cos 20° + i \sin 20°)]^3$

23. $\left[\sqrt{2}\left(\cos \frac{5\pi}{8} + i \sin \frac{5\pi}{8}\right)\right]^4$

24. $(1 - \sqrt{3}\,i)^6$

25. $(3 + 4i)^4$

26. Find all the complex cube roots of 27.

In Problems 27 and 28, use the figure to graph each of the following:

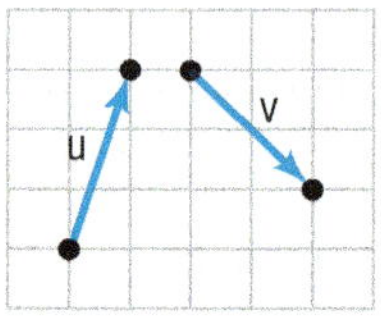

27. $\mathbf{u} + \mathbf{v}$

28. $2\mathbf{u} + 3\mathbf{v}$

In Problems 29 and 30, the vector $\mathbf{v}$ is represented by the directed line segment $\overrightarrow{PQ}$. Write $\mathbf{v}$ in the form $a\mathbf{i} + b\mathbf{j}$ and find $\|\mathbf{v}\|$.

29. $P = (1, -2)$; $Q = (3, -6)$

30. $P = (0, -2)$; $Q = (-1, 1)$

In Problems 31–35, use the vectors $\mathbf{v} = -2\mathbf{i} + \mathbf{j}$ and $\mathbf{w} = 4\mathbf{i} - 3\mathbf{j}$ to find:

31. $\mathbf{v} + \mathbf{w}$

32. $4\mathbf{v} - 3\mathbf{w}$

33. $\|\mathbf{v}\|$

34. $\|\mathbf{v}\| + \|\mathbf{w}\|$

35. Find a unit vector in the same direction as $\mathbf{v}$.

36. Find the vector $\mathbf{v}$ in the xy-plane with magnitude 3 if the direction angle of $\mathbf{v}$ is 60°.

37. Find the direction angle α of $\mathbf{v} = -\mathbf{i} + \sqrt{3}\,\mathbf{j}$.

In Problems 38 and 39, find the dot product $\mathbf{v} \cdot \mathbf{w}$ and the angle between $\mathbf{v}$ and $\mathbf{w}$.

38. $\mathbf{v} = -2\mathbf{i} + \mathbf{j}$, $\mathbf{w} = 4\mathbf{i} - 3\mathbf{j}$

39. $\mathbf{v} = \mathbf{i} - 3\mathbf{j}$, $\mathbf{w} = -\mathbf{i} + \mathbf{j}$

In Problems 40–42, determine whether $\mathbf{v}$ and $\mathbf{w}$ are parallel, orthogonal, or neither.

40. $\mathbf{v} = 2\mathbf{i} + 3\mathbf{j}$; $\mathbf{w} = -4\mathbf{i} - 6\mathbf{j}$

41. $\mathbf{v} = -2\mathbf{i} + 2\mathbf{j}$; $\mathbf{w} = -3\mathbf{i} + 2\mathbf{j}$

42. $\mathbf{v} = 3\mathbf{i} - 2\mathbf{j}$; $\mathbf{w} = 4\mathbf{i} + 6\mathbf{j}$

In Problems 43 and 44, decompose $\mathbf{v}$ into two vectors, one parallel to $\mathbf{w}$ and the other orthogonal to $\mathbf{w}$.

43. $\mathbf{v} = 2\mathbf{i} + \mathbf{j}$; $\mathbf{w} = -4\mathbf{i} + 3\mathbf{j}$

44. $\mathbf{v} = 2\mathbf{i} + 3\mathbf{j}$; $\mathbf{w} = 3\mathbf{i} + \mathbf{j}$

45. Actual Speed and Direction of a Swimmer A swimmer can maintain a constant speed of 5 miles per hour. If the swimmer heads directly across a river that has a current moving at the rate of 2 miles per hour, what is the actual speed of the swimmer? (See the figure.) If the river is 1 mile wide, how far downstream will the swimmer end up from the point directly across the river from the starting point?

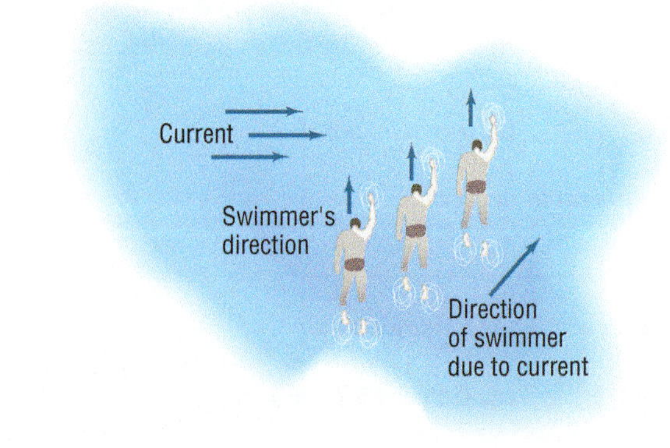

46. Static Equilibrium A weight of 2000 pounds is suspended from two cables, as shown in the figure. What are the tensions in each cable?

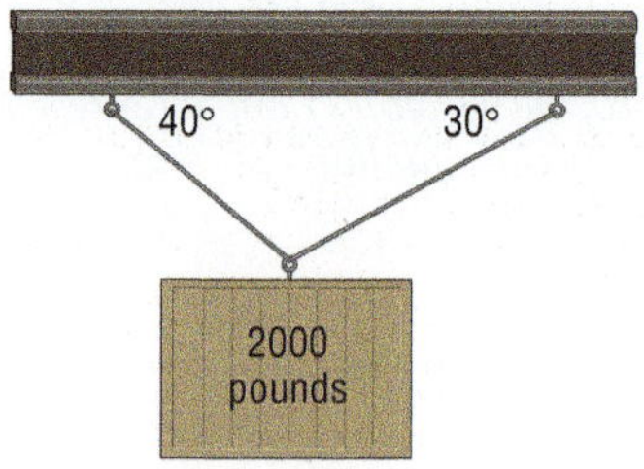

47. Computing Work Find the work done by a force of 5 pounds acting in the direction 60° to the horizontal in moving an object 20 feet from $(0, 0)$ to $(20, 0)$.

48. Braking Load A moving van with a gross weight of 8000 pounds is parked on a street with a 5° grade. Find the magnitude of the force required to keep the van from rolling down the hill. What is the magnitude of the force perpendicular to the hill?

Chapter Test

CHAPTER **Test Prep** VIDEOS

The Chapter Test Prep Videos are step-by-step solutions available in MyMathLab®, or on this text's YouTube™ Channel. Flip back to the Resources for Success page for a link to this text's YouTube channel.

In Problems 1–3, plot each point given in polar coordinates.

1. $\left(2, \frac{3\pi}{4}\right)$

2. $\left(3, -\frac{\pi}{6}\right)$

3. $\left(-4, \frac{\pi}{3}\right)$

4. Convert $(2, 2\sqrt{3})$ from rectangular coordinates to polar coordinates (r, θ), where $r > 0$ and $0 \le \theta < 2\pi$.

In Problems 5–7, convert the polar equation to a rectangular equation. Graph the equation.

5. $r = 7$

6. $\tan\theta = 3$

7. $r\sin^2\theta + 8\sin\theta = r$

In Problems 8 and 9, test the polar equation for symmetry with respect to the pole, the polar axis, and the line $\theta = \frac{\pi}{2}$.

8. $r^2\cos\theta = 5$

9. $r = 5\sin\theta\cos^2\theta$

In Problems 10–12, perform the given operation, where

$$z = 2(\cos 85° + i\sin 85°) \text{ and } w = 3(\cos 22° + i\sin 22°).$$

Write your answer in polar form.

10. $z \cdot w$

11. $\frac{w}{z}$

12. w^5

13. Find all the complex cube roots of $-8 + 8\sqrt{3}i$. Then plot them in the complex plane.

In Problems 14–18, $P_1 = (3\sqrt{2}, 7\sqrt{2})$ and $P_2 = (8\sqrt{2}, 2\sqrt{2})$.

14. Find the position vector $\mathbf{v}$ equal to $\overrightarrow{P_1P_2}$.

15. Find $\|\mathbf{v}\|$.

16. Find the unit vector in the direction of $\mathbf{v}$.

17. Find the direction angle of $\mathbf{v}$.

18. Decompose $\mathbf{v}$ into its vertical and horizontal components.

In Problems 19–22, $\mathbf{v}_1 = \langle 4, 6\rangle$, $\mathbf{v}_2 = \langle -3, -6\rangle$, $\mathbf{v}_3 = \langle -8, 4\rangle$, and $\mathbf{v}_4 = \langle 10, 15\rangle$.

19. Find the vector $\mathbf{v}_1 + 2\mathbf{v}_2 - \mathbf{v}_3$.

20. Which two vectors are parallel?

21. Which two vectors are orthogonal?

22. Find the angle between vectors $\mathbf{v}_1$ and $\mathbf{v}_2$.

23. A 1200-pound chandelier is to be suspended over a large ballroom; the chandelier will be hung on two cables of equal length whose ends will be attached to the ceiling, 16 feet apart. The chandelier will be free hanging so that the ends of the cable will make equal angles with the ceiling. If the top of the chandelier is to be 16 feet from the ceiling, what is the minimum tension each cable must be able to endure?

Cumulative Review

1. Find the real solutions, if any, of the equation $e^{x^2-9} = 1$.

2. Find an equation for the line containing the origin that makes an angle of 30° with the positive x-axis.

3. Find an equation for the circle with center at the point $(0, 1)$ and radius 3. Graph this circle.

4. What is the domain of the function $f(x) = \ln(1 - 2x)$?

5. Test the equation $x^2 + y^3 = 2x^4$ for symmetry with respect to the x-axis, the y-axis, and the origin.

6. Graph the function $y = |\ln x|$.

7. Graph the function $y = |\sin x|$.

8. Graph the function $y = \sin|x|$.

9. Find the exact value of $\sin^{-1}\left(-\frac{1}{2}\right)$.

10. Graph the equations $x = 3$ and $y = 4$ on the same set of rectangular coordinates.

11. Graph the equations $r = 2$ and $\theta = \frac{\pi}{3}$ on the same set of polar coordinates.

12. What is the amplitude and period of $y = -4\cos(\pi x)$?

Chapter Projects

I. Modeling Aircraft Motion Four aerodynamic forces act on an airplane in flight: lift, weight, thrust, and drag. While an aircraft is in flight, these four forces continuously battle each other. Weight opposes lift and drag opposes thrust. See the diagram below. In balanced flight at constant speed, the lift and weight are equal, and the thrust and drag are equal.

1. What will happen to the aircraft if the lift is held constant while the weight is decreased (say, from burning off fuel)?
2. What will happen to the aircraft if the lift is decreased while the weight is held constant?
3. What will happen to the aircraft if the thrust is increased while the drag is held constant?
4. What will happen to the aircraft if the drag is increased while the thrust is held constant?

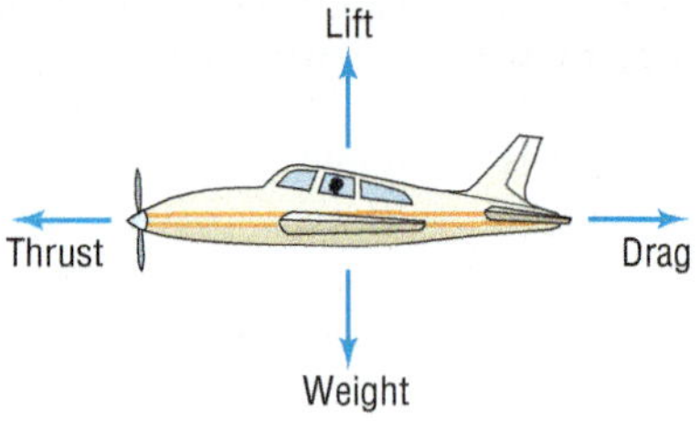

Source: *www.aeromuseum.org*

In 1903 the Wright brothers made the first controlled powered flight. The weight of their plane was approximately 700 pounds (lb). Newton's Second Law of motion states that force = mass × acceleration $(F = ma)$. If the mass is measured in kilograms (kg) and acceleration in meters per second squared (m/sec^2), then the force will be measured in newtons (N). [**Note:** $1\text{N} = 1\text{kg} \cdot \text{m/sec}^2$.]

5. If $1 \text{ kg} = 2.205 \text{ lb}$, convert the weight of the Wright brothers' plane to kilograms.
6. If acceleration due to gravity is $a = 9.80 \text{ m/sec}^2$, determine the force due to weight on the Wright brothers' plane.
7. What must be true about the lift force of the Wright brothers' plane for it to get off the ground?
8. The weight of a fully loaded Cessna 170B is 2200 lb. What lift force is required to get this plane off the ground?
9. The maximum gross weight of a Boeing 747 is 255,000 lb. What lift force is required to get this jet off the ground?

The following projects are available at the Instructors' Resource Center (IRC):

II. Project at Motorola ***Signal Fades Due to Interference*** Complex trigonometric functions are used to assure that a cellphone has optimal reception as the user travels up and down an elevator.

III. Compound Interest The effect of continuously compounded interest is analyzed using polar coordinates.

IV. Complex Equations Analysis of complex equations illustrates the connections between complex and real equations. At times, using complex equations is more efficient for proving mathematical theorems.

11 Analytic Geometry

The Orbit of the Hale-Bopp Comet

The orbits of the Hale-Bopp Comet and Earth can be modeled using *ellipses*, the subject of Section 11.3. The Internet-based Project at the end of this chapter explores the possibility of the Hale-Bopp Comet colliding with Earth.

—See the Internet-based Chapter Project I—

••• A Look Back

In Chapter 1, we introduced rectangular coordinates and showed how geometry problems can be solved algebraically. In Section 2.3, we defined a circle geometrically and then used the distance formula and rectangular coordinates to obtain an equation for a circle.

A Look Ahead •••

In this chapter, geometric definitions are given for the conics, and the distance formula and rectangular coordinates are used to obtain their equations.

Historically, Apollonius (200 BC) was among the first to study *conics* and discover some of their interesting properties. Today, conics are still studied because of their many uses. *Paraboloids of revolution* (parabolas rotated about their axes of symmetry) are used as signal collectors (the satellite dishes used with radar and dish TV, for example), as solar energy collectors, and as reflectors (telescopes, light projection, and so on). The planets circle the Sun in approximately *elliptical* orbits. Elliptical surfaces can be used to reflect signals such as light and sound from one place to another. A third conic, the *hyperbola,* can be used to determine the location of lightning strikes.

The Greeks used Euclidean geometry to study conics. However, we shall use the more powerful methods of analytic geometry, which employs both algebra and geometry, for our study of conics.

Outline

11.1 Conics

OBJECTIVE 1 Know the Names of the Conics (p. 792)

1 Know the Names of the Conics

The word *conic* derives from the word *cone,* which is a geometric figure that can be constructed in the following way: Let a and g be two distinct lines that intersect at a point V. Keep the line a fixed. Now rotate the line g about a, while maintaining the same angle between a and g. The collection of points swept out (generated) by the line g is called a **right circular cone**. See Figure 1. The fixed line a is called the **axis** of the cone; the point V is its **vertex**; the lines that pass through V and make the same angle with a as g are **generators** of the cone. Each generator is a line that lies entirely on the cone. The cone consists of two parts, called **nappes**, that intersect at the vertex.

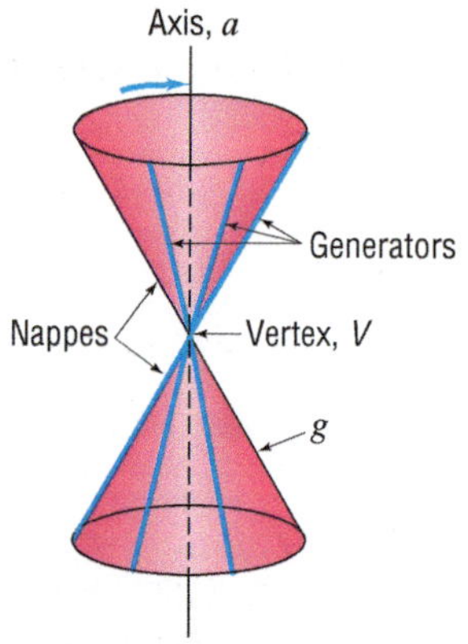

Figure 1 Right circular cone

Conics, an abbreviation for **conic sections**, are curves that result from the intersection of a right circular cone and a plane. The conics we shall study arise when the plane does not contain the vertex, as shown in Figure 2. These conics are **circles** when the plane is perpendicular to the axis of the cone and intersects each generator; **ellipses** when the plane is tilted slightly so that it intersects each generator, but intersects only one nappe of the cone; **parabolas** when the plane is tilted farther so that it is parallel to one (and only one) generator and intersects only one nappe of the cone; and **hyperbolas** when the plane intersects both nappes.

If the plane contains the vertex, the intersection of the plane and the cone is a point, a line, or a pair of intersecting lines. These are usually called **degenerate conics**.

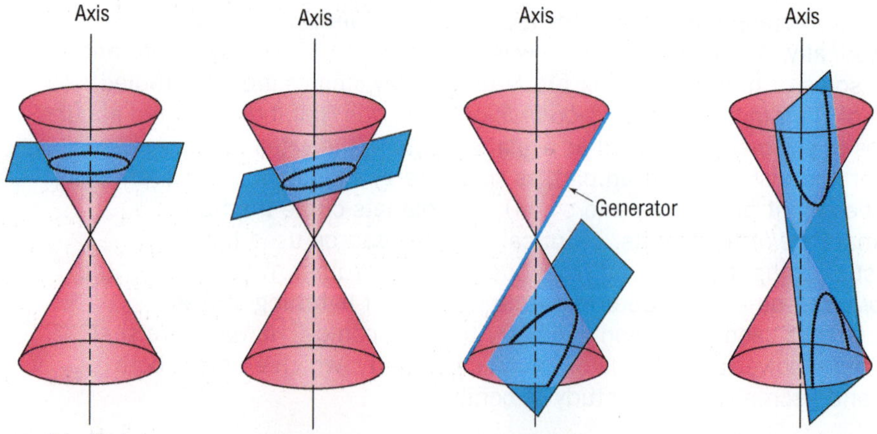

Figure 2 **(a)** Circle **(b)** Ellipse **(c)** Parabola **(d)** Hyperbola

Conic sections are used in modeling many different applications. For example, parabolas are used in describing searchlights and telescopes (see Figures 16 and 17 on page 799). Ellipses are used to model the orbits of planets and whispering chambers (see pages 810–811). And hyperbolas are used to locate lightning strikes and model nuclear cooling towers (see Problems 76 and 77 in Section 11.4).

11.2 The Parabola

PREPARING FOR THIS SECTION *Before getting started, review the following:*

- Distance Formula (Section 1.1, p. 85)
- Symmetry (Section 2.1, pp. 166–168)
- Square Root Method (Section 1.3, p. 112)
- Completing the Square (Chapter R, Review, Section R.5, p. 57)
- Graphing Techniques: Transformations (Section 3.5, pp. 256–264)

Now Work the **'Are You Prepared?'** problems on page 800.

OBJECTIVES
1 Analyze Parabolas with Vertex at the Origin (p. 793)
2 Analyze Parabolas with Vertex at (h, k) (p. 797)
3 Solve Applied Problems Involving Parabolas (p. 798)

In Section 4.3, we learned that the graph of a quadratic function is a parabola. In this section, we give a geometric definition of a parabola and use it to obtain an equation.

DEFINITION

A **parabola** is the collection of all points P in a plane that are the same distance d from a fixed point F as they are from a fixed line D. The point F is called the **focus** of the parabola, and the line D is its **directrix**. As a result, a parabola is the set of points P for which

$$d(F, P) = d(P, D) \tag{1}$$

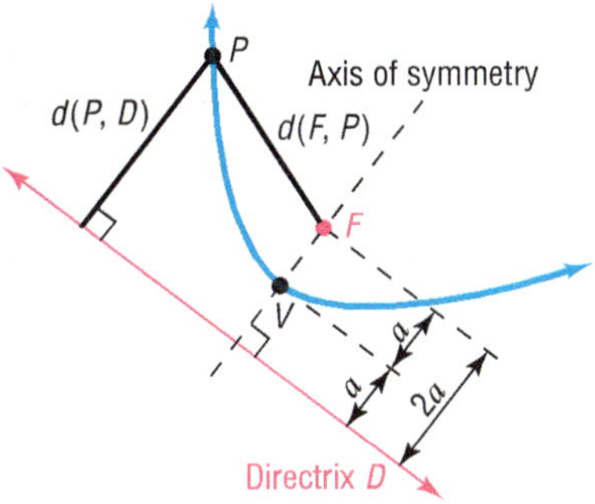

Figure 3 Parabola

Figure 3 shows a parabola (in blue). The line through the focus F and perpendicular to the directrix D is called the **axis of symmetry** of the parabola. The point of intersection of the parabola with its axis of symmetry is called the **vertex** V.

Because the vertex V lies on the parabola, it must satisfy equation (1): $d(F, V) = d(V, D)$. The vertex is midway between the focus and the directrix. We shall let a equal the distance $d(F, V)$ from F to V. Now we are ready to derive an equation for a parabola. To do this, we use a rectangular system of coordinates positioned so that the vertex V, focus F, and directrix D of the parabola are conveniently located.

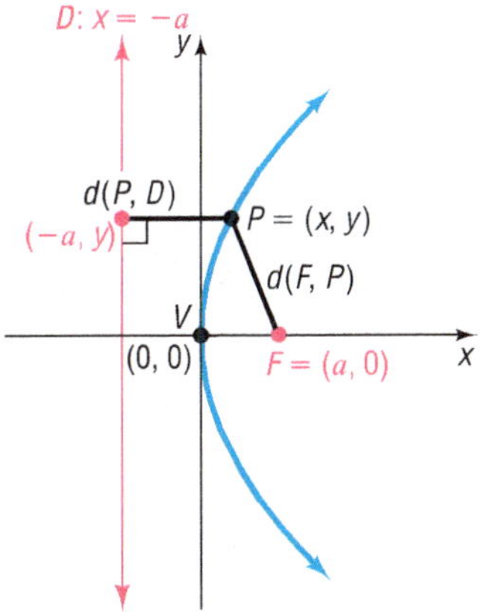

Figure 4

1 Analyze Parabolas with Vertex at the Origin

If we choose to locate the vertex V at the origin $(0, 0)$, we can conveniently position the focus F on either the x-axis or the y-axis. First, consider the case where the focus F is on the positive x-axis, as shown in Figure 4. Because the distance from F to V is a, the coordinates of F will be $(a, 0)$ with $a > 0$. Similarly, because the distance from V to the directrix D is also a, and because D must be perpendicular to the x-axis (since the x-axis is the axis of symmetry), the equation of the directrix D must be $x = -a$.

Now, if $P = (x, y)$ is any point on the parabola, P must satisfy equation (1):

$$d(F, P) = d(P, D)$$

So we have

$$\sqrt{(x-a)^2 + (y-0)^2} = \sqrt{(x-(-a))^2 + (y-y)^2} \quad \text{Use the Distance Formula.}$$
$$(x-a)^2 + y^2 = (x+a)^2 \quad \text{Square both sides.}$$
$$x^2 - 2ax + a^2 + y^2 = x^2 + 2ax + a^2 \quad \text{Multiply out.}$$
$$y^2 = 4ax \quad \text{Simplify.}$$

THEOREM

Equation of a Parabola: Vertex at (0, 0), Focus at (a, 0), $a > 0$

The equation of a parabola with vertex at $(0, 0)$, focus at $(a, 0)$, and directrix $x = -a$, $a > 0$, is

$$y^2 = 4ax \qquad (2)$$

Recall that a is the distance from the vertex to the focus of a parabola. When graphing the parabola $y^2 = 4ax$ it is helpful to determine the "opening" by finding the points that lie directly above or below the focus $(a, 0)$. This is done by letting $x = a$ in $y^2 = 4ax$, so $y^2 = 4a(a) = 4a^2$, or $y = \pm 2a$. The line segment joining the two points, $(a, 2a)$ and $(a, -2a)$, is called the **latus rectum**; its length is $4a$.

EXAMPLE 1 Finding the Equation of a Parabola and Graphing It

Find an equation of the parabola with vertex at $(0, 0)$ and focus at $(3, 0)$. Graph the equation.

Solution The distance from the vertex $(0, 0)$ to the focus $(3, 0)$ is $a = 3$. Based on equation (2), the equation of this parabola is

$$y^2 = 4ax$$
$$y^2 = 12x \quad a = 3$$

To graph this parabola, find the two points that determine the latus rectum by letting $x = 3$. Then

$$y^2 = 12x = 12(3) = 36$$
$$y = \pm 6 \quad \text{Solve for } y.$$

The points $(3, 6)$ and $(3, -6)$ determine the latus rectum. These points help in graphing the parabola because they determine the "opening." See Figure 5.

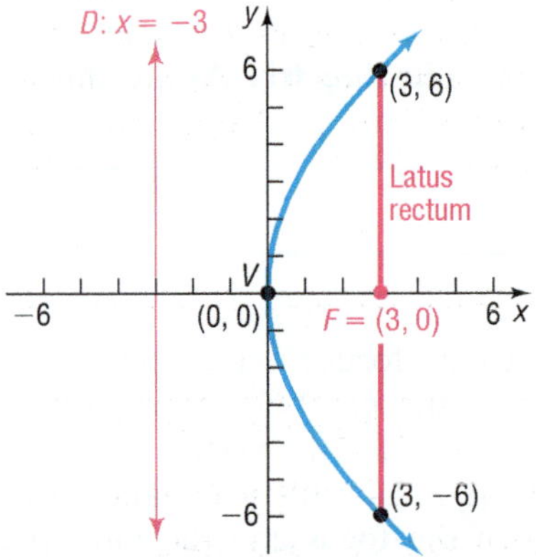

Figure 5 $y^2 = 12x$

Now Work PROBLEM 21

EXAMPLE 2 Graphing a Parabola Using a Graphing Utility

Graph the parabola $y^2 = 12x$.

Solution To graph the parabola $y^2 = 12x$, we need to graph the two functions $Y_1 = \sqrt{12x}$ and $Y_2 = -\sqrt{12x}$ on a square screen. Figure 6 shows the graph of $y^2 = 12x$. Notice that the graph fails the vertical line test, so $y^2 = 12x$ is not a function.

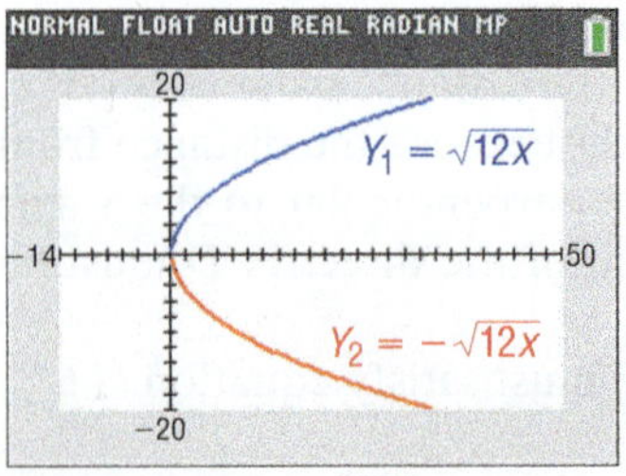

Figure 6 $Y_1 = \sqrt{12x}$; $Y_2 = -\sqrt{12x}$

By reversing the steps used to obtain equation (2), it follows that the graph of an equation of the form of equation (2), $y^2 = 4ax$, is a parabola; its vertex is at $(0, 0)$, its focus is at $(a, 0)$, its directrix is the line $x = -a$, and its axis of symmetry is the x-axis.

For the remainder of this section, the direction **"Analyze the equation"** will mean to find the vertex, focus, and directrix of the parabola and graph it.

EXAMPLE 3 Analyzing the Equation of a Parabola

Analyze the equation: $y^2 = 8x$

Solution The equation $y^2 = 8x$ is of the form $y^2 = 4ax$, where $4a = 8$, so $a = 2$. Consequently, the graph of the equation is a parabola with vertex at $(0, 0)$ and focus on the positive x-axis at $(a, 0) = (2, 0)$. The directrix is the vertical line $x = -2$. The two points that determine the latus rectum are obtained by letting $x = 2$. Then $y^2 = 16$, so $y = \pm 4$. The points $(2, -4)$ and $(2, 4)$ determine the latus rectum. See Figure 7(a) for the graph drawn by hand. Figure 7(b) shows the graph obtained using a graphing utility.

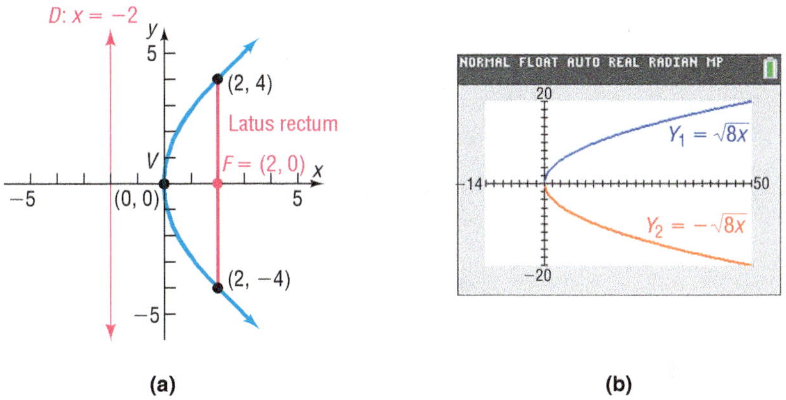

Figure 7 $y^2 = 8x$

Recall that we obtained equation (2) after placing the focus on the positive x-axis. If the focus is placed on the negative x-axis, positive y-axis, or negative y-axis, a different form of the equation for the parabola results. The four forms of the equation of a parabola with vertex at $(0, 0)$ and focus on a coordinate axis a distance a from $(0, 0)$ are given in Table 1, and their graphs are given in Figure 8. Notice that each graph is symmetric with respect to its axis of symmetry.

Table 1

Equations of a Parabola Vertex at (0, 0); Focus on an Axis; $a > 0$

Vertex	Focus	Directrix	Equation	Description
$(0, 0)$	$(a, 0)$	$x = -a$	$y^2 = 4ax$	Parabola, axis of symmetry is the x-axis, opens right
$(0, 0)$	$(-a, 0)$	$x = a$	$y^2 = -4ax$	Parabola, axis of symmetry is the x-axis, opens left
$(0, 0)$	$(0, a)$	$y = -a$	$x^2 = 4ay$	Parabola, axis of symmetry is the y-axis, opens up
$(0, 0)$	$(0, -a)$	$y = a$	$x^2 = -4ay$	Parabola, axis of symmetry is the y-axis, opens down

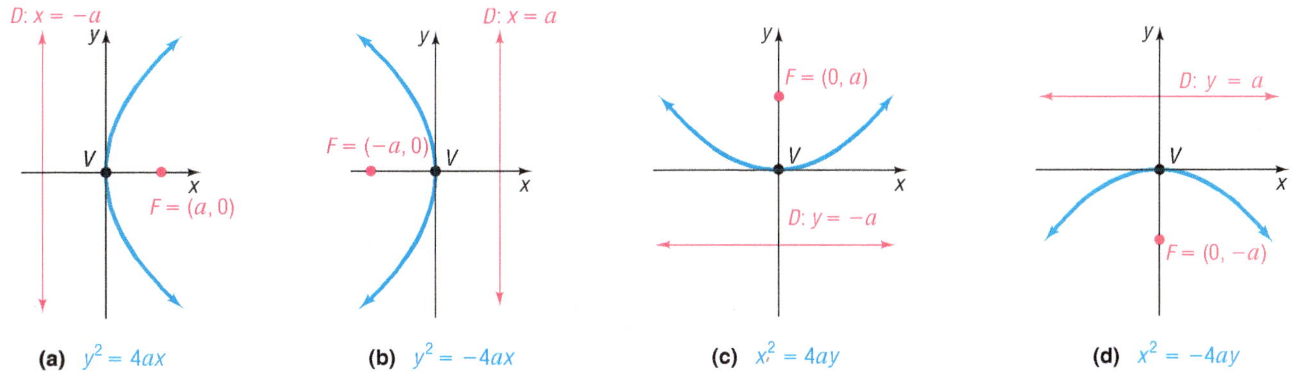

Figure 8

EXAMPLE 4 Analyzing the Equation of a Parabola

Analyze the equation: $x^2 = -12y$

Solution The equation $x^2 = -12y$ is of the form $x^2 = -4ay$, with $a = 3$. Consequently, the graph of the equation is a parabola with vertex at $(0, 0)$, focus at $(0, -3)$, and directrix the line $y = 3$. The parabola opens down, and its axis of symmetry is the y-axis. To obtain the points defining the latus rectum, let $y = -3$. Then $x^2 = 36$, so $x = \pm 6$. The points $(-6, -3)$ and $(6, -3)$ determine the latus rectum. See Figure 9(a) for the graph drawn by hand. Figure 9(b) shows the graph using a graphing utility.

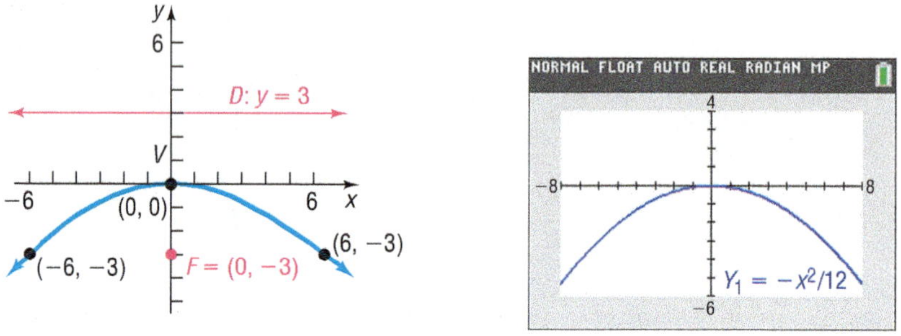

(a) (b)

Figure 9 $x^2 = -12y$

Now Work PROBLEM 41

EXAMPLE 5 Finding the Equation of a Parabola

Find the equation of the parabola with focus at $(0, 4)$ and directrix the line $y = -4$. Graph the equation.

Solution A parabola whose focus is at $(0, 4)$ and whose directrix is the horizontal line $y = -4$ will have its vertex at $(0, 0)$. (Do you see why? The vertex is midway between the focus and the directrix.) Since the focus is on the positive y-axis at $(0, 4)$, the equation of this parabola is of the form $x^2 = 4ay$, with $a = 4$; that is,

$$x^2 = 4ay = 4(4)y = 16y$$
$$\uparrow$$
$$a = 4$$

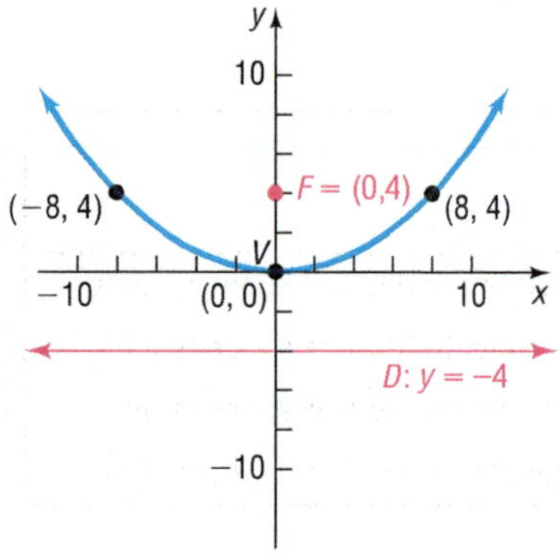

Figure 10 $x^2 = 16y$

Letting $y = 4$, we find $x^2 = 64$, so $x = \pm 8$. The points $(8, 4)$ and $(-8, 4)$ determine the latus rectum. Figure 10 shows the graph of $x^2 = 16y$.

Check: Verify the graph drawn in Figure 10 by graphing $Y_1 = \dfrac{x^2}{16}$ using a graphing utility.

EXAMPLE 6 Finding the Equation of a Parabola

Find the equation of a parabola with vertex at $(0, 0)$ if its axis of symmetry is the x-axis and its graph contains the point $\left(-\frac{1}{2}, 2\right)$. Find its focus and directrix, and graph the equation.

Solution The vertex is at the origin, the axis of symmetry is the x-axis, and the graph contains a point in the second quadrant, so the parabola opens to the left. From Table 1, note that the form of the equation is

$$y^2 = -4ax$$

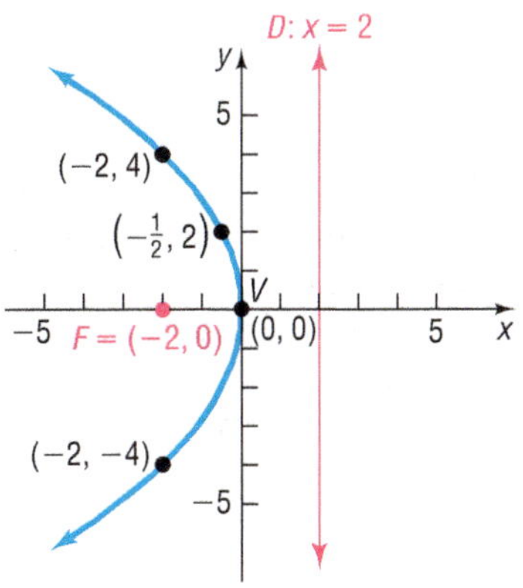

Figure 11 $y^2 = -8x$

Because the point $\left(-\frac{1}{2}, 2\right)$ is on the parabola, the coordinates $x = -\frac{1}{2}, y = 2$ must satisfy the equation $y^2 = -4ax$. Substituting $x = -\frac{1}{2}$ and $y = 2$ into the equation leads to

$$2^2 = -4a\left(-\frac{1}{2}\right) \qquad y^2 = -4ax;\ x = -\frac{1}{2},\ y = 2$$

$$a = 2$$

The equation of the parabola is

$$y^2 = -4(2)x = -8x$$

The focus is at $(-2, 0)$ and the directrix is the line $x = 2$. Letting $x = -2$ gives $y^2 = 16$, so $y = \pm 4$. The points $(-2, 4)$ and $(-2, -4)$ determine the latus rectum. See Figure 11. ■

Now Work PROBLEM 29

2 Analyze Parabolas with Vertex at (h, k)

If a parabola with vertex at the origin and axis of symmetry along a coordinate axis is shifted horizontally h units and then vertically k units, the result is a parabola with vertex at (h, k) and axis of symmetry parallel to a coordinate axis. The equations of such parabolas have the same forms as those in Table 1 but x is replaced by $x - h$ (the horizontal shift) and y is replaced by $y - k$ (the vertical shift). Table 2 gives the forms of the equations of such parabolas. Figures 12(a)–(d) illustrate the graphs for $h > 0, k > 0$.

Note: It is not recommended that Table 2 be memorized. Rather, use the ideas of transformations (shift horizontally h units, vertically k units), along with the fact that a represents the distance from the vertex to the focus, to determine the various components of a parabola. It is also helpful to remember that parabolas of the form "$x^2 =$ " open up or down, while parabolas of the form "$y^2 =$ " open left or right. ■

Table 2

Equations of a Parabola: Vertex at (h, k); Axis of Symmetry Parallel to a Coordinate Axis; $a > 0$

Vertex	Focus	Directrix	Equation	Description
(h, k)	$(h + a, k)$	$x = h - a$	$(y - k)^2 = 4a(x - h)$	Axis of symmetry is parallel to the x-axis, opens right
(h, k)	$(h - a, k)$	$x = h + a$	$(y - k)^2 = -4a(x - h)$	Axis of symmetry is parallel to the x-axis, opens left
(h, k)	$(h, k + a)$	$y = k - a$	$(x - h)^2 = 4a(y - k)$	Axis of symmetry is parallel to the y-axis, opens up
(h, k)	$(h, k - a)$	$y = k + a$	$(x - h)^2 = -4a(y - k)$	Axis of symmetry is parallel to the y-axis, opens down

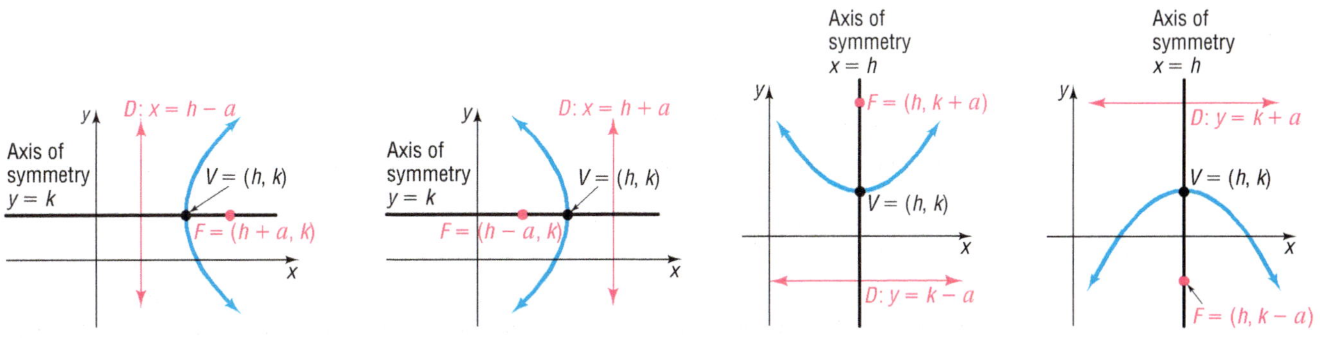

(a) $(y - k)^2 = 4a(x - h)$ (b) $(y - k)^2 = -4a(x - h)$ (c) $(x - h)^2 = 4a(y - k)$ (d) $(x - h)^2 = -4a(y - k)$

Figure 12

EXAMPLE 7 Finding the Equation of a Parabola, Vertex Not at the Origin

Find an equation of the parabola with vertex at $(-2, 3)$ and focus at $(0, 3)$. Graph the equation.

Solution The vertex $(-2, 3)$ and focus $(0, 3)$ both lie on the horizontal line $y = 3$ (the axis of symmetry). The distance a from the vertex $(-2, 3)$ to the focus $(0, 3)$ is $a = 2$. Also, because the focus lies to the right of the vertex, the parabola opens to the right. Consequently, the form of the equation is

$$(y - k)^2 = 4a(x - h)$$

where $(h, k) = (-2, 3)$ and $a = 2$. Therefore, the equation is

$$(y - 3)^2 = 4 \cdot 2[x - (-2)]$$
$$(y - 3)^2 = 8(x + 2)$$

To find the points that define the latus rectum, let $x = 0$, so that $(y - 3)^2 = 16$. Then $y - 3 = \pm 4$, so $y = -1$ or $y = 7$. The points $(0, -1)$ and $(0, 7)$ determine the latus rectum; the line $x = -4$ is the directrix. See Figure 13. ■

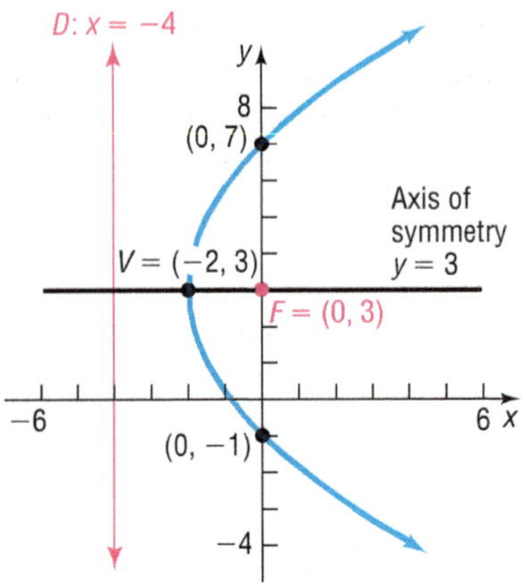

Figure 13 $(y - 3)^2 = 8(x + 2)$

Now Work PROBLEM 31

EXAMPLE 8 Using a Graphing Utility to Graph a Parabola, Vertex Not at Origin

Using a graphing utility, graph the equation $(y - 3)^2 = 8(x + 2)$.

Solution First, we must solve the equation for y.

$$(y - 3)^2 = 8(x + 2)$$
$$y - 3 = \pm\sqrt{8(x + 2)} \quad \text{Use the Square Root Method.}$$
$$y = 3 \pm \sqrt{8(x + 2)} \quad \text{Add 3 to both sides.}$$

Figure 14 shows the graphs of the equations $Y_1 = 3 + \sqrt{8(x + 2)}$ and $Y_2 = 3 - \sqrt{8(x + 2)}$. ■

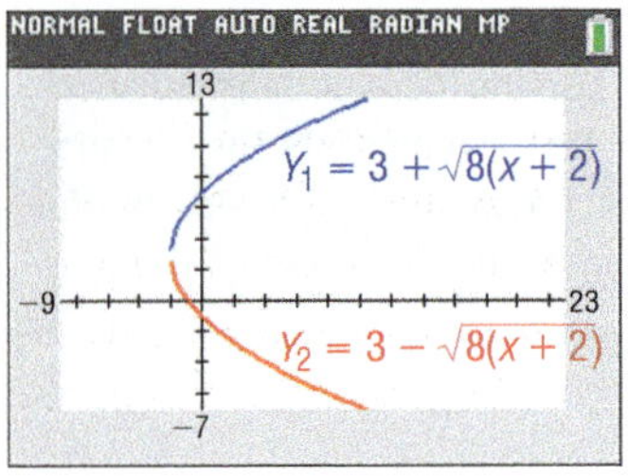

Figure 14

Polynomial equations define parabolas whenever they involve two variables that are quadratic in one variable and linear in the other.

EXAMPLE 9 Analyzing the Equation of a Parabola

Analyze the equation: $x^2 + 4x - 4y = 0$

Solution To analyze the equation $x^2 + 4x - 4y = 0$, complete the square involving the variable x.

$$x^2 + 4x - 4y = 0$$
$$x^2 + 4x = 4y \quad \text{Isolate the terms involving } x \text{ on the left side.}$$
$$x^2 + 4x + 4 = 4y + 4 \quad \text{Complete the square on the left side.}$$
$$(x + 2)^2 = 4(y + 1) \quad \text{Factor.}$$

This equation is of the form $(x - h)^2 = 4a(y - k)$, with $h = -2$, $k = -1$, and $a = 1$. The graph is a parabola with vertex at $(h, k) = (-2, -1)$ that opens up. The focus is $a = 1$ unit above the vertex at $(-2, 0)$, and the directrix is the line $y = -2$. See Figure 15. ■

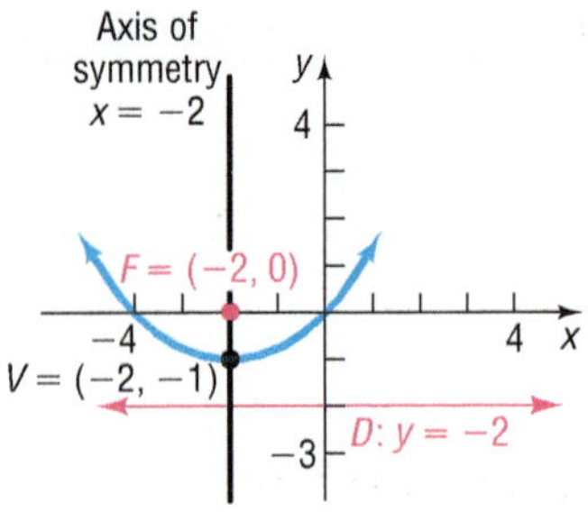

Figure 15 $x^2 + 4x - 4y = 0$

Now Work PROBLEM 49

3 Solve Applied Problems Involving Parabolas

Parabolas find their way into many applications. For example, as discussed in Section 4.4, suspension bridges have cables in the shape of a parabola. Another property of parabolas that is used in applications is their reflecting property.

Suppose that a mirror is shaped like a **paraboloid of revolution**, a surface formed by rotating a parabola about its axis of symmetry. If a light (or any other emitting source) is placed at the focus of the parabola, all the rays emanating from the light reflect off the mirror in lines parallel to the axis of symmetry. This principle is used in the design of searchlights, flashlights, certain automobile headlights, and other such devices. See Figure 16.

Conversely, suppose that rays of light (or other signals) emanate from a distant source so that they are essentially parallel. When these rays strike the surface of a parabolic mirror whose axis of symmetry is parallel to these rays, they are reflected to a single point at the focus. This principle is used in the design of some solar energy devices, satellite dishes, and the mirrors used in some types of telescopes. See Figure 17.

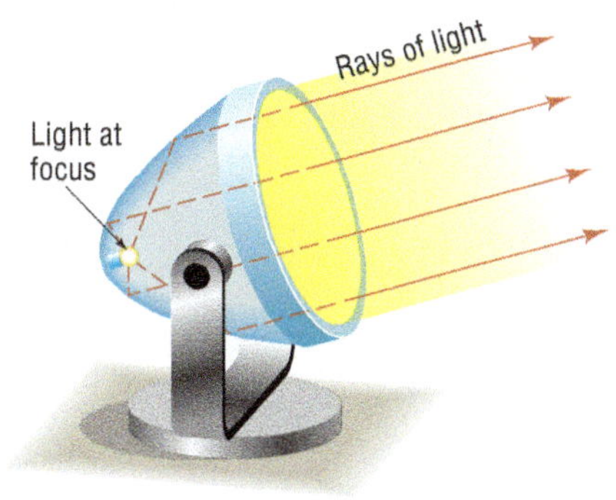

Figure 16 Searchlight

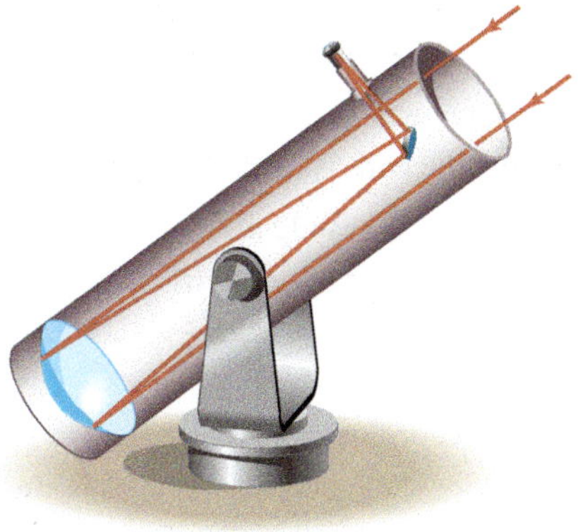

Figure 17 Telescope

EXAMPLE 10

Satellite Dish

A satellite dish is shaped like a paraboloid of revolution. The signals that emanate from a satellite strike the surface of the dish and are reflected to a single point, where the receiver is located. If the dish is 8 feet across at its opening and 3 feet deep at its center, at what position should the receiver be placed? That is, where is the focus?

Solution Figure 18(a) shows the satellite dish. On a rectangular coordinate system, draw the parabola used to form the dish so that the vertex of the parabola is at the origin and its focus is on the positive y-axis. See Figure 18(b).

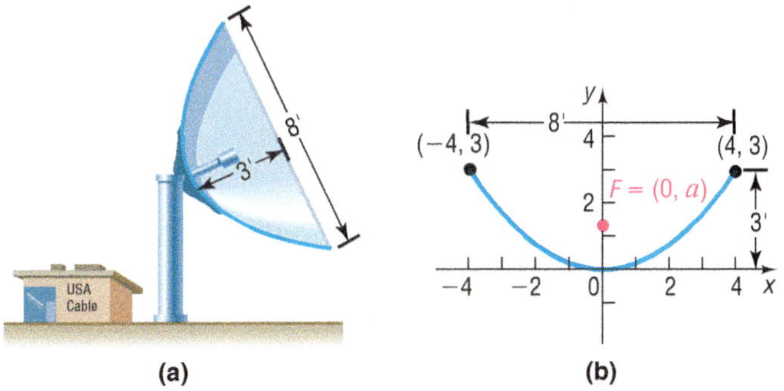

Figure 18

The form of the equation of the parabola is

$$x^2 = 4ay$$

and its focus is at $(0, a)$. Since $(4, 3)$ is a point on the graph, this gives

$$4^2 = 4a(3) \qquad x^2 = 4ay;\ x = 4,\ y = 3$$

$$a = \frac{4}{3} \qquad \text{Solve for } a.$$

The receiver should be located $1\frac{1}{3}$ feet (1 foot, 4 inches) from the base of the dish, along its axis of symmetry. ■

Now Work PROBLEM 65

11.2 Assess Your Understanding

'Are You Prepared?' *Answers are given at the end of these exercises. If you get a wrong answer, read the pages listed in red.*

1. The formula for the distance d from $P_1 = (x_1, y_1)$ to $P_2 = (x_2, y_2)$ is $d =$ ______________. (p. 85)
2. To complete the square of $x^2 - 4x$, add ____. (p. 57)
3. Use the Square Root Method to find the real solutions of $(x + 4)^2 = 9$. (p. 112)
4. The point that is symmetric with respect to the x-axis to the point $(-2, 5)$ is ________. (pp. 166–168)
5. To graph $y = (x - 3)^2 + 1$, shift the graph of $y = x^2$ to the right ____ units and then ______ 1 unit. (pp. 256–264)

Concepts and Vocabulary

6. A(n) __________ is the collection of all points in a plane that are the same distance from a fixed point as they are from a fixed line.
7. The line through the focus and perpendicular to the directrix is called the ______________ of the parabola.
8. For the parabola $y^2 = 4ax$, the line segment joining the two points $(a, 2a)$ and $(a, -2a)$ is called the __________.

Answer Problems 9–12 using the figure.

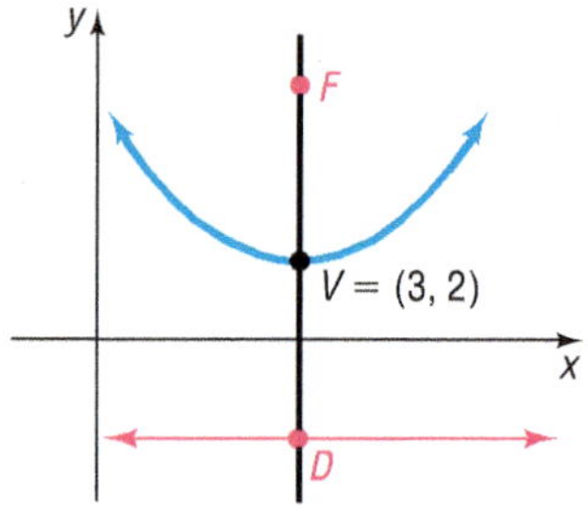

9. If $a > 0$, the equation of the parabola is of the form
   (a) $(y - k)^2 = 4a(x - h)$ (b) $(y - k)^2 = -4a(x - h)$
   (c) $(x - h)^2 = 4a(y - k)$ (d) $(x - h)^2 = -4a(y - k)$
10. The coordinates of the vertex are __________.
11. If $a = 4$, then the coordinates of the focus are ________.
    (a) $(-1, 2)$ (b) $(3, -2)$ (c) $(7, 2)$ (d) $(3, 6)$
12. If $a = 4$, then the equation of the directrix is ________.
    (a) $x = -3$ (b) $x = 3$ (c) $y = -2$ (d) $y = 2$

Skill Building

In Problems 13–20, the graph of a parabola is given. Match each graph to its equation.

(A) $y^2 = 4x$
(B) $x^2 = 4y$
(C) $y^2 = -4x$
(D) $x^2 = -4y$
(E) $(y - 1)^2 = 4(x - 1)$
(F) $(x + 1)^2 = 4(y + 1)$
(G) $(y - 1)^2 = -4(x - 1)$
(H) $(x + 1)^2 = -4(y + 1)$

13.
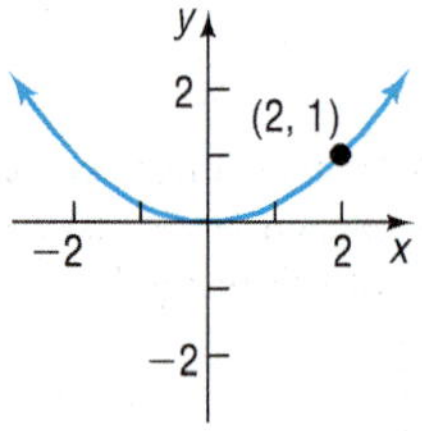

14.
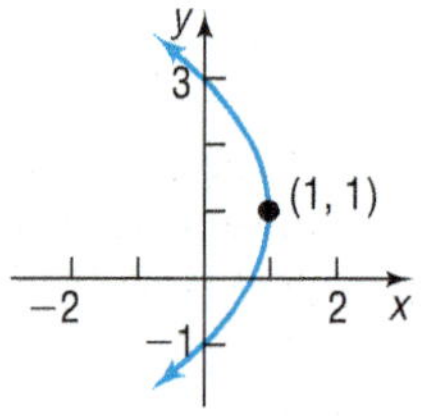

15.
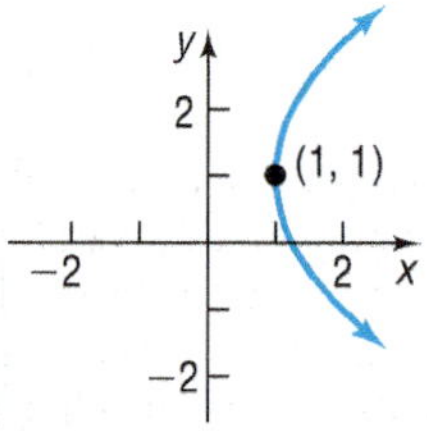

16.
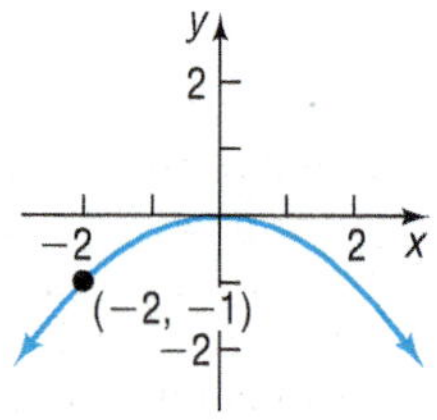

17.
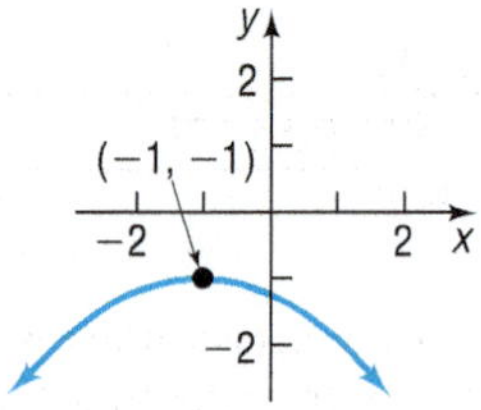

18.
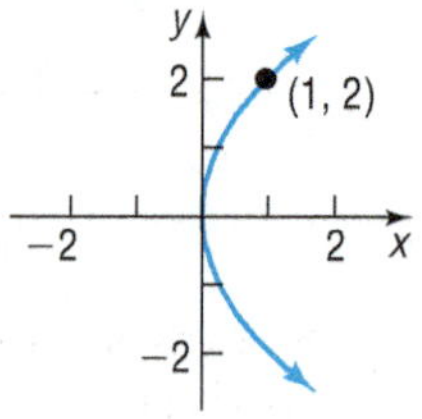

19.
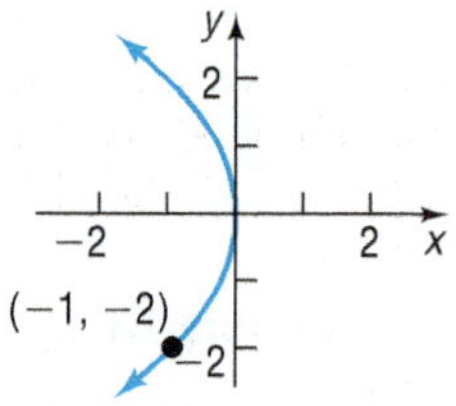

20.
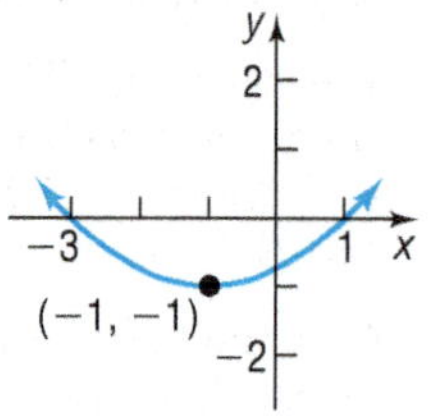

In Problems 21–38, find the equation of the parabola described. Find the two points that define the latus rectum, and graph the equation by hand.

21. Focus at $(4, 0)$; vertex at $(0, 0)$
22. Focus at $(0, 2)$; vertex at $(0, 0)$
23. Focus at $(0, -3)$; vertex at $(0, 0)$
24. Focus at $(-4, 0)$; vertex at $(0, 0)$
25. Focus at $(-2, 0)$; directrix the line $x = 2$
26. Focus at $(0, -1)$; directrix the line $y = 1$
27. Directrix the line $y = -\frac{1}{2}$; vertex at $(0, 0)$
28. Directrix the line $x = -\frac{1}{2}$; vertex at $(0, 0)$
29. Vertex at $(0, 0)$; axis of symmetry the y-axis; containing the point $(2, 3)$
30. Vertex at $(0, 0)$; axis of symmetry the x-axis; containing the point $(2, 3)$

31. Vertex at $(2, -3)$; focus at $(2, -5)$

32. Vertex at $(4, -2)$; focus at $(6, -2)$

33. Vertex at $(-1, -2)$; focus at $(0, -2)$

34. Vertex at $(3, 0)$; focus at $(3, -2)$

35. Focus at $(-3, 4)$; directrix the line $y = 2$

36. Focus at $(2, 4)$; directrix the line $x = -4$

37. Focus at $(-3, -2)$; directrix the line $x = 1$

38. Focus at $(-4, 4)$; directrix the line $y = -2$

In Problems 39–56, find the vertex, focus, and directrix of each parabola. Graph the equation by hand. Verify your graph using a graphing utility.

39. $x^2 = 4y$

40. $y^2 = 8x$

41. $y^2 = -16x$

42. $x^2 = -4y$

43. $(y - 2)^2 = 8(x + 1)$

44. $(x + 4)^2 = 16(y + 2)$

45. $(x - 3)^2 = -(y + 1)$

46. $(y + 1)^2 = -4(x - 2)$

47. $(y + 3)^2 = 8(x - 2)$

48. $(x - 2)^2 = 4(y - 3)$

49. $y^2 - 4y + 4x + 4 = 0$

50. $x^2 + 6x - 4y + 1 = 0$

51. $x^2 + 8x = 4y - 8$

52. $y^2 - 2y = 8x - 1$

53. $y^2 + 2y - x = 0$

54. $x^2 - 4x = 2y$

55. $x^2 - 4x = y + 4$

56. $y^2 + 12y = -x + 1$

In Problems 57–64, write an equation for each parabola.

57.

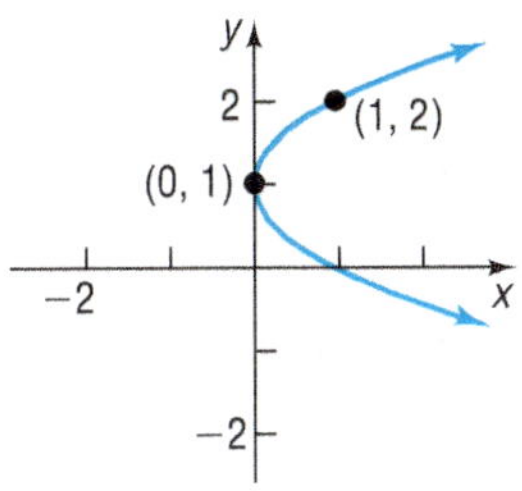

58.

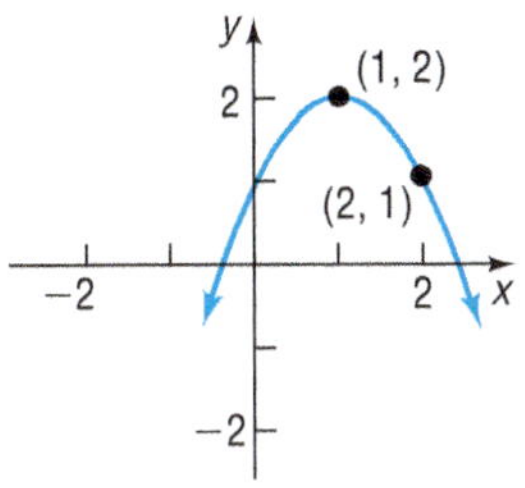

59.

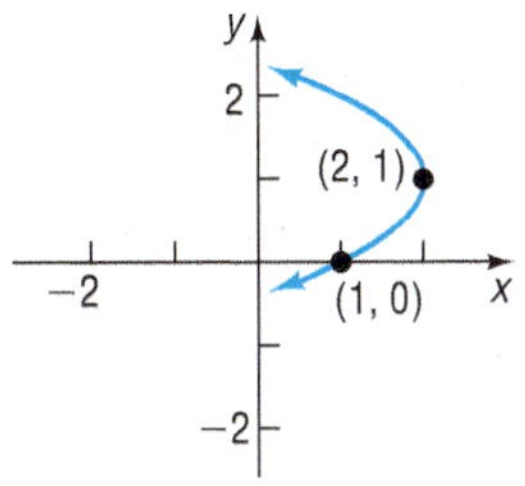

60.

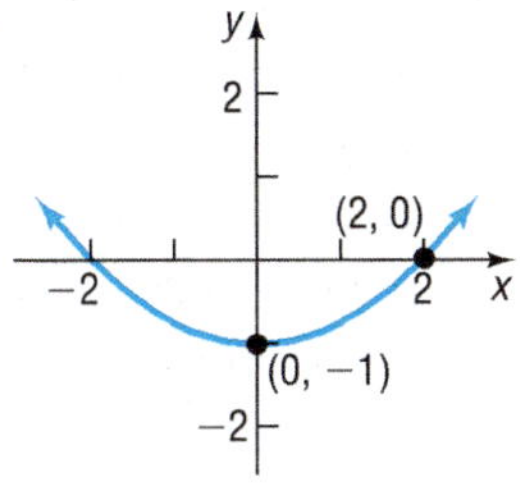

61.

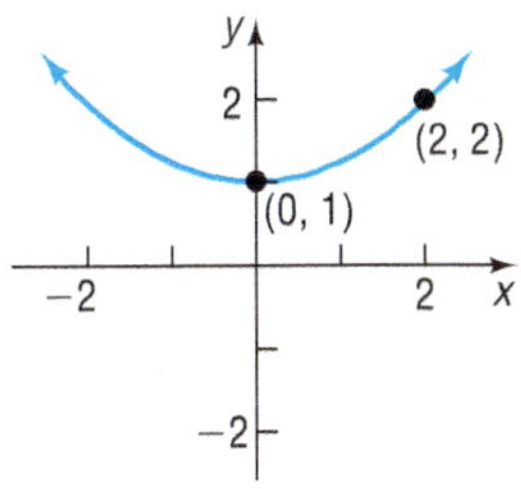

62.

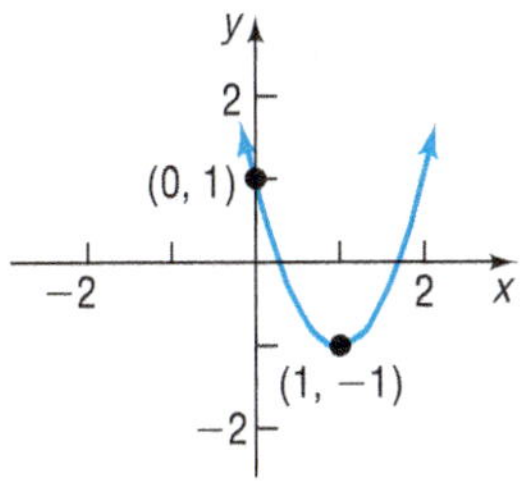

63.

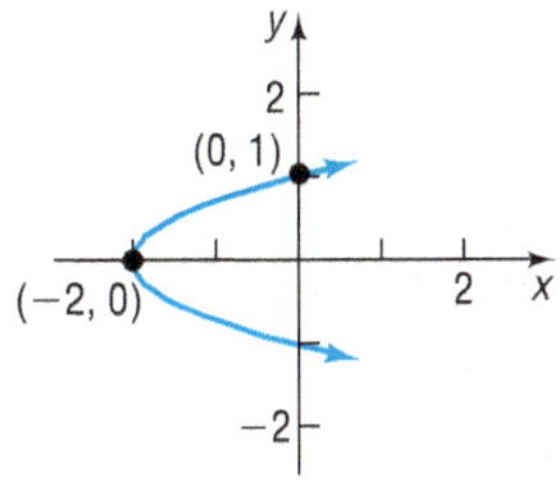

64.

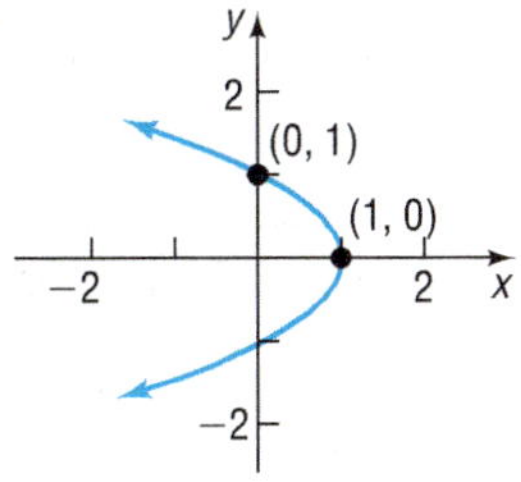

Applications and Extensions

65. **Satellite Dish** A satellite dish is shaped like a paraboloid of revolution. The signals that emanate from a satellite strike the surface of the dish and are reflected to a single point, where the receiver is located. If the dish is 10 feet across at its opening and 4 feet deep at its center, at what position should the receiver be placed?

66. **Constructing a TV Dish** A cable TV receiving dish is in the shape of a paraboloid of revolution. Find the location of the receiver, which is placed at the focus, if the dish is 6 feet across at its opening and 2 feet deep.

67. **Constructing a Flashlight** The reflector of a flashlight is in the shape of a paraboloid of revolution. Its diameter is 4 inches and its depth is 1 inch. How far from the vertex should the light bulb be placed so that the rays will be reflected parallel to the axis?

68. **Constructing a Headlight** A sealed-beam headlight is in the shape of a paraboloid of revolution. The bulb, which is placed at the focus, is 1 inch from the vertex. If the depth is to be 2 inches, what is the diameter of the headlight at its opening?

69. **Suspension Bridge** The cables of a suspension bridge are in the shape of a parabola, as shown in the figure. The towers supporting the cable are 600 feet apart and 80 feet high. If the cables touch the road surface midway between the towers, what is the height of the cable from the road at a point 150 feet from the center of the bridge?

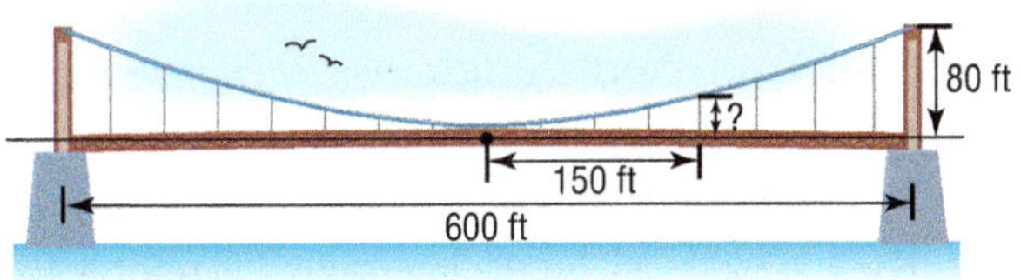

70. **Suspension Bridge** The cables of a suspension bridge are in the shape of a parabola. The towers supporting the cable are 400 feet apart and 100 feet high. If the cables are at a height of 10 feet midway between the towers, what is the height of the cable at a point 50 feet from the center of the bridge?

71. **Searchlight** A searchlight is shaped like a paraboloid of revolution. If the light source is located 2 feet from the base along the axis of symmetry and the opening is 5 feet across, how deep should the searchlight be?

72. **Searchlight** A searchlight is shaped like a paraboloid of revolution. If the light source is located 2 feet from the base along the axis of symmetry and the depth of the searchlight is 4 feet, what should the width of the opening be?

73. **Solar Heat** A mirror is shaped like a paraboloid of revolution and will be used to concentrate the rays of the sun at its focus, creating a heat source. See the figure. If the mirror is 20 feet across at its opening and is 6 feet deep, where will the heat source be concentrated?

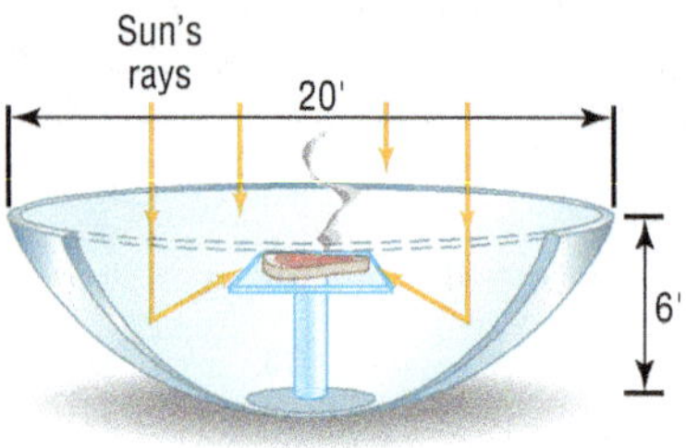

74. **Reflecting Telescope** A reflecting telescope contains a mirror shaped like a paraboloid of revolution. If the mirror is 4 inches across at its opening and is 3 inches deep, where will the collected light be concentrated?

75. **Parabolic Arch Bridge** A bridge is built in the shape of a parabolic arch. The bridge has a span of 120 feet and a maximum height of 25 feet. See the illustration. Choose a suitable rectangular coordinate system and find the height of the arch at distances of 10, 30, and 50 feet from the center.

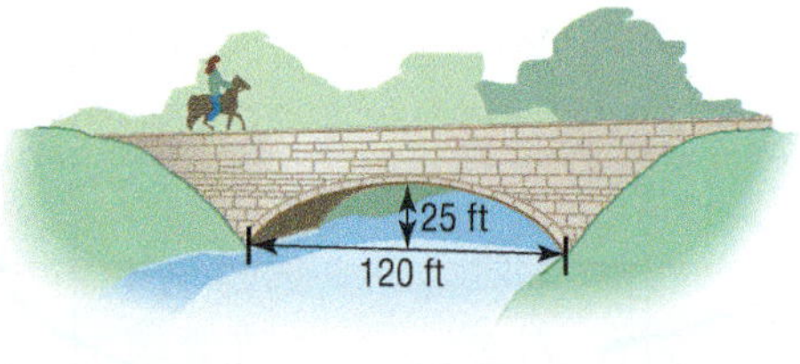

76. **Parabolic Arch Bridge** A bridge is to be built in the shape of a parabolic arch and is to have a span of 100 feet. The height of the arch a distance of 40 feet from the center is to be 10 feet. Find the height of the arch at its center.

77. **Gateway Arch** The Gateway Arch in St. Louis is often mistaken to be parabolic in shape. In fact, it is a *catenary*, which has a more complicated formula than a parabola. The Arch is 630 feet high and 630 feet wide at its base.

(a) Find the equation of a parabola with the same dimensions. Let x equal the horizontal distance from the center of the arch.

(b) The table below gives the height of the Arch at various widths; find the corresponding heights for the parabola found in (a).

Width (ft)	Height (ft)
567	100
478	312.5
308	525

(c) Do the data support the notion that the Arch is in the shape of a parabola?

Source: *gatewayarch.com*

78. Show that an equation of the form

$$Ax^2 + Ey = 0 \qquad A \neq 0, E \neq 0$$

is the equation of a parabola with vertex at $(0, 0)$ and axis of symmetry the y-axis. Find its focus and directrix.

79. Show that an equation of the form

$$Cy^2 + Dx = 0 \qquad C \neq 0, D \neq 0$$

is the equation of a parabola with vertex at $(0, 0)$ and axis of symmetry the x-axis. Find its focus and directrix.

80. Show that the graph of an equation of the form

$$Ax^2 + Dx + Ey + F = 0 \qquad A \neq 0$$

(a) Is a parabola if $E \neq 0$.
(b) Is a vertical line if $E = 0$ and $D^2 - 4AF = 0$.
(c) Is two vertical lines if $E = 0$ and $D^2 - 4AF > 0$.
(d) Contains no points if $E = 0$ and $D^2 - 4AF < 0$.

81. Show that the graph of an equation of the form

$$Cy^2 + Dx + Ey + F = 0 \qquad C \neq 0$$

(a) Is a parabola if $D \neq 0$.
(b) Is a horizontal line if $D = 0$ and $E^2 - 4CF = 0$.
(c) Is two horizontal lines if $D = 0$ and $E^2 - 4CF > 0$.
(d) Contains no points if $D = 0$ and $E^2 - 4CF < 0$.

Retain Your Knowledge

Problems 82–85 are based on material learned earlier in the course. The purpose of these problems is to keep the material fresh in your mind so that you are better prepared for the final exam.

82. For $x = 9y^2 - 36$, list the intercepts and test for symmetry.

83. Solve: $4^{x+1} = 8^{x-1}$

84. Given $\tan\theta = -\frac{5}{8}, \frac{\pi}{2} < \theta < \pi$, find the exact value of each of the remaining trigonometric functions.

85. Find the exact value: $\tan\left[\cos^{-1}\left(-\frac{3}{7}\right)\right]$

'Are You Prepared?' Answers

1. $\sqrt{(x_2 - x_1)^2 + (y_2 - y_1)^2}$ 2. 4 3. $\{-7, -1\}$ 4. $(-2, -5)$ 5. 3; up

11.3 The Ellipse

PREPARING FOR THIS SECTION *Before getting started, review the following:*

- Distance Formula (Section 1.1, p. 85)
- Completing the Square (Chapter R, Review, Section R.5, p. 57)
- Intercepts (Section 2.1, pp. 165–166)
- Symmetry (Section 2.1, pp. 166–168)
- Circles (Section 2.3, pp. 189–193)
- Graphing Techniques: Transformations (Section 3.5, pp. 256–264)

Now Work the 'Are You Prepared?' problems on page 811.

OBJECTIVES
1 Analyze Ellipses with Center at the Origin (p. 803)
2 Analyze Ellipses with Center at (h, k) (p. 808)
3 Solve Applied Problems Involving Ellipses (p. 810)

DEFINITION An **ellipse** is the collection of all points in a plane, the sum of whose distances from two fixed points, called the **foci**, is a constant.

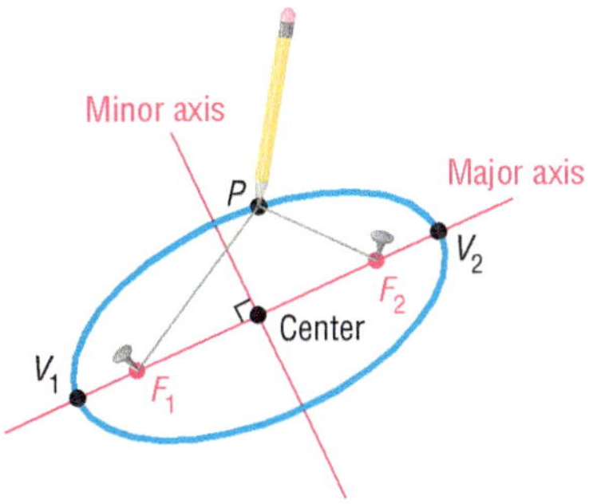

Figure 19 Ellipse

The definition contains within it a physical means for drawing an ellipse. Find a piece of string (the length of this string is the constant referred to in the definition). Then take two thumbtacks (the foci) and stick them into a piece of cardboard so that the distance between them is less than the length of the string. Now attach the ends of the string to the thumbtacks and, using the point of a pencil, pull the string taut. See Figure 19. Keeping the string taut, rotate the pencil around the two thumbtacks. The pencil traces out an ellipse, as shown in Figure 19.

In Figure 19, the foci are labeled F_1 and F_2. The line containing the foci is called the **major axis**. The midpoint of the line segment joining the foci is the **center** of the ellipse. The line through the center and perpendicular to the major axis is the **minor axis**.

The two points of intersection of the ellipse and the major axis are the **vertices**, V_1 and V_2, of the ellipse. The distance from one vertex to the other is the **length of the major axis**. The ellipse is symmetric with respect to its major axis, with respect to its minor axis, and with respect to its center.

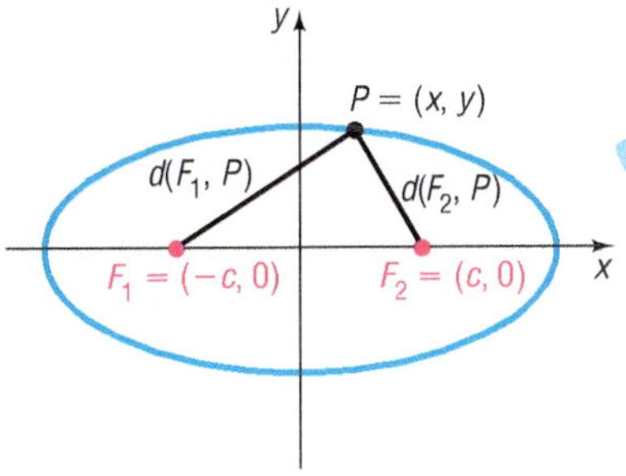

Figure 20

1 Analyze Ellipses with Center at the Origin

With these ideas in mind, we are ready to find the equation of an ellipse in a rectangular coordinate system. First, place the center of the ellipse at the origin. Second, position the ellipse so that its major axis coincides with a coordinate axis, say the x-axis, as shown in Figure 20. If c is the distance from the center to a focus, one focus is at $F_1 = (-c, 0)$ and the other at $F_2 = (c, 0)$. As we shall see, it is convenient to let $2a$ denote the constant distance referred to in the definition. Then, if $P = (x, y)$ is any point on the ellipse,

$$d(F_1, P) + d(F_2, P) = 2a \quad \text{Sum of the distances from } P \text{ to the foci equals a constant, } 2a.$$

$$\sqrt{(x - (-c))^2 + (y - 0)^2} + \sqrt{(x - c)^2 + (y - 0)^2} = 2a \quad \text{Use the Distance Formula.}$$

$$\sqrt{(x + c)^2 + y^2} = 2a - \sqrt{(x - c)^2 + y^2} \quad \text{Isolate one radical.}$$

$$(x + c)^2 + y^2 = 4a^2 - 4a\sqrt{(x - c)^2 + y^2} + (x - c)^2 + y^2 \quad \text{Square both sides.}$$

$$x^2 + 2cx + c^2 + y^2 = 4a^2 - 4a\sqrt{(x - c)^2 + y^2} + x^2 - 2cx + c^2 + y^2 \quad \text{Multiply out.}$$

$$4cx - 4a^2 = -4a\sqrt{(x - c)^2 + y^2} \quad \text{Simplify; isolate the radical.}$$

$$cx - a^2 = -a\sqrt{(x - c)^2 + y^2} \quad \text{Divide each side by 4.}$$

$$(cx - a^2)^2 = a^2[(x - c)^2 + y^2]$$ **Square both sides again.**

$$c^2x^2 - 2a^2cx + a^4 = a^2(x^2 - 2cx + c^2 + y^2)$$ **Multiply out.**

$$(c^2 - a^2)x^2 - a^2y^2 = a^2c^2 - a^4$$ **Rearrange the terms.**

$$(a^2 - c^2)x^2 + a^2y^2 = a^2(a^2 - c^2)$$ **Multiply each side by -1; factor a^2 on the right side.** **(1)**

To obtain points on the ellipse off the x-axis, it must be that $a > c$. To see why, look again at Figure 20. Then

$$d(F_1, P) + d(F_2, P) > d(F_1, F_2)$$ **The sum of the lengths of two sides of a triangle is greater than the length of the third side.**

$$2a > 2c$$ **$d(F_1, P) + d(F_2, P) = 2a$; $d(F_1, F_2) = 2c$**

$$a > c$$

Because $a > c > 0$, this means $a^2 > c^2$, so $a^2 - c^2 > 0$. Let $b^2 = a^2 - c^2$, $b > 0$. Then $a > b$ and equation (1) can be written as

$$b^2x^2 + a^2y^2 = a^2b^2$$

$$\frac{x^2}{a^2} + \frac{y^2}{b^2} = 1$$ **Divide each side by a^2b^2.**

As you can verify, the graph of this equation has symmetry with respect to the x-axis, y-axis, and origin.

Because the major axis is the x-axis, find the vertices of this ellipse by letting $y = 0$. The vertices satisfy the equation $\frac{x^2}{a^2} = 1$, the solutions of which are $x = \pm a$. Consequently, the vertices of this ellipse are $V_1 = (-a, 0)$ and $V_2 = (a, 0)$. The y-intercepts of the ellipse, found by letting $x = 0$, have coordinates $(0, -b)$ and $(0, b)$. These four intercepts, $(a, 0)$, $(-a, 0)$, $(0, b)$, and $(0, -b)$, are used to graph the ellipse.

THEOREM

Equation of an Ellipse: Center at (0, 0); Major Axis along the x-Axis

An equation of the ellipse with center at $(0, 0)$, foci at $(-c, 0)$ and $(c, 0)$, and vertices at $(-a, 0)$ and $(a, 0)$ is

$$\frac{x^2}{a^2} + \frac{y^2}{b^2} = 1, \quad \text{where } a > b > 0 \text{ and } b^2 = a^2 - c^2 \qquad \textbf{(2)}$$

The major axis is the x-axis. See Figure 21.

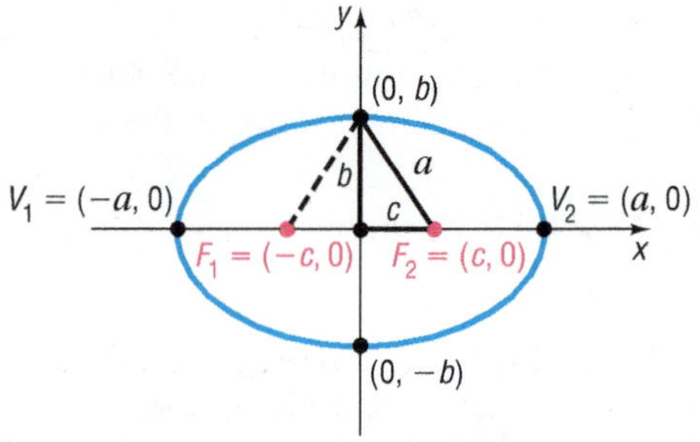

Figure 21

Notice in Figure 21 the right triangle formed by the points $(0, 0)$, $(c, 0)$, and $(0, b)$. Because $b^2 = a^2 - c^2$ (or $b^2 + c^2 = a^2$), the distance from the focus at $(c, 0)$ to the point $(0, b)$ is a.

This can be seen another way. Look at the two right triangles in Figure 21. They are congruent. Do you see why? Because the sum of the distances from the foci to a point on the ellipse is $2a$, it follows that the distance from $(c, 0)$ to $(0, b)$ is a.

EXAMPLE 1 Finding an Equation of an Ellipse

Find an equation of the ellipse with center at the origin, one focus at $(3, 0)$, and a vertex at $(-4, 0)$. Graph the equation.

Solution The ellipse has its center at the origin and, since the given focus and vertex lie on the x-axis, the major axis is the x-axis. The distance from the center, $(0, 0)$, to one of the foci, $(3, 0)$, is $c = 3$. The distance from the center, $(0, 0)$, to one of the vertices, $(-4, 0)$, is $a = 4$. From equation (2), it follows that

$$b^2 = a^2 - c^2 = 16 - 9 = 7$$

so an equation of the ellipse is

$$\frac{x^2}{16} + \frac{y^2}{7} = 1$$

Figure 22 shows the graph. ■

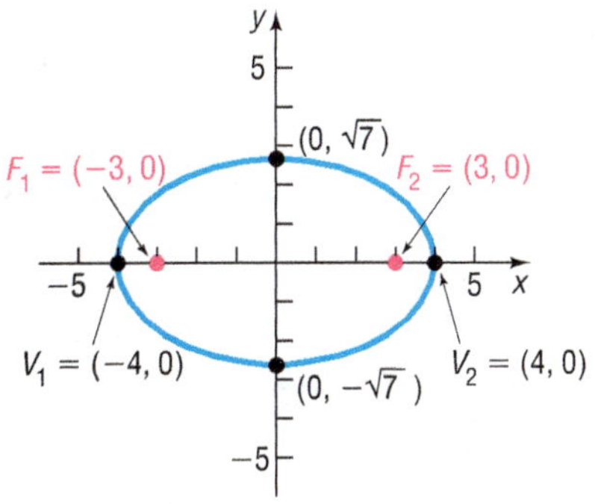

Figure 22 $\frac{x^2}{16} + \frac{y^2}{7} = 1$

In Figure 22, the intercepts of the equation are used to graph the ellipse. Following this practice will make it easier for you to obtain an accurate graph of an ellipse when graphing by hand. The intercepts also tell you how to set the initial viewing window when using a graphing utility.

Now Work PROBLEM 27

EXAMPLE 2 Graphing an Ellipse Using a Graphing Utility

Use a graphing utility to graph the ellipse $\frac{x^2}{16} + \frac{y^2}{7} = 1$.

Solution First, solve $\frac{x^2}{16} + \frac{y^2}{7} = 1$ for y.

$$\frac{y^2}{7} = 1 - \frac{x^2}{16} \qquad \text{Subtract } \frac{x^2}{16} \text{ from each side.}$$

$$y^2 = 7\left(1 - \frac{x^2}{16}\right) \qquad \text{Multiply both sides by 7.}$$

$$y = \pm\sqrt{7\left(1 - \frac{x^2}{16}\right)} \qquad \text{Apply the Square Root Method.}$$

Figure 23* shows the graphs of $Y_1 = \sqrt{7\left(1 - \frac{x^2}{16}\right)}$ and $Y_2 = -\sqrt{7\left(1 - \frac{x^2}{16}\right)}$. ■

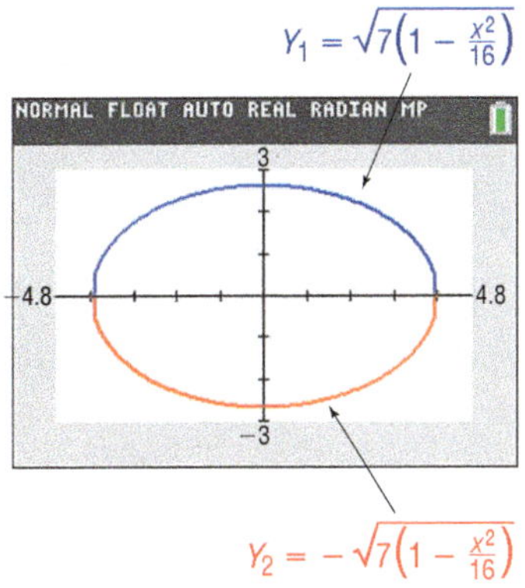

Figure 23

In Figure 23 a square screen is used. As with circles and parabolas, this is done to avoid a distorted view of the graph.

An equation of the form of equation (2), with $a > b$, is the equation of an ellipse with center at the origin, foci on the x-axis at $(-c, 0)$ and $(c, 0)$, where $c^2 = a^2 - b^2$, and major axis along the x-axis.

For the remainder of this section, the direction **"Analyze the equation"** will mean to find the center, major axis, foci, and vertices of the ellipse and graph it.

*The initial viewing window selected was $X\text{min} = -4$, $X\text{max} = 4$, $Y\text{min} = -3$, $Y\text{max} = 3$. Then we used the ZOOM-SQUARE option to obtain the window shown.

EXAMPLE 3 Analyzing the Equation of an Ellipse

Analyze the equation: $\frac{x^2}{25} + \frac{y^2}{9} = 1$

Solution The given equation is of the form of equation (2), with $a^2 = 25$ and $b^2 = 9$. The equation is that of an ellipse with center $(0, 0)$ and major axis along the x-axis. The vertices are at $(\pm a, 0) = (\pm 5, 0)$. Because $b^2 = a^2 - c^2$, this means

$$c^2 = a^2 - b^2 = 25 - 9 = 16$$

The foci are at $(\pm c, 0) = (\pm 4, 0)$. The y-intercepts are $(0, \pm b) = (0, \pm 3)$. Figure 24(a) shows the graph drawn by hand. Figure 24(b) shows the graph obtained using a graphing utility.

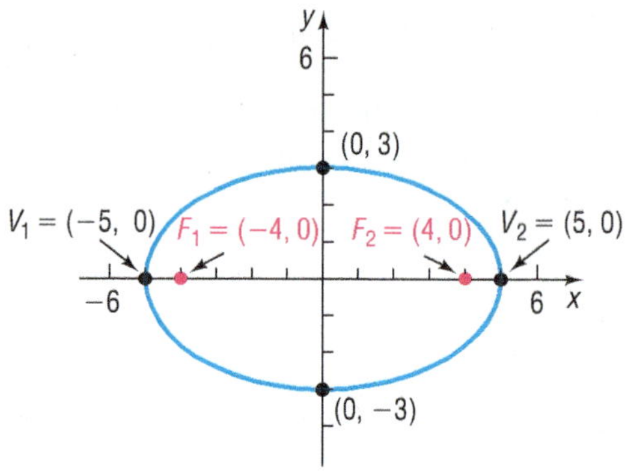

(a)

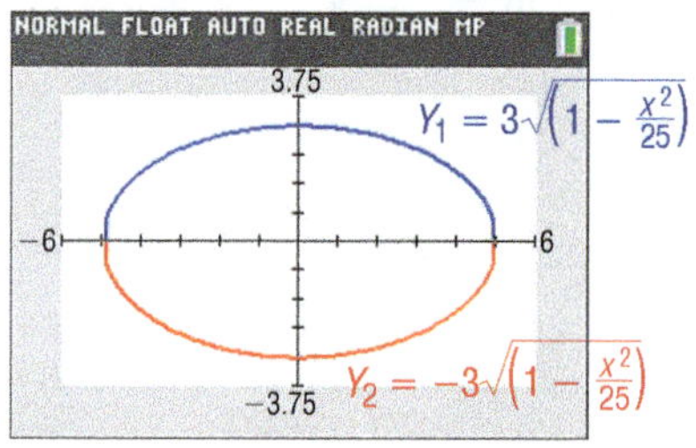

(b)

Figure 24 $\frac{x^2}{25} + \frac{y^2}{9} = 1$

Now Work PROBLEM 17

If the major axis of an ellipse with center at $(0, 0)$ lies on the y-axis, then the foci are at $(0, -c)$ and $(0, c)$. Using the same steps as before, the definition of an ellipse leads to the following result:

THEOREM

Equation of an Ellipse; Center at (0, 0); Major Axis along the *y*-Axis

An equation of the ellipse with center at $(0, 0)$, foci at $(0, -c)$ and $(0, c)$, and vertices at $(0, -a)$ and $(0, a)$ is

$$\frac{x^2}{b^2} + \frac{y^2}{a^2} = 1 \quad \text{where } a > b > 0 \text{ and } b^2 = a^2 - c^2 \qquad (3)$$

The major axis is the y-axis.

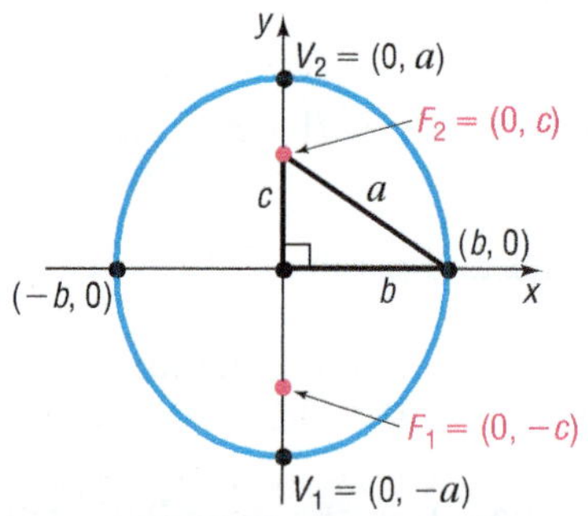

Figure 25 $\frac{x^2}{b^2} + \frac{y^2}{a^2} = 1, a > b > 0$

Figure 25 illustrates the graph of such an ellipse. Again, notice the right triangle formed by the points at $(0, 0)$, $(b, 0)$, and $(0, c)$, so that $a^2 = b^2 + c^2$ (or $b^2 = a^2 - c^2$).

Look closely at equations (2) and (3). Although they may look alike, there is a difference! In equation (2), the larger number, a^2, is in the denominator of the x^2-term, so the major axis of the ellipse is along the x-axis. In equation (3), the larger number, a^2, is in the denominator of the y^2-term, so the major axis is along the y-axis.

EXAMPLE 4 Analyzing the Equation of an Ellipse

Analyze the equation: $9x^2 + y^2 = 9$

Solution To put the equation in proper form, divide each side by 9.

$$x^2 + \frac{y^2}{9} = 1$$

The larger denominator, 9, is in the y^2-term so, based on equation (3), this is the equation of an ellipse with center at the origin and major axis along the y-axis. Also, $a^2 = 9$, $b^2 = 1$, and $c^2 = a^2 - b^2 = 9 - 1 = 8$. The vertices are at $(0, \pm a) = (0, \pm 3)$, and the foci are at $(0, \pm c) = (0, \pm 2\sqrt{2})$. The x-intercepts are at $(\pm b, 0) = (\pm 1, 0)$. Figure 26(a) shows the graph drawn by hand. Figure 26(b) shows the graph obtained using a graphing utility.

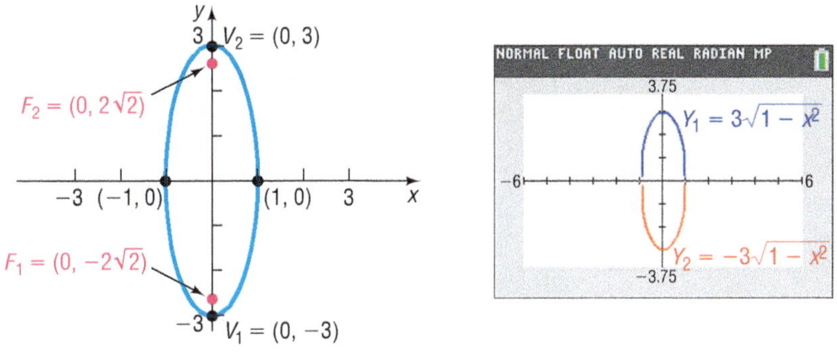

(a) (b)

Figure 26 $9x^2 + y^2 = 9$

Now Work PROBLEM 21

EXAMPLE 5 Finding an Equation of an Ellipse

Find an equation of the ellipse having one focus at $(0, 2)$ and vertices at $(0, -3)$ and $(0, 3)$. Graph the equation.

Solution By plotting the given focus and vertices, we find that the major axis is the y-axis. Because the vertices are at $(0, -3)$ and $(0, 3)$, the center of this ellipse is at their midpoint, the origin. The distance from the center, $(0, 0)$, to the given focus, $(0, 2)$, is $c = 2$. The distance from the center, $(0, 0)$, to one of the vertices, $(0, 3)$, is $a = 3$. So $b^2 = a^2 - c^2 = 9 - 4 = 5$. The form of the equation of this ellipse is given by equation (3).

$$\frac{x^2}{b^2} + \frac{y^2}{a^2} = 1$$

$$\frac{x^2}{5} + \frac{y^2}{9} = 1$$

Figure 27 shows the graph.

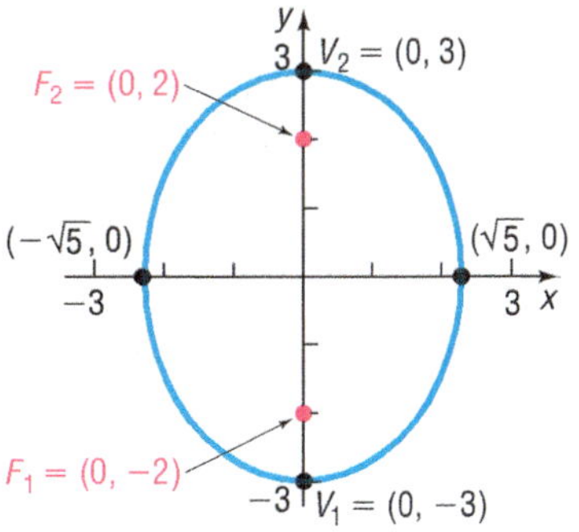

Figure 27 $\frac{x^2}{5} + \frac{y^2}{9} = 1$

Now Work PROBLEM 29

The circle may be considered a special kind of ellipse. To see why, let $a = b$ in equation (2) or (3). Then

$$\frac{x^2}{a^2} + \frac{y^2}{a^2} = 1$$

$$x^2 + y^2 = a^2$$

This is the equation of a circle with center at the origin and radius a. The value of c is

$$c^2 = a^2 - b^2 = 0$$
$$\uparrow$$
$$a = b$$

This indicates that the closer the two foci of an ellipse are to the center, the more the ellipse will look like a circle.

2 Analyze Ellipses with Center at (h, k)

If an ellipse with center at the origin and major axis coinciding with a coordinate axis is shifted horizontally h units and then vertically k units, the result is an ellipse with center at (h, k) and major axis parallel to a coordinate axis. The equations of such ellipses have the same forms as those given in equations (2) and (3), except that x is replaced by $x - h$ (the horizontal shift) and y is replaced by $y - k$ (the vertical shift). Table 3 gives the forms of the equations of such ellipses, and Figure 28 shows their graphs.

Note: It is not recommended that Table 3 be memorized. Rather, use transformations (shift horizontally h units, vertically k units), along with the fact that a represents the distance from the center to the vertices, c represents the distance from the center to the foci, and $b^2 = a^2 - c^2$ (or $c^2 = a^2 - b^2$). ■

Table 3

Equations of an Ellipse: Center at (h, k); Major Axis Parallel to a Coordinate Axis

Center	Major Axis	Foci	Vertices	Equation
(h, k)	Parallel to the x-axis	$(h + c, k)$	$(h + a, k)$	$\dfrac{(x-h)^2}{a^2} + \dfrac{(y-k)^2}{b^2} = 1,$
		$(h - c, k)$	$(h - a, k)$	$a > b > 0$ and $b^2 = a^2 - c^2$
(h, k)	Parallel to the y-axis	$(h, k + c)$	$(h, k + a)$	$\dfrac{(x-h)^2}{b^2} + \dfrac{(y-k)^2}{a^2} = 1,$
		$(h, k - c)$	$(h, k - a)$	$a > b > 0$ and $b^2 = a^2 - c^2$

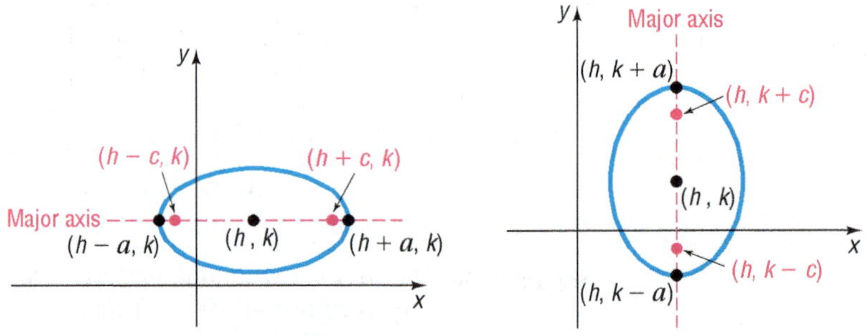

Figure 28 **(a)** $\dfrac{(x-h)^2}{a^2} + \dfrac{(y-k)^2}{b^2} = 1$ **(b)** $\dfrac{(x-h)^2}{b^2} + \dfrac{(y-k)^2}{a^2} = 1$

EXAMPLE 6 **Finding an Equation of an Ellipse, Center Not at the Origin**

Find an equation for the ellipse with center at $(2, -3)$, one focus at $(3, -3)$, and one vertex at $(5, -3)$. Graph the equation.

Solution The center is at $(h, k) = (2, -3)$, so $h = 2$ and $k = -3$. Plot the center, focus, and vertex, and note that the points all lie on the line $y = -3$. Therefore, the major axis is parallel to the x-axis. The distance from the center $(2, -3)$ to a focus $(3, -3)$ is $c = 1$; the distance from the center $(2, -3)$ to a vertex $(5, -3)$ is $a = 3$. Then $b^2 = a^2 - c^2 = 9 - 1 = 8$. The form of the equation is

$$\frac{(x-h)^2}{a^2} + \frac{(y-k)^2}{b^2} = 1 \quad h = 2, k = -3, a = 3, b = 2\sqrt{2}$$

$$\frac{(x-2)^2}{9} + \frac{(y+3)^2}{8} = 1$$

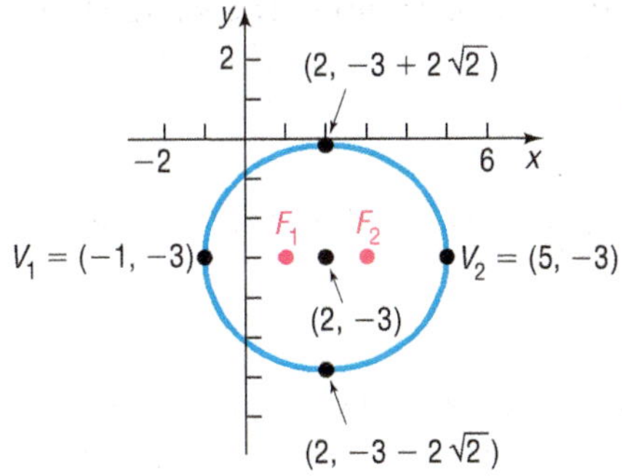

Figure 29 $\frac{(x-2)^2}{9} + \frac{(y+3)^2}{8} = 1$

To graph the equation, use the center $(h, k) = (2, -3)$ to locate the vertices. The major axis is parallel to the x-axis, so the vertices are $a = 3$ units left and right of the center $(2, -3)$. Therefore, the vertices are

$$V_1 = (2 - 3, -3) = (-1, -3) \quad \text{and} \quad V_2 = (2 + 3, -3) = (5, -3)$$

Since $c = 1$ and the major axis is parallel to the x-axis, the foci are 1 unit left and right of the center. Therefore, the foci are

$$F_1 = (2 - 1, -3) = (1, -3) \quad \text{and} \quad F_2 = (2 + 1, -3) = (3, -3)$$

Finally, use the value of $b = 2\sqrt{2}$ to find the two points above and below the center.

$$(2, -3 - 2\sqrt{2}) \quad \text{and} \quad (2, -3 + 2\sqrt{2})$$

Figure 29 shows the graph. ■

Now Work PROBLEM 55

EXAMPLE 7 Using a Graphing Utility to Graph an Ellipse, Center Not at the Origin

Using a graphing utility, graph the ellipse: $\frac{(x-2)^2}{9} + \frac{(y+3)^2}{8} = 1$

Solution First, solve the equation $\frac{(x-2)^2}{9} + \frac{(y+3)^2}{8} = 1$ for y.

$$\frac{(y+3)^2}{8} = 1 - \frac{(x-2)^2}{9} \qquad \text{Subtract } \frac{(x-2)^2}{9} \text{ from each side.}$$

$$(y+3)^2 = 8\left[1 - \frac{(x-2)^2}{9}\right] \qquad \text{Multiply each side by 8.}$$

$$y + 3 = \pm\sqrt{8\left[1 - \frac{(x-2)^2}{9}\right]} \qquad \text{Apply the Square Root Method.}$$

$$y = -3 \pm \sqrt{8\left[1 - \frac{(x-2)^2}{9}\right]} \qquad \text{Subtract 3 from each side.}$$

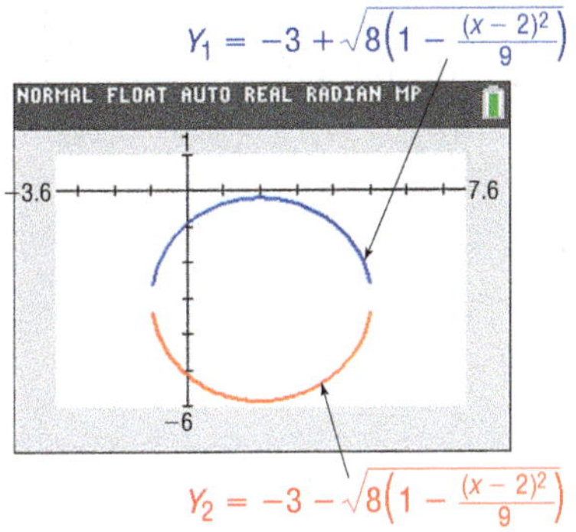

Figure 30

Figure 30 shows the graphs of $Y_1 = -3 + \sqrt{8\left[1 - \frac{(x-2)^2}{9}\right]}$ and $Y_2 = -3 - \sqrt{8\left[1 - \frac{(x-2)^2}{9}\right]}$. ■

EXAMPLE 8 Analyzing the Equation of an Ellipse

Analyze the equation: $4x^2 + y^2 - 8x + 4y + 4 = 0$

Solution Complete the squares in x and in y.

$$4x^2 + y^2 - 8x + 4y + 4 = 0$$

$$4x^2 - 8x + y^2 + 4y = -4 \qquad \text{Group like variables; place the constant on the right side.}$$

$$4(x^2 - 2x) + (y^2 + 4y) = -4 \qquad \text{Factor out 4 from the first two terms.}$$

$$4(x^2 - 2x + 1) + (y^2 + 4y + 4) = -4 + 4 + 4 \qquad \text{Complete each square.}$$

$$4(x-1)^2 + (y+2)^2 = 4 \qquad \text{Factor.}$$

$$(x-1)^2 + \frac{(y+2)^2}{4} = 1 \qquad \text{Divide each side by 4.}$$

This is the equation of an ellipse with center at $(1, -2)$ and major axis parallel to the y-axis. Since $a^2 = 4$ and $b^2 = 1$, we have $c^2 = a^2 - b^2 = 4 - 1 = 3$. The vertices are at $(h, k \pm a) = (1, -2 \pm 2)$ or $(1, -4)$ and $(1, 0)$. The foci are at $(h, k \pm c) = (1, -2 \pm \sqrt{3})$ or $(1, -2 - \sqrt{3})$ and $(1, -2 + \sqrt{3})$. Figure 31(a) shows the graph drawn by hand. Figure 31(b) shows the graph obtained using a graphing utility.

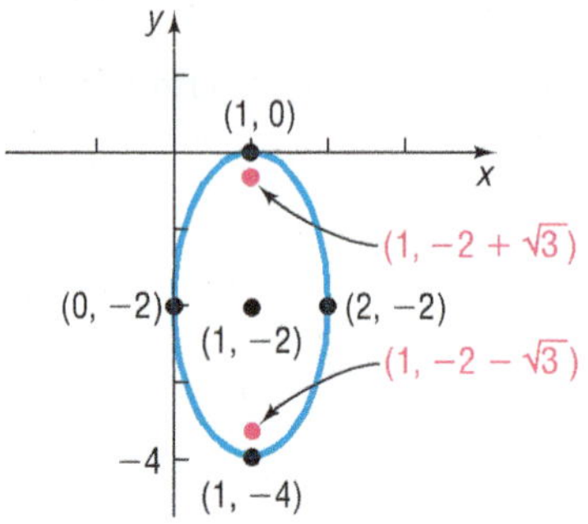

(a)

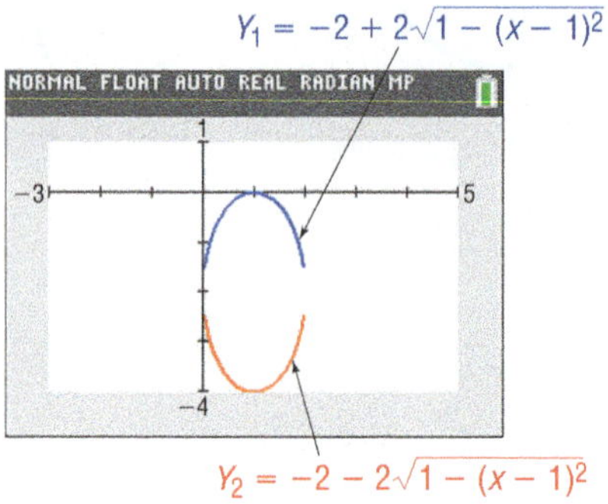

(b)

Figure 31 $4x^2 + y^2 - 8x + 4y + 4 = 0$

Now Work PROBLEM 47

3 Solve Applied Problems Involving Ellipses

Ellipses are found in many applications in science and engineering. For example, the orbits of the planets around the Sun are elliptical, with the Sun's position at a focus. See Figure 32.

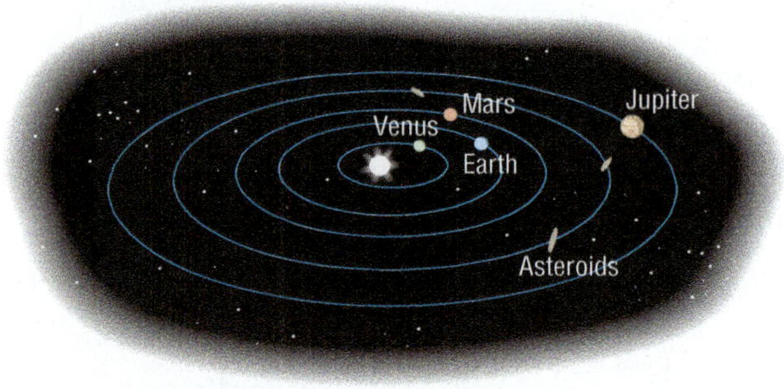

Figure 32 Planet orbits

Stone and concrete bridges are often shaped as semielliptical arches. Elliptical gears are used in machinery when a variable rate of motion is required.

Ellipses also have an interesting reflection property. If a source of light (or sound) is placed at one focus, the waves transmitted by the source reflect off the ellipse and concentrate at the other focus. This is the principle behind *whispering galleries,* which are rooms designed with elliptical ceilings. A person standing at one focus of the ellipse can whisper and be heard by a person standing at the other focus, because all the sound waves that reach the ceiling are reflected to the other person.

EXAMPLE 9 **A Whispering Gallery**

The whispering gallery in the Museum of Science and Industry in Chicago is 47.3 feet long. The distance from the center of the room to the foci is 20.3 feet. Find an equation that describes the shape of the room. How high is the room at its center?

Source: *Chicago Museum of Science and Industry Web site; www.msichicago.org*

Solution Set up a rectangular coordinate system so that the center of the ellipse is at the origin and the major axis is along the x-axis. The equation of the ellipse is

$$\frac{x^2}{a^2} + \frac{y^2}{b^2} = 1$$

Since the length of the room is 47.3 feet, the distance from the center of the room to each vertex (the end of the room) is $\frac{47.3}{2} = 23.65$ feet, so $a = 23.65$ feet. The distance from the center of the room to each focus is $c = 20.3$ feet. See Figure 33.

Since $b^2 = a^2 - c^2$, this means that $b^2 = 23.65^2 - 20.3^2 = 147.2325$. An equation that describes the shape of the room is given by

$$\frac{x^2}{23.65^2} + \frac{y^2}{147.2325} = 1$$

The height of the room at its center is $b = \sqrt{147.2325} \approx 12.1$ feet.

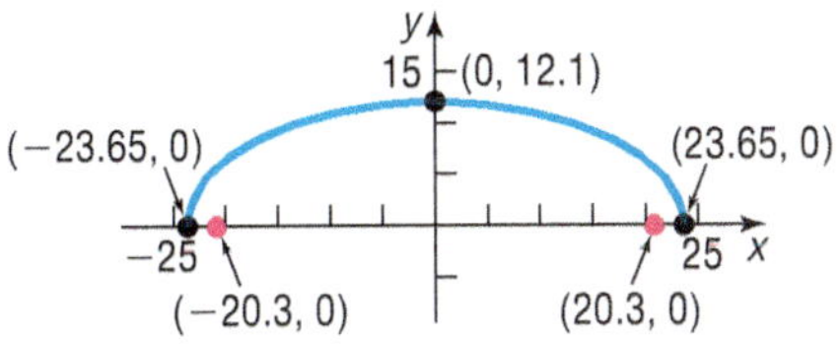

Figure 33

Now Work PROBLEM 71

11.3 Assess Your Understanding

'Are You Prepared?' *Answers are given at the end of these exercises. If you get a wrong answer, read the pages listed in red.*

1. The distance d from $P_1 = (2, -5)$ to $P_2 = (4, -2)$ is $d =$ ________. (p. 85)
2. To complete the square of $x^2 - 3x$, add ______. (p. 57)
3. Find the intercepts of the equation $y^2 = 16 - 4x^2$. (pp. 165–166)
4. The point that is symmetric with respect to the y-axis to the point $(-2, 5)$ is ________. (pp. 166–168)
5. To graph $y = (x + 1)^2 - 4$, shift the graph of $y = x^2$ to the (left/right) ______ unit(s) and then (up/down) ______ unit(s). (pp. 256–264)
6. The standard equation of a circle with center at $(2, -3)$ and radius 1 is ________. (pp. 189–193)

Concepts and Vocabulary

7. A(n) ________ is the collection of all points in a plane the sum of whose distances from two fixed points is a constant.
8. For an ellipse, the foci lie on a line called the ________.
   (a) minor axis (b) major axis
   (c) directrix (d) latus rectum
9. For the ellipse $\frac{x^2}{4} + \frac{y^2}{25} = 1$, the vertices are the points ________ and ________.
10. For the ellipse $\frac{x^2}{25} + \frac{y^2}{9} = 1$, the value of a is _____, the value of b is _____, and the major axis is the _____-axis.
11. If the center of an ellipse is $(2, -3)$, the major axis is parallel to the x-axis, and the distance from the center of the ellipse to its vertices is $a = 4$ units, then the coordinates of the vertices are ________ and ________.
12. If the foci of an ellipse are $(-4, 4)$ and $(6, 4)$, then the coordinates of the center of the ellipse are ________.
    (a) $(1, 4)$ (b) $(4, 1)$
    (c) $(1, 0)$ (d) $(5, 4)$

Skill Building

In Problems 13–16, the graph of an ellipse is given. Match each graph to its equation.

(A) $\frac{x^2}{4} + y^2 = 1$ (B) $x^2 + \frac{y^2}{4} = 1$ (C) $\frac{x^2}{16} + \frac{y^2}{4} = 1$ (D) $\frac{x^2}{4} + \frac{y^2}{16} = 1$

13.
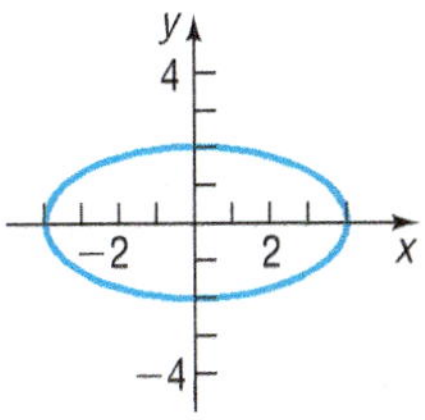

14.
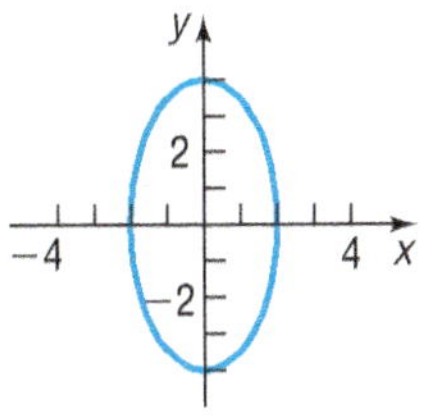

15.
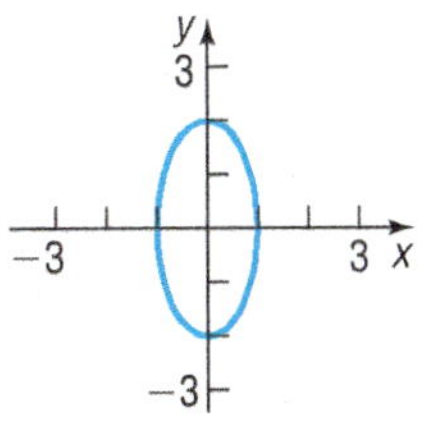

16.
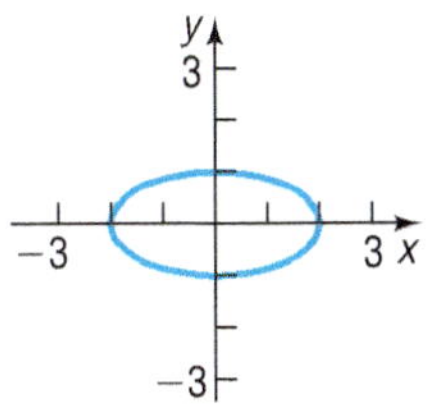

In Problems 17–26, find the vertices and foci of each ellipse. Graph each equation by hand. Verify your graph using a graphing utility.

17. $\frac{x^2}{25} + \frac{y^2}{4} = 1$ **18.** $\frac{x^2}{9} + \frac{y^2}{4} = 1$ **19.** $\frac{x^2}{9} + \frac{y^2}{25} = 1$ **20.** $x^2 + \frac{y^2}{16} = 1$

21. $4x^2 + y^2 = 16$ **22.** $x^2 + 9y^2 = 18$ **23.** $4y^2 + x^2 = 8$ **24.** $4y^2 + 9x^2 = 36$

25. $x^2 + y^2 = 16$ **26.** $x^2 + y^2 = 4$

In Problems 27–38, find an equation for each ellipse. Graph the equation by hand.

27. Center at $(0, 0)$; focus at $(3, 0)$; vertex at $(5, 0)$

28. Center at $(0, 0)$; focus at $(-1, 0)$; vertex at $(3, 0)$

29. Center at $(0, 0)$; focus at $(0, -4)$; vertex at $(0, 5)$

30. Center at $(0, 0)$; focus at $(0, 1)$; vertex at $(0, -2)$

31. Foci at $(\pm 2, 0)$; length of the major axis is 6

32. Foci at $(0, \pm 2)$; length of the major axis is 8

33. Focus at $(-4, 0)$; vertices at $(\pm 5, 0)$

34. Focus at $(0, -4)$; vertices at $(0, \pm 8)$

35. Foci at $(0, \pm 3)$; x-intercepts are ± 2

36. Vertices at $(\pm 4, 0)$; y-intercepts are ± 1

37. Center at $(0, 0)$; vertex at $(0, 4)$; $b = 1$

38. Vertices at $(\pm 5, 0)$; $c = 2$

In Problems 39–42, write an equation for each ellipse.

39.

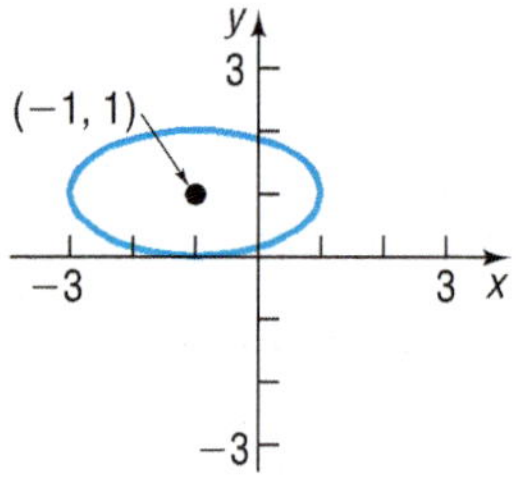

40.

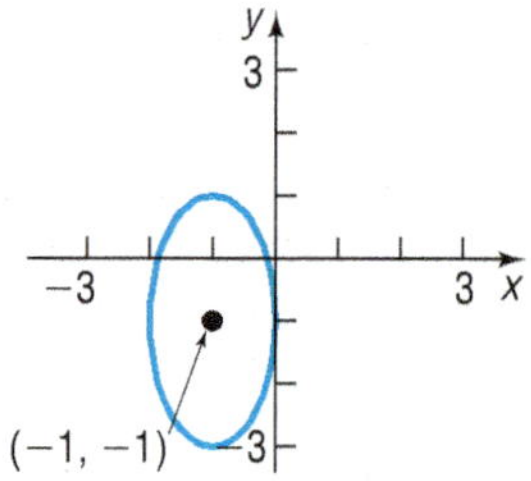

41.

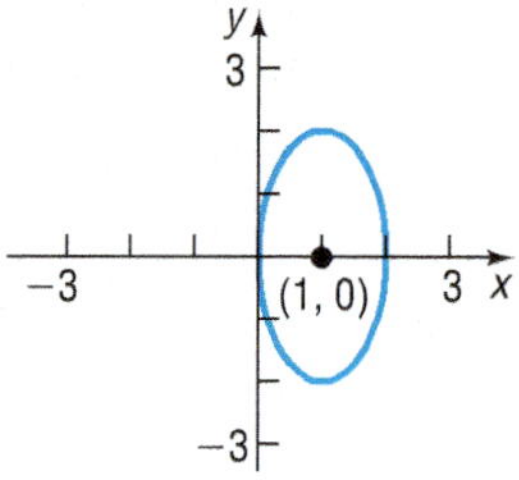

42.

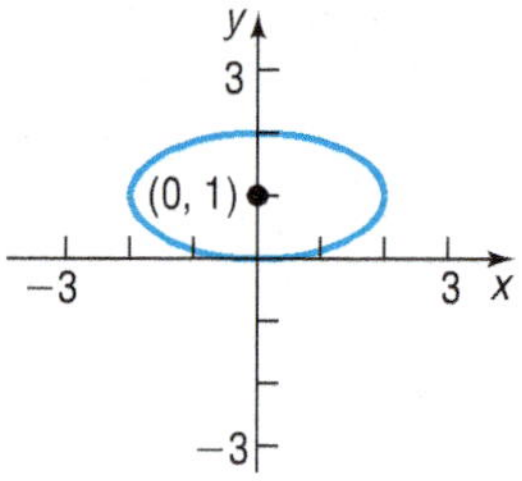

In Problems 43–54, analyze each equation; that is, find the center, foci, and vertices of each ellipse. Graph each equation by hand. Verify your graph using a graphing utility.

43. $\frac{(x-3)^2}{4} + \frac{(y+1)^2}{9} = 1$ **44.** $\frac{(x+4)^2}{9} + \frac{(y+2)^2}{4} = 1$ **45.** $(x+5)^2 + 4(y-4)^2 = 16$

46. $9(x-3)^2 + (y+2)^2 = 18$ **47.** $x^2 + 4x + 4y^2 - 8y + 4 = 0$ **48.** $x^2 + 3y^2 - 12y + 9 = 0$

49. $2x^2 + 3y^2 - 8x + 6y + 5 = 0$ **50.** $4x^2 + 3y^2 + 8x - 6y = 5$ **51.** $9x^2 + 4y^2 - 18x + 16y - 11 = 0$

52. $x^2 + 9y^2 + 6x - 18y + 9 = 0$ **53.** $4x^2 + y^2 + 4y = 0$ **54.** $9x^2 + y^2 - 18x = 0$

In Problems 55–64, find an equation for each ellipse. Graph the equation by hand.

55. Center at $(2, -2)$; vertex at $(7, -2)$; focus at $(4, -2)$

56. Center at $(-3, 1)$; vertex at $(-3, 3)$; focus at $(-3, 0)$

57. Vertices at $(4, 3)$ and $(4, 9)$; focus at $(4, 8)$

58. Foci at $(1, 2)$ and $(-3, 2)$; vertex at $(-4, 2)$

59. Foci at $(5, 1)$ and $(-1, 1)$; length of the major axis is 8

60. Vertices at $(2, 5)$ and $(2, -1)$; $c = 2$

61. Center at $(1, 2)$; focus at $(4, 2)$; contains the point $(1, 3)$

62. Center at $(1, 2)$; focus at $(1, 4)$; contains the point $(2, 2)$

63. Center at $(1, 2)$; vertex at $(4, 2)$; contains the point $(1, 5)$

64. Center at $(1, 2)$; vertex at $(1, 4)$; contains the point $(1 + \sqrt{3}, 3)$

In Problems 65–68, graph each function. Be sure to label all the intercepts. [**Hint:** *Notice that each function is half an ellipse.*]

65. $f(x) = \sqrt{16 - 4x^2}$ **66.** $f(x) = \sqrt{9 - 9x^2}$ **67.** $f(x) = -\sqrt{64 - 16x^2}$ **68.** $f(x) = -\sqrt{4 - 4x^2}$

Applications and Extensions

69. Semielliptical Arch Bridge An arch in the shape of the upper half of an ellipse is used to support a bridge that is to span a river 20 meters wide. The center of the arch is 6 meters above the center of the river. See the figure. Write an equation for the ellipse in which the x-axis coincides with the water level and the y-axis passes through the center of the arch.

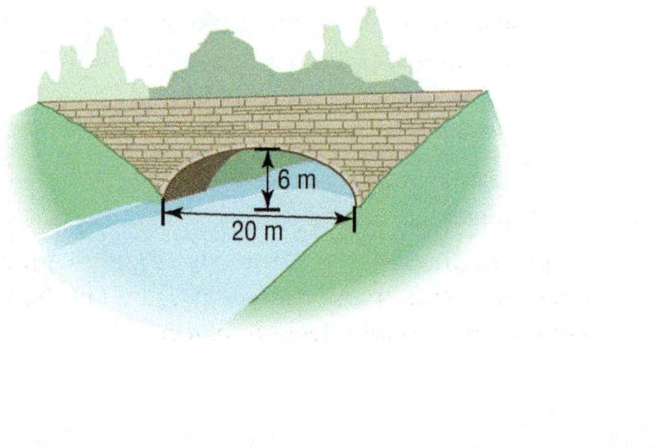

70. Semielliptical Arch Bridge The arch of a bridge is a semiellipse with a horizontal major axis. The span is 30 feet, and the top of the arch is 10 feet above the major axis. The roadway is horizontal and is 2 feet above the top of the arch. Find the vertical distance from the roadway to the arch at 5-foot intervals along the roadway.

71. Whispering Gallery A hall 100 feet in length is to be designed as a whispering gallery. If the foci are located 25 feet from the center, how high will the ceiling be at the center?

72. Whispering Gallery Jim, standing at one focus of a whispering gallery, is 6 feet from the nearest wall. His friend is standing at the other focus, 100 feet away. What is the length of this whispering gallery? How high is its elliptical ceiling at the center?

73. Semielliptical Arch Bridge A bridge is built in the shape of a semielliptical arch. The bridge has a span of 120 feet and a maximum height of 25 feet. Choose a suitable rectangular coordinate system and find the height of the arch at distances of 10, 30, and 50 feet from the center.

74. Semielliptical Arch Bridge A bridge is to be built in the shape of a semielliptical arch and is to have a span of 100 feet. The height of the arch, at a distance of 40 feet from the center, is to be 10 feet. Find the height of the arch at its center.

75. Racetrack Design Consult the figure. A racetrack is in the shape of an ellipse, 100 feet long and 50 feet wide. What is the width 10 feet from a vertex?

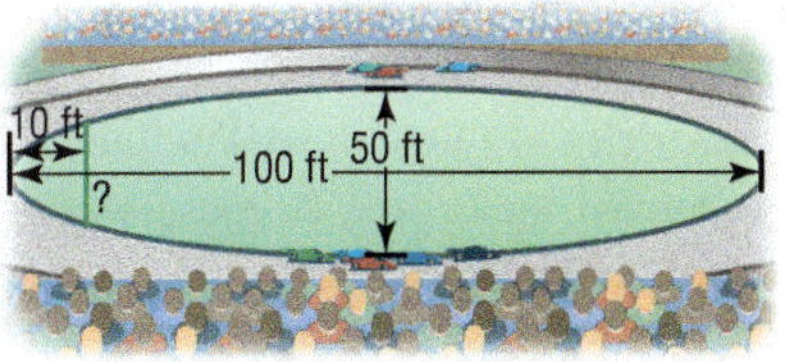

76. Semielliptical Arch Bridge An arch for a bridge over a highway is in the form of half an ellipse. The top of the arch is 20 feet above the ground level (the major axis). The highway has four lanes, each 12 feet wide; a center safety strip 8 feet wide; and two side strips, each 4 feet wide. What should the span of the bridge be (the length of its major axis) if the height 28 feet from the center is to be 13 feet?

77. Installing a Vent Pipe A homeowner is putting in a fireplace that has a 4-inch-radius vent pipe. He needs to cut an elliptical hole in his roof to accommodate the pipe. If the pitch of his roof is $\frac{5}{4}$, (a rise of 5, run of 4) what are the dimensions of the hole?

Source: *www.doe.virginia.gov*

78. Volume of a Football A football is in the shape of a **prolate spheroid,** which is simply a solid obtained by rotating an ellipse $\left(\frac{x^2}{a^2}+\frac{y^2}{b^2}=1\right)$ about its major axis. An inflated NFL football averages 11.125 inches in length and 28.25 inches in center circumference. If the volume of a prolate spheroid is $\frac{4}{3}\pi ab^2$, how much air does the football contain? (Neglect material thickness).

Source: *www.nfl.com*

In Problems 79–83, use the fact that the orbit of a planet about the Sun is an ellipse, with the Sun at one focus. The ***aphelion*** *of a planet is its greatest distance from the Sun, and the* ***perihelion*** *is its shortest distance. The* ***mean distance*** *of a planet from the Sun is the length of the semimajor axis of the elliptical orbit. See the illustration.*

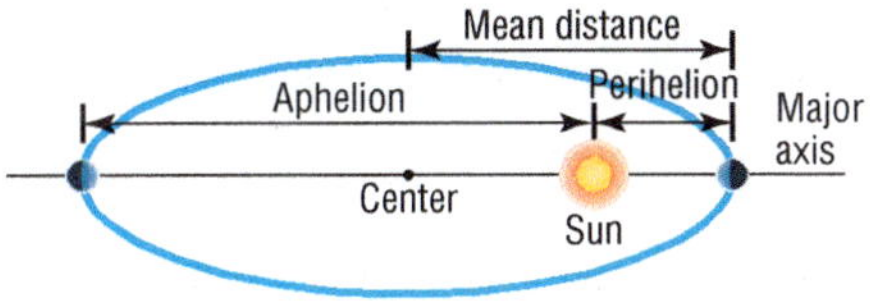

79. Earth The mean distance of Earth from the Sun is 93 million miles. If the aphelion of Earth is 94.5 million miles, what is the perihelion? Write an equation for the orbit of Earth around the Sun.

80. Mars The mean distance of Mars from the Sun is 142 million miles. If the perihelion of Mars is 128.5 million miles, what is the aphelion? Write an equation for the orbit of Mars about the Sun.

81. Jupiter The aphelion of Jupiter is 507 million miles. If the distance from the center of its elliptical orbit to the Sun is 23.2 million miles, what is the perihelion? What is the mean distance? Write an equation for the orbit of Jupiter around the Sun.

82. Pluto The perihelion of Pluto is 4551 million miles, and the distance from the center of its elliptical orbit to the Sun is 897.5 million miles. Find the aphelion of Pluto. What is the mean distance of Pluto from the Sun? Write an equation for the orbit of Pluto about the Sun.

83. Elliptical Orbit A planet orbits a star in an elliptical orbit with the star located at one focus. The perihelion of the planet is 5 million miles. The **eccentricity** e of a conic section is $e=\frac{c}{a}$. If the eccentricity of the orbit is 0.75, find the aphelion of the planet.†

84. A rectangle is inscribed in an ellipse with major axis of length 14 meters and minor axis of length 4 meters. Find the maximum area of a rectangle inscribed in the ellipse. Round your answer to two decimal places.†

85. Let D be the line given by the equation $x+5=0$. Let E be the conic section given by the equation $x^2+5y^2=20$. Let the point C be the vertex of E with the smaller x-coordinate, and let B be the endpoint of the minor axis of E with the larger y-coordinate. Determine the exact y-coordinate of the point M on D that is equidistant from points B and C.†

86. Show that an equation of the form

$$Ax^2+Cy^2+F=0, \qquad A\neq 0, C\neq 0, F\neq 0$$

where A and C are of the same sign and F is of opposite sign,
(a) is the equation of an ellipse with center at $(0,0)$ if $A\neq C$.
(b) is the equation of a circle with center $(0,0)$ if $A=C$.

87. Show that the graph of an equation of the form

$$Ax^2+Cy^2+Dx+Ey+F=0, \qquad A\neq 0, C\neq 0$$

where A and C are of the same sign,
(a) is an ellipse if $\frac{D^2}{4A}+\frac{E^2}{4C}-F$ is the same sign as A.
(b) is a point if $\frac{D^2}{4A}+\frac{E^2}{4C}-F=0$.
(c) contains no points if $\frac{D^2}{4A}+\frac{E^2}{4C}-F$ is of opposite sign to A.

†Courtesy of the Joliet Junior College Mathematics Department

Explaining Concepts: Discussion and Writing

88. The **eccentricity** e of an ellipse is defined as the number $\frac{c}{a}$, where a is the distance of a vertex from the center and c is the distance of a focus from the center. Because $a > c$, it follows that $e < 1$. Write a brief paragraph about the general shape of each of the following ellipses. Be sure to justify your conclusions.
(a) Eccentricity close to 0 (b) Eccentricity = 0.5 (c) Eccentricity close to 1

Retain Your Knowledge

Problems 89–92 are based on material learned earlier in the course. The purpose of these problems is to keep the material fresh in your mind so that you are better prepared for the final exam.

89. Find the zeros of the quadratic function $f(x) = (x - 5)^2 - 12$. What are the x-intercepts, if any, of the graph of the function?

90. Find the domain of the rational function $f(x) = \dfrac{2x - 3}{x - 5}$. Find any horizontal, vertical, or oblique asymptotes.

91. Find the work done by a force of 80 pounds acting in the direction of 50° to the horizontal in moving an object 12 feet from (0, 0) to (12, 0). Round to one decimal place.

92. Solve the right triangle shown.

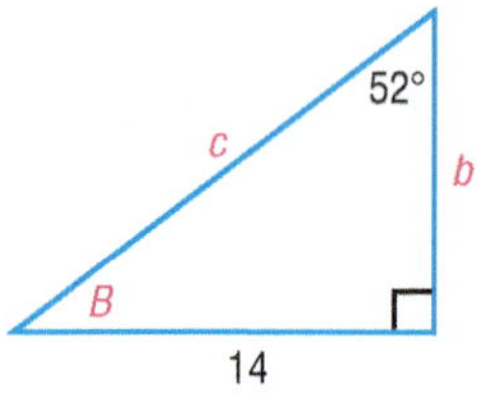

'Are You Prepared?' Answers

1. $\sqrt{13}$ **2.** $\frac{9}{4}$ **3.** $(-2, 0), (2, 0), (0, -4), (0, 4)$ **4.** $(2, 5)$ **5.** left 1; down 4 **6.** $(x - 2)^2 + (y + 3)^2 = 1$

11.4 The Hyperbola

PREPARING FOR THIS SECTION *Before getting started, review the following:*

- Distance Formula (Section 1.1, p. 85)
- Completing the Square (Chapter R, Review, Section R.5, p. 57)
- Intercepts (Section 2.1, pp. 165–166)
- Symmetry (Section 2.1, pp. 166–168)
- Asymptotes (Section 5.4, pp. 374–379)
- Graphing Techniques: Transformations (Section 3.5, pp. 256–264)
- Square Root Method (Section 1.3, p. 112)

Now Work the 'Are You Prepared?' problems on page 824.

OBJECTIVES
1 Analyze Hyperbolas with Center at the Origin (p. 815)
2 Find the Asymptotes of a Hyperbola (p. 819)
3 Analyze Hyperbolas with Center at (h, k) (p. 821)
4 Solve Applied Problems Involving Hyperbolas (p. 823)

DEFINITION A **hyperbola** is the collection of all points in a plane, the difference of whose distances from two fixed points, called the **foci**, is a constant.

Figure 34 illustrates a hyperbola with foci F_1 and F_2. The line containing the foci is called the **transverse axis**. The midpoint of the line segment joining the foci is the **center** of the hyperbola. The line through the center and perpendicular to the transverse axis is the **conjugate axis**. The hyperbola consists of two separate curves, called **branches**, that are symmetric with respect to the transverse axis, conjugate axis, and center. The two points of intersection of the hyperbola and the transverse axis are the **vertices**, V_1 and V_2, of the hyperbola.

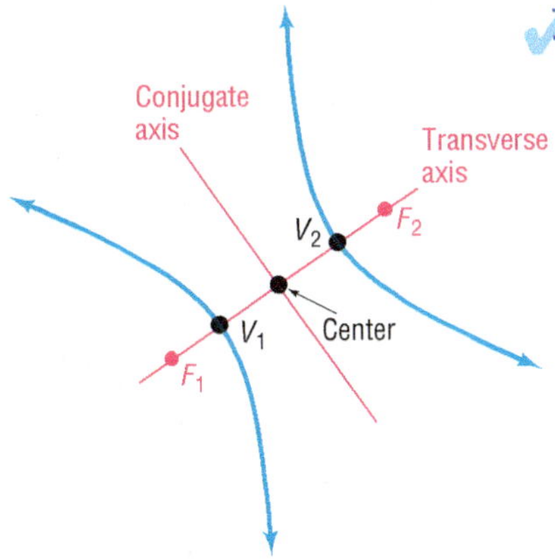

Figure 34 Hyperbola

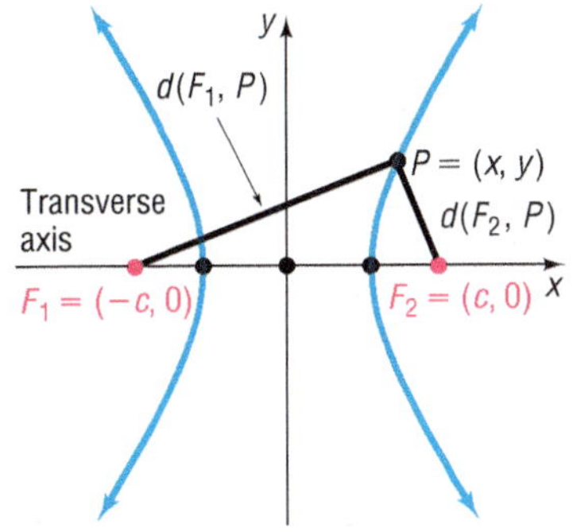

Figure 35
$d(F_1, P) - d(F_2, P) = \pm 2a$

1 Analyze Hyperbolas with Center at the Origin

With these ideas in mind, we are now ready to find the equation of a hyperbola in the rectangular coordinate system. First, place the center at the origin. Next, position the hyperbola so that its transverse axis coincides with a coordinate axis. Suppose that the transverse axis coincides with the x-axis, as shown in Figure 35.

If c is the distance from the center to a focus, one focus is at $F_1 = (-c, 0)$ and the other at $F_2 = (c, 0)$. Now we let the constant difference of the distances from any point $P = (x, y)$ on the hyperbola to the foci F_1 and F_2 be denoted by $\pm 2a$. (If P is on the right branch, the + sign is used; if P is on the left branch, the − sign is used.) The coordinates of P must satisfy the equation

$$d(F_1, P) - d(F_2, P) = \pm 2a \quad \text{Difference of the distances from } P \text{ to the foci equals } \pm 2a.$$

$$\sqrt{(x-(-c))^2 + y^2} - \sqrt{(x-c)^2 + y^2} = \pm 2a \quad \text{Use the Distance Formula.}$$

$$\sqrt{(x+c)^2 + y^2} = \pm 2a + \sqrt{(x-c)^2 + y^2} \quad \text{Isolate one radical.}$$

$$(x+c)^2 + y^2 = 4a^2 \pm 4a\sqrt{(x-c)^2 + y^2} + (x-c)^2 + y^2 \quad \text{Square both sides.}$$

Next multiply out.

$$x^2 + 2cx + c^2 + y^2 = 4a^2 \pm 4a\sqrt{(x-c)^2 + y^2} + x^2 - 2cx + c^2 + y^2$$

$$4cx - 4a^2 = \pm 4a\sqrt{(x-c)^2 + y^2} \quad \text{Simplify; isolate the radical.}$$

$$cx - a^2 = \pm a\sqrt{(x-c)^2 + y^2} \quad \text{Divide each side by 4.}$$

$$(cx - a^2)^2 = a^2[(x-c)^2 + y^2] \quad \text{Square both sides.}$$

$$c^2x^2 - 2ca^2x + a^4 = a^2(x^2 - 2cx + c^2 + y^2) \quad \text{Multiply out.}$$

$$c^2x^2 + a^4 = a^2x^2 + a^2c^2 + a^2y^2 \quad \text{Distribute and simplify.}$$

$$(c^2 - a^2)x^2 - a^2y^2 = a^2c^2 - a^4 \quad \text{Rearrange terms.}$$

$$(c^2 - a^2)x^2 - a^2y^2 = a^2(c^2 - a^2) \quad \text{Factor } a^2 \text{ on the right side.} \qquad (1)$$

To obtain points on the hyperbola off the x-axis, it must be that $a < c$. To see why, look again at Figure 35.

$$d(F_1, P) < d(F_2, P) + d(F_1, F_2) \quad \text{Use triangle } F_1PF_2.$$

$$d(F_1, P) - d(F_2, P) < d(F_1, F_2)$$

$$2a < 2c \quad P \text{ is on the right branch, so } d(F_1, P) - d(F_2, P) = 2a;\ d(F_1, F_2) = 2c.$$

$$a < c \quad \text{Divide each side by 2.}$$

Since $a < c$, we also have $a^2 < c^2$, so $c^2 - a^2 > 0$. Let $b^2 = c^2 - a^2, b > 0$. Then equation (1) can be written as

$$b^2x^2 - a^2y^2 = a^2b^2$$

$$\frac{x^2}{a^2} - \frac{y^2}{b^2} = 1 \quad \text{Divide each side by } a^2b^2.$$

To find the vertices of the hyperbola defined by this equation, let $y = 0$. The vertices satisfy the equation $\frac{x^2}{a^2} = 1$, the solutions of which are $x = \pm a$. Consequently, the vertices of the hyperbola are $V_1 = (-a, 0)$ and $V_2 = (a, 0)$. Notice that the distance from the center $(0, 0)$ to either vertex is a.

THEOREM

Equation of a Hyperbola: Center at (0, 0); Transverse Axis along the x-Axis

An equation of the hyperbola with center at $(0, 0)$, foci at $(-c, 0)$ and $(c, 0)$, and vertices at $(-a, 0)$ and $(a, 0)$ is

$$\frac{x^2}{a^2} - \frac{y^2}{b^2} = 1, \quad \text{where } b^2 = c^2 - a^2 \qquad \textbf{(2)}$$

The transverse axis is the x-axis.

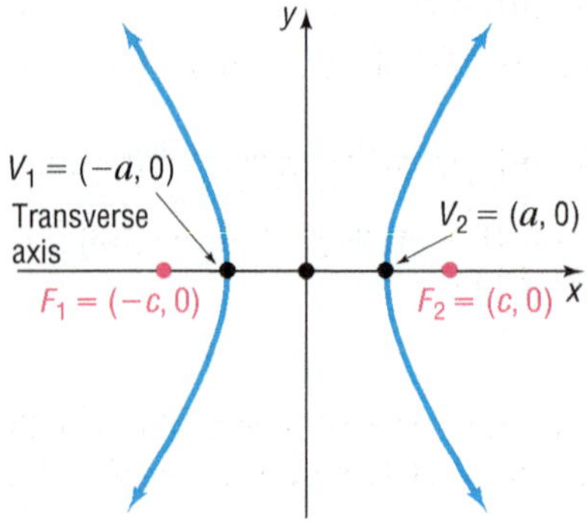

Figure 36 $\frac{x^2}{a^2} - \frac{y^2}{b^2} = 1, \quad b^2 = c^2 - a^2$

See Figure 36. As you can verify, the hyperbola defined by equation (2) is symmetric with respect to the x-axis, y-axis, and origin. To find the y-intercepts, if any, let $x = 0$ in equation (2). This results in the equation $\frac{y^2}{b^2} = -1$, which has no real solution, so the hyperbola defined by equation (2) has no y-intercepts. In fact, since $\frac{x^2}{a^2} - 1 = \frac{y^2}{b^2} \geq 0$, it follows that $\frac{x^2}{a^2} \geq 1$. There are no points on the graph for $-a < x < a$.

EXAMPLE 1

Finding and Graphing an Equation of a Hyperbola

Find an equation of the hyperbola with center at the origin, one focus at $(3, 0)$, and one vertex at $(-2, 0)$. Graph the equation.

Solution The hyperbola has its center at the origin. Plot the center, focus, and vertex. Since they all lie on the x-axis, the transverse axis coincides with the x-axis. One focus is at $(c, 0) = (3, 0)$, so $c = 3$. One vertex is at $(-a, 0) = (-2, 0)$, so $a = 2$. From equation (2), it follows that $b^2 = c^2 - a^2 = 9 - 4 = 5$, so an equation of the hyperbola is

$$\frac{x^2}{4} - \frac{y^2}{5} = 1$$

To graph a hyperbola, it is helpful to locate and plot other points on the graph. For example, to find the points above and below the foci, we let $x = \pm 3$. Then

$$\frac{x^2}{4} - \frac{y^2}{5} = 1$$

$$\frac{(\pm 3)^2}{4} - \frac{y^2}{5} = 1 \qquad x = \pm 3$$

$$\frac{9}{4} - \frac{y^2}{5} = 1$$

$$\frac{y^2}{5} = \frac{5}{4}$$

$$y^2 = \frac{25}{4}$$

$$y = \pm\frac{5}{2}$$

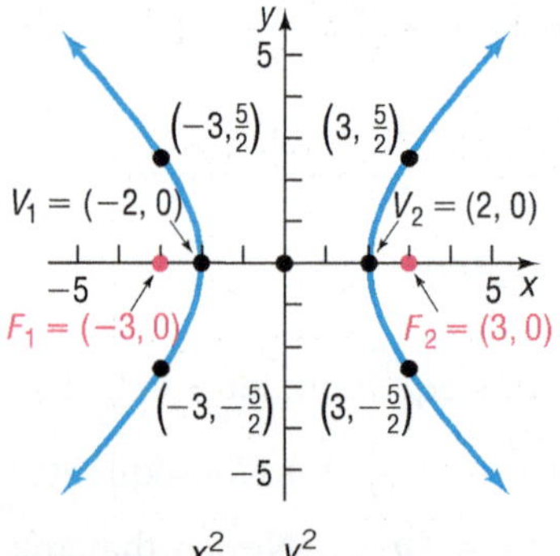

Figure 37 $\frac{x^2}{4} - \frac{y^2}{5} = 1$

The points above and below the foci are $\left(\pm 3, \frac{5}{2}\right)$ and $\left(\pm 3, -\frac{5}{2}\right)$. These points determine the "opening" of the hyperbola. See Figure 37.

EXAMPLE 2 Using a Graphing Utility to Graph a Hyperbola

Using a graphing utility, graph the hyperbola: $\dfrac{x^2}{4} - \dfrac{y^2}{5} = 1$

Solution To graph the hyperbola $\dfrac{x^2}{4} - \dfrac{y^2}{5} = 1$, we need to graph the two functions $Y_1 = \sqrt{5}\sqrt{\dfrac{x^2}{4} - 1}$ and $Y_2 = -\sqrt{5}\sqrt{\dfrac{x^2}{4} - 1}$. As with graphing circles, parabolas, and ellipses on a graphing utility, we use a square screen setting so that the graph is not distorted. Figure 38 shows the graph of the hyperbola. ■

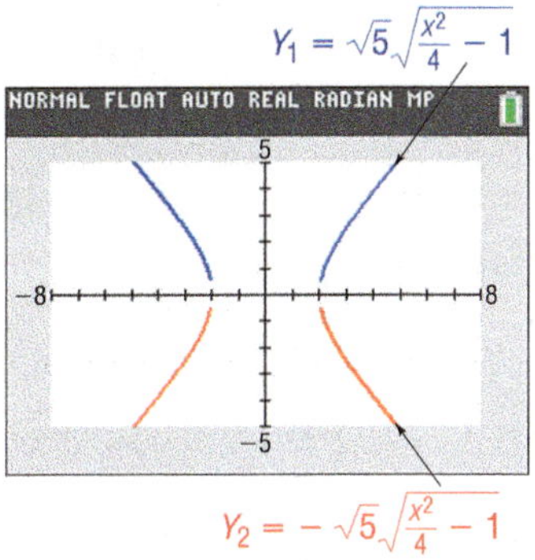

Figure 38

Now Work PROBLEM 19

An equation of the form of equation (2) is the equation of a hyperbola with center at the origin, foci on the x-axis at $(-c, 0)$ and $(c, 0)$, where $c^2 = a^2 + b^2$, and transverse axis along the x-axis.

For the next two examples of this section, the direction **"Analyze the equation"** will mean to find the center, transverse axis, vertices, and foci of the hyperbola and graph it.

EXAMPLE 3 Analyzing the Equation of a Hyperbola

Analyze the equation: $\dfrac{x^2}{16} - \dfrac{y^2}{4} = 1$

Solution The given equation is of the form of equation (2), with $a^2 = 16$ and $b^2 = 4$. The graph of the equation is a hyperbola with center at $(0, 0)$ and transverse axis along the x-axis. Also, $c^2 = a^2 + b^2 = 16 + 4 = 20$. The vertices are at $(\pm a, 0) = (\pm 4, 0)$, and the foci are at $(\pm c, 0) = (\pm 2\sqrt{5}, 0)$.

To locate the points on the graph above and below the foci, let $x = \pm 2\sqrt{5}$. Then

$$\begin{aligned}
\frac{x^2}{16} - \frac{y^2}{4} &= 1 \\
\frac{(\pm 2\sqrt{5})^2}{16} - \frac{y^2}{4} &= 1 \\
\frac{20}{16} - \frac{y^2}{4} &= 1 \\
\frac{5}{4} - \frac{y^2}{4} &= 1 \\
\frac{y^2}{4} &= \frac{1}{4} \\
y &= \pm 1
\end{aligned}$$

The points above and below the foci are $(\pm 2\sqrt{5}, 1)$ and $(\pm 2\sqrt{5}, -1)$. See Figure 39(a) for the graph drawn by hand. Figure 39(b) shows the graph obtained using a graphing utility, where $Y_1 = \sqrt{4\left(\dfrac{x^2}{16} - 1\right)}$ and $Y_2 = -\sqrt{4\left(\dfrac{x^2}{16} - 1\right)}$.

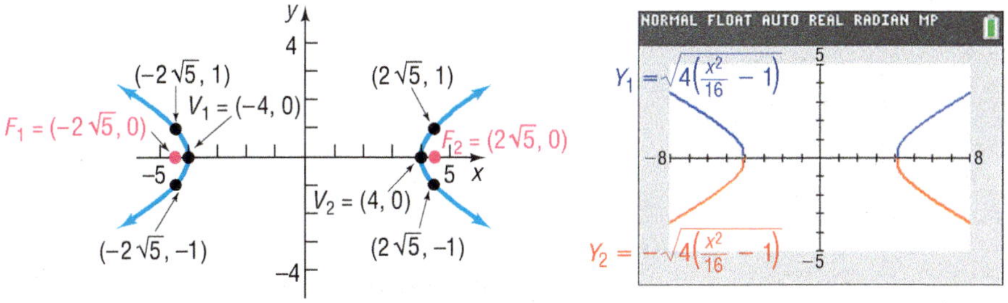

Figure 39 $\dfrac{x^2}{16} - \dfrac{y^2}{4} = 1$ (a) (b) ■

THEOREM

Equation of a Hyperbola; Center at (0, 0) Transverse Axis along the *y*-Axis

An equation of the hyperbola with center at $(0, 0)$, foci at $(0, -c)$ and $(0, c)$, and vertices at $(0, -a)$ and $(0, a)$ is

$$\frac{y^2}{a^2} - \frac{x^2}{b^2} = 1 \qquad \text{where } b^2 = c^2 - a^2 \qquad \textbf{(3)}$$

The transverse axis is the y-axis.

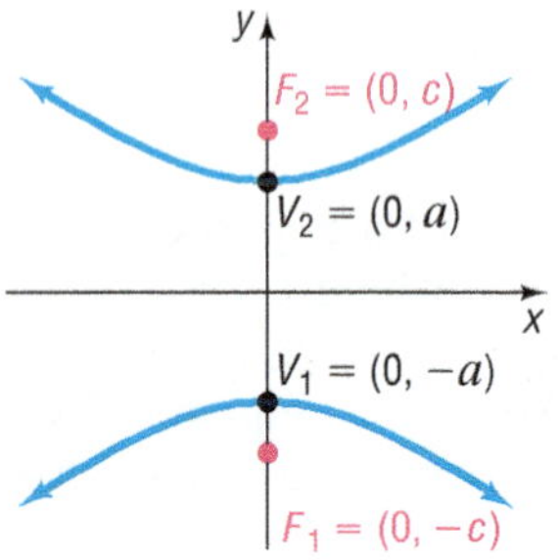

Figure 40
$\frac{y^2}{a^2} - \frac{x^2}{b^2} = 1,\ b^2 = c^2 - a^2$

Figure 40 shows the graph of a typical hyperbola defined by equation (3).

An equation of the form of equation (2), $\frac{x^2}{a^2} - \frac{y^2}{b^2} = 1$, is the equation of a hyperbola with center at the origin, foci on the x-axis at $(-c, 0)$ and $(c, 0)$, where $c^2 = a^2 + b^2$, and transverse axis along the x-axis.

An equation of the form of equation (3), $\frac{y^2}{a^2} - \frac{x^2}{b^2} = 1$, is the equation of a hyperbola with center at the origin, foci on the y-axis at $(0, -c)$ and $(0, c)$, where $c^2 = a^2 + b^2$, and transverse axis along the y-axis.

Notice the difference in the forms of equations (2) and (3). When the y^2-term is subtracted from the x^2-term, the transverse axis is along the x-axis. When the x^2-term is subtracted from the y^2-term, the transverse axis is along the y-axis.

EXAMPLE 4 **Analyzing the Equation of a Hyperbola**

Analyze the equation: $4y^2 - x^2 = 4$

Solution To put the equation in proper form, divide each side by 4:

$$y^2 - \frac{x^2}{4} = 1$$

Since the x^2-term is subtracted from the y^2-term, the equation is that of a hyperbola with center at the origin and transverse axis along the y-axis. Comparing the above equation to equation (3), note that $a^2 = 1$, $b^2 = 4$, and $c^2 = a^2 + b^2 = 5$. The vertices are at $(0, \pm a) = (0, \pm 1)$, and the foci are at $(0, \pm c) = (0, \pm\sqrt{5})$.

To locate points on the graph to the left and right of the foci, let $y = \pm\sqrt{5}$. Then

$$\begin{aligned} 4y^2 - x^2 &= 4 \\ 4(\pm\sqrt{5})^2 - x^2 &= 4 \quad y = \pm\sqrt{5} \\ 20 - x^2 &= 4 \\ x^2 &= 16 \\ x &= \pm 4 \end{aligned}$$

Four other points on the graph are $(\pm 4, \sqrt{5})$ and $(\pm 4, -\sqrt{5})$. See Figure 41(a) for the graph drawn by hand. Figure 41(b) shows the graph obtained using a graphing utility.

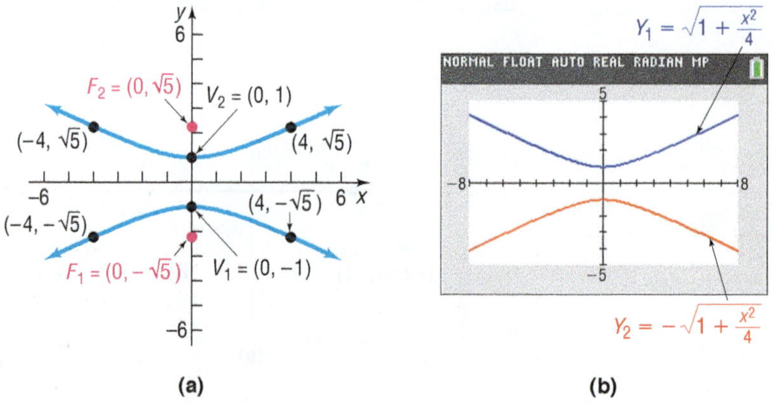

Figure 41 $4y^2 - x^2 = 4$

EXAMPLE 5 **Finding an Equation of a Hyperbola**

Find an equation of the hyperbola that has one vertex at $(0, 2)$ and foci at $(0, -3)$ and $(0, 3)$. Graph the equation.

Solution Since the foci are at $(0, -3)$ and $(0, 3)$, the center of the hyperbola, which is at their midpoint, is the origin. Also, the transverse axis is along the y-axis. The given information also reveals that $c = 3$, $a = 2$, and $b^2 = c^2 - a^2 = 9 - 4 = 5$. The form of the equation of the hyperbola is given by equation (3):

$$\frac{y^2}{a^2} - \frac{x^2}{b^2} = 1$$

$$\frac{y^2}{4} - \frac{x^2}{5} = 1$$

Let $y = \pm 3$ to obtain points on the graph on either side of the foci. See Figure 42.

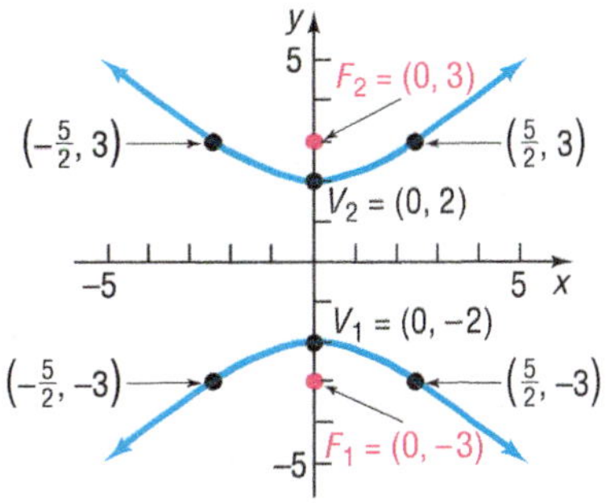

Figure 42 $\frac{y^2}{4} - \frac{x^2}{5} = 1$

Now Work PROBLEM 21

Look at the equations of the hyperbolas in Examples 3 and 5. For the hyperbola in Example 3, $a^2 = 16$ and $b^2 = 4$, so $a > b$; for the hyperbola in Example 5, $a^2 = 4$ and $b^2 = 5$, so $a < b$. We conclude that, for hyperbolas, there are no requirements involving the relative sizes of a and b. Contrast this situation to the case of an ellipse, in which the relative sizes of a and b dictate which axis is the major axis. Hyperbolas have another feature to distinguish them from ellipses and parabolas: Hyperbolas have asymptotes.

2 Find the Asymptotes of a Hyperbola

Recall from Section 5.4 that a horizontal or oblique asymptote of a graph is a line with the property that the distance from the line to points on the graph approaches 0 as $x \to -\infty$ or as $x \to \infty$. Asymptotes provide information about the end behavior of the graph of a hyperbola.

THEOREM

Asymptotes of a Hyperbola

The hyperbola $\frac{x^2}{a^2} - \frac{y^2}{b^2} = 1$ has the two oblique asymptotes

$$y = \frac{b}{a}x \quad \text{and} \quad y = -\frac{b}{a}x \tag{4}$$

Proof Begin by solving for y in the equation of the hyperbola.

$$\frac{x^2}{a^2} - \frac{y^2}{b^2} = 1$$

$$\frac{y^2}{b^2} = \frac{x^2}{a^2} - 1$$

$$y^2 = b^2\left(\frac{x^2}{a^2} - 1\right)$$

Since $x \neq 0$, the right side can be rearranged in the form

$$y^2 = \frac{b^2x^2}{a^2}\left(1 - \frac{a^2}{x^2}\right)$$

$$y = \pm\frac{bx}{a}\sqrt{1 - \frac{a^2}{x^2}}$$

Now, as $x \to -\infty$ or as $x \to \infty$, the term $\frac{a^2}{x^2}$ approaches 0, so the expression under the radical approaches 1. So, as $x \to -\infty$ or as $x \to \infty$, the value of y approaches $\pm\frac{bx}{a}$; that is, the graph of the hyperbola approaches the lines

$$y = -\frac{b}{a}x \quad \text{and} \quad y = \frac{b}{a}x$$

These lines are oblique asymptotes of the hyperbola. ■

The asymptotes of a hyperbola are not part of the hyperbola, but they do serve as a guide for graphing the hyperbola. For example, suppose that we want to graph the equation

$$\frac{x^2}{a^2} - \frac{y^2}{b^2} = 1$$

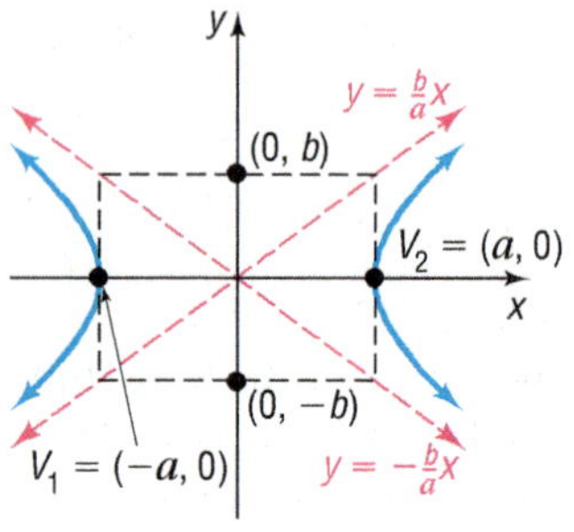

Figure 43 $\frac{x^2}{a^2} - \frac{y^2}{b^2} = 1$

Begin by plotting the vertices $(-a, 0)$ and $(a, 0)$. Then plot the points $(0, -b)$ and $(0, b)$ and use these four points to construct a rectangle, as shown in Figure 43. The diagonals of this rectangle have slopes $\frac{b}{a}$ and $-\frac{b}{a}$, and their extensions are the asymptotes $y = \frac{b}{a}x$ and $y = -\frac{b}{a}x$ of the hyperbola. If we graph the asymptotes, we can use them to establish the "opening" of the hyperbola and avoid plotting other points.

THEOREM

Asymptotes of a Hyperbola

The hyperbola $\frac{y^2}{a^2} - \frac{x^2}{b^2} = 1$ has the two oblique asymptotes:

$$y = \frac{a}{b}x \quad \text{and} \quad y = -\frac{a}{b}x \tag{5}$$

You are asked to prove this result in Problem 84.

For the remainder of this section, the direction **"Analyze the equation"** will mean to find the center, transverse axis, vertices, foci, and asymptotes of the hyperbola and graph it.

EXAMPLE 6

Analyzing the Equation of a Hyperbola

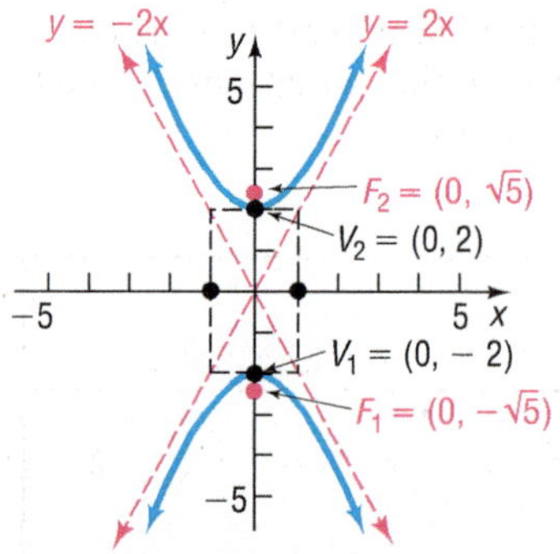

Figure 44 $\frac{y^2}{4} - x^2 = 1$

Analyze the equation: $\frac{y^2}{4} - x^2 = 1$

Solution Since the x^2-term is subtracted from the y^2-term, the equation is of the form of equation (3) and is a hyperbola with center at the origin and transverse axis along the y-axis. Comparing this equation to equation (3), note that $a^2 = 4$, $b^2 = 1$, and $c^2 = a^2 + b^2 = 5$. The vertices are at $(0, \pm a) = (0, \pm 2)$, and the foci are at $(0, \pm c) = (0, \pm\sqrt{5})$. Using equation (5) with $a = 2$ and $b = 1$, the asymptotes are the lines $y = \frac{a}{b}x = 2x$ and $y = -\frac{a}{b}x = -2x$. Form the rectangle containing the points $(0, \pm a) = (0, \pm 2)$ and $(\pm b, 0) = (\pm 1, 0)$. The extensions of the diagonals of this rectangle are the asymptotes. Now graph the asymptotes and the hyperbola. See Figure 44. ■

EXAMPLE 7 **Analyzing the Equation of a Hyperbola**

Analyze the equation: $9x^2 - 4y^2 = 36$

Solution Divide each side of the equation by 36 to put the equation in proper form.

$$\frac{x^2}{4} - \frac{y^2}{9} = 1$$

The center of the hyperbola is the origin. Since the x^2-term is first in the equation, the transverse axis is along the x-axis and the vertices and foci lie on the x-axis. Using equation (2), note that $a^2 = 4$, $b^2 = 9$, and $c^2 = a^2 + b^2 = 13$. The vertices are $a = 2$ units left and right of the center at $(\pm a, 0) = (\pm 2, 0)$, the foci are $c = \sqrt{13}$ units left and right of the center at $(\pm c, 0) = (\pm\sqrt{13}, 0)$, and the asymptotes have the equations

$$y = \frac{b}{a}x = \frac{3}{2}x \quad \text{and} \quad y = -\frac{b}{a}x = -\frac{3}{2}x$$

To graph the hyperbola by hand, form the rectangle containing the points $(\pm a, 0)$ and $(0, \pm b)$, that is, $(-2, 0)$, $(2, 0)$, $(0, -3)$, and $(0, 3)$. The extensions of the diagonals of this rectangle are the asymptotes. See Figure 45(a) for the graph drawn by hand. Figure 45(b) shows the graph obtained using a graphing utility.

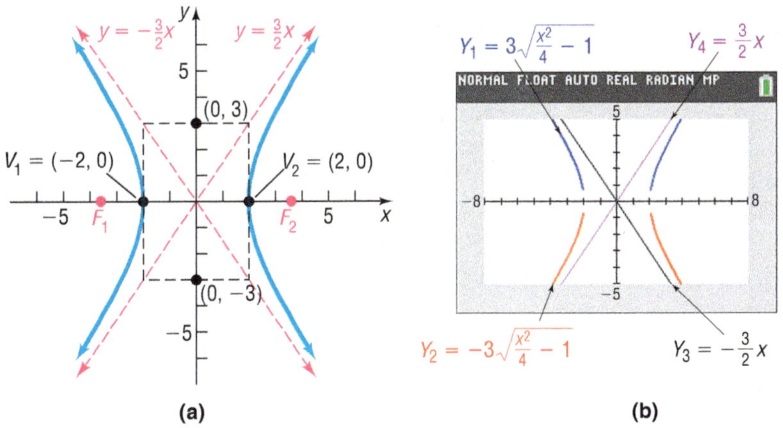

Figure 45 $9x^2 - 4y^2 = 36$

Seeing the Concept

Refer to Figure 45(b). Create a TABLE using Y_1 and Y_4 with $x = 10$, 100, 1000, and 10,000. Compare the values of Y_1 and Y_4. Repeat for Y_2 and Y_3. Now, create a TABLE using Y_1 and Y_3 with $x = -10, -100, -1000$, and $-10{,}000$. Repeat for Y_2 and Y_4.

Now Work PROBLEM 31

3 Analyze Hyperbolas with Center at (h, k)

If a hyperbola with center at the origin and transverse axis coinciding with a coordinate axis is shifted horizontally h units and then vertically k units, the result is a hyperbola with center at (h, k) and transverse axis parallel to a coordinate axis. The equations of such hyperbolas have the same forms as those given in equations (2) and (3), except that x is replaced by $x - h$ (the horizontal shift) and y is replaced by $y - k$ (the vertical shift). Table 4 gives the forms of the equations of such hyperbolas. See Figure 46 on the next page for typical graphs.

Table 4

Hyperbolas with Center at (h, k) and Transverse Axis Parallel to a Coordinate Axis

Center	Transverse Axis	Foci	Vertices	Equation	Asymptotes
(h, k)	Parallel to the x-axis	$(h \pm c, k)$	$(h \pm a, k)$	$\frac{(x-h)^2}{a^2} - \frac{(y-k)^2}{b^2} = 1, \quad b^2 = c^2 - a^2$	$y - k = \pm\frac{b}{a}(x - h)$
(h, k)	Parallel to the y-axis	$(h, k \pm c)$	$(h, k \pm a)$	$\frac{(y-k)^2}{a^2} - \frac{(x-h)^2}{b^2} = 1, \quad b^2 = c^2 - a^2$	$y - k = \pm\frac{a}{b}(x - h)$

Note: It is not recommended that Table 4 be memorized. Rather, use transformations (shift horizontally h units, vertically k units), along with the fact that a represents the distance from the center to the vertices, c represents the distance from the center to the foci, and $b^2 = c^2 - a^2$ (or $c^2 = a^2 + b^2$). ■

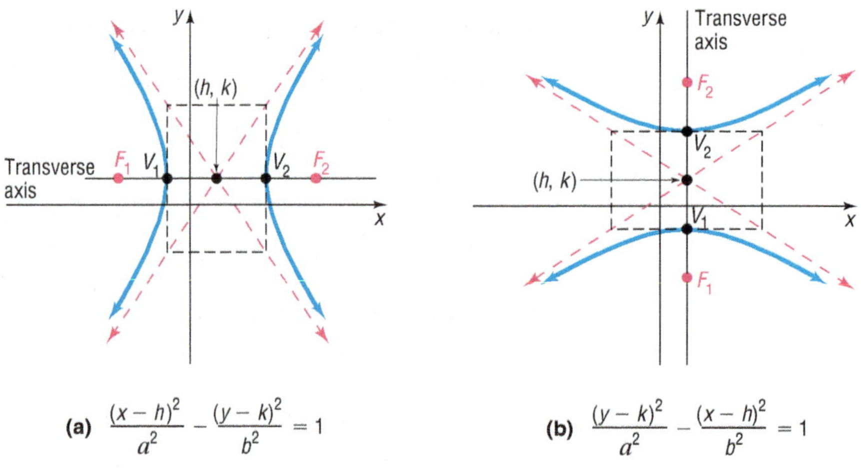

Figure 46

EXAMPLE 8 Finding an Equation of a Hyperbola, Center Not at the Origin

Find an equation for the hyperbola with center at $(1, -2)$, one focus at $(4, -2)$, and one vertex at $(3, -2)$. Graph the equation by hand.

Solution The center is at $(h, k) = (1, -2)$, so $h = 1$ and $k = -2$. Since the center, focus, and vertex all lie on the line $y = -2$, the transverse axis is parallel to the x-axis. The distance from the center $(1, -2)$ to the focus $(4, -2)$ is $c = 3$; the distance from the center $(1, -2)$ to the vertex $(3, -2)$ is $a = 2$. Thus, $b^2 = c^2 - a^2 = 9 - 4 = 5$. The equation is

$$\frac{(x-h)^2}{a^2} - \frac{(y-k)^2}{b^2} = 1$$

$$\frac{(x-1)^2}{4} - \frac{(y+2)^2}{5} = 1$$

See Figure 47. ■

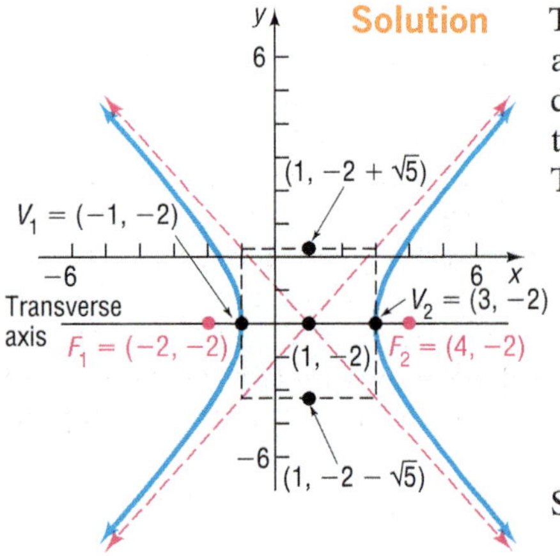

Figure 47 $\frac{(x-1)^2}{4} - \frac{(y+2)^2}{5} = 1$

Now Work PROBLEM 41

EXAMPLE 9 Analyzing the Equation of a Hyperbola

Analyze the equation: $-x^2 + 4y^2 - 2x - 16y + 11 = 0$

Solution Complete the squares in x and in y.

$$-x^2 + 4y^2 - 2x - 16y + 11 = 0$$

$$-(x^2 + 2x) + 4(y^2 - 4y) = -11 \quad \text{Group terms.}$$

$$-(x^2 + 2x + 1) + 4(y^2 - 4y + 4) = -11 - 1 + 16 \quad \text{Complete each square.}$$

$$-(x+1)^2 + 4(y-2)^2 = 4 \quad \text{Factor.}$$

$$(y-2)^2 - \frac{(x+1)^2}{4} = 1 \quad \text{Divide each side by 4.}$$

This is the equation of a hyperbola with center at $(-1, 2)$ and transverse axis parallel to the y-axis. Also, $a^2 = 1$ and $b^2 = 4$, so $c^2 = a^2 + b^2 = 5$. Since the transverse axis is parallel to the y-axis, the vertices and foci are located a and c units above and below the center, respectively. The vertices are at $(h, k \pm a) = (-1, 2 \pm 1)$, or $(-1, 1)$ and $(-1, 3)$. The foci are at $(h, k \pm c) = (-1, 2 \pm \sqrt{5})$. The asymptotes are

$y - 2 = \frac{1}{2}(x + 1)$ and $y - 2 = -\frac{1}{2}(x + 1)$. Figure 48(a) shows the graph drawn by hand. Figure 48(b) shows the graph obtained using a graphing utility.

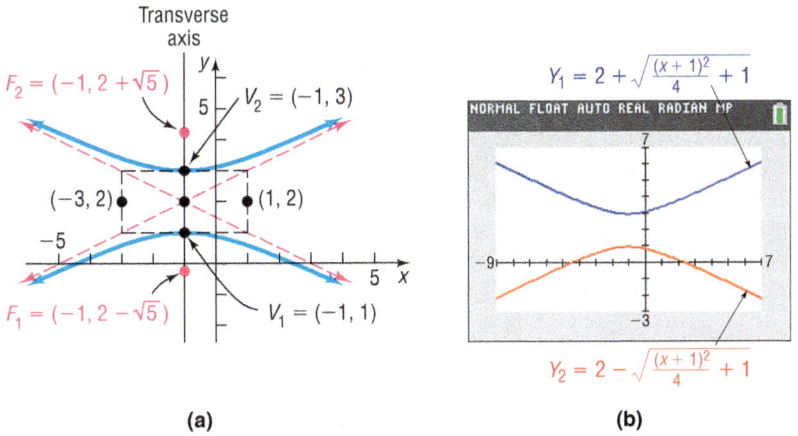

Figure 48 $-x^2 + 4y^2 - 2x - 16y + 11 = 0$

Now Work PROBLEM 55

4 Solve Applied Problems Involving Hyperbolas

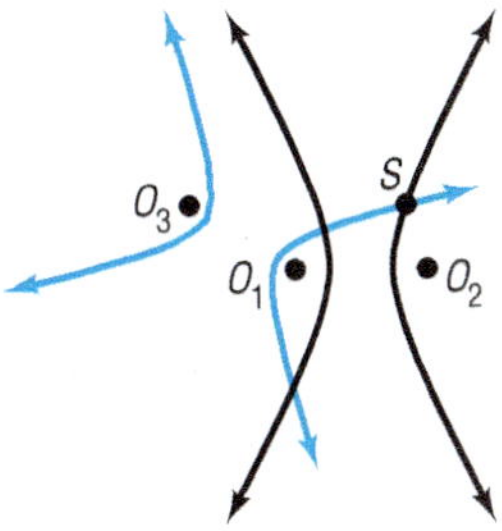

Figure 49

Look at Figure 49. Suppose that three microphones are located at points O_1, O_2, and O_3 (the foci of the two hyperbolas). In addition, suppose that a gun is fired at S and the microphone at O_1 records the gunshot 1 second after the microphone at O_2. Because sound travels at about 1100 feet per second, we conclude that the microphone at O_1 is 1100 feet farther from the gunshot than O_2. We can model this situation by saying that S lies on a branch of a hyperbola with foci at O_1 and O_2. (Do you see why? The difference of the distances from S to O_1 and from S to O_2 is the constant 1100.) If the third microphone at O_3 records the gunshot 2 seconds after O_1, then S lies on a branch of a second hyperbola with foci at O_1 and O_3. In this case, the constant difference is 2200. The intersection of the two hyperbolas identifies the location of S.

EXAMPLE 10 **Lightning Strikes**

Suppose that two people standing 1 mile apart both see a flash of lightning. After a period of time, the person standing at point A hears the thunder. One second later, the person standing at point B hears the thunder. If the person at B is due west of the person at A and the lightning strike is known to occur due north of the person standing at point A, where did the lightning strike occur?

Solution See Figure 50 in which the ordered pair (x, y) represents the location of the lightning strike. We know that sound travels at 1100 feet per second, so the person at point A is 1100 feet closer to the lightning strike than the person at point B. Since the difference of the distance from (x, y) to B and the distance from (x, y) to A is the constant 1100, the point (x, y) lies on a hyperbola whose foci are at A and B.

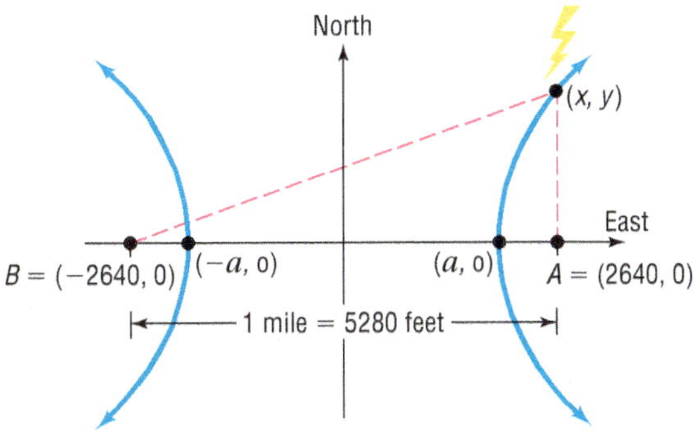

Figure 50

An equation of the hyperbola is

$$\frac{x^2}{a^2} - \frac{y^2}{b^2} = 1$$

where $2a = 1100$, so $a = 550$.

Because the distance between the two people is 1 mile (5280 feet) and each person is at a focus of the hyperbola, this means

$$2c = 5280$$
$$c = \frac{5280}{2} = 2640$$

Since $b^2 = c^2 - a^2 = 2640^2 - 550^2 = 6{,}667{,}100$, the equation of the hyperbola that describes the location of the lightning strike is

$$\frac{x^2}{550^2} - \frac{y^2}{6{,}667{,}100} = 1$$

Refer to Figure 50. Since the lightning strike occurred due north of the individual at the point $A = (2640, 0)$, let $x = 2640$ and solve the resulting equation.

$$\frac{2640^2}{550^2} - \frac{y^2}{6{,}667{,}100} = 1 \qquad x = 2640$$

$$-\frac{y^2}{6{,}667{,}100} = -22.04 \qquad \text{Subtract } \frac{2640^2}{550^2} \text{ from both sides.}$$

$$y^2 = 146{,}942{,}884 \qquad \text{Multiply both sides by } -6{,}667{,}100.$$

$$y = 12{,}122 \qquad y > 0 \text{ since the lightning strike occurred in quadrant I.}$$

The lightning strike occurred 12,122 feet north of the person standing at point A.

Check: The difference between the distance from $(2640, 12{,}122)$ to the person at the point $B = (-2640, 0)$ and the distance from $(2640, 12{,}122)$ to the person at the point $A = (2640, 0)$ should be 1100. Using the distance formula, the difference in the distances is

$$\sqrt{[2640 - (-2640)]^2 + (12{,}122 - 0)^2} - \sqrt{(2640 - 2640)^2 + (12{,}122 - 0)^2} = 1100$$

as required. ■

Now Work PROBLEM 75

11.4 Assess Your Understanding

'Are You Prepared?' *Answers are given at the end of these exercises. If you get a wrong answer, read the pages listed in red.*

1. The distance d from $P_1 = (3, -4)$ to $P_2 = (-2, 1)$ is $d =$ _____. (p. 85)
2. To complete the square of $x^2 + 5x$, add ____. (p. 57)
3. Find the intercepts of the equation $y^2 = 9 + 4x^2$. (pp. 165–166)
4. ***True or False*** The equation $y^2 = 9 + x^2$ is symmetric with respect to the x-axis, the y-axis, and the origin. (pp. 166–168)
5. To graph $y = (x - 5)^3 - 4$, shift the graph of $y = x^3$ to the (left/right) ________ unit(s) and then (up/down) ________ unit(s). (pp. 256–264)
6. Find the vertical asymptotes, if any, and the horizontal or oblique asymptote, if any, of $y = \frac{x^2 - 9}{x^2 - 4}$. (pp. 374–379)

Concepts and Vocabulary

7. A(n) __________ is the collection of points in a plane the difference of whose distances from two fixed points is a constant.

8. For a hyperbola, the foci lie on a line called the __________ _____.

Answer Problems 9–11 using the figure to the right.

9. The equation of the hyperbola is of the form

(a) $\dfrac{(x-h)^2}{a^2} - \dfrac{(y-k)^2}{b^2} = 1$

(b) $\dfrac{(y-k)^2}{a^2} - \dfrac{(x-h)^2}{b^2} = 1$

(c) $\dfrac{(x-h)^2}{a^2} + \dfrac{(y-k)^2}{b^2} = 1$

(d) $\dfrac{(x-h)^2}{b^2} + \dfrac{(y-k)^2}{a^2} = 1$

10. If the center of the hyperbola is $(2, 1)$ and $a = 3$, then the coordinates of the vertices are ______ and ________.

11. If the center of the hyperbola is $(2, 1)$ and $c = 5$, then the coordinates of the foci are ______ and ________.

12. In a hyperbola, if $a = 3$ and $c = 5$, then $b =$ _______.

(a) 1 (b) 2 (c) 4 (d) 8

13. For the hyperbola $\dfrac{x^2}{4} - \dfrac{y^2}{9} = 1$, the value of a is ____, the value of b is ____, and the transverse axis is the ____-axis.

14. For the hyperbola $\dfrac{y^2}{16} - \dfrac{x^2}{81} = 1$, the asymptotes are _________ and _______.

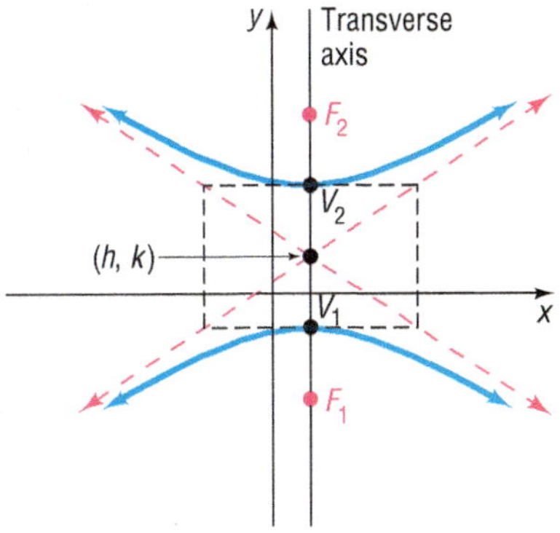

Skill Building

In Problems 15–18, the graph of a hyperbola is given. Match each graph to its equation.

(A) $\dfrac{x^2}{4} - y^2 = 1$ (B) $x^2 - \dfrac{y^2}{4} = 1$ (C) $\dfrac{y^2}{4} - x^2 = 1$ (D) $y^2 - \dfrac{x^2}{4} = 1$

15.

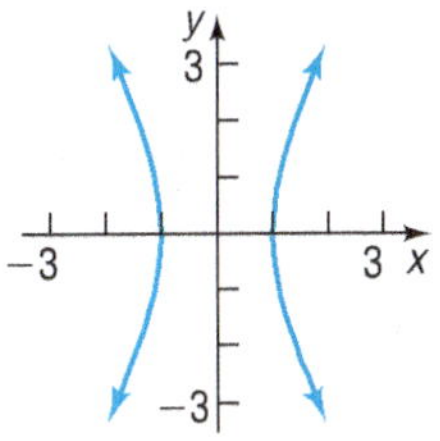

16.

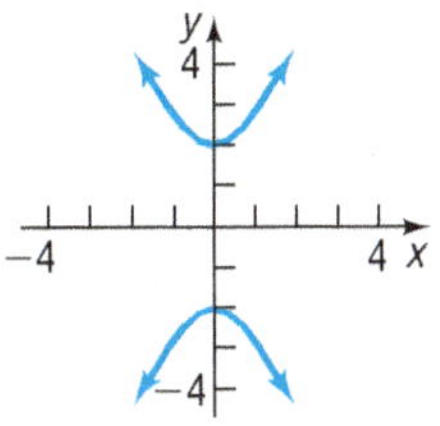

17.

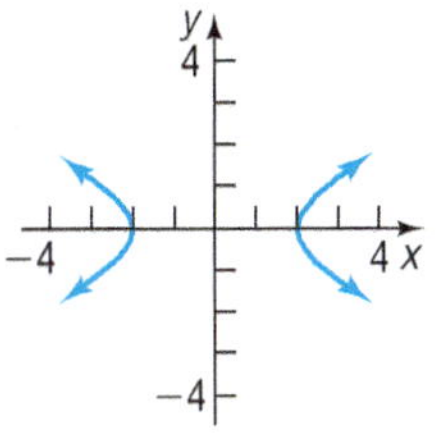

18.

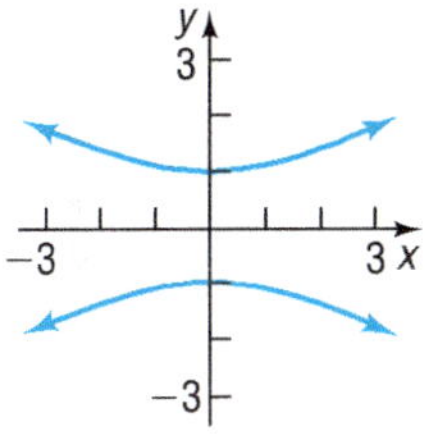

In Problems 19–28, find an equation for the hyperbola described. Graph the equation by hand.

19. Center at $(0, 0)$; focus at $(3, 0)$; vertex at $(1, 0)$

20. Center at $(0, 0)$; focus at $(0, 5)$; vertex at $(0, 3)$

21. Center at $(0, 0)$; focus at $(0, -6)$; vertex at $(0, 4)$

22. Center at $(0, 0)$; focus at $(-3, 0)$; vertex at $(2, 0)$

23. Foci at $(-5, 0)$ and $(5, 0)$; vertex at $(3, 0)$

24. Focus at $(0, 6)$; vertices at $(0, -2)$ and $(0, 2)$

25. Vertices at $(0, -6)$ and $(0, 6)$; asymptote the line $y = 2x$

26. Vertices at $(-4, 0)$ and $(4, 0)$; asymptote the line $y = 2x$

27. Foci at $(-4, 0)$ and $(4, 0)$; asymptote the line $y = -x$

28. Foci at $(0, -2)$ and $(0, 2)$; asymptote the line $y = -x$

In Problems 29–36, find the center, transverse axis, vertices, foci, and asymptotes. Graph each equation by hand. Verify your graph using a graphing utility.

29. $\dfrac{x^2}{25} - \dfrac{y^2}{9} = 1$

30. $\dfrac{y^2}{16} - \dfrac{x^2}{4} = 1$

31. $4x^2 - y^2 = 16$

32. $4y^2 - x^2 = 16$

33. $y^2 - 9x^2 = 9$

34. $x^2 - y^2 = 4$

35. $y^2 - x^2 = 25$

36. $2x^2 - y^2 = 4$

In Problems 37–40, write an equation for each hyperbola.

37.

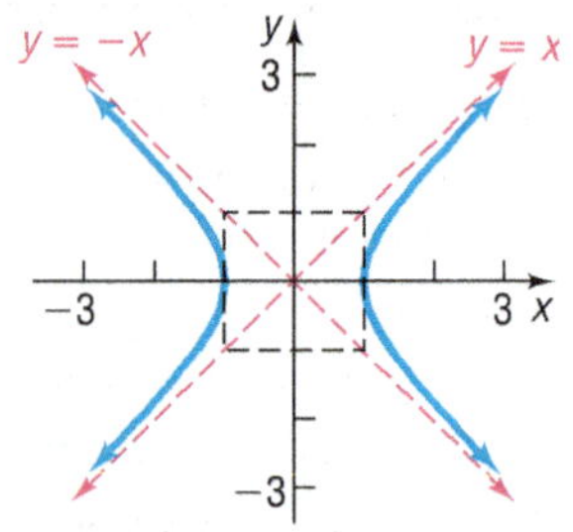

38.

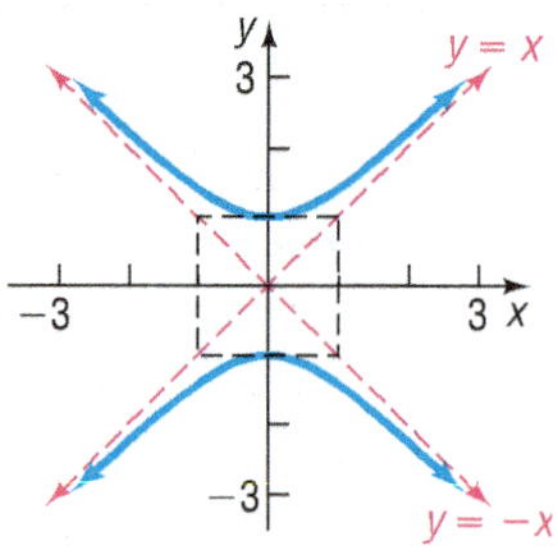

39.

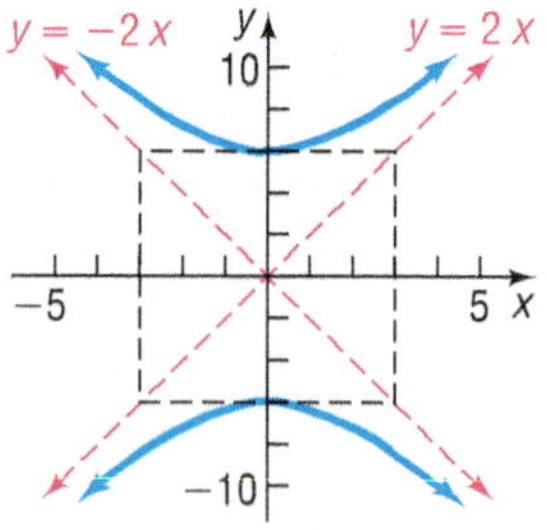

40.

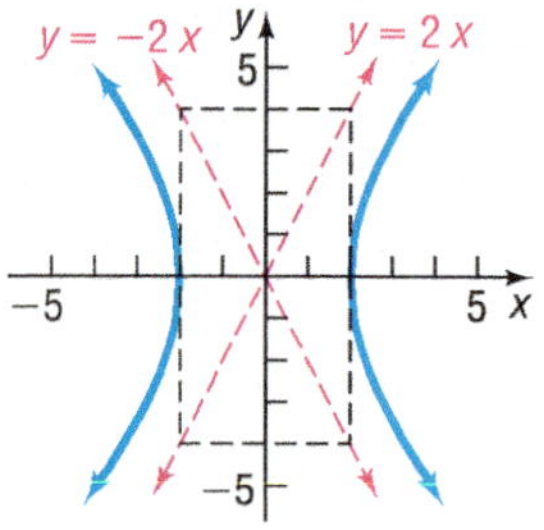

In Problems 41–48, find an equation for the hyperbola described. Graph the equation by hand.

41. Center at $(4, -1)$; focus at $(7, -1)$; vertex at $(6, -1)$

42. Center at $(-3, 1)$; focus at $(-3, 6)$; vertex at $(-3, 4)$

43. Center at $(-3, -4)$; focus at $(-3, -8)$; vertex at $(-3, -2)$

44. Center at $(1, 4)$; focus at $(-2, 4)$; vertex at $(0, 4)$

45. Foci at $(3, 7)$ and $(7, 7)$; vertex at $(6, 7)$

46. Focus at $(-4, 0)$ vertices at $(-4, 4)$ and $(-4, 2)$

47. Vertices at $(-1, -1)$ and $(3, -1)$; asymptote the line $y + 1 = \frac{3}{2}(x - 1)$

48. Vertices at $(1, -3)$ and $(1, 1)$; asymptote the line $y + 1 = \frac{3}{2}(x - 1)$

In Problems 49–62, find the center, transverse axis, vertices, foci, and asymptotes. Graph each equation by hand. Verify your graph using a graphing utility.

49. $\frac{(x-2)^2}{4} - \frac{(y+3)^2}{9} = 1$

50. $\frac{(y+3)^2}{4} - \frac{(x-2)^2}{9} = 1$

51. $(y-2)^2 - 4(x+2)^2 = 4$

52. $(x+4)^2 - 9(y-3)^2 = 9$

53. $(x+1)^2 - (y+2)^2 = 4$

54. $(y-3)^2 - (x+2)^2 = 4$

55. $x^2 - y^2 - 2x - 2y - 1 = 0$

56. $y^2 - x^2 - 4y + 4x - 1 = 0$

57. $y^2 - 4x^2 - 4y - 8x - 4 = 0$

58. $2x^2 - y^2 + 4x + 4y - 4 = 0$

59. $4x^2 - y^2 - 24x - 4y + 16 = 0$

60. $2y^2 - x^2 + 2x + 8y + 3 = 0$

61. $y^2 - 4x^2 - 16x - 2y - 19 = 0$

62. $x^2 - 3y^2 + 8x - 6y + 4 = 0$

In Problems 63–66, graph each function by hand. Be sure to label any intercepts. [**Hint:** *Notice that each function is half a hyperbola.*]

63. $f(x) = \sqrt{16 + 4x^2}$

64. $f(x) = -\sqrt{9 + 9x^2}$

65. $f(x) = -\sqrt{-25 + x^2}$

66. $f(x) = \sqrt{-1 + x^2}$

Mixed Practice

In Problems 67–74, analyze each equation.

67. $\frac{(x-3)^2}{4} - \frac{y^2}{25} = 1$

68. $\frac{(y+2)^2}{16} - \frac{(x-2)^2}{4} = 1$

69. $x^2 = 16(y - 3)$

70. $y^2 = -12(x + 1)$

71. $25x^2 + 9y^2 - 250x + 400 = 0$

72. $x^2 + 36y^2 - 2x + 288y + 541 = 0$

73. $x^2 - 6x - 8y - 31 = 0$

74. $9x^2 - y^2 - 18x - 8y - 88 = 0$

Applications and Extensions

75. **Fireworks Display** Suppose that two people standing 2 miles apart both see the burst from a fireworks display. After a period of time the first person, standing at point A, hears the burst. One second later the second person, standing at point B, hears the burst. If the person at point B is due west of the person at point A, and if the display is known to occur due north of the person at point A, where did the fireworks display occur?

76. **Lightning Strikes** Suppose that two people standing 1 mile apart both see a flash of lightning. After a period of time the first person, standing at point A, hears the thunder. Two seconds later the second person, standing at point B, hears the thunder. If the person at point B is due west of the person at point A, and if the lightning strike is known to occur due north of the person standing at point A, where did the lightning strike occur?

77. **Nuclear Power Plant** Some nuclear power plants utilize "natural draft" cooling towers in the shape of a **hyperboloid**, a solid obtained by rotating a hyperbola about its conjugate axis. Suppose that such a cooling tower has a base diameter of 400 feet and the diameter at its narrowest point, 360 feet above the ground, is 200 feet. If the diameter at the top of the tower is 300 feet, how tall is the tower?

Source: *Bay Area Air Quality Management District*

78. An Explosion Two recording devices are set 2400 feet apart, with the device at point A to the west of the device at point B. At a point between the devices 300 feet from point B, a small amount of explosive is detonated. The recording devices record the time until the sound reaches each. How far directly north of point B should a second explosion be done so that the measured time difference recorded by the devices is the same as that for the first detonation?

79. Rutherford's Experiment In May 1911, Ernest Rutherford published a paper in *Philosophical Magazine*. In this article, he described the motion of alpha particles as they are shot at a piece of gold foil 0.00004 cm thick. Before conducting this experiment, Rutherford expected that the alpha particles would shoot through the foil just as a bullet would shoot through snow. Instead, a small fraction of the alpha particles bounced off the foil. This led to the conclusion that the nucleus of an atom is dense, while the remainder of the atom is sparse. Only the density of the nucleus could cause the alpha particles to deviate from their path. The figure shows a diagram from Rutherford's paper that indicates that the deflected alpha particles follow the path of one branch of a hyperbola.

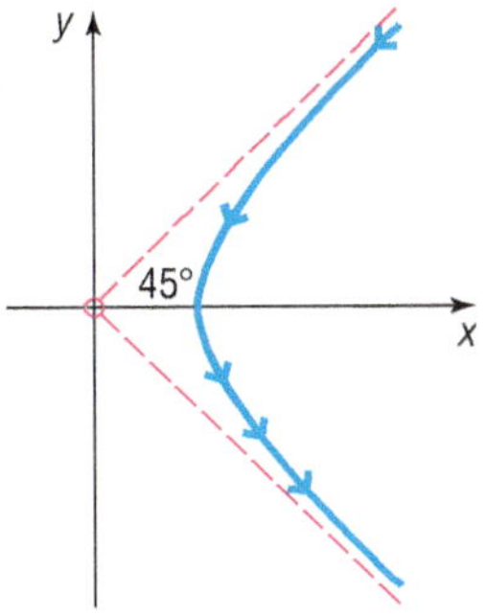

(a) Find an equation of the asymptotes under this scenario.

(b) If the vertex of the path of the alpha particles is 10 cm from the center of the hyperbola, find a model that describes the path of the particle.

80. Hyperbolic Mirrors Hyperbolas have interesting reflective properties that make them useful for lenses and mirrors. For example, if a ray of light strikes a convex hyperbolic mirror on a line that would (theoretically) pass through its rear focus, it is reflected through the front focus. This property, and that of the parabola, were used to develop the *Cassegrain* telescope in 1672. The focus of the parabolic mirror and the rear focus of the hyperbolic mirror are the same point. The rays are collected by the parabolic mirror, then are reflected toward the (common) focus, and thus are reflected by the hyperbolic mirror through the opening to its front focus, where the eyepiece is located. If the equation of the hyperbola is $\frac{y^2}{9} - \frac{x^2}{16} = 1$ and the focal length (distance from the vertex to the focus) of the parabola is 6, find the equation of the parabola.

Source: *www.enchantedlearning.com*

81. The **eccentricity** e of a hyperbola is defined as the number $\frac{c}{a}$, where a is the distance of a vertex from the center and c is the distance of a focus from the center. Because $c > a$, it follows that $e > 1$. Describe the general shape of a hyperbola whose eccentricity is close to 1. What is the shape if e is very large?

82. A hyperbola for which $a = b$ is called an **equilateral hyperbola**. Find the eccentricity e of an equilateral hyperbola.

[**Note:** The eccentricity of a hyperbola is defined in Problem 81.]

83. Two hyperbolas that have the same set of asymptotes are called **conjugate**. Show that the hyperbolas

$$\frac{x^2}{4} - y^2 = 1 \quad \text{and} \quad y^2 - \frac{x^2}{4} = 1$$

are conjugate. Graph each hyperbola on the same set of coordinate axes.

84. Prove that the hyperbola

$$\frac{y^2}{a^2} - \frac{x^2}{b^2} = 1$$

has the two oblique asymptotes

$$y = \frac{a}{b}x \quad \text{and} \quad y = -\frac{a}{b}x$$

85. Show that the graph of an equation of the form

$$Ax^2 + Cy^2 + F = 0 \qquad A \neq 0, C \neq 0, F \neq 0$$

where A and C are opposite in sign, is a hyperbola with center at $(0, 0)$.

86. Show that the graph of an equation of the form

$$Ax^2 + Cy^2 + Dx + Ey + F = 0 \qquad A \neq 0, C \neq 0$$

where A and C are opposite in sign,

(a) is a hyperbola if $\frac{D^2}{4A} + \frac{E^2}{4C} - F \neq 0$.

(b) is two intersecting lines if $\frac{D^2}{4A} + \frac{E^2}{4C} - F = 0$.

Retain Your Knowledge

Problems 87–90 are based on material learned earlier in the course. The purpose of these problems is to keep the material fresh in your mind so that you are better prepared for the final exam.

87. For $y = \frac{3}{2}\cos(6x + 3\pi)$, find the amplitude, period, and phase shift. Then graph the function, showing at least two periods.

88. Solve the triangle described: $a = 7$, $b = 10$, and $C = 100°$.

89. Find the rectangular coordinates of the point with the polar coordinates $\left(12, -\frac{\pi}{3}\right)$.

90. Transform the polar equation $r = 6 \sin \theta$ to an equation in rectangular coordinates. Then identify and graph the equation.

'Are You Prepared?' Answers

1. $5\sqrt{2}$ **2.** $\frac{25}{4}$ **3.** $(0, -3)$, $(0, 3)$ **4.** True **5.** right 5; down 4 **6.** Vertical: $x = -2$, $x = 2$; horizontal: $y = 1$

11.5 Rotation of Axes; General Form of a Conic

PREPARING FOR THIS SECTION *Before getting started, review the following:*

- Sum Formulas for Sine and Cosine (Section 8.5, pp. 652 and 655)
- Half-angle Formulas for Sine and Cosine (Section 8.6, p. 668)
- Double-angle Formulas for Sine and Cosine (Section 8.6, p. 665)

Now Work the 'Are You Prepared?' problems on page 834.

OBJECTIVES
1. Identify a Conic (p. 828)
2. Use a Rotation of Axes to Transform Equations (p. 829)
3. Analyze an Equation Using a Rotation of Axes (p. 831)
4. Identify Conics without a Rotation of Axes (p. 833)

In this section, we show that the graph of a general second-degree polynomial containing two variables x and y—that is, an equation of the form

$$Ax^2 + Bxy + Cy^2 + Dx + Ey + F = 0 \quad \textbf{(1)}$$

where A, B, and C are not simultaneously 0—is a conic. We shall not concern ourselves here with the degenerate cases of equation (1), such as $x^2 + y^2 = 0$, whose graph is a single point $(0, 0)$; or $x^2 + 3y^2 + 3 = 0$, whose graph contains no points; or $x^2 - 4y^2 = 0$, whose graph is two lines, $x - 2y = 0$ and $x + 2y = 0$.

We begin with the case where $B = 0$. In this case, the term containing xy is not present, so equation (1) has the form

$$Ax^2 + Cy^2 + Dx + Ey + F = 0$$

where either $A \neq 0$ or $C \neq 0$.

1 Identify a Conic

We have already discussed the procedure for identifying the graph of this kind of equation; we complete the squares of the quadratic expressions in x or y, or both. Once this has been done, the conic can be identified by comparing it to one of the forms studied in Sections 11.2 through 11.4.

In fact, though, the conic can be identified directly from the equation without completing the squares.

THEOREM

Identifying Conics without Completing the Squares

Excluding degenerate cases, the equation

$$Ax^2 + Cy^2 + Dx + Ey + F = 0 \quad \textbf{(2)}$$

where A and C are not both equal to zero:

(a) Defines a parabola if $AC = 0$.
(b) Defines an ellipse (or a circle) if $AC > 0$.
(c) Defines a hyperbola if $AC < 0$.

Proof

(a) If $AC = 0$, then either $A = 0$ or $C = 0$, but not both, so the form of equation (2) is either

$$Ax^2 + Dx + Ey + F = 0, \qquad A \neq 0$$

or

$$Cy^2 + Dx + Ey + F = 0, \qquad C \neq 0$$

Using the results of Problems 80 and 81 at the end of Section 11.2, it follows that, except for the degenerate cases, the equation is a parabola.

(b) If $AC > 0$, then A and C have the same sign. Using the results of Problem 87 at the end of Section 11.3, except for the degenerate cases, the equation is an ellipse.

(c) If $AC < 0$, then A and C are of opposite sign. Using the results of Problem 86 at the end of Section 11.4, except for the degenerate cases, the equation is a hyperbola. ■

We will not be concerned with the degenerate cases of equation (2). However, in practice, you should be alert to the possibility of degeneracy.

EXAMPLE 1 **Identifying a Conic without Completing the Squares**

Identify the graph of each equation without completing the squares.

(a) $3x^2 + 6y^2 + 6x - 12y = 0$ (b) $2x^2 - 3y^2 + 6y + 4 = 0$

(c) $y^2 - 2x + 4 = 0$

Solution (a) Compare the given equation to equation (2) and conclude that $A = 3$ and $C = 6$. Since $AC = 18 > 0$, the equation defines an ellipse.

(b) Here $A = 2$ and $C = -3$, so $AC = -6 < 0$. The equation defines a hyperbola.

(c) Here $A = 0$ and $C = 1$, so $AC = 0$. The equation defines a parabola. ■

Now Work PROBLEM 11

Although we can now identify the type of conic represented by any equation of the form of equation (2) without completing the squares, we still need to complete the squares if we desire additional information about the conic, such as its graph.

2 Use a Rotation of Axes to Transform Equations

Now let's turn our attention to equations of the form of equation (1), where $B \neq 0$. To discuss this case, we introduce a new procedure: *rotation of axes*.

In a **rotation of axes**, the origin remains fixed while the x-axis and y-axis are rotated through an angle θ to a new position; the new positions of the x-axis and the y-axis are denoted by x' and y', respectively, as shown in Figure 51(a).

Now look at Figure 51(b). There the point P has the coordinates (x, y) relative to the xy-plane, while the same point P has coordinates (x', y') relative to the $x'y'$-plane. We seek relationships that will enable us to express x and y in terms of x', y', and θ.

As Figure 51(b) shows, r denotes the distance from the origin O to the point P, and α denotes the angle between the positive x'-axis and the ray from O through P. Then, using the definitions of sine and cosine, we have

$$x' = r\cos\alpha \qquad y' = r\sin\alpha \tag{3}$$

$$x = r\cos(\theta + \alpha) \qquad y = r\sin(\theta + \alpha) \tag{4}$$

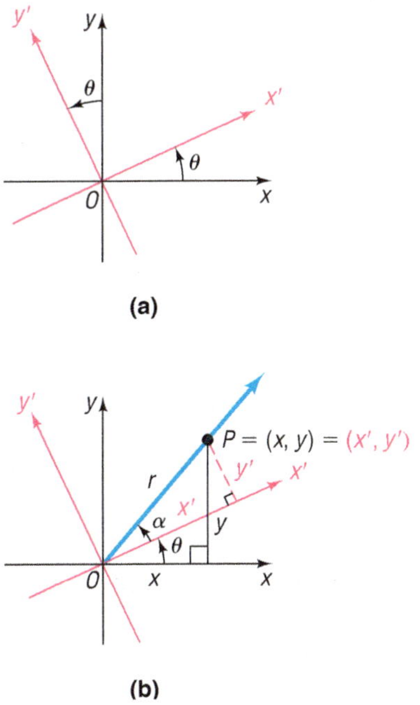

Figure 51 Rotation of axes

Now

$$\begin{aligned} x &= r\cos(\theta + \alpha) \\ &= r(\cos\theta\cos\alpha - \sin\theta\sin\alpha) && \text{Apply the Sum Formula for cosine.} \\ &= (r\cos\alpha)(\cos\theta) - (r\sin\alpha)(\sin\theta) \\ &= x'\cos\theta - y'\sin\theta && \text{By equation (3)} \end{aligned}$$

Similarly,

$$\begin{aligned} y &= r\sin(\theta + \alpha) \\ &= r(\sin\theta\cos\alpha + \cos\theta\sin\alpha) && \text{Apply the Sum Formula for sine.} \\ &= x'\sin\theta + y'\cos\theta && \text{By equation (3)} \end{aligned}$$

THEOREM

Rotation Formulas

If the x- and y-axes are rotated through an angle θ, the coordinates (x, y) of a point P relative to the xy-plane and the coordinates (x', y') of the same point relative to the new x'- and y'-axes are related by the formulas

$$x = x' \cos\theta - y' \sin\theta \qquad y = x' \sin\theta + y' \cos\theta \tag{5}$$

EXAMPLE 2 **Rotating Axes**

Express the equation $xy = 1$ in terms of new $x'y'$-coordinates by rotating the axes through a 45° angle. Discuss the new equation.

Solution Let $\theta = 45°$ in equation (5). Then

$$x = x' \cos 45° - y' \sin 45° = x' \frac{\sqrt{2}}{2} - y' \frac{\sqrt{2}}{2} = \frac{\sqrt{2}}{2}(x' - y')$$

$$y = x' \sin 45° + y' \cos 45° = x' \frac{\sqrt{2}}{2} + y' \frac{\sqrt{2}}{2} = \frac{\sqrt{2}}{2}(x' + y')$$

Substituting these expressions for x and y in $xy = 1$ gives

$$\left[\frac{\sqrt{2}}{2}(x' - y')\right]\left[\frac{\sqrt{2}}{2}(x' + y')\right] = 1$$

$$\frac{1}{2}(x'^2 - y'^2) = 1$$

$$\frac{x'^2}{2} - \frac{y'^2}{2} = 1$$

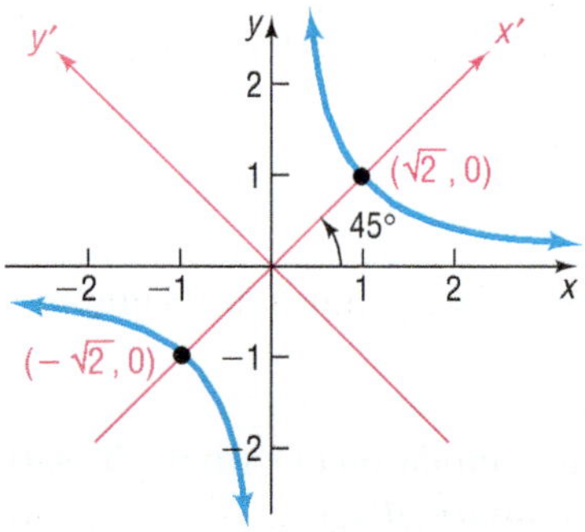

Figure 52 $xy = 1$

This is the equation of a hyperbola with center at $(0, 0)$ and transverse axis along the x'-axis. The vertices are at $(\pm\sqrt{2}, 0)$ on the x'-axis; the asymptotes are $y' = x'$ and $y' = -x'$ (which correspond to the original x- and y-axes). See Figure 52 for the graph.

As Example 2 illustrates, a rotation of axes through an appropriate angle can transform a second-degree equation in x and y containing an xy-term into one in x' and y' in which no $x'y'$-term appears. In fact, we will show that a rotation of axes through an appropriate angle transforms any equation of the form of equation (1) into an equation in x' and y' without an $x'y'$-term.

To find the formula for choosing an appropriate angle θ through which to rotate the axes, begin with equation (1),

$$Ax^2 + Bxy + Cy^2 + Dx + Ey + F = 0 \qquad B \neq 0$$

Next rotate through an angle θ using the rotation formulas (5).

$$\begin{aligned} A(x' \cos\theta - y' \sin\theta)^2 &+ B(x' \cos\theta - y' \sin\theta)(x' \sin\theta + y' \cos\theta) \\ &+ C(x' \sin\theta + y' \cos\theta)^2 + D(x' \cos\theta - y' \sin\theta) \\ &+ E(x' \sin\theta + y' \cos\theta) + F = 0 \end{aligned}$$

Expanding and collecting like terms gives

$$\begin{aligned} (A\cos^2\theta + B\sin\theta\cos\theta + C\sin^2\theta)x'^2 &+ [B(\cos^2\theta - \sin^2\theta) + 2(C - A)(\sin\theta\cos\theta)]x'y' \\ &+ (A\sin^2\theta - B\sin\theta\cos\theta + C\cos^2\theta)y'^2 \\ &+ (D\cos\theta + E\sin\theta)x' \\ &+ (-D\sin\theta + E\cos\theta)y' + F = 0 \end{aligned} \tag{6}$$

In equation (6), the coefficient of $x'y'$ is

$$B(\cos^2\theta - \sin^2\theta) + 2(C - A)(\sin\theta\cos\theta)$$

Since we want to eliminate the $x'y'$-term, we select an angle θ so that this coefficient is 0.

$$B(\cos^2\theta - \sin^2\theta) + 2(C - A)(\sin\theta\cos\theta) = 0$$

$$B\cos(2\theta) + (C - A)\sin(2\theta) = 0 \quad \text{Double-angle Formulas}$$

$$B\cos(2\theta) = (A - C)\sin(2\theta)$$

$$\cot(2\theta) = \frac{A - C}{B} \qquad B \neq 0$$

THEOREM

To transform the equation

$$Ax^2 + Bxy + Cy^2 + Dx + Ey + F = 0 \qquad B \neq 0$$

into an equation in x' and y' without an $x'y'$-term, rotate the axes through an angle θ that satisfies the equation

$$\cot(2\theta) = \frac{A - C}{B} \tag{7}$$

WARNING Be careful if you use a calculator to solve equation (7).

1. If $\cot(2\theta) = 0$, then $2\theta = 90°$ and $\theta = 45°$.
2. If $\cot(2\theta) \neq 0$, first find $\cos(2\theta)$. Then use the inverse cosine function key(s) to obtain 2θ, $0° < 2\theta < 180°$. Finally, divide by 2 to obtain the correct acute angle θ.

Equation (7) has an infinite number of solutions for θ. We shall adopt the convention of choosing the acute angle θ that satisfies (7). There are two possibilities:

If $\cot(2\theta) \geq 0$, then $0° < 2\theta \leq 90°$, so $0° < \theta \leq 45°$.

If $\cot(2\theta) < 0$, then $90° < 2\theta < 180°$, so $45° < \theta < 90°$.

Each of these results in a counterclockwise rotation of the axes through an acute angle θ.*

3 Analyze an Equation Using a Rotation of Axes

For the remainder of this section, the direction **"Analyze the equation"** will mean to transform the given equation so that it contains no xy-term and to graph the equation.

EXAMPLE 3

Analyzing an Equation Using a Rotation of Axes

Analyze the equation: $x^2 + \sqrt{3}xy + 2y^2 - 10 = 0$

Solution Since an xy-term is present, we must rotate the axes. Using $A = 1$, $B = \sqrt{3}$, and $C = 2$ in equation (7), the appropriate acute angle θ through which to rotate the axes satisfies the equation

$$\cot(2\theta) = \frac{A - C}{B} = \frac{-1}{\sqrt{3}} = -\frac{\sqrt{3}}{3} \qquad 0° < 2\theta < 180°$$

Since $\cot(2\theta) = -\frac{\sqrt{3}}{3}$, this means $2\theta = 120°$, so $\theta = 60°$. Using $\theta = 60°$ in the rotation formulas (5) yields

$$x = x'\cos 60° - y'\sin 60° = \frac{1}{2}x' - \frac{\sqrt{3}}{2}y' = \frac{1}{2}(x' - \sqrt{3}y')$$

$$y = x'\sin 60° + y'\cos 60° = \frac{\sqrt{3}}{2}x' + \frac{1}{2}y' = \frac{1}{2}(\sqrt{3}x' + y')$$

*Any rotation through an angle θ that satisfies $\cot(2\theta) = \frac{A - C}{B}$ will eliminate the $x'y'$-term. However, the final form of the transformed equation may be different (but equivalent), depending on the angle chosen.

Substituting these values into the original equation and simplifying gives

$$x^2 + \sqrt{3}xy + 2y^2 - 10 = 0$$

$$\frac{1}{4}(x' - \sqrt{3}y')^2 + \sqrt{3}\left[\frac{1}{2}(x' - \sqrt{3}y')\right]\left[\frac{1}{2}(\sqrt{3}x' + y')\right] + 2\left[\frac{1}{4}(\sqrt{3}x' + y')^2\right] = 10$$

Multiply both sides by 4 and expand to obtain

$$x'^2 - 2\sqrt{3}x'y' + 3y'^2 + \sqrt{3}(\sqrt{3}x'^2 - 2x'y' - \sqrt{3}y'^2) + 2(3x'^2 + 2\sqrt{3}x'y' + y'^2) = 40$$

$$10x'^2 + 2y'^2 = 40$$

$$\frac{x'^2}{4} + \frac{y'^2}{20} = 1$$

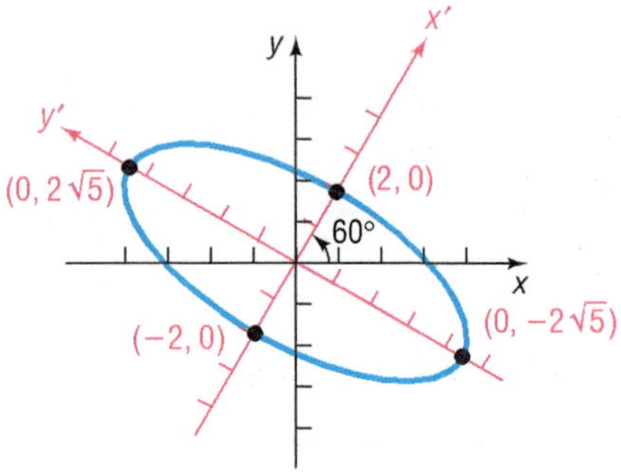

Figure 53 $\frac{x'^2}{4} + \frac{y'^2}{20} = 1$

This is the equation of an ellipse with center at $(0, 0)$ and major axis along the y'-axis. The vertices are at $(0, \pm 2\sqrt{5})$ on the y'-axis. See Figure 53 for the graph.

To graph the equation $x^2 + \sqrt{3}xy + 2y^2 - 10 = 0$ using a graphing utility, we need to solve the equation for y. Rearranging the terms, observe that the equation is quadratic in the variable y: $2y^2 + \sqrt{3}xy + (x^2 - 10) = 0$. Solve the equation for y using the quadratic formula with $a = 2$, $b = \sqrt{3}x$, and $c = x^2 - 10$.

$$Y_1 = \frac{-\sqrt{3}x + \sqrt{(\sqrt{3}x)^2 - 4(2)(x^2 - 10)}}{2(2)} = \frac{-\sqrt{3}x + \sqrt{-5x^2 + 80}}{4}$$

and

$$Y_2 = \frac{-\sqrt{3}x - \sqrt{(\sqrt{3}x)^2 - 4(2)(x^2 - 10)}}{2(2)} = \frac{-\sqrt{3}x - \sqrt{-5x^2 + 80}}{4}$$

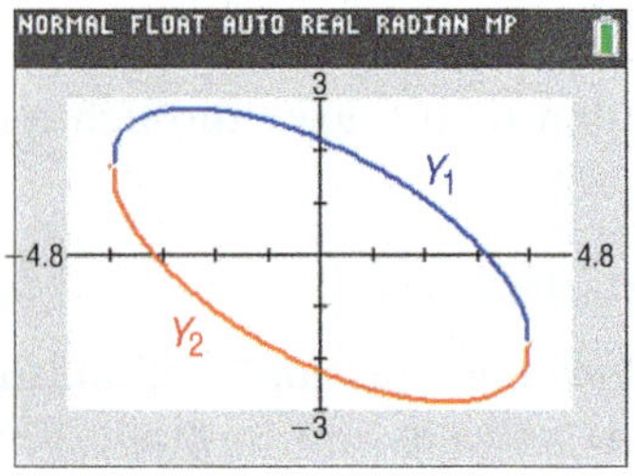

Figure 54

Figure 54 shows the graph of Y_1 and Y_2. ■

Now Work PROBLEM 31

In Example 3, the acute angle θ through which to rotate the axes was easy to find because of the numbers that we used in the given equation. In general, the equation $\cot(2\theta) = \frac{A - C}{B}$ will not have such a "nice" solution. As the next example shows, we can still find the appropriate rotation formulas without using a calculator approximation by applying Half-angle Formulas.

EXAMPLE 4 Analyzing an Equation Using a Rotation of Axes

Analyze the equation: $4x^2 - 4xy + y^2 + 5\sqrt{5}x + 5 = 0$

Solution Letting $A = 4$, $B = -4$, and $C = 1$ in equation (7), the appropriate angle θ through which to rotate the axes satisfies

$$\cot(2\theta) = \frac{A - C}{B} = \frac{3}{-4} = -\frac{3}{4}$$

To use the rotation formulas (5), we need to know the values of $\sin\theta$ and $\cos\theta$. Since we seek an acute angle θ, we know that $\sin\theta > 0$ and $\cos\theta > 0$. Use the Half-angle Formulas in the form

$$\sin\theta = \sqrt{\frac{1 - \cos(2\theta)}{2}} \qquad \cos\theta = \sqrt{\frac{1 + \cos(2\theta)}{2}}$$

Now we need to find the value of $\cos(2\theta)$. Since $\cot(2\theta) = -\frac{3}{4}$, then $90° < 2\theta < 180°$ (Do you know why?), so $\cos(2\theta) = -\frac{3}{5}$. Then

$$\sin\theta = \sqrt{\frac{1-\cos(2\theta)}{2}} = \sqrt{\frac{1-\left(-\frac{3}{5}\right)}{2}} = \sqrt{\frac{4}{5}} = \frac{2}{\sqrt{5}} = \frac{2\sqrt{5}}{5}$$

$$\cos\theta = \sqrt{\frac{1+\cos(2\theta)}{2}} = \sqrt{\frac{1+\left(-\frac{3}{5}\right)}{2}} = \sqrt{\frac{1}{5}} = \frac{1}{\sqrt{5}} = \frac{\sqrt{5}}{5}$$

With these values, the rotation formulas (5) are

$$x = \frac{\sqrt{5}}{5}x' - \frac{2\sqrt{5}}{5}y' = \frac{\sqrt{5}}{5}(x' - 2y')$$

$$y = \frac{2\sqrt{5}}{5}x' + \frac{\sqrt{5}}{5}y' = \frac{\sqrt{5}}{5}(2x' + y')$$

Substituting these values in the original equation and simplifying gives

$$4x^2 - 4xy + y^2 + 5\sqrt{5}x + 5 = 0$$

$$4\left[\frac{\sqrt{5}}{5}(x' - 2y')\right]^2 - 4\left[\frac{\sqrt{5}}{5}(x' - 2y')\right]\left[\frac{\sqrt{5}}{5}(2x' + y')\right]$$
$$+\left[\frac{\sqrt{5}}{5}(2x' + y')\right]^2 + 5\sqrt{5}\left[\frac{\sqrt{5}}{5}(x' - 2y')\right] = -5$$

Multiply both sides by 5 and expand to obtain

$$4(x'^2 - 4x'y' + 4y'^2) - 4(2x'^2 - 3x'y' - 2y'^2)$$
$$+ 4x'^2 + 4x'y' + y'^2 + 25(x' - 2y') = -25$$

$$25y'^2 - 50y' + 25x' = -25 \quad \text{Combine like terms.}$$

$$y'^2 - 2y' + x' = -1 \quad \text{Divide by 25.}$$

$$y'^2 - 2y' + 1 = -x' \quad \text{Complete the square in } y'.$$

$$(y' - 1)^2 = -x'$$

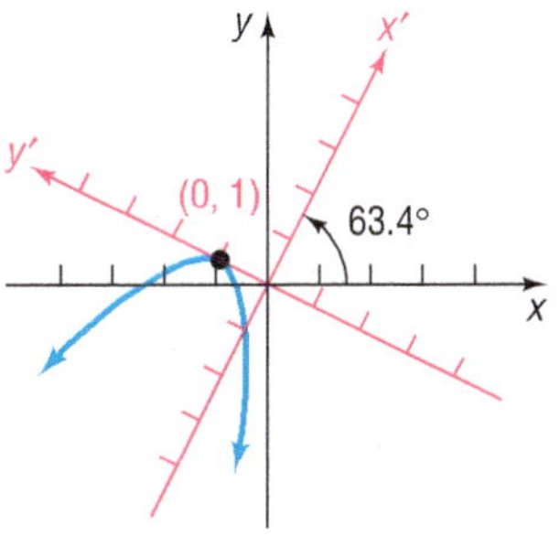

Figure 55 $(y' - 1)^2 = -x'$

This is the equation of a parabola with vertex at $(0, 1)$ in the $x'y'$-plane. The axis of symmetry is parallel to the x'-axis. Use a calculator to solve $\sin\theta = \frac{2\sqrt{5}}{5}$ and find that $\theta \approx 63.4°$. See Figure 55 for the graph. ■

Now Work PROBLEM 37

4 Identify Conics without a Rotation of Axes

Suppose that we are required only to identify (rather than analyze) the graph of an equation of the form

$$Ax^2 + Bxy + Cy^2 + Dx + Ey + F = 0 \qquad B \neq 0 \tag{8}$$

Applying the rotation formulas (5) to this equation gives an equation of the form

$$A'x'^2 + B'x'y' + C'y'^2 + D'x' + E'y' + F' = 0 \tag{9}$$

where A', B', C', D', E', and F' can be expressed in terms of A, B, C, D, E, F and the angle θ of rotation (see Problem 53). It can be shown that the value of $B^2 - 4AC$ in equation (8) and the value of $B'^2 - 4A'C'$ in equation (9) are equal no matter what angle θ of rotation is chosen (see Problem 55). In particular, if the angle θ of rotation satisfies equation (7), then $B' = 0$ in equation (9), and $B^2 - 4AC = -4A'C'$. Since equation (9) then has the form of equation (2),

$$A'x'^2 + C'y'^2 + D'x' + E'y' + F' = 0$$

we can identify its graph without completing the squares, as we did in the beginning of this section. In fact, now we can identify the conic described by any equation of the form of equation (8) without a rotation of axes.

THEOREM

Identifying Conics without a Rotation of Axes

Except for degenerate cases, the equation

$$Ax^2 + Bxy + Cy^2 + Dx + Ey + F = 0$$

(a) Defines a parabola if $B^2 - 4AC = 0$.
(b) Defines an ellipse (or a circle) if $B^2 - 4AC < 0$.
(c) Defines a hyperbola if $B^2 - 4AC > 0$.

You are asked to prove this theorem in Problem 56.

EXAMPLE 5 **Identifying a Conic without a Rotation of Axes**

Identify the graph of the equation: $8x^2 - 12xy + 17y^2 - 4\sqrt{5}x - 2\sqrt{5}y - 15 = 0$

Solution Here $A = 8, B = -12$, and $C = 17$, so $B^2 - 4AC = -400$. Since $B^2 - 4AC < 0$, the equation defines an ellipse.

Now Work PROBLEM 43

11.5 Assess Your Understanding

'Are You Prepared?' *Answers are given at the end of these exercises. If you get a wrong answer, read the pages listed in red.*

1. The sum formula for the sine function is $\sin(A + B) =$ ________. (p. 655)
2. The Double-angle Formula for the sine function is $\sin(2\theta) =$ ________. (p. 665)
3. If θ is acute, the Half-angle Formula for the sine function is $\sin\frac{\theta}{2} =$ ________. (p. 668)
4. If θ is acute, the Half-angle Formula for the cosine function is $\cos\frac{\theta}{2} =$ ________. (p. 668)

Concepts and Vocabulary

5. To transform the equation
$$Ax^2 + Bxy + Cy^2 + Dx + Ey + F = 0 \qquad B \neq 0$$
into one in x' and y' without an $x'y'$-term, rotate the axes through an acute angle θ that satisfies the equation ________.
6. Except for degenerate cases, the equation
$$Ax^2 + Bxy + Cy^2 + Dx + Ey + F = 0$$
defines a(n) ________ if $B^2 - 4AC = 0$.
(a) circle (b) ellipse (c) hyperbola (d) parabola
7. Except for degenerate cases, the equation
$$Ax^2 + Bxy + Cy^2 + Dx + Ey + F = 0$$
defines an ellipse if ________.
8. The equation $ax^2 + 6y^2 - 12y = 0$ defines an ellipse if ________.
(a) $a < 0$ (b) $a = 0$
(c) $a > 0$ (d) a is any real number
9. ***True or False*** The equation $3x^2 + Bxy + 12y^2 = 10$ defines a parabola if $B = -12$.
10. ***True or False*** To eliminate the xy-term from the equation $x^2 - 2xy + y^2 - 2x + 3y + 5 = 0$, rotate the axes through an angle θ, where $\cot\theta = B^2 - 4AC$.

Skill Building

In Problems 11–20, identify the graph of each equation without completing the squares.

11. $x^2 + 4x + y + 3 = 0$
12. $2y^2 - 3y + 3x = 0$
13. $6x^2 + 3y^2 - 12x + 6y = 0$
14. $2x^2 + y^2 - 8x + 4y + 2 = 0$

15. $3x^2 - 2y^2 + 6x + 4 = 0$

16. $4x^2 - 3y^2 - 8x + 6y + 1 = 0$

17. $2y^2 - x^2 - y + x = 0$

18. $y^2 - 8x^2 - 2x - y = 0$

19. $x^2 + y^2 - 8x + 4y = 0$

20. $2x^2 + 2y^2 - 8x + 8y = 0$

In Problems 21–30, determine the appropriate rotation formulas to use so that the new equation contains no xy-term.

21. $x^2 + 4xy + y^2 - 3 = 0$

22. $x^2 - 4xy + y^2 - 3 = 0$

23. $5x^2 + 6xy + 5y^2 - 8 = 0$

24. $3x^2 - 10xy + 3y^2 - 32 = 0$

25. $13x^2 - 6\sqrt{3}xy + 7y^2 - 16 = 0$

26. $11x^2 + 10\sqrt{3}xy + y^2 - 4 = 0$

27. $4x^2 - 4xy + y^2 - 8\sqrt{5}x - 16\sqrt{5}y = 0$

28. $x^2 + 4xy + 4y^2 + 5\sqrt{5}y + 5 = 0$

29. $25x^2 - 36xy + 40y^2 - 12\sqrt{13}x - 8\sqrt{13}y = 0$

30. $34x^2 - 24xy + 41y^2 - 25 = 0$

In Problems 31–42, rotate the axes so that the new equation contains no xy-term. Analyze and graph the new equation. Refer to Problems 21–30 for Problems 31–40.

31. $x^2 + 4xy + y^2 - 3 = 0$

32. $x^2 - 4xy + y^2 - 3 = 0$

33. $5x^2 + 6xy + 5y^2 - 8 = 0$

34. $3x^2 - 10xy + 3y^2 - 32 = 0$

35. $13x^2 - 6\sqrt{3}xy + 7y^2 - 16 = 0$

36. $11x^2 + 10\sqrt{3}xy + y^2 - 4 = 0$

37. $4x^2 - 4xy + y^2 - 8\sqrt{5}x - 16\sqrt{5}y = 0$

38. $x^2 + 4xy + 4y^2 + 5\sqrt{5}y + 5 = 0$

39. $25x^2 - 36xy + 40y^2 - 12\sqrt{13}x - 8\sqrt{13}y = 0$

40. $34x^2 - 24xy + 41y^2 - 25 = 0$

41. $16x^2 + 24xy + 9y^2 - 130x + 90y = 0$

42. $16x^2 + 24xy + 9y^2 - 60x + 80y = 0$

In Problems 43–52, identify the graph of each equation without applying a rotation of axes.

43. $x^2 + 3xy - 2y^2 + 3x + 2y + 5 = 0$

44. $2x^2 - 3xy + 4y^2 + 2x + 3y - 5 = 0$

45. $x^2 - 7xy + 3y^2 - y - 10 = 0$

46. $2x^2 - 3xy + 2y^2 - 4x - 2 = 0$

47. $9x^2 + 12xy + 4y^2 - x - y = 0$

48. $10x^2 + 12xy + 4y^2 - x - y + 10 = 0$

49. $10x^2 - 12xy + 4y^2 - x - y - 10 = 0$

50. $4x^2 + 12xy + 9y^2 - x - y = 0$

51. $3x^2 - 2xy + y^2 + 4x + 2y - 1 = 0$

52. $3x^2 + 2xy + y^2 + 4x - 2y + 10 = 0$

Applications and Extensions

In Problems 53–56, apply the rotation formulas (5) to

$$Ax^2 + Bxy + Cy^2 + Dx + Ey + F = 0$$

to obtain the equation

$$A'x'^2 + B'x'y' + C'y'^2 + D'x' + E'y' + F' = 0$$

53. Express A', B', C', D', E', and F' in terms of A, B, C, D, E, F, and the angle θ of rotation.

[**Hint:** Refer to equation (6).]

54. Show that $A + C = A' + C'$, which proves that $A + C$ is **invariant**; that is, its value does not change under a rotation of axes.

55. Refer to Problem 54. Show that $B^2 - 4AC$ is invariant.

56. Prove that, except for degenerate cases, the equation

$$Ax^2 + Bxy + Cy^2 + Dx + Ey + F = 0$$

(a) Defines a parabola if $B^2 - 4AC = 0$.
(b) Defines an ellipse (or a circle) if $B^2 - 4AC < 0$.
(c) Defines a hyperbola if $B^2 - 4AC > 0$.

57. Use the rotation formulas (5) to show that distance is invariant under a rotation of axes. That is, show that the distance from $P_1 = (x_1, y_1)$ to $P_2 = (x_2, y_2)$ in the xy-plane equals the distance from $P_1 = (x'_1, y'_1)$ to $P_2 = (x'_2, y'_2)$ in the $x'y'$-plane.

58. Show that the graph of the equation $x^{1/2} + y^{1/2} = a^{1/2}$ is part of the graph of a parabola.

Explaining Concepts: Discussion and Writing

59. Formulate a strategy for analyzing and graphing an equation of the form

$$Ax^2 + Cy^2 + Dx + Ey + F = 0$$

60. Explain how your strategy presented in Problem 59 changes if the equation is of the form

$$Ax^2 + Bxy + Cy^2 + Dx + Ey + F = 0$$

Retain Your Knowledge

Problems 61–64 are based on material learned earlier in the course. The purpose of these problems is to keep the material fresh in your mind so that you are better prepared for the final exam.

61. Solve the triangle described: $a = 7, b = 9$, and $c = 11$.

62. Find the area of the triangle described: $a = 14, b = 11$, and $C = 30°$.

63. Transform the equation $xy = 1$ from rectangular coordinates to polar coordinates.

64. Write the complex number $2 - 5i$ in polar form.

'Are You Prepared?' Answers

1. $\sin A \cos B + \cos A \sin B$ **2.** $2 \sin \theta \cos \theta$ **3.** $\sqrt{\dfrac{1 - \cos \theta}{2}}$ **4.** $\sqrt{\dfrac{1 + \cos \theta}{2}}$

11.6 Polar Equations of Conics

PREPARING FOR THIS SECTION *Before getting started, review the following:*

- Polar Coordinates (Section 10.1, pp. 731–738)

Now Work the 'Are You Prepared?' problems on page 841.

OBJECTIVES 1 Analyze and Graph Polar Equations of Conics (p. 836)
2 Convert the Polar Equation of a Conic to a Rectangular Equation (p. 840)

1 Analyze and Graph Polar Equations of Conics

In Sections 11.2 through 11.4, we gave separate definitions for the parabola, ellipse, and hyperbola based on geometric properties and the distance formula. This section presents an alternative definition that simultaneously defines all these conics. As we shall see, this approach is well suited to polar coordinate representation. (Refer to Section 10.1.)

DEFINITION

Let D denote a fixed line called the **directrix**; let F denote a fixed point called the **focus**, which is not on D; and let e be a fixed positive number called the **eccentricity**. A **conic** is the set of points P in a plane such that the ratio of the distance from F to P to the distance from D to P equals e. That is, a conic is the collection of points P for which

$$\frac{d(F, P)}{d(D, P)} = e \qquad (1)$$

If $e = 1$, the conic is a **parabola**.

If $e < 1$, the conic is an **ellipse**.

If $e > 1$, the conic is a **hyperbola**.

Observe that if $e = 1$ the definition of a parabola in equation (1) is exactly the same as the definition used earlier in Section 11.2.

In the case of an ellipse, the **major axis** is a line through the focus perpendicular to the directrix. In the case of a hyperbola, the **transverse axis** is a line through the focus perpendicular to the directrix. For both an ellipse and a hyperbola, the eccentricity e satisfies

$$e = \frac{c}{a} \tag{2}$$

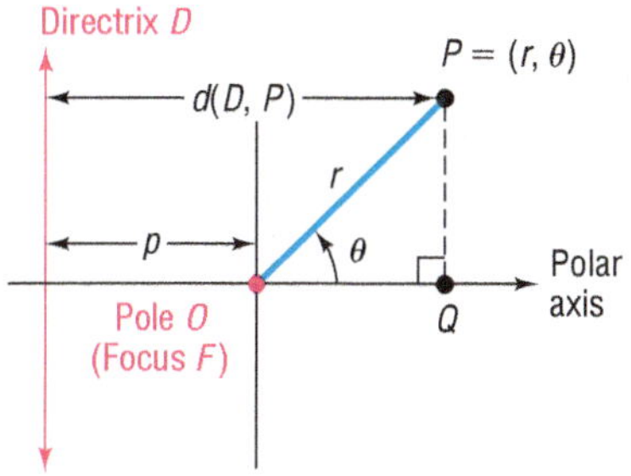

Figure 56

where c is the distance from the center to the focus and a is the distance from the center to a vertex.

Just as we did earlier using rectangular coordinates, we derive equations for the conics in polar coordinates by choosing a convenient position for the focus F and the directrix D. The focus F is positioned at the pole, and the directrix D is either parallel or perpendicular to the polar axis.

Suppose that we start with the directrix D perpendicular to the polar axis at a distance p units to the left of the pole (the focus F). See Figure 56.

If $P = (r, \theta)$ is any point on the conic, then, by equation (1),

$$\frac{d(F, P)}{d(D, P)} = e \quad \text{or} \quad d(F, P) = e \cdot d(D, P) \tag{3}$$

Now use the point Q obtained by dropping the perpendicular from P to the polar axis to calculate $d(D, P)$.

$$d(D, P) = p + d(O, Q) = p + r\cos\theta$$

Using this expression and the fact that $d(F, P) = d(O, P) = r$ in equation (3) gives

$$\begin{aligned} d(F, P) &= e \cdot d(D, P) \\ r &= e(p + r\cos\theta) \\ r &= ep + er\cos\theta \\ r - er\cos\theta &= ep \\ r(1 - e\cos\theta) &= ep \\ r &= \frac{ep}{1 - e\cos\theta} \end{aligned}$$

THEOREM

Polar Equation of a Conic; Focus at the Pole; Directrix Perpendicular to the Polar Axis a Distance p to the Left of the Pole

The polar equation of a conic with focus at the pole and directrix perpendicular to the polar axis at a distance p to the left of the pole is

$$r = \frac{ep}{1 - e\cos\theta} \tag{4}$$

where e is the eccentricity of the conic. ■

EXAMPLE 1 Analyzing and Graphing the Polar Equation of a Conic

Analyze and graph the equation: $r = \dfrac{4}{2 - \cos\theta}$

Solution The given equation is not quite in the form of equation (4), since the first term in the denominator is 2 instead of 1. Divide the numerator and denominator by 2 to obtain

$$r = \frac{2}{1 - \frac{1}{2}\cos\theta} \qquad r = \frac{ep}{1 - e\cos\theta}$$

This equation is in the form of equation (4), with

$$e = \frac{1}{2} \quad \text{and} \quad ep = 2$$

Then

$$\frac{1}{2}p = 2, \quad \text{so} \quad p = 4$$

Because $e = \frac{1}{2} < 1$, the conic is an ellipse. One focus is at the pole, and the directrix is perpendicular to the polar axis, a distance of $p = 4$ units to the left of the pole. It follows that the major axis is along the polar axis. To find the vertices, let $\theta = 0$ and $\theta = \pi$. The vertices of the ellipse are $(4, 0)$ and $\left(\frac{4}{3}, \pi\right)$. The midpoint of the vertices, $\left(\frac{4}{3}, 0\right)$ in polar coordinates, is the center of the ellipse. [Do you see why? The vertices $(4, 0)$ and $\left(\frac{4}{3}, \pi\right)$ in polar coordinates are $(4, 0)$ and $\left(-\frac{4}{3}, 0\right)$ in rectangular coordinates. The midpoint in rectangular coordinates is $\left(\frac{4}{3}, 0\right)$, which is also $\left(\frac{4}{3}, 0\right)$ in polar coordinates.] Then a = distance from the center to a vertex $= \frac{8}{3}$. Using $a = \frac{8}{3}$ and $e = \frac{1}{2}$ in equation (2), $e = \frac{c}{a}$, we find $c = ae = \frac{4}{3}$. Finally, using $a = \frac{8}{3}$ and $c = \frac{4}{3}$ in $b^2 = a^2 - c^2$ yields

$$b^2 = a^2 - c^2 = \frac{64}{9} - \frac{16}{9} = \frac{48}{9}$$

$$b = \frac{4\sqrt{3}}{3}$$

Figure 57(a) shows the graph drawn by hand. Figure 57(b) shows the graph of the equation obtained using a graphing utility in POLAR mode with $\theta\text{min} = 0$, $\theta\text{max} = 2\pi$, and $\theta\text{step} = \frac{\pi}{24}$. ■

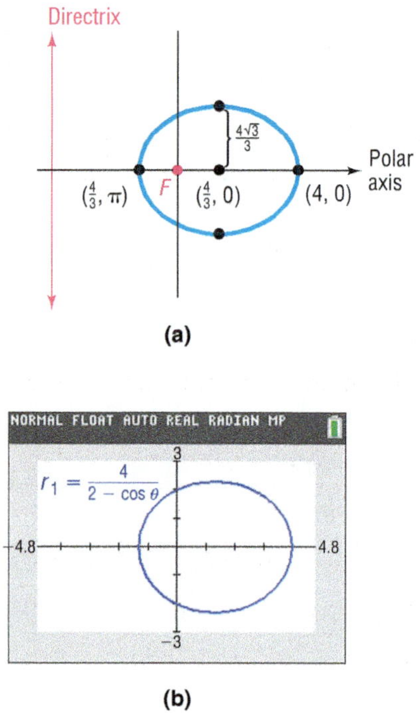

Figure 57 $r = \frac{4}{2 - \cos\theta}$

Exploration

Graph $r_1 = \frac{4}{2 + \cos\theta}$ and compare the result with Figure 57. What do you conclude? Clear the screen and graph $r_1 = \frac{4}{2 - \sin\theta}$ and then $r_1 = \frac{4}{2 + \sin\theta}$. Compare each of these graphs with Figure 57. What do you conclude? ■

Now Work PROBLEM 11

Equation (4) was obtained under the assumption that the directrix was perpendicular to the polar axis at a distance p units to the left of the pole. A similar derivation (see Problem 43), in which the directrix is perpendicular to the polar axis at a distance p units to the right of the pole, results in the equation

$$r = \frac{ep}{1 + e\cos\theta}$$

In Problems 44 and 45, you are asked to derive the polar equations of conics with focus at the pole and directrix parallel to the polar axis. Table 5 summarizes the polar equations of conics.

Table 5

Polar Equations of Conics (Focus at the Pole, Eccentricity e)	
Equation	**Description**
(a) $r = \dfrac{ep}{1 - e\cos\theta}$	Directrix is perpendicular to the polar axis at a distance p units to the left of the pole.
(b) $r = \dfrac{ep}{1 + e\cos\theta}$	Directrix is perpendicular to the polar axis at a distance p units to the right of the pole.
(c) $r = \dfrac{ep}{1 + e\sin\theta}$	Directrix is parallel to the polar axis at a distance p units above the pole.
(d) $r = \dfrac{ep}{1 - e\sin\theta}$	Directrix is parallel to the polar axis at a distance p units below the pole.
Eccentricity	
If $e = 1$, the conic is a parabola; the axis of symmetry is perpendicular to the directrix.	
If $e < 1$, the conic is an ellipse; the major axis is perpendicular to the directrix.	
If $e > 1$, the conic is a hyperbola; the transverse axis is perpendicular to the directrix.	

EXAMPLE 2 Analyzing and Graphing the Polar Equation of a Conic

Analyze and graph the equation: $r = \dfrac{6}{3 + 3\sin\theta}$

Solution To place the equation in proper form, divide the numerator and denominator by 3 to get

$$r = \frac{2}{1 + \sin\theta}$$

Referring to Table 5, conclude that this equation is in the form of equation (c) with

$$e = 1 \quad \text{and} \quad ep = 2$$
$$p = 2 \quad e = 1$$

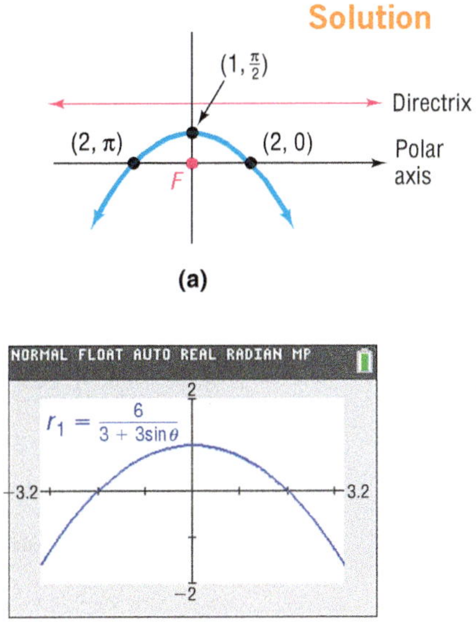

(a)

(b)

Figure 58 $r = \dfrac{6}{3 + 3\sin\theta}$

The conic is a parabola with focus at the pole. The directrix is parallel to the polar axis at a distance 2 units above the pole; the axis of symmetry is perpendicular to the polar axis. The vertex of the parabola, which is at the midpoint of the focus and directrix on the axis of symmetry, is $\left(1, \dfrac{\pi}{2}\right)$. See Figure 58(a) for the graph drawn by hand. Notice that we plotted two additional points, $(2, 0)$ and $(2, \pi)$, to assist in graphing. Figure 58(b) shows the graph of the equation using a graphing utility in POLAR mode with θmin $= 0$, θmax $= 2\pi$, and θstep $= \dfrac{\pi}{24}$. ■

Now Work PROBLEM 13

EXAMPLE 3 Analyzing and Graphing the Polar Equation of a Conic

Analyze and graph the equation: $r = \dfrac{3}{1 + 3\cos\theta}$

Solution This equation is in the form of equation (b) in Table 5. This means that

$$e = 3 \quad \text{and} \quad ep = 3$$
$$p = 1 \quad e = 3$$

This is the equation of a hyperbola with a focus at the pole. The directrix is perpendicular to the polar axis, $p = 1$ unit to the right of the pole. The transverse axis is along the polar axis. To find the vertices, let $\theta = 0$ and $\theta = \pi$. The vertices are $\left(\frac{3}{4}, 0\right)$ and $\left(-\frac{3}{2}, \pi\right)$. The center, which is at the midpoint of $\left(\frac{3}{4}, 0\right)$ and $\left(-\frac{3}{2}, \pi\right)$, is $\left(\frac{9}{8}, 0\right)$. Then $c =$ distance from the center to a focus $= \frac{9}{8}$. Since $e = 3$, it follows from equation (2), $e = \frac{c}{a}$, that $a = \frac{3}{8}$. Finally, using $a = \frac{3}{8}$ and $c = \frac{9}{8}$ in $b^2 = c^2 - a^2$ gives

$$b^2 = c^2 - a^2 = \frac{81}{64} - \frac{9}{64} = \frac{72}{64} = \frac{9}{8}$$

$$b = \frac{3}{2\sqrt{2}} = \frac{3\sqrt{2}}{4}$$

Figure 59(a) shows the graph drawn by hand. Notice that we plotted two additional points, $\left(3, \frac{\pi}{2}\right)$ and $\left(3, \frac{3\pi}{2}\right)$, on the left branch and used symmetry to obtain the right branch. The asymptotes of this hyperbola were found in the usual way by constructing the rectangle shown.

Figures 59(b) and (c) show the graph of the equation using a graphing utility in POLAR mode with $\theta\text{min} = 0$, $\theta\text{max} = 2\pi$, and $\theta\text{step} = \frac{\pi}{24}$, using both dot mode and connected mode. Notice the extraneous asymptotes in connected mode.

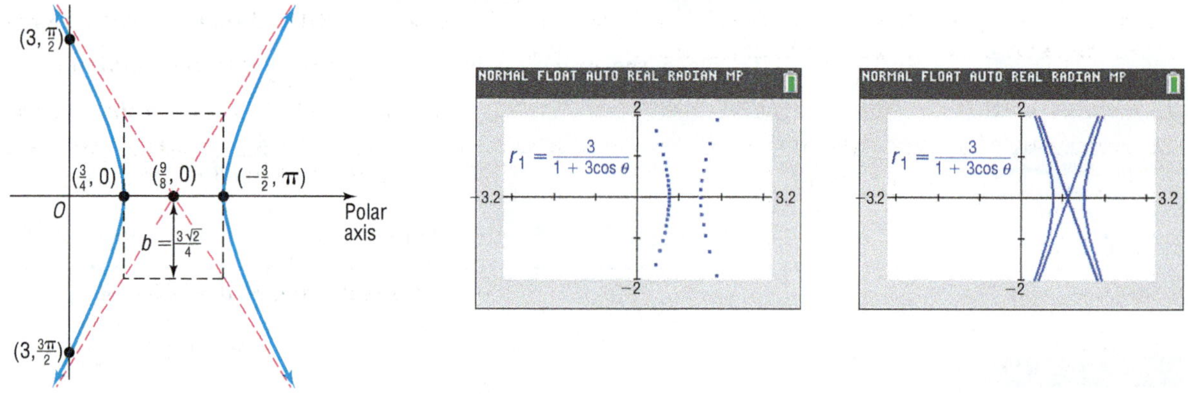

(a) **(b)** Dot Mode **(c)** Connected Mode

Figure 59 $r = \frac{3}{1 + 3\cos\theta}$

Now Work PROBLEM 17

2 Convert the Polar Equation of a Conic to a Rectangular Equation

EXAMPLE 4 **Converting a Polar Equation to a Rectangular Equation**

Convert the polar equation

$$r = \frac{1}{3 - 3\cos\theta}$$

to a rectangular equation.

Solution The strategy here is first to rearrange the equation and square each side before using the transformation equations.

$$r = \frac{1}{3 - 3\cos\theta}$$

$$3r - 3r\cos\theta = 1$$

$$3r = 1 + 3r\cos\theta \quad \text{Rearrange the equation.}$$

$$9r^2 = (1 + 3r\cos\theta)^2 \quad \text{Square each side.}$$

$$9(x^2 + y^2) = (1 + 3x)^2 \quad x^2 + y^2 = r^2;\ x = r\cos\theta$$

$$9x^2 + 9y^2 = 9x^2 + 6x + 1$$

$$9y^2 = 6x + 1$$

This is the equation of a parabola in rectangular coordinates. ■

Now Work PROBLEM 25

11.6 Assess Your Understanding

'Are You Prepared?' *Answers are given at the end of these exercises. If you get a wrong answer, read the pages listed in red.*

1. If (x, y) are the rectangular coordinates of a point P and (r, θ) are its polar coordinates, then $x =$ ______ and $y =$ ______. (pp. 733–734)

2. Transform the equation $r = 6\cos\theta$ from polar coordinates to rectangular coordinates. (pp. 737–738)

Concepts and Vocabulary

3. A ______ is the set of points P in a plane such that the ratio of the distance from a fixed point called the ______ to P to the distance from a fixed line called the ______ to P equals a constant e.

4. The eccentricity e of a parabola is ______, of an ellipse it is ______, and of a hyperbola it is ______.

5. ***True or False*** If (r, θ) are polar coordinates, the equation $r = \frac{2}{2 + 3\sin\theta}$ defines a hyperbola.

6. ***True or False*** The eccentricity e of any conic is $\frac{c}{a}$, where a is the distance of a vertex from the center and c is the distance of a focus from the center.

Skill Building

In Problems 7–12, identify the conic that each polar equation represents. Also, give the position of the directrix.

7. $r = \frac{1}{1 + \cos\theta}$

8. $r = \frac{3}{1 - \sin\theta}$

9. $r = \frac{4}{2 - 3\sin\theta}$

10. $r = \frac{2}{1 + 2\cos\theta}$

11. $r = \frac{3}{4 - 2\cos\theta}$

12. $r = \frac{6}{8 + 2\sin\theta}$

In Problems 13–24, analyze each equation and graph it.

13. $r = \frac{1}{1 + \cos\theta}$

14. $r = \frac{3}{1 - \sin\theta}$

15. $r = \frac{8}{4 + 3\sin\theta}$

16. $r = \frac{10}{5 + 4\cos\theta}$

17. $r = \frac{9}{3 - 6\cos\theta}$

18. $r = \frac{12}{4 + 8\sin\theta}$

19. $r = \frac{8}{2 - \sin\theta}$

20. $r = \frac{8}{2 + 4\cos\theta}$

21. $r(3 - 2\sin\theta) = 6$

22. $r(2 - \cos\theta) = 2$

23. $r = \frac{6\sec\theta}{2\sec\theta - 1}$

24. $r = \frac{3\csc\theta}{\csc\theta - 1}$

In Problems 25–36, convert each polar equation to a rectangular equation.

25. $r = \frac{1}{1 + \cos\theta}$

26. $r = \frac{3}{1 - \sin\theta}$

27. $r = \frac{8}{4 + 3\sin\theta}$

28. $r = \frac{10}{5 + 4\cos\theta}$

29. $r = \frac{9}{3 - 6\cos\theta}$

30. $r = \frac{12}{4 + 8\sin\theta}$

31. $r = \frac{8}{2 - \sin\theta}$

32. $r = \frac{8}{2 + 4\cos\theta}$

33. $r(3 - 2\sin\theta) = 6$

34. $r(2 - \cos\theta) = 2$

35. $r = \frac{6\sec\theta}{2\sec\theta - 1}$

36. $r = \frac{3\csc\theta}{\csc\theta - 1}$

In Problems 37–42, find a polar equation for each conic. Each has a focus at the pole.

37. $e = 1$; directrix is parallel to the polar axis 1 unit above the pole.

38. $e = 1$; directrix is parallel to the polar axis 2 units below the pole.

39. $e = \frac{4}{5}$; directrix is perpendicular to the polar axis 3 units to the left of the pole.

40. $e = \frac{2}{3}$; directrix is parallel to the polar axis 3 units above the pole.

41. $e = 6$; directrix is parallel to the polar axis 2 units below the pole.

42. $e = 5$; directrix is perpendicular to the polar axis 5 units to the right of the pole.

Applications and Extensions

43. Derive equation (b) in Table 5:

$$r = \frac{ep}{1 + e\cos\theta}$$

44. Derive equation (c) in Table 5:

$$r = \frac{ep}{1 + e\sin\theta}$$

45. Derive equation (d) in Table 5:

$$r = \frac{ep}{1 - e\sin\theta}$$

46. Orbit of Mercury The planet Mercury travels around the Sun in an elliptical orbit given approximately by

$$r = \frac{(3.442)10^7}{1 - 0.206\cos\theta}$$

where r is measured in miles and the Sun is at the pole. Find the distance from Mercury to the Sun at *aphelion* (greatest distance from the Sun) and at *perihelion* (shortest distance from the Sun). See the figure. Use the aphelion and perihelion to graph the orbit of Mercury using a graphing utility.

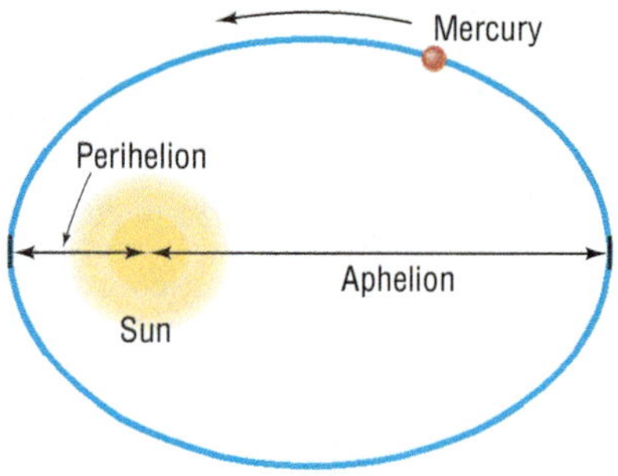

Retain Your Knowledge

Problems 47–50 are based on material learned earlier in the course. The purpose of these problems is to keep the material fresh in your mind so that you are better prepared for the final exam.

47. Find the area of the triangle described: $a = 7, b = 8$, and $c = 10$. Round to two decimal places.

48. Without graphing, determine the amplitude and period of $y = 4\cos\left(\frac{1}{5}x\right)$.

49. Solve $2\cos^2 x + \cos x - 1 = 0, 0 \le x < 2\pi$

50. For $\mathbf{v} = 10\mathbf{i} - 24\mathbf{j}$, find $\|\mathbf{v}\|$.

'Are You Prepared?' Answers

1. $r\cos\theta$; $r\sin\theta$

2. $x^2 + y^2 = 6x$ or $(x - 3)^2 + y^2 = 9$

11.7 Plane Curves and Parametric Equations

PREPARING FOR THIS SECTION *Before getting started, review the following:*

- Amplitude and Period of Sinusoidal Graphs (Section 7.6, pp. 576–578)

Now Work the **'Are You Prepared?'** problem on page 852.

OBJECTIVES
1 Graph Parametric Equations by Hand (p. 843)
2 Graph Parametric Equations Using a Graphing Utility (p. 843)
3 Find a Rectangular Equation for a Curve Defined Parametrically (p. 844)
4 Use Time as a Parameter in Parametric Equations (p. 846)
5 Find Parametric Equations for Curves Defined by Rectangular Equations (p. 849)

Equations of the form $y = f(x)$, where f is a function, have graphs that are intersected no more than once by any vertical line. The graphs of many of the conics and certain other, more complicated, graphs do not have this characteristic.

Yet each graph, like the graph of a function, is a collection of points (x, y) in the xy-plane; that is, each is a *plane curve*. In this section, we discuss another way of representing such graphs.

Let $x = f(t)$ and $y = g(t)$, where f and g are two functions whose common domain is some interval I. The collection of points defined by

$$(x, y) = (f(t), g(t))$$

is called a **plane curve**. The equations

$$x = f(t) \qquad y = g(t)$$

where t is in I, are called **parametric equations** of the curve. The variable t is called a **parameter**.

1 Graph Parametric Equations by Hand

Parametric equations are particularly useful in describing movement along a curve. Suppose that a curve is defined by the parametric equations

$$x = f(t), \qquad y = g(t), \qquad a \le t \le b$$

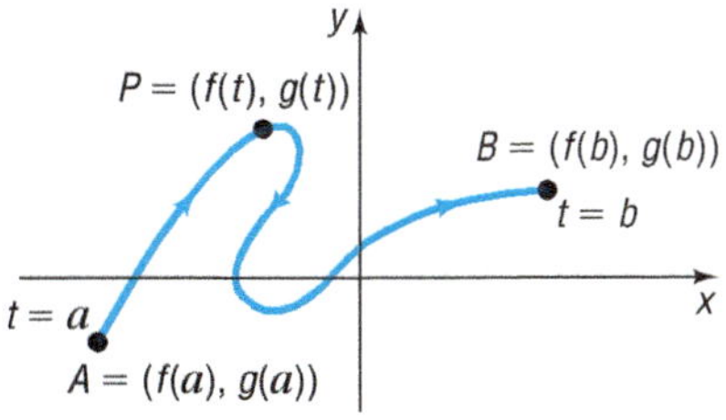

Figure 60

where f and g are each defined over the interval $a \le t \le b$. For a given value of t, we can find the value of $x = f(t)$ and $y = g(t)$, obtaining a point (x, y) on the curve. In fact, as t varies over the interval from $t = a$ to $t = b$, successive values of t give rise to a directed movement along the curve; that is, the curve is traced out in a certain direction by the corresponding succession of points (x, y). See Figure 60. The arrows show the direction, or **orientation**, along the curve as t varies from a to b.

EXAMPLE 1 **Graphing a Curve Defined by Parametric Equations**

Graph the curve defined by the parametric equations

$$x = 3t^2, \qquad y = 2t, \qquad -2 \le t \le 2$$

Solution For each number t, $-2 \le t \le 2$, there corresponds a number x and a number y. For example, when $t = -2$, then $x = 3(-2)^2 = 12$ and $y = 2(-2) = -4$. When $t = 0$, then $x = 0$ and $y = 0$. Set up a table listing various choices of the parameter t and the corresponding values for x and y, as shown in Table 6. Plotting these points and connecting them with a smooth curve leads to Figure 61. The arrows in Figure 61 are used to indicate the orientation.

Table 6

t	x	y	(x, y)
−2	12	−4	(12, −4)
−1	3	−2	(3, −2)
0	0	0	(0, 0)
1	3	2	(3, 2)
2	12	4	(12, 4)

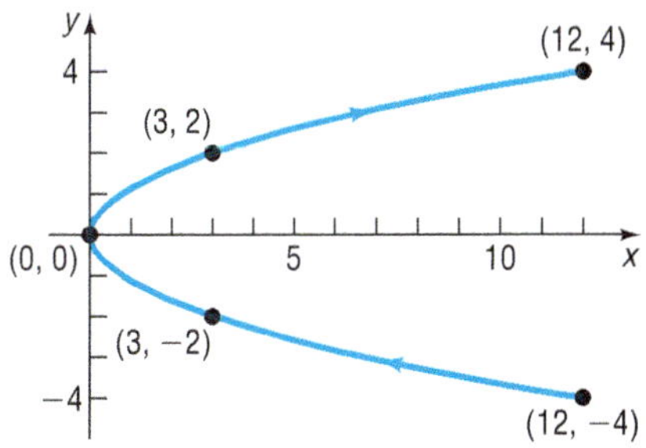

Figure 61 $x = 3t^2, y = 2t, -2 \le t \le 2$

2 Graph Parametric Equations Using a Graphing Utility

Most graphing utilities have the capability of graphing parametric equations. The following steps are usually required to obtain the graph of parametric equations. Check your owner's manual to see how yours works.

Graphing Parametric Equations Using a Graphing Utility

Step 1: Set the mode to PARAMETRIC. Enter $x(t)$ and $y(t)$.
Step 2: Select the viewing window. In addition to setting Xmin, Xmax, Xscl, and so on, the viewing window in parametric mode requires setting minimum and maximum values for the parameter t and an increment setting for t (Tstep).
Step 3: Graph.

EXAMPLE 2 Graphing a Curve Defined by Parametric Equations Using a Graphing Utility

Graph the curve defined by the parametric equations

$$x = 3t^2, \qquad y = 2t, \qquad -2 \le t \le 2 \tag{1}$$

Solution

Step 1: Enter the equations $x(t) = 3t^2$, $y(t) = 2t$ with the graphing utility in PARAMETRIC mode.

Step 2: Select the viewing window. The interval I is $-2 \le t \le 2$, so we select the following square viewing window:

$T\text{min} = -2$	$X\text{min} = -1$	$Y\text{min} = -5$
$T\text{max} = 2$	$X\text{max} = 15$	$Y\text{max} = 5$
$T\text{step} = 0.1$	$X\text{scl} = 1$	$Y\text{scl} = 1$

We choose $T\text{min} = -2$ and $T\text{max} = 2$ because $-2 \le t \le 2$. Finally, the choice for Tstep determines the number of points the graphing utility plots. For example, with Tstep at 0.1, the graphing utility evaluates x and y at $t = -2, -1.9, -1.8$, and so on. The smaller the Tstep, the more points the graphing utility plots. Experiment with different values of Tstep to see how the graph is affected.

Step 3: Graph. Notice the direction in which the graph is drawn as it appears. This direction shows the orientation of the curve. See Figure 62. ■

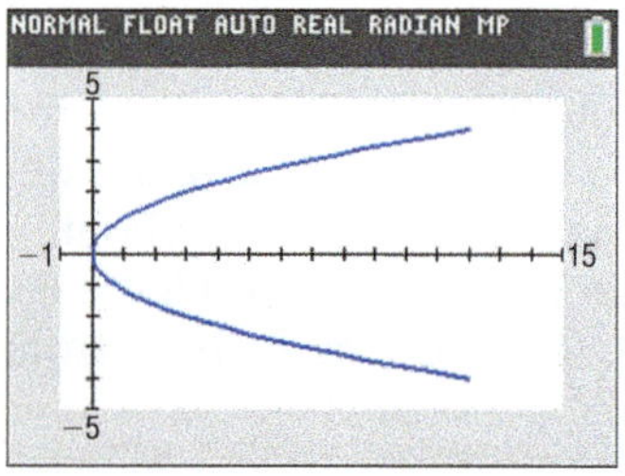

Figure 62
$x = 3t^2, y = 2t, -2 \le t \le 2$

Exploration

Graph the following parametric equations using a graphing utility with $X\text{min} = -1$, $X\text{max} = 15$, $Y\text{min} = -5$, $Y\text{max} = 5$, and $T\text{step} = 0.1$:

1. $x = \dfrac{3t^2}{4}$, $y = t$, $-4 \le t \le 4$
2. $x = 3t^2 + 12t + 12$, $y = 2t + 4$, $-4 \le t \le 0$
3. $x = 3t^{\frac{2}{3}}$, $y = 2\sqrt[3]{t}$, $-8 \le t \le 8$

Compare these graphs to the graph in Figure 62. Conclude that parametric equations defining a curve are not unique; that is, different parametric equations can represent the same graph. ■

3 Find a Rectangular Equation for a Curve Defined Parametrically

The curve given in Examples 1 and 2 should be familiar. To identify it accurately, find the corresponding rectangular equation by eliminating the parameter t from the parametric equations given in Example 1:

$$x = 3t^2, \quad y = 2t, \quad -2 \le t \le 2$$

Solve for t in $y = 2t$, obtaining $t = \dfrac{y}{2}$, and substitute this expression in the other equation to get

$$x = \underset{\uparrow\; t = \frac{y}{2}}{3t^2} = 3\left(\frac{y}{2}\right)^2 = \frac{3y^2}{4}$$

Exploration

In FUNCTION mode, graph $x = \dfrac{3y^2}{4}\left(Y_1 = \sqrt{\dfrac{4x}{3}} \text{ and } Y_2 = -\sqrt{\dfrac{4x}{3}}\right)$ with $X\text{min} = -1$, $X\text{max} = 15$, $Y\text{min} = -5$, $Y\text{max} = 5$. Compare this graph with Figure 62. Why do the graphs differ? ■

This equation, $x = \frac{3y^2}{4}$, is the equation of a parabola with vertex at $(0, 0)$ and axis of symmetry along the x-axis.

Note that the parameterized curve defined by equation (1) and shown in Figure 62 is only a part of the parabola $x = \frac{3y^2}{4}$. The graph of the rectangular equation obtained by eliminating the parameter in general contains more points than the original parameterized curve. Therefore, care must be taken when a parameterized curve is graphed after eliminating the parameter. Even so, the process of eliminating the parameter t of a parameterized curve to identify it accurately is sometimes a better approach than plotting points. However, the elimination process sometimes requires a little ingenuity.

EXAMPLE 3 Finding the Rectangular Equation of a Curve Defined Parametrically

Find the rectangular equation of the curve whose parametric equations are

$$x = a \cos t, \qquad y = a \sin t, \qquad -\infty < t < \infty$$

where $a > 0$ is a constant. Graph this curve, indicating its orientation.

Solution The presence of sines and cosines in the parametric equations suggests using a Pythagorean Identity. In fact, since

$$\cos t = \frac{x}{a} \qquad \sin t = \frac{y}{a}$$

this means that

$$\cos^2 t + \sin^2 t = 1$$

$$\left(\frac{x}{a}\right)^2 + \left(\frac{y}{a}\right)^2 = 1$$

$$x^2 + y^2 = a^2$$

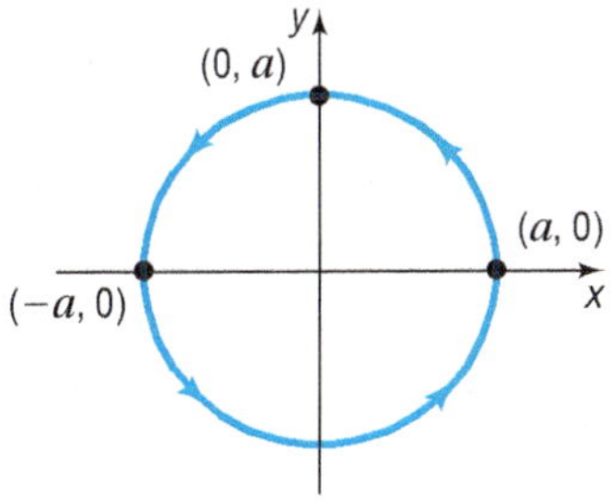

Figure 63 $x = a \cos t, y = a \sin t$

The curve is a circle with center at $(0, 0)$ and radius a. As the parameter t increases, say from $t = 0$ [the point $(a, 0)$] to $t = \frac{\pi}{2}$ [the point $(0, a)$] to $t = \pi$ [the point $(-a, 0)$], we see that the corresponding points are traced in a counterclockwise direction around the circle. The orientation is as indicated in Figure 63. ■

Now Work PROBLEMS 7 AND 19

Let's analyze the curve in Example 3 further. The domain of each parametric equation is $-\infty < t < \infty$. That means the graph in Figure 63 is actually being repeated each time that t increases by 2π.

If we wanted the curve to consist of exactly 1 revolution in the counterclockwise direction, we could write

$$x = a \cos t, \qquad y = a \sin t, \qquad 0 \le t \le 2\pi$$

This curve starts at $t = 0$ [the point $(a, 0)$] and, proceeding counterclockwise around the circle, ends at $t = 2\pi$ [also the point $(a, 0)$].

If we wanted the curve to consist of exactly three revolutions in the counterclockwise direction, we could write

$$x = a \cos t, \qquad y = a \sin t, \qquad -2\pi \le t \le 4\pi$$

or

$$x = a \cos t, \qquad y = a \sin t, \qquad 0 \le t \le 6\pi$$

or

$$x = a \cos t, \qquad y = a \sin t, \qquad 2\pi \le t \le 8\pi$$

EXAMPLE 4 Describing Parametric Equations

Find rectangular equations for the following curves defined by parametric equations. Graph each curve.

(a) $x = a \cos t, \quad y = a \sin t, \quad 0 \le t \le \pi, \quad a > 0$

(b) $x = -a \sin t, \quad y = -a \cos t, \quad 0 \le t \le \pi, \quad a > 0$

Solution (a) Eliminate the parameter t using a Pythagorean Identity.

$$\cos^2 t + \sin^2 t = 1$$
$$\left(\frac{x}{a}\right)^2 + \left(\frac{y}{a}\right)^2 = 1$$
$$x^2 + y^2 = a^2$$

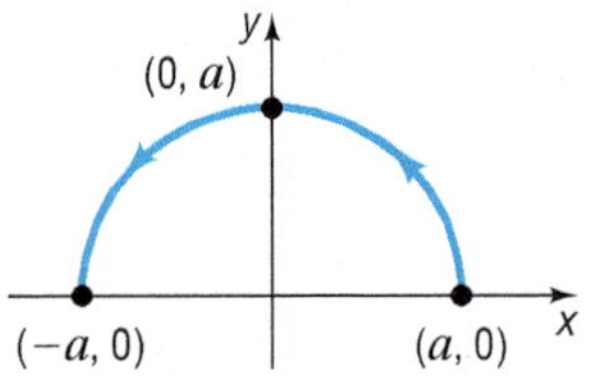

Figure 64
$x = a \cos t, y = a \sin t, 0 \le t \le \pi, a > 0$

The curve defined by these parametric equations lies on a circle, with radius a and center at $(0, 0)$. The curve begins at the point $(a, 0)$, when $t = 0$; passes through the point $(0, a)$, when $t = \frac{\pi}{2}$; and ends at the point $(-a, 0)$, when $t = \pi$.

The parametric equations define the upper semicircle of a circle of radius a with a counterclockwise orientation. See Figure 64. The rectangular equation is

$$y = \sqrt{a^2 - x^2} \qquad -a \le x \le a$$

(b) Eliminate the parameter t using a Pythagorean Identity.

$$\sin^2 t + \cos^2 t = 1$$
$$\left(\frac{x}{-a}\right)^2 + \left(\frac{y}{-a}\right)^2 = 1$$
$$x^2 + y^2 = a^2$$

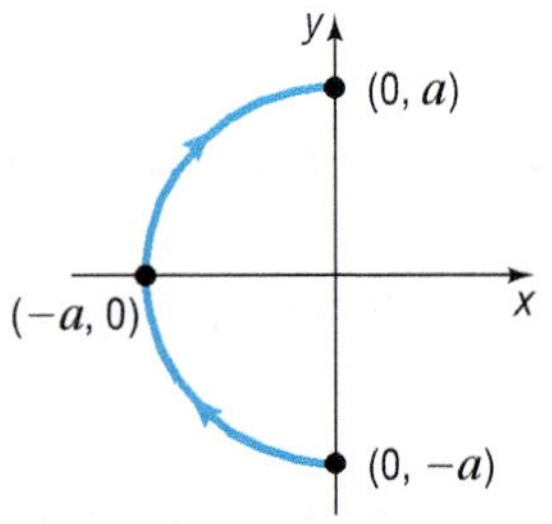

Figure 65
$x = -a \sin t, y = -a \cos t,$
$0 \le t \le \pi, a > 0$

The curve defined by these parametric equations lies on a circle with radius a and center at $(0, 0)$. The curve begins at the point $(0, -a)$, when $t = 0$; passes through the point $(-a, 0)$, when $t = \frac{\pi}{2}$; and ends at the point $(0, a)$, when $t = \pi$. The parametric equations define the left semicircle of a circle of radius a with a clockwise orientation. See Figure 65. The rectangular equation is

$$x = -\sqrt{a^2 - y^2} \qquad -a \le y \le a$$

■

Seeing the Concept

Graph $x = \cos t$, $y = \sin t$ for $0 \le t \le 2\pi$. Compare to Figure 63. Graph $x = \cos t$, $y = \sin t$ for $0 \le t \le \pi$. Compare to Figure 64. Graph $x = -\sin t$, $y = -\cos t$ for $0 \le t \le \pi$. Compare to Figure 65.

■

Example 4 illustrates the versatility of parametric equations for replacing complicated rectangular equations, while providing additional information about orientation. These characteristics make parametric equations very useful in applications, such as projectile motion.

4 Use Time as a Parameter in Parametric Equations

If we think of the parameter t as time, then the parametric equations $x = f(t)$ and $y = g(t)$ of a curve C specify how the x- and y-coordinates of a moving point vary with time.

For example, we can use parametric equations to model the motion of an object, sometimes referred to as **curvilinear motion**. Using parametric equations, we can specify not only where the object travels—that is, its location (x, y)—but also when it gets there—that is, the time t.

When an object is propelled upward at an inclination θ to the horizontal with initial speed v_0, the resulting motion is called **projectile motion**. See Figure 66(a).

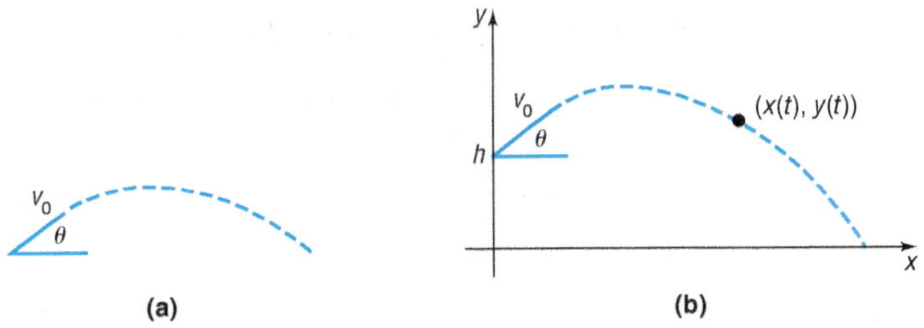

Figure 66

In calculus it is shown that the parametric equations of the path of a projectile fired at an inclination θ to the horizontal, with an initial speed v_0, from a height h above the horizontal are

$$x = (v_0 \cos \theta)t \qquad y = -\frac{1}{2}gt^2 + (v_0 \sin \theta)t + h \qquad \textbf{(2)}$$

where t is the time and g is the constant acceleration due to gravity (approximately 32 ft/sec/sec or 9.8 m/sec/sec). See Figure 66(b).

EXAMPLE 5 Projectile Motion

Suppose that Jim hit a golf ball with an initial velocity of 150 feet per second at an angle of 30° to the horizontal. See Figure 67.

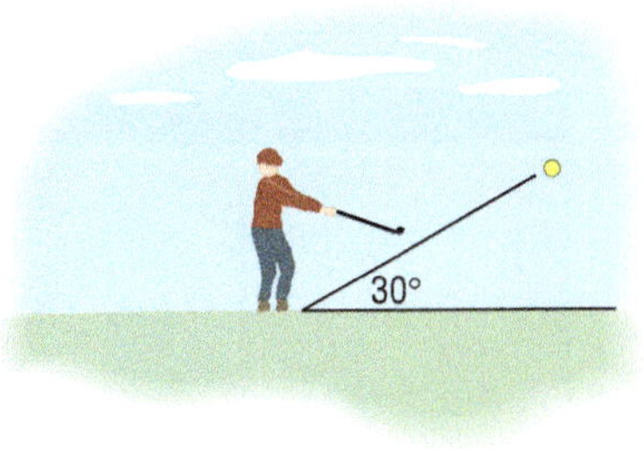

Figure 67

(a) Find parametric equations that describe the position of the ball as a function of time.
(b) How long was the golf ball in the air?
(c) When was the ball at its maximum height? Determine the maximum height of the ball.
(d) Determine the distance that the ball traveled.
(e) Using a graphing utility, simulate the motion of the golf ball by simultaneously graphing the equations found in part (a).

Solution

(a) We have $v_0 = 150$ ft/sec, $\theta = 30°$, $h = 0$ ft (the ball is on the ground), and $g = 32$ ft/sec^2 (since the units are in feet and seconds). Substitute these values into equations (2) to get

$$\begin{aligned} x &= (v_0 \cos \theta)t \\ &= (150 \cos 30°)t \\ &= 75\sqrt{3}\,t \end{aligned} \qquad \begin{aligned} y &= -\frac{1}{2}gt^2 + (v_0 \sin \theta)t + h \\ &= -\frac{1}{2}(32)t^2 + (150 \sin 30°)t + 0 \\ &= -16t^2 + 75t \end{aligned}$$

(b) To determine the length of time that the ball was in the air, solve the equation $y = 0$.

$$\begin{aligned} -16t^2 + 75t &= 0 \\ t(-16t + 75) &= 0 \end{aligned}$$

$$t = 0 \text{ sec} \quad \text{or} \quad t = \frac{75}{16} = 4.6875 \text{ sec}$$

The ball struck the ground after 4.6875 seconds.

(c) Notice that the height y of the ball is a quadratic function of t, so the maximum height of the ball can be found by determining the vertex of $y = -16t^2 + 75t$. The value of t at the vertex is

$$t = \frac{-b}{2a} = \frac{-75}{-32} = 2.34375 \text{ sec}$$

The ball was at its maximum height after 2.34375 seconds. The maximum height of the ball is found by evaluating the function y at $t = 2.34375$ seconds.

$$\text{Maximum height} = -16(2.34375)^2 + 75(2.34375) \approx 87.89 \text{ feet}$$

(d) Since the ball was in the air for 4.6875 seconds, the horizontal distance that the ball traveled is

$$x = (75\sqrt{3})4.6875 \approx 608.92 \text{ feet}$$

(e) Enter the equations from part (a) into a graphing utility with $T\text{min} = 0$, $T\text{max} = 4.7$, and $T\text{step} = 0.1$. Use ZOOM-SQUARE to avoid any distortion to the angle of elevation. See Figure 68.

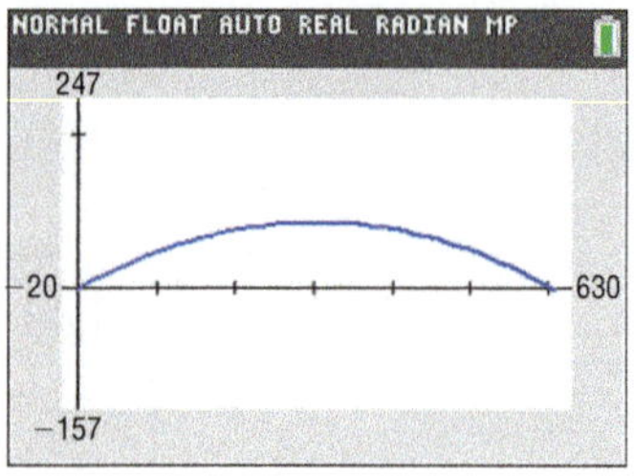

Figure 68

Exploration

Simulate the motion of a ball thrown straight up with an initial speed of 100 feet per second from a height of 5 feet above the ground. Use PARAMETRIC mode with $T\text{min} = 0$, $T\text{max} = 6.5$, $T\text{step} = 0.1$, $X\text{min} = 0$, $X\text{max} = 5$, $Y\text{min} = 0$, and $Y\text{max} = 180$. What happens to the speed with which the graph is drawn as the ball goes up and then comes back down? How do you interpret this physically? Repeat the experiment using other values for Tstep. How does this affect the experiment?

[**Hint:** In the projectile motion equations, let $\theta = 90°$, $v_0 = 100$, $h = 5$, and $g = 32$. Use $x = 3$ instead of $x = 0$ to see the vertical motion better.]

Result In Figure 69(a) the ball is going up. In Figure 69(b) the ball is near its highest point. Finally, in Figure 69(c) the ball is coming back down.

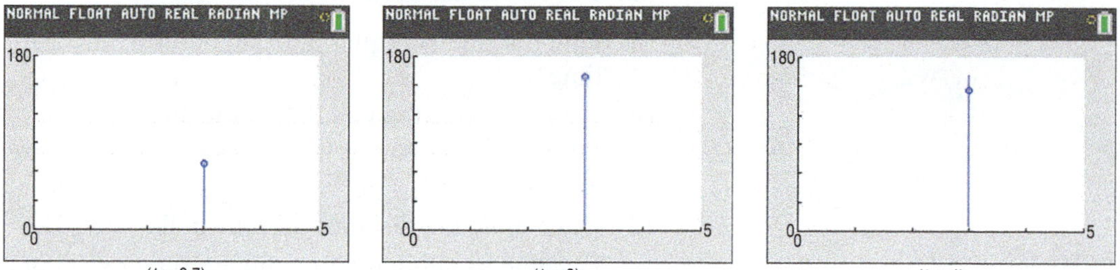

($t \approx 0.7$) **(a)** ($t \approx 3$) **(b)** ($t \approx 4$) **(c)**

Figure 69

Notice that, as the ball goes up, its speed decreases, until at the highest point it is zero. Then the speed increases as the ball comes back down.

Now Work PROBLEM 49

A graphing utility can be used to simulate other kinds of motion as well. Let's work again Example 5 from Section 1.6.

EXAMPLE 6

Simulating Motion

Tanya, who is a long distance runner, runs at an average speed of 8 miles per hour. Two hours after Tanya leaves your house, you leave in your Honda and follow the same route. If your average speed is 40 miles per hour, how long will it be before you catch up to Tanya? See Figure 70. Use a simulation of the two motions to verify the answer.

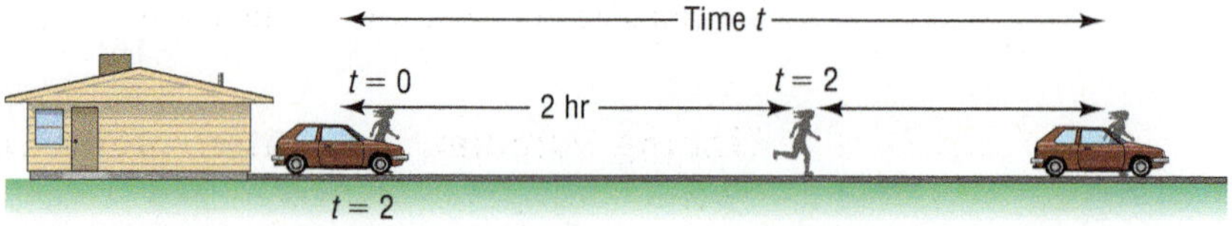

Figure 70

Solution Begin with two sets of parametric equations: one to describe Tanya's motion, the other to describe the motion of the Honda. We choose time $t = 0$ to be when Tanya leaves the house. If we choose $y_1 = 2$ as Tanya's path, then we can use $y_2 = 4$ as the parallel path of the Honda. The horizontal distances traversed in time t (Distance = Rate × Time) are

$$\text{Tanya: } x_1 = 8t \qquad \text{Honda: } x_2 = 40(t-2)$$

The Honda catches up to Tanya when $x_1 = x_2$.

$$\begin{aligned} 8t &= 40(t-2) \\ 8t &= 40t - 80 \\ -32t &= -80 \\ t &= \frac{-80}{-32} = 2.5 \end{aligned}$$

The Honda catches up to Tanya 2.5 hours after Tanya leaves the house.

In PARAMETRIC mode with Tstep $= 0.01$, simultaneously graph

$$\begin{aligned} &\text{Tanya: } x_1 = 8t \qquad &&\text{Honda: } x_2 = 40(t-2) \\ &\phantom{\text{Tanya: }} y_1 = 2 &&\phantom{\text{Honda: }} y_2 = 4 \end{aligned}$$

for $0 \le t \le 3$.

Figure 71 shows the relative position of Tanya and the Honda for $t = 0$, $t = 2$, $t = 2.25$, $t = 2.5$, and $t = 2.75$.

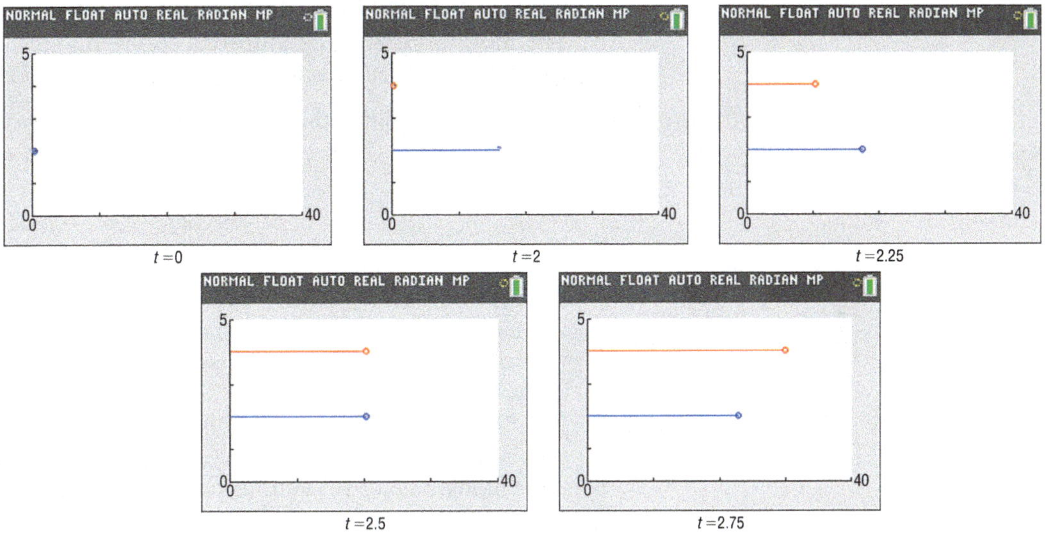

Figure 71

5 Find Parametric Equations for Curves Defined by Rectangular Equations

We now take up the question of how to find parametric equations of a given curve.

If a curve is defined by the equation $y = f(x)$, where f is a function, one way of finding parametric equations is to let $x = t$. Then $y = f(t)$ and

$$x = t, \quad y = f(t), \quad t \text{ in the domain of } f$$

are parametric equations of the curve.

EXAMPLE 7 **Finding Parametric Equations for a Curve Defined by a Rectangular Equation**

Find two different parametric equations for the equation $y = x^2 - 4$.

Solution For the first parametric equation, let $x = t$. Then the parametric equations are

$$x = t, \quad y = t^2 - 4, \qquad -\infty < t < \infty$$

A second parametric equation is found by letting $x = t^3$. Then the parametric equations become

$$x = t^3, \quad y = t^6 - 4, \qquad -\infty < t < \infty$$

Care must be taken when using the second approach in Example 7, since the substitution for x must be a function that allows x to take on all the values stipulated by the domain of f. For example, letting $x = t^2$ so that $y = t^4 - 4$ does not result in equivalent parametric equations for $y = x^2 - 4$, since only points for which $x \geq 0$ are obtained; yet the domain of $y = x^2 - 4$ is $\{x \mid x \text{ is any real number}\}$.

Now Work PROBLEM 33

EXAMPLE 8 Finding Parametric Equations for an Object in Motion

Find parametric equations for the ellipse

$$x^2 + \frac{y^2}{9} = 1$$

where the parameter t is time (in seconds) and

(a) The motion around the ellipse is clockwise, begins at the point $(0, 3)$, and requires 1 second for a complete revolution.

(b) The motion around the ellipse is counterclockwise, begins at the point $(1, 0)$, and requires 2 seconds for a complete revolution.

Solution (a) See Figure 72. Since the motion begins at the point $(0, 3)$, we want $x = 0$ and $y = 3$ when $t = 0$. Furthermore, since the given equation is an ellipse, we begin by letting

$$x = \sin(\omega t) \quad y = 3\cos(\omega t)$$

for some constant ω. These parametric equations satisfy the equation of the ellipse. Furthermore, with this choice, when $t = 0$, we have $x = 0$ and $y = 3$.

For the motion to be clockwise, the motion has to begin with the value of x increasing and y decreasing as t increases. This requires that $\omega > 0$. [Do you know why? If $\omega > 0$, then $x = \sin(\omega t)$ is increasing when $t > 0$ is near zero and $y = 3\cos(\omega t)$ is decreasing when $t > 0$ is near zero.] See the red part of the graph in Figure 72.

Finally, since 1 revolution requires 1 second, the period $\frac{2\pi}{\omega} = 1$, so $\omega = 2\pi$. Parametric equations that satisfy the conditions stipulated are

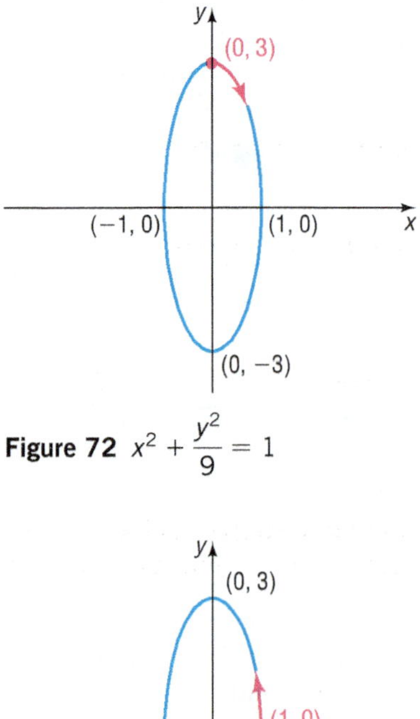

Figure 72 $x^2 + \frac{y^2}{9} = 1$

$$x = \sin(2\pi t), \qquad y = 3\cos(2\pi t), \qquad 0 \leq t \leq 1 \tag{3}$$

(b) See Figure 73. Since the motion begins at the point $(1, 0)$, we want $x = 1$ and $y = 0$ when $t = 0$. The given equation is an ellipse, so begin by letting

$$x = \cos(\omega t) \quad \text{and} \quad y = 3\sin(\omega t)$$

for some constant ω. These parametric equations satisfy the equation of the ellipse. Furthermore, with this choice, when $t = 0$ we have $x = 1$ and $y = 0$.

For the motion to be counterclockwise, the motion has to begin with the value of x decreasing and y increasing as t increases. This requires that $\omega > 0$. [Do you know why?] Finally, since 1 revolution requires 2 seconds, the period is $\frac{2\pi}{\omega} = 2$, so $\omega = \pi$. The parametric equations that satisfy the conditions stipulated are

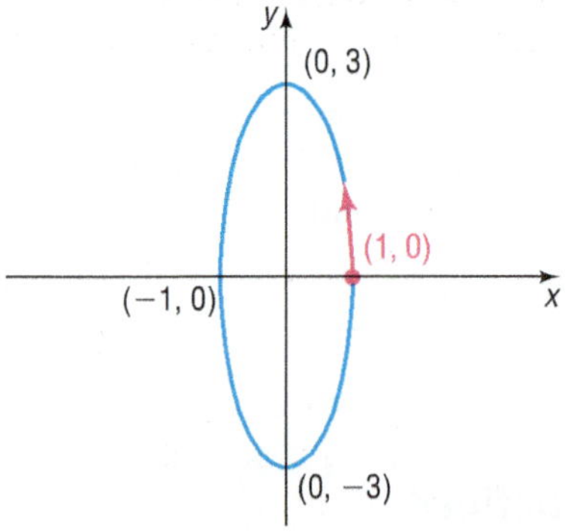

Figure 73 $x^2 + \frac{y^2}{9} = 1$

$$x = \cos(\pi t), \qquad y = 3\sin(\pi t), \qquad 0 \leq t \leq 2 \tag{4}$$

Either equations (3) or equations (4) can serve as parametric equations for the ellipse $x^2 + \frac{y^2}{9} = 1$ given in Example 8. The direction of the motion, the beginning point, and the time for 1 revolution give a particular parametric representation.

Now Work PROBLEM 39

The Cycloid

Suppose that a circle of radius a rolls along a horizontal line without slipping. As the circle rolls along the line, a point P on the circle will trace out a curve called a **cycloid** (see Figure 74). We now seek parametric equations* for a cycloid.

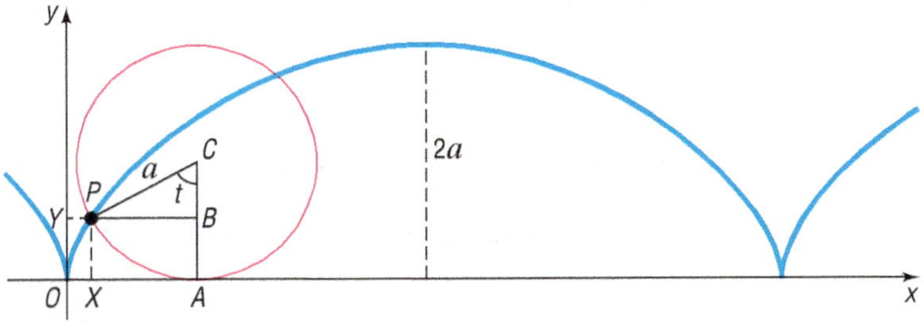

Figure 74 Cycloid

Begin with a circle of radius a and take the fixed line on which the circle rolls as the x-axis. Let the origin be one of the points at which the point P comes in contact with the x-axis. Figure 74 illustrates the position of this point P after the circle has rolled somewhat. The angle t (in radians) measures the angle through which the circle has rolled.

Since we require no slippage, it follows that

$$\text{Arc } AP = d(O, A)$$

The length of the arc AP is given by $s = r\theta$, where $r = a$ and $\theta = t$ radians. Then

$$at = d(O, A) \quad s = r\theta, \text{ where } r = a \text{ and } \theta = t$$

The x-coordinate of the point P is

$$d(O, X) = d(O, A) - d(X, A) = at - a \sin t = a(t - \sin t)$$

The y-coordinate of the point P is equal to

$$d(O, Y) = d(A, C) - d(B, C) = a - a \cos t = a(1 - \cos t)$$

The parametric equations of the cycloid are

$$x = a(t - \sin t) \qquad y = a(1 - \cos t) \tag{5}$$

Exploration

Graph $x = t - \sin t$, $y = 1 - \cos t$, $0 \le t \le 3\pi$, using your graphing utility with *T*step $= \frac{\pi}{36}$ and a square screen. Compare your results with Figure 74. ■

Applications to Mechanics

If a is negative in equation (5), we obtain an inverted cycloid, as shown in Figure 75(a). The inverted cycloid occurs as a result of some remarkable applications in the

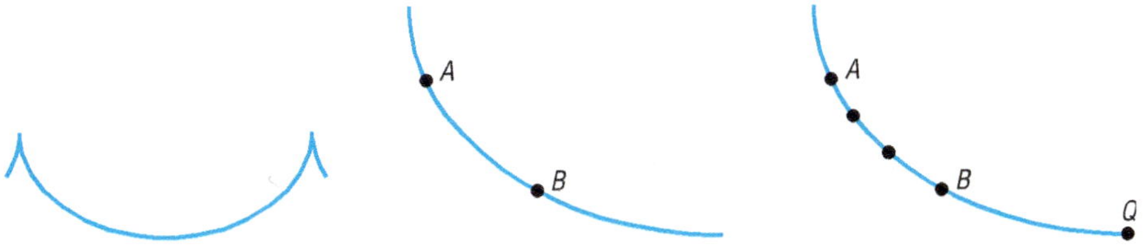

(a) Inverted cycloid **(b)** Curve of quickest descent **(c)** All reach Q at the same time

Figure 75

*Any attempt to derive the rectangular equation of a cycloid would soon demonstrate how complicated the task is.

field of mechanics. We shall mention two of them: the *brachistochrone* and the *tautochrone*.†

The **brachistochrone** is the curve of quickest descent. If a particle is constrained to follow some path from one point A to a lower point B (not on the same vertical line) and is acted on only by gravity, the time needed to make the descent is least if the path is an inverted cycloid. See Figure 75(b) on the previous page. This remarkable discovery, which is attributed to many famous mathematicians (including Johann Bernoulli and Blaise Pascal), was a significant step in creating the branch of mathematics known as the *calculus of variations*.

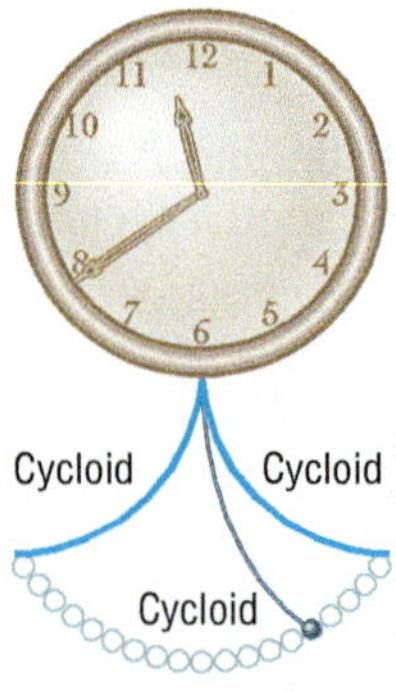

Figure 76

To define the **tautochrone**, let Q be the lowest point on an inverted cycloid. If several particles placed at various positions on an inverted cycloid simultaneously begin to slide down the cycloid, they will reach the point Q at the same time, as indicated in Figure 75(c). The tautochrone property of the cycloid was used by Christiaan Huygens (1629–1695), the Dutch mathematician, physicist, and astronomer, to construct a pendulum clock with a bob that swings along a cycloid (see Figure 76). In Huygen's clock, the bob was made to swing along a cycloid by suspending the bob on a thin wire constrained by two plates shaped like cycloids. In a clock of this design, the period of the pendulum is independent of its amplitude.

11.7 Assess Your Understanding

'Are You Prepared?' *Answers are given at the end of these exercises. If you get a wrong answer, read the pages listed in red.*

1. The function $f(x) = 3\sin(4x)$ has amplitude _____ and period __ . (pp. 576–578)

Concepts and Vocabulary

2. Let $x = f(t)$ and $y = g(t)$, where f and g are two functions whose common domain is some interval I. The collection of points defined by $(x, y) = (f(t), g(t))$ is called a(n) _____ _____. The variable t is called a(n) __________.

3. The parametric equations $x = 2\sin t$, $y = 3\cos t$ define a(n) _______.

(a) circle (b) ellipse
(c) hyperbola (d) parabola

4. If a circle rolls along a horizontal line without slippage, a fixed point P on the circle will trace out a curve called a(n) _______.

(a) cycloid (b) epitrochoid
(c) hyptrochoid (d) pendulum

5. ***True or False*** Parametric equations defining a curve are unique.

6. ***True or False*** Curves defined using parametric equations have an orientation.

Skill Building

In Problems 7–26, graph the curve whose parametric equations are given and show its orientation. Find the rectangular equation of each curve. Verify your graph using a graphing utility.

7. $x = 3t + 2, \quad y = t + 1; \quad 0 \le t \le 4$

8. $x = t - 3, \quad y = 2t + 4; \quad 0 \le t \le 2$

9. $x = t + 2, \quad y = \sqrt{t}; \quad t \ge 0$

10. $x = \sqrt{2t}, \quad y = 4t; \quad t \ge 0$

11. $x = t^2 + 4, \quad y = t^2 - 4; \quad -\infty < t < \infty$

12. $x = \sqrt{t} + 4, \quad y = \sqrt{t} - 4; \quad t \ge 0$

13. $x = 3t^2, \quad y = t + 1; \quad -\infty < t < \infty$

14. $x = 2t - 4, \quad y = 4t^2; \quad -\infty < t < \infty$

15. $x = 2e^t, \quad y = 1 + e^t; \quad t \ge 0$

16. $x = e^t, \quad y = e^{-t}; \quad t \ge 0$

17. $x = \sqrt{t}, \quad y = t^{3/2}; \quad t \ge 0$

18. $x = t^{3/2} + 1, \quad y = \sqrt{t}; \quad t \ge 0$

19. $x = 2\cos t, \quad y = 3\sin t; \quad 0 \le t \le 2\pi$

20. $x = 2\cos t, \quad y = 3\sin t; \quad 0 \le t \le \pi$

†In Greek, *brachistochrone* means "the shortest time" and *tautochrone* means "equal time."

21. $x = 2\cos t, \quad y = 3\sin t; \quad -\pi \le t \le 0$

22. $x = 2\cos t, \quad y = \sin t; \quad 0 \le t \le \frac{\pi}{2}$

23. $x = \sec t, \quad y = \tan t; \quad 0 \le t \le \frac{\pi}{4}$

24. $x = \csc t, \quad y = \cot t; \quad \frac{\pi}{4} \le t \le \frac{\pi}{2}$

25. $x = \sin^2 t, \quad y = \cos^2 t; \quad 0 \le t \le 2\pi$

26. $x = t^2, \quad y = \ln t; \quad t > 0$

In Problems 27–34, find two different parametric equations for each rectangular equation.

27. $y = 4x - 1$

28. $y = -8x + 3$

29. $y = x^2 + 1$

30. $y = -2x^2 + 1$

31. $y = x^3$

32. $y = x^4 + 1$

33. $x = y^{3/2}$

34. $x = \sqrt{y}$

In Problems 35–38, find parametric equations that define the curve shown.

35.

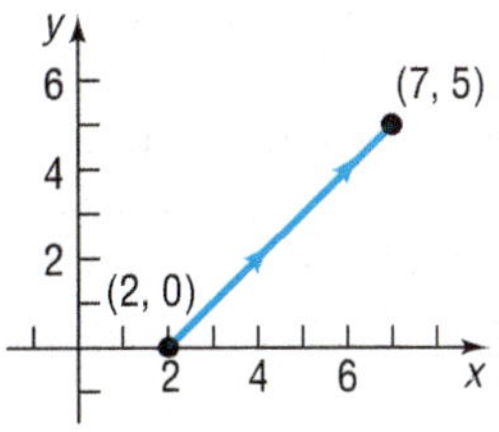

36.

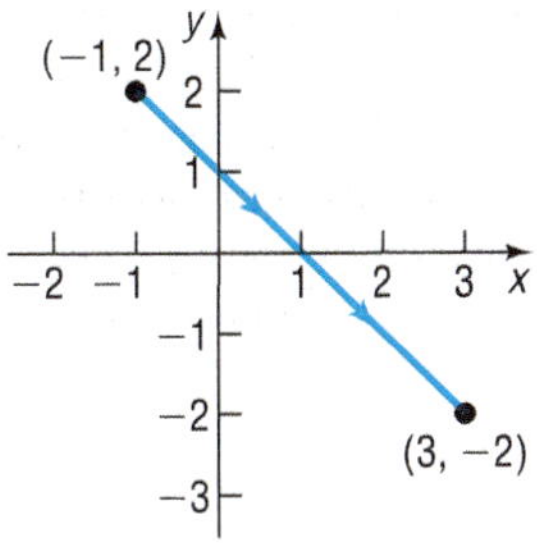

37.

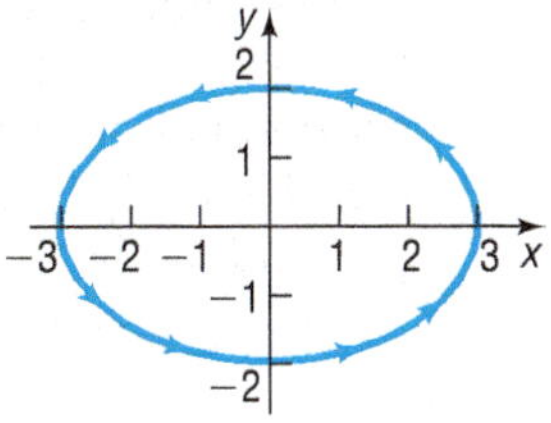

38.

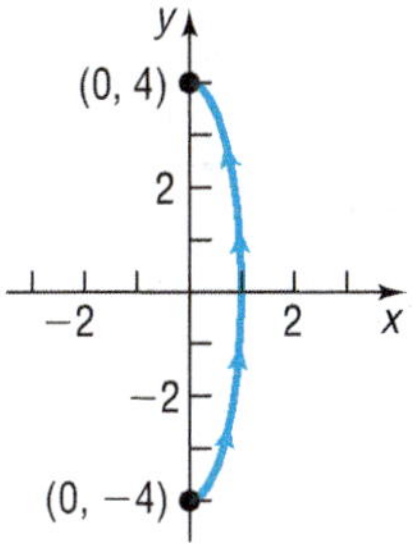

In Problems 39–42, find parametric equations for an object that moves along the ellipse $\frac{x^2}{4} + \frac{y^2}{9} = 1$ *with the motion described.*

39. The motion begins at $(2, 0)$, is clockwise, and requires 2 seconds for a complete revolution.

40. The motion begins at $(0, 3)$, is counterclockwise, and requires 1 second for a complete revolution.

41. The motion begins at $(0, 3)$, is clockwise, and requires 1 second for a complete revolution.

42. The motion begins at $(2, 0)$, is counterclockwise, and requires 3 seconds for a complete revolution.

In Problems 43 and 44, the parametric equations of four curves are given. Graph each of them, indicating the orientation.

43. C_1: $x = t, \quad y = t^2; \quad -4 \le t \le 4$
C_2: $x = \cos t, \quad y = 1 - \sin^2 t; \quad 0 \le t \le \pi$
C_3: $x = e^t, \quad y = e^{2t}; \quad 0 \le t \le \ln 4$
C_4: $x = \sqrt{t}, \quad y = t; \quad 0 \le t \le 16$

44. C_1: $x = t, \quad y = \sqrt{1 - t^2}; \quad -1 \le t \le 1$
C_2: $x = \sin t, \quad y = \cos t; \quad 0 \le t \le 2\pi$
C_3: $x = \cos t, \quad y = \sin t; \quad 0 \le t \le 2\pi$
C_4: $x = \sqrt{1 - t^2}, \quad y = t; \quad -1 \le t \le 1$

In Problems 45–48, use a graphing utility to graph the curve defined by the given parametric equations.

45. $x = t\sin t, \quad y = t\cos t, \quad t > 0$

46. $x = \sin t + \cos t, \quad y = \sin t - \cos t$

47. $x = 4\sin t - 2\sin(2t)$
$y = 4\cos t - 2\cos(2t)$

48. $x = 4\sin t + 2\sin(2t)$
$y = 4\cos t + 2\cos(2t)$

Applications and Extensions

49. Projectile Motion Bob throws a ball straight up with an initial speed of 50 feet per second from a height of 6 feet.
(a) Find parametric equations that model the motion of the ball as a function of time.
(b) How long is the ball in the air?
(c) When is the ball at its maximum height? Determine the maximum height of the ball.
(d) Simulate the motion of the ball by graphing the equations found in part (a).

50. Projectile Motion Alice throws a ball straight up with an initial speed of 40 feet per second from a height of 5 feet.
(a) Find parametric equations that model the motion of the ball as a function of time.
(b) How long is the ball in the air?
(c) When is the ball at its maximum height? Determine the maximum height of the ball.
(d) Simulate the motion of the ball by graphing the equations found in part (a).

51. Catching a Train Bill's train leaves at 8:06 AM and accelerates at the rate of 2 meters per second per second. Bill, who can run 5 meters per second, arrives at the train station 5 seconds after the train has left and runs for the train.
(a) Find parametric equations that model the motions of the train and Bill as a function of time.
[**Hint:** The position s at time t of an object having acceleration a is $s = \frac{1}{2}at^2$.]
(b) Determine algebraically whether Bill will catch the train. If so, when?
(c) Simulate the motion of the train and Bill by simultaneously graphing the equations found in part (a).

52. Catching a Bus Jodi's bus leaves at 5:30 PM and accelerates at the rate of 3 meters per second per second. Jodi, who can run 5 meters per second, arrives at the bus station 2 seconds after the bus has left and runs for the bus.

(a) Find parametric equations that model the motions of the bus and Jodi as a function of time.
[**Hint:** The position s at time t of an object having acceleration a is $s = \frac{1}{2}at^2$.]

(b) Determine algebraically whether Jodi will catch the bus. If so, when?

(c) Simulate the motion of the bus and Jodi by simultaneously graphing the equations found in part (a).

53. Projectile Motion Ichiro throws a baseball with an initial speed of 145 feet per second at an angle of 20° to the horizontal. The ball leaves Ichiro's hand at a height of 5 feet.

(a) Find parametric equations that model the position of the ball as a function of time.

(b) How long is the ball in the air?

(c) Determine the horizontal distance that the ball travels.

(d) When is the ball at its maximum height? Determine the maximum height of the ball.

(e) Using a graphing utility, simultaneously graph the equations found in part (a).

54. Projectile Motion Mark Texeira hit a baseball with an initial speed of 125 feet per second at an angle of 40° to the horizontal. The ball was hit at a height of 3 feet off the ground.

(a) Find parametric equations that model the position of the ball as a function of time.

(b) How long was the ball in the air?

(c) Determine the horizontal distance that the ball traveled.

(d) When was the ball at its maximum height? Determine the maximum height of the ball.

(e) Using a graphing utility, simultaneously graph the equations found in part (a).

55. Projectile Motion Suppose that Adam hits a golf ball off a cliff 300 meters high with an initial speed of 40 meters per second at an angle of 45° to the horizontal.

(a) Find parametric equations that model the position of the ball as a function of time.

(b) How long is the ball in the air?

(c) Determine the horizontal distance that the ball travels.

(d) When is the ball at its maximum height? Determine the maximum height of the ball.

(e) Using a graphing utility, simultaneously graph the equations found in part (a).

56. Projectile Motion Suppose that Karla hits a golf ball off a cliff 300 meters high with an initial speed of 40 meters per second at an angle of 45° to the horizontal on the Moon (gravity on the Moon is one-sixth of that on Earth).

(a) Find parametric equations that model the position of the ball as a function of time.

(b) How long is the ball in the air?

(c) Determine the horizontal distance that the ball travels.

(d) When is the ball at its maximum height? Determine the maximum height of the ball.

(e) Using a graphing utility, simultaneously graph the equations found in part (a).

57. Uniform Motion A Toyota Camry (traveling east at 40 mph) and a Chevy Impala (traveling north at 30 mph) are heading toward the same intersection. The Camry is 5 miles from the intersection when the Impala is 4 miles from the intersection. See the figure.

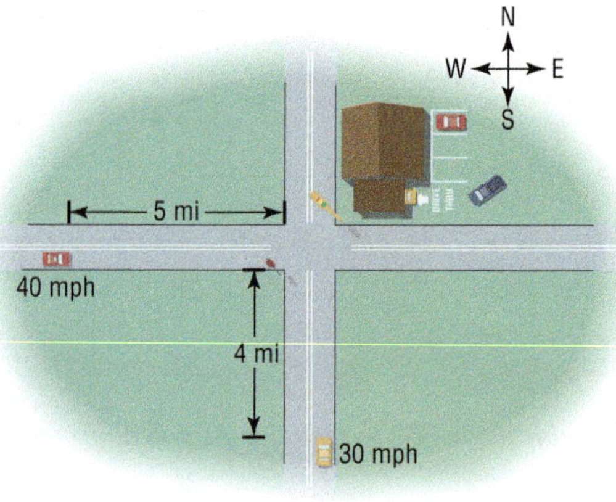

(a) Find parametric equations that model the motion of the Camry and Impala.

(b) Find a formula for the distance between the cars as a function of time.

(c) Graph the function in part (b) using a graphing utility.

(d) What is the minimum distance between the cars? When are the cars closest?

(e) Simulate the motion of the cars by simultaneously graphing the equations found in part (a).

58. Uniform Motion A Cessna (heading south at 120 mph) and a Boeing 747 (heading west at 600 mph) are flying toward the same point at the same altitude. The Cessna is 100 miles from the point where the flight patterns intersect, and the 747 is 550 miles from this intersection point. See the figure.

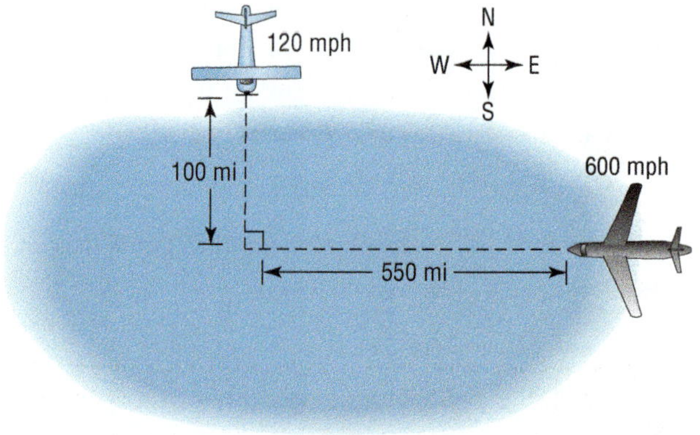

(a) Find parametric equations that model the motion of the Cessna and the 747.

(b) Find a formula for the distance between the planes as a function of time.

(c) Graph the function in part (b) using a graphing utility.

(d) What is the minimum distance between the planes? When are the planes closest?

(e) Simulate the motion of the planes by simultaneously graphing the equations found in part (a).

59. The Green Monster The left field wall at Fenway Park is 310 feet from home plate; the wall itself (affectionately named the Green Monster) is 37 feet high. A batted ball must clear the wall to be a home run. Suppose a ball leaves the bat 3 feet off the ground, at an angle of 45°. Use $g = 32$ feet per second^2 as the acceleration due to gravity and ignore any air resistance.

(a) Find parametric equations that model the position of the ball as a function of time.

(b) What is the maximum height of the ball if it leaves the bat with a speed of 90 miles per hour? Give your answer in feet.
(c) How far is the ball from home plate at its maximum height? Give your answer in feet.
(d) If the ball is hit straight down the left field wall, will it clear the Green Monster? If it does, by how much does it clear the wall?

Source: *The Boston Red Sox*

60. Projectile Motion The position of a projectile fired with an initial velocity v_0 feet per second and at an angle θ to the horizontal at the end of t seconds is given by the parametric equations

$$x = (v_0 \cos\theta)t \qquad y = (v_0 \sin\theta)t - 16t^2$$

See the illustration.

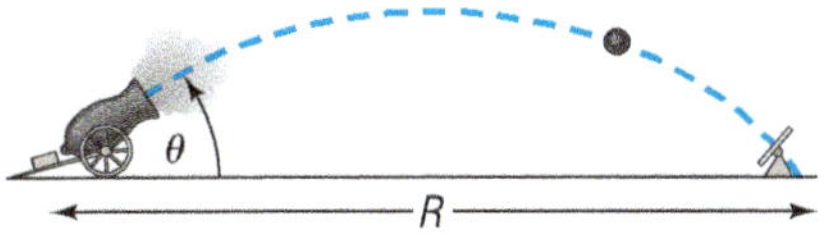

(a) Obtain the rectangular equation of the trajectory and identify the curve.
(b) Show that the projectile hits the ground $(y = 0)$ when $t = \frac{1}{16} v_0 \sin\theta$.
(c) How far has the projectile traveled (horizontally) when it strikes the ground? In other words, find the range R.
(d) Find the time t when $x = y$. Then find the horizontal distance x and the vertical distance y traveled by the projectile in this time. Then compute $\sqrt{x^2 + y^2}$. This is the distance R, the range, that the projectile travels up a plane inclined at 45° to the horizontal $(x = y)$. See the following illustration. (See also Problem 99 in Section 8.6.)

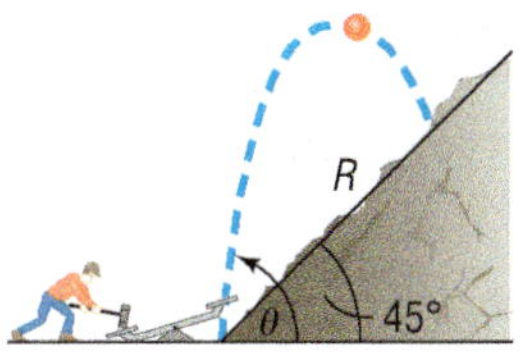

61. Show that the parametric equations for a line passing through the points (x_1, y_1) and (x_2, y_2) are

$$x = (x_2 - x_1)t + x_1$$
$$y = (y_2 - y_1)t + y_1, \quad -\infty < t < \infty$$

What is the orientation of this line?

62. Hypocycloid The hypocycloid is a curve defined by the parametric equations

$$x(t) = \cos^3 t, \quad y(t) = \sin^3 t, \quad 0 \le t \le 2\pi$$

(a) Graph the hypocycloid using a graphing utility.
(b) Find a rectangular equation of the hypocycloid.

Explaining Concepts: Discussion and Writing

63. In Problem 62, we graphed the hypocycloid. Now graph the rectangular equations of the hypocycloid. Did you obtain a complete graph? If not, experiment until you do.

64. Look up the curves called *hypocycloid* and *epicycloid*. Write a report on what you find. Be sure to draw comparisons with the cycloid.

Retain Your Knowledge

Problems 65–68 are based on material learned earlier in the course. The purpose of these problems is to keep the material fresh in your mind so that you are better prepared for the final exam.

65. Graph the equation $3x - 4y = 8$ on the xy-plane.

66. Graph $y = 2\cos(2x) + \sin\left(\frac{x}{2}\right)$ on the xy-plane.

67. The International Space Station (ISS) orbits Earth at a height of approximately 248 miles above the surface. What is the distance, in miles, on the surface of Earth that can be observed from the ISS? Assume that Earth's radius is 3960 miles.

Source: *nasa.gov*

68. The displacement d (in meters) of an object at time t (in seconds) is given by $d = 2\cos(4t)$.
(a) Describe the motion of the object.
(b) What is the maximum displacement of the object from its rest position?
(c) What is the time required for 1 oscillation?
(d) What is the frequency?

'Are You Prepared?' Answer

1. $3; \frac{\pi}{2}$

Chapter Review

Things to Know

Equations

Parabola (pp. 793–799)	See Tables 1 and 2 (pp. 795 and 797).
Ellipse (pp. 803–811)	See Table 3 (p. 808).
Hyperbola (pp. 814–824)	See Table 4 (p. 821).
General equation of a conic (p. 834)	$Ax^2 + Bxy + Cy^2 + Dx + Ey + F = 0$ Parabola if $B^2 - 4AC = 0$ Ellipse (or circle) if $B^2 - 4AC < 0$ Hyperbola if $B^2 - 4AC > 0$
Polar equations of a conic with focus at the pole (pp. 836–841)	See Table 5 (p. 839).
Parametric equations of a curve (p. 843)	$x = f(t), y = g(t)$, t is the parameter

Definitions

Parabola (p. 793)	Set of points P in a plane for which $d(F, P) = d(P, D)$, where F is the focus and D is the directrix
Ellipse (p. 803)	Set of points P in a plane, the sum of whose distances from two fixed points (the foci) is a constant
Hyperbola (p. 814)	Set of points P in a plane, the difference of whose distances from two fixed points (the foci) is a constant
Conic in polar coordinates (p. 836)	$\dfrac{d(F, P)}{d(P, D)} = e$ Parabola if $e = 1$ Ellipse if $e < 1$ Hyperbola if $e > 1$

Formulas

Rotation formulas (p. 830)	$x = x' \cos\theta - y' \sin\theta$ $y = x' \sin\theta + y' \cos\theta$
Angle θ of rotation that eliminates the $x'y'$-term (p. 831)	$\cot(2\theta) = \dfrac{A - C}{B}, \quad 0° < \theta < 90°$

Objectives

Section		You should be able to . . .	Examples	Review Exercises
11.1	1	Know the names of the conics (p. 792)		1–16
11.2	1	Analyze parabolas with vertex at the origin (p. 793)	1–6	1, 11
	2	Analyze parabolas with vertex at (h, k) (p. 797)	7–9	4, 6, 9, 14
	3	Solve applied problems involving parabolas (p. 798)	10	39
11.3	1	Analyze ellipses with center at the origin (p. 803)	1–5	3, 13
	2	Analyze ellipses with center at (h, k) (p. 808)	6–8	8, 10, 16, 38
	3	Solve applied problems involving ellipses (p. 810)	9	40
11.4	1	Analyze hyperbolas with center at the origin (p. 815)	1–5	2, 5, 12, 37
	2	Find the asymptotes of a hyperbola (p. 819)	6, 7	2, 5, 7
	3	Analyze hyperbolas with center at (h, k) (p. 821)	8, 9	7, 15, 17, 18
	4	Solve applied problems involving hyperbolas (p. 823)	10	41

Section		You should be able to . . .	Examples	Review Exercises
11.5	1	Identify a conic (p. 828)	1	19, 20
	2	Use a rotation of axes to transform equations (p. 829)	2	24–26
	3	Analyze an equation using a rotation of axes (p. 831)	3, 4	24–26, 44
	4	Identify conics without a rotation of axes (p. 833)	5	21–23
11.6	1	Analyze and graph polar equations of conics (p. 836)	1–3	27–29
	2	Convert the polar equation of a conic to a rectangular equation (p. 840)	4	30, 31
11.7	1	Graph parametric equations by hand (p. 843)	1	32–34
	2	Graph parametric equations using a graphing utility (p. 843)	2	32–34
	3	Find a rectangular equation for a curve defined parametrically (p. 844)	3, 4	32–34
	4	Use time as a parameter in parametric equations (p. 846)	5, 6	42, 43
	5	Find parametric equations for curves defined by rectangular equations (p. 849)	7, 8	35, 36

Review Exercises

In Problems 1–10, identify each equation. If it is a parabola, give its vertex, focus, and directrix; if it is an ellipse, give its center, vertices, and foci; if it is a hyperbola, give its center, vertices, foci, and asymptotes.

1. $y^2 = -16x$

2. $\dfrac{x^2}{25} - y^2 = 1$

3. $\dfrac{y^2}{25} + \dfrac{x^2}{16} = 1$

4. $x^2 + 4y = 4$

5. $4x^2 - y^2 = 8$

6. $x^2 - 4x = 2y$

7. $y^2 - 4y - 4x^2 + 8x = 4$

8. $4x^2 + 9y^2 - 16x - 18y = 11$

9. $4x^2 - 16x + 16y + 32 = 0$

10. $9x^2 + 4y^2 - 18x + 8y = 23$

In Problems 11–18, find an equation of the conic described. Graph the equation by hand.

11. Parabola; focus at $(-2, 0)$; directrix the line $x = 2$

12. Hyperbola; center at $(0, 0)$; focus at $(0, 4)$; vertex at $(0, -2)$

13. Ellipse; foci at $(-3, 0)$ and $(3, 0)$; vertex at $(4, 0)$

14. Parabola; vertex at $(2, -3)$; focus at $(2, -4)$

15. Hyperbola; center at $(-2, -3)$; focus at $(-4, -3)$; vertex at $(-3, -3)$

16. Ellipse; foci at $(-4, 2)$ and $(-4, 8)$; vertex at $(-4, 10)$

17. Center at $(-1, 2)$; $a = 3$; $c = 4$; transverse axis parallel to the x-axis

18. Vertices at $(0, 1)$ and $(6, 1)$; asymptote the line $3y + 2x = 9$

In Problems 19–23, identify each conic without completing the squares and without applying a rotation of axes.

19. $y^2 + 4x + 3y - 8 = 0$

20. $x^2 + 2y^2 + 4x - 8y + 2 = 0$

21. $9x^2 - 12xy + 4y^2 + 8x + 12y = 0$

22. $4x^2 + 10xy + 4y^2 - 9 = 0$

23. $x^2 - 2xy + 3y^2 + 2x + 4y - 1 = 0$

In Problems 24–26, rotate the axes so that the new equation contains no xy-term. Analyze and graph the new equation.

24. $2x^2 + 5xy + 2y^2 - \dfrac{9}{2} = 0$

25. $6x^2 + 4xy + 9y^2 - 20 = 0$

26. $4x^2 - 12xy + 9y^2 + 12x + 8y = 0$

In Problems 27–29, identify the conic that each polar equation represents and graph it by hand. Verify your graph using a graphing utility.

27. $r = \dfrac{4}{1 - \cos\theta}$

28. $r = \dfrac{6}{2 - \sin\theta}$

29. $r = \dfrac{8}{4 + 8\cos\theta}$

In Problems 30 and 31, convert each polar equation to a rectangular equation.

30. $r = \dfrac{4}{1 - \cos\theta}$

31. $r = \dfrac{8}{4 + 8\cos\theta}$

In Problems 32–34, graph the curve whose parametric equations are given by hand and show its orientation. Find the rectangular equation of each curve. Verify your graph using a graphing utility.

32. $x = 4t - 2,\quad y = 1 - t;\quad -\infty < t < \infty$

33. $x = 3\sin t,\quad y = 4\cos t + 2;\quad 0 \le t \le 2\pi$

34. $x = \sec^2 t,\quad y = \tan^2 t;\quad 0 \le t \le \dfrac{\pi}{4}$

35. Find two different parametric equations for $y = -2x + 4$.

36. Find parametric equations for an object that moves along the ellipse $\dfrac{x^2}{16} + \dfrac{y^2}{9} = 1$ where the motion begins at $(4, 0)$, is counterclockwise, and requires 4 seconds for a complete revolution.

37. Find an equation of the hyperbola whose foci are the vertices of the ellipse $4x^2 + 9y^2 = 36$ and whose vertices are the foci of this ellipse.

38. Describe the collection of points in a plane so that the distance from each point to the point $(3, 0)$ is three-fourths of its distance from the line $x = \dfrac{16}{3}$.

39. Searchlight A searchlight is shaped like a paraboloid of revolution. If a light source is located 1 foot from the vertex along the axis of symmetry and the opening is 2 feet across, how deep should the mirror be in order to reflect the light rays parallel to the axis of symmetry?

40. Semielliptical Arch Bridge A bridge is built in the shape of a semielliptical arch. The bridge has a span of 60 feet and a maximum height of 20 feet. Find the height of the arch at distances of 5, 10, and 20 feet from the center.

41. Calibrating Instruments In a test of their recording devices, a team of seismologists positioned two of the devices 2000 feet apart, with the device at point A to the west of the device at point B. At a point between the devices and 200 feet from point B, a small amount of explosive was detonated and a note made of the time at which the sound reached each device. A second explosion is to be carried out at a point directly north of point B. How far north should the site of the second explosion be chosen so that the measured time difference recorded by the devices for the second detonation is the same as that recorded for the first detonation?

42. Uniform Motion Mary's train leaves at 7:15 AM and accelerates at the rate of 3 meters per second per second. Mary, who can run 6 meters per second, arrives at the train station 2 seconds after the train has left.

(a) Find parametric equations that model the motion of the train and Mary as a function of time.
[**Hint:** The position s at time t of an object having acceleration a is $s = \dfrac{1}{2}at^2$.]

(b) Determine algebraically whether Mary will catch the train. If so, when?

(c) Simulate the motions of the train and Mary by simultaneously graphing the equations found in part (a).

43. Projectile Motion Drew Brees throws a football with an initial speed of 80 feet per second at an angle of 35° to the horizontal. The ball leaves Brees' hand at a height of 6 feet.

(a) Find parametric equations that model the position of the ball as a function of time.

(b) How long is the ball in the air?

(c) When is the ball at its maximum height? Determine the maximum height of the ball.

(d) Determine the horizontal distance that the ball travels.

(e) Using a graphing utility, simultaneously graph the equations found in part (a).

44. Formulate a strategy for discussing and graphing an equation of the form

$$Ax^2 + Bxy + Cy^2 + Dx + Ey + F = 0$$

Chapter Test

CHAPTER Test Prep VIDEOS

The Chapter Test Prep Videos are step-by-step solutions available in MyMathLab®, or on this text's YouTube™ Channel. Flip back to the Resources for Success page for a link to this text's YouTube channel.

In Problems 1–3, identify each equation. If it is a parabola, give its vertex, focus, and directrix; if it is an ellipse, give its center, vertices, and foci; if it is a hyperbola, give its center, vertices, foci, and asymptotes.

1. $\dfrac{(x+1)^2}{4} - \dfrac{y^2}{9} = 1$

2. $8y = (x - 1)^2 - 4$

3. $2x^2 + 3y^2 + 4x - 6y = 13$

In Problems 4–6, find an equation of the conic described; graph the equation.

4. Parabola: focus $(-1, 4.5)$, vertex $(-1, 3)$

5. Ellipse: center $(0, 0)$, vertex $(0, -4)$, focus $(0, 3)$

6. Hyperbola: center $(2, 2)$, vertex $(2, 4)$, contains the point $(2 + \sqrt{10}, 5)$

In Problems 7–9, identify each conic without completing the square or rotating axes.

7. $2x^2 + 5xy + 3y^2 + 3x - 7 = 0$ **8.** $3x^2 - xy + 2y^2 + 3y + 1 = 0$ **9.** $x^2 - 6xy + 9y^2 + 2x - 3y - 2 = 0$

10. Given the equation $41x^2 - 24xy + 34y^2 - 25 = 0$, rotate the axes so that there is no xy-term. Analyze and graph the new equation.

11. Identify the conic represented by the polar equation $r = \dfrac{3}{1 - 2\cos\theta}$. Find the rectangular equation.

12. Graph the curve whose parametric equations are given and show its orientation. Find the rectangular equation for the curve.

$$x = 3t - 2, \quad y = 1 - \sqrt{t}, \quad 0 \le t \le 9$$

13. A parabolic reflector (paraboloid of revolution) is used by TV crews at football games to pick up the referee's announcements, quarterback signals, and so on. A microphone is placed at the focus of the parabola. If a certain reflector is 4 feet wide and 1.5 feet deep, where should the microphone be placed?

Cumulative Review

1. For $f(x) = -3x^2 + 5x - 2$, find

$$\frac{f(x+h) - f(x)}{h} \qquad h \neq 0$$

2. In the complex number system, solve the equation

$$9x^4 + 33x^3 - 71x^2 - 57x - 10 = 0$$

3. For what numbers x is $6 - x \ge x^2$?

4. (a) Find the domain and range of $y = 3^x + 2$.
(b) Find the inverse of $y = 3^x + 2$ and state its domain and range.

5. $f(x) = \log_4(x - 2)$
(a) Solve $f(x) = 2$.
(b) Solve $f(x) \le 2$.

6. Find an equation for each of the following graphs.

(a) Line:

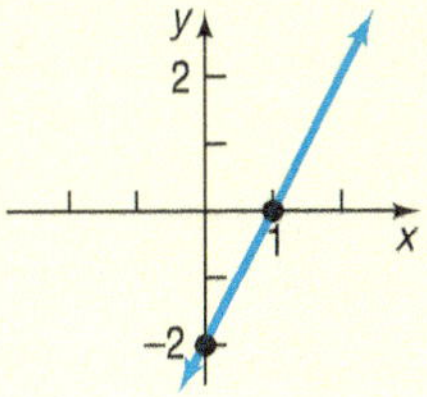

(b) Circle:

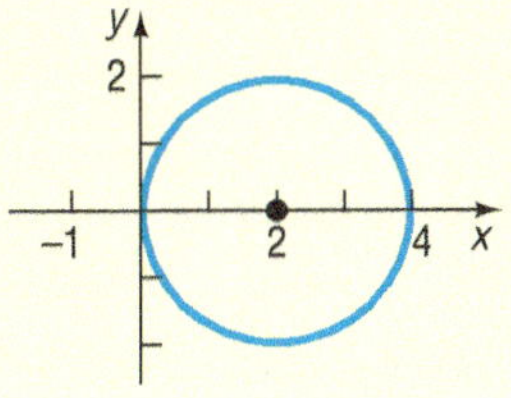

(c) Ellipse:

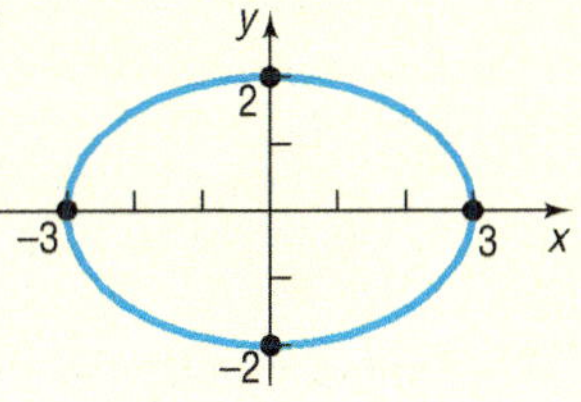

(d) Parabola:

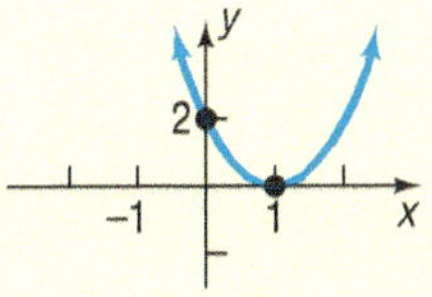

(e) Hyperbola:

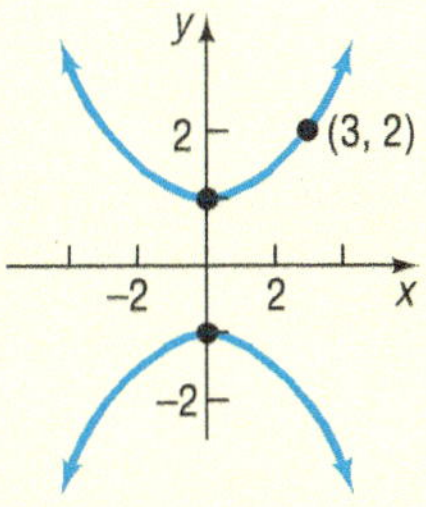

(f) Exponential:

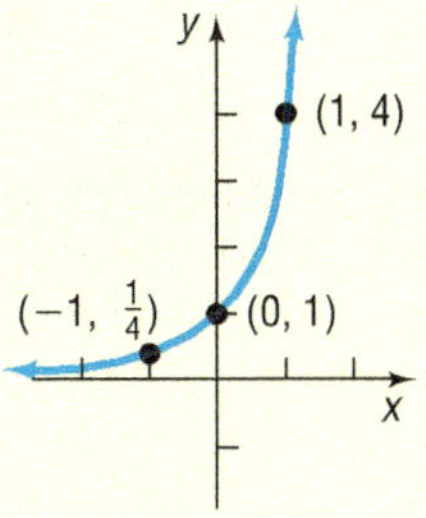

7. Find all the solutions of the equation $\sin(2\theta) = 0.5$.

8. Find a polar equation for the line containing the origin that makes an angle of 30° with the positive x-axis.

9. Find a polar equation for the circle with center at the point $(0, 4)$ and radius 4. Graph this circle.

10. What is the domain of the function $f(x) = \dfrac{3}{\sin x + \cos x}$?

11. Solve the equation $\cot(2\theta) = 1$, where $0° < \theta < 90°$.

12. Find the rectangular equation of the curve

$$x = 5\tan t, \; y = 5\sec^2 t, \; -\frac{\pi}{2} < t < \frac{\pi}{2}$$

Chapter Projects

I. Comet Hale-Bopp The orbits of planets and some comets about the Sun are ellipses, with the Sun at one focus. The **aphelion** of a planet is its greatest distance from the Sun, and the **perihelion** is its shortest distance. The **mean distance** of a planet from the Sun is the length of the semimajor axis of the elliptical orbit. See the illustration.

1. Research the history of Comet Hale-Bopp on the Internet. In particular, determine the aphelion and perihelion. Often these values are given in terms of astronomical units. What is an astronomical unit? What is it equivalent to in miles? In kilometers? What is the orbital period of Comet Hale-Bopp? When will it next be visible from Earth? How close does it come to Earth?
2. Find a model for the orbit of Comet Hale-Bopp around the Sun. Use the x-axis as the major axis.
3. Comet Hale-Bopp has an orbit that is roughly perpendicular to that of Earth. Find a model for the orbit of Earth using the y-axis as the major axis.
4. Use a graphing utility or some other graphing technology to graph the paths of the orbits. Based on the graphs, do the paths of the orbits intersect? Does this mean that Comet Hale-Bopp will collide with Earth?

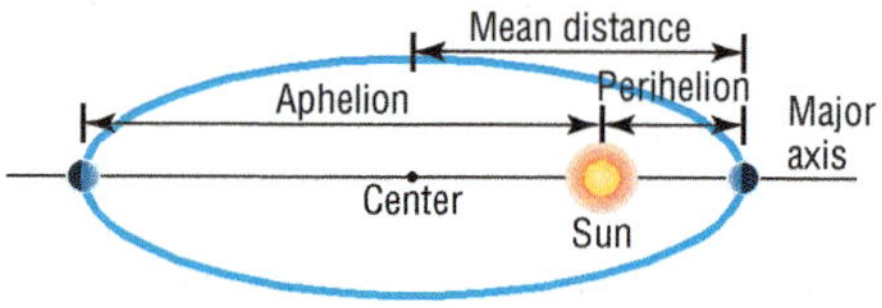

The following projects can be found at the Instructor's Resource Center (IRC):

II. The Orbits of Neptune and Pluto The astronomical body known as Pluto and the planet Neptune travel around the Sun in elliptical orbits. Pluto, at times, comes closer to the Sun than Neptune, the outermost planet. This project examines and analyzes the two orbits.

III. Project at Motorola ***Distorted Deployable Space Reflector Antennas*** An engineer designs an antenna that will deploy in space to collect sunlight.

IV. Constructing a Bridge over the East River The size of ships using a river and fluctuations in water height due to tides or flooding must be considered when designing a bridge that will cross a major waterway.

V. Systems of Parametric Equations Which approach to use when solving a system of equations depends on the form of the system and on the domains of the equations.

12 Systems of Equations and Inequalities

Economic Outcomes

Annual Earnings of Young Adults

For both males and females, earnings increase with education: full-time workers with at least a bachelor's degree have higher median earnings than those with less education. For example, in 2014, male college graduates earned 84% more than male high school completers. Females with a bachelor's or higher degree earned 81% more than female high school completers. Males and females who dropped out of high school earned 31% and 29% less, respectively, than male and female high school completers.

The median earnings of young adults who had at least a bachelor's degree declined in the 1970s relative to their counterparts who were high school completers, before increasing between 1980 and 2014. Males with a bachelor's degree or higher had earnings 19% higher than male high school completers in 1980 and had earnings 84% higher in 2014. Among females, those with at least a bachelor's degree had earnings 34% higher than female high school completers in 1980, compared with earnings 81% higher in 2014.

—See Chapter Project I—

••• A Look Back

In Chapters 1, 4, 5, 6, and 8 we solved various kinds of equations and inequalities involving a single variable.

A Look Ahead •••

In this chapter we take up the problem of solving equations and inequalities containing two or more variables. There are various ways to solve such problems.

The *method of substitution* for solving equations in several variables dates back to ancient times.

The *method of elimination*, although it had existed for centuries, was put into systematic order by Karl Friedrich Gauss (1777–1855) and by Camille Jordan (1838–1922).

The theory of *matrices* was developed in 1857 by Arthur Cayley (1821–1895), although only later were matrices used as we use them in this chapter. Matrices have become a very flexible instrument, useful in almost all areas of mathematics.

The method of *determinants* was invented by Takakazu Seki Kôwa (1642–1708) in 1683 in Japan and by Gottfried Wilhelm von Leibniz (1646–1716) in 1693 in Germany. *Cramer's Rule* is named after Gabriel Cramer (1704–1752) of Switzerland, who popularized the use of determinants for solving linear systems.

Section 12.5, on *partial fraction decomposition*, provides an application of systems of equations. This particular application is one that is used in integral calculus.

Section 12.8 introduces *linear programming*, a modern application of linear inequalities. This topic is particularly useful for students interested in operations research.

Outline

12.1 Systems of Linear Equations: Substitution and Elimination

PREPARING FOR THIS SECTION *Before getting started, review the following:*

- Linear Equations (Section 1.2, pp. 102–103)
- Lines (Section 2.2, pp. 173–184)

Now Work the 'Are You Prepared?' problems on page 872.

OBJECTIVES
1. Solve Systems of Equations by Substitution (p. 864)
2. Solve Systems of Equations by Elimination (p. 865)
3. Identify Inconsistent Systems of Equations Containing Two Variables (p. 867)
4. Express the Solution of a System of Dependent Equations Containing Two Variables (p. 867)
5. Solve Systems of Three Equations Containing Three Variables (p. 868)
6. Identify Inconsistent Systems of Equations Containing Three Variables (p. 870)
7. Express the Solution of a System of Dependent Equations Containing Three Variables (p. 871)

EXAMPLE 1 **Movie Theater Ticket Sales**

A movie theater sells tickets for \$10.00 each, with seniors receiving a discount of \$2.00. One evening the theater took in \$4630 in revenue. If x represents the number of tickets sold at \$10.00 and y the number of tickets sold at the discounted price of \$8.00, write an equation that relates these variables.

Solution Each nondiscounted ticket brings in \$10.00, so x tickets will bring in $10x$ dollars. Similarly, y discounted tickets bring in $8y$ dollars. Because the total brought in is \$4630, this means

$$10x + 8y = 4630$$

■

In Example 1, suppose it is also known that 525 tickets were sold that evening. Then a second equation relating the variables x and y is

$$x + y = 525$$

The two equations

$$\begin{cases} 10x + 8y = 4630 \\ x + y = 525 \end{cases}$$

form a *system* of equations.

In general, a **system of equations** is a collection of two or more equations, each containing one or more variables. Example 2 gives some illustrations of systems of equations.

EXAMPLE 2 **Examples of Systems of Equations**

(a) $\begin{cases} 2x + y = 5 & (1) \\ -4x + 6y = -2 & (2) \end{cases}$ Two equations containing two variables, x and y

(b) $\begin{cases} x + y^2 = 5 & (1) \\ 2x + y = 4 & (2) \end{cases}$ Two equations containing two variables, x and y

(c) $\begin{cases} x + y + z = 6 & (1) \\ 3x - 2y + 4z = 9 & (2) \\ x - y - z = 0 & (3) \end{cases}$ **Three equations containing three variables, x, y, and z**

(d) $\begin{cases} x + y + z = 5 & (1) \\ x - y = 2 & (2) \end{cases}$ **Two equations containing three variables, x, y, and z**

(e) $\begin{cases} x + y + z = 6 & (1) \\ 2x + 2z = 4 & (2) \\ y + z = 2 & (3) \\ x = 4 & (4) \end{cases}$ **Four equations containing three variables, x, y, and z**

We use a brace to remind us that we are dealing with a system of equations, and we number each equation in the system for convenient reference.

A **solution** of a system of equations consists of values for the variables that are solutions of each equation of the system. To **solve** a system of equations means to find all solutions of the system.

For example, $x = 2$, $y = 1$ is a solution of the system in Example 2(a), because

$$\begin{cases} 2x + y = 5 & (1) \\ -4x + 6y = -2 & (2) \end{cases} \qquad \begin{cases} 2(2) + 1 = 4 + 1 = 5 \\ -4(2) + 6(1) = -8 + 6 = -2 \end{cases}$$

This solution may also be written as the ordered pair $(2, 1)$.

A solution of the system in Example 2(b) is $x = 1$, $y = 2$, because

$$\begin{cases} x + y^2 = 5 & (1) \\ 2x + y = 4 & (2) \end{cases} \qquad \begin{cases} 1 + 2^2 = 1 + 4 = 5 \\ 2(1) + 2 = 2 + 2 = 4 \end{cases}$$

Another solution of the system in Example 2(b) is $x = \frac{11}{4}$, $y = -\frac{3}{2}$, which you can check for yourself.

A solution of the system in Example 2(c) is $x = 3$, $y = 2$, $z = 1$, because

$$\begin{cases} x + y + z = 6 & (1) \\ 3x - 2y + 4z = 9 & (2) \\ x - y - z = 0 & (3) \end{cases} \qquad \begin{cases} 3 + 2 + 1 = 6 \\ 3(3) - 2(2) + 4(1) = 9 - 4 + 4 = 9 \\ 3 - 2 - 1 = 0 \end{cases}$$

This solution may also be written as the ordered triplet $(3, 2, 1)$.

Note that $x = 3$, $y = 3$, $z = 0$ is not a solution of the system in Example 2(c).

$$\begin{cases} x + y + z = 6 & (1) \\ 3x - 2y + 4z = 9 & (2) \\ x - y - z = 0 & (3) \end{cases} \qquad \begin{cases} 3 + 3 + 0 = 6 \\ 3(3) - 2(3) + 4(0) = 3 \neq 9 \\ 3 - 3 - 0 = 0 \end{cases}$$

Although $x = 3$, $y = 3$, and $z = 0$ satisfy equations (1) and (3), they do not satisfy equation (2). Any solution of the system must satisfy *each* equation of the system.

Now Work PROBLEM 11

When a system of equations has at least one solution, it is said to be **consistent**. When a system of equations has no solution, it is called **inconsistent**.

An equation in n variables is said to be **linear** if it is equivalent to an equation of the form

$$a_1x_1 + a_2x_2 + \cdots + a_nx_n = b$$

where $x_1, x_2, \ldots, x_n$ are n distinct variables, $a_1, a_2, \ldots, a_n, b$ are constants, and at least one of the a's is not 0.

Some examples of linear equations are

$$2x + 3y = 2 \qquad 5x - 2y + 3z = 10 \qquad 8x + 8y - 2z + 5w = 0$$

If each equation in a system of equations is linear, the result is a **system of linear equations**. The systems in Examples 2(a), (c), (d), and (e) are linear, whereas the system in Example 2(b) is nonlinear. In this chapter we shall solve linear systems in Sections 12.1 to 12.3. Nonlinear systems are discussed in Section 12.6.

We begin by discussing a system of two linear equations containing two variables. The problem of solving such a system can be viewed as a geometry problem. The graph of each equation in such a system is a line. So a system of two linear equations containing two variables represents a pair of lines. The lines either (1) intersect, or (2) are parallel, or (3) are **coincident** (that is, identical).

1. If the lines intersect, the system of equations has one solution, given by the point of intersection. The system is **consistent** and the equations are **independent**. See Figure 1(a).
2. If the lines are parallel, the system of equations has no solution, because the lines never intersect. The system is **inconsistent**. See Figure 1(b).
3. If the lines are coincident (the lines lie on top of each other), the system of equations has infinitely many solutions, represented by the totality of points on the line. The system is **consistent** and the equations are **dependent**. See Figure 1(c).

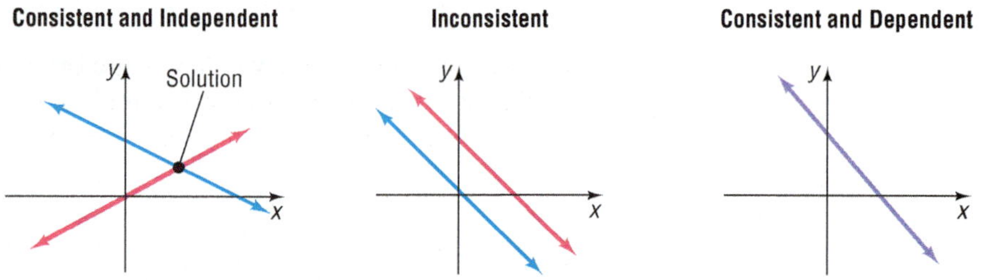

Figure 1 **(a)** Intersecting lines; system has one solution **(b)** Parallel lines; system has no solution **(c)** Coincident lines; system has infinitely many solutions

EXAMPLE 3 Solving a System of Linear Equations Using a Graphing Utility

Solve: $\begin{cases} 2x + y = -1 & (1) \\ -4x + 6y = 42 & (2) \end{cases}$

Solution First, solve each equation for y. This is equivalent to writing each equation in slope–intercept form. Equation (1) in slope–intercept form is $Y_1 = -2x - 1$. Equation (2) in slope–intercept form is $Y_2 = \frac{2}{3}x + 7$. Figure 2 shows the graphs using a TI-84 Plus C. The lines intersect, so the system is consistent and the equations are independent. Using INTERSECT gives the solution $(-3, 5)$. ■

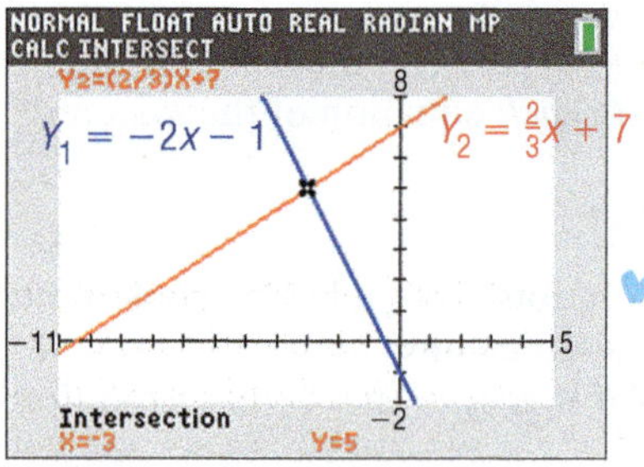

Figure 2

1 Solve Systems of Equations by Substitution

Most of the time algebraic methods must be used to obtain exact solutions. A number of methods are available for solving systems of linear equations algebraically. In this section, we introduce two methods: *substitution* and *elimination*. We illustrate the **method of substitution** by solving the system given in Example 3.

EXAMPLE 4 How to Solve a System of Linear Equations by Substitution

Solve: $\begin{cases} 2x + y = -1 & (1) \\ -4x + 6y = 42 & (2) \end{cases}$

Step-by-Step Solution

Step 1: Pick one of the equations, and solve for one of the variables in terms of the remaining variable(s).

Solve equation (1) for y.

$$\begin{aligned} 2x + y &= -1 && \text{Equation (1)} \\ y &= -2x - 1 && \text{Subtract } 2x \text{ from each side of (1).} \end{aligned}$$

Step 2: Substitute the result into the remaining equation(s).

Substitute $-2x - 1$ for y in equation (2). The result is an equation containing just the variable x, which we can solve.

$$\begin{aligned} -4x + 6y &= 42 && \text{Equation (2)} \\ -4x + 6(-2x - 1) &= 42 && \text{Substitute } -2x - 1 \text{ for } y \text{ in (2).} \end{aligned}$$

Step 3: If one equation in one variable results, solve this equation. Otherwise, repeat Steps 1 and 2 until a single equation with one variable remains.

$$\begin{aligned} -4x - 12x - 6 &= 42 && \text{Distribute.} \\ -16x - 6 &= 42 && \text{Combine like terms.} \\ -16x &= 48 && \text{Add 6 to both sides.} \\ x &= \frac{48}{-16} && \text{Divide both sides by } -16. \\ x &= -3 && \text{Simplify.} \end{aligned}$$

Step 4: Find the values of the remaining variables by back-substitution.

Because we know that $x = -3$, we can find the value of y by **back-substitution**, that is, by substituting -3 for x in one of the original equations. Equation (1) seems easier to work with, so we will back-substitute into equation (1).

$$\begin{aligned} 2x + y &= -1 && \text{Equation (1)} \\ 2(-3) + y &= -1 && \text{Substitute } x = -3 \text{ into equation (1).} \\ -6 + y &= -1 && \text{Simplify.} \\ y &= -1 + 6 && \text{Add 6 to both sides.} \\ y &= 5 \end{aligned}$$

Step 5: Check the solution found.

We have $x = -3$ and $y = 5$. Verify that both equations are satisfied (true) for these values.

$$\begin{cases} 2x + y = -1; & 2(-3) + 5 = -6 + 5 = -1 \\ -4x + 6y = 42; & -4(-3) + 6(5) = 12 + 30 = 42 \end{cases}$$

The solution of the system is $x = -3$ and $y = 5$. The solution can also be written as the ordered pair $(-3, 5)$. ■

Now Use Substitution to Work PROBLEM 21

2 Solve Systems of Equations by Elimination

A second method for solving a system of linear equations is the *method of elimination*. This method is usually preferred over substitution if substitution leads to fractions or if the system contains more than two variables. Elimination also provides the necessary motivation for solving systems using matrices (the subject of Section 12.2).

The idea behind the **method of elimination** is to replace the original system of equations by an equivalent system so that adding two of the equations eliminates a variable. The rules for obtaining equivalent equations are the same as those studied earlier. However, we may also interchange any two equations of the system and/or replace any equation in the system by the sum (or difference) of that equation and a nonzero multiple of any other equation in the system.

In Words

When using elimination, get the coefficients of one of the variables to be opposites of each other.

Rules for Obtaining an Equivalent System of Equations

1. Interchange any two equations of the system.
2. Multiply (or divide) each side of an equation by the same nonzero constant.
3. Replace any equation in the system by the sum (or difference) of that equation and a nonzero multiple of any other equation in the system.

An example will give you the idea. As you work through the example, pay particular attention to the pattern being followed.

EXAMPLE 5 How to Solve a System of Linear Equations by Elimination

Solve: $\begin{cases} 2x + 3y = 1 & (1) \\ -x + y = -3 & (2) \end{cases}$

Step-by-Step Solution

Step 1: Multiply both sides of one or both equations by a nonzero constant so that the coefficients of one of the variables are additive inverses.

Multiply both sides of equation (2) by 2 so that the coefficients of x in the two equations are additive inverses.

$$\begin{cases} 2x + 3y = 1 & (1) \\ -x + y = -3 & (2) \end{cases}$$

$$\begin{cases} 2x + 3y = 1 & (1) \\ 2(-x + y) = 2(-3) & (2) \quad \text{Multiply by 2.} \end{cases}$$

$$\begin{cases} 2x + 3y = 1 & (1) \\ -2x + 2y = -6 & (2) \end{cases}$$

Step 2: Add the equations to eliminate the variable. Solve the resulting equation for the remaining unknown.

$$\begin{cases} 2x + 3y = 1 & (1) \\ -2x + 2y = -6 & (2) \end{cases}$$

$$5y = -5 \quad \text{Add equations (1) and (2).}$$

$$y = -1 \quad \text{Divide both sides by 5.}$$

Step 3: Back-substitute the value of the variable found in Step 2 into one of the *original* equations to find the value of the remaining variable.

Back-substitute $y = -1$ into equation (1) and solve for x.

$$2x + 3y = 1 \quad \text{Equation (1)}$$

$$2x + 3(-1) = 1 \quad \text{Substitute } y = -1 \text{ into equation (1).}$$

$$2x - 3 = 1 \quad \text{Simplify.}$$

$$2x = 4 \quad \text{Add 3 to both sides.}$$

$$x = 2 \quad \text{Divide both sides by 2.}$$

Step 4: Check the solution found.

The check is left to you.

The solution of the system is $x = 2$ and $y = -1$. The solution also can be written as the ordered pair $(2, -1)$.

Now Use Elimination to Work PROBLEM 21

EXAMPLE 6 Movie Theater Ticket Sales

A movie theater sells tickets for \$10.00 each, with seniors receiving a discount of \$2.00. One evening the theater sold 525 tickets and took in \$4630 in revenue. How many of each type of ticket were sold?

Solution If x represents the number of tickets sold at \$10.00 and y the number of tickets sold at the discounted price of \$8.00, then the given information results in the system of equations

$$\begin{cases} 10x + 8y = 4630 & (1) \\ x + y = 525 & (2) \end{cases}$$

Using the method of elimination, first multiply the second equation by -8, and then add the equations.

$$\begin{cases} 10x + 8y = 4630 \\ -8x - 8y = -4200 \end{cases}$$
$$2x = 430 \quad \text{Add the equations.}$$
$$x = 215$$

Since $x + y = 525$, then $y = 525 - x = 525 - 215 = 310$. So 215 nondiscounted tickets and 310 senior discount tickets were sold.

3 Identify Inconsistent Systems of Equations Containing Two Variables

The previous examples dealt with consistent systems of equations that had a single solution. The next two examples deal with two other possibilities that may occur, the first being a system that has no solution.

EXAMPLE 7 An Inconsistent System of Linear Equations

Solve: $\begin{cases} 2x + y = 5 & (1) \\ 4x + 2y = 8 & (2) \end{cases}$

Solution We choose to use the method of substitution and solve equation (1) for y.

$$2x + y = 5 \quad (1)$$
$$y = -2x + 5 \quad \text{Subtract } 2x \text{ from each side.}$$

Now substitute $-2x + 5$ for y in equation (2) and solve for x.

$$4x + 2y = 8 \quad (2)$$
$$4x + 2(-2x + 5) = 8 \quad \text{Substitute } y = -2x + 5 \text{ in (2).}$$
$$4x - 4x + 10 = 8 \quad \text{Multiply out.}$$
$$0 = -2 \quad \text{Subtract 10 from both sides.}$$

This statement is false. Conclude that the system has no solution and is therefore inconsistent.

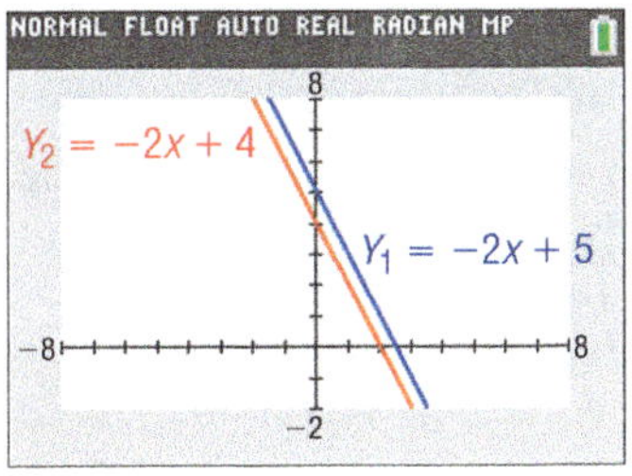

Figure 3

Figure 3 illustrates the pair of lines whose equations form the system in Example 7. Notice that the graphs of the two equations are lines, each with slope -2; one has a y-intercept of 5, the other a y-intercept of 4. The lines are parallel and have no point of intersection. This geometric statement is equivalent to the algebraic statement that the system has no solution.

4 Express the Solution of a System of Dependent Equations Containing Two Variables

EXAMPLE 8 Solving a System of Dependent Equations

Solve: $\begin{cases} 2x + y = 4 & (1) \\ -6x - 3y = -12 & (2) \end{cases}$

Solution We choose to use the method of elimination.

$$\begin{cases} 2x + y = 4 & \textbf{(1)} \\ -6x - 3y = -12 & \textbf{(2)} \end{cases}$$

$$\begin{cases} 6x + 3y = 12 & \textbf{(1)} \quad \textbf{Multiply each side of equation (1) by 3.} \\ -6x - 3y = -12 & \textbf{(2)} \end{cases}$$

$$0 = 0 \qquad \textbf{Add equations (1) and (2).}$$

The statement $0 = 0$ is true. This means the equation $6x + 3y = 12$ is equivalent to $-6x - 3y = -12$. Therefore, the original system is equivalent to a system containing one equation, so the equations are dependent. This means that any values of x and y that satisfy $6x + 3y = 12$ or, equivalently, $2x + y = 4$ are solutions. For example, $x = 2, y = 0$; $x = 0, y = 4$; $x = -2, y = 8$; $x = 4, y = -4$; and so on, are solutions. There are, in fact, infinitely many values of x and y for which $2x + y = 4$, so the original system has infinitely many solutions. We will write the solution of the original system either as

$$y = -2x + 4, \quad \text{where } x \text{ can be any real number}$$

or as

$$x = -\frac{1}{2}y + 2, \quad \text{where } y \text{ can be any real number.}$$

The solution can also be expressed as $\{(x, y) \mid y = -2x + 4, x \text{ is any real number}\}$ or as $\left\{(x, y) \,\middle|\, x = -\frac{1}{2}y + 2, y \text{ is any real number}\right\}$. ■

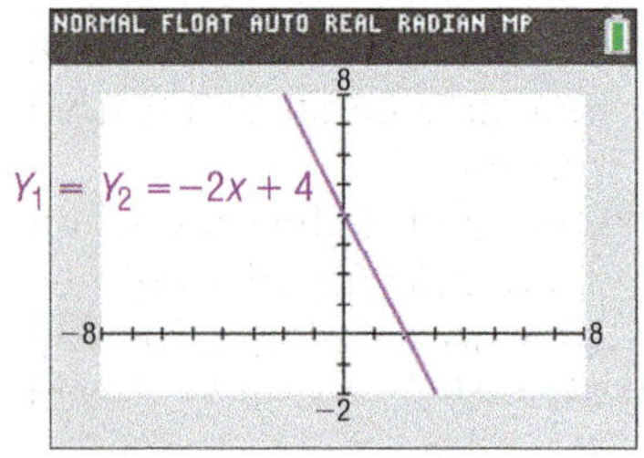

Figure 4 $y = -2x + 4$

Figure 4 illustrates the situation presented in Example 8. Notice that the graphs of the two equations are lines, each with slope -2 and each with y-intercept 4. The lines are coincident. Notice also that equation (2) in the original system is -3 times equation (1), indicating that the two equations are dependent.

For the system in Example 8, some of the infinite number of solutions can be written down by assigning values to x and then finding $y = -2x + 4$.

If $x = -1$, then $y = -2(-1) + 4 = 6$.

If $x = 0$, then $y = 4$.

If $x = 2$, then $y = 0$.

The ordered pairs $(-1, 6)$, $(0, 4)$, and $(2, 0)$ are three of the points on the line in Figure 4.

Now Work PROBLEMS 27 AND 31

5 Solve Systems of Three Equations Containing Three Variables

Just like a system of two linear equations containing two variables, a system of three linear equations containing three variables has (1) exactly one solution (a consistent system with independent equations), or (2) no solution (an inconsistent system), or (3) infinitely many solutions (a consistent system with dependent equations).

The problem of solving a system of three linear equations containing three variables can be viewed as a geometry problem. The graph of each equation in such a system is a plane in space. A system of three linear equations containing three variables represents three planes in space. Figure 5 illustrates some of the possibilities.

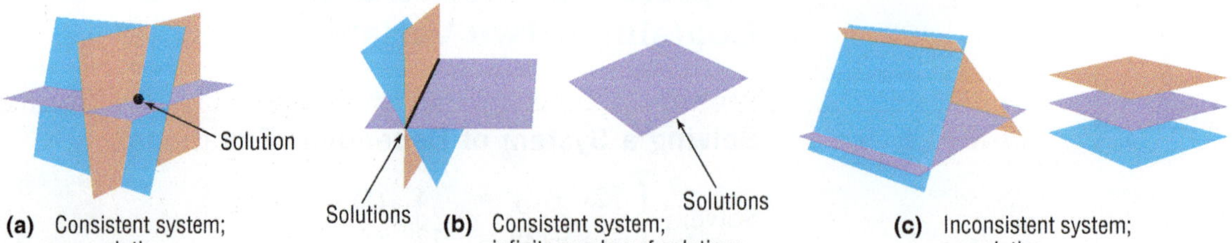

Figure 5 **(a)** Consistent system; one solution **(b)** Consistent system; infinite number of solutions **(c)** Inconsistent system; no solution

Recall that a **solution** to a system of equations consists of values for the variables that are solutions of each equation of the system. For example, $x = 3, y = -1, z = -5$ or, using an ordered triplet, $(3, -1, -5)$ is a solution to the system of equations

$$\begin{cases} x + y + z = -3 & (1) \\ 2x - 3y + 6z = -21 & (2) \\ -3x + 5y = -14 & (3) \end{cases} \qquad \begin{aligned} &3 + (-1) + (-5) = -3 \\ &2(3) - 3(-1) + 6(-5) = 6 + 3 - 30 = -21 \\ &-3(3) + 5(-1) = -9 - 5 = -14 \end{aligned}$$

because these values of the variables are solutions of each equation.

Typically, when solving a system of three linear equations containing three variables, we use the method of elimination. Recall that the idea behind the method of elimination is to form equivalent equations so that adding two of the equations eliminates a variable.

EXAMPLE 9 Solving a System of Three Linear Equations with Three Variables

Use the method of elimination to solve the system of equations.

$$\begin{cases} x + y - z = -1 & (1) \\ 4x - 3y + 2z = 16 & (2) \\ 2x - 2y - 3z = 5 & (3) \end{cases}$$

Solution For a system of three equations, attempt to eliminate one variable at a time, using pairs of equations, until an equation with a single variable remains. Our strategy for solving this system is to use equation (1) to eliminate the variable x from equations (2) and (3). We can then treat the new equations (2) and (3) as a system with two unknowns. Alternatively, we could use equation (1) to eliminate either y or z from equations (2) and (3). Try one of these approaches for yourself.

Begin by multiplying each side of equation (1) by -4 and adding the result to equation (2). (Do you see why? The coefficients of x are now negatives of one another.) Also multiply equation (1) by -2 and add the result to equation (3). Notice that these two procedures result in the elimination of the variable x from equations (2) and (3).

$$\begin{aligned} x + y - z &= -1 \quad (1) \quad \text{Multiply by } -4. \\ 4x - 3y + 2z &= 16 \quad (2) \end{aligned} \qquad \begin{aligned} -4x - 4y + 4z &= 4 \quad (1) \\ 4x - 3y + 2z &= 16 \quad (2) \\ \hline -7y + 6z &= 20 \quad \text{Add} \end{aligned}$$

$$\begin{aligned} x + y - z &= -1 \quad (1) \quad \text{Multiply by } -2. \\ 2x - 2y - 3z &= 5 \quad (3) \end{aligned} \qquad \begin{aligned} -2x - 2y + 2z &= 2 \quad (1) \\ 2x - 2y - 3z &= 5 \quad (3) \\ \hline -4y - z &= 7 \quad \text{Add} \end{aligned}$$

$$\begin{cases} x + y - z = -1 & (1) \\ -7y + 6z = 20 & (2) \\ -4y - z = 7 & (3) \end{cases}$$

Now concentrate on the new equations (2) and (3), treating them as a system of two equations containing two variables. It is easier to eliminate z. Multiply each side of equation (3) by 6, and add equations (2) and (3). The result is the new equation (3).

$$\begin{aligned} -7y + 6z &= 20 \quad (2) \\ -4y - z &= 7 \quad (3) \quad \text{Multiply by 6.} \end{aligned} \qquad \begin{aligned} -7y + 6z &= 20 \quad (2) \\ -24y - 6z &= 42 \quad (3) \\ \hline -31y &= 62 \quad \text{Add} \end{aligned} \qquad \begin{cases} x + y - z = -1 & (1) \\ -7y + 6z = 20 & (2) \\ -31y = 62 & (3) \end{cases}$$

Now solve equation (3) for y by dividing both sides of the equation by -31.

$$\begin{cases} x + y - z = -1 & (1) \\ -7y + 6z = 20 & (2) \\ y = -2 & (3) \end{cases}$$

Back-substitute $y = -2$ in equation (2) and solve for z.

$$\begin{aligned} -7y + 6z &= 20 && \text{(2)} \\ -7(-2) + 6z &= 20 && \text{Substitute } y = -2 \text{ in (2).} \\ 6z &= 6 && \text{Subtract 14 from both sides of the equation.} \\ z &= 1 && \text{Divide both sides of the equation by 6.} \end{aligned}$$

Finally, back-substitute $y = -2$ and $z = 1$ in equation (1) and solve for x.

$$\begin{aligned} x + y - z &= -1 && \text{(1)} \\ x + (-2) - 1 &= -1 && \text{Substitute } y = -2 \text{ and } z = 1 \text{ in (1).} \\ x - 3 &= -1 && \text{Simplify.} \\ x &= 2 && \text{Add 3 to both sides.} \end{aligned}$$

The solution of the original system is $x = 2, y = -2, z = 1$, or, using an ordered triplet, $(2, -2, 1)$. You should check this solution. ■

Look back over the solution given in Example 9. Note the pattern of removing one of the variables from two of the equations, followed by solving this system of two equations and two unknowns. Although which variables to remove is your choice, the methodology remains the same for all systems.

Now Work PROBLEM 45

6 Identify Inconsistent Systems of Equations Containing Three Variables

EXAMPLE 10 **Identify an Inconsistent System of Linear Equations**

Solve: $\begin{cases} 2x + y - z = -2 & \text{(1)} \\ x + 2y - z = -9 & \text{(2)} \\ x - 4y + z = 1 & \text{(3)} \end{cases}$

Solution Our strategy is the same as in Example 9. However, in this system, it seems easiest to eliminate the variable z first. Do you see why?

Multiply each side of equation (1) by -1, and add the result to equation (2). Also, add equations (2) and (3).

$$\begin{aligned} 2x + y - z &= -2 \ \text{(1)} \xrightarrow{\text{Multiply by } -1.} & -2x - y + z &= 2 \ \text{(1)} \\ x + 2y - z &= -9 \ \text{(2)} & x + 2y - z &= -9 \ \text{(2)} \\ & & \hline -x + y &= -7 \ \text{Add} \end{aligned}$$

$$\begin{aligned} x + 2y - z &= -9 \ \text{(2)} \\ x - 4y + z &= 1 \ \text{(3)} \\ \hline 2x - 2y &= -8 \ \text{Add} \end{aligned}$$

$$\begin{cases} 2x + y - z = -2 & \text{(1)} \\ -x + y = -7 & \text{(2)} \\ 2x - 2y = -8 & \text{(3)} \end{cases}$$

Now concentrate on the new equations (2) and (3), treating them as a system of two equations containing two variables. Multiply each side of equation (2) by 2, and add the result to equation (3).

$$\begin{aligned} -x + y &= -7 \ \text{(2)} \xrightarrow{\text{Multiply by 2.}} & -2x + 2y &= -14 \ \text{(2)} \\ 2x - 2y &= -8 \ \text{(3)} & 2x - 2y &= -8 \ \text{(3)} \\ & & \hline 0 &= -22 \ \text{Add} \end{aligned}$$

$$\begin{cases} 2x + y - z = -2 & \text{(1)} \\ -x + y = -7 & \text{(2)} \\ 0 = -22 & \text{(3)} \end{cases}$$

Equation (3) has no solution, so the system is inconsistent. ■

7 Express the Solution of a System of Dependent Equations Containing Three Variables

EXAMPLE 11 Solving a System of Dependent Equations

Solve: $\begin{cases} x - 2y - z = 8 & (1) \\ 2x - 3y + z = 23 & (2) \\ 4x - 5y + 5z = 53 & (3) \end{cases}$

Solution Our plan is to eliminate x from equations (2) and (3). Multiply each side of equation (1) by -2, and add the result to equation (2). Also, multiply each side of equation (1) by -4, and add the result to equation (3).

$$\begin{array}{lll} x - 2y - z = 8 \;(1) & \text{Multiply by } -2. & -2x + 4y + 2z = -16 \;(1) \\ 2x - 3y + z = 23 \;(2) & & \underline{2x - 3y + z = 23 \;(2)} \\ & & y + 3z = 7 \quad \text{Add} \end{array}$$

$$\begin{array}{lll} x - 2y - z = 8 \;(1) & \text{Multiply by } -4. & -4x + 8y + 4z = -32 \;(1) \\ 4x - 5y + 5z = 53 \;(3) & & \underline{4x - 5y + 5z = 53 \;(3)} \\ & & 3y + 9z = 21 \quad \text{Add} \end{array}$$

$$\begin{cases} x - 2y - z = 8 & (1) \\ y + 3z = 7 & (2) \\ 3y + 9z = 21 & (3) \end{cases}$$

Treat equations (2) and (3) as a system of two equations containing two variables, and eliminate the variable y by multiplying both sides of equation (2) by -3 and adding the result to equation (3).

$$\begin{array}{lll} y + 3z = 7 \;(2) & \text{Multiply by } -3. & -3y - 9z = -21 \\ 3y + 9z = 21 \;(3) & & \underline{3y + 9z = 21} \quad \text{Add} \\ & & 0 = 0 \end{array} \quad \begin{cases} x - 2y - z = 8 & (1) \\ y + 3z = 7 & (2) \\ 0 = 0 & (3) \end{cases}$$

The original system is equivalent to a system containing two equations, so the equations are dependent and the system has infinitely many solutions. If we solve equation (2) for y, we can express y in terms of z as $y = -3z + 7$. Substitute this expression into equation (1) to determine x in terms of z.

$$\begin{aligned} x - 2y - z &= 8 && (1) \\ x - 2(-3z + 7) - z &= 8 && \text{Substitute } y = -3z + 7 \text{ in (1).} \\ x + 6z - 14 - z &= 8 && \text{Multiply out.} \\ x + 5z &= 22 && \text{Combine like terms.} \\ x &= -5z + 22 && \text{Solve for } x. \end{aligned}$$

We write the solution to the system as

$$\begin{cases} x = -5z + 22 \\ y = -3z + 7 \end{cases} \quad \text{where } z \text{ can be any real number.}$$

This way of writing the solution makes it easier to find specific solutions of the system. To find specific solutions, choose any value of z and use the equations $x = -5z + 22$ and $y = -3z + 7$ to determine x and y. For example, if $z = 0$, then $x = 22$ and $y = 7$, and if $z = 1$, then $x = 17$ and $y = 4$.

Using ordered triplets, the solution is

$$\{(x, y, z) \mid x = -5z + 22, y = -3z + 7, z \text{ is any real number}\}$$

Now Work PROBLEM 47

Two distinct points in the Cartesian plane determine a unique line. Given three noncollinear points, we can find the (unique) quadratic function whose graph contains these three points.

EXAMPLE 12 **Curve Fitting**

Find real numbers a, b, and c so that the graph of the quadratic function $y = ax^2 + bx + c$ contains the points $(-1, -4)$, $(1, 6)$, and $(3, 0)$.

Solution The three points must satisfy the equation $y = ax^2 + bx + c$.

For the point $(-1, -4)$ we have: $-4 = a(-1)^2 + b(-1) + c$ $\quad -4 = a - b + c$

For the point $(1, 6)$ we have: $6 = a(1)^2 + b(1) + c$ $\quad 6 = a + b + c$

For the point $(3, 0)$ we have: $0 = a(3)^2 + b(3) + c$ $\quad 0 = 9a + 3b + c$

Determine a, b, and c so that each equation is satisfied. That is, solve the following system of three equations containing three variables:

$$\begin{cases} a - b + c = -4 & (1) \\ a + b + c = 6 & (2) \\ 9a + 3b + c = 0 & (3) \end{cases}$$

Solving this system of equations, we obtain $a = -2$, $b = 5$, and $c = 3$. So the quadratic function whose graph contains the points $(-1, -4)$, $(1, 6)$, and $(3, 0)$ is

$$y = -2x^2 + 5x + 3 \quad y = ax^2 + bx + c, \; a = -2, b = 5, c = 3$$

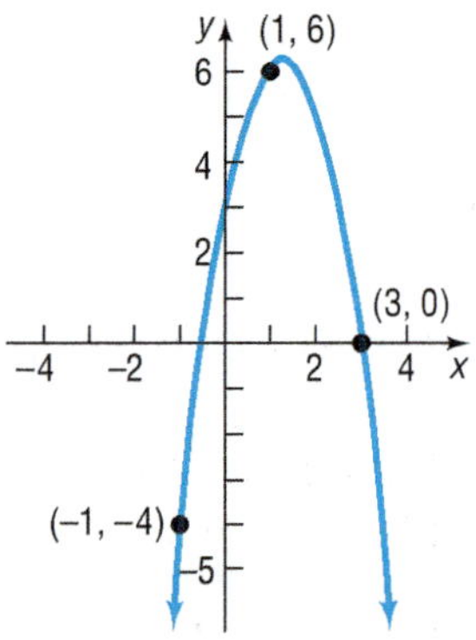

Figure 6 $y = -2x^2 + 5x + 3$

Figure 6 shows the graph of the function along with the three points. ■

Now Work PROBLEM 73

12.1 Assess Your Understanding

'Are You Prepared?' *Answers are given at the end of these exercises. If you get a wrong answer, read the pages listed in red.*

1. Solve the equation: $3x + 4 = 8 - x$. (pp. 102–103)

2. (a) Graph the line: $3x + 4y = 12$.
(b) What is the slope of a line parallel to this line?
(pp. 173–184)

Concepts and Vocabulary

3. If a system of equations has no solution, it is said to be __________.

4. If a system of equations has one solution, the system is __________ and the equations are __________.

5. If the solution to a system of two linear equations containing two unknowns is $x = 3$, $y = -2$, then the lines intersect at the point __________.

6. If the lines that make up a system of two linear equations are coincident, then the system is __________ and the equations are __________.

7. If a system of two linear equations in two variables is inconsistent, which of the following best describes the graphs of the lines in the system?
(a) intersecting (b) parallel
(c) coincident (d) perpendicular

8. If a system of dependent equations containing three variables has the general solution
$\{(x, y, z) \mid x = -z + 4, y = -2z + 5, z \text{ is any real number}\}$
which of the following is one of the infinite number of solutions of the system?
(a) $(1, -1, 3)$ (b) $(0, 4, 5)$ (c) $(4, -3, 0)$ (d) $(-1, 5, 7)$

Skill Building

In Problems 9–18, verify that the values of the variables listed are solutions of the system of equations.

9. $\begin{cases} 2x - y = 5 \\ 5x + 2y = 8 \end{cases}$
$x = 2, y = -1; (2, -1)$

10. $\begin{cases} 3x + 2y = 2 \\ x - 7y = -30 \end{cases}$
$x = -2, y = 4; (-2, 4)$

11. $\begin{cases} 3x - 4y = 4 \\ \frac{1}{2}x - 3y = -\frac{1}{2} \end{cases}$
$x = 2, y = \frac{1}{2}; \left(2, \frac{1}{2}\right)$

12. $\begin{cases} 2x + \frac{1}{2}y = 0 \\ 3x - 4y = -\frac{19}{2} \end{cases}$
$x = -\frac{1}{2}, y = 2; \left(-\frac{1}{2}, 2\right)$

13. $\begin{cases} x - y = 3 \\ \frac{1}{2}x + y = 3 \end{cases}$
$x = 4, y = 1; (4, 1)$

14. $\begin{cases} x - y = 3 \\ -3x + y = 1 \end{cases}$
$x = -2, y = -5; (-2, -5)$

15. $\begin{cases} 3x + 3y + 2z = 4 \\ x - y - z = 0 \\ 2y - 3z = -8 \end{cases}$
$x = 1, y = -1, z = 2;$
$(1, -1, 2)$

16. $\begin{cases} 4x - z = 7 \\ 8x + 5y - z = 0 \\ -x - y + 5z = 6 \end{cases}$
$x = 2, y = -3, z = 1;$
$(2, -3, 1)$

17. $\begin{cases} 3x + 3y + 2z = 4 \\ x - 3y + z = 10 \\ 5x - 2y - 3z = 8 \end{cases}$
$x = 2, y = -2, z = 2; (2, -2, 2)$

18. $\begin{cases} 4x - 5z = 6 \\ 5y - z = -17 \\ -x - 6y + 5z = 24 \end{cases}$
$x = 4, y = -3, z = 2; (4, -3, 2)$

In Problems 19–56, solve each system of equations. If the system has no solution, say that it is inconsistent. For Problems 19–30, graph the lines of the system.

19. $\begin{cases} x + y = 8 \\ x - y = 4 \end{cases}$

20. $\begin{cases} x + 2y = -7 \\ x + y = -3 \end{cases}$

21. $\begin{cases} 5x - y = 21 \\ 2x + 3y = -12 \end{cases}$

22. $\begin{cases} x + 3y = 5 \\ 2x - 3y = -8 \end{cases}$

23. $\begin{cases} 3x = 24 \\ x + 2y = 0 \end{cases}$

24. $\begin{cases} 4x + 5y = -3 \\ -2y = -8 \end{cases}$

25. $\begin{cases} 3x - 6y = 2 \\ 5x + 4y = 1 \end{cases}$

26. $\begin{cases} 2x + 4y = \frac{2}{3} \\ 3x - 5y = -10 \end{cases}$

27. $\begin{cases} 2x + y = 1 \\ 4x + 2y = 3 \end{cases}$

28. $\begin{cases} x - y = 5 \\ -3x + 3y = 2 \end{cases}$

29. $\begin{cases} 2x - y = 0 \\ 4x + 2y = 12 \end{cases}$

30. $\begin{cases} 3x + 3y = -1 \\ 4x + y = \frac{8}{3} \end{cases}$

31. $\begin{cases} x + 2y = 4 \\ 2x + 4y = 8 \end{cases}$

32. $\begin{cases} 3x - y = 7 \\ 9x - 3y = 21 \end{cases}$

33. $\begin{cases} 2x - 3y = -1 \\ 10x + y = 11 \end{cases}$

34. $\begin{cases} 3x - 2y = 0 \\ 5x + 10y = 4 \end{cases}$

35. $\begin{cases} 2x + 3y = 6 \\ x - y = \frac{1}{2} \end{cases}$

36. $\begin{cases} \frac{1}{2}x + y = -2 \\ x - 2y = 8 \end{cases}$

37. $\begin{cases} \frac{1}{2}x + \frac{1}{3}y = 3 \\ \frac{1}{4}x - \frac{2}{3}y = -1 \end{cases}$

38. $\begin{cases} \frac{1}{3}x - \frac{3}{2}y = -5 \\ \frac{3}{4}x + \frac{1}{3}y = 11 \end{cases}$

39. $\begin{cases} 3x - 5y = 3 \\ 15x + 5y = 21 \end{cases}$

40. $\begin{cases} 2x - y = -1 \\ x + \frac{1}{2}y = \frac{3}{2} \end{cases}$

41. $\begin{cases} \frac{1}{x} + \frac{1}{y} = 8 \\ \frac{3}{x} - \frac{5}{y} = 0 \end{cases}$

42. $\begin{cases} \frac{4}{x} - \frac{3}{y} = 0 \\ \frac{6}{x} + \frac{3}{2y} = 2 \end{cases}$

[**Hint:** Let $u = \frac{1}{x}$ and $v = \frac{1}{y}$, and solve for u and v. Then $x = \frac{1}{u}$ and $y = \frac{1}{v}$.]

43. $\begin{cases} x - y = 6 \\ 2x - 3z = 16 \\ 2y + z = 4 \end{cases}$

44. $\begin{cases} 2x + y = -4 \\ -2y + 4z = 0 \\ 3x - 2z = -11 \end{cases}$

45. $\begin{cases} x - 2y + 3z = 7 \\ 2x + y + z = 4 \\ -3x + 2y - 2z = -10 \end{cases}$

46. $\begin{cases} 2x + y - 3z = 0 \\ -2x + 2y + z = -7 \\ 3x - 4y - 3z = 7 \end{cases}$

47. $\begin{cases} x - y - z = 1 \\ 2x + 3y + z = 2 \\ 3x + 2y = 0 \end{cases}$

48. $\begin{cases} 2x - 3y - z = 0 \\ -x + 2y + z = 5 \\ 3x - 4y - z = 1 \end{cases}$

49. $\begin{cases} x - y - z = 1 \\ -x + 2y - 3z = -4 \\ 3x - 2y - 7z = 0 \end{cases}$

50. $\begin{cases} 2x - 3y - z = 0 \\ 3x + 2y + 2z = 2 \\ x + 5y + 3z = 2 \end{cases}$

51. $\begin{cases} 2x - 2y + 3z = 6 \\ 4x - 3y + 2z = 0 \\ -2x + 3y - 7z = 1 \end{cases}$

52. $\begin{cases} 3x - 2y + 2z = 6 \\ 7x - 3y + 2z = -1 \\ 2x - 3y + 4z = 0 \end{cases}$

53. $\begin{cases} x + y - z = 6 \\ 3x - 2y + z = -5 \\ x + 3y - 2z = 14 \end{cases}$

54. $\begin{cases} x - y + z = -4 \\ 2x - 3y + 4z = -15 \\ 5x + y - 2z = 12 \end{cases}$

55. $\begin{cases} x + 2y - z = -3 \\ 2x - 4y + z = -7 \\ -2x + 2y - 3z = 4 \end{cases}$

56. $\begin{cases} x + 4y - 3z = -8 \\ 3x - y + 3z = 12 \\ x + y + 6z = 1 \end{cases}$

Applications and Extensions

57. The perimeter of a rectangular floor is 90 feet. Find the dimensions of the floor if the length is twice the width.

58. The length of fence required to enclose a rectangular field is 3000 meters. What are the dimensions of the field if it is known that the difference between its length and width is 50 meters?

59. Orbital Launches In 2014 there was a total of 92 commercial and noncommercial orbital launches worldwide. In addition, the number of noncommercial orbital launches was three times the number of commercial orbital launches. Determine the number of commercial and noncommercial orbital launches in 2014.

Source: *Federal Aviation Administration*

60. Movie Theater Tickets A movie theater charges \$9.00 for adults and \$7.00 for senior citizens. On a day when 325 people paid for admission, the total receipts were \$2495. How many who paid were adults? How many were seniors?

61. Mixing Nuts A store sells cashews for \$5.00 per pound and peanuts for \$1.50 per pound. The manager decides to mix 30 pounds of peanuts with some cashews and sell the mixture for \$3.00 per pound. How many pounds of cashews should be mixed with the peanuts so that the mixture will produce the same revenue as selling the nuts separately?

62. Mixing a Solution A chemist wants to make 14 liters of a 40% acid solution. She has solutions that are 30% acid and 65% acid. How much of each must she mix?

63. Presale Order A wireless store owner takes presale orders for a new smartphone and tablet. He gets 340 preorders for the smartphone and 250 preorders for the tablet. The combined value of the preorders is \$270,500. If the price of a smartphone and tablet together is \$965, how much does each device cost?

64. Financial Planning A recently retired couple needs \$12,000 per year to supplement their Social Security. They have \$150,000 to invest to obtain this income. They have decided on two investment options: AA bonds yielding 10% per annum and a Bank Certificate yielding 5%.

(a) How much should be invested in each to realize exactly \$12,000?

(b) If, after 2 years, the couple requires \$14,000 per year in income, how should they reallocate their investment to achieve the new amount?

65. Computing Wind Speed With a tail wind, a small Piper aircraft can fly 600 miles in 3 hours. Against this same wind, the Piper can fly the same distance in 4 hours. Find the average wind speed and the average airspeed of the Piper.

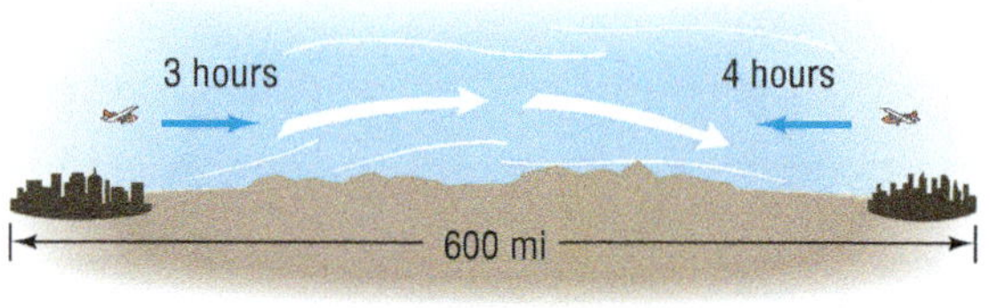

66. Computing Wind Speed The average airspeed of a single-engine aircraft is 150 miles per hour. If the aircraft flew the same distance in 2 hours with the wind as it flew in 3 hours against the wind, what was the wind speed?

67. Restaurant Management A restaurant manager wants to purchase 200 sets of dishes. One design costs \$25 per set, and another costs \$45 per set. If she has only \$7400 to spend, how many sets of each design should she order?

68. Cost of Fast Food One group of people purchased 10 hot dogs and 5 soft drinks at a cost of \$35.00. A second group bought 7 hot dogs and 4 soft drinks at a cost of \$25.25. What is the cost of a single hot dog? A single soft drink?

69. **Computing a Refund** The grocery store we use does not mark prices on its goods. My wife went to this store, bought three 1-pound packages of bacon and two cartons of eggs, and paid a total of \$13.45. Not knowing that she went to the store, I also went to the same store, purchased two 1-pound packages of bacon and three cartons of eggs, and paid a total of \$11.45. Now we want to return two 1-pound packages of bacon and two cartons of eggs. How much will be refunded?

70. **Finding the Current of a Stream** Pamela requires 3 hours to swim 15 miles downstream on the Illinois River. The return trip upstream takes 5 hours. Find Pamela's average speed in still water. How fast is the current? (Assume that Pamela's speed is the same in each direction.)

71. **Pharmacy** A doctor's prescription calls for a daily intake containing 40 milligrams (mg) of vitamin C and 30 mg of vitamin D. Your pharmacy stocks two liquids that can be used: One contains 20% vitamin C and 30% vitamin D, the other 40% vitamin C and 20% vitamin D. How many milligrams of each compound should be mixed to fill the prescription?

72. **Pharmacy** A doctor's prescription calls for the creation of pills that contain 12 units of vitamin B_{12} and 12 units of vitamin E. Your pharmacy stocks two powders that can be used to make these pills: One contains 20% vitamin B_{12} and 30% vitamin E, the other 40% vitamin B_{12} and 20% vitamin E. How many units of each powder should be mixed in each pill?

73. **Curve Fitting** Find real numbers a, b, and c so that the graph of the function $y = ax^2 + bx + c$ contains the points $(-1, 4)$, $(2, 3)$, and $(0, 1)$.

74. **Curve Fitting** Find real numbers a, b, and c so that the graph of the function $y = ax^2 + bx + c$ contains the points $(-1, -2)$, $(1, -4)$, and $(2, 4)$.

75. **IS–LM Model in Economics** In economics, the IS curve is a linear equation that represents all combinations of income Y and interest rates r that maintain an equilibrium in the market for goods in the economy. The LM curve is a linear equation that represents all combinations of income Y and interest rates r that maintain an equilibrium in the market for money in the economy. In an economy, suppose that the equilibrium level of income (in millions of dollars) and interest rates satisfy the system of equations

$$\begin{cases} 0.06Y - 5000r = 240 \\ 0.06Y + 6000r = 900 \end{cases}$$

Find the equilibrium level of income and interest rates.

76. **IS–LM Model in Economics** In economics, the IS curve is a linear equation that represents all combinations of income Y and interest rates r that maintain an equilibrium in the market for goods in the economy. The LM curve is a linear equation that represents all combinations of income Y and interest rates r that maintain an equilibrium in the market for money in the economy. In an economy, suppose that the equilibrium level of income (in millions of dollars) and interest rates satisfy the system of equations

$$\begin{cases} 0.05Y - 1000r = 10 \\ 0.05Y + 800r = 100 \end{cases}$$

Find the equilibrium level of income and interest rates.

77. **Electricity: Kirchhoff's Rules** An application of Kirchhoff's Rules to the circuit shown results in the following system of equations:

$$\begin{cases} I_2 = I_1 + I_3 \\ 5 - 3I_1 - 5I_2 = 0 \\ 10 - 5I_2 - 7I_3 = 0 \end{cases}$$

Find the currents I_1, I_2, and I_3.

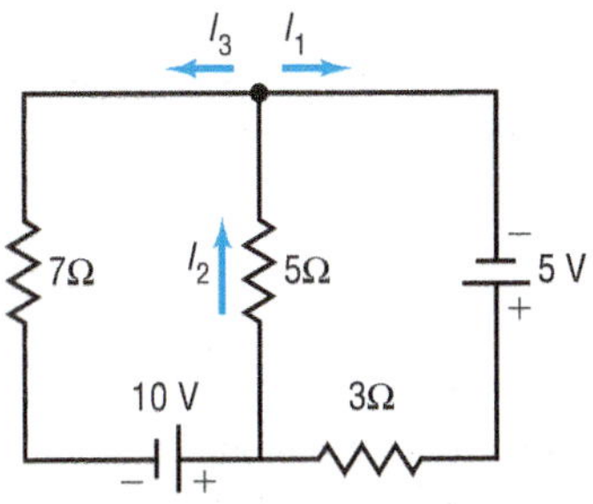

Source: *Physics for Scientists & Engineers,* 9th ed., by Serway. © 2013 Cengage Learning.

78. **Electricity: Kirchhoff's Rules** An application of Kirchhoff's Rules to the circuit shown below results in the following system of equations:

$$\begin{cases} I_3 = I_1 + I_2 \\ 8 = 4I_3 + 6I_2 \\ 8I_1 = 4 + 6I_2 \end{cases}$$

Find the currents I_1, I_2, and I_3.

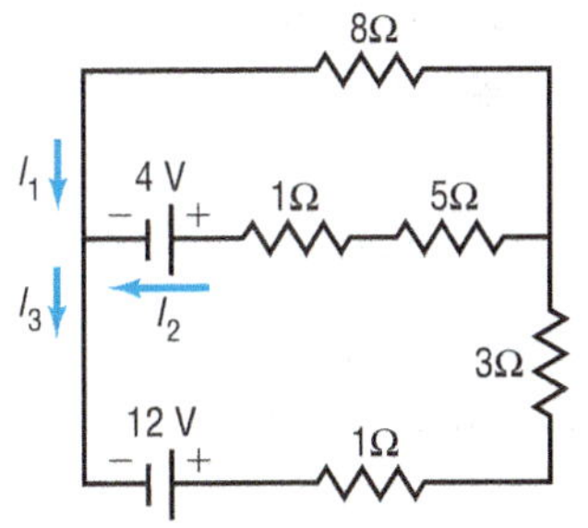

Source: *Physics for Scientists & Engineers,* 9th ed., by Serway. © 2013 Cengage Learning.

79. **Theater Revenues** A Broadway theater has 500 seats, divided into orchestra, main, and balcony seating. Orchestra seats sell for \$150, main seats for \$135, and balcony seats for \$110. If all the seats are sold, the gross revenue to the theater is \$64,250. If all the main and balcony seats are sold, but only half the orchestra seats are sold, the gross revenue is \$56,750. How many of each kind of seat are there?

80. **Theater Revenues** A movie theater charges \$11.00 for adults, \$6.50 for children, and \$9.00 for senior citizens. One day the theater sold 405 tickets and collected \$3315 in receipts. Twice as many children's tickets were sold as adult tickets. How many adults, children, and senior citizens went to the theater that day?

81. Nutrition A dietitian wishes a patient to have a meal that has 66 grams (g) of protein, 94.5 g of carbohydrates, and 910 milligrams (mg) of calcium. The hospital food service tells the dietitian that the dinner for today is chicken, corn, and 2% milk. Each serving of chicken has 30 g of protein, 35 g of carbohydrates, and 200 mg of calcium. Each serving of corn has 3 g of protein, 16 g of carbohydrates, and 10 mg of calcium. Each glass of 2% milk has 9 g of protein, 13 g of carbohydrates, and 300 mg of calcium. How many servings of each food should the dietitian provide for the patient?

82. Investments Kelly has $20,000 to invest. As her financial planner, you recommend that she diversify into three investments: Treasury bills that yield 5% simple interest, Treasury bonds that yield 7% simple interest, and corporate bonds that yield 10% simple interest. Kelly wishes to earn $1390 per year in income. Also, Kelly wants her investment in Treasury bills to be $3000 more than her investment in corporate bonds. How much money should Kelly place in each investment?

83. Prices of Fast Food One group of customers bought 8 deluxe hamburgers, 6 orders of large fries, and 6 large colas for $52.20. A second group ordered 10 deluxe hamburgers, 6 large fries, and 8 large colas and paid $63.20. Is there sufficient information to determine the price of each food item? If not, construct a table showing the various possibilities. Assume that the hamburgers cost between $3.50 and $4.50, the fries between $1.50 and $2.00, and the colas between $1.20 and $1.80.

84. Prices of Fast Food Use the information given in Problem 83. Suppose that a third group purchased 3 deluxe hamburgers, 2 large fries, and 4 large colas for $21.90. Now is there sufficient information to determine the price of each food item? If so, determine each price.

85. Painting a House Three painters (Beth, Bill, and Edie), working together, can paint the exterior of a home in 10 hours (h). Bill and Edie together have painted a similar house in 15 h. One day, all three worked on this same kind of house for 4 h, after which Edie left. Beth and Bill required 8 more hours to finish. Assuming no gain or loss in efficiency, how long should it take each person to complete such a job alone?

Explaining Concepts: Discussion and Writing

86. Make up a system of three linear equations containing three variables that has:
(a) No solution
(b) Exactly one solution
(c) Infinitely many solutions
Give the three systems to a friend to solve and critique.

87. Write a brief paragraph outlining your strategy for solving a system of two linear equations containing two variables.

88. Do you prefer the method of substitution or the method of elimination for solving a system of two linear equations containing two variables? Give your reasons.

Retain Your Knowledge

Problems 89–92 are based on material learned earlier in the course. The purpose of these problems is to keep the material fresh in your mind so that you are better prepared for the final exam.

89. Graph $f(x) = -3^{1-x} + 2$.

90. Factor each of the following:
(a) $4(2x-3)^3\cdot 2\cdot (x^3+5)^2 + 2(x^3+5)\cdot 3x^2\cdot (2x-3)^4$
(b) $\frac{1}{2}(3x-5)^{-\frac{1}{2}}\cdot 3\cdot (x+3)^{-\frac{1}{2}} - \frac{1}{2}(x+3)^{-\frac{3}{2}}(3x-5)^{\frac{1}{2}}$

91. Find the exact value of $\sin^{-1}\left[\sin\left(-\frac{10\pi}{9}\right)\right]$.

92. Write $-\sqrt{3}+i$ in polar form. Express each argument in degrees.

'Are You Prepared?' Answers

1. $\{1\}$ **2.** (a)

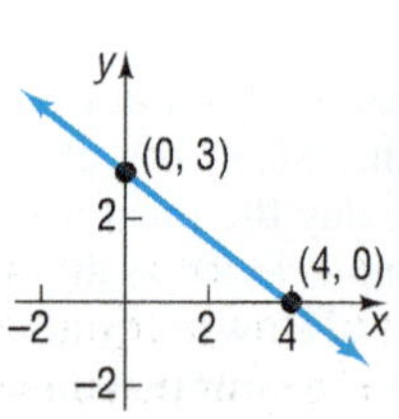

(b) $-\frac{3}{4}$

12.2 Systems of Linear Equations: Matrices

The systematic approach of the method of elimination for solving a system of linear equations provides another method of solution that involves a simplified notation.

Consider the following system of linear equations:

$$\begin{cases} x + 4y = 14 \\ 3x - 2y = 0 \end{cases}$$

If we choose not to write the symbols used for the variables, we can represent this system as

$$\left[\begin{array}{cc|c} 1 & 4 & 14 \\ 3 & -2 & 0 \end{array}\right]$$

where it is understood that the first column represents the coefficients of the variable x, the second column the coefficients of y, and the third column the constants on the right side of the equal signs. The vertical line serves as a reminder of the equal signs. The large square brackets are used to denote a *matrix* in algebra.

DEFINITION

A **matrix** is a rectangular array of numbers:

$$\begin{array}{c} \\ \text{Row 1} \\ \text{Row 2} \\ \vdots \\ \text{Row } i \\ \vdots \\ \text{Row } m \end{array} \begin{array}{c} \begin{array}{cccccc} \text{Column 1} & \text{Column 2} & & \text{Column } j & & \text{Column } n \end{array} \\ \begin{bmatrix} a_{11} & a_{12} & \cdots & a_{1j} & \cdots & a_{1n} \\ a_{21} & a_{22} & \cdots & a_{2j} & \cdots & a_{2n} \\ \vdots & \vdots & & \vdots & & \vdots \\ a_{i1} & a_{i2} & \cdots & a_{ij} & \cdots & a_{in} \\ \vdots & \vdots & & \vdots & & \vdots \\ a_{m1} & a_{m2} & \cdots & a_{mj} & \cdots & a_{mn} \end{bmatrix} \end{array} \quad (1)$$

Each number a_{ij} of the matrix has two indexes: the **row index** i and the **column index** j. The matrix shown in display (1) has m rows and n columns. The numbers a_{ij} are usually referred to as the **entries** of the matrix. For example, a_{23} refers to the entry in the second row, third column.

In Words
To augment means to increase or expand. An augmented matrix broadens the idea of matrices to systems of linear equations.

1 Write the Augmented Matrix of a System of Linear Equations

Now we will use matrix notation to represent a system of linear equations. The matrix used to represent a system of linear equations is called an **augmented matrix**. In writing the augmented matrix of a system, the variables of each equation must be on the left side of the equal sign and the constants on the right side. A variable that does not appear in an equation has a coefficient of 0.

EXAMPLE 1

Writing the Augmented Matrix of a System of Linear Equations

Write the augmented matrix of each system of equations.

(a) $\begin{cases} 3x - 4y = -6 & \text{(1)} \\ 2x - 3y = -5 & \text{(2)} \end{cases}$

(b) $\begin{cases} 2x - y + z = 0 & \text{(1)} \\ x + z - 1 = 0 & \text{(2)} \\ x + 2y - 8 = 0 & \text{(3)} \end{cases}$

Solution (a) The augmented matrix is

$$\left[\begin{array}{rr|r} 3 & -4 & -6 \\ 2 & -3 & -5 \end{array}\right]$$

(b) Care must be taken that the system be written so that the coefficients of all variables are present (if any variable is missing, its coefficient is 0). Also, all constants must be to the right of the equal sign. We need to rearrange the given system as follows:

$$\begin{cases} 2x - y + z = 0 & \text{(1)} \\ x + z - 1 = 0 & \text{(2)} \\ x + 2y - 8 = 0 & \text{(3)} \end{cases}$$

$$\begin{cases} 2x - y + z = 0 & \text{(1)} \\ x + 0 \cdot y + z = 1 & \text{(2)} \\ x + 2y + 0 \cdot z = 8 & \text{(3)} \end{cases}$$

The augmented matrix is

$$\left[\begin{array}{rrr|r} 2 & -1 & 1 & 0 \\ 1 & 0 & 1 & 1 \\ 1 & 2 & 0 & 8 \end{array}\right]$$

Caution Be sure variables and constants are lined up correctly before writing the augmented matrix.

If we do not include the constants to the right of the equal sign (that is, to the right of the vertical bar in the augmented matrix of a system of equations), the resulting matrix is called the **coefficient matrix** of the system. For the systems discussed in Example 1, the coefficient matrices are

$$\begin{bmatrix} 3 & -4 \\ 2 & -3 \end{bmatrix} \quad \text{and} \quad \begin{bmatrix} 2 & -1 & 1 \\ 1 & 0 & 1 \\ 1 & 2 & 0 \end{bmatrix}$$

Now Work PROBLEM 9

2 Write the System of Equations from the Augmented Matrix

EXAMPLE 2 **Writing the System of Linear Equations from the Augmented Matrix**

Write the system of linear equations that corresponds to each augmented matrix.

(a) $\left[\begin{array}{cc|c} 5 & 2 & 13 \\ -3 & 1 & -10 \end{array}\right]$ (b) $\left[\begin{array}{ccc|c} 3 & -1 & -1 & 7 \\ 2 & 0 & 2 & 8 \\ 0 & 1 & 1 & 0 \end{array}\right]$

Solution (a) The matrix has two rows and so represents a system of two equations. The two columns to the left of the vertical bar indicate that the system has two variables. If x and y are used to denote these variables, the system of equations is

$$\begin{cases} 5x + 2y = 13 & \textbf{(1)} \\ -3x + y = -10 & \textbf{(2)} \end{cases}$$

(b) Since the augmented matrix has three rows, it represents a system of three equations. Since there are three columns to the left of the vertical bar, the system contains three variables. If x, y, and z are the three variables, the system of equations is

$$\begin{cases} 3x - y - z = 7 & \textbf{(1)} \\ 2x + 2z = 8 & \textbf{(2)} \\ y + z = 0 & \textbf{(3)} \end{cases}$$

■

3 Perform Row Operations on a Matrix

Row operations on a matrix are used to solve systems of equations when the system is written as an augmented matrix. There are three basic row operations.

Row Operations

1. Interchange any two rows.
2. Replace a row by a nonzero multiple of that row.
3. Replace a row by the sum of that row and a constant nonzero multiple of some other row.

These three row operations correspond to the three rules given earlier for obtaining an equivalent system of equations. When a row operation is performed on a matrix, the resulting matrix represents a system of equations equivalent to the system represented by the original matrix.

For example, consider the augmented matrix

$$\left[\begin{array}{cc|c} 1 & 2 & 3 \\ 4 & -1 & 2 \end{array}\right]$$

Suppose that we want to apply a row operation to this matrix that results in a matrix whose entry in row 2, column 1 is a 0. The row operation to use is

Multiply each entry in row 1 by -4, and add the result to the corresponding entries in row 2. **(2)**

If we use R_2 to represent the new entries in row 2 and r_1 and r_2 to represent the original entries in rows 1 and 2, respectively, we can represent the row operation in statement (2) by

$$R_2 = -4r_1 + r_2$$

Then

$$\left[\begin{array}{rr|r} 1 & 2 & 3 \\ 4 & -1 & 2 \end{array}\right] \underset{R_2 = -4r_1 + r_2}{\rightarrow} \left[\begin{array}{cc|c} 1 & 2 & 3 \\ -4(1)+4 & -4(2)+(-1) & -4(3)+2 \end{array}\right] = \left[\begin{array}{rr|r} 1 & 2 & 3 \\ 0 & -9 & -10 \end{array}\right]$$

As desired, we now have the entry 0 in row 2, column 1.

EXAMPLE 3 Applying a Row Operation to an Augmented Matrix

Apply the row operation $R_2 = -3r_1 + r_2$ to the augmented matrix

$$\left[\begin{array}{rr|r} 1 & -2 & 2 \\ 3 & -5 & 9 \end{array}\right]$$

Solution The row operation $R_2 = -3r_1 + r_2$ says that the entries in row 2 are to be replaced by the entries obtained after multiplying each entry in row 1 by -3 and adding the result to the corresponding entries in row 2.

$$\left[\begin{array}{rr|r} 1 & -2 & 2 \\ 3 & -5 & 9 \end{array}\right] \underset{R_2 = -3r_1 + r_2}{\rightarrow} \left[\begin{array}{cc|c} 1 & -2 & 2 \\ -3(1)+3 & (-3)(-2)+(-5) & -3(2)+9 \end{array}\right] = \left[\begin{array}{rr|r} 1 & -2 & 2 \\ 0 & 1 & 3 \end{array}\right]$$

■

Now Work PROBLEM 19

EXAMPLE 4 Finding a Particular Row Operation

Find a row operation that results in the augmented matrix

$$\left[\begin{array}{rr|r} 1 & -2 & 2 \\ 0 & 1 & 3 \end{array}\right]$$

having a 0 in row 1, column 2.

Solution We want a 0 in row 1, column 2. Because there is a 1 in row 2, column 2, this result can be accomplished by multiplying row 2 by 2 and adding the result to row 1. That is, apply the row operation $R_1 = 2r_2 + r_1$.

$$\left[\begin{array}{rr|r} 1 & -2 & 2 \\ 0 & 1 & 3 \end{array}\right] \underset{R_1 = 2r_2 + r_1}{\rightarrow} \left[\begin{array}{cc|c} 2(0)+1 & 2(1)+(-2) & 2(3)+2 \\ 0 & 1 & 3 \end{array}\right] = \left[\begin{array}{rr|r} 1 & 0 & 8 \\ 0 & 1 & 3 \end{array}\right]$$

■

A word about the notation introduced here. A row operation such as $R_1 = 2r_2 + r_1$ changes the entries in row 1. Note also that for this type of row operation, we change the entries in a given row by multiplying the entries in some other row by an appropriate nonzero number and adding the results to the original entries of the row to be changed.

4 Solve a System of Linear Equations Using Matrices

To solve a system of linear equations using matrices, use row operations on the augmented matrix of the system to obtain a matrix that is in *row echelon form*.

DEFINITION

A matrix is in **row echelon form** when the following conditions are met:

1. The entry in row 1, column 1 is a 1, and only 0's appear below it.
2. The first nonzero entry in each row after the first row is a 1, only 0's appear below it, and the 1 appears to the right of the first nonzero entry in any row above.
3. Any rows that contain all 0's to the left of the vertical bar appear at the bottom.

For example, for a system of three equations containing three variables, x, y, and z, with a unique solution, the augmented matrix is in row echelon form if it is of the form

$$\left[\begin{array}{ccc|c} 1 & a & b & d \\ 0 & 1 & c & e \\ 0 & 0 & 1 & f \end{array}\right]$$

where a, b, c, d, e, and f are real numbers. The last row of this augmented matrix states that $z = f$. We can then determine the value of y using back-substitution with $z = f$, since row 2 represents the equation $y + cz = e$. Finally, x is determined using back-substitution again.

Two advantages of solving a system of equations by writing the augmented matrix in row echelon form are the following:

1. The process is algorithmic; that is, it consists of repetitive steps that can be programmed on a computer.
2. The process works on any system of linear equations, no matter how many equations or variables are present.

The next example shows how to solve a system of linear equations by writing its augmented matrix in row echelon form.

EXAMPLE 5 How to Solve a System of Linear Equations Using Matrices

Solve: $\begin{cases} 2x + 2y \phantom{{}+z} = 6 & (1) \\ x + y + z = 1 & (2) \\ 3x + 4y - z = 13 & (3) \end{cases}$

Step-by-Step Solution

Step 1: Write the augmented matrix that represents the system.

Write the augmented matrix of the system.

$$\left[\begin{array}{ccc|c} 2 & 2 & 0 & 6 \\ 1 & 1 & 1 & 1 \\ 3 & 4 & -1 & 13 \end{array}\right]$$

Step 2: Perform row operations that result in the entry in row 1, column 1 becoming 1.

To get a 1 in row 1, column 1, interchange rows 1 and 2. [Note that this is equivalent to interchanging equations (1) and (2) of the system.]

$$\left[\begin{array}{ccc|c} 1 & 1 & 1 & 1 \\ 2 & 2 & 0 & 6 \\ 3 & 4 & -1 & 13 \end{array}\right]$$

Step 3: Perform row operations that leave the entry in row 1, column 1 a 1, while causing the entries in column 1 below row 1 to become 0's.

Next, we want a 0 in row 2, column 1 and a 0 in row 3, column 1. Use the row operations $R_2 = -2r_1 + r_2$ and $R_3 = -3r_1 + r_3$ to accomplish this. Note that row 1 is unchanged using these row operations. Also, do you see that performing these row operations simultaneously is the same as doing one followed by the other?

$$\left[\begin{array}{ccc|c} 1 & 1 & 1 & 1 \\ 2 & 2 & 0 & 6 \\ 3 & 4 & -1 & 13 \end{array}\right] \rightarrow \left[\begin{array}{ccc|c} 1 & 1 & 1 & 1 \\ 0 & 0 & -2 & 4 \\ 0 & 1 & -4 & 10 \end{array}\right]$$

$R_2 = -2r_1 + r_2$
$R_3 = -3r_1 + r_3$

Step 4: Perform row operations that result in the entry in row 2, column 2 becoming 1 with 0's below it.

We want the entry in row 2, column 2 to be 1. We also want to have a 0 below the 1 in row 2, column 2. Interchanging rows 2 and 3 accomplishes both goals.

$$\left[\begin{array}{ccc|c} 1 & 1 & 1 & 1 \\ 0 & 0 & -2 & 4 \\ 0 & 1 & -4 & 10 \end{array}\right] \rightarrow \left[\begin{array}{ccc|c} 1 & 1 & 1 & 1 \\ 0 & 1 & -4 & 10 \\ 0 & 0 & -2 & 4 \end{array}\right]$$

Step 5: Repeat Step 4, placing a 1 in row 3, column 3.

To obtain a 1 in row 3, column 3, use the row operation $R_3 = -\frac{1}{2}r_3$. The result is

$$\left[\begin{array}{ccc|c} 1 & 1 & 1 & 1 \\ 0 & 1 & -4 & 10 \\ 0 & 0 & -2 & 4 \end{array}\right] \underset{R_3 = -\frac{1}{2}r_3}{\rightarrow} \left[\begin{array}{ccc|c} 1 & 1 & 1 & 1 \\ 0 & 1 & -4 & 10 \\ 0 & 0 & 1 & -2 \end{array}\right]$$

Step 6: The matrix on the right in Step 5 is the row echelon form of the augmented matrix. Use back-substitution to solve the original system.

The third row of the augmented matrix represents the equation $z = -2$. Using $z = -2$, back-substitute into the equation $y - 4z = 10$ (from the second row) and obtain

$$\begin{aligned} y - 4z &= 10 \\ y - 4(-2) &= 10 \quad z = -2 \\ y &= 2 \quad \text{Solve for } y. \end{aligned}$$

Finally, back-substitute $y = 2$ and $z = -2$ into the equation $x + y + z = 1$ (from the first row) and obtain

$$\begin{aligned} x + y + z &= 1 \\ x + 2 + (-2) &= 1 \quad y = 2, z = -2 \\ x &= 1 \quad \text{Solve for } x. \end{aligned}$$

The solution of the system is $x = 1, y = 2, z = -2$ or, using an ordered triplet, $(1, 2, -2)$. ■

In Words

To obtain an augmented matrix in row echelon form:

- Add rows, exchange rows, or multiply a row by a nonzero constant.
- Work from top to bottom and left to right.
- Get 1's in the main diagonal with 0's below the 1's.
- Once the entry in row 1, column 1 is 1 with 0's below it, do not use row 1 in your row operations.
- Once the entries in row 1, column 1 and row 2, column 2 are 1 with 0's below, do not use row 1 or 2 in your row operations (and so on).

Matrix Method for Solving a System of Linear Equations (Row Echelon Form)

STEP 1: Write the augmented matrix that represents the system.
STEP 2: Perform row operations that place the entry 1 in row 1, column 1.
STEP 3: Perform row operations that leave the entry 1 in row 1, column 1 unchanged, while causing 0's to appear below it in column 1.
STEP 4: Perform row operations that place the entry 1 in row 2, column 2, but leave the entries in columns to the left unchanged. If it is impossible to place a 1 in row 2, column 2, proceed to place a 1 in row 2, column 3. Once a 1 is in place, perform row operations to place 0's below it. (Place any rows that contain only 0's on the left side of the vertical bar, at the bottom of the matrix.)
STEP 5: Now repeat Step 4, placing a 1 in the next row, but one column to the right. Continue until the bottom row or the vertical bar is reached.
STEP 6: The matrix that results is the row echelon form of the augmented matrix. Analyze the system of equations corresponding to it to solve the original system.

EXAMPLE 6 **Solving a System of Linear Equations Using Matrices (Row Echelon Form)**

$$\text{Solve: } \begin{cases} x - y + z = 8 & (1) \\ 2x + 3y - z = -2 & (2) \\ 3x - 2y - 9z = 9 & (3) \end{cases}$$

Algebraic Solution

STEP 1: The augmented matrix of the system is

$$\left[\begin{array}{rrr|r} 1 & -1 & 1 & 8 \\ 2 & 3 & -1 & -2 \\ 3 & -2 & -9 & 9 \end{array}\right]$$

STEP 2: Because the entry 1 is already present in row 1, column 1, go to Step 3.

STEP 3: Perform the row operations $R_2 = -2r_1 + r_2$ and $R_3 = -3r_1 + r_3$. Each of these leaves the entry 1 in row 1, column 1 unchanged, while causing 0's to appear under it.

$$\left[\begin{array}{rrr|r} 1 & -1 & 1 & 8 \\ 2 & 3 & -1 & -2 \\ 3 & -2 & -9 & 9 \end{array}\right] \rightarrow \left[\begin{array}{rrr|r} 1 & -1 & 1 & 8 \\ 0 & 5 & -3 & -18 \\ 0 & 1 & -12 & -15 \end{array}\right]$$

$R_2 = -2r_1 + r_2$
$R_3 = -3r_1 + r_3$

STEP 4: The easiest way to obtain the entry 1 in row 2, column 2 without altering column 1 is to interchange rows 2 and 3 (another way would be to multiply row 2 by $\frac{1}{5}$, but this introduces fractions).

$$\left[\begin{array}{rrr|r} 1 & -1 & 1 & 8 \\ 0 & 1 & -12 & -15 \\ 0 & 5 & -3 & -18 \end{array}\right]$$

To get a 0 under the 1 in row 2, column 2, perform the row operation $R_3 = -5r_2 + r_3$.

$$\left[\begin{array}{rrr|r} 1 & -1 & 1 & 8 \\ 0 & 1 & -12 & -15 \\ 0 & 5 & -3 & -18 \end{array}\right] \rightarrow \left[\begin{array}{rrr|r} 1 & -1 & 1 & 8 \\ 0 & 1 & -12 & -15 \\ 0 & 0 & 57 & 57 \end{array}\right]$$

$R_3 = -5r_2 + r_3$

STEP 5: Continuing, obtain a 1 in row 3, column 3 by using $R_3 = \frac{1}{57}r_3$.

$$\left[\begin{array}{rrr|r} 1 & -1 & 1 & 8 \\ 0 & 1 & -12 & -15 \\ 0 & 0 & 57 & 57 \end{array}\right] \rightarrow \left[\begin{array}{rrr|r} 1 & -1 & 1 & 8 \\ 0 & 1 & -12 & -15 \\ 0 & 0 & 1 & 1 \end{array}\right]$$

$R_3 = \frac{1}{57}r_3$

STEP 6: The matrix on the right is the row echelon form of the augmented matrix. The system of equations represented by the matrix in row echelon form is

$$\begin{cases} x - y + z = 8 & (1) \\ y - 12z = -15 & (2) \\ z = 1 & (3) \end{cases}$$

Using $z = 1$, back-substitute to get

$$\begin{cases} x - y + 1 = 8 & (1) \\ y - 12(1) = -15 & (2) \end{cases} \xrightarrow{\text{Simplify.}} \begin{cases} x - y = 7 & (1) \\ y = -3 & (2) \end{cases}$$

Graphing Solution

The augmented matrix of the system is

$$\left[\begin{array}{rrr|r} 1 & -1 & 1 & 8 \\ 2 & 3 & -1 & -2 \\ 3 & -2 & -9 & 9 \end{array}\right]$$

Enter this matrix into a graphing utility and name it A. See Figure 7(a). Using the REF (Row Echelon Form) command on matrix A gives the results shown in Figure 7(b).

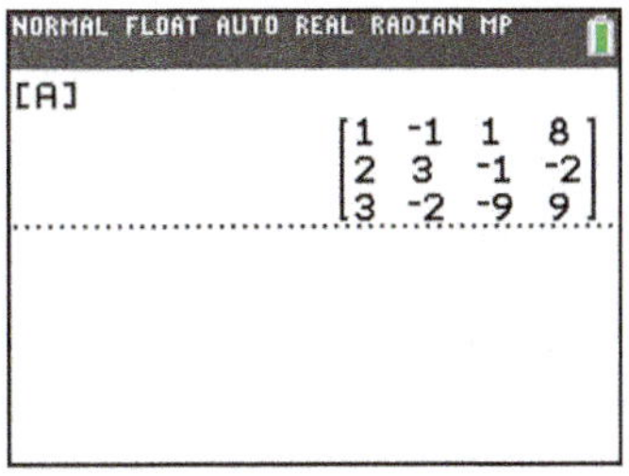

(a)

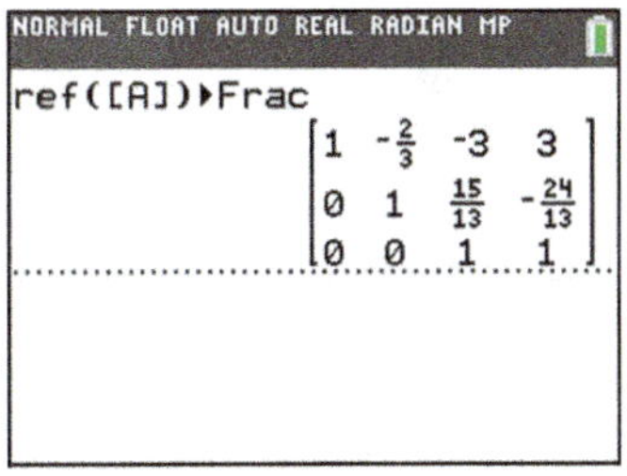

(b)

Figure 7 Row echelon form

The system of equations represented by the matrix in row echelon form is

$$\begin{cases} x - \frac{2}{3}y - 3z = 3 & (1) \\ y + \frac{15}{13}z = -\frac{24}{13} & (2) \\ z = 1 & (3) \end{cases}$$

Using $z = 1$, back-substitute to get

$$\begin{cases} x - \frac{2}{3}y - 3(1) = 3 & (1) \\ y + \frac{15}{13}(1) = -\frac{24}{13} & (2) \end{cases}$$

$$\begin{cases} x - \frac{2}{3}y = 6 & (1) \\ y = -\frac{39}{13} = -3 & (2) \end{cases}$$

Using $y = -3$ from the second equation, back-substitute into $x - \frac{2}{3}y = 6$ to get $x = 4$. The solution of the system is $x = 4, y = -3, z = 1$ or, using an ordered triplet, $(4, -3, 1)$. ■

Using $y = -3$, back-substitute into $x - y = 7$ to get $x = 4$. The solution of the system is $x = 4, y = -3, z = 1$ or, using an ordered triplet, $(4, -3, 1)$. ■

Notice that the row echelon form of the augmented matrix in the graphing solution differs from the row echelon form in the algebraic solution, yet both matrices provide the same solution! This is because the two solutions used different row operations to obtain the row echelon form. In all likelihood, the two solutions parted ways in Step 4 of the algebraic solution, where we avoided introducing fractions by interchanging rows 2 and 3.

Sometimes it is advantageous to write a matrix in **reduced row echelon form.** In this form, row operations are used to obtain entries that are 0 above (as well as below) the leading 1 in a row. For example, the row echelon form obtained in the algebraic solution to Example 6 is

$$\left[\begin{array}{rrr|r} 1 & -1 & 1 & 8 \\ 0 & 1 & -12 & -15 \\ 0 & 0 & 1 & 1 \end{array}\right]$$

To write this matrix in reduced row echelon form, proceed as follows:

$$\left[\begin{array}{rrr|r} 1 & -1 & 1 & 8 \\ 0 & 1 & -12 & -15 \\ 0 & 0 & 1 & 1 \end{array}\right] \underset{R_1 = r_2 + r_1}{\rightarrow} \left[\begin{array}{rrr|r} 1 & 0 & -11 & -7 \\ 0 & 1 & -12 & -15 \\ 0 & 0 & 1 & 1 \end{array}\right] \underset{\substack{R_1 = 11r_3 + r_1 \\ R_2 = 12r_3 + r_2}}{\rightarrow} \left[\begin{array}{rrr|r} 1 & 0 & 0 & 4 \\ 0 & 1 & 0 & -3 \\ 0 & 0 & 1 & 1 \end{array}\right]$$

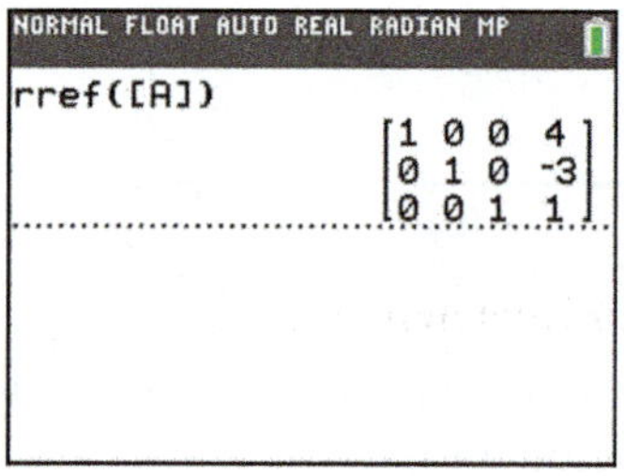

Figure 8 Reduced row echelon form

The matrix is now written in reduced row echelon form. The advantage of writing the matrix in this form is that the solution to the system, $x = 4$, $y = -3$, $z = 1$, is readily found, without the need to back-substitute. Another advantage will be seen in Section 12.4, where the inverse of a matrix is discussed. The method used to write a matrix in reduced row echelon form is called **Gauss–Jordan elimination.**

Most graphing utilities also have the ability to put a matrix in reduced row echelon form. Figure 8 shows the reduced row echelon form of the augmented matrix from Example 6 using the RREF command on a TI-84 Plus C graphing calculator.

Now Work PROBLEMS 39 AND 49

The matrix method for solving a system of linear equations also identifies systems that have infinitely many solutions and systems that are inconsistent.

EXAMPLE 7 Solving a Dependent System of Linear Equations Using Matrices

Solve: $\begin{cases} 6x - y - z = 4 & (1) \\ -12x + 2y + 2z = -8 & (2) \\ 5x + y - z = 3 & (3) \end{cases}$

Solution Start with the augmented matrix of the system and proceed to obtain a 1 in row 1, column 1 with 0's below.

$$\left[\begin{array}{rrr|r} 6 & -1 & -1 & 4 \\ -12 & 2 & 2 & -8 \\ 5 & 1 & -1 & 3 \end{array}\right] \underset{R_1 = -1r_3 + r_1}{\rightarrow} \left[\begin{array}{rrr|r} 1 & -2 & 0 & 1 \\ -12 & 2 & 2 & -8 \\ 5 & 1 & -1 & 3 \end{array}\right] \underset{\substack{R_2 = 12r_1 + r_2 \\ R_3 = -5r_1 + r_3}}{\rightarrow} \left[\begin{array}{rrr|r} 1 & -2 & 0 & 1 \\ 0 & -22 & 2 & 4 \\ 0 & 11 & -1 & -2 \end{array}\right]$$

Obtaining a 1 in row 2, column 2 without altering column 1 can be accomplished by $R_2 = -\frac{1}{22}r_2$, by $R_3 = \frac{1}{11}r_3$ and interchanging rows 2 and 3, or by $R_2 = \frac{23}{11}r_3 + r_2$. We shall use the first of these.

$$\left[\begin{array}{rrr|r} 1 & -2 & 0 & 1 \\ 0 & -22 & 2 & 4 \\ 0 & 11 & -1 & -2 \end{array}\right] \underset{R_2 = -\frac{1}{22}r_2}{\rightarrow} \left[\begin{array}{rrr|r} 1 & -2 & 0 & 1 \\ 0 & 1 & -\frac{1}{11} & -\frac{2}{11} \\ 0 & 11 & -1 & -2 \end{array}\right] \underset{R_3 = -11r_2 + r_3}{\rightarrow} \left[\begin{array}{rrr|r} 1 & -2 & 0 & 1 \\ 0 & 1 & -\frac{1}{11} & -\frac{2}{11} \\ 0 & 0 & 0 & 0 \end{array}\right]$$

This matrix is in row echelon form. Because the bottom row consists entirely of 0's, the system actually consists of only two equations.

$$\begin{cases} x - 2y = 1 & \textbf{(1)} \\ y - \frac{1}{11}z = -\frac{2}{11} & \textbf{(2)} \end{cases}$$

To make it easier to write down some of the solutions, we express both x and y in terms of z.

From the second equation, $y = \frac{1}{11}z - \frac{2}{11}$. Now back-substitute this solution for y into the first equation to get

$$x = 2y + 1 = 2\left(\frac{1}{11}z - \frac{2}{11}\right) + 1 = \frac{2}{11}z + \frac{7}{11}$$

The original system is equivalent to the system

$$\begin{cases} x = \frac{2}{11}z + \frac{7}{11} & \textbf{(1)} \\ y = \frac{1}{11}z - \frac{2}{11} & \textbf{(2)} \end{cases} \quad \text{where } z \text{ can be any real number.}$$

Let's look at the situation. The original system of three equations is equivalent to a system containing two equations. This means that any values of x, y, z that satisfy both

$$x = \frac{2}{11}z + \frac{7}{11} \quad \text{and} \quad y = \frac{1}{11}z - \frac{2}{11}$$

are solutions. For example, $z = 0, x = \frac{7}{11}, y = -\frac{2}{11}$; $z = 1, x = \frac{9}{11}, y = -\frac{1}{11}$; and $z = -1, x = \frac{5}{11}, y = -\frac{3}{11}$ are some of the solutions of the original system. There are, in fact, infinitely many values of x, y, and z for which the two equations are satisfied. That is, the original system has infinitely many solutions. We write the solution of the original system as

$$\begin{cases} x = \frac{2}{11}z + \frac{7}{11} \\ y = \frac{1}{11}z - \frac{2}{11} \end{cases} \quad \text{where } z \text{ can be any real number}$$

or, using ordered triplets, as

$$\left\{(x, y, z) \,\middle|\, x = \frac{2}{11}z + \frac{7}{11}, y = \frac{1}{11}z - \frac{2}{11}, z \text{ any real number}\right\}$$

■

The solution can also be found by writing the augmented matrix in reduced row echelon form. Begin with the row echelon form.

$$\left[\begin{array}{rrr|r} 1 & -2 & 0 & 1 \\ 0 & 1 & -\frac{1}{11} & -\frac{2}{11} \\ 0 & 0 & 0 & 0 \end{array}\right] \underset{\uparrow\; R_1 = 2r_2 + r_1}{\rightarrow} \left[\begin{array}{rrr|r} 1 & 0 & -\frac{2}{11} & \frac{7}{11} \\ 0 & 1 & -\frac{1}{11} & -\frac{2}{11} \\ 0 & 0 & 0 & 0 \end{array}\right]$$

The matrix on the right is in reduced row echelon form. The corresponding system of equations is

$$\begin{cases} x - \dfrac{2}{11}z = \dfrac{7}{11} & (1) \\ y - \dfrac{1}{11}z = -\dfrac{2}{11} & (2) \end{cases} \quad \text{where } z \text{ can be any real number}$$

or, equivalently,

$$\begin{cases} x = \dfrac{2}{11}z + \dfrac{7}{11} \\ y = \dfrac{1}{11}z - \dfrac{2}{11} \end{cases} \quad \text{where } z \text{ can be any real number}$$

Now Work PROBLEM 55

EXAMPLE 8 Solving an Inconsistent System of Linear Equations Using Matrices

Solve: $\begin{cases} x + y + z = 6 \\ 2x - y - z = 3 \\ x + 2y + 2z = 0 \end{cases}$

Solution Begin with the augmented matrix.

$$\left[\begin{array}{rrr|r} 1 & 1 & 1 & 6 \\ 2 & -1 & -1 & 3 \\ 1 & 2 & 2 & 0 \end{array}\right] \underset{\substack{\uparrow\\ R_2 = -2r_1 + r_2 \\ R_3 = -1r_1 + r_3}}{\rightarrow} \left[\begin{array}{rrr|r} 1 & 1 & 1 & 6 \\ 0 & -3 & -3 & -9 \\ 0 & 1 & 1 & -6 \end{array}\right] \underset{\substack{\uparrow\\ \text{Interchange rows 2 and 3.}}}{\rightarrow} \left[\begin{array}{rrr|r} 1 & 1 & 1 & 6 \\ 0 & 1 & 1 & -6 \\ 0 & -3 & -3 & -9 \end{array}\right] \underset{\substack{\uparrow\\ R_3 = 3r_2 + r_3}}{\rightarrow} \left[\begin{array}{rrr|r} 1 & 1 & 1 & 6 \\ 0 & 1 & 1 & -6 \\ 0 & 0 & 0 & -27 \end{array}\right]$$

This matrix is in row echelon form. The bottom row is equivalent to the equation

$$0x + 0y + 0z = -27$$

which has no solution. The original system is inconsistent. ■

Now Work PROBLEM 29

The matrix method is especially effective for systems of equations for which the number of equations and the number of variables are unequal. Here, too, such a system is either inconsistent or consistent. If it is consistent, it has either exactly one solution or infinitely many solutions.

EXAMPLE 9 Solving a System of Linear Equations Using Matrices

Solve: $\begin{cases} x - 2y + z = 0 & (1) \\ 2x + 2y - 3z = -3 & (2) \\ y - z = -1 & (3) \\ -x + 4y + 2z = 13 & (4) \end{cases}$

Solution Begin with the augmented matrix.

$$\left[\begin{array}{rrr|r} 1 & -2 & 1 & 0 \\ 2 & 2 & -3 & -3 \\ 0 & 1 & -1 & -1 \\ -1 & 4 & 2 & 13 \end{array}\right] \underset{\substack{R_2 = -2r_1 + r_2 \\ R_4 = r_1 + r_4}}{\rightarrow} \left[\begin{array}{rrr|r} 1 & -2 & 1 & 0 \\ 0 & 6 & -5 & -3 \\ 0 & 1 & -1 & -1 \\ 0 & 2 & 3 & 13 \end{array}\right] \underset{\text{Interchange rows 2 and 3.}}{\rightarrow} \left[\begin{array}{rrr|r} 1 & -2 & 1 & 0 \\ 0 & 1 & -1 & -1 \\ 0 & 6 & -5 & -3 \\ 0 & 2 & 3 & 13 \end{array}\right]$$

$$\underset{\substack{R_3 = -6r_2 + r_3 \\ R_4 = -2r_2 + r_4}}{\rightarrow} \left[\begin{array}{rrr|r} 1 & -2 & 1 & 0 \\ 0 & 1 & -1 & -1 \\ 0 & 0 & 1 & 3 \\ 0 & 0 & 5 & 15 \end{array}\right] \underset{R_4 = -5r_3 + r_4}{\rightarrow} \left[\begin{array}{rrr|r} 1 & -2 & 1 & 0 \\ 0 & 1 & -1 & -1 \\ 0 & 0 & 1 & 3 \\ 0 & 0 & 0 & 0 \end{array}\right]$$

We could stop here, since the matrix is in row echelon form, and back-substitute $z = 3$ to find x and y. Or we can continue and obtain the reduced row echelon form.

$$\left[\begin{array}{rrr|r} 1 & -2 & 1 & 0 \\ 0 & 1 & -1 & -1 \\ 0 & 0 & 1 & 3 \\ 0 & 0 & 0 & 0 \end{array}\right] \underset{R_1 = 2r_2 + r_1}{\rightarrow} \left[\begin{array}{rrr|r} 1 & 0 & -1 & -2 \\ 0 & 1 & -1 & -1 \\ 0 & 0 & 1 & 3 \\ 0 & 0 & 0 & 0 \end{array}\right] \underset{\substack{R_1 = r_3 + r_1 \\ R_2 = r_3 + r_2}}{\rightarrow} \left[\begin{array}{rrr|r} 1 & 0 & 0 & 1 \\ 0 & 1 & 0 & 2 \\ 0 & 0 & 1 & 3 \\ 0 & 0 & 0 & 0 \end{array}\right]$$

The matrix is now in reduced row echelon form, and we can see that the solution is $x = 1, y = 2, z = 3$ or, using an ordered triplet, $(1, 2, 3)$. ■

Now Work PROBLEM 71

EXAMPLE 10 Financial Planning

Adam and Michelle require an additional \$25,000 in annual income (beyond their pension benefits). They are rather risk averse and have narrowed their investment choices down to Treasury notes that yield 3%, Treasury bonds that yield 5%, and corporate bonds that yield 6%. They have \$600,000 to invest and want the amount invested in Treasury notes to equal the total amount invested in Treasury bonds and corporate bonds. How much should they place in each investment?

Solution Let n, b, and c represent the amounts invested in Treasury notes, Treasury bonds, and corporate bonds, respectively. There is a total of \$600,000 to invest, which means that the sum of the amounts invested in Treasury notes, Treasury bonds, and corporate bonds should equal \$600,000. The first equation is

$$n + b + c = 600{,}000 \quad (1)$$

If \$100,000 were invested in Treasury notes, the income would be 0.03 (\$100,000) = \$3000. In general, if n dollars were invested in Treasury notes, the income would be $0.03n$. Since the total income is to be \$25,000, the second equation is

$$0.03n + 0.05b + 0.06c = 25{,}000 \quad (2)$$

The amount invested in Treasury notes should equal the amount invested in Treasury bonds and corporate bonds, so the third equation is

$$n = b + c \quad \text{or} \quad n - b - c = 0 \quad (3)$$

We have the following system of equations:

$$\begin{cases} n + b + c = 600{,}000 & (1) \\ 0.03n + 0.05b + 0.06c = 25{,}000 & (2) \\ n - b - c = 0 & (3) \end{cases}$$

Begin with the augmented matrix and proceed as follows:

$$\left[\begin{array}{ccc|c} 1 & 1 & 1 & 600{,}000 \\ 0.03 & 0.05 & 0.06 & 25{,}000 \\ 1 & -1 & -1 & 0 \end{array}\right] \rightarrow \left[\begin{array}{ccc|c} 1 & 1 & 1 & 600{,}000 \\ 0 & 0.02 & 0.03 & 7000 \\ 0 & -2 & -2 & -600{,}000 \end{array}\right]$$

$R_2 = -0.03r_1 + r_2$
$R_3 = -r_1 + r_3$

$$\rightarrow \left[\begin{array}{ccc|c} 1 & 1 & 1 & 600{,}000 \\ 0 & 1 & 1.5 & 350{,}000 \\ 0 & -2 & -2 & -600{,}000 \end{array}\right] \rightarrow \left[\begin{array}{ccc|c} 1 & 1 & 1 & 600{,}000 \\ 0 & 1 & 1.5 & 350{,}000 \\ 0 & 0 & 1 & 100{,}000 \end{array}\right]$$

$R_2 = \dfrac{1}{0.02}r_2$ $\qquad$ $R_3 = 2r_2 + r_3$

The matrix is now in row echelon form. The final matrix represents the system

$$\begin{cases} n + b + c = 600{,}000 & (1) \\ b + 1.5c = 350{,}000 & (2) \\ c = 100{,}000 & (3) \end{cases}$$

From equation (3), note that Adam and Michelle should invest \$100,000 in corporate bonds. Back-substitute \$100,000 into equation (2) to find that $b = 200{,}000$, so Adam and Michelle should invest \$200,000 in Treasury bonds. Back-substitute these values into equation (1) and find that $n = 300{,}000$, so \$300,000 should be invested in Treasury notes. ■

12.2 Assess Your Understanding

Concepts and Vocabulary

1. An m by n rectangular array of numbers is called a(n) ________.

2. The matrix used to represent a system of linear equations is called a(n) ____________ matrix.

3. The notation a_{35} refers to the entry in the ________ row and ________ column of a matrix.

4. ***True or False*** The matrix $\left[\begin{array}{cc|c} 1 & 3 & -2 \\ 0 & 1 & 5 \\ 0 & 0 & 0 \end{array}\right]$ is in row echelon form.

5. Which of the following matrices is in reduced row echelon form?

(a) $\left[\begin{array}{cc|c} 1 & 2 & 9 \\ 3 & -1 & -1 \end{array}\right]$ (b) $\left[\begin{array}{cc|c} 1 & 0 & 1 \\ 0 & 1 & 4 \end{array}\right]$

(c) $\left[\begin{array}{cc|c} 1 & 2 & 9 \\ 0 & 0 & 28 \end{array}\right]$ (d) $\left[\begin{array}{cc|c} 1 & 2 & 9 \\ 0 & 1 & 4 \end{array}\right]$

6. Which of the following statements accurately describes the system represented by the matrix $\left[\begin{array}{ccc|c} 1 & 5 & -2 & 3 \\ 0 & 1 & 3 & -2 \\ 0 & 0 & 0 & 5 \end{array}\right]$?

(a) The system has one solution.
(b) The system has infinitely many solutions.
(c) The system has no solution.
(d) The number of solutions cannot be determined.

Skill Building

In Problems 7–18, write the augmented matrix of the given system of equations.

7. $\begin{cases} x - 5y = 5 \\ 4x + 3y = 6 \end{cases}$

8. $\begin{cases} 3x + 4y = 7 \\ 4x - 2y = 5 \end{cases}$

9. $\begin{cases} 2x + 3y - 6 = 0 \\ 4x - 6y + 2 = 0 \end{cases}$

10. $\begin{cases} 9x - y = 0 \\ 3x - y - 4 = 0 \end{cases}$

11. $\begin{cases} 0.01x - 0.03y = 0.06 \\ 0.13x + 0.10y = 0.20 \end{cases}$

12. $\begin{cases} \frac{4}{3}x - \frac{3}{2}y = \frac{3}{4} \\ -\frac{1}{4}x + \frac{1}{3}y = \frac{2}{3} \end{cases}$

13. $\begin{cases} x - y + z = 10 \\ 3x + 3y = 5 \\ x + y + 2z = 2 \end{cases}$

14. $\begin{cases} 5x - y - z = 0 \\ x + y = 5 \\ 2x - 3z = 2 \end{cases}$

15. $\begin{cases} x + y - z = 2 \\ 3x - 2y = 2 \\ 5x + 3y - z = 1 \end{cases}$

16. $\begin{cases} 2x + 3y - 4z = 0 \\ x - 5z + 2 = 0 \\ x + 2y - 3z = -2 \end{cases}$

17. $\begin{cases} x - y - z = 10 \\ 2x + y + 2z = -1 \\ -3x + 4y = 5 \\ 4x - 5y + z = 0 \end{cases}$

18. $\begin{cases} x - y + 2z - w = 5 \\ x + 3y - 4z + 2w = 2 \\ 3x - y - 5z - w = -1 \end{cases}$

In Problems 19–26, write the system of equations corresponding to each augmented matrix. Then perform the indicated row operation(s) on the given augmented matrix.

19. $\left[\begin{array}{cc|c} 1 & -3 & -2 \\ 2 & -5 & 5 \end{array}\right]$ $R_2 = -2r_1 + r_2$

20. $\left[\begin{array}{cc|c} 1 & -3 & -3 \\ 2 & -5 & -4 \end{array}\right]$ $R_2 = -2r_1 + r_2$

21. $\left[\begin{array}{ccc|c} 1 & -3 & 4 & 3 \\ 3 & -5 & 6 & 6 \\ -5 & 3 & 4 & 6 \end{array}\right]$ $R_2 = -3r_1 + r_2$, $R_3 = 5r_1 + r_3$

22. $\left[\begin{array}{ccc|c} 1 & -3 & 3 & -5 \\ -4 & -5 & -3 & -5 \\ -3 & -2 & 4 & 6 \end{array}\right]$ $R_2 = 4r_1 + r_2$, $R_3 = 3r_1 + r_3$

23. $\left[\begin{array}{ccc|c} 1 & -3 & 2 & -6 \\ 2 & -5 & 3 & -4 \\ -3 & -6 & 4 & 6 \end{array}\right]$ $R_2 = -2r_1 + r_2$, $R_3 = 3r_1 + r_3$

24. $\left[\begin{array}{ccc|c} 1 & -3 & -4 & -6 \\ 6 & -5 & 6 & -6 \\ -1 & 1 & 4 & 6 \end{array}\right]$ $R_2 = -6r_1 + r_2$, $R_3 = r_1 + r_3$

25. $\left[\begin{array}{ccc|c} 5 & -3 & 1 & -2 \\ 2 & -5 & 6 & -2 \\ -4 & 1 & 4 & 6 \end{array}\right]$ $R_1 = -2r_2 + r_1$, $R_3 = 2r_2 + r_3$

26. $\left[\begin{array}{ccc|c} 4 & -3 & -1 & 2 \\ 3 & -5 & 2 & 6 \\ -3 & -6 & 4 & 6 \end{array}\right]$ $R_1 = -r_2 + r_1$, $R_3 = r_2 + r_3$

In Problems 27–38, the reduced row echelon form of a system of linear equations is given. Write the system of equations corresponding to the given matrix. Use x, y; or x, y, z; or x_1, x_2, x_3, x_4 as variables. Determine whether the system is consistent or inconsistent. If it is consistent, give the solution.

27. $\left[\begin{array}{cc|c} 1 & 0 & 5 \\ 0 & 1 & -1 \end{array}\right]$

28. $\left[\begin{array}{cc|c} 1 & 0 & -4 \\ 0 & 1 & 0 \end{array}\right]$

29. $\left[\begin{array}{ccc|c} 1 & 0 & 0 & 1 \\ 0 & 1 & 0 & 2 \\ 0 & 0 & 0 & 3 \end{array}\right]$

30. $\left[\begin{array}{ccc|c} 1 & 0 & 0 & 0 \\ 0 & 1 & 0 & 0 \\ 0 & 0 & 0 & 2 \end{array}\right]$

31. $\left[\begin{array}{ccc|c} 1 & 0 & 2 & -1 \\ 0 & 1 & -4 & -2 \\ 0 & 0 & 0 & 0 \end{array}\right]$

32. $\left[\begin{array}{ccc|c} 1 & 0 & 4 & 4 \\ 0 & 1 & 3 & 2 \\ 0 & 0 & 0 & 0 \end{array}\right]$

33. $\left[\begin{array}{cccc|c} 1 & 0 & 0 & 0 & 1 \\ 0 & 1 & 0 & 1 & 2 \\ 0 & 0 & 1 & 2 & 3 \end{array}\right]$

34. $\left[\begin{array}{cccc|c} 1 & 0 & 0 & 0 & 1 \\ 0 & 1 & 0 & 2 & 2 \\ 0 & 0 & 1 & 3 & 0 \end{array}\right]$

35. $\left[\begin{array}{cccc|c} 1 & 0 & 0 & 4 & 2 \\ 0 & 1 & 1 & 3 & 3 \\ 0 & 0 & 0 & 0 & 0 \end{array}\right]$

36. $\left[\begin{array}{cccc|c} 1 & 0 & 0 & 0 & 1 \\ 0 & 1 & 0 & 0 & 2 \\ 0 & 0 & 1 & 2 & 3 \end{array}\right]$

37. $\left[\begin{array}{cccc|c} 1 & 0 & 0 & 1 & -2 \\ 0 & 1 & 0 & 2 & 2 \\ 0 & 0 & 1 & -1 & 0 \\ 0 & 0 & 0 & 0 & 0 \end{array}\right]$

38. $\left[\begin{array}{cccc|c} 1 & 0 & 0 & 0 & 1 \\ 0 & 1 & 0 & 0 & 2 \\ 0 & 0 & 1 & 0 & 3 \\ 0 & 0 & 0 & 1 & 0 \end{array}\right]$

In Problems 39–74, solve each system of equations using matrices (row operations). If the system has no solution, say that it is inconsistent.

39. $\begin{cases} x + y = 8 \\ x - y = 4 \end{cases}$

40. $\begin{cases} x + 2y = 5 \\ x + y = 3 \end{cases}$

41. $\begin{cases} 2x - 4y = -2 \\ 3x + 2y = 3 \end{cases}$

42. $\begin{cases} 3x + 3y = 3 \\ 4x + 2y = \dfrac{8}{3} \end{cases}$

43. $\begin{cases} x + 2y = 4 \\ 2x + 4y = 8 \end{cases}$

44. $\begin{cases} 3x - y = 7 \\ 9x - 3y = 21 \end{cases}$

45. $\begin{cases} 2x + 3y = 6 \\ x - y = \dfrac{1}{2} \end{cases}$

46. $\begin{cases} \dfrac{1}{2}x + y = -2 \\ x - 2y = 8 \end{cases}$

47. $\begin{cases} 3x - 5y = 3 \\ 15x + 5y = 21 \end{cases}$

48. $\begin{cases} 2x - y = -1 \\ x + \dfrac{1}{2}y = \dfrac{3}{2} \end{cases}$

49. $\begin{cases} x - y = 6 \\ 2x - 3z = 16 \\ 2y + z = 4 \end{cases}$

50. $\begin{cases} 2x + y = -4 \\ -2y + 4z = 0 \\ 3x - 2z = -11 \end{cases}$

51. $\begin{cases} x - 2y + 3z = 7 \\ 2x + y + z = 4 \\ -3x + 2y - 2z = -10 \end{cases}$

52. $\begin{cases} 2x + y - 3z = 0 \\ -2x + 2y + z = -7 \\ 3x - 4y - 3z = 7 \end{cases}$

53. $\begin{cases} 2x - 2y - 2z = 2 \\ 2x + 3y + z = 2 \\ 3x + 2y = 0 \end{cases}$

54. $\begin{cases} 2x - 3y - z = 0 \\ -x + 2y + z = 5 \\ 3x - 4y - z = 1 \end{cases}$

55. $\begin{cases} -x + y + z = -1 \\ -x + 2y - 3z = -4 \\ 3x - 2y - 7z = 0 \end{cases}$

56. $\begin{cases} 2x - 3y - z = 0 \\ 3x + 2y + 2z = 2 \\ x + 5y + 3z = 2 \end{cases}$

57. $\begin{cases} 2x - 2y + 3z = 6 \\ 4x - 3y + 2z = 0 \\ -2x + 3y - 7z = 1 \end{cases}$

58. $\begin{cases} 3x - 2y + 2z = 6 \\ 7x - 3y + 2z = -1 \\ 2x - 3y + 4z = 0 \end{cases}$

59. $\begin{cases} x + y - z = 6 \\ 3x - 2y + z = -5 \\ x + 3y - 2z = 14 \end{cases}$

60. $\begin{cases} x - y + z = -4 \\ 2x - 3y + 4z = -15 \\ 5x + y - 2z = 12 \end{cases}$

61. $\begin{cases} x + 2y - z = -3 \\ 2x - 4y + z = -7 \\ -2x + 2y - 3z = 4 \end{cases}$

62. $\begin{cases} x + 4y - 3z = -8 \\ 3x - y + 3z = 12 \\ x + y + 6z = 1 \end{cases}$

63. $\begin{cases} 3x + y - z = \dfrac{2}{3} \\ 2x - y + z = 1 \\ 4x + 2y = \dfrac{8}{3} \end{cases}$

64. $\begin{cases} x + y = 1 \\ 2x - y + z = 1 \\ x + 2y + z = \dfrac{8}{3} \end{cases}$

65. $\begin{cases} x + y + z + w = 4 \\ 2x - y + z = 0 \\ 3x + 2y + z - w = 6 \\ x - 2y - 2z + 2w = -1 \end{cases}$

66. $\begin{cases} x + y + z + w = 4 \\ -x + 2y + z = 0 \\ 2x + 3y + z - w = 6 \\ -2x + y - 2z + 2w = -1 \end{cases}$

67. $\begin{cases} x + 2y + z = 1 \\ 2x - y + 2z = 2 \\ 3x + y + 3z = 3 \end{cases}$

68. $\begin{cases} x + 2y - z = 3 \\ 2x - y + 2z = 6 \\ x - 3y + 3z = 4 \end{cases}$

69. $\begin{cases} x - y + z = 5 \\ 3x + 2y - 2z = 0 \end{cases}$

70. $\begin{cases} 2x + y - z = 4 \\ -x + y + 3z = 1 \end{cases}$

71. $\begin{cases} 2x + 3y - z = 3 \\ x - y - z = 0 \\ -x + y + z = 0 \\ x + y + 3z = 5 \end{cases}$

72. $\begin{cases} x - 3y + z = 1 \\ 2x - y - 4z = 0 \\ x - 3y + 2z = 1 \\ x - 2y = 5 \end{cases}$

73. $\begin{cases} 4x + y + z - w = 4 \\ x - y + 2z + 3w = 3 \end{cases}$

74. $\begin{cases} -4x + y = 5 \\ 2x - y + z - w = 5 \\ z + w = 4 \end{cases}$

Applications and Extensions

75. Curve Fitting Find the function $y = ax^2 + bx + c$ whose graph contains the points $(1, 2)$, $(-2, -7)$, and $(2, -3)$.

76. Curve Fitting Find the function $y = ax^2 + bx + c$ whose graph contains the points $(1, -1)$, $(3, -1)$, and $(-2, 14)$.

77. Curve Fitting Find the function $f(x) = ax^3 + bx^2 + cx + d$ for which $f(-3) = -112$, $f(-1) = -2$, $f(1) = 4$, and $f(2) = 13$.

78. Curve Fitting Find the function $f(x) = ax^3 + bx^2 + cx + d$ for which $f(-2) = -10$, $f(-1) = 3$, $f(1) = 5$, and $f(3) = 15$.

79. Nutrition A dietitian at Palos Community Hospital wants a patient to have a meal that has 78 grams (g) of protein, 59 g of carbohydrates, and 75 milligrams (mg) of vitamin A. The hospital food service tells the dietitian that the dinner for today is salmon steak, baked eggs, and acorn squash. Each serving of salmon steak has 30 g of protein, 20 g of carbohydrates, and 2 mg of vitamin A. Each serving of baked eggs contains 15 g of protein, 2 g of carbohydrates, and 20 mg of vitamin A. Each serving of acorn squash contains 3 g of protein, 25 g of carbohydrates, and 32 mg of vitamin A. How many servings of each food should the dietitian provide for the patient?

80. Nutrition A dietitian at General Hospital wants a patient to have a meal that has 47 grams (g) of protein, 58 g of carbohydrates, and 630 milligrams (mg) of calcium. The hospital food service tells the dietitian that the dinner for today is pork chops, corn on the cob, and 2% milk. Each serving of pork chops has 23 g of protein, 0 g of carbohydrates, and 10 mg of calcium. Each serving of corn on the cob contains 3 g of protein, 16 g of carbohydrates, and 10 mg of calcium. Each glass of 2% milk contains 9 g of protein, 13 g of carbohydrates, and 300 mg of calcium. How many servings of each food should the dietitian provide for the patient?

81. Financial Planning Carletta has \$10,000 to invest. As her financial consultant, you recommend that she invest in Treasury bills that yield 6%, Treasury bonds that yield 7%, and corporate bonds that yield 8%. Carletta wants to have an annual income of \$680, and the amount invested in corporate bonds must be half that invested in Treasury bills. Find the amount in each investment.

82. Landscaping A landscape company is hired to plant trees in three new subdivisions. The company charges the developer for each tree planted, an hourly rate to plant the trees, and a fixed delivery charge. In one subdivision it took 166 labor hours to plant 250 trees for a cost of \$7520. In a second subdivision it took 124 labor hours to plant 200 trees for a cost of \$5945. In the final subdivision it took 200 labor hours to plant 300 trees for a cost of \$8985. Determine the cost for each tree, the hourly labor charge, and the fixed delivery charge.
Source: *www.bx.org*

83. Production To manufacture an automobile requires painting, drying, and polishing. Epsilon Motor Company produces three types of cars: the Delta, the Beta, and the Sigma. Each Delta requires 10 hours (hr) for painting, 3 hr for drying, and 2 hr for polishing. A Beta requires 16 hr for painting, 5 hr for drying, and 3 hr for polishing, and a Sigma requires 8 hr for painting, 2 hr for drying, and 1 hr for polishing. If the company has 240 hr for painting, 69 hr for drying, and 41 hr for polishing per month, how many of each type of car are produced?

84. Production A Florida juice company completes the preparation of its products by sterilizing, filling, and labeling bottles. Each case of orange juice requires 9 minutes (min) for sterilizing, 6 min for filling, and 1 min for labeling. Each case of grapefruit juice requires 10 min for sterilizing, 4 min for filling, and 2 min for labeling. Each case of tomato juice requires 12 min for sterilizing, 4 min for filling, and 1 min for labeling. If the company runs the sterilizing machine for 398 min, the filling machine for 164 min, and the labeling machine for 58 min, how many cases of each type of juice are prepared?

85. Electricity: Kirchhoff's Rules An application of Kirchhoff's Rules to the circuit shown results in the following system of equations:

$$\begin{cases} -4 + 8 - 2I_2 = 0 \\ 8 = 5I_4 + I_1 \\ 4 = 3I_3 + I_1 \\ I_3 + I_4 = I_1 \end{cases}$$

Find the currents I_1, I_2, I_3, and I_4.

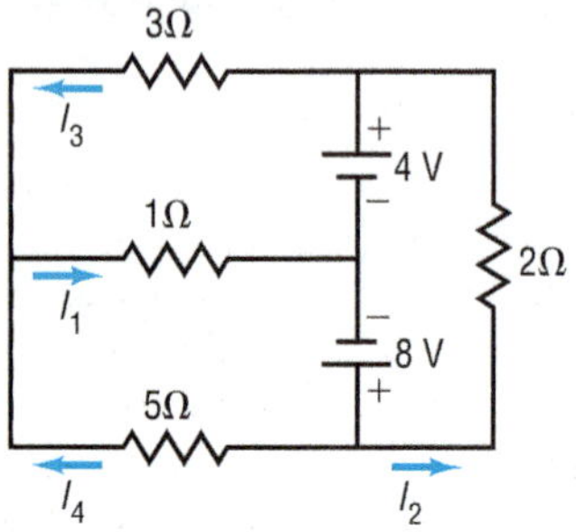

Source: *Based on Raymond Serway.* Physics, *3rd ed. (Philadelphia: Saunders, 1990), Prob. 34, p. 790.*

86. Electricity: Kirchhoff's Rules An application of Kirchhoff's Rules to the circuit shown results in the following system of equations:

$$\begin{cases} I_1 = I_3 + I_2 \\ 24 - 6I_1 - 3I_3 = 0 \\ 12 + 24 - 6I_1 - 6I_2 = 0 \end{cases}$$

Find the currents I_1, I_2, and I_3.

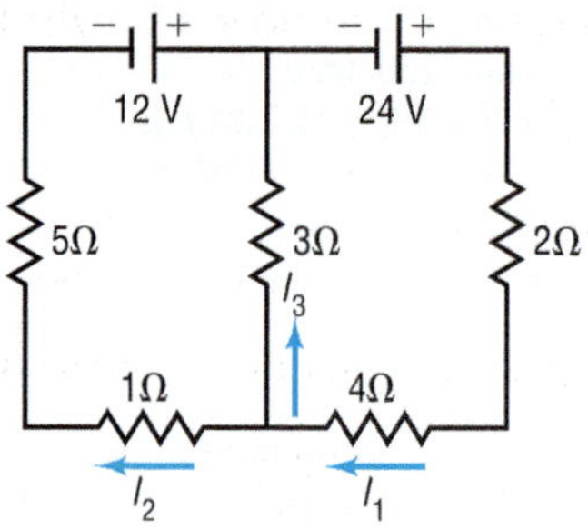

Source: Ibid., Prob. 38, p. 791.

87. Financial Planning Three retired couples each require an additional annual income of $2000 per year. As their financial consultant, you recommend that they invest some money in Treasury bills that yield 7%, some money in corporate bonds that yield 9%, and some money in "junk bonds" that yield 11%. Prepare a table for each couple showing the various ways that their goals can be achieved:
(a) If the first couple has $20,000 to invest.
(b) If the second couple has $25,000 to invest.
(c) If the third couple has $30,000 to invest.
(d) What advice would you give each couple regarding the amount to invest and the choices available?

[**Hint:** Higher yields generally carry more risk.]

88. Financial Planning A young couple has $25,000 to invest. As their financial consultant, you recommend that they invest some money in Treasury bills that yield 7%, some money in corporate bonds that yield 9%, and some money in junk bonds that yield 11%. Prepare a table showing the various ways that this couple can achieve the following goals:
(a) $1500 per year in income
(b) $2000 per year in income
(c) $2500 per year in income
(d) What advice would you give this couple regarding the income that they require and the choices available?

[**Hint:** Higher yields generally carry more risk.]

89. Pharmacy A doctor's prescription calls for a daily intake of a supplement containing 40 milligrams (mg) of vitamin C and 30 mg of vitamin D. Your pharmacy stocks three supplements that can be used: one contains 20% vitamin C and 30% vitamin D; a second, 40% vitamin C and 20% vitamin D; and a third, 30% vitamin C and 50% vitamin D. Create a table showing the possible combinations that could be used to fill the prescription.

90. Pharmacy A doctor's prescription calls for the creation of pills that contain 12 units of vitamin B_{12} and 12 units of vitamin E. Your pharmacy stocks three powders that can be used to make these pills: one contains 20% vitamin B_{12} and 30% vitamin E; a second, 40% vitamin B_{12} and 20% vitamin E; and a third, 30% vitamin B_{12} and 40% vitamin E. Create a table showing the possible combinations of these powders that could be mixed in each pill. Hint: 10 units of the first powder contains $10(0.2) = 2$ units of vitamin B_{12}.

Explaining Concepts: Discussion and Writing

91. Write a brief paragraph or two outlining your strategy for solving a system of linear equations using matrices.

92. When solving a system of linear equations using matrices, do you prefer to place the augmented matrix in row echelon form or in reduced row echelon form? Give reasons for your choice.

93. Make up a system of three linear equations containing three variables that has:
(a) No solution
(b) Exactly one solution
(c) Infinitely many solutions

Give the three systems to a friend to solve and critique.

Retain Your Knowledge

Problems 94–97 are based on material learned earlier in the course. The purpose of these problems is to keep the material fresh in your mind so that you are better prepared for the final exam.

94. Solve: $x^2 - 3x < 6 + 2x$

95. Graph: $f(x) = \dfrac{2x^2 - x - 1}{x^2 + 2x + 1}$

96. State the domain of $f(x) = -e^{x+5} - 6$.

97. Use a calculator to approximate $\cos^{-1}(-0.75)$ in radians, rounded to two decimal places.

12.3 Systems of Linear Equations: Determinants

OBJECTIVES
1. Evaluate 2 by 2 Determinants (p. 893)
2. Use Cramer's Rule to Solve a System of Two Equations Containing Two Variables (p. 894)
3. Evaluate 3 by 3 Determinants (p. 896)
4. Use Cramer's Rule to Solve a System of Three Equations Containing Three Variables (p. 898)
5. Know Properties of Determinants (p. 899)

The preceding section described a method of using matrices to solve a system of linear equations. This section deals with yet another method for solving systems of linear equations; however, it can be used only when the number of equations equals the number of variables. Although the method works for any system (provided that the number of equations equals the number of variables), it is most often used for systems of two equations containing two variables or three equations containing three variables. This method, called *Cramer's Rule,* is based on the concept of a *determinant.*

1 Evaluate 2 by 2 Determinants

DEFINITION

If $a, b, c,$ and d are four real numbers, the symbol

$$D = \begin{vmatrix} a & b \\ c & d \end{vmatrix}$$

is called a **2 by 2 determinant**. Its value is the number $ad - bc$; that is,

$$D = \begin{vmatrix} a & b \\ c & d \end{vmatrix} = ad - bc \qquad \textbf{(1)}$$

The following device may be helpful for remembering the value of a 2 by 2 determinant:

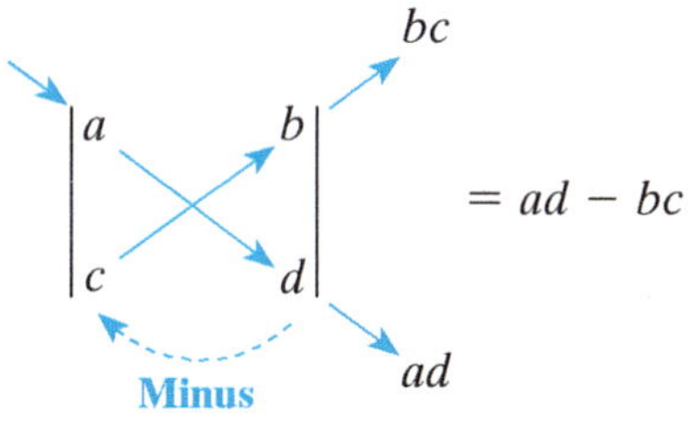

EXAMPLE 1 **Evaluating a 2 by 2 Determinant**

Evaluate: $\begin{vmatrix} 3 & -2 \\ 6 & 1 \end{vmatrix}$

Algebraic Solution

$$\begin{vmatrix} 3 & -2 \\ 6 & 1 \end{vmatrix} = (3)(1) - (-2)(6)$$
$$= 3 - (-12)$$
$$= 15$$

Graphing Solution

First, enter the matrix whose entries are those of the determinant into the graphing utility and name it A. Using the determinant command, obtain the result shown in Figure 9.

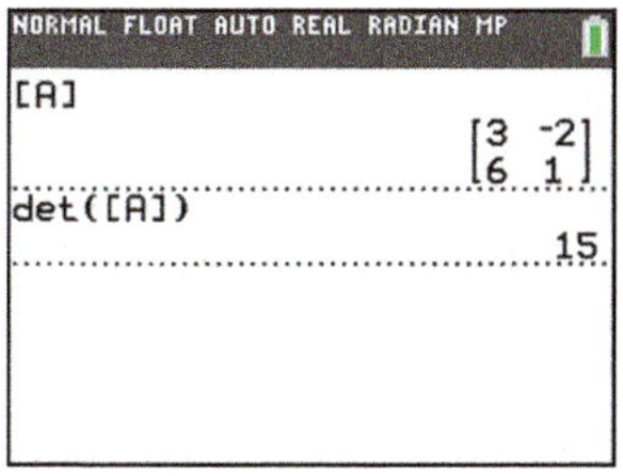

Figure 9

Now Work PROBLEM 7

2 Use Cramer's Rule to Solve a System of Two Equations Containing Two Variables

Let's see the role that a 2 by 2 determinant plays in the solution of a system of two equations containing two variables. Consider the system

$$\begin{cases} ax + by = s & \text{(1)} \\ cx + dy = t & \text{(2)} \end{cases} \qquad (2)$$

We use the method of elimination to solve this system.

Provided that $d \neq 0$ and $b \neq 0$, this system is equivalent to the system

$$\begin{cases} adx + bdy = sd & \text{(1) Multiply by } d. \\ bcx + bdy = tb & \text{(2) Multiply by } b. \end{cases}$$

Subtract the second equation from the first equation and obtain

$$\begin{cases} (ad - bc)x + 0 \cdot y = sd - tb & \text{(1)} \\ bcx + bdy = tb & \text{(2)} \end{cases}$$

Now the first equation can be rewritten using determinant notation.

$$\begin{vmatrix} a & b \\ c & d \end{vmatrix} x = \begin{vmatrix} s & b \\ t & d \end{vmatrix}$$

If $D = \begin{vmatrix} a & b \\ c & d \end{vmatrix} = ad - bc \neq 0$, solve for x to get

$$x = \frac{\begin{vmatrix} s & b \\ t & d \end{vmatrix}}{\begin{vmatrix} a & b \\ c & d \end{vmatrix}} = \frac{\begin{vmatrix} s & b \\ t & d \end{vmatrix}}{D} \qquad (3)$$

Return now to the original system (2). Provided that $a \neq 0$ and $c \neq 0$, the system is equivalent to

$$\begin{cases} acx + bcy = cs & \text{(1) Multiply by } c. \\ acx + ady = at & \text{(2) Multiply by } a. \end{cases}$$

Subtract the first equation from the second equation and obtain

$$\begin{cases} acx + bcy = cs & \text{(1)} \\ 0 \cdot x + (ad - bc)y = at - cs & \text{(2)} \end{cases}$$

The second equation can now be rewritten using determinant notation.

$$\begin{vmatrix} a & b \\ c & d \end{vmatrix} y = \begin{vmatrix} a & s \\ c & t \end{vmatrix}$$

If $D = \begin{vmatrix} a & b \\ c & d \end{vmatrix} = ad - bc \neq 0$, solve for y to get

$$y = \frac{\begin{vmatrix} a & s \\ c & t \end{vmatrix}}{\begin{vmatrix} a & b \\ c & d \end{vmatrix}} = \frac{\begin{vmatrix} a & s \\ c & t \end{vmatrix}}{D} \qquad (4)$$

Equations (3) and (4) lead to the following result, called **Cramer's Rule**.

THEOREM

Cramer's Rule for Two Equations Containing Two Variables

The solution to the system of equations

$$\begin{cases} ax + by = s & (1) \\ cx + dy = t & (2) \end{cases} \qquad (5)$$

is given by

$$x = \frac{\begin{vmatrix} s & b \\ t & d \end{vmatrix}}{\begin{vmatrix} a & b \\ c & d \end{vmatrix}} \qquad y = \frac{\begin{vmatrix} a & s \\ c & t \end{vmatrix}}{\begin{vmatrix} a & b \\ c & d \end{vmatrix}} \qquad (6)$$

provided that

$$D = \begin{vmatrix} a & b \\ c & d \end{vmatrix} = ad - bc \neq 0$$

In the derivation given for Cramer's Rule, we assumed that none of the numbers a, b, c, and d was 0. In Problem 65 you will be asked to complete the proof under the less stringent condition that $D = ad - bc \neq 0$.

Now look carefully at the pattern in Cramer's Rule. The denominator in the solution (6) is the determinant of the coefficients of the variables.

$$\begin{cases} ax + by = s \\ cx + dy = t \end{cases} \qquad D = \begin{vmatrix} a & b \\ c & d \end{vmatrix}$$

In the solution for x, the numerator is the determinant, denoted by D_x, formed by replacing the entries in the first column (the coefficients of x) of D by the constants on the right side of the equal sign.

$$D_x = \begin{vmatrix} s & b \\ t & d \end{vmatrix}$$

In the solution for y, the numerator is the determinant, denoted by D_y, formed by replacing the entries in the second column (the coefficients of y) of D by the constants on the right side of the equal sign.

$$D_y = \begin{vmatrix} a & s \\ c & t \end{vmatrix}$$

Cramer's Rule then states that if $D \neq 0$,

$$x = \frac{D_x}{D} \qquad y = \frac{D_y}{D} \qquad (7)$$

EXAMPLE 2 **Solving a System of Linear Equations Using Determinants**

Use Cramer's Rule, if applicable, to solve the system

$$\begin{cases} 3x - 2y = 4 & (1) \\ 6x + y = 13 & (2) \end{cases}$$

Algebraic Solution

The determinant D of the coefficients of the variables is

$$D = \begin{vmatrix} 3 & -2 \\ 6 & 1 \end{vmatrix} = (3)(1) - (-2)(6) = 15$$

Graphing Solution

Enter the coefficient matrix into the graphing utility. Call it A and evaluate det$[A]$. Since det$[A] \neq 0$, Cramer's Rule can be used. Enter the matrices D_x and D_y into the graphing utility and call them B and C, respectively. Finally, find x by

Because $D \neq 0$, Cramer's Rule (7) can be used.

$$x = \frac{D_x}{D} = \frac{\begin{vmatrix} 4 & -2 \\ 13 & 1 \end{vmatrix}}{15} = \frac{(4)(1) - (-2)(13)}{15} = \frac{30}{15} = 2 \qquad y = \frac{D_y}{D} = \frac{\begin{vmatrix} 3 & 4 \\ 6 & 13 \end{vmatrix}}{15} = \frac{(3)(13) - (4)(6)}{15} = \frac{15}{15} = 1$$

calculating $\frac{\det[B]}{\det[A]}$, and find y by calculating $\frac{\det[C]}{\det[A]}$. The results are shown in Figure 10.

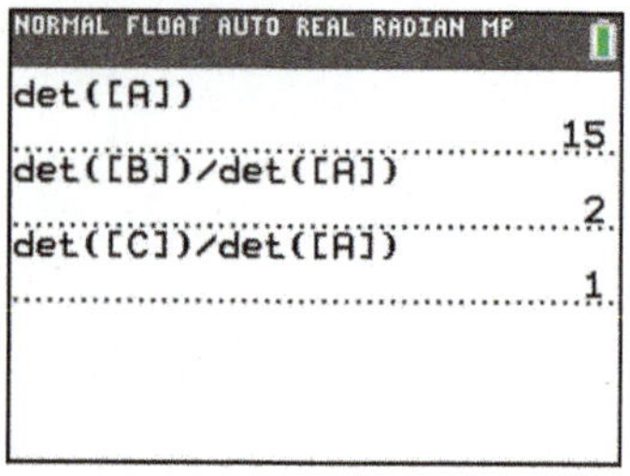

Figure 10

The solution is $x = 2$, $y = 1$ or, using an ordered pair, $(2, 1)$.

In attempting to use Cramer's Rule, if the determinant D of the coefficients of the variables is found to equal 0 (so that Cramer's Rule is not applicable), then the system either is consistent with dependent equations or is inconsistent. To determine whether the system has no solution or infinitely many solutions, solve the system using the methods of Section 12.1 or 12.2.

Now Work PROBLEM 15

3 Evaluate 3 by 3 Determinants

To use Cramer's Rule to solve a system of three equations containing three variables, we need to define a 3 by 3 determinant.

A **3 by 3 determinant** is symbolized by

$$\begin{vmatrix} a_{11} & a_{12} & a_{13} \\ a_{21} & a_{22} & a_{23} \\ a_{31} & a_{32} & a_{33} \end{vmatrix} \quad \textbf{(8)}$$

in which $a_{11}, a_{12}, \ldots,$ are real numbers.

As with matrices, we use a double subscript to identify an entry by indicating its row and column numbers. For example, the entry a_{23} is in row 2, column 3.

The value of a 3 by 3 determinant may be defined in terms of 2 by 2 determinants by the following formula:

$$\begin{vmatrix} a_{11} & a_{12} & a_{13} \\ a_{21} & a_{22} & a_{23} \\ a_{31} & a_{32} & a_{33} \end{vmatrix} = a_{11}\begin{vmatrix} a_{22} & a_{23} \\ a_{32} & a_{33} \end{vmatrix} - a_{12}\begin{vmatrix} a_{21} & a_{23} \\ a_{31} & a_{33} \end{vmatrix} + a_{13}\begin{vmatrix} a_{21} & a_{22} \\ a_{31} & a_{32} \end{vmatrix} \quad \textbf{(9)}$$

Minus (above $- a_{12}$) **Plus** (above $+ a_{13}$)

2 by 2 determinant left after removing the row and column containing a_{11} | **2 by 2 determinant left after removing the row and column containing a_{12}** | **2 by 2 determinant left after removing the row and column containing a_{13}**

In Words
The graphic below should help you visualize the minor of a_{11} in a 3 by 3 determinant:

$$\begin{vmatrix} a_{11} & a_{12} & a_{13} \\ a_{21} & a_{22} & a_{23} \\ a_{31} & a_{32} & a_{33} \end{vmatrix}$$

The 2 by 2 determinants shown in formula (9) are called **minors** of the 3 by 3 determinant. For an n by n determinant, the **minor** M_{ij} of entry a_{ij} is the determinant that results from removing the ith row and the jth column.

EXAMPLE 3 **Finding Minors of a 3 by 3 Determinant**

For the determinant $A = \begin{vmatrix} 2 & -1 & 3 \\ -2 & 5 & 1 \\ 0 & 6 & -9 \end{vmatrix}$, find: (a) M_{12} (b) M_{23}

Solution (a) M_{12} is the determinant that results from removing the first row and the second column from A.

$$A = \begin{vmatrix} 2 & -1 & 3 \\ -2 & 5 & 1 \\ 0 & 6 & -9 \end{vmatrix} \qquad M_{12} = \begin{vmatrix} -2 & 1 \\ 0 & -9 \end{vmatrix} = (-2)(-9) - (1)(0) = 18$$

(b) M_{23} is the determinant that results from removing the second row and the third column from A.

$$A = \begin{vmatrix} 2 & -1 & 3 \\ -2 & 5 & 1 \\ 0 & 6 & -9 \end{vmatrix} \qquad M_{23} = \begin{vmatrix} 2 & -1 \\ 0 & 6 \end{vmatrix} = (2)(6) - (-1)(0) = 12$$

Referring to formula (9), note that each element a_{ij} in the first row of the determinant is multiplied by its minor, but this term is sometimes added and other times subtracted. To determine whether to add or subtract a term, consider the *cofactor*.

DEFINITION

For an n by n determinant A, the **cofactor** of entry a_{ij}, denoted by A_{ij}, is given by

$$A_{ij} = (-1)^{i+j} M_{ij}$$

where M_{ij} is the minor of entry a_{ij}.

The exponent of $(-1)^{i+j}$ is the sum of the row and column of the entry a_{ij}, so if $i + j$ is even, $(-1)^{i+j}$ equals 1, and if $i + j$ is odd, $(-1)^{i+j}$ equals -1.

To find the value of a determinant, multiply each entry in any row or column by its cofactor and sum the results. This process is referred to as **expanding across a row or column**. For example, the value of the 3 by 3 determinant in formula (9) was found by expanding across row 1.

Expanding down column 2 gives

$$\begin{vmatrix} a_{11} & a_{12} & a_{13} \\ a_{21} & a_{22} & a_{23} \\ a_{31} & a_{32} & a_{33} \end{vmatrix} = (-1)^{1+2} a_{12} \begin{vmatrix} a_{21} & a_{23} \\ a_{31} & a_{33} \end{vmatrix} + (-1)^{2+2} a_{22} \begin{vmatrix} a_{11} & a_{13} \\ a_{31} & a_{33} \end{vmatrix} + (-1)^{3+2} a_{32} \begin{vmatrix} a_{11} & a_{13} \\ a_{21} & a_{23} \end{vmatrix}$$

Expand down column 2.

Expanding across row 3 gives

$$\begin{vmatrix} a_{11} & a_{12} & a_{13} \\ a_{21} & a_{22} & a_{23} \\ a_{31} & a_{32} & a_{33} \end{vmatrix} = (-1)^{3+1} a_{31} \begin{vmatrix} a_{12} & a_{13} \\ a_{22} & a_{23} \end{vmatrix} + (-1)^{3+2} a_{32} \begin{vmatrix} a_{11} & a_{13} \\ a_{21} & a_{23} \end{vmatrix} + (-1)^{3+3} a_{33} \begin{vmatrix} a_{11} & a_{12} \\ a_{21} & a_{22} \end{vmatrix}$$

Expand across row 3.

It can be shown that the value of a determinant does not depend on the choice of the row or column used in the expansion. However, expanding across a row or column that has an entry equal to 0 reduces the amount of work needed to compute the value of the determinant.

EXAMPLE 4 Evaluating a 3 by 3 Determinant

Find the value of the 3 by 3 determinant: $\begin{vmatrix} 3 & 0 & -1 \\ 4 & 6 & 2 \\ 8 & -2 & 3 \end{vmatrix}$

Solution Because of the 0 in row 1, column 2, it is easiest to expand across row 1 or down column 2. We choose to expand across row 1.

$$\begin{vmatrix} 3 & 0 & -1 \\ 4 & 6 & 2 \\ 8 & -2 & 3 \end{vmatrix} = (-1)^{1+1} \cdot 3 \cdot \begin{vmatrix} 6 & 2 \\ -2 & 3 \end{vmatrix} + (-1)^{1+2} \cdot 0 \cdot \begin{vmatrix} 4 & 2 \\ 8 & 3 \end{vmatrix} + (-1)^{1+3} \cdot (-1) \cdot \begin{vmatrix} 4 & 6 \\ 8 & -2 \end{vmatrix}$$

$$= 3(18 - (-4)) - 0 + (-1)(-8 - 48)$$

$$= 3(22) + (-1)(-56)$$

$$= 66 + 56 = 122$$

Now Work PROBLEM 11

4 Use Cramer's Rule to Solve a System of Three Equations Containing Three Variables

Consider the following system of three equations containing three variables.

$$\begin{cases} a_{11}x + a_{12}y + a_{13}z = c_1 \\ a_{21}x + a_{22}y + a_{23}z = c_2 \\ a_{31}x + a_{32}y + a_{33}z = c_3 \end{cases} \tag{10}$$

It can be shown that if the determinant D of the coefficients of the variables is not 0, that is, if

$$D = \begin{vmatrix} a_{11} & a_{12} & a_{13} \\ a_{21} & a_{22} & a_{23} \\ a_{31} & a_{32} & a_{33} \end{vmatrix} \neq 0$$

then the unique solution of system (10) is given by

THEOREM

Cramer's Rule for Three Equations Containing Three Variables

$$x = \frac{D_x}{D} \qquad y = \frac{D_y}{D} \qquad z = \frac{D_z}{D}$$

where

$$D_x = \begin{vmatrix} c_1 & a_{12} & a_{13} \\ c_2 & a_{22} & a_{23} \\ c_3 & a_{32} & a_{33} \end{vmatrix} \qquad D_y = \begin{vmatrix} a_{11} & c_1 & a_{13} \\ a_{21} & c_2 & a_{23} \\ a_{31} & c_3 & a_{33} \end{vmatrix} \qquad D_z = \begin{vmatrix} a_{11} & a_{12} & c_1 \\ a_{21} & a_{22} & c_2 \\ a_{31} & a_{32} & c_3 \end{vmatrix}$$

Do you see the similarity between this pattern and the pattern observed earlier for a system of two equations containing two variables?

EXAMPLE 5 **Using Cramer's Rule**

Use Cramer's Rule, if applicable, to solve the following system:

$$\begin{cases} 2x + y - z = 3 & (1) \\ -x + 2y + 4z = -3 & (2) \\ x - 2y - 3z = 4 & (3) \end{cases}$$

Solution The value of the determinant D of the coefficients of the variables is

$$D = \begin{vmatrix} 2 & 1 & -1 \\ -1 & 2 & 4 \\ 1 & -2 & -3 \end{vmatrix} = (-1)^{1+1} \cdot 2 \cdot \begin{vmatrix} 2 & 4 \\ -2 & -3 \end{vmatrix} + (-1)^{1+2} \cdot 1 \cdot \begin{vmatrix} -1 & 4 \\ 1 & -3 \end{vmatrix} + (-1)^{1+3}(-1) \begin{vmatrix} -1 & 2 \\ 1 & -2 \end{vmatrix}$$

$$= 2(2) - 1(-1) + (-1)(0)$$

$$= 4 + 1 = 5$$

Because $D \neq 0$, proceed to find the values of D_x, D_y, and D_z. To find D_x, replace the coefficients of x in D with the constants and then evaluate the determinant.

$$D_x = \begin{vmatrix} 3 & 1 & -1 \\ -3 & 2 & 4 \\ 4 & -2 & -3 \end{vmatrix} = (-1)^{1+1} \cdot 3 \cdot \begin{vmatrix} 2 & 4 \\ -2 & -3 \end{vmatrix} + (-1)^{1+2} \cdot 1 \cdot \begin{vmatrix} -3 & 4 \\ 4 & -3 \end{vmatrix} + (-1)^{1+3}(-1)\begin{vmatrix} -3 & 2 \\ 4 & -2 \end{vmatrix}$$

$$= 3(2) - 1(-7) + (-1)(-2) = 15$$

$$D_y = \begin{vmatrix} 2 & 3 & -1 \\ -1 & -3 & 4 \\ 1 & 4 & -3 \end{vmatrix} = (-1)^{1+1} \cdot 2 \cdot \begin{vmatrix} -3 & 4 \\ 4 & -3 \end{vmatrix} + (-1)^{1+2} \cdot 3 \cdot \begin{vmatrix} -1 & 4 \\ 1 & -3 \end{vmatrix} + (-1)^{1+3}(-1)\begin{vmatrix} -1 & -3 \\ 1 & 4 \end{vmatrix}$$

$$= 2(-7) - 3(-1) + (-1)(-1) = -10$$

$$D_z = \begin{vmatrix} 2 & 1 & 3 \\ -1 & 2 & -3 \\ 1 & -2 & 4 \end{vmatrix} = (-1)^{1+1} \cdot 2 \cdot \begin{vmatrix} 2 & -3 \\ -2 & 4 \end{vmatrix} + (-1)^{1+2} \cdot 1 \cdot \begin{vmatrix} -1 & -3 \\ 1 & 4 \end{vmatrix} + (-1)^{1+3} \cdot 3 \cdot \begin{vmatrix} -1 & 2 \\ 1 & -2 \end{vmatrix}$$

$$= 2(2) - 1(-1) + 3(0) = 5$$

As a result,

$$x = \frac{D_x}{D} = \frac{15}{5} = 3 \qquad y = \frac{D_y}{D} = \frac{-10}{5} = -2 \qquad z = \frac{D_z}{D} = \frac{5}{5} = 1$$

The solution is $x = 3$, $y = -2$, $z = 1$ or, using an ordered triplet, $(3, -2, 1)$. ■

Cramer's Rule cannot be used when the determinant of the coefficients on the variables, D, is 0. But can anything be learned about the system other than it is not a consistent and independent system if $D = 0$? The answer is yes!

Cramer's Rule with Inconsistent or Dependent Systems

- If $D = 0$ and at least one of the determinants D_x, D_y, or D_z is different from 0, then the system is inconsistent and the solution set is $\varnothing$, or $\{\ \}$.
- If $D = 0$ and all the determinants D_x, D_y, and D_z equal 0, then the system is consistent and dependent, so there are infinitely many solutions. The system must be solved using row reduction techniques.

Now Work PROBLEM 33

5 Know Properties of Determinants

Determinants have several properties that are sometimes helpful for obtaining their value. We list some of them here.

THEOREM

The value of a determinant changes sign if any two rows (or any two columns) are interchanged. **(11)**

■

Proof for 2 by 2 Determinants

$$\begin{vmatrix} a & b \\ c & d \end{vmatrix} = ad - bc \quad \text{and} \quad \begin{vmatrix} c & d \\ a & b \end{vmatrix} = bc - ad = -(ad - bc)$$ ■

EXAMPLE 6 **Demonstrating Theorem (11)**

$$\begin{vmatrix} 3 & 4 \\ 1 & 2 \end{vmatrix} = 6 - 4 = 2 \qquad \begin{vmatrix} 1 & 2 \\ 3 & 4 \end{vmatrix} = 4 - 6 = -2$$

THEOREM If all the entries in any row (or any column) equal 0, the value of the determinant is 0. **(12)**

Proof Expand across the row (or down the column) containing the 0's. ■

THEOREM If any two rows (or any two columns) of a determinant have corresponding entries that are equal, the value of the determinant is 0. **(13)**

In Problem 68, you are asked to prove this result for a 3 by 3 determinant in which the entries in column 1 equal the entries in column 3.

EXAMPLE 7 **Demonstrating Theorem (13)**

$$\begin{vmatrix} 1 & 2 & 3 \\ 1 & 2 & 3 \\ 4 & 5 & 6 \end{vmatrix} = (-1)^{1+1} \cdot 1 \cdot \begin{vmatrix} 2 & 3 \\ 5 & 6 \end{vmatrix} + (-1)^{1+2} \cdot 2 \cdot \begin{vmatrix} 1 & 3 \\ 4 & 6 \end{vmatrix} + (-1)^{1+3} \cdot 3 \cdot \begin{vmatrix} 1 & 2 \\ 4 & 5 \end{vmatrix}$$

$$= 1(-3) - 2(-6) + 3(-3) = -3 + 12 - 9 = 0$$

THEOREM If any row (or any column) of a determinant is multiplied by a nonzero number k, the value of the determinant is also changed by a factor of k. **(14)**

In Problem 67, you are asked to prove this result for a 3 by 3 determinant using row 2.

EXAMPLE 8 **Demonstrating Theorem (14)**

$$\begin{vmatrix} 1 & 2 \\ 4 & 6 \end{vmatrix} = 6 - 8 = -2$$

$$\begin{vmatrix} k & 2k \\ 4 & 6 \end{vmatrix} = 6k - 8k = -2k = k(-2) = k\begin{vmatrix} 1 & 2 \\ 4 & 6 \end{vmatrix}$$

THEOREM If the entries of any row (or any column) of a determinant are multiplied by a nonzero number k and the result is added to the corresponding entries of another row (or column), the value of the determinant remains unchanged. **(15)**

In Problem 69, you are asked to prove this result for a 3 by 3 determinant using rows 1 and 2.

EXAMPLE 9 **Demonstrating Theorem (15)**

$$\begin{vmatrix} 3 & 4 \\ 5 & 2 \end{vmatrix} = -14 \qquad \begin{vmatrix} 3 & 4 \\ 5 & 2 \end{vmatrix} \rightarrow \begin{vmatrix} -7 & 0 \\ 5 & 2 \end{vmatrix} = -14$$

Multiply row 2 by -2 and add to row 1.

Now Work PROBLEM 45

12.3 Assess Your Understanding

Concepts and Vocabulary

1. $D = \begin{vmatrix} a & b \\ c & d \end{vmatrix} =$ ______.

2. Using Cramer's Rule, the value of x that satisfies the system of equations $\begin{cases} 2x + 3y = 5 \\ x - 4y = -3 \end{cases}$ is $x = \dfrac{\text{\_\_\_\_\_\_}}{\begin{vmatrix} 2 & 3 \\ 1 & -4 \end{vmatrix}}$.

3. ***True or False*** A determinant can never equal 0.

4. ***True or False*** When using Cramer's Rule, if $D = 0$, then the system of linear equations is inconsistent.

5. ***True or False*** If any row (or any column) of a determinant is multiplied by a nonzero number k, the value of the determinant remains unchanged.

6. If any two rows of a determinant are interchanged, its value is best described by which of the following?
(a) changes sign (b) becomes zero (c) remains the same
(d) no longer relates to the original value

Skill Building

In Problems 7–14, find the value of each determinant.

7. $\begin{vmatrix} 6 & 4 \\ -1 & 3 \end{vmatrix}$ **8.** $\begin{vmatrix} 8 & -3 \\ 4 & 2 \end{vmatrix}$ **9.** $\begin{vmatrix} -3 & -1 \\ 4 & 2 \end{vmatrix}$ **10.** $\begin{vmatrix} -4 & 2 \\ -5 & 3 \end{vmatrix}$

11. $\begin{vmatrix} 3 & 4 & 2 \\ 1 & -1 & 5 \\ 1 & 2 & -2 \end{vmatrix}$ **12.** $\begin{vmatrix} 1 & 3 & -2 \\ 6 & 1 & -5 \\ 8 & 2 & 3 \end{vmatrix}$ **13.** $\begin{vmatrix} 4 & -1 & 2 \\ 6 & -1 & 0 \\ 1 & -3 & 4 \end{vmatrix}$ **14.** $\begin{vmatrix} 3 & -9 & 4 \\ 1 & 4 & 0 \\ 8 & -3 & 1 \end{vmatrix}$

In Problems 15–42, solve each system of equations using Cramer's Rule if it is applicable. If Cramer's Rule is not applicable, say so.

15. $\begin{cases} x + y = 8 \\ x - y = 4 \end{cases}$ **16.** $\begin{cases} x + 2y = 5 \\ x - y = 3 \end{cases}$ **17.** $\begin{cases} 5x - y = 13 \\ 2x + 3y = 12 \end{cases}$ **18.** $\begin{cases} x + 3y = 5 \\ 2x - 3y = -8 \end{cases}$

19. $\begin{cases} 3x = 24 \\ x + 2y = 0 \end{cases}$ **20.** $\begin{cases} 4x + 5y = -3 \\ -2y = -4 \end{cases}$ **21.** $\begin{cases} 3x - 6y = 24 \\ 5x + 4y = 12 \end{cases}$ **22.** $\begin{cases} 2x + 4y = 16 \\ 3x - 5y = -9 \end{cases}$

23. $\begin{cases} 3x - 2y = 4 \\ 6x - 4y = 0 \end{cases}$ **24.** $\begin{cases} -x + 2y = 5 \\ 4x - 8y = 6 \end{cases}$ **25.** $\begin{cases} 2x - 4y = -2 \\ 3x + 2y = 3 \end{cases}$ **26.** $\begin{cases} 3x + 3y = 3 \\ 4x + 2y = \dfrac{8}{3} \end{cases}$

27. $\begin{cases} 2x - 3y = -1 \\ 10x + 10y = 5 \end{cases}$ **28.** $\begin{cases} 3x - 2y = 0 \\ 5x + 10y = 4 \end{cases}$ **29.** $\begin{cases} 2x + 3y = 6 \\ x - y = \dfrac{1}{2} \end{cases}$ **30.** $\begin{cases} \dfrac{1}{2}x + y = -2 \\ x - 2y = 8 \end{cases}$

31. $\begin{cases} 3x - 5y = 3 \\ 15x + 5y = 21 \end{cases}$ **32.** $\begin{cases} 2x - y = -1 \\ x + \dfrac{1}{2}y = \dfrac{3}{2} \end{cases}$ **33.** $\begin{cases} x + y - z = 6 \\ 3x - 2y + z = -5 \\ x + 3y - 2z = 14 \end{cases}$ **34.** $\begin{cases} x - y + z = -4 \\ 2x - 3y + 4z = -15 \\ 5x + y - 2z = 12 \end{cases}$

35. $\begin{cases} x + 2y - z = -3 \\ 2x - 4y + z = -7 \\ -2x + 2y - 3z = 4 \end{cases}$ **36.** $\begin{cases} x + 4y - 3z = -8 \\ 3x - y + 3z = 12 \\ x + y + 6z = 1 \end{cases}$ **37.** $\begin{cases} x - 2y + 3z = 1 \\ 3x + y - 2z = 0 \\ 2x - 4y + 6z = 2 \end{cases}$ **38.** $\begin{cases} x - y + 2z = 5 \\ 3x + 2y = 4 \\ -2x + 2y - 4z = -10 \end{cases}$

39. $\begin{cases} x + 2y - z = 0 \\ 2x - 4y + z = 0 \\ -2x + 2y - 3z = 0 \end{cases}$ **40.** $\begin{cases} x + 4y - 3z = 0 \\ 3x - y + 3z = 0 \\ x + y + 6z = 0 \end{cases}$ **41.** $\begin{cases} x - 2y + 3z = 0 \\ 3x + y - 2z = 0 \\ 2x - 4y + 6z = 0 \end{cases}$ **42.** $\begin{cases} x - y + 2z = 0 \\ 3x + 2y = 0 \\ -2x + 2y - 4z = 0 \end{cases}$

In Problems 43–50, use properties of determinants to find the value of each determinant if it is known that

$$\begin{vmatrix} x & y & z \\ u & v & w \\ 1 & 2 & 3 \end{vmatrix} = 4$$

43. $\begin{vmatrix} 1 & 2 & 3 \\ u & v & w \\ x & y & z \end{vmatrix}$ **44.** $\begin{vmatrix} x & y & z \\ u & v & w \\ 2 & 4 & 6 \end{vmatrix}$ **45.** $\begin{vmatrix} x & y & z \\ -3 & -6 & -9 \\ u & v & w \end{vmatrix}$ **46.** $\begin{vmatrix} 1 & 2 & 3 \\ x - u & y - v & z - w \\ u & v & w \end{vmatrix}$

47. $\begin{vmatrix} 1 & 2 & 3 \\ x-3 & y-6 & z-9 \\ 2u & 2v & 2w \end{vmatrix}$ 48. $\begin{vmatrix} x & y & z-x \\ u & v & w-u \\ 1 & 2 & 2 \end{vmatrix}$ 49. $\begin{vmatrix} 1 & 2 & 3 \\ 2x & 2y & 2z \\ u-1 & v-2 & w-3 \end{vmatrix}$ 50. $\begin{vmatrix} x+3 & y+6 & z+9 \\ 3u-1 & 3v-2 & 3w-3 \\ 1 & 2 & 3 \end{vmatrix}$

Mixed Practice

In Problems 51–56, solve for x.

51. $\begin{vmatrix} x & x \\ 4 & 3 \end{vmatrix} = 5$ 52. $\begin{vmatrix} x & 1 \\ 3 & x \end{vmatrix} = -2$ 53. $\begin{vmatrix} x & 1 & 1 \\ 4 & 3 & 2 \\ -1 & 2 & 5 \end{vmatrix} = 2$

54. $\begin{vmatrix} 3 & 2 & 4 \\ 1 & x & 5 \\ 0 & 1 & -2 \end{vmatrix} = 0$ 55. $\begin{vmatrix} x & 2 & 3 \\ 1 & x & 0 \\ 6 & 1 & -2 \end{vmatrix} = 7$ 56. $\begin{vmatrix} x & 1 & 2 \\ 1 & x & 3 \\ 0 & 1 & 2 \end{vmatrix} = -4x$

Applications and Extensions

57. **Geometry: Equation of a Line** An equation of the line containing the two points (x_1, y_1) and (x_2, y_2) may be expressed as the determinant

$$\begin{vmatrix} x & y & 1 \\ x_1 & y_1 & 1 \\ x_2 & y_2 & 1 \end{vmatrix} = 0$$

Prove this result by expanding the determinant and comparing the result to the two-point form of the equation of a line.

58. **Geometry: Collinear Points** Using the result obtained in Problem 57, show that three distinct points (x_1, y_1), (x_2, y_2), and (x_3, y_3) are collinear (lie on the same line) if and only if

$$\begin{vmatrix} x_1 & y_1 & 1 \\ x_2 & y_2 & 1 \\ x_3 & y_3 & 1 \end{vmatrix} = 0$$

59. **Geometry: Area of a Triangle** A triangle has vertices (x_1, y_1), (x_2, y_2), and (x_3, y_3). The area of the triangle is given by the absolute value of D, where $D = \frac{1}{2}\begin{vmatrix} x_1 & x_2 & x_3 \\ y_1 & y_2 & y_3 \\ 1 & 1 & 1 \end{vmatrix}$.

Use this formula to find the area of a triangle with vertices $(2, 3)$, $(5, 2)$, and $(6, 5)$.

60. **Geometry: Area of a Polygon** The formula from Problem 59 can be used to find the area of a polygon. To do so, divide the polygon into non-overlapping triangular regions and find the sum of the areas. Use this approach to find the area of the given polygon.

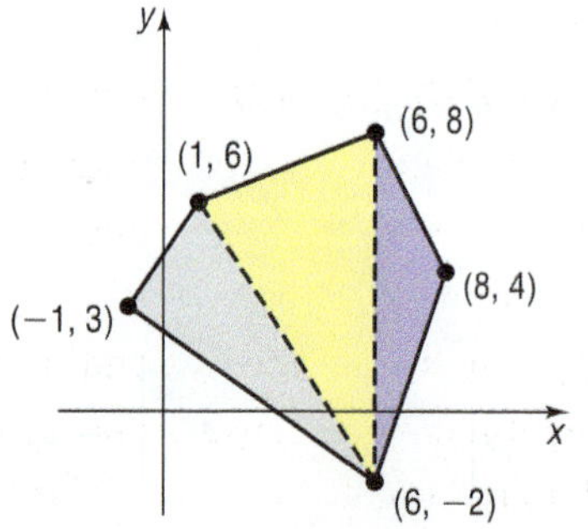

61. **Geometry: Area of a Polygon** Another approach for finding the area of a polygon by using determinants is to use the formula

$$A = \frac{1}{2}\left(\begin{vmatrix} x_1 & y_1 \\ x_2 & y_2 \end{vmatrix} + \begin{vmatrix} x_2 & y_2 \\ x_3 & y_3 \end{vmatrix} + \begin{vmatrix} x_3 & y_3 \\ x_4 & y_4 \end{vmatrix} + \cdots + \begin{vmatrix} x_n & y_n \\ x_1 & y_1 \end{vmatrix}\right)$$

where $(x_1, y_1), (x_2, y_2), \ldots, (x_n, y_n)$ are the n corner points in counterclockwise order. Use this formula to compute the area of the polygon from Problem 60 again. Which method do you prefer?

62. Show that the formula in Problem 61 yields the same result as the formula used in Problem 59.

63. **Geometry: Equation of a Circle** An equation of the circle containing the distinct points (x_1, y_1), (x_2, y_2), and (x_3, y_3) can be found using the following equation.

$$\begin{vmatrix} 1 & 1 & 1 & 1 \\ x & x_1 & x_2 & x_3 \\ y & y_1 & y_2 & y_3 \\ x^2+y^2 & x_1^2+y_1^2 & x_2^2+y_2^2 & x_3^2+y_3^2 \end{vmatrix} = 0$$

Find the equation of the circle containing the points $(7, -5)$, $(3, 3)$, and $(6, 2)$. Write the equation in standard form.

64. Show that $\begin{vmatrix} x^2 & x & 1 \\ y^2 & y & 1 \\ z^2 & z & 1 \end{vmatrix} = (y - z)(x - y)(x - z)$.

65. Complete the proof of Cramer's Rule for two equations containing two variables.

[**Hint:** In system (5), page 895, if $a = 0$, then $b \neq 0$ and $c \neq 0$, since $D = -bc \neq 0$. Now show that equation (6) provides a solution of the system when $a = 0$. Then three cases remain: $b = 0$, $c = 0$, and $d = 0$.]

66. Interchange columns 1 and 3 of a 3 by 3 determinant. Show that the value of the new determinant is -1 times the value of the original determinant.

67. Multiply each entry in row 2 of a 3 by 3 determinant by the number k, $k \neq 0$. Show that the value of the new determinant is k times the value of the original determinant.

68. Prove that a 3 by 3 determinant in which the entries in column 1 equal those in column 3 has the value 0.

69. Prove that if row 2 of a 3 by 3 determinant is multiplied by k, $k \neq 0$, and the result is added to the entries in row 1, there is no change in the value of the determinant.

Retain Your Knowledge

Problems 70–73 are based on material learned earlier in the course. The purpose of these problems is to keep the material fresh in your mind so that you are better prepared for the final exam.

70. For the points $P = (-4, 3)$ and $Q = (5, -1)$ write the vector $\mathbf{v}$ represented by the directed line segment $\overrightarrow{PQ}$ in the form $a\mathbf{i} + b\mathbf{j}$ and find $\|\mathbf{v}\|$.

71. List the potential rational zeros of the polynomial function $P(x) = 2x^3 - 5x^2 + x - 10$.

72. Graph $f(x) = (x + 1)^2 - 4$ using transformations (shifting, compressing, stretching, and/or reflecting).

73. Find the exact value of $\tan 42° - \cot 48°$ without using a calculator.

12.4 Matrix Algebra

OBJECTIVES

Section 12.2 defined a matrix as a rectangular array of real numbers and used an augmented matrix to represent a system of linear equations. There is, however, a branch of mathematics, called **linear algebra**, that deals with matrices in such a way that an algebra of matrices is permitted. This section provides a survey of how this **matrix algebra** is developed.

Before getting started, recall the definition of a matrix.

DEFINITION

A **matrix** is a rectangular array of numbers:

$$\begin{array}{c} \\ \text{Row 1} \\ \text{Row 2} \\ \vdots \\ \text{Row } i \\ \vdots \\ \text{Row } m \end{array} \begin{array}{c} \begin{array}{cccccc} \text{Column 1} & \text{Column 2} & & \text{Column } j & & \text{Column } n \end{array} \\ \begin{bmatrix} a_{11} & a_{12} & \cdots & a_{1j} & \cdots & a_{1n} \\ a_{21} & a_{22} & \cdots & a_{2j} & \cdots & a_{2n} \\ \vdots & \vdots & & \vdots & & \vdots \\ a_{i1} & a_{i2} & \cdots & a_{ij} & \cdots & a_{in} \\ \vdots & \vdots & & \vdots & & \vdots \\ a_{m1} & a_{m2} & \cdots & a_{mj} & \cdots & a_{mn} \end{bmatrix} \end{array}$$

Each number a_{ij} of the matrix has two indexes: the **row index** i and the **column index** j. The matrix shown here has m rows and n columns. The numbers a_{ij} are the **entries** of the matrix. For example, a_{23} refers to the entry in the second row, third column.

EXAMPLE 1 **Arranging Data in a Matrix**

In a survey of 900 people, the following information was obtained:

200 males	Thought federal defense spending was too high
150 males	Thought federal defense spending was too low
45 males	Had no opinion
315 females	Thought federal defense spending was too high
125 females	Thought federal defense spending was too low
65 females	Had no opinion

We can arrange these data in a rectangular array as follows:

	Too High	Too Low	No Opinion
Male	200	150	45
Female	315	125	65

or as the matrix

$$\begin{bmatrix} 200 & 150 & 45 \\ 315 & 125 & 65 \end{bmatrix}$$

This matrix has two rows (representing male and female) and three columns (representing "too high," "too low," and "no opinion").

The matrix developed in Example 1 has 2 rows and 3 columns. In general, a matrix with m rows and n columns is called an **m by n matrix.** The matrix developed in Example 1 is a 2 by 3 matrix and contains $2 \cdot 3 = 6$ entries. An m by n matrix contains $m \cdot n$ entries.

If an m by n matrix has the same number of rows as columns, that is, if $m = n$, then the matrix is a **square matrix**.

EXAMPLE 2 **Examples of Matrices**

(a) $\begin{bmatrix} 5 & 0 \\ -6 & 1 \end{bmatrix}$ A 2 by 2 square matrix (b) $\begin{bmatrix} 1 & 0 & 3 \end{bmatrix}$ A 1 by 3 matrix

(c) $\begin{bmatrix} 6 & -2 & 4 \\ 4 & 3 & 5 \\ 8 & 0 & 1 \end{bmatrix}$ A 3 by 3 square matrix

1 Find the Sum and Difference of Two Matrices

We begin our discussion of matrix algebra by first defining equivalent matrices and then defining the operations of addition and subtraction. It is important to note that these definitions require both matrices to have the same number of rows *and* the same number of columns as a condition for equality and for addition and subtraction.

Matrices usually are represented by capital letters, such as A, B, and C.

DEFINITION

Two matrices A and B are **equal**, written as

$$A = B$$

provided that A and B have the same number of rows and the same number of columns and each entry a_{ij} in A is equal to the corresponding entry b_{ij} in B.

For example,

$$\begin{bmatrix} 2 & 1 \\ 0.5 & -1 \end{bmatrix} = \begin{bmatrix} \sqrt{4} & 1 \\ \frac{1}{2} & -1 \end{bmatrix} \quad \text{and} \quad \begin{bmatrix} 3 & 2 & 1 \\ 0 & 1 & -2 \end{bmatrix} = \begin{bmatrix} \sqrt{9} & \sqrt{4} & 1 \\ 0 & 1 & \sqrt[3]{-8} \end{bmatrix}$$

$$\begin{bmatrix} 4 & 1 \\ 6 & 1 \end{bmatrix} \neq \begin{bmatrix} 4 & 0 \\ 6 & 1 \end{bmatrix}$$ Because the entries in row 1, column 2 are not equal

$$\begin{bmatrix} 4 & 1 & 2 \\ 6 & 1 & 2 \end{bmatrix} \neq \begin{bmatrix} 4 & 1 & 2 & 3 \\ 6 & 1 & 2 & 4 \end{bmatrix}$$ Because the matrix on the left has 3 columns and the matrix on the right has 4 columns

Suppose that A and B represent two m by n matrices. The **sum, $A + B$**, is defined as the m by n matrix formed by adding the corresponding entries a_{ij} of A and b_{ij} of B. The **difference, $A - B$**, is defined as the m by n matrix formed by subtracting the entries b_{ij} in B from the corresponding entries a_{ij} in A. Addition and subtraction of matrices are allowed only for matrices having the same number m of rows and the same number n of columns. For example, a 2 by 3 matrix and a 2 by 4 matrix cannot be added or subtracted.

EXAMPLE 3 Adding and Subtracting Matrices

Suppose that

$$A = \begin{bmatrix} 2 & 4 & 8 & -3 \\ 0 & 1 & 2 & 3 \end{bmatrix} \quad \text{and} \quad B = \begin{bmatrix} -3 & 4 & 0 & 1 \\ 6 & 8 & 2 & 0 \end{bmatrix}$$

Find: (a) $A + B$ (b) $A - B$

Algebraic Solution

(a) $$A + B = \begin{bmatrix} 2 & 4 & 8 & -3 \\ 0 & 1 & 2 & 3 \end{bmatrix} + \begin{bmatrix} -3 & 4 & 0 & 1 \\ 6 & 8 & 2 & 0 \end{bmatrix}$$

$$= \begin{bmatrix} 2+(-3) & 4+4 & 8+0 & -3+1 \\ 0+6 & 1+8 & 2+2 & 3+0 \end{bmatrix}$$ **Add corresponding entries.**

$$= \begin{bmatrix} -1 & 8 & 8 & -2 \\ 6 & 9 & 4 & 3 \end{bmatrix}$$

(b) $$A - B = \begin{bmatrix} 2 & 4 & 8 & -3 \\ 0 & 1 & 2 & 3 \end{bmatrix} - \begin{bmatrix} -3 & 4 & 0 & 1 \\ 6 & 8 & 2 & 0 \end{bmatrix}$$

$$= \begin{bmatrix} 2-(-3) & 4-4 & 8-0 & -3-1 \\ 0-6 & 1-8 & 2-2 & 3-0 \end{bmatrix}$$ **Subtract corresponding entries.**

$$= \begin{bmatrix} 5 & 0 & 8 & -4 \\ -6 & -7 & 0 & 3 \end{bmatrix}$$

Graphing Solution

Enter the matrices into a graphing utility. Name them $[A]$ and $[B]$. Figure 11 shows the results of adding and subtracting $[A]$ and $[B]$.

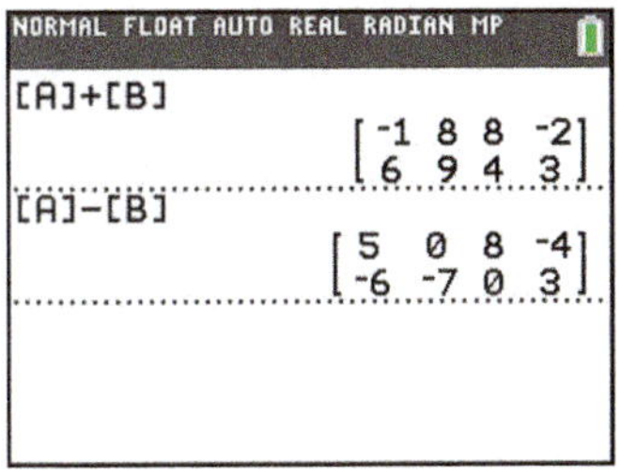

Figure 11 Matrix addition and subtraction

Now Work PROBLEM 9

Many of the algebraic properties of sums of real numbers are also true for sums of matrices. Suppose that A, B, and C are m by n matrices. Then matrix addition is **commutative**. That is,

Commutative Property of Matrix Addition

$$A + B = B + A$$

Matrix addition is also **associative**. That is,

Associative Property of Matrix Addition

$$(A + B) + C = A + (B + C)$$

Although we shall not prove these results, the proofs, as the following example illustrates, are based on the commutative and associative properties for real numbers.

EXAMPLE 4 Demonstrating the Commutative Property

$$\begin{bmatrix} 2 & 3 & -1 \\ 4 & 0 & 7 \end{bmatrix} + \begin{bmatrix} -1 & 2 & 1 \\ 5 & -3 & 4 \end{bmatrix} = \begin{bmatrix} 2+(-1) & 3+2 & -1+1 \\ 4+5 & 0+(-3) & 7+4 \end{bmatrix}$$

$$= \begin{bmatrix} -1+2 & 2+3 & 1+(-1) \\ 5+4 & -3+0 & 4+7 \end{bmatrix}$$

$$= \begin{bmatrix} -1 & 2 & 1 \\ 5 & -3 & 4 \end{bmatrix} + \begin{bmatrix} 2 & 3 & -1 \\ 4 & 0 & 7 \end{bmatrix}$$

A matrix whose entries are all equal to 0 is called a **zero matrix**. Each of the following matrices is a zero matrix.

$$\begin{bmatrix} 0 & 0 \\ 0 & 0 \end{bmatrix} \quad \text{2 by 2 square zero matrix} \qquad \begin{bmatrix} 0 & 0 & 0 \\ 0 & 0 & 0 \end{bmatrix} \quad \text{2 by 3 zero matrix} \qquad \begin{bmatrix} 0 & 0 & 0 \end{bmatrix} \quad \text{1 by 3 zero matrix}$$

Zero matrices have properties similar to the real number 0. If A is an m by n matrix and 0 is the m by n zero matrix, then

$$A + 0 = 0 + A = A$$

In other words, a zero matrix is the additive identity in matrix algebra.

2 Find Scalar Multiples of a Matrix

We can also multiply a matrix by a real number. If k is a real number and A is an m by n matrix, the matrix kA is the m by n matrix formed by multiplying each entry a_{ij} in A by k. The number k is sometimes referred to as a **scalar**, and the matrix kA is called a **scalar multiple** of A.

EXAMPLE 5 Operations Using Matrices

Suppose that

$$A = \begin{bmatrix} 3 & 1 & 5 \\ -2 & 0 & 6 \end{bmatrix} \qquad B = \begin{bmatrix} 4 & 1 & 0 \\ 8 & 1 & -3 \end{bmatrix} \qquad C = \begin{bmatrix} 9 & 0 \\ -3 & 6 \end{bmatrix}$$

Find: (a) $4A$ (b) $\frac{1}{3}C$ (c) $3A - 2B$

Algebraic Solution

(a) $$4A = 4\begin{bmatrix} 3 & 1 & 5 \\ -2 & 0 & 6 \end{bmatrix} = \begin{bmatrix} 4\cdot 3 & 4\cdot 1 & 4\cdot 5 \\ 4(-2) & 4\cdot 0 & 4\cdot 6 \end{bmatrix} = \begin{bmatrix} 12 & 4 & 20 \\ -8 & 0 & 24 \end{bmatrix}$$

(b) $$\frac{1}{3}C = \frac{1}{3}\begin{bmatrix} 9 & 0 \\ -3 & 6 \end{bmatrix} = \begin{bmatrix} \frac{1}{3}\cdot 9 & \frac{1}{3}\cdot 0 \\ \frac{1}{3}(-3) & \frac{1}{3}\cdot 6 \end{bmatrix} = \begin{bmatrix} 3 & 0 \\ -1 & 2 \end{bmatrix}$$

(c) $$3A - 2B = 3\begin{bmatrix} 3 & 1 & 5 \\ -2 & 0 & 6 \end{bmatrix} - 2\begin{bmatrix} 4 & 1 & 0 \\ 8 & 1 & -3 \end{bmatrix}$$

$$= \begin{bmatrix} 3\cdot 3 & 3\cdot 1 & 3\cdot 5 \\ 3(-2) & 3\cdot 0 & 3\cdot 6 \end{bmatrix} - \begin{bmatrix} 2\cdot 4 & 2\cdot 1 & 2\cdot 0 \\ 2\cdot 8 & 2\cdot 1 & 2(-3) \end{bmatrix}$$

$$= \begin{bmatrix} 9 & 3 & 15 \\ -6 & 0 & 18 \end{bmatrix} - \begin{bmatrix} 8 & 2 & 0 \\ 16 & 2 & -6 \end{bmatrix}$$

$$= \begin{bmatrix} 9-8 & 3-2 & 15-0 \\ -6-16 & 0-2 & 18-(-6) \end{bmatrix}$$

$$= \begin{bmatrix} 1 & 1 & 15 \\ -22 & -2 & 24 \end{bmatrix}$$

■

Graphing Solution

Enter the matrices $[A]$, $[B]$, and $[C]$ into a graphing utility. Figure 12 shows the required computations.

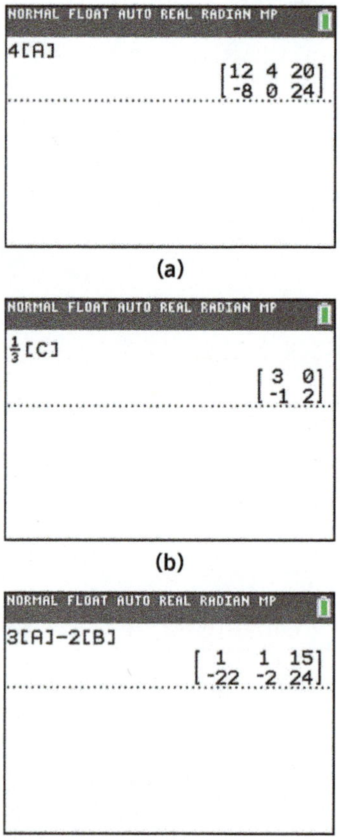

(a) (b) (c)

Figure 12

■

Now Work PROBLEM 13

Some of the algebraic properties of scalar multiplication are listed next. Let h and k be real numbers, and let A and B be m by n matrices. Then

Properties of Scalar Multiplication

$$k(hA) = (kh)A$$
$$(k + h)A = kA + hA$$
$$k(A + B) = kA + kB$$

3 Find the Product of Two Matrices

Unlike the straightforward definition for adding two matrices, the definition for multiplying two matrices is not what might be expected. In preparation for this definition, we need the following definitions:

DEFINITION

A **row vector** R is a 1 by n matrix

$$R = [r_1 \quad r_2 \quad \cdots \quad r_n]$$

A **column vector** C is an n by 1 matrix

$$C = \begin{bmatrix} c_1 \\ c_2 \\ \vdots \\ c_n \end{bmatrix}$$

The **product** RC of R times C is defined as the number

$$RC = [r_1 \quad r_2 \cdots r_n] \begin{bmatrix} c_1 \\ c_2 \\ \vdots \\ c_n \end{bmatrix} = r_1c_1 + r_2c_2 + \cdots + r_nc_n$$

Notice that a row vector and a column vector can be multiplied only if they contain the same number of entries.

EXAMPLE 6 The Product of a Row Vector and a Column Vector

If $R = [3 \quad -5 \quad 2]$ and $C = \begin{bmatrix} 3 \\ 4 \\ -5 \end{bmatrix}$, then

$$RC = [3 \quad -5 \quad 2] \begin{bmatrix} 3 \\ 4 \\ -5 \end{bmatrix} = 3 \cdot 3 + (-5)4 + 2(-5) = 9 - 20 - 10 = -21$$

EXAMPLE 7 Using Matrices to Compute Revenue

A clothing store sells men's shirts for \$40, silk ties for \$20, and wool suits for \$400. Last month, the store had sales consisting of 100 shirts, 200 ties, and 50 suits. What was the total revenue due to these sales?

Solution Set up a row vector R to represent the prices of these three items and a column vector C to represent the corresponding number of items sold. Then

$$\overset{\text{Prices}}{\underset{}{R = \overset{\text{Shirts Ties Suits}}{\begin{bmatrix} 40 & 20 & 400 \end{bmatrix}}}} \qquad C = \overset{\text{Number sold}}{\begin{bmatrix} 100 \\ 200 \\ 50 \end{bmatrix}} \begin{matrix} \text{Shirts} \\ \text{Ties} \\ \text{Suits} \end{matrix}$$

The total revenue obtained is the product RC. That is,

$$RC = \begin{bmatrix} 40 & 20 & 400 \end{bmatrix} \begin{bmatrix} 100 \\ 200 \\ 50 \end{bmatrix}$$

$$= \underbrace{40 \cdot 100}_{\text{Shirt revenue}} + \underbrace{20 \cdot 200}_{\text{Tie revenue}} + \underbrace{400 \cdot 50}_{\text{Suit revenue}} = \underbrace{\$28{,}000}_{\text{Total revenue}}$$

The definition for multiplying two matrices is based on the definition of a row vector times a column vector.

DEFINITION

Let A denote an m by r matrix and let B denote an r by n matrix. The **product** AB is defined as the m by n matrix whose entry in row i, column j is the product of the ith row of A and the jth column of B.

The definition of the product AB of two matrices A and B, in this order, requires that the number of columns of A equal the number of rows of B; otherwise, no product is defined.

In Words
To find the product *AB*, the number of columns in *A* must equal the number of rows in *B*.

A — m by r $\qquad$ B — r by n

Must be same for AB to be defined

AB is m by n.

An example will help to clarify the definition.

EXAMPLE 8 Multiplying Two Matrices

Find the product AB if

$$A = \begin{bmatrix} 2 & 4 & -1 \\ 5 & 8 & 0 \end{bmatrix} \quad \text{and} \quad B = \begin{bmatrix} 2 & 5 & 1 & 4 \\ 4 & 8 & 0 & 6 \\ -3 & 1 & -2 & -1 \end{bmatrix}$$

Solution First, observe that A is 2 by 3 and B is 3 by 4. The number of columns in A equals the number of rows in B, so the product AB is defined and will be a 2 by 4 matrix.

Algebraic Solution

Suppose that we want the entry in row 2, column 3 of AB. To find it, find the product of the row vector from row 2 of A and the column vector from column 3 of B.

$$\underset{\text{Row 2 of } A}{\begin{bmatrix} 5 & 8 & 0 \end{bmatrix}} \overset{\text{Column 3 of } B}{\begin{bmatrix} 1 \\ 0 \\ -2 \end{bmatrix}} = 5 \cdot 1 + 8 \cdot 0 + 0(-2) = 5$$

Graphing Solution

Enter the matrices A and B into a graphing utility. Figure 13 shows the product AB.

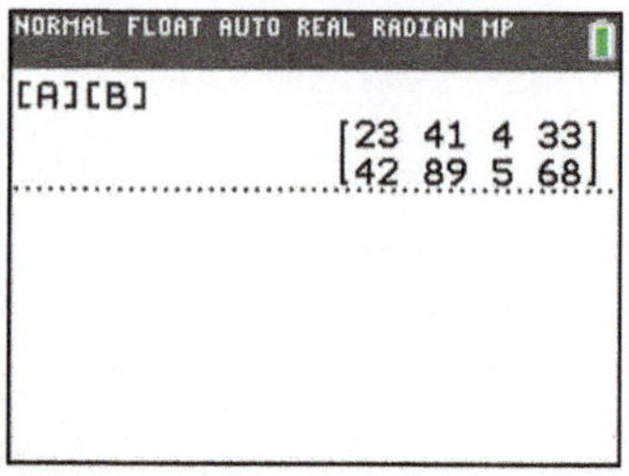

Figure 13 Matrix multiplication

So far, we have

$$AB = \begin{bmatrix} \_\_ & \_\_ & \_\_ & \_\_ \\ \_\_ & \_\_ & 5 & \_\_ \end{bmatrix} \leftarrow \text{Row 2}$$

(Column 3 ↓ points to the entry 5.)

Now, to find the entry in row 1, column 4 of AB, find the product of row 1 of A and column 4 of B.

$$\underset{\text{Row 1 of } A}{\begin{bmatrix} 2 & 4 & -1 \end{bmatrix}} \underset{\text{Column 4 of } B}{\begin{bmatrix} 4 \\ 6 \\ -1 \end{bmatrix}} = 2 \cdot 4 + 4 \cdot 6 + (-1)(-1) = 33$$

Continuing in this fashion, we find AB.

$$AB = \begin{bmatrix} 2 & 4 & -1 \\ 5 & 8 & 0 \end{bmatrix} \begin{bmatrix} 2 & 5 & 1 & 4 \\ 4 & 8 & 0 & 6 \\ -3 & 1 & -2 & -1 \end{bmatrix}$$

$$= \begin{bmatrix} \text{Row 1 of } A \text{ times column 1 of } B & \text{Row 1 of } A \text{ times column 2 of } B & \text{Row 1 of } A \text{ times column 3 of } B & \text{Row 1 of } A \text{ times column 4 of } B \\ \text{Row 2 of } A \text{ times column 1 of } B & \text{Row 2 of } A \text{ times column 2 of } B & \text{Row 2 of } A \text{ times column 3 of } B & \text{Row 2 of } A \text{ times column 4 of } B \end{bmatrix}$$

$$= \begin{bmatrix} 2 \cdot 2 + 4 \cdot 4 + (-1)(-3) & 2 \cdot 5 + 4 \cdot 8 + (-1)1 & 2 \cdot 1 + 4 \cdot 0 + (-1)(-2) & 33 \text{ (from earlier)} \\ 5 \cdot 2 + 8 \cdot 4 + 0(-3) & 5 \cdot 5 + 8 \cdot 8 + 0 \cdot 1 & 5 \text{ (from earlier)} & 5 \cdot 4 + 8 \cdot 6 + 0(-1) \end{bmatrix}$$

$$= \begin{bmatrix} 23 & 41 & 4 & 33 \\ 42 & 89 & 5 & 68 \end{bmatrix}$$

■

Now Work PROBLEM 27

Notice that for the matrices given in Example 8, the product BA is not defined because B is 3 by 4 and A is 2 by 3.

EXAMPLE 9 Multiplying Two Matrices

If

$$A = \begin{bmatrix} 2 & 1 & 3 \\ 1 & -1 & 0 \end{bmatrix} \quad \text{and} \quad B = \begin{bmatrix} 1 & 0 \\ 2 & 1 \\ 3 & 2 \end{bmatrix}$$

find: (a) AB (b) BA

Solution

(a) $$AB = \underset{2 \text{ by } 3}{\begin{bmatrix} 2 & 1 & 3 \\ 1 & -1 & 0 \end{bmatrix}} \underset{3 \text{ by } 2}{\begin{bmatrix} 1 & 0 \\ 2 & 1 \\ 3 & 2 \end{bmatrix}} = \underset{2 \text{ by } 2}{\begin{bmatrix} 13 & 7 \\ -1 & -1 \end{bmatrix}}$$

(b) $$BA = \underset{3 \text{ by } 2}{\begin{bmatrix} 1 & 0 \\ 2 & 1 \\ 3 & 2 \end{bmatrix}} \underset{2 \text{ by } 3}{\begin{bmatrix} 2 & 1 & 3 \\ 1 & -1 & 0 \end{bmatrix}} = \underset{3 \text{ by } 3}{\begin{bmatrix} 2 & 1 & 3 \\ 5 & 1 & 6 \\ 8 & 1 & 9 \end{bmatrix}}$$

■

Notice in Example 9 that AB is 2 by 2 and BA is 3 by 3. It is possible for both AB and BA to be defined and yet be unequal. In fact, even if A and B are both n by n matrices so that AB and BA are each defined and n by n, AB and BA will usually be unequal.

EXAMPLE 10 Multiplying Two Square Matrices

If

$$A = \begin{bmatrix} 2 & 1 \\ 0 & 4 \end{bmatrix} \quad \text{and} \quad B = \begin{bmatrix} -3 & 1 \\ 1 & 2 \end{bmatrix}$$

find: (a) AB (b) BA

Solution

(a) $AB = \begin{bmatrix} 2 & 1 \\ 0 & 4 \end{bmatrix}\begin{bmatrix} -3 & 1 \\ 1 & 2 \end{bmatrix} = \begin{bmatrix} -5 & 4 \\ 4 & 8 \end{bmatrix}$

(b) $BA = \begin{bmatrix} -3 & 1 \\ 1 & 2 \end{bmatrix}\begin{bmatrix} 2 & 1 \\ 0 & 4 \end{bmatrix} = \begin{bmatrix} -6 & 1 \\ 2 & 9 \end{bmatrix}$

The preceding examples demonstrate that an important property of real numbers, the commutative property of multiplication, is not shared by matrices. In general:

THEOREM

Matrix multiplication is not commutative.

Now Work PROBLEMS 15 AND 19

Next, consider two of the properties of real numbers that *are* shared by matrices. Assuming that each product and sum is defined, the following is true:

Associative Property of Matrix Multiplication

$$A(BC) = (AB)C$$

Distributive Property

$$A(B + C) = AB + AC$$

For an n by n square matrix, the entries located in row i, column i, $1 \le i \le n$, are called the **diagonal entries** or **the main diagonal.** The n by n square matrix whose diagonal entries are 1's, and all other entries are 0's, is called the **identity matrix** $\boldsymbol{I_n}$. For example,

$$I_2 = \begin{bmatrix} 1 & 0 \\ 0 & 1 \end{bmatrix} \qquad I_3 = \begin{bmatrix} 1 & 0 & 0 \\ 0 & 1 & 0 \\ 0 & 0 & 1 \end{bmatrix}$$

and so on.

EXAMPLE 11 Multiplication with an Identity Matrix

Let

$$A = \begin{bmatrix} -1 & 2 & 0 \\ 0 & 1 & 3 \end{bmatrix} \quad \text{and} \quad B = \begin{bmatrix} 3 & 2 \\ 4 & 6 \\ 5 & 2 \end{bmatrix}$$

Find: (a) AI_3 (b) I_2A (c) BI_2

Solution

(a) $AI_3 = \begin{bmatrix} -1 & 2 & 0 \\ 0 & 1 & 3 \end{bmatrix}\begin{bmatrix} 1 & 0 & 0 \\ 0 & 1 & 0 \\ 0 & 0 & 1 \end{bmatrix} = \begin{bmatrix} -1 & 2 & 0 \\ 0 & 1 & 3 \end{bmatrix} = A$

(b) $I_2A = \begin{bmatrix} 1 & 0 \\ 0 & 1 \end{bmatrix}\begin{bmatrix} -1 & 2 & 0 \\ 0 & 1 & 3 \end{bmatrix} = \begin{bmatrix} -1 & 2 & 0 \\ 0 & 1 & 3 \end{bmatrix} = A$

(c) $BI_2 = \begin{bmatrix} 3 & 2 \\ 4 & 6 \\ 5 & 2 \end{bmatrix}\begin{bmatrix} 1 & 0 \\ 0 & 1 \end{bmatrix} = \begin{bmatrix} 3 & 2 \\ 4 & 6 \\ 5 & 2 \end{bmatrix} = B$ ■

Example 11 demonstrates the following property:

Identity Property

If A is an m by n matrix, then

$$I_mA = A \quad \text{and} \quad AI_n = A$$

If A is an n by n square matrix,

$$AI_n = I_nA = A$$

An identity matrix has properties similar to those of the real number 1. In other words, the identity matrix is a multiplicative identity in matrix algebra.

4 Find the Inverse of a Matrix

DEFINITION

Let A be a square n by n matrix. If there exists an n by n matrix A^{-1} (read as "A inverse") for which

$$AA^{-1} = A^{-1}A = I_n$$

then A^{-1} is called the **inverse** of the matrix A. ■

Note: If the determinant of A is zero, A is singular. (Refer to Section 12.3.) ■

Not every square matrix has an inverse. When a matrix A does have an inverse A^{-1}, then A is said to be **nonsingular**. If a matrix A has no inverse, it is called **singular**.

EXAMPLE 12 **Multiplying a Matrix by Its Inverse**

Show that the inverse of

$$A = \begin{bmatrix} 3 & 1 \\ 2 & 1 \end{bmatrix} \quad \text{is} \quad A^{-1} = \begin{bmatrix} 1 & -1 \\ -2 & 3 \end{bmatrix}$$

Solution We need to show that $AA^{-1} = A^{-1}A = I_2$.

$$AA^{-1} = \begin{bmatrix} 3 & 1 \\ 2 & 1 \end{bmatrix}\begin{bmatrix} 1 & -1 \\ -2 & 3 \end{bmatrix} = \begin{bmatrix} 1 & 0 \\ 0 & 1 \end{bmatrix} = I_2$$

$$A^{-1}A = \begin{bmatrix} 1 & -1 \\ -2 & 3 \end{bmatrix}\begin{bmatrix} 3 & 1 \\ 2 & 1 \end{bmatrix} = \begin{bmatrix} 1 & 0 \\ 0 & 1 \end{bmatrix} = I_2$$ ■

The following discussion illustrates one way to find the inverse of

$$A = \begin{bmatrix} 3 & 1 \\ 2 & 1 \end{bmatrix}$$

Suppose that A^{-1} is given by

$$A^{-1} = \begin{bmatrix} x & y \\ z & w \end{bmatrix} \tag{1}$$

where x, y, z, and w are four variables. Based on the definition of an inverse, if A has an inverse, then

$$AA^{-1} = I_2$$

$$\begin{bmatrix} 3 & 1 \\ 2 & 1 \end{bmatrix}\begin{bmatrix} x & y \\ z & w \end{bmatrix} = \begin{bmatrix} 1 & 0 \\ 0 & 1 \end{bmatrix}$$

$$\begin{bmatrix} 3x + z & 3y + w \\ 2x + z & 2y + w \end{bmatrix} = \begin{bmatrix} 1 & 0 \\ 0 & 1 \end{bmatrix}$$

Because corresponding entries must be equal, it follows that this matrix equation is equivalent to two systems of linear equations.

$$\begin{cases} 3x + z = 1 \\ 2x + z = 0 \end{cases} \qquad \begin{cases} 3y + w = 0 \\ 2y + w = 1 \end{cases}$$

The augmented matrix of each system is

$$\left[\begin{array}{cc|c} 3 & 1 & 1 \\ 2 & 1 & 0 \end{array}\right] \quad \left[\begin{array}{cc|c} 3 & 1 & 0 \\ 2 & 1 & 1 \end{array}\right] \tag{2}$$

The usual procedure would be to transform each augmented matrix into reduced row echelon form. Notice, though, that the left sides of the augmented matrices are equal, so the same row operations (see Section 12.2) can be used to reduce each one. It is more efficient to combine the two augmented matrices (2) into a single matrix:

$$\left[\begin{array}{cc|cc} 3 & 1 & 1 & 0 \\ 2 & 1 & 0 & 1 \end{array}\right]$$

Next, use row operations to transform the matrix into reduced row echelon form.

$$\left[\begin{array}{cc|cc} 3 & 1 & 1 & 0 \\ 2 & 1 & 0 & 1 \end{array}\right] \underset{\uparrow}{\rightarrow} \left[\begin{array}{cc|cc} 1 & 0 & 1 & -1 \\ 2 & 1 & 0 & 1 \end{array}\right]$$

$R_1 = -1r_2 + r_1$

$$\underset{\uparrow}{\rightarrow} \left[\begin{array}{cc|cc} 1 & 0 & 1 & -1 \\ 0 & 1 & -2 & 3 \end{array}\right] \tag{3}$$

$R_2 = -2r_1 + r_2$

Matrix (3) is in reduced row echelon form.

Now reverse the earlier step of combining the two augmented matrices in (2), and write the single matrix (3) as two augmented matrices.

$$\left[\begin{array}{cc|c} 1 & 0 & 1 \\ 0 & 1 & -2 \end{array}\right] \text{ and } \left[\begin{array}{cc|c} 1 & 0 & -1 \\ 0 & 1 & 3 \end{array}\right]$$

The conclusion from these matrices is that $x = 1, z = -2$, and $y = -1, w = 3$. Substituting these values into matrix (1) results in

$$A^{-1} = \begin{bmatrix} 1 & -1 \\ -2 & 3 \end{bmatrix}$$

Notice in display (3) that the 2 by 2 matrix to the right of the vertical bar is, in fact, the inverse of A. Also notice that the identity matrix I_2 is the matrix that appears to the left of the vertical bar. These observations and the procedures used to get display (3) will work in general.

In Words

If A is nonsingular, begin with the matrix $[A|I_n]$, and after transforming it into reduced row echelon form, you end up with the matrix $[I_n|A^{-1}]$.

Procedure for Finding the Inverse of a Nonsingular Matrix*

To find the inverse of an n by n nonsingular matrix A, proceed as follows:

STEP 1: Form the matrix $[A|I_n]$.

STEP 2: Transform the matrix $[A|I_n]$ into reduced row echelon form.

STEP 3: The reduced row echelon form of $[A|I_n]$ contains the identity matrix I_n on the left of the vertical bar; the n by n matrix on the right of the vertical bar is the inverse of A.

EXAMPLE 13 Finding the Inverse of a Matrix

The matrix

$$A = \begin{bmatrix} 1 & 1 & 0 \\ -1 & 3 & 4 \\ 0 & 4 & 3 \end{bmatrix}$$

is nonsingular. Find its inverse.

Algebraic Solution

First, form the matrix

$$[A|I_3] = \left[\begin{array}{ccc|ccc} 1 & 1 & 0 & 1 & 0 & 0 \\ -1 & 3 & 4 & 0 & 1 & 0 \\ 0 & 4 & 3 & 0 & 0 & 1 \end{array}\right]$$

Next, use row operations to transform $[A|I_3]$ into reduced row echelon form.

$$\left[\begin{array}{ccc|ccc} 1 & 1 & 0 & 1 & 0 & 0 \\ -1 & 3 & 4 & 0 & 1 & 0 \\ 0 & 4 & 3 & 0 & 0 & 1 \end{array}\right] \rightarrow \left[\begin{array}{ccc|ccc} 1 & 1 & 0 & 1 & 0 & 0 \\ 0 & 4 & 4 & 1 & 1 & 0 \\ 0 & 4 & 3 & 0 & 0 & 1 \end{array}\right] \rightarrow \left[\begin{array}{ccc|ccc} 1 & 1 & 0 & 1 & 0 & 0 \\ 0 & 1 & 1 & \frac{1}{4} & \frac{1}{4} & 0 \\ 0 & 4 & 3 & 0 & 0 & 1 \end{array}\right]$$

$R_2 = r_1 + r_2$ $\qquad$ $R_2 = \frac{1}{4}r_2$

$$\rightarrow \left[\begin{array}{ccc|ccc} 1 & 0 & -1 & \frac{3}{4} & -\frac{1}{4} & 0 \\ 0 & 1 & 1 & \frac{1}{4} & \frac{1}{4} & 0 \\ 0 & 0 & -1 & -1 & -1 & 1 \end{array}\right] \rightarrow \left[\begin{array}{ccc|ccc} 1 & 0 & -1 & \frac{3}{4} & -\frac{1}{4} & 0 \\ 0 & 1 & 1 & \frac{1}{4} & \frac{1}{4} & 0 \\ 0 & 0 & 1 & 1 & 1 & -1 \end{array}\right]$$

$R_1 = -1r_2 + r_1$
$R_3 = -4r_2 + r_3$ $\qquad$ $R_3 = -1r_3$

$$\rightarrow \left[\begin{array}{ccc|ccc} 1 & 0 & 0 & \frac{7}{4} & \frac{3}{4} & -1 \\ 0 & 1 & 0 & -\frac{3}{4} & -\frac{3}{4} & 1 \\ 0 & 0 & 1 & 1 & 1 & -1 \end{array}\right]$$

$R_1 = r_3 + r_1$
$R_2 = -1r_3 + r_2$

Graphing Solution

Enter the matrix A into a graphing utility. Figure 14 shows A^{-1}.

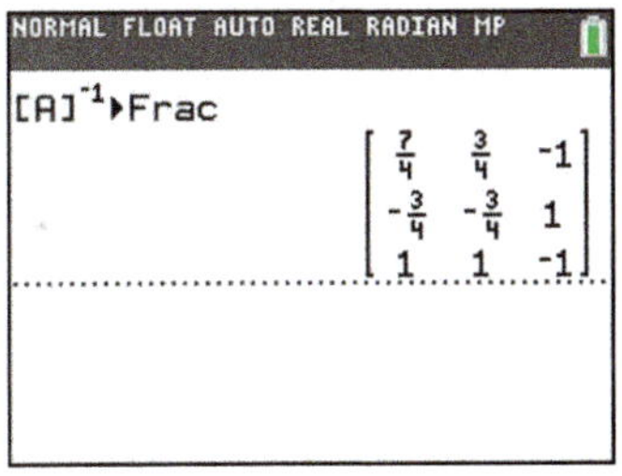

Figure 14 Inverse matrix

*For 2 by 2 matrices, there is a simple formula that can be used. See Problem 93.

The matrix $[A \mid I_3]$ is now in reduced row echelon form, and the identity matrix I_3 is on the left of the vertical bar. The inverse of A is

$$A^{-1} = \begin{bmatrix} \frac{7}{4} & \frac{3}{4} & -1 \\ -\frac{3}{4} & -\frac{3}{4} & 1 \\ 1 & 1 & -1 \end{bmatrix}$$

You should verify that this is the correct inverse by showing that

$$AA^{-1} = A^{-1}A = I_3.$$

Now Work PROBLEM 37

If transforming the matrix $[A \mid I_n]$ into reduced row echelon form does not result in the identity matrix I_n to the left of the vertical bar, A is singular and has no inverse.

EXAMPLE 14 **Showing That a Matrix Has No Inverse**

Show that the matrix $A = \begin{bmatrix} 4 & 6 \\ 2 & 3 \end{bmatrix}$ has no inverse.

Algebraic Solution

Begin by writing the matrix $[A \mid I_2]$.

$$[A \mid I_2] = \left[\begin{array}{cc|cc} 4 & 6 & 1 & 0 \\ 2 & 3 & 0 & 1 \end{array}\right]$$

Then use row operations to transform $[A \mid I_2]$ into reduced row echelon form.

$$[A \mid I_2] = \left[\begin{array}{cc|cc} 4 & 6 & 1 & 0 \\ 2 & 3 & 0 & 1 \end{array}\right] \rightarrow \left[\begin{array}{cc|cc} 1 & \frac{3}{2} & \frac{1}{4} & 0 \\ 2 & 3 & 0 & 1 \end{array}\right] \rightarrow \left[\begin{array}{cc|cc} 1 & \frac{3}{2} & \frac{1}{4} & 0 \\ 0 & 0 & -\frac{1}{2} & 1 \end{array}\right]$$

$R_1 = \frac{1}{4}r_1$ $\qquad$ $R_2 = -2r_1 + r_2$

The matrix $[A \mid I_2]$ is sufficiently reduced for it to be clear that the identity matrix cannot appear to the left of the vertical bar, so A is singular and has no inverse.

Graphing Solution

Enter the matrix A. Figure 15 shows the result of trying to find its inverse. The ERROR comes about because A is singular.

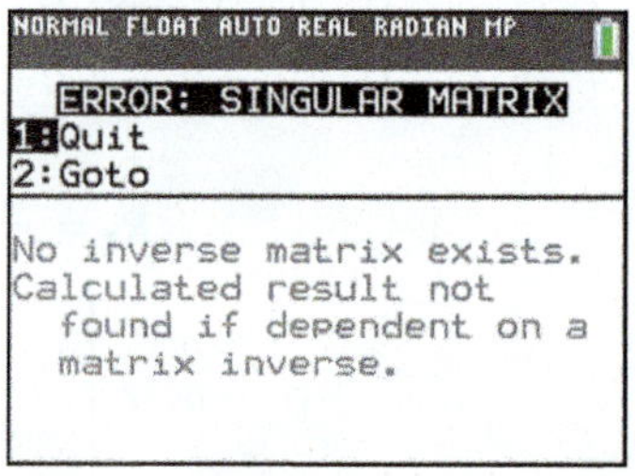

Figure 15

Seeing the Concept

Compute the determinant of A in Example 14 using a graphing utility. What is the result? Are you surprised?

Now Work PROBLEM 65

5 Solve a System of Linear Equations Using an Inverse Matrix

Inverse matrices can be used to solve systems of equations in which the number of equations is the same as the number of variables.

EXAMPLE 15 Using the Inverse Matrix to Solve a System of Linear Equations

Solve the system of equations: $\begin{cases} x + y = 3 \\ -x + 3y + 4z = -3 \\ 4y + 3z = 2 \end{cases}$

Solution Let

$$A = \begin{bmatrix} 1 & 1 & 0 \\ -1 & 3 & 4 \\ 0 & 4 & 3 \end{bmatrix} \quad X = \begin{bmatrix} x \\ y \\ z \end{bmatrix} \quad B = \begin{bmatrix} 3 \\ -3 \\ 2 \end{bmatrix}$$

Then the original system of equations can be written compactly as the matrix equation

$$AX = B \qquad \textbf{(4)}$$

From Example 13, the matrix A has the inverse A^{-1}. Multiply each side of equation (4) by A^{-1}.

$$\begin{aligned} AX &= B \\ A^{-1}(AX) &= A^{-1}B && \text{Multiply both sides by } A^{-1}. \\ (A^{-1}A)X &= A^{-1}B && \text{Associative Property of multiplication} \\ I_3X &= A^{-1}B && \text{Definition of an inverse matrix} \\ X &= A^{-1}B && \text{Property of the identity matrix} \end{aligned} \qquad \textbf{(5)}$$

Now use (5) to find $X = \begin{bmatrix} x \\ y \\ z \end{bmatrix}$. This can be done either algebraically or with a graphing utility.

Algebraic Solution

$$X = \begin{bmatrix} x \\ y \\ z \end{bmatrix} = A^{-1}B = \begin{bmatrix} \frac{7}{4} & \frac{3}{4} & -1 \\ -\frac{3}{4} & -\frac{3}{4} & 1 \\ 1 & 1 & -1 \end{bmatrix} \begin{bmatrix} 3 \\ -3 \\ 2 \end{bmatrix} = \begin{bmatrix} 1 \\ 2 \\ -2 \end{bmatrix}$$

↑ Example 13

Graphing Solution

Enter the matrices A and B into a graphing utility. Figure 16 shows the solution to the system of equations.

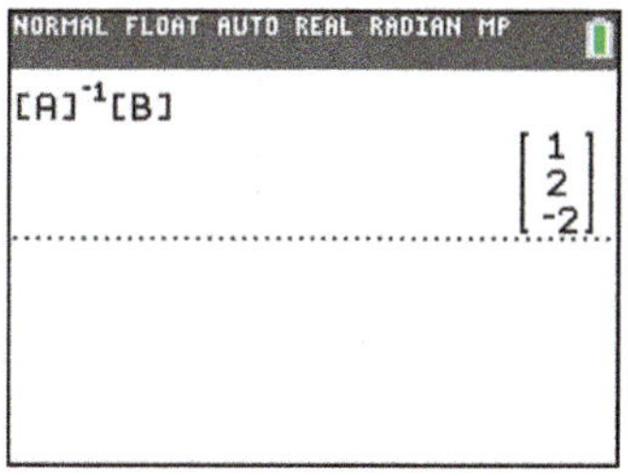

Figure 16

The solution is $x = 1, y = 2, z = -2$ or, using an ordered triplet, $(1, 2, -2)$. ■

The method used in Example 15 to solve a system of equations is particularly useful when it is necessary to solve several systems of equations in which the constants appearing to the right of the equal signs change, while the coefficients of the variables on the left side remain the same. See Problems 45–64 for some illustrations. Be careful; this method can be used only if the inverse exists. If it does not exist, row reduction must be used since the system is either inconsistent or dependent.

Now Work PROBLEM 49

Historical Feature

Arthur Cayley (1821–1895)

Matrices were invented in 1857 by Arthur Cayley (1821–1895) as a way of efficiently computing the result of substituting one linear system into another (see Historical Problem 3). The resulting system had incredible richness, in the sense that a wide variety of mathematical systems could be mimicked by the matrices. Cayley and his friend James J. Sylvester (1814–1897) spent much of the rest of their lives elaborating the theory. The torch was then passed to Georg Frobenius (1849–1917), whose deep investigations established a central place for matrices in modern mathematics. In 1924, rather to the surprise of physicists, it was found that matrices (with complex numbers in them) were exactly the right tool for describing the behavior of atomic systems. Today, matrices are used in a wide variety of applications.

Historical Problems

1. **Matrices and Complex Numbers** Frobenius emphasized in his research how matrices could be used to mimic other mathematical systems. Here, we mimic the behavior of complex numbers using matrices. Mathematicians call such a relationship an *isomorphism*.

Complex number $\longleftrightarrow$ Matrix

$$a + bi \longleftrightarrow \begin{bmatrix} a & b \\ -b & a \end{bmatrix}$$

Note that the complex number can be read off the top line of the matrix. Then

$$2 + 3i \longleftrightarrow \begin{bmatrix} 2 & 3 \\ -3 & 2 \end{bmatrix} \quad \text{and} \quad \begin{bmatrix} 4 & -2 \\ 2 & 4 \end{bmatrix} \longleftrightarrow 4 - 2i$$

(a) Find the matrices corresponding to $2 - 5i$ and $1 + 3i$.

(b) Multiply the two matrices.

(c) Find the corresponding complex number for the matrix found in part (b).

(d) Multiply $2 - 5i$ and $1 + 3i$. The result should be the same as that found in part (c).

The process also works for addition and subtraction. Try it for yourself.

2. Compute $(a + bi)(a - bi)$ using matrices. Interpret the result.

3. **Cayley's Definition of Matrix Multiplication** Cayley devised matrix multiplication to simplify the following problem:

$$\begin{cases} u = ar + bs \\ v = cr + ds \end{cases} \qquad \begin{cases} x = ku + lv \\ y = mu + nv \end{cases}$$

(a) Find x and y in terms of r and s by substituting u and v from the first system of equations into the second system of equations.

(b) Use the result of part (a) to find the 2 by 2 matrix A in

$$\begin{bmatrix} x \\ y \end{bmatrix} = A \begin{bmatrix} r \\ s \end{bmatrix}$$

(c) Now look at the following way to do it. Write the equations in matrix form.

$$\begin{bmatrix} u \\ v \end{bmatrix} = \begin{bmatrix} a & b \\ c & d \end{bmatrix} \begin{bmatrix} r \\ s \end{bmatrix} \qquad \begin{bmatrix} x \\ y \end{bmatrix} = \begin{bmatrix} k & l \\ m & n \end{bmatrix} \begin{bmatrix} u \\ v \end{bmatrix}$$

so

$$\begin{bmatrix} x \\ y \end{bmatrix} = \begin{bmatrix} k & l \\ m & n \end{bmatrix} \begin{bmatrix} a & b \\ c & d \end{bmatrix} \begin{bmatrix} r \\ s \end{bmatrix}$$

Do you see how Cayley defined matrix multiplication?

12.4 Assess Your Understanding

Concepts and Vocabulary

1. A matrix that has the same number of rows as columns is called a(n) _________ matrix.
2. ***True or False*** Matrix addition is commutative.
3. ***True or False*** If A and B are square matrices, then $AB = BA$,
4. Suppose that A is a square n by n matrix that is nonsingular. The matrix B for which $AB = BA = I_n$ is called the __________ of the matrix A.
5. ***True or False*** The identity matrix has properties similar to those of the real number 1.
6. If $AX = B$ represents a matrix equation where A is a nonsingular matrix, then we can solve the equation using $X =$ ______.
7. To find the product AB of two matrices A and B, which of the following must be true?
   (a) The number of columns in A must equal the number of rows in B.
   (b) The number of rows in A must equal the number of columns in B.
   (c) A and B must have the same number of rows and the same number of columns.
   (d) A and B must both be square matrices.
8. A matrix that has no inverse is called which of the following?
   (a) zero matrix (b) nonsingular matrix
   (c) identity matrix (d) singular matrix

Skill Building

In Problems 9–26, use the following matrices. Determine whether the given expression is defined. If it is defined, express the result as a single matrix; if it is not, say "not defined."

$$A = \begin{bmatrix} 0 & 3 & -5 \\ 1 & 2 & 6 \end{bmatrix} \quad B = \begin{bmatrix} 4 & 1 & 0 \\ -2 & 3 & -2 \end{bmatrix} \quad C = \begin{bmatrix} 4 & 1 \\ 6 & 2 \\ -2 & 3 \end{bmatrix}$$

9. $A + B$ **10.** $A - B$ **11.** $4A$ **12.** $-3B$

13. $3A - 2B$ **14.** $2A + 4B$ **15.** AC **16.** BC

17. AB **18.** BA **19.** CA **20.** CB

21. $C(A + B)$ **22.** $(A + B)C$ **23.** $AC - 3I_2$ **24.** $CA + 5I_3$

25. $CA - CB$ **26.** $AC + BC$

In Problems 27–34, determine whether the product is defined. If it is defined, find the product; if it is not, say "not defined."

27. $\begin{bmatrix} 2 & -2 \\ 1 & 0 \end{bmatrix}\begin{bmatrix} 2 & 1 & 4 & 6 \\ 3 & -1 & 3 & 2 \end{bmatrix}$ **28.** $\begin{bmatrix} 4 & 1 \\ 2 & 1 \end{bmatrix}\begin{bmatrix} -6 & 6 & 1 & 0 \\ 2 & 5 & 4 & -1 \end{bmatrix}$ **29.** $\begin{bmatrix} 1 & 2 & 3 \\ 0 & -1 & 4 \end{bmatrix}\begin{bmatrix} 1 & 2 \\ -1 & 0 \\ 2 & 4 \end{bmatrix}$ **30.** $\begin{bmatrix} 1 & -1 \\ -3 & 2 \\ 0 & 5 \end{bmatrix}\begin{bmatrix} 2 & 8 & -1 \\ 3 & 6 & 0 \end{bmatrix}$

31. $\begin{bmatrix} -4 \\ 2 \end{bmatrix}\begin{bmatrix} 1 & 0 \\ 3 & -1 \end{bmatrix}$ **32.** $\begin{bmatrix} 2 & -1 \\ 5 & 8 \\ -6 & 0 \end{bmatrix}\begin{bmatrix} 6 & 4 & 2 \\ -3 & 5 & -1 \\ 9 & 0 & 7 \end{bmatrix}$ **33.** $\begin{bmatrix} 1 & 0 & 1 \\ 2 & 4 & 1 \\ 3 & 6 & 1 \end{bmatrix}\begin{bmatrix} 1 & 3 \\ 6 & 2 \\ 8 & -1 \end{bmatrix}$ **34.** $\begin{bmatrix} 4 & -2 & 3 \\ 0 & 1 & 2 \\ -1 & 0 & 1 \end{bmatrix}\begin{bmatrix} 2 & 6 \\ 1 & -1 \\ 0 & 2 \end{bmatrix}$

In Problems 35–44, each matrix is nonsingular. Find the inverse of each matrix.

35. $\begin{bmatrix} 2 & 1 \\ 1 & 1 \end{bmatrix}$ **36.** $\begin{bmatrix} 3 & -1 \\ -2 & 1 \end{bmatrix}$ **37.** $\begin{bmatrix} 6 & 5 \\ 2 & 2 \end{bmatrix}$ **38.** $\begin{bmatrix} -4 & 1 \\ 6 & -2 \end{bmatrix}$ **39.** $\begin{bmatrix} 2 & 1 \\ a & a \end{bmatrix}$ $a \neq 0$

40. $\begin{bmatrix} b & 3 \\ b & 2 \end{bmatrix}$ $b \neq 0$ **41.** $\begin{bmatrix} 1 & -1 & 1 \\ 0 & -2 & 1 \\ -2 & -3 & 0 \end{bmatrix}$ **42.** $\begin{bmatrix} 1 & 0 & 2 \\ -1 & 2 & 3 \\ 1 & -1 & 0 \end{bmatrix}$ **43.** $\begin{bmatrix} 1 & 1 & 1 \\ 3 & 2 & -1 \\ 3 & 1 & 2 \end{bmatrix}$ **44.** $\begin{bmatrix} 3 & 3 & 1 \\ 1 & 2 & 1 \\ 2 & -1 & 1 \end{bmatrix}$

In Problems 45–64, use the inverses found in Problems 35–44 to solve each system of equations.

45. $\begin{cases} 2x + y = 8 \\ x + y = 5 \end{cases}$ **46.** $\begin{cases} 3x - y = 8 \\ -2x + y = 4 \end{cases}$ **47.** $\begin{cases} 2x + y = 0 \\ x + y = 5 \end{cases}$ **48.** $\begin{cases} 3x - y = 4 \\ -2x + y = 5 \end{cases}$

49. $\begin{cases} 6x + 5y = 7 \\ 2x + 2y = 2 \end{cases}$ **50.** $\begin{cases} -4x + y = 0 \\ 6x - 2y = 14 \end{cases}$ **51.** $\begin{cases} 6x + 5y = 13 \\ 2x + 2y = 5 \end{cases}$ **52.** $\begin{cases} -4x + y = 5 \\ 6x - 2y = -9 \end{cases}$

53. $\begin{cases} 2x + y = -3 \\ ax + ay = -a \end{cases}$ $a \neq 0$ **54.** $\begin{cases} bx + 3y = 2b + 3 \\ bx + 2y = 2b + 2 \end{cases}$ $b \neq 0$ **55.** $\begin{cases} 2x + y = \dfrac{7}{a} \\ ax + ay = 5 \end{cases}$ $a \neq 0$ **56.** $\begin{cases} bx + 3y = 14 \\ bx + 2y = 10 \end{cases}$ $b \neq 0$

57. $\begin{cases} x - y + z = 0 \\ -2y + z = -1 \\ -2x - 3y = -5 \end{cases}$ **58.** $\begin{cases} x + 2z = 6 \\ -x + 2y + 3z = -5 \\ x - y = 6 \end{cases}$ **59.** $\begin{cases} x - y + z = 2 \\ -2y + z = 2 \\ -2x - 3y = \dfrac{1}{2} \end{cases}$ **60.** $\begin{cases} x + 2z = 2 \\ -x + 2y + 3z = -\dfrac{3}{2} \\ x - y = 2 \end{cases}$

61. $\begin{cases} x + y + z = 9 \\ 3x + 2y - z = 8 \\ 3x + y + 2z = 1 \end{cases}$ **62.** $\begin{cases} 3x + 3y + z = 8 \\ x + 2y + z = 5 \\ 2x - y + z = 4 \end{cases}$ **63.** $\begin{cases} x + y + z = 2 \\ 3x + 2y - z = \dfrac{7}{3} \\ 3x + y + 2z = \dfrac{10}{3} \end{cases}$ **64.** $\begin{cases} 3x + 3y + z = 1 \\ x + 2y + z = 0 \\ 2x - y + z = 4 \end{cases}$

In Problems 65–70, show that each matrix has no inverse.

65. $\begin{bmatrix} 4 & 2 \\ 2 & 1 \end{bmatrix}$ **66.** $\begin{bmatrix} -3 & \frac{1}{2} \\ 6 & -1 \end{bmatrix}$ **67.** $\begin{bmatrix} 15 & 3 \\ 10 & 2 \end{bmatrix}$

68. $\begin{bmatrix} -3 & 0 \\ 4 & 0 \end{bmatrix}$ **69.** $\begin{bmatrix} -3 & 1 & -1 \\ 1 & -4 & -7 \\ 1 & 2 & 5 \end{bmatrix}$ **70.** $\begin{bmatrix} 1 & 1 & -3 \\ 2 & -4 & 1 \\ -5 & 7 & 1 \end{bmatrix}$

In Problems 71–74, use a graphing utility to find the inverse, if it exists, of each matrix. Round answers to two decimal places.

71. $\begin{bmatrix} 25 & 61 & -12 \\ 18 & -2 & 4 \\ 8 & 35 & 21 \end{bmatrix}$

72. $\begin{bmatrix} 18 & -3 & 4 \\ 6 & -20 & 14 \\ 10 & 25 & -15 \end{bmatrix}$

73. $\begin{bmatrix} 44 & 21 & 18 & 6 \\ -2 & 10 & 15 & 5 \\ 21 & 12 & -12 & 4 \\ -8 & -16 & 4 & 9 \end{bmatrix}$

74. $\begin{bmatrix} 16 & 22 & -3 & 5 \\ 21 & -17 & 4 & 8 \\ 2 & 8 & 27 & 20 \\ 5 & 15 & -3 & -10 \end{bmatrix}$

In Problems 75–78, use the idea behind Example 15 with a graphing utility to solve the following systems of equations. Round answers to two decimal places.

75. $\begin{cases} 25x + 61y - 12z = 10 \\ 18x - 12y + 7y = -9 \\ 3x + 4y - z = 12 \end{cases}$

76. $\begin{cases} 25x + 61y - 12z = 15 \\ 18x - 12y + 7z = -3 \\ 3x + 4y - z = 12 \end{cases}$

77. $\begin{cases} 25x + 61y - 12z = 21 \\ 18x - 12y + 7z = 7 \\ 3x + 4y - z = -2 \end{cases}$

78. $\begin{cases} 25x + 61y - 12z = 25 \\ 18x - 12y + 7z = 10 \\ 3x + 4y - z = -4 \end{cases}$

Mixed Practice

In Problems 79–86, algebraically solve each system of equations using any method you wish.

79. $\begin{cases} 2x + 3y = 11 \\ 5x + 7y = 24 \end{cases}$

80. $\begin{cases} 2x + 8y = -8 \\ x + 7y = -13 \end{cases}$

81. $\begin{cases} x - 2y + 4z = 2 \\ -3x + 5y - 2z = 17 \\ 4x - 3y = -22 \end{cases}$

82. $\begin{cases} 2x + 3y - z = -2 \\ 4x + 3z = 6 \\ 6y - 2z = 2 \end{cases}$

83. $\begin{cases} 5x - y + 4z = 2 \\ -x + 5y - 4z = 3 \\ 7x + 13y - 4z = 17 \end{cases}$

84. $\begin{cases} 3x + 2y - z = 2 \\ 2x + y + 6z = -7 \\ 2x + 2y - 14z = 17 \end{cases}$

85. $\begin{cases} 2x - 3y + z = 4 \\ -3x + 2y - z = -3 \\ -5y + z = 6 \end{cases}$

86. $\begin{cases} -4x + 3y + 2z = 6 \\ 3x + y - z = -2 \\ x + 9y + z = 6 \end{cases}$

Applications and Extensions

87. College Tuition Nikki and Joe take classes at a community college, LCCC, and a local university, SIUE. The number of credit hours taken and the cost per credit hour (2015–2016 academic year, tuition and approximate fees) are as follows:

	LCCC	SIUE
Nikki	6	9
Joe	3	12

	Cost per Credit Hour
LCCC	\$128.00
SIUE	\$341.60

(a) Write a matrix A for the credit hours taken by each student and a matrix B for the cost per credit hour.

(b) Compute AB and interpret the results.

Sources: *lc.edu, siue.edu*

88. School Loan Interest Jamal and Stephanie both have school loans issued from the same two banks. The amounts borrowed and the monthly interest rates are given next (interest is compounded monthly).

	Lender 1	Lender 2
Jamal	\$4000	\$3000
Stephanie	\$2500	\$3800

	Monthly Interest Rate
Lender 1	0.011 (1.1%)
Lender 2	0.006 (0.6%)

(a) Write a matrix A for the amounts borrowed by each student and a matrix B for the monthly interest rates.

(b) Compute AB and interpret the result.

(c) Let $C = \begin{bmatrix} 1 \\ 1 \end{bmatrix}$. Compute $A(C + B)$ and interpret the result.

89. Computing the Cost of Production The Acme Steel Company is a producer of stainless steel and aluminum containers. On a certain day, the following stainless steel containers were manufactured: 500 with 10-gallon (gal) capacity, 350 with 5-gal capacity, and 400 with 1-gal capacity. On the same day, the following aluminum containers were manufactured: 700 with 10-gal capacity, 500 with 5-gal capacity, and 850 with 1-gal capacity.

(a) Find a 2 by 3 matrix representing these data. Find a 3 by 2 matrix to represent the same data.

(b) If the amount of material used in the 10-gal containers is 15 pounds (lb), the amount used in the 5-gal containers is 8 lb, and the amount used in the 1-gal containers is 3 lb, find a 3 by 1 matrix representing the amount of material used.

(c) Multiply the 2 by 3 matrix found in part (a) and the 3 by 1 matrix found in part (b) to get a 2 by 1 matrix showing the day's usage of material.

(d) If stainless steel costs Acme \$0.10 per pound and aluminum costs \$0.05 per pound, find a 1 by 2 matrix representing cost.

(e) Multiply the matrices found in parts (c) and (d) to determine the total cost of the day's production.

90. Computing Profit Rizza Ford has two locations, one in the city and the other in the suburbs. In January, the city location sold 400 subcompacts, 250 intermediate-size cars, and 50 SUVs; in February, it sold 350 subcompacts, 100 intermediates, and 30 SUVs. At the suburban location in January, 450 subcompacts, 200 intermediates, and 140 SUVs were sold. In February, the suburban location sold 350 subcompacts, 300 intermediates, and 100 SUVs.

(a) Find 2 by 3 matrices that summarize the sales data for each location for January and February (one matrix for each month).

(b) Use matrix addition to obtain total sales for the 2-month period.
(c) The profit on each kind of car is \$100 per subcompact, \$150 per intermediate, and \$200 per SUV. Find a 3 by 1 matrix representing this profit.
(d) Multiply the matrices found in parts (b) and (c) to get a 2 by 1 matrix showing the profit at each location.

91. Cryptography One method of encryption is to use a matrix to encrypt the message and then use the corresponding inverse matrix to decode the message. The encrypted matrix, E, is obtained by multiplying the message matrix, M, by a key matrix, K. The original message can be retrieved by multiplying the encrypted matrix by the inverse of the key matrix. That is, $E = M \cdot K$ and $M = E \cdot K^{-1}$.

(a) Given the key matrix $K = \begin{bmatrix} 2 & 1 & 1 \\ 1 & 1 & 0 \\ 1 & 1 & 1 \end{bmatrix}$, find its inverse, K^{-1}. [**Note:** This key matrix is known as the Q_2^3 Fibonacci encryption matrix.]
(b) Use your result from part (a) to decode the encrypted matrix $E = \begin{bmatrix} 47 & 34 & 33 \\ 44 & 36 & 27 \\ 47 & 41 & 20 \end{bmatrix}$.
(c) Each entry in your result for part (b) represents the position of a letter in the English alphabet $(A = 1, B = 2, C = 3$, and so on$)$. What is the original message?

Source: *goldenmuseum.com*

92. Economic Mobility The relative income of a child (low, medium, or high) generally depends on the relative income of the child's parents. The matrix P, given by

$$\begin{array}{c} \text{Parent's Income} \\ \begin{array}{ccc} \text{L} & \text{M} & \text{H} \end{array} \end{array}$$
$$P = \begin{bmatrix} 0.4 & 0.2 & 0.1 \\ 0.5 & 0.6 & 0.5 \\ 0.1 & 0.2 & 0.4 \end{bmatrix} \begin{array}{l} \text{L} \\ \text{M} \\ \text{H} \end{array} \quad \text{Child's income}$$

is called a *left stochastic transition matrix.* For example, the entry $P_{21} = 0.5$ means that 50% of the children of low-relative-income parents will transition to the medium level of income. The diagonal entry P_{ii} represents the percent of children who remain in the same income level as their parents. Assuming that the transition matrix is valid from one generation to the next, compute and interpret P^2.

Source: *Understanding Mobility in America, April 2006*

93. Consider the 2 by 2 square matrix

$$A = \begin{bmatrix} a & b \\ c & d \end{bmatrix}$$

If $D = ad - bc \neq 0$, show that A is nonsingular and that

$$A^{-1} = \frac{1}{D}\begin{bmatrix} d & -b \\ -c & a \end{bmatrix}$$

Use the following discussion for Problems 94 and 95.

*In graph theory, an **adjacency matrix**, A, is a way of representing which nodes (or vertices) are connected. For a simple directed graph, each entry, a_{ij}, is either 1 (if a direct path exists from node i to node j) or 0 (if no direct path exists from node i to node j). For example, consider the following graph and corresponding adjacency matrix.*

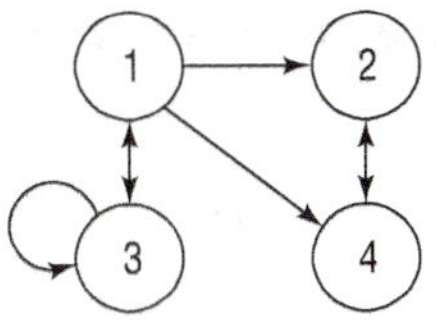

$$\begin{bmatrix} 0 & 1 & 1 & 1 \\ 0 & 0 & 0 & 1 \\ 1 & 0 & 1 & 0 \\ 0 & 1 & 0 & 0 \end{bmatrix}$$

The entry a_{14} is 1 because a direct path exists from node 1 to node 4. However, the entry a_{41} is 0 because no path exists from node 4 to node 1. The entry a_{33} is 1 because a direct path exists from node 3 to itself. The matrix $B_k = A + A^2 + \cdots + A^k$ indicates the number of ways to get from node i to node j within k moves (steps).

94. Website Map A content map can be used to show how different pages on a website are connected. For example, the following content map shows the relationship among the five pages of a certain website with links between pages represented by arrows.

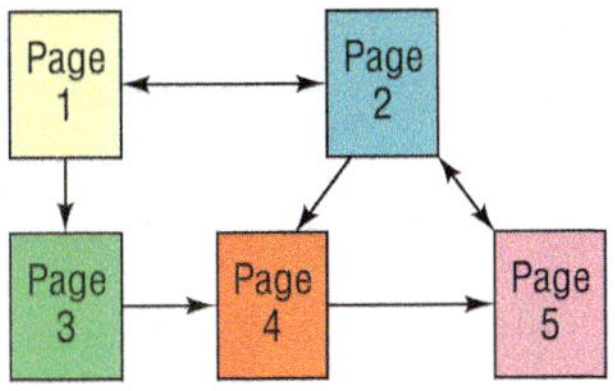

The content map can be represented by a 5 by 5 adjacency matrix where each entry, a_{ij}, is either 1 (if a link exists from page i to page j) or 0 (if no link exists from page i to page j).
(a) Write the 5 by 5 adjacency matrix that represents the given content map.
(b) Explain the significance of the entries on the main diagonal in your result from part (a).
(c) Find and interpret A^2.

95. Three-Click Rule An unofficial, and often contested, guideline for website design is to make all website content available to a user within three clicks. The webpage adjacency matrix for a certain website is given by

$$A = \begin{bmatrix} 0 & 1 & 1 & 0 & 0 \\ 1 & 0 & 0 & 1 & 1 \\ 1 & 0 & 0 & 1 & 0 \\ 0 & 0 & 1 & 0 & 1 \\ 0 & 1 & 0 & 0 & 0 \end{bmatrix}$$

(a) Find B_3. Does this website adhere to the Three-Click Rule?
(b) Which page can be reached the greatest number of ways from page 1 within three clicks?

96. Computer Graphics: Translating An important aspect of computer graphics is the ability to transform the coordinates of points within a graphic. For transformation purposes, a point (x, y) is represented as the column matrix $X = \begin{bmatrix} x \\ y \\ 1 \end{bmatrix}$.

To translate a point (x, y) horizontally h units and vertically k units, we use the translation matrix $S = \begin{bmatrix} 1 & 0 & h \\ 0 & 1 & k \\ 0 & 0 & 1 \end{bmatrix}$ and compute the matrix product SX. The translation is to the right for $h > 0$ and to the left for $h < 0$. Likewise, the translation is up for $k > 0$ and down for $k < 0$. The transformed coordinates are the first two entries in the resulting column matrix.

(a) Write the translation matrix needed to translate a point 3 units to the left and 5 units up.

(b) Find and interpret S^{-1}.

97. Computer Graphics: Rotating Besides translating a point, it is also important in computer graphics to be able to rotate a point. This is achieved by multiplying a point's column matrix (see Problem 96) by an appropriate rotation matrix R to form the matrix product RX. For example, to rotate a point 60°, the rotation matrix is $R = \begin{bmatrix} \frac{1}{2} & -\frac{\sqrt{3}}{2} & 0 \\ \frac{\sqrt{3}}{2} & \frac{1}{2} & 0 \\ 0 & 0 & 1 \end{bmatrix}$.

(a) Write the coordinates of the point $(6, 4)$ after it has been rotated 60°.

(b) Find and interpret R^{-1}.

Explaining Concepts: Discussion and Writing

98. Create a situation different from any found in the text that can be represented by a matrix.

99. Explain why the number of columns in matrix A must equal the number of rows in matrix B when finding the product AB.

100. If $a, b,$ and $c \neq 0$ are real numbers with $ac = bc$, then $a = b$. Does this same property hold for matrices? In other words, if $A, B,$ and C are matrices and $AC = BC$, must $A = B$?

101. What is the solution of the system of equations $AX = 0$ if A^{-1} exists? Discuss the solution of $AX = 0$ if A^{-1} does not exist.

Retain Your Knowledge

Problems 102–105 are based on material learned earlier in the course. The purpose of these problems is to keep the material fresh in your mind so that you are better prepared for the final exam.

102. Write a polynomial with minimum degree and leading coefficient 1 that has zeros $x = 3$ (multiplicity 2), $x = 0$ (multiplicity 3), and $x = -2$ (multiplicity 1).

103. Write $\cos(\csc^{-1} u)$ as an algebraic expression in u.

104. Solve: $\dfrac{5x}{x+2} = \dfrac{x}{x-2}$

105. For $\mathbf{v} = -2\mathbf{i} - \mathbf{j}$ and $\mathbf{w} = 2\mathbf{i} + \mathbf{j}$, find the dot product $\mathbf{v} \cdot \mathbf{w}$ and the angle between $\mathbf{v}$ and $\mathbf{w}$.

12.5 Partial Fraction Decomposition

PREPARING FOR THIS SECTION *Before getting started, review the following:*

- Identity (Section 1.2, p. 99)
- Proper and Improper Rational Functions (Section 5.4, p. 377)
- Factoring Polynomials (Chapter R, Review, Section R.5, pp. 51–56)
- Complex Zeros; Fundamental Theorem of Algebra (Section 5.3, pp. 366–370)

Now Work the **'Are You Prepared?'** problems on page 926.

OBJECTIVES

1 Decompose $\frac{P}{Q}$ Where Q Has Only Nonrepeated Linear Factors (p. 921)

2 Decompose $\frac{P}{Q}$ Where Q Has Repeated Linear Factors (p. 923)

3 Decompose $\frac{P}{Q}$ Where Q Has a Nonrepeated Irreducible Quadratic Factor (p. 925)

4 Decompose $\frac{P}{Q}$ Where Q Has a Repeated Irreducible Quadratic Factor (p. 926)

Consider the problem of adding two rational expressions:

$$\frac{3}{x+4} \quad \text{and} \quad \frac{2}{x-3}$$

The result is

$$\frac{3}{x+4} + \frac{2}{x-3} = \frac{3(x-3) + 2(x+4)}{(x+4)(x-3)} = \frac{5x-1}{x^2+x-12}$$

The reverse procedure, starting with the rational expression $\frac{5x-1}{x^2+x-12}$ and writing it as the sum (or difference) of the two simpler fractions $\frac{3}{x+4}$ and $\frac{2}{x-3}$, is referred to as **partial fraction decomposition**, and the two simpler fractions are called **partial fractions**. Decomposing a rational expression into a sum of partial fractions is important in solving certain types of calculus problems. This section presents a systematic way to decompose rational expressions.

Recall that a rational expression is the ratio of two polynomials, say P and $Q \neq 0$. Assume that P and Q have no common factors. Recall also that a rational expression $\frac{P}{Q}$ is called **proper** if the degree of the polynomial in the numerator is less than the degree of the polynomial in the denominator. Otherwise, the rational expression is called **improper**.

Because any improper rational expression can be reduced by long division to a mixed form consisting of the sum of a polynomial and a proper rational expression, we shall restrict the discussion that follows to proper rational expressions.

The partial fraction decomposition of the rational expression $\frac{P}{Q}$, in lowest terms, depends on the factors of the denominator Q. Recall from Section 5.3 that any polynomial whose coefficients are real numbers can be factored (over the real numbers) into products of linear and/or irreducible quadratic factors. This means that the denominator Q of the rational expression $\frac{P}{Q}$ will contain only factors of one or both of the following types:

- *Linear factors* of the form $x - a$, where a is a real number.
- *Irreducible quadratic factors* of the form $ax^2 + bx + c$, where a, b, and c are real numbers, $a \neq 0$, and $b^2 - 4ac < 0$ (which guarantees that $ax^2 + bx + c$ cannot be written as the product of two linear factors with real coefficients).

As it turns out, there are four cases to be examined. We begin with the case for which Q has only nonrepeated linear factors. Throughout we assume the rational expression $\frac{P}{Q}$ is in lowest terms.

1 Decompose $\frac{P}{Q}$ Where Q Has Only Nonrepeated Linear Factors

Case 1: Q has only nonrepeated linear factors.

Under the assumption that Q has only nonrepeated linear factors, the polynomial Q has the form

$$Q(x) = (x - a_1)(x - a_2) \cdot \cdots \cdot (x - a_n)$$

where no two of the numbers $a_1, a_2, \ldots, a_n$ are equal. In this case, the partial fraction decomposition of $\frac{P}{Q}$ is of the form

$$\frac{P(x)}{Q(x)} = \frac{A_1}{x - a_1} + \frac{A_2}{x - a_2} + \cdots + \frac{A_n}{x - a_n} \qquad (1)$$

where the numbers $A_1, A_2, \ldots, A_n$ are to be determined.

The example shows how to find these numbers.

EXAMPLE 1 Nonrepeated Linear Factors

Find the partial fraction decomposition of $\dfrac{x}{x^2 - 5x + 6}$.

Solution First, factor the denominator,

$$x^2 - 5x + 6 = (x - 2)(x - 3)$$

and notice that the denominator contains only nonrepeated linear factors. Then decompose the rational expression according to equation (1):

$$\frac{x}{x^2 - 5x + 6} = \frac{A}{x - 2} + \frac{B}{x - 3} \qquad \textbf{(2)}$$

where A and B are to be determined. To find A and B, clear the fractions by multiplying each side by $(x - 2)(x - 3) = x^2 - 5x + 6$. The result is

$$x = A(x - 3) + B(x - 2) \qquad \textbf{(3)}$$

or

$$x = (A + B)x + (-3A - 2B)$$

This equation is an identity in x. Equate the coefficients of like powers of x to get

$$\begin{cases} 1 = A + B & \text{Equate the coefficients of } x\text{: } 1x = (A + B)x. \\ 0 = -3A - 2B & \text{Equate the constants: } 0 = -3A - 2B. \end{cases}$$

This system of two equations containing two variables, A and B, can be solved using whatever method you wish. Solving it yields

$$A = -2 \qquad B = 3$$

From equation (2), the partial fraction decomposition is

$$\frac{x}{x^2 - 5x + 6} = \frac{-2}{x - 2} + \frac{3}{x - 3}$$

Check: The decomposition can be checked by adding the rational expressions.

$$\frac{-2}{x - 2} + \frac{3}{x - 3} = \frac{-2(x - 3) + 3(x - 2)}{(x - 2)(x - 3)} = \frac{x}{(x - 2)(x - 3)}$$
$$= \frac{x}{x^2 - 5x + 6}$$

The numbers to be found in the partial fraction decomposition can sometimes be found more readily by using suitable choices for x (which may include complex numbers) in the identity obtained after fractions have been cleared. In Example 1, the identity after clearing fractions is equation (3):

$$x = A(x - 3) + B(x - 2)$$

Let $x = 2$ in this expression, and the term containing B drops out, leaving $2 = A(-1)$, or $A = -2$. Similarly, let $x = 3$, and the term containing A drops out, leaving $3 = B$. As before, $A = -2$ and $B = 3$.

Now Work PROBLEM 13

2 Decompose $\frac{P}{Q}$ Where Q Has Repeated Linear Factors

Case 2: Q has repeated linear factors.

If the polynomial Q has a repeated linear factor, say $(x-a)^n, n \geq 2$ an integer, then, in the partial fraction decomposition of $\frac{P}{Q}$, allow for the terms

$$\frac{A_1}{x-a} + \frac{A_2}{(x-a)^2} + \cdots + \frac{A_n}{(x-a)^n}$$

where the numbers $A_1, A_2, \ldots, A_n$ are to be determined.

EXAMPLE 2 **Repeated Linear Factors**

Find the partial fraction decomposition of $\frac{x+2}{x^3-2x^2+x}$.

Solution First, factor the denominator,

$$x^3 - 2x^2 + x = x(x^2 - 2x + 1) = x(x-1)^2$$

and notice that the denominator has the nonrepeated linear factor x and the repeated linear factor $(x-1)^2$. By Case 1, the term $\frac{A}{x}$ must be in the decomposition; and by Case 2, the terms $\frac{B}{x-1} + \frac{C}{(x-1)^2}$ must be in the decomposition.

Now write

$$\frac{x+2}{x^3-2x^2+x} = \frac{A}{x} + \frac{B}{x-1} + \frac{C}{(x-1)^2} \quad \textbf{(4)}$$

Again, clear fractions by multiplying each side by $x^3 - 2x^2 + x = x(x-1)^2$. The result is the identity

$$x + 2 = A(x-1)^2 + Bx(x-1) + Cx \quad \textbf{(5)}$$

Let $x = 0$ in this expression and the terms containing B and C drop out, leaving $2 = A(-1)^2$, or $A = 2$. Similarly, let $x = 1$, and the terms containing A and B drop out, leaving $3 = C$. Then equation (5) becomes

$$x + 2 = 2(x-1)^2 + Bx(x-1) + 3x$$

Let $x = 2$ (any choice other than 0 or 1 will work as well). The result is

$$4 = 2(1)^2 + B(2)(1) + 3(2)$$
$$4 = 2 + 2B + 6$$
$$2B = -4$$
$$B = -2$$

Therefore, $A = 2$, $B = -2$, and $C = 3$.

From equation (4), the partial fraction decomposition is

$$\frac{x+2}{x^3-2x^2+x} = \frac{2}{x} + \frac{-2}{x-1} + \frac{3}{(x-1)^2}$$

■

EXAMPLE 3 **Repeated Linear Factors**

Find the partial fraction decomposition of $\dfrac{x^3-8}{x^2(x-1)^3}$.

Solution The denominator contains the repeated linear factors x^2 and $(x-1)^3$. The partial fraction decomposition takes the form

$$\frac{x^3-8}{x^2(x-1)^3}=\frac{A}{x}+\frac{B}{x^2}+\frac{C}{x-1}+\frac{D}{(x-1)^2}+\frac{E}{(x-1)^3} \quad (6)$$

As before, clear fractions and obtain the identity

$$x^3-8=Ax(x-1)^3+B(x-1)^3+Cx^2(x-1)^2+Dx^2(x-1)+Ex^2 \quad (7)$$

Let $x=0$. (Do you see why this choice was made?) Then

$$-8=B(-1)$$
$$B=8$$

Let $x=1$ in equation (7). Then

$$-7=E$$

Use $B=8$ and $E=-7$ in equation (7), and collect like terms.

$$x^3-8=Ax(x-1)^3+8(x-1)^3+Cx^2(x-1)^2+Dx^2(x-1)-7x^2$$
$$x^3-8-8(x^3-3x^2+3x-1)+7x^2=Ax(x-1)^3+Cx^2(x-1)^2+Dx^2(x-1)$$
$$-7x^3+31x^2-24x=x(x-1)[A(x-1)^2+Cx(x-1)+Dx]$$
$$x(x-1)(-7x+24)=x(x-1)[A(x-1)^2+Cx(x-1)+Dx]$$
$$-7x+24=A(x-1)^2+Cx(x-1)+Dx \quad (8)$$

Now work with equation (8). Let $x=0$. Then

$$24=A$$

Let $x=1$ in equation (8). Then

$$17=D$$

Use $A=24$ and $D=17$ in equation (8).

$$-7x+24=24(x-1)^2+Cx(x-1)+17x$$

Let $x=2$ and simplify.

$$-14+24=24+C(2)+34$$
$$-48=2C$$
$$-24=C$$

The numbers A, B, C, D, and E are all now known. So, from equation (6),

$$\frac{x^3-8}{x^2(x-1)^3}=\frac{24}{x}+\frac{8}{x^2}+\frac{-24}{x-1}+\frac{17}{(x-1)^2}+\frac{-7}{(x-1)^3}$$

Now Work Example 3 by solving the system of five equations containing five variables that the expansion of equation (7) leads to.

Now Work PROBLEM 19

The final two cases involve irreducible quadratic factors. A quadratic factor is irreducible if it cannot be factored into linear factors with real coefficients.

A quadratic expression $ax^2 + bx + c$ is irreducible whenever $b^2 - 4ac < 0$. For example, $x^2 + x + 1$ and $x^2 + 4$ are irreducible.

3 Decompose $\frac{P}{Q}$ Where Q Has a Nonrepeated Irreducible Quadratic Factor

Case 3: Q contains a nonrepeated irreducible quadratic factor.

If Q contains a nonrepeated irreducible quadratic factor of the form $ax^2 + bx + c$, then, in the partial fraction decomposition of $\frac{P}{Q}$, allow for the term

$$\frac{Ax + B}{ax^2 + bx + c}$$

where the numbers A and B are to be determined.

EXAMPLE 4 **Nonrepeated Irreducible Quadratic Factor**

Find the partial fraction decomposition of $\frac{3x - 5}{x^3 - 1}$.

Solution Factor the denominator,

$$x^3 - 1 = (x - 1)(x^2 + x + 1)$$

Notice that it has a nonrepeated linear factor $x - 1$ and a nonrepeated irreducible quadratic factor $x^2 + x + 1$. Allow for the term $\frac{A}{x - 1}$ by Case 1, and allow for the term $\frac{Bx + C}{x^2 + x + 1}$ by Case 3. Then

$$\frac{3x - 5}{x^3 - 1} = \frac{A}{x - 1} + \frac{Bx + C}{x^2 + x + 1} \qquad (9)$$

Multiply each side of equation (9) by $x^3 - 1 = (x - 1)(x^2 + x + 1)$ to obtain the identity

$$3x - 5 = A(x^2 + x + 1) + (Bx + C)(x - 1) \qquad (10)$$

Expand the identity in (10) to obtain

$$3x - 5 = (A + B)x^2 + (A - B + C)x + (A - C)$$

This identity leads to the system of equations

$$\begin{cases} A + B = 0 & (1) \\ A - B + C = 3 & (2) \\ A - C = -5 & (3) \end{cases}$$

The solution of this system is $A = -\frac{2}{3}, B = \frac{2}{3}, C = \frac{13}{3}$. Then, from equation (9),

$$\frac{3x - 5}{x^3 - 1} = \frac{-\frac{2}{3}}{x - 1} + \frac{\frac{2}{3}x + \frac{13}{3}}{x^2 + x + 1}$$

Now Work Example 4 using equation (10) and assigning values to x.

Now Work PROBLEM 21

4 Decompose $\frac{P}{Q}$ Where Q Has a Repeated Irreducible Quadratic Factor

Case 4: Q contains a repeated irreducible quadratic factor.

If the polynomial Q contains a repeated irreducible quadratic factor $(ax^2 + bx + c)^n$, $n \geq 2$, n an integer, then, in the partial fraction decomposition of $\frac{P}{Q}$, allow for the terms

$$\frac{A_1x + B_1}{ax^2 + bx + c} + \frac{A_2x + B_2}{(ax^2 + bx + c)^2} + \cdots + \frac{A_nx + B_n}{(ax^2 + bx + c)^n}$$

where the numbers $A_1, B_1, A_2, B_2, \ldots, A_n, B_n$ are to be determined.

EXAMPLE 5 **Repeated Irreducible Quadratic Factor**

Find the partial fraction decomposition of $\frac{x^3 + x^2}{(x^2 + 4)^2}$.

Solution The denominator contains the repeated irreducible quadratic factor $(x^2 + 4)^2$, so write

$$\frac{x^3 + x^2}{(x^2 + 4)^2} = \frac{Ax + B}{x^2 + 4} + \frac{Cx + D}{(x^2 + 4)^2} \qquad \textbf{(11)}$$

Clear fractions to obtain

$$x^3 + x^2 = (Ax + B)(x^2 + 4) + Cx + D$$

Collecting like terms yields the identity

$$x^3 + x^2 = Ax^3 + Bx^2 + (4A + C)x + 4B + D$$

Equating coefficients results in the system

$$\begin{cases} A = 1 \\ B = 1 \\ 4A + C = 0 \\ 4B + D = 0 \end{cases}$$

The solution is $A = 1, B = 1, C = -4, D = -4$. From equation (11),

$$\frac{x^3 + x^2}{(x^2 + 4)^2} = \frac{x + 1}{x^2 + 4} + \frac{-4x - 4}{(x^2 + 4)^2}$$

Now Work PROBLEM 35

12.5 Assess Your Understanding

'Are You Prepared?' *Answers are given at the end of these exercises. If you get a wrong answer, read the pages listed in red.*

1. ***True or False*** The equation $(x - 1)^2 - 1 = x(x - 2)$ is an example of an identity. (p. 99)
2. ***True or False*** The rational expression $\frac{5x^2 - 1}{x^3 + 1}$ is proper. (p. 377)
3. Factor completely: $3x^4 + 6x^3 + 3x^2$ (pp. 51–56)
4. ***True or False*** Every polynomial with real numbers as coefficients can be factored into products of linear and/or irreducible quadratic factors. (p. 369)

Skill Building

In Problems 5–12, tell whether the given rational expression is proper or improper. If improper, rewrite it as the sum of a polynomial and a proper rational expression.

5. $\dfrac{x}{x^2-1}$
6. $\dfrac{5x+2}{x^3-1}$
7. $\dfrac{x^2+5}{x^2-4}$
8. $\dfrac{3x^2-2}{x^2-1}$
9. $\dfrac{5x^3+2x-1}{x^2-4}$
10. $\dfrac{3x^4+x^2-2}{x^3+8}$
11. $\dfrac{x(x-1)}{(x+4)(x-3)}$
12. $\dfrac{2x(x^2+4)}{x^2+1}$

In Problems 13–46, find the partial fraction decomposition of each rational expression.

13. $\dfrac{4}{x(x-1)}$
14. $\dfrac{3x}{(x+2)(x-1)}$
15. $\dfrac{1}{x(x^2+1)}$
16. $\dfrac{1}{(x+1)(x^2+4)}$
17. $\dfrac{x}{(x-1)(x-2)}$
18. $\dfrac{3x}{(x+2)(x-4)}$
19. $\dfrac{x^2}{(x-1)^2(x+1)}$
20. $\dfrac{x+1}{x^2(x-2)}$
21. $\dfrac{1}{x^3-8}$
22. $\dfrac{2x+4}{x^3-1}$
23. $\dfrac{x^2}{(x-1)^2(x+1)^2}$
24. $\dfrac{x+1}{x^2(x-2)^2}$
25. $\dfrac{x-3}{(x+2)(x+1)^2}$
26. $\dfrac{x^2+x}{(x+2)(x-1)^2}$
27. $\dfrac{x+4}{x^2(x^2+4)}$
28. $\dfrac{10x^2+2x}{(x-1)^2(x^2+2)}$
29. $\dfrac{x^2+2x+3}{(x+1)(x^2+2x+4)}$
30. $\dfrac{x^2-11x-18}{x(x^2+3x+3)}$
31. $\dfrac{x}{(3x-2)(2x+1)}$
32. $\dfrac{1}{(2x+3)(4x-1)}$
33. $\dfrac{x}{x^2+2x-3}$
34. $\dfrac{x^2-x-8}{(x+1)(x^2+5x+6)}$
35. $\dfrac{x^2+2x+3}{(x^2+4)^2}$
36. $\dfrac{x^3+1}{(x^2+16)^2}$
37. $\dfrac{7x+3}{x^3-2x^2-3x}$
38. $\dfrac{x^3+1}{x^5-x^4}$
39. $\dfrac{x^2}{x^3-4x^2+5x-2}$
40. $\dfrac{x^2+1}{x^3+x^2-5x+3}$
41. $\dfrac{x^3}{(x^2+16)^3}$
42. $\dfrac{x^2}{(x^2+4)^3}$
43. $\dfrac{4}{2x^2-5x-3}$
44. $\dfrac{4x}{2x^2+3x-2}$
45. $\dfrac{2x+3}{x^4-9x^2}$
46. $\dfrac{x^2+9}{x^4-2x^2-8}$

Mixed Practice

In Problems 47–54, use the division algorithm to rewrite each improper rational expression as the sum of a polynomial and a proper rational expression. Find the partial fraction decomposition of the proper rational expression. Finally, express the improper rational expression as the sum of a polynomial and the partial fraction decomposition.

47. $\dfrac{x^3+x^2-3}{x^2+3x-4}$
48. $\dfrac{x^3-3x^2+1}{x^2+5x+6}$
49. $\dfrac{x^3}{x^2+1}$
50. $\dfrac{x^3+x}{x^2+4}$
51. $\dfrac{x^4-5x^2+x-4}{x^2+4x+4}$
52. $\dfrac{x^4+x^3-x+2}{x^2-2x+1}$
53. $\dfrac{x^5+x^4-x^2+2}{x^4-2x^2+1}$
54. $\dfrac{x^5-x^3+x^2+1}{x^4+6x^2+9}$

Retain Your Knowledge

Problems 55–58 are based on material learned earlier in the course. The purpose of these problems is to keep the material fresh in your mind so that you are better prepared for the final exam.

55. **Credit Card Balance** Nick has a credit card balance of \$4200. If the credit card company charges 18% interest compound daily, and Nick does not make any payments on the account, how long will it take for his balance to double? Round to two decimal places.
56. Given $f(x) = x + 4$ and $g(x) = x^2 - 3x$, find $(g \circ f)(-3)$.
57. Find the exact value of $\sec 52^\circ \cos 308^\circ$.
58. Plot the point given by the polar coordinates $\left(-1, \dfrac{5\pi}{4}\right)$ and find its rectangular coordinates.

'Are You Prepared?' Answers

1. True **2.** True **3.** $3x^2(x+1)^2$ **4.** True

12.6 Systems of Nonlinear Equations

PREPARING FOR THIS SECTION *Before getting started, review the following:*

- Lines (Section 2.2, pp. 173–184)
- Circles (Section 2.3, pp. 189–193)
- Parabolas (Section 11.2, pp. 793–799)
- Ellipses (Section 11.3, pp. 803–811)
- Hyperbolas (Section 11.4, pp. 814–824)

Now Work the **'Are You Prepared?'** problems on page 934.

OBJECTIVES 1 Solve a System of Nonlinear Equations Using Substitution (p. 928)
2 Solve a System of Nonlinear Equations Using Elimination (p. 929)

1 Solve a System of Nonlinear Equations Using Substitution

In Section 12.1 we observed that the solution to a system of linear equations could be found geometrically by determining the point(s) of intersection (if any) of the equations in the system. Similarly, in solving systems of nonlinear equations, the solution(s) also represent(s) the point(s) of intersection (if any) of the graphs of the equations.

There is no general methodology for solving a system of nonlinear equations. Sometimes substitution is best; at other times elimination is best; and sometimes neither of these methods works. Experience and a certain degree of imagination are your allies here.

Before we begin, two comments are in order.

- If the system contains two variables and if the equations in the system are easy to graph, then graph them. By graphing each equation in the system, you can get an idea of how many solutions a system has and approximately where they are located.
- Extraneous solutions can creep in when solving nonlinear systems, so it is imperative to check all apparent solutions.

EXAMPLE 1 **Solving a System of Nonlinear Equations**

Solve the following system of equations:

$$\begin{cases} 3x - y = -2 & \text{(1)} \quad \text{A line} \\ 2x^2 - y = 0 & \text{(2)} \quad \text{A parabola} \end{cases}$$

Algebraic Solution Using Substitution

First, notice that the system contains two variables and that we know how to graph each equation by hand. See Figure 17. The system apparently has two solutions.

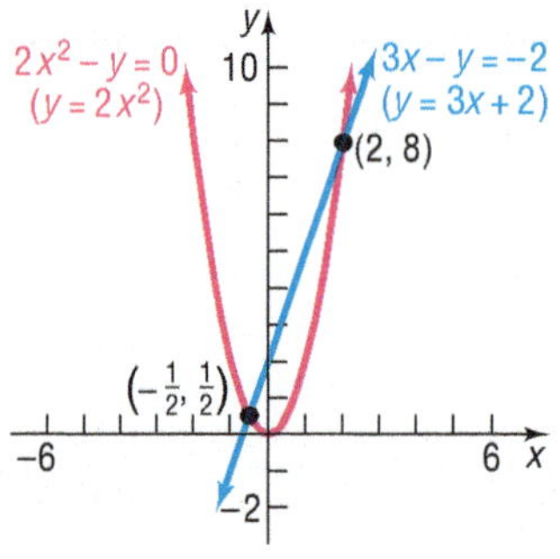

Figure 17

To use substitution to solve the system, we choose to solve equation (1) for y.

$$3x - y = -2 \quad \text{Equation (1)}$$
$$y = 3x + 2$$

Graphing Solution

Use a graphing utility to graph $Y_1 = 3x + 2$ and $Y_2 = 2x^2$. From Figure 18 observe that the system apparently has two solutions. Use INTERSECT to find that the solutions to the system of equations are $(-0.5, 0.5)$ and $(2, 8)$.

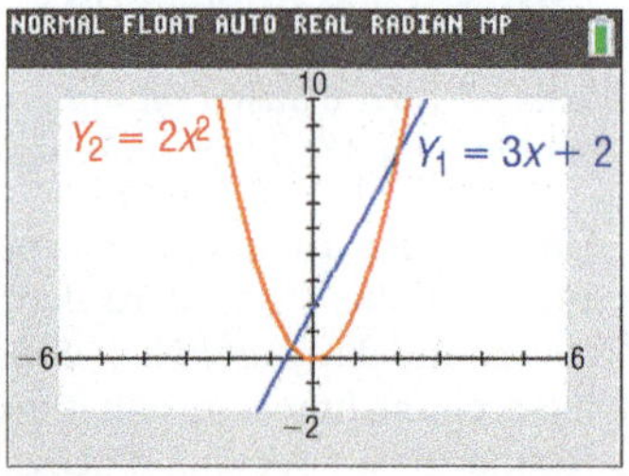

Figure 18

Substitute this expression for y in equation (2). The result is an equation containing just the variable x, which can then be solved.

$$2x^2 - y = 0 \quad \text{Equation (2)}$$
$$2x^2 - (3x + 2) = 0 \quad \text{Substitute } 3x + 2 \text{ for } y.$$
$$2x^2 - 3x - 2 = 0 \quad \text{Remove parentheses.}$$
$$(2x + 1)(x - 2) = 0 \quad \text{Factor.}$$
$$2x + 1 = 0 \quad \text{or} \quad x - 2 = 0 \quad \text{Apply the Zero-Product Property.}$$
$$x = -\frac{1}{2} \quad \text{or } x = 2$$

Use these values for x in $y = 3x + 2$ to find

$$y = 3\left(-\frac{1}{2}\right) + 2 = \frac{1}{2} \quad \text{or} \quad y = 3(2) + 2 = 8$$

The apparent solutions are $x = -\frac{1}{2}, y = \frac{1}{2}$ and $x = 2, y = 8$.

Check: For $x = -\frac{1}{2}, y = \frac{1}{2}$:

$$\begin{cases} 3\left(-\frac{1}{2}\right) - \frac{1}{2} = -\frac{3}{2} - \frac{1}{2} = -2 & (1) \\ 2\left(-\frac{1}{2}\right)^2 - \frac{1}{2} = 2\left(\frac{1}{4}\right) - \frac{1}{2} = 0 & (2) \end{cases}$$

For $x = 2, y = 8$:

$$\begin{cases} 3(2) - 8 = 6 - 8 = -2 & (1) \\ 2(2)^2 - 8 = 2(4) - 8 = 0 & (2) \end{cases}$$

Each solution checks. The graphs of the two equations intersect at the points $\left(-\frac{1}{2}, \frac{1}{2}\right)$ and $(2, 8)$, as shown in Figure 17. ■

Now Work PROBLEM 15 USING SUBSTITUTION

2 Solve a System of Nonlinear Equations Using Elimination

EXAMPLE 2 Solving a System of Nonlinear Equations

Solve: $\begin{cases} x^2 + y^2 = 13 & \text{(1) A circle} \\ x^2 - y = 7 & \text{(2) A parabola} \end{cases}$

Algebraic Solution Using Elimination

First graph each equation, as shown in Figure 19 on the next page. Based on the graph, four solutions are expected. Notice that subtracting equation (2) from equation (1) eliminates the variable x.

$$\begin{cases} x^2 + y^2 = 13 \\ x^2 - y = 7 \end{cases}$$
$$y^2 + y = 6 \quad \text{Subtract.}$$

Graphing Solution

Use a graphing utility to graph $x^2 + y^2 = 13$ and $x^2 - y = 7$. (Remember that to graph $x^2 + y^2 = 13$ requires two functions, $Y_1 = \sqrt{13 - x^2}$ and $Y_2 = -\sqrt{13 - x^2}$, and a square screen.) From Figure 20 on the next page, observe that the system apparently has four solutions.

This quadratic equation in y can be solved by factoring.

$$y^2 + y - 6 = 0$$
$$(y + 3)(y - 2) = 0$$
$$y = -3 \quad \text{or} \quad y = 2$$

Use these values for y in equation (2) to find x.
If $y = 2$, then $x^2 = y + 7 = 9$, so $x = 3$ or -3.
If $y = -3$, then $x^2 = y + 7 = 4$, so $x = 2$ or -2.

There are four solutions: $x = 3, y = 2$; $x = -3, y = 2$; $x = 2, y = -3$; and $x = -2, y = -3$.

You should verify that, in fact, these four solutions also satisfy equation (1), so all four are solutions of the system. The four points, $(3, 2)$, $(-3, 2)$, $(2, -3)$, and $(-2, -3)$, are the points of intersection of the graphs. Look again at Figure 19.

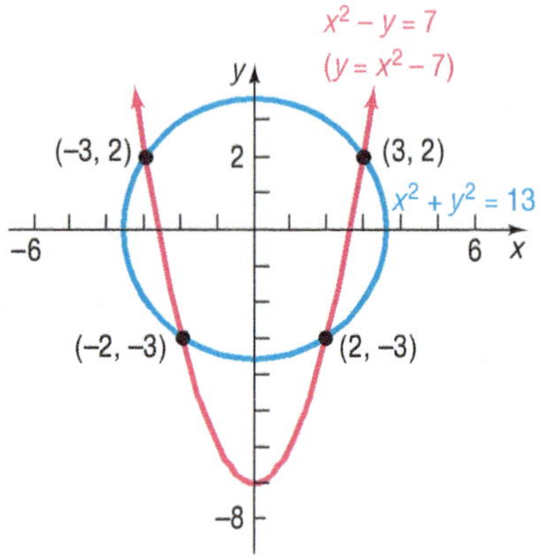

Figure 19

Use INTERSECT to find that the solutions to the system of equations are $(-3, 2)$, $(3, 2)$, $(-2, -3)$, and $(2, -3)$.

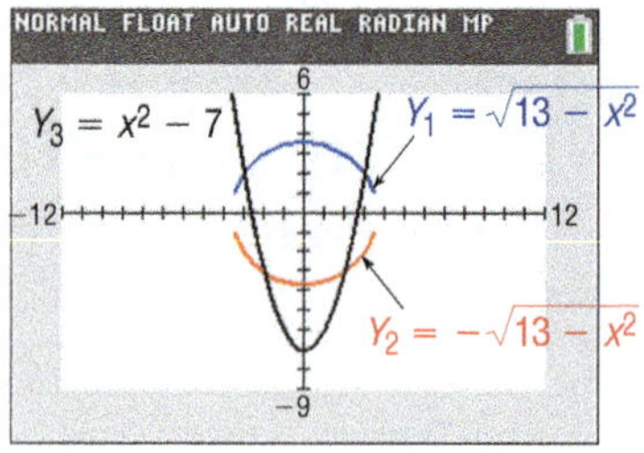

Figure 20

Now Work PROBLEM 13 USING ELIMINATION

EXAMPLE 3 Solving a System of Nonlinear Equations

Solve: $\begin{cases} x^2 - y^2 = 4 & \textbf{(1) A hyperbola} \\ y = x^2 & \textbf{(2) A parabola} \end{cases}$

Algebraic Solution Using Substitution

Either substitution or elimination can be used here. To use substitution, replace x^2 by y in equation (1).

$$x^2 - y^2 = 4 \quad \textbf{Equation (1)}$$
$$y - y^2 = 4 \quad y = x^2$$
$$y^2 - y + 4 = 0 \quad \textbf{Place in standard form.}$$

This is a quadratic equation whose discriminant is $(-1)^2 - 4 \cdot 1 \cdot 4 = 1 - 4 \cdot 4 = -15 < 0$. The equation has no real solutions, so the system is inconsistent. The graphs of these two equations do not intersect. See Figure 21.

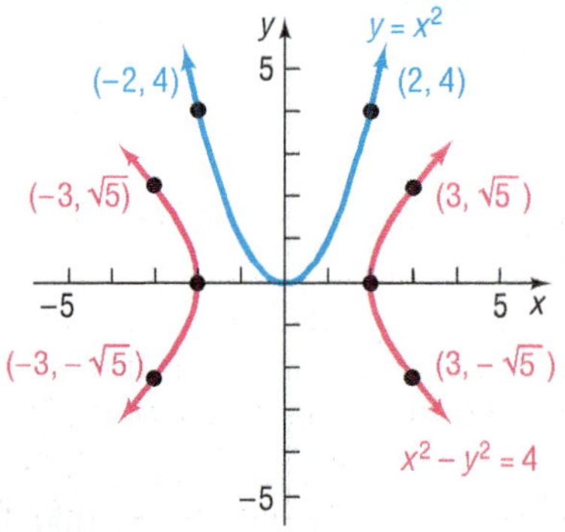

Figure 21

Graphing Solution

Graph $Y_1 = x^2$ and $x^2 - y^2 = 4$, as shown in Figure 22. To graph $x^2 - y^2 = 4$, use two functions:

$$Y_2 = \sqrt{x^2 - 4} \text{ and } Y_3 = -\sqrt{x^2 - 4}$$

From Figure 22, observe that the graphs of these two equations do not intersect. The system is inconsistent.

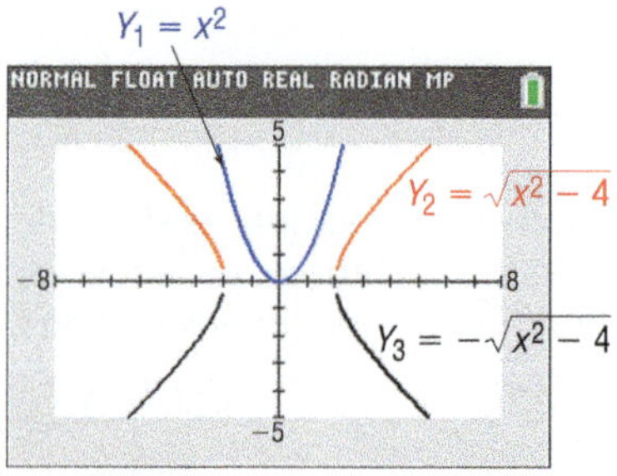

Figure 22

EXAMPLE 4 **Solving a System of Nonlinear Equations**

Solve: $\begin{cases} x^2 + x + y^2 - 3y + 2 = 0 & (1) \\ x + 1 + \dfrac{y^2 - y}{x} = 0 & (2) \end{cases}$

Algebraic Solution Using Elimination

Because it is not straightforward how to graph the equations in the system, we proceed directly to use the method of elimination.

First, multiply equation (2) by x to eliminate the fraction. The result is an equivalent system because x cannot be 0. [Look at the original equation (2) to see why.]

$$\begin{cases} x^2 + x + y^2 - 3y + 2 = 0 & (1) \\ x^2 + x + y^2 - y = 0 & (2) \quad x \neq 0 \end{cases}$$

Now subtract equation (2) from equation (1) to eliminate x. The result is

$$-2y + 2 = 0$$

$$y = 1 \quad \text{Solve for } y.$$

To find x, back-substitute $y = 1$ in equation (1).

$$x^2 + x + y^2 - 3y + 2 = 0 \quad \text{Equation (1)}$$

$$x^2 + x + (1)^2 - 3(1) + 2 = 0 \quad \text{Substitute 1 for } y \text{ in (1).}$$

$$x^2 + x = 0 \quad \text{Simplify.}$$

$$x(x + 1) = 0 \quad \text{Factor.}$$

$$x = 0 \quad \text{or} \quad x = -1 \quad \text{Apply the Zero-Product Property.}$$

Because x cannot be 0, the value $x = 0$ is extraneous, so discard it.

✓**Check:** Check $x = -1, y = 1$:

$$\begin{cases} (-1)^2 + (-1) + 1^2 - 3(1) + 2 = 1 - 1 + 1 - 3 + 2 = 0 & (1) \\ -1 + 1 + \dfrac{1^2 - 1}{-1} = 0 + \dfrac{0}{-1} = 0 & (2) \end{cases}$$

The solution is $x = -1, y = 1$. The point of intersection of the graphs of the equations is $(-1, 1)$. ■

Graphing Solution

First, multiply equation (2) by x to eliminate the fraction. The result is an equivalent system because x cannot be 0 [look at the original equation (2) to see why]:

$$\begin{cases} x^2 + x + y^2 - 3y + 2 = 0 & (1) \\ x^2 + x + y^2 - y = 0 & (2) \quad x \neq 0 \end{cases}$$

Solve each equation for y. First, solve equation (1) for y:

$$x^2 + x + y^2 - 3y + 2 = 0 \quad \text{Equation (1)}$$

$$y^2 - 3y = -x^2 - x - 2 \quad \text{Rearrange so that terms involving } y \text{ are on left side.}$$

$$y^2 - 3y + \frac{9}{4} = -x^2 - x - 2 + \frac{9}{4} \quad \text{Complete the square involving } y.$$

$$\left(y - \frac{3}{2}\right)^2 = -x^2 - x + \frac{1}{4} \quad \text{Factor and simplify.}$$

$$y - \frac{3}{2} = \pm\sqrt{-x^2 - x + \frac{1}{4}} \quad \text{Square Root Method}$$

$$y = \frac{3}{2} \pm \sqrt{-x^2 - x + \frac{1}{4}} \quad \text{Solve for } y.$$

Now solve equation (2) for y:

$$x^2 + x + y^2 - y = 0 \quad \text{Equation (2)}$$

$$y^2 - y = -x^2 - x \quad \text{Rearrange so that terms involving } y \text{ are on left side.}$$

$$y^2 - y + \frac{1}{4} = -x^2 - x + \frac{1}{4} \quad \text{Complete the square involving } y.$$

$$\left(y - \frac{1}{2}\right)^2 = -x^2 - x + \frac{1}{4} \quad \text{Factor.}$$

$$y - \frac{1}{2} = \pm\sqrt{-x^2 - x + \frac{1}{4}} \quad \text{Square Root Method}$$

$$y = \frac{1}{2} \pm \sqrt{-x^2 - x + \frac{1}{4}} \quad \text{Solve for } y.$$

Now graph each equation using a graphing utility. See Figure 23 on the next page.

Use INTERSECT to find that the points of intersection are $(-1, 1)$ and $(0, 1)$. Since $x \neq 0$ [look back at the original equation (2)], the graph of Y_3 has a hole at the point $(0, 1)$ and Y_4 has a hole at $(0, 0)$. The value $x = 0$ is extraneous, so discard it. The only solution is $x = -1, y = 1$.

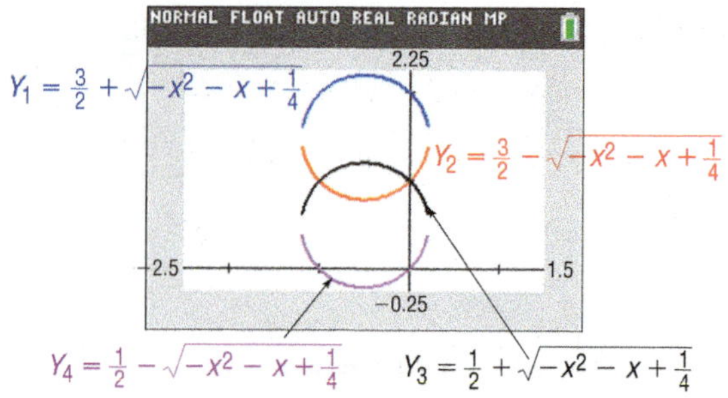

Figure 23

In Problem 55 you are asked to graph the equations given in Example 4 by hand. Be sure to show holes in the graph of equation (2) at $x = 0$.

Now Work PROBLEMS 29 AND 49

EXAMPLE 5 Solving a System of Nonlinear Equations

Solve: $\begin{cases} 3xy - 2y^2 = -2 & (1) \\ 9x^2 + 4y^2 = 10 & (2) \end{cases}$

Algebraic Solution

Multiply equation (1) by 2, and add the result to equation (2) to eliminate the y^2 terms.

$$\begin{cases} 6xy - 4y^2 = -4 & (1) \\ 9x^2 + 4y^2 = 10 & (2) \end{cases}$$

$$9x^2 + 6xy = 6 \quad \text{Add.}$$

$$3x^2 + 2xy = 2 \quad \text{Divide each side by 3.}$$

Since $x \neq 0$ (do you see why?), solve for y in this equation to get

$$y = \frac{2 - 3x^2}{2x} \quad x \neq 0 \quad (3)$$

Now substitute for y in equation (2) of the system.

$$9x^2 + 4y^2 = 10 \quad \text{Equation (2)}$$

$$9x^2 + 4\left(\frac{2 - 3x^2}{2x}\right)^2 = 10 \quad \text{Substitute } y = \frac{2 - 3x^2}{2x} \text{ in (2).}$$

$$9x^2 + \frac{4 - 12x^2 + 9x^4}{x^2} = 10$$

$$9x^4 + 4 - 12x^2 + 9x^4 = 10x^2 \quad \text{Multiply both sides by } x^2.$$

$$18x^4 - 22x^2 + 4 = 0 \quad \text{Subtract } 10x^2 \text{ from both sides.}$$

$$9x^4 - 11x^2 + 2 = 0 \quad \text{Divide both sides by 2.}$$

Graphing Solution

To graph $3xy - 2y^2 = -2$, solve for y. In this instance, it is easier to view the equation as a quadratic equation in the variable y.

$$3xy - 2y^2 = -2$$

$$2y^2 - 3xy - 2 = 0 \quad \text{Place in standard form.}$$

$$y = \frac{-(-3x) \pm \sqrt{(-3x)^2 - 4(2)(-2)}}{2(2)}$$

Use the quadratic formula with $a = 2, b = -3x, c = -2$.

$$y = \frac{3x \pm \sqrt{9x^2 + 16}}{4} \quad \text{Simplify.}$$

Using a graphing utility, graph

$$Y_1 = \frac{3x + \sqrt{9x^2 + 16}}{4} \text{ and } Y_2 = \frac{3x - \sqrt{9x^2 + 16}}{4}.$$

From equation (2), graph $Y_3 = \frac{\sqrt{10 - 9x^2}}{2}$ and $Y_4 = \frac{-\sqrt{10 - 9x^2}}{2}$. See Figure 24.

This quadratic equation (in x^2) can be factored:

$$(9x^2 - 2)(x^2 - 1) = 0$$

$$9x^2 - 2 = 0 \quad \text{or} \quad x^2 - 1 = 0$$

$$x^2 = \frac{2}{9} \qquad\qquad x^2 = 1$$

$$x = \pm\sqrt{\frac{2}{9}} = \pm\frac{\sqrt{2}}{3} \qquad\qquad x = \pm 1$$

To find y, use equation (3).

If $x = \dfrac{\sqrt{2}}{3}$: $y = \dfrac{2 - 3x^2}{2x} = \dfrac{2 - \frac{2}{3}}{2\left(\frac{\sqrt{2}}{3}\right)} = \dfrac{4}{2\sqrt{2}} = \sqrt{2}$

If $x = -\dfrac{\sqrt{2}}{3}$: $y = \dfrac{2 - 3x^2}{2x} = \dfrac{2 - \frac{2}{3}}{2\left(-\frac{\sqrt{2}}{3}\right)} = \dfrac{4}{-2\sqrt{2}} = -\sqrt{2}$

If $x = 1$: $y = \dfrac{2 - 3x^2}{2x} = \dfrac{2 - 3(1)^2}{2} = -\dfrac{1}{2}$

If $x = -1$: $y = \dfrac{2 - 3x^2}{2x} = \dfrac{2 - 3(-1)^2}{-2} = \dfrac{1}{2}$

The system has four solutions: $\left(\frac{\sqrt{2}}{3}, \sqrt{2}\right), \left(-\frac{\sqrt{2}}{3}, -\sqrt{2}\right), \left(1, -\frac{1}{2}\right)$, and $\left(-1, \frac{1}{2}\right)$. Check them for yourself. ■

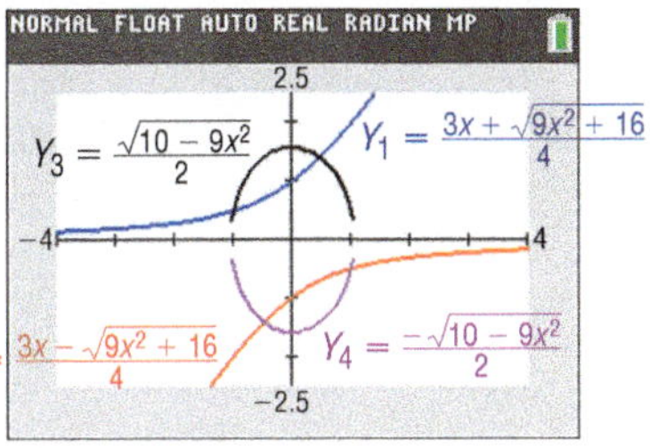

Figure 24

Use INTERSECT to find that the solutions to the system of equations are $(-1, 0.5)$, $(0.47, 1.41)$, $(1, -0.5)$, and $(-0.47, -1.41)$, each rounded to two decimal places. ■

Now Work PROBLEM 47

The next example illustrates an imaginative solution to a system of nonlinear equations.

EXAMPLE 6 Running a Long-Distance Race

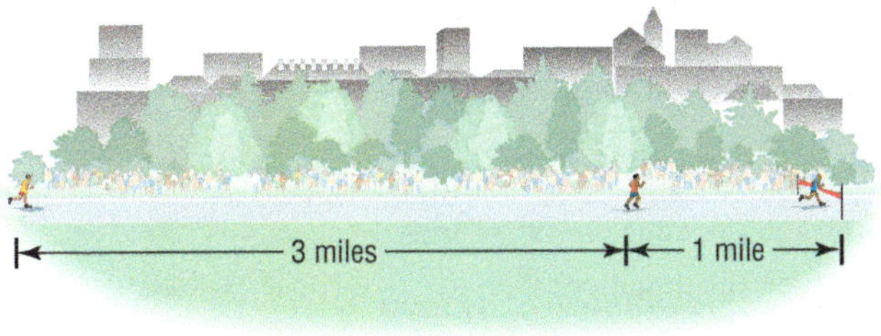

In a 50-mile race, the winner crosses the finish line 1 mile ahead of the second-place runner and 4 miles ahead of the third-place runner. Assuming that each runner maintains a constant speed throughout the race, by how many miles does the second-place runner beat the third-place runner?

Solution Let v_1, v_2, and v_3 denote the speeds of the first-, second-, and third-place runners, respectively. Let t_1 and t_2 denote the times (in hours) required for the first-place runner and the second-place runner to finish the race. Then the following system of equations results:

$$\begin{cases} 50 = v_1 t_1 & (1) \quad \text{First-place runner goes 50 miles in } t_1 \text{ hours.} \\ 49 = v_2 t_1 & (2) \quad \text{Second-place runner goes 49 miles in } t_1 \text{ hours.} \\ 46 = v_3 t_1 & (3) \quad \text{Third-place runner goes 46 miles in } t_1 \text{ hours.} \\ 50 = v_2 t_2 & (4) \quad \text{Second-place runner goes 50 miles in } t_2 \text{ hours.} \end{cases}$$

We seek the distance d of the third-place runner from the finish at time t_2. At time t_2, the third-place runner has gone a distance of $v_3 t_2$ miles, so the distance d remaining is $50 - v_3 t_2$. Now

$$\begin{aligned} d &= 50 - v_3 t_2 \\ &= 50 - v_3\left(t_1 \cdot \frac{t_2}{t_1}\right) \\ &= 50 - (v_3 t_1) \cdot \frac{t_2}{t_1} \\ &= 50 - 46 \cdot \frac{\frac{50}{v_2}}{\frac{50}{v_1}} && \begin{cases} \text{From (3), } v_3 t_1 = 46 \\ \text{From (4), } t_2 = \frac{50}{v_2} \\ \text{From (1), } t_1 = \frac{50}{v_1} \end{cases} \\ &= 50 - 46 \cdot \frac{v_1}{v_2} \\ &= 50 - 46 \cdot \frac{50}{49} && \text{From the quotient of (1) and (2).} \\ &\approx 3.06 \text{ miles} \end{aligned}$$

Historical Feature

In the beginning of this section, it was stated that imagination and experience are important in solving systems of nonlinear equations. Indeed, these kinds of problems lead into some of the deepest and most difficult parts of modern mathematics. Look again at the graphs in Examples 1 and 2 of this section (Figures 17 and 19). Example 1 has two solutions, and Example 2 has four solutions. We might conjecture that the number of solutions is equal to the product of the degrees of the equations involved. This conjecture was indeed made by Étienne Bézout (1730–1783), but working out the details took about 150 years. It turns out that arriving at the correct number of intersections requires counting not only the complex number intersections, but also those intersections that, in a certain sense, lie at infinity. For example, a parabola and a line lying on the axis of the parabola intersect at the vertex and at infinity. This topic is part of the study of algebraic geometry.

Historical Problem

A papyrus dating back to 1950 BC contains the following problem: "A given surface area of 100 units of area shall be represented as the sum of two squares whose sides are to each other as 1 is to $\frac{3}{4}$."

Solve for the sides by solving the system of equations

$$\begin{cases} x^2 + y^2 = 100 \\ x = \frac{3}{4}y \end{cases} \quad x = 6 \text{ units, } y = 8 \text{ units}$$

12.6 Assess Your Understanding

'Are You Prepared?' *Answers are given at the end of these exercises. If you get a wrong answer, read the pages listed in red.*

1. Graph the equation: $y = 3x + 2$ (pp. 173–184)
2. Graph the equation: $y + 4 = x^2$ (pp. 793–797)
3. Graph the equation: $y^2 = x^2 - 1$ (pp. 814–821)
4. Graph the equation: $x^2 + 4y^2 = 4$ (pp. 803–808)

Skill Building

In Problems 5–24, graph each equation of the system. Then solve the system to find the points of intersection.

5. $\begin{cases} y = x^2 + 1 \\ y = x + 1 \end{cases}$

6. $\begin{cases} y = x^2 + 1 \\ y = 4x + 1 \end{cases}$

7. $\begin{cases} y = \sqrt{36 - x^2} \\ y = 8 - x \end{cases}$

8. $\begin{cases} y = \sqrt{4 - x^2} \\ y = 2x + 4 \end{cases}$

9. $\begin{cases} y = \sqrt{x} \\ y = 2 - x \end{cases}$

10. $\begin{cases} y = \sqrt{x} \\ y = 6 - x \end{cases}$

11. $\begin{cases} x = 2y \\ x = y^2 - 2y \end{cases}$

12. $\begin{cases} y = x - 1 \\ y = x^2 - 6x + 9 \end{cases}$

13. $\begin{cases} x^2 + y^2 = 4 \\ x^2 + 2x + y^2 = 0 \end{cases}$

14. $\begin{cases} x^2 + y^2 = 8 \\ x^2 + y^2 + 4y = 0 \end{cases}$

15. $\begin{cases} y = 3x - 5 \\ x^2 + y^2 = 5 \end{cases}$

16. $\begin{cases} x^2 + y^2 = 10 \\ y = x + 2 \end{cases}$

17. $\begin{cases} x^2 + y^2 = 4 \\ y^2 - x = 4 \end{cases}$

18. $\begin{cases} x^2 + y^2 = 16 \\ x^2 - 2y = 8 \end{cases}$

19. $\begin{cases} xy = 4 \\ x^2 + y^2 = 8 \end{cases}$

20. $\begin{cases} x^2 = y \\ xy = 1 \end{cases}$

21. $\begin{cases} x^2 + y^2 = 4 \\ y = x^2 - 9 \end{cases}$

22. $\begin{cases} xy = 1 \\ y = 2x + 1 \end{cases}$

23. $\begin{cases} y = x^2 - 4 \\ y = 6x - 13 \end{cases}$

24. $\begin{cases} x^2 + y^2 = 10 \\ xy = 3 \end{cases}$

In Problems 25–54, solve each system. Use any method you wish.

25. $\begin{cases} 2x^2 + y^2 = 18 \\ xy = 4 \end{cases}$

26. $\begin{cases} x^2 - y^2 = 21 \\ x + y = 7 \end{cases}$

27. $\begin{cases} y = 2x + 1 \\ 2x^2 + y^2 = 1 \end{cases}$

28. $\begin{cases} x^2 - 4y^2 = 16 \\ 2y - x = 2 \end{cases}$

29. $\begin{cases} x + y + 1 = 0 \\ x^2 + y^2 + 6y - x = -5 \end{cases}$

30. $\begin{cases} 2x^2 - xy + y^2 = 8 \\ xy = 4 \end{cases}$

31. $\begin{cases} 4x^2 - 3xy + 9y^2 = 15 \\ 2x + 3y = 5 \end{cases}$

32. $\begin{cases} 2y^2 - 3xy + 6y + 2x + 4 = 0 \\ 2x - 3y + 4 = 0 \end{cases}$

33. $\begin{cases} x^2 - 4y^2 + 7 = 0 \\ 3x^2 + y^2 = 31 \end{cases}$

34. $\begin{cases} 3x^2 - 2y^2 + 5 = 0 \\ 2x^2 - y^2 + 2 = 0 \end{cases}$

35. $\begin{cases} 7x^2 - 3y^2 + 5 = 0 \\ 3x^2 + 5y^2 = 12 \end{cases}$

36. $\begin{cases} x^2 - 3y^2 + 1 = 0 \\ 2x^2 - 7y^2 + 5 = 0 \end{cases}$

37. $\begin{cases} x^2 + 2xy = 10 \\ 3x^2 - xy = 2 \end{cases}$

38. $\begin{cases} 5xy + 13y^2 + 36 = 0 \\ xy + 7y^2 = 6 \end{cases}$

39. $\begin{cases} 2x^2 + y^2 = 2 \\ x^2 - 2y^2 + 8 = 0 \end{cases}$

40. $\begin{cases} y^2 - x^2 + 4 = 0 \\ 2x^2 + 3y^2 = 6 \end{cases}$

41. $\begin{cases} x^2 + 2y^2 = 16 \\ 4x^2 - y^2 = 24 \end{cases}$

42. $\begin{cases} 4x^2 + 3y^2 = 4 \\ 2x^2 - 6y^2 = -3 \end{cases}$

43. $\begin{cases} \dfrac{5}{x^2} - \dfrac{2}{y^2} + 3 = 0 \\ \dfrac{3}{x^2} + \dfrac{1}{y^2} = 7 \end{cases}$

44. $\begin{cases} \dfrac{2}{x^2} - \dfrac{3}{y^2} + 1 = 0 \\ \dfrac{6}{x^2} - \dfrac{7}{y^2} + 2 = 0 \end{cases}$

45. $\begin{cases} \dfrac{1}{x^4} + \dfrac{6}{y^4} = 6 \\ \dfrac{2}{x^4} - \dfrac{2}{y^4} = 19 \end{cases}$

46. $\begin{cases} \dfrac{1}{x^4} - \dfrac{1}{y^4} = 1 \\ \dfrac{1}{x^4} + \dfrac{1}{y^4} = 4 \end{cases}$

47. $\begin{cases} x^2 - 3xy + 2y^2 = 0 \\ x^2 + xy = 6 \end{cases}$

48. $\begin{cases} x^2 - xy - 2y^2 = 0 \\ xy + x + 6 = 0 \end{cases}$

49. $\begin{cases} y^2 + y + x^2 - x - 2 = 0 \\ y + 1 + \dfrac{x - 2}{y} = 0 \end{cases}$

50. $\begin{cases} x^3 - 2x^2 + y^2 + 3y - 4 = 0 \\ x - 2 + \dfrac{y^2 - y}{x^2} = 0 \end{cases}$

51. $\begin{cases} \log_x y = 3 \\ \log_x(4y) = 5 \end{cases}$

52. $\begin{cases} \log_x(2y) = 3 \\ \log_x(4y) = 2 \end{cases}$

53. $\begin{cases} \ln x = 4 \ln y \\ \log_3 x = 2 + 2 \log_3 y \end{cases}$

54. $\begin{cases} \ln x = 5 \ln y \\ \log_2 x = 3 + 2 \log_2 y \end{cases}$

55. Graph the equations given in Example 4.

56. Graph the equations given in Problem 49.

In Problems 57–64, use a graphing utility to solve each system of equations. Express the solution(s) rounded to two decimal places.

57. $\begin{cases} y = x^{2/3} \\ y = e^{-x} \end{cases}$

58. $\begin{cases} y = x^{3/2} \\ y = e^{-x} \end{cases}$

59. $\begin{cases} x^2 + y^3 = 2 \\ x^3 y = 4 \end{cases}$

60. $\begin{cases} x^3 + y^2 = 2 \\ x^2 y = 4 \end{cases}$

61. $\begin{cases} x^4 + y^4 = 12 \\ xy^2 = 2 \end{cases}$

62. $\begin{cases} x^4 + y^4 = 6 \\ xy = 1 \end{cases}$

63. $\begin{cases} xy = 2 \\ y = \ln x \end{cases}$

64. $\begin{cases} x^2 + y^2 = 4 \\ y = \ln x \end{cases}$

Mixed Practice

In Problems 65–70, graph each equation and find the point(s) of intersection, if any.

65. The line $x + 2y = 0$ and the circle $(x - 1)^2 + (y - 1)^2 = 5$

66. The line $x + 2y + 6 = 0$ and the circle $(x + 1)^2 + (y + 1)^2 = 5$

67. The circle $(x - 1)^2 + (y + 2)^2 = 4$ and the parabola $y^2 + 4y - x + 1 = 0$

68. The circle $(x + 2)^2 + (y - 1)^2 = 4$ and the parabola $y^2 - 2y - x - 5 = 0$

69. $y = \dfrac{4}{x - 3}$ and the circle $x^2 - 6x + y^2 + 1 = 0$

70. $y = \dfrac{4}{x + 2}$ and the circle $x^2 + 4x + y^2 - 4 = 0$

Applications and Extensions

71. The difference of two numbers is 2 and the sum of their squares is 10. Find the numbers.

72. The sum of two numbers is 7 and the difference of their squares is 21. Find the numbers.

73. The product of two numbers is 4 and the sum of their squares is 8. Find the numbers.

74. The product of two numbers is 10 and the difference of their squares is 21. Find the numbers.

75. The difference of two numbers is the same as their product, and the sum of their reciprocals is 5. Find the numbers.

76. The sum of two numbers is the same as their product, and the difference of their reciprocals is 3. Find the numbers.

77. The ratio of a to b is $\frac{2}{3}$. The sum of a and b is 10. What is the ratio of $a + b$ to $b - a$?

78. The ratio of a to b is 4:3. The sum of a and b is 14. What is the ratio of $a - b$ to $a + b$?

79. Geometry The perimeter of a rectangle is 16 inches and its area is 15 square inches. What are its dimensions?

80. Geometry An area of 52 square feet is to be enclosed by two squares whose sides are in the ratio of 2:3. Find the sides of the squares.

81. Geometry Two circles have circumferences that add up to 12π centimeters and areas that add up to 20π square centimeters. Find the radius of each circle.

82. Geometry The altitude of an isosceles triangle drawn to its base is 3 centimeters, and its perimeter is 18 centimeters. Find the length of its base.

83. The Tortoise and the Hare In a 21-meter race between a tortoise and a hare, the tortoise leaves 9 minutes before the hare. The hare, by running at an average speed of 0.5 meter per hour faster than the tortoise, crosses the finish line 3 minutes before the tortoise. What are the average speeds of the tortoise and the hare?

84. Running a Race In a 1-mile race, the winner crosses the finish line 10 feet ahead of the second-place runner and 20 feet ahead of the third-place runner. Assuming that each runner maintains a constant speed throughout the race, by how many feet does the second-place runner beat the third-place runner?

85. Constructing a Box A rectangular piece of cardboard, whose area is 216 square centimeters, is made into an open box by cutting a 2-centimeter square from each corner and turning up the sides. See the figure. If the box is to have a volume of 224 cubic centimeters, what size cardboard should you start with?

86. Constructing a Cylindrical Tube A rectangular piece of cardboard, whose area is 216 square centimeters, is made into a cylindrical tube by joining together two sides of the rectangle. See the figure. If the tube is to have a volume of 224 cubic centimeters, what size cardboard should you start with?

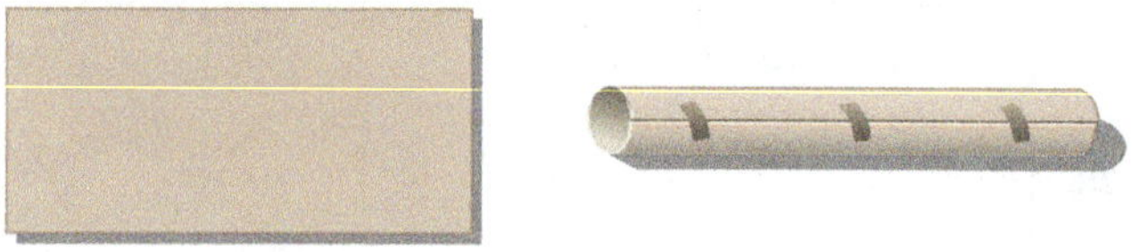

87. Fencing A farmer has 300 feet of fence available to enclose a 4500-square-foot region in the shape of adjoining squares, with sides of length x and y. See the figure. Find x and y.

88. Bending Wire A wire 60 feet long is cut into two pieces. Is it possible to bend one piece into the shape of a square and the other into the shape of a circle so that the total area enclosed by the two pieces is 100 square feet? If this is possible, find the length of the side of the square and the radius of the circle.

89. Geometry Find formulas for the length l and width w of a rectangle in terms of its area A and perimeter P.

90. Geometry Find formulas for the base b and one of the equal sides l of an isosceles triangle in terms of its altitude h and perimeter P.

91. Descartes' Method of Equal Roots Descartes' method for finding tangents depends on the idea that, for many graphs, the tangent line at a given point is the *unique* line that intersects the graph at that point only. We apply his method to find an equation of the tangent line to the parabola $y = x^2$ at the point $(2, 4)$. See the figure.

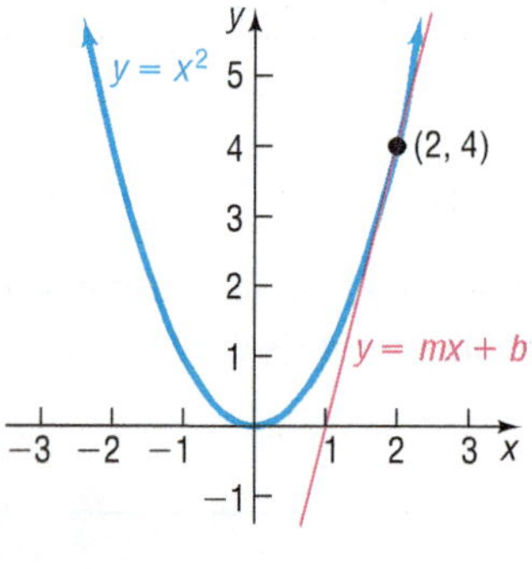

First, we know that the equation of the tangent line must be in the form $y = mx + b$. Using the fact that the point $(2, 4)$ is on the line, we can solve for b in terms of m and get the equation $y = mx + (4 - 2m)$. Now we want $(2, 4)$ to be the *unique* solution to the system

$$\begin{cases} y = x^2 \\ y = mx + 4 - 2m \end{cases}$$

From this system, we get $x^2 - mx + (2m - 4) = 0$. By using the quadratic formula, we get

$$x = \frac{m \pm \sqrt{m^2 - 4(2m - 4)}}{2}$$

To obtain a unique solution for x, the two roots must be equal; in other words, the discriminant $m^2 - 4(2m - 4)$ must be 0. Complete the work to get m, and write an equation of the tangent line.

In Problems 92–98, use Descartes' method from Problem 91 to find an equation of the line tangent to each graph at the given point.

92. $x^2 + y^2 = 10$; at $(1, 3)$

93. $y = x^2 + 2$; at $(1, 3)$

94. $x^2 + y = 5$; at $(-2, 1)$

95. $2x^2 + 3y^2 = 14$; at $(1, 2)$

96. $3x^2 + y^2 = 7$; at $(-1, 2)$

97. $x^2 - y^2 = 3$; at $(2, 1)$

98. $2y^2 - x^2 = 14$; at $(2, 3)$

99. If r_1 and r_2 are two solutions of a quadratic equation $ax^2 + bx + c = 0$, it can be shown that

$$r_1 + r_2 = -\frac{b}{a} \quad \text{and} \quad r_1 r_2 = \frac{c}{a}$$

Solve this system of equations for r_1 and r_2.

Explaining Concepts: Discussion and Writing

100. A circle and a line intersect at most twice. A circle and a parabola intersect at most four times. Deduce that a circle and the graph of a polynomial of degree 3 intersect at most six times. What do you conjecture about a polynomial of degree 4? What about a polynomial of degree n? Can you explain your conclusions using an algebraic argument?

101. Suppose that you are the manager of a sheet metal shop. A customer asks you to manufacture 10,000 boxes, each box being open on top. The boxes are required to have a square base and a 9-cubic-foot capacity. You construct the boxes by cutting out a square from each corner of a square piece of sheet metal and folding along the edges.

(a) What are the dimensions of the square to be cut if the area of the square piece of sheet metal is 100 square feet?

(b) Could you make the box using a smaller piece of sheet metal? Make a list of the dimensions of the box for various pieces of sheet metal.

Retain Your Knowledge

Problems 102–105 are based on material learned earlier in the course. The purpose of these problems is to keep the material fresh in your mind so that you are better prepared for the final exam.

102. Solve using the quadratic formula: $7x^2 = 8 - 6x$

103. Find an equation of the line with slope $-\frac{2}{5}$ that passes through the point $(10, -7)$.

104. Given $\cot\theta = \frac{24}{7}$ and $\cos\theta < 0$, find the exact value of each of the remaining trigonometric functions.

105. **Finding the Grade of a Mountain Trail** A straight trail with uniform inclination leads from a hotel, elevation 5300 feet, to a lake in the valley, elevation 4100 feet. The length of the trail is 4420 feet. What is the inclination (grade) of the trail?

'Are You Prepared?' Answers

1.

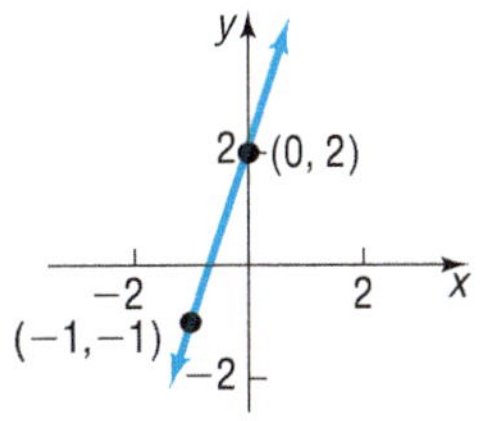

2.

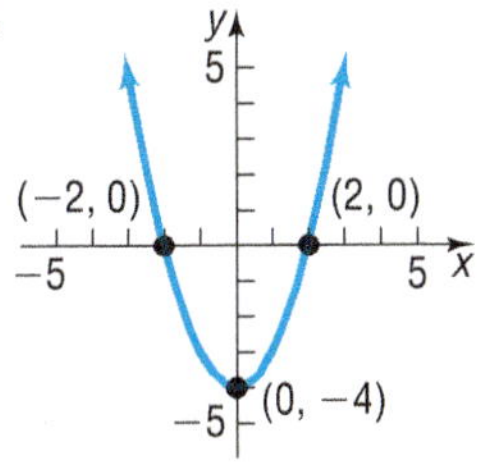

3.

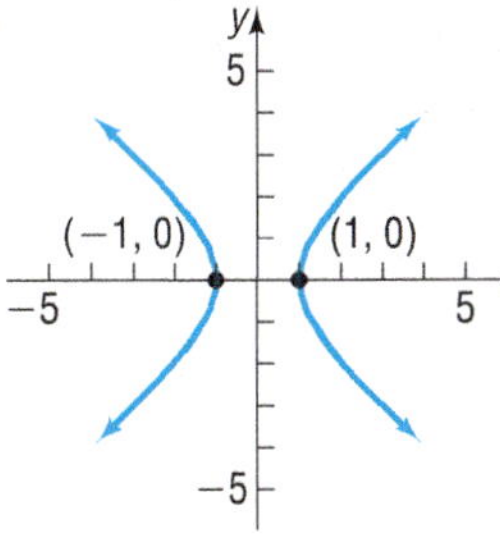

4.

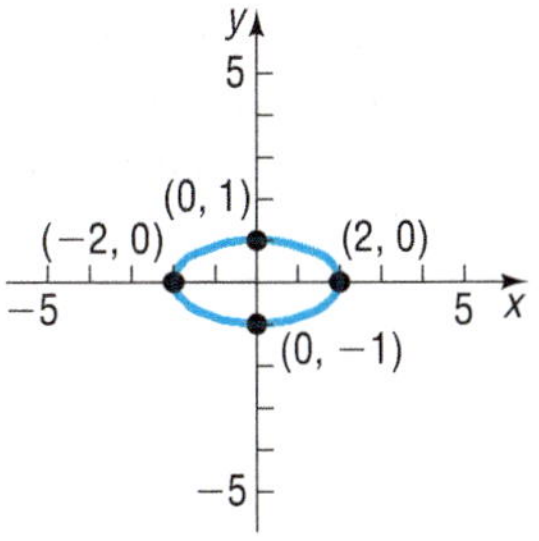

12.7 Systems of Inequalities

PREPARING FOR THIS SECTION *Before getting started, review the following:*

- Solving Linear Inequalities (Section 1.7, pp. 150–151)
- Lines (Section 2.2, pp. 173–184)
- Circles (Section 2.3, pp. 189–193)
- Graphing Techniques: Transformations (Section 3.5, pp. 256–264)

Now Work the 'Are You Prepared?' problems on page 944.

OBJECTIVES
1 Graph an Inequality by Hand (p. 938)
2 Graph an Inequality Using a Graphing Utility (p. 940)
3 Graph a System of Inequalities (p. 941)

Section 1.7 discussed inequalities in one variable. This section discusses inequalities in two variables.

EXAMPLE 1 **Examples of Inequalities in Two Variables**

(a) $3x + y \leq 6$ (b) $x^2 + y^2 < 4$ (c) $y^2 > x$ ■

1 Graph an Inequality by Hand

An inequality in two variables x and y is **satisfied** by an ordered pair (a, b) if, when x is replaced by a and y by b, a true statement results. The **graph of an inequality in two variables** x and y consists of all points (x, y) whose coordinates satisfy the inequality.

EXAMPLE 2 **Graphing an Inequality by Hand**

Graph the linear inequality: $3x + y \leq 6$

Solution Begin by graphing the equation

$$3x + y = 6$$

formed by replacing (for now) the $\leq$ symbol with an $=$ sign. The graph of the equation is a line. See Figure 25(a). This line is part of the graph of the inequality because the inequality is nonstrict, so the line is drawn as a solid line. (Do you see why? We are seeking points for which $3x + y$ is less than *or equal to* 6.)

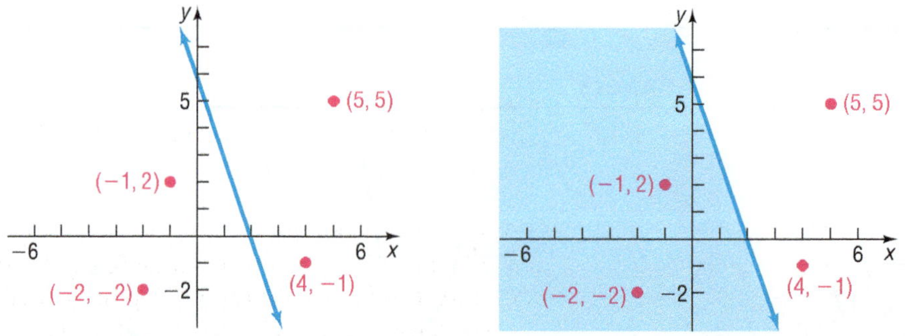

Figure 25 **(a)** $3x + y = 6$ **(b)** Graph of $3x + y \leq 6$

Now test a few randomly selected points to see whether they belong to the graph of the inequality.

	$3x + y \leq 6$	Conclusion
$(4, -1)$	$3(4) + (-1) = 11 > 6$	Does not belong to the graph
$(5, 5)$	$3(5) + 5 = 20 > 6$	Does not belong to the graph
$(-1, 2)$	$3(-1) + 2 = -1 \leq 6$	Belongs to the graph
$(-2, -2)$	$3(-2) + (-2) = -8 \leq 6$	Belongs to the graph

Look again at Figure 25(a). Notice that the two points that belong to the graph both lie on the same side of the line, and the two points that do not belong to the graph lie on the opposite side. As it turns out, all the points that satisfy the inequality will lie on one side of the line or on the line itself. All the points that do not satisfy the inequality will lie on the other side. The graph we seek consists of all points that lie on the line or on the same side of the line as $(-1, 2)$ and $(-2, -2)$. This graph is shown as the shaded region in Figure 25(b). ■

Now Work PROBLEM 15

The graph of any inequality in two variables may be obtained in a like way. The steps to follow are given next.

Note: The strict inequalities are $<$ and $>$. The nonstrict inequalities are $\leq$ and $\geq$. ■

Steps for Graphing an Inequality by Hand

STEP 1: Replace the inequality symbol by an equal sign, and graph the resulting equation. If the inequality is strict, use dashes; if it is nonstrict, use a solid mark. This graph separates the *xy*-plane into two or more regions.

STEP 2: In each region, select a test point P.

(a) If the coordinates of P satisfy the inequality, so do all the points in that region. Indicate this by shading the region.

(b) If the coordinates of P do not satisfy the inequality, none of the points in that region does.

EXAMPLE 3

Graphing an Inequality by Hand

Graph: $x^2 + y^2 \leq 4$

Solution

STEP 1: Graph the equation $x^2 + y^2 = 4$, a circle of radius 2, center at the origin. A solid circle is used because the inequality is not strict.

STEP 2: Use two test points, one inside the circle, the other outside.

Inside $(0, 0)$: $x^2 + y^2 = 0^2 + 0^2 = 0 \leq 4$ **Belongs to the graph**

Outside $(4, 0)$: $x^2 + y^2 = 4^2 + 0^2 = 16 > 4$ **Does not belong to the graph**

All the points inside and on the circle satisfy the inequality. See Figure 26. ■

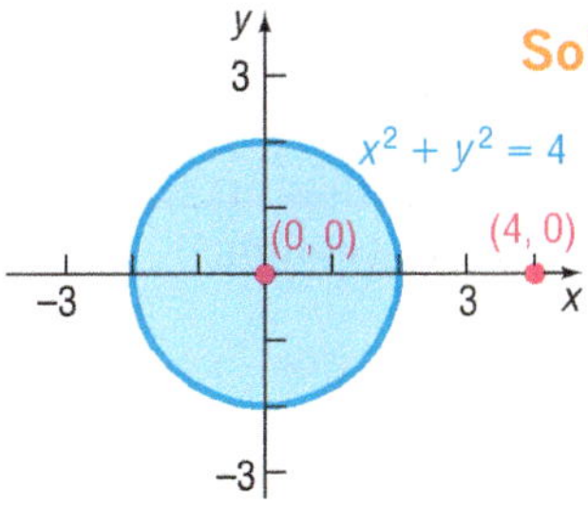

Figure 26 $x^2 + y^2 \leq 4$

Now Work PROBLEM 17

Linear Inequalities

Linear inequalities are inequalities equivalent to one of the forms

$$Ax + By < C \qquad Ax + By > C \qquad Ax + By \leq C \qquad Ax + By \geq C$$

where A and B are not both zero.

The graph of the corresponding equation of a linear inequality is a line that separates the *xy*-plane into two regions called **half-planes**. See Figure 27.

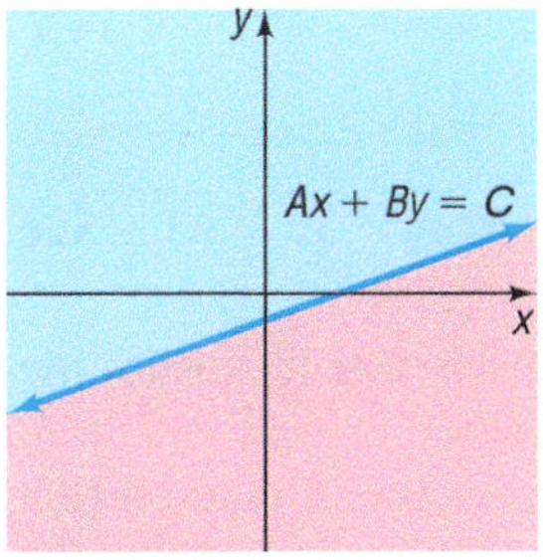

Figure 27

As shown, $Ax + By = C$ is the equation of the boundary line, and it divides the plane into two half-planes: one for which $Ax + By < C$ and the other for which $Ax + By > C$. Because of this, for linear inequalities, only one test point is required.

EXAMPLE 4 Graphing Linear Inequalities by Hand

Graph: (a) $y < 2$ (b) $y \geq 2x$

Solution (a) Points on the horizontal line $y = 2$ are not part of the graph of the inequality, so the graph is shown as a dashed line. Since $(0, 0)$ satisfies the inequality, the graph consists of the half-plane below the line $y = 2$. See Figure 28.

(b) Points on the line $y = 2x$ are part of the graph of the inequality, so the graph is shown as a solid line. Use $(3, 0)$ as a test point. It does not satisfy the inequality $[0 < 2 \cdot 3]$. Points in the half-plane on the opposite side of $(3, 0)$ satisfy the inequality. See Figure 29.

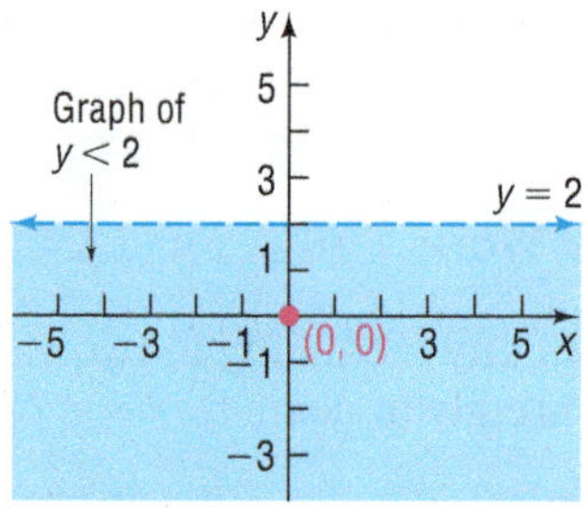

Figure 28 $y < 2$

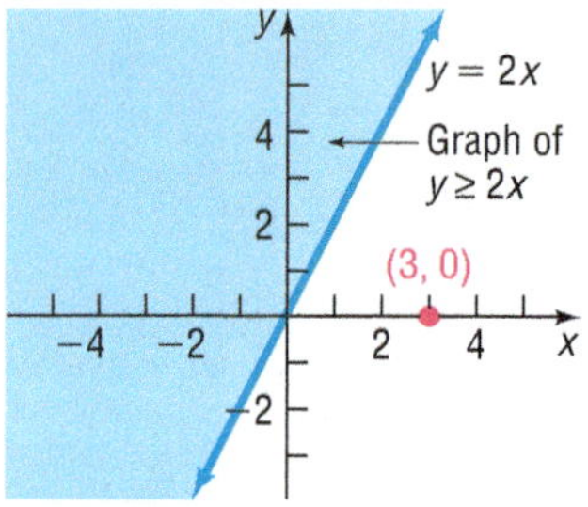

Figure 29 $y \geq 2x$

Now Work PROBLEM 13

2 Graph an Inequality Using a Graphing Utility

Graphing utilities can also be used to graph inequalities. The steps to follow are given next.

Steps for Graphing an Inequality Using a Graphing Utility

STEP 1: Replace the inequality symbol by an equal sign and graph the resulting equation. This graph separates the xy-plane into two or more regions.

STEP 2: Select a test point P in each region.

(a) Use a graphing utility to determine if the test point P satisfies the inequality. If the test point satisfies the inequality, then so do all the points in this region. Indicate this by using the graphing utility to shade the region.

(b) If the coordinates of P do not satisfy the inequality, then none of the points in that region does.

EXAMPLE 5 Graphing an Inequality Using a Graphing Utility

Use a graphing utility to graph $3x + y \leq 6$.

Solution

STEP 1: Begin by graphing the equation $3x + y = 6$ $(Y_1 = -3x + 6)$. See Figure 30.

STEP 2: Select a test point in one of the regions and determine whether it satisfies the inequality. To test the point $(-1, 2)$, for example, enter $3(-1) + 2 \leq 6$. See Figure 31(a). The 1 that appears indicates that the statement entered (the inequality) is true. When the point $(5, 5)$ is tested, a 0 appears, indicating that the statement entered is false. So $(-1, 2)$ is part of the graph of the inequality and

$(5, 5)$ is not. Shade the region containing the point $(-1, 2)$ that is below Y_1. Figure 31(b) shows the graph of the inequality on a TI-84 Plus C.

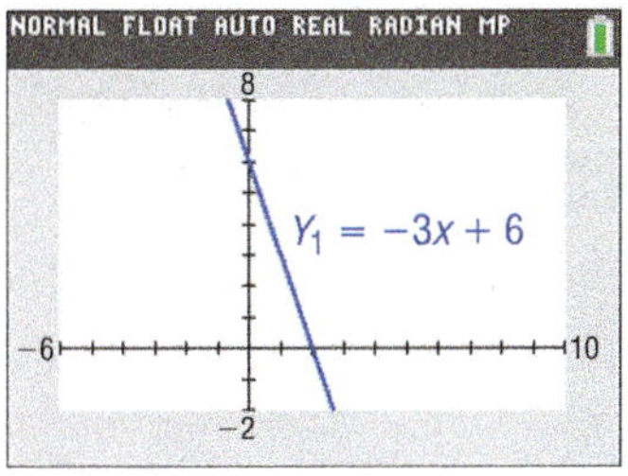

Figure 30 $3x + y = 6$

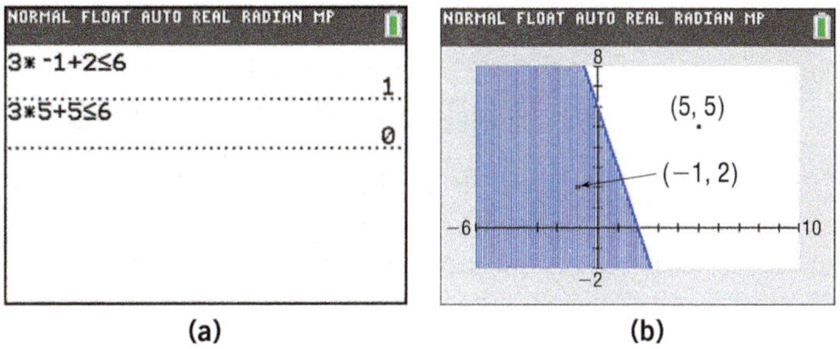

Figure 31 $3x + y \leq 6$ ■

Note: A second approach that could be used in Example 5 is to solve the inequality $3x + y \leq 6$ for y, $y \leq -3x + 6$, graph the corresponding line $y = -3x + 6$, and then shade below since the inequality is of the form $y <$ or $y \leq$ (shade above if of the form $y >$ or $y \geq$). ■

Now Work PROBLEM 15 USING A GRAPHING UTILITY

3 Graph a System of Inequalities

The **graph of a system of inequalities** in two variables x and y is the set of all points (x, y) that simultaneously satisfy *each* inequality in the system. The graph of a system of inequalities can be obtained by graphing each inequality individually and then determining where, if at all, they intersect.

EXAMPLE 6 Graphing a System of Linear Inequalities by Hand

Graph the system: $\begin{cases} x + y \geq 2 \\ 2x - y \leq 4 \end{cases}$

Solution Begin by graphing the lines $x + y = 2$ and $2x - y = 4$ using solid lines since the inequalities are nonstrict. Use the test point $(0, 0)$ on each inequality. For example, $(0, 0)$ does not satisfy $x + y \geq 2$, so shade above the line $x + y = 2$. See Figure 32(a). Also, $(0, 0)$ does satisfy $2x - y \leq 4$, so shade above the line $2x - y = 4$. See Figure 32(b). The intersection of the shaded regions (in purple) gives the result presented in Figure 32(c).

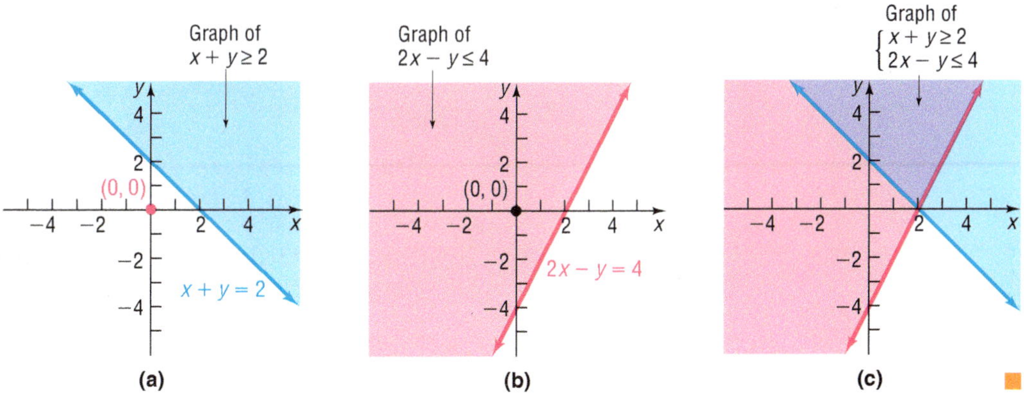

Figure 32 ■

EXAMPLE 7 Graphing a System of Linear Inequalities Using a Graphing Utility

Graph the system: $\begin{cases} x + y \geq 2 \\ 2x - y \leq 4 \end{cases}$

Solution First, graph the lines $x + y = 2$ $(Y_1 = -x + 2)$ and $2x - y = 4$ $(Y_2 = 2x - 4)$. See Figure 33 on the next page.

Notice that the graphs divide the viewing window into four regions. Select a test point for each region, and determine whether the point makes *both*

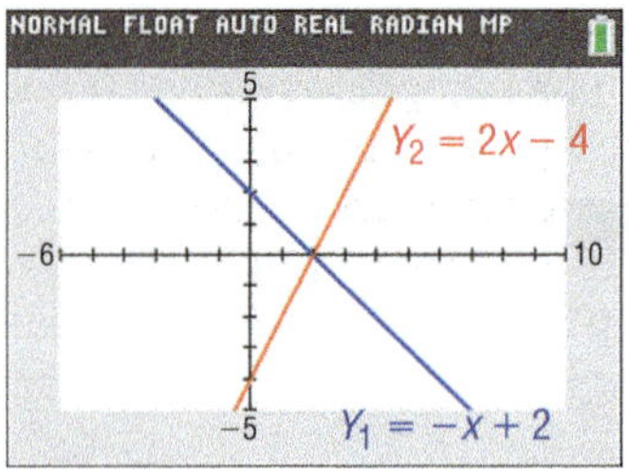

Figure 33

inequalities true. We choose to test $(0, 0)$, $(2, 3)$, $(4, 0)$, and $(2, -2)$. Figure 34(a) shows that $(2, 3)$ makes both inequalities true. The graph is shown in Figure 34(b).

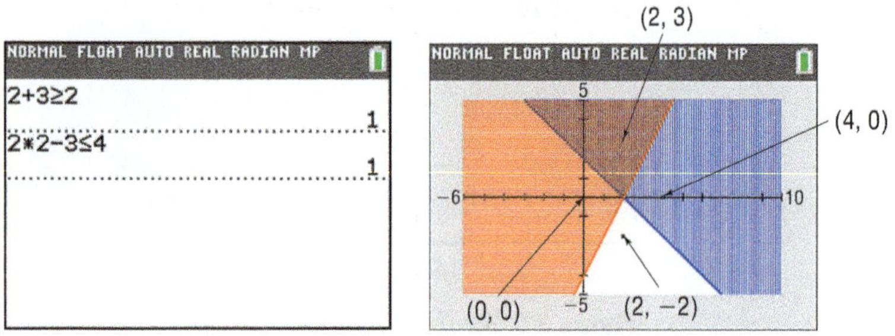

Figure 34 (a) (b)

Rather than testing four points, we could test just point $(0, 0)$ on each inequality. For example, $(0, 0)$ does not satisfy $x + y \geq 2$, so shade above the line $x + y = 2$. In addition, $(0, 0)$ does satisfy $2x - y \leq 4$, so shade above the line $2x - y = 4$. The intersection of the shaded regions gives the result presented in Figure 34(b).

Now Work PROBLEM 23

EXAMPLE 8 Graphing a System of Linear Inequalities by Hand

Graph the system: $\begin{cases} x + y \leq 2 \\ x + y \geq 0 \end{cases}$

Solution See Figure 35. The overlapping purple-shaded region between the two boundary lines is the graph of the system.

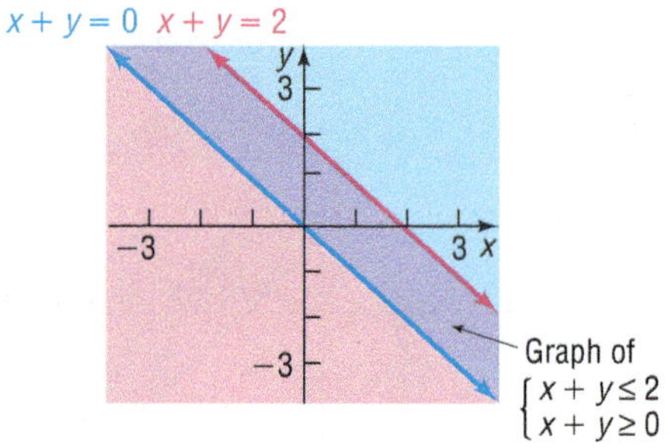

Figure 35

Now Work PROBLEM 29

EXAMPLE 9 Graphing a System of Linear Inequalities by Hand

Graph the systems:

(a) $\begin{cases} 2x - y \geq 0 \\ 2x - y \geq 2 \end{cases}$ (b) $\begin{cases} x + 2y \leq 2 \\ x + 2y \geq 6 \end{cases}$

Solution (a) See Figure 36. The overlapping purple-shaded region is the graph of the system. Note that the graph of the system is identical to the graph of the single inequality $2x - y \geq 2$.

(b) See Figure 37. Here, because no overlapping region results, there are no points in the xy-plane that simultaneously satisfy each inequality. The system has no solution.

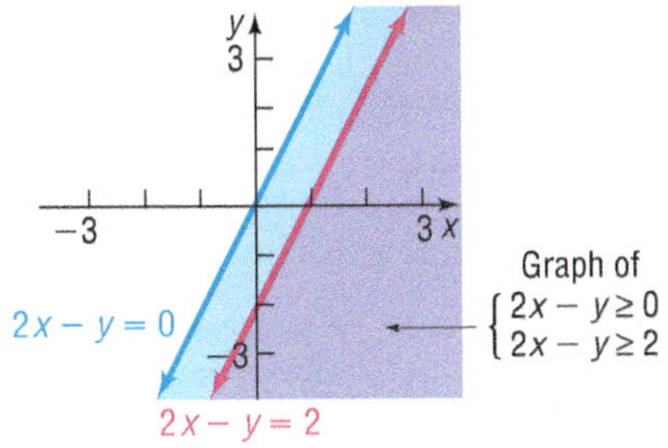

Figure 36

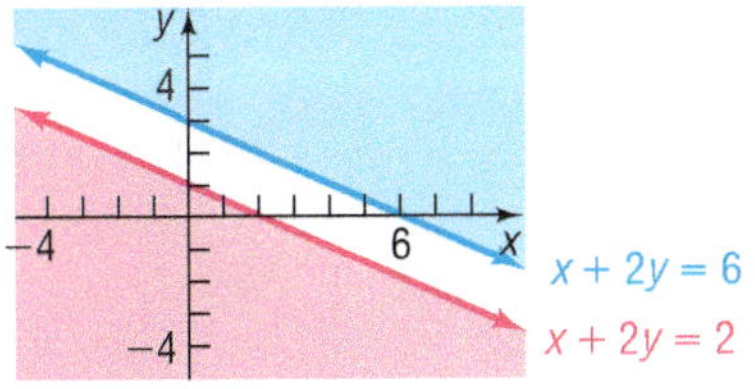

Figure 37

EXAMPLE 10 Graphing a System of Nonlinear Inequalities by Hand

Graph the region below the graph of $x + y = 2$ and above the graph of $y = x^2 - 4$ by graphing the system:

$$\begin{cases} y \geq x^2 - 4 \\ x + y \leq 2 \end{cases}$$

Label all points of intersection.

Solution Figure 38 shows the graph of the region above the graph of the parabola $y = x^2 - 4$ and below the graph of the line $x + y = 2$. The points of intersection are found by solving the system of equations

$$\begin{cases} y = x^2 - 4 \\ x + y = 2 \end{cases}$$

Use substitution to find

$$x + (x^2 - 4) = 2$$
$$x^2 + x - 6 = 0$$
$$(x + 3)(x - 2) = 0$$
$$x = -3 \quad \text{or} \quad x = 2$$

The two points of intersection are $(-3, 5)$ and $(2, 0)$.

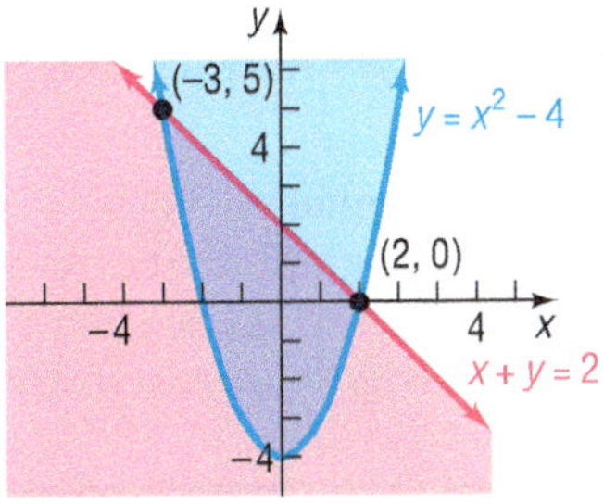

Figure 38

Now Work PROBLEM 37

EXAMPLE 11 Graphing a System of Four Linear Inequalities by Hand

Graph the system:
$$\begin{cases} x + y \geq 3 \\ 2x + y \geq 4 \\ x \geq 0 \\ y \geq 0 \end{cases}$$

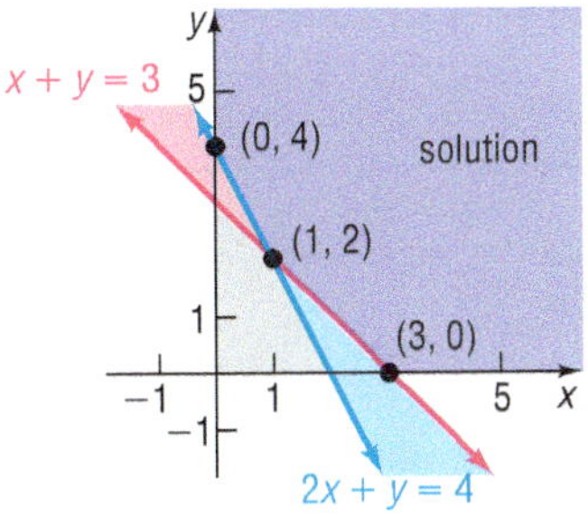

Figure 39

Solution See Figure 39. The two inequalities $x \geq 0$ and $y \geq 0$ require the graph of the system to be in quadrant I, which is shaded light gray. Concentrate on the remaining two inequalities. The intersection of the graphs of these two inequalities and quadrant I is shown in dark purple.

EXAMPLE 12 Financial Planning

A retired couple can invest up to \$25,000. As their financial adviser, you recommend that they place at least \$15,000 in Treasury bills yielding 2% and at most \$5000 in corporate bonds yielding 3%.

(a) Using x to denote the amount of money invested in Treasury bills and y the amount invested in corporate bonds, write a system of linear inequalities that describes the possible amounts of each investment. Assume that x and y are in thousands of dollars.

(b) Graph the system.

Solution (a) The system of linear inequalities is

$$\begin{cases} x \geq 0 \\ y \geq 0 \\ x + y \leq 25 \\ x \geq 15 \\ y \leq 5 \end{cases}$$

- $x \geq 0$, $y \geq 0$: **x and y are nonnegative variables since they represent money invested, in thousands of dollars.**
- $x + y \leq 25$: **The total of the two investments, $x + y$, cannot exceed \$25,000.**
- $x \geq 15$: **At least \$15,000 in Treasury bills**
- $y \leq 5$: **At most \$5000 in corporate bonds**

(b) See the shaded region in Figure 40. Note that the inequalities $x \geq 0$ and $y \geq 0$ require that the graph of the system be in quadrant I.

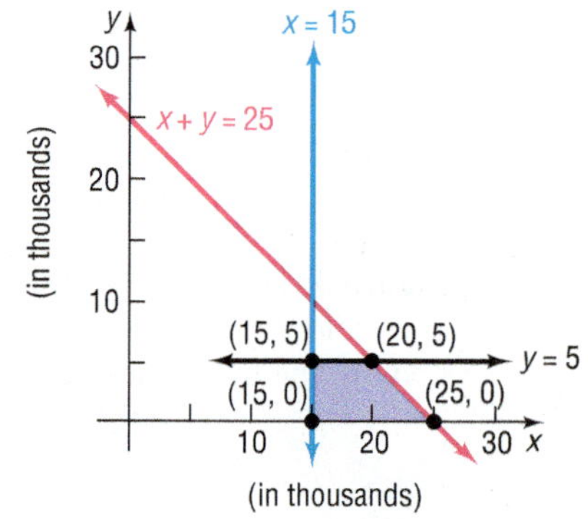

Figure 40

The graph of the system of linear inequalities in Figure 40 is **bounded**, because it can be contained within some circle of sufficiently large radius. A graph that cannot be contained in any circle is **unbounded**. For example, the graph of the system of linear inequalities in Figure 39 is unbounded, since it extends indefinitely in the positive x and positive y directions.

Notice in Figures 39 and 40 that those points that belong to the graph and are also points of intersection of boundary lines have been plotted. Such points are referred to as **vertices** or **corner points** of the graph. The system graphed in Figure 39 has three corner points: $(0, 4)$, $(1, 2)$, and $(3, 0)$. The system graphed in Figure 40 has four corner points: $(15, 0)$, $(25, 0)$, $(20, 5)$, and $(15, 5)$.

These ideas will be used in the next section in developing a method for solving linear programming problems, an important application of linear inequalities.

Now Work PROBLEM 45

12.7 Assess Your Understanding

'Are You Prepared?' *Answers are given at the end of these exercises. If you get a wrong answer, read the pages listed in red.*

1. Solve the inequality: $3x + 4 < 8 - x$ (pp. 150–151)
2. Graph the equation: $3x - 2y = 6$ (pp. 173–184)
3. Graph the equation: $x^2 + y^2 = 9$ (pp. 189–193)
4. Graph the equation: $y = x^2 + 4$ (pp. 256–264)
5. ***True or False*** The lines $2x + y = 4$ and $4x + 2y = 0$ are parallel. (pp. 181–182)
6. The graph of $y = (x - 2)^2$ may be obtained by shifting the graph of ________ to the (left/right) a distance of ________ units. (pp. 256–264)

Concepts and Vocabulary

7. When graphing an inequality in two variables, use ________ if the inequality is strict; if the inequality is nonstrict, use a ________ mark.

8. The graph of the corresponding equation of a linear inequality is a line that separates the xy-plane into two regions. The two regions are called ____________.

9. ***True or False*** The graph of a system of inequalities must have an overlapping region.

10. If a graph of a system of linear inequalities cannot be contained in any circle, then it is ____________.

Skill Building

In Problems 11–22, graph each inequality.

11. $x \geq 0$

12. $y \geq 0$

13. $x \geq 4$

14. $y \leq 2$

15. $2x + y \geq 6$

16. $3x + 2y \leq 6$

17. $x^2 + y^2 > 1$

18. $x^2 + y^2 \leq 9$

19. $y \leq x^2 - 1$

20. $y > x^2 + 2$

21. $xy \geq 4$

22. $xy \leq 1$

In Problems 23–34, graph each system of linear inequalities.

23. $\begin{cases} x + y \leq 2 \\ 2x + y \geq 4 \end{cases}$

24. $\begin{cases} 3x - y \geq 6 \\ x + 2y \leq 2 \end{cases}$

25. $\begin{cases} 2x - y \leq 4 \\ 3x + 2y \geq -6 \end{cases}$

26. $\begin{cases} 4x - 5y \leq 0 \\ 2x - y \geq 2 \end{cases}$

27. $\begin{cases} 2x - 3y \leq 0 \\ 3x + 2y \leq 6 \end{cases}$

28. $\begin{cases} 4x - y \geq 2 \\ x + 2y \geq 2 \end{cases}$

29. $\begin{cases} x - 2y \leq 6 \\ 2x - 4y \geq 0 \end{cases}$

30. $\begin{cases} x + 4y \leq 8 \\ x + 4y \geq 4 \end{cases}$

31. $\begin{cases} 2x + y \geq -2 \\ 2x + y \geq 2 \end{cases}$

32. $\begin{cases} x - 4y \leq 4 \\ x - 4y \geq 0 \end{cases}$

33. $\begin{cases} 2x + 3y \geq 6 \\ 2x + 3y \leq 0 \end{cases}$

34. $\begin{cases} 2x + y \geq 0 \\ 2x + y \geq 2 \end{cases}$

In Problems 35–42, graph each system of inequalities.

35. $\begin{cases} x^2 + y^2 \leq 9 \\ x + y \geq 3 \end{cases}$

36. $\begin{cases} x^2 + y^2 \geq 9 \\ x + y \leq 3 \end{cases}$

37. $\begin{cases} y \geq x^2 - 4 \\ y \leq x - 2 \end{cases}$

38. $\begin{cases} y^2 \leq x \\ y \geq x \end{cases}$

39. $\begin{cases} x^2 + y^2 \leq 16 \\ y \geq x^2 - 4 \end{cases}$

40. $\begin{cases} x^2 + y^2 \leq 25 \\ y \leq x^2 - 5 \end{cases}$

41. $\begin{cases} xy \geq 4 \\ y \geq x^2 + 1 \end{cases}$

42. $\begin{cases} y + x^2 \leq 1 \\ y \geq x^2 - 1 \end{cases}$

In Problems 43–52, graph each system of linear inequalities. Tell whether the graph is bounded or unbounded, and label the corner points.

43. $\begin{cases} x \geq 0 \\ y \geq 0 \\ 2x + y \leq 6 \\ x + 2y \leq 6 \end{cases}$

44. $\begin{cases} x \geq 0 \\ y \geq 0 \\ x + y \geq 4 \\ 2x + 3y \geq 6 \end{cases}$

45. $\begin{cases} x \geq 0 \\ y \geq 0 \\ x + y \geq 2 \\ 2x + y \geq 4 \end{cases}$

46. $\begin{cases} x \geq 0 \\ y \geq 0 \\ 3x + y \leq 6 \\ 2x + y \leq 2 \end{cases}$

47. $\begin{cases} x \geq 0 \\ y \geq 0 \\ x + y \geq 2 \\ 2x + 3y \leq 12 \\ 3x + y \leq 12 \end{cases}$

48. $\begin{cases} x \geq 0 \\ y \geq 0 \\ x + y \geq 1 \\ x + y \leq 7 \\ 2x + y \leq 10 \end{cases}$

49. $\begin{cases} x \geq 0 \\ y \geq 0 \\ x + y \geq 2 \\ x + y \leq 8 \\ 2x + y \leq 10 \end{cases}$

50. $\begin{cases} x \geq 0 \\ y \geq 0 \\ x + y \geq 2 \\ x + y \leq 8 \\ x + 2y \geq 1 \end{cases}$

51. $\begin{cases} x \geq 0 \\ y \geq 0 \\ x + 2y \geq 1 \\ x + 2y \leq 10 \end{cases}$

52. $\begin{cases} x \geq 0 \\ y \geq 0 \\ x + 2y \geq 1 \\ x + 2y \leq 10 \\ x + y \geq 2 \\ x + y \leq 8 \end{cases}$

In Problems 53–56, write a system of linear inequalities for the given graph.

53.

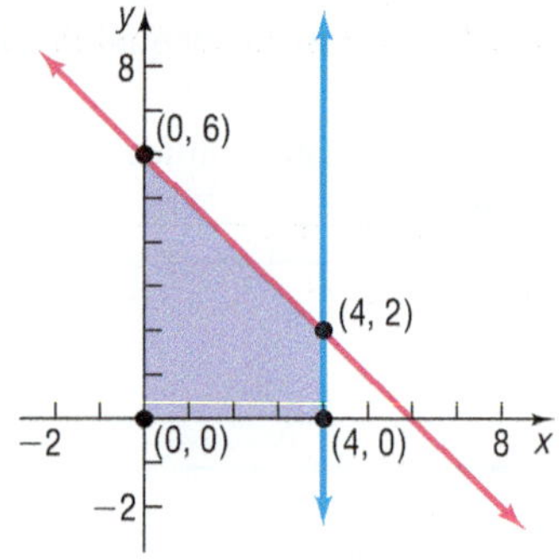

54.

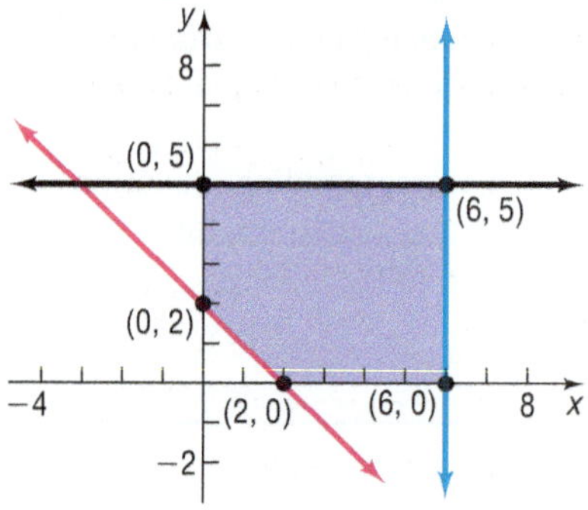

55.

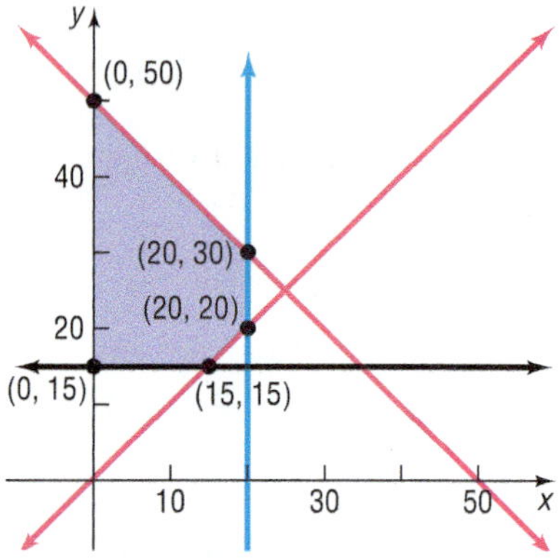

56.

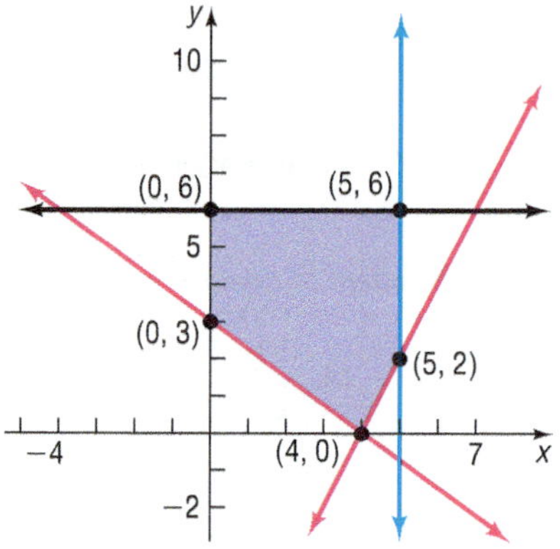

Applications and Extensions

57. Financial Planning A retired couple has up to \$50,000 to invest. As their financial adviser, you recommend that they place at least \$35,000 in Treasury bills yielding 1% and at most \$10,000 in corporate bonds yielding 3%.

(a) Using x to denote the amount of money invested in Treasury bills and y to denote the amount invested in corporate bonds, write a system of linear inequalities that describes the possible amounts of each investment.

(b) Graph the system and label the corner points.

58. Manufacturing Trucks Mike's Toy Truck Company manufactures two models of toy trucks, a standard model and a deluxe model. Each standard model requires 2 hours (h) for painting and 3 h for detail work; each deluxe model requires 3 h for painting and 4 h for detail work. Two painters and three detail workers are employed by the company, and each works 40 h per week.

(a) Using x to denote the number of standard-model trucks and y to denote the number of deluxe-model trucks, write a system of linear inequalities that describes the possible numbers of each model of truck that can be manufactured in a week.

(b) Graph the system and label the corner points.

59. Blending Coffee Bill's Coffee House, a store that specializes in coffee, has available 75 pounds (lb) of A grade coffee and 120 lb of B grade coffee. These will be blended into 1-lb packages as follows: an economy blend that contains 4 ounces (oz) of A grade coffee and 12 oz of B grade coffee, and a superior blend that contains 8 oz of A grade coffee and 8 oz of B grade coffee.

(a) Using x to denote the number of packages of the economy blend and y to denote the number of packages of the superior blend, write a system of linear inequalities that describes the possible numbers of packages of each kind of blend.

(b) Graph the system and label the corner points.

60. Mixed Nuts Nola's Nuts, a store that specializes in selling nuts, has available 90 pounds (lb) of cashews and 120 lb of peanuts. These are to be mixed in 12-ounce (oz) packages as follows: a lower-priced package containing 8 oz of peanuts and 4 oz of cashews, and a quality package containing 6 oz of peanuts and 6 oz of cashews.

(a) Use x to denote the number of lower-priced packages, and use y to denote the number of quality packages. Write a system of linear inequalities that describes the possible numbers of each kind of package.

(b) Graph the system and label the corner points.

61. Transporting Goods A small truck can carry no more than 1600 pounds (lb) of cargo and no more than 150 cubic feet (ft^3) of cargo. A printer weighs 20 lb and occupies 3 ft^3 of space. A microwave oven weighs 30 lb and occupies 2 ft^3 of space.

(a) Using x to represent the number of microwave ovens and y to represent the number of printers, write a system of linear inequalities that describes the number of ovens and printers that can be hauled by the truck.

(b) Graph the system and label the corner points.

Retain Your Knowledge

Problems 62–65 are based on material learned earlier in the course. The purpose of these problems is to keep the material fresh in your mind so that you are better prepared for the final exam.

62. Solve $2(x+1)^2 + 8 = 0$ in the complex number system.

63. Write the polar equation $3r = \sin\theta$ as an equation in rectangular coordinates. Identify the equation and graph it.

64. Use the Intermediate Value Theorem to show that $f(x) = 6x^2 + 5x - 6$ has a real zero on the interval $[-1, 2]$.

65. Solve the equation $2\cos^2\theta - \cos\theta - 1 = 0$ on the interval $0 \le \theta < 2\pi$.

'Are You Prepared?' Answers

1. $\{x \mid x < 1\}$ or $(-\infty, 1)$

2.

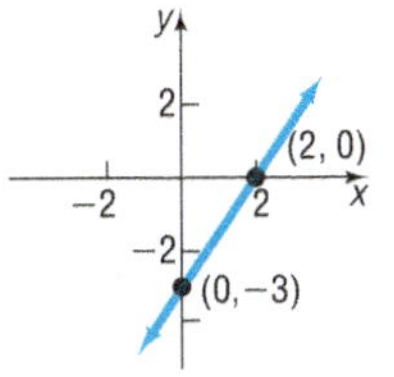

3.

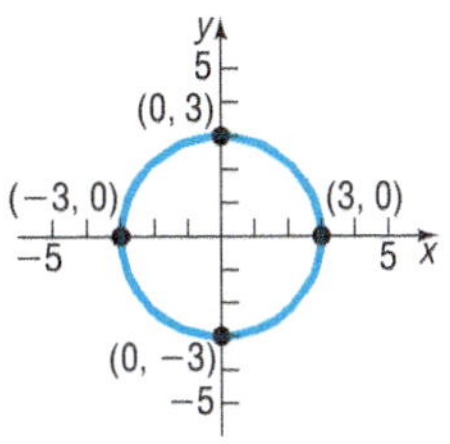

4.

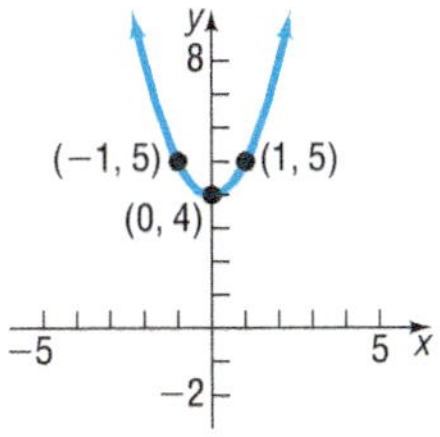

5. True

6. $y = x^2$; right; 2

12.8 Linear Programming

OBJECTIVES 1 Set Up a Linear Programming Problem (p. 947)
2 Solve a Linear Programming Problem (p. 948)

Historically, linear programming evolved as a technique for solving problems involving resource allocation of goods and materials for the U.S. Air Force during World War II. Today, linear programming techniques are used to solve a wide variety of problems, such as optimizing airline scheduling and establishing telephone lines. Although most practical linear programming problems involve systems of several hundred linear inequalities containing several hundred variables, we limit our discussion to problems containing only two variables, because we can solve such problems using graphing techniques.*

1 Set Up a Linear Programming Problem

Let's begin by returning to Example 12 of the previous section.

EXAMPLE 1 Financial Planning

A retired couple has up to \$25,000 to invest. As their financial adviser, you recommend that they place at least \$15,000 in Treasury bills yielding 2% and at most \$5000 in corporate bonds yielding 3%. Develop a model that can be used to determine how much money they should place in each investment so that income is maximized.

*The **simplex method** is a way to solve linear programming problems involving many inequalities and variables. This method was developed by George Dantzig in 1946 and is particularly well suited for computerization. In 1984, Narendra Karmarkar of Bell Laboratories discovered a way of solving large linear programming problems that improves on the simplex method.

Solution The problem is typical of a *linear programming problem*. The problem requires that a certain linear expression, the income, be maximized. If I represents income, x the amount invested in Treasury bills at 2%, and y the amount invested in corporate bonds at 3%, then

$$I = 0.02x + 0.03y$$

Assume, as before, that I, x, and y are in thousands of dollars.

The linear expression $I = 0.02x + 0.03y$ is called the **objective function**. Further, the problem requires that the maximum income be achieved under certain conditions, or **constraints**, each of which is a linear inequality involving the variables. (See Example 12 in Section 12.7.) The linear programming problem may be modeled as

$$\text{Maximize} \qquad I = 0.02x + 0.03y$$

subject to the conditions that

$$\begin{cases} x \geq 0 \\ y \geq 0 \\ x + y \leq 25 \\ x \geq 15 \\ y \leq 5 \end{cases}$$

In general, every linear programming problem has two components:

1. A linear objective function that is to be maximized or minimized
2. A collection of linear inequalities that must be satisfied simultaneously

DEFINITION

A **linear programming problem** in two variables x and y consists of maximizing (or minimizing) a linear objective function

$$z = Ax + By \qquad A \text{ and } B \text{ are real numbers, not both } 0$$

subject to certain conditions, or constraints, expressible as linear inequalities in x and y.

2 Solve a Linear Programming Problem

To maximize (or minimize) the quantity $z = Ax + By$, we need to identify points (x, y) that make the expression for z the largest (or smallest) possible. But not all points (x, y) are eligible; only those that also satisfy each linear inequality (constraint) can be used. Each point (x, y) that satisfies the system of linear inequalities (the constraints) is a **feasible point**. Linear programming problems seek the feasible point(s) that maximizes (or minimizes) the objective function.

Look again at the linear programming problem in Example 1.

EXAMPLE 2 **Analyzing a Linear Programming Problem**

Consider the linear programming problem

$$\text{Maximize} \qquad I = 0.02x + 0.03y$$

subject to the conditions that

$$\begin{cases} x \geq 0 \\ y \geq 0 \\ x + y \leq 25 \\ x \geq 15 \\ y \leq 5 \end{cases}$$

Graph the constraints. Then graph the objective function for $I = 0, 0.3, 0.45, 0.55,$ and 0.6.

Solution Figure 41 shows the graph of the constraints. We superimpose on this graph the graph of the objective function for the given values of I.

For $I = 0$, the objective function is the line $0 = 0.02x + 0.03y$.
For $I = 0.3$, the objective function is the line $0.3 = 0.02x + 0.03y$.
For $I = 0.45$, the objective function is the line $0.45 = 0.02x + 0.03y$.
For $I = 0.55$, the objective function is the line $0.55 = 0.02x + 0.03y$.
For $I = 0.6$, the objective function is the line $0.6 = 0.02x + 0.03y$.

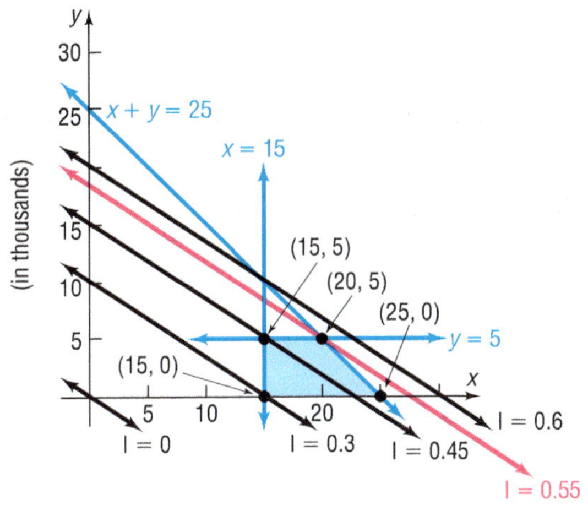

Figure 41

DEFINITION A **solution** to a linear programming problem consists of a feasible point that maximizes (or minimizes) the objective function, together with the corresponding value of the objective function.

One condition for a linear programming problem in two variables to have a solution is that the graph of the feasible points be bounded. (Refer to page 944.)

If none of the feasible points maximizes (or minimizes) the objective function or if there are no feasible points, the linear programming problem has no solution.

Consider the linear programming problem posed in Example 2, and look again at Figure 41. The feasible points are the points that lie in the shaded region. For example, $(20, 3)$ is a feasible point, as are $(15, 5)$, $(20, 5)$, $(18, 4)$, and so on. To find the solution of the problem requires finding a feasible point (x, y) that makes $I = 0.02x + 0.03y$ as large as possible. Notice that as I increases in value from $I = 0$ to $I = 0.3$ to $I = 0.45$ to $I = 0.55$ to $I = 0.6$, the result is a collection of parallel lines. Further, notice that the largest value of I that can be obtained using feasible points is $I = 0.55$, which corresponds to the line $0.55 = 0.02x + 0.03y$. Any larger value of I results in a line that does not pass through any feasible points. Finally, notice that the feasible point that yields $I = 0.55$ is the point $(20, 5)$, a corner point. These observations form the basis of the following results, which are stated without proof.

THEOREM

Location of the Solution of a Linear Programming Problem

If a linear programming problem has a solution, it is located at a corner point of the graph of the feasible points.

If a linear programming problem has multiple solutions, at least one of them is located at a corner point of the graph of the feasible points.

In either case, the corresponding value of the objective function is unique. ■

We shall not consider linear programming problems that have no solution. As a result, we can outline the procedure for solving a linear programming problem as follows:

Procedure for Solving a Linear Programming Problem

STEP 1: Write an expression for the quantity to be maximized (or minimized). This expression is the objective function.

STEP 2: Write all the constraints as a system of linear inequalities, and graph the system.

STEP 3: List the corner points of the graph of the feasible points.

STEP 4: List the corresponding values of the objective function at each corner point. The largest (or smallest) of these is the solution.

EXAMPLE 3 Solving a Minimum Linear Programming Problem

Minimize the expression

$$z = 2x + 3y$$

subject to the constraints

$$y \le 5 \qquad x \le 6 \qquad x + y \ge 2 \qquad x \ge 0 \qquad y \ge 0$$

Solution

STEP 1: The objective function is $z = 2x + 3y$.

STEP 2: We seek the smallest value of z that can occur if x and y are solutions of the system of linear inequalities

$$\begin{cases} y \le 5 \\ x \le 6 \\ x + y \ge 2 \\ x \ge 0 \\ y \ge 0 \end{cases}$$

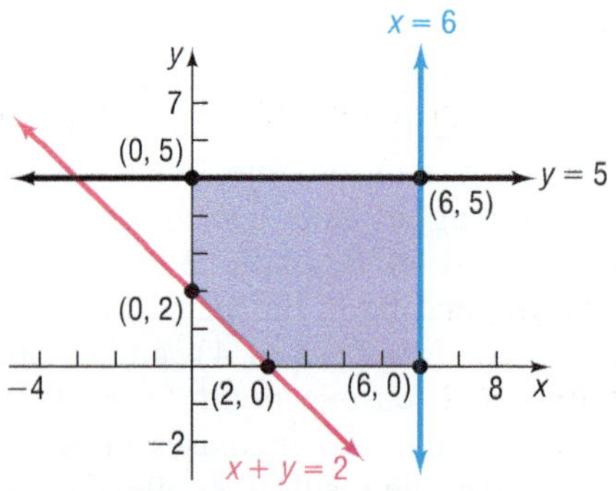

Figure 42

STEP 3: The graph of this system (the set of feasible points) is shown as the shaded region in Figure 42. The corner points have also been plotted.

STEP 4: Table 1 lists the corner points and the corresponding values of the objective function. From the table, the minimum value of z is 4, and it occurs at the point $(2, 0)$.

Table 1

Corner Point (x, y)	Value of the Objective Function $z = 2x + 3y$
(0, 2)	$z = 2(0) + 3(2) = 6$
(0, 5)	$z = 2(0) + 3(5) = 15$
(6, 5)	$z = 2(6) + 3(5) = 27$
(6, 0)	$z = 2(6) + 3(0) = 12$
(2, 0)	$z = 2(2) + 3(0) = 4$

Now Work PROBLEMS 5 AND 11

EXAMPLE 4 Maximizing Profit

At the end of every month, after filling orders for its regular customers, a coffee company has some pure Colombian coffee and some special-blend coffee remaining. The practice of the company has been to package a mixture of the two coffees into 1-pound (lb) packages as follows: a low-grade mixture containing 4 ounces (oz) of Colombian coffee and 12 oz of special-blend coffee, and a high-grade mixture containing 8 oz of Colombian and 8 oz of special-blend coffee. A profit of \$0.30 per package is made on the low-grade mixture, whereas a profit of \$0.40 per package is made on the high-grade mixture. This month, 120 lb of special-blend coffee and 100 lb of pure Colombian coffee remain. How many packages of each mixture should be prepared to achieve a maximum profit? Assume that all packages prepared can be sold.

Solution

STEP 1: Begin by assigning symbols for the two variables.

x = Number of packages of the low-grade mixture

y = Number of packages of the high-grade mixture

If P denotes the profit, then

$$P = \$0.30x + \$0.40y \quad \text{Objective function}$$

STEP 2: The goal is to maximize P subject to certain constraints on x and y. Because x and y represent numbers of packages, the only meaningful values for x and y are nonnegative integers. This yields the two constraints

$$x \geq 0 \qquad y \geq 0 \quad \text{Nonnegative constraints}$$

There is only so much of each type of coffee available. For example, the total amount of Colombian coffee used in the two mixtures cannot exceed 100 lb, or 1600 oz. Because 4 oz are used in each low-grade package and 8 oz are used in each high-grade package, this leads to the constraint

$$4x + 8y \leq 1600 \quad \text{Colombian coffee constraint}$$

Similarly, the supply of 120 lb, or 1920 oz, of special-blend coffee leads to the constraint

$$12x + 8y \leq 1920 \quad \text{Special-blend coffee constraint}$$

The linear programming problem may be stated as

$$\text{Maximize} \qquad P = 0.3x + 0.4y$$

subject to the constraints

$$x \geq 0 \qquad y \geq 0 \qquad 4x + 8y \leq 1600 \qquad 12x + 8y \leq 1920$$

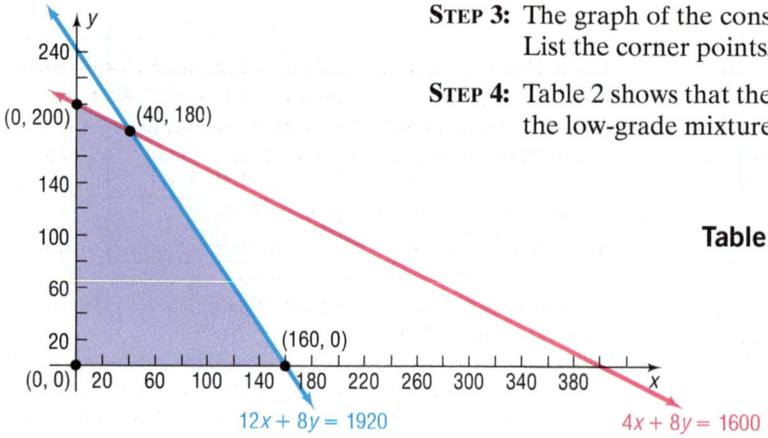

Figure 43

STEP 3: The graph of the constraints (the feasible points) is illustrated in Figure 43. List the corner points, and evaluate the objective function at each point.

STEP 4: Table 2 shows that the maximum profit, \$84, is achieved with 40 packages of the low-grade mixture and 180 packages of the high-grade mixture.

Table 2

Corner Point (x, y)	Value of Profit $P = 0.3x + 0.4y$
(0, 0)	$P = 0$
(0, 200)	$P = 0.3(0) + 0.4(200) = \$80$
(40, 180)	$P = 0.3(40) + 0.4(180) = \$84$
(160, 0)	$P = 0.3(160) + 0.4(0) = \$48$

Now Work PROBLEM 19

12.8 Assess Your Understanding

Concepts and Vocabulary

1. A linear programming problem requires that a linear expression, called the __________ __________, be maximized or minimized.

2. ***True or False*** If a linear programming problem has a solution, it is located at a corner point of the graph of the feasible points.

Skill Building

In Problems 3–8, find the maximum and minimum value of the given objective function of a linear programming problem. The figure illustrates the graph of the feasible points.

3. $z = x + y$

4. $z = 2x + 3y$

5. $z = x + 10y$

6. $z = 10x + y$

7. $z = 5x + 7y$

8. $z = 7x + 5y$

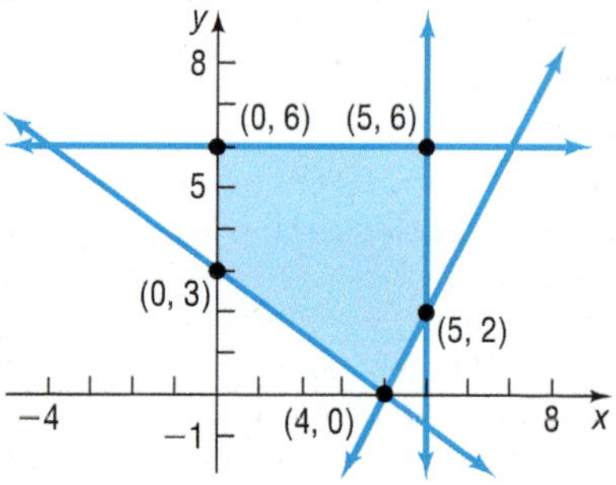

In Problems 9–18, solve each linear programming problem.

9. Maximize $z = 2x + y$ subject to $x \geq 0$, $y \geq 0$, $x + y \leq 6$, $x + y \geq 1$

10. Maximize $z = x + 3y$ subject to $x \geq 0$, $y \geq 0$, $x + y \geq 3$, $x \leq 5$, $y \leq 7$

11. Minimize $z = 2x + 5y$ subject to $x \geq 0$, $y \geq 0$, $x + y \geq 2$, $x \leq 5$, $y \leq 3$

12. Minimize $z = 3x + 4y$ subject to $x \geq 0$, $y \geq 0$, $2x + 3y \geq 6$, $x + y \leq 8$

13. Maximize $z = 3x + 5y$ subject to $x \geq 0$, $y \geq 0$, $x + y \geq 2$, $2x + 3y \leq 12$, $3x + 2y \leq 12$

14. Maximize $z = 5x + 3y$ subject to $x \geq 0$, $y \geq 0$, $x + y \geq 2$, $x + y \leq 8$, $2x + y \leq 10$

15. Minimize $z = 5x + 4y$ subject to $x \geq 0$, $y \geq 0$, $x + y \geq 2$, $2x + 3y \leq 12$, $3x + y \leq 12$

16. Minimize $z = 2x + 3y$ subject to $x \geq 0$, $y \geq 0$, $x + y \geq 3$, $x + y \leq 9$, $x + 3y \geq 6$

17. Maximize $z = 5x + 2y$ subject to $x \geq 0$, $y \geq 0$, $x + y \leq 10$, $2x + y \geq 10$, $x + 2y \geq 10$

18. Maximize $z = 2x + 4y$ subject to $x \geq 0$, $y \geq 0$, $2x + y \geq 4$, $x + y \leq 9$

Applications and Extensions

19. **Maximizing Profit** A manufacturer of skis produces two types: downhill and cross-country. Use the following table to determine how many of each kind of ski should be produced to achieve a maximum profit. What is the maximum profit? What would the maximum profit be if the time available for manufacturing were increased to 48 hours?

	Downhill	Cross-country	Time Available
Manufacturing time per ski	2 hours	1 hour	40 hours
Finishing time per ski	1 hour	1 hour	32 hours
Profit per ski	$70	$50	

20. **Farm Management** A farmer has 70 acres of land available for planting either soybeans or wheat. The cost of preparing the soil, the workdays required, and the expected profit per acre planted for each type of crop are given in the following table.

	Soybeans	Wheat
Preparation cost per acre	$60	$30
Workdays required per acre	3	4
Profit per acre	$180	$100

The farmer cannot spend more than $1800 in preparation costs and cannot use a total of more than 120 workdays. How many acres of each crop should be planted to maximize the profit? What is the maximum profit? What is the maximum profit if the farmer is willing to spend no more than $2400 on preparation?

21. **Banquet Seating** A banquet hall offers two types of tables for rent: 6-person rectangular tables at a cost of $28 each and 10-person round tables at a cost of $52 each. Kathleen would like to rent the hall for a wedding banquet and needs tables for 250 people. The hall can have a maximum of 35 tables, and the hall has only 15 rectangular tables available. How many of each type of table should be rented to minimize cost and what is the minimum cost?

Source: *facilities.princeton.edu*

22. **Spring Break** The student activities department of a community college plans to rent buses and vans for a spring-break trip. Each bus has 40 regular seats and 1 special seat designed to accommodate travelers with disabilities. Each van has 8 regular seats and 3 special seats. The rental cost is $350 for each van and $975 for each bus. If 320 regular and 36 special seats are required for the trip, how many vehicles of each type should be rented to minimize cost?

Source: *www.busrates.com*

23. **Return on Investment** An investment broker is instructed by her client to invest up to $20,000, some in a junk bond yielding 9% per annum and some in Treasury bills yielding 7% per annum. The client wants to invest at least $8000 in T-bills and no more than $12,000 in the junk bond.

(a) How much should the broker recommend that the client place in each investment to maximize income if the client insists that the amount invested in T-bills must equal or exceed the amount placed in the junk bond?

(b) How much should the broker recommend that the client place in each investment to maximize income if the client insists that the amount invested in T-bills must not exceed the amount placed in the junk bond?

24. **Production Scheduling** In a factory, machine 1 produces 8-inch (in.) pliers at the rate of 60 units per hour (h) and 6-in. pliers at the rate of 70 units/h. Machine 2 produces 8-in. pliers at the rate of 40 units/h and 6-in. pliers at the rate of 20 units/h. It costs $50/h to operate machine 1, and machine 2 costs $30/h to operate. The production schedule requires that at least 240 units of 8-in. pliers and at least 140 units of 6-in. pliers be produced during each 10-h day. Which combination of machines will cost the least money to operate?

25. **Managing a Meat Market** A meat market combines ground beef and ground pork in a single package for meat loaf. The ground beef is 75% lean (75% beef, 25% fat) and costs the market $0.75 per pound (lb). The ground pork is 60% lean and costs the market $0.45/lb. The meat loaf must be at least 70% lean. If the market wants to use at least 50 lb of its available pork, but no more than 200 lb of its available ground beef, how much ground beef should be mixed with ground pork so that the cost is minimized?

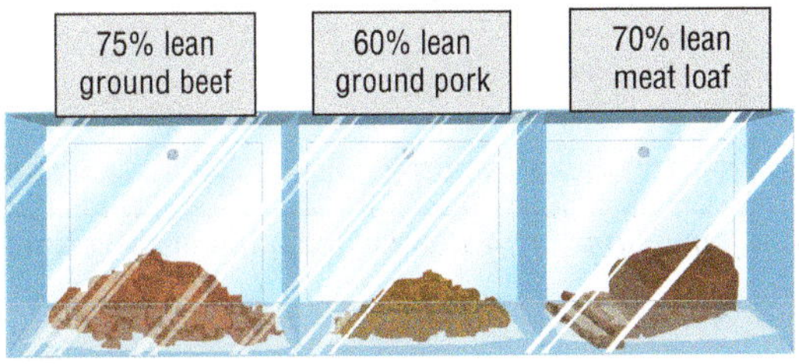

26. **Ice Cream** The Mom and Pop Ice Cream Company makes two kinds of chocolate ice cream: regular and premium. The properties of 1 gallon (gal) of each type are shown in the table:

	Regular	Premium
Flavoring	24 oz	20 oz
Milk-fat products	12 oz	20 oz
Shipping weight	5 lbs	6 lbs
Profit	$0.75	$0.90

In addition, current commitments require the company to make at least 1 gal of premium for every 4 gal of regular. Each day, the company has available 725 pounds (lb) of flavoring and 425 lb of milk-fat products. If the company can ship no more than 3000 lb of product per day, how many gallons of each type should be produced daily to maximize profit?

Source: *www.scitoys.com/ingredients/ice_cream.html*

27. **Maximizing Profit on Ice Skates** A factory manufactures two kinds of ice skates: racing skates and figure skates. The racing skates require 6 work-hours in the fabrication department, whereas the figure skates require 4 work-hours there. The racing skates require 1 work-hour in the finishing

department, whereas the figure skates require 2 work-hours there. The fabricating department has available at most 120 work-hours per day, and the finishing department has no more than 40 work-hours per day available. If the profit on each racing skate is \$10 and the profit on each figure skate is \$12, how many of each should be manufactured each day to maximize profit? (Assume that all skates made are sold.)

28. **Financial Planning** A retired couple have up to \$50,000 to place in fixed-income securities. Their financial adviser suggests two securities to them: one is an AAA bond that yields 8% per annum; the other is a certificate of deposit (CD) that yields 4%. After careful consideration of the alternatives, the couple decide to place at most \$20,000 in the AAA bond and at least \$15,000 in the CD. They also instruct the financial adviser to place at least as much in the CD as in the AAA bond. How should the financial adviser proceed to maximize the return on their investment?

29. **Product Design** An entrepreneur is having a design group produce at least six samples of a new kind of fastener that he wants to market. It costs \$9.00 to produce each metal fastener and \$4.00 to produce each plastic fastener. He wants to have at least two of each version of the fastener and needs to have all the samples 24 hours (h) from now. It takes 4 h to produce each metal sample and 2 h to produce each plastic sample. To minimize the cost of the samples, how many of each kind should the entrepreneur order? What will be the cost of the samples?

30. **Animal Nutrition** Kevin's dog Amadeus likes two kinds of canned dog food. Gourmet Dog costs 40 cents a can and has 20 units of a vitamin complex; the calorie content is 75 calories. Chow Hound costs 32 cents a can and has 35 units of vitamins and 50 calories. Kevin likes Amadeus to have at least 1175 units of vitamins a month and at least 2375 calories during the same time period. Kevin has space to store only 60 cans of dog food at a time. How much of each kind of dog food should Kevin buy each month to minimize his cost?

31. **Airline Revenue** An airline has two classes of service: first class and coach. Management's experience has been that each aircraft should have at least 8 but no more than 16 first-class seats and at least 80 but no more than 120 coach seats.
   (a) If management decides that the ratio of first-class seats to coach seats should never exceed 1:12, with how many of each type of seat should an aircraft be configured to maximize revenue?
   (b) If management decides that the ratio of first-class seats to coach seats should never exceed 1:8, with how many of each type of seat should an aircraft be configured to maximize revenue?
   (c) If you were management, what would you do?
   [**Hint:** Assume that the airline charges \$$C$ for a coach seat and \$$F$ for a first-class seat; $C > 0$, $F > C$.]

Explaining Concepts: Discussion and Writing

32. Explain in your own words what a linear programming problem is and how it can be solved.

Retain Your Knowledge

Problems 33–36 are based on material learned earlier in the course. The purpose of these problems is to keep the material fresh in your mind so that you are better prepared for the final exam.

33. Solve: $2m^{2/5} - m^{1/5} = 1$

34. Graph $y = -\tan\left(x - \frac{\pi}{2}\right)$ for at least two periods. Use the graph to determine the domain and range.

35. **Radioactive Decay** The half-life of titanium-44 is 63 years. How long will it take 200 grams to decay to 75 grams? Round to one decimal place.

36. Find the equation of the line that is parallel to $y = 3x + 11$ and passes through the point $(-2, 1)$.

Chapter Review

Things to Know

Systems of equations (pp. 862–864)

Systems with no solutions are inconsistent. Systems with a solution are consistent.

Consistent systems of linear equations have either a unique solution (independent) or an infinite number of solutions (dependent).

Matrix (p. 877) Rectangular array of numbers, called entries

Augmented matrix (p. 877)

Row operations (p. 879)

Row echelon form (p. 881)

Determinants and Cramer's Rule (pp. 893, 895, 896–897, and 898)

Matrix (p. 903)

m by n matrix (p. 904)	Matrix with m rows and n columns
Identity matrix I_n (p. 910)	An n by n square matrix whose diagonal entries are 1's, while all other entries are 0's
Inverse of a matrix (p. 911)	A^{-1} is the inverse of A if $AA^{-1} = A^{-1}A = I_n$.
Nonsingular matrix (p. 911)	A square matrix that has an inverse

Linear programming problem (p. 948)

Maximize (or minimize) a linear objective function, $z = Ax + By$, subject to certain conditions, or constraints, expressible as linear inequalities in x and y. A feasible point (x, y) is a point that satisfies the constraints (linear inequalities) of a linear programming problem.

Location of solution (p. 950)

If a linear programming problem has a solution, it is located at a corner point of the graph of the feasible points. If a linear programming problem has multiple solutions, at least one of them is located at a corner point of the graph of the feasible points. In either case, the corresponding value of the objective function is unique.

Objectives

Section		You should be able to ...	Example(s)	Review Exercises
12.1	1	Solve systems of equations by substitution (p. 864)	4	1–7, 56, 59, 60
	2	Solve systems of equations by elimination (p. 865)	5, 6	1–7, 56, 59, 60
	3	Identify inconsistent systems of equations containing two variables (p. 867)	7	5, 54
	4	Express the solution of a system of dependent equations containing two variables (p. 867)	8	7, 53
	5	Solve systems of three equations containing three variables (p. 868)	9	8–10, 55, 57
	6	Identify inconsistent systems of equations containing three variables (p. 870)	10	10
	7	Express the solution of a system of dependent equations containing three variables (p. 871)	11	9
12.2	1	Write the augmented matrix of a system of linear equations (p. 877)	1	20–25
	2	Write the system of equations from the augmented matrix (p. 879)	2	11, 12
	3	Perform row operations on a matrix (p. 879)	3, 4	20–25
	4	Solve a system of linear equations using matrices (p. 880)	5–10	20–25
12.3	1	Evaluate 2 by 2 determinants (p. 893)	1	26
	2	Use Cramer's Rule to solve a system of two equations containing two variables (p. 894)	2	29, 30
	3	Evaluate 3 by 3 determinants (p. 896)	4	27, 28
	4	Use Cramer's Rule to solve a system of three equations containing three variables (p. 898)	5	31
	5	Know properties of determinants (p. 899)	6–9	32, 33
12.4	1	Find the sum and difference of two matrices (p. 904)	3, 4	13
	2	Find scalar multiples of a matrix (p. 906)	5	14
	3	Find the product of two matrices (p. 907)	6–11	15, 16
	4	Find the inverse of a matrix (p. 911)	12–14	17–19
	5	Solve a system of linear equations using an inverse matrix (p. 914)	15	20–25
12.5	1	Decompose $\frac{P}{Q}$, where Q has only nonrepeated linear factors (p. 921)	1	34
	2	Decompose $\frac{P}{Q}$, where Q has repeated linear factors (p. 923)	2, 3	35
	3	Decompose $\frac{P}{Q}$, where Q has a nonrepeated irreducible quadratic factor (p. 925)	4	36, 38
	4	Decompose $\frac{P}{Q}$, where Q has a repeated irreducible quadratic factor (p. 926)	5	37

Section	You should be able to ...	Example(s)	Review Exercises
12.6	1 Solve a system of nonlinear equations using substitution (p. 928)	1, 3	39–43
	2 Solve a system of nonlinear equations using elimination (p. 929)	2, 4, 5	39–43
12.7	1 Graph an inequality by hand (p. 938)	2–4	44, 45
	2 Graph an inequality using a graphing utility (p. 940)	5, 7	44, 45
	3 Graph a system of inequalities (p. 941)	6–12	46–50, 58
12.8	1 Set up a linear programming problem (p. 947)	1	61
	2 Solve a linear programming problem (p. 948)	2–4	51, 52, 61

Review Exercises

In Problems 1–10, solve each system of equations using the method of substitution or the method of elimination. If the system has no solution, say that it is inconsistent.

1. $\begin{cases} 2x - y = 5 \\ 5x + 2y = 8 \end{cases}$

2. $\begin{cases} 3x - 4y = 4 \\ x - 3y = \dfrac{1}{2} \end{cases}$

3. $\begin{cases} x - 2y - 4 = 0 \\ 3x + 2y - 4 = 0 \end{cases}$

4. $\begin{cases} y = 2x - 5 \\ x = 3y + 4 \end{cases}$

5. $\begin{cases} x - 3y + 4 = 0 \\ \dfrac{1}{2}x - \dfrac{3}{2}y + \dfrac{4}{3} = 0 \end{cases}$

6. $\begin{cases} 2x + 3y - 13 = 0 \\ 3x - 2y = 0 \end{cases}$

7. $\begin{cases} 2x + 5y = 10 \\ 4x + 10y = 20 \end{cases}$

8. $\begin{cases} x + 2y - z = 6 \\ 2x - y + 3z = -13 \\ 3x - 2y + 3z = -16 \end{cases}$

9. $\begin{cases} 2x - 4y + z = -15 \\ x + 2y - 4z = 27 \\ 5x - 6y - 2z = -3 \end{cases}$

10. $\begin{cases} x - 4y + 3z = 15 \\ -3x + y - 5z = -5 \\ -7x - 5y - 9z = 10 \end{cases}$

In Problems 11 and 12, write the system of equations that corresponds to the given augmented matrix.

11. $\left[\begin{array}{cc|c} 3 & 2 & 8 \\ 1 & 4 & -1 \end{array}\right]$

12. $\left[\begin{array}{ccc|c} 1 & 2 & 5 & -2 \\ 5 & 0 & -3 & 8 \\ 2 & -1 & 0 & 0 \end{array}\right]$

In Problems 13–16, use the following matrices to compute each expression.

$$A = \begin{bmatrix} 1 & 0 \\ 2 & 4 \\ -1 & 2 \end{bmatrix} \quad B = \begin{bmatrix} 4 & -3 & 0 \\ 1 & 1 & -2 \end{bmatrix} \quad C = \begin{bmatrix} 3 & -4 \\ 1 & 5 \\ 5 & 2 \end{bmatrix}$$

13. $A + C$ **14.** $6A$ **15.** AB **16.** BC

In Problems 17–19, find the inverse, if there is one, of each matrix. If there is not an inverse, say that the matrix is singular.

17. $\begin{bmatrix} 4 & 6 \\ 1 & 3 \end{bmatrix}$

18. $\begin{bmatrix} 1 & 3 & 3 \\ 1 & 2 & 1 \\ 1 & -1 & 2 \end{bmatrix}$

19. $\begin{bmatrix} 4 & -8 \\ -1 & 2 \end{bmatrix}$

In Problems 20–25, solve each system of equations using matrices. If the system has no solution, say that it is inconsistent.

20. $\begin{cases} 3x - 2y = 1 \\ 10x + 10y = 5 \end{cases}$

21. $\begin{cases} 5x - 6y - 3z = 6 \\ 4x - 7y - 2z = -3 \\ 3x + y - 7z = 1 \end{cases}$

22. $\begin{cases} 2x + y + z = 5 \\ 4x - y - 3z = 1 \\ 8x + y - z = 5 \end{cases}$

23. $\begin{cases} x - 2z = 1 \\ 2x + 3y = -3 \\ 4x - 3y - 4z = 3 \end{cases}$

24. $\begin{cases} x - y + z = 0 \\ x - y - 5z - 6 = 0 \\ 2x - 2y + z - 1 = 0 \end{cases}$

25. $\begin{cases} x - y - z - t = 1 \\ 2x + y - z + 2t = 3 \\ x - 2y - 2z - 3t = 0 \\ 3x - 4y + z + 5t = -3 \end{cases}$

In Problems 26–28, find the value of each determinant.

26. $\begin{vmatrix} 3 & 4 \\ 1 & 3 \end{vmatrix}$

27. $\begin{vmatrix} 1 & 4 & 0 \\ -1 & 2 & 6 \\ 4 & 1 & 3 \end{vmatrix}$

28. $\begin{vmatrix} 2 & 1 & -3 \\ 5 & 0 & 1 \\ 2 & 6 & 0 \end{vmatrix}$

In Problems 29–31, use Cramer's Rule, if applicable, to solve each system.

29. $\begin{cases} x - 2y = 4 \\ 3x + 2y = 4 \end{cases}$

30. $\begin{cases} 2x + 3y - 13 = 0 \\ 3x - 2y = 0 \end{cases}$

31. $\begin{cases} x + 2y - z = 6 \\ 2x - y + 3z = -13 \\ 3x - 2y + 3z = -16 \end{cases}$

In Problems 32 and 33, use properties of determinants to find the value of each determinant if it is known that $\begin{vmatrix} x & y \\ a & b \end{vmatrix} = 8$.

32. $\begin{vmatrix} 2x & y \\ 2a & b \end{vmatrix}$

33. $\begin{vmatrix} y & x \\ b & a \end{vmatrix}$

In Problems 34–38, write the partial fraction decomposition of each rational expression.

34. $\dfrac{6}{x(x-4)}$

35. $\dfrac{x-4}{x^2(x-1)}$

36. $\dfrac{x}{(x^2+9)(x+1)}$

37. $\dfrac{x^3}{(x^2+4)^2}$

38. $\dfrac{x^2}{(x^2+1)(x^2-1)}$

In Problems 39–43, solve each system of equations.

39. $\begin{cases} 2x + y + 3 = 0 \\ x^2 + y^2 = 5 \end{cases}$

40. $\begin{cases} 2xy + y^2 = 10 \\ 3y^2 - xy = 2 \end{cases}$

41. $\begin{cases} x^2 + y^2 = 6y \\ x^2 = 3y \end{cases}$

42. $\begin{cases} 3x^2 + 4xy + 5y^2 = 8 \\ x^2 + 3xy + 2y^2 = 0 \end{cases}$

43. $\begin{cases} x^2 - 3x + y^2 + y = -2 \\ \dfrac{x^2 - x}{y} + y + 1 = 0 \end{cases}$

In Problems 44 and 45, graph each inequality by hand. Verify your results using a graphing utility.

44. $3x + 4y \le 12$

45. $y \le x^2$

In Problems 46–48, graph each system of inequalities by hand. Tell whether the graph is bounded or unbounded, and label the corner points.

46. $\begin{cases} -2x + y \le 2 \\ x + y \ge 2 \end{cases}$

47. $\begin{cases} x \ge 0 \\ y \ge 0 \\ x + y \le 4 \\ 2x + 3y \le 6 \end{cases}$

48. $\begin{cases} x \ge 0 \\ y \ge 0 \\ 2x + y \le 8 \\ x + 2y \ge 2 \end{cases}$

In Problems 49 and 50, graph each system of inequalities by hand.

49. $\begin{cases} x^2 + y^2 \le 16 \\ x + y \ge 2 \end{cases}$

50. $\begin{cases} x^2 + y^2 \le 25 \\ xy \le 4 \end{cases}$

In Problems 51 and 52, solve each linear programming problem.

51. Maximize $z = 3x + 4y$ subject to $x \ge 0, y \ge 0, 3x + 2y \ge 6, x + y \le 8$

52. Minimize $z = 3x + 5y$ subject to $x \ge 0, y \ge 0, x + y \ge 1, 3x + 2y \le 12, x + 3y \le 12$

---

53. Find A so that the system of equations has infinitely many solutions.

$$\begin{cases} 2x + 5y = 5 \\ 4x + 10y = A \end{cases}$$

54. Find A so that the system in Problem 53 is inconsistent.

55. Curve Fitting Find the quadratic function $y = ax^2 + bx + c$ that passes through the three points $(0, 1)$, $(1, 0)$, and $(-2, 1)$.

56. Blending Coffee A coffee distributor is blending a new coffee that will cost \$6.90 per pound. It will consist of a blend of \$6.00-per-pound coffee and \$9.00-per-pound coffee. What amounts of each type of coffee should be mixed to achieve the desired blend?

[**Hint:** Assume that the weight of the blended coffee is 100 pounds.]

57. Cookie Orders A cookie company makes three kinds of cookies (oatmeal raisin, chocolate chip, and shortbread) packaged in small, medium, and large boxes. The small box contains 1 dozen oatmeal raisin and 1 dozen chocolate chip; the medium box has 2 dozen oatmeal raisin, 1 dozen chocolate chip, and 1 dozen shortbread; the large box contains 2 dozen oatmeal raisin, 2 dozen chocolate chip, and 3 dozen shortbread. If you require exactly 15 dozen oatmeal raisin, 10 dozen chocolate chip, and 11 dozen shortbread, how many of each size box should you buy?

58. Mixed Nuts A store that specializes in selling nuts has available 72 pounds (lb) of cashews and 120 lb of peanuts. These are to be mixed in 12-ounce (oz) packages as follows: a lower-priced package containing 8 oz of peanuts and 4 oz of cashews, and a quality package containing 6 oz of peanuts and 6 oz of cashews.

(a) Use x to denote the number of lower-priced packages, and use y to denote the number of quality packages. Write a system of linear inequalities that describes the possible numbers of each kind of package.

(b) Graph the system and label the corner points.

59. Determining the Speed of the Current of the Aguarico River On a recent trip to the Cuyabeno Wildlife Reserve in the Amazon region of Ecuador, Mike took a 100-kilometer trip by speedboat down the Aguarico River from Chiritza to the Flotel Orellana. As Mike watched the Amazon unfold, he wondered how fast the speedboat was going and how fast the current of the white-water Aguarico River was. Mike timed the trip downstream at 2.5 hours and the return trip at 3 hours. What were the two speeds?

60. Constant Rate Jobs If Bruce and Bryce work together for 1 hour and 20 minutes, they will finish a certain job. If Bryce and Marty work together for 1 hour and 36 minutes, the same job can be finished. If Marty and Bruce work together, they can complete this job in 2 hours and 40 minutes. How long would it take each of them, working alone, to finish the job?

61. Minimizing Production Cost A factory produces gasoline engines and diesel engines. Each week the factory is obligated to deliver at least 20 gasoline engines and at least 15 diesel engines. Due to physical limitations, however, the factory cannot make more than 60 gasoline engines or more than 40 diesel engines in any given week. Finally, to prevent layoffs, a total of at least 50 engines must be produced. If gasoline engines cost \$450 each to produce and diesel engines cost \$550 each to produce, how many of each should be produced per week to minimize the cost? What is the excess capacity of the factory? That is, how many of each kind of engine are being produced in excess of the number that the factory is obligated to deliver?

62. Describe four ways of solving a system of three linear equations containing three variables. Which method do you prefer? Why?

Chapter Test

CHAPTER **Test Prep** VIDEOS

The Chapter Test Prep Videos are step-by-step solutions available in MyMathLab®, or on this text's YouTube Channel. Flip back to the Resources for Success page for a link to this text's YouTube channel.

In Problems 1–4, solve each system of equations using the method of substitution or the method of elimination. If the system has no solution, say that it is inconsistent.

1. $\begin{cases} -2x + y = -7 \\ 4x + 3y = 9 \end{cases}$

2. $\begin{cases} \frac{1}{3}x - 2y = 1 \\ 5x - 30y = 18 \end{cases}$

3. $\begin{cases} x - y + 2z = 5 \\ 3x + 4y - z = -2 \\ 5x + 2y + 3z = 8 \end{cases}$

4. $\begin{cases} 3x + 2y - 8z = -3 \\ -x - \frac{2}{3}y + z = 1 \\ 6x - 3y + 15z = 8 \end{cases}$

5. Write the augmented matrix corresponding to the system of equations: $\begin{cases} 4x - 5y + z = 0 \\ -2x - y + 6 = -19 \\ x + 5y - 5z = 10 \end{cases}$

6. Write the system of equations corresponding to the augmented matrix: $\left[\begin{array}{ccc|c} 3 & 2 & 4 & -6 \\ 1 & 0 & 8 & 2 \\ -2 & 1 & 3 & -11 \end{array}\right]$

In Problems 7–10, use the given matrices to compute each expression.

$$A = \begin{bmatrix} 1 & -1 \\ 0 & -4 \\ 3 & 2 \end{bmatrix} \quad B = \begin{bmatrix} 1 & -2 & 5 \\ 0 & 3 & 1 \end{bmatrix} \quad C = \begin{bmatrix} 4 & 6 \\ 1 & -3 \\ -1 & 8 \end{bmatrix}$$

7. $2A + C$ **8.** $A - 3C$

9. CB **10.** BA

In Problems 11 and 12, find the inverse of each nonsingular matrix.

11. $A = \begin{bmatrix} 3 & 2 \\ 5 & 4 \end{bmatrix}$

12. $B = \begin{bmatrix} 1 & -1 & 1 \\ 2 & 5 & -1 \\ 2 & 3 & 0 \end{bmatrix}$

In Problems 13–16, solve each system of equations using matrices. If the system has no solution, say that it is inconsistent.

13. $\begin{cases} 6x + 3y = 12 \\ 2x - y = -2 \end{cases}$

14. $\begin{cases} x + \frac{1}{4}y = 7 \\ 8x + 2y = 56 \end{cases}$

15. $\begin{cases} x + 2y + 4z = -3 \\ 2x + 7y + 15z = -12 \\ 4x + 7y + 13z = -10 \end{cases}$

16. $\begin{cases} 2x + 2y - 3z = 5 \\ x - y + 2z = 8 \\ 3x + 5y - 8z = -2 \end{cases}$

In Problems 17 and 18, find the value of each determinant.

17. $\begin{vmatrix} -2 & 5 \\ 3 & 7 \end{vmatrix}$

18. $\begin{vmatrix} 2 & -4 & 6 \\ 1 & 4 & 0 \\ -1 & 2 & -4 \end{vmatrix}$

In Problems 19 and 20, use Cramer's Rule, if possible, to solve each system.

19. $\begin{cases} 4x + 3y = -23 \\ 3x - 5y = 19 \end{cases}$

20. $\begin{cases} 4x - 3y + 2z = 15 \\ -2x + y - 3z = -15 \\ 5x - 5y + 2z = 18 \end{cases}$

In Problems 21 and 22, solve each system of equations.

21. $\begin{cases} 3x^2 + y^2 = 12 \\ y^2 = 9x \end{cases}$

22. $\begin{cases} 2y^2 - 3x^2 = 5 \\ y - x = 1 \end{cases}$

23. Graph the system of inequalities: $\begin{cases} x^2 + y^2 \le 100 \\ 4x - 3y \ge 0 \end{cases}$

In Problems 24 and 25, write the partial fraction decomposition of each rational expression.

24. $\dfrac{3x + 7}{(x + 3)^2}$

25. $\dfrac{4x^2 - 3}{x(x^2 + 3)^2}$

26. Graph the system of inequalities. Tell whether the graph is bounded or unbounded, and label all corner points.

$$\begin{cases} x \ge 0 \\ y \ge 0 \\ x + 2y \ge 8 \\ 2x - 3y \ge 2 \end{cases}$$

27. Maximize $z = 5x + 8y$ subject to $x \ge 0, 2x + y \le 8$, and $x - 3y \le -3$.

28. Megan went clothes shopping and bought 2 pairs of flare jeans, 2 camisoles, and 4 T-shirts for \$90.00. At the same store, Paige bought one pair of flare jeans and 3 T-shirts for \$42.50, while Kara bought 1 pair of flare jeans, 3 camisoles, and 2 T-shirts for \$62.00. Determine the price of each clothing item.

Cumulative Review

In Problems 1–6, solve each equation.

1. $2x^2 - x = 0$

2. $\sqrt{3x + 1} = 4$

3. $2x^3 - 3x^2 - 8x - 3 = 0$

4. $3^x = 9^{x+1}$

5. $\log_3(x - 1) + \log_3(2x + 1) = 2$

6. $3^x = e$

7. Determine whether the function $g(x) = \dfrac{2x^3}{x^4 + 1}$ is even, odd, or neither. Is the graph of g symmetric with respect to the x-axis, y-axis, or origin?

8. Find the center and radius of the circle $x^2 + y^2 - 2x + 4y - 11 = 0$. Graph the circle.

9. Graph $f(x) = 3^{x-2} + 1$ using transformations. What is the domain, range, and horizontal asymptote of f?

10. The function $f(x) = \dfrac{5}{x + 2}$ is one-to-one. Find f^{-1}. Find the domain and the range of f and the domain and the range of f^{-1}.

11. Graph each equation.
(a) $y = 3x + 6$
(b) $x^2 + y^2 = 4$
(c) $y = x^3$
(d) $y = \dfrac{1}{x}$
(e) $y = \sqrt{x}$
(f) $y = e^x$
(g) $y = \ln x$
(h) $2x^2 + 5y^2 = 1$
(i) $x^2 - 3y^2 = 1$
(j) $x^2 - 2x - 4y + 1 = 0$

12. $f(x) = x^3 - 3x + 5$
(a) Using a graphing utility, graph f and approximate the zero(s) of f.
(b) Using a graphing utility, approximate the local maxima and local minima.
(c) Determine the intervals on which f is increasing.

Chapter Projects

Highest Educational Level of Parents	Maximum Education That Children Achieve		
	College	High School	Elementary
College	80%	18%	2%
High school	40%	50%	10%
Elementary	20%	60%	20%

I. Markov Chains A **Markov chain** (or process) is one in which future outcomes are determined by a current state. Future outcomes are based on probabilities. The probability of moving to a certain state depends only on the state previously occupied and does not vary with time. An example of a Markov chain is the maximum education achieved by children based on the highest educational level attained by their parents, where the states are (1) earned college degree, (2) high school diploma only, (3) elementary school only. If p_{ij} is the probability of moving from state i to state j, the **transition matrix** is the $m \times m$ matrix

$$P = \begin{bmatrix} p_{11} & p_{12} & \cdots & p_{1m} \\ \vdots & \vdots & & \vdots \\ p_{m1} & p_{m2} & \cdots & p_{mm} \end{bmatrix}$$

The table represents the probabilities for the highest educational level of children based on the highest educational level of their parents. For example, the table shows that the probability p_{21} is 40% that parents with a high-school education (row 2) will have children with a college education (column 1).

1. Convert the percentages to decimals.
2. What is the transition matrix?
3. Sum across the rows. What do you notice? Why do you think that you obtained this result?
4. If P is the transition matrix of a Markov chain, the (i, j)th entry of P^n (nth power of P) gives the probability of passing from state i to state j in n stages. What is the probability that the grandchild of a college graduate is a college graduate?
5. What is the probability that the grandchild of a high school graduate finishes college?
6. The row vector $v^{(0)} = [0.319 \quad 0.564 \quad 0.117]$ represents the proportion of the U.S. population 25 years or older that has college, high school, and elementary school, respectively, as the highest educational level in 2014.* In a Markov chain the probability distribution $v^{(k)}$ after k stages is $v^{(k)} = v^{(0)}P^k$, where P^k is the kth power of the transition matrix. What will be the distribution of highest educational attainment of the grandchildren of the current population?
7. Calculate $P^3, P^4, P^5, \ldots$. Continue until the matrix does not change. This is called the long-run or steady-state distribution. What is the long-run distribution of highest educational attainment of the population?

*__*Source:*__ *U.S. Census Bureau.*

The following projects are available at the Instructor's Resource Center (IRC).

II. Project at Motorola: ***Error Control Coding*** The high-powered engineering needed to ensure that wireless communications are transmitted correctly is analyzed using matrices to control coding errors.

III. Using Matrices to Find the Line of Best Fit Have you wondered how our calculators get a line of best fit? See how to find the line by solving a matrix equation.

IV. CBL Experiment Simulate two people walking toward each other at a constant rate. Then solve the resulting system of equations to determine when and where they will meet.

13 Sequences; Induction; the Binomial Theorem

UN Projects World Population Will Hit 9.6 Billion by 2050

The population of the planet is expected to reach 9.6 billion by 2050, according to a new UN report—a slightly larger number than anticipated, because fertility projections have increased in nations where women have the most children.

More than half of this projected demographic growth will be in Africa, which continues to add people even as population growth in the world at large slows down.

"Although population growth has slowed for the world as a whole, this report reminds us that some developing countries, especially in Africa, are still growing rapidly," said Under-Secretary-General for Economic and Social Affairs Wu Hongbo, in the UN report.

World Population Prospects: The 2012 Revision notes that the population in the developed regions of the world should remain stable at around 1.3 billion until 2050 thanks to trends of low fertility, while the world's 49 least-developed countries are projected to double in population size.

The new figures do *not* mean that world population growth has started to speed up again. As countries industrialize, they tend to undergo "demographic transition," wherein high death and birth rates are slowly replaced with low birth and death rates. It's a demographic shift often helped along by an increase in the granting of rights for women.

Some experts suspect that the world population will plateau around 2060, and there's a possibility that—after a transitional period of a higher death rate than birth rate—world birth and death rates could actually even out, keeping the human population stable.

Source: Faine Greenwood, June 14, 2013, 11:36. GlobalPost®(www.globalpost.com/dispatch/news/science/130614/un-projects-world-population-will-hit-96-billion-2050)

—See the Internet-based Chapter Project I—

••• A Look Back, A Look Ahead •••

This chapter may be divided into three independent parts: Sections 13.1–13.3, Section 13.4, and Section 13.5.

In Chapter 3, we defined a function and its domain, which was usually some set of real numbers. In Sections 13.1–13.3, we discuss a sequence, which is a function whose domain is the set of positive integers.

Throughout this text, where it seemed appropriate, we have given proofs of many of the results. In Section 13.4, a technique for proving theorems involving natural numbers is discussed.

In Chapter R, Review, Section R.4, there are formulas for expanding $(x + a)^2$ and $(x + a)^3$. In Section 13.5, we discuss the Binomial Theorem, a formula for the expansion of $(x + a)^n$, where n is any positive integer.

The topics introduced in this chapter are covered in more detail in courses titled *Discrete Mathematics*. Applications of these topics can be found in the fields of computer science, engineering, business and economics, the social sciences, and the physical and biological sciences.

Outline

13.1 Sequences

PREPARING FOR THIS SECTION *Before getting started, review the following:*

- Functions (Section 3.1, pp. 207–213)
- Compound Interest (Section 6.7, pp. 475–480)

Now Work the 'Are You Prepared?' problems on page 971.

OBJECTIVES
1. Write the First Several Terms of a Sequence (p. 962)
2. Write the Terms of a Sequence Defined by a Recursive Formula (p. 965)
3. Use Summation Notation (p. 966)
4. Find the Sum of a Sequence Algebraically and Using a Graphing Utility (p. 967)
5. Solve Annuity and Amortization Problems (p. 969)

When you hear the word *sequence* as it is used in the phrase "a sequence of events," you probably think of a collection of events, one of which happens first, another second, and so on. In mathematics, the word *sequence* also refers to outcomes that are first, second, and so on.

DEFINITION A **sequence** is a function whose domain is the set of positive integers.

In a sequence, then, the inputs are 1, 2, 3, ... Because a sequence is a function, it will have a graph. Figure 1(a) shows the graph of the function $f(x) = \frac{1}{x}, x > 0$. If all the points on this graph were removed except those whose x-coordinates are positive integers—that is, if all points were removed except $(1, 1), \left(2, \frac{1}{2}\right), \left(3, \frac{1}{3}\right)$, and so on—the remaining points would be the graph of the sequence $f(n) = \frac{1}{n}$, as shown in Figure 1(b). Note that n is used to represent the independent variable in a sequence. This serves to remind us that n is a positive integer.

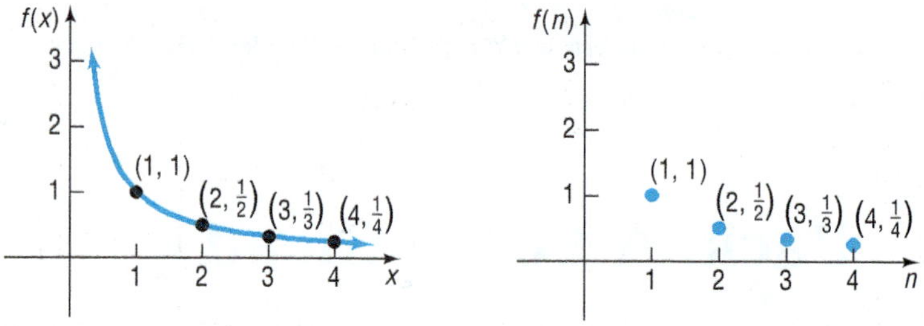

Figure 1 (a) $f(x) = \frac{1}{x}, x > 0$ (b) $f(n) = \frac{1}{n}$, n a positive integer

1 Write the First Several Terms of a Sequence

A sequence is usually represented by listing its values in order. For example, the sequence whose graph is given in Figure 1(b) might be represented as

$$f(1), f(2), f(3), f(4), \ldots \quad \text{or} \quad 1, \frac{1}{2}, \frac{1}{3}, \frac{1}{4}, \ldots$$

The list never ends, as the ellipsis indicates. The numbers in this ordered list are called the **terms** of the sequence.

In dealing with sequences, subscripted letters are used such as a_1 to represent the first term, a_2 for the second term, a_3 for the third term, and so on.

For the sequence $f(n) = \frac{1}{n}$, this means

$$a_1 = f(1) = 1 \quad a_2 = f(2) = \frac{1}{2} \quad a_3 = f(3) = \frac{1}{3} \quad a_4 = f(4) = \frac{1}{4} \ldots a_n = f(n) = \frac{1}{n} \ldots$$

first term — second term — third term — fourth term — nth term

In other words, the traditional function notation $f(n)$ is not used for sequences. For this particular sequence, we have a rule for the nth term, which is $a_n = \frac{1}{n}$, so it is easy to find any term of the sequence.

When a formula for the nth term (sometimes called the **general term**) of a sequence is known, the entire sequence can be represented by placing braces around the formula for the nth term. For example, the sequence whose nth term is $b_n = \left(\frac{1}{2}\right)^n$ may be represented as

$$\{b_n\} = \left\{\left(\frac{1}{2}\right)^n\right\}$$

or by

$$b_1 = \frac{1}{2}, \quad b_2 = \frac{1}{4}, \quad b_3 = \frac{1}{8}, \ldots, \quad b_n = \left(\frac{1}{2}\right)^n, \ldots$$

EXAMPLE 1 Writing the First Several Terms of a Sequence

Write down the first six terms of the following sequence and graph it.

$$\{a_n\} = \left\{\frac{n-1}{n}\right\}$$

Algebraic Solution

The first six terms of the sequence are

$$a_1 = \frac{1-1}{1} = 0, \quad a_2 = \frac{1}{2}, \quad a_3 = \frac{2}{3},$$

$$a_4 = \frac{3}{4}, \quad a_5 = \frac{4}{5}, \quad a_6 = \frac{5}{6}$$

See Figure 2 for the graph.

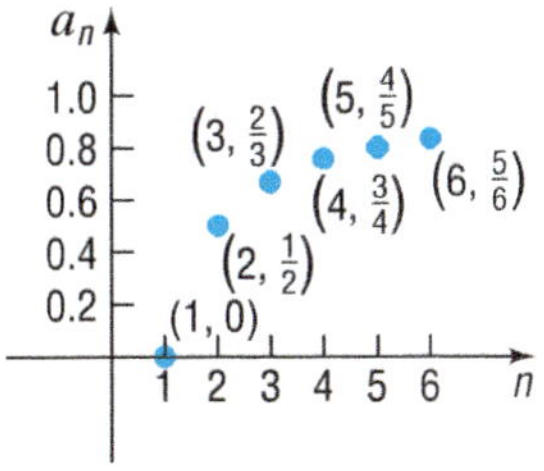

Figure 2 $\{a_n\} = \left\{\frac{n-1}{n}\right\}$

Graphing Solution

Figure 3 shows the sequence generated on a TI-84 Plus C graphing calculator. We can see the first few terms of the sequence on the screen.

We could also obtain the terms of the sequence using the TABLE feature. First, put the graphing utility in SEQuence mode. Press $Y=$ and enter the formula for the sequence into the graphing utility. See Figure 4. Set up the table with TblStart = 1 and ΔTbl = 1. See Table 1. Finally, we can graph the sequence. See Figure 5. Notice that the first term of the sequence is hard to see since it lies on the x-axis. TRACEing the graph allows you to determine the terms of the sequence.

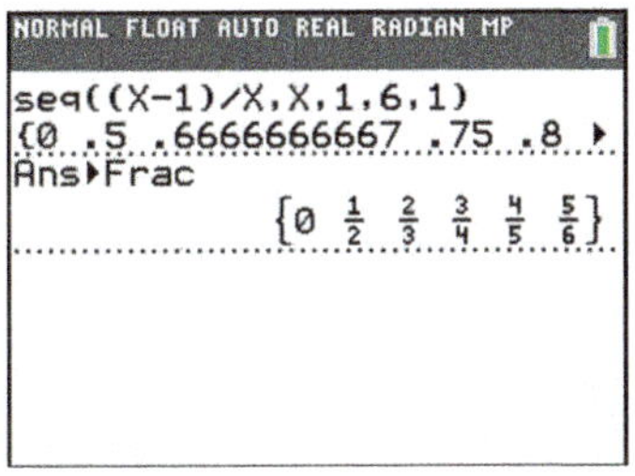

Figure 3

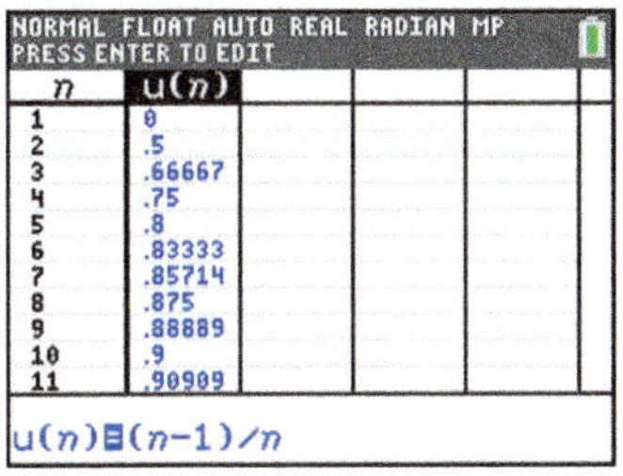

n	u(n)
1	0
2	.5
3	.66667
4	.75
5	.8
6	.83333
7	.85714
8	.875
9	.88889
10	.9
11	.90909

Table 1

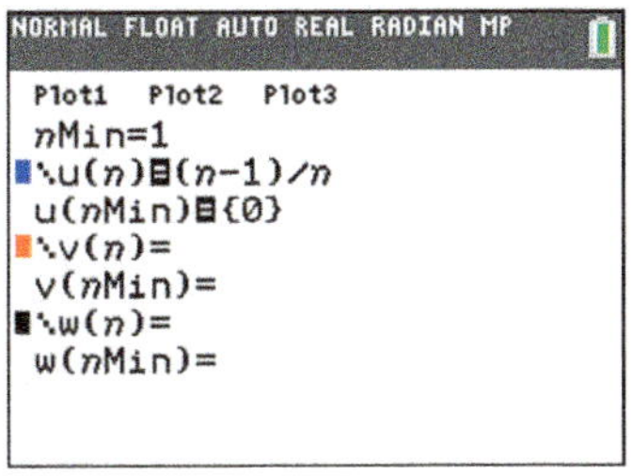

Figure 4

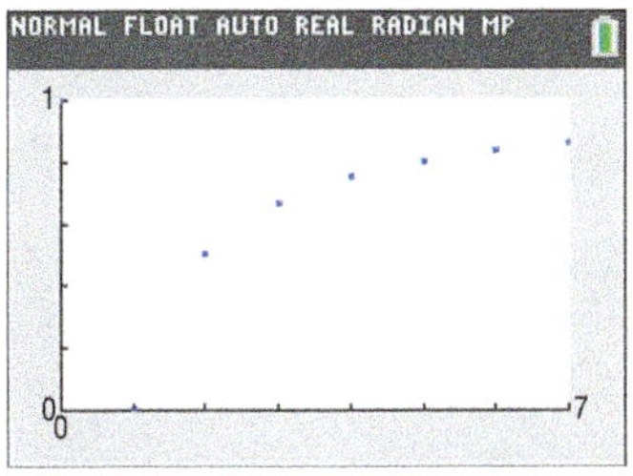

Figure 5

Now Work PROBLEM 19

We will usually provide solutions done by hand. The reader is encouraged to verify solutions using a graphing utility.

EXAMPLE 2 Writing the First Several Terms of a Sequence

Write down the first six terms of the following sequence and graph it.

$$\{b_n\} = \left\{(-1)^{n+1}\left(\frac{2}{n}\right)\right\}$$

Solution The first six terms of the sequence are

$$b_1 = (-1)^{1+1}\left(\frac{2}{1}\right) = 2 \quad b_2 = (-1)^{2+1}\left(\frac{2}{2}\right) = -1 \quad b_3 = (-1)^{3+1}\left(\frac{2}{3}\right) = \frac{2}{3}$$

$$b_4 = -\frac{1}{2} \qquad b_5 = \frac{2}{5} \qquad b_6 = -\frac{1}{3}$$

See Figure 6 for the graph. ■

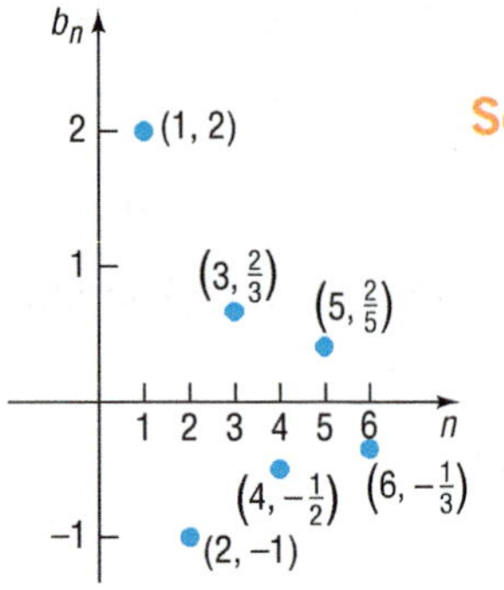

Figure 6 $\{b_n\} = \left\{(-1)^{n+1}\left(\frac{2}{n}\right)\right\}$

Notice in the sequence $\{b_n\}$ in Example 2 that the signs of the terms **alternate**. This occurs when we use factors such as $(-1)^{n+1}$, which equals 1 if n is odd and -1 if n is even, or $(-1)^n$, which equals -1 if n is odd and 1 if n is even.

EXAMPLE 3 Writing the First Several Terms of a Sequence

Write down the first six terms of the following sequence and graph it.

$$\{c_n\} = \left\{\begin{array}{ll} n & \text{if } n \text{ is even} \\ \frac{1}{n} & \text{if } n \text{ is odd} \end{array}\right\}$$

Solution The first six terms of the sequence are

$$c_1 = 1 \quad c_2 = 2 \quad c_3 = \frac{1}{3} \quad c_4 = 4 \quad c_5 = \frac{1}{5} \quad c_6 = 6$$

See Figure 7 for the graph. ■

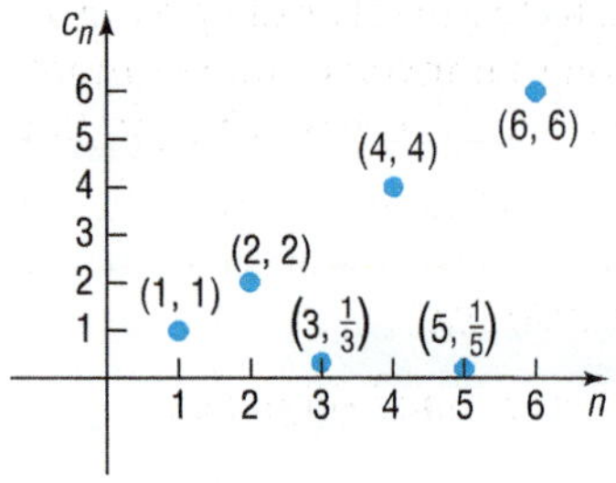

Figure 7 $\{c_n\} = \left\{\begin{array}{ll} n & \text{if } n \text{ is even} \\ \frac{1}{n} & \text{if } n \text{ is odd} \end{array}\right\}$

Now Work PROBLEM 21

Note that the formulas that generate the terms of a sequence are not unique. For example, the terms of the sequence in Example 3 could also be found using

$$\{d_n\} = \{n^{(-1)^n}\}$$

Sometimes a sequence is indicated by an observed pattern in the first few terms that makes it possible to infer the makeup of the nth term. In the example that follows, a sufficient number of terms of the sequence is given so that a natural choice for the nth term is suggested.

EXAMPLE 4 Determining a Sequence from a Pattern

(a) $e, \frac{e^2}{2}, \frac{e^3}{3}, \frac{e^4}{4}, \ldots \qquad a_n = \frac{e^n}{n}$

(b) $1, \frac{1}{3}, \frac{1}{9}, \frac{1}{27}, \ldots \qquad b_n = \frac{1}{3^{n-1}}$

(c) $1, 3, 5, 7, \ldots \qquad c_n = 2n - 1$

(d) $1, 4, 9, 16, 25, \ldots \qquad d_n = n^2$

(e) $1, -\frac{1}{2}, \frac{1}{3}, -\frac{1}{4}, \frac{1}{5}, \ldots \qquad e_n = (-1)^{n+1}\left(\frac{1}{n}\right)$ ■

Now Work PROBLEM 29

The Factorial Symbol

Some sequences in mathematics involve a special product called a *factorial.*

DEFINITION

If $n \geq 0$ is an integer, the **factorial symbol $n!$** is defined as follows:

$$0! = 1 \qquad 1! = 1$$
$$n! = n(n-1)\cdot\cdots\cdot 3\cdot 2\cdot 1 \qquad \text{if } n \geq 2$$

Table 2

n	$n!$
0	1
1	1
2	2
3	6
4	24
5	120
6	720

For example, $2! = 2\cdot 1 = 2, 3! = 3\cdot 2\cdot 1 = 6, 4! = 4\cdot 3\cdot 2\cdot 1 = 24$, and so on. Table 2 lists the values of $n!$ for $0 \leq n \leq 6$.

Because

$$n! = n\underbrace{(n-1)(n-2)\cdot\ldots\cdot 3\cdot 2\cdot 1}_{(n-1)!}$$

the formula

$$n! = n(n-1)!$$

can be used to find successive factorials. For example, because $6! = 720$,

$$7! = 7\cdot 6! = 7(720) = 5040$$

and

$$8! = 8\cdot 7! = 8(5040) = 40{,}320$$

Exploration

Use your calculator's factorial key to see how fast factorials increase in value. Find the value of 69!. What happens when you try to find 70!? In fact, 70! is larger than 10^{100} (a googol), the largest number most calculators can display.

Now Work PROBLEM 13

2 Write the Terms of a Sequence Defined by a Recursive Formula

A second way of defining a sequence is to assign a value to the first (or the first few) term(s) and specify the nth term by a formula or equation that involves one or more of the terms preceding it. Such sequences are said to be defined **recursively**, and the rule or formula is called a **recursive formula**.

EXAMPLE 5 **Writing the Terms of a Recursively Defined Sequence**

Write down the first five terms of the following recursively defined sequence.

$$s_1 = 1, \qquad s_n = ns_{n-1}$$

Algebraic Solution

The first term is given as $s_1 = 1$. To get the second term, use $n = 2$ in the formula $s_n = ns_{n-1}$ to get $s_2 = 2s_1 = 2\cdot 1 = 2$. To get the third term, use $n = 3$ in the formula to get $s_3 = 3s_2 = 3\cdot 2 = 6$. To get a new term requires knowing the value of the preceding term. The first five terms are

$$s_1 = 1$$
$$s_2 = 2\cdot 1 = 2$$
$$s_3 = 3\cdot 2 = 6$$
$$s_4 = 4\cdot 6 = 24$$
$$s_5 = 5\cdot 24 = 120$$

Do you recognize this sequence? $s_n = n!$.

Graphing Solution

First, put the graphing utility into SEQuence mode. Press $Y=$ and enter the recursive formula into the graphing utility to generate the desired sequence. See Figure 8(a). Next, set up the viewing window. Finally, graph the recursive relation and use TRACE to determine the terms in the sequence. See Figure 8(b). For example, the fourth term of the sequence is 24. Table 3 also shows the terms of the sequence.

Table 3

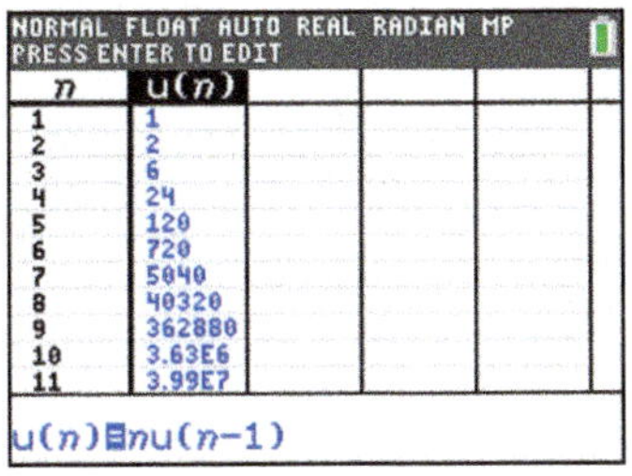
NORMAL FLOAT AUTO REAL RADIAN MP
PRESS ENTER TO EDIT

n	u(n)
1	1
2	2
3	6
4	24
5	120
6	720
7	5040
8	40320
9	362880
10	3.63E6
11	3.99E7

u(n)=nu(n−1)

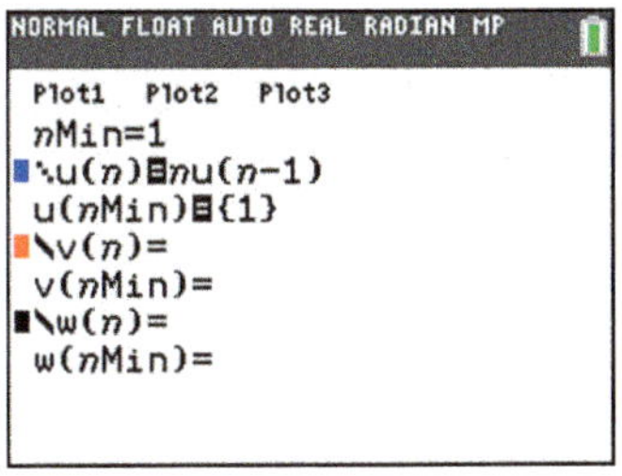

(a)

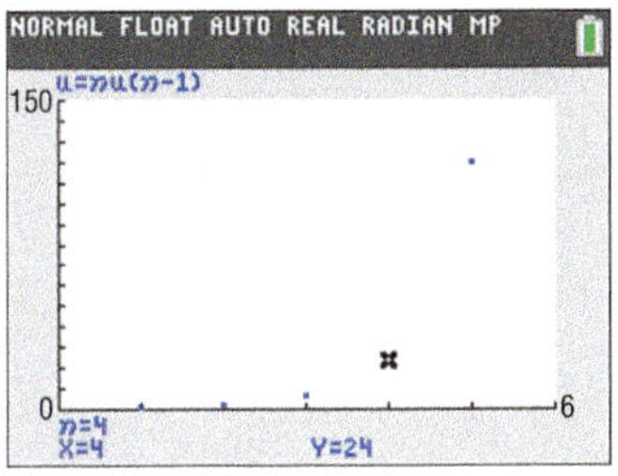

(b)

Figure 8

EXAMPLE 6

Writing the Terms of a Recursively Defined Sequence

Write down the first five terms of the following recursively defined sequence.

$$u_1 = 1 \qquad u_2 = 1 \qquad u_n = u_{n-2} + u_{n-1}$$

Solution The first two terms are given. Finding each successive term requires knowing the previous two terms. That is,

$$u_1 = 1$$
$$u_2 = 1$$
$$u_3 = u_1 + u_2 = 1 + 1 = 2$$
$$u_4 = u_2 + u_3 = 1 + 2 = 3$$
$$u_5 = u_3 + u_4 = 2 + 3 = 5$$

■

Now Work PROBLEMS 37 AND 45

The sequence given in Example 6 is called the **Fibonacci sequence**, and the terms of this sequence are called **Fibonacci numbers**. These numbers appear in a wide variety of applications (see Problems 91–94).

3 Use Summation Notation

It is often important to find the sum of the first n terms of a sequence $\{a_n\}$—that is,

$$a_1 + a_2 + a_3 + \cdots + a_n$$

Rather than writing down all these terms, we can use **summation notation** to express the sum more concisely:

$$a_1 + a_2 + a_3 + \cdots + a_n = \sum_{k=1}^{n} a_k$$

The symbol Σ (the Greek letter sigma, which is an S in our alphabet) is simply an instruction to sum, or add up, the terms. The integer k is called the **index** of the sum; it tells where to start the sum and where to end it. The expression

$$\sum_{k=1}^{n} a_k$$

is an instruction to add the terms a_k of the sequence $\{a_n\}$ starting with $k = 1$ and ending with $k = n$. The expression is read as "the sum of a_k from $k = 1$ to $k = n$."

EXAMPLE 7

Expanding Summation Notation

Write out each sum.

(a) $\sum_{k=1}^{n} \frac{1}{k}$ (b) $\sum_{k=1}^{n} k!$

Solution (a) $\sum_{k=1}^{n} \frac{1}{k} = 1 + \frac{1}{2} + \frac{1}{3} + \cdots + \frac{1}{n}$ (b) $\sum_{k=1}^{n} k! = 1! + 2! + \cdots + n!$ ■

Now Work PROBLEM 53

EXAMPLE 8

Writing a Sum in Summation Notation

Express each sum using summation notation.

(a) $1^2 + 2^2 + 3^2 + \cdots + 9^2$ (b) $1 + \frac{1}{2} + \frac{1}{4} + \frac{1}{8} + \cdots + \frac{1}{2^{n-1}}$

Solution (a) The sum $1^2 + 2^2 + 3^2 + \cdots + 9^2$ has 9 terms, each of the form k^2, and starts at $k = 1$ and ends at $k = 9$:

$$1^2 + 2^2 + 3^2 + \cdots + 9^2 = \sum_{k=1}^{9} k^2$$

(b) The sum

$$1 + \frac{1}{2} + \frac{1}{4} + \frac{1}{8} + \cdots + \frac{1}{2^{n-1}}$$

has n terms, each of the form $\frac{1}{2^{k-1}}$, and starts at $k = 1$ and ends at $k = n$:

$$1 + \frac{1}{2} + \frac{1}{4} + \frac{1}{8} + \cdots + \frac{1}{2^{n-1}} = \sum_{k=1}^{n} \frac{1}{2^{k-1}}$$

Now Work PROBLEM 63

The index of summation need not always begin at 1 or end at n; for example, the sum in Example 8(b) could also be expressed as

$$\sum_{k=0}^{n-1} \frac{1}{2^k} = 1 + \frac{1}{2} + \frac{1}{4} + \cdots + \frac{1}{2^{n-1}}$$

Letters other than k may be used as the index. For example,

$$\sum_{j=1}^{n} j! \quad \text{and} \quad \sum_{i=1}^{n} i!$$

both represent the same sum given in Example 7(b).

4 Find the Sum of a Sequence Algebraically and Using a Graphing Utility

The following theorem lists some properties of sequences using summation notation. These properties are useful for adding the terms of a sequence algebraically.

THEOREM

Properties of Sequences

If $\{a_n\}$ and $\{b_n\}$ are two sequences and c is a real number, then:

$$\sum_{k=1}^{n} (ca_k) = ca_1 + ca_2 + \cdots + ca_n = c(a_1 + a_2 + \cdots + a_n) = c\sum_{k=1}^{n} a_k \quad (1)$$

$$\sum_{k=1}^{n} (a_k + b_k) = \sum_{k=1}^{n} a_k + \sum_{k=1}^{n} b_k \quad (2)$$

$$\sum_{k=1}^{n} (a_k - b_k) = \sum_{k=1}^{n} a_k - \sum_{k=1}^{n} b_k \quad (3)$$

$$\sum_{k=j+1}^{n} a_k = \sum_{k=1}^{n} a_k - \sum_{k=1}^{j} a_k \quad \text{where } 0 < j < n \quad (4)$$

The proof of property (1) follows from the distributive property of real numbers. The proofs of properties 2 and 3 are based on the commutative and associative properties of real numbers. Property (4) states that the sum from $j + 1$ to n equals the sum from 1 to n minus the sum from 1 to j. It can be helpful to employ this property when the index of summation begins at a number larger than 1.

THEOREM **Formulas for Sums of Sequences**

$$\sum_{k=1}^{n} c = \underbrace{c + c + \cdots + c}_{n \text{ terms}} = cn \qquad c \text{ is a real number} \tag{5}$$

$$\sum_{k=1}^{n} k = 1 + 2 + 3 + \cdots + n = \frac{n(n+1)}{2} \tag{6}$$

$$\sum_{k=1}^{n} k^2 = 1^2 + 2^2 + 3^2 + \cdots + n^2 = \frac{n(n+1)(2n+1)}{6} \tag{7}$$

$$\sum_{k=1}^{n} k^3 = 1^3 + 2^3 + 3^3 + \cdots + n^3 = \left[\frac{n(n+1)}{2}\right]^2 \tag{8}$$

The proof of formula (5) follows from the definition of summation notation. You are asked to prove formula (6) in Problem 98. The proofs of formulas (7) and (8) require mathematical induction, which is discussed in Section 13.4.

Notice the difference between formulas (5) and (6). In (5), the constant c is being summed from 1 to n, while in (6) the index of summation k is being summed from 1 to n.

EXAMPLE 9 Finding the Sum of a Sequence

Find the sum of each sequence.

(a) $\sum_{k=1}^{5} (3k)$ (b) $\sum_{k=1}^{10} (k^3 + 1)$

(c) $\sum_{k=1}^{24} (k^2 - 7k + 2)$ (d) $\sum_{k=6}^{20} (4k^2)$

Algebraic Solution

(a)
$$\begin{aligned} \sum_{k=1}^{5} (3k) &= 3\sum_{k=1}^{5} k && \text{Property (1)} \\ &= 3\left(\frac{5(5+1)}{2}\right) && \text{Formula (6)} \\ &= 3(15) \\ &= 45 \end{aligned}$$

(b)
$$\begin{aligned} \sum_{k=1}^{10} (k^3 + 1) &= \sum_{k=1}^{10} k^3 + \sum_{k=1}^{10} 1 && \text{Property (2)} \\ &= \left(\frac{10(10+1)}{2}\right)^2 + 1(10) && \text{Formulas (8) and (5)} \\ &= 3025 + 10 \\ &= 3035 \end{aligned}$$

Graphing Solution

(a) Figure 9 shows the solution using a TI-84 Plus C graphing calculator. So $\sum_{k=1}^{5} (3k) = 45$.

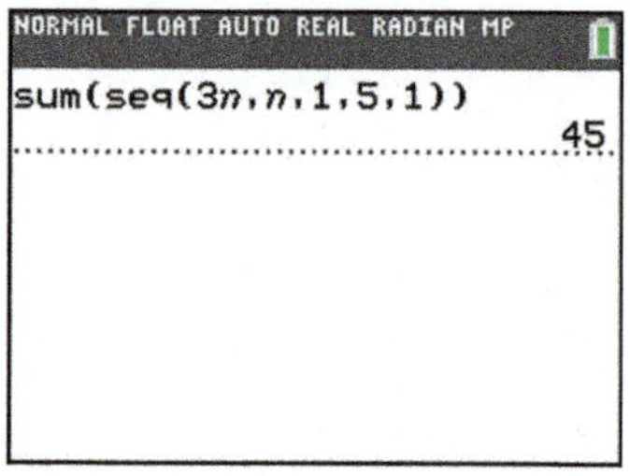

Figure 9

(b) Figure 10 shows the solution using a TI-84 Plus C graphing calculator. So $\sum_{k=1}^{10} (k^3 + 1) = 3035$.

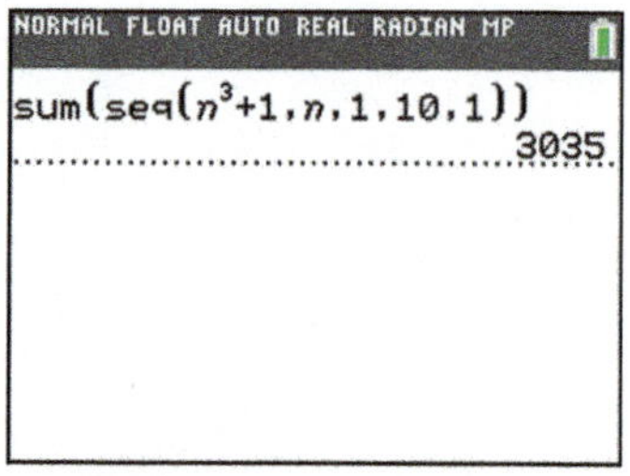

Figure 10

(c) $\sum_{k=1}^{24}(k^2-7k+2)=\sum_{k=1}^{24}k^2-\sum_{k=1}^{24}(7k)+\sum_{k=1}^{24}2$ Properties (2) and (3)

$=\sum_{k=1}^{24}k^2-7\sum_{k=1}^{24}k+\sum_{k=1}^{24}2$ Property (1)

$=\frac{24(24+1)(2\cdot24+1)}{6}-7\left(\frac{24(24+1)}{2}\right)+2(24)$ Formulas (7), (6), (5)

$=4900-2100+48$

$=2848$

(c) Figure 11 shows the solution using a TI-84 Plus C graphing calculator.

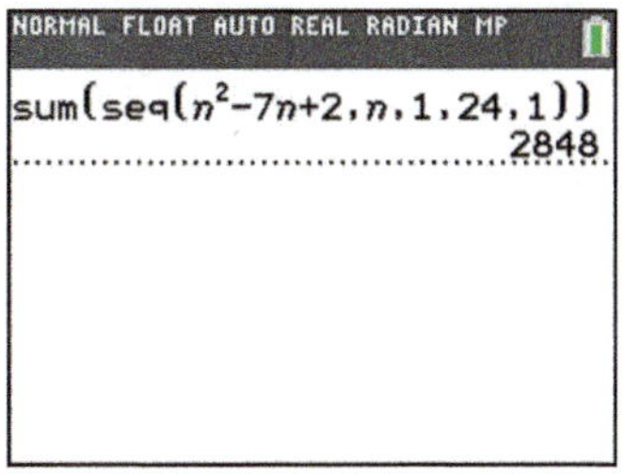

Figure 11

(d) Notice that the index of summation starts at 6. Use property (4) as follows:

$$\sum_{k=6}^{20}(4k^2)=4\sum_{k=6}^{20}k^2=4\left[\sum_{k=1}^{20}k^2-\sum_{k=1}^{5}k^2\right]=4\left[\frac{20(21)(41)}{6}-\frac{5(6)(11)}{6}\right]$$

Property (1) Property (4) Formula (7)

$=4[2870-55]=11{,}260$

(d) Figure 12 shows the solution using a TI-84 Plus C graphing calculator.

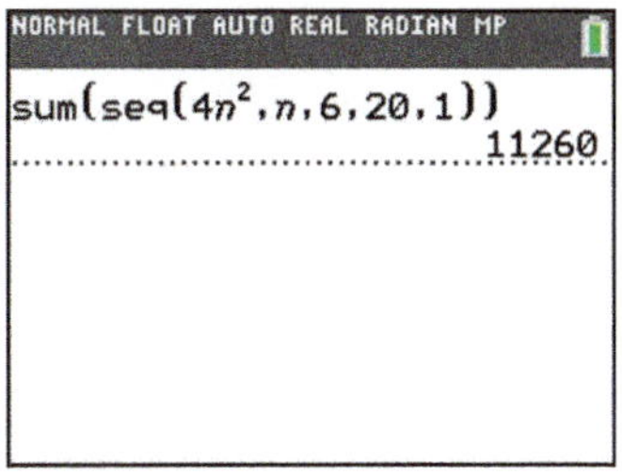

Figure 12

Now Work PROBLEM 75

5 Solve Annuity and Amortization Problems

In Section 6.7 we developed the compound interest formula, which gives the future value when a fixed amount of money is deposited in an account that pays interest compounded periodically. Often, though, money is invested in small amounts at periodic intervals. An **annuity** is a sequence of equal periodic deposits. The periodic deposits may be made annually, quarterly, monthly, or daily.

When deposits are made at the same time that the interest is credited, the annuity is called **ordinary**. We will only deal with ordinary annuities here. The **amount of an annuity** is the sum of all deposits made plus all interest paid.

Suppose that the initial amount deposited in an annuity is $\$M$, the periodic deposit is $\$P$, and the per annum rate of interest is $r\%$ (expressed as a decimal) compounded N times per year. The periodic deposit is made at the same time that the interest is credited, so N deposits are made per year. The amount A_n of the annuity after n deposits will equal A_{n-1}, the amount of the annuity after $n-1$ deposits, plus the interest earned on this amount, plus P, the periodic deposit. That is,

$$A_n = A_{n-1} + \frac{r}{N}A_{n-1} + P = \left(1+\frac{r}{N}\right)A_{n-1} + P$$

Amount after n deposits; Amount in previous period; Interest earned; Periodic deposit

We have established the following result:

THEOREM

Annuity Formula

If $A_0 = M$ represents the initial amount deposited in an annuity that earns $r\%$ per annum compounded N times per year, and if P is the periodic deposit made at each payment period, then the amount A_n of the annuity after n deposits is given by the recursive sequence

$$A_0 = M \qquad A_n = \left(1+\frac{r}{N}\right)A_{n-1} + P \qquad n \ge 1 \qquad \textbf{(9)}$$

Formula (9) may be explained as follows: the money in the account initially, A_0, is \$$M$; the money in the account after $n-1$ payments, A_{n-1}, earns interest $\frac{r}{N}A_{n-1}$ during the nth period; so when the periodic payment of P dollars is added, the amount after n payments, A_n, is obtained.

EXAMPLE 10

Saving for Spring Break

A trip to Cancun during spring break will cost \$450 and full payment is due March 2. To have the money, a student, on September 1, deposits \$100 in a savings account that pays 4% per annum compounded monthly. On the first of each month, the student deposits \$50 in this account.

(a) Find a recursive sequence that explains how much is in the account after n months.

(b) Use the TABLE feature to list the amounts of the annuity for the first 6 months.

(c) After the deposit on March 1 is made, is there enough in the account to pay for the Cancun trip?

(d) If the student deposits \$60 each month, will there be enough for the trip after the March 1 deposit?

Solution

(a) The initial amount deposited in the account is $A_0 = \$100$. The monthly deposit is $P = \$50$, and the per annum rate of interest is $r = 0.04$ compounded $N = 12$ times per year. The amount A_n in the account after n monthly deposits is given by the recursive sequence

$$A_0 = 100 \qquad A_n = \left(1 + \frac{r}{N}\right)A_{n-1} + P = \left(1 + \frac{0.04}{12}\right)A_{n-1} + 50$$

Table 4

NORMAL FLOAT AUTO REAL RADIAN MP
PRESS ENTER TO EDIT

n	u(n)
0	100
1	150.33
2	200.83
3	251.5
4	302.34
5	353.35
6	404.53
7	455.88
8	507.4
9	559.09
10	610.95

u(n)■(1+0.04/12)u(n−1)+50

(b) In SEQuence mode on a TI-84 Plus C, enter the sequence $\{A_n\}$ and create Table 4. On September 1 $(n = 0)$, there is \$100 in the account. After the first payment on October 1, the value of the account is \$150.33. After the second payment on November 1, the value of the account is \$200.83. After the third payment on December 1, the value of the account is \$251.50, and so on.

(c) On March 1 $(n = 6)$, there is only \$404.53, not enough to pay for the trip to Cancun.

(d) If the periodic deposit, P, is \$60, then on March 1, there is \$465.03 in the account, enough for the trip. See Table 5.

Table 5

NORMAL FLOAT AUTO REAL RADIAN MP
PRESS ENTER TO EDIT

n	u(n)
0	100
1	160.33
2	220.87
3	281.6
4	342.54
5	403.68
6	465.03
7	526.58
8	588.34
9	650.3
10	712.46

u(n)■(1+0.04/12)u(n−1)+60

Now Work PROBLEM 85

Recursive sequences can also be used to compute information about loans. When equal periodic payments are made to pay off a loan, the loan is said to be **amortized**.

THEOREM

Amortization Formula

If \$$B$ is borrowed at an interest rate of r% (expressed as a decimal) per annum compounded monthly, the balance A_n due after n monthly payments of \$$P$ is given by the recursive sequence

$$A_0 = B \qquad A_n = \left(1 + \frac{r}{12}\right)A_{n-1} - P \qquad n \geq 1 \qquad \textbf{(10)}$$

Formula (10) may be explained as follows: The initial loan balance is \$$B$. The balance due A_n after n payments will equal the balance due previously, A_{n-1}, plus the interest charged on that amount, reduced by the periodic payment P.

EXAMPLE 11 Mortgage Payments

John and Wanda borrowed \$180,000 at 7% per annum compounded monthly for 30 years to purchase a home. Their monthly payment is determined to be \$1197.54.

(a) Find a recursive formula that represents their balance after each payment of \$1197.54 has been made.

(b) Determine their balance after the first payment is made.

(c) When will their balance be below \$170,000?

Table 6

NORMAL FLOAT AUTO REAL RADIAN MP
PRESS ENTER TO EDIT

n	u(n)
0	180000
1	179852
2	179704
3	179555
4	179405
5	179254
6	179102
7	178949
8	178795
9	178641
10	178485

u(n)=(1+0.07/12)u(n−1)−119

Table 7

NORMAL FLOAT AUTO REAL RADIAN MP
PRESS ENTER TO EDIT

n	u(n)
48	171854
49	171659
50	171463
51	171266
52	171067
53	170868
54	170667
55	170465
56	170262
57	170057
58	169852

u(n)=(1+0.07/12)u(n−1)−119

Solution

(a) Use formula (10) with $A_0 = 180{,}000$, $r = 0.07$, and $P = \$1197.54$. Then

$$A_0 = 180{,}000 \qquad A_n = \left(1 + \frac{0.07}{12}\right)A_{n-1} - 1197.54$$

(b) In SEQuence mode on a TI-84 Plus C, enter the sequence $\{A_n\}$ and create Table 6. After the first payment is made, the balance is $A_1 = \$179{,}852$.

(c) Scroll down until the balance is below \$170,000. See Table 7. After the fifty-eighth payment is made $(n = 58)$, or 4 years, 10 months, the balance is below \$170,000. ■

Now Work PROBLEM 87

13.1 Assess Your Understanding

'Are You Prepared?' *Answers are given at the end of these exercises. If you get a wrong answer, read the pages listed in red.*

1. For the function $f(x) = \dfrac{x-1}{x}$, find $f(2)$ and $f(3)$. (pp. 210–213)

2. ***True or False*** A function is a relation between two sets D and R so that each element x in the first set D is related to exactly one element y in the second set R. (pp. 207–210)

3. If \$1000 is invested at 4% per annum compounded semiannually, how much is in the account after 2 years? (pp. 475–477)

4. How much do you need to invest now at 5% per annum compounded monthly so that in 1 year you will have \$10,000? (p. 479)

Concepts and Vocabulary

5. A(n) ________ is a function whose domain is the set of positive integers.

6. ***True or False*** The notation a_5 represents the fifth term of a sequence.

7. If $n \geq 0$ is an integer, then $n! =$ ________________ when $n \geq 2$.

8. The sequence $a_1 = 5$, $a_n = 3a_{n-1}$ is an example of a(n) ________ sequence.

(a) alternating (b) recursive
(c) Fibonacci (d) summation

9. The notation $a_1 + a_2 + a_3 + \cdots + a_n = \sum_{k=1}^{n} a_k$ is an example of ________ notation.

10. $\sum_{k=1}^{n} k = 1 + 2 + 3 + \cdots + n =$ ________.

(a) $n!$ (b) $\dfrac{n\,(n+1)}{2}$

(c) nk (d) $\dfrac{n\,(n+1)\,(2n+1)}{6}$

Skill Building

In Problems 11–16, evaluate each factorial expression.

11. $10!$ **12.** $9!$ **13.** $\dfrac{9!}{6!}$ **14.** $\dfrac{12!}{10!}$ **15.** $\dfrac{3!\,7!}{4!}$ **16.** $\dfrac{5!\,8!}{3!}$

In Problems 17–28, write down the first five terms of each sequence.

17. $\{s_n\} = \{n\}$ **18.** $\{s_n\} = \{n^2 + 1\}$ **19.** $\{a_n\} = \left\{\dfrac{n}{n+2}\right\}$ **20.** $\{b_n\} = \left\{\dfrac{2n+1}{2n}\right\}$

21. $\{c_n\} = \{(-1)^{n+1}n^2\}$ **22.** $\{d_n\} = \left\{(-1)^{n-1}\left(\frac{n}{2n-1}\right)\right\}$ **23.** $\{s_n\} = \left\{\frac{2^n}{3^n+1}\right\}$ **24.** $\{s_n\} = \left\{\left(\frac{4}{3}\right)^n\right\}$

25. $\{t_n\} = \left\{\frac{(-1)^n}{(n+1)(n+2)}\right\}$ **26.** $\{a_n\} = \left\{\frac{3^n}{n}\right\}$ **27.** $\{b_n\} = \left\{\frac{n}{e^n}\right\}$ **28.** $\{c_n\} = \left\{\frac{n^2}{2^n}\right\}$

In Problems 29–36, the given pattern continues. Write down the nth term of a sequence $\{a_n\}$ suggested by the pattern.

29. $\frac{1}{2}, \frac{2}{3}, \frac{3}{4}, \frac{4}{5}, \dots$ **30.** $\frac{1}{1\cdot 2}, \frac{1}{2\cdot 3}, \frac{1}{3\cdot 4}, \frac{1}{4\cdot 5}, \dots$ **31.** $1, \frac{1}{2}, \frac{1}{4}, \frac{1}{8}, \dots$ **32.** $\frac{2}{3}, \frac{4}{9}, \frac{8}{27}, \frac{16}{81}, \dots$

33. $1, -1, 1, -1, 1, -1, \dots$ **34.** $1, \frac{1}{2}, 3, \frac{1}{4}, 5, \frac{1}{6}, 7, \frac{1}{8}, \dots$ **35.** $1, -2, 3, -4, 5, -6, \dots$ **36.** $2, -4, 6, -8, 10, \dots$

In Problems 37–50, a sequence is defined recursively. Write down the first five terms.

37. $a_1 = 2;\quad a_n = 3 + a_{n-1}$ **38.** $a_1 = 3;\quad a_n = 4 - a_{n-1}$ **39.** $a_1 = -2;\quad a_n = n + a_{n-1}$

40. $a_1 = 1;\quad a_n = n - a_{n-1}$ **41.** $a_1 = 5;\quad a_n = 2a_{n-1}$ **42.** $a_1 = 2;\quad a_n = -a_{n-1}$

43. $a_1 = 3;\quad a_n = \frac{a_{n-1}}{n}$ **44.** $a_1 = -2;\quad a_n = n + 3a_{n-1}$ **45.** $a_1 = 1;\quad a_2 = 2;\quad a_n = a_{n-1}\cdot a_{n-2}$

46. $a_1 = -1;\quad a_2 = 1;\quad a_n = a_{n-2} + na_{n-1}$ **47.** $a_1 = A;\quad a_n = a_{n-1} + d$ **48.** $a_1 = A;\quad a_n = ra_{n-1},\quad r \neq 0$

49. $a_1 = \sqrt{2};\quad a_n = \sqrt{2 + a_{n-1}}$ **50.** $a_1 = \sqrt{2};\quad a_n = \sqrt{\frac{a_{n-1}}{2}}$

In Problems 51–60, write out each sum.

51. $\sum_{k=1}^{n}(k+2)$ **52.** $\sum_{k=1}^{n}(2k+1)$ **53.** $\sum_{k=1}^{n}\frac{k^2}{2}$ **54.** $\sum_{k=1}^{n}(k+1)^2$ **55.** $\sum_{k=0}^{n}\frac{1}{3^k}$

56. $\sum_{k=0}^{n}\left(\frac{3}{2}\right)^k$ **57.** $\sum_{k=0}^{n-1}\frac{1}{3^{k+1}}$ **58.** $\sum_{k=0}^{n-1}(2k+1)$ **59.** $\sum_{k=2}^{n}(-1)^k \ln k$ **60.** $\sum_{k=3}^{n}(-1)^{k+1}2^k$

In Problems 61–70, express each sum using summation notation.

61. $1 + 2 + 3 + \cdots + 20$ **62.** $1^3 + 2^3 + 3^3 + \cdots + 8^3$

63. $\frac{1}{2} + \frac{2}{3} + \frac{3}{4} + \cdots + \frac{13}{13+1}$ **64.** $1 + 3 + 5 + 7 + \cdots + [2(12) - 1]$

65. $1 - \frac{1}{3} + \frac{1}{9} - \frac{1}{27} + \cdots + (-1)^6\left(\frac{1}{3^6}\right)$ **66.** $\frac{2}{3} - \frac{4}{9} + \frac{8}{27} - \cdots + (-1)^{12}\left(\frac{2}{3}\right)^{11}$

67. $3 + \frac{3^2}{2} + \frac{3^3}{3} + \cdots + \frac{3^n}{n}$ **68.** $\frac{1}{e} + \frac{2}{e^2} + \frac{3}{e^3} + \cdots + \frac{n}{e^n}$

69. $a + (a+d) + (a+2d) + \cdots + (a+nd)$ **70.** $a + ar + ar^2 + \cdots + ar^{n-1}$

In Problems 71–82, find the sum of each sequence.

71. $\sum_{k=1}^{40} 5$ **72.** $\sum_{k=1}^{50} 8$ **73.** $\sum_{k=1}^{40} k$ **74.** $\sum_{k=1}^{24}(-k)$

75. $\sum_{k=1}^{20}(5k+3)$ **76.** $\sum_{k=1}^{26}(3k-7)$ **77.** $\sum_{k=1}^{16}(k^2+4)$ **78.** $\sum_{k=0}^{14}(k^2-4)$

79. $\sum_{k=10}^{60}(2k)$ **80.** $\sum_{k=8}^{40}(-3k)$ **81.** $\sum_{k=5}^{20} k^3$ **82.** $\sum_{k=4}^{24} k^3$

Applications and Extensions

83. Trout Population A pond currently has 2000 trout in it. A fish hatchery decides to add an additional 20 trout each month. In addition, it is known that the trout population is growing 3% per month. The size of the population after n months is given by the recursively defined sequence

$$p_0 = 2000, \qquad p_n = 1.03p_{n-1} + 20$$

(a) How many trout are in the pond at the end of the second month? That is, what is p_2?
(b) Using a graphing utility, determine how long it will be before the trout population reaches 5000.

84. Environmental Control The Environmental Protection Agency (EPA) determines that Maple Lake has 250 tons of pollutants as a result of industrial waste and that 10% of the pollutant present is neutralized by solar oxidation every year. The EPA imposes new pollution control laws that result in 15 tons of new pollutant entering the lake each year. The amount of pollutant in the lake at the end of each year is given by the recursively defined sequence

$$p_0 = 250, \qquad p_n = 0.9p_{n-1} + 15$$

(a) Determine the amount of pollutant in the lake at the end of the second year. That is, determine p_2.
(b) Using a graphing utility, provide pollutant amounts for the next 20 years.
(c) What is the equilibrium level of pollution in Maple Lake? That is, what is $\lim_{n\to\infty} p_n$?

85. Roth IRA On January 1, Liam deposits $1500 into a Roth individual retirement account (IRA) and decides to deposit an additional $750 at the end of each quarter into the account.

(a) Find a recursive formula that represents Liam's balance at the end of each quarter if the rate of return is assumed to be 5% per annum compounded quarterly.
(b) Use a graphing utility to determine how long it will be before the value of the account exceeds $150,000.
(c) What will be the value of the account in 30 years, when Liam retires?

86. Education Savings Account On January 1, Aubrey's parents deposit $4000 in an education savings account and decide to place an additional $75 into the account at the end of each month.

(a) Find a recursive formula that represents the balance at the end of each month if the rate of return is assumed to be 1.5% per annum compounded monthly.
(b) Use a graphing utility to determine how long it will be before the value of the account exceeds $10,000.
(c) What will be the value of the account in 16 years when Aubrey goes to college?

87. Credit Card Debt John has a balance of $3000 on his Discover card that charges 1% interest per month on any unpaid balance. John can afford to pay $100 toward the balance each month. His balance each month after making a $100 payment is given by the recursively defined sequence

$$B_0 = \$3000, \qquad B_n = 1.01B_{n-1} - 100$$

(a) Determine John's balance after making the first payment. That is, determine B_1.
(b) Using a graphing utility, determine when John's balance will be below $2000. How many payments of $100 have been made?
(c) Using a graphing utility, determine when John will pay off the balance. What is the total of all the payments?
(d) What was John's interest expense?

88. Car Loans Phil bought a car by taking out a loan for $18,500 at 0.5% interest per month. Phil's normal monthly payment is $434.47 per month, but he decides that he can afford to pay $100 extra toward the balance each month. His balance each month is given by the recursively defined sequence

$$B_0 = \$18{,}500, \qquad B_n = 1.005B_{n-1} - 534.47$$

(a) Determine Phil's balance after making the first payment. That is, determine B_1.
(b) Using a graphing utility, determine when Phil's balance will be below $10,000. How many payments of $534.47 have been made?
(c) Using a graphing utility, determine when Phil will pay off the balance. What is the total of all the payments?
(d) What was Phil's interest expense?

89. Home Loan Bill and Laura borrowed $150,000 at 6% per annum compounded monthly for 30 years to purchase a home. Their monthly payment is determined to be $899.33.
(a) Find a recursive formula for their balance after each monthly payment has been made.
(b) Determine Bill and Laura's balance after the first payment.
(c) Using a graphing utility, create a table showing Bill and Laura's balance after each monthly payment.
(d) Using a graphing utility, determine when Bill and Laura's balance will be below $140,000.
(e) Using a graphing utility, determine when Bill and Laura will pay off the balance.
(f) Determine Bill and Laura's interest expense when the loan is paid.
(g) Suppose that Bill and Laura decide to pay an additional $100 each month on their loan. Answer parts (a) to (f) under this scenario.
(h) Is it worthwhile for Bill and Laura to pay the additional $100? Explain.

90. Home Loan Jodi and Jeff borrowed $120,000 at 6.5% per annum compounded monthly for 30 years to purchase a home. Their monthly payment is determined to be $758.48.
(a) Find a recursive formula for their balance after each monthly payment has been made.
(b) Determine Jodi and Jeff's balance after the first payment.
(c) Using a graphing utility, create a table showing Jodi and Jeff's balance after each monthly payment.
(d) Using a graphing utility, determine when Jodi and Jeff's balance will be below $100,000.
(e) Using a graphing utility, determine when Jodi and Jeff will pay off the balance.
(f) Determine Jodi and Jeff's interest expense when the loan is paid.
(g) Suppose that Jodi and Jeff decide to pay an additional $100 each month on their loan. Answer parts (a) to (f) under this scenario.
(h) Is it worthwhile for Jodi and Jeff to pay the additional $100? Explain.

91. Growth of a Rabbit Colony A colony of rabbits begins with one pair of mature rabbits, which will produce a pair of offspring (one male, one female) each month. Assume that all rabbits mature in 1 month and produce a pair of offspring (one male, one female) after 2 months. If no rabbits ever die, how many pairs of mature rabbits are there after 7 months?

[**Hint:** The Fibonacci sequence models this colony. Do you see why?]

92. Fibonacci Sequence Let

$$u_n = \frac{(1+\sqrt{5})^n - (1-\sqrt{5})^n}{2^n\sqrt{5}}$$

define the nth term of a sequence.

(a) Show that $u_1 = 1$ and $u_2 = 1$.
(b) Show that $u_{n+2} = u_{n+1} + u_n$.
(c) Draw the conclusion that $\{u_n\}$ is the Fibonacci sequence.

93. The Pascal Triangle Divide the triangular array shown (called the Pascal triangle) using diagonal lines as indicated. Find the sum of the numbers in each of these diagonal rows. Do you recognize this sequence?

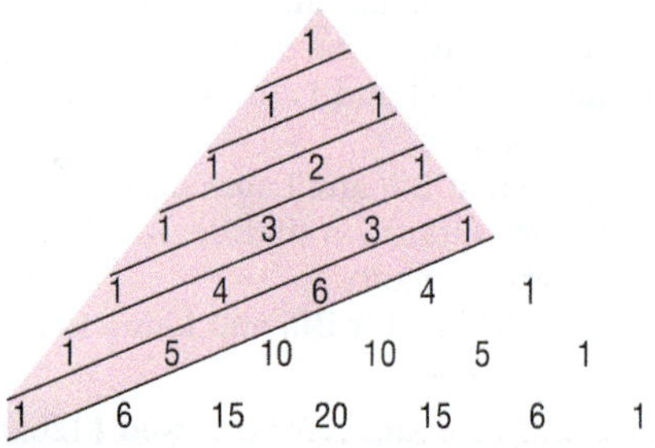

94. Fibonacci Sequence Use the result of Problem 92 to do the following problems:

(a) Write the first 10 terms of the Fibonacci sequence.
(b) Compute the ratio $\frac{u_{n+1}}{u_n}$ for the first 10 terms.
(c) As n gets large, what number does the ratio approach? This number is referred to as the **golden ratio**. Rectangles whose sides are in this ratio were considered pleasing to the eye by the Greeks. For example, the facade of the Parthenon was constructed using the golden ratio.
(d) Compute the ratio $\frac{u_n}{u_{n+1}}$ for the first 10 terms.
(e) As n gets large, what number does the ratio approach? This number is also referred to as the **conjugate golden ratio**. This ratio is believed to have been used in the construction of the Great Pyramid in Egypt. The ratio equals the sum of the areas of the four face triangles divided by the total surface area of the Great Pyramid.

95. Approximating $f(x) = e^x$ In calculus, it can be shown that

$$f(x) = e^x = \sum_{k=0}^{\infty} \frac{x^k}{k!}$$

We can approximate the value of $f(x) = e^x$ for any x using the following sum:

$$f(x) = e^x \approx \sum_{k=0}^{n} \frac{x^k}{k!}$$

for some n.

(a) Approximate $f(1.3)$ with $n = 4$.
(b) Approximate $f(1.3)$ with $n = 7$.
(c) Use a calculator to approximate $f(1.3)$.
(d) Using trial and error along with a graphing utility's SEQuence mode, determine the value of n required to approximate $f(1.3)$ correct to eight decimal places.

96. Approximating $f(x) = e^x$ Refer to Problem 95.

(a) Approximate $f(-2.4)$ with $n = 3$.
(b) Approximate $f(-2.4)$ with $n = 6$.
(c) Use a calculator to approximate $f(-2.4)$.
(d) Using trial and error along with a graphing utility's SEQuence mode, determine the value of n required to approximate $f(-2.4)$ correct to eight decimal places.

97. Bode's Law In 1772, Johann Bode published the following formula for predicting the mean distances, in astronomical units (AU), of the planets from the sun:

$$a_1 = 0.4 \qquad \{a_n\} = \{0.4 + 0.3 \cdot 2^{n-2}\}, n \ge 2$$

where n is the number of the planet from the sun.

(a) Determine the first eight terms of this sequence.
(b) At the time of Bode's publication, the known planets were Mercury (0.39 AU), Venus (0.72 AU), Earth (1 AU), Mars (1.52 AU), Jupiter (5.20 AU), and Saturn (9.54 AU). How do the actual distances compare to the terms of the sequence?
(c) The planet Uranus was discovered in 1781, and the asteroid Ceres was discovered in 1801. The mean orbital distances from the sun to Uranus and Ceres* are 19.2 AU and 2.77 AU, respectively. How well do these values fit within the sequence?
(d) Determine the ninth and tenth terms of Bode's sequence.
(e) The planets Neptune and Pluto* were discovered in 1846 and 1930, respectively. Their mean orbital distances from the sun are 30.07 AU and 39.44 AU, respectively. How do these actual distances compare to the terms of the sequence?
(f) On July 29, 2005, NASA announced the discovery of a dwarf planet $(n = 11)$, which has been named Eris.* Use Bode's Law to predict the mean orbital distance of Eris from the sun. Its actual mean distance is not yet known, but Eris is currently about 97 astronomical units from the sun.

Source: *NASA*

98. Show that

$$1 + 2 + \cdots + (n-1) + n = \frac{n(n+1)}{2}$$

(continued)

*Ceres, Haumea, Makemake, Pluto, and Eris are referred to as dwarf planets.

[**Hint:** Let

$$S = 1 + 2 + \cdots + (n-1) + n$$
$$S = n + (n-1) + (n-2) + \cdots + 1$$

Add these equations. Then

$$2S = \underbrace{[1+n] + [2 + (n-1)] + \cdots + [n+1]}_{n \text{ terms in brackets}}$$

Now complete the derivation.]

Computing Square Roots *A method for approximating $\sqrt{p}$ can be traced back to the Babylonians. The formula is given by the recursively defined sequence*

$$a_0 = k \qquad a_n = \frac{1}{2}\left(a_{n-1} + \frac{p}{a_{n-1}}\right)$$

where k is an initial guess as to the value of the square root. Use this recursive formula to approximate the following square roots by finding a_5. Compare this result to the value provided by your calculator.

99. $\sqrt{5}$

100. $\sqrt{8}$

101. $\sqrt{21}$

102. $\sqrt{89}$

103. Triangular Numbers A **triangular number** is a term of the sequence

$$u_1 = 1 \qquad u_{n+1} = u_n + (n+1)$$

Write down the first seven triangular numbers.

104. For the sequence given in Problem 103, show that

$$u_{n+1} = \frac{(n+1)(n+2)}{2}.$$

105. For the sequence given in Problem 103, show that

$$u_{n+1} + u_n = (n+1)^2.$$

Explaining Concepts: Discussion and Writing

106. Investigate various applications that lead to a Fibonacci sequence, such as art, architecture, or financial markets. Write an essay on these applications.

107. Write a paragraph that explains why the numbers found in Problem 103 are called triangular.

Retain Your Knowledge

Problems 108–111 are based on material learned earlier in the course. The purpose of these problems is to keep the material fresh in your mind so that you are better prepared for the final exam.

108. If \$2500 is invested at 3% compounded monthly, find the amount that results after a period of 2 years.

109. Write the complex number $-1 - i$ in polar form. Express the argument in degrees.

110. For $\mathbf{v} = 2\mathbf{i} - \mathbf{j}$ and $\mathbf{w} = \mathbf{i} + 2\mathbf{j}$, find the dot product $\mathbf{v} \cdot \mathbf{w}$.

111. Find an equation of the parabola with vertex $(-3, 4)$ and focus $(1, 4)$.

'Are You Prepared?' Answers

1. $f(2) = \frac{1}{2}; f(3) = \frac{2}{3}$ **2.** True **3.** \$1082.43 **4.** \$9513.28

13.2 Arithmetic Sequences

OBJECTIVES
1 Determine Whether a Sequence Is Arithmetic (p. 975)
2 Find a Formula for an Arithmetic Sequence (p. 976)
3 Find the Sum of an Arithmetic Sequence (p. 977)

1 Determine Whether a Sequence Is Arithmetic

When the difference between successive terms of a sequence is always the same number, the sequence is called **arithmetic**.

DEFINITION

An **arithmetic sequence*** may be defined recursively as $a_1 = a$, $a_n - a_{n-1} = d$, or as

$$a_1 = a \qquad a_n = a_{n-1} + d \qquad \textbf{(1)}$$

where $a_1 = a$ and d are real numbers. The number a is the first term, and the number d is called the **common difference**.

*Sometimes called an **arithmetic progression**.

The terms of an arithmetic sequence with first term a_1 and common difference d follow the pattern

$$a_1 \quad a_1 + d \quad a_1 + 2d \quad a_1 + 3d \ldots$$

EXAMPLE 1 **Determining Whether a Sequence Is Arithmetic**

The sequence

$$4, \quad 6, \quad 8, \quad 10, \ldots$$

is arithmetic since the difference of successive terms is 2. The first term is $a_1 = 4$, and the common difference is $d = 2$.

EXAMPLE 2 **Determining Whether a Sequence Is Arithmetic**

Show that the following sequence is arithmetic. Find the first term and the common difference.

$$\{s_n\} = \{3n + 5\}$$

Solution The first term is $s_1 = 3 \cdot 1 + 5 = 8$. The nth term and the $(n-1)$st term of the sequence $\{s_n\}$ are

$$s_n = 3n + 5 \quad \text{and} \quad s_{n-1} = 3(n-1) + 5 = 3n + 2$$

Their difference d is

$$d = s_n - s_{n-1} = (3n + 5) - (3n + 2) = 5 - 2 = 3$$

Since the difference of any two successive terms is the constant 3, the sequence $\{s_n\}$ is arithmetic, and the common difference is 3.

EXAMPLE 3 **Determining Whether a Sequence Is Arithmetic**

Show that the sequence $\{t_n\} = \{4 - n\}$ is arithmetic. Find the first term and the common difference.

Solution The first term is $t_1 = 4 - 1 = 3$. The nth term and the $(n-1)$st term are

$$t_n = 4 - n \quad \text{and} \quad t_{n-1} = 4 - (n-1) = 5 - n$$

Their difference d is

$$d = t_n - t_{n-1} = (4 - n) - (5 - n) = 4 - 5 = -1$$

Since the difference of any two successive terms is the constant -1, $\{t_n\}$ is an arithmetic sequence whose common difference is -1.

Now Work PROBLEM 9

2 Find a Formula for an Arithmetic Sequence

Suppose that a is the first term of an arithmetic sequence whose common difference is d. We seek a formula for the nth term, a_n. To see the pattern, consider the first few terms.

$$\begin{aligned}
a_1 &= a \\
a_2 &= a_1 + d = a_1 + 1 \cdot d \\
a_3 &= a_2 + d = (a_1 + d) + d = a_1 + 2 \cdot d \\
a_4 &= a_3 + d = (a_1 + 2 \cdot d) + d = a_1 + 3 \cdot d \\
a_5 &= a_4 + d = (a_1 + 3 \cdot d) + d = a_1 + 4 \cdot d \\
&\vdots \\
a_n &= a_{n-1} + d = [a_1 + (n-2)d] + d = a_1 + (n-1)d
\end{aligned}$$

This leads to the following result:

THEOREM

nth Term of an Arithmetic Sequence

For an arithmetic sequence $\{a_n\}$ whose first term is a_1 and whose common difference is d, the nth term is determined by the formula

$$a_n = a_1 + (n - 1)d \quad \textbf{(2)}$$

EXAMPLE 4 **Finding a Particular Term of an Arithmetic Sequence**

Find the 41st term of the arithmetic sequence: 2, 6, 10, 14, 18, . . .

Solution The first term of this arithmetic sequence is $a_1 = 2$, and the common difference is $d = 4$. By formula (2), the nth term is

$$a_n = 2 + (n - 1)4 \quad a_n = a_1 + (n - 1)d;\ a_1 = 2,\ d = 4$$

The 41st term is

$$a_{41} = 2 + (41 - 1) \cdot 4 = 162$$

Now Work PROBLEM 25

EXAMPLE 5 **Finding a Recursive Formula for an Arithmetic Sequence**

The 8th term of an arithmetic sequence is 75, and the 20th term is 39.

(a) Find the first term and the common difference.
(b) Give a recursive formula for the sequence.
(c) What is the nth term of the sequence?

Solution (a) Formula (2) states that $a_n = a_1 + (n - 1)d$. As a result,

$$\begin{cases} a_8 = a_1 + 7d = 75 \\ a_{20} = a_1 + 19d = 39 \end{cases}$$

This is a system of two linear equations containing two variables, a_1 and d, which can be solved by elimination. Subtracting the second equation from the first gives

$$-12d = 36$$
$$d = -3$$

With $d = -3$, use $a_1 + 7d = 75$ to find that $a_1 = 75 - 7d = 75 - 7(-3) = 96$. The first term is $a_1 = 96$, and the common difference is $d = -3$.

(b) Using formula (1), a recursive formula for this sequence is

$$a_1 = 96 \qquad a_n = a_{n-1} - 3$$

(c) Using formula (2), a formula for the nth term of the sequence $\{a_n\}$ is

$$a_n = a_1 + (n - 1)d = 96 + (n - 1)(-3) = 99 - 3n$$

Exploration

Graph the recursive formula from Example 5, $a_1 = 96$, $a_n = a_{n-1} - 3$, using a graphing utility. Conclude that the graph of the recursive formula behaves like the graph of a linear function. How is d, the common difference, related to m, the slope of a line?

Now Work PROBLEMS 17 AND 31

3 Find the Sum of an Arithmetic Sequence

The next result gives two formulas for finding the sum of the first n terms of an arithmetic sequence.

THEOREM

Sum of the First n Terms of an Arithmetic Sequence

Let $\{a_n\}$ be an arithmetic sequence with first term a_1 and common difference d. The sum S_n of the first n terms of $\{a_n\}$ may be found in two ways:

$$S_n = a_1 + a_2 + a_3 + \cdots + a_n = \sum_{k=1}^{n} [a_1 + (k-1)d]$$

$$= \frac{n}{2}[2a_1 + (n-1)d] \quad \textbf{(3)}$$

$$= \frac{n}{2}(a_1 + a_n) \quad \textbf{(4)}$$

Proof

$$S_n = a_1 + a_2 + a_3 + \cdots + a_n \quad \text{Sum of first } n \text{ terms}$$

$$= a_1 + (a_1 + d) + (a_1 + 2d) + \cdots + [a_1 + (n-1)d] \quad \text{Formula (2)}$$

$$= \underbrace{(a_1 + a_1 + \cdots + a_1)}_{n \text{ terms}} + [d + 2d + \cdots + (n-1)d] \quad \text{Rearrange terms.}$$

$$= na_1 + d[1 + 2 + \cdots + (n-1)]$$

$$= na_1 + d\left[\frac{(n-1)n}{2}\right] \quad \text{Formula 6, Section 13.1}$$

$$= na_1 + \frac{n}{2}(n-1)d$$

$$= \frac{n}{2}[2a_1 + (n-1)d] \quad \text{Factor out } \frac{n}{2}\text{; this is formula (3).}$$

$$= \frac{n}{2}[a_1 + a_1 + (n-1)d]$$

$$= \frac{n}{2}(a_1 + a_n) \quad \text{Use formula (2); this is formula (4).}$$

■

Notice that formula (3) involves the first term and common difference, whereas formula (4) involves the first term and the nth term. Use whichever form is easier.

EXAMPLE 6

Finding the Sum of an Arithmetic Sequence

Find the sum S_n of the first n terms of the sequence $\{a_n\} = \{3n + 5\}$; that is, find

$$8 + 11 + 14 + \cdots + (3n + 5) = \sum_{k=1}^{n} (3k + 5)$$

Solution The sequence $\{a_n\} = \{3n + 5\}$ is an arithmetic sequence with first term $a_1 = 8$ and nth term $a_n = 3n + 5$. To find the sum S_n, use formula (4).

$$S_n = \sum_{k=1}^{n} (3k + 5) \underset{\uparrow}{=} \frac{n}{2}[8 + (3n + 5)] = \frac{n}{2}(3n + 13)$$

$$S_n = \frac{n}{2}(a_1 + a_n)$$

Now Work PROBLEM 39

EXAMPLE 7

Finding the Sum of an Arithmetic Sequence

Find the sum: $60 + 64 + 68 + 72 + \cdots + 120$

Solution This is the sum S_n of an arithmetic sequence $\{a_n\}$ whose first term is $a_1 = 60$ and whose common difference is $d = 4$. The nth term is $a_n = 120$. Use formula (2) to find n.

$$a_n = a_1 + (n-1)d \quad \text{Formula (2)}$$
$$120 = 60 + (n-1)\cdot 4 \quad a_n = 120, a_1 = 60, d = 4$$
$$60 = 4(n-1) \quad \text{Simplify.}$$
$$15 = n - 1 \quad \text{Simplify.}$$
$$n = 16 \quad \text{Solve for } n.$$

Now use formula (4) to find the sum S_{16}.

$$60 + 64 + 68 + \cdots + 120 = S_{16} = \frac{16}{2}(60 + 120) = 1440$$
$$\uparrow\ S_n = \frac{n}{2}(a_1 + a_n)$$

Now Work PROBLEM 43

EXAMPLE 8

Creating a Floor Design

A ceramic tile floor is designed in the shape of a trapezoid 20 feet wide at the base and 10 feet wide at the top. See Figure 13. The tiles, which measure 12 inches by 12 inches, are to be placed so that each successive row contains one fewer tile than the preceding row. How many tiles will be required?

Figure 13

Solution The bottom row requires 20 tiles and the top row, 10 tiles. Since each successive row requires one fewer tile, the total number of tiles required is

$$S = 20 + 19 + 18 + \cdots + 11 + 10$$

This is the sum of an arithmetic sequence; the common difference is -1. The number of terms to be added is $n = 11$, with the first term $a_1 = 20$ and the last term $a_{11} = 10$. The sum S is

$$S = \frac{n}{2}(a_1 + a_{11}) = \frac{11}{2}(20 + 10) = 165$$

In all, 165 tiles will be required.

13.2 Assess Your Understanding

Concepts and Vocabulary

1. In a(n) ________ sequence, the difference between successive terms is a constant.
2. ***True or False*** For an arithmetic sequence $\{a_n\}$ whose first term is a_1 and whose common difference is d, the nth term is determined by the formula $a_n = a_1 + nd$.
3. If the 5th term of an arithmetic sequence is 12 and the common difference is 5, then the 6th term of the sequence is ___.
4. ***True or False*** The sum S_n of the first n terms of an arithmetic sequence $\{a_n\}$ whose first term is a_1 can be found using the formula $S_n = \frac{n}{2}(a_1 + a_n)$.
5. An arithmetic sequence can always be expressed as a(n) ________ sequence.
   (a) Fibonacci (b) alternating
   (c) geometric (d) recursive
6. If $a_n = -2n + 7$ is the nth term of an arithmetic sequence, the first term is ________.
   (a) -2 (b) 0
   (c) 5 (d) 7

Skill Building

In Problems 7–16, show that each sequence is arithmetic. Find the common difference and write out the first four terms.

7. $\{s_n\} = \{n + 4\}$
8. $\{s_n\} = \{n - 5\}$
9. $\{a_n\} = \{2n - 5\}$
10. $\{b_n\} = \{3n + 1\}$
11. $\{c_n\} = \{6 - 2n\}$
12. $\{a_n\} = \{4 - 2n\}$
13. $\{t_n\} = \left\{\frac{1}{2} - \frac{1}{3}n\right\}$
14. $\{t_n\} = \left\{\frac{2}{3} + \frac{n}{4}\right\}$
15. $\{s_n\} = \{\ln 3^n\}$
16. $\{s_n\} = \{e^{\ln n}\}$

In Problems 17–24, find the nth term of the arithmetic sequence $\{a_n\}$ whose initial term a_1 and common difference d are given. What is the 51st term?

17. $a_1 = 2;\quad d = 3$
18. $a_1 = -2;\quad d = 4$
19. $a_1 = 5;\quad d = -3$
20. $a_1 = 6;\quad d = -2$
21. $a_1 = 0;\quad d = \frac{1}{2}$
22. $a_1 = 1;\quad d = -\frac{1}{3}$
23. $a_1 = \sqrt{2};\quad d = \sqrt{2}$
24. $a_1 = 0;\quad d = \pi$

In Problems 25–30, find the indicated term in each arithmetic sequence.

25. 100th term of 2, 4, 6, ...
26. 80th term of $-1, 1, 3, \ldots$
27. 90th term of $1, -2, -5, \ldots$
28. 80th term of $5, 0, -5, \ldots$
29. 80th term of $2, \frac{5}{2}, 3, \frac{7}{2}, \ldots$
30. 70th term of $2\sqrt{5}, 4\sqrt{5}, 6\sqrt{5}, \ldots$

In Problems 31–38, find the first term and the common difference of the arithmetic sequence described. Give a recursive formula for the sequence. Find a formula for the nth term.

31. 8th term is 8; 20th term is 44
32. 4th term is 3; 20th term is 35
33. 9th term is -5; 15th term is 31
34. 8th term is 4; 18th term is -96
35. 15th term is 0; 40th term is -50
36. 5th term is -2; 13th term is 30
37. 14th term is -1; 18th term is -9
38. 12th term is 4; 18th term is 28

In Problems 39–56, find each sum.

39. $1 + 3 + 5 + \cdots + (2n - 1)$
40. $2 + 4 + 6 + \cdots + 2n$
41. $7 + 12 + 17 + \cdots + (2 + 5n)$
42. $-1 + 3 + 7 + \cdots + (4n - 5)$
43. $2 + 4 + 6 + \cdots + 70$
44. $1 + 3 + 5 + \cdots + 59$
45. $5 + 9 + 13 + \cdots + 49$
46. $2 + 5 + 8 + \cdots + 41$
47. $73 + 78 + 83 + 88 + \cdots + 558$
48. $7 + 1 - 5 - 11 - \cdots - 299$
49. $4 + 4.5 + 5 + 5.5 + \cdots + 100$
50. $8 + 8\frac{1}{4} + 8\frac{1}{2} + 8\frac{3}{4} + 9 + \cdots + 50$
51. $\sum_{n=1}^{80} (2n - 5)$
52. $\sum_{n=1}^{90} (3 - 2n)$
53. $\sum_{n=1}^{100} \left(6 - \frac{1}{2}n\right)$
54. $\sum_{n=1}^{80} \left(\frac{1}{3}n + \frac{1}{2}\right)$
55. The sum of the first 120 terms of the sequence
    14, 16, 18, 20, ...
56. The sum of the first 46 terms of the sequence
    $2, -1, -4, -7, \ldots$

Applications and Extensions

57. Find x so that $x + 3, 2x + 1,$ and $5x + 2$ are consecutive terms of an arithmetic sequence.

58. Find x so that $2x, 3x + 2,$ and $5x + 3$ are consecutive terms of an arithmetic sequence.

59. How many terms must be added in an arithmetic sequence whose first term is 11 and whose common difference is 3 to obtain a sum of 1092?

60. How many terms must be added in an arithmetic sequence whose first term is 78 and whose common difference is -4 to obtain a sum of 702?

61. Drury Lane Theater The Drury Lane Theater has 25 seats in the first row and 30 rows in all. Each successive row contains one additional seat. How many seats are in the theater?

62. Football Stadium The corner section of a football stadium has 15 seats in the first row and 40 rows in all. Each successive row contains two additional seats. How many seats are in this section?

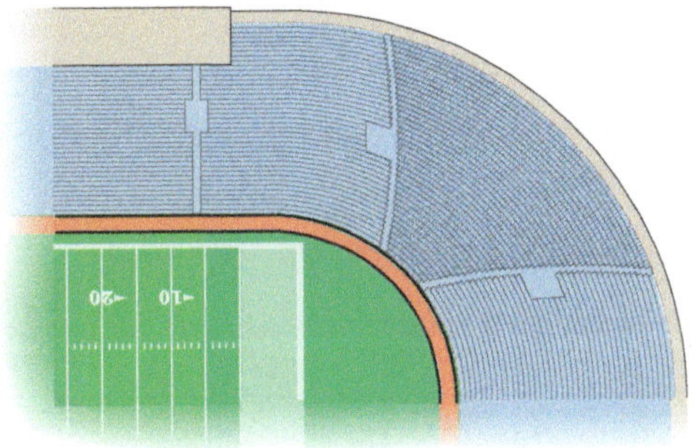

63. Creating a Mosaic A mosaic is designed in the shape of an equilateral triangle, 20 feet on each side. Each tile in the mosaic is in the shape of an equilateral triangle, 12 inches to a side. The tiles are to alternate in color as shown in the illustration. How many tiles of each color will be required?

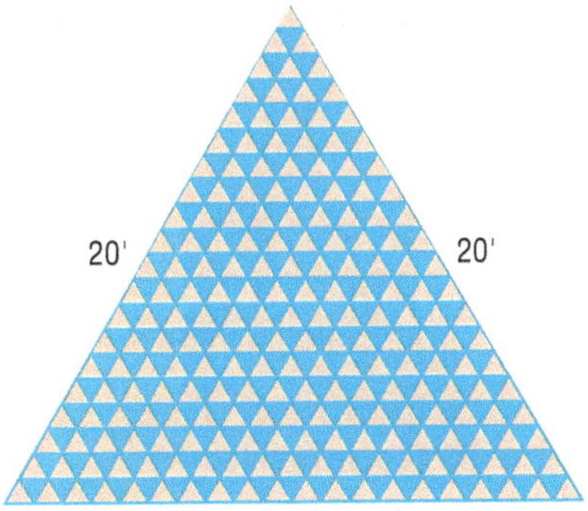

64. Constructing a Brick Staircase A brick staircase has a total of 30 steps. The bottom step requires 100 bricks. Each successive step requires two fewer bricks than the prior step.

(a) How many bricks are required for the top step?

(b) How many bricks are required to build the staircase?

65. Cooling Air As a parcel of air rises (for example, as it is pushed over a mountain), it cools at the *dry adiabatic lapse rate* of 5.5°F per 1000 feet until it reaches its dew point. If the ground temperature is 67°F, write a formula for the sequence of temperatures, $\{T_n\}$, of a parcel of air that has risen n thousand feet. What is the temperature of a parcel of air if it has risen 5000 feet?

Source: *National Aeronautics and Space Administration*

66. Citrus Ladders Ladders used by fruit pickers are typically tapered with a wide bottom for stability and a narrow top for ease of picking. If the bottom rung of such a ladder is 49 inches wide and the top rung is 24 inches wide, how many rungs does the ladder have if each rung is 2.5 inches shorter than the one below it? How much material would be needed to make the rungs for the ladder described?

Source: *www.stokesladders.com*

67. Seats in an Amphitheater An outdoor amphitheater has 35 seats in the first row, 37 in the second row, 39 in the third row, and so on. There are 27 rows altogether. How many can the amphitheater seat?

68. Stadium Construction How many rows are in the corner section of a stadium containing 2040 seats if the first row has 10 seats and each successive row has 4 additional seats?

69. Salary If you take a job with a starting salary of \$35,000 per year and a guaranteed raise of \$1400 per year, how many years will it be before your aggregate salary is \$280,000?

[**Hint:** Remember that your aggregate salary after 2 years is \$35,000 + (\$35,000 + \$1400).]

Explaining Concepts: Discussion and Writing

70. Make up an arithmetic sequence. Give it to a friend and ask for its 20th term.

71. Describe the similarities and differences between arithmetic sequences and linear functions.

Retain Your Knowledge

Problems 72–75 are based on material learned earlier in the course. The purpose of these problems is to keep the material fresh in your mind so that you are better prepared for the final exam.

72. If a credit card charges 15.3% interest compounded monthly, find the effective rate of interest.

73. The vector $\mathbf{v}$ has initial point $P = (-1, 2)$ and terminal point $Q = (3, -4)$. Write $\mathbf{v}$ in the form $a\mathbf{i} + b\mathbf{j}$; that is, find its position vector.

74. Analyze and graph the equation: $25x^2 + 4y^2 = 100$

75. Find the inverse of the matrix $\begin{bmatrix} 2 & 0 \\ 3 & -1 \end{bmatrix}$, if there is one; otherwise, state that the matrix is singular.

13.3 Geometric Sequences; Geometric Series

OBJECTIVES
1 Determine Whether a Sequence Is Geometric (p. 982)
2 Find a Formula for a Geometric Sequence (p. 983)
3 Find the Sum of a Geometric Sequence (p. 984)
4 Determine Whether a Geometric Series Converges or Diverges (p. 985)

1 Determine Whether a Sequence Is Geometric

When the ratio of successive terms of a sequence is always the same nonzero number, the sequence is called **geometric**.

DEFINITION

A **geometric sequence*** may be defined recursively as $a_1 = a$, $\dfrac{a_n}{a_{n-1}} = r$, or as

$$a_1 = a \qquad a_n = ra_{n-1} \qquad \textbf{(1)}$$

where $a_1 = a$ and $r \neq 0$ are real numbers. The number a_1 is the first term, and the nonzero number r is called the **common ratio.**

The terms of a geometric sequence with first term a_1 and common ratio r follow the pattern

$$a_1 \quad a_1r \quad a_1r^2 \quad a_1r^3 \ldots$$

EXAMPLE 1 Determining Whether a Sequence Is Geometric

The sequence

$$2, 6, 18, 54, 162, \ldots$$

is geometric because the ratio of successive terms is 3; $\left(\dfrac{6}{2} = \dfrac{18}{6} = \dfrac{54}{18} = \cdots = 3\right)$. The first term is $a_1 = 2$, and the common ratio is 3.

EXAMPLE 2 Determining Whether a Sequence Is Geometric

Show that the following sequence is geometric.

$$\{s_n\} = \{2^{-n}\}$$

Find the first term and the common ratio.

*Sometimes called a **geometric progression**.

Solution The first term is $s_1 = 2^{-1} = \frac{1}{2}$. The nth term and the $(n-1)$st term of the sequence $\{s_n\}$ are

$$s_n = 2^{-n} \quad \text{and} \quad s_{n-1} = 2^{-(n-1)}$$

Their ratio is

$$\frac{s_n}{s_{n-1}} = \frac{2^{-n}}{2^{-(n-1)}} = 2^{-n+(n-1)} = 2^{-1} = \frac{1}{2}$$

Because the ratio of successive terms is the nonzero constant $\frac{1}{2}$, the sequence $\{s_n\}$ is geometric with common ratio $\frac{1}{2}$. ■

EXAMPLE 3 **Determining Whether a Sequence Is Geometric**

Show that the following sequence is geometric.

$$\{t_n\} = \{3 \cdot 4^n\}$$

Find the first term and the common ratio.

Solution The first term is $t_1 = 3 \cdot 4^1 = 12$. The nth term and the $(n-1)$st term are

$$t_n = 3 \cdot 4^n \quad \text{and} \quad t_{n-1} = 3 \cdot 4^{n-1}$$

Their ratio is

$$\frac{t_n}{t_{n-1}} = \frac{3 \cdot 4^n}{3 \cdot 4^{n-1}} = 4^{n-(n-1)} = 4$$

The sequence, $\{t_n\}$, is a geometric sequence with common ratio 4. ■

Now Work PROBLEM 11

2 Find a Formula for a Geometric Sequence

Suppose that a_1 is the first term of a geometric sequence with common ratio $r \neq 0$. We seek a formula for the nth term, a_n. To see the pattern, consider the first few terms:

$$\begin{aligned}
a_1 &= a_1 \cdot 1 = a_1 r^0 \\
a_2 &= r a_1 = a_1 r^1 \\
a_3 &= r a_2 = r(a_1 r) = a_1 r^2 \\
a_4 &= r a_3 = r(a_1 r^2) = a_1 r^3 \\
a_5 &= r a_4 = r(a_1 r^3) = a_1 r^4 \\
&\vdots \\
a_n &= r a_{n-1} = r(a_1 r^{n-2}) = a_1 r^{n-1}
\end{aligned}$$

This leads to the following result:

THEOREM ***n*th Term of a Geometric Sequence**

For a geometric sequence $\{a_n\}$ whose first term is a_1 and whose common ratio is r, the nth term is determined by the formula

$$a_n = a_1 r^{n-1} \qquad r \neq 0 \qquad \textbf{(2)}$$

■

EXAMPLE 4 **Finding a Particular Term of a Geometric Sequence**

(a) Find the nth term of the geometric sequence: $10, 9, \frac{81}{10}, \frac{729}{100}, \ldots$
(b) Find the 9th term of this sequence.
(c) Find a recursive formula for this sequence.

Solution (a) The first term of this geometric sequence is $a_1 = 10$, and the common ratio is $\frac{9}{10}$. (Use $\frac{9}{10}$ or $\frac{\frac{81}{10}}{9} = \frac{9}{10}$ or any two successive terms.) Then, by formula (2), the nth term is

$$a_n = 10\left(\frac{9}{10}\right)^{n-1} \qquad a_n = a_1 r^{n-1};\ a_1 = 10,\ r = \frac{9}{10}$$

(b) The 9th term is

$$a_9 = 10\left(\frac{9}{10}\right)^{9-1} = 10\left(\frac{9}{10}\right)^{8} = 4.3046721$$

(c) The first term in the sequence is 10, and the common ratio is $r = \frac{9}{10}$. Using formula (1), the recursive formula is $a_1 = 10,\ a_n = \frac{9}{10}a_{n-1}$. ■

Exploration

Use a graphing utility to find the ninth term of the sequence given in Example 4. Use it to find the 20th and 50th terms. Now use a graphing utility to graph the recursive formula found in Example 4(c). Conclude that the graph of the recursive formula behaves like the graph of an exponential function. How is r, the common ratio, related to a, the base of the exponential function $y = a^x$? ■

Now Work PROBLEMS 19, 27, AND 35

3 Find the Sum of a Geometric Sequence

THEOREM

Sum of the First *n* Terms of a Geometric Sequence

Let $\{a_n\}$ be a geometric sequence with first term a_1 and common ratio r, where $r \neq 0, r \neq 1$. The sum S_n of the first n terms of $\{a_n\}$ is

$$S_n = a_1 + a_1 r + a_1 r^2 + \cdots + a_1 r^{n-1} = \sum_{k=1}^{n} a_1 r^{k-1}$$
$$= a_1 \cdot \frac{1 - r^n}{1 - r} \qquad r \neq 0, 1 \qquad \textbf{(3)}$$

■

Proof The sum S_n of the first n terms of $\{a_n\} = \{a_1 r^{n-1}\}$ is

$$S_n = a_1 + a_1 r + \cdots + a_1 r^{n-1} \qquad \textbf{(4)}$$

Multiply each side by r to obtain

$$rS_n = a_1 r + a_1 r^2 + \cdots + a_1 r^n \qquad \textbf{(5)}$$

Now, subtract (5) from (4). The result is

$$S_n - rS_n = a_1 - a_1 r^n$$
$$(1 - r)S_n = a_1(1 - r^n)$$

Since $r \neq 1$, solve for S_n.

$$S_n = a_1 \cdot \frac{1 - r^n}{1 - r}$$ ■

EXAMPLE 5 **Finding the Sum of the First *n* Terms of a Geometric Sequence**

Find the sum S_n of the first n terms of the sequence $\left\{\left(\frac{1}{2}\right)^n\right\}$; that is, find

$$\frac{1}{2} + \frac{1}{4} + \frac{1}{8} + \cdots + \left(\frac{1}{2}\right)^n = \sum_{k=1}^{n} \frac{1}{2}\left(\frac{1}{2}\right)^{k-1}$$

Solution The sequence $\left\{\left(\frac{1}{2}\right)^n\right\}$ is a geometric sequence with $a_1 = \frac{1}{2}$ and $r = \frac{1}{2}$. Use formula (3) to get

$$S_n = \sum_{k=1}^{n} \frac{1}{2}\left(\frac{1}{2}\right)^{k-1} = \frac{1}{2} + \frac{1}{4} + \frac{1}{8} + \cdots + \left(\frac{1}{2}\right)^n$$

$$= \frac{1}{2}\left[\frac{1 - \left(\frac{1}{2}\right)^n}{1 - \frac{1}{2}}\right] \quad \text{Formula (3); } a_1 = \frac{1}{2}, r = \frac{1}{2}$$

$$= \frac{1}{2}\left[\frac{1 - \left(\frac{1}{2}\right)^n}{\frac{1}{2}}\right]$$

$$= 1 - \left(\frac{1}{2}\right)^n$$

Now Work PROBLEM 41

EXAMPLE 6 Using a Graphing Utility to Find the Sum of a Geometric Sequence

Use a graphing utility to find the sum of the first 15 terms of the sequence $\left\{\left(\frac{1}{3}\right)^n\right\}$; that is, find

$$\frac{1}{3} + \frac{1}{9} + \frac{1}{27} + \cdots + \left(\frac{1}{3}\right)^{15} = \sum_{k=1}^{15} \frac{1}{3}\left(\frac{1}{3}\right)^{k-1}$$

Solution Figure 14 shows the result using a TI-84 Plus C graphing calculator. The sum of the first 15 terms of the sequence $\left\{\left(\frac{1}{3}\right)^n\right\}$ is approximately 0.4999999652.

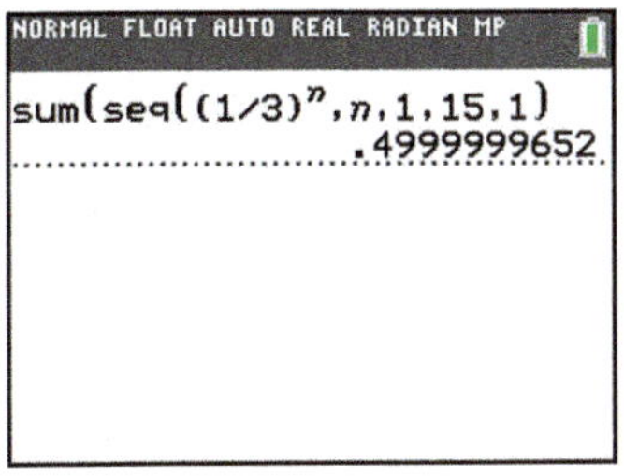

Figure 14

Now Work PROBLEM 47

4 Determine Whether a Geometric Series Converges or Diverges

DEFINITION

An infinite sum of the form

$$a_1 + a_1 r + a_1 r^2 + \cdots + a_1 r^{n-1} + \cdots$$

with first term a_1 and common ratio r, is called an **infinite geometric series** and is denoted by

$$\sum_{k=1}^{\infty} a_1 r^{k-1}$$

Based on formula (3), the sum S_n of the first n terms of a geometric series is

$$S_n = a_1 \cdot \frac{1 - r^n}{1 - r} = \frac{a_1}{1 - r} - \frac{a_1 r^n}{1 - r} \qquad (6)$$

Note: In calculus, limit notation is used, and the sum is written

$$L = \lim_{n\to\infty} S_n = \lim_{n\to\infty} \sum_{k=1}^{n} a_1 r^{k-1} = \sum_{k=1}^{\infty} a_1 r^{k-1}$$

If this finite sum S_n approaches a number L as $n \to \infty$, then the infinite geometric series $\sum_{k=1}^{\infty} a_1 r^{k-1}$ **converges** to L and L is called the **sum of the infinite geometric series**. The sum is written as

$$L = \sum_{k=1}^{\infty} a_1 r^{k-1}$$

A series that does not converge is called a **divergent series.**

THEOREM

Convergence of an Infinite Geometric Series

If $|r| < 1$, the infinite geometric series $\sum_{k=1}^{\infty} a_1 r^{k-1}$ converges. Its sum is

$$\sum_{k=1}^{\infty} a_1 r^{k-1} = \frac{a_1}{1-r} \qquad \textbf{(7)}$$

Intuitive Proof Since $|r| < 1$, it follows that $|r^n|$ approaches 0 as $n \to \infty$. Then, based on formula (6), the term $\frac{a_1 r^n}{1-r}$ approaches 0, so the sum S_n approaches $\frac{a_1}{1-r}$ as $n \to \infty$.

EXAMPLE 7 Determining Whether a Geometric Series Converges or Diverges

Determine whether the geometric series

$$\sum_{k=1}^{\infty} 2\left(\frac{2}{3}\right)^{k-1} = 2 + \frac{4}{3} + \frac{8}{9} + \cdots$$

converges or diverges. If it converges, find its sum.

Solution Comparing $\sum_{k=1}^{\infty} 2\left(\frac{2}{3}\right)^{k-1}$ to $\sum_{k=1}^{\infty} a_1 r^{k-1}$, the first term is $a_1 = 2$ and the common ratio is $r = \frac{2}{3}$. Since $|r| < 1$, the series converges. Use formula (7) to find its sum:

$$\sum_{k=1}^{\infty} 2\left(\frac{2}{3}\right)^{k-1} = 2 + \frac{4}{3} + \frac{8}{9} + \cdots = \frac{2}{1-\frac{2}{3}} = 6$$

Now Work PROBLEM 53

EXAMPLE 8 Repeating Decimals

Show that the repeating decimal 0.999 . . . equals 1.

Solution The decimal $0.999\ldots = 0.9 + 0.09 + 0.009 + \cdots = \frac{9}{10} + \frac{9}{100} + \frac{9}{1000} + \cdots$ is an infinite geometric series. Write it in the form $\sum_{k=1}^{\infty} a_1 r^{k-1}$ and use formula (7).

$$0.999\ldots = \frac{9}{10} + \frac{9}{100} + \frac{9}{1000} + \cdots = \sum_{k=1}^{\infty} \frac{9}{10^k} = \sum_{k=1}^{\infty} \frac{9}{10 \cdot 10^{k-1}} = \sum_{k=1}^{\infty} \frac{9}{10}\left(\frac{1}{10}\right)^{k-1}$$

Compare this series to $\sum_{k=1}^{\infty} a_1 r^{k-1}$ and note that $a_1 = \frac{9}{10}$ and $r = \frac{1}{10}$. Since $|r| < 1$, the series converges and its sum is

$$0.999\ldots = \frac{\frac{9}{10}}{1 - \frac{1}{10}} = \frac{\frac{9}{10}}{\frac{9}{10}} = 1$$

The repeating decimal 0.999... equals 1. ■

EXAMPLE 9 Pendulum Swings

Initially, a pendulum swings through an arc of 18 inches. See Figure 15. On each successive swing, the length of the arc is 0.98 of the previous length.

(a) What is the length of the arc of the 10th swing?
(b) On which swing is the length of the arc first less than 12 inches?
(c) After 15 swings, what total distance will the pendulum have swung?
(d) When it stops, what total distance will the pendulum have swung?

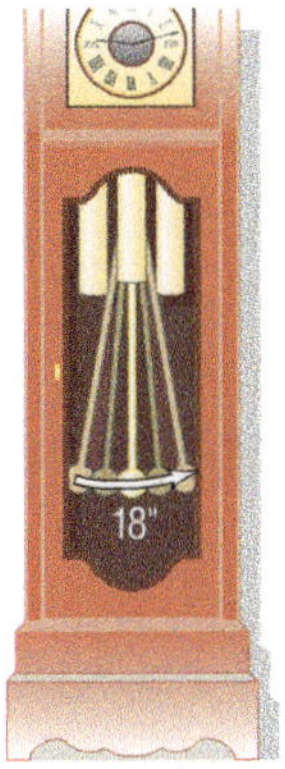

Figure 15

Solution

(a) The length of the first swing is 18 inches.
The length of the second swing is $0.98(18)$ inches.
The length of the third swing is $0.98(0.98)(18) = 0.98^2(18)$ inches.
The length of the arc of the 10th swing is

$$(0.98)^9(18) \approx 15.007 \text{ inches}$$

(b) The length of the arc of the nth swing is $(0.98)^{n-1}(18)$. For this to be exactly 12 inches requires that

$$(0.98)^{n-1}(18) = 12$$

$$(0.98)^{n-1} = \frac{12}{18} = \frac{2}{3} \qquad \text{Divide both sides by 18.}$$

$$n - 1 = \log_{0.98}\left(\frac{2}{3}\right) \qquad \text{Express as a logarithm.}$$

$$n = 1 + \frac{\ln\left(\frac{2}{3}\right)}{\ln 0.98} \approx 1 + 20.07 = 21.07 \qquad \text{Solve for } n\text{; use the Change of Base Formula.}$$

The length of the arc of the pendulum exceeds 12 inches on the 21st swing and is first less than 12 inches on the 22nd swing.

(c) After 15 swings, the pendulum will have swung the following total distance L:

$$L = \underbrace{18}_{\text{1st}} + \underbrace{0.98(18)}_{\text{2nd}} + \underbrace{(0.98)^2(18)}_{\text{3rd}} + \underbrace{(0.98)^3(18)}_{\text{4th}} + \cdots + \underbrace{(0.98)^{14}(18)}_{\text{15th}}$$

This is the sum of a geometric sequence. The common ratio is 0.98; the first term is 18. The sum has 15 terms, so

$$L = 18 \cdot \frac{1 - 0.98^{15}}{1 - 0.98} \approx 18(13.07) \approx 235.3 \text{ inches}$$

The pendulum will have swung through approximately 235.3 inches after 15 swings.

(d) When the pendulum stops, it will have swung the following total distance T:

$$T = 18 + 0.98(18) + (0.98)^2(18) + (0.98)^3(18) + \cdots$$

This is the sum of an infinite geometric series. The common ratio is $r = 0.98$; the first term is $a_1 = 18$. Since $|r| < 1$, the series converges. Its sum is

$$T = \frac{a_1}{1 - r} = \frac{18}{1 - 0.98} = 900$$

The pendulum will have swung a total of 900 inches when it finally stops. ■

Now Work PROBLEM 87

Historical Feature

Fibonacci

Sequences are among the oldest objects of mathematical investigation, having been studied for over 3500 years. After the initial steps, however, little progress was made until about 1600.

Arithmetic and geometric sequences appear in the Rhind papyrus, a mathematical text containing 85 problems copied around 1650 BC by the Egyptian scribe Ahmes from an earlier work (see Historical Problem 1). Fibonacci (AD 1220) wrote about problems similar to those found in the Rhind papyrus, leading one to suspect that Fibonacci may have had material available that is now lost. This material would have been in the non-Euclidean Greek tradition of Heron (about AD 75) and Diophantus (about AD 250). One problem, again modified slightly, is still with us in the familiar puzzle rhyme "As I was going to St. Ives . . . " (see Historical Problem 2).

The Rhind papyrus indicates that the Egyptians knew how to add up the terms of an arithmetic or geometric sequence, as did the Babylonians. The rule for summing up a geometric sequence is found in Euclid's *Elements* (Book IX, 35, 36), where, like all Euclid's algebra, it is presented in a geometric form.

Investigations of other kinds of sequences began in the 1500s, when algebra became sufficiently developed to handle the more complicated problems. The development of calculus in the 1600s added a powerful new tool, especially for finding the sum of an infinite series, and the subject continues to flourish today.

Historical Problems

1. *Arithmetic sequence problem from the Rhind papyrus (statement modified slightly for clarity)* One hundred loaves of bread are to be divided among five people so that the amounts that they receive form an arithmetic sequence. The first two together receive one-seventh of what the last three receive. How many loaves does each receive?

   [*Partial answer:* First person receives $1\frac{2}{3}$ loaves.]

2. The following old English children's rhyme resembles one of the Rhind papyrus problems.

   As I was going to St. Ives
   I met a man with seven wives
   Each wife had seven sacks
   Each sack had seven cats
   Each cat had seven kits [kittens]
   Kits, cats, sacks, wives
   How many were going to St. Ives?

   (a) Assuming that the speaker and the cat fanciers met by traveling in opposite directions, what is the answer?

   (b) How many kittens are being transported?

   (c) Kits, cats, sacks, wives; how many?

13.3 Assess Your Understanding

Concepts and Vocabulary

1. The formula for the nth term of a geometric sequence is ______________.

2. If the eighth term of a geometric sequence is $-\frac{8}{9}$ and the common ratio is $\frac{3}{4}$, then the ninth term of the sequence is _____.

3. In a geometric sequence, the ________ of successive terms is a constant.
   (a) difference (b) product (c) ratio (d) sum

4. If $|r| < 1$, the sum of the geometric series $\sum_{k=1}^{\infty} ar^{k-1}$ is ________.

5. If a series does not converge, it is called________.
   (a) arithmetic (b) divergent (c) geometric (d) insurgent

6. ***True or False*** A geometric sequence may be defined recursively.

7. ***True or False*** In a geometric sequence, the common ratio is always a positive number.

8. ***True or False*** For a geometric sequence with first term a_1 and common ratio r, where $r \neq 0, r \neq 1$, the sum of the first n terms is $S_n = a_1 \cdot \frac{1 - r^n}{1 - r}$.

Skill Building

In Problems 9–18, show that each sequence is geometric. Then find the common ratio and write out the first four terms.

9. $\{s_n\} = \{3^n\}$

10. $\{s_n\} = \{(-5)^n\}$

11. $\{a_n\} = \left\{-3\left(\frac{1}{2}\right)^n\right\}$

12. $\{b_n\} = \left\{\left(\frac{5}{2}\right)^n\right\}$

13. $\{c_n\} = \left\{\frac{2^{n-1}}{4}\right\}$

14. $\{d_n\} = \left\{\frac{3^n}{9}\right\}$

15. $\{e_n\} = \{2^{n/3}\}$

16. $\{f_n\} = \{3^{2n}\}$

17. $\{t_n\} = \left\{\frac{3^{n-1}}{2^n}\right\}$

18. $\{u_n\} = \left\{\frac{2^n}{3^{n-1}}\right\}$

In Problems 19–26, find the fifth term and the nth term of the geometric sequence whose initial term a_1 and common ratio r are given.

19. $a_1 = 2;\quad r = 3$

20. $a_1 = -2;\quad r = 4$

21. $a_1 = 5;\quad r = -1$

22. $a_1 = 6;\quad r = -2$

23. $a_1 = 0;\quad r = \frac{1}{2}$

24. $a_1 = 1;\quad r = -\frac{1}{3}$

25. $a_1 = \sqrt{2};\quad r = \sqrt{2}$

26. $a_1 = 0;\quad r = \frac{1}{\pi}$

In Problems 27–32, find the indicated term of each geometric sequence.

27. 7th term of $1, \frac{1}{2}, \frac{1}{4}, \ldots$

28. 8th term of $1, 3, 9, \ldots$

29. 9th term of $1, -1, 1, \ldots$

30. 10th term of $-1, 2, -4, \ldots$

31. 8th term of $0.4, 0.04, 0.004, \ldots$

32. 7th term of $0.1, 1.0, 10.0, \ldots$

In Problems 33–40, find the nth term a_n of each geometric sequence. When given, r is the common ratio.

33. $7, 14, 28, 56, \ldots$

34. $5, 10, 20, 40, \ldots$

35. $-3, 1, -\frac{1}{3}, \frac{1}{9}, \ldots$

36. $4, 1, \frac{1}{4}, \frac{1}{16}, \ldots$

37. $a_6 = 243;\quad r = -3$

38. $a_2 = 7;\quad r = \frac{1}{3}$

39. $a_2 = 7;\quad a_4 = 1575$

40. $a_3 = \frac{1}{3};\quad a_6 = \frac{1}{81}$

In Problems 41–46, find each sum.

41. $\frac{1}{4} + \frac{2}{4} + \frac{2^2}{4} + \frac{2^3}{4} + \cdots + \frac{2^{n-1}}{4}$

42. $\frac{3}{9} + \frac{3^2}{9} + \frac{3^3}{9} + \cdots + \frac{3^n}{9}$

43. $\sum_{k=1}^{n} \left(\frac{2}{3}\right)^k$

44. $\sum_{k=1}^{n} 4 \cdot 3^{k-1}$

45. $-1 - 2 - 4 - 8 - \cdots - (2^{n-1})$

46. $2 + \frac{6}{5} + \frac{18}{25} + \cdots + 2\left(\frac{3}{5}\right)^{n-1}$

For Problems 47–52, use a graphing utility to find the sum of each geometric sequence.

47. $\frac{1}{4} + \frac{2}{4} + \frac{2^2}{4} + \frac{2^3}{4} + \cdots + \frac{2^{14}}{4}$

48. $\frac{3}{9} + \frac{3^2}{9} + \frac{3^3}{9} + \cdots + \frac{3^{15}}{9}$

49. $\sum_{n=1}^{15} \left(\frac{2}{3}\right)^n$

50. $\sum_{n=1}^{15} 4 \cdot 3^{n-1}$

51. $-1 - 2 - 4 - 8 - \cdots - 2^{14}$

52. $2 + \frac{6}{5} + \frac{18}{25} + \cdots + 2\left(\frac{3}{5}\right)^{15}$

In Problems 53–68, determine whether each infinite geometric series converges or diverges. If it converges, find its sum.

53. $1 + \frac{1}{3} + \frac{1}{9} + \cdots$

54. $2 + \frac{4}{3} + \frac{8}{9} + \cdots$

55. $8 + 4 + 2 + \cdots$

56. $6 + 2 + \frac{2}{3} + \cdots$

57. $2 - \frac{1}{2} + \frac{1}{8} - \frac{1}{32} + \cdots$

58. $1 - \frac{3}{4} + \frac{9}{16} - \frac{27}{64} + \cdots$

59. $8 + 12 + 18 + 27 + \cdots$

60. $9 + 12 + 16 + \frac{64}{3} + \cdots$

61. $\sum_{k=1}^{\infty} 5\left(\frac{1}{4}\right)^{k-1}$

62. $\sum_{k=1}^{\infty} 8\left(\frac{1}{3}\right)^{k-1}$

63. $\sum_{k=1}^{\infty} \frac{1}{2} \cdot 3^{k-1}$

64. $\sum_{k=1}^{\infty} 3\left(\frac{3}{2}\right)^{k-1}$

65. $\sum_{k=1}^{\infty} 6\left(-\frac{2}{3}\right)^{k-1}$

66. $\sum_{k=1}^{\infty} 4\left(-\frac{1}{2}\right)^{k-1}$

67. $\sum_{k=1}^{\infty} 3\left(\frac{2}{3}\right)^k$

68. $\sum_{k=1}^{\infty} 2\left(\frac{3}{4}\right)^k$

Mixed Practice

In Problems 69–82, determine whether the given sequence is arithmetic, geometric, or neither. If the sequence is arithmetic, find the common difference; if it is geometric, find the common ratio. If the sequence is arithmetic or geometric, find the sum of the first 50 terms.

69. $\{n + 2\}$

70. $\{2n - 5\}$

71. $\{4n^2\}$

72. $\{5n^2 + 1\}$

73. $\left\{3 - \frac{2}{3}n\right\}$

74. $\left\{8 - \frac{3}{4}n\right\}$

75. $1, 3, 6, 10, \ldots$

76. $2, 4, 6, 8, \ldots$

77. $\left\{\left(\frac{2}{3}\right)^n\right\}$

78. $\left\{\left(\frac{5}{4}\right)^n\right\}$

79. $-1, 2, -4, 8, \ldots$

80. $1, 1, 2, 3, 5, 8, \ldots$

81. $\{3^{n/2}\}$

82. $\{(-1)^n\}$

Applications and Extensions

83. Find x so that $x, x + 2$, and $x + 3$ are consecutive terms of a geometric sequence.

84. Find x so that $x - 1, x$, and $x + 2$ are consecutive terms of a geometric sequence.

85. Salary Increases If you have been hired at an annual salary of \$42,000 and expect to receive annual increases of 3%, what will your salary be when you begin your fifth year?

86. Equipment Depreciation A new piece of equipment cost a company \$15,000. Each year, for tax purposes, the company depreciates the value by 15%. What value should the company give the equipment after 5 years?

87. Pendulum Swings Initially, a pendulum swings through an arc of 2 feet. On each successive swing, the length of the arc is 0.9 of the previous length.

(a) What is the length of the arc of the 10th swing?

(b) On which swing is the length of the arc first less than 1 foot?

(c) After 15 swings, what total length will the pendulum have swung?

(d) When it stops, what total length will the pendulum have swung?

88. Bouncing Balls A ball is dropped from a height of 30 feet. Each time it strikes the ground, it bounces up to 0.8 of the previous height.

(a) What height will the ball bounce up to after it strikes the ground for the third time?

(b) How high will it bounce after it strikes the ground for the nth time?

(c) How many times does the ball need to strike the ground before its bounce is less than 6 inches?

(d) What total vertical distance does the ball travel before it stops bouncing?

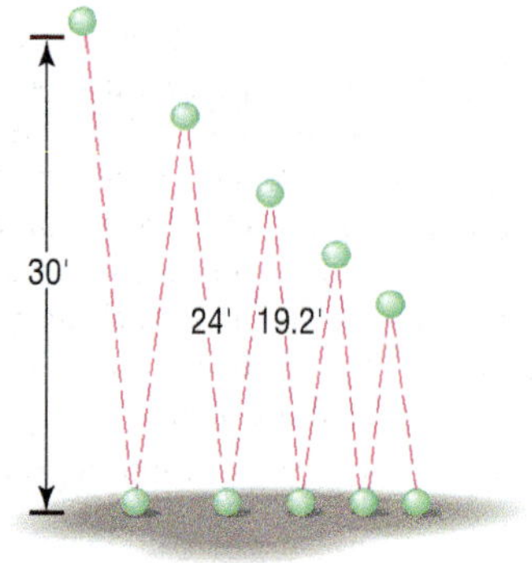

Amount of an Annuity *Suppose that P is the deposit in dollars made at the end of each payment period for an annuity (see Section 13.1) paying* $i = \frac{r}{N}$ *percent interest per payment period. The amount A of the annuity after n deposits is given by*

$$A = P\frac{(1 + i)^n - 1}{i}$$

Use this result to answer Problems 89–94.

89. Retirement Christine contributes \$100 each month to her 401(k). What will be the value of Christine's 401(k) after the 360th deposit (30 years) if the per annum rate of return is assumed to be 12% compounded monthly?

90. Saving for a Home Jolene wants to purchase a new home. Suppose that she invests \$400 per month into a mutual fund. If the per annum rate of return of the mutual fund is assumed to be 10% compounded monthly, how much will Jolene have for a down payment after the 36th deposit (3 years)?

91. Tax-Sheltered Annuity Don contributes \$500 at the end of each quarter to a tax-sheltered annuity (TSA). What will the value of the TSA be after the 80th deposit (20 years) if the per annum rate of return is assumed to be 8% compounded quarterly?

92. Retirement Ray contributes \$1000 to an individual retirement account (IRA) semiannually. What will the value of the IRA be when Ray makes his 30th deposit (after 15 years) if the per annum rate of return is assumed to be 10% compounded semiannually?

93. Sinking Fund Scott and Alice want to purchase a vacation home in 10 years and need \$50,000 for a down payment. How much should they place in a savings account each month if the per annum rate of return is assumed to be 6% compounded monthly?

94. Sinking Fund For a child born in 2015, the cost of a 4-year college education at a public university is projected to be \$200,000. Assuming an 8% per annum rate of return compounded monthly, how much must be contributed to a college fund every month to have \$200,000 in 18 years when the child begins college?

95. Grains of Wheat on a Chess Board In an old fable, a commoner who had saved the king's life was told he could ask the king for any just reward. Being a shrewd man, the commoner said, "A simple wish, sire. Place one grain of wheat on the first square of a chessboard, two grains on the second square, four grains on the third square, continuing until you have filled the board. This is all I seek." Compute the total number of grains needed to do this to see why the request, seemingly simple, could not be granted. (A chessboard consists of $8 \times 8 = 64$ squares.)

96. Look at the figure. What fraction of the square is eventually shaded if the indicated shading process continues indefinitely?

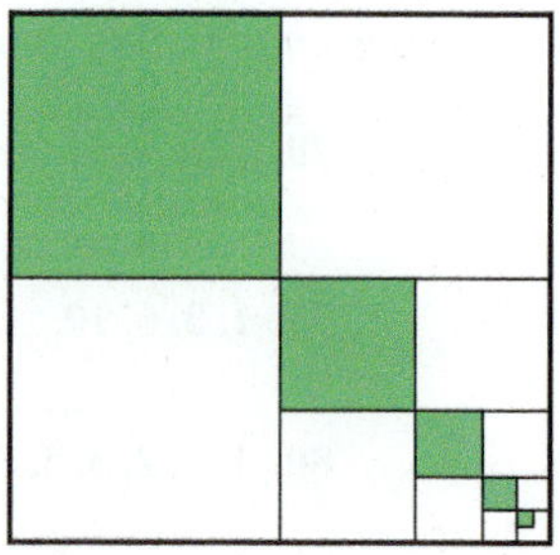

97. **Multiplier** Suppose that, throughout the U.S. economy, individuals spend 90% of every additional dollar that they earn. Economists would say that an individual's **marginal propensity to consume** is 0.90. For example, if Jane earns an additional dollar, she will spend $0.9(1) = \$0.90$ of it. The individual who earns \$0.90 (from Jane) will spend 90% of it, or \$0.81. This process of spending continues and results in an infinite geometric series as follows:

$$1, 0.90, 0.90^2, 0.90^3, 0.90^4, \ldots$$

The sum of this infinite geometric series is called the **multiplier**. What is the multiplier if individuals spend 90% of every additional dollar that they earn?

98. **Multiplier** Refer to Problem 97. Suppose that the marginal propensity to consume throughout the U.S. economy is 0.95. What is the multiplier for the U.S. economy?

99. **Stock Price** One method of pricing a stock is to discount the stream of future dividends of the stock. Suppose that a stock pays $\$P$ per year in dividends, and historically, the dividend has been increased $i\%$ per year. If you desire an annual rate of return of $r\%$, this method of pricing a stock states that the price that you should pay is the present value of an infinite stream of payments:

$$\text{Price} = P + P\cdot\frac{1+i}{1+r} + P\cdot\left(\frac{1+i}{1+r}\right)^2 + P\cdot\left(\frac{1+i}{1+r}\right)^3 + \cdots$$

The price of the stock is the sum of an infinite geometric series. Suppose that a stock pays an annual dividend of \$4.00, and historically, the dividend has been increased 3% per year. You desire an annual rate of return of 9%. What is the most you should pay for the stock?

100. **Stock Price** Refer to Problem 99. Suppose that a stock pays an annual dividend of \$2.50, and historically, the dividend has increased 4% per year. You desire an annual rate of return of 11%. What is the most that you should pay for the stock?

101. **A Rich Man's Promise** A rich man promises to give you \$1000 on September 1, 2015. Each day thereafter he will give you $\frac{9}{10}$ of what he gave you the previous day. What is the first date on which the amount you receive is less than 1¢? How much have you received when this happens?

102. Show that the "Amount of an Annuity" formula that you used in Problems 89–94 results from summing a geometric sequence.

103. **Seating Revenue** A special section in the end zone of a football stadium has 2 seats in the first row and 14 rows total. Each successive row has 2 seats more than the row before. In this particular section, the first seat is sold for 1 cent, and each following seat sells for 5% more than the previous seat. Find the total revenue generated if every seat in the section is sold. Round only the final answer, and state the final answer in dollars rounded to two decimal places. (JJC)[†]

[†]Courtesy of the Joliet Junior College Mathematics Department.

Explaining Concepts: Discussion and Writing

104. **Critical Thinking** You are interviewing for a job and receive two offers for a five-year contract:

   A: \$40,000 to start, with guaranteed annual increases of 6% for the first 5 years

   B: \$44,000 to start, with guaranteed annual increases of 3% for the first 5 years

   Which offer is better if your goal is to be making as much as possible after 5 years? Which is better if your goal is to make as much money as possible over the contract (5 years)?

105. **Critical Thinking** Which of the following choices, *A* or *B*, results in more money?

   A: To receive \$1000 on day 1, \$999 on day 2, \$998 on day 3, with the process to end after 1000 days

   B: To receive \$1 on day 1, \$2 on day 2, \$4 on day 3, for 19 days

106. **Critical Thinking** You have just signed a 7-year professional football league contract with a beginning salary of \$2,000,000 per year. Management gives you the following options with regard to your salary over the 7 years.

   1. A bonus of \$100,000 each year
   2. An annual increase of 4.5% per year beginning after 1 year
   3. An annual increase of \$95,000 per year beginning after 1 year

   Which option provides the most money over the 7-year period? Which the least? Which would you choose? Why?

107. **Critical Thinking** Suppose you were offered a job in which you would work 8 hours per day for 5 workdays per week for 1 month at hard manual labor. Your pay the first day would be 1 penny. On the second day your pay would be two pennies; the third day 4 pennies. Your pay would double on each successive workday. There are 22 workdays in the month. There will be no sick days. If you miss a day of work, there is no pay or pay increase. How much do you get paid if you work all 22 days? How much do you get paid for the 22nd workday? What risks do you run if you take this job offer? Would you take the job?

108. Can a sequence be both arithmetic and geometric? Give reasons for your answer.

109. Make up a geometric sequence. Give it to a friend and ask for its 20th term.

110. Make up two infinite geometric series, one that has a sum and one that does not. Give them to a friend and ask for the sum of each series.

111. Describe the similarities and differences between geometric sequences and exponential functions.

Retain Your Knowledge

Problems 112–115 are based on material learned earlier in the course. The purpose of these problems is to keep the material fresh in your mind so that you are better prepared for the final exam.

112. Use the Change-of-Base Formula and a calculator to evaluate $\log_7 62$. Round the answer to three decimal places.

113. Find the unit vector in the same direction as $\mathbf{v} = 8\mathbf{i} - 15\mathbf{j}$.

114. Find the equation of the hyperbola with vertices at $(-2, 0)$ and $(2, 0)$ and a focus at $(4, 0)$.

115. Find the value of the determinant: $\begin{vmatrix} 3 & 1 & 0 \\ 0 & -2 & 6 \\ 4 & -1 & -2 \end{vmatrix}$.

13.4 Mathematical Induction

OBJECTIVE 1 Prove Statements Using Mathematical Induction (p. 992)

1 Prove Statements Using Mathematical Induction

Mathematical induction is a method for proving that statements involving natural numbers are true for all natural numbers.*

For example, the statement "$2n$ is always an even integer" can be proved for all natural numbers n by using mathematical induction. Also, the statement "the sum of the first n positive odd integers equals n^2," that is,

$$1 + 3 + 5 + \cdots + (2n - 1) = n^2 \qquad \textbf{(1)}$$

can be proved for all natural numbers n by using mathematical induction.

Before stating the method of mathematical induction, let's try to gain a sense of the power of the method. We shall use the statement in equation (1) for this purpose by restating it for various values of $n = 1, 2, 3, \ldots$.

$n = 1$ The sum of the first positive odd integer is 1^2; $1 = 1^2$.

$n = 2$ The sum of the first 2 positive odd integers is 2^2; $1 + 3 = 4 = 2^2$.

$n = 3$ The sum of the first 3 positive odd integers is 3^2; $1 + 3 + 5 = 9 = 3^2$.

$n = 4$ The sum of the first 4 positive odd integers is 4^2; $1 + 3 + 5 + 7 = 16 = 4^2$.

Although from this pattern we might conjecture that statement (1) is true for any choice of n, can we really be sure that it does not fail for some choice of n? The method of proof by mathematical induction will, in fact, prove that the statement is true for all n.

THEOREM

The Principle of Mathematical Induction

Suppose that the following two conditions are satisfied with regard to a statement about natural numbers:

CONDITION I: The statement is true for the natural number 1.

CONDITION II: If the statement is true for some natural number k, it is also true for the next natural number $k + 1$.

Then the statement is true for all natural numbers.

*Recall that the natural numbers are the numbers $1, 2, 3, 4, \ldots$. In other words, the terms *natural numbers* and *positive integers* are synonymous.

Figure 16

We shall not prove this principle. However, the following physical interpretation illustrates why the principle works. Think of a collection of natural numbers obeying a statement as a collection of infinitely many dominoes. See Figure 16.

Now, suppose that two facts are given:

1. The first domino is pushed over.
2. If one domino falls over, say the kth domino, so will the next one, the $(k + 1)$st domino.

Is it safe to conclude that *all* the dominoes fall over? The answer is yes, because if the first one falls (Condition I), the second one does also (by Condition II); and if the second one falls, so does the third (by Condition II); and so on.

EXAMPLE 1 Using Mathematical Induction

Show that the following statement is true for all natural numbers n.

$$1 + 3 + 5 + \cdots + (2n - 1) = n^2 \qquad \textbf{(2)}$$

Solution First show that statement (2) holds for $n = 1$. Because $1 = 1^2$, statement (2) is true for $n = 1$. Condition I holds.

Next, show that Condition II holds. From statement (2), assume that

$$1 + 3 + 5 + \cdots + (2k - 1) = k^2 \qquad \textbf{(3)}$$

is true for some natural number k.

Now show that, based on equation (3), statement (2) holds for $k + 1$. Look at the sum of the first $k + 1$ positive odd integers to determine whether this sum equals $(k + 1)^2$.

$$\begin{aligned} 1 + 3 + 5 + \cdots + [2(k + 1) - 1] &= \underbrace{[1 + 3 + 5 + \cdots + (2k - 1)]}_{= k^2 \text{ by equation (3)}} + (2k + 1) \\ &= k^2 + (2k + 1) \\ &= k^2 + 2k + 1 = (k + 1)^2 \end{aligned}$$

Conditions I and II are satisfied; by the Principle of Mathematical Induction, statement (2) is true for all natural numbers n. ■

EXAMPLE 2 Using Mathematical Induction

Show that the following statement is true for all natural numbers n.

$$2^n > n$$

Solution First, show that the statement $2^n > n$ holds when $n = 1$. Because $2^1 = 2 > 1$, the inequality is true for $n = 1$. Condition I holds.

Next, assume, for some natural number k, that $2^k > k$. Now show that the formula holds for $k + 1$; that is, show that $2^{k+1} > k + 1$.

$$2^{k+1} = 2 \cdot 2^k \underset{\substack{\uparrow \\ \text{We know that} \\ 2^k > k.}}{>} 2 \cdot k = k + k \underset{\substack{\uparrow \\ k \geq 1}}{\geq} k + 1$$

If $2^k > k$, then $2^{k+1} > k + 1$, so Condition II of the Principle of Mathematical Induction is satisfied. The statement $2^n > n$ is true for all natural numbers n. ■

EXAMPLE 3 Using Mathematical Induction

Show that the following formula is true for all natural numbers n.

$$1 + 2 + 3 + \cdots + n = \frac{n(n+1)}{2} \tag{4}$$

Solution First, show that formula (4) is true when $n = 1$. Because

$$\frac{1(1+1)}{2} = \frac{1(2)}{2} = 1$$

Condition I of the Principle of Mathematical Induction holds.

Next, assume that formula (4) holds for some k, and determine whether the formula then holds for $k + 1$. Assume that

$$1 + 2 + 3 + \cdots + k = \frac{k(k+1)}{2} \quad \text{for some } k \tag{5}$$

Now show that

$$1 + 2 + 3 + \cdots + (k+1) = \frac{(k+1)[(k+1)+1]}{2} = \frac{(k+1)(k+2)}{2}$$

as follows:

$$\begin{aligned}1 + 2 + 3 + \cdots + (k+1) &= \underbrace{[1 + 2 + 3 + \cdots + k]}_{} + (k+1)\\ &= \frac{k(k+1)}{2} \quad \text{by equation (5)}\\ &= \frac{k(k+1)}{2} + (k+1)\\ &= \frac{k^2 + k + 2k + 2}{2}\\ &= \frac{k^2 + 3k + 2}{2} = \frac{(k+1)(k+2)}{2}\end{aligned}$$

Condition II also holds. As a result, formula (4) is true for all natural numbers n. ■

Now Work PROBLEM 1

EXAMPLE 4 Using Mathematical Induction

Show that $3^n - 1$ is divisible by 2 for all natural numbers n.

Solution First, show that the statement is true when $n = 1$. Because $3^1 - 1 = 3 - 1 = 2$ is divisible by 2, the statement is true when $n = 1$. Condition I is satisfied.

Next, assume that the statement holds for some k, and determine whether the statement holds for $k + 1$. Assume that $3^k - 1$ is divisible by 2 for some k. Now show that $3^{k+1} - 1$ is divisible by 2.

$$\begin{aligned}3^{k+1} - 1 &= 3^{k+1} - 3^k + 3^k - 1 \quad \text{Subtract and add } 3^k.\\ &= 3^k(3-1) + (3^k - 1) = 3^k \cdot 2 + (3^k - 1)\end{aligned}$$

Because $3^k \cdot 2$ is divisible by 2 and $3^k - 1$ is divisible by 2, it follows that $3^k \cdot 2 + (3^k - 1) = 3^{k+1} - 1$ is divisible by 2. Condition II is also satisfied. As a result, the statement "$3^n - 1$ is divisible by 2" is true for all natural numbers n. ■

Now Work PROBLEM 19

WARNING The conclusion that a statement involving natural numbers is true for all natural numbers is made only after *both* Conditions I and II of the Principle of Mathematical Induction have been satisfied. Problem 28 demonstrates a statement for which only Condition I holds, and the statement is *not* true for all natural numbers. Problem 29 demonstrates a statement for which only Condition II holds, and the statement is *not* true for any natural number. ■

13.4 Assess Your Understanding

Skill Building

In Problems 1–22, use the Principle of Mathematical Induction to show that the given statement is true for all natural numbers n.

1. $2 + 4 + 6 + \cdots + 2n = n(n + 1)$
2. $1 + 5 + 9 + \cdots + (4n - 3) = n(2n - 1)$
3. $3 + 4 + 5 + \cdots + (n + 2) = \frac{1}{2}n(n + 5)$
4. $3 + 5 + 7 + \cdots + (2n + 1) = n(n + 2)$
5. $2 + 5 + 8 + \cdots + (3n - 1) = \frac{1}{2}n(3n + 1)$
6. $1 + 4 + 7 + \cdots + (3n - 2) = \frac{1}{2}n(3n - 1)$
7. $1 + 2 + 2^2 + \cdots + 2^{n-1} = 2^n - 1$
8. $1 + 3 + 3^2 + \cdots + 3^{n-1} = \frac{1}{2}(3^n - 1)$
9. $1 + 4 + 4^2 + \cdots + 4^{n-1} = \frac{1}{3}(4^n - 1)$
10. $1 + 5 + 5^2 + \cdots + 5^{n-1} = \frac{1}{4}(5^n - 1)$
11. $\frac{1}{1 \cdot 2} + \frac{1}{2 \cdot 3} + \frac{1}{3 \cdot 4} + \cdots + \frac{1}{n(n + 1)} = \frac{n}{n + 1}$
12. $\frac{1}{1 \cdot 3} + \frac{1}{3 \cdot 5} + \frac{1}{5 \cdot 7} + \cdots + \frac{1}{(2n - 1)(2n + 1)} = \frac{n}{2n + 1}$
13. $1^2 + 2^2 + 3^2 + \cdots + n^2 = \frac{1}{6}n(n + 1)(2n + 1)$
14. $1^3 + 2^3 + 3^3 + \cdots + n^3 = \frac{1}{4}n^2(n + 1)^2$
15. $4 + 3 + 2 + \cdots + (5 - n) = \frac{1}{2}n(9 - n)$
16. $-2 - 3 - 4 - \cdots - (n + 1) = -\frac{1}{2}n(n + 3)$
17. $1 \cdot 2 + 2 \cdot 3 + 3 \cdot 4 + \cdots + n(n + 1) = \frac{1}{3}n(n + 1)(n + 2)$
18. $1 \cdot 2 + 3 \cdot 4 + 5 \cdot 6 + \cdots + (2n - 1)(2n) = \frac{1}{3}n(n + 1)(4n - 1)$
19. $n^2 + n$ is divisible by 2.
20. $n^3 + 2n$ is divisible by 3.
21. $n^2 - n + 2$ is divisible by 2.
22. $n(n + 1)(n + 2)$ is divisible by 6.

Applications and Extensions

In Problems 23–27, prove each statement.

23. If $x > 1$, then $x^n > 1$.
24. If $0 < x < 1$, then $0 < x^n < 1$.
25. $a - b$ is a factor of $a^n - b^n$.
    [Hint: $a^{k+1} - b^{k+1} = a(a^k - b^k) + b^k(a - b)$]
26. $a + b$ is a factor of $a^{2n+1} + b^{2n+1}$.
27. $(1 + a)^n \geq 1 + na$, for $a > 0$
28. Show that the statement "$n^2 - n + 41$ is a prime number" is true for $n = 1$ but is not true for $n = 41$.
29. Show that the formula
    $$2 + 4 + 6 + \cdots + 2n = n^2 + n + 2$$
    obeys Condition II of the Principle of Mathematical Induction. That is, show that if the formula is true for some k, it is also true for $k + 1$. Then show that the formula is false for $n = 1$ (or for any other choice of n).
30. Use mathematical induction to prove that if $r \neq 1$, then
    $$a + ar + ar^2 + \cdots + ar^{n-1} = a\frac{1 - r^n}{1 - r}$$
31. Use mathematical induction to prove that
    $$a + (a + d) + (a + 2d) + \cdots + [a + (n - 1)d] = na + d\frac{n(n - 1)}{2}$$
32. **Extended Principle of Mathematical Induction** The Extended Principle of Mathematical Induction states that if Conditions I and II hold, that is,
    (I) A statement is true for a natural number j.
    (II) If the statement is true for some natural number $k \geq j$, then it is also true for the next natural number $k + 1$.
    then the statement is true for all natural numbers $\geq j$. Use the Extended Principle of Mathematical Induction to show that the number of diagonals in a convex polygon of n sides is $\frac{1}{2}n(n - 3)$.
    [**Hint:** Begin by showing that the result is true when $n = 4$ (Condition I).]
33. **Geometry** Use the Extended Principle of Mathematical Induction to show that the sum of the interior angles of a convex polygon of n sides equals $(n - 2) \cdot 180°$.

Explaining Concepts: Discussion and Writing

34. How would you explain the Principle of Mathematical Induction to a friend?

Retain Your Knowledge

Problems 35–38 are based on material learned earlier in the course. The purpose of these problems is to keep the material fresh in your mind so that you are better prepared for the final exam.

35. Solve: $\log_2 \sqrt{x+5} = 4$

36. A mass of 500 kg is suspended from two cables, as shown in the figure. What are the tensions in the two cables?

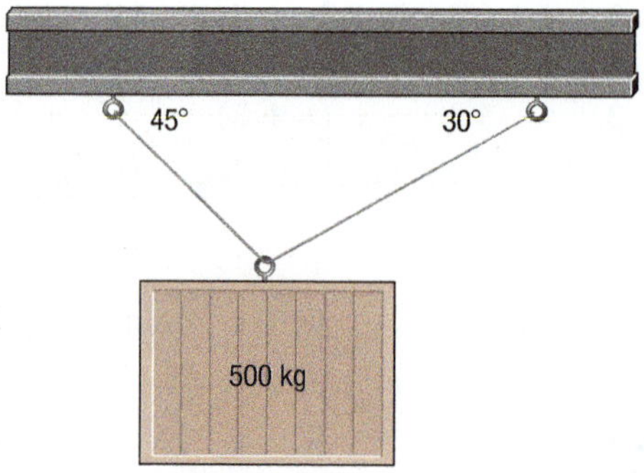

37. Solve the system: $\begin{cases} 4x + 3y = -7 \\ 2x - 5y = 16 \end{cases}$

38. For $A = \begin{bmatrix} 1 & 2 & -1 \\ 0 & 1 & 4 \end{bmatrix}$ and $B = \begin{bmatrix} 3 & -1 \\ 1 & 0 \\ -2 & 2 \end{bmatrix}$, find $A \cdot B$.

13.5 The Binomial Theorem

OBJECTIVES **1** Evaluate $\binom{n}{j}$ (p. 996)

2 Use the Binomial Theorem (p. 998)

Formulas have been given for expanding $(x+a)^n$ for $n = 2$ and $n = 3$. The *Binomial Theorem** is a formula for the expansion of $(x+a)^n$ for any positive integer n. If $n = 1, 2, 3$, and 4, the expansion of $(x+a)^n$ is straightforward.

$(x+a)^1 = x + a$ — **Two terms, beginning with x^1 and ending with a^1**

$(x+a)^2 = x^2 + 2ax + a^2$ — **Three terms, beginning with x^2 and ending with a^2**

$(x+a)^3 = x^3 + 3ax^2 + 3a^2x + a^3$ — **Four terms, beginning with x^3 and ending with a^3**

$(x+a)^4 = x^4 + 4ax^3 + 6a^2x^2 + 4a^3x + a^4$ — **Five terms, begining with x^4 and ending with a^4**

Notice that each expansion of $(x+a)^n$ begins with x^n and ends with a^n. From left to right, the powers of x are decreasing by 1, while the powers of a are increasing by 1. Also, the number of terms equals $n + 1$. Notice, too, that the degree of each monomial in the expansion equals n. For example, in the expansion of $(x+a)^3$, each monomial $(x^3, 3ax^2, 3a^2x, a^3)$ is of degree 3. As a result, it is reasonable to conjecture that the expansion of $(x+a)^n$ would look like this:

$$(x+a)^n = x^n + \_\_\_\_ ax^{n-1} + \_\_\_\_ a^2x^{n-2} + \cdots + \_\_\_\_ a^{n-1}x + a^n$$

where the blanks are numbers to be found. This is in fact the case, as will be seen shortly.

COMMENT On a graphing calculator, the symbol $\binom{n}{j}$ may be denoted by the key [nCr]. ■

Before we can fill in the blanks, we need to introduce the symbol $\binom{n}{j}$.

1 Evaluate $\binom{n}{j}$

The symbol $\binom{n}{j}$, read "n taken j at a time," is defined next.

*The name *binomial* is derived from the fact that $x + a$ is a binomial; that is, it contains two terms.

DEFINITION

If j and n are integers with $0 \le j \le n$, the symbol $\binom{n}{j}$ is defined as

$$\binom{n}{j} = \frac{n!}{j!(n-j)!} \qquad \textbf{(1)}$$

EXAMPLE 1

Evaluating $\binom{n}{j}$

Find:

(a) $\binom{3}{1}$ (b) $\binom{4}{2}$ (c) $\binom{8}{7}$ (d) $\binom{65}{15}$

Solution

(a) $\binom{3}{1} = \frac{3!}{1!(3-1)!} = \frac{3!}{1!\,2!} = \frac{3\cdot 2\cdot 1}{1(2\cdot 1)} = \frac{6}{2} = 3$

(b) $\binom{4}{2} = \frac{4!}{2!(4-2)!} = \frac{4!}{2!\,2!} = \frac{4\cdot 3\cdot 2\cdot 1}{(2\cdot 1)(2\cdot 1)} = \frac{24}{4} = 6$

(c) $\binom{8}{7} = \frac{8!}{7!(8-7)!} = \frac{8!}{7!\,1!} = \frac{8\cdot \cancel{7!}}{\cancel{7!}\cdot 1!} = \frac{8}{1} = 8$

$8! = 8\cdot 7!$

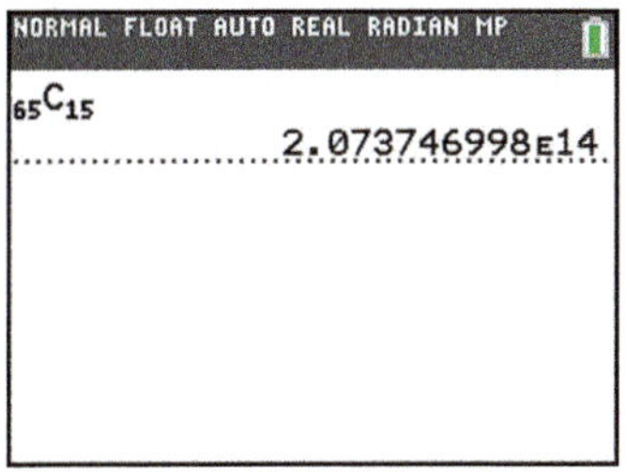

Figure 17

(d) Figure 17 shows the solution using a TI-84 Plus C graphing calculator. So

$$\binom{65}{15} \approx 2.073746998 \times 10^{14}$$

Now Work PROBLEM 5

Four useful formulas involving the symbol $\binom{n}{j}$ are

$$\binom{n}{0} = 1 \qquad \binom{n}{1} = n \qquad \binom{n}{n-1} = n \qquad \binom{n}{n} = 1$$

Proof $\binom{n}{0} = \frac{n!}{0!(n-0)!} = \frac{\cancel{n!}}{0!\,\cancel{n!}} = \frac{1}{1} = 1$

$$\binom{n}{1} = \frac{n!}{1!(n-1)!} = \frac{n!}{(n-1)!} = \frac{n\cancel{(n-1)!}}{\cancel{(n-1)!}} = n$$

You are asked to prove the remaining two formulas in Problem 45.

Suppose that the values of the symbol $\binom{n}{j}$ are arranged in a triangular display, as shown next and in Figure 18.

$$\binom{0}{0}$$

$$\binom{1}{0} \quad \binom{1}{1}$$

$$\binom{2}{0} \quad \binom{2}{1} \quad \binom{2}{2}$$

$$\binom{3}{0} \quad \binom{3}{1} \quad \binom{3}{2} \quad \binom{3}{3}$$

$$\binom{4}{0} \quad \binom{4}{1} \quad \binom{4}{2} \quad \binom{4}{3} \quad \binom{4}{4}$$

$$\binom{5}{0} \quad \binom{5}{1} \quad \binom{5}{2} \quad \binom{5}{3} \quad \binom{5}{4} \quad \binom{5}{5}$$

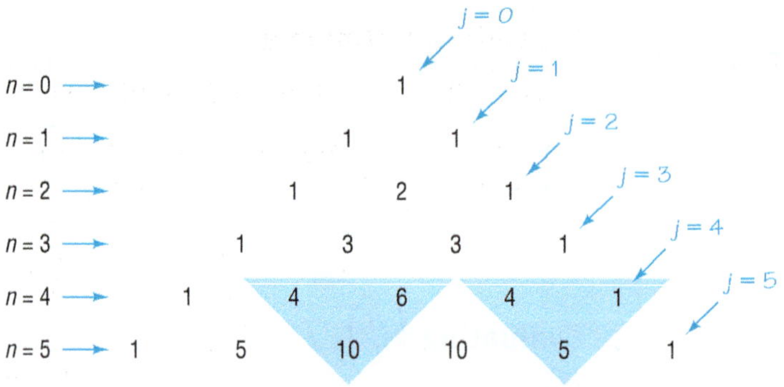

Figure 18 The Pascal Triangle

This display is called the **Pascal triangle**, named after Blaise Pascal (1623–1662), a French mathematician.

The Pascal triangle has 1's down the sides. To get any other entry, add the two nearest entries in the row above it. The shaded triangles in Figure 18 illustrate this feature of the Pascal triangle. Based on this feature, the row corresponding to $n = 6$ is found as follows:

$n = 5 \rightarrow$ 1 5 10 10 5 1

$n = 6 \rightarrow$ 1 6 15 20 15 6 1

This addition always works (see the theorem on page 1000).

Although the Pascal triangle provides an interesting and organized display of the symbol $\binom{n}{j}$, in practice it is not all that helpful. For example, if you wanted to know the value of $\binom{12}{5}$, you would need to produce 13 rows of the triangle before seeing the answer. It is much faster to use definition (1).

2 Use the Binomial Theorem

THEOREM

Binomial Theorem

Let x and a be real numbers. For any positive integer n, we have

$$(x+a)^n = \binom{n}{0}x^n + \binom{n}{1}ax^{n-1} + \cdots + \binom{n}{j}a^jx^{n-j} + \cdots + \binom{n}{n}a^n$$

$$= \sum_{j=0}^{n}\binom{n}{j}x^{n-j}a^j \qquad \textbf{(2)}$$

Now you know why it was necessary to introduce the symbol $\binom{n}{j}$; these symbols are the numerical coefficients that appear in the expansion of $(x + a)^n$. Because of this, the symbol $\binom{n}{j}$ is called a **binomial coefficient**.

EXAMPLE 2 **Expanding a Binomial**

Use the Binomial Theorem to expand $(x + 2)^5$.

Solution In the Binomial Theorem, let $a = 2$ and $n = 5$. Then

$$(x+2)^5 = \binom{5}{0}x^5 + \binom{5}{1}2x^4 + \binom{5}{2}2^2x^3 + \binom{5}{3}2^3x^2 + \binom{5}{4}2^4x + \binom{5}{5}2^5$$

↑ Use equation (2).

$$= 1\cdot x^5 + 5\cdot 2x^4 + 10\cdot 4x^3 + 10\cdot 8x^2 + 5\cdot 16x + 1\cdot 32$$

↑ Use row $n = 5$ of the Pascal triangle or definition (1) for $\binom{n}{j}$.

$$= x^5 + 10x^4 + 40x^3 + 80x^2 + 80x + 32$$

EXAMPLE 3

Expanding a Binomial

Expand $(2y - 3)^4$ using the Binomial Theorem.

Solution First, rewrite the expression $(2y - 3)^4$ as $[2y + (-3)]^4$. Now use the Binomial Theorem with $n = 4$, $x = 2y$, and $a = -3$.

$$[2y + (-3)]^4 = \binom{4}{0}(2y)^4 + \binom{4}{1}(-3)(2y)^3 + \binom{4}{2}(-3)^2(2y)^2 + \binom{4}{3}(-3)^3(2y) + \binom{4}{4}(-3)^4$$

$$= 1 \cdot 16y^4 + 4(-3)8y^3 + 6 \cdot 9 \cdot 4y^2 + 4(-27)2y + 1 \cdot 81$$

↑ Use row $n = 4$ of the Pascal triangle or definition (1) for $\binom{n}{j}$.

$$= 16y^4 - 96y^3 + 216y^2 - 216y + 81$$

In this expansion, note that the signs alternate because $a = -3 < 0$. ■

Now Work PROBLEM 21

EXAMPLE 4

Finding a Particular Coefficient in a Binomial Expansion

Find the coefficient of y^8 in the expansion of $(2y + 3)^{10}$.

Solution Write out the expansion using the Binomial Theorem.

$$(2y + 3)^{10} = \binom{10}{0}(2y)^{10} + \binom{10}{1}(2y)^9(3)^1 + \binom{10}{2}(2y)^8(3)^2 + \binom{10}{3}(2y)^7(3)^3 + \binom{10}{4}(2y)^6(3)^4 + \cdots + \binom{10}{9}(2y)(3)^9 + \binom{10}{10}(3)^{10}$$

From the third term in the expansion, the coefficient of y^8 is

$$\binom{10}{2}(2)^8(3)^2 = \frac{10!}{2!\,8!} \cdot 2^8 \cdot 9 = \frac{10 \cdot 9 \cdot \cancel{8!}}{2 \cdot \cancel{8!}} \cdot 2^8 \cdot 9 = 103{,}680$$ ■

As this solution demonstrates, the Binomial Theorem can be used to find a particular term in an expansion without writing the entire expansion.

Based on the expansion of $(x + a)^n$, the term containing x^j is

$$\binom{n}{n - j} a^{n-j} x^j \quad \textbf{(3)}$$

Example 4 can be solved by using formula (3) with $n = 10$, $a = 3$, $x = 2y$, and $j = 8$. Then the term containing y^8 is

$$\binom{10}{10 - 8} 3^{10-8}(2y)^8 = \binom{10}{2} \cdot 3^2 \cdot 2^8 \cdot y^8 = \frac{10!}{2!\,8!} \cdot 9 \cdot 2^8 y^8$$

$$= \frac{10 \cdot 9 \cdot \cancel{8!}}{2 \cdot \cancel{8!}} \cdot 9 \cdot 2^8\, y^8 = 103{,}680y^8$$

EXAMPLE 5 Finding a Particular Term in a Binomial Expansion

Find the 6th term in the expansion of $(x + 2)^9$.

Solution A Expand using the Binomial Theorem until the 6th term is reached.

$$(x+2)^9 = \binom{9}{0}x^9 + \binom{9}{1}x^8 \cdot 2 + \binom{9}{2}x^7 \cdot 2^2 + \binom{9}{3}x^6 \cdot 2^3 + \binom{9}{4}x^5 \cdot 2^4 + \binom{9}{5}x^4 \cdot 2^5 + \cdots$$

The 6th term is

$$\binom{9}{5}x^4 \cdot 2^5 = \frac{9!}{5!\,4!} \cdot x^4 \cdot 32 = 4032x^4$$

Solution B The 6th term in the expansion of $(x + 2)^9$, which has 10 terms total, contains x^4. (Do you see why?) By formula (3), the 6th term is

$$\binom{9}{9-4}2^{9-4}\,x^4 = \binom{9}{5}2^5\,x^4 = \frac{9!}{5!\,4!} \cdot 32x^4 = 4032x^4$$

■

Now Work PROBLEMS 29 AND 35

The following theorem shows that the *triangular addition* feature of the Pascal triangle illustrated in Figure 18 always works.

THEOREM

If n and j are integers with $1 \le j \le n$, then

$$\binom{n}{j-1} + \binom{n}{j} = \binom{n+1}{j} \tag{4}$$

■

Proof

$$\binom{n}{j-1} + \binom{n}{j} = \frac{n!}{(j-1)!\,[n-(j-1)]!} + \frac{n!}{j!\,(n-j)!}$$

$$= \frac{n!}{(j-1)!\,(n-j+1)!} + \frac{n!}{j!\,(n-j)!}$$

$$= \frac{jn!}{j(j-1)!\,(n-j+1)!} + \frac{(n-j+1)n!}{j!\,(n-j+1)\,(n-j)!}$$

Multiply the first term by $\frac{j}{j}$ and the second term by $\frac{n-j+1}{n-j+1}$ to make the denominators equal.

$$= \frac{jn!}{j!\,(n-j+1)!} + \frac{(n-j+1)n!}{j!\,(n-j+1)!}$$

$$= \frac{jn! + (n-j+1)n!}{j!\,(n-j+1)!}$$

$$= \frac{n!\,(j+n-j+1)}{j!\,(n-j+1)!}$$

$$= \frac{n!\,(n+1)}{j!\,(n-j+1)!} = \frac{(n+1)!}{j!\,[(n+1)-j]!} = \binom{n+1}{j}$$

■

Historical Feature

Omar Khayyám (1048–1131)

The case $n = 2$ of the Binomial Theorem, $(a + b)^2$, was known to Euclid in 300 BC, but the general law seems to have been discovered by the Persian mathematician and astronomer Omar Khayyám (1048–1131), who is also well known as the author of the *Rubáiyát*, a collection of four-line poems making observations on the human condition. Omar Khayyám did not state the Binomial Theorem explicitly, but he claimed to have a method for extracting third, fourth, and fifth roots, and so on. A little study shows that one must know the Binomial Theorem to create such a method.

The heart of the Binomial Theorem is the formula for the numerical coefficients, and, as we saw, they can be written in a symmetric triangular form. The Pascal triangle appears first in the books of Yang Hui (about 1270) and Chu Shih-chieh (1303). Pascal's name is attached to the triangle because of the many applications he made of it, especially to counting and probability. In establishing these results, he was one of the earliest users of mathematical induction.

Many people worked on the proof of the Binomial Theorem, which was finally completed for all n (including complex numbers) by Niels Abel (1802–1829).

13.5 Assess Your Understanding

Concepts and Vocabulary

1. The ________ ________ is a triangular display of the binomial coefficients.

2. $\binom{n}{0} =$ __ and $\binom{n}{1} =$ __.

3. ***True or False*** $\binom{n}{j} = \dfrac{j!}{(n-j)!\,n!}$

4. The ________ ________ can be used to expand expressions like $(2x + 3)^6$.

Skill Building

In Problems 5–16, evaluate each expression.

5. $\binom{5}{3}$ **6.** $\binom{7}{3}$ **7.** $\binom{7}{5}$ **8.** $\binom{9}{7}$

9. $\binom{50}{49}$ **10.** $\binom{100}{98}$ **11.** $\binom{1000}{1000}$ **12.** $\binom{1000}{0}$

13. $\binom{55}{23}$ **14.** $\binom{60}{20}$ **15.** $\binom{47}{25}$ **16.** $\binom{37}{19}$

In Problems 17–28, expand each expression using the Binomial Theorem.

17. $(x + 1)^5$ **18.** $(x - 1)^5$ **19.** $(x - 2)^6$ **20.** $(x + 3)^5$

21. $(3x + 1)^4$ **22.** $(2x + 3)^5$ **23.** $(x^2 + y^2)^5$ **24.** $(x^2 - y^2)^6$

25. $(\sqrt{x} + \sqrt{2})^6$ **26.** $(\sqrt{x} - \sqrt{3})^4$ **27.** $(ax + by)^5$ **28.** $(ax - by)^4$

In Problems 29–42, use the Binomial Theorem to find the indicated coefficient or term.

29. The coefficient of x^6 in the expansion of $(x + 3)^{10}$

30. The coefficient of x^3 in the expansion of $(x - 3)^{10}$

31. The coefficient of x^7 in the expansion of $(2x - 1)^{12}$

32. The coefficient of x^3 in the expansion of $(2x + 1)^{12}$

33. The coefficient of x^7 in the expansion of $(2x + 3)^9$

34. The coefficient of x^2 in the expansion of $(2x - 3)^9$

35. The 5th term in the expansion of $(x + 3)^7$

36. The 3rd term in the expansion of $(x - 3)^7$

37. The 3rd term in the expansion of $(3x - 2)^9$

38. The 6th term in the expansion of $(3x + 2)^8$

39. The coefficient of x^0 in the expansion of $\left(x^2 + \dfrac{1}{x}\right)^{12}$

40. The coefficient of x^0 in the expansion of $\left(x - \dfrac{1}{x^2}\right)^9$

41. The coefficient of x^4 in the expansion of $\left(x - \dfrac{2}{\sqrt{x}}\right)^{10}$

42. The coefficient of x^2 in the expansion of $\left(\sqrt{x} + \dfrac{3}{\sqrt{x}}\right)^8$

Applications and Extensions

43. Use the Binomial Theorem to find the numerical value of $(1.001)^5$ correct to five decimal places.
[**Hint:** $(1.001)^5 = (1 + 10^{-3})^5$]

44. Use the Binomial Theorem to find the numerical value of $(0.998)^6$ correct to five decimal places.

45. Show that $\binom{n}{n-1} = n$ and $\binom{n}{n} = 1$.

46. Show that if n and j are integers with $0 \le j \le n$, then,

$$\binom{n}{j} = \binom{n}{n-j}$$

Conclude that the Pascal triangle is symmetric with respect to a vertical line drawn from the topmost entry.

47. If n is a positive integer, show that

$$\binom{n}{0} + \binom{n}{1} + \cdots + \binom{n}{n} = 2^n$$

[**Hint:** $2^n = (1+1)^n$; now use the Binomial Theorem.]

48. If n is a positive integer, show that

$$\binom{n}{0} - \binom{n}{1} + \binom{n}{2} - \cdots + (-1)^n\binom{n}{n} = 0$$

49. $\binom{5}{0}\left(\frac{1}{4}\right)^5 + \binom{5}{1}\left(\frac{1}{4}\right)^4\left(\frac{3}{4}\right) + \binom{5}{2}\left(\frac{1}{4}\right)^3\left(\frac{3}{4}\right)^2$
$+ \binom{5}{3}\left(\frac{1}{4}\right)^2\left(\frac{3}{4}\right)^3 + \binom{5}{4}\left(\frac{1}{4}\right)\left(\frac{3}{4}\right)^4 + \binom{5}{5}\left(\frac{3}{4}\right)^5 = ?$

50. Stirling's Formula An approximation for $n!$, when n is large, is given by

$$n! \approx \sqrt{2n\pi}\left(\frac{n}{e}\right)^n\left(1 + \frac{1}{12n-1}\right)$$

Calculate 12!, 20!, and 25! on your calculator. Then use Stirling's formula to approximate 12!, 20!, and 25!.

Retain Your Knowledge

Problems 51–54 are based on material learned earlier in the course. The purpose of these problems is to keep the material fresh in your mind so that you are better prepared for the final exam.

51. Solve $6^x = 5^{x+1}$. Express the answer both in exact form and as a decimal rounded to three decimal places.

52. For $\mathbf{v} = 2\mathbf{i} + 3\mathbf{j}$ and $\mathbf{w} = 3\mathbf{i} - 2\mathbf{j}$:
(a) Find the dot product $\mathbf{v}\cdot\mathbf{w}$.
(b) Find the angle between $\mathbf{v}$ and $\mathbf{w}$.
(c) Are the vectors parallel, orthogonal, or neither?

53. Solve the system of equations:

$$\begin{cases} x - y - z = 0 \\ 2x + y + 3z = -1 \\ 4x + 2y - z = 12 \end{cases}$$

54. Graph the system of inequalities. Tell whether the graph is bounded or unbounded, and label the corner points.

$$\begin{cases} x \ge 0 \\ y \ge 0 \\ x + y \le 6 \\ 2x + y \le 10 \end{cases}$$

Chapter Review

Things to Know

Sequence (p. 962) — A function whose domain is the set of positive integers

Factorials (p. 965) — $0! = 1, 1! = 1, n! = n(n-1)\cdot\cdots\cdot 3\cdot 2\cdot 1$ if $n \ge 2$ is an integer

Arithmetic sequence (pp. 975 and 977) — $a_1 = a,\ a_n = a_{n-1} + d$, where $a_1 = a$ = first term, d = common difference
$a_n = a_1 + (n-1)d$

Sum of the first n terms of an arithmetic sequence (p. 978) — $S_n = \frac{n}{2}[2a_1 + (n-1)d] = \frac{n}{2}(a_1 + a_n)$

Geometric sequence (pp. 982 and 983) — $a_1 = a,\ \ a_n = ra_{n-1}$, where $a_1 = a$ = first term, r = common ratio
$a_n = a_1 r^{n-1} \quad r \ne 0$

Sum of the first n terms of a geometric sequence (p. 984) — $S_n = a_1\frac{1-r^n}{1-r} \quad r \ne 0, 1$

Infinite geometric series (p. 985) — $a_1 + a_1 r + \cdots + a_1 r^{n-1} + \cdots = \sum_{k=1}^{\infty} a_1 r^{k-1}$

Sum of a convergent infinite geometric series (p. 986) — If $|r| < 1$, $\sum_{k=1}^{\infty} a_1 r^{k-1} = \frac{a_1}{1-r}$

Principle of Mathematical Induction (p. 992)	If the following two conditions are satisfied, Condition I: The statement is true for the natural number 1. Condition II: If the statement is true for some natural number k, it is also true for $k + 1$. then the statement is true for all natural numbers.
Binomial coefficient (p. 997)	$\binom{n}{j} = \dfrac{n!}{j!(n-j)!}$
The Pascal triangle (p. 998)	See Figure 18.
Binomial Theorem (p. 998)	$(x+a)^n = \binom{n}{0}x^n + \binom{n}{1}ax^{n-1} + \cdots + \binom{n}{j}a^j x^{n-j} + \cdots + \binom{n}{n}a^n = \sum_{j=0}^{n}\binom{n}{j}x^{n-j}a^j$

Objectives

Section		You should be able to . . .	Examples	Review Exercises
13.1	1	Write the first several terms of a sequence (p. 962)	1–4	1, 2
	2	Write the terms of a sequence defined by a recursive formula (p. 965)	5, 6	3, 4
	3	Use summation notation (p. 966)	7, 8	5, 6
	4	Find the sum of a sequence algebraically and using a graphing utility (p. 967)	9	13, 14
	5	Solve annuity and amortization problems (p. 969)	10, 11	37
13.2	1	Determine whether a sequence is arithmetic (p. 975)	1–3	7–12
	2	Find a formula for an arithmetic sequence (p. 976)	4, 5	17, 19–21, 34(a)
	3	Find the sum of an arithmetic sequence (p. 977)	6–8	7, 10, 14, 34(b), 35
13.3	1	Determine whether a sequence is geometric (p. 982)	1–3	7–12
	2	Find a formula for a geometric sequence (p. 983)	4	11, 18, 36(a)–(c), 38
	3	Find the sum of a geometric sequence (p. 984)	5, 6	9, 11, 15, 16
	4	Determine whether a geometric series converges or diverges (p. 985)	7–9	22–25, 36(d)
13.4	1	Prove statements using mathematical induction (p. 992)	1–4	26–28
13.5	1	Evaluate $\binom{n}{j}$ (p. 996)	1	29
	2	Use the Binomial Theorem (p. 998)	2–5	30–33

Review Exercises

In Problems 1–4, write down the first five terms of each sequence.

1. $\{a_n\} = \left\{(-1)^n\left(\dfrac{n+3}{n+2}\right)\right\}$ **2.** $\{c_n\} = \left\{\dfrac{2^n}{n^2}\right\}$ **3.** $a_1 = 3;\ a_n = \dfrac{2}{3}a_{n-1}$ **4.** $a_1 = 2;\ a_n = 2 - a_{n-1}$

5. Write out $\sum_{k=1}^{4}(4k+2)$.

6. Express $1 - \dfrac{1}{2} + \dfrac{1}{3} - \dfrac{1}{4} + \cdots + \dfrac{1}{13}$ using summation notation.

In Problems 7–12, determine whether the given sequence is arithmetic, geometric, or neither. If the sequence is arithmetic, find the common difference and the sum of the first n terms. If the sequence is geometric, find the common ratio and the sum of the first n terms.

7. $\{a_n\} = \{n+5\}$ **8.** $\{c_n\} = \{2n^3\}$ **9.** $\{s_n\} = \{2^{3n}\}$

10. $0, 4, 8, 12, \ldots$ **11.** $3, \dfrac{3}{2}, \dfrac{3}{4}, \dfrac{3}{8}, \dfrac{3}{16}, \ldots$ **12.** $\dfrac{2}{3}, \dfrac{3}{4}, \dfrac{4}{5}, \dfrac{5}{6}, \ldots$

In Problems 13–16, find each sum.

13. $\sum_{k=1}^{30} (k^2 + 2)$ **14.** $\sum_{k=1}^{40} (-2k + 8)$ **15.** $\sum_{k=1}^{7} \left(\frac{1}{3}\right)^k$ **16.** $\sum_{k=1}^{10} (-2)^k$

In Problems 17–19, find the indicated term in each sequence. [**Hint:** *Find the general term first.*]

17. 9th term of 3, 7, 11, 15, ... **18.** 11th term of $1, \frac{1}{10}, \frac{1}{100}, \ldots$ **19.** 9th term of $\sqrt{2}, 2\sqrt{2}, 3\sqrt{2}, \ldots$

In Problems 20 and 21, find a general formula for each arithmetic sequence.

20. 7th term is 31; 20th term is 96

21. 10th term is 0; 18th term is 8

In Problems 22–25, determine whether each infinite geometric series converges or diverges. If it converges, find its sum.

22. $3 + 1 + \frac{1}{3} + \frac{1}{9} + \cdots$ **23.** $2 - 1 + \frac{1}{2} - \frac{1}{4} + \cdots$

24. $\frac{1}{2} + \frac{3}{4} + \frac{9}{8} + \cdots$ **25.** $\sum_{k=1}^{\infty} 4\left(\frac{1}{2}\right)^{k-1}$

In Problems 26–28, use the Principle of Mathematical Induction to show that the given statement is true for all natural numbers.

26. $3 + 6 + 9 + \cdots + 3n = \frac{3n}{2}(n + 1)$ **27.** $2 + 6 + 18 + \cdots + 2 \cdot 3^{n-1} = 3^n - 1$

28. $1^2 + 4^2 + 7^2 + \cdots + (3n - 2)^2 = \frac{1}{2}n(6n^2 - 3n - 1)$

29. Evaluate: $\binom{5}{2}$

In Problems 30 and 31, expand each expression using the Binomial Theorem.

30. $(x + 2)^5$ **31.** $(3x - 4)^4$

---

32. Find the coefficient of x^7 in the expansion of $(x + 2)^9$.

33. Find the coefficient of x^2 in the expansion of $(2x + 1)^7$.

34. Constructing a Brick Staircase A brick staircase has a total of 25 steps. The bottom step requires 80 bricks. Each step thereafter requires three fewer bricks than the prior step.

(a) How many bricks are required for the top step?

(b) How many bricks are required to build the staircase?

35. Creating a Floor Design A mosaic tile floor is designed in the shape of a trapezoid 30 feet wide at the base and 15 feet wide at the top. The tiles, 12 inches by 12 inches, are to be placed so that each successive row contains one fewer tile than the row below. How many tiles will be required?

36. Bouncing Balls A ball is dropped from a height of 20 feet. Each time it strikes the ground, it bounces up to three-quarters of the height of the previous bounce.

(a) What height will the ball bounce up to after it strikes the ground for the 3rd time?

(b) How high will it bounce after it strikes the ground for the nth time?

(c) How many times does the ball need to strike the ground before its bounce is less than 6 inches?

(d) What total distance does the ball travel before it stops bouncing?

37. Retirement Planning Chris gets paid once a month and contributes \$200 each pay period into his 401(k). If Chris plans on retiring in 20 years, what will be the value of his 401(k) if the per annum rate of return of the 401(k) is 10% compounded monthly?

38. Salary Increases Your friend has just been hired at an annual salary of \$50,000. If she expects to receive annual increases of 4%, what will be her salary as she begins her 5th year?

Chapter Test

CHAPTER **Test Prep** VIDEOS

The Chapter Test Prep Videos are step-by-step solutions available in MyMathLab®, or on this text's YouTube Channel. Flip back to the Resources for Success page for a link to this text's YouTube channel.

In Problems 1 and 2, write down the first five terms of each sequence.

1. $\{s_n\} = \left\{\dfrac{n^2-1}{n+8}\right\}$

2. $a_1 = 4,\ a_n = 3a_{n-1} + 2$

In Problems 3 and 4, write out each sum. Evaluate each sum.

3. $\sum_{k=1}^{3} (-1)^{k+1}\left(\dfrac{k+1}{k^2}\right)$

4. $\sum_{k=1}^{4}\left[\left(\dfrac{2}{3}\right)^k - k\right]$

5. Write the following sum using summation notation.

$$-\frac{2}{5} + \frac{3}{6} - \frac{4}{7} + \cdots + \frac{11}{14}$$

In Problems 6–11, determine whether the given sequence is arithmetic, geometric, or neither. If the sequence is arithmetic, find the common difference and the sum of the first n terms. If the sequence is geometric, find the common ratio and the sum of the first n terms.

6. $6, 12, 36, 144, \ldots$

7. $\left\{-\dfrac{1}{2}\cdot 4^n\right\}$

8. $-2, -10, -18, -26, \ldots$

9. $\left\{-\dfrac{n}{2} + 7\right\}$

10. $25, 10, 4, \dfrac{8}{5}, \ldots$

11. $\left\{\dfrac{2n-3}{2n+1}\right\}$

12. Determine whether the infinite geometric series

$$256 - 64 + 16 - 4 + \cdots$$

converges or diverges. If it converges, find its sum.

13. Expand $(3m+2)^5$ using the Binomial Theorem.

14. Use the Principle of Mathematical Induction to show that the given statement is true for all natural numbers.

$$\left(1+\frac{1}{1}\right)\left(1+\frac{1}{2}\right)\left(1+\frac{1}{3}\right)\cdots\left(1+\frac{1}{n}\right) = n+1$$

15. A new car sold for \$31,000. If the vehicle loses 15% of its value each year, how much will it be worth after 10 years?

16. A weightlifter begins his routine by benching 100 pounds and increases the weight by 30 pounds for each set. If he does 10 repetitions in each set, what is the total weight lifted after 5 sets?

Cumulative Review

1. Find all the solutions, real and complex, of the equation

$$|x^2| = 9$$

2. (a) Graph the circle $x^2 + y^2 = 100$ and the parabola $y = 3x^2$.

(b) Solve the system of equations: $\begin{cases} x^2 + y^2 = 100 \\ y = 3x^2 \end{cases}$

(c) Where do the circle and the parabola intersect?

3. Solve the equation: $2e^x = 5$

4. Find an equation of the line with slope 5 and x-intercept 2.

5. Find the standard equation of the circle whose center is the point $(-1, 2)$ if $(3, 5)$ is a point on the circle.

6. $f(x) = \dfrac{3x}{x-2}$ and $g(x) = 2x + 1$

Find:

(a) $(f \circ g)(2)$

(b) $(g \circ f)(4)$

(c) $(f \circ g)(x)$

(d) The domain of $(f \circ g)(x)$

(e) $(g \circ f)(x)$

(f) The domain of $(g \circ f)(x)$

(g) The function g^{-1} and its domain

(h) The function f^{-1} and its domain

7. Find the equation of an ellipse with center at the origin, a focus at $(0, 3)$, and a vertex at $(0, 4)$.

8. Find the equation of a parabola with vertex at $(-1, 2)$ and focus at $(-1, 3)$.

9. Find the polar equation of a circle with center at $(0, 4)$ that passes through the pole. What is the rectangular equation?

10. Solve the equation

$$2\sin^2 x - \sin x - 3 = 0, \quad 0 \le x < 2\pi$$

11. Find the exact value of $\cos^{-1}(-0.5)$.

12. If $\sin\theta = \dfrac{1}{4}$ and θ is in the second quadrant, find:

(a) $\cos\theta$

(b) $\tan\theta$

(c) $\sin(2\theta)$

(d) $\cos(2\theta)$

(e) $\sin\left(\dfrac{1}{2}\theta\right)$

Chapter Projects

I. Population Growth The size of the population of the United States essentially depends on its current population, the birth and death rates of the population, and immigration. Let b represent the birth rate of the U.S. population, and let d represent its death rate. Then $r = b - d$ represents the growth rate of the population, where r varies from year to year. The U.S. population after n years can be modeled using the recursive function

$$p_n = (1 + r)p_{n-1} + I$$

where I represents net immigration into the United States.

1. Using data from the CIA World Factbook at *https://www.cia.gov/library/publications/the-world-factbook/* determine the birth and death rates in the United States for the most recent year that data are available. Birth rates and death rates are given as the number of live births per 1000 population. Each must be computed as the number of births (deaths) per individual. For example, in 2014, the birth rate was 13.42 per 1000 and the death rate was 8.15 per 1000, so

$$b = \frac{13.42}{1000} = 0.01342, \quad \text{and} \quad d = \frac{8.15}{1000} = 0.00815.$$

   Next, using data from the Immigration and Naturalization Service at *www.fedstats.gov*, determine the net immigration into the United States for the same year used to obtain b and d.
2. Determine the value of r, the growth rate of the population.
3. Find a recursive formula for the population of the United States.
4. Use the recursive formula to predict the population of the United States in the following year. In other words, if data are available for the year 2015, predict the U.S. population in 2016.
5. Does your prediction seem reasonable? Explain.
6. Repeat Problems 1–5 for Uganda using the CIA World Factbook (in 2014, the birth rate was 47.17 per 1000 and the death rate was 10.97 per 1000).
7. Do your results for the United States (a developed country) and Uganda (a developing country) seem in line with the article in the chapter opener? Explain.
8. Do you think the recursive formula found in Problem 3 will be useful in predicting future populations? Why or why not?

The following projects are available at the Instructor's Resource Center (IRC):

II. Project at Motorola ***Digital Wireless Communication*** Cell phones take speech and change it into digital code using only zeros and ones. See how the code length can be modeled using a mathematical sequence.

III. Economics Economists use the current price of a good and a recursive model to predict future consumer demand and to determine future production.

IV. Standardized Tests Many tests of intelligence, aptitude, and achievement contain questions asking for the terms of a mathematical sequence.

Answers

CHAPTER R Review

R.1 Assess Your Understanding (page 16)

1. rational **2.** 31 **3.** Distributive **4.** $5(x+3)=6$ **5.** a **6.** b **7.** T **8.** F **9.** F **10.** T **11.** $\{1,2,3,4,5,6,7,8,9\}$ **13.** $\{4\}$ **15.** $\{1,3,4,6\}$ **17.** $\{0,2,6,7,8\}$ **19.** $\{0,1,2,3,5,6,7,8,9\}$ **21.** $\{0,1,2,3,5,6,7,8,9\}$ **23. (a)** $\{2,5\}$ **(b)** $\{-6,2,5\}$ **(c)** $\left\{-6,\frac{1}{2},-1.333\ldots,2,5\right\}$ **(d)** $\{\pi\}$ **(e)** $\left\{-6,\frac{1}{2},-1.333\ldots,\pi,2,5\right\}$ **25. (a)** $\{1\}$ **(b)** $\{0,1\}$ **(c)** $\left\{0,1,\frac{1}{2},\frac{1}{3},\frac{1}{4}\right\}$ **(d)** None **(e)** $\left\{0,1,\frac{1}{2},\frac{1}{3},\frac{1}{4}\right\}$ **27. (a)** None **(b)** None **(c)** None **(d)** $\left\{\sqrt{2},\pi,\sqrt{2}+1,\pi+\frac{1}{2}\right\}$ **(e)** $\left\{\sqrt{2},\pi,\sqrt{2}+1,\pi+\frac{1}{2}\right\}$ **29. (a)** 18.953 **(b)** 18.952 **31. (a)** 28.653 **(b)** 28.653 **33. (a)** 0.063 **(b)** 0.062 **35. (a)** 9.999 **(b)** 9.998 **37. (a)** 0.429 **(b)** 0.428 **39. (a)** 34.733 **(b)** 34.733 **41.** $3+2=5$ **43.** $x+2=3\cdot 4$ **45.** $3y=1+2$ **47.** $x-2=6$ **49.** $\frac{x}{2}=6$ **51.** 7 **53.** 6 **55.** 1 **57.** $\frac{13}{3}$ **59.** -11 **61.** 11 **63.** -4 **65.** 1 **67.** 6 **69.** $\frac{2}{7}$ **71.** $\frac{4}{45}$ **73.** $\frac{23}{20}$ **75.** $\frac{79}{30}$ **77.** $\frac{13}{36}$ **79.** $-\frac{16}{45}$ **81.** $\frac{1}{60}$ **83.** $\frac{15}{22}$ **85.** 1 **87.** $\frac{15}{8}$ **89.** $6x+24$ **91.** x^2-4x **93.** $\frac{3}{2}x-1$ **95.** x^2+6x+8 **97.** x^2-x-2 **99.** $x^2-10x+16$ **101.** $2x+3x=(2+3)x=5x$ **103.** $2(3\cdot 4)=2\cdot 12=24;\ (2\cdot 3)\cdot(2\cdot 4)=6\cdot 8=48$ **105.** No; $2-3\neq 3-2$ **107.** No; $2\div 3\neq 3\div 2$ **109.** Symmetric Property **111.** No; no

R.2 Assess Your Understanding (page 27)

1. variable **2.** origin **3.** strict **4.** base; exponent, or power **5.** 1.2345678×10^3 **6.** T **7.** T **8.** F **9.** F **10.** F **11.** d **12.** b **13.** 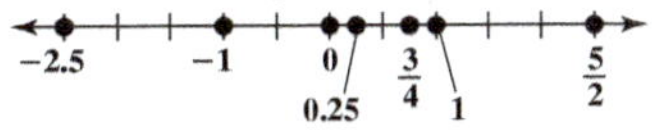

15. > **17.** > **19.** > **21.** = **23.** < **25.** $x>0$ **27.** $x<2$ **29.** $x\le 1$ **31.** 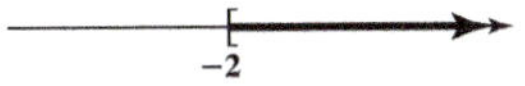

33. 

35. 1 **37.** 2 **39.** 6 **41.** 4 **43.** -28 **45.** $\frac{4}{5}$ **47.** 0 **49.** 1 **51.** 5 **53.** 1 **55.** 22 **57.** 2 **59.** $x=0$ **61.** $x=3$ **63.** None **65.** $x=0, x=1, x=-1$ **67.** $\{x\mid x\neq 5\}$ **69.** $\{x\mid x\neq -4\}$ **71.** 0°C **73.** 25°C **75.** 16 **77.** $\frac{1}{16}$ **79.** $\frac{1}{9}$ **81.** 9 **83.** 5 **85.** 4 **87.** $64x^6$ **89.** $\frac{x^4}{y^2}$ **91.** $\frac{x}{y}$ **93.** $-\frac{8x^3z}{9y}$ **95.** $\frac{16x^2}{9y^2}$ **97.** -4 **99.** 5 **101.** 4 **103.** 2 **105.** $\sqrt{5}$ **107.** $\frac{1}{2}$ **109.** 10; 0 **111.** 81 **113.** 304,006.671 **115.** 0.004 **117.** 481.890 **119.** 0.000 **121.** 4.542×10^2 **123.** 1.3×10^{-2} **125.** 3.2155×10^4 **127.** 4.23×10^{-4} **129.** 61,500 **131.** 0.001214 **133.** 110,000,000 **135.** 0.081 **137.** $A=lw$ **139.** $C=\pi d$ **141.** $A=\frac{\sqrt{3}}{4}x^2$ **143.** $V=\frac{4}{3}\pi r^3$ **145.** $V=x^3$ **147. (a)** \$6000 **(b)** \$8000 **149.** $|x-4|\ge 6$ **151. (a)** $2\le 5$ **(b)** $6>5$ **153. (a)** Yes **(b)** No **155.** 400,000,000 m **157.** 0.0000005 m **159.** 5×10^{-4} in. **161.** 5.865696×10^{12} mi **163.** No; $\frac{1}{3}$ is larger; 0.000333... **165.** No

R.3 Assess Your Understanding (page 37)

1. right; hypotenuse **2.** $A=\frac{1}{2}bh$ **3.** $C=2\pi r$ **4.** similar **5.** c **6.** b **7.** T **8.** T **9.** F **10.** T **11.** T **12.** F **13.** 13 **15.** 26 **17.** 25 **19.** Right triangle; 5 **21.** Not a right triangle **23.** Right triangle; 25 **25.** Not a right triangle **27.** 8 in.^{2} **29.** 4 in.^{2} **31.** $A=25\pi$ m^2; $C=10\pi$ m **33.** $V=224$ ft^3; $S=232$ ft^2 **35.** $V=\frac{256}{3}\pi$ cm^3; $S=64\pi$ cm^2 **37.** $V=648\pi$ in.^{3}; $S=306\pi$ in.^{2} **39.** π square units **41.** 2π square units **43.** $x=4$ units; $A=90°$; $B=60°$; $C=30°$ **45.** $x=67.5$ units; $A=60°$; $B=95°$; $C=25°$ **47.** About 16.8 ft **49.** 64 ft^2 **51.** $24+2\pi\approx 30.28$ ft^2; $16+2\pi\approx 22.28$ ft **53.** 160 paces **55.** About 5.477 mi **57.** From 100 ft: 12.2 mi; From 150 ft: 15.0 mi

R.4 Assess Your Understanding (page 48)

1. 4; 3 **2.** x^4-16 **3.** x^3-8 **4.** a **5.** c **6.** F **7.** T **8.** F **9.** Monomial; variable: x; coefficient: 2; degree: 3 **11.** Not a monomial; the exponent of the variable is not a nonnegative integer **13.** Monomial; variables: x, y; coefficient: -2; degree: 3 **15.** Not a monomial; the exponent of one of the variables is not a nonnegative integer **17.** Not a monomial; it has more than one term **19.** Yes; 2 **21.** Yes; 0 **23.** No; the exponent of the variable of one of the terms is not a nonnegative integer **25.** Yes; 3 **27.** No; the polynomial of the denominator has a degree greater than 0 **29.** x^2+7x+2 **31.** x^3-4x^2+9x+7 **33.** $6x^5+5x^4+3x^2+x$ **35.** $7x^2-x-7$ **37.** $-2x^3+18x^2-18$ **39.** $2x^2-4x+6$ **41.** $15y^2-27y+30$ **43.** x^3+x^2-4x **45.** $-8x^5-10x^2$ **47.** x^3+3x^2-2x-4 **49.** x^2+6x+8 **51.** $2x^2+9x+10$ **53.** x^2-2x-8 **55.** x^2-5x+6 **57.** $2x^2-x-6$ **59.** $-2x^2+11x-12$ **61.** $2x^2+8x+8$ **63.** $x^2-xy-2y^2$ **65.** $-6x^2-13xy-6y^2$ **67.** x^2-49 **69.** $4x^2-9$ **71.** $x^2+8x+16$ **73.** $x^2-8x+16$ **75.** $9x^2-16$ **77.** $4x^2-12x+9$ **79.** x^2-y^2 **81.** $9x^2-y^2$ **83.** $x^2+2xy+y^2$ **85.** $x^2-4xy+4y^2$ **87.** $x^3-6x^2+12x-8$ **89.** $8x^3+12x^2+6x+1$ **91.** $4x^2-11x+23$; remainder -45 **93.** $4x-3$; remainder $x+1$ **95.** $5x^2-13$; remainder $x+27$ **97.** $2x^2$; remainder $-x^2+x+1$ **99.** $x^2-2x+\frac{1}{2}$; remainder $\frac{5}{2}x+\frac{1}{2}$ **101.** $-4x^2-3x-3$; remainder -7 **103.** x^2-x-1; remainder $2x+2$ **105.** x^2+ax+a^2; remainder 0

R.5 Assess Your Understanding (page 58)

1. $3x(x-2)(x+2)$ **2.** prime **3.** c **4.** b **5.** d **6.** c **7.** T **8.** F **9.** $3(x+2)$ **11.** $a(x^2+1)$ **13.** $x(x^2+x+1)$ **15.** $2x(x-1)$

17. $3xy(x - 2y + 4)$ **19.** $(x + 1)(x - 1)$ **21.** $(2x + 1)(2x - 1)$ **23.** $(x + 4)(x - 4)$ **25.** $(5x + 2)(5x - 2)$ **27.** $(x + 1)^2$ **29.** $(x + 2)^2$ **31.** $(x - 5)^2$ **33.** $(2x + 1)^2$ **35.** $(4x + 1)^2$ **37.** $(x - 3)(x^2 + 3x + 9)$ **39.** $(x + 3)(x^2 - 3x + 9)$ **41.** $(2x + 3)(4x^2 - 6x + 9)$ **43.** $(x + 2)(x + 3)$ **45.** $(x + 6)(x + 1)$ **47.** $(x + 5)(x + 2)$ **49.** $(x - 8)(x - 2)$ **51.** $(x - 8)(x + 1)$ **53.** $(x + 8)(x - 1)$ **55.** $(x + 2)(2x + 3)$ **57.** $(x - 2)(2x + 1)$ **59.** $3(2x + 3)(3x + 2)$ **61.** $(3x + 1)(x + 1)$ **63.** $(z + 1)(2z + 3)$ **65.** $(x + 2)(3x - 4)$ **67.** $(x - 2)(3x + 4)$ **69.** $4x^2(x + 4)(3x + 2)$ **71.** $(x + 4)(3x - 2)$ **73.** 25; $(x + 5)^2$ **75.** 9; $(y - 3)^2$ **77.** $\frac{1}{16}$; $\left(x - \frac{1}{4}\right)^2$ **79.** $(x + 6)(x - 6)$ **81.** $2(1 + 2x)(1 - 2x)$ **83.** $8(x + 1)(x + 10)$ **85.** $(x - 7)(x - 3)$ **87.** $4(x^2 - 2x + 8)$ **89.** Prime **91.** $-(x - 5)(x + 3)$ **93.** $3(x + 2)(x - 6)$ **95.** $y^2(y + 5)(y + 6)$ **97.** $2x^3(2x + 3)^2$ **99.** $2(3x + 1)(x + 1)$ **101.** $(x - 3)(x + 3)(x^2 + 9)$ **103.** $(x - 1)^2(x^2 + x + 1)^2$ **105.** $x^5(x - 1)(x + 1)$ **107.** $(4x + 3)^2$ **109.** $-(4x - 5)(4x + 1)$ **111.** $(2y - 5)(2y - 3)$ **113.** $-(3x - 1)(3x + 1)(x^2 + 1)$ **115.** $(x + 3)(x - 6)$ **117.** $(x + 2)(x - 3)$ **119.** $(3x - 5)(9x^2 - 3x + 7)$ **121.** $(x + 5)(3x + 11)$ **123.** $(x - 1)(x + 1)(x + 2)$ **125.** $(x - 1)(x + 1)(x^2 - x + 1)$ **127.** $2(3x + 4)(9x + 13)$ **129.** $2x(3x + 5)$ **131.** $5(x + 3)(x - 2)^2(x + 1)$ **133.** $3(4x - 3)(4x - 1)$ **135.** $6(3x - 5)(2x + 1)^2(5x - 4)$ **137.** The possibilities are $(x \pm 1)(x \pm 4) = x^2 \pm 5x + 4$ or $(x \pm 2)(x \pm 2) = x^2 \pm 4x + 4$, none of which equals $x^2 + 4$.

R.6 Assess Your Understanding *(page 62)*

1. quotient; divisor; remainder **2.** $-3\overline{)2\ 0\ -5\ 1}$ **3.** d **4.** a **5.** T **6.** T **7.** $x^2 + x + 4$; remainder 12 **9.** $3x^2 + 11x + 32$; remainder 99 **11.** $x^4 - 3x^3 + 5x^2 - 15x + 46$; remainder -138 **13.** $4x^5 + 4x^4 + x^3 + x^2 + 2x + 2$; remainder 7 **15.** $0.1x^2 - 0.11x + 0.321$; remainder -0.3531 **17.** $x^4 + x^3 + x^2 + x + 1$; remainder 0 **19.** No **21.** Yes **23.** Yes **25.** No **27.** Yes **29.** -9

R.7 Assess Your Understanding *(page 71)*

1. lowest terms **2.** least common multiple **3.** d **4.** a **5.** T **6.** F **7.** $\frac{3}{x - 3}$ **9.** $\frac{x}{3}$ **11.** $\frac{4x}{2x - 1}$ **13.** $\frac{y + 5}{2(y + 1)}$ **15.** $\frac{x + 5}{x - 1}$ **17.** $-(x + 7)$ **19.** $\frac{3}{5x(x - 2)}$ **21.** $\frac{2x(x^2 + 4x + 16)}{x + 4}$ **23.** $\frac{8}{3x}$ **25.** $\frac{x - 3}{x + 7}$ **27.** $\frac{4x}{(x - 2)(x - 3)}$ **29.** $\frac{4}{5(x - 1)}$ **31.** $-\frac{(x - 4)^2}{4x}$ **33.** $\frac{(x + 3)^2}{(x - 3)^2}$ **35.** $\frac{(x - 4)(x + 3)}{(x - 1)(2x + 1)}$ **37.** $\frac{x + 5}{2}$ **39.** $\frac{(x - 2)(x + 2)}{2x - 3}$ **41.** $\frac{3x - 2}{x - 3}$ **43.** $\frac{x + 9}{2x - 1}$ **45.** $\frac{4 - x}{x - 2}$ **47.** $\frac{2(x + 5)}{(x - 1)(x + 2)}$ **49.** $\frac{3x^2 - 2x - 3}{(x + 1)(x - 1)}$ **51.** $\frac{-(11x + 2)}{(x + 2)(x - 2)}$ **53.** $\frac{2(x^2 - 2)}{x(x - 2)(x + 2)}$ **55.** $(x - 2)(x + 2)(x + 1)$ **57.** $x(x - 1)(x + 1)$ **59.** $x^3(2x - 1)^2$ **61.** $x(x - 1)^2(x + 1)(x^2 + x + 1)$ **63.** $\frac{5x}{(x - 6)(x - 1)(x + 4)}$ **65.** $\frac{2(2x^2 + 5x - 2)}{(x - 2)(x + 2)(x + 3)}$ **67.** $\frac{5x + 1}{(x - 1)^2(x + 1)^2}$ **69.** $\frac{-x^2 + 3x + 13}{(x - 2)(x + 1)(x + 4)}$ **71.** $\frac{x^3 - 2x^2 + 4x + 3}{x^2(x + 1)(x - 1)}$ **73.** $\frac{-1}{x(x + h)}$ **75.** $\frac{x + 1}{x - 1}$ **77.** $\frac{(x - 1)(x + 1)}{2x(2x + 1)}$ **79.** $\frac{2(5x - 1)}{(x - 2)(x + 1)^2}$ **81.** $\frac{-2x(x^2 - 2)}{(x + 2)(x^2 - x - 3)}$ **83.** $\frac{-1}{x - 1}$ **85.** $\frac{3x - 1}{2x + 1}$ **87.** $\frac{19}{(3x - 5)^2}$ **89.** $\frac{(x + 1)(x - 1)}{(x^2 + 1)^2}$ **91.** $\frac{x(3x + 2)}{(3x + 1)^2}$ **93.** $-\frac{(x + 3)(3x - 1)}{(x^2 + 1)^2}$ **95.** $f = \frac{R_1 \cdot R_2}{(n - 1)(R_1 + R_2)}$; $\frac{2}{15}$ m

R.8 Assess Your Understanding *(page 79)*

3. index **4.** cube root **5.** b **6.** d **7.** c **8.** c **9.** T **10.** F **11.** 3 **13.** -2 **15.** $2\sqrt{2}$ **17.** $10\sqrt{7}$ **19.** $2\sqrt[3]{4}$ **21.** $-2x\sqrt[3]{x}$ **23.** $3\sqrt[4]{3}$ **25.** x^3y^2 **27.** x^2y **29.** $6\sqrt{x}$ **31.** $3x^2y^3\sqrt[4]{2x}$ **33.** $6x\sqrt{x}$ **35.** $15\sqrt[3]{3}$ **37.** $12\sqrt{3}$ **39.** $7\sqrt{2}$ **41.** $\sqrt{2}$ **43.** $2\sqrt{3}$ **45.** $-\sqrt[3]{2}$ **47.** $x - 2\sqrt{x} + 1$ **49.** $(2x - 1)\sqrt[3]{2x}$ **51.** $(2x - 15)\sqrt{2x}$ **53.** $-(x + 5y)\sqrt[3]{2xy}$ **55.** $\frac{\sqrt{2}}{2}$ **57.** $-\frac{\sqrt{15}}{5}$ **59.** $\frac{(5 + \sqrt{2})\sqrt{3}}{23}$ **61.** $\frac{8\sqrt{5} - 19}{41}$ **63.** $5\sqrt{2} + 5$ **65.** $\frac{5\sqrt[3]{4}}{2}$ **67.** $\frac{2x + h - 2\sqrt{x^2 + xh}}{h}$ **69.** 4 **71.** -3 **73.** 64 **75.** $\frac{1}{27}$ **77.** $\frac{27\sqrt{2}}{32}$ **79.** $\frac{27\sqrt{2}}{32}$ **81.** $-\frac{1}{10}$ **83.** $\frac{25}{16}$ **85.** $x^{7/12}$ **87.** xy^2 **89.** $x^{2/3}y$ **91.** $\frac{8x^{5/4}}{y^{3/4}}$ **93.** $\frac{3x + 2}{(1 + x)^{1/2}}$ **95.** $\frac{x(3x^2 + 2)}{(x^2 + 1)^{1/2}}$ **97.** $\frac{22x + 5}{10\sqrt{(x - 5)(4x + 3)}}$ **99.** $\frac{2 + x}{2(1 + x)^{3/2}}$ **101.** $\frac{4 - x}{(x + 4)^{3/2}}$ **103.** $\frac{1}{x^2(x^2 - 1)^{1/2}}$ **105.** $\frac{1 - 3x^2}{2\sqrt{x}(1 + x^2)^2}$ **107.** $\frac{1}{2}(5x + 2)(x + 1)^{1/2}$ **109.** $2x^{1/2}(3x - 4)(x + 1)$ **111.** $(x^2 + 4)^{1/3}(11x^2 + 12)$ **113.** $(3x + 5)^{1/3}(2x + 3)^{1/2}(17x + 27)$ **115.** $\frac{3(x + 2)}{2x^{1/2}}$ **117.** 1.41 **119.** 1.59 **121.** 4.89 **123.** 2.15 **125.** **(a)** 15,660.4 gal **(b)** 390.7 gal **127.** $2\sqrt{2}\pi \approx 8.89$ sec

CHAPTER 1 Graphs, Equations, and Inequalities

1.1 Assess Your Understanding *(page 94)*

7. x-coordinate or abscissa; y-coordinate or ordinate **8.** quadrants **9.** midpoint **10.** F **11.** F **12.** T **13.** d **14.** c

15. (a) quadrant II **(b)** x-axis **(c)** quadrant III **(d)** quadrant I **(e)** y-axis **(f)** quadrant IV

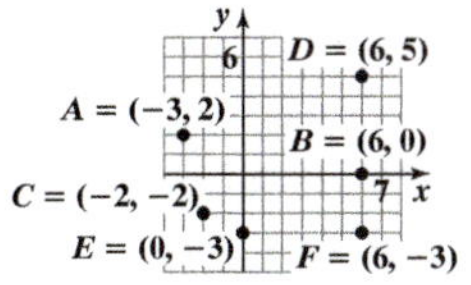

17. The points will be on a vertical line that is 2 units to the right of the y-axis.

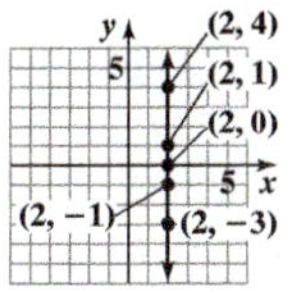

19. $(-1, 4)$; quadrant II **21.** $(3, 1)$; quadrant I **23.** $Xmin = -11$, $Xmax = 5$, $Xscl = 1$, $Ymin = -3$, $Ymax = 6$, $Yscl = 1$ **25.** $Xmin = -30$, $Xmax = 50$, $Xscl = 10$, $Ymin = -90$, $Ymax = 50$, $Yscl = 10$ **27.** $Xmin = -10$, $Xmax = 110$, $Xscl = 10$, $Ymin = -10$, $Ymax = 160$, $Yscl = 10$ **29.** $Xmin = -6$, $Xmax = 6$, $Xscl = 2$, $Ymin = -4$, $Ymax = 4$, $Yscl = 2$

31. $Xmin = -6$, $Xmax = 6$, $Xscl = 2$, $Ymin = -1$, $Ymax = 3$, $Yscl = 1$ **33.** $Xmin = 3$, $Xmax = 9$, $Xscl = 1$, $Ymin = 2$, $Ymax = 10$, $Yscl = 2$

35. $\sqrt{5}$ **37.** $\sqrt{10}$ **39.** $2\sqrt{17}$ **41.** 20 **43.** $\sqrt{53}$ **45.** $\sqrt{a^2 + b^2}$ **47.** $4\sqrt{10}$ **49.** $2\sqrt{65}$

51. $d(A, B) = \sqrt{13}$
$d(B, C) = \sqrt{13}$
$d(A, C) = \sqrt{26}$
$(\sqrt{13})^2 + (\sqrt{13})^2 = (\sqrt{26})^2$
$\text{Area} = \frac{13}{2}$ square units

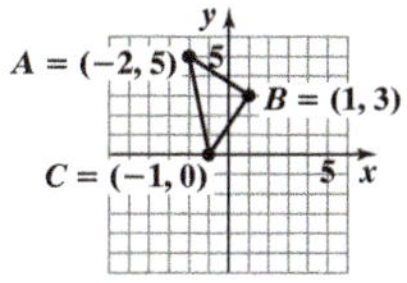

53. $d(A, B) = \sqrt{130}$
$d(B, C) = \sqrt{26}$
$d(A, C) = 2\sqrt{26}$
$(\sqrt{26})^2 + (2\sqrt{26})^2 = (\sqrt{130})^2$
Area = 26 square units

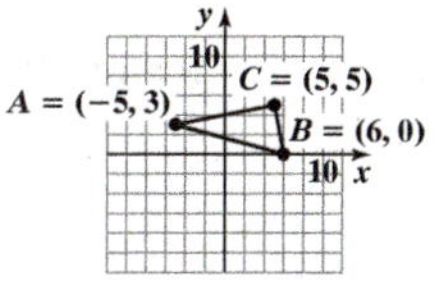

55. $d(A, B) = 4$
$d(B, C) = \sqrt{41}$
$d(A, C) = 5$
$4^2 + 5^2 = (\sqrt{41})^2$
Area = 10 square units

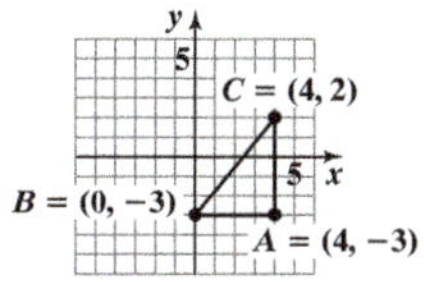

57. $(4, 0)$ **59.** $(3, 3)$ **61.** $(5, -1)$ **63.** $\left(\frac{a}{2}, \frac{b}{2}\right)$ **65.** $(0, 0)$ is on the graph. **67.** $(0, 3)$ is on the graph. **69.** $(0, 2)$ and $(\sqrt{2}, \sqrt{2})$ are on the graph. **71.** $(-1, 0), (1, 0)$ **73.** $\left(-\frac{\pi}{2}, 0\right), (0, 1), \left(\frac{\pi}{2}, 0\right)$ **75.** $(0, 2), (1, 0), (0, -2)$ **77.** $(-4, 0), (-1, 0), (0, -3), (4, 0)$

79.

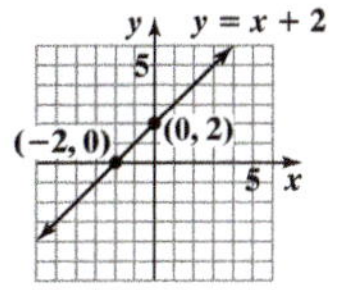

81.

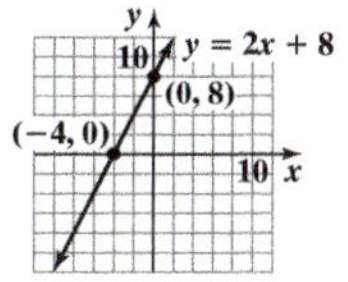

83.

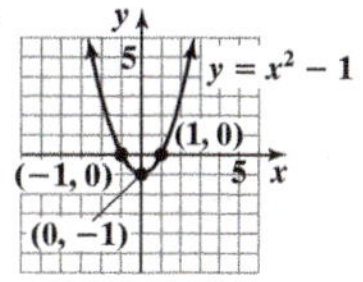

85.

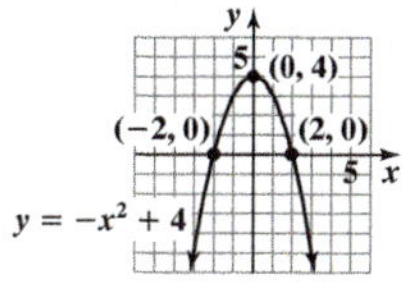

87.

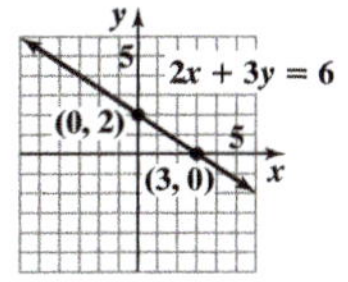

89.

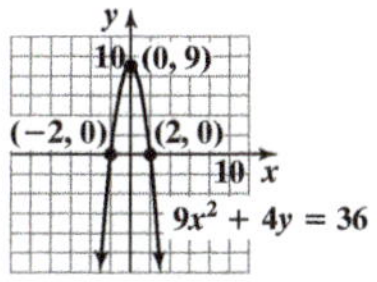

91. x-intercept: 6.5
y-intercept: -13

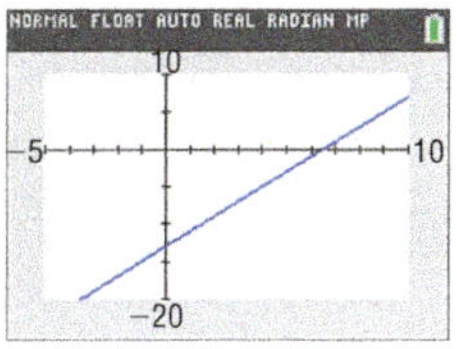

93. x-intercepts: $-2.74, 2.74$
y-intercept: -15

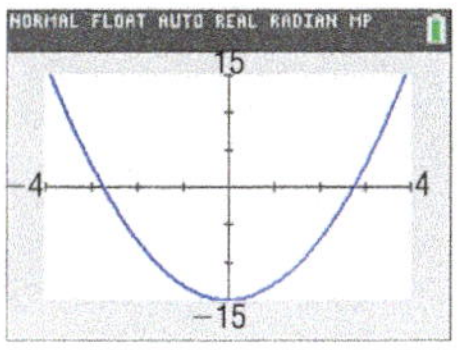

95. x-intercept: 14.33
y-intercept: -21.5

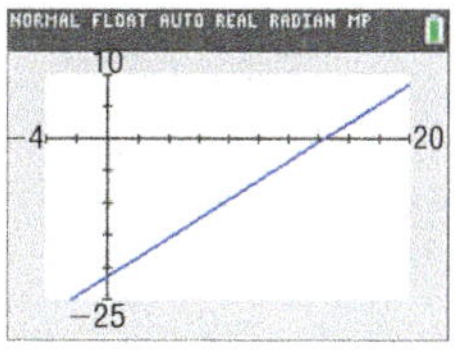

97. x-intercepts: $-2.72, 2.72$
y-intercept: 12.33

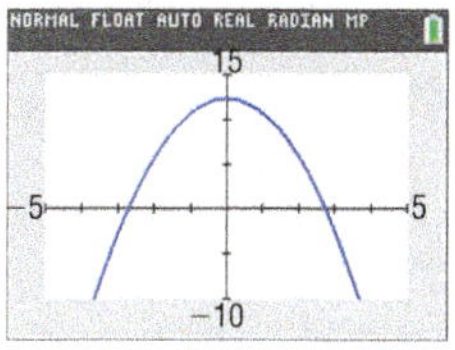

99. $(5, 3)$ **101.** $\sqrt{17}; 2\sqrt{5}; \sqrt{29}$ **103.** $d(P_1, P_2) = 6$; $d(P_2, P_3) = 4$; $d(P_1, P_3) = 2\sqrt{13}$; right triangle **105.** $d(P_1, P_2) = 2\sqrt{17}$; $d(P_2, P_3) = \sqrt{34}$; $d(P_1, P_3) = \sqrt{34}$; isosceles right triangle **107.** $(5, -2)$ **109.** $90\sqrt{2} \approx 127.28$ ft **111. (a)** $(90, 0), (90, 90), (0, 90)$ **(b)** $5\sqrt{2161} \approx 232.43$ ft **(c)** $30\sqrt{149} \approx 366.20$ ft **113.** $d = 50t$ mi **115. (a)** $(2.65, 1.6)$ **(b)** ≈ 1.285 units **117.** \$21,582.50

1.2 Assess Your Understanding *(page 107)*

5. b **6.** identity **7.** linear; first **8.** d **9.** T **10.** F **11.** $\{3\}$ **13.** $\{-5\}$ **15.** $\left\{\frac{3}{2}\right\}$ **17.** $\left\{\frac{5}{4}\right\}$ **19.** $\{-2.21, 0.54, 1.68\}$ **21.** $\{-1.55, 1.15\}$ **23.** $\{-1.12, 0.36\}$ **25.** $\{-2.69, -0.49, 1.51\}$ **27.** $\{-2.86, -1.34, 0.20, 1.00\}$ **29.** No real solutions **31.** $\{-2\}$ **33.** $\{3\}$ **35.** $\{-1\}$ **37.** $\{-18\}$

39. $\{-4\}$ **41.** $\left\{-\frac{5}{4}\right\}$ **43.** $\left\{-\frac{3}{4}\right\}$ **45.** $\{-2\}$ **47.** $\{0.5\}$ **49.** $\left\{\frac{29}{10}\right\}$ **51.** $\{2\}$ **53.** $\{8\}$ **55.** $\{2\}$ **57.** $\{-1\}$ **59.** $\{3\}$ **61.** No solution **63.** No solution **65.** $\{-6\}$ **67.** $\{34\}$ **69.** $\left\{-\frac{20}{39}\right\}$ **71.** $\{-1\}$ **73.** $\left\{-\frac{11}{6}\right\}$ **75.** $\{-6\}$ **77.** $-\frac{2}{5}$ **79.** $2a + 3b = 6$ **81.** $x = \frac{b+c}{a}$ **83.** $x = \frac{abc}{a+b}$ **85.** $x = a^2$ **87.** $a = 3$ **89.** $R = \frac{R_1R_2}{R_1 + R_2}$ **91.** $R = \frac{mv^2}{F}$ **93.** $r = \frac{S-a}{S}$ **95.** \$11,500 will be invested in bonds and \$8500 in CDs. **97.** Scott will receive \$400,000, Alice \$300,000, and Tricia \$200,000. **99.** The regular hourly rate is \$11.50. **101.** Brooke needs a score of 85. **103.** The original price was \$200,000; purchasing the model saves \$30,000. **105.** The theater paid \$0.80 for each box. **107.** There were 400 adults. **109.** The length is 19 ft; the width is 11 ft. **111.** Approximately 184 million people **113.** In obtaining step (7) we divided by $x - 2$. Since $x = 2$ from step (1), we actually divided by 0.

Historical Problems *(page 117)*

1. The area of each shaded square is 9, so the larger square will have area $85 + 4(9) = 121$. The area of the larger square is also given by the expression $(x+6)^2$, so $(x+6)^2 = 121$. Taking the positive square root of each side, $x + 6 = 11$ or $x = 5$.
2. Let $z = -6$, so $z^2 + 12z - 85 = -121$. We get the equation $u^2 - 121 = 0$ or $u^2 = 121$. Thus $u = \pm 11$, so $x = \pm 11 - 6$. $x = -17$ or $x = 5$.

3.
$$\left(x + \frac{b}{2a}\right)^2 = \left(\frac{\sqrt{b^2-4ac}}{2a}\right)^2$$
$$\left(x + \frac{b}{2a}\right)^2 - \left(\frac{\sqrt{b^2-4ac}}{2a}\right)^2 = 0$$
$$\left(x + \frac{b}{2a} - \frac{\sqrt{b^2-4ac}}{2a}\right)\left(x + \frac{b}{2a} + \frac{\sqrt{b^2-4ac}}{2a}\right) = 0$$
$$\left(x + \frac{b - \sqrt{b^2-4ac}}{2a}\right)\left(x + \frac{b + \sqrt{b^2-4ac}}{2a}\right) = 0$$
$$x = \frac{-b+\sqrt{b^2-4ac}}{2a} \text{ or } x = \frac{-b-\sqrt{b^2-4ac}}{2a}$$

1.3 Assess Your Understanding *(page 117)*

6. repeated; multiplicity 2 **7.** discriminant; negative **8.** F **9.** b **10.** d **11.** $\{0, 9\}$ **13.** $\{-5, 5\}$ **15.** $\{-3, 2\}$ **17.** $\left\{-\frac{1}{2}, 3\right\}$ **19.** $\{-4, 4\}$ **21.** $\{2, 6\}$ **23.** $\left\{\frac{3}{2}\right\}$ **25.** $\left\{-\frac{2}{3}, \frac{3}{2}\right\}$ **27.** $\left\{-\frac{2}{3}, \frac{3}{2}\right\}$ **29.** $\left\{-\frac{3}{4}, 2\right\}$ **31.** $\{-5, 5\}$ **33.** $\{-1, 3\}$ **35.** $\{-3, 0\}$ **37.** $\{-7, 3\}$ **39.** $\left\{-\frac{1}{4}, \frac{3}{4}\right\}$ **41.** $\left\{\frac{-1-\sqrt{7}}{6}, \frac{-1+\sqrt{7}}{6}\right\}$ **43.** $\{2-\sqrt{2}, 2+\sqrt{2}\}$ **45.** $\{2-\sqrt{5}, 2+\sqrt{5}\}$ **47.** $\left\{1, \frac{3}{2}\right\}$ **49.** No real solution **51.** $\left\{\frac{-1-\sqrt{5}}{4}, \frac{-1+\sqrt{5}}{4}\right\}$ **53.** $\left\{0, \frac{9}{4}\right\}$ **55.** $\left\{\frac{1}{3}\right\}$ **57.** $\left\{-\frac{2}{3}, 1\right\}$ **59.** $\left\{\frac{3-\sqrt{29}}{10}, \frac{3+\sqrt{29}}{10}\right\}$ **61.** $\left\{\frac{-2-\sqrt{10}}{2}, \frac{-2+\sqrt{10}}{2}\right\}$ **63.** $\left\{\frac{1-\sqrt{33}}{8}, \frac{1+\sqrt{33}}{8}\right\}$ **65.** $\left\{\frac{9-\sqrt{73}}{2}, \frac{9+\sqrt{73}}{2}\right\}$ **67.** No real solution **69.** Repeated real solution **71.** Two unequal real solutions **73.** $\{-\sqrt{5}, \sqrt{5}\}$ **75.** $\left\{\frac{1}{4}\right\}$ **77.** $\left\{-\frac{3}{5}, \frac{5}{2}\right\}$ **79.** $\left\{-\frac{1}{2}, \frac{2}{3}\right\}$ **81.** $\left\{\frac{-\sqrt{2}-2}{2}, \frac{-\sqrt{2}+2}{2}\right\}$ **83.** $\left\{\frac{-1-\sqrt{17}}{2}, \frac{-1+\sqrt{17}}{2}\right\}$ **85.** $\left\{-\frac{1}{4}, \frac{2}{3}\right\}$ **87.** $\{5\}$ **89.** 2; 5 meters, 12 meters, 13 meters; 20 meters, 21 meters, 29 meters **91.** The dimensions are 11 ft by 13 ft. **93.** The dimensions are 5 m by 8 m. **95.** The dimensions should be 4 ft by 4 ft. **97. (a)** The ball strikes the ground after 6 sec. **(b)** The ball passes the top of the building on its way down after 5 sec. **99.** The dimensions should be 11.55 cm by 6.55 cm by 3 cm. **101.** The border will be 2.71 ft wide. **103.** The border will be 2.56 ft wide. **105.** The screen of an iPad Air in 4:3 format has an area of 45.16 square inches; the screen of the Google Nexus in 16:10 format has an area of 44.94 square inches. The iPad Air has a larger screen. **107.** 1.1 ft **109.** 29 hours **111.** 36 consecutive integers must be added.
113. $\frac{-b+\sqrt{b^2-4ac}}{2a} + \frac{-b-\sqrt{b^2-4ac}}{2a} = \frac{-2b}{2a} = -\frac{b}{a}$ **115.** $k = \frac{1}{2}$ or $k = -\frac{1}{2}$ **117.** $ax^2 + bx + c = 0, x = \frac{-b \pm \sqrt{b^2-4ac}}{2a}$; $ax^2 - bx + c = 0, x = \frac{b \pm \sqrt{(-b)^2-4ac}}{2a} = \frac{b \pm \sqrt{b^2-4ac}}{2a} = -\frac{-b \pm \sqrt{b^2-4ac}}{2a}$ **119.** b

1.4 Assess Your Understanding *(page 128)*

4. real; imaginary; imaginary unit **5.** F **6.** T **7.** F **8.** b **9.** a **10.** c **11.** $8 + 5i$ **13.** $-7 + 6i$ **15.** $-6 - 11i$ **17.** $6 - 18i$ **19.** $6 + 4i$ **21.** $10 - 5i$ **23.** 37 **25.** $\frac{6}{5} + \frac{8}{5}i$ **27.** $1 - 2i$ **29.** $\frac{5}{2} - \frac{7}{2}i$ **31.** $-\frac{1}{2} + \frac{\sqrt{3}}{2}i$ **33.** $2i$ **35.** $-i$ **37.** i **39.** -6 **41.** $-10i$ **43.** $-2 + 2i$ **45.** 0 **47.** 0 **49.** $2i$ **51.** $5i$ **53.** $2\sqrt{3}i$ **55.** $10\sqrt{2}i$ **57.** $5i$ **59.** $\{-2i, 2i\}$ **61.** $\{-4, 4\}$ **63.** $\{3-2i, 3+2i\}$ **65.** $\{3-i, 3+i\}$ **67.** $\left\{\frac{1}{4} - \frac{1}{4}i, \frac{1}{4} + \frac{1}{4}i\right\}$ **69.** $\left\{\frac{1}{5} - \frac{2}{5}i, \frac{1}{5} + \frac{2}{5}i\right\}$ **71.** $\left\{-\frac{1}{2} - \frac{\sqrt{3}}{2}i, -\frac{1}{2} + \frac{\sqrt{3}}{2}i\right\}$ **73.** $\{2, -1-\sqrt{3}i, -1+\sqrt{3}i\}$ **75.** $\{-2, 2, -2i, 2i\}$ **77.** $\{-3i, -2i, 2i, 3i\}$ **79.** Two complex solutions that are conjugates of each other **81.** Two unequal real solutions **83.** A repeated real solution **85.** $2 - 3i$ **87.** 6 **89.** 25 **91.** $2 + 3i$ ohms **93.** $z + \overline{z} = (a + bi) + (a - bi) = 2a$; $z - \overline{z} = (a + bi) - (a - bi) = 2bi$
95. $\overline{z + w} = \overline{(a + bi) + (c + di)} = \overline{(a + c) + (b + d)i} = (a + c) - (b + d)i = (a - bi) + (c - di) = \overline{z} + \overline{w}$

1.5 Assess Your Understanding *(page 135)*

6. F **7.** quadratic in form **8.** T **9.** a **10.** c **11.** $\{22\}$ **13.** $\{1\}$ **15.** No real solution **17.** $\{-13\}$ **19.** $\{4\}$ **21.** $\{-1\}$ **23.** $\{0, 64\}$ **25.** $\{3\}$ **27.** $\{2\}$
29. $\left\{-\frac{8}{5}\right\}$ **31.** $\{8\}$ **33.** $\{-1, 3\}$ **35.** $\{1, 5\}$ **37.** $\{1\}$ **39.** $\{5\}$ **41.** $\{2\}$ **43.** $\{-4, 4\}$ **45.** $\{-2, 2\}$ **47.** $\{-2, -1, 1, 2\}$ **49.** $\{-1, 1\}$ **51.** $\{-2, 1\}$
53. $\{-6, -5\}$ **55.** $\left\{-\frac{3}{2}, 2\right\}$ **57.** $\{0, 16\}$ **59.** $\{16\}$ **61.** $\{1\}$ **63.** $\left\{\left(\frac{9-\sqrt{17}}{8}\right)^4, \left(\frac{9+\sqrt{17}}{8}\right)^4\right\}$ **65.** $\left\{-2, -\frac{1}{2}\right\}$ **67.** $\left\{-\frac{3}{2}, \frac{1}{3}\right\}$ **69.** $\left\{-\frac{1}{8}, 27\right\}$
71. $\{-4, 1\}$ **73.** $\left\{-1, \frac{3}{2}\right\}$ **75.** $\{-4, 4\}$ **77.** $\left\{-\frac{1}{2}, \frac{1}{2}\right\}$ **79.** $\left\{-\frac{27}{2}, \frac{27}{2}\right\}$ **81.** $\left\{-\frac{36}{5}, \frac{24}{5}\right\}$ **83.** No real solution **85.** $\{-3, 3\}$ **87.** $\{-1, 3\}$
89. $\{-3, 0, 3\}$ **91.** $\{-5, 0, 4\}$ **93.** $\{-1, 1\}$ **95.** $\{-2, 2, 3\}$ **97.** $\left\{-2, \frac{1}{2}, 2\right\}$ **99.** $\{0.34, 11.66\}$ **101.** $\{-1.03, 1.03\}$ **103.** $\left\{-4, \frac{5}{3}\right\}$
105. $\left\{-3, -\frac{2}{5}, 3\right\}$ **107.** $\left\{-\frac{1}{5}, 1\right\}$ **109.** $\{5\}$ **111.** $\left\{-2, -\frac{4}{5}\right\}$ **113.** $\{-2, 6\}$ **115.** $\{2\}$ **117.** $\left\{\frac{-3-\sqrt{6}}{3}, \frac{-3+\sqrt{6}}{3}\right\}$ **119.** $\{-2, -1, 0, 1\}$
121. $\{\sqrt{2}, \sqrt{3}\}$ **123.** $\{-2, 2, -2i, 2i\}$ **125.** $\left\{1, 2, -1-\sqrt{3}i, -1+\sqrt{3}i, -\frac{1}{2}-\frac{\sqrt{3}}{2}i, -\frac{1}{2}+\frac{\sqrt{3}}{2}i\right\}$ **127.** $\left\{\frac{3}{2}, 5\right\}$
129. $(2, 2); (2, -4)$ **131.** $(0, 0); (8, 0)$ **133.** $\{1, 3\}$ **135.** The distance is approximately 229.94 ft. **137.** approx. 221 feet

1.6 Assess Your Understanding *(page 143)*

1. mathematical modeling **2.** interest **3.** uniform motion **4.** F **5.** T **6.** a **7.** b **8.** c **9.** $A = \pi r^2$; r = radius, A = area
11. $A = s^2$; A = area, s = length of a side **13.** $F = ma$; F = force, m = mass, a = acceleration **15.** $W = Fd$; W = work, F = force, d = distance
17. $C = 150x$; C = total variable cost, x = number of dishwashers **19.** Invest \$31,250 in bonds and \$18,750 in CDs. **21.** \$11,600 was loaned out at 8%.
23. Mix 75 lb of Earl Grey tea with 25 lb of Orange Pekoe tea. **25.** Mix 160 lb of cashews with the almonds. **27.** The speed of the current is 2.286 mi/h.
29. The speed of the current is 5 mi/h. **31.** Karen walked at 4.05 ft/sec. **33.** A doubles tennis court is 78 feet long and 36 feet wide.
35. Working together, it takes 12 min. **37. (a)** The dimensions are 10 ft by 5 ft. **(b)** The area is 50 sq ft. **(c)** The dimensions would be 7.5 ft by 7.5 ft.
(d) The area would be 56.25 sq ft. **39.** The defensive back catches up to the tight end at the tight end's 45-yd line. **41.** Add $\frac{2}{3}$ gal of water.
43. Evaporate 10.67 oz of water. **45.** 40 g of 12-karat gold should be mixed with 20 g of pure gold.
47. Mike passes Dan $\frac{1}{3}$ mile from the start, 2 min from the time Mike started to run. **49.** Start the auxiliary pump at 9:45 AM. **51.** The tub will fill in 1 h.
53. Run: 12 miles; bicycle: 75 miles **55.** Bolt would beat Burke by 19.25 m. **57.** Set the original price at \$40. At 50% off, there will be no profit.
61. The tail wind was 91.47 knots.

1.7 Assess Your Understanding *(page 154)*

4. negative **5.** closed interval **6.** $-a \le u \le a$ **7.** $(-\infty, a]$ **8.** T **9.** T **10.** F **11.** a **12.** c **13.** $[0, 2]; 0 \le x \le 2$
15. $[2, \infty); x \ge 2$ **17.** $[0, 3); 0 \le x < 3$ **19. (a)** $6 < 8$ **(b)** $-2 < 0$ **(c)** $9 < 15$ **(d)** $-6 > -10$
21. (a) $2x + 4 < 5$ **(b)** $2x - 4 < -3$ **(c)** $6x + 3 < 6$ **(d)** $-4x - 2 > -4$

23. $[0, 4]$

25. $[4, 6)$

27. $[4, \infty)$

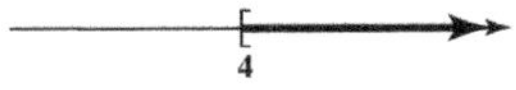

29. $(-\infty, -4)$

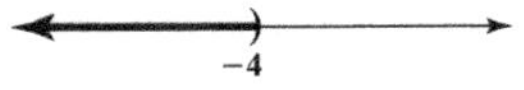

31. $2 \le x \le 5$

33. $-3 < x < -2$

35. $x \ge 4$

37. $x < -3$

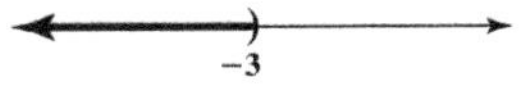

39. $<$ **41.** $>$ **43.** $\ge$ **45.** $<$ **47.** $<$ **49.** $\ge$

51. $\{x \mid x > 3\}$ or $(3, \infty)$

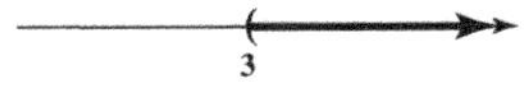

53. $\{x \mid x \ge -1\}$ or $[-1, \infty)$

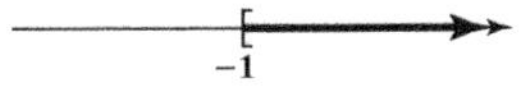

55. $\{x \mid x \ge 2\}$ or $[2, \infty)$

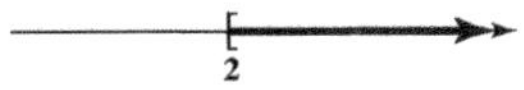

57. $\{x \mid x > -7\}$ or $(-7, \infty)$

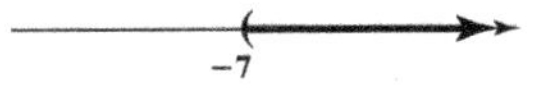

59. $\left\{x \,\middle|\, x \le \frac{2}{3}\right\}$ or $\left(-\infty, \frac{2}{3}\right]$

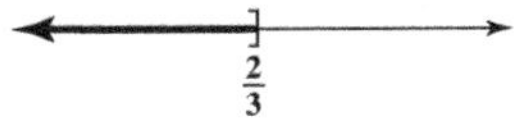

61. $\{x \mid x < -20\}$ or $(-\infty, -20)$

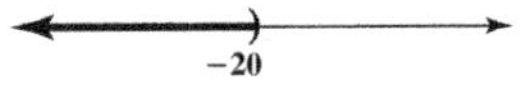

63. $\{x \mid 3 \le x \le 5\}$ or $[3, 5]$

65. $\left\{x \,\middle|\, \frac{2}{3} \le x \le 3\right\}$ or $\left[\frac{2}{3}, 3\right]$

67. $\left\{x \,\middle|\, -\frac{11}{2} < x < \frac{1}{2}\right\}$ or $\left(-\frac{11}{2}, \frac{1}{2}\right)$

69. $\{x \mid -6 < x < 0\}$ or $(-6, 0)$

71. $\{x \mid x < -5\}$ or $(-\infty, -5)$

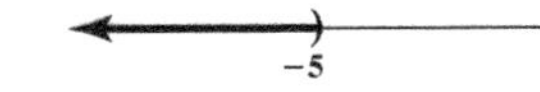

73. $\{x \mid -4 < x < 4\}$; $(-4, 4)$

75. $\{x \mid x < -4 \text{ or } x > 4\}$; $(-\infty, -4) \cup (4, \infty)$

77. $\left\{t \,\middle|\, -\frac{2}{3} \le t \le 2\right\}$; $\left[-\frac{2}{3}, 2\right]$

79. $\{x \mid 1 < x < 3\}$; $(1, 3)$

81. $\{x \mid x \le 1 \text{ or } x \ge 5\}$; $(-\infty, 1] \cup [5, \infty)$

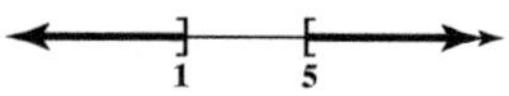

83. $\{x \mid x < -1 \text{ or } x > 2\}$; $(-\infty, -1)$ or $(2, \infty)$

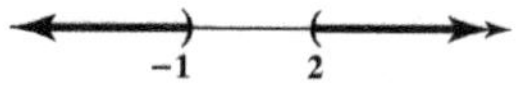

85. $\left\{x \middle| -1 < x < \frac{3}{2}\right\}$; $\left(-1, \frac{3}{2}\right)$

87. No solution

89. $\{x \mid x > -2\}$; $(-2, \infty)$

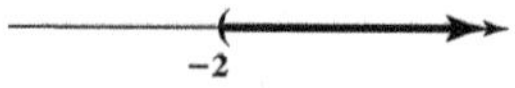

91. $\left\{x \middle| x \le -\frac{5}{2} \text{ or } x \ge \frac{3}{2}\right\}$; $\left(-\infty, -\frac{5}{2}\right] \cup \left[\frac{3}{2}, \infty\right)$

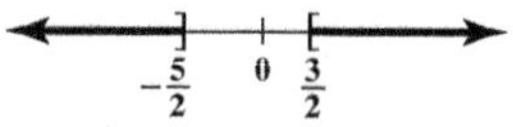

93. $\left\{x \middle| x \ge \frac{4}{3}\right\}$; $\left[\frac{4}{3}, \infty\right)$

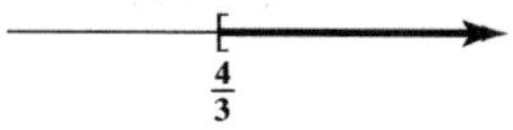

95. $\{x \mid -3 \le x < 7\}$; $[-3, 7)$

97. $\{x \mid x \ge -1\}$; $[-1, \infty)$;

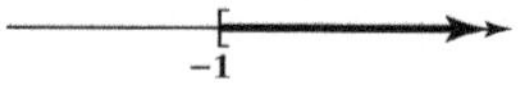

99. $\left\{x \middle| \frac{17}{6} < x < \frac{19}{6}\right\}$; $\left(\frac{17}{6}, \frac{19}{6}\right)$

101. $\{x \mid -3 \le x < 4\}$; $[-3, 4)$

103. $\{x \mid -2 < x < 4\}$; $(-2, 4)$

105. $\{x \mid x > 5\}$; $(5, \infty)$

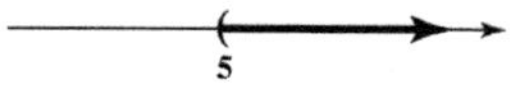

107. $\{x \mid x > 3\}$; $(3, \infty)$

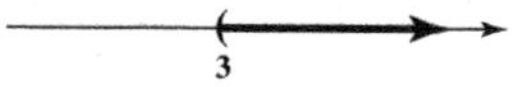

109. $|x - 2| < \frac{1}{2}$; $\left\{x \middle| \frac{3}{2} < x < \frac{5}{2}\right\}$

111. $|x + 3| > 2$; $\{x \mid x < -5 \text{ or } x > -1\}$ **113.** $21 < \text{age} < 30$ **115.** $|x - 98.6| \ge 1.5$; $\{x \mid x \le 97.1 \text{ or } x \ge 100.1\}$
117. (a) Male ≥ 81.9 **(b)** Female ≥ 85.6 **(c)** A female can expect to live at least 3.7 years longer. **119.** The agent's commission ranges from \$45,000 to \$95,000, inclusive. As a percent of selling price, the commission ranges from 5% to approximately 8.6%, inclusive.
121. The amount withheld varies from \$134.50 to \$184.50, inclusive. **123.** The usage varies from approximately 700 to 2700 kilowatt-hours, inclusive.
125. The dealer's cost varies from \$7457.63 to \$7857.14, inclusive. **127. (a)** You need at least a 74 on the last test.
(b) You need at least a 77 on the last test. **129.** $|x - 13.6| < 1.8$; between 11.8 and 15.4 books per year are read, on average.

131. $\frac{a+b}{2} - a = \frac{a+b-2a}{2} = \frac{b-a}{2} > 0$; therefore, $a < \frac{a+b}{2}$

$b - \frac{a+b}{2} = \frac{2b-a-b}{2} = \frac{b-a}{2} > 0$; therefore, $b > \frac{a+b}{2}$.

133. $(\sqrt{ab})^2 - a^2 = ab - a^2 = a(b-a) > 0$; thus, $(\sqrt{ab})^2 > a^2$ and $\sqrt{ab} > a$.

$b^2 - (\sqrt{ab})^2 = b^2 - ab = b(b-a) > 0$; thus, $b^2 > (\sqrt{ab})^2$ and $b > \sqrt{ab}$.

135. $\frac{1}{h} = \frac{1}{2}\left(\frac{1}{a} + \frac{1}{b}\right) = \frac{1}{2}\left(\frac{b+a}{ab}\right)$; $h = \frac{2ab}{b+a}$

$h - a = \frac{2ab}{b+a} - a = \frac{2ab - a(b+a)}{b+a} = \frac{2ab - ab - a^2}{b+a} = \frac{ab - a^2}{b+a} = \frac{a(b-a)}{b+a} > 0$, since $0 < a < b$.

$b - h = b - \frac{2ab}{b+a} = \frac{b(b+a) - 2ab}{b+a} = \frac{b^2 + ab - 2ab}{b+a} = \frac{b^2 - ab}{b+a} = \frac{b(b-a)}{b+a} > 0$, since $0 < a < b$.

Review Exercises *(page 159)*

1. $\{-18\}$ **2.** $\{6\}$ **3.** $\left\{\frac{1}{5}\right\}$ **4.** $\{6\}$ **5.** No real solution **6.** $\left\{-\frac{27}{13}\right\}$ **7.** $\left\{-2, \frac{3}{2}\right\}$ **8.** $\left\{\frac{1-\sqrt{13}}{4}, \frac{1+\sqrt{13}}{4}\right\}$ **9.** $\{-3, 3\}$ **10.** No real solution

11. No real solution **12.** $\{-2, -1, 1, 2\}$ **13.** $\{2\}$ **14.** $\left\{\frac{13}{2}\right\}$ **15.** $\left\{\frac{\sqrt{5}}{2}\right\}$ **16.** $\left\{\frac{9}{4}\right\}$ **17.** $\left\{-1, \frac{1}{2}\right\}$ **18.** $\left\{\frac{m}{1-n}, \frac{m}{1+n}\right\}$ **19.** $\left\{-\frac{9b}{5a}, \frac{2b}{a}\right\}$

20. $\left\{-\frac{9}{5}\right\}$ **21.** $\{-5, 2\}$ **22.** $\left\{-\frac{5}{3}, 3\right\}$ **23.** $\left\{0, \frac{3}{2}\right\}$ **24.** $\left\{-\frac{5}{2}, -2, 2\right\}$ **25.** $\{3\}$ **26.** $\{-1, 5\}$ **27.** $\{-2.49, 0.66, 1.83\}$ **28.** $\{-1.14, 1.64\}$

29. $\{x \mid x \ge 14\}$; $[14, \infty)$

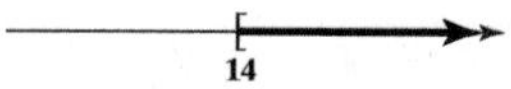

30. $\left\{x \middle| -\frac{31}{2} \le x \le \frac{33}{2}\right\}$; $\left[-\frac{31}{2}, \frac{33}{2}\right]$

31. $\{x \mid -23 < x < -7\}$; $(-23, -7)$

32. $\left\{x \middle| -\frac{3}{2} < x < -\frac{7}{6}\right\}$; $\left(-\frac{3}{2}, -\frac{7}{6}\right)$

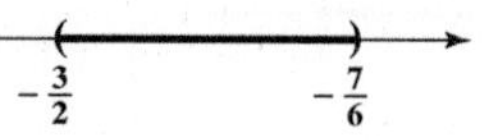

33. $\{x \mid x \le -2 \text{ or } x \ge 7\}$; $(-\infty, -2] \cup [7, \infty)$

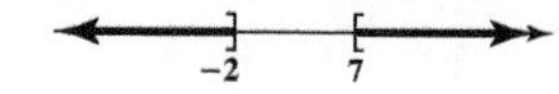

34. $\left\{x \middle| 0 \le x \le \frac{4}{3}\right\}$; $\left[0, \frac{4}{3}\right]$

35. $\left\{x \middle| x < -1 \text{ or } x > \frac{7}{3}\right\}; \left(-\infty, -1\right) \cup \left(\frac{7}{3}, \infty\right)$ **36.** $4 + 7i$ **37.** $-3 + 2i$ **38.** $\frac{9}{10} - \frac{3}{10}i$ **39.** -1 **40.** $-46 + 9i$ **41.** $\left\{-\frac{1}{2} - \frac{\sqrt{3}}{2}i, -\frac{1}{2} + \frac{\sqrt{3}}{2}i\right\}$

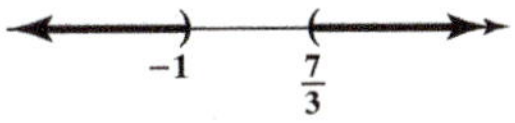

42. $\left\{\frac{-1 - \sqrt{17}}{4}, \frac{-1 + \sqrt{17}}{4}\right\}$ **43.** $\left\{\frac{1}{2} - \frac{\sqrt{11}}{2}i, \frac{1}{2} + \frac{\sqrt{11}}{2}i\right\}$ **44.** $\{-2, 1\}$ **45. (a)** $2\sqrt{5}$ **(b)** $(2, 1)$ **46. (a)** 5 **(b)** $\left(-\frac{1}{2}, 1\right)$ **47. (a)** 12 **(b)** $(4, 2)$

48.

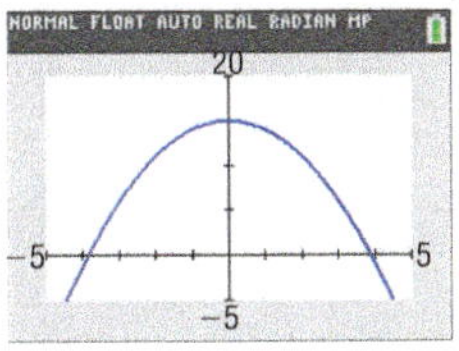

49. $(-4, 0), (0, 2), (0, 0), (0, -2), (2, 0)$ **50.**

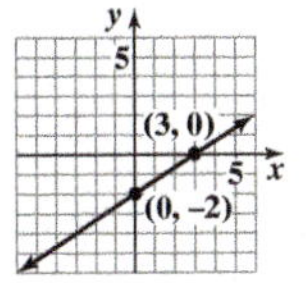

51.

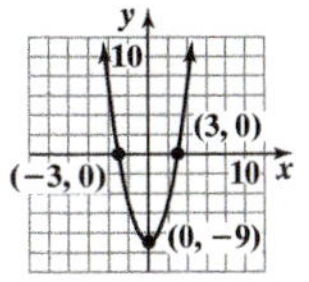

52.

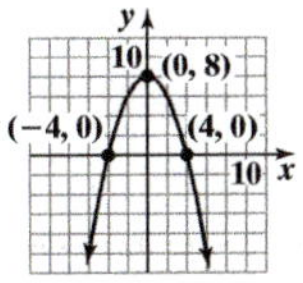

53. $d(A, B) = \sqrt{13}$; $d(B, C) = \sqrt{13}$ **54.** -4 and 8 **55.** $p = 2l + 2w$ **56.** The interest is \$630. **57.** He should invest \$50,000 in A-rated bonds and \$20,000 in the CD. **58.** The storm is 3300 ft away. **59.** The range of distances is from 0.5 to 0.75 m, $0.5 \le x \le 0.75$. **60.** The search plane can go as far as 616 mi. **61.** The helicopter will reach the life raft in a little less than 1 hr 35 min. **62.** The bees meet for the first time in 18.75 seconds. The bees meet for the second time 37.5 seconds later. **63. (a)** The object will strike the ground in 8 seconds. **(b)** The height is 896 ft. **64.** It takes Clarissa 10 days by herself.

65. Add $6\frac{2}{3}$ lb of \$8/lb coffee to get $26\frac{2}{3}$ lb of \$5/lb coffee. **66.** Evaporate 51.2 oz of water. **67.** 5 cm and 12 cm **68.** The freight train is 190.67 ft long.

69. (a) 6.5 in. by 6.5 in.; 12.5 in. by 12.5 in. **(b)** $8\frac{2}{3}$ in. by $4\frac{1}{3}$ in.; $14\frac{2}{3}$ in. by $10\frac{1}{3}$ in. **70.** It will take the smaller pump 2 h.

71. The length should be approximately 6.47 ft. **72.** 36 seniors went on the trip; each paid \$13.40.

73. (a) No **(b)** Todd wins again. **(c)** Todd wins by $\frac{1}{4}$ m. **(d)** Todd should line up 5.26 m behind the start line. **(e)** Yes

Chapter Test *(page 162)*

1. (a) $d = 10$ **(b)** $M = (1, 1)$ **2.** $\left\{-\frac{3}{2}, -1\right\}$ **3.** $\{2\}$ **4.** $\left\{-\frac{3}{2}, 2\right\}$ **5.** $\{2\}$ **6.** $\{0, 3\}$ **7.** $\{-2, 2\}$ **8.** $\{2 - \sqrt{2}, 2 + \sqrt{2}\}$ **9.** $\{3 - 2\sqrt{3}, 3 + 2\sqrt{3}\}$

10.

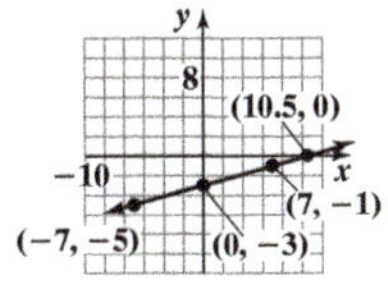

11.

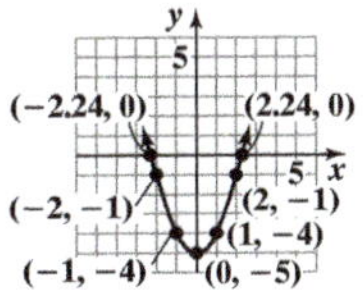

12. $\{-1, 0.5, 1\}$
13. $\{-2.50, 2.50\}$
14. $\{-2.46, -0.24, 1.70\}$
15. $\left\{x \middle| x < -\frac{11}{2}\right\}; \left(-\infty, -\frac{11}{2}\right)$

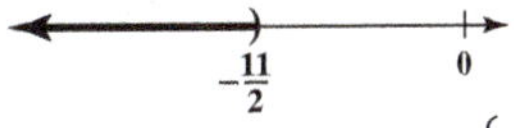

16. $\{x | x \le -5 \text{ or } x \ge 2\}$; $(\infty, -5] \cup [2, \infty)$

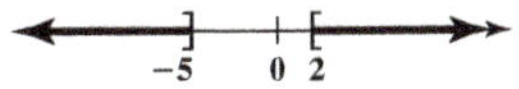

17. $\{x | -1 \le x < 2\}$; $[-1, 2)$

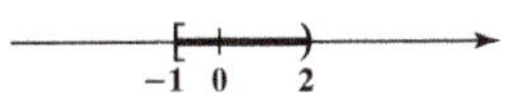

18. $\left\{x \middle| -4 < x < \frac{4}{3}\right\}; \left(-4, \frac{4}{3}\right)$

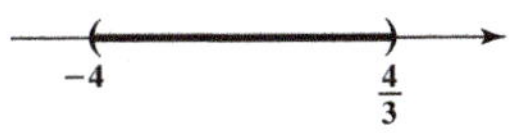

19. $2 - 25i$ **20.** $14 + 83i$ **21.** $\frac{7}{34} + \frac{11}{34}i$ **22.** $\left\{\frac{1}{2} - i, \frac{1}{2} + i\right\}$ **23.** About 204.63 min (3.41 h) **24.** 23.75 lb of banana chips

25. The sale price is \$159.50. **26.** Glenn will earn \$100 in interest after 3 months.

CHAPTER 2 Graphs

2.1 Assess Your Understanding *(page 170)*

3. intercepts **4.** y-axis **5.** 4 **6.** $(-3, 4)$ **7.** T **8.** F **9.** a **10.** c

11. $(-2, 0), (0, 2)$

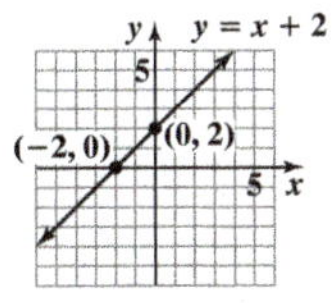

13. $(-4, 0), (0, 8)$

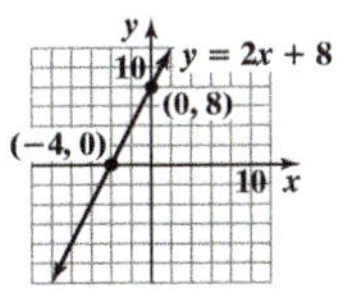

15. $(-1, 0), (1, 0), (0, -1)$

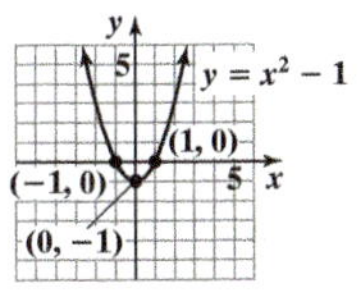

17. $(-2, 0), (2, 0), (0, 4)$

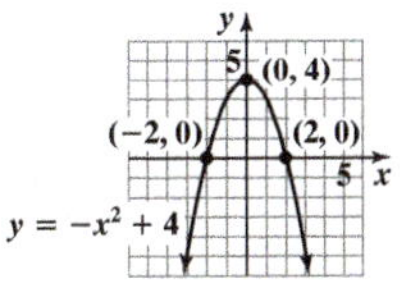

19. $(3, 0), (0, 2)$

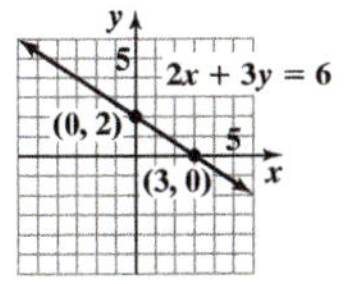

21. $(-2, 0), (2, 0), (0, 9)$

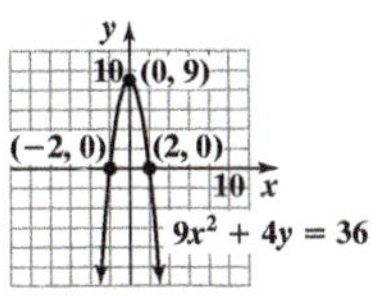

23.

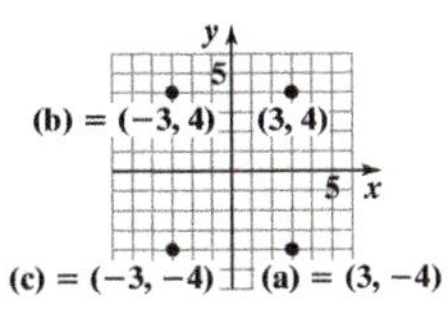

25.

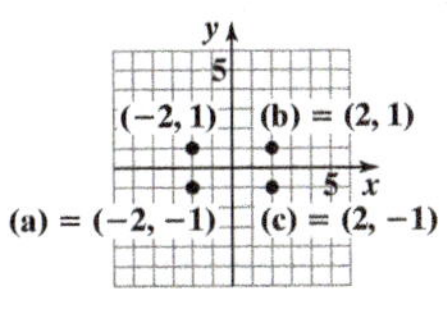

27.

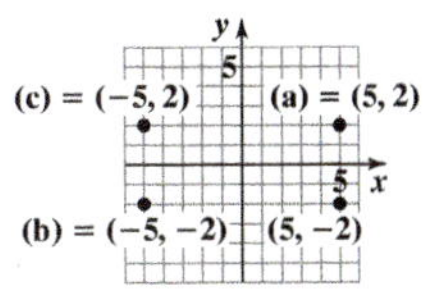

29.

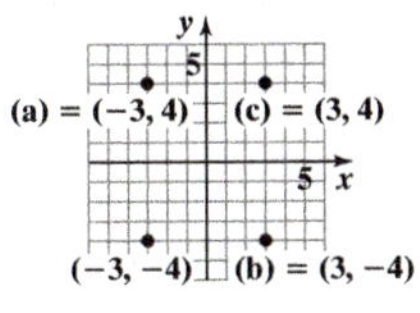

31.

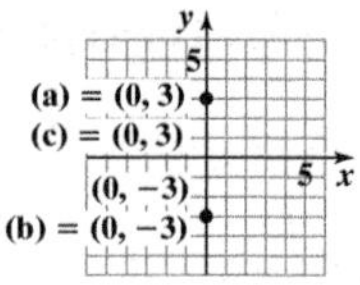

33. (a) $(-1, 0), (1, 0)$ **(b)** Symmetric with respect to the x-axis, the y-axis, and the origin **35. (a)** $\left(-\frac{\pi}{2}, 0\right), (0, 1), \left(\frac{\pi}{2}, 0\right)$ **(b)** Symmetric with respect to the y-axis **37. (a)** $(0, 0)$ **(b)** Symmetric with respect to the x-axis **39. (a)** $(-2, 0), (0, 0), (2, 0)$ **(b)** Symmetric with respect to the origin **41. (a)** $(x, 0), -2 \le x \le 1$ **(b)** No Symmetry **43. (a)** No intercepts **(b)** Symmetric with respect to the origin

45.

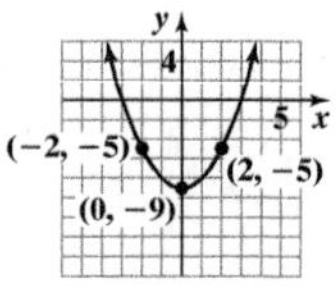

47.

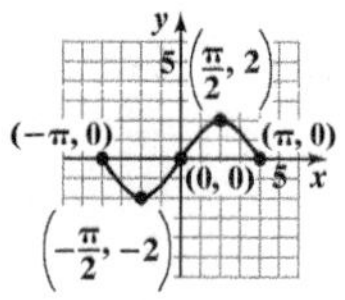

49. $(-4, 0), (0, -2), (0, 2)$; symmetric with respect to the x-axis **51.** $(0, 0)$; symmetric with respect to the origin **53.** $(0, -9), (3, 0), (-3, 0)$; symmetric with respect to the y-axis **55.** $(-2, 0), (2, 0), (0, -3), (0, 3)$; symmetric with respect to the x-axis, y-axis, and origin **57.** $(0, -27), (3, 0)$; no symmetry **59.** $(0, -4), (4, 0), (-1, 0)$; no symmetry **61.** $(0, 0)$; symmetric with respect to the origin **63.** $(0, 0)$; symmetric with respect to the origin

65.

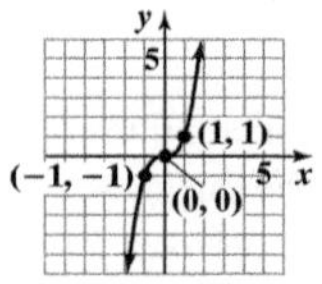

67.

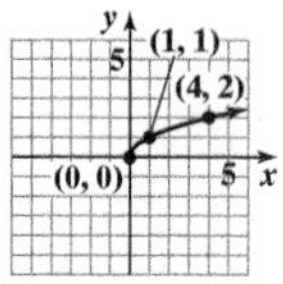

69. $b = 13$ **71.** $a = -4$ or $a = 1$
73. (a) $(0, -5), (-\sqrt{5}, 0), (\sqrt{5}, 0)$ **(b)** Symmetric with respect to the y-axis **(c)**

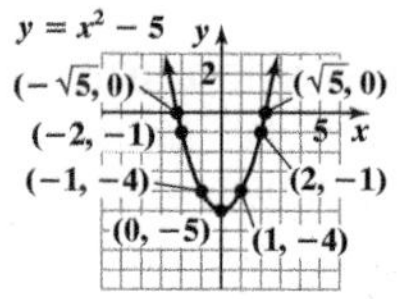

75. (a) $(-9, 0), (0, -3), (0, 3)$ **(b)** Symmetric with respect to the x-axis **(c)**

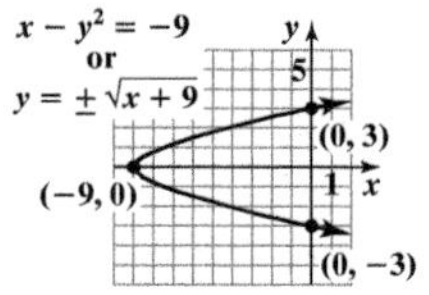

77. (a) $(0, 3), (0, -3), (-3, 0), (3, 0)$ **(b)** Symmetric with respect to the x-axis, y-axis, and origin **(c)**

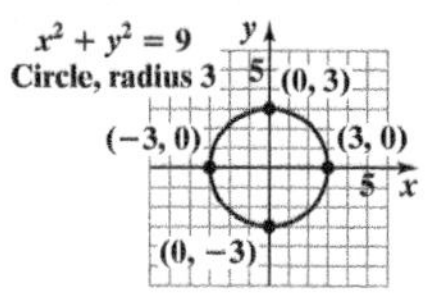

79. (a) $(0, 0), (-2, 0), (2, 0)$ **(b)** Symmetric with respect to the origin **(c)**

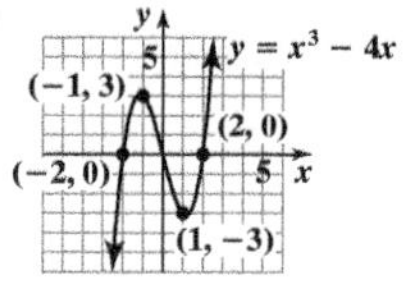

81. $(-1, -2)$ **83.** 4 **85. (a)** $(0, 0), (2, 0), (0, 1), (0, -1)$ **(b)** x-axis symmetry **87. (a)** $y = \sqrt{x^2}$ and $y = |x|$ have the same graph. **(b)** $\sqrt{x^2} = |x|$ **(c)** $x \ge 0$ for $y = (\sqrt{x})^2$, while x can be any real number for $y = x$. **(d)** $y \ge 0$ for $y = \sqrt{x^2}$
94. $\frac{1}{2}$ **95.** $3(x-5)^2$ **96.** $14i$ **97.** $\{4 - 2\sqrt{3}, 4 + 2\sqrt{3}\}$

2.2 Assess Your Understanding *(page 184)*

1. undefined; 0 **2.** 3; 2 **3.** T **4.** F **5.** T **6.** $m_1 = m_2$; y-intercepts; $m_1 m_2 = -1$ **7.** 2 **8.** $-\frac{1}{2}$ **9.** F **10.** d **11.** c **12.** b
13. (a) Slope $= \frac{1}{2}$ **(b)** If x increases by 2 units, y will increase by 1 unit. **15. (a)** Slope $= -\frac{1}{3}$ **(b)** If x increases by 3 units, y will decrease by 1 unit.

17. Slope $= -\frac{3}{2}$

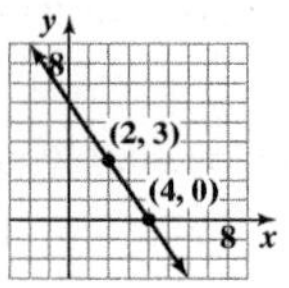

19. Slope $= -\frac{1}{2}$

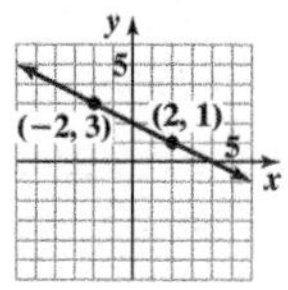

21. Slope $= 0$

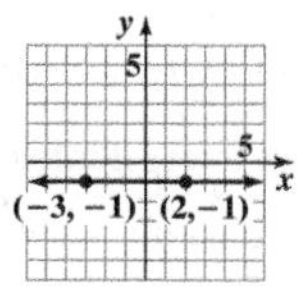

23. Slope undefined

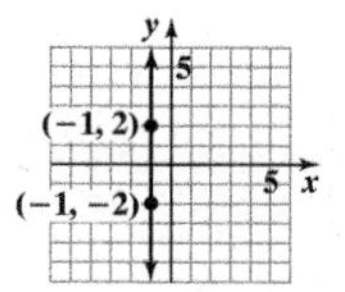

25.

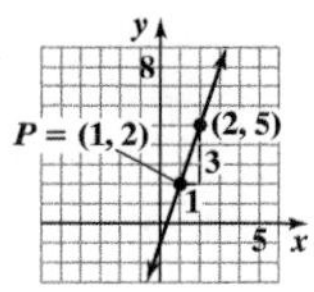

27.

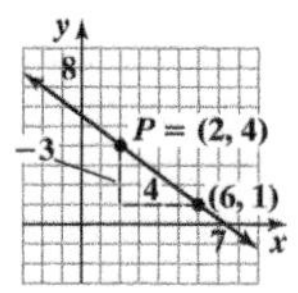

29.

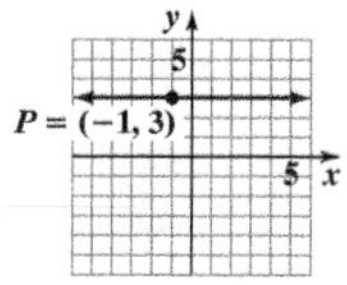

31.

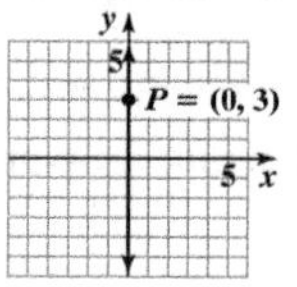

33. $(2, 6); (3, 10); (4, 14)$ **35.** $(4, -7); (6, -10); (8, -13)$ **37.** $(-1, -5); (0, -7); (1, -9)$ **39.** $x - 2y = 0$ or $y = \frac{1}{2}x$
41. $x + y = 2$ or $y = -x + 2$ **43.** $2x - y = 3$ or $y = 2x - 3$ **45.** $x + 2y = 5$ or $y = -\frac{1}{2}x + \frac{5}{2}$ **47.** $3x - y = -9$ or $y = 3x + 9$
49. $2x + 3y = -1$ or $y = -\frac{2}{3}x - \frac{1}{3}$ **51.** $3x + y = 3$ or $y = -3x + 3$ **53.** $x - 2y = -5$ or $y = \frac{1}{2}x + \frac{5}{2}$ **55.** $x - 2y = 2$ or $y = \frac{1}{2}x - 1$
57. $x = 2$; no slope–intercept form **59.** $y = 2$ **61.** $4x - y = -6$ or $y = 4x + 6$ **63.** $5x - y = 0$ or $y = 5x$ **65.** $x = 4$; no slope–intercept form
67. $6x + y = 0$ or $y = -6x$ **69.** $5x - 2y = -3$ or $y = \frac{5}{2}x + \frac{3}{2}$ **71.** $y = 4$

73. Slope = 2; y-intercept = 3

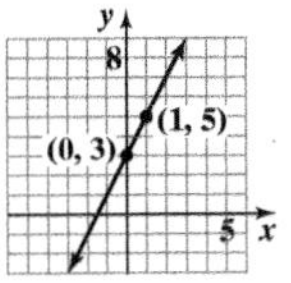

75. Slope = 4; y-intercept = −4

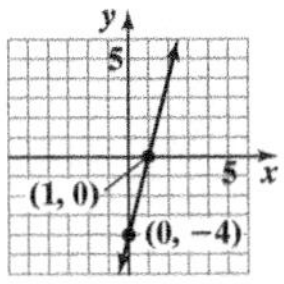

77. Slope = $\frac{1}{2}$; y-intercept = 2

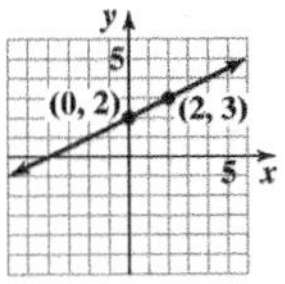

79. Slope = $-\frac{1}{4}$; y-intercept = 1

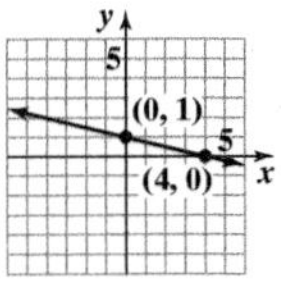

81. Slope = $\frac{2}{3}$; y-intercept = −2

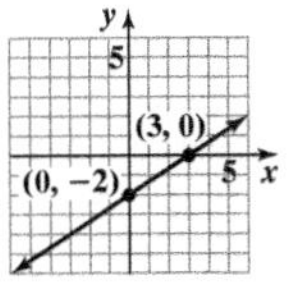

83. Slope = −1; y-intercept = 1

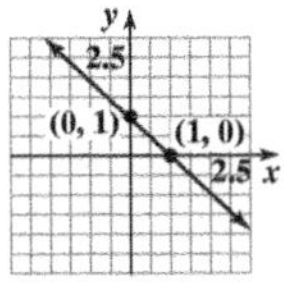

85. Slope undefined; no y-intercept

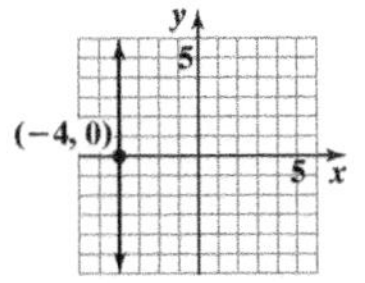

87. Slope = 0; y-intercept = 5

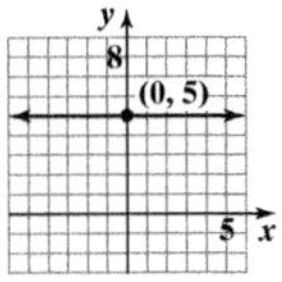

89. Slope = 1; y-intercept = 0

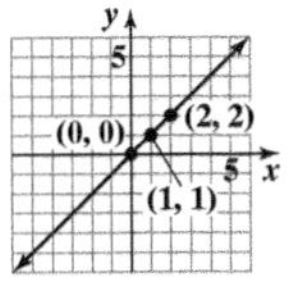

91. Slope = $\frac{3}{2}$; y-intercept = 0

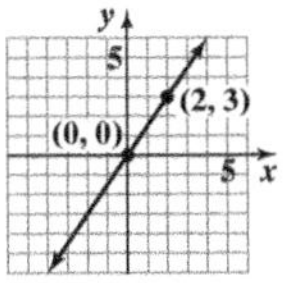

93. (a) x-intercept: 3; y-intercept: 2
(b)

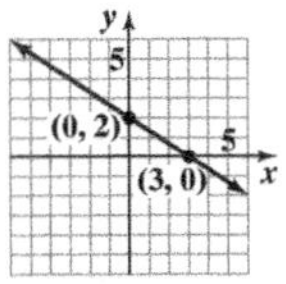

95. (a) x-intercept: −10; y-intercept: 8
(b)

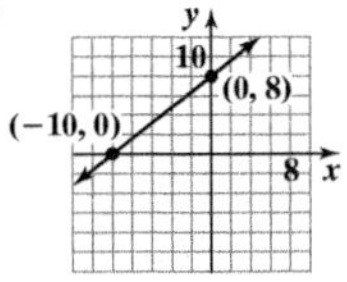

97. (a) x-intercept: 3; y-intercept: $\frac{21}{2}$
(b)

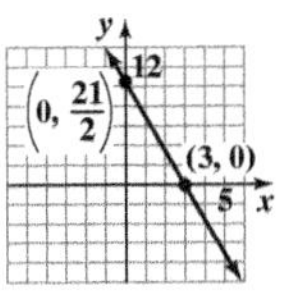

99. (a) x-intercept: 2; y-intercept: 3
(b)

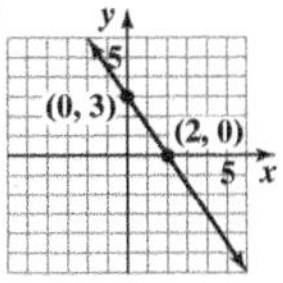

101. (a) x-intercept: 5; y-intercept: −2
(b)

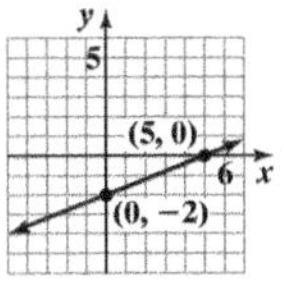

103. $y = 0$
105. Parallel
107. Neither
109. $x - y = -2$ or $y = x + 2$
111. $x + 3y = 3$ or $y = -\frac{1}{3}x + 1$

113. $P_1 = (-2, 5), P_2 = (1, 3), m_1 = -\frac{2}{3}$; $P_2 = (1, 3), P_3 = (-1, 0), m_2 = \frac{3}{2}$; because $m_1 m_2 = -1$, the lines are perpendicular and the points $(-2, 5)$, $(1, 3)$, and $(-1, 0)$ are the vertices of a right triangle.
115. $P_1 = (-1, 0), P_2 = (2, 3), m = 1$; $P_3 = (1, -2), P_4 = (4, 1), m = 1$; $P_1 = (-1, 0), P_3 = (1, -2), m = -1$; $P_2 = (2, 3), P_4 = (4, 1), m = -1$; opposite sides are parallel, and adjacent sides are perpendicular; the points are the vertices of a rectangle.
117. $C = 0.60x + 39$; \$105; \$177 **119.** $C = 0.16x + 1461$

121. (a) $C = 0.0757x + 15.14, 0 \le x \le 800$
(b)

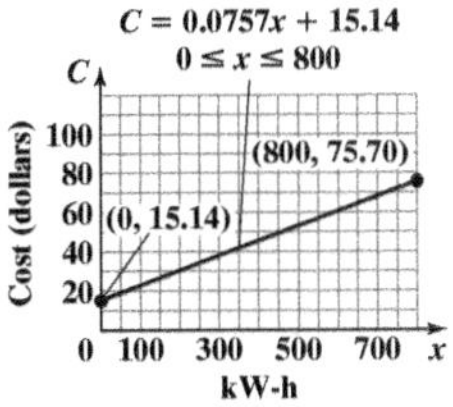

(c) \$30.28 **(d)** \$52.99
(e) Each additional kW-h used adds \$0.0757 to the bill.

123. °C = $\frac{5}{9}$(°F − 32); approximately 21.1°C **125. (a)** $y = -\frac{2}{25}x + 30$ **(b)** x-intercept: 375; The ramp meets the floor 375 in. (31.25 ft) from the base of the platform. **(c)** The ramp does not meet design requirements. It has a run of 31.25 ft. **(d)** The only slope possible for the ramp to comply with the requirement is for it to drop 1 in. for every 12-in. run.

127. (a) $A = \frac{1}{5}x + 20{,}000$ **(b)** \$80,000 **(c)** Each additional box sold requires an additional \$0.20 in advertising. **129.** All have the same slope, 2; the lines are parallel.

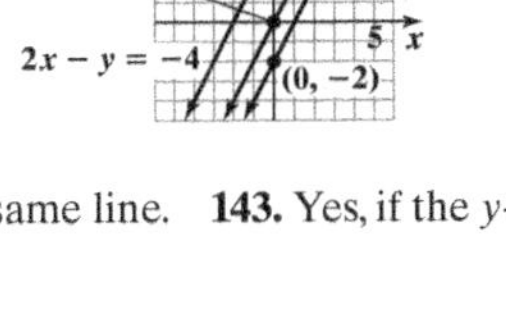

131. (b), (c), (e), (g) **133.** (c) **139.** No; no **141.** They are the same line. **143.** Yes, if the y-intercept is 0. **146.** x^4y^{16} **147.** 17
148. $\{3 - 2\sqrt{6}, 3 + 2\sqrt{6}\}$ **149.** $\{x \mid 1 < x < 4\}$ or $(1, 4)$

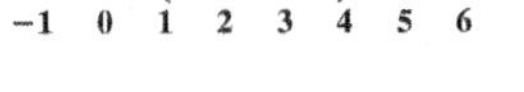

2.3 Assess Your Understanding *(page 193)*

3. F **4.** radius **5.** T **6.** F **7.** d **8.** a **9.** Center $(2, 1)$; radius = 2; $(x - 2)^2 + (y - 1)^2 = 4$
11. Center $\left(\frac{5}{2}, 2\right)$; radius = $\frac{3}{2}$; $\left(x - \frac{5}{2}\right)^2 + (y - 2)^2 = \frac{9}{4}$

13. $x^2 + y^2 = 4$; $x^2 + y^2 - 4 = 0$

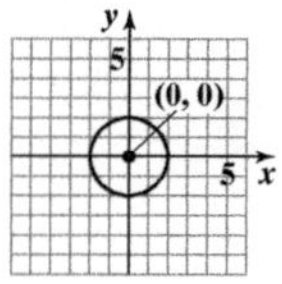

15. $x^2 + (y - 2)^2 = 4$; $x^2 + y^2 - 4y = 0$

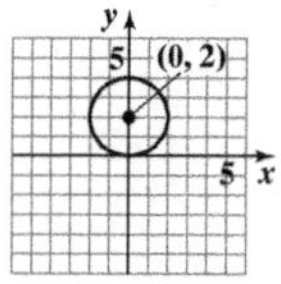

17. $(x - 4)^2 + (y + 3)^2 = 25$; $x^2 + y^2 - 8x + 6y = 0$

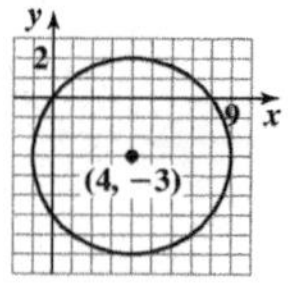

19. $(x + 2)^2 + (y - 1)^2 = 16$; $x^2 + y^2 + 4x - 2y - 11 = 0$

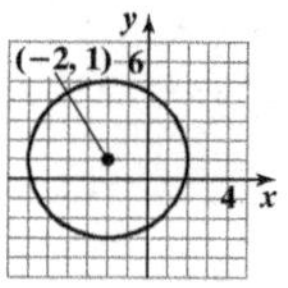

21. $\left(x - \frac{1}{2}\right)^2 + y^2 = \frac{1}{4}$; $x^2 + y^2 - x = 0$

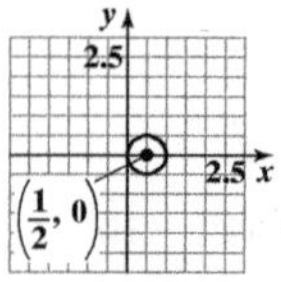

23. (a) $(h, k) = (0, 0); r = 2$ **(b)**

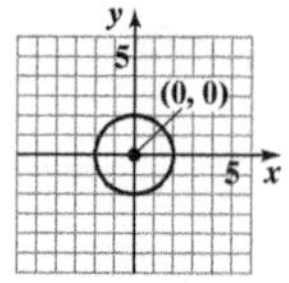

(c) $(\pm 2, 0)$; $(0, \pm 2)$

25. (a) $(h, k) = (3, 0); r = 2$ **(b)**

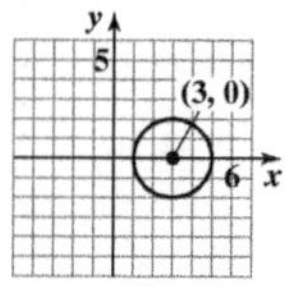

(c) $(1, 0)$; $(5, 0)$

27. (a) $(h, k) = (1, 2); r = 3$ **(b)**

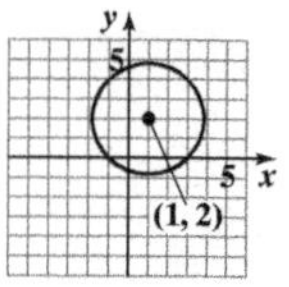

(c) $(1 \pm \sqrt{5}, 0)$; $(0, 2 \pm 2\sqrt{2})$

29. (a) $(h, k) = (-2, 2); r = 3$ **(b)**

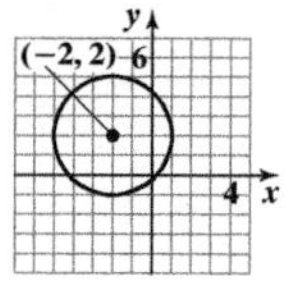

(c) $(-2 \pm \sqrt{5}, 0)$; $(0, 2 \pm \sqrt{5})$

31. (a) $(h, k) = \left(\frac{1}{2}, -1\right); r = \frac{1}{2}$ **(b)**

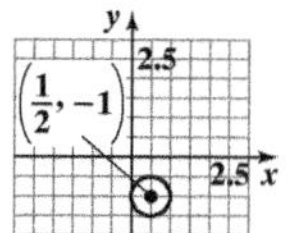

(c) $(0, -1)$

33. (a) $(h, k) = (3, -2); r = 5$ **(b)**

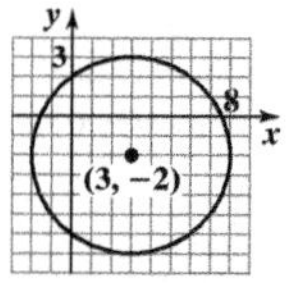

(c) $(3 \pm \sqrt{21}, 0)$; $(0, -6)$, $(0, 2)$

35. (a) $(h, k) = (-2, 0); r = 2$ **(b)**

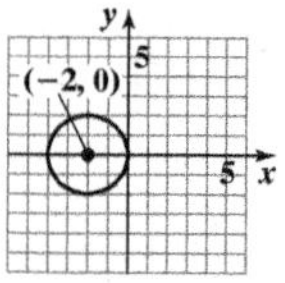

(c) $(0, 0)$, $(-4, 0)$

37. $x^2 + y^2 = 13$ **39.** $(x - 2)^2 + (y - 3)^2 = 9$ **41.** $(x + 1)^2 + (y - 3)^2 = 5$ **43.** $(x + 1)^2 + (y - 3)^2 = 1$ **45.** (c) **47.** (b) **49.** 18 units^2 **51.** $x^2 + (y - 139)^2 = 15{,}625$ **53.** $x^2 + y^2 + 2x + 4y - 4168.16 = 0$ **55.** $\sqrt{2}x + 4y - 9\sqrt{2} = 0$ **57.** $(1, 0)$ **59.** $y = 2$ **61.** (b), (c), (e), (g) **65.** $A = 169\pi \text{cm}^2$; $C = 26\pi \text{cm}$ **66.** $3x^3 - 8x^2 + 13x - 6$ **67.** {1} **68.** 12.32 min

2.4 Assess Your Understanding *(page 199)*

1. $y = kx$ **2.** F **3.** b **4.** c **5.** $y = \frac{1}{5}x$ **7.** $A = \pi x^2$ **9.** $F = \frac{250}{d^2}$ **11.** $z = \frac{1}{5}(x^2 + y^2)$ **13.** $M = \frac{9d^2}{2\sqrt{x}}$ **15.** $T^2 = \frac{8a^3}{d^2}$ **17.** $V = \frac{4\pi}{3}r^3$ **19.** $A = \frac{1}{2}bh$ **21.** $F = 6.67 \times 10^{-11}\left(\frac{mM}{d^2}\right)$ **23.** $p = 0.00649B$; \$941.05 **25.** 144 ft; 2 sec **27.** 2.25 **29.** $R = 3.95g$; \$41.48 **31. (a)** $D = \frac{429}{p}$ **(b)** 143 bags **33.** 450 cm^3 **35.** 124.76 lb **37.** $V = \pi r^2 h$ **39.** 0.012 foot-candle **41.** $\sqrt[3]{6} \approx 1.82$ in. **43.** 2812.5 joules **45.** 384 psi **51.** $(3x + 25)(x + 2)(x - 2)$ **52.** $\frac{6}{x + 4}$ **53.** $\frac{8}{125}$ **54.** $\sqrt{7} + 2$

Review Exercises *(page 203)*

1. (a) $\frac{1}{2}$ **(b)** For each run of 2, there is a rise of 1. **2. (a)** $-\frac{4}{3}$ **(b)** For each run of 3, there is a rise of -4. **3. (a)** undefined **(b)** no change in x **4. (a)** 0 **(b)** no change in y **5.** $(0, 0)$; symmetric with respect to the x-axis **6.** $(\pm 4, 0)$, $(0, \pm 2)$; symmetric with respect to the x-axis, y-axis, and origin **7.** $(\pm 2, 0)$, $(0, -4)$; symmetric with respect to the y-axis **8.** $(0, 0)$, $(\pm 1, 0)$; symmetric with respect to the origin **9.** $(0, 0)$, $(-1, 0)$, $(0, -2)$; no symmetry **10.** $(x + 2)^2 + (y - 3)^2 = 16$ **11.** $(x + 1)^2 + (y + 2)^2 = 1$

12. Center $(0, 1)$; radius $= 2$

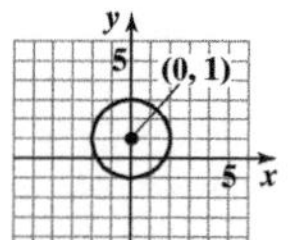

Intercepts: $(-\sqrt{3}, 0)$, $(\sqrt{3}, 0)$, $(0, -1)$, $(0, 3)$

13. Center $(1, -2)$; radius $= 3$

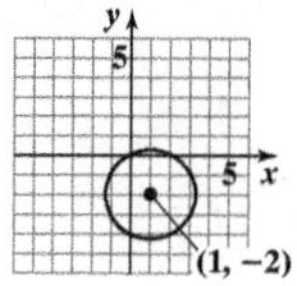

Intercepts: $(1 - \sqrt{5}, 0)$, $(1 + \sqrt{5}, 0)$, $(0, -2 - 2\sqrt{2})$, $(0, -2 + 2\sqrt{2})$

14. Center $(1, -2)$; radius $= \sqrt{5}$

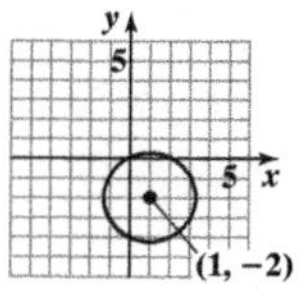

Intercepts: $(0, 0)$, $(2, 0)$, $(0, -4)$

15. $2x + y = 5$ or $y = -2x + 5$ **16.** $y = 4$ **17.** $x = -3$; no slope–intercept form **18.** $5x + 2y = 10$ or $y = -\frac{5}{2}x + 5$ **19.** $x + 5y = -10$ or $y = -\frac{1}{5}x - 2$ **20.** $5x + y = 11$ or $y = -5x + 11$ **21.** $2x - 3y = -19$ or $y = \frac{2}{3}x + \frac{19}{3}$ **22.** $x + 3y = 10$ or $y = -\frac{1}{3}x + \frac{10}{3}$

23. Slope $= \frac{4}{5}$; y-intercept $= 4$

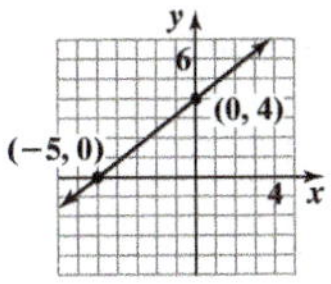

24. Slope $= \frac{3}{2}$; y-intercept $= \frac{1}{2}$

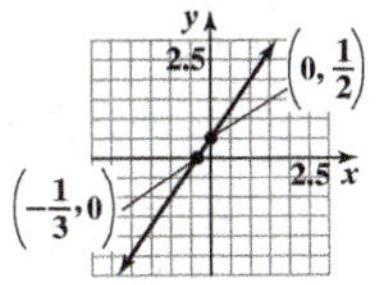

25. Intercepts: $(6, 0)$, $(0, -4)$

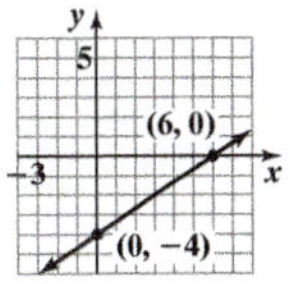

26. Intercepts: $(4, 0)$, $(0, 6)$

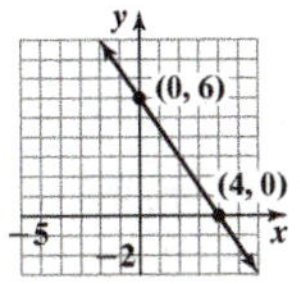

27.

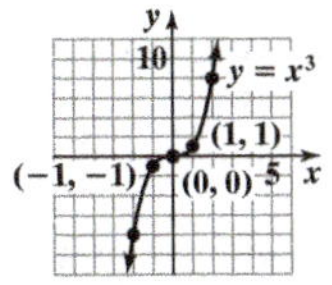

28.

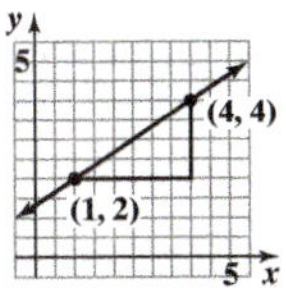

29. Slope from A to B is -2; slope from A to C is $\frac{1}{2}$. Since $(-2)\left(\frac{1}{2}\right) = -1$, the lines are perpendicular. **30.** Center: $(1, -2)$; radius: $4\sqrt{2}$; $(x - 1)^2 + (y + 2)^2 = 32$ **31.** Slope from A to B is -1; slope from A to C is -1. **32.** $p = \frac{854}{130{,}000}B$; \$1083.92 **33.** 199.9 lb **34.** 189 Btu

Chapter Test *(page 204)*

1. (a) $m = -\frac{2}{3}$ **(b)** For every 3-unit change in x, y will change by -2 units.

2.

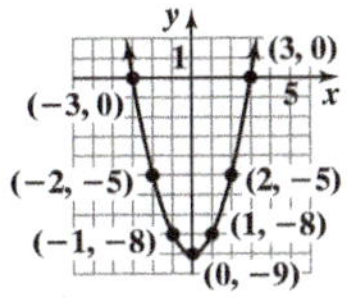

3.

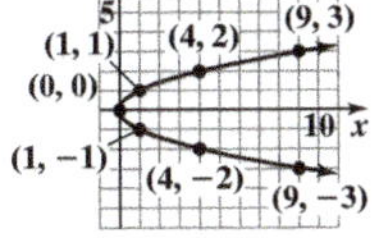

4. Intercepts: $(-3, 0)$, $(3, 0)$, $(0, 9)$; symmetric with respect to the y-axis

5. $y = -2x + 2$

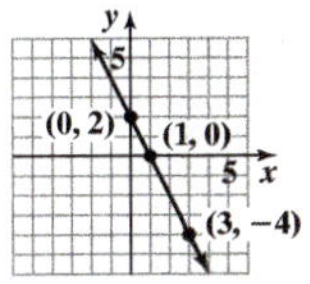

6. Slope $= -\frac{2}{3}$; y-intercept $= 3$

7.

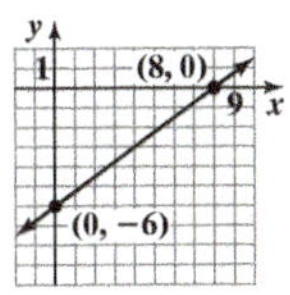

8. $x^2 + y^2 - 8x + 6y = 0$

9. Center: $(-2, 1)$; radius: 3

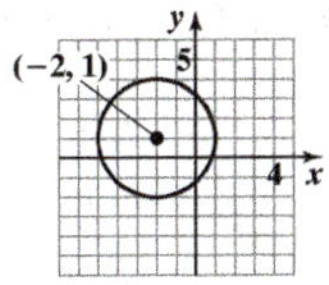

10. Parallel line: $y = -\frac{2}{3}x - \frac{1}{3}$; perpendicular line: $y = \frac{3}{2}x + 3$ **11.** 14.69 ohms

Cumulative Review *(page 204)*

1. $\left\{\frac{5}{3}\right\}$ **2.** $\{-3, 4\}$ **3.** $\left\{-\frac{1}{2}, 3\right\}$ **4.** $\{1 - \sqrt{3}, 1 + \sqrt{3}\}$ **5.** No real solution

6. $\{4\}$ **7.** $\{1, 3\}$ **8.** $\{-2 - 2\sqrt{2}, -2 + 2\sqrt{2}\}$ **9.** $\{-3i, 3i\}$ **10.** $\{1 - 2i, 1 + 2i\}$ **11.** $\{x \mid x \le 5\}$ or $(\infty, 5]$;

12. $\{x \mid -5 < x < 1\}$ or $(-5, 1)$;

13. $\{x \mid 1 \le x \le 3\}$ or $[1, 3]$;

14. $\{x \mid x < -5 \text{ or } x > 1\}$ or $(\infty, -5) \cup (1, \infty)$;

15. $5\sqrt{2}$; $\left(\frac{3}{2}, \frac{1}{2}\right)$ **16.** (a), (b)

17.

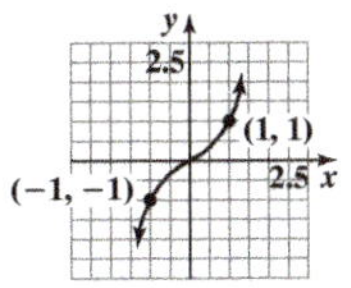

18. $y = -2x + 2$ **19.** $y = -\frac{1}{2}x + \frac{13}{2}$;

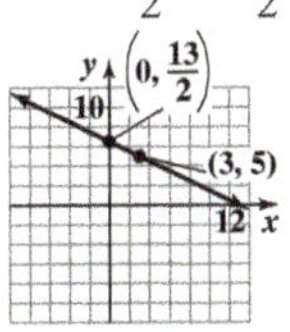

20.

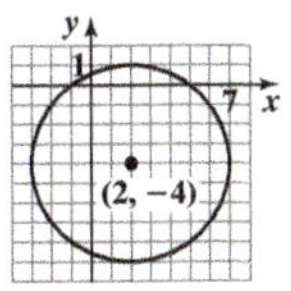

CHAPTER 3 Functions and Their Graphs

3.1 Assess Your Understanding *(page 218)*

7. independent; dependent **8.** $[0, 5]$ **9.** $\neq$; f; g **10.** $(g - f)(x)$ **11.** F **12.** T **13.** F **14.** F **15.** a **16.** c **17.** d **18.** a **19.** Function; Domain: {Elvis, Colleen, Kaleigh, Marissa}; Range: {January 8, March 15, September 17} **21.** Not a function; Domain: {20, 30, 40}; Range: {200, 300, 350, 425} **23.** Not a function; Domain: $\{-3, 2, 4\}$; Range: {6, 9, 10} **25.** Function; Domain: {1, 2, 3, 4}; Range: {3} **27.** Not a function; Domain: $\{-2, 0, 3\}$; Range: {3, 4, 6, 7} **29.** Function; Domain: $\{-2, -1, 0, 1\}$; Range: {0, 1, 4} **31.** Function **33.** Function

35. Not a function **37.** Not a function **39.** Function **41.** Not a function **43. (a)** -4 **(b)** 1 **(c)** -3 **(d)** $3x^2 - 2x - 4$ **(e)** $-3x^2 - 2x + 4$ **(f)** $3x^2 + 8x + 1$ **(g)** $12x^2 + 4x - 4$ **(h)** $3x^2 + 6xh + 3h^2 + 2x + 2h - 4$ **45. (a)** 0 **(b)** $\frac{1}{2}$ **(c)** $-\frac{1}{2}$ **(d)** $\frac{-x}{x^2+1}$ **(e)** $\frac{-x}{x^2+1}$ **(f)** $\frac{x+1}{x^2+2x+2}$ **(g)** $\frac{2x}{4x^2+1}$ **(h)** $\frac{x+h}{x^2+2xh+h^2+1}$ **47. (a)** 4 **(b)** 5 **(c)** 5 **(d)** $|x| + 4$ **(e)** $-|x| - 4$ **(f)** $|x+1| + 4$ **(g)** $2|x| + 4$ **(h)** $|x+h| + 4$ **49. (a)** $-\frac{1}{5}$ **(b)** $-\frac{3}{2}$ **(c)** $\frac{1}{8}$ **(d)** $\frac{2x-1}{3x+5}$ **(e)** $\frac{-2x-1}{3x-5}$ **(f)** $\frac{2x+3}{3x-2}$ **(g)** $\frac{4x+1}{6x-5}$ **(h)** $\frac{2x+2h+1}{3x+3h-5}$ **51.** All real numbers **53.** All real numbers **55.** $\{x \mid x \neq -4, x \neq 4\}$ **57.** $\{x \mid x \neq 0\}$ **59.** $\{x \mid x \geq 4\}$ **61.** $\{x \mid x > 1\}$ **63.** $\{x \mid x > 4\}$ **65.** $\{t \mid t \geq 4, t \neq 7\}$
67. (a) $(f+g)(x) = 5x + 1$; All real numbers **(b)** $(f-g)(x) = x + 7$; All real numbers **(c)** $(f \cdot g)(x) = 6x^2 - x - 12$; All real numbers **(d)** $\left(\frac{f}{g}\right)(x) = \frac{3x+4}{2x-3}; \left\{x \middle| x \neq \frac{3}{2}\right\}$ **(e)** 16 **(f)** 11 **(g)** 10 **(h)** -7 **69. (a)** $(f+g)(x) = 2x^2 + x - 1$; All real numbers **(b)** $(f-g)(x) = -2x^2 + x - 1$; All real numbers **(c)** $(f \cdot g)(x) = 2x^3 - 2x^2$; All real numbers **(d)** $\left(\frac{f}{g}\right)(x) = \frac{x-1}{2x^2}; \{x \mid x \neq 0\}$ **(e)** 20 **(f)** -29 **(g)** 8 **(h)** 0
71. (a) $(f+g)(x) = \sqrt{x} + 3x - 5; \{x \mid x \geq 0\}$ **(b)** $(f-g)(x) = \sqrt{x} - 3x + 5; \{x \mid x \geq 0\}$ **(c)** $(f \cdot g)(x) = 3x\sqrt{x} - 5\sqrt{x}; \{x \mid x \geq 0\}$ **(d)** $\left(\frac{f}{g}\right)(x) = \frac{\sqrt{x}}{3x-5}; \left\{x \middle| x \geq 0, x \neq \frac{5}{3}\right\}$ **(e)** $\sqrt{3} + 4$ **(f)** -5 **(g)** $\sqrt{2}$ **(h)** $-\frac{1}{2}$ **73. (a)** $(f+g)(x) = 1 + \frac{2}{x}; \{x \mid x \neq 0\}$ **(b)** $(f-g)(x) = 1; \{x \mid x \neq 0\}$ **(c)** $(f \cdot g)(x) = \frac{1}{x} + \frac{1}{x^2}; \{x \mid x \neq 0\}$ **(d)** $\left(\frac{f}{g}\right)(x) = x + 1; \{x \mid x \neq 0\}$ **(e)** $\frac{5}{3}$ **(f)** 1 **(g)** $\frac{3}{4}$ **(h)** 2
75. (a) $(f+g)(x) = \frac{6x+3}{3x-2}; \left\{x \middle| x \neq \frac{2}{3}\right\}$ **(b)** $(f-g)(x) = \frac{-2x+3}{3x-2}; \left\{x \middle| x \neq \frac{2}{3}\right\}$ **(c)** $(f \cdot g)(x) = \frac{8x^2+12x}{(3x-2)^2}; \left\{x \middle| x \neq \frac{2}{3}\right\}$ **(d)** $\left(\frac{f}{g}\right)(x) = \frac{2x+3}{4x}; \left\{x \middle| x \neq 0, x \neq \frac{2}{3}\right\}$ **(e)** 3 **(f)** $-\frac{1}{2}$ **(g)** $\frac{7}{2}$ **(h)** $\frac{5}{4}$ **77.** $g(x) = 5 - \frac{7}{2}x$ **79.** 4 **81.** $2x + h$ **83.** $2x + h - 1$
85. $\frac{-(2x+h)}{x^2(x+h)^2}$ **87.** $\frac{6}{(x+3)(x+h+3)}$ **89.** $\frac{1}{\sqrt{x+h-2} + \sqrt{x-2}}$ **91.** $\{-2, 4\}$ **93.** $A = -\frac{7}{2}$ **95.** $A = -4$ **97.** $A(x) = \frac{1}{2}x^2$ **99.** $G(x) = 14x$
101. (a) P is the dependent variable; a is the independent variable. **(b)** $P(20) = 231.427$ million; In 2012, there were 231.427 million people 20 years of age or older. **(c)** $P(0) = 327.287$ million; In 2012, there were 327.287 million people.
103. (a) 15.1 m, 14.071 m, 12.944 m, 11.719 m **(b)** 1.01 sec, 1.43 sec, 1.75 sec **(c)** 2.02 sec **105. (a)** \$222 **(b)** \$225 **(c)** \$220 **(d)** \$230
107. $R(x) = \frac{L(x)}{P(x)}$ **109.** $H(x) = P(x) \cdot I(x)$ **111. (a)** $P(x) = -0.05x^3 + 0.8x^2 + 155x - 500$ **(b)** $P(15) = \$1836.25$ **(c)** When 15 hundred cellphones are sold, the profit is \$1836.25. **113. (a)** $D(v) = 0.05v^2 + 2.6v - 15$ **(b)** 321 feet **(c)** The car will need 321 feet to stop once the impediment is observed. **115.** No; domain of f is all real numbers; domain of g is $\{x \mid x \neq -1\}$
117. $H(x) = \frac{3x - x^3}{\text{age}}$ **118.** Intercepts: $(-16, 0), (-8, 0)$; x-axis symmetry **119.** $(4, 32)$ **120.** 7.5 lb **121.** $\{-3, 2, 3\}$

3.2 Assess Your Understanding *(page 226)*

3. vertical **4.** 5; -3 **5.** $a = -2$ **6.** F **7.** F **8.** T **9.** c **10.** a **11. (a)** $f(0) = 3; f(-6) = -3$ **(b)** $f(6) = 0; f(11) = 1$ **(c)** Positive **(d)** Negative **(e)** -3, 6, and 10 **(f)** $-3 < x < 6; 10 < x \leq 11$ **(g)** $\{x \mid -6 \leq x \leq 11\}$ **(h)** $\{y \mid -3 \leq y \leq 4\}$ **(i)** $-3, 6, 10$ **(j)** 3 **(k)** 3 times **(l)** Once **(m)** 0, 4 **(n)** $-5, 8$ **13.** Not a function **15.** Function **(a)** Domain: $\{x \mid -\pi \leq x \leq \pi\}$; Range: $\{y \mid -1 \leq y \leq 1\}$ **(b)** $\left(-\frac{\pi}{2}, 0\right), \left(\frac{\pi}{2}, 0\right), (0, 1)$ **(c)** y-axis **17.** Not a function **19.** Function **(a)** Domain: $\{x \mid 0 < x < 3\}$; Range: $\{y \mid y < 2\}$ **(b)** $(1, 0)$ **(c)** None
21. Function **(a)** Domain: all real numbers; Range: $\{y \mid y \leq 2\}$ **(b)** $(-3, 0), (3, 0), (0, 2)$ **(c)** y-axis **23.** Function **(a)** Domain: all real numbers; Range: $\{y \mid y \geq -3\}$ **(b)** $(1, 0), (3, 0), (0, 9)$ **(c)** None **25. (a)** Yes **(b)** $f(-2) = 9; (-2, 9)$ **(c)** $0, \frac{1}{2}; (0, -1), \left(\frac{1}{2}, -1\right)$ **(d)** All real numbers **(e)** $-\frac{1}{2}, 1$ **(f)** -1 **27. (a)** No **(b)** $f(4) = -3; (4, -3)$ **(c)** 14; $(14, 2)$ **(d)** $\{x \mid x \neq 6\}$ **(e)** -2 **(f)** $-\frac{1}{3}$ **29. (a)** Yes **(b)** $f(2) = \frac{8}{17}; \left(2, \frac{8}{17}\right)$ **(c)** $-1, 1; (-1, 1), (1, 1)$ **(d)** All real numbers **(e)** 0 **(f)** 0
31. (a) 3 **(b)** -2 **(c)** -1 **(d)** 1 **(e)** 2 **(f)** $-\frac{1}{3}$

33. (a) Approximately 10.4 ft high
(b) Approximately 9.9 ft high
(c)

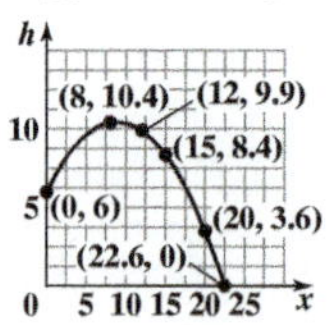

(d) The ball will not go through the hoop; $h(15) \approx 8.4$ ft. If $v = 30$ ft/sec, $h(15) = 10$ ft.

35. (a) About 81.07 ft **(b)** About 129.59 ft **(c)** About 26.63 ft
(d) About 528.13 ft
(e)

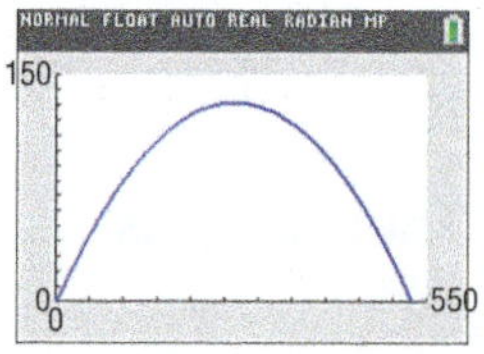

(f) About 115.07 ft and 413.05 ft
(g) 275 ft; maximum height shown in the table is 131.8 ft **(h)** 264 ft

37. (a) \$223; \$220 **(b)** $\{x \mid x > 0\}$

(c)

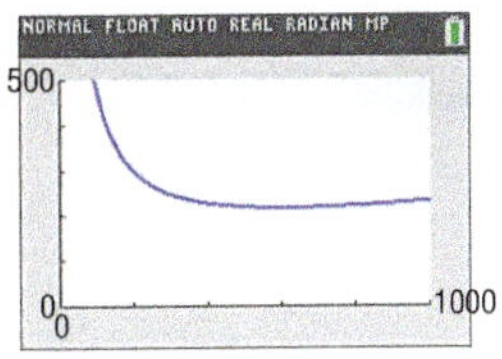

(d)

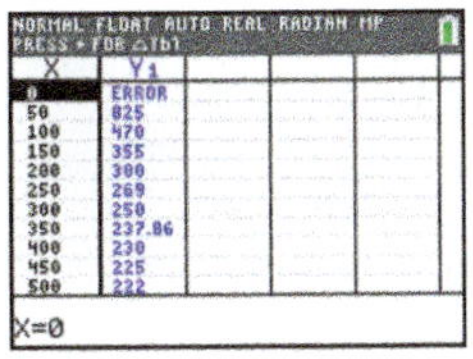

(e) 600 mi/h

39. (a) \$30; It costs \$30 if you use 0 gigabytes. **(b)** \$30; It costs \$30 if you use 5 gigabytes. **(c)** \$90; It costs \$90 if you use 15 gigabytes. **(d)** $\{g \mid 0 \le g \le 60\}$. There are at most 60 gigabytes in a month.

41. The x-intercepts can number anywhere from 0 to infinitely many. There is at most one y-intercept.

43. (a) III **(b)** IV **(c)** I **(d)** V **(e)** II

45.

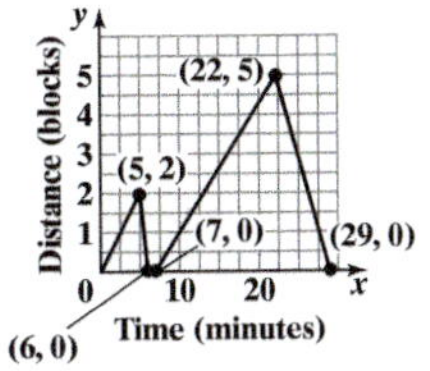

47. (a) 2 hr elapsed during which Kevin was between 0 and 3 mi from home **(b)** 0.5 hr elapsed during which Kevin was 3 mi from home **(c)** 0.3 hr elapsed during which Kevin was between 0 and 3 mi from home **(d)** 0.2 hr elapsed during which Kevin was 0 mi from home **(e)** 0.9 hr elapsed during which Kevin was between 0 and 2.8 mi from home **(f)** 0.3 hr elapsed during which Kevin was 2.8 mi from home **(g)** 1.1 hr elapsed during which Kevin was between 0 and 2.8 mi from home **(h)** 3 mi **(i)** Twice

49. No points whose x-coordinate is 5 or whose y-coordinate is 0 can be on the graph.

52. $2(2x+3)^2$ **53.** $2\sqrt{10}$ **54.** $y = \frac{2}{3}x + 8$ **55.** $4x^3 - 8x^2 - 5x + 4$

3.3 Assess Your Understanding *(page 240)*

6. increasing **7.** even; odd **8.** T **9.** T **10.** F **11.** c **12.** d **13.** Yes **15.** No **17.** $[-8,-2]$; $[0,2]$; $[5,7]$ **19.** Yes; 10 **21.** $-2, 2$; 6, 10 **23.** $f(-8) = -4$ **25. (a)** $(-2,0)$, $(0,3)$, $(2,0)$ **(b)** Domain: $\{x \mid -4 \le x \le 4\}$ or $[-4,4]$; Range: $\{y \mid 0 \le y \le 3\}$ or $[0,3]$ **(c)** Increasing on $[-2,0]$ and $[2,4]$; Decreasing on $[-4,-2]$ and $[0,2]$ **(d)** Even **27. (a)** $(0,1)$ **(b)** Domain: all real numbers; Range: $\{y \mid y > 0\}$ or $(0,\infty)$ **(c)** Increasing on $(-\infty,\infty)$ **(d)** Neither **29. (a)** $(-\pi,0)$, $(0,0)$, $(\pi,0)$ **(b)** Domain: $\{x \mid -\pi \le x \le \pi\}$ or $[-\pi,\pi]$; Range: $\{y \mid -1 \le y \le 1\}$ or $[-1,1]$ **(c)** Increasing on $\left[-\frac{\pi}{2},\frac{\pi}{2}\right]$; Decreasing on $\left[-\pi,-\frac{\pi}{2}\right]$ and $\left[\frac{\pi}{2},\pi\right]$ **(d)** Odd **31. (a)** $\left(0,\frac{1}{2}\right)$, $\left(\frac{1}{3},0\right)$, $\left(\frac{5}{2},0\right)$ **(b)** Domain: $\{x \mid -3 \le x \le 3\}$ or $[-3,3]$; Range: $\{y \mid -1 \le y \le 2\}$ or $[-1,2]$ **(c)** Increasing on $[2,3]$; Decreasing on $[-1,1]$; Constant on $[-3,-1]$ and $[1,2]$ **(d)** Neither **33. (a)** 0; 3 **(b)** $-2, 2$; 0, 0

35. (a) $\frac{\pi}{2}$; 1 **(b)** $-\frac{\pi}{2}$; -1 **37.** Odd **39.** Even **41.** Odd **43.** Neither **45.** Even **47.** Odd

49. Absolute maximum: $f(1) = 4$; absolute minimum: $f(5) = 1$; local maximum: $f(3) = 3$; local minimum: $f(2) = 2$

51. Absolute maximum: $f(3) = 4$; absolute minimum: $f(1) = 1$; local maximum: $f(3) = 4$; local minimum: $f(1) = 1$

53. Absolute maximum: none; absolute minimum: $f(0) = 0$: local maximum: $f(2) = 3$; local minimum: $f(0) = 0$ and $f(3) = 2$

55. Absolute maximum: none; absolute minimum: none: local maximum: none; local minimum: none

57.

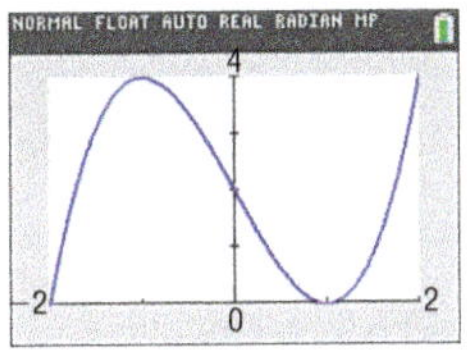

Increasing: $[-2,-1]$, $[1,2]$
Decreasing: $[-1,1]$
Local maximum: $f(-1) = 4$
Local minimum: $f(1) = 0$

59.

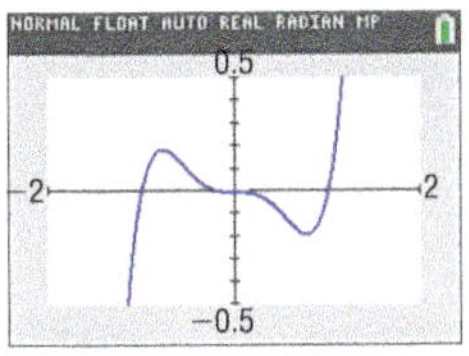

Increasing: $[-2,-0.77]$, $[0.77,2]$
Decreasing: $[-0.77,0.77]$
Local maximum: $f(-0.77) = 0.19$
Local minimum: $f(0.77) = -0.19$

61.

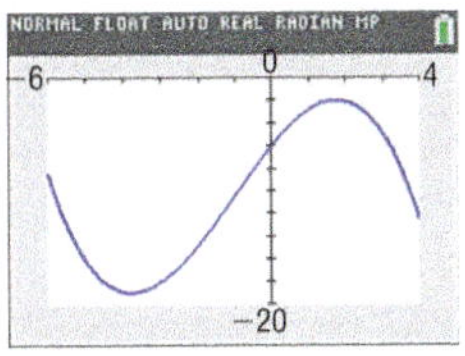

Increasing: $[-3.77, 1.77]$
Decreasing: $[-6,-3.77]$, $[1.77,4]$
Local maximum: $f(1.77) = -1.91$
Local minimum: $f(-3.77) = -18.89$

63.

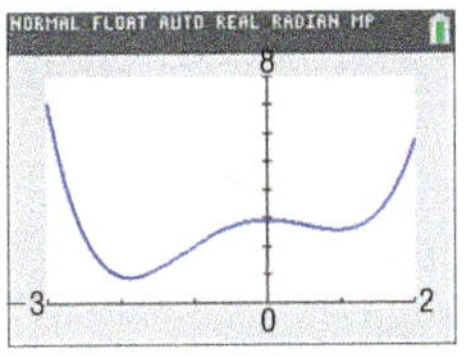

Increasing: $[-1.87,0]$, $[0.97,2]$
Decreasing: $[-3,-1.87]$, $[0,0.97]$
Local maximum: $f(0) = 3$
Local minima: $f(-1.87) = 0.95$, $f(0.97) = 2.65$

65. (a) -4 **(b)** -8 **(c)** -10

(d)

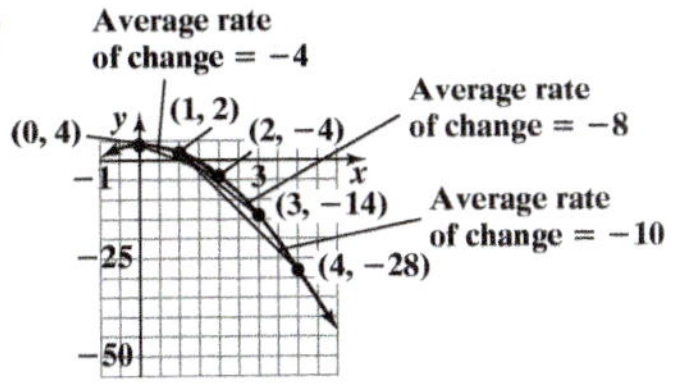

67. (a) 17 **(b)** -1 **(c)** 11 **69. (a)** 5 **(b)** $y = 5x - 2$

71. (a) -1 **(b)** $y = -x$

(c)

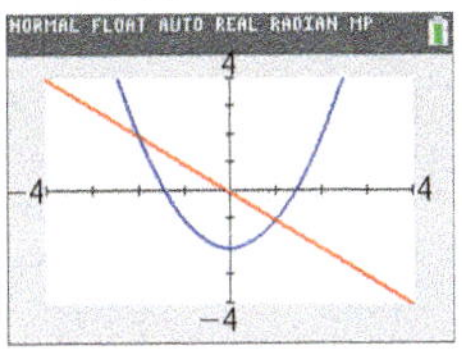

73. (a) 4 **(b)** $y = 4x - 8$

(c)

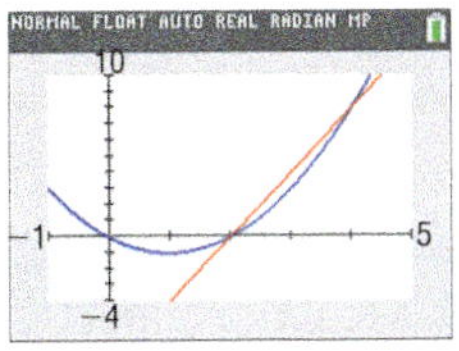

75. (a) Odd **(b)** Local maximum value: 54 at $x = -3$ **77. (a)** Even **(b)** Local maximum value: 25 at $x = -2$ **(c)** 50.4 sq. units

79. (a)
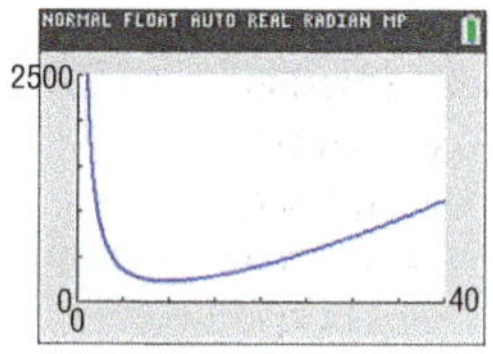

(b) 10 riding lawn mowers
(c) \$239/mower

81. (a), (b)
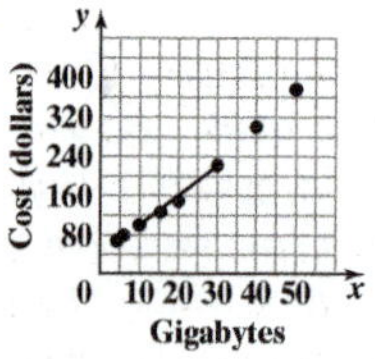

(c) \$5/gigabyte
(d) \$6.25/gigabyte
(e) \$7.50/gigabyte
(f) The average rate of change is increasing as the number of gigabytes increases.

83. (a) On average, the population is increasing at a rate of 0.036 g/h from 0 to 2.5 h. **(b)** On average, from 4.5 to 6 h, the population is increasing at a rate of 0.1 g/h. **(c)** The average rate of change is increasing over time.

85. (a) 1 **(b)** 0.5 **(c)** 0.1 **(d)** 0.01
(e). 0001
(f)
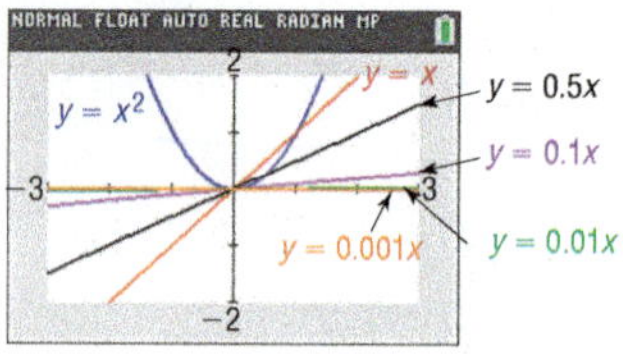

(g) They are getting closer to the tangent line at $(0, 0)$.
(h) They are getting closer to 0.

87. (a) 2
(b) 2; 2; 2; 2
(c) $y = 2x + 5$
(d)
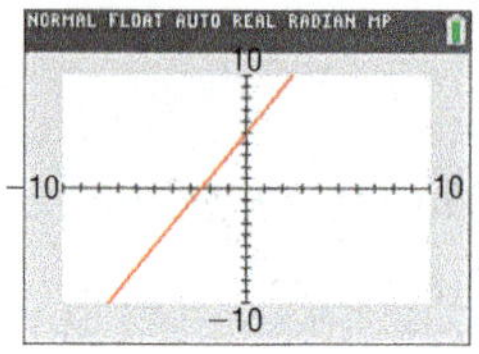

89. (a) $2x + h + 2$
(b) 4.5; 4.1; 4.01; 4
(c) $y = 4.01x - 1.01$
(d)
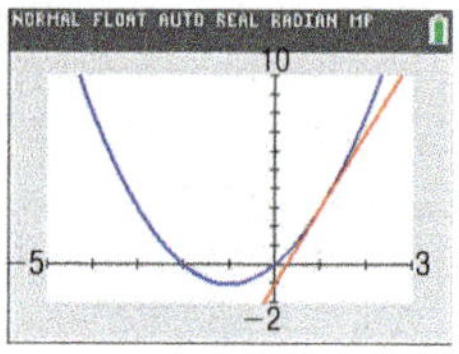

91. (a) $4x + 2h - 3$
(b) 2; 1.2; 1.02; 1
(c) $y = 1.02x - 1.02$
(d)
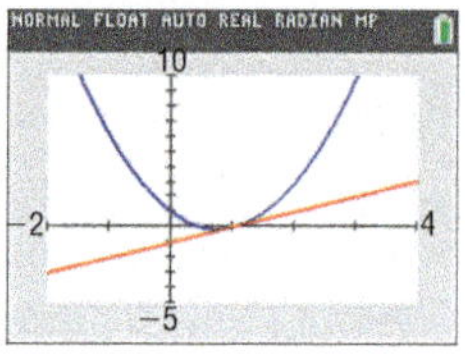

93. (a) $-\dfrac{1}{(x + h)x}$
(b) $-\dfrac{2}{3}; -\dfrac{10}{11}; -\dfrac{100}{101}; -1$
(c) $y = -\dfrac{100}{101}x + \dfrac{201}{101}$
(d)
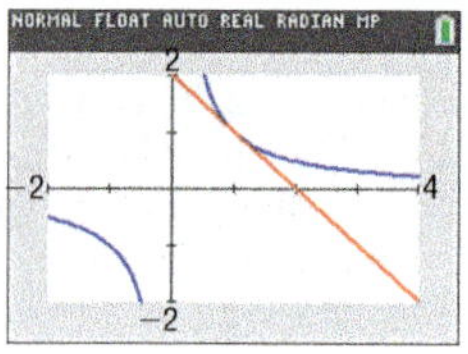

97. At most one **99.** Yes; the function $f(x) = 0$ is both even and odd. **101.** Not necessarily. It just means $f(5) > f(2)$.
103. (a) 7.01×10^{-6} **(b)** 2.305×10^{9} **104.** $6\sqrt{15}$ **105.** $[-8, -3)$ **106.** 8.25 days

3.4 Assess Your Understanding *(page 252)*

4. $(-\infty, 0]$ **5.** piecewise-defined **6.** T **7.** F **8.** F **9.** b **10.** a **11.** C **13.** E **15.** B **17.** F

19.
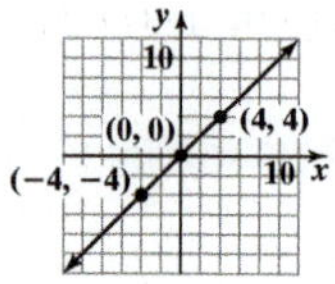

21.
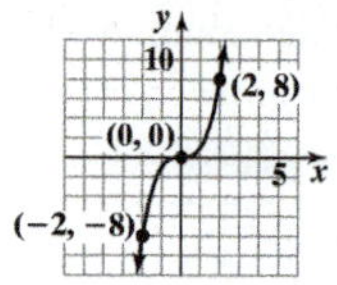

23.
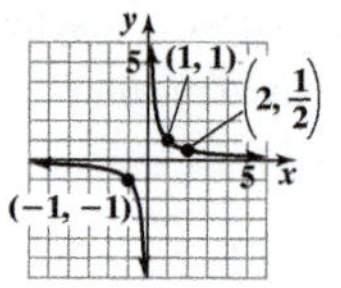

25.
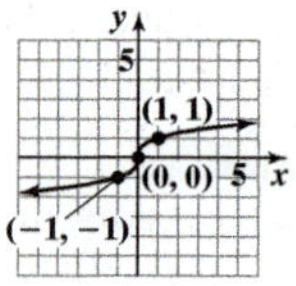

27. (a) 4 **(b)** 2 **(c)** 5 **29. (a)** −4 **(b)** −2 **(c)** 0 **(d)** 25

31. (a) All real numbers
(b) $(0, 1)$
(c)
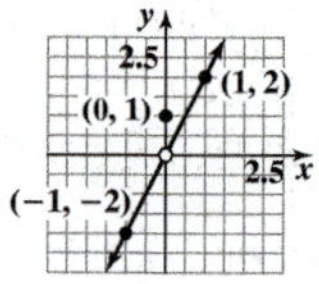

(d) $\{y \mid y \neq 0\}$; $(-\infty, 0) \cup (0, \infty)$

33. (a) All real numbers
(b) $(0, 3)$
(c)
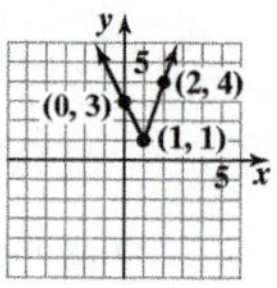

(d) $\{y \mid y \geq 1\}$; $[1, \infty)$

35. (a) $\{x \mid x \geq -2\}$; $[-2, \infty)$
(b) $(0, 3)$, $(2, 0)$
(c)
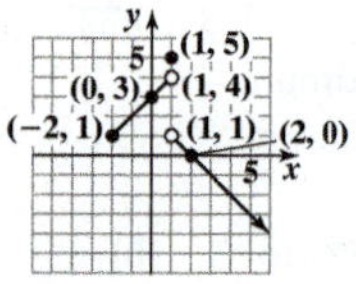

(d) $\{y \mid y < 4, y = 5\}$; $(-\infty, 4) \cup \{5\}$

37. (a) All real numbers
(b) $(-1, 0)$, $(0, 0)$
(c)
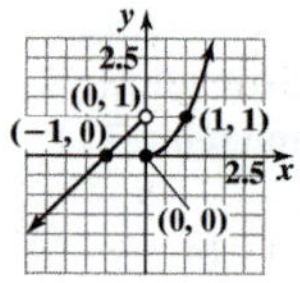

(d) All real numbers

39. (a) $\{x \mid x \geq -2, x \neq 0\}$; $[-2, 0) \cup (0, \infty)$
(b) No intercepts
(c)
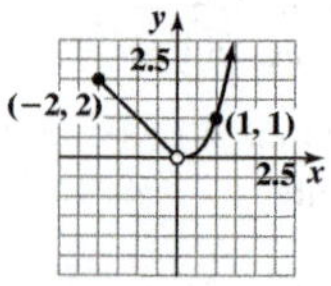

(d) $\{y \mid y > 0\}$; $(0, \infty)$

41. (a) All real numbers
(b) $(x, 0)$ for $0 \leq x < 1$
(c)
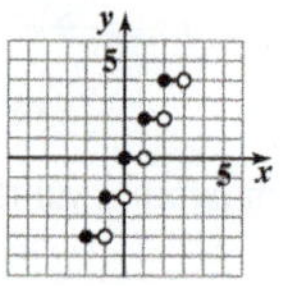

(d) Set of even integers

43. $f(x) = \begin{cases} -x & \text{if } -1 \leq x \leq 0 \\ \frac{1}{2}x & \text{if } 0 < x \leq 2 \end{cases}$ (Other answers are possible.)

45. $f(x) = \begin{cases} -x & \text{if } x \leq 0 \\ -x + 2 & \text{if } 0 < x \leq 2 \end{cases}$ (Other answers are possible.)

47. (a) 2 **(b)** 3 **(c)** −4

49. (a)
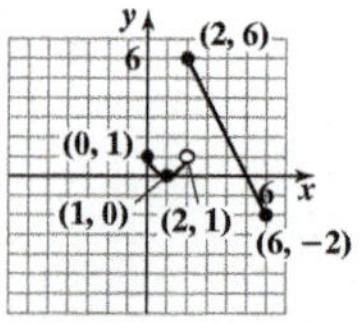

(b) $[0, 6]$
(c) Absolute maximum: $f(2) = 6$; absolute minimum: $f(6) = -2$
(d) Local maximum: $f(2) = 6$; local minimum: $f(1) = 0$

51. (a) \$34.99 **(b)** \$64.99 **(c)** \$184.99

53. (a) \$44.80 **(b)** \$128.51

(c) $C(x) = \begin{cases} 1.26486x + 19.50 & \text{if } 0 \le x \le 30 \\ 0.5922x + 39.6798 & \text{if } x > 30 \end{cases}$

(d)

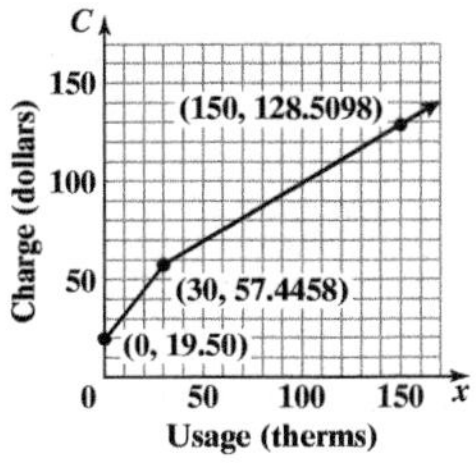

55. $f(x) = \begin{cases} 0.10x & \text{if } 0 < x \le 9{,}225 \\ 922.50 + 0.15(x - 9{,}225) & \text{if } 9{,}225 < x \le 37{,}450 \\ 5{,}156.25 + 0.25(x - 37{,}450) & \text{if } 37{,}450 < x \le 90{,}750 \\ 18{,}481.25 + 0.28(x - 90{,}750) & \text{if } 90{,}750 < x \le 189{,}300 \\ 46{,}075.25 + 0.33(x - 189{,}300) & \text{if } 189{,}300 < x \le 411{,}500 \\ 119{,}401.25 + 0.35(x - 411{,}500) & \text{if } 411{,}500 < x \le 413{,}200 \\ 119{,}996.25 + 0.396(x - 413{,}200) & \text{if } x > 413{,}200 \end{cases}$

57. (a)

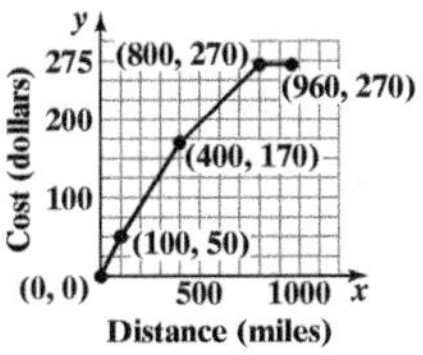

(b) $C(x) = 10 + 0.4x$ **(c)** $C(x) = 70 + 0.25x$

59. (a) $C(s) = \begin{cases} 9000 & \text{if } s \le 659 \\ 7500 & \text{if } 660 \le s \le 679 \\ 5250 & \text{if } 680 \le s \le 699 \\ 3000 & \text{if } 700 \le s \le 719 \\ 1500 & \text{if } 720 \le s \le 739 \\ 750 & \text{if } s \ge 740 \end{cases}$ **(b)** \$1500 **(c)** \$7500

61. (a) 10°C **(b)** 4°C **(c)** −3°C **(d)** −4°C
(e) The wind chill is equal to the air temperature.
(f) At wind speed greater than 20 m/sec, the wind chill factor depends only on the air temperature.

63. $C(x) = \begin{cases} 0.98 & \text{if } 0 < x \le 1 \\ 1.20 & \text{if } 1 < x \le 2 \\ 1.42 & \text{if } 2 < x \le 3 \\ 1.64 & \text{if } 3 < x \le 4 \\ 1.86 & \text{if } 4 < x \le 5 \\ 2.08 & \text{if } 5 < x \le 6 \\ 2.30 & \text{if } 6 < x \le 7 \\ 2.52 & \text{if } 7 < x \le 8 \\ 2.74 & \text{if } 8 < x \le 9 \\ 2.96 & \text{if } 9 < x \le 10 \\ 3.18 & \text{if } 10 < x \le 11 \\ 3.40 & \text{if } 11 < x \le 12 \\ 3.62 & \text{if } 12 < x \le 13 \end{cases}$

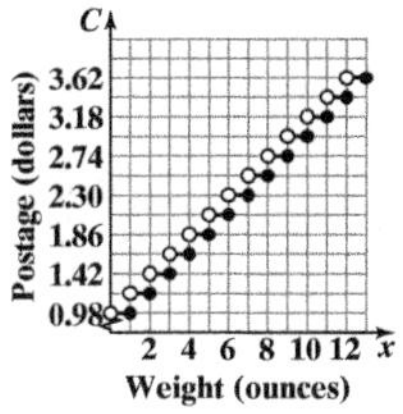

65. Each graph is that of $y = x^2$, but shifted horizontally. If $y = (x - k)^2$, $k > 0$, the shift is right k units; if $y = (x + k)^2$, $k > 0$, the shift is left k units. **67.** The graph of $y = -f(x)$ is the reflection about the x-axis of the graph of $y = f(x)$. **69.** Yes. The graph of $y = (x - 1)^3 + 2$ is the graph of $y = x^3$ shifted right 1 unit and up 2 units. **71.** They all have the same general shape. All three go through the points $(-1, -1)$, $(0, 0)$, and $(1, 1)$. As the exponent increases, the steepness of the curve increases (except near $x = 0$). **74.** $22 - 7i$ **75.** $(h, k) = (0, 3)$; $r = 5$ **76.** $\{-8\}$ **77.** CD: \$22,000; Mutual fund: \$38,000

3.5 Assess Your Understanding *(page 264)*

1. horizontal; right **2.** y **3.** F **4.** T **5.** d **6.** a **7.** B **9.** H **11.** I **13.** L **15.** F **17.** G **19.** $y = (x - 4)^3$ **21.** $y = x^3 + 4$ **23.** $y = -x^3$ **25.** $y = 4x^3$ **27.** $y = -(\sqrt{-x} + 2)$ **29.** $y = -\sqrt{x + 3} + 2$ **31.** c **33.** c **35. (a)** −7 and 1 **(b)** −3 and 5 **(c)** −5 and 3 **(d)** −3 and 5 **37. (a)** $[-3, 3]$ **(b)** $[4, 10]$ **(c)** Decreasing on $[-1, 5]$ **(d)** Decreasing on $[-5, 1]$

39.

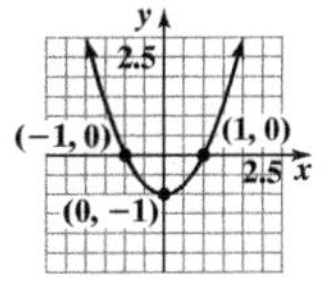

Domain: $(-\infty, \infty)$; Range: $[-1, \infty)$

41.

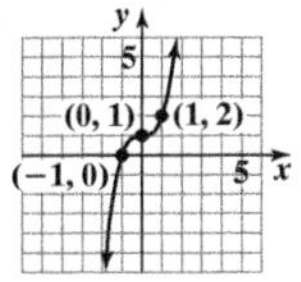

Domain: $(-\infty, \infty)$; Range: $(-\infty, \infty)$

43.

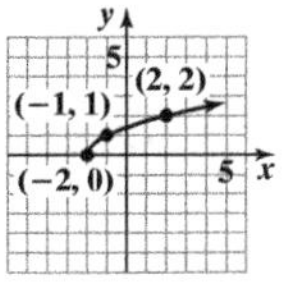

Domain: $[-2, \infty)$; Range: $[0, \infty)$

45.

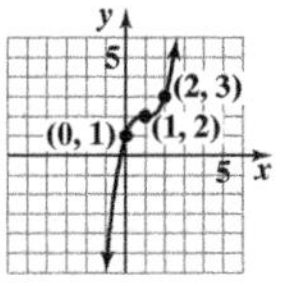

Domain: $(-\infty, \infty)$; Range: $(-\infty, \infty)$

47.

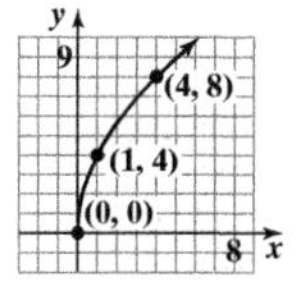

Domain: $[0, \infty)$; Range: $[0, \infty)$

49.

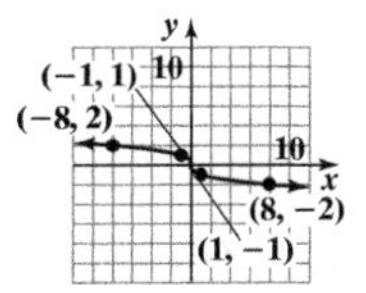

Domain: $(-\infty, \infty)$; Range: $(-\infty, \infty)$

51.

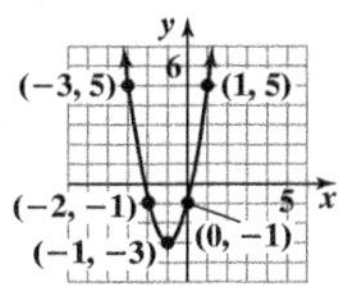

Domain: $(-\infty, \infty)$; Range: $[-3, \infty)$

53.

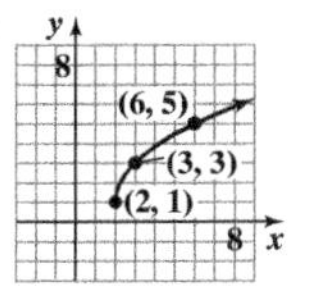

Domain: $[2, \infty)$; Range: $[1, \infty)$

55.

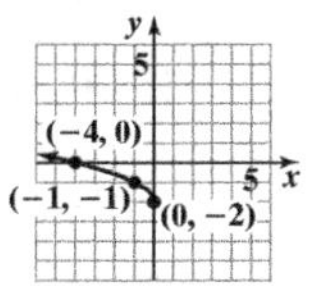

Domain: $(-\infty, 0]$; Range: $[-2, \infty)$

57.

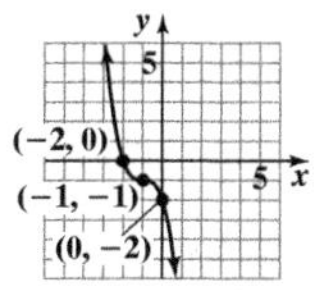

Domain: $(-\infty, \infty)$; Range: $(-\infty, \infty)$

59.

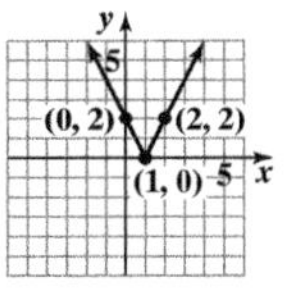

Domain: $(-\infty, \infty)$; Range: $[0, \infty)$

61.

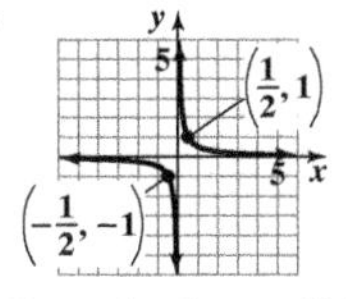

Domain: $(-\infty, 0) \cup (0, \infty)$; Range: $(-\infty, 0) \cup (0, \infty)$

63. (a) $F(x) = f(x) + 3$

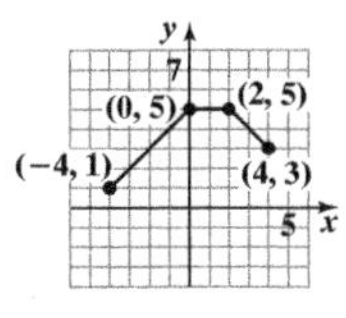

(b) $G(x) = f(x + 2)$

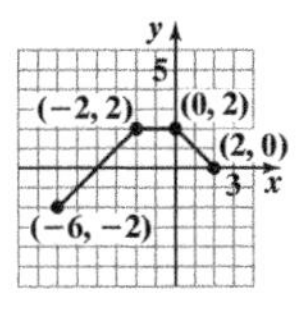

(c) $P(x) = -f(x)$

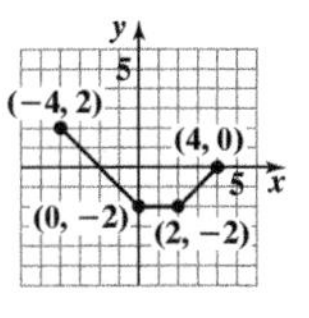

(d) $H(x) = f(x+1) - 2$

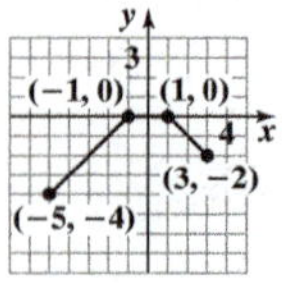

(e) $Q(x) = \frac{1}{2}f(x)$

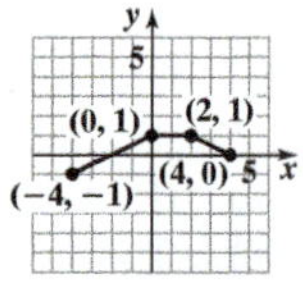

(f) $g(x) = f(-x)$

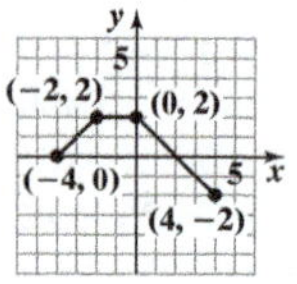

(g) $h(x) = f(2x)$

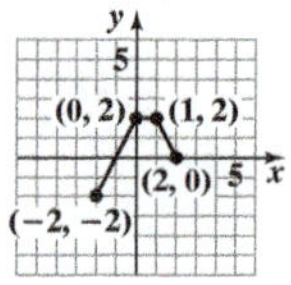

65. (a) $F(x) = f(x) + 3$

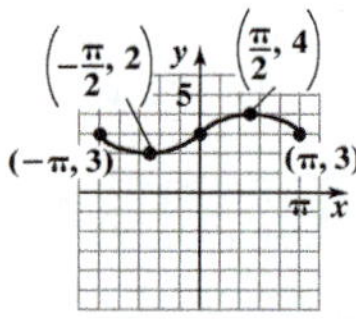

(b) $G(x) = f(x+2)$

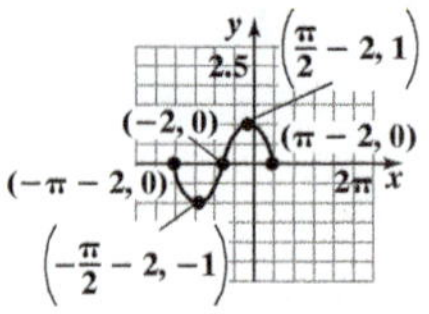

(c) $P(x) = -f(x)$

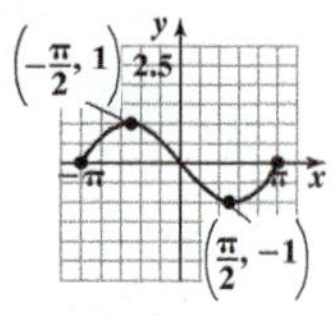

(d) $H(x) = f(x+1) - 2$

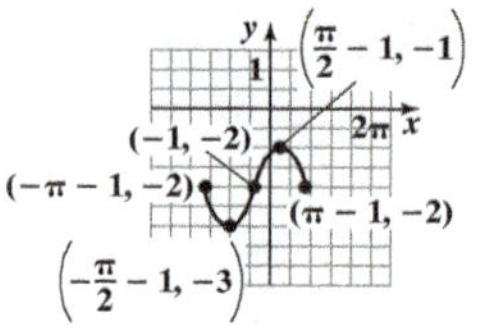

(e) $Q(x) = \frac{1}{2}f(x)$

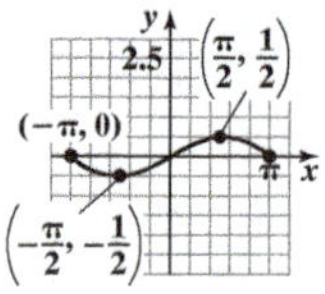

(f) $g(x) = f(-x)$

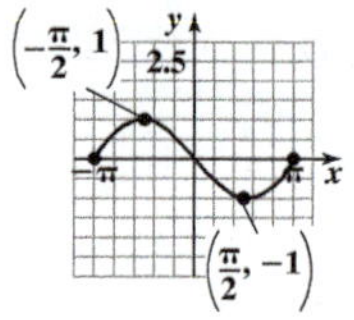

(g) $h(x) = f(2x)$

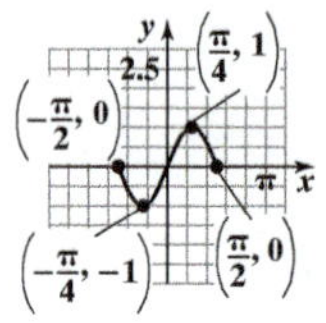

67. (a)

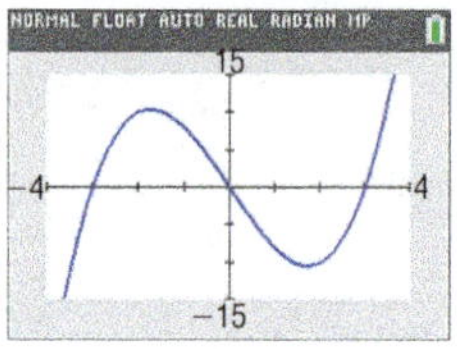

(b) −3, 0, 3 **(c)** Local maximum: 10.39 at $x = -1.73$; local minimum: −10.39 at $x = 1.73$ **(d)** Increasing: [−4, −1.73], [1.73, 4]; decreasing: [−1.73, 1.73] **(e)** Intercepts: −5, −2, 1; local maximum: 10.39 at $x = -3.73$; local minimum: −10.39 at $x = -0.27$; increasing: [−6, −3.73], [−0.27, 2]; decreasing: [−3.73, −0.27]
(f) Intercepts: −3, 0, 3; Local maximum: 20.78 at $x = -1.73$; local minimum: −20.78 at $x = 1.73$; increasing: [−4 ,−1.73], [1.73, 4]; decreasing: [−1.73, 1.73] **(g)** Intercepts: −3, 0, 3; local maximum: 10.39 at $x = 1.73$; local minimum: −10.39 at $x = -1.73$; increasing: [−1.73, 1.73]; decreasing: [−4, −1.73], [1.73, 4]

69. $f(x) = (x+1)^2 - 1$

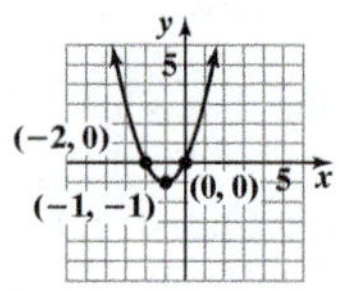

71. $f(x) = (x-4)^2 - 15$

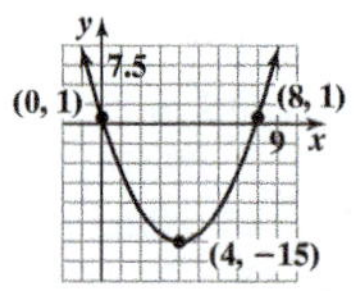

73. $f(x) = 2(x-3)^2 + 1$

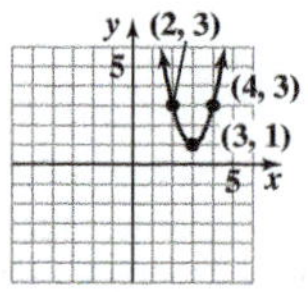

75. $f(x) = -3(x+2)^2 - 5$

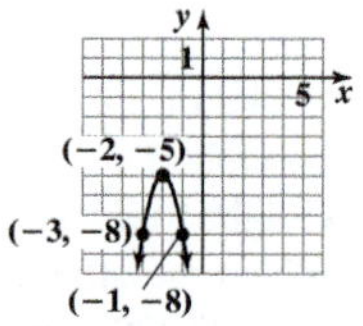

77. (a)

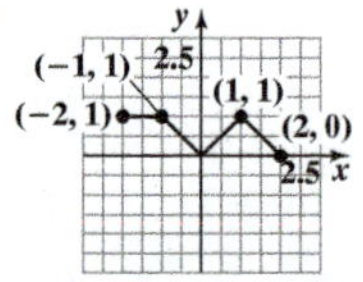

(b)

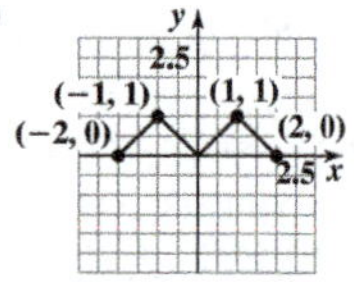

79. (a) (−2, 2) **(b)** (3, −5) **(c)** (−1, 3)

81. (a)

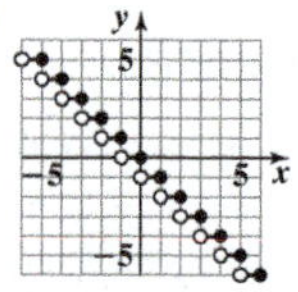

(b)

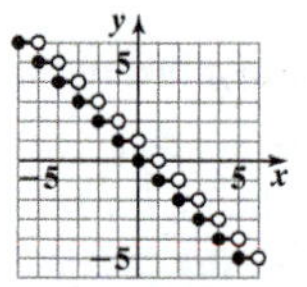

83. (a)

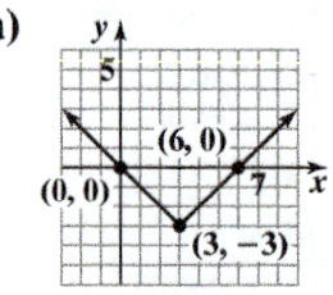

(b) 9 square units

85. (a) 72°F; 65°F
(b) The temperature decreases by 2° to 70°F during the day and 63°F overnight.

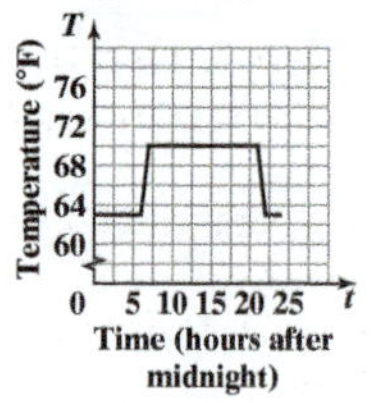

(c) The time at which the temperature adjusts between the daytime and overnight settings is moved to 1 hr sooner. It begins warming up at 5:00 AM instead of 6:00 AM, and it begins cooling down at 8:00 PM instead of 9:00 PM.

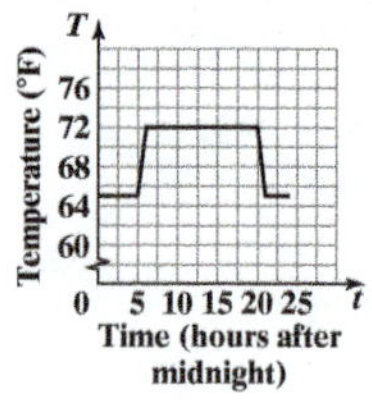

87.

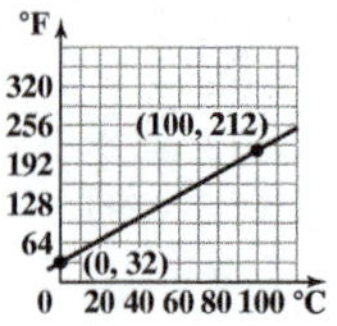

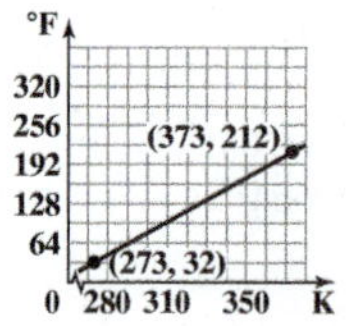

89.

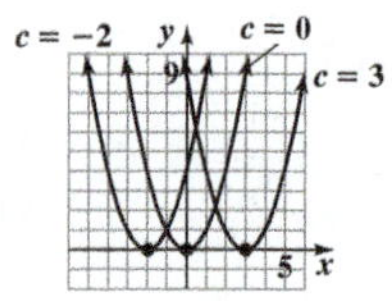

91. The graph of $y = 4f(x)$ is a vertical stretch by a factor of 4. The graph of $y = f(4x)$ is a horizontal compression by a factor of $\frac{1}{4}$.

93. $\frac{16}{3}$ sq. units **95.** The domain of $g(x) = \sqrt{x}$ is $[0, \infty)$. The graph of $g(x-k)$ is the graph of g shifted k units to the right, so the domain of $g(x-k)$ is $[k, \infty)$. **96.** $m = \frac{3}{5}$; $b = -6$ **97.** $\frac{y^2}{x^4}$ **98.** 15.75 gal **99.** Intercepts: (0, −2), (0, 2), (−4, 0); x-axis symmetry

3.6 Assess Your Understanding *(page 270)*

1. (a) $d(x) = \sqrt{x^4 - 15x^2 + 64}$
(b) $d(0) = 8$
(c) $d(1) = \sqrt{50} \approx 7.07$
(d)
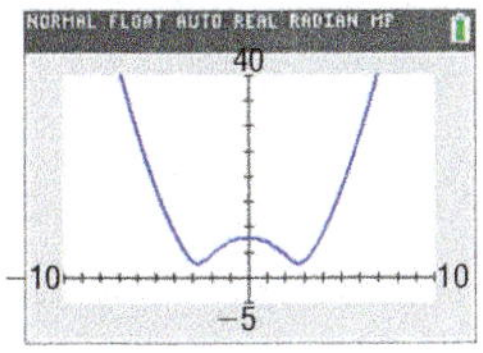

(e) d is smallest when $x \approx -2.74$ or $x \approx 2.74$

3. (a) $d(x) = \sqrt{x^2 - x + 1}$
(b)
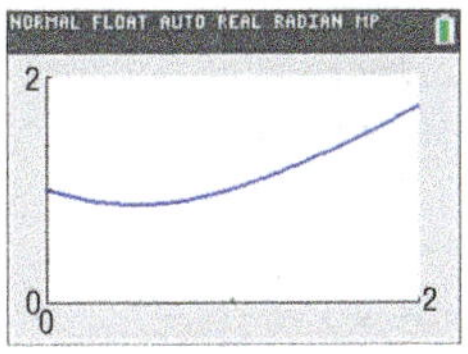

(c) d is smallest when $x = 0.5$.

5. $A(x) = \frac{1}{2}x^4$

7. (a) $A(x) = x(16 - x^2)$
(b) Domain: $\{x|0 < x < 4\}$
(c) The area is largest when $x \approx 2.31$.
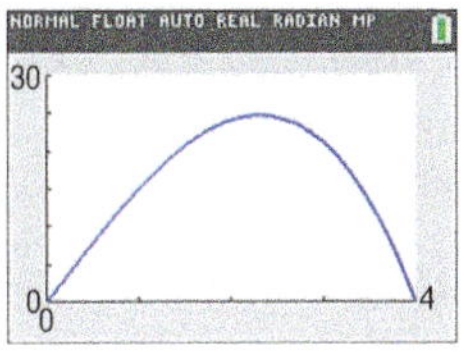

9. (a) $A(x) = 4x\sqrt{4 - x^2}$ **(b)** $p(x) = 4x + 4\sqrt{4 - x^2}$
(c) A is largest when $x \approx 1.41$. **(d)** p is largest when $x \approx 1.41$.
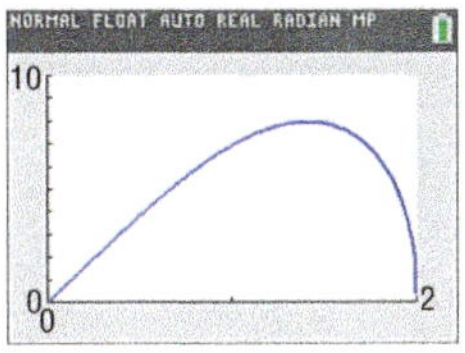

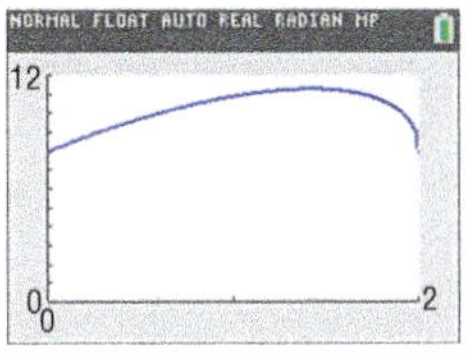

11. (a) $A(x) = x^2 + \dfrac{25 - 20x + 4x^2}{\pi}$
(b) Domain: $\{x|0 < x < 2.5\}$
(c) A is smallest when $x \approx 1.40$ m.
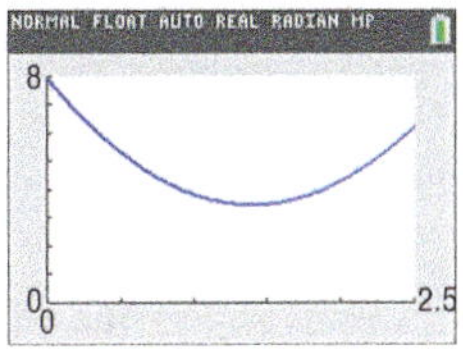

13. (a) $C(x) = x$ **(b)** $A(x) = \dfrac{x^2}{4\pi}$ **15. (a)** $A(r) = 2r^2$ **(b)** $p(r) = 6r$ **17.** $A(x) = \left(\dfrac{\pi}{3} - \dfrac{\sqrt{3}}{4}\right)x^2$

19. (a) $d(t) = \sqrt{2500t^2 - 360t + 13}$
(b) d is smallest when $t \approx 0.07$ hr.
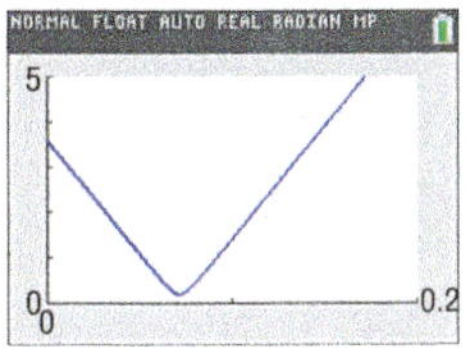

21. $V(r) = \dfrac{\pi H(R - r)r^2}{R}$

23. (a) $T(x) = \dfrac{12 - x}{5} + \dfrac{\sqrt{x^2 + 4}}{3}$
(b) $\{x|0 \le x \le 12\}$
(c) 3.09 hr **(d)** 3.55 hr

25. (a) $V(x) = x(24 - 2x)^2$
(b) 972 in.^{3} **(c)** 160 in.^{3}
(d) V is largest when $x = 4$.
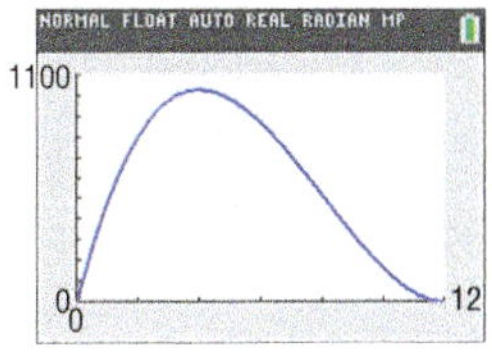

27. $\{0, 3\}$ **28.** 64 mi/h **29.** $m = -4$ **30.** 5.6

Review Exercises *(page 275)*

1. Function; domain $\{-1, 2, 4\}$, range $\{0, 3\}$ **2.** Not a function **3. (a)** 2 **(b)** -2 **(c)** $-\dfrac{3x}{x^2 - 1}$ **(d)** $-\dfrac{3x}{x^2 - 1}$ **(e)** $\dfrac{3(x - 2)}{x^2 - 4x + 3}$ **(f)** $\dfrac{6x}{4x^2 - 1}$

4. (a) 0 **(b)** 0 **(c)** $\sqrt{x^2 - 4}$ **(d)** $-\sqrt{x^2 - 4}$ **(e)** $\sqrt{x^2 - 4x}$ **(f)** $2\sqrt{x^2 - 1}$ **5. (a)** 0 **(b)** 0 **(c)** $\dfrac{x^2 - 4}{x^2}$ **(d)** $-\dfrac{x^2 - 4}{x^2}$ **(e)** $\dfrac{x(x - 4)}{(x - 2)^2}$ **(f)** $\dfrac{x^2 - 1}{x^2}$

6. $\{x|x \ne -3, x \ne 3\}$ **7.** $\{x|x \le 2\}$ **8.** $\{x|x \ne 0\}$ **9.** $\{x|x \ne -3, x \ne 1\}$ **10.** $[-1, 2) \cup (2, \infty)$ **11.** $\{x|x > -8\}$

12. $(f + g)(x) = 2x + 3$; Domain: all real numbers
$(f - g)(x) = -4x + 1$; Domain: all real numbers
$(f \cdot g)(x) = -3x^2 + 5x + 2$; Domain: all real numbers
$\left(\dfrac{f}{g}\right)(x) = \dfrac{2 - x}{3x + 1}$; Domain: $\left\{x \middle| x \ne -\dfrac{1}{3}\right\}$

13. $(f + g)(x) = 3x^2 + 4x + 1$; Domain: all real numbers
$(f - g)(x) = 3x^2 - 2x + 1$; Domain: all real numbers
$(f \cdot g)(x) = 9x^3 + 3x^2 + 3x$; Domain: all real numbers
$\left(\dfrac{f}{g}\right)(x) = \dfrac{3x^2 + x + 1}{3x}$; Domain: $\{x|x \ne 0\}$

14. $(f + g)(x) = \dfrac{x^2 + 2x - 1}{x(x - 1)}$; Domain: $\{x|x \ne 0, x \ne 1\}$
$(f - g)(x) = \dfrac{x^2 + 1}{x(x - 1)}$; Domain: $\{x|x \ne 0, x \ne 1\}$
$(f \cdot g)(x) = \dfrac{x + 1}{x(x - 1)}$; Domain: $\{x|x \ne 0, x \ne 1\}$
$\left(\dfrac{f}{g}\right)(x) = \dfrac{x(x + 1)}{x - 1}$; Domain: $\{x|x \ne 0, x \ne 1\}$

15. $-4x + 1 - 2h$ **16. (a)** Domain: $\{x|-4 \le x \le 3\}$; Range: $\{y|-3 \le y \le 3\}$
(b) $(0, 0)$ **(c)** -1 **(d)** -4 **(e)** $\{x|0 < x \le 3\}$
(f)
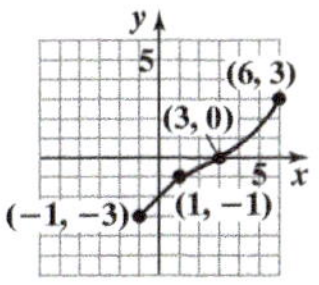

(g)
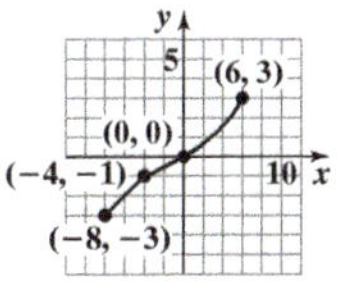

(h)
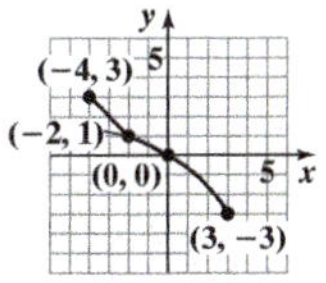

17. (a) Domain: $\{x|x \le 4\}$ or $(-\infty, 4]$
Range: $\{y|y \le 3\}$ or $(-\infty, 3]$
(b) Increasing on $(-\infty, -2]$ and $[2, 4]$; Decreasing on $[-2, 2]$
(c) Local maximum value is 1 and occurs at $x = -2$.
Local minimum value is -1 and occurs at $x = 2$.
(d) Absolute maximum: $f(4) = 3$
Absolute minimum: none
(e) No symmetry
(f) Neither
(g) x-intercepts: $-3, 0, 3$ y-intercept: 0

18. Odd **19.** Even **20.** Neither **21.** Odd

22.

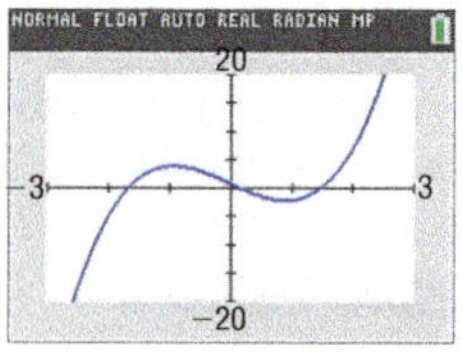

Local maximum value: 4.04 at $x = -0.91$
Local minimum value: -2.04 at $x = 0.91$
Increasing: $[-3, -0.91]$; $[0.91, 3]$
Decreasing: $[-0.91, 0.91]$

23.

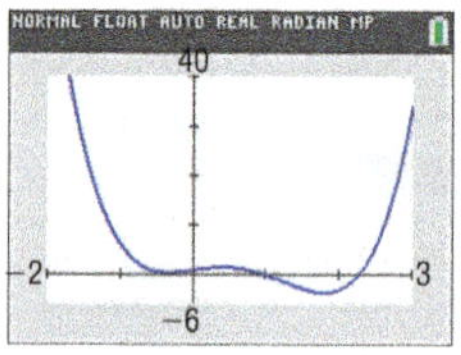

Local maximum value: 1.53 at $x = 0.41$
Local minima values: 0.54 at $x = -0.34$ and -3.56 at $x = 1.80$
Increasing: $[-0.34, 0.41]$; $[1.80, 3]$
Decreasing: $[-2, -0.34]$; $[0.41, 1.80]$

24. (a) 23 **(b)** 7 **(c)** 47 **25.** -5 **26.** -17 **27.** No **28.** Yes

29.

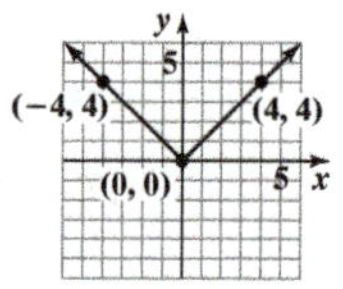

30.

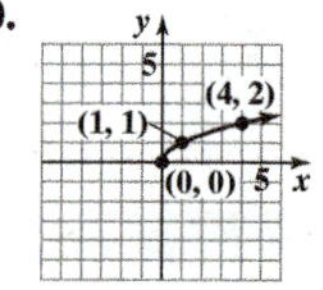

31.

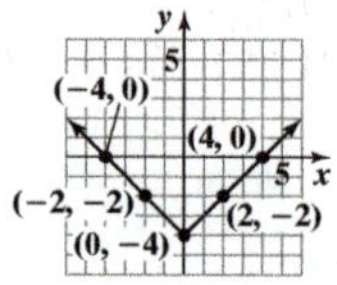

Intercepts: $(-4, 0)$, $(4, 0)$, $(0, -4)$
Domain: all real numbers
Range: $\{y \mid y \geq -4\}$ or $[-4, \infty)$

32.

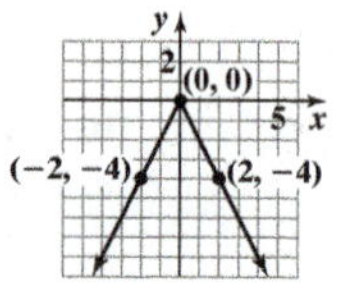

Intercept: $(0, 0)$
Domain: all real numbers
Range: $\{y \mid y \leq 0\}$ or $(-\infty, 0]$

33.

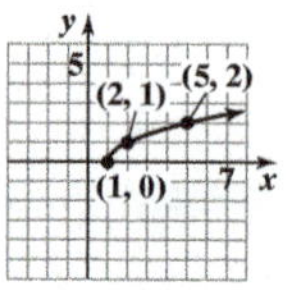

Intercept: $(1, 0)$
Domain: $\{x \mid x \geq 1\}$ or $[1, \infty)$
Range: $\{y \mid y \geq 0\}$ or $[0, \infty)$

34.

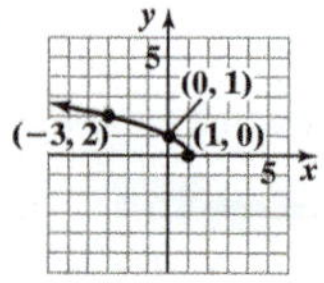

Intercepts: $(0, 1)$, $(1, 0)$
Domain: $\{x \mid x \leq 1\}$ or $(-\infty, 1]$
Range: $\{y \mid y \geq 0\}$ or $[0, \infty)$

35.

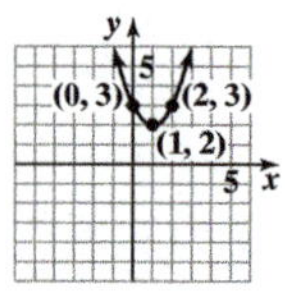

Intercept: $(0, 3)$
Domain: all real numbers
Range: $\{y \mid y \geq 2\}$ or $[2, \infty)$

36.

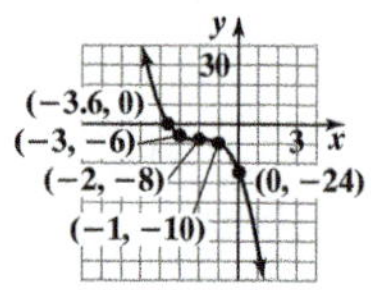

Intercepts: $(0, -24)$, $(-2 - \sqrt[3]{4}, 0)$ or about $(-3.6, 0)$
Domain: all real numbers
Range: all real numbers

37. (a) $\{x \mid x > -2\}$ or $(-2, \infty)$
(b) $(0, 0)$
(c)

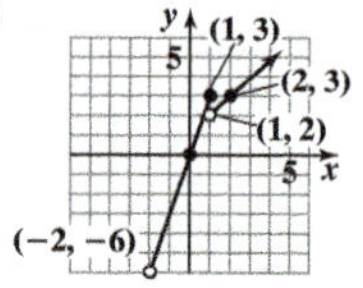

(d) $\{y \mid y > -6\}$ or $(-6, \infty)$

38. (a) $\{x \mid x \geq -4\}$ or $[-4, \infty)$
(b) $(0, 1)$
(c)

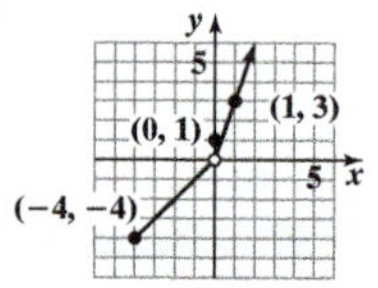

(d) $\{y \mid -4 \leq y < 0 \text{ or } y > 0\}$ or $[-4, 0) \cup (0, \infty)$

39. $A = 11$

40. (a) $A(x) = 2x^2 + \dfrac{40}{x}$
(b) 42 ft^2
(c) 28 ft^2
(d)

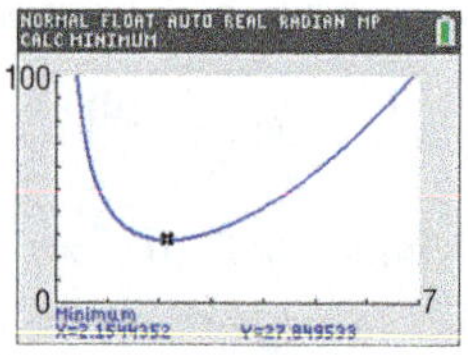

A is smallest when $x \approx 2.15$ ft.

41. (a) $A(x) = 10x - x^3$
(b) The largest area that can be enclosed by the rectangle is approximately 12.17 square units.

Chapter Test *(page 277)*

1. (a) Function; Domain: $\{2, 4, 6, 8\}$; Range: $\{5, 6, 7, 8\}$ **(b)** Not a function **(c)** Not a function **(d)** Function; Domain: all real numbers; Range: $\{y \mid y \geq 2\}$

2. Domain: $\left\{x \,\middle|\, x \leq \dfrac{4}{5}\right\}$; $f(-1) = 3$ **3.** Domain: $\{x \mid x \neq -2\}$; $g(-1) = 1$ **4.** Domain: $\{x \mid x \neq -9, x \neq 4\}$; $h(-1) = \dfrac{1}{8}$

5. (a) Domain: $\{x \mid -5 \leq x \leq 5\}$; Range: $\{y \mid -3 \leq y \leq 3\}$ **(b)** $(0, 2)$, $(-2, 0)$, and $(2, 0)$ **(c)** $f(1) = 3$ **(d)** $x = -5$ and $x = 3$
(e) $\{x \mid -5 \leq x < -2 \text{ or } 2 < x \leq 5\}$ or $[-5, -2) \cup (2, 5]$ **6.** Local maxima values: $f(-0.85) \approx -0.86$; $f(2.35) \approx 15.55$; local minimum value: $f(0) = -2$; the function is increasing on the intervals $[-5, -0.85]$ and $[0, 2.35]$ and decreasing on the intervals $[-0.85, 0]$ and $[2.35, 5]$.

7. (a)

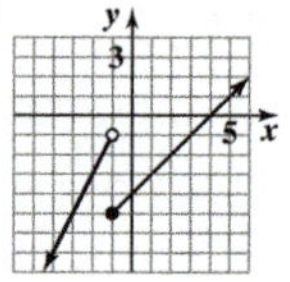

(b) $(0, -4)$, $(4, 0)$
(c) $g(-5) = -9$
(d) $g(2) = -2$

8. 19

9. (a) $(f - g)(x) = 2x^2 - 3x + 3$
(b) $(f \cdot g)(x) = 6x^3 - 4x^2 + 3x - 2$
(c) $f(x + h) - f(x) = 4xh + 2h^2$

10. (a)

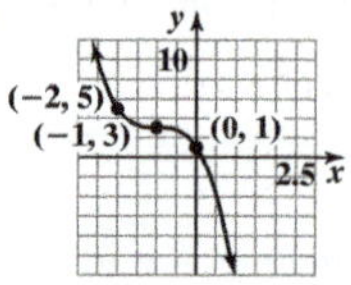

(b)

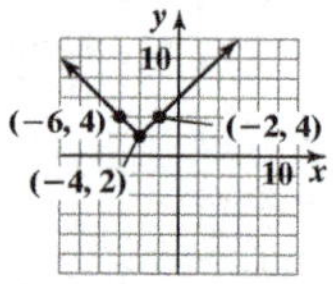

11. (a) 8.67% occurring in 1997 ($x \approx 5$) **(b)** The model predicts that the interest rate will be -10.343%. This is not reasonable.

12. (a) $V(x) = \dfrac{x^2}{8} - \dfrac{5x}{4} + \dfrac{\pi x^2}{64}$ **(b)** 1297.61 ft^3

Cumulative Review *(page 278)*

1. {6} **2.** $\left\{0, \frac{1}{3}\right\}$ **3.** $\{-1, 9\}$ **4.** $\left\{\frac{1}{3}, \frac{1}{2}\right\}$ **5.** $\left\{-\frac{7}{2}, \frac{1}{2}\right\}$ **6.** $\left\{\frac{1}{2}\right\}$

7. $\left\{x \middle| x < -\frac{4}{3}\right\}; \left(-\infty, -\frac{4}{3}\right)$

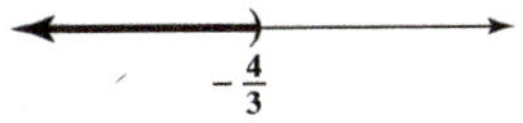

8. $\{x|1 < x < 4\}$; $(1, 4)$

9. $\left\{x \middle| x \le -2 \text{ or } x \ge \frac{3}{2}\right\}; (-\infty, -2] \cup \left[\frac{3}{2}, \infty\right)$

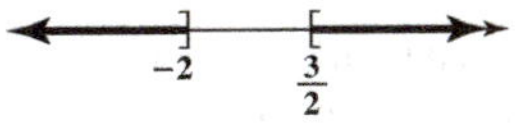

10. (a) distance: $\sqrt{29}$ **(b)** midpoint: $\left(\frac{1}{2}, -4\right)$ **(c)** slope: $-\frac{2}{5}$

11.

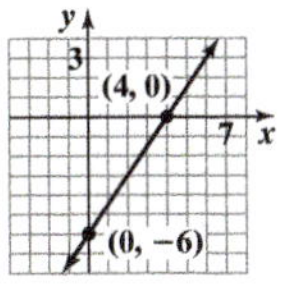

12.

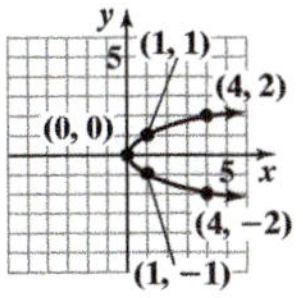

13.

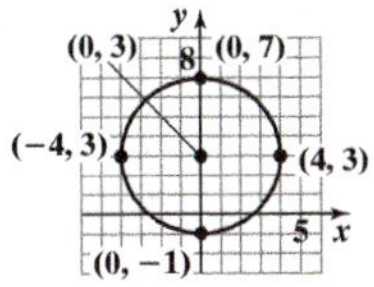

14.

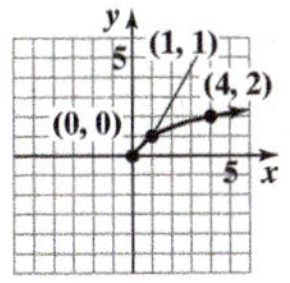

15. Intercepts: $(0, -3)$, $(-2, 0)$, $(2, 0)$; symmetry with respect to the y-axis

16. $y = \frac{1}{2}x + 5$

17.

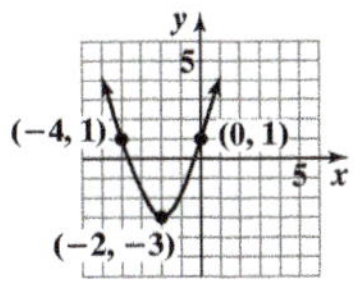

18.

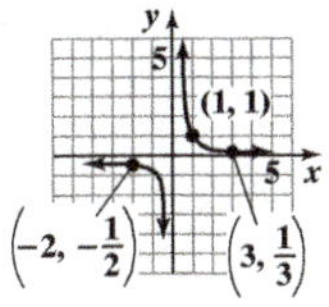

19.

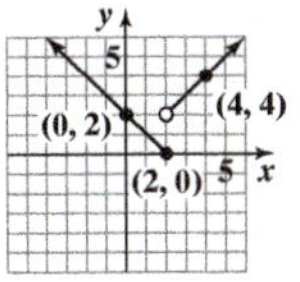

CHAPTER 4 Linear and Quadratic Functions

4.1 Assess Your Understanding *(page 287)*

7. slope; y-intercept **8.** positive **9.** T **10.** F **11.** a **12.** d

13. (a) $m = 2$; $b = 3$
(b)

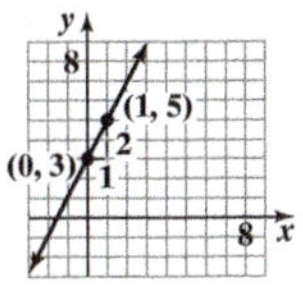

(c) 2 **(d)** Increasing

15. (a) $m = -3$; $b = 4$
(b)

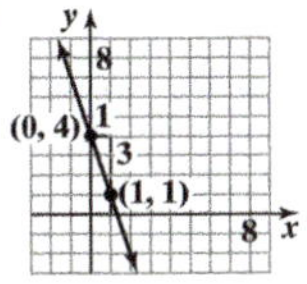

(c) −3 **(d)** Decreasing

17. (a) $m = \frac{1}{4}$; $b = -3$
(b)

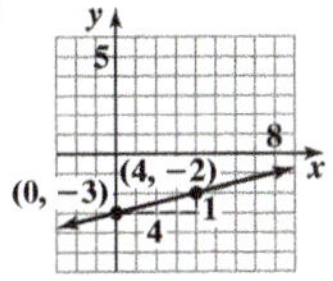

(c) $\frac{1}{4}$ **(d)** Increasing

19. (a) $m = 0$; $b = 4$
(b)

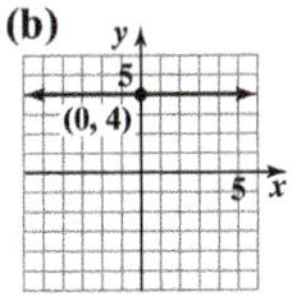

(c) 0 **(d)** Constant

21. Linear; $f(x) = -3x - 2$
23. Nonlinear
25. Nonlinear
27. Linear; $f(x) = 8$

29. (a) $\frac{1}{4}$ **(b)** $\left\{x \middle| x > \frac{1}{4}\right\}$ or $\left(\frac{1}{4}, \infty\right)$
(c) 1 **(d)** $\{x|x \le 1\}$ or $(-\infty, 1]$
(e)

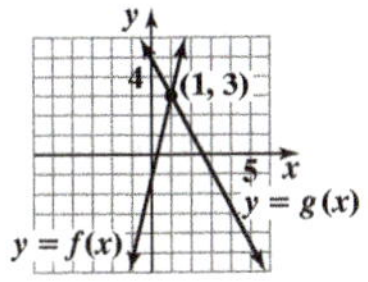

31. (a) 40 **(b)** 88 **(c)** −40 **(d)** $\{x|x > 40\}$ or $(40, \infty)$ **(e)** $\{x|x \le 88\}$ or $(-\infty, 88]$ **(f)** $\{x|-40 < x < 88\}$ or $(-40, 88)$ **33. (a)** −4 **(b)** $\{x|x < -4\}$ or $(-\infty, -4)$ **35. (a)** −6 **(b)** $\{x|-6 \le x < 5\}$ or $[-6, 5)$ **37. (a)** \$59 **(b)** 180 mi **(c)** 300 mi **(d)** $\{x|x \ge 0\}$ or $[0, \infty)$ **(e)** The cost of renting the car for a day increases \$0.35 for each mile driven, or there is a charge of \$0.35 per mile to rent the car in addition to a fixed charge of \$45. **(f)** It costs \$45 to rent the car if 0 miles are driven, or there is a fixed charge of \$45 to rent the car in addition to a charge that depends on mileage. **39. (a)** \$24; 600 T-shirts **(b)** $\$0 \le p < \$24$ **(c)** The price will increase.

41. (a) $\{x|9225 \le x \le 37{,}450\}$ or $[9225, 37{,}450]$
(b) \$2538.75
(c) The independent variable is adjusted gross income, x. The dependent variable is the tax bill, T.
(d)

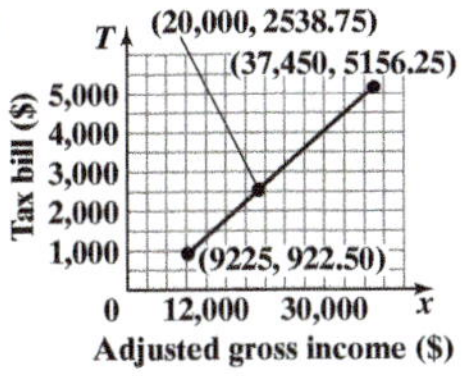

(e) \$27,500
(f) For each additional dollar of taxable income between \$9225 and \$37,450, the tax bill of a single person in 2015 increased by \$0.15.

43. (a) $x = 5000$
(b) $x > 5000$

45. (a) $V(x) = -1000x + 3000$
(b) $\{x|0 \le x \le 3\}$ or $[0, 3]$
(c)

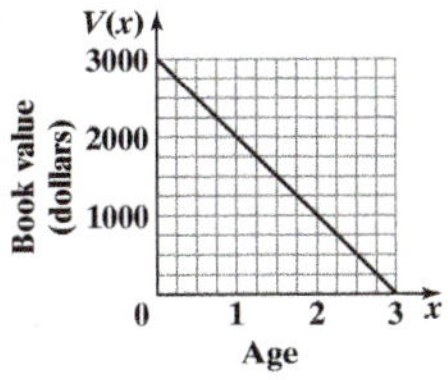

(d) \$1000
(e) After 1 year

47. (a) $C(x) = 90x + 1800$
(b)

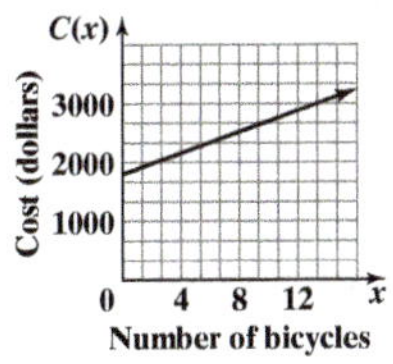

(c) \$3060 **(d)** 22 bicycles

49. (a) $C(x) = 0.89x + 39.95$ **(b)** \$137.85; \$244.65

51. (a)

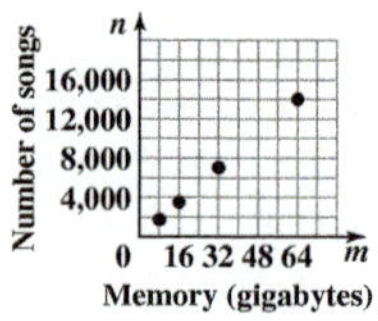

(b) Since each input (memory) corresponds to a single output (number of songs), we know that number of songs is a function of memory. Also, because the average rate of change is a constant 218.75 songs per gigabyte, the function is linear.

(c) $n(m) = 218.75m$ **(d)** $\{m \mid m \geq 0\}$ or $[0, \infty)$

(e)

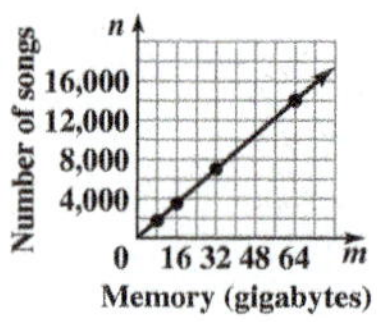

(f) If memory increases by 1 GB, then the number of songs increases by 218.75.

53. (d), (e) **55.** $b = 0$; yes, $f(x) = b$

57.

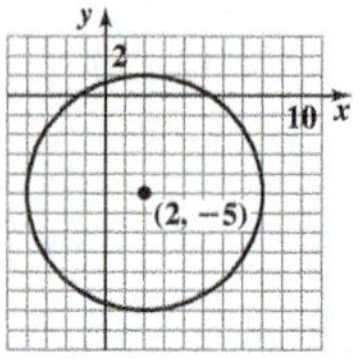

58. 6 **59.** 7 **60.**

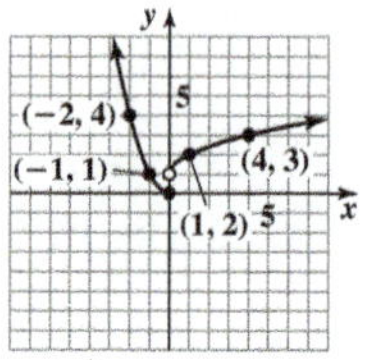

4.2 Assess Your Understanding *(page 294)*

3. scatter diagram **4.** decrease; 0.008 **5.** Linear relation, $m > 0$ **7.** Linear relation, $m < 0$ **9.** Nonlinear relation

11. (a)

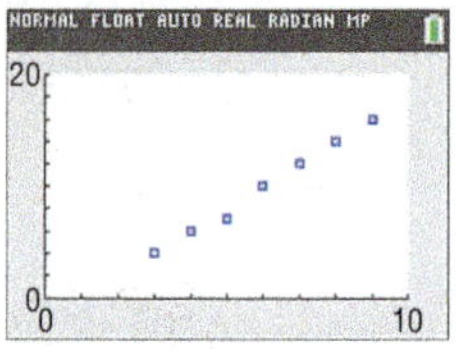

(b) Answers will vary. Using (4, 6) and (8, 14), $y = 2x - 2$.

(c)

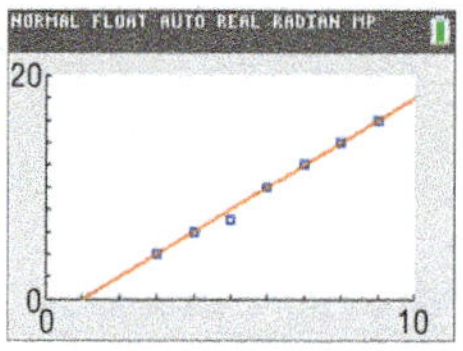

(d) $y = 2.0357x - 2.3571$

(e)

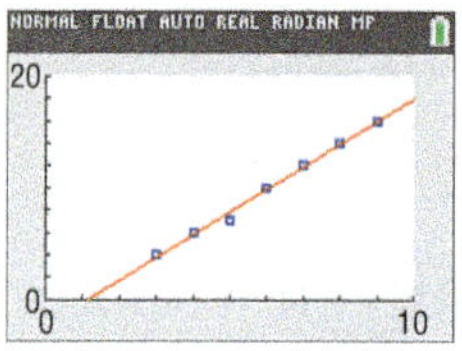

13. (a)

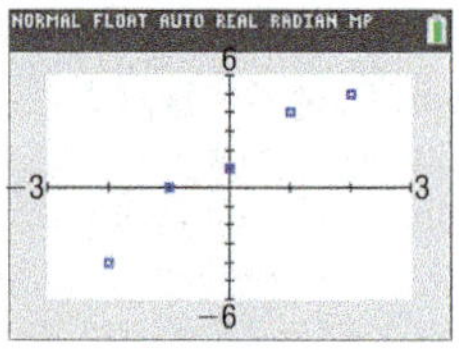

(b) Answers will vary. Using $(-2, -4)$ and $(2, 5)$, $y = \frac{9}{4}x + \frac{1}{2}$.

(c)

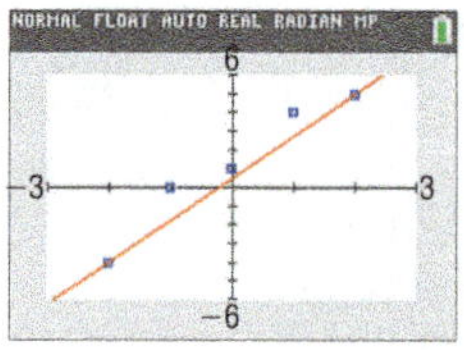

(d) $y = 2.2x + 1.2$

(e)

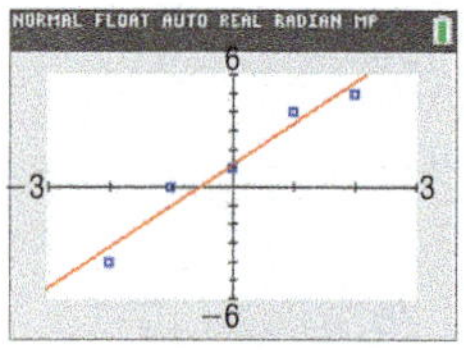

15. (a)

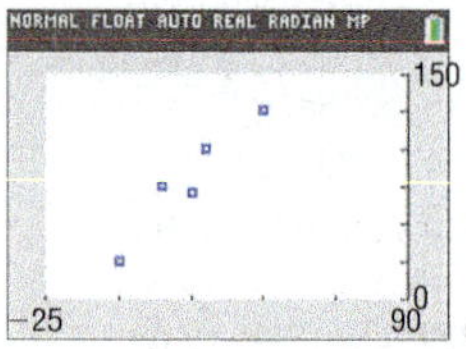

(b) Answers will vary. Using $(-20, 100)$ and $(-10, 140)$, $y = 4x + 180$.

(c)

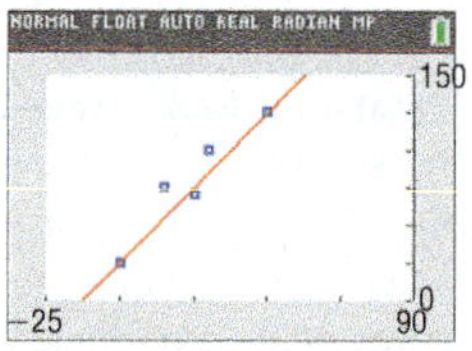

(d) $y = 3.8613x + 180.2920$

(e)

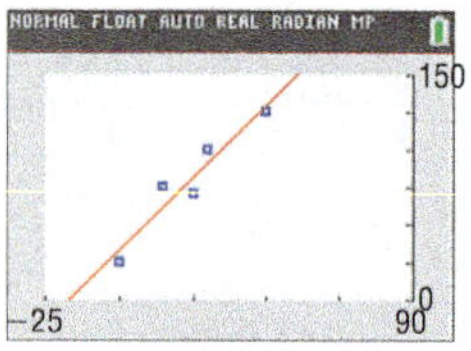

17. (a)

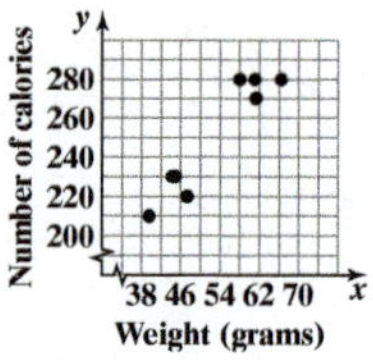

(b) Linear with positive slope

(c) Answers will vary. Using the points (39.52, 210) and (66.45, 280), $y = 2.599x + 107.288$.

(d)

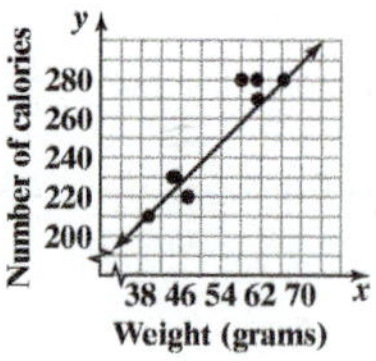

(e) 269 calories

(f) If the weight of a candy bar is increased by 1 gram, the number of calories will increase by 2.599, on average.

19. (a) The independent variable is the number of hours spent playing video games, and cumulative grade-point average is the dependent variable, because we are using number of hours playing video games to predict (or explain) cumulative grade-point average.

(b)

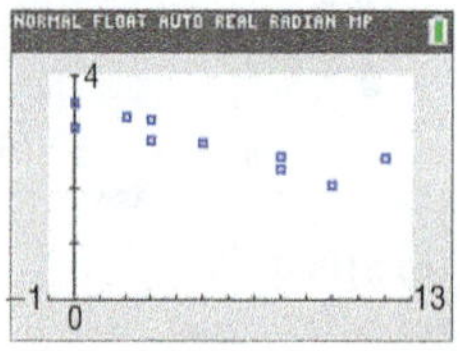

(c) $G(h) = -0.0942h + 3.2763$

(d) If the number of hours playing video games in a week increases by 1 hour, the cumulative grade-point average decreases 0.09, on average.

(e) 2.52

(f) Approximately 9.3 hours

21. (a) No

(b)

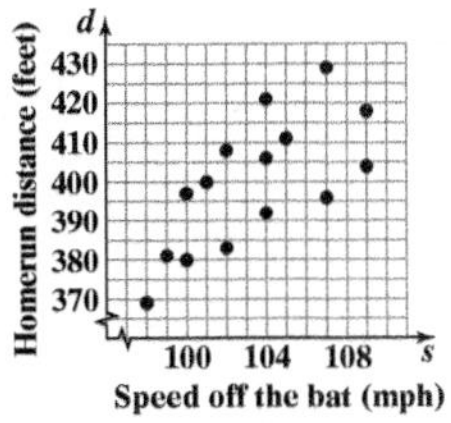

(c) $d = 3.3641s + 51.8233$
(d) If the speed off the bat increases by 1 mile per hour, the homerun distance increases by 3.3641 feet, on average.
(e) $d(s) = 3.3641s + 51.8233$
(f) $\{s \mid s > 0\}$ or $(0, \infty)$
(g) Approximately 398 feet

23.

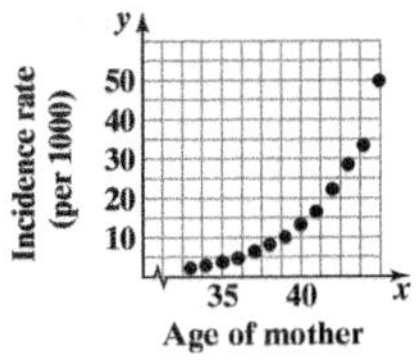

No, the data do not follow a linear pattern.

25. No linear relation **27.** 34.8 hours; A student whose GPA is 0 spends 34.8 hours each week playing video games. $G(0) = 3.28$. The average GPA of a student who does not play video games is 3.28. **28.** $2x + y = 3$ or $y = -2x + 3$ **29.** $\{x \mid x \neq -5, x \neq 5\}$ **30.** $(g - f)(x) = x^2 - 8x + 12$ **31.** $y = (x + 3)^2 - 4$

4.3 Assess Your Understanding *(page 306)*

5. parabola **6.** axis or axis of symmetry **7.** $-\dfrac{b}{2a}$ **8.** T **9.** T **10.** T **11.** a **12.** d **13.** *C* **15.** *F* **17.** *G* **19.** *H*

21.

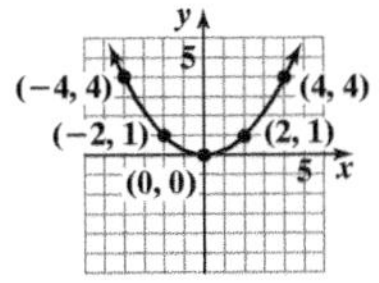

23.

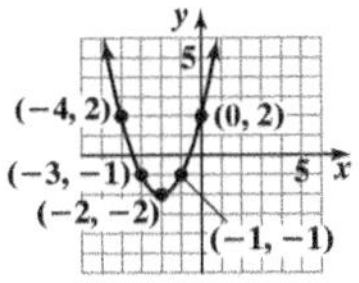

25. $f(x) = (x + 2)^2 - 2$

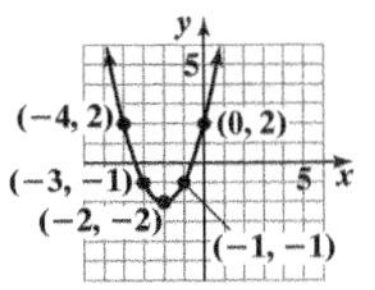

27. $f(x) = 2(x - 1)^2 - 1$

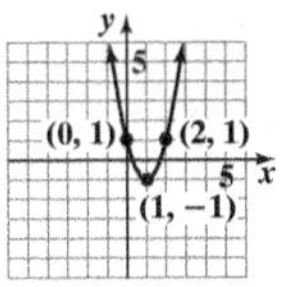

29. $f(x) = -(x + 1)^2 + 1$

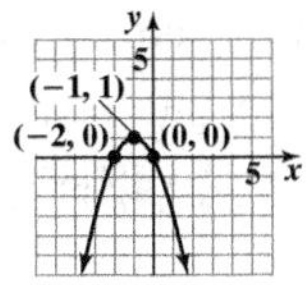

31. $f(x) = \dfrac{1}{2}(x + 1)^2 - \dfrac{3}{2}$

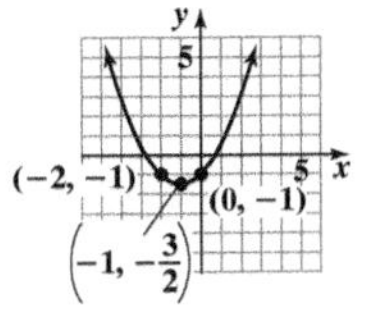

33. (a)

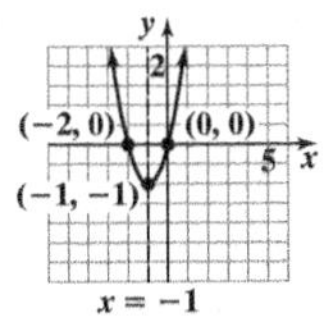

(b) Domain: $(-\infty, \infty)$
Range: $[-1, \infty)$
(c) Decreasing: $(-\infty, -1]$
Increasing: $[-1, \infty)$

35. (a)

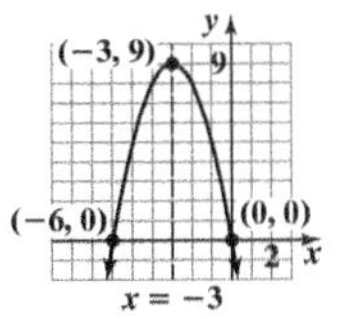

(b) Domain: $(-\infty, \infty)$
Range: $(-\infty, 9]$
(c) Increasing: $(-\infty, -3]$
Decreasing: $[-3, \infty)$

37. (a)

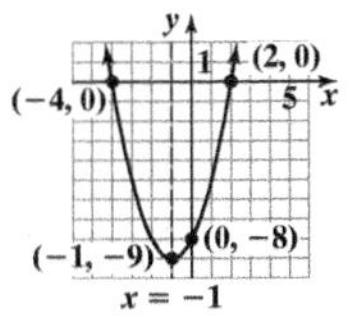

(b) Domain: $(-\infty, \infty)$
Range: $[-9, \infty)$
(c) Decreasing: $(-\infty, -1]$
Increasing: $[-1, \infty)$

39. (a)

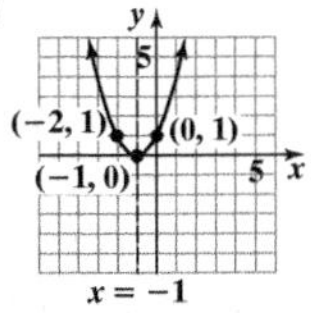

(b) Domain: $(-\infty, \infty)$
Range: $[0, \infty)$
(c) Decreasing: $(-\infty, -1]$
Increasing: $[-1, \infty)$

41. (a)

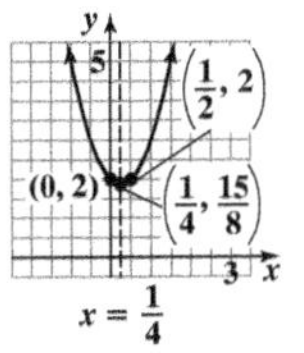

(b) Domain: $(-\infty, \infty)$
Range: $\left[\dfrac{15}{8}, \infty\right)$
(c) Decreasing: $\left(-\infty, \dfrac{1}{4}\right]$
Increasing: $\left[\dfrac{1}{4}, \infty\right)$

43. (a)

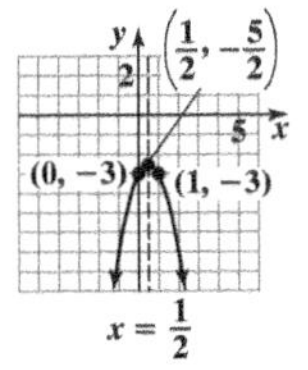

(b) Domain: $(-\infty, \infty)$
Range: $\left(-\infty, -\dfrac{5}{2}\right]$
(c) Increasing: $\left(-\infty, \dfrac{1}{2}\right]$
Decreasing: $\left[\dfrac{1}{2}, \infty\right)$

45. (a)

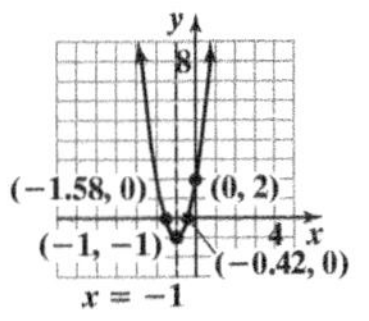

(b) Domain: $(-\infty, \infty)$
Range: $[-1, \infty)$
(c) Decreasing: $(-\infty, -1]$
Increasing: $[-1, \infty)$

47. (a)

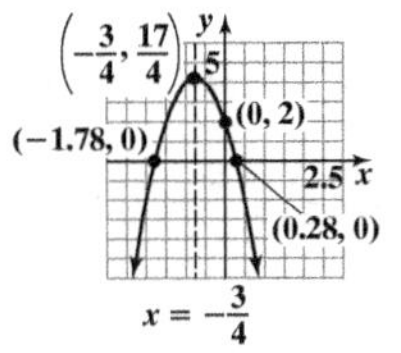

(b) Domain: $(-\infty, \infty)$
Range: $\left(-\infty, \dfrac{17}{4}\right]$
(c) Increasing: $\left(-\infty, -\dfrac{3}{4}\right]$
Decreasing: $\left[-\dfrac{3}{4}, \infty\right)$

49. $f(x) = (x + 1)^2 - 2 = x^2 + 2x - 1$ **51.** $f(x) = -(x + 3)^2 + 5 = -x^2 - 6x - 4$ **53.** $f(x) = 2(x - 1)^2 - 3 = 2x^2 - 4x - 1$
55. Minimum value; -18 **57.** Minimum value; -21 **59.** Maximum value; 21 **61.** Maximum value; 13

63. (a)

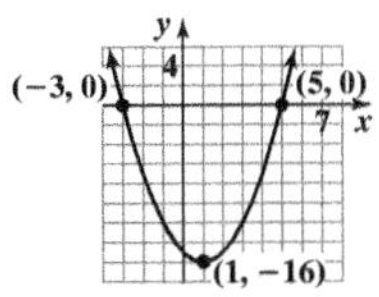

(b) Domain: $(-\infty, \infty)$
Range: $[-16, \infty)$
(c) Decreasing: $(-\infty, 1]$
Increasing: $[1, \infty)$

65. (a)

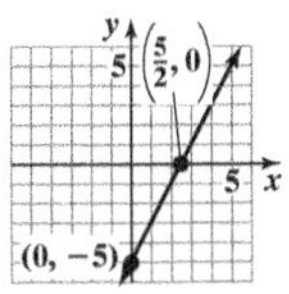

(b) Domain: $(-\infty, \infty)$
Range: $(-\infty, \infty)$
(c) Increasing: $(-\infty, \infty)$

67. (a)

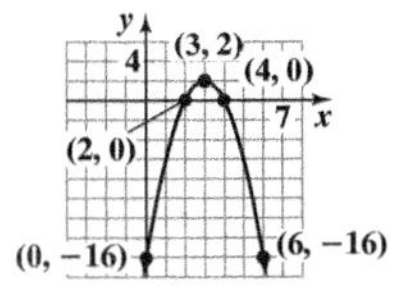

(b) Domain: $(-\infty, \infty)$
Range: $(-\infty, 2]$
(c) Increasing: $(-\infty, 3]$
Decreasing: $[3, \infty)$

69. (a)

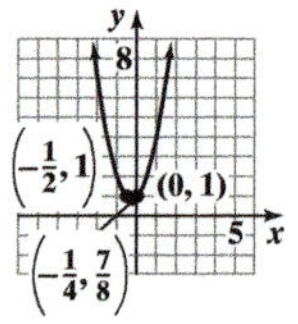

(b) Domain: $(-\infty, \infty)$

Range: $\left[\frac{7}{8}, \infty\right)$

(c) Decreasing: $\left(-\infty, -\frac{1}{4}\right]$

Increasing: $\left[-\frac{1}{4}, \infty\right)$

71. (a)

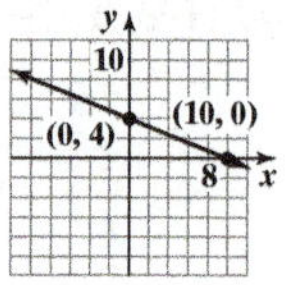

(b) Domain: $(-\infty, \infty)$

Range: $(-\infty, \infty)$

(c) Decreasing: $(-\infty, \infty)$

73. (a)

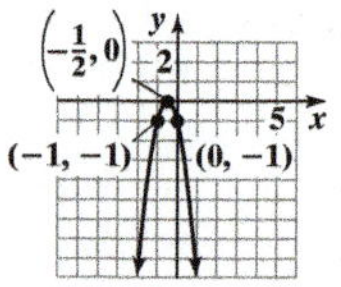

(b) Domain: $(-\infty, \infty)$

Range: $(-\infty, 0]$

(c) Increasing: $\left(-\infty, -\frac{1}{2}\right]$

Decreasing: $\left[-\frac{1}{2}, \infty\right)$

75. $a = 6, b = 0, c = 2$

77. (a), (c), (d)

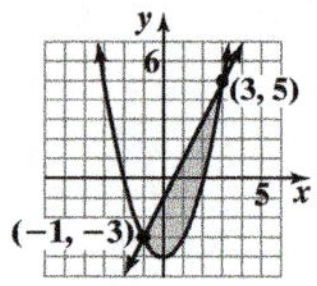

(b) $\{-1, 3\}$

79. (a), (c), (d)

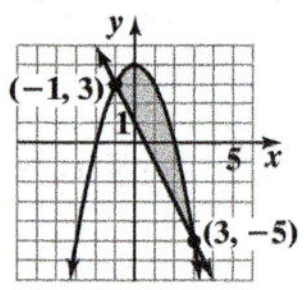

(b) $\{-1, 3\}$

81. (a), (c), (d)

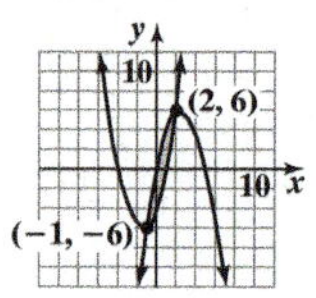

(b) $\{-1, 2\}$

83. (a) $a = 1: f(x) = (x + 3)(x - 1) = x^2 + 2x - 3$
$a = 2: f(x) = 2(x + 3)(x - 1) = 2x^2 + 4x - 6$
$a = -2: f(x) = -2(x + 3)(x - 1) = -2x^2 - 4x + 6$
$a = 5: f(x) = 5(x + 3)(x - 1) = 5x^2 + 10x - 15$

(b) The value of a does not affect the x-intercepts, but it changes the y-intercept by a factor of a.

(c) The value of a does not affect the axis of symmetry. It is $x = -1$ for all values of a.

(d) The value of a does not affect the x-coordinate of the vertex. However, the y-coordinate of the vertex is multiplied by a.

(e) The mean of the x-intercepts is the x-coordinate of the vertex.

85. (a) $(-2, -25)$
(b) $-7, 3$
(c) $-4, 0; (-4, -21), (0, -21)$
(d)

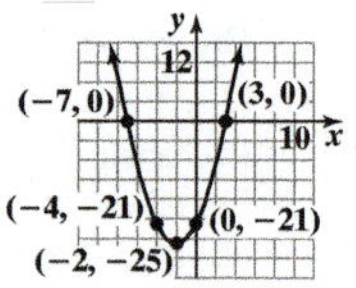

87. $(2, 2)$ **89.** \$500; \$1,000,000 **91. (a)** 70,000 digital music players **(b)** \$2500
93. (a) 187 or 188 watches; \$7031.20 **(b)** $P(x) = -0.2x^2 + 43x - 1750$
(c) 107 or 108 watches; \$561.20 **95. (a)** 171 ft **(b)** 49 mph **(c)** Reaction time
97. If x is even, then ax^2 and bx are even and $ax^2 + bx$ is even, which means that $ax^2 + bx + c$ is odd. If x is odd, then ax^2 and bx are odd and $ax^2 + bx$ is even, which means that $ax^2 + bx + c$ is odd. In either case, $f(x)$ is odd.

99.

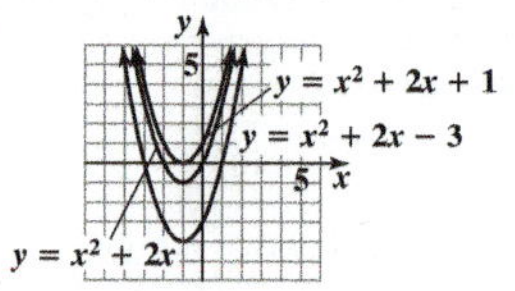

101. $b^2 - 4ac < 0$ **103.** No
105. Symmetric with respect to the x-axis, the y-axis, and the origin.
106. $\{x | x \le 4\}$ or $(-\infty, 4]$ **107.** Center $(5, -2)$; radius $= 3$
108. $y = \sqrt{-x}$

4.4 Assess Your Understanding *(page 315)*

3. (a) $R(x) = -\frac{1}{6}x^2 + 100x$ **(b)** $\{x | 0 \le x \le 600\}$ **(c)** \$13,333.33 **(d)** 300; \$15,000 **(e)** \$50 **5. (a)** $R(x) = -\frac{1}{5}x^2 + 20x$ **(b)** \$255
(c) 50; \$500 **(d)** \$10 **(e)** Between \$8 and \$12 **7. (a)** $A(w) = -w^2 + 200w$ **(b)** A is largest when $w = 100$ yd. **(c)** 10,000 yd^2 **9.** 2,000,000 m^2
11. (a) $\frac{625}{16} \approx 39$ ft **(b)** $\frac{7025}{32} \approx 219.5$ ft **(c)** About 170 ft
(d)

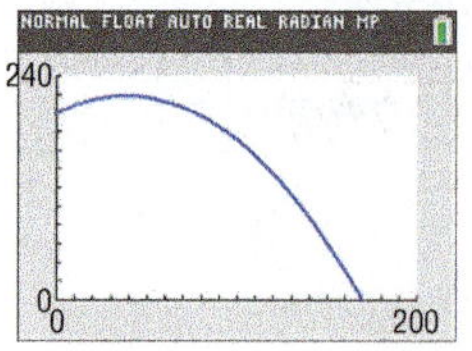

(f) When the height is 100 ft, the projectile is about 135.7 ft from the cliff.

13. 18.75 m **15. (a)** 3 in. **(b)** Between 2 in. and 4 in.
17. $\frac{750}{\pi} \approx 238.73$ m by 375 m
19. $x = \frac{a}{2}$ **21.** $\frac{38}{3}$ **23.** $\frac{248}{3}$

25. (a)

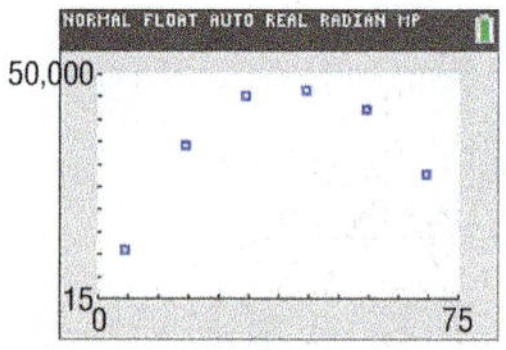

The data appear to follow a quadratic relation with $a < 0$.

(b) $I(x) = -44.759x^2 + 4295.356x - 55{,}045.418$
(c) About 48.0 years of age
(d) Approximately \$48,007

(e)

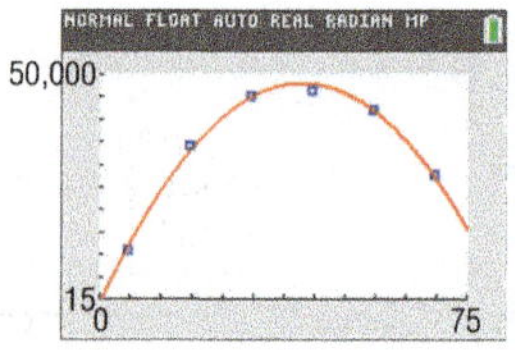

27. (a)

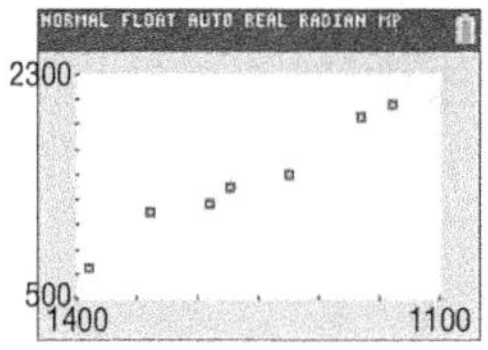

The data appear to be linearly related with positive slope.
(b) $R(x) = 1.229x + 917.385$ **(c)** \$1993

29. (a)

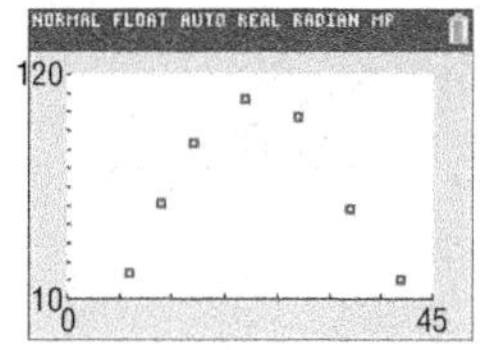

The data appear to follow a quadratic relation with $a < 0$.
(b) $B(a) = -0.547a^2 + 31.190a - 342.218$ **(c)** 79.357

32. $15i$ **33.** 13 **34.** $(x + 6)^2 + y^2 = 7$ **35.** $\left\{\frac{-4 - \sqrt{31}}{5}, \frac{-4 + \sqrt{31}}{5}\right\}$

4.5 Assess Your Understanding *(page 322)*

3. (a) $\{x|x < -2 \text{ or } x > 2\}; (-\infty, -2) \cup (2, \infty)$ **(b)** $\{x|-2 \le x \le 2\}; [-2, 2]$
5. (a) $\{x|-2 \le x \le 1\}; [-2, 1]$ **(b)** $\{x|x < -2 \text{ or } x > 1\}; (-\infty, -2) \cup (1, \infty)$ **7.** $\{x|-2 < x < 5\}; (-2, 5)$
9. $\{x|x < 0 \text{ or } x > 4\}; (-\infty, 0) \cup (4, \infty)$ **11.** $\{x|-3 < x < 3\}; (-3, 3)$ **13.** $\{x|x < -4 \text{ or } x > 3\}; (-\infty, -4) \cup (3, \infty)$
15. $\left\{x\middle|-\frac{1}{2} < x < 3\right\}; \left(-\frac{1}{2}, 3\right)$ **17.** No real solution **19.** No real solution **21.** $\left\{x\middle| x < -\frac{2}{3} \text{ or } x > \frac{3}{2}\right\}; \left(-\infty, -\frac{2}{3}\right) \cup \left(\frac{3}{2}, \infty\right)$
23. $\{x|x \le -4 \text{ or } x \ge 4\}; (-\infty, -4] \cup [4, \infty)$
25. (a) $\{-1, 1\}$ **(b)** $\{-1\}$ **(c)** $\{-1, 4\}$ **(d)** $\{x|x < -1 \text{ or } x > 1\}; (-\infty, -1) \cup (1, \infty)$ **(e)** $\{x|x \le -1\}; (-\infty, -1]$
(f) $\{x|x < -1 \text{ or } x > 4\}; (-\infty, -1) \cup (4, \infty)$ **(g)** $\{x|x \le -\sqrt{2} \text{ or } x \ge \sqrt{2}\}; (-\infty, -\sqrt{2}] \cup [\sqrt{2}, \infty)$
27. (a) $\{-1, 1\}$ **(b)** $\left\{-\frac{1}{4}\right\}$ **(c)** $\{-4, 0\}$ **(d)** $\{x|-1 < x < 1\}; (-1, 1)$ **(e)** $\left\{x\middle|x \le -\frac{1}{4}\right\}; \left(-\infty, -\frac{1}{4}\right]$ **(f)** $\{x|-4 < x < 0\}; (-4, 0)$ **(g)** $\{0\}$
29. (a) $\{-2, 2\}$ **(b)** $\{-2, 2\}$ **(c)** $\{-2, 2\}$ **(d)** $\{x|x < -2 \text{ or } x > 2\}; (-\infty, -2) \cup (2, \infty)$ **(e)** $\{x|x \le -2 \text{ or } x \ge 2\}; (-\infty, -2] \cup [2, \infty)$
(f) $\{x|x < -2 \text{ or } x > 2\}; (-\infty, -2) \cup (2, \infty)$ **(g)** $\{x|x \le -\sqrt{5} \text{ or } x \ge \sqrt{5}\}; (-\infty, -\sqrt{5}] \cup [\sqrt{5}, \infty)$
31. (a) $\{-1, 2\}$ **(b)** $\{-2, 1\}$ **(c)** $\{0\}$ **(d)** $\{x|x < -1 \text{ or } x > 2\}; (-\infty, -1) \cup (2, \infty)$ **(e)** $\{x|-2 \le x \le 1\}; [-2, 1]$ **(f)** $\{x|x < 0\}; (-\infty, 0)$
(g) $\left\{x\middle|x \le \frac{1 - \sqrt{13}}{2} \text{ or } x \ge \frac{1 + \sqrt{13}}{2}\right\}; \left(-\infty, \frac{1 - \sqrt{13}}{2}\right] \cup \left[\frac{1 + \sqrt{13}}{2}, \infty\right)$
33. (a) 5 sec **(b)** The ball is more than 96 ft above the ground for time t between 2 and 3 sec, $2 < t < 3$.
35. (a) \$0, \$1000 **(b)** The revenue is more than \$800,000 for prices between \$276.39 and \$723.61, $\$276.39 < p < \$723.61$.
37. (a) $\{c \mid 0.112 < c < 81.907\}; (0.112, 81.907)$ **(b)** It is possible to hit a target 75 km away if $c = 0.651$ or $c = 1.536$. **44.** $\{x|x \le 5\}$
45. (a) $(9, 0), (0, -6)$
(b)

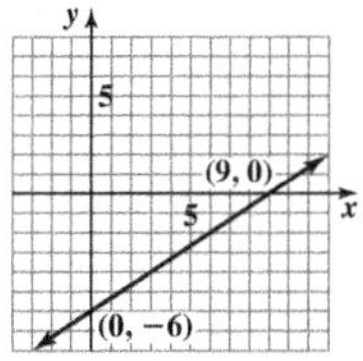

46. Odd
47. $19 - 26i$

Review Exercises *(page 325)*

1. (a) $m = 2; b = -5$ **(b)** 2
(c)

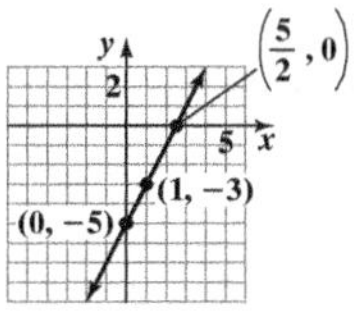

(d) Increasing

2. (a) $m = -\frac{1}{3}; b = 1$ **(b)** $-\frac{1}{3}$
(c)

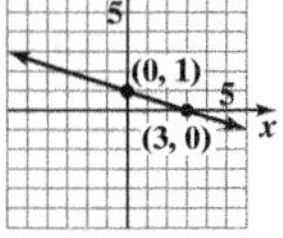

(d) Decreasing

3. (a) $m = 0; b = 4$ **(b)** 0
(c)

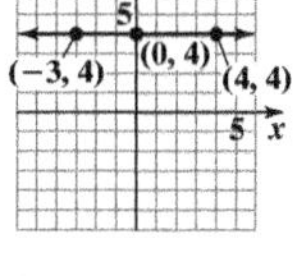

(d) Constant
4. Linear; $f(x) = 5x + 3$
5. Nonlinear

6.

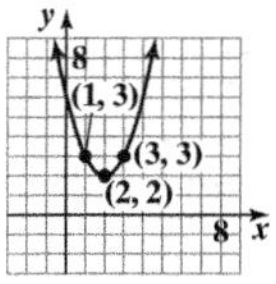

7.

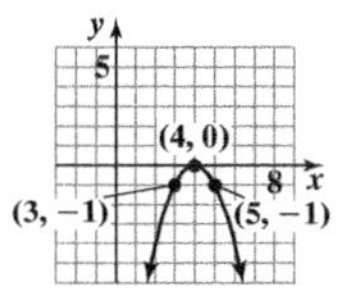

8.

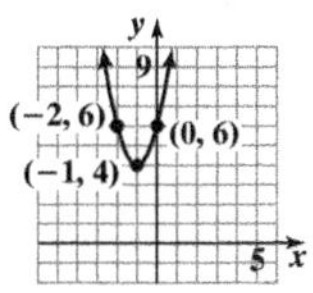

9. (a)

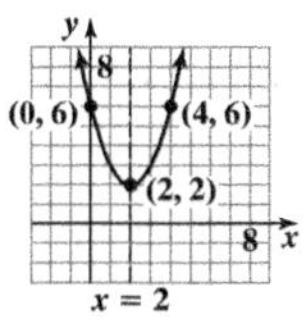

(b) Domain: $(-\infty, \infty)$
Range: $[2, \infty)$
(c) Decreasing: $(-\infty, 2]$
Increasing: $[2, \infty)$

10. (a)

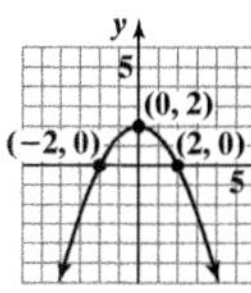

(b) Domain: $(-\infty, \infty)$
Range: $(-\infty, 2]$
(c) Increasing: $(-\infty, 0]$
Decreasing: $[0, \infty)$

11. (a)

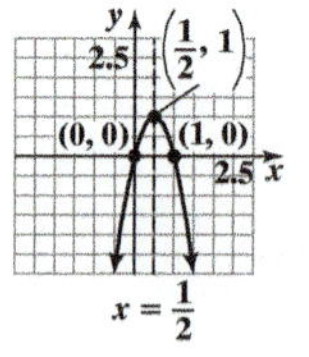

(b) Domain: $(-\infty, \infty)$
Range: $(-\infty, 1]$

(c) Increasing: $\left(-\infty, \frac{1}{2}\right]$
Decreasing: $\left[\frac{1}{2}, \infty\right)$

12. (a)

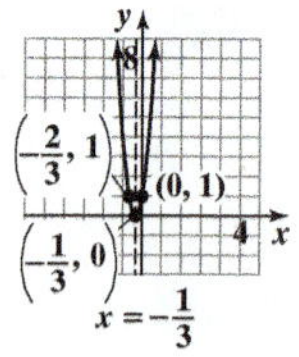

(b) Domain: $(-\infty, \infty)$
Range: $[0, \infty)$

(c) Decreasing: $\left(-\infty, -\frac{1}{3}\right]$
Increasing: $\left[-\frac{1}{3}, \infty\right)$

13. (a)

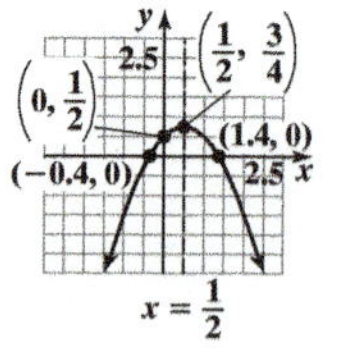

(b) Domain: $(-\infty, \infty)$
Range: $\left(-\infty, \frac{3}{4}\right]$

(c) Increasing: $\left(-\infty, \frac{1}{2}\right]$
Decreasing: $\left[\frac{1}{2}, \infty\right)$

14. (a)

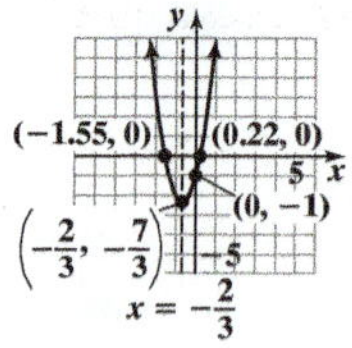

(b) Domain: $(-\infty, \infty)$
Range: $\left[-\frac{7}{3}, \infty\right)$

(c) Decreasing: $\left(-\infty, -\frac{2}{3}\right]$
Increasing: $\left[-\frac{2}{3}, \infty\right)$

15. Minimum value; 1 **16.** Maximum value; 12 **17.** Maximum value; 4 **18.** $\{x \mid -8 < x < 2\}$; $(-8, -2)$

19. $\left\{x \,\middle|\, x \le -\frac{1}{3} \text{ or } x \ge 5\right\}$; $\left(-\infty, -\frac{1}{3}\right] \cup [5, \infty)$ **20.** $y = x^2 + 2x + 3$ **21.** $y = x^2 - 4x + 5$

22. (a) $S(x) = 0.01x + 25{,}000$ **(b)** \$35,000 **(c)** \$7,500,000 **(d)** $x >$ \$12,500,000 **23. (a)** $R(x) = -\frac{1}{10}x^2 + 150x$ **(b)** \$14,000 **(c)** 750; \$56,250 **(d)** \$75

24. 4,166,666.7 m^2 **25. (a)** 63 clubs **(b)** \$151.90 **26. (a)** $A(x) = -x^2 + 10x$; 25 units^2 **27.** 3.6 ft

28. (a)

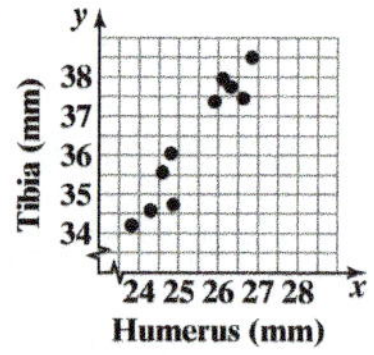

(b) Yes
(c) $y = 1.3092x + 1.1140$
(d) 37.95 mm

29. (a) Quadratic, $a < 0$

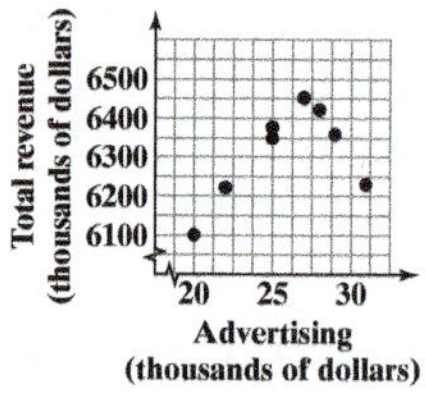

(b) About \$26.5 thousand
(c) \$6408 thousand

(e)

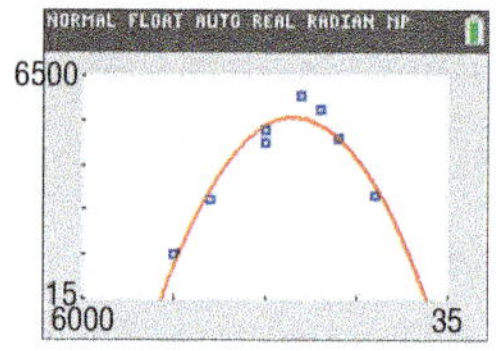

Chapter Test *(page 327)*

1. (a) Slope: −4; y-intercept: 3
(b) Decreasing
(c)

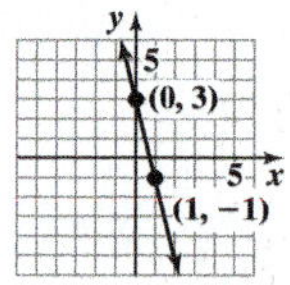

2. Linear; $y = -5x + 2$

3.

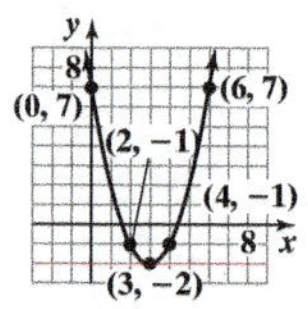

4. (a) Opens up **(b)** $(2, -8)$ **(c)** $x = 2$
(d) x-intercepts: $\frac{6 - 2\sqrt{6}}{3}, \frac{6 + 2\sqrt{6}}{3}$;
y-intercept: 4
(e)

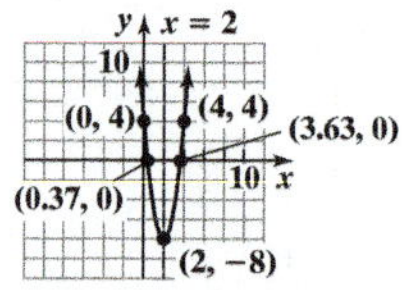

(f) Domain: All real numbers; $(-\infty, \infty)$
Range: $\{y \mid y \ge -8\}$; $[-8, \infty)$
(g) Decreasing: $(-\infty, 2]$; Increasing: $[2, \infty)$

5. (a) Opens down **(b)** $(1, -3)$ **(c)** $x = 1$
(d) No x-intercepts: y-intercept: −5

(e)

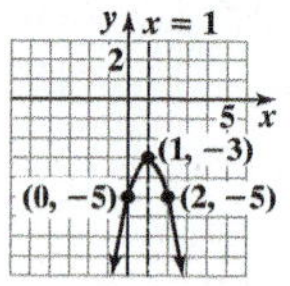

(f) Domain: All real numbers; $(-\infty, \infty)$
Range: $\{y \mid y \le -3\}$; $(-\infty, -3]$
(g) Increasing: $(-\infty, 1]$; Decreasing: $[1, \infty)$

6. $f(x) = 2x^2 - 4x - 30$ **7.** Maximum value; 21 **8.** $\{x \mid x \le 4 \text{ or } x \ge 6\}$; $(-\infty, 4] \cup [6, \infty)$ **9. (a)** $C(m) = 0.15m + 129.50$ **(b)** \$258.50

(c) 562 miles **10. (a)** $R(x) = -\frac{1}{10}x^2 + 1000x$ **(b)** \$384,000 **(c)** 5000 units; \$2,500,000 **(d)** \$500

11. (a)

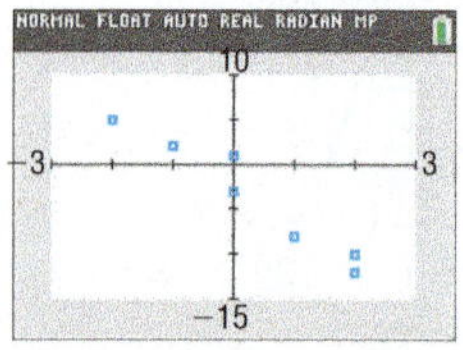

Linear with negative slope

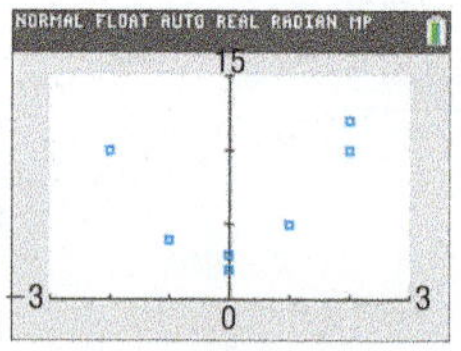

Quadratic that opens up

(b) $y = -4.234x - 2.362$ **(c)** $y = 1.993x^2 + 0.289x + 2.503$

Cumulative Review *(page 328)*

1. $5\sqrt{2}; \left(\frac{3}{2}, \frac{1}{2}\right)$ **2.** $(-2, -1)$ and $(2, 3)$ are on the graph.

3. $\left\{x \middle| x \geq -\frac{3}{5}\right\}$ or $\left[-\frac{3}{5}, \infty\right)$

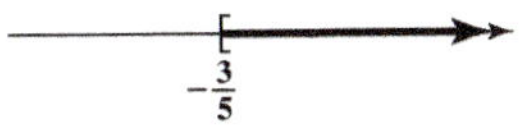

4. $y = -2x + 2$

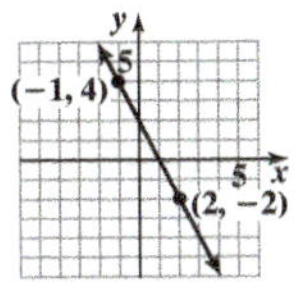

5. $y = -\frac{1}{2}x + \frac{13}{2}$

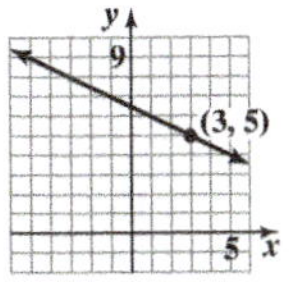

6. $(x - 2)^2 + (y + 4)^2 = 25$

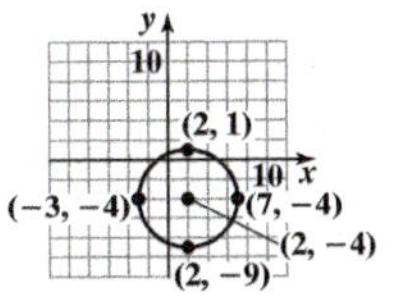

7. Yes **8. (a)** -3 **(b)** $x^2 - 4x - 2$ **(c)** $x^2 + 4x + 1$ **(d)** $-x^2 + 4x - 1$ **(e)** $x^2 - 3$ **(f)** $2x + h - 4$ **9.** $\left\{z \middle| z \neq \frac{7}{6}\right\}$ **10.** Yes
11. (a) No **(b)** -1; $(-2, -1)$ is on the graph. **(c)** -8; $(-8, 2)$ is on the graph. **12.** Neither **13.** Local maximum value is 5.30 and occurs at $x = -1.29$. Local minimum value is -3.30 and occurs at $x = 1.29$. Increasing: $[-4, -1.29]$ and $[1.29, 4]$; Decreasing: $[-1.29, 1.29]$
14. (a) -4 **(b)** $\{x | x > -4\}$ or $(-4, \infty)$ **15. (a)** Domain: $\{x | -4 \leq x \leq 4\}$; Range: $\{y | -1 \leq y \leq 3\}$ **(b)** $(-1, 0), (0, -1), (1, 0)$ **(c)** y-axis **(d)** 1 **(e)** -4 and 4 **(f)** $\{x | -1 < x < 1\}$

(g)

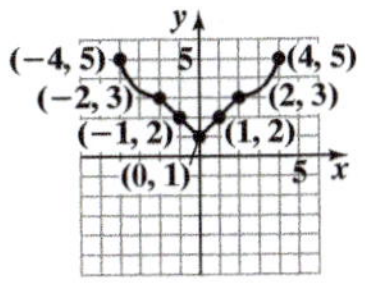

(h)

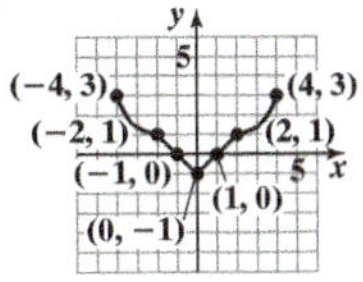

(i)

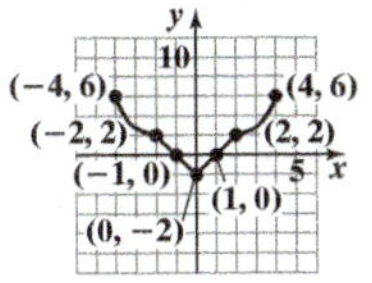

(j) Even **(k)** $[0, 4]$

CHAPTER 5 Polynomial and Rational Functions

5.1 Assess Your Understanding *(page 346)*

7. smooth; continuous **8.** touches **9.** $(-1, 1)$; $(0, 0)$; $(1, 1)$ **10.** r is a real zero of f; r is an x-intercept of the graph of f; $x - r$ is a factor of f.
11. turning points **12.** ∞; ∞ **13.** ∞; $-\infty$ **14.** As x increases in the positive direction, $f(x)$ decreases without bound. **15.** b **16.** d
17. Yes; degree 3; $f(x) = x^3 + 4x$; leading term: x^3; constant term: 0 **19.** Yes; degree 2; $g(x) = -\frac{1}{2}x^2 + \frac{1}{2}$; leading term: $-\frac{1}{2}x^2$; constant term: $\frac{1}{2}$
21. No; x is raised to the -1 power **23.** No; x is raised to the $\frac{3}{2}$ power **25.** Yes; degree 4; $F(x) = 5x^4 - \pi x^3 + \frac{1}{2}$; leading term: $5x^4$; constant term: $\frac{1}{2}$
27. Yes; degree 4; $G(x) = 2x^4 - 4x^3 + 4x^2 - 4x + 2$; leading term: $2x^4$; constant term: 2

29.

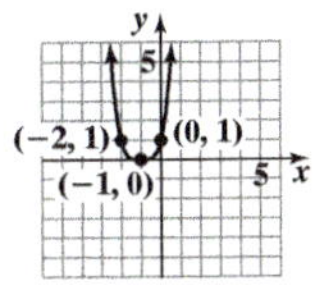

31.

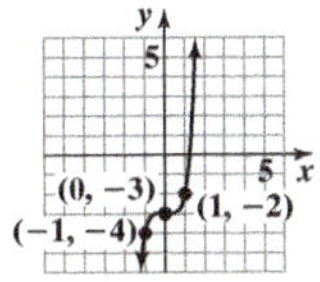

33.

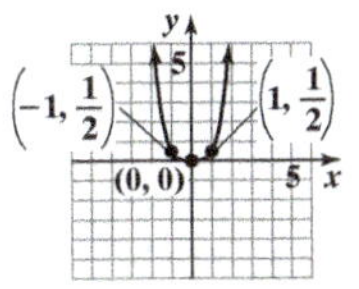

35.

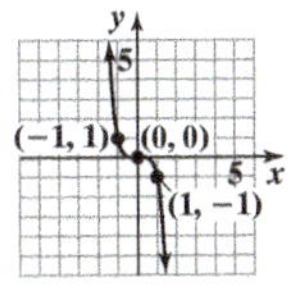

37.

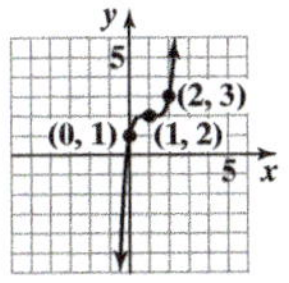

39.

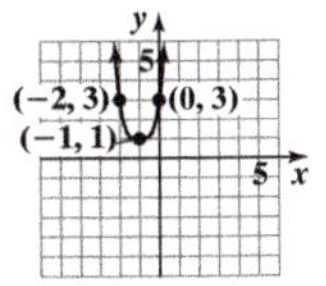

41.

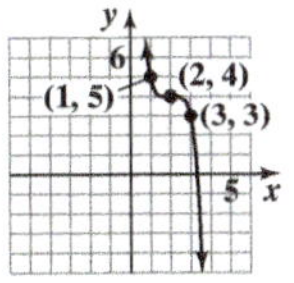

43. $f(x) = (x + 1)(x - 1)(x - 3)$
$= x^3 - 3x^2 - x + 3$ for $a = 1$

45. $f(x) = x(x + 3)(x - 4)$
$= x^3 - x^2 - 12x$ for $a = 1$

47. $f(x) = (x + 4)(x + 1)(x - 2)(x - 3)$
$= x^4 - 15x^2 + 10x + 24$ for $a = 1$

49. $f(x) = (x + 1)(x - 3)^2$
$= x^3 - 5x^2 + 3x + 9$ for $a = 1$

51. $f(x) = 2(x + 3)(x - 1)(x - 4)$
$= 2x^3 - 4x^2 - 22x + 24$

53. $f(x) = 16x(x + 1)(x - 2)(x - 4)$
$= 16x^4 - 80x^3 + 32x^2 + 128x$

55. $f(x) = 5(x + 1)^2(x - 1)^2$
$= 5x^4 - 10x^2 + 5$

57. (a) 7, multiplicity 1; -3, multiplicity 2 **(b)** Graph touches the x-axis at -3 and crosses it at 7. **(c)** 2 **(d)** $y = 3x^3$
59. (a) 2, multiplicity 3 **(b)** Graph crosses the x-axis at 2. **(c)** 4 **(d)** $y = 4x^5$ **61. (a)** $-\frac{1}{2}$, multiplicity 2; -4, multiplicity 3
(b) Graph touches the x-axis at $-\frac{1}{2}$ and crosses at -4. **(c)** 4 **(d)** $y = -2x^5$ **63. (a)** 5, multiplicity 3; -4, multiplicity 2
(b) Graph touches the x-axis at -4 and crosses it at 5. **(c)** 4 **(d)** $y = x^5$ **65. (a)** No real zeros **(b)** Graph neither crosses nor touches the x-axis.
(c) 5 **(d)** $y = 3x^6$ **67. (a)** 0, multiplicity 2; $-\sqrt{2}, \sqrt{2}$, multiplicity 1 **(b)** Graph touches the x-axis at 0 and crosses at $-\sqrt{2}$ and $\sqrt{2}$. **(c)** 3
(d) $y = -2x^4$ **69.** Could be; zeros: $-1, 1, 2$; Least degree is 3. **71.** Cannot be the graph of a polynomial; gap at $x = -1$ **73.** $f(x) = x(x - 1)(x - 2)$
75. $f(x) = -\frac{1}{2}(x + 1)(x - 1)^2(x - 2)$ **77.** $f(x) = 0.2(x + 4)(x + 1)^2(x - 3)$ **79.** $f(x) = -x(x + 3)^2(x - 3)^2$

81. Step 1: $y = x^3$
Step 2: x-intercepts: 0, 3; y-intercept: 0
Step 3: 0: multiplicity 2; touches; 3: multiplicity 1; crosses
Step 4:

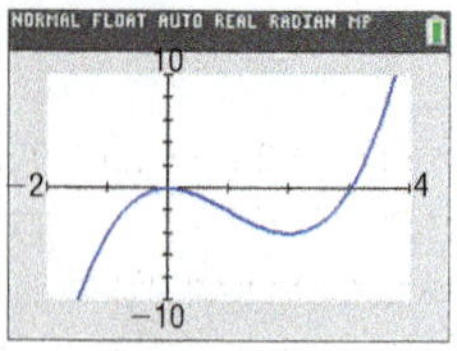

Step 5: $(2, -4)$; $(0, 0)$
Step 6:

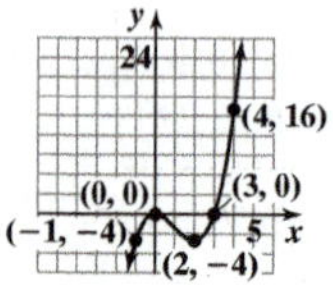

Step 7: Domain: $(-\infty, \infty)$; Range: $(-\infty, \infty)$
Step 8: Increasing on $(-\infty, 0]$ and $[2, \infty)$
Decreasing on $[0, 2]$

83. Step 1: $y = -x^3$
Step 2: x-intercepts: $-4, 1$; y-intercept: 16
Step 3: -4: multiplicity 2, touches; 1: multiplicity 1, crosses
Step 4:

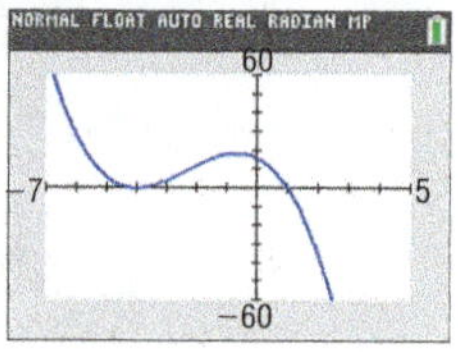

Step 5: $(-4, 0)$; $(-0.67, 18.52)$
Step 6:

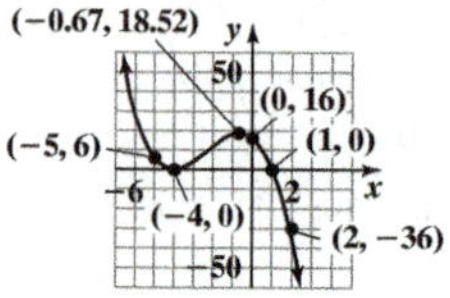

Step 7: Domain: $(-\infty, \infty)$; Range: $(-\infty, \infty)$
Step 8: Increasing on $[-4, -0.67]$
Decreasing on $(-\infty, -4]$ and $[-0.67, \infty)$

85. Step 1: $y = -2x^4$
Step 2: x-intercepts: $-2, 2$; y-intercept: 32
Step 3: -2: multiplicity 1, crosses; 2: multiplicity 3, crosses
Step 4:

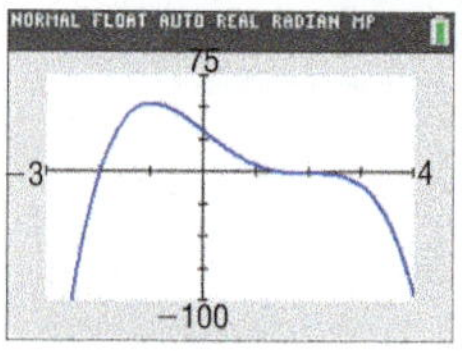

Step 5: $(-1, 54)$
Step 6:

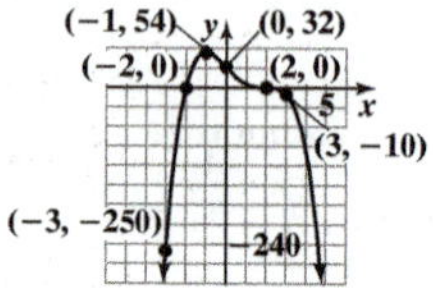

Step 7: Domain: $(-\infty, \infty)$; Range: $(-\infty, 54]$
Step 8: Increasing on $(-\infty, -1]$
Decreasing on $[-1, \infty)$

87. Step 1: $y = x^3$
Step 2: x-intercepts: $-4, -1, 2$; y-intercept: -8
Step 3: $-4, -1, 2$: multiplicity 1, crosses
Step 4:

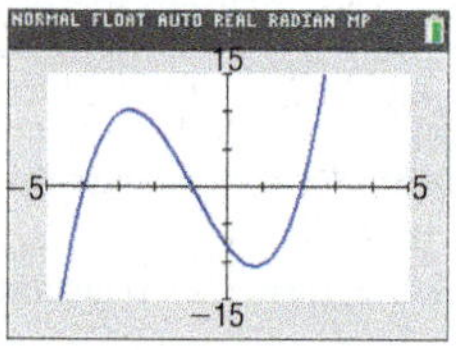

Step 5: $(-2.73, 10.39)$; $(0.73, -10.39)$
Step 6:

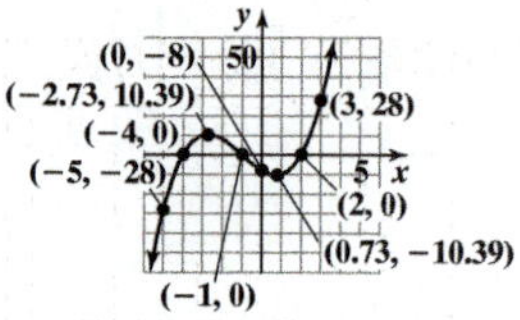

Step 7: Domain: $(-\infty, \infty)$; Range: $(-\infty, \infty)$
Step 8: Increasing on $(-\infty, -2.73]$ and $[0.73, \infty)$
Decreasing on $[-2.73, 0.73]$

89. Step 1: $y = x^4$
Step 2: x-intercepts: $-2, 0, 2$; y-intercept: 0
Step 3: $-2, 2$: multiplicity 1, crosses; 0: multiplicity 2, touches
Step 4:

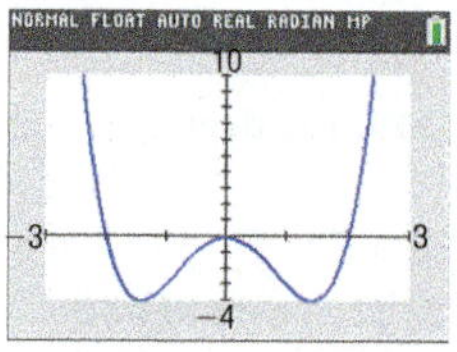

Step 5: $(-1.41, -4)$; $(1.41, -4)$; $(0, 0)$
Step 6:

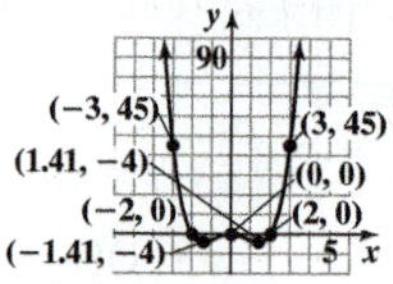

Step 7: Domain: $(-\infty, \infty)$; Range: $[-4, \infty)$
Step 8: Increasing on $[-1.41, 0]$ and $[1.41, \infty)$
Decreasing on $(-\infty, -1.41]$, and $[0, 1.41]$

91. Step 1: $y = x^4$
Step 2: x-intercepts: $-1, 2$; y-intercept: 4
Step 3: $-1, 2$: multiplicity 2, touches
Step 4:

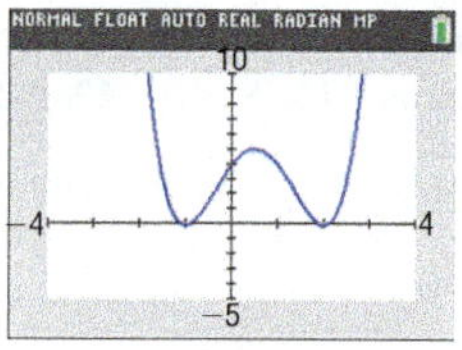

Step 5: $(-1, 0)$; $(2, 0)$; $(0.5, 5.06)$
Step 6:

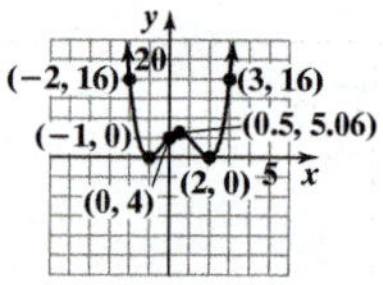

Step 7: Domain: $(-\infty, \infty)$; Range: $[0, \infty)$
Step 8: Increasing on $[-1, 0.5]$ and $[2, \infty)$
Decreasing on $(-\infty, -1]$ and $[0.5, 2]$

93. Step 1: $y = x^4$
Step 2: x-intercepts: $-1, 0, -3$; y-intercept: 0
Step 3: $-1, -3$; multiplicity 1, crosses; 0: multiplicity 2, touches
Step 4:

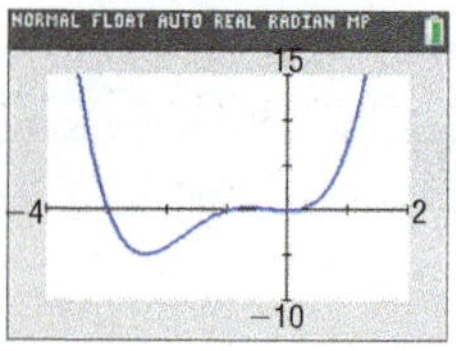

Step 5: $(-2.37, -4.85)$; $(-0.63, 0.35)$; $(0, 0)$
Step 6:

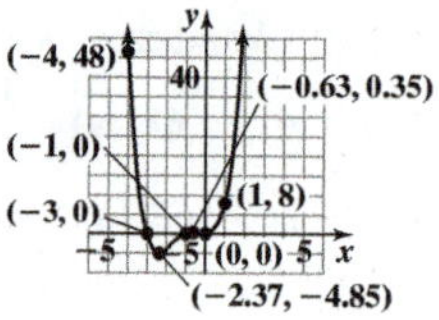

Step 7: Domain: $(-\infty, \infty)$; Range: $[-4.85, \infty)$
Step 8: Increasing on $[-2.37, -0.63]$ and $[0, \infty)$
Decreasing on $(-\infty, -2.37]$ and $[-0.63, 0]$

95. Step 1: $y = 5x^4$
Step 2: x-intercepts: $-3, -2, 0, 2$; y-intercept: 0
Step 3: $-3, -2, 0, 2$: multiplicity 1, crosses
Step 4:

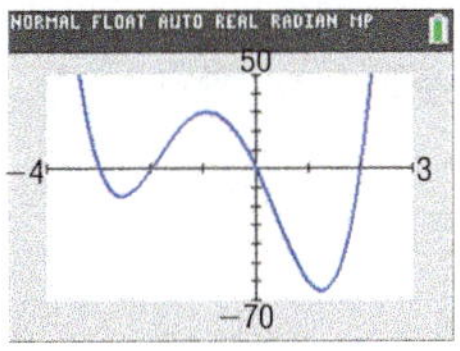

Step 5:
$(-2.57, -14.39)$; $(-0.93, 30.18)$; $(1.25, -64.75)$
Step 6:

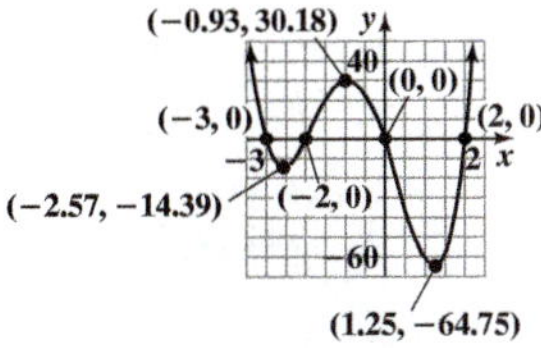

Step 7: Domain: $(-\infty, \infty)$; Range: $[-64.75, \infty)$
Step 8: Increasing on $[-2.57, -0.93]$ and $[1.25, \infty)$
Decreasing on $(-\infty, -2.57]$ and $[-0.93, 1.25]$

97. Step 1: $y = x^5$
Step 2: x-intercepts: 0, 2; y-intercept: 0
Step 3: 0: multiplicity 2, touches; 2: multiplicity 1, crosses
Step 4:

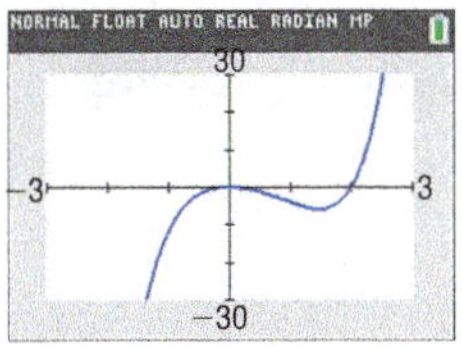

Step 5: $(0, 0)$; $(1.48, -5.91)$
Step 6:

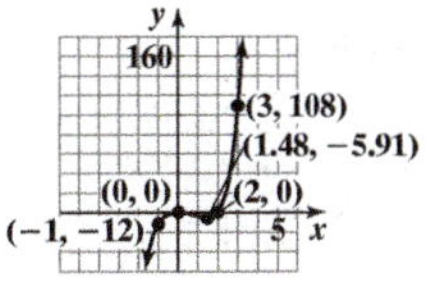

Step 7: Domain: $(-\infty, \infty)$;
Range: $(-\infty, \infty)$
Step 8: Increasing on $(-\infty, 0]$ and $[1.48, \infty)$
Decreasing on $[0, 1.48]$

99. Step 1: $y = x^3$
Step 2:

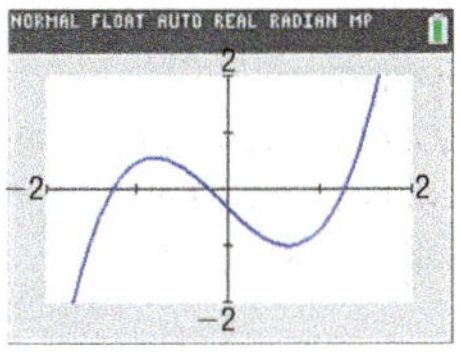

Step 3: x-intercepts: $-1.26, -0.20, 1.26$; y-intercept: -0.31752
Step 4:

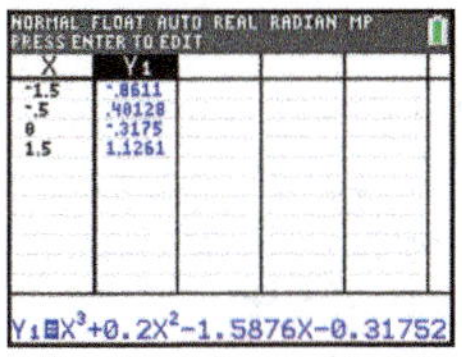

Step 5: $(-0.80, 0.57)$; $(0.66, -0.99)$
Step 6:

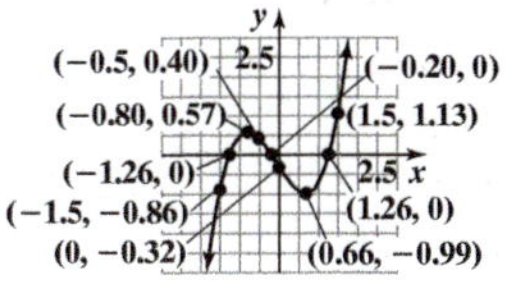

Step 7: Domain: $(-\infty, \infty)$; Range: $(-\infty, \infty)$
Step 8: Increasing on $(-\infty, -0.80]$ and $[0.66, \infty)$
Decreasing on $[-0.80, 0.66]$

101. Step 1: $y = x^3$
Step 2:

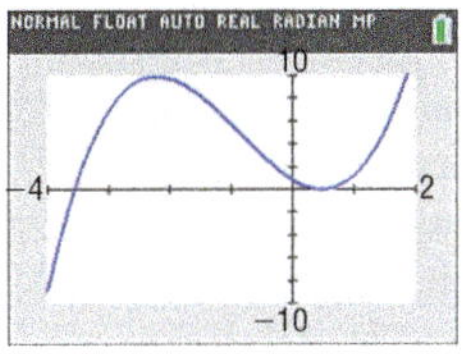

Step 3: x-intercepts: $-3.56, 0.50$; y-intercept: 0.89
Step 4:

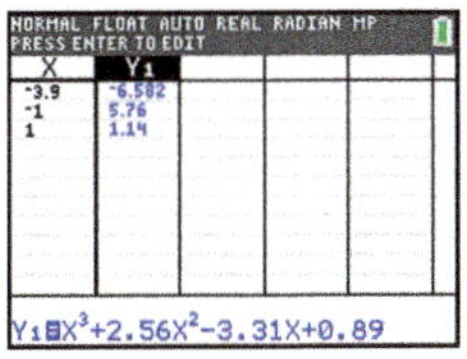

Step 5: $(-2.21, 9.91)$; $(0.50, 0)$
Step 6:

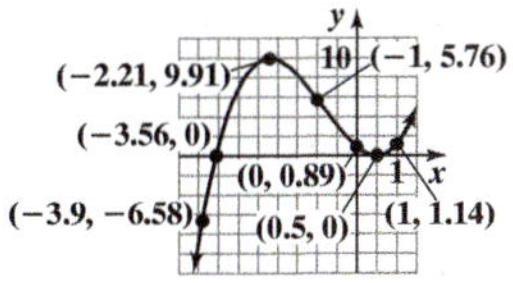

Step 7: Domain: $(-\infty, \infty)$;
Range: $(-\infty, \infty)$
Step 8: Increasing on $(-\infty, -2.21]$ and $[0.50, \infty)$
Decreasing on $[-2.21, 0.50]$

103. Step 1: $y = x^4$
Step 2:

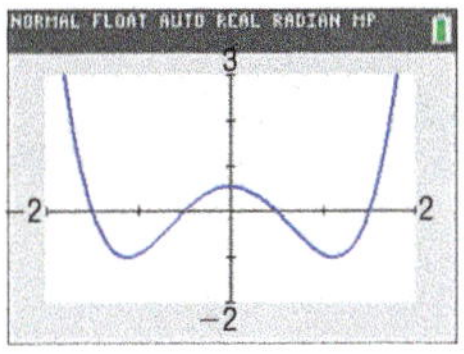

Step 3: x-intercepts: $-1.5, -0.5, 0.5, 1.5$; y-intercept: 0.5625
Step 4:

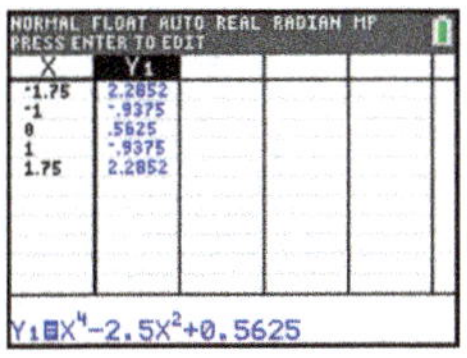

Step 5: $(-1.12, -1)$; $(1.12, -1)$, $(0, 0.56)$
Step 6:

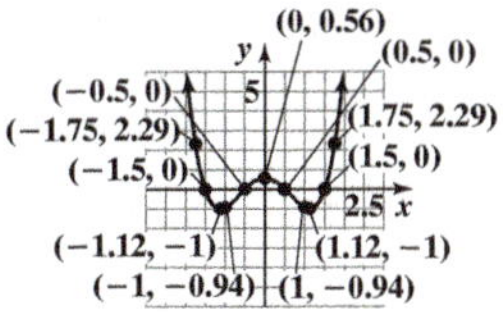

Step 7: Domain: $(-\infty, \infty)$;
Range: $[-1, \infty)$
Step 8: Increasing on $[-1.12, 0]$ and $[1.12, \infty)$
Decreasing on $(-\infty, -1.12]$ and $[0, 1.12]$

105. Step 1: $y = 2x^4$
Step 2:

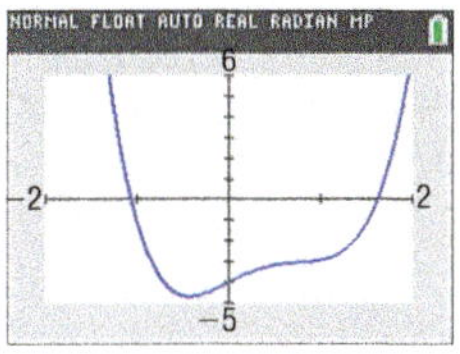

Step 3: x-intercepts: $-1.07, 1.62$; y-intercept: -4
Step 4:

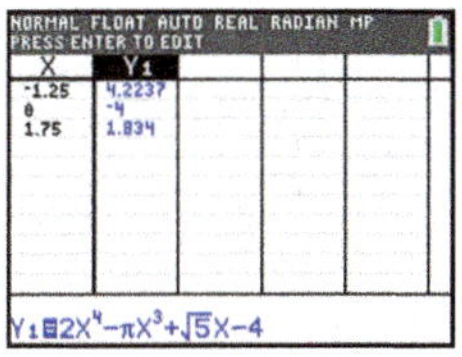

Step 5: $(-0.42, -4.64)$
Step 6:

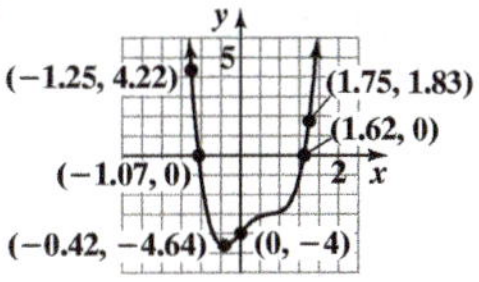

Step 7: Domain: $(-\infty, \infty)$;
Range: $[-4.64, \infty)$
Step 8: Increasing on $[-0.42, \infty)$
Decreasing on $(-\infty, -0.42]$

107. $f(x) = -x(x+2)(x-2)$
Step 1: $y = -x^3$
Step 2: x-intercepts: $-2, 0, 2$; y-intercept: 0
Step 3: $-2, 0, 2$: multiplicity 1, crosses
Step 4:

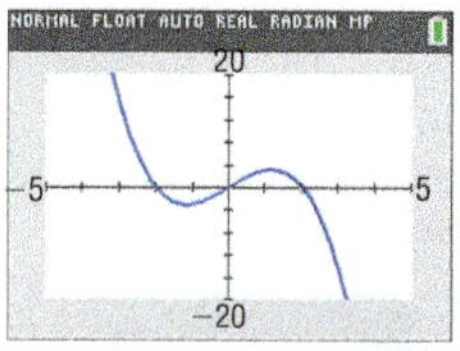

Step 5: $(-1.15, -3.08)$; $(1.15, 3.08)$
Step 6:

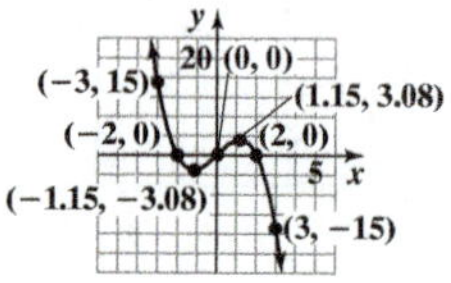

Step 7: Domain: $(-\infty, \infty)$; Range: $(-\infty, \infty)$
Step 8: Increasing on $[-1.15, 1.15]$
Decreasing on $(-\infty, -1.15]$ and $[1.15, \infty)$

109. $f(x) = x(x+4)(x-3)$
Step 1: $y = x^3$
Step 2: x-intercepts: $-4, 0, 3$; y-intercept: 0
Step 3: $-4, 0, 3$: multiplicity 1, crosses
Step 4:

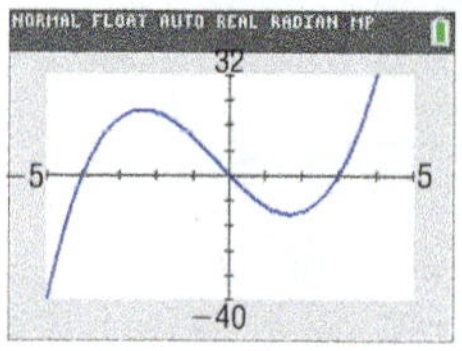

Step 5: $(-2.36, 20.75)$; $(1.69, -12.60)$
Step 6:

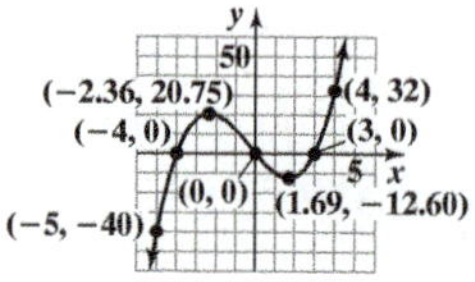

Step 7: Domain: $(-\infty, \infty)$; Range: $(-\infty, \infty)$
Step 8: Increasing on $(-\infty, -2.36]$ and $[1.69, \infty)$
Decreasing on $[-2.36, 1.69]$

111. $f(x) = 2x(x+6)(x-2)(x+2)$
Step 1: $y = 2x^4$
Step 2: x-intercepts: $-6, -2, 0, 2$; y-intercept: 0
Step 3: $-6, -2, 0, 2$: multiplicity 1, crosses
Step 4:

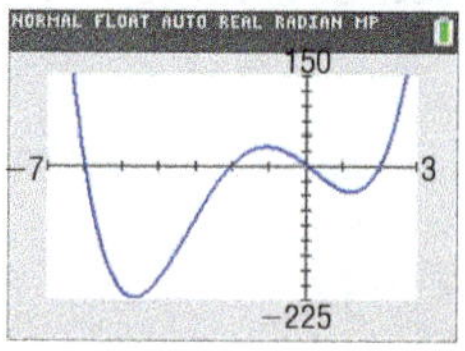

Step 5: $(-4.65, -221.25)$; $(-1.06, 30.12)$; $(1.21, -44.25)$
Step 6:

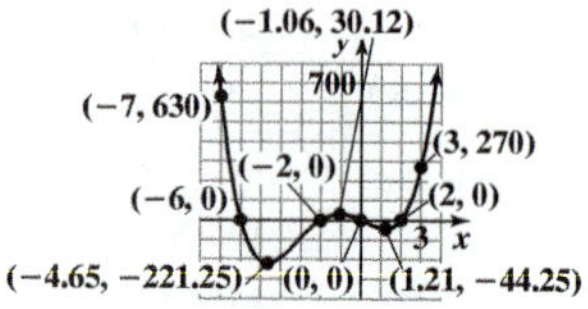

Step 7: Domain: $(-\infty, \infty)$; Range: $[-221.25, \infty)$
Step 8: Decreasing on $(-\infty, -4.65]$ and $[-1.06, 1.21]$
Increasing on $[-4.65, -1.06]$ and $[1.21, \infty)$

113. $f(x) = -x^2(x+1)^2(x-1)$
Step 1: $y = -x^5$
Step 2: x-intercepts: $-1, 0, 1$; y-intercept: 0
Step 3: 1: multiplicity 1, crosses; -1, 0: multiplicity 2, touches
Step 4:

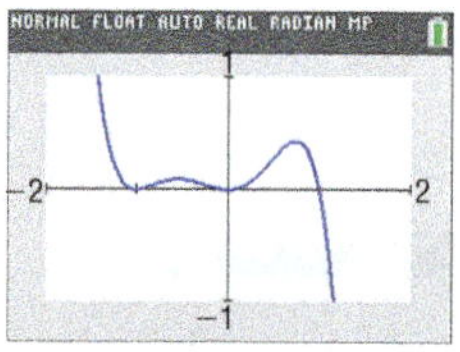

Step 5: $(-1, 0)$; $(-0.54, 0.10)$; $(0, 0)$; $(0.74, 0.43)$
Step 6:

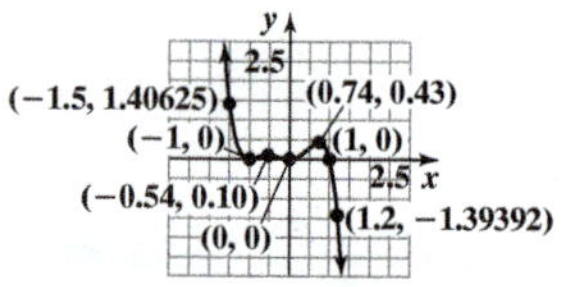

Step 7: Domain: $(-\infty, \infty)$; Range: $(-\infty, \infty)$
Step 8: Increasing on $[-1, -0.54]$ and $[0, 0.74]$
Decreasing on $(-\infty, -1]$, $[-0.54, 0]$, and $[0.74, \infty)$

115. $f(x) = 3(x+3)(x-1)(x-4)$ **117.** $f(x) = -2(x+5)^2(x-2)(x-4)$ **119. (a)** $-3, 2$ **(b)** $-6, -1$

121. (a)

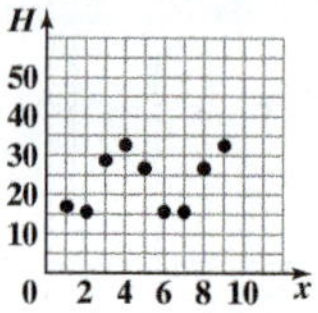

The relation appears to be cubic.
(b) $H(x) = 0.3948x^3 - 5.9563x^2 + 26.1965x - 7.4127$ **(c)** ≈ 24
(d)

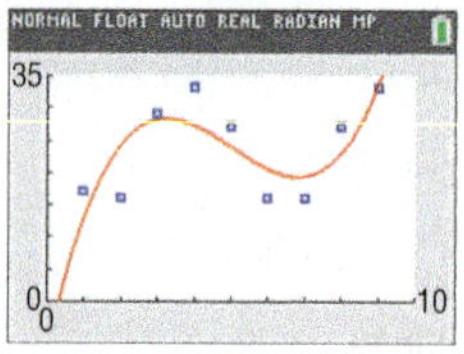

(e) ≈ 54; no. The end behavior of the model indicates that as time goes on, the number of major hurricanes will continue to increase each decade without limit. This is unrealistic. End behavior should not be used to make predictions too far outside the data used to create the model.

123. (a)

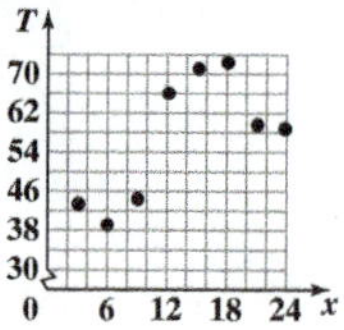

The relation appears to be cubic.
(b) 6°/h **(c)** 0.17°/h
(d) $T(x) = -0.01992x^3 + 0.6745x^2 - 4.4360x + 48.4643$; 70.1°F
(e)

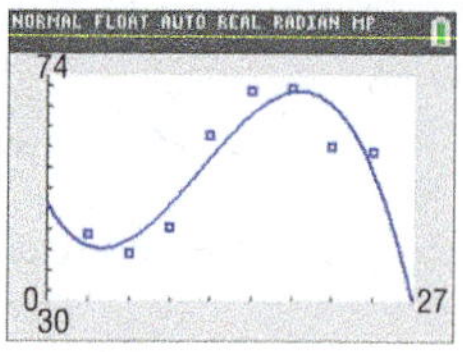

(f) The predicted temperature at midnight is 48.5°F.

125. (a)

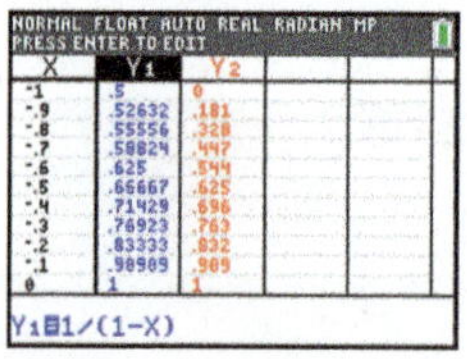

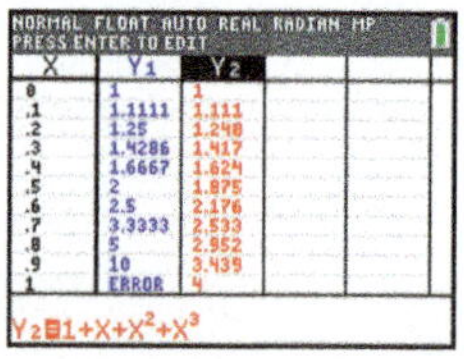

(b)

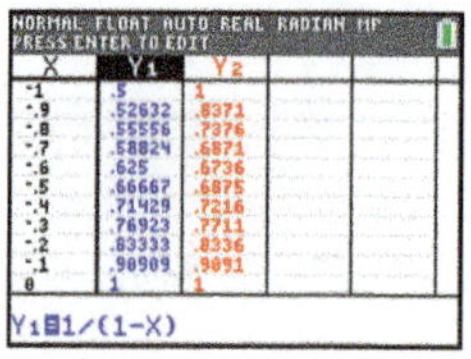

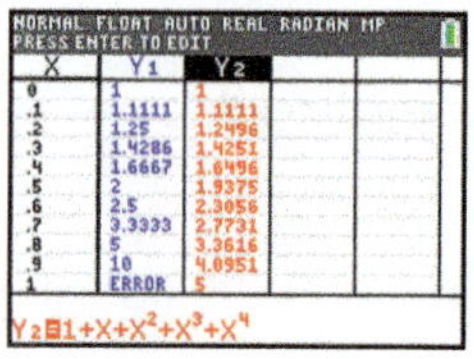

(c)

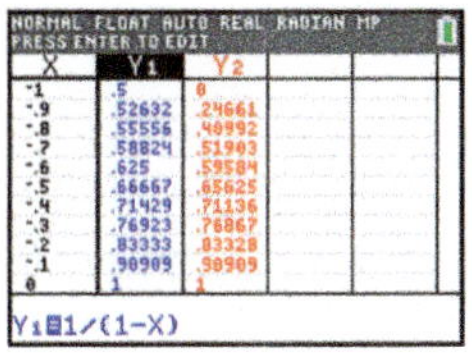

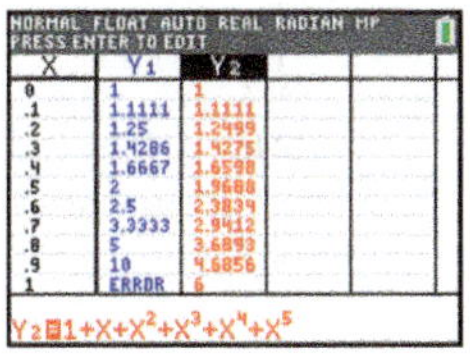

(d) As more terms are added, the values of the polynomial function get closer to the values of f. The approximations near 0 are better than those near -1 or 1.

131. (a) − (d) **135.** $y = -\frac{2}{5}x - \frac{11}{5}$ **136.** $\{x \mid x \neq -5\}$ **137.** $\frac{-2-\sqrt{7}}{2}, \frac{-2+\sqrt{7}}{2}$ **138.** $\left\{-\frac{4}{5}, 2\right\}$

Historical Problems *(page 363)*

1.
$$\left(x-\frac{b}{3}\right)^3 + b\left(x-\frac{b}{3}\right)^2 + c\left(x-\frac{b}{3}\right) + d = 0$$
$$x^3 - bx^2 + \frac{b^2x}{3} - \frac{b^3}{27} + bx^2 - \frac{2b^2x}{3} + \frac{b^3}{9} + cx - \frac{bc}{3} + d = 0$$
$$x^3 + \left(c - \frac{b^2}{3}\right)x + \left(\frac{2b^3}{27} - \frac{bc}{3} + d\right) = 0$$

Let $p = c - \frac{b^2}{3}$ and $q = \frac{2b^3}{27} - \frac{bc}{3} + d$. Then $x^3 + px + q = 0$.

2.
$$(H+K)^3 + p(H+K) + q = 0$$
$$H^3 + 3H^2K + 3HK^2 + K^3 + pH + pK + q = 0$$

Let $3HK = -p$.

$$H^3 - pH - pK + K^3 + pH + pK + q = 0, \quad H^3 + K^3 = -q$$

3.
$$3HK = -p$$
$$K = -\frac{p}{3H}$$
$$H^3 + \left(-\frac{p}{3H}\right)^3 = -q$$
$$H^3 - \frac{p^3}{27H^3} = -q$$
$$27H^6 - p^3 = -27qH^3$$
$$27H^6 + 27qH^3 - p^3 = 0$$
$$H^3 = \frac{-27q \pm \sqrt{(27q)^2 - 4(27)(-p^3)}}{2 \cdot 27}$$
$$H^3 = \frac{-q}{2} \pm \sqrt{\frac{27^2q^2}{2^2(27^2)} + \frac{4(27)p^3}{2^2(27^2)}}$$
$$H^3 = \frac{-q}{2} \pm \sqrt{\frac{q^2}{4} + \frac{p^3}{27}}$$
$$H = \sqrt[3]{\frac{-q}{2} + \sqrt{\frac{q2}{4} + \frac{p3}{27}}}$$

Choose the positive root for now.

4.
$$H^3 + K^3 = -q$$
$$K^3 = -q - H^3$$
$$K^3 = -q - \left[\frac{-q}{2} + \sqrt{\frac{q^2}{4} + \frac{p^3}{27}}\right]$$
$$K^3 = \frac{-q}{2} - \sqrt{\frac{q^2}{4} + \frac{p^3}{27}}$$
$$K = \sqrt[3]{\frac{-q}{2} - \sqrt{\frac{q^2}{4} + \frac{p^3}{27}}}$$

5. $x = H + K$
$$x = \sqrt[3]{\frac{-q}{2} + \sqrt{\frac{q^2}{4} + \frac{p^3}{27}}} + \sqrt[3]{\frac{-q}{2} - \sqrt{\frac{q^2}{4} + \frac{p^3}{27}}}$$

(Note that had we used the negative root in 3 the result would be the same.)

6. $x = 3$ **7.** $x = 2$ **8.** $x = 2$

5.2 Assess Your Understanding *(page 363)*

5. a **6.** $f(c)$ **7.** b **8.** F **9.** 0 **10.** T **11.** $R = f(2) = 8$; no **13.** $R = f(2) = 0$; yes **15.** $R = f(-3) = 0$; yes **17.** $R = f(-4) = 1$; no **19.** $R = f\left(\frac{1}{2}\right) = 0$; yes **21.** 7; 3 or 1 positive; 2 or 0 negative **23.** 6; 2 or 0 positive; 2 or 0 negative **25.** 3; 2 or 0 positive; 1 negative **27.** 4; 2 or 0 positive; 2 or 0 negative **29.** 5; 0 positive; 3 or 1 negative **31.** 6; 1 positive; 1 negative **33.** 4; $\pm 1, \pm\frac{1}{3}$ **35.** 5; $\pm 1, \pm 3$ **37.** 3; $\pm 1, \pm 2, \pm\frac{1}{4}, \pm\frac{1}{2}$ **39.** 4; $\pm 1, \pm 3, \pm 9, \pm\frac{1}{2}, \pm\frac{1}{3}, \pm\frac{1}{6}, \pm\frac{3}{2}, \pm\frac{9}{2}$ **41.** 5; $\pm 1, \pm 2, \pm 3, \pm 4, \pm 6, \pm 12, \pm\frac{1}{2}, \pm\frac{3}{2}$ **43.** 4; $\pm 1, \pm 2, \pm 4, \pm 5, \pm 10, \pm 20, \pm\frac{1}{2}, \pm\frac{5}{2}, \pm\frac{1}{3}, \pm\frac{2}{3}, \pm\frac{4}{3}, \pm\frac{5}{3}, \pm\frac{10}{3}, \pm\frac{20}{3}, \pm\frac{1}{6}, \pm\frac{5}{6}$

45. -1 and 1

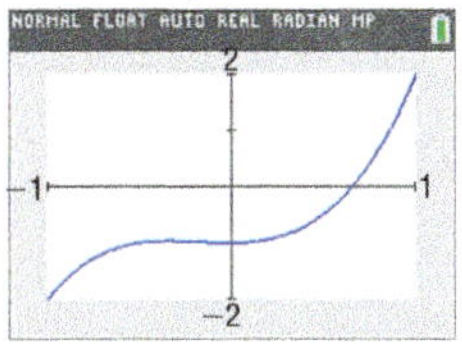

47. -11 and 11

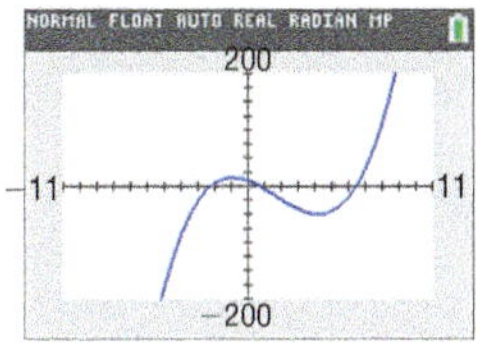

49. -9 and 3

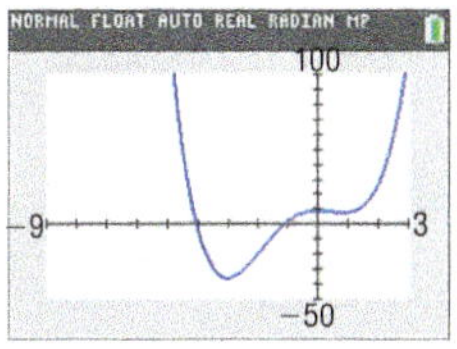

51. $-3, -1, 2; f(x) = (x+3)(x+1)(x-2)$ **53.** $\frac{1}{2}, 3, 3; f(x) = (2x-1)(x-3)^2$ **55.** $-\frac{1}{3}; f(x) = (3x+1)(x^2+x+1)$

57. $4, 3-\sqrt{5}, 3+\sqrt{5}; f(x) = (x-4)(x-3+\sqrt{5})(x-3-\sqrt{5})$ **59.** $-2, -1, 1, 1; f(x) = (x+2)(x+1)(x-1)^2$

61. $-\frac{7}{3}, -1, \frac{2}{7}, 2; f(x) = (x-2)(x+1)(3x+7)(7x-2)$ **63.** $3, \frac{5+\sqrt{17}}{2}, \frac{5-\sqrt{17}}{2}; f(x) = (x-3)\left(x - \frac{5+\sqrt{17}}{2}\right)\left(x - \frac{5-\sqrt{17}}{2}\right)$

65. $-\frac{1}{2}, \frac{1}{2}; f(x) = (2x+1)(2x-1)(x^2+2)$ **67.** $\frac{\sqrt{2}}{2}, -\frac{\sqrt{2}}{2}, 2; f(x) = (x-2)(\sqrt{2}x-1)(\sqrt{2}x+1)(2x^2+1)$ **69.** $-5.9, -0.3, 3$

71. $-3.8, 4.5$ **73.** $-43.5, 1, 23$ **75.** $\{-1, 2\}$ **77.** $\left\{\frac{2}{3}, -1+\sqrt{2}, -1-\sqrt{2}\right\}$ **79.** $\left\{\frac{1}{3}, \sqrt{5}, -\sqrt{5}\right\}$ **81.** $\{-3, -2\}$ **83.** $\left\{-\frac{1}{3}\right\}$

85. $f(0) = -1; f(1) = 10$; Zero: 0.21 **87.** $f(-5) = -58; f(-4) = 2$; Zero: -4.04 **89.** $f(1.4) = -0.17536; f(1.5) = 1.40625$; Zero: 1.41

91. Step 1: $y = x^3$
Step 2: x-intercepts: $-3, -1, 2$; y-intercept: -6
Step 3: $-3, -1, 2$: multiplicity 1, crosses
Step 4:

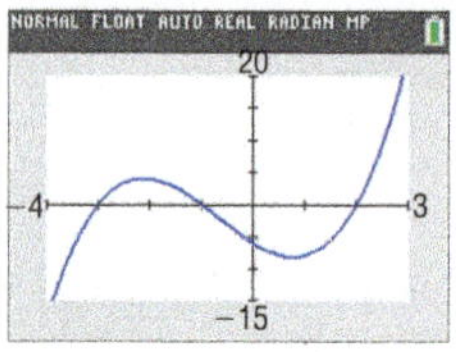

Step 5: $(-2.12, 4.06); (0.79, -8.21)$
Step 6:

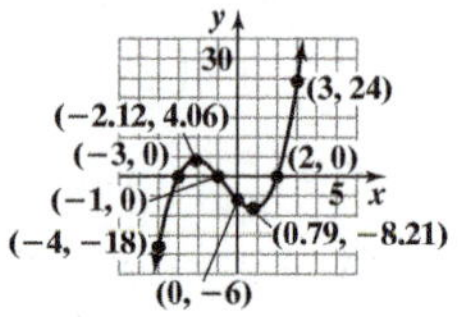

Step 7: Domain and range: $(-\infty, \infty)$
Step 8: Increasing: $(-\infty, -2.12], [0.79, \infty)$
Decreasing: $[-2.12, 0.79]$

93. Step 1: $y = x^4$
Step 2: x-intercepts: $-2, -1, 1$; y-intercept: 2
Step 3: $-2, -1$: multiplicity 1, crosses; 1: multiplicity 2, touches
Step 4:

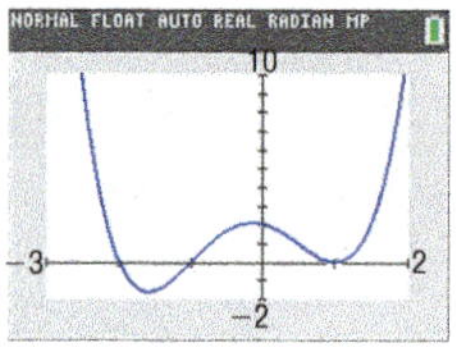

Step 5: $(-1.59, -1.62), (-0.16, 2.08), (1, 0)$
Step 6:

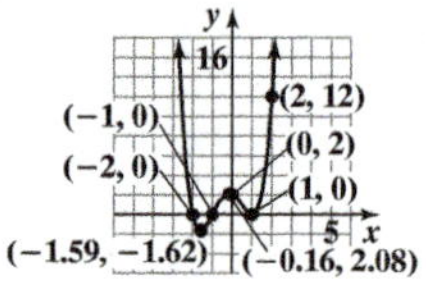

Step 7: Domain: $(-\infty, \infty)$; Range: $[-1.62, \infty)$
Step 8: Increasing: $[-1.59, -0.16], [1, \infty)$
Decreasing: $(-\infty, -1.59], [-0.16, 1]$

95. Step 1: $y = 4x^5$
Step 2: x-intercepts: $-\frac{\sqrt{2}}{2}, \frac{\sqrt{2}}{2}, 2$; y-intercept: 2
Step 3: $-\frac{\sqrt{2}}{2}, \frac{\sqrt{2}}{2}, 2$: multiplicity 1, crosses
Step 4:

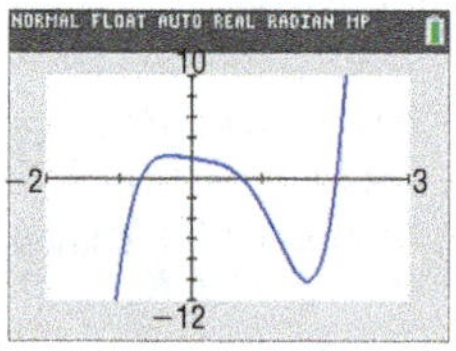

Step 5: $(-0.30, 2.23); (1.61, -10.09)$
Step 6:

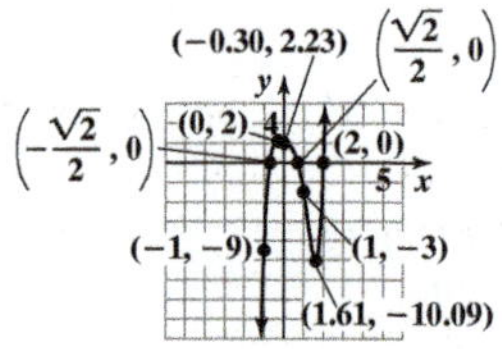

Step 7: Domain and range: $(-\infty, \infty)$
Step 8: Increasing: $(-\infty, -0.30], [1.61, \infty)$
Decreasing: $[-0.30, 1.61]$

97. Step 1: $y = 6x^4$
Step 2: x-intercepts: $-\frac{1}{2}, \frac{2}{3}, 3$; y-intercept: -18
Step 3: $-\frac{1}{2}, \frac{2}{3}$: multiplicity 1, crosses; 3: multiplicity 2, touches
Step 4:

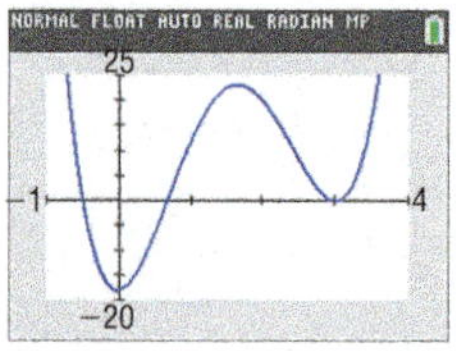

Step 5: $(-0.03, -18.04), (1.65, 23.12), (3, 0)$
Step 6:

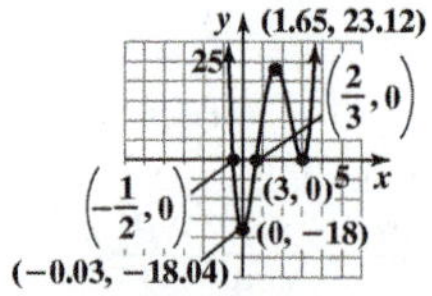

Step 7: Domain: $(-\infty, \infty)$; Range: $[-18.04, \infty)$
Step 8: Increasing: $[-0.03, 1.65], [3, \infty)$;
Decreasing: $(-\infty, -0.03], [1.65, 3]$

99. $k = 5$ **101.** -7 **103.** If $f(x) = x^n - c^n$, then $f(c) = c^n - c^n = 0$; so $x - c$ is a factor of f. **105.** 5 **107.** 7 in.
109. All the potential rational zeros are integers, so r is either an integer or is not a rational zero (and is, therefore, irrational). **111.** 0.215
113. No; by the Rational Zeros Theorem, $\frac{1}{3}$ is not a potential rational zero. **115.** No; by the Rational Zeros Theorem, $\frac{2}{3}$ is not a potential rational zero.
116. $y = \frac{2}{5}x - \frac{3}{5}$ **117.** $[3, 8)$ **118.** $(0, -2\sqrt{3}), (0, 2\sqrt{3}), (4, 0)$ **119.** $[-3, 2]$ and $[5, \infty)$

5.3 Assess Your Understanding *(page 370)*

3. one **4.** $3 - 4i$ **5.** T **6.** F **7.** $4 + i$ **9.** $-i, 1 - i$ **11.** $-i, -2i$ **13.** $-i$ **15.** $2 - i, -3 + i$ **17.** $f(x) = x^4 - 14x^3 + 77x^2 - 200x + 208; a = 1$

19. $f(x) = x^5 - 4x^4 + 7x^3 - 8x^2 + 6x - 4; a = 1$ **21.** $f(x) = x^4 - 6x^3 + 10x^2 - 6x + 9; a = 1$ **23.** $-2i, 4$ **25.** $2i, -3, \frac{1}{2}$ **27.** $3 + 2i, -2, 5$

29. $4i, -\sqrt{11}, \sqrt{11}, -\frac{2}{3}$ **31.** $1, -\frac{1}{2} - \frac{\sqrt{3}}{2}i, -\frac{1}{2} + \frac{\sqrt{3}}{2}i; f(x) = (x-1)\left(x + \frac{1}{2} + \frac{\sqrt{3}}{2}i\right)\left(x + \frac{1}{2} - \frac{\sqrt{3}}{2}i\right)$

33. $2, 3-2i, 3+2i; f(x) = (x-2)(x-3+2i)(x-3-2i)$ **35.** $-i, i, -2i, 2i; f(x) = (x+i)(x-i)(x+2i)(x-2i)$

37. $-5i, 5i, -3, 1; f(x) = (x + 5i)(x - 5i)(x + 3)(x - 1)$ **39.** $-4, \frac{1}{3}, 2 - 3i, 2 + 3i; f(x) = 3(x + 4)\left(x - \frac{1}{3}\right)(x - 2 + 3i)(x - 2 - 3i)$

41. 130 **43. (a)** $f(x) = (x^2 - \sqrt{2}x + 1)(x^2 + \sqrt{2}x + 1)$ **(b)** $-\frac{\sqrt{2}}{2} - \frac{\sqrt{2}}{2}i, -\frac{\sqrt{2}}{2} + \frac{\sqrt{2}}{2}i, \frac{\sqrt{2}}{2} - \frac{\sqrt{2}}{2}i, \frac{\sqrt{2}}{2} + \frac{\sqrt{2}}{2}i$

45. Zeros that are complex numbers must occur in conjugate pairs; or a polynomial with real coefficients of odd degree must have at least one real zero.
47. If the remaining zero were a complex number, its conjugate would also be a zero, creating a polynomial of degree 5.

49.

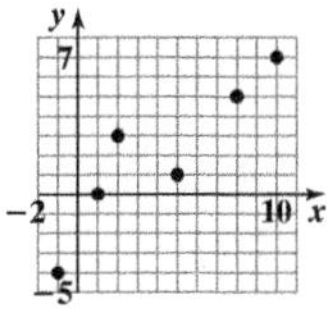

50. $\{-22\}$ **51.** $6x^3 - 13x^2 - 13x + 20$ **52.** $A = 9\pi$ ft^2 (about 28.274 ft^2); $C = 6\pi$ ft (about 18.850 ft)

5.4 Assess Your Understanding *(page 379)*

5. F **6.** horizontal asymptote **7.** vertical asymptote **8.** proper **9.** T **10.** F **11.** $y = 0$ **12.** T **13.** d **14.** a

15. All real numbers except 3; $\{x \mid x \neq 3\}$ **17.** All real numbers except 2 and -4; $\{x \mid x \neq 2, x \neq -4\}$

19. All real numbers except $-\frac{1}{2}$ and 3; $\left\{x \,\middle|\, x \neq -\frac{1}{2}, x \neq 3\right\}$ **21.** All real numbers except 2; $\{x \mid x \neq 2\}$ **23.** All real numbers

25. All real numbers except -3 and 3; $\{x \mid x \neq -3, x \neq 3\}$
27. (a) Domain: $\{x \mid x \neq 2\}$; range: $\{y \mid y \neq 1\}$ **(b)** $(0, 0)$ **(c)** $y = 1$ **(d)** $x = 2$ **(e)** None

29. (a) Domain: $\{x \mid x \neq 0\}$; range: all real numbers **(b)** $(-1, 0), (1, 0)$ **(c)** None **(d)** $x = 0$ **(e)** $y = 2x$

31. (a) Domain: $\{x \mid x \neq -2, x \neq 2\}$; range: $\{y \mid y \leq 0, y > 1\}$ **(b)** $(0, 0)$ **(c)** $y = 1$ **(d)** $x = -2, x = 2$ **(e)** None

33. (a)

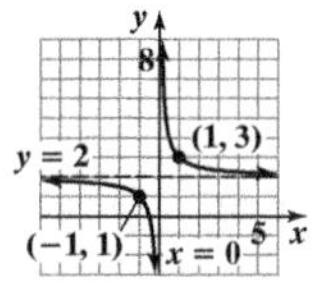

(b) Domain: $\{x \mid x \neq 0\}$; range: $\{y \mid y \neq 2\}$
(c) Vertical asymptote: $x = 0$;
horizontal asymptote: $y = 2$

35. (a)

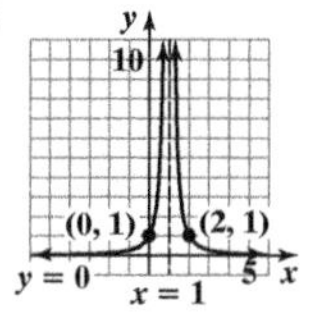

(b) Domain: $\{x \mid x \neq 1\}$; range: $\{y \mid y > 0\}$
(c) Vertical asymptote: $x = 1$;
horizontal asymptote: $y = 0$

37. (a)

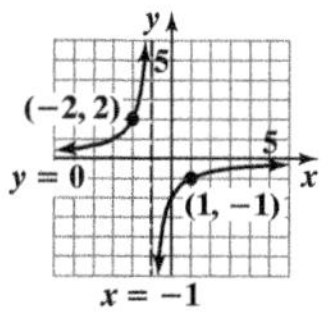

(b) Domain: $\{x \mid x \neq -1\}$; range: $\{y \mid y \neq 0\}$
(c) Vertical asymptote: $x = -1$;
horizontal asymptote: $y = 0$

39. (a)

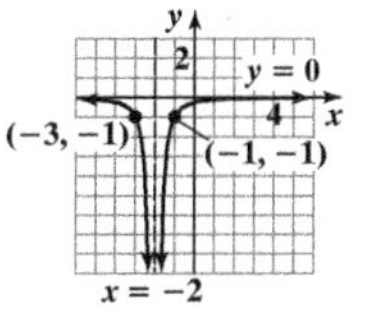

(b) Domain: $\{x \mid x \neq -2\}$; range: $\{y \mid y < 0\}$
(c) Vertical asymptote: $x = -2$;
horizontal asymptote: $y = 0$

41. (a)

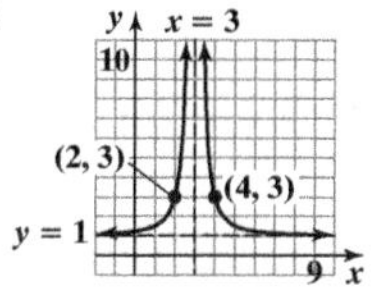

(b) Domain: $\{x \mid x \neq 3\}$; range: $\{y \mid y > 1\}$
(c) Vertical asymptote: $x = 3$;
horizontal asymptote: $y = 1$

43. (a)

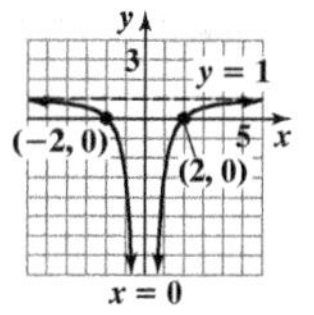

(b) Domain: $\{x \mid x \neq 0\}$; range: $\{y \mid y < 1\}$
(c) Vertical asymptote: $x = 0$;
horizontal asymptote: $y = 1$

45. Vertical asymptote: $x = -4$; horizontal asymptote: $y = 3$ **47.** Vertical asymptote: $x = 3$; oblique asymptote: $y = x + 5$
49. Vertical asymptotes: $x = 1, x = -1$; horizontal asymptote: $y = 0$ **51.** Vertical asymptote: $x = -\frac{1}{3}$; horizontal asymptote: $y = \frac{2}{3}$
53. No asymptotes **55.** Vertical asymptote: $x = 0$; no horizontal or oblique asymptote

57. (a) 9.8208 m/sec^2 **(b)** 9.8195 m/sec^2 **(c)** 9.7936 m/sec^2 **(d)** $y = 0$ **(e)** $\varnothing$

59. (a)

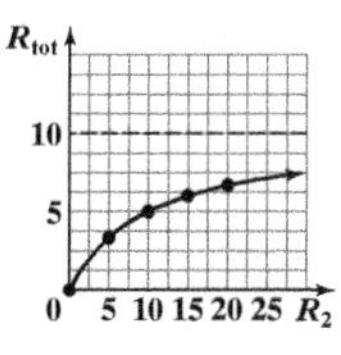

(b) Horizontal: $R_{tot} = 10$; as the resistance of R_2 increases without bound, the total resistance approaches 10 ohms, the resistance R_1.
(c) $R_1 \approx 103.5$ ohms

61. (a) $R(x) = 2 + \frac{5}{x - 1} = 5\left(\frac{1}{x - 1}\right) + 2$

(b)

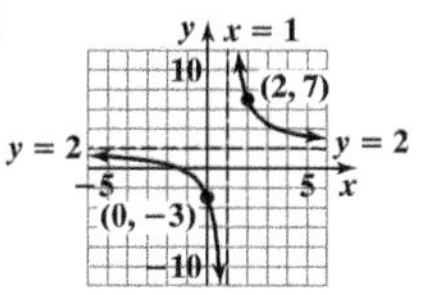

(c) Vertical asymptote: $x = 1$;
horizontal asymptote: $y = 2$

67. $x = 5$ **68.** $\left\{-\frac{4}{19}\right\}$

69. x-axis symmetry

70. $(-3, 11), (2, -4)$

5.5 Assess Your Understanding *(page 390)*

2. False **3.** c **4.** a **5. (a)** $\{x \mid x \neq 2\}$ **(b)** 0 **6.** True

7. Step 1: Domain: $\{x \mid x \neq 0, x \neq -4\}$
Step 2: R is in lowest terms
Step 3: no y-intercept; x-intercept: -1
Step 4: R is in lowest terms; vertical asymptotes: $x = 0$, $x = -4$
Step 5: Horizontal asymptote: $y = 0$, intersected at $(-1, 0)$
Step 6:

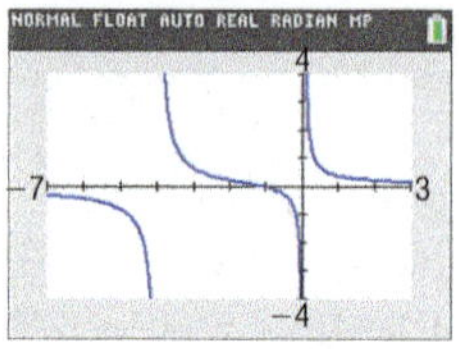

Step 7:

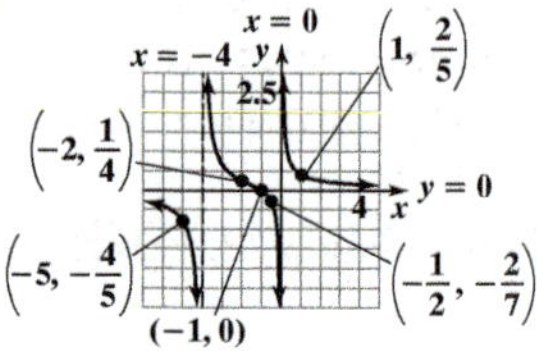

9. Step 1: $R(x) = \dfrac{3(x+1)}{2(x+2)}$; domain: $\{x \mid x \neq -2\}$
Step 2: R is in lowest terms
Step 3: y-intercept: $\frac{3}{4}$; x-intercept: -1
Step 4: R is in lowest terms; vertical asymptote: $x = -2$
Step 5: Horizontal asymptote: $y = \frac{3}{2}$, not intersected
Step 6:

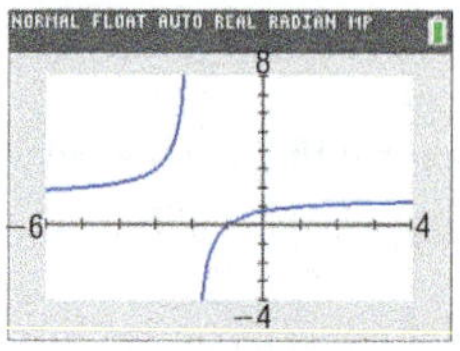

Step 7:

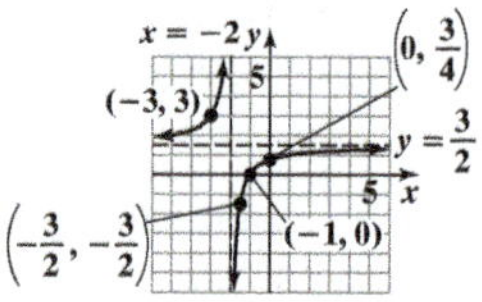

11. Step 1: $R(x) = \dfrac{3}{(x+2)(x-2)}$; domain: $\{x \mid x \neq -2, x \neq 2\}$
Step 2: R is in lowest terms
Step 3: y-intercept: $-\frac{3}{4}$; no x-intercept
Step 4: R is in lowest terms; vertical asymptotes: $x = 2, x = -2$
Step 5: Horizontal asymptote: $y = 0$, not intersected
Step 6:

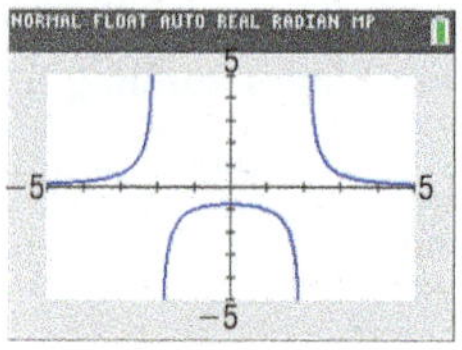

Step 7:

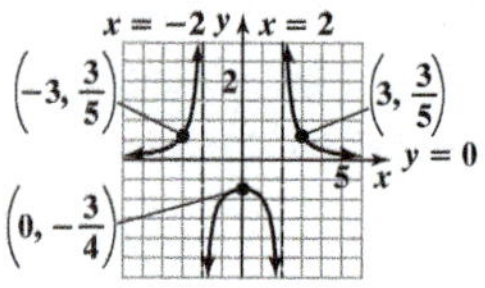

13. Step 1: $P(x) = \dfrac{(x^2+x+1)(x^2-x+1)}{(x+1)(x-1)}$; domain: $\{x \mid x \neq -1, x \neq 1\}$
Step 2: P is in lowest terms
Step 3: y-intercept: -1; no x-intercept
Step 4: P is in lowest terms; vertical asymptotes: $x = -1, x = 1$
Step 5: No horizontal or oblique asymptote
Step 6:

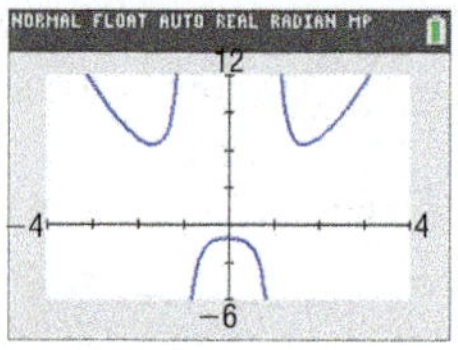

Step 7:

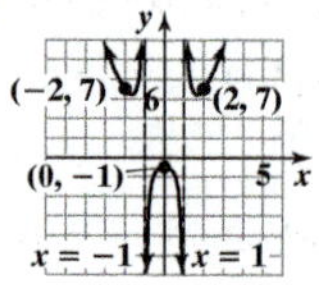

15. Step 1: $H(x) = \dfrac{(x-1)(x^2+x+1)}{(x+3)(x-3)}$; domain: $\{x \mid x \neq -3, x \neq 3\}$
Step 2: H is in lowest terms
Step 3: y-intercept: $\frac{1}{9}$; x-intercept: 1
Step 4: H is in lowest terms; vertical asymptotes: $x = 3, x = -3$
Step 5: Oblique asymptote: $y = x$, intersected at $\left(\frac{1}{9}, \frac{1}{9}\right)$
Step 6:

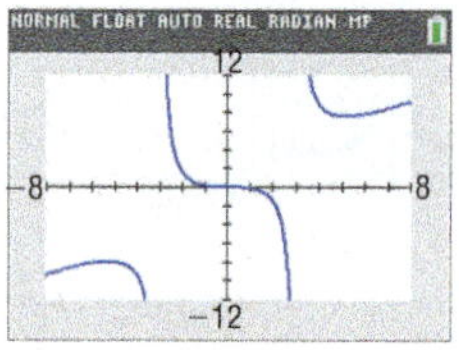

Step 7:

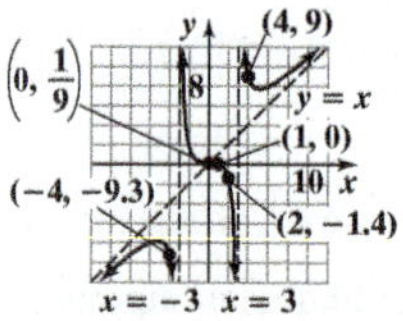

17. Step 1: $R(x) = \dfrac{x^2}{(x+3)(x-2)}$; domain: $\{x \neq -3, x \neq 2\}$
Step 2: R is in lowest terms
Step 3: y-intercept: 0; x-intercept: 0
Step 4: R is in lowest terms; vertical asymptotes: $x = 2, x = -3$
Step 5: Horizontal asymptote: $y = 1$, intersected at $(6, 1)$
Step 6:

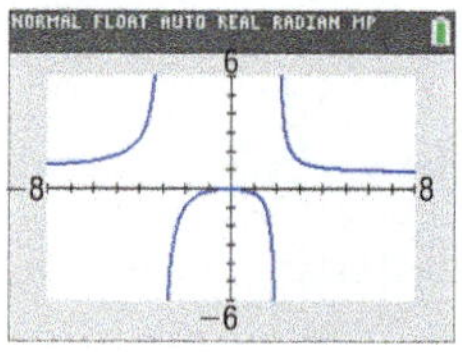

Step 7:

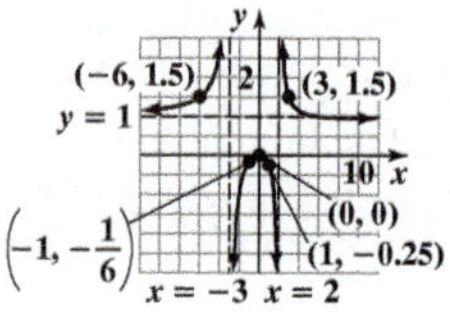

19. Step 1: $G(x) = \dfrac{x}{(x+2)(x-2)}$; domain: $\{x \mid x \neq -2, x \neq 2\}$
Step 2: G is in lowest terms
Step 3: y-intercept: 0; x-intercept: 0
Step 4: G is in lowest terms; vertical asymptotes: $x = -2, x = 2$
Step 5: Horizontal asymptote: $y = 0$, intersected at $(0, 0)$

Step 6:

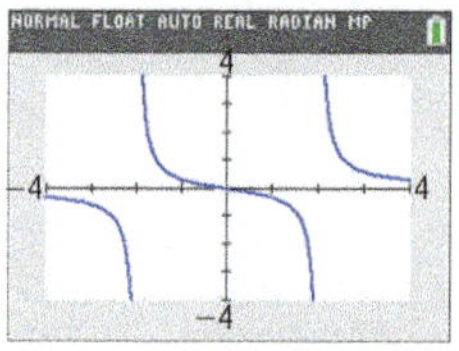

Step 7:

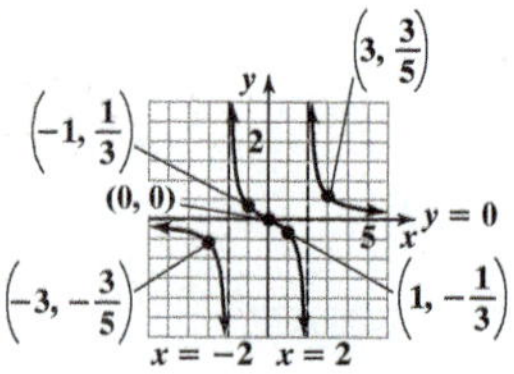

21. Step 1: $R(x) = \dfrac{3}{(x-1)(x+2)(x-2)}$; domain: $\{x \mid x \neq 1, x \neq -2, x \neq 2\}$
Step 2: R is in lowest terms
Step 3: y-intercept: $\dfrac{3}{4}$; no x-intercept
Step 4: R is in lowest terms; vertical asymptotes: $x = -2, x = 1, x = 2$
Step 5: Horizontal asymptote: $y = 0$, not intersected
Step 6:

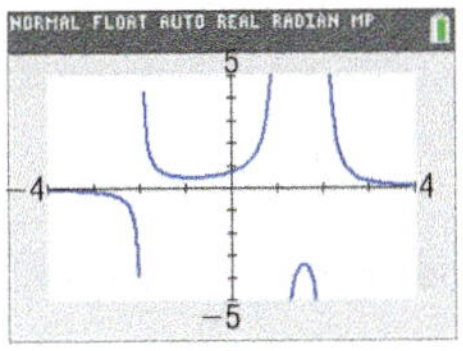

Step 7:

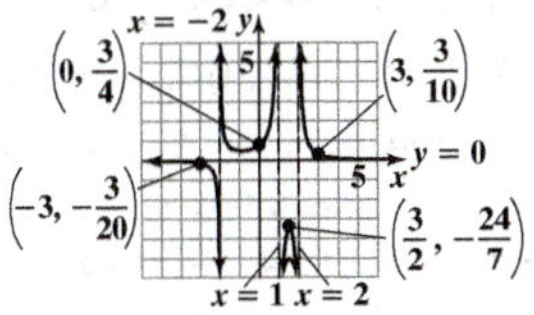

23. Step 1: $H(x) = \dfrac{(x+1)(x-1)}{(x^2+4)(x+2)(x-2)}$; domain: $\{x \mid x \neq -2, x \neq 2\}$
Step 2: H is in lowest terms
Step 3: y-intercept: $\dfrac{1}{16}$; x-intercepts: $-1, 1$
Step 4: H is in lowest terms; vertical asymptotes: $x = -2, x = 2$
Step 5: Horizontal asymptote: $y = 0$, intersected at $(-1, 0)$ and $(1, 0)$
Step 6:

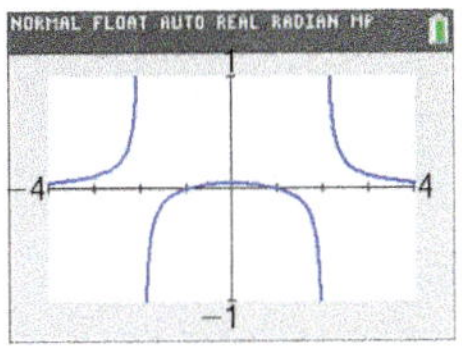

Step 7:

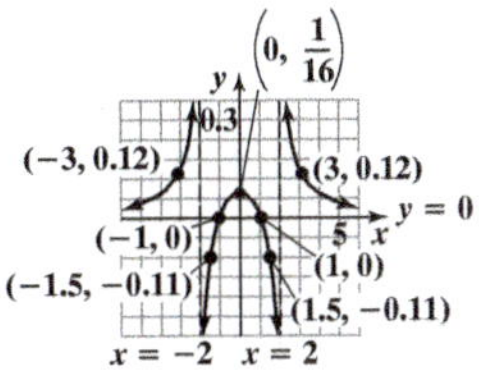

25. Step 1: $F(x) = \dfrac{(x+1)(x-4)}{x+2}$; domain: $\{x \mid x \neq -2\}$
Step 2: F is in lowest terms
Step 3: y-intercept: -2; x-intercepts: $-1, 4$
Step 4: F is in lowest terms; vertical asymptote: $x = -2$
Step 5: Oblique asymptote: $y = x - 5$, not intersected
Step 6:

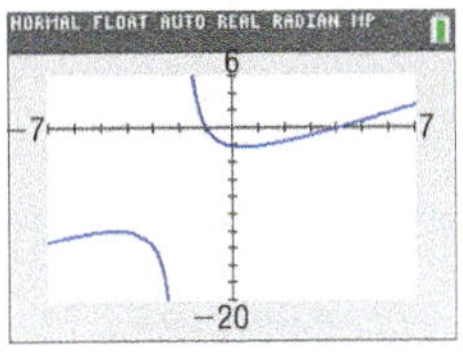

Step 7:

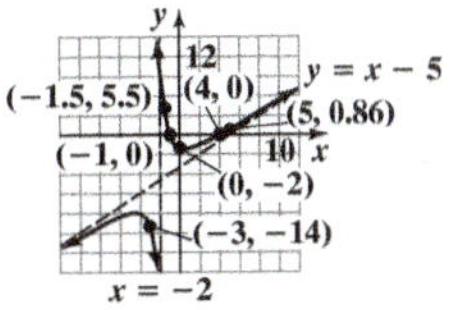

27. Step 1: $R(x) = \dfrac{(x+4)(x-3)}{x-4}$; domain: $\{x \mid x \neq 4\}$
Step 2: R is in lowest terms
Step 3: y-intercept: 3; x-intercepts: $-4, 3$
Step 4: R is in lowest terms; vertical asymptote: $x = 4$
Step 5: Oblique asymptote: $y = x + 5$, not intersected
Step 6:

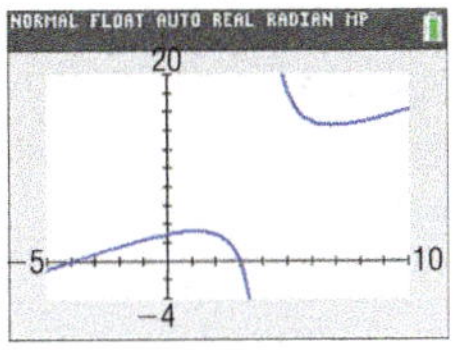

Step 7:

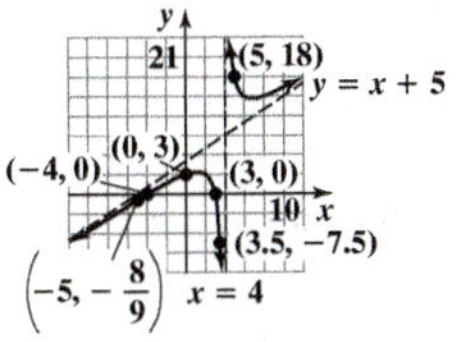

29. Step 1: $F(x) = \dfrac{(x+4)(x-3)}{x+2}$; domain: $\{x \mid x \neq -2\}$
Step 2: F is in lowest terms
Step 3: y-intercept: -6; x-intercepts: $-4, 3$
Step 4: F is in lowest terms; vertical asymptote: $x = -2$
Step 5: Oblique asymptote: $y = x - 1$, not intersected
Step 6:

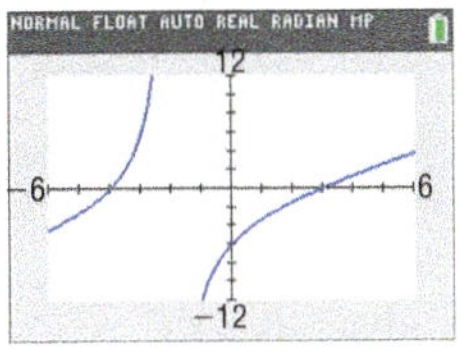

Step 7:

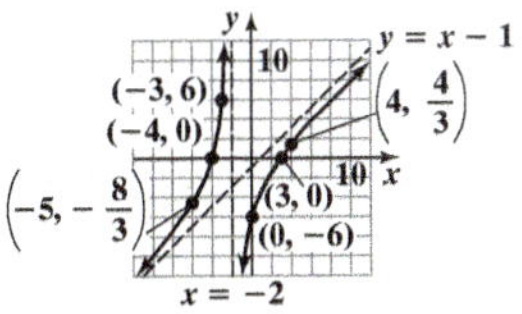

31. Step 1: Domain: $\{x \mid x \neq -3\}$
Step 2: R is in lowest terms
Step 3: y-intercept: 0; x-intercepts: 0, 1
Step 4: Vertical asymptote: $x = -3$
Step 5: Horizontal asymptote: $y = 1$, not intersected
Step 6:

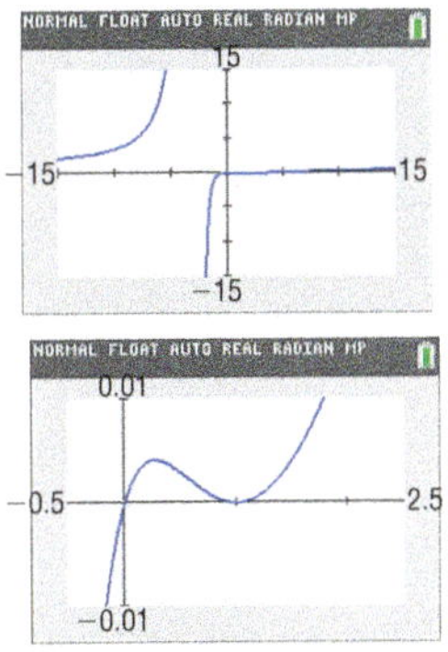

Step 7:

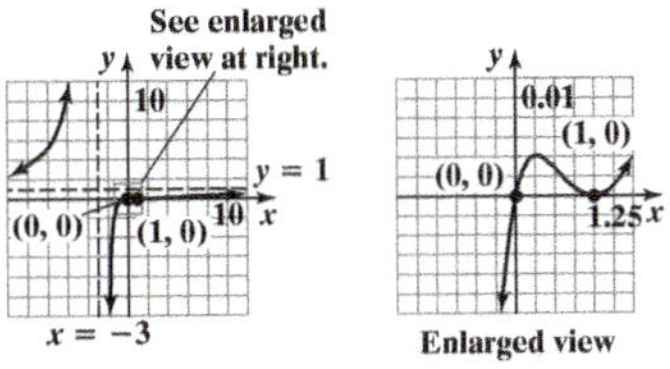

33. Step 1: $R(x) = \dfrac{(x+4)(x-3)}{(x+2)(x-3)}$; domain: $\{x \mid x \neq -2, x \neq 3\}$
Step 2: In lowest terms, $R(x) = \dfrac{x+4}{x+2}$
Step 3: y-intercept: 2; x-intercept: -4
Step 4: Vertical asymptote: $x = -2$; hole at $\left(3, \dfrac{7}{5}\right)$
Step 5: Horizontal asymptote: $y = 1$, not intersected
Step 6:

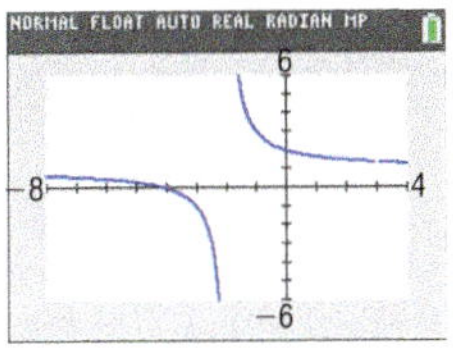

Step 7:

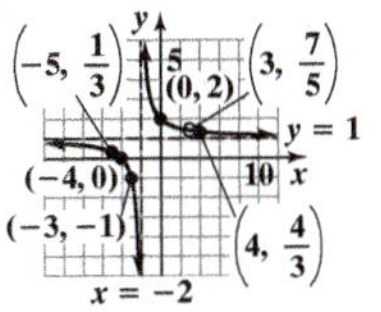

35. Step 1: $R(x) = \dfrac{(3x+1)(2x-3)}{(x-2)(2x-3)}$;
domain: $\left\{x \,\middle|\, x \neq \dfrac{3}{2}, x \neq 2\right\}$

Step 2: In lowest terms, $R(x) = \dfrac{3x+1}{x-2}$

Step 3: y-intercept: $-\dfrac{1}{2}$; x-intercept: $-\dfrac{1}{3}$

Step 4: Vertical asymptote: $x = 2$; hole at $\left(\dfrac{3}{2}, -11\right)$

Step 5: Horizontal asymptote: $y = 3$, not intersected

Step 6:

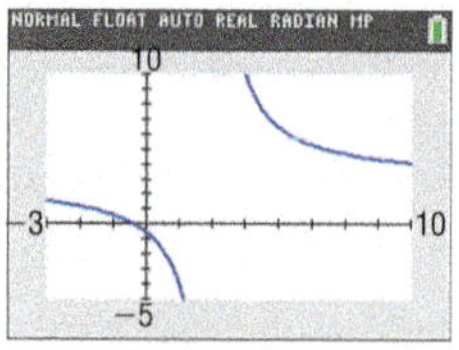

Step 7:

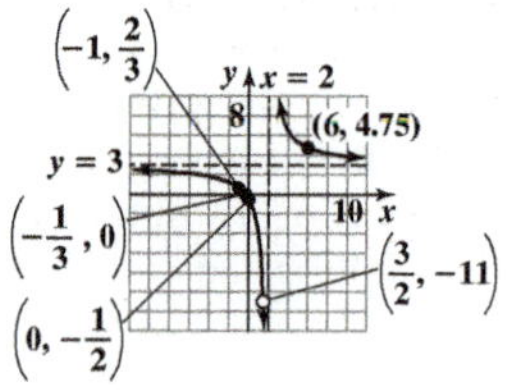

37. Step 1: $R(x) = \dfrac{(x+3)(x+2)}{x+3}$;
domain: $\{x \mid x \neq -3\}$

Step 2: In lowest terms, $R(x) = x + 2$

Step 3: y-intercept: 2; x-intercept: -2

Step 4: Vertical asymptote: none; hole at $(-3, -1)$

Step 5: No horizontal or oblique asymptote

Step 6:

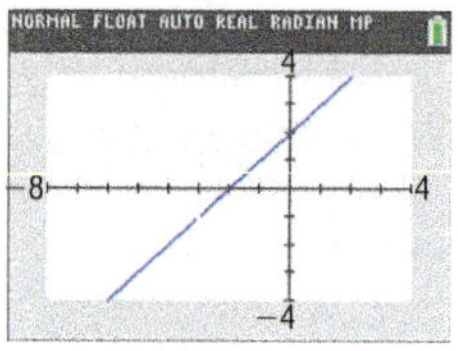

Step 7:

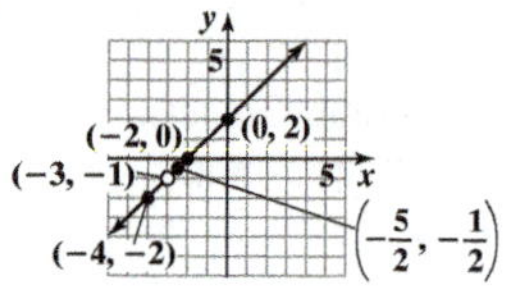

39. 1. $H(x) = \dfrac{-3(x-2)}{(x-2)(x+2)}$; domain: $\{x \mid x \neq -2, x \neq 2\}$ **2.** In lowest terms, $H(x) = \dfrac{-3}{x+2}$ **3.** y-intercept: $-\dfrac{3}{2}$; no x-intercept

4. Vertical asymptote: $x = -2$; hole at $\left(2, -\dfrac{3}{4}\right)$ **5.** Horizontal asymptote: $y = 0$; not intersected

6.

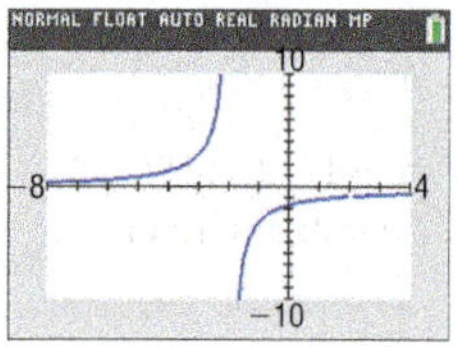

7.

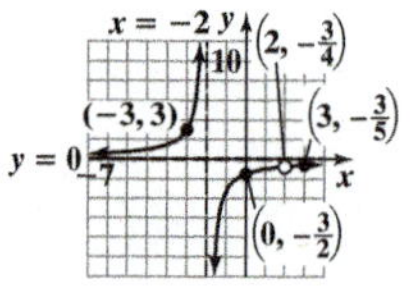

41. 1. $F(x) = \dfrac{(x-1)(x-4)}{(x-1)^2}$; domain: $\{x \mid x \neq 1\}$ **2.** In lowest terms, $F(x) = \dfrac{x-4}{x-1}$ **3.** y-intercept: 4; x-intercept: 4

4. Vertical asymptote: $x = 1$ **5.** Horizontal asymptote: $y = 1$; not intersected

6.

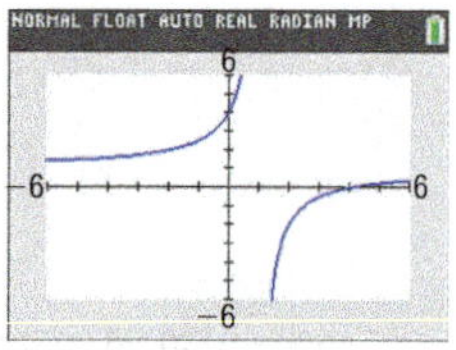

7.

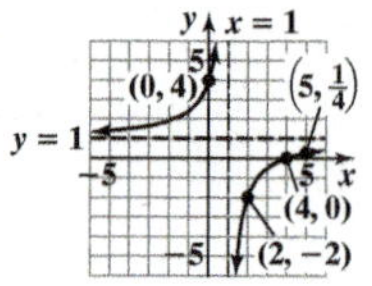

43. 1. $G(x) = \dfrac{x}{(x+2)^2}$; domain: $\{x \mid x \neq -2\}$ **2.** G is in lowest terms **3.** y-intercept: 0; x-intercept: 0 **4.** Vertical asymptote: $x = -2$

5. Horizontal asymptote: $y = 0$; intersected at $(0, 0)$

6.

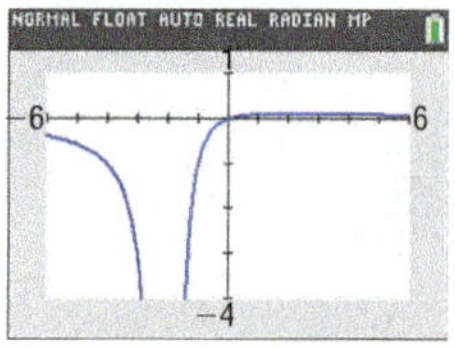

7.

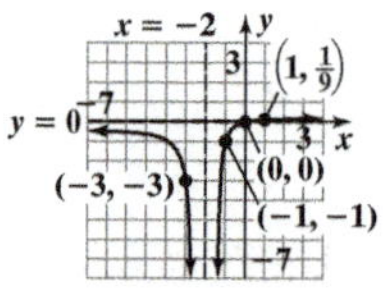

45. Step 1: $f(x) = \dfrac{x^2+1}{x}$; domain: $\{x \mid x \neq 0\}$

Step 2: f is in lowest terms

Step 3: no y-intercept; no x-intercepts

Step 4: f is in lowest terms; vertical asymptote: $x = 0$

Step 5: Oblique asymptote: $y = x$, not intersected

Step 6:

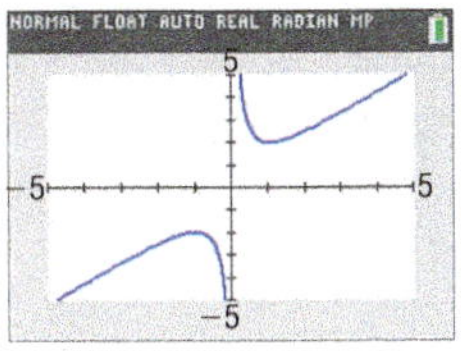

Step 7:

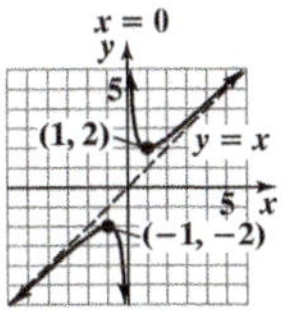

47. Step 1: $f(x) = \dfrac{x^3+1}{x} = \dfrac{(x+1)(x^2-x+1)}{x}$; domain: $\{x \mid x \neq 0\}$

Step 2: f is in lowest terms

Step 3: no y-intercept; x-intercept: -1

Step 4: f is in lowest terms; vertical asymptote: $x = 0$

Step 5: No horizontal or oblique asymptote

Step 6:

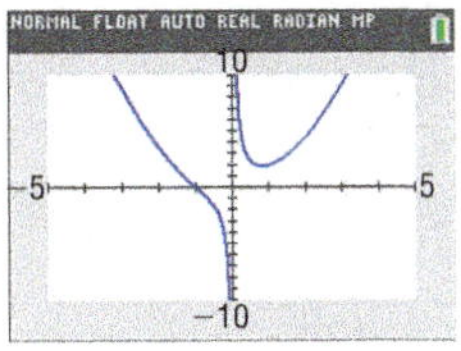

Step 7:

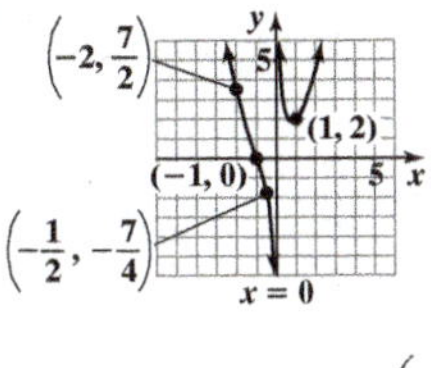

49. Step 1: $f(x) = \dfrac{x^4+1}{x^3}$; domain: $\{x \mid x \neq 0\}$

Step 2: f is in lowest terms

Step 3: no y-intercept; no x-intercepts

Step 4: f is in lowest terms; vertical asymptote: $x = 0$

Step 5: Oblique asymptote: $y = x$, not intersected

Step 6:

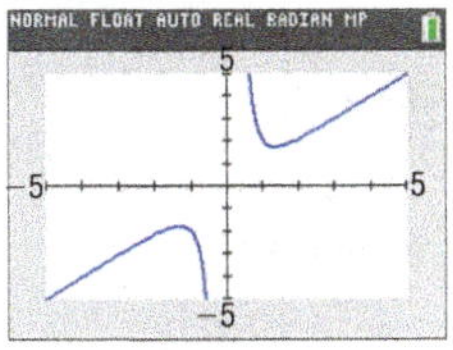

Step 7:

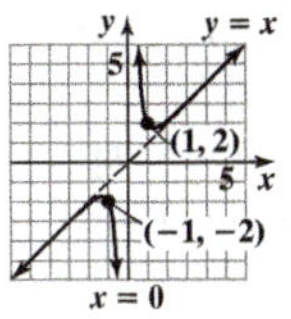

51. One possibility: $R(x) = \dfrac{x^2}{x^2-4}$ **53.** One possibility: $R(x) = \dfrac{(x-1)(x-3)\left(x^2+\frac{4}{3}\right)}{(x+1)^2(x-2)^2}$

55.

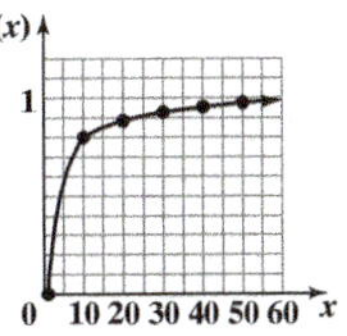

The likelihood of your ball being chosen decreases very quickly and approaches 0 as the number of attendees, x, increases.

57. (a) t-axis; $C(t) \to 0$

(b)

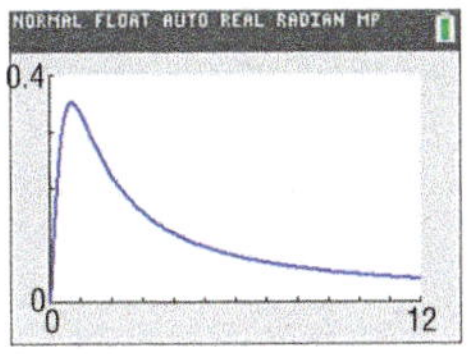

(c) 0.71 h after injection

59. (a) $C(x) = 16x + \dfrac{5000}{x} + 100$ **(b)** $x > 0$

(c)

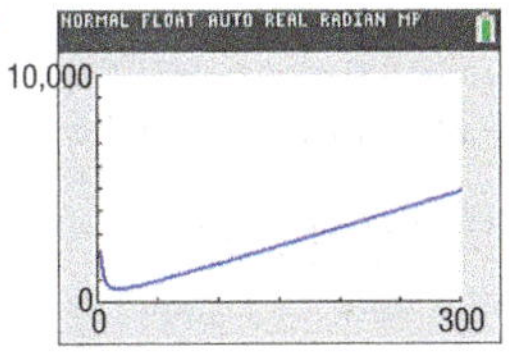

(d) Approximately 17.7 ft by 56.6 ft (longer side parallel to river)

61. (a) $S(x) = 2x^2 + \dfrac{40{,}000}{x}$

(b)

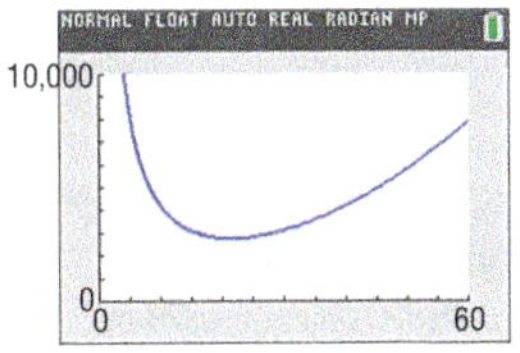

(c) 2784.95 in.^{2}

(d) 21.54 in. × 21.54 in. × 21.54 in.

(e) To minimize the cost of materials needed for construction

63. (a) $C(r) = 12\pi r^2 + \dfrac{4000}{r}$

(b)

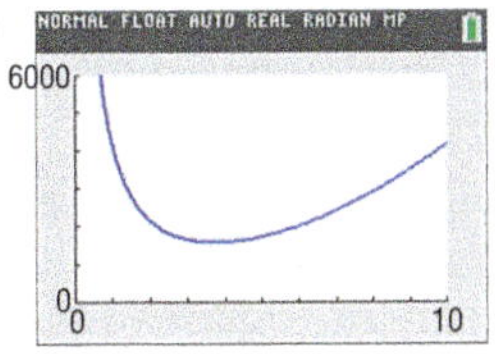

The cost is smallest when $r = 3.76$ cm.

65. No. Each function is a quotient of polynomials, but it is not written in lowest terms. Each function is undefined for $x = 1$; each graph has a hole at $x = 1$. **71.** If there is a common factor between the numerator and the denominator, and the factor yields a real zero, then the graph will have a hole. **72.** $4x^3 - 5x^2 + 2x + 2$ **73.** $\left\{-\dfrac{1}{10}\right\}$ **74.** $\dfrac{17}{2}$ **75.** ≈ -0.164

5.6 Assess Your Understanding *(page 398)*

3. c **4.** F **5. (a)** $\{x \mid 0 < x < 1 \text{ or } x > 2\}$; $(0,1) \cup (2,\infty)$ **(b)** $\{x \mid x \leq 0 \text{ or } 1 \leq x \leq 2\}$; $(-\infty,0] \cup [1,2]$
7. (a) $\{x \mid -1 < x < 0 \text{ or } x > 1\}$; $(-1,0) \cup (1,\infty)$ **(b)** $\{x \mid x < -1 \text{ or } 0 \leq x < 1\}$; $(-\infty,-1) \cup [0,1)$
9. $\{x \mid x < 0 \text{ or } 0 < x < 3\}$; $(-\infty,0) \cup (0,3)$ **11.** $\{x \mid x \leq 1\}$; $(-\infty,1]$ **13.** $\{x \mid x \leq -2 \text{ or } x \geq 2\}$; $(-\infty,-2] \cup [2,\infty)$
15. $\{x \mid -4 < x < -1 \text{ or } x > 0\}$; $(-4,-1) \cup (0,\infty)$ **17.** $\{x \mid -2 < x \leq -1\}$; $(-2,-1]$ **19.** $\{x \mid x < -2\}$; $(-\infty,-2)$ **21.** $\{x \mid x > 4\}$; $(4,\infty)$
23. $\{x \mid -4 < x < 0 \text{ or } x > 0\}$; $(-4,0) \cup (0,\infty)$ **25.** $\{x \mid x \leq 1 \text{ or } 2 \leq x \leq 3\}$; $(-\infty,1] \cup [2,3]$ **27.** $\{x \mid -1 < x < 0 \text{ or } x > 3\}$; $(-1,0) \cup (3,\infty)$
29. $\{x \mid x < -1 \text{ or } x > 1\}$; $(-\infty,-1) \cup (1,\infty)$ **31.** $\{x \mid x < -1 \text{ or } x > 1\}$; $(-\infty,-1) \cup (1,\infty)$ **33.** $\{x \mid x < -1 \text{ or } x > 1\}$; $(-\infty,-1) \cup (1,\infty)$
35. $\{x \mid x \leq -1 \text{ or } 0 < x \leq 1\}$; $(-\infty,-1] \cup (0,1]$ **37.** $\{x \mid x < -1 \text{ or } x > 1\}$; $(-\infty,-1) \cup (1,\infty)$ **39.** $\{x \mid x < 2\}$; $(-\infty,2)$
41. $\{x \mid -2 < x \leq 9\}$; $(-2,9]$ **43.** $\{x \mid x < 2 \text{ or } 3 < x < 5\}$; $(-\infty,2) \cup (3,5)$

45. $\{x|x < -5 \text{ or } -4 \le x \le -3 \text{ or } x = 0 \text{ or } x > 1\}$; $(-\infty, -5) \cup [-4, -3] \cup \{0\} \cup (1, \infty)$

47. $\left\{x\middle|-\frac{1}{2} < x < 1 \text{ or } x > 3\right\}$; $\left(-\frac{1}{2}, 1\right) \cup (3, \infty)$ **49.** $\{x|-1 < x < 3 \text{ or } x > 5\}$; $(-1, 3) \cup (5, \infty)$

51. $\left\{x\middle|x \le -4 \text{ or } x \ge \frac{1}{2}\right\}$; $(-\infty, -4] \cup \left[\frac{1}{2}, \infty\right)$ **53.** $\{x|x < 3 \text{ or } x \ge 7\}$; $(-\infty, 3) \cup [7, \infty)$ **55.** $\{x|x < 2\}$; $(-\infty, 2)$

57. $\left\{x\middle|x < -\frac{2}{3} \text{ or } 0 < x < \frac{3}{2}\right\}$; $\left(-\infty, -\frac{2}{3}\right) \cup \left(0, \frac{3}{2}\right)$ **59.** $\{x|x \le -3 \text{ or } 0 \le x \le 3\}$; $(-\infty, -3] \cup [0, 3]$

61. (a) $-4, -\frac{1}{2}, 3$
(b) $f(x) = (x + 4)^2(2x + 1)(x - 3)$
(c)

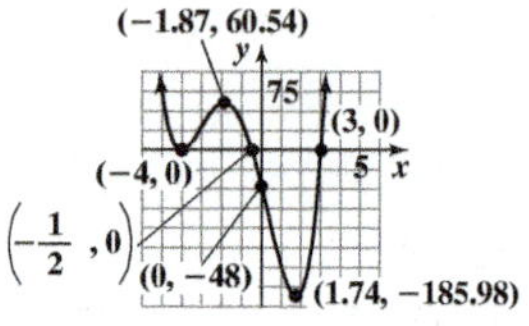

(d) $\left(-\frac{1}{2}, 3\right)$

63. (a)

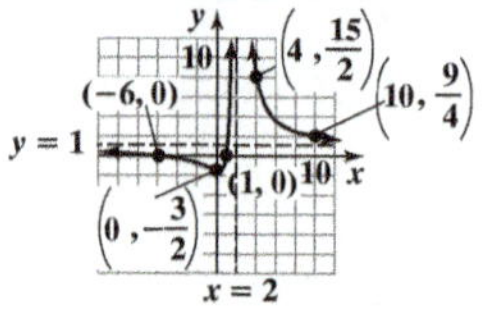

(b) $(-\infty, -6] \cup [1, 2) \cup (2, \infty)$

65. (a)

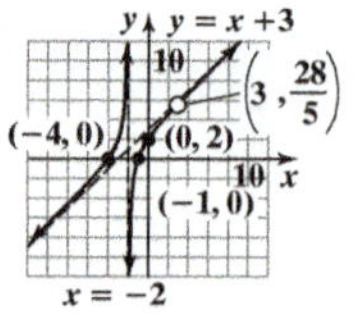

(b) $[-4, -2) \cup [-1, 3) \cup (3, \infty)$

67. $\{x|x > 4\}$; $(4, \infty)$
69. $\{x|x \le -2 \text{ or } x \ge 2\}$; $(-\infty, -2] \cup [2, \infty)$
71. $\{x|x < -4 \text{ or } x \ge 2\}$; $(-\infty, -4) \cup [2, \infty)$

73.

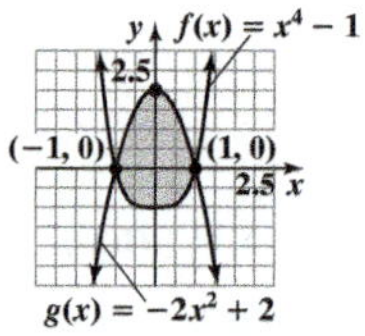

$f(x) \le g(x)$ if $-1 \le x \le 1$

75.

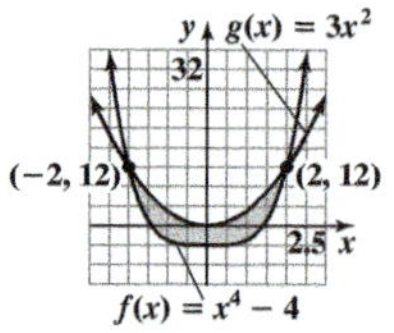

$f(x) \le g(x)$ if $-2 \le x \le 2$

77. Produce at least 250 bicycles
79. (a) The stretch is less than 39 ft.
(b) The ledge should be at least 84 ft above the ground for a 150-lb jumper.
81. At least 50 students must attend. **86.** $\left[\frac{4}{3}, \infty\right)$
87. $x^2 - x - 4$ **88.** $3x^2y^4(x + 2y)(2x - 3y)$ **89.** $y = \frac{2}{3}\sqrt{x}$

Review Exercises *(page 402)*

1. Polynomial of degree 5 **2.** Rational **3.** Neither **4.** Polynomial of degree 0

5.

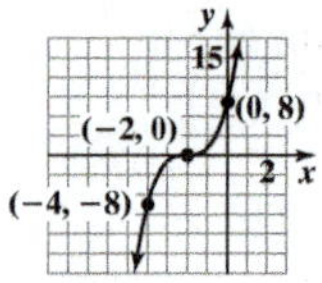

6.

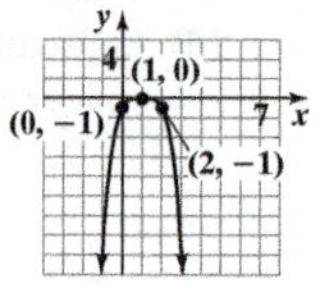

7.

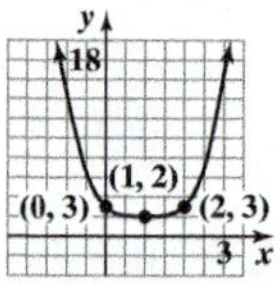

8. Step 1: $y = x^3$
Step 2: x-intercepts: $-4, -2, 0$; y-intercept: 0
Step 3: $-4, -2, 0$: multiplicity 1; crosses
Step 4:

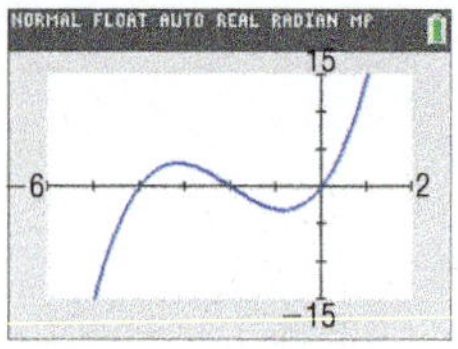

Step 5: $(-3.15, 3.08)$, $(-0.85, -3.08)$
Step 6:

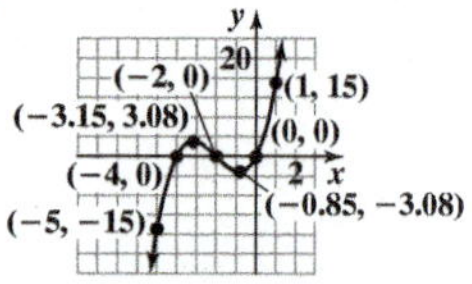

Step 7: Domain: $(-\infty, \infty)$; Range: $(-\infty, \infty)$
Step 8: Increasing on $(-\infty, -3.15]$ and $[-0.85, \infty)$
Decreasing on $[-3.15, -0.85]$

9. Step 1: $y = x^3$
Step 2: x-intercepts: $-4, 2$; y-intercept: 16
Step 3: -4: multiplicity 1; crosses;
2: multiplicity 2; touches
Step 4:

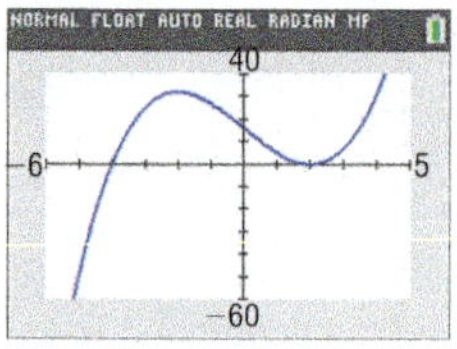

Step 5: $(-2, 32)$, $(2, 0)$
Step 6:

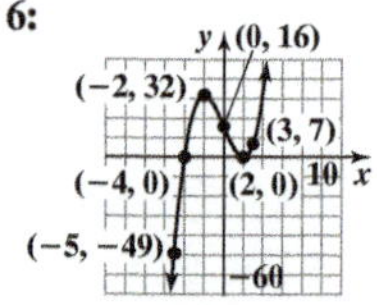

Step 7: Domain: $(-\infty, \infty)$; Range: $(-\infty, \infty)$
Step 8: Increasing on $(-\infty, -2]$ and $[2, \infty)$
Decreasing on $[-2, 2]$

10. Step 1: $y = -2x^3$
Step 2: x-intercepts: 0, 2; y-intercept: 0
Step 3: 0: multiplicity 2; touches;
2: multiplicity 1; crosses
Step 4:

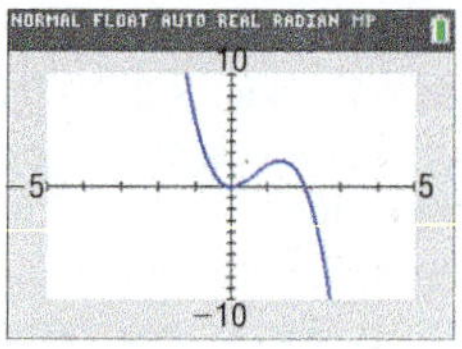

Step 5: $(0, 0)$, $(1.33, 2.37)$
Step 6 :

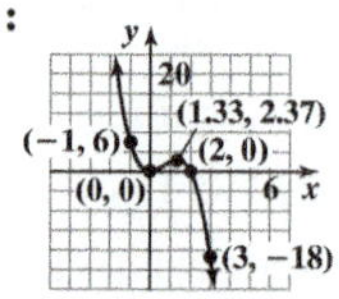

Step 7: Domain: $(-\infty, \infty)$; Range: $(-\infty, \infty)$
Step 8: Increasing on $[0, 1.33]$
Decreasing on $(-\infty, 0]$ and $[1.33, \infty)$

11. Step 1: $y = x^4$
Step 2: x-intercepts: $-3, -1, 1$; y-intercept: 3
Step 3: $-3, -1$: multiplicity 1; crosses; 1: multiplicity 2; touches

Step 4:

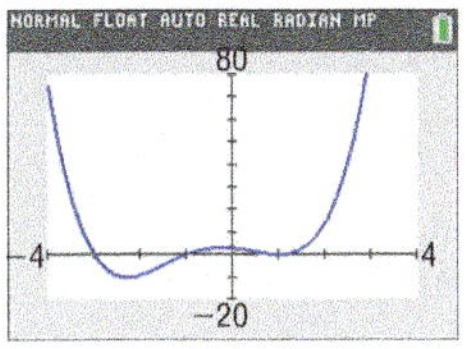

Step 5: $(-2.28, -9.91)$, $(-0.22, 3.23)$, $(1, 0)$

Step 6:

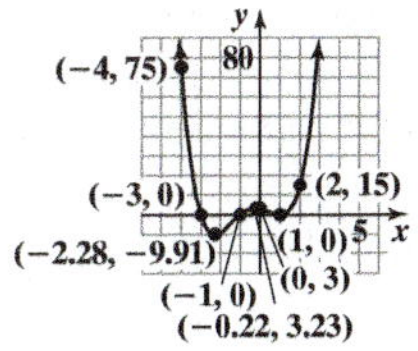

Step 7: Domain: $(-\infty, \infty)$; Range: $[-9.91, \infty)$
Step 8: Increasing on $[-2.28, -0.22]$ and $[1, \infty)$
Decreasing on $(-\infty, -2.28]$ and $[-0.22, 1]$

12. $R = 10$; g is not a factor of f. **13.** $R = 0$; g is a factor of f. **14.** $f(4) = 47{,}105$ **15.** 4, 2, or 0 positive; 2 or 0 negative **16.** 1 positive; 2 or 0 negative

17. $\pm 1, \pm 3, \pm\frac{1}{2}, \pm\frac{3}{2}, \pm\frac{1}{3}, \pm\frac{1}{4}, \pm\frac{3}{4}, \pm\frac{1}{6}, \pm\frac{1}{12}$ **18.** $-2, 1, 4$; $f(x) = (x + 2)(x - 1)(x - 4)$ **19.** $\frac{1}{2}$, multiplicity 2; -2; $f(x) = 4\left(x - \frac{1}{2}\right)^2(x + 2)$

20. 2, multiplicity 2; $f(x) = (x - 2)^2(x^2 + 5)$ **21.** $\{-3, 2\}$ **22.** $\left\{-3, -1, -\frac{1}{2}, 1\right\}$

23. -2 and 3

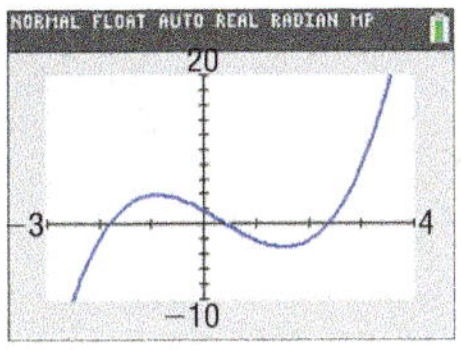

24. -5 and 5

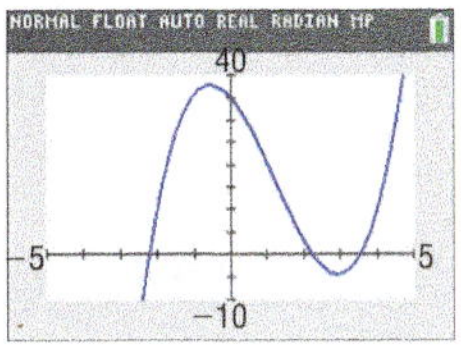

25. $f(0) = -1$; $f(1) = 1$ **26.** $f(0) = -1$; $f(1) = 1$ **27.** 1.52 **28.** 0.93
29. $4 - i$; $f(x) = x^3 - 14x^2 + 65x - 102$ **30.** $-i, 1 - i$; $f(x) = x^4 - 2x^3 + 3x^2 - 2x + 2$
31. $-2, 1, 4$; $f(x) = (x + 2)(x - 1)(x - 4)$
32. $-2, \frac{1}{2}$ (multiplicity 2); $f(x) = 4(x + 2)\left(x - \frac{1}{2}\right)^2$

33. 2 (multiplicity 2), $-\sqrt{5}i, \sqrt{5}i$; $f(x) = (x + \sqrt{5}i)(x - \sqrt{5}i)(x - 2)^2$ **34.** $-3, 2, -\frac{\sqrt{2}}{2}i, \frac{\sqrt{2}}{2}i$; $f(x) = 2(x + 3)(x - 2)\left(x + \frac{\sqrt{2}}{2}i\right)\left(x - \frac{\sqrt{2}}{2}i\right)$

35. Domain: $\{x \mid x \neq -3, x \neq 3\}$; horizontal asymptote: $y = 0$; vertical asymptotes: $x = -3, x = 3$
36. Domain: $\{x \mid x \neq -2\}$; horizontal asymptote: $y = 1$; vertical asymptote: $x = -2$

37. Step 1: $R(x) = \dfrac{2(x - 3)}{x}$; domain: $\{x \mid x \neq 0\}$
Step 2: R is in lowest terms
Step 3: no y-intercept; x-intercept: 3
Step 4: R is in lowest terms; vertical asymptote: $x = 0$
Step 5: Horizontal asymptote: $y = 2$; not intersected
Step 6:

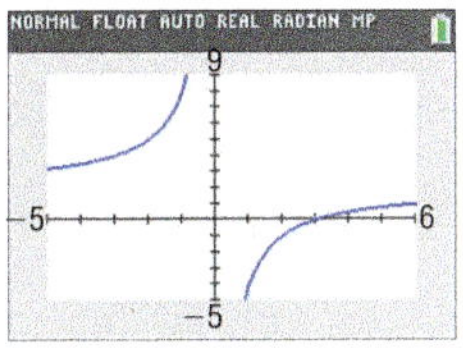

Step 7:

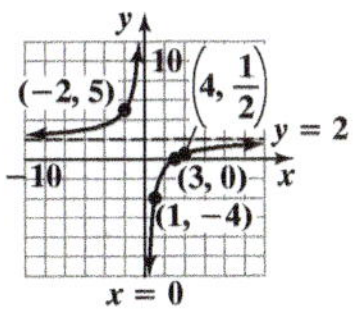

38. Step 1: Domain: $\{x \mid x \neq 0, x \neq 2\}$
Step 2: H is in lowest terms
Step 3: no y-intercept; x-intercept: -2
Step 4: H is in lowest terms; vertical asymptote: $x = 0, x = 2$
Step 5: Horizontal asymptote: $y = 0$; intersected at $(-2, 0)$
Step 6:

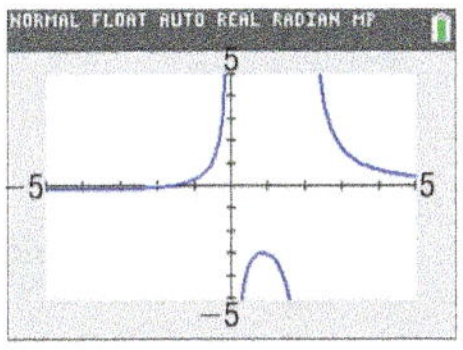

Step 7:

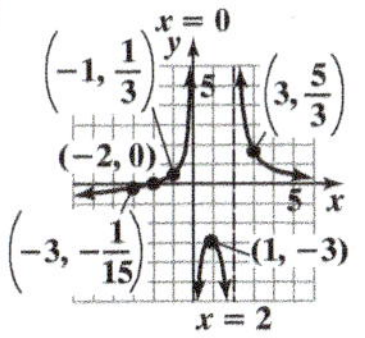

39. Step 1: $R(x) = \dfrac{(x + 3)(x - 2)}{(x - 3)(x + 2)}$; domain: $\{x \mid x \neq -2, x \neq 3\}$
Step 2: R is in lowest terms
Step 3: y-intercept: 1; x-intercepts: $-3, 2$
Step 4: R is in lowest terms; vertical asymptotes: $x = -2, x = 3$
Step 5: Horizontal asymptote: $y = 1$; intersected at $(0, 1)$
Step 6:

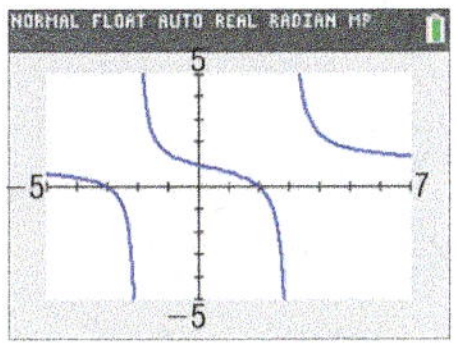

Step 7:

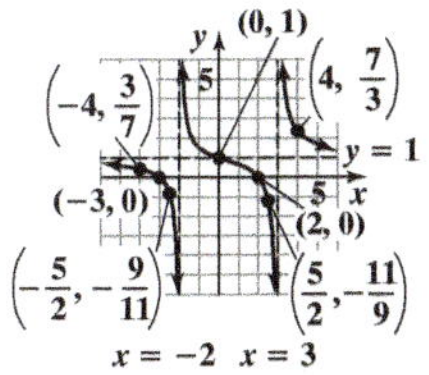

40. Step 1: $F(x) = \dfrac{x^3}{(x + 2)(x - 2)}$; domain: $\{x \mid x \neq -2, x \neq 2\}$
Step 2: F is in lowest terms
Step 3: y-intercept: 0; x-intercept: 0
Step 4: F is in lowest terms; vertical asymptotes: $x = -2, x = 2$
Step 5: Oblique asymptote: $y = x$; intersected at $(0, 0)$

Step 6:

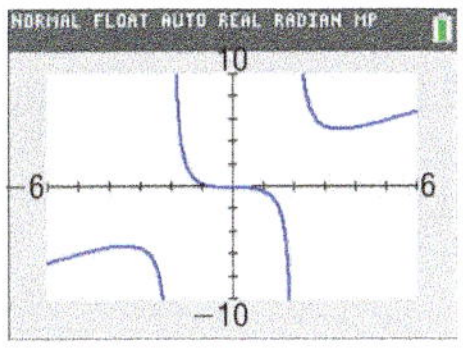

Step 7:

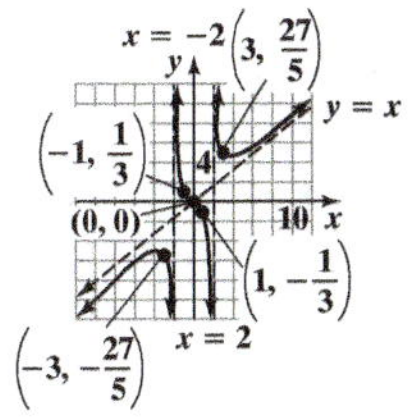

41. Step 1: Domain: $\{x \mid x \neq 1\}$
Step 2: R is in lowest terms
Step 3: y-intercept: 0; x-intercept: 0
Step 4: R is in lowest terms;
vertical asymptote: $x = 1$
Step 5: No oblique or horizontal asymptote

Step 6:

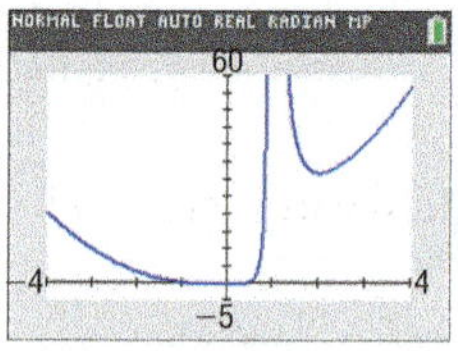

Step 7:

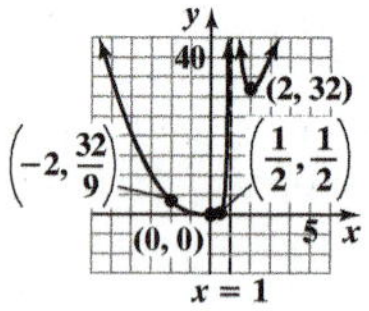

42. Step 1: $G(x) = \dfrac{(x + 2)(x - 2)}{(x + 1)(x - 2)}$;
domain: $\{x \mid x \neq -1, x \neq 2\}$
Step 2: In lowest terms, $G(x) = \dfrac{x + 2}{x + 1}$
Step 3: y-intercept: 2; x-intercept: -2
Step 4: Vertical asymptote: $x = -1$;
hole at $\left(2, \dfrac{4}{3}\right)$
Step 5: Horizontal asymptote: $y = 1$,
not intersected

Step 6:

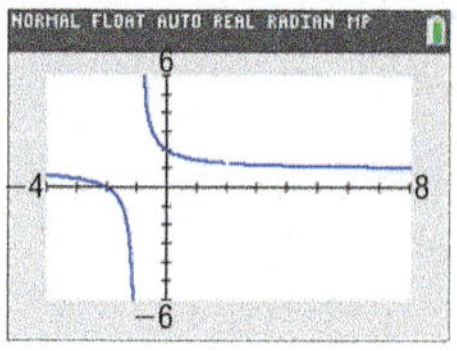

Step 7:

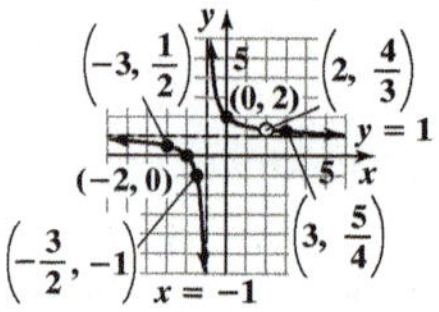

43. **(a)** $\{-3, 2\}$ **(b)** $(-3, 2) \cup (2, \infty)$ **(c)** $(-\infty, -3] \cup \{2\}$ **(d)** $f(x) = (x - 2)^2(x + 3)$

44. **(a)** $y = 0.25$ **(b)** $x = -2$, $x = 2$ **(c)** $(-3, -2) \cup (-1, 2)$ **(d)** $(-\infty, -3] \cup (-2, -1] \cup (2, \infty)$ **(e)** $R(x) = \dfrac{x^2 + 4x + 3}{4x^2 - 16}$

45. $\{x \mid x < -2 \text{ or } -1 < x < 2\}$; $(-\infty, -2) \cup (-1, 2)$

46. $\{x \mid -4 \le x \le -1 \text{ or } x \ge 1\}$; $[-4, -1] \cup [1, \infty)$

47. $\{x \mid x < 1 \text{ or } x > 2\}$; $(-\infty, 1) \cup (2, \infty)$

48. $\{x \mid 1 \le x \le 2 \text{ or } x > 3\}$; $[1, 2] \cup (3, \infty)$

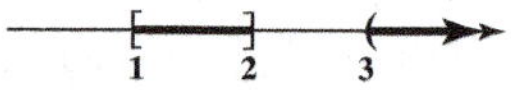

49. $\{x \mid x < -4 \text{ or } 2 < x < 4 \text{ or } x > 6\}$; $(-\infty, -4) \cup (2, 4) \cup (6, \infty)$

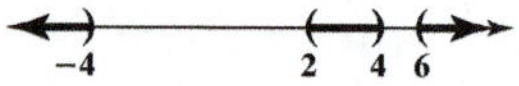

50. **(a)** $A(r) = 2\pi r^2 + \dfrac{500}{r}$
(b) 223.22 cm^2
(c) 257.08 cm^2
(d)

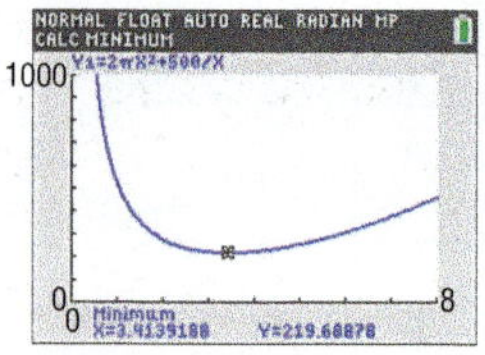

A is smallest when $r \approx 3.41$ cm.

51. **(a)**

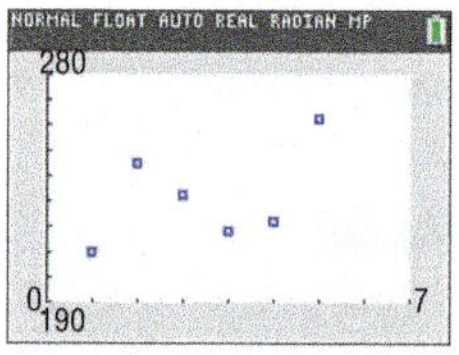

The relation appears to be cubic.
(b) $P(t) = 4.4926t^3 - 45.5294t^2 + 136.1209t + 115.4667$; $\approx \$928{,}000$
(c)

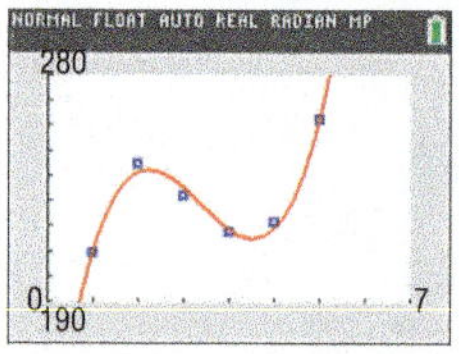

52. **(a)** Even **(b)** Positive **(c)** Even **(d)** The graph touches the x-axis at $x = 0$, but does not cross it there. **(e)** 8

Chapter Test *(page 404)*

1.

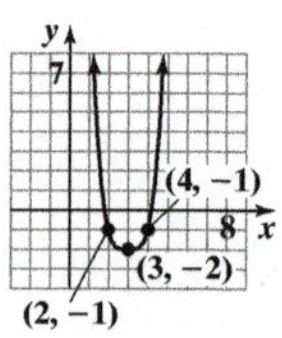

2. **(a)** 3 **(b)** $\dfrac{p}{q}$: $\pm\dfrac{1}{2}, \pm 1, \pm\dfrac{3}{2}, \pm\dfrac{5}{2}, \pm 3, \pm 5, \pm\dfrac{15}{2}, \pm 15$ **(c)** $-5, -\dfrac{1}{2}, 3$; $g(x) = (x + 5)(2x + 1)(x - 3)$

(d) y-intercept: -15; x-intercepts: $-5, -\dfrac{1}{2}, 3$ **(e)** Crosses at $-5, -\dfrac{1}{2}, 3$ **(f)** $y = 2x^3$

(g) $(-3.15, 60.30)$, $(1.48, -39.00)$ **(h)**

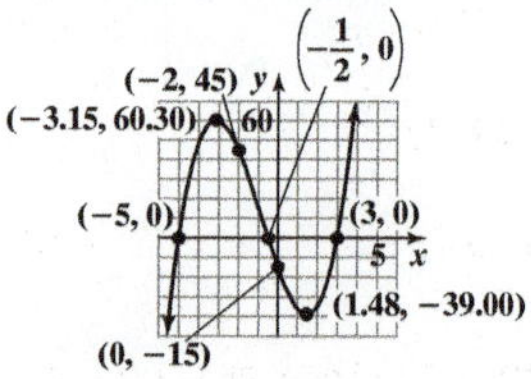

3. $4, -5i, 5i$ **4.** $\left\{1, \frac{5-\sqrt{61}}{6}, \frac{5+\sqrt{61}}{6}\right\}$ **5.** Domain: $\{x|x \neq -10, x \neq 4\}$; asymptotes: $x = -10, y = 2$
6. Domain: $\{x|x \neq -1\}$; asymptotes: $x = -1, y = x + 1$
7.

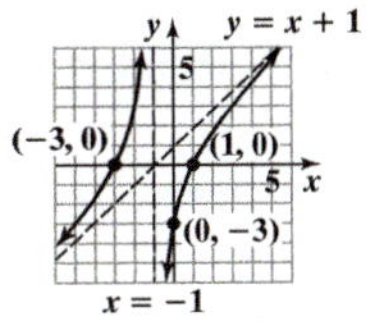

8. Answers may vary. One possibility is $f(x) = x^4 - 4x^3 - 2x^2 + 20x$.
9. Answers may vary. One possibility is $r(x) = \frac{2(x-9)(x-1)}{(x-4)(x-9)}$.
10. $f(0) = 8; f(4) = -36$; Since $f(0) = 8 > 0$ and $f(4) = -36 < 0$, the Intermediate Value Theorem guarantees that there is at least one real zero between 0 and 4. **11.** $\{x|x < 3 \text{ or } x > 8\}$; $(-\infty, 3) \cup (8, \infty)$

Cumulative Review *(page 404)*

1. $\sqrt{26}$ **2.** $\{x|x \leq 0 \text{ or } x \geq 1\}$; $(-\infty, 0]$ or $[1, \infty)$

3. $\{x|-1 < x < 4\}$; $(-1, 4)$

4. $f(x) = -3x + 1$

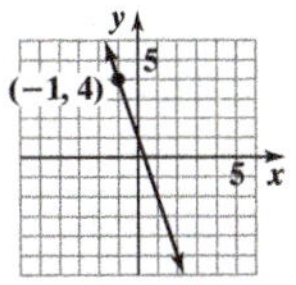

5. $y = 2x - 1$

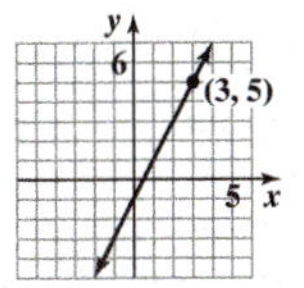

6.

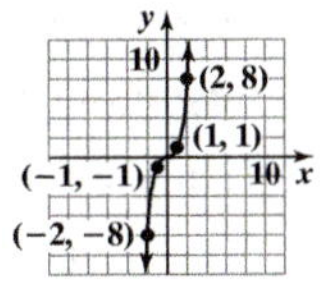

7. Not a functions; 3 has two images. **8.** $\{0, 2, 4\}$ **9.** $\left\{x \middle| x \geq \frac{3}{2}\right\}$; $\left[\frac{3}{2}, \infty\right)$

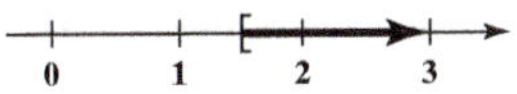

10. Center: $(-2, 1)$; radius: 3

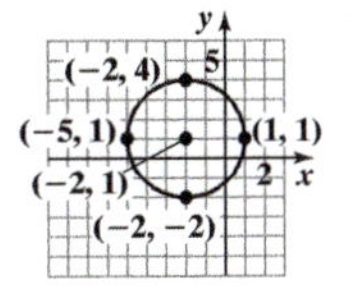

11. x-intercepts: $-3, 0, 3$; y-intercept: 0; symmetric with respect to the origin **12.** $y = -\frac{2}{3}x + \frac{17}{3}$
13. Not a function; it fails the Vertical Line Test. **14. (a)** 22 **(b)** $x^2 - 5x - 2$ **(c)** $-x^2 - 5x + 2$
(d) $9x^2 + 15x - 2$ **(e)** $2x + h + 5$ **15. (a)** $\{x|x \neq 1\}$ **(b)** No; (2,7) is on the graph. **(c)** 4; (3, 4) is on the graph.
(d) $\frac{7}{4}$; $\left(\frac{7}{4}, 9\right)$ is on the graph. **(e)** Rational

16.

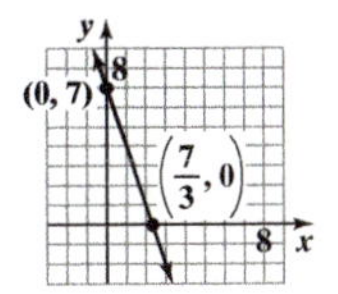

17.

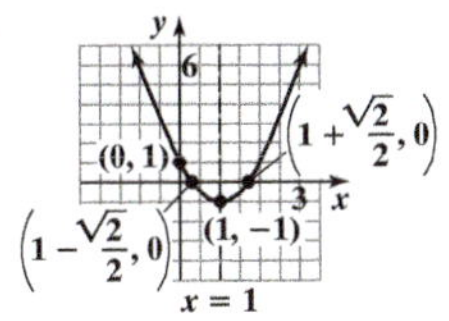

18. 6; $y = 6x - 1$ **19. (a)** x-intercepts: $-5, -1, 5$; y-intercept: -3 **(b)** No symmetry
(c) Neither **(d)** Increasing: $(-\infty, -3]$ and $[2, \infty)$; decreasing: $[-3, 2]$
(e) Local maximum value is 5 and occurs at $x = -3$.
(f) Local minimum value is -6 and occurs at $x = 2$. **20.** Odd

21. (a) Domain: $\{x|x > -3\}$ or $(-3, \infty)$
(b) x-intercept: $-\frac{1}{2}$; y-intercept: 1
(c)

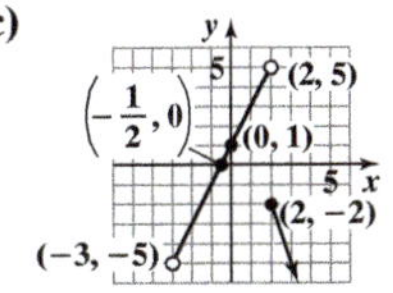

(d) Range: $\{y|y < 5\}$ or $(-\infty, 5)$

22.

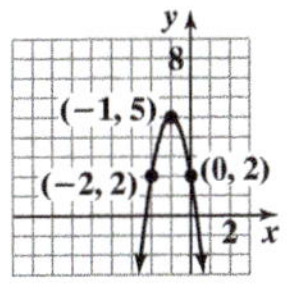

23. (a) $(f + g)(x) = x^2 - 9x - 6$; domain: all real numbers
(b) $\left(\frac{f}{g}\right)(x) = \frac{x^2 - 5x + 1}{-4x - 7}$; domain: $\left\{x \middle| x \neq -\frac{7}{4}\right\}$
24. (a) $R(x) = -\frac{1}{10}x^2 + 150x$ **(b)** \$14,000 **(c)** 750; \$56,250 **(d)** \$75

CHAPTER 6 Exponential and Logarithmic Functions

6.1 Assess Your Understanding *(page 413)*

4. composite function; $f(g(x))$ **5.** F **6.** c **7.** a **8.** F **9. (a)** -1 **(b)** -1 **(c)** 8 **(d)** 0 **(e)** 8 **(f)** -7 **11. (a)** 4 **(b)** 5 **(c)** -1 **(d)** -2
13. (a) 98 **(b)** 49 **(c)** 4 **(d)** 4 **15. (a)** 97 **(b)** $-\frac{163}{2}$ **(c)** 1 **(d)** $-\frac{3}{2}$ **17. (a)** $2\sqrt{2}$ **(b)** $2\sqrt{2}$ **(c)** 1 **(d)** 0 **19. (a)** $\frac{1}{17}$ **(b)** $\frac{1}{5}$ **(c)** 1 **(d)** $\frac{1}{2}$
21. (a) $\frac{3}{\sqrt[3]{4} + 1}$ **(b)** 1 **(c)** $\frac{6}{5}$ **(d)** 0 **23. (a)** $(f \circ g)(x) = 6x + 3$; all real numbers **(b)** $(g \circ f)(x) = 6x + 9$; all real numbers
(c) $(f \circ f)(x) = 4x + 9$; all real numbers **(d)** $(g \circ g)(x) = 9x$; all real numbers **25. (a)** $(f \circ g)(x) = 3x^2 + 1$; all real numbers
(b) $(g \circ f)(x) = 9x^2 + 6x + 1$; all real numbers **(c)** $(f \circ f)(x) = 9x + 4$; all real numbers **(d)** $(g \circ g)(x) = x^4$; all real numbers
27. (a) $(f \circ g)(x) = x^4 + 8x^2 + 16$; all real numbers **(b)** $(g \circ f)(x) = x^4 + 4$; all real numbers **(c)** $(f \circ f)(x) = x^4$; all real numbers
(d) $(g \circ g)(x) = x^4 + 8x^2 + 20$; all real numbers **29. (a)** $(f \circ g)(x) = \frac{3x}{2 - x}$; $\{x|x \neq 0, x \neq 2\}$ **(b)** $(g \circ f)(x) = \frac{2(x-1)}{3}$; $\{x|x \neq 1\}$
(c) $(f \circ f)(x) = \frac{3(x-1)}{4 - x}$; $\{x|x \neq 1, x \neq 4\}$ **(d)** $(g \circ g)(x) = x$; $\{x|x \neq 0\}$ **31. (a)** $(f \circ g)(x) = \frac{4}{4 + x}$; $\{x|x \neq -4, x \neq 0\}$
(b) $(g \circ f)(x) = \frac{-4(x-1)}{x}$; $\{x|x \neq 0, x \neq 1\}$ **(c)** $(f \circ f)(x) = x$; $\{x|x \neq 1\}$ **(d)** $(g \circ g)(x) = x$; $\{x|x \neq 0\}$

33. (a) $(f \circ g)(x) = \sqrt{2x+3}; \left\{x \middle| x \geq -\frac{3}{2}\right\}$ **(b)** $(g \circ f)(x) = 2\sqrt{x} + 3; \{x|x \geq 0\}$ **(c)** $(f \circ f)(x) = \sqrt[4]{x}; \{x|x \geq 0\}$ **(d)** $(g \circ g)(x) = 4x + 9$; all real numbers **35. (a)** $(f \circ g)(x) = x; \{x|x \geq 1\}$ **(b)** $(g \circ f)(x) = |x|$; all real numbers **(c)** $(f \circ f)(x) = x^4 + 2x^2 + 2$; all real numbers **(d)** $(g \circ g)(x) = \sqrt{\sqrt{x-1} - 1}; \{x|x \geq 2\}$ **37. (a)** $(f \circ g)(x) = -\frac{4x-17}{2x-1}; \left\{x \middle| x \neq 3; x \neq \frac{1}{2}\right\}$ **(b)** $(g \circ f)(x) = -\frac{3x-3}{2x+8}; \{x|x \neq -4; x \neq -1\}$ **(c)** $(f \circ f)(x) = -\frac{2x+5}{x-2}; \{x|x \neq -1; x \neq 2\}$ **(d)** $(g \circ g)(x) = -\frac{3x-4}{2x-11}; \left\{x \middle| x \neq \frac{11}{2}; x \neq 3\right\}$

39. $(f \circ g)(x) = f(g(x)) = f\left(\frac{1}{2}x\right) = 2\left(\frac{1}{2}x\right) = x; (g \circ f)(x) = g(f(x)) = g(2x) = \frac{1}{2}(2x) = x$

41. $(f \circ g)(x) = f(g(x)) = f(\sqrt[3]{x}) = (\sqrt[3]{x})^3 = x; (g \circ f)(x) = g(f(x)) = g(x^3) = \sqrt[3]{x^3} = x$

43. $(f \circ g)(x) = f(g(x)) = f\left(\frac{1}{2}(x+6)\right) = 2\left[\frac{1}{2}(x+6)\right] - 6 = x + 6 - 6 = x; (g \circ f)(x) = g(f(x)) = g(2x-6) = \frac{1}{2}(2x-6+6) = \frac{1}{2}(2x) = x$

45. $(f \circ g)(x) = f(g(x)) = f\left(\frac{1}{a}(x-b)\right) = a\left[\frac{1}{a}(x-b)\right] + b = x; (g \circ f)(x) = g(f(x)) = g(ax+b) = \frac{1}{a}(ax+b-b) = x$

47. $f(x) = x^4; g(x) = 2x + 3$ (Other answers are possible.) **49.** $f(x) = \sqrt{x}; g(x) = x^2 + 1$ (Other answers are possible.) **51.** $f(x) = |x|; g(x) = 2x + 1$ (Other answers are possible.) **53.** $(f \circ g)(x) = 11; (g \circ f)(x) = 2$ **55.** $-3, 3$ **57. (a)** $(f \circ g)(x) = acx + ad + b$ **(b)** $(g \circ f)(x) = acx + bc + d$ **(c)** The domains of both $f \circ g$ and $g \circ f$ are all real numbers. **(d)** $f \circ g = g \circ f$ when $ad + b = bc + d$

59. $S(t) = \frac{16}{9}\pi t^6$ **61.** $C(t) = 15{,}000 + 800{,}000t - 40{,}000t^2$ **63.** $C(p) = \frac{2\sqrt{100-p}}{25} + 600, 0 \leq p \leq 100$ **65.** $V(r) = 2\pi r^3$

67. (a) $f(x) = 0.9428x$ **(b)** $g(x) = 126.457x$ **(c)** $g(f(x)) = g(0.9428x) = 119.2236596x$ **(d)** 119,223.6596 yen **69. (a)** $f(p) = p - 200$ **(b)** $g(p) = 0.8p$ **(c)** $(f \circ g)(p) = 0.8p - 200; (g \circ f)(p) = 0.8p - 160$; The 20% discount followed by the $200 rebate is the better deal. **71.** 15 **73.** f is an odd function, so $f(-x) = -f(x)$. g is an even function, so $g(-x) = g(x)$. Then $(f \circ g)(-x) = f(g(-x)) = f(g(x)) = (f \circ g)(x)$. So $f \circ g$ is even. Also, $(g \circ f)(-x) = g(f(-x)) = g(-f(x)) = g(f(x)) = (g \circ f)(x)$, so $g \circ f$ is even.

74. $(f + g)(x) = 4x + 3$; Domain: all real numbers
$(f - g)(x) = 2x + 13$; Domain: all real numbers
$(f \cdot g)(x) = 3x^2 - 7x - 40$; Domain: all real numbers
$\left(\frac{f}{g}\right)(x) = \frac{3x+8}{x-5}$; Domain: $\{x|x \neq 5\}$

75. $\frac{1}{4}, 4$

76.

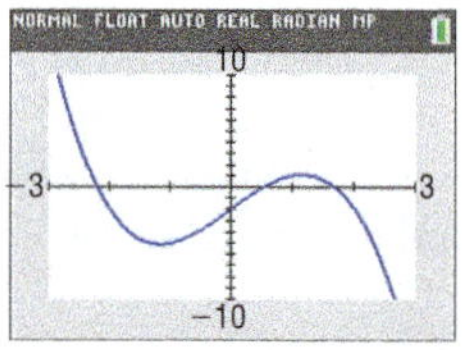

Local minimum: -5.08 at $x = -1.15$
Local maximum: 1.08 at $x = 1.15$
Decreasing: $[-3, -1.15]; [1.15, 3]$
Increasing: $[-1.15, 1.15]$

77. Domain: $\{x|x \neq 3\}$
Vertical asymptote: $x = 3$
Oblique asymptote: $y = x + 9$

6.2 Assess Your Understanding *(page 424)*

5. $f(x_1) \neq f(x_2)$ **6.** one-to-one **7.** 3 **8.** $y = x$ **9.** $[4, \infty)$ **10.** T **11.** a **12.** d **13.** one-to-one **15.** not one-to-one **17.** not one-to-one **19.** one-to-one **21.** one-to-one **23.** not one-to-one **25.** one-to-one

27.

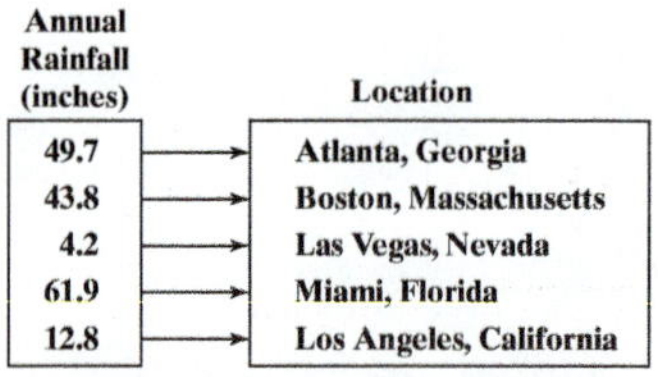

Domain: {49.7, 43.8, 4.2, 61.9, 12.8}
Range: {Atlanta, Boston, Las Vegas, Miami, Los Angeles}

29.

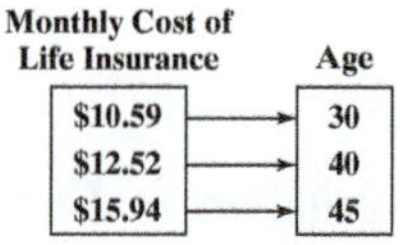

Domain: {$10.59, $12.52, $15.94}
Range: {30, 40, 45}

31. $\{(5, -3), (9, -2), (2, -1), (11, 0), (-5, 1)\}$
Domain: $\{5, 9, 2, 11, -5\}$
Range: $\{-3, -2, -1, 0, 1\}$

33. $\{(1, -2), (2, -3), (0, -10), (9, 1), (4, 2)\}$
Domain: $\{1, 2, 0, 9, 4\}$
Range: $\{-2, -3, -10, 1, 2\}$

35. $f(g(x)) = f\left(\frac{1}{3}(x-4)\right) = 3\left[\frac{1}{3}(x-4)\right] + 4 = (x-4) + 4 = x$
$g(f(x)) = g(3x+4) = \frac{1}{3}[(3x+4) - 4] = \frac{1}{3}(3x) = x$

37. $f(g(x)) = f\left(\frac{x}{4} + 2\right) = 4\left[\frac{x}{4} + 2\right] - 8 = (x+8) - 8 = x$
$g(f(x)) = g(4x-8) = \frac{4x-8}{4} + 2 = (x-2) + 2 = x$

39. $f(g(x)) = f(\sqrt[3]{x+8}) = (\sqrt[3]{x+8})^3 - 8 = (x+8) - 8 = x$
$g(f(x)) = g(x^3 - 8) = \sqrt[3]{(x^3-8) + 8} = \sqrt[3]{x^3} = x$

41. $f(g(x)) = f\left(\frac{1}{x}\right) = \frac{1}{\left(\frac{1}{x}\right)} = x; x \neq 0, g(f(x)) = g\left(\frac{1}{x}\right) = \frac{1}{\left(\frac{1}{x}\right)} = x, x \neq 0$

43. $f(g(x)) = f\left(\dfrac{4x-3}{2-x}\right) = \dfrac{2\left(\dfrac{4x-3}{2-x}\right)+3}{\dfrac{4x-3}{2-x}+4}$

$= \dfrac{2(4x-3)+3(2-x)}{4x-3+4(2-x)} = \dfrac{5x}{5} = x, x \neq 2$

$g(f(x)) = g\left(\dfrac{2x+3}{x+4}\right) = \dfrac{4\left(\dfrac{2x+3}{x+4}\right)-3}{2-\dfrac{2x+3}{x+4}}$

$= \dfrac{4(2x+3)-3(x+4)}{2(x+4)-(2x+3)} = \dfrac{5x}{5} = x, x \neq -4$

45.

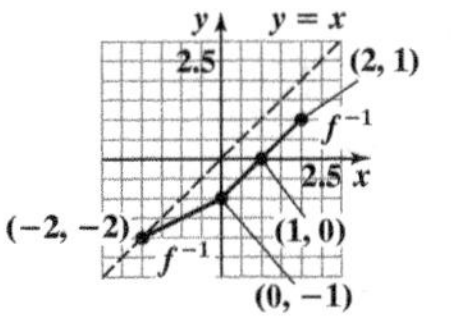

47.

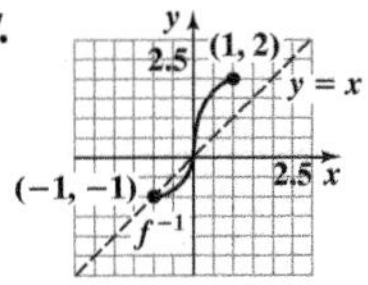

49.

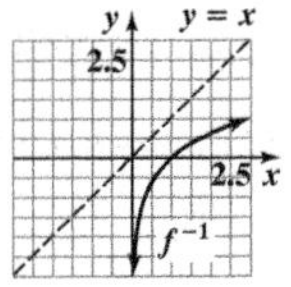

51. (a) $f^{-1}(x) = \frac{1}{3}x$

$f(f^{-1}(x)) = f\left(\frac{1}{3}x\right) = 3\left(\frac{1}{3}x\right) = x$

$f^{-1}(f(x)) = f^{-1}(3x) = \frac{1}{3}(3x) = x$

(b) Domain of f = Range of f^{-1} = All real numbers; Range of f = Domain of f^{-1} = All real numbers

(c)

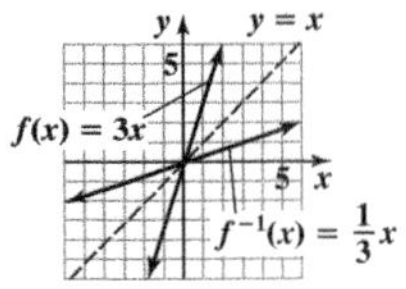

53. (a) $f^{-1}(x) = \dfrac{x}{4} - \dfrac{1}{2}$

$f(f^{-1}(x)) = f\left(\dfrac{x}{4} - \dfrac{1}{2}\right) = 4\left(\dfrac{x}{4} - \dfrac{1}{2}\right) + 2$

$= (x-2) + 2 = x$

$f^{-1}(f(x)) = f^{-1}(4x+2) = \dfrac{4x+2}{4} - \dfrac{1}{2}$

$= \left(x + \dfrac{1}{2}\right) - \dfrac{1}{2} = x$

(b) Domain of f = Range of f^{-1} = All real numbers; Range of f = Domain of f^{-1} = All real numbers

(c)

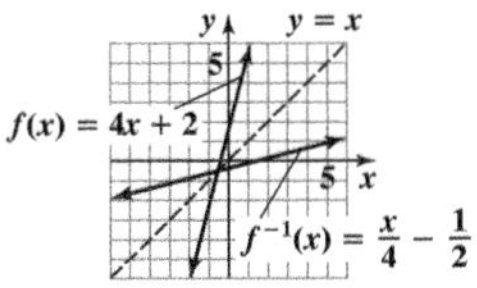

55. (a) $f^{-1}(x) = \sqrt[3]{x+1}$

$f(f^{-1}(x)) = f(\sqrt[3]{x+1})$

$= (\sqrt[3]{x+1})^3 - 1 = x$

$f^{-1}(f(x)) = f^{-1}(x^3-1)$

$= \sqrt[3]{(x^3-1)+1} = x$

(b) Domain of f = Range of f^{-1} = All real numbers; Range of f = Domain of f^{-1} = All real numbers

(c)

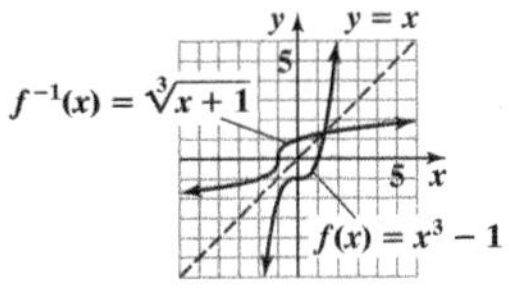

57. (a) $f^{-1}(x) = \sqrt{x-4}, x \geq 4$

$f(f^{-1}(x)) = f(\sqrt{x-4}) = (\sqrt{x-4})^2 + 4 = x$

$f^{-1}(f(x)) = f^{-1}(x^2+4) = \sqrt{(x^2+4)-4} = \sqrt{x^2} = x, x \geq 0$

(b) Domain of f = Range of f^{-1} = $\{x \mid x \geq 0\}$; Range of f = Domain of f^{-1} = $\{x \mid x \geq 4\}$

(c)

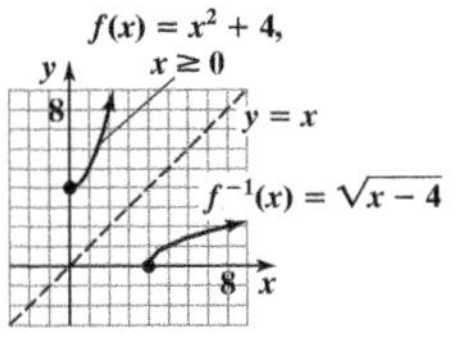

59. (a) $f^{-1}(x) = \dfrac{4}{x}$

$f(f^{-1}(x)) = f\left(\dfrac{4}{x}\right) = \dfrac{4}{\left(\dfrac{4}{x}\right)} = x$

$f^{-1}(f(x)) = f^{-1}\left(\dfrac{4}{x}\right) = \dfrac{4}{\left(\dfrac{4}{x}\right)} = x$

(b) Domain of f = Range of f^{-1} = $\{x \mid x \neq 0\}$; Range of f = Domain of f^{-1} = $\{x \mid x \neq 0\}$

(c)

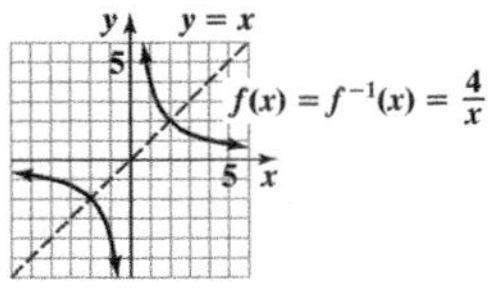

61. (a) $f^{-1}(x) = \dfrac{2x+1}{x}$

$f(f^{-1}(x)) = f\left(\dfrac{2x+1}{x}\right) = \dfrac{1}{\dfrac{2x+1}{x} - 2} = \dfrac{x}{(2x+1) - 2x} = x$

$f^{-1}(f(x)) = f^{-1}\left(\dfrac{1}{x-2}\right) = \dfrac{2\left(\dfrac{1}{x-2}\right) + 1}{\dfrac{1}{x-2}} = \dfrac{2 + (x-2)}{1} = x$

(b) Domain of f = Range of f^{-1} = $\{x \mid x \neq 2\}$; Range of f = Domain of f^{-1} = $\{x \mid x \neq 0\}$

(c)

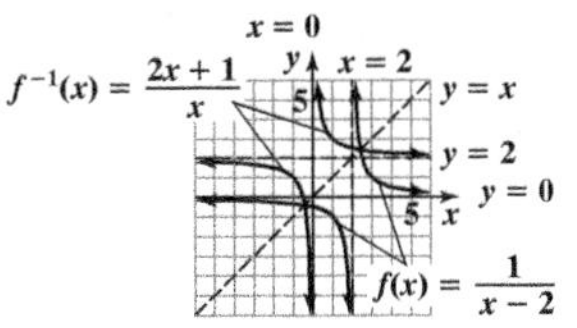

63. (a) $f^{-1}(x) = \dfrac{2-3x}{x}$

$$f(f^{-1}(x)) = f\left(\frac{2-3x}{x}\right) = \frac{2}{3+\dfrac{2-3x}{x}} = \frac{2x}{3x+2-3x} = \frac{2x}{2} = x$$

$$f^{-1}(f(x)) = f^{-1}\left(\frac{2}{3+x}\right) = \frac{2-3\left(\dfrac{2}{3+x}\right)}{\dfrac{2}{3+x}} = \frac{2(3+x)-3\cdot 2}{2} = \frac{2x}{2} = x$$

(b) Domain of f = Range of $f^{-1} = \{x \mid x \neq -3\}$; Range of f = Domain of $f^{-1} = \{x \mid x \neq 0\}$

65. (a) $f^{-1}(x) = \dfrac{-2x}{x-3}$

$$f(f^{-1}(x)) = f\left(\frac{-2x}{x-3}\right) = \frac{3\left(\dfrac{-2x}{x-3}\right)}{\dfrac{-2x}{x-3}+2} = \frac{3(-2x)}{-2x+2(x-3)} = \frac{-6x}{-6} = x$$

$$f^{-1}(f(x)) = f^{-1}\left(\frac{3x}{x+2}\right) = \frac{-2\left(\dfrac{3x}{x+2}\right)}{\dfrac{3x}{x+2}-3} = \frac{-2(3x)}{3x-3(x+2)} = \frac{-6x}{-6} = x$$

(b) Domain of f = Range of $f^{-1} = \{x \mid x \neq -2\}$; Range of f = Domain of $f^{-1} = \{x \mid x \neq 3\}$

67. (a) $f^{-1}(x) = \dfrac{x}{3x-2}$

$$f(f^{-1}(x)) = f\left(\frac{x}{3x-2}\right) = \frac{2\left(\dfrac{x}{3x-2}\right)}{3\left(\dfrac{x}{3x-2}\right)-1} = \frac{2x}{3x-(3x-2)} = \frac{2x}{2} = x$$

$$f^{-1}(f(x)) = f^{-1}\left(\frac{2x}{3x-1}\right) = \frac{\dfrac{2x}{3x-1}}{3\left(\dfrac{2x}{3x-1}\right)-2} = \frac{2x}{6x-2(3x-1)} = \frac{2x}{2} = x$$

(b) Domain of f = Range of $f^{-1} = \left\{x \,\middle|\, x \neq \frac{1}{3}\right\}$; Range of f = Domain of $f^{-1} = \left\{x \,\middle|\, x \neq \frac{2}{3}\right\}$

69. (a) $f^{-1}(x) = \dfrac{3x+4}{2x-3}$

$$f(f^{-1}(x)) = f\left(\frac{3x+4}{2x-3}\right) = \frac{3\left(\dfrac{3x+4}{2x-3}\right)+4}{2\left(\dfrac{3x+4}{2x-3}\right)-3} = \frac{3(3x+4)+4(2x-3)}{2(3x+4)-3(2x-3)} = \frac{17x}{17} = x$$

$$f^{-1}(f(x)) = f^{-1}\left(\frac{3x+4}{2x-3}\right) = \frac{3\left(\dfrac{3x+4}{2x-3}\right)+4}{2\left(\dfrac{3x+4}{2x-3}\right)-3} = \frac{3(3x+4)+4(2x-3)}{2(3x+4)-3(2x-3)} = \frac{17x}{17} = x$$

(b) Domain of f = Range of $f^{-1} = \left\{x \,\middle|\, x \neq \frac{3}{2}\right\}$; Range of f = Domain of $f^{-1} = \left\{x \,\middle|\, x \neq \frac{3}{2}\right\}$

71. (a) $f^{-1}(x) = \dfrac{-2x+3}{x-2}$

$$f(f^{-1}(x)) = f\left(\frac{-2x+3}{x-2}\right) = \frac{2\left(\dfrac{-2x+3}{x-2}\right)+3}{\dfrac{-2x+3}{x-2}+2} = \frac{2(-2x+3)+3(x-2)}{-2x+3+2(x-2)} = \frac{-x}{-1} = x$$

$$f^{-1}(f(x)) = f^{-1}\left(\frac{2x+3}{x+2}\right) = \frac{-2\left(\dfrac{2x+3}{x+2}\right)+3}{\dfrac{2x+3}{x+2}-2} = \frac{-2(2x+3)+3(x+2)}{2x+3-2(x+2)} = \frac{-x}{-1} = x$$

(b) Domain of f = Range of $f^{-1} = \{x \mid x \neq -2\}$; Range of f = Domain of $f^{-1} = \{x \mid x \neq 2\}$

73. (a) $f^{-1}(x) = \dfrac{2}{\sqrt{1-2x}}$

$$f(f^{-1}(x)) = f\left(\frac{2}{\sqrt{1-2x}}\right) = \frac{\dfrac{4}{1-2x} - 4}{2 \cdot \dfrac{4}{1-2x}} = \frac{4 - 4(1-2x)}{2 \cdot 4} = \frac{8x}{8} = x$$

$$f^{-1}(f(x)) = f^{-1}\left(\frac{x^2-4}{2x^2}\right) = \frac{2}{\sqrt{1 - 2\left(\dfrac{x^2-4}{2x^2}\right)}} = \frac{2}{\sqrt{\dfrac{4}{x^2}}} = \sqrt{x^2} = x, \text{ since } x > 0$$

(b) Domain of f = Range of $f^{-1} = \{x | x > 0\}$; Range of f = Domain of $f^{-1} = \left\{x \middle| x < \dfrac{1}{2}\right\}$

75. (a) 0 **(b)** 2 **(c)** 0 **(d)** 1 **77.** 7 **79.** Domain of f^{-1}: $[-2, \infty)$; range of f^{-1}: $[5, \infty)$ **81.** Domain of g^{-1}: $[0, \infty)$; range of g^{-1}: $(-\infty, 0]$

83. Increasing on the interval $[f(0), f(5)]$ **85.** $f^{-1}(x) = \dfrac{1}{m}(x - b), m \neq 0$ **87.** Quadrant I

89. Possible answer: $f(x) = |x|, x \geq 0$, is one-to-one; $f^{-1}(x) = x, x \geq 0$

91. (a) $r(d) = \dfrac{d + 90.39}{6.97}$

(b) $r(d(r)) = \dfrac{6.97r - 90.39 + 90.39}{6.97} = \dfrac{6.97r}{6.97} = r$

$d(r(d)) = 6.97\left(\dfrac{d + 90.39}{6.97}\right) - 90.39 = d + 90.39 - 90.39 = d$

(c) 56 miles per hour

93. (a) 77.6 kg

(b) $h(W) = \dfrac{W - 50}{2.3} + 60 = \dfrac{W + 88}{2.3}$

(c) $h(W(h)) = \dfrac{50 + 2.3(h - 60) + 88}{2.3} = \dfrac{2.3h}{2.3} = h$

$W(h(W)) = 50 + 2.3\left(\dfrac{W + 88}{2.3} - 60\right)$

$= 50 + W + 88 - 138 = W$

(d) 73 inches

95. (a) $\{g | 37{,}450 \leq g \leq 90{,}750\}$

(b) $\{T | 5156.25 \leq T \leq 18{,}481.25\}$

(c) $g(T) = \dfrac{T - 5156.25}{0.25} + 37{,}450$

Domain: $\{T | 5156.25 \leq T \leq 18{,}481.25\}$

Range: $\{g | 37{,}450 \leq g \leq 90{,}750\}$

97. (a) t represents time, so $t \geq 0$.

(b) $t(H) = \sqrt{\dfrac{H - 100}{-4.9}} = \sqrt{\dfrac{100 - H}{4.9}}$

(c) 2.02 seconds

99. $f^{-1}(x) = \dfrac{-dx + b}{cx - a}$; $f = f^{-1}$ if $a = -d$ **103.** No

107.

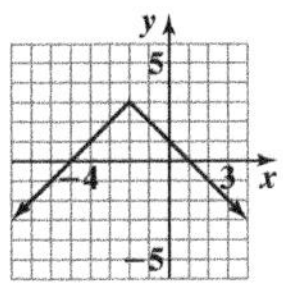

108. Zeros: $\dfrac{-5 - \sqrt{13}}{6}, \dfrac{-5 + \sqrt{13}}{6}$, x-intercepts: $\dfrac{-5 - \sqrt{13}}{6}, \dfrac{-5 + \sqrt{13}}{6}$

109. Domain: $\left\{x \middle| x \neq -\dfrac{3}{2}, x \neq 2\right\}$; Vertical asymptote: $x = -\dfrac{3}{2}$,

Horizontal asymptote: $y = 3$

110. $6xh + 3h^2 - 7h$

6.3 Assess Your Understanding *(page 439)*

6. Exponential function; growth factor; initial value **7.** a **8.** T **9.** T **10.** $\left(-1, \dfrac{1}{a}\right)$; $(0, 1)$; $(1, a)$ **11.** 4 **12.** F **13.** b **14.** c

15. (a) 8.815 **(b)** 8.821 **(c)** 8.824 **(d)** 8.825 **17. (a)** 21.217 **(b)** 22.217 **(c)** 22.440 **(d)** 22.459 **19.** 1.265 **21.** 0.347 **23.** 3.320 **25.** 149.952

27. Neither **29.** Exponential; $H(x) = 4^x$ **31.** Exponential; $f(x) = 3(2^x)$ **33.** Linear; $H(x) = 2x + 4$ **35.** B **37.** D **39.** A **41.** E

43.

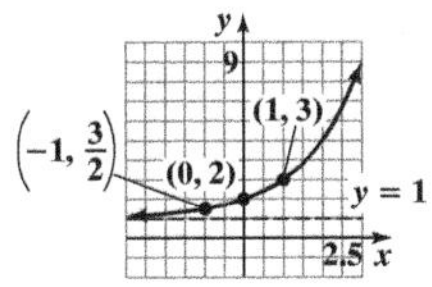

Domain: All real numbers

Range: $\{y | y > 1\}$ or $(1, \infty)$

Horizontal asymptote: $y = 1$

45.

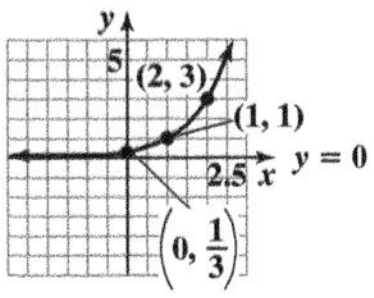

Domain: All real numbers

Range: $\{y | y > 0\}$ or $(0, \infty)$

Horizontal asymptote: $y = 0$

47.

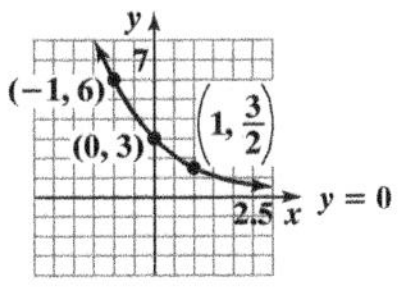

Domain: All real numbers

Range: $\{y | y > 0\}$ or $(0, \infty)$

Horizontal asymptote: $y = 0$

49.

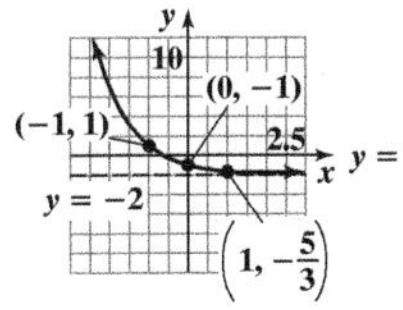

Domain: All real numbers

Range: $\{y | y > -2\}$ or $(-2, \infty)$

Horizontal asymptote: $y = -2$

51.

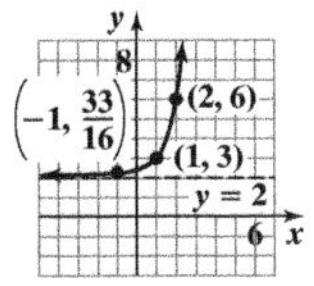

Domain: All real numbers

Range: $\{y | y > 2\}$ or $(2, \infty)$

Horizontal asymptote: $y = 2$

53.

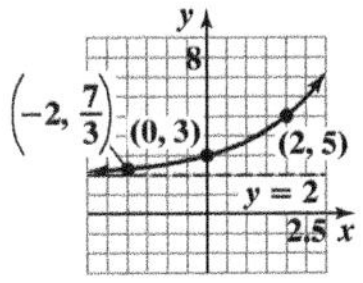

Domain: All real numbers

Range: $\{y | y > 2\}$ or $(2, \infty)$

Horizontal asymptote: $y = 2$

55.

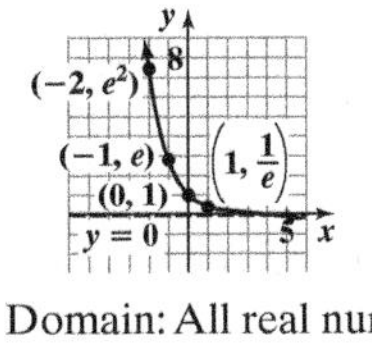

Domain: All real numbers

Range: $\{y | y > 0\}$ or $(0, \infty)$

Horizontal asymptote: $y = 0$

57.

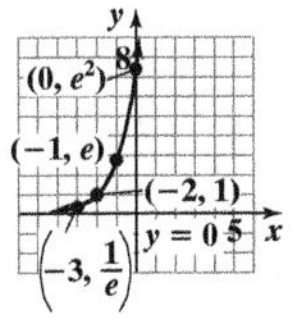

Domain: All real numbers

Range: $\{y | y > 0\}$ or $(0, \infty)$

Horizontal asymptote: $y = 0$

59.

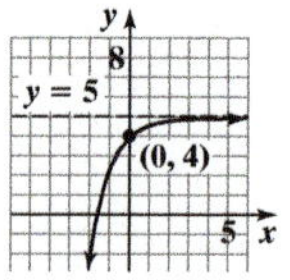

Domain: All real numbers
Range: $\{y \mid y < 5\}$ or $(-\infty, 5)$
Horizontal asymptote: $y = 5$

61.

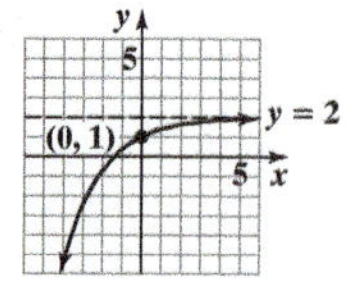

Domain: All real numbers
Range: $\{y \mid y < 2\}$ or $(-\infty, 2)$
Horizontal asymptote: $y = 2$

63. $\{3\}$ **65.** $\{-4\}$ **67.** $\{2\}$ **69.** $\left\{\frac{3}{2}\right\}$ **71.** $\{-\sqrt{2}, 0, \sqrt{2}\}$ **73.** $\{6\}$ **75.** $\{-1, 7\}$ **77.** $\{-4, 2\}$ **79.** $\{-4\}$ **81.** $\{1, 2\}$ **83.** $\frac{1}{49}$ **85.** $\frac{1}{4}$ **87.** 5 **89.** $f(x) = 3^x$ **91.** $f(x) = -6^x$ **93.** $f(x) = 3^x + 2$ **95.** **(a)** 16; $(4, 16)$ **(b)** -4; $\left(-4, \frac{1}{16}\right)$ **97.** **(a)** $\frac{9}{4}$; $\left(-1, \frac{9}{4}\right)$ **(b)** 3; $(3, 66)$ **99.** **(a)** 60; $(-6, 60)$ **(b)** -4; $(-4, 12)$ **(c)** -2

101.

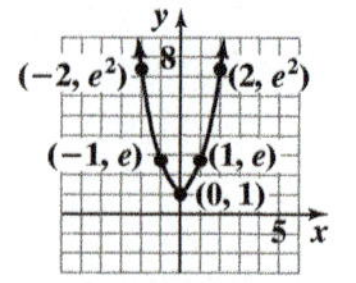

Domain: $(-\infty, \infty)$
Range: $[1, \infty)$
Intercept: $(0, 1)$

103.

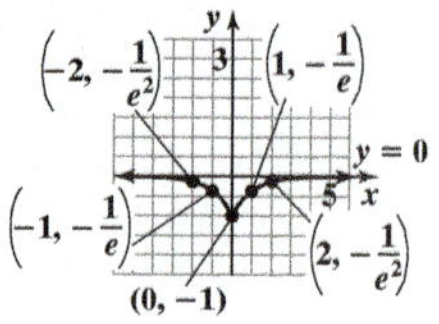

Domain: $(-\infty, \infty)$
Range: $[-1, 0)$
Intercept: $(0, -1)$

105. **(a)** 74% **(b)** 47% **(c)** Each pane allows only 97% of light to pass through. **107.** **(a)** \$16,231 **(b)** \$8626 **(c)** As each year passes, the sedan is worth 90% of its value the previous year. **109.** **(a)** 30% **(b)** 9% **(c)** Each year only 30% of the previous survivors survive again. **111.** 3.35 mg; 0.45 mg **113.** **(a)** 0.632 **(b)** 0.982 **(c)** 1 **(d)**

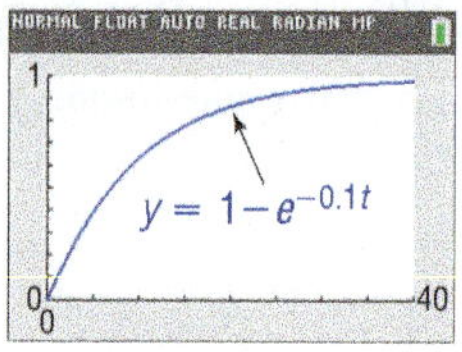

(e) About 7 min

115. **(a)** 0.0516 **(b)** 0.0888 **117.** **(a)** 70.95% **(b)** 72.62% **(c)** 100% **119.** **(a)** 5.41 amp, 7.59 amp, 10.38 amp **(b)** 12 amp **(c)**, **(f)**

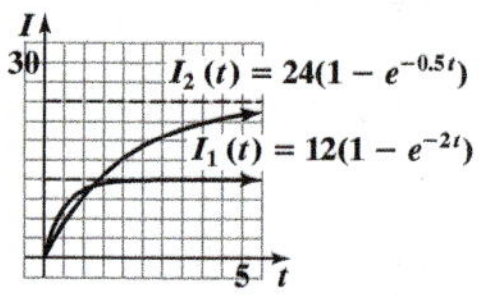

(d) 3.34 amp, 5.31 amp, 9.44 amp
(e) 24 amp

121. 36 **123.**

Final Denominator	Value of Expression	Compare Value to $e \approx 2.718281828$
1 + 1	2.5	$2.5 < e$
2 + 2	2.8	$2.8 > e$
3 + 3	2.7	$2.7 < e$
4 + 4	2.721649485	$2.721649485 > e$
5 + 5	2.717770035	$2.717770035 < e$
6 + 6	2.718348855	$2.718348855 > e$

125. $f(A + B) = a^{A+B} = a^A \cdot a^B = f(A) \cdot f(B)$ **127.** $f(\alpha x) = a^{\alpha x} = (a^x)^\alpha = [f(x)]^\alpha$

129. **(a)**
$$f(-x) = \frac{1}{2}(e^{-x} + e^{-(-x)}) = \frac{1}{2}(e^{-x} + e^{x}) = \frac{1}{2}(e^{x} + e^{-x}) = f(x)$$

(b)

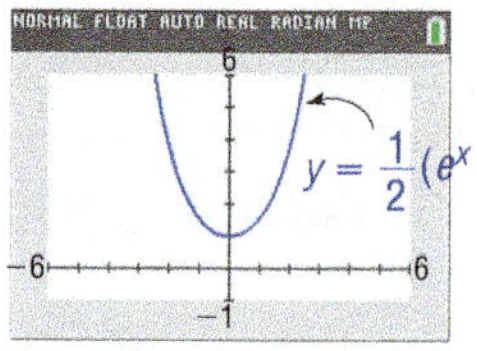

(c)
$$(\cosh x)^2 - (\sinh x)^2 = \left[\frac{1}{2}(e^x + e^{-x})\right]^2 - \left[\frac{1}{2}(e^x - e^{-x})\right]^2 = \frac{1}{4}[e^{2x} + 2 + e^{-2x} - e^{2x} + 2 - e^{-2x}] = \frac{1}{4}(4) = 1$$

131. 59 minutes **135.** $a^{-x} = (a^{-1})^x = \left(\frac{1}{a}\right)^x$ **136.** $(-\infty, -5] \cup [-2, 2]$ **137.** $(2, \infty)$ **138.** $f(x) = -2x^2 + 12x - 13$

139. **(a)**

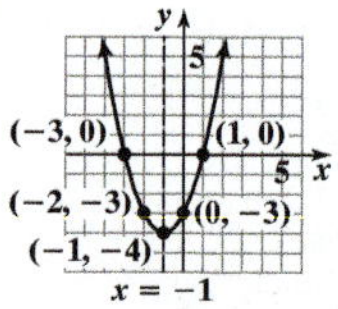

(b) Domain: $(-\infty, \infty)$; Range: $[-4, \infty)$
(c) Decreasing: $(-\infty, -1]$; Increasing: $[-1, \infty)$

6.4 Assess Your Understanding *(page 454)*

4. $\{x \mid x > 0\}$ or $(0, \infty)$ **5.** $\left(\frac{1}{a}, -1\right)$, $(1, 0)$, $(a, 1)$ **6.** 1 **7.** F **8.** T **9.** a **10.** c **11.** $\log_3 9 = 2$ **13.** $\log_a 1.6 = 2$ **15.** $\log_2 7.2 = x$ **17.** $\ln 8 = x$ **19.** $2^3 = 8$ **21.** $a^6 = 3$ **23.** $3^x = 2$ **25.** $e^x = 4$ **27.** 0 **29.** 2 **31.** -4 **33.** $\frac{1}{2}$ **35.** 4 **37.** $\frac{1}{2}$ **39.** $\{x \mid x > 3\}$; $(3, \infty)$ **41.** All real numbers except 0; $\{x \mid x \neq 0\}$; $(-\infty, 0) \cup (0, \infty)$ **43.** $\{x \mid x > 10\}$; $(10, \infty)$ **45.** $\{x \mid x > -1\}$; $(-1, \infty)$ **47.** $\{x \mid x < -1 \text{ or } x > 0\}$; $(-\infty, -1) \cup (0, \infty)$ **49.** $\{x \mid x \geq 1\}$; $[1, \infty)$ **51.** 0.511 **53.** 30.099 **55.** 2.303 **57.** -53.991 **59.** $\sqrt{2}$

61.

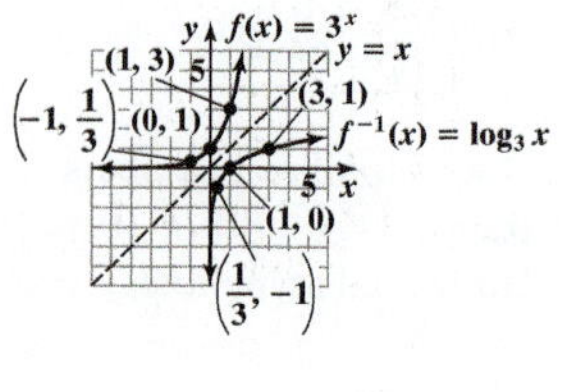

63.

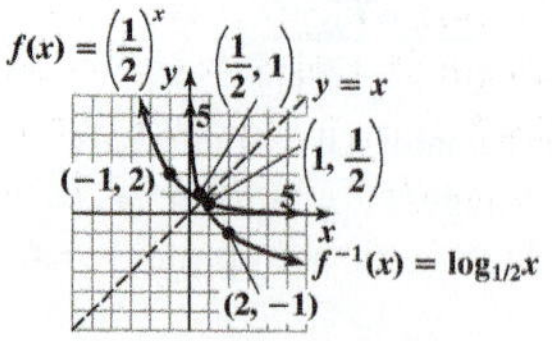

65. B **67.** D **69.** A **71.** E

73. **(a)** Domain: $(-4, \infty)$

(b)

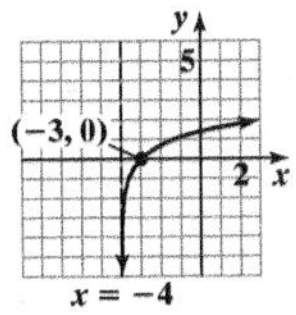

(c) Range: $(-\infty, \infty)$
Vertical asymptote: $x = -4$
(d) $f^{-1}(x) = e^x - 4$
(e) Domain of f^{-1}: $(-\infty, \infty)$
Range of f^{-1}: $(-4, \infty)$

(f)

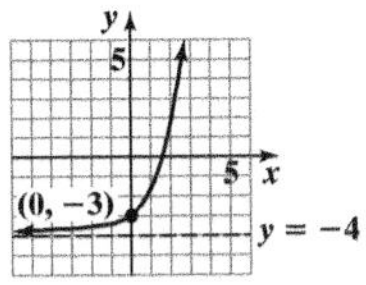

75. **(a)** Domain: $(0, \infty)$

(b)

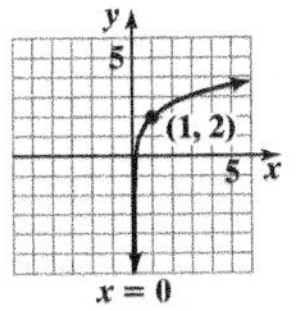

(c) Range: $(-\infty, \infty)$
Vertical asymptote: $x = 0$
(d) $f^{-1}(x) = e^{x-2}$
(e) Domain of f^{-1}: $(-\infty, \infty)$
Range of f^{-1}: $(0, \infty)$

(f)

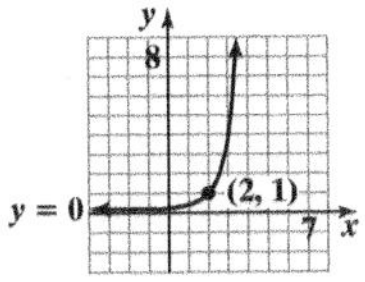

77. **(a)** Domain: $(0, \infty)$

(b)

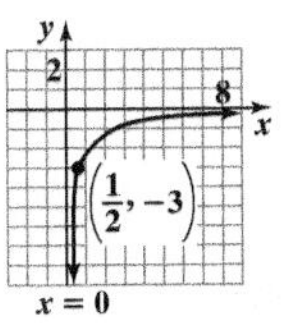

(c) Range: $(-\infty, \infty)$
Vertical asymptote: $x = 0$
(d) $f^{-1}(x) = \frac{1}{2}e^{x+3}$
(e) Domain of f^{-1}: $(-\infty, \infty)$
Range of f^{-1}: $(0, \infty)$

(f)

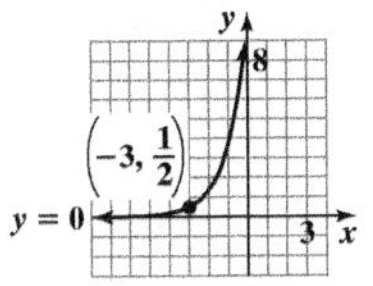

79. **(a)** Domain: $(4, \infty)$

(b)

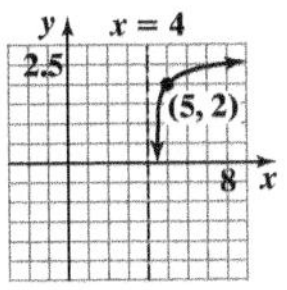

(c) Range: $(-\infty, \infty)$
Vertical asymptote: $x = 4$
(d) $f^{-1}(x) = 10^{x-2} + 4$
(e) Domain of f^{-1}: $(-\infty, \infty)$
Range of f^{-1}: $(4, \infty)$

(f)

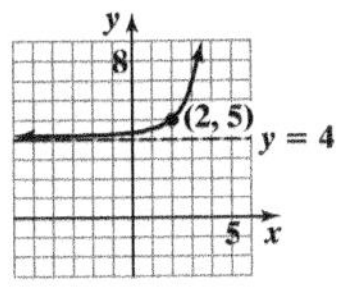

81. **(a)** Domain: $(0, \infty)$

(b)

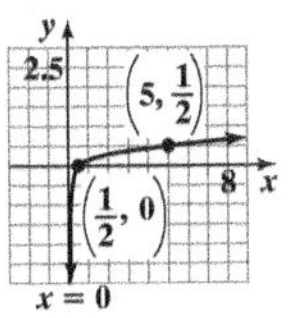

(c) Range: $(-\infty, \infty)$
Vertical asymptote: $x = 0$
(d) $f^{-1}(x) = \frac{1}{2} \cdot 10^{2x}$
(e) Domain of f^{-1}: $(-\infty, \infty)$
Range of f^{-1}: $(0, \infty)$

(f)

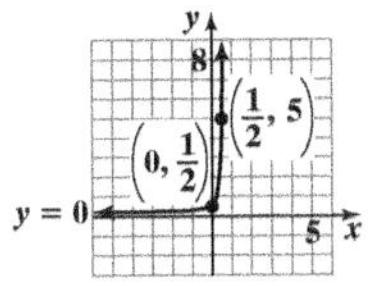

83. **(a)** Domain: $(-2, \infty)$

(b)

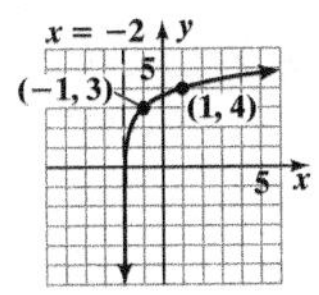

(c) Range: $(-\infty, \infty)$
Vertical asymptote: $x = -2$
(d) $f^{-1}(x) = 3^{x-3} - 2$
(e) Domain of f^{-1}: $(-\infty, \infty)$
Range of f^{-1}: $(-2, \infty)$

(f)

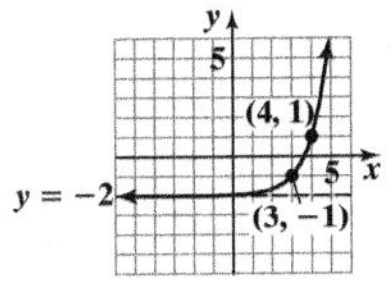

85. **(a)** Domain: $(-\infty, \infty)$

(b)

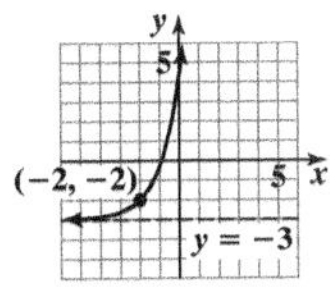

(c) Range: $(-3, \infty)$
Horizontal asymptote: $y = -3$
(d) $f^{-1}(x) = \ln(x + 3) - 2$
(e) Domain of f^{-1}: $(-3, \infty)$
Range of f^{-1}: $(-\infty, \infty)$

(f)

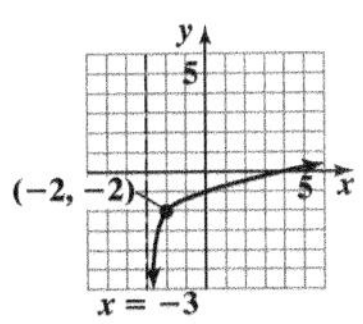

87. **(a)** Domain: $(-\infty, \infty)$

(b)

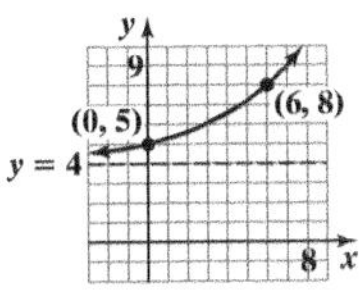

(c) Range: $(4, \infty)$
Horizontal asymptote: $y = 4$
(d) $f^{-1}(x) = 3\log_2(x - 4)$
(e) Domain of f^{-1}: $(4, \infty)$
Range of f^{-1}: $(-\infty, \infty)$

(f)

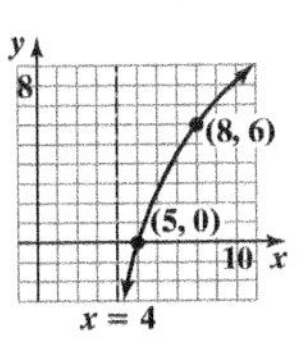

89. $\{9\}$ **91.** $\left\{\frac{7}{2}\right\}$ **93.** $\{2\}$ **95.** $\{5\}$ **97.** $\{3\}$

99. $\{2\}$ **101.** $\left\{\frac{\ln 10}{3}\right\}$ **103.** $\left\{\frac{\ln 8 - 5}{2}\right\}$

105. $\{-2\sqrt{2}, 2\sqrt{2}\}$

107. $\{-1\}$

109. $\left\{5 \ln \frac{7}{5}\right\}$

111. $\left\{2 - \log \frac{5}{2}\right\}$

113. **(a)** $\left\{x \middle| x > -\frac{1}{2}\right\}; \left(-\frac{1}{2}, \infty\right)$
(b) 2; (40, 2) **(c)** 121; (121, 3) **(d)** 4

115.

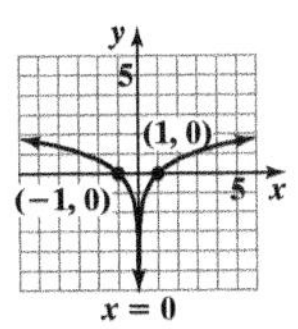

Domain: $\{x | x \neq 0\}$
Range: $(-\infty, \infty)$
Intercepts: $(-1, 0)$, $(1, 0)$

117.

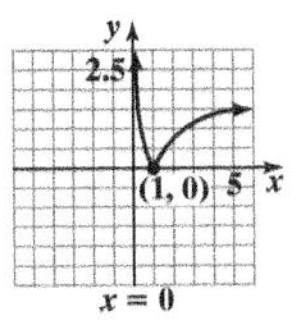

Domain: $\{x | x > 0\}$
Range: $\{y | y \geq 0\}$
Intercept: $(1, 0)$

119. **(a)** 1 **(b)** 2 **(c)** 3
(d) It increases. **(e)** 0.000316
(f) 3.981×10^{-8}

121. **(a)** 5.97 km **(b)** 0.90 km

123. **(a)** 6.93 min **(b)** 16.09 min

125. $h \approx 2.29$, so the time between injections is about 2 h, 17 min.

127. 0.2695 s
0.8959 s

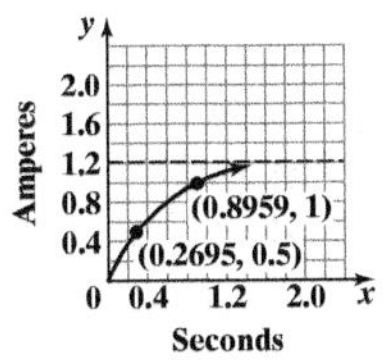

129. 50 decibels (dB) **131.** 90 dB **133.** 8.1 **135. (a)** $k \approx 11.216$ **(b)** 6.73 **(c)** 0.41% **(d)** 0.14%

137. Because $y = \log_1 x$ means $1^y = 1 = x$, which cannot be true for $x \neq 1$ **139.** Zeros: $-3, -\frac{1}{2}, \frac{1}{2}, 3$; x-intercepts: $-3, -\frac{1}{2}, \frac{1}{2}, 3$
140. 12 **141.** $f(1) = -5; f(2) = 17$ **142.** $3 + i; f(x) = x^4 - 7x^3 + 14x^2 + 2x - 20; a = 1$

6.5 Assess Your Understanding *(page 465)*

1. 0 **2.** M **3.** r **4.** $\log_a M; \log_a N$ **5.** $\log_a M; \log_a N$ **6.** $r \log_a M$ **7.** 7 **8.** F **9.** F **10.** F **11.** b **12.** b **13.** 71 **15.** −4 **17.** 7 **19.** 1 **21.** 1 **23.** 3 **25.** $\frac{5}{4}$ **27.** 4 **29.** $a + b$ **31.** $b - a$ **33.** $3a$ **35.** $\frac{1}{5}(a + b)$ **37.** $2 + \log_5 x$ **39.** $3 \log_2 z$ **41.** $1 + \ln x$ **43.** $\ln x - x$ **45.** $2 \log_a u + 3 \log_a v$ **47.** $2 \ln x + \frac{1}{2}\ln(1 - x)$ **49.** $3 \log_2 x - \log_2(x - 3)$ **51.** $\log x + \log(x + 2) - 2\log(x + 3)$ **53.** $\frac{1}{3}\ln(x - 2) + \frac{1}{3}\ln(x + 1) - \frac{2}{3}\ln(x + 4)$ **55.** $\ln 5 + \ln x + \frac{1}{2}\ln(1 + 3x) - 3\ln(x - 4)$ **57.** $\log_5(u^3v^4)$ **59.** $\log_3\left(\frac{1}{x^{5/2}}\right)$ **61.** $\log_4\left[\frac{x - 1}{(x + 1)^4}\right]$ **63.** $-2\ln(x - 1)$ **65.** $\log_2[x(3x - 2)^4]$ **67.** $\log_a\left(\frac{25x^6}{\sqrt{2x + 3}}\right)$ **69.** $\log_2\left[\frac{(x + 1)^2}{(x + 3)(x - 1)}\right]$ **71.** 2.771 **73.** −3.880 **75.** 5.615 **77.** 0.874

79. $y = \frac{\log x}{\log 4}$

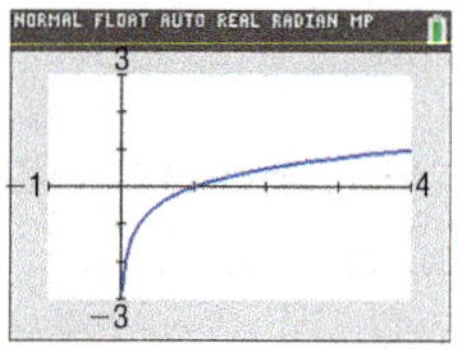

81. $y = \frac{\log(x + 2)}{\log 2}$

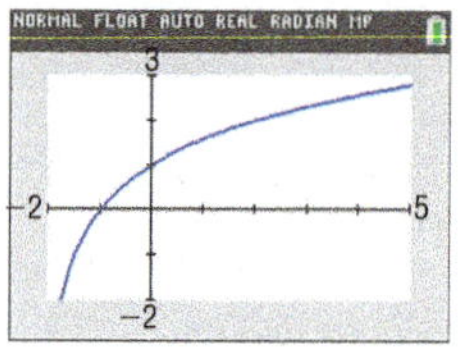

83. $y = \frac{\log(x + 1)}{\log(x - 1)}$

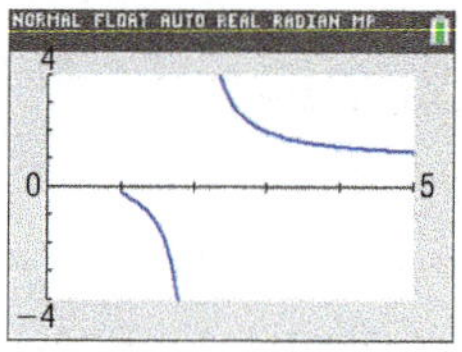

85. (a) $(f \circ g)(x) = x$; $\{x \mid x \text{ is any real number}\}$ or $(-\infty, \infty)$
(b) $(g \circ f)(x) = x$; $\{x \mid x > 0\}$ or $(0, \infty)$ **(c)** 5
(d) $(f \circ h)(x) = \ln x^2$; $\{x \mid x \neq 0\}$ or $(-\infty, 0) \cup (0, \infty)$ **(e)** 2

87. $y = Cx$ **89.** $y = Cx(x + 1)$ **91.** $y = Ce^{3x}$ **93.** $y = Ce^{-4x} + 3$

95. $y = \frac{\sqrt[3]{C}(2x + 1)^{1/6}}{(x + 4)^{1/9}}$ **97.** 3 **99.** 1

101. $\log_a(x + \sqrt{x^2 - 1}) + \log_a(x - \sqrt{x^2 - 1}) = \log_a[(x + \sqrt{x^2 - 1})(x - \sqrt{x^2 - 1})] = \log_a[x^2 - (x^2 - 1)] = \log_a 1 = 0$
103. $\ln(1 + e^{2x}) = \ln[e^{2x}(e^{-2x} + 1)] = \ln e^{2x} + \ln(e^{-2x} + 1) = 2x + \ln(1 + e^{-2x})$

105. $y = f(x) = \log_a x$; $a^y = x$ implies $a^y = \left(\frac{1}{a}\right)^{-y} = x$, so $-y = \log_{1/a} x = -f(x)$.

107. $f(x) = \log_a x$; $f\left(\frac{1}{x}\right) = \log_a \frac{1}{x} = \log_a 1 - \log_a x = -f(x)$

109. $\log_a \frac{M}{N} = \log_a(M \cdot N^{-1}) = \log_a M + \log_a N^{-1} = \log_a M - \log_a N$, since $a^{\log_a N^{-1}} = N^{-1}$ implies $a^{-\log_a N^{-1}} = N$; that is, $\log_a N = -\log_a N^{-1}$.

115. $\{-1.78, 1.29, 3.49\}$ **116.** A repeated real solution (double root) **117.** $-2, \frac{1}{5}, \frac{-5 - \sqrt{21}}{2}, \frac{-5 + \sqrt{21}}{2}$
118.

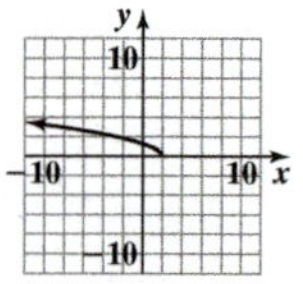

Domain: $\{x \mid x \leq 2\}$ or $(-\infty, 2]$
Range: $\{y \mid y \geq 0\}$ or $[0, \infty)$

6.6 Assess Your Understanding *(page 472)*

5. $\{16\}$ **7.** $\left\{\frac{16}{5}\right\}$ **9.** $\{6\}$ **11.** $\{16\}$ **13.** $\left\{\frac{1}{3}\right\}$ **15.** $\{3\}$ **17.** $\{5\}$ **19.** $\left\{\frac{21}{8}\right\}$ **21.** $\{-6\}$ **23.** $\{-2\}$ **25.** $\{-1 + \sqrt{1 + e^4}\} \approx \{6.456\}$ **27.** $\left\{\frac{-5 + 3\sqrt{5}}{2}\right\} \approx \{0.854\}$ **29.** $\{2\}$ **31.** $\left\{\frac{9}{2}\right\}$ **33.** $\{7\}$ **35.** $\{-2 + 4\sqrt{2}\}$ **37.** $\{-\sqrt{3}, \sqrt{3}\}$ **39.** $\left\{\frac{1}{3}, 729\right\}$ **41.** $\{8\}$

43. $\{\log_2 10\} = \left\{\frac{\ln 10}{\ln 2}\right\} \approx \{3.322\}$ **45.** $\{-\log_8 1.2\} = \left\{-\frac{\ln 1.2}{\ln 8}\right\} \approx \{-0.088\}$ **47.** $\left\{\frac{1}{3}\log_2 \frac{8}{5}\right\} = \left\{\frac{\ln \frac{8}{5}}{3 \ln 2}\right\} \approx \{0.226\}$

49. $\left\{\frac{\ln 3}{2\ln 3 + \ln 4}\right\} \approx \{0.307\}$ **51.** $\left\{\frac{\ln 7}{\ln 0.6 + \ln 7}\right\} \approx \{1.356\}$ **53.** $\{0\}$ **55.** $\left\{\frac{\ln \pi}{1 + \ln \pi}\right\} \approx \{0.534\}$ **57.** $\left\{\frac{\ln 3}{\ln 2}\right\} \approx \{1.585\}$

59. $\{0\}$ **61.** $\left\{\log_4(-2 + \sqrt{7})\right\} \approx \{-0.315\}$ **63.** $\{\log_5 4\} \approx \{0.861\}$ **65.** No real solution **67.** $\{\log_4 5\} \approx \{1.161\}$ **69.** $\{2.79\}$

71. $\{-0.57\}$ **73.** $\{-0.70\}$ **75.** $\{0.57\}$ **77.** $\{0.39, 1.00\}$ **79.** $\{1.32\}$ **81.** $\{1.31\}$ **83.** $\{1\}$ **85.** $\{16\}$ **87.** $\left\{-1, \frac{2}{3}\right\}$ **89.** $\{0\}$

91. $\{\ln(2+\sqrt{5})\} \approx \{1.444\}$ **93.** $\left\{e^{\frac{\ln 5 \cdot \ln 3}{\ln 15}}\right\} \approx \{1.921\}$ **95.(a)** $\{5\}$; $(5, 3)$ **(b)** $\{5\}$; $(5, 4)$ **(c)** $\{1\}$; yes, at $(1, 2)$ **(d)** $\{5\}$ **(e)** $\left\{-\frac{1}{11}\right\}$

97. (a), (b)

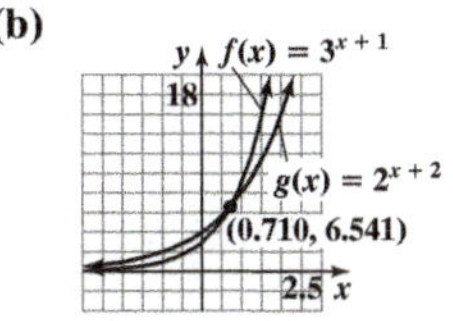

(c) $\{x \mid x > 0.710\}$ or $(0.710, \infty)$

99. (a), (b), (c)

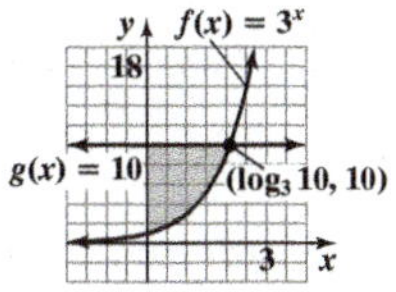

101. (a), (b), (c)

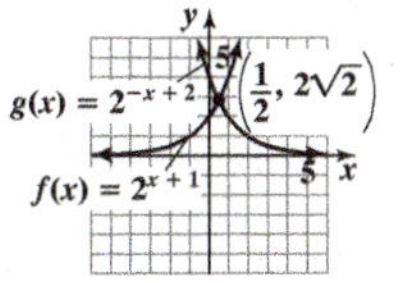

103. (a)

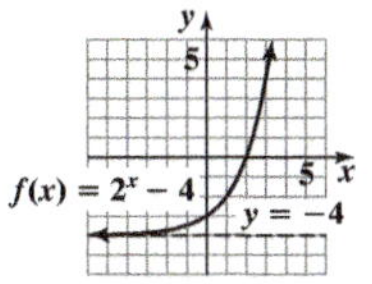

(b) 2 **(c)** $\{x \mid x < 2\}$ or $(-\infty, 2)$

105. (a) 2047 **(b)** 2059

107. (a) After 4.2 yr
(b) After 6.5 yr
(c) After 12.8 yr

110. $\left\{-3, \frac{1}{4}, 2\right\}$

111. one-to-one

112. $(f \circ g)(x) = \dfrac{x+5}{-x+11}$; $\{x \mid x \neq 3, x \neq 11\}$

113. $\{x \mid x \geq 1\}$, or $[1, \infty)$

6.7 Assess Your Understanding *(page 481)*

3. principal **4.** *I; Prt;* simple interest **5.** 4 **6.** effective rate of interest **7.** \$108.29 **9.** \$609.50 **11.** \$697.09 **13.** \$1246.08 **15.** \$88.72 **17.** \$860.72 **19.** \$554.09 **21.** \$59.71 **23.** 5.095% **25.** 5.127% **27.** $6\frac{1}{4}$% compounded annually **29.** 9% compounded monthly **31.** 25.992% **33.** 24.573% **35. (a)** About 8.69 yr **(b)** About 8.66 yr **37.** 6.823% **39.** 10.15 yr; 10.14 yr **41.** 15.27 yr or 15 yr, 3 mo **43.** \$104,335 **45.** \$12,910.62 **47.** About \$30.17 per share or \$3017 **49.** Not quite. Jim will have \$1057.60. The second bank gives a better deal, since Jim will have \$1060.62 after 1 yr. **51.** Will has \$11,632.73; Henry has \$10,947.89. **53.(a)** \$64,589 **(b)** \$45,062 **55.** About \$1020 billion; about \$233 billion **57.** \$940.90 **59.** 2.53% **61.** 34.31 yr **63. (a)** \$3686.45 **(b)** \$3678.79 **65.** \$6439.28

67. (a) 11.90 yr **(b)** 22.11 yr **(c)**

$$mP = P\left(1 + \frac{r}{n}\right)^{nt}$$
$$m = \left(1 + \frac{r}{n}\right)^{nt}$$
$$\ln m = \ln\left(1 + \frac{r}{n}\right)^{nt} = nt \ln\left(1 + \frac{r}{n}\right)$$
$$t = \frac{\ln m}{n \ln\left(1 + \frac{r}{n}\right)}$$

69. (a) 1.99% **(b)** In 2026 or after 17 yr **71.** 22.7 yr **76.** $R = 0$; yes **77.** $f^{-1}(x) = \dfrac{2x}{x-1}$ **78.** $-2, 5$; $f(x) = (x+2)^2(x-5)(x^2+1)$ **79.** $\{6\}$

6.8 Assess Your Understanding *(page 492)*

1. (a) 500 insects **(b)** $0.02 = 2\%$ per day **(c)**

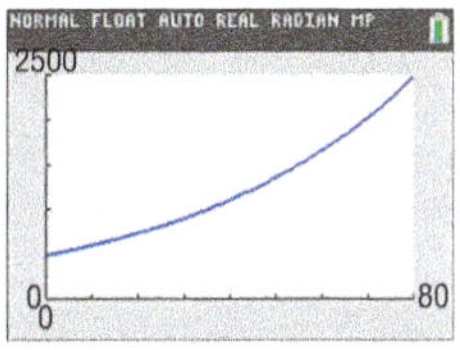

(d) About 611 insects **(e)** After about 23.5 days **(f)** After about 34.7 days

3. (a) $-0.0244 = -2.44\%$ per year **(b)**

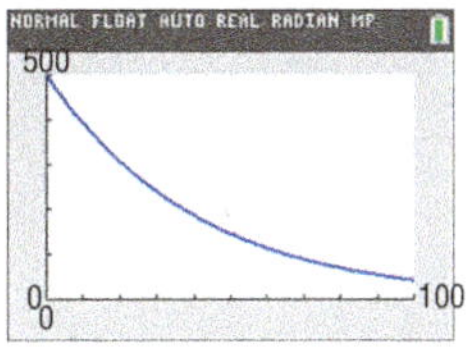

(c) About 391.7 g **(d)** After about 9.1 yr **(e)** 28.4 yr

5. (a) $N(t) = N_0 e^{kt}$ **(b)** 5832 **(c)** 3.9 days

7. (a) $N(t) = N_0 e^{kt}$ **(b)** 25,198 **9.** 9.797 g

11. (a) 9953 years ago
(b)

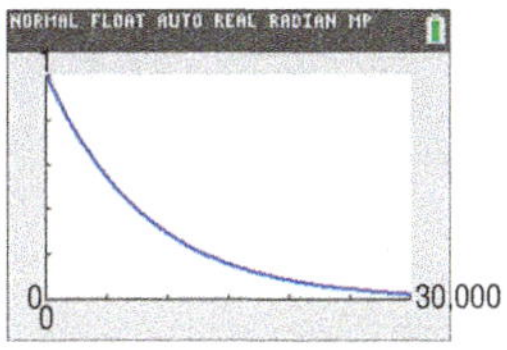

(c) 5730 yr

13. (a) 5:18 PM
(b)

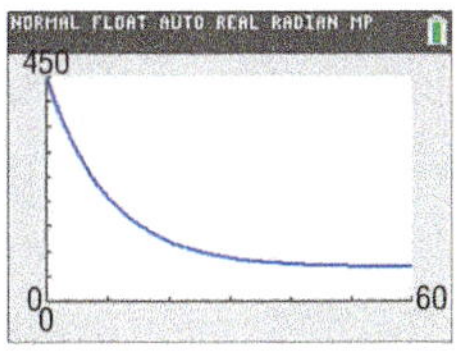

(c) About 14.3 min **(d)** The temperature of the pizza approaches 70°F.

15. (a) 18.63°C; 25.07°C
(b)

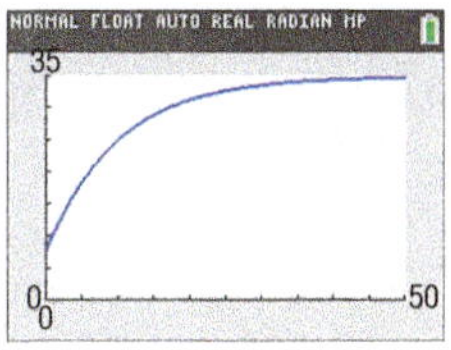

17. 1.7 ppm; 7.17 days, or 172 h **19.** 0.26 M; 6.58 hr, or 395 min **21.** 26.6 days

23. (a) 1000 **(b)** 43.9% **(c)** 30 g

(d)

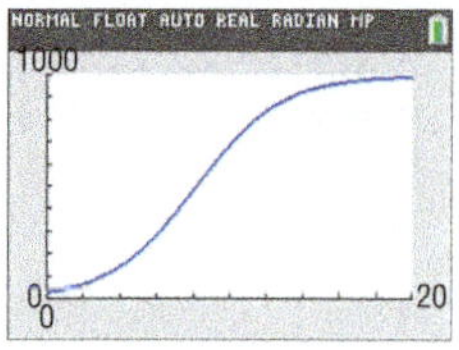

(e) 616.6 g **(f)** After 9.85 h **(g)** About 7.9 h

25. (a) $P(0) \approx 48$; In 1900, about 48 invasive species were present in the Great Lakes. **(b)** 1.7%

(c)

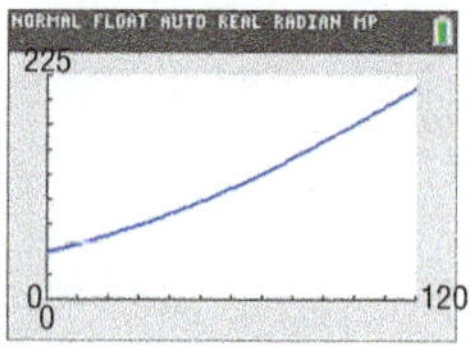

(d) About 176 **(e)** During 1999

27. (a) In 1984, 91.8% of households did not own a personal computer.

(b)

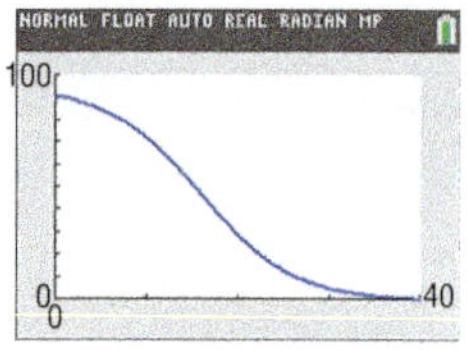

(c) 70.6% **(d)** During 2011

29. (a)

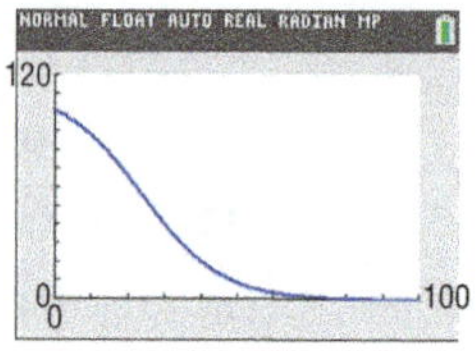

(b) 0.78, or 78%
(c) 50 people
(d) As n increases, the probability decreases.

31. (a) $P(t) = 25{,}000(2)^{t/8}$

(b)

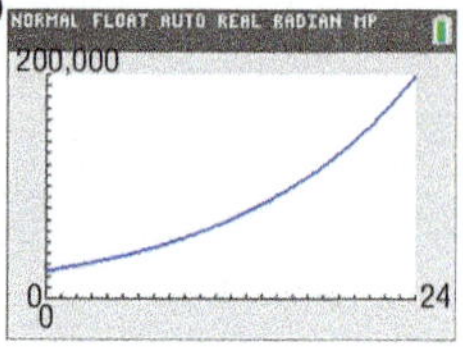

(c) 32,421 people **(d)** In about 13.42 yr
(e) $P(t) = 25{,}000e^{0.087t}$

33. $f(x) = -\frac{3}{2}x + 7$ **34.** Neither **35.** $2 \ln x + \frac{1}{2} \ln y - \ln z$ **36.** $2\sqrt[3]{5}$

6.9 Assess Your Understanding *(page 500)*

1. (a)

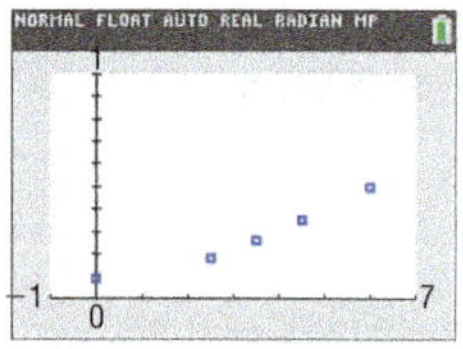

(b) $y = 0.0903(1.3384)^x$
(c) $N(t) = 0.0903e^{0.2915t}$

(d)

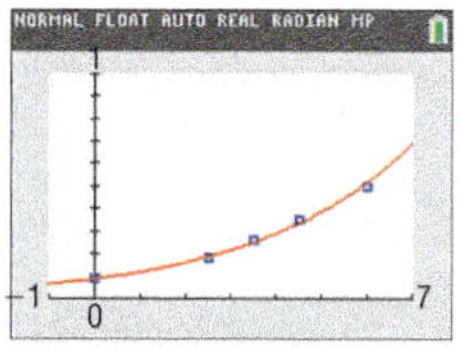

(e) 0.69
(f) After about 7.26 hr

3. (a)

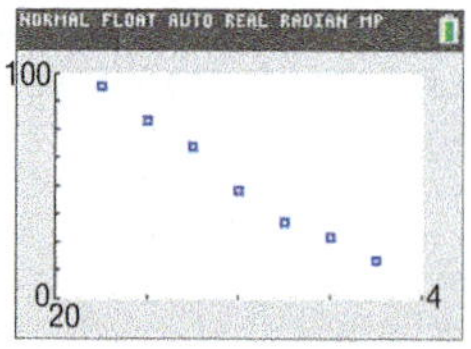

(b) $y = 118.7226(0.7013)^x$
(c) $A(t) = 118.7226e^{-0.3548t}$

(d)

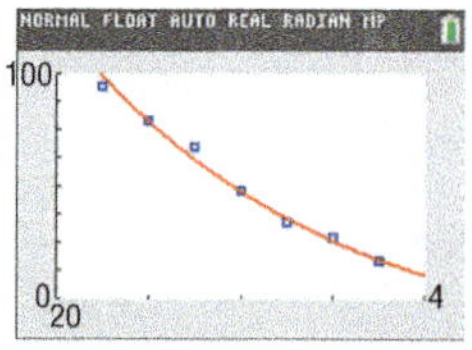

(e) 28.7%
(f) $k = -0.3548 = -35.48\%$ is the exponential growth rate. It represents the rate at which the percent of patients surviving advanced-stage breast cancer is decreasing.

5. (a)

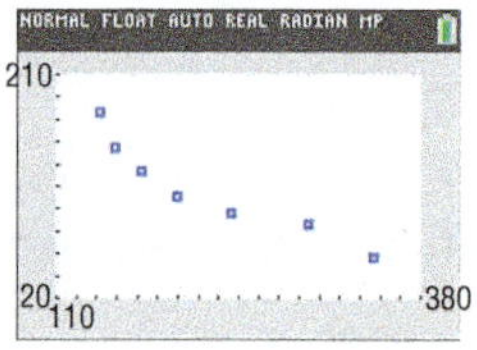

(b) $y = 330.0549 - 34.5008 \ln x$

(c)

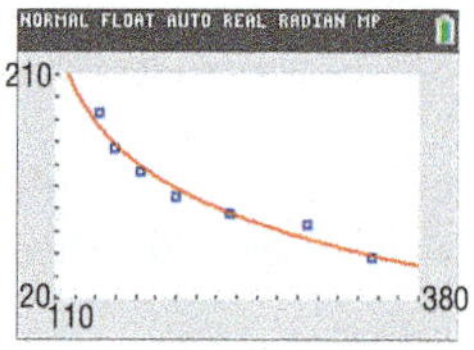

(d) 185 billion pounds
(e) Under by 5 billion pounds

7. (a)

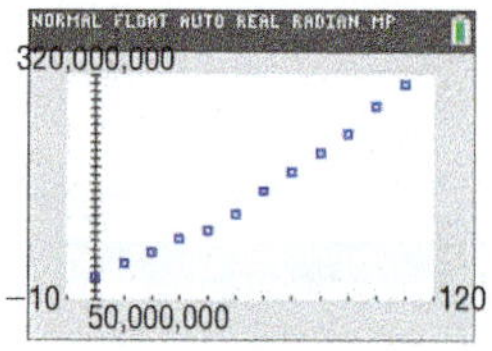

(b) $y = \dfrac{762{,}176{,}844.4}{1 + 8.7428e^{-0.0162x}}$

(c)

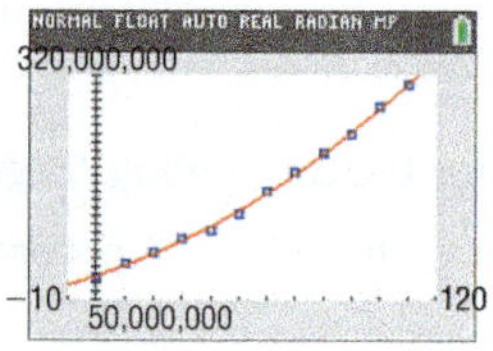

(d) 762,176,844
(e) Approximately 315,203,288 **(f)** 2023

9. (a)

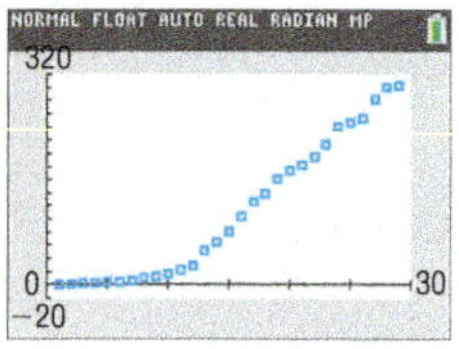

(b) $y = \dfrac{321.0384}{1 + 135.3081e^{-0.2516x}}$

(c)

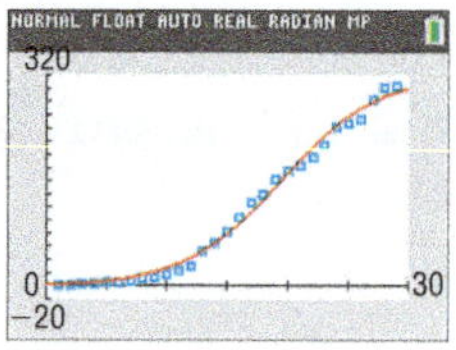

(d) 321.0 thousand cell sites
(e) 314.7 thousand cell sites

11. (a)

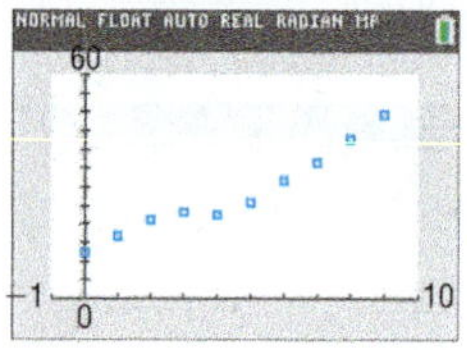

(b) Cubic
(c) $y = 0.0691x^3 - 0.6538x^2 + 4.4323x + 13.0352$

(d)

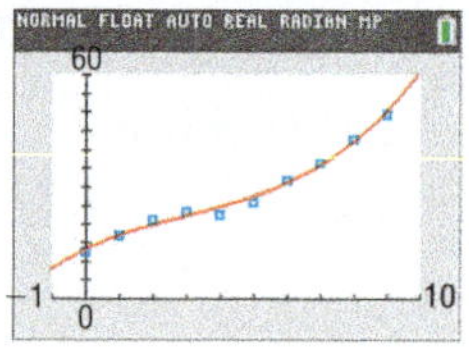

(e) About $74.6 billion

13. (a)

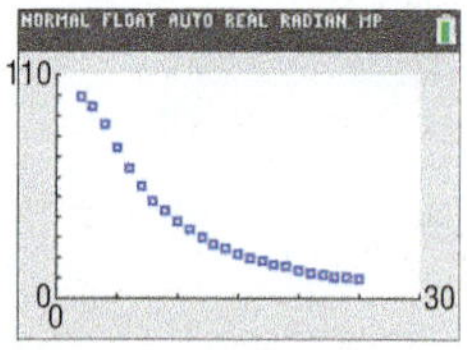

(b) Exponential
(c) $y = 115.5779(0.9012)^x$

(d)

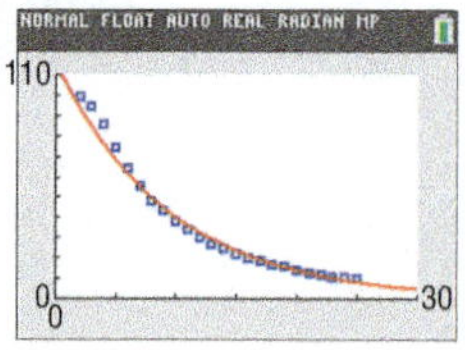

(e) 5.1%

14. $f(x) = \frac{1}{3}(x + 3)(x + 1)^2(x - 2)$

15. $\frac{3\sqrt{2}}{2}$ **16.** $\sqrt{3}$ **17.**

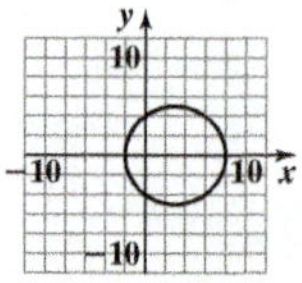

Review Exercises *(page 506)*

1. (a) -4 **(b)** 1 **(c)** -6 **(d)** -6 **2. (a)** -26 **(b)** -241 **(c)** 16 **(d)** -1 **3. (a)** $\sqrt{11}$ **(b)** 1 **(c)** $\sqrt{\sqrt{6}+2}$ **(d)** 19

4. (a) e^4 **(b)** $3e^{-2}-2$ **(c)** e^{e^4} **(d)** -17

5. $(f \circ g)(x) = 1 - 3x$, all real numbers; $(g \circ f)(x) = 7 - 3x$, all real numbers; $(f \circ f)(x) = x$, all real numbers; $(g \circ g)(x) = 9x + 4$, all real numbers

6. $(f \circ g)(x) = \sqrt{3 + 3x + 3x^2}$, all real numbers; $(g \circ f)(x) = 1 + \sqrt{3x} + 3x$, $\{x | x \geq 0\}$;
$(f \circ f)(x) = \sqrt{3\sqrt{3x}}$, $\{x | x \geq 0\}$; $(g \circ g)(x) = 3 + 3x + 4x^2 + 2x^3 + x^4$, all real numbers

7. $(f \circ g)(x) = \dfrac{1+x}{1-x}$, $\{x | x \neq 0, x \neq 1\}$; $(g \circ f)(x) = \dfrac{x-1}{x+1}$, $\{x | x \neq -1, x \neq 1\}$; $(f \circ f)(x) = x$, $\{x | x \neq 1\}$; $(g \circ g)(x) = x$, $\{x | x \neq 0\}$

8. (a) one-to-one **(b)** $\{(2, 1), (5, 3), (8, 5), (10, 6)\}$ **9.**

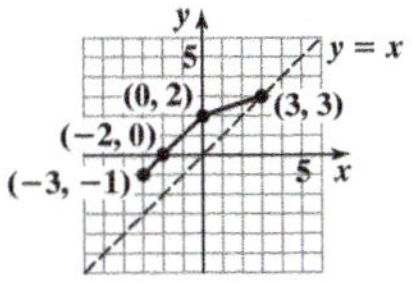

10. $f^{-1}(x) = \dfrac{2x+3}{5x-2}$

$$f(f^{-1}(x)) = \frac{2\left(\frac{2x+3}{5x-2}\right)+3}{5\left(\frac{2x+3}{5x-2}\right)-2} = x$$

$$f^{-1}(f(x)) = \frac{2\left(\frac{2x+3}{5x-2}\right)+3}{5\left(\frac{2x+3}{5x-2}\right)-2} = x$$

Domain of f = range of f^{-1} = all real numbers except $\frac{2}{5}$

Range of f = domain of f^{-1} = all real numbers except $\frac{2}{5}$

11. $f^{-1}(x) = \dfrac{x+1}{x}$

$$f(f^{-1}(x)) = \frac{1}{\frac{x+1}{x} - 1} = x$$

$$f^{-1}(f(x)) = \frac{\frac{1}{x-1}+1}{\frac{1}{x-1}} = x$$

Domain of f = range of f^{-1} = all real numbers except 1
Range of f = domain of f^{-1} = all real numbers except 0

12. $f^{-1}(x) = x^2 + 2, x \geq 0$
$f(f^{-1}(x)) = \sqrt{x^2+2-2} = |x| = x, x \geq 0$
$f^{-1}(f(x)) = (\sqrt{x-2})^2 + 2 = x$
Domain of f = range of f^{-1} = $[2, \infty)$
Range of f = domain of f^{-1} = $[0, \infty)$

13. $f^{-1}(x) = (x-1)^3$;
$f(f^{-1}(x)) = ((x-1)^3)^{1/3} + 1 = x$
$f^{-1}(f(x)) = (x^{1/3} + 1 - 1)^3 = x$
Domain of f = range of f^{-1} = $(-\infty, \infty)$
Range of f = domain of f^{-1} = $(-\infty, \infty)$

14. (a) 81 **(b)** 2 **(c)** $\frac{1}{9}$ **(d)** -3
15. $\log_5 z = 2$ **16.** $5^{13} = u$ **17.** $\left\{x \middle| x > \frac{2}{3}\right\}$; $\left(\frac{2}{3}, \infty\right)$
18. $\{x | x < 1 \text{ or } x > 2\}$; $(-\infty, 1) \cup (2, \infty)$
19. -3 **20.** $\sqrt{2}$ **21.** 0.4
22. $\log_3 u + 2\log_3 v - \log_3 w$ **23.** $8\log_2 a + 2\log_2 b$

24. $2\log x + \frac{1}{2}\log(x^3+1)$ **25.** $2\ln(2x+3) - 2\ln(x-1) - 2\ln(x-2)$ **26.** $\frac{25}{4}\log_4 x$ **27.** $-2\ln(x+1)$ **28.** $\ln\left[\dfrac{16\sqrt{x^2+1}}{\sqrt{x(x-4)}}\right]$ **29.** 2.124

30.

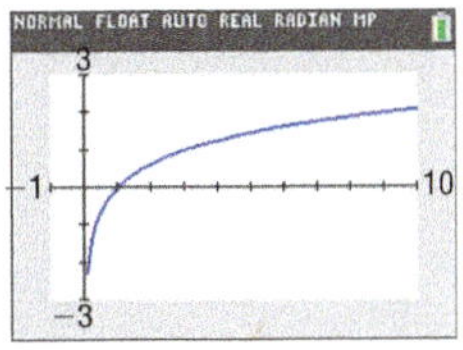

31. (a) Domain of f: $(-\infty, \infty)$
(b)

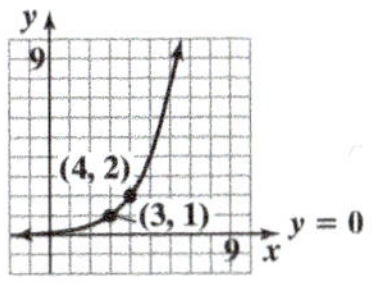

(c) Range of f: $(0, \infty)$
Horizontal asymptote: $y = 0$
(d) $f^{-1}(x) = 3 + \log_2 x$
(e) Domain of f^{-1}: $(0, \infty)$
Range of f^{-1}: $(-\infty, \infty)$
(f)

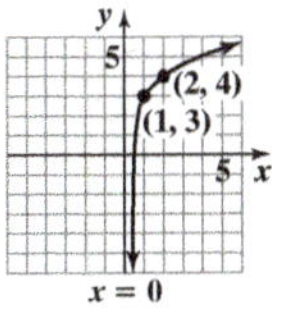

32. (a) Domain of f: $(-\infty, \infty)$
(b)

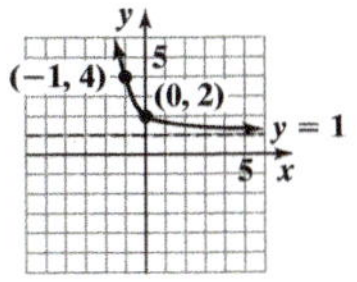

(c) Range of f: $(1, \infty)$
Horizontal asymptote: $y = 1$
(d) $f^{-1}(x) = -\log_3(x-1)$
(e) Domain of f^{-1}: $(1, \infty)$
Range of f^{-1}: $(-\infty, \infty)$
(f)

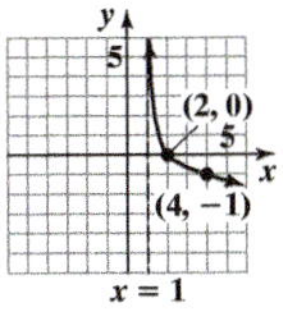

33. (a) Domain of f: $(-\infty, \infty)$
(b)

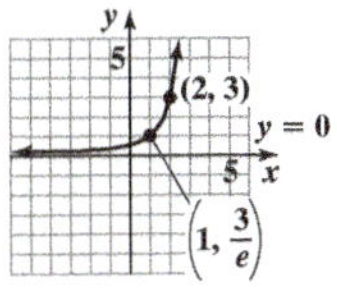

(c) Range of f: $(0, \infty)$
Horizontal asymptote: $y = 0$
(d) $f^{-1}(x) = 2 + \ln\left(\dfrac{x}{3}\right)$
(e) Domain of f^{-1}: $(0, \infty)$
Range of f^{-1}: $(-\infty, \infty)$
(f)

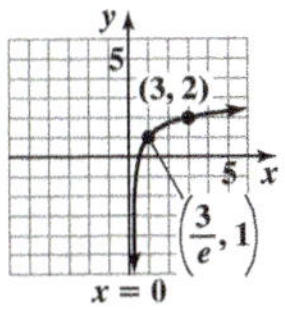

34. (a) Domain of f: $(-3, \infty)$
(b)

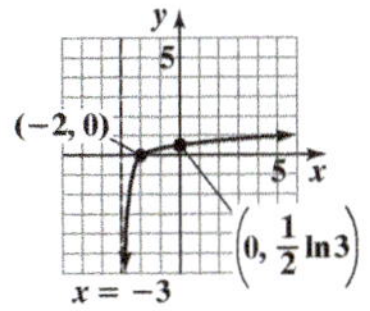

(c) Range of f: $(-\infty, \infty)$
Vertical asymptote: $x = -3$
(d) $f^{-1}(x) = e^{2x} - 3$
(e) Domain of f^{-1}: $(-\infty, \infty)$
Range of f^{-1}: $(-3, \infty)$
(f)

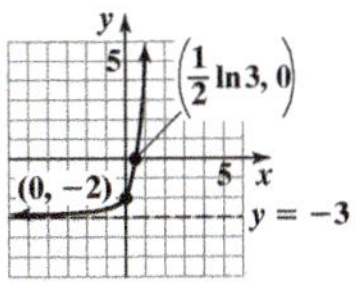

35. $\left\{-\frac{16}{9}\right\}$ **36.** $\left\{\frac{-1-\sqrt{3}}{2}, \frac{-1+\sqrt{3}}{2}\right\} \approx \{-1.366, 0.366\}$ **37.** $\left\{\frac{1}{4}\right\}$ **38.** $\left\{\frac{2\ln 3}{\ln 5 - \ln 3}\right\} \approx \{4.301\}$ **39.** $\{-2, 6\}$ **40.** $\{83\}$ **41.** $\left\{\frac{1}{2}, -3\right\}$

42. $\{1\}$ **43.** $\{-1\}$ **44.** $\{1 - \ln 5\} \approx \{-0.609\}$ **45.** $\left\{\log_3(-2+\sqrt{7})\right\} = \left\{\frac{\ln(-2+\sqrt{7})}{\ln 3}\right\} \approx \{-0.398\}$

46. (a), (e)

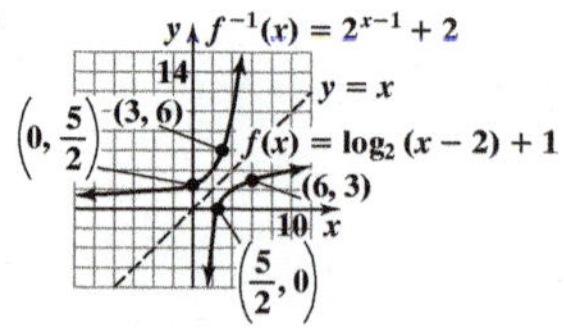

(b) 3; (6, 3)
(c) 10; (10, 4)
(d) $\left\{x \middle| x > \frac{5}{2}\right\}$ or $\left(\frac{5}{2}, \infty\right)$
(e) $f^{-1}(x) = 2^{x-1} + 2$

47. (a) 37.3 W **(b)** 6.9 dB **48. (a)** 11.77 **(b)** 9.56 in. **49. (a)** 9.85 yr **(b)** 4.27 yr **50.** \$20,398.87; 4.04%; 17.5 yr **51.** \$41,668.97 **52.** 24,765 yr ago **53.** 55.22 min, or 55 min, 13 sec **54.** 7,615,278,125 **55.** 7.204 g; 0.519 g

56. (a) 0.3 **(b)** 0.8
(c)

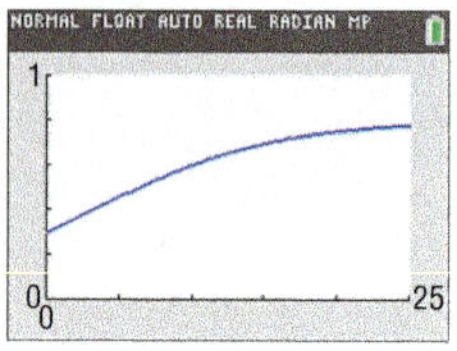

(d) In 2026

57. (a)

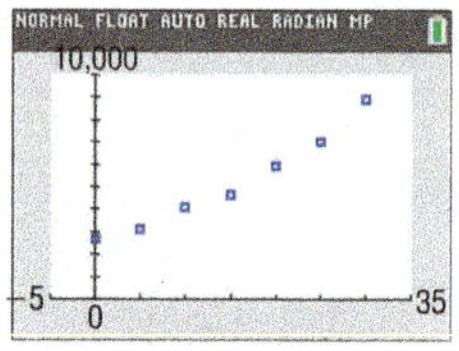

(b) $y = 2638.26(1.0407)^x$
(c) $A(t) = 2638.26e^{0.0399x}$
(d)

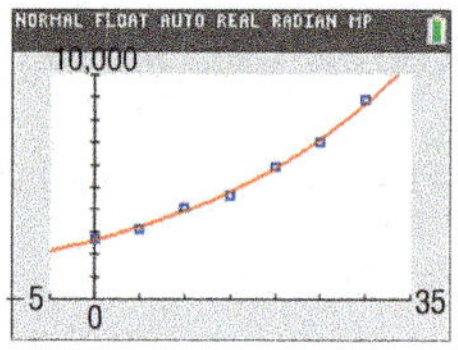

(e) 2021–22

58. (a)

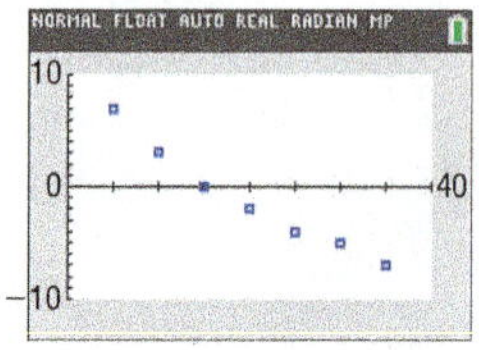

(b) $y = 18.921 - 7.096 \ln x$
(c)

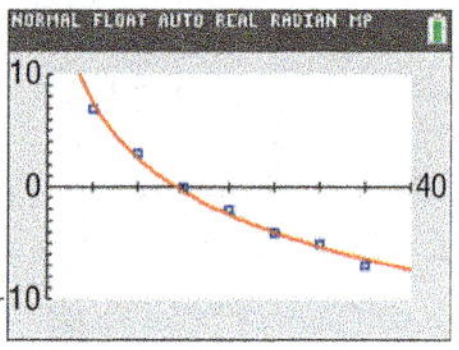

(d) Approximately −3°F

59. (a)

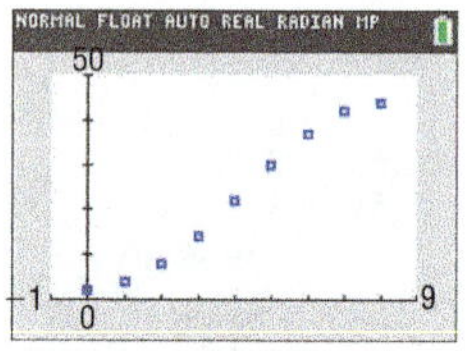

(b) $C = \frac{46.9292}{1 + 21.2733e^{-0.7306t}}$
(c)

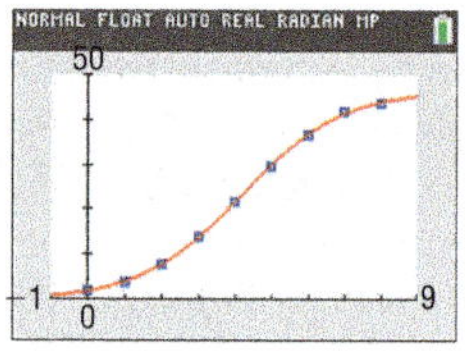

(d) About 47 people; 50 people
(e) 2.4 days; during the tenth hour of day 3
(f) 9.5 days

Chapter Test *(page 509)*

1. (a) $f \circ g = \frac{2x+7}{2x+3}$; domain: $\left\{x \middle| x \neq -\frac{3}{2}\right\}$ **(b)** $(g \circ f)(-2) = 5$ **(c)** $(f \circ g)(-2) = -3$

2. (a) The function is not one-to-one. **(b)** The function is one-to-one.

3. $f^{-1}(x) = \frac{2+5x}{3x}$; domain of $f = \left\{x \middle| x \neq \frac{5}{3}\right\}$, range of $f = \{y | y \neq 0\}$; domain of $f^{-1} = \{x | x \neq 0\}$; range of $f^{-1} = \left\{y \middle| y \neq \frac{5}{3}\right\}$

4. The point $(-5, 3)$ must be on the graph of f^{-1}. **5.** $\{5\}$ **6.** $\{4\}$ **7.** $\{625\}$ **8.** $e^3 + 2 \approx 22.086$ **9.** $\log 20 \approx 1.301$

10. $\log_3 21 = \frac{\ln 21}{\ln 3} \approx 2.771$ **11.** $\ln 133 \approx 4.890$

12. (a) Domain of *f*: $\{x | -\infty < x < \infty\}$ or $(-\infty, \infty)$
(b)

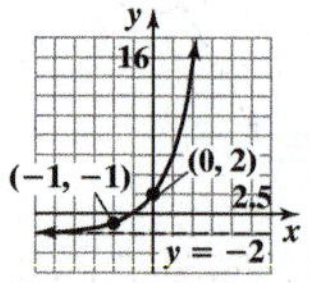

(c) Range of *f*: $\{y | y > -2\}$ or $(-2, \infty)$;
Horizontal asymptote: $y = -2$
(d) $f^{-1}(x) = \log_4(x+2) - 1$
(e) Domain of f^{-1}: $\{x | x > -2\}$ or $(-2, \infty)$
Range of f^{-1}: $\{y | -\infty < y < \infty\}$ or $(-\infty, \infty)$
(f)

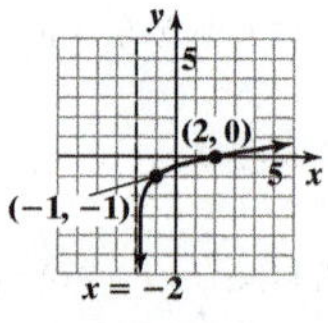

13. (a) Domain of *f*: $\{x | x > 2\}$ or $(2, \infty)$
(b)

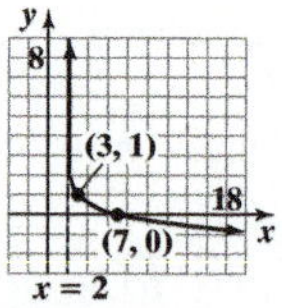

(c) Range of *f*: $\{y | -\infty < y < \infty\}$ or $(-\infty, \infty)$;
vertical asymptote: $x = 2$
(d) $f^{-1}(x) = 5^{1-x} + 2$
(e) Domain of f^{-1}: $\{x | -\infty < x < \infty\}$ or $(-\infty, \infty)$
Range of f^{-1}: $\{y | y > 2\}$ or $(2, \infty)$
(f)

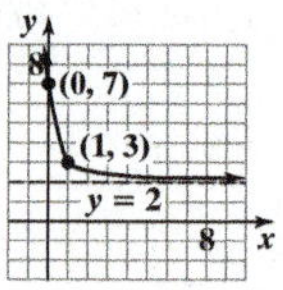

14. $\{1\}$ **15.** $\{91\}$ **16.** $\{-\ln 2\} \approx \{-0.693\}$ **17.** $\left\{\frac{1-\sqrt{13}}{2}, \frac{1+\sqrt{13}}{2}\right\} \approx \{-1.303, 2.303\}$ **18.** $\left\{\frac{3\ln 7}{1 - \ln 7}\right\} \approx \{-6.172\}$

19. $\{2\sqrt{6}\} \approx \{4.899\}$ **20.** $2 + 3\log_2 x - \log_2(x-6) - \log_2(x+3)$ **21.** About 250.39 days **22. (a)** \$1033.82 **(b)** \$963.42 **(c)** 11.9 yr

23. (a) About 83 dB **(b)** The pain threshold will be exceeded if 31,623 people shout at the same time.

Cumulative Review *(page 510)*

1. (a) Yes; no **(b)** Polynomial; the graph is smooth and continuous. **2. (a)** 10 **(b)** $2x^2 + 3x + 1$ **(c)** $2x^2 + 4xh + 2h^2 - 3x - 3h + 1$

3. $\left(\frac{1}{2}, \frac{\sqrt{3}}{2}\right)$ is on the graph. **4.** $\{-26\}$

5.

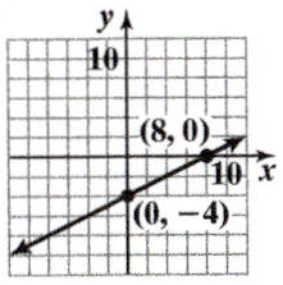

6. (a)

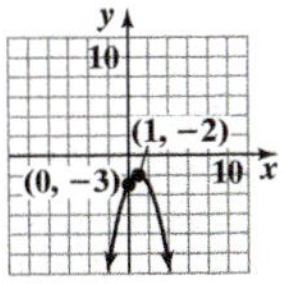

(b) All real numbers; $(-\infty, \infty)$ **7.** $f(x) = 2(x-4)^2 - 8 = 2x^2 - 16x + 24$

8. Exponential; $f(x) = 2 \cdot 3^x$ **9.**

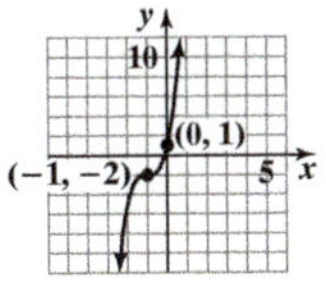

10. (a) $f(g(x)) = \dfrac{4}{(x-3)^2} + 2$; domain: $\{x \mid x \neq 3\}$; 3

(b) $f(g(x)) = \log_2 x + 2$; domain: $\{x \mid x > 0\}$; $2 + \log_2 14$

11. (a) Zeros: $-4, -\frac{1}{4}, 2$

(b) x-intercepts: $-4, -\frac{1}{4}, 2$; y-intercept: -8

(c) Local maximum value of 60.75 occurs at $x = -2.5$.
Local minimum value of -25 occurs at $x = 1$.

(d)

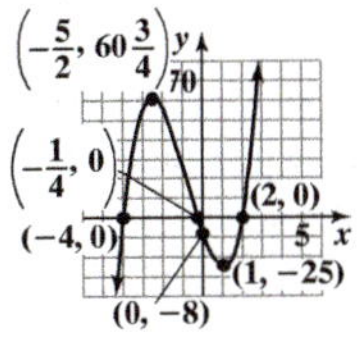

12. (a), (c)

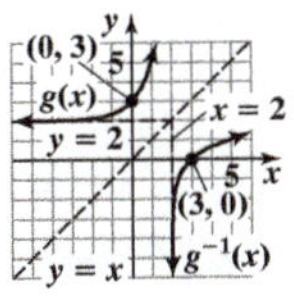

Domain g = range $g^{-1} = (-\infty, \infty)$
Range g = domain $g^{-1} = (2, \infty)$

(b) $g^{-1}(x) = \log_3(x - 2)$

13. $\left\{-\frac{3}{2}\right\}$ **14.** $\{2\}$

15. (a) $\{-1\}$ **(b)** $\{x \mid x > -1\}$ or $(-1, \infty)$ **(c)** $\{25\}$

16. (a)

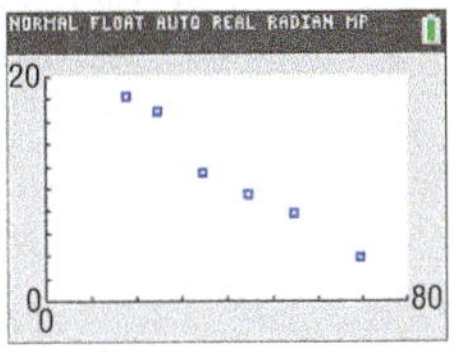

(b) Logarithmic; $y = 49.293 - 10.563 \ln x$
(c) Highest value of $|r|$

CHAPTER 7 Trigonometric Functions

7.1 Assess Your Understanding *(page 522)*

3. standard position **4.** central angle **5.** d **6.** $r\theta$; $\frac{1}{2}r^2\theta$ **7.** b **8.** $\frac{s}{t}$; $\frac{\theta}{t}$ **9.** T **10.** F

11.

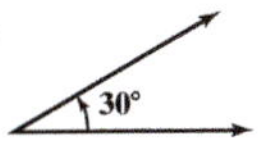

13.

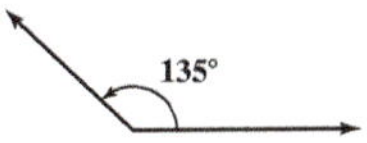

15.

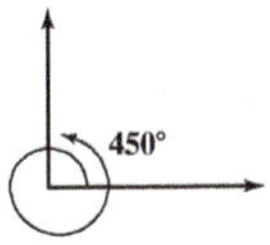

17.

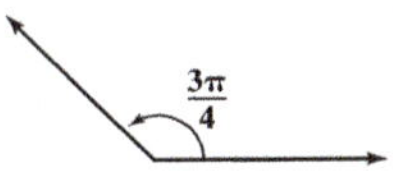

19.

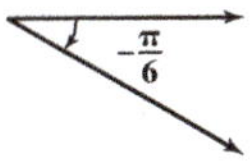

21.

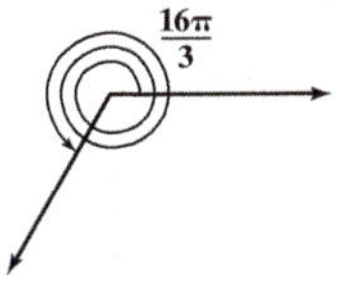

23. $\frac{\pi}{6}$ **25.** $\frac{4\pi}{3}$ **27.** $-\frac{\pi}{3}$ **29.** π **31.** $-\frac{3\pi}{4}$ **33.** $-\frac{\pi}{2}$ **35.** 60° **37.** −225° **39.** 90° **41.** 15° **43.** −90° **45.** −30° **47.** 0.30 **49.** −0.70 **51.** 2.18 **53.** 179.91° **55.** 114.59° **57.** 362.11° **59.** 40.17° **61.** 50.24° **63.** 9.15° **65.** 40°19′12″ **67.** 18°15′18″ **69.** 19°59′24″ **71.** 5 m **73.** 6 ft **75.** 0.6 radian **77.** $\frac{\pi}{3} \approx 1.047$ in. **79.** 25 m² **81.** $2\sqrt{3} \approx 3.464$ ft **83.** 0.24 radian **85.** $\frac{\pi}{3} \approx 1.047$ in.² **87.** $s = 2.094$ ft; $A = 2.094$ ft² **89.** $s = 14.661$ yd; $A = 87.965$ yd² **91.** $3\pi \approx 9.42$ in; $5\pi \approx 15.71$ in. **93.** $2\pi \approx 6.28$ m² **95.** $\frac{675\pi}{2} \approx 1060.29$ ft² **97.** $\frac{1075\pi}{3} \approx 1125.74$ in.² **99.** $\omega = \frac{1}{60}$ radian/s; $v = \frac{1}{12}$ cm/s **101.** ≈23.2 mph **103.** ≈120.6 km/h **105.** ≈452.5 rpm **107.** ≈359 mi **109.** ≈898 mi/h **111.** ≈2292 mi/h **113.** $\frac{3}{4}$ rpm **115.** ≈2.86 mi/h **117.** ≈31.47 rpm **119.** 63π ft² **121.** ≈1037 mi/h **123.** Radius ≈ 3979 mi; circumference ≈ 25,000 mi **125.** $v_1 = r_1\omega_1$, $v_2 = r_2\omega_2$, and $v_1 = v_2$, so $r_1\omega_1 = r_2\omega_2$ and $\frac{r_1}{r_2} = \frac{\omega_2}{\omega_1}$. **134.** $x = -\frac{7}{3}$ **135.** $\left\{-3, \frac{1}{5}\right\}$ **136.** $y = -|x+3| - 4$ **137.** HA: $y = 3$; VA: $x = 7$

7.2 Assess Your Understanding *(page 535)*

3. complementary **4.** cosine **5.** 62° **6.** c **7.** T **8.** a **9.** T **10.** F

11. $\sin\theta = \frac{5}{13}$; $\cos\theta = \frac{12}{13}$; $\tan\theta = \frac{5}{12}$; $\csc\theta = \frac{13}{5}$; $\sec\theta = \frac{13}{12}$; $\cot\theta = \frac{12}{5}$ **13.** $\sin\theta = \frac{2\sqrt{13}}{13}$; $\cos\theta = \frac{3\sqrt{13}}{13}$; $\tan\theta = \frac{2}{3}$; $\csc\theta = \frac{\sqrt{13}}{2}$; $\sec\theta = \frac{\sqrt{13}}{3}$; $\cot\theta = \frac{3}{2}$ **15.** $\sin\theta = \frac{\sqrt{3}}{2}$; $\cos\theta = \frac{1}{2}$; $\tan\theta = \sqrt{3}$; $\csc\theta = \frac{2\sqrt{3}}{3}$; $\sec\theta = 2$; $\cot\theta = \frac{\sqrt{3}}{3}$ **17.** $\sin\theta = \frac{\sqrt{6}}{3}$; $\cos\theta = \frac{\sqrt{3}}{3}$; $\tan\theta = \sqrt{2}$; $\csc\theta = \frac{\sqrt{6}}{2}$; $\sec\theta = \sqrt{3}$; $\cot\theta = \frac{\sqrt{2}}{2}$ **19.** $\sin\theta = \frac{\sqrt{5}}{5}$; $\cos\theta = \frac{2\sqrt{5}}{5}$; $\tan\theta = \frac{1}{2}$; $\csc\theta = \sqrt{5}$; $\sec\theta = \frac{\sqrt{5}}{2}$; $\cot\theta = 2$ **21.** $\tan\theta = \frac{\sqrt{3}}{3}$; $\csc\theta = 2$; $\sec\theta = \frac{2\sqrt{3}}{3}$; $\cot\theta = \sqrt{3}$ **23.** $\tan\theta = \frac{2\sqrt{5}}{5}$; $\csc\theta = \frac{3}{2}$; $\sec\theta = \frac{3\sqrt{5}}{5}$; $\cot\theta = \frac{\sqrt{5}}{2}$ **25.** $\cos\theta = \frac{\sqrt{2}}{2}$; $\tan\theta = 1$; $\csc\theta = \sqrt{2}$; $\sec\theta = \sqrt{2}$; $\cot\theta = 1$

27. $\sin\theta = \frac{2\sqrt{2}}{3}$; $\tan\theta = 2\sqrt{2}$; $\csc\theta = \frac{3\sqrt{2}}{4}$; $\sec\theta = 3$; $\cot\theta = \frac{\sqrt{2}}{4}$ **29.** $\sin\theta = \frac{\sqrt{5}}{5}$; $\cos\theta = \frac{2\sqrt{5}}{5}$; $\csc\theta = \sqrt{5}$; $\sec\theta = \frac{\sqrt{5}}{2}$; $\cot\theta = 2$

31. $\sin\theta = \frac{2\sqrt{2}}{3}$; $\cos\theta = \frac{1}{3}$; $\tan\theta = 2\sqrt{2}$; $\csc\theta = \frac{3\sqrt{2}}{4}$; $\cot\theta = \frac{\sqrt{2}}{4}$ **33.** $\sin\theta = \frac{\sqrt{6}}{3}$; $\cos\theta = \frac{\sqrt{3}}{3}$; $\csc\theta = \frac{\sqrt{6}}{2}$; $\sec\theta = \sqrt{3}$; $\cot\theta = \frac{\sqrt{2}}{2}$

35. $\sin\theta = \frac{1}{2}$; $\cos\theta = \frac{\sqrt{3}}{2}$; $\tan\theta = \frac{\sqrt{3}}{3}$; $\sec\theta = \frac{2\sqrt{3}}{3}$; $\cot\theta = \sqrt{3}$ **37.** 1 **39.** 1 **41.** 0 **43.** 0 **45.** 1 **47.** 0 **49.** 0 **51.** 1 **53.** 1

55. (a) $\frac{1}{2}$ **(b)** $\frac{3}{4}$ **(c)** 2 **(d)** 2 **57. (a)** 17 **(b)** $\frac{1}{4}$ **(c)** 4 **(d)** $\frac{17}{16}$ **59. (a)** $\frac{1}{4}$ **(b)** 15 **(c)** 4 **(d)** $\frac{16}{15}$ **61. (a)** 0.78 **(b)** 0.79 **(c)** 1.27 **(d)** 1.28 **(e)** 1.61 **(f)** 0.78 **(g)** 0.62 **(h)** 1.27 **63.** 0.6 **65.** 20° **67.** $5\cos\theta$ **69.** $7\sec\theta$ **71.** $6\tan\theta$

73. (a) 10 min
(b) 20 min
(c) $T(\theta) = 5\left(1 - \frac{1}{3\tan\theta} + \frac{1}{\sin\theta}\right)$
(d) Approximately 15.8 min
(e) Approximately 10.4 min
(f) 70.5°; 177 ft; 9.7 min

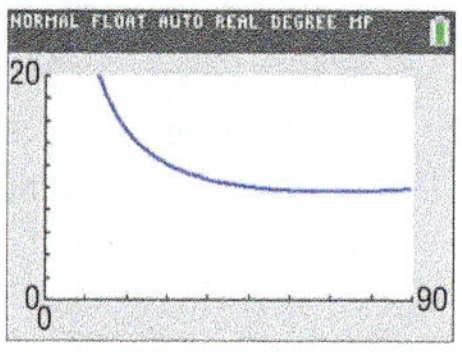

75. (a) $Z = 200\sqrt{13} \approx 721.1$ ohms **(b)** $\tan\phi = \frac{2}{3}$; $\sin\phi = \frac{2\sqrt{13}}{13}$; $\cos\phi = \frac{3\sqrt{13}}{13}$; $\cot\phi = \frac{3}{2}$; $\csc\phi = \frac{\sqrt{13}}{2}$; $\sec\phi = \frac{\sqrt{13}}{3}$

77. (a) $|OA| = |OC| = 1$; angle OAC = angle OCA; angle OAC + angle $OAC + 180° - \theta = 180°$; angle $OAC = \frac{\theta}{2}$
(b) $\sin\theta = \frac{|CD|}{|OC|} = |CD|$; $\cos\theta = \frac{|OD|}{|OC|} = |OD|$ **(c)** $\tan\frac{\theta}{2} = \frac{|CD|}{|AD|} = \frac{\sin\theta}{1 + |OD|} = \frac{\sin\theta}{1 + \cos\theta}$

79. $h = x\tan\theta$ and $h = (1 - x)\tan n\theta$; thus, $x\tan\theta = (1 - x)\tan n\theta$, so $x = \frac{\tan n\theta}{\tan\theta + \tan n\theta}$.

81. (a) Area $\triangle OAC = \frac{1}{2}|AC|\,|OC| = \frac{1}{2}\cdot\frac{|AC|}{1}\cdot\frac{|OC|}{1} = \frac{1}{2}\sin\alpha\cos\alpha$

(b) Area $\triangle OCB = \frac{1}{2}|BC|\,|OC| = \frac{1}{2}|OB|^2\frac{|BC|}{|OB|}\cdot\frac{|OC|}{|OB|} = \frac{1}{2}|OB|^2\sin\beta\cos\beta$

(c) Area $\triangle OAB = \frac{1}{2}|BD|\,|OA| = \frac{1}{2}|OB|\frac{|BD|}{|OB|} = \frac{1}{2}|OB|\sin(\alpha + \beta)$

(d) $\frac{\cos\alpha}{\cos\beta} = \frac{\frac{|OC|}{1}}{\frac{|OC|}{|OB|}} = |OB|$

(e)
$$\text{Area}\,\triangle OAB = \text{area}\,\triangle OAC + \text{area}\,\triangle OCB$$
$$\frac{1}{2}|OB|\sin(\alpha + \beta) = \frac{1}{2}\sin\alpha\cos\alpha + \frac{1}{2}|OB|^2\sin\beta\cos\beta$$
$$\sin(\alpha + \beta) = \frac{\sin\alpha\cos\alpha + |OB|^2\sin\beta\cos\beta}{|OB|}$$
$$\sin(\alpha + \beta) = \frac{\sin\alpha\,(|OB|\cos\beta) + |OB|^2\sin\beta\left(\frac{\cos\alpha}{|OB|}\right)}{|OB|}$$
$$\sin(\alpha + \beta) = \sin\alpha\cos\beta + \cos\alpha\sin\beta$$

83. $\sin\alpha = \tan\alpha\cos\alpha = \cos\beta\cos\alpha = \cos\beta\tan\beta = \sin\beta$;
$$\sin^2\alpha + \cos^2\alpha = 1$$
$$\sin^2\alpha + \tan^2\beta = 1$$
$$\sin^2\alpha + \frac{\sin^2\beta}{\cos^2\beta} = 1$$
$$\sin^2\alpha + \frac{\sin^2\alpha}{1 - \sin^2\alpha} = 1$$
$$\sin^2\alpha - \sin^4\alpha + \sin^2\alpha = 1 - \sin^2\alpha$$
$$\sin^4\alpha - 3\sin^2\alpha + 1 = 0$$
$$\sin^2\alpha = \frac{3 \pm \sqrt{5}}{2}$$
$$\sin^2\alpha = \frac{3 - \sqrt{5}}{2}$$
$$\sin\alpha = \sqrt{\frac{3 - \sqrt{5}}{2}}$$

85. Since $a^2 + b^2 = c^2$, $a > 0$, $b > 0$, then $0 < a^2 < c^2$ or $0 < a < c$. Thus $0 < \frac{a}{c} < 1$ and $0 < \cos\theta < 1$ or $\frac{\cos\theta}{\cos\theta} < \frac{1}{\cos\theta}$; therefore, $\sec\theta > 1$.

89. $\left\{x \middle| x > -\frac{2}{5}\right\}$ or $\left(-\frac{2}{5}, \infty\right)$ **90.** $4 - 3i, -5$ **91.** $R = 134$ **92.** 81π ft^2

7.3 Assess Your Understanding *(page 545)*

1. $\frac{3}{2}$ **2.** 0.84 **3.** T **4.** F **5.** b **6.** c **7.** $\sin 45° = \frac{\sqrt{2}}{2}$; $\cos 45° = \frac{\sqrt{2}}{2}$; $\tan 45° = 1$; $\csc 45° = \sqrt{2}$; $\sec 45° = \sqrt{2}$; $\cot 45° = 1$ **9.** $\frac{\sqrt{3}}{2}$ **11.** $\frac{1}{2}$ **13.** $\frac{3}{4}$

15. $\sqrt{3}$ **17.** $\frac{\sqrt{3}}{4}$ **19.** $\sqrt{2}$ **21.** 2 **23.** $\sqrt{2} + \frac{4\sqrt{3}}{3}$ **25.** $-\frac{8}{3}$ **27.** $\frac{1}{2}$ **29.** 0 **31.** 0.47 **33.** 0.38 **35.** 1.33 **37.** 0.31 **39.** 3.73 **41.** 1.04 **43.** 0.84

45. 0.02 **47.** 0.31 **49.** $\frac{1+\sqrt{3}}{2}$ **51.** $\frac{1}{2}$ **53.** $\frac{\sqrt{3}}{2}$ **55.** $\frac{\sqrt{2}}{4}$ **57. (a)** $\frac{\sqrt{2}}{2}$; $\left(\frac{\pi}{4}, \frac{\sqrt{2}}{2}\right)$ **(b)** $\left(\frac{\sqrt{2}}{2}, \frac{\pi}{4}\right)$ **(c)** $\left(\frac{\pi}{4}, -2\right)$ **59.** 4.59 in.; 6.55 in.

61. (a) 5.52 in. or 11.83 in. **63.** $R \approx 310.56$ ft; $H \approx 77.64$ ft **65.** $R \approx 19{,}541.95$ m; $H \approx 2278.14$ m **67. (a)** 1.20 s **(b)** 1.12 s **(c)** 1.20 s

69. 1 **71.** $\frac{\sqrt{2}}{2}$ **73.** 70.02 ft **75.** 985.91 ft **77.** $50\frac{\sqrt{3}}{3} \approx 28.87$ m **79.** 1978.09 ft **81.** 60.27 ft **83.** $190\sqrt{2} \approx 268.70$ ft **85.** 3.83 mi

87. (a) $T(\theta) = 1 + \frac{2}{3\sin\theta} - \frac{1}{4\tan\theta}$
(b) 1.9 h; 0.57 h
(c) 1.69 h; 0.75 h
(d) 1.63 h; 0.86 h
(e) 1.67 h
(f) 2.75 h
(g)

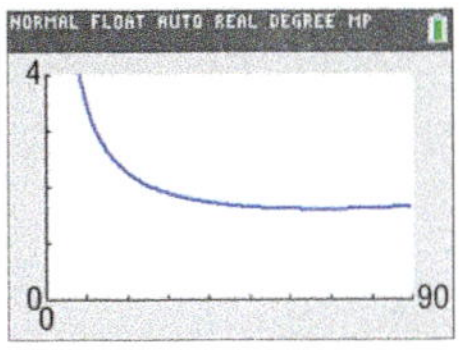

67.98°; 1.62 h; 0.9 h

89. The white ball should hit the top cushion 4.125 ft from the upper-left corner.

91. (a) $L(\theta) = \frac{15}{\cos\theta} + 5\sqrt{40 - 36\tan\theta + 9\tan^2\theta}$
(b) $0° < \theta < 63.4°$
(c) $L(45°) = 15\sqrt{2} + 5\sqrt{13}$ ft ≈ 39.2 ft; 15 ft
(d)

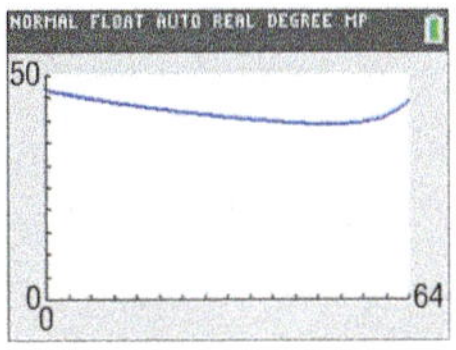

$\theta = 50.2°$
(e) $L_{\min} = 39.05$ ft; ≈18 ft from the 15-ft pole

93.

θ	0.5	0.4	0.2	0.1	0.01	0.001	0.0001	0.00001
$f(\theta) = \frac{\sin\theta}{\theta}$	0.9589	0.9735	0.9933	0.9983	1.0000	1.0000	1.0000	1.0000

$\frac{\sin\theta}{\theta}$ approaches 1 as $\theta \to 0$.

97. All real numbers **98.** Vertex: $(3, 5)$; axis of symmetry: $x = 3$ **99.** $\{\ln 6 + 4\}$ **100.** 9

7.4 Assess Your Understanding *(page 558)*

1. tangent; cotangent **2.** coterminal **3.** 60° **4.** F **5.** T **6.** T **7.** 60° **8.** b **9.** $\frac{\pi}{2}, \frac{3\pi}{2}$ **10.** d **11.** $\sin\theta = \frac{4}{5}$; $\cos\theta = -\frac{3}{5}$; $\tan\theta = -\frac{4}{3}$; $\csc\theta = \frac{5}{4}$; $\sec\theta = -\frac{5}{3}$; $\cot\theta = -\frac{3}{4}$ **13.** $\sin\theta = -\frac{3\sqrt{13}}{13}$; $\cos\theta = \frac{2\sqrt{13}}{13}$; $\tan\theta = -\frac{3}{2}$; $\csc\theta = -\frac{\sqrt{13}}{3}$; $\sec\theta = \frac{\sqrt{13}}{2}$; $\cot\theta = -\frac{2}{3}$

15. $\sin\theta = -\frac{\sqrt{2}}{2}$; $\cos\theta = -\frac{\sqrt{2}}{2}$; $\tan\theta = 1$; $\csc\theta = -\sqrt{2}$; $\sec\theta = -\sqrt{2}$; $\cot\theta = 1$ **17.** $\sin\theta = \frac{1}{2}$; $\cos\theta = \frac{\sqrt{3}}{2}$; $\tan\theta = \frac{\sqrt{3}}{3}$; $\csc\theta = 2$; $\sec\theta = \frac{2\sqrt{3}}{3}$; $\cot\theta = \sqrt{3}$ **19.** $\sin\theta = -\frac{\sqrt{2}}{2}$; $\cos\theta = \frac{\sqrt{2}}{2}$; $\tan\theta = -1$; $\csc\theta = -\sqrt{2}$; $\sec\theta = \sqrt{2}$; $\cot\theta = -1$ **21.** $\frac{\sqrt{2}}{2}$ **23.** 1 **25.** 1 **27.** $\sqrt{3}$

29. $\frac{\sqrt{2}}{2}$ **31.** 0 **33.** II **35.** IV **37.** IV **39.** III **41.** 30° **43.** 60° **45.** 60° **47.** $\frac{\pi}{4}$ **49.** $\frac{\pi}{3}$ **51.** 45° **53.** $\frac{\pi}{3}$ **55.** 80° **57.** $\frac{\pi}{4}$ **59.** $\frac{1}{2}$ **61.** $\frac{1}{2}$ **63.** $\frac{\sqrt{2}}{2}$

65. -2 **67.** $-\sqrt{3}$ **69.** $\frac{\sqrt{2}}{2}$ **71.** $-\frac{\sqrt{2}}{2}$ **73.** $-\frac{\sqrt{3}}{2}$ **75.** $-\sqrt{3}$ **77.** 0 **79.** 0 **81.** -1

83. $\cos\theta = -\frac{5}{13}$; $\tan\theta = -\frac{12}{5}$; $\csc\theta = \frac{13}{12}$; $\sec\theta = -\frac{13}{5}$; $\cot\theta = -\frac{5}{12}$ **85.** $\sin\theta = -\frac{3}{5}$; $\tan\theta = \frac{3}{4}$; $\csc\theta = -\frac{5}{3}$; $\sec\theta = -\frac{5}{4}$; $\cot\theta = \frac{4}{3}$

87. $\cos\theta = -\frac{12}{13}$; $\tan\theta = -\frac{5}{12}$; $\csc\theta = \frac{13}{5}$; $\sec\theta = -\frac{13}{12}$; $\cot\theta = -\frac{12}{5}$ **89.** $\sin\theta = -\frac{2\sqrt{2}}{3}$; $\tan\theta = 2\sqrt{2}$; $\csc\theta = -\frac{3\sqrt{2}}{4}$; $\sec\theta = -3$; $\cot\theta = \frac{\sqrt{2}}{4}$

91. $\cos\theta = -\frac{\sqrt{5}}{3}$; $\tan\theta = -\frac{2\sqrt{5}}{5}$; $\csc\theta = \frac{3}{2}$; $\sec\theta = -\frac{3\sqrt{5}}{5}$; $\cot\theta = -\frac{\sqrt{5}}{2}$ **93.** $\sin\theta = -\frac{\sqrt{3}}{2}$; $\cos\theta = \frac{1}{2}$; $\tan\theta = -\sqrt{3}$; $\csc\theta = -\frac{2\sqrt{3}}{3}$; $\cot\theta = -\frac{\sqrt{3}}{3}$ **95.** $\sin\theta = -\frac{3}{5}$; $\cos\theta = -\frac{4}{5}$; $\csc\theta = -\frac{5}{3}$; $\sec\theta = -\frac{5}{4}$; $\cot\theta = \frac{4}{3}$ **97.** $\sin\theta = \frac{\sqrt{10}}{10}$; $\cos\theta = -\frac{3\sqrt{10}}{10}$; $\csc\theta = \sqrt{10}$; $\sec\theta = -\frac{\sqrt{10}}{3}$; $\cot\theta = -3$ **99.** $\sin\theta = -\frac{1}{2}$; $\cos\theta = -\frac{\sqrt{3}}{2}$; $\tan\theta = \frac{\sqrt{3}}{3}$; $\sec\theta = -\frac{2\sqrt{3}}{3}$; $\cot\theta = \sqrt{3}$

101. 0 **103. (a)** $-\frac{\sqrt{2}}{2}$; $\left(315°, -\frac{\sqrt{2}}{2}\right)$ **(b)** $\sqrt{2}$; $(315°, \sqrt{2})$ **(c)** -1; $(315°, -1)$ **105. (a)** $-\frac{\sqrt{3}}{2}$; $\left(\frac{7\pi}{6}, -\frac{\sqrt{3}}{2}\right)$ **(b)** -2; $\left(\frac{7\pi}{6}, -2\right)$ **(c)** 1; $(-315°, 1)$
107. -0.2 **109.** 3 **111.** -5 **113.** 0 **115. (a)** Approximately 16.6 ft **(b)**

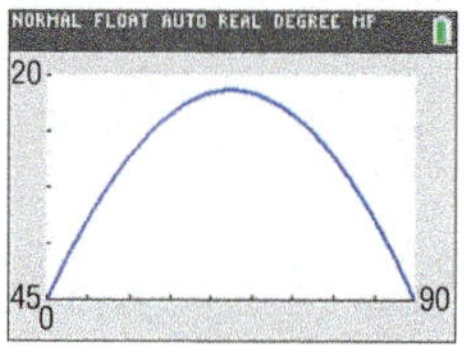

(c) 67.5° **117.** 54.7° **121.** $\{-10\}$
122. 2,476,000 units **123.** $f(12) = 5$ **124.** $g(x) = x - 4$

7.5 Assess Your Understanding *(page 569)*

4. 2π; π **5.** b; a **6.** $\frac{b}{r}$; $\frac{a}{r}$ **7.** -0.2; 0.2 **8.** T **9.** b **10.** a **11.** $\sin t = -\frac{1}{2}$; $\cos t = \frac{\sqrt{3}}{2}$; $\tan t = -\frac{\sqrt{3}}{3}$; $\csc t = -2$; $\sec t = \frac{2\sqrt{3}}{3}$; $\cot t = -\sqrt{3}$

13. $\sin t = -\frac{\sqrt{2}}{2}$; $\cos t = -\frac{\sqrt{2}}{2}$; $\tan t = 1$; $\csc t = -\sqrt{2}$; $\sec t = -\sqrt{2}$; $\cot t = 1$

15. $\sin t = \frac{2}{3}$; $\cos t = \frac{\sqrt{5}}{3}$; $\tan t = \frac{2\sqrt{5}}{5}$; $\csc t = \frac{3}{2}$; $\sec t = \frac{3\sqrt{5}}{5}$; $\cot t = \frac{\sqrt{5}}{2}$ **17.** $\sin\theta = -\frac{4}{5}$; $\cos\theta = \frac{3}{5}$; $\tan\theta = -\frac{4}{3}$; $\csc\theta = -\frac{5}{4}$; $\sec\theta = \frac{5}{3}$; $\cot\theta = -\frac{3}{4}$

19. $\sin\theta = \frac{3\sqrt{13}}{13}$; $\cos\theta = -\frac{2\sqrt{13}}{13}$; $\tan\theta = -\frac{3}{2}$; $\csc\theta = \frac{\sqrt{13}}{3}$; $\sec\theta = -\frac{\sqrt{13}}{2}$; $\cot\theta = -\frac{2}{3}$

21. $\sin\theta = -\frac{\sqrt{2}}{2}$; $\cos\theta = -\frac{\sqrt{2}}{2}$; $\tan\theta = 1$; $\csc\theta = -\sqrt{2}$; $\sec\theta = -\sqrt{2}$; $\cot\theta = 1$ **23.** $\frac{\sqrt{2}}{2}$ **25.** 1 **27.** 1 **29.** $\sqrt{3}$ **31.** $\frac{\sqrt{2}}{2}$ **33.** 0 **35.** $\sqrt{2}$

37. $\frac{\sqrt{3}}{3}$ **39.** $-\frac{\sqrt{3}}{2}$ **41.** $-\frac{\sqrt{3}}{3}$ **43.** 2 **45.** -1 **47.** -1 **49.** $\frac{\sqrt{2}}{2}$ **51.** 0 **53.** $-\sqrt{2}$ **55.** $\frac{2\sqrt{3}}{3}$ **57.** -1 **59.** -2 **61.** $\frac{2-\sqrt{2}}{2}$ **63.** $(-\infty, \infty)$

65. At odd multiples of $\frac{\pi}{2}$ **67.** At odd multiples of $\frac{\pi}{2}$ **69.** $[-1, 1]$ **71.** $(-\infty, \infty)$ **73.** $(-\infty, -1] \cup [1, \infty)$ **75.** Odd; yes; origin **77.** Odd; yes; origin

79. Even; yes; y-axis **81.** 0.9 **83.** 9 **85. (a)** $-\frac{1}{3}$ **(b)** 1 **87. (a)** -2 **(b)** 6 **89. (a)** -4 **(b)** -12

91. (a) Using graph: $\sin 1 \approx 0.8$, $\cos 1 \approx 0.5$, $\tan 1 \approx 1.6$, $\csc 1 \approx 1.3$, $\sec 1 \approx 2$, $\cot 1 \approx 0.6$;
Using calculator: $\sin 1 \approx 0.8$, $\cos 1 \approx 0.5$, $\tan 1 \approx 1.6$, $\csc 1 \approx 1.2$, $\sec 1 \approx 1.9$, $\cot 1 \approx 0.6$
(b) Using graph: $\sin 5.1 \approx -0.9$, $\cos 5.1 \approx 0.4$, $\tan 5.1 \approx -2.3$, $\csc 5.1 \approx -1.1$, $\sec 5.1 \approx 2.5$, $\cot 5.1 \approx -0.4$;
Using calculator: $\sin 5.1 \approx -0.9$, $\cos 5.1 \approx 0.4$, $\tan 5.1 \approx -2.4$, $\csc 5.1 \approx -1.1$, $\sec 5.1 \approx 2.6$, $\cot 5.1 \approx -0.4$

93. Suppose that there is a number p, $0 < p < 2\pi$, for which $\sin(\theta + p) = \sin\theta$ for all θ. If $\theta = 0$, then $\sin(0 + p) = \sin p = \sin 0 = 0$, so $p = \pi$. If $\theta = \frac{\pi}{2}$, then $\sin\left(\frac{\pi}{2} + p\right) = \sin\left(\frac{\pi}{2}\right)$. But $p = \pi$. Thus $\sin\left(\frac{3\pi}{2}\right) = -1 = \sin\left(\frac{\pi}{2}\right) = 1$. This is impossible. Therefore, the smallest positive number p for which $\sin(\theta + p) = \sin\theta$ for all θ is $p = 2\pi$. **95.** $\sec\theta = \frac{1}{\cos\theta}$; since $\cos\theta$ has period 2π, so does $\sec\theta$.

97. If $P = (a, b)$ is the point on the unit circle corresponding to θ, then $Q = (-a, -b)$ is the point on the unit circle corresponding to $\theta + \pi$. Thus $\tan(\theta + \pi) = \frac{(-b)}{(-a)} = \frac{b}{a} = \tan\theta$. If there exists a number p, $0 < p < \pi$, for which $\tan(\theta + p) = \tan\theta$ for all θ, then if $\theta = 0$, $\tan(p) = \tan 0 = 0$. But this means that p is a multiple of π. Since no multiple of π exists in the interval $(0, \pi)$, this is impossible. Therefore, the fundamental period of $f(\theta) = \tan\theta$ is π.

99. Let $P = (x, y)$ be the point on the unit circle that corresponds to t. Consider the equation $\tan t = \frac{y}{x} = a$. Then $y = ax$. But $x^2 + y^2 = 1$, so $x^2 + a^2x^2 = 1$. Thus $x = \pm\frac{1}{\sqrt{1 + a^2}}$ and $y = \pm\frac{a}{\sqrt{1 + a^2}}$; that is, for any real number a, there is a point $P = (x, y)$ on the unit circle for which $\tan t = a$. In other words, $-\infty < \tan t < \infty$, and the range of the tangent function is the set of all real numbers.

101. $m = \frac{\sin\theta - 0}{\cos\theta - 0} = \frac{\sin\theta}{\cos\theta} = \tan\theta$ **106.** $\frac{5 - \sqrt{10}}{3}$, $\frac{5 + \sqrt{10}}{3}$ **107.** $y = \frac{1}{4}x + \frac{31}{4}$ **108.** 6.12% **109.** Vertical: $x = -3$; horizontal: $y = 5$

7.6 Assess Your Understanding *(page 582)*

3. 1; $\frac{\pi}{2}$ **4.** 3; π **5.** 3; $\frac{\pi}{3}$ **6.** T **7.** F **8.** T **9.** d **10.** d **11. (a)** 0 **(b)** $-\frac{\pi}{2} \le x \le \frac{\pi}{2}$ **(c)** 1 **(d)** $0, \pi, 2\pi$

(e) $f(x) = 1$ for $x = -\frac{3\pi}{2}, \frac{\pi}{2}$; $f(x) = -1$ for $x = -\frac{\pi}{2}, \frac{3\pi}{2}$ **(f)** $-\frac{5\pi}{6}, -\frac{\pi}{6}, \frac{7\pi}{6}, \frac{11\pi}{6}$ **(g)** $\{x \mid x = k\pi, k \text{ an integer}\}$ **13.** Amplitude $= 2$; period $= 2\pi$

15. Amplitude $= 4$; period $= \pi$ **17.** Amplitude $= 6$; period $= 2$ **19.** Amplitude $= \frac{1}{2}$; period $= \frac{4\pi}{3}$ **21.** Amplitude $= \frac{5}{3}$; period $= 3$ **23.** F **25.** A
27. H **29.** C **31.** J

33.

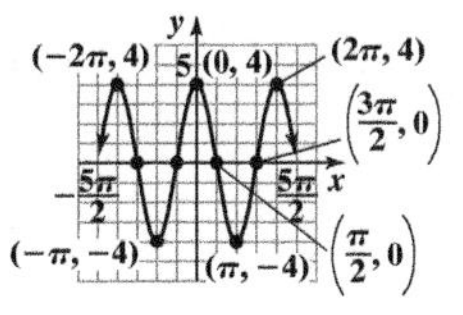

Domain: $(-\infty, \infty)$
Range: $[-4, 4]$

35.

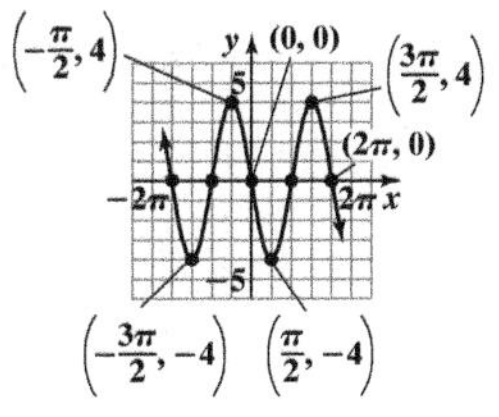

Domain: $(-\infty, \infty)$
Range: $[-4, 4]$

37.

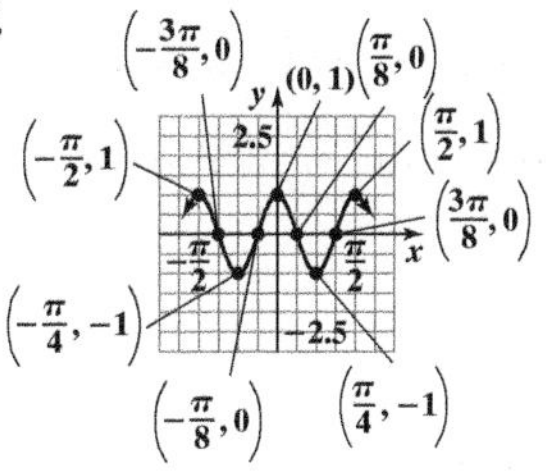

Domain: $(-\infty, \infty)$
Range: $[-1, 1]$

39.

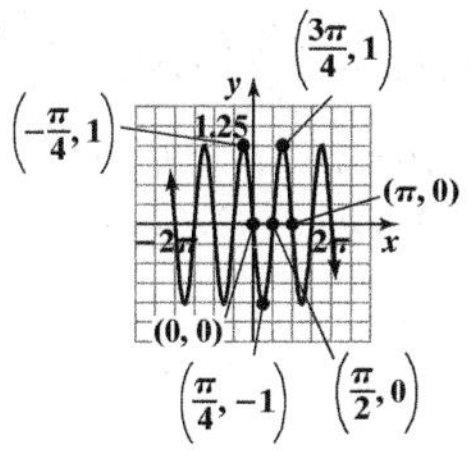

Domain: $(-\infty, \infty)$
Range: $[-1, 1]$

41.

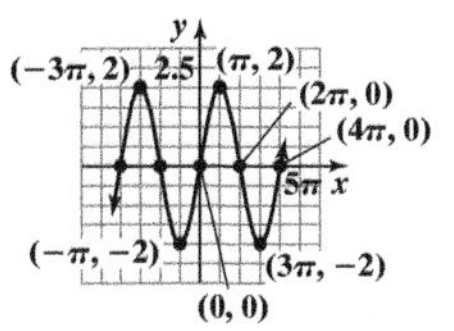

Domain: $(-\infty, \infty)$
Range: $[-2, 2]$

43.

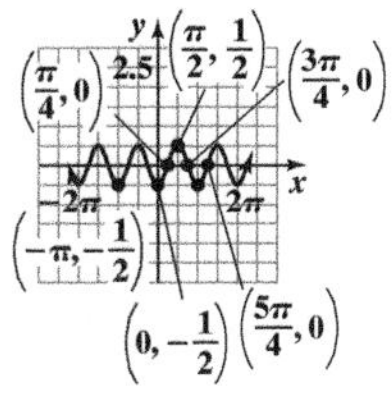

Domain: $(-\infty, \infty)$
Range: $\left[-\frac{1}{2}, \frac{1}{2}\right]$

45.

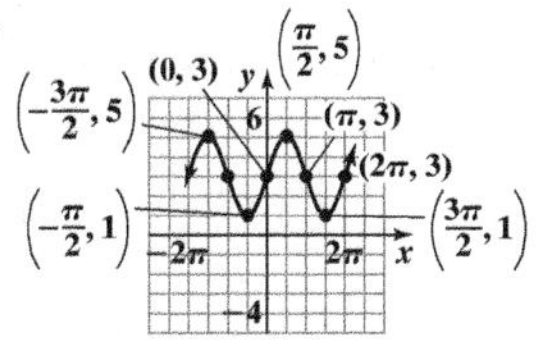

Domain: $(-\infty, \infty)$
Range: $[1, 5]$

47.

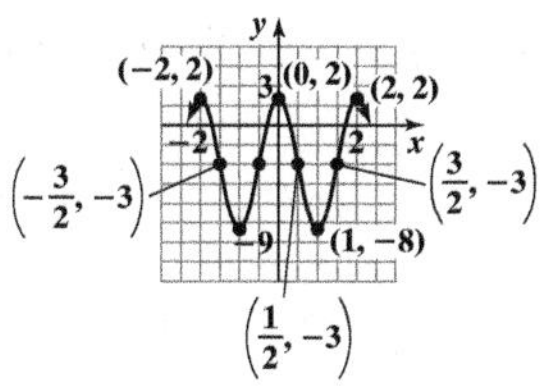

Domain: $(-\infty, \infty)$
Range: $[-8, 2]$

49.

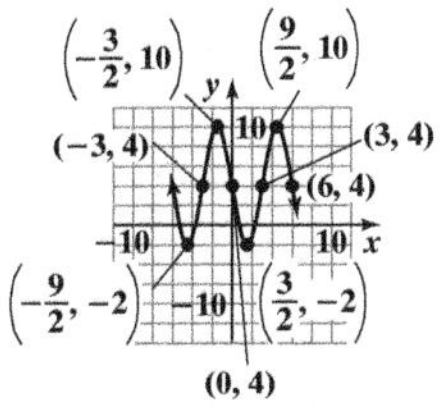

Domain: $(-\infty, \infty)$
Range: $[-2, 10]$

51.

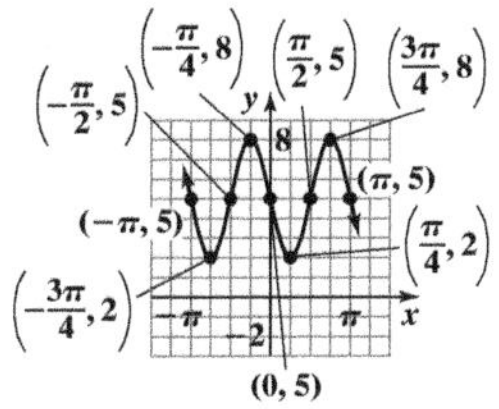

Domain: $(-\infty, \infty)$
Range: $[2, 8]$

53.

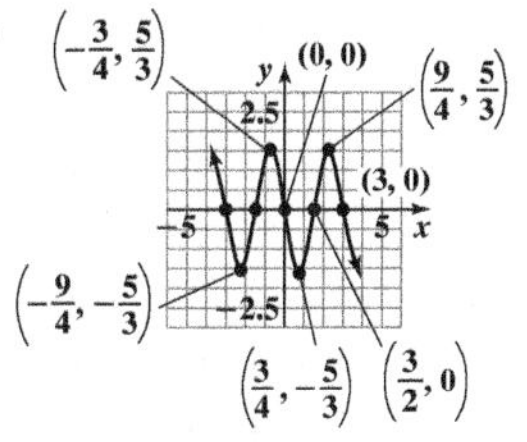

Domain: $(-\infty, \infty)$
Range: $\left[-\frac{5}{3}, \frac{5}{3}\right]$

55.

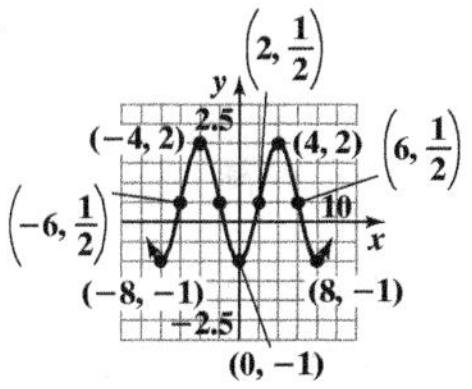

Domain: $(-\infty, \infty)$
Range: $[-1, 2]$

57. $y = \pm 3 \sin(2x)$ **59.** $y = \pm 3 \sin(\pi x)$ **61.** $y = 5 \cos\left(\frac{\pi}{4}x\right)$ **63.** $y = -3 \cos\left(\frac{1}{2}x\right)$ **65.** $y = \frac{3}{4} \sin(2\pi x)$ **67.** $y = -\sin\left(\frac{3}{2}x\right)$

69. $y = -\cos\left(\frac{4\pi}{3}x\right) + 1$ **71.** $y = 3 \sin\left(\frac{\pi}{2}x\right)$ **73.** $y = -4 \cos(3x)$ **75.** $\frac{2}{\pi}$ **77.** $\frac{\sqrt{2}}{\pi}$

79. $(f \circ g)(x) = \sin(4x)$

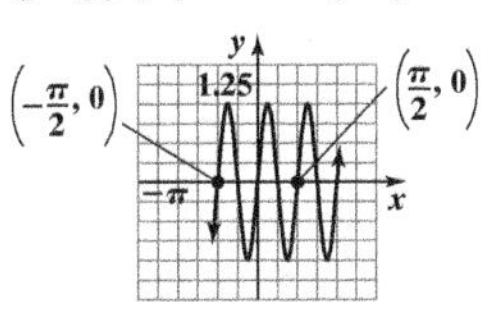

$(g \circ f)(x) = 4 \sin x$

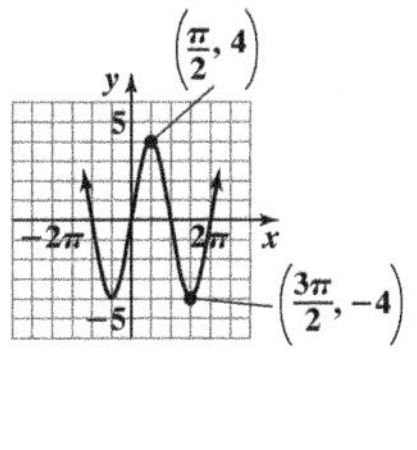

81. $(f \circ g)(x) = -2 \cos x$

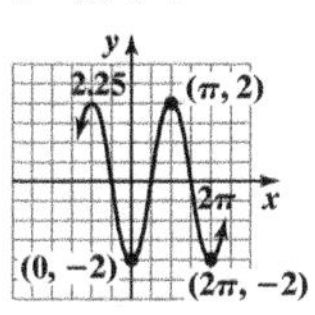

$(g \circ f)(x) = \cos(-2x)$

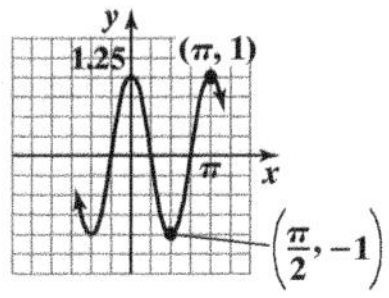

83.

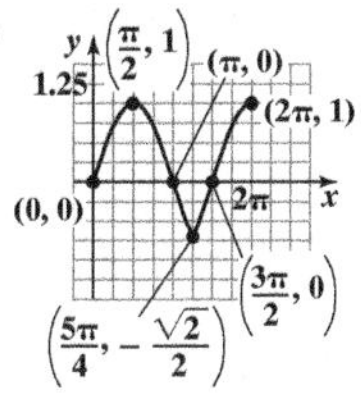

85. Period $= \frac{1}{30}$ s

Amplitude $= 220$ amp

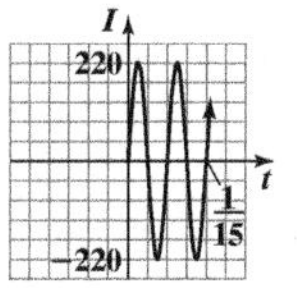

87. (a) Amplitude $= 220$ V

Period $= \frac{1}{60}$ s

(b), (e)

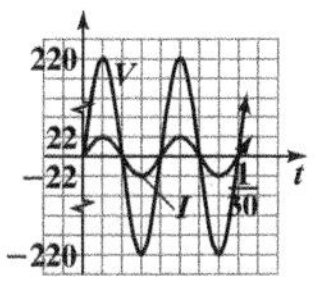

(c) $I(t) = 22 \sin(120\pi t)$
(d) Amplitude $= 22$ amp

Period $= \frac{1}{60}$ s

89. (a) $P(t) = \frac{[V_0 \sin(2\pi f t)]^2}{R} = \frac{V_0^2}{R} \sin^2(2\pi f t)$ **(b)** Since the graph of P has amplitude $\frac{V_0^2}{2R}$ and period $\frac{1}{2f}$ and is of the form $y = A\cos(\omega t) + B$, then $A = -\frac{V_0^2}{2R}$ and $B = \frac{V_0^2}{2R}$. Since $\frac{1}{2f} = \frac{2\pi}{\omega}$, then $\omega = 4\pi f$. Therefore, $P(t) = -\frac{V_0^2}{2R}\cos(4\pi f t) + \frac{V_0^2}{2R} = \frac{V_0^2}{2R}[1 - \cos(4\pi f t)]$.

91.

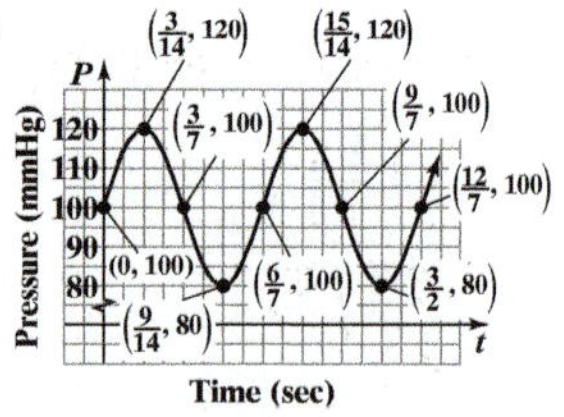

93.

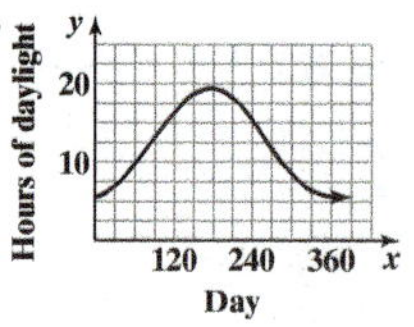

95. (a) Physical potential: $\omega = \frac{2\pi}{23}$; emotional potential: $\omega = \frac{\pi}{14}$; intellectual potential: $\omega = \frac{2\pi}{33}$

(b)

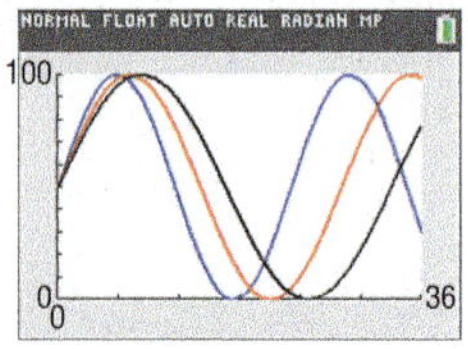

(c) No **(d)** Physical potential peaks at 15 days after 20th birthday. Emotional potential is 50% at 17 days, with a maximum at 10 days and a minimum at 24 days. Intellectual potential starts fairly high, drops to a minimum at 13 days, and rises to a maximum at 29 days.

97.

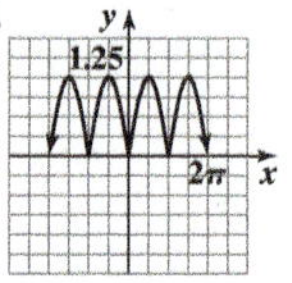

99. Answers may vary. $\left(-\frac{5\pi}{3}, \frac{1}{2}\right), \left(-\frac{\pi}{3}, \frac{1}{2}\right), \left(\frac{\pi}{3}, \frac{1}{2}\right), \left(\frac{5\pi}{3}, \frac{1}{2}\right)$ **101.** Answers may vary. $\left(-\frac{3\pi}{4}, 1\right), \left(\frac{\pi}{4}, 1\right), \left(\frac{5\pi}{4}, 1\right), \left(\frac{9\pi}{4}, 1\right)$ **107.** $2x + h - 5$

108. $(2, 5)$ **109.** $(0, 5), \left(-\frac{5}{3}, 0\right), \left(-\frac{7}{3}, 0\right)$ **110.** $\{-8\}$

7.7 Assess Your Understanding *(page 593)*

3. origin; odd multiples of $\frac{\pi}{2}$ **4.** y-axis; odd multiples of $\frac{\pi}{2}$ **5.** b **6.** T **7.** 0 **9.** 1

11. $\sec x = 1$ for $x = -2\pi, 0, 2\pi$; $\sec x = -1$ for $x = -\pi, \pi$ **13.** $-\frac{3\pi}{2}, -\frac{\pi}{2}, \frac{\pi}{2}, \frac{3\pi}{2}$ **15.** $-\frac{3\pi}{2}, -\frac{\pi}{2}, \frac{\pi}{2}, \frac{3\pi}{2}$

17.

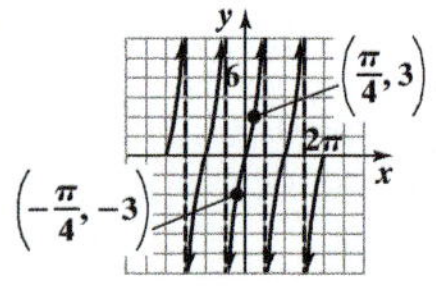

Domain: $\left\{x \middle| x \neq \frac{k\pi}{2}, \ k \text{ is an odd integer}\right\}$

Range: $(-\infty, \infty)$

19.

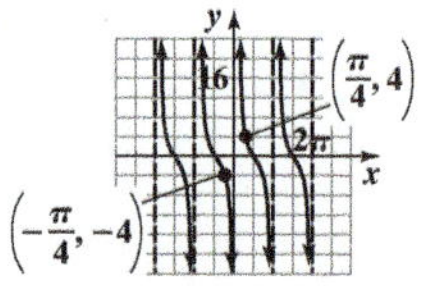

Domain: $\{x \mid x \neq k\pi, \ k \text{ is an integer}\}$

Range: $(-\infty, \infty)$

21.

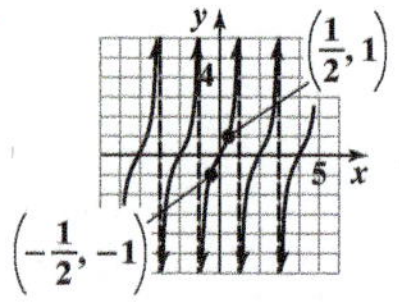

Domain: $\{x \mid x \text{ does not equal an odd integer}\}$

Range: $(-\infty, \infty)$

23.

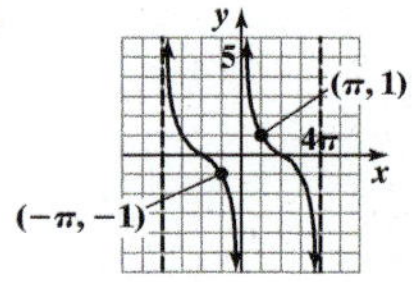

Domain: $\{x \mid x \neq 4k\pi, k \text{ is an integer}\}$

Range: $(-\infty, \infty)$

25.

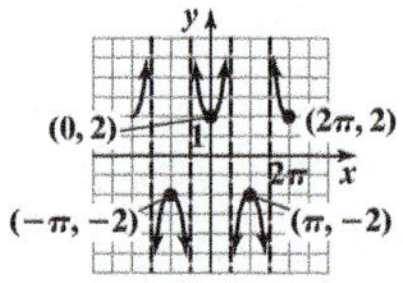

Domain: $\left\{x \middle| x \neq \frac{k\pi}{2}, \ k \text{ is an odd integer}\right\}$

Range: $\{y \mid y \leq -2 \text{ or } y \geq 2\}$

27.

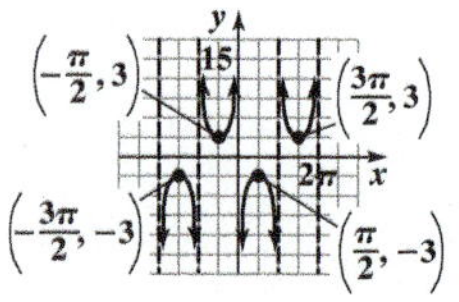

Domain: $\{x \mid x \neq k\pi, k \text{ is an integer}\}$

Range: $\{y \mid y \leq -3 \text{ or } y \geq 3\}$

29.

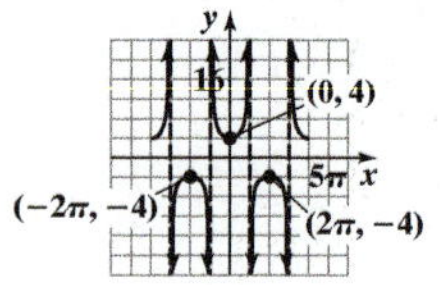

Domain: $\{x \mid x \neq k\pi, k \text{ is an odd integer}\}$

Range: $\{y \mid y \leq -4 \text{ or } y \geq 4\}$

31.

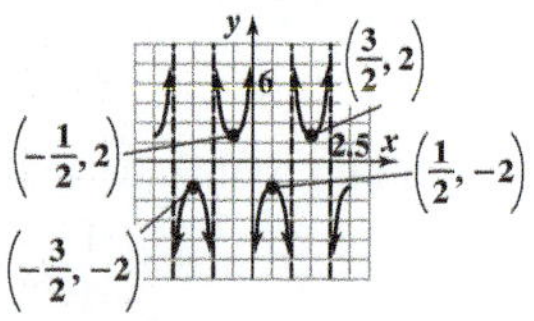

Domain: $\{x \mid x \text{ does not equal an integer}\}$

Range: $\{y \mid y \leq -2 \text{ or } y \geq 2\}$

33.

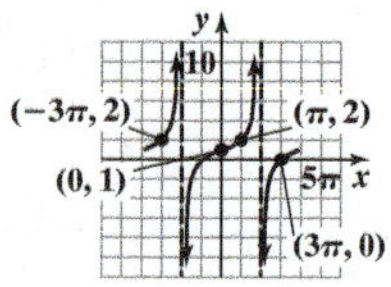

Domain: $\{x \mid x \neq 2\pi k, k \text{ is an odd integer}\}$

Range: $(-\infty, \infty)$

35.

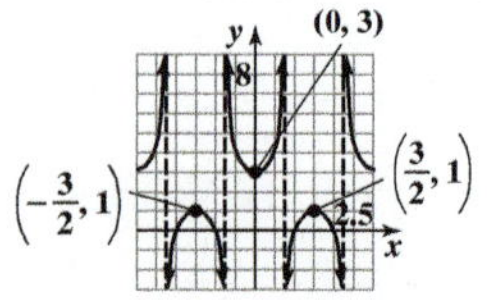

Domain: $\left\{x \middle| x \neq \frac{3}{4}k, k \text{ is an odd integer}\right\}$

Range: $\{y \mid y \leq 1 \text{ or } y \geq 3\}$

37.

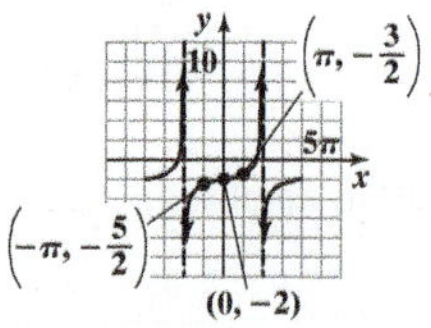

Domain: $\{x \mid x \neq 2\pi k, k \text{ is an odd integer}\}$

Range: $(-\infty, \infty)$

39.

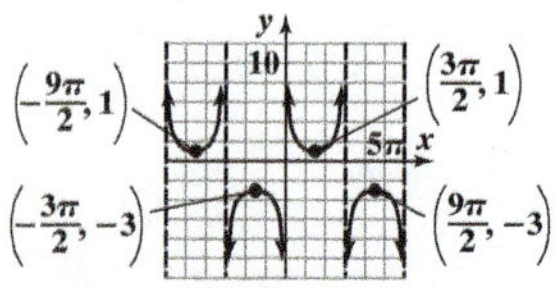

Domain: $\{x \mid x \neq 3\pi k, k \text{ is an integer}\}$

Range: $\{y \mid y \leq -3 \text{ or } y \geq 1\}$

41. $\frac{2\sqrt{3}}{\pi}$ **43.** $\frac{6\sqrt{3}}{\pi}$ **45.** $(f \circ g)(x) = \tan(4x)$

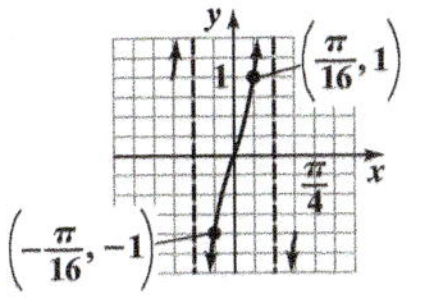

$(g \circ f)(x) = 4\tan x$

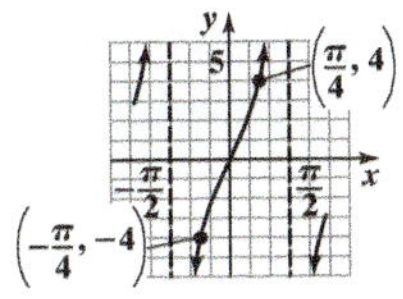

47. $(f \circ g)(x) = -2\cot x$

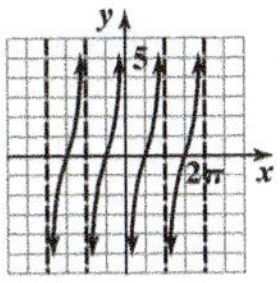

$(g \circ f)(x) = \cot(-2x)$

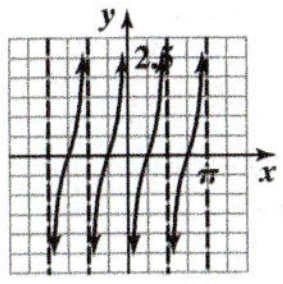

49.

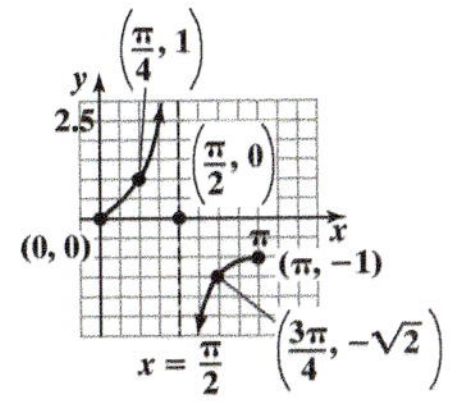

51. (a) $L(\theta) = \frac{3}{\cos\theta} + \frac{4}{\sin\theta} = 3\sec\theta + 4\csc\theta$
(c) ≈ 0.83 **(d)** ≈ 9.86 ft

(b)

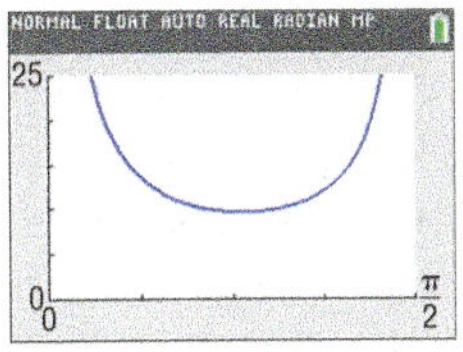

53.

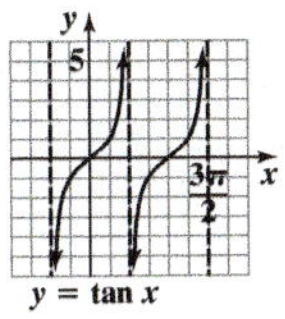

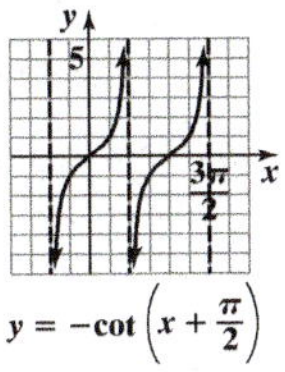

54. $(5p - 2q^2)(25p^2 + 10pq^2 + 4q^4)$ **55.** Hazel: 4 hours; Gwyneth: 6 hours **56.** $\{-1, 3\}$ **57.**

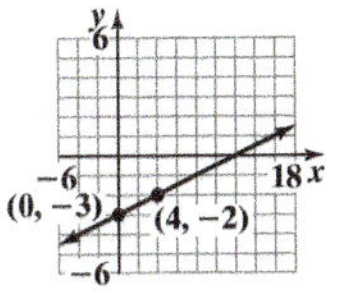

7.8 Assess Your Understanding *(page 603)*

1. phase shift **2.** False

3. Amplitude = 4
Period = π
Phase shift = $\frac{\pi}{2}$

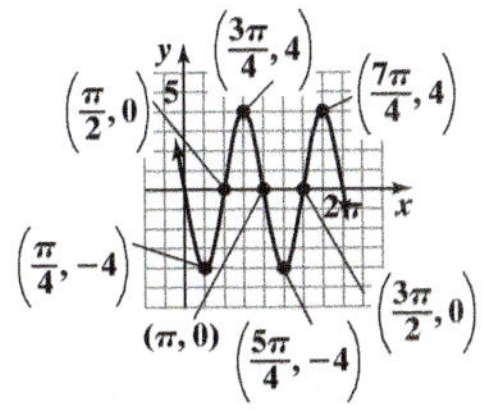

5. Amplitude = 2
Period = $\frac{2\pi}{3}$
Phase shift = $-\frac{\pi}{6}$

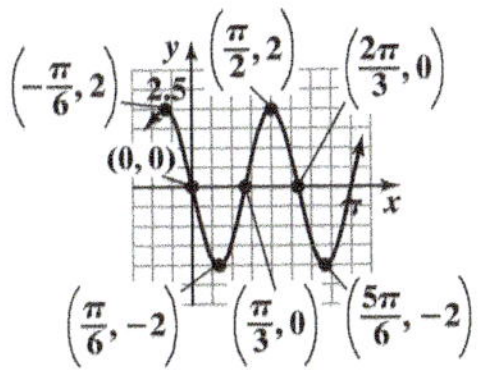

7. Amplitude = 3
Period = π
Phase shift = $-\frac{\pi}{4}$

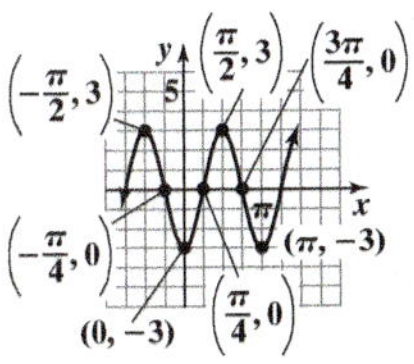

9. Amplitude = 4
Period = 2
Phase shift = $-\frac{2}{\pi}$

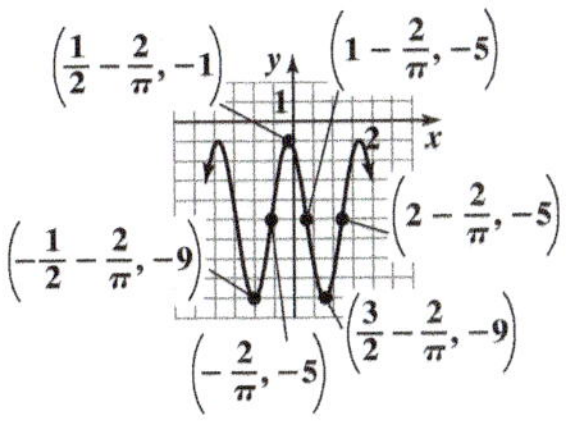

11. Amplitude = 3
Period = 2
Phase shift = $\frac{2}{\pi}$

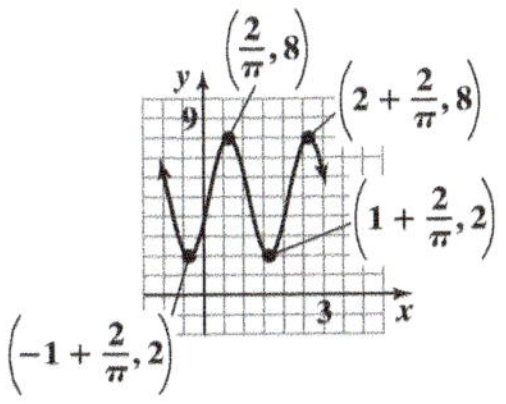

13. Amplitude = 3
Period = π
Phase shift = $\frac{\pi}{4}$

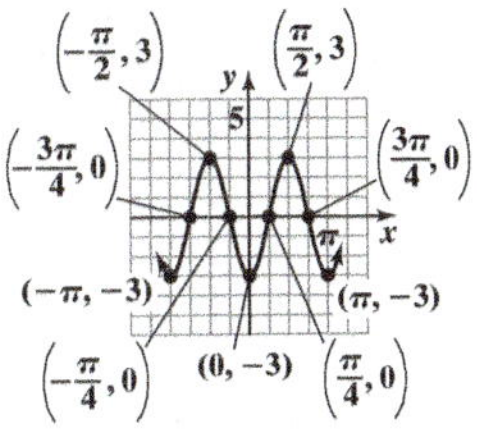

15. $y = 2\sin\left[2\left(x - \frac{1}{2}\right)\right]$ or $y = 2\sin(2x - 1)$

17. $y = 3\sin\left[\frac{2}{3}\left(x + \frac{1}{3}\right)\right]$ or $y = 3\sin\left(\frac{2}{3}x + \frac{2}{9}\right)$

19.

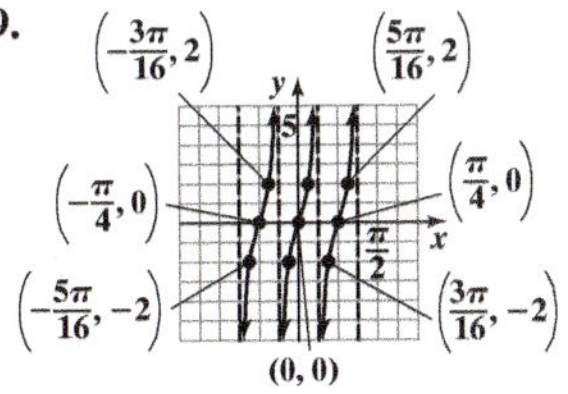

21.

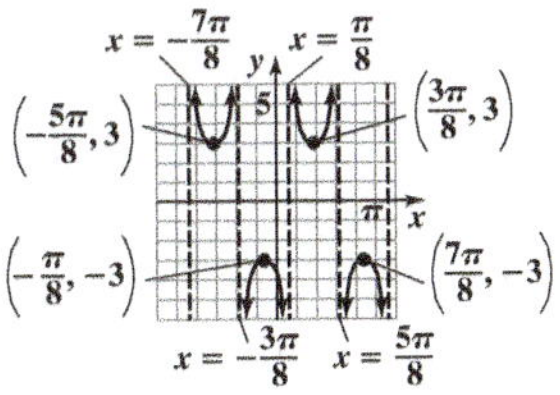

23.

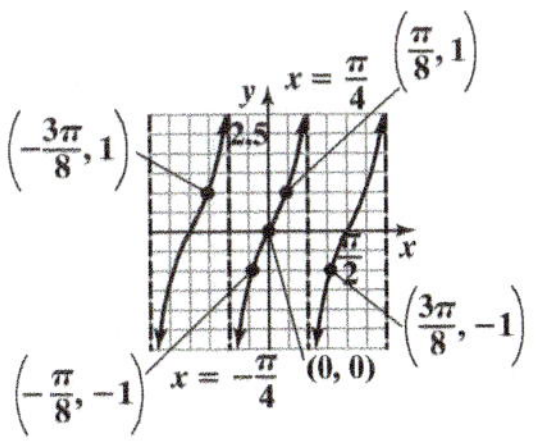

25.

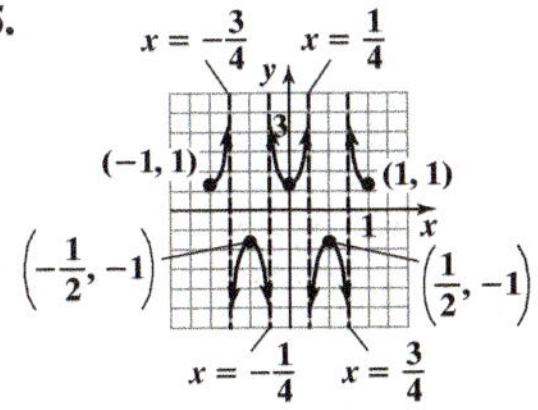

27. Period = $\frac{1}{15}$ s
Amplitude = 120 amp
Phase shift = $\frac{1}{90}$ s

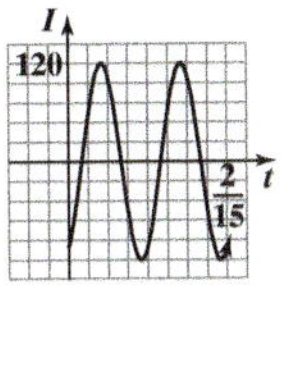

29. (a)

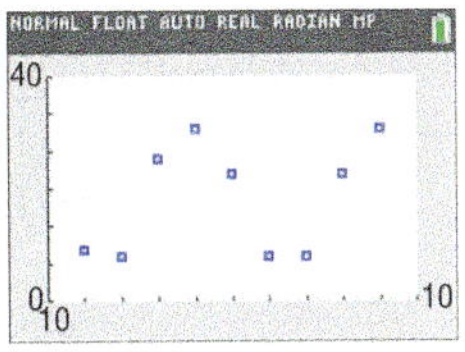

(b) $y = 8.5\sin\left[\frac{2\pi}{5}\left(x - \frac{11}{4}\right)\right] + 24.5$
or
$y = 8.5\sin\left(\frac{2\pi}{5}x - \frac{11\pi}{10}\right) + 24.5$

(c)

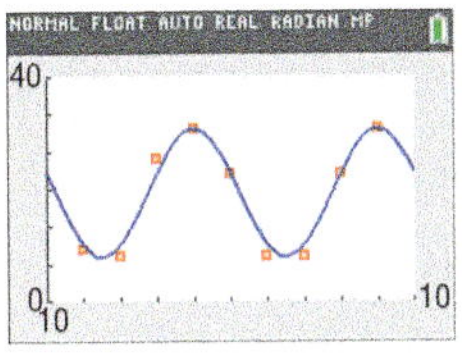

(d) $y = 9.46\sin(1.247x + 2.096) + 24.088$

(e)

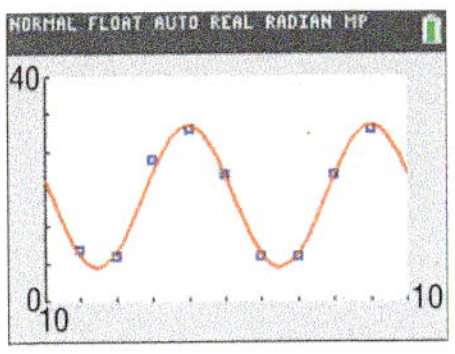

31. (a) 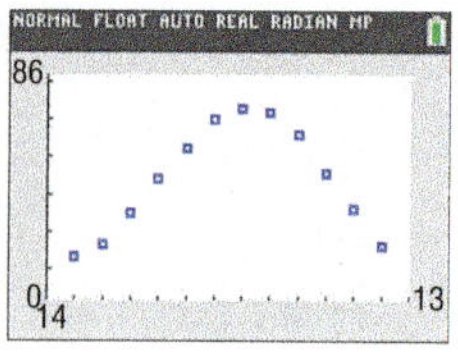

(b) $y = 23.65\sin\left[\frac{\pi}{6}(x-4)\right] + 51.75$ or $y = 23.65\sin\left(\frac{\pi}{6}x - \frac{2\pi}{3}\right) + 51.75$

(c) 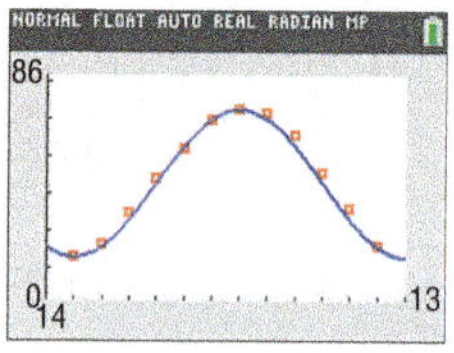

(d) $y = 24.25\sin(0.493x - 1.927) + 51.61$

(e) 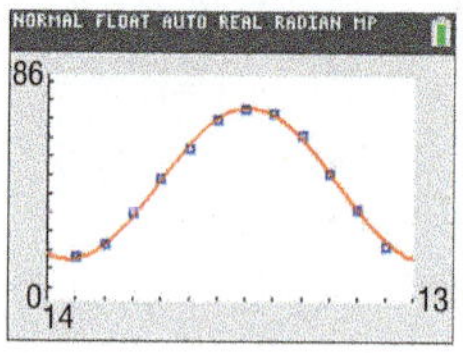

33. (a) 2:37 PM **(b)** $y = 2.2\sin\left[\frac{24\pi}{149}(x + 0.9042)\right] + 3.07$ or $y = 2.2\sin\left[\frac{24\pi}{149}x + 0.4575\right] + 3.07$ **(c)** 2.51 ft

35. (a) $y = 1.615\sin\left(\frac{2\pi}{365}x - 1.39\right) + 12.135$

(b) 12.42 h

(c) 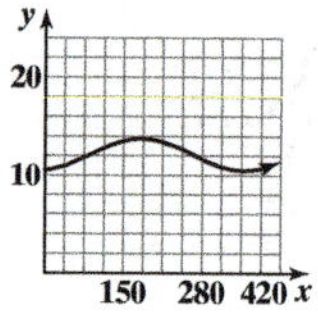

(d) The actual hours of sunlight on April 1, 2015, were 12.43 hours. This is close to the predicted amount of 12.42 hours.

37. (a) $y = 6.96\sin\left(\frac{2\pi}{365}x - 1.39\right) + 12.41$

(b) 13.63 h

(c) 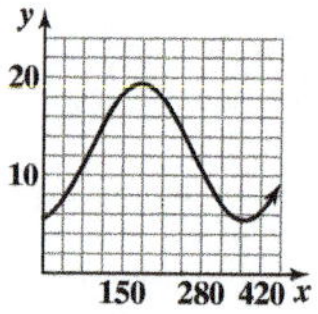

(d) The actual hours of sunlight on April 1, 2015, were 13.35 hours. This is close to the predicted amount of 13.63 hours.

41. $f^{-1}(x) = \frac{2x-9}{4}$ **42.** $\left\{\frac{7}{3}\right\}$ **43.** $64x^2 + 240xy + 225y^2$ **44.** $2\sqrt{13}$

Review Exercises *(page 609)*

1. $\frac{3\pi}{4}$ **2.** $\frac{\pi}{12}$ **3.** 120° **4.** −450° **5.** $\frac{1}{2}$ **6.** $\frac{3\sqrt{2}}{2} - \frac{4\sqrt{3}}{3}$ **7.** $-3\sqrt{2} - 2\sqrt{3}$ **8.** 3 **9.** 0 **10.** 0 **11.** 1 **12.** 1 **13.** 1 **14.** −1 **15.** 1

16. $\cos\theta = \frac{3}{5}$; $\tan\theta = \frac{4}{3}$; $\csc\theta = \frac{5}{4}$; $\sec\theta = \frac{5}{3}$; $\cot\theta = \frac{3}{4}$ **17.** $\sin\theta = -\frac{12}{13}$; $\cos\theta = -\frac{5}{13}$; $\csc\theta = -\frac{13}{12}$; $\sec\theta = -\frac{13}{5}$; $\cot\theta = \frac{5}{12}$

18. $\sin\theta = \frac{3}{5}$; $\cos\theta = -\frac{4}{5}$; $\tan\theta = -\frac{3}{4}$; $\csc\theta = \frac{5}{3}$; $\cot\theta = -\frac{4}{3}$ **19.** $\sin\theta = -\frac{4}{5}$; $\tan\theta = \frac{4}{3}$; $\csc\theta = -\frac{5}{4}$; $\sec\theta = -\frac{5}{3}$; $\cot\theta = \frac{3}{4}$

20. $\cos\theta = \frac{12}{13}$; $\tan\theta = -\frac{5}{12}$; $\csc\theta = -\frac{13}{5}$; $\sec\theta = \frac{13}{12}$; $\cot\theta = -\frac{12}{5}$ **21.** $\sin\theta = -\frac{\sqrt{10}}{10}$; $\cos\theta = -\frac{3\sqrt{10}}{10}$; $\csc\theta = -\sqrt{10}$; $\sec\theta = -\frac{\sqrt{10}}{3}$; $\cot\theta = 3$

22. $\sin\theta = -\frac{1}{4}$; $\cos\theta = -\frac{\sqrt{15}}{4}$; $\tan\theta = \frac{\sqrt{15}}{15}$; $\sec\theta = -\frac{4\sqrt{15}}{15}$; $\cot\theta = \sqrt{15}$ **23.** $\sin\theta = \frac{\sqrt{5}}{5}$; $\cos\theta = -\frac{2\sqrt{5}}{5}$; $\tan\theta = -\frac{1}{2}$; $\csc\theta = \sqrt{5}$; $\sec\theta = -\frac{\sqrt{5}}{2}$

24. 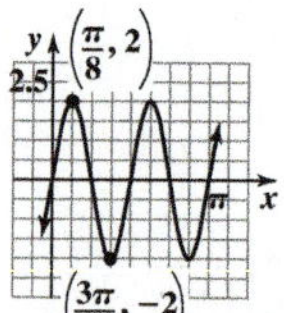

Domain: $(-\infty, \infty)$
Range: $[-2, 2]$

25. 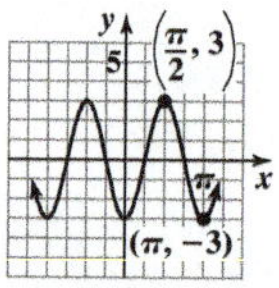

Domain: $(-\infty, \infty)$
Range: $[-3, 3]$

26. 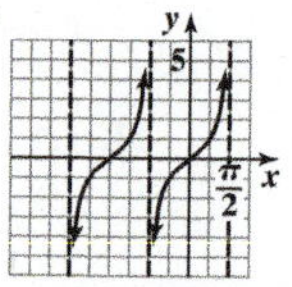

Domain: $\left\{x \middle| x \neq \frac{k\pi}{2}, k \text{ is an odd integer}\right\}$
Range: $(-\infty, \infty)$

27. 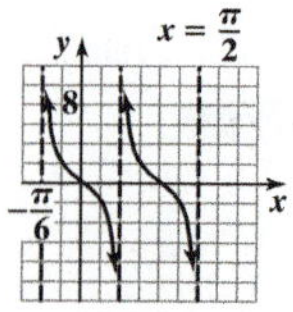

Domain: $\left\{x \middle| x \neq \frac{\pi}{6} + k\cdot\frac{\pi}{3}, k \text{ is an integer}\right\}$
Range: $(-\infty, \infty)$

28. 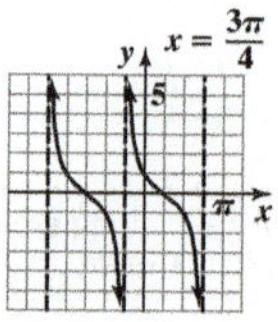

Domain: $\left\{x \middle| x \neq -\frac{\pi}{4} + k\pi, k \text{ is an integer}\right\}$
Range: $(-\infty, \infty)$

29. 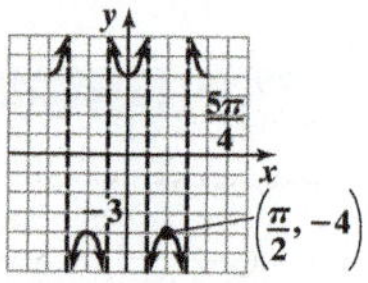

Domain: $\left\{x \middle| x \neq \frac{k\pi}{4}, k \text{ is an odd integer}\right\}$
Range: $\{y | y \leq -4 \text{ or } y \geq 4\}$

30.

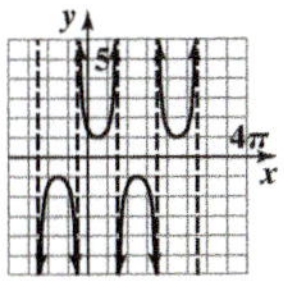

Domain: $\left\{x \middle| x \neq -\frac{\pi}{4} + k\pi, k \text{ is an integer}\right\}$

Range: $\{y | y \leq -1 \text{ or } y \geq 1\}$

31.

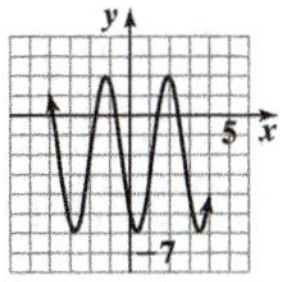

Domain: $(-\infty, \infty)$
Range: $[-6, 2]$

32.

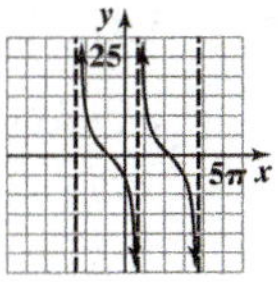

Domain: $\left\{x \middle| x \neq \frac{3\pi}{4} + k \cdot 3\pi, k \text{ is an integer}\right\}$

Range: $(-\infty, \infty)$

33. 1; π **34.** 2; $\frac{2}{3}$

35. Amplitude $= 4$

Period $= \frac{2\pi}{3}$

Phase shift $= 0$

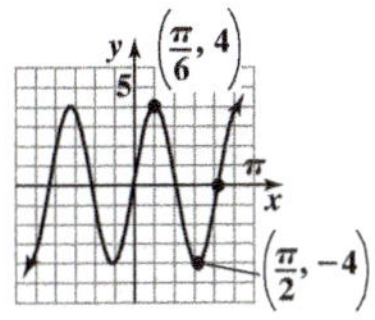

36. Amplitude $= 1$
Period $= 4\pi$
Phase shift $= -\pi$

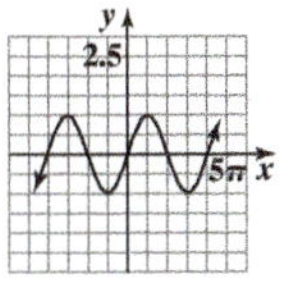

37. Amplitude $= \frac{1}{2}$

Period $= \frac{4\pi}{3}$

Phase shift $= \frac{2\pi}{3}$

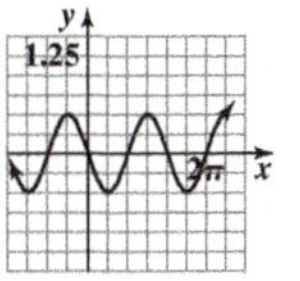

38. Amplitude $= \frac{2}{3}$

Period $= 2$

Phase shift $= \frac{6}{\pi}$

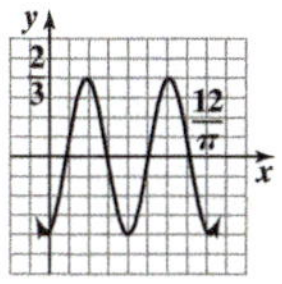

39. $y = 5\cos\frac{x}{4}$ **40.** $y = -7\sin\left(\frac{\pi}{4}x\right)$ **41.** $\sin\theta = \frac{3}{7}$; $\cos\theta = \frac{2\sqrt{10}}{7}$; $\tan\theta = \frac{3\sqrt{10}}{20}$; $\csc\theta = \frac{7}{3}$; $\sec\theta = \frac{7\sqrt{10}}{20}$; $\cot\theta = \frac{2\sqrt{10}}{3}$ **42.** 0.38; 1.02

43. Sine, cosine, cosecant and secant: negative; tangent and cotangent: positive **44.** 30°

45. $\sin\theta = \frac{2\sqrt{2}}{3}$; $\cos\theta = -\frac{1}{3}$; $\tan\theta = -2\sqrt{2}$; $\csc\theta = \frac{3\sqrt{2}}{4}$; $\sec\theta = -3$; $\cot\theta = -\frac{\sqrt{2}}{4}$ **46.** $\sin t = \frac{5\sqrt{29}}{29}$; $\cos t = -\frac{2\sqrt{29}}{29}$; $\tan t = -\frac{5}{2}$

47. Domain: $\left\{x \middle| x \neq \text{odd multiple of } \frac{\pi}{2}\right\}$; range: $\{y \,|\, |y| \geq 1\}$; period $= 2\pi$ **48. (a)** 32.34° **(b)** 63°10′48″ **49.** $\frac{\pi}{3} \approx 1.05$ ft; $\frac{\pi}{3} \approx 1.05$ ft^2

50. $8\pi \approx 25.13$ in.; $\frac{16\pi}{3} \approx 16.76$ in. **51.** Approximately 114.59 revolutions/h **52.** 0.1 revolution/sec $= \frac{\pi}{5}$ radian/sec **53.** 839.10 ft

54. 23.32 ft **55.** 2.15 mi **56. (a)** $\frac{1}{15}$ **(b)** 220 **(c)** $-\frac{1}{180}$ **(d)**

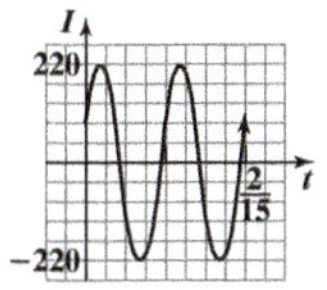

57. (a)

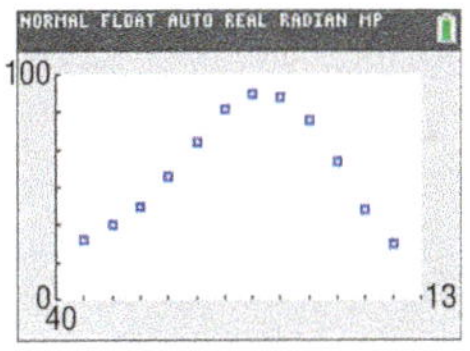

(b) $y = 20\sin\left[\frac{\pi}{6}(x-4)\right] + 75$ or

$y = 20\sin\left(\frac{\pi}{6}x - \frac{2\pi}{3}\right) + 75$

(c)

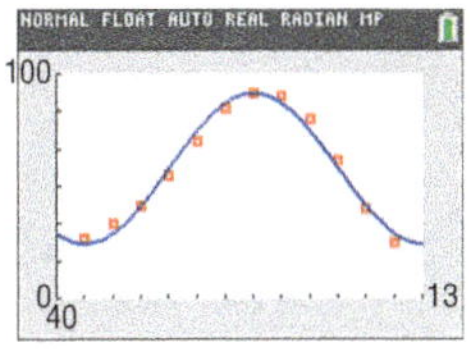

(d) $y = 19.81\sin(0.543x - 2.296) + 75.66$

(e)

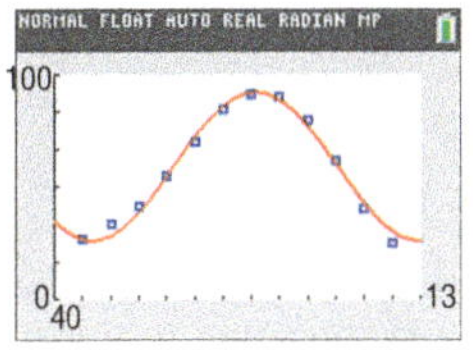

Chapter Test *(page 611)*

1. $\frac{13\pi}{9}$ **2.** $-\frac{20\pi}{9}$ **3.** $\frac{13\pi}{180}$ **4.** $-22.5°$ **5.** 810° **6.** 135° **7.** $\frac{1}{2}$ **8.** 0 **9.** $-\frac{1}{2}$ **10.** $-\frac{\sqrt{3}}{3}$ **11.** 2 **12.** $\frac{3(1-\sqrt{2})}{2}$ **13.** 0.292 **14.** 0.309

15. -1.524 **16.** 2.747 **17.**

	$\sin\theta$	$\cos\theta$	$\tan\theta$	$\sec\theta$	$\csc\theta$	$\cot\theta$
θ in QI	+	+	+	+	+	+
θ in QII	+	−	−	−	+	−
θ in QIII	−	−	+	−	−	+
θ in QIV	−	+	−	+	−	−

18. $-\frac{3}{5}$ **19.** $\cos\theta = -\frac{2\sqrt{6}}{7}$; $\tan\theta = -\frac{5\sqrt{6}}{12}$; $\csc\theta = \frac{7}{5}$; $\sec\theta = -\frac{7\sqrt{6}}{12}$; $\cot\theta = -\frac{2\sqrt{6}}{5}$ **20.** $\sin\theta = -\frac{\sqrt{5}}{3}$; $\tan\theta = -\frac{\sqrt{5}}{2}$; $\csc\theta = -\frac{3\sqrt{5}}{5}$; $\sec\theta = \frac{3}{2}$; $\cot\theta = -\frac{2\sqrt{5}}{5}$ **21.** $\sin\theta = \frac{12}{13}$; $\cos\theta = -\frac{5}{13}$; $\csc\theta = \frac{13}{12}$; $\sec\theta = -\frac{13}{5}$; $\cot\theta = -\frac{5}{12}$ **22.** $\frac{7\sqrt{53}}{53}$ **23.** $-\frac{5\sqrt{146}}{146}$ **24.** $-\frac{1}{2}$

25.

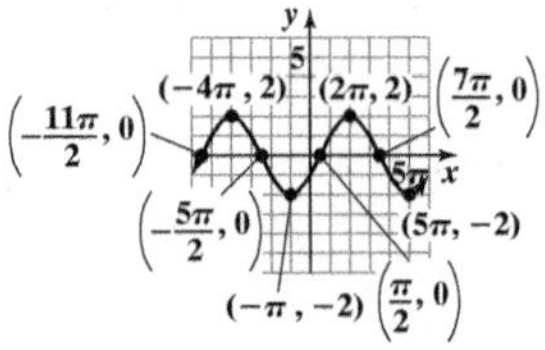

26.

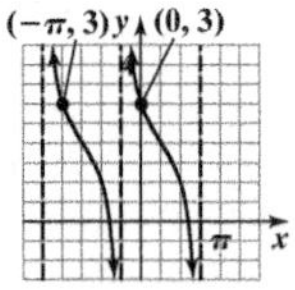

27. $y = -3\sin\left(3x + \frac{3\pi}{4}\right)$

28. 78.93 ft^2

29. 143.5 rpm

30. The ship is about 838 ft from the statue.

31. The building is 122.4 ft tall.

Cumulative Review *(page 612)*

1. $\left\{-1, \frac{1}{2}\right\}$ **2.** $y - 5 = -3(x + 2)$ or $y = -3x - 1$ **3.** $x^2 + (y + 2)^2 = 16$

4. A line; slope $\frac{2}{3}$; intercepts $(6, 0)$ and $(0, -4)$

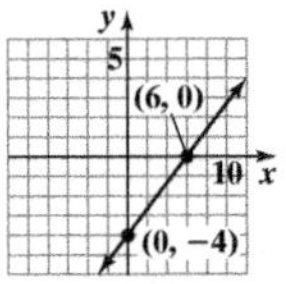

5. A circle; center $(1, -2)$; radius 3

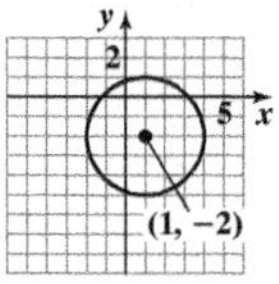

6.

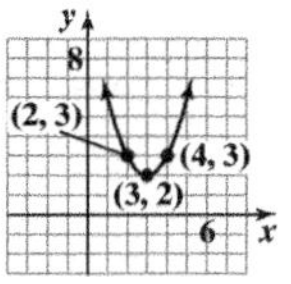

7. (a)

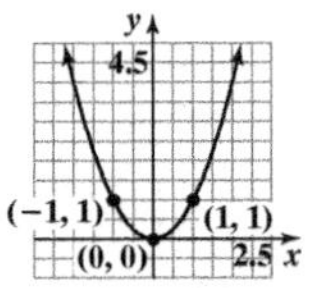

(b)

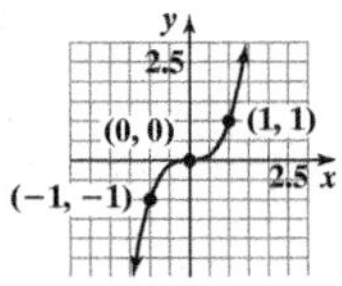

(c)

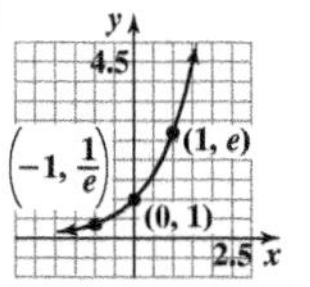

(d)

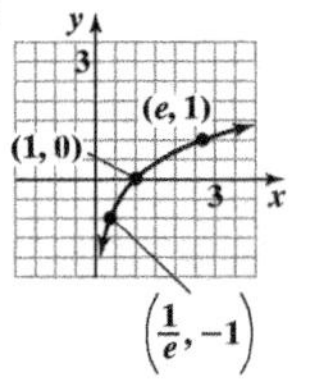

(e)

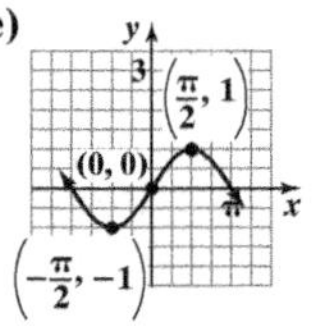

(f)

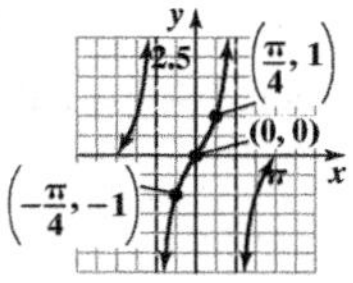

8. $f^{-1}(x) = \frac{1}{3}(x + 2)$ **9.** -2

10.

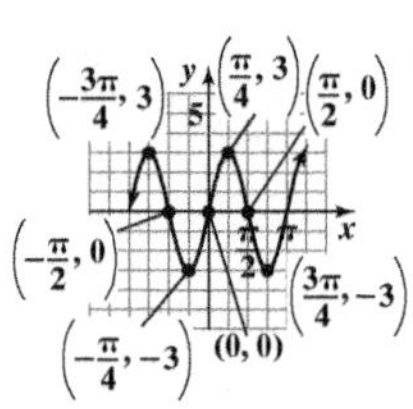

11. $3 - \frac{3\sqrt{3}}{2}$ **12.** $y = 2(3^x)$ **13.** $y = 3\cos\left(\frac{\pi}{6}x\right)$

14. (a) $f(x) = -3x - 3$;
$m = -3$; $(-1, 0)$, $(0, -3)$

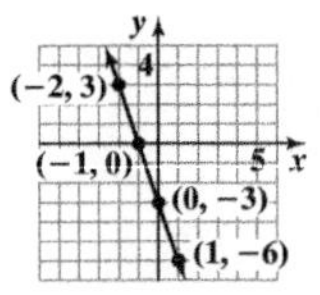

(b) $f(x) = (x - 1)^2 - 6$; $(0, -5)$, $(-\sqrt{6} + 1, 0)$, $(\sqrt{6} + 1, 0)$

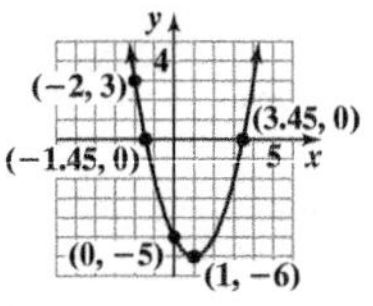

(c) We have that $y = 3$ when $x = -2$ and $y = -6$ when $x = 1$. Both points satisfy $y = ae^x$. Therefore, for $(-2, 3)$ we have $3 = ae^{-2}$, which implies that $a = 3e^2$. But for $(1, -6)$ we have $-6 = ae^1$, which implies that $a = -6e^{-1}$. Therefore, there is no exponential function $y = ae^x$ that contains $(-2, 3)$ and $(1, -6)$.

15. (a) $f(x) = \frac{1}{6}(x + 2)(x - 3)(x - 5)$

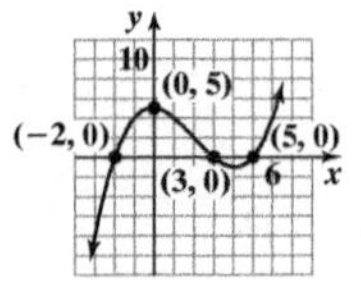

(b) $R(x) = -\frac{(x + 2)(x - 3)(x - 5)}{3(x - 2)}$

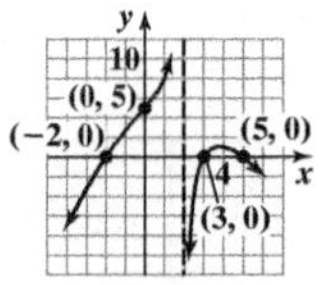

CHAPTER 8 Analytic Trigonometry

8.1 Assess Your Understanding *(page 625)*

7. $x = \sin y$ **8.** $0 \le x \le \pi$ **9.** $-\infty < x < \infty$ **10.** F **11.** T **12.** T **13.** d **14.** a **15.** 0 **17.** $-\frac{\pi}{2}$ **19.** 0 **21.** $\frac{\pi}{4}$ **23.** $\frac{\pi}{3}$ **25.** $\frac{5\pi}{6}$ **27.** 0.10 **29.** 1.37 **31.** 0.51 **33.** -0.38 **35.** -0.12 **37.** 1.08 **39.** $\frac{4\pi}{5}$ **41.** $-\frac{3\pi}{8}$ **43.** $-\frac{\pi}{8}$ **45.** $\frac{\pi}{3}$ **47.** $-\frac{\pi}{5}$ **49.** $\frac{\pi}{3}$ **51.** $\frac{\pi}{4}$ **53.** Not defined **55.** $\frac{1}{4}$ **57.** 4 **59.** Not defined **61.** π

63. $f^{-1}(x) = \sin^{-1}\frac{x-2}{5}$
Range of f = Domain of $f^{-1} = [-3, 7]$
Range of $f^{-1} = \left[-\frac{\pi}{2}, \frac{\pi}{2}\right]$

65. $f^{-1}(x) = \frac{1}{3}\cos^{-1}\left(-\frac{x}{2}\right)$
Range of f = Domain of $f^{-1} = [-2, 2]$
Range of $f^{-1} = \left[0, \frac{\pi}{3}\right]$

67. $f^{-1}(x) = -\tan^{-1}(x+3) - 1$
Range of f = Domain of $f^{-1} = (-\infty, \infty)$
Range of $f^{-1} = \left(-1 - \frac{\pi}{2}, \frac{\pi}{2} - 1\right)$

69. $f^{-1}(x) = \frac{1}{2}\left[\sin^{-1}\left(\frac{x}{3}\right) - 1\right]$
Range of f = Domain of $f^{-1} = [-3, 3]$
Range of $f^{-1} = \left[-\frac{1}{2} - \frac{\pi}{4}, -\frac{1}{2} + \frac{\pi}{4}\right]$

71. $\left\{\frac{\sqrt{2}}{2}\right\}$ **73.** $\left\{-\frac{1}{4}\right\}$ **75.** $\{\sqrt{3}\}$ **77.** $\{-1\}$

79. (a) 13.92 h or 13 h, 55 min **(b)** 12 h **(c)** 13.85 h or 13 h, 51 min
81. (a) 13.3 h or 13 h, 18 min **(b)** 12 h **(c)** 13.26 h or 13 h, 15 min
83. (a) 12 h **(b)** 12 h **(c)** 12 h **(d)** It is 12 h. **85.** 3.35 min

87. (a) $\frac{\pi}{3}$ square units **(b)** $\frac{5\pi}{12}$ square units **89.** 4250 mi **91.** $\left[-\frac{2}{3}, 2\right]$ **92.** The graph passes the horizontal-line test.

93. $f^{-1}(x) = \log_2(x-1)$ **94.** $(2x+1)^{-\frac{1}{2}}(x^2+3)^{-\frac{3}{2}}(-x^2-x+3)$

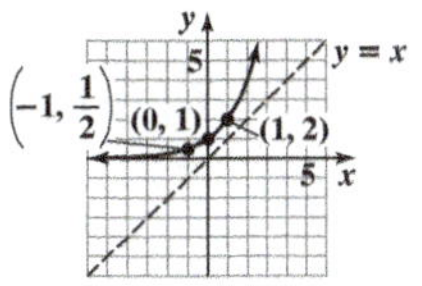

8.2 Assess Your Understanding *(page 632)*

4. $x = \sec y$; ≥ 1; 0; π **5.** cosine **6.** F **7.** T **8.** T **9.** $\frac{\sqrt{2}}{2}$ **11.** $-\frac{\sqrt{3}}{3}$ **13.** 2 **15.** $\sqrt{2}$ **17.** $-\frac{\sqrt{2}}{2}$ **19.** $\frac{2\sqrt{3}}{3}$ **21.** $\frac{3\pi}{4}$ **23.** $-\frac{\pi}{3}$ **25.** $\frac{\sqrt{2}}{4}$ **27.** $\frac{\sqrt{5}}{2}$ **29.** $-\frac{\sqrt{14}}{2}$ **31.** $-\frac{3\sqrt{10}}{10}$ **33.** $\sqrt{5}$ **35.** $-\frac{\pi}{4}$ **37.** $\frac{\pi}{6}$ **39.** $-\frac{\pi}{2}$ **41.** $\frac{\pi}{6}$ **43.** $\frac{2\pi}{3}$ **45.** 1.32 **47.** 0.46 **49.** -0.34 **51.** 2.72 **53.** -0.73 **55.** 2.55 **57.** $\frac{1}{\sqrt{1+u^2}}$ **59.** $\frac{u}{\sqrt{1-u^2}}$ **61.** $\frac{\sqrt{u^2-1}}{|u|}$ **63.** $\frac{\sqrt{u^2-1}}{|u|}$ **65.** $\frac{1}{u}$ **67.** $\frac{5}{13}$ **69.** $\frac{3\pi}{4}$ **71.** $-\frac{3}{4}$ **73.** $\frac{5}{13}$ **75.** $\frac{5\pi}{6}$ **77.** $-\sqrt{15}$

79. (a) $\theta = 31.89°$ **(b)** 54.64 ft in diameter **(c)** 37.96 ft high **81. (a)** $\theta = 22.3°$ **(b)** $v_0 = 2940.23$ ft/s

83.

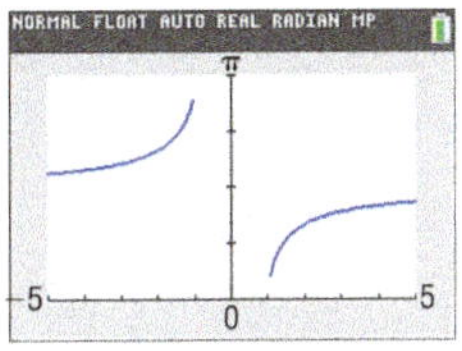

87. $-5i, 5i, -2, 2$ **88.** Neither **89.** $\frac{7\pi}{4}$ **90.** $\frac{5\pi}{2} \approx 7.85$ in.

8.3 Assess Your Understanding *(page 639)*

7. F **8.** T **9.** T **10.** F **11.** d **12.** a **13.** $\left\{\frac{7\pi}{6}, \frac{11\pi}{6}\right\}$ **15.** $\left\{\frac{7\pi}{6}, \frac{11\pi}{6}\right\}$ **17.** $\left\{\frac{3\pi}{4}, \frac{7\pi}{4}\right\}$ **19.** $\left\{\frac{2\pi}{3}, \frac{4\pi}{3}\right\}$ **21.** $\left\{\frac{3\pi}{4}, \frac{5\pi}{4}\right\}$ **23.** $\left\{\frac{\pi}{3}, \frac{2\pi}{3}, \frac{4\pi}{3}, \frac{5\pi}{3}\right\}$ **25.** $\left\{\frac{\pi}{4}, \frac{3\pi}{4}, \frac{5\pi}{4}, \frac{7\pi}{4}\right\}$ **27.** $\left\{\frac{\pi}{2}, \frac{7\pi}{6}, \frac{11\pi}{6}\right\}$ **29.** $\left\{\frac{\pi}{3}, \frac{2\pi}{3}, \frac{4\pi}{3}, \frac{5\pi}{3}\right\}$ **31.** $\left\{\frac{4\pi}{9}, \frac{8\pi}{9}, \frac{16\pi}{9}\right\}$ **33.** $\left\{\frac{3\pi}{4}, \frac{7\pi}{4}\right\}$ **35.** $\left\{\frac{11\pi}{6}\right\}$

37. $\left\{\theta \middle| \theta = \frac{\pi}{6} + 2k\pi, \theta = \frac{5\pi}{6} + 2k\pi\right\}$; $\frac{\pi}{6}, \frac{5\pi}{6}, \frac{13\pi}{6}, \frac{17\pi}{6}, \frac{25\pi}{6}, \frac{29\pi}{6}$ **39.** $\left\{\theta \middle| \theta = \frac{5\pi}{6} + k\pi\right\}$; $\frac{5\pi}{6}, \frac{11\pi}{6}, \frac{17\pi}{6}, \frac{23\pi}{6}, \frac{29\pi}{6}, \frac{35\pi}{6}$

41. $\left\{\theta \middle| \theta = \frac{\pi}{2} + 2k\pi, \theta = \frac{3\pi}{2} + 2k\pi\right\}$; $\frac{\pi}{2}, \frac{3\pi}{2}, \frac{5\pi}{2}, \frac{7\pi}{2}, \frac{9\pi}{2}, \frac{11\pi}{2}$ **43.** $\left\{\theta \middle| \theta = \frac{\pi}{3} + k\pi, \theta = \frac{2\pi}{3} + k\pi\right\}$; $\frac{\pi}{3}, \frac{2\pi}{3}, \frac{4\pi}{3}, \frac{5\pi}{3}, \frac{7\pi}{3}, \frac{8\pi}{3}$

45. $\left\{\theta \middle| \theta = \frac{8\pi}{3} + 4k\pi, \theta = \frac{10\pi}{3} + 4k\pi\right\}$; $\frac{8\pi}{3}, \frac{10\pi}{3}, \frac{20\pi}{3}, \frac{22\pi}{3}, \frac{32\pi}{3}, \frac{34\pi}{3}$ **47.** $\{0.41, 2.73\}$ **49.** $\{1.37, 4.51\}$ **51.** $\{2.69, 3.59\}$ **53.** $\{1.82, 4.46\}$

55. $\{2.08, 5.22\}$ **57.** $\{0.73, 2.41\}$ **59.** $\left\{\frac{\pi}{2}, \frac{2\pi}{3}, \frac{4\pi}{3}, \frac{3\pi}{2}\right\}$ **61.** $\left\{\frac{\pi}{2}, \frac{7\pi}{6}, \frac{11\pi}{6}\right\}$ **63.** $\left\{0, \frac{\pi}{4}, \frac{5\pi}{4}\right\}$ **65.** $\left\{\frac{\pi}{2}, \frac{2\pi}{3}, \frac{4\pi}{3}, \frac{3\pi}{2}\right\}$ **67.** $\{\pi\}$ **69.** $\left\{\frac{\pi}{4}, \frac{5\pi}{4}\right\}$

71. $\left\{0, \frac{\pi}{3}, \pi, \frac{5\pi}{3}\right\}$ **73.** $\left\{\frac{\pi}{6}, \frac{5\pi}{6}, \frac{3\pi}{2}\right\}$ **75.** $\left\{\frac{\pi}{2}\right\}$ **77.** $\{0\}$ **79.** $\left\{\frac{\pi}{3}, \frac{5\pi}{3}\right\}$ **81.** No real solution **83.** $\{-1.31, 1.98, 3.84\}$ **85.** $\{0.52\}$

87. $\{1.26\}$ **89.** $\{-1.02, 1.02\}$ **91.** $\{0, 2.15\}$ **93.** $\{0.76, 1.35\}$ **95.** $\frac{\pi}{3}, \frac{2\pi}{3}, \frac{4\pi}{3}, \frac{5\pi}{3}$

97. (a) $-2\pi, -\pi, 0, \pi, 2\pi, 3\pi, 4\pi$ **(b)**

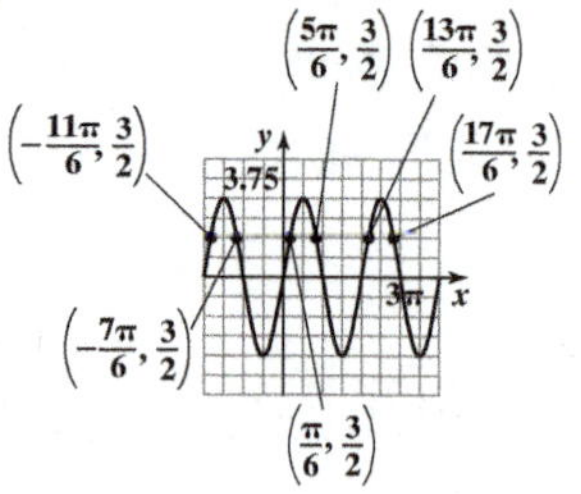

(c) $\left\{-\frac{11\pi}{6}, -\frac{7\pi}{6}, \frac{\pi}{6}, \frac{5\pi}{6}, \frac{13\pi}{6}, \frac{17\pi}{6}\right\}$

(d) $\left\{x \middle| -\frac{11\pi}{6} < x < -\frac{7\pi}{6} \text{ or } \frac{\pi}{6} < x < \frac{5\pi}{6} \text{ or } \frac{13\pi}{6} < x < \frac{17\pi}{6}\right\}$

99. (a) $\left\{x \middle| x = -\frac{\pi}{4} + k\pi, k \text{ is any integer}\right\}$ **(b)** $-\frac{\pi}{2} < x < -\frac{\pi}{4}$ or $\left(-\frac{\pi}{2}, -\frac{\pi}{4}\right)$

101. (a), (d)

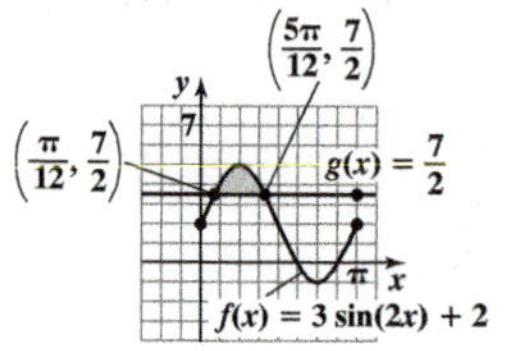

(b) $\left\{\frac{\pi}{12}, \frac{5\pi}{12}\right\}$

(c) $\left\{x \middle| \frac{\pi}{12} < x < \frac{5\pi}{12}\right\}$ or $\left(\frac{\pi}{12}, \frac{5\pi}{12}\right)$

103. (a), (d)

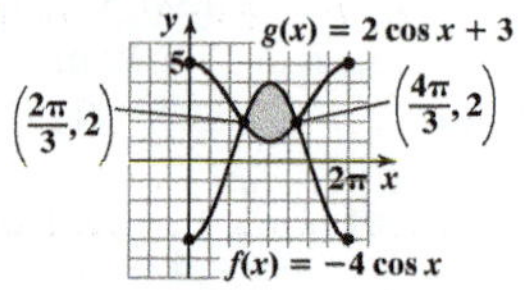

(b) $\left\{\frac{2\pi}{3}, \frac{4\pi}{3}\right\}$

(c) $\left\{x \middle| \frac{2\pi}{3} < x < \frac{4\pi}{3}\right\}$ or $\left(\frac{2\pi}{3}, \frac{4\pi}{3}\right)$

105. (a) 0 s, 0.43 s, 0.86 s **(b)** 0.21 s **(c)** $[0, 0.03] \cup [0.39, 0.43] \cup [0.86, 0.89]$ **107. (a)** 150 mi **(b)** 6.06, 8.44, 15.72, 18.11 min **(c)** Before 6.06 min, between 8.44 and 15.72 min, and after 18.11 min **(d)** No **109.** 2.03, 4.91

111. (a) 30°, 60° **(b)** 123.6 m **113.** 28.90° **115.** Yes; it varies from 1.25 to 1.34. **117.** 1.47

(c)

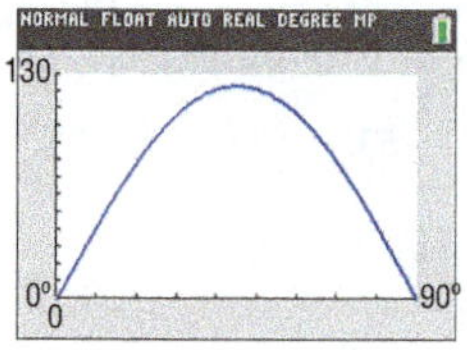

119. If θ is the original angle of incidence and ϕ is the angle of refraction, then $\frac{\sin\theta}{\sin\phi} = n_2$. The angle of incidence of the emerging beam is also ϕ, and the index of refraction is $\frac{1}{n_2}$. Thus, θ is the angle of refraction of the emerging beam.

123. $x = \log_6 y$ **124.** $\frac{9 - \sqrt{17}}{4}, \frac{9 + \sqrt{17}}{4}$ **125.** $\tan\theta = -\frac{1}{3}$; $\csc\theta = -\sqrt{10}$; $\sec\theta = \frac{\sqrt{10}}{3}$; $\cot\theta = -3$ **126.** Amplitude: 2
Period: π
Phase shift: $\frac{\pi}{2}$

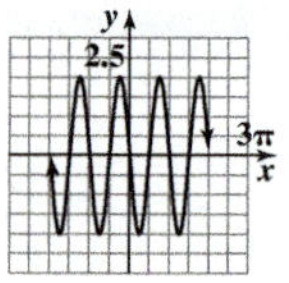

8.4 Assess Your Understanding *(page 649)*

3. identity; conditional **4.** -1 **5.** 0 **6.** T **7.** F **8.** T **9.** c **10.** b **11.** $\frac{1}{\cos\theta}$ **13.** $\frac{1 + \sin\theta}{\cos\theta}$ **15.** $\frac{1}{\sin\theta\cos\theta}$ **17.** 2 **19.** $\frac{3\sin\theta + 1}{\sin\theta + 1}$

21. $\csc\theta \cdot \cos\theta = \frac{1}{\sin\theta} \cdot \cos\theta = \frac{\cos\theta}{\sin\theta} = \cot\theta$ **23.** $1 + \tan^2(-\theta) = 1 + (-\tan\theta)^2 = 1 + \tan^2\theta = \sec^2\theta$

25. $\cos\theta(\tan\theta + \cot\theta) = \cos\theta\left(\frac{\sin\theta}{\cos\theta} + \frac{\cos\theta}{\sin\theta}\right) = \cos\theta\left(\frac{\sin^2\theta + \cos^2\theta}{\cos\theta\sin\theta}\right) = \cos\theta\left(\frac{1}{\cos\theta\sin\theta}\right) = \frac{1}{\sin\theta} = \csc\theta$

27. $\tan u \cot u - \cos^2 u = \tan u \cdot \frac{1}{\tan u} - \cos^2 u = 1 - \cos^2 u = \sin^2 u$ **29.** $(\sec\theta - 1)(\sec\theta + 1) = \sec^2\theta - 1 = \tan^2\theta$

31. $(\sec\theta + \tan\theta)(\sec\theta - \tan\theta) = \sec^2\theta - \tan^2\theta = 1$ **33.** $\cos^2\theta(1 + \tan^2\theta) = \cos^2\theta\sec^2\theta = \cos^2\theta \cdot \frac{1}{\cos^2\theta} = 1$

35. $(\sin\theta + \cos\theta)^2 + (\sin\theta - \cos\theta)^2 = \sin^2\theta + 2\sin\theta\cos\theta + \cos^2\theta + \sin^2\theta - 2\sin\theta\cos\theta + \cos^2\theta$
$= \sin^2\theta + \cos^2\theta + \sin^2\theta + \cos^2\theta = 1 + 1 = 2$

37. $\sec^4\theta - \sec^2\theta = \sec^2\theta(\sec^2\theta - 1) = (1 + \tan^2\theta)\tan^2\theta = \tan^4\theta + \tan^2\theta$

39. $\sec u - \tan u = \frac{1}{\cos u} - \frac{\sin u}{\cos u} = \frac{1 - \sin u}{\cos u} \cdot \frac{1 + \sin u}{1 + \sin u} = \frac{1 - \sin^2 u}{\cos u(1 + \sin u)} = \frac{\cos^2 u}{\cos u(1 + \sin u)} = \frac{\cos u}{1 + \sin u}$

41. $3\sin^2\theta + 4\cos^2\theta = 3\sin^2\theta + 3\cos^2\theta + \cos^2\theta = 3(\sin^2\theta + \cos^2\theta) + \cos^2\theta = 3 + \cos^2\theta$

43. $1 - \frac{\cos^2\theta}{1 + \sin\theta} = 1 - \frac{1 - \sin^2\theta}{1 + \sin\theta} = 1 - \frac{(1 + \sin\theta)(1 - \sin\theta)}{1 + \sin\theta} = 1 - (1 - \sin\theta) = \sin\theta$

45. $\dfrac{1+\tan v}{1-\tan v}=\dfrac{1+\dfrac{1}{\cot v}}{1-\dfrac{1}{\cot v}}=\dfrac{\dfrac{\cot v+1}{\cot v}}{\dfrac{\cot v-1}{\cot v}}=\dfrac{\cot v+1}{\cot v-1}$ **47.** $\dfrac{\sec\theta}{\csc\theta}+\dfrac{\sin\theta}{\cos\theta}=\dfrac{\dfrac{1}{\cos\theta}}{\dfrac{1}{\sin\theta}}+\tan\theta=\dfrac{\sin\theta}{\cos\theta}+\tan\theta=\tan\theta+\tan\theta=2\tan\theta$

49. $\dfrac{1+\sin\theta}{1-\sin\theta}=\dfrac{1+\dfrac{1}{\csc\theta}}{1-\dfrac{1}{\csc\theta}}=\dfrac{\dfrac{\csc\theta+1}{\csc\theta}}{\dfrac{\csc\theta-1}{\csc\theta}}=\dfrac{\csc\theta+1}{\csc\theta-1}$

51. $\dfrac{1-\sin v}{\cos v}+\dfrac{\cos v}{1-\sin v}=\dfrac{(1-\sin v)^2+\cos^2 v}{\cos v(1-\sin v)}=\dfrac{1-2\sin v+\sin^2 v+\cos^2 v}{\cos v(1-\sin v)}=\dfrac{2-2\sin v}{\cos v(1-\sin v)}=\dfrac{2(1-\sin v)}{\cos v(1-\sin v)}=\dfrac{2}{\cos v}=2\sec v$

53. $\dfrac{\sin\theta}{\sin\theta-\cos\theta}=\dfrac{1}{\dfrac{\sin\theta-\cos\theta}{\sin\theta}}=\dfrac{1}{1-\dfrac{\cos\theta}{\sin\theta}}=\dfrac{1}{1-\cot\theta}$

55. $(\sec\theta-\tan\theta)^2=\sec^2\theta-2\sec\theta\tan\theta+\tan^2\theta=\dfrac{1}{\cos^2\theta}-\dfrac{2\sin\theta}{\cos^2\theta}+\dfrac{\sin^2\theta}{\cos^2\theta}=\dfrac{1-2\sin\theta+\sin^2\theta}{\cos^2\theta}=\dfrac{(1-\sin\theta)^2}{1-\sin^2\theta}=\dfrac{(1-\sin\theta)^2}{(1-\sin\theta)(1+\sin\theta)}$

$=\dfrac{1-\sin\theta}{1+\sin\theta}$

57. $\dfrac{\cos\theta}{1-\tan\theta}+\dfrac{\sin\theta}{1-\cot\theta}=\dfrac{\cos\theta}{1-\dfrac{\sin\theta}{\cos\theta}}+\dfrac{\sin\theta}{1-\dfrac{\cos\theta}{\sin\theta}}=\dfrac{\cos\theta}{\dfrac{\cos\theta-\sin\theta}{\cos\theta}}+\dfrac{\sin\theta}{\dfrac{\sin\theta-\cos\theta}{\sin\theta}}=\dfrac{\cos^2\theta}{\cos\theta-\sin\theta}+\dfrac{\sin^2\theta}{\sin\theta-\cos\theta}$

$=\dfrac{\cos^2\theta-\sin^2\theta}{\cos\theta-\sin\theta}=\dfrac{(\cos\theta-\sin\theta)(\cos\theta+\sin\theta)}{\cos\theta-\sin\theta}=\sin\theta+\cos\theta$

59. $\tan\theta+\dfrac{\cos\theta}{1+\sin\theta}=\dfrac{\sin\theta}{\cos\theta}+\dfrac{\cos\theta}{1+\sin\theta}=\dfrac{\sin\theta(1+\sin\theta)+\cos^2\theta}{\cos\theta(1+\sin\theta)}=\dfrac{\sin\theta+\sin^2\theta+\cos^2\theta}{\cos\theta(1+\sin\theta)}=\dfrac{\sin\theta+1}{\cos\theta(1+\sin\theta)}=\dfrac{1}{\cos\theta}=\sec\theta$

61. $\dfrac{\tan\theta+\sec\theta-1}{\tan\theta-\sec\theta+1}=\dfrac{\tan\theta+(\sec\theta-1)}{\tan\theta-(\sec\theta-1)}\cdot\dfrac{\tan\theta+(\sec\theta-1)}{\tan\theta+(\sec\theta-1)}=\dfrac{\tan^2\theta+2\tan\theta(\sec\theta-1)+\sec^2\theta-2\sec\theta+1}{\tan^2\theta-(\sec^2\theta-2\sec\theta+1)}$

$=\dfrac{\sec^2\theta-1+2\tan\theta(\sec\theta-1)+\sec^2\theta-2\sec\theta+1}{\sec^2\theta-1-\sec^2\theta+2\sec\theta-1}=\dfrac{2\sec^2\theta-2\sec\theta+2\tan\theta(\sec\theta-1)}{-2+2\sec\theta}$

$=\dfrac{2\sec\theta(\sec\theta-1)+2\tan\theta(\sec\theta-1)}{2(\sec\theta-1)}=\dfrac{2(\sec\theta-1)(\sec\theta+\tan\theta)}{2(\sec\theta-1)}=\tan\theta+\sec\theta$

63. $\dfrac{\tan\theta-\cot\theta}{\tan\theta+\cot\theta}=\dfrac{\dfrac{\sin\theta}{\cos\theta}-\dfrac{\cos\theta}{\sin\theta}}{\dfrac{\sin\theta}{\cos\theta}+\dfrac{\cos\theta}{\sin\theta}}=\dfrac{\dfrac{\sin^2\theta-\cos^2\theta}{\cos\theta\sin\theta}}{\dfrac{\sin^2\theta+\cos^2\theta}{\cos\theta\sin\theta}}=\dfrac{\sin^2\theta-\cos^2\theta}{1}=\sin^2\theta-\cos^2\theta$

65. $\dfrac{\tan u-\cot u}{\tan u+\cot u}+1=\dfrac{\dfrac{\sin u}{\cos u}-\dfrac{\cos u}{\sin u}}{\dfrac{\sin u}{\cos u}+\dfrac{\cos u}{\sin u}}+1=\dfrac{\dfrac{\sin^2 u-\cos^2 u}{\cos u\sin u}}{\dfrac{\sin^2 u+\cos^2 u}{\cos u\sin u}}+1=\sin^2 u-\cos^2 u+1=\sin^2 u+(1-\cos^2 u)=2\sin^2 u$

67. $\dfrac{\sec\theta+\tan\theta}{\cot\theta+\cos\theta}=\dfrac{\dfrac{1}{\cos\theta}+\dfrac{\sin\theta}{\cos\theta}}{\dfrac{\cos\theta}{\sin\theta}+\cos\theta}=\dfrac{\dfrac{1+\sin\theta}{\cos\theta}}{\dfrac{\cos\theta+\cos\theta\sin\theta}{\sin\theta}}=\dfrac{1+\sin\theta}{\cos\theta}\cdot\dfrac{\sin\theta}{\cos\theta(1+\sin\theta)}=\dfrac{\sin\theta}{\cos\theta}\cdot\dfrac{1}{\cos\theta}=\tan\theta\sec\theta$

69. $\dfrac{1-\tan^2\theta}{1+\tan^2\theta}+1=\dfrac{1-\tan^2\theta+1+\tan^2\theta}{1+\tan^2\theta}=\dfrac{2}{1+\tan^2\theta}=\dfrac{2}{\sec^2\theta}=2\cos^2\theta$

71. $\dfrac{\sec\theta-\csc\theta}{\sec\theta\csc\theta}=\dfrac{\sec\theta}{\sec\theta\csc\theta}-\dfrac{\csc\theta}{\sec\theta\csc\theta}=\dfrac{1}{\csc\theta}-\dfrac{1}{\sec\theta}=\sin\theta-\cos\theta$

73. $\sec\theta-\cos\theta=\dfrac{1}{\cos\theta}-\cos\theta=\dfrac{1-\cos^2\theta}{\cos\theta}=\dfrac{\sin^2\theta}{\cos\theta}=\sin\theta\cdot\dfrac{\sin\theta}{\cos\theta}=\sin\theta\tan\theta$

75. $\dfrac{1}{1-\sin\theta}+\dfrac{1}{1+\sin\theta}=\dfrac{1+\sin\theta+1-\sin\theta}{(1+\sin\theta)(1-\sin\theta)}=\dfrac{2}{1-\sin^2\theta}=\dfrac{2}{\cos^2\theta}=2\sec^2\theta$

77. $\dfrac{\sec\theta}{1-\sin\theta}=\dfrac{\sec\theta}{1-\sin\theta}\cdot\dfrac{1+\sin\theta}{1+\sin\theta}=\dfrac{\sec\theta(1+\sin\theta)}{1-\sin^2\theta}=\dfrac{\sec\theta(1+\sin\theta)}{\cos^2\theta}=\dfrac{1+\sin\theta}{\cos^3\theta}$

79. $\dfrac{(\sec v-\tan v)^2+1}{\csc v(\sec v-\tan v)}=\dfrac{\sec^2 v-2\sec v\tan v+\tan^2 v+1}{\dfrac{1}{\sin v}\left(\dfrac{1}{\cos v}-\dfrac{\sin v}{\cos v}\right)}=\dfrac{2\sec^2 v-2\sec v\tan v}{\dfrac{1}{\sin v}\left(\dfrac{1-\sin v}{\cos v}\right)}=\dfrac{\dfrac{2}{\cos^2 v}-\dfrac{2\sin v}{\cos^2 v}}{\dfrac{1-\sin v}{\sin v\cos v}}=\dfrac{2-2\sin v}{\cos^2 v}\cdot\dfrac{\sin v\cos v}{1-\sin v}$

$=\dfrac{2(1-\sin v)}{\cos v}\cdot\dfrac{\sin v}{1-\sin v}=\dfrac{2\sin v}{\cos v}=2\tan v$

81. $\dfrac{\sin\theta+\cos\theta}{\cos\theta}-\dfrac{\sin\theta-\cos\theta}{\sin\theta}=\dfrac{\sin\theta}{\cos\theta}+1-1+\dfrac{\cos\theta}{\sin\theta}=\dfrac{\sin^2\theta+\cos^2\theta}{\cos\theta\sin\theta}=\dfrac{1}{\cos\theta\sin\theta}=\sec\theta\csc\theta$

83. $\dfrac{\sin^3\theta+\cos^3\theta}{\sin\theta+\cos\theta}=\dfrac{(\sin\theta+\cos\theta)(\sin^2\theta-\sin\theta\cos\theta+\cos^2\theta)}{\sin\theta+\cos\theta}=\sin^2\theta+\cos^2\theta-\sin\theta\cos\theta=1-\sin\theta\cos\theta$

85. $\dfrac{\cos^2\theta-\sin^2\theta}{1-\tan^2\theta}=\dfrac{\cos^2\theta-\sin^2\theta}{1-\dfrac{\sin^2\theta}{\cos^2\theta}}=\dfrac{\cos^2\theta-\sin^2\theta}{\dfrac{\cos^2\theta-\sin^2\theta}{\cos^2\theta}}=\cos^2\theta$

87. $\dfrac{(2\cos^2\theta-1)^2}{\cos^4\theta-\sin^4\theta}=\dfrac{[2\cos^2\theta-(\sin^2\theta+\cos^2\theta)]^2}{(\cos^2\theta-\sin^2\theta)(\cos^2\theta+\sin^2\theta)}=\dfrac{(\cos^2\theta-\sin^2\theta)^2}{\cos^2\theta-\sin^2\theta}=\cos^2\theta-\sin^2\theta=(1-\sin^2\theta)-\sin^2\theta=1-2\sin^2\theta$

89. $\dfrac{1+\sin\theta+\cos\theta}{1+\sin\theta-\cos\theta}=\dfrac{(1+\sin\theta)+\cos\theta}{(1+\sin\theta)-\cos\theta}\cdot\dfrac{(1+\sin\theta)+\cos\theta}{(1+\sin\theta)+\cos\theta}=\dfrac{1+2\sin\theta+\sin^2\theta+2(1+\sin\theta)\cos\theta+\cos^2\theta}{1+2\sin\theta+\sin^2\theta-\cos^2\theta}$

$=\dfrac{1+2\sin\theta+\sin^2\theta+2(1+\sin\theta)(\cos\theta)+(1-\sin^2\theta)}{1+2\sin\theta+\sin^2\theta-(1-\sin^2\theta)}=\dfrac{2+2\sin\theta+2(1+\sin\theta)(\cos\theta)}{2\sin\theta+2\sin^2\theta}$

$=\dfrac{2(1+\sin\theta)+2(1+\sin\theta)(\cos\theta)}{2\sin\theta(1+\sin\theta)}=\dfrac{2(1+\sin\theta)(1+\cos\theta)}{2\sin\theta(1+\sin\theta)}=\dfrac{1+\cos\theta}{\sin\theta}$

91. $(a\sin\theta+b\cos\theta)^2+(a\cos\theta-b\sin\theta)^2=a^2\sin^2\theta+2ab\sin\theta\cos\theta+b^2\cos^2\theta+a^2\cos^2\theta-2ab\sin\theta\cos\theta+b^2\sin^2\theta$
$=a^2(\sin^2\theta+\cos^2\theta)+b^2(\cos^2\theta+\sin^2\theta)=a^2+b^2$

93. $\dfrac{\tan\alpha+\tan\beta}{\cot\alpha+\cot\beta}=\dfrac{\tan\alpha+\tan\beta}{\dfrac{1}{\tan\alpha}+\dfrac{1}{\tan\beta}}=\dfrac{\tan\alpha+\tan\beta}{\dfrac{\tan\beta+\tan\alpha}{\tan\alpha\tan\beta}}=(\tan\alpha+\tan\beta)\cdot\dfrac{\tan\alpha\tan\beta}{\tan\alpha+\tan\beta}=\tan\alpha\tan\beta$

95. $(\sin\alpha+\cos\beta)^2+(\cos\beta+\sin\alpha)(\cos\beta-\sin\alpha)=(\sin^2\alpha+2\sin\alpha\cos\beta+\cos^2\beta)+(\cos^2\beta-\sin^2\alpha)$
$=2\cos^2\beta+2\sin\alpha\cos\beta=2\cos\beta(\cos\beta+\sin\alpha)=2\cos\beta(\sin\alpha+\cos\beta)$
97. $\ln|\sec\theta|=\ln|\cos\theta|^{-1}=-\ln|\cos\theta|$
99. $\ln|1+\cos\theta|+\ln|1-\cos\theta|=\ln(|1+\cos\theta||1-\cos\theta|)=\ln|1-\cos^2\theta|=\ln|\sin^2\theta|=2\ln|\sin\theta|$

101. $g(x)=\sec x-\cos x=\dfrac{1}{\cos x}-\cos x=\dfrac{1}{\cos x}-\dfrac{\cos^2 x}{\cos x}=\dfrac{1-\cos^2 x}{\cos x}=\dfrac{\sin^2 x}{\cos x}=\sin x\cdot\dfrac{\sin x}{\cos x}=\sin x\cdot\tan x=f(x)$

103. $f(\theta)=\dfrac{1-\sin\theta}{\cos\theta}-\dfrac{\cos\theta}{1+\sin\theta}=\dfrac{1-\sin\theta}{\cos\theta}\cdot\dfrac{1+\sin\theta}{1+\sin\theta}-\dfrac{\cos\theta}{1+\sin\theta}\cdot\dfrac{\cos\theta}{\cos\theta}=\dfrac{1-\sin^2\theta}{\cos\theta(1+\sin\theta)}-\dfrac{\cos^2\theta}{\cos\theta(1+\sin\theta)}$

$=\dfrac{\cos^2\theta}{\cos\theta(1+\sin\theta)}-\dfrac{\cos^2\theta}{\cos\theta(1+\sin\theta)}=0=g(\theta)$

105. $\sqrt{16+16\tan^2\theta}=\sqrt{16}\sqrt{1+\tan^2\theta}=4\sqrt{\sec^2\theta}=4\sec\theta$, since $\sec\theta>0$ for $-\dfrac{\pi}{2}<\theta<\dfrac{\pi}{2}$

107. $1200\sec\theta\,(2\sec^2\theta-1)=1200\dfrac{1}{\cos\theta}\left(\dfrac{2}{\cos^2\theta}-1\right)=1200\dfrac{1}{\cos\theta}\left(\dfrac{2}{\cos^2\theta}-\dfrac{\cos^2\theta}{\cos^2\theta}\right)=1200\dfrac{1}{\cos\theta}\left(\dfrac{2-\cos^2\theta}{\cos^2\theta}\right)=\dfrac{1200\,(1+1-\cos^2\theta)}{\cos^3\theta}$

$=\dfrac{1200\,(1+\sin^2\theta)}{\cos^3\theta}$

113. Maximum, 1250 **114.** $(f\circ g)(x)=\dfrac{x-1}{x-2}$ **115.** $\sin\theta=\dfrac{5}{13}$; $\cos\theta=-\dfrac{12}{13}$; $\tan\theta=-\dfrac{5}{12}$; $\csc\theta=\dfrac{13}{5}$; $\sec\theta=-\dfrac{13}{12}$; $\cot\theta=-\dfrac{12}{5}$ **116.** $-\dfrac{2}{\pi}$

8.5 Assess Your Understanding *(page 661)*

5. − **6.** − **7.** F **8.** F **9.** F **10.** T **11.** a **12.** d **13.** $-\frac{1}{4}(\sqrt{2}+\sqrt{6})$ **15.** $2-\sqrt{3}$ **17.** $\frac{1}{4}(\sqrt{6}+\sqrt{2})$ **19.** $\frac{1}{4}(\sqrt{2}-\sqrt{6})$

21. $-\frac{1}{4}(\sqrt{6}+\sqrt{2})$ **23.** $\sqrt{6}-\sqrt{2}$ **25.** $\frac{1}{2}$ **27.** 0 **29.** 1 **31.** −1 **33.** $\frac{1}{2}$ **35. (a)** $\frac{2\sqrt{5}}{25}$ **(b)** $\frac{11\sqrt{5}}{25}$ **(c)** $\frac{2\sqrt{5}}{5}$ **(d)** 2

37. (a) $\frac{4-3\sqrt{3}}{10}$ **(b)** $\frac{-3-4\sqrt{3}}{10}$ **(c)** $\frac{4+3\sqrt{3}}{10}$ **(d)** $\frac{25\sqrt{3}+48}{39}$ **39. (a)** $-\frac{5+12\sqrt{3}}{26}$ **(b)** $\frac{12-5\sqrt{3}}{26}$ **(c)** $\frac{-5+12\sqrt{3}}{26}$ **(d)** $\frac{-240+169\sqrt{3}}{69}$

41. (a) $-\frac{2\sqrt{2}}{3}$ **(b)** $\frac{-2\sqrt{2}+\sqrt{3}}{6}$ **(c)** $\frac{-2\sqrt{2}+\sqrt{3}}{6}$ **(d)** $\frac{9-4\sqrt{2}}{7}$ **43.** $\frac{1-2\sqrt{6}}{6}$ **45.** $\frac{\sqrt{3}-2\sqrt{2}}{6}$ **47.** $\frac{8\sqrt{2}-9\sqrt{3}}{5}$

49. $\sin\left(\dfrac{\pi}{2}+\theta\right)=\sin\dfrac{\pi}{2}\cos\theta+\cos\dfrac{\pi}{2}\sin\theta=1\cdot\cos\theta+0\cdot\sin\theta=\cos\theta$

51. $\sin(\pi-\theta)=\sin\pi\cos\theta-\cos\pi\sin\theta=0\cdot\cos\theta-(-1)\sin\theta=\sin\theta$
53. $\sin(\pi+\theta)=\sin\pi\cos\theta+\cos\pi\sin\theta=0\cdot\cos\theta+(-1)\sin\theta=-\sin\theta$

55. $\tan(\pi-\theta)=\dfrac{\tan\pi-\tan\theta}{1+\tan\pi\tan\theta}=\dfrac{0-\tan\theta}{1+0\cdot\tan\theta}=-\tan\theta$

57. $\sin\left(\dfrac{3\pi}{2}+\theta\right)=\sin\dfrac{3\pi}{2}\cos\theta+\cos\dfrac{3\pi}{2}\sin\theta=(-1)\cos\theta+0\cdot\sin\theta=-\cos\theta$

59. $\sin(\alpha+\beta)+\sin(\alpha-\beta)=\sin\alpha\cos\beta+\cos\alpha\sin\beta+\sin\alpha\cos\beta-\cos\alpha\sin\beta=2\sin\alpha\cos\beta$

61. $\dfrac{\sin(\alpha+\beta)}{\sin\alpha\cos\beta}=\dfrac{\sin\alpha\cos\beta+\cos\alpha\sin\beta}{\sin\alpha\cos\beta}=\dfrac{\sin\alpha\cos\beta}{\sin\alpha\cos\beta}+\dfrac{\cos\alpha\sin\beta}{\sin\alpha\cos\beta}=1+\cot\alpha\tan\beta$

63. $\dfrac{\cos(\alpha+\beta)}{\cos\alpha\cos\beta}=\dfrac{\cos\alpha\cos\beta-\sin\alpha\sin\beta}{\cos\alpha\cos\beta}=\dfrac{\cos\alpha\cos\beta}{\cos\alpha\cos\beta}-\dfrac{\sin\alpha\sin\beta}{\cos\alpha\cos\beta}=1-\tan\alpha\tan\beta$

65. $\dfrac{\sin(\alpha+\beta)}{\sin(\alpha-\beta)}=\dfrac{\sin\alpha\cos\beta+\cos\alpha\sin\beta}{\sin\alpha\cos\beta-\cos\alpha\sin\beta}=\dfrac{\dfrac{\sin\alpha\cos\beta+\cos\alpha\sin\beta}{\cos\alpha\cos\beta}}{\dfrac{\sin\alpha\cos\beta-\cos\alpha\sin\beta}{\cos\alpha\cos\beta}}=\dfrac{\dfrac{\sin\alpha\cos\beta}{\cos\alpha\cos\beta}+\dfrac{\cos\alpha\sin\beta}{\cos\alpha\cos\beta}}{\dfrac{\sin\alpha\cos\beta}{\cos\alpha\cos\beta}-\dfrac{\cos\alpha\sin\beta}{\cos\alpha\cos\beta}}=\dfrac{\tan\alpha+\tan\beta}{\tan\alpha-\tan\beta}$

67. $\cot(\alpha+\beta)=\dfrac{\cos(\alpha+\beta)}{\sin(\alpha+\beta)}=\dfrac{\cos\alpha\cos\beta-\sin\alpha\sin\beta}{\sin\alpha\cos\beta+\cos\alpha\sin\beta}=\dfrac{\dfrac{\cos\alpha\cos\beta-\sin\alpha\sin\beta}{\sin\alpha\sin\beta}}{\dfrac{\sin\alpha\cos\beta+\cos\alpha\sin\beta}{\sin\alpha\sin\beta}}=\dfrac{\dfrac{\cos\alpha\cos\beta}{\sin\alpha\sin\beta}-\dfrac{\sin\alpha\sin\beta}{\sin\alpha\sin\beta}}{\dfrac{\sin\alpha\cos\beta}{\sin\alpha\sin\beta}+\dfrac{\cos\alpha\sin\beta}{\sin\alpha\sin\beta}}=\dfrac{\cot\alpha\cot\beta-1}{\cot\beta+\cot\alpha}$

69. $\sec(\alpha+\beta)=\dfrac{1}{\cos(\alpha+\beta)}=\dfrac{1}{\cos\alpha\cos\beta-\sin\alpha\sin\beta}=\dfrac{\dfrac{1}{\sin\alpha\sin\beta}}{\dfrac{\cos\alpha\cos\beta-\sin\alpha\sin\beta}{\sin\alpha\sin\beta}}=\dfrac{\dfrac{1}{\sin\alpha}\cdot\dfrac{1}{\sin\beta}}{\dfrac{\cos\alpha\cos\beta}{\sin\alpha\sin\beta}-\dfrac{\sin\alpha\sin\beta}{\sin\alpha\sin\beta}}=\dfrac{\csc\alpha\csc\beta}{\cot\alpha\cot\beta-1}$

71. $\sin(\alpha-\beta)\sin(\alpha+\beta)=(\sin\alpha\cos\beta-\cos\alpha\sin\beta)(\sin\alpha\cos\beta+\cos\alpha\sin\beta)=\sin^2\alpha\cos^2\beta-\cos^2\alpha\sin^2\beta$
$=(\sin^2\alpha)(1-\sin^2\beta)-(1-\sin^2\alpha)(\sin^2\beta)=\sin^2\alpha-\sin^2\beta$

73. $\sin(\theta+k\pi)=\sin\theta\cos k\pi+\cos\theta\sin k\pi=(\sin\theta)(-1)^k+(\cos\theta)(0)=(-1)^k\sin\theta$, k any integer

75. $\dfrac{\sqrt{3}}{2}$ **77.** $-\dfrac{24}{25}$ **79.** $-\dfrac{33}{65}$ **81.** $\dfrac{63}{65}$ **83.** $\dfrac{48+25\sqrt{3}}{39}$ **85.** $\dfrac{4}{3}$ **87.** $u\sqrt{1-v^2}-v\sqrt{1-u^2}$; $-1\le u\le 1$; $-1\le v\le 1$

89. $\dfrac{u\sqrt{1-v^2}-v}{\sqrt{1+u^2}}$; $-\infty<u<\infty$; $-1\le v\le 1$ **91.** $\dfrac{uv-\sqrt{1-u^2}\sqrt{1-v^2}}{v\sqrt{1-u^2}+u\sqrt{1-v^2}}$; $-1\le u\le 1$; $-1\le v\le 1$ **93.** $\left\{\dfrac{\pi}{2},\dfrac{7\pi}{6}\right\}$ **95.** $\left\{\dfrac{\pi}{4}\right\}$ **97.** $\left\{\dfrac{11\pi}{6}\right\}$

99. Let $\alpha=\sin^{-1}v$ and $\beta=\cos^{-1}v$. Then $\sin\alpha=\cos\beta=v$, and since $\sin\alpha=\cos\left(\dfrac{\pi}{2}-\alpha\right)$, $\cos\left(\dfrac{\pi}{2}-\alpha\right)=\cos\beta$.

If $v\ge 0$, then $0\le\alpha\le\dfrac{\pi}{2}$, so $\left(\dfrac{\pi}{2}-\alpha\right)$ and β both lie on $\left[0,\dfrac{\pi}{2}\right]$. If $v<0$, then $-\dfrac{\pi}{2}\le\alpha<0$, so $\left(\dfrac{\pi}{2}-\alpha\right)$ and β both lie on $\left(\dfrac{\pi}{2},\pi\right]$.

Either way, $\cos\left(\dfrac{\pi}{2}-\alpha\right)=\cos\beta$ implies $\dfrac{\pi}{2}-\alpha=\beta$, or $\alpha+\beta=\dfrac{\pi}{2}$.

101. Let $\alpha=\tan^{-1}\dfrac{1}{v}$ and $\beta=\tan^{-1}v$. Because $v\ne 0$, $\alpha,\beta\ne 0$. Then $\tan\alpha=\dfrac{1}{v}=\dfrac{1}{\tan\beta}=\cot\beta$, and since

$\tan\alpha=\cot\left(\dfrac{\pi}{2}-\alpha\right)$, $\cot\left(\dfrac{\pi}{2}-\alpha\right)=\cot\beta$. Because $v>0$, $0<\alpha<\dfrac{\pi}{2}$, and so $\left(\dfrac{\pi}{2}-\alpha\right)$ and β both lie on $\left(0,\dfrac{\pi}{2}\right)$.

Then $\cot\left(\dfrac{\pi}{2}-\alpha\right)=\cot\beta$ implies $\dfrac{\pi}{2}-\alpha=\beta$, or $\alpha=\dfrac{\pi}{2}-\beta$.

103. $\sin(\sin^{-1}v+\cos^{-1}v)=\sin(\sin^{-1}v)\cos(\cos^{-1}v)+\cos(\sin^{-1}v)\sin(\cos^{-1}v)=(v)(v)+\sqrt{1-v^2}\sqrt{1-v^2}=v^2+1-v^2=1$

105. $\dfrac{\sin(x+h)-\sin x}{h}=\dfrac{\sin x\cos h+\cos x\sin h-\sin x}{h}=\dfrac{\cos x\sin h-\sin x(1-\cos h)}{h}=\cos x\cdot\dfrac{\sin h}{h}-\sin x\cdot\dfrac{1-\cos h}{h}$

107. (a) $\tan(\tan^{-1}1+\tan^{-1}2+\tan^{-1}3)=\tan((\tan^{-1}1+\tan^{-1}2)+\tan^{-1}3)=\dfrac{\tan(\tan^{-1}1+\tan^{-1}2)+\tan(\tan^{-1}3)}{1-\tan(\tan^{-1}1+\tan^{-1}2)\tan(\tan^{-1}3)}$

$$=\frac{\dfrac{\tan(\tan^{-1}1)+\tan(\tan^{-1}2)}{1-\tan(\tan^{-1}1)\tan(\tan^{-1}2)}+3}{1-\dfrac{\tan(\tan^{-1}1)+\tan(\tan^{-1}2)}{1-\tan(\tan^{-1}1)\tan(\tan^{-1}2)}\cdot 3}=\frac{\dfrac{1+2}{1-1\cdot 2}+3}{1-\dfrac{1+2}{1-1\cdot 2}\cdot 3}=\frac{\dfrac{3}{-1}+3}{1-\dfrac{3}{-1}\cdot 3}=\frac{-3+3}{1+9}=\frac{0}{10}=0$$

(b) From the definition of the inverse tangent function, $0<\tan^{-1}1<\dfrac{\pi}{2}$, $0<\tan^{-1}2<\dfrac{\pi}{2}$, and $0<\tan^{-1}3<\dfrac{\pi}{2}$,

so $0<\tan^{-1}1+\tan^{-1}2+\tan^{-1}3<\dfrac{3\pi}{2}$.

On the interval $\left(0,\dfrac{3\pi}{2}\right)$, $\tan\theta=0$ if and only if $\theta=\pi$. Therefore, from part (a), $\tan^{-1}1+\tan^{-1}2+\tan^{-1}3=\pi$.

109. $\tan\theta=\tan(\theta_2-\theta_1)=\dfrac{\tan\theta_2-\tan\theta_1}{1+\tan\theta_1\tan\theta_2}=\dfrac{m_2-m_1}{1+m_1m_2}$

111. $2\cot(\alpha-\beta)=\dfrac{2}{\tan(\alpha-\beta)}=2\left(\dfrac{1+\tan\alpha\tan\beta}{\tan\alpha-\tan\beta}\right)=2\left(\dfrac{1+(x+1)(x-1)}{(x+1)-(x-1)}\right)=2\left(\dfrac{1+x^2-1}{x+1-x+1}\right)=\dfrac{2x^2}{2}=x^2$

113. $\tan\frac{\pi}{2}$ is not defined; $\tan\left(\frac{\pi}{2}-\theta\right) = \dfrac{\sin\left(\frac{\pi}{2}-\theta\right)}{\cos\left(\frac{\pi}{2}-\theta\right)} = \dfrac{\cos\theta}{\sin\theta} = \cot\theta.$

114. $\left(-\frac{1}{3}, -\frac{5}{9}\right), (-5, 1)$ **115.** $510°$ **116.** $\frac{9\pi}{2}$ cm$^2 \approx 14.14$ cm^2 **117.** $\sin\theta = -\frac{2\sqrt{5}}{5}$; $\cos\theta = \frac{\sqrt{5}}{5}$; $\csc\theta = -\frac{\sqrt{5}}{2}$; $\sec\theta = \sqrt{5}$; $\cot\theta = -\frac{1}{2}$

8.6 Assess Your Understanding *(page 671)*

1. $\sin^2\theta$; $2\cos^2\theta$; $2\sin^2\theta$ **2.** $1-\cos\theta$ **3.** $\sin\theta$ **4.** T **5.** F **6.** F **7.** b **8.** c

9. (a) $\frac{24}{25}$ **(b)** $\frac{7}{25}$ **(c)** $\frac{\sqrt{10}}{10}$ **(d)** $\frac{3\sqrt{10}}{10}$ **(e)** $\frac{24}{7}$ **(f)** $\frac{1}{3}$

11. (a) $\frac{24}{25}$ **(b)** $-\frac{7}{25}$ **(c)** $\frac{2\sqrt{5}}{5}$ **(d)** $-\frac{\sqrt{5}}{5}$ **(e)** $-\frac{24}{7}$ **(f)** -2

13. (a) $-\frac{2\sqrt{2}}{3}$ **(b)** $\frac{1}{3}$ **(c)** $\sqrt{\frac{3+\sqrt{6}}{6}}$ **(d)** $\sqrt{\frac{3-\sqrt{6}}{6}}$ **(e)** $-2\sqrt{2}$ **(f)** $\sqrt{\frac{3+\sqrt{6}}{3-\sqrt{6}}}$ or $\sqrt{5+2\sqrt{6}}$

15. (a) $\frac{4\sqrt{2}}{9}$ **(b)** $-\frac{7}{9}$ **(c)** $\frac{\sqrt{3}}{3}$ **(d)** $\frac{\sqrt{6}}{3}$ **(e)** $-\frac{4\sqrt{2}}{7}$ **(f)** $\frac{\sqrt{2}}{2}$

17. (a) $-\frac{4}{5}$ **(b)** $\frac{3}{5}$ **(c)** $\sqrt{\frac{5+2\sqrt{5}}{10}}$ **(d)** $\sqrt{\frac{5-2\sqrt{5}}{10}}$ **(e)** $-\frac{4}{3}$ **(f)** $\sqrt{\frac{5+2\sqrt{5}}{5-2\sqrt{5}}}$ or $\sqrt{9+4\sqrt{5}}$

19. (a) $-\frac{3}{5}$ **(b)** $-\frac{4}{5}$ **(c)** $\frac{1}{2}\sqrt{\frac{10-\sqrt{10}}{5}}$ **(d)** $-\frac{1}{2}\sqrt{\frac{10+\sqrt{10}}{5}}$ **(e)** $\frac{3}{4}$ **(f)** $-\sqrt{\frac{10-\sqrt{10}}{10+\sqrt{10}}}$ or $-\frac{\sqrt{11-2\sqrt{10}}}{3}$

21. $\frac{\sqrt{2-\sqrt{2}}}{2}$ **23.** $1-\sqrt{2}$ **25.** $-\frac{\sqrt{2+\sqrt{3}}}{2}$ **27.** $\frac{2}{\sqrt{2+\sqrt{2}}} = (2-\sqrt{2})\sqrt{2+\sqrt{2}}$ **29.** $-\frac{\sqrt{2-\sqrt{2}}}{2}$ **31.** $-\frac{4}{5}$ **33.** $\frac{\sqrt{10(5-\sqrt{5})}}{10}$

35. $\frac{4}{3}$ **37.** $-\frac{7}{8}$ **39.** $\frac{\sqrt{10}}{4}$ **41.** $-\frac{\sqrt{15}}{3}$ **43.** $\sin^4\theta = (\sin^2\theta)^2 = \left(\frac{1-\cos(2\theta)}{2}\right)^2 = \frac{1}{4}[1-2\cos(2\theta)+\cos^2(2\theta)] = \frac{1}{4}-\frac{1}{2}\cos(2\theta)+\frac{1}{4}\cos^2(2\theta)$

$= \frac{1}{4}-\frac{1}{2}\cos(2\theta)+\frac{1}{4}\left(\frac{1+\cos(4\theta)}{2}\right) = \frac{1}{4}-\frac{1}{2}\cos(2\theta)+\frac{1}{8}+\frac{1}{8}\cos(4\theta) = \frac{3}{8}-\frac{1}{2}\cos(2\theta)+\frac{1}{8}\cos(4\theta)$

45. $\cos(3\theta) = 4\cos^3\theta - 3\cos\theta$ **47.** $\sin(5\theta) = 16\sin^5\theta - 20\sin^3\theta + 5\sin\theta$ **49.** $\cos^4\theta - \sin^4\theta = (\cos^2\theta+\sin^2\theta)(\cos^2\theta-\sin^2\theta) = \cos(2\theta)$

51. $$\cot(2\theta) = \frac{1}{\tan(2\theta)} = \frac{1-\tan^2\theta}{2\tan\theta} = \frac{1-\frac{1}{\cot^2\theta}}{2\left(\frac{1}{\cot\theta}\right)} = \frac{\frac{\cot^2\theta-1}{\cot^2\theta}}{\frac{2}{\cot\theta}} = \frac{\cot^2\theta-1}{\cot^2\theta}\cdot\frac{\cot\theta}{2} = \frac{\cot^2\theta-1}{2\cot\theta}$$

53. $$\sec(2\theta) = \frac{1}{\cos(2\theta)} = \frac{1}{2\cos^2\theta-1} = \frac{1}{\frac{2}{\sec^2\theta}-1} = \frac{1}{\frac{2-\sec^2\theta}{\sec^2\theta}} = \frac{\sec^2\theta}{2-\sec^2\theta}$$

55. $\cos^2(2u) - \sin^2(2u) = \cos[2(2u)] = \cos(4u)$

57. $$\frac{\cos(2\theta)}{1+\sin(2\theta)} = \frac{\cos^2\theta-\sin^2\theta}{1+2\sin\theta\cos\theta} = \frac{(\cos\theta-\sin\theta)(\cos\theta+\sin\theta)}{\sin^2\theta+\cos^2\theta+2\sin\theta\cos\theta} = \frac{(\cos\theta-\sin\theta)(\cos\theta+\sin\theta)}{(\sin\theta+\cos\theta)(\sin\theta+\cos\theta)} = \frac{\cos\theta-\sin\theta}{\cos\theta+\sin\theta}$$

$$= \frac{\frac{\cos\theta-\sin\theta}{\sin\theta}}{\frac{\cos\theta+\sin\theta}{\sin\theta}} = \frac{\frac{\cos\theta}{\sin\theta}-\frac{\sin\theta}{\sin\theta}}{\frac{\cos\theta}{\sin\theta}+\frac{\sin\theta}{\sin\theta}} = \frac{\cot\theta-1}{\cot\theta+1}$$

59. $$\sec^2\frac{\theta}{2} = \frac{1}{\cos^2\left(\frac{\theta}{2}\right)} = \frac{1}{\frac{1+\cos\theta}{2}} = \frac{2}{1+\cos\theta}$$

61. $$\cot^2\frac{v}{2} = \frac{1}{\tan^2\left(\frac{v}{2}\right)} = \frac{1}{\frac{1-\cos v}{1+\cos v}} = \frac{1+\cos v}{1-\cos v} = \frac{1+\frac{1}{\sec v}}{1-\frac{1}{\sec v}} = \frac{\frac{\sec v+1}{\sec v}}{\frac{\sec v-1}{\sec v}} = \frac{\sec v+1}{\sec v}\cdot\frac{\sec v}{\sec v-1} = \frac{\sec v+1}{\sec v-1}$$

63. $$\frac{1-\tan^2\left(\frac{\theta}{2}\right)}{1+\tan^2\left(\frac{\theta}{2}\right)} = \frac{1-\frac{1-\cos\theta}{1+\cos\theta}}{1+\frac{1-\cos\theta}{1+\cos\theta}} = \frac{\frac{1+\cos\theta-(1-\cos\theta)}{1+\cos\theta}}{\frac{1+\cos\theta+1-\cos\theta}{1+\cos\theta}} = \frac{2\cos\theta}{1+\cos\theta}\cdot\frac{1+\cos\theta}{2} = \cos\theta$$

65. $\dfrac{\sin(3\theta)}{\sin\theta} - \dfrac{\cos(3\theta)}{\cos\theta} = \dfrac{\sin(3\theta)\cos\theta - \cos(3\theta)\sin\theta}{\sin\theta\cos\theta} = \dfrac{\sin(3\theta - \theta)}{\frac{1}{2}(2\sin\theta\cos\theta)} = \dfrac{2\sin(2\theta)}{\sin(2\theta)} = 2$

67. $\tan(3\theta) = \tan(\theta + 2\theta) = \dfrac{\tan\theta + \tan(2\theta)}{1 - \tan\theta\tan(2\theta)} = \dfrac{\tan\theta + \dfrac{2\tan\theta}{1-\tan^2\theta}}{1 - \dfrac{\tan\theta(2\tan\theta)}{1-\tan^2\theta}} = \dfrac{\tan\theta - \tan^3\theta + 2\tan\theta}{1 - \tan^2\theta - 2\tan^2\theta} = \dfrac{3\tan\theta - \tan^3\theta}{1 - 3\tan^2\theta}$

69. $\frac{1}{2}(\ln|1 - \cos(2\theta)| - \ln 2) = \ln\left(\dfrac{|1-\cos(2\theta)|}{2}\right)^{1/2} = \ln|\sin^2\theta|^{1/2} = \ln|\sin\theta|$ **71.** $\left\{\frac{\pi}{3}, \frac{2\pi}{3}, \frac{4\pi}{3}, \frac{5\pi}{3}\right\}$ **73.** $\left\{0, \frac{2\pi}{3}, \frac{4\pi}{3}\right\}$

75. $\left\{0, \frac{\pi}{3}, \frac{\pi}{2}, \frac{2\pi}{3}, \pi, \frac{4\pi}{3}, \frac{3\pi}{2}, \frac{5\pi}{3}\right\}$ **77.** No real solution **79.** $\left\{0, \frac{\pi}{3}, \pi, \frac{5\pi}{3}\right\}$ **81.** $\frac{\sqrt{3}}{2}$ **83.** $\frac{7}{25}$ **85.** $\frac{24}{7}$ **87.** $\frac{24}{25}$ **89.** $\frac{1}{5}$ **91.** $\frac{25}{7}$ **93.** $0, \frac{\pi}{3}, \pi, \frac{5\pi}{3}$

95. $\frac{\pi}{2}, \frac{3\pi}{2}$ **97. (a)** $W = 2D(\csc\theta - \cot\theta) = 2D\left(\dfrac{1}{\sin\theta} - \dfrac{\cos\theta}{\sin\theta}\right) = 2D\dfrac{1-\cos\theta}{\sin\theta} = 2D\tan\dfrac{\theta}{2}$ **(b)** $\theta = 24.45°$

99. (a) $R = \dfrac{v_0^2\sqrt{2}}{16}\cos\theta(\sin\theta - \cos\theta)$

$= \dfrac{v_0^2\sqrt{2}}{32}(2\cos\theta\sin\theta - 2\cos^2\theta)$

$= \dfrac{v_0^2\sqrt{2}}{32}[\sin(2\theta) - \cos(2\theta) - 1]$

(b) $\frac{3\pi}{8}$ or 67.5° **(c)** $32(2 - \sqrt{2}) \approx 18.75$ ft

(d) 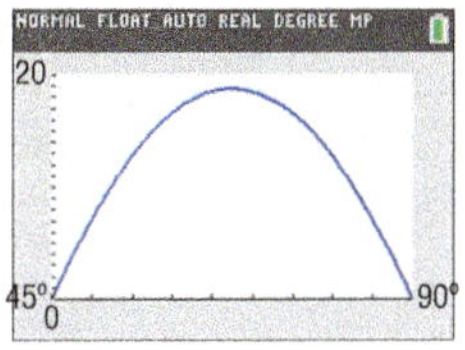

$\theta = 67.5°$ $\left(\frac{3\pi}{8}\text{ radians}\right)$ makes R largest. $R = 18.75$ ft

101. $A = \frac{1}{2}h(\text{base}) = h\left(\frac{1}{2}\text{base}\right) = s\cos\frac{\theta}{2}\cdot s\sin\frac{\theta}{2} = \frac{1}{2}s^2\sin\theta$

103. (a) $150 + 75\sqrt{3}$ cm² **(b)** $\dfrac{10 + 5\sqrt{3}}{2}$ cm **(c)** $\dfrac{175\pi + 100\pi\sqrt{3}}{4}$ cm² **(d)** $\dfrac{600 + 300\sqrt{3} - 175\pi - 100\pi\sqrt{3}}{4}$ cm² **105.** $-\frac{1}{4}$

107. $\dfrac{2z}{1+z^2} = \dfrac{2\tan\left(\frac{\alpha}{2}\right)}{1 + \tan^2\left(\frac{\alpha}{2}\right)} = \dfrac{2\tan\left(\frac{\alpha}{2}\right)}{\sec^2\left(\frac{\alpha}{2}\right)} = \dfrac{\dfrac{2\sin\left(\frac{\alpha}{2}\right)}{\cos\left(\frac{\alpha}{2}\right)}}{\dfrac{1}{\cos^2\left(\frac{\alpha}{2}\right)}} = 2\sin\left(\frac{\alpha}{2}\right)\cos\left(\frac{\alpha}{2}\right) = \sin\left(2\cdot\frac{\alpha}{2}\right) = \sin\alpha$ **109.**

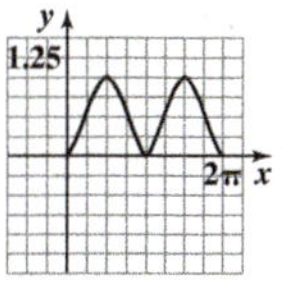

111. $\sin\dfrac{\pi}{24} = \dfrac{\sqrt{2}}{4}\sqrt{4 - \sqrt{6} - \sqrt{2}}$

$\cos\dfrac{\pi}{24} = \dfrac{\sqrt{2}}{4}\sqrt{4 + \sqrt{6} + \sqrt{2}}$

113. $\sin^3\theta + \sin^3(\theta + 120°) + \sin^3(\theta + 240°) = \sin^3\theta + (\sin\theta\cos120° + \cos\theta\sin120°)^3 + (\sin\theta\cos240° + \cos\theta\sin240°)^3$

$= \sin^3\theta + \left(-\frac{1}{2}\sin\theta + \frac{\sqrt{3}}{2}\cos\theta\right)^3 + \left(-\frac{1}{2}\sin\theta - \frac{\sqrt{3}}{2}\cos\theta\right)^3$

$= \sin^3\theta + \frac{1}{8}(3\sqrt{3}\cos^3\theta - 9\cos^2\theta\sin\theta + 3\sqrt{3}\cos\theta\sin^2\theta - \sin^3\theta) - \frac{1}{8}(\sin^3\theta + 3\sqrt{3}\sin^2\theta\cos\theta + 9\sin\theta\cos^2\theta + 3\sqrt{3}\cos^3\theta)$

$= \frac{3}{4}\sin^3\theta - \frac{9}{4}\cos^2\theta\sin\theta = \frac{3}{4}[\sin^3\theta - 3\sin\theta(1 - \sin^2\theta)] = \frac{3}{4}(4\sin^3\theta - 3\sin\theta) = -\frac{3}{4}\sin(3\theta)$ (from Example 2)

115. $-\frac{1}{2}$ **117.** $y = \frac{1}{2}x - 4$ **118.**

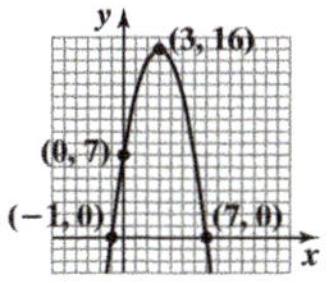

119. $\dfrac{\sqrt{3}+1}{2}$ **120.**

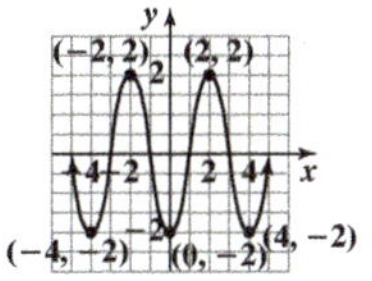

8.7 Assess Your Understanding *(page 677)*

1. $\frac{1}{2}\left(\frac{\sqrt{3}}{2} - 1\right)$ **3.** $-\frac{1}{2}\left(\frac{\sqrt{3}}{2} + 1\right)$ **5.** $\frac{\sqrt{2}}{2}$ **7.** $\frac{1}{2}[\cos(2\theta) - \cos(6\theta)]$ **9.** $\frac{1}{2}[\sin(6\theta) + \sin(2\theta)]$ **11.** $\frac{1}{2}[\cos(2\theta) + \cos(8\theta)]$

13. $\frac{1}{2}[\cos\theta - \cos(3\theta)]$ **15.** $\frac{1}{2}[\sin(2\theta) + \sin\theta]$ **17.** $2\sin\theta\cos(3\theta)$ **19.** $2\cos(3\theta)\cos\theta$ **21.** $2\sin(2\theta)\cos\theta$ **23.** $2\sin\theta\sin\frac{\theta}{2}$

25. $\dfrac{\sin\theta + \sin(3\theta)}{2\sin(2\theta)} = \dfrac{2\sin(2\theta)\cos\theta}{2\sin(2\theta)} = \cos\theta$ **27.** $\dfrac{\sin(4\theta) + \sin(2\theta)}{\cos(4\theta) + \cos(2\theta)} = \dfrac{2\sin(3\theta)\cos\theta}{2\cos(3\theta)\cos\theta} = \dfrac{\sin(3\theta)}{\cos(3\theta)} = \tan(3\theta)$

29. $\dfrac{\cos\theta - \cos(3\theta)}{\sin\theta + \sin(3\theta)} = \dfrac{2\sin(2\theta)\sin\theta}{2\sin(2\theta)\cos\theta} = \dfrac{\sin\theta}{\cos\theta} = \tan\theta$

31. $\sin\theta[\sin\theta + \sin(3\theta)] = \sin\theta[2\sin(2\theta)\cos\theta] = \cos\theta[2\sin(2\theta)\sin\theta] = \cos\theta\left[2\cdot\frac{1}{2}[\cos\theta - \cos(3\theta)]\right] = \cos\theta[\cos\theta - \cos(3\theta)]$

33. $\dfrac{\sin(4\theta) + \sin(8\theta)}{\cos(4\theta) + \cos(8\theta)} = \dfrac{2\sin(6\theta)\cos(2\theta)}{2\cos(6\theta)\cos(2\theta)} = \dfrac{\sin(6\theta)}{\cos(6\theta)} = \tan(6\theta)$

35. $\dfrac{\sin(4\theta) + \sin(8\theta)}{\sin(4\theta) - \sin(8\theta)} = \dfrac{2\sin(6\theta)\cos(-2\theta)}{2\sin(-2\theta)\cos(6\theta)} = \dfrac{\sin(6\theta)}{\cos(6\theta)}\cdot\dfrac{\cos(2\theta)}{-\sin(2\theta)} = \tan(6\theta)[-\cot(2\theta)] = -\dfrac{\tan(6\theta)}{\tan(2\theta)}$

37. $\dfrac{\sin\alpha + \sin\beta}{\sin\alpha - \sin\beta} = \dfrac{2\sin\frac{\alpha+\beta}{2}\cos\frac{\alpha-\beta}{2}}{2\sin\frac{\alpha-\beta}{2}\cos\frac{\alpha+\beta}{2}} = \dfrac{\sin\frac{\alpha+\beta}{2}}{\cos\frac{\alpha+\beta}{2}}\cdot\dfrac{\cos\frac{\alpha-\beta}{2}}{\sin\frac{\alpha-\beta}{2}} = \tan\dfrac{\alpha+\beta}{2}\cot\dfrac{\alpha-\beta}{2}$

39. $\dfrac{\sin\alpha + \sin\beta}{\cos\alpha + \cos\beta} = \dfrac{2\sin\frac{\alpha+\beta}{2}\cos\frac{\alpha-\beta}{2}}{2\cos\frac{\alpha+\beta}{2}\cos\frac{\alpha-\beta}{2}} = \dfrac{\sin\frac{\alpha+\beta}{2}}{\cos\frac{\alpha+\beta}{2}} = \tan\dfrac{\alpha+\beta}{2}$

41. $1 + \cos(2\theta) + \cos(4\theta) + \cos(6\theta) = [1 + \cos(6\theta)] + [\cos(2\theta) + \cos(4\theta)] = 2\cos^2(3\theta) + 2\cos(3\theta)\cos(-\theta)$
$= 2\cos(3\theta)[\cos(3\theta) + \cos\theta] = 2\cos(3\theta)[2\cos(2\theta)\cos\theta] = 4\cos\theta\cos(2\theta)\cos(3\theta)$

43. $\left\{0, \frac{\pi}{3}, \frac{\pi}{2}, \frac{2\pi}{3}, \pi, \frac{4\pi}{3}, \frac{3\pi}{2}, \frac{5\pi}{3}\right\}$

45. $\left\{0, \frac{\pi}{5}, \frac{2\pi}{5}, \frac{3\pi}{5}, \frac{4\pi}{5}, \pi, \frac{6\pi}{5}, \frac{7\pi}{5}, \frac{8\pi}{5}, \frac{9\pi}{5}\right\}$

47. (a) $y = 2\sin(2061\pi t)\cos(357\pi t)$
(b) $y_{\text{max}} = 2$
(c)

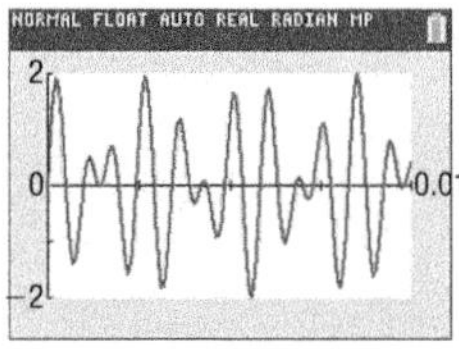

49. $I_u = I_x\cos^2\theta + I_y\sin^2\theta - 2I_{xy}\sin\theta\cos\theta = I_x\cos^2\theta + I_y\sin^2\theta - I_{xy}\sin 2\theta$
$= I_x\left(\dfrac{\cos 2\theta + 1}{2}\right) + I_y\left(\dfrac{1 - \cos 2\theta}{2}\right) - I_{xy}\sin 2\theta$
$= \dfrac{I_x}{2}\cos 2\theta + \dfrac{I_x}{2} + \dfrac{I_y}{2} - \dfrac{I_y}{2}\cos 2\theta - I_{xy}\sin 2\theta$
$= \dfrac{I_x + I_y}{2} + \dfrac{I_x - I_y}{2}\cos 2\theta - I_{xy}\sin 2\theta$

$I_v = I_x\sin^2\theta + I_y\cos^2\theta + 2I_{xy}\sin\theta\cos\theta = I_x\left(\dfrac{1 - \cos 2\theta}{2}\right) + I_y\left(\dfrac{\cos 2\theta + 1}{2}\right) + I_{xy}\sin 2\theta$
$= \dfrac{I_x}{2} - \dfrac{I_x}{2}\cos 2\theta + \dfrac{I_y}{2}\cos 2\theta + \dfrac{I_y}{2} + I_{xy}\sin 2\theta$
$= \dfrac{I_x + I_y}{2} - \dfrac{I_x - I_y}{2}\cos 2\theta + I_{xy}\sin 2\theta$

51. $\sin(2\alpha) + \sin(2\beta) + \sin(2\gamma) = 2\sin(\alpha + \beta)\cos(\alpha - \beta) + \sin(2\gamma) = 2\sin(\alpha + \beta)\cos(\alpha - \beta) + 2\sin\gamma\cos\gamma$
$= 2\sin(\pi - \gamma)\cos(\alpha - \beta) + 2\sin\gamma\cos\gamma = 2\sin\gamma\cos(\alpha - \beta) + 2\sin\gamma\cos\gamma = 2\sin\gamma[\cos(\alpha - \beta) + \cos\gamma]$
$= 2\sin\gamma\left(2\cos\dfrac{\alpha - \beta + \gamma}{2}\cos\dfrac{\alpha - \beta - \gamma}{2}\right) = 4\sin\gamma\cos\dfrac{\pi - 2\beta}{2}\cos\dfrac{2\alpha - \pi}{2} = 4\sin\gamma\cos\left(\dfrac{\pi}{2} - \beta\right)\cos\left(\alpha - \dfrac{\pi}{2}\right)$
$= 4\sin\gamma\sin\beta\sin\alpha = 4\sin\alpha\sin\beta\sin\gamma$

53.
$$\begin{aligned}\sin(\alpha - \beta) &= \sin\alpha\cos\beta - \cos\alpha\sin\beta\\ \sin(\alpha + \beta) &= \sin\alpha\cos\beta + \cos\alpha\sin\beta\\ \sin(\alpha - \beta) + \sin(\alpha + \beta) &= 2\sin\alpha\cos\beta\\ \sin\alpha\cos\beta &= \frac{1}{2}[\sin(\alpha + \beta) + \sin(\alpha - \beta)]\end{aligned}$$

55. $2\cos\dfrac{\alpha + \beta}{2}\cos\dfrac{\alpha - \beta}{2} = 2\cdot\dfrac{1}{2}\left[\cos\left(\dfrac{\alpha + \beta}{2} + \dfrac{\alpha - \beta}{2}\right) + \cos\left(\dfrac{\alpha + \beta}{2} - \dfrac{\alpha - \beta}{2}\right)\right] = \cos\dfrac{2\alpha}{2} + \cos\dfrac{2\beta}{2} = \cos\alpha + \cos\beta$

57. $\{13\}$ **58.** Amplitude: 5; Period: $\frac{\pi}{2}$; Phase shift: $\frac{\pi}{4}$ **59.** $\frac{2\sqrt{6}}{7}$

60. $f^{-1}(x) = \sin^{-1}\left(\dfrac{x + 5}{3}\right)$; Range of f = Domain of $f^{-1} = [-8, -2]$; Range of $f^{-1} = \left[-\frac{\pi}{2}, \frac{\pi}{2}\right]$

Review Exercises *(page 680)*

1. $\frac{\pi}{2}$ **2.** $\frac{\pi}{2}$ **3.** $\frac{\pi}{4}$ **4.** $-\frac{\pi}{6}$ **5.** $\frac{5\pi}{6}$ **6.** $-\frac{\pi}{3}$ **7.** $\frac{\pi}{4}$ **8.** $\frac{3\pi}{4}$ **9.** $\frac{3\pi}{8}$ **10.** $\frac{3\pi}{4}$ **11.** $-\frac{\pi}{3}$ **12.** $\frac{\pi}{7}$ **13.** $-\frac{\pi}{9}$ **14.** 0.9 **15.** 0.6 **16.** 5 **17.** Not defined

18. $-\frac{\pi}{6}$ **19.** π **20.** $-\sqrt{3}$ **21.** $\frac{2\sqrt{3}}{3}$ **22.** $\frac{4}{5}$ **23.** $-\frac{4}{3}$ **24.** $f^{-1}(x) = \frac{1}{3}\sin^{-1}\left(\frac{x}{2}\right)$; Range of f = Domain of $f^{-1} = [-2, 2]$; Range of $f^{-1} = \left[-\frac{\pi}{6}, \frac{\pi}{6}\right]$

25. $f^{-1}(x) = \cos^{-1}(3 - x)$; Range of f = Domain of $f^{-1} = [2, 4]$; Range of $f^{-1} = [0, \pi]$ **26.** $\sqrt{1 - u^2}$ **27.** $\dfrac{|u|}{u\sqrt{u^2 - 1}}$

28. $\tan\theta\cot\theta - \sin^2\theta = 1 - \sin^2\theta = \cos^2\theta$ **29.** $\sin^2\theta(1 + \cot^2\theta) = \sin^2\theta\csc^2\theta = 1$

30. $5\cos^2\theta + 3\sin^2\theta = 2\cos^2\theta + 3(\cos^2\theta + \sin^2\theta) = 3 + 2\cos^2\theta$

31. $\dfrac{1 - \cos\theta}{\sin\theta} + \dfrac{\sin\theta}{1 - \cos\theta} = \dfrac{(1 - \cos\theta)^2 + \sin^2\theta}{\sin\theta(1 - \cos\theta)} = \dfrac{1 - 2\cos\theta + \cos^2\theta + \sin^2\theta}{\sin\theta(1 - \cos\theta)} = \dfrac{2(1 - \cos\theta)}{\sin\theta(1 - \cos\theta)} = 2\csc\theta$

32. $\dfrac{\cos\theta}{\cos\theta - \sin\theta} = \dfrac{\dfrac{\cos\theta}{\cos\theta}}{\dfrac{\cos\theta - \sin\theta}{\cos\theta}} = \dfrac{1}{1 - \dfrac{\sin\theta}{\cos\theta}} = \dfrac{1}{1 - \tan\theta}$ **33.** $\dfrac{\csc\theta}{1 + \csc\theta} = \dfrac{\dfrac{1}{\sin\theta}}{1 + \dfrac{1}{\sin\theta}} = \dfrac{1}{1 + \sin\theta} = \dfrac{1}{1 + \sin\theta}\cdot\dfrac{1 - \sin\theta}{1 - \sin\theta} = \dfrac{1 - \sin\theta}{1 - \sin^2\theta} = \dfrac{1 - \sin\theta}{\cos^2\theta}$

34. $\csc\theta - \sin\theta = \dfrac{1}{\sin\theta} - \sin\theta = \dfrac{1 - \sin^2\theta}{\sin\theta} = \dfrac{\cos^2\theta}{\sin\theta} = \cos\theta\cdot\dfrac{\cos\theta}{\sin\theta} = \cos\theta\cot\theta$

35. $\dfrac{1 - \sin\theta}{\sec\theta} = \cos\theta(1 - \sin\theta)\cdot\dfrac{1 + \sin\theta}{1 + \sin\theta} = \dfrac{\cos\theta(1 - \sin^2\theta)}{1 + \sin\theta} = \dfrac{\cos^3\theta}{1 + \sin\theta}$ **36.** $\cot\theta - \tan\theta = \dfrac{\cos\theta}{\sin\theta} - \dfrac{\sin\theta}{\cos\theta} = \dfrac{\cos^2\theta - \sin^2\theta}{\sin\theta\cos\theta} = \dfrac{1 - 2\sin^2\theta}{\sin\theta\cos\theta}$

37. $\dfrac{\cos(\alpha + \beta)}{\cos\alpha\sin\beta} = \dfrac{\cos\alpha\cos\beta - \sin\alpha\sin\beta}{\cos\alpha\sin\beta} = \dfrac{\cos\alpha\cos\beta}{\cos\alpha\sin\beta} - \dfrac{\sin\alpha\sin\beta}{\cos\alpha\sin\beta} = \cot\beta - \tan\alpha$

38. $\dfrac{\cos(\alpha - \beta)}{\cos\alpha\cos\beta} = \dfrac{\cos\alpha\cos\beta + \sin\alpha\sin\beta}{\cos\alpha\cos\beta} = \dfrac{\cos\alpha\cos\beta}{\cos\alpha\cos\beta} + \dfrac{\sin\alpha\sin\beta}{\cos\alpha\cos\beta} = 1 + \tan\alpha\tan\beta$

39. $(1 + \cos\theta)\left(\tan\dfrac{\theta}{2}\right) = (1 + \cos\theta)\cdot\dfrac{\sin\theta}{1 + \cos\theta} = \sin\theta$

40. $2\cot\theta\cot 2\theta = 2\left(\dfrac{\cos\theta}{\sin\theta}\right)\left(\dfrac{\cos 2\theta}{\sin 2\theta}\right) = \dfrac{2\cos\theta(\cos^2\theta - \sin^2\theta)}{2\sin^2\theta\cos\theta} = \dfrac{\cos^2\theta - \sin^2\theta}{\sin^2\theta} = \cot^2\theta - 1$

41. $1 - 8\sin^2\theta\cos^2\theta = 1 - 2(2\sin\theta\cos\theta)^2 = 1 - 2\sin^2(2\theta) = \cos(4\theta)$

42. $\dfrac{\sin(3\theta)\cos\theta - \sin\theta\cos(3\theta)}{\sin(2\theta)} = \dfrac{\sin(2\theta)}{\sin(2\theta)} = 1$ **43.** $\dfrac{\sin(2\theta) + \sin(4\theta)}{\cos(2\theta) + \cos(4\theta)} = \dfrac{2\sin(3\theta)\cos(-\theta)}{2\cos(3\theta)\cos(-\theta)} = \tan(3\theta)$

44. $\dfrac{\cos(2\theta) - \cos(4\theta)}{\cos(2\theta) + \cos(4\theta)} - \tan\theta\tan(3\theta) = \dfrac{-2\sin(3\theta)\sin(-\theta)}{2\cos(3\theta)\cos(-\theta)} - \tan\theta\tan(3\theta) = \tan(3\theta)\tan\theta - \tan\theta\tan(3\theta) = 0$ **45.** $\dfrac{1}{4}(\sqrt{6} - \sqrt{2})$

46. $-2 - \sqrt{3}$ **47.** $\dfrac{1}{4}(\sqrt{6} - \sqrt{2})$ **48.** $\dfrac{1}{4}(\sqrt{2} - \sqrt{6})$ **49.** $\dfrac{1}{2}$ **50.** $\dfrac{1}{2}$ **51.** $\sqrt{2} - 1$ **52.** $\dfrac{\sqrt{2 + \sqrt{2}}}{2}$

53. (a) $-\dfrac{33}{65}$ **(b)** $-\dfrac{56}{65}$ **(c)** $-\dfrac{63}{65}$ **(d)** $\dfrac{33}{56}$ **(e)** $\dfrac{24}{25}$ **(f)** $\dfrac{119}{169}$ **(g)** $\dfrac{5\sqrt{26}}{26}$ **(h)** $\dfrac{2\sqrt{5}}{5}$

54. (a) $-\dfrac{16}{65}$ **(b)** $-\dfrac{63}{65}$ **(c)** $-\dfrac{56}{65}$ **(d)** $\dfrac{16}{63}$ **(e)** $\dfrac{24}{25}$ **(f)** $\dfrac{119}{169}$ **(g)** $\dfrac{\sqrt{26}}{26}$ **(h)** $-\dfrac{\sqrt{10}}{10}$

55. (a) $-\dfrac{63}{65}$ **(b)** $\dfrac{16}{65}$ **(c)** $\dfrac{33}{65}$ **(d)** $-\dfrac{63}{16}$ **(e)** $\dfrac{24}{25}$ **(f)** $-\dfrac{119}{169}$ **(g)** $\dfrac{2\sqrt{13}}{13}$ **(h)** $-\dfrac{\sqrt{10}}{10}$

56. (a) $\dfrac{-\sqrt{3} - 2\sqrt{2}}{6}$ **(b)** $\dfrac{1 - 2\sqrt{6}}{6}$ **(c)** $\dfrac{-\sqrt{3} + 2\sqrt{2}}{6}$ **(d)** $\dfrac{8\sqrt{2} + 9\sqrt{3}}{23}$ **(e)** $-\dfrac{\sqrt{3}}{2}$ **(f)** $-\dfrac{7}{9}$ **(g)** $\dfrac{\sqrt{3}}{3}$ **(h)** $\dfrac{\sqrt{3}}{2}$

57. (a) 1 **(b)** 0 **(c)** $-\dfrac{1}{9}$ **(d)** Not defined **(e)** $\dfrac{4\sqrt{5}}{9}$ **(f)** $-\dfrac{1}{9}$ **(g)** $\dfrac{\sqrt{30}}{6}$ **(h)** $-\dfrac{\sqrt{6}\sqrt{3 - \sqrt{5}}}{6}$ **58.** $\dfrac{4 + 3\sqrt{3}}{10}$ **59.** $\dfrac{33}{65}$

60. $-\dfrac{48 + 25\sqrt{3}}{39}$ **61.** $-\dfrac{\sqrt{2}}{10}$ **62.** $-\dfrac{24}{25}$ **63.** $-\dfrac{7}{25}$ **64.** $\left\{\dfrac{\pi}{3}, \dfrac{5\pi}{3}\right\}$ **65.** $\left\{\dfrac{2\pi}{3}, \dfrac{5\pi}{3}\right\}$ **66.** $\left\{\dfrac{3\pi}{4}, \dfrac{7\pi}{4}\right\}$ **67.** $\left\{0, \dfrac{\pi}{2}, \pi, \dfrac{3\pi}{2}\right\}$ **68.** $\left\{\dfrac{\pi}{3}, \dfrac{2\pi}{3}, \dfrac{4\pi}{3}, \dfrac{5\pi}{3}\right\}$

69. $\{0.25, 2.89\}$ **70.** $\left\{0, \dfrac{2\pi}{3}, \pi, \dfrac{4\pi}{3}\right\}$ **71.** $\left\{0, \dfrac{\pi}{6}, \dfrac{5\pi}{6}\right\}$ **72.** $\left\{\dfrac{\pi}{6}, \dfrac{\pi}{2}, \dfrac{5\pi}{6}\right\}$ **73.** $\left\{\dfrac{\pi}{3}, \dfrac{5\pi}{3}\right\}$ **74.** $\left\{\dfrac{\pi}{4}, \dfrac{\pi}{2}, \dfrac{3\pi}{4}, \dfrac{3\pi}{2}\right\}$ **75.** $\left\{\dfrac{\pi}{2}, \pi\right\}$ **76.** 0.78

77. -1.11 **78.** 1.77 **79.** 1.23 **80.** 2.90 **81.** $\{1.11\}$ **82.** $\{0.87\}$ **83.** $\{2.22\}$ **84.** $\left\{-\dfrac{\sqrt{3}}{2}\right\}$ **85.** $\{0\}$

86. $\sin 15° = \sqrt{\dfrac{1 - \cos 30°}{2}} = \sqrt{\dfrac{1 - \dfrac{\sqrt{3}}{2}}{2}} = \sqrt{\dfrac{2 - \sqrt{3}}{4}} = \dfrac{\sqrt{2 - \sqrt{3}}}{2}$;

$\sin 15° = \sin(45° - 30°) = \sin 45°\cos 30° - \cos 45°\sin 30° = \dfrac{\sqrt{2}}{2}\cdot\dfrac{\sqrt{3}}{2} - \dfrac{\sqrt{2}}{2}\cdot\dfrac{1}{2} = \dfrac{\sqrt{6}}{4} - \dfrac{\sqrt{2}}{4} = \dfrac{\sqrt{6} - \sqrt{2}}{4}$;

$\left[\dfrac{\sqrt{2 - \sqrt{3}}}{2}\right]^2 = \dfrac{2 - \sqrt{3}}{4} = \dfrac{4(2 - \sqrt{3})}{4\cdot 4} = \dfrac{8 - 4\sqrt{3}}{16} = \dfrac{6 - 2\sqrt{12} + 2}{16} = \left(\dfrac{\sqrt{6} - \sqrt{2}}{4}\right)^2$

Chapter Test *(page 682)*

1. $\frac{\pi}{6}$ **2.** $-\frac{\pi}{4}$ **3.** $\frac{\pi}{5}$ **4.** $\frac{7}{3}$ **5.** 3 **6.** $-\frac{4}{3}$ **7.** 0.39 **8.** 0.78 **9.** 1.25 **10.** 0.20

11. $\frac{\csc\theta+\cot\theta}{\sec\theta+\tan\theta}=\frac{\csc\theta+\cot\theta}{\sec\theta+\tan\theta}\cdot\frac{\csc\theta-\cot\theta}{\csc\theta-\cot\theta}=\frac{\csc^2\theta-\cot^2\theta}{(\sec\theta+\tan\theta)(\csc\theta-\cot\theta)}=\frac{1}{(\sec\theta+\tan\theta)(\csc\theta-\cot\theta)}$
$=\frac{1}{(\sec\theta+\tan\theta)(\csc\theta-\cot\theta)}\cdot\frac{\sec\theta-\tan\theta}{\sec\theta-\tan\theta}=\frac{\sec\theta-\tan\theta}{(\sec^2\theta-\tan^2\theta)(\csc\theta-\cot\theta)}=\frac{\sec\theta-\tan\theta}{\csc\theta-\cot\theta}$

12. $\sin\theta\tan\theta+\cos\theta=\sin\theta\cdot\frac{\sin\theta}{\cos\theta}+\cos\theta=\frac{\sin^2\theta}{\cos\theta}+\frac{\cos^2\theta}{\cos\theta}=\frac{\sin^2\theta+\cos^2\theta}{\cos\theta}=\frac{1}{\cos\theta}=\sec\theta$

13. $\tan\theta+\cot\theta=\frac{\sin\theta}{\cos\theta}+\frac{\cos\theta}{\sin\theta}=\frac{\sin^2\theta}{\sin\theta\cos\theta}+\frac{\cos^2\theta}{\sin\theta\cos\theta}=\frac{\sin^2\theta+\cos^2\theta}{\sin\theta\cos\theta}=\frac{1}{\sin\theta\cos\theta}=\frac{2}{2\sin\theta\cos\theta}=\frac{2}{\sin(2\theta)}=2\csc(2\theta)$

14. $\frac{\sin(\alpha+\beta)}{\tan\alpha+\tan\beta}=\frac{\sin\alpha\cos\beta+\cos\alpha\sin\beta}{\frac{\sin\alpha}{\cos\alpha}+\frac{\sin\beta}{\cos\beta}}=\frac{\sin\alpha\cos\beta+\cos\alpha\sin\beta}{\frac{\sin\alpha\cos\beta}{\cos\alpha\cos\beta}+\frac{\cos\alpha\sin\beta}{\cos\alpha\cos\beta}}=\frac{\sin\alpha\cos\beta+\cos\alpha\sin\beta}{\frac{\sin\alpha\cos\beta+\cos\alpha\sin\beta}{\cos\alpha\cos\beta}}$
$=\frac{\sin\alpha\cos\beta+\cos\alpha\sin\beta}{1}\cdot\frac{\cos\alpha\cos\beta}{\sin\alpha\cos\beta+\cos\alpha\sin\beta}=\cos\alpha\cos\beta$

15. $\sin(3\theta)=\sin(\theta+2\theta)=\sin\theta\cos(2\theta)+\cos\theta\sin(2\theta)=\sin\theta\cdot(\cos^2\theta-\sin^2\theta)+\cos\theta\cdot 2\sin\theta\cos\theta=\sin\theta\cos^2\theta-\sin^3\theta+2\sin\theta\cos^2\theta$
$=3\sin\theta\cos^2\theta-\sin^3\theta=3\sin\theta(1-\sin^2\theta)-\sin^3\theta=3\sin\theta-3\sin^3\theta-\sin^3\theta=3\sin\theta-4\sin^3\theta$

16. $\frac{\tan\theta-\cot\theta}{\tan\theta+\cot\theta}=\frac{\frac{\sin\theta}{\cos\theta}-\frac{\cos\theta}{\sin\theta}}{\frac{\sin\theta}{\cos\theta}+\frac{\cos\theta}{\sin\theta}}=\frac{\frac{\sin^2\theta-\cos^2\theta}{\sin\theta\cos\theta}}{\frac{\sin^2\theta+\cos^2\theta}{\sin\theta\cos\theta}}=\frac{\sin^2\theta-\cos^2\theta}{\sin^2\theta+\cos^2\theta}=\frac{(1-\cos^2\theta)-\cos^2\theta}{1}=1-2\cos^2\theta$ **17.** $\frac{1}{4}(\sqrt{6}+\sqrt{2})$

18. $2+\sqrt{3}$ **19.** $\frac{\sqrt{5}}{5}$ **20.** $\frac{12\sqrt{85}}{49}$ **21.** $\frac{2\sqrt{13}(\sqrt{5}-3)}{39}$ **22.** $\frac{2+\sqrt{3}}{4}$ **23.** $\frac{\sqrt{6}}{2}$ **24.** $\frac{\sqrt{2}}{2}$ **25.** $\left\{\frac{\pi}{3},\frac{2\pi}{3},\frac{4\pi}{3},\frac{5\pi}{3}\right\}$ **26.** $\{0, 1.911, \pi, 4.373\}$

27. $\left\{\frac{3\pi}{8},\frac{7\pi}{8},\frac{11\pi}{8},\frac{15\pi}{8}\right\}$ **28.** $\{0.285, 3.427\}$ **29.** $\{0.253, 2.889\}$

Cumulative Review *(page 682)*

1. $\left\{\frac{-1-\sqrt{13}}{6},\frac{-1+\sqrt{13}}{6}\right\}$ **2.** $y+1=-1(x-4)$ or $x+y=3$; $6\sqrt{2}$; $(1,2)$ **3.** x-axis symmetry; $(0,-3)$, $(0,3)$, $(3,0)$

4.

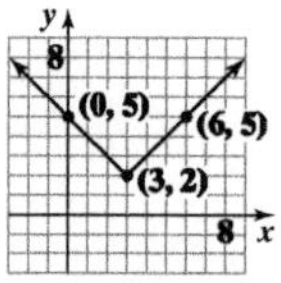

5.

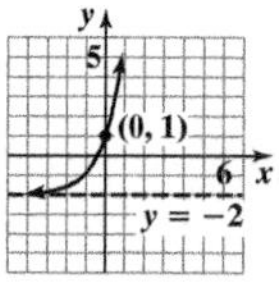

6.

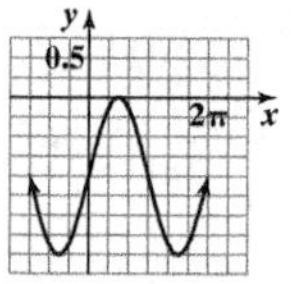

7. (a)

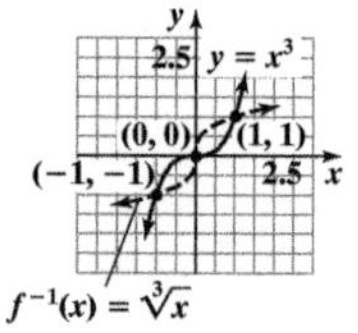

(b)

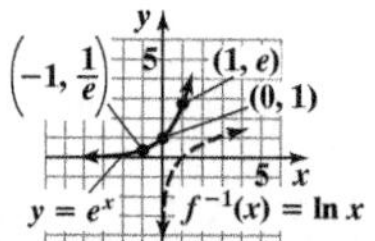

(c)

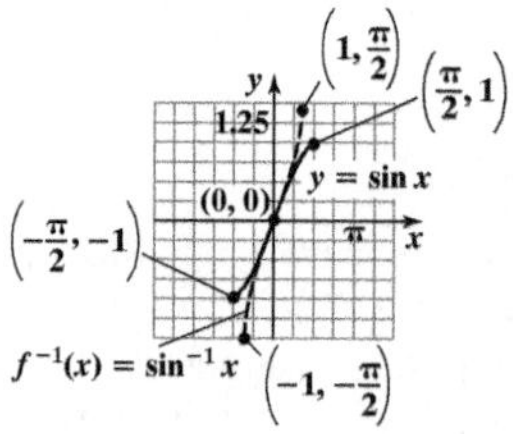

(d)

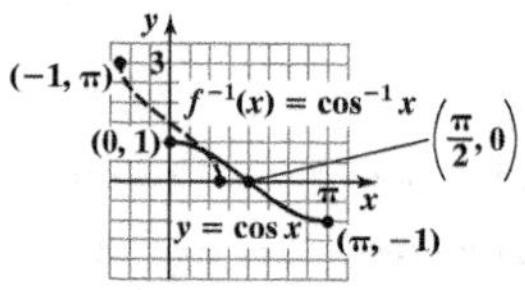

8. (a) $-\frac{2\sqrt{2}}{3}$ **(b)** $\frac{\sqrt{2}}{4}$ **(c)** $\frac{4\sqrt{2}}{9}$ **(d)** $\frac{7}{9}$ **(e)** $\sqrt{\frac{3+2\sqrt{2}}{6}}$ **(f)** $-\sqrt{\frac{3-2\sqrt{2}}{6}}$ **9.** $\frac{\sqrt{5}}{5}$ **10. (a)** $-\frac{2\sqrt{2}}{3}$ **(b)** $-\frac{2\sqrt{2}}{3}$ **(c)** $\frac{7}{9}$ **(d)** $\frac{4\sqrt{2}}{9}$ **(e)** $\frac{\sqrt{6}}{3}$

11. (a) $f(x)=(2x-1)(x-1)^2(x+1)^2$; $\frac{1}{2}$ multiplicity 1; 1 and -1 multiplicity 2
(b) $(0,-1)$; $\left(\frac{1}{2},0\right)$; $(-1,0)$; $(1,0)$ **(c)** $y=2x^5$
(d)

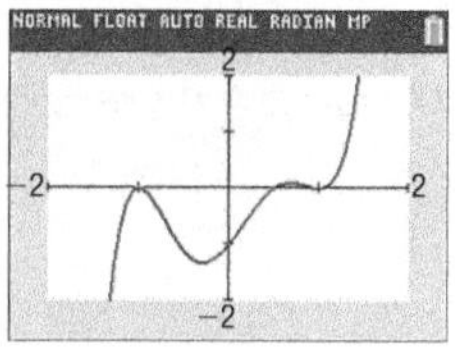

(e) Local minimum value -1.33 at $x=-0.29$,
Local minimum value 0 at $x=1$
Local maximum value 0 at $x=-1$,
Local maximum value 0.10 at $x=0.69$

(f)

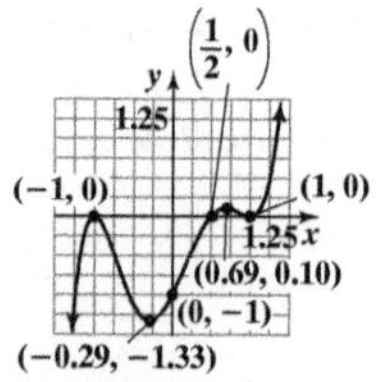

(g) Increasing: $(-\infty,-1]$, $[-0.29, 0.69]$, $[1,\infty)$
Decreasing: $[-1,-0.29]$, $[0.69, 1]$

12. (a) $\left\{-1,-\frac{1}{2}\right\}$
(b) $\{-1, 1\}$
(c) $(-\infty,-1)\cup\left(-\frac{1}{2},\infty\right)$
(d) $(-\infty,-1]\cup[1,\infty)$

CHAPTER 9 Applications of Trigonometric Functions

9.1 Assess Your Understanding *(page 688)*

5. b **6.** direction; bearing **7.** T **8.** F **9.** $a \approx 13.74$, $c \approx 14.62$, $A = 70°$ **11.** $b \approx 5.03$, $c \approx 7.83$, $A = 50°$ **13.** $a \approx 0.71$, $c \approx 4.06$, $B = 80°$ **15.** $b \approx 10.72$, $c \approx 11.83$, $B = 65°$ **17.** $b \approx 3.08$, $a \approx 8.46$, $A = 70°$ **19.** $c \approx 5.83$, $A \approx 59.0°$, $B \approx 31.0°$ **21.** $b \approx 4.58$, $A \approx 23.6°$, $B \approx 66.4°$ **23.** 23.6° and 66.4° **25.** 4.59 in.; 6.55 in. **27.** 80.5° **29.** **(a)** 111.96 ft/s or 76.3 mi/h **(b)** 82.42 ft/s or 56.2 mi/h **(c)** Under 18.8° **31.** **(a)** 2.4898×10^{13} miles **(b)** 0.000214° **33.** S76.6°E **35.** The embankment is 30.5 m high. **37.** The buildings are 7984 ft apart. **39.** 69.0° **41.** 38.9° **43.** 76.94 in. **46.** Yes **47.** $\dfrac{\sqrt{6}-\sqrt{2}}{4}$ or $\dfrac{\sqrt{2-\sqrt{3}}}{2}$ **48.** 0.236, 0.243, 0.248 **49.** $\left\{\dfrac{\pi}{2}, \dfrac{7\pi}{6}, \dfrac{11\pi}{6}\right\}$

9.2 Assess Your Understanding *(page 697)*

4. a **5.** $\dfrac{\sin A}{a} = \dfrac{\sin B}{b} = \dfrac{\sin C}{c}$ **6.** F **7.** F **8.** ambiguous case **9.** $a \approx 3.23$, $b \approx 3.55$, $A = 40°$ **11.** $a \approx 3.25$, $c \approx 4.23$, $B = 45°$ **13.** $C = 95°$, $c \approx 9.86$, $a \approx 6.36$ **15.** $A = 40°$, $a = 2$, $c \approx 3.06$ **17.** $C = 120°$, $b \approx 1.06$, $c \approx 2.69$ **19.** $A = 100°$, $a \approx 5.24$, $c \approx 0.92$ **21.** $B = 40°$, $a \approx 5.64$, $b \approx 3.86$ **23.** $C = 100°$, $a \approx 1.31$, $b \approx 1.31$ **25.** One triangle; $B \approx 30.7°$, $C \approx 99.3°$, $c \approx 3.86$ **27.** One triangle; $C \approx 36.2°$, $A \approx 43.8°$, $a \approx 3.51$ **29.** No triangle **31.** Two triangles; $C_1 \approx 30.9°$, $A_1 \approx 129.1°$, $a_1 \approx 9.07$ or $C_2 \approx 149.1°$, $A_2 \approx 10.9°$, $a_2 \approx 2.20$ **33.** No triangle **35.** Two triangles; $A_1 \approx 57.7°$, $B_1 \approx 97.3°$, $b_1 \approx 2.35$ or $A_2 \approx 122.3°$, $B_2 \approx 32.7°$, $b_2 \approx 1.28$ **37.** 1490.48 ft **39.** 335.16 ft **41.** 153.42 ft; 136.59 ft **43.** The tree is 39.4 ft high. **45.** Adam receives 100.6 more frequent flyer miles. **47.** **(a)** Station Able is about 143.33 mi from the ship: Station Baker is about 135.58 mi from the ship. **(b)** Approximately 41 min **49.** 84.7°; 183.72 ft **51.** 2.64 mi **53.** 38.5 in. **55.** 449.36 ft **57.** 187,600,000 km or 101,440,000 km **59.** The diameter is 252 ft.

61. $\dfrac{a-b}{c} = \dfrac{a}{c} - \dfrac{b}{c} = \dfrac{\sin A}{\sin C} - \dfrac{\sin B}{\sin C} = \dfrac{\sin A - \sin B}{\sin C} = \dfrac{2\sin\left(\dfrac{A-B}{2}\right)\cos\left(\dfrac{A+B}{2}\right)}{2\sin\dfrac{C}{2}\cos\dfrac{C}{2}} = \dfrac{\sin\left(\dfrac{A-B}{2}\right)\cos\left(\dfrac{\pi}{2}-\dfrac{C}{2}\right)}{\sin\dfrac{C}{2}\cos\dfrac{C}{2}} = \dfrac{\sin\left(\dfrac{A-B}{2}\right)}{\cos\dfrac{C}{2}}$

63. $\dfrac{a-b}{a+b} = \dfrac{\dfrac{a-b}{c}}{\dfrac{a+b}{c}} = \dfrac{\dfrac{\sin\left[\frac{1}{2}(A-B)\right]}{\cos\frac{C}{2}}}{\dfrac{\cos\left[\frac{1}{2}(A-B)\right]}{\sin\frac{C}{2}}} = \dfrac{\tan\left[\frac{1}{2}(A-B)\right]}{\cot\frac{C}{2}} = \dfrac{\tan\left[\frac{1}{2}(A-B)\right]}{\tan\left(\frac{\pi}{2}-\frac{C}{2}\right)} = \dfrac{\tan\left[\frac{1}{2}(A-B)\right]}{\tan\left[\frac{1}{2}(A+B)\right]}$

69. $\left\{-3, -\dfrac{4}{3}, 3\right\}$ **70.** $3\sqrt{5} \approx 6.71$ **71.** $-\dfrac{\sqrt{15}}{7}$ **72.**

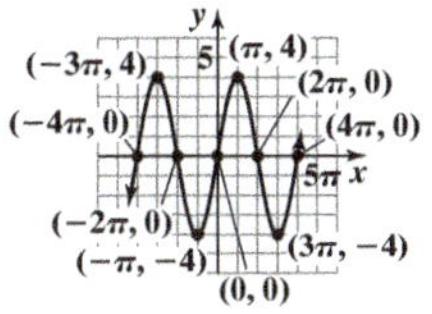

9.3 Assess Your Understanding *(page 704)*

3. Cosines **4.** a **5.** b **6.** F **7.** F **8.** T **9.** $b \approx 2.95$, $A \approx 28.7°$, $C \approx 106.3°$ **11.** $c \approx 3.75$, $A \approx 32.1°$, $B \approx 52.9°$ **13.** $A \approx 48.5°$, $B \approx 38.6°$, $C \approx 92.9°$ **15.** $A \approx 127.2°$, $B \approx 32.1°$, $C \approx 20.7°$ **17.** $c \approx 2.57$, $A \approx 48.6°$, $B \approx 91.4°$ **19.** $a \approx 2.99$, $B \approx 19.2°$, $C \approx 80.8°$ **21.** $b \approx 4.14$, $A \approx 43.0°$, $C \approx 27.0°$ **23.** $c \approx 1.69$, $A = 65.0°$, $B = 65.0°$ **25.** $A \approx 67.4°$, $B = 90°$, $C \approx 22.6°$ **27.** $A = 60°$, $B = 60°$, $C = 60°$ **29.** $A \approx 33.6°$, $B \approx 62.2°$, $C \approx 84.3°$ **31.** $A \approx 97.9°$, $B \approx 52.4°$, $C \approx 29.7°$ **33.** $A = 85°$, $a = 14.56$, $c = 14.12$ **35.** $A = 40.8°$, $B = 60.6°$, $C = 78.6°$ **37.** $A = 80°$, $b = 8.74$, $c = 13.80$ **39.** Two triangles: $B_1 = 35.4°$, $C_1 = 134.6°$, $c_1 = 12.29$; $B_2 = 144.6°$, $C_2 = 25.4°$, $c_2 = 7.40$ **41.** $B = 24.5°$, $C = 95.5°$, $a = 10.44$ **43.** 165 yd **45.** **(a)** 26.4° **(b)** 30.8 h **47.** **(a)** 63.7 ft **(b)** 66.8 ft **(c)** 92.8° **49.** **(a)** 492.6 ft **(b)** 269.3 ft **51.** **(a)** 59.2 mm **(b)** male **53.** **(a)** $\alpha \approx 9.9°$; $\beta \approx 8.3°$ **(b)** 21.73 yd **(c)** 0.36 yd or 13 in. **55.** 342.33 ft **57.** The footings should be 7.65 ft apart.

59. Suppose $0 < \theta < \pi$. Then, by the Law of Cosines, $d^2 = r^2 + r^2 - 2r^2\cos\theta = 4r^2\left(\dfrac{1-\cos\theta}{2}\right) \Rightarrow d = 2r\sqrt{\dfrac{1-\cos\theta}{2}} = 2r\sin\dfrac{\theta}{2}$.

Since, for any angle in $(0, \pi)$, d is strictly less than the length of the arc subtended by θ, that is, $d < r\theta$, then $2r\sin\dfrac{\theta}{2} < r\theta$, or $2\sin\dfrac{\theta}{2} < \theta$.

Since $\cos\dfrac{\theta}{2} < 1$, then, for $0 < \theta < \pi$, $\sin\theta = 2\sin\dfrac{\theta}{2}\cos\dfrac{\theta}{2} < 2\sin\dfrac{\theta}{2} < \theta$. If $\theta \geq \pi$, then, since $\sin\theta \leq 1$, $\sin\theta < \theta$. Thus $\sin\theta < \theta$ for all $\theta > 0$.

61. $\sin\frac{C}{2} = \sqrt{\frac{1-\cos C}{2}} = \sqrt{\frac{1-\frac{a^2+b^2-c^2}{2ab}}{2}} = \sqrt{\frac{2ab-a^2-b^2+c^2}{4ab}} = \sqrt{\frac{c^2-(a-b)^2}{4ab}} = \sqrt{\frac{(c+a-b)(c+b-a)}{4ab}}$
$= \sqrt{\frac{(2s-2b)(2s-2a)}{4ab}} = \sqrt{\frac{(s-a)(s-b)}{ab}}$

68.

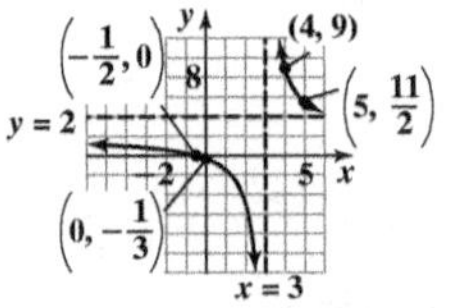

69. $\left\{\frac{\ln 3}{\ln 4 - \ln 3}\right\} \approx \{3.819\}$

70. $\sin\theta = \frac{2\sqrt{6}}{7}$; $\csc\theta = \frac{7\sqrt{6}}{12}$; $\sec\theta = -\frac{7}{5}$; $\cot\theta = -\frac{5\sqrt{6}}{12}$

71. $y = -3\sin(4x)$

9.4 Assess Your Understanding *(page 711)*

2. $\frac{1}{2}ab\sin C$ **3.** $\sqrt{s(s-a)(s-b)(s-c)}$; $\frac{1}{2}(a+b+c)$ **4.** T **5.** c **6.** c **7.** 2.83 **9.** 2.99 **11.** 14.98 **13.** 9.56 **15.** 3.86 **17.** 1.48 **19.** 2.82

21. 30 **23.** 1.73 **25.** 19.90 **27.** $K = \frac{1}{2}ab\sin C = \frac{1}{2}a\sin C\left(\frac{a\sin B}{\sin A}\right) = \frac{a^2\sin B\sin C}{2\sin A}$ **29.** 0.92 **31.** 2.27 **33.** 5.44 **35.** 9.03 sq ft **37.** $5446.38

39. The area of the lot is about 8376 sq ft. **41.** 15.27 in.^{2} **43.** The lake area is 1645.14 ft^2. **45.** The ground area is 7517.4 ft^2.

47. (a) Area $\Delta OAC = \frac{1}{2}|OC||AC| = \frac{1}{2}\cdot\frac{|OC|}{1}\cdot\frac{|AC|}{1} = \frac{1}{2}\sin\alpha\cos\alpha$

(b) Area $\Delta OCB = \frac{1}{2}|BC||OC| = \frac{1}{2}|OB|^2\frac{|BC|}{|OB|}\cdot\frac{|OC|}{|OB|} = \frac{1}{2}|OB|^2\sin\beta\cos\beta$

(c) Area $\Delta OAB = \frac{1}{2}|BD||OA| = \frac{1}{2}|OB|\frac{|BD|}{|OB|} = \frac{1}{2}|OB|\sin(\alpha+\beta)$

(d) $\frac{\cos\alpha}{\cos\beta} = \frac{\frac{|OC|}{1}}{\frac{|OC|}{|OB|}} = |OB|$

(e) Area ΔOAB = Area ΔOAC + Area ΔOCB

$\frac{1}{2}|OB|\sin(\alpha+\beta) = \frac{1}{2}\sin\alpha\cos\alpha + \frac{1}{2}|OB|^2\sin\beta\cos\beta$

$\sin(\alpha+\beta) = \frac{1}{|OB|}\sin\alpha\cos\alpha + |OB|\sin\beta\cos\beta$

$\sin(\alpha+\beta) = \frac{\cos\beta}{\cos\alpha}\sin\alpha\cos\alpha + \frac{\cos\alpha}{\cos\beta}\sin\beta\cos\beta$

$\sin(\alpha+\beta) = \sin\alpha\cos\beta + \cos\alpha\sin\beta$

49. 31,145 ft^2 **51. (a)** The perimeter and area are both 36. **(b)** The perimeter and area are both 60.

53. $K = \frac{1}{2}ah = \frac{1}{2}ab\sin C \Rightarrow h = b\sin C = \frac{a\sin B\sin C}{\sin A}$

55. $\angle POQ = 180^\circ - \left(\frac{A}{2}+\frac{B}{2}\right) = 180^\circ - \frac{1}{2}(180^\circ - C) = 90^\circ + \frac{C}{2}$, and $\sin\left(90^\circ + \frac{C}{2}\right) = \cos\left(-\frac{C}{2}\right) = \cos\frac{C}{2}$, since cosine is an even function.

Therefore, $r = \frac{c\sin\frac{A}{2}\sin\frac{B}{2}}{\sin\left(90^\circ+\frac{C}{2}\right)} = \frac{c\sin\frac{A}{2}\sin\frac{B}{2}}{\cos\frac{C}{2}}$.

57. $\cot\frac{A}{2} + \cot\frac{B}{2} + \cot\frac{C}{2} = \frac{s-a}{r} + \frac{s-b}{r} + \frac{s-c}{r} = \frac{3s-(a+b+c)}{r} = \frac{3s-2s}{r} = \frac{s}{r}$ **59.** $\frac{\sqrt{2+\sqrt{3}}}{4}$ or $\frac{\sqrt{2}+\sqrt{6}}{8}$ square units

63. Maximum value; 17 **64.** $(-\infty, -3) \cup [-1, 3)$ **65.** $\sin t = \frac{\sqrt{2}}{3}$, $\cos t = -\frac{\sqrt{7}}{3}$, $\tan t = -\frac{\sqrt{14}}{7}$, $\csc t = \frac{3\sqrt{2}}{2}$, $\sec t = -\frac{3\sqrt{7}}{7}$, $\cot t = -\frac{\sqrt{14}}{2}$

66. $\csc\theta - \sin\theta = \frac{1}{\sin\theta} - \sin\theta = \frac{1-\sin^2\theta}{\sin\theta} = \frac{\cos^2\theta}{\sin\theta} = \cos\theta\cdot\frac{\cos\theta}{\sin\theta} = \cos\theta\cot\theta$

9.5 Assess Your Understanding *(page 721)*

2. Simple harmonic; amplitude **3.** Simple harmonic; damped **4.** T **5.** $d = -5\cos(\pi t)$ **7.** $d = -6\cos(2t)$ **9.** $d = -5\sin(\pi t)$

11. $d = -6\sin(2t)$ **13. (a)** Simple harmonic **(b)** 5 m **(c)** $\frac{2\pi}{3}$ sec **(d)** $\frac{3}{2\pi}$ oscillation/sec **15. (a)** Simple harmonic **(b)** 6 m **(c)** 2 sec

(d) $\frac{1}{2}$ oscillation/sec **17. (a)** Simple harmonic **(b)** 3 m **(c)** 4π sec **(d)** $\frac{1}{4\pi}$ oscillation/sec **19. (a)** Simple harmonic **(b)** 2 m

(c) 1 sec **(d)** 1 oscillation/sec

21.

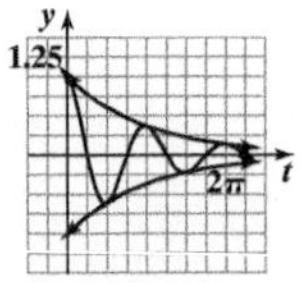

23.

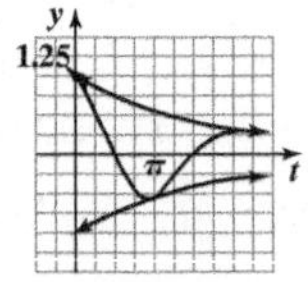

25.

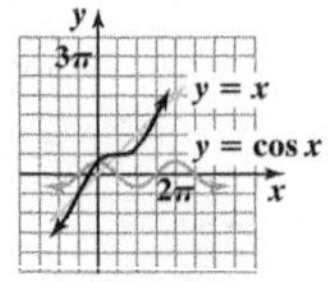

27.

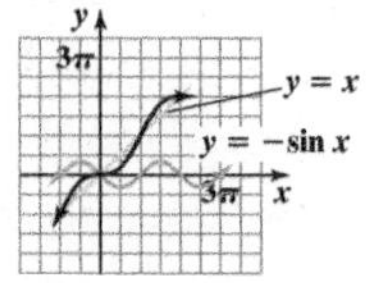

29.

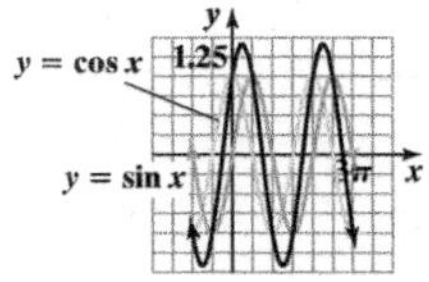

31.

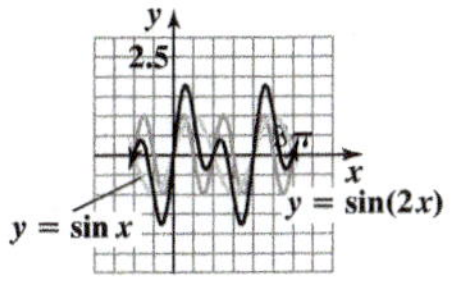

33. (a) $f(x) = \frac{1}{2}[\cos x - \cos(3x)]$

(b)

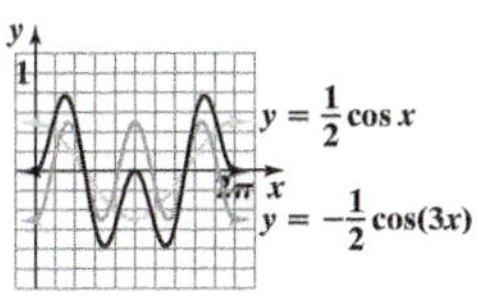

35. (a) $G(x) = \frac{1}{2}[\cos(6x) + \cos(2x)]$

(b)

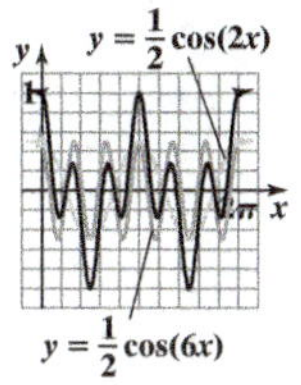

37. (a) $H(x) = \sin(4x) + \sin(2x)$

(b)

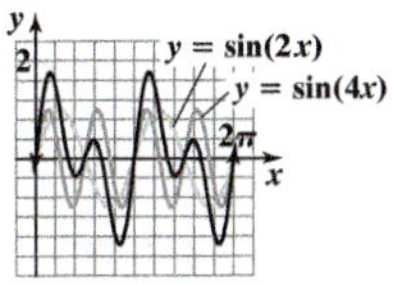

39. (a) $d = -10e^{-0.7t/50}\cos\left(\sqrt{\frac{4\pi^2}{25} - \frac{0.49}{2500}}\,t\right)$

(b)

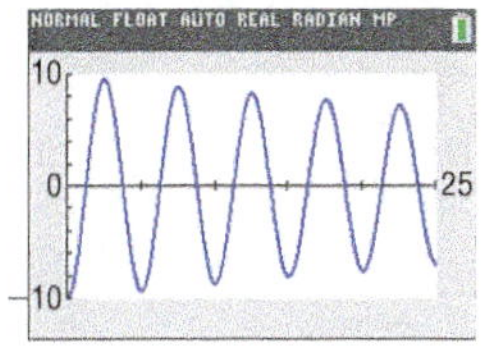

41. (a) $d = -18e^{-0.6t/60}\cos\left(\sqrt{\frac{\pi^2}{4} - \frac{0.36}{3600}}\,t\right)$

(b)

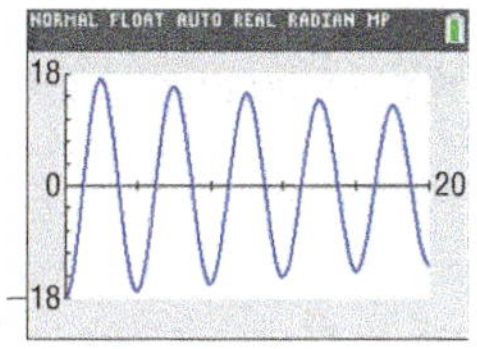

43. (a) $d = -5e^{-0.8t/20}\cos\left(\sqrt{\frac{4\pi^2}{9} - \frac{0.64}{400}}\,t\right)$

(b)

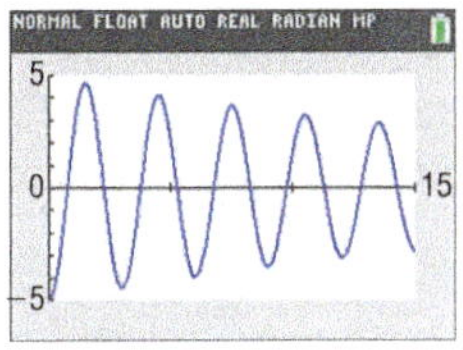

45. (a) The motion is damped. The bob has mass $m = 20$ kg with a damping factor of 0.7 kg/sec.

(b) 20 m leftward

(c)

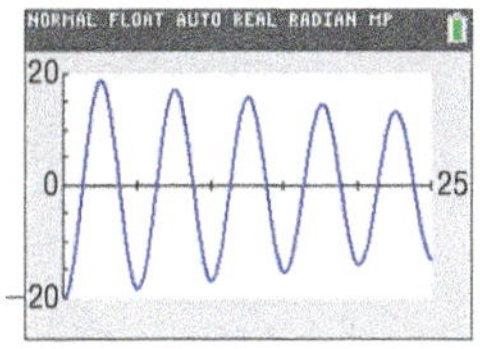

(d) 18.33 m leftward **(e)** $d \to 0$

47. (a) The motion is damped. The bob has mass $m = 40$ kg with a damping factor of 0.6 kg/sec.

(b) 30 m leftward

(c)

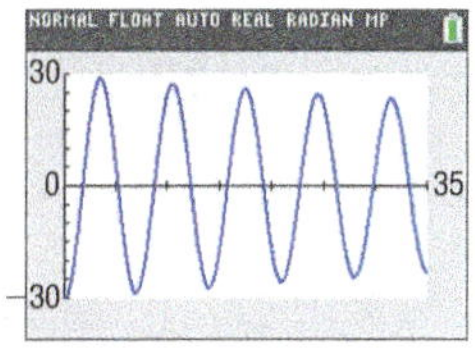

(d) 28.47 m leftward **(e)** $d \to 0$

49. (a) The motion is damped. The bob has mass $m = 15$ kg with a damping factor of 0.9 kg/sec.

(b) 15 m leftward

(c)

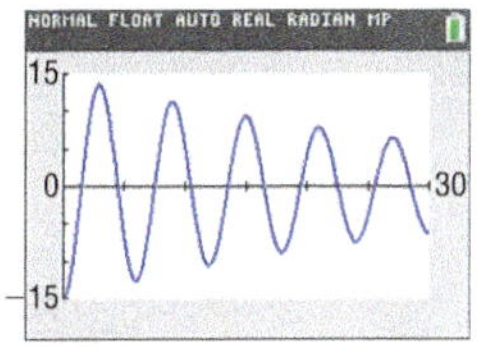

(d) 12.53 m leftward **(e)** $d \to 0$

51. $\omega = 1040\pi$; $d = 0.80\cos(1040\pi t)$

53. $\omega = 880\pi$; $d = 0.01\sin(880\pi t)$

55. (a)

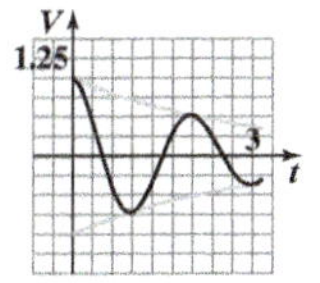

(b) At $t = 0$, $t = 2$; at $t = 1$, $t = 3$

(c) During the approximate intervals $0.35 < t < 0.67$, $1.29 < t < 1.75$, and $2.19 < t \le 3$

57.

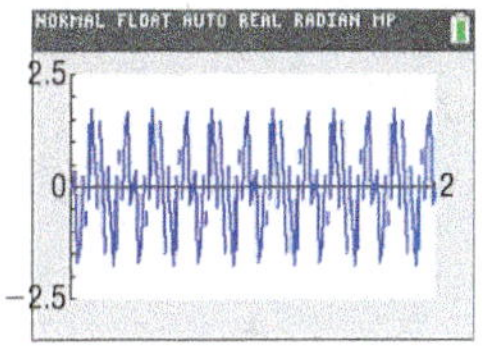

61.

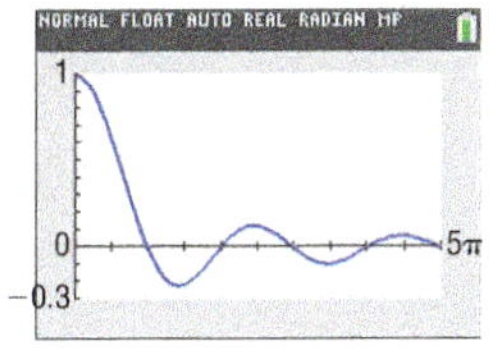

63. $y = \frac{1}{x}\sin x$

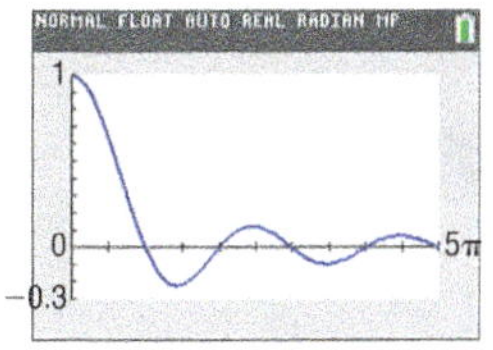

$y = \frac{1}{x^2}\sin x$

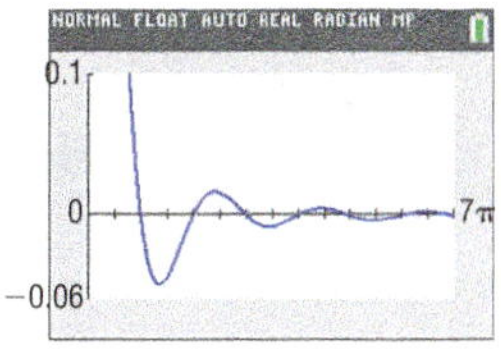

$y = \frac{1}{x^3}\sin x$

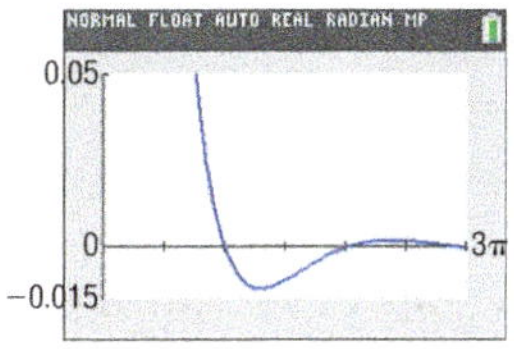

65. $f^{-1}(x) = \frac{4x - 3}{x - 1}$ **66.** $\log_7\left(\frac{xy^3}{x + y}\right)$ **67.** $\{4\}$ **68. (a)** $\frac{3\sqrt{10}}{10}$ **(b)** $\frac{\sqrt{10}}{10}$ **(c)** $\frac{1}{3}$

Review Exercises *(page 724)*

1. $A = 70°$, $b \approx 3.42$, $a \approx 9.40$ **2.** $a \approx 4.58$, $A \approx 66.4°$, $B \approx 23.6°$ **3.** $C = 100°$, $b \approx 0.65$, $c \approx 1.29$ **4.** $B \approx 56.8°$, $C \approx 23.2°$, $b \approx 4.25$ **5.** No triangle **6.** $b \approx 3.32$, $A \approx 62.8°$, $C \approx 17.2°$ **7.** $A \approx 36.2°$, $C \approx 63.8°$, $c \approx 4.55$ **8.** No triangle **9.** $A \approx 83.3°$, $B \approx 44.0°$, $C \approx 52.6°$ **10.** $c \approx 2.32$, $A \approx 16.1°$, $B \approx 123.9°$ **11.** $B \approx 36.2°$, $C \approx 63.8°$, $c \approx 4.55$ **12.** $A \approx 39.6°$, $B \approx 18.6°$, $C \approx 121.9°$ **13.** Two triangles: $B_1 \approx 13.4°$, $C_1 \approx 156.6°$, $c_1 \approx 6.86$ or $B_2 \approx 166.6°$, $C_2 \approx 3.4°$, $c_2 \approx 1.02$ **14.** $b \approx 11.52$, $c \approx 10.13$, $C \approx 60°$ **15.** $a \approx 5.23$, $B \approx 46.0°$, $C \approx 64.0°$ **16.** 1.93 **17.** 18.79 **18.** 6 **19.** 3.80 **20.** 0.32 **21.** 12.7° **22.** 29.97 ft **23.** 6.22 mi **24. (a)** 131.8 mi **(b)** 23.1° **(c)** 0.21 hr **25.** 8798.67 ft^2 **26.** \$222,983.51 **27.** 1.92 in.^{2} **28.** S4.0°E **29.** $d = -3\cos\left(\frac{\pi}{2}t\right)$ **30. (a)** simple harmonic **(b)** 6 ft **(c)** π s **(d)** $\frac{1}{\pi}$ oscillation/sec **31. (a)** simple harmonic **(b)** 2 ft **(c)** 2 s **(d)** $\frac{1}{2}$ oscillation/sec

32. (a) $d = -15e^{-0.75t/80}\cos\left(\sqrt{\dfrac{4\pi^2}{25} - \dfrac{0.5625}{6400}}\,t\right)$ **(b)**

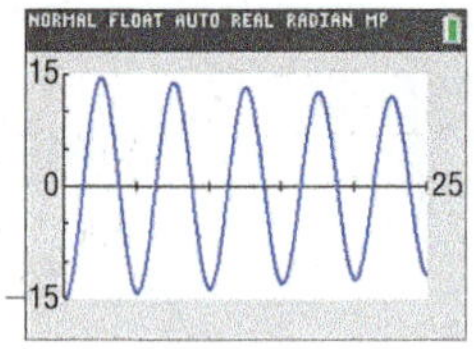

33. (a) The motion is damped. The bob has mass $m = 20$ kg with a damping factor of 0.6 kg/s. **(b)** 15 m leftward **(c)**

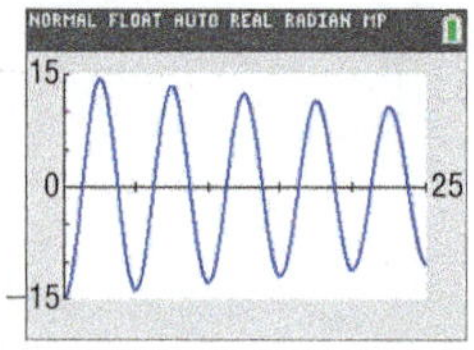

(d) 13.92 m leftward **(e)** $d \to 0$

34.

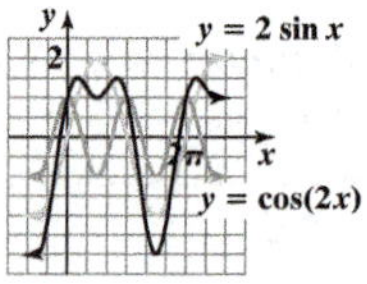

Chapter Test *(page 726)*

1. 61.0° **2.** 1.3° **3.** $a = 15.88$, $B \approx 57.5°$, $C \approx 70.5°$ **4.** $b \approx 6.85$, $C = 117°$, $c \approx 16.30$ **5.** $A \approx 52.4°$, $B \approx 29.7°$, $C \approx 97.9°$ **6.** $b \approx 4.72$, $c \approx 1.67$, $B = 105°$ **7.** No triangle **8.** $c \approx 7.62$, $A \approx 80.5°$, $B \approx 29.5°$ **9.** 15.04 square units **10.** 19.81 square units **11.** The area of the shaded region is 9.26 cm^2. **12.** 54.15 square units **13.** Madison will have to swim about 2.23 miles. **14.** 12.63 square units **15.** The lengths of the sides are 15, 18, and 21. **16.** $d = 5(\sin 42°)\sin\left(\dfrac{\pi t}{3}\right)$ or $d \approx 3.346\sin\left(\dfrac{\pi t}{3}\right)$

Cumulative Review *(page 727)*

1. $\left\{\dfrac{1}{3}, 1\right\}$ **2.** $(x + 5)^2 + (y - 1)^2 = 9$

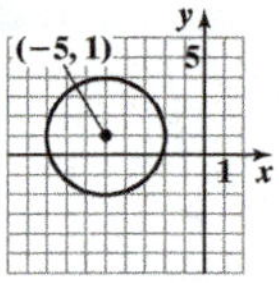

3. $\{x \mid x \le -1 \text{ or } x \ge 4\}$ **4.**

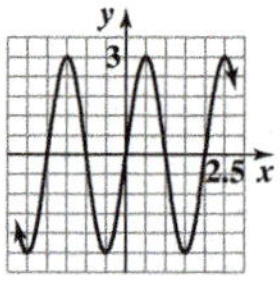

5.

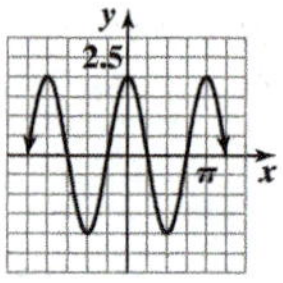

6. (a) $-\dfrac{2\sqrt{5}}{5}$ **(b)** $\dfrac{\sqrt{5}}{5}$ **(c)** $-\dfrac{4}{5}$ **(d)** $-\dfrac{3}{5}$ **(e)** $\sqrt{\dfrac{5 - \sqrt{5}}{10}}$ **(f)** $-\sqrt{\dfrac{5 + \sqrt{5}}{10}}$

7. (a)

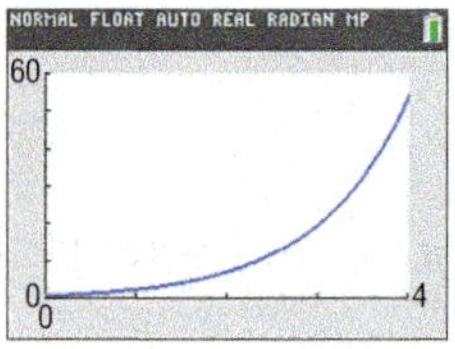

(b)

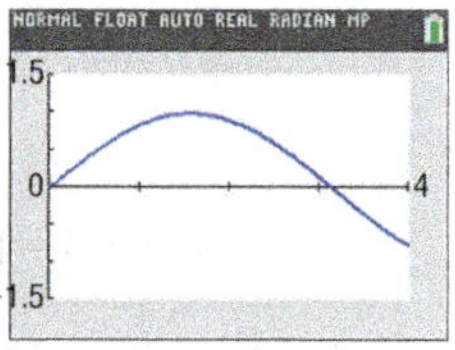

(c)

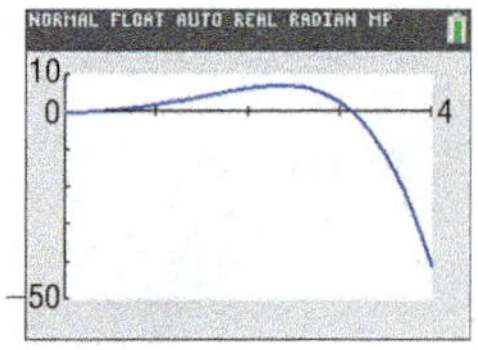

(d)

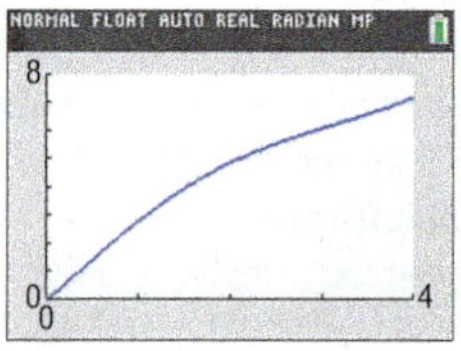

8. (a)

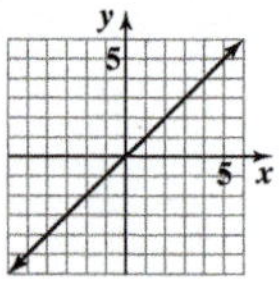

(b)

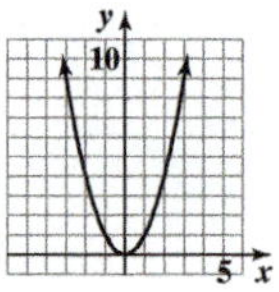

(c)

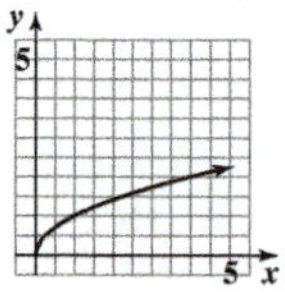

(d)

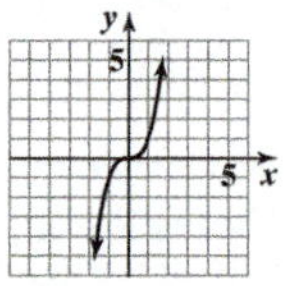

(e)

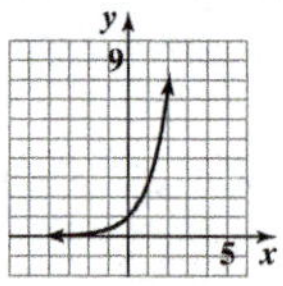

(f)

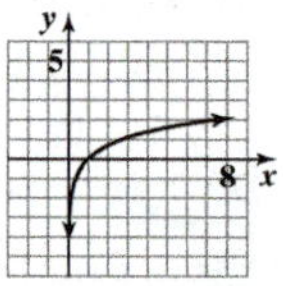

(g)

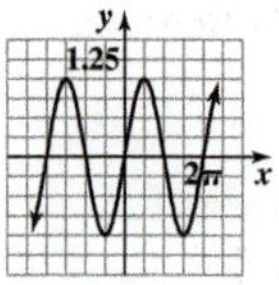

(h)

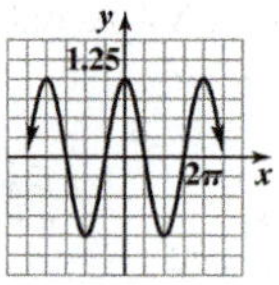

(i)

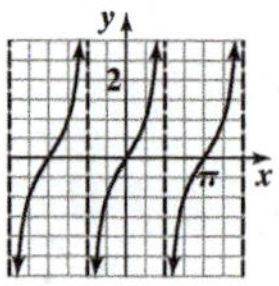

9. Two triangles: $A_1 \approx 59.0°$, $B_1 \approx 81.0°$, $b_1 \approx 23.05$ or $A_2 \approx 121.0°$, $B_2 \approx 19.0°$, $b_2 \approx 7.59$

10. $\left\{-2i, 2i, \dfrac{1}{3}, 1, 2\right\}$

11. $R(x) = \dfrac{(2x + 1)(x - 4)}{(x + 5)(x - 3)}$; domain: $\{x \mid x \ne -5, x \ne 3\}$

Intercepts: $\left(-\dfrac{1}{2}, 0\right)$, $(4, 0)$, $\left(0, \dfrac{4}{15}\right)$

No symmetry

Vertical asymptotes: $x = -5$, $x = 3$

Horizontal asymptote: $y = 2$

Intersects: $\left(\dfrac{26}{11}, 2\right)$

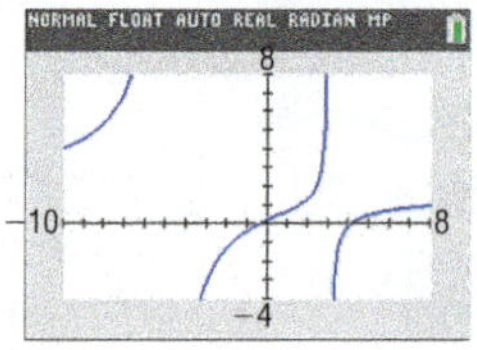

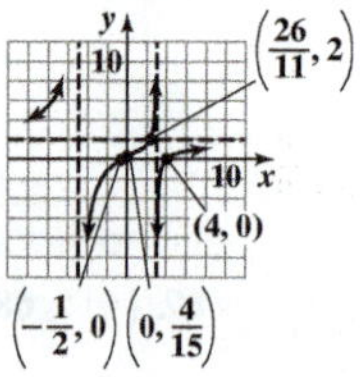

12. {2.26} **13.** {1} **14. (a)** $\left\{-\dfrac{5}{4}\right\}$ **(b)** {2} **(c)** $\left\{\dfrac{-1 - 3\sqrt{13}}{2}, \dfrac{-1 + 3\sqrt{13}}{2}\right\}$ **(d)** $\left\{x \,\middle|\, x > -\dfrac{5}{4}\right\}$ or $\left(-\dfrac{5}{4}, \infty\right)$

(e) $\{x \mid -8 \le x \le 3\}$ or $[-8, 3]$ **(f)**

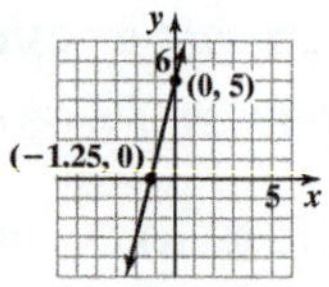

(g)

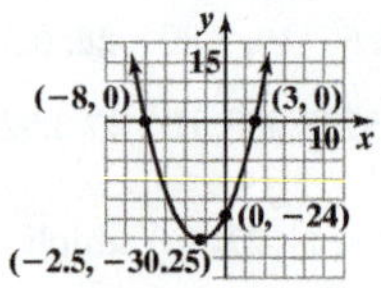

CHAPTER 10 Polar Coordinates; Vectors

10.1 Assess Your Understanding *(page 738)*

5. pole; polar axis **6.** $r\cos\theta$; $r\sin\theta$ **7.** b **8.** d **9.** T **10.** F **11.** *A* **13.** *C* **15.** *B* **17.** *A*

19.
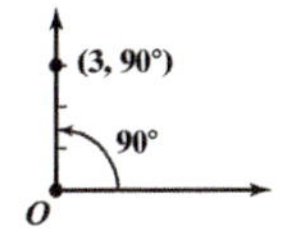

21.

23.
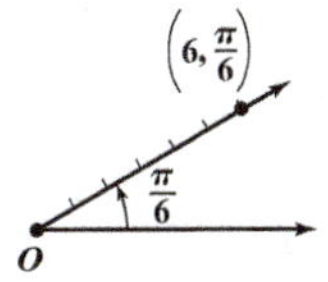

25.
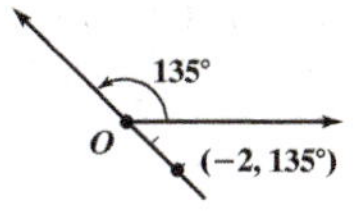

27.
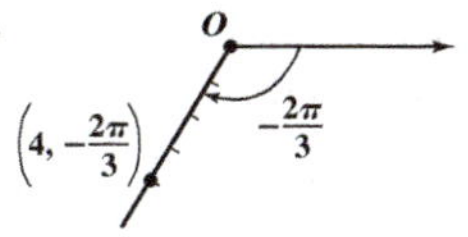

29.
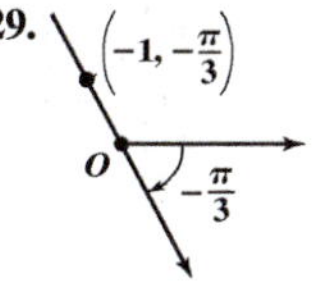

31.
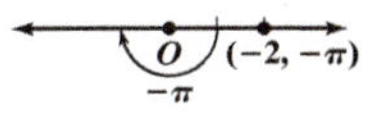

33.
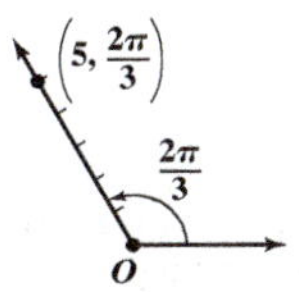

(a) $\left(5, -\frac{4\pi}{3}\right)$ **(b)** $\left(-5, \frac{5\pi}{3}\right)$ **(c)** $\left(5, \frac{8\pi}{3}\right)$

35.
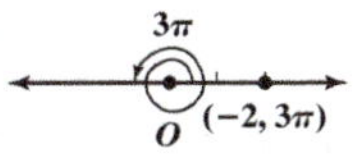

(a) $(2, -2\pi)$
(b) $(-2, \pi)$
(c) $(2, 2\pi)$

37.
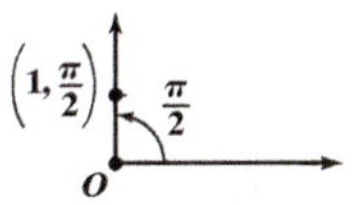

(a) $\left(1, -\frac{3\pi}{2}\right)$
(b) $\left(-1, \frac{3\pi}{2}\right)$
(c) $\left(1, \frac{5\pi}{2}\right)$

39.
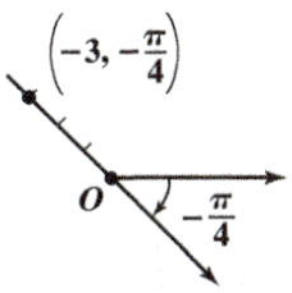

(a) $\left(3, -\frac{5\pi}{4}\right)$
(b) $\left(-3, \frac{7\pi}{4}\right)$
(c) $\left(3, \frac{11\pi}{4}\right)$

41. $(0, 3)$ **43.** $(-2, 0)$ **45.** $(-3\sqrt{3}, 3)$ **47.** $(\sqrt{2}, -\sqrt{2})$ **49.** $\left(-\frac{1}{2}, \frac{\sqrt{3}}{2}\right)$ **51.** $(2, 0)$

53. $(-2.57, 7.05)$ **55.** $(-4.98, -3.85)$ **57.** $(3, 0)$ **59.** $(1, \pi)$ **61.** $\left(\sqrt{2}, -\frac{\pi}{4}\right)$

63. $\left(2, \frac{\pi}{6}\right)$ **65.** $(2.47, -1.02)$ **67.** $(9.30, 0.47)$ **69.** $r^2 = \frac{3}{2}$ or $r = \frac{\sqrt{6}}{2}$

71. $r^2\cos^2\theta - 4r\sin\theta = 0$ **73.** $r^2\sin 2\theta = 1$ **75.** $r\cos\theta = 4$

77. $x^2 + y^2 - x = 0$ or $\left(x - \frac{1}{2}\right)^2 + y^2 = \frac{1}{4}$ **79.** $(x^2 + y^2)^{3/2} - x = 0$

81. $x^2 + y^2 = 4$ **83.** $y^2 = 8(x + 2)$

85. (a) $(-10, 36)$ **(b)** $\left(2\sqrt{349}, 180° + \tan^{-1}\left(-\frac{18}{5}\right)\right) \approx (37.36, 105.5°)$ **(c)** $(-3, -35)$ **(d)** $\left(\sqrt{1234}, 180° + \tan^{-1}\left(\frac{35}{3}\right)\right) \approx (35.13, 265.1°)$

90. $\left\{\frac{19}{15}\right\}$ **91.** 2 or 0 positive real zeros; 1 negative real zero **92.** $\left(-\frac{5}{4}, \frac{9}{2}\right)$ **93.** $(0, -11)$

10.2 Assess Your Understanding *(page 753)*

7. polar equation **8.** F **9.** $-\theta$ **10.** $\pi - \theta$ **11.** T **12.** $2n$; n **13.** c **14.** b

15. $x^2 + y^2 = 16$; circle, radius 4, center at pole

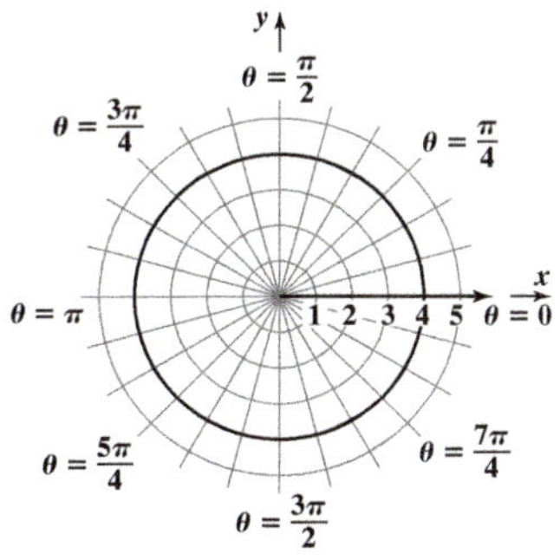

17. $y = \sqrt{3}x$; line through pole, making an angle of $\frac{\pi}{3}$ with polar axis

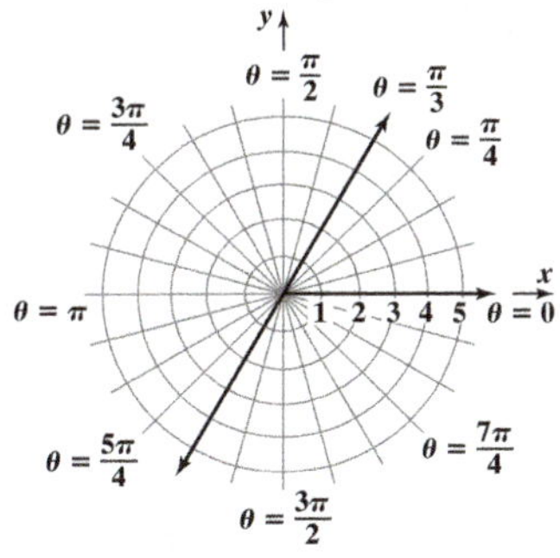

19. $y = 4$; horizontal line 4 units above the pole

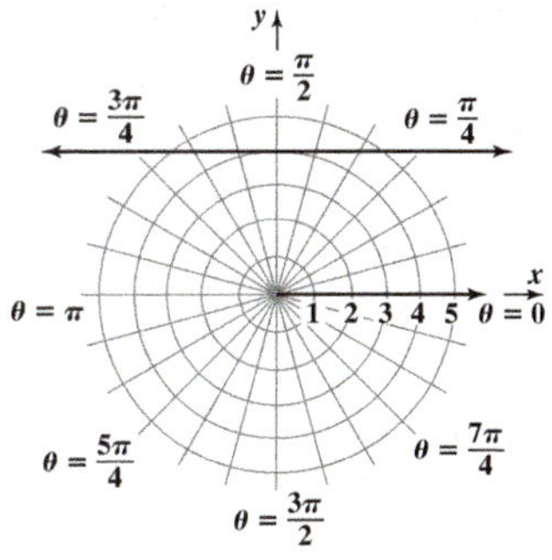

21. $x = -2$; vertical line 2 units to the left of the pole

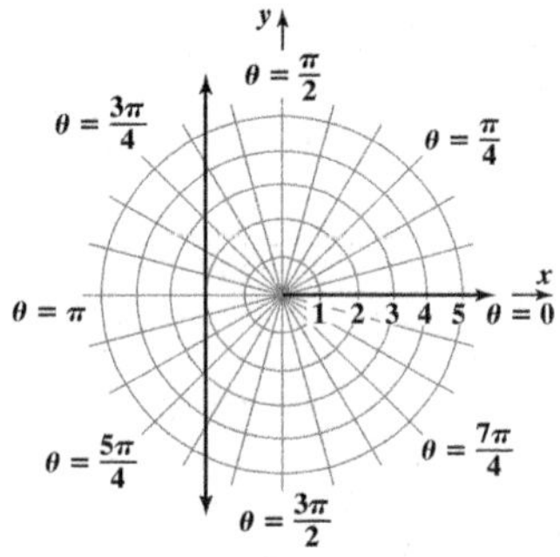

23. $(x - 1)^2 + y^2 = 1$; circle, radius 1, center $(1, 0)$ in rectangular coordinates

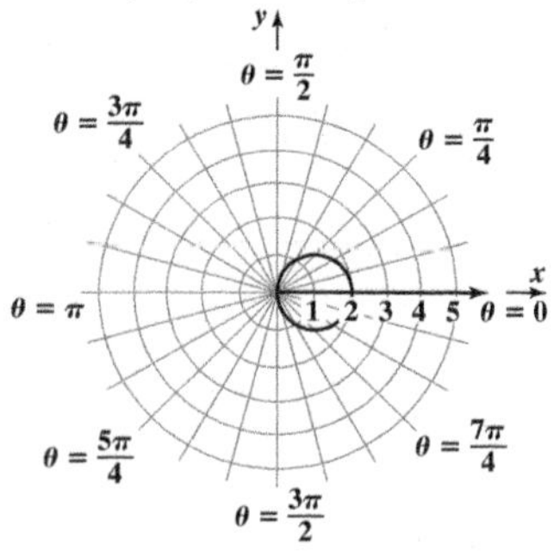

25. $x^2 + (y + 2)^2 = 4$; circle, radius 2, center at $(0, -2)$ in rectangular coordinates

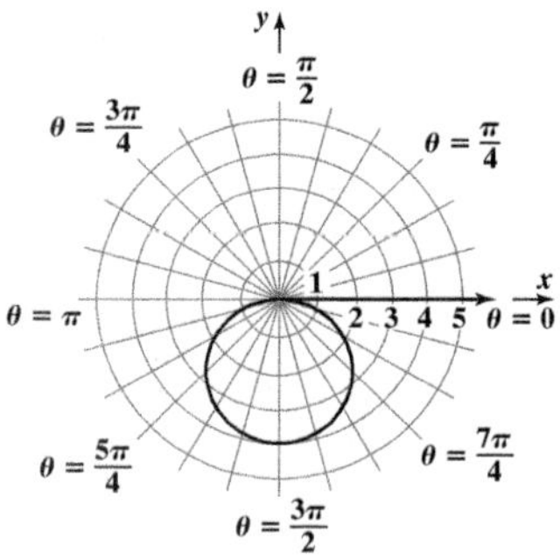

27. $(x - 2)^2 + y^2 = 4, x \neq 0$; circle, radius 2, center at $(2, 0)$ in rectangular coordinates, hole at $(0, 0)$

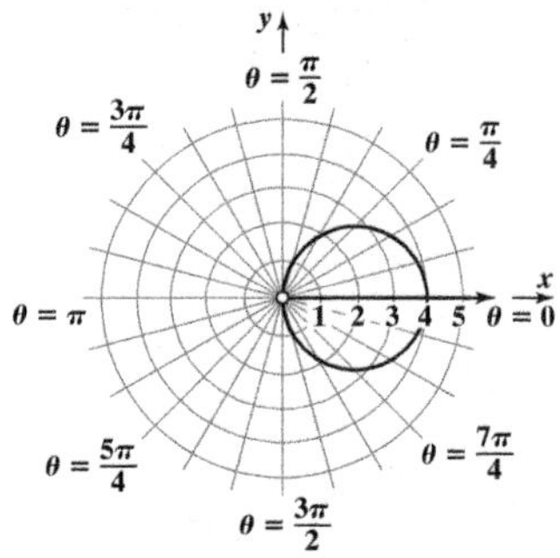

29. $x^2 + (y + 1)^2 = 1, x \neq 0$; circle, radius 1, center at $(0, -1)$ in rectangular coordinates, hole at $(0, 0)$

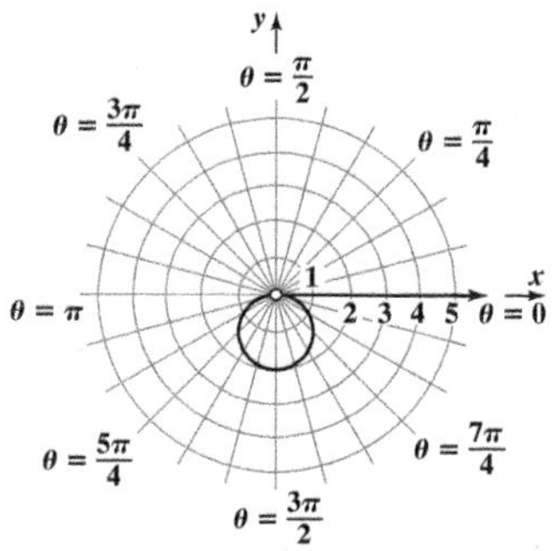

31. *E* **33.** *F* **35.** *H* **37.** *D*

39. Cardioid

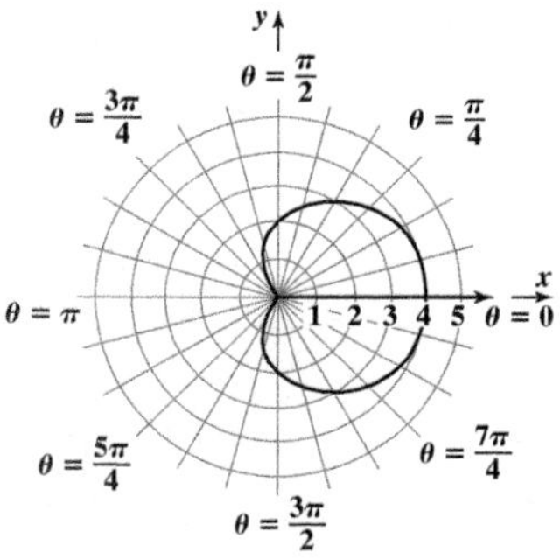

41. Cardioid

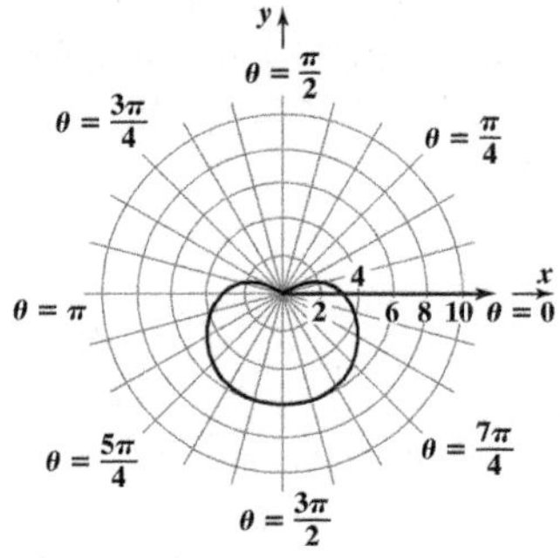

43. Limaçon without inner loop

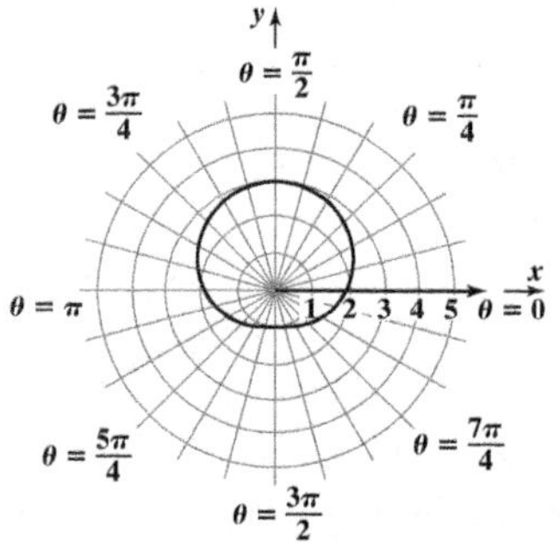

45. Limaçon without inner loop

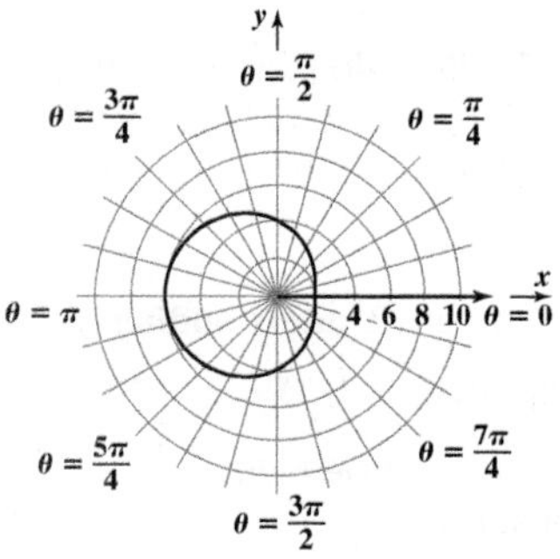

47. Limaçon with inner loop

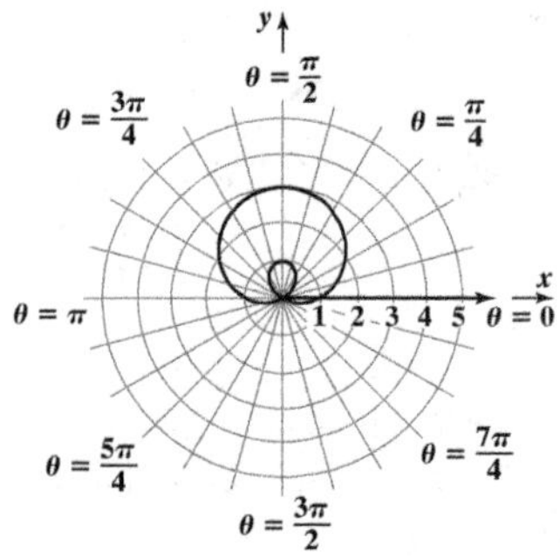

49. Limaçon with inner loop

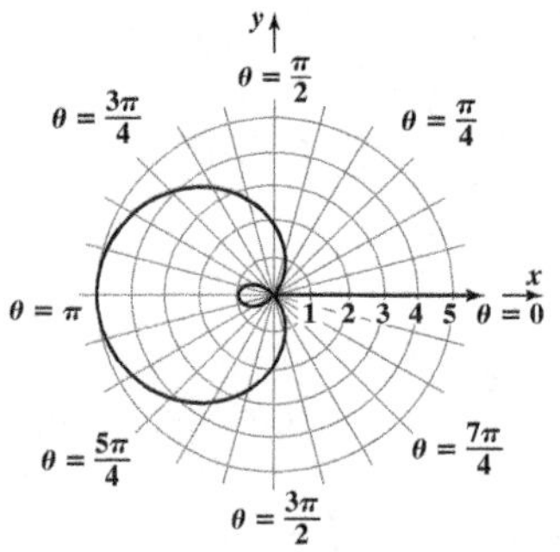

51. Rose

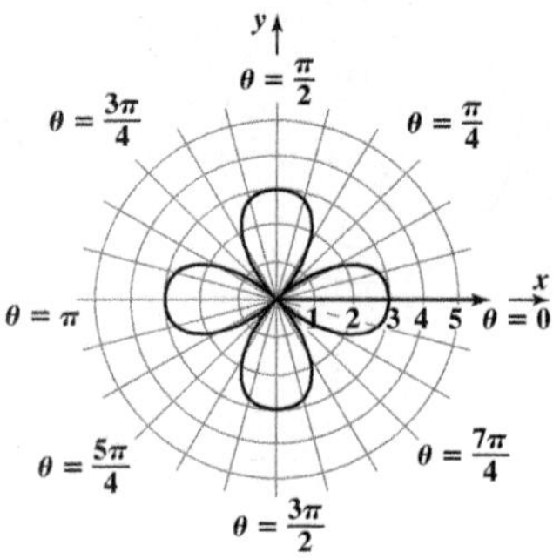

53. Rose

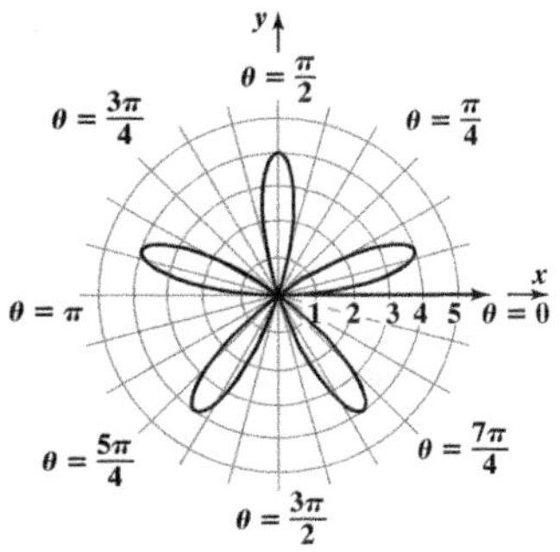

55. Lemniscate

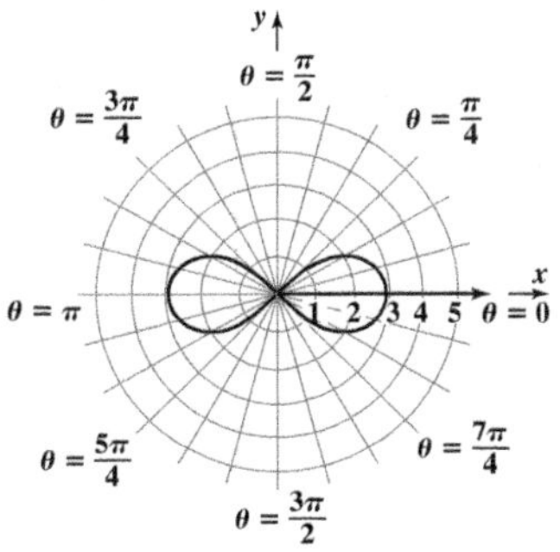

57. Spiral

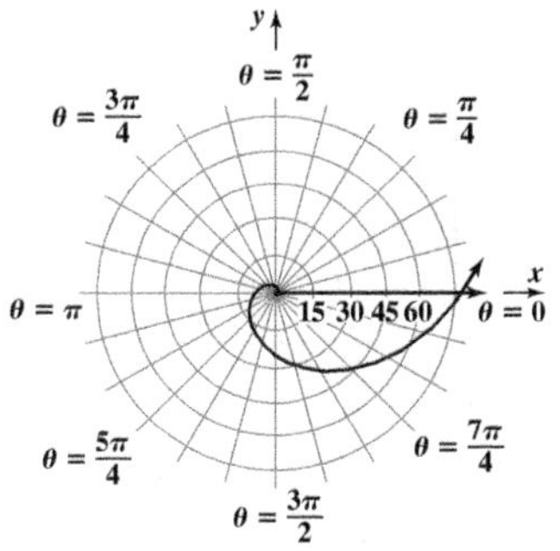

59. Cardioid

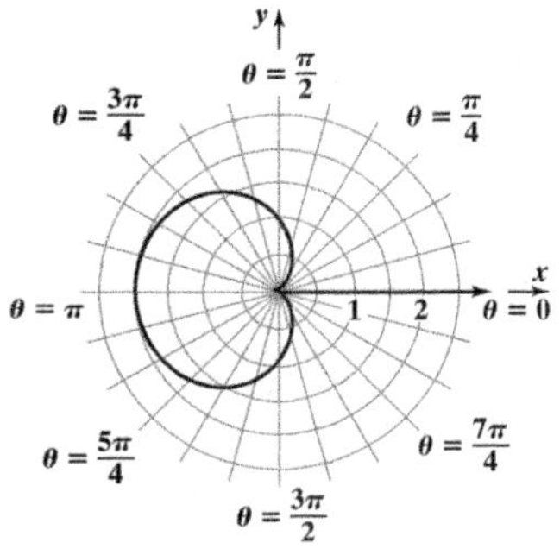

61. Limaçon with inner loop

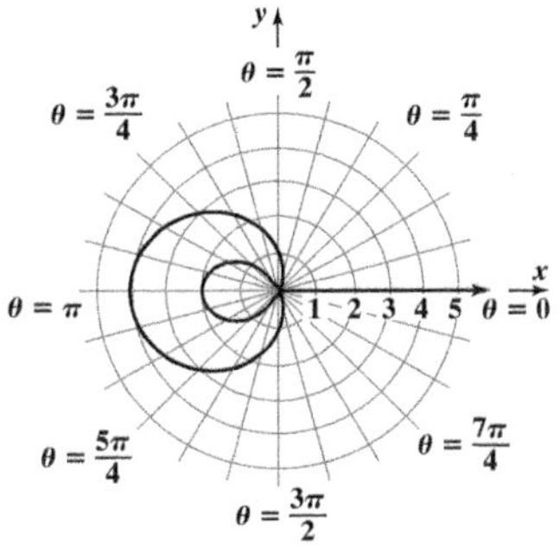

63.

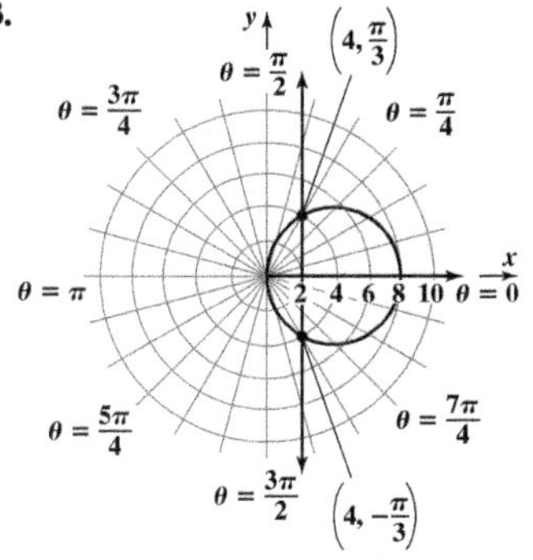

65.

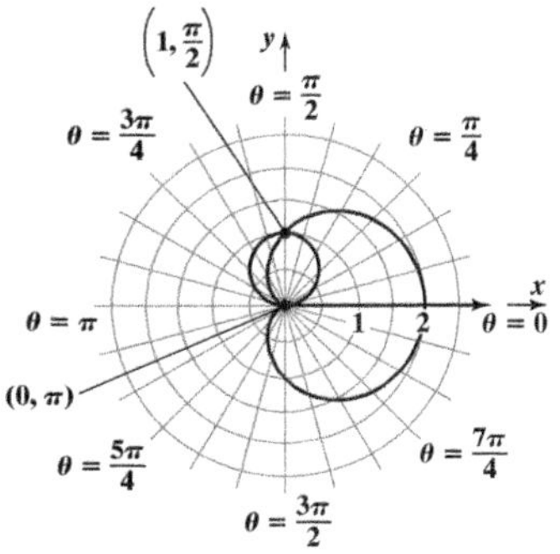

67.

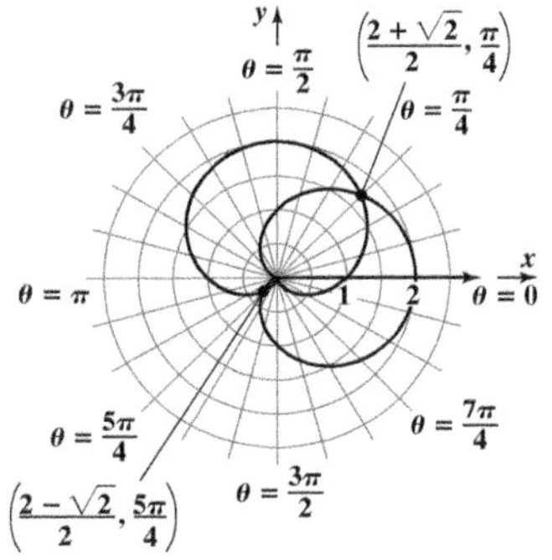

69. $r = 3 + 3\cos\theta$ **71.** $r = 4 + \sin\theta$

73.

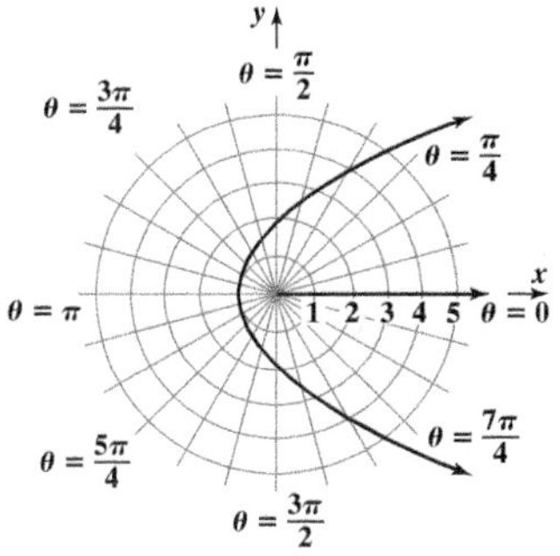

75.

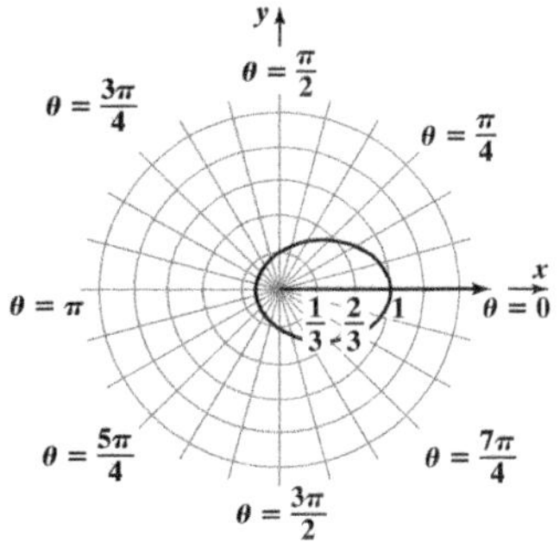

77.

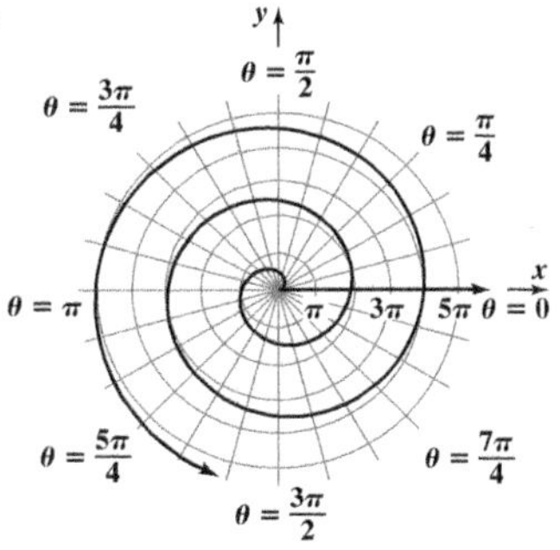

79.

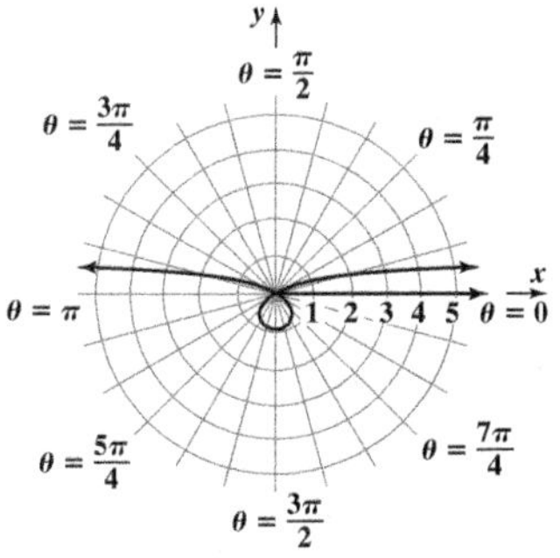

81.

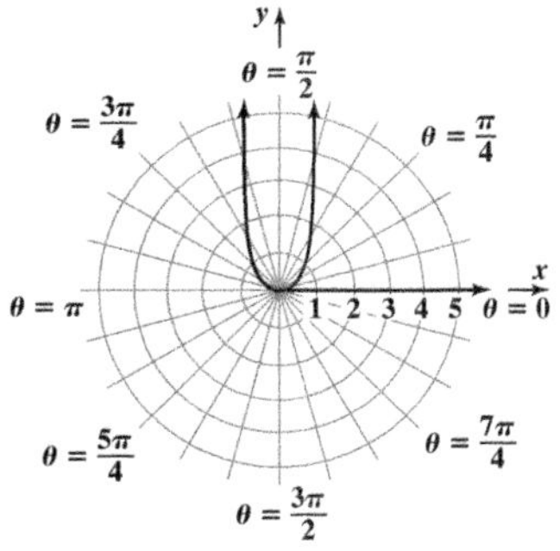

83. $r\sin\theta = a$

$y = a$

85.

$$r = 2a\sin\theta$$
$$r^2 = 2ar\sin\theta$$
$$x^2 + y^2 = 2ay$$
$$x^2 + y^2 - 2ay = 0$$
$$x^2 + (y - a)^2 = a^2$$

Circle, radius a, center at $(0, a)$ in rectangular coordinates

87.

$$r = 2a\cos\theta$$
$$r^2 = 2ar\cos\theta$$
$$x^2 + y^2 = 2ax$$
$$x^2 - 2ax + y^2 = 0$$
$$(x - a)^2 + y^2 = a^2$$

Circle, radius a, center at $(a, 0)$ in rectangular coordinates

89. (a) $r^2 = \cos\theta$; $r^2 = \cos(\pi - \theta)$

$r^2 = -\cos\theta$

Not equivalent; test fails.

$(-r)^2 = \cos(-\theta)$

$r^2 = \cos\theta$

New test works.

(b) $r^2 = \sin\theta$: $r^2 = \sin(\pi - \theta)$

$r^2 = \sin\theta$

Test works.

$(-r)^2 = \sin(-\theta)$

$r^2 = -\sin\theta$

Not equivalent; new test fails.

93. $\{x|3 < x \leq 8\}$, or $(3, 8]$ **94.** 420° **95.** Amplitude $= 2$; period $= \frac{2\pi}{5}$ **96.** Horizontal asymptote: $y = 0$
Vertical asymptote: $x = 4$

Historical Problems *(page 762)*

1. (a) $1 + 4i, 1 + i$ **(b)** $-1, 2 + i$

10.3 Assess Your Understanding *(page 762)*

5. real; imaginary **6.** magnitude; modulus; argument **7.** $r_1 r_2$; $\theta_1 + \theta_2$; $\theta_1 + \theta_2$ **8.** r^n; $n\theta$; $n\theta$ **9.** three **10.** T **11.** c **12.** a

13.

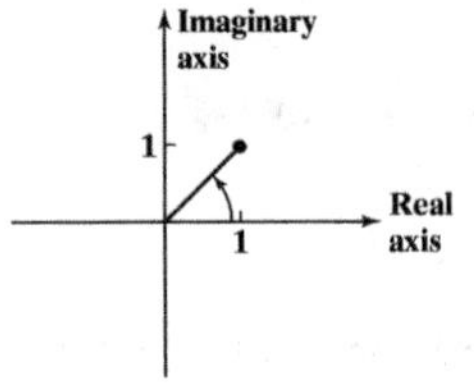

$\sqrt{2}\,(\cos 45° + i \sin 45°)$

15.

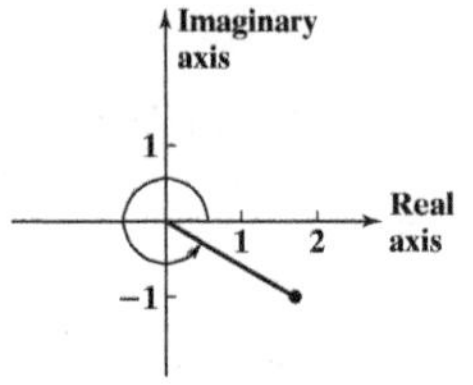

$2(\cos 330° + i \sin 330°)$

17.

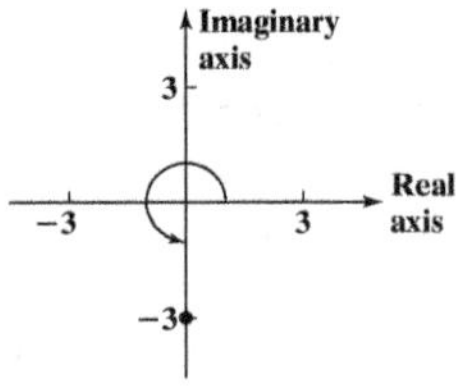

$3(\cos 270° + i \sin 270°)$

19.

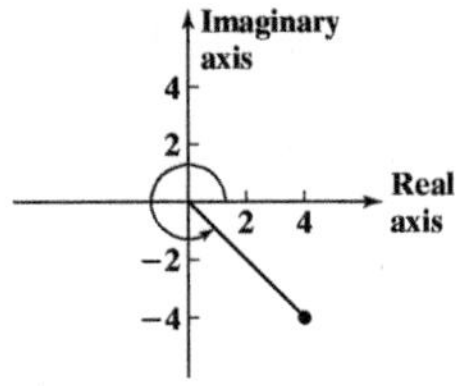

$4\sqrt{2}\,(\cos 315° + i \sin 315°)$

21.

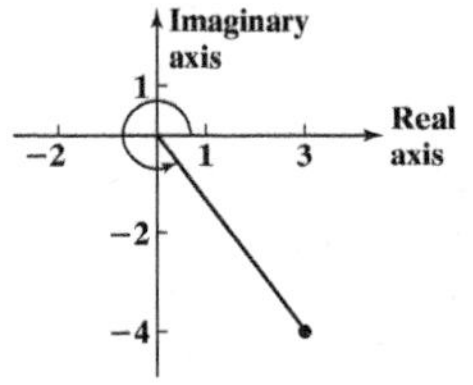

$5(\cos 306.9° + i \sin 306.9°)$

23.

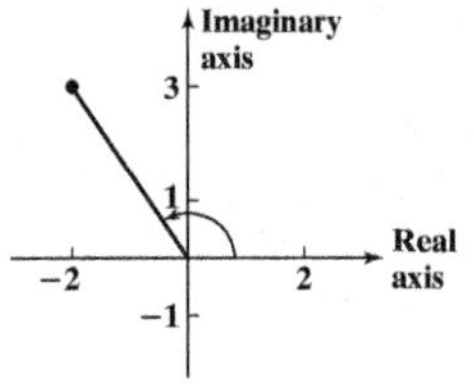

$\sqrt{13}\,(\cos 123.7° + i \sin 123.7°)$

25. $-1 + \sqrt{3}i$ **27.** $2\sqrt{2} - 2\sqrt{2}i$ **29.** $-3i$ **31.** $-0.035 + 0.197i$ **33.** $1.970 + 0.347i$

35. $zw = 8(\cos 60° + i \sin 60°)$; $\frac{z}{w} = \frac{1}{2}(\cos 20° + i \sin 20°)$ **37.** $zw = 12(\cos 40° + i \sin 40°)$; $\frac{z}{w} = \frac{3}{4}(\cos 220° + i \sin 220°)$

39. $zw = 4\left(\cos \frac{9\pi}{40} + i \sin \frac{9\pi}{40}\right)$; $\frac{z}{w} = \cos \frac{\pi}{40} + i \sin \frac{\pi}{40}$ **41.** $zw = 4\sqrt{2}\,(\cos 15° + i \sin 15°)$; $\frac{z}{w} = \sqrt{2}\,(\cos 75° + i \sin 75°)$

43. $-32 + 32\sqrt{3}i$ **45.** $32i$ **47.** $\frac{27}{2} + \frac{27\sqrt{3}}{2}i$ **49.** $-\frac{25\sqrt{2}}{2} + \frac{25\sqrt{2}}{2}i$ **51.** $-4 + 4i$ **53.** $-23 + 14.142i$

55. $\sqrt[6]{2}(\cos 15° + i \sin 15°)$, $\sqrt[6]{2}(\cos 135° + i \sin 135°)$, $\sqrt[6]{2}(\cos 255° + i \sin 255°)$

57. $\sqrt[4]{8}(\cos 75° + i \sin 75°)$, $\sqrt[4]{8}(\cos 165° + i \sin 165°)$, $\sqrt[4]{8}(\cos 255° + i \sin 255°)$, $\sqrt[4]{8}(\cos 345° + i \sin 345°)$

59. $2(\cos 67.5° + i \sin 67.5°)$, $2(\cos 157.5° + i \sin 157.5°)$, $2(\cos 247.5° + i \sin 247.5°)$, $2(\cos 337.5° + i \sin 337.5°)$

61. $\cos 18° + i \sin 18°$, $\cos 90° + i \sin 90°$, $\cos 162° + i \sin 162°$, $\cos 234° + i \sin 234°$, $\cos 306° + i \sin 306°$

63. $1, i, -1, -i$

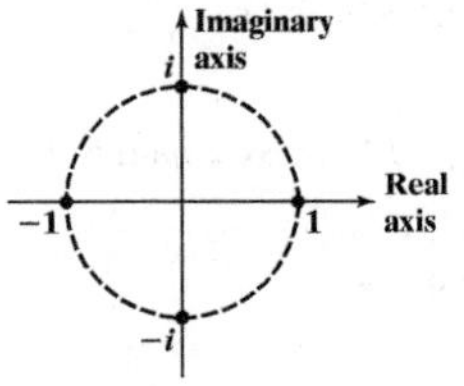

65. Look at formula (8). $|z_k| = \sqrt[n]{r}$ for all k.

67. Look at formula (8). The z_k are spaced apart by an angle of $\frac{2\pi}{n}$.

69. Assume the theorem is true for $n \geq 1$.

For $n = 0$:

$z^0 = r^0[\cos(0 \cdot \theta) + i \sin(0 \cdot \theta)]$
$1 = 1 \cdot [\cos(0) + i \sin(0)]$
$1 = 1 \cdot [1 + 0]$
$1 = 1$ True

For negative integers:

$$z^{-n} = (z^n)^{-1} = (r^n[\cos(n\theta) + i \sin(n\theta)])^{-1} \quad \text{with } n \geq 1$$
$$= \frac{1}{r^n[\cos(n\theta) + i \sin(n\theta)]}$$
$$= \frac{1}{r^n[\cos(n\theta) + i \sin(n\theta)]} \cdot \frac{\cos(n\theta) - i \sin(n\theta)}{\cos(n\theta) - i \sin(n\theta)}$$
$$= \frac{\cos(n\theta) - i \sin(n\theta)}{r^n(\cos^2(n\theta) + \sin^2(n\theta))}$$
$$= \frac{\cos(n\theta) - i \sin(n\theta)}{r^n}$$
$$= r^{-n}[\cos(n\theta) - i \sin(n\theta)]$$
$$= r^{-n}[\cos(-n\theta) + i \sin(-n\theta)]$$

Thus, De Moivre's Theorem is true for all integers.

71. ≈ 40.50 **72.** $\frac{4}{3}\pi$ **73.** $2y\sqrt[3]{3x^2y^2}$ **74.** Minimum: $f\left(\frac{6}{5}\right) = -\frac{16}{5}$

10.4 Assess Your Understanding *(page 775)*

1. vector **2.** 0 **3.** unit **4.** position **5.** horizontal; vertical **6.** resultant **7.** T **8.** F **9.** a **10.** b

11.

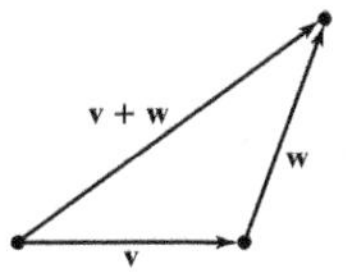

13.

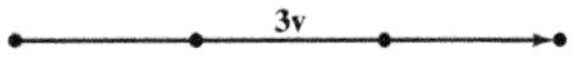

15.

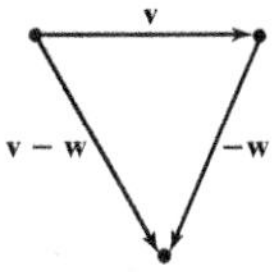

17.

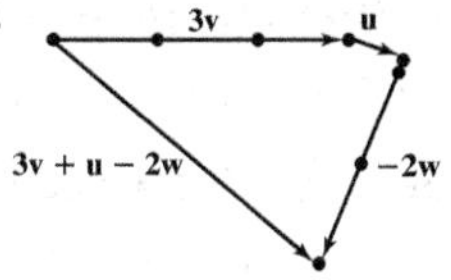

19. T **21.** F **23.** F **25.** T **27.** 12 **29.** $\mathbf{v} = 3\mathbf{i} + 4\mathbf{j}$ **31.** $\mathbf{v} = 2\mathbf{i} + 4\mathbf{j}$ **33.** $\mathbf{v} = 8\mathbf{i} - \mathbf{j}$ **35.** $\mathbf{v} = -\mathbf{i} + \mathbf{j}$ **37.** 5 **39.** $\sqrt{2}$ **41.** $\sqrt{13}$ **43.** $-\mathbf{j}$

45. $\sqrt{89}$ **47.** $\sqrt{34} - \sqrt{13}$ **49.** $\mathbf{i}$ **51.** $\frac{3}{5}\mathbf{i} - \frac{4}{5}\mathbf{j}$ **53.** $\frac{\sqrt{2}}{2}\mathbf{i} - \frac{\sqrt{2}}{2}\mathbf{j}$ **55.** $\mathbf{v} = \frac{8\sqrt{5}}{5}\mathbf{i} + \frac{4\sqrt{5}}{5}\mathbf{j}$, or $\mathbf{v} = -\frac{8\sqrt{5}}{5}\mathbf{i} - \frac{4\sqrt{5}}{5}\mathbf{j}$

57. $\{-2 + \sqrt{21}, -2 - \sqrt{21}\}$ **59.** $\mathbf{v} = \frac{5}{2}\mathbf{i} + \frac{5\sqrt{3}}{2}\mathbf{j}$ **61.** $\mathbf{v} = -7\mathbf{i} + 7\sqrt{3}\mathbf{j}$ **63.** $\mathbf{v} = \frac{25\sqrt{3}}{2}\mathbf{i} - \frac{25}{2}\mathbf{j}$ **65.** 45° **67.** 150° **69.** 333.4° **71.** 258.7°

73. $\mathbf{F} = 20\sqrt{3}\mathbf{i} + 20\mathbf{j}$ **75.** $\mathbf{F} = (20\sqrt{3} + 30\sqrt{2})\mathbf{i} + (20 - 30\sqrt{2})\mathbf{j}$

77. (a) $\mathbf{v}_a = 550\mathbf{j}$; $\mathbf{v}_w = 50\sqrt{2}\mathbf{i} + 50\sqrt{2}\mathbf{j}$
(b) $\mathbf{v}_g = 50\sqrt{2}\mathbf{i} + (550 + 50\sqrt{2})\mathbf{j}$
(c) $\|\mathbf{v}_g\| = 624.7$ mph; N6.5°E

79. $\mathbf{v} = (250\sqrt{2} - 30)\mathbf{i} + (250\sqrt{2} + 30\sqrt{3})\mathbf{j}$; 518.8 km/h; N38.6°E

81. Approximately 4031 lb **83.** 8.6° left of direct heading across the river; 1.52 min

85. (a) N7.05°E **(b)** 12 min **87.** Tension in right cable: 1000 lb; tension in left cable: 845.2 lb

89. Tension in right part: 1088.4 lb; tension in left part: 1089.1 lb **91.** $\mu = 0.36$

93. 13.68 lb **95.** The truck must pull with a force of 4635.2 lb.

97. (a) $(-1, 4)$
(b)

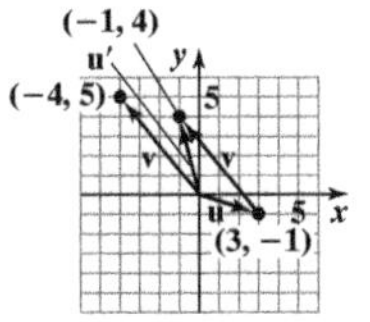

99.

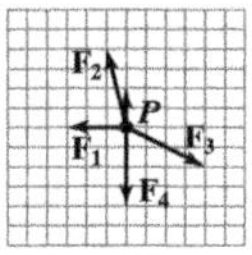

103. $\{29\}$ **104.** $-3x(x + 2)(x - 6)$ **105.** $\sqrt{3}$ **106.** Amplitude $= \frac{3}{2}$; period $= \frac{\pi}{3}$

Phase shift $= -\frac{\pi}{2}$

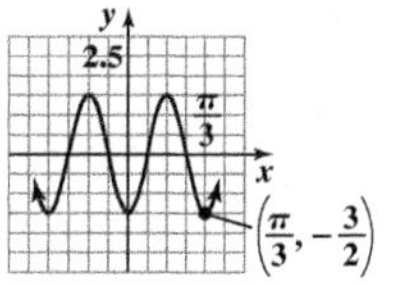

Historical Problem *(page 784)*

$(a\mathbf{i} + b\mathbf{j}) \cdot (c\mathbf{i} + d\mathbf{j}) = ac + bd$

Real part $[(\overline{a + bi})(c + di)]$ = real part$[(a - bi)(c + di)]$ = real part$[ac + adi - bci - bdi^2] = ac + bd$

10.5 Assess Your Understanding *(page 784)*

2. dot product **3.** orthogonal **4.** parallel **5.** T **6.** F **7.** d **8.** b **9. (a)** 0 **(b)** 90° **(c)** orthogonal **11. (a)** 0 **(b)** 90° **(c)** orthogonal

13. (a) $\sqrt{3} - 1$ **(b)** 75° **(c)** neither **15. (a)** -50 **(b)** 180° **(c)** parallel **17. (a)** 0 **(b)** 90° **(c)** orthogonal

19. $\frac{2}{3}$ **21.** $\mathbf{v}_1 = \frac{5}{2}\mathbf{i} - \frac{5}{2}\mathbf{j}$, $\mathbf{v}_2 = -\frac{1}{2}\mathbf{i} - \frac{1}{2}\mathbf{j}$ **23.** $\mathbf{v}_1 = -\frac{1}{5}\mathbf{i} - \frac{2}{5}\mathbf{j}$, $\mathbf{v}_2 = \frac{6}{5}\mathbf{i} - \frac{3}{5}\mathbf{j}$ **25.** $\mathbf{v}_1 = \frac{14}{5}\mathbf{i} + \frac{7}{5}\mathbf{j}$, $\mathbf{v}_2 = \frac{1}{5}\mathbf{i} - \frac{2}{5}\mathbf{j}$ **27.** Approximately 1.353

29. 9 ft-lb

31. (a) $\|\mathbf{I}\| \approx 0.022$; the intensity of the sun's rays is approximately 0.022 W/cm². $\|\mathbf{A}\| = 500$; the area of the solar panel is 500 cm².
(b) $W = 10$; ten watts of energy is collected. **(c)** Vectors **I** and **A** should be parallel with the solar panels facing the sun.

33. Force required to keep the Sienna from rolling down the hill: 737.6 lb; force perpendicular to the hill: 5248.4 lb

35. Timmy must exert 85.5 lb. **37.** 60° **39.** Let $\mathbf{v} = a\mathbf{i} + b\mathbf{j}$. Then $\mathbf{0} \cdot \mathbf{v} = 0a + 0b = 0$.

41. $\mathbf{v} = \cos\alpha\mathbf{i} + \sin\alpha\mathbf{j}$, $0 \le \alpha \le \pi$; $\mathbf{w} = \cos\beta\mathbf{i} + \sin\beta\mathbf{j}$, $0 \le \beta \le \pi$. If θ is the angle between **v** and **w**, then $\mathbf{v} \cdot \mathbf{w} = \cos\theta$, since $\|\mathbf{v}\| = 1$ and $\|\mathbf{w}\| = 1$. Now $\theta = \alpha - \beta$ or $\theta = \beta - \alpha$. Since the cosine function is even, $\mathbf{v} \cdot \mathbf{w} = \cos(\alpha - \beta)$. Also, $\mathbf{v} \cdot \mathbf{w} = \cos\alpha\cos\beta + \sin\alpha\sin\beta$. So $\cos(\alpha - \beta) = \cos\alpha\cos\beta + \sin\alpha\sin\beta$.

43. (a) If $\mathbf{u} = a_1\mathbf{i} + b_1\mathbf{j}$ and $\mathbf{v} = a_2\mathbf{i} + b_2\mathbf{j}$, then, since $\|\mathbf{u}\| = \|\mathbf{v}\|$, $a_1^2 + b_1^2 = \|\mathbf{u}\|^2 = \|\mathbf{v}\|^2 = a_2^2 + b_2^2$,
$(\mathbf{u} + \mathbf{v}) \cdot (\mathbf{u} - \mathbf{v}) = (a_1 + a_2)(a_1 - a_2) + (b_1 + b_2)(b_1 - b_2) = (a_1^2 + b_1^2) - (a_2^2 + b_2^2) = 0.$
(b) The legs of the angle can be made to correspond to vectors $\mathbf{u} + \mathbf{v}$ and $\mathbf{u} - \mathbf{v}$.

45. $(\|\mathbf{w}\|\mathbf{v} + \|\mathbf{v}\|\mathbf{w}) \cdot (\|\mathbf{w}\|\mathbf{v} - \|\mathbf{v}\|\mathbf{w}) = \|\mathbf{w}\|^2\mathbf{v}\cdot\mathbf{v} - \|\mathbf{w}\|\|\mathbf{v}\|\mathbf{v}\cdot\mathbf{w} + \|\mathbf{v}\|\|\mathbf{w}\|\mathbf{w}\cdot\mathbf{v} - \|\mathbf{v}\|^2\mathbf{w}\cdot\mathbf{w} = \|\mathbf{w}\|^2\mathbf{v}\cdot\mathbf{v} - \|\mathbf{v}\|^2\mathbf{w}\cdot\mathbf{w} = \|\mathbf{w}\|^2\|\mathbf{v}\|^2 - \|\mathbf{v}\|^2\|\mathbf{w}\|^2 = 0$

47. $\|\mathbf{u} + \mathbf{v}\|^2 - \|\mathbf{u} - \mathbf{v}\|^2 = (\mathbf{u} + \mathbf{v}) \cdot (\mathbf{u} + \mathbf{v}) - (\mathbf{u} - \mathbf{v}) \cdot (\mathbf{u} - \mathbf{v}) = (\mathbf{u}\cdot\mathbf{u} + \mathbf{u}\cdot\mathbf{v} + \mathbf{v}\cdot\mathbf{u} + \mathbf{v}\cdot\mathbf{v}) - (\mathbf{u}\cdot\mathbf{u} - \mathbf{u}\cdot\mathbf{v} - \mathbf{v}\cdot\mathbf{u} + \mathbf{v}\cdot\mathbf{v})$
$= 2(\mathbf{u}\cdot\mathbf{v}) + 2(\mathbf{v}\cdot\mathbf{u}) = 4(\mathbf{u}\cdot\mathbf{v})$

49. 12 **50.** $\frac{9}{2}$ **51.** $(1 - \sin^2\theta)(1 + \tan^2\theta) = (\cos^2\theta)(\sec^2\theta)$
$= \cos^2\theta \cdot \frac{1}{\cos^2\theta}$
$= 1$

52. $V(x) = x(19 - 2x)(13 - 2x)$, or $V(x) = 4x^3 - 64x^2 + 247x$

Review Exercises *(page 787)*

1. $\left(\frac{3\sqrt{3}}{2}, \frac{3}{2}\right)$

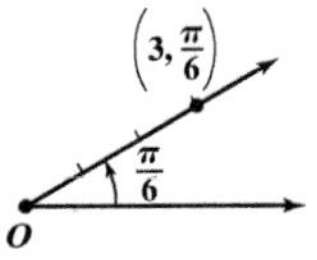

2. $(1, \sqrt{3})$

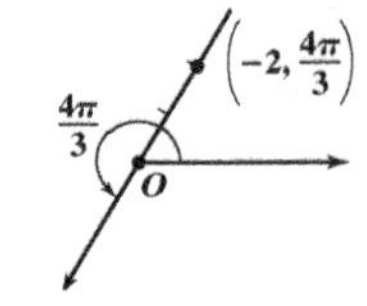

3. $(0, 3)$

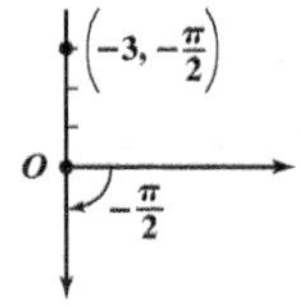

4. $\left(3\sqrt{2}, \frac{3\pi}{4}\right), \left(-3\sqrt{2}, -\frac{\pi}{4}\right)$

5. $\left(2, -\frac{\pi}{2}\right), \left(-2, \frac{\pi}{2}\right)$

6. $(5, 0.93), (-5, 4.07)$

7. (a) $x^2 + (y-1)^2 = 1$ **(b)** circle, radius 1, center $(0, 1)$ in rectangular coordinates

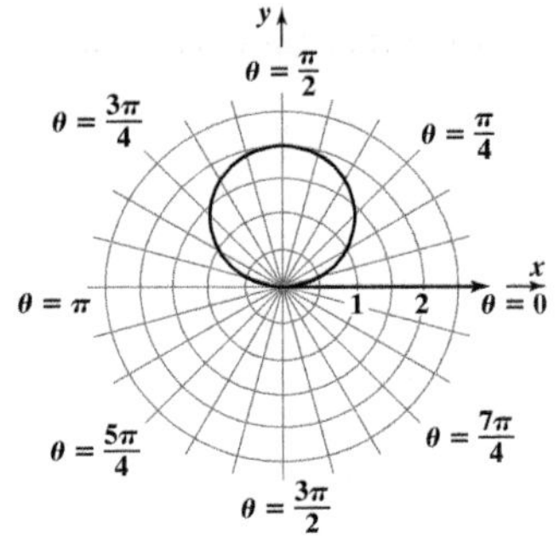

8. (a) $x^2 + y^2 = 25$ **(b)** circle, radius 5, center at pole

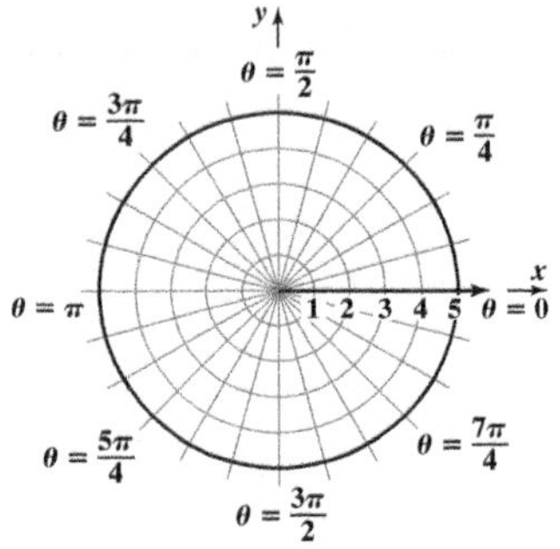

9. (a) $x - y = 0$ **(b)** line through pole, making an angle of $\frac{\pi}{4}$ with polar axis

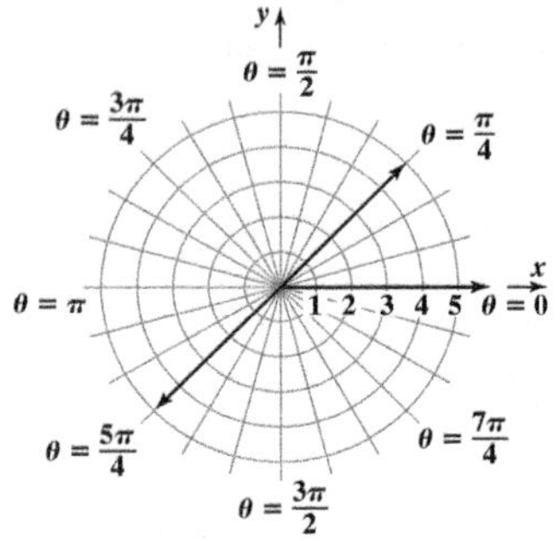

10. (a) $(x-4)^2 + (y+2)^2 = 25$ **(b)** circle, radius 5, center $(4, -2)$ in rectangular coordinates

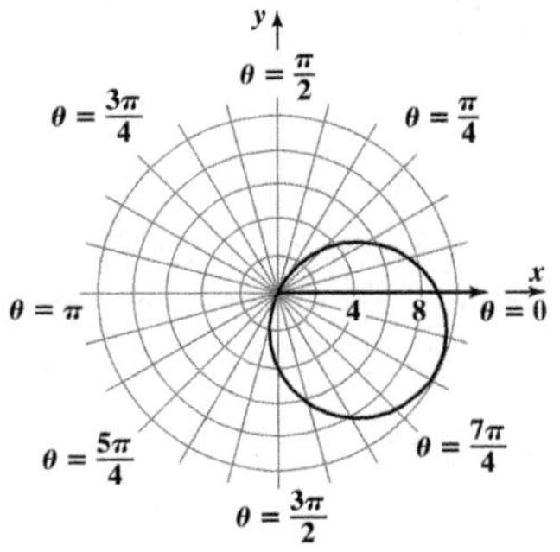

11. Circle; radius 2, center at $(2, 0)$ in rectangular coordinates; symmetric with respect to the polar axis

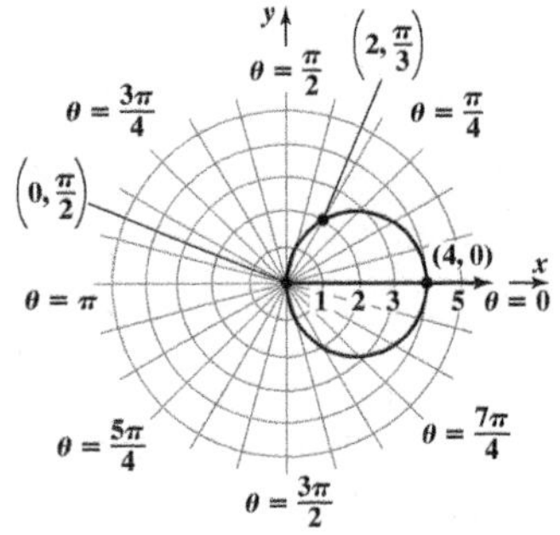

12. Cardioid; symmetric with respect to the line $\theta = \frac{\pi}{2}$

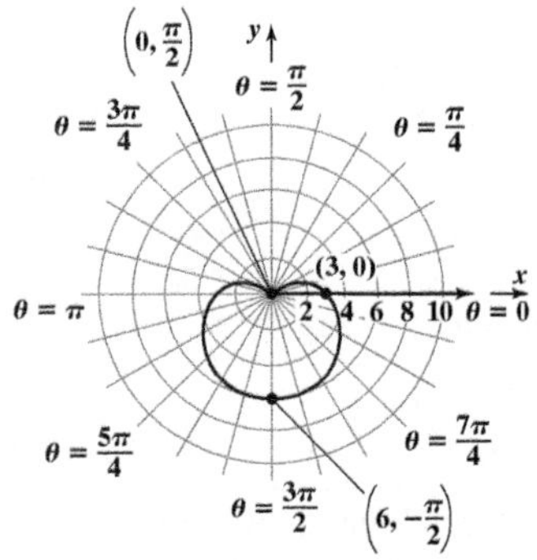

13. Limaçon without inner loop; symmetric with respect to the polar axis

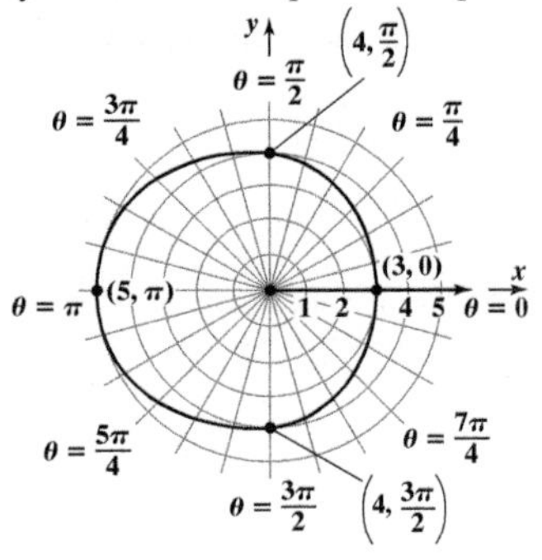

14. $\sqrt{2}(\cos 225° + i \sin 225°)$

15. $5(\cos 323.1° + i \sin 323.1°)$

16. $-\sqrt{3} + i$

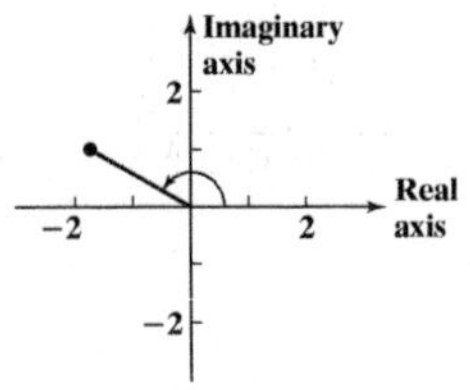

17. $-\frac{3}{2} + \frac{3\sqrt{3}}{2}i$

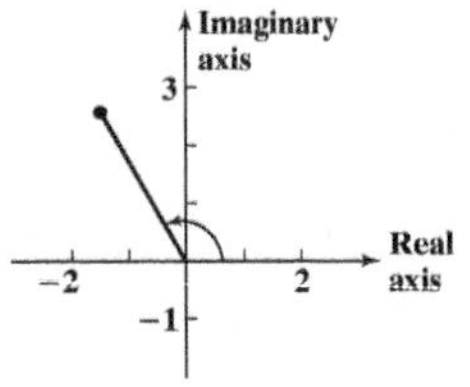

18. $0.10 - 0.02i$

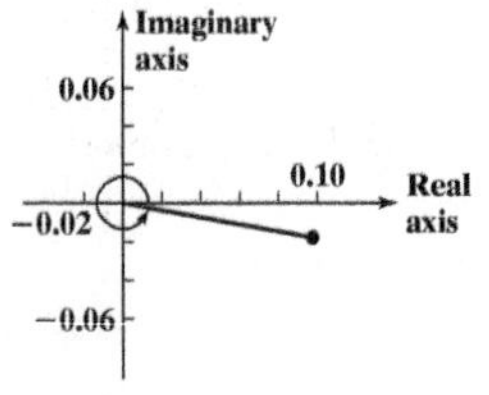

19. $zw = \cos 130° + i \sin 130°;\ \dfrac{z}{w} = \cos 30° + i \sin 30°$ **20.** $zw = 6(\cos 0 + i \sin 0) = 6;\ \dfrac{z}{w} = \dfrac{3}{2}\left(\cos \dfrac{8\pi}{5} + i \sin \dfrac{8\pi}{5}\right)$

21. $zw = 5(\cos 5° + i \sin 5°);\ \dfrac{z}{w} = 5(\cos 15° + i \sin 15°)$ **22.** $\dfrac{27}{2} + \dfrac{27\sqrt{3}}{2} i$ **23.** $4i$ **24.** 64 **25.** $-527 - 336i$

26. $3,\ 3(\cos 120° + i \sin 120°),\ 3(\cos 240° + i \sin 240°)$ or $3,\ -\dfrac{3}{2} + \dfrac{3\sqrt{3}}{2} i,\ -\dfrac{3}{2} - \dfrac{3\sqrt{3}}{2} i$

27.

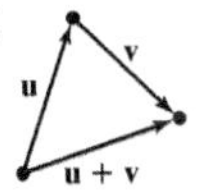

28.

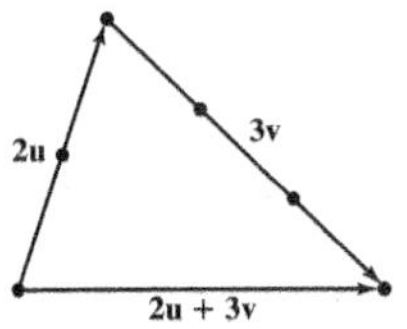

29. $\mathbf{v} = 2\mathbf{i} - 4\mathbf{j};\ \|\mathbf{v}\| = 2\sqrt{5}$ **30.** $\mathbf{v} = -\mathbf{i} + 3\mathbf{j};\ \|\mathbf{v}\| = \sqrt{10}$ **31.** $2\mathbf{i} - 2\mathbf{j}$ **32.** $-20\mathbf{i} + 13\mathbf{j}$

33. $\sqrt{5}$ **34.** $\sqrt{5} + 5 \approx 7.24$ **35.** $-\dfrac{2\sqrt{5}}{5}\mathbf{i} + \dfrac{\sqrt{5}}{5}\mathbf{j}$ **36.** $\mathbf{v} = \dfrac{3}{2}\mathbf{i} + \dfrac{3\sqrt{3}}{2}\mathbf{j}$

37. 120° **38.** $\mathbf{v} \cdot \mathbf{w} = -11;\ \theta \approx 169.7°$ **39.** $\mathbf{v} \cdot \mathbf{w} = -4;\ \theta \approx 153.4°$ **40.** Parallel **41.** Neither

42. Orthogonal **43.** $\mathbf{v}_1 = \dfrac{4}{5}\mathbf{i} - \dfrac{3}{5}\mathbf{j};\ \mathbf{v}_2 = \dfrac{6}{5}\mathbf{i} + \dfrac{8}{5}\mathbf{j}$ **44.** $\mathbf{v}_1 = \dfrac{9}{10}(3\mathbf{i} + \mathbf{j});\ \mathbf{v}_2 = -\dfrac{7}{10}\mathbf{i} + \dfrac{21}{10}\mathbf{j}$

45. $\sqrt{29} \approx 5.39$ mph; 0.4 mi **46.** Left cable: 1843.21 lb; right cable: 1630.41 lb **47.** 50 ft-lb **48.** A force of 697.2 lb is needed to keep the van from rolling down the hill. The magnitude of the force on the hill is 7969.6 lb.

Chapter Test *(page 789)*

1–3.

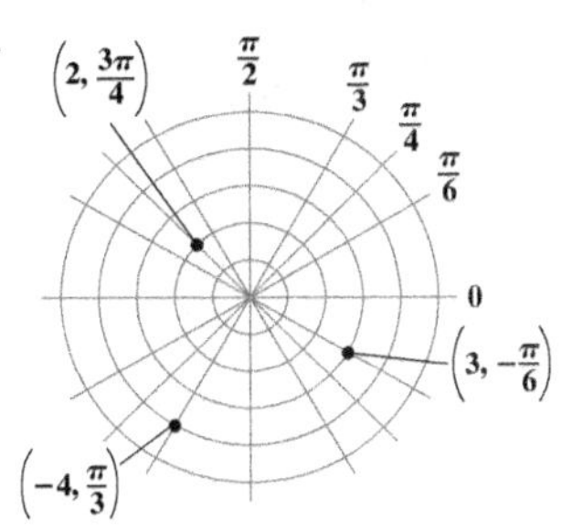

4. $\left(4, \dfrac{\pi}{3}\right)$

5. $x^2 + y^2 = 49$

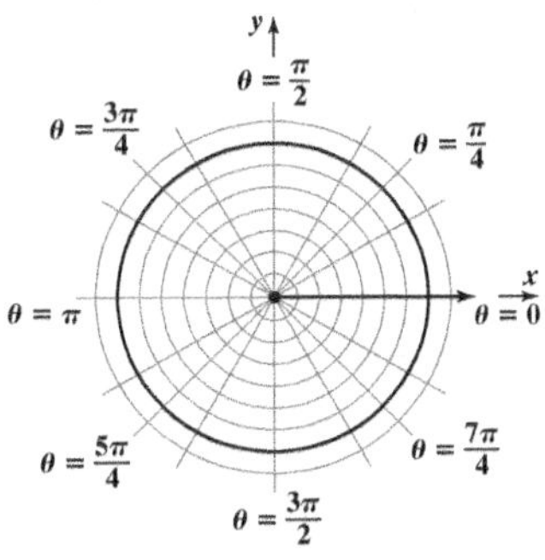

6. $\dfrac{y}{x} = 3$ or $y = 3x$

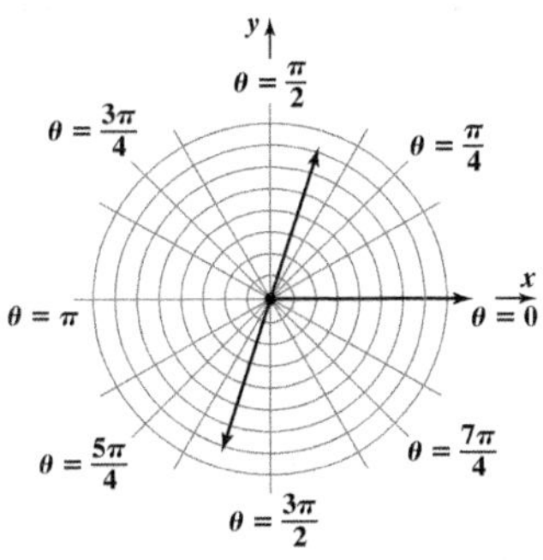

7. $8y = x^2$

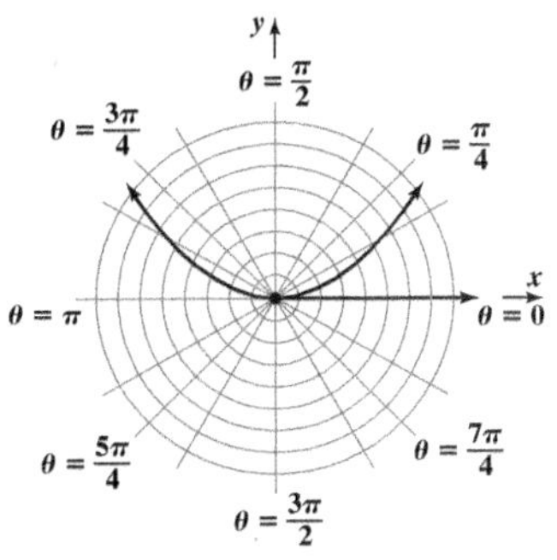

8. $r^2 \cos \theta = 5$ is symmetric about the pole, the polar axis, and the line $\theta = \dfrac{\pi}{2}$.

9. $r = 5 \sin \theta \cos^2 \theta$ is symmetric about the line $\theta = \dfrac{\pi}{2}$. The tests for symmetry about the pole and the polar axis fail, so the graph of $r = 5 \sin \theta \cos^2 \theta$ may or may not be symmetric about the pole or polar axis.

10. $z \cdot w = 6(\cos 107° + i \sin 107°)$ **11.** $\dfrac{w}{z} = \dfrac{3}{2}(\cos 297° + i \sin 297°)$ **12.** $w^5 = 243(\cos 110° + i \sin 110°)$

13. $z_0 = 2\sqrt[3]{2}(\cos 40° + i \sin 40°),\ z_1 = 2\sqrt[3]{2}(\cos 160° + i \sin 160°),\ z_2 = 2\sqrt[3]{2}(\cos 280° + i \sin 280°)$

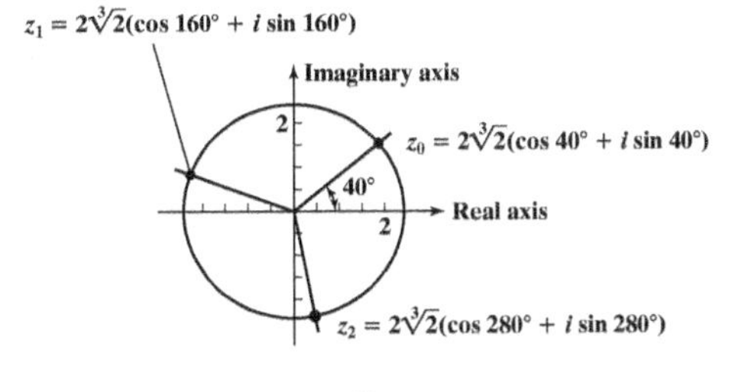

14. $\mathbf{v} = \langle 5\sqrt{2}, -5\sqrt{2} \rangle$ **15.** $\|\mathbf{v}\| = 10$ **16.** $\mathbf{u} = \dfrac{\mathbf{v}}{\|\mathbf{v}\|} = \left\langle \dfrac{\sqrt{2}}{2}, -\dfrac{\sqrt{2}}{2} \right\rangle$ **17.** 315° off the positive x-axis **18.** $\mathbf{v} = 5\sqrt{2}\mathbf{i} - 5\sqrt{2}\mathbf{j}$

19. $\mathbf{v}_1 + 2\mathbf{v}_2 - \mathbf{v}_3 = \langle 6, -10 \rangle$ **20.** Vectors $\mathbf{v}_1$ and $\mathbf{v}_4$ are parallel. **21.** Vectors $\mathbf{v}_2$ and $\mathbf{v}_3$ are orthogonal. **22.** 172.87°
23. The cable must be able to endure a tension of approximately 670.82 lb.

Cumulative Review *(page 789)*

1. $\{-3, 3\}$ **2.** $y = \dfrac{\sqrt{3}}{3}x$ **3.** $x^2 + (y - 1)^2 = 9$ **4.** $\left\{x \middle| x < \dfrac{1}{2}\right\}$ or $\left(-\infty, \dfrac{1}{2}\right)$ **5.** Symmetry with respect to the y-axis **6.**

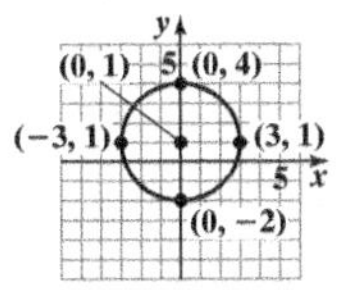

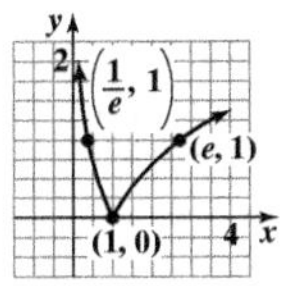

7.

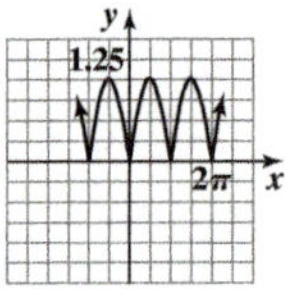

8.

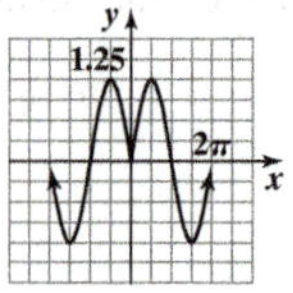

9. $-\frac{\pi}{6}$ **10.**

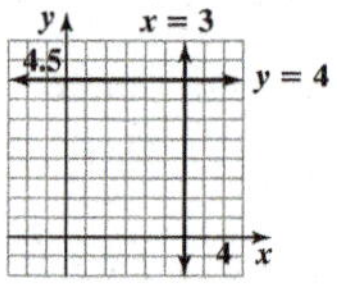

11.

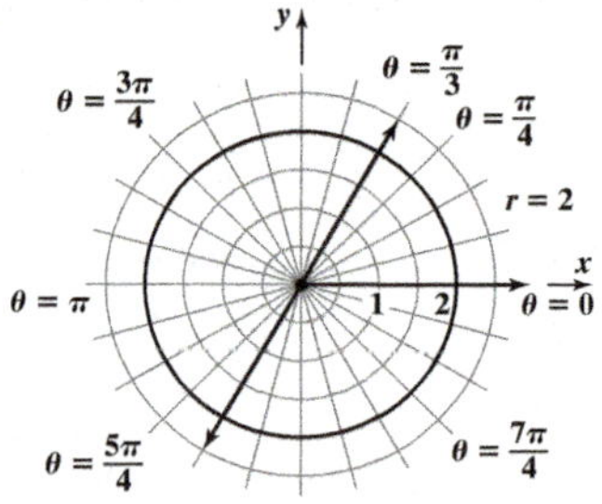

12. Amplitude: 4; period: 2

CHAPTER 11 Analytic Geometry

11.2 Assess Your Understanding *(page 800)*

6. parabola **7.** axis of symmetry **8.** latus rectum **9.** c **10.** $(3, 2)$ **11.** d **12.** c **13.** B **15.** E **17.** H **19.** C

21. $y^2 = 16x$

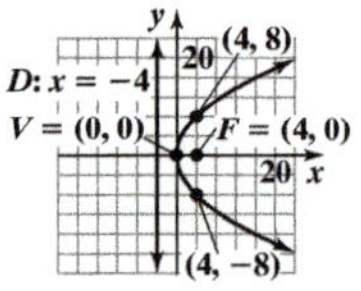

23. $x^2 = -12y$

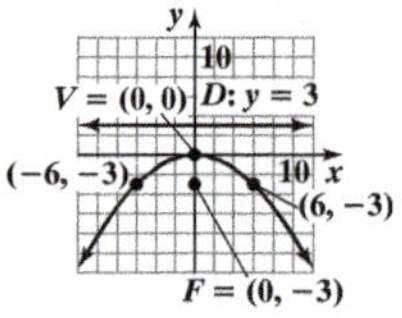

25. $y^2 = -8x$

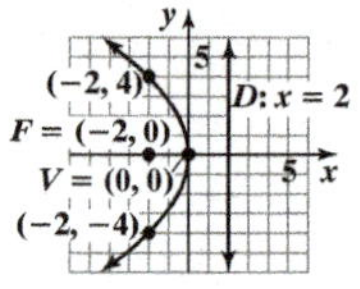

27. $x^2 = 2y$

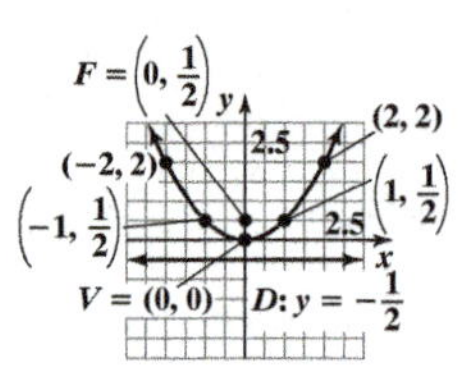

29. $x^2 = \frac{4}{3}y$

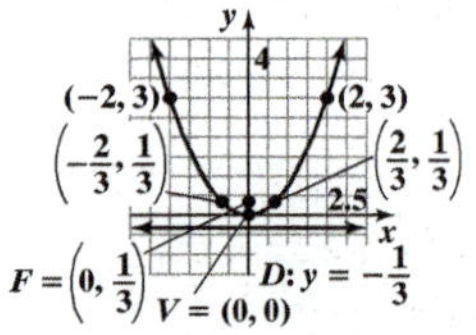

31. $(x - 2)^2 = -8(y + 3)$

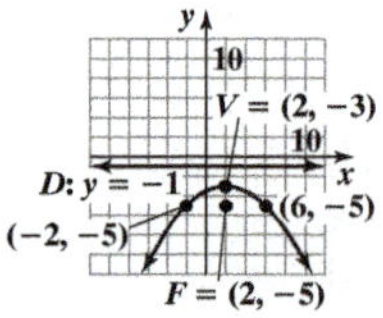

33. $(y + 2)^2 = 4(x + 1)$

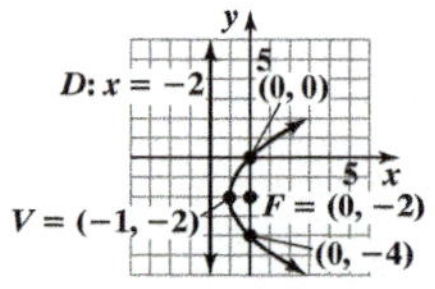

35. $(x + 3)^2 = 4(y - 3)$

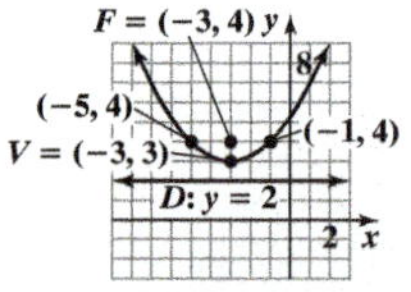

37. $(y + 2)^2 = -8(x + 1)$

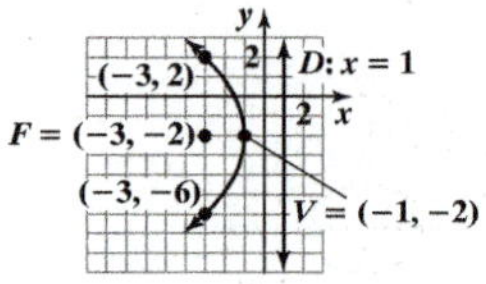

39. Vertex: $(0, 0)$; focus: $(0, 1)$; directrix: $y = -1$

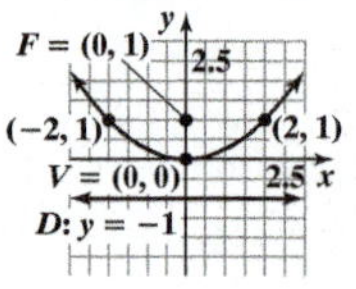

41. Vertex: $(0, 0)$; focus: $(-4, 0)$; directrix: $x = 4$

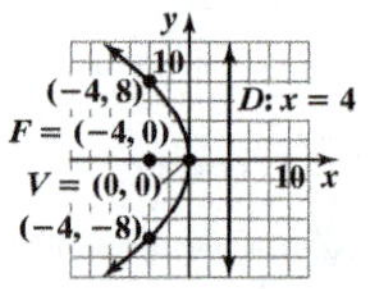

43. Vertex: $(-1, 2)$; focus: $(1, 2)$; directrix: $x = -3$

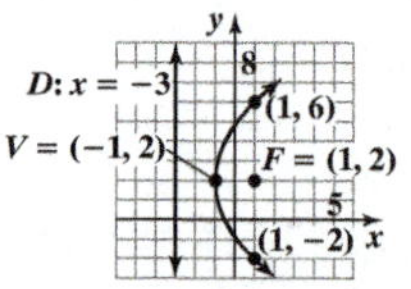

45. Vertex: $(3, -1)$; focus: $\left(3, -\frac{5}{4}\right)$; directrix: $y = -\frac{3}{4}$

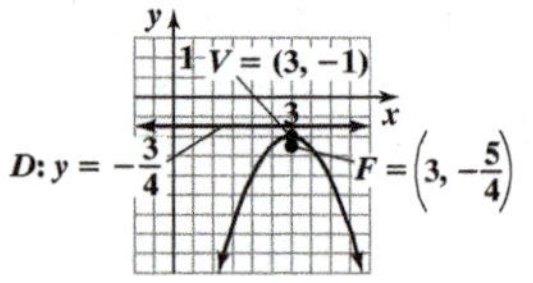

47. Vertex: $(2, -3)$; focus: $(4, -3)$; directrix: $x = 0$

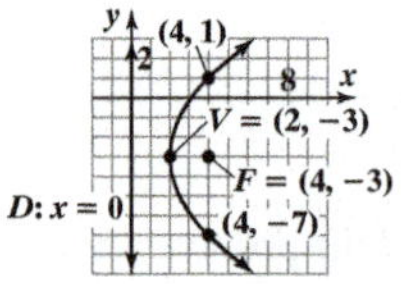

49. Vertex: $(0, 2)$; focus: $(-1, 2)$; directrix: $x = 1$

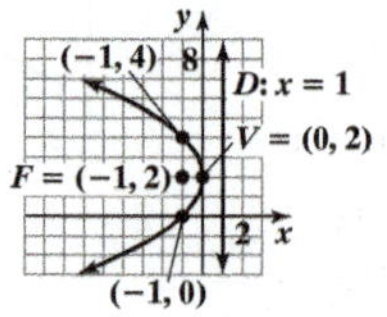

51. Vertex: $(-4, -2)$; focus: $(-4, -1)$; directrix: $y = -3$

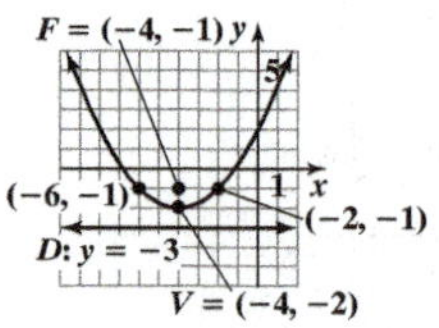

53. Vertex: $(-1, -1)$; focus: $\left(-\frac{3}{4}, -1\right)$; directrix: $x = -\frac{5}{4}$

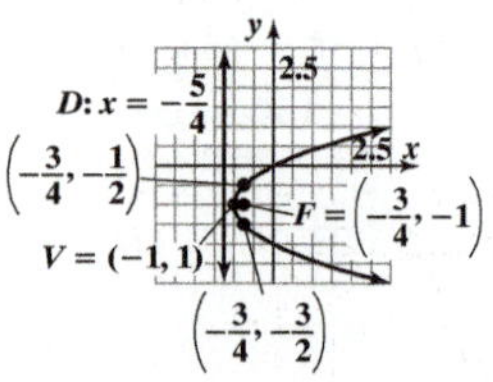

55. Vertex: $(2, -8)$; focus: $\left(2, -\frac{31}{4}\right)$; directrix: $y = -\frac{33}{4}$

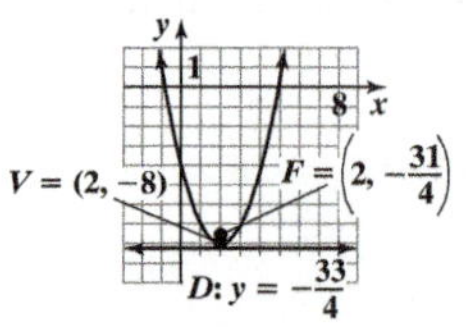

57. $(y - 1)^2 = x$ **59.** $(y - 1)^2 = -(x - 2)$

61. $x^2 = 4(y - 1)$ **63.** $y^2 = \frac{1}{2}(x + 2)$

65. 1.5625 ft from the base of the dish, along the axis of symmetry

67. 1 in. from the vertex, along the axis of symmetry

69. 20 ft **71.** 0.78125 ft

73. 4.17 ft from the base, along the axis of symmetry

75. 24.31 ft, 18.75 ft, 7.64 ft

77. **(a)** $y = -\frac{2}{315}x^2 + 630$

(b) 567 ft: 119.7 ft; 478 ft: 267.3 ft; 308 ft: 479.4 ft **(c)** No

79. $Cy^2 + Dx = 0, C \neq 0, D \neq 0$

$Cy^2 = -Dx$

$y^2 = -\frac{D}{C}x$

This is the equation of a parabola with vertex at $(0, 0)$ and axis of symmetry the x-axis. The focus is $\left(-\frac{D}{4C}, 0\right)$; the directrix is the line $x = \frac{D}{4C}$. The parabola opens to the right if $-\frac{D}{C} > 0$ and to the left if $-\frac{D}{C} < 0$.

81. $Cy^2 + Dx + Ey + F = 0, C \neq 0$

$Cy^2 + Ey = -Dx - F$

$y^2 + \frac{E}{C}y = -\frac{D}{C}x - \frac{F}{C}$

$\left(y + \frac{E}{2C}\right)^2 = -\frac{D}{C}x - \frac{F}{C} + \frac{E^2}{4C^2}$

$\left(y + \frac{E}{2C}\right)^2 = -\frac{D}{C}x + \frac{E^2 - 4CF}{4C^2}$

(a) If $D \neq 0$, then the equation may be written as $\left(y + \frac{E}{2C}\right)^2 = -\frac{D}{C}\left(x - \frac{E^2 - 4CF}{4CD}\right)$. This is the equation of a parabola with vertex at $\left(\frac{E^2 - 4CF}{4CD}, -\frac{E}{2C}\right)$ and axis of symmetry parallel to the x-axis.

(b)–(d) If $D = 0$, the graph of the equation contains no points if $E^2 - 4CF < 0$, is a single horizontal line if $E^2 - 4CF = 0$, and is two horizontal lines if $E^2 - 4CF > 0$.

82. $(0, 2)$, $(0, -2)$, $(-36, 0)$; symmetric with respect to the x-axis **83.** $\{5\}$

84. $\sin\theta = \frac{5\sqrt{89}}{89}$; $\cos\theta = -\frac{8\sqrt{89}}{89}$; $\csc\theta = \frac{\sqrt{89}}{5}$; $\sec\theta = -\frac{\sqrt{89}}{8}$; $\cot\theta = -\frac{8}{5}$ **85.** $-\frac{2\sqrt{10}}{3}$

11.3 Assess Your Understanding *(page 811)*

7. ellipse **8.** b **9.** $(0, -5)$; $(0, 5)$ **10.** 5; 3; x **11.** $(-2, -3)$; $(6, -3)$ **12.** a **13.** C **15.** B

17. Vertices: $(-5, 0)$, $(5, 0)$; foci: $(-\sqrt{21}, 0)$, $(\sqrt{21}, 0)$

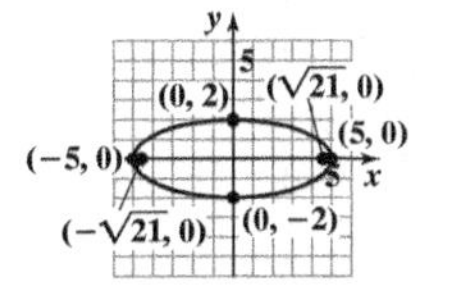

19. Vertices: $(0, -5)$, $(0, 5)$; foci: $(0, -4)$, $(0, 4)$

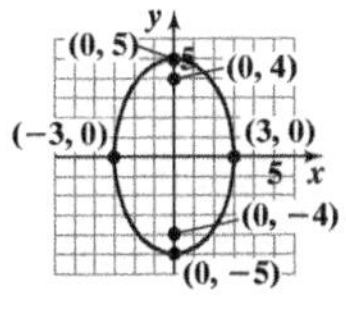

21. $\frac{x^2}{4} + \frac{y^2}{16} = 1$
Vertices: $(0, -4)$, $(0, 4)$; foci: $(0, -2\sqrt{3})$, $(0, 2\sqrt{3})$

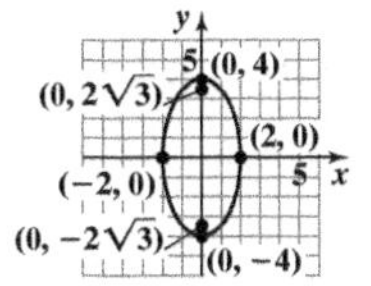

23. $\frac{x^2}{8} + \frac{y^2}{2} = 1$
Vertices: $(-2\sqrt{2}, 0)$, $(2\sqrt{2}, 0)$; foci: $(-\sqrt{6}, 0)$, $(\sqrt{6}, 0)$

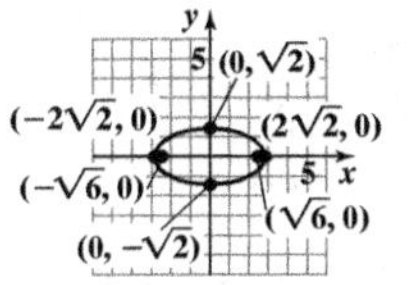

25. $\frac{x^2}{16} + \frac{y^2}{16} = 1$
Vertices: $(-4, 0)$, $(4, 0)$, $(0, -4)$, $(0, 4)$; focus: $(0, 0)$

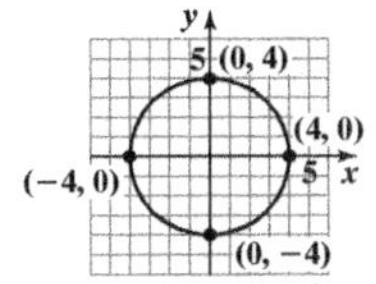

27. $\frac{x^2}{25} + \frac{y^2}{16} = 1$

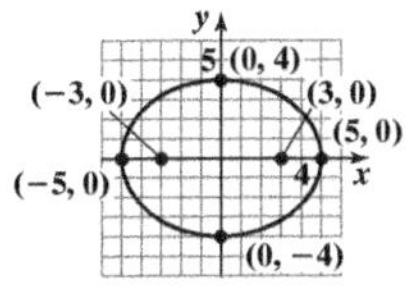

29. $\frac{x^2}{9} + \frac{y^2}{25} = 1$

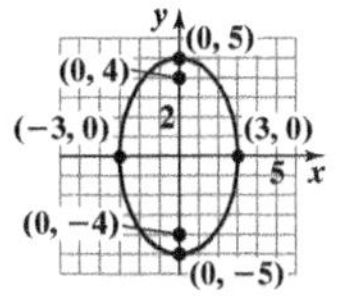

31. $\frac{x^2}{9} + \frac{y^2}{5} = 1$

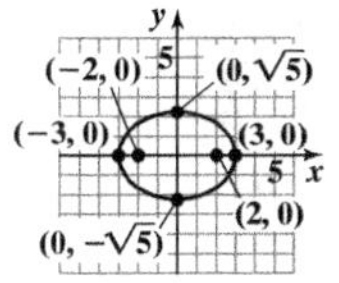

33. $\frac{x^2}{25} + \frac{y^2}{9} = 1$

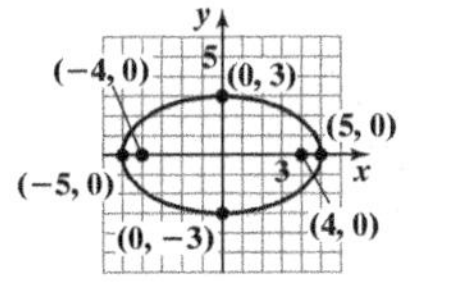

35. $\frac{x^2}{4} + \frac{y^2}{13} = 1$

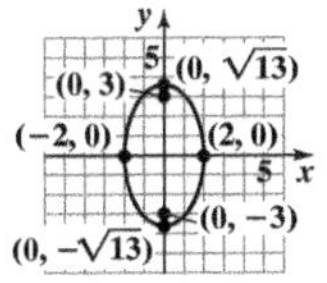

37. $x^2 + \frac{y^2}{16} = 1$

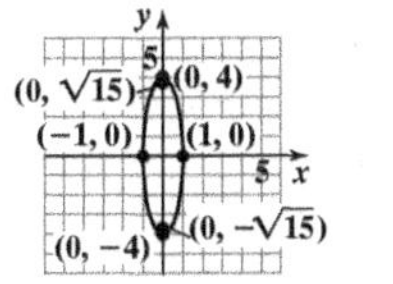

39. $\frac{(x + 1)^2}{4} + (y - 1)^2 = 1$

41. $(x - 1)^2 + \frac{y^2}{4} = 1$

43. Center: $(3, -1)$; vertices: $(3, -4)$, $(3, 2)$; foci: $(3, -1 - \sqrt{5})$, $(3, -1 + \sqrt{5})$

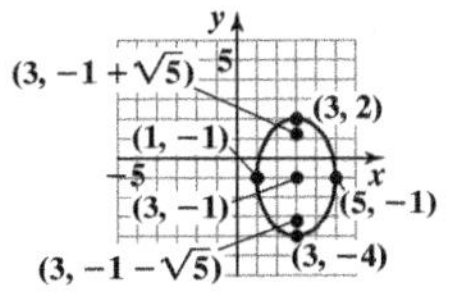

45. $\frac{(x + 5)^2}{16} + \frac{(y - 4)^2}{4} = 1$
Center: $(-5, 4)$; vertices: $(-9, 4)$, $(-1, 4)$; foci: $(-5 - 2\sqrt{3}, 4)$, $(-5 + 2\sqrt{3}, 4)$

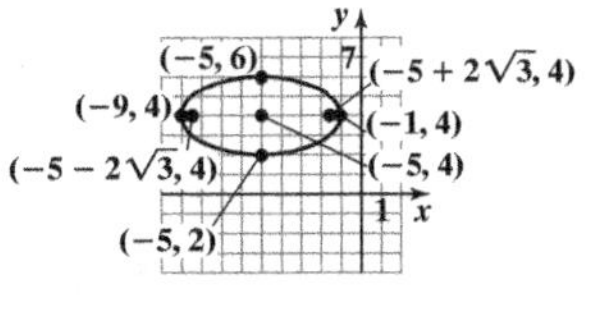

47. $\frac{(x + 2)^2}{4} + (y - 1)^2 = 1$
Center: $(-2, 1)$; vertices: $(-4, 1)$, $(0, 1)$; foci: $(-2 - \sqrt{3}, 1)$, $(-2 + \sqrt{3}, 1)$

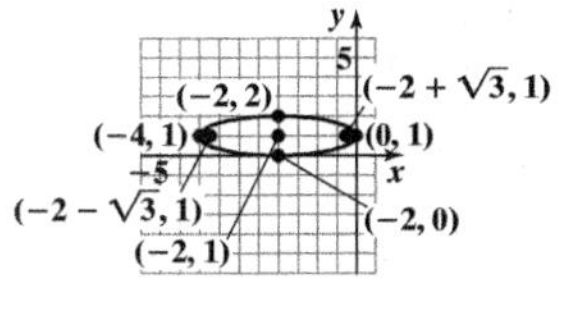

49. $\frac{(x-2)^2}{3}+\frac{(y+1)^2}{2}=1$
Center: $(2,-1)$; vertices: $(2-\sqrt{3},-1)$, $(2+\sqrt{3},-1)$; foci: $(1,-1)$, $(3,-1)$

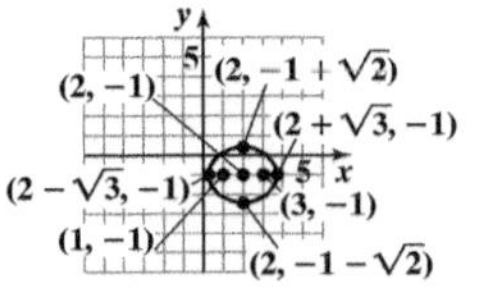

51. $\frac{(x-1)^2}{4}+\frac{(y+2)^2}{9}=1$
Center: $(1,-2)$; vertices: $(1,-5)$, $(1,1)$; foci: $(1,-2-\sqrt{5})$, $(1,-2+\sqrt{5})$

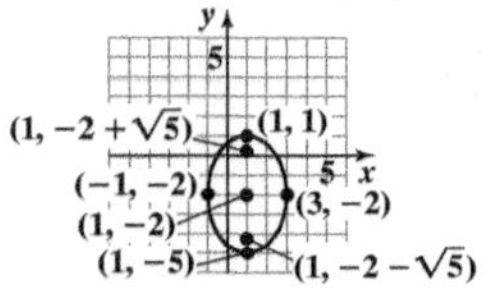

53. $x^2+\frac{(y+2)^2}{4}=1$
Center: $(0,-2)$; vertices: $(0,-4)$, $(0,0)$; foci: $(0,-2-\sqrt{3})$, $(0,-2+\sqrt{3})$

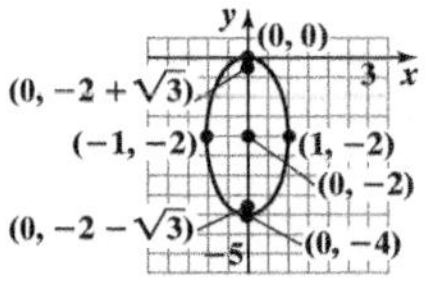

55. $\frac{(x-2)^2}{25}+\frac{(y+2)^2}{21}=1$

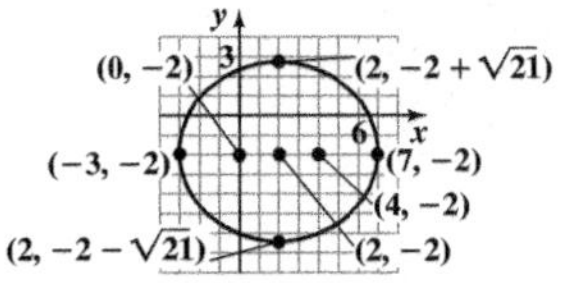

57. $\frac{(x-4)^2}{5}+\frac{(y-6)^2}{9}=1$

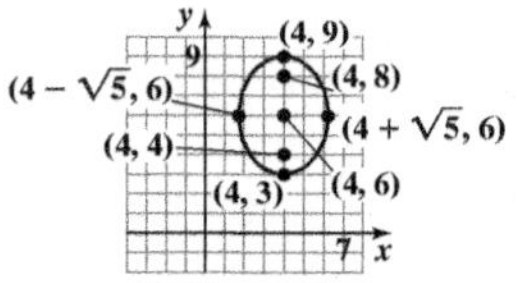

59. $\frac{(x-2)^2}{16}+\frac{(y-1)^2}{7}=1$

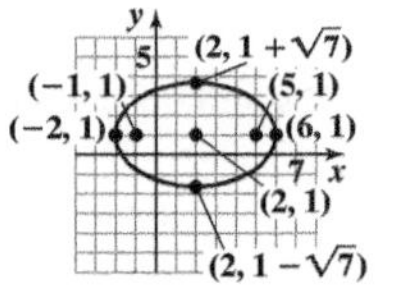

61. $\frac{(x-1)^2}{10}+(y-2)^2=1$

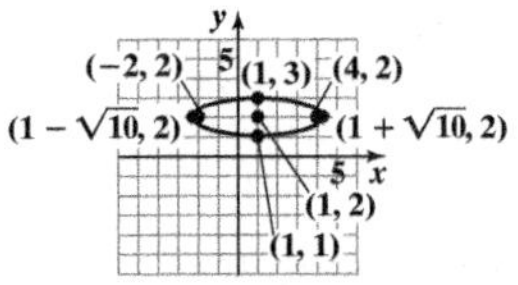

63. $\frac{(x-1)^2}{9}+\frac{(y-2)^2}{9}=1$

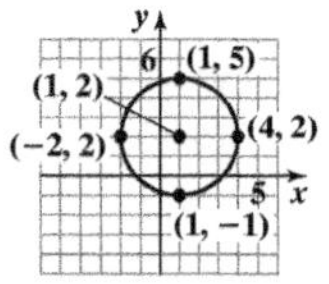

65.

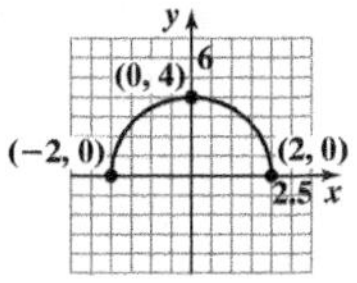

67.

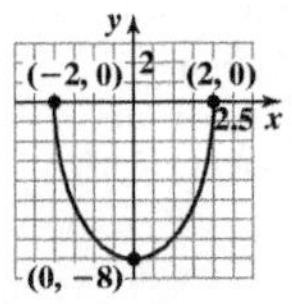

69. $\frac{x^2}{100}+\frac{y^2}{36}=1$ **71.** 43.3 ft **73.** 24.65 ft, 21.65 ft, 13.82 ft **75.** 30 ft **77.** The elliptical hole will have a major axis of length $2\sqrt{41}$ in. and a minor axis of length 8 in. **79.** 91.5 million mi; $\frac{x^2}{(93)^2}+\frac{y^2}{8646.75}=1$

81. Perihelion: 460.6 million mi; mean distance: 483.8 million mi; $\frac{x^2}{(483.8)^2}+\frac{y^2}{233{,}524.2}=1$

83. 35 million mi **85.** $5\sqrt{5}-4$
87. $Ax^2+Cy^2+Dx+Ey+F=0 \quad A\neq 0, C\neq 0$

$$Ax^2+Dx+Cy^2+Ey=-F$$

$$A\left(x^2+\frac{D}{A}x\right)+C\left(y^2+\frac{E}{Cy}\right)=-F$$

$$A\left(x+\frac{D}{2A}\right)^2+C\left(y+\frac{E}{2C}\right)^2=-F+\frac{D^2}{4A}+\frac{E^2}{4C}$$

(a) If $\frac{D^2}{4A}+\frac{E^2}{4C}-F$ is of the same sign as A (and C), this is the equation of an ellipse with center at $\left(-\frac{D}{2A},-\frac{E}{2C}\right)$.

(b) If $\frac{D^2}{4A}+\frac{E^2}{4C}-F$, the graph is the single point $\left(-\frac{D}{2A},-\frac{E}{2C}\right)$.

(c) If $\frac{D^2}{4A}+\frac{E^2}{4C}-F$ is of the sign opposite that of A (and C), the graph contains no points, because in this case, the left side has the sign opposite that of the right side.

89. Zeros: $5-2\sqrt{3}, 5+2\sqrt{3}$; x-intercepts: $5-2\sqrt{3}, 5+2\sqrt{3}$ **90.** Domain: $\{x \mid x\neq 5\}$; Horizontal asymptote: $y=2$; Vertical asymptote: $x=5$
91. 617.1 ft-lb **92.** $b\approx 10.94$, $c\approx 17.77$, $B=38°$

11.4 Assess Your Understanding *(page 824)*

7. hyperbola **8.** transverse axis **9.** b **10.** $(2,4)$; $(2,-2)$ **11.** $(2,6)$; $(2,-4)$ **12.** c **13.** 2; 3; x **14.** $y=-\frac{4}{9}x$; $y=\frac{4}{9}x$ **15.** B **17.** A

19. $x^2 - \frac{y^2}{8} = 1$

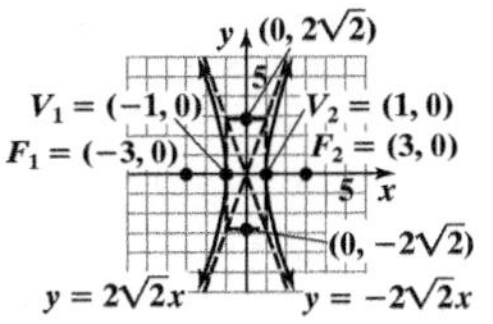

21. $\frac{y^2}{16} - \frac{x^2}{20} = 1$

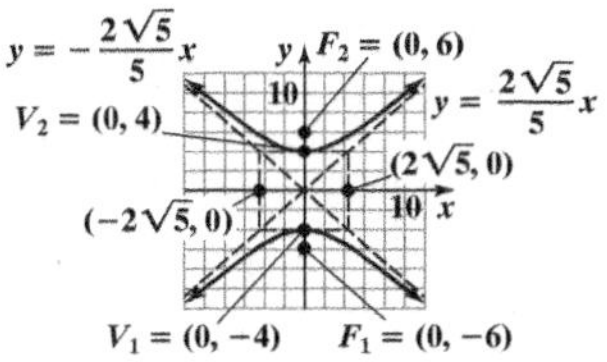

23. $\frac{x^2}{9} - \frac{y^2}{16} = 1$

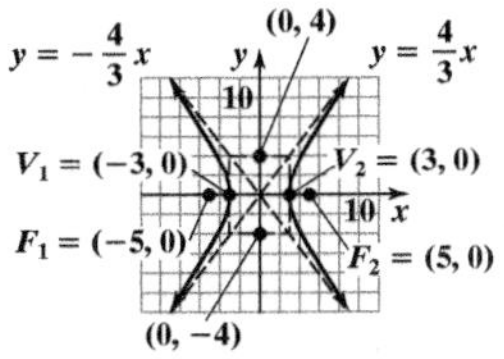

25. $\frac{y^2}{36} - \frac{x^2}{9} = 1$

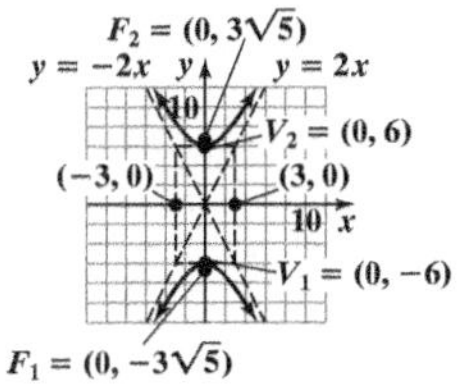

27. $\frac{x^2}{8} - \frac{y^2}{8} = 1$

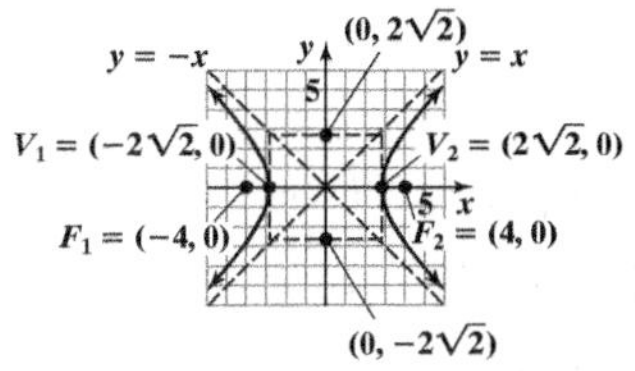

29. $\frac{x^2}{25} - \frac{y^2}{9} = 1$

Center: $(0, 0)$
Transverse axis: x-axis
Vertices: $(-5, 0)$, $(5, 0)$
Foci: $(-\sqrt{34}, 0)$, $(\sqrt{34}, 0)$
Asymptotes: $y = \pm\frac{3}{5}x$

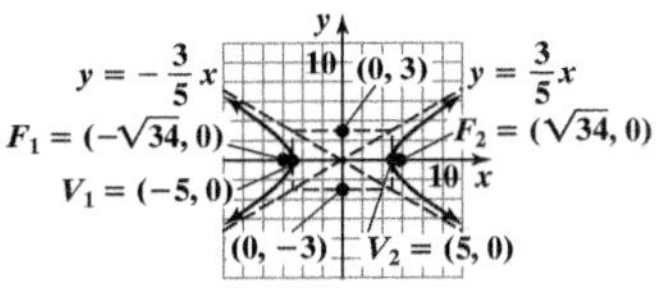

31. $\frac{x^2}{4} - \frac{y^2}{16} = 1$

Center: $(0, 0)$
Transverse axis: x-axis
Vertices: $(-2, 0)$, $(2, 0)$
Foci: $(-2\sqrt{5}, 0)$, $(2\sqrt{5}, 0)$
Asymptotes: $y = \pm 2x$

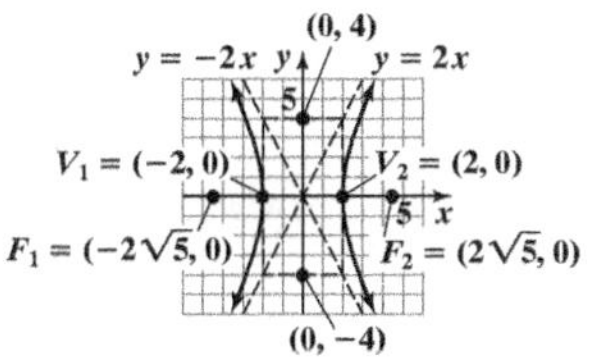

33. $\frac{y^2}{9} - x^2 = 1$

Center: $(0, 0)$
Transverse axis: y-axis
Vertices: $(0, -3)$, $(0, 3)$
Foci: $(0, -\sqrt{10})$, $(0, \sqrt{10})$
Asymptotes: $y = \pm 3x$

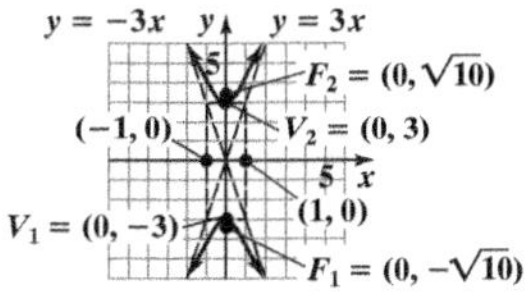

35. $\frac{y^2}{25} - \frac{x^2}{25} = 1$

Center: $(0, 0)$
Transverse axis: y-axis
Vertices: $(0, -5)$, $(0, 5)$
Foci: $(0, -5\sqrt{2})$, $(0, 5\sqrt{2})$
Asymptotes: $y = \pm x$

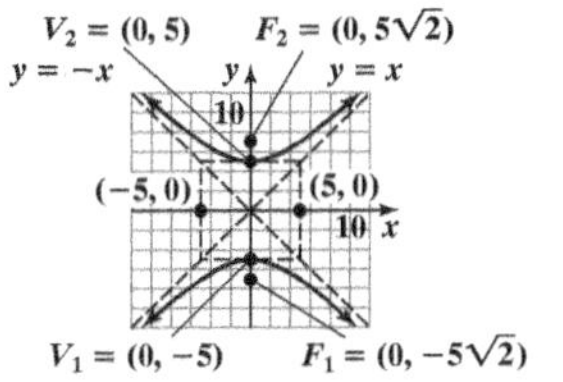

37. $x^2 - y^2 = 1$

39. $\frac{y^2}{36} - \frac{x^2}{9} = 1$

41. $\frac{(x-4)^2}{4} - \frac{(y+1)^2}{5} = 1$

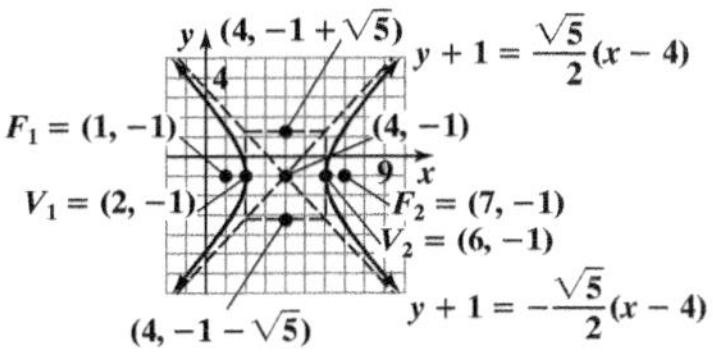

43. $\frac{(y+4)^2}{4} - \frac{(x+3)^2}{12} = 1$

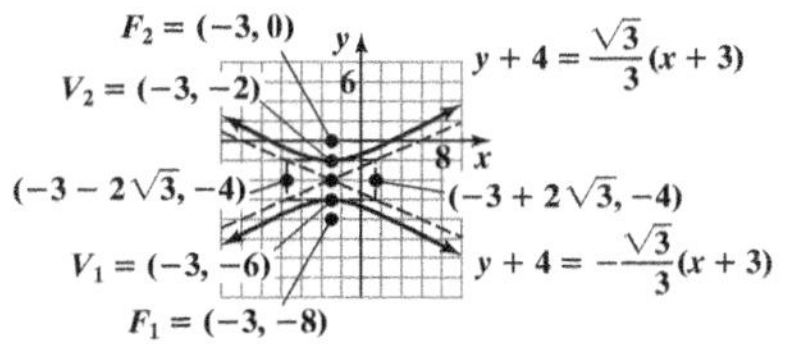

45. $(x-5)^2 - \frac{(y-7)^2}{3} = 1$

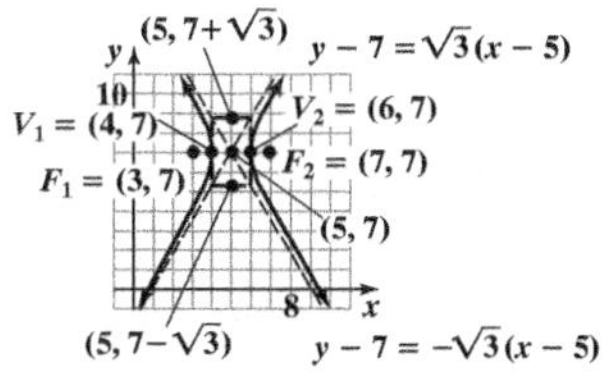

47. $\frac{(x-1)^2}{4} - \frac{(y+1)^2}{9} = 1$

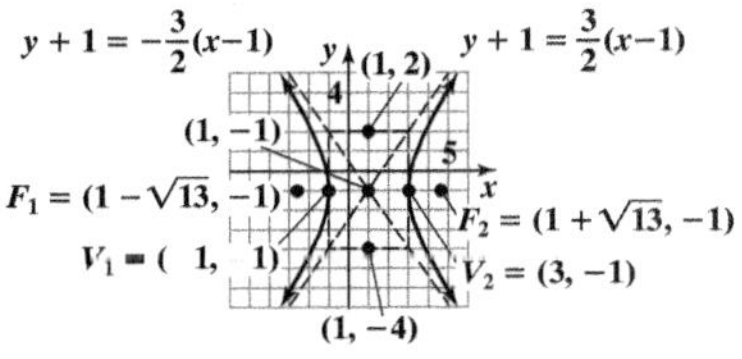

49. $\frac{(x-2)^2}{4} - \frac{(y+3)^2}{9} = 1$

Center: $(2, -3)$
Transverse axis: parallel to x-axis
Vertices: $(0, -3)$, $(4, -3)$
Foci: $(2 - \sqrt{13}, -3)$, $(2 + \sqrt{13}, -3)$
Asymptotes: $y + 3 = \pm\frac{3}{2}(x - 2)$

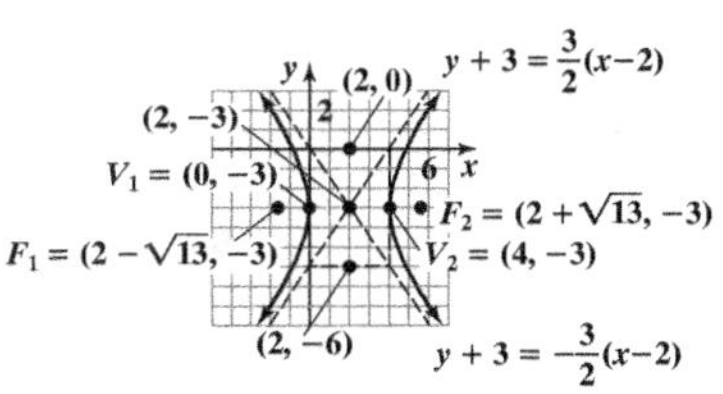

51. $\dfrac{(y-2)^2}{4} - (x+2)^2 = 1$
Center: $(-2, 2)$
Transverse axis: parallel to y-axis
Vertices: $(-2, 0)$, $(-2, 4)$
Foci: $(-2, 2 - \sqrt{5})$, $(-2, 2 + \sqrt{5})$
Asymptotes: $y - 2 = \pm 2(x + 2)$

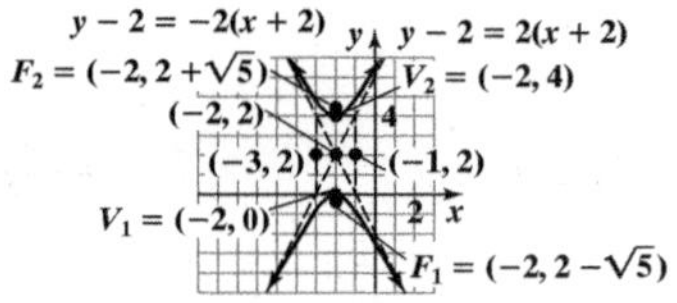

53. $\dfrac{(x+1)^2}{4} - \dfrac{(y+2)^2}{4} = 1$
Center: $(-1, -2)$
Transverse axis: parallel to x-axis
Vertices: $(-3, -2)$, $(1, -2)$
Foci: $(-1 - 2\sqrt{2}, -2)$, $(-1 + 2\sqrt{2}, -2)$
Asymptotes: $y + 2 = \pm(x + 1)$

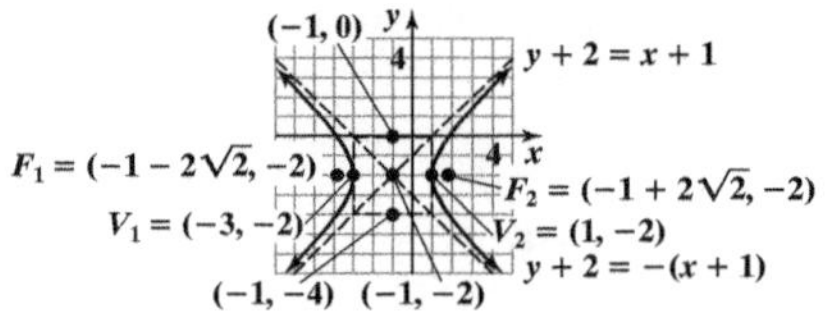

55. $(x-1)^2 - (y+1)^2 = 1$
Center: $(1, -1)$
Transverse axis: parallel to x-axis
Vertices: $(0, -1)$, $(2, -1)$
Foci: $(1 - \sqrt{2}, -1)$, $(1 + \sqrt{2}, -1)$
Asymptotes: $y + 1 = \pm(x - 1)$

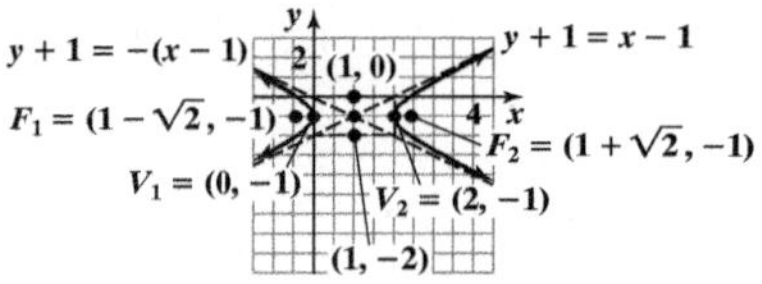

57. $\dfrac{(y-2)^2}{4} - (x+1)^2 = 1$
Center: $(-1, 2)$
Transverse axis: parallel to y-axis
Vertices: $(-1, 0)$, $(-1, 4)$
Foci: $(-1, 2 - \sqrt{5})$, $(-1, 2 + \sqrt{5})$
Asymptotes: $y - 2 = \pm 2(x + 1)$

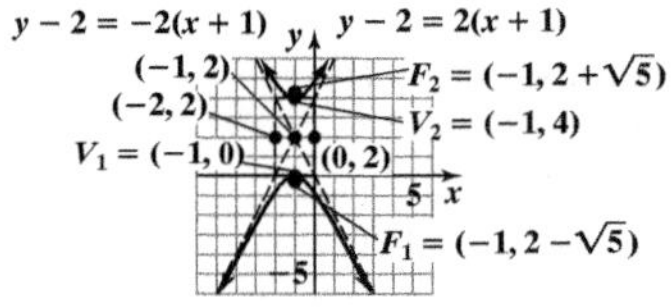

59. $\dfrac{(x-3)^2}{4} - \dfrac{(y+2)^2}{16} = 1$
Center: $(3, -2)$
Transverse axis: parallel to x-axis
Vertices: $(1, -2)$, $(5, -2)$
Foci: $(3 - 2\sqrt{5}, -2)$, $(3 + 2\sqrt{5}, -2)$
Asymptotes: $y + 2 = \pm 2(x - 3)$

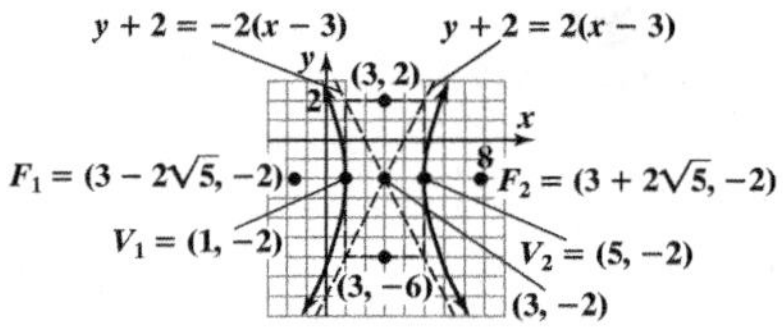

61. $\dfrac{(y-1)^2}{4} - (x+2)^2 = 1$
Center: $(-2, 1)$
Transverse axis: parallel to y-axis
Vertices: $(-2, -1)$, $(-2, 3)$
Foci: $(-2, 1 - \sqrt{5})$, $(-2, 1 + \sqrt{5})$
Asymptotes: $y - 1 = \pm 2(x + 2)$

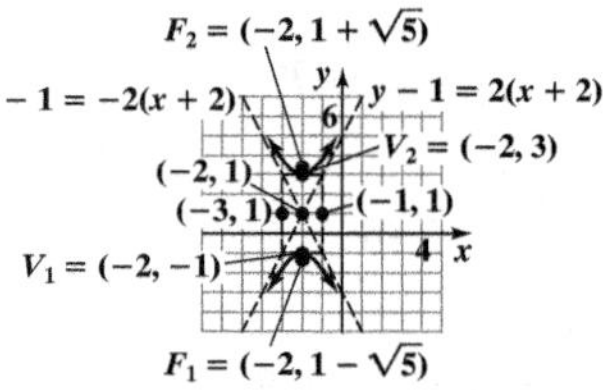

63.

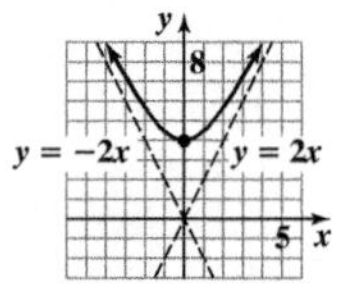

65.

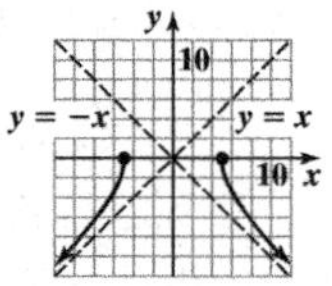

67. Center: $(3, 0)$
Transverse axis: parallel to x-axis
Vertices: $(1, 0)$, $(5, 0)$
Foci: $(3 - \sqrt{29}, 0)$, $(3 + \sqrt{29}, 0)$
Asymptotes: $y = \pm\dfrac{5}{2}(x - 3)$

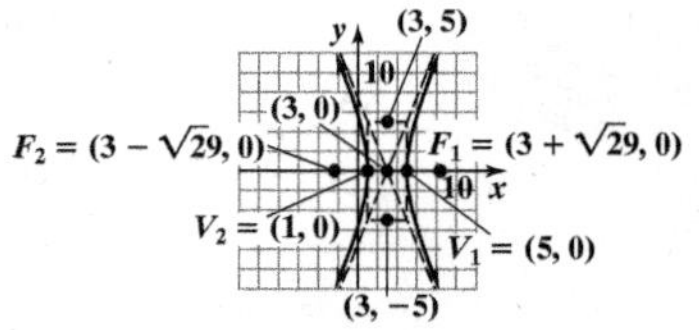

69. Vertex: $(0, 3)$; focus: $(0, 7)$; directrix: $y = -1$

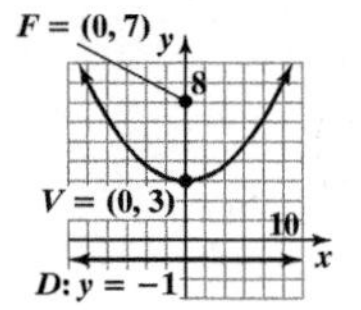

71. $\dfrac{(x-5)^2}{9} + \dfrac{y^2}{25} = 1$
Center: $(5, 0)$; vertices: $(5, 5)$, $(5, -5)$; foci: $(5, -4)$, $(5, 4)$

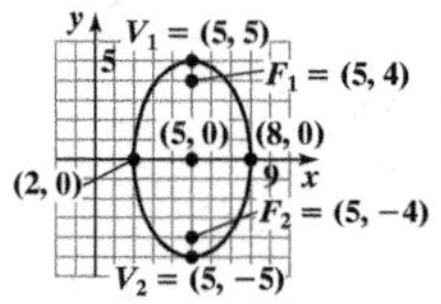

73. $(x-3)^2 = 8(y+5)$
Vertex: $(3, -5)$; focus: $(3, -3)$; directrix: $y = -7$

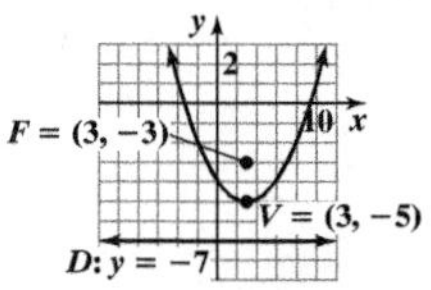

75. The fireworks display is 50,138 ft north of the person at point A. **77.** The tower is 592.4 ft tall. **79. (a)** $y = \pm x$ **(b)** $\dfrac{x^2}{100} - \dfrac{y^2}{100} = 1, x \geq 0$
81. If the eccentricity is close to 1, the "opening" of the hyperbola is very small. As e increases, the opening gets bigger.

83. $\dfrac{x^2}{4} - y^2 = 1$; asymptotes: $y = \pm\dfrac{1}{2}x$

$y^2 - \dfrac{x^2}{4} = 1$; asymptotes: $y = \pm\dfrac{1}{2}x$

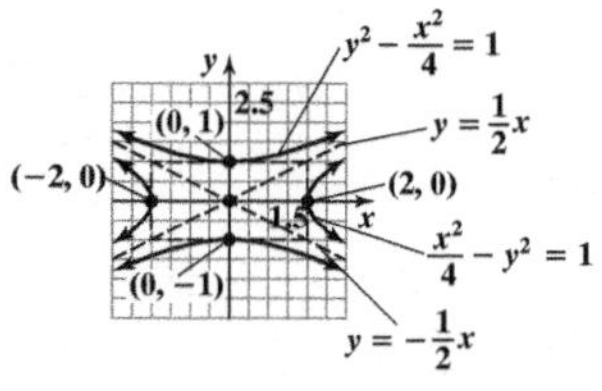

85. $Ax^2 + Cy^2 + F = 0$
$Ax^2 + Cy^2 = -F$

If A and C are opposite in sign and $F \neq 0$, this equation may be written as $\dfrac{x^2}{\left(-\dfrac{F}{A}\right)} + \dfrac{y^2}{\left(-\dfrac{F}{C}\right)} = 1$, where $-\dfrac{F}{A}$ and $-\dfrac{F}{C}$ are opposite in sign. This is the equation of a hyperbola with center $(0, 0)$. The transverse axis is the x-axis if $-\dfrac{F}{A} > 0$; the transverse axis is the y-axis if $-\dfrac{F}{A} < 0$.

87. Amplitude $= \frac{3}{2}$; Period $= \frac{\pi}{3}$; Phase shift $= -\frac{\pi}{2}$

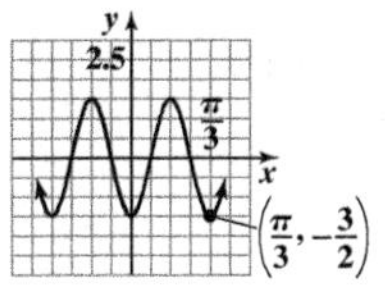

88. $c \approx 13.16, A \approx 31.6°, B = 48.4°$ **89.** $(6, -6\sqrt{3})$

90. $x^2 + (y-3)^2 = 9$; circle, radius 3, center at $(0, 3)$ in rectangular coordinates

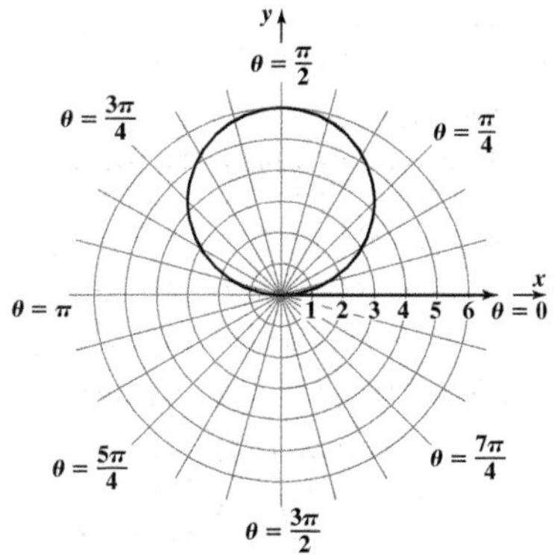

11.5 Assess Your Understanding *(page 834)*

5. $\cot(2\theta) = \frac{A-C}{B}$ **6.** d **7.** $B^2 - 4AC < 0$ **8.** c **9.** T **10.** F **11.** Parabola **13.** Ellipse **15.** Hyperbola **17.** Hyperbola **19.** Circle **21.** $x = \frac{\sqrt{2}}{2}(x' - y'), y = \frac{\sqrt{2}}{2}(x' + y')$ **23.** $x = \frac{\sqrt{2}}{2}(x' - y'), y = \frac{\sqrt{2}}{2}(x' + y')$ **25.** $x = \frac{1}{2}(x' - \sqrt{3}y'), y = \frac{1}{2}(\sqrt{3}x' + y')$ **27.** $x = \frac{\sqrt{5}}{5}(x' - 2y'), y = \frac{\sqrt{5}}{5}(2x' + y')$ **29.** $x = \frac{\sqrt{13}}{13}(3x' - 2y'), y = \frac{\sqrt{13}}{13}(2x' + 3y')$

31. $\theta = 45°$ (see Problem 21)
$x'^2 - \frac{y'^2}{3} = 1$
Hyperbola
Center at origin
Transverse axis is the x'-axis.
Vertices at $(\pm 1, 0)$

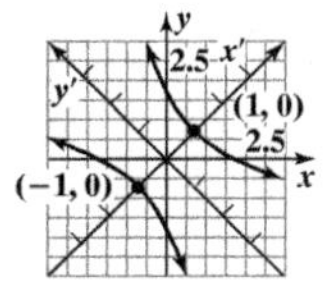

33. $\theta = 45°$ (see Problem 23)
$x'^2 + \frac{y'^2}{4} = 1$
Ellipse
Center at $(0, 0)$
Major axis is the y'-axis.
Vertices at $(0, \pm 2)$

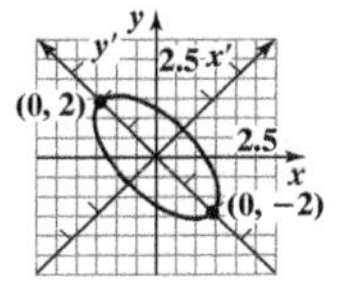

35. $\theta = 60°$ (see Problem 25)
$\frac{x'^2}{4} + y'^2 = 1$
Ellipse
Center at $(0, 0)$
Major axis is the x'-axis.
Vertices at $(\pm 2, 0)$

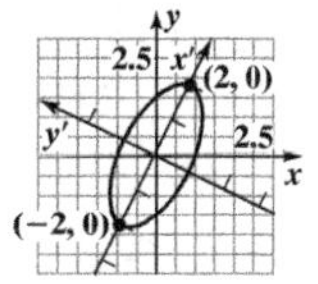

37. $\theta \approx 63°$ (see Problem 27)
$y'^2 = 8x'$
Parabola
Vertex at $(0, 0)$
Focus at $(2, 0)$

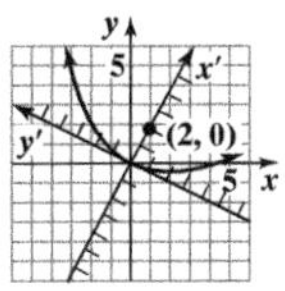

39. $\theta \approx 34°$ (see Problem 29)
$\frac{(x'-2)^2}{4} + y'^2 = 1$
Ellipse
Center at $(2, 0)$
Major axis is the x'-axis.
Vertices at $(4, 0)$ and $(0, 0)$

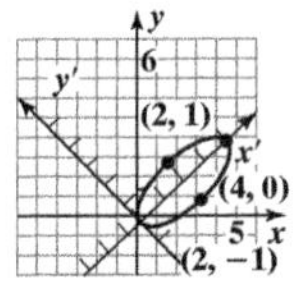

41. $\cot(2\theta) = \frac{7}{24}$;
$\theta = \sin^{-1}\left(\frac{3}{5}\right) \approx 37°$
$(x' - 1)^2 = -6\left(y' - \frac{1}{6}\right)$
Parabola
Vertex at $\left(1, \frac{1}{6}\right)$
Focus at $\left(1, -\frac{4}{3}\right)$

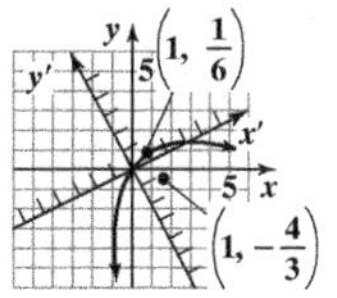

43. Hyperbola **45.** Hyperbola **47.** Parabola **49.** Ellipse **51.** Ellipse

53. Refer to equation (6): $A' = A\cos^2\theta + B\sin\theta\cos\theta + C\sin^2\theta$

$$B' = B(\cos^2\theta - \sin^2\theta) + 2(C - A)(\sin\theta\cos\theta)$$
$$C' = A\sin^2\theta - B\sin\theta\cos\theta + C\cos^2\theta$$
$$D' = D\cos\theta + E\sin\theta$$
$$E' = -D\sin\theta + E\cos\theta$$
$$F' = F$$

55. Use Problem 53 to find $B'^2 - 4A'C'$. After much cancellation, $B'^2 - 4A'C' = B^2 - 4AC$.

57. The distance between P_1 and P_2 in the $x'y'$-plane equals $\sqrt{(x_2' - x_1')^2 + (y_2' - y_1')^2}$.
Assuming that $x' = x\cos\theta - y\sin\theta$ and $y' = x\sin\theta + y\cos\theta$, then

$$(x_2' - x_1')^2 = (x_2\cos\theta - y_2\sin\theta - x_1\cos\theta + y_1\sin\theta)^2$$
$$= \cos^2\theta(x_2 - x_1)^2 - 2\sin\theta\cos\theta(x_2 - x_1)(y_2 - y_1) + \sin^2\theta(y_2 - y_1)^2, \text{ and}$$
$$(y_2' - y_1')^2 = (x_2\sin\theta + y_2\cos\theta - x_1\sin\theta - y_1\cos\theta)^2 = \sin^2\theta(x_2 - x_1)^2 + 2\sin\theta\cos\theta(x_2 - x_1)(y_2 - y_1) + \cos^2\theta(y_2 - y_1)^2.$$

Therefore, $(x_2' - x_1')^2 + (y_2' - y_1')^2 = \cos^2\theta(x_2 - x_1)^2 + \sin^2\theta(x_2 - x_1)^2 + \sin^2\theta(y_2 - y_1)^2 + \cos^2\theta(y_2 - y_1)^2$
$$= (x_2 - x_1)^2(\cos^2\theta + \sin^2\theta) + (y_2 - y_1)^2(\sin^2\theta + \cos^2\theta) = (x_2 - x_1)^2 + (y_2 - y_1)^2.$$

61. $A \approx 39.4°, B \approx 54.7°, C \approx 85.9°$ **62.** 38.5 **63.** $r^2\cos\theta\sin\theta = 1$ **64.** $\sqrt{29}(\cos 291.8° + i\sin 291.8°)$

11.6 Assess Your Understanding *(page 841)*

3. conic; focus; directrix **4.** 1; <1; >1 **5.** T **6.** T **7.** Parabola; directrix is perpendicular to the polar axis, 1 unit to the right of the pole. **9.** Hyperbola; directrix is parallel to the polar axis, $\frac{4}{3}$ units below the pole. **11.** Ellipse; directrix is perpendicular to the polar axis, $\frac{3}{2}$ units to the left of the pole.

13. Parabola; directrix is perpendicular to the polar axis, 1 unit to the right of the pole; vertex is at $\left(\frac{1}{2}, 0\right)$.

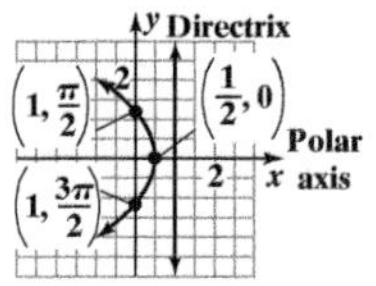

15. Ellipse; directrix is parallel to the polar axis, $\frac{8}{3}$ units above the pole; vertices are at $\left(\frac{8}{7}, \frac{\pi}{2}\right)$ and $\left(8, \frac{3\pi}{2}\right)$.

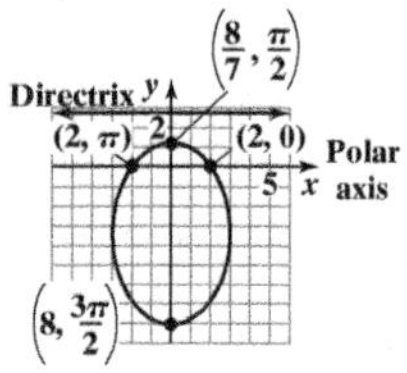

17. Hyperbola; directrix is perpendicular to the polar axis, $\frac{3}{2}$ units to the left of the pole; vertices are at $(-3, 0)$ and $(1, \pi)$.

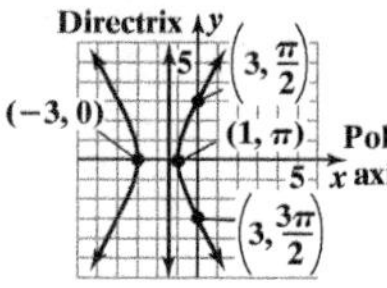

19. Ellipse; directrix is parallel to the polar axis, 8 units below the pole; vertices are at $\left(8, \frac{\pi}{2}\right)$ and $\left(\frac{8}{3}, \frac{3\pi}{2}\right)$.

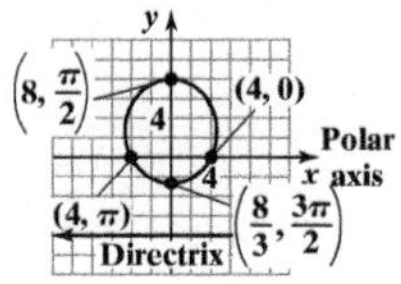

21. Ellipse; directrix is parallel to the polar axis, 3 units below the pole; vertices are at $\left(6, \frac{\pi}{2}\right)$ and $\left(\frac{6}{5}, \frac{3\pi}{2}\right)$.

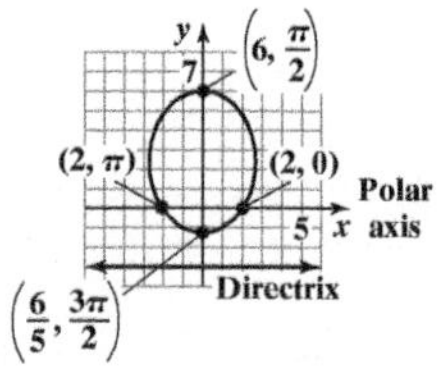

23. Ellipse; directrix is perpendicular to the polar axis, 6 units to the left of the pole; vertices are at $(6, 0)$ and $(2, \pi)$.

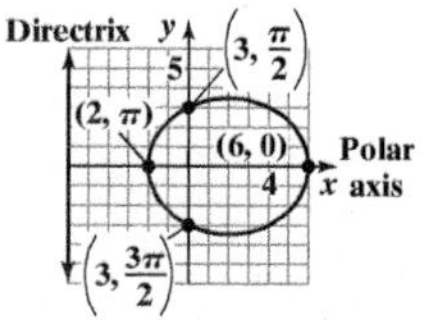

25. $y^2 + 2x - 1 = 0$ **27.** $16x^2 + 7y^2 + 48y - 64 = 0$ **29.** $3x^2 - y^2 + 12x + 9 = 0$ **31.** $4x^2 + 3y^2 - 16y - 64 = 0$
33. $9x^2 + 5y^2 - 24y - 36 = 0$ **35.** $3x^2 + 4y^2 - 12x - 36 = 0$ **37.** $r = \frac{1}{1 + \sin\theta}$ **39.** $r = \frac{12}{5 - 4\cos\theta}$ **41.** $r = \frac{12}{1 - 6\sin\theta}$
43. Use $d(D, P) = p - r\cos\theta$ in the derivation of equation (a) in Table 5. **45.** Use $d(D, P) = p + r\sin\theta$ in the derivation of equation (a) in Table 5.
47. 27.81 **48.** Amplitude = 4; Period = 10π **49.** $\left\{\frac{\pi}{3}, \pi, \frac{5\pi}{3}\right\}$ **50.** 26

11.7 Assess Your Understanding *(page 852)*

2. plane curve; parameter **3.** b **4.** a **5.** F **6.** T

7.
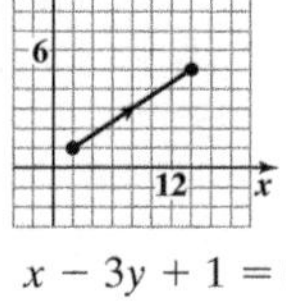

$x - 3y + 1 = 0$

9.
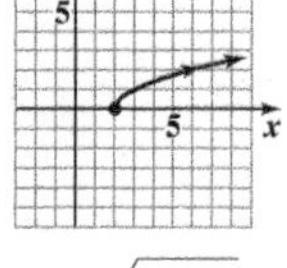

$y = \sqrt{x - 2}$

11.
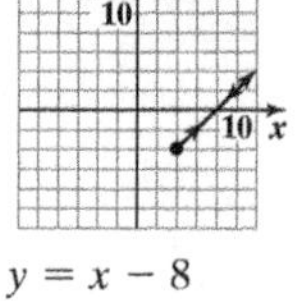

$y = x - 8$

13.
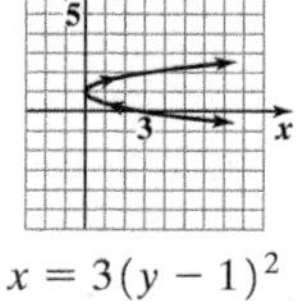

$x = 3(y - 1)^2$

15.
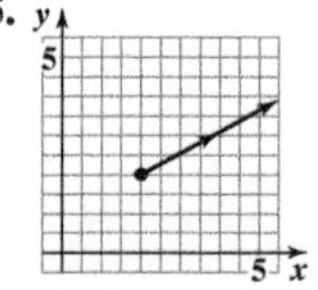

$2y = 2 + x$

17.
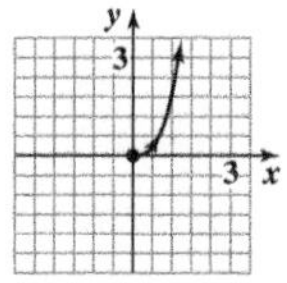

$y = x^3$

19.
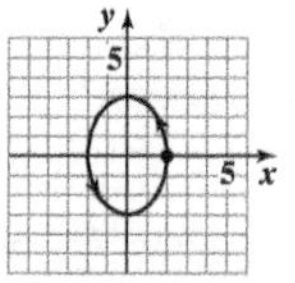

$\frac{x^2}{4} + \frac{y^2}{9} = 1$

21.
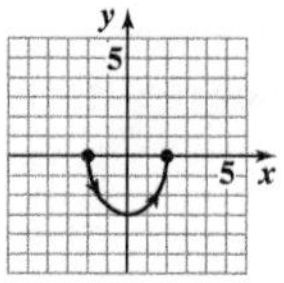

$\frac{x^2}{4} + \frac{y^2}{9} = 1$

23.

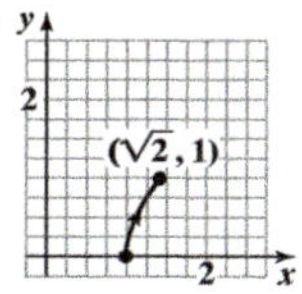

$x^2 - y^2 = 1$

25.

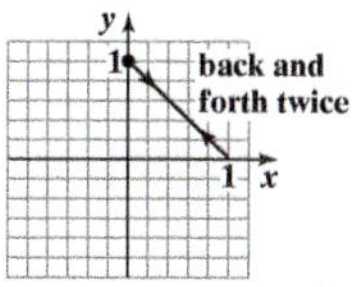

$x + y = 1$

27. $x = t$ or $x = \frac{t+1}{4}$
$y = 4t - 1$ $\quad y = t$

29. $x = t$ or $x = t^3$
$y = t^2 + 1$ $\quad y = t^6 + 1$

31. $x = t$ or $x = \sqrt[3]{t}$
$y = t^3$ $\quad y = t$

33. $x = t$ or $x = t^3$
$y = t^{2/3}, t \geq 0$ $\quad y = t^2, t \geq 0$

35. $x = t + 2, y = t, 0 \leq t \leq 5$ **37.** $x = 3\cos t, y = 2\sin t, 0 \leq t \leq 2\pi$

39. $x = 2\cos(\pi t), y = -3\sin(\pi t), 0 \leq t \leq 2$

41. $x = 2\sin(2\pi t), y = 3\cos(2\pi t), 0 \leq t \leq 1$

43.

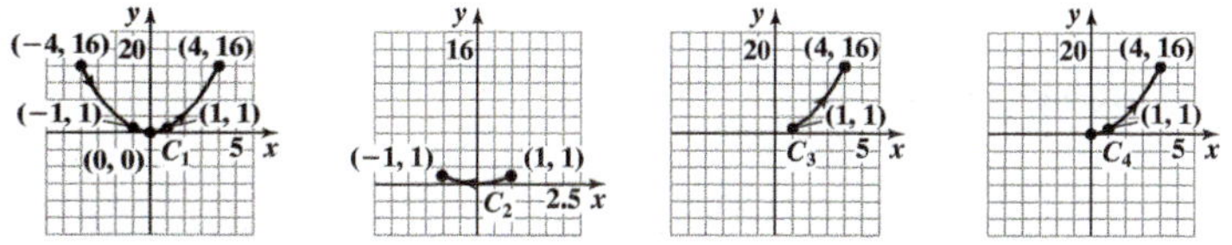

45.

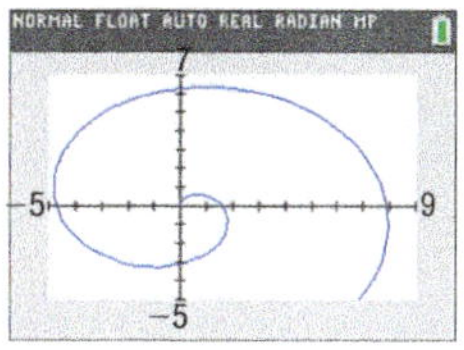

47.

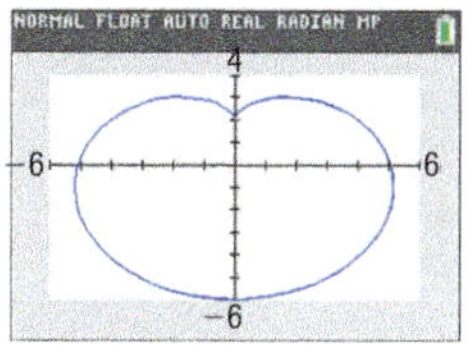

49. (a) $x = 3$
$y = -16t^2 + 50t + 6$
(b) 3.24 s
(c) 1.56 s; 45.06 ft
(d)

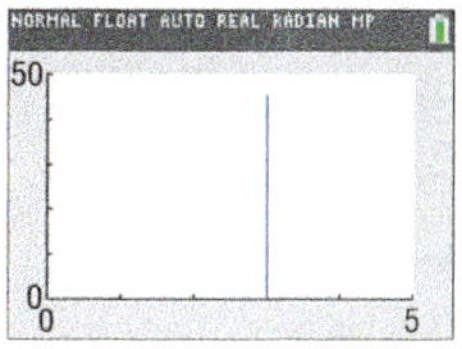

51. (a) Train: $x_1 = t^2, y_1 = 1$;
Bill: $x_2 = 5(t - 5), y_2 = 3$
(b) Bill won't catch the train.
(c)

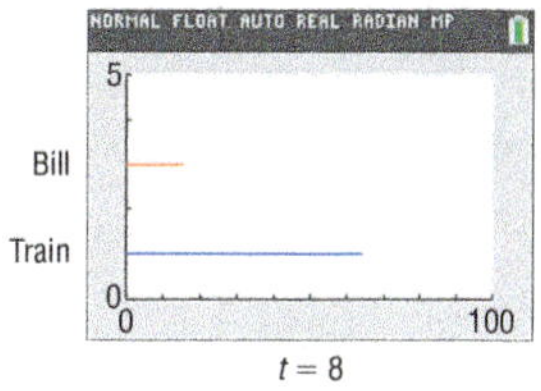

53. (a) $x = (145\cos 20°)t$
$y = -16t^2 + (145\sin 20°)t + 5$
(b) 3.20 s
(c) 435.65 ft
(d) 1.55 s; 43.43 ft
(e)

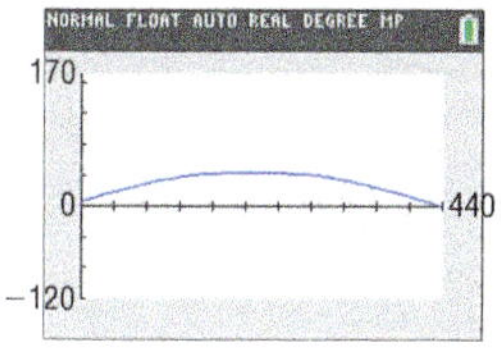

55. (a) $x = (40\cos 45°)t$
$y = -4.9t^2 + (40\sin 45°)t + 300$
(b) 11.23 s
(c) 317.52 m
(d) 2.89 s; 340.82 m
(e)

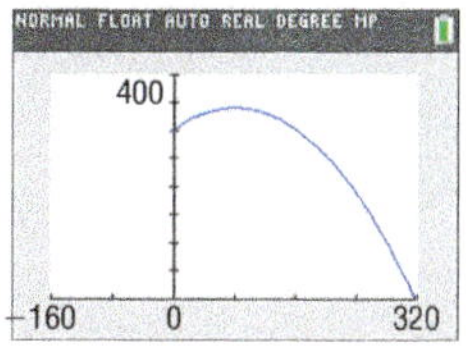

57. (a) Camry: $x = 40t - 5, y = 0$; Chevy Impala: $x = 0, y = 30t - 4$
(b) $d = \sqrt{(40t - 5)^2 + (30t - 4)^2}$
(c)

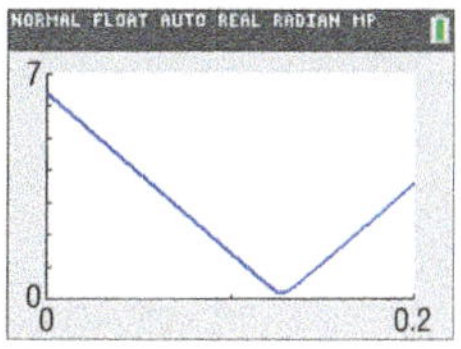

(d) 0.2 mi; 7.68 min
(e)

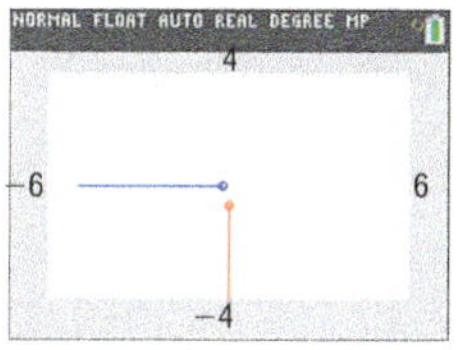

59. (a) $x = \frac{\sqrt{2}}{2}v_0 t, y = -16t^2 + \frac{\sqrt{2}}{2}v_0 t + 3$ **(b)** Maximum height is 139.1 ft. **(c)** The ball is 272.25 ft from home plate. **(d)** Yes, the ball will clear the wall by about 99.5 ft. **61.** The orientation is from (x_1, y_1) to (x_2, y_2).

65.

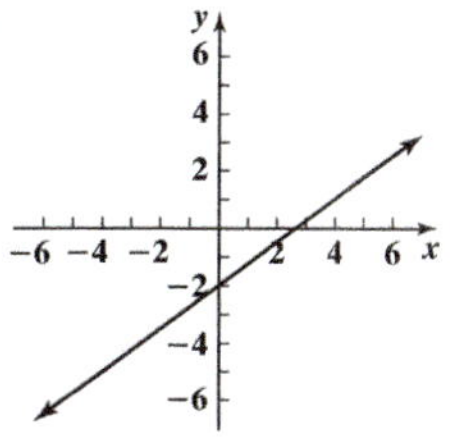

66.

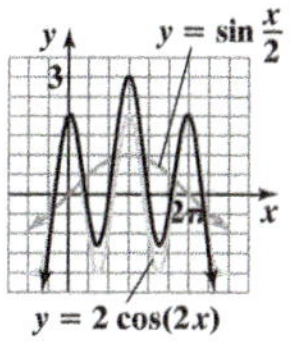

67. Approximately 2733 miles **68. (a)** Simple harmonic **(b)** 2 m **(c)** $\frac{\pi}{2}$ s **(d)** $\frac{2}{\pi}$ oscillations/s

Review Exercises *(page 857)*

1. Parabola; vertex: $(0,0)$; focus: $(-4,0)$; directrix: $x = 4$ **2.** Hyperbola; center: $(0,0)$; vertices: $(5,0)$ and $(-5,0)$; foci: $(\sqrt{26},0)$ and $(-\sqrt{26},0)$; asymptotes: $y = \frac{1}{5}x$ and $y = -\frac{1}{5}x$ **3.** Ellipse; center: $(0,0)$; vertices: $(0,5)$ and $(0,-5)$; foci: $(0,3)$ and $(0,-3)$

4. $x^2 = -4(y-1)$: Parabola; vertex: $(0,1)$; focus: $(0,0)$; directrix: $y = 2$ **5.** $\frac{x^2}{2} - \frac{y^2}{8} = 1$: Hyperbola; center: $(0,0)$; vertices: $(\sqrt{2},0)$ and $(-\sqrt{2},0)$; foci: $(\sqrt{10},0)$ and $(-\sqrt{10},0)$; asymptotes: $y = 2x$ and $y = -2x$ **6.** $(x-2)^2 = 2(y+2)$: Parabola; vertex: $(2,-2)$; focus: $\left(2,-\frac{3}{2}\right)$; directrix: $y = -\frac{5}{2}$

7. $\frac{(y-2)^2}{4} - (x-1)^2 = 1$: Hyperbola; center: $(1,2)$; vertices: $(1,4)$ and $(1,0)$; foci: $(1, 2+\sqrt{5})$ and $(1, 2-\sqrt{5})$; asymptotes: $y - 2 = \pm 2(x-1)$

8. $\frac{(x-2)^2}{9} + \frac{(y-1)^2}{4} = 1$: Ellipse; center: $(2,1)$; vertices: $(5,1)$ and $(-1,1)$; foci: $(2+\sqrt{5},1)$ and $(2-\sqrt{5},1)$

9. $(x-2)^2 = -4(y+1)$: Parabola; vertex: $(2,-1)$; focus: $(2,-2)$; directrix: $y = 0$

10. $\frac{(x-1)^2}{4} + \frac{(y+1)^2}{9} = 1$: Ellipse; center: $(1,-1)$; vertices: $(1,2)$ and $(1,-4)$; foci: $(1,-1+\sqrt{5})$ and $(1,-1-\sqrt{5})$

11. $y^2 = -8x$

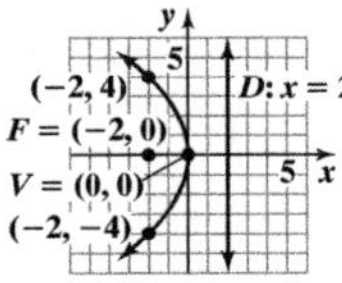

12. $\frac{y^2}{4} - \frac{x^2}{12} = 1$

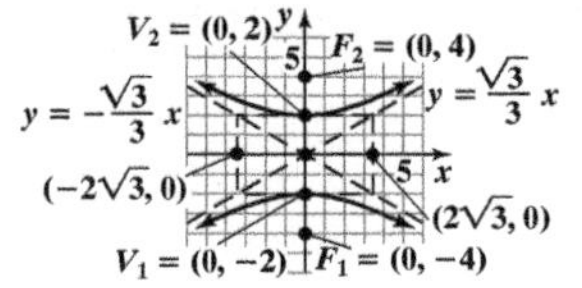

13. $\frac{x^2}{16} + \frac{y^2}{7} = 1$

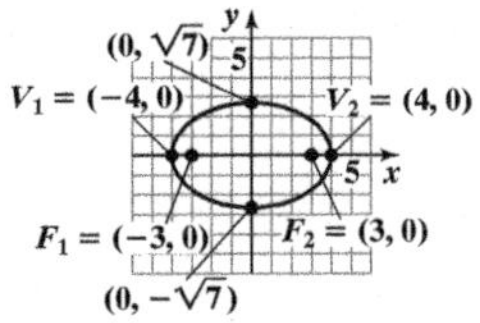

14. $(x-2)^2 = -4(y+3)$

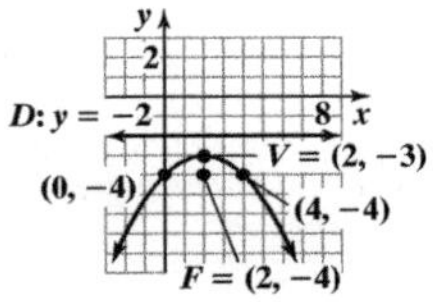

15. $(x+2)^2 - \frac{(y+3)^2}{3} = 1$

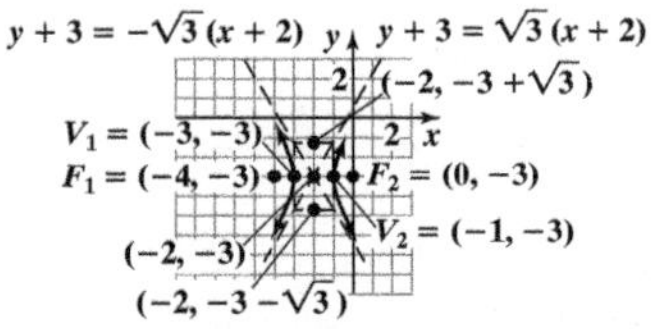

16. $\frac{(x+4)^2}{16} + \frac{(y-5)^2}{25} = 1$

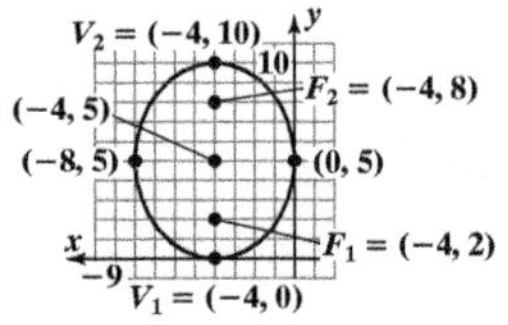

17. $\frac{(x+1)^2}{9} - \frac{(y-2)^2}{7} = 1$

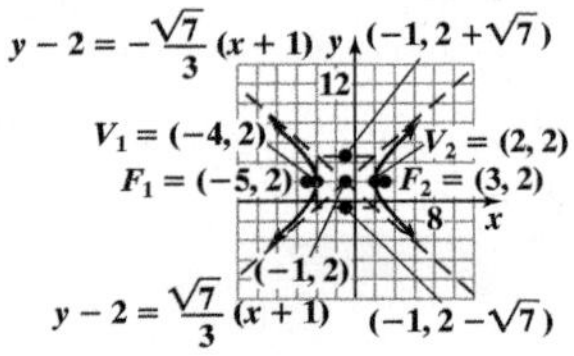

18. $\frac{(x-3)^2}{9} - \frac{(y-1)^2}{4} = 1$

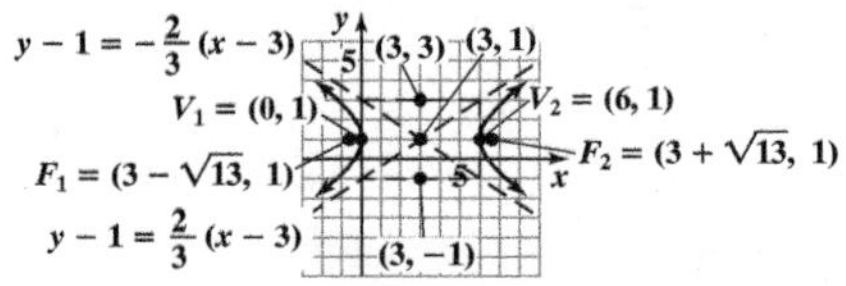

19. Parabola **20.** Ellipse
21. Parabola **22.** Hyperbola
23. Ellipse

24. $x'^2 - \frac{y'^2}{9} = 1$

Hyperbola
Center at the origin
Transverse axis the x'-axis
Vertices at $(\pm 1, 0)$

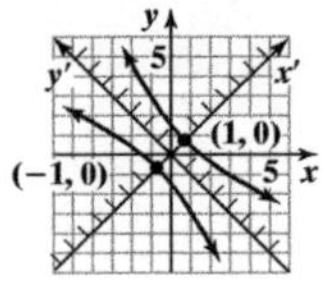

25. $\frac{x'^2}{2} + \frac{y'^2}{4} = 1$

Ellipse
Center at origin
Major axis the y'-axis
Vertices at $(0, \pm 2)$

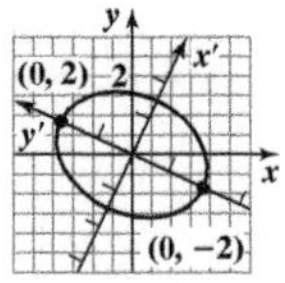

26. $y'^2 = -\frac{4\sqrt{13}}{13}x'$

Parabola
Vertex at the origin
Focus on the x'-axis at $\left(-\frac{\sqrt{13}}{13}, 0\right)$

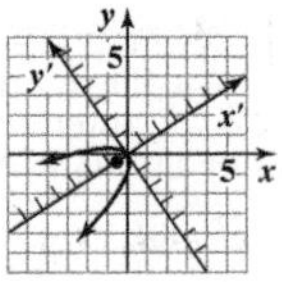

27. Parabola; directrix is perpendicular to the polar axis, 4 units to the left of the pole; vertex is $(2, \pi)$.

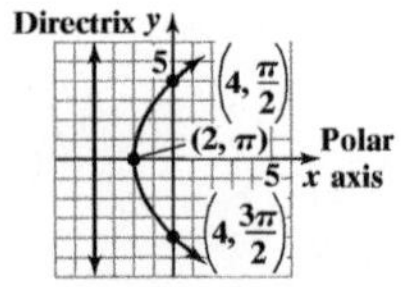

28. Ellipse; directrix is parallel to the polar axis, 6 units below the pole; vertices are $\left(6, \frac{\pi}{2}\right)$ and $\left(2, \frac{3\pi}{2}\right)$.

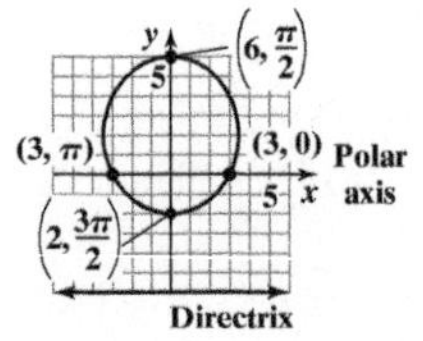

29. Hyperbola; directrix is perpendicular to the polar axis, 1 unit to the right of the pole; vertices are $\left(\frac{2}{3}, 0\right)$ and $(-2, \pi)$.

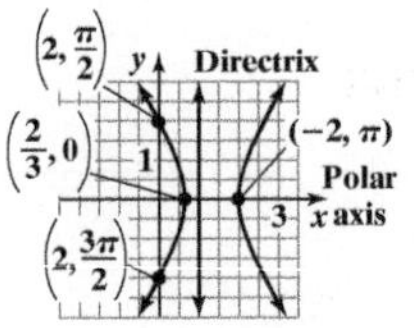

30. $y^2 - 8x - 16 = 0$ **31.** $3x^2 - y^2 - 8x + 4 = 0$

32.

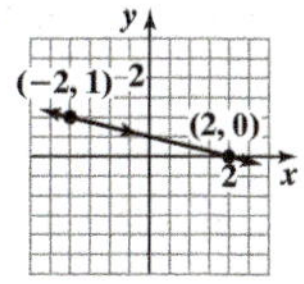

$x + 4y = 2$

33.

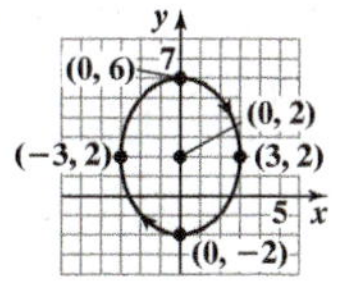

$\dfrac{x^2}{9} + \dfrac{(y-2)^2}{16} = 1$

34.

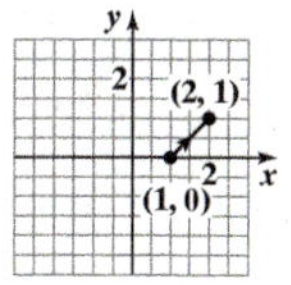

$1 + y = x$

35. $x = t, y = -2t + 4, -\infty < t < \infty$

$x = \dfrac{t-4}{-2}, y = t, -\infty < t < \infty$

36. $x = 4\cos\left(\dfrac{\pi}{2}t\right), y = 3\sin\left(\dfrac{\pi}{2}t\right), 0 \le t \le 4$ **37.** $\dfrac{x^2}{5} - \dfrac{y^2}{4} = 1$ **38.** The ellipse $\dfrac{x^2}{16} + \dfrac{y^2}{7} = 1$

39. $\dfrac{1}{4}$ ft or 3 in. **40.** 19.72 ft, 18.86 ft, 14.91 ft **41.** 450 ft

42. (a) Train: $x_1 = \dfrac{3}{2}t^2, y_1 = 1$
Mary: $x_2 = 6(t-2), y_2 = 3$
(b) Mary won't catch the train.

(c)

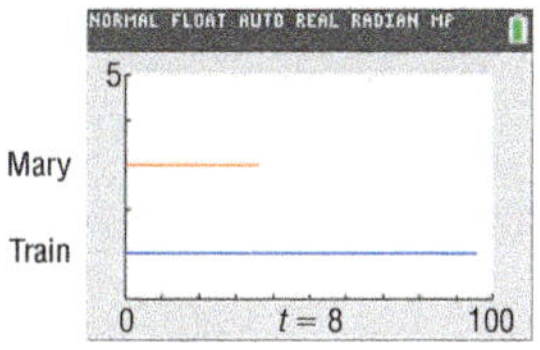

43. (a) $x = (80\cos 35°)t$
$y = -16t^2 + (80\sin 35°)t + 6$
(b) 2.9932 s
(c) 1.4339 s; 38.9 ft
(d) 196.15 ft

(e)

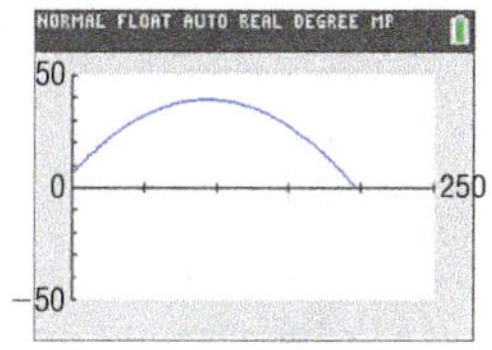

Chapter Test *(page 858)*

1. Hyperbola; center: $(-1, 0)$; vertices: $(-3, 0)$ and $(1, 0)$; foci: $(-1 - \sqrt{13}, 0)$ and $(-1 + \sqrt{13}, 0)$; asymptotes: $y = -\dfrac{3}{2}(x+1)$ and $y = \dfrac{3}{2}(x+1)$

2. Parabola; vertex: $\left(1, -\dfrac{1}{2}\right)$; focus: $\left(1, \dfrac{3}{2}\right)$; directrix: $y = -\dfrac{5}{2}$

3. Ellipse; center: $(-1, 1)$; foci: $(-1 - \sqrt{3}, 1)$ and $(-1 + \sqrt{3}, 1)$; vertices: $(-4, 1)$ and $(2, 1)$

4. $(x+1)^2 = 6(y-3)$

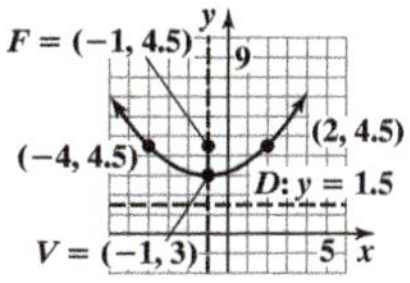

5. $\dfrac{x^2}{7} + \dfrac{y^2}{16} = 1$

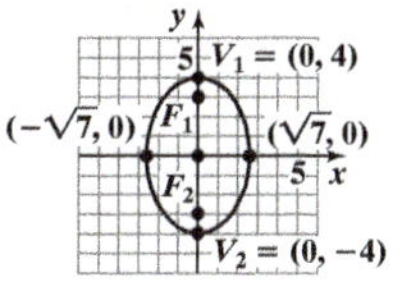

6. $\dfrac{(y-2)^2}{4} - \dfrac{(x-2)^2}{8} = 1$

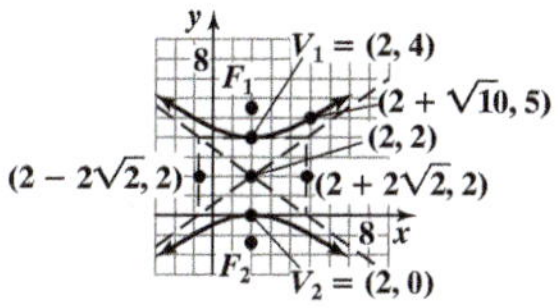

7. Hyperbola **8.** Ellipse **9.** Parabola

10. $x'^2 + 2y'^2 = 1$. This is the equation of an ellipse with center at $(0, 0)$ in the $x'y'$-plane. The vertices are at $(-1, 0)$ and $(1, 0)$ in the $x'y'$-plane.

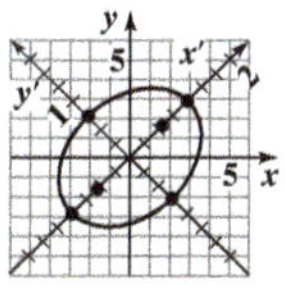

11. Hyperbola; $(x+2)^2 - \dfrac{y^2}{3} = 1$ **12.** $y = 1 - \sqrt{\dfrac{x+2}{3}}$

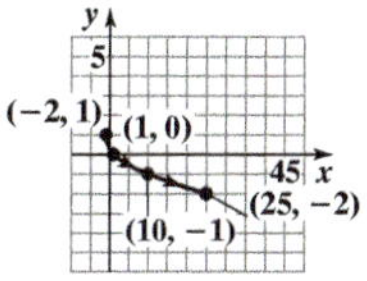

13. The microphone should be located $\dfrac{2}{3}$ ft from the base of the reflector, along its axis of symmetry.

Cumulative Review *(page 859)*

1. $-6x + 5 - 3h$ **2.** $\left\{-5, -\dfrac{1}{3}, 2\right\}$ **3.** $\{x \mid -3 \le x \le 2\}$ or $[-3, 2]$ **4. (a)** Domain: $(-\infty, \infty)$; range: $(2, \infty)$
(b) $y = \log_3(x - 2)$; domain: $(2, \infty)$; range: $(-\infty, \infty)$ **5. (a)** $\{18\}$ **(b)** $(2, 18]$

6. (a) $y = 2x - 2$ **(b)** $(x-2)^2 + y^2 = 4$ **(c)** $\dfrac{x^2}{9} + \dfrac{y^2}{4} = 1$ **(d)** $y = 2(x-1)^2$ **(e)** $y^2 - \dfrac{x^2}{3} = 1$ **(f)** $y = 4^x$

7. $\theta = \dfrac{\pi}{12} \pm \pi k$, k is any integer; $\theta = \dfrac{5\pi}{12} \pm \pi k$, k is any integer **8.** $\theta = \dfrac{\pi}{6}$

9. $r = 8\sin\theta$

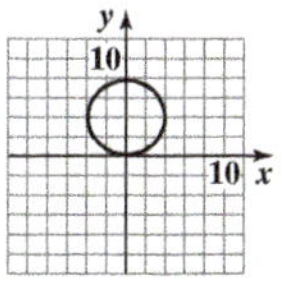

10. $\left\{x \,\middle|\, x \ne \dfrac{3\pi}{4} \pm \pi k, k \text{ is an integer}\right\}$ **11.** $\{22.5°\}$ **12.** $y = \dfrac{x^2}{5} + 5$

CHAPTER 12 Systems of Equations and Inequalities

12.1 Assess Your Understanding *(page 872)*

3. inconsistent **4.** consistent; independent **5.** $(3, -2)$ **6.** consistent; dependent **7.** b **8.** a **9.** $\begin{cases} 2(2) - (-1) = 5 \\ 5(2) + 2(-1) = 8 \end{cases}$

11. $\begin{cases} 3(2) - 4\left(\frac{1}{2}\right) = 4 \\ \frac{1}{2}(2) - 3\left(\frac{1}{2}\right) = -\frac{1}{2} \end{cases}$ **13.** $\begin{cases} 4 - 1 = 3 \\ \frac{1}{2}(4) + 1 = 3 \end{cases}$ **15.** $\begin{cases} 3(1) + 3(-1) + 2(2) = 4 \\ 1 - (-1) - 2 = 0 \\ 2(-1) - 3(2) = -8 \end{cases}$ **17.** $\begin{cases} 3(2) + 3(-2) + 2(2) = 4 \\ 2 - 3(-2) + 2 = 10 \\ 5(2) - 2(-2) - 3(2) = 8 \end{cases}$

19. $x = 6, y = 2; (6, 2)$

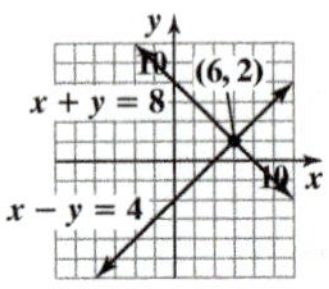

21. $x = 3, y = -6; (3, -6)$

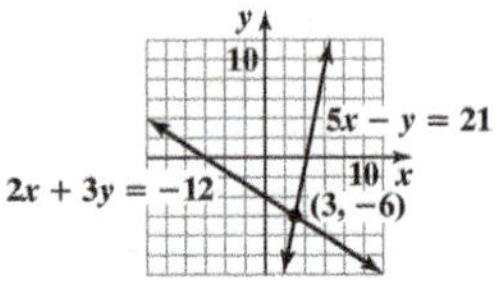

23. $x = 8, y = -4; (8, -4)$

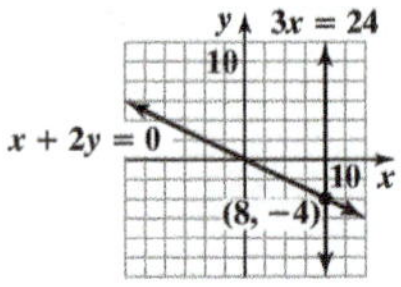

25. $x = \frac{1}{3}, y = -\frac{1}{6}; \left(\frac{1}{3}, -\frac{1}{6}\right)$

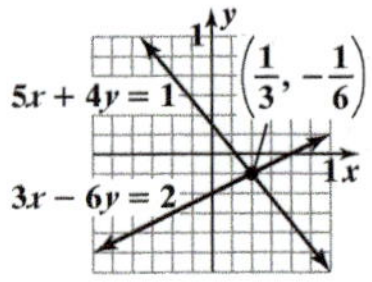

27. Inconsistent

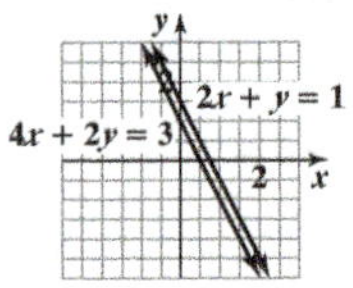

29. $x = \frac{3}{2}, y = 3; \left(\frac{3}{2}, 3\right)$

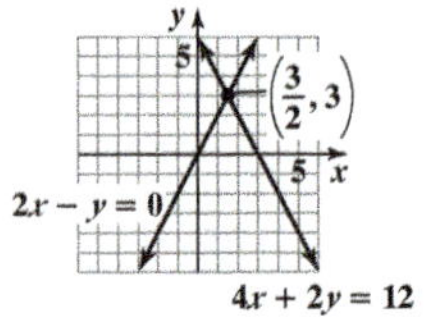

31. $\{(x, y) \mid x = 4 - 2y, y \text{ is any real number}\}$, or $\left\{(x, y) \,\middle|\, y = \frac{4 - x}{2}, x \text{ is any real number}\right\}$ **33.** $x = 1, y = 1; (1, 1)$ **35.** $x = \frac{3}{2}, y = 1; \left(\frac{3}{2}, 1\right)$

37. $x = 4, y = 3; (4, 3)$ **39.** $x = \frac{4}{3}, y = \frac{1}{5}; \left(\frac{4}{3}, \frac{1}{5}\right)$ **41.** $x = \frac{1}{5}, y = \frac{1}{3}; \left(\frac{1}{5}, \frac{1}{3}\right)$ **43.** $x = 8, y = 2, z = 0; (8, 2, 0)$

45. $x = 2, y = -1, z = 1; (2, -1, 1)$ **47.** Inconsistent **49.** $\{(x, y, z) \mid x = 5z - 2, y = 4z - 3; z \text{ is any real number}\}$ **51.** Inconsistent

53. $x = 1, y = 3, z = -2; (1, 3, -2)$ **55.** $x = -3, y = \frac{1}{2}, z = 1; \left(-3, \frac{1}{2}, 1\right)$ **57.** Length 30 ft; width 15 ft

59. 23 commercial launches and 69 noncommercial launches **61.** 22.5 lb **63.** Smartphone: \$325; tablet: \$640

65. Average wind speed 25 mph; average airspeed 175 mph **67.** 80 \$25 sets and 120 \$45 sets **69.** \$9.96

71. Mix 50 mg of first compound with 75 mg of second. **73.** $a = \frac{4}{3}, b = -\frac{5}{3}, c = 1$ **75.** $Y = 9000, r = 0.06$ **77.** $I_1 = \frac{10}{71}, I_2 = \frac{65}{71}, I_3 = \frac{55}{71}$

79. 100 orchestra, 210 main, and 190 balcony seats **81.** 1.5 chicken, 1 corn, 2 milk

83. If x = price of hamburgers, y = price of fries, and z = price of colas, then $x = 5.5 - z$, $y = \frac{41}{30} + \frac{1}{3}z$, $\$1.20 \le z \le \$1.80$. There is not sufficient information:

x	\$4.26	\$4.02	\$3.72
y	\$1.78	\$1.86	\$1.96
z	\$1.24	\$1.48	\$1.78

85. It will take Beth 30 hr, Bill 24 hr, and Edie 40 hr.

89.

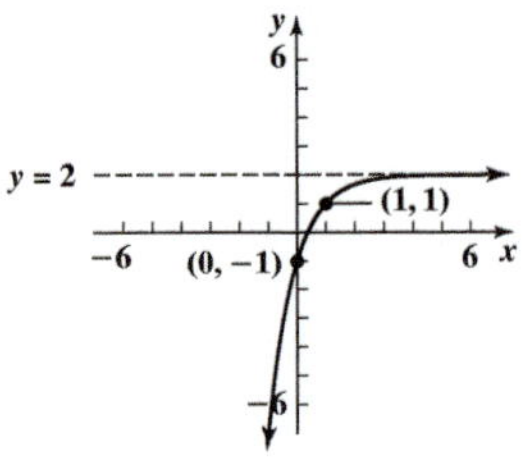

90. **(a)** $2(2x - 3)^3(x^3 + 5)(10x^3 - 9x^2 + 20)$ **(b)** $7(3x - 5)^{-1/2}(x + 3)^{-3/2}$

91. $\frac{\pi}{9}$ **92.** $2(\cos 150° + i \sin 150°)$

12.2 Assess Your Understanding *(page 888)*

1. matrix **2.** augmented **3.** third; fifth **4.** T **5.** b **6.** c **7.** $\left[\begin{array}{rr|r} 1 & -5 & 5 \\ 4 & 3 & 6 \end{array}\right]$ **9.** $\left[\begin{array}{rr|r} 2 & 3 & 6 \\ 4 & -6 & -2 \end{array}\right]$ **11.** $\left[\begin{array}{rr|r} 0.01 & -0.03 & 0.06 \\ 0.13 & 0.10 & 0.20 \end{array}\right]$

13. $\left[\begin{array}{rrr|r} 1 & -1 & 1 & 10 \\ 3 & 3 & 0 & 5 \\ 1 & 1 & 2 & 2 \end{array}\right]$ **15.** $\left[\begin{array}{rrr|r} 1 & 1 & -1 & 2 \\ 3 & -2 & 0 & 2 \\ 5 & 3 & -1 & 1 \end{array}\right]$ **17.** $\left[\begin{array}{rrr|r} 1 & -1 & -1 & 10 \\ 2 & 1 & 2 & -1 \\ -3 & 4 & 0 & 5 \\ 4 & -5 & 1 & 0 \end{array}\right]$ **19.** $\begin{cases} x - 3y = -2 & (1) \\ 2x - 5y = 5 & (2) \end{cases}$; $\left[\begin{array}{rr|r} 1 & -3 & -2 \\ 0 & 1 & 9 \end{array}\right]$

21. $\begin{cases} x - 3y + 4z = 3 & (1) \\ 3x - 5y + 6z = 6 & (2) \\ -5x + 3y + 4z = 6 & (3) \end{cases}$; $\left[\begin{array}{rrr|r} 1 & -3 & 4 & 3 \\ 0 & 4 & -6 & -3 \\ 0 & -12 & 24 & 21 \end{array}\right]$ **23.** $\begin{cases} x - 3y + 2z = -6 & (1) \\ 2x - 5y + 3z = -4 & (2) \\ -3x - 6y + 4z = 6 & (3) \end{cases}$; $\left[\begin{array}{rrr|r} 1 & -3 & 2 & -6 \\ 0 & 1 & -1 & 8 \\ 0 & -15 & 10 & -12 \end{array}\right]$

25. $\begin{cases} 5x - 3y + z = -2 & (1) \\ 2x - 5y + 6z = -2 & (2); \\ -4x + y + 4z = 6 & (3) \end{cases} \left[\begin{array}{ccc|c} 1 & 7 & -11 & 2 \\ 2 & -5 & 6 & -2 \\ 0 & -9 & 16 & 2 \end{array}\right]$

27. $\begin{cases} x = 5 \\ y = -1 \end{cases}$
Consistent; $x = 5, y = -1$ or $(5, -1)$

29. $\begin{cases} x = 1 \\ y = 2 \\ 0 = 3 \end{cases}$
Inconsistent

31. $\begin{cases} x + 2z = -1 \\ y - 4z = -2 \\ 0 = 0 \end{cases}$
Consistent:
$\begin{cases} x = -1 - 2z \\ y = -2 + 4z \\ z \text{ is any real number or} \end{cases}$
$\{(x, y, z) \mid x = -1 - 2z, y = -2 + 4z,$ z is any real number$\}$

33. $\begin{cases} x_1 = 1 \\ x_2 + x_4 = 2 \\ x_3 + 2x_4 = 3 \end{cases}$
Consistent:
$\begin{cases} x_1 = 1, x_2 = 2 - x_4 \\ x_3 = 3 - 2x_4 \\ x_4 \text{ is any real number or} \end{cases}$
$\{(x_1, x_2, x_3, x_4) \mid x_1 = 1, x_2 = 2 - x_4, x_3 = 3 - 2x_4,$ x_4 is any real number$\}$

35. $\begin{cases} x_1 + 4x_4 = 2 \\ x_2 + x_3 + 3x_4 = 3 \\ 0 = 0 \end{cases}$
Consistent:
$\begin{cases} x_1 = 2 - 4x_4 \\ x_2 = 3 - x_3 - 3x_4 \\ x_3, x_4 \text{ are any real numbers or} \end{cases}$
$\{(x_1, x_2, x_3, x_4) \mid x_1 = 2 - 4x_4, x_2 = 3 - x_3 - 3x_4, x_3, x_4$ are any real numbers$\}$

37. $\begin{cases} x_1 + x_4 = -2 \\ x_2 + 2x_4 = 2 \\ x_3 - x_4 = 0 \\ 0 = 0 \end{cases}$
Consistent:
$\begin{cases} x_1 = -2 - x_4 \\ x_2 = 2 - 2x_4 \\ x_3 = x_4 \\ x_4 \text{ is any real number or} \end{cases}$
$\{(x_1, x_2, x_3, x_4) \mid x_1 = -2 - x_4, x_2 = 2 - 2x_4, x_3 = x_4, x_4$ is any real number$\}$

39. $x = 6, y = 2; (6, 2)$ **41.** $x = \frac{1}{2}, y = \frac{3}{4}; \left(\frac{1}{2}, \frac{3}{4}\right)$ **43.** $x = 4 - 2y$, y is any real number; $\{(x, y) \mid x = 4 - 2y, y$ is any real number$\}$
45. $x = \frac{3}{2}, y = 1; \left(\frac{3}{2}, 1\right)$ **47.** $x = \frac{4}{3}, y = \frac{1}{5}; \left(\frac{4}{3}, \frac{1}{5}\right)$ **49.** $x = 8, y = 2, z = 0; (8, 2, 0)$ **51.** $x = 2, y = -1, z = 1; (2, -1, 1)$ **53.** Inconsistent
55. $x = 5z - 2, y = 4z - 3$, where z is any real number; $\{(x, y, z) \mid x = 5z - 2, y = 4z - 3, z$ is any real number$\}$ **57.** Inconsistent
59. $x = 1, y = 3, z = -2; (1, 3, -2)$ **61.** $x = -3, y = \frac{1}{2}, z = 1; \left(-3, \frac{1}{2}, 1\right)$ **63.** $x = \frac{1}{3}, y = \frac{2}{3}, z = 1; \left(\frac{1}{3}, \frac{2}{3}, 1\right)$
65. $x = 1, y = 2, z = 0, w = 1; (1, 2, 0, 1)$ **67.** $y = 0, z = 1 - x$, x is any real number; $\{(x, y, z) \mid y = 0, z = 1 - x, x$ is any real number$\}$
69. $x = 2, y = z - 3$, z is any real number; $\{(x, y, z) \mid x = 2, y = z - 3, z$ is any real number$\}$ **71.** $x = \frac{13}{9}, y = \frac{7}{18}, z = \frac{19}{18}; \left(\frac{13}{9}, \frac{7}{18}, \frac{19}{18}\right)$
73. $x = \frac{7}{5} - \frac{3}{5}z - \frac{2}{5}w, y = -\frac{8}{5} + \frac{7}{5}z + \frac{13}{5}w$, where z and w are any real numbers; $\left\{(x, y, z, w) \middle| x = \frac{7}{5} - \frac{3}{5}z - \frac{2}{5}w, y = -\frac{8}{5} + \frac{7}{5}z + \frac{13}{5}w,\right.$ z and w are any real numbers$\}$ **75.** $y = -2x^2 + x + 3$ **77.** $f(x) = 3x^3 - 4x^2 + 5$ **79.** 1.5 salmon steak, 2 baked eggs, 1 acorn squash
81. \$4000 in Treasury bills, \$4000 in Treasury bonds, \$2000 in corporate bonds **83.** 8 Deltas, 5 Betas, 10 Sigmas **85.** $I_1 = \frac{44}{23}, I_2 = 2, I_3 = \frac{16}{23}, I_4 = \frac{28}{23}$

87. (a)

Amount Invested At		
7%	9%	11%
0	10,000	10,000
1000	8000	11,000
2000	6000	12,000
3000	4000	13,000
4000	2000	14,000
5000	0	15,000

(b)

Amount Invested At		
7%	9%	11%
12,500	12,500	0
14,500	8500	2000
16,500	4500	4000
18,750	0	6250

(c) All the money invested at 7% provides \$2100, more than what is required.

89.

First Supplement	Second Supplement	Third Supplement
50 mg	75 mg	0 mg
36 mg	76 mg	8 mg
22 mg	77 mg	16 mg
8 mg	78 mg	24 mg

94. $\{x \mid -1 < x < 6\}$, or $(-1, 6)$
95.

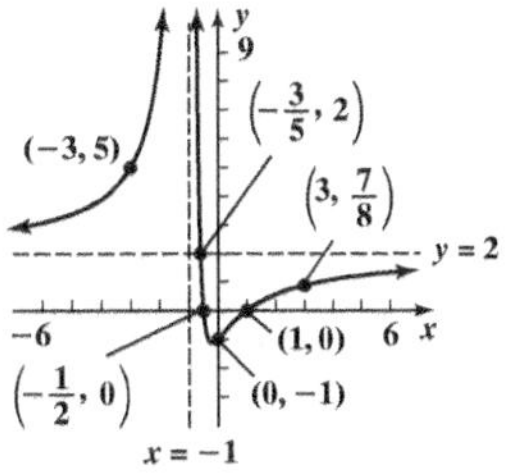

96. $\{x \mid x$ is any real nunber$\}$, or $(-\infty, \infty)$ **97.** 2.42

12.3 Assess Your Understanding *(page 901)*

1. $ad - bc$ **2.** $\begin{vmatrix} 5 & 3 \\ -3 & -4 \end{vmatrix}$ **3.** F **4.** F **5.** F **6.** a **7.** 22 **9.** -2 **11.** 10 **13.** -26 **15.** $x = 6, y = 2; (6, 2)$ **17.** $x = 3, y = 2; (3, 2)$

19. $x = 8, y = -4; (8, -4)$ **21.** $x = 4, y = -2; (4, -2)$ **23.** Not applicable **25.** $x = \frac{1}{2}, y = \frac{3}{4}; \left(\frac{1}{2}, \frac{3}{4}\right)$ **27.** $x = \frac{1}{10}, y = \frac{2}{5}; \left(\frac{1}{10}, \frac{2}{5}\right)$

29. $x = \frac{3}{2}, y = 1; \left(\frac{3}{2}, 1\right)$ **31.** $x = \frac{4}{3}, y = \frac{1}{5}; \left(\frac{4}{3}, \frac{1}{5}\right)$ **33.** $x = 1, y = 3, z = -2; (1, 3, -2)$ **35.** $x = -3, y = \frac{1}{2}, z = 1; \left(-3, \frac{1}{2}, 1\right)$

37. Not applicable **39.** $x = 0, y = 0, z = 0; (0, 0, 0)$ **41.** Not applicable **43.** -4 **45.** 12 **47.** 8 **49.** 8 **51.** -5 **53.** $\frac{13}{11}$ **55.** 0 or -9

57.
$$\begin{aligned}
(y_1 - y_2)x - (x_1 - x_2)y + (x_1y_2 - x_2y_1) &= 0 \\
(y_1 - y_2)x + (x_2 - x_1)y &= x_2y_1 - x_1y_2 \\
(x_2 - x_1)y - (x_2 - x_1)y_1 &= (y_2 - y_1)x + x_2y_1 - x_1y_2 - (x_2 - x_1)y_1 \\
(x_2 - x_1)(y - y_1) &= (y_2 - y_1)x - (y_2 - y_1)x_1 \\
y - y_1 &= \frac{y_2 - y_1}{x_2 - x_1}(x - x_1)
\end{aligned}$$

59. The triangle has an area of 5 square units. **61.** 50.5 square units **63.** $(x - 3)^2 + (y + 2)^2 = 25$

65. If $a = 0$, we have

$$by = s$$
$$cx + dy = t$$

Thus, $y = \frac{s}{b}$ and

$$x = \frac{t - dy}{c} = \frac{tb - ds}{bc}$$

Using Cramer's Rule, we get

$$x = \frac{sd - tb}{-bc} = \frac{tb - sd}{bc}$$
$$y = \frac{-sc}{-bc} = \frac{s}{b}$$

If $b = 0$, we have

$$ax = s$$
$$cx + dy = t$$

Since $D = ad \neq 0$, then $a \neq 0$ and $d \neq 0$.

Thus, $x = \frac{s}{a}$ and

$$y = \frac{t - cx}{d} = \frac{ta - cs}{ad}$$

Using Cramer's Rule, we get

$$x = \frac{sd}{ad} = \frac{s}{a}$$
$$y = \frac{ta - cs}{ad}$$

If $c = 0$, we have

$$ax + by = s$$
$$dy = t$$

Since $D = ad \neq 0$, then $a \neq 0$ and $d \neq 0$.

Thus, $y = \frac{t}{d}$ and

$$x = \frac{s - by}{a} = \frac{sd - bt}{ad}$$

Using Cramer's Rule, we get

$$x = \frac{sd - bt}{ad}$$
$$y = \frac{at}{ad} = \frac{t}{d}$$

If $d = 0$, we have

$$ax + by = s$$
$$cx = t$$

Since $D = -bc \neq 0$, then $b \neq 0$ and $c \neq 0$.

Thus, $x = \frac{t}{c}$ and

$$y = \frac{s - ax}{b} = \frac{sc - at}{bc}$$

Using Cramer's Rule, we get

$$x = \frac{-tb}{-bc} = \frac{t}{c}$$
$$y = \frac{at - sc}{-bc} = \frac{sc - at}{bc}$$

67.
$$\begin{aligned}
\begin{vmatrix} a_{11} & a_{12} & a_{13} \\ ka_{21} & ka_{22} & ka_{23} \\ a_{31} & a_{32} & a_{33} \end{vmatrix} &= -ka_{21}(a_{12}a_{33} - a_{32}a_{13}) + ka_{22}(a_{11}a_{33} - a_{31}a_{13}) - ka_{23}(a_{11}a_{32} - a_{31}a_{12}) \\
&= k[-a_{21}(a_{12}a_{33} - a_{32}a_{13}) + a_{22}(a_{11}a_{33} - a_{31}a_{13}) - a_{23}(a_{11}a_{32} - a_{31}a_{12})] = k\begin{vmatrix} a_{11} & a_{12} & a_{13} \\ a_{21} & a_{22} & a_{23} \\ a_{31} & a_{32} & a_{33} \end{vmatrix}
\end{aligned}$$

69.
$$\begin{aligned}
\begin{vmatrix} a_{11} + ka_{21} & a_{12} + ka_{22} & a_{13} + ka_{23} \\ a_{21} & a_{22} & a_{23} \\ a_{31} & a_{32} & a_{33} \end{vmatrix} &= (a_{11} + ka_{21})(a_{22}a_{33} - a_{32}a_{23}) - (a_{12} + ka_{22})(a_{21}a_{33} - a_{31}a_{23}) + (a_{13} + ka_{23})(a_{21}a_{32} - a_{31}a_{22}) \\
&= a_{11}a_{22}a_{33} - a_{11}a_{32}a_{23} + \cancel{ka_{21}a_{22}a_{33}} - \cancel{ka_{21}a_{32}a_{23}} - a_{12}a_{21}a_{33} + a_{12}a_{31}a_{23} \\
&\quad - \cancel{ka_{22}a_{21}a_{33}} + \cancel{ka_{22}a_{31}a_{23}} + a_{13}a_{21}a_{32} - a_{13}a_{31}a_{22} + \cancel{ka_{23}a_{21}a_{32}} - \cancel{ka_{23}a_{31}a_{22}} \\
&= a_{11}a_{22}a_{33} - a_{11}a_{32}a_{23} - a_{12}a_{21}a_{33} + a_{12}a_{31}a_{23} + a_{13}a_{21}a_{32} - a_{13}a_{31}a_{22} \\
&= a_{11}(a_{22}a_{33} - a_{32}a_{23}) - a_{12}(a_{21}a_{33} - a_{31}a_{23}) + a_{13}(a_{21}a_{32} - a_{31}a_{22}) \\
&= \begin{vmatrix} a_{11} & a_{12} & a_{13} \\ a_{21} & a_{22} & a_{23} \\ a_{31} & a_{32} & a_{33} \end{vmatrix}
\end{aligned}$$

70. $\mathbf{v} = 9\mathbf{i} - 4\mathbf{j}; \sqrt{97}$ **71.** $\pm\frac{1}{2}, \pm\frac{5}{2}, \pm 1, \pm 2, \pm 5, \pm 10$

72.

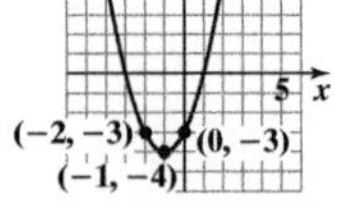

73. 0

Historical Problems *(page 916)*

1. (a) $2 - 5i \longleftrightarrow \begin{bmatrix} 2 & -5 \\ 5 & 2 \end{bmatrix}, 1 + 3i \longleftrightarrow \begin{bmatrix} 1 & 3 \\ -3 & 1 \end{bmatrix}$ **(b)** $\begin{bmatrix} 2 & -5 \\ 5 & 2 \end{bmatrix}\begin{bmatrix} 1 & 3 \\ -3 & 1 \end{bmatrix} = \begin{bmatrix} 17 & 1 \\ -1 & 17 \end{bmatrix}$ **(c)** $17 + i$ **(d)** $17 + i$

2. $\begin{bmatrix} a & b \\ -b & a \end{bmatrix}\begin{bmatrix} a & -b \\ b & a \end{bmatrix} = \begin{bmatrix} a^2+b^2 & 0 \\ 0 & b^2+a^2 \end{bmatrix}$; the product is a real number.

3. (a) $x = k(ar + bs) + l(cr + ds) = r(ka + lc) + s(kb + ld)$
$y = m(ar + bs) + n(cr + ds) = r(ma + nc) + s(mb + nd)$ **(b)** $A = \begin{bmatrix} ka + lc & kb + ld \\ ma + nc & mb + nd \end{bmatrix}$

12.4 Assess Your Understanding *(page 916)*

1. square **2.** T **3.** F **4.** inverse **5.** T **6.** $A^{-1}B$ **7.** a **8.** d **9.** $\begin{bmatrix} 4 & 4 & -5 \\ -1 & 5 & 4 \end{bmatrix}$ **11.** $\begin{bmatrix} 0 & 12 & -20 \\ 4 & 8 & 24 \end{bmatrix}$ **13.** $\begin{bmatrix} -8 & 7 & -15 \\ 7 & 0 & 22 \end{bmatrix}$

15. $\begin{bmatrix} 28 & -9 \\ 4 & 23 \end{bmatrix}$ **17.** Not defined **19.** $\begin{bmatrix} 1 & 14 & -14 \\ 2 & 22 & -18 \\ 3 & 0 & 28 \end{bmatrix}$ **21.** $\begin{bmatrix} 15 & 21 & -16 \\ 22 & 34 & -22 \\ -11 & 7 & 22 \end{bmatrix}$ **23.** $\begin{bmatrix} 25 & -9 \\ 4 & 20 \end{bmatrix}$ **25.** $\begin{bmatrix} -13 & 7 & -12 \\ -18 & 10 & -14 \\ 17 & -7 & 34 \end{bmatrix}$ **27.** $\begin{bmatrix} -2 & 4 & 2 & 8 \\ 2 & 1 & 4 & 6 \end{bmatrix}$

29. $\begin{bmatrix} 5 & 14 \\ 9 & 16 \end{bmatrix}$ **31.** Not defined **33.** $\begin{bmatrix} 9 & 2 \\ 34 & 13 \\ 47 & 20 \end{bmatrix}$ **35.** $\begin{bmatrix} 1 & -1 \\ -1 & 2 \end{bmatrix}$ **37.** $\begin{bmatrix} 1 & -\frac{5}{2} \\ -1 & 3 \end{bmatrix}$ **39.** $\begin{bmatrix} 1 & -\frac{1}{a} \\ -1 & \frac{2}{a} \end{bmatrix}$ **41.** $\begin{bmatrix} 3 & -3 & 1 \\ -2 & 2 & -1 \\ -4 & 5 & -2 \end{bmatrix}$ **43.** $\begin{bmatrix} -\frac{5}{7} & \frac{1}{7} & \frac{3}{7} \\ \frac{9}{7} & \frac{1}{7} & -\frac{4}{7} \\ \frac{3}{7} & -\frac{2}{7} & \frac{1}{7} \end{bmatrix}$

45. $x = 3, y = 2; (3, 2)$ **47.** $x = -5, y = 10; (-5, 10)$ **49.** $x = 2, y = -1; (2, -1)$ **51.** $x = \frac{1}{2}, y = 2; \left(\frac{1}{2}, 2\right)$ **53.** $x = -2, y = 1; (-2, 1)$

55. $x = \frac{2}{a}, y = \frac{3}{a}; \left(\frac{2}{a}, \frac{3}{a}\right)$ **57.** $x = -2, y = 3, z = 5; (-2, 3, 5)$ **59.** $x = \frac{1}{2}, y = -\frac{1}{2}, z = 1; \left(\frac{1}{2}, -\frac{1}{2}, 1\right)$

61. $x = -\frac{34}{7}, y = \frac{85}{7}, z = \frac{12}{7}; \left(-\frac{34}{7}, \frac{85}{7}, \frac{12}{7}\right)$ **63.** $x = \frac{1}{3}, y = 1, z = \frac{2}{3}; \left(\frac{1}{3}, 1, \frac{2}{3}\right)$ **65.** $\left[\begin{array}{cc|cc} 4 & 2 & 1 & 0 \\ 2 & 1 & 0 & 1 \end{array}\right] \to \left[\begin{array}{cc|cc} 1 & \frac{1}{2} & \frac{1}{4} & 0 \\ 2 & 1 & 0 & 1 \end{array}\right] \to \left[\begin{array}{cc|cc} 1 & \frac{1}{2} & \frac{1}{4} & 0 \\ 0 & 0 & -\frac{1}{2} & 1 \end{array}\right]$

67. $\left[\begin{array}{cc|cc} 15 & 3 & 1 & 0 \\ 10 & 2 & 0 & 1 \end{array}\right] \to \left[\begin{array}{cc|cc} 1 & \frac{1}{5} & \frac{1}{15} & 0 \\ 10 & 2 & 0 & 1 \end{array}\right] \to \left[\begin{array}{cc|cc} 1 & \frac{1}{5} & \frac{1}{15} & 0 \\ 0 & 0 & -\frac{2}{3} & 1 \end{array}\right]$

69. $\left[\begin{array}{ccc|ccc} -3 & 1 & -1 & 1 & 0 & 0 \\ 1 & -4 & -7 & 0 & 1 & 0 \\ 1 & 2 & 5 & 0 & 0 & 1 \end{array}\right] \to \left[\begin{array}{ccc|ccc} 1 & 2 & 5 & 0 & 0 & 1 \\ 1 & -4 & -7 & 0 & 1 & 0 \\ -3 & 1 & -1 & 1 & 0 & 0 \end{array}\right] \to \left[\begin{array}{ccc|ccc} 1 & 2 & 5 & 0 & 0 & 1 \\ 0 & -6 & -12 & 0 & 1 & -1 \\ 0 & 7 & 14 & 1 & 0 & 3 \end{array}\right]$

$\to \left[\begin{array}{ccc|ccc} 1 & 2 & 5 & 0 & 0 & 1 \\ 0 & 1 & 2 & 0 & -\frac{1}{6} & \frac{1}{6} \\ 0 & 1 & 2 & \frac{1}{7} & 0 & \frac{3}{7} \end{array}\right] \to \left[\begin{array}{ccc|ccc} 1 & 2 & 5 & 0 & 0 & 1 \\ 0 & 1 & 2 & 0 & -\frac{1}{6} & \frac{1}{6} \\ 0 & 0 & 0 & \frac{1}{7} & \frac{1}{6} & \frac{11}{42} \end{array}\right]$

71. $\begin{bmatrix} 0.01 & 0.05 & -0.01 \\ 0.01 & -0.02 & 0.01 \\ -0.02 & 0.01 & 0.03 \end{bmatrix}$ **73.** $\begin{bmatrix} 0.02 & -0.04 & -0.01 & 0.01 \\ -0.02 & 0.05 & 0.03 & -0.03 \\ 0.02 & 0.01 & -0.04 & 0.00 \\ -0.02 & 0.06 & 0.07 & 0.06 \end{bmatrix}$

75. $x = 4.57, y = -6.44, z = -24.07; (4.57, -6.44, -24.07)$ **77.** $x = -1.19, y = 2.46, z = 8.27; (-1.19, 2.46, 8.27)$

79. $x = -5, y = 7; (-5, 7)$ **81.** $x = -4, y = 2, z = \frac{5}{2}; \left(-4, 2, \frac{5}{2}\right)$

83. Inconsistent; $\varnothing$ **85.** $x = -\frac{1}{5}z + \frac{1}{5}, y = \frac{1}{5}z - \frac{6}{5}$, where z is any real number; $\left\{(x, y, z) \,\middle|\, x = -\frac{1}{5}z + \frac{1}{5}, y = \frac{1}{5}z - \frac{6}{5}, z \text{ is any real number}\right\}$

87. (a) $A = \begin{bmatrix} 6 & 9 \\ 3 & 12 \end{bmatrix}; B = \begin{bmatrix} 128.00 \\ 341.60 \end{bmatrix}$ **(b)** $AB = \begin{bmatrix} 3842.40 \\ 4483.20 \end{bmatrix}$; Nikki's total tuition is \$3842.40, and Joe's total tuition is \$4483.20.

89. (a) $\begin{bmatrix} 500 & 350 & 400 \\ 700 & 500 & 850 \end{bmatrix}; \begin{bmatrix} 500 & 700 \\ 350 & 500 \\ 400 & 850 \end{bmatrix}$ **(b)** $\begin{bmatrix} 15 \\ 8 \\ 3 \end{bmatrix}$ **(c)** $\begin{bmatrix} 11{,}500 \\ 17{,}050 \end{bmatrix}$ **(d)** $[0.10 \quad 0.05]$ **(e)** \$2002.50

91. (a) $K^{-1} = \begin{bmatrix} 1 & 0 & -1 \\ -1 & 1 & 1 \\ 0 & -1 & 1 \end{bmatrix}$ **(b)** $M = \begin{bmatrix} 13 & 1 & 20 \\ 8 & 9 & 19 \\ 6 & 21 & 14 \end{bmatrix}$ **(c)** Math is fun.

93. If $D = ad - bc \neq 0$, then $a \neq 0$ and $d \neq 0$, or $b \neq 0$ and $c \neq 0$. Assuming the former,

$$\left[\begin{array}{cc|cc} a & b & 1 & 0 \\ c & d & 0 & 1 \end{array}\right] \rightarrow \left[\begin{array}{cc|cc} 1 & \frac{b}{a} & \frac{1}{a} & 0 \\ c & d & 0 & 1 \end{array}\right] \rightarrow \left[\begin{array}{cc|cc} 1 & \frac{b}{a} & \frac{1}{a} & 0 \\ 0 & \frac{D}{a} & -\frac{c}{a} & 1 \end{array}\right] \rightarrow \left[\begin{array}{cc|cc} 1 & \frac{b}{a} & \frac{1}{a} & 0 \\ 0 & 1 & -\frac{c}{D} & \frac{a}{D} \end{array}\right] \rightarrow \left[\begin{array}{cc|cc} 1 & 0 & \frac{d}{D} & -\frac{b}{D} \\ 0 & 1 & -\frac{c}{D} & \frac{a}{D} \end{array}\right]$$

$$R_1 = \frac{1}{a} r_1 \qquad R_2 = -c r_1 + r_2 \qquad R_2 = \frac{a}{D} r_2 \qquad R_1 = -\frac{b}{a} r_2 + r_1$$

95. (a) $B_3 = A + A^2 + A^3 = \begin{bmatrix} 2 & 4 & 5 & 2 & 3 \\ 5 & 3 & 2 & 5 & 4 \\ 4 & 2 & 2 & 4 & 2 \\ 2 & 2 & 3 & 2 & 3 \\ 1 & 3 & 2 & 1 & 2 \end{bmatrix}$; Yes, all pages can reach every other page within 3 clicks. **(b)** Page 3

97. (a) $(3 - 2\sqrt{3}, 3\sqrt{3} + 2)$ **(b)** $R^{-1} = \begin{bmatrix} \frac{1}{2} & \frac{\sqrt{3}}{2} & 0 \\ -\frac{\sqrt{3}}{2} & \frac{1}{2} & 0 \\ 0 & 0 & 1 \end{bmatrix}$; This is the rotation matrix needed to get the translated coordinates back to the original coordinates.

102. $x^6 - 4x^5 - 3x^4 + 18x^3$ **103.** $\frac{\sqrt{u^2 - 1}}{|u|}$ **104.** $\{0, 3\}$ **105.** $\mathbf{v} \cdot \mathbf{w} = -5; \theta = 180°$

12.5 Assess Your Understanding *(page 926)*

5. Proper **7.** Improper; $1 + \frac{9}{x^2 - 4}$ **9.** Improper; $5x + \frac{22x - 1}{x^2 - 4}$ **11.** Improper; $1 + \frac{-2(x - 6)}{(x + 4)(x - 3)}$ **13.** $\frac{-4}{x} + \frac{4}{x - 1}$ **15.** $\frac{1}{x} + \frac{-x}{x^2 + 1}$

17. $\frac{-1}{x - 1} + \frac{2}{x - 2}$ **19.** $\frac{\frac{1}{4}}{x + 1} + \frac{\frac{3}{4}}{x - 1} + \frac{\frac{1}{2}}{(x - 1)^2}$ **21.** $\frac{\frac{1}{12}}{x - 2} + \frac{-\frac{1}{12}(x + 4)}{x^2 + 2x + 4}$ **23.** $\frac{\frac{1}{4}}{x - 1} + \frac{\frac{1}{4}}{(x - 1)^2} + \frac{-\frac{1}{4}}{x + 1} + \frac{\frac{1}{4}}{(x + 1)^2}$

25. $\frac{-5}{x + 2} + \frac{5}{x + 1} + \frac{-4}{(x + 1)^2}$ **27.** $\frac{\frac{1}{4}}{x} + \frac{1}{x^2} + \frac{-\frac{1}{4}(x + 4)}{x^2 + 4}$ **29.** $\frac{\frac{2}{3}}{x + 1} + \frac{\frac{1}{3}(x + 1)}{x^2 + 2x + 4}$ **31.** $\frac{\frac{2}{7}}{3x - 2} + \frac{\frac{1}{7}}{2x + 1}$ **33.** $\frac{\frac{3}{4}}{x + 3} + \frac{\frac{1}{4}}{x - 1}$

35. $\frac{1}{x^2 + 4} + \frac{2x - 1}{(x^2 + 4)^2}$ **37.** $\frac{-1}{x} + \frac{2}{x - 3} + \frac{-1}{x + 1}$ **39.** $\frac{4}{x - 2} + \frac{-3}{x - 1} + \frac{-1}{(x - 1)^2}$ **41.** $\frac{x}{(x^2 + 16)^2} + \frac{-16x}{(x^2 + 16)^3}$

43. $\frac{-\frac{8}{7}}{2x + 1} + \frac{\frac{4}{7}}{x - 3}$ **45.** $\frac{-\frac{2}{9}}{x} + \frac{\frac{1}{3}}{x^2} + \frac{\frac{1}{6}}{x - 3} + \frac{\frac{1}{18}}{x + 3}$ **47.** $x - 2 + \frac{10x - 11}{x^2 + 3x - 4}; \frac{\frac{51}{5}}{x + 4} + \frac{-\frac{1}{5}}{x - 1}; x - 2 + \frac{\frac{51}{5}}{x + 4} - \frac{\frac{1}{5}}{x - 1}$

49. $x - \frac{x}{x^2 + 1}$ **51.** $x^2 - 4x + 7 + \frac{-11x - 32}{x^2 + 4x + 4}; \frac{-11}{x + 2} + \frac{-10}{(x + 2)^2}; x^2 - 4x + 7 - \frac{11}{x + 2} - \frac{10}{(x + 2)^2}$

53. $x + 1 + \frac{2x^3 + x^2 - x + 1}{x^4 - 2x^2 + 1}; \frac{1}{x + 1} + \frac{\frac{1}{4}}{(x + 1)^2} + \frac{1}{x - 1} + \frac{\frac{3}{4}}{(x - 1)^2}; x + 1 + \frac{1}{x + 1} + \frac{\frac{1}{4}}{(x + 1)^2} + \frac{1}{x - 1} + \frac{\frac{3}{4}}{(x - 1)^2}$

55. 3.85 years **56.** -2 **57.** 1

58. $\left(\frac{\sqrt{2}}{2}, \frac{\sqrt{2}}{2}\right)$;

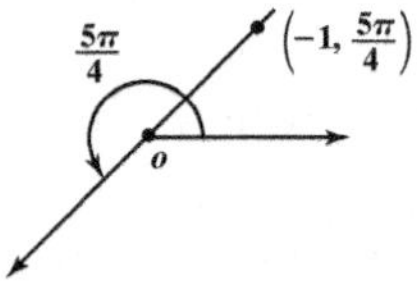

Historical Problem *(page 934)*

$x = 6$ units, $y = 8$ units

12.6 Assess Your Understanding *(page 934)*

5.

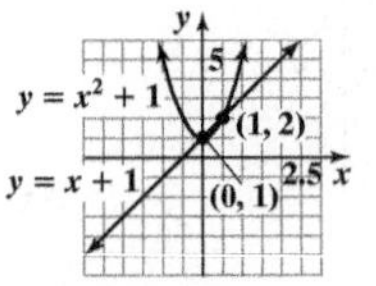

7.

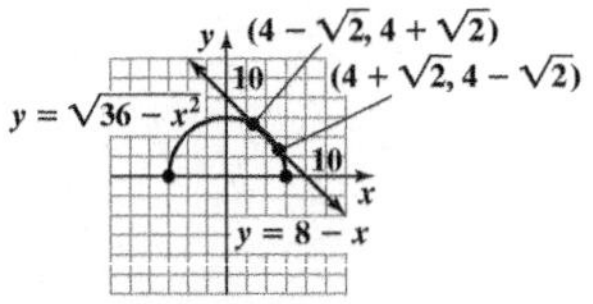

9.

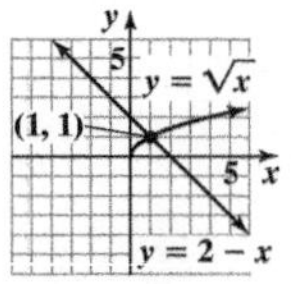

11.

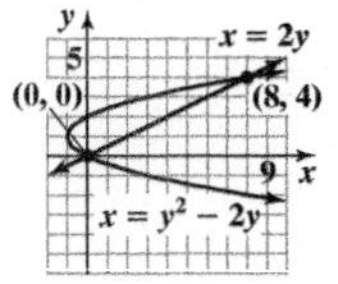

13.

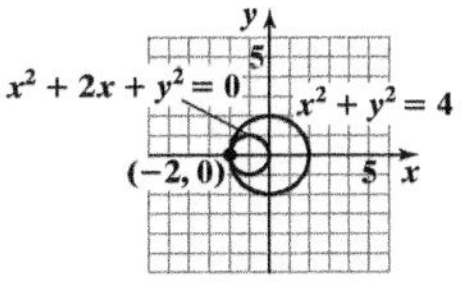

15.

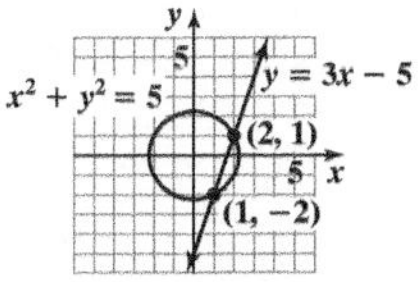

17.

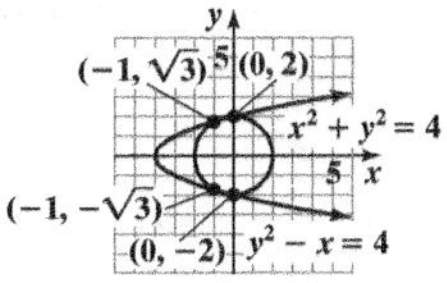

19.

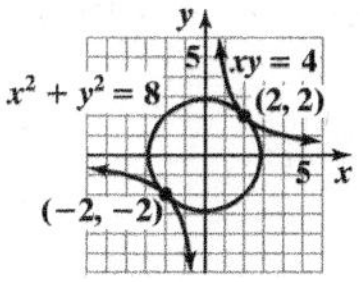

21. No points of intersection

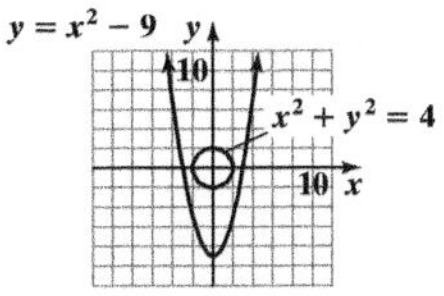

23.

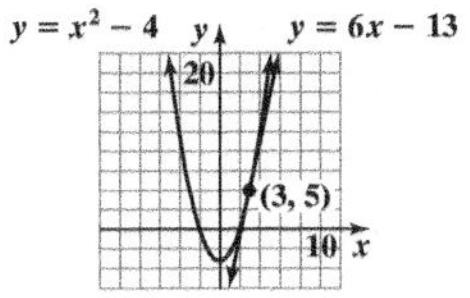

25. $x = 1, y = 4; x = -1, y = -4; x = 2\sqrt{2}, y = \sqrt{2};\ x = -2\sqrt{2}, y = -\sqrt{2}$ or $(1, 4), (-1, -4), (2\sqrt{2}, \sqrt{2}), (-2\sqrt{2}, -\sqrt{2})$

27. $x = 0, y = 1; x = -\frac{2}{3}, y = -\frac{1}{3}$ or $(0, 1), \left(-\frac{2}{3}, -\frac{1}{3}\right)$ **29.** $x = 0, y = -1; x = \frac{5}{2}, y = -\frac{7}{2}$ or $(0, -1), \left(\frac{5}{2}, -\frac{7}{2}\right)$

31. $x = 2, y = \frac{1}{3}; x = \frac{1}{2}, y = \frac{4}{3}$ or $\left(2, \frac{1}{3}\right), \left(\frac{1}{2}, \frac{4}{3}\right)$ **33.** $x = 3, y = 2; x = 3, y = -2; x = -3, y = 2; x = -3, y = -2$ or $(3, 2), (3, -2), (-3, 2), (-3, -2)$ **35.** $x = \frac{1}{2}, y = \frac{3}{2}; x = \frac{1}{2}, y = -\frac{3}{2}; x = -\frac{1}{2}, y = \frac{3}{2}; x = -\frac{1}{2}, y = -\frac{3}{2}$ or $\left(\frac{1}{2}, \frac{3}{2}\right), \left(\frac{1}{2}, -\frac{3}{2}\right), \left(-\frac{1}{2}, \frac{3}{2}\right), \left(-\frac{1}{2}, -\frac{3}{2}\right)$

37. $x = \sqrt{2}, y = 2\sqrt{2}; x = -\sqrt{2}, y = -2\sqrt{2}$ or $(\sqrt{2}, 2\sqrt{2}), (-\sqrt{2}, -2\sqrt{2})$ **39.** No real solution exists.

41. $x = \frac{8}{3}, y = \frac{2\sqrt{10}}{3}; x = -\frac{8}{3}, y = \frac{2\sqrt{10}}{3}; x = \frac{8}{3}, y = -\frac{2\sqrt{10}}{3}; x = -\frac{8}{3}, y = -\frac{2\sqrt{10}}{3}$ or $\left(\frac{8}{3}, \frac{2\sqrt{10}}{3}\right), \left(-\frac{8}{3}, \frac{2\sqrt{10}}{3}\right), \left(\frac{8}{3}, -\frac{2\sqrt{10}}{3}\right), \left(-\frac{8}{3}, -\frac{2\sqrt{10}}{3}\right)$

43. $x = 1, y = \frac{1}{2}; x = -1, y = \frac{1}{2}; x = 1, y = -\frac{1}{2}; x = -1, y = -\frac{1}{2}$ or $\left(1, \frac{1}{2}\right), \left(-1, \frac{1}{2}\right), \left(1, -\frac{1}{2}\right), \left(-1, -\frac{1}{2}\right)$ **45.** No real solution exists.

47. $x = \sqrt{3}, y = \sqrt{3}; x = -\sqrt{3}, y = -\sqrt{3}; x = 2, y = 1; x = -2, y = -1$ or $(\sqrt{3}, \sqrt{3}), (-\sqrt{3}, -\sqrt{3}), (2, 1), (-2, -1)$

49. $x = 0, y = -2; x = 0, y = 1;\ x = 2, y = -1$ or $(0, -2), (0, 1), (2, -1)$ **51.** $x = 2, y = 8$ or $(2, 8)$ **53.** $x = 81, y = 3$ or $(81, 3)$

55.

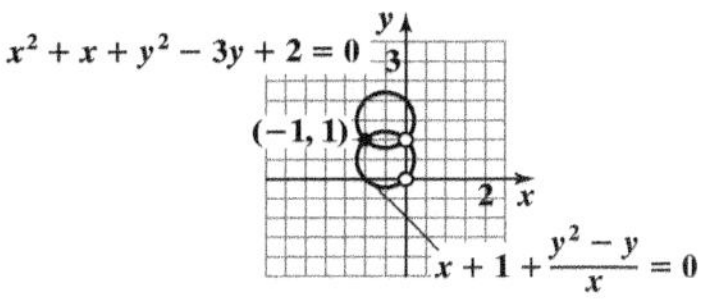

57. $x = 0.48, y = 0.62$ **59.** $x = -1.65, y = -0.89$

61. $x = 0.58, y = 1.86; x = 1.81, y = 1.05; x = 0.58, y = -1.86; x = 1.81, y = -1.05$

63. $x = 2.35, y = 0.85$

65.

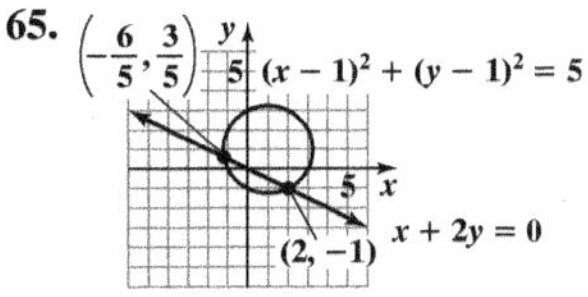

67.

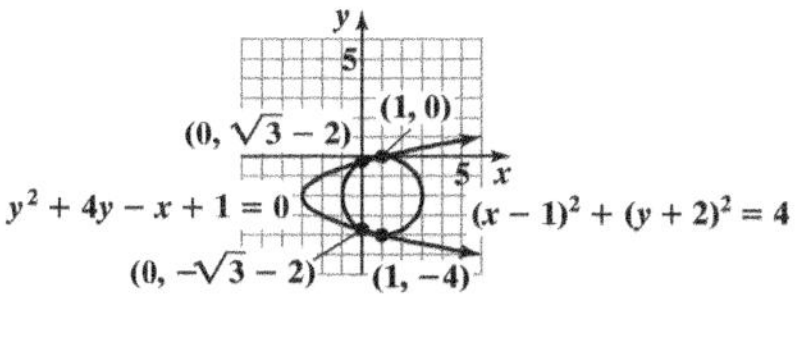

69.

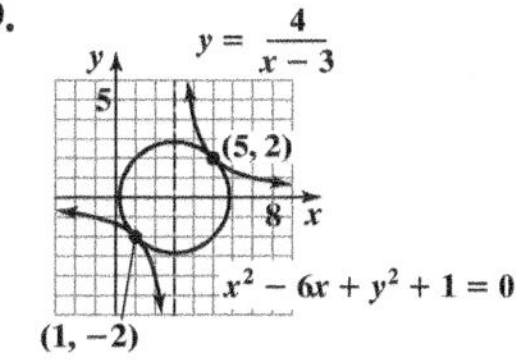

71. 3 and 1; −1 and −3 **73.** 2 and 2; −2 and −2 **75.** $\frac{1}{2}$ and $\frac{1}{3}$ **77.** 5 **79.** 5 in. by 3 in. **81.** 2 cm and 4 cm **83.** Tortoise: 7 m/h, hare: $7\frac{1}{2}$ m/h

85. 12 cm by 18 cm **87.** $x = 60$ ft; $y = 30$ ft **89.** $l = \frac{P + \sqrt{P^2 - 16A}}{4}$; $w = \frac{P - \sqrt{P^2 - 16A}}{4}$ **91.** $y = 4x - 4$ **93.** $y = 2x + 1$

95. $y = -\frac{1}{3}x + \frac{7}{3}$ **97.** $y = 2x - 3$ **99.** $r_1 = \frac{-b + \sqrt{b^2 - 4ac}}{2a}$; $r_2 = \frac{-b - \sqrt{b^2 - 4ac}}{2a}$ **101. (a)** 4.274 ft by 4.274 ft or 0.093 ft by 0.093 ft

102. $\left\{\frac{-3 - \sqrt{65}}{7}, \frac{-3 + \sqrt{65}}{7}\right\}$ **103.** $y = -\frac{2}{5}x - 3$ **104.** $\sin\theta = -\frac{7}{25}$; $\cos\theta = -\frac{24}{25}$; $\tan\theta = \frac{7}{24}$; $\csc\theta = -\frac{25}{7}$; $\sec\theta = -\frac{25}{24}$ **105.** $\approx 15.8°$

12.7 Assess Your Understanding *(page 944)*

7. dashes; solid **8.** half-planes **9.** F **10.** unbounded

11.

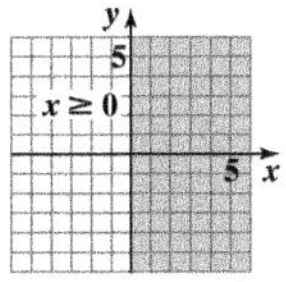

13.

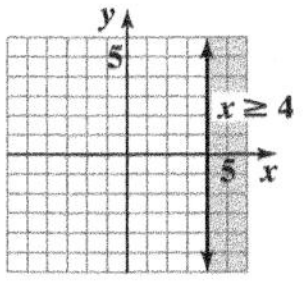

15.

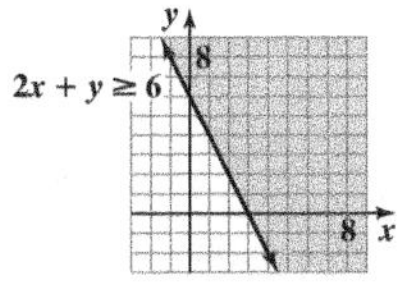

17.

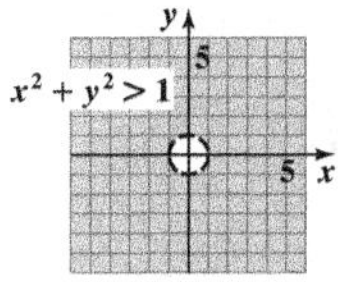

19.

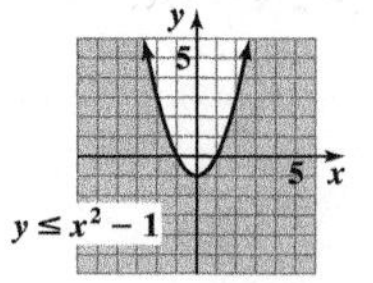

21.

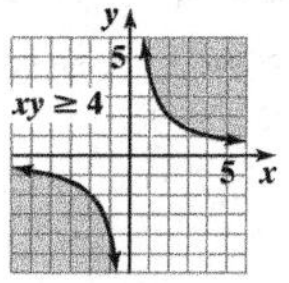

23.

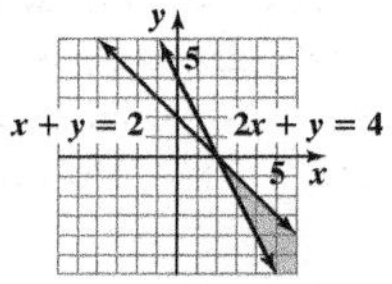

25.

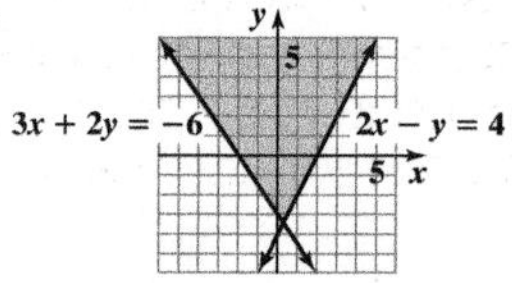

27.

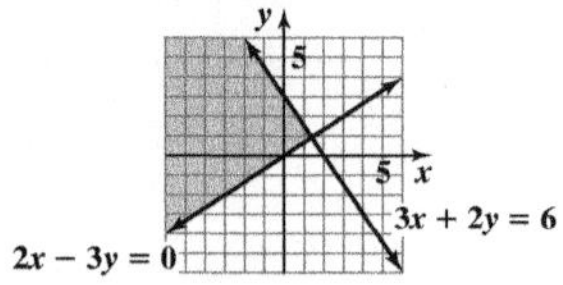

29.

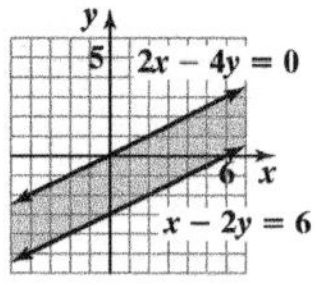

31.

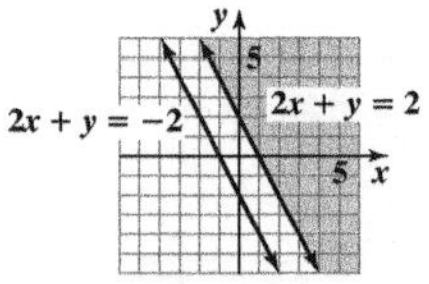

33. No solution

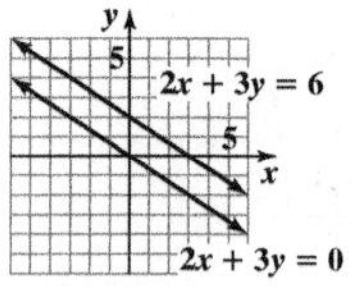

35.

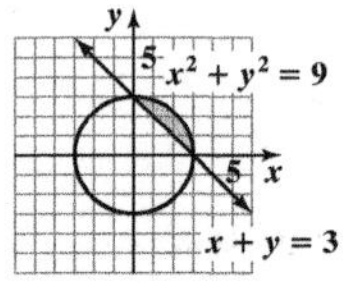

37.

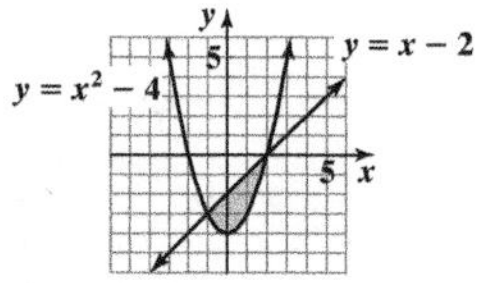

39.

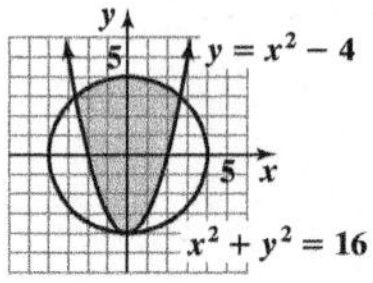

41.

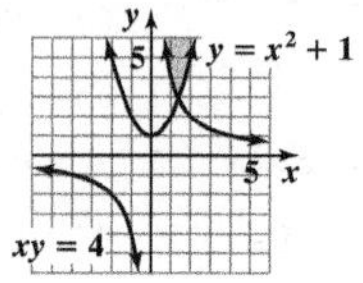

43. Bounded

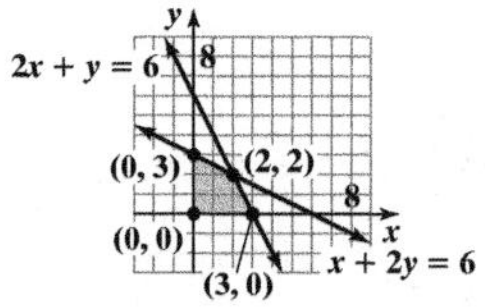

45. Unbounded

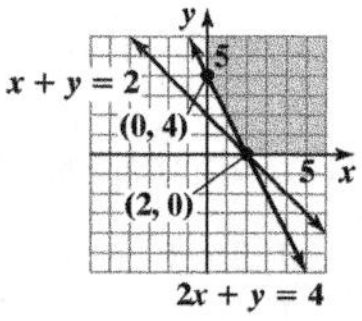

47. Bounded

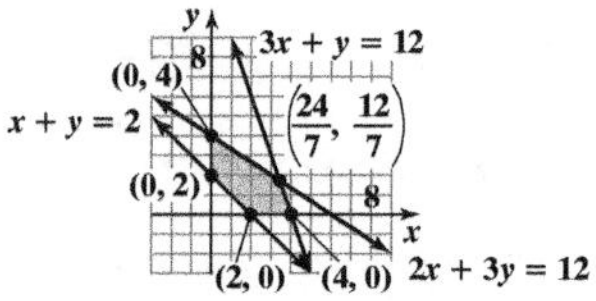

49. Bounded

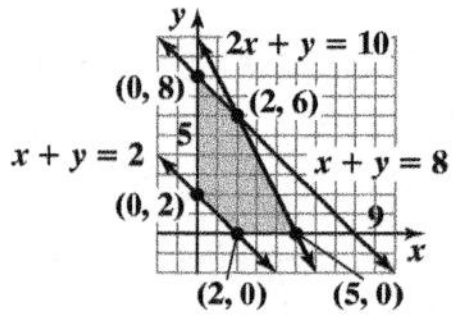

51. Bounded

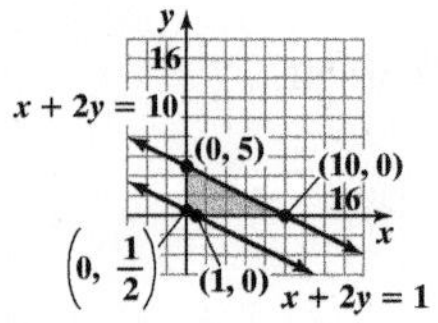

53. $\begin{cases} x \le 4 \\ x + y \le 6 \\ x \ge 0 \\ y \ge 0 \end{cases}$

55. $\begin{cases} x \le 20 \\ y \ge 15 \\ x + y \le 50 \\ x - y \le 0 \\ x \ge 0 \end{cases}$

57. (a) $\begin{cases} x + y \le 50{,}000 \\ x \ge 35{,}000 \\ y \le 10{,}000 \\ y \ge 0 \end{cases}$

(b)

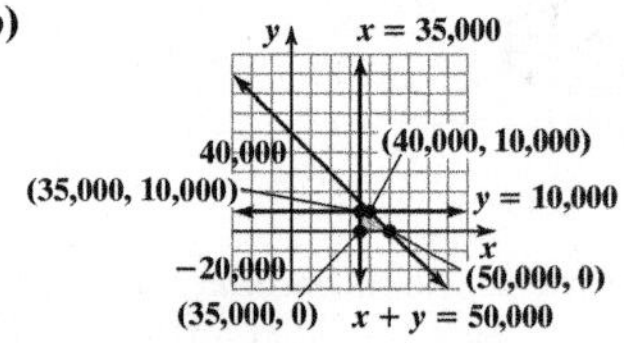

59. (a) $\begin{cases} x \ge 0 \\ y \ge 0 \\ x + 2y \le 300 \\ 3x + 2y \le 480 \end{cases}$

(b)

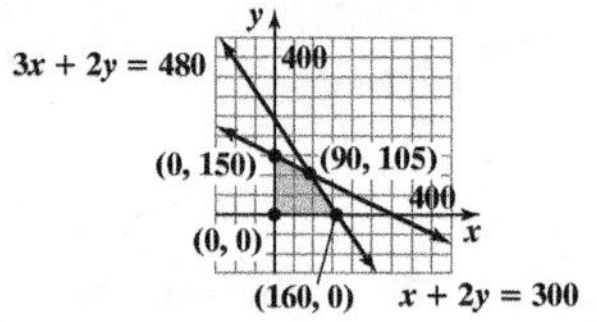

61. (a) $\begin{cases} 3x + 2y \le 160 \\ 2x + 3y \le 150 \\ x \ge 0 \\ y \ge 0 \end{cases}$

(b)

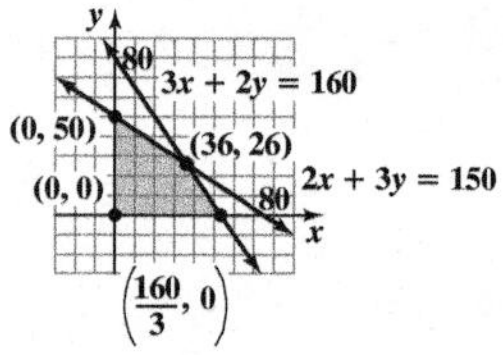

62. $\{-1 - 2i, -1 + 2i\}$

63. $x^2 + \left(y - \frac{1}{6}\right)^2 = \frac{1}{36}$; circle with radius $\frac{1}{6}$ and center $\left(0, \frac{1}{6}\right)$;

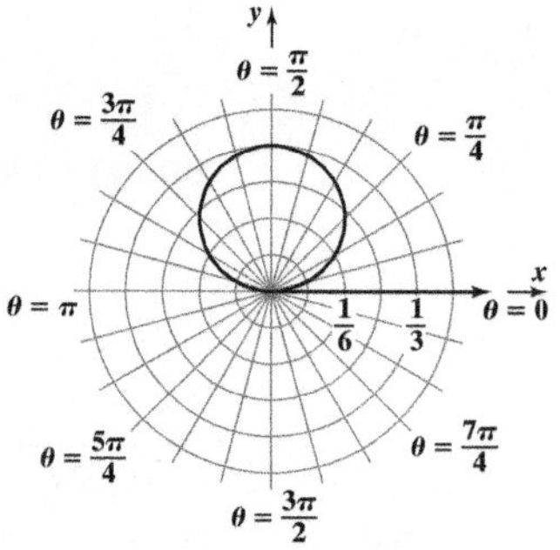

64. $f(-1) = -5$; $f(2) = 28$

65. $\left\{0, \frac{2\pi}{3}, \frac{4\pi}{3}\right\}$

12.8 Assess Your Understanding *(page 952)*

1. objective function **2.** T **3.** Maximum value is 11; minimum value is 3. **5.** Maximum value is 65; minimum value is 4.
7. Maximum value is 67; minimum value is 20. **9.** The maximum value of z is 12, and it occurs at the point $(6, 0)$.
11. The minimum value of z is 4, and it occurs at the point $(2, 0)$. **13.** The maximum value of z is 20, and it occurs at the point $(0, 4)$.
15. The minimum value of z is 8, and it occurs at the point $(0, 2)$. **17.** The maximum value of z is 50, and it occurs at the point $(10, 0)$.
19. Produce 8 downhill and 24 cross-country; \$1760; \$1920 which is the profit when producing 16 downhill and 16 cross-country.

21. Rent 15 rectangular tables and 16 round tables for a minimum cost of \$1252. **23. (a)** \$10,000 in a junk bond and \$10,000 in Treasury bills **(b)** \$12,000 in a junk bond and \$8000 in Treasury bills **25.** 100 lb of ground beef should be mixed with 50 lb of pork. **27.** Manufacture 10 racing skates and 15 figure skates. **29.** Order 2 metal samples and 4 plastic samples; \$34 **31. (a)** Configure with 10 first-class seats and 120 coach seats. **(b)** Configure with 15 first-class seats and 120 coach seats.

33. $\left\{-\frac{1}{32}, 1\right\}$

34.

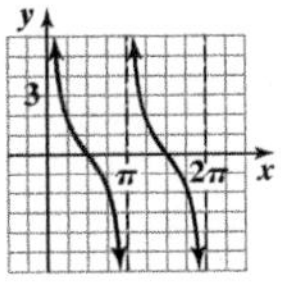

Domain: $\{x \mid x \neq k\pi, k \text{ is an integer}\}$; range: $(-\infty, \infty)$

35. 89.1 years **36.** $y = 3x + 7$

Review Exercises *(page 956)*

1. $x = 2, y = -1$ or $(2, -1)$ **2.** $x = 2, y = \frac{1}{2}$ or $\left(2, \frac{1}{2}\right)$ **3.** $x = 2, y = -1$ or $(2, -1)$ **4.** $x = \frac{11}{5}, y = -\frac{3}{5}$ or $\left(\frac{11}{5}, -\frac{3}{5}\right)$ **5.** Inconsistent

6. $x = 2, y = 3$ or $(2, 3)$ **7.** $y = -\frac{2}{5}x + 2$, where x is any real number, or $\left\{(x, y) \mid y = -\frac{2}{5}x + 2, x \text{ is any real number}\right\}$

8. $x = -1, y = 2, z = -3$ or $(-1, 2, -3)$

9. $x = \frac{7}{4}z + \frac{39}{4}, y = \frac{9}{8}z + \frac{69}{8}$, where z is any real number, or $\left\{(x, y, z) \,\middle|\, x = \frac{7}{4}z + \frac{39}{4}, y = \frac{9}{8}z + \frac{69}{8}, z \text{ is any real number}\right\}$ **10.** Inconsistent

11. $\begin{cases} 3x + 2y = 8 \\ x + 4y = -1 \end{cases}$ **12.** $\begin{cases} x + 2y + 5z = -2 \\ 5x \quad\quad - 3z = 8 \\ 2x - y \quad\quad = 0 \end{cases}$ **13.** $\begin{bmatrix} 4 & -4 \\ 3 & 9 \\ 4 & 4 \end{bmatrix}$ **14.** $\begin{bmatrix} 6 & 0 \\ 12 & 24 \\ -6 & 12 \end{bmatrix}$ **15.** $\begin{bmatrix} 4 & -3 & 0 \\ 12 & -2 & -8 \\ -2 & 5 & -4 \end{bmatrix}$ **16.** $\begin{bmatrix} 9 & -31 \\ -6 & -3 \end{bmatrix}$

17. $\begin{bmatrix} \frac{1}{2} & -1 \\ -\frac{1}{6} & \frac{2}{3} \end{bmatrix}$ **18.** $\begin{bmatrix} -\frac{5}{7} & \frac{9}{7} & \frac{3}{7} \\ \frac{1}{7} & \frac{1}{7} & -\frac{2}{7} \\ \frac{3}{7} & -\frac{4}{7} & \frac{1}{7} \end{bmatrix}$ **19.** Singular **20.** $x = \frac{2}{5}, y = \frac{1}{10}$ or $\left(\frac{2}{5}, \frac{1}{10}\right)$ **21.** $x = 9, y = \frac{13}{3}, z = \frac{13}{3}$ or $\left(9, \frac{13}{3}, \frac{13}{3}\right)$

22. Inconsistent **23.** $x = -\frac{1}{2}, y = -\frac{2}{3}, z = -\frac{3}{4}$, or $\left(-\frac{1}{2}, -\frac{2}{3}, -\frac{3}{4}\right)$

24. $z = -1, x = y + 1$, where y is any real number, or $\{(x, y, z) \mid x = y + 1, z = -1, y \text{ is any real number}\}$

25. $x = 4, y = 2, z = 3, t = -2$ or $(4, 2, 3, -2)$ **26.** 5 **27.** 108 **28.** -100 **29.** $x = 2, y = -1$ or $(2, -1)$ **30.** $x = 2, y = 3$ or $(2, 3)$

31. $x = -1, y = 2, z = -3$ or $(-1, 2, -3)$ **32.** 16 **33.** -8 **34.** $\frac{-\frac{3}{2}}{x} + \frac{\frac{3}{2}}{x - 4}$ **35.** $\frac{-3}{x - 1} + \frac{3}{x} + \frac{4}{x^2}$ **36.** $\frac{-\frac{1}{10}}{x + 1} + \frac{\frac{1}{10}x + \frac{9}{10}}{x^2 + 9}$

37. $\frac{x}{x^2 + 4} + \frac{-4x}{(x^2 + 4)^2}$ **38.** $\frac{\frac{1}{2}}{x^2 + 1} + \frac{\frac{1}{4}}{x - 1} + \frac{-\frac{1}{4}}{x + 1}$ **39.** $x = -\frac{2}{5}, y = -\frac{11}{5}; x = -2, y = 1$ or $\left(-\frac{2}{5}, -\frac{11}{5}\right), (-2, 1)$

40. $x = 2\sqrt{2}, y = \sqrt{2}; x = -2\sqrt{2}, y = -\sqrt{2}$ or $(2\sqrt{2}, \sqrt{2}), (-2\sqrt{2}, -\sqrt{2})$

41. $x = 0, y = 0; x = -3, y = 3; x = 3, y = 3$ or $(0, 0), (-3, 3), (3, 3)$

42. $x = \sqrt{2}, y = -\sqrt{2}; x = -\sqrt{2}, y = \sqrt{2}; x = \frac{4}{3}\sqrt{2}, y = -\frac{2}{3}\sqrt{2}; x = -\frac{4}{3}\sqrt{2}, y = \frac{2}{3}\sqrt{2}$ or $(\sqrt{2}, -\sqrt{2}), (-\sqrt{2}, \sqrt{2}), \left(\frac{4}{3}\sqrt{2}, -\frac{2}{3}\sqrt{2}\right)$, $\left(-\frac{4}{3}\sqrt{2}, \frac{2}{3}\sqrt{2}\right)$ **43.** $x = 1, y = -1$ or $(1, -1)$

44.

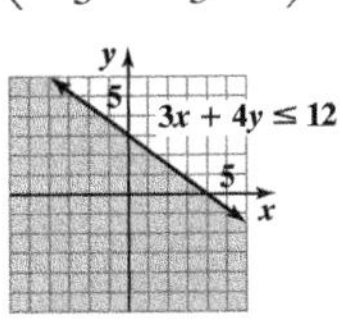

45.

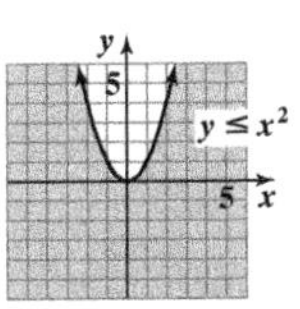

46. Unbounded

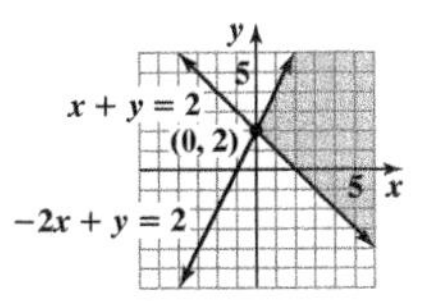

47. Bounded

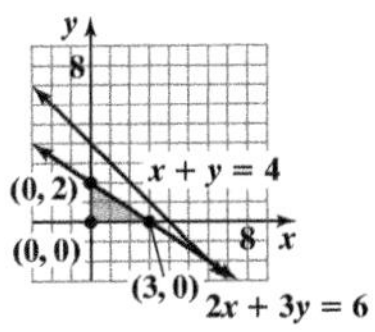

48. Bounded

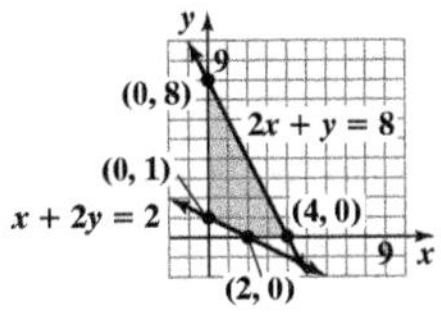

49.

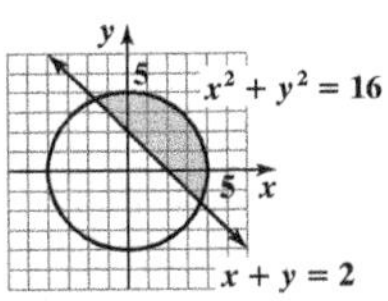

50.

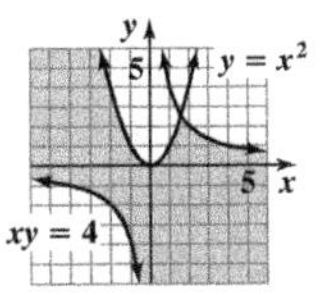

51. The maximum value is 32 when $x = 0$ and $y = 8$. **52.** The minimum value is 3 when $x = 1$ and $y = 0$. **53.** 10 **54.** A is any real number, $A \neq 10$.

55. $y = -\frac{1}{3}x^2 - \frac{2}{3}x + 1$ **56.** Mix 70 lb of \$6.00 coffee and 30 lb of \$9.00 coffee. **57.** Buy 1 small, 5 medium, and 2 large.

58. (a) $\begin{cases} x \geq 0 \\ y \geq 0 \\ 4x + 3y \leq 960 \\ 2x + 3y \leq 576 \end{cases}$

(b)

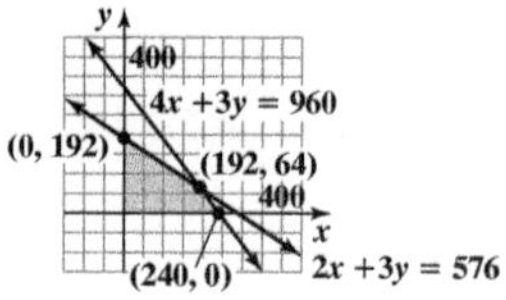

59. Speedboat: 36.67 km/hr; Aguarico River: 3.33 km/hr
60. Bruce: 4 hr; Bryce: 2 hr; Marty: 8 hr
61. Produce 35 gasoline engines and 15 diesel engines; the factory is producing an excess of 15 gasoline engines and 0 diesel engines.

Chapter Test *(page 958)*

1. $x = 3, y = -1$ or $(3, -1)$ **2.** Inconsistent **3.** $x = -z + \frac{18}{7}, y = z - \frac{17}{7}$, where z is any real number, or $\left\{(x, y, z) \,\middle|\, x = -z + \frac{18}{7}, y = z - \frac{17}{7}, z \text{ is any real number}\right\}$ **4.** $x = \frac{1}{3}, y = -2, z = 0$ or $\left(\frac{1}{3}, -2, 0\right)$ **5.** $\left[\begin{array}{ccc|c} 4 & -5 & 1 & 0 \\ -2 & -1 & 0 & -25 \\ 1 & 5 & -5 & 10 \end{array}\right]$

6. $\begin{cases} 3x + 2y + 4z = -6 \\ 1x + 0y + 8z = 2 \\ -2x + 1y + 3z = -11 \end{cases}$ or $\begin{cases} 3x + 2y + 4z = -6 \\ x + 8z = 2 \\ -2x + y + 3z = -11 \end{cases}$ **7.** $\begin{bmatrix} 6 & 4 \\ 1 & -11 \\ 5 & 12 \end{bmatrix}$ **8.** $\begin{bmatrix} -11 & -19 \\ -3 & 5 \\ 6 & -22 \end{bmatrix}$ **9.** $\begin{bmatrix} 4 & 10 & 26 \\ 1 & -11 & 2 \\ -1 & 26 & 3 \end{bmatrix}$ **10.** $\begin{bmatrix} 16 & 17 \\ 3 & -10 \end{bmatrix}$

11. $\begin{bmatrix} 2 & -1 \\ -\frac{5}{2} & \frac{3}{2} \end{bmatrix}$ **12.** $\begin{bmatrix} 3 & 3 & -4 \\ -2 & -2 & 3 \\ -4 & -5 & 7 \end{bmatrix}$ **13.** $x = \frac{1}{2}, y = 3$ or $\left(\frac{1}{2}, 3\right)$ **14.** $x = -\frac{1}{4}y + 7$, where y is any real number, or $\left\{(x, y) \,\middle|\, x = -\frac{1}{4}y + 7, y \text{ is any real number}\right\}$ **15.** $x = 1, y = -2, z = 0$ or $(1, -2, 0)$ **16.** Inconsistent **17.** -29 **18.** -12

19. $x = -2, y = -5$ or $(-2, -5)$ **20.** $x = 1, y = -1, z = 4$ or $(1, -1, 4)$ **21.** $(1, -3)$ and $(1, 3)$ **22.** $(3, 4)$ and $(1, 2)$

23.

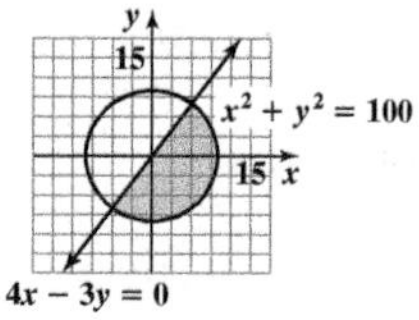

24. $\frac{3}{x + 3} + \frac{-2}{(x + 3)^2}$

25. $\frac{-\frac{1}{3}}{x} + \frac{\frac{1}{3}x}{(x^2 + 3)} + \frac{5x}{(x^2 + 3)^2}$

26. Unbounded

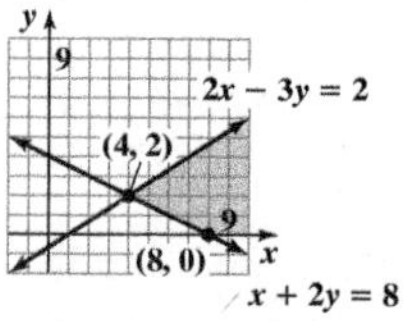

27. The maximum value of z is 64, and it occurs at the point $(0, 8)$. **28.** Flare jeans cost \$24.50, camisoles cost \$8.50, and T-shirts cost \$6.00.

Cumulative Review *(page 959)*

1. $\left\{0, \frac{1}{2}\right\}$ **2.** $\{5\}$ **3.** $\left\{-1, -\frac{1}{2}, 3\right\}$ **4.** $\{-2\}$ **5.** $\left\{\frac{5}{2}\right\}$ **6.** $\left\{\frac{1}{\ln 3}\right\}$ **7.** Odd; symmetric with respect to the origin

8. Center: $(1, -2)$; radius $= 4$

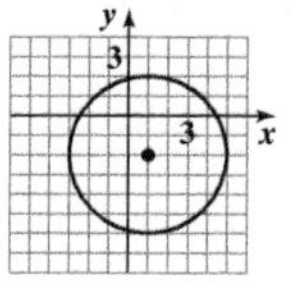

9. Domain: all real numbers
Range: $\{y \mid y > 1\}$
Horizontal asymptote: $y = 1$

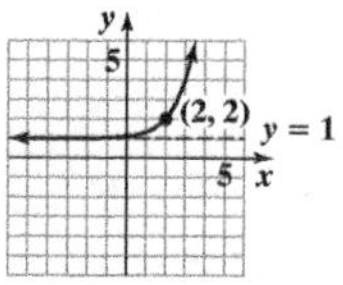

10. $f^{-1}(x) = \frac{5}{x} - 2$

Domain of f: $\{x \mid x \neq -2\}$
Range of f: $\{y \mid y \neq 0\}$
Domain of f^{-1}: $\{x \mid x \neq 0\}$
Range of f^{-1}: $\{y \mid y \neq -2\}$

11. (a)

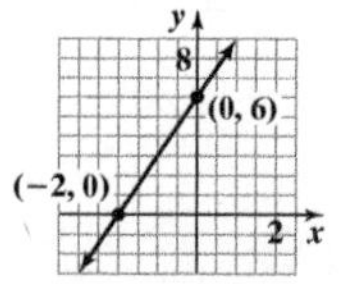

(b)

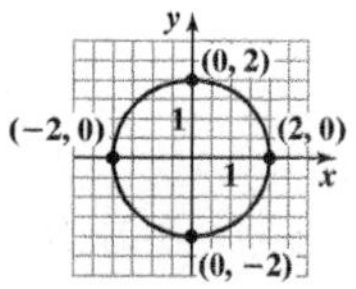

(c)

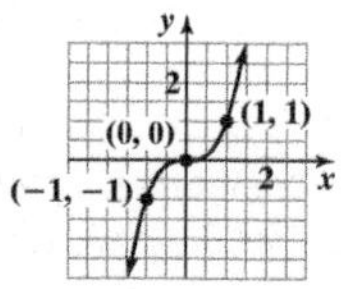

(d)

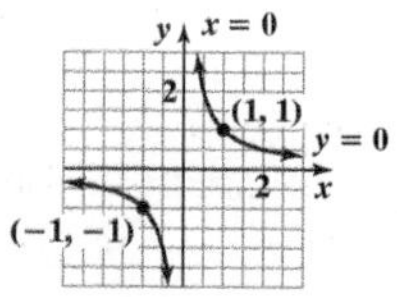

(e)

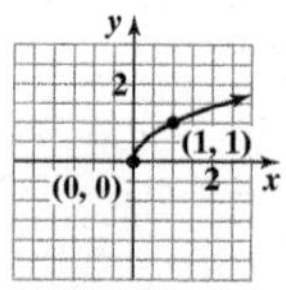

(f) 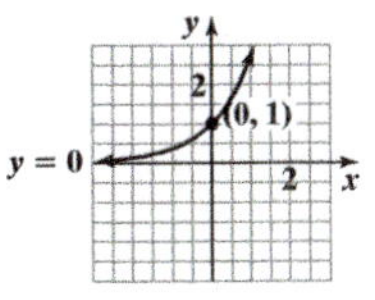

(g) 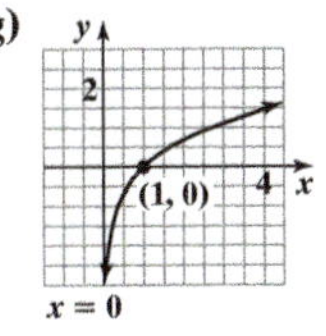

(h) 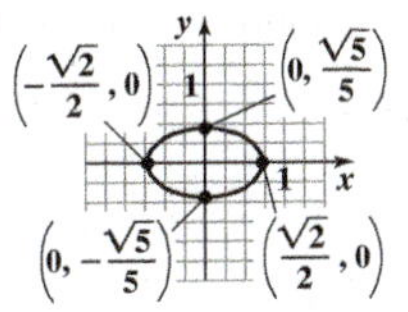

(i) 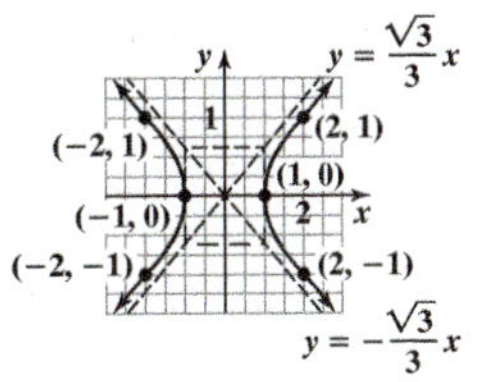

(j) 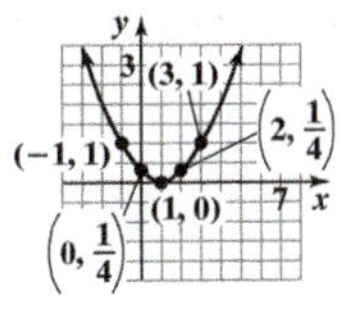

12. (a) 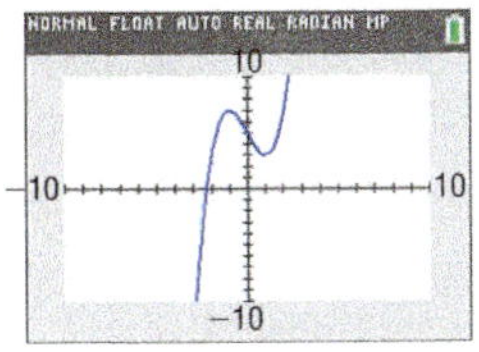

; Zero: −2.28

(b) Local maximum of 7 at $x = -1$; local minimum of 3 at $x = 1$
(c) $(-\infty, -1], [1, \infty)$

CHAPTER 13 Sequences; Induction; the Binomial Theorem

13.1 Assess Your Understanding *(page 971)*

5. sequence **6.** True **7.** $n(n-1)\cdot\cdot\cdot 3\cdot 2\cdot 1$ **8.** b **9.** summation **10.** b **11.** 3,628,800 **13.** 504 **15.** 1260 **17.** $s_1 = 1, s_2 = 2, s_3 = 3, s_4 = 4, s_5 = 5$
19. $a_1 = \frac{1}{3}, a_2 = \frac{1}{2}, a_3 = \frac{3}{5}, a_4 = \frac{2}{3}, a_5 = \frac{5}{7}$ **21.** $c_1 = 1, c_2 = -4, c_3 = 9, c_4 = -16, c_5 = 25$ **23.** $s_1 = \frac{1}{2}, s_2 = \frac{2}{5}, s_3 = \frac{2}{7}, s_4 = \frac{8}{41}, s_5 = \frac{8}{61}$
25. $t_1 = -\frac{1}{6}, t_2 = \frac{1}{12}, t_3 = -\frac{1}{20}, t_4 = \frac{1}{30}, t_5 = -\frac{1}{42}$ **27.** $b_1 = \frac{1}{e}, b_2 = \frac{2}{e^2}, b_3 = \frac{3}{e^3}, b_4 = \frac{4}{e^4}, b_5 = \frac{5}{e^5}$ **29.** $a_n = \frac{n}{n+1}$ **31.** $a_n = \frac{1}{2^{n-1}}$
33. $a_n = (-1)^{n+1}$ **35.** $a_n = (-1)^{n+1}n$ **37.** $a_1 = 2, a_2 = 5, a_3 = 8, a_4 = 11, a_5 = 14$ **39.** $a_1 = -2, a_2 = 0, a_3 = 3, a_4 = 7, a_5 = 12$
41. $a_1 = 5, a_2 = 10, a_3 = 20, a_4 = 40, a_5 = 80$ **43.** $a_1 = 3, a_2 = \frac{3}{2}, a_3 = \frac{1}{2}, a_4 = \frac{1}{8}, a_5 = \frac{1}{40}$ **45.** $a_1 = 1, a_2 = 2, a_3 = 2, a_4 = 4, a_5 = 8$
47. $a_1 = A, a_2 = A + d, a_3 = A + 2d, a_4 = A + 3d, a_5 = A + 4d$
49. $a_1 = \sqrt{2}, a_2 = \sqrt{2 + \sqrt{2}}, a_3 = \sqrt{2 + \sqrt{2 + \sqrt{2}}},$
$a_4 = \sqrt{2 + \sqrt{2 + \sqrt{2 + \sqrt{2}}}},$
$a_5 = \sqrt{2 + \sqrt{2 + \sqrt{2 + \sqrt{2 + \sqrt{2}}}}}$
51. $3 + 4 + \cdots + (n + 2)$ **53.** $\frac{1}{2} + 2 + \frac{9}{2} + \cdots + \frac{n^2}{2}$ **55.** $1 + \frac{1}{3} + \frac{1}{9} + \cdots + \frac{1}{3^n}$
57. $\frac{1}{3} + \frac{1}{9} + \cdots + \frac{1}{3^n}$ **59.** $\ln 2 - \ln 3 + \ln 4 - \cdots + (-1)^n \ln n$ **61.** $\sum_{k=1}^{20} k$
63. $\sum_{k=1}^{13} \frac{k}{k+1}$ **65.** $\sum_{k=0}^{6} (-1)^k\left(\frac{1}{3^k}\right)$ **67.** $\sum_{k=1}^{n} \frac{3^k}{k}$ **69.** $\sum_{k=0}^{n} (a + kd)$ or $\sum_{k=1}^{n+1} [a + (k-1)d]$ **71.** 200 **73.** 820 **75.** 1110 **77.** 1560 **79.** 3570
81. 44,000 **83. (a)** 2162 **(b)** After 26 months **85. (a)** $A_0 = 1500, A_n = (1.0125)A_{n-1} + 750$ **(b)** After 99 quarters **(c)** \$213,073.11
87. (a) \$2930 **(b)** 14 payments have been made. **(c)** 36 payments; \$3584.62 **(d)** \$584.62
89. (a) $a_0 = 150{,}000, a_n = (1.005)a_{n-1} - 899.33$
(b) \$149,850.67 **(c)** 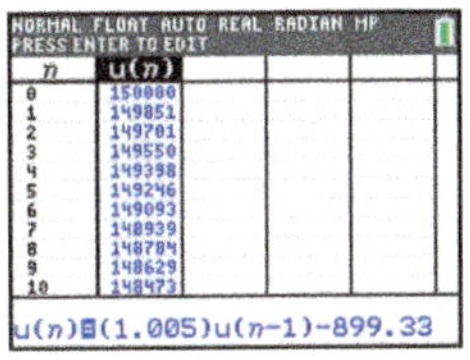

(d) After 58 payments or 4 years and 10 months later
(e) After 359 payments of \$899.33, plus last payment of \$895.10
(f) \$173,754.57

(g) (a) $a_0 = 150{,}000, a_n = (1.005)a_{n-1} - 999.33$
(b) \$149,750.67 **(c)** 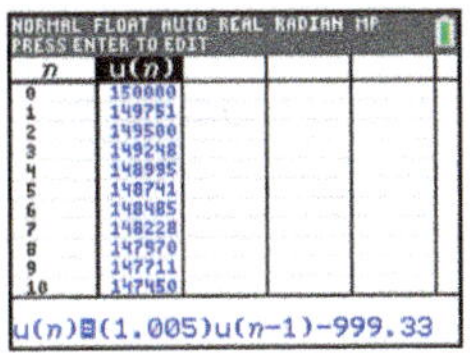

(d) After 37 payments or 3 years and 1 month later
(e) After 278 payments of \$999.33, plus last payment of \$353.69(1.005) = \$355.46
(f) \$128,169.20

91. 21 pairs **93.** Fibonacci sequence **95. (a)** 3.630170833 **(b)** 3.669060828 **(c)** 3.669296668 **(d)** 12
97. (a) $a_1 = 0.4; a_2 = 0.7; a_3 = 1; a_4 = 1.6; a_5 = 2.8; a_6 = 5.2; a_7 = 10; a_8 = 19.6$
(b) Except for term 5, which has no match, Bode's formula provides excellent approximations for the mean distances of the planets from the sun.
(c) The mean distance of Ceres from the sun is approximated by $a_5 = 2.8$, and that of Uranus is $a_8 = 19.6$.
(d) $a_9 = 38.8; a_{10} = 77.2$
(e) Pluto's distance is approximated by a_9, but no term approximates Neptune's mean distance from the sun.
(f) According to Bode's Law, the mean orbital distance of Eris will be 154 AU from the sun.
99. $a_0 = 2; a_5 = 2.236067977; 2.236067977$ **101.** $a_0 = 4; a_5 = 4.582575695; 4.582575695$ **103.** 1, 3, 6, 10, 15, 21, 28

105. $u_n = 1 + 2 + 3 + \cdots + n = \sum_{k=1}^{n} k = \dfrac{n(n+1)}{2}$, and from Problem 104, $u_{n-1} = \dfrac{(n+1)(n+2)}{2}$.

Thus, $u_{n+1} + u_n = \dfrac{(n+1)(n+2)}{2} + \dfrac{n(n+1)}{2} = \dfrac{(n+1)[(n+2)+n]}{2} = (n+1)^2$.

108. \$2654.39 **109.** $\sqrt{2}(\cos 225° + i \sin 225°)$ **110.** 0 **111.** $(y-4)^2 = 16(x+3)$

13.2 Assess Your Understanding *(page 980)*

1. arithmetic **2.** F **3.** 17 **4.** T **5.** d **6.** c **7.** $s_n - s_{n-1} = (n+4) - [(n-1)+4] = n + 4 - (n+3) = n + 4 - n - 3 = 1$, a constant; $d = 1; s_1 = 5, s_2 = 6, s_3 = 7, s_4 = 8$
9. $a_n - a_{n-1} = (2n-5) - [2(n-1)-5] = 2n - 5 - (2n-2-5) = 2n - 5 - (2n-7) = 2n - 5 - 2n + 7 = 2$, a constant; $d = 2; a_1 = -3, a_2 = -1, a_3 = 1, a_4 = 3$

11. $c_n - c_{n-1} = (6-2n) - [6-2(n-1)] = 6 - 2n - (6-2n+2) = 6 - 2n - (8-2n) = 6 - 2n - 8 + 2n = -2$, a constant; $d = -2; c_1 = 4, c_2 = 2, c_3 = 0, c_4 = -2$

13. $t_n - t_{n-1} = \left(\frac{1}{2} - \frac{1}{3}n\right) - \left[\frac{1}{2} - \frac{1}{3}(n-1)\right] = \frac{1}{2} - \frac{1}{3}n - \left(\frac{1}{2} - \frac{1}{3}n + \frac{1}{3}\right) = \frac{1}{2} - \frac{1}{3}n - \left(\frac{5}{6} - \frac{1}{3}n\right) = \frac{1}{2} - \frac{1}{3}n - \frac{5}{6} + \frac{1}{3}n = -\frac{1}{3}$, a constant; $d = -\frac{1}{3}; t_1 = \frac{1}{6}, t_2 = -\frac{1}{6}, t_3 = -\frac{1}{2}, t_4 = -\frac{5}{6}$

15. $s_n - s_{n-1} = \ln 3^n - \ln 3^{n-1} = n \ln 3 - (n-1)\ln 3 = n \ln 3 - (n \ln 3 - \ln 3) = n \ln 3 - n \ln 3 + \ln 3 = \ln 3$, a constant; $d = \ln 3; s_1 = \ln 3, s_2 = 2\ln 3, s_3 = 3 \ln 3, s_4 = 4 \ln 3$

17. $a_n = 3n - 1; a_{51} = 152$ **19.** $a_n = 8 - 3n; a_{51} = -145$ **21.** $a_n = \frac{1}{2}(n-1); a_{51} = 25$ **23.** $a_n = \sqrt{2}n; a_{51} = 51\sqrt{2}$ **25.** 200 **27.** −266 **29.** $\frac{83}{2}$
31. $a_1 = -13; d = 3; a_n = a_{n-1} + 3; a_n = -16 + 3n$ **33.** $a_1 = -53; d = 6; a_n = a_{n-1} + 6; a_n = -59 + 6n$
35. $a_1 = 28; d = -2; a_n = a_{n-1} - 2; a_n = 30 - 2n$ **37.** $a_1 = 25; d = -2; a_n = a_{n-1} - 2; a_n = 27 - 2n$ **39.** n^2 **41.** $\frac{n}{2}(9 + 5n)$ **43.** 1260 **45.** 324
47. 30,919 **49.** 10,036 **51.** 6080 **53.** −1925 **55.** 15,960 **57.** $-\frac{3}{2}$ **59.** 24 terms **61.** 1185 seats **63.** 210 beige and 190 blue
65. $\{T_n\} = \{-5.5n + 67\}; T_5 = 39.5°F$ **67.** The amphitheater has 1647 seats. **69.** 8 yr **72.** 16.42% **73.** $\mathbf{v} = 4\mathbf{i} - 6\mathbf{j}$

74. Ellipse: Center: (0, 0); Vertices: $(0, -5), (0, 5)$; Foci: $(0, -\sqrt{21}), (0, \sqrt{21})$,

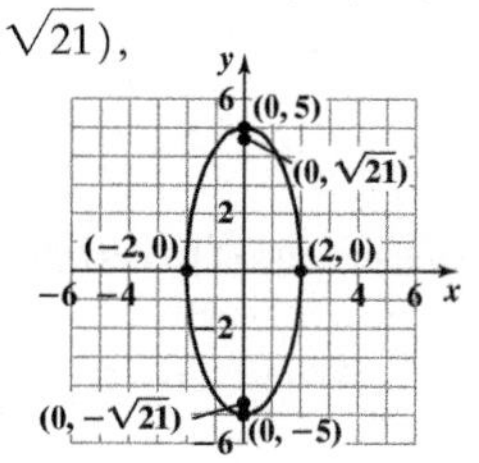

75. $\begin{bmatrix} \frac{1}{2} & 0 \\ \frac{3}{2} & -1 \end{bmatrix}$

Historical Problems *(page 988)*

1. $1\frac{2}{3}$ loaves, $10\frac{5}{6}$ loaves, 20 loaves, $29\frac{1}{6}$ loaves, $38\frac{1}{3}$ loaves **2. (a)** 1 person **(b)** 2401 kittens **(c)** 2800

13.3 Assess Your Understanding *(page 988)*

1. $a_n = a_1 \cdot r^{n-1}$ **2.** $-\frac{2}{3}$ **3.** c **4.** $\frac{a}{1-r}$ **5.** b **6.** T **7.** F **8.** T **9.** $r = 3; s_1 = 3, s_2 = 9, s_3 = 27, s_4 = 81$
11. $r = \frac{1}{2}; a_1 = -\frac{3}{2}, a_2 = -\frac{3}{4}, a_3 = -\frac{3}{8}, a_4 = -\frac{3}{16}$ **13.** $r = 2; c_1 = \frac{1}{4}, c_2 = \frac{1}{2}, c_3 = 1, c_4 = 2$ **15.** $r = 2^{1/3}; e_1 = 2^{1/3}, e_2 = 2^{2/3}, e_3 = 2, e_4 = 2^{4/3}$
17. $r = \frac{3}{2}; t_1 = \frac{1}{2}, t_2 = \frac{3}{4}, t_3 = \frac{9}{8}, t_4 = \frac{27}{16}$ **19.** $a_5 = 162; a_n = 2 \cdot 3^{n-1}$ **21.** $a_5 = 5; a_n = 5 \cdot (-1)^{n-1}$ **23.** $a_5 = 0; a_n = 0$
25. $a_5 = 4\sqrt{2}; a_n = (\sqrt{2})^n$ **27.** $a_7 = \frac{1}{64}$ **29.** $a_9 = 1$ **31.** $a_8 = 0.00000004$ **33.** $a_n = 7 \cdot 2^{n-1}$ **35.** $a_n = -3 \cdot \left(-\frac{1}{3}\right)^{n-1} = \left(-\frac{1}{3}\right)^{n-2}$
37. $a_n = -(-3)^{n-1}$ **39.** $a_n = \frac{7}{15}(15)^{n-1} = 7 \cdot 15^{n-2}$ **41.** $-\frac{1}{4}(1 - 2^n)$ **43.** $2\left[1 - \left(\frac{2}{3}\right)^n\right]$ **45.** $1 - 2^n$

47.
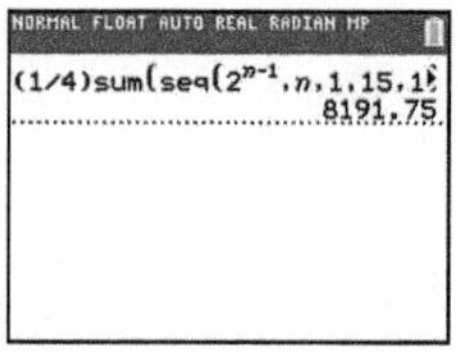

49.
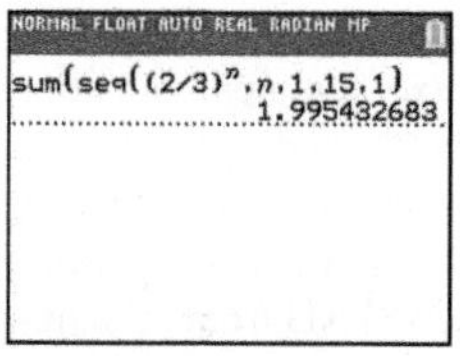

51.
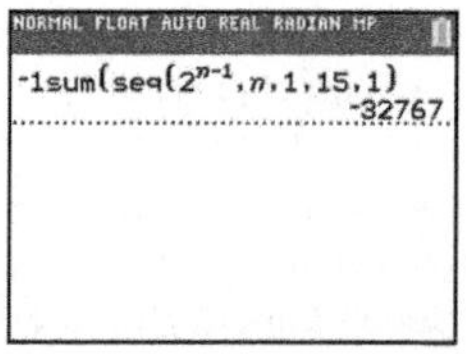

53. Converges; $\frac{3}{2}$ **55.** Converges; 16 **57.** Converges; $\frac{8}{5}$ **59.** Diverges **61.** Converges; $\frac{20}{3}$ **63.** Diverges **65.** Converges; $\frac{18}{5}$ **67.** Converges; 6

69. Arithmetic; $d = 1$; 1375 **71.** Neither **73.** Arithmetic; $d = -\frac{2}{3}$; -700 **75.** Neither **77.** Geometric; $r = \frac{2}{3}$; $2\left[1 - \left(\frac{2}{3}\right)^{50}\right]$

79. Geometric; $r = -2$; $-\frac{1}{3}[1 - (-2)^{50}]$ **81.** Geometric; $r = 3^{1/2}$; $-\frac{\sqrt{3}}{2}(1 + \sqrt{3})(1 - 3^{25})$

83. -4 **85.** \$47,271.37 **87. (a)** 0.775 ft **(b)** 8th **(c)** 15.88 ft **(d)** 20 ft **89.** \$349,496.41 **91.** \$96,885.98 **93.** \$305.10 **95.** 1.845×10^{19} **97.** 10 **99.** \$72.67 per share **101.** December 20, 2015; \$9999.92 **103.** \$5633.36 **105.** Option B results in more money (\$524,287 versus \$500,500).

107. Total pay: \$41,943.03; pay on day 22: \$20,971.52 **112.** 2.121 **113.** $\frac{8}{17}\mathbf{i} - \frac{15}{17}\mathbf{j}$ **114.** $\frac{x^2}{4} - \frac{y^2}{12} = 1$ **115.** 54

13.4 Assess Your Understanding *(page 995)*

1. (I) $n = 1: 2(1) = 2$ and $1(1 + 1) = 2$
(II) If $2 + 4 + 6 + \cdots + 2k = k(k + 1)$, then $2 + 4 + 6 + \cdots + 2k + 2(k + 1) = (2 + 4 + 6 + \cdots + 2k) + 2(k + 1)$ $= k(k + 1) + 2(k + 1) = k^2 + 3k + 2 = (k + 1)(k + 2) = (k + 1)[(k + 1) + 1]$.

3. (I) $n = 1: 1 + 2 = 3$ and $\frac{1}{2}(1)(1 + 5) = \frac{1}{2}(6) = 3$
(II) If $3 + 4 + 5 + \cdots + (k + 2) = \frac{1}{2}k(k + 5)$, then $3 + 4 + 5 + \cdots + (k + 2) + [(k + 1) + 2]$
$= [3 + 4 + 5 + \cdots + (k + 2)] + (k + 3) = \frac{1}{2}k(k + 5) + k + 3 = \frac{1}{2}(k^2 + 7k + 6) = \frac{1}{2}(k + 1)(k + 6) = \frac{1}{2}(k + 1)[(k + 1) + 5]$.

5. (I) $n = 1: 3(1) - 1 = 2$ and $\frac{1}{2}(1)[3(1) + 1] = \frac{1}{2}(4) = 2$
(II) If $2 + 5 + 8 + \cdots + (3k - 1) = \frac{1}{2}k(3k + 1)$, then $2 + 5 + 8 + \cdots + (3k - 1) + [3(k + 1) - 1]$
$= [2 + 5 + 8 + \cdots + (3k - 1)] + (3k + 2) = \frac{1}{2}k(3k + 1) + (3k + 2) = \frac{1}{2}(3k^2 + 7k + 4) = \frac{1}{2}(k + 1)(3k + 4)$
$= \frac{1}{2}(k + 1)[3(k + 1) + 1]$.

7. (I) $n = 1: 2^{1-1} = 1$ and $2^1 - 1 = 1$
(II) If $1 + 2 + 2^2 + \cdots + 2^{k-1} = 2^k - 1$, then $1 + 2 + 2^2 + \cdots + 2^{k-1} + 2^{(k+1)-1} = (1 + 2 + 2^2 + \cdots + 2^{k-1}) + 2^k$
$= 2^k - 1 + 2^k = 2(2^k) - 1 = 2^{k+1} - 1$.

9. (I) $n = 1: 4^{1-1} = 1$ and $\frac{1}{3}(4^1 - 1) = \frac{1}{3}(3) = 1$
(II) If $1 + 4 + 4^2 + \cdots + 4^{k-1} = \frac{1}{3}(4^k - 1)$, then $1 + 4 + 4^2 + \cdots + 4^{k-1} + 4^{(k+1)-1} = (1 + 4 + 4^2 + \cdots + 4^{k-1}) + 4^k$
$= \frac{1}{3}(4^k - 1) + 4^k = \frac{1}{3}[4^k - 1 + 3(4^k)] = \frac{1}{3}[4(4^k) - 1] = \frac{1}{3}(4^{k+1} - 1)$.

11. (I) $n = 1: \frac{1}{1 \cdot 2} = \frac{1}{2}$ and $\frac{1}{1 + 1} = \frac{1}{2}$
(II) If $\frac{1}{1 \cdot 2} + \frac{1}{2 \cdot 3} + \frac{1}{3 \cdot 4} + \cdots + \frac{1}{k(k + 1)} = \frac{k}{k + 1}$, then $\frac{1}{1 \cdot 2} + \frac{1}{2 \cdot 3} + \frac{1}{3 \cdot 4} + \cdots + \frac{1}{k(k + 1)} + \frac{1}{(k + 1)[(k + 1) + 1]}$
$= \left[\frac{1}{1 \cdot 2} + \frac{1}{2 \cdot 3} + \frac{1}{3 \cdot 4} + \cdots + \frac{1}{k(k + 1)}\right] + \frac{1}{(k + 1)(k + 2)} = \frac{k}{k + 1} + \frac{1}{(k + 1)(k + 2)} = \frac{k(k + 2) + 1}{(k + 1)(k + 2)}$
$= \frac{k^2 + 2k + 1}{(k + 1)(k + 2)} = \frac{(k + 1)^2}{(k + 1)(k + 2)} = \frac{k + 1}{k + 2} = \frac{k + 1}{(k + 1) + 1}$.

13. (I) $n = 1: 1^2 = 1$ and $\frac{1}{6} \cdot 1 \cdot 2 \cdot 3 = 1$
(II) If $1^2 + 2^2 + 3^2 + \cdots + k^2 = \frac{1}{6}k(k + 1)(2k + 1)$, then $1^2 + 2^2 + 3^2 + \cdots + k^2 + (k + 1)^2$
$= (1^2 + 2^2 + 3^2 + \cdots + k^2) + (k + 1)^2 = \frac{1}{6}k(k + 1)(2k + 1) + (k + 1)^2 = \frac{1}{6}(2k^3 + 9k^2 + 13k + 6)$
$= \frac{1}{6}(k + 1)(k + 2)(2k + 3) = \frac{1}{6}(k + 1)[(k + 1) + 1][2(k + 1) + 1]$.

15. (I) $n = 1: 5 - 1 = 4$ and $\frac{1}{2}(1)(9 - 1) = \frac{1}{2} \cdot 8 = 4$
(II) If $4 + 3 + 2 + \cdots + (5 - k) = \frac{1}{2}k(9 - k)$, then $4 + 3 + 2 + \cdots + (5 - k) + [5 - (k + 1)]$
$= [4 + 3 + 2 + \cdots + (5 - k)] + 4 - k = \frac{1}{2}k(9 - k) + 4 - k = \frac{1}{2}(9k - k^2 + 8 - 2k) = \frac{1}{2}(-k^2 + 7k + 8)$
$= \frac{1}{2}(k + 1)(8 - k) = \frac{1}{2}(k + 1)[9 - (k + 1)]$.

17. **(I)** $n = 1: 1 \cdot (1 + 1) = 2$ and $\frac{1}{3} \cdot 1 \cdot 2 \cdot 3 = 2$

(II) If $1 \cdot 2 + 2 \cdot 3 + 3 \cdot 4 + \cdots + k(k + 1) = \frac{1}{3}k(k + 1)(k + 2)$, then $1 \cdot 2 + 2 \cdot 3 + 3 \cdot 4 + \cdots + k(k + 1)$
$+ (k + 1)[(k + 1) + 1] = [1 \cdot 2 + 2 \cdot 3 + 3 \cdot 4 + \cdots + k(k + 1)] + (k + 1)(k + 2)$
$= \frac{1}{3}k(k + 1)(k + 2) + \frac{1}{3} \cdot 3(k + 1)(k + 2) = \frac{1}{3}(k + 1)(k + 2)(k + 3) = \frac{1}{3}(k + 1)[(k + 1) + 1][(k + 1) + 2].$

19. **(I)** $n = 1: 1^2 + 1 = 2$, which is divisible by 2.
(II) If $k^2 + k$ is divisible by 2, then $(k + 1)^2 + (k + 1) = k^2 + 2k + 1 + k + 1 = (k^2 + k) + 2k + 2$. Since $k^2 + k$ is divisible by 2 and $2k + 2$ is divisible by 2, $(k + 1)^2 + (k + 1)$ is divisible by 2.

21. **(I)** $n = 1: 1^2 - 1 + 2 = 2$, which is divisible by 2.
(II) If $k^2 - k + 2$ is divisible by 2, then $(k + 1)^2 - (k + 1) + 2 = k^2 + 2k + 1 - k - 1 + 2 = (k^2 - k + 2) + 2k$. Since $k^2 - k + 2$ is divisible by 2 and $2k$ is divisible by 2, $(k + 1)^2 - (k + 1) + 2$ is divisible by 2.

23. **(I)** $n = 1$: If $x > 1$, then $x^1 = x > 1$.
(II) Assume, for an arbitrary natural number k, that if $x > 1$ then $x^k > 1$. Multiply both sides of the inequality $x^k > 1$ by x. If $x > 1$, then $x^{k+1} > x > 1$.

25. **(I)** $n = 1$: $a - b$ is a factor of $a^1 - b^1 = a - b$.
(II) If $a - b$ is a factor of $a^k - b^k$, then $a^{k+1} - b^{k-1} = a(a^k - b^k) + b^k(a - b)$.
Since $a - b$ is a factor of $a^k - b^k$ and $a - b$ is a factor of $a - b$, then $a - b$ is a factor of $a^{k+1} - b^{k+1}$.

27. **(a)** $n = 1$: $(1 + a)^1 = 1 + a \geq 1 + 1 \cdot a$
(b) Assume that there is an integer k for which the inequality holds. So $(1 + a)^k \geq 1 + ka$. We need to show that $(1 + a)^{k+1} \geq 1 + (k + 1)a$.
$(1 + a)^{k+1} = (1 + a)^k(1 + a) \geq (1 + ka)(1 + a) = 1 + ka^2 + a + ka = 1 + (k + 1)a + ka^2 \geq 1 + (k + 1)a$.

29. If $2 + 4 + 6 + \cdots + 2k = k^2 + k + 2$, then $2 + 4 + 6 + \cdots + 2k + 2(k + 1)$
$= (2 + 4 + 6 + \cdots + 2k) + 2k + 2 = k^2 + k + 2 + 2k + 2 = k^2 + 3k + 4 = (k^2 + 2k + 1) + (k + 1) + 2$
$= (k + 1)^2 + (k + 1) + 2$.
But $2 \cdot 1 = 2$ and $1^2 + 1 + 2 = 4$. The fact is that $2 + 4 + 6 + \cdots + 2n = n^2 + n$, not $n^2 + n + 2$ (Problem 1).

31. **(I)** $n = 1$: $[a + (1 - 1)d] = a$ and $1 \cdot a + d\frac{1 \cdot (1 - 1)}{2} = a$.

(II) If $a + (a + d) + (a + 2d) + \cdots + [a + (k - 1)d] = ka + d\frac{k(k - 1)}{2}$, then

$a + (a + d) + (a + 2d) + \cdots + [a + (k - 1)d] + [a + ((k + 1) - 1)d] = ka + d\frac{k(k - 1)}{2} + a + kd$

$= (k + 1)a + d\frac{k(k - 1) + 2k}{2} = (k + 1)a + d\frac{(k + 1)(k)}{2} = (k + 1)a + d\frac{(k + 1)[(k + 1) - 1]}{2}.$

33. **(I)** $n = 3$: The sum of the angles of a triangle is $(3 - 2) \cdot 180° = 180°$.
(II) Assume that for some $k \geq 3$, the sum of the angles of a convex polygon of k sides is $(k - 2) \cdot 180°$.
A convex polygon of $k + 1$ sides consists of a convex polygon of k sides plus a triangle (see the illustration).
The sum of the angles is $(k - 2) \cdot 180° + 180° = (k - 1) \cdot 180° = [(k + 1) - 2] \cdot 180°$.

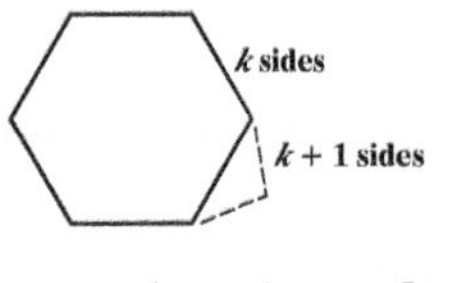

35. {251} **36.** Left: 448.3 kg; right: 366.0 kg **37.** $x = \frac{1}{2}, y = -3; \left(\frac{1}{2}, -3\right)$ **38.** $\begin{bmatrix} 7 & -3 \\ -7 & 8 \end{bmatrix}$

13.5 Assess Your Understanding *(page 1001)*

1. Pascal triangle **2.** 1; n **3.** F **4.** Binomial Theorem **5.** 10 **7.** 21 **9.** 50 **11.** 1 **13.** $\approx 1.8664 \times 10^{15}$ **15.** $\approx 1.4834 \times 10^{13}$
17. $x^5 + 5x^4 + 10x^3 + 10x^2 + 5x + 1$ **19.** $x^6 - 12x^5 + 60x^4 - 160x^3 + 240x^2 - 192x + 64$ **21.** $81x^4 + 108x^3 + 54x^2 + 12x + 1$
23. $x^{10} + 5x^8y^2 + 10x^6y^4 + 10x^4y^6 + 5x^2y^8 + y^{10}$ **25.** $x^3 + 6\sqrt{2}x^{5/2} + 30x^2 + 40\sqrt{2}x^{3/2} + 60x + 24\sqrt{2}x^{1/2} + 8$
27. $a^5x^5 + 5a^4bx^4y + 10a^3b^2x^3y^2 + 10a^2b^3x^2y^3 + 5ab^4xy^4 + b^5y^5$ **29.** 17,010 **31.** −101,376 **33.** 41,472 **35.** $2835x^3$ **37.** $314{,}928x^7$ **39.** 495

41. 3360 **43.** 1.00501 **45.** $\binom{n}{n-1} = \frac{n!}{(n - 1)!\,[n - (n - 1)]!} = \frac{n!}{(n - 1)!\,1!} = \frac{n \cdot (n - 1)!}{(n - 1)!} = n$; $\binom{n}{n} = \frac{n!}{n!\,(n - n)!} = \frac{n!}{n!\,0!} = \frac{n!}{n!} = 1$

47. $2^n = (1 + 1)^n = \binom{n}{0}1^n + \binom{n}{1}(1)^{n-1}(1) + \cdots + \binom{n}{n}1^n = \binom{n}{0} + \binom{n}{1} + \cdots + \binom{n}{n}$ **49.** 1

51. $\left\{\frac{\ln 5}{\ln 6 - \ln 5}\right\} \approx \{8.827\}$ **52.** **(a)** 0 **(b)** 90° **(c)** Orthogonal **53.** $x = 1, y = 3, z = -2; (1, 3, -2)$

54. Bounded

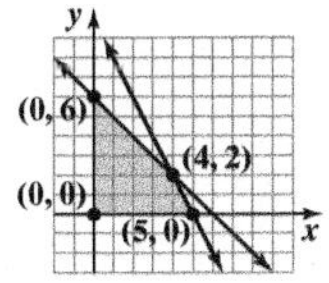

Review Exercises *(page 1003)*

1. $a_1 = -\frac{4}{3}, a_2 = \frac{5}{4}, a_3 = -\frac{6}{5}, a_4 = \frac{7}{6}, a_5 = -\frac{8}{7}$ **2.** $c_1 = 2, c_2 = 1, c_3 = \frac{8}{9}, c_4 = 1, c_5 = \frac{32}{25}$ **3.** $a_1 = 3, a_2 = 2, a_3 = \frac{4}{3}, a_4 = \frac{8}{9}, a_5 = \frac{16}{27}$

4. $a_1 = 2, a_2 = 0, a_3 = 2, a_4 = 0, a_5 = 2$ **5.** $6 + 10 + 14 + 18 = 48$ **6.** $\sum_{k=1}^{13} (-1)^{k+1}\frac{1}{k}$ **7.** Arithmetic; $d = 1$; $S_n = \frac{n}{2}(n + 11)$ **8.** Neither

9. Geometric; $r = 8$; $S_n = \frac{8}{7}(8^n - 1)$ **10.** Arithmetic; $d = 4$; $S_n = 2n(n - 1)$ **11.** Geometric; $r = \frac{1}{2}$; $S_n = 6\left[1 - \left(\frac{1}{2}\right)^n\right]$ **12.** Neither

13. 9515 **14.** –1320 **15.** $\frac{1093}{2187} \approx 0.49977$ **16.** 682 **17.** 35 **18.** $\frac{1}{10^{10}}$ **19.** $9\sqrt{2}$ **20.** $\{a_n\} = \{5n - 4\}$ **21.** $\{a_n\} = \{n - 10\}$ **22.** Converges; $\frac{9}{2}$

23. Converges; $\frac{4}{3}$ **24.** Diverges **25.** Converges; 8

26. (I) $n = 1: 3 \cdot 1 = 3$ and $\frac{3 \cdot 1}{2}(1 + 1) = 3$

(II) If $3 + 6 + 9 + \cdots + 3k = \frac{3k}{2}(k + 1)$, then $3 + 6 + 9 + \cdots + 3k + 3(k + 1) = (3 + 6 + 9 + \cdots + 3k) + (3k + 3)$

$= \frac{3k}{2}(k + 1) + (3k + 3) = \frac{3k^2}{2} + \frac{3k}{2} + \frac{6k}{2} + \frac{6}{2} = \frac{3}{2}(k^2 + 3k + 2) = \frac{3}{2}(k + 1)(k + 2) = \frac{3(k + 1)}{2}[(k + 1) + 1].$

27. (I) $n = 1: 2 \cdot 3^{1-1} = 2$ and $3^1 - 1 = 2$

(II) If $2 + 6 + 18 + \cdots + 2 \cdot 3^{k-1} = 3^k - 1$, then $2 + 6 + 18 + \cdots + 2 \cdot 3^{k-1} + 2 \cdot 3^{(k+1)-1} = (2 + 6 + 18 + \cdots + 2 \cdot 3^{k-1}) + 2 \cdot 3^k$
$= 3^k - 1 + 2 \cdot 3^k = 3 \cdot 3^k - 1 = 3^{k+1} - 1.$

28. (I) $n = 1: (3 \cdot 1 - 2)^2 = 1$ and $\frac{1}{2} \cdot 1 \cdot [6(1)^2 - 3(1) - 1] = 1$

(II) If $1^2 + 4^2 + 7^2 + \cdots + (3k - 2)^2 = \frac{1}{2}k(6k^2 - 3k - 1)$, then $1^2 + 4^2 + 7^2 + \cdots + (3k - 2)^2 + [3(k + 1) - 2]^2$

$= [1^2 + 4^2 + 7^2 + \cdots + (3k - 2)^2] + (3k + 1)^2 = \frac{1}{2}k(6k^2 - 3k - 1) + (3k + 1)^2 = \frac{1}{2}(6k^3 - 3k^2 - k) + (9k^2 + 6k + 1)$

$= \frac{1}{2}(6k^3 + 15k^2 + 11k + 2) = \frac{1}{2}(k + 1)(6k^2 + 9k + 2) = \frac{1}{2}(k + 1)[6(k + 1)^2 - 3(k + 1) - 1].$

29. 10 **30.** $x^5 + 10x^4 + 40x^3 + 80x^2 + 80x + 32$ **31.** $81x^4 - 432x^3 + 864x^2 - 768x + 256$ **32.** 144 **33.** 84

34. (a) 8 bricks **(b)** 1100 bricks **35.** 360 **36. (a)** $20\left(\frac{3}{4}\right)^3 = \frac{135}{16}$ ft **(b)** $20\left(\frac{3}{4}\right)^n$ ft **(c)** 13 times **(d)** 140 ft **37.** \$151,873.77 **38.** \$58,492.93

Chapter Test *(page 1005)*

1. $0, \frac{3}{10}, \frac{8}{11}, \frac{5}{4}, \frac{24}{13}$ **2.** 4, 14, 44, 134, 404 **3.** $2 - \frac{3}{4} + \frac{4}{9} = \frac{61}{36}$ **4.** $-\frac{1}{3} - \frac{14}{9} - \frac{73}{27} - \frac{308}{81} = -\frac{680}{81}$ **5.** $\sum_{k=1}^{10} (-1)^k\left(\frac{k + 1}{k + 4}\right)$ **6.** Neither

7. Geometric; $r = 4$; $S_n = \frac{2}{3}(1 - 4^n)$ **8.** Arithmetic: $d = -8$; $S_n = n(2 - 4n)$ **9.** Arithmetic; $d = -\frac{1}{2}$; $S_n = \frac{n}{4}(27 - n)$

10. Geometric; $r = \frac{2}{5}$; $S_n = \frac{125}{3}\left[1 - \left(\frac{2}{5}\right)^n\right]$ **11.** Neither **12.** Converges; $\frac{1024}{5}$ **13.** $243m^5 + 810m^4 + 1080m^3 + 720m^2 + 240m + 32$

14. First we show that the statement holds for $n = 1$. $\left(1 + \frac{1}{1}\right) = 1 + 1 = 2$. The equality is true for $n = 1$, so Condition I holds. Next we assume that $\left(1 + \frac{1}{1}\right)\left(1 + \frac{1}{2}\right)\left(1 + \frac{1}{3}\right) \cdots \left(1 + \frac{1}{n}\right) = n + 1$ is true for some k, and we determine whether the formula then holds for $k + 1$. We assume that $\left(1 + \frac{1}{1}\right)\left(1 + \frac{1}{2}\right)\left(1 + \frac{1}{3}\right) \cdots \left(1 + \frac{1}{k}\right) = k + 1$. Now we need to show that $\left(1 + \frac{1}{1}\right)\left(1 + \frac{1}{2}\right)\left(1 + \frac{1}{3}\right) \cdots \left(1 + \frac{1}{k}\right)\left(1 + \frac{1}{k + 1}\right)$ $= (k + 1) + 1 = k + 2$. We do this as follows:

$$\left(1 + \frac{1}{1}\right)\left(1 + \frac{1}{2}\right)\left(1 + \frac{1}{3}\right) \cdots \left(1 + \frac{1}{k}\right)\left(1 + \frac{1}{k + 1}\right) = \left[\left(1 + \frac{1}{1}\right)\left(1 + \frac{1}{2}\right)\left(1 + \frac{1}{3}\right) \cdots \left(1 + \frac{1}{k}\right)\right]\left(1 + \frac{1}{k + 1}\right)$$

$$= (k + 1)\left(1 + \frac{1}{k + 1}\right)(\text{induction assumption}) = (k + 1) \cdot 1 + (k + 1) \cdot \frac{1}{k + 1} = k + 1 + 1 = k + 2$$

Condition II also holds. Thus, the formula holds true for all natural numbers.

15. After 10 years, the Durango will be worth \$6103.11. **16.** The weightlifter will have lifted a total of 8000 pounds after 5 sets.

Cumulative Review *(page 1005)*

1. $\{-3, 3, -3i, 3i\}$ **2. (a)**

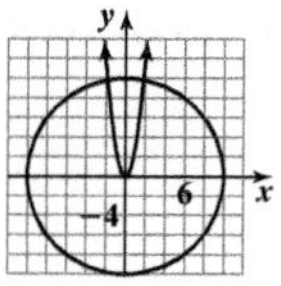

(b) $\left\{\left(\sqrt{\dfrac{-1+\sqrt{3601}}{18}}, \dfrac{-1+\sqrt{3601}}{6}\right), \left(-\sqrt{\dfrac{-1+\sqrt{3601}}{18}}, \dfrac{-1+\sqrt{3601}}{6}\right)\right\}$

(c) The circle and the parabola intersect at

$\left(\sqrt{\dfrac{-1+\sqrt{3601}}{18}}, \dfrac{-1+\sqrt{3601}}{6}\right), \left(-\sqrt{\dfrac{-1+\sqrt{3601}}{18}}, \dfrac{-1+\sqrt{3601}}{6}\right).$

3. $\left\{\ln\left(\dfrac{5}{2}\right)\right\}$ **4.** $y = 5x - 10$ **5.** $(x+1)^2 + (y-2)^2 = 25$ **6. (a)** 5 **(b)** 13 **(c)** $\dfrac{6x+3}{2x-1}$ **(d)** $\left\{x \middle| x \neq \dfrac{1}{2}\right\}$ **(e)** $\dfrac{7x-2}{x-2}$ **(f)** $\{x|x \neq 2\}$ **(g)** $g^{-1}(x) = \dfrac{1}{2}(x-1)$; all reals **(h)** $f^{-1}(x) = \dfrac{2x}{x-3}$; $\{x|x \neq 3\}$ **7.** $\dfrac{x^2}{7} + \dfrac{y^2}{16} = 1$ **8.** $(x+1)^2 = 4(y-2)$

9. $r = 8 \sin \theta$; $x^2 + (y-4)^2 = 16$ **10.** $\left\{\dfrac{3\pi}{2}\right\}$ **11.** $\dfrac{2\pi}{3}$ **12. (a)** $-\dfrac{\sqrt{15}}{4}$ **(b)** $-\dfrac{\sqrt{15}}{15}$ **(c)** $-\dfrac{\sqrt{15}}{8}$ **(d)** $\dfrac{7}{8}$ **(e)** $\sqrt{\dfrac{1+\dfrac{\sqrt{15}}{4}}{2}} = \dfrac{\sqrt{4+\sqrt{15}}}{2\sqrt{2}}$

Credits

Photos

Cover Leigh Prather/Shutterstock

Endpaper 3 Pearson Education, Inc.

Chapter R Page 1, Aslysun/Shutterstock; Page 32, Feraru Nicolae/Shutterstock.

Chapter 1 Pages 82 and 163, Andres Rodriguez/Fotolia; Page 109, A.R. Monko/Shutterstock; Page 120 (L), Stanca Sanda/Alamy; Page 120 (R), Jeff Chiu/AP Images; Page 156, Nancy R. Cohen/Getty Images; Page 160, Hemera Technologies/Photos.com/Getty Images.

Chapter 2 Pages 164 and 205, Andrey Popov/Shutterstock; Page 166, Snapper/Fotolia; Page 172 (L), Stockyimages/Fotolia; Page 172 (R), Department of Energy; Page 187, Tetra Images/Alamy; Page 195, Tom Donoghue/Polaris/Newscom.

Chapter 3 Pages 206 and 278, Jeff Metzger/Shutterstock; Page 220, NASA; Page 228, Exactostock-1557/SuperStock; Page 267, Dreamstime LLC.

Chapter 4 Pages 280 and 329, 123RF; Page 317, Geno EJ Sajko Photography/Shutterstock.

Chapter 5 Pages 330 and 405, Muzhik/Shutterstock; Page 392, Corbis.

Chapter 6 Pages 407 and 511, Andrey Lobachev/Fotolia; Page 436 and 464, Pearson Education, Inc.; Page 474 (L), Stockbyte/Getty Images; Page 474 (R), Pascal Saez/Alamy; Page 479, Lacroix Serge/iStock/Getty Images; Page 494, Jupiterimages/Getty Images.

Chapter 7 Pages 512 and 613, Northern Lights Software; Page 524, Ryan McVay/Digital Vision/Getty Images; Page 548, PhotoSerg/Shutterstock; Page 586, Srdjan Draskovic/Dreamstime LLC.

Chapter 8 Pages 614 and 683, Pankajstock123/Fotolia.

Chapter 9 Pages 684 and 728, Jennifer Thermes/Photodisc/Getty Images; Page 713, Alexandre Fagundes De Fagundes/Dreamstime LLC; Page 715, Anton Ignatenco/iStock/Getty Images.

Chapter 10 Pages 730 and 790, Andrii Gorulko/Shutterstock; Page 753, Pearson Education, Inc.; Page 762, Science & Society Picture Library/SSPL/Getty Images; Page 774, Hulton Archive/Getty Images.

Chapter 11 Pages 791 and 860, MarcelClemens/Shutterstock; Page 811, Thomas Barrat/Shutterstock.

Chapter 12 Pages 861 and 960, Rawpixel/Shutterstock; Page 916, Pearson Education, Inc.

Chapter 13 Pages 961 and 1006, Arthimedes/Shutterstock; Pages 988 and 1001, Pearson Education, Inc.

Text

TI 84 Plus C screenshots courtesy of Texas Instruments. Screenshots from Microsoft® Excel®. Used by permission of Microsoft Corporation. Chapter 3, Page 209: Diamond price. Used with permission of Diamonds.com. Copyright © Martin Rapaport. All rights reserved. Chapter 6, Page 502: Cell Phone Towers by CTIA-The Wireless Association. Copyright © 2013 by CTIA-The Wireless Association. Used by permission of CTIA-The Wireless Association®. Chapter 8, Page 644: Excerpt from *Fundamentals of Physics*, 10th ed. by David Halliday, Robert Resnick and Jearl Walker. Published by Wiley, © 2014.

Subject Index

CONICS

Parabola

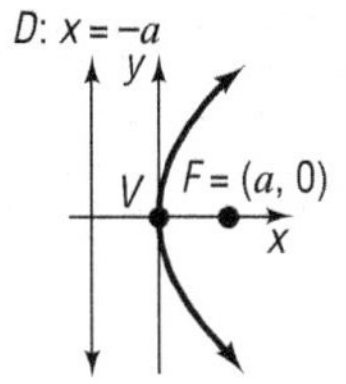

$y^2 = 4ax$

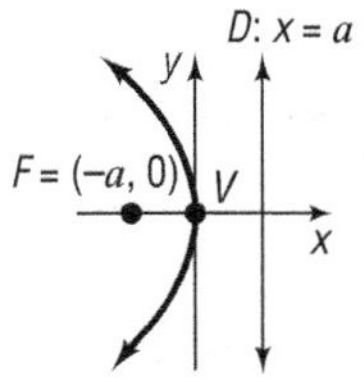

$y^2 = -4ax$

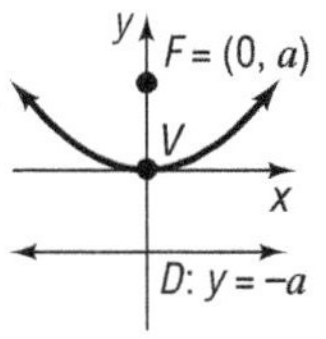

$x^2 = 4ay$

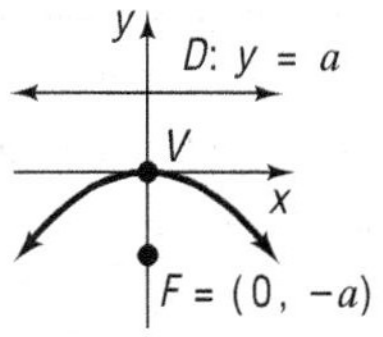

$x^2 = -4ay$

Ellipse

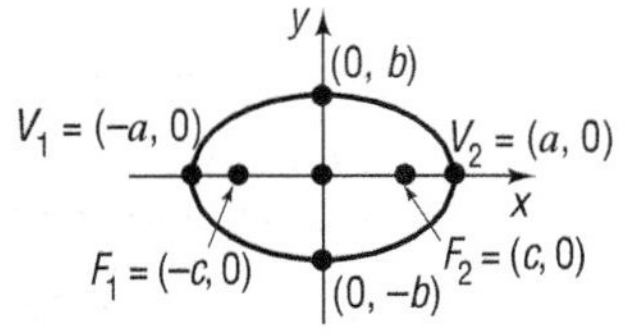

$$\frac{x^2}{a^2} + \frac{y^2}{b^2} = 1, \quad a > b, \quad c^2 = a^2 - b^2$$

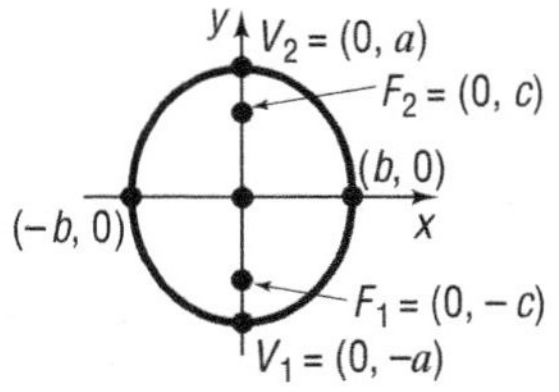

$$\frac{x^2}{b^2} + \frac{y^2}{a^2} = 1, \quad a > b, \quad c^2 = a^2 - b^2$$

Hyperbola

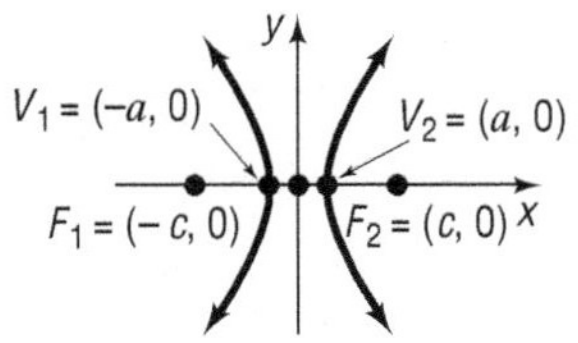

$$\frac{x^2}{a^2} - \frac{y^2}{b^2} = 1, \quad c^2 = a^2 + b^2$$

Asymptotes: $y = \frac{b}{a}x, \quad y = -\frac{b}{a}x$

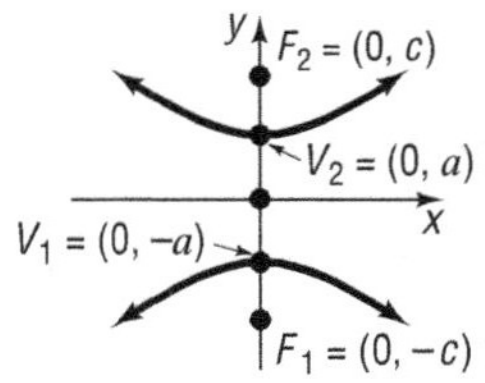

$$\frac{y^2}{a^2} - \frac{x^2}{b^2} = 1, \quad c^2 = a^2 + b^2$$

Asymptotes: $y = \frac{a}{b}x, \quad y = -\frac{a}{b}x$

PROPERTIES OF LOGARITHMS

$\log_a(MN) = \log_a M + \log_a N$

$\log_a\left(\frac{M}{N}\right) = \log_a M - \log_a N$

$\log_a M^r = r\log_a M$

$\log_a M = \frac{\log M}{\log a} = \frac{\ln M}{\ln a}$

$a^r = e^{r \ln a}$

PERMUTATIONS/COMBINATIONS

$0! = 1 \qquad 1! = 1$

$n! = n(n-1) \cdot \ldots \cdot (3)(2)(1)$

$P(n, r) = \frac{n!}{(n-r)!}$

$C(n, r) = \binom{n}{r} = \frac{n!}{(n-r)!\,r!}$

BINOMIAL THEOREM

$$(a+b)^n = a^n + \binom{n}{1}ba^{n-1} + \binom{n}{2}b^2a^{n-2} + \cdots + \binom{n}{n-1}b^{n-1}a + b^n$$

ARITHMETIC SEQUENCE

$$a_1 + (a_1 + d) + (a_1 + 2d) + \cdots + [a_1 + (n-1)d] = \frac{n}{2}[2a_1 + (n-1)d] = \frac{n}{2}[a_1 + a_n]$$

GEOMETRIC SEQUENCE

$$a_1 + a_1r + a_1r^2 + \cdots + a_1r^{n-1} = a_1 \cdot \frac{1 - r^n}{1 - r}$$

GEOMETRIC SERIES

If $|r| < 1$, $a_1 + a_1r + a_1r^2 + \cdots = \sum_{k=1}^{\infty} a_1r^{k-1} = \frac{a_1}{1-r}$

LIBRARY OF FUNCTIONS

Identity Function

$f(x) = x$

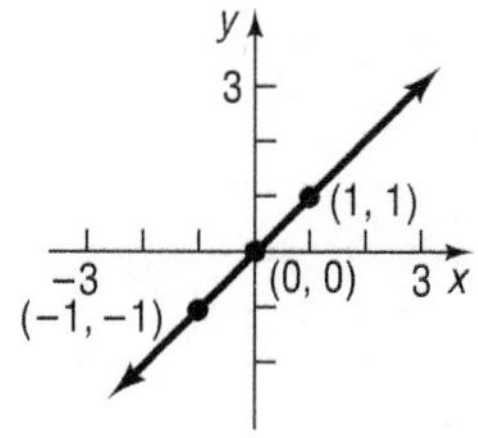

Square Function

$f(x) = x^2$

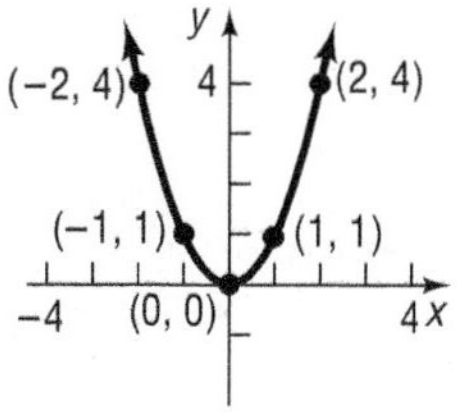

Cube Function

$f(x) = x^3$

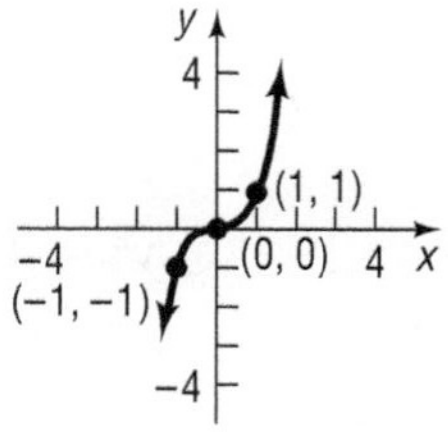

Square Root Function

$f(x) = \sqrt{x}$

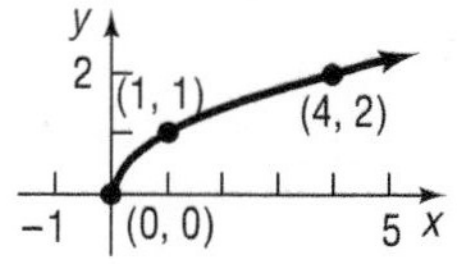

Reciprocal Function

$f(x) = \dfrac{1}{x}$

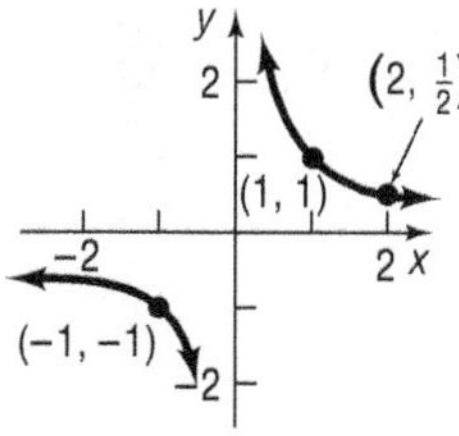

Cube Root Function

$f(x) = \sqrt[3]{x}$

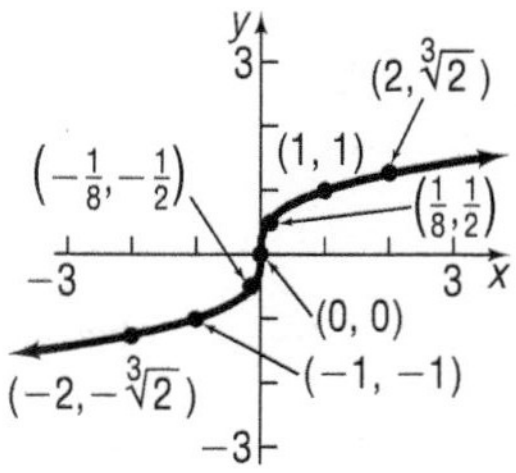

Absolute Value Function

$f(x) = |x|$

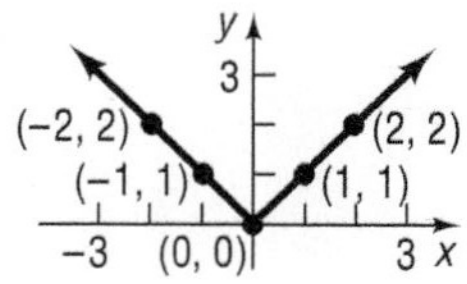

Exponential Function

$f(x) = e^x$

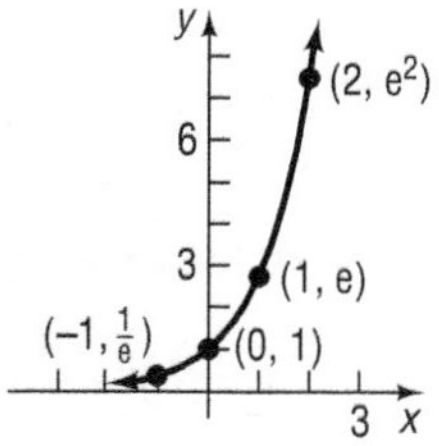

Natural Logarithm Function

$f(x) = \ln x$

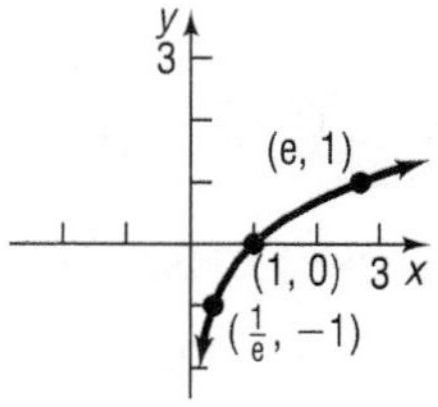

Sine Function

$f(x) = \sin x$

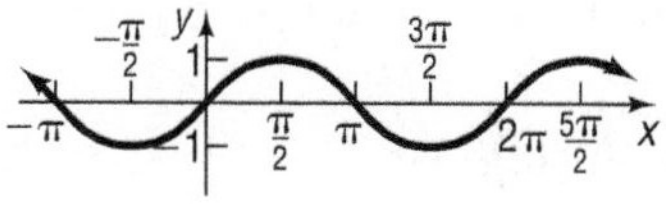

Cosine Function

$f(x) = \cos x$

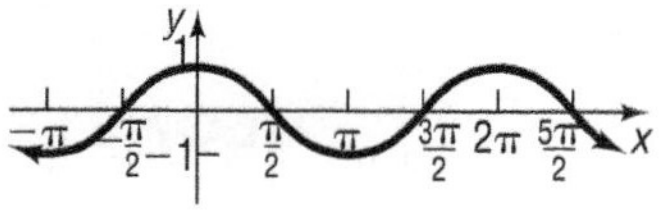

Tangent Function

$f(x) = \tan x$

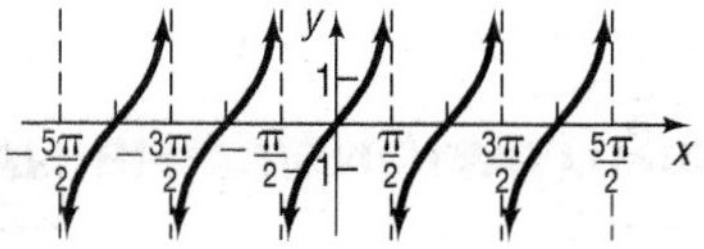

Cosecant Function

$f(x) = \csc x$

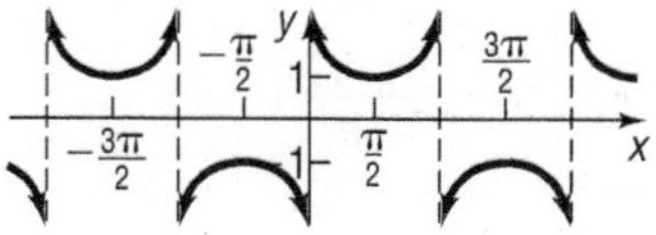

Secant Function

$f(x) = \sec x$

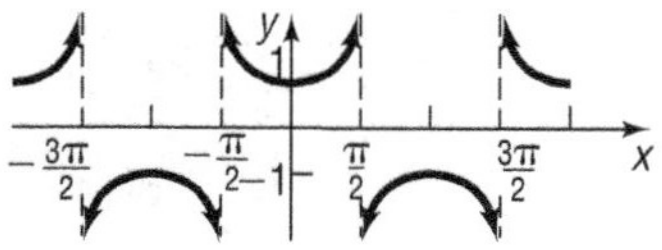

Cotangent Function

$f(x) = \cot x$

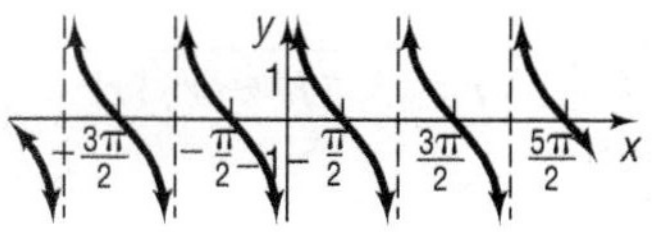

FORMULAS/EQUATIONS

Distance Formula

If $P_1 = (x_1, y_1)$ and $P_2 = (x_2, y_2)$, the distance from P_1 to P_2 is

$$d(P_1, P_2) = \sqrt{(x_2 - x_1)^2 + (y_2 - y_1)^2}$$

Standard Equation of a Circle

The standard equation of a circle of radius r with center at (h, k) is

$$(x - h)^2 + (y - k)^2 = r^2$$

Slope Formula

The slope m of the line containing the points $P_1 = (x_1, y_1)$ and $P_2 = (x_2, y_2)$ is

$$m = \frac{y_2 - y_1}{x_2 - x_1} \qquad \text{if } x_1 \neq x_2$$

$$m \text{ is undefined} \qquad \text{if } x_1 = x_2$$

Point–Slope Equation of a Line

The equation of a line with slope m containing the point (x_1, y_1) is

$$y - y_1 = m(x - x_1)$$

Slope–Intercept Equation of a Line

The equation of a line with slope m and y-intercept b is

$$y = mx + b$$

Quadratic Formula

The solutions of the equation $ax^2 + bx + c = 0, a \neq 0$, are

$$x = \frac{-b \pm \sqrt{b^2 - 4ac}}{2a}$$

If $b^2 - 4ac > 0$, there are two unequal real solutions.
If $b^2 - 4ac = 0$, there is a repeated real solution.
If $b^2 - 4ac < 0$, there are two complex solutions that are not real.

GEOMETRY FORMULAS

Circle

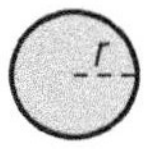

r = Radius, A = Area, C = Circumference

$$A = \pi r^2 \qquad C = 2\pi r$$

Triangle

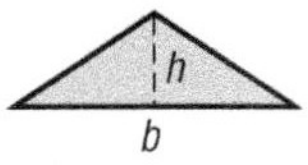

b = Base, h = Altitude (Height), A = area

$$A = \frac{1}{2} bh$$

Rectangle

l = Length, w = Width, A = area, P = perimeter

$$A = lw \qquad P = 2l + 2w$$

Rectangular Box (closed)

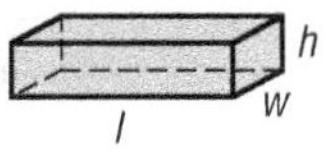

l = Length, w = Width, h = Height, V = Volume, S = Surface area

$$V = lwh \qquad S = 2lw + 2lh + 2wh$$

Sphere

r = Radius, V = Volume, S = Surface area

$$V = \frac{4}{3}\pi r^3 \qquad S = 4\pi r^2$$

Right Circular Cylinder (closed)

r = Radius, h = Height, V = Volume, S = Surface area

$$V = \pi r^2 h \qquad S = 2\pi r^2 + 2\pi rh$$

TRIGONOMETRIC FUNCTIONS

Of an Acute Angle

$$\sin\theta = \frac{b}{c} = \frac{\text{Opposite}}{\text{Hypotenuse}} \qquad \cos\theta = \frac{a}{c} = \frac{\text{Adjacent}}{\text{Hypotenuse}} \qquad \tan\theta = \frac{b}{a} = \frac{\text{Opposite}}{\text{Adjacent}}$$

$$\csc\theta = \frac{c}{b} = \frac{\text{Hypotenuse}}{\text{Opposite}} \qquad \sec\theta = \frac{c}{a} = \frac{\text{Hypotenuse}}{\text{Adjacent}} \qquad \cot\theta = \frac{a}{b} = \frac{\text{Adjacent}}{\text{Opposite}}$$

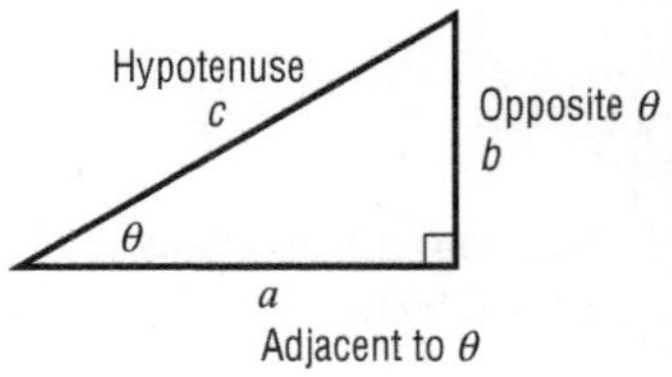

Of a General Angle

$$\sin\theta = \frac{b}{r} \qquad \cos\theta = \frac{a}{r} \qquad \tan\theta = \frac{b}{a} \quad a \neq 0$$

$$\csc\theta = \frac{r}{b} \quad b \neq 0 \qquad \sec\theta = \frac{r}{a} \quad a \neq 0 \qquad \cot\theta = \frac{a}{b} \quad b \neq 0$$

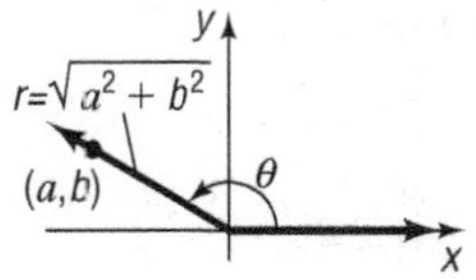

TRIGONOMETRIC IDENTITIES

Fundamental Identities

$$\tan\theta = \frac{\sin\theta}{\cos\theta} \qquad \cot\theta = \frac{\cos\theta}{\sin\theta}$$

$$\csc\theta = \frac{1}{\sin\theta} \qquad \sec\theta = \frac{1}{\cos\theta} \qquad \cot\theta = \frac{1}{\tan\theta}$$

$$\sin^2\theta + \cos^2\theta = 1$$

$$\tan^2\theta + 1 = \sec^2\theta$$

$$\cot^2\theta + 1 = \csc^2\theta$$

Half-Angle Formulas

$$\sin\frac{\theta}{2} = \pm\sqrt{\frac{1-\cos\theta}{2}}$$

$$\cos\frac{\theta}{2} = \pm\sqrt{\frac{1+\cos\theta}{2}}$$

$$\tan\frac{\theta}{2} = \frac{1-\cos\theta}{\sin\theta}$$

Double-Angle Formulas

$$\sin(2\theta) = 2\sin\theta\cos\theta$$

$$\cos(2\theta) = \cos^2\theta - \sin^2\theta$$

$$\cos(2\theta) = 2\cos^2\theta - 1$$

$$\cos(2\theta) = 1 - 2\sin^2\theta$$

$$\tan(2\theta) = \frac{2\tan\theta}{1-\tan^2\theta}$$

Even-Odd Identities

$$\sin(-\theta) = -\sin\theta \qquad \csc(-\theta) = -\csc\theta$$

$$\cos(-\theta) = \cos\theta \qquad \sec(-\theta) = \sec\theta$$

$$\tan(-\theta) = -\tan\theta \qquad \cot(-\theta) = -\cot\theta$$

Product-to-Sum Formulas

$$\sin\alpha\sin\beta = \tfrac{1}{2}[\cos(\alpha-\beta) - \cos(\alpha+\beta)]$$

$$\cos\alpha\cos\beta = \tfrac{1}{2}[\cos(\alpha-\beta) + \cos(\alpha+\beta)]$$

$$\sin\alpha\cos\beta = \tfrac{1}{2}[\sin(\alpha+\beta) + \sin(\alpha-\beta)]$$

Sum and Difference Formulas

$$\sin(\alpha+\beta) = \sin\alpha\cos\beta + \cos\alpha\sin\beta$$

$$\sin(\alpha-\beta) = \sin\alpha\cos\beta - \cos\alpha\sin\beta$$

$$\cos(\alpha+\beta) = \cos\alpha\cos\beta - \sin\alpha\sin\beta$$

$$\cos(\alpha-\beta) = \cos\alpha\cos\beta + \sin\alpha\sin\beta$$

$$\tan(\alpha+\beta) = \frac{\tan\alpha + \tan\beta}{1 - \tan\alpha\tan\beta}$$

$$\tan(\alpha-\beta) = \frac{\tan\alpha - \tan\beta}{1 + \tan\alpha\tan\beta}$$

Sum-to-Product Formulas

$$\sin\alpha + \sin\beta = 2\sin\frac{\alpha+\beta}{2}\cos\frac{\alpha-\beta}{2}$$

$$\sin\alpha - \sin\beta = 2\sin\frac{\alpha-\beta}{2}\cos\frac{\alpha+\beta}{2}$$

$$\cos\alpha + \cos\beta = 2\cos\frac{\alpha+\beta}{2}\cos\frac{\alpha-\beta}{2}$$

$$\cos\alpha - \cos\beta = -2\sin\frac{\alpha+\beta}{2}\sin\frac{\alpha-\beta}{2}$$

SOLVING TRIANGLES

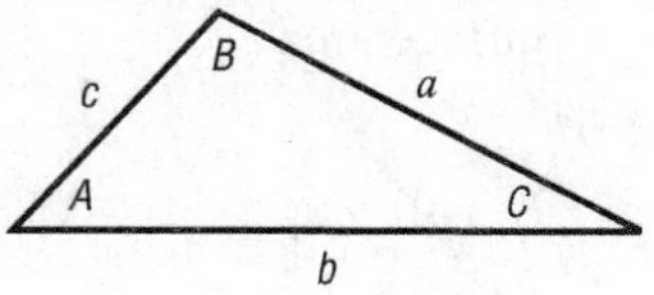

Law of Sines

$$\frac{\sin A}{a} = \frac{\sin B}{b} = \frac{\sin C}{c}$$

Law of Cosines

$$a^2 = b^2 + c^2 - 2bc\cos A$$

$$b^2 = a^2 + c^2 - 2ac\cos B$$

$$c^2 = a^2 + b^2 - 2ab\cos C$$